All–in–One Electronics Simplified

Dr Anil K. Maini
Nakul Maini

All-in-One

ELECTRONICS SIMPLIFIED

The Q & A Book

DR ANIL K. MAINI

NAKUL MAINI

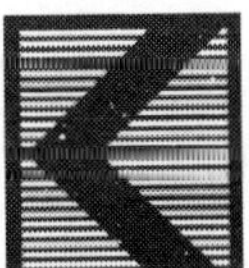

KHANNA BOOK PUBLISHING CO. (P) LTD.

Publisher of Engineering and Computer Books

4C/4344, Ansari Road, Darya Ganj, New Delhi-110002

Phone: 011-23244447-48 **Mobile:** +91-9910909320

E-mail: contact@khannabooks.com

Website: www.khannabooks.com

Price : Rs. 725.00

ALL-IN-ONE ELECTRONICS SIMPLIFIED
Dr Anil K. Maini, Nakul Maini

Copyright © Khanna Book Publishing Co. (P) Ltd.

This book is sold subject to the condition that it shall not, by way of trade or otherwise, be lent, resold, hired out, or otherwise circulated without the publisher's prior consent in any form of binding or cover other than that in which it is published and without a similar condition including this condition being imposed on the subsequent purchaser and without limiting the rights under copyright reserved above, no part of this publication may be reproduced, stored in or introduced into retrieval system, or transmitted in any form or by any means (electronic, mechanical, photocopying, recording or otherwise), without the prior written permission of both the copyright owner and the above mentioned publisher of this book.

ISBN: 978-93-86173-39-3

First Edition: 2018
Reprint: 2021

Published by:

KHANNA BOOK PUBLISHING CO. (P) LTD.
4C/4344, Ansari Road, Darya Ganj, New Delhi-110 002
Phone: 011-23244447-48 **Mobile:** +91-9910909320
E-mail: contact@khannabooks.com

Printed in India by:
India Book Printers & Binders, Delhi

Dedicated to

Little Sweetheart KIAAN

Dr Anil K. Maini
Nakul Maini

Preface

All-in-One Electronics Simplified – The Q & A Book is a comprehensive treatise on the whole gamut of topics in Electronics covering as many as six major subjects taught at the undergraduate level. The subjects covered include **Electronic Components, Devices and Circuits, Networks, Operational Amplifiers and Linear Integrated Circuits, Digital Electronics, Feedback and Control systems and Electrical Measurements and Instrumentation**. Different subjects are exhaustively covered in thirty chapters in the form of probing **Questions and Answers**. The Questions and Answers in each chapter are followed up by a large number of solved numerical problems. A comprehensive **Self-Evaluation Exercise** comprising multiple choice questions and other forms of objective type exercises with answers concludes each chapter. Each topic is covered in entirety beginning with basic concepts and then gradually moving on to more complex concepts. The choice of questions and the order in which they are presented in each chapter makes it a step-by-step learning exercise.

The book is broadly divided into eight sections namely **Passive Electronic Components, Discrete Semiconductor Devices, Electronic Circuitry, Operational Amplifiers, Linear Integrated Circuits, Digital Electronics** including **Programmable Logic Devices, Feedback and Control Systems, Electrical Measurements and Electronic Instrumentation.**

The section on **Passive Electronic Components** comprises four chapters namely *Resistors, Capacitors, Inductors and Transformers and Hardware Electronic Components*. In each of these four chapters, the discussion is mainly centred on role of these components in electronic circuits, fundamentals around these components, selection criteria and other application related topics. For example, the first three chapters on resistors, capacitors and inductors are replete with topics such as standard values of these components, colour coding schemes, advantages and disadvantages of different types etc. The fourth chapter on hardware components covers cables, connectors, relays, switches, fuses and batteries describing fundamentals, types and characteristic features of each one of these components.

The section on **Networks** also comprises four chapters namely *Network Fundamentals, Network Theorems, Fourier Series Analysis of Networks and Two-Port Networks and State Equations*. Major topics covered in *Network Fundamentals* include network graphs and matrices associated with network graphs, Kirchhoff's laws, mesh current and node voltage analysis, voltage and current sources, source transformations and RLC transients. Different theorems discussed under *Network Theorems* include Thevenin's and Norton's theorems, maximum power transfer theorem, superposition theorem, reciprocity theorem, compensation theorem, substitution theorem, Tellegen's theorem, Millman's theorem, star-delta and delta-star transformations and application of network theorems to solving networks. In the chapter on *Fourier Series Analysis of Networks*, are covered odd and even functions, Dirichlet conditions, trigonometric and exponential Fourier series, Fourier series expansion of important waveforms, solving networks using Fourier series expansion and two-port networks and state equations.

The section on **Discrete Semiconductor Devices** has four chapters namely *semiconductor diodes and transistors, field effect transistors* including junction FETs and MOSFETs, *uni-junction transistors and Thyristors and optoelectronic devices*. Device fundamentals, operational principle, current-voltage characteristics, biasing techniques, characteristic features, performance specifications, circuit configurations and applications are the major topics covered for each of these devices.

The section on **Electronic Circuitry** has five chapters namely *amplifiers, oscillators, linear power supplies, switched mode power supplies* and *wave shaping circuits*. Basic circuit configurations, biasing techniques, stability considerations, small signal analysis, frequency response and large signal amplifiers are the major topics discussed under *amplifiers. Sinusoidal oscillators* including R-C phase shift oscillator, Wein bridge oscillator, twin-T oscillator, Bubba oscillator, Hartley, Colpitt and Clapp oscillators, crystal oscillator, Piercing oscillator and *non-sinusoidal oscillators* including Bistable, Monostable, Astable multivibrators and Schmitt trigger are covered under oscillators. In the chapter on *linear power supplies*, major topics covered in the chapter include different types of rectifier and filter circuits, different types of regulator circuits including emitter-follower regulators, series pass regulators and shunt regulators and power supply protection features. The chapter on switched mode power supplies includes discussion on comparison of linear power supplies versus switched mode power supplies, constituents of a switched mode power supply, types of switched mode power supplies, different types of DC-to-DC converters such as forward converters, push-pull converters and fly back converters and their design, switching regulators including buck, boost and inverting regulators, off-line switched mode power supplies and series and parallel connection of DC-to-DC converters. The last chapter in this section covers various wave shaping circuits including R-C integrator and differentiator circuits, R-L integrator and differentiator circuits, diode clipping circuits and clamping circuits.

The next section focuses on **operational amplifiers and linear integrated circuits**. The first of the two chapters on operational amplifiers cover device fundamentals as regards the internal circuitry of operational amplifier, characteristic parameters and types of operational amplifiers such as general purpose opamps, Norton opamps, opamp comparators, video opamps, isolation opamps, low noise opamps and instrumentation opamps. The second chapter covers operational aspects and design guidelines of a large number of opamp circuits, which includes opamp as voltage follower, opamp as inverting and non-inverting amplifiers, inverting and non-inverting summing amplifiers, inverting and non-inverting comparators, window comparator, subtractor, logarithmic and anti-logarithmic amplifiers, peak detector, half wave and full wave rectifiers, low pass, high pass and band pass filters, voltage regulators, sinusoidal oscillators, clipper and clamper circuits, current-to-voltage and voltage-to-current converters, averager circuit and opamp as instrumentation amplifier. The next two chapters in this section cover IC timers and phase locked loops. Again, discussion is mainly centred on operational basics and application circuits. Application circuits built around IC timer 555 include different variants of monostable and astable multivibrator circuits, pulse width and pulse time modulator circuits. Phase locked loop application circuits covered in the chapter include frequency multiplier and divider circuits, AM and FM demodulator circuits.

Following operational amplifiers and linear integrated circuits, different topics of **digital electronics** are covered next. The section begins with fundamental topics of digital electronics including number systems, number codes and binary arithmetic. Logic gates and logic families are discussed next. This is followed by a discussion on different laws and theorems of Boolean algebra and techniques used for simplification of complex Boolean expressions. Combinational logic circuits including arithmetic circuits, multiplexers, de-multiplexers, magnitude comparators and parity generators and checkers are covered next. Latches and flip flops, counters and registers are extensively covered under sequential logic circuits. Other topics exhaustively covered under digital electronics include characteristic parameters, types and applications of digital-to-analog and analog-to-digital converters, different types of primary and secondary memory devices, memory expansion, and programmable logic devices including PROM, PAL, PLA, HAL, CPLD and FPGA.

Feedback and control systems are discussed next. Major topics covered in this chapter include control system fundamentals, stability criteria for absolute stability including Routh, Hurwitz and continued fraction stability criteria, block diagram algebra, signal flow graphs, concept of relative stability, stability analysis, error constants, root locus analysis, Nyquist analysis, Bode analysis, control system compensators including lag, lead and lag-lead compensators, controllers including proportional, integral, differential, proportional-integral and proportional-integral-differential controllers and state variable analysis of control systems.

The concluding section is on **electrical measurements and instrumentation**. There are three chapters in this part of the book covering electrical measurement fundamentals including basic meter movements, use of series resistances and shunts to convert the basic movement into voltmeters and ammeters respectively, DC and AC bridges used for measurement of unknown resistances, inductances and capacitances in the first chapter; electronic instruments including digital multimeters, oscilloscopes, frequency synthesisers, synthesised function generators, lock-in amplifiers, logic analysers and frequency counters, concept of virtual instrumentation in the second chapter and sensors and transducers in the third chapter. The concluding chapter of this part and also of the book covers operational fundamentals and applications of a wide range of sensors and transducers. The chapter begins with brief description of difference between a sensor, a transducer and an actuator and then goes on to discuss different types of actuators, transducer parameters, different categories of transducers, which includes strain gauges including resistive, semiconductor and vibrating wire strain gauges, potentiometer transducers, inductive, capacitive and induction-type transducers, Hall effect transducers, Linear Variable Differential Transformer (LVDT), thermocouples, thermopiles and thermistors, resistance and thermocouple thermometry, synchro transducers, shaft encoders including absolute and incremental shaft encoders, hot wire gauges, Pirani gauge and proximity sensors.

The book is primarily intended for undergraduate students of **Electronics and Communication Engineering (ECE)** and **Instrumentation Engineering.** The subjects of semiconductor devices and digital electronics will also interest undergraduate students of **Computer Science (CS) and Information Technology (IT)**. It is aimed at providing to them in one book concise and yet complete coverage of as many as six major subjects taught at the undergraduate level. The treatment of different topics in each of these six subjects is such as to cover them in entirety in simple and lucid language. The **Question and Answer format** of presentation material and inclusion of a large number of **solved problems** and **self-evaluation exercises** in each chapter is also aimed at comprehensively preparing them for the examination; which is an important part of academics. We, the authors, have made a sincere attempt towards achieving the intended objective of bringing out this book and we do hope that the book will be well received by the students. We shall look forward to receiving suggestions from students and also faculty; which will enable us improve upon the book in future editions. You may send your suggestions to the publishers.

Dr Anil K. Maini

Nakul Maini

Content

Chapter 1

Resistors and Resistive Networks

Learning Objectives

After completing the chapter, you will learn the following topics:

1. Concept of potential difference
2. Resistance and its purpose
3. Ohm's law
4. Resistance in DC circuits
5. Resistance in AC circuits
6. Specifications of Resistor
7. Resistors in series
8. Resistors in parallel
9. Resistors in series-parallel
10. Resistor colour coding
11. Types of resistors
12. Variable resistors
13. Voltage dependent resistors or Varistors
14. Thermistors

Q.1. What is a RESISTOR? How do you represent a resistance in an electrical circuit and what are its units of measurement?

Ans. A RESISTOR is a passive electronic component that when connected in a circuit offers a specific amount of electrical resistance to the flow of current. A fixed resistance is represented in a circuit by the circuit symbol shown in Fig. 1.1. It is measured in OHMS (represented by symbol Ω). Ohm however is a very small unit. Most of the practical resistors will have resistances in thousands or hundreds of thousands of ohms. Therefore, resistance is often measured in KILO-OHMS and MEGA-OHMS, with 1 Kilo-ohm = 1000 ohms and 1 Mega-ohm = 1000 Kilo-ohms = 1000000 (or 10^6) ohms.

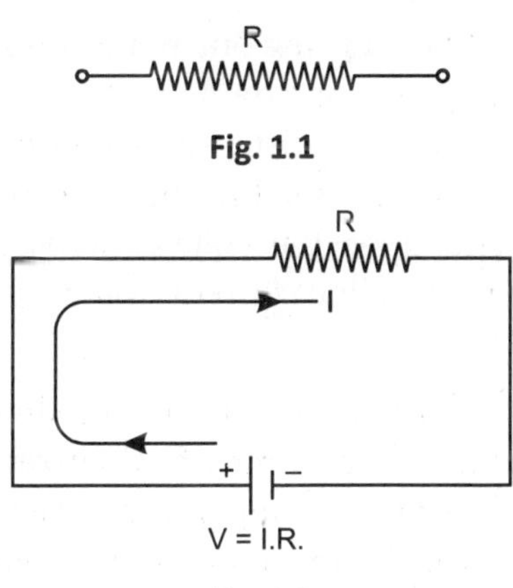

Fig. 1.1

Fig. 1.2

Q.2. How do you define POTENTIAL DIFFERENCE between two points in an electrical circuit? Can the

magnitude of potential difference be greater than the magnitude of the potential of either of the two points?

Ans. The POTENTIAL DIFFERENCE between two points in an electrical circuit is given by the difference in potentials of the two points with respect to a common reference point, which is generally ground potential or zero volt point. As an example, if the potentials of two given points in an electrical circuit were 5V and –3V, the potential difference across these points would then be 5–(–3) = 8V.

The magnitude of potential difference is greater than that of either of the involved potentials if they had opposite polarities. This is evident from the example given in the previous paragraph.

Q.3. What quantum of resistance does one ohm represent? Which law of electricity is associated with the definition of an ohm?

Or

State and briefly explain Ohm's Law.

Ans. The resistance offered by a resistor is 1 ohm when a voltage of 1 volt when impressed across it results in a current of 1 Amp through it (Fig. 1.2). Definition of resistance is an outcome of the famous Ohm's Law, which states that the resistance (*R*) equals the ratio of Voltage (*V*) appearing across (*R*) to the current (*I*) flowing through it, *i.e.* $R = V/I$.

According to the Ohm's law, the current (*I*) flowing through a linear circuit is directly proportional to the potential difference (*V*) applied across it. That is, $I \propto V$ the constant of proportionality here is equal to (*I/R*) where (*R*) is the resistance of the circuit. That is, $I = \frac{V}{R}$ or $V = IR$. A linear circuit is the one that comprises of linear components. In a linear component, the value of the component such as its resistance value (in the case of a resistor) or its capacitance value (in the case of a capacitor) does not depend upon the voltage across it or current flowing through it. A linear circuit follows superposition principle.

Q.4. Write expressions for power dissipated in a resistor of value (*R*) in terms of (*a*) current (*I*) flowing through it and voltage (*V*) appearing across it (*b*) voltage (*V*) and resistance (*R*) and (*c*) current (*I*) and resistance (*R*).

Ans.(*a*) Power dissipated (*P*) in a resistor (*R*) is given by product of voltage (*V*) across it and current (*I*) flowing through it.

That is, $P = VI$

(*b*) According to Ohm's law, $V = IR$ or $I = V/R$

Therefore, $P = V^2/R$

(*c*) Substituting $V = IR$, we get $P = I^2R$

Q.5. How is the resistance (*R*) of a cylindrical conductor related to its length (*l*), cross-section (*A*) and the material resistivity (ρ)?

Ans. Resistance (*R*) of a cylindrical conductor of length (*l*) and cross section (*A*) is given by

$$R = \rho \frac{l}{A} \text{ ohms}$$

where

l = Length in m

A = Cross-section in m^2

ρ = Material resistivity in ohm-m

Q.6. How do you find total resistance of a series connection of more than one resistance? Which law of electrical networks has been used to deduce this relationship?

Ans. Total Resistance (*R*) of a series connection of (*n*) resistors R_1, R_2, R_3 ..... R_n is expressed by

$$R = R_1 + R_2 + R_3 + \text{........} + R_n$$

Refer to Fig. 1.3. According to Kirchoff's Voltage Law (*KVL*), the impressed voltage equals the algebraic sum of all voltages in a closed mesh[1].

Therefore

$$V = IR_1 + IR_2 + IR_3 + \text{.....} + IR_n$$

or $$V = I(R_1 + R_2 + R_3 + \text{....} + R_n) \quad \text{...(1)}$$

If (*R*) is the total resistance offered by a series combination of different resistors, then

$$V = IR \quad \text{...(2)}$$

Comparing equations (1) and (2) $R = R_1 + R_2 + R_3 +$ .... $+ R_n$.

The equivalent resistance in a series connection is always greater than the largest resistance value in the series connection.

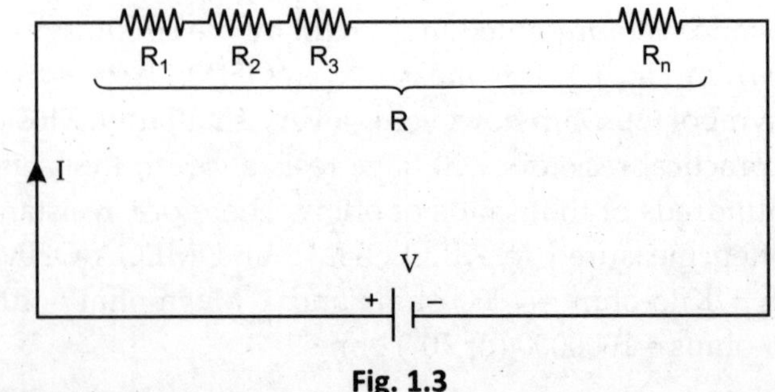

Fig. 1.3

[1] KVL is discussed in detail in chapter 5.

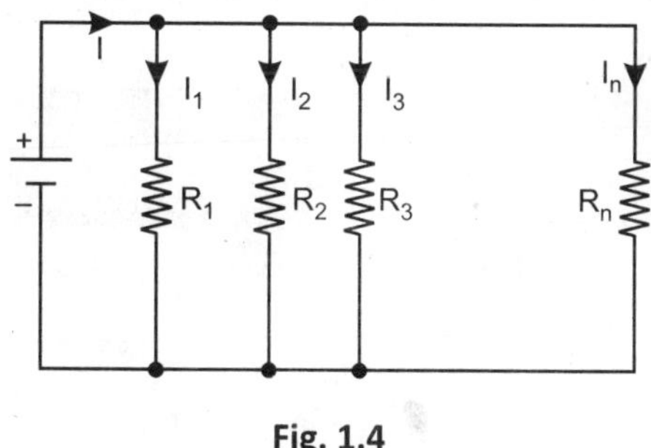

Fig. 1.4

Q.7. How is the total current flowing through a series connection of multiple resistors related to the currents flowing through individual resistors in the series connection? How total voltage is related to the voltage drops across each of the series connected resistors?

Ans. In a series connection of multiple resistors, total current is the same as the current flowing through each of the resistors. The voltage across series connected resistors is equal to the sum of the voltage drops across individual resistors. Voltage drop across any given resistor in the series connection is equal to the product of the current flowing through it and the resistance value.

Q.8. How do you find the total resistance offered by a parallel connection of more than one resistance? Which law of electrical network can be used to deduce relationship between equivalent resistance and individual resistance values?

Ans. The total resistance (R) of a parallel connection of resistances R_1, R_2, R_3,.....R_n is given by

$$\frac{1}{R} = \left(\frac{1}{R_1} + \frac{1}{R_2} + \frac{1}{R_3} \cdots + \frac{1}{R_n}\right)$$

The above relationship is an outcome of Kirchoff's Current Law (*KCL*)[2]. Refer to Fig. 1.4. As per Kirchoff's Current Law, algebraic sum of currents at a node is always zero. *i.e.* total current entering a node or moving towards a node must equal the total current leaving the node or flowing away from the node. From Fig. 1.4.

$$I = I_1 + I_2 + I_3 + \ldots\ldots + I_n$$

or $$\frac{V}{R} = \frac{V}{R_1} + \frac{V}{R_2} + \frac{V}{R_3} + \ldots + \frac{V}{R_n}$$

or $$\frac{1}{R} = \frac{1}{R_1} + \frac{1}{R_2} + \frac{1}{R_3} + \frac{1}{R_4} + \ldots + \frac{1}{R_n}$$

The equivalent resistance in a parallel connection of multiple resistors is always less than the smallest value resistance.

[2] Kirchoff's laws are discussed in detail in Chapter-5.

Q.9. In the case of parallel connection of two resistors (R_1) and (R_2), if one of the resistors were much larger than the other, the equivalent resistance will be approximately equal to the smaller resistance. Justify!

Ans. In this case $R = 1/[(1/R_1) + (1/R_2)]$

Let us assume that $R_2 >> R_1$, then $(1/R_1) >> (1/R_2)$

That is, $R \cong 1/(1/R_1) = R_1$ and hence the answer

Q.10. Write expressions for equivalent DELTA CONNECTED RESISTANCE for given STAR-CONNECTED RESISTANCES (Fig. 1.5).

Ans. Refer to Fig. 1.5

$$R_{AB} = R_A + R_B + \frac{R_A R_B}{R_C};$$

$$R_{BC} = R_B + R_C + \frac{R_B R_C}{R_A};$$

$$R_{CA} = R_C + R_A + \frac{R_C R_A}{R_B}$$

These expressions are derived in chapter-5.

Q.11. Write expressions for equipvalent STAR-CONNECTED RESISTANCES for given DELTA CONNECTED RESESTANCES (Fig. 1.5)

Ans. Refer Fig. 1.5.

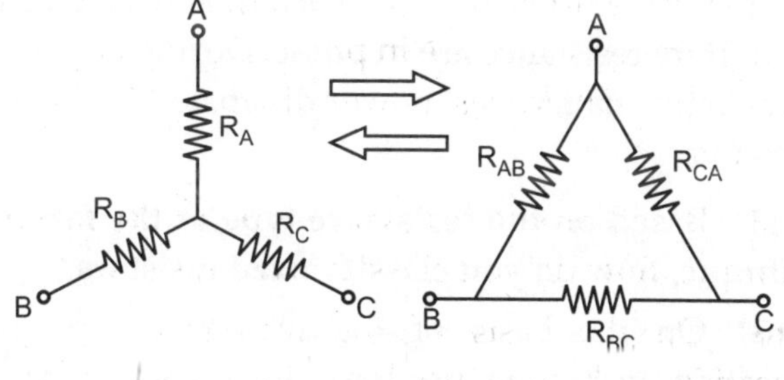

Fig. 1.5

$$R_A = \left(\frac{R_{AB} \cdot R_{CA}}{R_{AB} + R_{BC} + R_{CA}}\right);$$

$$R_B = \left(\frac{R_{BC} \cdot R_{AB}}{R_{AB} + R_{BC} + R_{CA}}\right);$$

$$R_C = \left(\frac{R_{CA} \cdot R_{BC}}{R_{AB} + R_{BC} + R_{CA}}\right).$$

These expressions are also derived in chapter-5.

Q.12. How does a resistor behave in an *AC* circuit? How is the voltage impressed across the resistor, current flowing through it and the resistance related? What would be the magnitude of power dissipated in the resistor?

Ans. A pure resistance (the one that has no stray inductance and stray capacitance) behaves in the same manner in an *AC* circuit as it would do in a *DC* circuit. Ohm's law is equally valid in this case too. If a voltage $v = V_m \sin\omega t$ is impressed across a resistance (R), then the current (i) flowing through it is given by $i = I_m \sin\omega t$ where $I_m = V_m/R$. Power dissipated in (R) in this case is given by the product of *RMS* values of voltage and current.

That is, $P = V_{rms} \times I_{rms}$ where $V_{rms} = V_m/\sqrt{2}$ and $I_{rms} = I_m/\sqrt{2}$

Also, $P = V_{rms}^2/R = I_{rms}^2 R$

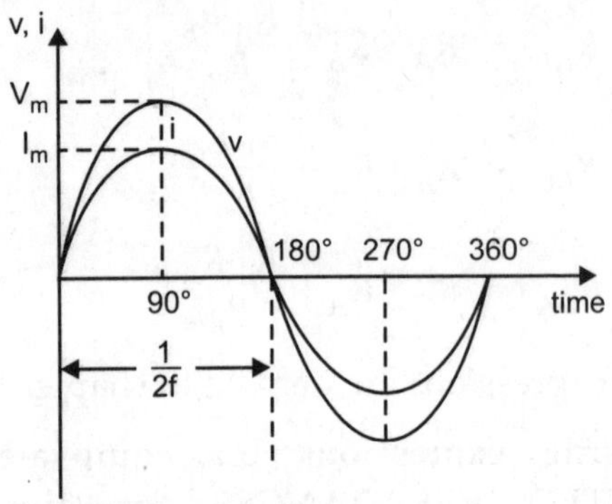

Fig. 1.6

Fig. 1.6 shows the relevant waveforms. It is evident from the waveforms also that current and voltage in the case of pure resistance are in phase. During both positive and negative half cycles, power dissipated in the resistor is positive.

Q.13. Based on the resistance type or the fabrication technique, how do you classify fixed resistors?

Ans. On the basis of the resistance type or the fabrication technique used for its manufacture, fixed resistors are classified as:

1. Carbon composition resistors
2. Carbon film resistors
3. Metal film resistors
4. Wire wound resistors

The resistance element in carbon composition resistors is a slug of resistive material (carbon in this case) whereas in film resistors (both carbon film and metal-film), it is a relatively thin layer of resistive material painted or deposited on a ceramic or a glass-tube. The resistive material is carbon in carbon-film and metal, metal-alloy or metal oxide in case of metal-film resistors. Resistive element in a wire wound resistor is a length of wire wound on an insulating core in one or more than one layers. Fig. 1.7 shows the different types of resistors.

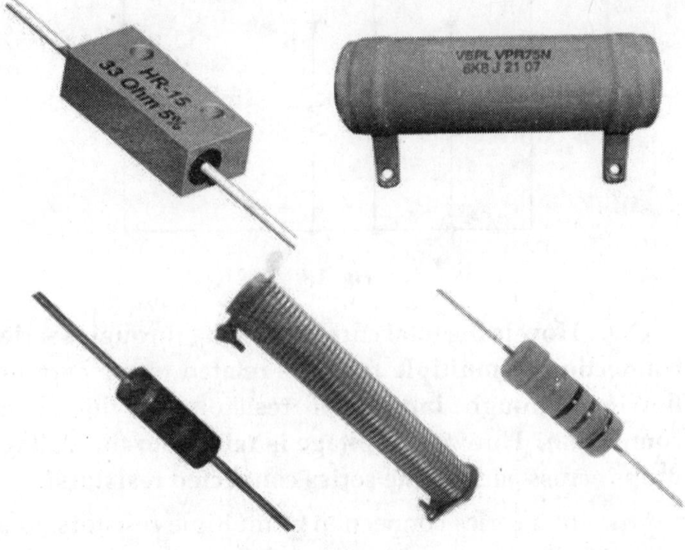

Fig. 1.7

Q.14. Briefly describe the major electrical specifications of a resistor?

Ans. The major electrical specifications of a resistor are:

1. Resistance value
2. Tolerance
3. Wattage rating
4. Voltage rating
5. Temperature co-efficient of resistance

The RESISTANCE VALUE is the resistance in ohms offered by the resistor. It is either marked or coded on the resistor.

TOLERANCE is the permissible plus or minus resistance deviation in the nominal resistance value. For instance, the actual resistance value of a 100 $K \pm$ 10% resistor may be anywhere between 90 K and 110 K. Fixed resistors are available in wide range of tolerance specifications form ± 20% to ± 0.01%.

WATTAGE RATING or the POWER RATING is the maximum power in watts that the resistance can safely dissipate at the specified ambient temperature. The wattage rating must be derated as per the curve supplied by the manufacturer in case the operational temperature is higher than the one for which the power rating has been specified. Carbon composition and film resistors are available in the wattage ratings of 1/8 W, 1/4 W, 1/2 W, 1 W and 2 W. Wire wound resistors are available in relatively much higher power ratings.

VOLTAGE RATING or the RATED CONTINUOUS WORKING VOLTAGE of a resistor is given by $\sqrt{W.R}$

where (W) is the Wattage rating in watts and (R) is resistance value in ohms. Also, there is an upper limit for the voltage rating in case of resistors of different power ratings and the resistor types. In case of general purpose carbon composition resistors, this limit is 150V (for 1/8 W,) 250 V (for 1/4 W), 350 V (for 1/2 W,) 500 V (for 1 W) and 750 V (for 2 W). For general purpose carbon film resistors, the values are more or less identical. Thus the voltage rating of a given resistor is given by $\sqrt{W.R}$ or the specified upper limit whichever is less. For instance, a 1/4 W, 100 K carbon composition resistor can sustain a voltage of about 160 volts whereas a 1 M resistor of the same wattage rating has a maximum continuous working voltage of 250 volts.

This implies that a 1 watt resistor with its wattage rating specified at an ambient of 25°C would be capable of safely dissipating a power of less than 1 watt at all ambient temperatures of greater than 25°C. Another interesting observation is that a 2 watt resistor may have a lower power dissipation capability when operated at 70°C than a 1 watt resistor operated at 30°C if their power ratings have been specified at an ambient temperature of 25°C.

TEMPERATURE CO-EFFICIENT OF RESISTANCE is a parameter indicating the rate of change of nominal resistance value as a function of temperature. It is expressed as ppm/°C or % per °C.

Q.15. How does a typical power derating curve for a resistor look like? Mark the important points on this curve and briefly explain their significance.

Ans. Fig. 1.8 shows the typical power derating curve (*i.e.* graph showing variation of maximum power dissipation capability of the resistor as a function of ambient temperature). As is clear from the curve, the power dissipation capability remains equal to the specified value upto a certain temperature and then it falls linearly till it becomes zero at a temperature called the HOT SPOT TEMPERATURE. You should always remember that at any given temperature a resistor can safely dissipate only that much power whose heat equivalent equals the heat loss to the ambient by mechanisms like conduction, convection and radiation. In the limiting case, if the ambient temperature equals the hot spot temperature, the power dissipation capability of the resistor becomes zero. The hot-spot temperature is typically 150°C for film and composition resistors. It is in the range of 250°C to 350°C for wire wound types.

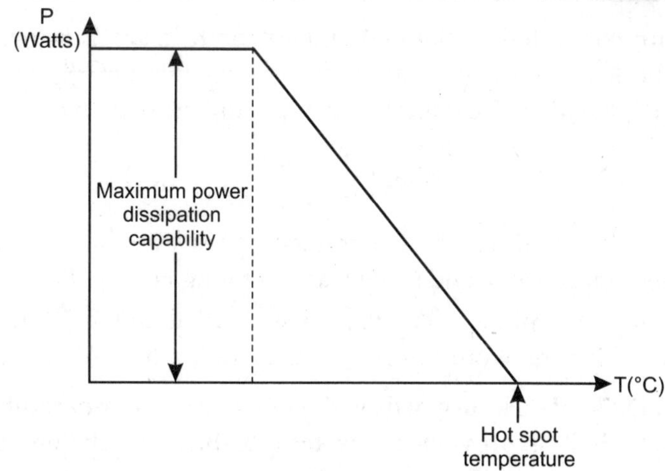

Fig. 1.8

Q.16. Resistors are often connected in series or in parallel to get an overall resistance of desired value and also sometimes the desired wattage rating. How would you know the wattage rating of a series or parallel combination of more than one resistor if the wattage ratings of individual resistors were known to you?

Ans. When resistors R_1, R_2, R_3, ... R_n with wattage ratings W_1, W_2, W_3..... Wn respectively are connected in series, the effective wattage rating of the series combination can be computed from:

$$W = \frac{W_m}{R_m}(R_1 + R_2 + R_3 + ... + R_n)$$

where $1 \leq m \leq n$ so that (W_m/R_m) has the least value.

For instance, if R_1, R_2, R_3 and R_4 are connected in series and we find that (W_3/R_3) has the least value amongst (W_1/R_1), (W_2/R_2), (W_3/R_3) and (W_4R_4), then overall wattage rating would be given by

$$W = \frac{W_3}{R_3}(R_1 + R_2 + R_3 + R_4)$$

An important corollary of this expression is that if all series connected resistors have same resistance value and same wattage rating, then the overall wattage rating of (n) such series connected resistors would be (nW) where $W_1 = W_2 = W_3 = W_4 ...... = W_n = W$.

In case of parallel connection of more than one resistors, the effective wattage rating is given by:

$$W = W_m R_m \left(\frac{1}{R_1} + \frac{1}{R_2} + \frac{1}{R_3} + ... + \frac{1}{R_n}\right)$$

where (m) is an integer $(1 \leq m \leq n)$ so that $(W_m \cdot R_m)$ has the least value. As an example, if R_1, R_2, R_3, R_4, are connected in parallel and we find that $(W_2 R_2)$ has the least value amongst $(W_1 R_1)$, $(W_2 R_2)$, $(W_3 R_3)$ and $(W_4 R_4)$, then the overall wattage rating is given by:

$$W = W_2 R_2\left(\frac{1}{R_1} + \frac{1}{R_2} + \frac{1}{R_3} + \frac{1}{R_4}\right)$$

Again, if all parallel connected resistors have same resistance value and same wattage rating i.e. $R_1 = R_2 = \ldots\ldots = R_n = R$ and $W_1 = W_2 = \ldots\ldots = W_n = W$, then the parallel combination would have a wattage rating of (nW).

Q.17. Resistance value depends upon temperature, which is expressed as its temperature co-efficient of resistance briefly described earlier. This variation in resistance occurs due to temperature dependence of the resistivity of the material of which the resistance is made. Which mathematical expression explains the temperature dependence of the resistivity of conducting materials?

Ans. Resistivity (ρ) varies as a function of temperature according to

$$\rho = \rho_o (1 + \alpha \Delta T)$$

where

α = Temperature coefficient of resistivity

ΔT = Change in temperature

ρ_o = Initial resistivity

Q.18. When two resistors with known tolerance specifications are connected in series or in parallel, how do we compute the tolerance specification of the series or parallel combination?

Ans. In case of series connection of resistors R_1, R_2 with respective tolerance specifications as T_1, T_2, the tolerance (T) of the series combination is given by:

$$T = \pm\left[\frac{T_1 R_1 + T_2 R_2}{R_1 + R_2}\right] \times 100,$$

T_1, T_2 are expressed as fraction.

In case of parallel connection,

$$T = \pm\left[\frac{(R_1 + R_2).(1 + T_1).(1 + T_2)}{R_1(1 + T_1) + R_2(1 + T_2)} - 1\right] \times 100,$$

T_1, T_2 are expressed as fraction.

The generalised expression for series connection of (n) resistors is given by:

$$T = \pm\left[\frac{T_1 R_1 + T_2 R_2 + T_3 R_3 + \ldots\ldots + T_n R_n}{R_1 + R_2 + \ldots\ldots + R_n}\right] \times 100$$

The expression for (n) parallel connected resistors is too cumbersome to be given here. For knowing the tolerance specification of series or parallel connected resistors, following inferences can be drawn from the above given expressions.

1. In case the series or parallel connected resistors have identical tolerance specification; the tolerance specification of the equivalent resistance remains unchanged.
2. In the case of series connected resistors with unequal tolerance specifications; the tolerance specification of the equivalent resistance is closer to that of the largest resistance value. As an illustration, if (R_1) (with tolerance T_1) is connected in series with (R_2) (with tolerance T_2) and $R_2 >> R_1$; then tolerance of equivalent series resistance $(= R_1 + R_2)$ will be approximately equal to (T_2).
3. In the case of parallel connected resistors with unequal tolerance specifications; the tolerance specification of the equivalent resistance is closer to that of the smallest resistance value. As an illustration, if (R_1) (with tolerance T_1) is connected in parallel with (R_2) (with tolerance T_2) and $R_2 >> R_1$; Then tolerance of equivalent parallel resistance $[= R_1 . R_2/(R_1 + R_2)]$ will be approximately equal to (T_1).
4. In case the series or parallel connected resistors have same resistance values but different tolerance specifications; the tolerance of the equivalent resistance is equal to the average of the tolerance values of individual resistors. As an illustration, if 100K, ± 10% resistor was connected in parallel with a 100K, ± 20% resistor; the equivalent resistance of 50K resistance value shall have tolerance of ± 15%.

Q.19. Based on the resistor's tolerance specification, how do you classify resistors?

Or

Distinguish between General purpose resistors, Semi-precision resistors, Precision resistors and Ultra-precision resistors.

Ans. Based on the tolerance specification, resistors are classified as:

1. General purpose resistors having a tolerance specification range of ± 20% to ± 5%.

2. Semi-precision resistors having a tolerance specification range of ± 5% to ± 1%.
3. Precision resistors having a tolerance specification range of ± 1% to ± 0.1%.
4. Ultra-precision resistors having a tolerance specification range of ± 0.1% to ± 0.01%.

General purpose resistors are available in all types (carbon composition, carbon-film, metal-film, wire wound). Semi-precision resistors are available in carbon-film, metal-film and wire wound varieties while precision and ultra-precision resistors are made only in metal-film and wire wound varieties.

Q.20. Briefly describe the advantages and disadvantages of carbon composition resistors.

Ans. The resistance element in a carbon composition resistor is a compressed slug of a finely ground mixture of carbon and silica. The significant advantages of composition resistors include:

1. Extremely small inherent inductance and capacitance
2. Do not fail catastrophically
3. Can withstand higher voltages than the film resistors of standard configuration

The Disadvantages include :

1. These are comparatively noisier
2. Less stable (*i.e.* these have a higher temperature co-efficient of resistance)
3. Their *AC* resistance reduces with increase in frequency and it drops at a faster rate for higher value of resistors. For instance, at 1 MHz, a 100 K resistance is typically 99 K. while a 1 M resistance is approximately 770 K and a 10 M resistance may be as small as 3.8 M.

Q.21. Outline advantages and disadvantages of carbon-film resistors. How do they compare with carbon composition resistors?

Ans. In carbon-film resistors, the resistance element is a thin layer of carbon on the surface of a ceramic or glass rod or tube. Except for low value resistances (upto 1K), higher value resistances are fabricated by forming a film in a spiral shape.

Carbon-film resistors have the following advantages

1. They have better stability as compared to their carbon composition counterparts.
2. They have comparatively better tolerance specifications (± 5.0%).
3. They have a wide resistance range (1 ohm to 100 mega ohms)

Carbon-film resistors have the following disadvantages

1. They are prone to catastrophic failure. The thin film is fragile and has no ability to withstand electrical over loads.
2. They have comparatively higher inherent inductances.

Q.22. What are wire wound resistors? For which applications are they preferred over carbon-film or carbon composition resistors? What are non-inductive type of wire wound resistors?

In the answer to the above question, you will also find answers to the following.

(*a*) **Difference between inductive and non-inductive type of wire wound resistors and**

(*b*) **Different classes of wire wound resistors on the basis of precision of resistance value.**

Ans. Wire wound resistors are fabricated by winding one or more than one layer of a high resistance wire. The fragile windings are protected against mechanical and environmental hazards by dipped, sprayed, molded or rigid cover of high temperature silicone, inorganic cement, vitreous enamel, plastic or ceramic materials.

Wire wound resistors are preferred over carbon composition resistors in low resistance, low noise applications. Precision wire wound resistors are preferred over film resistors when there is a need to handle transient overloads. In fact, for a power rating of more than 2 watts, there is hardly any other choice.

These are manufactured in a wide variety of resistance values, tolerances and power ratings. Due to the very nature of fabrication technique, wire wound resistors are likely to have a large inductance and shunt capacitance. It may not pose any serious problem upto 25 kHz. For high frequency applications, non-inductive type wire wound resistors should be used. These are fabricated by having two windings with same number of turns wound in opposite directions and connected in parallel (Fig. 1.9).

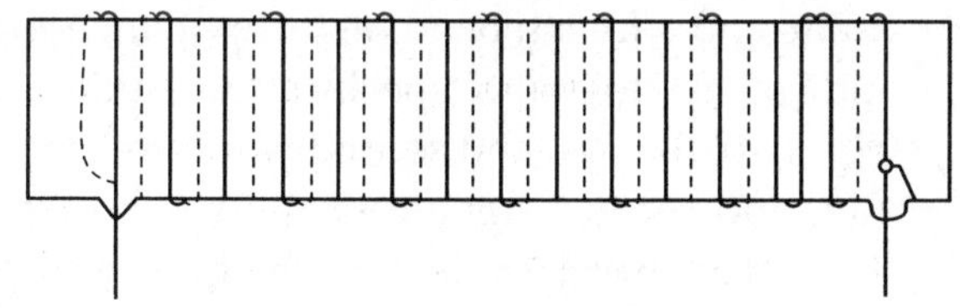

Fig. 1.9

General purpose wirewound resistors available in power rating of 1 watt to 10 watts, have a resistance range of 1Ω to 200 kΩ, tolerance specification of ± 5%, ± 10%, maximum operating temperature of upto 350°C (in ceramic flameproof molded type) and a temperature co-efficient of ± 20 ppm/°C to 200 ppm/°C.

Semiprecision wire wound resistors available in power rating of 1/4 watt to 20 watt, have a resistance range of 0.1Ω to 300 kΩ, tolerance specification of ± 5% to ± 0.05%, maximum operating temperature of upto + 350°C (in case of silicone coated conformal) and a temperature co-efficient of ± 20 ppm/°C.

Precision and ultra-precision wire wound resistors are usually available in the power ratings ranging from 1/10 watt to 100 watt, have a resistance of 0.1Ω to 20 M Ω and a tolerance as good as ±0.05%. The temperature co-efficient could be as small as ± 1 ppm°C (in case of Bobbin type molded Epoxy) and the maximum operating temperature as high as 275°C. In case of Bobbin molded epoxy type, maximum operating temperature is typically 150°C.

The power rating of high power wire wound resistors could be as high as 500 W with a resistance range of 0.1Ω to 100kΩ.

Q.23 Outline the salient characteristics of metal film resistors.

Ans. These are similar to carbon-film resistors in construction except that the film is of a metal or a metal alloy or a metal oxide. These have the advantages of both carbon-film and precision wire wound resistors without having the disadvantages of either of the two types.

In some metal film resistor types, the tolerance specification could be as good as ± 0.01% and temperature co-efficient as small as ± 2 ppm/°C.

Q.24. Outline the procedure for colour coding of

(*i*) General Propose Fixed Resistors,

(*ii*) Semi-precision Fixed Resistors.

Ans. (*i*) The general purpose carbon composition resistors are colour coded by using three bands on the extreme left to indicate the resistance value and a fourth band on the extreme right to indicate the tolerance specification. As shown in Fig. 1.10, the first and second bands indicate the first and second significant digits respectively. The third band gives the decimal multiplier factor. Gold and Silver coloured tolerance bands indicate tolerance specifications of ± 5% and ± 10% respectively whereas an absence of tolerance band indicates ± 20% tolerance. Fig. 1.10 and the accompanying table are self explanatory.

Colour	*Digit*	*Multiplier*	*Tolerance*
Black	0	1	–
Brown	1	10	–
Red	2	100	–
Orange	3	1000	–
Yellow	4	10000	–
Green	5	100000	–
Blue	6	1000000	–
Violet	7	10000000	–
Grey	8	–	–
White	9	–	–
Gold	–	0.1	±5%
Silver	–	–	± 10%
No Colour	–	–	± 20%

(*ii*) Fig. 1.11 gives the colour code scheme for ± 1% and better tolerance resistors. The resistance value in this case is represented by first four bands with the first three bands giving three significant digits and the fourth colour band providing the decimal multiplier. The digit

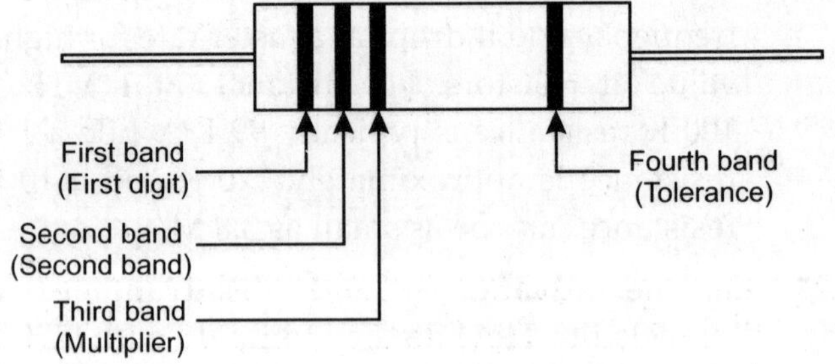

Fig. 1.10

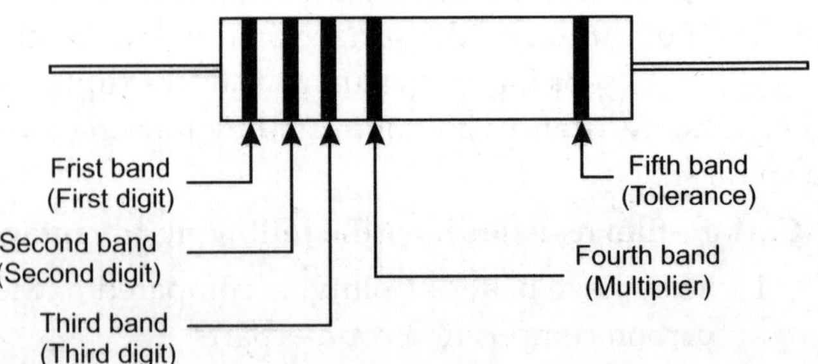

Fig. 1.11

and the multiplier codes in this case are the same as the ones just outlined for general purpose resistors and depicted in Fig. 1.10. The tolerance code is given in Fig. 1.12 First band if wider than the other bands indicates a wire wound resistor.

Black	–	Blue	±0.25%
Brown	±1%	Violet	±0.1%
Red	±2%	Grey	±0.05%
Orange	–	White	–
Yellow	–	Gold	±5%
Green	±0.5%	Silver	±10%

Fig. 1.12

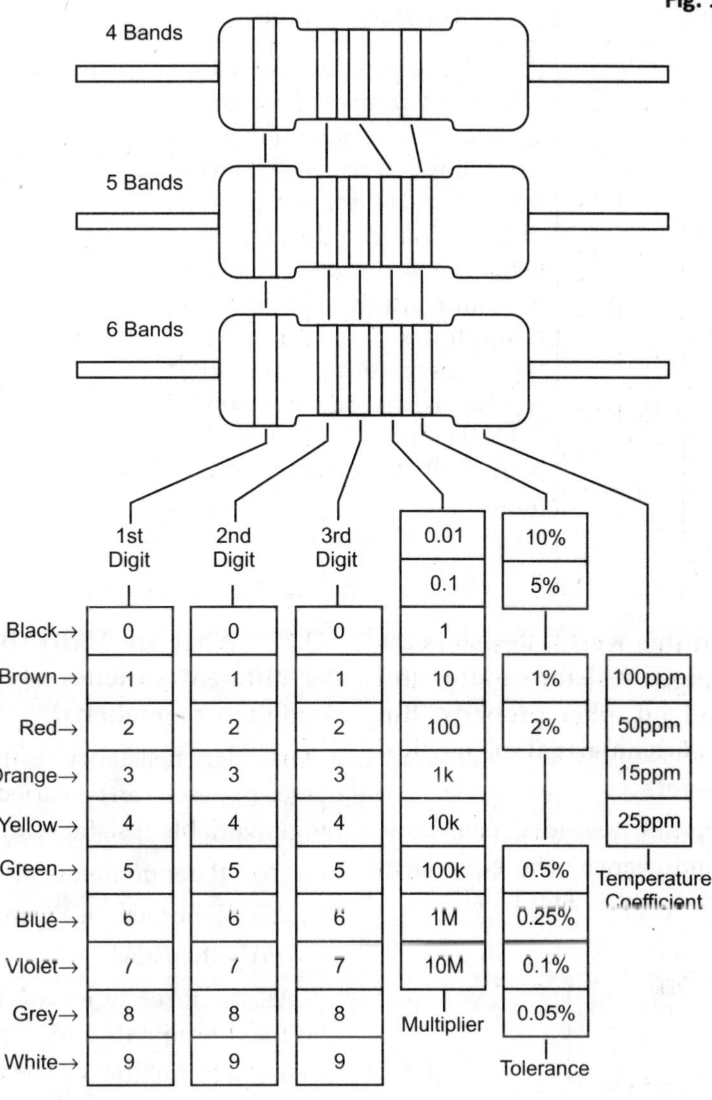

Fig. 1.13

A six-band colour code is also used in the case of precision resistors. The sixth band on the extreme right in this case represents the temperature co-efficient of resistance in ppm/°C. Temperature co-efficients of 100 ppm/°C, 50 ppm/°C, 15 ppm/°C and 25 ppm/°C are respectively represented by brown, red, orange and yellow coloured sixth band. Four, five and six-band colour coding schemes of fixed resistors are summarized in Fig. 1.13.

Q.25. Sometimes the resistors are coded with a certain numerical code known as Industrial type designation. Explain with the help of an illustration the industrial type designation of fixed resistors.

Ans. In this type of coding or type designation, starting from left, the first two alphabet represent the power rating (for instance BB indicates 1/8 watt) at 70°C, the three digits after the first two alphabet indicate the resistance value with first two digits giving the two significant digits and the third providing the decimal multiplier or the number of 0s to be added and finally another digit after these three indicates the tolerance specification.

Take case of a resistor with the industrial type designation CB 5645. Interpretation of the specifications of this resistor are illustrated in the accomparying table. The resistance in this case is a 560K, +5% resistor.

CB 564 5

Power Rating At 70°C				**Resistance Value**	**Tolerance Code**		
Type No.	*Power Rating (Watts)*	*MIL-R-11 Style*	*MIL-R-39008 Style*	First two digits represent the significant figures. Last digit specifies the number of zeroes to follow. For resistances less than 10 ohms, in the industrial type designation, letter (G) subtituted in place of third digit signifies a multiplier factor of 0.1. In MIL type designation, letter (R) represents the decimal point. For instance, 27 G and 2R7 both mean a resistance of 2.7 ohms	*Industrial Type Designation*	*Tolerance*	*MIL Designation*
BB	1/8	RC05	RCR05		5	± 5%	J
CB	1/4	RC07	RCR07		2	± 20%	M
EB	1/2	RC20	RCR20		1	± 10%	K
GB	1	RC32	RCR32		–	± 2%	G
HB	2	RC42	RCR42		–	± 1%	F
GM	3	–	–		–	± 0.5%	D
HM	4	–	–		–	± 0.25%	C
					–	± 0.1%	B

Q.26. Nothing is perfect in this world. Resistors are no exception. Ideally speaking, a resistor should offer only electrical resistance and no other property. But that is never the case. How does an actual or practical resistor differ from an ideal one?

Ans. For all practical purposes, resistors have some associated capacitance and inductance. The equivalent circuit of a practical resistor is shown in Fig. 1.14.

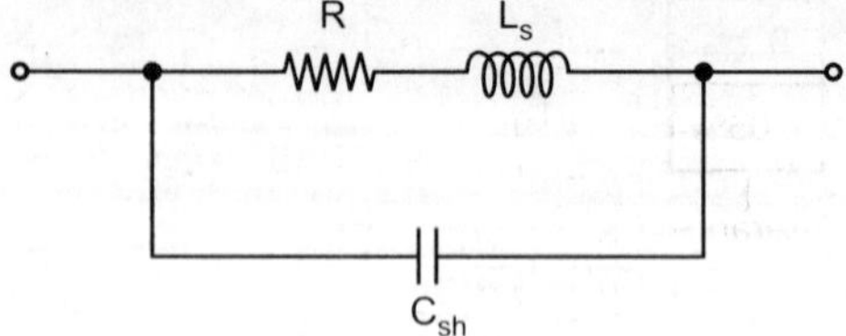

Fig. 1.14

The values of capacitance and inductance depend upon the type and dimensions of the resistor. Cylindrical film resistors and wire wound resistors have high series inductance. Composition resistors have negligible inductance and capacitance. In general, inductance and capacitance values increase with increase in dimensions of the resistor. Surface mount varieties have negligible inductance and capacitance.

Q.27. What are VARIABLE RESISTORS? What are the different varieties of variable resistors and their circuit representations?

Ans. The resistance value of variable resistors, as the name suggests, can be varied over the specified resistance range. Variable resistors can be classified as

(*i*) Potentiometers or simply Pots

(*ii*) Trimmers or Presets

(*iii*) Rheostats

All the three types are three terminal resistors with two fixed terminals and a third terminal that is attached to a movable tap that can be moved along the length of the resistive element thus varying the resistance between the movable terminal and the fixed terminal at either of the two ends of the resistance element. For all practical purposes, these three types are simply and collectively known as potentiometers. Specifically, a POTENTIOMETER is a variable resistance designed for frequent and sometimes continuous movement of adjustable terminal. TRIMMERS (or Presets) are designed for occasional movement of the tap. When connected in a circuit, these are usually varied and then set for optimum performances. RHEOSTAT is a wire wound variable resistance of higher wattage and is primarily used for current limiting applications. Fig. 1.15

(*a*), (*b*) and (*c*) respectively show circuit representations of potentiometers, presets and rheostats.

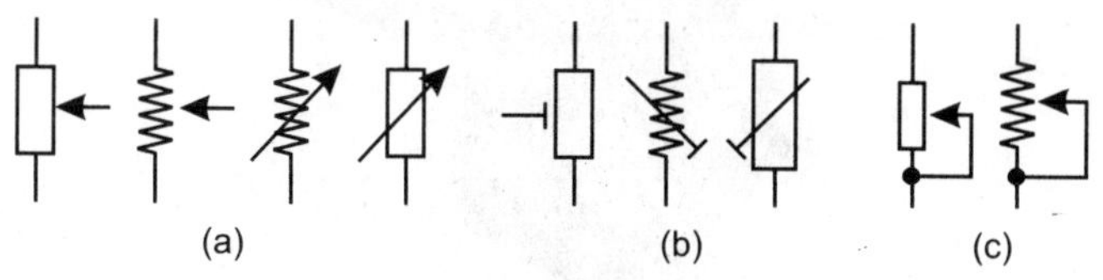

Fig. 1.15

Q.28. Define the following with the reference to variable resistors:

(*a*) Total resistance

(*b*) Resolution

(*c*) Linearity

(*d*) Precision

(*e*) Equivalent Noise Resistance

(*f*) Temperature co-efficient

Ans. (*a*) TOTAL RESISTANCE is the resistance measured between the end terminals of the potentiometer.

(*b*) RESOLUTION is the smallest incremental resistance change that is possible in a wire wound potentiometer. It is a function of diameter and the inter-turn spacing of the wire used for making resistance element. Carbon composition, conductive plastic and cermet potentiometers for all practical purposes have infinite resolution.

(*c*) LINEARITY is the deviation of the output vs. rotation characteristics from a straight line. If the straight line considered is drawn between zero voltage and full voltage points, the linearity is termed as ABSOLUTE LINEARITY. If the deviation is measured around the best possible straight line drawn from zero through output vs. rotation characteristics curve, it is the ZERO BASED LINEARITY. It is expressed as a percentage.

(*d*) PRECISION is an indication of linearity. Better the linearity, more precise the potentiometer is. Precision potentiometers have a linearity of 1% or better whereas in case of semi precision potentiometers, it is 1% to 3%. Linearity of 3% or worse is an indication of non-precision potentiometer.

(*e*) EQUIVALENT NOISE RESISTANCE gives an idea about the electrical noise generated while adjusting a wire wound potentiometer. It is expressed in terms of an equivalent maximum resistance in series with the wiper.

(*f*) TEMPERATURE CO-EFFICIENT is the variation in the total resistance of the potentiometer as a function of temperature. It is expressed as a change in the resistance per degree change in temperature.

Q.29. Write short notes on the following:

(*a*) Carbon Composition Potentiometers

(*b*) Conductive Plastic Potentiometers

(*c*) Cermet Potentiometers

(*d*) Wire wound Potentiometers

(*e*) Rheostats.

Ans. (*a*) Carbon Composition Potentiometers (Fig. 1.16)

These potentiometers have two types of constructions (*i*) coated film type in which a mixture of carbon, some filler material and a binder is coated on the periphery of an insulating material and (*ii*) the molded type in which the mixture is molded in the form of a cavity in a plastic base.

Molded composition potentiometers, due to the fact that these are totally enclosed and sealed against moisture and other atmospheric contamination, are more rugged than film type carbon composition potentiometers. These are available in both linear as well as logarithmic varieties. Due to low inductance and capacitance, they are suitable for high frequency applications. Maximum operating temperature is typically at 120°C.

Figs 1.16 to 1.20 show photographs of some popular types of potentiometers and rheostats.

Fig. 1.16

Fig. 1.17

(b) **Conductive Plastic Potertiometers (Fig. 1.17)**

The resistance element in conductive plastic potentiometers is a carbon resin mixture. They are also of the molded type or film type. Their characteristics are more or less similar to those of carbon composition potentiometers. Maximum operating temperature is again typically 120°C. They have comparatively lower power ratings. They are available in power rating of upto 1/2 W against 2 W in case of carbon composition potentiometers.

(c) **Cermet Potentiometers (Fig. 1.18)**

Cermet potentiometers, where the resistance element is a hybrid mixture of a ceramic and a metal, are immune to humidity and have better temperature co-efficient than either carbon composition or the conductive plastic types. Maximum operating temperature is typically ± 175°C and are available in power ratings of upto 5 watts.

(d) **Wire wound Potentiometers (Fig. 1.19)**

In wire wound potentionmeters, the resistive element is usually a single layer of bare resistance wire wound on an insulating core. The resistance of a potentiometer is a function of the wire diameter, wire material, core cross-section, core length and the adjacent turn spacing. It may be mentioned here that although fixed wire wound resistors can be wound to get a non-inductive variety, it is not so in case of wire wound potentiometers. So, these are basically high resistance inductors. Their strong points are better tolerance, better temperature coefficient (especially in low resistance values). Moreover, if you are looking for a very low resistance potentiometer (few ohms) wire wound potentiometer is perhaps the only choice. These are no good for high frequency applications and are seldom used beyond several tens of kilohertz.

Fig. 1.18

Fig. 1.19

(e) **Rheostats (Fig. 1.20)**

A rheostat is a high wattage potentiometer of wire or ribbon wound type generally used as a two terminal variable resistor for current limiting applications. Rheostats are available in the power rating specifications of 5 W to more than 50 W. The size of the wire or ribbon to be used for making the resistive element is governed by the desired power rating.

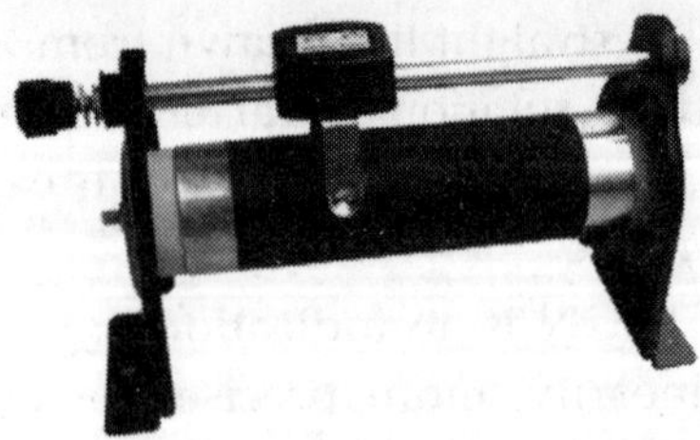

Fig. 1.20

Q.30. What are TRIMMER POTENTIOMETERS? Distinguish between MULTI-TURN TRIMMERS and MULTI-TURN POTS.

Ans. Trimmer potentiometers are variable resistors that are varied and then set for best performance. Trimmer pots are not meant for frequent adjustment as in the case of regular potentiometers. They are of two basic types (*i*) Single turn trimmers (*ii*) Multi-turn trimmers. Multi-turn trimmer differs from a multiturn potentiometer in the sense that here the multi-turn does not mean a long resistance wire wound in a spiral form. In multi-turn trimmers, multi-turn means that the slider is actuated either by a threaded screw or a worm gear. Fig. 1.21 shows multi-turn trimmers (Fig. 1.21 *a*) and multiturn potentiometers (Fig. 1.21 *b*).

Q.31. What are non-linear resistors? Both Thermistors and Varistors are non-linear resistors, How do they differ from each other?

Fig. 1.21

Ans. In a non-linear resistor, the current through it does not vary according to Ohm's law. It varies as some function of voltage in case of VOLTAGE DEPENDENT RESISTORS (also called VARISTORS) or the body temperature in case of THERMISTORS. There are thermistors with negative temperature co-efficient called NTC thermistors as well as thermistors with positive temperature co-efficient called PTC thermistors. VDRs are essentially non-linear resistors with negative coefficients.

Q.32. Define the terms TIME CONSTANT, DISSIPATION CONSTANT and BREAK POINT for thermistors.

Ans. When a step change in temperature is impressed across a thermistor, the time taken by the thermistor to indicate that change equal to 63% of the temperature difference between the step function impressed new temperature and the original temperature is called the **Time Constant.**

Dissipation Constant is the power in milliwatts that will raise the thermistor temperature by 1°C. The measurement is normally made at 25°C in a specified mounting arrangement.

Break Point is defined only with reference to PTC thermistors. It is a point on the rapidly rising portion of the resistance-temperature characteristics above the curie point at which the resistance of the PTC thermistor is 5 times the base resistance.

Q.33. Draw the typical Resistance vs Temperature characteristics of NTC and PTC thermistors.

Ans. Fig. 1.22 (*a*) shows the typical Resistance *vs.* Temperature characteristics of NTC thermistors. NTC thermistors, as is clear from the curve, have a smooth negative temperature-resistance characteristics that is usable over a wide temperature range. On the other hand, a PTC thermistor is usable over a much narrower temperature range but with a much greater sensitivity. Fig. 1.22 (*b*) shows typical Resistance *vs.* Temperature characteristics of PTC thermistors. In fact, PTC thermistors can be fabricated and are available for operation over selected temperature bands.

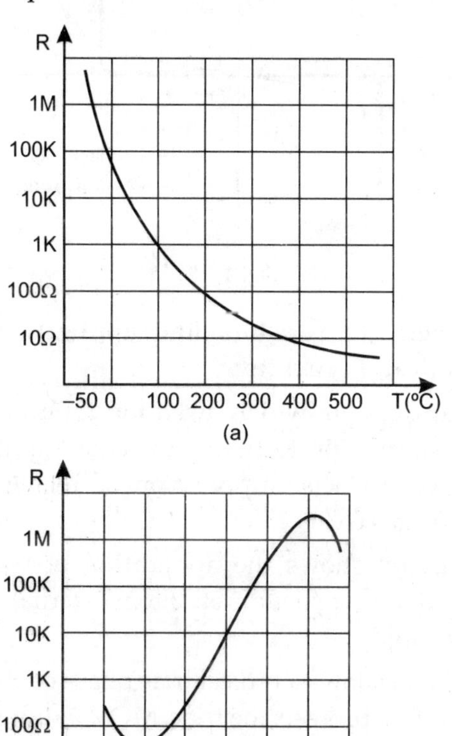

Fig. 1.22

Q.34. How do the Volt-Ampere characteristics of Silicon Carbide Varistors differ from those of Metal Oxide Varistors (MOV)?

Ans. Volt-Ampere characteristics of a varistor are expressed by

$i = K \cdot e^n$

e = instantaneous voltage

k = constant (in Amp/Volt)

n = an exponent whose value depends upon the type of varistor.

There are two types of varistors, namely the SILICON CARBIDE VARISTORS and the METAL OXIDE

VARISTORS. The value of exponent (n) for silicon carbide varistors varies between 1 and 6. It is 25 or more for metal oxide varistors. Higher value of exponent for metal oxide varistors make their I-V characteristics highly non-linear and thus result in a much sharper change from the high resistance region to the low resistance region compared to the silicon carbide varistors. The I-V characteristics for the two types are shown in Fig. 1.23.

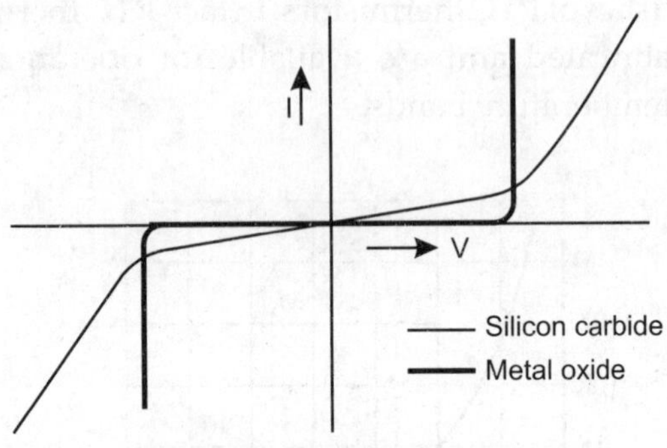

Fig. 1.23

Q.35. Where do varistors find application ? Give typical application circuits.

Ans. Varistors are widely used for applications such as power supply protection, protection of thyristors, protection of contacts, suppression of relay/contractor switching surges etc.

Fig. 1.24 (*a*) shows the application of varistors to protect the diodes, capacitors etc. From external transients in a power supply.

Fig. 1.24 (*b*) shows a typical triac phase control circuit in which MOV-1 protects the triac, MOV-2 suppresses the transients generated by the load and MOV-3 protects the circuits from power line transjents and surges.

In the circuit application of Fig. 1.24 (*c*), the varistors provide protection to the relay contacts. It has been observed that the contact resistance of the relays remain essentially unaltered even after million operations.

Fig. 1.24 (*d*) shows the use of a varistor to suppress surges generated by relays and contractors during switch-off operation.

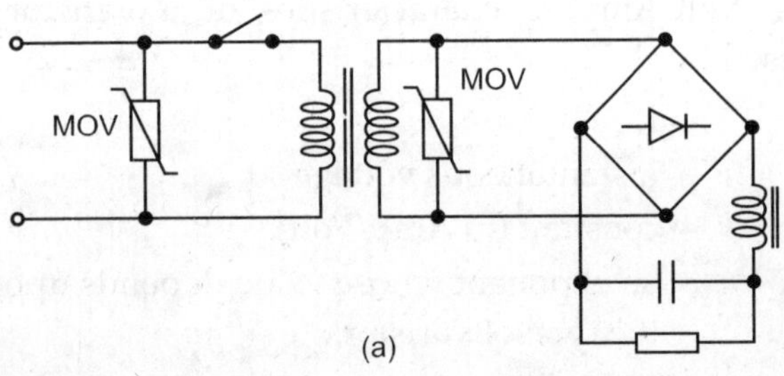

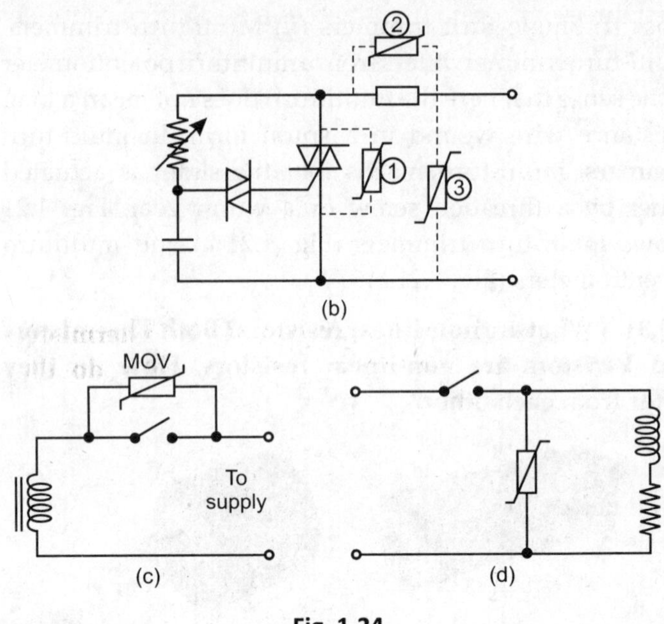

Fig. 1.24

Q.36. What are surface mount resistors?

Ans. Surface mount resistors or SMD resistors or chip resistors are small rectangular shaped resistors that can be soldered directly on to the surface of the printed circuit board (Fig. 1.25). They comprise of a thick layer of metal layer of metal oxide deposited on a ceramic substrate. The resistance value is governed by its length and thickness of metal oxide layer and type of metal oxide film. SMD resistors are precision resistors with very tight tolerance specifications. Tolerance levels down to ± 0.1% are feasible.

Fig. 1.25 Surface mount resistor

Q.37. How are SMD resistors coded?

Ans. SMD resistors are generally coded using a 3-digit or 4-digit numerical code. The first two digits in the 3-digit code represent the first two significant digits and the third digit represents the decimal multiplier. Resistance values less than 10 ohms have a letter 'R' to indicate the position of decimal point. As an illustration, 223, 4R7 and R23 would respectively represent 22K, 4.7Ω and 0.23Ω. In the 4-digit code, first three digits represent first three significant digits and the fourth digit represents the decimal multiplier. 3-digit and 4-digit codes are respectively used for ± 5% and ± 1% tolerance resistors.

Q.38. What are different sizes (or packages) the SMD resistors are fabricated in? Illustrate with the help of examples the commonly used codes for representing package size.

Ans. The package size of an SMD resistor is represented by its length (l), width (w) and height (h) as shown in Fig. 1.26. The size depends upon wattage rating. Two commonly used codes for representing package size are the Imperial Code indicating various dimensions in inches and the Metric Code indicating dimensions in millimeters. Both codes are 4-digit codes. In both codes, first two digits represent the length and the next two digits represent the width. In the case of Imperial Code, while deciphering the code, a decimal point is placed before the first two digits to indicate length in inches. A decimal point is placed before the next two digits to indicate width in inches. For example, 0603 package would be an SMD resistor with length equal to 0.06 inch and width equal to 0.03 inch. In the case of metric code designation, decimal point is placed between first and second digits to indicate length in mm and a decimal point is placed between third and fourth digits to indicate width in mm. For example, metric code designation 1005 implies a length of 1.0 mm and width of 0.5 mm. Imperial Code and Metric Code designations and their features are shown in the following table.

Fig. 1.26 SMD resistor package size

Code Imperial	*Code Metric*	*Length Inches*	*Length mm*	*Width Inches*	*Width mm*	*Height Inches*	*Height mm*	*Power Watt*
0201	0603	0.02	0.6	0.01	0.3	0.01	0.25	1/20
0402	1005	0.04	1.0	0.02	0.5	0.014	0.35	1/16
0603	1608	0.06	1.55	0.03	0.85	0.018	0.45	1/10
0805	2012	0.08	2.0	0.05	1.2	0.018	0.45	1/8
1206	3216	0.12	3.2	0.06	1.6	0.022	0.55	1/4
1210	3225	0.12	3.2	0.10	2.5	0.022	0.55	1/2
1218	3246	0.12	3.2	0.18	4.6	0.022	0.55	1
2010	5025	0.20	5.0	0.10	2.5	0.024	0.6	3/4
2512	6332	0.25	6.3	0.12	3.2	0.024	0.6	1

Q.38. What are the major sources of noise in resistors? Give relevant mathematical expressions to illustrate your answer?

Ans. There are two main sources of electrical noise associated with resistors. These are:

1. Thermal Noise (also called Johnson Noise or White Noise)
2. Current Noise

Thermal noise is due to thermal agitation of the charge carriers. RMS value of thermal noise voltage is given by:

$$V_{th}(RMS) = \sqrt{4\,KTRB}$$

where

T = Absolute temperature

R = Resistance value

B = Bandwidth over which noise is measured

K = Boltzmann's constant

Current noise is a low frequency noise caused by flow of charge carriers through the non-homogeneous material of the resistor. RMS value of the current noise voltage is given by:

$$V_i\,(RMS) = cf^{-\gamma}\,V^{\alpha}\,B$$

where

V = Applied DC voltage across the resistor

B = Bandwidth over which noise is measured

f = frequency

c, γ, α = constants

Value of (γ) is approximately unity due to which current noise is also referred to as ($1/f$) noise. It is insignificant for frequencies above 100 kHz.

Q.39. Which standard or preferred values the General purpose, semiprecision, precision and ultra-precision fixed resistors are available in?

Ans. The standard or preferred resistance values available in different power ratings and tolerance specifications in case of general purpose, semi-precision, precision and ultra-precision resistors are listed below. The values given below are as per EIA/MIL standards. The number of preferred resistance values depends upon the tolerance specification, these values (for one decade) for ± 20% tolerance (E6 series), ± 10% tolerance (E-12 series), ± 5% tolerance (E24 series) and ± 1% tolerance (E96 series) are given as under. Preferred resistance value list for one decade for ± 0.5%, ± 0.25% and ± 0.1% is also presented. The value given are for only one decade. All order decades are the same with appropriate multiplier.

General Purpose Resistors (± 20%)

1.0 Ω, 1.5 Ω, 2.2 Ω, 3.3 Ω, 4.7 Ω, 6.8 Ω

General Purpose Resistors (± 10%)

1 Ω, 1.2 Ω, 1.5 Ω, 1.8 Ω, 2.2 Ω, 2.7 Ω, 3.3 Ω, 3.9 Ω, 4.7 Ω, 5.6 Ω, 6.8 Ω, 8.2 Ω

Semiprecision Resistors (± 5%)

1 Ω, 1.1 Ω, 1.2 Ω, 1.3 Ω, 1.5 Ω, 1.6 Ω, 1.8 Ω, 2.0 Ω, 2.2 Ω, 2.4 Ω, 2.4 Ω, 2.7 Ω, 3.0 Ω, 3.3 Ω, 3.6 Ω, 3.9 Ω, 4.3 Ω, 4.7 Ω, 5.1 Ω, 5.6 Ω, 6.2 Ω, 6.8 Ω, 7.5 Ω, 8.2 Ω, 9.1 Ω

Precision Resistors (± 1%)

1.00 Ω, 1.02 Ω, 1.05 Ω, 1.07 Ω, 1.10 Ω, 1.13, 1.15, 1.18, 1.21, 1,24, 1.27, 1.30, 1.22, 1.37, 1.40, 1.43, 1.50, 1.54, 1.58, 1.62, 1.65. 1.69, 1.74, 1.78, 1.82, 1.91, 1.96, 2.00, 2.05, 2.10, 2.15, 2.21, 2.21, 2.26,2.32, 2.37, 2.43, 2.49, 2.55. 2.61, 2.67, 2.71, 2.74, 2.80, 2.87, 2.94, 3.01, 3.09, 3.16, 3.24, 3.32, 3.40, 3.48, 3.57, 3.65, 3.74, 3.83, 3.92, 4.02, 4.12, 4.22, 4.32, 4.42, 4.53, 4.64, 4.75, 4.87. 4.99, 5.11, 5.23, 5.36, 5.49, 5.62, 5.76, 5.90, 6.04, 6.09, 6.34, 6.49, 6.65, 6.81, 6.98, 7.15, 7.32, 7.50, 7.68, 7.87, 8.06, 8.25, 8.45, 8.66, 8.87, 9.09, 9.31, 9.53, 9.76

Ultra precision Resistors (± 0.5%, ± 0.25%, ± 0.1%)

1.00 Ω, 1.01 Ω, 1.02 Ω, 1.04 Ω, 1.05 Ω, 1,06, 1.07, 1.09, 1.10, 1.11, 1.13, 1.14, 1.15, 1.17, 1.18, 1.20, 1.21, 1.23, 1.24, 1.26, 1.27, 1.29, 1.30, 1.32, 1.33, 1.35, 1.37, 1.38, 1.40, 1.42, 1.43, 1.45, 1.47, 1.49, 1.50, 1.52, 1.54, 1.56, 1.58, 1.60, 1.62, 1.64, 1.65, 1.67, 1.69, 1.72, 1.74, 1.76, 1.78, 1.80, 1.82, 1.84, 1.87, 1.89, 1.91, 1.93, 1.96, 1.98, 2.00, 2.03, 2.05, 2.08, 2.10, 2.13, 2.15, 2.18, 2.21, 2.23, 2.26, 2.29,2.32, 2.34, 2.37, 2.40, 2.43, 2.46, 2.49, 2.52, 2.55, 2.58, 2.61, 2.64, 2.67, 2.71, 2.74, 2.77, 2.80, 2.84, 2.87, 2.91, 2.94, 2.98, 3.01, 3.05, 3.09, 3.12, 3.16, 3.20, 3.24, 3.28, 3.32, 3.36, 3.40, 3.44, 3.48, 3.52, 3.57, 3.61, 3.65, 3.70, 3.74, 3.79, 3.83, 3.88, 3.92, 3.97, 4.02, 4.07, 4.12, 4.22, 4.27, 4.32, 4.37, 4.42, 4.48, 4.53, 4.64, 4.75, 4.87, 4.99, 5.11, 5.23, 5.36, 5.49, 5.62, 5.76, 5.90, 5.97, 6.04, 6.12, 6.19, 6.26, 6.34, 6.42, 6.49, 6.57, 6.65, 6.73, 6.81, 6.90, 6.98, 7.06, 7.15, 7.23, 7.32, 7.41, 7.50, 7.59, 7.68, 7.77, 7.87, 8.16, 8.25, 8.25, 8.35, 8.45, 8.56, 8.66, 8.76, 8.87, 8.98, 9.09, 9.20, 9.31, 9.42, 9.53, 9.65, 9.76, 9.88

SOLVED PROBLEMS

Problem 1.1. *A given cylindrical conductor has a resistance of 10 ohms. What will be the resistance of another cylindrical conductor whose material resistivity is twice that of the first conductor. The length and area of cross-section are twice and one-fourth respectively as compared to the first conductor.*

Solution. Let us assume that (l_1 A_1, ρ_1) and (l_2, A_2, ρ_2) are respectively the length, area of cross-section and material resistivity of first and second conductors.

From the given data, therefore,

$$R_1 = \left(\frac{\rho_1 l_1}{A_1}\right) = 10; \text{ Also, } \rho_2 = 2\rho_1, l_2 = 2l_1 \text{ and}$$

$$A_2 = \frac{1}{4}A_1$$

Therefore

$$R_2 = \rho_2\left(\frac{l_2}{A_2}\right) = 2\rho_1\left(\frac{2l_1}{A_1/4}\right) = 16\rho_1\frac{l_1}{A_1} = 160\ \Omega$$

Problem 1.2. *Find the total resistance offered by various series parallel combinations shown in Fig. 1.27. All resistance values are in ohms.*

Solution: Fig. 1.27 (*a*): Total resistance

$$= 10 + 10 + \frac{10 \times 10}{10 + 10} = 25 \text{ ohms.}$$

Fig. 1.27 (*b*): Total resistance

$$= \frac{10 \times 10}{10+10} + \frac{10 \times 10}{10+10} + \frac{10 \times 10}{10+10} = 15 \text{ ohms.}$$

Fig. 1.27 (*c*): Total resistance

$$= \frac{1}{\left(\frac{1}{300} + \frac{1}{300} + \frac{1}{300}\right)} = \frac{1}{\left(\frac{1}{100}\right)}$$

$= 100$ ohms.

Fig. 1.27 (*d*): Total resistance

$$= \frac{1}{\left(\frac{1}{20} + \frac{1}{10} + \frac{1}{10}\right)} = \frac{1}{\left(\frac{5}{20}\right)} = 4 \text{ ohms.}$$

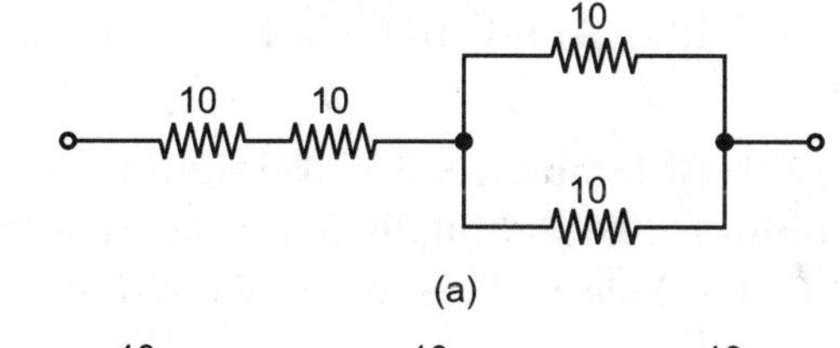

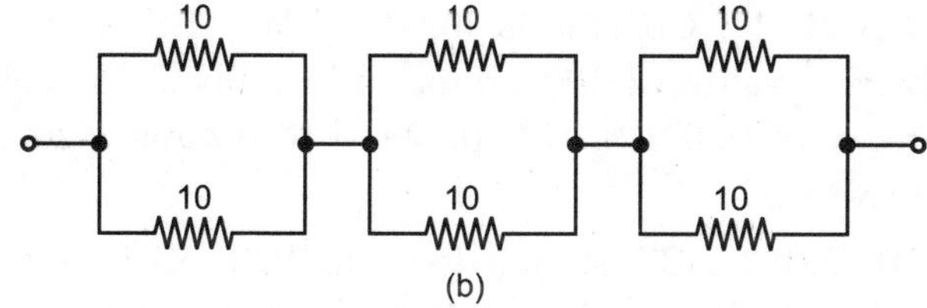

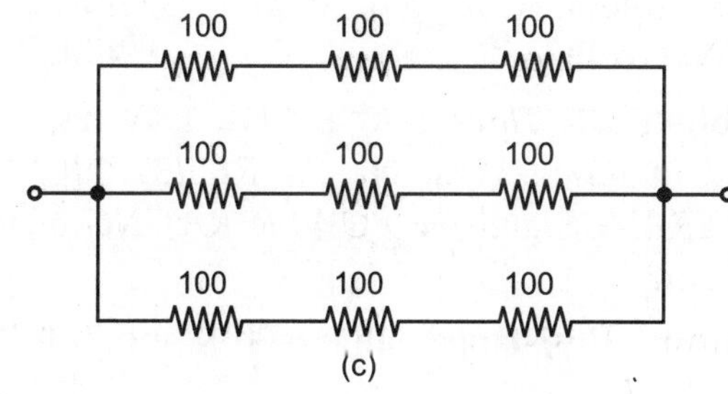

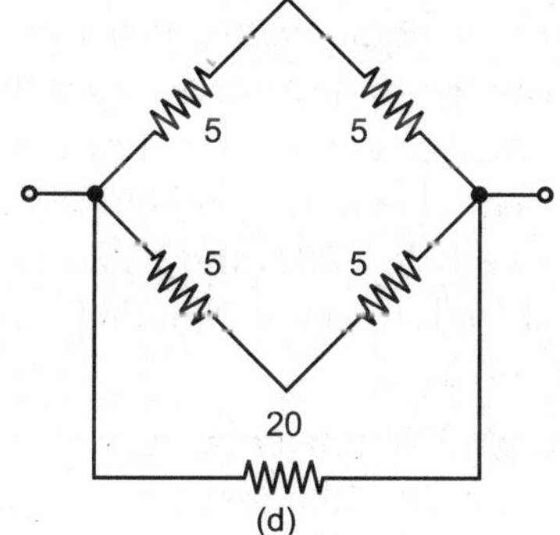

Fig. 1.27

Problem 1.3. *Justify that when more than one resistor are connected in parallel, the overall resistance is less than the one that has the least value.*

Solution. Let us take the case of two resistors R_1 and R_2 connected in parallel. The overall resistance (R) of this parallel combination is given by

$$R = \frac{R_1 R_2}{R_1 + R_2} = R_1\left(\frac{R_2}{R_1 + R_2}\right) = R_2\left(\frac{R_1}{R_1 + R_2}\right)$$

Now both $\left(\frac{R_2}{R_1 + R_2}\right)$ and $\left(\frac{R_1}{R_1 + R_2}\right)$ are less than 1.

Therefore R is less than R_1 as well as R_2 and hence the result.

Similarly, if R_1, R_2 and R_3 are connected in parallel, we can write the expression for overall resistance (R) as

$$R = \frac{R_1 R_2 R_3}{R_1 R_2 + R_2 R_3 + R_3 R_1}$$

$$= R_1\left(\frac{R_2 R_3}{Rl\, R_2 + R_2 R_3 + R_3 R_1}\right)$$

$$= R_2\left(\frac{R_3 R_1}{R_1 R_2 + R_2 R_3 + R_3 R_1}\right)$$

$$= R_3\left(\frac{R_1 R_2}{R_1 R_2 + R_2 R_3 + R_3 R_1}\right)$$

Thus R is less than R_1, R is less than R_2, R is less than R_3.

Therefore, we can conclude the following. When more than one resistances are connected in parallel, the overall resistance is always less than each one of the individual resistances. That it is less than the least resistance is implied.

Problem 1.4. *Give the colour coding arrangement for following resistors.*

(*a*)	*220 ohms, ± 5%*	(*b*)	*15 K, ± 10%*
(*c*)	*390 K, ± 10%*	(*d*)	*IM, ± 5%*
(*e*)	*330 ohms, ± 1%*	(*f*)	*1.5 K, ± 0.1%*

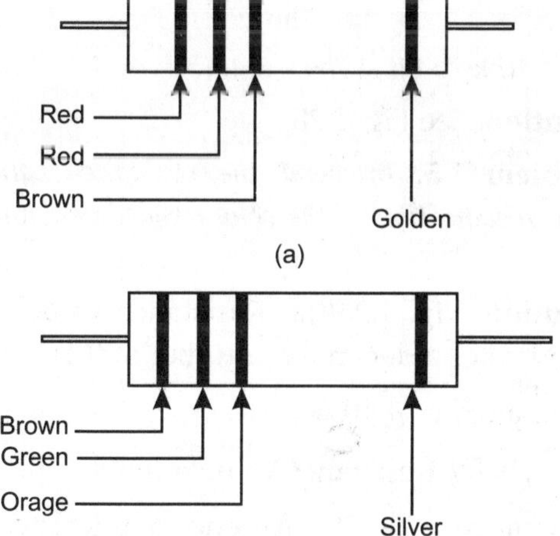

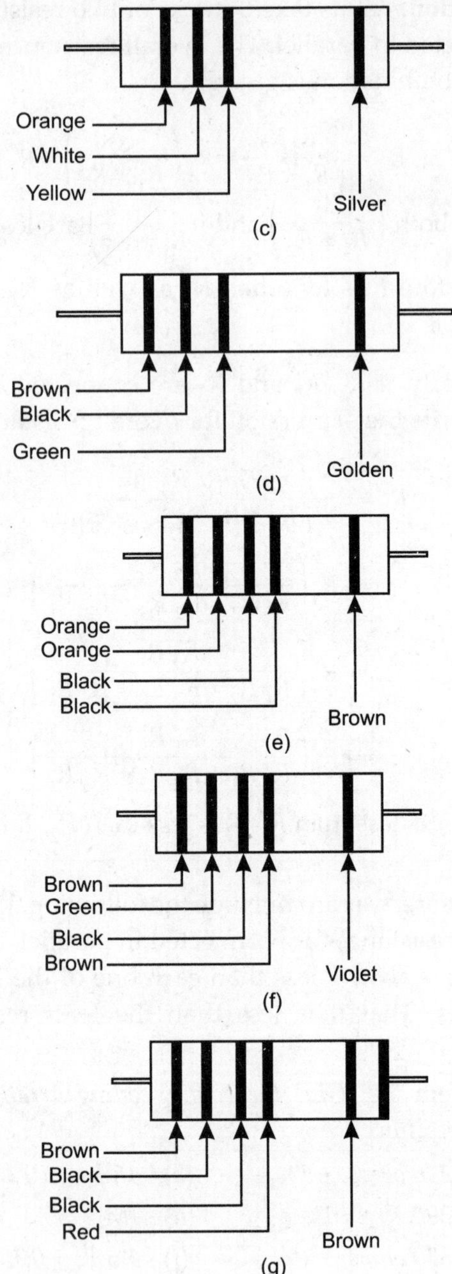

Fig. 1.28

(*g*) *10 K, ± 1%, Wire wound*

Solution. See Fig. 1.28

Problem 1.5. *Interpret the resistance value and the tolerance specification of the colour coded resistors shown in Fig. 1.29.*

Solution. Fig. 1.2.9 (*a*) Resistance value = 4.7 ohms (Gold indicates a decimal multiplier of 0.1)

Tolerance = ± 10%

Fig. 1.29 (*b*)-Resistance value = 100 *K*

Tolerance = ± 20% (Absence of tolerance band indicates ± 20%).

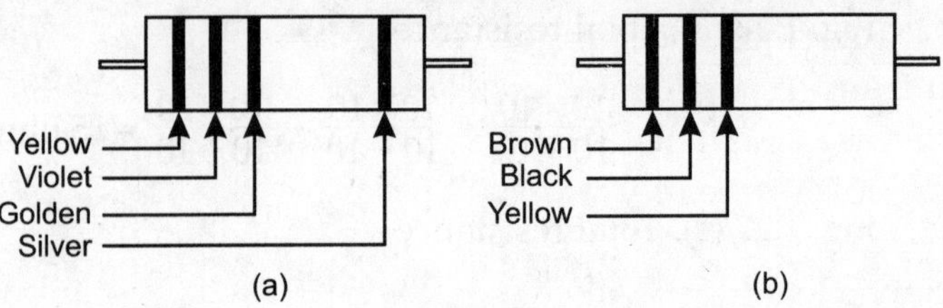

Fig. 1.29

Problem 1.6. *Decode the resistance value, the power rating and the tolerance of following resistors having an industrial type designation. EB 1041, GM 3935, CB 1012, BB 33G5.*

Solution.

EB 1041 : EB stands for 1/2 watt, 104 signifies a resistance of 100000 ohms = 100*K* and finally 1 indicates a tolerance of ± 10%. So EB 1041 is a 100*K*, ± 10%, 1/2 Watt resistor.

GM 3935 : GM stands for 3 W, 393 signifies a resistance of 39000 ohms = 39*K* and finally 5 indicates a tolerance of ± 5%. So, GM 3935 is a 39*K*, ± 5%, 3 W resistor.

CB 1012 : CB stands for 1/4 Watt, 101 signifies a resistance value of 100 ohms and finally 2 indicates a tolerance of ± 20%. So, CB 1012 is a 100 Ohms, ± 20%, 1/4 watt resistor.

BB 33G5 : BB stands for 1/8 Watt, 33 G signifies resistance value of 33 × 0.1 = 3.3 ohms and finally 5 indicates a tolerance of ± 5%. So, BB 33G5 is a 3.3 ohms, ± 5%, 1/8 Watt resistor.

Problem 1.7. *Three 10K, ± 10%, 1 W resistances are connected in series. What are the RESISTANCE VALUE, the TOLERANCE and the POWER RATING of the series combination?*

Solution. Resistance value = 30 *K* and Tolerance = ± 10%

In the case of equal value resistors, the tolerance specification can be determined by taking the average value of individual tolerance specifications. For instance, if the three equal value resistances had tolerance specifications of ± 10%, ± 20% and ± 5%, then the tolerance specification of the series combination would have been approximately ± 12%.

Power rating = 3 W. Power rating is equal to the sum of power ratings of individual resistors connected in series.

Problem 1.8. *Three 100K, ± 10%, 1 W resistance are connected in parallel. What are the RESISTANCE VALUE, the TOLERANCE and the POWER RATING of this parallel combination.*

Solution. Resistance value = 33 *K* (approx.) and Tolerance = ± 10%

Tolerance specification is again the average of the tolerance of individual resistors connected in parallel. In the case of parallel connection of equal value resistors, the tolerance specification of equipment resistor is the average of the tolerance specifications of parallel connected resistors.

Power rating = 3W. Power rating is the sum of power ratings of individual resistors.

An important inference of practical interest from problems 7 and 8 is that if you are looking for a resistor of a certain value and wattage rating and you don't have that right away, look for an appropriate series or preferably parallel combination. For instance, a 33 K, 5 W requirement could be easily met with by connecting three 100 K, 2 W resistances in Parallel. Series and parallel combinations also assume significance when you are looking for higher wattage (higher than 2 Watts) composition and film resistors and when you cannot afford to use wire wound resistors for some reasons like large associated inductances and capacitances.

Problem 1.9. *Two 100 K, ± 10%, 1 Watt resistors are connected in parallel. Determine the resistance value, the tolerance and the wattage rating of the equivalent parallel resistance.*

Solution. Resistance value of parallel combination

$$= \frac{100}{2} = 50K$$

Tolerance of parallel combination

$$= \pm\left[\frac{(R_1 + R_2)(1 + T_1)(1 + T_2)}{R_1(1 + T_1) + R_2(1 + T_2)} - 1\right] \times 100$$

$$= \pm\left[\frac{(100 + 100)(1.1)(1.1)}{100(1.1) + 100(1.1)} - 1\right] \times 100$$

$$= \pm 10\%$$

Wattage rating of parallel combination

$$= W_m R_m \left(\frac{1}{R_1} + \frac{1}{R_2}\right)$$

Now

$$W_1 R_1 = 100 \text{ k}\Omega\text{–watt}$$

$$W_2 R_2 = 100 \text{ k}\Omega\text{–watt}$$

Therefore, wattage rating

$$= 100\left(\frac{1}{100} + \frac{1}{100}\right) = 2 \text{ watts.}$$

Problem 1.10. *Four resistors of (10 K, 0.5 W), (20 K, 1 W), (30 K, 2 W) and (60 K, 4 W) are connected in parallel. Find the effective wattage rating of the parallel combination.*

Solution.

$$W_1 R_1 = 5 \text{ k}\Omega\text{-watt};$$

$$W_2 R_2 = 20 \text{ k}\Omega\text{-watt};$$

$$W_3 R_3 = 60 \text{ k}\Omega\text{-watt};$$

$$W_4 R_4 = 240 \text{ k}\Omega\text{-watt}$$

$(W_1 R_1)$ is the least of all. Therefore, effective wattage rating of parallel combination is given by:

$$W = 5\left(\frac{1}{10} + \frac{1}{20} + \frac{1}{30} + \frac{1}{60}\right) = \frac{5 \times 12}{60}$$

$$= 1 \text{ Watt.}$$

Problem 1.11. *R_1 = 10 K/0.5 W, R_2 = 20 K/l W, R_3 = 30 K/2 W and R_4 = 40 K/4 W are connected in series. Determine the wattage rating of the series combination.*

Solution. If W_1, W_2, W_3 and W_4 are wattage ratings of R_1, R_2, R_3 and R_4 respectively, then

$$\frac{W_1}{R_1} = \frac{0.5}{10} = 0.50 \text{ W/k}\Omega; \quad \frac{W_2}{R_2} = \frac{1}{20} = 0.05 \text{ W/k}\Omega$$

$$\frac{W_3}{R_3} = \frac{2}{30} = 0.067 \text{ W/k}\Omega; \quad \frac{W_4}{R_4} = \frac{4}{40} = 0.1 \text{ W/k}\Omega$$

$$\text{Least value} = \frac{W_1}{R_1} = \frac{W_2}{R_2} = 0.05 \text{ W/k}\Omega$$

Therefore, effective wattage rating,

$$W = \frac{W_1}{R_1}(R_1 + R_2 + R_3 + R_4)$$

$$= 0.05\,(10 + 20 + 30 + 40) = 5 \text{ Watt.}$$

Problem 1.12. *Fig. 1.30 shows cross-sections of two conductors having same length and material. The conductor (B) fits snugly within conductor (A). Determine the ratio (R_A/ R_B) if R_A and R_B are the resistance values of the two conductors.*

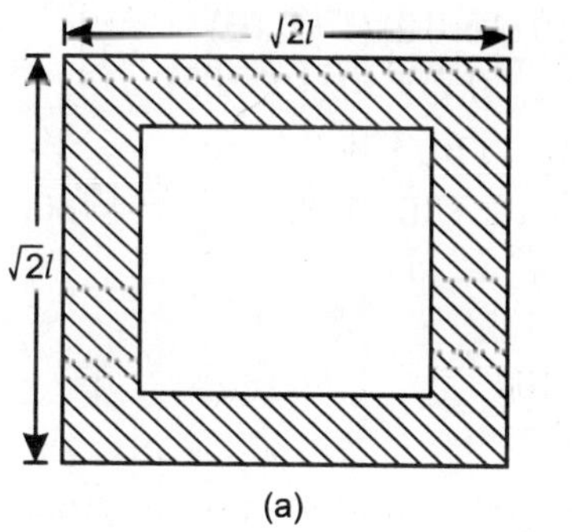

(a)

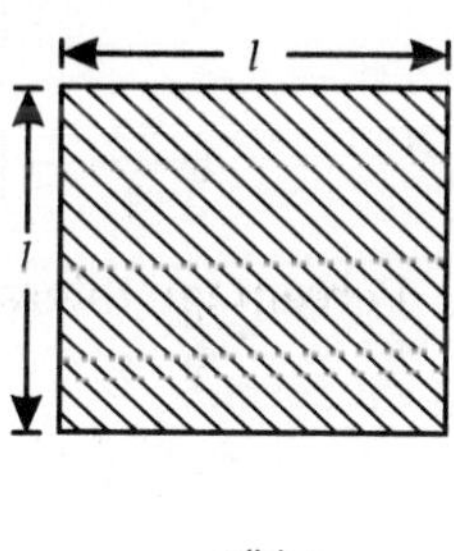

(b)

Fig. 1.30

Solution.

Cross-sectional area of conductor $(A) = (\sqrt{2}\,l)^2 - l^2 = l^2$

Cross-sectional area of conductor $(B) = l^2$

Now Resistance = $\rho\,(l/A)$

where

ρ = Resistivity

l = Length

A = Cross-sectional area

Since resistivity (ρ) is a material property, all three parameters turn out to be the same for the two conductors.

Therefore,

$$R_A = R_B \text{ i.e. } (R_A/R_B) = 1.$$

Problem 1.13. *Resistors $R_1 = 100\ K \pm 10\%$ and $R_2 = 10\ K \pm 1\%$ are connected in series. Determine the percentage tolerance of the combination. What do you deduce from the result?*

Solution. Tolerance of the series combination is given by:

$$T = \pm\frac{100 \times 0.1 + 10 \times 0.01}{100 + 10} = \pm\frac{10 + 0.1}{110}$$

$$= \pm\frac{10.1}{110}$$

Percentage tolerance

$$= \pm\frac{10.1 \times 100}{110} \cong \pm 9.2\%$$

Deduction: The tolerance of series combination is closer to that of the larger of the two resistance values.

Problem 1.14. *When R_1 = 100 K, ± 20% is connected across R_2 = 10 K, ± 5% what will be the tolerance of the parallel combination ? What do you deduce from the result?*

Solution. Tolerance of parallel combination is given by:

$$T = \pm\left[\frac{(R_1 + R_2)(1 + T_1)(1 + T_2)}{R_1(1 + T_1) + R_2(1 + T_2)} - 1\right]$$

$$= \pm\left[\frac{110\,(1 + 0.2)\,(1 + 0.05)}{100\,(1 + 0.2) + 10\,(1 + 0.05)} - 1\right]$$

$$= \pm\left[\frac{110 \times 1.2 \times 1{,}05}{100 \times 1.2 + 10 \times 1.05} - 1\right] = \pm\left(\frac{8.1}{130.5}\right)$$

Percentage tolerance

$$= \pm\frac{8.1 \times 100}{130.5} = \pm 6.2\%$$

Deduction: Tolerance of parallel combination of two resistors is closer to that of the smaller of the two resistance values.

Problem 1.15. *Three resistors R_1, R_2, R_3 are connected in series and a voltage (V) is applied across the series combination. If voltage across (R_1) is 20 volts, power dissipated in (R_2) is 25 watts and (R_3) has a resistance of 2Ω, determine (V) if the current flowing through the series combination is 5A.*

Solution.

$$R_1 = \frac{20}{5} = 4\Omega;\ R_2 = \frac{25}{(5)^2} = 1\Omega;\ R_3 = 2\Omega \text{ (Given)}$$

Total resistance of series combination = 4 + 1 + 2 = 7Ω

Therefore.

$$V = 7 \times 5 = 35 \text{ volts}$$

Problem 1.16. *Refer to Fig. 1.31. Determine (V)*

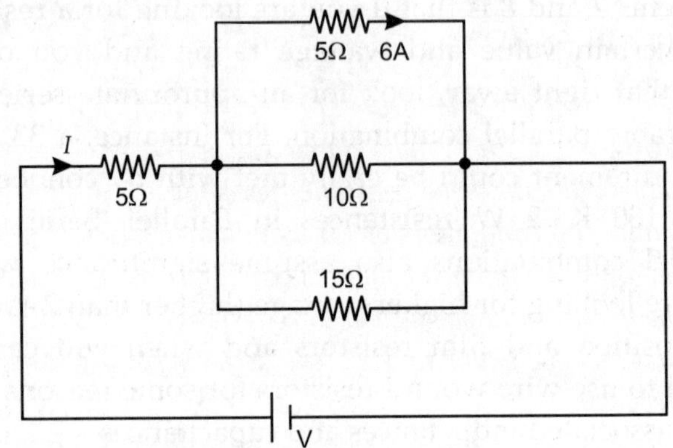

Fig. 1.31

Solution. 6 A through 5 Ω resistor means a voltage of 30 volts across it and also across 10 Ω and 15 Ω resistors.

This gives current through 10 Ω resistor = 3 *A* and current through 15 Ω resistor = 2 *A*

This gives

$$I = 6 + 3 + 2 = 11A$$

which gives

$$V = 11 \times 5 + 30 = 85 \text{ volts.}$$

Problem 1.17. *A copper transmission line consists of 24 strands of copper wire, each wire 1.6 mm in diameter. Find resistance of 10 km of cable length given that resistivity (ρ) of copper is 1.72 × 10^{-8} Ω-m.*

Solution. Diameter of each strand of wire, d = 1.6 mm = 1.6 × 10^{-3} m

Length, l = 10 km = 10,000 m and Resistivity, ρ = 1.72 × 10^{-8} Ω-m

Therefore, resistance of one strand of 10 km length

$$= \rho \times \frac{1}{A} = \rho\frac{l}{(\pi d^2/4)} = \frac{4\rho l}{\pi d^2} = \frac{4 \times 1.72 \times 10^{-8} \times 10000}{3.14 \times (16 \times 10^{-3})^2}$$

$$= 85.6\ \Omega$$

Since the transmission line has 24 strands running parallel throughout the length, the resistance of the line would therefore be equal to 24 resistances of 85.6 Ω connected in parallel.

Resistance of the line = 85.6/24 = 3.56 Ω

Problem 1.18. *A stranded cable consists of 8 copper conductors, each having a diameter of 1 mm. Twisting the wire to form cable requires the length of each conductor to be 3 percent longer than the length of the cable when finished. Determine the resistance of 1 km length of this cable (Given : Resistivity of copper = 1.72 × 10^{-8} Ω-m).*

Solution. Each strand length = 1000 + 1000 × 3/100 = 1030 m. Resistance offered by each strand is given by :

$$\frac{1.72\times10^{-8}\times1030}{(\pi\times10^{-3}\times10^{-3})}=\frac{1.72\times1030\times4\times10^{-8}}{\pi\times10^{-6}}=22.57\ \Omega$$

Resistance of cable = 22.57/8 = 2.82 Ω

Problem 1.19. *What diameter of the solid copper conductor will have same resistance per km as a stranded cable of 6 aluminium conductors of 1.3 mm diameter each assuming that making the stranded cable requires 3 percent extra length than the length of finished cable. Take resistivities of copper and aluminium to be equal to 1.72 × 10^{-8} Ω-m and 2.83 × 10^{-8} Ω–m respectively.*

Solution. Resistance of stranded cable is given by:

$$\frac{2.83\times10^{-8}\times1030\times4}{3.14\times(1.3)^2\times10^{-6}\times6}=3.65\ \Omega$$

If (*D*) is the diameter of the solid copper conductor, its resistance per km length is given by:

$$\frac{1.72\times10^{-8}\times1000\times4}{3.14\times D^2}=\frac{2.19\times10^{-5}}{D^2}$$

(D) can be computed from : $\dfrac{2.19\times10^{-5}}{D^2}=3.65$

or $$D=\sqrt{\frac{2.19\times10^{-5}}{3.65}}=2.46\times10^{-3}\ \text{m}$$

= 2.46 mm

Problem 1.20. *The resistance of a carbon rod is 100 Ω at 20°C. What would be its resistance at 70°C if the temperature co-efficient of resistance has a magnitude of 0.0005/°C at 20°C.*

Solution. If (R_1) and (R_2) are resistance values of the carbon rod at 20°C and 70°C respectively, then

$$R_2 = R_1(1+\alpha\Delta T)$$

where

α = Temperature co-efficient of resistance = 0.0005/°C

ΔT = Change in temperature = 70°C – 20°C = 50°C

Carbon has a negative temperature co-efficient of resistance.

Therefore, $R_2 = R_1(1 - 0.0005\times50) = 100\,(1-0.025) =$ 97.5 Ω

Problem 1.21. *A copper conductor has a resistance of 12 Ω. at 20°C. Determine its resistance at 100°C given that the temperature co-efficient of resistance for copper at 20°C is 0.004/°C. At what temperature will the resistance be 6 Ω?*

Solution. If (R_1) and (R_2) are the resistance values at 20°C and 100°C respectively, then

$$R_2 = R_1(1+\alpha\Delta T)$$

where α = 0.004/°C ΔT = 100 – 20 = 80°C

Therefore,

$$R_2 = 12\,(1+0.004\times80) = 12\,(1.32) = 15.84\ \Omega$$

Temperature at which resistance would be 6 Ω can be computed from :

$$6 = 12\,[1+\alpha(T-20)] \quad\text{or}\quad \alpha(T-20) = -0.5$$

or $$T = \frac{-0.5}{0.004}+20 = -105°\text{C}$$

Problem 1.22. *A certain conductor has resistance of 10 Ω. and 11.35 Ω. at 20°C and 50°C respectively. Of what material should this conductor be made? Take the help of data given in the accompanying table.*

Conductor	*Temp. Co-eff. at 20° C*
Silver	0.0038
Copper	0.00393
Aluminium	0.0039
Tungsten	0.0045
Constantan	0.000008
Nickel	0.006

Solution. Resistance at 20°C = 10 Ω, Resistance at 50°C = 11.35 Ω

Now

R(at 50°C) = R(at 20°C) [1 + α Δ T].

Here ΔT = 50 – 20 = 30°

Therefore, 11.35 = 10 (1 + 30 α)

or 1+ 30 α = 11.35/10 = 1.135 or α = 0.135/30 = 0.0045

Referring to the table, the material is tungsten.

Problem 1.23. *By how many degrees must the temperature of an aluminium rod be to increase its resistance by 10% of its value at room temperature given that temperature co-efficient of resistance of aluminium at room temperature is 0.0039°C.*

Solution. $R_2 = R_1(1+\alpha\Delta T)$ or $1+\alpha\Delta T = R_2/R_1$

or $$\Delta T = \frac{(R_2/R_1)-1}{\alpha} = \frac{\frac{110}{100}-1}{0.0039} = \frac{1}{0.039}$$

= 25.64° C

Thus the temperature must be increased by 25.64°C above the room temperature to increase the resistance by 10%.

Problem 1.24. *What would be the at resistance of a copper conductor at – 20°C if its resistance at +30°C is 1.3 Ω. Temperature co-efficient of copper is 0.00393/°C at 30°C.*

Solution. If (R_1) and (R_2) are the resistance values of the copper conductor at + 30°C and – 20°C, then

$$R_2 = R_1 (1 + \alpha \Delta T); \text{ where } \alpha = 0.00393/°C$$

$$\Delta T = -20 - 30 = -50°C$$

$$R_2 = 1.3 (1 - 0.00393 \times 50) = 1.3 (1 - 0.1965)$$

$$= 1.3 \times 0.8035 = 1.04$$

Problem 1. 25. *An electric motor drawing 10 A is to be operated at a distance of 200 m from 220 V source. This current raises the temperature of copper conductor feeding the motor by 50°C. What is the minimum diameter of the wire that can be used without the voltage drop in the conductor exceeding 10% of applied voltage ? (Resistivity of copper = 1.72 x 10^{-8} Ω-m)*

Solution. Allowable voltage drop

$$= 220 \times 0.1 = 22V, \text{ Current} = 10A$$

Therefore, permissible maximum conductor resistance = 22/10 = 2.2 Ω

Length of conductor = 200 m ΔT = 50°C, Conductor diameter = D

Cross-sectional area = $\pi D^2/4$

$$\frac{1.72 \times 10^{-8} \times 200 \times 4}{3.14\, D^2}(1 + \alpha \Delta T) = 2.2$$

$$\text{or} \quad \frac{438.2 \times 10^{-8}}{D^2}(1 + 0.00393 \times 50) = 2.2$$

$$\text{or} \quad \frac{438.2 \times 10^{-8}}{D^2}(1.1965) = 2.2$$

$$\text{or} \quad D^2 = \frac{438.2 \times 10^{-8} \times 1.1965}{2.2} = 238.32 \times 10^{-8}$$

$$D = \sqrt{238.32 \times 10^{-8}} = 15.44 \times 10^{-4} \text{ m}$$

$$= 1.544 \text{ mm}$$

Problem 1.26. *A 10K, 20W resistor, 80K, 100W resistor and a 20K, 200W resistor are connected in series. What maximum voltage can be applied to the network without exceeding the power rating of any resistor?*

Solution. Fig 1.32 shows the arrangement

$$\text{Now,} \quad I = \frac{V}{110 \times 10^3}$$

If dissipation in (R_1) is not to exceed 20 watts (its rated wattage), $\left(\frac{V_{max}}{110 \times 10^3}\right)^2 \times 10 \times 10^3 = 20$

$$\text{or} \quad V_{max}^2 = \frac{20 \times 110 \times 110 \times 10^6}{10^4} = 242 \times 10^5$$

$$\text{or} \quad V_{max} = \sqrt{242 \times 10^5} = 4919.3 \text{ volts}$$

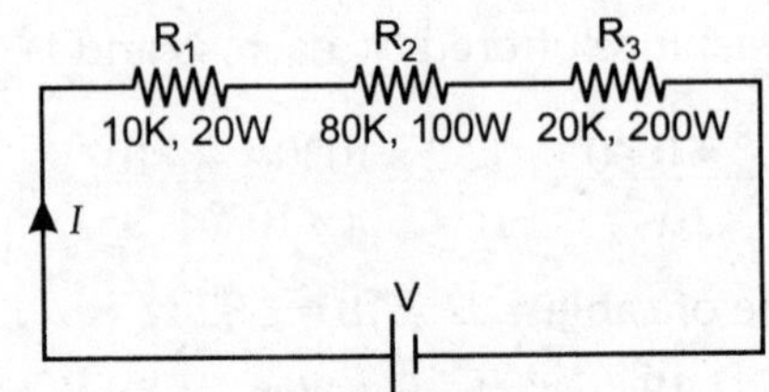

Fig. 1.32

If dissipation in (R_2) is not to exceed 100 watts (its rated wattage), then

$$\left(\frac{V_{max}}{110 \times 10^3}\right)^2 \times 80 \times 10^3 = 100$$

$$\text{or} \quad V_{max}^2 = \frac{100 \times 110 \times 110 \times 10^6}{80 \times 10^3} = \frac{121 \times 10^6}{8}$$

$$= 15.128 \times 10^6$$

$$V_{max} = \sqrt{15.125 \times 10^6} = 3889 \text{ volts}$$

If dissipation in (R_3) is not to exceed 200 watts (its rated wattage), then

$$\left(\frac{V_{max}}{110 \times 10^3}\right)^2 \times 20 \times 10^3 = 200$$

$$\text{or} \quad V_{max}^2 = \frac{200 \times 110 \times 110 \times 10^6}{20 \times 10^3} = 121 \times 10^6$$

$$V_{max} = \sqrt{121 \times 10^6} = 11000 \text{ volts.}$$

From the above results, the voltage that can be safely applied across the series combination so that the wattage ratings of individual resistors are not exceeded is 3889 volts.

Problem 1.27. *Three resistors (R_1), (R_2) and (R_3) connected in parallel (Fig.1.33) draw a total current of 0.6 A from the source. (R_1) is 400 Ω, (R_2) draws a current of 60 mA and (R_3) has a voltage of 150 V across it. Calculate resistance values of (R_2) and (R_3).*

Solution. $R_2 = \frac{150}{60 \times 10^{-3}} = 2.5 \text{ K}$

Current through $R_1 = \frac{150}{400} = 375$ mA

Therefore, current through

$$R_3 = 600 - (375 + 60) = 165 \text{ mA}$$

$R_3 = 150/0.165 = 909/1\ \Omega$

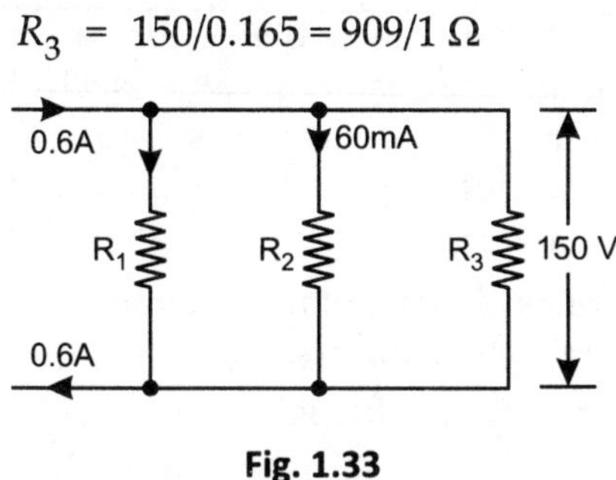

Fig. 1.33

Problem 1.28. *Three resistors (R_1), (R_2) and (R_3) connected in parallel draw energy from a source at a total rate of 840 watts. If (R_1) draws 5A, (R_2) drops 60V across it and (R_3) dissipates 360 watts, find (R_1), (R_2) and (R_3).*

Solution. (Refer to Fig. 1.34)

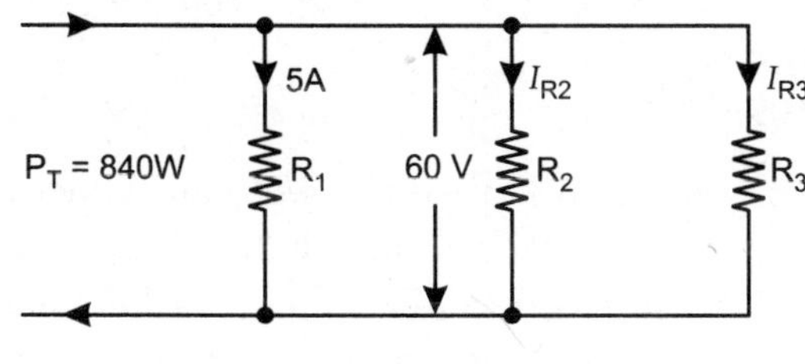

Fig. 1.34

$P_T = 840$ watts

$P_{R3} = 360$ watts

$R_1 = 60/5 = 12\ \Omega$

$I_{R2} = 60/R_2$ and $I_{R3} = 60/R_3$

Also,

$(60)^2/R_3 = 360$

or $R_3 = 3600/360 = 10\ \Omega$, which gives $I_{R3} = 60/10 = 6\ A$

Now $60\ (I_{R1} + I_{R2} + I_{R3}) = 840$

or $I_{R1} + I_{R2} + I_{R3} = 840/60 = 14\ A$

Therefore, $I_{R2} = 14 - 6 - 5 = 3\ A$

which gives $R_2 = 60/3 = 20\ \Omega$

Problem 1.29. *12 V source with an internal resistance of 2 Ω, a 0 to 100 Ω potentiometer and a fixed 10 Ω load are connected in series. What setting of potentiometer connected as a variable resistance will produce maximum power in a fixed 10 Ω load? What is the maximum power transferred?*

Solution. Refer to Fig. 1.35. For maximum power transfer, source resistance should equal load resistance.[3]

Load resistance, $R_L = 10\ \Omega$

Therefore, total source resistance = $10\ \Omega$

for maximum power transfer

Potentiometer setting = $8\ \Omega$

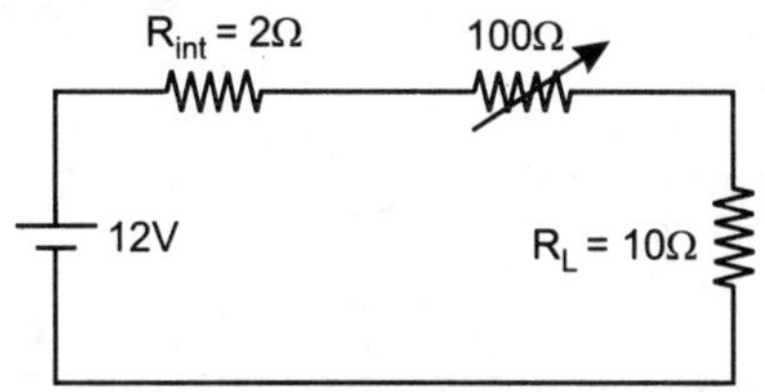

Fig. 1.35

Maximum power transferred $= \left(\frac{12}{10+10}\right)^2 \times 10$

$= 3.6$ watts

Problem 1.30. *A generator whose internal resistance is 1 Ω feeds a load connected with a pair of wires having a total resistance of 4 Ω. The generator voltage is adjusted to produce 120 V across load. If the total dissipation in the wire is not to exceed 10% of the total generator power, what is the maximum power that can be delivered to the load?*

Solution. (Refer to Fig. 1.36)

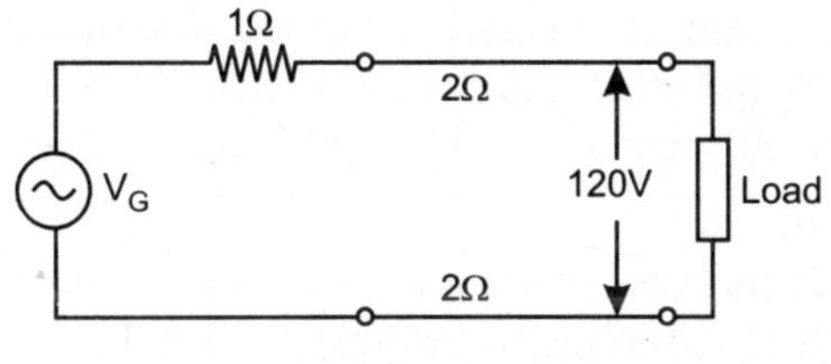

Fig. 1.36

$$\frac{V_G R_L}{R_L + 5} = 120$$

Also, $V_G I_L = 10\ I_L^2 \times 4 = 40\ I_L^2$

or $V_G = 40\ I_L = \frac{40 \times 120}{R_L}$

Therefore, $\frac{(4800/R_L) \times R_L}{R_L + 5} = 120$

or $(R_L + 5) = \frac{4800}{120} = 40$

which gives $R_L = 35\ \Omega$

$$P_L\ (\text{max}) = \frac{(120)^2}{35}$$

$= 411.4$ watts

Problem 1.31. *Three resistors (R_1), (R_2) and (R_3) are connected in series across a 120 V generator (Fig. 1.37). R_1 = 50 Ω, current through R_2 equals 0.5 A and voltage drop across (R_3) is 50 V, determine (R_2) and (R_3).*

[3] Maximum power transfer theorem is discused in chapter 5

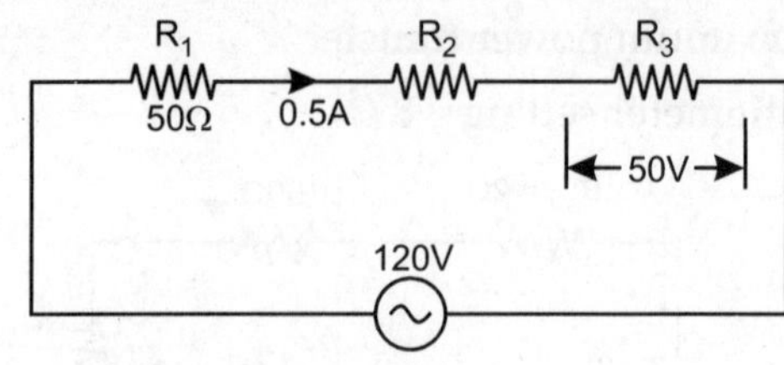

Fig. 1.37

Solution. Drop across $(R_1) = 0.5 \times 50 = 25$ V

Therefore drop across $(R_2) = 120 - (25 + 50) = 45$ V

$$R_2 = \frac{45}{0.5} = 90\ \Omega \text{ and}$$

$$R_3 = \frac{50}{0.5} = 100\ \Omega.$$

Problem 1.32. *Two resistors of 100 K, ± 10% each are connected in series. What will be the percentage tolerance of the series combination? Also, find the resulting percentage tolerance if the two resistances were connected in parallel.*

Solution. ± 10% tolerance implies that the resistance could be anywhere in 90 K to 110 K range. When the two resistances are connected in series, the minimum and maximum possible values of resistance for the series combination would be 180 K and 220 K. This implies a percentage tolerance of ± (20/200) × 100 = ± 10%.

When the two resistances are connected in parallel, the nominal value of the resistance combination is 50 K. The minimum and maximum possible values of resistance for the parallel combination would be 45 K and 55 K respectively. This implies that the percentage tolerance of the parallel combination is ± (5/50) × 100 = ± 10%.

From the above results, we can safely conclude that the tolerance specification of the equivalent resistance of several series connected or parallel connected resistors is the same as that of respective individual resistors if the resistors connected in series or parallel have same tolerance specification.

Problem 1.33. *Two resistors of 10 K ± 10% each are connected in series. What will be the percentage tolerance of the series combination? Also find out the percentage tolerance of the parallel combination. What do you deduce from the results?*

Solution. This problem can be solved on the lines of the solution to problem 1.32. The tolerance in the two cases turns out to be again ± 10%. The deduction is also the same.

Problem 1.34. *A resistor of 100 K, ± 10% is connected in series with a 10 K, ± 1% resistor. What would be the percentage tolerance of the series combination? Also determine percentage tolerance of the parallel combination. What do you deduce from the results?*

Solution. When the two resistors are connected in series, the nominal resistance value is 110 K. Minimum and maximum values of resistance for the series combination would be 90 + 9.9 = 99.9 K and 110 + 10.1 = 120.1 K respectively

This implies a percentage tolerance of ± (10.1/110) × 100 = ± 9.2% when the two resistors are connected in parallel, the nominal value is

$$\frac{100 \times 10}{110} = 0.09 \text{ K}$$

$$\text{Minimum possible resistance value} = \frac{90 \times 9.9}{99.9} = 8.92 \text{ K}$$

$$\text{Maximum possible resistance value} = \frac{110 \times 10.1}{120.1} = 9.25 \text{ K}$$

This gives a percentage tolerance of approximately

$$\pm \frac{0.16}{9.09} \times 100 = \pm 1.75\%$$

From the above results, we can conclude that when two unidentical resistors with different tolerance specifications are connected in series, the tolerance of the series combination is closer to that of resistor having larger resistance value. If the ratio of the two resistance values is very large (say greater than 10), the tolerance of the series combination becomes almost equal to that of larger resistance value resistor.[4]

In case of parallel combination, the tolerance of the combination approaches that of the smaller resistance value resistor.

Problem 1.35. *Derive an expression to prove that when R_1, ± x% is connected in series or in parallel with R_2, ± y%, the series or parallel combination will have a tolerance of ± x% or ± y% provided that x = y.*

Solution. When (R_1) and (R_2) are connected in series,

$$\text{Nominal resistance} = (R_1 + R_2)\ \Omega$$

$$\text{Minimum resistor} = \left(R_1 - \frac{R_1 x}{100}\right) + \left(R_2 - \frac{R_2 y}{100}\right)$$

$$= (R_1 + R_2) - \frac{1}{100}(R_1 x + R_2 y)$$

$$= (R_1 + R_2) - \frac{x}{100}(R_1 + R_2)...$$

$$\text{for } x = y = (R_1 + R_2)\left(1 - \frac{x}{100}\right)\Omega$$

On similar lines, Maximum resistance

$$= (R_1 + R_2)\left(1 + \frac{x}{100}\right)\Omega$$

[4] Problems 1.32 and 1.33 can also solved with the help of mathematical expressions given in this chapter.

The difference between maximum resistance and nominal resistance equals the difference between nominal resistance and the minimum resistance and this difference is given by:

$$\frac{(R_1 + R_2)\,x}{100}\,\Omega$$

Therefore, percentage tolerance of the series combination is given by:

$$\frac{(R_1 + R_2)\,x/100}{(R_1 + R_2)} \times 100 = x\%$$

When the two resistors are connected in parallel,

$$\text{Nominal resistance} = \left(\frac{R_1 R_2}{R_1 + R_2}\right)\Omega$$

$$\text{Minimum resistance} = \frac{R_1 (1 - x/100) \times R_2 (1 - y/100)}{R_1 (1 - x/100) + R_2 (1 - y/100)}$$

$$= \frac{R_1 R_2 (1 - x/100)^2}{(R_1 + R_2)(1 - x/100)} \text{ ... for } x = y$$

$$= \left(\frac{R_1 R_2}{R_1 + R_2}\right)\left(1 - \frac{x}{100}\right)\Omega$$

$$\text{Maximum resistance} = \frac{R_1 (1 + x/100) \times R_2 (1 + y/100)}{R_1 (1 + x/100) + R_2 (1 + y/100)}$$

$$= \frac{R_1 R_2 (1 + x/100)^2}{(R_1 + R_2)(1 + x/100)} \text{ ... for } x = y$$

$$= \left(\frac{R_1 R_2}{R_1 + R_2}\right)\left(1 + \frac{x}{100}\right)\Omega$$

Difference between maximum resistance and nominal resistance = Difference between nominal resistance and minimum resistance

$$= \left(\frac{R_1 R_2}{R_1 + R_2}\right)\left(1 + \frac{x}{100}\right) - \left(\frac{R_1 R_2}{R_1 + R_2}\right)$$

$$= \left(\frac{R_1 R_2}{R_1 + R_2}\right) - \left(\frac{R_1 R_2}{R_1 + R_2}\right)\left(1 - \frac{x}{100}\right)$$

$$= \frac{R_1 R_2\, x}{(R_1 + R_2) \times 100}$$

Therefore, percentage tolerance

$$= \frac{R_1 R_2 x/(R_1 + R_2) + 100}{R_1 R_2 / (R_1 + R_2)} \times 100$$

$$= x\%$$

Problem 1.36. *Prove that for a series connection of $R_1 \pm x\%$ and R_2, $\pm y\%$, the equivalent resistance would have a tolerance approaching $\pm y\%$ for $R_2 >> R_1$ and $\pm x\%$ for $R_1 >> R_2$.*

Solution. (From Problem - 19)

$$\%\text{ tolerance} = \pm \frac{1/100\,(R_1 x + R_2 y)}{(R_1 + R_2)} \times 100$$

$$= \pm\left(\frac{R_1 x + R_2 y}{R_1 + R_2}\right)$$

When $R_2 >> R_1$, $\frac{R_1}{R_2} << 1$

$$\%\text{ tolerance} = \pm\left(\frac{R_1 x + R_2 y}{R_1 + R_2}\right)$$

$$= \pm \frac{R_2 (y + R_1 . R_2 x)}{R_2 (1 + R_1/R_2)}$$

$$= \pm \frac{y + (R_1/R_2)x}{(1 + R_1/R_2)} = \pm y$$

Also $$\%\text{ tolerance} = \pm \frac{R_1 (x + R_2/R_1 y)}{R_1 (1 + R_2/R_1)}$$

for $R_1 >> R_2$, $\frac{R_2}{R_1} << 1$ % tolerance = $\pm x$.

Problem 1.37. *What are the resistance values and tolerance specifications for the resistors colour coded as (a) three orange bands (b) Gray, Red, Black, Silver.*

Solution.

(*a*) Three orange bands stand for $33 \times 10^3\,\Omega$ i.e. 33K. Since there is no tolerance band, it indicates a tolerance of ± 20%

(*b*) Gray, Red and Black stand for $82 \times 10^0\,\Omega$ i.e. 82 Ω. Silver indicates ± 10% tolerance.

The two cases are shown in Figs. 1.38 (*a*) and (*b*).

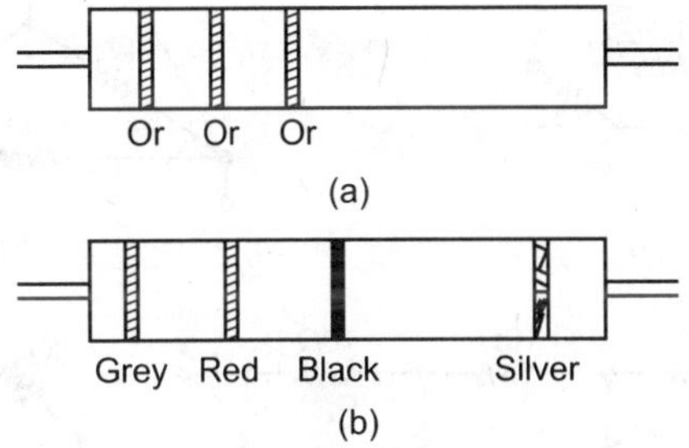

Fig. 1.38

Problem 1.39. *Determine colour code for the following resistors (a) 680 K, ± 20% (b) 0.2 Ω, ± 5% (c) 18 K, ± 1%.*

Solution.

(*a*) 680 K would be represented by Blue, Gray and Yellow as

$680\text{ K} = 680 \times 10^3\,\Omega = 68 \times 10^4\,\Omega$

± 20% can be expressed by an absence of colour band.

(*b*) $0.2\,\Omega = 20 \times 10^{-2}\,\Omega$

The colour bands representing 0.2 Ω will therefore be Red, Black, Silver, ± 5% is represented by Golden band

(*c*) $18\text{ K} = 180 \times 10^2\,\Omega$

18 K can therefore be represented by Brown, Gray, Black, Red. Here, the resistance value has been represented by four colour bands instead of three. Four colour bands are used in case of resistors with a tolerance specification better than ± 5%.

± 1% tolerance is represented by Brown. The three cases of Problem 1.39 are shown in Figs. 1.39 (*a*), (*b*) and (*c*).

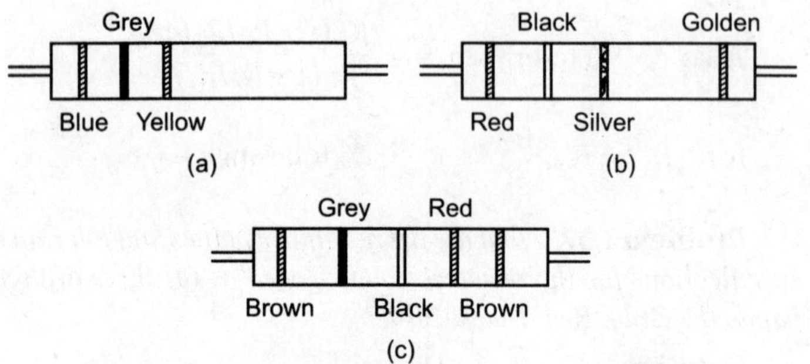

Fig. 1.39

Problem 1.40. *Determine equivalent resistance between X-X for the networks shown in Figs. 1.40 (a) to (d).*

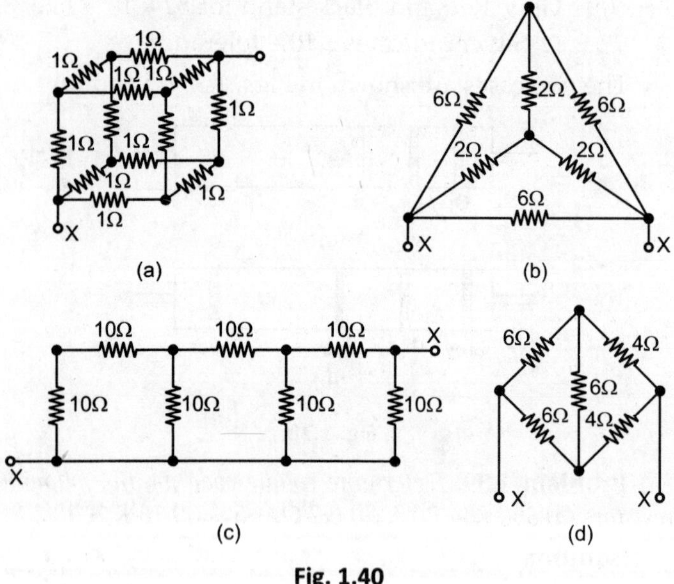

Fig. 1.40

Solution. Fig. 1.40 (*a*): By visual inspection, the network can be redrawn as shown in Fig. 1.41.

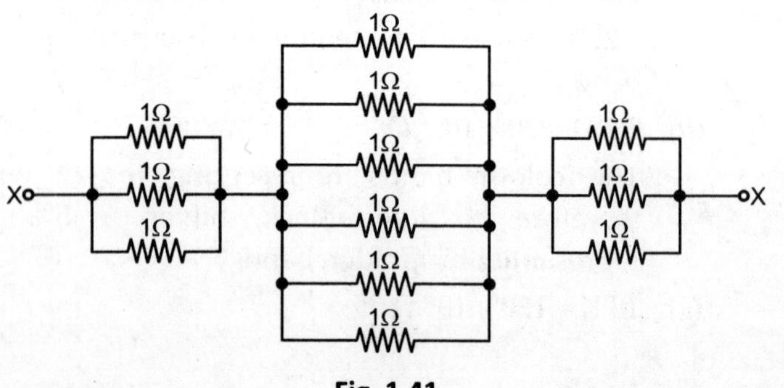

Fig. 1.41

The equivalent resistance between X-X is given by:

$$\frac{1}{3}+\frac{1}{6}+\frac{1}{3} = \frac{5}{6}\ \Omega$$

Fig. 1.40 (*b*): Converting the inner Star-network into an equivalent Delta, the network takes the shape of Fig. 1.42 (*a*).

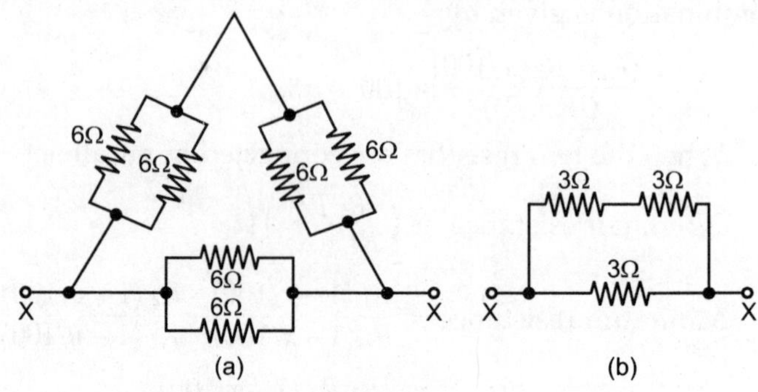

Fig. 1.42

Each of the equivalent delta resistances for the inner star would be:

$$2+2+\frac{2\times 2}{2} = 6\ \Omega$$

The network of Fig. 1.42 (*a*) can be redrawn as shown in Fig. 1.42 (*b*).

$$\text{The equivalent resistance between } X-X = \frac{3\times 6}{3+6} = 2\ \Omega$$

Fig. 1.40 (*c*): Different steps of network simplification are shown in Fig. 1.43.

The equivalent resistance between X – X is given by:

$$\frac{65/4\times 10}{65/4+10} = \frac{130}{21}\ \Omega$$

Fig. 1.43

Fig 1.40 (*d*): Converting the left hand side Delta consisting of 6 Ω each into an equivalent star, we get the network of Fig. 1.44 (*a*). Each of the star network resistances is given by:

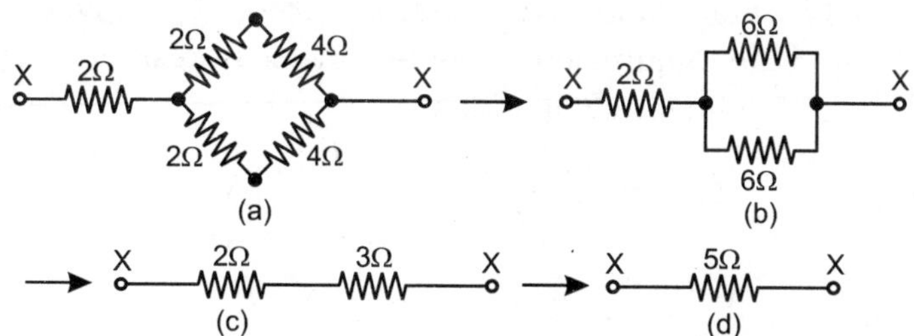

Fig. 1.44

$$\frac{6 \times 6}{6+6+6} = 2\ \Omega$$

Subsequent steps of simplification are show in Figs 1.44 (*b*), (*c*) and (*d*).

The equivalent resistance between $X - X = 5\ \Omega$.

Summary of Important Formulas

1. Resistance (R) of a cylindrical conductor of length (l), cross-section (A) and resistivity (ρ) is given by:

   $R = \rho \times (l/A)$

2. If a voltage (V) is applied across a resistance (R), then the current (I) flowing through the resistance is given by:

   $V = IR$

   The above expression is also known by Ohm's law. Ohm's law is equally valid for both DC as well as AC excitation.

3. Current (i) flowing through a pure resistance (R) due to a voltage (v) impressed across it is always in phase with the applied voltage. If the applied voltage is given by $v = V \sin\omega t$; then the resultant current would be given by $i = I \sin\omega t$ where $I = V/R$

4. Equivalent resistance of a series connection of (n) resistors (R_1), (R_2), (R_3)...... (R_n) is given by:

   $R = R_1 + R_2 + R_3 + \ldots\ldots + R_n$

5. Equivalent resistance of a parallel connection of (n) resistors (R_1), (R_2), (R_3)...... (R_n) is given by:

   $1/R = 1/R_1 + 1/R_2 + 1/R_3 + \ldots\ldots + 1/R_n$

6. When resistors (R_1), (R_2), (R_3)......(R_n) with wattage ratings (W_1), (W_2), (W_3)......(W_n) respectively are connected in series; the effective wattage rating of the equivalent series resistance is given by:

   $W = W_m/R_m \times (R_1 + R_2 + R_3 + \ldots\ldots + R_n)$

   Where $1 \le m \le n$ and (W_m/R_m) has the least value

7. When resistors (R_1), (R_2), (R_3)......(R_n) with wattage ratings (W_1), (W_2), (W_3)......(W_n) respectively are connected in parallel; the effective wattage rating of the equivalent parallel resistance is given by:

   $W = W_m \times R_m\ (1/R_1 + 1/R_2 + 1/R_3 + \ldots\ldots + 1/R_n)$

   Where $1 \le m \le n$ and ($W_m \times R_m$) has the least value

8. When resistors (R_1), (R_2), (R_3)......(R_n) with tolerance specifications (T_1), (T_2), (T_3)......(T_n) respectively are connected in series; the effective tolerance (T) of the equivalent series resistance is given by:

   $$T = \pm[(T_1R_1 + T_2R_2 + T_3R_3 + \ldots\ldots + T_nR_n)/(R_1 + R_2 + R_3 + \ldots\ldots + R_n)] \times 100$$

   (T_1, T_2, T_3......T_n) are expressed as fraction.

9. In the case of parallel connection of resistors (R_1) and (R_2) with respective tolerances of (T_1) and (T_2), the effective tolerance (T) of the effective parallel resistance (R) is given by:

   $$T = \pm[\{(R_1 + R_2) \times (1 + T_1) \times (1 + T_2)\}/\{R_1 \times (1 + T_1) + R_2 \times (1 + T_2)\} - 1] \times 100$$

   (T_1, T_2, T_3 ......T_n) are expressed as fraction.

10. *V-I* characteristics of a Varistor are given by:

    $i = K.v^n$

    v – Instantaneous voltage in Volts, K = Constant in Amp/Volt and n = an exponent whose value depends upon type of varistor. (n) is between 1 and 6 for silicon carbide Varistors and greater than equal to 25 for metal oxide varistors.

11. RMS value of thermal noise voltage also called Johnson noise voltage is given by:

    $V_{TH}\ (RMS) = \sqrt{(4KTRB)}$

    R = Resistance value

    K = Boltzmann's constant

    T = Absolute temperature

    B = Bandwidth over which noise is measured

12. RMS value of the noise voltage due to charge carriers flowing through the non-homogeneous material of resistor is given by:

$$V_i\,(RMS) = c \times f^{-\gamma} \times V^{\alpha} \times B$$

c, γ, α = Constants

f = frequency

B = Bandwidth over which noise is measured

V = DC voltage applied across the resistor

13. Delta-connected resistors can be converted to equivalent star-connected resistors by following equations.

$$R_A = R_{AB} \times R_{CA} / (R_{AB} + R_{BC} + R_{CA})$$

$$R_B = R_{BC} \times R_{AB} / (R_{AB} + R_{BC} + R_{CA})$$

$$R_C = R_{CA} \times R_{BC} / (R_{AB} + R_{BC} + R_{CA})$$

14. Star-connected resistors can be converted to equivalent delta-connected resistors by following equations.

$$R_{AB} = R_A + R_B + (R_A \times R_B/R_C)$$

$$R_{BC} = R_B + R_C + (R_B \times R_C/R_A)$$

$$R_{AB} = R_C + R_A + (R_C \times R_A/R_B)$$

15. Temperature dependence of resistivity (ρ) is given by the following formula.

$$\rho = \rho_o \times (1 + \alpha\Delta T)$$

Where ρ_o = Initial resistivity

α = Temperature coefficient of resistivity

ΔT = Temperature difference

SELF EVALUATION EXERCISE

1. Say True or False in case of the following statements:

(a) The equivalent resistance of a parallel combination of R_1, R_2 amd R_3 is given by $\dfrac{R_1 . R_2 + R_2 . R_3 + R_3 . R_1}{R_1 R_2 R_3}$

(b) A resistance R dissipates a power of 100 mW when a voltage of 10 VDC is applied across it. The power dissipated will be 200 mW if the voltage is increased to 14.14 volts.

(c) Parallel combination of 100K and 100 ohms yields a resistance of 100 K.

(d) A 100 K, 100 ppm/°C resistor is more stable than a 100 K, 0.01% per °C resistor.

(e) If we talk about power factor in case of resistors as we do in case of inductors and capacitors, it will be zero in case of resistors.

(f) Multi-turn potentiometers have a better resolution than single turn potentiometers.

(g) A 100 ohms, 1/4W general purpose carbon composition or carbon film resistor can withstand a maximum of 250 Volts.

(h) Of all the general purpose fixed resistor types, the one with the smallest associated inductance are the carbon composition resistors.

(i) Series combination of 2.2 M and 47Ω yields 2.2 M.

(j) The current flowing through a resistance is always in phase with the voltage across it irrespective of waveshape of voltage applied.

2. List all standard values for general purpose (± 10%, ± 20%) carbon composition or carbon-film resistors from 10 ohms to 100 ohms.

3. Identify the resistor type from the circuit symbols shown in Fig. 1.45. Choose from the following words. FIXED RESISTOR, VARIABLE RESISTOR, GANGED RESISTOR, PRESET RESISTOR, NONLINEAR RESISTOR, THERMISTOR

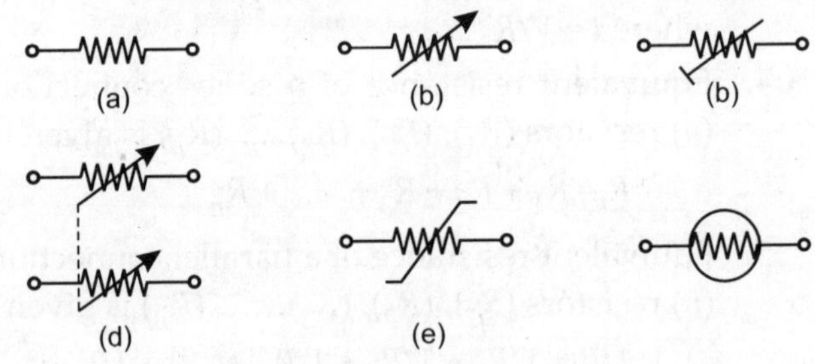

Fig. 1.45

4. Find out the resistance values and tolerance of the coded resistors shown in Fig. 1.46.

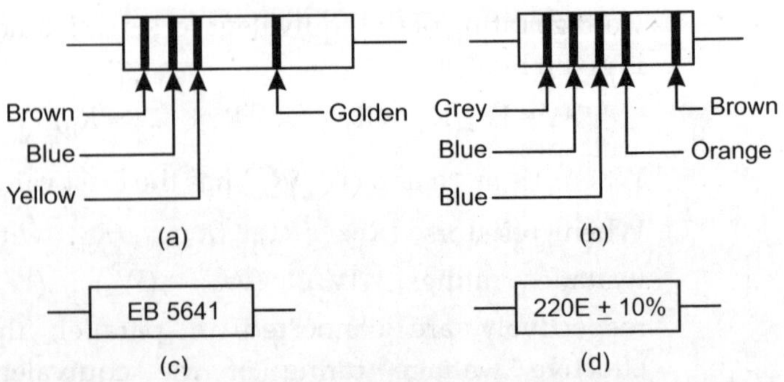

Fig. 1.46

Multichoice Questions

1. The general purpose fixed resistor type that does not fail catastrophically under fault conditions is the
   (*a*) carbon film type
   (*b*) carbon composition type
   (*c*) metal film type
   (*d*) wirewound type
2. Out of metal-film type and carbon-film type, the one with better tolerance specification is
   (*a*) carbon-film type (*b*) metal film type
3. The colour band on the extreme left in general purpose fixed resistors represents,
   (*a*) first significant digit (*b*) tolerance
   (*c*) wattage rating (*d*) voltage rating
4. The colour coding of a resistor shows 100 K, ± 10%. When the resistance was measured with an ohmmeter, it read
   (*a*) 95 K (*b*) 80 K
   (*c*) 120 K (*d*) 85 K
5. In a 5.6 ohms, ±10% carbon composition resistor, the colour of the third strip from extreme left is
   (*a*) golden (*b*) silver
   (*c*) green (*d*) black
6. Volume control found on the front panel of television receivers or transistor receivers is
   (*a*) A ganged resistor
   (*b*) A variable resistor of logarithmic type
   (*c*) A variable resistor of linear type
   (*d*) A trimmer potentiometer
7. In a varistor, as the voltage increases beyond a certain value, its resistance
   (*a*) rapidly increases.
   (*b*) rapidly decreases,
   (*c*) becomes negative *i.e.* varistor goes into negative resistance region
   (*d*) None of these
8. When three 100 K, ± 10% fixed resistors of 1/4 watt power rating are connected in parallel, the resulting 33 K resistor will have a tolerance specification and power rating of
   (*a*) ± 10% and 3/4 W respectively
   (*b*) ± 10% and 1/12 W respectively
   (*c*) ± 3.3% and 3/4 W respectively
   (*d*) ± 30% and 3/4 W respectively
9. When a 10 K, ± 10%, 1/4 W resistor is connected in series with a 12K, ± 1%, 1/4 W resistor, the resulting combination has the following specifications :
   (*a*) 22 K, ± 11%, 1/2 W
   (*b*) 22 K, ± 5.5%, 1/2 W
   (*c*) 22 K, ± 5.5%, 1/8 W
   (*d*) None of these.
10. In the 1/4W general purpose fixed resistor category (*i.e.* ± 10%, ± 20% tolerance), the standard values between 56 K and 100 K are
    (*a*) 70 K, 80 K (*b*) 68 K, 82 K, 89 K
    (*c*) 68K, 82 K (*d*) 82K.
11. An AC voltage of 9.9 sin $2\pi \times 10^4$ t is applied across a series connection of 27Ω and 22Ω resistor. Power dissipated in 22Ω and 27Ω resistors respectively would be
    (*a*) $\frac{22}{49}$ watt and $\frac{27}{49}$ watt
    (*b*) $\frac{27}{49}$ watt and $\frac{22}{49}$ watt
    (*c*) 0.5 watt each
    (*d*) Indeterminate from given data
12. One of the following types of fixed resistors has the least lead inductance.
    (*a*) Carbon film
    (*b*) Carbon composition
    (*c*) Metal film
    (*d*) Wirewound
13. A 90K resistance is constructed by using one of the following combinations. The one with least tolerance specification would be
    (*a*) Series connection of 1K ± 1% and 89 K, 10% resistance
    (*b*) Parallel connection of two 180 K, ± 5% resistor
    (*c*) Series connection of 1K, ± 10% and 89 K, ± 1% resistor
    (*d*) Parallel connection of 100 K, ± 10% and 900 K, ± 10% resistors
14. A fixed resistor is coded as CB1041. Identify the resistor from the following
    (*a*) 100 K, ± 10%, $\frac{1}{2}$ watt
    (*b*) 100 K, ± 10%, $\frac{1}{4}$ watt

(c) 100 K, ± 1%, $\frac{1}{4}$ watt

(d) 10 K, ± 10%, $\frac{1}{2}$ watt

15. What are the preferred resistance values between 10Ω and 100Ω in ±10% tolerance category.
    (a) 12Ω, 15Ω, 18Ω, 22Ω, 27Ω, 33Ω, 39Ω, 47Ω, 56Ω, 68Ω, and 82Ω
    (b) 15Ω, 22Ω, 33, 47Ω, 56Ω, 68Ω, and 82Ω
    (c) 15Ω, 33Ω, 47Ω, 68Ω, 68Ω and 82Ω
    (d) 22Ω, 33Ω, 47Ω, 56Ω, 68Ω and 82Ω
16. If thermal noise voltage in a resistor ie
    (a) directly proportional to square roof of absolute temperature
    (b) directly proportional to square roof of resistance value
    (c) directly proportional to square roof of bandwidth
    (d) all of the above
17. Parallel combination of 1M, 100K and 1K would yield
    (a) 1M (b) 100K
    (c) 500Ω (d) 1K
18. Series combination of 1M, ± 10%, 10K, ± 10% and 1K, 10% would yield
    (a) 1 M, ± 10 % (b) 1.1 M, ± 10%
    (c) 1K, ± 10% (d) 1.011 M, ± 3.33%
19. Refer to the resistance network of fig. 1.47. Power dissipated in (R_2) is

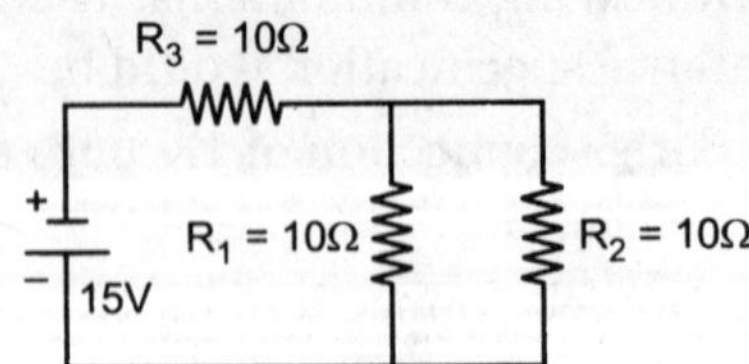

Fig. 1.47

    (a) 2.5 watt (b) 10 watt
    (c) 5 watt (d) None of these
20. An AC voltage $v = 10 \sin 6.28 \times 10^3 t$ is applied across a parallel combination of R_1 = 9Ω and R_2 = 1Ω. Expression for instantaneous current through (R_2) would then be given by
    (a) $\sin 6.28 \times 10^3 t$
    (b) $10 \sin (6.28 \times 10^3 t - \frac{\pi}{2})$
    (c) $10 \sin 6.28 \times 10^3 t$ (d) None of these
21. A 10K, 1W resistor is connected in series with a 1K, 5W resistor Maximum voltage the series combination can dissipate is
    (a) 1.1 W (b) 6W
    (c) 5W (d) 1W
22. A 10K, 1W resistor is connected in parallel with a 1K, 5W resistor. Maximum power the parallel combination can safely dissipate would be
    (a) 1 W (b) 5 W
    (c) 5.5 W (d) 6W
23. The *V-I* characteristic of a 1.732 Ω resistance were plotted by taking *I* (in mA) along *y*-axis and *V* (in volts) along x-axis. The straight line representing the characteristics makes an angle (θ) with x-axis. θ is
    (a) 60° (b) 30°
    (c) 45° (d) None of these
24. When an AC voltage was applied across a resistor, the current through it was observed to be lagging behind the voltage by a small angle. The resistance
    (a) is a pure resistance
    (b) has some associated stray inductance
    (c) has some associated stray capacitance
    (d) None of these
25. An SMD resistor is marked '000'.
    (a) It has a tolerance 0%
    (b) It has a temperature coefficient of 0 PPM/°C
    (c) It has a negligible failure rate
    (d) It is a zero ohm shorting link

ANSWERS TO SELF EVALUATION EXERCISE

1. (a) False (b) True (c) False (d) False
   (e) False (f) True (g) False (h) True
   (i) True (j) True
2. 10Ω, 12Ω, 15Ω, 18Ω, 22Ω, 27Ω, 33Ω, 39Ω, 47Ω, 56Ω, 68Ω, 82Ω, 100Ω
3. Fig. 1.45 (a) - Fixed resistor

Fig. 1.45 (*b*) - Variable resistor
Fig. 1.45 (*c*) - Preset resistor
Fig. 1.45 (*d*) - Ganged resistor
Fig. 1.45 (*e*) - Non-linear resistor
Fig. 1.45 (*f*) - Thermistor

4. Fig. 1.46 (*a*) - 160 K, ± 5%
Fig. 1.46 (*b*) - 866 K, ± 1%
Fig. 1.46 (*c*) - 560 K, ± 10%
Fig. 1.46 (*d*) - 220 Ω, ± 10%

MULTICHOICE QUESTIONS

1. (*b*)	2. (*b*)	3. (*a*)	4. (*a*)
5. (*a*)	6. (*b*)	7. (*b*)	8. (*a*)
9. (*b*)	10. (*c*)	11. (*a*)	12. (*b*)
13. (*c*)	14. (*b*)	15. (*a*)	16. (*d*)
17. (*d*)	18. (*a*)	19. (*a*)	20. (*c*)
21. (*a*)	22. (*c*)	23. (*b*)	24. (*b*)
25. (*d*)			

Chapter 2

Capacitors and Capacitive Networks

Learning Objectives

After completing the chapter, you will learn the following topics:

1. Capacitor basics
2. Equivalent circuit of a capacitor and useful frequency range of operation
3. Important specifications
4. Capacitors in series
5. Capacitors in parallel
6. Star-Delta and Delta-Star conversion
7. Standard or preferred values of capacitors
8. Marking or coding of capacitors
9. Types of capacitors and their application areas
10. Variable capacitors

Q.1. What are capacitors ? Outline some of the basic electrical laws governing the functional aspects of capacitors.

Ans. A capacitor is an electronic component that can store energy in the form of electrostatic field. In the most basic form, a capacitor is constituted by two parallel plates separated by an insulating medium. In this form it is called a parallel plate capacitor. The circuit representation of a fixed capacitor is as shown in Fig. 2.1.

The basic unit of measurement of capacitance is a Farad (F). The capacitance of a capacitor is one Farad when it requires a charge of one coulomb to establish a potential difference of one volt across it. Another way of explaining this is as follows. The charge on a capacitor is directly proportional to the potential difference or voltage across it. Capacitance in this case is the constant of proportionality expressed as ratio of charge to voltage. $1F = 1C/V$. Capacitance of practical capacitors is measured in micro-farads (μF), nano-farads (nF) and pico-farads (pF).

$1\ \mu F = 1000\ nF = 1000000\ pF$. $1F = 1000\ mF$ (milli-farad) $= 1000000\ \mu F$.

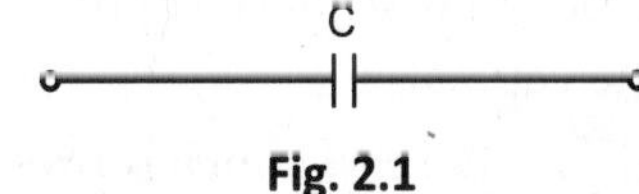

Fig. 2.1

Some of the basic laws governing the functioning of a capacitor are as follows :

1. When a capacitor (C) is charged to a DC voltage (V), the charge (q) stored in it is given by

$$q = CV$$

When (C) is in Farads, (V) is in volts; then charge (q) is in Coulombs.

2. The electrostatic energy (E) stored in a capacitor (C) charged to a DC voltage (V) is given by

$$E = \frac{1}{2}CV^2$$

When (C) is in Farads, (V) is in volts; then energy (E) is in Joules

3. Reactance (X_C) offered by a capacitor (C) to an AC signal of radian frequency (ω) is given by

$$X_C = \frac{1}{\omega C}$$

When (C) is in Farads, (ω) is in radian/s; then reactance (X_C) is in ohms.

4. Instantaneous voltage (v) across a capacitor (C) is related to instantaneous current (i) through it by

$$v = \frac{1}{C}\int i\,dt \text{ or } i = C\frac{dv}{dt}$$

An ideal capacitor does not dissipate any energy. A practical capacitor however does dissipate some energy with the quantum of energy loss depending upon its departure from ideal characteristics.

Q.2. How is instantaneous current (*i*) related to instantaneous voltage (*v*) in an ideal capacitor? How does a practical capacitor differ from an ideal one?

Ans. In an ideal capacitor, the instantaneous current (i) leads the instantaneous voltage (v) by 90° (Fig. 2.2 *a*). In a practical capacitor, (i) leads (v) by an angle that is slightly less than 90° as shown in Fig. 2.2 (*b*). The angle by which the phase difference between (i) and (v) differs from 90° is called the *Loss Angle* usually denoted by (δ).

A practical capacitor will have associated inductances and resistances. The electrical equivalent circuit of a practical capacitor is as shown in Fig. 2.2 (*c*). In the equivalent circuit shown,

C = Capacitance

L = Associated series inductance

R_s = Associated series resistance representing losses in the capacitor

R_p = Associated parallel resistance representing insulation resistance of the capacitor.

In the ideal capacitor, $L = R_s = 0$, R_p = infinity.

An electrolytic capacitor is often represented by the electrical equivalent circuit of Fig. 2.2 (*d*). Here ESR stands for *Equivalent Series Resistance*.

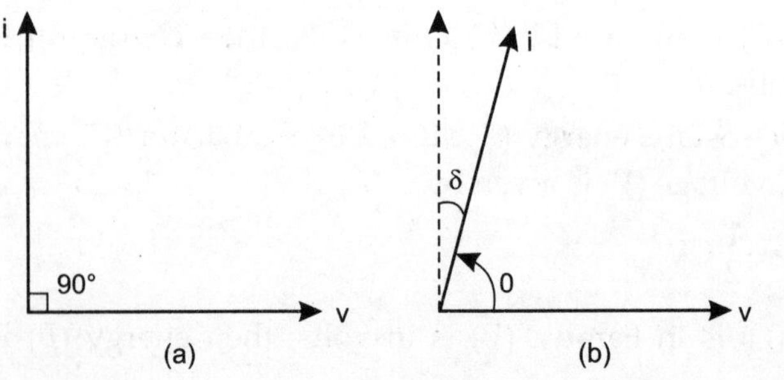

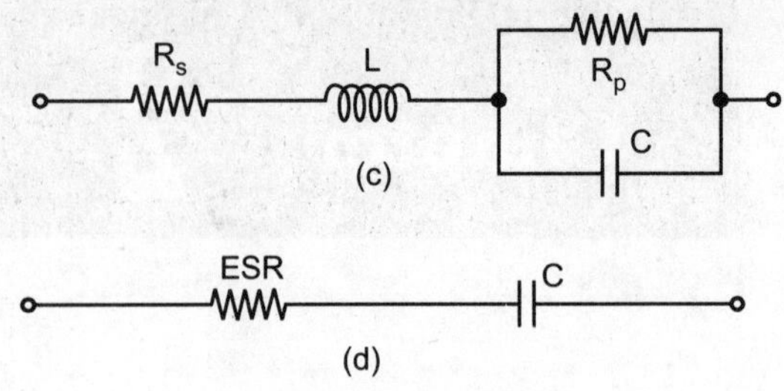

Fig. 2.2

Q.3. How does the response of a capacitor to *AC* differ from its response to *DC*?

Ans. When a *DC* voltage is applied across a capacitor, it is charged to the value of the applied voltage. In the case of an ideal capacitor, this charge is held indefinitely even if the charging voltage is removed. In a practical capacitor, it will self-discharge slowly through its normally very high insulation resistance. The charge is held even in a practical capacitor if the charging voltage stood connected as any loss of charge due to self-discharge would be replenished by charging voltage. During the charging process, a charging current proportional to rate of change of voltage across the capacitor will flow. The magnitude of charging current is given by *Cdv*/*dt*. Once the capacitor is fully charged; the rate of change of capacitor voltage becomes zero reducing the charging current also to zero.

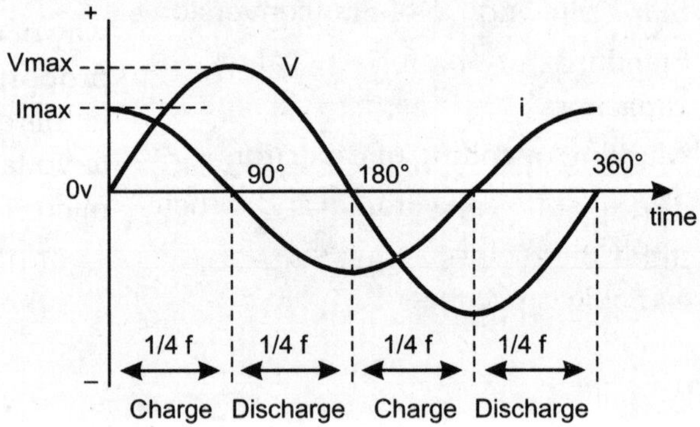

Fig. 2.3

When the capacitor is connected across an *AC* supply; the capacitor will alternately charge and discharge at a rate determined by the frequency of the supply. Since the voltage across the capacitor is constantly changing with respect to time; there is a constant flow of current (= *Cdv*/*dt*) in this case as against a zero charging current in the case of *DC* excitation when the capacitor is fully charged. Also, in the case of *AC* excitation, rate of change of voltage is at maximum during the cross-over points from negative half cycles to positive half cycles (0° point) and positive half cycles to negative half cycles (180° point). This leads to current maxima at these points. It is at minimum during the positive (90°) and negative peaks (270°) causing the current minima at these points. Figure 2.3 shows the voltage and current waveforms, which are self-explanatory. If a voltage $v = V_m \sin\omega t$ is applied across a pure capacitor; current flowing through it will then be given by $i = I_m \sin(\omega t + \pi/2)$. $I_m = V_m/X_C$ where (X_C) is capacitive reactance.

Q.4. Define the following terms with reference to capacitors.

1. **Nominal capacitance**
2. **Working voltage**
3. **Tolerance**
4. **Equivalent series resistance (ESR)**
5. **Insulation Resistance**
6. **Temperature coefficient**
7. **Power factor**
8. **Loss angle and loss tangent**
9. **Dissipation factor**
10. **Quality factor**
11. **Polarization**
12. **Time constant**

Ans. NOMINAL CAPACITANCE is the rated value of capacitance of the capacitor. The real value of capacitance may not be the same as the rated value as it varies with operating frequency and temperature. The variation with operating frequency and temperature is quite significant in some cases. The nominal capacitance is usually measured in pico-Farads (*pF*), nano-Farads (*nF*) and micro-Farads (μF). It is marked on to the body of capacitors as numbers, letters or coloured bands.

WORKING VOLTAGE is the maximum continuous *DC* or *AC* voltage that can be applied across the capacitor without causing any damage or failure during its working life. Capacitor manufactures usually mark the DC working voltage on the capacitor. This working voltage rating is valid for a specified temperature range. In the case of AC voltage care should be taken to make sure that peak value of applied AC voltage is well within the specified DC working voltage rating. Common DC working voltage ratings are 10V, 16V, 25V, 35V, 50V, 100V, 160V, 250V, 400V and 1000V. Some special varieties of high voltage capacitors are available in DC working voltage rating of several tens of kilo-volts.

TOLERANCE is the plus or minus deviation from the nominal capacitance value of a capacitor that is likely to occur. For instance, in a 100 pF ± 10% capacitor, the actual value of the capacitance may be anywhere between 90 pF and 110 pF. Tolerance of capacitors less than 100 pF is usually specified as ± pF rather than ± %. Tolerance specification can range anywhere from –20% to +80%.

The EQUIVALENT SERIES RESISTANCE (ESR) is the sum of all internal series resistances concentrated or lumped at one point in the electrical equivalent circuit and treated as a single resistance. Higher ESR signifies higher power loss in the capacitor.

INSULATION RESISTANCE is the direct current resistance measured across the capacitor terminals. Greater the value of this resistance, lesser is the leakage current.

TEMPERATURE CO-EFFICIENT is the change in capacitance of a capacitor per degree change in temperature. It may be positive or negative or even zero and is usually expressed in PPM/°C. The temperature coefficient can be either positive or negative. A capacitor marked or specified as 'N100' has a temperature coefficient of – 100 PPM/°C. 'P200' signifies a temperature coefficient of +200PPM/°C. 'NPO' signifies zero temperature coefficient.

The POWER FACTOR is always defined for a non-ideal capacitor. It is given by cos θ where (θ) is the angle by which instantaneous current leads the instantaneous voltage.

In case of an ideal capacitor, θ = 90°, which implies that the power factor of an ideal capacitor is zero. The power factor for a non-ideal capacitor represented by a series combination of a resistance (ESR) and the capacitance is also given by (R/Z) where (R) is the Equivalent Series resistance and (Z) is the impedance equal to $\sqrt{R^2 + X_C^2}$.

The LOSS ANGLE is denoted by δ and is given by (90 – θ). LOSS TANGENT is given by tan δ.

The DISSIPATION FACTOR is given by tan δ where (δ) is the loss angle. Smaller the (δ), Smaller is the dissipation factor.

Dissipation factor is given by $\left(\frac{R}{X_C}\right)$

The QUALITY FACTOR is defined as the ratio of energy stored to the energy dissipated. It is given by $\left(\frac{X_c}{R}\right)$ = ($1/\omega CR$) where (R) is the Equivalent Series Resistance of the capacitor.

It is reciprocal of Dissipation factor.

POLARIZATION refers to electrolytic capacitor, aluminium electrolytic capacitors in particular. These capacitors have polarity marked on them. Voltage should be applied to these capacitors with correct polarity. That is, positive should be connected to the positive and negative should be connected to the negative. Incorrect polarity causes capacitor's oxide layer to break down resulting in flow of large current thereby damaging the capacitor.

TIME CONSTANT is the product of the capacitance value and the insulation resistance of the capacitor. It is usually defined for large value capacitors meant for charge/discharge applications and is specified as Mega-ohm. μF (or seconds). This specification is indicative of the self discharge characteristics of the capacitor once it is charged and left unloaded with the source of charging removed. For instance, a 100 μF capacitor with a time constant specification of 500 seconds will self discharge to 36.8 percent of its initial voltage in 500 seconds. Also, such a capacitor has an insulation resistance of 5 MΩ.

Q.5. With the help of the equivalent circuit of the capacitor as shown in Fig. 2.2 (c), briefly describe the impedance characteristics of the capacitor as a function of frequency. Comment on the role of different parasitic components in the equivalent circuit in determining the usable frequency range of the capacitor.

Ans. The equivalent circuit of Fig. 2.2 (c) is reproduced below in Fig. 2.3. The expression for the impedance can be written as :

$$Z = R_s + j\omega L + \frac{R_p}{1 + j\omega R_p C}$$

Fig. 2.4

As the frequency increases, the inductive reactance increases and above a certain frequency, the capacitor behaves more like an inductor. This puts an upper limit on the frequency upto which the capacitor can be usefully employed as a capacitor. Without going into too much of mathematics, it can be verified that for condition given by $\left[\frac{R_s}{2}\sqrt{\frac{L}{C}} < 1\right]$, which is achievable for low loss capacitors characterised by small values of (R_S) and (L), the usable frequency range is between ($1/R_pC$) and ($1/\sqrt{LC}$). For $\left[\frac{R_s}{2}\sqrt{\frac{L}{C}} > 1\right]$, which is the case for high loss capacitors, the usable frequency range is between ($1/R_pC$) and ($1/R_SC$). In the frequency range for which $\omega^2 R_p^2 C >> 1$, the expression for impedance can be simplified as :

$$Z = \left[R_s + \frac{1}{R_p\,\omega^2 C^2}\right] + j\left(\omega L - \frac{1}{\omega C}\right)$$

$$= R_{ESR} + j(X_L - X_C)$$

Fig. 2.5

The simplified equivalent circuit is then reduced to what is shown in Fig. 2.5 (R_{ESR}) represents capacitor's energy losses in terms of an equivalent series resistance. In case of *DC* or very low frequency operation, the equivalent of circuit further reduces to a series connection of (R_{ESR}) and (C).

The generalised frequency response of a capacitor would look like the curve of Fig. 2.6.

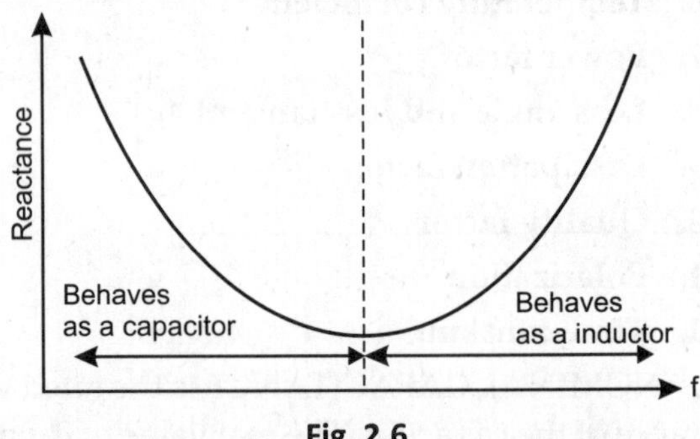

Fig. 2.6

Q.6. How do you calculate the total capacitance of a number of capacitors connected in series?

Ans. Refer to Fig. 2.7. The equivalent series capacitance (C_T) for a number of series connected capacitors C_1, C_2, C_3, C_4,..C_n is given by

$$\frac{1}{C_T} = \frac{1}{C_1} + \frac{1}{C_2} + \frac{1}{C_3} + \frac{1}{C_4} + \ldots\ldots + \frac{1}{C_n}$$

$$\text{or } C_T = \frac{1}{\left(\frac{1}{C_1}\right) + \left(\frac{1}{C_2}\right) + \left(\frac{1}{C_3}\right) + \left(\frac{1}{C_4}\right) + \ldots\ldots + \left(\frac{1}{C_n}\right)}$$

Fig. 2.7

The concept behind this result is that whenever a string of series connected capacitors is connected across a voltage source, this applied voltage is equal to the sum of voltages across individual capacitors. Also, the charge that would be stored on the equivalent capacitor due to the applied voltage will be the same as the charge stored on each one of the individual capacitors. This forms the basis of the derivation of the above expression.

Q.7. How do you find out the equivalent capacitance (C_T) of a number of capacitors C_1, C_2, C_3, C_n connected in parallel?

Ans. Refer to Fig. 2.8. The equivalent capacitance (C_T) for a number of capacitors C_1, C_2, C_3.....C_n connected in parallel is given by

$$C_T = C_1 + C_2 + C_3 + ... + C_n$$

Fig. 2.8

The concept behind this result is that in case of parallel connected capacitors, individual capacitors have the same potential difference across them and the charge that would be stored on the equivalent capacitance would be equal to sum of charges stored on individual capacitors. This forms the basis of derivation of above expression.

Q.8. How do you convert Star connected capacitors into equivalent Delta connected capacitors ?

Ans. Refer to Fig. 2.9. The equivalent delta connected capacitors are given by

$$C_{AB} = \frac{C_A \cdot C_B}{C_A + C_B + C_C}; \; C_{BC} = \frac{C_B \cdot C_C}{C_A + C_B + C_C};$$

$$C_{CA} = \frac{C_A \cdot C_C}{C_A + C_B + C_C}$$

Q.9. How do you convert given Delta connected capacitors into equivalent Star-connected capacitors ?

Ans. Refer to Fig. 2.9. The equivalent star-connected capacitors are given by

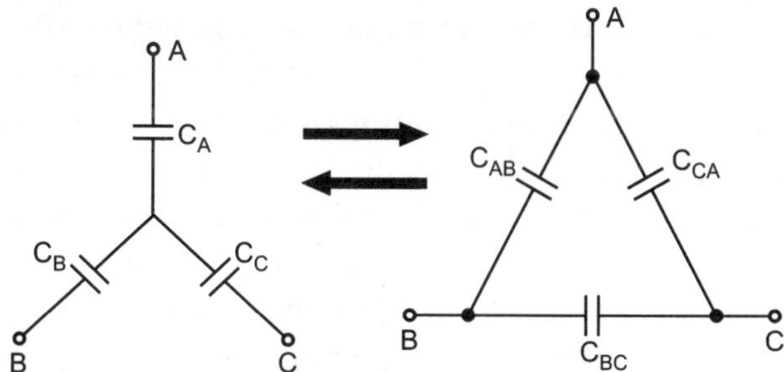

Fig. 2.9

$$C_A = C_{AB} + C_{CA} + \frac{C_{AB} \cdot C_{CA}}{C_{BC}}; \; CB = C_{BC} + C_{AB} + \frac{C_{BC} \cdot C_{AB}}{C_{CA}}; \; C_C = C_{CA} + C_{BC} + \frac{C_{CA} \cdot C_{BC}}{C_{AB}}$$

Q.10. There can be various types of conductor-dielectric configurations to form capacitors. Some of these include

(*a*) Parallel plate capacitor with air dielectric as shown in Fig. 2.10 (*a*)

(*b*) Parallel plate capacitor with a dielectric material with relative permittivity (ε_r) as shown in Fig. 2.10 (*b*).

(*c*) Parallel plate capacitor with a composite dielectric as shown in Fig. 2.10 (*c*).

(*d*) Capacitor with multiple plates as shown in Fig. 2.10 (*d*).

(*e*) Capacitance between two wires separated by a dielectric as shown in Fig. 2.10 (*e*)

(*f*) Capacitance formed by two concentric cylinders as shown in Fig. 2.10 (*f*).

(g) Capacitance formed by two concentric spheres as shown in Fig. 2.10 (g).

(*h*) Capacitance of an isolated sphere.

Write expressions for the capacitance value in each of the above cases.

Ans. (*a*) $C = \frac{\varepsilon_0 A}{d}$

where (A) is area of each plate and (d) is inter-plate separation.

(*b*) $C = \frac{\varepsilon_0 \varepsilon_r A}{d}$

where ε_0 = Free space permittivity = 8.85×10^{-12} F/m

ε_r = Relative permittivity of dielectric.

Relative permittivity of commonly used capacitor dielectric materials is given in Table 2.1.

(*c*) $C = \frac{\varepsilon_0 A}{(d_1/\varepsilon_{r1}) + (d_2/\varepsilon_{r2})}$

(a) Parallel plate capacitor with air dielectric

(b) Parallel plate capacitor with a dielectric (ε_r)

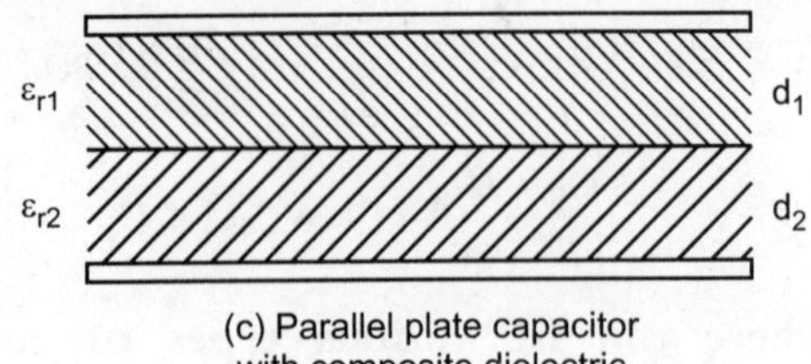

(c) Parallel plate capacitor with composite dielectric

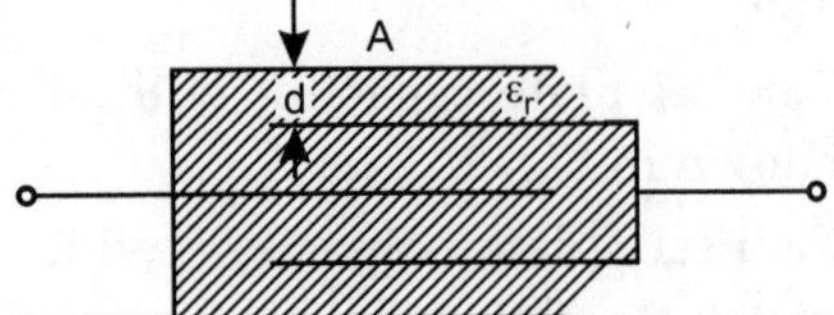

(d) Parallel plate capacitor with multiple plates.

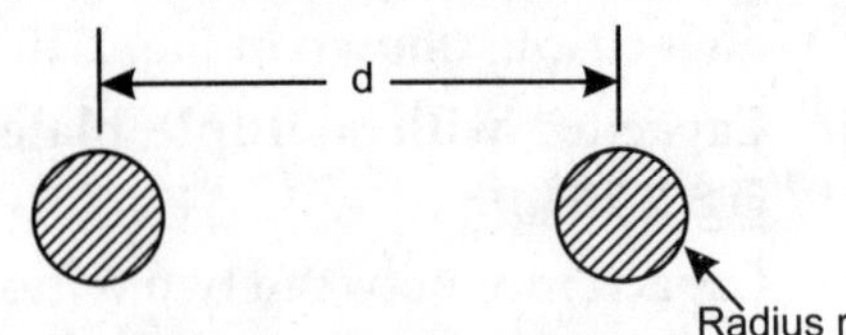

(e) Two wires separated by a dielectric of dielectric strength (ε_r)

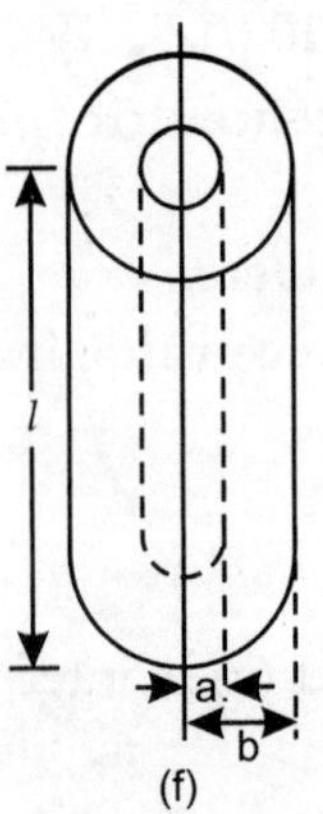

(f)

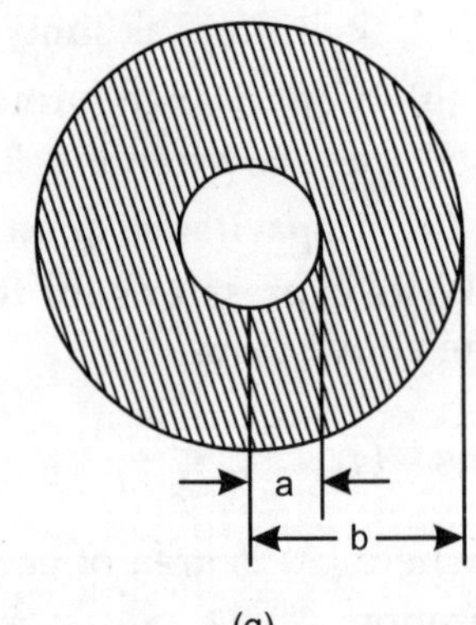

(g)

Fig. 2.10

(*d*) $C = \dfrac{(n-1)\,\varepsilon_0 \varepsilon_r\, A}{d}$ where (*n*) is number of plates

(*e*) $C = \dfrac{\pi\,\varepsilon_0 \varepsilon_r\, l}{2.3 \log_{10}(d/r)}$ where *l* = length of each wire

(*f*) $C = \dfrac{2\,\pi\,\varepsilon_0\,\varepsilon_r\, l}{2.3 \log_{10}(b/a)}$

(g) $C = \dfrac{4\pi\,\varepsilon_0\,\varepsilon_r\, ab}{(b-a)}$

(*h*) $C = 4\pi\,\varepsilon_0\, R$ where *R* = Radius of the sphere.

Q.11. Do capacitors also have a preferred set of capacitance values like we have in the case of resistors? If yes, what are the standard or preferred capacitance values in which different types of capacitors are generally manufactured?

Ans. Like resistors and inductors, capacitors too have a set of standard values. These standard values between 1 pF and 10 pF in the case of ceramic, film and tantalum capacitors are 1.2, 1.3, 1.5, 1.6, 1.8, 2.0, 2.2, 2.4, 2.7, 3.0, 3.3, 3.6, 3.9, 4.3, 4.7, 5.1, 5.6, 6.2, 6.8, 7.2, 8.5 and 9.1 (all in pF). Higher standard capacitance values can be obtained by multiplying these values by 10, 100 and 1000. For example, 1.2 pF, 12 pF, 120 pF and 1200 pF are all standard values. Some of these values including 1.5, 2.2, 3.3, 4.7 and 6.8 are commonly available up to nine decades. For example, 2.2, 22, 220, 2200 (all in pF), 0.022, 0.22, 2.2, 22, 220, 2200 (all in μF) are standard values. Aluminium electrolytic and double layer capacitors are available in still higher capacitance values. While aluminium electrolytic capacitors are available in capacitance range of 0.1 μF to 0.1 F; double layer capacitors also called supercapacitors or ultracapacitors are available in capacitance range of 0.1 F to 100 F.

Q.12. Based on the type of dielectric used, fixed capacitors can be categorized as

(*i*) Paper capacitors
(*ii*) Plastic film capacitors, like Polyester Capacitors, Polystyrene capacitors etc.
(*iii*) Mica capacitors
(*iv*) Ceramic capacitors
(*v*) Electrolytic capacitors
(*vi*) Oil filled capacitors
(*vii*) Air dielectric capacitors and so on.

Do a comparative study of these types as regards their advantages and disadvantages.

Ans. Paper capacitors

The dielectric used in paper capacitors is paper or oil impregnated paper (in case of energy storage capacitors). Paper capacitors are extensively used for high voltage high discharge current applications.

However, paper capacitors have a very small capacitance to volume ratio. Also since paper is extremely hygroscopic, it absorbs moisture from atmosphere despite plastic enclosures and impregnants thus increasing its power factor and decreasing its insulation resistance. Metalized paper capacitors are comparatively smaller in size and have higher voltage ratings. But then these are suitable for low current applications only. Fig. 2. 11 shows construction and pictures of a variety of paper capacitors.

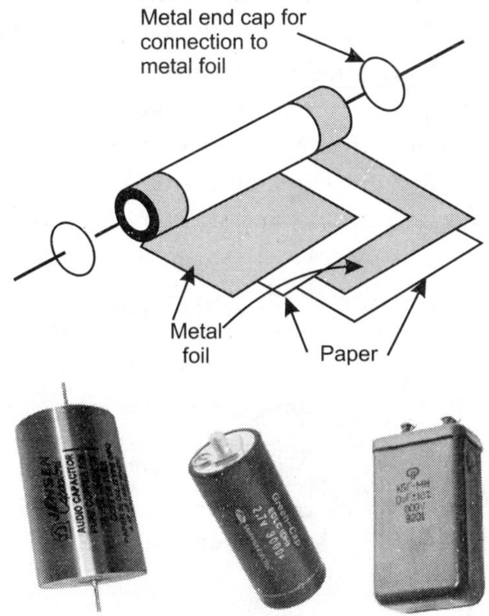

Fig. 2.11

Plastic Film Capacitors

Depending upon the type of plastic film used, these capacitors are further classified as Polyester (or MYLAR) capacitors (Polyester or Mylar is the dielectric used), Polystyrene capacitors (dielectric used in this case is Polystyrene), Polycarbonate capacitors

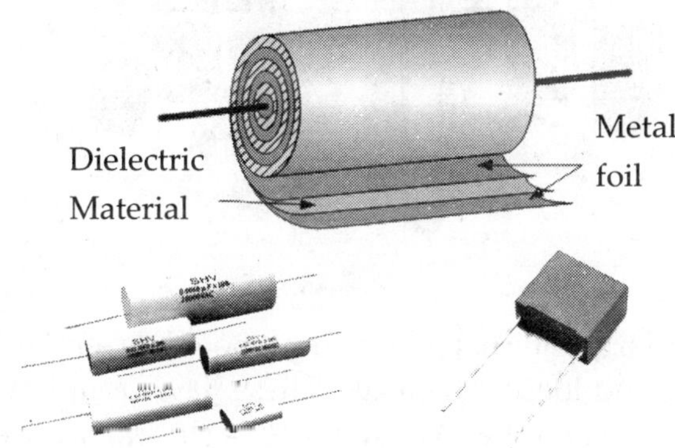

Fig. 2.12

(dielectric used in Polycarbonate), Polypropylene capacitors (Polypropylene is the dielectric), PTFE-fluorocarbon (Teflon) capacitors and so on. Metalized film capacitors are also made in polyester and polycarbonate types.

Polyester (or Mylar) capacitors are smaller in size as compared to paper capacitors of comparable specifications and have almost completely replaced paper capacitors for most of DC applications with operating voltages up to 2000 volts or so and an operating temperature of + 125°C (Max). Moisture pickup is less. The temperature stability of mylar capacitors is poorer than that of paper capacitors. Mylar capacitors are the most extensively used polyester film capacitors.

Polystyrene capacitors on the other hand have excellent stability, low moisture pick-up and a slightly negative temperature coefficient of capacitance that can be used to advantage to match the positive temperature coefficient of other components. These are comparatively bigger in size and their maximum operating temperature is limited to + 85°C.

Polycarbonate capacitors have superior insulation resistance, dielectric absorption and dissipation factor as compared to polyester capacitors. Moisture pick up is also small and have nearly zero temperature coefficient.

Polypropylene capacitor are known for extremely small dissipation factor, very high dielectric strength and low moisture pick-up.

PTFE capacitors have operating temperature of upto +170°C and an extremely high insulation resistance. These are very expensive and have large size.

Metalized plastic film capacitors are reliable and significantly smaller in size but like metalized paper capacitors are good only for low current applications.

Fig. 2.12. shows construction and pictures of the whole range of plastic film capacitors.

Mica capacitors

Mica capacitor use Mica as the dielectric material. It is one of the oldest dielectric materials used in capacitors. Mica is very stable electrically, mechanically and chemically. There are two types of mica capacitors namely clamped mica capacitors and silver mica capacitors. Silver mica capacitors due to their superior characteristics have rendered as obsolete the clamped mica capacitors. Silver mica capacitors are fabricated by sandwiching mica sheets coated with metal on both sides. The assembly is dipped in epoxy. Silver mica capacitor are high stability, high precision and high reliability capacitors having large capacitance to volume ratio. They are low loss capacitors, which makes them particularly attractive for high frequency applications. They are usually available in small capacitance values ranging form 1fF to 100 nF.

Ceramic Capacitors

In ceramic capacitors, we have two major categories

(*i*) *Class-I:* Temperature compensating ceramic capacitors using mixtures of complex titanate compounds as dielectric and (*ii*) *Class-II:* High dielectric strength type ceramic capacitors using Barium titanate based dielectrics. We also have class-III and class-IV and barrier layer capacitors. These are not standardized any more.

Temperature compensating ceramic capacitors are low cost, small size capacitors. These have excellent high frequency characteristics and have good reliability. These have very low tolerances, low losses and high insulation resistance. These have a predictable linear capacitance change with operating temperature. The disadvantages include change in capacitance with change in applied voltage and ageing effects. These are particularly suitable for tuned circuits, timing and other precision circuits.

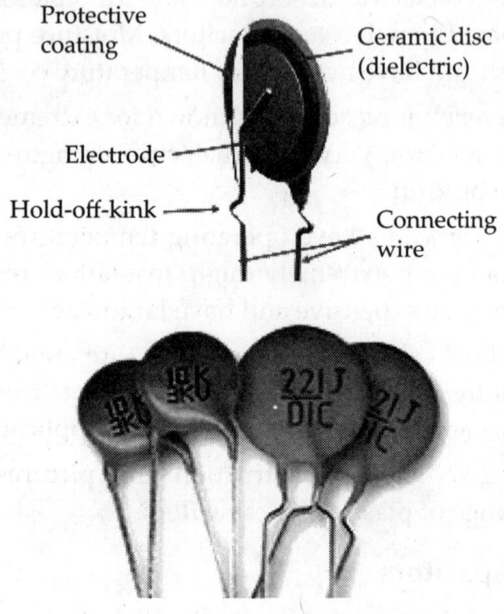

Fig. 2.13

Class-II ceramic capacitors are comparatively much smaller than class-I ceramic capacitors due to higher dielectric strength of the ceramics used. However, these are comparatively less stable. However, Class-II ceramic capacitors have relatively higher temperature coefficient and loss factor as compared to Class-I ceramic capacitors. These are commonly used as bypass, and decoupling capacitors. Fig. 2.13 shows construction and pictures of typical ceramic capacitors.

Electrolytic Capacitors

The Electrolytic Capacitors use some form of electrolyte as the dielectric. It is aluminium oxide in case of aluminium electrolytic capacitors and tantalum oxide in case of Tantalum electrolytic capacitors. An aluminium electrolytic capacitor consists of two strips of aluminium foil (anode and cathode) with inleaved paper. The foils and paper are impregnated with electrolyte. Aluminium oxide formed on anode foil surface acts as dielectric. Electrolytic capacitors in general have a large capacitance to volume ratio. Aluminium electrolytic capacitors are inexpensive and are extensively used in power supply filtering and decoupling applications. Disadvantages of aluminium electrolytics include higher direct current leakage, large internal inductances, poor low temperature stability and loose tolerances. These have been found to burst open when overloaded. Fig. 2.14. shows construction and pictures of a cross-section of aluminium electrolytic capacitors.

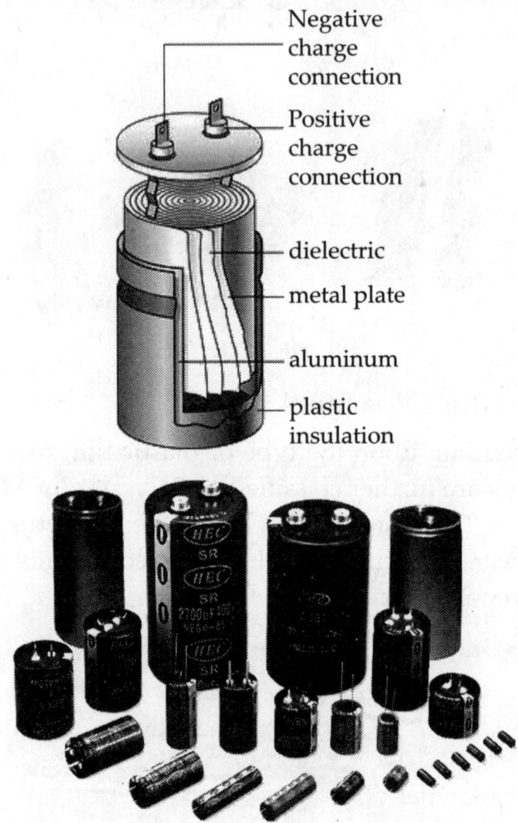

Fig. 2.14

Fig. 2.15

Tantalum Capacitors on the other hand have good stability, wide operating temperature range and reliable operating life. Aluminium electrolytic capacitors are available only in ' polarised variety whereas tantalum capacitors are available in both polarised as well as non-polarised varieties. Tantalum capacitors are extensively used in miniaturised electronic equipment and computers. Fig. 2.15 shows pictures of typical tantalum capacitors.

Oil Filled Capacitors

In this category, there are oil filled capacitors for AC applications and there are oil filled capacitors for DC. applications. Alternating oil filled capacitors use oil impregnated paper as dielectric, and are primarily designed to provide very large capacitances for industrial AC applications where these are required to withstand large current and high peak voltages at power line frequencies. Applications include AC motor starting and running, phase splitting, power factor correction, voltage regulation etc. Direct current oil filled capacitors use paper or paper-polyester film combination as dielectric. These are primarily designed for DC applications such as Filtering, Bypassing, Coupling, Arc suppression, Voltage multiplication etc. Fig. 2.16. shows pictures of some oil filled capacitors.

Fig. 2.16

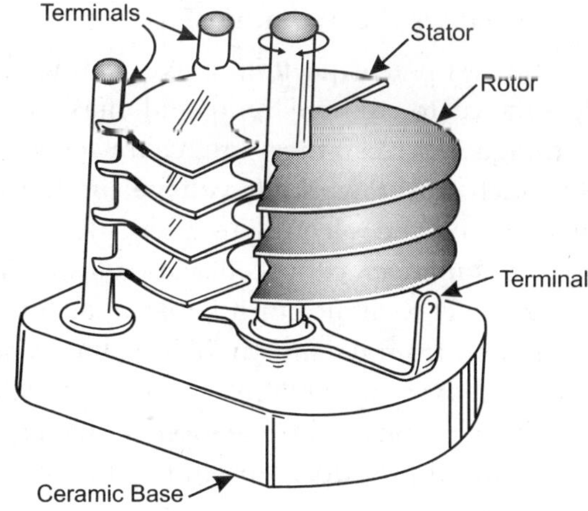

Fig. 2.17

Air Dielectric Capacitors

These usually belong to the class of variable capacitors. Though air as dielectric has lower insulation resistance and dielectric constant, these capacitors exhibit excellent stability. Fig. 2.17 shows the construction and photograph of a typical air dielectric capacitor.

Q.13. What are supercapacitors? How do they differ from conventional capacitors in their construction, operation and characteristics?

Ans. In the simplest form, a conventional capacitor consists of two plates separated by an insulator. The capacitance of such a capacitor is given by $C = \varepsilon A/d$ where (A) is the area of each plate, (d) is separation between plates and (ε) is the permittivity of insulating medium. The insulating medium in a conventional capacitor is made of a solid and dry dielectric material such as paper, mica, polyester, polystyrene, polyethylene, polycarbonate, ceramic, Teflon etc. The dielectric material in this case has a finite thickness ranging from a few microns to a milli-meter in some cases. Due to an upper limit on the surface area of the plates for an acceptable capacitor size and a lower limit on the plate separation due to finite thickness of dielectric used, it is not feasible to realise capacitors with very high capacitance values approaching Farads. **Supercapacitors** also called **ultracapacitors** or **double layer capacitors** or **electric double layer capacitors** (EDLC) offer a solution to overcome this shortcoming.

A **supercapacitor** differs from a conventional or ordinary capacitor in two important ways namely relatively much larger plate surface area and much smaller inter-plate distance. The plates or electrodes are made of metal foil coated on both sides with a porous substance such as powdery, activated carbon, which effectively gives them a larger area that contributes towards achieving a higher capacitance value. The activated carbon for example may be electrochemically etched on the metal foil leading to an effective surface area that would be typically 100,000 times the surface area of a smooth surface.

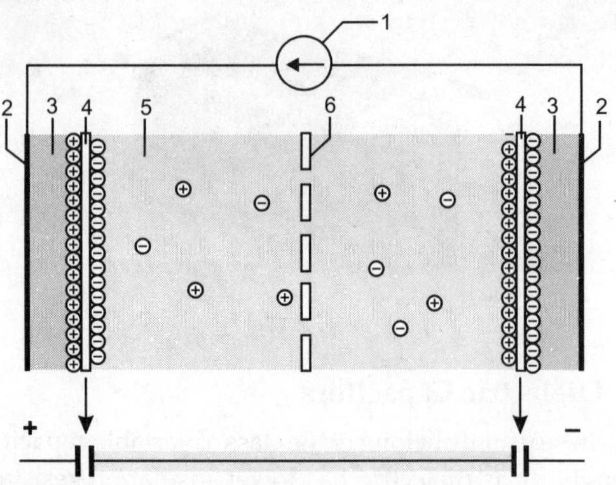

1. Voltage source 2. Metal foil 3. Activated carbon electrodes 4. Electric double layers 5. Electrolyte with positive and negative ions 6. Ion-permeable separator

Fig. 2.18

Also, supercapacitors use a wet dielectric as against a solid dry dielectric used in case of conventional capacitors. Both plates are soaked in an electrolyte, which establishes an ionic connection between the two electrodes. The electrodes are separated by a very thin ion-permeable membrane used as an insulator that acts as a separator to prevent electrodes from getting short circuited. The electrodes along with the separator are rolled or folded into a cylindrical or rectangular shape and then stacked into an aluminium can or a suitable rectangular housing. In recent times, new nano-structured electrode materials such as carbon nano-tubes, metal oxides and conducting polymers have been developed. These electrode materials have demonstrated superior electrochemical capacitance properties that will lead to high performance supercapacitors with improved specific capacitance, rate capability and cyclic stability.

Fig. 2.19

When a conventional capacitor is charged, one of the plates is positively charged and the other plate is negatively charged. This creates an electric field between the plates, which polarises the dielectric to oppose the electric field. That means the plates or electrodes can store more charge at a given voltage. In the case of a supercapacitor, when the electrodes are polarised by an applied voltage, an electric double layer is formed at the interface of electrode and electrolyte on either side of the separator having a polarity opposite to that of the electrodes as shown in Fig. 2.18. For example, in the case of positively polarised electrodes, a layer of negative ions is formed at the electrode/electrolyte interface along with a charge-balancing layer of positive ions adsorbing onto the negative layer. The opposite is true for the negatively polarised electrode. This double layer is very thin of the order of thickness of one molecule as compared to a dielectric whose thickness may vary from a few microns to a millimeter or more in a conventional capacitor. The two electrodes form an equivalent of a series connection of two individual capacitors C_1 and C_2. The capacitance of the supercapacitor cell is then given by $C_1 . C_2/ (C_1 + C_2)$. Fig. 2.19 shows photographs of some typical supercapacitors.

Q.14. Briefly describe salient features of super-capacitors. What are the limitations?

Ans. While super capacitors have extraordinarily high capacitance to volume ratio and make excellent energy storage devices; they are characterised by a relatively much lower working voltage limited by the decomposition of the electrolyte used. The working voltage of a supercapacitor cell varies from 1 to 3 volts. While the capacitance of a supercapacitor can be further increased from its already high value by using a parallel connection of more than one supercapacitor; its working voltage rating can be enhanced by using a series connection of multiple supercapacitors. Supercapacitors also suffer from relatively lower energy density (energy storage capability per unit weight) as compared to batteries.

Supercapacitors offer a combination of high power, high energy as against conventional capacitors that have enormous power but store small amounts of energy. Supercapacitors fill the gap between conventional capacitors and batteries/fuel cells in applications requiring high power and high energy.

Supercapacitors can be charged and discharged almost unlimited number of times without suffering any serious degradation. They are capable of very fast charges and discharges. They can be discharged in time frames ranging from milli-seconds to minutes. They can be charged in time frames ranging from seconds to minutes. They don't release any thermal heat during discharge and are free from the risk of overcharging. Unlike chemical batteries, they are not affected by deep discharge. They have a long shelf life, which reduces maintenance costs. The lifetime is estimated to be about 20 years. Supercapacitors offer *DC*-to-*DC* round trip efficiency in the range of 80 – 95%.

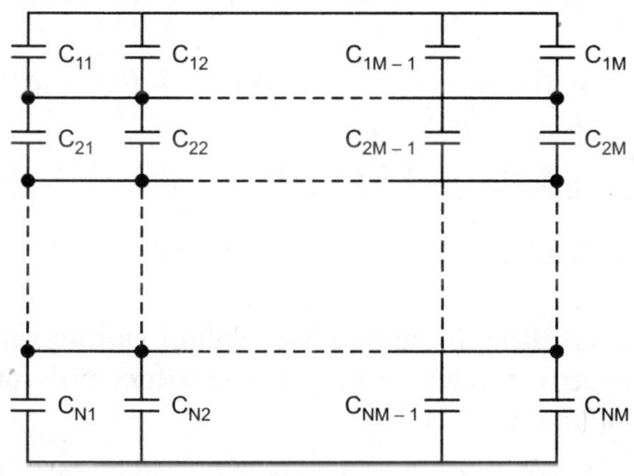

Fig. 2.20

Q.15. How can we enhance both the capacitance value and the working voltage of a supercapacitor?

Ans. Capacitance value can be enhanced, as outlined earlier, by using a parallel connection of capacitors. Working voltage rating can be increased by using a series connection of supercapacitors. A 2-dimensional array of supercapacitors involving both series and parallel connections can be used to increase both specifications. Fig. 2.20 shows the arrangement. The arrangement has (*M*) columns, which means (*M*) capacitors are connected in parallel. It has (*N*) rows, which means (*N*) parallel blocks (each having '*M*' parallel connected supercapacitors) are connected in series. If each supercapacitor has a capacitance of (*C*) and working voltage of (*V*); the array arrangement will have capacitance of $C \times (M/N)$ and a working voltage of (*NV*).

Q.16. What are the important application areas of supercapacitors?

Ans. Supercapacitors are commonly used for energy storage applications, which includes load levelling, uninterrupted power supplies (UPS) or power bridging and pulse power applications demanding a high pulse power for a relatively shorter period of time. If you need to store a reasonable amount of energy for a relatively short period of time ranging from a few seconds to a few minutes, it may be too much of energy to store in a capacitor and the available time may be little to charge a battery. In such situations, a supercapacitor offers an optimal solution. Due to their many benefits as outlined in answer to Q-14, while supercapacitors are currently being used in thousands of different applications in providing back-up power, regenerative power, burst power, quick charge and cold starting; they are being considered in an equally diverse range of future applications.

Supercapacitors have been widely used as energy reservoirs to smooth out power supplies to electrical and electronic equipment. Supercapacitors can also be connected to batteries to regulate the power they supply. One common application is use of large supercapacitors to smooth out intermittent power supplied by wind in the operation of wind turbines. In UPS or power bridging applications, supercapacitors are used to maintain power until power back-up can be established. One such application could be to maintain power while battery is being replaced. They can also be used to maintain power quality to computers and other electronic equipment that are sensitive to voltage dips and surges.

Another common application of supercapacitors is as a temporary energy storage device in the regenerative braking system of electric hybrid vehicles. In the regenerative braking system, the energy that would have otherwise gone waste during braking action is briefly stored and then reused when it starts moving again. However, supercapacitors suitable for this application would need to have a working voltage of hundreds of volts as the motors that drive electric vehicles run off power supplies rated in that range. This is made possible by a series connection of hundreds of supercapacitors.

Supercapacitors are increasingly replacing batteries in a large number of applications ranging from small ones used in cellular phones to large ones used in automobiles. Though supercapacitors have relatively lower energy density as compared to chemical batteries; their advantages such as large number of charge/discharge cycles, faster charge/discharge capability, no parasitic thermal release that heats up the battery and long life make them an attractive choice. Another application where supercapacitors have either replaced or being explored to replace batteries is in engine starting applications including large diesel generators, tank, submarine and locomotive engines. In such applications, as compared to batteries, supercapacitors offer smaller size and weight, excellent cold weather starting capability, long life and low maintenance. Supercapacitors have also replaced batteries in missiles in some cases because of their insensitivity to extreme temperatures, longer shelf life and higher reliability.

Q.17. How do you calculate the working voltage rating of

(i) a parallel combination of capacitors and

(ii) series combination of capacitors ?

Ans.

(i) The working voltage of a parallel combination of capacitors is always limited by the smallest working voltage of an individual capacitor. For instance, if you connected 0.1 μF, 100 V and 0.22 μF, 40 V in parallel, the overall capacitance of 0.32 μF will have an effective working voltage of 40 V only.

(ii) The working voltage of a series combination of identical capacitors is equal to sum of voltage ratings of individual capacitors provided equalising resistors are used to ensure equal voltage division (Fig. 2.21.) The resistance value of equalising resistors should be much smaller than the insulation resistance of capacitors.

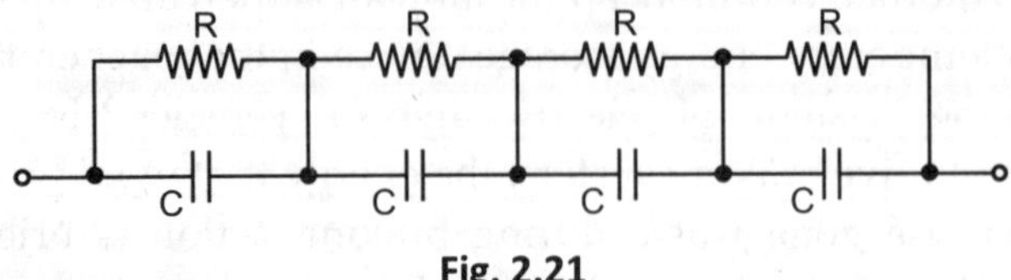

Fig. 2.21

Q.18. When a certain number of capacitors are in turn connected in series first and then in parallel, do you think there is a relationship between the equivalent series and equivalent parallel capacitors. If yes, what is it. Elaborate with the help of an illustration.

Ans. When (n) capacitors are connected in series to get an equivalent series capacitance (C_s) and when the same set of (n) capacitors are connected in parallel to get an equivalent capacitor (C_p), then

$$\left[\frac{C_p}{C_s}\right] > n^2$$

In a special case where all the (n) capacitors have same capacitance, then

$$\left[\frac{C_p}{C_s}\right] = n^2$$

As an illustration, let us say that we have connected four capacitors of capacitance value 2 μF, 4 μF, 8 μF and 16 μF in series, The equivalent series capacitance is given by

$$\frac{1}{C_s} = \frac{1}{2}+\frac{1}{4}+\frac{1}{8}+\frac{1}{16} = \frac{15}{16}$$

$$\text{or} \quad C_s = \frac{16}{15} \cong 1.07\ \mu\text{F}$$

If these capacitors, were connected in parallel, the equivalent parallel capacitance would be given by

$$C_p = 2+4+8+16 = 30\ \mu\text{F}$$

Now $n = 4$ which gives $n^2 = 16$

Also C_p (= 30 μF in this case) > 16 C_s (where C_s = 1.07 μF)

As another illustration, let us say all four capacitors are of 4 μF each. In that case,

$$C_s = 1\ \mu\text{F} \text{ and } C_p = 16\ \mu\text{F}$$

Which gives $\frac{C_p}{C_s} = 16 = n^2$ (Since $n = 4$).

Q.19. Outline the methods of colour coding ceramic disc capacitors, tubular ceramic capacitors and molded mica capacitors.

Ans. Colour coding of capacitors today is not as popular as it was in early days. Today, majority of capacitor types are plainly marked with important capacitor specifications like the nominal capacitance value, the tolerance and the working voltage. The polarity is also marked in case of polarised capacitors.

The ceramic discs of nominal value of less than 1000 pF are usually plain marked. For instance, for 220 pF capacitors, it will be marked 220 only. For capacitance values of 1000 pF or more, a three digit code is used. The first two digits represent the two significant digits and the third digit represents a decimal multiplier, for instance, 102 represents a capacitance of $10 \times 10^2 = 1000$ pF and 10^4 represents a capacitance of 10×10^4 pF = 100 kpF = 0.1 μF (Fig. 2.22)

Fig. 2.22

Colour coding of tubular ceramic disc capacitors is given in Fig. 2.23 and Fig. 2.24 and the accompanying data in Tables 2.2, 2.3 and 2.4. Colour coding of molded Mica Capacitors is given in Fig. 2.25 and the accompanying data in Table 2.5. Information given in Figs. 2.23, 2.24 and 2.25 is self explanatory.

Table 2.1: Dielectric constants of commonly used dielectric Materials

S.No.	*Dielectric Material*	*Dielectric Strength*
1	Air	1.0001
2	Vacuum	1.0000
3	Impregnated paper	4.6
4	Polyester	3.0
5	Polystyrene	2.5 to 2.7
6	Polycarbonate	3.0
7	Polypropylene	2.0
8	Polysulphone	3.0
9	Teflon	2.0
10	Polyimide	3.5
11	Mica	4.5 to 7.5
12	Glass	6.7
13	Green Glass	8.3
14	Quartz	4.2
15	Steatite	5.5 to 6.5
16	Titanium Dioxide	80 to 120
17	Barium Titanate	200 to 16000
18	Aluminium Oxide	10.0
19	Titanium Oxide	11.0

Table 2.2: (Capacitance value & Tolerance for 5 & 6 band codes)

Colour	*Digit (C&D)*	*Multiplier factors (E)*	*Tolerance (F)*	
			C < 10 pF	C . 10 pF
Black	0	10^0 or 1	± 2.0 pF	± 20%
Brown	1	10^1	± 0.1 pF	± 10%
Red	2	10^2	–	± 2%
Orange	3	10^3	–	± 3%
Yellow	4	10^4	+ 100% – 0%	+ 100% – 0%
Green	5	–	± 0.5 pF	± 5%
Blue	6	–	–	–
Violet	7	–	–	–
Grey	8	10^{-2} or 0.01	± 0.25 pF	+ 80– 20%
White	9	10^{-1} or 0.1	± 1.0 pF	± 10

Table 2.3 (Temp, coeff. for 5 band code)

A	*T.C.*
Black	NP0
Brown	N030
Red	N080
Orange	N150
Yellow	N220
Green	N330
Blue	N470
Violet	N750
Gold	P100
White	–
Grey	–

Table 2.4 (Temp, coeff. for 6 band code)

Digit		*T.C.*
A	**B**	
Grey	Black	
Orange	Orange	N 1500

Cont....

Yellow	Orange	N 2200
Green	Orange	N 3300
Blue	Orange	N 4700
Red	Violet	P 100
Green	Blue	P 030
Gold	Orange	X5F
Brown	Orange	Z5F
Gold	Yellow	X5P
Brown	Yellow	Z5P
Gold	Blue	X5S
Brown	Blue	Z5S
Gold	Grey	X5U
Brown	Grey	Z5U

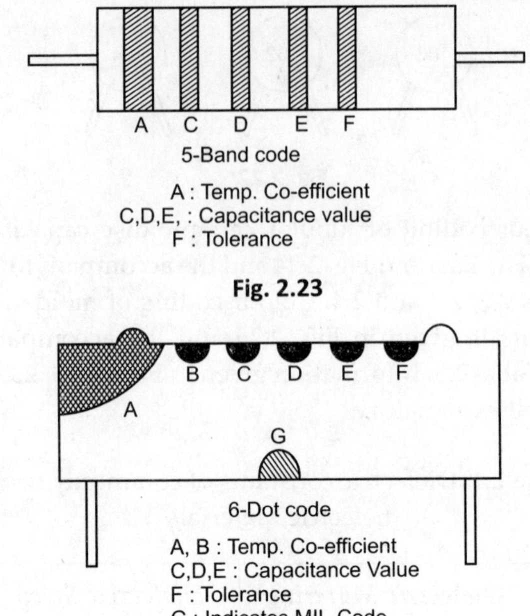

Fig. 2.23

Fig. 2.24

Table 2.5

Colour	*Digits*	*Multiplier*	*Tolerance*	*D.C. Working*	*Operating*
Black	0	100 or 1	± 20 %	–	– 55°C to + 70°C
Brown	1	10^1	± 1 %	–	–
Red	2	102	± 1 %	–	– 55°C to + 85°C
Orange	3	103	–	300	–
Yellow	4	104	–	–	– 55°C to + 125°C
Green	5	–	± 5 %	500	–
Blue	6	–	–	–	– 55°C to + 150°C
Violet	7	–	–	–	–
Grey	8	–	–	–	–
White	9	–	–	–	–
Gold	–	–	± 0.5 %	1000	–
Silver	–	–	± 10%	–	–

Q.20. Capacitors are made in a large variety depending upon the type of dielectric used. What are the typical capacitance ranges available in each of the popular capacitor varieties?

Ans. Capacitor ranges available in different capacitor types are outlined in Table 2.6.

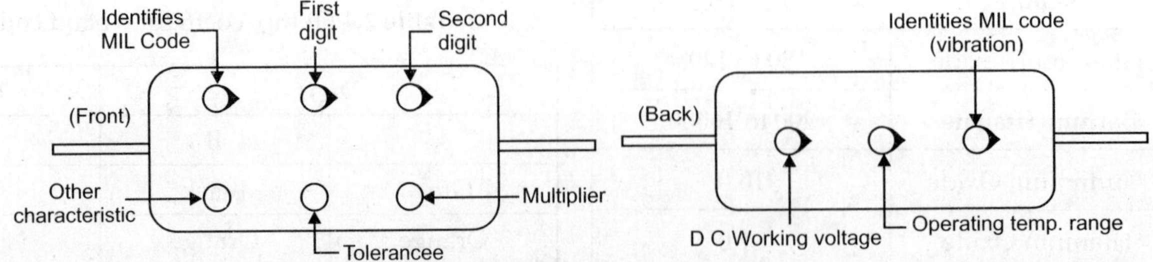

Fig. 2.25

Table 2.6: Capacitance ranges available in different capacitor types.

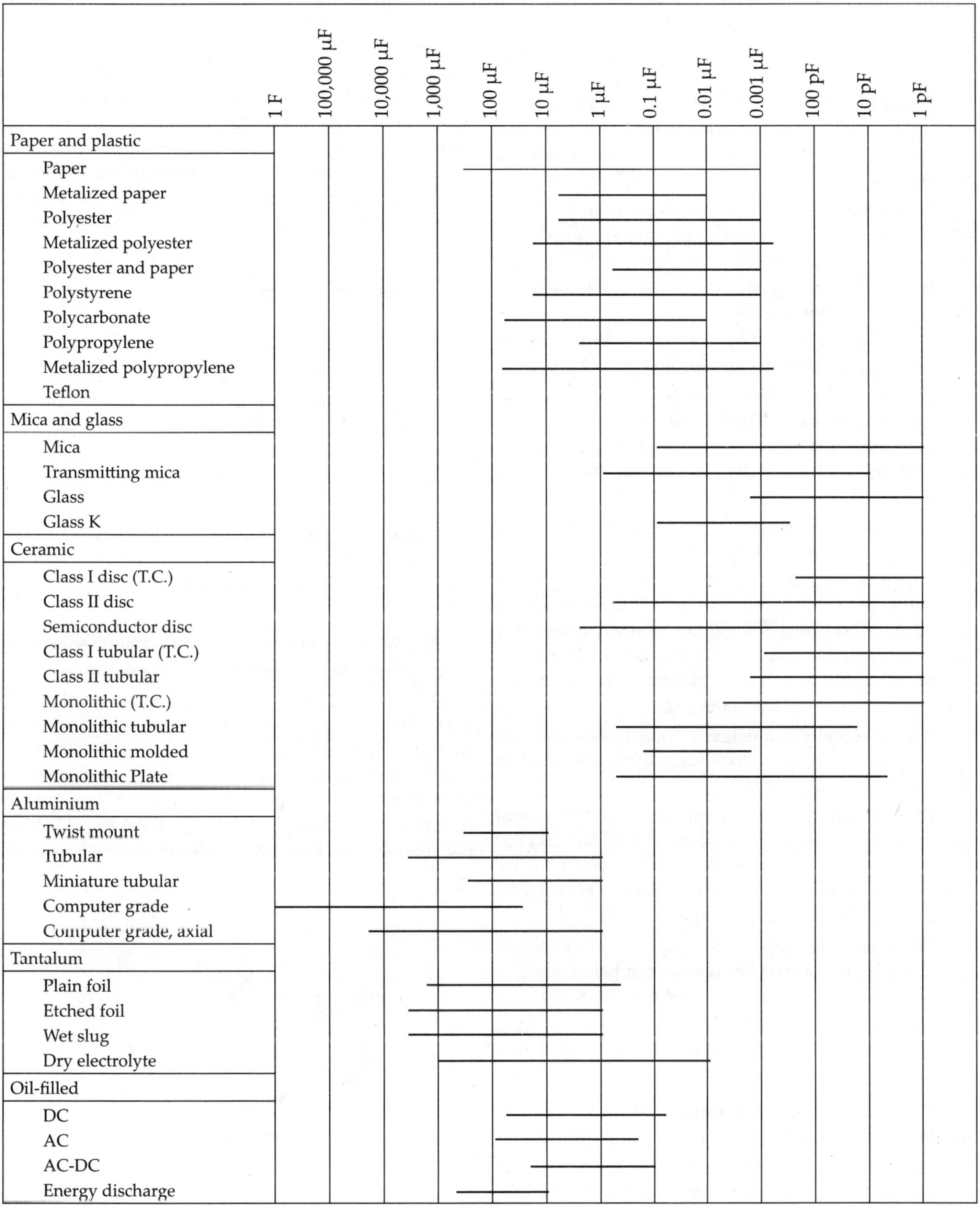

	1 F	100,000 μF	10,000 μF	1,000 μF	100 μF	10 μF	1 μF	0.1 μF	0.01 μF	0.001 μF	100 pF	10 pF	1 pF
Paper and plastic													
Paper													
Metalized paper													
Polyester													
Metalized polyester													
Polyester and paper													
Polystyrene													
Polycarbonate													
Polypropylene													
Metalized polypropylene													
Teflon													
Mica and glass													
Mica													
Transmitting mica													
Glass													
Glass K													
Ceramic													
Class I disc (T.C.)													
Class II disc													
Semiconductor disc													
Class I tubular (T.C.)													
Class II tubular													
Monolithic (T.C.)													
Monolithic tubular													
Monolithic molded													
Monolithic Plate													
Aluminium													
Twist mount													
Tubular													
Miniature tubular													
Computer grade													
Computer grade, axial													
Tantalum													
Plain foil													
Etched foil													
Wet slug													
Dry electrolyte													
Oil-filled													
DC													
AC													
AC-DC													
Energy discharge													

Q.21. Briefly compare salient features of supercapacitors and batteries as an energy storage device.

Ans. A brief comparison of supercapacitors and batteries on the basis of energy storage, charging method, charge/discharge time, form factor, energy density, power density, working voltage, lifetime and operating temperature range is presented in the following paragraphs.

Energy storage: While supercapacitors store energy of the order of watt-seconds employing surface charge storage mechanism; batteries have much higher energy storage capability of the order of watt-hours and employ chemical storage mechanism.

Charging method: Supercapacitors use constant voltage charging. Batteries employ both constant voltage charging and constant current charging.

Charge/discharge Times: Supercapacitors have charge/discharge times in the range of milli-seconds to seconds. Batteries have charge/discharge times of the order of hours.

Form factor: Supercapacitors have a much smaller form factor as compared batteries.

Energy density: Batteries offer highest energy density capability of conventional capacitors, supercapacitors and batteries. Supercapacitors offer energy density in the range 1 – 10 watt-hour/kg. The energy density in the case of batteries could be anywhere between few watt-hours/kg to several hundred watt-hours/kg.

Power density: Supercapacitors have relatively higher power density (> 5000 watt/kg) as compared to batteries (100 – 3000 watt/kg).

Working voltage: working voltage of a supercapacitor cell is anywhere between 2.3 volts and 2.75 volts. The same for a battery cell is between 1.2 volts and 4.2 volts. The working voltage in both cases can be enhanced by connecting individual cells in series.

Lifetime: Supercapacitors have a much longer lifetime in excess of 100,000 cycles as compared batteries where it is usually limited to a few thousand cycles.

Operating temperature range: Supercapacitors have a much wider operating temperature range of – 40°C to + 85°C as compared to – 20°C to + 65°C in case of batteries.

Q.22. What are variable capacitors? How are they represented in circuits? Where do they find application?

Ans. A variable capacitor is the one whose capacitance can be varied either electronically or mechanically. Like resistors, they are also classified as variable capacitors whose capacitance can be continuously or frequently varied and preset capacitors also called trimmer capacitors whose capacitance is usually varied and set for desired performance. There is a third type known as ganged capacitor. A gang capacitor is generally a set of two or more variable capacitors mounted on a common shaft to allow capacitance value adjustment by a common control. Fig. 2.26 shows circuit representations of variable capacitor, preset or trimmer capacitor and a gang capacitor.

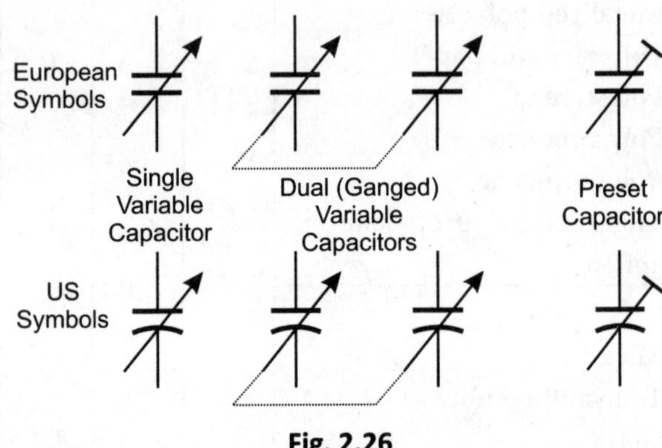

Fig. 2.26

Variable capacitors are commonly used for tuning resonant frequency of LC circuits and as a variable reactance for impedance matching in antenna tuning applications.

Q.23. What are the commonly used mechanisms for varying capacitance value in variable capacitors?

Ans. Two most commonly used mechanisms employed in variable capacitors are the mechanical control and electronic control. In the case of mechanically controlled variable capacitors, in most cases, variable capacitance is accomplished by varying the distance between the parallel plates in a capacitor or by varying the cross-sectional area at which the plates overlap one another.

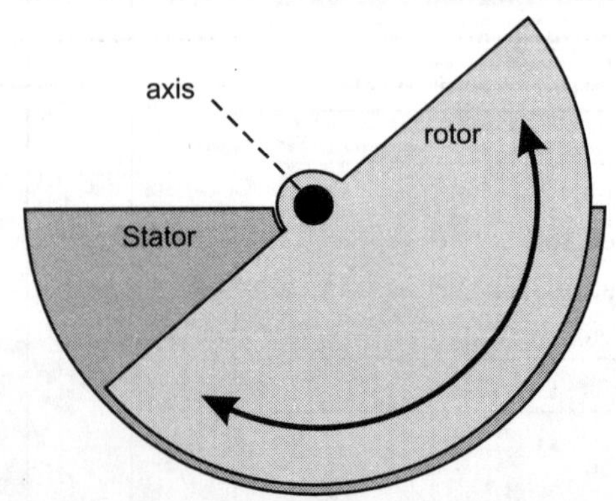

Fig. 2.27

A common form of mechanically controlled variable capacitor consists of two sets of conducting plates intermeshed with each other. One of the sets of conducting plates called stator is fixed and the other set of conducting plates called rotor is fixed to a rotary shaft. The intermeshed plates are separated by a dielectric medium that is either air or some kind of solid dielectric. The capacitance is varied by rotating the rotor shaft. Angular rotation of the shaft alters the overlapping cross-section of stator and rotor plates thereby changing the capacitance. Rotary plates of different shapes are used to get the desired capacitance versus angle characteristics to suit the intended application. Semi-circular plates are commonly used. Fig. 2.27 illustrates the operating principle. Gear reduction mechanisms may be used to achieve finer control of capacitance variation. Fig. 2.28 shows one such mechanically controlled variable capacitor with air dielectric.

Fig. 2.28

Another type of mechanically controlled variable capacitor is the piston capacitor. It comprises of a metal piston that is made to move in and out of a cylindrical metal shell. The two are separated by vacuum, air or ceramic dielectric (Fig. 2.29). Capacitance is varied by moving the piston. The capacitance increases as the piston extends further into the metal shell. The piston is usually attached to a screw shaft to enable piston movement and consequent capacitance value adjustment. This mechanism is commonly used in trimmer capacitors.

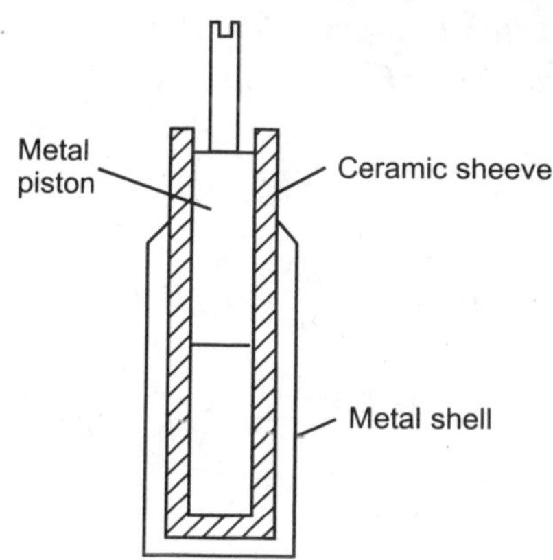

Fig. 2.29

Another common type of mechanically controlled variable capacitor is the compression capacitor. It consists of multiple sets of metal plates separated by sheets of mica dielectric. The entire assembly is mounted on a ceramic or some other form of holder frame. The capacitance is varied by altering either the area of metal plates and dielectric or by changing the number of alternating metal plate-mica layers. The capacitance value variation is enabled by a screw shaft. Force is usually present on both sides of a pair of plates surrounding a mica sheet in order to preserve unit integrity. The capacitor assembly is typically mounted whole on a holder made of ceramic or similar material. This mechanism is commonly used in trimmer capacitors. Fig. 2.30 shows constructional features of a typical compression capacitor.

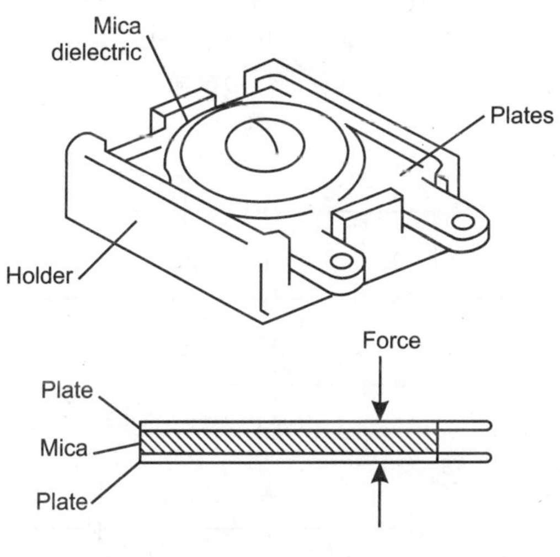

Fig. 2.30

An electronically controlled variable capacitor makes use of capacitance available across a reverse biased P-N junction diode. The capacitance value is varied by applying a variable reverse bias voltage, which in turn varies the depletion width of the junction thereby varying the capacitance. These variable capacitors are known as varactors or varicaps. P-N junction diodes intended for variable capacitance applications are fabricated with large junction area and doping profile to maximize capacitance. Fig. 2.31 shows different circuit representations of a varactor. Varactors are commonly used in frequency modulation of oscillators, high frequency voltage controlled oscillators (VCO) such as those used in phase locked loops (PLL) and tuning of RF filters.

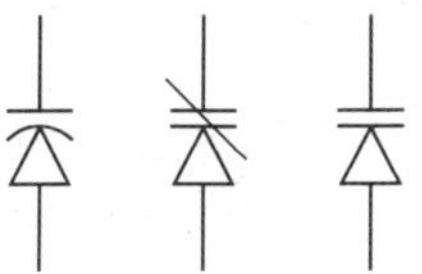

Fig. 2.31

Solved Problems

Problem 2.1. *Find the total equivalent capacitance of the series parallel combinations of capacitors shown in (Fig. 2.32.)*

Solution.

Fig. 2.32 (*a*) – Equivalent capacitance = 0.05 + 0.05

$= 0.1\ \mu$

Fig. 2.32 (*b*) – Equivalent capacitance = 1 + 1 + 1 = 3μ

Fig. 2.32 (*c*) – Equivalent capacitance = $\dfrac{1}{\left(\dfrac{1}{3}+\dfrac{1}{3}+\dfrac{1}{3}\right)}$

$= 1\ \mu F.$

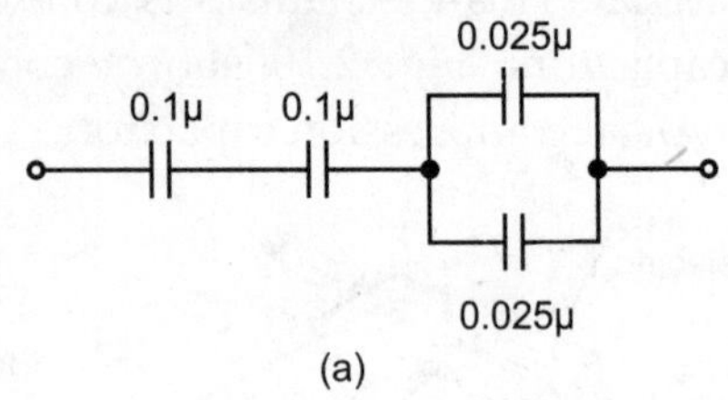

(a)

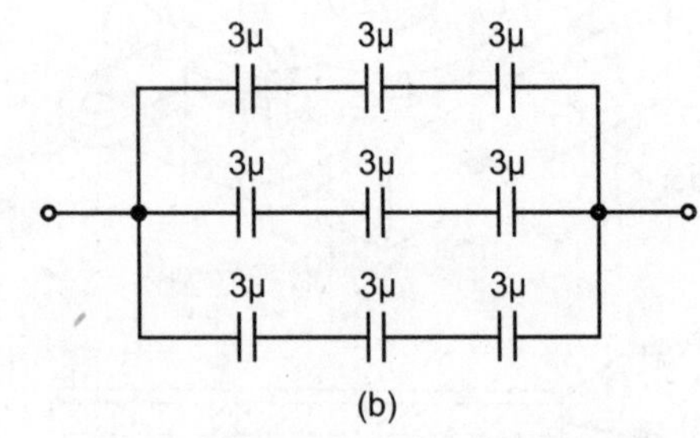

(b)

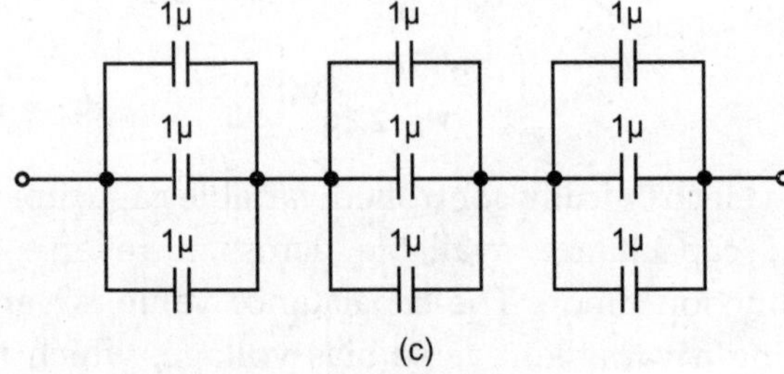

(c)

Fig. 2.32

Problem 2.2. *A given capacitor has a power factor of 0.5. Calculate the Dissipation Factor.*

Solution. Power Factor = 0.5

If the instantaneous current leads the instantaneous voltage by angle (θ).

then cos θ = Power factor = 0.5, that is θ = 60°

If (δ) is the loss angle, then δ = 30°. Dissipation factor $= \tan\delta = \tan 30° = \dfrac{1}{\sqrt{3}}$.

Problem 2.3. *Calculate the Dissipation Factor of a 10 μF capacitor at 1 kHz if its equivalent Series Resistance (ESR) is 100 Ω. Also calculate its loss angle.*

Solution. Dissipation factor is given by tan δ where δ is loss angle. It is also expressed as Dissipation factor = $\dfrac{ESR}{X_C}$, where X_c = Reactance

Reactance (X_c) is given by 1/ωC where ω is radian frequency.

$$\omega = 2\pi f = 6.28 \times 1000 = 6280$$

Therefore

$$X_C = \frac{1}{6280 \times 10 \times 10^{-6}}$$

$$= 16\ \Omega \text{ (approx.)}$$

Dissipation factor is then 100/16 = 6.25. Also

$$\tan\delta = 6.25$$

Therefore $\delta = \tan^{-1} 6.25 = 80.9°$

Problem 2.4. *Temperature co-efficients of some given samples of ceramic capacitors are expressed as N200, P150 and NPO. What do these inscriptions indicate?*

Solution.

N200 on a capacitor indicates that it has a negative temperature co-efficient of 200 ppm/°C.

P150 implies a positive temperature co-efficient of 150 ppm/°C.

NPO means a zero temperature coefficient of capacitance.

Problem 2.5. *The power factor of a certain capacitor is 0.5. Determine its Dissipation factor and loss angle.*

Solution. If (θ) is angle between (*i*) and (*v*), then, Power factor = cos θ = 0.5 which gives θ = 60°

If (δ) is the loss angle, then, δ = 90° – 60° – 30°

Dissipation factor = $\tan\delta = \tan 30° = \dfrac{1}{\sqrt{3}}$.

Problem 2.6. *Fig 2.33. shows the Impedance triangle of a practical capacitor. Determine the Quality factor, Power factor and Dissipation factor for this capacitor.*

Solution. Quality factor = $\dfrac{X_C}{R} = \dfrac{10}{1} = 10$

Dissipation factor = $\dfrac{R}{X_C} = \dfrac{1}{10} = 0.1$

Power factor = $\dfrac{R}{Z} = \cos\theta = \dfrac{R}{\sqrt{R^2 + X_C^2}} = \dfrac{1}{\sqrt{1^2 + 10^2}}$

$= 0.0996.$

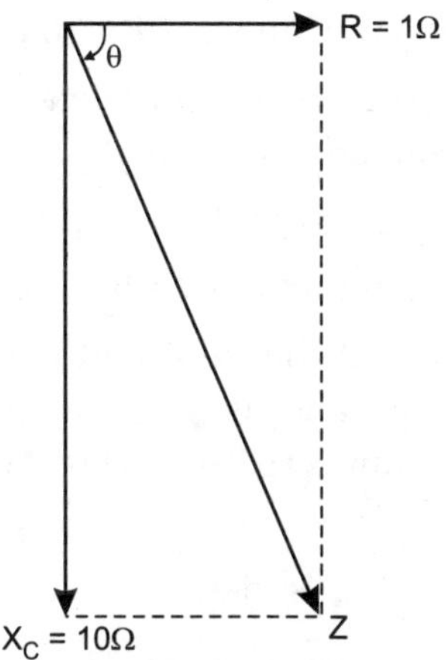

Fig. 2.33

Problem 2.7. *Fig. 2.34 (a) shows the construction of a parallel plate capacitor using a composite dielectric. The dielectric between half of the areas of the two plates has a relative permittivity of (ε_{r1}). The other half has a dielectric with a relative permittivity of (ε_{r2}). Show that the capacitance of this capacitor is* $\frac{\varepsilon_0 A}{2d}(\varepsilon_{r1}+\varepsilon_{r2})$.

Solution. The given capacitor is equivalent to two capacitors (C_1) and (C_2) in parallel as shown in Fig. 2.35 (*b*) where

$$C_1 = \frac{\varepsilon_0 \varepsilon_{r1} A}{2d} \text{ and } C_2 = \frac{\varepsilon_0 \varepsilon_{r2} A}{2d}$$

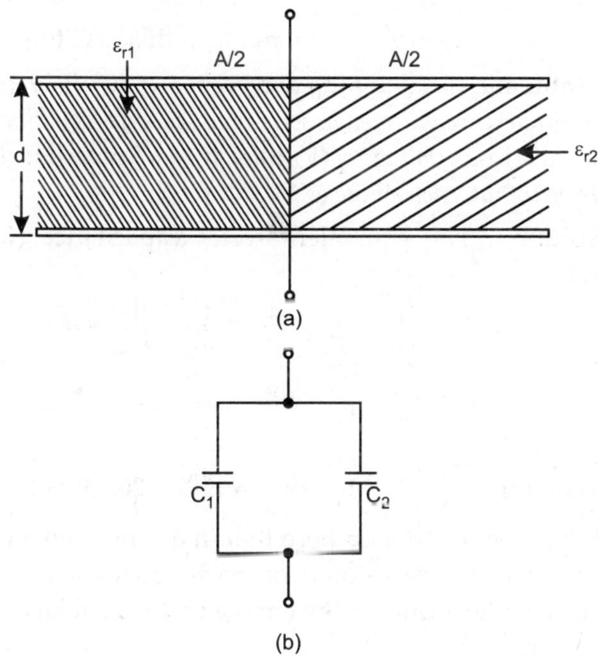

Fig. 2.34

The capacitor of the given capacitor = $C_1 + C_2$ =

$$\frac{\varepsilon_0 \varepsilon_{e1} A}{2d} + \frac{\varepsilon_0 \varepsilon_{r2} A}{2d} = \frac{\varepsilon_0 A}{2d}(\varepsilon_{r1}+\varepsilon_{r2})$$

Problem 2.8. *Determine the capacitance of the parallel plate capacitor with composite dielectric as shown in Fig. 2.35(a).*

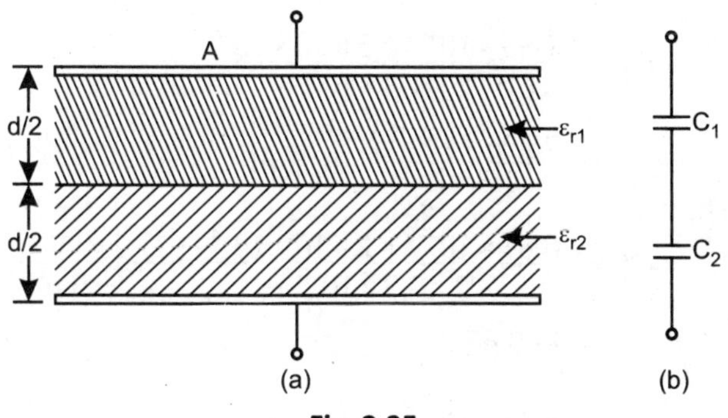

Fig. 2.35

Solution. The given capacitor is equivalent to two capacitors (C_1) and (C_2) in series as shown in Fig. 2.35 (*b*) where

$$C_1 = \frac{\varepsilon_0 \varepsilon_{r1} A}{d/2} = \frac{2\varepsilon_0 \varepsilon r_1 A}{d} \text{ and } C_2 = \frac{\varepsilon_0 \varepsilon_{r2} A}{d/2} = \frac{2\varepsilon_0 \varepsilon r_2 A}{d}$$

The capacitance of given capacitor, $C = \frac{C_1 C_2}{C_1 + C_2}$

$$= \frac{\left(\frac{2\varepsilon_0 A}{d}\right)^2 \varepsilon_{r1}\varepsilon_{r2}}{(2\varepsilon_0 \varepsilon_{r1} A/d) + (2\varepsilon_0 \varepsilon_{r2} A/d)} = \frac{2\varepsilon_0 A}{d}\left(\frac{\varepsilon_{r1}\varepsilon_{r2}}{\varepsilon_{r1}+\varepsilon_{r2}}\right)$$

Problem 2.9. *Refer to Fig. 2.36 (a). Determine the capacitance value of the capacitor represented by structural arrangement of Fig. 2.36. (a)*

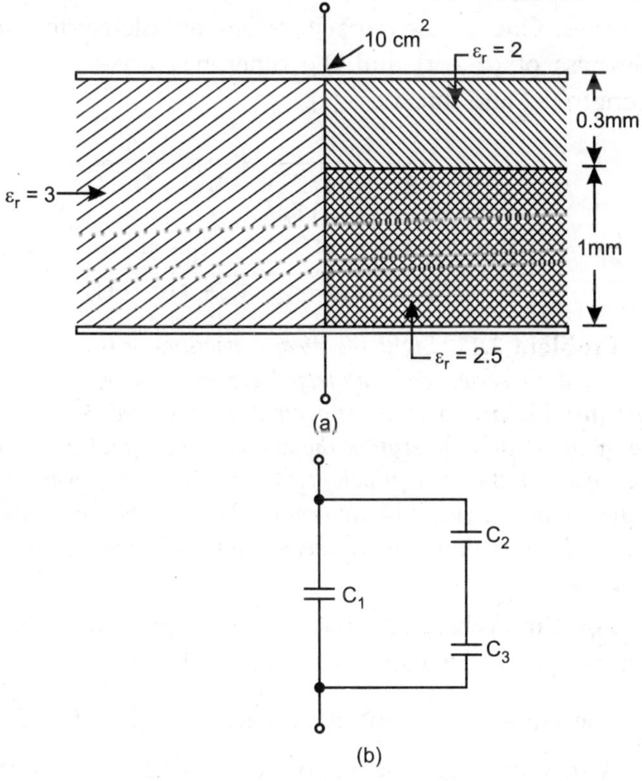

Fig. 2.36

Solution. The given capacitor is a combination of three capacitors C_1, C_2 and C_3 having an equivalent circuit of Fig. 2.37 (*b*).

$$C_1 = \frac{8.85\times10^{-12}\times3\times10\times10^{-4}}{2\times1.3\times10^{-3}}$$

$$= \frac{26.55\times10^{-12}}{1.3\times2} = 10.2 \text{ pF}$$

$$C_2 = \frac{8.85\times10^{-12}\times2\times10\times10^{-4}}{2\times0.3\times10^{-3}} = \frac{17.7\times10^{-12}}{2\times0.3}$$

$$= 29.5 \text{ pF}$$

$$C_3 = \frac{8.85\times10^{-12}\times2.5\times10\times2.5\times10^{-4}}{2\times1\times10^{-3}} = 11 \text{ pF}$$

$$C = C_1 + \frac{C_2C_3}{C_2+C_3} = 10.2 + \frac{29.5\times11}{29.5+11}$$

$$= 10.2 + \frac{324.5}{40.5} = 10.2 + 8 = 18.5 \text{ pF.}$$

Problem 2.10. *A parallel plate capacitor with an air dielectric has a plate area of (A) and plate separation of (d). A dielectric slab of thickness (τ) is inserted between the plates. If the dielectric strength of slab is (ε_r), prove that*

$$C = \frac{\varepsilon_0\varepsilon_r A}{\varepsilon_r d - t(\varepsilon_r - 1)}$$

Solution. When the dielectric slab is inserted, the given capacitor becomes equivalent to two capacitors in series. One of the capacitors has air dielectric with thickness of (d – τ) and the other has a dielectric of strength (ε_r) and thickness (τ).

$$C = \frac{\varepsilon_0 A}{\left(\frac{d-\tau}{1}\right)+\left(\frac{\tau}{\varepsilon_r}\right)} = \frac{\varepsilon_0\varepsilon_r A}{\varepsilon_r(d-\tau)+\tau}$$

$$= \frac{\varepsilon_0\varepsilon_r A}{\varepsilon_r d - \varepsilon_r\tau + \tau} = \frac{\varepsilon_0\varepsilon_r A}{\varepsilon_{rd} - \tau(\varepsilon_r - 1)}$$

Problem 2.11. *Eight identical capacitors of (C) each are connected in series and connected across voltage (V). Each capacitor in the series arrangement is observed to store a charge of 10 μC. Determine the charge that would be stored if a capacitor that is equivalent to the series arrangement is connected across (V). Also determine the total charge stored if those eight capacitors were connected in parallel and connected across (V).*

Solution. When the capacitors are connected in series, voltage across each capacitor equals (V/8).

Therefore, charge on each capacitor = $\frac{CV}{8}$ = 10 μC

Also, equivalent capacitance of series arrangement = C/8

Charge stored on (C/8) = $CV/8$ = 10 μC

When the capacitors are connected in parallel, each capacitor has a voltage (V) across it.

Therefore, charge on each capacitor = CV = 80 μC

Total charge stored = 80 × 8 = 640 μC

We can notice that charge stored when the capacitors are connected in parallel is 64 times the charge stored when they were connected in series. We can generalise and say that

$$q \text{ (parallel)} = n^2 q \text{ (series)},$$

where n = number of capacitors.

This is valid if all capacitors are identical. In case, They are not, Then q (parallel) > n^2 q (series).

Problem 2.12. *In case of data given in Problem 2.11, determine the ratio of total energy stored when the capacitors are connected in series and when connected in parallel*

Solution. Energy stored when they are in series is

$$8\times\frac{1}{2}C\left(\frac{V}{8}\right)^2 = \frac{4CV^2}{64} = \frac{CV^2}{16}$$

Energy stored when they are connected in parallel is

$$8\times\frac{1}{2}CV^2 = 4CV^2$$

Again E (parallel) = 64 E (series) = n^2 E (series)

Similar deductions can be made in this case too.

Problem 2.13. *Four unequal capacitors of 10 μF, 20 μF, 30 μF and 40 μF are connected in series and the series combination connected across a voltage of (V) volts. If 10 μF capacitor stores charge of 100 μC, determine (V).*

Solution. The equivalent series capacitance (C) is given by :

$$\frac{1}{C} = \frac{1}{10}+\frac{1}{20}+\frac{1}{30}+\frac{1}{40} = \frac{25}{120} = \frac{5}{24}$$

$$C = \frac{24}{5} \text{ μF,}$$

Therefore, $\frac{24}{5}\times V = 100$ or $V = \frac{500}{24} = 20.83$ volts.

It may be mentioned here that in a series connection of capacitors, charge stored on each capacitor is same, which is further equal to the charge that would be stored equivalent series capacitance.

Problem 2.13. *Series combination of three capacitors C, 2C and 4C (all in μF) produces an equivalent capacitance of 4 μF. What would be the equivalent capacitance of they were connected in parallel?*

Solution. Equivalent series capacitance (C_s) is given by

$$\frac{1}{C_s} = \frac{1}{C} + \frac{1}{2C} + \frac{1}{4C} = \frac{7}{4C}$$

$$C = \frac{4C}{7} = 4\ \mu F. \text{ This gives } C = 7\ \mu F$$

Equivalent parallel capacitance, $C_p = C + 2C + 4C = 7C = 7 \times 7 = 49\ \mu F$.

Problem 2.15. *Fig. 2.37 shows voltage (v) vs. time (t) and current (i) vs. time (t) graphs of a certain circuit element. Determine whether the circuit element is a Resistor, a Capacitor or an Inductor. Also determine the value of the element.*

Solution. It is clear from the graphs that the element can not be a Resistance as the element does not obey ohm's law. The current (i) here is not proportional to the voltage (v). It can not be an Inductor too because in case of an inductance, $v = (L\ di/dt)$ which implies that a constant current should lead to a zero voltage.

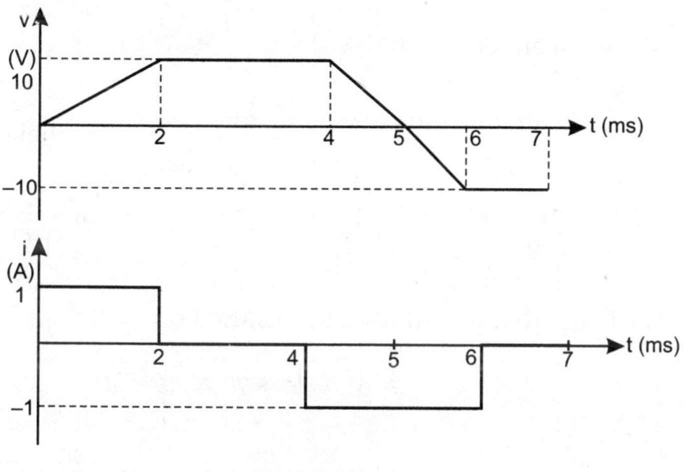

Fig 2.37

In the given graphs, a linearly rising voltage is producing a constant positive current $\left(i = C\frac{dv}{dt}\right)$. A constant voltage is producing a zero current $\left(as\ \frac{dv}{dt} = 0\right)$ and a linearly decreasing voltage is producing a constant negative current ([dv/dt]) is negative]. All this leads to the conclusion that the circuit elements is a capacitor.

Also, $\quad i = C\frac{dv}{dt}$ or $1 = C \times \frac{10}{2 \times 10^{-3}} = 5 \times 10^3\ C$

or $\quad C = \frac{1}{5 \times 10^3} F = \frac{10^6}{5 \times 10^3}\ \mu F = 200\ \mu F$

Problem 2.16. *Refer to Fig. 2.38. Capacitors (C_1) and (C_2) are initially to charged to 100 V but with opposite polarity as shown and switches (S_1) and (S_2) are open. What would be the potential difference across X – X′ after the two switches are closed and the steady state is reached.*

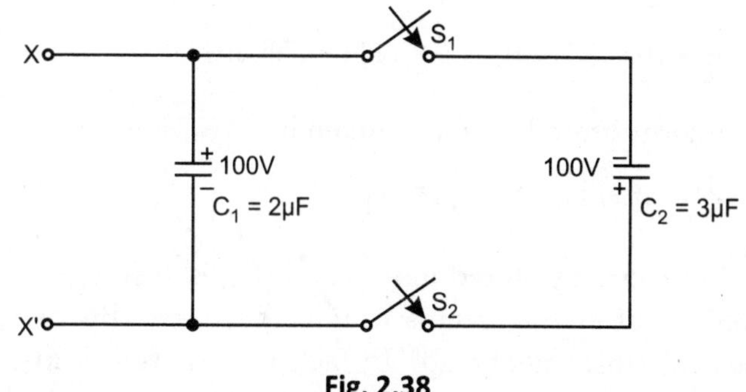

Fig. 2.38

Solution. When the two switches are open, net charge stored is given by:

$100 \times 2 - 100 \times 3 = -100\ \mu C$

After the switches are closed and steady state is reached, if (V) is the potential difference across X – X′, then

$V(2 + 3) = -100$ or $V = -20$ V.

Problem 2.17. *Refer to Fig. 2.39. (C_1) is charged to 100 V while switch (S_1) is open and (S_2) is closed. Then, switch (S_2) is opened and switch (S_1) closed at the same time. What would be the potential difference across each capacitor when equilibrium is reached?*

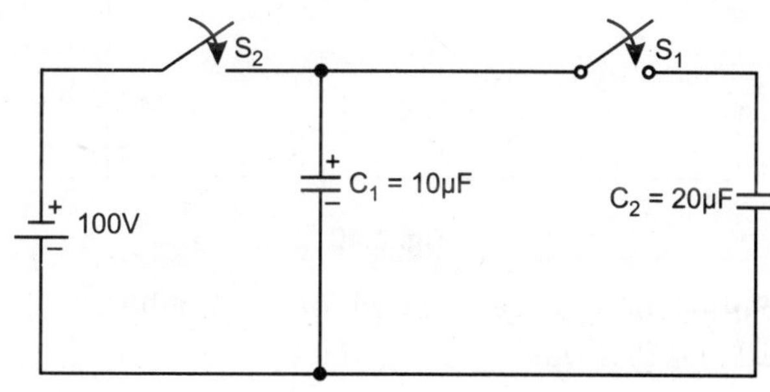

Fig. 2.39

Solution. When (C_1) is charged to 100 V, charge on C_1 = 10 x 100 = 1000 μC.

When switch (S_1) is closed, this charge is shared between (C_1) and (C_2) till both capacitors acquire equal potential difference. If (V) is this new potential difference,

$V(C_1 + C_2) = 1000\ \mu C$

or $\quad V = \frac{1000 \times 10^{-6}}{(10 + 20) \times 10^{-6}} = 33.33$ Volts

We can make a very interesting observation from the results obtained. When the switch (S_1) is open, the energy stored in (C_1) is given by :

$$\frac{1}{2} \times 10 \times (100)^2 = 5 \times 10^4\ \mu J = 50{,}000\ \mu J$$

When (S_1) is closed and the charge is redistributed. Energy stored in (C_1) is given by:

$$\frac{1}{2} \times 10 \times (33.33)^2 \cong 5 \times 1111 = 5555\ \mu J$$

Energy stored in (C_2) is given by:

$$\frac{1}{2} \times 20 \times (33.33)^2 = 11110\ \mu J$$

Total energy stored in (C_1) and (C_2) is 16665 μJ. Thus 33335 μJ of energy seems to have been lost. But where does this lost energy go? In fact, this energy is lost as heat energy in the wire connecting (C_1) and (C_2) during redistribution of charge and consequent flow of current. It can be proved that the fraction of stored energy that is lost is $[C_2/(C_1 + C_2)]$.

Problem 2.18. *Refer to Fig. 2.40. Initially (S_1) is closed and (S_2) is open. After steady state is reached and (C_1) is fully charged, (S_1) is opened and (S_2) is closed at the same time instant.*

Determine the potential difference across (C_1) after a new steady state is reached. Also determine charge on (C_2) and (C_3).

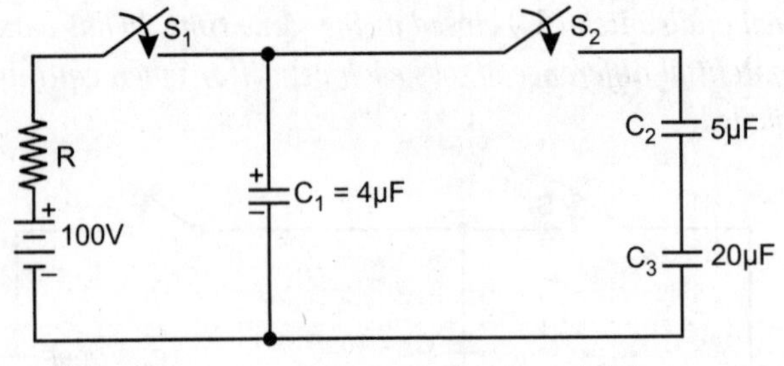

Fig. 2.40

Solution. Charged stored on (C_1) when (S_1) was closed and (S_2) was open is given by :

$$4 \times 100 = 400\ \mu C = 400 \times 10^{-6}\ C$$

when (S_1) is opened and (S_2) is closed, this charge is redistributed. Now (C_2) and (C_3) are in series and therefore have an equivalent capacitance of:

$$\frac{5 \times 20}{5 + 20} = 4\ \mu F$$

Therefore, after redistribution of charge, if (V) is the new potential difference across (C_1), then

$$V(4 + 4) \times 10^{-6} = 400 \times 10^{-6} \text{ or } V = \frac{400}{8} = 50 \text{ volts}$$

Thus after redistribution, (C_1) has 200 μC of charge and the equivalent capacitance of series combination of (C_2) and (C_3) also has 200 μC (= 400 – 200) of charge. As (C_2) and (C_3) are in series, charge on, C_2 = charge on C_3 = 200 μC.

Problem 2.19. Refer to Fig. 2.41. Determine charge on each capacitor when *(a)* only (S_1) is closed and *(b)* both (S_1) and (S_2) are closed.

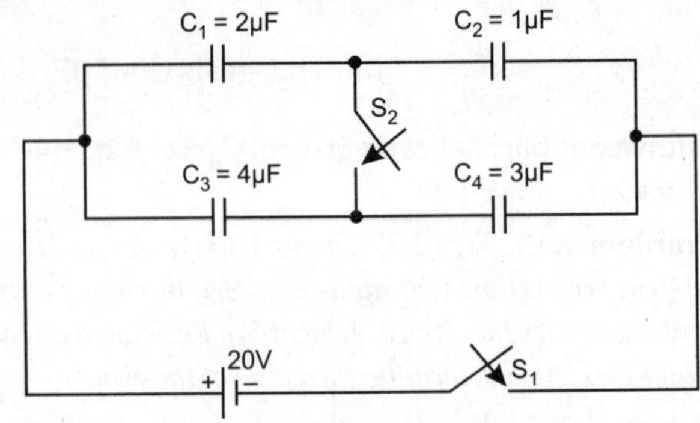

Fig. 2.41

Solution. *(a)* When only (S_1) is closed, (C_1) and (C_2) are in series and (C_3) and (C_4) are in series.

$$\text{Equivalent of } (C_1) \text{ and } (C_2) = \frac{1 \times 2}{1 + 2} = \frac{2}{3}\ \mu F$$

$$\text{Equivalent of } (C_3) \text{ and } (C_4) = \frac{3 \times 4}{3 + 4} = \frac{12}{7}\ \mu F$$

Charge that would have been stored on

$$\frac{2}{3}\mu F = \frac{2}{3} \times 20 = \frac{40}{3}\ \mu C$$

$$\text{Therefore charge on } (C_1) = \text{Charge on } (C_2) = \frac{40}{3}\ \mu C$$

Charge that would have been stored on

$$\frac{12}{7}\ \mu F = \frac{12}{7} \times 20 = \frac{240}{7}\ \mu C$$

$$\text{Therefore charge on } (C_3) = \text{charge on } (C_4) = \frac{240}{7}\ \mu C$$

(b) When both (S_1) and (S_2) are closed, (C_1) and (C_3) are in parallel and (C_2) and (C_4) are in parallel.

The two parallel combinations are connected in series.

Parallel combination of (C_1) and (C_3) = 1 + 3 = 4 μF

Parallel combination of (C_2) and (C_4) = 2 + 4 = 6 μF

Series combination of 4 μF and 6 μF yields

$$\frac{4 \times 6}{4 + 6} = 2.4\ \mu F$$

Charge stored = 2.4 × 20 = 48 μC

Therefore charge stored on (C_1) and (C_3) = charge stored on (C_2) and (C_4) = 48 μC which means

Voltage across (C_1) – (C_3) parallel combination = (48/4) = 12 V

and voltage across (C_2) – (C_4) parallel combination = (48/6) = 8 V

Therefore charge on $(C_1) = 12 \times 1 = 12$ μC

Charge on $(C_3) = 12 \times 3 = 36$ μC

Charge on $(C_2) = 8 \times 2 = 16$ μC

Charge on $(C_4) = 8 \times 4 = 32$ μC.

Problem 2.20. *Refer to Fig. 2.42 All capacitors are initially uncharged. The switch 'S' is first switched to position-1 and then after some time to position-2. Find the charge stored in each of the three capacitors.*

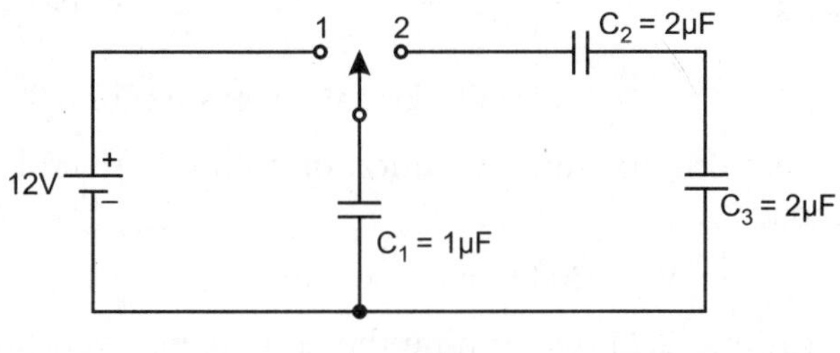

Fig. 2.42

Solution. When the switch is in position-1, C_1 (= 1 μF) is charged to 12 volts. The switch is then thrown to position-2. The charge stored on (C_1) in then re-distributed between (C_1), (C_2) and (C_3) so that voltage across (C_1) is same as that across series combination of (C_2) and (C_3). Also, sum of the new charge stored on (C_1) and equivalent of series combination of (C_2) and (C_3) equals the charge originally stored on (C_1). If (V) is the new voltage across (C_1).

Then

$$12 \times 1 = V \times 1 + V \times \left(\frac{2 \times 2}{2+2}\right) = V + V = 2V$$

or
$$V = \frac{12}{2} = 6 \text{ volts}$$

Therefore, charge stored on $C_1 = 6 \times 1 = 6$ μJ

Charge stored on C_2 = Charge stored on C_3 = 12 – 6 – 5 μJ

Problem 2.21. A capacitor (C_1) is initially charged to a voltage (V_1) and the source of voltage is removed. Then, an uncharged capacitor (C_2) is put across an already charged (C_1). There is redistribution of charge and it is observed that both (C_1) and (C_2) acquire a potential difference of (V_2) volts. Prove that during redistribution, an energy quantum equal to $\left(\frac{C_2}{C_1+C_2}\right) \times 100$ percent of energy originally stored in (C_1) is lost.

Solution. Energy stored in $C_1 = \frac{1}{2}C_1V_1^2$

After redistribution,

Energy stored in $C_1 = \frac{1}{2}C_1V_2^2$ and Energy stored in

$$C_2 = \frac{1}{2}C_2V_2^2$$

That energy stored after redistribution $= \frac{1}{2}V_2^2(C_1 + C_2)$

Also, due to conservation of charge,

$$C_1V_1 = V_2(C_1 + C_2) \text{ or } V_2 = \left(\frac{C_1}{C_1+C_2}\right)V_1$$

Substituting for V_2,

Total energy stored after redistribution =

$$\frac{1}{2}(C_1 + C_2) \times \frac{C_1^2V_1^2}{(C_1+C_2)} = \frac{C_1^2V_1^2}{2(C_1+C_2)}$$

Energy stored initially $= \frac{1}{2}C_1V_1^2$

Therefore, fraction of initial energy still retained is given by

$$\frac{C_1^2V_1^2}{2(C_1+C_2)} \times \frac{2}{C_1V_1^2} = \frac{C_1}{(C_1+C_2)}$$

Therefore, percentage of initial energy lost

$$= \left[1 - \frac{C_1}{C_1+C_2}\right] \times 100 = \left(\frac{C_2}{C_1+C_2}\right) \times 100.$$

Problem 2.22. *Three capacitors of 10μF, 20μF and 40μF are connected in series across 500 V source. Determine (a) voltage across each capacitor (b) magnitude of charge stored by each capacitor (c) equivalent capacitance. When the capacitors are fully charged, they are disconnected from the source and from each other. They are then connected in parallel. What is the new voltage across each capacitor?*

Solution. The equivalent capacitance (C) of the series combination is given by:

$$\frac{1}{C} = \frac{1}{10} + \frac{1}{20} + \frac{1}{40} = \frac{7}{40}$$

Therefore, $C = \frac{40}{7}$ μF

Charge stored in each of the series connected capacitors is the same as charge stored in (40/7) μF given by;

$$Q = \frac{40}{7} \times 500 = \frac{20000}{7} \ \mu C$$

Now charge stored in 10μF = charge stored in 20μF =

charge stored in 40μF = $\frac{20000}{7}$ μC

Therefore, voltage across 10μF capacitor =

$$\frac{20,000}{7} \times \frac{1}{10} = \frac{2000}{7} V = 285.7 \text{ V}$$

Voltage across 20μF capacitor $= \frac{20000}{7} \times \frac{1}{20} = \frac{1000}{7}$

$= 142.85$ V

Voltage across 40μF capacitor $= \frac{20000}{7} \times \frac{1}{40}$

$= 71.425$ V

When these charged capacitors are disconnected from the source and connected in parallel, there will be redistribution of charge so as to have a common voltage across each capacitor. If this new voltage is (V), then

$$V(C_1 + C_2 + C_3) = \frac{3 \times 20000}{7} \mu C \text{ or } 70V = \frac{60000}{7} \mu C$$

$$V = \frac{60000}{7 \times 70} = 122.44 \text{ Volts.}$$

Problem 2.23. *Given three capacitors of 0.01 μF, 0.02 μF and 0.05 and a rated working voltage of 400 volts for each capacitor. Determine the highest voltage that can safely be applied across a series combination of these capacitors.*

Solution. The equivalent capacitance of the series combination is given by:

$$\frac{1}{(1/0.01) + (1/0.02) + (1/0.05)} = \frac{1}{170} = 0.00588 \ \mu F$$

If (V) is the maximum voltage across the series combination, then ensure safe operation of 0.01 μF,

$$0.00588\, V = 0.01 \times 400; V = \frac{0.01 \times 400}{0.00588} = 680.27 \text{ Volts}$$

To ensure save operation of 0.02 μF,

$$0.00588\, V = 0.02 \times 400; V = \frac{0.02 \times 400}{0.00588} = 1360.54 \text{ Volts}$$

To ensure safe operation of 0.05 μF,

$$0.00588\, V = 0.05 \times 400; V = \frac{0.05 \times 400}{0.00588} = 3401.36 \text{ Volts}$$

To ensure that each capacitor in the series combination gets a voltage equal to or less than the rated working voltage, maximum voltage across the series combination = 680.27 volts.

Problem 2.24. *A 0.1 μF, 400 volt capacitor is connected in series with a 0.05 μF 200 volt capacitor. What is the maximum voltage that can be safely applied across this combination?*

Solution. The equivalent capacitance of the series combination is given by;

$$\frac{0.1 \times 0.05}{0.1 + 0.05} = 0.033 \ \mu F$$

If (V) is the maximum voltage that can be applied across the series combination, then

$$0.033V = 0.1 \times 400$$

or $$V = \frac{0.1 \times 400}{0.033}$$

= 1212.12 volts for safe operation of 0.1 μF

and $$V = \frac{0.05 \times 200}{0.033}$$

= 303 volts for safe operation of 0.05 μF

Therefore, for safe operation of both 0.1 μF and 0.05 μF capacitors,

$$V = 303 \text{ volts}$$

Problem 2.25. Determine the equivalent capacitance across X-X terminals for the networks shown in Figs. 2.43 (*a*) to (*c*).

Solution. Fig. 2.43 (*a*). the network is redrawn in Fig. 2.44 (*a*). The subsequent steps of simplification are shown in Figs. 2.44 (*b*) and (*c*).

Fig. 2.43. (*b*) : Converting the inner star into an equivalent delta, the circuit takes the form of Fig. 2.45. (*a*). the subsequent simplifications are shown in Figs. 2.45 (*b*) and (*c*). While transforming inner start into an equivalent delta, each of the equivalent delta capacitances is given by:

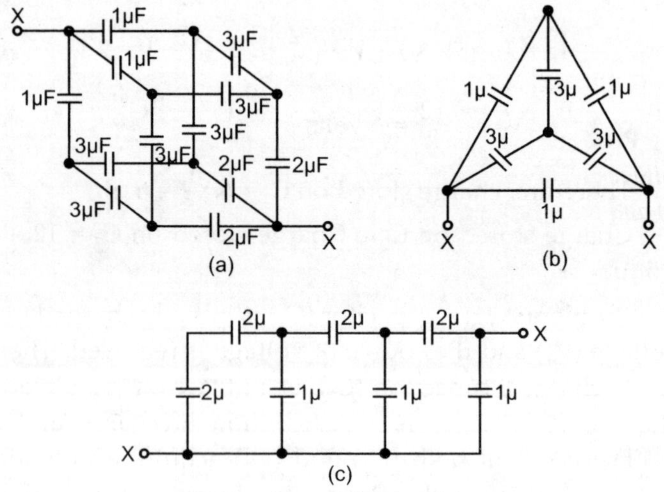

Fig. 2.43

The equivalent capacitance between $X - X = 2 + 1 = 3$ μF

Fig. 2.43 (*c*) : Different steps of network simplification are show in Fig. 2.46. The steps are self explanatory.

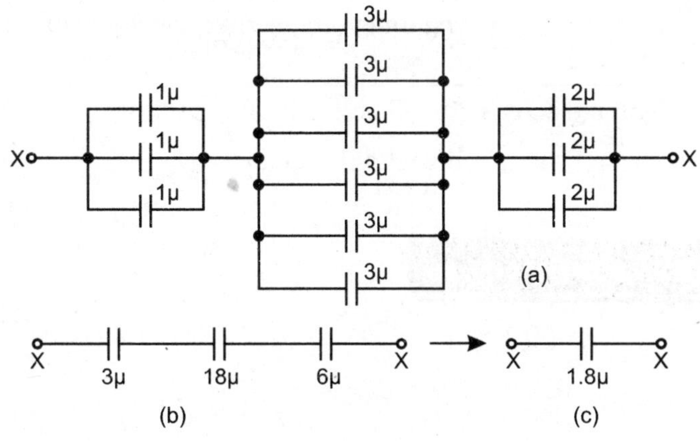

Fig. 2.44

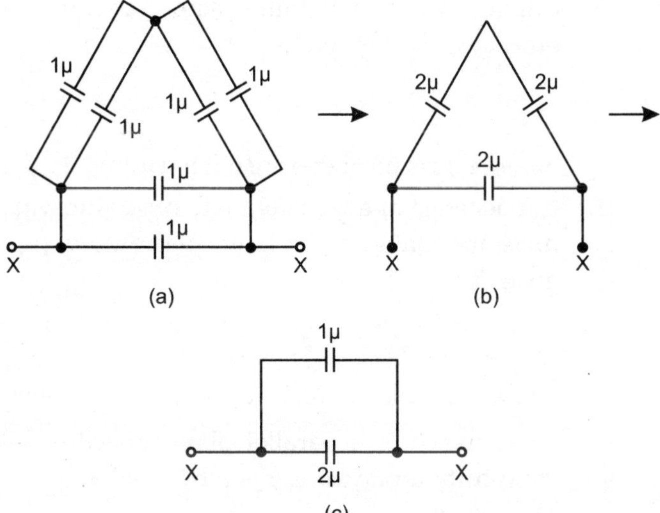

Fig. 2.45

Problem 2.26. *Find the equivalent capacitance and the voltage across each of the capacitors in the network of Fig. 2.47. (a) and (b).*

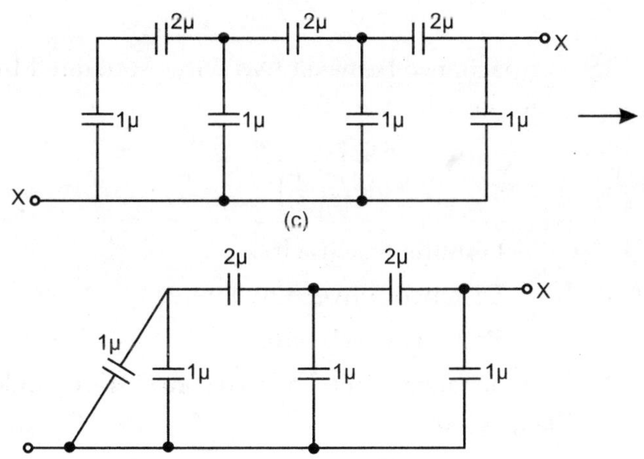

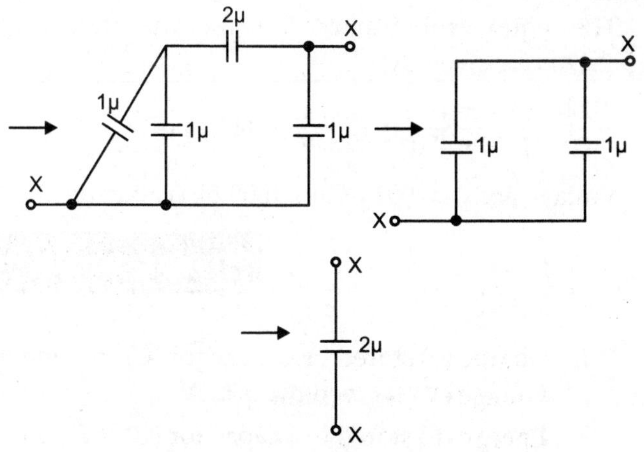

Fig. 2.46

Solution. The equivalent capacitance for the network of Fig. 2.47 (*a*) is given by:

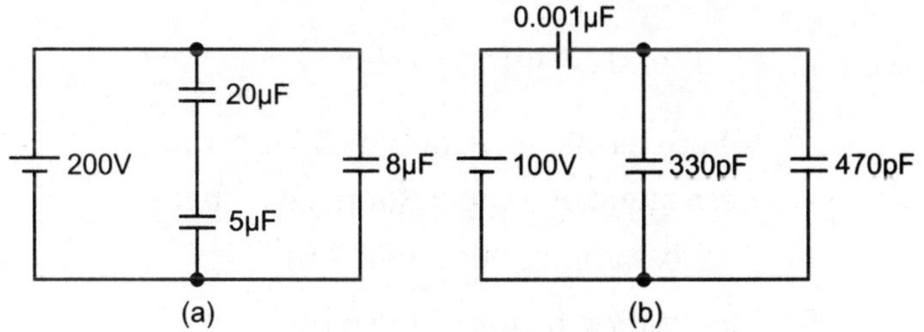

Fig. 2.47

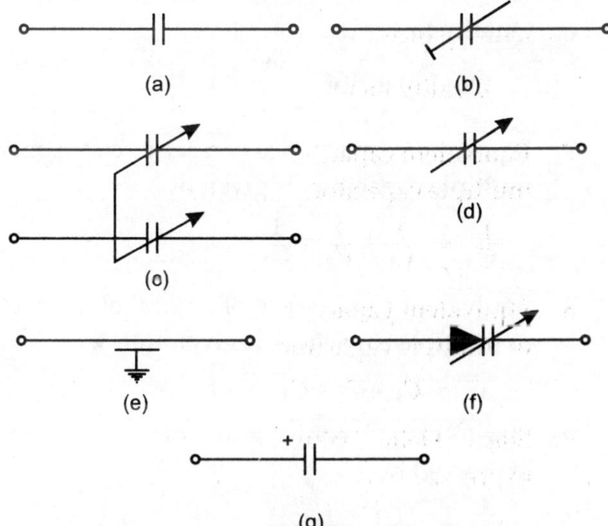

Fig. 2.48.

$$8 + \frac{20 \times 5}{20 + 5} = 8 + 4 = 12 \ \mu F$$

Voltage across 8 μF capacitor = 200 Volts

Voltage across 20 μF capacitor = $\frac{200 \times 5}{25}$ = 40 Volts

Voltage across 5 μF capacitor = $\frac{200 \times 20}{25}$ = 160 Volts

The equivalent capacitance for the network of Fig. 2.47 (*b*) is given by:

$$\frac{1000 \times (330 + 470)}{1000 + 330 + 470} = \frac{80000}{1800} = 444.44\ \mu F$$

Voltage across 0.001 μF (or 1000 pF) capacitor

$$= \frac{100 \times 800}{1800} = 44.44 \text{ Volts}$$

Voltage across 300 pF and 470 pF capacitors

$$= \frac{100 \times 1000}{1800} = 55.56 \text{ Volts}$$

Summary of Important Formulas

1. Charge (q) stored in a capacitor (C) charged to a voltage (V) is given by: q = CV
2. Energy (E) stored in a capacitor (C) charged to a voltage (V) is given by:
   $$E = \frac{1}{2} CV^2$$
3. Power factor of a capacitor is given by:
   $$\text{Power factor} = \cos\theta = \frac{R}{Z} = \frac{R}{\sqrt{R^2 + X_C^2}}$$
   where terms have their usual meaning
4. Loss angle of a capacitor is given by:
   $$\text{Loss angle,}\quad \delta = (90° - \theta)$$
5. Dissipation factor is given by:
   $$\text{Dissipation factor} = \frac{R}{X_C}$$
6. Quality factor is given by
   $$\text{Quality factor} = \frac{X_C}{R}$$
7. Equivalent capacitance of a series connection of multiple capacitors is given by:
   $$\frac{1}{C} = \frac{1}{C_1} + \frac{1}{C_2} + \frac{1}{C_3} + ... + \frac{1}{C_n}$$
8. Equivalent capacitance of a parallel connection of multiple capacitors is given by:
   $$C = C_1 + C_2 + C_3 + ... + C_n$$
9. Star-to-Delta conversion of capacitors is expressed by:
   $$C_{AB} = \frac{C_A . C_B}{C_A + C_B + C_C}$$
   $$C_{BC} = \frac{C_B . C_C}{C_A + C_B + C_C}$$
   $$C_{CA} = \frac{C_C . C_A}{C_A + C_B + C_C}$$
10. Delta-to-Star conversion of capacitor is expressed by:
    $$C_A = C_{AB} + C_{CA} + \frac{C_{AB} . C_{CA}}{C_{BC}}$$
    $$C_B = C_{BC} + C_{AB} + \frac{C_{BC} . C_{AB}}{C_{CA}}$$
    $$C_C = C_{CA} + C_{BC} + \frac{C_{CA} . C_{BC}}{C_{AB}}$$
11. Capacitance of a parallel plate capacitor with air dielectric is given by:
    $$C = \frac{\varepsilon_0 A}{d}$$
    where terms have their usual meaning
12. Capacitance of a parallel plate capacitor with a dielectric material of relative permittivity (ε_r) is given by:
    $$C = \frac{\varepsilon_0 \varepsilon_r A}{d}$$
    where terms have their usual meaning.
13. Capacitance of a parallel plate capacitor with composite capacitor is given by:
    $$C = \frac{\varepsilon_0 A}{(d_1/\varepsilon_{r1}) + (d_2/\varepsilon_{r2})}$$
    where terms have their usual meaning.
14. Capacitance of a parallel plate capacitor with multiple plates is given by:
    $$C = \frac{(n-1)\,\varepsilon_0\,\varepsilon_r A}{d}$$
15. Capacitance between two wires separated by a dielectric is given by:
    $$C = \frac{\pi\,\varepsilon_0\,\varepsilon_r l}{2.3 \log_{10}\left(\frac{d}{r}\right)}$$
    l = Length of each wire.
    d = Distance between two wires
    r = Radius of each wire
16. Capacitance formed by two concentric cylinders is given by:
    $$C = \frac{2\pi\,\varepsilon_0 \varepsilon_r l}{2.3 \log_{10}\left(\frac{b}{a}\right)}$$
    l = Length of each cylinder

a = Radius of inner cylinder

b = radius of outer cylinder

17. Capacitance formed by two concentric spheres is given by:

$$C = \frac{4\pi \varepsilon_0 \varepsilon_r ab}{(b-a)}$$

a = Radius of inner sphere

b = Radius of outer sphere

18. Capacitance of an isolated sphere is given by:

$$C = 4\pi \varepsilon_0 R$$

R = Radius of sphere

Self Evaluation Exercise

1. Name the capacitors from the circuit symbols shown in Fig. 2.48 Choose from : FIXED CAPACITORS.

   VARIABLE CAPACITORS, PRESET CAPACITORS, GANGED CAPACITORS, VARAC-TORS POLARIZED CAPACITORS, NON-POLARIZED CAPACITORS, FEED THROUGH CAPACITORS.

2. We give you five false statements regarding capacitors. Give the corresponding correct statements.

(*a*) The voltage across an ideal capacitor leads the current through it by 90°.

(*b*) Since a capacitor (C) when charged to a DC voltage (V) stores an energy equal to $\left(\frac{1}{2}CV^2\right)$, therefore the same capacitor when charged to a voltage of ($2V$) will store an energy (CV^2).

(*c*) The power factor in case of non-ideal capacitors is defined as the ratio of resistance to reactance.

(*d*) The overall capacitance of series combination of more than one capacitors is always greater than the capacitance of any of the individual capacitors.

(*e*) For an ideal capacitor, dissipation factor is reciprocal of the power factor.

3. Table 2.6 lists some of the important specifications of a capacitor. In column-2 of the table, are given some statements in a haphazard way. For each of the specifications mentioned in column-1, find out the relevant statement from column-2.

Table 2.6

S.N.	*Specification*	*Statement*
1.	TOLERANCE	The prefixes P, N or NP0 are used with reference to this specification
2.	INSULATION RESISTANCE	Greater the value of this resistance, larger is the dissipation factor
3.	EQUIVALENT SERIES RESISTANCE (ESR)	Greater the value of this parameter, lesser is the leakage current. Its value decreases with increase in temperature.
4.	DISSIPATION FACTOR	It is the ratio of resistance to impedance expressed as a percentage
5.	POWER FACTOR	It is the ratio of resistance (ESR) to reactance expressed as a percentage.
6.	Q-FACTOR	It is the reciprocal of the Dissipation Factor.
7.	TEMPERATURE CO-EFFICIENT	If it is ± 10% for a capacitor of nominal value 1 μF, the actual value may lie anywhere between 0.9 μF and 1.1 μF.

4. We are looking for the most appropriate capacitor type for the kind of requirements mentioned below.

   (*a*) For high frequency decoupling of the DC supply being used for integrated circuits.

   (*b*) a 10 μF coupling capacitor that is highly stable.

   (*c*) a 1000 μF capacitor to be used for filtering purpose in a conventional Transformer-Rectifier Filter arrangement.

   (*d*) 0.1 μF and 22 μF capacitors to be used separately as timing capacitors in timing circuit configurations built around 555 timer IC.

   (*e*) a high voltage trimmer capacitor in the capacitance range of say 5 pF to 20 pF to be used for tuning purposes in an *RF* amplifier.

(f) a 10 μF, 2000 V capacitor to be used for repetitive charge/discharge applications.

Choose from the following words :

Ceramic Disc, Solid Tantalum, Electrolytic, Polystyrene, Energy Storage Capacitor Using Oil Impregnated Paper As Dielectric, Vacuum Capacitor, Ceramic Disc Trimmer, Mica Trimmers.

5. For each of the capacitive networks shown in Fig. 2.49 find out equivalent capacitance between X – X. All capacitances are in microfarads.

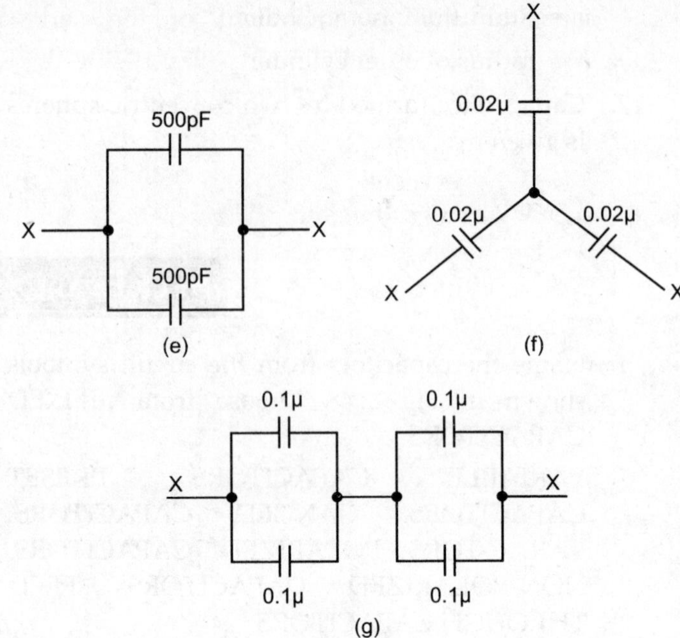

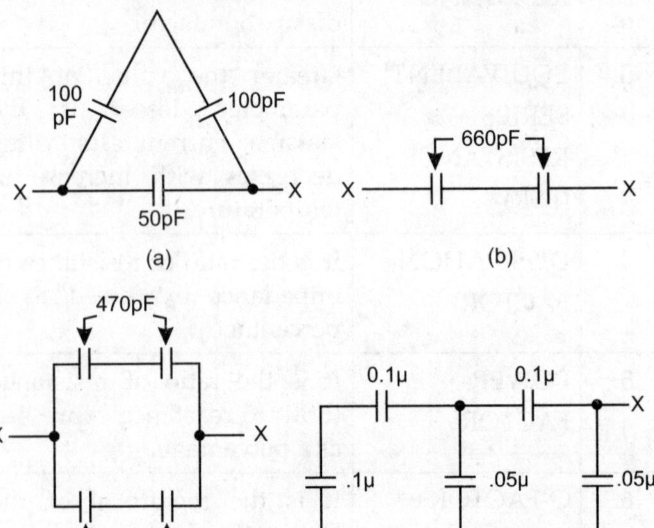

Fig. 2.49

Multichoice Questions

1. When a 20 μF capacitor is storing an energy of 10 Joules, voltage across it must be
   (a) 1000 volts (b) 1414 volts
   (c) 707 volts (d) very high
2. The circuit element in which the phase difference between instantaneous voltage and instantaneous current is 90° is
   (a) necessarily a resistor
   (b) necessarily a capacitor
   (c) necessarily an inductor
   (d) either an inductor or capacitor
3. A 100 pF air dielectric capacitor is charged to a voltage of 100 volts and then the source of voltage removed. A dielectric with a relative permittivity of 2.5 is inserted to fill the entire air space. What will be the new capacitance and voltage.
   (a) 100 pF, 40 V (b) 100 pF, 250 V
   (c) 250 pF, 100 V (d) 250 pF, 40 V
4. A constant current source when connected across a 0.1 μF capacitor for a given time charged it to 10 V. The same current source when connected across an unknown capacitor for the same time charges it to 22 V. The capacitance of the unknown capacitor is
   (a) 0.22 μF
   (b) 0.045 μF
   (c) Indeterminate from given data
   (d) none of these
5. The charge stored on each of the four series connected capacitors is observed to be 10 μJ when the series combination was connected across 100 VDC. The charge stored on the

capacitor that is equivalent of the series combination of capacitors when connected across same 100 VDC will be

(*a*) 40 μJ (*b*) 2.5 μJ
(*c*) 10 μJ (*d*) none of these

6. Whenever a series combination of more than one capacitors is connected across a source of potential difference, then

(*a*) largest capacitor always stores highest energy
(*b*) largest capacitor always stores lowest energy
(*c*) smallest capacitor always stores highest energy
(*d*) all capacitors store same energy
(*e*) both (*b*) and (*c*) are correct

7. The series combination of three equal value capacitors yields an equivalent capacitance of 5 μF. Their parallel combination would have yielded

(*a*) 45μF (*b*) 15μF
(*c*) 30μF (*d*) none of these

8. The number of 1 μF capacitors required to be connected in parallel to store a charge to 1 coulomb with 100 volts connected across the parallel combination is

(*a*) 1000 (*b*) 9999
(*c*) 10,000 (*d*) 100

9. A capacitor C_1 = 10 μF is charged till it stores an energy of 18 Joules. The source of voltage is then removed and another capacitor C_2 = 20 μF is connected across the charged capacitor. The total energy stored in the two capacitors would now be

(*a*) 18 J (*b*) 36 J
(*c*) 9 J (*d*) 6 J

10. Fig. 2.50 shows voltage versus time characteristics of four capacitors when they were individually connected to the same constant current source. Arrange the capacitors C_1, C_2, C_3 and C_4 is ascending order of their capacitance values.

(*a*) C_1, C_2, C_3, C_4 (*b*) C_4, C_3, C_2, C_1
(*c*) C_1, C_3, C_4, C_2 (*d*) C_4, C_3, C_1, C_2

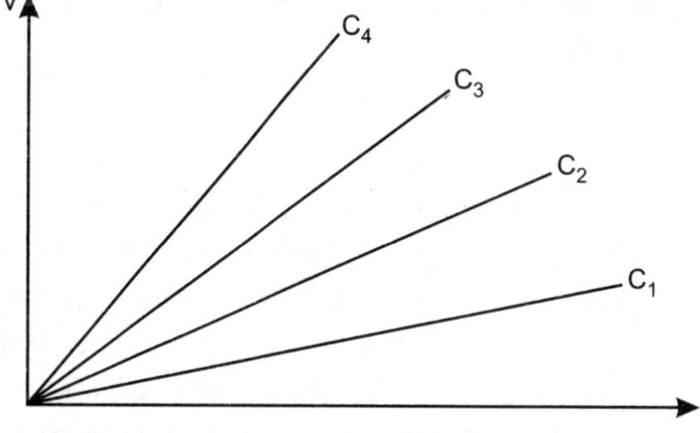

Fig. 2.50

11. A ceramic disc capacitor is marked 104. Its capacitance value is

(*a*) 100 pF (*b*) 100 kpF
(*c*) 10000 μF (*d*) 0.1 μF

12. A supercapacitor achieves its extraordinarily large capacitance value due to its

(*a*) extremely large area of plates
(*b*) extremely small dielectric thickness
(*c*) both (*a*) and (*b*)
(*d*) none of these

13. The capacitance value and working voltage of a supercapacitor can be further enhanced by

(*a*) using a two-dimensional array of series and parallel connection of individual supercapacitor cells.
(*b*) using parallel connection of individual supercapacitor cell
(*c*) using series connection of individual supercapacitor cells
(*d*) none of these

14. When (*N*) equal value capacitor were connected in series and a voltage (*V*) was applied across the series connection; the charge stored on each capacitor was observe to be equal to (*Q*). When these capacitors are connected in parallel and same voltage is applied across the parallel combination; charge stored on each capacitor would be

(*a*) *Q* (*b*) N^2Q
(*c*) *Q/N* (*d*) *NQ*

15. When two equal value capacitor were connected in series and a certain voltage was applied across the series combination; total energy stored was observed to be 16 Joules. When the voltage source was removed and the charged capacitors were connected in parallel; total energy stored in the capacitors after redistribution of charge would be

(*a*) 8 Joules (*b*) 16 Joules
(*c*) 4 Joules (*d*) 64 Joules

16. A capacitor when connected across a certain constant current source for a certain time stored an energy of 10 Joules. If the capacitor instead had been connected to a current source with the magnitude of constant current equal to twice the amplitude of the previous current source for the same time duration; the energy stored would be

(*a*) 10 Joules (*b*) 20 Joules
(*c*) 5 Joules (*d*) 40 Joules

17. A capacitor is marked P200. It means that

(*a*) it has working voltage of + 200 volts
(*b*) it has a temperature coefficient of + 200 PPM/°C
(*c*) it has a temperature coefficient of + 2% per°C
(*d*) it has a tolerance of + 200 pF

18. One of the following types of variable capacitors in an electronically controlled one.

(*a*) compression capacitor
(*b*) piston capacitor
(*c*) varactor
(*d*) none of these

19. Five equal value capacitors when connected in series produce an equivalent series capacitor of 1 μF. The such capacitors when connected in parallel would produce an equivalent capacitance of

(*a*) 25 μF (*b*) 2 μF
(c) 10 μF (d) 0.5 μF

20. Comparing supercapacitors with chemical batteries as energy storage device, one of the following statements is false.

(*a*) Supercapacitors are capable of relatively much faster charge/discharge rates.
(*b*) Supercapacitors have relatively much longer lifetime
(*c*) Supercapacitors have relatively higher energy density
(*d*) Supercapacitors have much smaller form factor than chemical batteries

Answers to Self Evaluation Exercises

1. Fig 2.49 (*a*) – Fixed Capacitor
Fig 2.49 (*b*) – Preset Capacitor
Fig 2.49 (*c*) – Ganged Capacitor
Fig 2.49 (*d*) – Variable Capacitor
Fig 2.49 (*e*) – Feed through Capacitor
Fig 2.49 (*f*) – Varactor
Fig 2.49 (*g*) – Polarised Capacitor

2. (*a*) The voltage across an ideal capacitors lags the current through it by 90°.
(*b*) Since a capacitor (C) when charged to a DC voltage (*V*) stores an energy equal to (1/2 CV^2), therefore the same capacitor when charged to a voltage (2*V*) will store an energy equal to ($2CV^2$).
(*c*) The power factor in case of non-ideal capacitors is defined as the ratio of resistance to impedance.
(*d*) The overall capacitance of series combination of more than one capacitors is always smaller than the capacitance of any of the individuals capacitors.
(*e*) For an ideal capacitor, Dissipation Factor is reciprocal of the *Q*-Factor.

3. (1-7), (2-3), (3-2), (4-5), (5-4), (6-6), (7-1)

4. (*a*) Ceramic Disc (*b*) Solid Tantalum
(*c*) Electrolytic (*d*) Polystyrene
(*e*) Ceramic Disc Trimmer
(*f*) Energy Storage Capacitors using oil impregnated paper as dielectric.

5. Fig 2.50 (*a*) – 100 pF Fig 2.50 (*b*) – 330 pF
Fig 2.50 (*c*) – 470 pF Fig 2.50 (*d*) – 0.1 μF
Fig 2.50 (*e*) – 1 kpF Fig 2.50 (*f*) – 0.01 μF
Fig 2.50 (*g*) – 0.1 μF

Multichoice Questions

1. (*a*)	2. (*d*)	3. (*d*)	4. (*b*)
5. (*c*)	6. (*e*)	7. (*a*)	8. (*c*)
9. (*d*)	10. (*b*)	11. (*b*)	12. (*c*)
13. (*a*)	14. (*d*)	15. (*c*)	16. (*d*)
17. (*b*)	18. (*c*)	19. (*a*)	20. (*c*)

Chapter 3

Inductors and Transformers

Learning Objectives

After completing the chapter, you will learn the following topics:

1. Inductor and transformer basics
2. Equivalent circuit of an inductor
3. Important inductor and transformer parameters
4. Winding configurations of inductors
5. Inductors in series
6. Inductors in parallel
7. Standard or preferred values of inductors
8. Types of magnetic materials used in inductors and transformers
9. Variable Inductors
10. Types of transformers
11. Design of transformers
12. Inductor and transformer applications

Q.1. How do you define inductance? What is an inductor?

Ans. Inductance is the property of an electric conductor or circuit due to which an electromotive force (EMF) is induced in it by a changing current flowing through it. This is based on the principle that an electric current flowing through an electric conductor causes a magnetic field to surround the conductor. A changing current causes a changing magnetic field. An EMF is induced in the conductor if it links with a changing magnetic flux and the magnitude of induced EMF is directly proportional to rate of change of magnetic flux. Also, the induced EMF is directly proportional to the number of turns if the conductor were wound in the form of a coil or solenoid (Fig. 3.1).

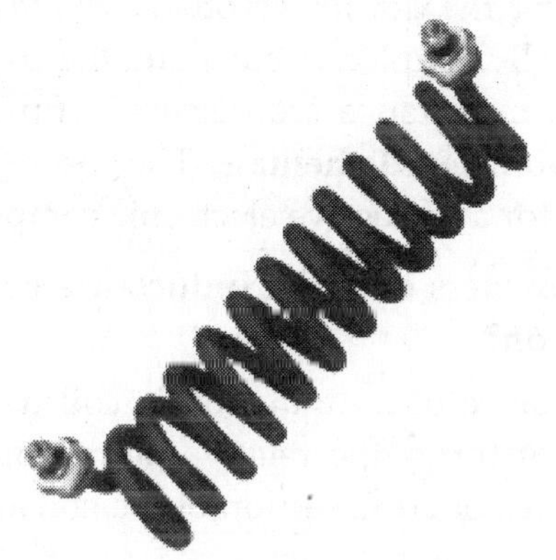

Fig. 3.1

An inductor is a passive electronic component that is designed to possess inductance. An inductor opposes changes in electrical current flowing through it. It stores electrical energy in a magnetic field. SI unit of inductance is Henry. Inductance is measured in Henries (H), milli-Henries (mH), micro-Henries (μH) and nano-Henries (nH). 1 H = 10^3 mH = 10^6 μH = 10^9 nH. The inductance is 1.0 H if a rate of change of current of 1 Ampere/sec induces an EMF of 1.0 volt. Induced EMF = L× di/dt. Also, 1.0 H = 1.0 Weber/Ampere for a single-turn coil as L = dϕ/di where (dϕ) is change in magnetic flux. Weber (Wb) is the SI unit of measurement of magnetic flux.

Q.2. What is the magnetic energy stored in an inductor (L) for a DC current (I) flowing through it?

Ans. Magnetic energy stored in an inductor is given by:

$E = (1/2) \times LI^2$

E = Energy in Joules

L = Inductance in Henries

I = Current in Amperes

Q.3. What do you understand by self-inductance and mutual inductance?

Ans. Self-inductance is the property of an inductor due to which a changing current induces a voltage across it. Self-inductance always refers to a single inductor or coil. **Mutual inductance** refers to two or more inductors or coils that are magnetically coupled. It is an inductive effect due to which a changing current in one circuit causes an induced voltage across a second circuit as a result of the magnetic field linking both circuits. This effect is used in transformers. Magnitude of mutual inductance between two inductors (L_1) and (L_2) is given by $M = K \times \sqrt{L_1 . L_2}$ where (K) is the coefficient of coupling. The polarity of mutual inductance is usually expressed by a dot convention. When the current is either entering or leaving both the dots; the mutual inductance is taken as positive. If the current were entering the dot in one inductor and leaving the dot in the other; it is taken as negative. Magnetically coupled circuits and the concepts of self and mutual inductance are discussed further in chapter-5 on **Network Fundamentals.** The focus in this chapter is on inductor as a passive electronic component.

Q.4. What parameters does the inductance of a coil in general depend on?

Ans. The inductance of an inductor or coil depends upon the number of turns, the relative permeability of the core material, area of cross section of the coil and the length of the coil.

Mathematically, Inductance (L) of a coil having (N) Number of turns, cross sectional area (A), length (l) and wound on a magnetic material with a relative permeability of (μ_r) is given by:

$$L = \frac{\mu_o . \mu_r . N^2 . A}{l}$$

where,

L = Coil inductance in Henries

μ_o = Permeability of free space = $4\pi \times 10^{-7}$ H/m

μ_r = Relative permeability of core material

N = Number of turns

A = Arrea of crossection of the coil in (m^2)

l = Lenght of the coil in (m).

Q.5. How is the voltage vector related to the current vector in an ideal and a practical (or non-ideal) inductor of inductance (L). The practical inductor can be assumed to have an equivalent series resistance (R).

Ans. The voltage across an ideal inductor leads the current through it by 90° (Fig. 3.2 *a*). The voltage leads the current by an angle (θ) that is less than 90° in case of a practical inductor (Fig. 3.2 *b*). Angle (θ) is given by:

$$\theta = \tan^{-1} (\omega L/R),$$

(ω) is the radian frequency of the signal.

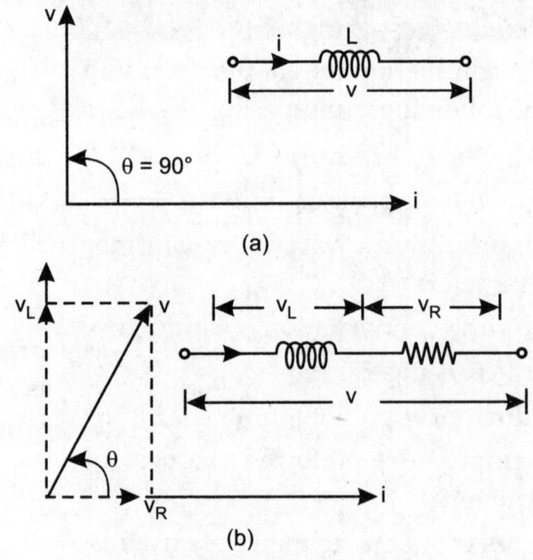

Fig. 3.2

Q.6. Draw the equivalent circuit of a practical inductor and briefly describe the sources and implications of the undesirable components in the electrical equivalent circuit.

Ans. The electrical equivalent circuit of a practical inductor is shown in Fig. 3.3 The associated series resistance (R) in equivalent circuit comes from the DC resistance of the wire used to wind the coil, the hysterisis and eddy current losses due to the core material. These losses become more pronounced at higher frequencies. The effect of a large associated resistance is the reduction in the coil's Q-factor. The capacitance (C) shown in the equivalent circuit is the interwinding capacitance (Fig. 3.4) also known as the distributed capacitance. Larger the interwinding capacitance, smaller is the Q-factor of the coil. The interwinding capacitance together with the inductance of the coil determines its self resonance frequency.

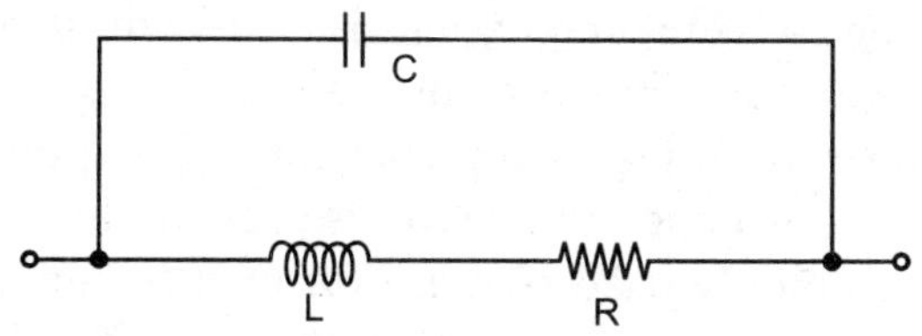

Fig. 3.3

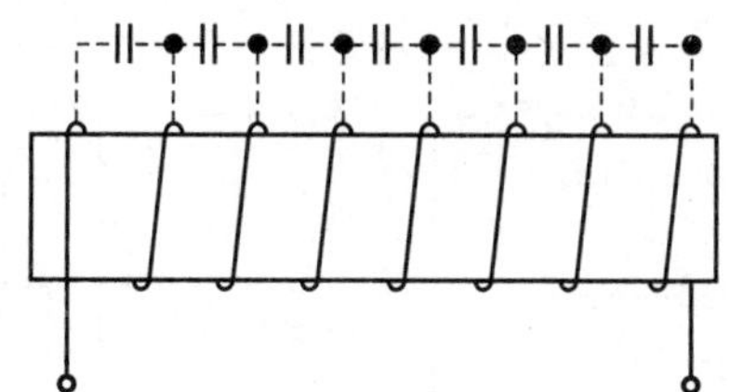

Fig. 3.4

Q.7. Define the *Q*-factor of a practical inductor. What should be its magnitude for an ideal inductor?

Ans. The *Q*-factor of a practical inductor is given by:

$$Q = (1/R)\sqrt{L/C}$$

where *L*, *R* and *C* are the inductance, series resistance and the interwinding capacitance of the coil in question.

The quality factor of an inductor is given by

$$Q = \frac{\omega_0 L}{R}$$

where

ω_0 = self resonant frequency of the coil

The self resonance frequency of this coil is given by:

$$\omega_0 = \frac{1}{\sqrt{LC}}$$

Substituting the value of (C) in the expression for *Q*-factor, we get,

$$Q = \frac{1}{R}\sqrt{\frac{L}{C}}$$

Thus *Q*-factor may also be defined as ratio of inductive reactance at the self resonance frequency to the series resistance. *Q*-factor is also known as the efficiency factor. It is also a direct measure of the coil's selectivity.

Q.8. Briefly explain as to what you understand by:

(*a*) Fixed inductors

(*b*) Variable inductors

(*c*) Preset inductors

(*d*) Chokes

Ans. Fixed Inductors: Fixed inductors offer a fixed inductance when connected in an electrical circuit. There is no method to alter the inductance. Fig. 3.5 (*a*) shows the circuit representation of fixed inductors. Fixed inductors are wound in such a fashion that the number of turns do not change and if it uses any core of a magnetic material, iron core or ferrite, its position with respect to the winding remains fixed. Coils wound on toroidal cores are an example of fixed inductors (Fig. 3.5 *b*).

Variable Inductors: A variable inductor is the one whose inductance can be changed over a specified range as and when required. Fig. 3.5 (*c*) shows the circuit representation of a variable inductor. The inductance is usually varied by changing the position of the core with respect to the winding.

Preset Inductors: A preset inductor is a variable inductor that is varied and then set for the most optimum performance. A typical application of a preset inductor is in Intermediate Frequency Transformers (IFTs) used in radio and television receivers where the inductance is varied by moving the core at the factory premises to get the desired IF response curve. Fig. 3.5 (*d*) shows the circuit representation of a preset inductor

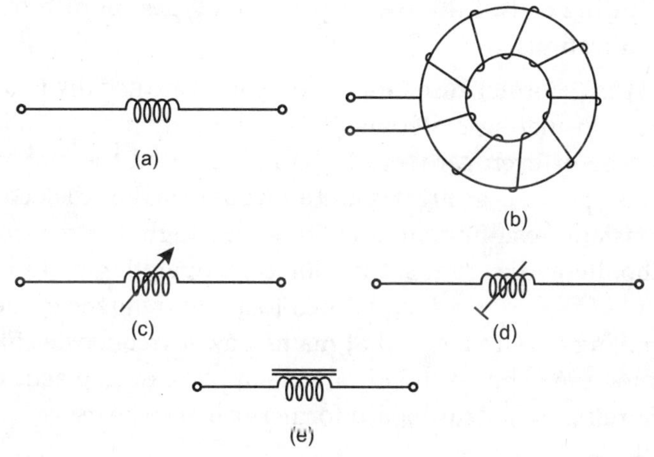

Fig. 3.5

Chokes: A choke is usually an inductor wound on a core (Iron core or Ferrites) that is used to impede the flow of alternating current or a pulsating direct current. Such chokes are also known as the smoothing chokes. Chokes are also used as coupling component for coupling the signal from one stage to another. A choke differs from a conventional coil in the respect that it usually carries through it a sizable amount of DC in addition to AC and as a result it is often necessary to have an air gap in the core to prevent the choke from the ill effects of core saturation. Fig. 3.5 (*e*) shows the circuit representation of a choke.

Q.9. One of the applications of a choke is in the choke-input filter where it is used in conjunction with a

capacitor to smoothen the ripple in a DC power supply. What do you understand by smoothing factor in a choke input filter?

Ans. The smoothing factor of a choke in a choke input filter (Fig. 3.6) is defined as the ratio of capacitive reactance to inductive reactance. If the smoothing factor is say 10, the ripple in the filtered output will be one tenth of what it would have been across the smoothing capacitor in the absence of the choke. Thus for the known values of smoothing capacitor and the acceptable ripple, one can calculate the value of the smoothing choke.

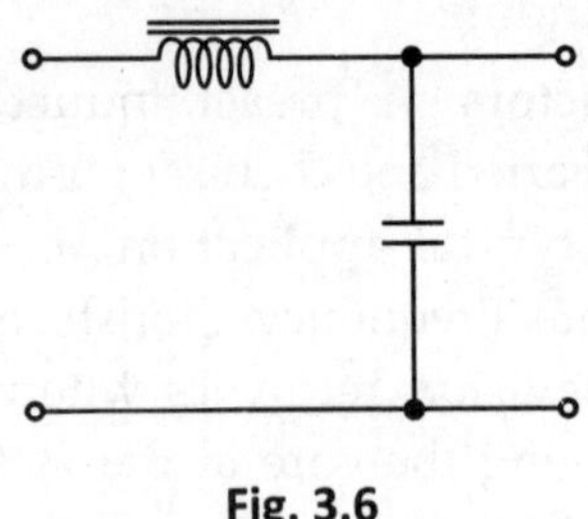

Fig. 3.6

Q.10. What are the types of formers used for winding coils. Indicate the preferred types for different applications.

Ans. Formers used for winding coils are usually made from insulating materials like ceramic, ebonite, paxolin, plastic or even cardboard. Ceramic is used for making coils operating at 30 MHz and above as it has low dielectric constant resulting in low losses at high frequencies. Ebonite or paxolin is used for coils operating at lower end of radio frequencies down to audio frequencies. For making coils to be used at mains power frequencies like smoothing chokes in DC power supplies, even plastic or more economic cardboard formers can also be used.

Q.11. Different types of core materials are used for making coils and transformers. Briefly describe the following magnetic materials for different applications.

(*a*) **Iron cores**

(*b*) **Powdered iron cores**

(*c*) **Ferrites**

(*d*) **Permallory cores. shiko**

Ans. Iron Cores: Iron is a ferromagnetic material and is commonly used as the core material for making mains power transformers. Iron as a core material has its own problems mainly arising from the low resistivity of iron. Low resistivity of iron results in prohibitively large amount of eddy current losses as we tend to operate at even slightly elevated frequencies. Infact, at the power line frequency of 50 Hz too, the losses tend to be very high. The overcome this, iron core is used in the form of laminations so as to offer high resistance paths to the flow of eddy currents. However in the form of laminations also, iron core is just acceptable for use at power line frequency, it is unacceptable for use beyond that. Fig. 3.7 shows commonly used shapes of iron cove laminations. These include shell-type laminations of *E-I* and *E-E* type (Fig. 3.7*a*) and core type laminations of *L-L* and *U-I* type (Fig. 3.7 *b*).

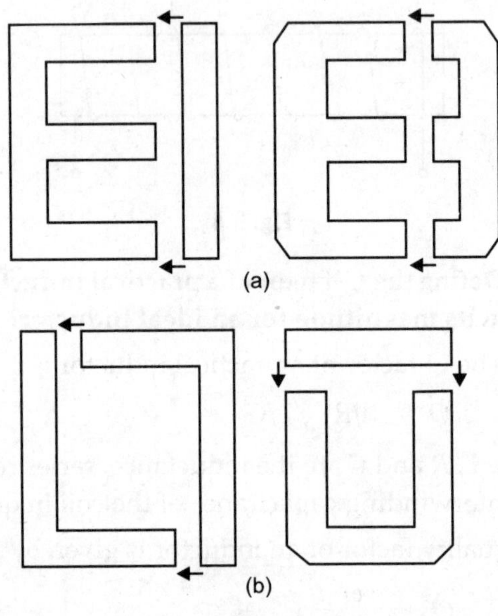

Fig. 3.7

Powdered Iron Cores: Powdered iron cores are made by subjecting a mixture of powdered iron dust and an insulating cement to high pressure in moulds. The result is a magnetic material that has relatively much smaller eddy current losses. Powdered iron cores can be satisfactorily used for making coils for use upto 500 kHz.

Ferrites: Ferrite is a magnetic material produced from a powdered mixture of iron oxide and an oxide of divalent metal ion. The general chemical formula for a ferrite is MFe_2O_4 where M is the divalent metal ion. Mg, Mn, Fe, Co, Ni, Zn, Cu and Cd are commonly used metal ions. If only one metal ion is used besides iron oxide, the result is a simple ferrite. But not all simple ferrites are magnetic. For instance, $ZnFe_2O_4$ (Zinc Ferrite) and $CdFe_2O_4$ (Cadmium Ferrite) are non-magnetic. Most of the practical ferrites use more than one of the above listed divalent oxides, of which Ni-Zn and Mn-Zn ferrites are the most popular.

Wide use of ferrites as magnetic core material at higher frequencies is attributed to extremely high resistivity (as high as 10^{11} ohm-m in some cases) and consequent extremely low eddy current losses, stable magnetic properties and availability of ferrites in a large variety of performance specifications. While ferrites with a linear B-H curve are preferred for making coils, pulse and wideband transformers, antenna rods and loading coils; ferrites with a nonlinear B-H curve and a high flux density are extensively used for making flyback transformers, deflection yokes, choke coils etc. Memory cores and switching cores are made from ferrites that have a nonlinear B-H curve and predominantly square hysterisis loop. Ferrite cover are made in a vide variety of core shapes and sizes as shown in Fig. 3.8.

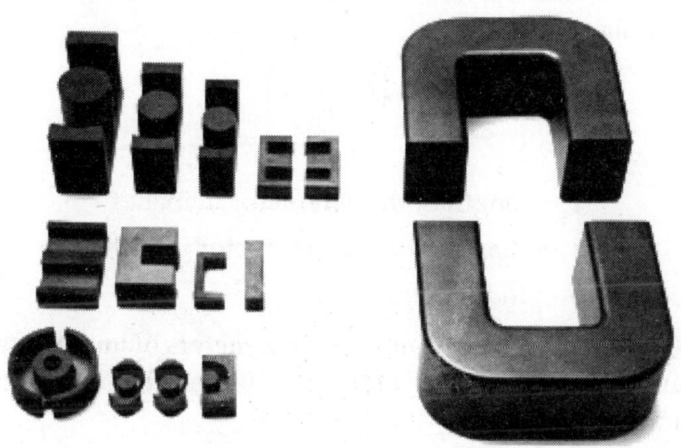

Fig. 3.8

Permalloy Cores: Permalloy is a magnetic material that is an alloy of nickel and iron with nickel content varying from 36 percent to 80 percent. Elements like Chromium, Copper or Molybdenum are also added to get the desired properties. Permalloys are characterised by a high permeability. For instance permalloy with 78.5 per cent of Nickel and about 4 percent of Molybdenum has an initial permeability of 25000 and a maximum permeability as high as 1,50000.

Permalloy cores are available in the form of ribbons in thicknesses from a few microns to 0.5 mm, in the form of rods in diameters ranging from a few mm to 50 mm and in the form of sheets of 1 to 2 mm thickness. Fig. 3.9 shows some common permalloy cove shapes.

Fig. 3.9

Q.12. Distinguish between soft and hard magnetic materials.

Ans. Soft magnetic materials are the ones that lose their magnetism upon removal of magnetising force. These are characterised by low loss, high initial and maximum permeability and low coercive force. The coercive force is usually less than 40 A/m. The magnetic properties of soft magnetic materials depend on the chemical purity and degree of distortion in the crystal lattice of the material.

Hard magnetic materials retain their magnetism even after removal of magnetising force. As their name suggests, these are hard to magnetise but once magnetised, these can retain their magnetism for years. These have high coercive force (usually greater than 40 A/m) and remanent flux densitv (typically 0.1 tesla). These are widely used for making permanent magnets.

Q.13. Sketch the typical HYSTERISIS LOOP for a magnetic material and define the following terms.

(a) Initial permeability, μ_o
(b) Maximum permeability, μ_m
(c) Coercive force, H_c
(d) Remanent flux densityh, B_r
(e) Saturation flux density, B_s
(f) Squareness ratio.

Ans. Initial Permeability, μ_o: Refer to Fig. 3.10. Initial permeability (μ_o) is defined as the slope of B-H curve at zero value of the magnetising force (H).

Maximum Permeability, μ_m: In case of magnetic materials having a non-linear B-H hysterisis loop, the permeability of the material increases as the magnetising force increases from zero. The permeability reaches a maximum just before the core begins to saturate (Fig. 3.10). The maximum and initial permeability are almost the same in case of materials having a linear B-H curve.

Coercive Force, H_c: Coercive force is the negative magnetic force (Fig. 3.10) that must be applied to the material once saturated to reduce the magnetic flux density or magnetic induction back to zero.

Remanent Flux Density, B_r: As the magnetising force H is increased, the magnetic induction or flux density also increases either linearly or non-linearly depending upon the material till it reaches saturation. If the magnetising force is now reduced, it is observed that the magnetic flux density does not follow the same curve. Infact, there is some residual magnetism even when force is reduced to zero. This residual magnetism is known as remanent flux density, B_r.

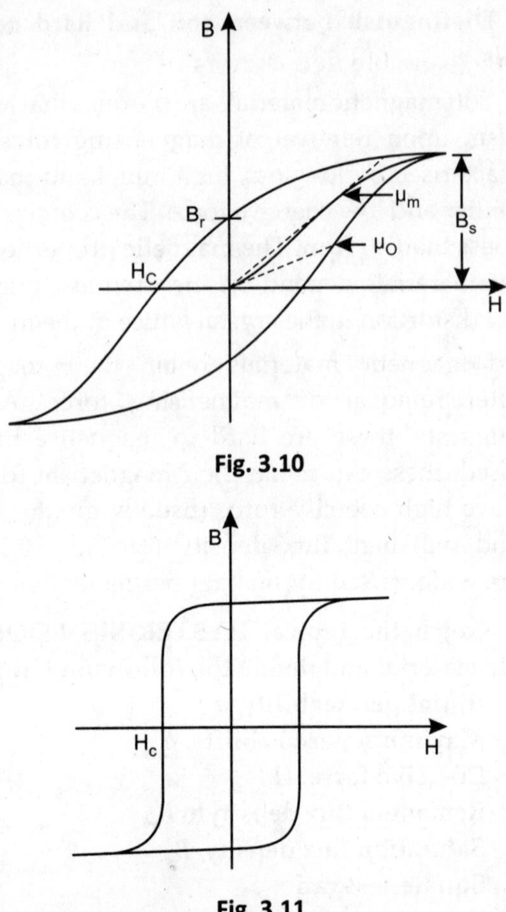

Fig. 3.10

Fig. 3.11

Saturation Flux Density, B_s or B_{max}: It is the maximum flux density. Once reached, further increase in magnetising force does not produce an increased flux density. In saturation, the permeability falls rapidly and the inductance of the coil decreases in the same proportion.

Squareness Ratio: It indicates the squareness of the B-H loop. It is defined as the ratio of remanent flux density (B_r) to the saturation flux density (B_s). In case of square loop cores used for memory and switching applications, the Squareness ratio is typically 0.98 (Fig. 3.11).

Q.14. What do you understand by μ.Q product of a magnetic material?

Ans. While designing inductors, μ.Q product of the core material has been found to be a useful measure of the quality of the material: (μ) here is the initial permeability of the material and (Q) is the quality factor of the inductor. Larger the value of μ.Q product, better is the material. Ferrites have the highest $\mu.Q$ product of all the commercially known magnetic materials. Iron dust has a $\mu.Q$ of about 2000, permalloy powder has a $\mu.Q$ of 10,000, the same for 12.5 μm thick molybdenum permalloy tape is 100,000 and for MnZn ferrite it is as high as 2,50,000.

These figures have been obtained from measurements done at 20kHz.

Q.15. While working at high frequencies, it is often important to know the effects of lead inductances as even small values of lead inductances offer significant reactances at high frequencies. For known wire diameter and length, how do you determine the inductance of a straight wire conductor?

Ans. The inductance of a straight wire conductor is mainly a function of length and is given by the empirical formula

$$L = 5.081\left(Ln\frac{4l}{d}-1\right)$$

where

l = length of the conductor in inches

d = diameter of the conductor in inches

L = Inductance in nH.

Q.16. For known geometric parameters of an air cored single layer coil (Fig. 3.12), how do you determine the inductance ?

Ans. Inductance (L) is expressed by

$$L = \frac{r^2N^2}{(9r+10l)}$$

where

L = Inductance in μH

r = Mean radius of the coil in inches

l = Length of the coil in inches

N = Number of turns.

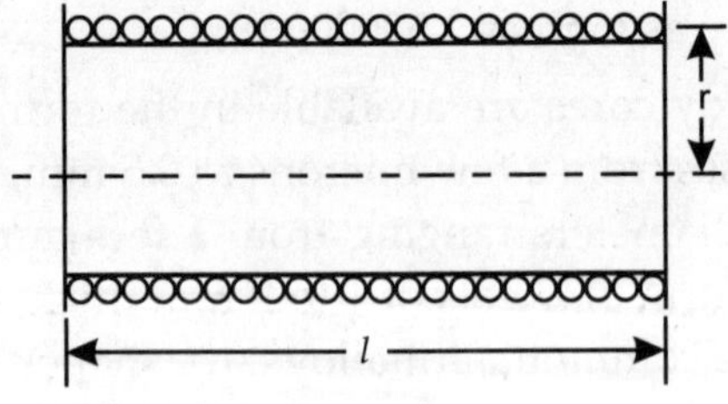

Fig. 3.12

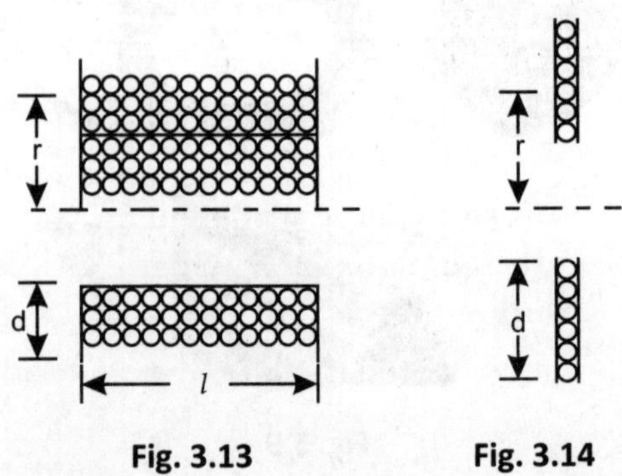

Fig. 3.13 Fig. 3.14

Q.17. For known geometric parameters of a multilayered air core coil (Fig. 3.13) how do you determine the inductance ?

Ans. Inductance (L) is given by

$$L = \frac{0.8\,r^2N^2}{(6r + 9l + 10d)}$$

where

L = Inductance in μH

r = Mean radius of the coil in inches

l = Length of coil in inches

d = Depth of the coil in inches

N = Number of turns.

Q.18. How do you calculate the inductance of a spiral coil (Fig. 3.14) of given geometric parameters ?

Ans. Inductance (L) is given by

$$L = \frac{r^2N^2}{(6r + 11d)}$$

where

r = Mean radius in inches

d = Depth of the coil in inches

N = Number of turns

L = Inductance in μH.

Q.19. How do you compute the effective inductance of a series or a parallel connection of more than one inductor?

Ans. When inductors L_1, L_2, L_3.....L_n are connected in series, the equivalent inductance of this series connection of inductors is given by:

$$L_s = L_1 + L_2 + L_3 + \ldots.. + L_n$$

The above expression is valid only when the individual inductors are uncoupled.

When inductors L_1, L_2, L_3.....L_n are connected in parallel, the equivalent inductance of this parallel connection of inductors is given by :

$$\frac{1}{L_P} = \frac{1}{L_1} + \frac{1}{L_2} + \frac{1}{L_3} + \ldots.. + \frac{1}{L_n}$$

The above expression is also valid only when the individual parallel connected inductors are uncoupled.

Q.20. How do you compute the energy stored in an inductor of inductance value (L) when a DC current of (I) is flowing through it ?

Ans. The energy stored can be computed from :

$$E = \frac{1}{2}LI^2.$$

Q.21. What are the standard values of inductance commonly available in inductors?

Ans. Standard values of inductance commonly available in inductors are 1.0, 1.1, 1.2, 1.3, 1.5, 1.6, 1.8, 2.0, 2.2, 2.4, 2.7, 3.0, 3.3, 3.6, 3.9, 4.3, 4.7, 5.1, 5.6, 6.2, 6.8, 7.5, 8.2, 8.7 and 9.1 (all in nH and μH). These are the standard values for one decade. Higher values can be obtained by multiplying these values by 10, 100 and 1000. For example, 1.0, 10, 100 and 1000 (all in nH) and 1.0, 10, 100 and 1000 (all in μH) are standard values. Also, 9.1, 91, 910, 9100 (all in nH) and 9.1, 91, 910 and 9100 (all in μH) are standard values.

Q.22. Do we have a colour coding standard for marking of inductors like the way we have it for resistors and capacitors? If yes, what is it?

Ans. We do have a colour coding scheme for marking of inductors. There is the 4-band code similar to what we have for resistors. There is a 5-band code for MIL standard inductors.

In the case of 4-band code, first three bands give inductance value in μH. The first two bands identify the significant digits and the third band identifies the decimal multiplier. On the extreme right is the tolerance band. Significant digits 0, 1, 2, 3, 4, 5, 6, 7, 8 and 9 are respectively represented by Black, Brown, Red, Orange, Yellow, Green, Blue, Violet, Grey and White. Decimal multipliers of 10^0, 10^1, 10^2, 10^3, 10^4, 10^5, 10^6, 10^7, 10^8, 10^9, 10^{-1} and 10^{-2} are respectively represented by Black, Brown, Red, Orange, Yellow, Green, Blue, Violet, Grey, Gold and Silver. Tolerances of ± 5%, ± 10% and ± 20% are respectively represented by Gold, Silver and Black (or no band). For example, 470 μH, ± 10% inductor will have Yellow, Violet and Brown as the first three bands in the same order. A Silver band on the extreme right will indicate tolerance of ± 10%.

In the case of 5-band code, first band is always silver and it indicates the MIL standard. The width of this band is double the width of other bands. Bands 2, 3 and 4 give inductance value in μH. Band-5 gives tolerance. For inductance values less than 10 μH, a Gold band as the second or third band signifies decimal point. For example, 0.39 μH inductor will respectively have Gold, Orange and White as bands 2, 3 and 4. As another example, bands 2, 3 and 4 in the case of 4.7 μH inductor will respectively be Yellow, Gold and Violet. Fig. 3.15 summarizes 4-band and MIL-STD 5-band inductor colour code.

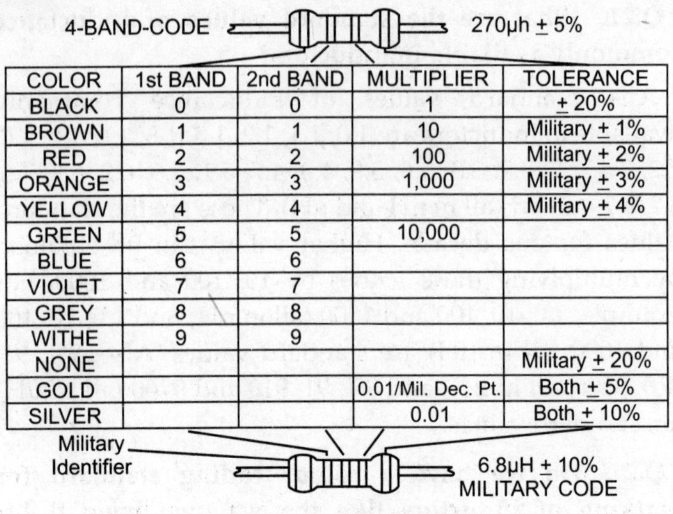

COLOR	1st BAND	2nd BAND	MULTIPLIER	TOLERANCE
BLACK	0	0	1	± 20%
BROWN	1	1	10	Military ± 1%
RED	2	2	100	Military ± 2%
ORANGE	3	3	1,000	Military ± 3%
YELLOW	4	4		Military ± 4%
GREEN	5	5	10,000	
BLUE	6	6		
VIOLET	7	7		
GREY	8	8		
WITHE	9	9		
NONE				Military ± 20%
GOLD			0.01/Mil. Dec. Pt.	Both ± 5%
SILVER			0.01	Both ± 10%

Fig. 3.15

Q.23. What essentially is a transformer ? What are the broad areas of applications of transformers ?

Ans. Essentially, a transformer is a component that consists of two or more than two coils, wound on an air core or a magnetic material, that are electrically isolated but magnetically coupled. It does the job of transforming one or more electrical parameters from one alternating circuit to another. The parameters to be transformed could be voltage (as in a Voltage Transformer), current (as in a Current Transformer), impedance (as in an Impedance Transformer) phase and so on. It may even provide isolation between two alternating current circuits (Isolation Transformer) or may be used for precise measurements of current and voltage. Fig. 3.16 shows the circuit representations of air core (Fig. 3.16) and iron or ferrite core (Fig. 3.16) transformers.

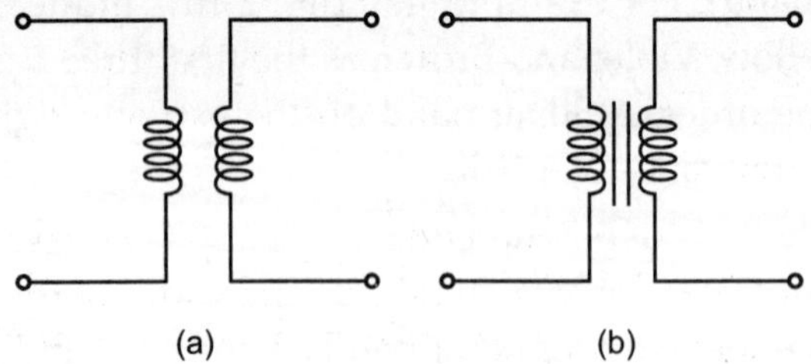

Fig. 3.16

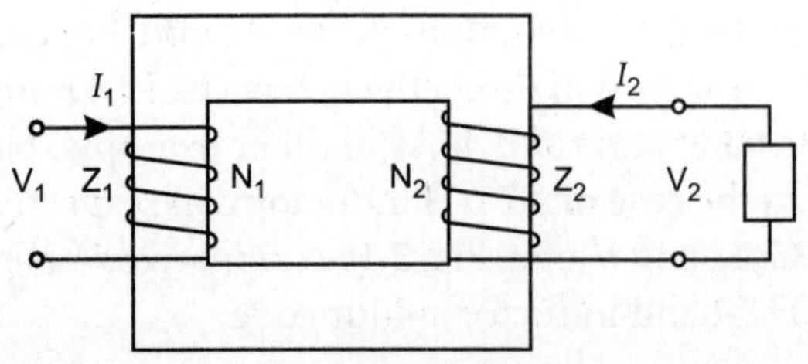

Fig. 3.17

Q.24. What is the basic principle underlying the operation of practically all transformers ? How are current, voltage, impedance and number of turns related to one another in the primary and secondary of the transformer?

Ans. The primary and secondary windings of a transformer are magnetically coupled, though the co-efficient of coupling can vary. In an ideal transformer (the one that has no losses), a time varying voltage signal (V_1) applied to the primary of the transformer causes a time varying current (I_1) to flow through the. primary (Fig. 3.17). The magnetic field thus produced is linearly proportional to the current. This magnetic field is coupled to the secondary winding and induces a voltage (V_2) across it. If the secondary circuit is closed, a current (I_2) flows. Following mathematical relationships relate different electrical parameters in the primary with their counterparts in the secondary.

$$(V_2/V_1) = (N_2/N_1) = (I_1/I_2) = n$$
$$(Z_2/Z_1) = (V_2/V_1)^2 = (I_1/I_2)^2 = n^2$$

Q.25. Classsify transformers on the basis of

(*a*) the nature of function performed

(*b*) the frequency of the signal being handled.

Ans. (*a*) On the basis of the function performed by the transformers, they are classified as:

Voltage Transformers (Fig. 3.18 *a*) used to change the magnitude of voltage.

Current Transformers (Fig. 3.18 *b*) used to change the magnitude of current.

Impedance Transformers (Fig. 3.18 *c*) used to transform impedance level. They usually transform the output impedance of one circuit (the source circuit) in such a way that it matches the input impedance of another circuit (the load circuit).

Isolation Transformers (Fig. 3.18 *d*) used to provide isolation between two circuits without changing the electrical parameters.

Instrument Transformers (Fig. 3.18 *e*) used to accurately measure current and voltage.

Trigger Transformers (Fig. 3.18 *f*) used to generate trigger pulses for components like thyristors, flash lamps, spark gaps etc.

(*b*) On the basis of frequency of the signals being handled, transformers are classified as

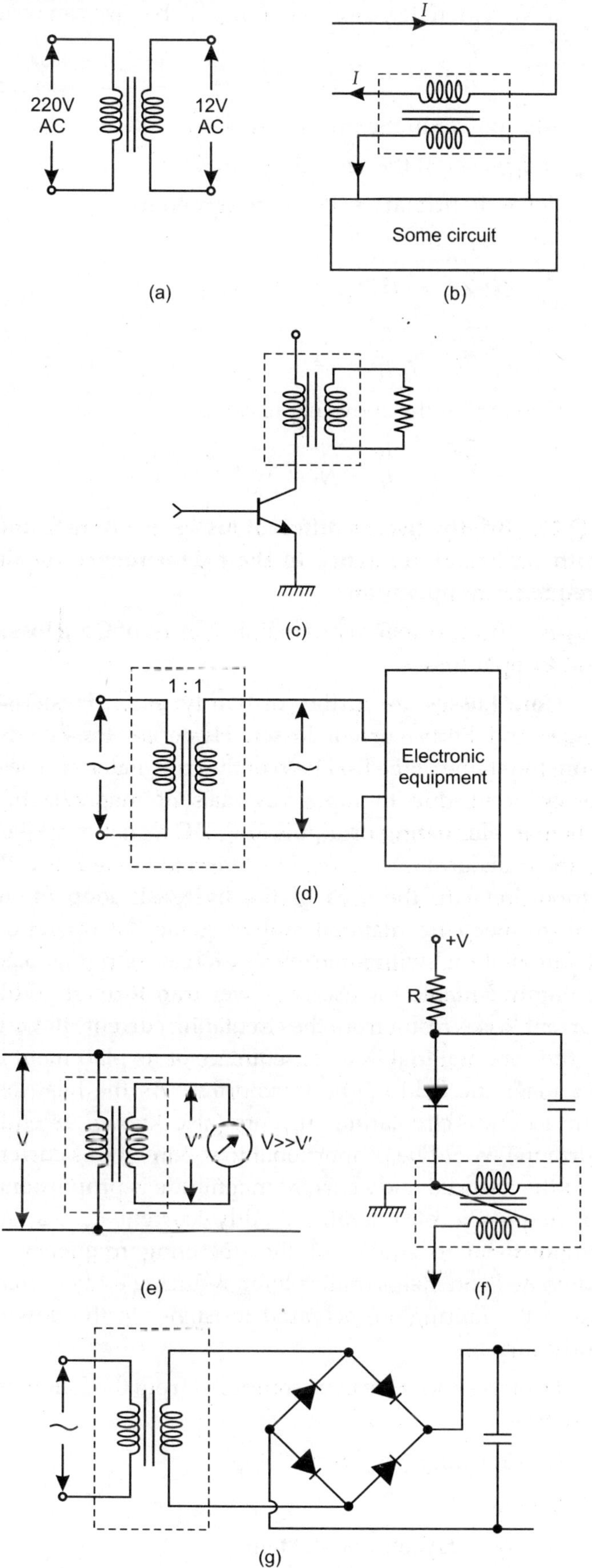

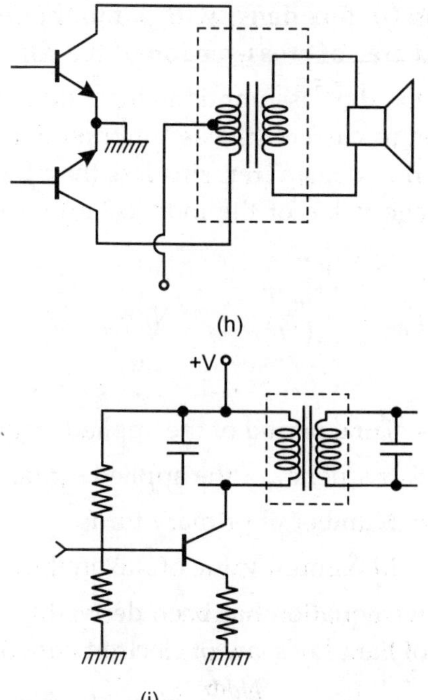

Fig. 3.18

Power Transformers that operate at power line frequency of 50 Hz. These find extensive applications in AC/DC power supplies and for generating AC voltages other than the available AC mains voltage. Fig. 3.18 (*g*) shows a typical application.

Audio Transformers used to transform electrical signals spread out in frequency in the audio frequency range from one circuit to another. A typical application is that of coupling the audio signal present at the output of the final stage of audio amplifier to the speaker (Fig. 3.17*h*). Here the transformer provides impedance matching between relatively high output impedance of the amplifier circuit and a much lower impedance of the speaker's voice coil to achieve optimum energy transfer.

Video Transformers are also wideband transformers like audio transformers. A video transformer is typically required to couple signal voltages spread over more than five decades of frequency range from one circuit to another.

While audio and video transformers are broadband transformers, **RF** and **IF** transformers are narrow band transformers that couple signal voltages in a narrow band around a center frequency from one stage to another (Fig. 3.18 *i*).

Q.26. Derive from first principles the basic transformer equations that relate the induced voltages in the primary and secondary windings to the frequency

of operation (*f*), flux density (B_m), number of turns (N_p and N_S) and area of cross-section of the core (*A*).

Ans. With the assumption that flux is linearly proportional to current (the assumption is valid as long as the core flux density remains less than the saturation limit), average value of the induced primary voltage is given by :

$$E_{av} = \frac{N_p \cdot \Phi_m}{(T/4)} = 4f\,N_p\,\phi_m$$

where

T = Time period of the applied signal

f = frequency of the applied signal

N_p = Number of primary turns

Φ_m = Maximum value of flux in the core

The above equation has been derived from the basic expression of Faraday's law of electromagnetic induction according to which $E = \frac{Nd\Phi}{dt}$. Here, the flux (Φ) reaches the maximum value of (Φ_m) in quarter of one full cycle of applied signal.

RMS value of the primary voltage is given by

$$E_p = 1.11\,(4fN_p\,\Phi_m)$$
$$= 4.44\,(fN_p\Phi_m) = 4.44\,fN_p\,B_m A$$

where

B_m = Maximum flux density

A = Area of cross section of the core

RMS value of the secondary voltage is given by

$$E_s = 4.44\,fN_s\Phi_m$$

From the above two epressions for E_P and E_s.

$$\frac{Ep}{Es} = \frac{Np}{Ns}$$

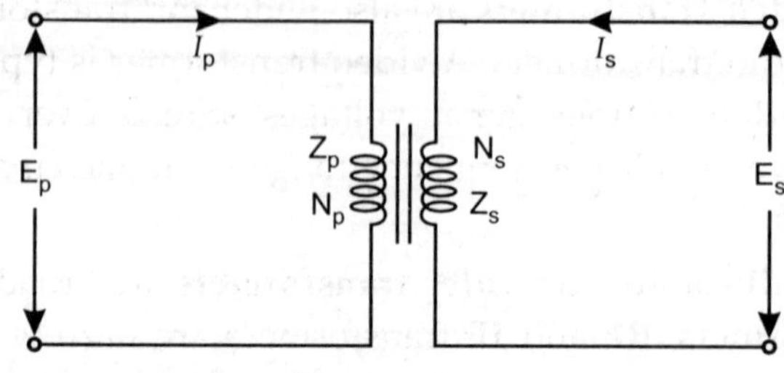

Fig. 3.19

Also in an ideal transformer, power in the primary equals the power in the secondary.

Therefore,

$$E_p \cdot I_P = E_s \cdot I_s \text{ or } \frac{E_P}{E_S} = \frac{I_S}{I_P}$$

If N_S/N_P = n (transformation ratio), than we can write

$$\frac{E_P}{E_S} = \frac{I_S}{I_P} = \frac{N_P}{N_S} = \frac{1}{n}$$

Also power in the primary = $I_P^2.Z_P$

and power in the secondary = $I_p^2.Z_S$

Power in primary = Power in secondary

or

$$I_P^2.\,Zp = I_s^2.Z_s$$

$$\frac{Z_P}{Z_S} = \frac{I_S^2}{I_P^2} = \frac{1}{n^2}$$

Combining the above ezpressions

$$\frac{E_P}{E_S} = \frac{I_S}{I_P} = \frac{N_P}{N_S} = \sqrt{Z_P/Z_S} = \frac{1}{n}.$$

Q.27. Briefly discuss different losses in a transformer with particular reference to their dependence on the frequency of operation.

Ans. Transformer losses mainly consist of **Core losses** and **Copper losses.**

Core Losses are further of two types *i.e.* Hysterisis losses and Eddy current losses. Hysterisis losses arise from the shape of the B-H loop of the core material. These losses occur due to rapid reversals of magnetization when an alternating or a pulsating DC signal is applied to the transformer primary. Hysterisis losses are directly proportional to the area of the hysterisis loop of the transformer core material and frequency of operation. A core material with a narrow hysterisis loop (Fig. 3.20) is highly suitable for use in power transformers. Eddy current losses result from the circulating currents flowing in the core material as a consequence of its placement in the magnetic field of the transformer. As the heat loss due to these circulating currents (also known as eddy currents) would be proportional to square of this current and the induced eddy current magnitude is proportional to frequency of operation, eddy current losses are proportional to square of the operating frequency. A laminated core helps quite a lot in reducing eddy current losses by offering an increased resistance to the flow of eddy currents.

Hysteresis losses can be computed from the following equation.

$$W_h \text{ (in watts)} = K_h f\,(B_m)^{1.6}$$

Where

K_h = Hysteresis constant

f = Operating frequency

B_m = Maimum flux density

Eddy current losses can be computed from the following equation.

W_e (in watts) = $K_e f^2 k_f B_m^2$

Where

K_e = Eddy current constant

K_f = form constant

Copper losses result from the DC resistances of the transformer windings and the currents flowing through them.

Copper losses can be computed from the following equation.

We (in watts) = $I_L^2 R_2 t$ + stray losses

where

I_L = Load current

R_2 = Transformer resistance referred to secondary

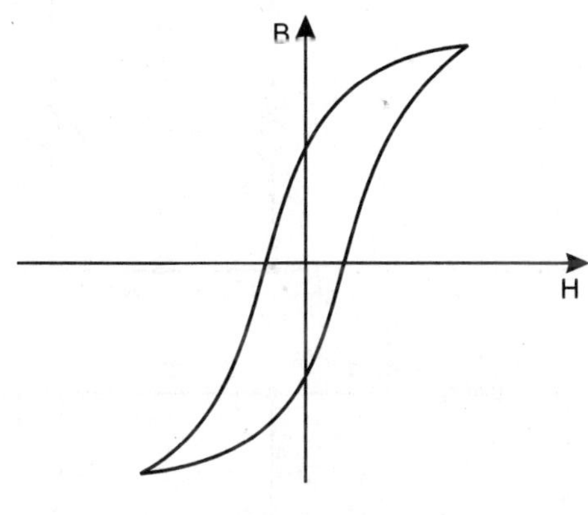

Fig. 3.20

Q.28. How do the laminated iron cores used for designing power transformers help in drastically reducing eddy current losses and make it a practical core material for use at power line frequencies. Why is it that even after lamination, iron core is not suitable for use at higher frequencies due to rapidly increasing eddy current losses.

Ans. As explained in the last question, eddy current losses in the core material result from induced e.m.f. in the core due to the fact that the core is placed in the magnetic field. This is a situation which is essential to the operation of the transformer and can not be avoided. The induced e.m.f. gives rise to the circulating current in the core whose magnitude depends upon the resistivity of the core material and the length and area of cross-section of the core as seen by the circulating eddy currents. By using a laminated iron core with all the laminations electrically insulated from ench other, the magnitude of eddy currents is significantly reduced and the effect of low resistivity of iron core is compensated for to some extent.

However, as the frequency increases, eddy current losses which are proportional to square of frequency, become prohibitively large and render the laminated iron core useless at frequencies beyond a few hundreds of hertz.

Q.29. Outline different steps involved in designing a power transformer of given specifications.

Ans. 1. As a first step, ensure that the transformer core remains away from the saturation limit for the desired load current or the output power requirement. This requirement can be met with by choosing an appropriate core cross-section. The optimum core cross-section (A_C) can be determined with sufficient accuracy from the empirical formula

$$A_c = \frac{\sqrt{W}}{5.6}$$

Where

W = Wattage required to be delivered by the transformer

A_c = Core cross section in sq. inches.

2. The ratio of the stack thickness to the width of the center limb should be chosen to be between 1.1 and 1.5 (Refer to Fig. 3.21)
3. The turns per volt for the windings can be calculated from:

$$\text{Turns per Volt} = \frac{10^8}{4.44 \times f \times A_c \times B}$$

where

f – frequency – 50 Hz

B = flux density in lines/sq. inch

A_c = core cross-section in sq. inch

In this formula if (A_c) is in sq. cm, then B should be substituted in Gauss. For STALLOY or other similar core materials used for winding power transformers, value of (B) can be taken to be 50,000 lines/sq. inch

4. Having computed Turns/Volt, both primary as well as secondary turns can be calculated from

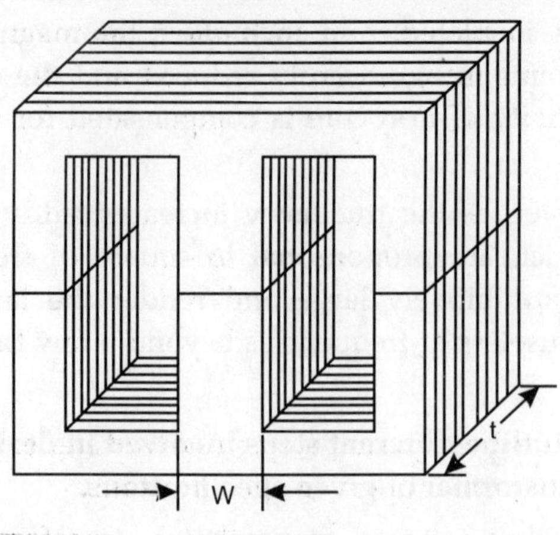

Fig. 3.21

the known values of primary and secondary voltages.

5. Primary current can be calculated from Primary current = $\frac{\text{Output Wattage}}{\text{Efficiency} \times \text{Primary Voltage}}$

The transformer efficiency can be taken to be 85 to 90 percent.

6. Secondary current can be calculated from Secondary current = $\frac{\text{Primary current}}{n}$

7. Primary and secondary wire sizes can be determined for the calculated values of primary and secondary currents from the Standard Wire Gauge Table (Refer Table 3.1). The current handling capability of different wire sizes as shown in Table 3.1 is with the assumption of a current density of 1000 amp per sq. inch. For power transformers, it is reasonably safe to assume a current density of 2000 amp per sq. inch. Thus for selecting a wire size for a particular current handling requirement in a power transformer, choose a wire size that corresponds to half of the desired current requirement.

Q.30. What is a Pulse Transformer ? Briefly describe the factors to be considered and the steps to be followed while designing a pulse transformer of desired specifications.

Ans. A pulse transformer as the name suggests transforms pulses from one circuit (the source circuit) to another (the load circuit). The transformed or the coupled pulse is either the exact replica of the original pulse or it has undergone a change in its amplitude or polarity or both. The fact that a pulse transformer handles pulses, it has got to be a wideband transformer as a rectangular pulse is supposed to comprise frequency components from almost DC to a frequency determined by the rise time of the pulse to be transformed.

A pulse transformer or for that matter any transformer in general can be represented by an equivalent circuit of Fig. 3.22 (*a*). Here, R_1 is the sum of source resistance and the transformer primary's DC resistance, (σ) is the leakage inductance and is given by transformer's primary inductance with its secondary short circuited. (L) is the magnetising inductance. It is the primary inductance with secondary open circuited. (C) is the total shunt capacitance including the inter-winding capacitance and any load capacitance reflected onto the primary

$$C = C_d + n^2 C_L$$

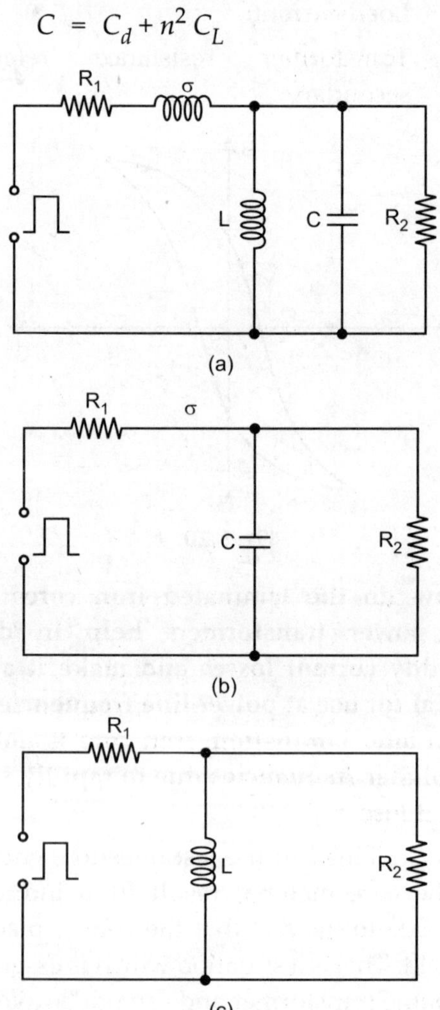

Fig 3.22

(R_2) is the total load resistance reflected onto the primary side. It is given by

$$R_2 = \frac{(R_{sec} + R_L)}{n^2}$$

In order to determine the response of the transformer to an applied rectangular pulse, the entire pulse can be divided into three different parts i.e. (*i*) the leading edge of the pulse, (*ii*) the flat top and (*iii*) the trailing edge.

Quite obviously, how truly the leading edge of the pulse is reproduced across the load depends upon the high frequency response of the transformer. High frequency equivalent circuit is shown in Fig. 3.22 (*b*). Magnetising inductance has been ignored as at high frequencies, its reactance can be considered to be nearly an open circuit. The equivalent circuit is reduced to a series *R-L-C* circuit. When a step input is applied to such a circuit. For critical damping, it responds to the step input with a rise time given by

$$t_r = 3.35 \sqrt{\left(\frac{R_2}{R_1 + R_2}\right) \cdot \sigma C}$$

Improvement in rise time can be achieved at the cost of a small overshoot. An overshoot of about 10 percent is usually acceptable and the rise time in such a case is given by

$$t_r = 1.67 \sqrt{\left(\frac{R_2}{R_1 + R_2}\right) \cdot \sigma C}$$

The conclusion from the above discussion is that a fast rise time and a good leading edge response is obtained when the pulse transformer has extremely small leakage inductance and inter-winding capacitance. All good pulse transformers have a coupling co-efficient of greater than 99 percent.

A small leakage inductance and a small inter-winding capacitance are conflicting requirements. An attempt to reduce leakage inductance by having closely would turns increases the inter-winding capacitance and vice versa. One way to overcome this problem and meet both the requirements is to use a very high permeability core material so that very few turns are needed to get the desired magnetising inductance. Tape wound permalloy cores where the permeability approaches 80,000 make very good core material for making pulse transformers. It may be mentioned here that the effective permeability during the leading edge is much less than the maximum value mentioned above due to eddy currents that flow. Ferrite cores on the other hand have a smaller value of maximum permeability as compared to permalloy cores but have a comparatively greater effective permeability during the rising edge as the eddy current effects in this case are far less severe due to a much larger resistivity. Therefore, ferrite cores, particularly the *Mn-Zn* ferrites, are also very useful for making pulse transformers. **Either toroids or pot cores are used due to their inherent close coupling and self-shielding features.**

A large step-up ratio should also be avoided as it increases the magnitude of load capacitance reflected onto the primary. Increase in rise time is directly proportional to the step-up ratio.

The flat top response and the back swing are dictated by the low frequency response of the transformer. From the low frequency equivalent circuit of Fig. 3.22 (*c*), the droop in the flat top for a pulse of width (τ) is given by

$$= R\tau/L_p$$

where

$$\tau = \text{pulse width and } R = \frac{R_1 . R_2}{R_1 + R_2}$$

A large magnetising inductance is desirable for a smaller droop. For good pulse transformers, L_p = 50 Rτ giving only a 2 percent droop in the flat top. Smaller the droop, smaller is the back swing on the trailing edge of the pulse.

The points to be borne in mind while designing a pulse transformer can be summarised as under:

1. For the given pulse width, calculate the magnetising inductance so that it is at least 50 times the (Rτ) product. This will give a 2 per cent droop. Other wise if the acceptable droop is given, the required magnetising inductance can be calculated from

$$L = \frac{R.\tau}{\%\ droop} \times 100$$

2. Choose a high permeability core so that minimum number of turns are required to get the desired magnetising inductance.
3. Verify that the core does not saturate for the calculated number of turns (*N*), given pulse amplitude (*V*), width (τ) and cross-sectional area (*A*) of the chosen core.
4. Use a winding technique to achieve a close coupling and avoid as far as possible a step-up ratio (*n*) of more than 10.

Q.31. Briefly describe the following (or Write short notes on)

1. **Isolation transformer**
2. **Autotransformer**
3. **Variac**
4. **Instrument transformer**

Ans. Isolation Transformer. An isolation transformer is used to electrically isolate two AC circuits without changing their current and voltage levels. It is very useful when it is desired to isolate an equipment from the power lines. In some cases, it may be required to ground the chassis of the equipment for improved performance. Connecting such an equipment to power lines without an isolation transformer could cause a catastrophe unless it is ensured that ground side of the equipment is connected to grounded side of power lines. An isolation transformer between the equipment and the power lines avoids the possibility of any damage to the equipment that must be grounded.

Autotransformer. Unlike the conventional transformers where primary and secondary windings are only magnetically coupled, in an autotransformer, (Fig. 3.23) they are both electrically and magnetically coupled. Infact, secondary winding is a continuation of the primary winding.

Auto transformers are used in applications where primary and secondary voltages do not differ by a large amount. It is seldom used where the step-down ratio equals or exceeds 10. In such a case, the power transferred directly to the load reduces proportionately and most of the power is transformed by transformer action. As a result, there is no significant saving in the copper loss.

Variac also called a power stat is a variable auto transformer (Fig. 3. 24) The coil in a variac is a single layer winding on a toroidal core of high permeability. It has a movable tap which can be set by a knob or a dial attached to a sliding brush contact anywhere on the circumference of the winding. The voltage output can thus be varied from zero to slightly above the input voltage. While using a variac, one important point to be borne in mind is that the rated current of the variac should never be allowed to exceed for any output voltage setting.

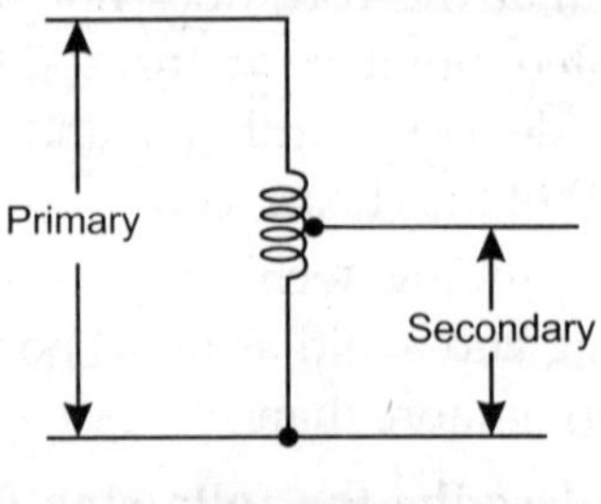

Fig. 3.23

Unlike a conventional step-down transformer where the secondary current is greater than the primary current, the maximum current that can be drawn from an a variac secondary even at lower output voltages is the same as that can be drawn at full line voltage. This maximum current is nothing but the rated current of the variac. A fuse for the rated current connected in a variac can protect it from any accidental damage.

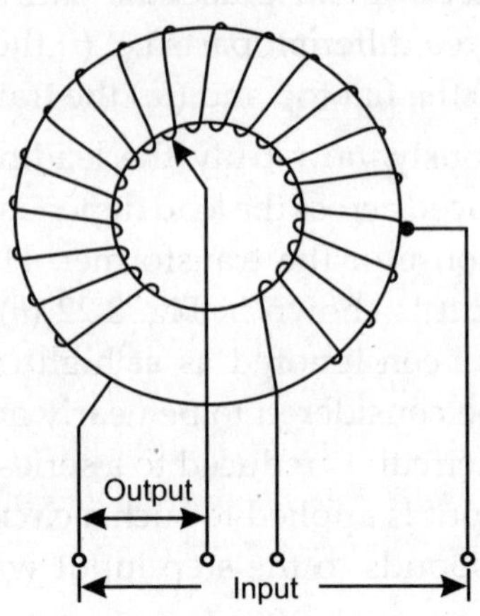

Fig. 3.24

Instrument transformers are used in conjunction with measuring instruments like Voltmeters, Ammeters or Control devices like Relays and are designed to maintain a specific magnitude and phase relationship between primary and secondary currents and voltages.

Potential transformers (Fig. 3.25*a*) enable measurement of extremely high voltages with a small low range meter. The meter can be calibrated to directly read the main voltage. Similarly, a current transformer (Fig. 3.25 *b*) can be used to step down a high current to a low value and thus make the measurement possible with a low range ammeter.

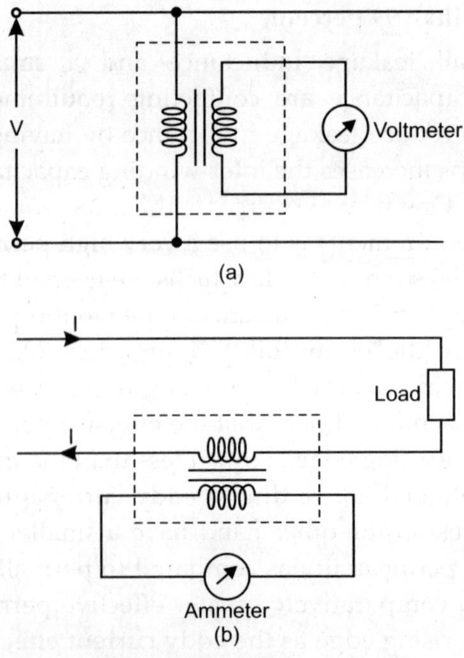

Fig. 3.25

Solved Problems

Problem 3.1. *What is the Q-factor of a coil having a self resonant frequency of 10 MHz, an inductance of 10 μH and an associated series resistance of 10 ohms?*

Solution. Q-factor is given by the epression

$$Q = \frac{2\pi fL}{R} = \frac{6.28 \times 10 \times 10^6 \times 10 \times 10^{-6}}{10} = 6.28$$

Problem 3.2. *An air core coil having a magnetic path length of 10 cm, an area of cross-section of 2 sq. cm. and 100 turns has an inductance of 50 μH. What will be the inductance μf the same coil was wound on a former with a core having a permeability of 1000? The magnetic path length, the area of cross section and the number of turns remain the same.*

Solution. The inductance in general is given by

$$L = \frac{\mu_o \mu_r N^2 A}{l}$$

The inductance in the second case will therefore be 1000 times the inductance in case of air core coil (μ_r for air is approximately 1). That is, inductance of the coil with core is 50 mH.

Problem 3.3 *An AC/DC power supply consists of a transformer, a full wave rectifier and a capacitor filter of 100 μF. The DC output has a peak to peak ripple voltage of 100 mV, If the filter is changed to a choke input one, what should be the inductance of the choke to reduce the ripple by a factor of 10 ?*

Solution. As the ripple is to be reduced by a factor of 10, the smoothing factor in this case is 1/10. Also, the smoothing factor in case of a choke input filter in terms of capacitive and inductive reactances is given by : (X_C/X_L) Now,

$$X_C = 1/(2\pi fC)$$

where f = 100 Hz (Ripple Frequency)

$$X_C = 1/(2\pi \times 100 \times 100 \times 10^{-6}) = 16 \text{ ohms}$$

Now,

$$X_C/X_L = 1/10$$

or $$X_L = 10X_C = 10 \times 16 = 160 \text{ ohms}$$

Therefore,

$$L = X_L/2\pi f = 160/6.28 \times 100 = 255 \text{ mH.}$$

Problem 3.4. *Determine the Q-factor of a coil whose electrical equivalent circuit is shown in Fig. 3.26*

Solution. The Q-factor of a coil in terms of its inductance, series resistance and interwinding capacitance is given by:

$(1/R)\sqrt{L/C}$

100 nH   10 Ω   10 pH

Fig. 3.26

Therefore,

$$Q = (1/10)\sqrt{100 \times 10^{-9}/10 \times 10^{-12}} = 10.$$

Problem 3.5. *What will be the reactance offered by a 0.01 inch thick lead length of a quarter of an inch when you are working at 1 GHz?*

Solution.

Inductance (in nH)

$$= 5.081\left(Ln\frac{(4l)}{d} - 1\right) = 5.081\left(Ln\frac{4 \times 1/4}{0.01} - 1\right)$$

$$= 5.081\,(Ln\,100 - 1)$$

$$= 5.081\,(4.6 - 1) = 18\,nH$$

Reactance = 2πfL = 6.28 × 10^9 × 18 × 10^{-9} = 113 ohms.

Problem 3.6. *A 220 volts, 50 Hz supplies a load of 22 Amp. The total resistance of the transmission line joining the source to the load is 2 ohms. A step-up transformer at the generator site and a step-down transformer at the load site have been used as shown in Fig. 3.27. Calculate*

(*a*) *Power loss in the line*

(*b*) *Current drawn from the generator*

Solution. (*a*) If I_L is the current flowing in the line, then

$$\frac{I_L}{22} = \frac{220}{2200} = \frac{1}{10} \text{ Hence } I_L = 2.2 \text{ amp}$$

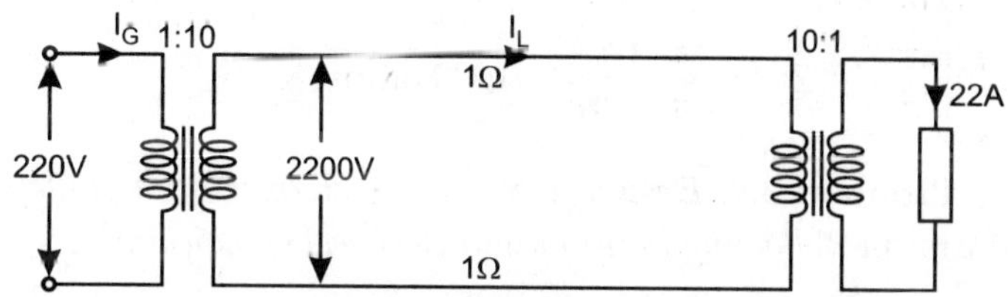

Fig. 3.27

Therefore power loss in the line = $(2.2)^2 \times 2 = 4.84 \times 2 = 9.68$ W

(b) If I_G is the current drawn from the generator, then

$$\frac{I_L}{22} = \frac{2200}{220} = 10 \text{ or } I_G = 22 \text{ amp.}$$

Problem 3.7. *A power transformer has 1000 primary turns and 100 secondary turns. The cross-sectional area of* *the core is 6 sq.cm and the maximum flux density while in operation is 10,000 Gauss. Calculate TURNS PER VOLT for the primary and secondary windings.*

Solution. The RMS value of the primary voltage is given by

$$E_p = 4.44 f N_p B_m A$$
$$= 4.44 \times 50 \times 1000 \times 1 \times 6 \times 10^{-4}$$
$$(10000\ G = 1\ \text{tesla})$$
$$= 133\ \text{volts}$$

The RMS value of the secondary voltage is given by

$$E_s = 4.44 f N_S B_m A$$
$$= 4.44 \times 50 \times 100 \times 1 \times 6 \times 10^{-4} = 13.3\ \text{volts}$$

$$\text{Turns per Volt of the primary} = \frac{1000}{133} = 7.5$$

$$\text{Turns per Volt of the secondary} = \frac{100}{13.3} = 7.5$$

Problem 3.8. *Calculate the impedance of the transformer looking into the primary for the loaded transformer shown in Fig. 3.28*

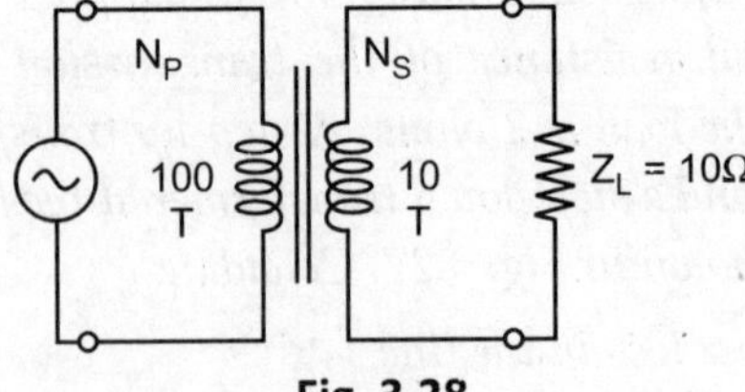

Fig. 3.28

Solution. Impedance looking into the primary of the transformer is given by

$$Z_P = \left(\frac{Z_L}{n^2}\right) \text{ where } n = \frac{N_S}{N_P}$$

There fore,

$$Z_P = \frac{10}{(1/10)^2} = 1000\ \text{ohms.}$$

Problem 3.9. *Design a power transformer on a stalloy core for the following input-output electrical specifications.*

1. Primary voltage = 220 VAC,
2. Secondary Voltage = 1000 VAC and
3. Load current = 100 mA

Assume B = 60,000 lines per sq. inch for the chosen core material and an efficiency of 90 percent.

Solution. Core cross-section in given by

$$A_C = \frac{\sqrt{W}}{5.6} = \frac{1000 \times 0.1}{5.6} = \frac{10}{5.6}$$
$$= 1.785\ \text{sq. inch}$$

Turns/Volt

$$= \frac{10^8}{4.44 \times f \times B \times A_C}$$
$$= \frac{10^8}{4.44 \times 50 \times 600000 \times 1.785} = 4.2$$

Number of primary turns,

$$N_p = 4.2 \times 220 = 924$$

Number of secondary turns,

$$N_S = 4.2 \times 1000 = 4200$$

Primary current can be calculated from

$$I_p = n \times I_S$$
$$\text{where } n = \frac{N_S}{N_P} = \frac{1000}{220} = 4.545$$

Therefore

$$I_p = 4.545 \times 0.1 = 0.454\ \text{amp}$$

Primary and secondary wire sizes can be selected form Table 3.1

Primary wire size = 26 SWG assuming a current density of 2000 amp per sq. inch secondary wire size = 35 SWG

The above design assumes that the chosen core has sufficient window area to accommodate both windings with appropriate inter-layer insulation.

Problem 3.10. *Design a power transformer with a multioutput secondary and the following input-output electrical specifications.*

Primary = 220 V, 50 Hz Secondary windings (i) Center tapped 12-0-12 V at 100 mA and (ii) 5 V at 1A.

Solution.

The wattage to be delivered by the transformer secondary is given by

Output watts = 12 × 0.1 + 5 × 1 = 6.2

$$\text{Core cross-section} = \frac{\sqrt{6.2}}{5.6} = 0.444\ \text{sq. inch}$$

$$\text{Turns per volt} = \frac{10^8}{4.44 \times 50 \times 60{,}000 \times 0.444} = 16.9$$

Primary turns = 16.9 × 220 = 3719

Secondary turns for 12-0-12 volt winding = 16.9 × 12 = 202.

Therefore 12-0-12 volt secondary winding has 202-0-202 turns.

Secondary turns for the 5 V winding = 16.9 × 5 = 84.5

$$\text{Primary current} = \frac{6.2}{0.9 \times 220} = 0.031$$ (assuming 90% efficiency)

Primary wire size = 40 SWG

Secondary wire size for 12-0-12 V winding = 35 SWG

Secondary wire size for 5V winding = 23 SWG

The wire sizes have been selected assuming a current density of 2000 amp per sq. inch.

Table 3.1

SWG No.	*Diameter (inch)*	*Wire Cross-section (sq. inch)*	*Current at 1000 per sq. inch*	*SWG No.*	*Diameter (inch)*	*W ire Cross-section (Sq. inch)*	*Current at 1000A per SQ. inch*
1	0.30	0.07	71.0	2 6	0.00025	0.018	0.25
2	0.27	0.06	60.0	27	0.00020	0.017	0.21
3	0.25	0.05	50.0	28	0.00017	0.015	017
4	0.23	0.04	42.0	29	0.00014	0.013	0.15
5	0.21	0.03	35.0	30	0.00012	0.012	012
6	0.19	0.028	29.0	31	0.00011	0.011	0.10
7	0.17	0.024	24.0	32	0.00009	0.0108	0.09
8	0.16	0.020	20.0	33	0.00008	0.0100	0.07
9	0.14	0.016	16.0	34	0.00007	0.0090	0.06
10	0.12	0.013	13.0	35	0.00006	0.0080	0.05
11	0.11	0.010	11.0	36	0.00005	0.0070	0.04
12	0.10	0.008	9.0	37	0.00004	0.0068	0.03
13	0.09	0.007	7.0	38	0.00003	0.0060	0.028
14	0.08	0.005	5.0	39	0.00002	0.0050	0.021
15	0.07	0.0047	4.0	40	0.000018	0.0048	0.018
16	0.06	0.0030	3.0	41	0.000015	0.0044	0.015
17	0.05	0.0025	2.0	42	0.000012	0.0040	0.012
18	0.048	0.0020	1.8	43	0.000010	0.0036	0.010
19	0.040	0.0013	1.3	44	---	0.0032	0.008
20	0.036	0.0010	1.0	45	---	0.0028	0.006
21	0.034	0.0008	0.8	46	...	0.0024	0.004
22	0.028	0.0006	0.6	47	---	0.0020	0.003
23	0.024	0.0005	0.5	48	...	0.0016	0.002
24	0.022	0.0004	0.4	49	---	0.0012	0.001
25	0.020	0.0003	0.3	50	---	0.0010	0.0007

Problem 3.11. *Fig. 3.29 shows an impedance transformer with a tapped secondary winding. Calculate the magnitude of the missing impedance.*

Solution.

The three possible impedances of the tapped secondary winding are related to each other by the expression

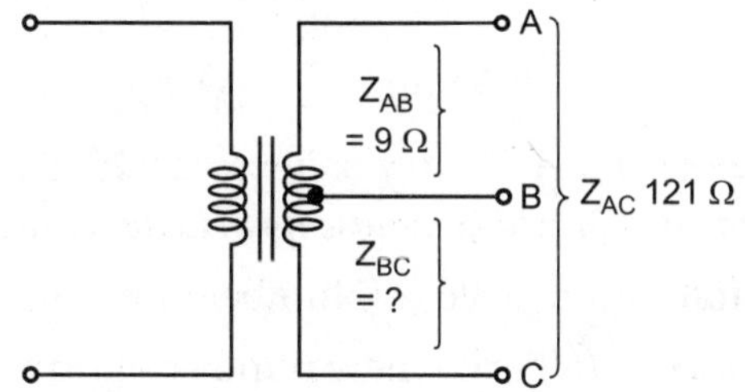

Fig. 3.29

$$\sqrt{Z_{AC}} = \sqrt{Z_{AB}} + \sqrt{Z_{BC}}$$

Substituting the values of (Z_{AB}) and (Z_{AC}) in the above expression, we get

$$Z_{BC} = (\sqrt{Z_{AC}} - \sqrt{Z_{AB}})^2 = 64 \text{ ohms.}$$

Problem 3.12. *Fig. 3.30 shows the waveform of current (i) flowing through a 10 mH inductance. Determine the magnitude of induced E.M.F. across the inductance at time t = 2 ms 4 ms, 8 ms and 14 ms.*

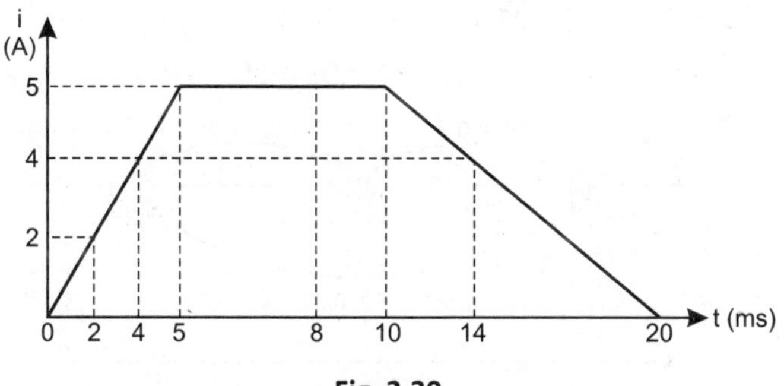

Fig. 3.30

Solution.

(*a*) For t = 2 ms

Rate of change of current = 1 A/s = 10^3 A/s

Therefore, induced

$$\text{E.M.F.} = L\frac{di}{dt} = 10\times 10^{-3} \times 1 \times 10^3 = 10 \text{ Volts}$$

(*b*) For t = 4 ms

Rate of change of current = 1 A/ms = 10^3 A/s

$$\text{Induced E.M.F.} = L\frac{di}{dt} = 10 \times 10^{-3} \times 10^3$$

$$= 10 \text{ volts}$$

(*c*) For t = 8 ms

Rate of change of current = 0

$$\text{Induced E.M.F.} = L\frac{di}{dt} = 0$$

(*d*) For $t = 14$ ms

Rate of change of current = 0.25×10^3 A/ms

Therefore,

$$\text{induced E.M.F.} = L\frac{di}{dt} = -10 \times 10^{-3} \times 0.25 \times 10^3$$

$$= -2.5 \text{ Volts.}$$

Problem 3.13. *A 2 henry inductor has 1200 turns. How many turns must be added to raise its inductance to 3 henries ?*

Solution. Inductance $\propto$ (Number of turns)^2

Therefore, if (N) is the new number of turns, then

$$\left(\frac{N}{1200}\right)^2 = \frac{3}{2} = 1.5 \text{ or } N = 1200\sqrt{1.5} = 1200 \times 1.224$$

$$= 1470 \text{ turns}$$

Number of turns to be added = 1470 –1200 = 270 turns.

Problem 3.14 *Refer to Fig. 3.31 What voltage would be induced across the series combination for current (i) changing at a rate of 10 A/s.*

Solution. The equivalent inductance of the series combination is given by :

$$(3 + 12 + 27) + 2 \times 0.5\sqrt{3 \times 12}$$

$$-2 \times 0.2\sqrt{12 \times 27} - 2 \times 0.1\sqrt{3 \times 27}$$

$$= 42 + 6 - 7.2 - 1.8 = 39 \text{ mH}$$

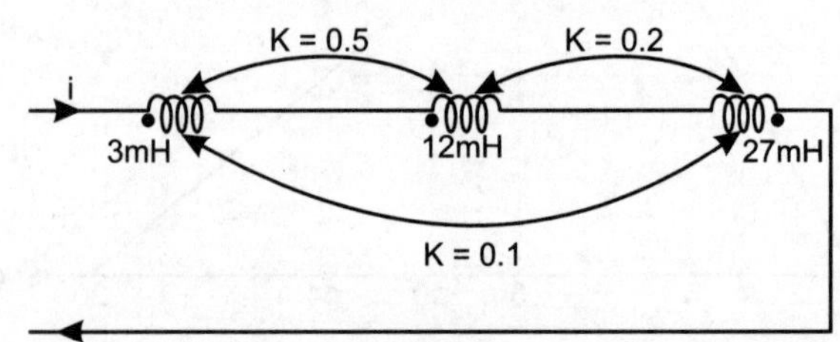

Fig. 3.31

Rate of change of current = 10 A / s

Therefore induced voltage

$$= 39 \times 10^{-3} \times 10 = 0.39 \text{ Volt}$$

Problem 3.15. *A 50 μF capacitor is subjected to a voltage of 120 cos 400 t. What is the current flowing through the capacitor?*

Solution. Voltage across capacitor = 120 cos 400 t,

$C = 50\ \mu\text{F}$

Current through capacitor

$$= C\frac{dv}{dt} = 50 \times 10^{-6}\frac{d}{dt}(120 \cos 400\, t)$$

$$= 50 \times 10^{-6} \times (-120 \times 400 \sin 400\, t)$$

$$= -50 \times 120 \times 400 \times 10^{-6} \sin 400\, t$$

$$= -2.4 \sin 400\, t.$$

Problem 3.16 *If current through a 2H inductor is given by the waveform of Fig. 3.32, sketck the waveform of voltage across the inductor. What would be the energy stored in the inductor at t = 9 ms?*

Solution. The instantaneous voltage across the inductor is given by:

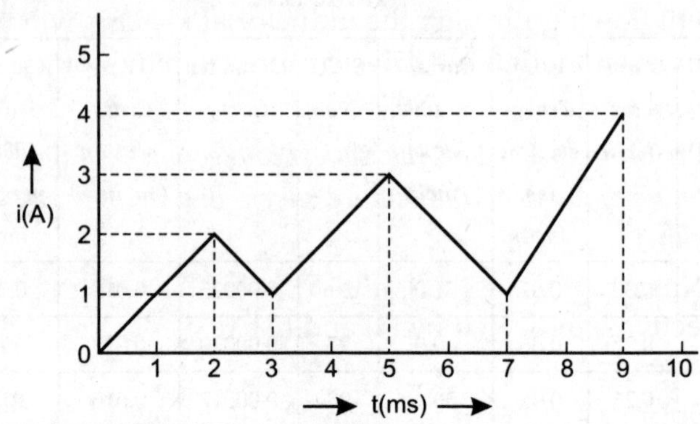

Fig. 3.32

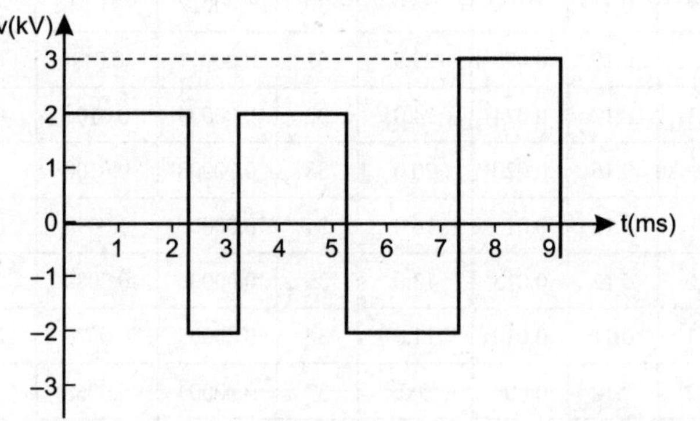

Fig. 3.33

$$v = L\frac{di}{dt} \text{ where } L = 2\ H$$

Now $\frac{di}{dt}$ for $t = 0$ to $t = 2$ ms is given by

$$\frac{2}{2} = 1\text{A/ms}$$

$\frac{di}{dt}$ for $t = 2$ ms to $t = 3$ ms is given by

$$\frac{-1}{1} = -1 \text{ A/ms}$$

$\frac{di}{dt}$ for t = 3 ms to t = 5 ms is given by $\frac{2}{2} = 1$ A/ms

$\frac{di}{dt}$ for t = 5 ms to t = 7ms is given by $\frac{-2}{2} = -1$ A/ms

$\frac{di}{dt}$ for t = 5 ms to t = 7 ms is given by $\frac{-2}{2} = -1$A/ms

$\frac{di}{dt}$ for t = 7 ms to t = 9 ms is given by $\frac{3}{2} = 1.5$ A/ms

The voltage waveform is shown in Fig. 3.33

Energy stored in the inductor at $t = 9$ ms is given by:

$\frac{1}{2} \times 2 \times 4^2 = 16$ Joules

We should remember that current through an inductor cannot change instantaneously. As a result, current flowing through the inductor at $t = 9$ ms will be 4 A only even though the drive current from the source has reduced to zero.

Problem 3.17. *For the series connection of Fig. 3.34, determine equivalent inductance given that $K_{12} = 0.8$, $K_{13} = 0.6$ and $K_{23} = 0.5$.*

Note that (L_1 – L_2) and (L_1 – L_3) couplings are respectively indicated by (•) and (Δ). ($L_2 - L_3$) coupling is indicated by (□)

Solution. As is clear from the figure, (L_1) and (L_2) have a positive sign of mutual inductance (indicated by position of•), (L_1) and (L_3) also have a positive sign of mutual inductance (indicated by position of Δ) whereas (L_2) and (L_3) have a negative sign of mutual inductance (indicated by location of □).

Also, mutual inductance between any two given coils

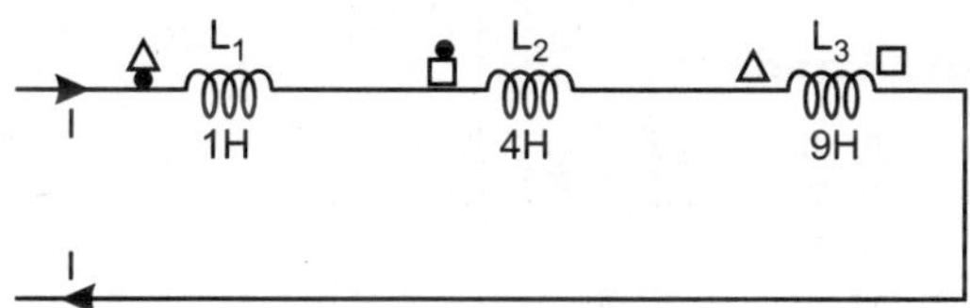

Fig. 3.34

$= K \sqrt{\text{Product of their self inductances}}$

Also, mutual inductance between (L_1) and (L_2) is same as that between (L_2) and (L_1). With this background, equivalent inductance is given by:

$1 + 4 + 9 + 0.8 \sqrt{1 \times 4} + 0.6 \sqrt{1 \times 9} + 0.8 \sqrt{1 \times 4} - 0.5 \sqrt{4 \times 9} - 0.5 \sqrt{4 \times 9} + 0.6 \sqrt{1 \times 9}$

$= 14 + 1.6 + 1.8 + 1.6 - 3 - 3 + 1.8$

$= 20.8 - 6 = 14.8$ H.

Problem 3.18 *What is the equivalent inductance of the network shown in Fig. 3.35.*

Note that ($L_1 - L_2$), ($L_1 - L_3$) and ($L_1 - L_4$) couplings are respectively indicated by (•), (□) and (Δ). Also, ($L_2 - L_3$) and ($L_2 - L_4$) couplings are respectively indicated by (◊) and (•). ($L_3 - L_4$) coupling is indicated by (o)

$K_{12} = 0.92$, $K_{13} = 0.86$, $K_{14} = 0.72$, $K_{23} = 0.66$,

$K_{24} = 0.875$ and $K_{34} = 0.782$

Solution. The equivalent inductance is given by:

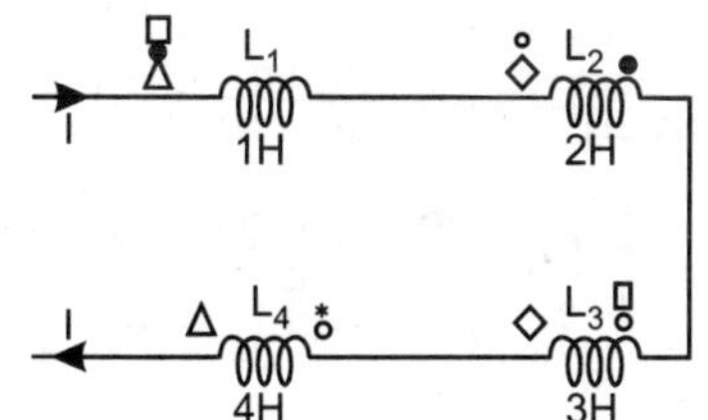

Fig. 3.35

$$
\begin{aligned}
L_{eq} &= 1 + 2 + 3 + 4 - 2 \times 0.92 \sqrt{1 \times 2} \\
&\quad + 2 \times 0.86 \sqrt{1 \times 3} - 2 \times 0.72 \sqrt{1 \times 4} \\
&\quad - 2 \times 0.66 \sqrt{2 \times 3} + 2 \times 0.875 \sqrt{2 \times 4} \\
&\quad + 2 \times 0.782 \sqrt{3 \times 4} \\
&= 10 - 1.84 \times 1.414 + 1.72 \times 1.732 - 1.44 \times 2 \\
&\quad - 1.32 \times 2.45 \\
&\quad + 1.75 \times 2.82 + 1.564 \times 3.464 \\
&= 10 - 2.6 + 2.98 - 2.88 - 3.234 + 4.95 + 5.417 \\
&= 14.633 \text{ H.}
\end{aligned}
$$

Problem 3.19. *What turns ratio in an audio output transformer (an ideal one) would be required to transfer maximum power from an amplifier with an output impedance of 4000 Ω to a 16 Ω speaker?*

Solution. Let the primary to secondary turns ratio be (n).

Then for maximum power transfer,

$$16\,n^2 = 4000$$

$$n = \sqrt{4000/16} = 15.8$$

Here, ($16\,n^2$) is the load impedance reflected onto the primary side. For maximum power transfer, source impedance equals the load impedance.

Problem 3.20. *In the network' shown in Fig. 3.36 determine (n) to make 4 Ω. resistor dissipate 225 watts.*

Solution. For 225 watts dissipated in 4 Ω. resistor, voltage across the resistor = secondary voltage = $\sqrt{225 \times 4}$ = 30 volts Voltage reflected to the primary = 30 n volts Also, impedance reflected to the primary = $4n^2$

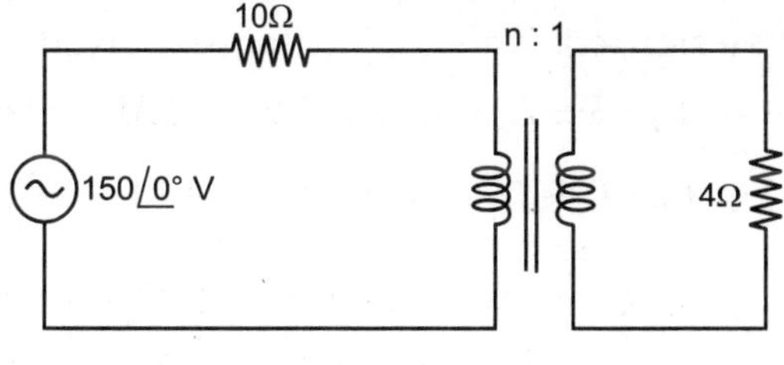

Fig. 3.36

Therefore, $\dfrac{105\angle 0° \times 4\,n^2}{10 + 4n^2} = 30\text{n}$

or $\frac{420\,n^2}{4\,n^2+10} = 30\,n$ *or* $\frac{420\,n}{4n^2+10} = 30$

or $120\,n^2 - 420n + 300 = 0$ *or* $2n^2 - 7n + 5 = 0$

$$n = \frac{7 \pm \sqrt{49-40}}{4} = \frac{7 \pm 3}{4} = 1{,}2.5$$

Therefore, turns ratio is 1 or 2.5.

Problem 3.21. *Two mutually coupled coils (L_1) and (L_2) connected in series have an equivalent inductance of (L_x). The terminals of one of the coils are reversed and reconnected with the result that equivalent inductance changes to (L_y). Determine mutual inductance (M) in terms of (L_x) and (L_y).*

Solution. In the first case, when the equivalent inductance is (L_x), if the mutual inductance has a positive sign, then in the second case with the terminals of one of the coils as reversed, the mutual inductance would have a negative sign.

Therefore,

$$L_1 + L_2 + 2\,M = L_x \quad \text{... (1)}$$

and

$$L_1 + L_2 - 2M = L_y \quad \text{... (2)}$$

Subtracting (2) from (1) , we get $4M = L_x - L_y$

which gives,

$$M = 1/4\,(L_x - L_y).$$

Problem 3.22. *Three coils (L_1), (L_2), (L_3), mutually coupled to each other, are connected in series. They are so oriented that mutual inductances M_{12} (between L_1 and L_2), M_{13} (between L_1 and L_3), M_{23} (between L_2 and L_3) all have a positive sign. The equivalent inductance in this case is 20 H. When the terminals of (L_1) are reversed, the equivalent inductance changes to 15 H. With the terminals of (L_1) restored and terminals of (L_2) reversed instead, the equivalent inductance changes to 10 H. And when both (L_1) and (L_2) are in the original orientation and terminals of (L_3) are reversed, the equivalent inductance becomes 8 H. Determine M_{12}, M_{23} and M_{13}.*

Solution. In the first case, when all mutual inductances have a + ve sign, equivalent inductance is given by:

$$L_1 + L_2 + L_3 + 2M_{12} + 2M_{23} + 2M_{13} = 20$$

or $\quad L_1 + L_2 + L_3 + 2\,(M_{12} + M_{23} + M_{13}) = 20 \quad \text{... (1)}$

In the second case, when terminals of (L_1) are reversed,

$$L_1 + L_2 + L_3 - 2M_{12} + 2M_{23} - 2M_{13} = 15$$

or $\quad L_1 + L_2 + L_3 + 2\,(-M_{12} + M_{23} - M_{13}) = 15 \quad \text{... (2)}$

In the third case, when terminals of (L_2) are reversed,

$$L_1 + L_2 + L_3 - 2M_{12} - 2M_{23} + 2M_{13} = 10$$

or $\quad L_1 + L_2 + L_3 + 2\,(-M_{12} - M_{23} + M_{13}) = 10 \quad \text{... (3)}$

In the fourth case, when terminals of (L_3) are reversed,

$$L_1 + L_2 + L_3 + 2M_{12} - 2M_{23} - 2M_{13} = 8$$

or $\quad L_1 + L_2 + L_3 + 2\,(M_{12} - M_{23} + M_{13}) = 8 \quad \text{... (4)}$

Subtracting (2) from (1), we get

$$4\,(M_{12} + M_{13}) = 5$$

$$\text{or } M_{12} + M_{13} = 5/4 \quad \text{... (5)}$$

Subtracting (3) from (1), we get

$$4\,(M_{12} + M_{23}) = 10$$

$$\text{or } M_{12} + M_{23} = 5/2 \quad \text{... (6)}$$

Subtracting (4) from (1), we get

$$4\,(M_{23} + M_{13}) = 12$$

$$\text{or } M_{23} + M_{13} = 3 \quad \text{... (7)}$$

Adding (5), (6) and (7), we get

$$2\,(M_{12} + M_{23} + M_{13}) = 27/4$$

$$\text{or } M_{12} + M_{23} + M_{13} = 27/8 \quad \text{... (8)}$$

From (7) and (8),

$$M_{12} = \frac{27}{8} - 3 = \frac{3}{8}$$

From (6),

$$M_{23} = \frac{5}{2} - \frac{3}{8} = \frac{17}{8}$$

From (7),

$$M_{13} = 3 - \frac{17}{8} = \frac{7}{8}$$

So,

$$M_{12} = \frac{3}{8}\text{ H},\ M_{23} = \frac{17}{8}\text{H},\ M_{13} = \frac{7}{8}\ H.$$

Problem 3.23 *Fig. 3.37 shows an impedance trans former. Determine the impedance looking into primary of the transformer.*

Solution. Impedance across CD is given by:

$(\sqrt{100} + \sqrt{25})^2 = 225\ \Omega$

Therefore, primary impedance

$= 225 \times (10)^2 = 22500\ \Omega = 22.5$ K

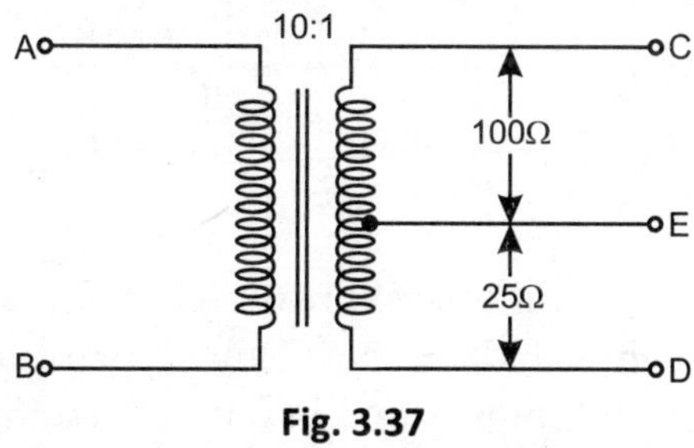

Fig. 3.37

Problem 3.24 *Determine equivalent inductance across (X – X) for the network of Fig. 3.38 The three coils have no mutual coupling with one another.*

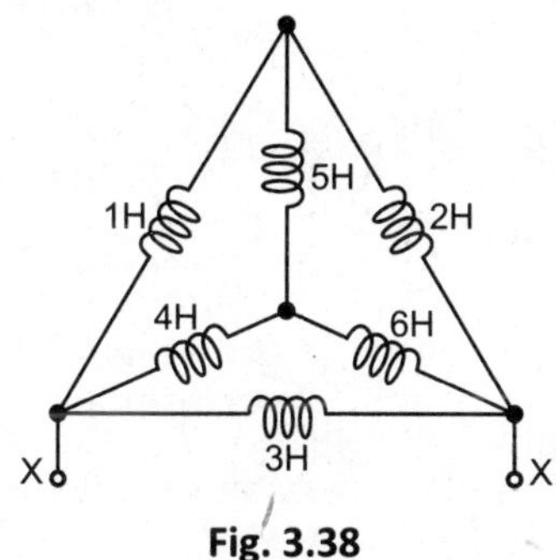

Fig. 3.38

Solution. Converting the inner star into an equivalent delta, the delta inductance, that would appear across 1 H inductance, is given by:

$$\frac{5 \times 4 + 5 \times 6 + 6 \times 4}{6} = \frac{74}{6} = \frac{37}{3}\text{H}$$

Similarly, inductance appearing across 2H inductance is given by $\frac{74}{4} = \frac{37}{2}$ H and the inductance appearing across 3 H inductance is given by $\frac{74}{5}$ H. The network then transforms to the one shown in Fig. 3.39

The subsequent transformations are also shown in Fig. 3.39. The equivalent inductance across $X - X$ is given by:

$$\frac{2.73 \times 2.5}{2.73 + 2.5} = \frac{6.825}{5.23} = 1.3 \text{ H}.$$

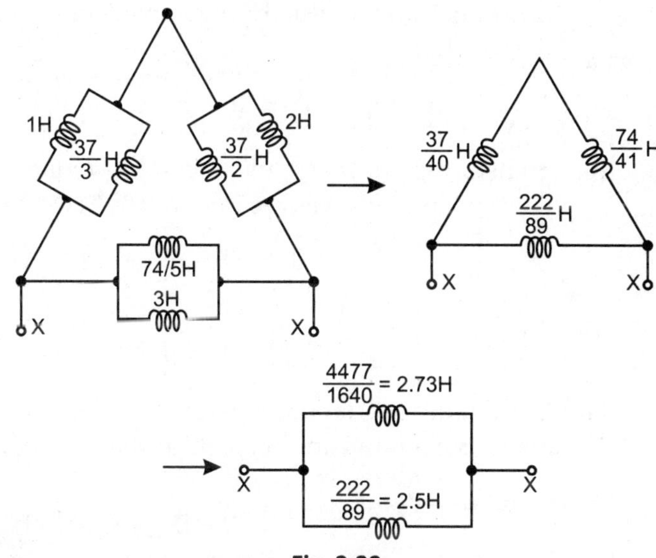

Fig. 3.39

SUMMARY OF IMPORTANT FORMULAS

1. Energy stored in an inductor is given by:

$$E = \frac{1}{2} LI^2$$

where the terms have their usual meaning.

2. Inductance of an inductor is given by:

$$L = \frac{\mu_o \mu_r N^2 A}{l}$$

where the terms have their usual meaning.

3. Quality factor of an inductor is given by:

$$Q = \frac{\omega_o L}{R}$$

where ω_o = Self resonant frequency = $\frac{1}{\sqrt{LC}}$

which gives $Q = \frac{1}{R}\sqrt{\frac{L}{C}}$

Here (*R*) and (*C*) are the equivalent series resistance and inter winding capacitance of the inductor.

4. The inductance of a straight wire conductor is given by:

$$L = 5.081 \left[L_n \frac{4l}{d} - 1 \right]$$

where length (*l*), diameter (*d*) and inductance (*L*) respectively are in inches, inches and *nH*.

5. Inductance of an air are single layered coil is given by:

$$L = \frac{r^2 N^2}{(9r + 10l)}$$

where (*L*) is in μH; radius (*r*) and length (*l*) are in inches.

6. Inductance of a multilayered coil is given by;

$$L = \frac{0.8\, r^2 N^2}{(6r + 9l + 10d)}$$

where mean radius (*r*) of the coil, length (*l*) of the coil and depth (*d*) of the coil are in inches. Inductance (*L*) is in μH.

7. Inductance of a spiral coil is given by:

$$L = \frac{r^2 N^2}{(6r + 11d)}$$

where mean radius (r) and depth (d) of the coil are in inches. Inductance (L) is in µH.

8. Equivalent inductance (L) of multiple series connected and mutually uncoupled inductors is given by:

$$L = L_1 + L_2 + L_3 + ... + L_n$$

9. Equivalent inductance (L) of multiple parallel connected and mutually uncoupled inductors is given by:

$$\frac{1}{L} = \frac{1}{L_1} + \frac{1}{L_2} + \frac{1}{L_3} + ... + \frac{1}{L_n}$$

10. Transformer primary and secondary voltages, currents, impedances are inter related by:

$$\frac{V_S}{V_P} = \frac{I_P}{I_S} = \frac{N_S}{N_P} = n$$

and

$$\frac{Z_S}{Z_P} = \left(\frac{V_S}{V_P}\right)^2 = \left(\frac{I_P}{I_S}\right)^2 = n^2$$

11. Importance equations governing design of mains power transformer include the following.

$$A_C = \frac{\sqrt{W}}{5.6}$$

$$\text{Turns per volt} = \frac{10^8}{4.44\, f\, A_C B}$$

$$\text{Primary current} = \frac{\text{Output wallage, W}}{\text{Efficiency} \times \text{Primary Voltage}}$$

$$\text{Secondary current} = \frac{\text{Primary current}}{n}$$

where $n = \frac{N_S}{N_P}$

In the above expressions, (A_C) is in square inches and (B) is in lines per square inch. If (A_C) is substituted in (cm^2), then (B) should be substituted in Gauss.

12. Important equations governing design of tube transformers include the following.

$$\text{Rise time, } t_r = 3.35\sqrt{\left(\frac{R_2}{R_1 + R_2}\right).\sigma C}$$

If 10 percent overshoot is acceptable, then

$$\text{Rise time, } t_r = 3.35\sqrt{\left(\frac{R_2}{R_1 + R_2}\right).\sigma C}$$

R_1 = Sum of source resistance and transformer primary's DC resistance

R_2 = Told resistance reflected on to primary $= \frac{(R_{sec} + R_L)}{n^2}$

σ = Leakage inductance

C = Total shunt capacitance including inter winding capacitance and any load capacitance reflected onto primary

The droop in the flat of the pulse of width (τ) is given by:

$$\text{Percentage droop} = \frac{R\tau}{L} \times 100$$

where

$$R = \frac{R_1 R_2}{R_1 + R_2}$$

SELF EVALUATION EXERCISE

1. Name the coils and transformers from the circuit symbols shown in Fig. 3.40. Choose from the following words:

Iron cored inductor, Fixed inductor, Saturable core transformer, Autotransformer, Preset inductor, Ganged inductor, Variable inductor.

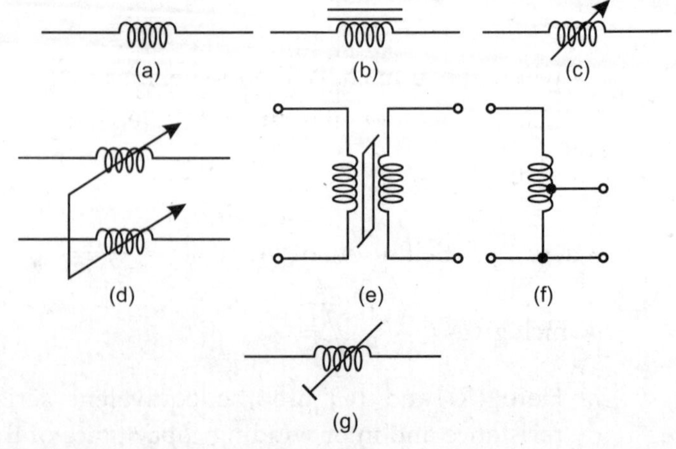

Fig. 3.40

2. Find the *Q*-factor for the *R–L* circuit arrangements shown in Fig. 3. 41

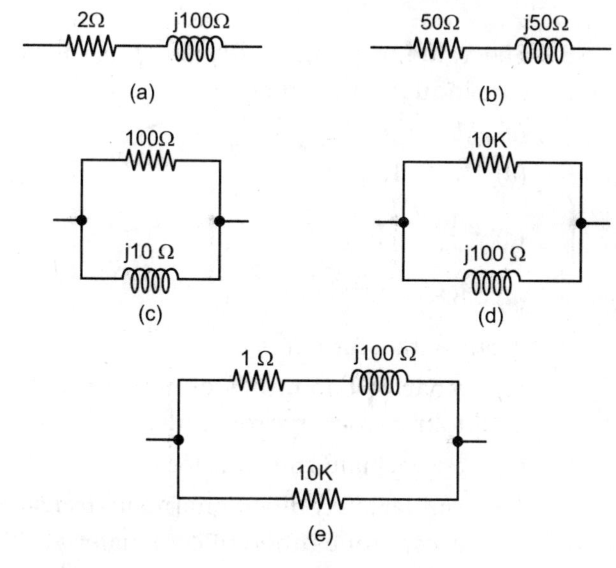

Fig. 3.41

3. On the left hand side of Table 3.2 are listed some inductor and transformer terms. Each one of the listed terms has a relevant statement on the right hand side. You have to match the terms with the corresponding relevant statements.

Table 3.2

S.No.	*Term*	*Statement*
1	Choke	It is that magnetic property which tells us that there is a time lag; between the application of a magnetizing force and the actual magnetization of the material.
2	Hysterisis	It is actually a wideband transformer and while designing the same, utmost importance is attached to see that it has least leakage inductance and stray capacitance.
3	Permeability	It is an inductor whose prime function is to impede the flow of alternating current or pulsating DC.
4	Pulse Transformer	Exceeding this parameter in case of an inductor leads to fast reduction of its inductance.
5	Auto transformer	Essentially a single winding transformer where a part of the primary winding acts as secondary winding.
6	Maximum flux density	It is that magnetic property that gives us an idea about the resistance offered by the material to the magnetic lines of force.

4. Fig. 3.42 shows E-E type silicon steel laminations to construct the core for a certain mains power transformer. What are the magnitudes of (*i*) core cross-section (*ii*) Cross window area (*iii*) Effective magnetic path length.

5. For the impedance transformer shown in Fig. 3.43 (*a*), what is the impedance seen by the source looking into the primary of the transformer. In case of Fig. 3.43 (*b*) what should be the turns ratio if the source has to see in impedance of 1000 ohms.

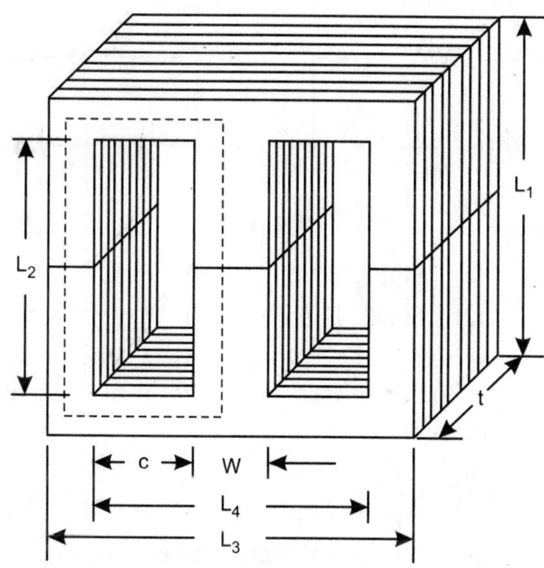

Fig. 3.42

Multichoice Questions

1. An inductance is a circuit element that
   (*a*) helps a change in voltage
   (*b*) opposes a change in voltage
   (*c*) opposes a change in impedance
   (*d*) opposes a change in current

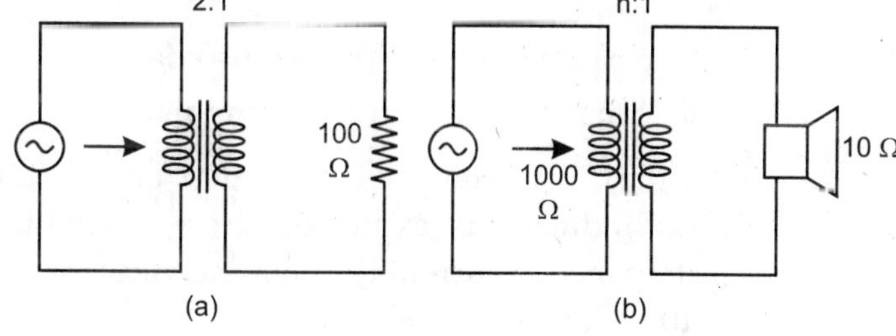

Fig. 3.43

2. One of the circuit representations of Fig. 3. 44 can be regarded as the equivalent circuit of a non-ideal inductor.

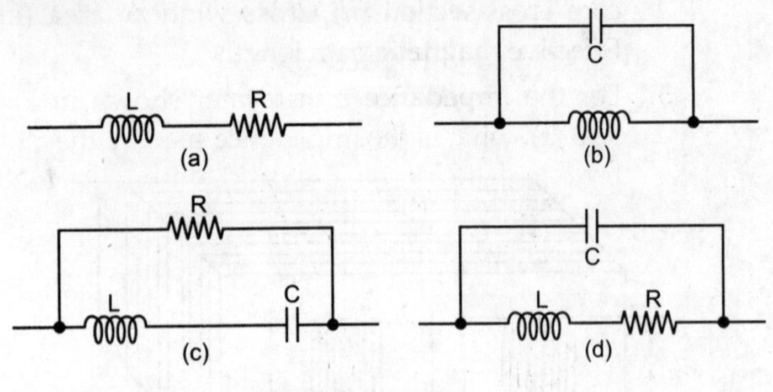

Fig. 3.44

(a) Fig. 3.44 (d) (b) Fig. 3.44 (b)
(c) Fig. 3. 44 (c) (d) Fig. 3.44 (a)

3. When a DC voltage is applied across an inductor, what would determine its steady state current value?
(a) Inductance
(b) Applied voltage
(c) Applied voltage and inductance
(d) Applied voltage and the series resistance associated with it
(e) Inductance and DC resistance of the winding

4. When a constant DC voltage (V) is impressed across a pure inductance (L), the instantaneous current (i) flowing through the inductance as a function of time (t) can be expressed by
(a) $i = L\frac{dV}{dt}$ (b) $i = \frac{V}{L}t$
(c) $V = L\frac{d^2i}{dt^2}$ (d) none of these

5. When the current flowing through an ideal inductance is a ramp, the voltage across it will be
(a) a ramp too (b) sinusoidal
(c) constant (d) triangular

6. If the instantaneous voltage across an ideal inductor is expressed by $v = 10t$, the instantaneous current through the inductor at $t = 10$ μs for $L = 10$ μH will be
(a) 50 μA (b) 50 mA
(c) 5 mA (d) 5 A

7. Mark the only correct expression. The terms have their usual meaning
(a) $L = \frac{\mu_o N^2 A^2}{l}$ (b) $L = \frac{\mu_r N^2 A}{l}$
(c) $L = \frac{\mu_o \mu_r N^2 A}{l}$ (d) $L = \frac{\mu_o \mu_r N^2 l}{A}$

8. The value of permeability $\mu = \mu_o \mu_r$ for an air core inductor would be
(a) 1
(b) $4\pi \times 10^{-7}$ H/m
(c) $\frac{10^{-7}}{4\pi}$ H/m
(d) 8.85×10^{-12} H/m

9. A pulse transformer is
(a) a narrowband transformer so as to be able to transform narrow pulses
(b) a wideband transformer
(c) similar to a conventional mains transformer except for a different core material
(d) wound on a core of very low permeability

10. Fig. 3.45 shows *v-i* relationship for a certain inductor. The *Q*-factor for this inductor would be
(a) $\sqrt{3}$ (b) $1/\sqrt{3}$
(c) 2 (d) 0.5

11. An inductor is colour coded using a 4 - band code. Starting from left, the four bands are red, violet, brown and black. Identify the inductance value and percentage tolerance.
(a) 270 μH, ± 20 % (b) 270 μH, ± 10 %
(c) 27 μH, ± 20 % (d) 270 μH, ± 5 %

12. An inductor is colour coded using MIL-STD 5 - band code. Starting from extreme left, it has wide silver band followed by yellow, gold, violet and red bands. Identify the inductance value and tolerance.
(a) 47 μH, ± 2 % (b) 4.7 μH, ± 5 %
(c) 4.7 μH, ± 2 % (d) 4.7 μH, ± 10 %

13. It it is given that 1.5 n H is a standard inductance value then one of the following sets of inductance value would also be standard values.
(a) 15 nH, 150 nH, 1.5 μH, 15 μH
(b) 3nH, 4.5 nH, 6 nH, 7.5 nH
(c) 2 nH, 2.5 nH, 3 nH, 3.5 nH
(d) None of these

14. The core materials preferred for inductors meant for memory and switching applications are characterised by.

(*a*) square ness ratio of less than 0.5
(*b*) a linear B-H loop
(*c*) square ness ratio of near unity
(*d*) None of these

15. Hysteresis losses and eddy current losses are directly proportional to.

(*a*) (f) and (f^2) respectively
(*b*) (f^2) and (f) respectively
(*c*) ($f^{1.6}$) and (f^2) respectively
(*d*) (f^2) and ($f^{1.6}$) respectively
(f) is the operating frequency

ANSWERS TO SELF EVALUATION EXERCISE

1. Fig. 3.40 (*a*) Fixed Inductor
   Fig. 3.40 (*b*) Iron Core Inductor
   Fig. 3.40 (*c*) Variable Inductor
   Fig. 3.40 (*d*) Ganged Inductor
   Fig. 3.40 (*e*) Saturable Core Transformer
   Fig. 3.40 (f) Autotransformer
   Fig. 3.40 (g) Preset Inductor

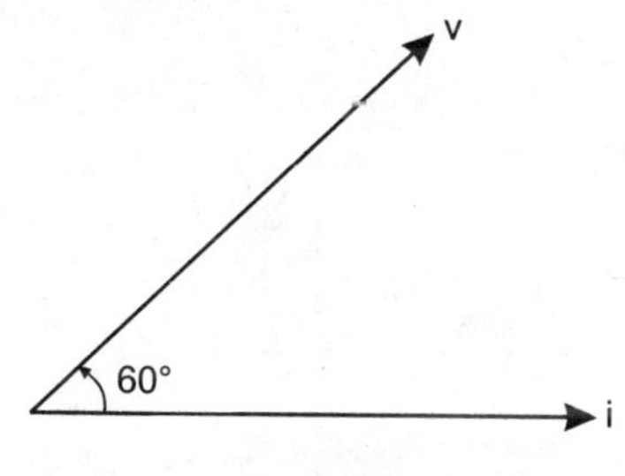

Fig. 3.45

2. Fig. 3. 41 (*a*) 50 Fig. 3.41 (*b*) 1 Fig. 3.41 (*c*) 10 Fig. 3.41 (*d*) 100 Fig. 3.41 (*e*) 100.
3. (1–3) (2–1) (3–6) (4–2) (5–5) (6–4)
4. Core Cross-section = $w.t$
   Window Area = $L_2.C$
   Effective Magnetic path Length
   $= L_1 + L_2 + \frac{L_3 + L_4}{2}$
5. Fig. 3. 43 (a) 400 ohms
   Fig. 3.43 (b) 10 : 1

MULTICHOICE QUESTIONS

1. (*d*)	2. (*a*)	3. (*d*)	4. (*b*)
5. (*c*)	6. (*a*)	7. (*c*)	8. (*b*)
9. (*b*)	10. (*a*)	11. (*a*)	12. (*c*)
13. (*a*)	14. (*c*)	15. (*a*)	

Chapter 4

Hardware Electronic Components

Learning Objectives

After completing the chapter, you will learn the following topics:

1. Role of switches in electronics circuits
2. Different types of switches including toggle switches, slide switches, rocker switches, snap acting switches and rotary switches
3. Operating parameters of switches
4. Types of switch contacts
5. Relays and types of relay contacts
6. Types of relays including time delay relays, power relays, latching relays, crystal can relays, coaxial relays, meter relays, reed relays, hybrid relays and solid state relays
7. Operate and release times of relays
8. Different types of connectors including coaxial *RF* connectors, *BNC* connectors, *TNC* connectors, *N* type connectors, *UHF* connectors, *SMA* connectors and D-type connectors
9. Bayonet coupled and Crimp connectors
10. Coaxial cables
11. Characteristic impedance of a coaxial cable
12. Flexible, semi-rigid and rigid cables
13. Fuse and its primary function
14. Major performance specifications of a fuse
15. Fast blow, normal blow and slow blow types of fuses
16. Batteries
17. Basic operation of a battery
18. Primary and secondary batteries
19. Important performance parameters of a battery
20. Alkaline batteries
21. Lead acid batteries
22. Nickel-cadmium (Ni-Cd) batteries
23. Nickel metal hydride (NiMH) batteries
24. Lithium ion batteries
25. Fuel cells

Q.1. What is an electrical switch? Distinguish between a Mechanical, an Electromagnetic and an Electronic switch.

Ans. An electrical switch is a component that can be used either for opening and closing an electrical circuit (Fig. 4.1*a*) or for rerouting the electrical signal through a circuit as shown in Fig. 4.1 (*b*). Mechanical switches are the manually operated ones. Typical examples are the Toggle switches, Rotary switches, the Snap acting switches, Slide switches and so on. Switches driven by an electric motor or a solenoid are termed as Electromagnetic switches. Electronic devices used for switching action constitute Electronic switches. These devices mainly include semiconductor transistors, FETs and MOSFETs, thyristors etc.

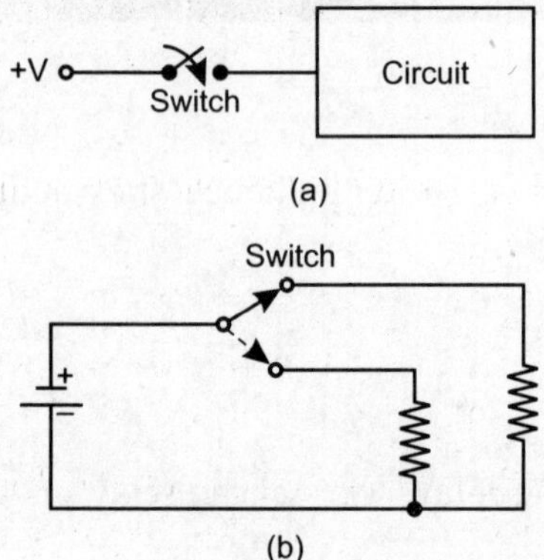

Fig. 4.1

Fig. 4.2 illustrates the use of a bipolar transistor as a switch. Whenever a transistor is used as a switch, it is either in the cut-off region (corresponding to the open condition of the switch) or in the saturation region (corresponding to the closed position of the switch). It is never operated in the active region. It may be mentioned here that electronic switches such as bipolar transistors, MOSFETs, IGBTs constitute one of the most vital components of switched mode power supplies.

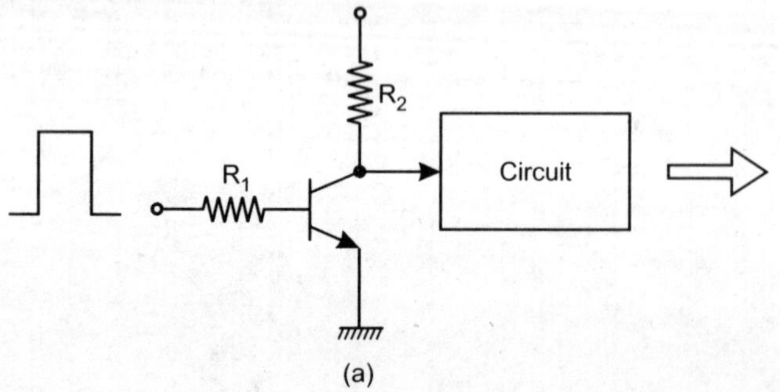

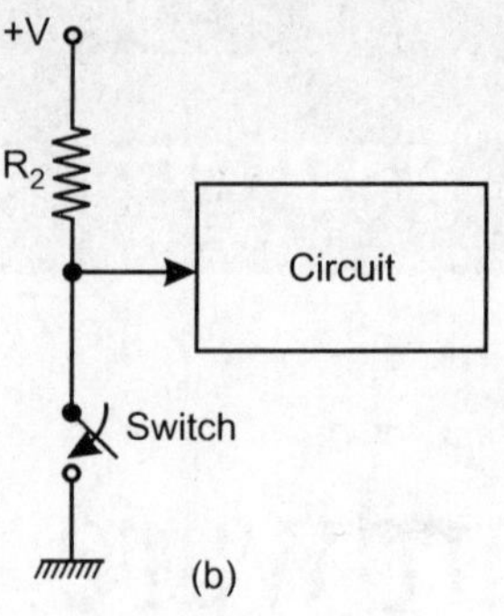

Fig. 4.2

Q.2. What is a toggle switch? What are the contact types usually available in toggle switches?

Ans. A toggle switch (Fig. 4.3) is a manually operated mechanical switch in which the lever operates a spring loaded toggle linkage to make or break contacts. Fig. 4.4(*a*) shows the basic construction. The contact types usually available with toggle switches are:

(*i*) Single Pole Single Throw (SPST) known as an ON/OFF switch,

(*ii*) Single Pole Double Throw (SPDT) and

(*iii*) Double Pole Double Throw (DPDT)

Fig. 4.3

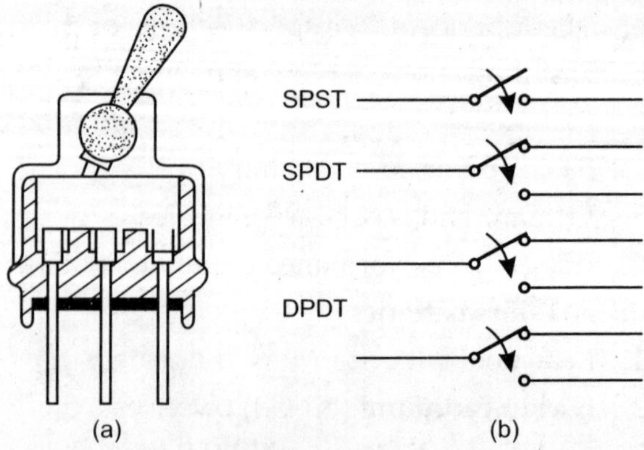

Fig. 4.4

The contact arrangements for these types of contacts are shown in Fig. 4.4 (*b*).

Toggle switches are made in a variety of sizes and shapes. In the SPST and SPDT toggle switches, there are two distinct varieties of contact types. In one of the types, there are two ON positions only (Fig. 4.5 *a*) while in the other, there is an OFF position in the center in addition to the two ON positions on the extremes (Fig. 4.5 *b*).

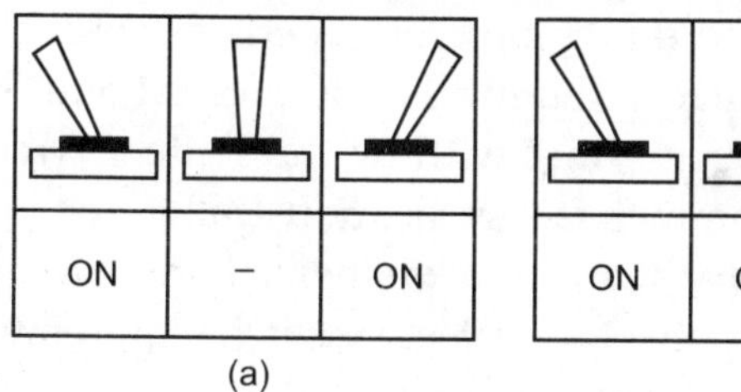

Fig. 4.5

Q.3. How do SLIDE and ROCKER switches differ from the basic toggle switch?

Ans. The moving contact in a slide switch is a spring loaded copper alloy strip that moves back and forth across the stud contacts fixed in a plastic insulator base. Slide switches are also made in a variety of sizes and shapes (Fig. 4.6). Fig. 4.7 shows the basic construction of a slide switch. The contact forms usually available with slide switches are the SPST, SPDT, DPDT, SP3T and DP4T.

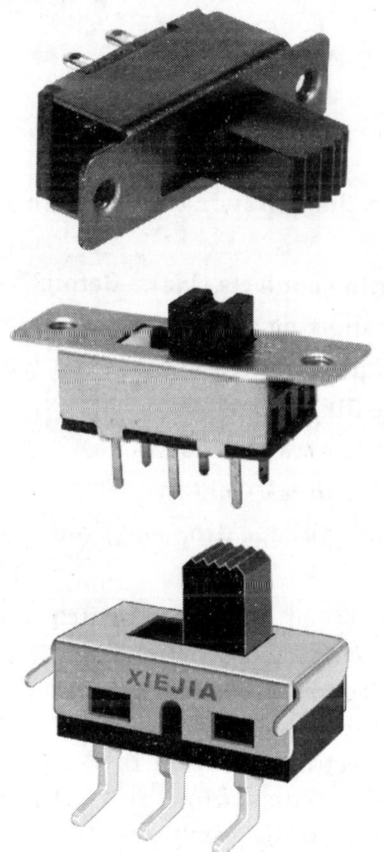

Fig. 4.6

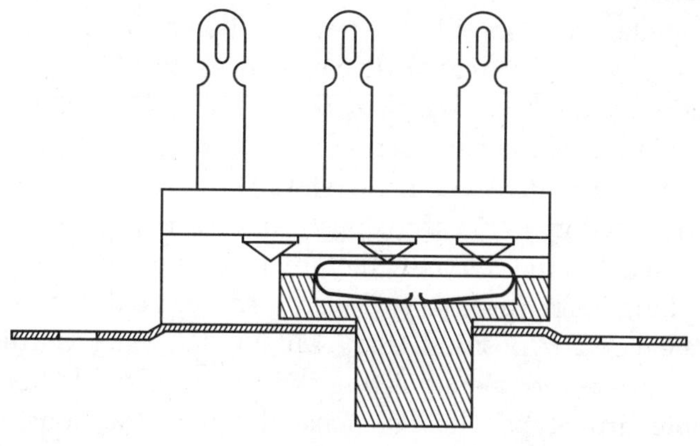

Fig. 4.7

When a rocker handle is attached to a toggle, slide or a snap acting switch (snap acting switches are discussed in the next question), it becomes a rocker switch (Fig. 4.8). Rocker switches are also available in a variety of sizes and shapes and for different contact current and voltage specifications like toggle and slide switches. The contact forms available with rocker switches are more or less identical to the ones available in case of toggle and slide switches. A popular type of rocker switch is the lighted rocker switch. A lighted rocker switch has either an opaque plastic rocker handle with a lamp embedded in it or a translucent handle illuminated from behind by a lamp that is not a part of the switch.

Fig. 4.8

Q.4. What is a snap acting switch and how does it function?

Ans. A snap acting switch (Fig. 4.9) is also a manually operated electrical switch usually having a push button operated actuating plunger. These switches are available in the form of basic switch alone or a switch with an actuator or a switch with an actuator and an enclosure. The moving contact arm of a snap acting switch (Fig. 4.10*a*) has two positions of equilibrium. When the actuating plunger is made to deflect a part of the moving contact arm, the contact arm, moves rapidly in almost snap action from one position of equilibrium to the other.

The contact arm stays in the new position as long as the plunger remains pressed. It comes back to its original position when the plunger is released. The release operation is also a snap action only. In fact, if you have used such a switch earlier, you would have observed that the snap action occurs as the push button is being pressed or released as the case may be. Any further movement of the plunger, forwards or backwards, makes no difference to the contact position. Fig. 4.10 (*b*) shows a circuit representation of a snap acting switch of the single pole double throw type. These switches also go by the name of micro-switches. One typical application of a micro-switch is in digital circuits using flip-flops, counters, registers etc. where at times you are interested to send a SET or a CLEAR pulse. A micro-switch along with a debouncing circuit does the job well. Each press and release operation of the switch produces one positive going pulse at the designated output (Fig. 4.11).

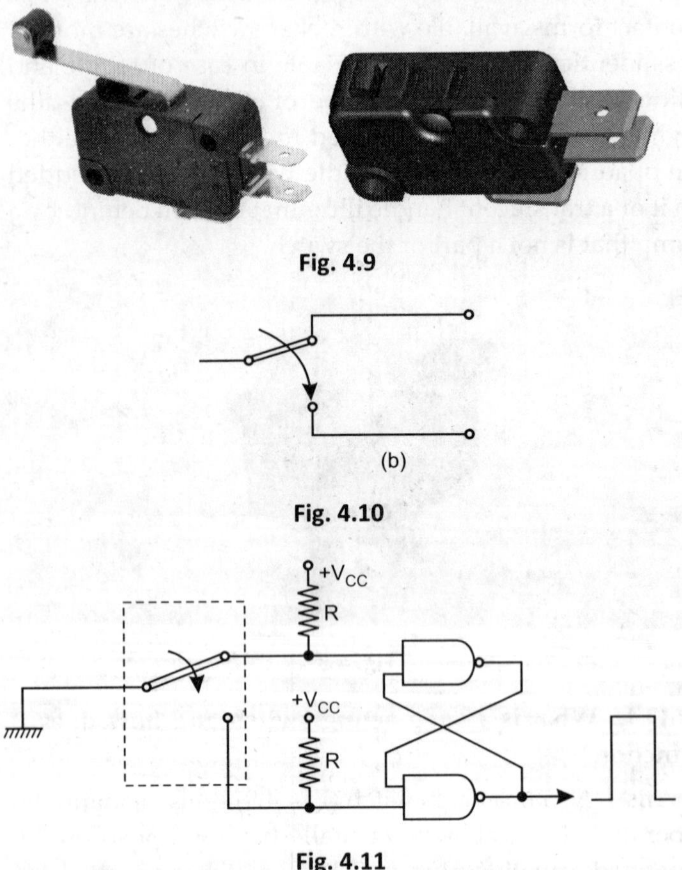

Fig. 4.9

Fig. 4.10

Fig. 4.11

Q.5. What are rotary switches and what are their main applications?

Ans. Rotary switches are also manually operated switches usually having 1, 2 or 3 poles and multiple throws. They get their name from the circular or elliptical configuration in which the switch contacts are fabricated (Fig. 4.12). Based on the size, they are made in heavy duty, miniature, sub-miniature and micro-miniature categories. PCB mounting rotary switches are also available.

A rotary switch is undoubtedly the most popular choice when you are looking for a selector switch. These are available in a variety of contact current and voltage specifications. Based on the type of materials chosen for fabricating the contacts, the insulation material and the quality standards being met with by the switches, they are classified as commercial grade, industrial grade and military grade rotary switches. Commercial grade rotary switches are used to meet the selector switch requirement of various consumer gadgets. Industrial grade switches are used in large numbers in electronics test equipment and automotive industries. Military grade rotary switches are used in all kinds of military systems.

Fig. 4.12

Q.6. Define the following terms with reference to switches.

(*a*) Shorting contacts (Make-Before-Break)

(*b*) Non-shorting contacts (Break-Before-Make)

(*c*) Operating point and operating force

(*d*) Force differential

(*e*) Double Make Contacts

(*f*) Double Break Contacts

(*g*) Contact voltage drop and Contact Resistance

Ans.

(*a*) In case of **shorting switch contacts,** the moving contact arm makes contact with the new operated position before it breaks contact with the previous position. In other words, the electrical function of the selected switch position settles before the electrical function of the previously established position had been interrupted.

(*b*) In case of **non shorting switch contacts,** the moving contact arm breaks contact with the existing position before it makes contact with the new operated position. In other words, the electrical function of the selected position settles after the electrical function of the previously established position had been interrupted.

(*c*) **Operating point** is the position of the plunger of a snap acting switch at which the contact arm moves from its normal contact position to the operated contact position. **Operating force** is the force that need to be applied to reach the operating point.

(*d*) **Force Differential** is the difference between the operating force and the release force of a snap switch.

(*e*) A contact combination in which a pair of contacts on the same conductive support simultaneously close an electrical circuit connected to an independent pair of contacts is known as a **Double Make Contact.**

(*f*) A contact combination in which a pair of contacts on the same conductive support simultaneously open an electrical circuit connected to an independent pair of contacts is known as **Double Break Contact.**

(*g*) **Contact resistance** is the resistance across a pair of closed contacts. **Contact voltage** drop is the voltage across a pair of closed contacts. It is obviously equal to contact resistance multiplied by the current flowing through the contacts.

Q.7. Draw the circuit representations of the following types of contact forms usually available with snap acting switches:

1. **Single Pole-Single Throw-Single Break**
2. **Single Pole-Single Throw-Double Break**
3. **Single Pole-Double Throw-Single Break**
4. **Single Pole-Double Throw-Double Break**
5. **Double Pole-Single Throw-Single Break**
6. **Double Pole-Single Throw-Double Break**
7. **Double Pole-Double Throw-Single Break**
8. **Double Pole-Double Throw-Double Break.**

Ans. The circuit representations of the above types of contact forms are shown in Figs. 4.13 (*a*) to (*h*).

Q.8. What is a RELAY? How does an electromagnetic relay differ from an electromagnetic switch?

Ans. A relay is an electrically operated switch. The relay contacts can be made to operate in the pre-arranged fashion. For instance, normally open contacts close and normally closed contacts open. In electromagnetic relays, the contacts however complex they might be, they have only two positions *i.e.* OPEN and CLOSED whereas in case of electromagnetic switches, the contacts can have multiple positions.

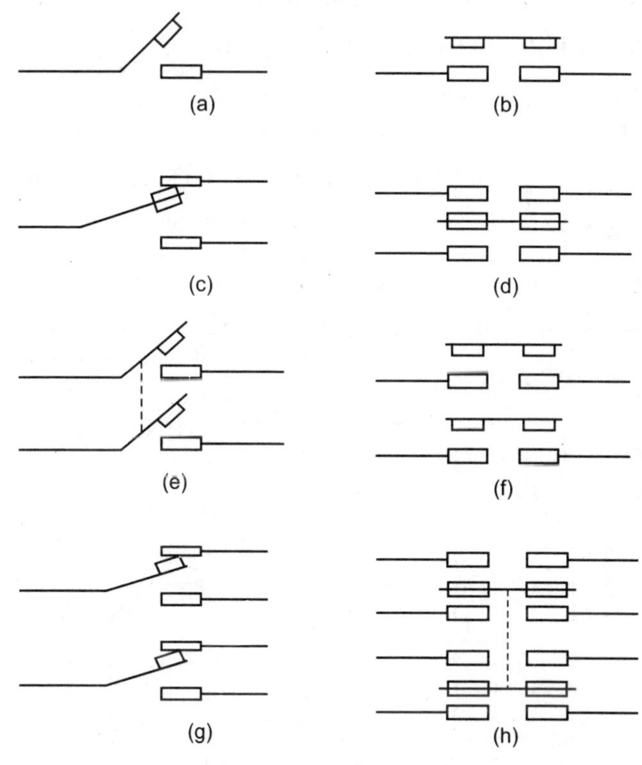

Fig. 4.13

Q.9. How does the basic electromagnetic relay function? What are the different contact forms commonly available with electromagnetic relays?

Ans. An electromagnetic relay in its simplest form consists of a coil, a DC current passing through which produces a magnetic field. This magnetic field attracts an armature which in turn operates the contacts. Normally open contacts close and normally closed contacts open. Fig. 4.14 shows the constructional features of a simple electromagnetic relay.

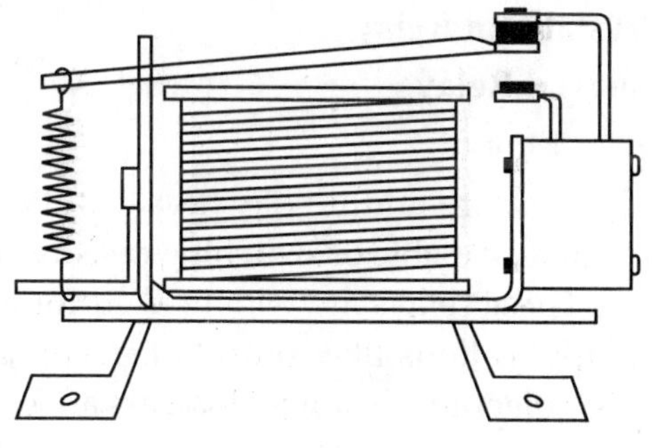

Fig. 4.14

Electromagnetic relays are made in a large variety of contact forms. Some of the popular contact forms are shown in Fig. 4.15.

Fig. 4.15 (*a*): SPST (Normally Open)

Fig. 4.15 (*b*): SPST (Normally Closed)

Fig. 4.15 (*c*): SPDT (Break Before Make)

Fig. 4.15 (*d*): SPDT (Make Before Break)

Fig. 4.15 (*e*): SPST (Normally Open)

Fig. 4.15 (*f*): SPST (Normally Closed)

Fig. 4.15 (*g*): SPST (Normally Open, Double Make)

Fig. 4.15 (*h*): SPST (Normally Closed, Double Break

Fig. 4.15 (*i*): SPDT (Double Make-Double Break)

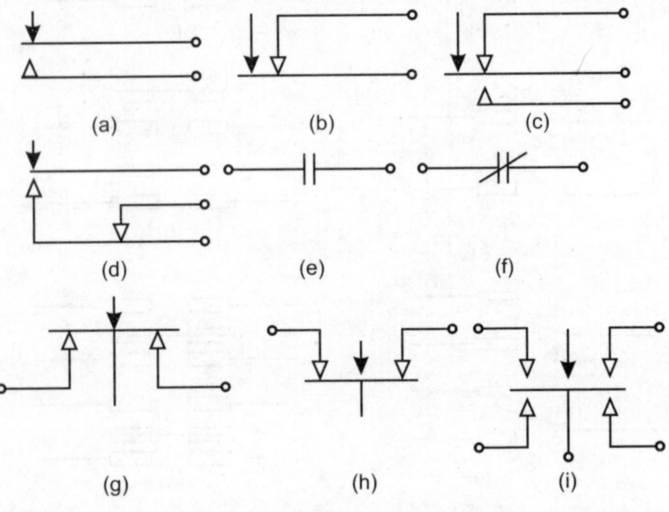

Fig. 4.15

Q.10. Differentiate a CONTACTOR from a RELAY.

Ans. An electromagnetic relay having a contact current of say more than 25 Amp is normally classified as a contactor. The operating principle is however the same.

Q.11. Briefly describe as to what do you understand by:

1. **Power Relay**
2. **Time Delay Relay**
3. **Latching Relay**
4. **Crystal Can Relay**
5. **Co-axial Relay**

Ans. Power Relay

Power Relays are multi-pole heavy duty clapper type relays that are capable of switching resistive loads of upto 25 Amp. These relays are widely used for a variety of industrial applications like control of fractional horse power motors, solenoids, heating elements and so on.

These relays usually have button like silver or silver alloy contacts and the contact welding due to heavy in–rush of current is avoided by wiping action of the contacts. In some relays, a permanent magnet is placed near the contacts to quench the arc during high voltage DC switching thus avoiding contact welding.

Time Delay Relay

A **Time Delay Relay** is the one in which there is a desired amount of time delay between the application of the actuating signal and operation of the load switching device. There are four functional classes of time delay relays. In the first class of time delay relays known as TIME DELAY RELAY, the load circuit is energised a specific time delay interval after the relay control circuit is energised (Fig. 4.16 *a*). In the second class of such relays known as DELAY TIMER, the load circuit gets energised at the time of energising the relay control circuit and remains energised for a specific period of time after the relay control circuit is de-energised (Fig. 4.16 *b*). In the third class of time delay relays called INTERVAL TIMER, the load circuit is energised at the time of energising the relay control circuit and remains so for a specified interval of time (Fig. 4.16 *c*). In the fourth class of these relays knows as REPEAT CYCLE TIMER, the load circuit is repeatedly energised and de-energised for predetermined time periods as long as the relay control circuit remains energised (Fig. 4.16*d*).

Latching Relay

In a **Latching Relay**, the relay contacts remain in the last energised position even after removal of signal in the relay control circuit. The contacts are held in the last relay energised position after removal of energisation either electrically or magnetically. The contacts can be released to the normal position electrically or mechanically.

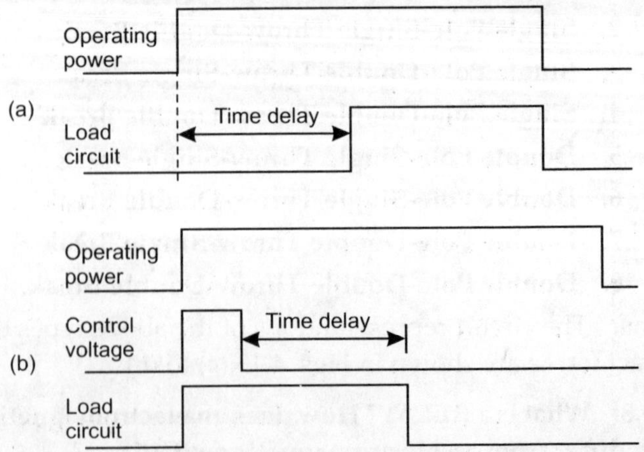

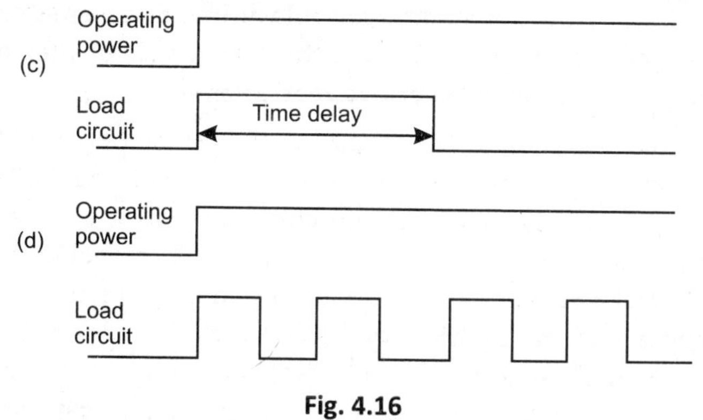

Fig. 4.16

Crystal Can Relay

Crystal Can Relays (Fig. 4.17) are so called as they resemble quartz crystals in external shape. These are high performance hermetically sealed miniature and sub-miniature relays widely used in aerospace and military applications. These relays usually have gold plated contacts and thus have extremely low contact resistance. Due to the low moment of inertia of the armature and also due to statically and dynamically balanced nature of the armature, these relays switch quite reliably even under extreme conditions of shock and vibration. The hermeticity protects the relay from any moisture penetration.

Fig. 4.17

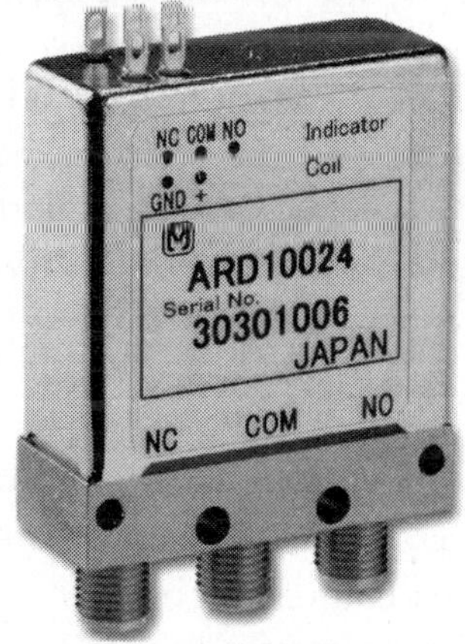

Fig. 4.18

Co-axial Relay

A **Co-axial Relay** (Fig. 4.18) has two basic parts, an actuator which is nothing but some kind of a coil and a cavity housing the relay contacts. One type of co-axial relay has RF connector as an integral part of the relay. These relays are terminated in a wide selection of RF connectors like BNC, TNC, N, UHF. The relay cavity housing the contacts is so configured that it provides a very close match to the characteristic impedance of the interconnecting cables. In the other style, the shielded cables are directly connected to the cavity.

The co-axial relays are extensively used for radio frequency switching operations of equipment interconnected by cables.

Meter Relay

Meter Relays also knows as **Instrument Relays** are extremely sensitive relays extensively used in industrial process control and monitoring applications. There are three broad categories of meter relays *i.e.* LOCKING COIL METER RELAYS, MAGNETIC METER RELAYS and OPTICAL METER RELAYS. All the three types use modified D' Arsonval type meter movements as the actuator for the relay contacts. In the locking coil type meter relay, the meter pointer is in physical contact with the contacts. In the magnetic meter relay too, the meter pointer physically closes the contacts but it is held in that position by a magnet and needs to be reset either mechanically or electromagnetically. In the optical meter relays, there is no physical contact between the meter movement and the switching contacts.

Q.12. What is a Reed Relay? What are the different contact forms available with reed relays?

Ans. A **Reed Relay** basically consists of one or more reed switches and a coil. The number of reed switches usually decides the number of poles. A reed switch that forms the contact in a reed relay basically consists of ferromagnetic reeds enclosed in a sealed glass capsule which is either evacuated or filled with an inert gas. In case of multiple reed relays, more than one such capsule is used. Reed relays are fast acting and have a long operating life. As they are hermetically sealed, they perform well in stringent environmental conditions.

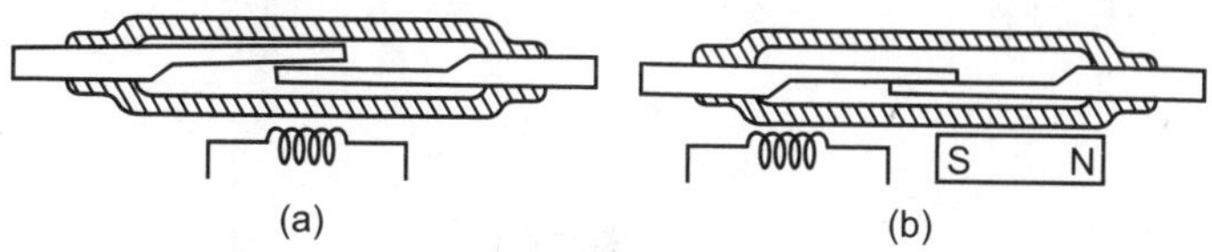

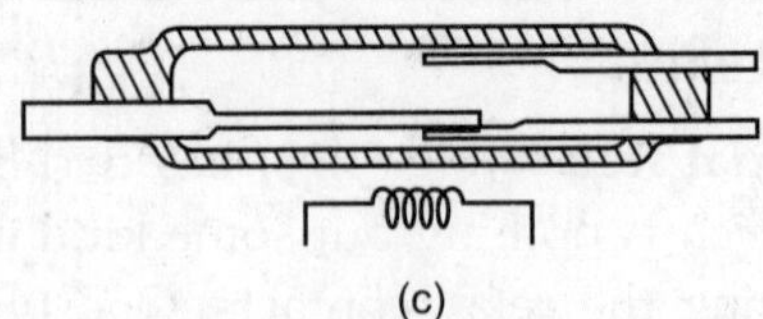

(c)

Fig. 4.19

Reed relays are available in one of the following four contact forms: FORM-A contact (Fig. 4.19*a*) which is a SPST-NO contact. When the coil is energised, the normally open contacts close. FORM-B contact (Fig. 4.19*b*) is a normally closed contact (SPST-NC). The contacts are kept in the closed state by the biasing action of the permanent magnet. When the coil is energised in the proper direction so that its magnetic field overcomes the biasing field of the permanent magnet, the contacts open. As the current through the coil must flow in a particular direction for the operation of the relay, it is a polarized relay. FORM-C contact represents a changeover contact. There are two types of Form-C contacts. In the mechanically biased Form-C contacts, the movable reed is so positioned that the force is exerted on the normally closed contact (Fig. 4.19*c*) and to limit the magnetic attraction between the two, a pad of a non-magnetic material is brazed or welded to the stationary normally closed contact. When the coil is energised, the contacts change over. In the polarised magnetically biased Form-C contact form, the movable reed is center stable and it is held on to the normally closed stationary contact by means of a permanent magnet. When the relay coil is energised in the appropriate direction, the contacts change over. Reed relays are available in hermetically sealed D.I.L. packages (Fig. 4.20).

Fig. 4.20

Q.13. How does a dry read relay differ from a mercury wetted reed relay? What are the advantages of mercury wetted reed relays over dry reed relays?

Ans. In mercury wetted reed relays, the hermetically sealed glass capsules contain mercury at the bottom (Fig. 4.21). Some of the mercury rises up the reed by capillary action to wet both fixed and movable reed contact surfaces. These are also available in the same contact forms as the dry reed relays. The advantages of wetted reed relays include low contact resistance, reduced contact wear, no contact bounce and consistent operate and release times.

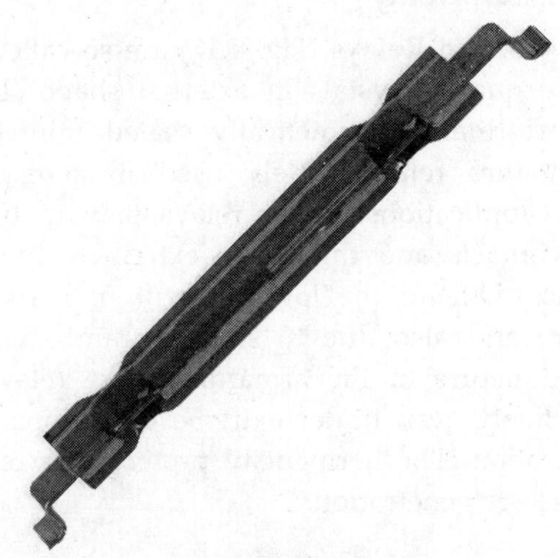

Fig. 4.21

Q.14. Write short notes on the following:

1. **Hybrid Relays**
2. **Solid-state Relays**

Ans. HYBRID RELAYS

Hybrid Relays are formed by a combination of conventional electromagnetic relays and solid-state devices. There are two possible types of hybrid relays. In the first type, an electromagnetic relay drives a solid state device that acts as the load switching circuit (Fig. 4.22*a*). The load switch may be a bipolar transistor, a FET, an SCR or a triac depending upon the end performance specifications one is looking for. In the second type, a solid-state amplifier drives a conventional electromagnetic relay (Fig. 4.22*b*).

SOLID-STATE RELAYS

A **Solid-State Relay** has a solid-state switching device driven by an LED or a solid-state amplifier. The solid-state switch could be a bipolar transistor, and FET or a thyristor depending upon nature of load circuit.

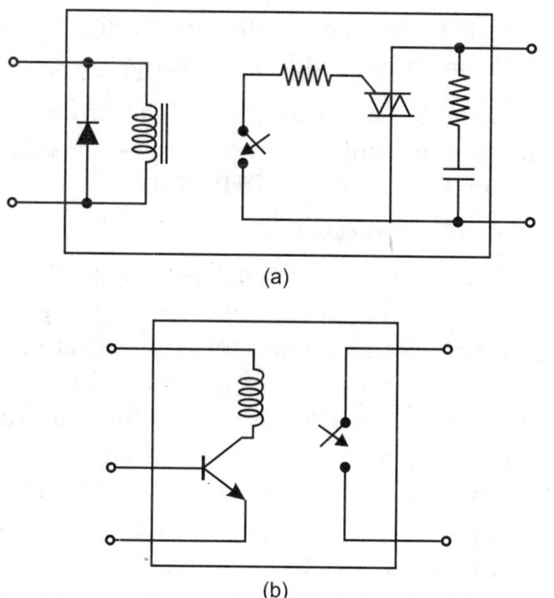

Fig. 4.22

While bipolar transistors are a better choice for high current DC switching, triacs are invariably used for AC switching. Fig. 4.23 shows the internal arrangement of some of the popular solid-state relay types.

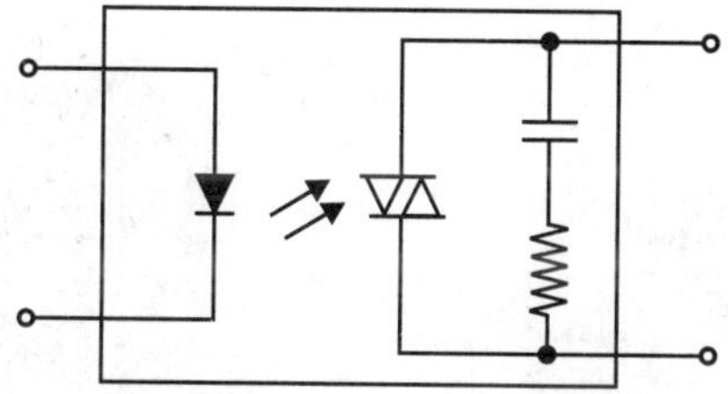

Fig. 4.23

Solid-state relays are reliable and have longer life. Since there is no physical opening and closing of contacts, operation is silent and free from transients.

Q.15. Define OPERATE and RELEASE TIMES in case of electromagnetic relays. On what factors do these parameters depend?

Ans. The **Operate Time** of a relay is the time interval that elapses from the instant the relay coil is energised till the closing of the normally open contacts. The **Release Time** is the time interval between the instant the relay coil is de-energised and the instant the normally open contacts begin to open from the closed position.

The operate time of a coil actuated relay depends primarily on the resistance and inductance of the coil, relay adjustment and the source voltage. The operate time reduces with voltage or current overdrive but this is at the cost of increased contact bounce. The operate and release times may increase as much as ten times the nominal value if R-C networks have been used to protect the relay contacts. The release time also deteriorates as much as by hundred times in case the coil has been protected by the means of a transient suppression diode.

Q.16. Briefly describe the following terms with reference to connectors.

Bayonet Coupling, Crimp Connectors, Receptacle, Plug, Polarisation, Threaded Coupling, Socket Contact.

Ans. Bayonet coupling is the coupling technique where pin type contacts on one connector shell mate with spiralled grooves in the other shell. The locking of the contacts is achieved by applying axial force against a latching spring and giving a partial turn to the bayonet ring. Bayonet coupling is very common in circular connectors. BNC connectors belong to this class. Most of the circular connectors used for military and aerospace applications have bayonet coupling.

Crimp connectors are the ones that use crimping type contacts rather than solder type contacts. A crimp contact is produced by either indenting or compressing the contact tail against or around the wire to lock or retain the wire within the contact tail (Fig. 4.24). Crimp type connectors are extensively used for military and aerospace applications due to their environment worthiness and greater joint strength.

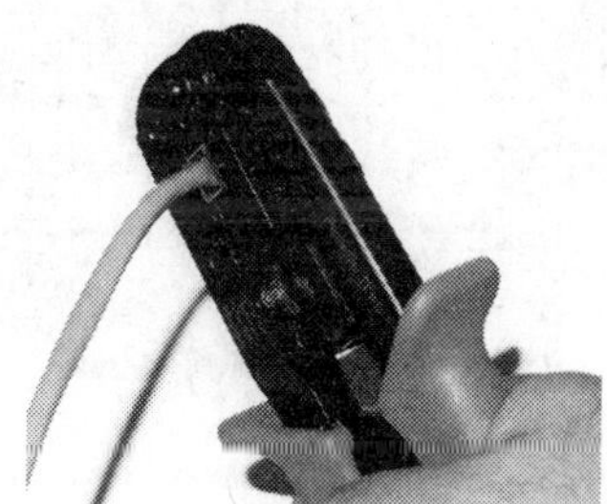

Fig. 4.24

A **Receptacle** is a connector having either pin or socket type contacts. It is permanently mounted on the panel chassis. Fig. 4.25 shows both pin-contact and socket-contact receptacles.

Fig. 4.25

A connector attached to a cable is called a **plug**. It can have pin type or socket type contacts (Fig. 4.26).

Fig. 4.26

Polarisation means contact arrangement or shell configuration of a connector that ensures that the mating connectors are guided into proper alingment during mating.

Threaded coupling is used on circular connectors and contact engagement is achieved by several turns on the threaded part. In some circular connectors, the engagement takes place by applying an axial force and the connector locking is done by giving several turns on the threaded receptacle (Fig. 4.27).

Fig. 4.27

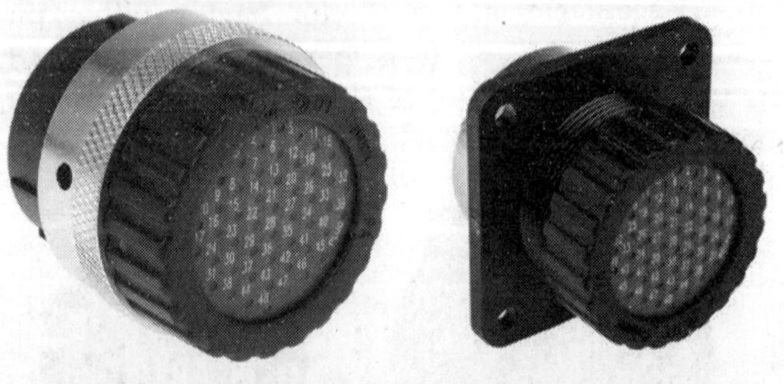

Fig. 4.28

A socket contact is a hollow and usually cylindrical contact that has been designed to mate with a pin contact. Socket contacts are preferably used on a connector used on the source side of the circuit. Fig. 4.28 shows a receptacle and a plug with socket contacts.

Q.17. Briefly describe the characteristics and applications of the following types of co-axial RF connectors: BNC, TNC, N, UHF, SMA.

Ans. BNC Connector

The BNC connectors (Fig. 4.29) are the quick disconnect type RF connectors having a bayonet coupling. These are the most widely used connector type and are invariably found on the front panel of majority of electronic test and measuring equipment, telecommunication and aerospace equipment. These have a nominal impedance of 50 ohms and a working voltage of 500 volts (RMS). The BNC plugs are available with both captive as well as non-captive center contacts. For making the cable, the center contact is soldered to the center conductor of the cable and the outer shield is either clamped or crimped. These connectors when terminated in RG 55/U, RG 58/U, RG 141 A/U, RG 142 A/U, RG 223/U, RG 303/U give excellent performance upto 4 GHz with a VSWR between 1 and 1.3.

Fig. 4.29

TNC CONNECTORS

TNC type RF connectors (Fig. 4.30) resemble BNC connectors except that the coupling mechanism used here is the threaded coupling instead of a bayonet coupling used in BNC connectors. These are comparatively smaller and weatherproof and are thus ideally suited for aircraft and missile applications and other such applications where one is likely to encounter extreme vibrations. These have a nominal impedance of 50 ohms and a usable frequency range extended upto 11 GHz. These are available in captive as well as non-captive center contact varieties. The center conductor of the cable is soldered to the center contact and the outer conductor is secured by clamping.

Fig. 4.30

N-Type CONNECTORS

N-type connectors (Fig. 4.31) use threaded coupling, have a nominal impedance of 50 ohms and a maximum working voltage of 1000 V (RMS). The usable frequency range is extended upto 11 GHz. These are widely used in RF test and measurement equipment. With RG 8A/U and RG 9 B/U cables, these offer excellent electrical performance upto 11 GHz.

UHF CONNECTORS

UHF connectors are general purpose, low cost threaded coupling type RF connectors that offer satisfactory electrical performance upto 300 MHz. These are used in applications where mismatch is not a serious problem. These have fixed center contact which is soldered to the center conductor of the cable. Crimp type UHF connectors are also available. Their applications include electronic test and measuring instruments, communication equipment, radar, television and many other applications.

Fig. 4.31

Fig. 4.32

SMA CONNECTORS

Of all the RF connectors, SMA connectors have the best RF performance. SMA connectors (Fig. 4.32) are precision connectors designed to give a low VSWR at frequencies as large as 20 GHz. These have a nominal impedance of 50 ohms and a maximum working voltage of 1000 V (RMS). SMA connectors find extensive use in RF test and measurement, radar and communication equipment, microwave communication systems etc.

Q.18. Write short notes on the following :

1. **PC board edge type connectors**
2. **Sub-miniature D-type connectors**
3. **Multipin circular connectors.**

Ans. PC BOARD CONNECTORS

PC board connectors are available in two types. One of the PC card connectors are basically receptacles that mate directly with the edge of the printed circuit card. The contact can be made with one or both sides of the printed circuit card. These are available in a variety of contact spacings, sizes and configurations (Fig. 4.33). These are extensively used in commercial and industrial applications.

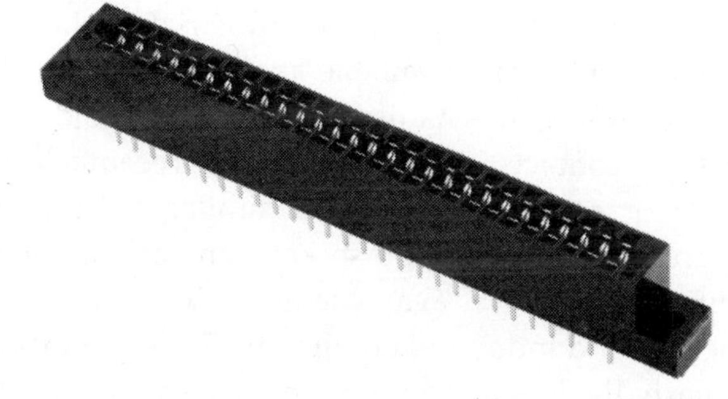

Fig. 4.33

The other type of PC card connectors is the two piece connector assembly in which one of the mating connectors is hard wired to the printed circuit board (Fig. 4.34).

SUB-MINIATURE D-TYPE CONNECTORS

These are sub-miniature type of Rack and Panel connectors. They derive their name from the shape of the connector. These connectors are made by several manufacturers under different trade names. Upto 50 contacts with normal spacing and a current carrying capability of 5A per contact are available in the general purpose category. Double density versions may have upto 100 contacts.

Fig. 4.34

The miniature size and light weight of these connectors make them ideally suitable for a wide variety of applications which do not require stringent environmental capabilities. The applications include computers and data processing equipment, instrumentation, telecommunication equipment and so on.

MULTIPIN CIRCULAR CONNECTORS

Multipin circular connectors (Fig. 4.35) are used in a variety of military, aerospace and industrial applications. These connectors are available in both threaded and bayonet coupling mechanisms. Also, both solder and crimp type contacts are available. The receptacle may have pin or socket contacts and similarly the mating plug can also have pin or socket contacts. Hundreds of configurations are available for use in consumer, commercial and industrial equipment. These connectors, particularly the bayonet coupling type crimp connectors, due to their excellent mechanical and environmental capabilities find extensive use in defence applications like aerospace and missile systems.

Fig. 4.35

Q.19. What is a co-axial cable and what are its advantages?

Ans. A co-axial cable is a two conductor transmission line with a center conductor surrounded by a braided shield. The inner conductor is supported by some form of dielectric insulation.

The most significant feature of a co-axial cable is its shielded structure. In a shielded structure like this, the electromagnetic fields associated with the two conductors remain confined within the space between the inner conductor and the outer shield. As the frequency of the signal carried by the cable increases, the current concentrates itself on the inside of the outer conductor (skin effect). As a result of this, a co-axial structure is a self-shielding one with the shielding characteristics improving with increase in frequency. The major application of co-axial cable is in transmission of high frequency broadband signals. These are rarely used at lower frequencies as their shielding properties become poor at low frequencies and they tend to be more expensive than the twisted pairs for the same transmission loss. Fig. 4.36 shows conductor arrangement in different co-axial cables.

Fig. 4.36

Q.20. How do you determine characteristic impedance of a co-axial cable?

Ans. The characteristic impedance of a lossless air insulated co-axial cable is given by

$$Z_0 = 138 \log_{10} (D/d)$$

The characteristic impedance of a lossless dielectric insulated co-axial cable having a dielectric strength of (ε) is given by:

$$Z_0 = 138/\sqrt{\varepsilon} \log_{10} (D/d), \text{ refer to Fig. 4.37}(a)$$

The characteristic impedance of a lossless co-axial cable with inner and outer conductors separated by insulating discs at definite intervals (Fig. 4.37 *b*) is given by:

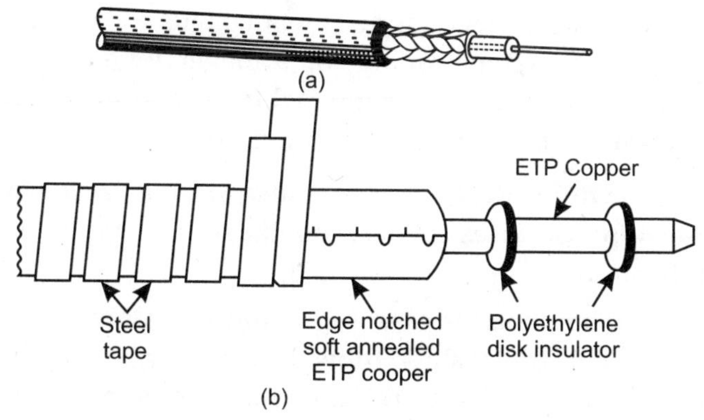

Fig. 4.37

$$Z_0 = 138 \log_{10} \frac{(D/d)}{\sqrt{1+(\varepsilon-1)W/S}}$$

where

d = Outer diameter of inner conductor

D = Inner diameter of outer conductor

W = Thickness of insulating disc

S = Center-center spacing of adjacent discs

Q.21. Briefly describe FLEXIBLE, SEMIRIGID and RIGID cables?

Ans. A flexible cable (Fig. 4.38*a*) has a solid or stranded inner conductor, a braided outer conductor with a solid or expanded plastic separating the two. The electrical characteristics of flexible co-axial cables vary over a large length of the cable. These cables are mainly used for shorter runs like making interconnections between high frequency circuits in an equipment or interconnecting one sub-system to another within a large system.

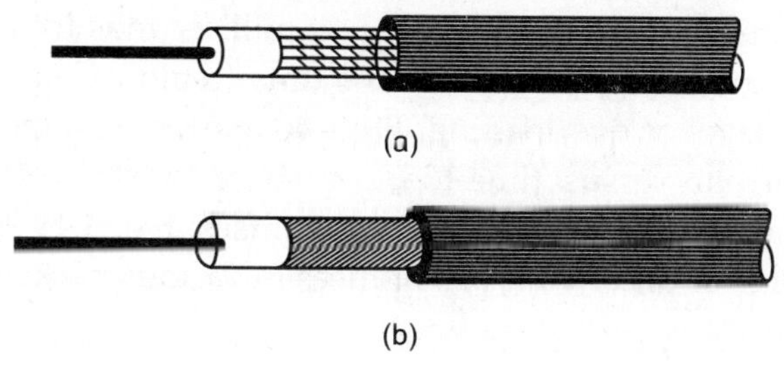

Fig. 4.38

A semirigid cable (Fig. 4.38*b*) has a thin walled copper tubing as the outer conductor and a solid dielectric. The dielectric can be either insulating discs or expanding plastic insulation. One of the popular applications of semirigid cables is in closed circuit television installations.

A rigid cable has an inner conductor that is either a solid wire or a hollow tube that can be liquid cooled in high power applications. The outer conductor is heavy and rigid. Rigid cables are used in places where extremely good electrical performance is desired and no bending is needed to install the cable. The dielectric used in usually semisolid like discs for instance and the rigid outer conductor allows the dielectric supports to be spaced far apart. This helps minimize dielectric losses.

Dielectric Constants

Material	*Dielectric Constant*
Air	1.0001
Vacuum	1.0000
Teflon	2
Impregnated Paper	4.6
Aluminium Oxide	10
Polyster	3
Mica	4.5 to 7.5
Glass	6.7
Tantalum Oxide	11

Q.22. What is the primary function of a fuse element? What is the fundamental principle underlying the functioning of a fuse?

Ans. The primary function of fuse is to automatically disconnect individual circuits, components or equipment from the source of power on the occurance of a potentially damaging fault in the system. The fault could either be moderate overload or even a short circuit.

The fuse element is made of a low melting point alloy or of aluminium. The current flowing through the fusible element causes an increase in its temperature. When the temperature reaches the melting point, the element volatilizes and opens the circuit if the resulting arc is self extinguishing.

The operation of a fuse is based on a time element principle. According to this, on the occurance of a short circuit, the fuse operates almost instantaneously and in an overload condition, it operates after a certain definite time lag. This time lag varies inversely with the magnitude of overload as shown in Fig. 4.39. The time-to-blow depends upon the amount of energy that must be dissipated in the fusible element to cause the required temperature rise to reach the melting point. The energy quantum in turn depends upon the characteristics of the fusible element including the material, its mass and specific heat. For a given resistance of the fuse element, the time-to-blow varies as inverse square of the current, which explains the shape of the curve.

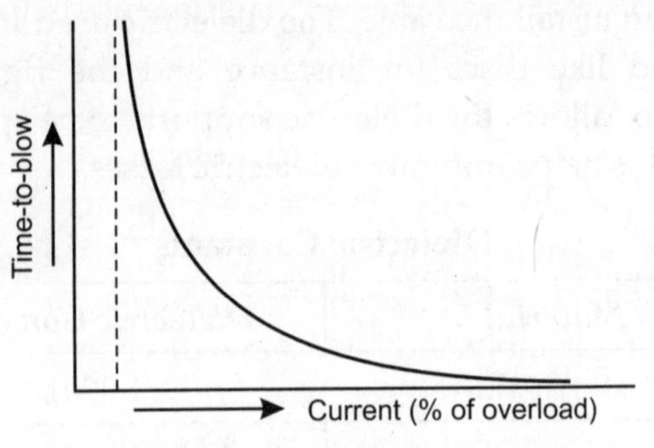

Fig. 4.39

Q.23. Briefly describe major specifications of a fuse.

Ans. Major specifications of a fuse include:

1. Current Rating,
2. Voltage Rating and
3. (I^2t) Rating also known as 'Let through' current.

1. **Current Rating** is the maximum nominal amount of DC or RMS AC current a fuse can carry indefinitely without blowing. This means that the current rating of the fuse must exceed the maximum DC or RMS AC current drawn by the circuit or device to be protected.
2. **Voltage Rating** is the maximum nominal DC or AC RMS voltage for which the fuse is designed. It is indicative of the ability of the fuse to extinguish the arc generated across the fuse terminals as it melts during fault conditions. It is not necessarily equal to the supply voltage as the voltage generated across the fuse not only depends upon the supply voltage but the nature of the circuit connected in series with the fuse. For example, if the fuse feeds an inductive circuit, the voltage generated across it as it melts will be much larger than the supply voltage.
3. **(I^2t) Rating** is the amount of energy that must be dissipated in the fuse element to cause it to melt. It is also sometimes referred to as pre-arcing let through current. On the basis of (I^2t) rating, the fuses are usually categorized as fast blow, normal and slow blow types.

Q.24. What can be the possible implications of a fuse with inadequate voltage rating?

Ans. As already mentioned during the answer to the previous question, the voltage that actually appears across the fuse terminals during the clearance transient not only depends upon the supply voltage but also on the type of circuit it is protecting. In case of inductive circuits, this voltage could become much higher than the supply voltage. If the voltage rating is inadequate, it could lead to sustained arcing. Arcing for a longer time leads to increase in let through energy, which in the worst case could cause fuse cartridge explosion.

Q.25. Briefly describe fast blow, normal and slow blow type of fuses. Where do these fuse types find application?

Ans. The time-to-blow for a **normal slow** fuse is between that of a **slow blow** (also known as a time delay fuse) and a **fast blow fuse**. These fuses can usually handle short term high current transients. These are generally meant for short circuit protection. The time-to-blow typically varies from a few milliseconds to several tens of milliseconds for a normal blow fuse of given current rating depending upon quantum of overloading. A higher overload means a shorter clearance time.

A slow blow or a time delay fuse on the other hand can provide large currents for relatively longer periods of time without getting ruptured. These are used to protect equipment or circuits that draw a high initial current that later drops a lower normal operating current. Typical examples are high in-rush current as compared to the normal running current of an electric motor or the high initial surge current drawn by a capacitor. The time-to-blow typically varies from hundreds of milliseconds to a few seconds for a fuse of given current rating again depending upon extent of overloading. Fast blow fuses are used for protection of semiconductor devices. These fuses clear very rapidly even under short term current transients. The time-to-blow typically varies from a fraction of a millisecond to a few milliseconds for a fuse of given current rating with the time-to-blow depending upon level of overloading. Fig 4.40 shows a comparison of current versus time-to-blow characteristics of slow blow, normal blow and fast blow fuses. Fig. 4.41 shows pictures of different types of fuses in various packages.

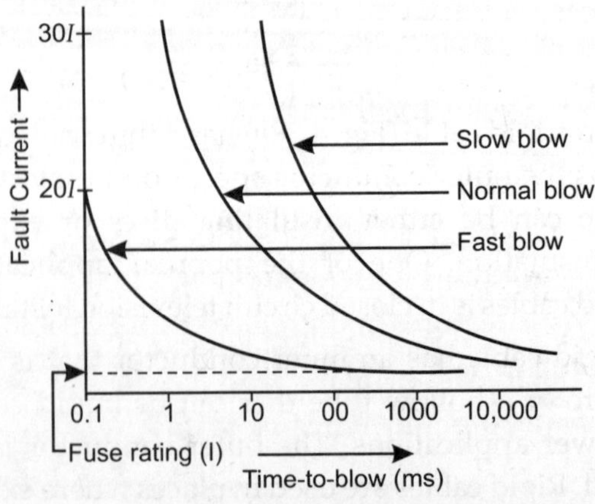

Fig. 4.40

Fig. 4.41

Q.26. What is a battery? How does it work?

Ans. A **battery** is essentially an electro-chemical cell that converts stored chemical energy into electrical energy. It comprises of a positive electrode called cathode, a negative electrode called anode, electrolyte that brings into contact two different chemical compositions of anode and cathode materials and an electrolyte separator that prevents flow of electrons from anode to cathode within the battery. The electrolyte may be aqueous or non-aqueous, in liquid, paste, or solid form. The chemical reaction in the battery causes build-up of electrons at the anode thereby creating a kind of potential difference between anode and cathode. This is unstable build-up of electrons and electrons have a tendency to re-arrange themselves and move to a place with fewer electrons. If the external circuit is closed; the electrons move towards the cathode thereby providing flow of electric current through the external circuit. Depending upon electrode compositions, type of electrolyte used and battery chemistry, there is a large variety of batteries. Common types include carbon-zinc, alkaline, lead acid, nickel-cadmium, nickel metal hydride and lithium ion batteries. Each one of these is briefly described in the following questions with regards to their basic chemistry, characteristic features and preferred application areas.

Q.27. What are primary and secondary batteries? Name some common types of primary and secondary batteries.

Ans. Primary batteries are one-time use batteries that need to be discarded after they have run out of stored chemical energy. Primary batteries cannot be reliably recharged as the chemical reactions involved in their operation are not easily reversible and the active materials may not return to their original forms. Primary batteries are capable of delivering electrical current immediately after their assembly and have a low discharge rate. These are generally preferred for use in those portable devices with low current or are used well away from an alternative power source. Some common types of primary batteries are carbon-zinc batteries also known as dry cells, alkaline batteries and Lithium cells. Carbon zinc primary batteries are zinc–acidic manganese dioxide batteries while alkaline primary batteries are zinc–alkaline manganese dioxide batteries. Silver oxide and zinc-air cells are the other primary batteries. Primary batteries are used in low current drain applications such as in clocks, remote controls, hearing aids, calculators and so on. They have long shelf life. In the case of lithium primary cells, it may be well above 10 years.

Secondary batteries are re-chargeable batteries. In contrast with the primary batteries that can deliver electrical current immediately after assembly; secondary batteries are usually assembled with active materials in the discharged state. Secondary batteries can be recharged by applying electric current, which reverses the chemical reactions that occur during its use. Some common types of secondary batteries are re-chargeable alkaline batteries, lead acid batteries, nickel-cadmium batteries, nickel metal hydride batteries and lithium ion batteries. Based on the cathode oxide material used, lithium ion batteries are further of three types namely lithium-ion cobalt, lithium-ion manganese and lithium ion phosphate. Lithium-ion polymer is another re-chargeable lithium ion battery that has a distinctly different architecture than the three above mentioned lithium ion battery types. Lithium metal is another emerging re-chargeable lithium battery.

Q.28. Briefly describe important specifications of a battery.

Ans. Major specifications of batteries include nominal voltage, cut-off voltage, capacity, energy capacity, cycle life, specific energy, specific power, energy density, power density, maximum continuous discharge current, charging voltage, float voltage recommended charging current, maximum 30 second discharge pulse current and maximum internal resistance. Each of these specifications is briefly described in the following paragraphs.

Nominal voltage of a cell or battery is its terminal voltage when it is fully charged and is delivering rated capacity at a specific discharge rate. It is 2.0 volts for a lead acid cell, 1.2 volts for nickel cadmium and nickel metal hydride cells and 3.6 volts for a lithium ion cell.

Cut-off voltage is the minimum allowable voltage and is the voltage that generally defines the empty state of the battery. It is the prescribed lower-limit voltage at which battery discharge is considered complete. The significance of the cut-off voltage of a battery is that maximum useful capacity of the battery is considered to have been utilized when it reaches the cut-off voltage. It is 0.9 volt for an alkaline cell, 1.0 volt in the case of nickel cadmium and nickel metal hydride cells and approximately 3.3 volts in the case of lithium ion cell.

Nominal Capacity: Nominal capacity is the total Ampere-hours (Ah) available with the battery for a specified discharge current as indicated by its C-rate when it is discharged from 100 per cent state-of-charge to the cut-off voltage. Nominal capacity in Ampere-hours is given by the product of discharge current in Amperes and discharge time in hours. Capacity decreases with increase in C-rate. Rated capacity of secondary batteries is generally specified at 1C rate. In the case of primary batteries, typical rated capacities of AAA, A, C and D cells and 9 volt pack respectively are 1150 mAh (for rated load of 75 ohms), 2870 mAh (for rated load of 75 ohms), 7800 mAh (for rated load of 39 ohms), 17000 mAh (for rated load of 39 ohms) and 570 mAh (for rated load of 620 ohms). The mAh rating depends upon the discharge rate. Capacity is also specified as **Energy Capacity** measured in watt-hour (Wh). It is the total watt-hours available when the battery is discharged at a specified discharge current from 100 per cent state-of-charge to cut-off voltage. It is given by product of discharge voltage in watts and discharge current in hours and like Ampere-hour capacity, watt-hour capacity also decreases with increase in C-rate.

Specific Energy and Specific Power: Specific energy measured in Wh/kg is the battery energy per unit mass. It depends upon the battery chemistry and packaging. **Specific power** measured in W/kg is the maximum available power per unit mass. Like specific energy, specific power too depends upon battery chemistry and packaging. Amongst major types of primary and secondary batteries, alkaline primary, lithium-ion and lithium polymer secondary batteries have the highest specific energy in the range of 100–150 Wh/kg followed by nickel metal hydride batteries with specific energy of 70–80 Wh/kg. Lead acid and nickel cadmium batteries have specific energy in the range of 30–35 Wh/kg.

Energy Density and Power Density: Energy density is the nominal energy per unit volume and **power density** is the nominal power unit volume. These are respectively measured in Wh/litre and W/litre. Amongst major types of primary and secondary batteries, alkaline primary and lithium-ion secondary batteries have the highest energy density of about 400 Wh/litre followed by nickel metal hydride and lithium polymer batteries with energy density of about 250 Wh/litre. Lead acid and nickel cadmium batteries have energy density of about 75 and 100 Wh/litre respectively.

Cycle Life: It is the number of complete charge-discharge cycles the battery can support before its capacity falls below 80 per cent of rated capacity. It depends upon depth-of-discharge. The cycle life reduces with increasing depth-of-discharge. It is also affected by temperature and humidity conditions. It is important here to underline the words '*complete charge-discharge*'. If a battery were discharged to 50 percent and then charged to 70 percent; it wouldn't be considered as a complete cycle. Typical cycle life figures for lead-acid, nickel-cadmium, nickel metal hydride, lithium-ion cobalt, lithium-ion manganese and lithium-ion phosphate batteries for 80 per cent depth-of-discharge are 200–300, 1000, 300–500, 500–1000, 500–1000 and 1000–2000 respectively.

C-Rate and E-Rate: C-rate is used in batteries to indicate the discharge current. A 1C rate would imply a discharge current that would discharge the battery in one hour. It is a measure of the rate at which battery is discharged relative to its maximum capacity. As an example, a battery of 100 Ah capacity and 1C rate implies a discharge current of 100 Amperes. If this battery were specified to have a maximum discharge current of 5C; then maximum discharge current would be 500 Amperes. If two batteries with different capacity are specified to have same maximum discharge current in terms of C-rate; then the one with greater capacity will have higher maximum discharge current. In other words, if two batteries had the same capacity; higher C-rate would signify higher maximum discharge current. E-rate describes the discharge power. A 1E rate similarly is the discharge power that would discharge the entire battery in 1 hour.

Maximum Continuous Discharge Current: It is the maximum continuous discharge current the battery can support without affecting its capacity or causing any damage to the battery. It is specified as a C-rate.

Maximum continuous discharge current can be computed from the capacity of the battery and the C-rate. As an example, a 100 Ah battery with maximum discharge current of 5C shall have maximum discharge current of 500 Amperes. Lead acid, nickel-cadmium, nickel metal hydride, lithium-ion (cobalt), lithium-ion (manganese) and lithium-ion (phosphate) batteries have maximum continuous discharge current ratings of 5C, 20C, 5C, > 3C, > 30C and > 30C respectively.

Maximum Pulse Discharge Current: It is the maximum discharge current the battery can support in the pulsed discharge mode. It is usually specified for a 30 second discharge time period. It is specified by the battery manufacturers to prevent excessive discharge rates that can cause damage to the battery.

State-of-Charge (SOC): State-of-charge is the present capacity of the battery expressed as a percentage of its maximum capacity.

Depth-of-Discharge (DOD): It is the battery capacity that has been utilized or discharged expressed as a percentage of maximum capacity. A DOD of equal to and greater than 80 percent is considered as deep discharge.

Open Circuit Voltage: It is the voltage across battery terminals under no load conditions. It depends upon state-of-charge. Higher state-of-charge leads to increased open circuit voltage.

Terminal Voltage: It is the voltage across battery terminals with load connected across it. It depends upon state-of-charge and also the discharge current.

Internal Resistance: Internal resistance is the resistance offered by the battery to flow of current within the battery. It has two main components namely the electronic resistance and ionic resistance. Electronic resistance is due to actual materials used in making the battery and also how well these materials make contact with each other. Ionic resistance is opposition to flow of current due to various electrochemical factors such as conductivity of electrolyte, electrode surface area and ionic mobility. It is in the range of few tens of milli-ohms to few hundreds of milli-ohms. Primary batteries have relatively much larger internal resistance. Amongst lithium, carbon-zinc and alkaline primary batteries, alkaline batteries have the lowest internal resistance of the order of 1.0 ohm. It is an indicative of efficiency of the battery. Higher internal resistance causes lower efficiency.

Self-Discharge Rate: Self-discharge is the drop in the capacity of the battery when it is lying in the shelf unused. It is caused by electrochemical processes within the battery and is equivalent to the application of a small external load. Amongst secondary batteries, it is 5 percent/month for lead-acid batteries, 20 percent/month for nickel-cadmium batteries, 30 percent/month for nickel metal hydride batteries and less than 10 percent/month for lithium batteries. In the case of nickel-cadmium, nickel metal hydride and lithium batteries, self-discharge rate is higher immediately after full charge in the first twenty four hours and then stabilizes to above mentioned figures. Self-discharge rate increases with temperature. The above figures are typical of room temperature operation.

Q.29. What are primary batteries? Briefly describe carbon-zinc, alkaline and lithium primary batteries as regards their operation, battery chemistry, advantages and disadvantages.

Ans. As outlined in an earlier question, a primary battery is the one that is not rechargeable. The electrochemical reaction in a primary battery is not reversible. A primary battery is used once and then discarded. Common types of primary batteries include carbon-zinc, alkaline and lithium batteries. Each of three types is briefly described in the following paragraphs.

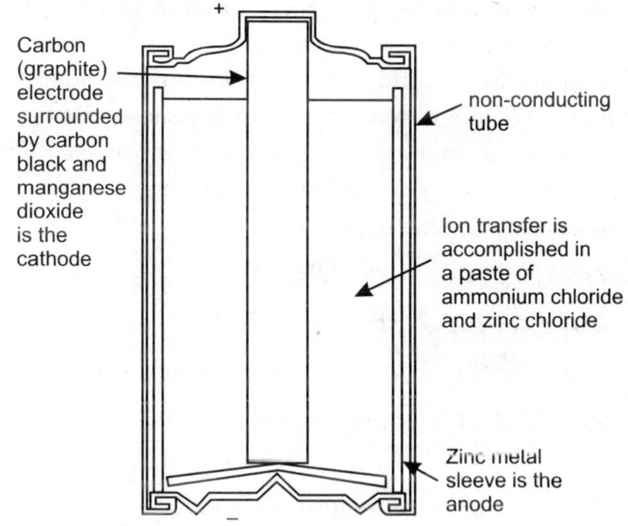

Fig. 4.42

Carbon-Zinc Battery

There are two types of carbon-zinc batteries namely carbon-zinc battery and zinc chloride battery. A **carbon-zinc battery** also called **Leclanche battery** uses zinc as the anode (negative terminal), carbon rod surrounded by a mix of manganese dioxide and carbon powder, usually graphite powder, as the cathode (positive terminal) and ammonium chloride along with some quantity of zinc chloride as the electrolyte. Zinc chloride battery also uses same anode and cathode materials but the electrolyte in this case is zinc chloride. Fig. 4.42 shows the constructional details of a carbon-zinc battery. The chemistry of both types of carbon-zinc batteries is given as follows in the form of different chemical reactions involved therein.

During discharge, zinc anode involves in oxidation reaction and each zinc atom involved in this reaction releases two electrons.

$$Zn \rightarrow Zn^{+2} + 2e$$

In a Leclanche battery, MnO_2 in the cathode is reduced to Mn_2O_3 in reaction with ammonium ion (NH_4^+) present in NH_4Cl electrolyte to produce Mn_2O_3. In addition, this reaction also produces ammonia (NH_3) and water (H_2O) as expressed by the following reaction.

$$2\,NH_4^+ + 2\,MnO_2 + 2\,e \rightarrow 2\,e + Mn_2O_3 + H_2O + 2\,NH_3$$

During this chemical process, some of ammonium ions (NH_4^+) are directly reduced by electrons and form gaseous ammonia (NH_3) and hydrogen (H_2) as expressed by the following chemical reactions.

$$2\,NH_4{+} + 2\,e \rightarrow 2\,NH_3 + H_2$$

In carbon-zinc battery, the ammonia gas generated in the above chemical reaction further reacts with zinc chloride to form solid zinc ammonium chloride. Also, gaseous hydrogen reacts with manganese dioxide to form solid di-manganese trioxide and water. These two reactions as given below prevent formation of gas pressure during discharging of battery.

$$ZnCl_2 + 2\,NH_3 \rightarrow Zn\,(NH_3)_2\,Cl_2$$

$$2\,MnO_2 + H_2 \rightarrow Mn_2O_3 + H_2O$$

The overall reaction can be written as follows.

$$Zn + 2\,MnO_2 + 2\,NH_4Cl \rightarrow Mn_2O_3 + Zn\,(NH_3)_2Cl_2 + H_2O$$

To sum up the operation of a carbon-zinc primary battery, the electrons leave the zinc, which is the anode and the negative terminal; go to the ammonium chloride, which is the electrolyte. The electrons from a hydrogen by product leave the ammonium chloride and go to the manganese dioxide, which is the cathode and the positive terminal. The electrons therefore go from anode, to cathode.

Standard voltage rating of a carbon-zinc battery depends upon electrode potentials of zinc (= – 0.76 volt) and manganese dioxide (= 1.23 volts). Theoretical voltage therefore should be equal to – (– 0.76) + 1.23 = 1.99 volts. Due a number of practical considerations, the voltage rating of a carbon-zinc battery is typically 1.5 volts. A low discharge carbon zinc battery has energy density of 75 Ah/kg whereas a heavy-duty, intermittent discharge battery has energy density of 35 Ah/kg. Carbon-zinc batteries are available in a number of different sizes including AAA, AA, C, D and 9 volt batteries. Nominal voltage and Ampere-hour (Ah) capacity of these batteries respectively are 1.5 volts and 320 mAh (AAA); 1.5 volts and 591 mAh (AA); 1.5 volts and 2172 mAh (C); 1.5 volts and 4733 mAh (D) and 9 volt and 400 mAh (9 V). The average voltage under load conditions is typically 1.1 volts. Advantages of carbon zinc batteries include low cost, reliability and availability in different shapes and sizes. Disadvantages include low energy density, high leakage resistance, poor low temperature performance, steady drop in terminal voltage with discharge and low shelf-life. Zinc chloride batteries on the other hand perform better than Leclanche batteries in terms of leakage resistance, energy density and operation at heavy discharge conditions and at low temperatures. Zinc chloride batteries are however more oxygen sensitive and have higher gassing rate.

Alkaline Battery

Alkaline battery also called **alkaline-manganese battery** is an improved version of the carbon-zinc battery. In an alkaline battery, powder zinc serves as the anode; manganese dioxide mixed with coal dust is the cathode and potassium hydroxide is the electrolyte. It derives its name from the alkaline nature of its electrolyte. Fig. 4.43 shows the constructional features of an alkaline battery. It comprises of a hollow steel drum whose inner peripheral surface is deposited with fine grained powder of manganese dioxide mixed with coal dust acting as the cathode. The cathode (or positive) terminal of the battery is projected from the top of this drum. The inner surface of the thick layer of cathode mixture is covered with paper separator. The central space, inside this paper separator is filled by zinc powder with potassium hydroxide electrolyte. Zinc powder serves as the anode (or negative terminal). Its powder form increases the contact surface. The paper separator soaked with potassium hydroxide holds the electrolyte in between cathode and anode. A metallic pin (preferably made of brass) called negative collector pin is inserted along the central axis of the alkaline battery to collect negative charge.

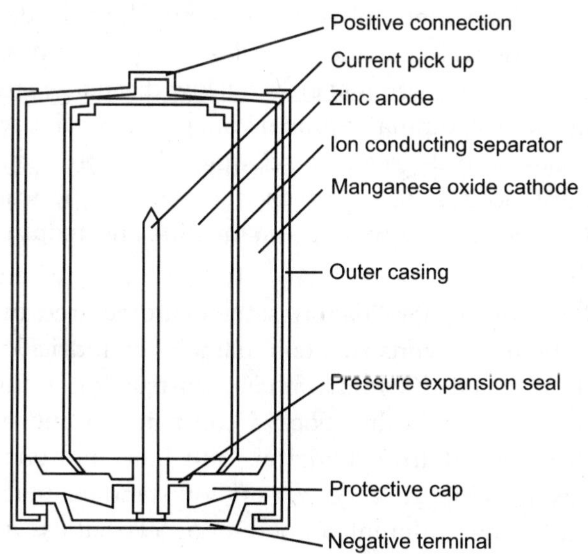

Fig. 4.43

Different chemical reactions involved in the operation of the alkaline battery are given as under.

The first reaction is expressed as:

$Zn + 2\ HO^- \rightarrow ZnO + H_2O + 2\ e$

The second half reaction is expressed as:

$2\ MnO_2 + H_2O + 2\ e \rightarrow Mn_2O_3 + 2\ HO^-$

Overall reaction is expressed as:

$Zn + 2\ MnO_2 \leftrightarrow Mn_2O_3 + ZnO$

The nominal voltage is 1.5 volts. The average voltage under load conditions is between 1.1 and 1.3 volts. Based on type of active materials used, other common alkaline batteries are nickel-iron, nickel-cadmium and silver-zinc batteries. Like carbon-zinc batteries, alkaline batteries are also available in AAA, AA, C, D and 9 volt varieties. Different packages of alkaline batteries have ampere-hour capacity of 1150 mAh (AAA), 2122 mAh (AA), 7800 mAh (C), 17000 mAh (Δ) and 600 mAh (9 volt). The ampere-hour capacity reduces with increase in discharge current.

The advantages of alkaline batteries as compared to carbon-zinc or Leclanche batteries are the following: relatively higher energy density, better performance at higher discharge currents and low temperatures, lower leakage and self-discharge, longer shelf life, lower internal resistance, better dimensional stability. The only disadvantage vis-à-vis carbon-zinc battery is its comparatively higher cost.

Lithium Primary Battery

Lithium primary battery is a family of a large variety of primary batteries each having a different chemistry. It should not be confused with lithium-ion secondary batteries. Common types of lithium primary batteries are Lithium Manganese Dioxide (Li-MnO_2), Lithium Iron Disulphide (Li-Fe-S2), Lithium Thionyl Chloride (Li-$SOCl_2$), Lithium Sulpher Dioxide (Li-SO_2)and Lithium Iodine (LiI_2) batteries. All lithium primary batteries use lithium metal or its compounds as the anode. Different types use different cathode and electrolyte combinations.

Lithium manganese dioxide battery is the most common consumer grade lithium primary battery. It uses lithium foil as the anode and manganese dioxide as the cathode. Lithium perchlorate in propylene carbonate and dimethoxyethane is used as the electrolyte. It uses inexpensive materials and is particularly suitable for low drain, long life and low cost applications. The nominal voltage is 3.0 volts and power density is typically 280 Wh/kg. The internal resistance increases leading to decrease in terminal voltage with increase in discharge current. It also suffers from self-discharge at elevated temperatures.

The chemical reactions involved in a lithium manganese dioxide battery are given as under.

The first half reaction (oxidation) is expressed as:

$Li \rightarrow Li^+ + e^-$

The second half reaction (reduction) is expressed as:

$MnO_2 + Li^+ + e^- \rightarrow MnOOLi$

Overall reaction is expressed as:

$MnO_2 + Li \rightarrow MnOOLi$

In the case of **Lithium iron disulphide battery,** the cathode material is iron disulphide. The electrolyte is a lithium salt in an organic solvent blend. The involved chemical reactions are as follows.

First half reaction (oxidation) is expressed as:

$Li \rightarrow Li^+ + e^-$

Second half reaction (reduction) is expressed as:

$FeS_2 + 4\ Li^+ + 4\ e^- \rightarrow Fe + 2\ Li_2S$

The overall reaction is given by:

$4\ Li + FeS_2 \rightarrow Fe + 2Li_2S$

With a nominal voltage of 1.5 volts, this lithium battery can work as a replacement for alkaline batteries offering higher shelf life due to lower self-discharge, better performance at extreme low temperatures and higher power density, typically 300 Wh/kg.

Lithium Thionyl Chloride batteries use Thionyl Chloride as the liquid cathode material and lithium tetrachloroaluminate in Thionyl chloride as the electrolyte. The involved chemical reactions are given as under.

First half reaction (oxidation) is expressed as:

$Li \rightarrow Li^+ + e^-$

Second half reaction (reduction) is expressed as:

$2\,SOCl_2 + 4\,Li + + 4\,e^- \rightarrow 4\,LiCl + S + SO_2$

The overall reaction is expressed as under.

$2\,SOCl_2 + 4\,Li \rightarrow 4\,LiCl + S + SO_2$

The nominal voltage is 3.5 volts. These batteries have the highest energy density of all lithium primary batteries, typically 500 Wh/kg. When kept in storage for a long time, a passivation layer is formed on the anode leading to a temporary delay in appearance of voltage after it is pressed into service. Safety concerns limit its use in civilian applications. These batteries can explode under short circuit conditions.

Lithium Sulpher Dioxide Batteries employ sulpher dioxide on Teflon-bonded carbon as the cathode material. The electrolyte is lithium bromide in sulpher dioxide with small amount of acetonitrile. The involved chemical reactions are given as under.

$2\,Li + 2\,SO_2 \rightarrow Li_2S_2O_4$

$Li_2S_2O_4 \rightarrow SO_2 + S + 2\,Li_2SO_3$

The nominal voltage and power density values are typically 2.85 volts and 250 Wh/kg respectively. It also has safety concerns and therefore needs to have a safety vent. It performs well over a wide temperature range of – 55°C to + 70°C. At high discharge currents and low temperatures, it offers superior performance over lithium-manganese dioxide batteries.

In the case of **lithium iodine battery**, the cathode is iodine that has been mixed and heated with poly-2-vinylpyridine (P2VP) to form a solid organic charge transfer complex. The electrolyte is a solid monomolecular layer of crystalline Lithium iodide. The involved chemical reactions are as follows.

The first half reaction (oxidation) is expressed as under.

$Li \rightarrow Li^+ + e^-$

The second half reaction (reduction) is expressed as under.

$\frac{1}{2}\,I_2 + e^- \rightarrow I^-$

The overall reaction is expressed as under.

$Li + \frac{1}{2}\,I_2 \rightarrow LiI$

Lithium iodine batteries are characterised by high reliability and low self-discharge rate that results in their having a long life. Unlike lithium sulpher dioxide and lithium Thionyl chloride batteries, there are no safety issues. These are suitable only for low discharge current applications. Nominal terminal voltage is 2.8 volts, which decreases with degree of discharge due to precipitation of lithium iodide. These are used in medical applications such as pacemakers that need a long life and require low current.

Lithium primary batteries are manufactured in two basic types of constructions namely cylindrical cells and button cells. Cylindrical cell construction is further of two types including bobbin construction and spiral construction. Both cylindrical cells and button cells are available in a large variety of standard package sizes offering different ampere-hour capacities. As an illustration, Figs. 4.44 and 4.45 show constructional features of bobbin and spiral type of lithium manganese dioxide cells. Fig. 4.46 shows the construction of a typical lithium manganese dioxide button cell.

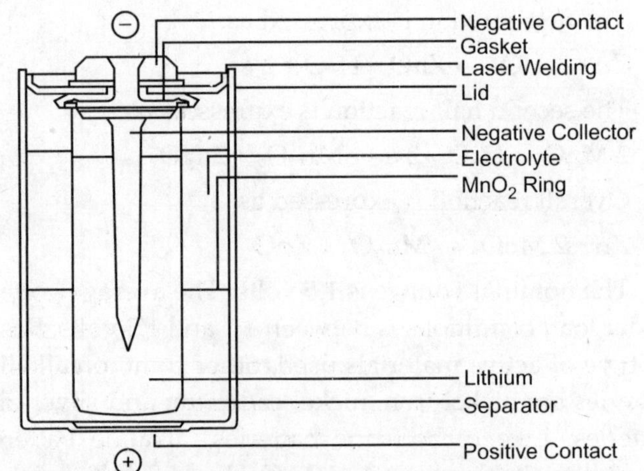

Fig. 4.44

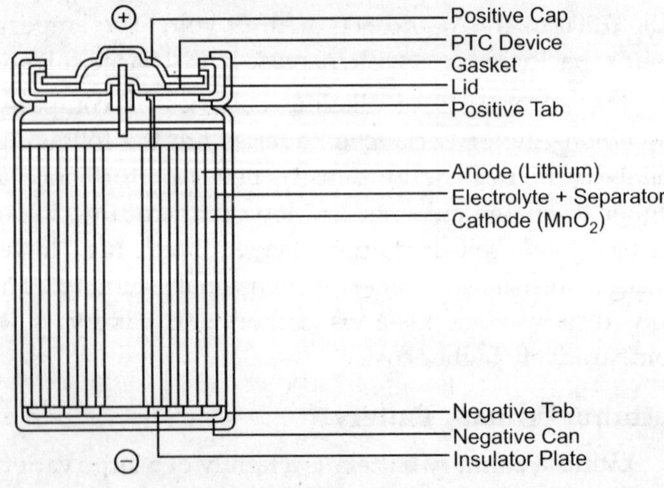

Fig. 4.45

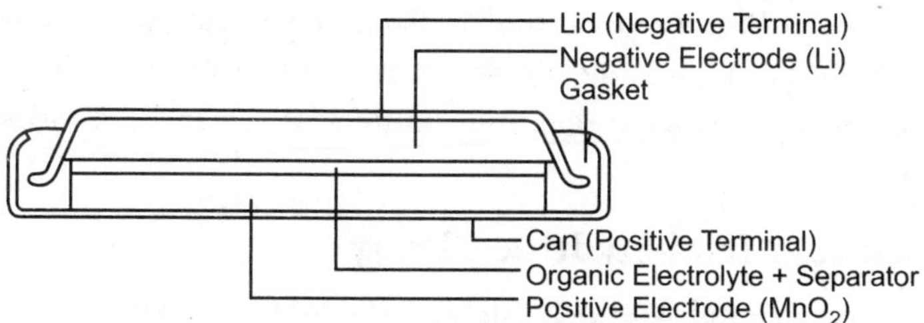

Fig. 4.46

Q.30. What are secondary batteries? Briefly describe lead-acid, Nickel-Cadmium, Nickel Metal Hydride and Lithium Ion secondary batteries. Compare them in terms of their performance features and advantages and disadvantages.

Ans. A secondary battery is a rechargeable battery. It can be electrically recharged after use to their original pre-discharge condition by passing current through the circuit in the opposite direction to the current during discharge. The electrochemical reactions occurring in the case of a secondary battery are reversible. Commonly used secondary batteries include the following.

1. Lead acid battery
2. Nickel-cadmium battery
3. Nickel-metal-hydride battery
4. Lithium-ion battery

Fig. 4.47

Lead Acid Battery

Lead acid battery is the oldest rechargeable battery system. Main active materials used in a lead acid battery are lead peroxide (PbO_2), sponge lead (Pb) and dilute sulphuric acid (H_2SO_4). Lead peroxide is the cathode; lead in soft sponge condition is the anode and dilute sulphuric acid in the ratio of water: acid of 3:1 is the electrolyte. Lead acid battery is made by dipping lead peroxide, sponge lead plates in dilute sulphuric acid. Relevant chemical reactions are the given as under. It may be mentioned here that hydrogen ions produced due to splitting of sulphuric acid molecules receive electrons at the lead peroxide cathode to produce hydrogen atoms.

$$H_2SO_4 \rightarrow H^+ + HSO_4^-$$

$$PbO_2 + 2\,H \rightarrow PbO + H_2O$$

$$PbO + H_2SO_4 \rightarrow PbSO_4 + H_2O$$

$$PbO_2 + H_2SO_4 + 2\,H \rightarrow PbSO_4 + 2\,H_2O$$

Sealed lead acid batteries also called maintenance free batteries use an electrolyte that is impregnated into a moistened separator instead of being in liquid form as is the case with flooded type lead acid batteries. It may be mentioned here that even sealed lead acid batteries too cannot be totally sealed. A valve is added to allow venting of gases during stressful charge and rapid discharge. Sealed lead acid batteries can be operated in any physical orientation without leakage. Fig. 4.47 shows photographs of a typical sealed lead acid battery.

Lead acid batteries are simple to manufacture. Typical values of important performance parameters are nominal voltage of 2.0 volts, self-discharge (3 – 4 %), specific energy (30 – 50 Wh/kg), specific power (180 – 200 W/kg), peak discharge current (5C), long charging time (8 – 16 hours) and cycle life (200 – 300 charge/discharge cycles). Main advantages of lead acid batteries include low cost per watt-hour, low self-discharge, high specific power, high peak discharge current capability and good performance in low and high temperature conditions. Disadvantages or more precisely the limitations include low specific energy, poor weight-to-energy ratio, and large charging time; need to store the battery in charged condition to prevent sulfation and limited cycle life with repeated deep cycling reducing battery life. In addition, flooded type batteries require watering and have transportation restrictions.

Nickel-Cadmium Battery

Nickel cadmium is one of the most rugged and high endurance batteries and is preferred in applications where long service life, high discharge current and extreme operating temperature conditions are a requirement. The battery chemistry allows ultrafast charging with minimal stress. Cadmium (Cd) is used as the anode material; nickel hydroxide [Ni $(OH)_2$] is used as the cathode and alkaline potassium hydroxide (KOH) is the electrolyte. Fig. 4.48 shows construction of a typical nickel-cadmium cell.

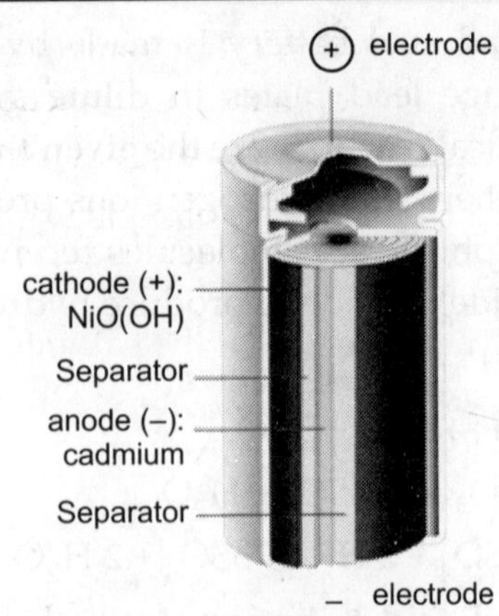

Fig. 4.48

The battery chemistry is explained by following chemical reactions.

First half reaction (oxidation) occurring at the anode is expressed as follows.

$Cd + 2\,OH^- \rightarrow Cd\,(OH)_2 + 2\,e^-$

The second half reaction (reduction) occurring at the cathode is given as follows.

$2\,NiO\,(OH) + 2\,H_2O + 2\,e^- \rightarrow 2\,Ni\,(OH)_2 + 2\,OH^-$

The net reaction during discharge is given as under.

$2\,NiO\,(OH) + Cd + 2\,H_2O \rightarrow 2\,Ni\,(OH)_2 + Cd\,(OH)_2$

Important performance parameters of nickel-cadmium batteries are nominal cell voltage of 1.2 volts, specific energy of 45 – 80 Wh/kg, specific power of 150 W/kg, cycle life of 1000 charge/discharge cycles, self-discharge of 20%, charging time of 1 – 2 hours, peak discharge current of 20C and operational temperature range of – 20°C to + 65°C. Major applications include their use in power tools, aircraft batteries, uninterrupted power supplies and medical devices. Due to toxicity of cadmium, they are being replaced by other battery chemistries.

Major advantages of nickel-cadmium batteries include fast and simple charging, high number of charge/discharge cycles, long shelf life and ability to be stored in discharged state, good low temperature performance, low cost per cycle and availability in a wide range of sizes and performance options. Disadvantages or limitations include low specific energy as compared to newer systems such as lithium-ion batteries, memory effect necessitating periodic full discharges; high self-discharge requiring recharging after storage and hazards due to toxicity of cadmium. Nickel cadmium batteries are also available in a wide variety shapes and sized including button cells, cylindrical cells and battery packs using more than one cell to produce higher terminal voltages. One such common battery pack that is designed to replace 9 volt battery pack uses six or seven nickel-cadmium cells in series to produce a terminal voltage of 7.2 and 8.4 volts respectively.

Nickel-Metal-Hydride Battery

A **Nickel-Metal-Hydride (NiMH) battery** is a practical replacement for Nickel-Cadmium (Ni-Cd) battery being more environmental friendly due to use of mild toxic metals. Though NiMH batteries are also susceptible to memory effect like nickel-cadmium batteries; the effect is less pronounced in NiMH batteries. Like a nickel-cadmium battery, it uses a cathode of Nickel-hydroxide. Unlike a nickel-cadmium battery that uses a cadmium anode; NiMH battery uses an anode of hydrogen absorbing alloys. Anode is made from a metal hydride, usually alloys of Lanthanum and rare earths that serve as a solid source of reduced hydrogen that can be oxidized to form protons. The electrolyte in both batteries is Potassium-hydroxide (KOH). It may also be mentioned here that the anode, the cathode and the electrolyte used in NiMH batteries are collectively more benign than the active chemicals used in rival Lithium batteries. The battery chemistry is described in the following chemical reactions.

At the negative electrode, the hydrogen is desorbed and combines with a hydroxyl ion to form water while also contributing an electron to the circuit. The first half reaction is given by:

$Alloy\,(H) + OH^- \leftrightarrow Alloy + H_2O + e^-$

At the positive electrode, nickel oxy-hydroxide is reduced to its lower valence state, nickel hydroxide. The second half reaction is given by:

$NiOOH + H_2O + e^- \leftrightarrow Ni\,(OH)_2 + OH^-$

Important performance parameters of nickel-metal-hydride batteries are nominal cell voltage of 1.2 volts, specific energy of 60 – 120 Wh/kg, specific power of 250 – 1000 W/kg, cycle life of 300 – 500 charge/discharge cycles, self-discharge of 30%, charging time of 2 – 4 hours, peak discharge current of 5C and operational temperature range of – 20°C to + 65°C.

Major advantages of nickel-metal-hydride batteries include 30 – 40 percent higher capacity than a standard nickel-cadmium battery, being less prone to memory effect, being environmentally friendly due to use of only mild toxins, simple storage and transportation not subject to any regulatory control and profitable recycling due to nickel content. The shortcomings include limited service life with deep discharge reducing it further, requirement of complex charging algorithm, memory effect though not as pronounced as it is in case of nickel-cadmium batteries, low tolerance to over-charge necessitating trickle charge to be kept low, heat generation during fast charge and high load discharge and a high self-discharge.

Nickel-metal-hydride batteries largely replaced nickel-cadmium batteries in low cost consumer applications, electronic gadgets such as cameras, mobile phones, pagers, camcorders, electric razors etc. and medical instruments and equipment. Other applications are use of high energy NiMH batteries in automotive industry, in uninterrupted power supplies and in telecommunications equipment.

Lithium-Ion Battery

Among all secondary batteries, **lithium-ion battery** has shown the greatest promise. Lithium-ion battery and its many variants are fast replacing nickel-based rechargeable batteries in many applications previously served by nickel-cadmium and nickel-metal-hydride batteries. Though lithium-ion batteries are more expensive than most other rechargeable batteries; its high cycle count and low maintenance more than compensate for its high initial cost.

Depending upon type of cathode material used, lithium-ion battery has many variants. Common cathode materials are Lithium Cobalt Oxide or Lithium Cobaltate ($LiCoO_2$), Lithium Manganese Oxide also known as spinel ($LiMn_2O_4$), Lithium Iron Phosphate ($LiFePO_4$), Lithium Nickel Manganese Cobalt Oxide ($LiNiMnCoO_2$), Lithium Nickel Cobalt Aluminum Oxide (LiNiCoAlO2) and Lithium Titanate ($Li_4Ti_5O_{12}$). Each cathode material gives the corresponding lithium-ion battery its unique performance characteristics.

Graphite is used as the anode material. Earlier, coke was used but it has now been replaced by graphite due to latter producing a flatter discharge characteristics. Carbon nano tubes are also being explored. Silicon is another emerging anode material for use in lithium ion batteries due to their potential for offering higher specific energy. It may be mentioned here that while six graphite (or carbon) atoms bind a single lithium ion; a single silicon atom can bind to four lithium ions. Lithium salts such as $LiPF_6$, $LiBF_4$, $LiClO_4$ in an organic solvent such as ether serves as the electrolyte. During the discharge process, lithium ions move from anode to cathode. Anode undergoes oxidation or loss of electrons while cathode sees a reduction or gain of electrons. During charging process, direction of movement of lithium ions is reversed. The following chemical reactions given here for a lithium cobalt oxide battery explain the chemistry.

The first half reaction at the anode is given by:

$LiC_6 \leftrightarrow 6C + Li^+ + e^-$

The second half reaction at the cathode is given by:

$Li_{0.5}CoO_2 + Li^+ + e^- \leftrightarrow LiCoO_2$

Overall reaction on a Li-ion cell is given by:

$C + LiCoO_2 \leftrightarrow LiC_6 + Li_{0.5}CoO_2$

Lithium cobalt oxide battery as outlined earlier uses cobalt oxide cathode and graphite carbon anode. Its key asset is its high specific energy, which makes it a preferred choice for use in cell phones, laptops and digital cameras. The shortcomings include relatively short life span, low thermal stability and limited discharge current capabilities. Li-cobalt cannot be charged and discharged at a current higher than its rating. Charge/discharge currents higher than recommended ratings shortens life. Major performance parameters are nominal voltage of 3.6 volts, specific energy of 150 – 200 Wh/kg, charge C-rate of 0.7 – 1C, discharge C-rate of 1C, cycle life of 500 – 1000 charge/discharge cycles (dependent on depth-of-discharge, load current and temperature) and thermal runaway at 150°C.

Lithium manganese oxide battery consists of manganese oxide cathode and graphite carbon anode. Low internal resistance, fast charge, high current discharge, high thermal stability and enhanced safety are the major advantages. The disadvantages include relatively lower specific energy as compared to lithium cobalt batteries and limited cycle life. Lithium manganese oxide battery is used in power tools, medical instrumentation and electric and hybrid vehicles. Major performance parameters are nominal voltage of 3.7 volts, specific energy of 100 – 150 Wh/kg, charge C-rate of 0.7 – 1C (3C maximum), discharge C-rate of 1C (10C maximum and 30C for 5 second pulse), cycle life of 300 – 700 charge/discharge cycles (dependent on depth-of-discharge and temperature) and thermal runaway at 250°C.

Lithium iron phosphate battery uses a phosphate cathode. The key advantages of $LiFePO_4$ batteries include high current rating and long cycle life, good thermal stability, enhanced safety and high tolerance to abuse. Disadvantages include low specific energy, high self-discharge and performance deterioration at low and elevated temperatures. Lithium phosphate batteries can be used to replace lead acid starter battery. A series connection of four lithium phosphate batteries produces the same full charge voltage of 14.4 volts as the series connection of six lead acid batteries. It may be mentioned here that full charge voltage of lithium phosphate batteries is 3.6 volts. Major performance parameters are nominal voltage of 3.2 volts, specific energy of 90 – 120 Wh/kg, charge C-rate of 1C, discharge C-rate of 1C (25C maximum), cycle life of 1000 – 2000 charge/discharge cycles (dependent on depth-of-discharge and temperature) and thermal runaway at 270°C.

Lithium nickel manganese cobalt oxide battery uses nickel-manganese-cobalt cathode. The cathode combination of typically one-third nickel, one-third manganese and one-third cobalt offers a unique blend that also lowers raw material cost due to reduced cobalt content. The three elements in other proportions have also been used. Key advantages of this battery result from combination of nickel and manganese. The combination of nickel and manganese enhances each other's strengths including high specific energy of nickel and low internal resistance offered by spinel structure of manganese. It is the preferred battery for power tools, e-bikes and electric power trains. Major performance parameters are nominal voltage of 3.6 volts, specific energy of 150 – 220 Wh/kg, charge C-rate of 0.7 – 1C, discharge C-rate of 1C (2C maximum), cycle life of 1000 – 2000 charge/discharge cycles (dependent on depth-of-discharge and temperature) and thermal runaway at 210°C.

Lithium Nickel Cobalt Aluminum Oxide battery is characterised by high specific energy and reasonably good specific power and a long life span. Major performance parameters are nominal voltage of 3.6 volts, specific energy of 200 – 260 Wh/kg, charge C-rate of 0.7C, discharge C-rate of 1C, cycle life of 500 charge/discharge cycles (dependent on depth-of-discharge and temperature) and thermal runaway at 150°C.

In the case of **lithium titanate batteries**, Li-titanate replaces the graphite in the anode of a typical lithium-ion battery and the material forms into a spinel structure. The cathode is graphite and resembles the architecture of a typical lithium-metal battery. The battery is characterised by excellent safety, low temperature performance and life span. Major performance parameters are nominal voltage of 2.4 volts, specific energy of 70 – 80 Wh/kg, charge C-rate of 1C (5C maximum), discharge C-rate of 10C (maximum), pulsed discharge rate of 30C (5 second pulse) and cycle life of 3000 – 7000 charge/discharge cycles. Thermal runaway is not specified here as lithium titanate battery is one of the safest lithium-ion batteries.

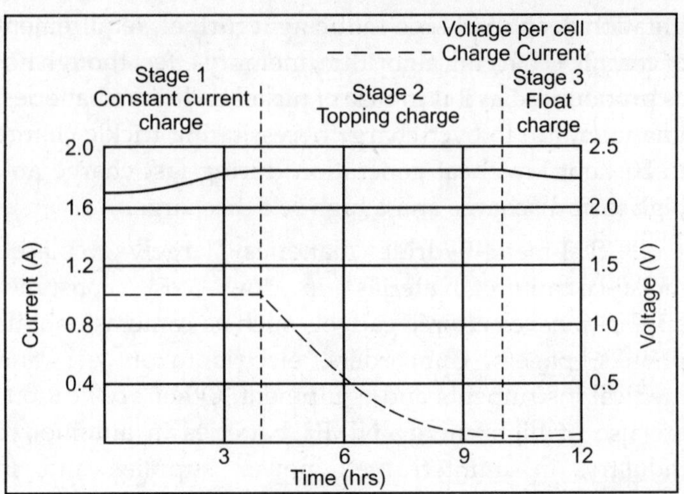

Fig. 4.49

Q.31. What is MEMORY EFFECT with reference to rechargeable batteries?

Ans. Memory effect in rechargeable batteries refers to a phenomenon where the battery remembers its usual discharge point before it is charged again. As a result, the battery refuses to run at its rated capacity. This effect occurs when rechargeable batteries are not fully discharged between charge cycles. The effect is observed in nickel cadmium and nickel–metal hydride rechargeable batteries. It causes them to hold less charge. As an example, if a nickel-cadmium or nickel-metal-hydride battery were always discharged to only 60 percent of its capacity before it is charged again; it will eventually not run below 60 percent mark. The source of the effect is changes in the characteristics of the underused active materials of the cell.

Q.32. What are the charging requirements of lead acid batteries?

Ans. Lead acid batteries use a voltage-based charge algorithm and are charged in three stages. The three stages of charging include (*a*) **constant-current charge** (*b*) **topping charge and** (*c*) **float charge**. Fig. 4.49 illustrates the three stages of the charge cycle of a lead acid battery. In the first stage of constant-current charge, the battery charges to about 70 percent in 5–8 hours. The remaining 30 percent is filled during the second stage of charging with the slower topping charge that lasts another 7–10 hours. The topping charge continues at a lower charge current and provides saturation. The float charge in the third stage maintains the battery at full charge by compensating the loss of charge due to self-discharge. The charge time of a sealed lead acid battery is typically 12 to 16 hours. It could be up to 36 to 48 hours for large stationary batteries. The charge voltage/cell selection is critical and varies from 2.30 volts to 2.45 volts. While it is desirable to have a fully charged battery to get maximum capacity and avoid sulfation on negative plate; over-saturated condition causes grid corrosion on the positive plate and induces gassing. During the second stage, the charging current begins to drop as the battery moves towards saturation. When the current reaches about three percent of rated current; the battery is considered to be fully charged. The float voltage must be reduced at full charge.

Q.33. What are the charging requirements of nickel-cadmium batteries?

Ans. While lead acid batteries can be charged using a voltage-based charge algorithm and an overcharge can be controlled by setting a maximum charge voltage; it is not so in the case of nickel-based batteries such as nickel cadmium and nickel metal hydride batteries as they don't have a float charge voltage. Charging in this case is based on forcing current through the battery. If nickel cadmium battery packs were to be charged in parallel; either separate charging circuit should be used for each string in the parallel pack or current balancing resistors should be used for current control. There are two common methods of determining end-of-charge. These are delta voltage and delta temperature methods. When the battery is fully charged; oxygen begins to be generated at the nickel electrode. The oxygen diffuses through the separator and reacts with cadmium electrode to form cadmium hydroxide. This lowers the cell voltage, which can be used to detect end-of-charge. The chemical reaction also generates heat, which can be measured by a thermistor to detect end-of-charge. In the case of temperature sensing, there is a likelihood of overcharge as the core of a cell may be several degrees warmer than the skin where the temperature is measured. A temperature cut-off at 50°C is generally recommended. Advanced chargers sense the rate of temperature rise (dT/dt) and charge termination occurs when the temperature rises 1°C per minute. If the battery cannot achieve the pace of temperature rise, an absolute temperature cut-off set to 60°C terminates the charge. Fig. 4.50 shows the charge cycle of a nickel-cadmium battery. Most rechargeable cells include a re-sealable safety vent to release excess pressure if incorrectly charged. The vent is usually designed to open at a pressure of 150–200 psi. Though pressure release through a re-sealable vent causes no damage; the battery should preferably be never stressed to the point of venting. Multiple venting may eventually lead to a dry-out condition due to leakage of some electrolyte each time vent opens.

The cheapest way to charge a nickel cadmium battery is to charge at C/10 rate for 16 hours. This method does not require an end-of-charge sensor and ensures a full charge. Cells can be charged at this rate no matter what the initial state of charge is. The minimum voltage required for a full charge varies with temperature and is at least 1.41 volts per cell at 20°C. A timer may be used to prevent overcharging to continue past 16 hours. Nickel-cadmium batteries can be charged faster at C/3 rate for five hours or C/5 rate for eight hours. In this case, battery should be fully discharged before charging. By using a temperature or voltage monitor to detect end-of-charge, the batteries can be charged up to 1C rate for 1.5 hours. In a standby mode, the battery may be kept topped up by using trickle charge at 0.05C to 0.06C rate.

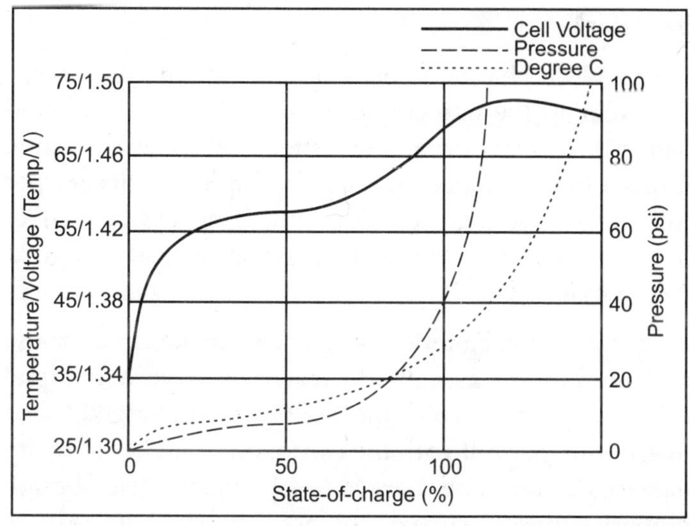

Fig. 4.50

Q.34. What are charging requirements of nickel metal hydride batteries?

Ans. The charge algorithm in the case of nickel metal hydride (NiMH) batteries is similar to the one described in the case of nickel-cadmium (Ni-Cd) batteries in the previous question. There are though a few issues that make charging of NiMH batteries more complex than it is for Ni-Cd batteries. The first and foremost constraint is that the negative delta voltage (NDV) used for full-charge detection is not as pronounced as it is in the case of Ni-Cd batteries. It is more so if the charging rate is less than 0.5 C. The NDV in the case of NiMH batteries is about 5 mV per cell, which puts stringent requirements on the charger. NiMH chargers include NDV, delta temperature (dT/dt), and temperature threshold and time-out timer as a part of full charge algorithm. All these features are actually ORed to detect full-charge condition. A number of NiMH chargers include a 30 minute topping charge of 0.1 C to add a bit of extra charge.

A common algorithm used in nickel-based batteries is the step-differential-charge in which there is an initial fast charge at 1C followed by charging at lower current. The charging current is further reduced with progression of charging process. Also, NiMH batteries don't like over-charge. Trickle charge in this case is restricted to 0.05C against 0.1C in case of Ni-Cd batteries. Due to differences in trickle charge and need to have more sensitive full-charge detection, Ni-Cd battery chargers are not suitable for NiMH batteries. A NiMH battery if charged from a Ni-Cd battery charger will overheat; though a Ni-Cd battery can be charged from a NiMH battery charger.

Q. 35. What are charging requirements of lithium-ion batteries?

Ans. The charge algorithm of a lithium-ion battery is similar to the one employed in the case of lead-acid batteries except for a few differences. These include lithium-ion batteries having a higher voltage per cell, tighter voltage tolerance implying intolerance of overcharge and the absence of trickle or float charge at full charge.

Lithium-ion batteries using cobalt, nickel, manganese and aluminium as cathode materials have a nominal cell voltage of 3.6 volts and are typically charged to 4.1 to 4.2 volts per cell with the exception of higher capacity batteries that may be charged to 4.3 volts per cell. Though higher voltage achieves higher capacity; it reduces service life. Lithium phosphate batteries have a nominal voltage of 3.2 volts per cell and are charged to 3.65 volts. Lithium titanate with a nominal cell voltage of 2.4 volts is charged to 2.85 volts. As a result of difference in voltage, 3.6 volt lithium batteries (lithium cobalt, lithium nickel, and lithium manganese and lithium aluminium) cannot be charged from lithium phosphate or lithium titanate battery chargers. The reverse is also not true as it would lead to overcharge. Chargers designed for multi-use are equipped with a provision for voltage adjustment.

Fig. 4.51 shows charge algorithm for a 3.6 volt lithium-ion battery. The charge process comprises of four stages in general. In the first stage, the charging occurs at constant current. The charge rate is in the range 0.5 C to 1C. In the second stage, which is the saturation stage, the voltage peaks and the current decreases.

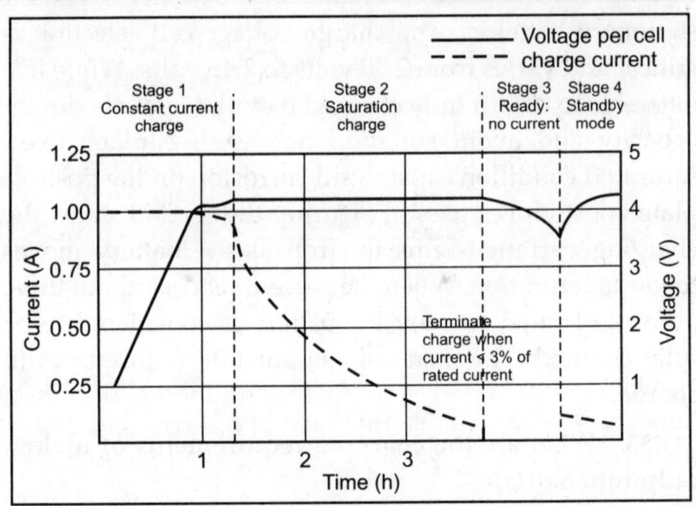

Fig. 4.51

The battery is considered to be fully charged when the battery reaches the voltage threshold and the current has reduced to three percent of rated current or when the current has levelled off and cannot reduce further. Though an increase in charge current during the constant current charging phase hastens the completion of first stage; the second stage takes longer. Also, lowering the voltage threshold or eliminating the saturation stage altogether increases battery life but it does so at the cost of reduced capacity or run time. In the third stage, charge is terminated and stage four is the standby mode for occasional topping charge.

Q. 36. Why do we need to connect batteries in series, parallel or series-parallel? What are the recommended procedures?

Ans. Individual cells or batteries are connected in series to increase the voltage rating. If (n) cells each having voltage rating of (V) and internal resistance (r) are connected in series; resultant voltage and internal resistance of the battery pack would be (nV) and (nr) respectively. Series connection of two batteries would double its operating voltage; the current rating remains unchanged. Some common examples include series connection of four nickel-cadmium cells to produce 4.8 volts, six lead-acid cells to produce 12 volts and four lithium-ion cells to produce 14.4 volts. Fig. 4.52 shows a series connection of two 6 volt batteries to get a battery pack of 12 volts with current rating equal to that of individual batteries. A higher voltage keeps the size of conductor small by reducing the required current for a given power. Series connection of batteries has the disadvantage that opening of one cell, which is likely in a long string of cells, can break the circuit and shorting of one cell lowers the overall voltage. Also, higher voltage battery packs require careful cell matching.

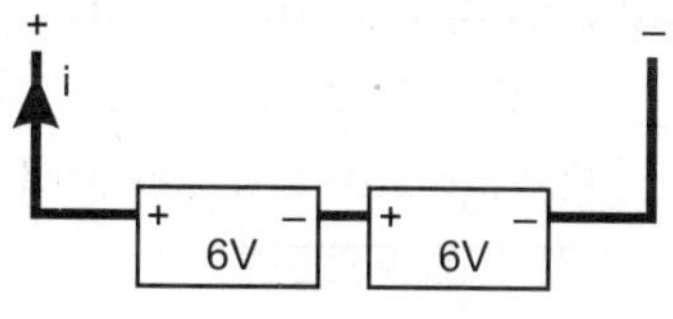

Fig. 4.52

On the other hand, parallel connection increases the current rating with operating voltage remaining unaltered. As an example, parallel connection of two 1.2 volt/1000 mA nickel-cadmium cells produce a 1.2 volt/2000 mA battery. If (*n*) cells each having voltage rating of (*V*) and internal resistance (*r*) are connected in parallel; resultant voltage and internal resistance of the battery pack would be (*V*) and (r/n) respectively. Fig. 4.53 shows parallel connection of two 12 volt batteries to get a 12 volt battery pack and a current rating twice the current rating of individual batteries. Parallel connection is used when single cells with desired current rating are not available or when larger cells don't fit the design constraint. An open cell in a parallel connection is less critical than it is in case of series connection though it reduces the load capability. A shorted cell is more serious here. Large battery packs using parallel connection generally include a fusing system to disconnect the faulty cell.

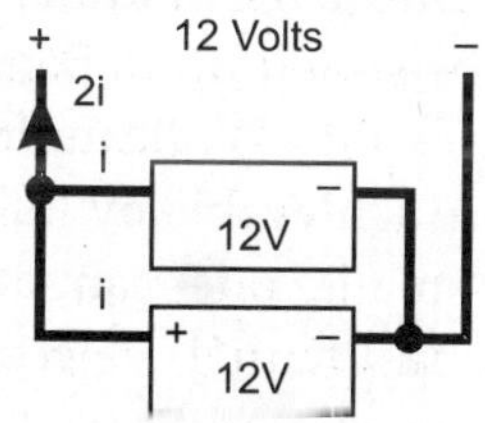

Fig. 4.53

Series-parallel connection of cells is used to achieve desired voltage and current ratings from a standard cell. Series-parallel connection lends itself well to lithium-ion batteries. Various series-parallel configurations such as 2S2P and 4S2P are available. Laptop batteries commonly use 4S2P configuration of lithium-ion cells. Fig. 4.54 shows 2S2P and 6S2P series-parallel configurations. Fig. 4.54 (*a*) shows 2S2P connection of 6 volt batteries producing a 12 volt battery pack with current rating twice that of individual battery. Fig. 4.54 (*b*) shows 6S2P connection of 2 volt lead acid cells again producing a 12 volt battery pack with current rating twice that of individual battery. In the case of nSmP battery pack, resultant voltage and internal resistance of battery pack would be (nV) and (nr/m) respectively.

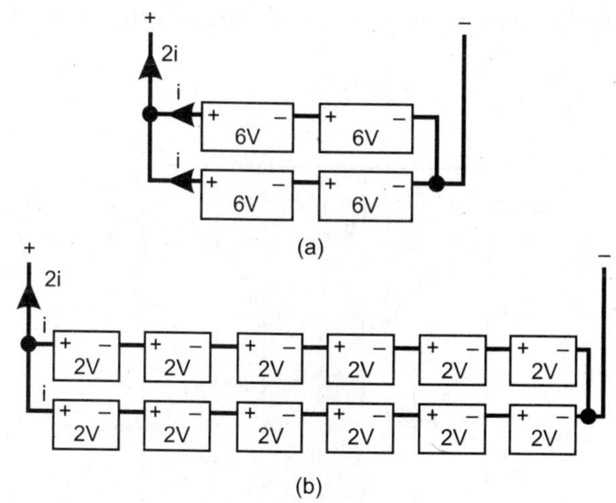

Fig. 4.54

Q.37. What is smart battery?

Ans. A smart battery is the one in which there is some level of communication occurring between the battery, the equipment and the user. Most basic smart battery may contain nothing more than a chip that sets the charger to the correct charge algorithm. Providing state-of-charge (SoC) indication and battery safety are the other key objectives of building intelligence into the battery pack. Battery packs used for medical and computing applications use smart batteries.

Q. 38. What is a fuel cell? Briefly describe the operational principle of a fuel cell.

Ans. A fuel cell operates on the principle of conversion of chemical energy of a fuel into electrical energy by a chemical reaction. The fuel used here is hydrogen and the chemical reaction takes place between the positively charged hydrogen ions and oxygen or another oxidising agent. Basic fuel cell comprises of two electrodes namely anode and cathode, electrolyte and sources of supply of fuel i.e. hydrogen and oxygen. Fuel cells are classified on the basis of electrolyte used and the associated start-up time. There are proton exchange membrane fuel cells (PEMFC) with a start-up time of typically one second and solid oxide fuel cells (SOFC) with a star-up time of 10 minutes. The electrodes have catalysts that enable chemical reactions at the electrodes. The basic fuel cell operates as follows. An oxidation reaction at the anode splits hydrogen atoms in the hydrogen fuel into positively charged hydrogen ions (or protons) and electrons. The hydrogen ions move through the electrolyte to the other electrode of the fuel cell where they combine with oxygen and electrons in a chemical reaction at the cathode to form water. The electrons move from anode to cathode through an external circuit generating electrical energy.

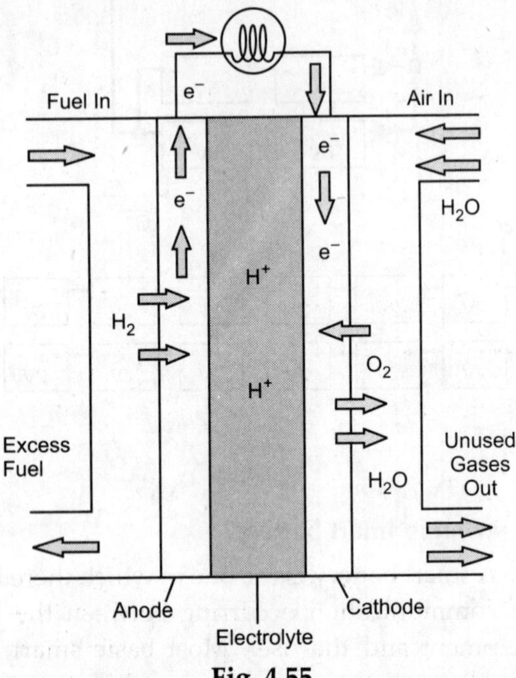

Fig. 4.55

Fig. 4.55 illustrates the basic operation of a fuel cell. Individual fuel cell produces an electric potential of typically 0.7 volt, which necessitates stacking of a large number of cells in series to generate the desired voltage for the intended application requirement. Fuel cells in addition to generating electrical energy also produce water, gaseous emissions such as nitrogen dioxide and heat as by products. Energy conversion efficiency is typically in the range 40 to 60 percent. Efficiency approaching 85 percent is achievable if waste heat energy were captured and used.

Q.39. How are fuel cells different from batteries?

Ans. Both batteries and fuel cells convert chemical energy into electrical energy. In the case of batteries, chemicals contained in the battery react to generate an electromotive force. Once the battery is discharged or used up; it must be thrown away in the case of a primary battery or recharged in the case of secondary batteries. Fuel cells on the other hand require a continuous supply of fuel and oxidiser stored externally to sustain chemical reactions occurring at the two electrodes. Electrical energy is available from the fuel cell as long as fuel and oxidiser inputs are supplied.

Q. 40. What are the key advantages and disadvantages of fuel cells as an energy source?

Ans. Key advantages of fuel cells as an energy source are the following.

1. Fuel cells offer higher conversion efficiency as compared to internal combustion engines, more so when the waste heat is captured and utilized in cogeneration application.
2. Fuel cells offer a higher power density, which allows having compact source of electrical power for a given application.
3. Most fuel cells, due to their nature of operation, are extremely silent in operation. This feature is particularly desirable in applications where noise pollution is not acceptable or desirable.
4. Fuel cells are responsive to changing electrical loads.
5. Fuel cells are far less polluting than other incumbent energy sources. Fuel cells using pure hydrogen have practically zero emission.
6. Scalability is another major benefit offered by fuel cells. Due to the modular nature of fuel cells, they can be scaled up depending on the power needs of a facility. While multi-megawatt output powers are achievable by linking together large fuel cells; smaller ones can satisfy specific power needs at residential, telecommunications, or small commercial facilities.

The key disadvantages include the following.

1. Fuel cells are relatively more expensive.
2. Fuel cell operation requires continuous fuel supply. Fuelling cells is still an issue as production, transportation, distribution and storage of hydrogen is difficult.

Solved Problems

Problem 4.1. *Calculate the characteristic impedance of the co-axial cable shown in Fig. 4.56.*

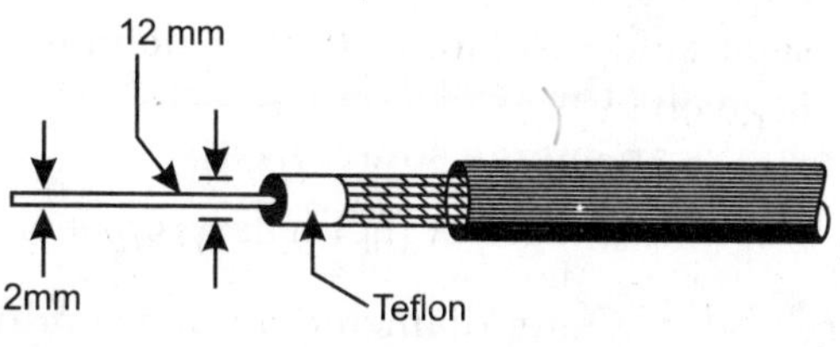

Fig. 4.56

Solution. From Fig. 4.56,

Outer diameter of the inner conductor = 2 mm

Inner diameter of the outer conductor = 12 mm

Permittivity of teflon = 2

The characteristic impedance of the cable is therefore given by:

$$= (138/\sqrt{2}) \log_{10} 6 = 75 \text{ ohms}$$

Problem 4.2. *A co-axial cable has an inductance per unit length of 10 nH per meter and a capacitance per unit length of 4 pF per meter. What should be the characteristic impedance of the cable?*

Solution. Induction per unit length = 10 nH

Capacitance per unit length = 4 pF

Characteristics impedance = $\sqrt{10000/4}$ = 50 ohms.

Problem 4.3. *Four 1.2 volt/1000 mAh nickel-cadmium cells are connected in series. Determine the resultant voltage and capacity of the battery pack. If the individual cells were connected in parallel, what would be resultant voltage and capacity?*

Solution: Series connection:

Resultant voltage = 1.2 × 4 = 4.8 volts

Resultant capacity = 1000 mAh

Parallel connection:

Resultant voltage = 1.2 volts

Resultant capacity = 1000 × 4 = 4000 mAh

Problem 4.4. *Eight 3.6 volt/2000 mA lithium cobalt cells are wired in 4S2P configuration. What is the resultant voltage and the load current capability of the battery pack?*

Solution: In a 4S2P battery configuration, there are two parallel connected strings of cells, each string having four cells connected in series.

Resultant voltage of each string = 3.6 × 4 = 14.4 volts

Resultant voltage of battery pack = 14.4 volts as parallel connection doesn't alter the voltage rating. Load current capability of each string = 2000 mA as series connection doesn't alter the current rating.

Load current capability of battery pack = 2000 × 2 = 4000 mA

Problem 4.5. *What are the charging voltage requirements of?*

(*a*) *4S2P battery configuration of 3.6 volt/1000 mA lithium-ion cells with recommended charging voltage of 4.2 volts per cell.*

(*b*) *6S4P battery configuration of 2 volt/2000 mA lead-acid cells with recommended charging voltage of 2.35 volts per cell*

Solution:

(*a*) Resultant voltage of battery pack = 3.6 × 4 = 14.4 volts

Charge voltage per cell = 4.2 volts

Charging voltage for the battery pack = 4.2 × 4 = 18.8 volts

(*b*) Resultant voltage of battery pack = 2 × 6 = 12 volts

Charging voltage = 2.35 × 6 = 14.1 volts

Problem 4.6. *If the individual cells of problem 4.4 had internal resistance of 1.0 ohm; what would be the internal resistance of the battery pack?*

Solution: If (*r*) is the internal resistance of individual cells; internal resistance of 4S2P battery pack = 4r/2 = 2r = 2.0 ohms

Problem 4.7. *The discharge rate of a 12 volt/3.5 Ah battery is specified as 0.5C. What is the discharge current in amperes?*

Solution: Discharge rate = 0.5 C

Battery capacity = 3.5 Ah

At 1C rate, discharge current = 3.5 Amperes

Discharge current at 0.5 C = 1.75 Amperes

Summary of Important Formulas

1. The characteristic impedance of a lossless air insulated co-axial cable is given by:

$$Z_0 = 138 \log_{10}\left(\frac{D}{d}\right)$$

where d = Outer diameter of inner conductor
D = Inner diameter of outer conductor

2. The characteristic impedance of a lossless dielectric insulated co-axial cable having a dielectric strength of (ε) is given by:

$$Z_0 = \frac{138}{\sqrt{\varepsilon}} \log_{10}\left(\frac{D}{d}\right)$$

where d = Outer diameter of inner conductor
D = Inner diameter outer conductor

3. The characteristic impedance of a lossless co-axial cable with inner and outer conductors separated insulating discs at definite intervals is given by:

$$Z_0 = 138 \log_{10} \frac{(D/d)}{\sqrt{1+(\varepsilon-1)\,W/S}}$$

where d = Outer diameter of inner conductor
D = Inner diameter of outer conductor
ε = Dielectric strength of insulating discs
W = Thickness of the insulating disc.
S = Center-to-center spacing between adjacent discs

Self Evaluation Exercise

I. Do as directed:

(*a*) It is low cost co-axial RF connector that can be satisfactorily used upto a frequency of 500 MHz and has a peak voltage rating of about 500 volts. It has an impedance that is not constant. Which connector is being referred to?

(UHF, BNC, N TNC)

(*b*) It is most extensively used co-axial RF connector. It is rated at 500 V (RMS) and can be terminated with cables like RG 55/U, 58/U, 141 A/U, 223/U etc. It is .... connector.

(BNC, N, SMA, UHF)

(*c*) It is high frequency connector (0 to 20 GHz), has a small size and threaded coupling. It is .... connector.

(SMA, N, TNC, BNC)

(*d*) A coupling technique used on circular connectors where pins on one connector shell engage in spiralled grooves in the mating connector shell.

(Bayonet coupling, Threaded coupling)

II. Identify True and False Statements:

(*a*) In a coil actuated relay, the operate time is a function of coil inductance, coil resistance, effective source voltage and steady state current.

(*b*) The operate time usually increases further in case of voltage overdrive or current overdrive.

(*c*) Locking coil is found in a type of instrument relay.

(*d*) Operate and release times increase manifold if you are using R-C networks for protecting relay contacts in case of inductive surges.

(*e*) Coil actuated relay usually has a diode connected across the coil. This is done to protect the drive circuit from inductive transient surges.

(*f*) The operate and release times in case of general purpose coil actuated relays are of the order of hundreds of milliseconds.

(*g*) The operate and release time in case of general purpose coil actuated relays are of the order of tens of milliseconds.

(*h*) A relay in which contacts are mechanically or magnetically held in the last relay energised position after the energisation is removed is removed is called a latching relay.

(*i*) When a relay is feeding an incandescent lamp, its contacts must be rated for a current that is significantly higher than the continuous load requirement.

(*j*) A relay in which the contacts operate after a preset time delay after the relay is energised is called a clock work timer.

III. Complete the following statements with appropriate words

1. Characteristics of a fuse are expressed in terms of its current rating, voltage rating and its ....
2. The preferred fuse type for overload protection of semiconductor devices is the ....

3. We have the slow blow fuses, normal blow fuses and the .... fuses.
4. A high frequency (0 to 20 GHz), small size, thread coupled co-axial connector is ....
5. A coupling technique used with circular connectors where pins on one connector shell engage in spiraled grooves in the mating connector shell is ....

Multichoice Questions

1. A12 V/4AH battery is discharged at 1C rate. The discharge current is
   (*a*) 4.0 A
   (*b*) 2.0 A
   (*c*) 1.0 A
   (*d*) Indeterminate from given data
2. The internal resistance of parallel connection of six 2 V/1Ah cells each having internal resistance of 0.6 ohm is
   (*a*) 0.6 ohm
   (*b*) 0.1 ohm
   (*c*) 3.6 ohms
   (*d*) None of these
3. The load current capability of 4S2P connection of 3.6 V/1000 mA lithium-ion cells would be
   (*a*) 1000 mA
   (*b*) 6000 mA
   (*c*) 2000 mA
   (*d*) None of these
4. The internal resistance of a battery comprising of 6S3P configuration of 2V/1000 mA/0.5 ohm lead acid cells would be
   (*a*) 0.5 ohm
   (*b*) 3 ohms
   (*c*) 1.0 ohm
   (*d*) None of these
5. The present capacity of a battery expressed as a percentage of maximum capacity is called
   (*a*) Depth-of-discharge
   (*b*) State-of-charge
   (*c*) Self-discharge rate
   (*d*) None of these
6. Deep discharge is considered to be equal to depth-of- discharge of
   (*a*) 50%
   (*b*) 25%
   (*c*) $\geq$ 80%
   (*d*) None of these
7. A battery has E-rate of 0.5 E. If the battery is specified as 12 V/1.5 Ah, discharge power is
   (*a*) 9.0 watts
   (*b*) 18 watts
   (*c*) 4.5 watts
   (*d*) None of these
8. One of the following is type of primary battery.
   (*a*) Alkaline battery
   (*b*) lithium-ion battery
   (*c*) Lead-acid battery
   (*d*) Nickel-cadmium battery
9. One of the following is type of secondary battery
   (*a*) Lithium manganese dioxide battery
   (*b*) Carbon-zinc battery
   (*c*) Alkaline battery
   (*d*) Nickel-metal-hydride battery
10. Memory effect is associated with one of the following battery types.
    (*a*) Lithium-ion batteries
    (*b*) Nickel-cadmium batteries
    (*c*) Carbon-zinc batteries
    (*d*) Lithium primary batteries
11. As compared to batteries, fuel cells
    (*a*) Produce higher voltage per cell
    (*b*) Offer lower power density
    (*c*) Are relatively quieter and less polluting
    (*d*) Relatively cheaper
12. Lithium batteries are classified on the basis of
    (*a*) Anode materials used
    (*b*) Cathode materials used
    (*c*) Electrolyte used
    (*d*) None of these

Answers to Self Evaluation Exercise

I.

(*a*) UHF connector (*b*) BNC connector
(*c*) SMA connector (*d*) Bayonet coupling

II.

(*a*) True (*b*) False (*c*) True (*d*) True
(*e*) True (*f*) False (*g*) False (*h*) True
(*i*) True (*j*) False

III.

1. I^2t or 'Let through current' rating
2. Fast acting fuses
3. Quick-blow
4. SMA-connector
5. Bayonet coupling

Multichoice Questions

1. (*a*)	2. (*b*)	3. (*c*)	4. (*c*)	5. (*b*)	6. (*c*)
7. (*a*)	8. (*a*)	9. (*d*)	10. (*b*)	11. (*c*)	12. (*b*)

Chapter 5

Network Fundamentals

Network Graphs – Mesh Current Analysis – Node Voltage Analysis – RLC Transients – Voltage and Current Sources – Coupled Circuits

Learning Objectives

After completing the chapter, you will learn the following topics:

1. Network Graphs
2. Matrices Associated with Network Graphs
3. Kirchhoff's Laws
4. Mesh Current Analysis
5. Node Voltage Analysis
6. Voltage and Current Sources
7. Source Transformations
8. RLC Transients
9. Coupled Circuits

Q.1. What do you understand by Linear Graph with reference to a network?

Ans. A **Linear Graph** of a network comprises of a collection of nodes or vertices (points) and branches (line segments), with the different nodes connected together by branches. It is drawn by representing the voltage and current sources by their internal impedances (note that the internal impedance of an ideal voltage source is zero, hence it is denoted as a short circuit and that of an ideal current source is infinite, hence it is denoted as an open circuit) and keeping all the nodes and branches of the network. Figure 5.1(*b*) shows the graph corresponding to the network of Figure 5.1(*a*).

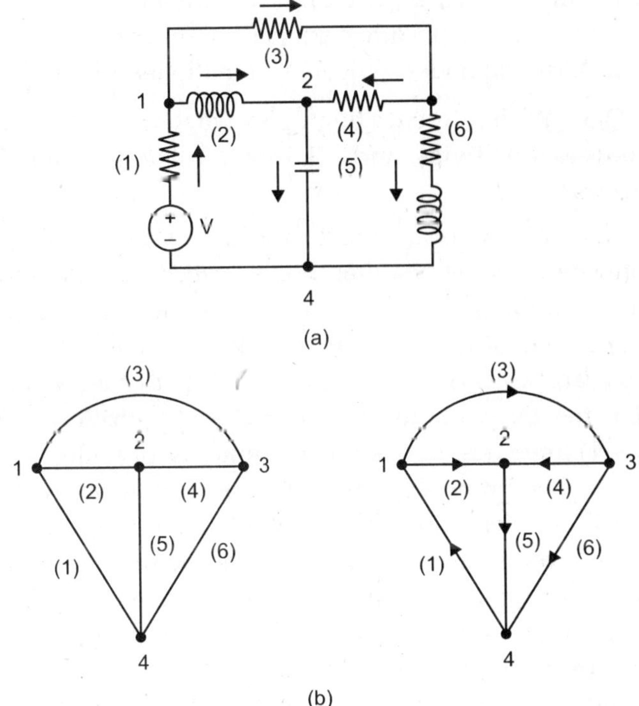

Fig. 5.1

Q.2. What are *directed* and *undirected* graphs? What is a *connected* graph?

Ans. A graph whose branches carry an arrow to indicate their orientation are referred to as *Directed* or *Oriented* graphs, else they are referred to as *Undirected* graphs. A *Connected* graph is one in which there is a continuous path through all the branches (any of which may be traversed more than once) which touches all the nodes.

Q.3. What do you understand by rank of an undirected graph?

Ans. In graph theory, there are two unrelated definitions of rank of an undirected graph. According to the *Matrix Theory,* The rank of a graph is defined as the rank of its adjacency matrix. According to the *Matroid Theory,* the rank of an undirected graph is $(n - c)$ where (n) is the number of nodes and (c) is the number of connected components of the graph. Equivalently, the rank of a graph is the rank of the oriented incidence matrix associated with the graph.

Q.4. What is a sub-graph?

Ans. A *sub-graph* is a subset of the branches and nodes of a graph. A sub-graph is said to be proper if it consists of strictly less than all the branches and nodes of the graph. A path is a particular sub-graph having the following properties: 1. One branch of the sub-graph is incident at its two terminal nodes. 2. At the other nodes, referred to as the internal nodes, two branches are incident. 3. Also, there should be no other proper sub-graph having the same terminal nodes with above mentioned properties.

Q.5. With reference to graphs, what are (*a*) Treesand Co-trees (*c*) Twigs and Chords (*d*) Forests and Co-forests?

Ans. A *tree* is a set of branches with each node connected to every other node, directly or indirectly such that the removal of any single branch destroys this property. In other words, a tree of the graph is a subset of the branches such that all graph nodes are connected by branches but without forming a closed path. A tree has $(n - 1)$ branches, where n is the number of nodes of the graph. The branches of the tree are referred to as *twigs* and those branches which are not the part of the tree are referred to as links or *chords*. All the links combined together form a *co-tree* and it represents the complement of the tree. Figure. 5.2 shows the possible trees, its twigs and corresponding co-tree for the graph shown in Figure. 5.1 (*a*). If the graph is not-connected, then a set of trees, one for each separate part is referred to as a *forest*. The complement of a forest is a *co-forest*.

Tree	Twigs of Tree	Links of Cotree
(2), (5), (6)	{2, 5, 6}	{1, 3, 4}
(2), (4), (5)	{2, 4, 5}	{1, 3, 6}
(4), (1), (6)	{1, 4, 6}	{2, 3, 5}

Fig. 5.2

Q.6. What do understand by *Complete Incidence Matrix* and *Reduced Incidence Matrix* with reference to graphs?

Ans. For a graph with n nodes and b branches, the complete incidence matrix A_a is a rectangular matrix of the order $n \times b$, with number of rows equal to the number of nodes in the graph and number of columns equal to the number of branches of the graph and has a rank of $(n - 1)$.

The element a_{mk} of the matrix is given by:

$$a_{mk} = 0, \text{ if branch } k \text{ is not associated with node } m$$
$$= 1, \text{ if branch } k \text{ is associated with node } m \text{ and oriented away from node } m$$
$$= -1, \text{ if branch } k \text{ is associated with node } m \text{ and oriented towards node } m$$

The *complete incidence matrix* for the network graph shown of Figure 5.1 (b) is given in Table 5.1. Each row in the complete incidence matrix corresponds to the node of the graph and has non-zero entries in all those columns of the braches which are associated with that node and zero entries in all other columns. The sum of the entries in any column of the complete incidence matrix is zero and its determinant for a closed loop is zero. The entries highlighted in grey are the elements of the incidence matrix A_a.

Two different graphs having the same number of nodes and branches and one-to-one correspondence between their nodes as well their branches have the same complete incidence matrix. Such graphs are referred to as isomorphic graphs.

Table 5.1

Nodes ↓	*Branches* →					
	1	2	3	4	5	6
1	−1	1	1	0	0	0
2	0	−1	0	−1	1	0
3	0	0	−1	1	0	1
4	1	0	0	0	−1	−1

$$A = \begin{bmatrix} -1 & 1 & 1 & 0 & 0 & 0 \\ 0 & -1 & 0 & -1 & 1 & 0 \\ 0 & 0 & -1 & 1 & 0 & 1 \\ 1 & 0 & 0 & 0 & -1 & -1 \end{bmatrix}$$

It may be noted here, that all the rows are not independent and at least one row can be obtained by the algebraic manipulation of other rows. The matrix obtained from the complete incidence matrix by eliminating one of the rows is referred to as the *incidence matrix* or the *reduced incidence matrix* (denoted by A). For the incidence matrix given in Table-5.1, the reduced incidence matrix formed by eliminating the last row is given by the following matrix.

$$A = \begin{bmatrix} -1 & 1 & 1 & 0 & 0 & 0 \\ 0 & -1 & 0 & -1 & 1 & 0 \\ 0 & 0 & -1 & 1 & 0 & 1 \end{bmatrix}$$

Number of trees of a graph can be computed from its incidence matrix and is given by $T = \det\{[A][A]^T$ where, $[A]$ and $[A]^T$ are the reduced incidence matrix and its transpose respectively.

Q.7. What is an Adjacency Matrix? Give two examples to illustrate your answer.

Ans. The *Adjacency Matrix*, also called the *Connection Matrix*, of a simple labeled graph is a matrix with rows and columns labeled by graph vertices or nodes, with a 1 or 0 in position (v_i, v_j) according to whether v_i and v_j are adjacent or not.

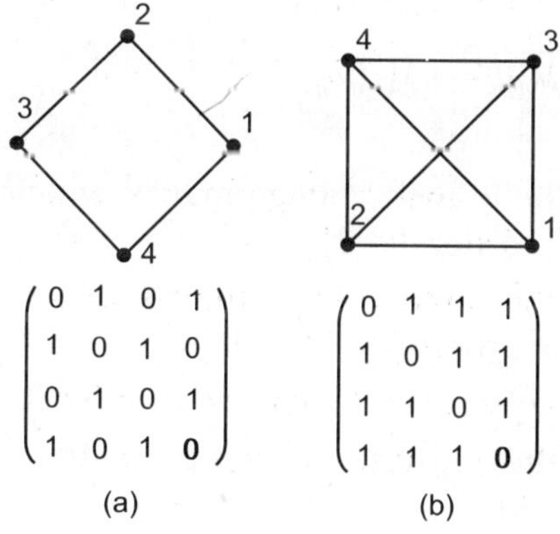

Fig. 5.3

Following deductions can be made: 1. For a simple graph with no self-loops; adjacency matrix must have 0s on the diagonal and 2. For an undirected graph, the adjacency matrix is symmetric. As an illustration, Figures 5.3 (*a*) and (*b*) show two network graphs and corresponding adjacency matrices.

Q.8. What are Cut Sets? What is a Cut Set Matrix?

Ans. A *cut-set* is the minimum set of branches of a connected graph provided that the removal of these branches cuts the graphs into two parts. In order words, cut-set is the minimum set of branches such that the removal of these branches reduces the rank of the graph by one.

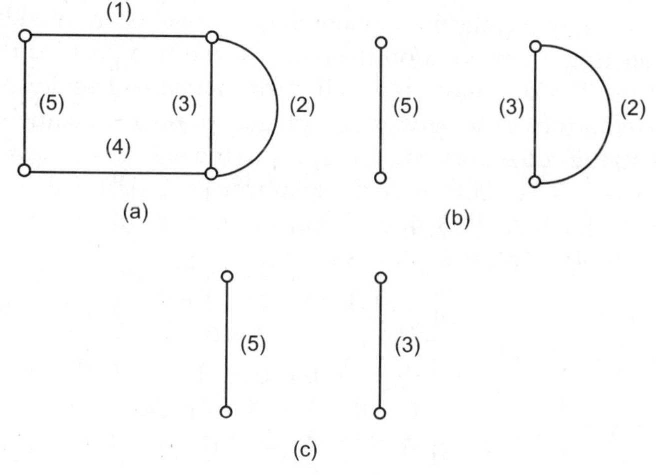

Fig. 5.4

As an example, let us consider a network graph shown in Figure 5.4 (*a*). As is evident from the figure; the rank of the graph is 3. Removal of braches 1 and 4 breaks the graph into two connected sub-graphs as shown in Figure 5.4 (*b*). Therefore, [1, 4] may be a cut-set. Also, removing the branches 1, 2 and 4 reduces the graph into two connected sub-graphs as shown in Figure 5.4 (*c*). Therefore [1, 2, 4] may be a cut-set. Since, [1, 4] is a subset of [1, 2, 4], therefore [1, 4] is a proper cut-set and not [1, 2, 4] as it contains the minimum set of edges.

Each branch of a cut-set has one of its terminals incident at a node in one group and its other end at a node in the other group. The *cut-set matrix* has its rows corresponding to the cut-sets and the columns are the branches of the graph and is given by $Q_a = [q_{ij}]$,

Where,

q_{ij} = 1, if branch (j) is in the cut-set (i) and the orientations coincide

q_{ij} = −1, if branch (j) is in the cut-set (i) and the orientations do not coincide

q_{ij} = 0, if the branch (j) is not in the cut-set (i)

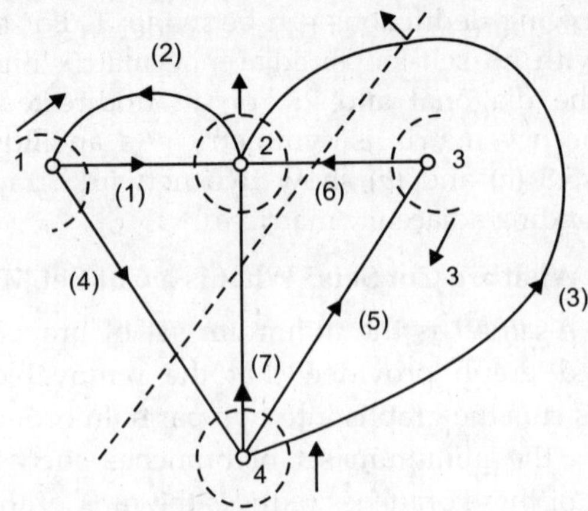

Fig. 5.5

It may be mentioned here that a cut-set is oriented by selecting an orientation from one of the two parts to the other. The orientations of a branch in a cut-set as per the orientation of the network graph may or may not coincide with the cut-set orientation. Refer to the network graph of Figure 5.5. The cut-sets of the graph are [1, 2, 4], [1, 2, 3, 6, 7], [5, 6], [3, 4, 5, 7], [3, 4, 6, 7] and so on. The cut-set matrix Q_a can be written as follows.

$$Q_a = \begin{bmatrix} 1 & -1 & 0 & 1 & 0 & 0 & 0 \\ -1 & 1 & -1 & 0 & 0 & -1 & -1 \\ 0 & 0 & 0 & 0 & -1 & 1 & 0 \\ 0 & 0 & 1 & -1 & 1 & 0 & 1 \\ 0 & 0 & 1 & -1 & 0 & 1 & 1 \end{bmatrix}$$

Those graphs whose complete incidence matrices are contained within the cut-set matrix are referred to as non-separable.

Q.9. What do you understand by a fundamental cut set and a fundamental cut set matrix?

Ans. For a network graph, select a tree and focus on the twig b_k of the tree that disconnects the tree into two pieces. All the links which go from one part of this disconnected tree to the other along with the twig b_k will constitute a cut-set. This is referred to as *fundamental cut-set* or *f-cut set*. Therefore, for each branch of the tree, there will be an *f*-cut set. The direction of an *f*-cut set is chosen to be that of the twig defining the cut set. For a network graph with (n) nodes, there are (n – 1) twigs in a tree and therefore there are (n – 1) number of *f*-cut-sets. The fundamental or the *f*-cut-set matrix is defined as a matrix whose each row represents an *f*-cut-set with respect to a given tree of the graph. The rank of the matrix Q is (n – 1). The *f*-cut-set $Q = [q_{ij}]$ has one row for each *f*-cut-set of the graph and one column for each edge, where

q_{ij} = 1, if edge (j) is in *f*-cut-set (i) and the direction of edge (j) coincides with that of the *f*-cut-set (i)

= –1, if edge (j) is in *f*-cut-set (i) and the direction of edge (j) does not coincide with that of the *f*-Cut set (i)

= 0, if edge (j) is not in *f*-cut-set (i)

For the network graph shown in Figure 5.5, consider a tree shown in Figure 5.6 (a). The *f*-cut sets are shown in Figure 5.6 (b).

The *f*-cut-set matrix (Q) is as follows.

$$Q = \begin{bmatrix} 1 & -1 & 0 & 1 & 0 & 0 & 0 \\ 0 & 0 & 0 & 0 & 1 & -1 & 0 \\ 0 & 0 & 1 & -1 & 0 & 1 & 1 \end{bmatrix}$$

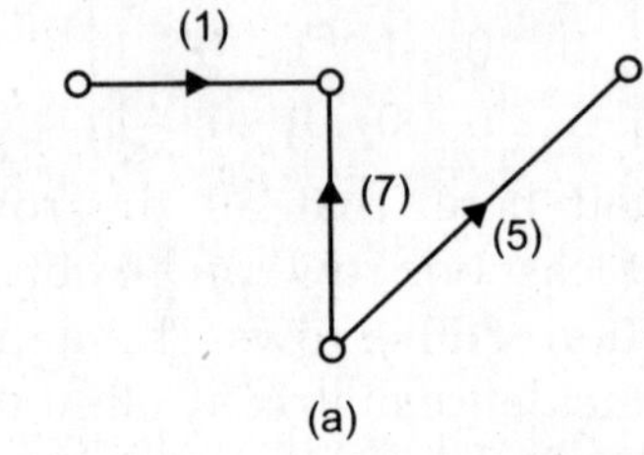

(a)

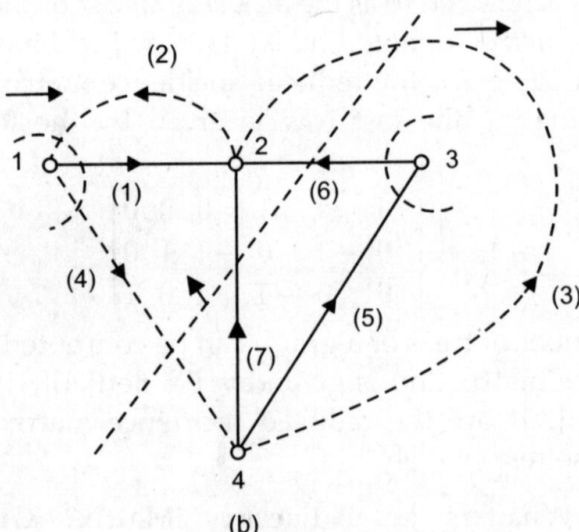

(b)

Fig. 5.6

The rows correspond to the *f*-cut-sets 1, 5 and 7 and the columns correspond to the branches 1, 2, 3, 4, 5, 6 and 7. The branch voltage matrix $[V_b]$ is expressed in terms of node voltage matrix $[V_n]$ as follows.

$$[V_b] = [Q]^T [V_n]$$

$$[V_b] = [A]^T [V_n]$$

Where,

$[V_b]$ is the branch voltage matrix which is a column matrix of order ($b \times 1$)

$[V_n]$ is the node voltage matrix which is a column matrix of order $[(n-1) \times 1]$

$[Q]$ is the *f*-cut-set matrix of the order $[(n-1) \times b]$

[A] is the reduced incidence matrix of the order $[(n-1) \times b]$.

Q.10. What is circuit or tie-set matrix? What is a fundamental tie-set matrix?

Ans. The circuit or the tie set matrix (represented by B_a) of a graph with (n) nodes and (b) branches is a rectangular matrix with (b) columns and as many rows as there are loops. The rank of the circuit matrix is $[b - (n - 1)]$. The element of mth row and k^{th} column is given by:

$$b_{mk} = 0, \text{ if branch k is not in loop } m$$
$$= +1, \text{ if branch } k \text{ is in loop } m \text{ and their orientations coincide}$$
$$= -1, \text{ if branch } k \text{ is in loop } m \text{ and their orientations do not coincide}$$

Tie-set is a set of branches contained in a loop such that each loop contains one link and the remainder is tree branches. *Fundamental tie-set* (f-tie set) is referred to as a group of branches contacting only one link and minimum number of twigs. Number of tie sets of a graph is equal to number of links or the number of KVL mesh equations. It may be mentioned here that the KVL equations can be obtained from the rows of fundamental tie-set matrix and the branch currents can be obtained from the columns of the f-tie set matrix. The fundamental tie-set matrix is formed by using the following procedure.

1. Draw the directed graph of a network and choose a tree.
2. Each link forms an independent loop. The direction of the loop is the same as that of the corresponding link.
3. Elements b_{ij} of the tie-set matrix are given by $b_{ij} = 1$, when the branch (b_j) is in loop (i) and is directed in the same direction as the loop current; equal to (– 1) when the branch (b_j) is in loop (i) and is directed in the opposite direction as the loop current and equal to 0 when the branch (b_j) is not in loop (i). The tie-set matrix is an $(b - n + 1) \times b$ matrix

The relation between the branch current matrix $[I_b]$ and the loop current matrix $[I_L]$ is given by $[I_b] = [B]^T[I_L]$.

Q.11. How are the Incidence Matrix, Fundamental Cut-set Matrix and Fundamental Tie-set matrices interrelated?

Ans. If A is the incidence matrix, B the fundamental circuit matrix and Q the fundamental cut-set matrix then

$$A = [A_t : A_l]$$

Where,

At is a square matrix of twigs of order $(n - 1) \times (n - 1)$ and rank $(n - 1)$

A_l is the matrix of links of order of $(n - 1) \times (b - n + 1)$

$$B = [B_t : B_l] = [B_t : U]$$

Where,

B_t is a matrix of twigs of order $(b - n + 1) \times (n - 1)$

B_l is the matrix of links of order of $(b - n + 1) \times (b - n + 1)$ and is an identity matrix

$$Q = [Q_t : Q_l] = [U : Q_l]$$

Where,

Q_t is identity matrix of twigs of order $(n - 1) \times (n - 1)$ and is an identity matrix

B_l is the matrix of links of order of $(n - 1) \times (b - n + 1)$

The inter-relationships between different matrices are expressed as under.

$$B_t = [A_t^{-1} A_l]^T$$
$$Q_l = A_t^{-1} A_l$$
$$BQ^T = QB^T = 0$$

Q.12. Write expressions for instantaneous voltage (v) instantaneous current (i) and instantaneous power (p) in case of PURE RESISTANCE (R), PURE INDUCTANCE (L) and PURE CAPACITANCE (C) as depicted in Fig. 5.7 (a), (b) and (c) respectively.

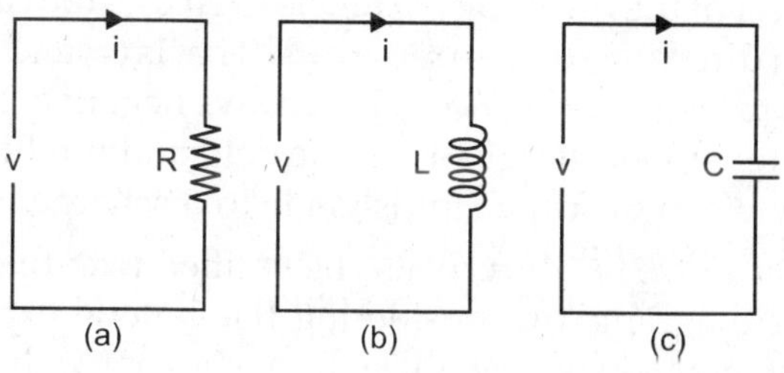

Fig. 5.7

Ans. In case of resistance, R

$$v = iR$$
$$i = v/R$$
$$p = vi = i^2R = \frac{v^2}{R}$$

In case of inductance, L

$$v = L\, di/dt$$
$$p = vi = iL\frac{di}{dt}; \quad i = \frac{1}{L}\int vdt + k_1$$

In case of capacitance, C

$$v = \frac{1}{C}\int idt + k_2; \; i = C\, dv/dt; \; vi = vC\, dv/dt$$

Q.13. In the light of the expressions derived in case of pure *R*, *L* and *C* for the instantaneous power (*p*), how do you justify the well known belief that whereas resistive elements consume power from the driving source, pure *L* and pure *C* are energy storage elements.

Ans. In case of a resistance, instantaneous power is given by $p = (i^2R)$ or $p = (v^2/R)$. In either case, it is proportional to square of a quantity (i or v). Therefore, (p) will be positive irrespective of whether (v) and (i) are respectively negative or positive. This indicates that resistances dissipate power. And actually they do dissipate power (equal in magnitude to i^2R) and this dissipated power or energy appears in the form of heat.

Let us take the case of inductance (L). The expression for instantaneous power (p) is ($p = iL\, di/dt$). In this expression, ($L\, di/dt$) is the voltage induced across inductance (L) when current (i) flows through it. This induced voltage will have a polarity so as to send a current in a direction opposite to that of (i). Therefore, product ($iL\, di/dt$) is always negative which in other words could be interpreted as saying that a pure inductance does not consume any power. It, infact, stores energy in the magnetic field and then returns it as the field is made to collapse.

Similarly, in case of a capacitance too, it is well known, when the voltage across the capacitor is building up, the current through the same is decaying. In other words, when a voltage (v) is impressed across a capacitance (C), the effect of the resulting current is to charge the capacitor (C) in a direction so as to oppose (v). That is (v) and (i) have opposite sense and hence (p) is always negative. Infact, a capacitor stores energy in electric field and it returns this energy when the capacitor is made to discharge.

The above results can also be verified with the help of phasor diagrams by considering the general expression for instantaneous power given by $p = (vi \cos \theta)$ where (θ) is phase angle between (v) and (i). In case of pure R, θ equals zero *i.e.* (v) is in phase with (i). In case of pure capacitance, (v) lags (i) by 90° whereas in case of a pure inductance, (v) leads (i) by 90°. The conclusions are obvious.

Q.14. State Kirchoff's Current Law (abbreviated as *KCL*) and explain its significance from the viewpoint of simplification of complex networks. Why should sum of all currents leaving a node or a junction always equal sum of all currents entering a node or a junction?

Ans. Kirchoff's Current Law (*KCL*) states that the algebraic sum of currents at a node (or a junction) is always zero. This implies that the sum of currents flowing towards a node (which may be considered to be positive) must equal sum of currents flowing away from it (which then have to be considered as negative). Infact, the basis of *KCL* is CHARGE CONSERVATION which means charge cannot accumulate at a junction. As current is nothing but rate of flow of charge ($i = dq/dt$), the total outgoing current must be equal to total incoming current. See Fig. 5.8 for illustration.

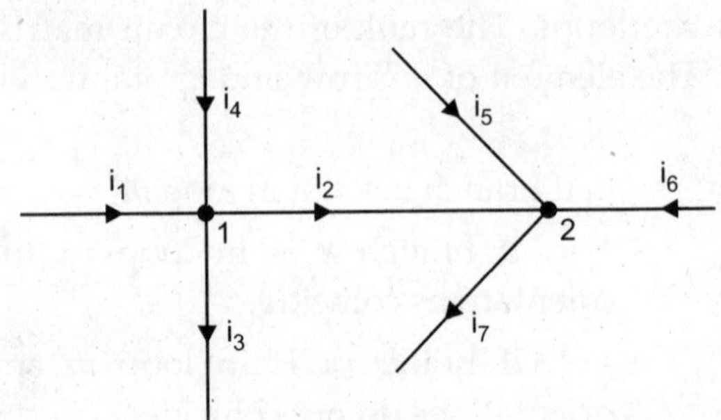

Fig. 5.8

At node : 1 $\quad i_1 - i_2 - i_3 + i_4 = 0$ or $(i_1 + i_4) = (i_2 + i_3)$

At node : 2 $\quad i_2 + i_5 + i_6 - i_7 = 0$ or $i_2 + i_5 + i_6 = i_7$

Kirchoff's current law finds frequent application in the simplification of networks, in particular, those haying more number of nodes or parallel branches. The simplest example of use of *KCL* is in deriving an expression for total equivalent resistance or inductance or capacitance when a number of resistances or inductances or capacitances have been connected in parallel.

Q.15. State Kirchoff's Voltage Law (abbreviated as *KVL*). Which type of network configurations is preferably simplified by applying *KVL*?

Ans. Kirchoff's Voltage Law (*KVL*) states that algebraic sum of voltages around any closed path (or loop) is zero. That is, sum of all potential drops must equal sum of all potential rises around any closed path. *KVL* as applied to Mesh 1 and 2 of the network of Fig. 5.9 gives following expressions:

For Mesh 1

$$V_1 - I_1R_1 + V_2 - I_2R_2 - I_1R_3 = 0 \;.....$$

Algebraic sum of voltages = 0

or $V_1 + V_2 = I_1 R_1 + I_1 R_3 + I_2 R_2 \;.....$

Potential rises = Potential drops

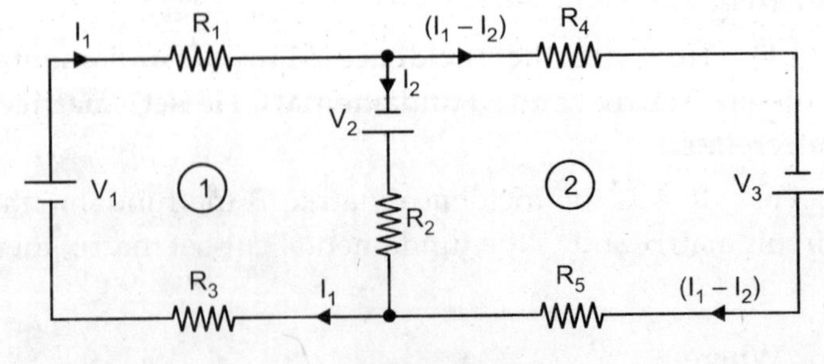

Fig. 5.9

For Mesh 2 :

$$-V_2-(I_1-I_2)R_4+V_3-(I_1-I_2)R_5+I_2R_2=0 \; .....$$

Algebraic sum of voltage = 0

or $V_3+I_2R_2=V_2+(I_1-I_2)R_4+(I_1-I_2)R_5$

or $V_3+I_2R_2=V_2+(I_1-I_2)(R_4+R_5)$ .....

Potential rises = Potential drops.

Q.16. Use Kirchoff's current law (KCL) to prove that

(i) $\frac{1}{R}=\frac{1}{R_1}+\frac{1}{R_2}+\frac{1}{R_3}$ **for Fig. 5.10 (a)**

(ii) $C=C_1+C_2+C_3$ **for Fig. 5.10 (b)**

(iii) $1/L=1/L_1+1/L_2+1/L_3$ **for Fig. 5.4 (c).**

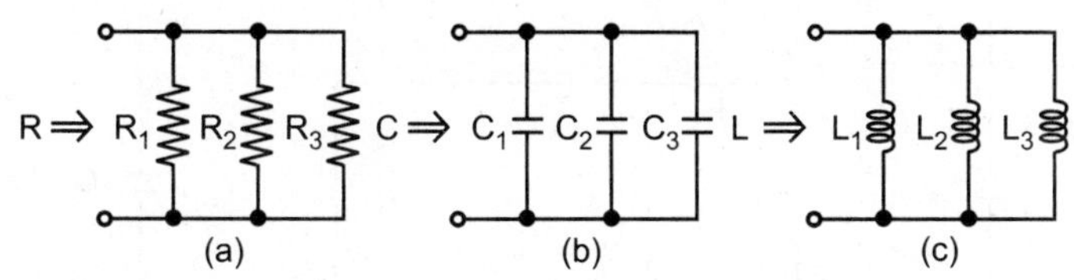

Fig. 5.10

Ans. (*i*) With reference to Fig. 5.11 (*a*), applying *KCL* at node (*X*),

$$I = I_1+I_2+I_3$$

or $$V/R = V/R_1+V/R_2+V/R_3$$

or $$1/R = 1/R_1+1/R_2+1/R_3$$

(*ii*) With reference to Fig. 5.11 (*b*), applying *KCL* at node (*X*),

$$I = I_1+I_2+I_3$$

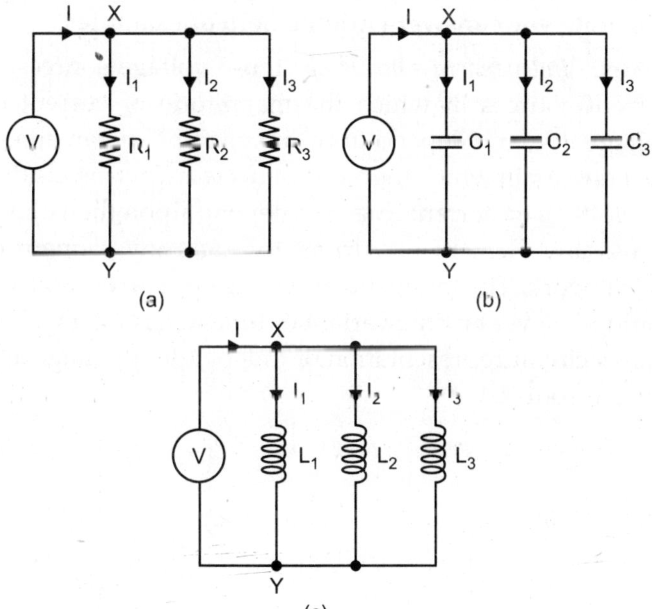

Fig. 5.11

or $$C\frac{dv}{dt} = C_1\frac{dv}{dt}+C_2\frac{dv}{dt}+C_3\frac{dv}{dt}$$

$$C = C_1+C_2+C_3$$

(*iii*) With reference to Fig. 5.11 (*c*), applying *KCL* at node *X*,

$$I = I_1+I_2+I_3$$

or $$\frac{1}{L}\int vdt = \frac{1}{L_1}\int vdt+\frac{1}{L_2}\int vdt+\frac{1}{L_3}\int vdt$$

$$\frac{1}{L} = \frac{1}{L_1}+\frac{1}{L_2}+\frac{1}{L_3}$$

Q.17. Use Kirchoff's Voltage Law (KVL) to prove

(*i*) $R=R_1+R_2+R_3$ for Fig. 5.12 (*a*)

(*ii*) $L=L_1+L_2+L_3$ for Fig. 5.12. (*b*)

(*iii*) $1/C=1/C_1+1/C_2+1/C_3$ for Fig. 5.12 (*c*).

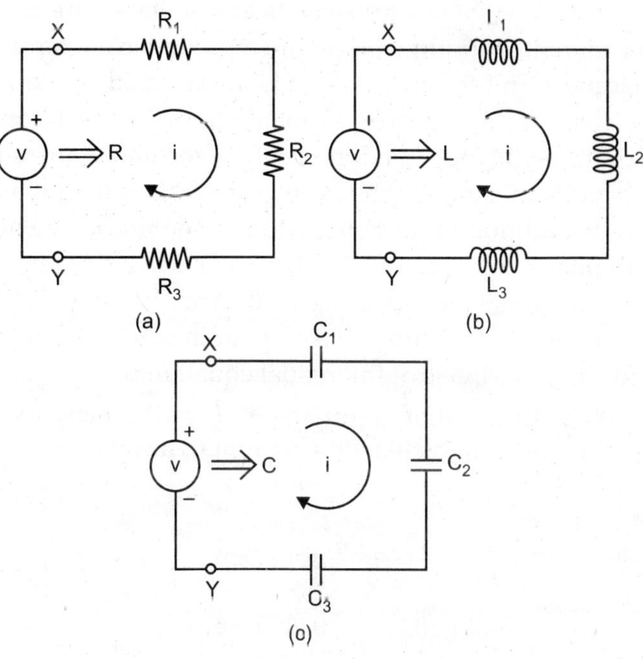

Fig. 5.12

Ans. (*i*) Applying (*KVL*) to the network of Fig. 5.12 (*a*)

$$v = iR_1+iR_2+iR_3=i(R_1+R_2+R_3)$$

$$\frac{v}{i} = (R_1+R_2+R_3)-R$$

(*ii*) Applying (*KVL*) to the network of Fig. 5.12 (*b*)

$$v = L_1\,di/dt+L_2\,di/dt+L_3\,di/dt$$

If *L* was the total effective inductance across *X–Y*, then

$$v = Ldi/dt$$

Therefore,

$$Ldi/dt = L_1\,di/dt+L_2\,di/dt+L_3\,di/dt$$

$$L = L_1+L_2+L_3$$

(iii) Applying (KVL) to the network of Fig. 5.12 (c)

$$v = \frac{1}{C_1}\int idt + \frac{1}{C_2}\int idt + \frac{1}{C_3}\int idt$$

or $$\frac{1}{C}\int idt = \frac{1}{C_1}\int idt + \frac{1}{C_2}\int idt + \frac{1}{C_3}\int idt$$

or $$1/C = 1/C_1 + 1/C_2 + 1/C_3.$$

Q.18. Are Kirchoff's laws (KCL and KVL) valid for steady state AC analysis and transient analysis? If yes, with what modifications?

Ans. Kirchoff's laws are as well valid for steady state AC analysis and transient analysis as these are in case of steady state DC analysis. In case of steady state AC analysis, all voltages and currents must be expressed as complex quantities and resistance should be replaced by complex impedances. In case of transient analysis, instantaneous values of currents and voltages are to be considered while impedances are represented as complex quantities. In other words, while doing steady state AC analysis, we must equate phasor addition of all out-going currents with phasor addition of all incoming currents of a particular node while applying KCL and similarly phasor addition of all potential rises should be equated with phasor addition of all potential falls in a closed path of a network while applying *KVL*. Also, writing *KCL* at a node or *KVL* around a closed path in case of transient analysis gives integro-differential equations.

As an illustration, applying *KCL* to the network of Fig. 5.13, we can calculate the current (*i*) as follows :

$$i = i_1 + i_2 + i_3$$

$$i_1 = \frac{10\angle 45°}{10} = 1\angle 45°$$

$$= (0.707 + j\,0.707) \text{ Amp.}$$

Now, $(7.07 + j\,7.07) = 10\angle 45°$, which gives

$$i_2 = \frac{10\angle 45°}{10\angle 45°} = 1\angle 0° \text{ Amp.}$$

Also, $(7.07 - j\,7.07) = 10\angle -45°$, which gives

$$i_3 = \frac{10\angle 45°}{10\angle -45°} = 1\angle 90° = (0 + j\,1) \text{ Amp.}$$

$$i = 0.707 + j\,0.707 + 1 + 0 + j1$$

$$= 1.707 + j\,1.707 = 2.4\angle 45° \text{ Amp.}$$

It may be mentioned that a simple addition of magnitudes of three currents would have yielded $i = 3$ Amp which is not correct.

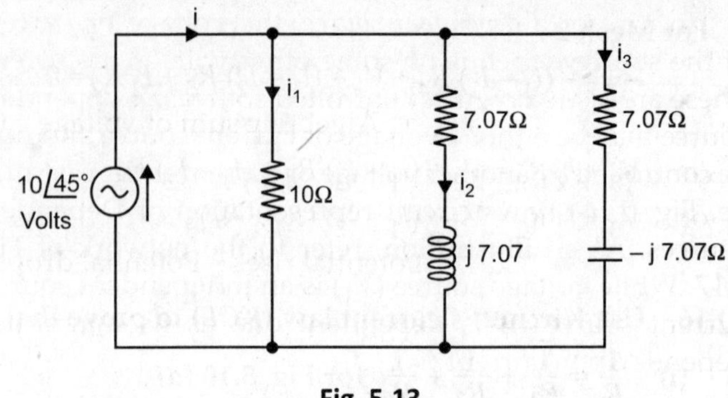

Fig. 5.13

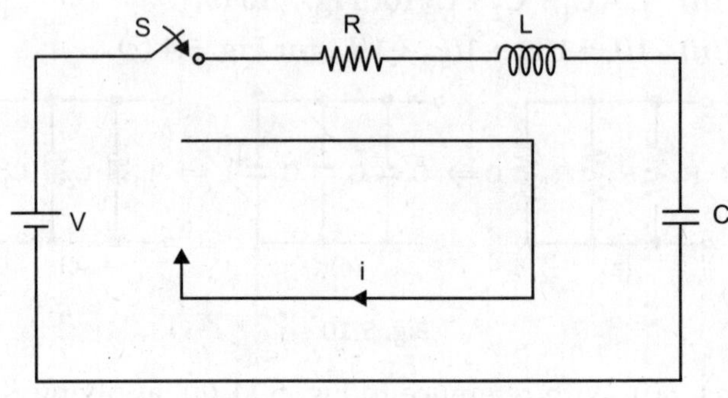

Fig. 5.14

Writing *KVL* for the *R-L-C* network of Fig. 5.14, while carrying transient analysis after the switch *S* is closed at $t = 0$, we get the following expression:

$$V = Ri + L\frac{di}{dt} + \frac{1}{C}\int idt$$

or $$Ld^2i/dt^2 + Rdi/dt + i/C = 0$$

Q.19. What are Independent and Dependent sources? Illustrate your answer with the help of examples.

Ans. *Independent sources* are those voltage sources or current sources in which the magnitude of current or voltage does not depend upon any external parameter of the network in which they are connected. This means that the voltage or current does not depend upon the voltage across any element or current through any element of the network. The independent voltage or current sources could however be time variant or time invariant. Fig. 5.15 shows circuit representation of independent voltage and current sources.

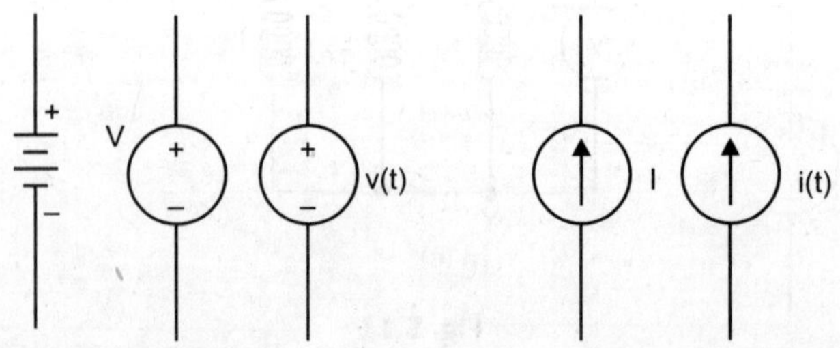

Fig. 5.15

In case of a *dependent source,* the voltage or current of the source depends on some other voltage or current. These are also known as controlled sources. A dependent source may be either a voltage or current source and may be controlled by another voltage or current as the case may be. Fig. 5.16 shows circuit representation of Dependent sources. As an illustration, refer to the network of Fig. 5.17. While voltage source (V_1) is an independent source, current source (i) is dependent one as its magnitude depends upon current (i_2).

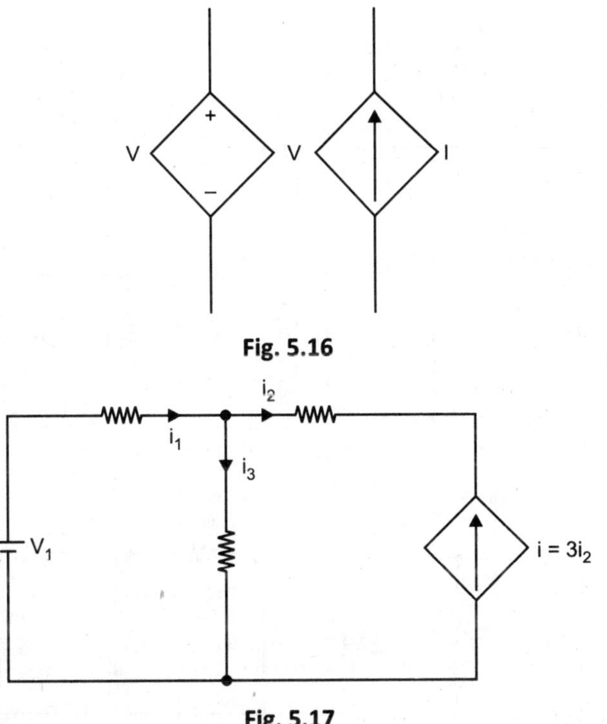

Fig. 5.16

Fig. 5.17

Q.20. What are constant voltage and constant current sources?

Ans. In case of a *constant voltage source,* the voltage across the source terminals remains constant irrespective of the current drawn from the source. Fig. 5.18 (*a*) shows the equivalent circuit of a generalized voltage source. Fig. 5.18 (*b*) shows its *V-I* characteristics. In case of an ideal voltage source or a true constant voltage source, $R_0 = 0$ and the *V-I* characteristics take the form of Fig. 5.18 (*c*).

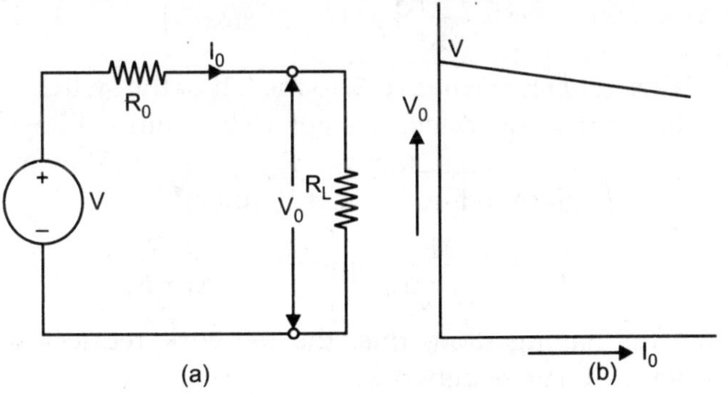

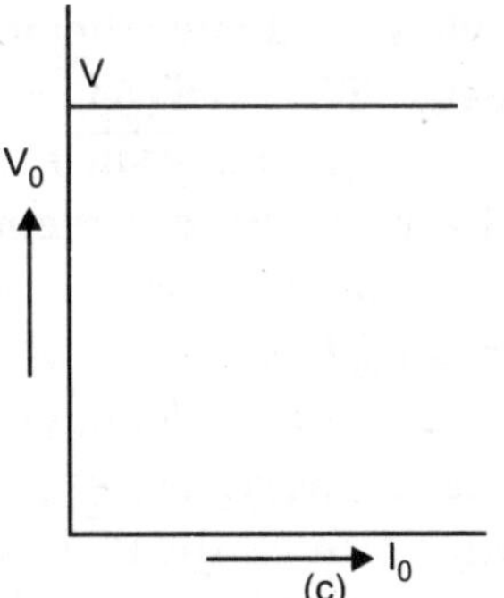

Fig. 5.18

In case of a *constant current source,* the current available from the source is constant irrespective of the voltage across its terminals. In other words, it supplies a constant current to varying load resistance. Fig. 5.19 (*a*) shows the equivalent circuit of a generalized current source. Fig. 5.19 (*b*) shows the *V-I* characteristics. In case of a true constant current source, R_0 = infinity and the *V-I* characteristics take the shape of Fig. 5.19 (*c*),

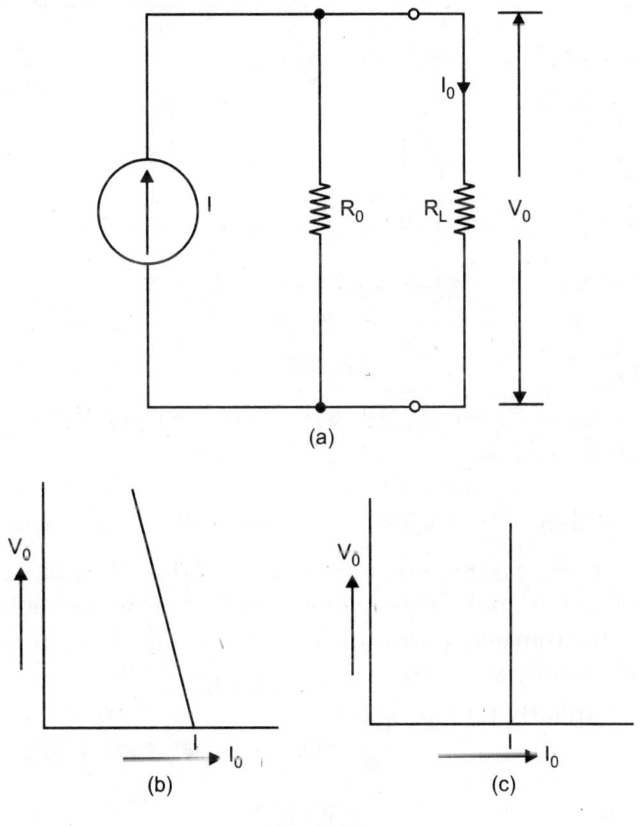

Fig. 5.19

Q.21. "Whenever a network is switched from one condition to another either by change in applied source magnitude or by change in circuit elements, there is always a transition period during which the branch currents and node voltages change from earlier values to newer ones. This transition phase is known

as TRANSIENT phase and once the newer values have been attained, we say STEADYSTATE has been achieved. We also say that solution of any such switched network comprises of two parts (*i*) complementary solution corresponding to transient part and (*ii*) particular integral corresponding to steadystate part. Now in this light, state what are the complementary solutions and particular integrals for the networks of Figs. 5.20 (*a*), (*b*) and (*c*) when switch *S* in all the three cases is closed at $t = 0$. Note that we are looking for i_L in case of Fig. 5.20 (*a*), v_c in case of Figs. 5.20 (*b*) and (*c*).

Ans. In case of Fig. 5.20 (*a*), we have already seen that

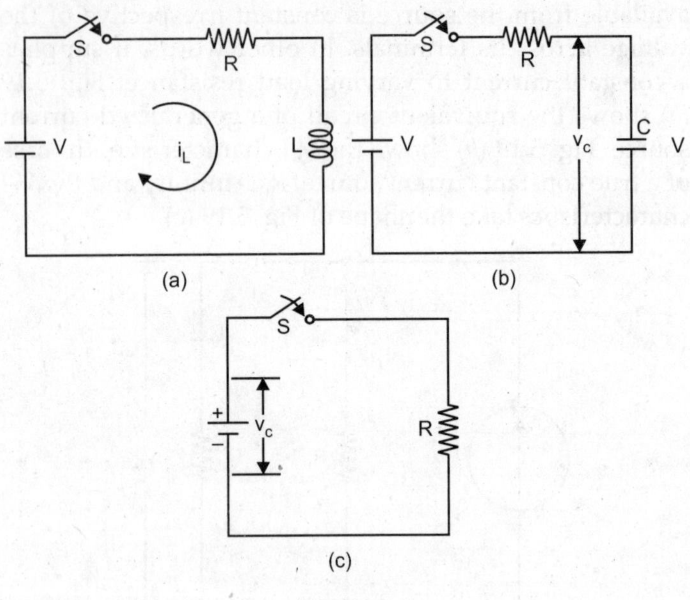

Fig. 5.20

$$i_L = V/R\,(1 - e^{-R\,t/L}) = (V/R) - (V/R\, e^{-R\,t/L})$$

$$= i_s + i_t$$

where $i_s = V/R$ and $i_t = (-V/R)\, e^{-R\,t/L}$

Here (i_s) is the steady state part and (i_t) is the transient part. If you look at these expressions, you would notice that the transient part vanishes at $t = \infty$ and what remains is the steady state part.

Similarly for Fig. 5.20 (*b*),

$$v_c = V\,(1 - e^{-t/RC}) = (V - V\,e^{-t/RC}) = (v_s + v_t)$$

where $v_s = V,\ v_t = -V\,e^{-t/RC}$

Again the transient part ($-V\,e^{-t/RC}$) vanishes at $t = \infty$ and what is left is the steady state part. In practice, one doesn't wait upto $t = \infty$. After a time equal to that of approximately five time constants, steady state is supposed to have been achieved.

For the network of Fig. 5.20 (*c*).

$$v_c = V\,e^{-t/RC} = v_s + v_t$$

where $v_s = 0,\ v_t = V\,e^{-t/RC}$

A steadystate value of zero indicates that the capacitor voltage is zero in the steadystate.

Q.22. For the networks of Figs. 5.21 (*a*) to 5.21 (*h*), write the expressions for the variables marked? Do not derive the expressions. Give brief justification wherever needed.

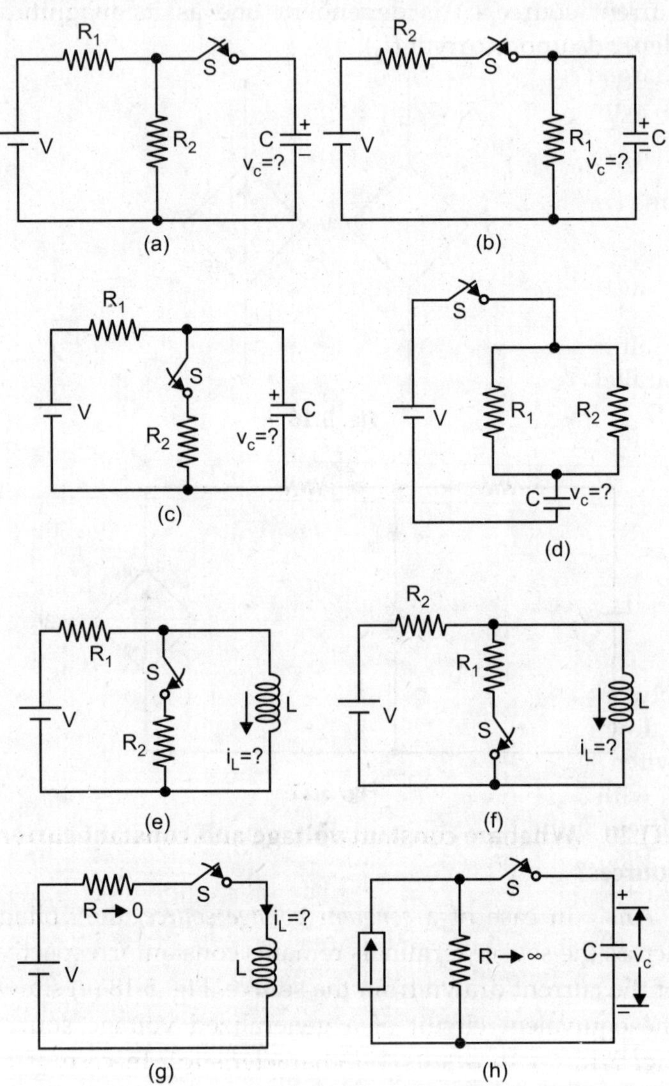

Fig. 5.21

Ans. (*a*) $$v_c = \frac{VR_2}{R_1 + R_2}\left[1 - e^{-\frac{(R_1 + R_2)\,t}{R_1 R_2 C}}\right]$$

If we replace the network to the left of the switch *S* by its Theverrin's equivalent, we get a Thevenin's voltage of V_{th} and Thevenin's resistance of R_{th} given by (Thevenin's theorem is discussed in the next chapter)

$$V_{th} = \frac{VR_2}{R_1 + R_2} \text{ and } R_{th} = \frac{R_1 R_2}{R_1 + R_2}$$

After having done this, the network reduces to a simple *R—C* series network.

(*b*) The network is similar to the one in Fig. 5.21 (*a*) except that R_2 is in place of R_1 and R_1 is in place of R_2.

$$v_c = \frac{VR_1}{R_1 + R_2}\left[1 - e^{\frac{(R_1+R_2)t}{R_1 R_2 C}}\right]$$

(*c*) $$v_c = \frac{VR_2}{R_1 + R_2}\left[1 - e^{-\frac{(R_1+R_2)t}{R_1 R_2 C}}\right] + V\,e^{-\frac{t}{R_2 C}}$$

Initially, when switch is open, the capacitor gets charged to V through R_1. When S is closed, an already charged capacitor (charged to V volts) discharges through R_2 with a time constant R_2C whereas it is also getting charged from an effective voltage of $\left(\frac{VR_2}{R_1 + R_2}\right)$ with a time constant $\left(\frac{R_1 R_2}{R_1 + R_2}\right) C$.

(*d*) $$v_c = V\left[1 - e^{\frac{-(R_1+R_2)t}{R_1 R_2 C}}\right]$$

It is a case of simple R–C network with R given by a parallel combination of R_1 and R_2.

(*e*) $$i_L = \left[\frac{VR_2}{(R_1 + R_2)} / \frac{R_1 R_2}{(R_1 + R_2)}\right]\left[1 - e^{-\frac{R_1 R_2 t}{(R_1+R_2)L}}\right] + \frac{V}{R_1}e^{-\frac{R_2 t}{L}}$$

$$= \frac{V}{R_1}\left[1 - e^{-\frac{R_1 R_2 t}{(R_1+R_2)L}} + e^{-\frac{R_2 t}{L}}\right]$$

The situation is similar to the one seen in case of network of Fig. 5.21 (*c*). A steady state current of (V/R_1) is flowing through L before switch is closed. When the switch is closed, this current (= V/R_1) decays through R_2 with a time constant (L/R_2) and at the same time, the current through L builds up from an effective voltage of $\left(\frac{VR_2}{R_1 + R_2}\right)$ through a resistance $\left(\frac{R_1 R_2}{R_1 + R_2}\right)$.

(*f*) It is just the same as in case of the network of Fig. 5.21 (*e*) except that R_1 and R_2 have interchanged their positions.

Therefore,

$$i_L = \frac{V}{R_2}\left[1 - e^{-\frac{R_1 R_2 t}{(R_1+R_2)L}} + e^{-\frac{R_1 t}{L}}\right]$$

(*g*) $$i_L = \frac{V}{L}t$$

(*h*) $$V_c = \frac{I}{C}t.$$

Q.23. The STEP-WAVEFORM shown in Fig. 5.22 (*a*) is applied as an input to various networks shown in Fig, 5.22 (*b*) to 5.22 (*x*). The step signal could be an input current (I_i) or an input voltage (V_i). The output signal waveform (I_o if the input is V_i and V_o if the input is I_i) is all we are interested in. Please justify briefly how you have arrived at these waveforms only in cases where these are not self explanatory.

Ans. Fig. 5.22 (*b*), (*c*), (*d*) : The derivative of a step is an impulse. Current through a capacitor C is $C\frac{dv}{dt}$. Voltage across a capacitor = $\frac{I}{C}\int idt$ and when $i = I$, then $v_a = \frac{I}{C}t$. See Fig. 5.23 to 5.25.

Fig. 5.22 (*e*) : Current through inductance $I_0 = \frac{1}{L}\int vdt$ and for $v = V$, $I_o = \frac{V}{L}t$. See Fig. 5.26.

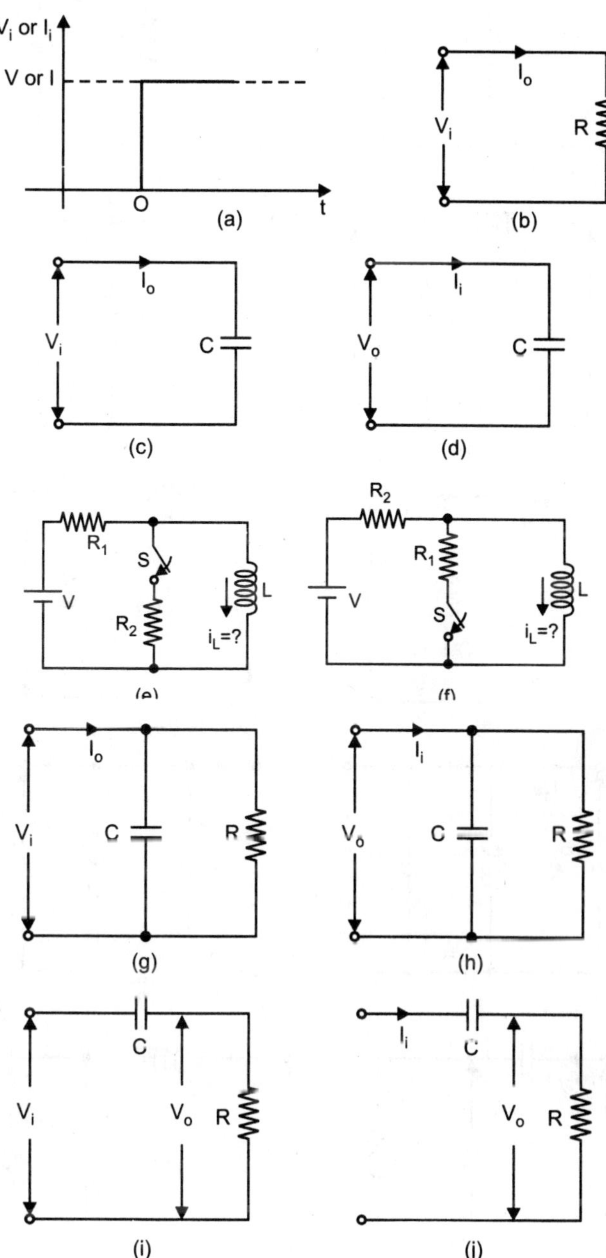

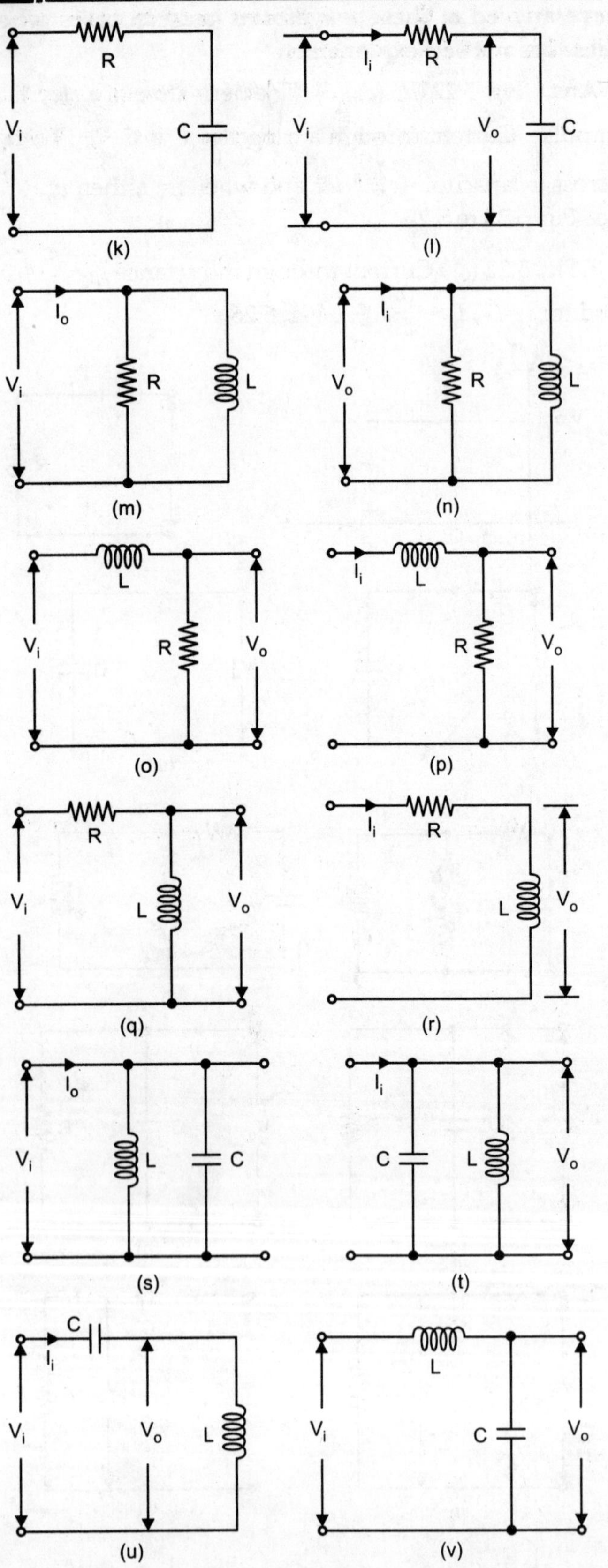

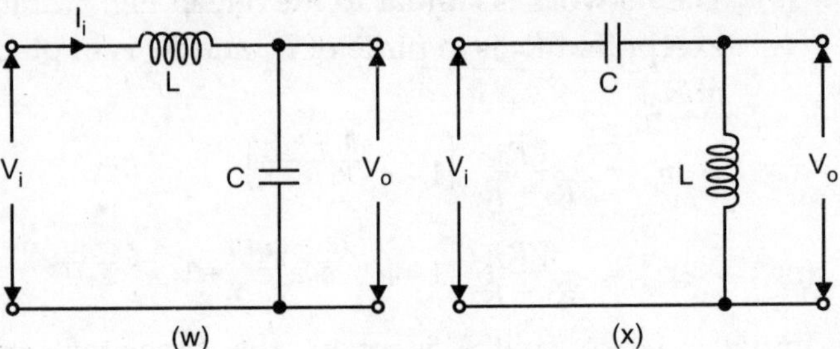

Fig. 5.22

Fig. 5.22 (*f*) : Now $v = L\, di/dt$ and when (*i*) is a step, di/dt is in impulse. See Fig. 5.27.

Fig. 5.22 (*h*) : Quite obviously, V_o is *R* multiplied by current through *R*. Now current through *C* is an exponential decay and current through *R* is an exponential rise (Refer to Fig. 5.29).

Fig. 5.22 (*n*) : Again current through *R* is initially *I* as current through an inductance cannot change instantaneously. But then current through *L* grows exponentially as $I\,(I - e^{-Rt/L})$ so that current through *R* is $I - I\,(1 - e^{-Rt/L}) = Ie^{-Rt/L}$ and hence $V_o = (I_i\, R) = IRe^{-Rt/L}$ (Refer Fig. 5.35).

Fig. 5.22 (*q*) : Now $V_o = L\dfrac{di}{dt}$. (*i*) in this case is given by $\dfrac{V}{R}\,(1 - e^{-Rt/L})$ and hence the result. See Fig. 5.38.

Fig. 5.22 (*t*) : Writing *KCL* in Laplace transform form,

$$\frac{I}{s} = CsV_o\,(s) + \frac{Vo\,(s)}{Ls}$$

where $V_o\,(s)$ is Laplace transform of $V_o\,(t)$. Simplifying on R.H.S.

$$\frac{I}{s} = V_o\,(s)\left[\frac{s^2\,LC+1}{Ls}\right]$$

$$\text{or} \quad V_o\,(s) = \frac{I/s \times sL}{s^2\,LC+1} = \frac{IL/LC}{s^2 + \dfrac{1}{LC}} = \frac{I/C}{s^2 + \dfrac{1}{LC}}$$

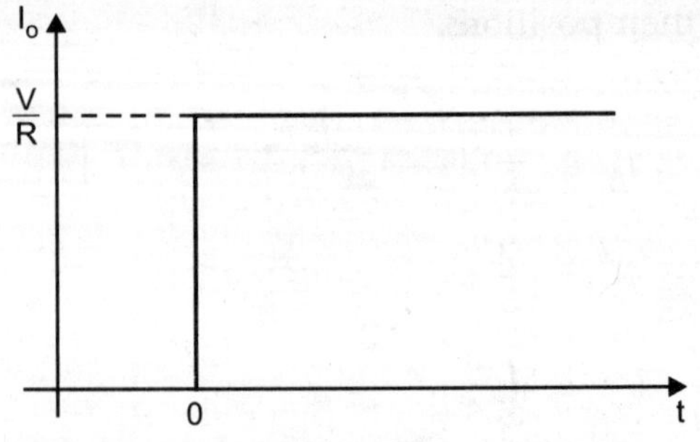

Fig. 5.23

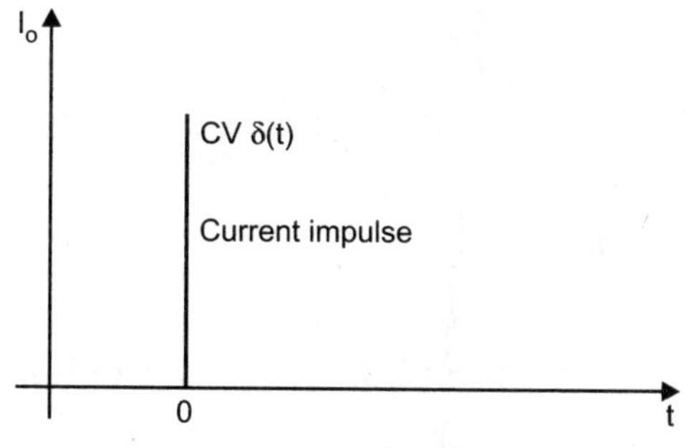

Fig. 5.24

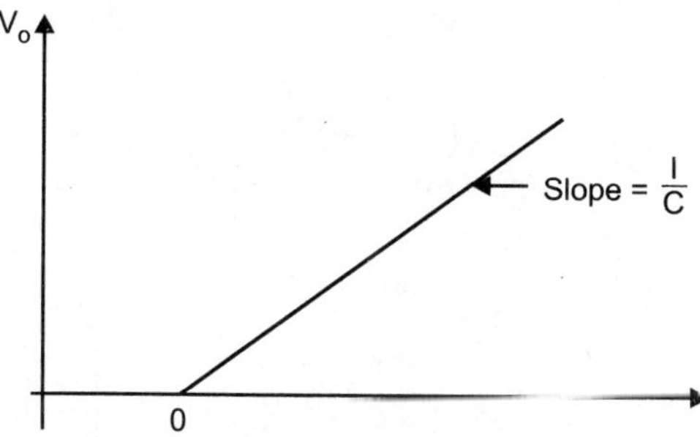

Fig. 5.25

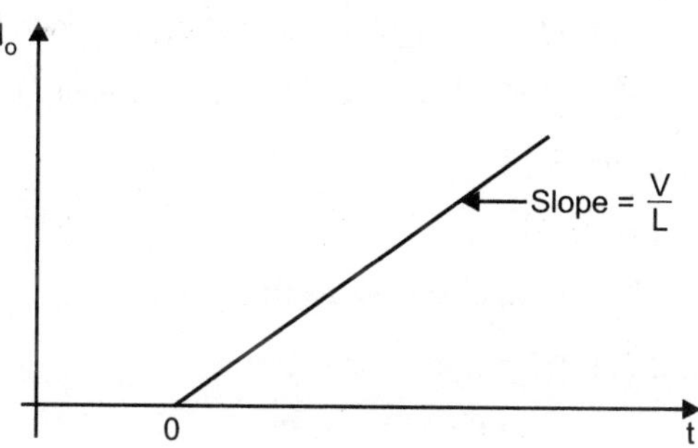

Fig. 5.26

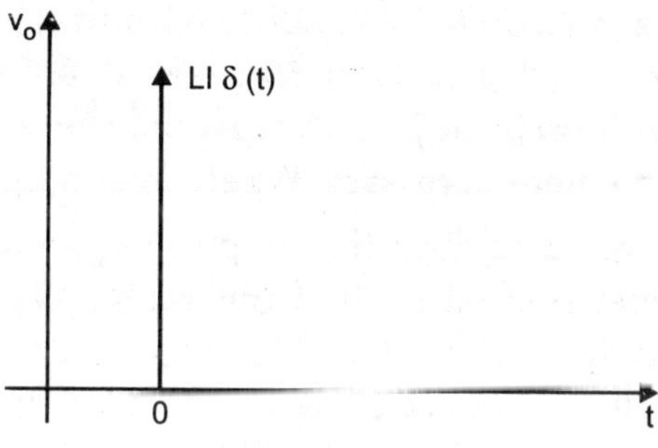

Fig. 5.27

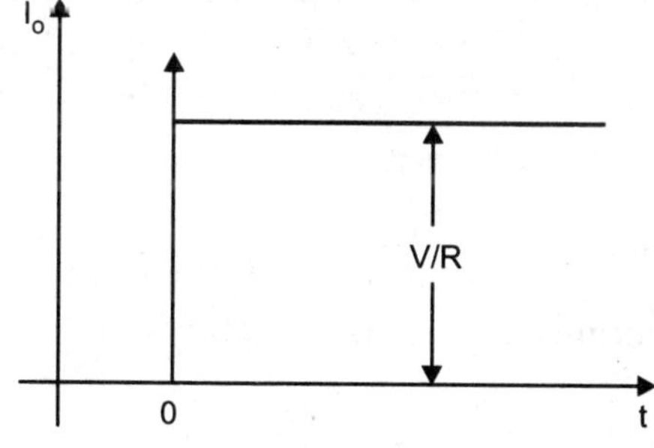

Fig. 5.28

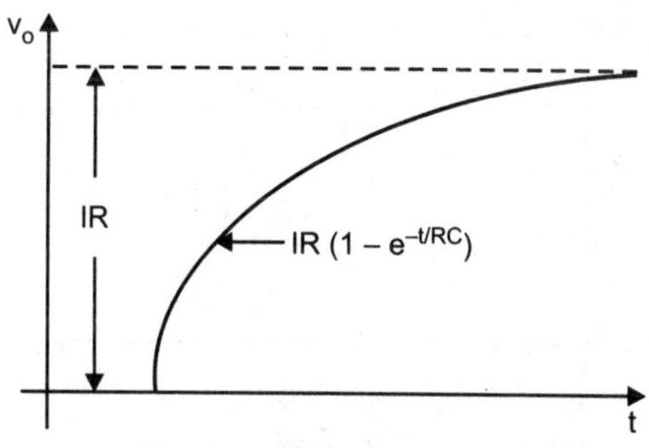

Fig. 5.29

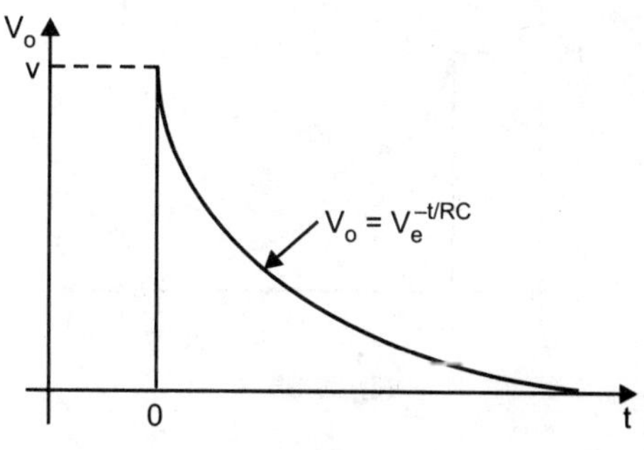

Fig. 5.30

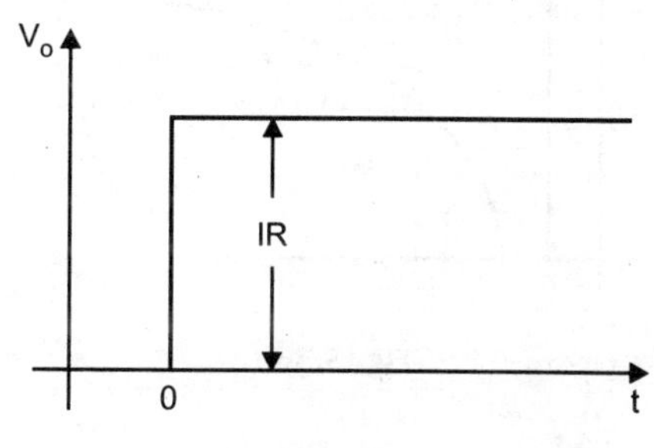

Fig. 5.31

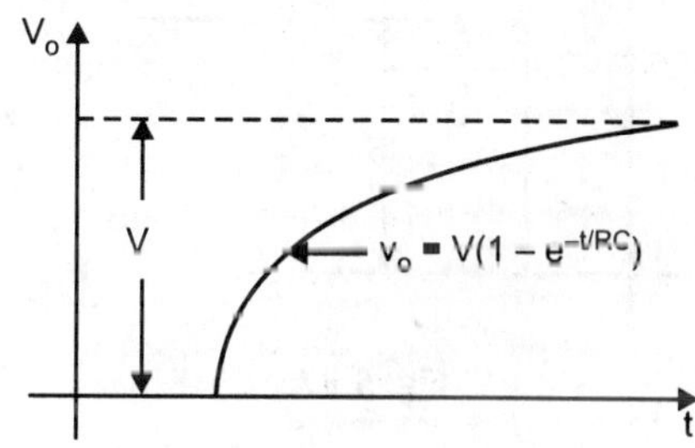

Fig. 5.32

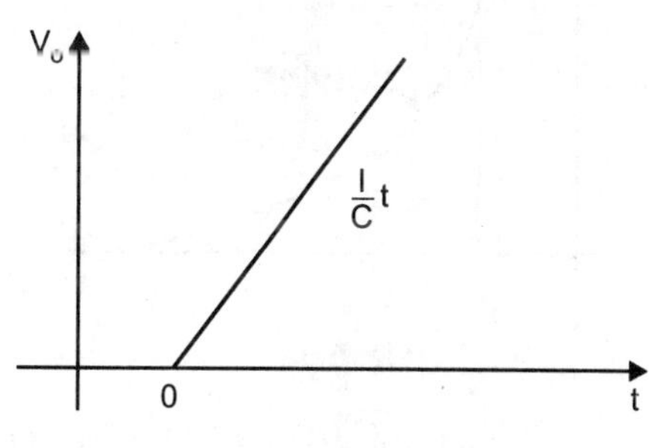

Fig. 5.33

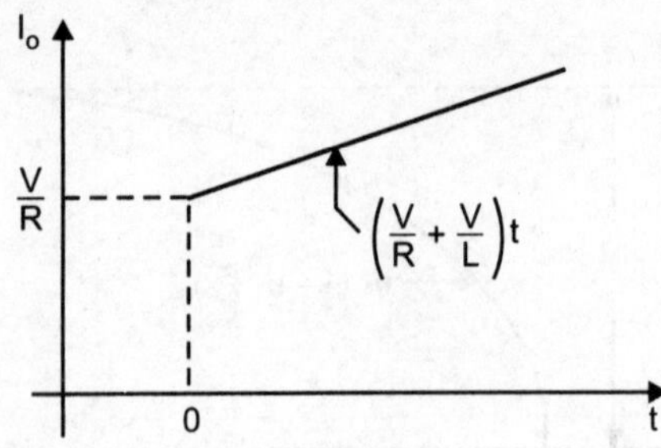

Fig. 5.34

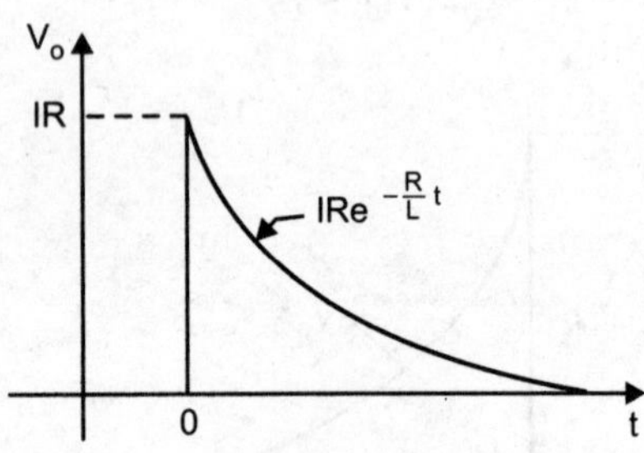

Fig. 5.35

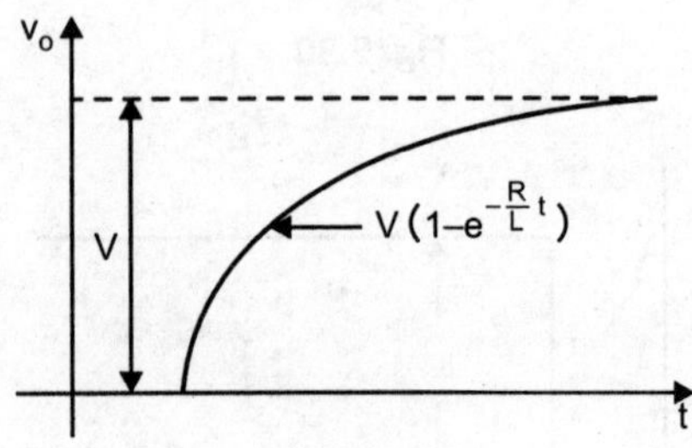

Fig. 5.36

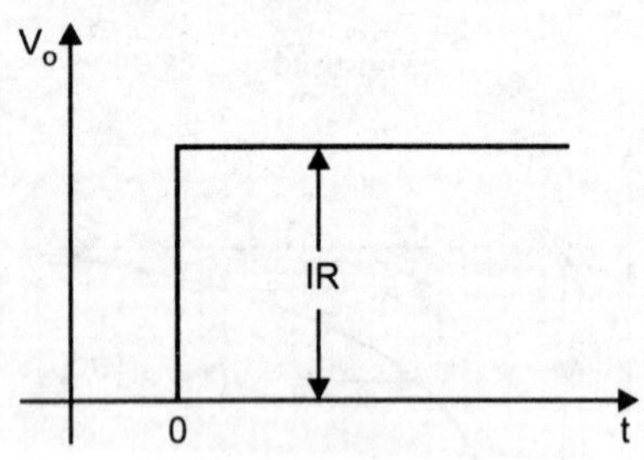

Fig. 5.37

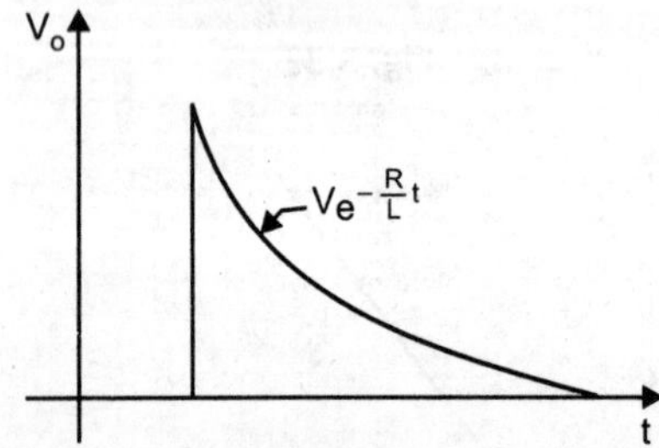

Fig. 5.38

$$\text{or} \quad V_O(s) = \frac{I/C\,(\sqrt{LC}/\sqrt{LC})}{s^2+1/LC} = \frac{1\sqrt{L/C}\times\omega_r}{s^2+\omega_r^2}$$

$$\text{where } \omega_r = \frac{1}{\sqrt{LC}}$$

$$V_o(t) = I\sqrt{L/C}\,\sin\omega_r t$$

Refer to Fig. 5.41

Fig. 5.22 (u):

$$\frac{V}{s} = sLI + \frac{1}{Cs}I(s) \quad \text{in Laplace transform}$$

This gives, after simplification,

$$I(s) = \frac{V/L}{s^2+\omega_r^2} \quad \text{where } \omega_r = \frac{l}{\sqrt{LC}}$$

$$I(s) = \frac{V/L}{s^2+\omega_r^2} = \frac{\frac{V}{\sqrt{C}/\sqrt{L}}\cdot\omega_r}{s^2+\omega_r^2}$$

$$\text{or} \quad i(t) = \frac{V\sqrt{C}}{\sqrt{L}}\sin\omega_r t$$

$$\therefore \quad V_o = \frac{L\,di}{dt} = \left[\frac{V\sqrt{C}}{\sqrt{L}}\omega_r\cos\omega_r t\right].L = V\cos\omega_r t$$

Refer to Fig. 5.42

Fig. 5.22 (v) : If the output was taken across C,

$$V_o = V - V\cos\omega_r t = V(1-\cos\omega_r t).$$

Refer to Fig. 5.43

Fig. 5.22 (w) : Here,

$$V_o = It/C. \text{ Refer to Fig. 5.44}$$

Fig. 5.22 (x) : Here,

$$V_o = LI\,\delta(t)$$

Q.25. The networks are once again the same [the ones given in Fig. 5.22 (*b*) to 5.22 (*x*).] The input (V_i or I_i) is a pulse (Shown in Fig. 5.46) instead of a step. Sketch the output waveform (I_o or V_o as the case may be) along with justification where necessary. Where ever necessary.

Ans. If we remember that a pulse is formed by a positive going step (at $t = 0$ in this case) followed by a negative going step (at $t = T$ in this case) the above waveforms shown in Figs. 5.47 to 5.52 corresponding to the networks of Figs. 5.22 (*b*) to 5.22 (*g*) will be clear.

Look at Fig. 5.22 (*h*), for time t between $t = 0$ and $t = T$

$$I = C\frac{dV_o}{dt} + \frac{V_o}{R}$$

with simple mathematics,

$$V_o(s) = \frac{RI}{s(1+RCs)} = \frac{IC/C}{s(s+1/RC)}$$

Using method of partial of fractions,

$$V_o(s) = \frac{IR}{s} + \frac{(-IR)}{(s+1/RC)}$$

Taking Inverse Laplace transform, (The Laplace transform and Inverse Laplace transform tables are given at the end of this chapter).

$$v_o(t) = IR - IR\,e^{-t/RC} = IR(1 - e^{-t/RC})$$

$$V_o(t) = IR(1 - e^{-T/RC}) \text{ (at } t = T)$$

Now at t = T, I_i goes to zero. As result, the capacitor C begins to discharge through resistance R from its value of v_T.

$v_o(t)$ (for $T < t\ \infty$) $= v_T\, e^{-(t-T)/RC}$

where $v_T = IR(1 - e^{-T/RC})$ Fig. 5.53 shows the waveform.

When the input voltage goes from zero to V at t = 0, output voltage follows that change as voltage across capacitor cannot change instantaneously [Refer to the network of Fig. 5.22 (*i*)]. But after that, the voltage across capacitor builds up exponentially so that V_o decays exponentially. At $t = T$, $V_o = (V\,e^{-T/RC})$. Now V_i, goes to zero at $t = T$. Again as voltage across a capacitor can not change instantaneously, this change appears across R so that at $t = T$, V_o changes from $V\,e^{-T/RC}$ to $V\,e^{-T/RC} - V = \{-V(1 - e^{-T/RC})\}$. Since $V_i = 0$ for $t \geq T$ the capacitor discharges through R and the output voltage is given by

$$V_o = [-V(1 - e^{-T/RC})]e^{-(t-T)/RC}$$

Fig. 5.54 shows the waveform.

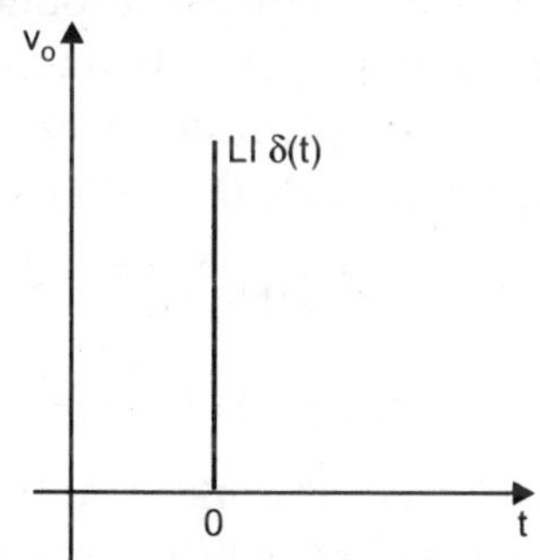

Fig. 5.39

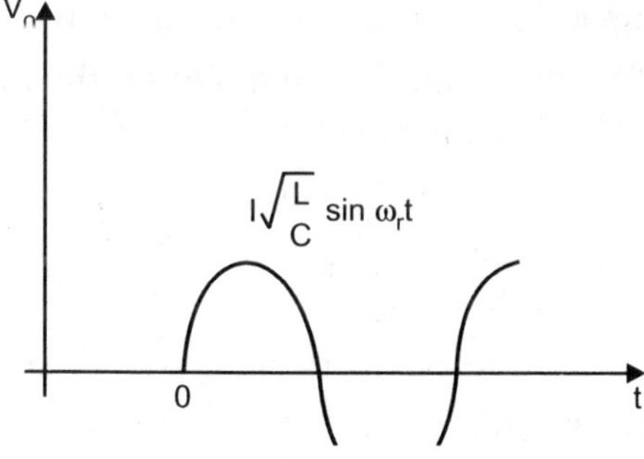

Fig. 5.40

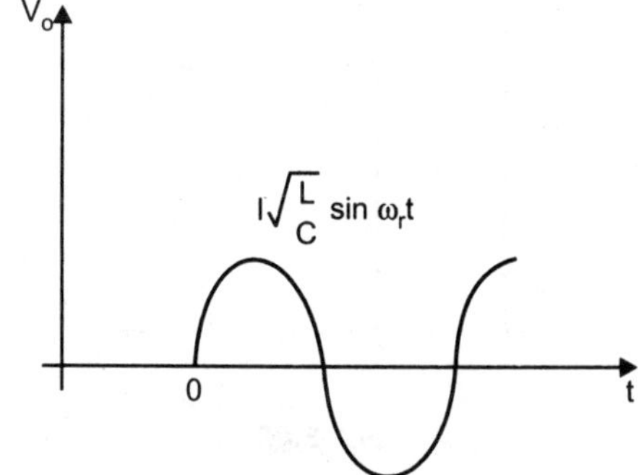

Fig. 5.41

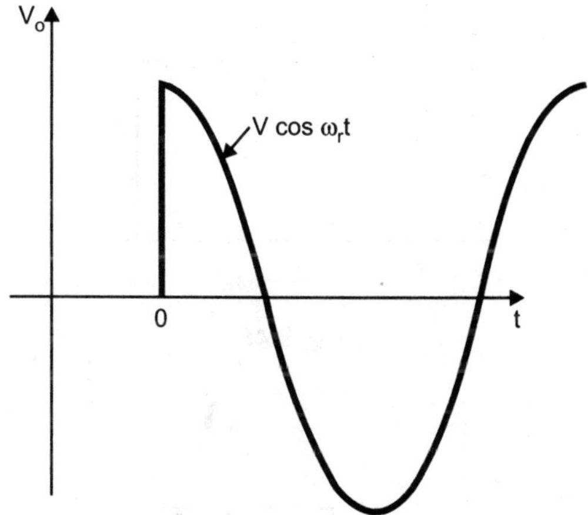

Fig. 5.42

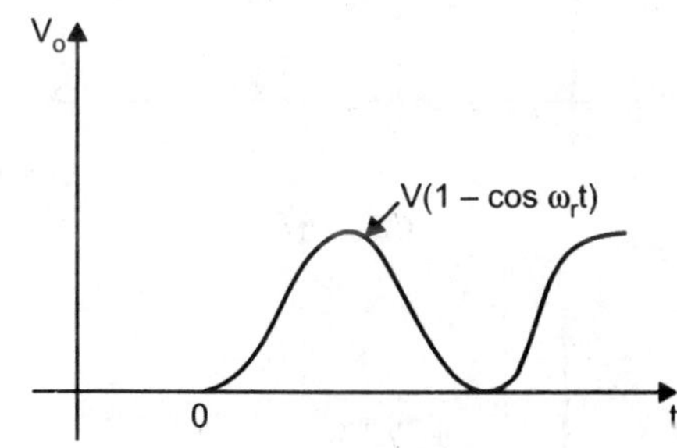

Fig. 5.43

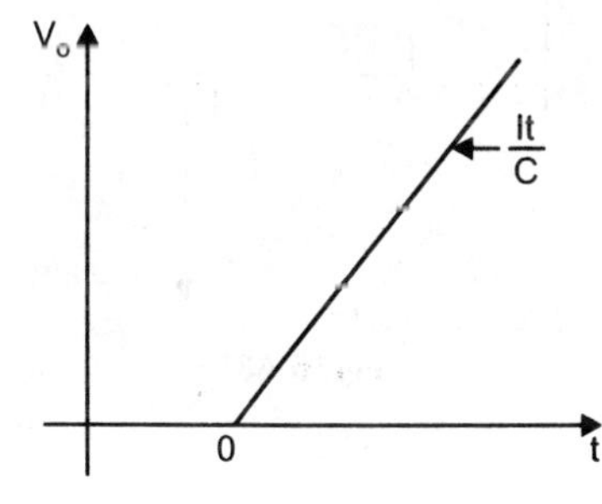

Fig. 5.44

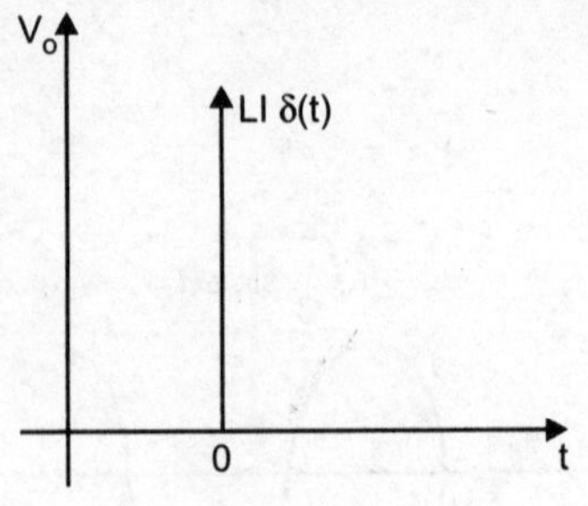

Fig. 5.45

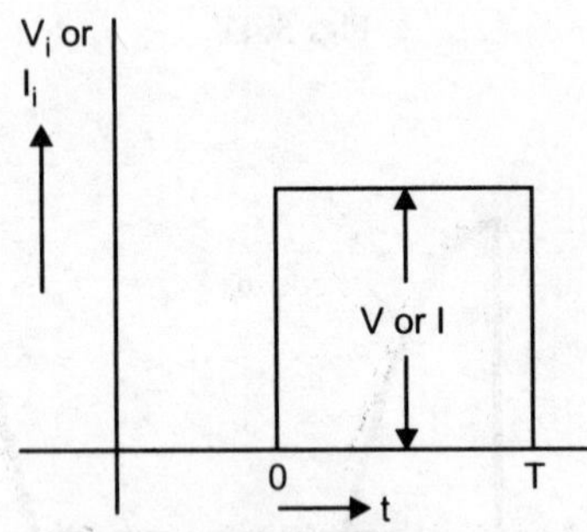

Fig. 5.46

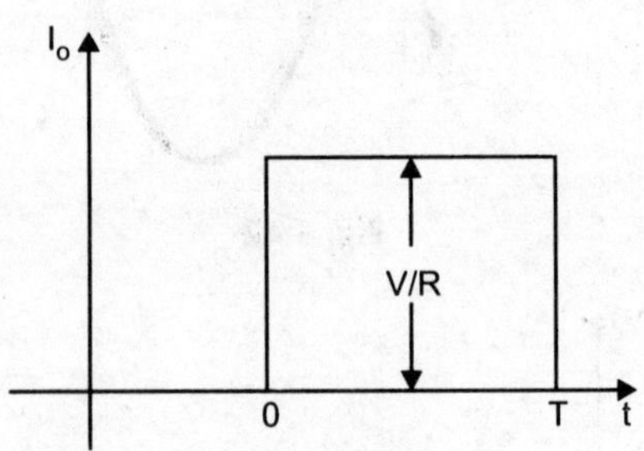

Fig. 5.47

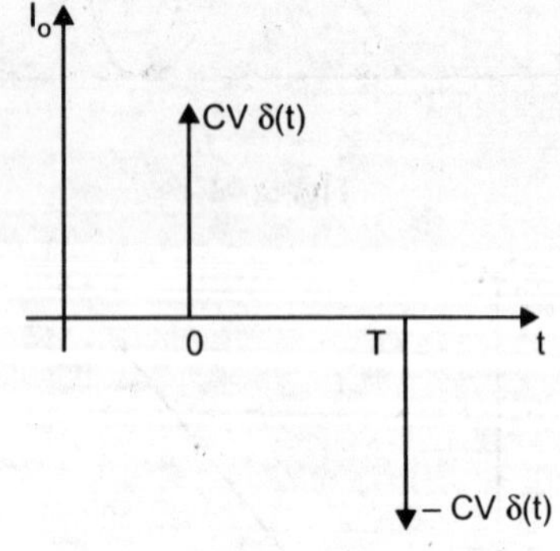

Fig. 5.48

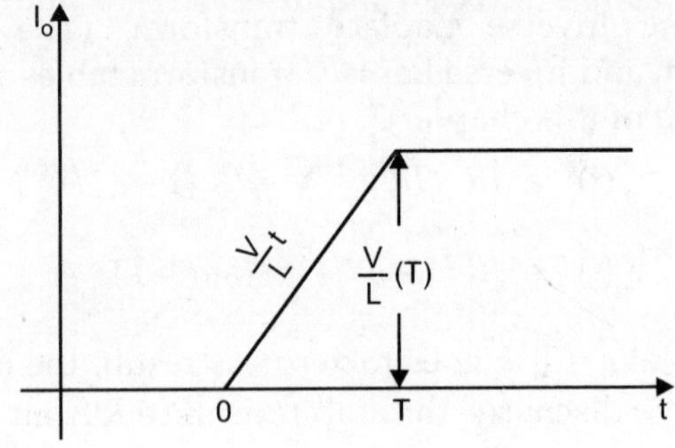

Fig. 5.49

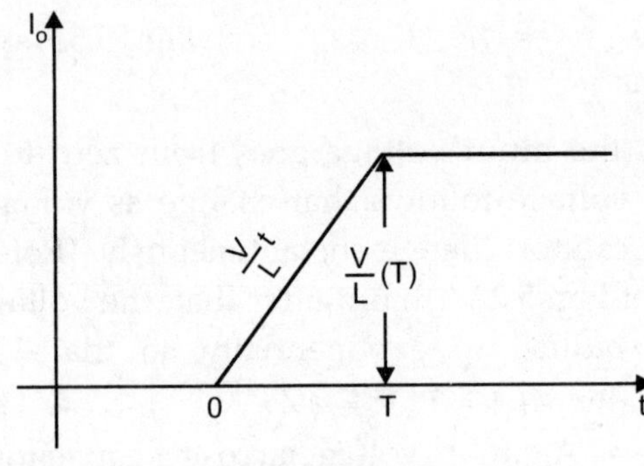

Fig. 5.50

Figs. 5.55 to 5.57 shows the output waveform corresponding to network of figs. 5.22 (*j*) to 5.22 (*l*). These are self explanatory.

Refer to Fig. 5.22 (*m*). At $t = 0$, input voltage goes to V. At that instant (*i.e.* $t = 0$), current through R is V/R and that through L is zero because current through an inductance cannot change instantaneously. Therefore, I_o experiences a step rise equal to V/R only. Current through R remains V/R as long as $V_i = V$ (*i.e.*, for $0 \le t \le T$). Current through L rises $\left(\frac{V}{L}\right)t$ for $0 \le t \le T$. That is, $\frac{V}{R} + \frac{V}{L}(t) = \frac{V}{L}\left(1 + \frac{R}{L}t\right)$ for $0 \le t \le T$. At $t = T$, $V_i = 0$. As a result, current through R becomes zero. Current through L at that instant is V/L (T) and since it cannot change instantaneously, the output current abruptly falls by (V/R). For $t > T$, the output current remains constant at $V/L(T)$ as there is a short circuit across the coil (due to $V_i = 0$). Fig. 5.58 shows the waveform

Fig. 5.22 (*n*) : It can be explained on the lines of case (*i*). Fig. 5.59 shows the waveform.

Figs. 5.60 to 5.62 show output waveform corresponding to networks of Fig. 5.22 (*o*) to 5.22 (*q*). These are self explanatory.

Fig. 5.31 (*r*) : Output voltage V_o is $\frac{L\,di}{dt}$ and (*i*) in this case is a pulse. This pulse can be considered equivalent to a current step (at $t = 0$) which is positive going and a negative going step at $t = T$. Derivative of step is an impulse. Hence V_o is an impulse at $t = 0$ and another impulse (negative going in this case) at $t = T$ (Refer Fig. 5.63)

Fig. 5.31 (*s*): The output current is sum of inductor and capacitor currents. The firm lines represent capacitor current whereas lines with small dots represent inductor current variation with time (Fig. 5.64).

Figs. 5.65. to 5.69 show relevant output waveforms corresponding to networks of Figs. 5.22 (*t*) to 5.22 (*x*). These are self-explanatory.

Q.26. Look at the networks (simple *R-C* networks) shown in Fig. 5.70 (*a*) to 5.70 (*c*). Find out s-domain transfer functions for each of them. Also draw *T(s)* versus (ω) plots for these networks. What conclusions do you draw from the results.

Ans. The *s*-domain transfer function for the network of Fig. 5.70 (*a*) is given by

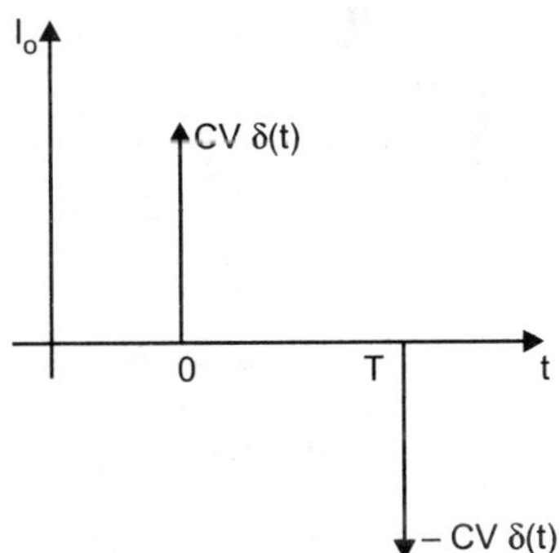

Fig. 5.51

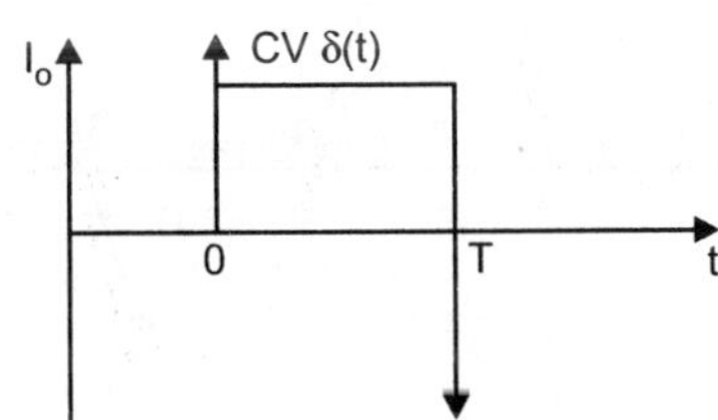

Fig. 5.52

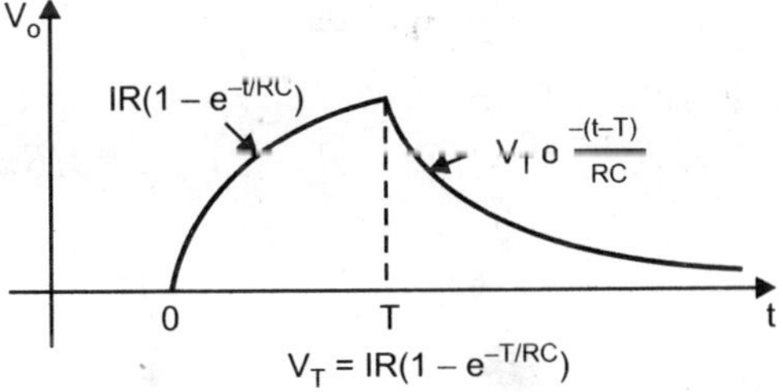

Fig. 5.53

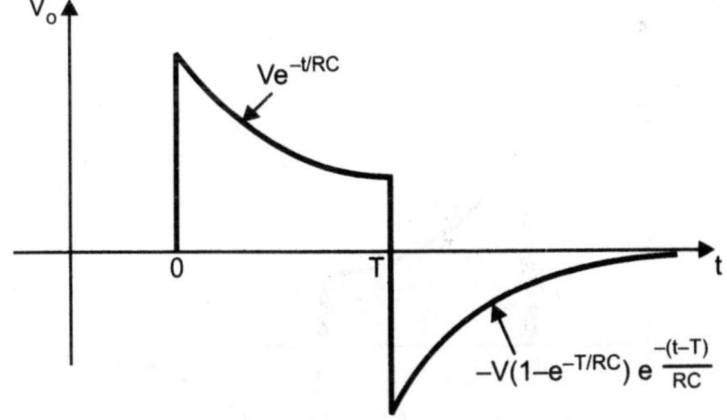

Fig. 5.54

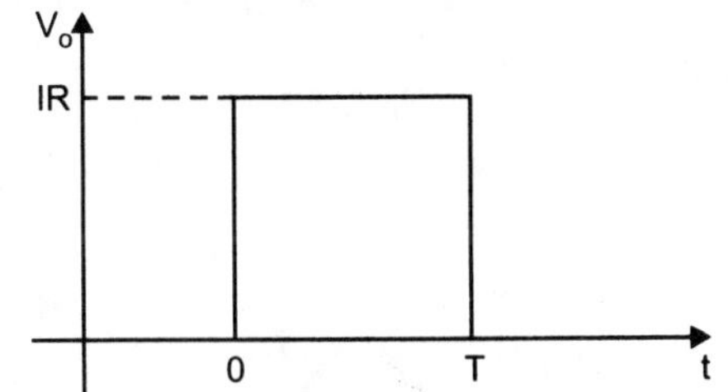

Fig. 5.55

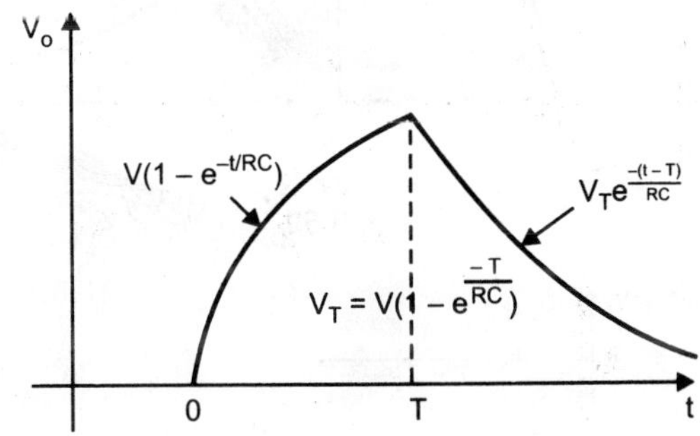

Fig. 5.56

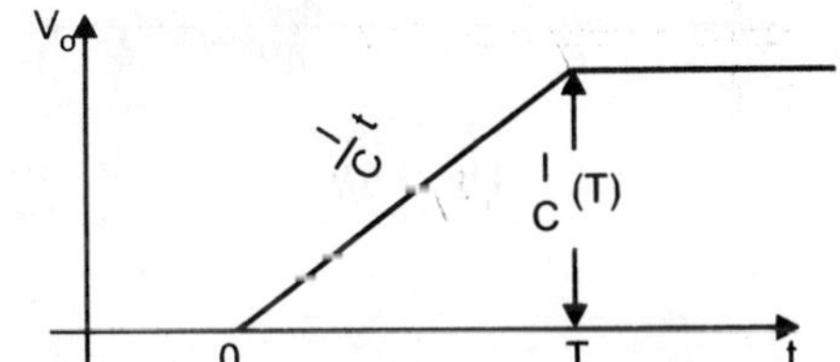

Fig. 5.57

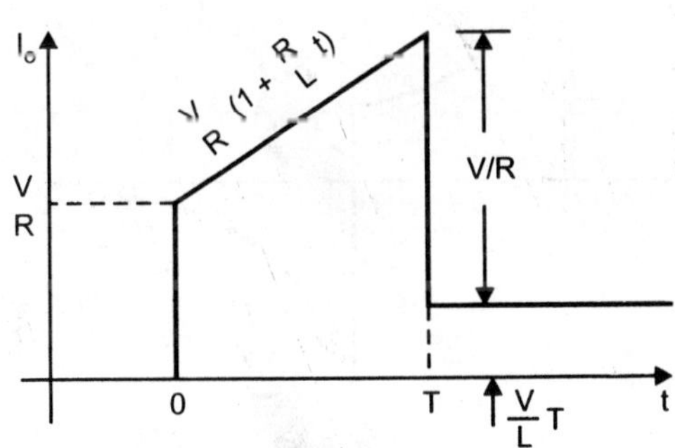

Fig. 5.58

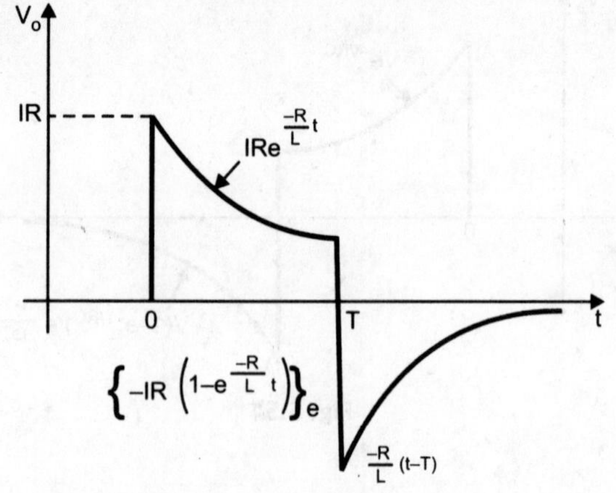

Fig. 5.59

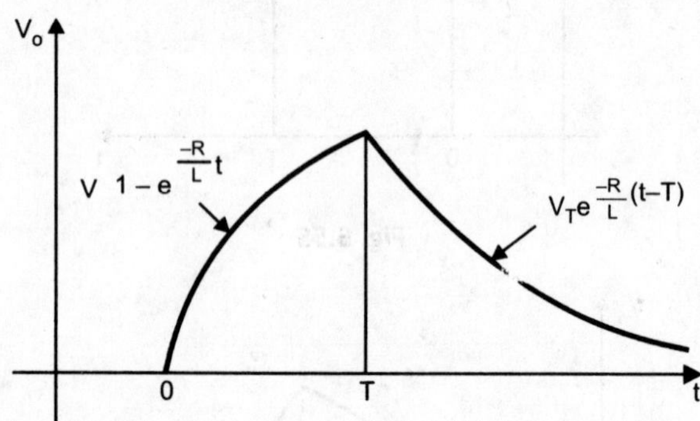

Fig. 5.60

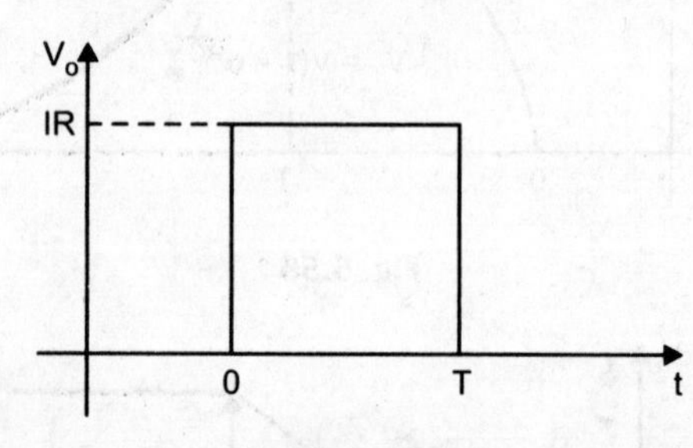

Fig. 5.61

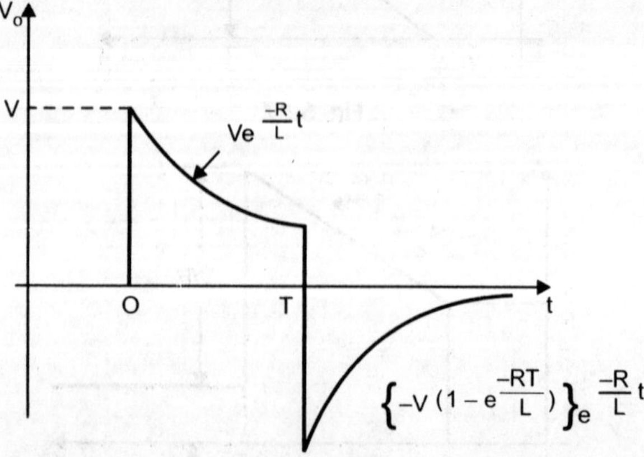

Fig. 5.62

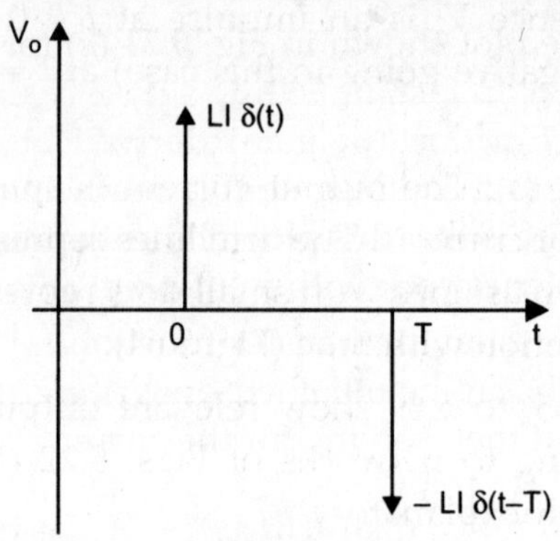

Fig. 5.63

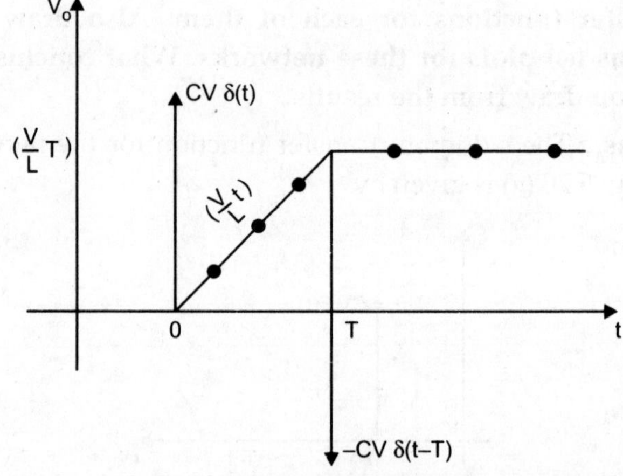

Fig. 5.64

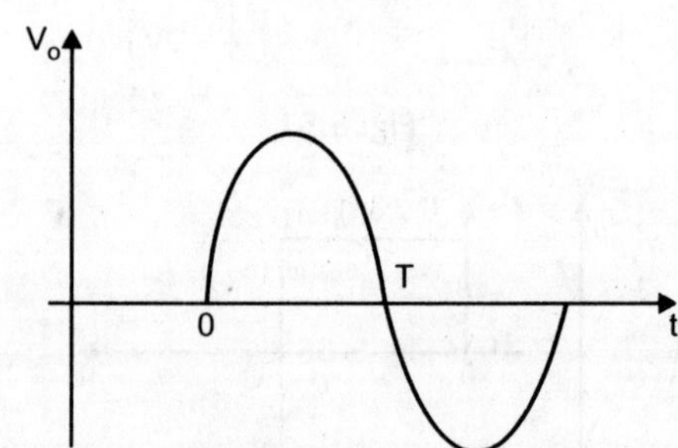

$V_o = I\,(L/C)^{1/2}\ \sin \omega_r t\ \ 0 \le t \le T$

$\quad = I\,(L/C)^{1/2}\ \{\sin \omega_r t - \sin \omega_r\,(t - T)\}\ \ t \ge T$

Fig. 5.65

$$\frac{V_O(s)}{V_i(s)} = \frac{R}{R + 1/Cs} = \frac{RCs}{RCs + 1} = \frac{s}{s + 1/RC}$$

$$= \frac{s}{s + \omega_0}$$

where $\omega_0 = \dfrac{1}{RC}$

The gain vs-frequency plot for this transfer function is given in Fig. 5.71 (*a*).

Avoid going into mathematical details to prove the

validly of the plot shown in Fig. 5.71 (*a*) for the network of Fig. 5.70 (*a*). In fact, by letting $s = j\,\omega$ in the expression for transfer function (or gain) derived above, one can easily get or/to that using simple mathematics. What is important here is that this plot tells us that this network is a HIGH PASS network with a cut-off at $\omega_o = 1/RC$.

The expression for transfer function for the network of Fig. 5.70 (*b*) can be obtained by writing the two mesh equations and then solving for the current I_2 as follows :

$$\left(R+\frac{1}{Cs}\right)I_1(s) - RI_2(s) = V_i(s)$$

$$\text{and } -RI_1(s)+\left(2R+\frac{1}{Cs}\right)I_2(s) = 0$$

$$I_2(s) = \frac{\Delta_2}{\Delta}$$

$$\text{where } \Delta_2 = \begin{vmatrix} R+1/Cs & V_i(s) \\ -R & 0 \end{vmatrix}$$

$$\text{and } \Delta = \begin{vmatrix} R+1/Cs & -R \\ -R & (2R+1/Cs) \end{vmatrix}$$

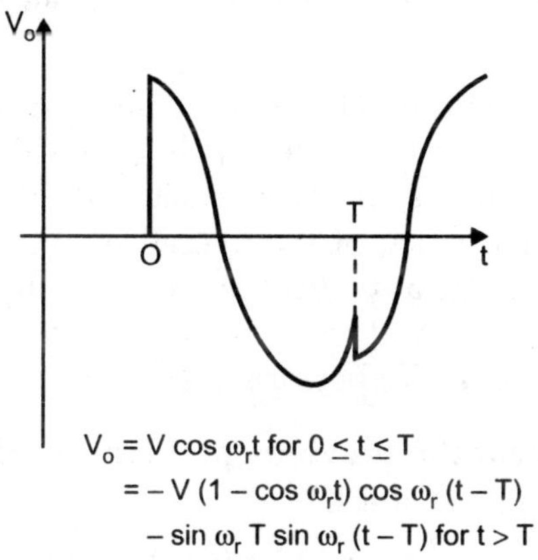

Fig. 5.66

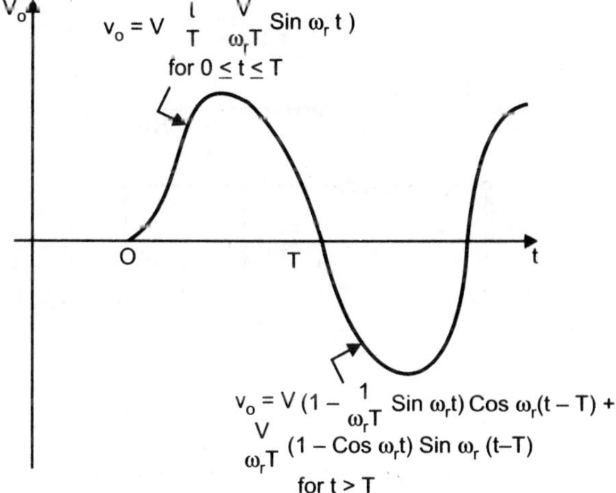

Fig. 5.67

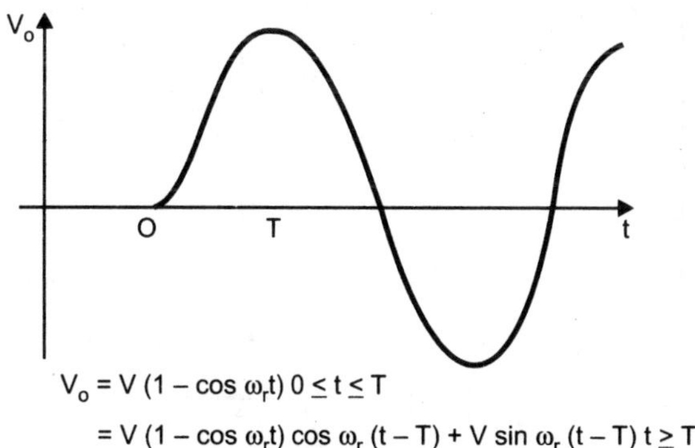

Fig. 5.68

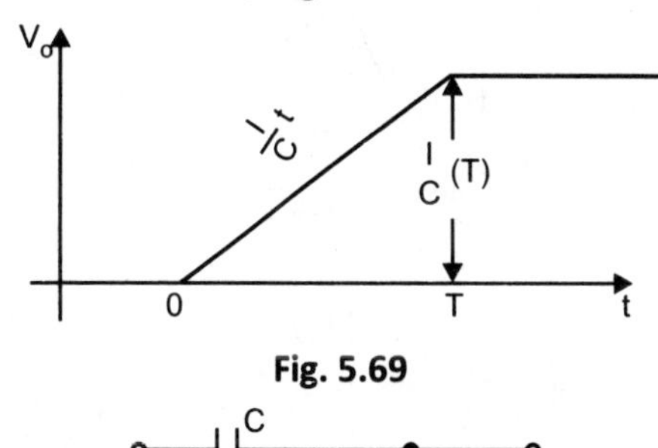

Fig. 5.69

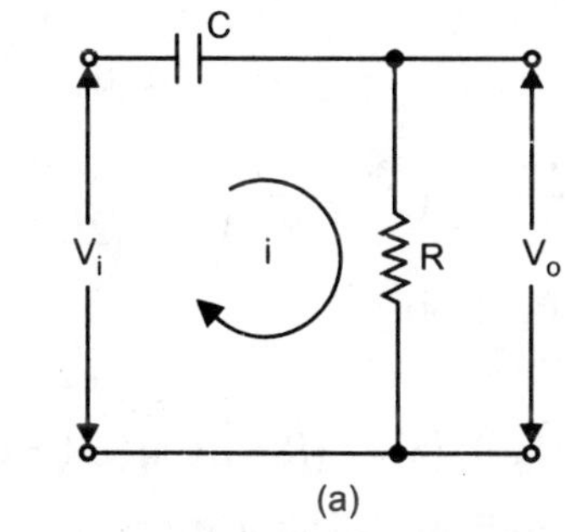

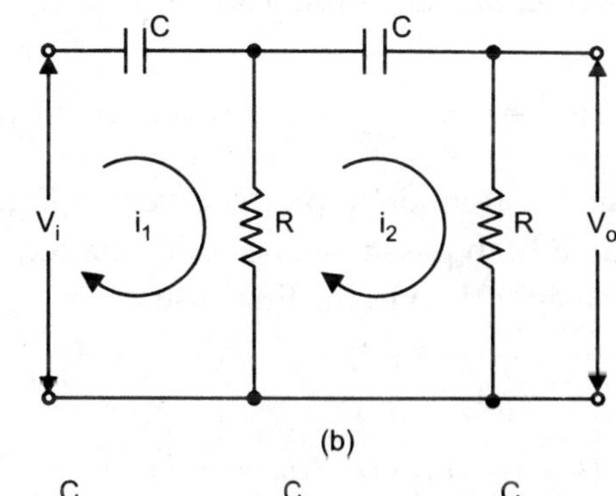

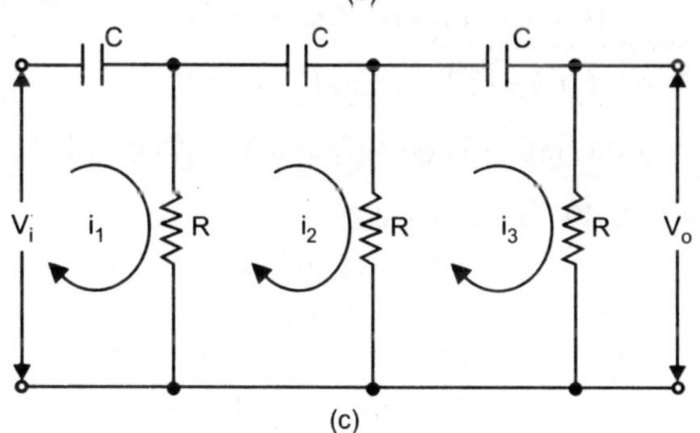

Fig. 5.70

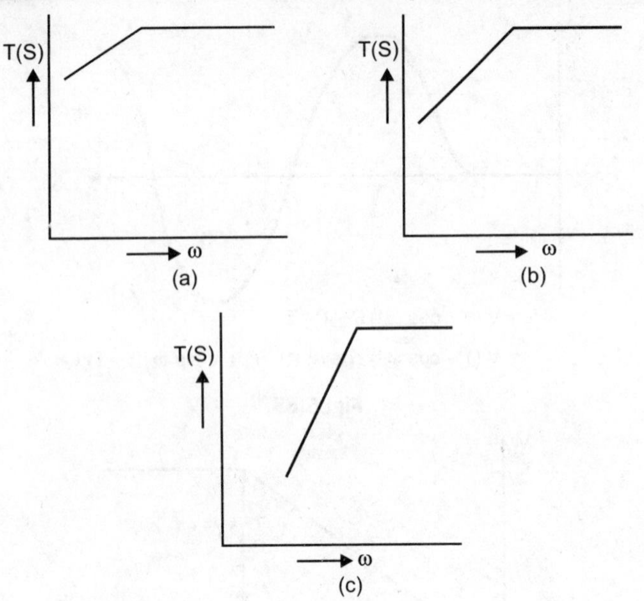

Fig. 5.71

Therefore $$I_2(s) = \frac{RV_i(s)}{(R+1/Cs)(2R+1/Cs)-R^2}$$

$$V_o(s) = RI_2(s) = \frac{V_i(s)R^2}{(R+1/Cs)(2R+1/Cs)-R^2}$$

or Transfer function,

$$T(s) = \frac{V_O(s)}{V_i(s)} = \frac{R^2}{(R+1/Cs)(2R+1Cs)-R^2}$$

After mathematical simplification,

$$T(s) = \frac{s^2}{s^2+3\omega_0 s+\omega_0^2} \text{ where } \omega_0 = \left(\frac{1}{RC}\right).$$

The gain *vs.* frequency plot is given in Fig. 5.71 (*b*). This is again a high pass response with the roll-off in the stop-band steeper than in the first case.

Similarly, writing mesh equations for the network of Fig. 5.70 (*c*).

$$(R+1/Cs)\, I_1(s) - RI_2(s) + 0\, I_3(s) = V_i(s)$$

$$-R\, I_1(s) + (2R+1/Cs)\, I_2(s) - RI_3(s) = 0$$

$$0\, I_1(s) - RI_2(s) + (2R+1/Cs)\, I_3(s) = 0$$

Solving for $I_3(s)$, we get

$$I_s(s) = \frac{\begin{vmatrix} R+1Cs & -R & V_i(s) \\ -R & (2R+1/Cs) & 0 \\ 0 & -R & 0 \end{vmatrix}}{\begin{vmatrix} R+1/Cs & -R & 0 \\ -R & (2R+1/Cs) & -R \\ 0 & R & (2R+1/Cs) \end{vmatrix}}$$

Solving for these determinants and doing mathematical simplification, we get

$$T(s) = \frac{RI_3(s)}{V_1(s)} = \frac{sC}{(s^3+6\omega_0 s^2+5\omega_0^2 s+1)}$$

Some conclusions. Now if we look at the expressions or the transfer functions for all the three cases, we would notice that the transfer function for the network of Fig. 5.70 (*b*) is not equal to the square of the transfer function for the network of Fig. 5.70 (*a*) although network of Fig. 5.70 (*b*) is nothing but two stages of Fig. 5.70 (*a*) cascaded together Similarly, the transfer function of the network of Fig. 5.70 (*c*) is not the cube of that of Fig. 5.70 (*a*) network. The reasons for this are as follows: when two or more than two identical stages (or even nonidentical stages) are cascaded together, each stage loads onto the stage immediately preceding it and when we determine the overall transfer function by simple multiplication of the transfer functions of individual stages determined independently, we are overlooking this interstage dependence. It is for these reasons that we get false results.

Secondly, we have also noticed that by cascading more number of high pass stages with identical components and hence same cut-off frequencies, we do not change the cut-off frequency of the overall network. What gets affected is the rate at which the output to input ratio falls in the stop-band. And from that angle, more the number of stages, better is the performance.

Q.27. Repeat the previous problem for the networks of Fig. 5.72 (*a*), (*b*) and (*c*).

Ans. The transfer function for the network of Fig. 5.72 (*a*) is given by

$$T(s) = \frac{V_o(s)}{V_i(s)} = \frac{1/Cs}{R+1/Cs} = \frac{1}{RCs+1} = \frac{1/RC}{s+1/RC}$$

$$= \frac{\omega_0}{s+\omega_0}; \text{ where } \omega_0 = \frac{1}{RC}$$

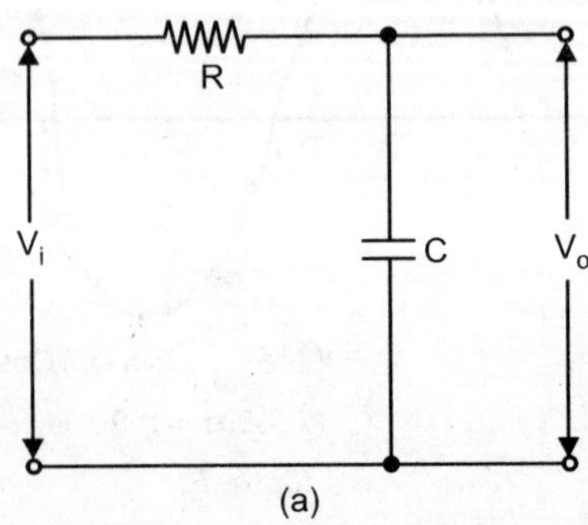

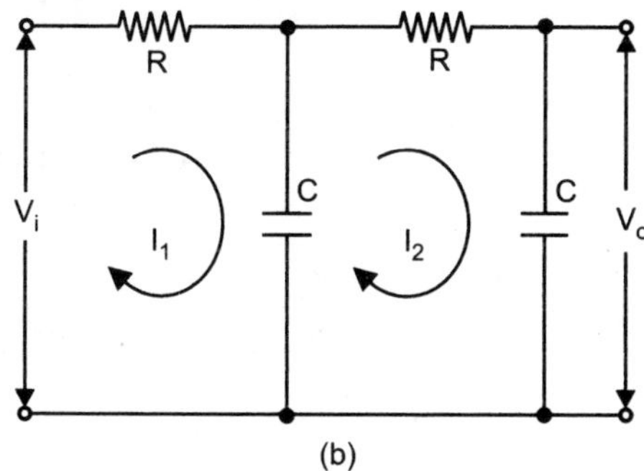

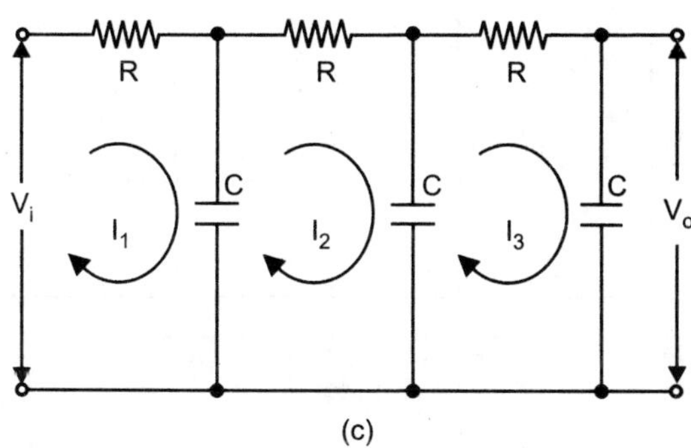

Fig. 5.72

The gain vs. frequency plot is given in Fig. 5.73 (*a*). The transfer function for the network of Fig. 5.72 (*b*) can be obtained by writing mesh equation and then solving for I_2 as follows:

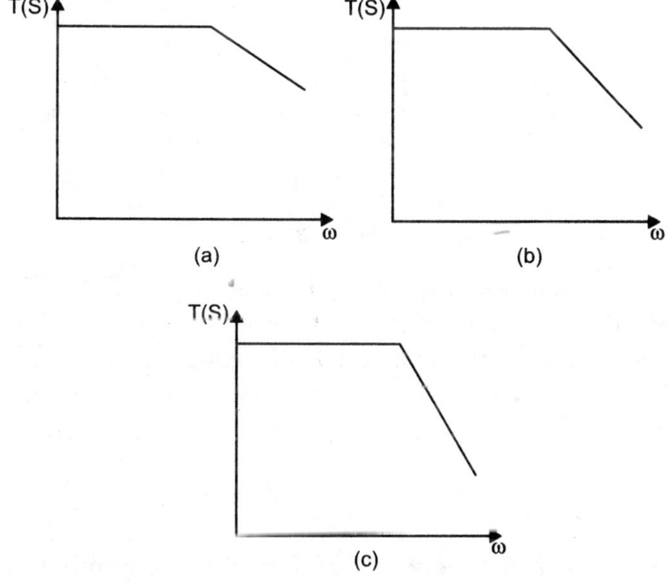

Fig. 5.73

$$\left(R+\frac{1}{Cs}\right)I_1(s)-\left(\frac{1}{Cs}\right)I_2(s) = \text{Vi}(s)$$

and $$\left(-\frac{1}{Cs}\right)I_1(s)+\left(R+\frac{2}{Cs}\right)I_2(s) = 0$$

$$\text{Is}(s) = \frac{\begin{vmatrix} R+\frac{1}{Cs} & V_i(s) \\ \frac{-1}{Cs} & 0 \end{vmatrix}}{\begin{vmatrix} R+\frac{1}{Cs} & \frac{-1}{Cs} \\ \frac{-1}{Cs} & R+\frac{2}{Cs} \end{vmatrix}}$$

$$= \frac{V_i(s)/Cs}{\left(R+\frac{1}{Cs}\right)\left(R+\frac{2}{Cs}\right)-\frac{1}{c^2s^2}}$$

$$V_0(s) = I_2(s)\times\frac{1}{Cs}$$

$$= \frac{V_i(s)/(Cs)^2}{\left(R+\frac{1}{Cs}\right)\left(R+\frac{2}{Cs}\right)-\frac{1}{C^2s^2}}$$

or $$T(s) = \frac{V_0(s)}{V_i(s)} = \frac{1/C^2s^2}{(R+1/Cs)(R+2/Cs)-1/C^2s^2}$$

After mathematical simplification,

$$T(s) = \frac{\omega_0^2}{s^2+3\omega_0+\omega_0^2}; \quad \text{where } \omega_0 = \left(\frac{1}{RC}\right)$$

Similarly for the network of Fig. 5.73 (*c*).

$$\left(R+\frac{1}{Cs}\right)I_1(s)-\left(\frac{1}{Cs}\right)I_2(s)+0\,I_3(s) = V_i(s)$$

$$\left(-\frac{1}{Cs}\right)I_1(s)+\left(R+\frac{2}{Cs}\right)I_2(s)-\left(\frac{1}{Cs}\right)I_3(s) = 0$$

$$0\,I_1(s)-\left(\frac{1}{Cs}\right)I_2(s)+\left(R+\frac{2}{Cs}\right)I_3(s) = 0$$

$$I_3(s) = \frac{\begin{vmatrix} \left(R+\frac{1}{Cs}\right) & \left(\frac{-1}{Cs}\right) & V_i \\ \left(\frac{-1}{Cs}\right) & \left(R+\frac{2}{Cs}\right) & 0 \\ 0 & \left(-\frac{1}{Cs}\right) & 0 \end{vmatrix}}{\begin{vmatrix} \left(R+\frac{1}{Cs}\right) & \left(\frac{-1}{Cs}\right) & 0 \\ \left(\frac{-1}{Cs}\right) & \left(R+\frac{2}{Cs}\right) & \left(\frac{-1}{Cs}\right) \\ 0 & \left(\frac{-1}{Cs}\right) & \left(R+\frac{2}{Cs}\right) \end{vmatrix}}$$

$$V_0(s) = I_3(s)\times\frac{1}{Cs}$$

Solving for $V_0(s)$ and determining $\frac{V_0(s)}{V_i(s)}$, we get

$$T(s) = \frac{V_0(s)}{V_i(s)} = \frac{\omega_0^3}{s^3+5\omega_0 s^2+6\omega_0^2 s+\omega_0^3}$$

The gain us. frequency plot is shown in Fig. 5.73 (*c*)

The observations are similar to the ones made in case of previous question. Firstly, these are low pass networks, the attenuation in the stop band becoming more and more steep with addition of sections. Secondly, once again the overall transfer function cannot be determined by simple multiplication of the transfer functions of the individual stages determined independently without taking into consideration interdependence of various stages.

Q.28. Find the s-domain transfer function for the network of Fig. 5.74. Draw qualitative gain vs. frequency plot and state the condition under which the given network might behave as a band pass network.

Ans. The two mesh equations are

$$\left(R_1 + \frac{1}{C_1 s}\right) I_1(s) - \left(\frac{1}{C_1 s}\right) I_2(s) = V_i(s)$$

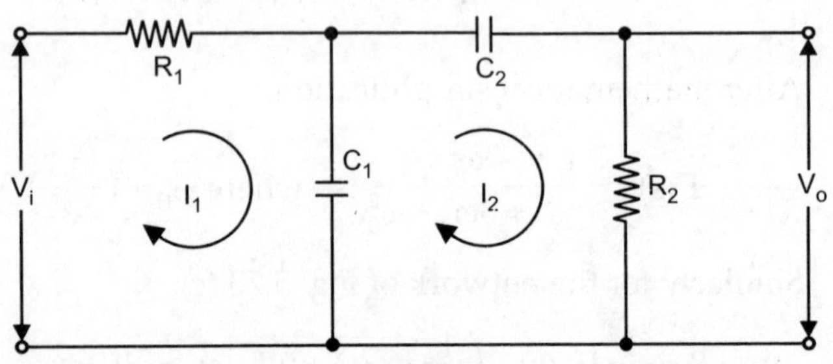

Fig. 5.74

$$-\left(\frac{1}{C_1 s}\right) I_1(s) \left(R_2 + \frac{1}{C_1 s} + \frac{1}{C_2 s}\right) I_2(s) = 0$$

$$I_2(s) = \frac{\begin{vmatrix} R + \dfrac{1}{C_1 s} & V_i(s) \\ \dfrac{-1}{C_1 s} & 0 \end{vmatrix}}{\begin{vmatrix} R_1 + \dfrac{1}{C_1 s} & -\dfrac{1}{C_1 s} \\ \dfrac{-1}{C_1 s} & R_2 + \dfrac{1}{C_1 s} + \dfrac{1}{C_2 s} \end{vmatrix}}$$

$$= \frac{V_i(s) \times 1/C_1 s}{R_1 R_2 + \dfrac{R_1}{C_1 s} + \dfrac{R_2}{C_1 s} + \dfrac{1}{(C_1 s)^2} + \dfrac{1}{C_1 C_2 s^2}}$$

$$V_0(s) = I_2(s) R_2$$

$$= \frac{V_i(s) \times R_2/C_1 s}{R_1 R_2 + \dfrac{R_1 + R_2}{C_1 s} + \dfrac{1}{(C_1 s)^2} + \dfrac{1}{C_1 C_2 s^2}}$$

$$T(s) = \frac{V_0(s)}{V_i(s)}$$

$$= \frac{R_2/C_{1s}}{R_1 R_2 + \dfrac{R_1 + R_2}{C_1 s} + \dfrac{1}{(C_1 s)^2} + \dfrac{1}{C_1 C_2 s^2}}$$

$$T(s) = \frac{s}{R_1 C_1 s^2 + \left(\dfrac{R_1 C_1 + R_2 C_2 + R_1 C_2}{R_2 C_2}\right) s + \dfrac{1}{R_2 C_2}}$$

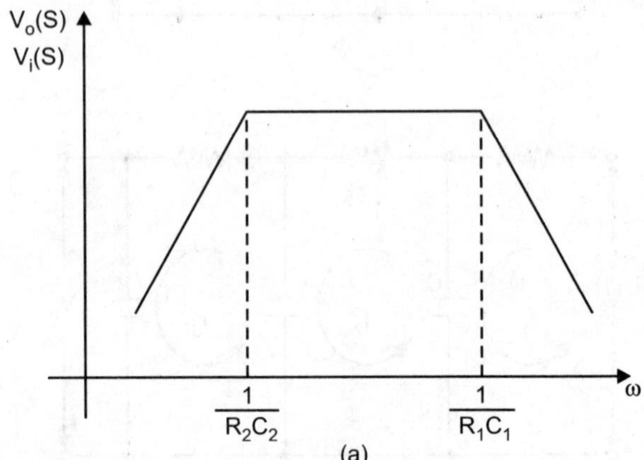

(a)

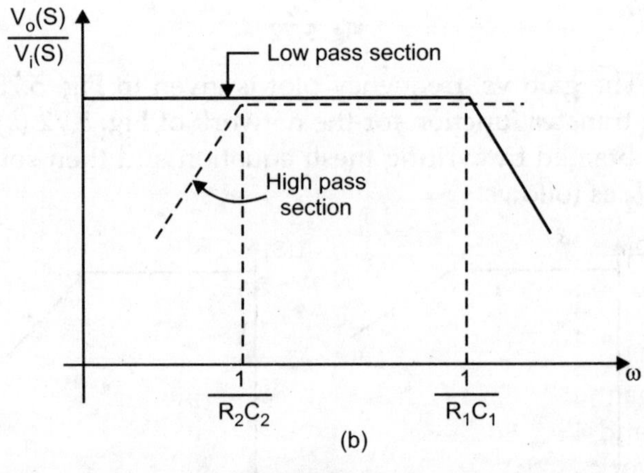

(b)

Fig. 5.75

The gain vs. frequency plot is shown in Fig. 5.75. It will have these band pass characteristics provided that the low pass section has a cut-off frequency greater than the cut-off frequency of the high pass section following the low pass section. See Fig. 5.75 (*b*). Mathematically,

$$\frac{1}{R_1 C_1} > \frac{1}{R_2 C_2} \text{ or } R_1 C_1 < R_2 C_2$$

Q.29. For the network of Fig. 5.76, determine the s-domain transfer function. Show that for $R_1 = R_2 = R$, $C_1 = C_2 = C$ and $\omega = 1/RC$, the transfer function is real with magnitude equal to 0.33. Also, such a network finds use in one of the commonly used audio frequency oscillators. Can you name that ?

Ans. The s-domain transfer function is given by

$$T(s) = \frac{\frac{R_2/C_2 s}{R_2 + 1/C_2 s}}{\left[R_1 + \frac{1}{C_1 s}\right] + \left[\frac{R_2/C_2 s}{R_2 + 1/C_2 s}\right]}$$

After mathematical simplification and substituting $R_1 = R_2 = R$ and $C_1 = C_2 = C$

$$T(s) = \frac{s/Rc}{\left\{s^2 + \left(\frac{3}{Rc}\right)s + \frac{1}{R^2 + 1/C_2 s}\right\}}$$

$$= \frac{j\omega/RC}{\left(\frac{1}{R^2C^2} - \omega^2\right) + j\,\omega\,(3/RC)}$$

Substituting $\omega = \frac{1}{RC}$ we get

$$T(s) = \frac{(j/R^2C^2)}{0 + j3/R^2C^2} = \frac{1/R^2C^2}{3/R^2C^2} = \frac{1}{3} = 0.33$$

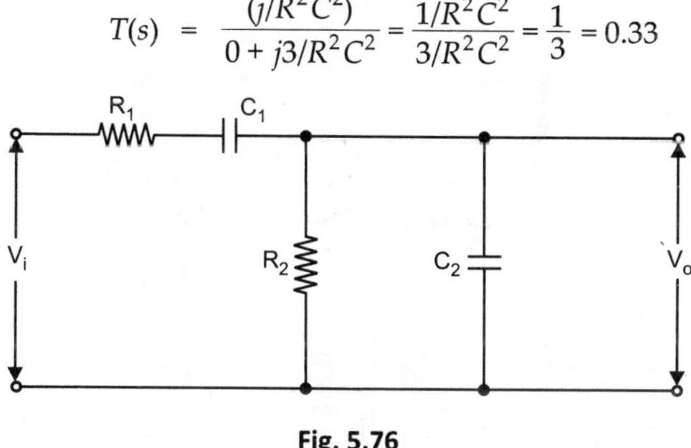

Fig. 5.76

Thus for $\omega = 1/RC$, the transfer function is real and its magnitude is 0.33. In the above network when $R_1 = R_2 = R$ and $C_1 = C_2 = C$, the network becomes the feedback network of a Wien Bridge Oscillator whose frequency of oscillation is then $= \left(\frac{1}{2\pi RC}\right)$. The series R-C and the shunt R C arms of the above network also form the two arms of a WIEN BRIDGE used for impedance measurements.

Q.30. Two coils with self-inductances L_1 and L_2 are closely coupled with a coupling co-efficient of k. What is the magnitude of mutual inductance between the two coils. How did you arrive at these results.

Ans. The mutual inductance M is given by $k\sqrt{L_1 L_2}$

Derivation

Let us assume that a current (i_1) flowing through the coil of self inductance L_1 produces a flux (ϕ_1). Then voltage induce across L_1 is

$$N_1 \frac{d\varphi_1}{dt} = \frac{L_1 di_1}{dt}$$

i.e., $$N_1\, d\varphi_1 = L_1\, di_1 \text{ or } \frac{N_1 d\varphi_1}{di_1} \quad ...(1)$$

If the co-efficient of coupling is (k), then the flux linking the second coil is $(k\phi_1)$ and the voltage induced is given by

$$N_2\, d\,\frac{(k\varphi_1)}{dt} = M\frac{di_1}{dt} \text{ i.e., } N_2\, dk\, \varphi_1 = M\, di_1$$

or $$M = \frac{N_2\, dk\varphi_1}{di_1} = \frac{kN_2\, d\varphi_1}{di_1} \quad ...(2)$$

If we had assumed a current (i_2) flowing in the second coil we would at

$$\frac{N_2 d\varphi_2}{di_2} = L_2 \quad ...(3)$$

and $$\frac{kN_1 d\varphi_2}{di_2} = M \quad ...(4)$$

From eqns. 2 and 4

$$M_2 = \left(\frac{kN_2 d\varphi_1}{di_1}\right)\left(\frac{kN_1\, d\varphi_2}{di_2}\right)$$

$$= \frac{k^2 N_1 d\varphi_1}{di_1} \times \frac{N_2 d\varphi_2}{di_2} = k^2$$

$$= k_2\, L_1\, L_2 \text{ from eqns. 1 and 3}$$

Therefore, $$M = k\sqrt{L_1 L_2}$$

Q.31. Write down the expressions for the time variant parameter indicated by (?) in the circuits of Fig. 5.77 (*a*) to (*f*).

Ans. Fig. 5.77 (*a*) : By visual Inspection itself, one can write

$$v_2\,(t) = M\frac{dI_i}{dt}$$

We have taken (+) sign for M because current would enter the dots in both the windings.

Fig. 5.77. (*b*) : Again one can write expression of $v_2\,(t)$ as

$$v_2\,(t) = L_2\frac{di_2\,(t)}{dt} - M\frac{di_1\,(t)}{dt}$$

Now why minus sign for M, the reason is while $i_1\,(t)$ is leaving the dot, $i_2\,(t)$ is entering the dot.

Fig. 5.77 (c) : Now $v_1\,(t) = M\frac{d}{dt}$ (current through L_2)

Current through

$$L_2 = \frac{1}{L_2}\int v_2(t)\, dt$$

$$\therefore \quad v_1(t) = M\frac{d}{dt}\left\{\frac{1}{L_2}(t)dt\right\} = \frac{M}{L_2}v_2(t)$$

Fig. 5.77. (*d*) : This is straight forward

$$v_1(t) = M\frac{di_2(t)}{dt}$$

For 5.77 (*e*) : Now $-L_1\frac{di_1(t)}{dt}$ = voltage induced across $(L_1) = \frac{Mdi_2(t)}{dt}$

or $\quad di_1(t) = \frac{M}{L_1}di_2(t)$ or $i_1(t) = -\frac{M}{L_1}i_2(t)$

Fig. 5.77 (*f*) : $v(t) = Ri(t) + L_2\frac{di(t)}{dt} - M\frac{di(t)}{dt} + L_1\frac{di(t)}{dt}$

$$-M\frac{di(t)}{dt} + \frac{1}{C}\int i(t)dt$$

$$= R_i(t) + L_1\frac{di(t)}{dt} + L_2\frac{di(t)}{dt} - 2M\frac{di(t)}{dt}$$

$$+\frac{1}{C}\int i(t)dt$$

It may be mentioned that the mutually induced e.m.f has been taken to be of minus sign in both the cases because while the coil L_2 has been wound in the clockwise fashion, the coil L_1 has been wound anti-clockwise.

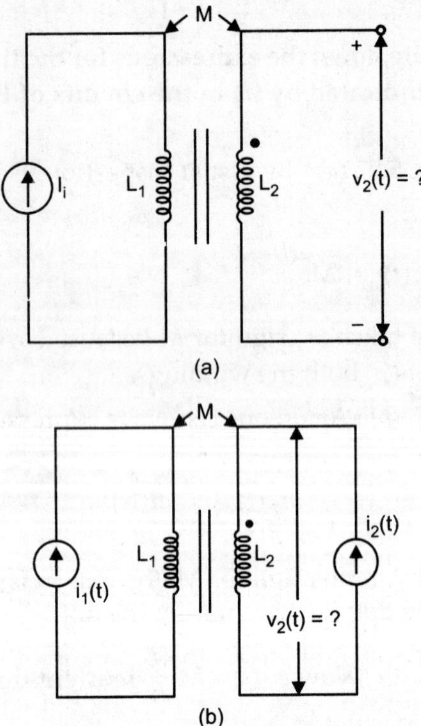

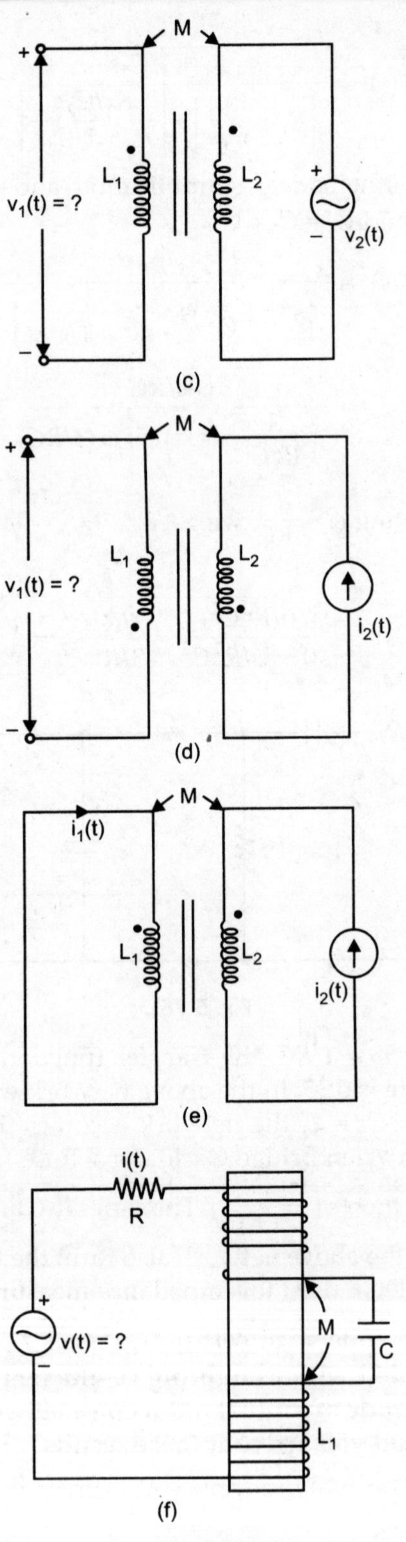

Fig. 5.77

Solved Problems

Problem 5.1. *A network has 7 nodes and 5 independent loops. Determine of the number of branches in the network.*

Solution: $e = b - (n - 1)$, where (b) is the number of branches, (n) is number of nodes and e is the number of independent loops. Therefore, $b = e + (n - 1) = 5 + (7 - 1) = 5 + 6 = 11$.

Problem 5.2. *For the network shown in Figure 5.78, determine the maximum number of possible trees. Assume all active sources to be ideal.*

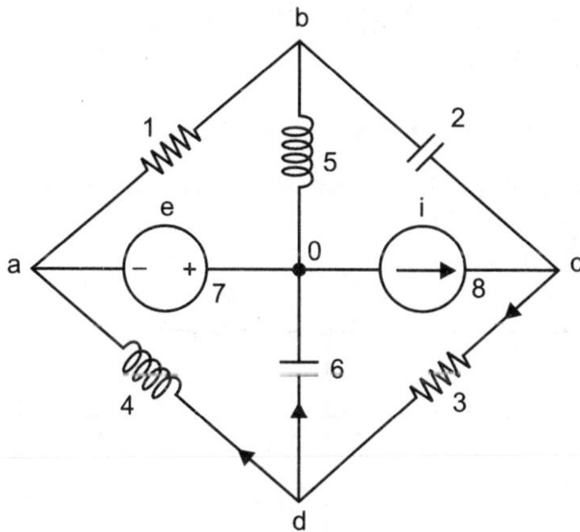

Fig. 5.78

Solution. Considering the ideal current source as open-circuit and the ideal voltage source as short circuit, the oriented network graph can be drawn as shown in Figure-5.79. Node (d) is taken as the reference node.

The incidence matrix is given by:

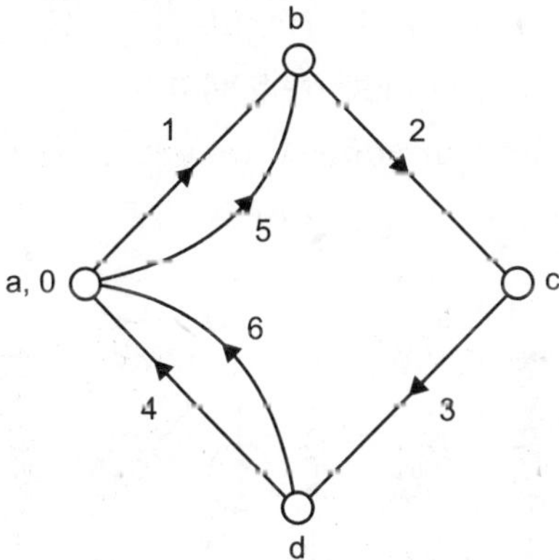

Fig. 5.79

$$A = \begin{bmatrix} 1 & 0 & 0 & -1 & 1 & -1 \\ -1 & 1 & 0 & 1 & -1 & 0 \\ 0 & -1 & 1 & 0 & 0 & 0 \end{bmatrix}$$

The transpose of the matrix A is given by:

$$A^T = \begin{bmatrix} 1 & -1 & 0 \\ 0 & 1 & -1 \\ 0 & 0 & 1 \\ -1 & 1 & 0 \\ 1 & -1 & 0 \\ -1 & 0 & 0 \end{bmatrix}$$

$$\det \quad AA^T = \begin{bmatrix} 1 & 0 & 0 & -1 & 1 & -1 \\ -1 & 1 & 0 & 1 & -1 & 0 \\ 0 & -1 & 1 & 0 & 0 & 0 \end{bmatrix} \begin{bmatrix} 1 & -1 & 0 \\ 0 & 1 & -1 \\ 0 & 0 & 1 \\ -1 & 1 & 0 \\ 1 & -1 & 0 \\ -1 & 0 & 0 \end{bmatrix} = 12$$

Therefore, possible number of trees = 12

Problem 5.3. *For the following reduced incidence matrix [A], determine the number of branches and nodes in the corresponding network.*

$$\begin{bmatrix} -1 & 0 & 0 & +1 & 0 & +1 & 0 & 0 \\ +1 & +1 & 0 & 0 & 0 & 0 & +1 & 0 \\ 0 & -1 & +1 & 0 & -1 & 0 & 0 & 0 \\ 0 & 0 & -1 & -1 & 0 & 0 & 0 & -1 \\ 0 & 0 & 0 & 0 & +1 & -1 & -1 & +1 \end{bmatrix}$$

Solution. For a reduced incidence matrix with r rows, the number of nodes is $(r + 1)$. Therefore, the number of nodes for the given matrix = 6.

For the reduced incidence matrix having (c) columns, the network has (c) branches. Therefore, the number of branches = 8.

Problem 5.4. *For the network shown in Figure 5.80, determine the incidence matrix.*

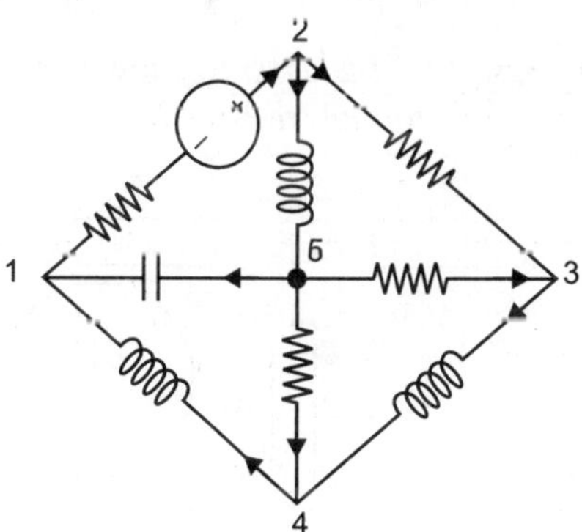

Fig. 5.80

Solution. The network shown in the Figure 5.80 has five nodes and eight branches. The graph for the network is shown in Fig. 5.81.

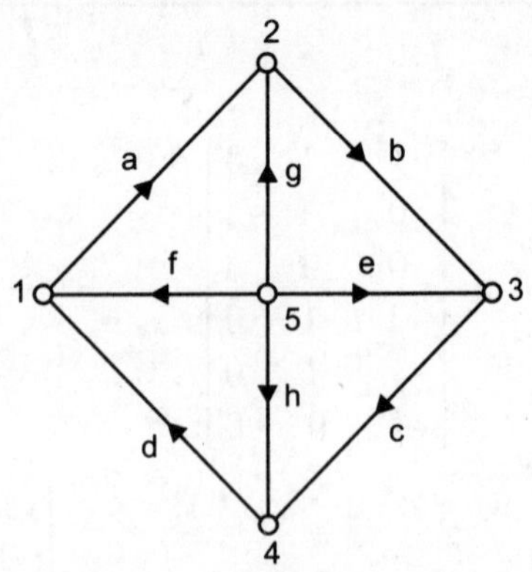

Fig. 5.81

The incidence matrix A_a is comprised of elements a_{ij} such that

a_{ij} = + 1, when the current in branch j leaves node (i)

= – 1, when the current in branch (j) enters node (i)

= 0, when branch (j) is not connected to node (i)

Therefore, incidence matrix is given by:

Nodes	*Branch Numbers*							
	a	*b*	*c*	*d*	*e*	*f*	*g*	*h*
1	+1	0	0	–1	0	–1	0	0
2	–1	+1	0	0	0	0	–1	0
3	0	–1	+1	0	– 1	0	0	0
4	0	0	–1	+1	0	0	0	–1
5	0	0	0	0	+1	+1	+1	+1

The incidence matrix can also be represented as follows.

$$A = \begin{bmatrix} +1 & 0 & 0 & -1 & 0 & -1 & 0 & 0 \\ -1 & +1 & 0 & 0 & 0 & 0 & -1 & 0 \\ 0 & -1 & +1 & 0 & -1 & 0 & 0 & 0 \\ 0 & 0 & -1 & +1 & 0 & 0 & 0 & -1 \\ 0 & 0 & 0 & 0 & +1 & +1 & +1 & +1 \end{bmatrix}$$

Problem. 5.5. *For the circuit shown in Figure 5.82, determine the current through resistor R_1.*

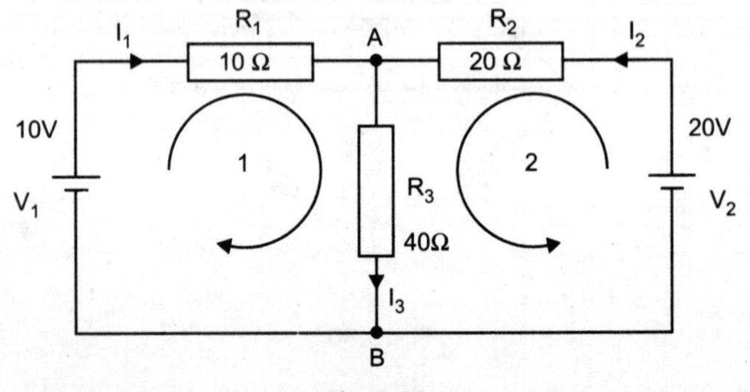

Fig. 5.82

Solution. Applying KCL at node *A*, we get $I_3 = I_1 + I_2$ ...(1)

Applying KVL in left loop, we get $10 - 10I_1 - 40I_3 = 0$ ...(2)

Applying KVL in right loop, we get $20 - 20I_2 - 40I_3 = 0$ ...(3)

Multiplying the second equation by 2 and adding to third equation, we get

$40 - 20\,(I_1 + I_2) - 120I_3 = 0$

Substituting the value of $I_1 + I_2 = I_3$ in the above equation, we get $40 - 140I_3 = 0$

Therefore, $I_3 = 40/140 = 2/7A$

Substituting the value of I_3 in second equation, we get $10 - 10I_1 - 80/7 = 0$

Therefore, $I_1 = -1/7\text{A} = -143\text{mA}$

Problem. 5.6. *Refer to the network of Figure. 5.83. Determine (V).*

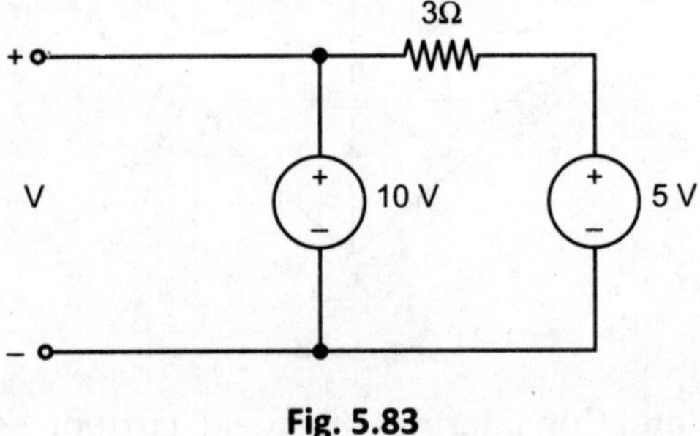

Fig. 5.83

Solution. The voltage (V) in the Figure 5.83 is equal to the voltage across the terminals marked as (+) and (–) which is equal to 10V.

Problem 5.7. *Refer to the network of Figure 5.84 (a). Determine current (i_4) in the circuit.*

Solution. Refer to Figure 5.84 (*b*).

Applying KCL at node-*A*, we get $i_0 + i_1 + i_4 = 0$

Therefore, $7 + 5 + i_4 = 0$ or $i_4 = -12\text{A}$

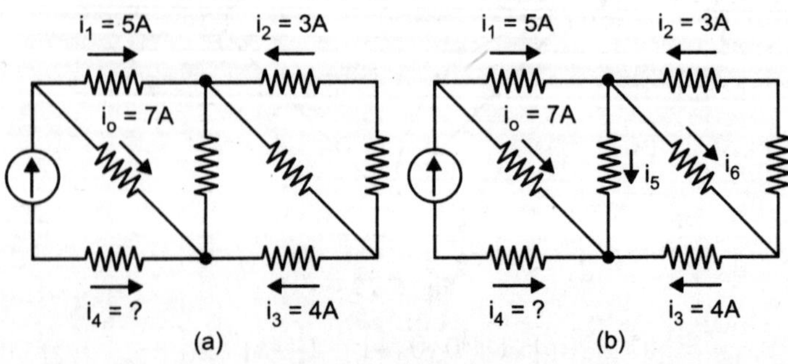

Fig. 5.84

Problem. 5.8. *For the circuit shown in Figure 5.85, determine the power delivered by element D.*

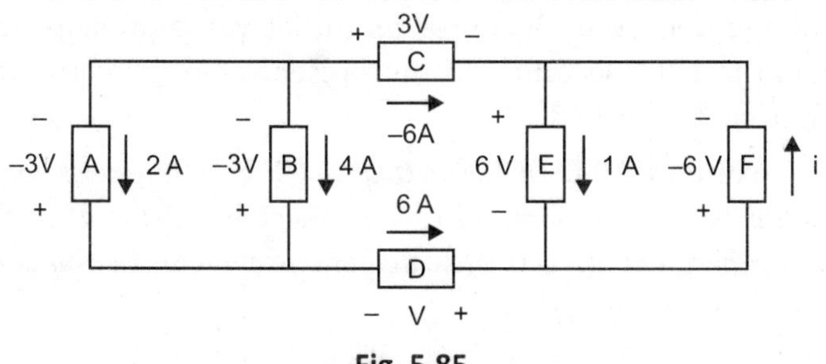

Fig. 5.85

Solution. Applying KVL to loop having elements C, E, D and B, we get $3 + 6 + v + (-3) = 0$

Therefore, $v = -6$V.

Therefore, power supplied by element (D) is $P_D = (-6) \times (6) = -36$W. In other words, the element receives a power of 36W or in theoretical terms, delivers a power of – 36W

Problem 5.9. Apply phasor form of KCL to prove that the voltage (v) in Fig. 5.86 is 100 volts.

Solution. If v_R, v_L and v_C are the voltages across R, L and C respectively.

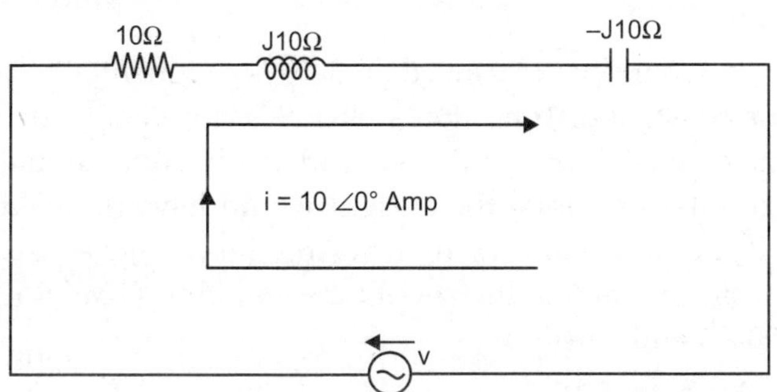

Fig. 5.86

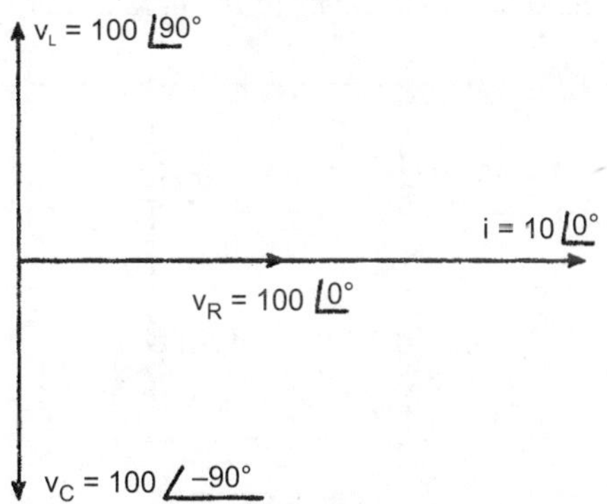

Fig. 5.87

Where the addition on the right hand side has to be addition of phasors. The phasor diagram is given in Fig. 5.87.

The phasor diagram of Fig. 5.87 is self explanatory. Quite obviously, phasor sum of v_R, v_L and v_C equals v_R.

$$v = v_R = 100 \text{ Volts.}$$

Problem 5.10. *Write the expressions for the instantaneous current (i) for t > 0 for networks of Fig. 5.88 (a) 5.88 (d) when the switch S is closed in case of networks of Fig. 5.88 (a) and (b) and is opened in case of network of Fig. 5.88. (c) and (d).*

Solution. (*a*) $\quad i = \frac{10}{10}(1 - e^{-10t}) = (1 - e^{-10t})$

In general, the current grows exponentially in an R-L circuit to which a DC voltage of V volts has been applied at $t = 0$. The generalized expression is given by $i = V/R\ (1 - e^{-Rt/L})$.

(*b*) $\quad i = (1 - e^{-5t})$

(*c*) In general, when a current I is flowing through an inductance L and if it is shorted with a resistance R at time $t = 0$ cutting off the DC voltage source V, then the current I decays exponentially with instantaneous current (i) given by

$$i = I e^{-(R/L)t}$$

Keeping this in mind, for the network of Fig. 5.88 (c), $i = 2e^{-10t}$

(*d*) $\quad i = e^{-20t}$

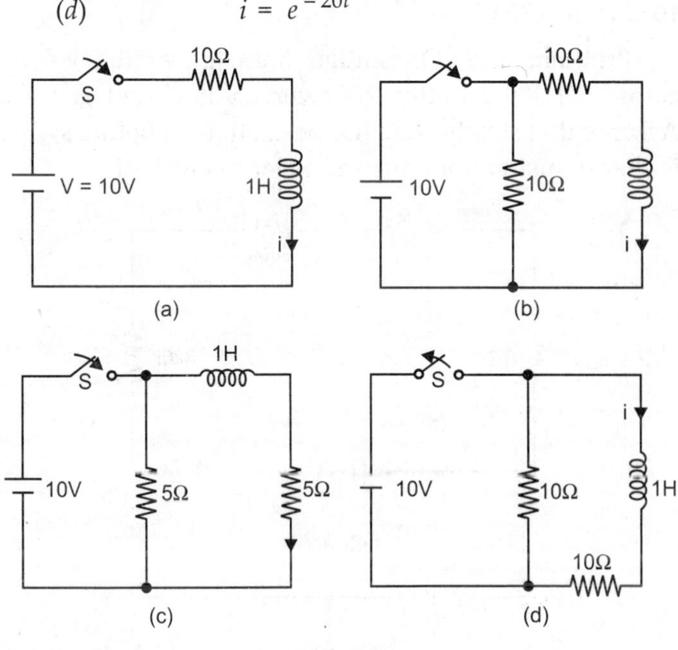

Fig. 5.88

Problem 5.11. *Write expressions for (v) in the networks of Fig. 5.88 (a) and (b). In case (a), S is closed at t = 0. In case (b) too, S is closed at t = 0 but at the same time, you can assume that steady state has been achieved. Also write expressions for instantaneous charge (q) in the capacitor.*

Solution. (*a*) $\quad v = V(1 - e^{-t/RC})$

(*b*) $\quad v = V e^{-t/RC}$

The instantaneous charge (q) for the network of Fig. 5.89 (*a*) is

$$q = CV(1 - e^{-t/RC}) \text{ as } Q = Cv$$

and that for the network of Fig. 5.89 (*b*) is $q = CV e^{-t/RC}$ as $Q = CV$.

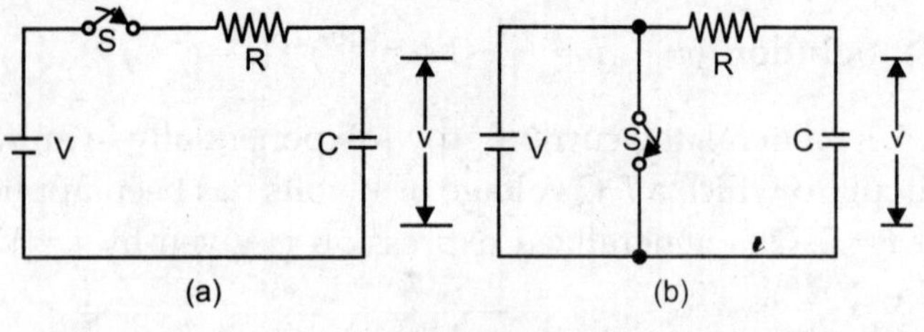

Fig. 5.89

Problem 5.12. *Look at the network of Fig. 5.-S0. Guess what should be the voltage across the capacitor after 1 second if switch S is closed at t = 0. No mathematics please ! But do write the basis of your guess.*

Solution. 6.32 volts.

The network has a time constant of 1 second ($\tau = RC$). In a time equal to time constant, the voltage across *C* in an *R-C* network rises to 63.2% of the steadystate value which is 10 volts in this case. Hence the answer (63.2% of 10 volts is 6.32 volts).

Problem 5.13. On similar lines, guess the voltage across (*C*) 100 μs after the switch *S* is closed at *t* = 0. Assume that steady state has been attained before switch is closed. Justify your answer. Refer to Fig. 5.91.

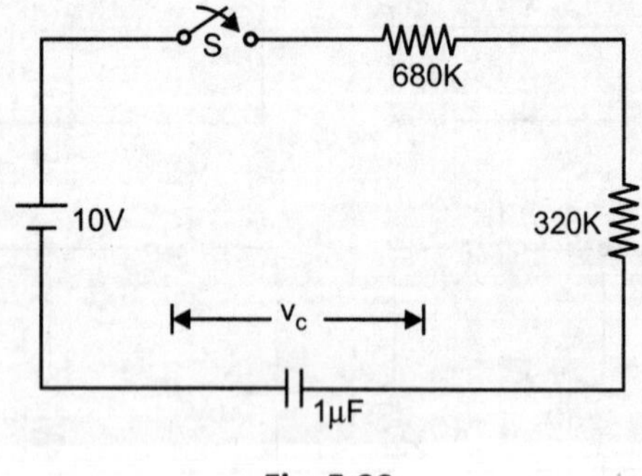

Fig. 5.90

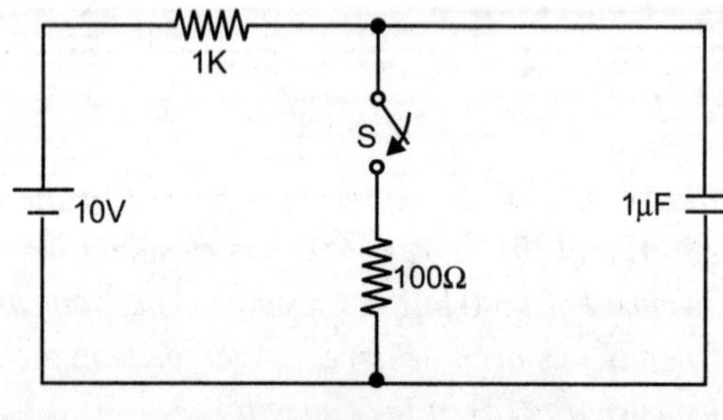

Fig. 5.91

Solution. If a capacitor *C* charged to a voltage *V* initially is made to discharge through a resistance *R*, its voltage decays to 36.8% of its initial value after a time equal to (*RC*) seconds. In the present case *RC* = 100 μs. Therefore, $v_c = 3.68\ V$.

Problem 5.14. *In the network of Fig. 5.92, guess what should be voltage across 10 ohms resistance 2 ms after the switch is closed at t = 0. No calculations allowed. Justify your answer.*

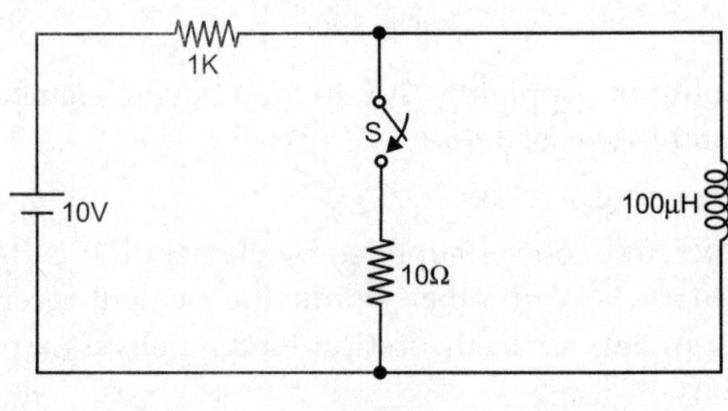

Fig. 5.92

Solution. 13.5 mV.

When the steady state has been attained before the switch S is closed, current through inductance is 10/1 = 10 mA.

This current of 10 mA then decays exponentially with time constant of 1 ms after switch S is closed at *t* – 0. In a time equal to 2 ms (which is incidentally equal to that of two time constants) the current would have decayed to ($1/e^2$) of initial value of 10 mA. And the voltage across 10 Ω resistance at that instant of time would be ($10/e^2 \times 10$) = ($100/e^2$) = 13.5 mV.

Problem 5.15. *What should be the magnitudes of currents i_1, i_2, i_3, and i at the time instants t = 0^+ and t = 10 secs if the switch S is closed at t – 0. Refer to the network of Fig. 5.93.*

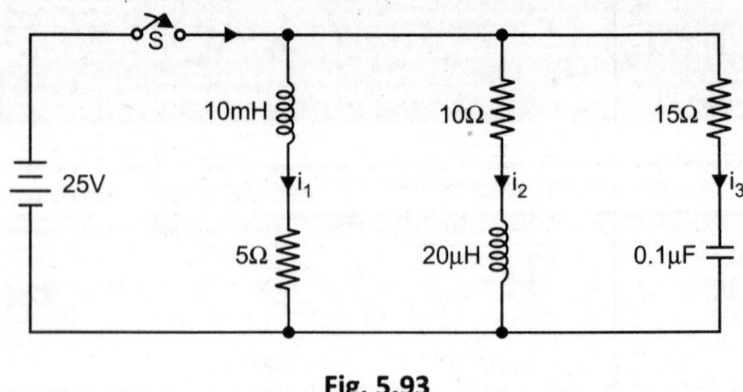

Fig. 5.93

Solution. As the current through an inductance and voltage across a capacitance cannot change instantaneously, the inductor will thus remain an open circuit and the capacitor a short circuit at the instant $t = 0^+$ if that was so at $t = 0$.

Therefore, at $t = 0^+$

$$i_1 = 0, i_2 = 0, = \frac{5}{3} A \text{ and } i = \frac{5}{3} A.$$

The time constants of the three branches are respectively 2 ms, 2μs and 1.5 μs. As the steady state is supposed to be attained in a time equal to five times the time constants, we can safely say that after 10 seconds, all the three branches will be in the steady state.

Therefore at $t = 10$ secs.

$$i = \frac{25}{5} = 5 A, i_2 = \frac{25}{10} = 2.5 A, i_3 = 0$$

and $\quad i = 5 + 2.5 + 0 = 7.5$ A

Problem 5.16. *Refer to the network of Fig. 5.94 Determine potential difference between points A and B.*

Solution. The two loops here are independent and the 2V battery interconnecting the two has no role in supplying current to either of the two loops.

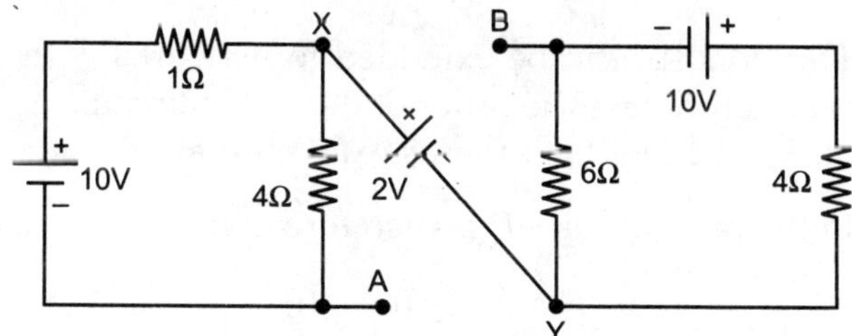

Fig. 5.94

Now $V_{AB} = V_{AX} + V_{XY} + V_{YB}$

Now $V_{AX} = \left(\frac{10}{1+4}\right) \times 4 = 8$ V (Potential rise)

$V_{XY} = -2\, V$ (Potential fall)

$V_{YB} = \frac{10}{(6+4)} \times 6 = -6$V (Potential fall)

Therefore,

$$V_{AB} = 8 - 2 - 6 = 0 \text{ V.}$$

Problem 5.17. *Refer to the bridge circuit of Fig. 5.95. Using mesh current analysis and the Cramer's rule, determine the magnitude and direction of current through the detector assembly.*

Solution. The network of Fig. 5.95 is redrawn in Fig. 5.96 with chosen mesh currents. Mesh currents are so chosen as to have only one current flowing through the detector assembly. This simplifies computation.

The loop equations can be written in the matrix form by network inspection as follows :

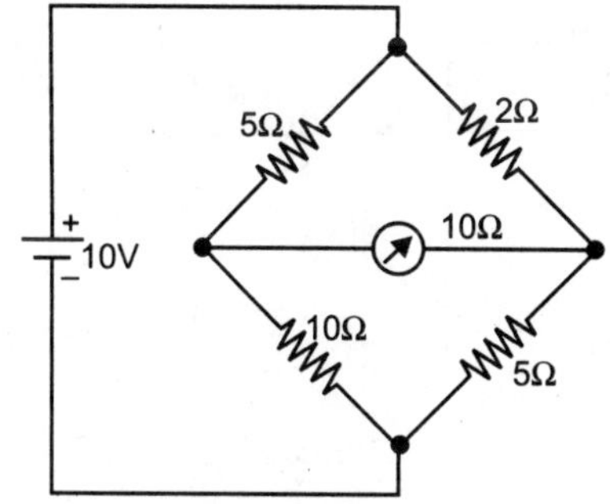

Fig. 5.95

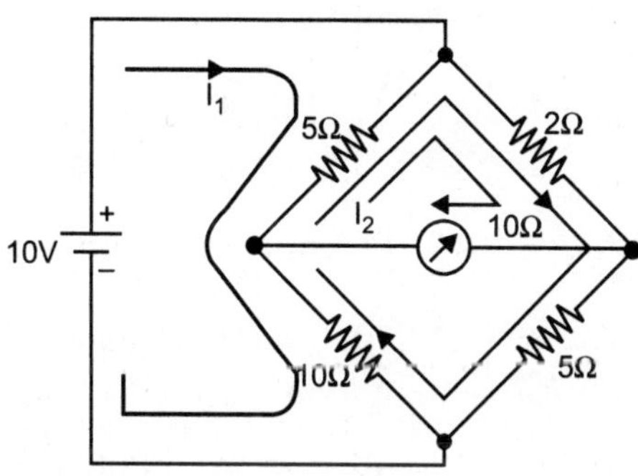

Fig. 5.96

The loop equations can be written in the matrix form by network inspection as follows

$$\begin{bmatrix} 15 & -5 & -15 \\ -5 & 17 & 7 \\ -15 & 7 & 22 \end{bmatrix} \begin{bmatrix} I_1 \\ I_2 \\ I_3 \end{bmatrix} = \begin{bmatrix} 2 \\ 0 \\ 0 \end{bmatrix}$$

Using Cramer's role $i_2 = \frac{\Delta_2}{\Delta_z}$

where $\Delta_2 = \begin{vmatrix} 15 & 2 & -15 \\ -5 & 0 & 7 \\ -15 & 0 & 22 \end{vmatrix}$ and $\Delta_z = \begin{vmatrix} 15 & -5 & -15 \\ -5 & 17 & 7 \\ -15 & 7 & 22 \end{vmatrix}$

Now $\quad \Delta_2 = -2\,(-110 + 105) = 10$

$\Delta_z = 15\,(374 - 49) + 5\,(-110 + 105) \quad 15\,(-35 + 255)$

$= 15 \times 325 - 25 - 15 \times 220 = 4875 - 25 - 3300 = 1550$

This gives $I_2 = \frac{10}{1550}$ A = 0.00645 A = 6.45 mA.

Problem 5.18. *Determine the voltage V_{AB} in the network of Fig. 5.97*

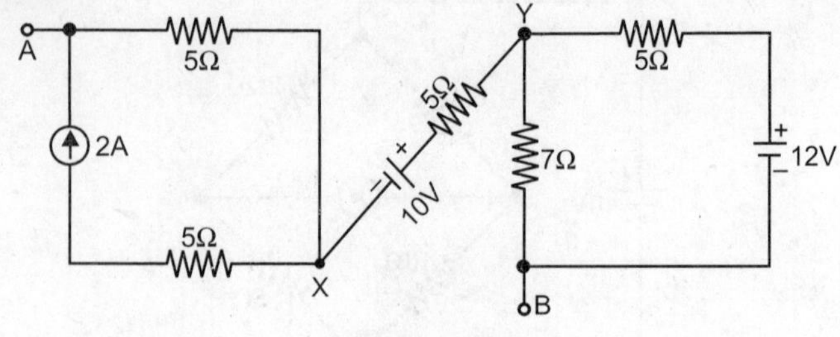

Fig. 5.97

Solution. The two closed loops are independent and therefore the current flowing through the interconnecting branch is zero.

$$V_{AB} = V_{AX} + V_{XY} + V_{YB}$$

$$V_{AX} = 2 \times 5 = 10 \text{ V (Potential rise from } A \text{ to } X)$$

$$V_{XY} = 10 \text{ V (Potential rise from } X \text{ to } Y)$$

$$V_{YB} = \left(\frac{12}{7+5}\right) \times 7 = 7 \text{ V (Potential fall from } Y \text{ to } B)$$

Therefore $V_{AB} = 10 + 10 - 7 = 13$ V (Potential rise from A to B). Point B is at a higher potential than point A.

Problem 5.19. *One of the methods to solve for various mesh equations in a given network is to make use of DETERMINANTS. For the network of Fig. 5.98 (a) write the expression for mesh current I_1 in the determinant form.*

Solution. We shall first explain the basis of writing mesh equations in the determinant form. If we write the two mesh equations, we get

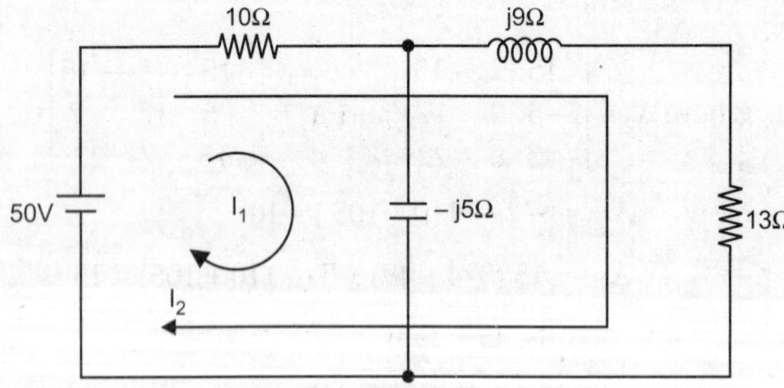

Fig. 5.98

Fig. 5.99

$$(10 - j5)\, I_1 + 10\, I_2 = 50$$

and $$10\, I_1 + (23 + j\,9)\, I_2 = 50$$

The determinant formed by the coefficients of I_1 and I_2 is given by

$$\Delta = \begin{vmatrix} (10 - j5) & 10 \\ 10 & (23 + j9) \end{vmatrix}$$

If in the above determinant, first column is replaced (column constituted by co-efficients of I_1 in the two equations) by the column on the R.H.S. of the two equations, we get

Similarly, $$\Delta_1 = \begin{vmatrix} 50 & 10 \\ 50 & (23 + j9) \end{vmatrix}$$

$$\Delta_2 = \begin{vmatrix} 10 - j5 & 50 \\ 10 & 50 \end{vmatrix}$$

Then $$I_2 = \frac{\Delta_1}{\Delta} \text{ and } I_2 = \frac{\Delta_2}{\Delta}$$

This concept can be extended to networks with any number of meshes. The order of the determinants (Δ_1, Δ_2, Δ_3 ....Δ_n and Δ) is 'n' if there are 'n' meshes.

In the present case, I_1 is therefore given by

$$I_1 = \frac{\begin{vmatrix} 50 & 10 \\ 50 & (23 + j9) \end{vmatrix}}{\begin{vmatrix} (10 - j5) & 10 \\ 10 & (23 + j9) \end{vmatrix}}$$

Problem 5.20. *On similar lines, write expression for node voltage (V_2) in the determinant form. Justify your answer by writing node equations. Refer to the network of Fig. 5.99.*

Solution. The node equations are as under :

$$2V_1 - V_2 - V_3 = 1$$

$$-V_1 + 3V_2 - V_3 = 0$$

$$-V_1 - V_2 + 2V_3 = 1$$

Therefore, $$\Delta = \begin{vmatrix} 2 & -1 & -1 \\ -1 & 3 & -1 \\ -1 & -1 & 2 \end{vmatrix} \quad \Delta_2 = \begin{vmatrix} 2 & 1 & -1 \\ -1 & 0 & -1 \\ -1 & 1 & -2 \end{vmatrix}$$

Hence $$V_2 = \frac{\begin{vmatrix} 2 & 1 & -1 \\ -1 & 0 & -1 \\ -1 & 1 & 2 \end{vmatrix}}{\begin{vmatrix} 2 & -1 & -1 \\ -1 & 3 & -1 \\ -1 & -1 & 2 \end{vmatrix}}$$

Problem 5.21. *Look at the network shown in Fig. 5.100 (a). The given network has three meshes. Let the three mesh* *currents be I_1, I_2 and I_3. The mesh currents were so chosen (not indicated in figure) that the following set of mesh equations was obtained. From the given set of mesh equations, indicate the mesh currents in the network.*

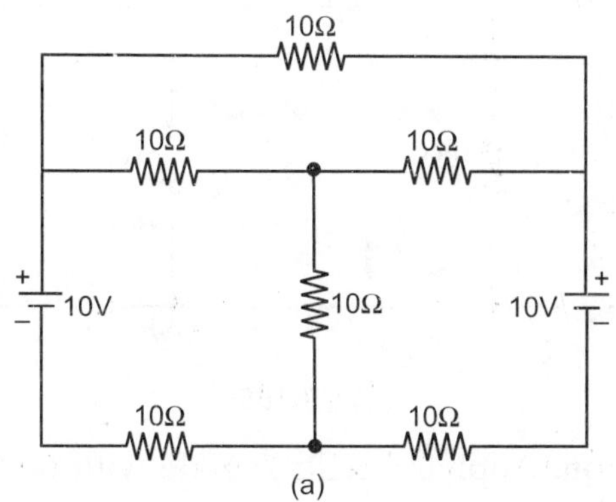

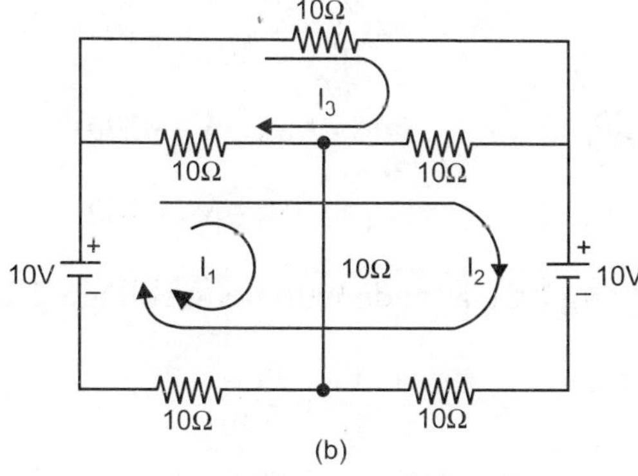

Fig. 5.100

$$30\,I_1 + 20\,I_2 - 10\,I_3 = 10$$
$$20\,I_1 + 30\,I_2 - 20\,I_3 = 0$$
$$-10\,I_1 - 20\,I_2 + 30\,I_3 = 0$$

Solution. The mesh currents are shown in Fig. 5.100 (*b*).

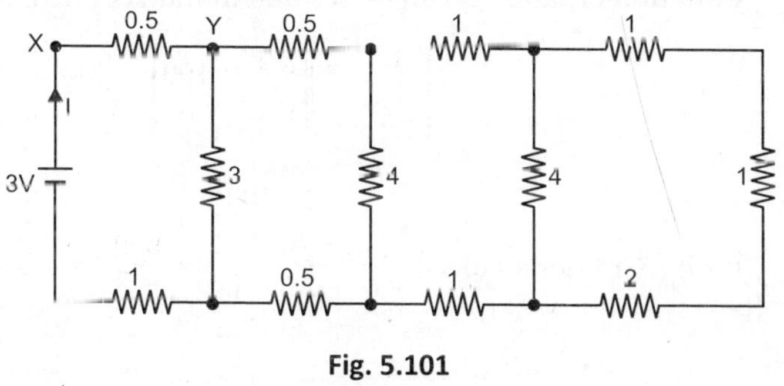

Fig. 5.101

Problem 5.22. *Find current (I) in the network of Fig. 5. 101. All resistances are in ohms.*

Solution. The network of Fig. 5. 101 can be simplified step by step as depicted in Figs. 5.102 (*a*) to 5.102 (*c*). From Fig. 5.102 (*c*).

$$I = 1A$$

If we had made an attempt solving this problem using mesh current analysis, we would have written four mesh equations and solving them simultaneously would have filled at least a couple of pages.

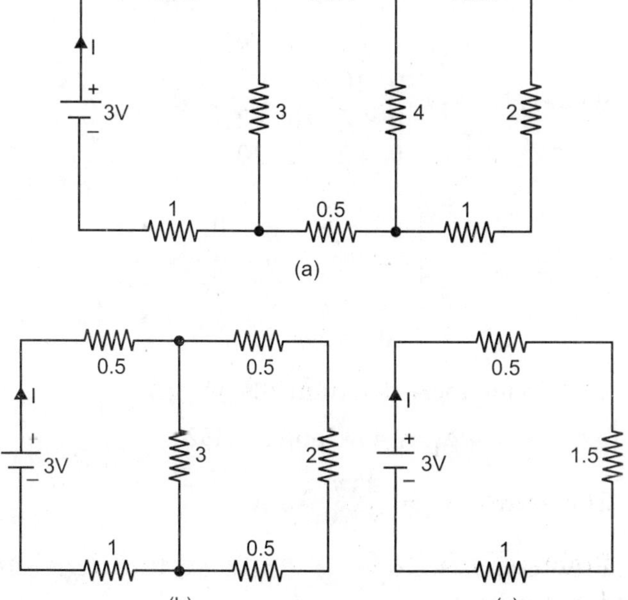

Fig. 5.102

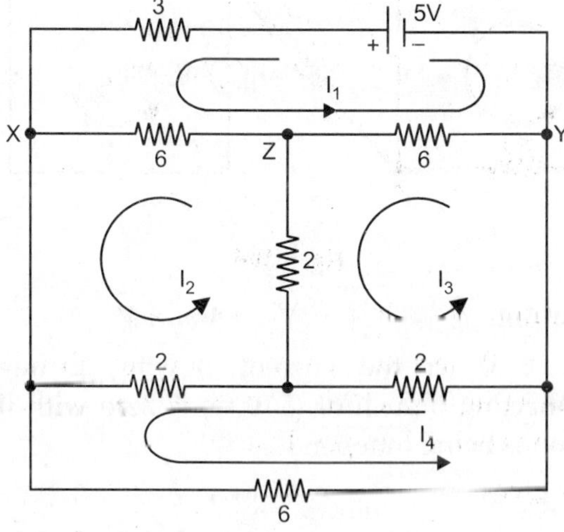

Fig. 5.103

Problem 5.23. *Find current (I) in the network of Fig. 5.103 by Mesh current analysis. What conclusions do you draw ? All resistances are in ohms.*

Solution. (*i*) From the network of Fig. 5.103 and the mesh currents chosen, following mesh-current expressions can be written on application of KVL in all the meshes individually.

$$15\,I_1 - 6\,I_2 - 6\,I_3 + 0\,I_4 = 5$$

$$-6I_1 + 10I_2 - 2I_3 - 2I_4 = 0$$
$$-6I_1 - 2I_2 + 10I_3 - 2I_4 = 0$$
$$0I_1 - 2I_2 - 2I_3 + 10I_4 = 0$$

From determinant method, $I_1 = \dfrac{\Delta_1}{\Delta}$

where $\Delta_1 = \begin{vmatrix} 5 & -6 & -6 & 0 \\ 0 & 10 & -2 & -2 \\ 0 & -2 & 10 & -2 \\ 0 & -2 & -2 & 10 \end{vmatrix}$ and

$$\Delta = \begin{vmatrix} 15 & -6 & -6 & 0 \\ -6 & -10 & -2 & -2 \\ -6 & -2 & 10 & -2 \\ 0 & -2 & -2 & 10 \end{vmatrix}$$

On solving these determinants, we get

$$\Delta_1 = 4320 \text{ and } \Delta = 4320$$

Therefore $I_1 = \dfrac{4320}{4320} = 1\text{ A}.$

Problem 5.24. *Refer to the network of Fig. 5.104. Determine voltage V_{AB}.*

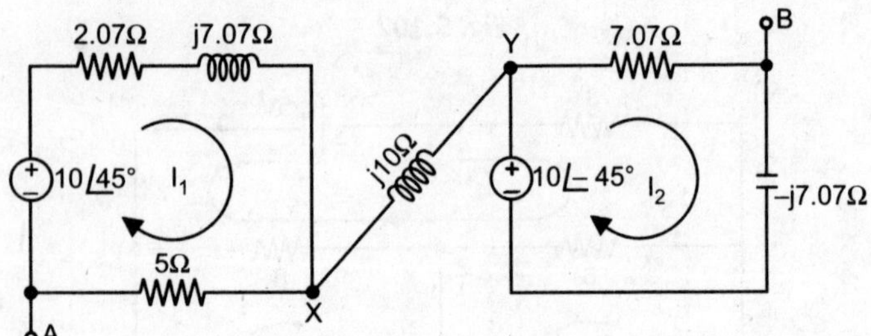

Fig. 5.104

Solution. $V_{AB} = V_{AX} + V_{XY} + V_{YB}$

$V_{XY} = 0$ as the current flowing through the interconnecting branch of (j 10 Ω) is zero with the two closed loops being independent.

This gives $V_{AB} = V_{AX} + V_{YB}$

$$V_{AX} = -5I_1 \text{ and } V_{YB} = 7.07\,I_2$$

$$I_1 = \frac{10}{(7.07 + j\,7.07)} = \frac{10\angle -45^\circ}{10\angle -45^\circ} = 1\angle 0^\circ$$

$$I_2 = \frac{10\angle -45^\circ}{(7.07 - j\,7.07)} = \frac{10\angle -45^\circ}{10\angle -45^\circ} = 1\angle 0^\circ$$

Therefore $V_{AX} = -5 \times 1 + 7.07 \times 1 = 2.07$ volts

Problem 5.25. *Fig. 5.105 shows a network with a dependent current source. Determine current (I_2).*

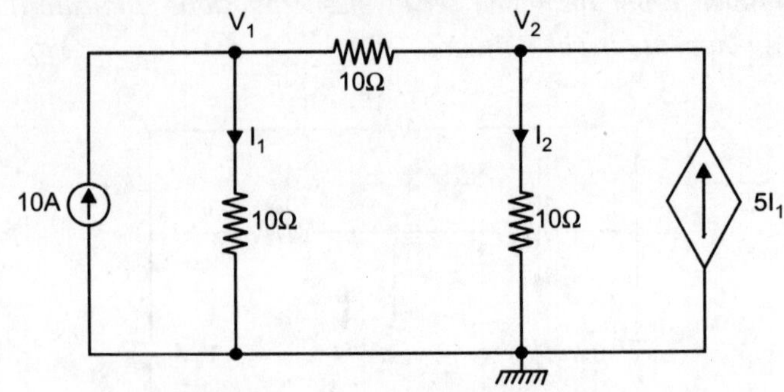

Fig. 5.105

Solution. Applying KCL at node with node voltage (V_1),

$$\frac{V_1}{10} + \frac{V_1 - V_2}{10} = 10$$

or $$V_1 + V_1 - V_2 = 100$$

or $$2V_1 - V_2 = 100 \quad \text{...(1)}$$

Applying KCL at node with node voltage (V_2),

$$\frac{V_2}{10} + \frac{V_2 - V_1}{10} = 5\,I_1$$

or $$V_2 + V_2 - V_1 = 50\,I_1$$

$$= \frac{50\,V_1}{10} = 5\,V_1$$

or $$-6V_1 + 2\,V_2 = 0 \quad \text{...(2)}$$

Equation (1) and (2) can be written in matrix form as

$$\begin{bmatrix} 2 & -1 \\ -6 & 2 \end{bmatrix}\begin{bmatrix} V_1 \\ V_2 \end{bmatrix} = \begin{bmatrix} 100 \\ 0 \end{bmatrix}$$

Using Cramer's rule, $V_2 = \dfrac{\begin{vmatrix} 2 & 100 \\ -6 & 6 \end{vmatrix}}{\begin{vmatrix} 2 & -1 \\ -6 & 2 \end{vmatrix}} = \dfrac{600}{-2} = -300$ volts

This gives $I_2 = \dfrac{-300}{10} = -30A.$

Problem 5.26. *Refer to the network of Fig. 5.106. It uses a controlled source. Determine the power in the controlled source. It is given that voltage across the 5 Ω resistor is 10 volts.*

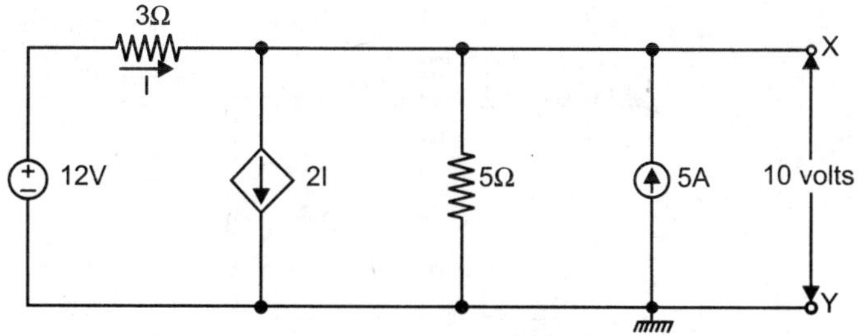

Fig. 5.106

Solution. Applying KCL at node X,

$$\frac{V_X - 1_2}{3} + 2I + \frac{10}{5} = 5$$

or
$$-\frac{2}{3} + 2I + 2 = 5$$

or
$$I = 11/6 \text{ A}$$

Current delivered by controlled source

$$= -2I = -11/3 \text{ A}$$

Therefore, power in the controller source

$$= -11/3 \times 10 = -36.67 \text{ V.}$$

The negative sign here indicates that the source absorbs power and does not deliver power.

Problem 5.27. *In the network of 5.107, find the value of constant (k) for which the loop current (i) will be zero.*

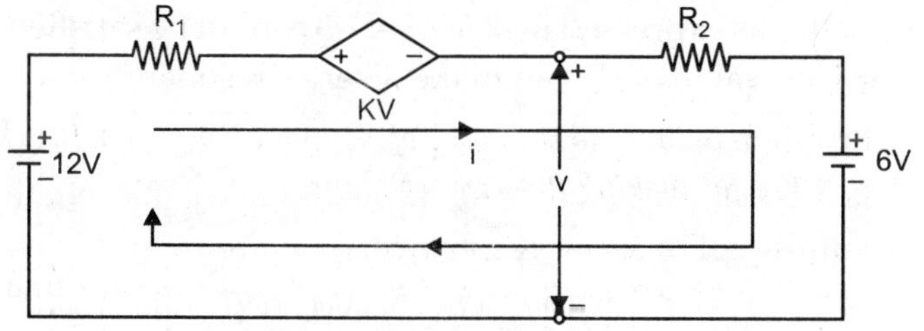

Fig. 5.107

Solution. Applying KVL around the loop,

$$1_2 - iR_1 - kV - iR_2 - 6 = 0$$

or
$$i(R_1 + R_2) = kV - 6$$

or
$$i = \frac{kV - 6}{(R_1 + R_2)}$$

For $i = 0$, $kV - 6 = 0$, which gives $k = \frac{6}{V}$

Also, $V = i\,R_2 + 6 = 6$ (for $i = 0$) Therefore, $k = \frac{6}{6} = 1.$

Problem 5.28. *Refer to the network of Fig. 5.108. Use Nodal analysis method to determine the condition for which voltage $V_2 = 0$. (V_2) is the voltage at node-2 with respect to the reference node.*

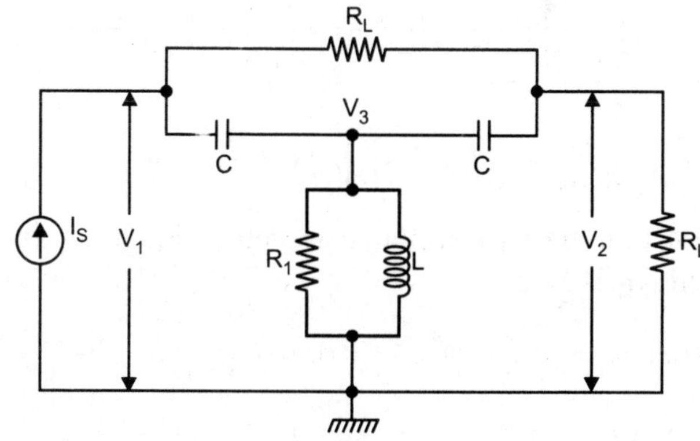

Fig. 5.108

Solution. The three nodal equations at nodes represented by voltages V_1, V_2 and V_3 can be written in the matrix form by inspection as follows:

$$\begin{bmatrix} \left(\frac{1}{R_1} + j\omega C\right) & \left(\frac{1}{R_1}\right) & (-j\omega C) \\ \left(-\frac{1}{R_1}\right) & \left(\frac{1}{R_1} + j\omega C + \frac{1}{R_L}\right) & (-j\omega C) \\ (-j\omega C) & (-j\omega C) & \left(\frac{1}{R_2} + j2\omega C - \frac{j}{\omega L}\right) \end{bmatrix}$$

$$\times \begin{bmatrix} V_1 \\ V_2 \\ V_3 \end{bmatrix} = \begin{bmatrix} I_S \\ 0 \\ 0 \end{bmatrix}$$

Now V_2 can be computed from:

$$V_2 = \frac{\begin{vmatrix} \left(\frac{1}{R_1} + j\omega C\right) & I_S & -j\omega C \\ -\frac{1}{R_1} & 0 & -j\omega C \\ -j\omega C & 0 & \left(\frac{1}{R_2} + j2\omega C - \frac{j}{\omega L}\right) \end{vmatrix}}{\begin{vmatrix} \left(\frac{1}{R_1} + j\omega C\right) & -\frac{1}{R_1} & -j\omega C \\ -\frac{1}{R_1} & \left(\frac{1}{R_1} + j\omega C + \frac{1}{R_L}\right) & -j\omega C \\ -j\omega C & -j\omega C & \left(\frac{1}{R_2} + j2\omega C - \frac{i}{\omega L}\right) \end{vmatrix}}$$

For $V_2 = 0$, we get

$$\begin{vmatrix} \left(\frac{1}{R_1} + j\omega C\right) & I_S & -j\omega C \\ -\frac{1}{R_1} & 0 & -j\omega C \\ -j\omega C & 0 & \left(\frac{1}{R_2} + j\omega C - \frac{j}{\omega L}\right) \end{vmatrix} = 0$$

This leads to

$$-I_S\left[-\frac{1}{R_1}\left\{\left(\frac{1}{R_2}+j2\omega C-\frac{j}{\omega L}\right)\right\}-\{(-j\omega C)^2\}\right]=0$$

or $$-\frac{1}{R_1R_2}-\frac{j}{R_1}\left(2\omega C-\frac{1}{\omega L}\right)+\omega^2C^2=0$$

For V_2 to be zero, both real as well as imaginary parts should equal zero.

That is, $-\frac{1}{R_1R_2}+\omega^2C^2=0$ or $R_1R_2=\frac{1}{\omega^2C^2}$

and $2\omega C-\frac{1}{\omega L}=0$ or $2\omega C=\frac{1}{\omega L}$ or $\omega^2LC=\frac{1}{2}$

Problem 5.29. *For the networks of Fig. 5.109 (a), (b) and (c), determine the co-efficient of coupling (k) for which these network are in series resonance.*

Solution. Fig. 5.109 (*a*).

For series resonance, total inductive reactance = total capacitance reactance,

i.e., $$j3+j27+2j\omega M=j\,33.6$$

$$2j\omega M=j\,3.6$$

$$2j\omega k\sqrt{L_1L_2}=k\,3.6 \text{ or } jk\sqrt{\omega L_1\cdot\omega L_2}=\frac{j\,3.6}{2}=j\,1.8$$

Substituting $2L_1=3$ and $\omega L_2=27$,

we get, $$k=\frac{1.8}{\sqrt{27\times3}}=0.2$$

We have taken (+) sign for ($j\omega M$) because the current through the network will either enter or leave the dots in both the coils.

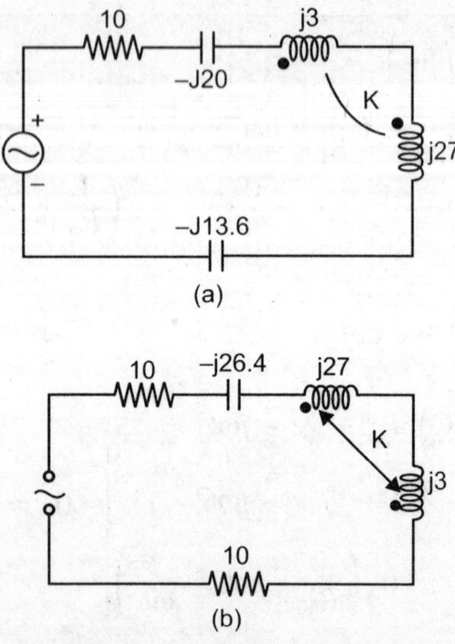

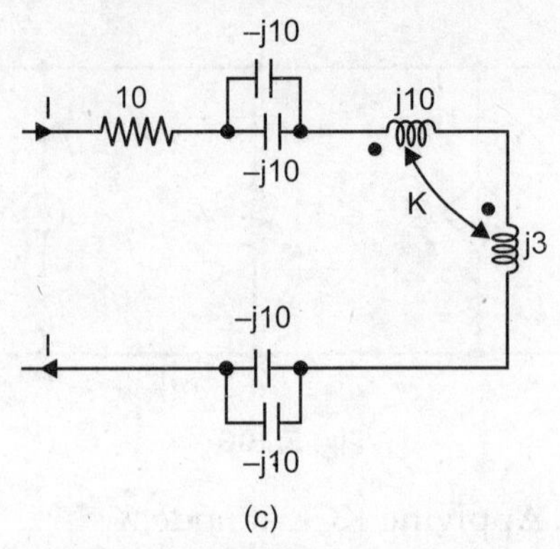

Fig. 5.109

Fig. 5.109 (*b*); $j27+j3-2j\omega M=j\,26.4$..... for series resonance

or $2j\,\omega M=j\,3.6$ which gives $K=0.2$.

A (–) sign for $j\omega M$ has been taken because for any assumed direction of current, if the current enters the dot in one coil, it would leave the dot in the other.

Fig. 5.109 (*c*)

Total capacity reactance $=-j\,10$

For series resonance $j10+j3+2j\omega M-j10=0$; $2j\omega M=-j\,3$

$$j\omega k\sqrt{L_1L_2}=-j3/2...\ \omega L_1=10,\ \omega L_2=3$$

As per dot placing, the sign for mutual reactance will be positive. As a consequence, the condition of series resonance as expressed by $2\,j\omega M=-j3$ cannot be satisfied. Hence, no value of (*k*) can make it series resonant.

Problem 5.30. *Find the value of X_L in the coupled network of Fig. 5.110 for making it series resonant.*

Solution. For series resonance,

$$j10+jXL-2j\,\omega M=j40$$

or $$10+X_L-2\omega M=40$$

or $$2\omega M=10+X_L-40=X_L-30$$

or $$\omega k\sqrt{L_1L_2}=\left(\frac{X_L-30}{2}\right)...\omega L_1=10,\ \omega L_2=XL$$

or $$K\sqrt{\omega L_1\cdot\omega L_2}=\left(\frac{X_L-30}{2}\right)$$

or $$0.1\sqrt{10\times X_L}=\frac{X_L-30}{2}$$

or $$0.01\times10\,X_L=\left(\frac{X_L-30}{2}\right)^2$$

or $$X_L^2-60X_L+900=0.4\,X_L$$

or $\quad X_L = \dfrac{60.4 \pm \sqrt{(-60.4)^2 - 3600}}{2}$

$$= \frac{60.4 \pm \sqrt{3612 - 3600}}{2}$$

or $\quad X_L = \dfrac{60.4 \pm \sqrt{12}}{2} = \dfrac{+60.4 \pm 3.46}{2}$

Fig. 5.110

Fig. 5.111

Problem 5.31. *For the networks of Fig. 5.111 (a) and (b), place dots appropriately so as to make them resonant.*

Solution. This problem can be solved by visual inspection itself. For resonance, the total inductive reactance must equal total capacitive reactance. In case of Fig. 5.111 (*a*) the mutual coupling reactance must have a minus sign to make the two reactances equal. As a result, the dots can be placed either like shown in Fig. 5.112 (*a*) or like shown in Fig. 5.112 (*b*).

Similarly, for the network of Fig. 5.111 (*b*), the overall mutual reactance has to be (– *j*16) to achieve series resonance and hence the dot placing of Fig. 5.112 (*c*) and (*d*). Both these dot placings are valid for the network of Fig. 5.112. (*b*) to make it series resonant.

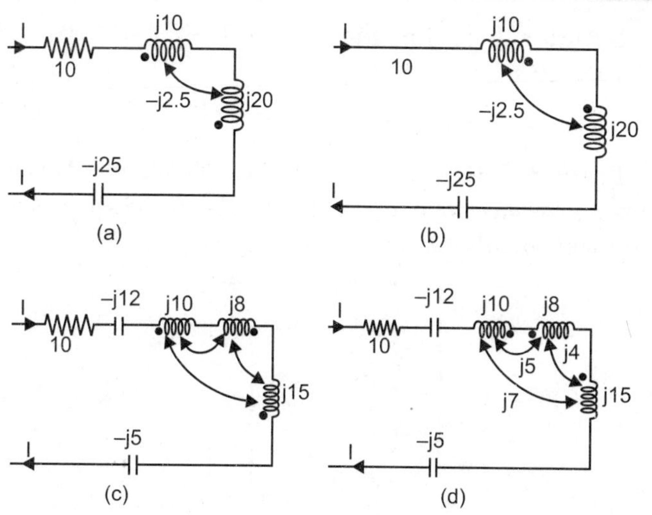

Fig. 5.112

Problem 5.32. *Is it possible to achieve series resonance in the network of Fig. 5.113 for any value of coupling co-efficient. If yes, what is that value of (k) and if no, why not.*

Solution. For series resonance, total inductive reactance must be (*j*40) in this case. That implies : 2 *jωM* equals *j*15 (*jωM* has + sign here).

Now $2j\omega M \;=\; j\,15 = 2k\sqrt{j\omega L_1 \,.\, j\omega L_2}$

where, $j\omega L_1 = j16$ and $j2L_2 = j9 = 2k\sqrt{j16\,.\,j9} = j24\text{K}$

or $\quad k = \dfrac{j\,15}{j\,24} = 0.625$

Fig. 5.113

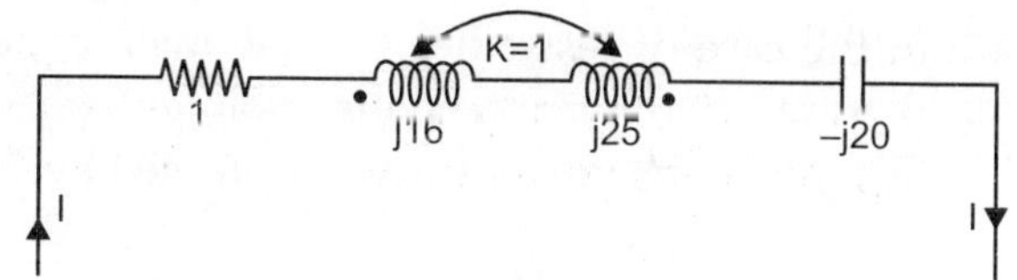

Fig. 5.114

Yes, it is possible. Any value of *k* given by $0 \le k \le 1$ is achieveable.

Problem 5.33. *Find the input impedance for the network of Fig. 5.114.*

Solution. Input impedance $= 1 + j16 + j25 - 2\sqrt{j16 \times j25} - j20 = 1\ j19$

$$|Z_{in}| = \sqrt{1^2 + 19^2} \cong 19\Omega$$

Problem 5.34. *The coupled coils shown in Fig. 5.115 have coupling co-efficients of 0.6. What are the two possible dotted equivalent circuits for the coupled coils of Fig. 5.115.*

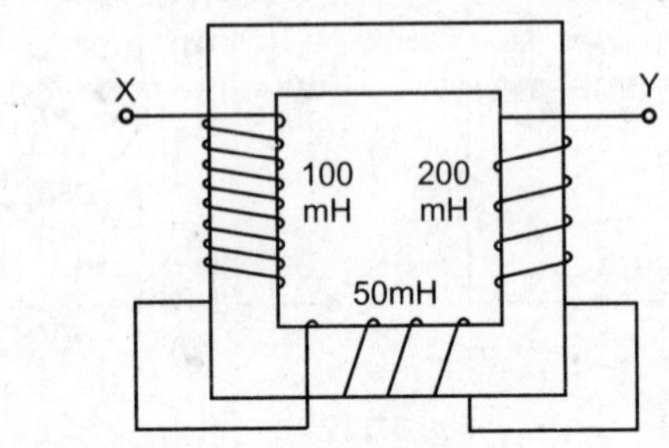

Fig. 5.115

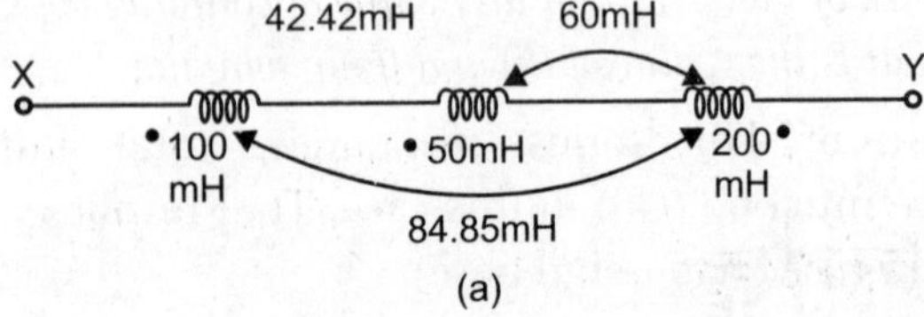

(a)

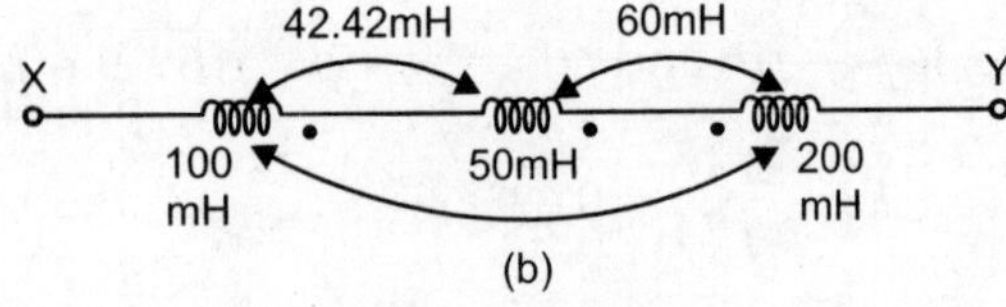

(b)

Fig. 5.116

Solution. The equivalent inductance between XY is given by

$$100 + 50 + 200 + 2 \times 0.6\sqrt{100 \times 50} - 2 \times 0.6\sqrt{50 \times 200} - 2 \times 0.6\sqrt{200 \times 100} = 350 + 84.84 - 120 - 169.70 = 145.15$$ mH.

The two possible dotted equivalent circuits are shown in Fig. 5.116.

Problem 5.35. *Refer to the network of Fig. 5.117. Determine voltage V_{XY} as marked in the network. Use mesh current analysis technique and Cramer's rule.*

Solution. The matrix form of mesh equation can be written by inspection as follows

$$\begin{bmatrix} (7+j3) & -j5 & -5 \\ -j5 & (10+j5) & -10 \\ -5 & -10 & 25 \end{bmatrix} \begin{bmatrix} I_1 \\ I_2 \\ I_3 \end{bmatrix} = \begin{bmatrix} 10 \\ -5 \\ 0 \end{bmatrix}$$

Using Cramer's rule, $I_3 = \dfrac{\begin{vmatrix} (7+j3) & -j5 & 10 \\ -j5 & (10+j5) & -5 \\ -5 & -10 & 0 \end{vmatrix}}{\begin{vmatrix} (7+j3) & -j5 & -5 \\ -j5 & (10+j5) & -10 \\ -5 & -10 & 25 \end{vmatrix}}$

$$= \frac{(7+j3)(-50) + j5(-25) + 10[(j\,50) + 5(10+j5)]}{(7+j3)[25(10+j5)-100] + j5[-125-50] - 5[j50 + 5(10+j5)]}$$

$$= \frac{-350 - j\,150 - j\,125 + j\,500 + j\,250 + 50}{(7-j3)(250 + j125 - 100) + 625 - 250 - j\,375 - 250}$$

$$= \frac{150 + j\,475}{(7+j3)(150 + j\,125) + 375 - j\,625}$$

$$= \frac{150 + j475}{675 + j\,1325 + 375 - j\,625}$$

$$= \frac{150 + j\,475}{1050 + j\,700}$$

$$= \frac{498\angle 72.5^\circ}{1262\angle 33.7^\circ}$$

Now $V_{XY} = -I_3 \times 10$

$= -39.4\angle 38.8^\circ$ volts.

Summary of Important Formulas

1. In the case of resistance (R), the instantaneous current (i), instantaneous voltage (v) and instantaneous power (p) are expressed by:

$$v = i \times R,\ i = \frac{v}{R},\ p = vi = i^2R = \frac{v^2}{R}$$

2. In the case of the an inductance (L), the instantaneous current (i), the instantaneous voltage (v) and the instantaneous power (p) are expressed:

$$v = L\frac{di}{dt},\ p = vi = i \times L\frac{di}{dt},\ i = \frac{1}{L}\int v dt$$

3. In the case of a capacitance (C), the instantaneous, current (i), the instantaneous voltage (v) and the instantaneous power (p) are expressed by:

$$v = \frac{1}{C}\int idt,\ i = C\frac{dv}{dt},\ p = vi = vc\frac{dv}{dt}$$

4. In a series R-L circuit, when a DC voltage (V) is applied across the circuit at time $t = 0$, the current (i_L) for $t \geq 0$ can be expressed by:

$$i_L = \frac{V}{R}\left(1 - e^{\frac{-Rt}{L}}\right)$$

5. In a series *R-C* circuit, when a DC voltage (*V*) is applied across the circuit at time $t = 0$; the current (v_C) for $t \geq 0$ can be expressed by:

$$v_C = V(1 - e^{-t/RC})$$

6. If a steady state current (*I*) is flowing through series *R-L* circuit and the source of DC voltage is replaced by a short circuit at $t = 0$; the current (*I*) will decay for $t \geq 0$ according to:

$$i_L = Ie^{-\frac{R}{L}t}$$

7. If a capacitor (*C*) is charged to a DC voltage (*V*) through a resistance (*R*) and the DC voltage (*V*) is replaced by a short circuit at $t = 0$; then the voltage (v_C) across capacitor (*C*) will decay according to:

$$v_C = Ve^{-\frac{t}{RC}}$$

8. A capacitor (*C*) when charged from a constant current source (*I*) will have voltage (v_c) across capacitor expressed by:

$$v_C = \frac{It}{C}$$

9. The current (i_L) through an inductance (*L*) when connected across a DC voltage (*V*) will have (i_L) expressed as a function of time by:

$$i_L = \frac{V}{L}t$$

10. The time constant (*C*) of a series *R-L* circuit is expressed by:

$$C = \frac{L}{R}$$

11. The time constant (*C*) of series *R-C* circuit is expressed by:

$$C = RC$$

Self Evaluation Exercise

Multiple Choice Questions

1. An electrical network having six independent branches and three nodes including the reference node should be analyzed using
   (*a*) Mesh-current analysis technique
   (*b*) Node voltage analysis technique
   (*c*) Kirchoffs voltage Law
   (*d*) Impedance matrix technique
2. An electrical network with seven independent branches and five independent nodes excluding the reference node should preferably be solved by
   (*a*) Mesh current analysis
   (*b*) Node voltage analysis
   (*c*) Kirchoffs current law
   (*d*) None of these
3. When we talk about mesh current analysis using matrix formulation, we talk about
   (*a*) Admittance matrix
   (*b*) Z-parameters and Impedance matrix
   (*c*) Y-parameters
   (*d*) None of these
4. When we talk about node voltage analysis using matrix formulation, we talk about
   (*a*) Impedance matrix
   (*b*) Admittance matrix
   (*c*) Laplace transform
   (*d*) Z-parameters
5. The number of loop equations required to completely analyze a network comprising of nine independent branches and five nodes including the reference node shall be
   (*a*) 5 (*b*) 4
   (*c*) 9 (*d*) None of these
6. One of the following laws of electrical networks is used in node voltage analysis technique
   (*a*) Kirchoffs voltage law
   (*b*) Faraday's laws
   (*c*) Ohm's law
   (*c*) Kirchoffs current law
   (*e*) none of these
7. Bring out the correct statement amongst the following

(a) In the node voltage analysis, all outgoing currents must always be taken as negative and all incoming currents taken as positive.

(b) In the mesh current analysis, all loop currents should either be all clockwise or all anticlockwise

(c) In the node voltage analysis, if outgoing currents are taken as positive, then incoming currents must be taken as negative and vice-versa.

(d) In the Mesh-current analysis, the directions of the loop currents should be so chosen that no branch should have more than two currents otherwise the results are likely to be incorrect.

8. The series combination of R_1 Ω. and another resistance which is further a parallel combination of R_1 and R_2 Ω is given by

(a) $\dfrac{R_1^2 + 2R_1 R_2}{(R_1 + R_2)}$ (b) $\dfrac{R_1 R_2}{R_1 + R_2}$

(a) $2 R_1 + R_2$ (b) $\dfrac{2R_1 R_2}{2R_1 + R_2}$

9. Identify which of the following is NOT a tree of the graph shown in the Figure 5.118

(a) begh (b) defg

(c) adhg (d) aegh

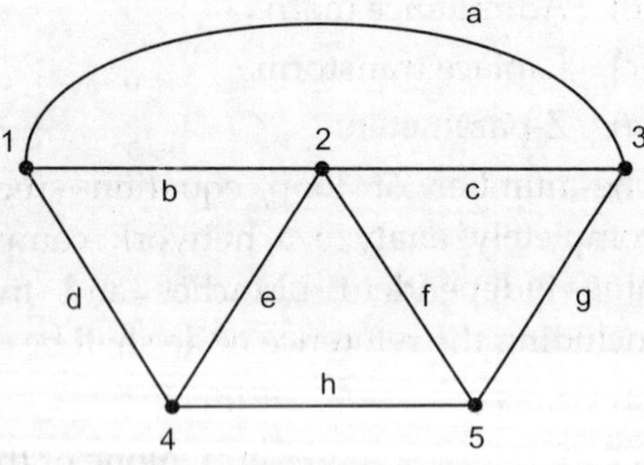

Fig. 5.118

10. Relative to a given fixed tree of a network,

(a) Link currents form an independent set

(b) Branch currents form an independent set

(c) Link voltages form an independent set

(d) Branch voltages form an independent

11. For the incidence matrix given below, the corresponding network graph is shown in

$$A = \begin{bmatrix} 1 & 0 & 0 & 0 & 1 & 0 & 0 & 1 \\ 0 & 1 & 0 & 0 & -1 & 1 & 0 & 0 \\ 0 & 0 & 1 & 0 & 0 & -1 & 1 & -1 \\ 0 & 0 & 0 & 1 & 1 & 0 & -1 & 0 \end{bmatrix}$$

(a) Figure 5.119 (a) (b) Figure 5.119 (b)

(c) Figure 5.119 (c) (d) Figure 5.119 (d)

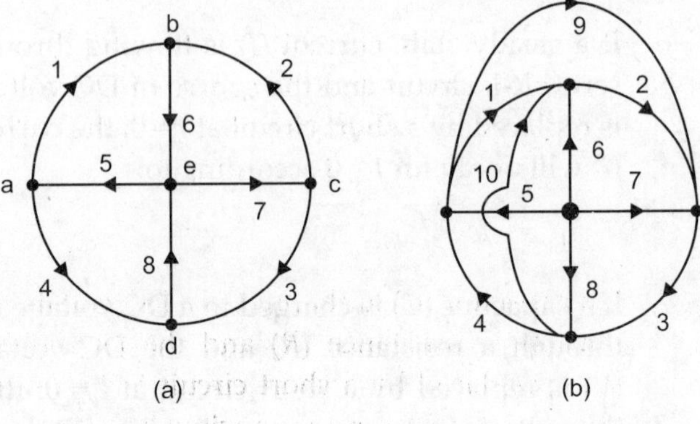

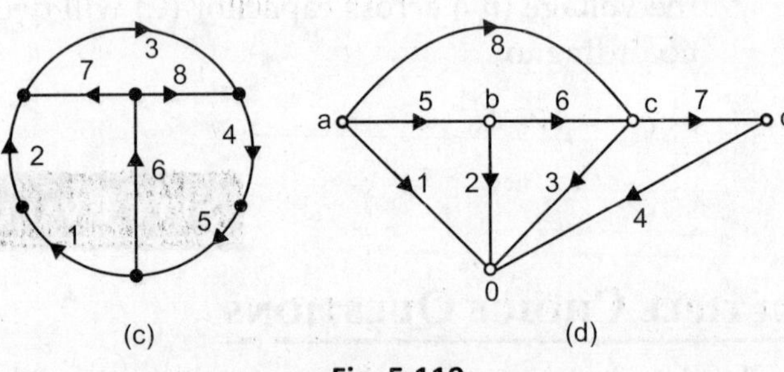

Fig. 5.119

12. The following matrix identity represents

$$\begin{bmatrix} Z_{11} & Z_{12} & Z_{13} & \cdots & Z_{1m} \\ Z_{21} & Z_{22} & Z_{23} & \cdots & Z_{2m} \\ \cdot & \cdot & \cdot & \cdots & \cdot \\ \cdot & \cdot & \cdot & & \cdot \\ Z_{m1} & Z_{m2} & Z_{m3} & \cdots & Zmm \end{bmatrix} \begin{bmatrix} I_1 \\ I_2 \\ \cdot \\ \cdot \\ I_m \end{bmatrix} = \begin{bmatrix} V_1 \\ V_2 \\ \cdot \\ \cdot \\ V_m \end{bmatrix}$$

(a) Mesh current analysis

(b) Node voltage analysis

(c) Application of Kirchhoff's Voltage Law

(d) None of these

13. The following matrix identity represents

$$\begin{bmatrix} Y_{11} & Y_{12} & Y_{13} & \cdots & Y_{1m} \\ Y_{21} & Y_{22} & Y_{23} & \cdots & Y_{2m} \\ \cdot & \cdot & \cdot & \cdots & \cdot \\ \cdot & \cdot & \cdot & \cdots & \cdot \\ Y_{m1} & Y_{m2} & Y_{m3} & \cdots & Y_{mm} \end{bmatrix} \begin{bmatrix} V_1 \\ V_2 \\ \cdot \\ \cdot \\ V_m \end{bmatrix} = \begin{bmatrix} I_1 \\ I_2 \\ \cdot \\ \cdot \\ I_m \end{bmatrix}$$

(a) Mesh current analysis

(b) Node voltage analysis

(c) Kirchhoff's Current Law

(d) None of these

14. For the network shown in Fig. 5.120, the current through resistor R_6 is
    (*a*) 420 mA
    (*b*) 310 mA
    (*c*) 750 mA
    (*d*) 350 mA

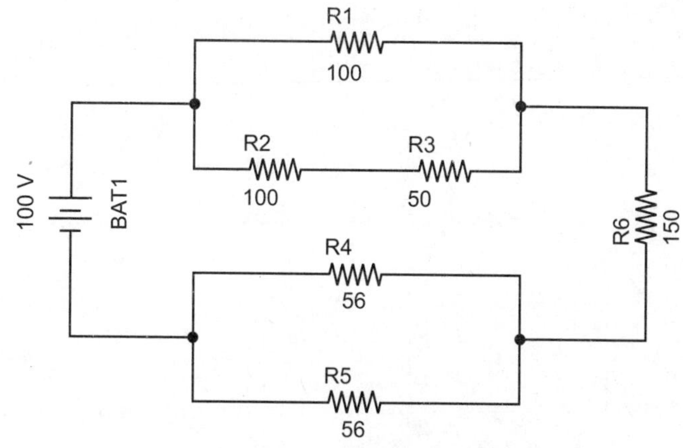

Fig. 5.120

Answers to Self Evaluation Exercise

1. (*b*)	2. (*a*)	3. (*b*)	4. (*b*)	5. (*a*)
6. (*d*)	7. (*c*)	8. (*a*)	9. (*c*)	10. (*a*)
11. (*d*)	12. (*a*)	13. (*b*)	14. (*a*)	

Chapter 6

Network Theorems

Learning Objectives

After completing the chapter, you will learn the following topics:

1. Thevenin's Theorem
2. Norton's Theorem
3. Maximum Power Transfer Theorem
4. Superposition Theorem
5. Reciprocity Theorem
6. Compensation Theorem
7. Substitution Theorem
8. Tellegen's Theorem
9. Millman's Theorem
10. Star-Delta Transformation
11. Delta-Star Transformation
12. Application of Network Theorems to Solving Networks

Q.1. State and explain Thevenin's. Theorem. Given an illustration to prove effectiveness of the theorem in solving circuit problems.

Ans. As per Thevenin's theorem, "A linear, active network with output terminals *X-Y* is electricallly equivalent to a voltage source (V_{th}) in series with an impedance (Z_{th}) as shown in Fig. 6.1. V_{th} is the open

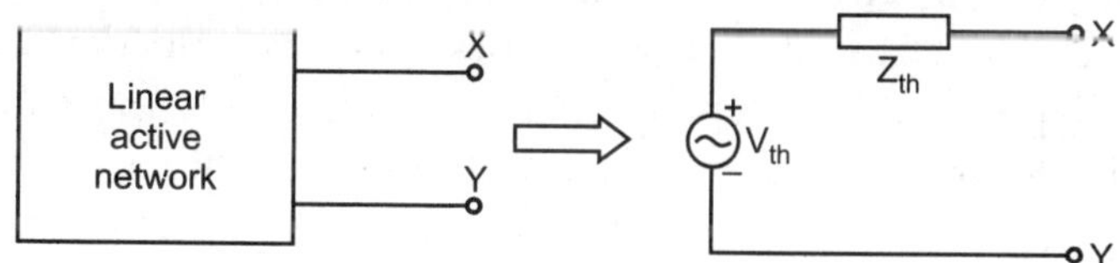

Fig. 6.1

circuit voltage across *X-Y* and Z_{th} is the driving point impedance across *X-Y* with all voltage sources replaced by their internal impedances (short circuits in case of ideal voltage sources) and all current sources replaced by their output impedances (open circuits in case of ideal currents sources).

At this point, it is important to define a linear active network. A linear network has only linear circuit elements; the ones having linear voltage-current relationships. As long as the impedance of a circuit element is fixed, it can be termed as linear. Resistances are linear. Inductances and capacitances are also linear as long as frequency remains fixed during operation. An active network is the one that contains both passive as well as active elements. Voltage and current sources are active elements.

Let us take up an example of finding current (i) through a resistance (R) connected across terminals *X-Y* in the network of Fig. 6.2 for the three cases (a) $R = 3\ \Omega$. (b) $R = 6\ \Omega$. (c) $R = 9\ \Omega$.

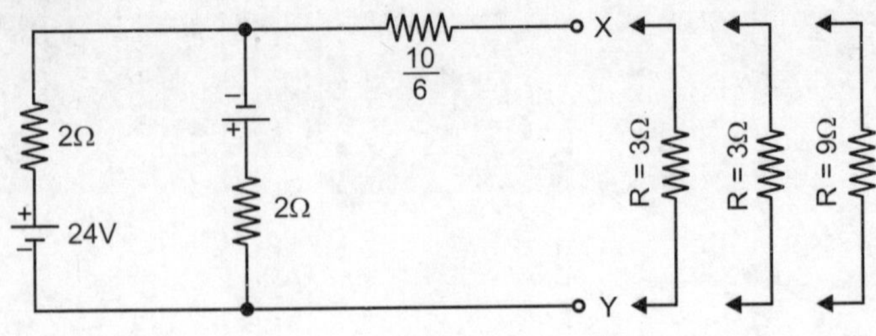

Fig. 6.2

If we start solving this network problem by writing mesh equations; it would be a tedious job to do. And moreover, the set of loop equations is going to be different for the three cases and alternative approach is to first determine the Thevenin's equivalent of this network across X-Y and then calculate current through R for different values of R. And moreover, once the Thevenin's equivalent is determined, there is practically not much do be done whether you have to find the current for one value of R or three values of R. Even if you have to determine the current through R for a single value, solving through Thevenin's equivalent approach quite often turns out to be simper and convenient. However, it is for you to decide which circuit analysis technique is appropriate for a particular network.

Fig. 6.3 gives Thevenin's equivalent for the network of Fig. 6.3. Quite obviously,

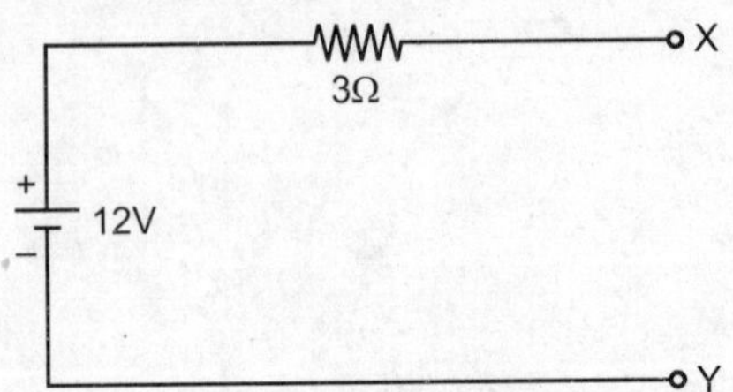

Fig. 6.3

Current (*i*) when $R = 3\ \Omega$ = 12/6 = 2 A

Current (*i*) when $R = 6\ \Omega$ = 12/9 = 1.33 A.

Current (*i*) when $R = 9\ \Omega$ = 12/12 = 1 A.

Q.2. What are the different steps involved in computing the Thevenin's impedance while applying the Thevenin's theorem?

Ans. The impedance Z_{TH} can be calculated as given below.

1. If the circuit contains only independent sources and resistors; deactivate the sources and find Z_{TH} by circuit reduction technique. Independent ideal current sources are deactivated by opening them while independent ideal voltage sources are deactivated by shorting them. Non-ideal voltage and current sources are replaced by their internal impedances.
2. If the circuit contains resistors, dependent and independent sources; determine the open circuit voltage with the sources activated and the short circuit current when a short circuit is applied to the terminals. Z_{TH} is the ratio of the open circuit voltage to the short circuit current.
3. If the circuit contains impedances and only dependent sources, then connect 1A current source to the desired terminals, where Z_{TH} has to be found and determine the terminal voltage. Z_{TH} in this case is the ratio of the terminal voltage to 1A current.

Q.3. State and explain Norton's theorem. How are Thevenin's and Norton's equivalent circuits interrelated?

Ans. A linear active network with output terminals X-Y is equivalent to a current source (I_n) in parallel with an impedance (Z_n) as shown in Fig. 6.4 (I_n) is the short circuit current through X-Y; (Z_n) has the same meaning as Z_{th} of the Thevenins equivalent circuit. Thevenin's equivalent can be derived from Norton's equivalent and vice versa as follows :

$$I_n = \frac{V_{th}}{Z_{th}} \ ; \ Z_n = Z_{th}$$

$$V_{th} = I_n Z_n \ ; \ Z_{th} = Z_n$$

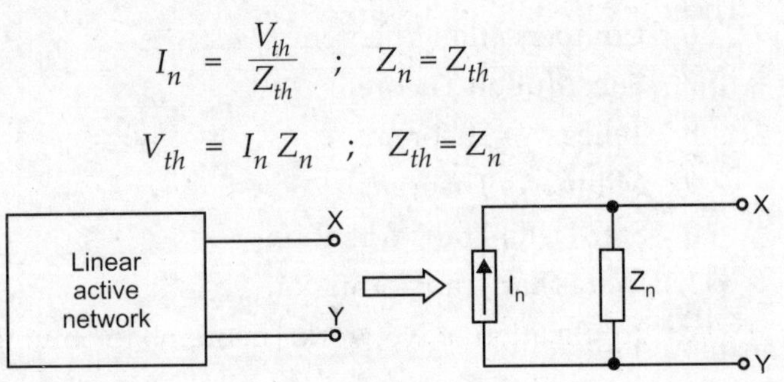

Fig. 6.4

Q.4. How do you go about determining the Norton's current source (I_n) and Norton's impedance (Z_n) for any given network?

Ans. If the network contains impedances and independent sources, then deactivate the sources and find (Z_n) by circuit reduction techniques and find (I_n) with sources activated. If the network contains impedances, independent and dependent sources, then determine the short-circuit current (I_n) with all sources activated and find the open-circuit voltage (V_{OC}). (Z_n) is the ratio of the open circuit voltage (V_{OC}) to the short-circuit current (IN). If the network contains only impedances and dependent sources, then connect 1A current source to the terminals and find the terminal voltage. (Z_n) is the ratio of the open circuit terminal voltage to the 1A current.

Q.5. A linear and active network which can be represented by an equivalent network comprising of a voltage source (V_s) in series with an impedance (Z_s) across two terminals X-Y, then under what conditions will there be maximum power transfer to a load impedance Z_L connected across X-Y. What will be the transfer efficiency in that case? Justify your answer.

Ans. The power transferred from the network in question is going to be the maximum when the load impedance is a complex conjugate of the equivalent series impedance of the equivalent voltage source. See Fig. 6.5 for illustration.

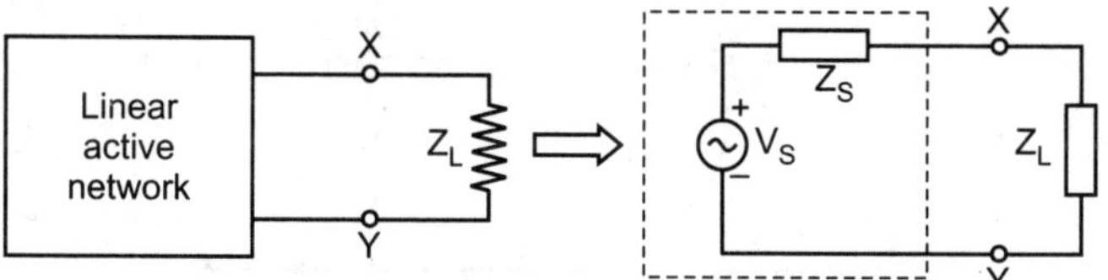

Fig. 6.5

For maximum power transfer,

If (Z_S) is in general equal to ($R_S \pm jX_S$), then (Z_L) must equal ($R_L \mp jX_L$) with $R_L = R_S$ and $X_L = X_S$.

These are essential conditions for maximum transfer of power from source to load. Incidentally, these conditions are also known by the name of MAXIMUM POWER TRANSFER THEOREM.

It can be mathematically verified that the transfer efficiency in case of optimum power transfer is 50 per cent.

From Fig. 6.5 when conditions for maximum power transfer are being met with, we can write an expression for power transferred to the load as follows :

$$\text{Power delivered to load} = \left(\frac{V}{R_S + R_L}\right)^2 R_L$$

$$\text{Power drawn from source} = \frac{V^2}{(R_S + R_L)}$$

$$\text{Therefore, transfer efficiency} = \frac{[V/(R_S + R_L)]^2 R_L}{V^2/(R_S + R_L)}$$

$$= \frac{V^2 R_L}{(R_S + R_L)^2} \times \frac{(R_S + R_L)}{V^2}$$

$$= \frac{R_L}{R_S + R_L} = 0.5 \text{ (for } R_S = R_L)$$

Q.6. Derive an expression to establish the conditions of maximum power transfer for a Thevenin's equivalent representation.

Ans. From answer to Q-4, expression for power delivered to the load in the case of a linear active network represented by an equivalent Thevenin's circuit is given by:

$$P_L = [V/(R_S + R_L)]^2 \times R_L$$

Conditions for maximum power transfer are expressed by $dP_L/R_L = 0$

That is, $[V^2 \times (R_S + R_L)^2 - 2VR_L \times 2 \times (R_S + R_L)]/(R_S + R_L)^4 = 0$

Simplifying this equation gives $[V^2 \times (R_S + R_L)^2 - V^2R_L \times 2 \times (R_S + R_L)] = 0$, which further simplifies to

$$R_S = R_L.$$

Q.7. Fig. 6.6 shows the Norton's equivalent of a linear, active network. When a load resistance R_L is. connected across X-Y, under what conditions will there be maximum transfer of power?

Ans. From Fig. 6.6

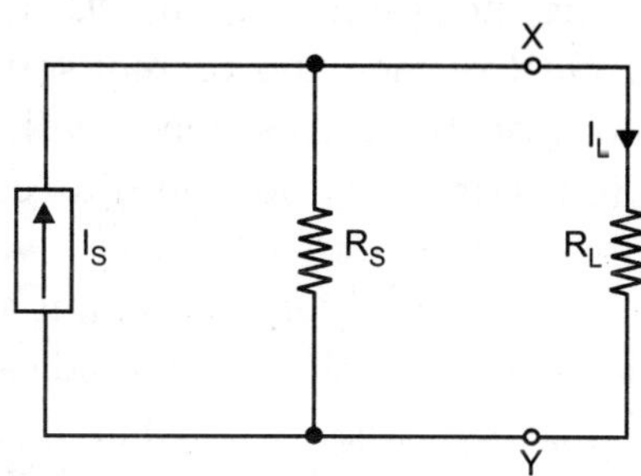

Fig. 6.6

$$I_L = \frac{I_S R_S}{(R_S + R_L)}$$

Power delivered to $R_L = I_L^2 R_L$

$$= \frac{I_S^2 R_S^2 R_L}{(R_S + R_L)} = P_L \text{ (say)}$$

For deriving condition for maximum power transfer put $\frac{dP_L}{dR_L} = 0$

That is, $\frac{d}{dR_L}\left(\frac{I_S^2 R_L R_S^2}{(R_S + R_L)^2}\right) = 0$

or $\frac{I_S^2 R_S^2 (R_S + R_L)^2 - 2(R_S + R_L) I_S^2 R_L R_S^2}{(R_S + R_L)^4} = 0$

or $I_S^2 R_S^2 (R_S + R_L)^2 = I_S^2 R_S^2 (R_S + R_L) \times 2R_L$

or $(R_S + R_L) = 2 R_L.$

or $R_L = R_S$

We can draw a very useful conclusion from the outcome of this small mathematical treatment that we have just concluded and that is "Power transferred from a source to a load is always the maximum when load resistance equals source resistance irrespective of the fact whether the source is a current source or a voltage source."

In fact, we can extend this statement to include sources having complex impedances. In that case, for maximum power transfer, load impedance should be complex conjugate of source impedance as stated before.

Q.8. What is superposition theorem? Under what conditions is this theorem valid Illustrate with the help of simple examples the application of superposition theorem.

Ans. In a linear bilateral active network containing two or more than two sources, a branch current or a node voltage in any part of the network can be determined by considering only one source at a time with all other sources removed (*i.e.* all voltage sources replaced by short circuits AND all current sources replaced by open circuits) and then superposing the results. Let me clarify here that I am assuming ideal voltage and current sources. In case of practical (nonideal) sources, these will be replaced by the series impedance (for voltage sources) and shunt impedance (for current sources). When we talk about sources vis-a-vis superposition theorem, it also includes initial condition sources such as a charged capacitor.

Now the question arises : whether this theorem is valid without any limitations! Infact, is valid only in cases where the network has INDEPENDENT sources *i.e.* the characteristics of the sources do not vary with variation in circuit elements. In case the sources are dependent ones and the controlling elements are external to the network, even then the theorem is valid. Also, the network should be linear and bilateral.

Let us see how we can make use of superposition theorem to calculate current through R_3 in the network of Fig. 6.7 (*a*).

Considering 15 volts source alone and replacing 7.5 volts source by a short circuit, we get a total current of 1A (= 15/15) drawn from the source. This 1 ampere of current then gets divided between R_2 and R_3, Therefore current through R_3 as a result of only 15V source is given by;

$I = \frac{1 \times 10}{20}$ = 0.5 A in a direction given by dotted line arrow.

Similarly when only 7.5V source is considered, the total current drawn from 7.5V source would be 0.5 A (= 7.5/15). This would then get divided between R_1 and R_3.

Therefore, current through R_3 as a result of only 7.5 V source is given by:

$I' = \frac{0.5 \times 10}{20}$ = 0.25 A in a direction given by firm line arrow.

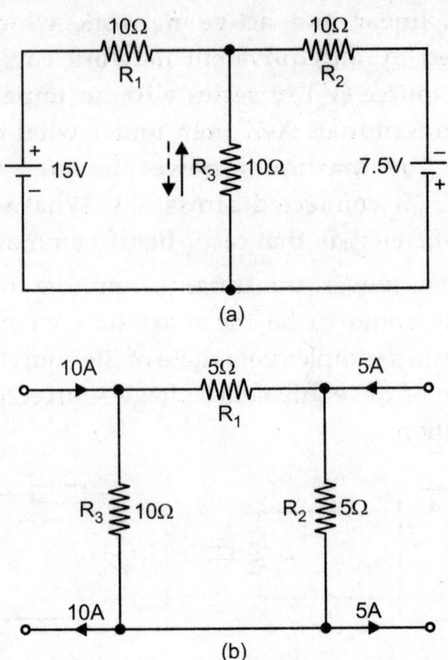

Fig. 6.7

The current that would flow through R_3 when both 15 V as well as 7.5 V sources are active is given by 0.5 – 0.25 = 0.25 in a direction given by dotted line arrow (Note that it is important to take into consideration the directions of flow of constituent currents.)

Let us take another example. Look at the network of Fig. 6.7 (*b*) and try to use superposition concept to calculate potential difference across R_2.

Ignoring 10 A current source, current through

$$R_2 = \frac{5 \times 15}{20} = \frac{15}{4} A$$

Therefore, potential difference across

$$R_3 = \frac{15}{4} \times 5 = \frac{75}{4} \text{ volts}$$

Ignoring 5 A current source, current through

$$R_2 = \frac{10 \times 10}{20} = 5 \text{ A}$$

Therefore potential difference across

$$R_2 = 5 \times 5 = 25 \text{ volts}$$

Note that both these potential differences have same polarities; overall potential difference when both the current sources are active is therefore,

$$75/4 + 25 = 175/4 = 43.75 \text{ volts}$$

Q.9. Is superposition theorem valid for computation of power in a circuit element or a network for which it is otherwise valid. Whatever be your reply, justify.

Ans. It is NOT VALID for computation of power. It is

valid for computation of a node voltage for given current excitations because voltage is linearly related to excitation (current in this case) for a linear network. Similarly, it can be used for computation a branch current for give voltage excitations again because for a linear network, they are linearly related. In case of power the relationship with current as well as voltage is QUADRATIC, *i.e.* power is proportional square of current or voltage even for linear circuit elements. That is why superposition theorem cannot be used for computation of power in a circuit element of a linear active network with two or more than two active sources.

Q.10. What do you understand by ADDITIVE PROPERTY and HOMOGENEITY of linear networks?

Ans. According to the additive property of linear networks, a given response in network voltage across a circuit element or current through it, resulting from a number of independent sources including the initial condition sources can be computed from taking the sum of responses from individual sources considering them one at a time with other sources made inoperative. The sum of all responses is taken with due regards to the polarity. According to homogeneity property of linear networks, if all sources are multiplied by a certain constant, the response is also multiplied by the same constant. For instance, in case of network of Fig. 6.7 (*a*), if 15V source was changed to 30 V and 7.5 V hanged to 15 V, the current through R_3 would be 0.5 A a direction given by dotted line.

Additivity and Homogeneity properties are the basis of superposition theorem.

Q.11. State Reciprocity theorem and illustrate its application with the help of an example.

Ans. In a linear, active and bilateral network with a single source, the ratio of excitation to response remains unchanged when the two (excitation and response) interchange their positions, For example, if a voltage V applied between terminals X-Y of the network produces a current I in a branch A-B, then the voltage V applied between A-B will result in a current I through X-Y. Similarly if a current I injected between terminals X-Y produces a voltage V across A-B, then a current I injected between A-B will produce a voltage V across X-Y. The networks Where this holds are called reciprocal networks.

Refere to Fig. 6.8,

Voltage V produced across R_4 as a result of a 3 A source connected across R_2 is given as

$$V = \left(\frac{3 \times 10}{30}\right) \times 5 = 5 \text{ volts. [Fig. 6.8 (a)]}$$

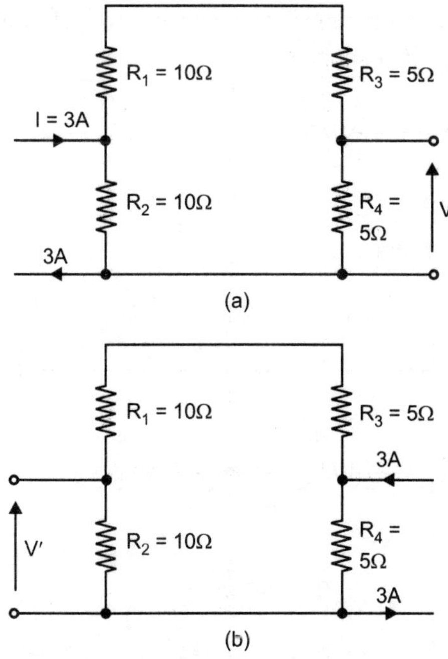

Fig. 6.8

Voltage V' produced across R_2 as a result of a 3 A source connected across R_4 is given as

$$V' = \left(\frac{3 \times 5}{30}\right) \times 10 = 5 \text{ volts. [Fig. 6.8 (b)]}$$

Hence $V = V'$ and this proves reciprocity theorem.

Q.12. What are the conditions under which Reciprocity theorem might be applied to a network?

Ans. There should be a single source. There should not be any initial condition sources. This implies that the network must be initially relaxed or in the zero state. The network should have only linear elements and should not contain dependent sources.

Q.13. Which network theorem you would like to employ in order to find change in current produced is an element of a linear active network as a result of change in the impedance of that element. State in simple words the theorem in question and explain with the help of a simple illustration as to how you can put this concept to use.

Ans. It is the COMPENSATION THEOREM. According to this theorem, if in a linear active network, a current I flows through an impedance Z, then the change in current ΔI produced as a result of change in impedance ΔZ is given by the current that would flow if a compensation source equal in magnitude and phase to ($I \Delta Z$) is connected in series with Z with all other sources removed *i.e.* all voltage sources replaced by short circuits and all current sources replaced by open circuits (assuming ideal sources).

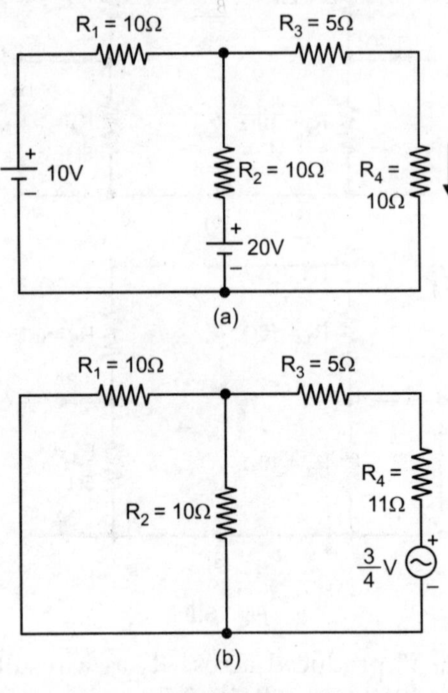

Fig. 6.9

One can use any of the theorems discussed in the preceding questions (Thevenin or superposition for example) to say that current flowing through R_4 in the network of Fig. 6.9 (*a*) is (3/4) Amp. Now let us find change in current flowing through R_4 when its resistance changes to 11. In order to do that, a compensation source equal to $I\,\Delta\,Z$ (= 3/4 × 1) has been connected as in Fig. 6.9 (*b*) and all voltage sources have been replaced by short circuits. Let me point out here that the polarity of compensation source would have been opposite, had resistance R_4 changed from 10 Ω to 9Ω. Obviously increase in resistance value of R_4 would tend to reduce current through it. This decides the polarity of the compensation source.

Quite clearly, current flowing in R_4 in the network of Fig. 6.9 (*b*) is given as:

– (3/4) / (5 + 5 + 11) = – 3/84 amperes

Minus sign indicates that the change in current is in a direction opposite to that of *I i.e.*, current through R_4 decreases as a result of this impedances change.

Q.14. State and explain substitution theorem.

Ans. According to substitution theorem, any branch in a network may be substituted by a different branch or any combination of elements without disturbing the voltages and currents in the entire network; provided that the new branch has the same set of terminal voltages and current as the original branch. In other words, according to the substitution theorem, for branch equivalence, the terminal voltage and current must be same. The concept of substitution theorem is illustrated with the following example. Refer to the circuit diagrams of Fig. 6.10 (*a*), (*b*) and (*c*). The circuit diagram of Fig. 6.10 (*b*) is same as the circuit diagram of Fig. 6.10 (*a*) with resistance (R_3) replaced by a voltage source (V_3), which is same as voltage across (R_3). Also, the circuit diagram of Fig. 6.10 (*c*) is same as the circuit diagram of Fig. 6.10 (*a*) with resistance (R_3) replaced by a current source (I_3), which is same as current flowing through (R_3). In this case all three networks have identical branch voltages and currents.

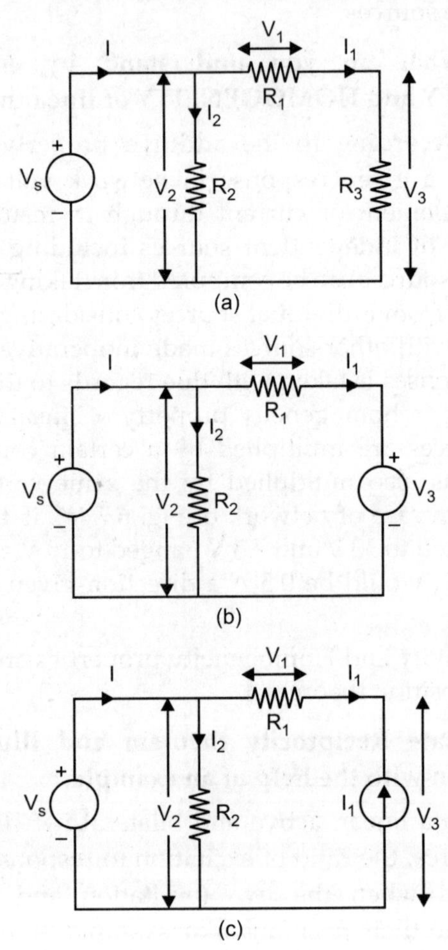

Fig. 6.10

Q.15. State Millman's Theorem for a number of voltage sources connected in parallel and also for a number of current sources connected parallel.

Ans. As per Millman's theorem, voltage sources V_1, V_2, V_3,.......V_n with respective internal impedances Z_1, Z_2, Z_3,.......Z_n when connected in parallel are electrically equivalent to a single voltage source V with internal impedance Z (Fig. 6.11).

$$\text{where} \quad V = \left[\frac{(V_1/Z_1) + (V_2/Z_2) + (V_3/Z_3) .... + (V_n/Z_n)}{(1/Z_1) + (1/Z_2) + (1/Z_3) + ... (1/Z_n)}\right]$$

$$Z = \left[\frac{1}{(1/Z_1)+(1/Z_2)+(1/Z_3)+\ldots\ldots(1/Z_n)}\right]$$

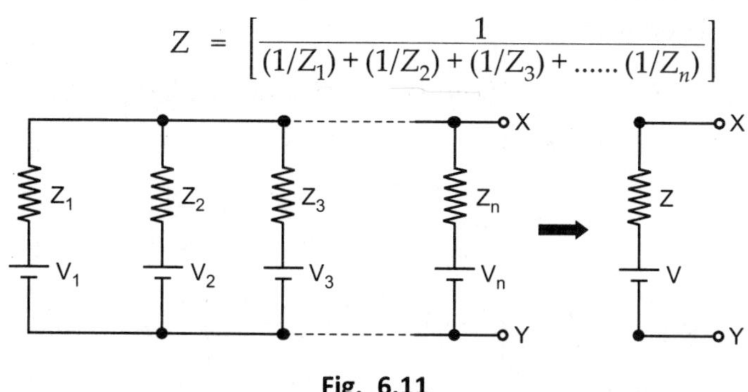

Fig. 6.11

Similarly, a number of current sources $I_1, I_2, I_3 \ldots\ldots I_n$ with respective shunt impedances $Z_1, Z_2, Z_3 \ldots\ldots Z_n$ when connected in parallel are electrically equivalent to a single current source I with shunt impedance Z (Fig. 6.12).

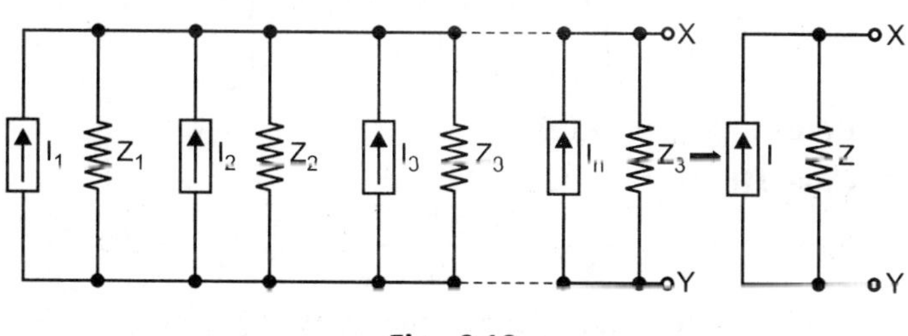

Fig. 6.12

where $I = I_1 + I_2 + I_3 + \ldots\ldots\ldots I_n$

$$Z = \frac{1}{(1/Z_1)+((1/Z_2)+(1/Z_3)+\ldots\ldots(1/Z_n)}$$

Q.16. State and explain Tellegen's Theorem.

Ans. Tellegen's theorem states that for any lumped network whose network graph has (*b*) branches and (*n*) nodes, with branch voltages and currents represented by (V_k) and (i_k) respectively (k = 1, 2, ...b) measured with respect to arbitrarily chosen associated reference directions and satisfying all constraints of KVL and KCL respectively are represented by the following expression.

$$\sum_{k=1}^{b} v_k\, i_k = 0$$

It may be mentioned here that the Tellegen's theorem is applicable for any lumped network having linear or non-linear, active or passive and time-varying or time-invariant elements.

Q.17. What are the conversion, formulae for representing three STAR connected impedances Z_A, Z_B and Z_C by equivalent DELTA-connected impedances Z_{AB}, Z_{BC} and Z_{CA}. In other words, write the expressions for Z_{AB}, Z_{BC} and Z_{CA} in terms of Z_A, Z_B and Z_C and write the expressions for Z_A, Z_B and Z_C in terms of Z_{AB}, Z_{AC} and Z_{CA}. Yon need not derive these expressions.

Ans. With reference to Fig. 6.13 (*a*)

$$Z_{AB} = \frac{Z_A Z_B + Z_B Z_C + Z_C Z_A}{Z_C}$$

$$Z_{BC} = \frac{Z_A Z_B + Z_B Z_C + Z_C Z_A}{Z_A};$$

$$Z_{CA} = \frac{Z_A Z_B + Z_B Z_C + Z_C Z_A}{Z_B}$$

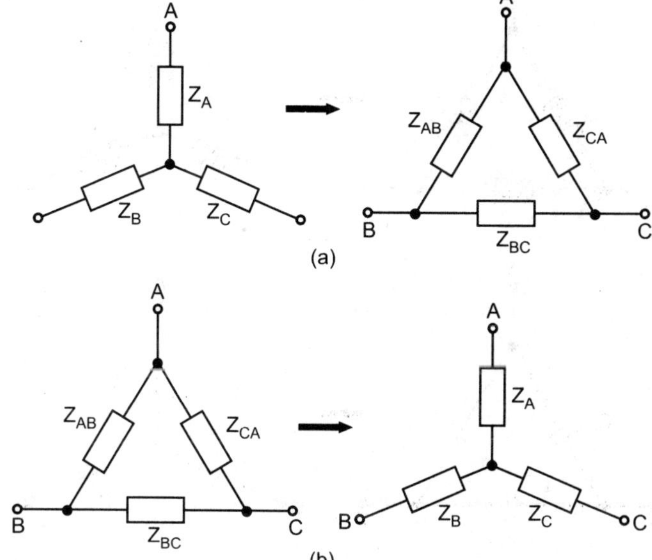

Fig. 6.13

With reference to Fig. 6.13 (*b*)

$$Z_A = \frac{Z_{AB} Z_{CA}}{Z_{AB} + Z_{BC} + Z_{CA}}$$

$$Z_B = \frac{Z_{AB} Z_{BC}}{Z_{AB} + Z_{BC} + Z_{CA}};$$

$$Z_C = \frac{Z_{BC} Z_{CA}}{Z_{AB} + Z_{BC} + Z_{CA}}.$$

Q.18. Refer to Fig. 6.13 (*a*). What would be corresponding DELTA impedances if all STAR impedances were either PURE RESISTORS or PURE INDUCTORS or PURE CAPACITORS. That is, you have to find expressions for Z_{AB}, Z_{BC} and Z_{CA} for three different cases namely (*i*) $Z_A = R_A$, $Z_B = R_R$, $Z_C = R_C$ (*ii*) Z_A = impedance of an inductor L_A, Z_B = impedance of an inductor L_B, Z_C = impedance of an inductor L_C (*iii*) Z_A = impedance of a capacitor C_A, Z_B = impedance of a capacitor C_B, Z_C = impedance of a capacitor C_C.

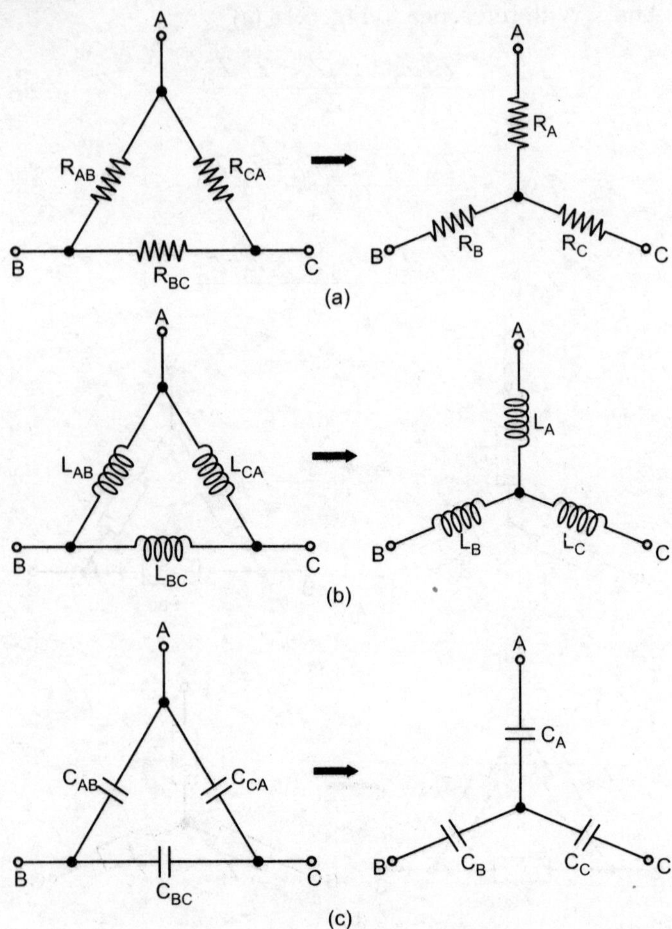

Fig. 6.14

Ans. Case I. We have already said that

$$Z_{AB} = \frac{Z_A Z_B + Z_B Z_C + Z_C Z_A}{Z_C}$$

$$Z_{BC} = \frac{Z_A Z_B + Z_B Z_C + Z_C Z_A}{Z_A};$$

$$Z_{CA} = \frac{Z_A Z_B + Z_B Z_C + Z_C Z_A}{Z_B}$$

Substituting for Z_A, Z_B, Z_C, we get

$$Z_{AB} = \frac{R_A R_B + R_B R_C + R_C R_A}{R_A} = R_{AB}$$

$$Z_{BC} = \frac{R_A R_B + R_B R_C + R_C R_A}{R_A} = R_{BC}$$

$$Z_{CA} = \frac{R_A R_B + R_B R_C + R_C R_A}{R_B} = R_{CA}$$

Refer to Fig. 6.14 (*a*)

Case II

Now $Z_A = j\omega L_A,\ Z_B = j\omega L_B,\ Z_C = j\omega L_C$

Therefore,

$$Z_{AB} = \left[\frac{(j\omega)^2 (L_A L_B + L_B L_C + L_C L_A)}{j\omega L_C}\right]$$

$$= j\omega\left\{\frac{L_A L_B + L_B L_C + L_C L_A)}{L_C}\right\}$$

or $$L_{AB} = \left\{\frac{L_A L_B + L_B L_C + L_C L_A}{L_A}\right\}$$

Similarly,

$$L_{BC} = \left\{\frac{L_A L_B + L_B L_C + L_C L_A}{L_A}\right\}$$

and $$L_{CA} = \left\{\frac{L_A L_B + L_B L_C + L_C L_A}{L_B}\right\}$$

Refer to Fig. 6.14 (*b*).

Case III

Now $Z_A = -j/\omega C_A,\ Z_B = -j/\omega C_B,\ Z_C = -j/\omega C_C$

$$Z_{AB} = \frac{(-j/\omega C_A)(-j/\omega C_B) + (-j/\omega C_B)(-j/\omega C_C) + (-j/\omega C_C)(-j\omega C_A)}{-(j/\omega C_C)}$$

$$= \left(-\frac{j}{\omega}\right)\left\{\frac{1/C_A C_B + 1/C_B C_C + 1/C_C C_A}{1/C_C}\right\}$$

$$= -j/\omega\left\{\frac{(C_A + C_B + C_C)}{C_A C_B C_C} \times C_C\right\}$$

$$= \left(-\frac{j}{\omega}\right)\left\{\frac{(C_A + C_B + C_C)}{C_A C_B}\right\}$$

$$= \frac{-j}{\omega C_A C_B/(C_A + C_B + C_C)} = -j/C_{AB}$$

where $C_{AB} = \dfrac{C_A C_B}{C_A + C_B + C_C}$

Similarly,

$$C_{BC} = \frac{C_B C_C}{C_A + C_B + C_C} \quad \text{and}$$

$$C_{CA} = \frac{C_C C_A}{C_A + C_B + C_C}$$

Refre to Fig. 6.14 (*c*).

Q.19. What would be corresponding STAR-impedances if all DELTA-impedances were either all RESISTORS or all INDUCTORS or all CAPACITORS.

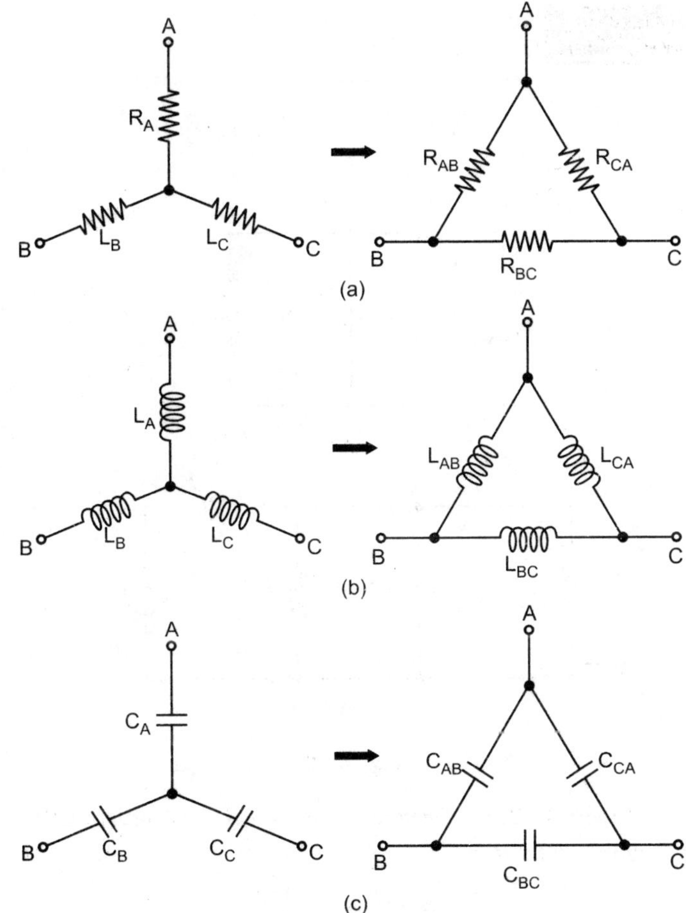

Fig. 6.15

Ans. Case I $Z_{AB} = R_{AB}, Z_{BC} = R_{BC}, Z_{CA} = R_{CA}$.

Substituting for Z_{AB}, Z_{BC} and Z_{CA} in the expressions written earlier.

$$Z_A = R_A = \frac{R_{AB}\,R_{CA}}{R_{AB}+R_{BC}+R_{CA}}$$

$$Z_B = R_B = \frac{R_{AB}\,R_{BC}}{R_{AB}+R_{BC}+R_{CA}}$$

and
$$Z_C = \frac{R_{BC}\,R_{CA}}{R_{AB}+R_{BC}+R_{CA}}$$

Refer to Fig. 6.15 (*a*).

Case II

$$Z_A = j\omega L_A = \frac{(j\omega\, L_{AB})\,(J\,\omega\, L_{CA})}{j\omega\,(L_{AB}+L_{BC}+L_{CA})}$$

$$= j\,\omega\,\frac{L_{AB}\,L_{CA}}{(L_{AB}+L_{BC}+L_{CA})}$$

or
$$L_A = \frac{L_{AB}\,L_{CA}}{L_{AB}+L_{BC}+L_{CA}}$$

Similarly,

$$L_B = \frac{L_{AB}\,L_{BC}}{L_{AB}+L_{BC}+L_{CA}} \quad \text{and}$$

$$L_C = \frac{L_{BC}\,L_{CA}}{L_{AB}+L_{BC}+L_{CA}}.$$

Refer to Fig. 6.15 (*b*)

Case III

$$Z_A = \frac{-j}{\omega\,C_A} = \frac{(j/\omega\,C_{AB})\,(-\,j/\omega\,C_{CA})}{j/\omega\left(\dfrac{1}{C_{AB}}+\dfrac{1}{C_{BC}}+\dfrac{1}{C_{CA}}\right)}$$

$$= \frac{-\,(j/\omega)\,(1/C_{AB}\,C_{CA})}{\left\{\dfrac{C_{AB}\,C_{BC}+C_{BC}\,C_{CA}+C_{CA}\,C_{AB}}{C_{AB}\,C_{BC}\,C_{CA}}\right\}}$$

$$= \frac{j}{\left(\omega\,\dfrac{C_{AB}\,C_{BC}+C_{BC}\,C_{CA}+C_{CA}\,C_{AB}}{C_{BC}}\right)}$$

or
$$C_A = \left(\frac{C_{AB}\,C_{BC}+C_{BC}\,C_{CA}+C_{CA}\,C_{AB}}{C_{BC}}\right)$$

Similarly,

$$C_B = \left(\frac{C_{AB}\,C_{BC}+C_{BC}\,C_{CA}+C_{CA}\,C_{AB}}{C_{CA}}\right) \quad \text{and}$$

$$C_C = \left(\frac{C_{AB}\,C_{BC}+C_{BC}\,C_{CA}+C_{CA}\,C_{AB}}{C_{AB}}\right)$$

Refer to Fig. 6.15 (*c*)

Q.20. Prove that for a DELTA CONNECTED network with a pure Resistor R Ω in one arm, a pure inductive reactance (jX) Ω in the second arm and a pure capacitive reactance (– jX) Ω in the third arm, the corresponding STAR connected elements will also be a pure resistance R Ω, a pure inductive reactance (JX) Ω and a pure capacitive reactance (– JX) Ω. provided that $R = X$.

Ans. The Delta network is given in Fig. 6.16 (*a*). The corresponding STAR elements are given by

$$\frac{(jX)\,(-\,jX)}{(R+jX-jX)} = \frac{X^2}{R} = R \;\ldots\ldots\; (\text{as } R = X) \qquad (\text{element } AO)$$

$$\frac{R\,(jX)}{(R+jX-jX)} = (jX) \qquad (\text{element } BO)$$

$$\frac{R\,(-\,jX)}{(R+j\,X-jX)} = (-\,jX) \qquad (\text{element } CO)$$

(a) (b)

Fig. 6.16

SOLVED PROBLEMS

Problem 6.1. *For the circuit in Figure-6.17, use superposition theorem to determine voltage $v_0(t)$ given that $R_1 = 1\Omega$, $R_2 = 1\Omega$, $C_1 = 1F$, $L_1 = 1H$, $v(t) = 2t$, $i(t) = e^{-2t}$. The initial conditions may be assumed to be zero).*

Solution. The output voltage can be found using the superposition theorem. It is given that the initial conditions are zero, therefore, the initial voltage across the capacitor is zero and the initial current through the inductor is zero.

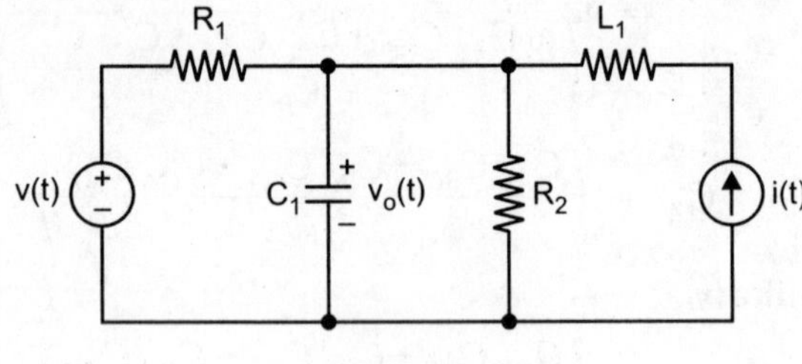

Fig. 6.17

Considering the effect of voltage source $v(t)$ and making the current source $i(t)$ inoperative, the equivalent circuit is shown in Fig. 6.18 (*a*).

Therefore,

$$V'_0(s) = \frac{\left(\dfrac{1}{1/R_2 + sC_1}\right)}{\left[\left(\dfrac{1}{1/R_2 + sC_1}\right) + R_1\right]} \times V(s) = \frac{V(s)}{s+2} = \frac{2}{s^2(s+2)}$$

Taking inverse Laplace transform, we get

$$v_0'(t) = \left(t + \frac{1}{2}e^{-2t} - \frac{1}{2}\right)u(t)$$

Considering the effect of current source $i(t)$ and making the voltage source $v(t)$ inoperative, the equivalent circuit is shown in Fig. 6.18 (*b*). Therefore,

$$V_0''(s) = \left(\frac{1}{1/R_1 + 1/R_2 + 1/C_1 s}\right) \times I(s)$$

$$= \frac{I(s)}{(s+1)} = \frac{1}{(s+1)(s+2)} = \frac{1}{(s+1)} - \frac{1}{(s+2)}$$

Taking inverse Laplace transform, we get

$$v_0''(t) = (e^{-t} - e^{-2t})\,u(t)$$

Applying superposition theorem, total output $v_o(t)$ is the sum of $v_o'(t)$ and $v_o''(t)$.

Therefore,

$$v_o(t) = \left(t + \frac{1}{2}e^{-2t} - \frac{1}{2}\right)u(t) + (e^{-t} - e^{-2t})\,u(t)$$

$$= \left(e^{-t} - \frac{1}{2}e^{-2t} + t - \frac{1}{2}\right)u(t)$$

(a)

(b)

Fig. 6.18

Problem 6.2. *For the network of Fig. 6.19, use superposition theorem to compute i_o.*

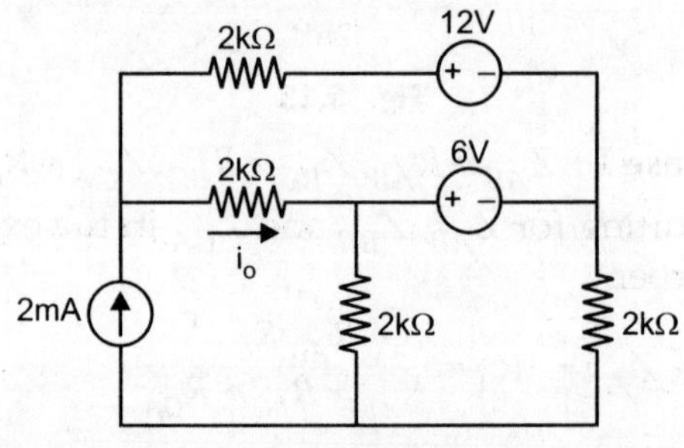

Fig. 6.19

Solution. We will apply the superposition theorem to find out the current i_o. Keeping the 12V source active and making the 2mA current source as an open circuit and the 6V voltage source as a short circuit, the resultant circuit is shown in Fig. 6.20 (*a*).

Current

$$i_0' = \frac{12}{(2+2)\times 10^3} = 3\text{mA}$$

(a)

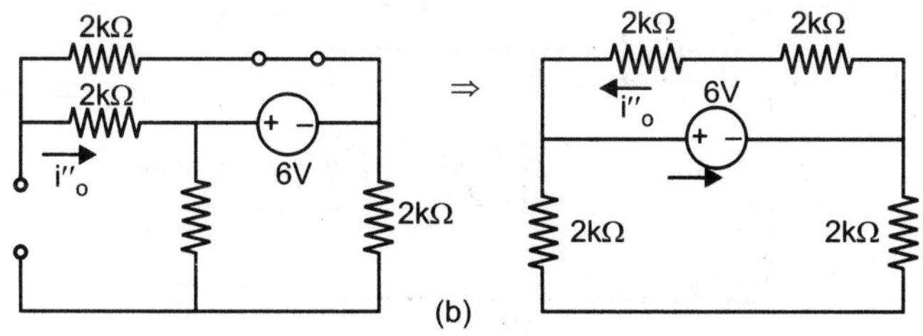

Fig. 6.20

Keeping the 6V source active and making the 2mA current source as an open circuit and the 12V voltage source as a short circuit, the resultant circuit is shown in Fig. 6.20 (*b*).

Applying KVL to the upper loop, we get $2 \times 10^3 i_o'' + 2 \times 10^3 i_o'' + 6 = 0$

Therefore, $i_o'' = -1.5\text{mA}$

Keeping the 2mA current source active and replacing the voltage sources of 6V and 12V as short circuits, the resultant circuit is shown in Fig. 6.21.

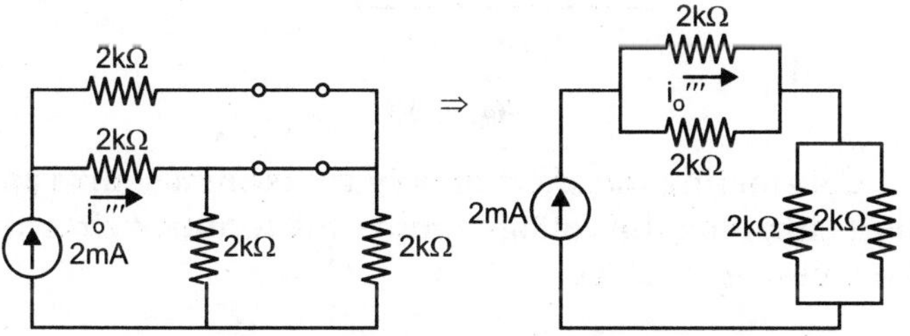

Fig. 6.21

Current of 2mA gets divided equally. Therefore, $i_o''' = 1\text{mA}$

Applying superposition theorem, we get $i_o = i_o' + i_o'' + i_o''' = 3 - 1.5 + 1 = 2.5\text{mA}$

Problem 6.3. *Determine the Thevenin's equivalent voltage V_{th} appearing between the terminals A and B of the network shown in the Fig. 6.22. Assume the voltage source to be ideal.*

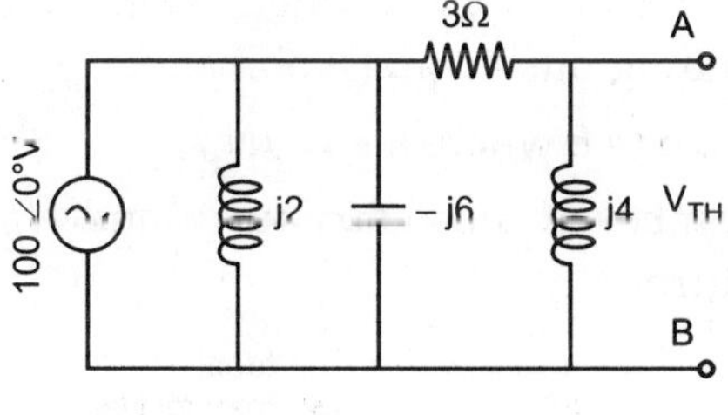

Fig. 6.22

Solution: Thevenin's equivalent voltage can be determined by replacing all voltage sources by short circuits and all current sources by open circuits assuming they are ideal sources.

$$V_{TH} = 100\angle 0° \frac{4j}{3+4j}$$

$$= \frac{100 \times 4j\,(3-4j)}{25} = 16j\,(3-4j)$$

Therefore, Thevenin's equivalent voltage

Problem 6.4. *Apply superposition theorem to determine i_o in the network of Fig. 6.23.*

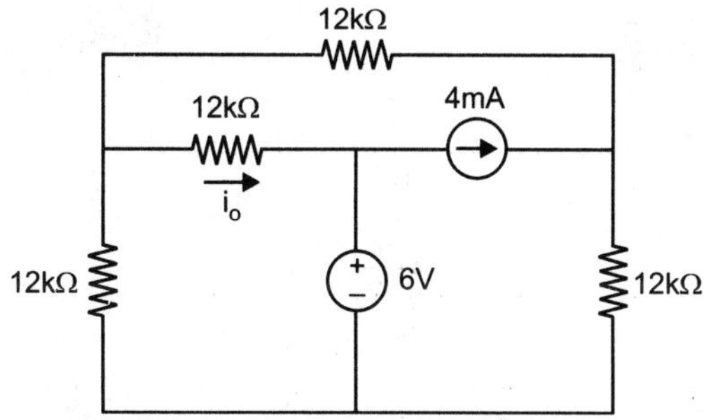

Fig. 6.23

Solution: The network shown in Fig. 6.23 has two sources, one voltage source of 6V and one current source of 4mA. Considering only the voltage source and replacing the current source by open circuit, the equivalent circuit is shown in Fig. 6.24 (*a*). Circuit in Fig. 6.24 (*a*) can be further simplified to the one shown in Fig. 6.24 (*b*).

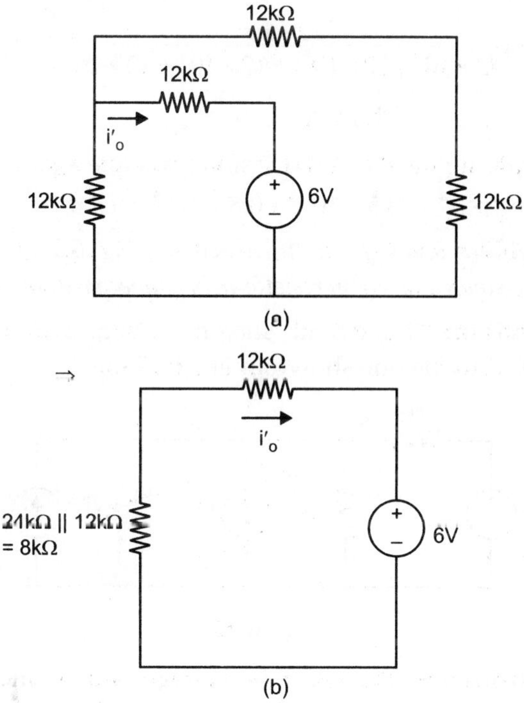

Fig. 6.24

$$i_o' = \frac{-6}{(8+12)\times 10^3} = -0.3\text{ mA}$$

Considering only the current source and replacing the voltage source by short circuit, the equivalent circuit is shown in Fig. 6.25 (*a*). Circuit in Fig. 6.25 (*a*) can be further simplified to the one shown in Fig. 6.25 (*b*).

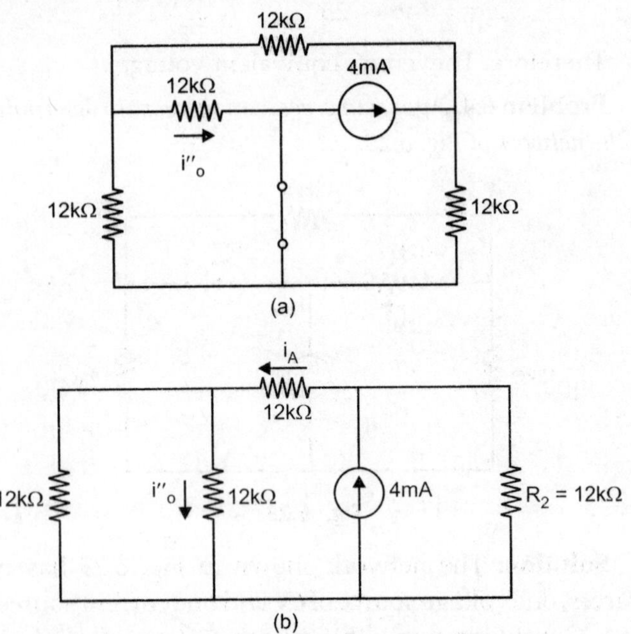

Fig. 6.25

Applying the current division principle, current iA is given by:

$$i_A = \left[\frac{12 \times 10^3}{(2 \times 10^3 \parallel 12 \times 10^3) + 12 \times 10^3 + 12 \times 10^3}\right] \times 4 \times 10^{-3}$$

$$= 1.6\text{mA}$$

Applying the current division principle again, we get $i_o = i_o' + i_o'' = -0.3 \times 10^{-3} + 0.8 \times 10^{-3} = 0.5$ mA

Problem 6.5. *Refer to the network of Fig. 6.26. Compute power dissipated in 2 ohm resistor using superposition theorem.*

Solution: The circuit shown in Fig. 6.26 can be simplified to the one shown in Fig. 6.27 (*a*).

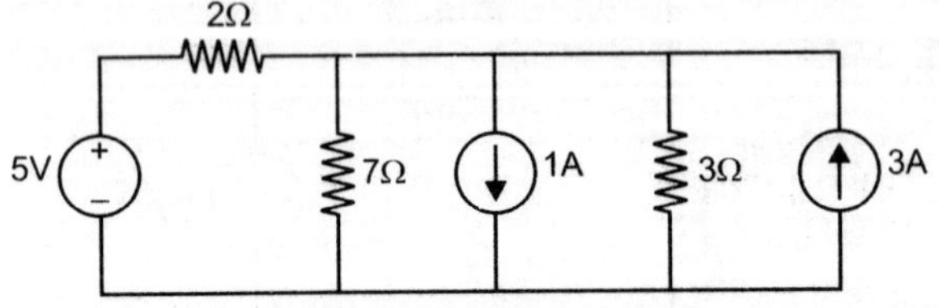

Fig. 6.26

Considering the effect of voltage source and open circuiting the current source, the resultant circuit is shown in Fig. 6.27 (*b*).

$$i'_1 = \frac{5}{2 + 2.1} = 1.22 \text{ A}$$

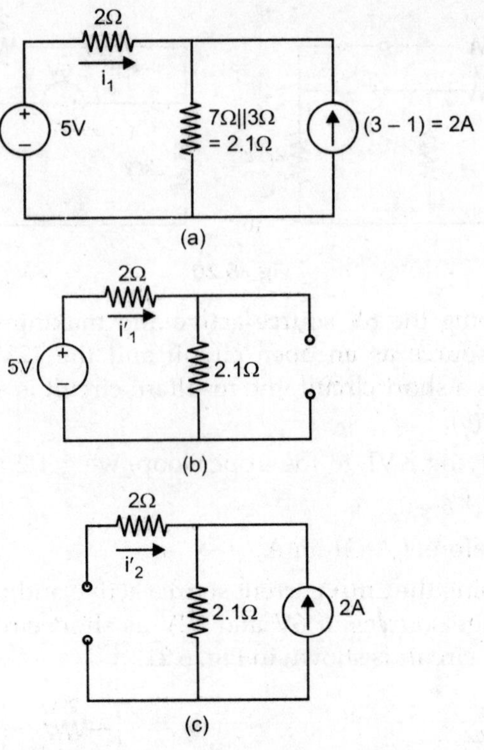

Fig. 6.27

Considering the effect of only the current source and short circuiting the voltage source, the resultant circuit is shown in Fig. 6.27 (*c*).

$$i'_2 = \frac{2 \times 2.1}{2 + 2.1} = -1.024 \text{ A}$$

Total current $i = i'_1 + i'_2 = 1.22 - 1.024 = 0.196$A

Power dissipated in the 2Ω resistor = $i^2R = (0.196)^2 \times 2 = 76.8$ mW.

Problem 6.6. *If the secondary winding of the ideal transformer shown in the circuit of figure has 40 turns, compute the number of turns in the primary winding for maximum power transfer to the 2 Ω resistor.*

Solution: For maximum power transfer to 2 ohm resistor, the 2 ohm resistor when reflected back to the primary winding must equal 8 ohms.

That is, $2/n^2 = 8$ where $n = N_S/N_P$

N_S = Number of secondary turns and N_P = Number of primary turns

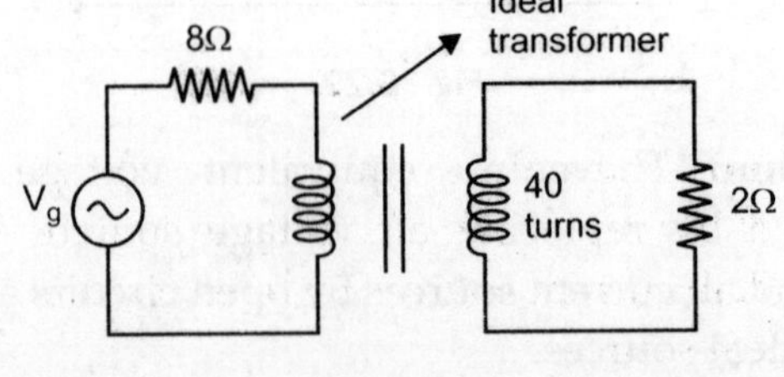

Fig. 6.28

$n^2 = 2/8 = 1/4$

Therefore, $n = 1/2$ or $N_P = 80$ turns

Problem 6.7. *In the network of Fig. 6.29, what should be the resistance value of (R_L) for maximum power transfer? You may use one or more network theorems to solve the problem.*

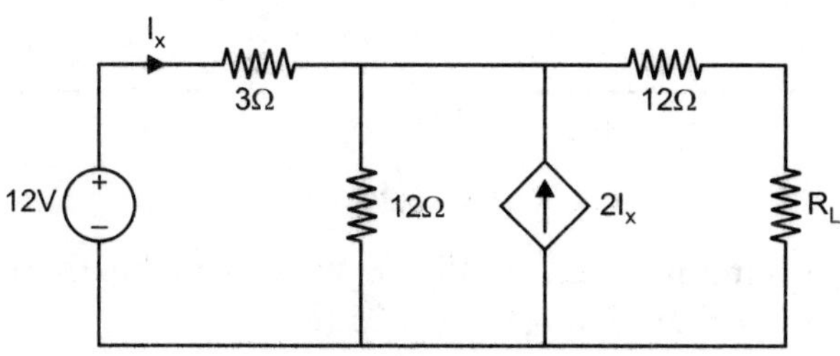

Fig. 6.29

Solution: Open circuiting the load resistor R_L, the equivalent circuit is shown in Fig. 6.30 (*a*).

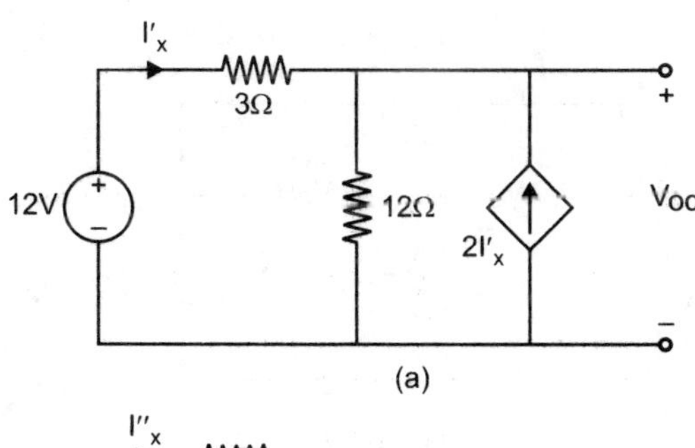

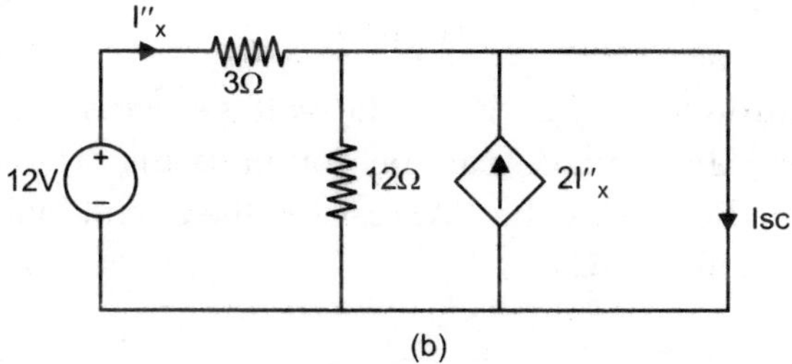

Fig. 6.30

For the circuit shown above $\frac{(V_{oc} - 12)}{3} + \frac{V_{oc}}{12} - 2I'_x = 0$

and $I'_x = \frac{(12 - V_{oc})}{3}$.

Solving the two equations we get $V_{OC} = 144/13$V

Short circuiting the load resistor RL, the equivalent circuit is shown in Fig. 6.30 (*b*).

$I''_x = 12/3 = 4A$

$I_{SC} = I''_x + 2I''_x = 12A$

Thevenin's equivalent resistance $R_{TH} = V_{OC}/I_{SC} = 12/13\Omega$.

The value of load resistor R_L for maximum power transfer is $R_L = 12 + 12/13 = 12.92\Omega \approx 13\Omega$

Problem 6.8. *Refer to Fig. 6.31. The value of the resistance (R) connected across the terminals A and B, which will absorb the maximum power, is*

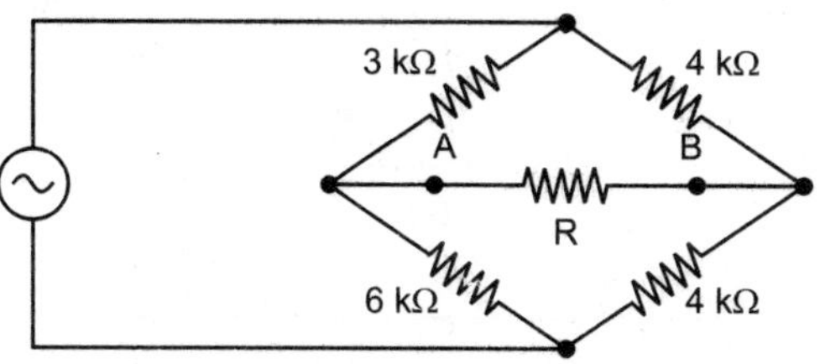

Fig. 6.31

Solution: As a first step, we shall find Thevenin's resistance across *A* and *B* terminals. The equivalent Thevenin's resistance must then be equal to the resistance (*R*) for maximum power transfer. Thevenin's equivalent resistance can be determined by replacing voltage source by a short circuit. This puts 3 kΩ and 6 kΩ in parallel and also the two 4 kΩ resistors in parallel. The two resulting resistances will then be in series.

Therefore,

$R_{AB} = R_{TH} = (3 \times 6/9) + (4 \times 4/8)\ k\Omega = 4\ k\Omega$

Therefore, $R = 4\ k\Omega$

Problem 6.9. *For the network shown in Fig. 6.32, compute the value of R_L that will achieve maximum power transfer. Also compute the magnitude of maximum power transferred.*

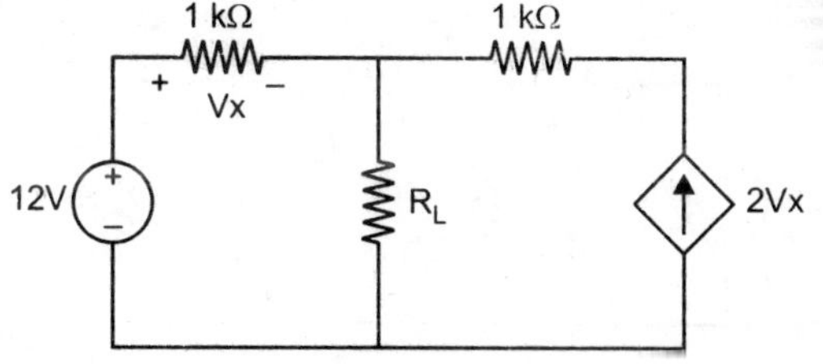

Fig. 6.32

Solution: Considering the load resistance R_L as an open circuit, the equivalent circuit is shown in Fig. 6.33 (*a*).

Applying KVL to the outer loop

$12 - 1 \times 10^3 I - 1 \times 10^3 I - 2V'_x = 0$

Also $V'_X = 1 \times 10^3 I$. Therefore, $I = 3$mA.

Applying KVL to the left loop, we get

$12 - V'_X - V_{OC} = 0$

Solving the above equations, we get

$V_{OC} = 12 - V'_X = 12 - 1 \times 10^3 I = 9V$

Considering the load resistance R_L as short circuit, the equivalent circuit is shown in Fig. 6.33 (*b*).

$V''_X = 12V$

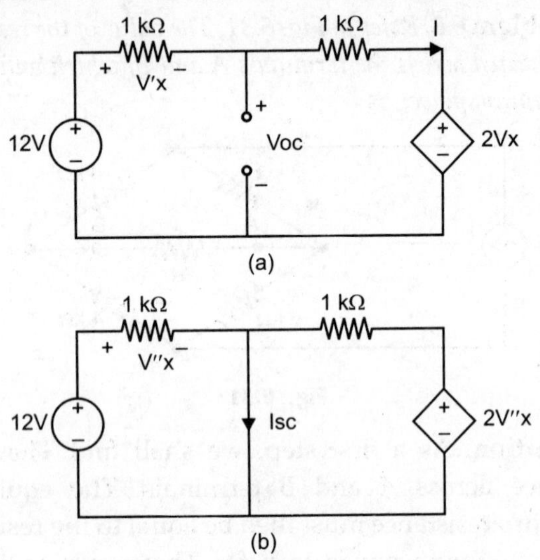

Fig. 6.33

$$I_{SC} = 12/1 \times 10^3 + 24/1 \times 10^3 = 36\text{mA}$$

Thevenin's equivalent resistance

$$R_{TH} = V_{OC}/I_{SC} = 9/36 \times 10^{-3} = 1/4\text{k}\Omega$$

The load resistance value that lets maximum power transfer is 1/4kΩ.

Maximum transferred power = $(1/4)[(9)^2/(1/4)]$mW = 81mW

Problem 6.10. *Fig. 6.34 shows a network. Compute the current flowing through resistor R.*

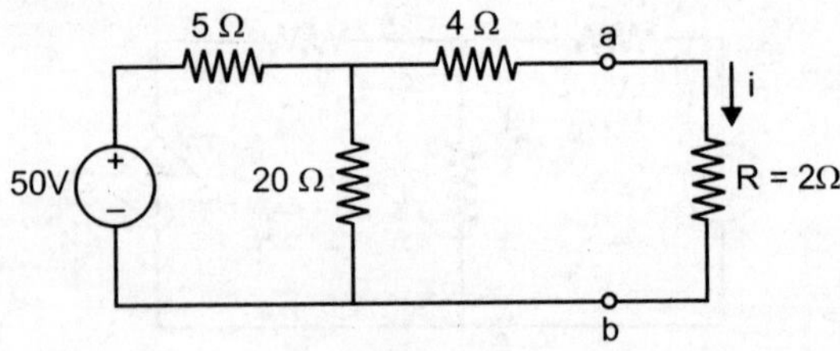

Fig. 6.34

Solution: The current can be found out using the Thevenin's theorem.

The resistor *R* sees the circuit shown in Fig. 6.35.

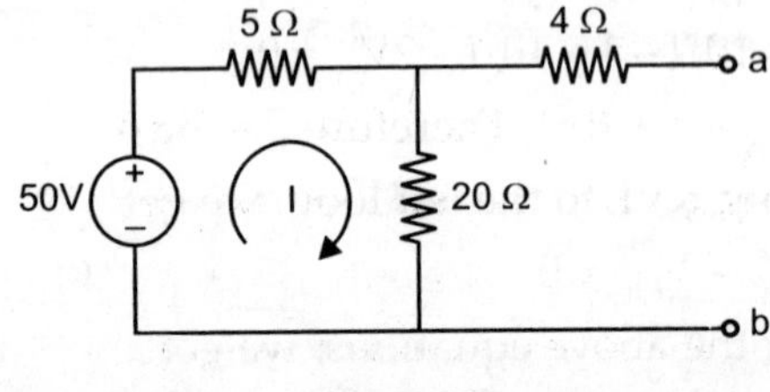

Fig. 6.35

The Thevenin's equivalent resistance is obtained by replacing the voltage source by short-circuit. Therefore,

The Thevenin's equivalent voltage is given by V_{TH} = 50 × 20/25 = 40V

The Thevenin's equivalent circuit as seen by resistor *R* is shown in Fig. 6.36 (*a*).

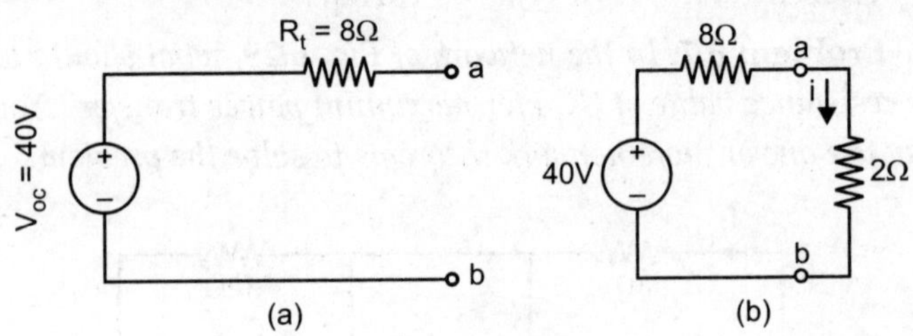

Fig. 6.36

The circuit given in Fig. 6.36 (*a*) can be therefore redrawn as that given in Fig. 6.36 (*b*).

The current (*i*) through resistor (*R*) is given by *i* = 40/(8 + 2) = 4A

Problem 6.11. *For the circuit shown in Fig. 6.37, compute the voltage V_O.*

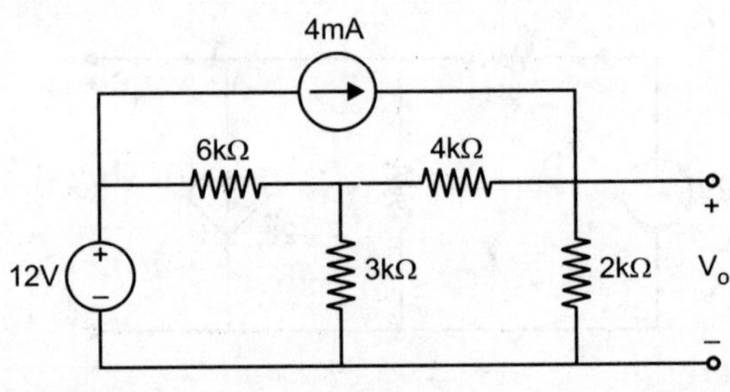

Fig. 6.37

Solution: Voltage V_O is the voltage across the 2kΩ resistor. The voltage can be found using Thevenin's theorem. Removing the 2kΩ resistor, the circuit looks like as shown in Fig. 6.38.

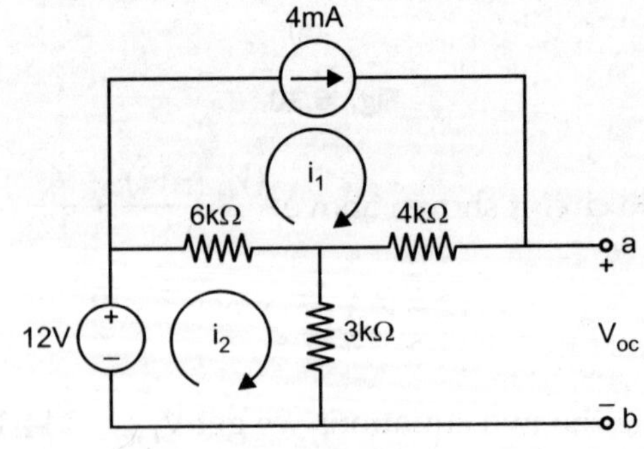

Fig. 6.38

From the given figure, i_1 = 4mA.

Applying KVL in the bottom loop, we get $12 - 6 \times 10^3 (i_2 - i_1) - 3 \times 10^3 (i_2) = 0$

Substituting the value of i_1 in the above equation and solving it we get i_2 = 4 mA

The voltage V_{OC} can be found by using KVL, $3 \times 10^3 i_2 + 4 \times 10^3 i_1 - V_{OC} = 0$

Solving the above equation we get $V_{OC} = 28V$

The Thevenin's equivalent resistance can be found by replacing the voltage source by short circuit and the current source by open circuit as shown in Fig. 6.39.

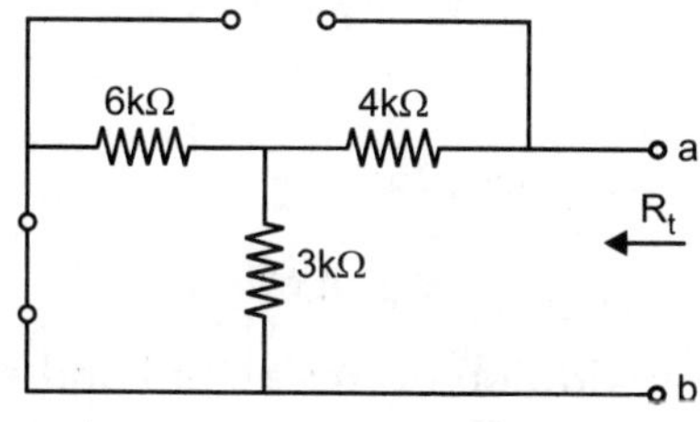

Fig. 6.39

$$R_{TH} = (6 \times 10^3 \| 3 \times 10^3) + 4 \times 10^3 = 6k\Omega$$

The Thevenin's equivalent of the circuit in Figure-6.39 is shown in Fig. 6.40.

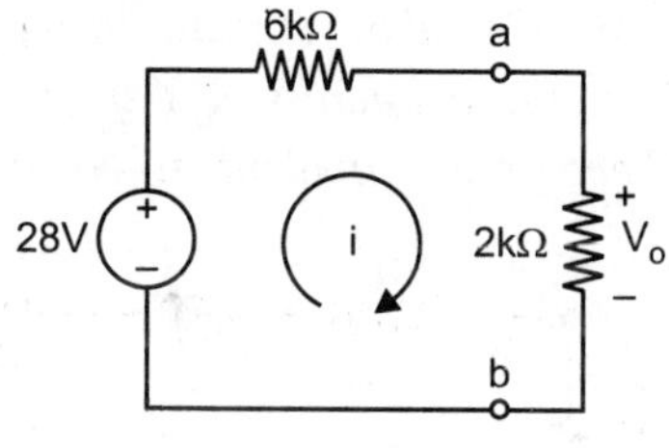

Fig. 6.40

Therefore,

$$V_o = i \times 2 \times 10^3$$

$$= [28/(6 \times 10^3 + 2 \times 10^3)] \times 2 \times 10^3 = 7V$$

Problem 6.12. *In the network of Fig. 6.41, in the absence of the current source of 10mA, the current i_a is equal to 2 mA. Compute the value of i_a (in mA) when the current source of 10mA is attached.*

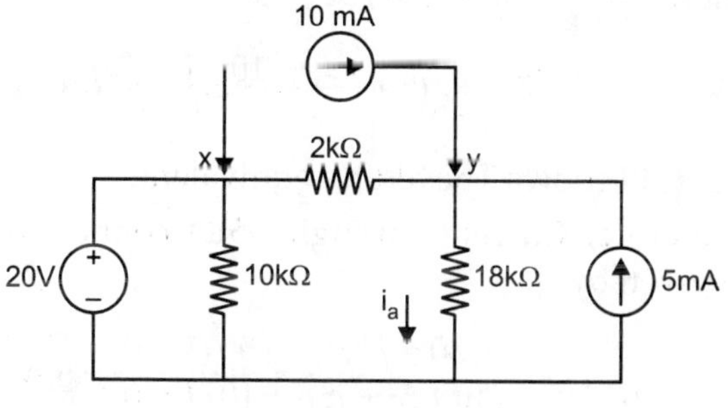

Fig. 6.41

Solution: The total current i_a can be found using superposition theorem.

The given current i_a of 2mA is that considering the effect of 20V voltage source and 5mA current source. Considering the effect of new current source of 10mA and deactivating the other two sources, we get the resultant circuit as shown in Fig. 6.42.

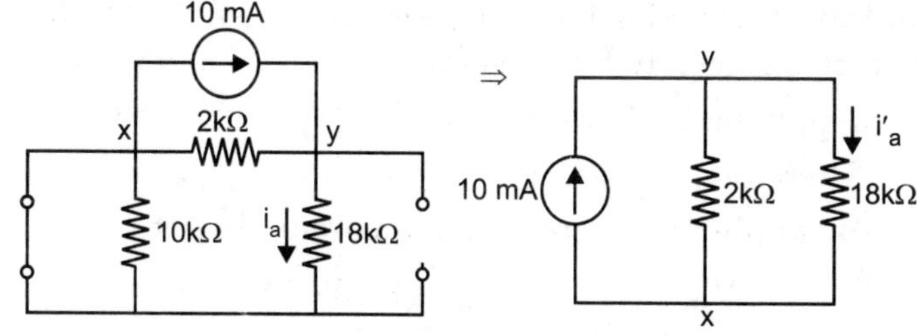

Fig. 6.42

Current i'_a is given by

$$i'_a = 10 \times 10^{-3} \left(\frac{2 \times 10^3}{18 \times 10^3 + 2 \times 10^3} \right) = 1 \text{ mA}$$

Therefore total current $i_a = 2\text{mA} + 1\text{mA} = 3\text{mA}$

Problem 6.13. *Calculate the voltage V_{ab} for the circuit shown in Fig. 6.43.*

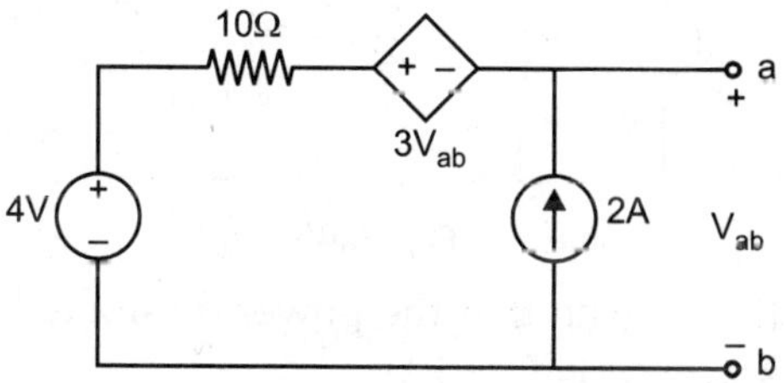

Fig. 6.43

Solution: The voltage V_{ab} can be calculated using the superposition theorem.

Considering the effect of only the voltage source of 4V and open circuiting the current source of 2A, the resultant circuit is shown in Fig. 6.44 (*a*).

Applying KVL, we get $4 - 10 \times 0 - 3V_{ab1} - V_{ab1} = 0$

Therefore, $V_{ab1} = 1V$

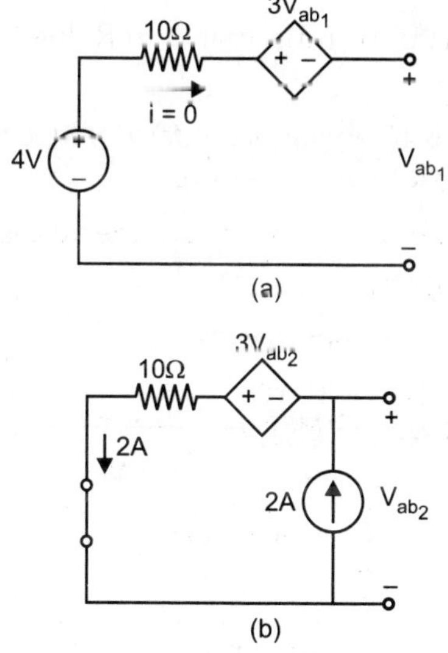

Fig. 6.44

Considering the effect of only the current source of 2A

and short circuiting the voltage source of 4V, the resultant circuit is shown in Fig. 6.44 (*b*).

Applying KVL, we get $10 \times 2 - 3V_{ab2} - V_{ab2} = 0$

Therefore, $V_{ab2} = 5V$

By applying superposition theorem, we get

$$V_{ab} = V_{ab1} + V_{ab2} = 1 + 5 = 6V$$

Problem 6.14. *In the circuit of Fig. 6.45, the power dissipated in the resistor R is 1 W when only source-1 is present and source-2 is replaced by a short. The power dissipated in the same resistor R is 4 W when only source-2 is present and source-1 is replaced by a short. Compute the power dissipated when both sources are present.*

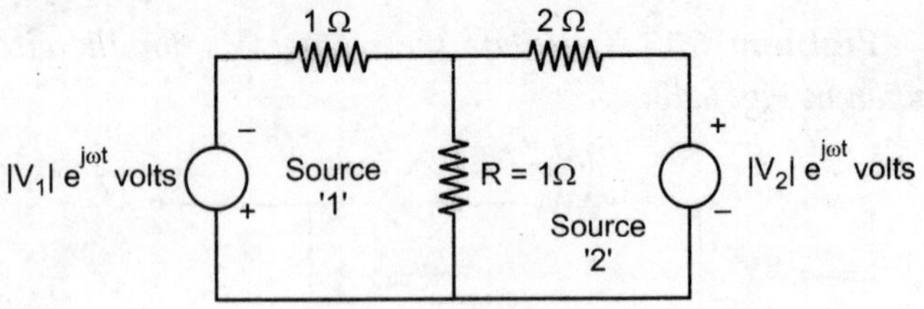

Fig. 6.45

Solution: Given that the power dissipated in resistor R due to source-1 is $P_1 = 1$ W

Given that the power dissipated in resistor *R* due to source-2 is $P_2 = 4$ W

Therefore, current (i_1) through (*R*) due to source-1 only is equal to $\sqrt{(P_1 \times R)} = 1$A and

Current (i_2) through (*R*) due to source-2 only is equal to $-\sqrt{(P_2 \times R)} = 3$A

Since the polarities of both the sources are different, therefore the total current when both the sources are present is given by (2 – 1) = 1A.

Therefore, power dissipated in R due to both sources $= 1^2 \times 1 = 1$W

Problem 6.15. *Apply phasor form of KCL to prove that the voltage (v) in Fig. 6.46 is 100 volts.*

Solution: If v_R, v_L and v_C are the voltages across *R*, *L* and *C* respectively.

$$v = v_R + v_L + v_C$$

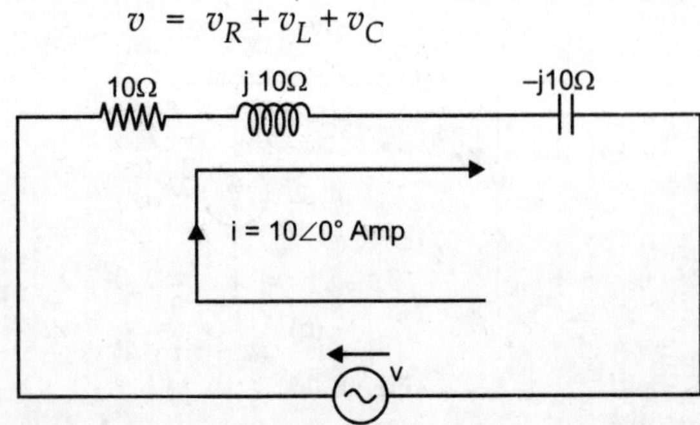

Fig. 6.46

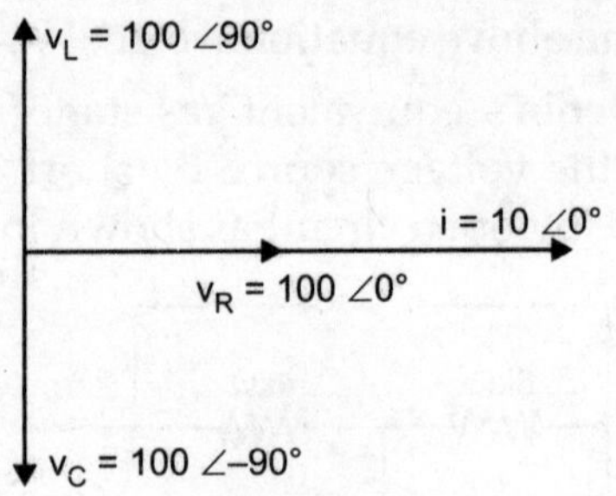

Fig. 6.47

Where the addition on the right hand side has to be addition of phasors. The phasor diagram is given in Fig. 6.47.

The phasor diagram of Fig. 6.47 is self explanatory. Quite obviously, phasor sum of v_R, v_L and v_C equals v_R.

$$v = v_R = 100 \text{ Volts.}$$

Problem 6.16. *Calculate current through an inductive reactance of j5Ω connected across X-Y in the network of Fig. 6.48 using (i) Thevenin's equivalent (ii) Norton's equivalent.*

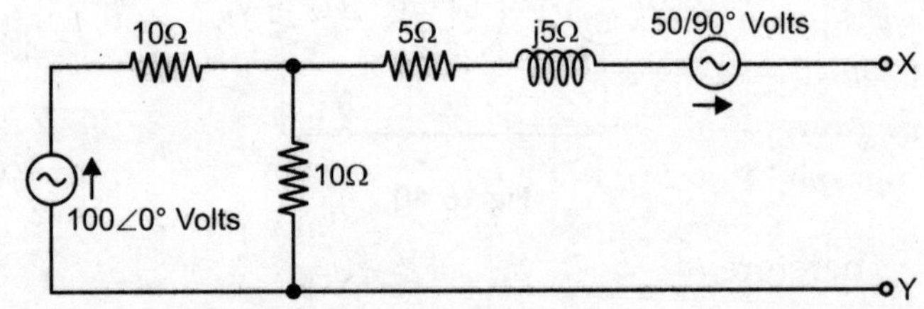

Fig. 6.48

Solution: (*i*) The open circuit voltage across *X-Y* is quite clearly given by

$$\frac{100\angle 0° \times 10}{20} + 50\angle 90° = (50 + j\,0) + j\,50 = (50 + j\,50)$$

Therefore, $V_{th} = (50\, j\, 50)$ Volts

Also $Z_{th} = (5 + j\,5) + \dfrac{10 \times 10}{20} = 10 + j\,5$

Fig. 6.49 shows Thevenin's equivalent.

Therefore, Current through (*j*5)Ω connected across *X-Y* is given by

$$i = \frac{50 + j50}{(10 + 5j + j5)} = \frac{50\,(1 + j)}{10\,(1 + j)} = 5 \text{ A}$$

(*ii*) Norton's equivalent can be derived from Thevenin's equivalent of Fig. 6.49. It is shown in Fig. 6.50.

Current through (*j* 5) Ω connected across *X-Y* is given by

$$\frac{(6 + j\,2) \times (10 + j\,5)}{(10 + j\,10)} = \frac{(50 + j\,50)}{(10 + j\,10)} = 5 \text{ A.}$$

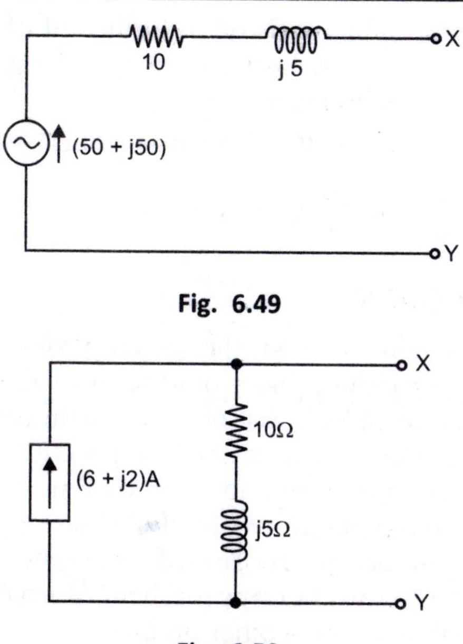

Fig. 6.49

Fig. 6.50

Problem 6.17. *In the network of Fig. 6.51, find the current through 3 Ω resistance using (i) Thevenin theorem (ii) Superposition theorem (iii) using KCL or Mesh current analysis (iv) Millman's theorem if possible. Then try to reach at a conclusion as to which approach in this case in the most convenient one.*

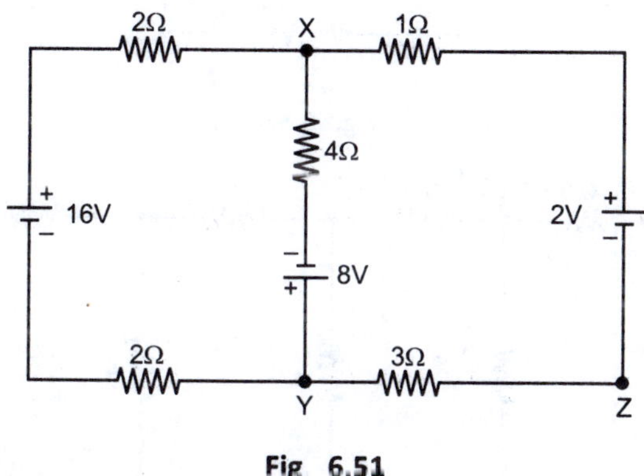

Fig 6.51

Solution:

(*i*) Let us find Thevenin's equivalent across 3 Ω resistance. The network is redrawn Fig. 6.52 (*a*).

Quite clearly, open circuit voltage V_{ZY} is given by $\{(-8) + (3 \times 4)\} - 2 = 2$ volts

and Thevenin impedance $= 1 + \dfrac{4 \times 4}{8} = 3\ \Omega$.

The Thevenin equivalent is shown in Fig. 6.52 (*b*).

Current through 3 Ω resistance is

$$\left(\frac{2}{3+3}\right) = 1/3 \text{ A (from Z to Y).}$$

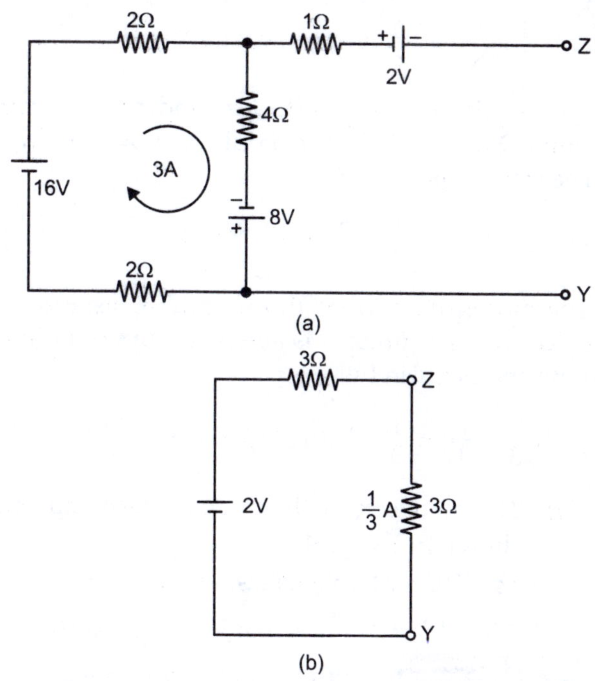

Fig. 6.52

(*ii*) Let us find out current due to 16 V source alone. The network with the remaining two sources replaced by short circuits is shown in Fig. 6.53 (*a*). Quite clearly, current through 3 Ω resistance is

$$\frac{16}{6 \times 2} = \frac{4}{3} \text{ A from Z to Y.}$$

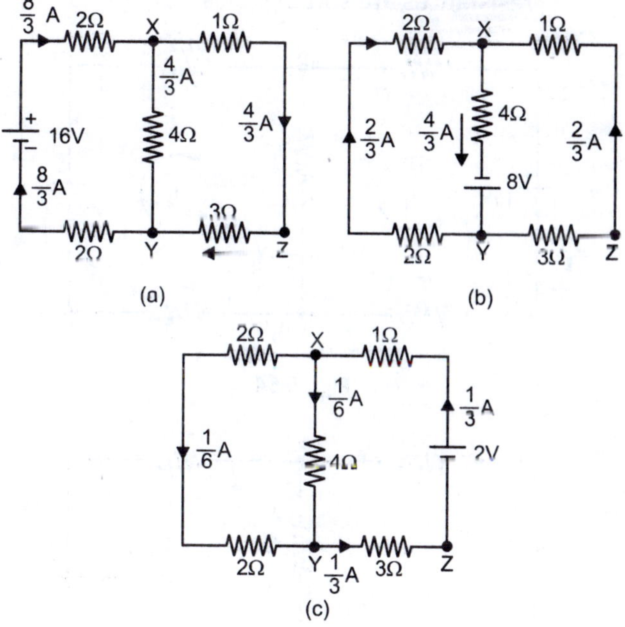

Fig. 6.53

The equivalent network for finding out current through 3Ω due to 8 V source only is given by Fig. 6.53 (*b*). The current is given by

$$\frac{4}{3} \times \frac{4}{8} = \frac{2}{3} \text{ A (from } Y \text{ to } Z)$$

The equivalent network for finding out current through 3Ω due to 2 V source only is shown in Fig. 6.53 (*c*). The current is given by

$$\frac{1}{3} \text{ A. (from } Y \text{ to } Z)$$

Therefore, the current through 3Ω resistance when all the sources are present is algebraic sum of individual currents and is given by

$$\left(\frac{4}{3} - \frac{2}{3} - \frac{1}{3}\right) = \frac{1}{3} \text{ A (from } Z \text{ to } Y)$$

(*iii*) The network with chosen Mesh currents is shown in Fig. 6.54.

Writing KVL for two meshes, we get

$$8 I_1 - 4 I_2 = 24 \quad \text{or} \quad 2 I_1 - I_2 = 6$$

and $-4 I_1 - 8 I_2 = -10$ or $2 I - 4 I_2 = 5$

From above equation,

$$I_2 = \frac{1}{3} \text{ A (From } Z \text{ to } Y)$$

(*iv*) Look at Fig. 6.55. It is nothing but network of Fig. 6.55 redrawn with the two resistances to the left of X-Y rearranged. It has not made any difference to the network as both these resistances are still in series.

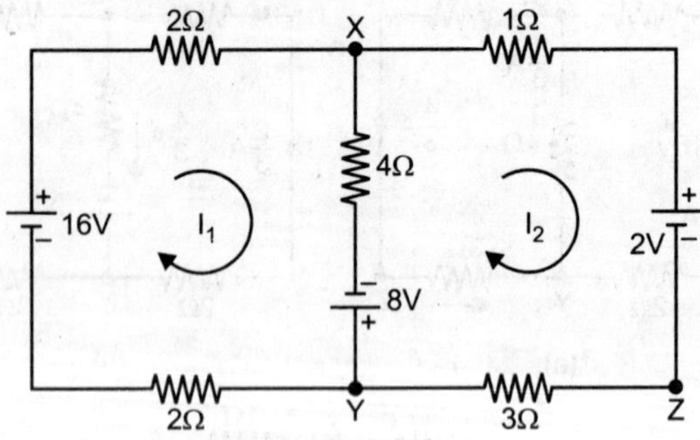

Fig. 6.54

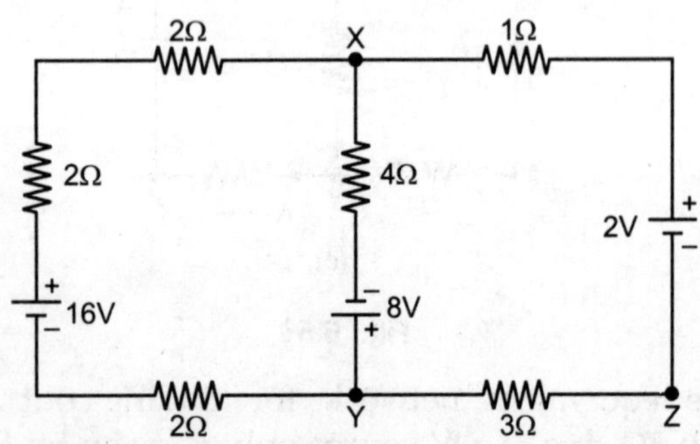

Fig. 6.55

Applying Millman theorem to the left of *X-Y* in the network of Fig. 6.55, the network of Fig. 6.55, gets reduced to the one shown in Fig. 6.56.

Current through 3Ω. resistance is thus

$$\frac{4-2}{6} = \frac{1}{3} \text{ A (From } Z \text{ to } Y)$$

CONCLUSION

After having solved the given problem by four different approaches, you would have seen; yourself by now that this problem was solved in no time by applying Millman theorem. It was just one-step solution. Solutions by Thevenin's approach and mesh current analysis were also:-obtained quite conveniently while application of superposition theorem consumed lot of time. But it is not always so. Superposition approach may turn out to be the best alternative in some other problem.

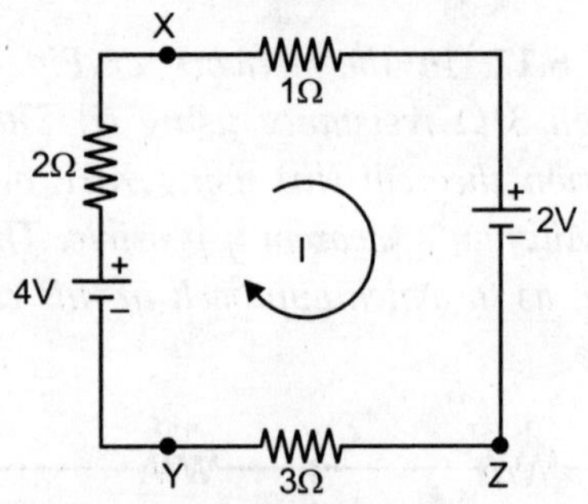

Fig. 6.56

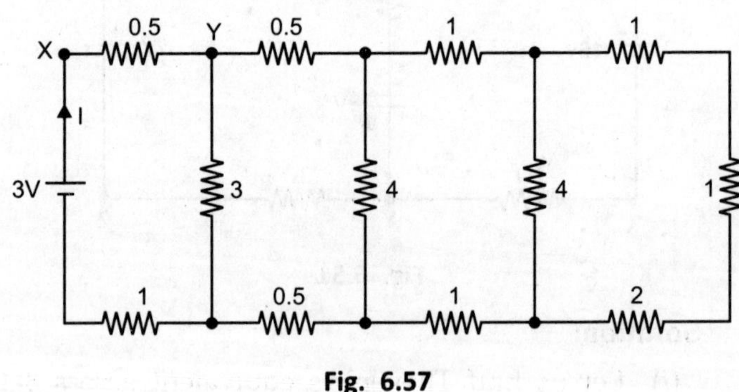

Fig. 6.57

Problem 6.18. *Find current (I) in the network of Fig. 6.57. All resistances are in ohms.*

Solution: The network of Fig. 6.57 can be simplified step by step as depicted in Fig. 6.58 (*a*) to 6.58 (*c*). From Fig. 6.58 (*c*).

$$I = 1\text{A}$$

If we had made an attempt solving this problem using mesh current analysis, we would have written four mesh equations and solving them simultaneously would have filled at least a couple of pages.

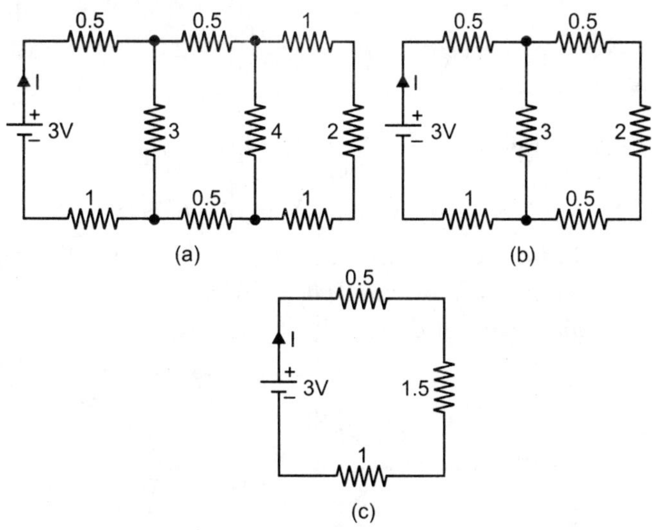

Fig. 6.58

On the other hand, finding Thevenin's equivalent across X-Y (see Fig. 6.57), we get the network of Fig. 6.59.

$$I = \frac{3}{(2.5 + 0.5)} = 1 \text{ A}.$$

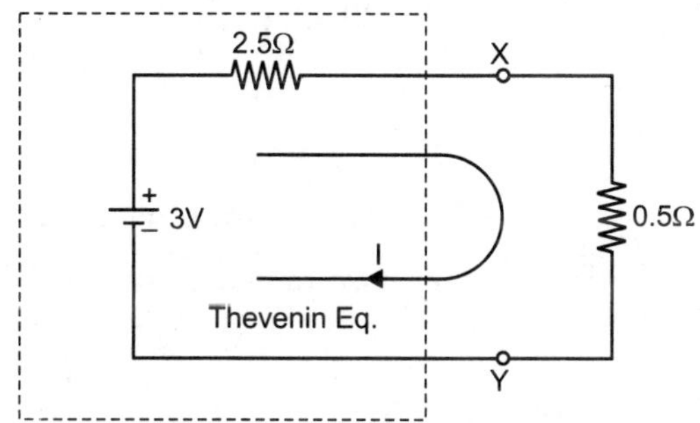

Fig. 6.59

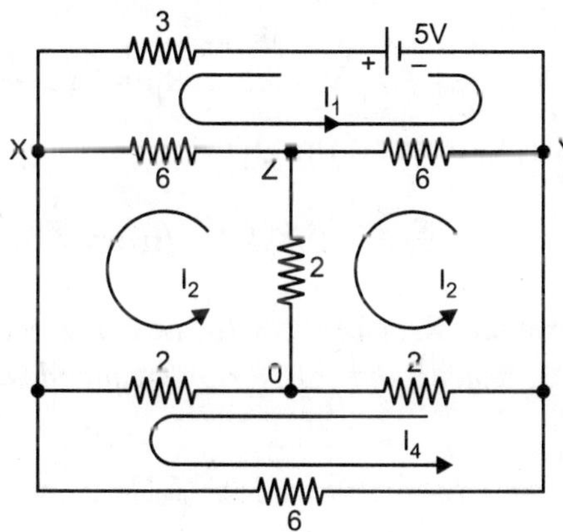

Fig. 6.60

Problem 6.19. *Find current (I) in the network of Fig. 6.60 by (i) Mesh current analysis (ii) Star-delta transformation. What conclusions do you draw? All resistances are in ohms.*

Solution: (*i*) From the network of Fig. 6.60 and the mesh currents chosen, following mesh-current expressions can be written on application of KVL in all the meshes individually.

$$15\,I_1 - 6\,I_2 - 6\,I_3 + 0\,I_4 = 5$$

$$-6\,I_1 + 10\,I_2 - 2\,I_3 - 2\,I_4 = 0$$

$$-6\,I_1 - 2\,I_2 + 10\,I_3 - 2\,I_4 = 0$$

$$0\,I_1 - 2\,I_2 - 2\,I_3 + 10\,I_4 = 0$$

From determinant method, $I_1 = \dfrac{\Delta_1}{\Delta}$

where $\Delta_1 = \begin{vmatrix} 5 & -6 & -6 & 0 \\ 0 & 10 & -2 & -2 \\ 0 & -2 & 10 & -2 \\ 0 & -2 & -2 & 10 \end{vmatrix}$ and $\Delta = \begin{vmatrix} 15 & -6 & -6 & 0 \\ -6 & 10 & -2 & -2 \\ -6 & -2 & 10 & -2 \\ 0 & -2 & -2 & 10 \end{vmatrix}$

On solving these determinants, we get

$$\Delta_1 = 4320 \quad \text{and} \quad \Delta = 4320$$

Therefore

$$I_1 = \frac{4320}{4320} = 1 \text{ A}.$$

(*ii*) The network of Fig. 6.61 (*a*) is nothing but network of Fig. 6.60 redrawn to bring out hidden information.

Converting the inner star into equivalent delta, we get network of Fig. 6.61 (*b*). The network of Fig. 6.61 (*b*) is clearly reduced to the network of Fig. 6.61 (*c*).

From Fig. 6.61 (*c*), then $I = 1$ A.

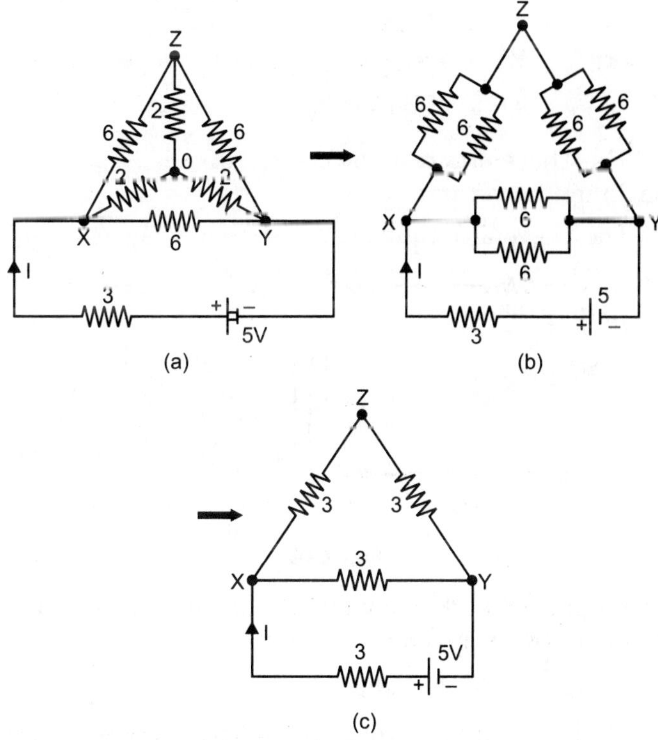

Fig. 6.61

One can clearly see that the second approach has been much more convenient than the first. What is required is a good look at the network configuration to be analyzed before a decision is made regarding the approach to be adopted.

Problem 6.20. *The networks of Fig. 6.62 (a) and 6.62 (b) are such that they are transferring maximum power to the load. If the power transferred in case of the network of Fig. 6.62 (a) is 10 watts, then draw the Thevenin's equivalent for the network of Fig. 6.62 (a) and Norton's equivalent for the network of Fig. 6.62 (b).*

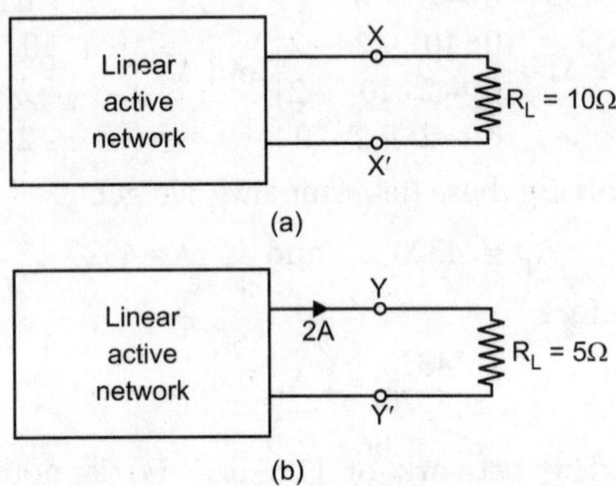

Fig. 6.62

Solution: For the network of Fig. 6.62 (*a*), as the power transferred to the load resistance (= 10 Ω) is maximum, we have

$$\frac{V^2}{4R_L} = 10 \text{ where V is the thevenin voltage}$$

or $V^2 = 4R_L \times 10 = 4 \times 10 \times 10 = 400$

$V = \sqrt{400} = 20$ V.

The Thevenin's resistance = R_L 10Ω .... for maximum power transfer.

The Thevenin's equivalent is shown in Fig. 6.63 (*a*).

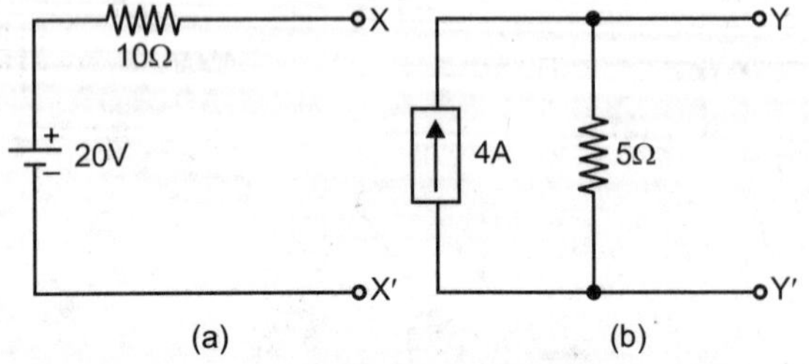

Fig. 6.63

For the network of Fig. 6.62 (*b*), the power in the load resistance (R_L) is given by

$(2)^2 \times 5 = 20$ watts.

Norton's equivalent resistance = 5Ω....... for maximum power transfer.

Norton's current = 4 A as $\dfrac{I_n \times R_n}{R_n + R_L} = 2$

which gives $I_n = \dfrac{2(R_n + R_L)}{R_n} = 4\ (R_n = R_L)$.

Norton's equivalent is given in Fig. 6.63 (*b*).

Problem 6.21. *For the network of Fig. 6.64, find the value R that receives a maximum power. Mark the direction and magnitude of current through R.*

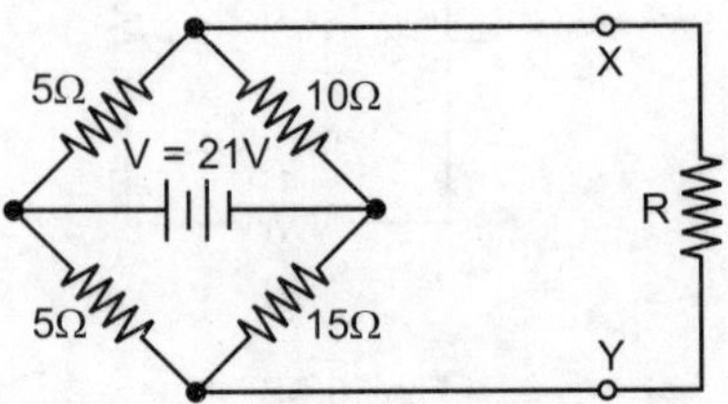

Fig. 6.64

Solution: If we try to find thevenin's equivalent across X-Y. The Thevenin's resistance is quite clearly given by

$$\frac{10 \times 5}{15} + \frac{15 \times 5}{20} = \frac{50}{15} + \frac{75}{20}$$

$$\frac{425}{60} \cong 7\ \Omega.$$

Therefore, $R = 7\ \Omega$ for maximum power transfer.

Let us find open circuit voltage across *X-Y*. It is given by

$$\frac{21 \times 5}{20} - \frac{21 \times 5}{15} = -1.75 \text{ V}.$$

Thus we have both the components of Thevenin's equivalent circuit with

$$V_{th} = -1.75 \text{ V} \quad \text{and} \quad R_{th} = 7\ \Omega.$$

Therefore, Current through

$$R = \frac{1.75}{14} = 0.124 \text{ A} \quad (\text{from } Y \text{ to } X).$$

Problem 6.22. *Replace the network of Fig. 6.65 by its (a) Thevenin's equivalent and (b) Norton's equivalent across the terminals X – Y.*

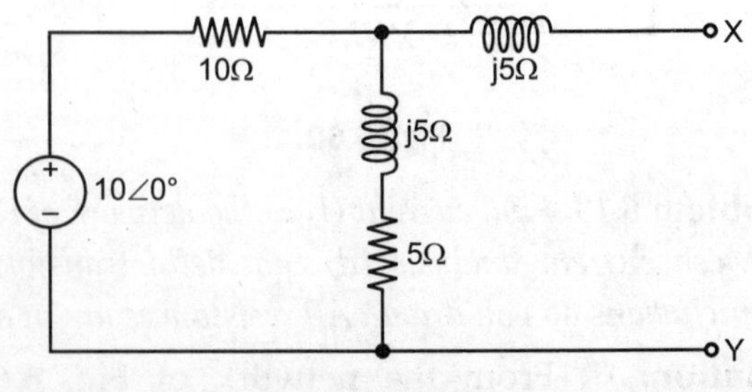

Fig. 6.65

Solution: (*a*) The Thevenin's voltage (V_{th}) is the open circuit voltage across XY. It is given by:

$$V_{th} = \frac{10\,(5+j5)}{(15+j5)} = \frac{(50+j50)}{(15+j5)}$$

$$= \frac{70.7\angle 45^\circ}{15.8\angle 18.43^\circ} = 4.47\angle 26.57^\circ$$

The Thevenin's impedance (Z_{th}) is given by :

$$Z_{th} = j5 + \frac{10\,(5+j5)}{(15+j5)} = j5 + \frac{(50+j50)}{(15+j5)}$$

$$= \frac{-25+j75+50+j50}{(15+j5)}$$

$$= \frac{25+j\,125}{(15+j5)} = \frac{127.5\angle 78.7^\circ}{15.8\angle 18.43^\circ} = 8\angle 60.27^\circ$$

The Thevenin's equivalent circuit is shown in Fig. 6.66 (*a*).

(*b*) Norton's current,

$$I_n = \frac{4.47\angle 26.57^\circ}{8\angle 60.27^\circ} = 0.558\angle -33.7^\circ$$

Norton's impedance, $Z_n = Z_{th}$

Norton's equivalent circuit is shown in Fig. 6.66 (*b*)

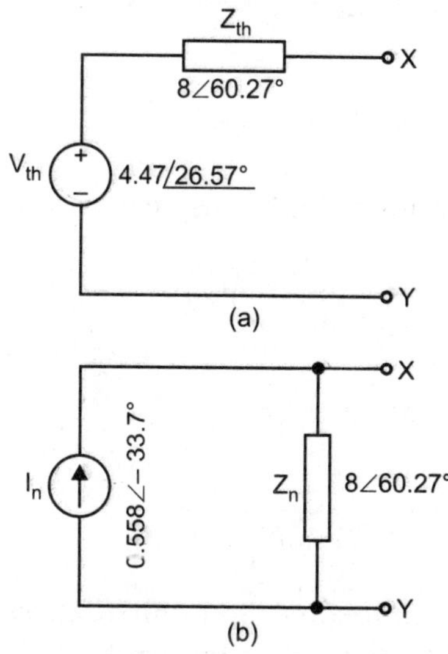

Fig. 6.66

Summary of Important Formulas

1. Thevenin's voltage (V_{th}), Thevenin's impedance (Z_{th}), Norton's current (I_n) and Norton's impedance (Z_n) are interrelated by following expressions.

$$I_n = \frac{V_{th}}{Z_{th}},\; Z_{th} = Z_n,\; V_{th} = I_n Z_n,\; Z_{th} = Z_n$$

2. For maximum power transfer from a source with devies impedance $Z_S = R_S + jX$ to a load $Z_L = R_L + jX_L$,

$$R_L = R_S \text{ and } X_L = -X_S$$

Transfer efficiency for maximum power transfer = 0.5

3. Voltage source $V_1, V_2, V_3 \ldots\ldots V_n$ with respective internal impedances $Z_1, Z_2, Z_3 \ldots\ldots\ldots Z_n$ when connected in parallel are electrically equivalent to a single voltage source (V) with internal impedance (Z) given by :

$$V = \left[\frac{(V_1/Z_1) + (V_2/Z_2) + (V_3/Z_3) + \ldots\ldots (V_n/Z_n)}{(1/Z_1) + (1/Z_2) + (1/23) + \ldots\ldots + (1/Z_n)}\right]$$

$$Z = \left[\frac{1}{(1/Z_1) + (1/Z_2) + (1/Z_3) + \ldots\ldots + (1/Z_n)}\right]$$

4. Current sources $I_1, I_2, I_3, \ldots\ldots I_n$ with respective shunt impedances $Z_1, Z_2, Z_3, \ldots\ldots, Z_n$ are electrically equivalent to a single curren sources (I) with shunt impedance (Z) given by :

$$I = I_1 + I_2 + I_3 + \ldots\ldots + I_n$$

$$Z = \left[\frac{1}{(1/Z_1) + (1/Z_2) + (1/Z_3) + \ldots\ldots + (1/Z_n)}\right]$$

5. The equivalent delta impedances Z_{AB}, Z_{BC} and Z_{CA} for star impedances Z_A, Z_B and Z_C are given by :

$$Z_{AB} = Z_A + Z_B + \frac{Z_A Z_B}{Z_C}$$

$$Z_{BC} = Z_B + Z_C + \frac{Z_B Z_C}{Z_A}$$

$$Z_{CA} = Z_C + Z_A + \frac{Z_C Z_A}{Z_B}$$

The same when translated to resistors,

$$R_{AB} = R_A + R_B + \frac{R_A R_B}{R_C}$$

$$R_{BC} = R_B + R_C + \frac{R_B R_C}{R_A}$$

$$R_{CA} = R_C + R_A + \frac{R_C R_A}{R_B}$$

The same when translated to inductors,

$$L_{AB} = L_A + L_B + \frac{L_A L_B}{L_C}$$

$$L_{BC} = L_B + L_C + \frac{L_B L_C}{L_A}$$

$$L_{CA} = L_C + L_A + \frac{L_C L_A}{L_B}$$

The same when translated to capacitors,

$$C_{AB} = \frac{C_A C_B}{C_A + C_B + C_C}$$

$$C_{BC} = \frac{C_B + C_C}{C_A + C_B + C_C}$$

$$C_{CA} = \frac{C_C + C_A}{C_A + C_B + C_C}$$

6. The equivalent star impedances Z_A, Z_B and Z_C for delta impedances Z_{AB}, Z_{BC} and Z_{CA} are given by :

$$Z_A = \frac{Z_{AB} Z_{CA}}{Z_{AB} + Z_{BC} + Z_{CA}}$$

$$Z_B = \frac{Z_{BC} Z_{AB}}{Z_{AB} + Z_{BC} + Z_{CA}}$$

$$Z_C = \frac{Z_{BC} Z_{CA}}{Z_{AB} + Z_{BC} + Z_{CA}}$$

In the case of resistors, inductors and capacitors, corresponding expressions are :

$$R_A = \frac{R_{AB} R_{AC}}{R_{AB} + R_{BC} + R_{CA}}$$

$$R_B = \frac{R_{BC} R_{AB}}{R_{AB} + R_{BC} + R_{CA}}$$

$$R_C = \frac{R_{CA} R_{BC}}{R_{AB} + R_{BC} + R_{CA}}$$

$$L_A = \frac{L_{AB} L_{CA}}{L_{AB} + L_{BC} + L_{CA}}$$

$$L_B = \frac{L_{BC} L_{AB}}{L_{AB} + L_{BC} + L_{CA}}$$

$$L_C = \frac{L_{CA} L_{BC}}{L_{AB} + L_{BC} + L_{CA}}$$

$$C_A = C_{AB} + C_{CA} + \frac{C_{AB} C_{CA}}{C_{BC}}$$

$$C_B = C_{BC} + C_{AB} + \frac{C_{AB} C_{BC}}{C_{CA}}$$

$$C_C = C_{CA} + C_{BC} + \frac{C_{CA} C_{BC}}{C_{AB}}$$

SELF EVALUATION EXERCISE

MULTIPLE CHOICE QUESTIONS

1. The Thevenin's equivalent voltage and equivalent impedance across a certain set of terminals in a given network are 10 volts and 100 ohms respectively. The Norton's impedance across same terminals in the same network would be:
   (*a*) 100 ohms
   (*b*) 10 ohms
   (*c*) 1000 ohms
   (*d*) Cannot be determined from given data
2. In a given network, across a certain set of terminals, the Thevenin's voltage and the Norton's current were determined to be equal to 100 Volts and 1 Ampere respectively. The Thevenin's impedance in this case would be:
   (*a*) 10 ohms
   (*b*) 100 ohms
   (*c*) Cannot be determined from given data
   (*d*) None of these
3. In Q-2, Norton's impedance will be:
   (*a*) 100 ohms (*b*) 10 ohms
   (*c*) 1000 ohms (*d*) None of these
4. According to maximum power transfer theorem, for maximum power transfer, the load impedance will be (10 + *j*50) ohms for a source impedance of:
   (*a*) (10 – *j*50) ohms (*b*) (10 + *j*50) ohms
   (*c*) (10 – *j*10) ohms (*d*) (50 – *j*10) ohms
5. A certain delta network has equal value resistors

equal to (R) ohms in three arms. The three resistors in the equivalent star network will

(a) Be unequal

(b) Have same value equal to (R/3)

(c) Have same value equal to (3R)

(d) None of these

6. A certain star network has equal value resistors equal to (R) ohms in three arms. The three resistors in the equivalent delta network will

(e) Be unequal

(f) Have same value equal to (R/3)

(g) Have same value equal to (3R)

(h) None of these

7. A certain network has two active sources. Power dissipated in a given resistive element (R) is 2 watts when only one source is present. The power dissipated due to second source alone is 4.5 watts. If the two sources result in currents through that element in opposite directions; then the power dissipated with both sources present will be:

(a) 3 watts (b) 2 watts

(c) 6.5 watts (d) 0.5 watt

8. In Q-7, if the currents flowing through the resistive element due to the two sources were in the same direction; then the dissipated power will be:

(a) 12.5 watts (b) 0.5 watt

(c) 3 watts (d) 6.5 watts

9. Three resistances R_1, R_2 and R_3 have been connected in the DELTA fashion. The equivalent STAR resistance connected to $R_1 - R_2$ node of the delta is given by

(a) $\frac{R_1 R_2}{R_1 + R_2 + R_3}$ (b) $\frac{R_2 R_3}{R_1 + R_2 + R_3}$

(c) $\frac{R_3 R_1}{R_1 + R_2 + R_3}$ (d) $R_1 + R_2 + \frac{R_1 R_2}{R_3}$

10. In Q. 9 the equivalent Star resistances that would originate from ($R_2 - R_3$) node and ($R_3 - R_1$) onode can be determined by multiplying the equivalent STAR resistance of ($R_1 - R_2$) node by

(a) R_3/R_2 and R_3/R_1 Respectively

(b) R_3/R_1 and R_3/R_2 Respectively

(c) R_2/R_1 and R_3/R_1 Respectively

(d) can not be determined like that

(e) None of these.

11. The resistances R_A, R_B and R_C have been connected in a STAR fashion with R_A connected between a point A and a common point, R_B connected between a point B and a common point and R_C connected between a point C and a common point. The equivalent DELTA resistance R_{AB} between points A and B is given by

(a) $R_B + R_C + \frac{R_B R_C}{R_A}$

(b) $\frac{R_A R_B}{R_A + R_B + R_C}$

(c) $R_A + R_B + \frac{R_A R_B}{R_C}$

(d) $\frac{R_A R_B + R_B R_C + R_C R_A}{R_C}$

12. In Q.11 equivalent star resistances R_{BC} and R_{CA} are respectively obtained from R_{AB} as:

(a) $\frac{R_{AB} R_C}{R_A}$ and $\frac{R_{AB} R_C}{R_B}$

(b) $\frac{R_{AB} R_C}{R_B}$ and $\frac{R_{AB} R_C}{R_A}$

(c) $\frac{R_{AB} R_A}{R_C}$ and $\frac{R_{AB} R_B}{R_C}$

(d) can not be determined from R_{AB}

(e) None of these.

13. In Q.9, the resistances R_1, R_2 and R_3 are replaced by pure inductances L_1, L_2 and L_3. The statement remaining the same, the equivalent STAR, inductance between L_1 and L_2 is

(a) $\frac{L_1 L_2}{L_1 + L_2 + L_3}$

(b) $\frac{L_2 L_3}{L_1 + L_2 + L_3}$

(c) $\frac{L_1 L_2 + L_2 L_3 + L_3 L_1}{L_3}$

(d) None of these.

14. In Q.11, resistances R_A, R_B and R_C care replaced by pure inductances, L_A, L_B and L_C respectively, the statement remaining the same, the equivalent DELTA inductance between the points A and B is given by

(a) $\frac{L_A L_B}{L_A + L_B + L_C}$

(b) $\frac{L_A L_B + L_B L_C + L_C L_A}{L_C}$

(c) $\frac{L_A L_B + L_B L_C + L_C L_A}{L_A}$

(d) $LA + LA + \frac{L_B L_C}{L_A}$.

15. In Q. 9 resistances R_1, R_2 and R_3 are replaced by pure capacitors, C_1, C_2 and C_3 respectively be :

(a) $\frac{C_1 C_2 + C_2 C_3 + C_3 C_1}{C_1}$

(b) $\frac{C_1 C_2 + C_2 C_3 + C_3 C_1}{C_2}$

(c) $\frac{C_1 C_2 + C_2 C_3 + C_3 C_1}{C_3}$

(d) $\frac{C_1 C_2}{C_1 + C_2 + C_3}$.

16. In Q. 11, Resistors R_A, R_B and R_C are replaced by pure capacitors C_A, C_B and C_C respectively. The staterment remaining same, the equivalent DELTA capacitance C_{BC} is given by

(a) $\frac{C_B C_C}{C_A + C_B + C_C}$

(b) $\frac{C_A + C_B + C_C}{C_A C_B}$

(c) $\frac{C_A + C_B + C_C}{C_B C_C}$

(d) $C_B + C_C + \frac{C_B C_C}{C_A}$.

17. "A DLETA - connection consists of a resistance of 5Ω, a reactive impedance of + j 5Ω and another reactive impedance of – j 5Ω. The equivalent STAR impedance will also be 5 Ω, + j5 Ω and – j 5 Ω. Comment on the statement

(a) The statement is false

(b) The statement is true but it is a pure coincidence

(c) The statement is true and it is true for all those cases where DELTA consists of R,+ jX and – jX with $R = X$.

(d) The statement is meaningless.

18. Two voltage sources of magnitude 12 volts and 18 volts and respective series resistances 1Ω and 2Ω are equivalent to a single source of magnitude (V) volts and series resistance (R)Ω where

(a) V = 18 volts, R = 2Ω

(b) V = 14 volts, R = 2/3Ω

(c) V = 12 volts, R = 1Ω

(d) V = 14 volts, R = 1Ω

19. Two current sources of magnitude 10 Amperes and 15 Amperes and respective shunt resistances 100Ω and 50Ω are equivalent to a single current source of magnitude (I) Amperes and series resistance (R)Ω where

(a) I = 15A, R = 50Ω

(b) I = 10 volts, R = 100Ω

(c) I = 25 A, R = 150Ω

(d) I = 25 A, R = 33.33Ω

20. A network has two voltage sources and some passive components. Power dissipated in a certain resistance due to one voltage source alone is 5 watts and the power dissipated in the same resistance due to the other voltage source alone is 10 watts. Power dissipated in the resistance when both voltage sources are present will be equal to

(a) 10 watts

(b) 15 watts

(c) Cannot be determined from given data

(d) √125 watts

ANSWERS TO SELF EVALUATION EXERCISE

1. (a)	**2.** (b)	**3.** (a)	**4.** (a)	**5.** (b)	**6.** (c)
7. (d)	**8.** (a)	**9.** (a)	**10.** (b)	**11.** (c) and (d)	**12.** (a)
13. (a)	**14.** (b)	**15.** (c)	**16.** (a)	**17.** (c)	**18.** (b)
19. (d)	**20.** (c)				

Chapter 7

Fourier Analysis of Waveforms

Learning Objectives

After completing the chapter, you will learn the following topics:

1. Fourier Series Analysis of Networks
2. Odd and Even Functions
3. Half Wave Symmetry
4. Dirichlet Conditions
5. Trigonometric Fourier series
6. Exponential Fourier Series
7. Fourier Series Expansion of Important Waveforms
8. Fourier Integral for Non-recurring Pulse

Q.1. Where does the Fourier method of waveform analysis find application in analyzing linear networks?

Ans. The steadystate response of linear networks to input DC or sinusoidal input can be expressed by a single expression and this expression is valid for all values of time (t). This is so because a single expression describes these input functions for all values of (t) *e.g.*, v = constant (for DC) and $v = V_m \sin \omega t$ (for sinusoidal) for all values of (t). Certain periodic waveforms, sawtooth, triangular, square being some of them, which are of considerable importance in electronics circuitry, can be described by a single expression only within an interval (the interval being equal to the time period after which the waveform repeats itself). As a result, the steadystate response of linear networks to such periodic waveforms cannot be satisfactorily determined for all values of (t). If there existed a method by which a periodic non-sinusoidal waveform could be expressed as a sum of finite or infinite number of sinusoidal functions, the response of a linear network to a non-sinusoidal excitation could be easily found out by determining response to individual sinusoidal functions and then making use of superposition principle.

Fourier method of waveform analysis does exactly this. While FOURIER SERIES can be used to express non-sinusoidal periodic waveforms as a sum of sinusoidal and cosinusoidal functions, FOURIER INTEGRAL can be used to represent a non-periodic non-sinusoidal waveform (*i.e.* a pulse) in terms of sinusoidal and cosinusoidal functions.

Q.2. Can Fourier series be used to express all periodic waveforms or is it that waveforms have to satisfy certain conditions, if yes, what are these conditions and what are they popularly called?

Ans. The periodic waveforms must satisfy the following conditions so that they can be expressed by a Fourier series :

(*i*) If the waveform is discontinuous, there should be finite number of discontinuities in the period T.

(*ii*) It should have a finite average value for the period T.

(*iii*) It should have a finite number of positive and negative maxima in the period T.

These are called DIRICHLET CONDITIONS.

Q.3. How can a generalised periodic function be represented by trigonometric Fourier series? How does trignometric Fourier series differ from exponential Fourier series? How do you convert a trignometric Fourier series expression into corresponding exponential Fourier series?

Ans. A periodic function $f(t)$ so that $f(t) = f(t + T)$ can be represented by trignometric Fourier series as

$$f(t) = \frac{a_0}{2} + a_1 \cos \omega t + a_2 \cos 2\omega t + a_3 \cos 3\omega t + \ldots.. a_n \cos n\omega t + b_1 \sin \omega t + b_2 \sin 2\omega t + b_3 \sin 3\omega t + \ldots.. b_n \sin n\omega t$$

where $a_n = \frac{1}{\pi}\int_0^{2\pi} f(t) \cos n\omega t \, d(\omega t) = \frac{2}{T}\int_0^T f(t) \cos n\omega t \, dt$

$$b_n = \frac{1}{\pi}\int_0^{2\pi} f(t) \sin n\omega t \, d(\omega t) = \frac{2}{T}\int_0^T f(t) \sin n\omega t \, dt$$

and $\frac{a_0}{2}$ = average value of the function.

It can be calculated by putting $n = 0$ in the expression for a_n or merely by inspection of the given function.

Exponential Fourier series can be obtained from trignometric Fourier series by replacing sine and cosine components by their corresponding exponential equivalents like

$$\sin \omega t = \frac{e^{j\omega t} - e^{-j\omega t}}{2j}$$

$$\sin 2\omega t = \frac{e^{j2\omega t} - e^{-j2\omega t}}{2j} \text{ and so on.}$$

And similarly,

$$\cos \omega t = \frac{e^{j\omega t} + e^{-j\omega t}}{2}$$

$$\cos 2\omega t = \frac{e^{j2\omega t} + e^{-j1\omega t}}{2} \text{ and so on.}$$

Substituting for sinusoidal and cosinusoidal components in the Fourier series of trignometric form and doing a little bit of mathematics, we get

$$f(t) = \ldots.. + A_{-2} e^{-j2\omega t} + A_{-1} e^{-j\omega t} + A_0 + A_1 e^{j\omega t} + A_2 e^{j2\omega t} + \ldots..$$

where $A_n = \frac{1}{T}\int_0^T f(t) e^{-jn\omega t} dt$

$$= \frac{1}{2\pi}\int_0^{2\pi} f(t) e^{-jn\omega t} d(\omega t)$$

This is the exponential form of Fourier series.

Q.4. What is an EVEN function? Tell whether the functions shown in Fig. 7.1 (*a*), (*b*) and (*c*) are even. Is the resulting function obtained by additions of the functions shown in Fig. 7.1 (*a*) and (*b*) even?

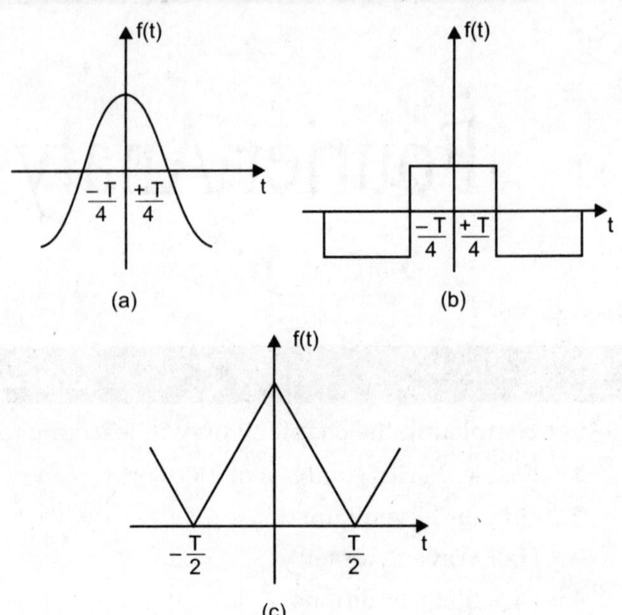

Fig. 7.1

Ans. A function $f(t)$ is even when $f(t) = f(-t)$ for all values of (t). All the three functions are even as $f(t) = f(-t)$ in all the three cases. Sum of two even functions is also an even function. Therefore, the function obtained by addition of the functions shown in figures (*a*) and (*b*) will also be an even function.

Q.5. What is an ODD function? Look at the waveforms shown in Fig. 7.2 (*a*), (*b*) and (*c*) and tell whether they are represented by odd-functions. Is the sum of two odd-functions an odd function. Illustrate with an example.

Ans. A function $f(t)$ is ODD provided

$$f(t) = -f(-t)$$

This condition is valid in case of all the three waveforms shown in Fig. 7.2 (*a*) to (*c*). Hence all of them are odd functions. The sum of two odd functions is an odd function.

Let us consider two odd functions given by

$$f_1(t) = V_m \sin \omega_1 t$$

and $f_2(t) = V_m \sin \omega_2 t$

Then $f(t) = f_1(t) + f_2(t) = V_m (\sin \omega_1 t + \sin \omega_2 t)$

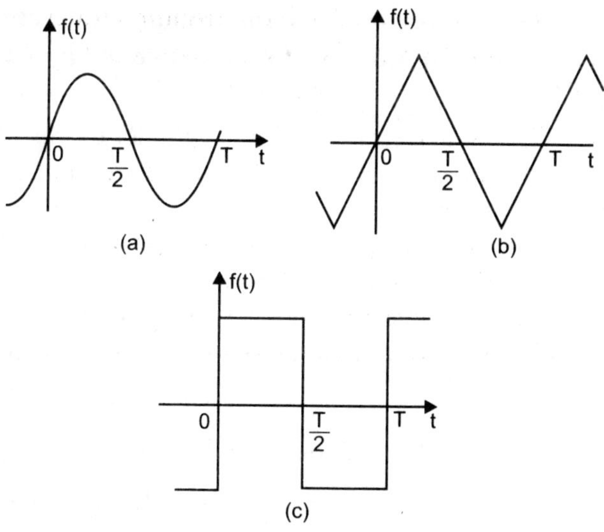

Fig. 7.2

$$= 2\,V_m\left[\sin\left(\frac{\omega_1-\omega_2}{2}\right)t\cos\left(\frac{\omega_1-\omega_2}{2}\right)t\right]$$

Now $f(-t)$ =

$$2\,V_m\left[\sin\left(\frac{\omega_1+\omega_2}{2}\right)(-t)\cos\left(\frac{\omega_1-\omega_2}{2}\right)(-t)\right]$$

$$= -2\,V_m\left[\sin\left(\frac{\omega_1+\omega_2}{2}\right)t\cos\left(\frac{\omega_1-\omega_2}{2}\right)t\right]$$

$$= -f(t)$$

and hence the result.

Q.6. What do you understand by HALF WAVE SYMMETRY of a periodic waveform? Justify in case of the waveform shown in Fig. 7.3 that it is half wave symmetric.

Ans. A waveform represented by function $f(t)$ is half-wave symmetric provided

$$f(t) = -f(t+T/2)$$

where, T = Time period of the waveforms.

$f(t)$ can be mathematically represented as

$$f(t) = \frac{2\,|f(t)|_{\max}}{T}(t) \text{ for } 0 \le t \le \frac{T}{2}$$

$$f(t) = \frac{-2\,|f(t)|_{\max}}{T}\left(t-\frac{T}{2}\right) \text{ for } \frac{T}{2} \le t \le T$$

We will notice that for any given (t), the time $(t + T/2)$ will always lie in the following half cycle of the function.

Thus replacing (t) by $(t + T/2)$, we get

$$f\left(t+\frac{T}{2}\right) = \frac{-2\,|f(t)|_{\max}}{T}\left(t+\frac{T}{2}-\frac{T}{2}\right)$$

$$= \frac{-2\,|f(t)|_{\max}}{T}(t) = -f(t)$$

This proves that the given waveform is half-wave symmetric over first complete cycle. It can similarly be proved for the succeeding cycles by writing mathematical expressions for the cycles in question.

The preceding discussion might have been a bit confusing. I will put it in simple words.

Let us assume $t = T/4$

Now since the time instant $t = T/4$ lies in the first half of the time period T, we will make use of relevant expression (1) for $f(t)$ to calculate the value of the function at $t = T/4$

$$|f(t)|_{t=T/4} = \frac{2\,|f(t)|_{\max}}{T}\times\frac{T}{4} = \frac{|f(t)|_{\max}}{2}$$

Now we have to calculate $f(t + T/2)$ *i.e.* $f(T/4 + T/2) = f(3T/4)$ or in other words $|f(t)|_{t=\frac{3T}{4}}$.

As the time instant $t = 3T/4$ lies in the second half,

$$|f(t)|_{t=\frac{3T}{4}} = -\frac{2\,|f(t)|_{\max}}{T}\left(\frac{3T}{4}-\frac{T}{2}\right)$$

$$= -\frac{2\,|f(t)|_{\max}}{T}\times\frac{T}{4} = -\frac{|f(t)|_{\max}}{2}$$

$$|f(t)|_{t=3T/4} = -\frac{|f(t)|_{\max}}{2}$$

Therefore, $f(t) = -f(t + T/2)$ and hence the result.

Q.7. When we are doing Fourier series expansion of a periodic waveform, how does the knowledge of waveform symmetry *i.e.*, whether the function is odd or even or half wave symmetric or combination of these, help us in solving the problem.

Ans. If the periodic waveform is represented by an odd function, its fourier series expansion contains only the sine terms with or without a constant depending upon whether the function is odd after omitting the constant or the function is odd and does not contain any constant. Thus once we know that the waveform is represented by an odd function, we need not evaluate the co-efficients (a_n) in the trignometric fourier series expansion. (A_n) values will be pure imaginaries and $A_n = 0$ for n even and $A_n \neq 0$ for n odd in the exponential Fourier series expansion.

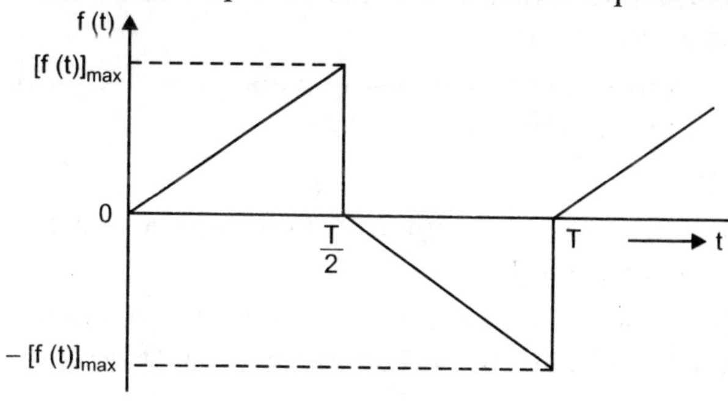

Fig. 7.3

If the periodic waveform is represented by an even function, its Fourier series expansion contains only cosine terms plus a constant. Thus in this case, we need not evaluate co-efficients (b_n) in the Trigonometric fourier series expansion and (A_n) will be pure real and also $A_n = 0$ for 'n' odd and $A_n \neq 0$ for n even in the exponential series expansion.

A half-wave symmetric waveform contains both sine and cosine terms with only odd harmonics provided it is neither odd nor even. It actual practice, the waveform may also be either even or odd if it is also half wave symmetric. In that case, the trignometric Fourier series expansion contains odd harmonics of sine terms or odd harmonics of cosine terms depending upon whether the waveform is odd and half wave symmetric or whether the waveform is even and half wave symmetric. If the waveform has a finite average value, it would also have a constant term.

Thus we have seen in the preceding paragraphs that the knowledge of waveform symmetry considerably simplifies the problem of doing Fourier series expansion of a given periodic waveform. Infact, knowing waveform symmetry is the first step towards the solution of a Fourier series expansion problem.

Q.8. Let us see if you can tell how would the trignometrie Fourier series expansions of the waveforms shown in Fig. 7.1 (*c*), 7.2 (*b*) and 7.3 would look like. No detailed expressions please ! Just give the composition like whether you are going to have both sine and cosine or only sine or only cosine terms and also whether there will be a constant or not.

Ans. Fig. 7.1 (*c*) :

It is an even function. It has a finite average value.

Therefore, Fourier series expansion will contain only cosine terms and a constant.

Fig. 7.2 (*b*) :

It is an odd function with zero average value. It is also half-wave symmetric.

Therefore, Fourier series expansion would contain only odd harmonics of sine terms.

Fig. 7.3 :

It is half wave symmetric with zero average. It is neither even nor odd.

Therefore, Fourier series expansion contains odd harmonics of both sine and cosine terms with no constant term.

Q.9. Find the expressions for trignometric Fourier series expansion for the waveforms shown in Fig. 7.1 (*b*) and 7.1 (*c*).

Ans. For the waveform of Fig. 7.1 (*b*), let us consider one complete cycle from (– $T/4$) to ($3T/4$) and let (A_{max}) be the peak value.

Since the waveform is represented by an EVEN-function, it will contain only the cosine terms and hence there is no need to evaluate the (b_n) co-efficients.

Trignometric Fourier series expansion, in general, is given by

$$f(t) = \frac{a_0}{2} + a_1 \cos \omega t + a_2 \cos 2\omega t + ... a_n \cos n\omega t + b_1 \sin \omega t + b_2 \sin 2\omega t + ... + b_n \sin n\omega t$$

In this case, the expression gets reduced to

$$f(t) = a_1 \cos \omega t + a_2 \cos 2\omega t + ... a_n \cos n\omega t$$

where
$$a_n = \frac{2}{T}\int_0^T f(t) \cos n\omega t\, dt$$

$$= \frac{2}{T}\left\{\int_{-T/4}^{+T/4} A_{max} \cos n\omega dt + \int_{+T/4}^{+3T/4} - A_{max} \cos n\omega t\, dt\right\}$$

$$= \frac{2A_{max}}{T}\left\{\left|\frac{\sin n\omega t}{\omega n}\right|_{-T/4}^{+T/4} - \left|\frac{\sin n\omega t}{\omega n}\right|_{T/4}^{3T/4}\right\}$$

$$= \frac{2A_{max}}{T\omega n}\left[\left\{\sin\left(\frac{n\omega T}{4}\right) + \sin\left(\frac{n\omega T}{4}\right)\right\} - \left\{\sin\left(\frac{3n\omega T}{4}\right) - \sin\left[\frac{n\omega T}{4}\right]\right\}\right]$$

$$= \frac{2A_{max}}{T\omega n}\left(3 \sin \frac{n\omega T}{4} - \sin \frac{3n\omega T}{4}\right)$$

$$a_1 = \frac{2A_{max}}{\omega T}(3+1) = \frac{8A_{max}}{\omega T}$$

$$a_2 = \frac{A_{max}}{\omega T}(0-0) = 0$$

$$a_3 = \frac{2A_{max}}{3\omega T}(-3-1) = \frac{-8A_{max}}{3\omega T}$$

$$f(t) = \frac{8A_{max}}{\omega T} \cos \omega t - \frac{8A_{max}}{3\omega T} \cos 3\omega t + ...$$

$$= \frac{8A_{max}}{\omega T}\left(\cos \omega t - \frac{1}{3} \cos 3\,\omega t + \frac{1}{5} \cos 5\,\omega t - ...\right)$$

The waveform of Fig. 7.1 (*c*) is also an even function with an average value of $\left(\frac{A_{max}}{2}\right)$ where A_{max} is the peak value

$$a_n = \frac{2}{T}\int_0^T f(t)\cos n\omega t\,dt$$

$$= \frac{2}{T}\left\{\int_{-T/2}^{0}\left(\frac{2A_{max}}{T}t + A_{max}\right)\cos n\omega t\,dt + \int_0^{T/2}\left(-\frac{2A_{max}}{T}t + A_{max}\right)\cos n\omega t\,dt\right\}$$

$$= \frac{2}{T}\int_{-T/2}^{0} A_{max}\cos n\omega t\,dt + \int_{-T/2}^{0}\frac{2A_{max}}{T}t\cos n\omega t\,dt + \int_0^{T/2} A_{max}\cos n\omega t\,dt - \int_0^{T/2}\frac{2A_{max}}{T}t\cos n\omega t\,dt\} \quad ...(1)$$

Now $\int_{-T/2}^{0} A_{max}\cos n\omega t\,dt = A_{max}\left|\frac{\sin n\omega t}{\omega n}\right|_{-T/2}^{0}$

$A_{max}\left(\frac{\sin n\pi}{\omega n}\right) = 0, \quad$ for $n = 1, 2, 3 ....$

$$\int_{-T/2}^{0}\frac{2A_{max}}{T}t\cos n\omega t\,dt = 2\frac{A_{max}}{T}\left[\left|\frac{\sin n\omega t}{\omega n}\right|_{-T/2}^{0} - \int_{-T/2}^{0}\frac{\sin n\omega t}{\omega n}dt\right]$$

$$= \frac{2A_{max}}{T}\left[\frac{\sin \omega nt}{\omega n}.t + \frac{\cos n\omega t}{(\omega n)^2}\right]_{-T/2}^{0}$$

$$= \frac{2A_{max}}{T}\left[\frac{1}{(\omega n)^2} - \frac{\cos n\omega T/2}{(\omega n)^2}\right]$$

Similarly, $\int_0^{T/2} A_{max}\cos n\omega t\,dt = 0$

and $\int_0^{T2}\frac{2A_{max}}{T}t\cos n\omega t\,dt$

$$= \frac{2A_{max}}{T}\left|\frac{\sin n\omega t}{\omega n}.t + \frac{\cos n\omega t}{(\omega n)^2}\right|_0^{T/2}$$

$$= \frac{2A_{max}}{T}\left[\frac{\cos\frac{n\omega T}{2}}{(\omega n)^2} - \frac{1}{(\omega n)^2}\right]$$

Substituting in (1)

$$a_n = \frac{4A_{max}}{T^2}\left[\left\{\frac{1}{(\omega n)^2} - \frac{\cos n\omega T/2}{(\omega n)^2}\right\} - \left\{\frac{-1}{(\omega n)^2} + \frac{\cos n\omega T/2}{(n\omega)^2}\right\}\right]$$

$$= \frac{4A_{max}}{T^2}\left(\frac{2}{\omega^2 n^2} - \frac{2\cos n\omega T/2}{\omega^2 n^2}\right)$$

$$= \frac{8A_{max}}{\omega^2 T^2 n^2}\left(\frac{1-\cos n\omega T}{2}\right)$$

$$a_1 = \frac{8A_{max}}{\omega^2 T^2}(2) = \frac{16A_{max}}{\omega^2 T^2}$$

$$= \left(\frac{4A_{max}}{\pi^2}\right) \text{ for } \omega T = 2\pi$$

$$a_2 = \frac{8A_{max}}{4\omega^2 T^2}(0) = 0$$

$$a_3 = \frac{8A_{max}}{9\omega^2 T^2}(2) = \frac{16A_{max}}{9\omega^2 T^2} = \frac{4A_{max}}{9\pi^2}$$

$$a_4 = 0$$

$$a_5 = \frac{8A_{max}}{25\omega^2 T^2}(2) = \frac{16A_{max}}{25\omega^2 T^2} = \frac{4A_{max}}{25\pi^2}$$

$$f(t) = \left(\frac{A_{max}}{2}\right) + \left(\frac{4A_{max}}{\pi^2}\right)\cos\omega t + \left(\frac{4A_{max}}{9\pi^2}\right)\cos 3\omega t + ...$$

$$= \left(\frac{A_{max}}{2}\right) + \frac{4A_{max}}{\pi^2}\left\{\cos\omega t + \frac{1}{9}\cos 3\omega t + \frac{1}{25}\cos 5\omega t + ...\right\}$$

$$= \left(\frac{A_{max}}{2}\right) + \frac{4A_{max}}{\pi^2}\left\{\cos\omega t + \frac{1}{(3)^2}\cos 3\omega t + \frac{1}{(5)^2}\cos 5\omega t + ...\right\}$$

Q.10. Find trignometric fourier series expansion for the waveform of Fig. 7.2 (*b*).

Ans. Let A_{max} be the peak amplitude.

Now $f(t) = \frac{4A_{max}}{T}.t$ for $\frac{-T}{4} \le t \le \frac{T}{4}$

$= -\frac{4A_{max}}{T}.t + 2A_{max}$ for $\frac{T}{4} \le t \le \frac{3T}{4}$

$$b_n = \frac{2}{T}\left\{\int_{-T/4}^{T/4}\frac{4A_{max}}{T}.t\sin n\omega t dt + \int_{T/4}^{3T/4}\left(-\frac{4A_{max}}{T}.t + 2A_{max}\right)\sin n\omega t\,dt\right\}$$

$$= \frac{2}{T}\left\{\int_{-T/4}^{T/4}\frac{4A_{max}}{T}.t\sin n\omega t dt - \int_{T/4}^{3T/4}\frac{4A_{max}}{T}.t\sin n\omega t\,dt + \int_{T/4}^{3T/4} 2A_{max}\sin n\omega t\,dt\right\}$$

Now $\int_{-T/4}^{T/4}\frac{4A_{max}}{T}.t\sin n\omega t\,dt$

$$= \frac{4A_{max}}{T}\left\{\left|\frac{\cos n\omega t}{\omega n}.\right|_{-T/4}^{T/4} + \int_{-T/4}^{T/4} + \frac{\cos\omega nt}{\omega n}dt\right\}$$

$$= \frac{4 A_{max}}{T}\left\{\left|\frac{-\cos n\omega t}{\omega n}.t\right|_{-T/4}^{T/4} + \left|\frac{\sin n\omega t}{(\omega n)^2}\right|_{-T/4}^{T/4}\right\}$$

$$= \frac{4 A_{max}}{T}\left(\frac{\sin n\omega T/4}{\omega^2 n^2} + \frac{\sin n\omega T/4}{\omega^2 n^2}\right)$$

$$= \frac{8 A_{max}}{T}\left(\frac{\sin n\omega T/4}{\omega^2 n^2}\right)$$

Similarly, $\int_{T/4}^{3T/4} \frac{4 A_{max}}{T}.t \sin n\omega t\, dt$

$$= \frac{4 A_{max}}{T\omega^2 n^2}\left(\frac{3n\omega T}{4} - \sin\frac{n\omega T}{4}\right)$$

$$\int_{T/4}^{3T/4} 2 A_{max} \sin n\,\omega t\, dt = -2A_{max}\left|\frac{\cos n\omega t}{\omega n}\right|_{T/4}^{3T/4} = 0.$$

Therefore,

$$b_n = \frac{2}{T}\left[\frac{8 A_{max}}{T\omega^2 n^2}\sin\frac{n\omega T}{4} - \frac{4 A_{max}}{\omega^2 n^2 T}\left\{\frac{3n\omega T}{4} - \sin\frac{n\omega T}{4}\right\}\right]$$

$a_n = 0$ (As the waveforms is represented by an odd function)

$$f(t) = b_1 \sin\omega t + b_2 \sin 2\omega t + b_3 \sin 3\omega t + ...$$

$$b_1 = \frac{2}{T}\left[\frac{8 A_{max}}{\omega^2 T} - \frac{4 A_{max}}{\omega^2 T}(-1-1)\right]$$

$$= \frac{2}{T}\left[\frac{8 A_{max}}{\omega^2 T} + \frac{8 A_{max}}{\omega^2 T}\right] = \frac{32 A_{max}}{\omega^2 T^2}$$

$$= \frac{8 A_{max}}{\pi^2} \qquad \omega T = 2\pi$$

$$b_2 = \frac{2}{T}\left[\frac{8 A_{max}}{4\omega^2 T}\times 0 - \frac{4 A_{max}}{\omega^2 n^2 T}(0)\right] = 0$$

$$b_3 = \frac{2}{T}\left[\frac{8 A_{max}}{9\omega^2 T}(-1) - \frac{4 A_{max}}{9\omega^2 T}(1+1)\right]$$

$$= \frac{2}{T}\left(-\frac{8 A_{max}}{9\omega^2 T} - \frac{8 A_{max}}{9\omega^2 T}\right)$$

$$= \frac{-32 A_{max}}{9\omega^2 T^2} = -\frac{8 A_{max}}{9\pi^2}$$

$$b_4 = 0.$$

Therefore,

$$f(t) = \frac{8 A_{max}}{\pi^2}\sin\omega t - \frac{8 A_{max}}{9\pi^2}\sin 2\omega t + \frac{8 A_{max}}{25\pi^2}\sin 5\omega t ....$$

$$= \frac{8 A_{max}}{\pi^2}\left(\sin\omega t - \frac{1}{(3)^2}\sin 3\omega t + \frac{1}{(5)^2}\sin 5\omega t ....\right)$$

Q.11. Find Trignometric fourier series expansion for the waveform shown in Fig. 7.3.

Ans. $f(t) = \frac{2|f(t)|_{max}}{T}(t)\ 0 \le t \le \frac{T}{2}$

$$= -\frac{2|f(t)|_{max}}{T}(t) + |f(t)|_{max}\ \frac{T}{2} \le t \le T.$$

The waveform is half wave symmetric; therefore, it will have only odd harmonics. The average value is zero.

Now $a_n = \frac{2}{T}\int_0^{T/2} f(t)\cos n\omega t\, dt$

$$= \frac{2}{T}\left\{\int_0^{T/2} 2|f(t)|_{max}\cos n\omega t\, dt + \int_{T/2}^{T}\left(-\frac{2|f(t)|_{max}}{T}.t + |f(t)|_{max}\right)\cos n\omega t\, dt\right\}$$

Now $\int_0^{T/2}\frac{2|f(t)|_{max}}{T}.t\cos n\omega t\, dt$

$$= \frac{2|f(t)|_{max}}{T}\int_0^{T/2} t\cos n\omega t\, dt$$

$$= \frac{2|f(t)|_{max}}{T}\left\{\left|\frac{\sin n\omega t}{\omega n}.t\right|_0^{T/2} - \int_0^{T/2}\frac{\sin n\omega t}{\omega n}dt\right\}$$

$$= \frac{2|f(t)|_{max}}{T}\left\{\left|\frac{\sin n\omega t}{\omega n}t + \frac{\cos n\omega t}{\omega^2 n^2}\right|_0^{T/2}\right\}$$

$$= \frac{2|f(t)|_{max}}{T}\left\{\frac{\cos n\omega T/2}{\omega^2 n^2} - \frac{1}{\omega^2 n^2}\right\}$$

Also, $\int_{T/2}^{T} -\frac{2|f(t)|_{max}}{T}.t\cos n\omega t\, dt$

$$= -\frac{2|f(t)|_{max}}{T}\left|\frac{\sin n\omega t}{\omega n}.t + \frac{\cos n\omega t}{\omega^2 n^2}\right|_{T/2}^{T}$$

$$= -\frac{2|f(t)|_{max}}{T}\left\{\frac{\cos n\omega T}{\omega^2 n^2} - \frac{\cos n\omega T/2}{\omega^2 n^2}\right\}$$

and, $\int_{T/2}^{T}|f(t)|_{max}\cos n\omega t\, dt$

$$= |f(t)|_{max}\left|\frac{\sin n\omega t}{\omega n}\right|_{T/2}^{T} = 0$$

Therefore, $a_n = \frac{2}{T}\left[\frac{2|f(t)|_{max}}{T}\left\{\frac{\cos n\omega T}{\omega^2 n^2} - \frac{1}{\omega^2 n^2}\right\} - \frac{2|f(t)|_{max}}{T}\left\{\frac{\cos n\omega T}{\omega^2 n^2} - \frac{\cos n\omega T/2}{\omega^2 n^2}\right\}\right]$

$$= \frac{4|f(t)|_{max}}{T^2}\left[\frac{\frac{\cos n\omega T}{2} - 1}{\omega^2 n^2}\right.$$

$$\left. - \frac{\cos n\omega T - \cos \omega \frac{T}{2}}{\omega^2 n^2}\right]$$

$$= \frac{4|f(t)|_{max}}{\omega^2 n^2 T^2}\left[2\cos\frac{n\omega T}{2} - \cos n\omega T - 1\right]$$

$$a_1 = \frac{4|f(t)|_{max}}{\omega^2 n^2 T^2} \cdot (-4) = \frac{-16|f(t)|_{max}}{\omega^2 T^2}$$

$$= \frac{-4|f(t)|_{max}}{\pi^2}$$

$$a_3 = \frac{4|f(t)|_{max}}{9\omega^2 T^2} \cdot (-4) = \frac{-16|f(t)|_{max}}{9\omega^2 T^2}$$

$$= \frac{-4|f(t)|_{max}}{9\pi^2}$$

$$a_5 = \frac{4|f(t)|_{max}}{25\pi^2} \text{ and so on}$$

Now $b_n = \frac{2}{T}\int_0^T f(t)\sin n\omega t\, dt$

$$= \frac{2}{T}\left\{\int_0^{T/2}\frac{2|f(t)|_{max}}{T}(t)\sin n\omega t dt\right.$$

$$+ \int_{T/2}^{T}\left(\frac{-2|f(t)|_{max}}{T}(t) + |f(t)|_{max}\right)$$

$$\left.\sin n\omega t\, dt\right\}$$

Now $\int_0^{T/2}\frac{2|f(t)|_{max}}{T}(t)\, ain\, n\,\omega t$

$$= \frac{2|f(t)|_{max}}{T}\left\{\left|\frac{-\cos n\omega t}{\omega n}\right|(t)\Big|_0^{T/2} + \int_0^{T/2}\frac{\cos n\omega t}{\omega n}dt\right\}$$

$$= \frac{2|f(t)|_{max}}{T}\left|-\frac{\cos n\omega t}{\omega n}(t) + \frac{\sin n\omega t}{\omega^2 n^2}\right|_0^{T/2}$$

$$= \frac{2|f(t)|_{max}}{T}\left(-\frac{\cos n\omega T/2}{\omega n}\left(\frac{T}{2}\right)\right)$$

$$= \frac{-|f(t)|_{max}}{\omega n}\cos\frac{n\omega T}{2}$$

and $\int_{T/2}^{T}\left(\frac{-2|f(t)|_{max}}{T}(t) + |f(t)|_{max}\right)\sin n\omega t\, dt$

$$= \frac{-2|f(t)|_{max}}{T}\left|\frac{-\cos n\omega t}{\omega t}(t) + \frac{\sin n\omega t}{\omega^2 n^2}\right|_{T/2}^{T}$$

$$- \left|\frac{|f(t)|_{max}\cos n\omega t}{\omega n}\right|_{T/2}^{T}$$

$$= \frac{-2|f(t)|_{max}}{T}\left(\frac{-\cos n\omega T}{\omega n}(T) + \frac{\cos n\omega T/2}{\omega n}\left(\frac{T}{2}\right)\right) - \frac{|f(t)|_{max}}{\omega n}$$

$$\left(\cos n\omega T - \cos\frac{n\omega T}{2}\right)$$

$$b_n = \frac{2}{T}\left[\left(\frac{-|f(t)|_{max}}{\omega n}\cos\frac{n\omega T}{2}\right)\right.$$

$$- \frac{2|f(t)|_{max}}{T}$$

$$\left(-\frac{\cos n\omega T}{\omega n}(T) + \frac{\cos n\omega T/2}{\omega n}\left(\frac{T}{2}\right)\right)$$

$$\left. - \frac{|f(t)|_{max}}{\omega n}\left(\cos n\omega T - \cos\frac{n\omega T}{2}\right)\right]$$

$$= \frac{2}{T}\left[-\frac{|f(t)|_{max}}{\omega n}\cos\frac{n\omega T}{2} + \frac{2|f(t)|_{max}}{\omega n}\cos n\omega T - \frac{|f(t)|_{max}}{\omega n}\cos\frac{n\omega T}{2}\right.$$

$$\left. - \frac{|f(t)|_{max}}{\omega n}\cos n\omega T + \frac{|f(t)|_{max}}{\omega n}\cos\frac{n\omega T}{2}\right]$$

$$= \frac{2}{T}\left[\frac{-|f(t)|_{max}}{\omega n}\cos n\omega T - \frac{|f(t)|_{max}}{\omega n}\cos\frac{n\omega T}{2}\right]$$

$$b_1 = \frac{2}{T}\left[\frac{|f(t)|_{max}}{\omega} + \frac{|f(t)|_{max}}{\omega}\right]$$

$$= \frac{4|f(t)|_{max}}{\omega T} = \frac{2|f(t)|_{max}}{\pi}$$

$$b_3 = \frac{2}{\omega T}\left[\frac{|f(t)|_{max}}{3} + \frac{|f(t)|_{max}}{3}\right]$$

$$= \frac{4|f(t)|_{max}}{3\omega T} - \frac{2|f(t)|_{max}}{3\pi}$$

Similarly, $b_5 = \frac{2|f(t)|_{max}}{5\pi}$ and so on.

Therefore,

$$f(t) = a_1\cos\omega t + a_2\cos 2\omega t + \ldots + a_n\cos n\omega t + b_1\sin\omega t + b_2\sin 2\omega t + \ldots + b_n\sin n\omega t$$

In this case

$$a_2 = a_4 = a_6 = a_8 \ldots = 0$$

$$b_2 = b_4 = b_6 = b_8 \ldots = 0$$

$$\therefore \quad f(t) = a_1\cos\omega t + a_3\cos 3\omega t + \ldots + b_1\sin\omega t + b_3\sin 3\omega t + \ldots$$

$$= \frac{4|f(t)|_{max}}{\pi^2}\cos\omega t - \frac{4|f(t)|_{max}}{9\pi^2}\cos 3\omega t - \frac{4|f(t)|_{max}}{25\pi^2}\cos 5\omega t - \ldots$$

$$+ \frac{2|f(t)|_{max}}{\pi} \sin \omega t + \frac{2|f(t)|_{max}}{3\pi} \sin 3\omega t + \frac{2|f(t)|_{max}}{5\pi} \sin 5\omega t + ...$$

$$= -\frac{4|f(t)|_{max}}{\pi^2}\left(\cos \omega t + \frac{1}{9}\cos 3\omega t + \frac{1}{25}\cos 5\omega t + ...\right) + \frac{2|f(t)|_{max}}{\pi}\left(\sin \omega t + \frac{1}{3}\sin 3\omega t + \frac{1}{5}\sin 5\omega t + ...\right)$$

Q.12. How do you convert the trignometric series expansion of a given function into the exponential series expansion of the same function without actually solving for exponential Fourier series expansion evaluation integrals. Similarly, how do you convert the exponential Fourier series expansion of a given function into the corresponding trignometric series expansion.

Ans. Exponential to Trignometric

The trignometric Fourier series co-efficients are given by

$$a_n = A_n + A_{-n} \qquad ...(1)$$

$$b_n = j(A_n - A_{-n}) \qquad ...(2)$$

where (A_n) and (A_{-n}) are exponential fourier series co-efficients.

From (1) and (2)

$$a_n = A_n + A_{-n} \qquad ...(3)$$

and $$jb_n = j^2(A_n - A_{-n}) = -A_n + A_{-n} \qquad ...(4)$$

Trignometric to Exponential

From (3 and 4)

$$A_n = \frac{1}{2}(a_n - jb_n) \qquad ...(5)$$

$$A_{-n} = \frac{1}{2}(a_n + jb_n) \qquad ...(6)$$

Q.13. You have seen in one of the previous questions that a periodic waveform mathematically expressed by a function $f(t)$ satisfying the condition $f(t) = -f(t + T/2)$ will have its Fourier series expansion contain only odd harmonics. Tell what would the Fourier series expansion look like for a waveform that satisfied $f(t) = f(t + T/2)$? Will it contain only even harmonics? If yes, do all waveforms, whose Fourier series expansions contain only even harmonies satisfy $f(t) = f(t + T/2)$?

Ans. Yes, if we have a periodic waveform with period (T) expressed by a function $f(t)$ satisfying $f(t) = F(t + T/2)$, its Fourier series expansion would certainly contain even harmonics only. But this condition of $f(t) = f(t + T/2)$ is SUFFICIENT but not NECESSARY. That means: there can be a waveform that contains only even harmonics but does not satisfy $f(t + T/2)$ whereas all waveforms that satisfy $f(t) = f(t + T/2)$ would contain only even harmonics. A similar explanation holds good for the waveforms satisfying $f(t) = -f(t + T/2)$. These waveforms contain only odd harmonics.

Q.14. What do you understand by LINE SPECTRUM in reference to Fourier series expansion of periodic waveforms? What characteristic information regarding the waveform do you get from the line spectrum?

Ans. As discussed in a number of previous questions, Fourier series expansion (trignometric or exponential) of a periodic waveform consists of a summation of fundamental frequency and its harmonics. Line spectrum is a plot of harmonic amplitudes present in the waveform. Figures 7.4 (*a*) and (*b*) shows the line spectrum of the same waveform., one obtained from the trignometric series expansion while the other obtained from exponential series expansion. When the two plots are observed closely, you would notice that they convey same sense ; for instance the amplitude of third harmonic in case of Fig. 7.4 (*a*) is ($4V/3\pi$) and in case of (*b*) also, it is

$$\left(\frac{2V}{3\pi} + \frac{2V}{3\pi} = \frac{4V}{3\pi}\right)$$

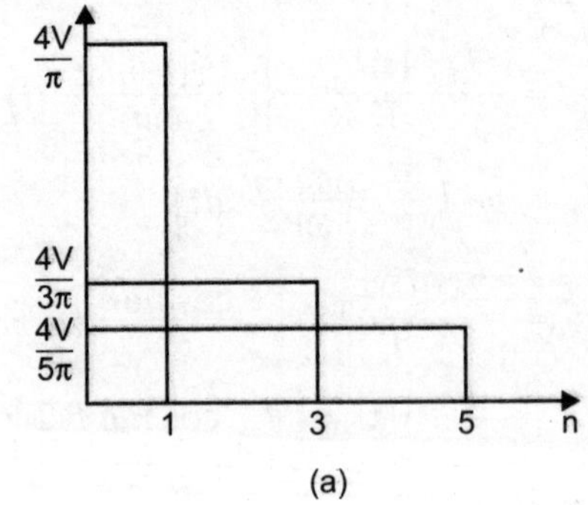

(a)

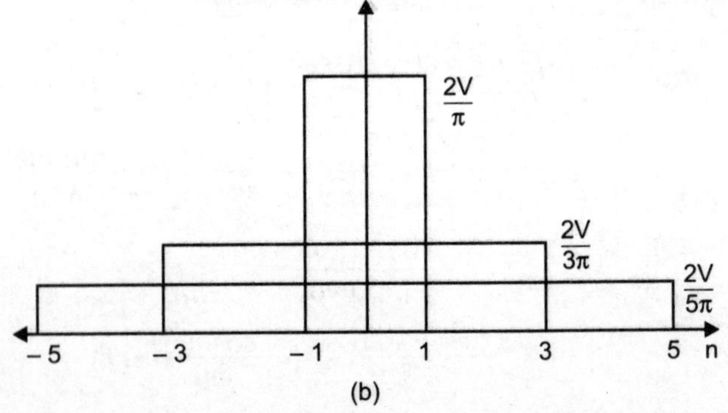

(b)

Fig. 7.4

Quite obviously, the line spectrum is the discrete frequency spectrum of a periodic waveform. It tells us about the frequency components present in a given periodic waveform alongwith respective amplitudes. This gives us an idea as to whether the waveform is fast converging (which is the case when the waveform has sharp discontinuities) or slow converging (in case of waveforms with more gradual variations). All these become important when the periodic waveforms are used as excitations to networks.

Q.15. How can the principles of Fourier series be extended to determine the frequency spectrum of a single non-recurring pulse? Illustrate with the help of an example.

Ans. You would notice that when the period (T) of a periodic waveform increases, the line spectrum spacing = ($2\pi/T$) reduces and in the limit when $T \rightarrow \infty$, the spacing reduces to zero and as a result, the discrete line spectrum becomes a continuous spectrum. A single non-recurring pulse in a way is periodic waveform with $T = \infty$. Therefore, the frequency spectrum of a single pulse is an extended case of discrete line spectrum with $T = \infty$.

If we recall the Fourier series expansion, we wrote an expression for co-efficients (A_n) in case of exponential series expansion as

$$A_n = \frac{1}{T}\int_0^T f(t)\, e^{-j\omega nt}\, dt$$

Let us say $g_n = \frac{A_n}{\omega}$ and g_n becomes $g(\omega)$ in the limit. Then we can write

$$g(\omega) = \frac{1}{2\pi}\int_{-\infty}^{\infty} f(t)\, e^{-j\omega t}\, dt$$

or $$f(t) = \int_{-\infty}^{\infty} g(\omega)\, e^{j\omega t}\, dt$$

These are the expressions for Fourier integral. Infact, the discrete line spectrum (of a periodic waveform) represented by a summation gets replaced by a continuous frequency spectrum (in case of a single non-recurring pulse) represented by a Fourier integral.

$$g(\omega) = \frac{1}{2\pi}\int_{-\infty}^{\infty} f(t)\, e^{-j\omega t}\, dt$$

As an illustration, for the single pulse of Fig. 7.5.

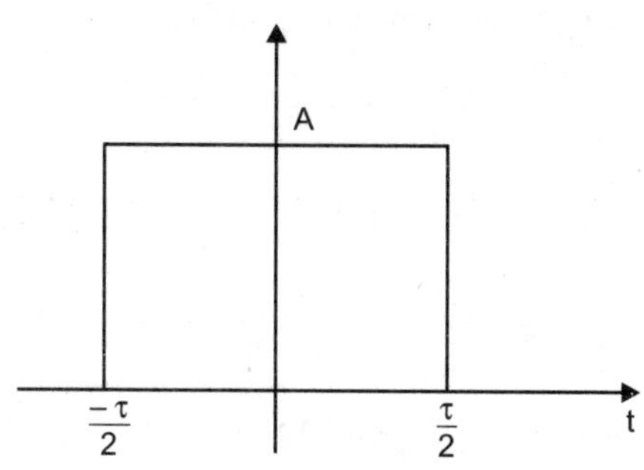

Fig. 7.5

$$g(\omega) = \frac{1}{2\pi}\int_{-\infty}^{\infty} f(t)\, e^{-j\omega t}\, dt$$

$$= \frac{1}{2\pi}\int_{-\tau/2}^{\tau/2} A\, e^{-j\omega t}\, dt$$

$$= \frac{A}{2\pi}\left|\frac{e^{-j\omega t}}{-j\omega}\right|^{\tau/2}$$

$$= \frac{A}{\omega\pi}\left(\frac{e^{j\omega\tau/2} - e^{-j\omega\tau/2}}{2j}\right)$$

$$= \frac{A}{\omega\pi}\sin\left(\frac{\omega\tau}{2}\right) = \frac{A}{2\pi f.\pi}\sin(\pi f\tau)$$

$$= \frac{A\tau}{2\pi}\left(\frac{\sin \pi f\tau}{\pi f\tau}\right)$$

The frequency spectrum is of $\left(\frac{\sin x}{x}\right)$ form with $x = (\pi f\tau)$.

Solved Problems

Problem 7.1. *Refer to the square waveform of Fig. 7.6. Determine the trignoemtric Fourier series expansion of this waveform. Also determine exponential Fourier series expansion for the same.*

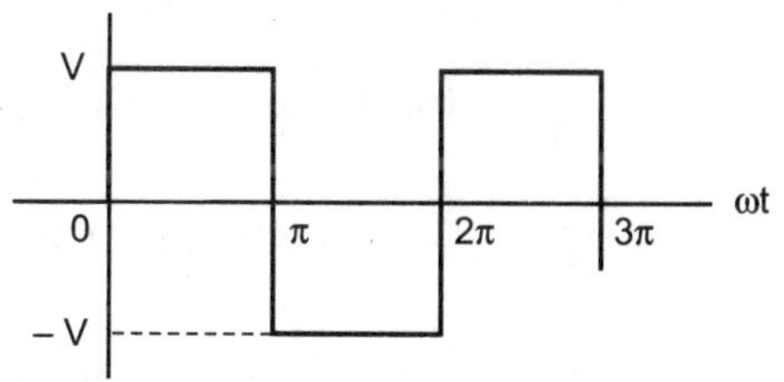

Fig. 7.6

Solution. As is evident from the waveform of Fig. 7.6,

$f(t) = V$ for $0 < \omega t < \pi$ and

$f(t) = -V$ for $\pi < \omega t < 2\pi$

Also, the average value of the given waveform = 0, which implies $\frac{a_0}{2} = 0$ in the given waveform, $f(t) = -f(-t)$.

Therefore, it is an odd function. As a consequence its Fourier serious expansion will not have any cosine terms. It will have only sine terms. The co-efficients (b_n) of the sine terms are given by:

$$b_n = \frac{1}{\pi}\left\{\int_0^{\pi} V \sin n\,\omega t\; d(\omega t)\right.$$

$$+ \int_{\pi}^{2\pi} - V \sin n\omega t \, d(\omega t)\Big\}$$

$$= \frac{V}{\pi}\left\{\left[-\frac{1}{n}\cos n\omega t\right]_0^{\pi} + \left[\frac{1}{n}\cos n\omega t\right]_{\pi}^{2\pi}\right\}$$

This simplifies to

$$b_n = \frac{2V}{n\pi}(1 - \cos n\pi)$$

$$b_n = 0 \quad \text{for } n = 2, 4, 6, \ldots$$

and $$b_n = \frac{4V}{n\pi} \quad \text{for } n = 1, 3, 5, \ldots$$

The Fourier series expansion in trignometric form is then given by,

$$f(t) = \frac{4V}{\pi}\sin \omega t + \frac{4V}{3\pi}\sin 3\omega t + \frac{4V}{5\pi}\sin 5\omega t + \ldots$$

In the case of exponential fourier series expansion of this waveform,

$$A_0 = 0 \text{ and } A_n = \frac{1}{2\pi}\left\{\int_{-\pi}^{0} - V e^{-jn\omega t} d(wt)\right.$$

$$\left. + \int_0^{\pi} V e^{-jn\omega t} d(\omega t)\right\}$$

$$= \frac{V}{2\pi}\left\{-\left[\frac{1}{(-jn)} e^{-j\omega nt}\right]_{-\pi}^{0} + \left[\frac{1}{(-jn)} e^{-j\omega nt}\right]_0^{\pi}\right\}$$

$$= j\frac{V}{n\pi}(e^{jn\pi} - 1)$$

For $n = 2, 4, 6, \ldots \; e^{jn\pi} = 1$ and $A_n = 0$

and for $n = 1, 3, 5, \ldots \; e^{jn\pi} = -1$ and $A_n = -j\left(\frac{2V}{n\pi}\right)$

The Fourier series expansions in exponential form is therefore given by:

$$f(t) = +j\frac{2V}{3\pi}e^{-j3\omega t} + j\frac{2V}{\pi}e^{-j\omega t} - j\frac{2V}{\pi}e^{j\omega t}$$

$$- j\frac{2V}{3\pi}e^{j3\omega t} \; \ldots\ldots$$

Problem 7.2. *Sketch the line spectra for the square waveform of Fig. 7.6 for both trignometric as well as exponential forms of Fourier series expansion.*

Solution. The line spectra can be drawn from the Fourier series expansion expressions in the two cases as derived in Problem 7.1. Fig. 7.7 shows in line spectra. The figures are self explanatory. Fig. 7.7(*a*) shows line spectrum for thetrignometric fourier series expansion and Fig. 7.7(*b*) shows line spectrum for exponential fourier series expansion.

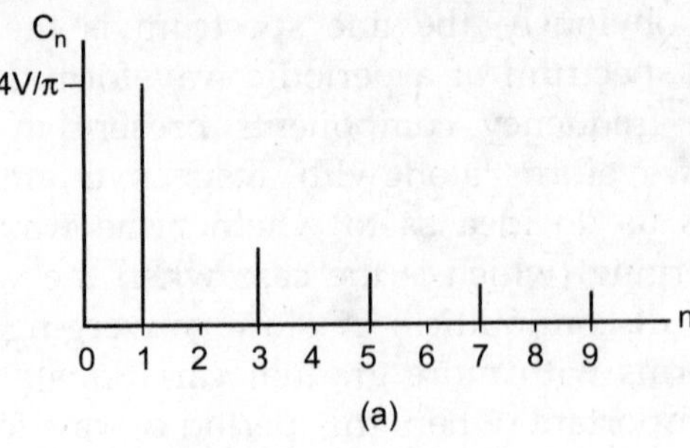

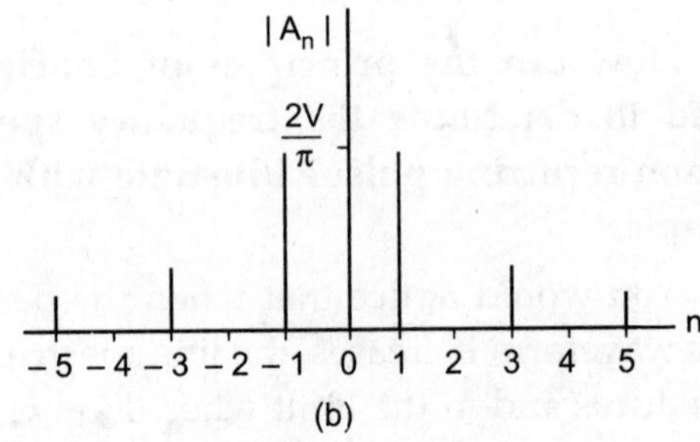

Fig. 7.7

Problem 7.3. *Find the trignometric and exponential fourier series expansions of the triangular waveform shown in Fig. 7.8.*

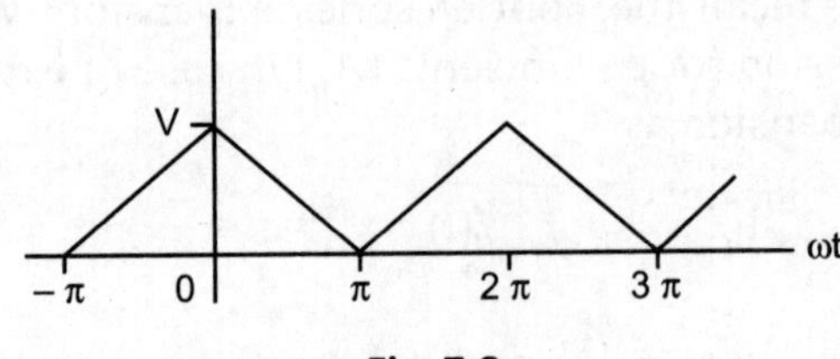

Fig. 7.8

Solution. Following observations can be made from the given waveform.

1. It has an average value equal to $\left(\frac{V}{2}\right)$
2. It is an even function as $f(t) = f(-t)$
3. It is also halfwave symmetric if average value subtracted. In that case, $f(t) = -f(t + \pi)$
4. The trignometric Fourier series expansion will have only cosine terms i.e. $b_n = 0$
5. The exponential Fourier series expansion will have real (A_n) coefficients
6. $f(t) = V + \frac{V}{\pi}\omega t$ for $-\pi < \omega t < 0$
7. $f(t) = V - \frac{V}{\pi}\omega t$ for $0 < \omega t < \pi$

Now $$a_n = \frac{1}{\pi}\int_{-\pi}^{0}\left[V + \frac{V}{\pi}\omega t\right]\cos n\omega t \, d(\omega t)$$

$$+ \frac{1}{\pi}\int_0^{\pi}\left[V - \frac{V}{\pi}\omega t\right]\cos n\omega t \, d(\omega t)$$

$$= \frac{V}{\pi}\left\{\int_{-\pi}^{0} \cos n\omega t \, d(\omega t) + \int_{-\pi}^{0} \frac{\omega t}{\pi} \cos n\omega t \, d(\omega t) - \int_{0}^{\pi} \frac{\omega t}{\pi} \cos n\omega t \, d(\omega t)\right\}$$

This simplifies to

$$a_n = \frac{V}{\pi^2}\left\{\left[\frac{1}{n^2}\cos n\omega t + \frac{\omega t}{\pi}\sin n\omega t\right]_{-\pi}^{0} - \left[\frac{1}{n^2}\cos n\omega t + \frac{\omega t}{\pi}\sin n\omega t\right]_{0}^{\pi}\right\}$$

$$= \frac{2V}{\pi^2 n^2}(1 - \cos n\pi)$$

$$a_n = 0 \quad \text{for} \quad n = 2, 4, 6, \ldots$$

(from halfwave symmetry)

$$a_n = \frac{4V}{\pi^2 n^2} \quad \text{for} \quad n = 1, 3, 5, \ldots$$

Therefore, trignometric Fourier series expansion is given by:

$$f(t) = \frac{V}{2} + \frac{4V}{\pi^2}\cos \omega t + \frac{4V}{(3\pi)^2}\cos 3\omega t + \frac{4V}{(5\pi)^2}\cos 5\omega t + \ldots$$

For exponential fourier series expansion,

$$A_n = \frac{1}{2\pi}\left\{\int_{-\pi}^{0}\left[V + \frac{V}{\pi}\omega t\right]e^{-jn\omega t}\, d(\omega t) + \int_{0}^{\pi}\left[V - \frac{V}{\pi}\omega t\right]e^{-jn\omega t}\, d(\omega t)\right\}$$

$$= \frac{V}{2\pi^2}\left\{\int_{-\pi}^{0}\omega t\, e^{-jn\omega t}\, d(\omega t) + \int_{0}^{\pi}(-\omega t)\, e^{-jn\omega t}\, d(\omega t) + \int_{-\pi}^{\pi}\pi\, e^{-jn\omega t}\, d(\omega t)\right\}$$

$$= \frac{V}{2\pi^2}\left\{\left[\frac{e^{-jn\omega t}}{(-jn)^2}(-jn\omega t - 1)\right]_{-\pi}^{0} - \left[\frac{e^{-jn\omega t}}{(-jn)^2}(-jn\omega t - 1)\right]_{0}^{\pi}\right\}$$

$$= \frac{V}{\pi^2 n^2}\left[1 - e^{jn\pi}\right]$$

For n = even, $e^{jn\pi} = +1$ and $A_n = 0$

For n = odd, $A_n = \frac{2V}{\pi^2 n^2}$

Therefore,

$$f(t) = \ldots + \frac{2V}{(-3\pi)^2}e^{-j3\omega t} + \frac{2V}{(-\pi)^2}e^{-j\omega t} + \frac{V}{2} + \frac{2V}{\pi^2}e^{j\omega t} + \frac{2V}{(3\pi)^2}e^{j3\omega t} + \ldots$$

Problem 7.4. *Draw the line spectra for the trignometric and exponential fourier series expansions of the triangular waveform of Problem 7.3.*

Solution. The line spectra can be drawn from the magnitudes of different terms in the two expansions derived as a part of solution to problem 7.3. As will be evident from the two expressions, both have the same line spectrum. It is shown in Figure 7.9, which is self explanatory. It may be mentioned here that line spectrum shows that the series converges very rapidly due to presence of $\left(\frac{1}{n^2}\right)$ term in the coefficients.

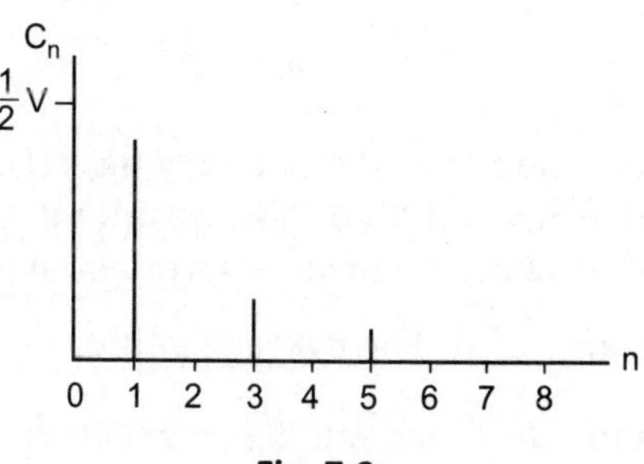

Fig. 7.9

Problem 7.5. *Below are given five different expressions representing five different waveforms. Identify whether these waveforms are odd, even, halfwave symmetric and also whether they have a non-zero average value.*

(a) $f(t) = \left(\frac{V}{\pi}\right)\omega t$

(b) $f(t) = \left(\frac{V}{\pi}\right)\omega t$ *for* $0 < \omega t < \pi$

$= 0$ *for* $\pi < \omega t < 2\pi$

(c) $f(t) = V \sin \omega t$

(d) $f(t) = V \sin \omega t$ *for* $0 < \omega t < \pi$

$= 0$ *for* $\pi < \omega t < 2\pi$

(e) $f(t) = V - \frac{V}{2\pi}\omega t$

Solution.

(a) $f(t) = \frac{V}{\pi}(\omega t)$

$f(-t) = -\frac{V}{\pi}(\omega t)$ or $f(t) = -f(-t)$

Therefore, the waveform represented by this function has odd symmetry. Also, the average value is zero.

(*b*) For $0 < \omega t < \pi$, $f(t) = -f(-t)$

For $\pi < \omega t < 2\pi$, $f(t) = 0$

Therefore, this function's neither even or odd.

Also, average value $= \dfrac{V}{4}$

The waveform represented by this function has no symmetry.

(*c*) $f(t) = V \sin \omega t$

$f(-t) = -V \sin \omega t$

or $f(t) = -f(-t)$

The waveform has odd symmetry. Also, it has a single frequency component and average value is zero.

(*d*) It is a half wave rectified sinusoidal waveform. It has no symmetry. The average value is $\left(\dfrac{V}{\pi}\right)$.

(*e*) $f(t) = V - \dfrac{V}{2\pi}(\omega t)$

The function has odd symmetry (*i.e.* $f(t) = -f(-t)$ only if constant terms omitted. The resultant Fourier series expansion will have a constant term equal to the average value and sine term. The average value is $\left(\dfrac{V}{2}\right)$.

Problem 7.6. *Determine the trignometric Fourier series expansion for the half-wave rectified sinusoidal waveform of Fig. 7.10. Also, sketch its line spectrum.*

Solution. $f(t) = V_m \sin \omega t$ for $0 < \omega t < \pi$

$= 0$ for $\pi < \omega t < 2\pi$

Evidently, the waveform exhibits no symmetry, which means that the trignometric Fourier series expansion will have both sine and cosine terms.

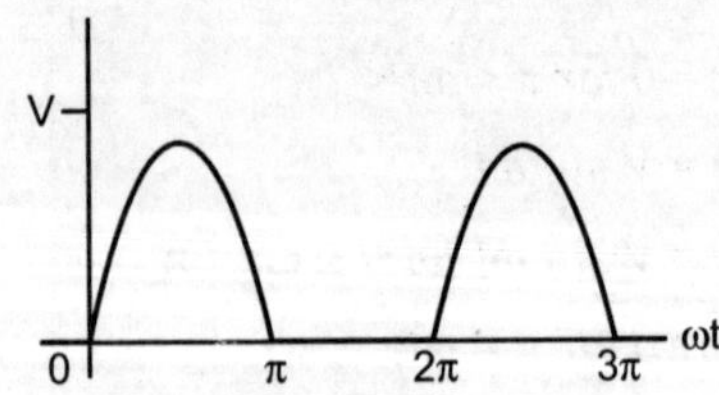

Fig. 7.10.

Now, $$a_0 = \frac{1}{\pi}\int_0^{\pi} V_m \sin \omega t \, d(\omega t)$$

$$= \frac{V_m}{\pi}[-\cos \omega t]_0^{\pi} = \frac{2V_m}{\pi}$$

Therefore,

$$\frac{a_0}{2} = \frac{V_m}{\pi}$$

Coefficient, a_n are expressed by

$$a_n = \frac{1}{\pi}\int_0^{\pi} V_m \sin \omega t \cos n\omega t \, d(\omega t)$$

$$= \frac{V_m}{\pi}\left[\frac{-n \sin \omega t \sin n\omega t - \cos n\omega t \cos \omega t}{-n^2+1}\right]_0^{\pi}$$

$$= \frac{V_m}{\pi(1-n^2)}(\cos n\pi + 1)$$

when (n) is odd, $a_n = 0$

when (n) is even, $a_n = \dfrac{2V_m}{\pi(1-n^2)}$

The expression is indeterminate form $n = 1$. Coefficient (a_1) is determined separately.

$$a_1 = \frac{1}{\pi}\int_0^{\pi} V_m \sin \omega t \cos \omega t \, d(\omega t)$$

$$= \frac{V_m}{\pi}\int_0^{\pi} \frac{1}{2} \sin 2\omega t \, d(\omega t) = 0$$

Coefficients, b_n are expressed by :

$$b_n = \frac{1}{\pi}\int_0^{\pi} V_m \sin \omega t \, n\omega t \, d(\omega t)$$

$$= \frac{V_m}{\pi}\left[\frac{n \sin \omega t \cos n\omega t - \sin n\omega t \cos n\omega t}{(-n^2+1)}\right]_0^{\pi}$$

$$= 0$$

However, expression is indeterminate for $n = 1$. Therefore, coefficient (b_1) is determined separately.

$$b_1 = \frac{1}{\pi}\int_0^{\pi} V_m \sin^2 \omega t \, d(\omega t)$$

$$= \frac{V_m}{\pi}\left[\frac{\omega t}{2} - \frac{\sin 2\omega t}{4}\right]_0^{\pi}$$

$$= \frac{V_m}{2}$$

The trignometric Fourier series expansion can be written as:

$$f(t) = \frac{V}{\pi}\Big[1 + \frac{\pi}{2}\sin \omega t - \frac{2}{3}\cos 2\omega t - \frac{2}{15}\cos 4\omega t - \frac{2}{35}\cos 6\omega t - \ldots\Big]$$

Fig. 7.11 shows line spectrum, which is self-explanatory.

Problem 7.7. *Find trignometric Fourier series expansion for the waveform of Fig. 7.12.*

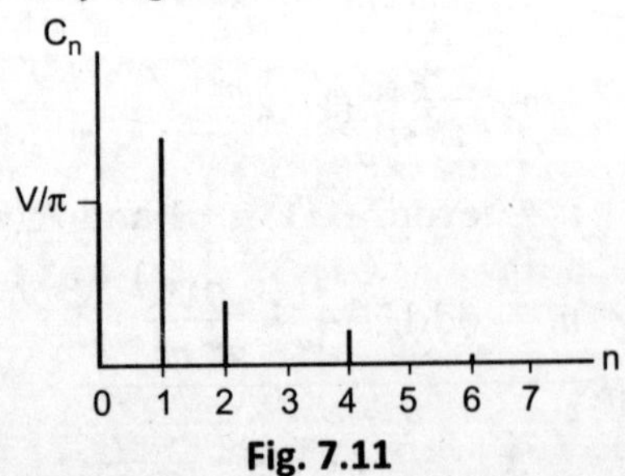

Fig. 7.11

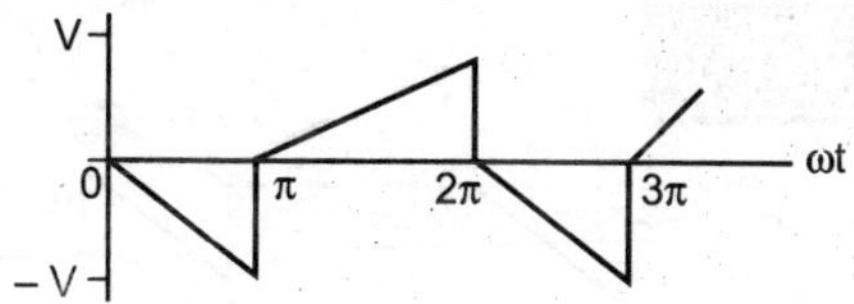

Fig. 7.12

Solution. The waveform shows neither odd nor even symmetry. However, it exhibits half-wave symmetry as quite evidently, $f(t) = -f\left(t + \frac{T}{2}\right)$. The resultant trignometric Fourier series expansion will have odd harmonics of both sine and cosine terms. The waveform shown in Fig. 7.12 is the same as the one shown earlier in Fig. 7.3. The trignometric Fourier series expansion for the waveform was derived earlier in answer to Q.11. Substituting $|f(t)|_{max} = V$, we get the solution.

Therefore,

$$f(t) = \frac{4V}{\pi^2}\left(\cos \omega t + \frac{1}{9}\cos 3\omega t + \frac{1}{25}\cos 5\omega t + ...\right) - \frac{2V}{\pi}\left(\sin \omega t + \frac{1}{3}\sin 3\omega t + \frac{1}{5}\sin 5\omega t + ...\right)$$

Summary of Important Formulas

1. For a periodic function $f(t)$ defined by $f(t) = f(t+T)$ to be an even function,

$$f(t) = f(-t) \text{ for all values of } (t)$$

2. For a periodic function $f(t)$ defined by $f(t) = f(t+T)$ to be an odd function,

$$f(t) = -f(-t)$$

3. For a periodic function $f(t)$ defined by $f(t) = f(t+T)$ to be half wave symmetric,

$$f(t) = -f\left(t + \frac{T}{2}\right)$$

4. A periodic function $f(t)$ defined by $f(t) = f(t+T)$ can be expressed by trignometric Fourier series by:

$$f(t) = \frac{a_0}{2} + a_1 \cos \omega t + a_2 \cos 2\omega t + ... + a_n \cos \omega nt + b_1 \sin \omega t + b_2 \sin 2\omega t + ... + b_n \sin \omega nt$$

where
$$a_n = \frac{1}{\pi}\int_0^{2\pi} f(t) \cos n\omega t \, d(\omega t) = \frac{2}{T}\int_0^T f(t) \cos n\omega t \, dt$$

$$b_n = \frac{1}{\pi}\int_0^{2\pi} f(t) \sin n\omega t \, d(\omega t) = \frac{2}{T}\int_0^T f(t) \sin n\omega t \, dt$$

$$\frac{a_0}{2} = \text{Average value of the function}$$

5. A periodic function $f(t)$ defined by $f(t) = f(t+T)$ can be expressed by exponential Fourier series by:

$$f(t) = ... + A_{-2} e^{-j2\omega t} + A_{-1} e^{-j\omega t} + A_0 + A_1 e^{j\omega t} + A_2 e^{j2\omega t} + ...$$

where
$$A_n = \frac{1}{T}\int_0^T f(t) e^{-jn\omega t} dt = \frac{1}{2\pi}\int_0^{2\pi} f(t) e^{-jn\omega t} d(\omega t)$$

6. Trignometric Fourier series expansion of a periodic function can be converted into an exponential Fourier series expansion by using following transformations.

$$A_n = \frac{1}{2}(a_n - jb_n)$$

$$A_{-n} = \frac{1}{2}(a_n + jb_n)$$

7. Exponential fourier series expansion of a periodic function can be converted into a trignometric fourier series expansion by using following transformations.

$$a_n = A_n + A_{-n}$$

$$b_n = j(A_n - A_{-n})$$

8. A single non-recurring pulse can be expressed by the following fourier integral.

$$g(w) = \frac{1}{2\pi}\int_{-\infty}^{\infty} f(t) e^{-j\omega t} dt$$

Self Evaluation Exercise

Multiple Choice Questions

1. A certain waveform is expressed by sum of a constant term and a time dependent term. The wave form shows odd symmetry if the constant term were omitted. The trigonometric Fourier series expansion of this waveform will have
   (*a*) Constant term and sine terms
   (*b*) Constant term and cosine terms
   (*c*) Constant term and both sine and cosine terms
   (*d*) Constant term and odd harmonics of sine terms

2. A function *f* (*t*) is an EVEN function if for all values of (*t*)
   (*a*) $f(t) = f(-t)$
   (*b*) $f(t) = -f(-t)$
   (*c*) $f(t) = f(t+T)$
   (*d*) $f(t) = -f\left(t+\frac{T}{2}\right)$
   (*e*) $f(t) = -f(t+T)$

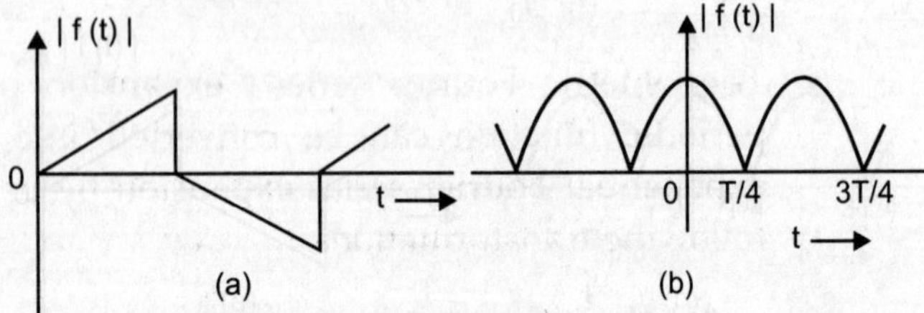

Fig. 7.13

3. A function *f* (*t*) is an ODD function if for all values of (*t*)
   (*a*) $f(t) = f(-t)$
   (*b*) $f(t) = -f(-t)$
   (*c*) $f(t) = f(t+T)$
   (*d*) $f(t) = f(t-T)$
   (*e*) $f(t) = f(t+T/2)$

4. For a periodic waveform to be half-wave symmetric, it must be represented by a function satisfying
   (*a*) $f(t) = f(t+T/2)$
   (*b*) $f(t) = -f(t+T/2)$
   (*c*) $f(t) = -f(-t)$
   (d) $f(t) = f(-t)$
   (e) None of these

5. A sinusoidal waveform is mathematically represented by
   (*a*) an EVEN function
   (*b*) an ODD function
   (*c*) a function that is even as well as half wave symmetric
   (*d*) a function that is neither even or odd
   (*e*) None of these

6. The trignometric Fourier Series expansion of an odd-function shall have
   (*a*) only sine terms
   (*b*) only cosine terms
   (*c*) odd harmonics of both sine as well as cosine terms
   (*d*) even harmonics of both sine as well as cosine terms
   (*e*) None of these

7. The trignometric Fourier series expansion of an even function that is also half-wave symmetric shall contain
   (*a*) odd harmonics of sine terms only
   (*b*) both sine and cosine terms
   (*c*) only cosine terms
   (*d*) only odd harmonics of cosine terms
   (*e*) None of these

8. A waveform with discontinuities is always characterised by
   (*a*) fast converging line fourier spectra
   (*b*) strong harmonic content
   (*c*) half-wave symmetry
   (*d*) quarter-wave symmetry
   (*e*) None of these

9. The waveform that results from addition of 10 sin ω*t*, 5 sin 2 ω*t*, 3.33 sin 3 ω*t*, 2.5 sin 4 ω*t*, 1.66 sin 6 ω*t*, 1.43 sin 7 ω*t*, 1.25 sin 8 ω*t* and 1.11 sin 9 ω*t* is
   (*a*) a sawtooth waveform of frequency (ω) having a finite average value
   (*b*) a sawtooth waveform of frequency (ω) having zero average value
   (*c*) a square waveform of frequency (ω)

(*d*) a complex waveform with no standard shape
(*e*) None of these.

10. A function that is neither even nor odd but has half-wave symmetry can be expressed in terms of Fourier series expansion as a sum of
(*a*) odd harmonics of both sine and cosine terms
(*b*) even harmonics of both sine and cosine terms
(*c*) odd harmonics of sine terms
(*d*) even harmonics of cosine terms

11. The exponential Fourier coefficients of function *f* (*t*), whose trignometric Fourier expansion is found to contain only cosine terms, are
(*a*) always pure imaginary
(*b*) always real numbers
(*c*) partly real and partly imaginary
(*d*) None of these

12. When the period *T* of a periodic waveform increases, the adjacent lines of the line spectrum of that waveform
(*a*) will be spaced further apart
(*b*) will come closer
(*c*) will not be affected
(*d*) None of these.

13. The frequency spectrum of a single non-recurring pulse will be
(a) a line spectrum containing all harmonics very closely spaced
(b) a single line
(c) a continuous frequency spectrum defined by Fourier integral rather than a discrete line specturum defined by Fourier series
(*d*) None of these

14. Look at the waveform shown in Fig. 7.13 (*a*). It falls in one of the following categories
(*a*) Even
(*b*) Odd
(*c*) Even and half wave symmetric
(*d*) Odd and half wave symmetric
(*e*) Quarter-wave symmetric

15. How would the trignometric Fourier series expansion of the waveform of Fig. 7.13 (*b*) look like? Just make a guess. Options are given to you below.
(*a*) It would be a summation of a constant term and even harmonics of cosine terms
(*b*) It would be a series consisting of a constant term and both sine and cosine terms.
(*c*) It would be series of cosine terms only plus a constant term
(*d*) None of these

Answers to Self Evaluation Exercise

1. (*a*)	2. (*a*)	3. (*b*)	4. (*b*)	5. (*b*)	6. (*a*)	7. (*d*)	8. (*b*)
9. (*b*)	10. (*a*)	11. (*b*)	12. (*b*)	13. (*c*)	14. (*d*)	15. (*a*)	

Chapter 8

Two-Port Networks and State Equations

Learning Objectives

After completing the chapter, you will learn the following topics:

1. Two Port Networks
2. Impedance Parameters
3. Admittance Parameters
4. Transmission Parameters
5. Inverse Transmission Parameters
6. Hybrid Parameters
7. Inverse Hybrid or 'g' Parameters
8. Interrelationship between Different Network Parameters
9. Interconnection of Two Port Networks
10. Network Functions
11. Analysis of Networks Using State Equations

Q.1. Define a two-port network. What are network parameters?

Ans. A two-port network is an electrical network with two separate ports for input and output. Figure. 8.1 shows a two-port network. The port labeled 1-1′ is the input port while the port labeled 2-2′ is the output port. The port variables are port currents and port voltages as shown in the figure. Current I_1 is the input current; I_2 the output current; V_1 is the input voltage and V_2 is the output voltage. Two set of linear equations are required for describing the relationships between the port voltages and currents of a two-port network. The various terms that relate these voltages and currents are referred to as network parameters. The choice of two independent and two dependent parameters results in different parameters namely the impedance, admittance, transmission, inverse transmission, hybrid and inverse hybrid parameters. These parameters are discussed in the following questions.

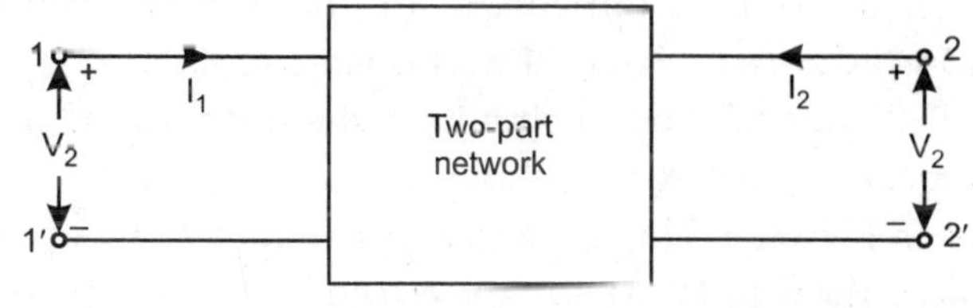

Fig. 8.1

Q.2. What are IMPEDANCE PARAMETERS with reference to a 2-port network ? Why are these parameters also called open circuit impedance parameters ?

Ans. Fig. 8.2 shows block schematic representation of a 2-port network. If we express input and output voltages in the terms of input and output currents, the following generalised set of two equations can be written.

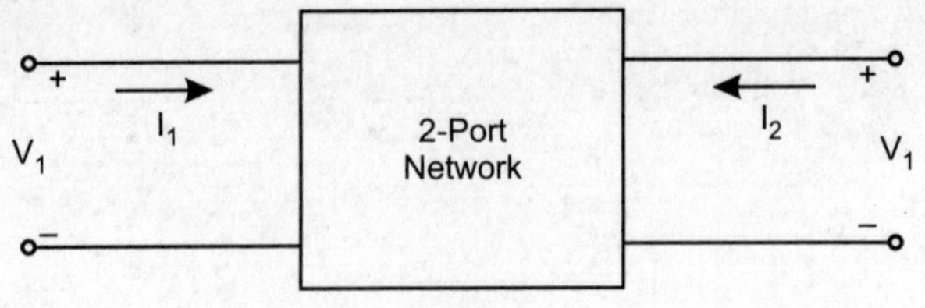

Fig. 8.2

$$V_1 = Z_{11} I_1 + Z_{12} I_2$$
$$V_2 = Z_{21} I_1 + Z_{22} I_2$$

Obviously, Z_{11}, Z_{12}, Z_{21} and Z_{22} have to be impedances in order that both sides of these two equations have identical dimensions. From the above set of equations, we can write expressions for Z_{11}, Z_{12}, Z_{21} and Z_{22} as follows :

$$Z_{11} = \left.\frac{V_1}{V_1}\right|_{I_2=0} ; Z_{12} = \left.\frac{V_1}{I_2}\right|_{I_1=0}$$

$$Z_{21} = \left.\frac{V_2}{I_1}\right|_{I_2=0} ; Z_{22} = \left.\frac{V_2}{I_2}\right|_{I_1=0}$$

Z_{11}, Z_{12}, Z_{21} and Z_{22} are the impedance parameters of this 2-port network. We will notice that for each of these four parameters, either input port or the output port is open circuited implied by $I_1 = 0$ and $I_2 = 0$ respectively. This explains why these parameters are also called open circuit impedance parameters.

Q. 3. What are different impedance parameters called? How are these computed?

Ans. z_{11} is the open circuit input impedance or the input driving-point impedance with output port open-circuited. z_{12} is the open circuit transfer impedance from port 1 to port 2 or the reverse transfer impedance with input port open-circuited. z_{21} is the open circuit transfer impedance from port 2 to port 1 or the forward transfer impedance with output port open-circuited. z_{22} is the open circuit output impedance or the output driving point impedance with input port open-circuited. z_{11} and z_{21} are obtained by open-circuiting the output port and input port is excited by a known voltage source V_1 as shown in Figure-8.3 (a). z_{12} and z_{22} are obtained by open-circuiting the input port and the output port is excited by a known voltage source V_2 as shown in Figure-8.3 (*b*).

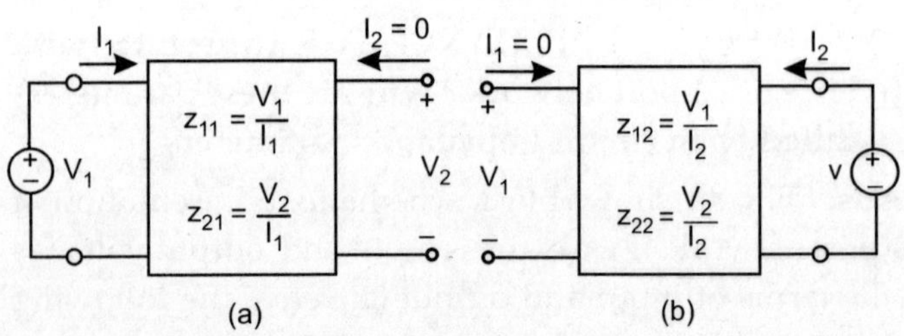

Fig. 8.3

Q. 4. What do you understand by reciprocal networks? What is the condition to be satisfied for the impedance parameters if a two-port network is to be reciprocal?

Ans. A network is said to be reciprocal if the ratio of the response transform to the excitation transform is invariant to an interchange of the positions of excitation and response in the network. If the transfer impedances z_{12} and z_{21} are equal, then the network is reciprocal. Therefore, for a two-port network to be reciprocal, $z_{12} = z_{21}$.

Q. 5. What do you understand by network symmetry? What is the condition to be satisfied in the case of impedance parameters for symmetry of a two-port network?

Ans. A two-port network is said to be symmetric, if the ports can be interchanged without changing the port voltages and currents. For a two-port network to be symmetric, $z_{11} = z_{22}$.

Q. 6. What are ADMITTANCE PARAMETERS with reference to 2-port networks ? Why are these parameters also ailed short circuit admittance parameters ?

Ans. With reference to the 2-port network of Fig. 8.2, if we express the input and output currents (I_1, I_2) in terms of input and output voltages (V_1, V_2), we can write the following set of generalized equations,

$$I_1 = y_{11} V_1 + y_{12} V_2$$
$$I_2 = y_{21} V_1 + y_{22} V_2$$

From the above set of equations, we can write the following expressions for the admittance parameters y_{11}, y_{12}, y_{13}, y_{21} and y_{22}.

$$y_{11} = \left.\frac{I_1}{V_1}\right|_{V_2=0} ; y_{12} = \left.\frac{I_1}{V_2}\right|_{V_1=0}$$

$$y_{21} = \left.\frac{I_2}{V_1}\right|_{V_2=0} \text{ and } y_{22} = \left.\frac{I_2}{V_2}\right|_{V_1=0}$$

y_{11} and y_{22} are respectively the input and output admittances. y_{12} and y_{21} are transfer admittances. We can notice that for each of these parameters, either the input port or the output port is short circuited implied by $V_1 = 0$ and $V_2 = 0$ respectively. This explains why these parameters are also known as short circuit admittance parameters.

Q. 7. What are different admittance parameters called? How are these computed?

Ans. y_{11} is the short circuit input admittance or the input driving-point admittance with output port short-circuited. y_{12} is the short circuit transfer admittance from port 1 to port 2 or the reverse transfer admittance with input port short circuited. y_{21} is the short circuit transfer admittance from port 2 to port 1 or the forward transfer admittance with output port short circuited. y_{22} is the short circuit output admittance or the output driving point admittance with input port short circuited. y_{11} and y_{21} are obtained by short-circuiting the output port and input port is excited by a known current source I_1 as shown in Figure-8.4 (*a*). y_{12} and y_{22} are obtained by short-circuiting the input port and the output port is excited by a known current source I_2 as shown in Figure-8.4 (b).

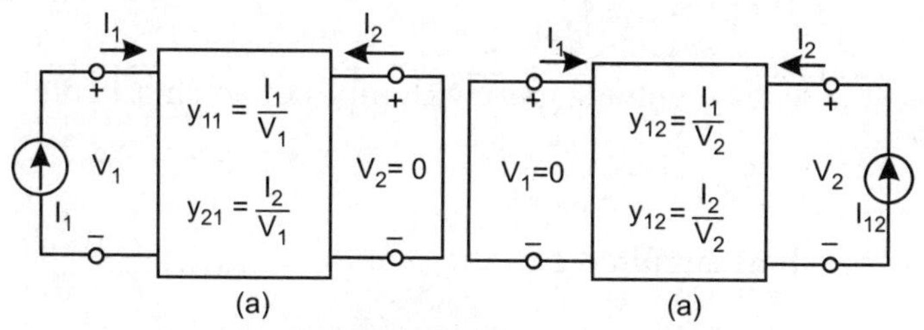

Fig. 8.4

Q. 8. What is the condition to be satisfied for the admittance parameters if a two-port network is to be reciprocal?

Ans. If the transfer admittances y_{12} and y_{21} are equal, then the network is reciprocal. Therefore, for a two-port network to be reciprocal, $y_{12} = y_{21}$.

Q. 9. What is the condition to be satisfied in the case of admittance parameters for symmetry of a two-port network?

Ans. For a two-port network to be symmetric, $y_{11} = y_{22}$.

Q. 10. What are TRANSMISSION PARAMETERS with reference to a 2-port network? What are INVERSE TRANSMISSION PARAMETERS ?

Ans. The transmission parameters relate input voltage and input current to output voltage and output current The set of equations is as follows:

$$V_1 = AV_2 - BI_2 \quad ; \; I_1 = CV_2 - DI_2$$

A, B, C, D here are the TRANSMISSION PARAMETERS also known as *ABCD* parameters. The negative sign for the second term in the above set of equations is due to two different conventions used for assigning a positive direction to the current I_2. While in case of impedance and admittance parameters, the positive direction of I_2 was considered as the one shown in Fig. 8.2, in case of transmission and inverse transmission parameters, the positive direction for I_2 is considered as the one that is opposite to that shown in Fig. 8.2 The negative sign therefore is for I_2 and not for parameters *B* and *D*. The expressions for *A,B,C,* and *D* can be written as

$$\frac{1}{A} = \left.\frac{V_2}{V_1}\right|_{I_2=0} ; -\frac{1}{B} = \left.\frac{I_2}{V_1}\right|_{V_2=0}$$

$$\frac{1}{C} = \left.\frac{V_2}{I_1}\right|_{I_2=0} \text{ and } -\frac{1}{D} = \left.\frac{I_2}{I_1}\right|_{V_2=0}$$

In case of INVERSE TRANSMISSION PARAMETERS, output voltage and output currents are expressed in terms of input voltage and input current as

$$V_2 = A' V_1 - B' I_1 ; I_2 = C' V_1 - D' I_1$$

which gives

$$\frac{1}{A'} = \left.\frac{V_1}{V_2}\right|_{I_1=0} \text{ and } -\frac{1}{B'} = \left.\frac{I_1}{V_2}\right|_{V_1=0}$$

$$\frac{1}{C'} = \left.\frac{V_1}{I_2}\right|_{I_1=0} \text{ and } -\frac{1}{D'} = \left.\frac{I_1}{I_2}\right|_{V_1=0}.$$

For reciprocal networks,

AD – BC = 1 and A'D' – B'C' = 1

Q. 11. What are A, B, D, A', B', C' and D' are called?

Ans. *A* is the reverse voltage ratio with the receiving end open-circuited or the open-circuit voltage ratio, *B* is the transfer impedance with the receiving end short-circuited or the negative short-circuit transfer impedance, *C* is the transfer admittance with the receiving end open-circuited or open circuit transfer admittance and *D* is the reverse current ratio with the receiving end short-circuited or the negative short-circuit current ratio.

A′ is the forward voltage ratio with the sending end open-circuited or the open-circuit voltage gain, *B′* is the transfer impedance with the sending end short-circuited or the negative short-circuit transfer impedance, *C′* is the transfer admittance with the sending end open-circuited or open circuit transfer admittance and *D′* is the forward current ratio with the sending end short-circuited or the negative short-circuit current gain.

Q. 12. What are the conditions for transmission parameters for reciprocal and symmetric networks?

Ans. For reciprocal networks, AD – BC = 1. For a two-port network to be symmetric, A = D.

Q. 13. What are the conditions for inverse transmission parameters for reciprocal and symmetric networks?

Ans. For a two-port network to be reciprocal, A′D′ – B′C′ = 1. For a two-port network to be symmetric, A′ = D′.

Q. 14. What type of 2-port network problems the transmission parameters are particularly applicable to ?

Ans. Transmission parameters are particularly useful

in describing 2-port networks that have been connected in cascade. It can be shown that if two 2-port networks with transmission parameters as $A_1\ B_1$, $C_1\ D_1$ and $A_2\ B_2$ $C_2\ D_2$ are cascaded, the, transmission parameters *A, B, C, D* of the cascaded arrangement can be expressed as;

$$\begin{bmatrix} A & B \\ C & D \end{bmatrix} = \begin{bmatrix} A_1 & B_1 \\ C_1 & D_1 \end{bmatrix} \begin{bmatrix} A_2 & B_2 \\ C_1 & D_1 \end{bmatrix}$$

Q. 15 How are different types of network parameters such as the Impedance parameters, Admittance parameters and Transmission parameters interrelated ?

Ans. Impedance and Admittance parameters are interrelated by :

$$Z_{11} Y_{11} = Z_{22} Y_{22} \text{ and } \Delta_z \Delta_y = 1$$

where $\Delta_z = \begin{vmatrix} Z_{11} & Z_{12} \\ Z_{21} & Z_{22} \end{vmatrix}$ and $\Delta_y = \begin{vmatrix} Y_{11} & Y_{12} \\ Y_{21} & Y_{22} \end{vmatrix}$

Transmission parameters can be expressed in terms of impedance parameters and admittance parameters as:

$$A = \frac{Z_{11}}{Z_{21}} = -\frac{Y_{22}}{Y_{21}}\ ;\ B = \frac{\Delta z}{Z_{21}} = -\frac{1}{Y_{21}}$$

$$C = \frac{1}{Z_{21}} = -\frac{\Delta_y}{Y_{21}} \text{ and } D = \frac{Z_{22}}{Z_{21}} = -\frac{Y_{11}}{Y_{21}}$$

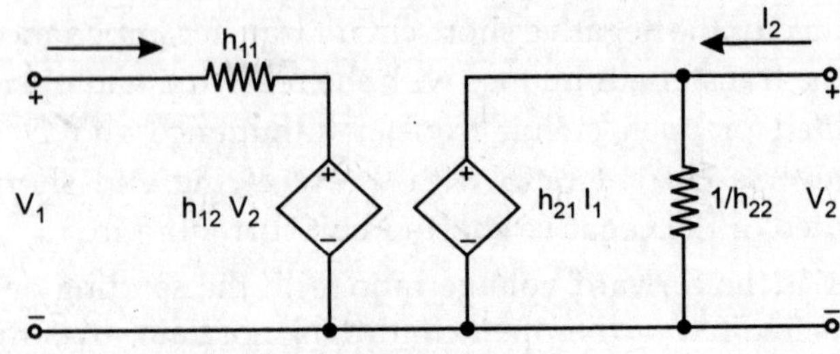

Fig. 8.5

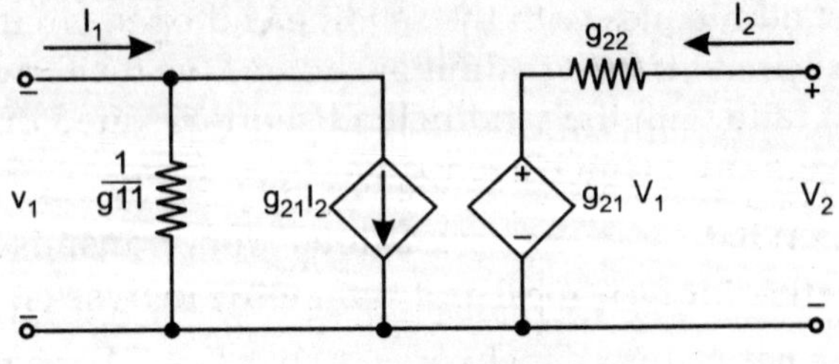

Fig. 8.6

In Inverse transmission parameters can be expressed in terms of impedance and admittance parameters as :

$$A' = \frac{Z_{22}}{Z_{12}} = -\frac{Y_{11}}{Y_{12}}\ ;\ B' = \frac{\Delta_z}{Z_{12}} = -\frac{1}{Y_{12}}$$

$$C' = \frac{1}{Z_{12}} = -\frac{\Delta_Y}{Y_{12}} \text{ and } D' = \frac{Z_{11}}{Z_{12}} = -\frac{Y_{22}}{Y_{12}}$$

Q.16. Briefly describe Hybrid parameters with reference to a 2-port network.

Ans. Refer to Fig 8.5. The HYBRID PARAMETERS also known as *h*-parameters are defined by the following set of equations written for a generalized 2-port network

$$V_1 = h_{11} I_1 + h_{12} V_2$$

$$I_2 = h_{21} I_1 + h_{22} V_2$$

The h-parameters h_{11}, h_{12}, h_{21} and h_{22} can be defined as:

$$h_{11} = \left.\frac{V_1}{I_1}\right|_{V_2=0}$$

It is the input impedance with output short circuited.

$$h_{12} = \left.\frac{V_1}{V_2}\right|_{I_1=0}$$

It is reverse voltage gain with input open circuited

$$h_{21} = \left.\frac{I_2}{I_1}\right|_{V_2=0}$$

It is short circuited current gain

$$h_{22} = \left.\frac{I_2}{V_2}\right|_{I_1=0}$$

It is the open circuited output admittance.

These parameters are called hybrid parameters as these are dimensionally mixed. These parameters are particularly used in constructing transistor models.

Q. 17. Briefly describe inverse Hybrid Parameters.

Ans. Refer to fig. 8.6 The Inverse Hybrid Parameters are governed by equations:

$$I_1 = g_{11} V_1 + g_{12} \cdot I_2$$

$$V_1 = g_{21} V_1 + g_{22} I_2$$

where g_{11} = open circuited input admittance = $\left.\frac{I_1}{V_1}\right|_{I_2=0}$

g_{12} = short circuited current ratio = $\left.\frac{I_1}{I_2}\right|_{V_1=0}$

g_{21} = Open circuited voltage ratio = $\left.\frac{V_2}{V_1}\right|_{I_2=0}$

and g_{22} = Short circuited output impedance = $\left.\frac{V_2}{I_1}\right|_{V_1=0}$

Q. 18. Outline the conditions for reciprocity and symmetry for hybrid and inverse hybrid parameters.

Ans. For Hybrid Parameters: For a two-port network to be reciprocal,

$$h_{12} = h_{21}.$$

For a two-port network to be symmetric

$$\begin{vmatrix} h_{11} & h_{12} \\ h_{21} & h_{22} \end{vmatrix} = 1$$

which simplifies to $h_{11}h_{22} - h_{12}h_{21} = 1$

For **Inverse Hybrid Parameters:**

For a two-port network to be reciprocal,

$$g_{12} = g_{21}.$$

For a two-port network to be symmetric,

$$\begin{vmatrix} g_{11} & g_{12} \\ g_{21} & g_{22} \end{vmatrix} = 1$$

which simplifies to $g_{11}g_{22} - g_{12}g_{21} = 1$

Q. 19. What type of 2-port network parameters are particularly suited to analysis of more than one 2-port networks connected in parallel ?

Ans. Short circuit admittance parameters are ideally suited for this purpose. The overall short circuit admittance function of parallel connected networks can be determined from addition of individual short circuit admittance functions. For example, if two 2-port networks with admittance parameters $y_{11a}, y_{12a}, y_{21a}, y_{22a}$ and $y_{11b}, y_{12b}, y_{21b}$ and y_{22b} are connected in parallel, then the admittance parameters of the parallel combination $y_{11}, y_{12}, y_{21}, y_{22}$ are given by :

$$y_{11} = y_{11a} + y_{11b};\ y_{12} = y_{12a} + y_{12b}$$

$$y_{21} = y_{21a} + y_{21b} \text{ and } y_{22} = y_{22a} + y_{22b}$$

Q. 20. Briefly describe cascade, series and parallel connection of two-port networks.

Ans. Cascade Connection: Figure.8.7 (*a*) shows the cascade connection of two networks N_a and N_b with transmission parameters A_a, B_a, C_a and D_a and A_b, B_b, C_b and D_b respectively. The transmission parameters A, B, C, D of the cascaded network is

$$\begin{bmatrix} A & B \\ C & D \end{bmatrix} = \begin{bmatrix} A_a & B_a \\ C_a & D_a \end{bmatrix}\begin{bmatrix} A_b & B_b \\ C_b & D_b \end{bmatrix}$$

Series Connection: Figure.8.7 (*b*) shows the series connection of two networks N_a and N_b with z- parameters Z_{11a}, Z_{12a}, Z_{21a} and Z_{22a} and Z_{11b}, Z_{12b}, Z_{21b} and Z_{22b} respectively. The z-parameters Z_{11}, Z_{12}, Z_{21} and Z_{22} of the series network is

$$\begin{bmatrix} Z_{11} & Z_{12} \\ Z_{21} & Z_{22} \end{bmatrix} = \begin{bmatrix} Z_{11a} + Z_{11b} & Z_{12a} + Z_{12b} \\ Z_{21a} + Z_{21b} & Z_{22a} + Z_{22b} \end{bmatrix}$$

Parallel Connection: Figure. 8.7 (*c*) shows the parallel connection of two networks N_a and N_b with *y*- parameters Y_{11a}, Y_{12a}, Y_{21a} and Y_{22a} and Y_{11b}, Y_{12b}, Y_{21b} and Y_{22b} respectively. The y-parameters Y_{11}, Y_{12}, Y_{21} and Y_{22} of the parallel network is

$$\begin{bmatrix} Y_{11} & Y_{12} \\ Y_{21} & Y_{22} \end{bmatrix} = \begin{bmatrix} Y_{11a} + Y_{11b} & Y_{12a} + Y_{12b} \\ Y_{21a} + Y_{21b} & Y_{22a} + Y_{22b} \end{bmatrix}$$

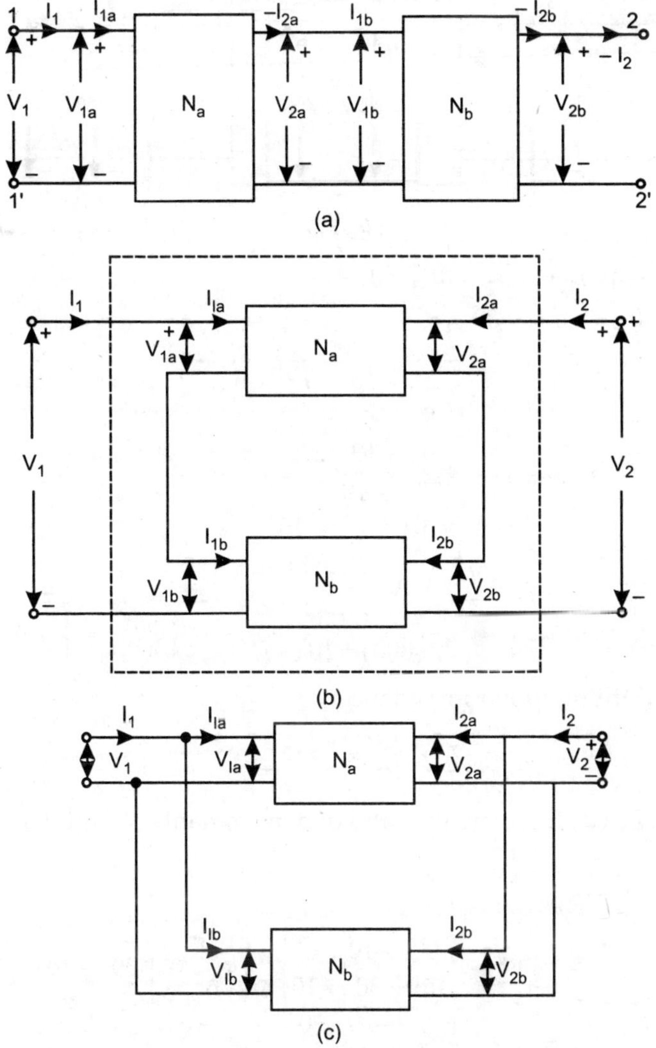

Fig. 8.7

Q. 21. What do you understand by DRIVING POINT IMPEDANCE of a network ? Calculate the driving point impedance of the network shown in Fig. 8.8 Explain how you could have arrived at the same result using the method of determinants.

Ans. Driving point impedance is in a way the input impedance. It is the impedance seen across the input terminals by setting all sources equal to zero *i.e.* all voltage sources shorted but their internal impedances retained and all current sources replaced by open circuits but their output impedance if any, retained. It can be calculated by driving the input with a voltage *V* and calculating the resulting current I and then finding the ratio (V/I).

In the network of Fig. 8.8, driving point impedance is given by $(20/I_1)$.

Writing the three mesh equations:

$$20\,I_1 - 10\,I_2 + 0I_3 = 20$$

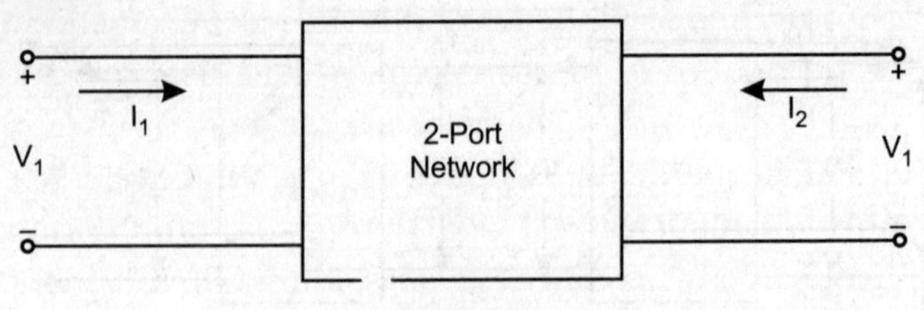

Fig. 8.8

$$-10\,I_1 + 30\,I_2 - 10I_3 = 0$$
$$0I_1 - 10\,I_2 + 30\,I_3 = 0$$

$$\text{This gives } I_1 = \frac{\begin{vmatrix} 20 & -10 & 0 \\ 0 & 30 & -10 \\ 0 & -10 & 30 \end{vmatrix}}{\begin{vmatrix} 20 & -10 & 0 \\ -10 & 30 & -10 \\ 0 & -10 & 30 \end{vmatrix}}$$

$$= \frac{16000}{20\,(800) + 10\,(-300)} = \frac{16000}{13000} = \frac{16}{13}\,A.$$

Driving point impedance

$$= 20 \times \frac{13}{16} = \frac{65}{4}\,\Omega$$

Let us calculate the ratio of determinants Δ_z and Δ_{11}

where

$$\Delta_z = \begin{vmatrix} 20 & -10 & 0 \\ -10 & 30 & -10 \\ 0 & -10 & 30 \end{vmatrix} \text{ and } \Delta_{11} = \begin{vmatrix} 30 & -10 \\ -10 & 30 \end{vmatrix}$$

Determinant obtained by ignoring first row and first column

Now Δ_z = 20 (900 – 100) + 10 (– 300) = 16000 – 3000 = 13000

and Δ_{11} = 900 – 100 = 800

$$\text{Therefore, } \frac{\Delta_z}{\Delta_{11}} = \frac{13000}{800} = \frac{65}{4}\,\Omega$$

which is same as the driving point impedance. Thus the driving point impedance can also be calculated by the method of determinants as explained above. The proof for the validity of above approach can be found in any of the standard texts on NETWORKS.

Q. 22. Define the term TRANSFER IMPEDANCE. Using the method of determinants, calculate the transfer impedances Z_{13} and Z_{12} for the network shown in Fig. 8.8.

Ans. Transfer impedance $\underline{Zpq}$ is the ratio of voltage impressed (V_p) in a mesh (p) to the resulting current (I_q) in mesh (q). As per this terminology, transfer impedance Z_{12} means (V_1/I_2) and transfer impedance Z_{13} is (V_1/I_3). Similarly, transfer impedance Z_{23} will be (V_2/I_3). Here V_2 is the voltage impressed in the mesh 2 and I_3 is the resulting current in mesh 3. In the network of Fig. 8.8, V_2 = 0 and therefore Z_{23} will also be zero. Another important point, while calculating transfer impedance, to be borne in mind is that for transfer impedance Z_{23}, you will consider V_2 only and set all other sources to zero and similarly for Z_{13}, it is only V_1, which is considered. Such a situation arises in networks with more than one active sources.

$$Z_{12} = \frac{\Delta_z}{\Delta_{12}}$$

Where Δ_z is the determinant as defined earlier, Δ_{12} is the determinant left after ignoring first row and second column.

$$Z_{12} = \frac{\begin{vmatrix} 20 & -10 & 0 \\ -10 & 30 & -10 \\ 0 & -10 & 30 \end{vmatrix}}{\begin{vmatrix} -10 & -10 \\ 0 & 30 \end{vmatrix}} = \frac{13000}{-300} = -\frac{130}{2}$$

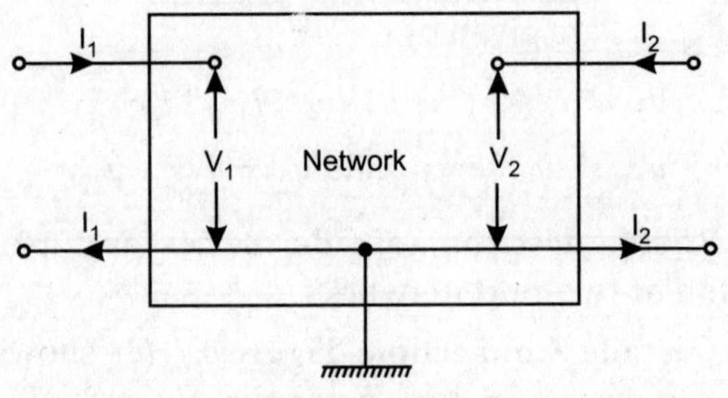

Fig. 8.9

$$\text{or} \quad |Z_{12}| = \frac{130}{3}\,\Omega.$$

Similarly,

$$Z_{13} = \frac{\Delta_z}{\Delta_{13}} = \frac{13000}{\begin{vmatrix} -10 & 30 \\ 0 & -10 \end{vmatrix}} = \frac{13000}{100} = 130\,\Omega.$$

Q. 23. Define the terms TRANSFER ADMITTANCE and DRIVING POINT ADMITTANCE. A network with three nodes with node voltages V_1, V_2 and V_3 driven by a current source of 10 A at the input (current injected into node 1) has transfer admittance $|Y_{13}|$ = 0.1. W What should be the node voltage V_3.

Ans. The DRIVING POINT ADMITTANCE is defined as the ratio of driving current of a single current source injected between two nodes to the resulting voltage drop across the same two nodes. In case of networks with more than one source, the driving point admittance or input admittance is again the ratio of (I) to (V) where (I) is the current injected between terminals in question and (V) is the voltage between the same terminals with all other sources set equal to zero. (*i.e.* voltage sources replaced by shorts and current sources replaced by open circuits Internal impedances of voltage sources and output impedances of current sources are however retained. Look at the block schematic of Fig. 8.9. The driving point admittance $Y_{input\ (1)}$ is giver by

$$Y_{input\ (1)} = \frac{I_1}{V_1} \text{ with } I_2 \text{ set equal to zero.}$$

$$\text{and } Y_{input\ (2)} = \frac{I_2}{V_2} \text{ with } I_1 \text{ set equal to zero.}$$

Mathematically,

$$Y_{input\ (1)} = \frac{\Delta Y}{\Delta_{11}} \text{ and } Y_{input\ (2)} = \frac{\Delta Y}{\Delta_{22}}$$

Where ΔY = Admittance determinant which can be obtained by first writing node equations and Δ_{11} is the determinant left over after ignoring first row and first column of ΔY and Δ_{22} is the determinant left over after ignoring second row and second column of ΔY.

The TRANSFER ADMITTANCE is the ratio of current driving in at one node to the resulting voltage at another node, all other sources being set equal to zero.

From Fig. 8.9

$$Y_{input\ (12)} = \frac{I_1}{V_2} \text{ With } I_2 = 0 \text{ and } Y_{input\ (21)} = \frac{I_2}{V_1}, \text{ with } I_1 = 0$$

It is given that $Y_{transfer\ (13)} = 0.1\ \Omega$

That implies

$$\frac{I_1}{V_3} = 0.1$$

$$V_3 = \frac{I_1}{0.1} = 10\ I_1 = 10 \times 10 = 100 \text{ V.}$$

Q. 24. What are network functions? What are the four network transfer functions?

Ans. For a two-port network, the ratio of one voltage to another voltage is referred to as the voltage transfer function, the ratio of one current to another current is referred to as the current transfer function and the ratio of one current to another voltage or one voltage to another current is referred to as the transfer admittance or the transfer impedance functions respectively.

For a two-port network shown in Fig. 8.10, there are four transfer functions as mentioned below.

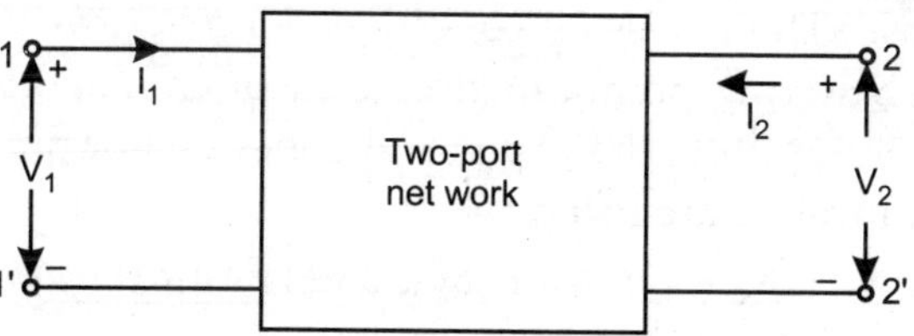

Fig. 8.10

The voltage transfer function is expressed as:

$$G_{21}(s) = \frac{V_2(s)}{V_1(s)}$$

The current transfer function is expressed as:

$$\alpha_{21}(s) = \frac{I_2(s)}{I_1(s)}$$

The admittance transfer function is expressed as:

$$Y_{21}(s) = \frac{I_2(s)}{V_1(s)}$$

The impedance transfer function is expressed as:

$$Z_{21}(s) = \frac{V_2(s)}{I_1(s)}$$

Q. 25. What are poles and zeros of network functions?

Ans. It may be mentioned here that all network functions, mentionedin the previous questioncan be expressed in the generalized form as:

$$N(s) = \frac{p(s)}{q(s)} = \frac{a_o s^n + a_1 s^{n-1} + ... + a_{n-1}s + a_n}{b_o s^m + b_1 s^{m-1} + ... + b_{m-1}s + b_m}$$

Where, a's and b's are coefficients of real positive value. The degree of the numerator polynomial is (n) and that of the denominator polynomial is (m).

Let the numerator p(s) has (n) roots $z_1, z_2, z_3.......z_n$ and the denominator has (m) roots $p_1, p_2, ...p_m$. Then the network function N(s) can be written as

$$N(s) = C\frac{(s-z_1)(s-z_2)......(s-z_n)}{(s-p_1)(s-p_2)......(s-p_m)}$$

Where,C is the scale factor and is given by $C = a_o/b_o$ For $s = z_1, z_2....z_n$, the network function becomes zero and these values of (s) are referred to as the zeros of the network function. For $s = p_1, p_2....p_n$, the network function becomes infinite and these values of (s) are referred to as the poles of the network function. A network function is completely specified in terms of its zeros, poles and the scale factor. Stable network functions with all left-hand plane zeros are classified as minimum-phase functions and those with any zero on the right-half plane are non-minimum phase functions.

Q. 26. What are the necessary conditions for network functions to be driving point functions?

Ans. The necessary conditions for a network function to be a driving-point function with common factors in the numerator polynomial $p(s)$ and denominator polynomial $q(s)$ cancelled are listed as

1. The coefficients in the polynomials $p(s)$ and $q(s)$ should be real and positive
2. Complex and imaginary poles and zeros must be conjugate
3. The real part of all the poles and zeros must not be positive. If the real part is zero, then the pole and zero must be simple.
4. The polynomials $p(s)$ and $q(s)$ must not have missing terms between the highest and lowest degree, unless all the even or odd terms are missing
5. The degree of $p(s)$ and $q(s)$ may differ by zero or one only
6. The terms of lowest degree in $p(s)$ and $q(s)$ should differ in degree by one at the most

Q. 27. What are the necessary conditions for network functions to be transfer functions?

Ans. The necessary conditions for a network function to be a transfer function with common factors in the numerator polynomial $p(s)$ and denominator polynomial $q(s)$ cancelled are listed as

1. The coefficients in the polynomials $p(s)$ and $q(s)$ should be real and positive
2. Complex and imaginary poles and zeros must be conjugate
3. The real part of all the poles and zeros must not be positive. If the real part is zero, then the pole and zero must be simple.
4. The polynomial $q(s)$ must not have missing terms between the highest and lowest degree, unless all the even or odd terms are missing
5. The polynomial $p(s)$ may have terms missing between the terms of lowest and highest degree and some of the coefficients may be negative
6. The degree of $p(s)$ may be as small as zero, independent of the degree of $q(s)$
7. For G_{12} and α_{12}, the maximum degree of $p(s)$ is the degree of $q(s)$. For Z_{12} and Y_{12}, the maximum degree of $p(s)$ is the degree of $q(s)$ plus one.

Q. 28. What is state variable analysis used for? Briefly explain state equations of networks in normal form and matrix differential form.

Ans. State variable analysis is used for solution of first order, second order and higher order systems. The nth order system is described in terms of set of (n) simultaneous first-order equations obtained either directly or from the nth-order equation of the system. Then the set is replaced by a single first-order matrix equation. The solution of the matrix equation is obtained using standard numerical and computer techniques.

State Equations in Normal Form

State equations are written making use of the concept of graph theory. Graph theory was discussed in detail in chapter-5.

The steps to be followed to formulate first-order state equations for a given network are as follows.

1. Draw the network graph for the given circuit
2. In the graph, choose a tree that contains all voltage sources and maximum possible numbers of capacitors. All the current sources and the inductors are left for the co-tree. The maximum number of control voltages should be in the tree and the control currents in the co-tree. The resistors can be either in the tree or in the co-tree.
3. Assign a voltage to each capacitor and mark its polarity. Similarly for an inductor, assign a current and specify its direction. These capacitor voltages and the inductor currents are the state variables.
4. Using KVL, write the loop equation for each loop comprising of an inductor link and a path in the tree.
5. Using KCL, write a node equation at each capacitor.
6. If the resistor voltages occur in the KVL equations, use KCL to formulate v_R/R to a sum of link currents. If resistor currents occur in the KCL equations, use KVL to set $i_R R$ equal to a sum of loop voltages. Substitute these values in the equations obtained in steps 3 and 4. The set of equations obtained are the normal state equations.

State Matrix Differential Equation

The response of passive electrical networks can be described in the form of a differential equation in the time-domain as given below.

$$\frac{d^n y}{dt_n} + b_{n-1}\frac{d^{n-1} y}{dt^{n-1}} + \dots + b_2\frac{d^2 y}{dt^2} + b_1\frac{dy}{dt} + b_o y = f(t)$$

Where, $f(t)$ is the forcing function

$$y \equiv x_1, \frac{dy}{dt} \equiv x_2, \frac{d^2y}{dt^2} \equiv x_3 ....., \frac{d^{n-1}y}{dt^{n-1}} \equiv x_n.$$

The equation can be replaced by first order system as given below.

$$\dot{x}_1 = 0x_1 + 1x_2 + 0x_3 + 0x_4 + ...+ 0x_n$$

$$\dot{x}_2 = 0x_1 + 0x_2 + 1x_3 + 0x_4 + ...+ 0x_n$$

$$\dot{x}_3 = 0x_1 + 0x_2 + 0x_3 + 1x_4 + ...+ 0x_n$$

....

$$\dot{x}_{n-1} = 0x_1 + 0x_2 + 0x_3 + 0x_4 + ...+ 1x_n$$

$$\dot{x}_n = -b_o x_1 - b_1 x_2 - b_2 x_3 - b_3 x_4 + ...-b_{n-1} x_n + f(t)$$

The coefficients of x's can be denoted by a (n × n) matrix A. f(t) is a (n × 1) column matrix having function f(t) in the last row and zero otherwise. Therefore, the above equation for ($\dot{x}_n$) can be rewritten as a single first-order matrix differential equation for the state vector $x(t)$ as:

$$\dot{x} = Ax(t) + f(t)$$

The matrix (A) is the characteristic or the system matrix and $f(t)$ is the forcing vector.

SOLVED PROBLEMS

Problem 8.1 *For the single π network of Fig. 8.11 determine input admittance (y_{11}), output admittance (y_{22}) and the two transfer admittance parameters (y_{12}) and (y_{21}).*

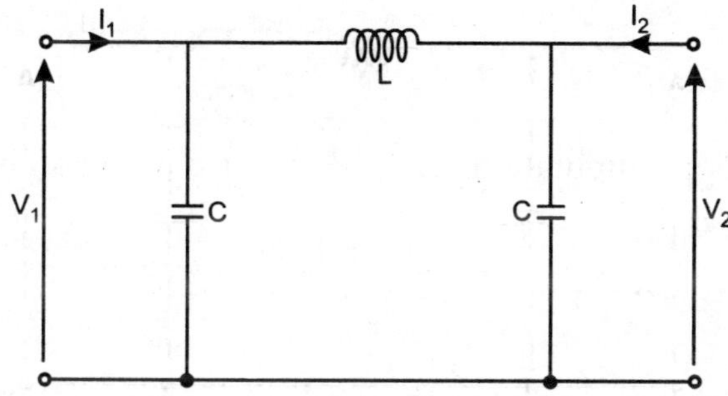

Fig. 8.11

Solution. The input admittance (y_{11}) is given by

$$y_{11} = \left.\frac{I_1}{V_1}\right|_{(V_2=0)}$$

The condition $V_2 = 0$ reduces the network to a parallel connection of (L) and (C) acrosS (V_1). This gives

$$I_1 = V_1 (j\omega C - j/\omega L)$$

or $$y_{11} = (j\omega C - j/\omega L) = j\left(wC - \frac{1}{\omega L}\right)$$

The output admittance (y_{22}) is given by

$$y_{22} = \left.\frac{I_2}{V_2}\right|_{V_1=0}$$

The condition $V_1 = 0$ reduces the network to a parallel connection of (L) and (C) across (V_2). This gives

$$I_2 = V_2\left(j\omega C - \frac{j}{\omega L}\right) \text{ or } y_{22} = j\left(\omega C - \frac{1}{\omega L}\right)$$

The transfer admittance (y_{12}) is given by

$$y_{12} = \left.\frac{I_1}{V_2}\right|_{V_1=0}$$

For the condition $V_1 = 0$, current (I_1) is same as the current through inductance (L). Under this condition, voltage across (L) is (V_2).

Therefore,

$$I_1 = -V_2\left(\frac{-j}{\omega L}\right)$$

This leads to

$$y_{12} = \frac{I_1}{V_2} = \left(\frac{-j}{\omega L}\right) = \frac{j}{\omega L}$$

Transfer admittance (y_{21}) is given by

$$y_{21} = \left.\frac{I_2}{V_1}\right|_{V_2=0}$$

Under the condition $V_2 = 0$, we can write

$$I_2 = V_1\left(\frac{-j}{\omega L}\right)$$

This gives $y_{21} = \frac{I_2}{V_1} = -\left(\frac{-j}{\omega L}\right) = \frac{j}{\omega L}$.

Problem 8.2 *Determine all four impedance parameters for the T-network of fig.* 8.12

Solution.

The input impedance (Z_{11}) is given by;

$$Z_{11} = \left.\frac{V_1}{I_1}\right|_{I_2=0}$$

This gives

$$Z_{11} = j\,\omega\,L - \frac{j}{\omega\,C}$$

$$= j\left(\omega\,L - \frac{1}{\omega\,C}\right)$$

The output impedance (Z_{22}) is given by:

$$Z_{22} = \left.\frac{V_2}{I_2}\right|_{I_1=0}$$

This gives

$$Z_{22} = j\,\omega\,L - \frac{j}{\omega\,C}$$

$$= j\left(\omega L - \frac{1}{\omega\,C}\right)$$

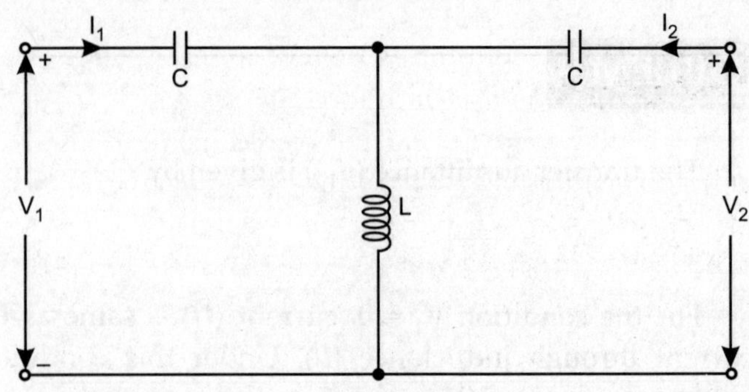

Fig 8.12

Transfer impedance (Z_{12}) is given by $Z_{12} = \left.\frac{V_1}{I_2}\right|_{I_1=0}$

This gives V_1 = voltae across (L) = $j\,\omega\,L \times I_2$

Therefore,

$$Z_{12} = \frac{V_1}{I_2} = j\,\omega L$$

Transfer impedance (Z_{21}) is given by $Z_{21} = \left.\frac{V_2}{I_1}\right|_{I_2=0}$

This given V_2 = Voltage across (L) = $j\,\omega\,L \times I_1$

Therefore,

$$Z_{21} = \frac{V2}{I1} = j\,\omega\,L.$$

Problem 8.3 *For the network of Fig 8.13 find impedance parameters Z_{11}, Z_{12}, Z_{21} and Z_{22} and admitance parameters y_{11}, y_{12}, y_{21} and y_{22}.*

Solution.

Z-Parameters

For computation of $Z_{11}\left(=\left.\frac{V_1}{I_1}\right|_{I_2=0}\right)$, output port is open

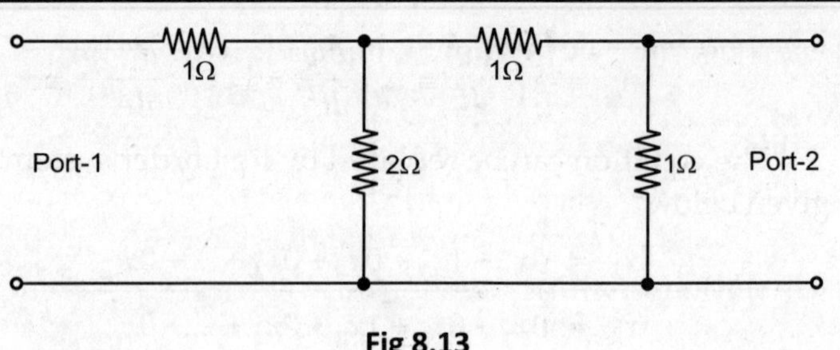

Fig 8.13

This given

$$Z_{11} = 1 + \frac{2\times 2}{2+2} = 2\,\Omega$$

For computation of $Z_{12}\left(\left.\frac{V_1}{I_2}\right|_{I_1=0}\right)$, input port is open.

This gives $Z_{12} = \frac{V_1}{I_2}$ with (V_1) given by

$$V_1 = 2\times\frac{I_2\times 1}{1+3} = \frac{2I_2}{4} = \frac{I_2}{2}$$

This gives

$$Z_{12} = \frac{1}{2}\,\Omega$$

For computation of $Z_{21}\left(\left.\frac{V_2}{I_2}\right|_{I_2=0}\right)$, output port is open.

$$V_2 = \frac{I_1\times 2}{2+2} = \frac{I_1}{2} \text{ This given } Z_{21} = \frac{V_2}{I_1} = \frac{1}{2}\text{ W}$$

For computation of $Z_{22}\left(=\left.\frac{V_2}{I_2}\right|_{I_1=0}\right)$, input port is open.

This gives $Z_{22} = \frac{1\times 3}{1+3} = \frac{3}{4}\,\Omega$

y- Parameters

For computation of $y_{11}\left(=\left.\frac{I_1}{V_1}\right|_{V_2=0}\right)$, output port is shorted.

The network reducces to the one shown in Fig. 8.14 (b).

$$I_1 = \frac{V_1}{1+\frac{2}{3}} = \frac{V_1}{5/3} = \frac{3V_1}{5}$$

This gives $y_{11} = \frac{I_1}{V_1} = \frac{3}{5} = 0.6$ Siemen

For Computation of $y_{12}\left(=\left.\frac{I_1}{V_2}\right|_{V_1=0}\right)$, intput port is shorted. This reduces the network to the one shown in Fig. 8.14 (*a*).

$$I_1 = \frac{-V_2 \times (2/3)}{\left(1+\frac{2}{3}\right)} = -V_2\left(\frac{2}{3}\times\frac{3}{5}\right) = \frac{-2V_2}{5}$$

or $$y_{12} = \frac{I_1}{V_2} = -\frac{2}{5} = -0.4 \text{ Siemen}$$

For computation of $y_{21}\left(\frac{I_2}{V_1}\Big|_{V_2=0}\right)$, output port is shorted. This reduces the network to the one shown in Fig. 8.14 (*b*).

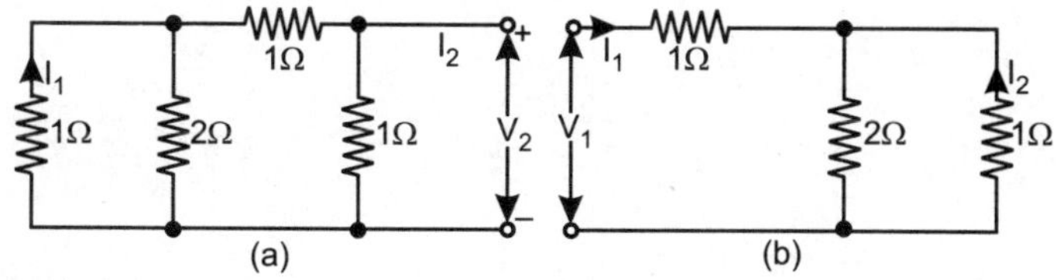

Fig. 8.14

$$I_2 = \frac{V_1 \times (2/3)}{\left(1+\frac{2}{3}\right)} = -\frac{2V_1}{5}$$

or $$y_{21} = \frac{I_2}{V_1} = -\frac{2}{5} = -0.4 \text{ Siemen}$$

For computation of $y_{22}\left(=\frac{I_2}{V_2}\Big|_{V_1=0}\right)$, input port is shorted. The network reduces to the one shown in Fig 8.14 (*a*).

$$I_2 = \frac{V_2}{(1\times 5/3)\left(1+\frac{5}{3}\right)} = \frac{V_2}{(5/3)/(8/3)} = \frac{8V_2}{5}$$

This gives

$$y_{22} = \frac{I_2}{V_2} = \frac{8}{5} = 1.6 \text{ Siemen.}$$

Problem 8.4 *Determine admittance parameters for the bridged-T network of Fig 8.15*

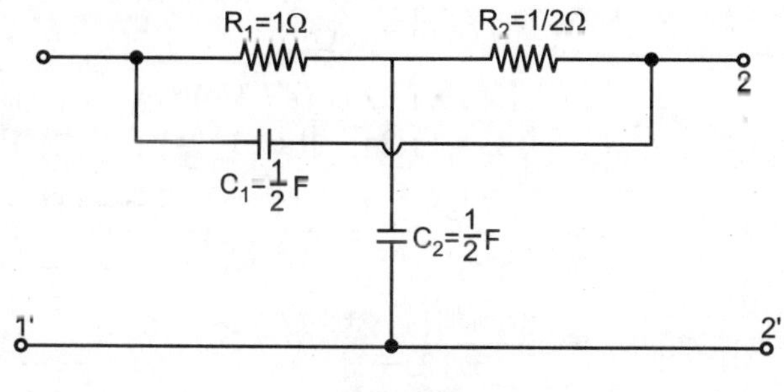

Fig. 8.15

Solution. For computation of $y_{11}\left(=\frac{I_1}{V_1}\Big|_{V_2=0}\right)$ and $y_{21}\left(\frac{I_2}{V_1}\Big|_{V_2=0}\right)$, the given network reduces to what is shown in Fig. 8.16 (*a*). For computation of $y_{22}\left(=\frac{I_2}{V_1}\Big|_{V_1=0}\right)$ and $y_{12}\left(=\frac{I_1}{V_2}\Big|_{V_1=0}\right)$, The given network reduces to what is shown in Fig 8.16 (*b*)

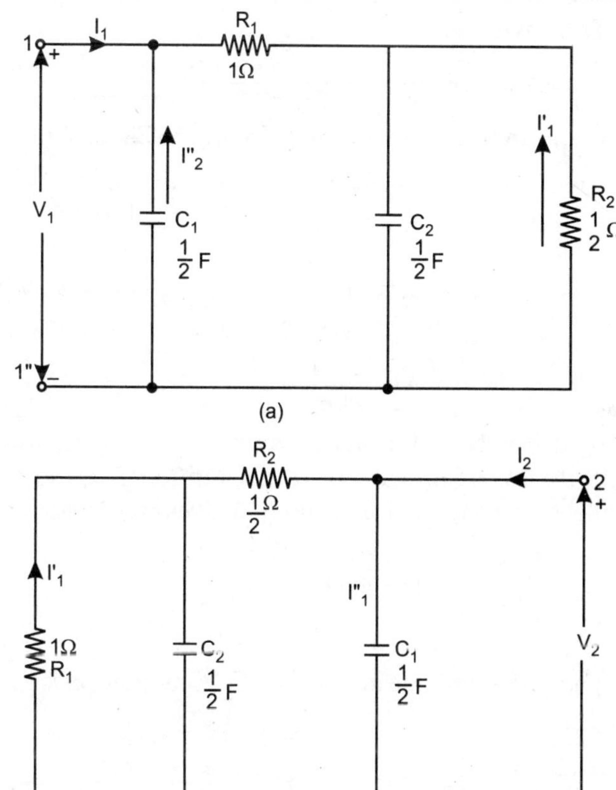

Fig. 8.16

$$y_{11} = \frac{s}{2} + \left[\frac{1\times(s/2+2)}{1+\frac{s}{2}+2}\right] = \frac{s}{2} + \left[\frac{(s+4)}{(s+6)}\right]$$

$$= \frac{s(s+6)+2(s+4)}{2(s+6)} = \frac{s^2+6s+8}{2(s+6)}$$

(y_{21}) can also be computed by referring to Fig. 8.16 (*a*). When the output port is shorted, we can visualize that the current (I_2) in this case would be sum of currents flowing through (R_2) and (C_1). These currents are marked (I'_2) and (I''_2) in Fig. 8.16 (*a*).

Now,

$$I'_2 = -\left[\frac{V_1\times 2}{(s+4)} \div \left(1+\frac{2}{s+4}\right)\right]\times 2$$

$$= -\left[\frac{2V_1}{(s+4)}\times\frac{(s+4)}{(s+6)}\right]\times 2$$

$$= \frac{-4V_1}{(s+6)} \text{ and } I''_2 = \frac{-V_1 s}{2}$$

Therefore,

$$I_2 = (I'_2 + I''_2) = -\left[\frac{4V_1}{(s+6)} + \frac{V_1 s}{2}\right]$$

$$= -\left[\frac{8V_1 + V_1 s(s+6)}{2(s+6)}\right] = -\frac{V_1(s^2+6s+8)}{2(s+6)}$$

This gives

$$y_{21} = \frac{I_2}{V_1} = \frac{-(s^2+6s+8)}{2(s+6)}$$

(y_{22}) can be computed by referring to Fig 8.16 (*b*).

$$y_{22} = \frac{s}{2} + \left[\frac{2\times[1+s/2]}{2+1+\frac{s}{2}}\right] = \frac{s}{2} + \left[\frac{2(s+2)}{(s+6)}\right]$$

$$= \frac{s(s+6)+4(s+2)}{2(s+6)} = \frac{s^2+6s+4s+8}{2(s+6)}$$

$$= \frac{s^2+10s+8}{2(s+6)}$$

(y_{12}) can be computed form (I_1/V_2) by referring to Fig. 8.16. (*b*). Again current (I'_1) is sum of (I''_1) and (I''_1). Parallel combination of (R_1) and (C_2) gives an impedance of

$$\frac{\frac{1\times 2}{s}}{1+\frac{2}{s}} = \frac{2V_2/(s+2)}{(s+6)/2(s+2)}$$

Therefore, voltage across $R_1 - C_2$ combination is given by :

$$\frac{\frac{V_2\times 2}{(s+2)}}{\frac{1}{2}+\frac{2}{(s+2)}} = \frac{2V_2/(s+2)}{(s+6)/2(s+2)} = \frac{2V_2}{(s+2)} \times \frac{2(s+2)}{(s+6)}$$

$$= \frac{4V_2}{(s+6)}$$

This gives

$$I'_1 = \frac{-4V_2}{(s+6)} \text{ Also } I''_1 = \frac{-V_2 s}{2}$$

Therefore,

$$I_1 = -\left[\frac{V_2 s}{2} + \frac{4V_2}{(s+6)}\right] = -V_2\left[\frac{s(s+6)+8}{2(s+6)}\right]$$

$$= -V_2\left[\frac{s^2+6s+8}{2(s+6)}\right]$$

This gives

$$y_{12} = \frac{I_1}{V_2} = -\left[\frac{s^2+6s+8}{2(s+6)}\right].$$

Problem 8.5 *For the Twin-T network of Fig. 8.17, find the impedance parameters, Z_{11}, Z_{12}, Z_{21} and Z_{22}.*

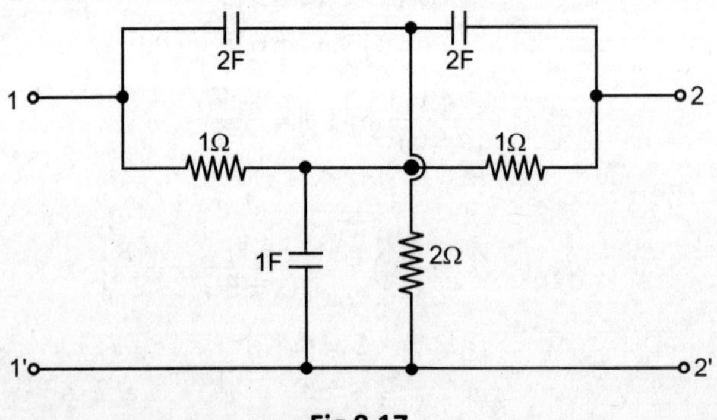

Fig 8.17

Solution. The twin -T network of Fig. 8.17 is nothing but a parallel connection of two T-networks shown in Fig 8.18 (*a*) and (*b*). If we can find z-parameters for these individual networks, the desired z-parameters can then be computed from the parallel equivalent of individual values.

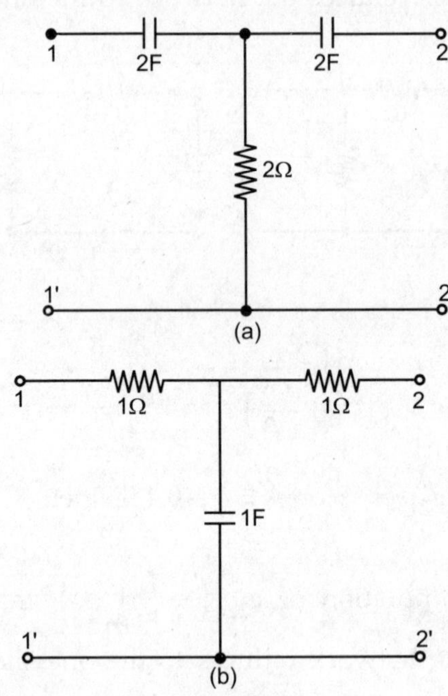

Fig. 8.18

$$Z_{11}(a) = 2 + \frac{1}{2s} = \frac{4s+1}{2s}$$

$$Z_{11}(b) = 1 + \frac{1}{s} = \frac{s+1}{2s}$$

Therefore,

$$Z_{11} = \frac{Z_{11}(a)\times Z_{11}(b)}{Z_{11}(a)+Z_{11}(b)}$$

$$= \frac{[(4s+1)2s]\times(s+1/s)]}{[(4s+1)/2s]+[(s+1)/s]}$$

$$= \frac{(4s+1)(s+1)}{s(4s+1)+2s(s+1)} = \frac{(4s+1)(s+1)}{4s^2+s+2s^2+2s}$$

$$= \frac{(4s+1)(s+1)}{3s(2s+1)}$$

Since each of the two networks is symmetrical, therefore,

$$Z_{22}(a) = Z_{11}(a) \text{ and } Z_{22}(b) = Z_{11}(b),$$

which gives

$$Z_{22} = Z_{11} = \frac{(4s+1)(s+1)}{3s(2s+1)}$$

Now

$$Z_{12}(a) = 2 \text{ and } Z_{12}(b) = 1/s$$

Therefore,

$$Z_{12} = \frac{2 \times 1/s}{2 + 1/s} = \frac{2}{(2s+1)}$$

$$\text{Again } Z_{21} = Z_{12} = \frac{2}{(2s+1)}$$

Problem 8.6. *Express the series impedance (Z) and the shunt admittances (Y_1) and (Y_2) of the unsymmetrical π-network of Fig. 8.19 in terms of the ABCD parameters.*

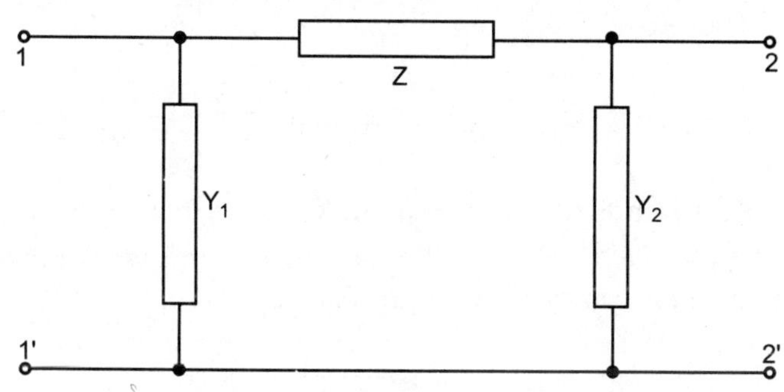

Fig. 8.19

Solution. ABCD parameters are given by :

$$\frac{1}{A} = \left.\frac{V_2}{V_1}\right|_{I_2=0} ; -\frac{1}{B} = \left.\frac{I_2}{V_1}\right|_{V_2=0}$$

$$\frac{1}{C} = \left.\frac{V_2}{I_1}\right|_{I_2=0} ; -\frac{1}{D} = \left.\frac{I_2}{I_1}\right|_{V_2=0}$$

Remember that minus sign for (1/B) and (1/D) only indicates that the direction of (I_2) in case of transmission parameters is chosen to be opposite to that chosen for defining other parameters such as Impedance and Admittance parameters.

Now when the output port is open circuited i.e. $I_2 = 0$, (V_2) is nothing but voltage across (Y_2).

$$V_2 = \frac{V_1 \times \frac{1}{Y_2}}{Z + \frac{1}{Y_2}} - \frac{V_1}{1 + ZY_2} \text{. This gives,}$$

$$A = \frac{V_1}{V_2} = 1 + ZY_2$$

Also, with $I_2 = 0$. the impedance across 1 – 1' port is given by:

$$\frac{\frac{1}{Y_1} \times \left(Z + \frac{1}{Y_2}\right)}{\frac{1}{Y_1} + Z + \frac{1}{Y_2}} = \frac{1 + ZY_2}{Y_1 + Y_2 + z}$$

This gives

$$V_1 = I_1\left(\frac{1 + Z\,Y_2}{Y_1 + Y_2 + Z}\right) = V_2(1 + ZY_2)$$

Therefore

$$C = \frac{(1 + ZY_2)(Y_1 + Y_2 + Z)}{(1 + ZY_2)} = (Y_1 + Y_2 + Z)$$

For $V_2 = 0$,

$$I_2 = \frac{I_1 \times \frac{1}{Y_1}}{Z + \frac{1}{Y_1}} = \frac{I_1}{1 + ZY_1}$$

This gives $D = (1 + ZY_1)$. Alos, I_2 = Current through (z) $= \frac{V_1}{Z}$

This gives $B = Z$. Now $D = (1 + Z\,Y_1)$

Substituting for Z, $D = 1 + BY_1$, which gives

$$Y_1 = \left(\frac{D-1}{B}\right)$$

Also substituting for (Z) in $A = 1 + ZY_2$ and $A = 1 + B\,Y_2$

or $$Y_2 = \left(\frac{A-1}{B}\right); \; Z = B$$

$$Y_1 = \left(\frac{D-1}{B}\right) \text{ and } Y_2 = \left(\frac{A-1}{B}\right)$$

Problem-8.7: For the network shown in Figure 8.20, compute parameters z_{11} and z_{21}.

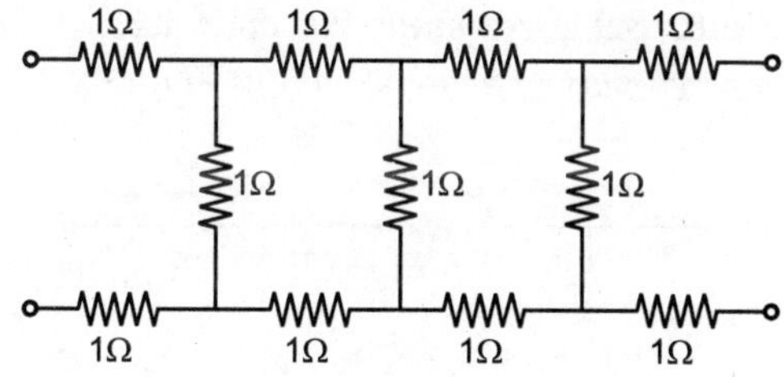

Fig. 8.20

Solution: To calculate z_{11} and z_{21} connect a current source I_1 at the left side terminals as shown in Figure. 8.21.

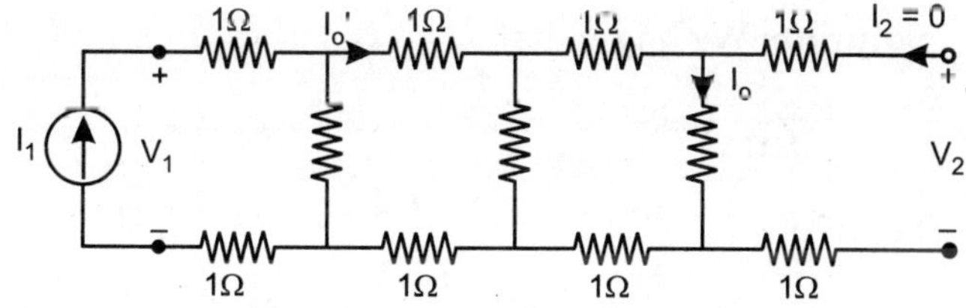

Fig. 8.21

$$Z_{11} = \left.\frac{V_1}{I_1}\right|_{I_2=0}$$

$$= 2 + 1 \,\|\, [(1+1) + 1 \,\|\, (1+1+1)] = 2.733\ \Omega$$

From the above figure, we can see that $I_0 = 1/4I_0'$, $I_0' = [1/(1+11/4)]I_1 = 4/15I_1$

Therefore, $I_0 = 1/15I_1$

$$V_2 = 2I_0 = 2/15I_1$$

Therefore,

$$z_{21} = \left.\frac{V_2}{I_1}\right|_{I_2=0} = 2/15 = 0.133\Omega$$

Problem.8.8 *For the network shown in Figure.8.22, compute parameters z_{12} and z_{22}.*

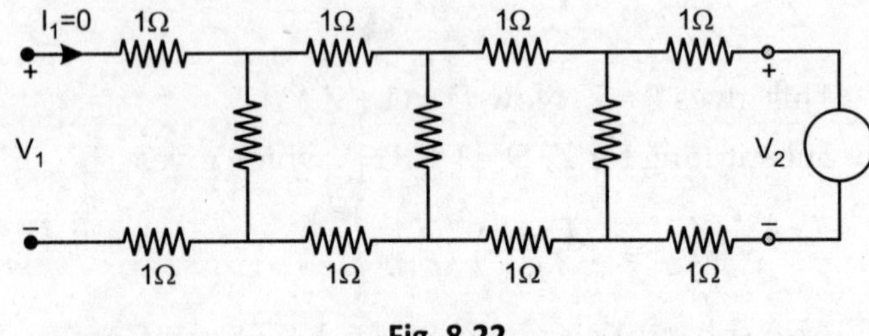

Fig. 8.22

Solution: For the given network $z_{12} = z_{21} = 0.133\Omega$

To calculate z_{22}, connect voltage source V_2 across the right most terminals as shown in Figure 8.22.

$$z_{22} = \left.\frac{V_2}{I_2}\right|_{I_1=0} = 2 + 1 \,\|[(1+1)+1\,\|\,(1+1+1)]$$

$$= 2.733\ \Omega$$

Problem. 8.9 *Determine the open circuit impedance matrix of the two-part network shown in Figure 8.23.*

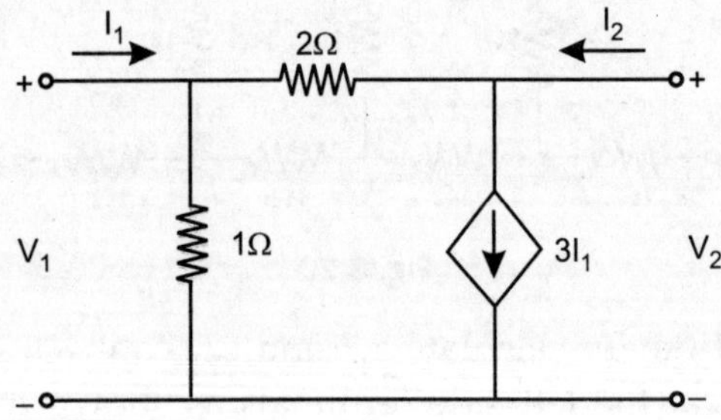

Fig. 8.23

Solution: We know that

$$Z_{11} = \left.\frac{V_1}{I_1}\right|_{I_2=0}$$

Therefore,

$$Z_{11} = \frac{-2I_1 \times 1}{I_1} = -2$$

$$Z_{21} = \left.\frac{V_2}{I_2}\right|_{I_2=0}$$

Therefore, $$Z_{21} = \frac{-6I_1 + V_1}{I_1} = \frac{-6I_1 - 2I_1}{I_1} = -8$$

$$Z_{12} = \left.\frac{V_1}{I_2}\right|_{I_1=0}$$

Therefore, $$Z_{12} = \frac{1 \times I_2}{I_2} = 1\ \Omega$$

$$Z_{22} = \left.\frac{V_2}{I_2}\right|_{I_1=0}$$

Therefore, $$Z_{22} = \frac{2I_2 + 1I_2}{I_2} = 3\ \Omega$$

Hence the impedance matrix is $\begin{bmatrix} -2 & 1 \\ -8 & 3 \end{bmatrix}$

Problem 8.10. *For the circuit shown in the Figure-8.24, the z-parameters are $z_{11} = 10$, $z_{12} = -6$, $z_{21} = -4$ and $z_{22} = 12$. Compute the values of current I_1 and I_2.*

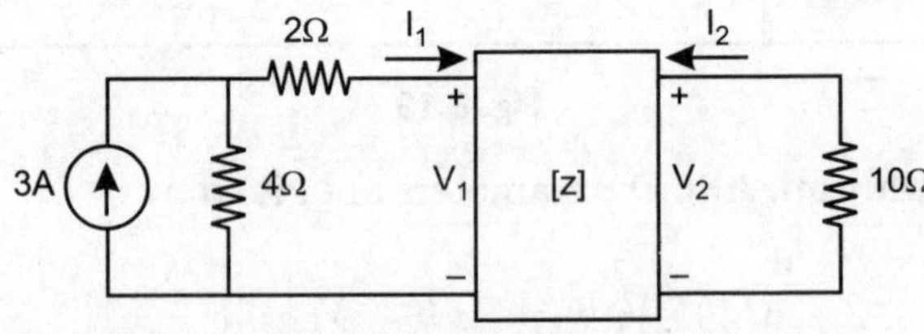

Fig. 8.24

Solution: The current source of 3A can be converted into an equivalent voltage source as shown in Figure 8.25.

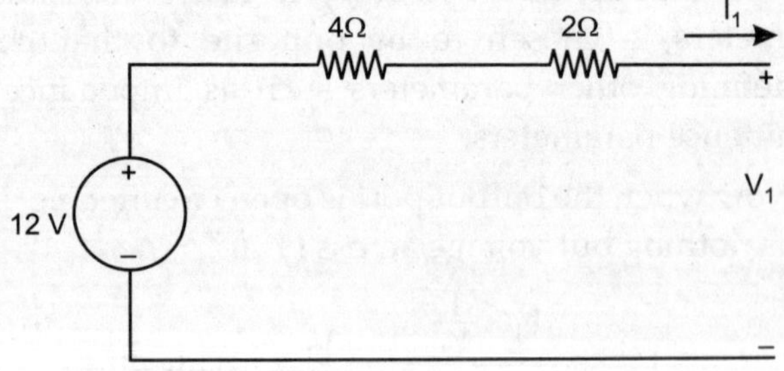

Fig. 8.25

Applying KVL, We get $-12 + 6I_1 + V_1 = 0$. Therefore, $V_1 = 12 - 6I_1$

Form the given values of z-parameters,we get $V_1 = 10I_1 - 6I_2$ and $V_2 = -4I_1 + 12I_2$

Form Figure8.24, $V_2 = -10I_2$

Solving the above four equation, we get $I_2 = 0.15$A, $I_1 = 0.81$ A

Problem 8.11: *For the two-port network in Figure 8.26, compute the ABCD parameters.*

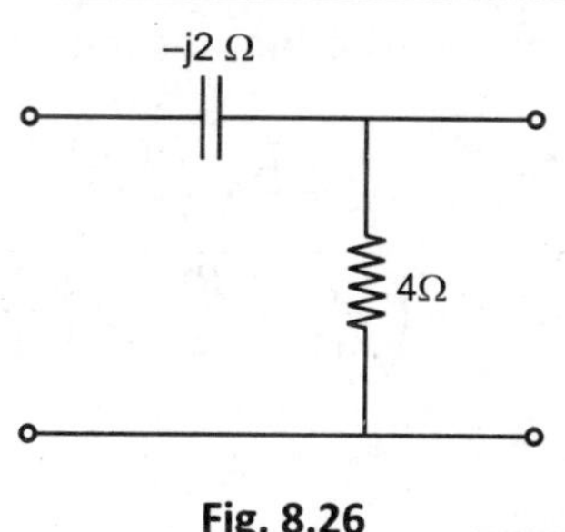

Fig. 8.26

Solution: To determine A and C parameters, connect source V_1 to the left termainals as shown in Figure 8.27 (*a*).

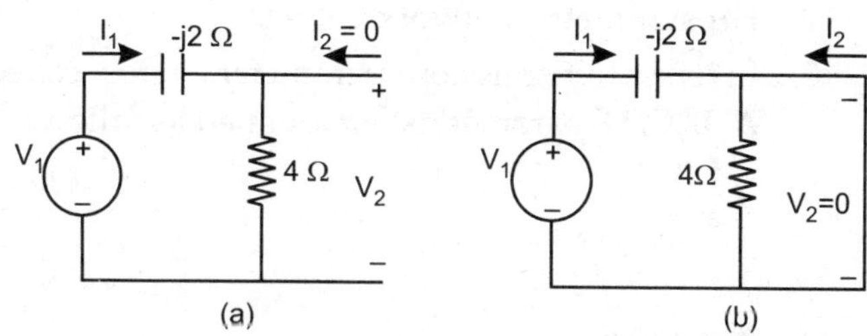

Fig. 8.27

From the figure 8.27 (*a*), applying KVL we get $V_1 = (4 \times 2j)I_1$ and $V_2 = 4I_1$

Parameter

$$A = V_1/V_2 = (4 - 2j)/4 = 1 - 0.5j$$

Parameter

$$C = I_1/V_2 = I_1/4I_1 = 0.25$$

To find the parameters B and D, consider the Figure 8.27 (b)

From the figure, we see that $I_1 = -I_2$. Now parameter $D = -I_1/I_2 = 1$

$$V_1 = -j2I_1 = j2I_2$$

Now, parameter $B = -V_1/I_2 = -2j\Omega$

The ABCD parameter matrix is

$$\begin{bmatrix} A & B \\ C & D \end{bmatrix} = \begin{bmatrix} 1 - j0.5 & -j2\Omega \\ 0.25S & 1 \end{bmatrix}$$

Problem 8.12. *Compute the admittance parameter y_{12} (in milli- mhos) in the 2-port network of Figure 8.28.*

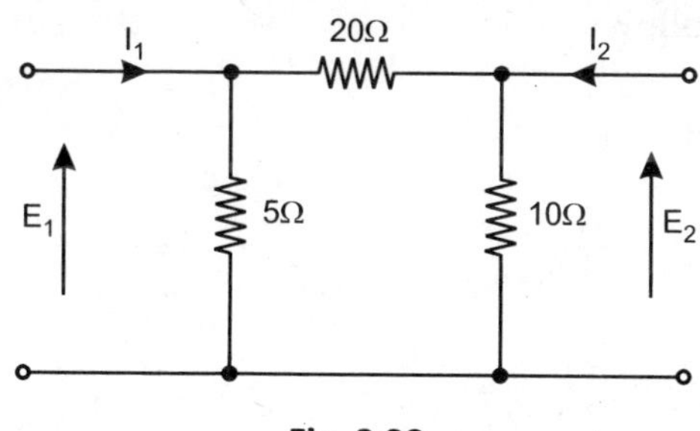

Fig. 8.28

Solution: Figure 8.29 shows the circuit given in the problem.

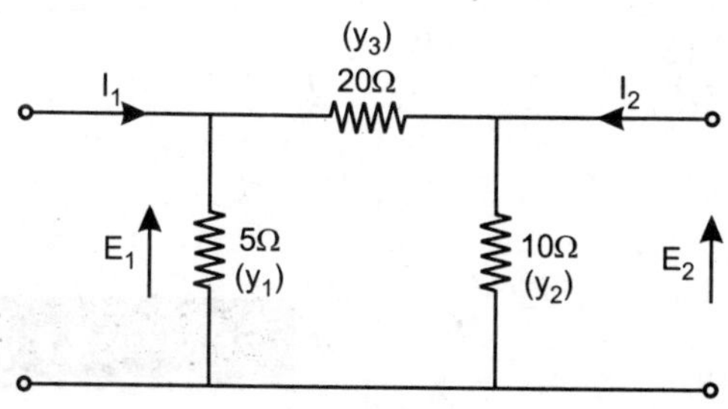

Fig. 8.29

The admittance matrix is given for the network in Figure 8.29 is

$$\begin{bmatrix} y_{11} & y_{12} \\ y_{21} & y_{22} \end{bmatrix} = \begin{bmatrix} y_1 + y_3 & -y_3 \\ -y_3 & y_2 + y_3 \end{bmatrix}$$

Therefore,

$$y_{12} = -y_3$$

Hence $y_{12} = -\frac{1}{20} = -0.05 \text{ mho} = -50 \text{ milli-mhos}$

Problem 8.13: *For the network shown in Figure 8.30, what is the driving point impedance as a function of (s)?*

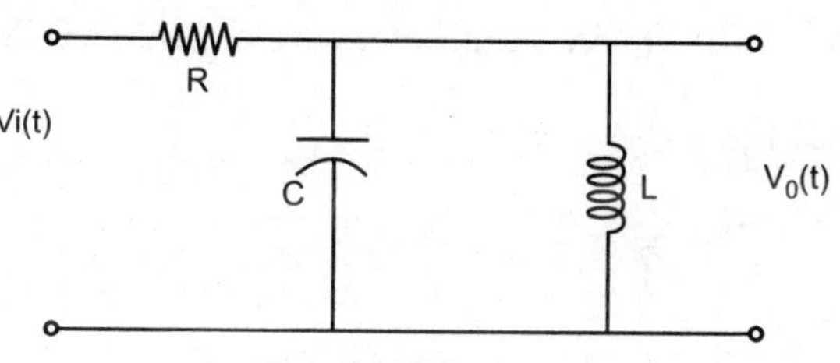

Fig. 8.30

Solution: Driving point impedance is given by the following equation.

$$Z(s) = R + \left(\frac{1}{sC} \parallel sL\right) = R + \frac{\left(\frac{1}{sC}\right)(sL)}{\left(\frac{1}{sC} + sL\right)}$$

Solving the above equation, we get

$$Z(s) = \frac{s^2 LCR + sL + R}{s^2 LC + 1}$$

Problem 8.14 *For the network in Figure 8.31, determine the transfer function as a function of (s).*

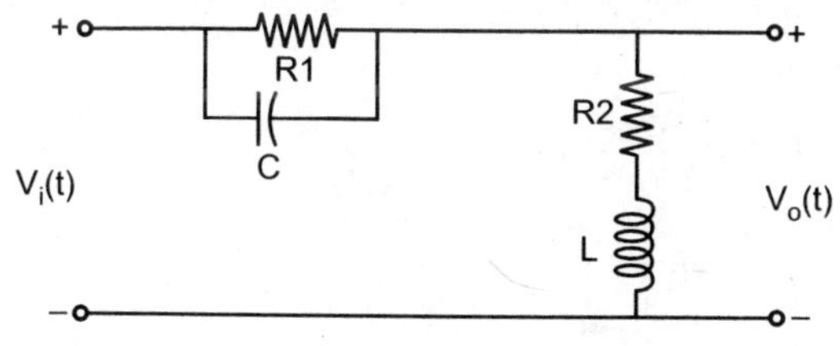

Fig. 8.31

Solution: Transfer fucntion is given by the following equation.

$$\frac{V_o(S)}{V_i(S)} = \frac{(R_2+sL)}{(R_2+sL)+\dfrac{R_1/sC}{R_1+\dfrac{1}{sC}}}$$

Solving the above equation, we get

$$\frac{V_o(s)}{V_i(s)} = \frac{\left(s+\dfrac{R_2}{L}\right)\left(s+\dfrac{1}{CR_1}\right)}{s^2+\left(\dfrac{1}{R_1C}+\dfrac{R_2}{L}\right)s+\left(\dfrac{R_1+R_2}{LCR_1}\right)}$$

Summary of Important Formulas

1. z-parameters are defined as follows.

$$z_{11} = \left.\frac{V_1}{I_1}\right|_{I_2=0}$$

$$z_{21} = \left.\frac{V_2}{I_1}\right|_{I_2=0}$$

$$z_{12} = \left.\frac{V_1}{I_2}\right|_{I_1=0}$$

$$z_{22} = \left.\frac{V_2}{I_2}\right|_{I_2=0}$$

2. For reciproal networks, $z_{12} = z_{21}$
3. For symmetric networks, $z_{11} = z_{22}$
4. Admiltance parameters are defind as follows.

$$y_{11} = \left.\frac{I_1}{V_1}\right|_{V_2=0}$$

$$y_{21} = \left.\frac{I_2}{V_1}\right|_{V_2=0}$$

$$y_{12} = \left.\frac{I_1}{V_2}\right|_{V_1=0}$$

$$y_{22} = \left.\frac{I_2}{V_2}\right|_{V_1=0}$$

5. For reciprocal networks, $y_{12} = y_{21}$
6. For symmetric networks, $y_{11} = y_{22}$
7. Transmission parameters (also called ABCD parametess) are defined as follows.

$$A = \left.\frac{V_1}{V_2}\right|_{I_2=0}$$

$$B = \left.\frac{-V_1}{I_2}\right|_{V_2=0}$$

$$C = \left.\frac{I_1}{V_2}\right|_{I_2=0}$$

$$D = \left.\frac{-I_1}{I_2}\right|_{V_2=0}$$

8. For reciprocal networks, $AD - BC = 1$
9. For symmetric networks, $A = D$
10. Inverse transmission parameters (alos called A',B',C,'D' parameters) are defined as follows.

$$A' = \left.\frac{V_2}{V_1}\right|_{I_1=0}$$

$$B' = \left.\frac{-V_2}{I_1}\right|_{V_1=0}$$

$$C' = \left.\frac{I_2}{V_1}\right|_{I_1=0}$$

$$D' = -\left.\frac{-I_2}{I_1}\right|_{V_1=0}$$

11. For reciprocal networks, $A'D' - B'C' = 1$
12. For symmetric networks , $A' = D'$
13. Hybrid parameters are defined as follows.

$$h_{11} = \left.\frac{V_1}{I_1}\right|_{V_2=0}$$

$$h_{21} = \left.\frac{I_2}{I_1}\right|_{V_2=0}$$

$$h_{12} = \left.\frac{V_1}{V_2}\right|_{I_1=0}$$

$$h_{22} = \left.\frac{I_2}{V_2}\right|_{I_1=0}$$

14. For reciporal networks, $h_{12} = -h_{21}$
15. For symmetric networks, $h_{11}h_{22} - h_{12}h_{21} = 1$
16. Inverse hybrid parameters are defined as follows.

$$g_{11} = \left.\frac{I_1}{V_1}\right|_{I_2=0}$$

$$g_{21} = \left.\frac{V_2}{V_1}\right|_{I_2=0}$$

$$g_{12} = \left.\frac{I_1}{I_2}\right|_{V_1=0}$$

$$g_{22} = \frac{V_2}{I_2}\bigg|_{V_1=0}$$

17. For reciprocal networks, $g_{12} = -g_{21}$
18. For symmetric networks, $g_{11}g_{22} - g_{12}g_{21} = 1$
19. when two 2-port networks are cascaded; the transmission parameters A,B,C,D of cascaded arrangement are expressed by:

$$\begin{bmatrix} A & B \\ C & D \end{bmatrix} = \begin{bmatrix} A_1 & B_1 \\ C_1 & D_1 \end{bmatrix} \begin{bmatrix} A_2 & B_2 \\ C_2 & D_2 \end{bmatrix}$$

20. When two 2 -port networks are connected in series; the impedance parametess of the series connected aor angement are expressed by:

$$\begin{bmatrix} Z_{11} & Z_{12} \\ Z_{21} & Z_{22} \end{bmatrix} = \begin{bmatrix} z_{11}(1) + z_{11}(2) & z_{12}(1) + z_{12}(2) \\ z_{21}(1) + z_{21}(2) & z_{22}(1) + z_{22}(2) \end{bmatrix}$$

$z_{11}(1), z_{12}(1), z_{21}(1)$ and $z_{22}(1)$ are the impedance parameters of first network. $z_{11}(2), z_{12}(2), z_{21}(2)$ and $z_{22}(2)$ are the inpedance parameters of the second network.

21. When two 2-port networks are connected in parallel; the adinittance parameters of parallel arrangement are exfressed by:

$$\begin{bmatrix} y_{11} & y_{12} \\ y_{21} & y_{22} \end{bmatrix} = \begin{bmatrix} y_{11}(1) + y_{11}(2) & y_{12}(1) + y_{12}(2) \\ y_{21}(1) + y_{21}(2) & y_{22}(1) + y_{22}(2) \end{bmatrix}$$

$y_{11}(1), y_{12}(1), y_{21}(1)$ and $y_{22}(1)$ are admittance parameters of first network. $y_{11}(2), y_{12}(2), y_{21}(2)$ and $y_{22}(2)$ are admittance parmeters of second network.

Self Evaluation Exercise

Multiple Choice Questions

1. Two two-port networks are connected in parallel. The combination is to be represented as a single two-port network. The parameters of this network are obtained by addition of the individual
   (a) z- parameters
   (b) h- parameters
   (c) y-parameters
   (d) ABCD parameters
2. The condition AD – BC = 1 for a two-port network implies that the network is a:
   (a) Reciprocal network
   (b) Lumped element network
   (c) Lossless network
   (d) Unilateral element network
3. Two two-port networks are connected in cascade. The combination is to be represented as a single two-port network. The parameters of the network that are obtained by multiplying the individual parameter matrices are
   (a) z-parameter
   (b) h-parameter
   (c) y-parameter
   (d) ABCD parameter
4. Which of the following statements are TRUE?

   S1: If n number of two-port networks with z-parameters $[Z]_1, [Z]_2, [Z]_3, \ldots, [Z]_n$ are connected in series, then the equivalent z-parameters of the equivalent two-port network is given by:

   $[Z]_{eq} = [Z]_1 + [Z]_2 + [Z]_3 \ldots\ldots + [Z]_n$

   S2: If n two-port networks with y-parameters $[Y]_1, [Y]_2, [Y]_3, \ldots, [Y]_n$ are connected in parallel, then the equivalent y-parameters of the equivalent two-port network is given by:

   $[Y]_{eq} = [Y]_1 + [Y]_2 + [Y]_3 \ldots\ldots + [Y]_n$

   S3: If n two-port networks with transmission parameters $[A]_1, [A]_2, [A]_3, \ldots, [A]_n$ are connected in cascade, then the equivalent transmission parameters of the equivalent two-port network is given by:

   $[A]_{eq} = [A]_1 * [A]_2 * [A]_3 \ldots\ldots * [A]_n$

   (a) S_1 and S2 (b) S2 and S3
   (c) S3 and S1 (d) S1, S2 and S3
5. For a 2-port network to be reciprocal,
   (a) $z_{11} = z_{22}$ (b) $y_{21} = y_{12}$
   (c) $h_{21} = -h_{12}$ (d) AD – BC = 0
6. The network in Figure-8.32 is a
   (a) High pass filter
   (b) Low pass filter
   (c) Band pass filter
   (d) Band reject filter

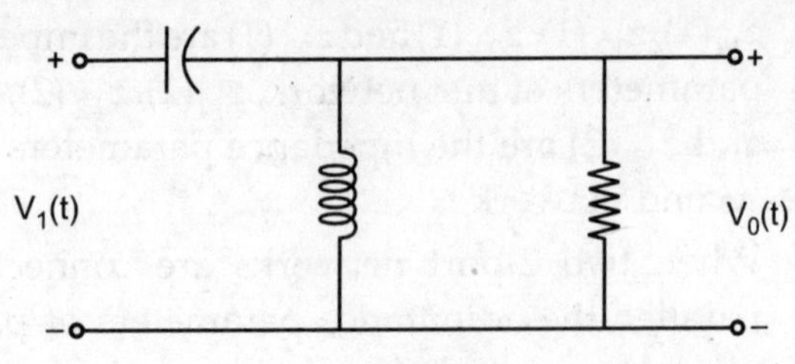

Fig 8.32

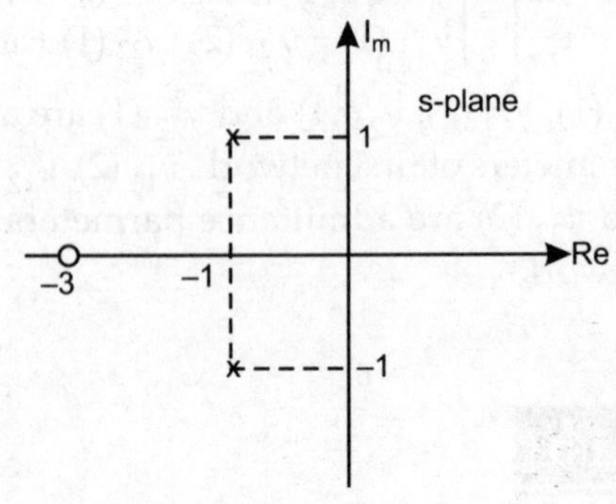

Fig 8.33

7. The first and the last critical frequency of an RC-driving point impedance function must respectively be
   (a) A zero and a pole
   (b) A zero and a zero
   (c) A pole and a pole
   (d) A pole and a zero
8. The driving-point impedance Z(s) of a network has the pole-zero locations as shown in Figure 8.33. If Z(0) = 3, then Z(s) is:
   (a) $\frac{3(s+3)}{s^2+2s+3}$ (b) $\frac{2(s+3)}{s^2+2s+2}$
   (c) $\frac{3(s-3)}{s^2-2s-2}$ (d) $\frac{2(s-3)}{s^2-2s-3}$
9. For a symmetric network, one of the following conditions should be satisfied for impedance parameters.
   (a) $z_{11} = z_{22}$ (b) $z_{12} = z_{21}$
   (c) $z_{11} = -z_{22}$ (d) $z_{12} = -z_{21}$
10. For a symmetric network, one of the following conditions should be satisfied for hybrid parameters.
   (a) $h_{11}h_{22} + h_{12}h_{21} = 1$
   (b) $h_{11}h_{22} - h_{12}h_{21} = 1$
   (c) $h_{11}h_{12} - h_{22}h_{21} = 1$
   (d) None of these
11. The Driving Point Impedance of a network can be calculated from (the terms have their usual meaning)
   (a) Δ_z/Δ_{11} (b) Δ_{11}/Δ_z
   (c) Δ_y/Δ_{11} (d) Δ_{11}/Δ_y
12. The TransferAdmittance (y_{23}) of a network can be calculated from:
   (a) Δ_y/Δ_{11} (b) Δ_{23}/Δ_y
   (c) Δ_y/Δ_{23} (d) None of these
13. When we talk about mesh current analysis using matrix formulation; we talk about
   (a) Admittance matrix
   (b) Z-parameters and impedance matrix
   (c) Y-parameters
   (d) None of these
14. When we talk about node voltage analysis using matrix; we talk about
   (a) Impedance matrix
   (b) Admittance matrix
   (c) Z-parameters
   (d) Laplace transform
15. For the two-port network in Figure-8.34, the z-parameter matrix is
   (a) $\begin{bmatrix} 4.8\Omega & -0.4\Omega \\ 0.4\Omega & 4.2\Omega \end{bmatrix}$ (b) $\begin{bmatrix} 4.8\Omega & 0.4\Omega \\ 0.4\Omega & 4.2\Omega \end{bmatrix}$
   (c) $\begin{bmatrix} 4.8\Omega & -0.4\Omega \\ -0.4\Omega & -4.2\Omega \end{bmatrix}$ (d) $\begin{bmatrix} 4.8\Omega & -0.4\Omega \\ -0.4\Omega & 4.2\Omega \end{bmatrix}$

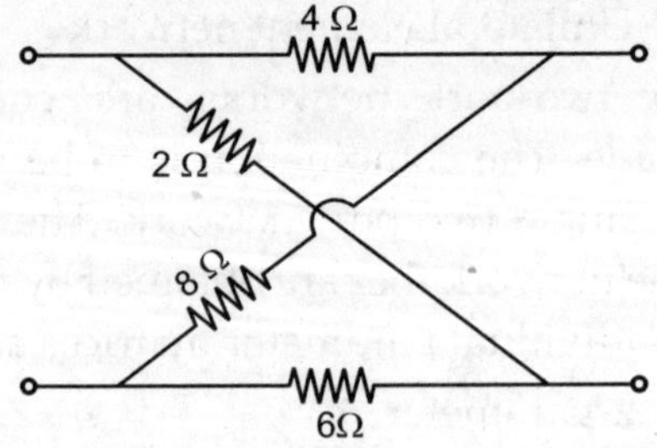

fig. 8.34

ANSWERS TO MULTIPLE CHOICE QUESTIONS

1. (c)	2. (a)	3. (d)	4. (d)	5. (b) and (c)	6. (a)
7. (d)	8. (b)	9. (a)	10. (b)	11. (a)	12. (c)
13. (b)	14. (b)	15. (d)			

Chapter 9

Diodes and Transistors

Learning Objectives

After completing the chapter, you will learn the following topics:

1. Insulators and Conductors
2. Semiconductor materials
3. Types of semiconductors
4. Law of mass action
5. Drift and diffusion currents
6. Biasing in P-N Junction
7. Ideal and practical diodes
8. V-I characteristics of a Diode
9. Temperature dependence of the V-I characteristics
10. Diode resistances
11. Diode junction capacitances
12. Diode equivalent circuits
13. DC and AC load line analysis of a diode circuit
14. Breakdown diodes – Avalanche and Zener diodes
15. Varactor diodes
16. Tunnel diodes
17. Schottky diodes
18. Point contact diodes
19. Power diodes
20. Transistor types – NPN and PNP
21. Transistor operation
22. Transistor configurations – Common Base, Common Emitter and Common Collector

Q.1. What is an insulator?

Ans. Insulators are materials that offer a large resistance to the flow of current through them. The typical resistivity level of an insulator is of the order of 10^{10} to 10^{12} Ω-cm. Hence, the application of voltage across the insulator results in negligible flow of current. Some of the popular insulator materials are mica, glass, quartz etc.

Q.2. What is a semiconductor material? How does it differ from a conductor?

Ans. A semiconductor is a material whose conductivity lies somewhere between that of a conductor and an insulator. Resistivity of pure silicon is about 2.5 × 10^3 ohm-m as compared to a figure of 1.7 × 10^{-8} ohm-m in for copper (conductor) and 10^{10} to 10^{14} ohm-m for glass (insulator) Conductivity is nothing hut reciprocal of resistivity and is expressed as mhos per metre. Germanium and silicon are two most commonly known semiconductor materials. Gallium arsenide is another semiconductor material becoming increasingly popular particularly for microwave semiconductor devices.

Q.3. What is the difference between intrinsic and extrinsic semiconductors?

Ans. Intrinsic semiconductors are semiconductors with very low level of impurity concentration. They are as pure as can be available through modern technology. The impurity is of the order of 1 part in 10 billion. Conduction in intrinsic semiconductors is either due to thermal excitation or due to crystal defects. Silicon and germanium are the two most important semiconductors used. Compared to extrinsic semiconductors they conduct a very small amount of current.

The electrical characteristics of an intrinsic semiconductor can be significantly changed by adding impurity atoms to the pure semiconductor material. The impurities added are of the order of 1 part in 10^5. Even though the alteration might seem to be small, it results in a large increase in conductivity. This process of addition of impurities is called doping.

Q.4. What are '*N*'-type' and '*P*-type' semiconductors?

Ans. Pure semiconductors have got poo r conductivity. However, when certain impurities are added, they are found to be exhibiting very important characteristics.

When a trivalent impurity (atoms having three electronsintheoutermostorbit)isaddedtoasemiconductor material, it becomes a *P*-type semiconductor. Frequently used *P*-type impurities are Indium, Boron. When a pentavalent impurity (atoms having five electrons in the outermost orbit) is added to a semiconductor material, it becomes an *N*-type semiconductor. Frequently used *N*-type impurities are Phosphorus, Antimony, Arsenic.

Q.5. What are 'donor' and 'acceptor' impurities?

Ans. Pentavalent impurities (Arsenic, Antimony) are also called DONOR IMPURITIES as they have got the ability to donate one electron from each impurity atom added to the pure semiconductor material.

Trivalent impurities (indium. Boron) have got three electrons in their outermost orbit and thus have a tendency to accept one electron from adjacent atoms to complete lattice structure. That is why, they are also coiled ACCEPTOR IMPURITIES.

Q.6. What are 'majority carriers' and 'minority carriers'?

Ans. Presence of pentavalent impurities in N-type materials causes excess of electrons which are available as charge carriers. These electrons are called as MAJORITY CARRIERS as they are in majority (in abundance) depending upon impurity concentration. In addition to these electrons, thermal agitation may also free some electrons from the valence-band. These electrons (a few in number; join the excess electrons (large in number) to become the majority carries. But the holes left behind by these thermally agitated electrons form the MINORITY CARRIERS. When a *P-N* junction is forward biased, the current is due to majority carriers only whereas a minority carrier current flows in the reverse biased condition.

Similarly, holes in the *P*-type constitute majority carriers whereas electrons left behind by the thermally agitated holes are called minority carriers.

Q.7. 'An *N*-type semiconductor has excess of electrons whereas a *P*-type semiconductor has excess of holes," Comment on it.

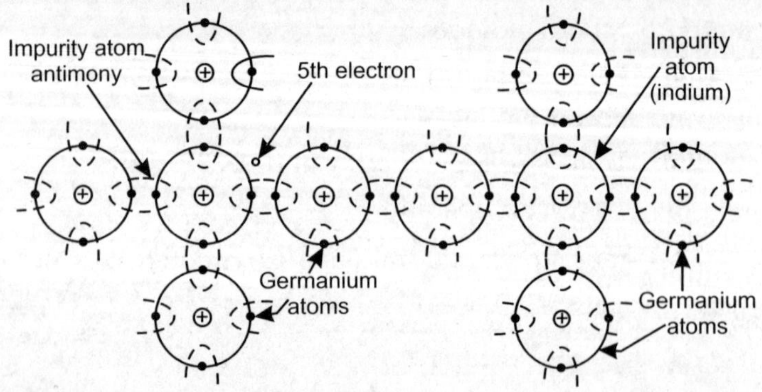

Fig. 9.1

Ans. Both, Germanium and Silicon have four electrons in their outermost orbits. In a germanium or a silicon crystal, different atoms form a sort of a lattice as shown in Fig. 9.1

When a pentavalent impurity is added, four of the five electrons present in the outermost orbit of impurity atom from bonds with the adjacent atoms whereas the fifth electron is available as free electron for conduction. This fifth electron in the outermost shell of the impurity atom is available as a free electron because it typically requires only 0.01 eV of energy to get dislodged. This much energy is available at the room temperature. Also remember that the escape of fifth electron leaves behind a bound positive charge, which can not move. Recombination is also not possible due to the shielding effect of the remaining electrons. Thus an *N*-type material has got excess of electrons depending upon impurity concentration, which are available as free charge carriers.

Similarly, when a trivalent impurity such as Aluminium, Boron, Indium is added to Germanium or Silicon, there is a deficiency of one electron per impurity atom for completing the covalent bonds. This allows the impurity atom to accept one electron from an adjoining semiconductor atom thus creating a vacancy of electron there. Vacancy of an electron is considered as creation of a hole. Thus as the trivalent impurity concentration increases, there are a large number of holes available for conduction. As electrons jump from one semiconductor atom to another to fill the vacancies, the corresponding holes move in the opposite direction. Most of the current that flows in a *P*-type semiconductor material is due to electrons filling vacancies (or holes) in the valence band. This mechanism is different from electrons moving in the conduction band of an N-type semiconductor material.

Q.8. 'An *N*-type semiconductor has excess of electrons and a *P*-type excess of holes'. Does that mean, *N*-type materials have a net negative charge and *P* type materials have a net positive charge ? Comment.

Ans. In an *N*-type semiconductor, when one electron is available for conduction from a pentavalent impurity atom, it leaves the impurity atom as positively charged by the same amount as the negative charge carried by an electron so that semiconductor as a whole remains neutral. Same is the situation in *P*-type semiconductors also.

Q.9. What is doping?

Ans. Addition of controlled amounts of impurities to the pure semiconductor materials is called doping.

Q.10. What is the law of mass action?

Ans. According to the law of mass action, the product of free-electron concentration and hole concentration in any semiconductor is constant. The expression for the law of mass action is given by:

$$np = n_i^2$$

(n) is the free electron concentration (negatively charged carriers), (p) is the hole concentration (positively charged carriers) and (n_i)is the intrinsic concentration.

Q.11. Explain the current flow in semiconductors.

Ans. The current flow in semiconductors is the sum of the drift and diffusion currents whereas the current flow in conductors is only due to the drift phenomenon.

Drift current is due to the potential gradient in the semiconductor and is given by the following expression.

$$J = (n\mu_n + p\mu_p)q\varepsilon$$

where J is the current density, n is the free electron concentration in the material, p is the hole concentration in the material, μ_n is the mobility of an electron in the material, μ_p is the mobility of a hole in the material, q is the charge of an electron ($-$ 1.6 × 10^{-19} C) and ε is the applied electric field.

Diffusion current is caused by the concentration gradient in the semiconductor when there is non-uniform concentration of charge particles in a semiconductor. The hole diffusion current density is given by the following expression:

$$J_p = -qD_p \frac{dp}{dx}$$

D_p is the diffusion constant of holes and $\frac{dp}{dx}$ is the variation in hole concentration with distance x, which is positive when the hole concentration increases with distance and is negative when hole concentration deceases with distance.

$$J_n = qD_n \frac{dn}{dx}$$

Dn is the diffusion constant of electrons and $\frac{dn}{dx}$ is the variation in electron concentration with distance x, which is positive when the electron concentration increases with distance and is negative when electron concentration deceases with distance.

Q.12. Explain the formation of a *P*-N Junction with no applied bias. What is barrier potential?

Ans. As already discussed, the *N*-type semiconductor is formed by introducing pentavalent dopant impurity atoms while the *P*-type semiconductor is formed by introducing trivalent dopant impurity atoms into the intrinsic semiconductor material. Moreover, electrons are the majority carriers in an *N*-type semiconductor while holes are the majority carriers in a *P*-type semiconductor.

P-N junction is formed by introducing the donor impurities on one side and acceptor impurities on the other side of a single crystal of a semiconductor.

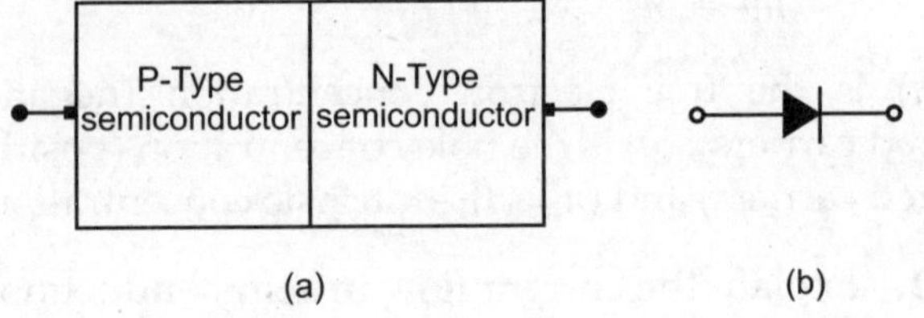

Fig. 9.2

Electrons (majority carriers) in the *N*-region and negatively charged ions in the *P*-region, near the junction repel each other. Similarly, holes in the *P*-region (majority carriers) and positively charged ions in the *N*-region, near the junction also repel each other. An effective potential of the order of few tenths of a volt, referred to as the contact potential or the barrier potential is developed across the depletion region. However, some of these holes and electrons have sufficient kinetic energy to overcome the contact potential and be able to pass through the depletion region. This results in a flow of electrons from the *N*-region to the *P*-region and flow of holes from the *P*-region to the *N*-region. This constitutes the majority carrier flow vector.

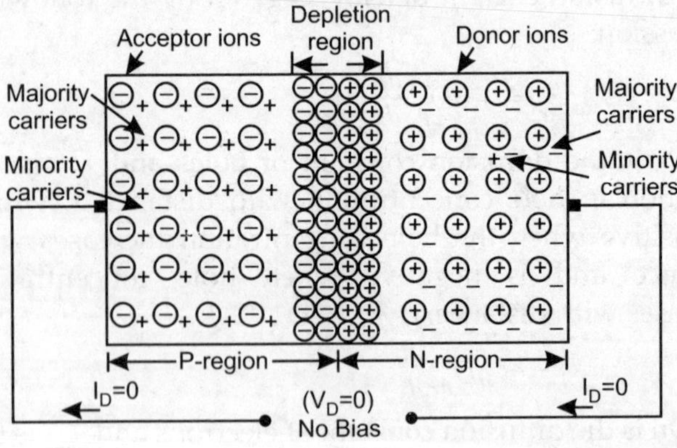

Fig. 9.3

Also, holes (minority carriers) that are present in the depletion region of the *N*-region will pass to the *P*-region. Similarly, electrons (minority carriers) that are present in the depletion region of the *P*-region will pass to the *N*-region. Figure-9.3 shows the *P*-N junction with no applied bias.

Q.13. Show the basic diode action diagrammatically when it is : (*i*) Unbiased, (*ii*) Reverse biased, (*iii*) Forward biased.

Ans. (*i*) Fig. 9.4 a shows a *P-N* junction diode when it is unbiased.

P-type material contains excess of holes (+ve charges) and negatively charged impurity atoms whereas *N*-type material contains excess of electrons and positively charged impurity atoms.

Under the conditions of no external voltage, electrons in *N*-type and negatively charged atoms in *P*-type repel each other.

Similarly, holes in *P*-type and positively charged atoms in *N*-type also repel each other. This results in electrons as well as holes moving away from junction, producing a region near the junction that is depleted of the mobile charges. This region is called DEPLETION REGION. Thus a sort of potential barrier appears across the junction representing the extent of depletion region.

(*ii*) When an external voltage is applied with *P*-type material negative and *N*-type positive [Fig. 9.4 b], the electrons which were originally repelled away from the junction, due to negatively charged atoms on *P*-side, are repelled further due to attraction force between positive voltage (applied externally) and electrons. Thus electrons as well as holes move further away from junction, further reducing the possibility of any conduction. This is termed as reverse bias.

(*iii*) When an external voltage applied with *P*-side (+ve) and *N*-side (-ve) is sufficient to overcome the barrier potential, [Fig.9.4 c], electrons and holes cross the junction and recombine. For every recombination of an electron and a hole, one electron enters the *N*-side from negative terminal of battery and one electron from an electron pair bond breaks up the bond to enter the positive terminal, creating a hole. Thus, current flow is due to flow of electrons in the outside circuit, flow of holes in *P*-type material and flow of electrons in the *N*-type material.

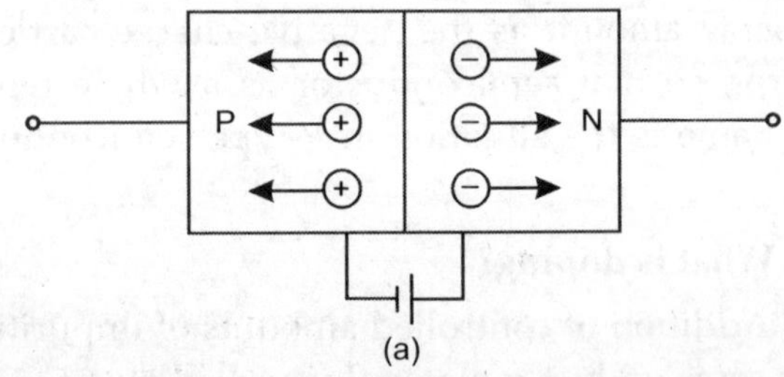

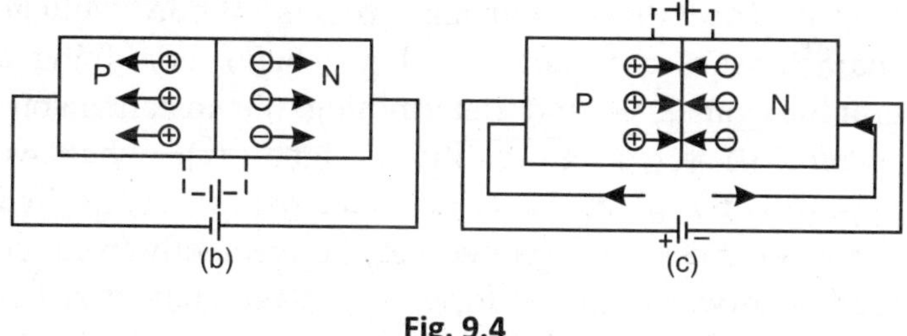

Fig. 9.4

Q.14. Briefly differentiate between ideal and practical diodes.

Ans. An ideal diode behaves like a switch that conducts current only in one direction from anode to cathode. An ideal diode acts as a short circuit when forward biased and as an open circuit when reverse biased. Thus, the resistance of the forward-biased diode is zero and the resistance of the reverse-biased diode is infinite.

In contrast to the ideal diodes, practical diodes do not conduct until a certain value of forward voltage is applied to them during forward bias state. This voltage referred to as the *cut-in voltage* or *knee voltage* or *threshold voltage* is of the order of less than 1V for semiconductor diodes. Also, the ON-resistance of the practical diode is not zero and varies from few ohms to few hundreds of ohms. In the reverse bias state, the practical diode differs from the ideal open switch as in this condition a small amount of current referred to as the reverse saturation current flows through the diode. Moreover, there is sharp increase in the reverse current when the applied reverse-bias voltage exceeds the reverse breakdown voltage.

Q.15. Sketch I_F–V_F characteristics (I_F and V_F are forward current and forward voltage respectively) of a semiconductor junction diode, both for Silicon and Germanium diodes. Indicate differences if any in the characteristic of the two types.

Ans. The (I_F) vs. (V_F) characteristics of a real (not the ideal) semiconductor *P-N* junction diode are shown in Fig. 9.5 (*b*) for a silicon diode and Fig. 9.5 (*c*) for a germanium diode.

When the diode is forward biased, there is always a minimum voltage that must be exceeded before there is sufficient conduction of current through the diode. This is known as the CUT-IN VOLTAGE and is 0.6 V for Silicon and 0.2 V for Germanium diodes. For all forward voltages greater than the cut-in voltage, there is a sharp rise in forward current *i.e.* a very small change in forward voltage causes a very large change in forward current. Infact, for all practical purposes, the voltage across the diode can be considered to be equal to the cut-in voltage irrespective of the forward current flowing through the diode.

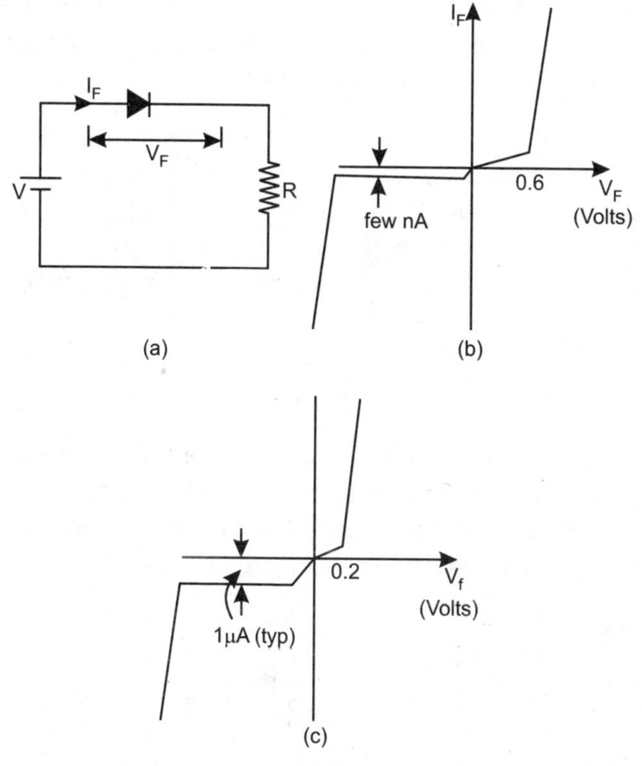

Fig. 9.5

In the reverse biased condition, the current that flows through the diode is the reverse leakage current. It may be a few nanoamperes in case of Silicon diodes and typically 1 μA for Germanium diodes. As the reverse bias voltage is increased further, a point comes where the junction breaks down and there is a steep rise in current which ultimately culminates in device getting damaged. This reverse bias voltage at which the break down occurs is known as PEAK INVERSE VOLTAGE of the diode.

Q.16. Draw the *I-V* characteristics and the superimposed DC load line for the semiconductor diode feeding a load resistance R_L as shown in Fig. 9.6 (*a*). Define STATIC and DYNAMIC resistance of the diode.

Ans. The *I-V* characteristics and the DC load line are shown in Fig. 9.6 (*b*). The equation of the DC load line is given by

$$V = V_F + IR_L \quad \text{or} \quad I = \left(-\frac{V_F}{R_L} + \frac{V}{R_L}\right)$$

This is the equation of a straight line with slope equal to $(-1/R_L)$ and Y-axis intercept of (V/R_L). Also if the voltage (V) is increased or decreased, the new load line, shifts upwards or downwards respectively parallel to the existing line as indicated by the dotted lines. The DC load line for a given load resistance determines the quiescent point. It tells us about the DC current, through the diode and the voltage (V_F) across the diode for a given input voltage (V) and lead resistance (R_L).

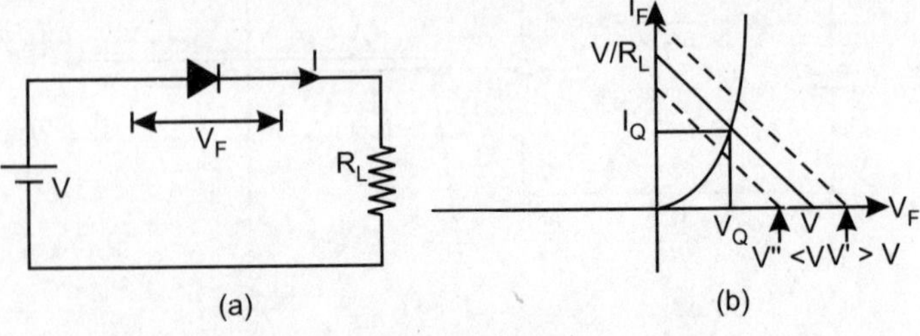

Fig. 9.6

The STATIC RESISTANCE (R) is the resistance offered by the diode to DC current. It is given by the ratio of (V_F) to (I_F) at the point where the DC load line intersects the characteristics. From Fig. 9.6 (*b*), $R = (V_Q/I_Q)$.

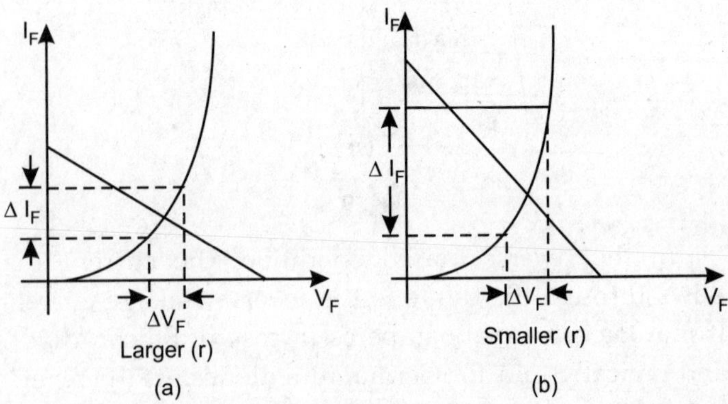

Fig. 9.7

The DYNAMIC RESISTANCE (r) is the resistance offered by the diode to AC signals. It is therefore given by ratio of a small change in diode's forward voltage to the corresponding change in diode's forward current around the operating point. We will notice that for larger values of average forward current, the dynamic resistance (r) reduces as the characteristics become more and more steep as the forward current increases (Fig. 9.7). The dynamic resistance (r) as a function of forward current is expressed by

$$r = \left(\frac{26}{I_F}\right) \text{ for silicon diodes ; } r = \left(\frac{52}{I_F}\right) \text{ for}$$

germanium diodes.

Q.17. With the help of diode's *I-V* characteristics, describe graphically the use of a semiconductor diode as a rectifier. Refer to Fig. 9.8. (*a*). (*V*) is the sinusoidal input signal with a peak to peak voltage of (2 V_m) and (R_L) is the load resistance, Assume that diode's dynamic resistance is negligible as compared to (R_L).

Ans. The diode current versus diode voltage characteristics are shown in Fig. 9.8 (*b*). For doing a graphical analysis and determining the instantaneous current (i) for a given input signal (v), what we require is the (i) versus (V) characteristics. The desired characteristics can be plotted very conveniently from the diode's forward current-forward voltage characteristics and the given value of (R_L)- The procedure is as follows :

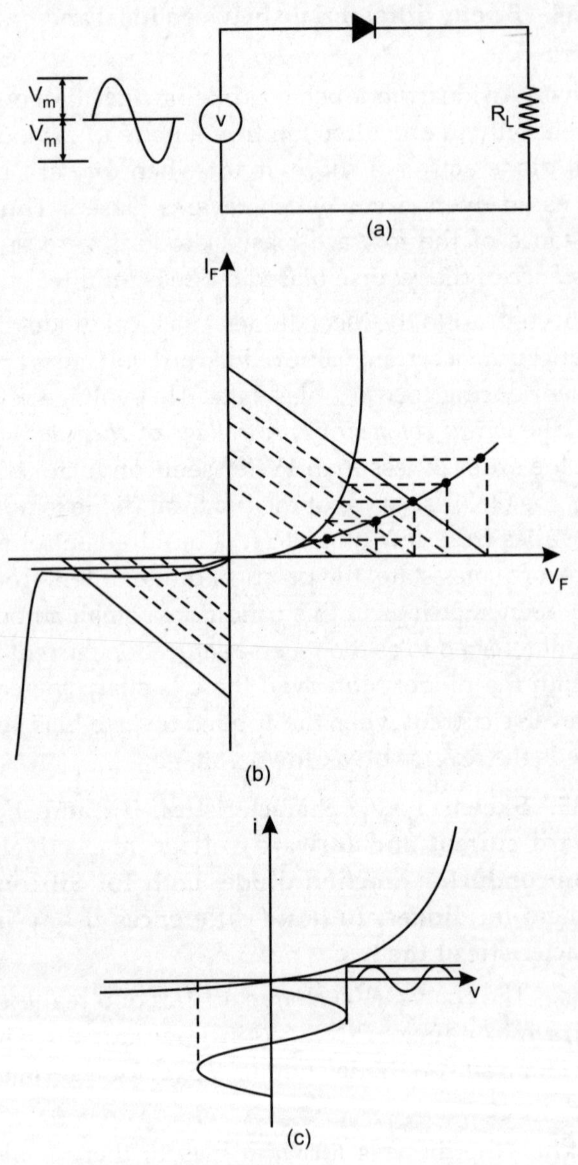

Fig. 9.8

Draw the load line for $v = V_m$ (the positive peak of the input signal). The *X*-axis and *Y*-axis co-ordinates are (V_m, 0) and (0, V_m/R_L) respectively. The intersection of the load line with the diode's forward characteristics gives the magnitude of current (I_m) corresponding to an input of (V_m). The point (V_m, I_m) thus lies on the (i) vs. (v) curve. Similarly, more points can be located by considering (v) to be within range $-V_m \le v \le V_m$ and drawing the corresponding load lines.

All load lines will have the same slope (= $-1/R_L$). These points then can be joined to get (*i*) vs (*v*) curve. Determination of output current waveform for a given input voltage waveform graphically is demonstrated in Fig. 9.8 (*c*). The diagram is self-explanatory.

Q.18. How do changes in ambient temperature affect the *I-V* characteristics of semiconductor junction diode? When changes in temperature affect both Silicon and Germanium diodes in the same proportion, why is a Germanium diode then said to be more strongly temperature dependent.

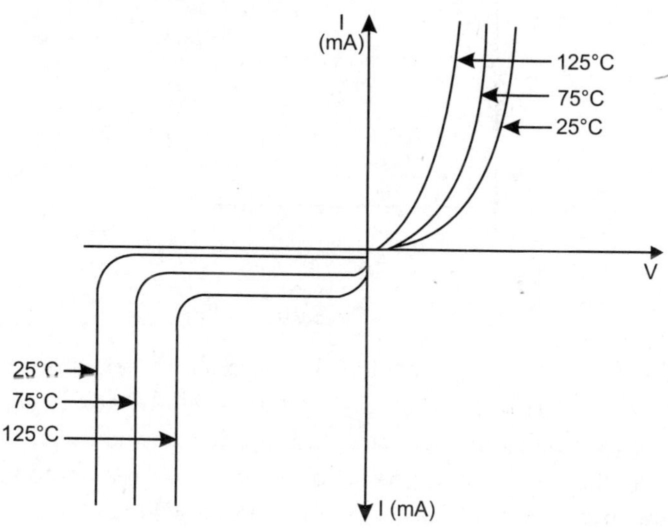

Fig. 9.9

Ans. Figure 9.9 shows the *I-V* characteristics of a semiconductor junction diode at an ambient temperature of 25°C. The cut-in voltage ($V\gamma$) is 0.6 volt if it is a Silicon diode and 0.2 volt if it is a Germanium diode. The reverse saturation current (I_0) is typically 1 nA for Silicon diode and 1 μA for a Germanium diode. Effect of temperature D on the diode current (*I*) is expressed by

$$I = I_0\left[e^{\frac{(11600\,V)}{T}} - 1\right]$$

for Germanium diodes

$$I = I_0\left[e^{\frac{(5800\,V)}{T}} - 1\right] \text{ for silicon diodes}$$

where (*V*) is voltage across the diode in volts and (*T*) is ambient temperature in degrees Kelvin. In fact, the diode's current-voltage relationship can be expressed by :

$$I = I_0 \times \left[e^{\frac{V}{\eta V_T}} - 1\right]$$

where $V_T = \frac{T}{11600}$, a quantity dependent upon temperature and measured in volts

η = constant. It is 1 for Germanium and 2 for Silicon

Substituting the values of (V_T), and (η), we get the expression for Silicon and Germanium diodes.

Effect on forward biased characteristics

The forward voltage changes with temperature at a rate of – 2.5 m *V*/°*C*.

That is, as the temperature increases, the forward voltage decreases for a constant current. Or in other words, for a constant forward voltage, the forward current increases with temperature. This appears to be in contrast with what is indicated by the expressions given above.

Although, the factors $\left[e^{\frac{(11600\,V)}{T}} - 1\right]$ and $\left[e^{\frac{(5800\,V)}{T}} - 1\right]$ reduce with increase in (T) for a constant (*V*) for a constant (*V*), there is always an overall increase in current as (I_0) increases with temperature by a comparatively larger proportion.

Effect on reverse biased characteristics

It is clear from the expressions given above that for (*V*) more negative than – 0.2 Volt or so, the reverse current remains practically constant at (I_0) called the reverse saturation current. (I_0) is strongly affected by temperature changes. It doubles for every 10°C rise in temperature for both Silicon and Germanium diodes. The reverse saturation current (I_0) at 75°C is thus 32 times of what it is at 25°C. So, an (I_0) of 1 nA (typical of Silicon diodes) becomes 32 nA which is still very small and thus ignorable. On the other hand, (I_0) of 1 μA (typical of Germanium diodes) becomes 32 μA which makes germanium diodes unusable at elevated temperatures.

Q.19. Define the PEAK INVERSE VOLTAGE (PIV) or (PEAK REVERSE VOLTAGE (PRV) REVERSE RECOVERY TIME (t_{rr}) FORWARD CURRENT, SURGE CURRENT and I^2t RATING.

Ans. The PEAK REVERSE VOLTAGE (abbreviated as PRV) or the PEAK INVERSE VOLTAGE (abbreviated as PIV) is the maximum allowable reverse voltage that can appear across the diode without causing it any damage. There are usually two different PIV ratings usually specified in case of diodes. One is the repetitive peak inverse voltage denoted by (V_R) or (V_{RRM}) and the other is the non-repetitive PIV denoted by (V_{RSM}). The non-repetitive PIV is obviously larger than the repetitive PIV. Diodes with PIV rating of 50 V to 1500 V are commercially available. High voltage diodes having a stack of semiconductor junctions are available in PIV ratings of tens of kilovolts.

The REVERSE RECOVERY TIME (t_{rr}) is the time that elapses before a forward biased diode returns to the reverse biased state after an input signal abruptly reverses the bias voltage. The (t_{rr}) specification infact tells about the switching speed of the diode. For instance, diode IN 3899 with a (t_{rr}) specification of 150 ns is a high speed diode also known as fast recovery diode.

FORWARD CURRENT is also specified in two different ways. One is the maximum average rectified current that can pass through the diode keeping its power dissipation within limits. This is denoted by I_F (av) or (I_F). The other is the peak forward current (I_{FSM}) or the non-repetitive forward current the diode can withstand without getting damaged. Also referred to as SURGE CURRENT, it is significantly higher than the maximum average rectified current.

(I^2t) RATING (AMPERE SQUARED SECONDS) tells us about the subcycle surge current; capability of the diode when used as rectifier. It is usually specified for one complete cycle of 50 Hz operation. It can be calculated from one cycle peak surge current rating of the diode from

$$I^2\,t \text{ rating} = \frac{(\text{One cycle peak surger rating})^2}{2} \times 20{\times}10^{-3}$$

The full cycle time period here is taken to be 20 ms corresponding to AC line frequency of 50 Hz.

Q.20. Why do we need to connect diodes in series or in parallel and what are the precautions to be observed while doing so?

Ans. Diodes are connected in series to enhance the PIV rating beyond the capability of individual diodes. If (N identical diodes each having a PIV rating of (V) volts are connected in series, the overall PIV rating of the series stack in the ideal case would be (N.V) volts. However, this assumes that the applied reverse voltage is equally divided amongst the series connected diodes. To ensure an equal division of voltage, which is desirable for a series stack, the individual diodes to be connected in series should have closely matched reverse characteristics. Another technique to force an equal division of voltage is to use R-C networks across each of the series connected diodes as shown in Fig. 9.10 (a). Resistors could however be dispensed with in case of high frequency operation, for instance, the fast recovery rectifiers used in switched mode power supplies. In case of low frequency operation, both resistor and capacitor would control the division of voltage.

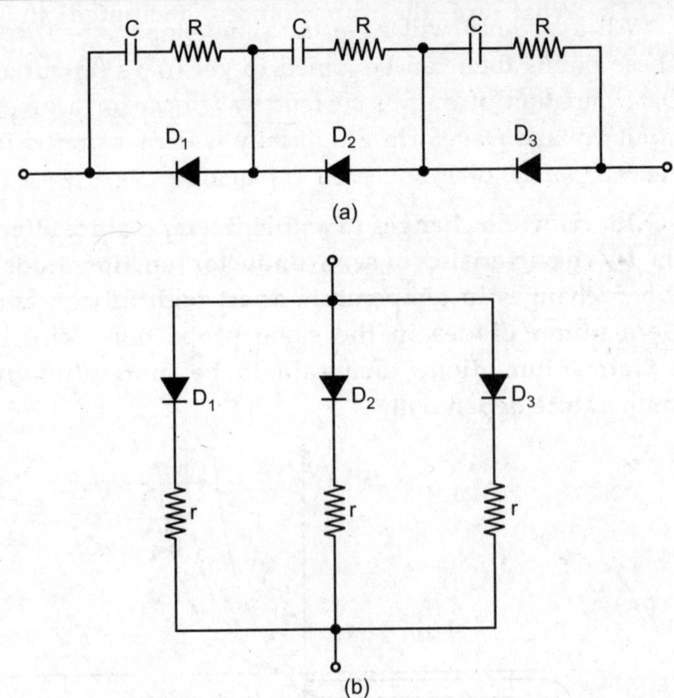

Fig. 9.10

Diodes are connected in parallel to increase the current carrying capability of the parallel stack beyond the capability of the individual diodes. In this case, it is essential to have an equal division of current amongst the individual diodes. An unequal division of current primarily due to mismatch in the forward characteristics of the parallel connected diodes can lead to destruction of one and then subsequently the other diodes in the stack. The diode that has the least forward voltage drop draws the largest current initially. The diode heats up and this further reduces the forward voltage drop. The current drawn by this diode increases further leading to more heating of the diode junction. This process is regenerative and culminates in destruction of the diode. Destruction of one diode puts more strain on the other diodes of the stack and they too become vulnerable to the undesired regenerative effect described above. Thus, the diodes to be connected in parallel should have closely matched forward characteristics. An equal division of current can be forced by using series resistors or balancing inductors with each of the parallel connected diodes as shown in Fig. 9.10 (b).

In case of series connection, the resistance (R) should be much smaller than the reverse biased resistance of the individual diodes in the stack. In case of parallel connection, the resistance (r) should be much larger than the forward biased resistance of the individual diodes.

Q.21. What are the capacitances attributed to a semiconductor P-N junction diode. Which capacitance predominantly determines the switching speed of the diode?

Ans. A semiconductor P-N junction diode is associated with capacitive effects when reverse biased and also when it is forward biased. However the two capacitive effects are different. When the diode is reverse biased, the capacitance of the diode originates from the concentrations of positive and negative charge carriers on the *P* and *N* sides constituting the two parallel plates separated by the depletion region acting as the dielectric. This is known as TRANSITION CAPACITANCE or SPACE CHARGE CAPACITANCE and typically varies from 20 pF for no external bias to about 5 pF for a reverse bias of about 5 volts. The voltage variable nature of the transition capacitance is made use of in varactor diodes.

In a forward biased diode, the predominant capacitance effect is due to minority carriers outside the depletion region. This is known as DIFFUSION CAPACITANCE and is comparatively much larger as a small change in applied forward bias voltage causes a large change in stored minority carrier charge. Typically, it is few hundred pFs and could be even as large as few microfarads for heavy forward bias. Transition capacitance is negligible as compared to diffusion capacitance in forward biased diode where as it is the transition capacitance that is predominant when the diode is reverse biased. Both diffusion and transition capacitances affect the switching speed of the diode When the diode is used at very high switching speeds, the .mail toward biased dynamic resistance of the diode together with very large diffusion capacitance could also adversely affect the performance.

Q.22. What is Avalanche-multiplication?

Ans. The forward current flowing in a. P N junction diode consists mainly of majority carriers and smaller number of minority carriers. The current contributed by minority carriers saturates after sometime amen there are no more minority carriers left. But if the reverse voltage across the diode is increased to a large value, a stage may come where a number of charge carriers from valence band break up their covalont bonds and reach the conduction band to contribute current. This breaking up of bonds is so violent that it knocks out more charge carriers from adjacent atoms. These newly freed charge carriers further knock out more of them the process being regenerative. Thus, a stage comes where minority carriers exceed the majority carriers There is a sudden increase in current. If diode is not able to sustain this heavy current, it. may be spoiled. The voltage at which the current starts rising all of a sudden is called avalanche breakdown voltage and process being termed as avalanche multiplication.

Q.23. What is zener breakdown?

Ans. As explained before, when the reverse voltage across a P-N diode is increased, a stage comes when valence electrons break up their covalent bonds and reach the conduction band to constitute current resulting in a sudden increase in current. This is called zener breakdown. It is different from avalanche breakdown in the respect that in the latter case, the voltage is strong enough to cause such a violent breaking up of bonds that the freed electrons may knock out more electrons from adjacent atoms, the process being regenerative.

Q.24. What is zener diode? Draw its circuit symbol and characteristics?

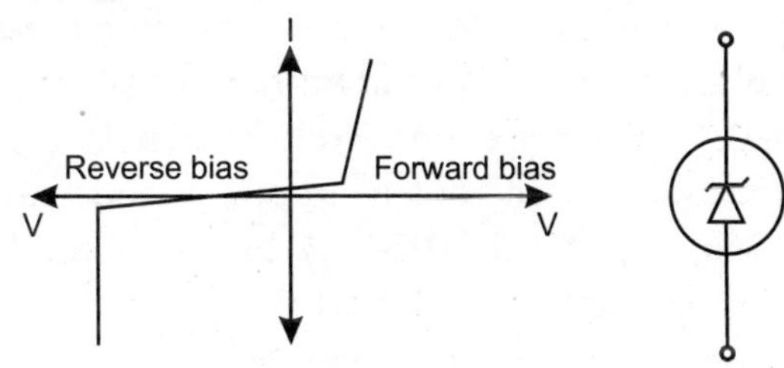

Fig. 9.11

Ans. A *P-N* junction diode which is made capable of sustaining heavy current of the zener breakdown region is called a zener diode. Zener diodes are operated in the reverse biased region.

It is clear from the characteristics of Fig. 9.11 that when a diode is operated in the reverse region, at a certain voltage V_{BO}, the current increases all of a sudden. This voltage is called zener breakdown voltage.

It is the characteristic of the zener diode that as the current increases, its resistance decreases keeping almost a constant voltage across its terminals.

Q.25. Shown how a zener diode may be used to regulate the voltage?

Ans. A simple zener diode voltage regulator circuit is shown in Fig. 9.12

The output voltage in the zener diode regulator circuit of Fig. 9.12 is the zener breakdown voltage of the zener diode irrespective of input voltage fluctuations provided that the zener diode always remains biased by current that is greater than a certain specified minimum value that keeps (typically 10 mA) and less than the maximum value that keeps the zener diode dissipation within acceptable limits.

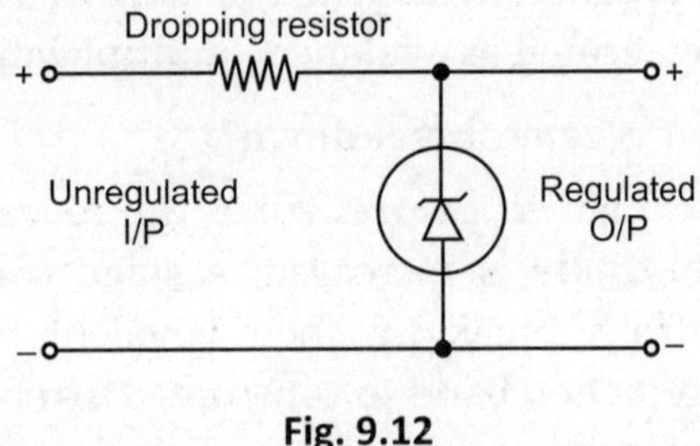

Fig. 9.12

Whenever input voltage fluctuates, change in input voltage appears across the dropping resistor (R) and the output voltage remains constable.

Q.26. What is a tunnel diode? How does it differ from an ordinary diode?

Ans. A funnel diode is also a *P-N* junction diode but with the difference in the impurity concentration in *P* and *N*-type materials. In an ordinary semiconductor diode, the impurity concentration may be about 1 part in 10^8 parts. The depletion width, which is inversely proportional to square root of impurity concentration, is about 5 microns (= 5 × 10^{-6} m). In tunnel diode, this impurity concentration is 1 part in 10^3 parts and the depletion width is about 10^{-6} cm.

Q.27. What is tunneling phenomenon?

Ans. In an ordinary diode (*P-N* junction diode), in order to make the diode conduct in the forward direction, a voltage sufficient to overcome the barrier potential is needed to be applied. This voltage is about 0.6 V in case of Silicon diodes. It has been observed that if the depiction region is too narrow, the charge carriers instead of climbing up the potential barrier may pierce through the junction barrier with little or no energy applied. This phenomenon is called tunneling. Thus a tunnel diode may act as a very good conductor in the forward direction for a very small applied voltage, of the order 0.1 V or less.

Q.28. Draw the tunnel diode circuit symbol and its *V-I* characteristic. Explain in brief the behaviour of the tunnel diode from the characteristics.

Ans. From the characteristics (Fig. 9.13), it is clear that both in the forward as well as backward direction, the tunnel diode responds with a huge current for a very small applied voltage. This is due to tunnelling phenomenon.

In the forward direction, the current reaches maximum value I_P (called peak point) at a voltage V_P (peak voltage).

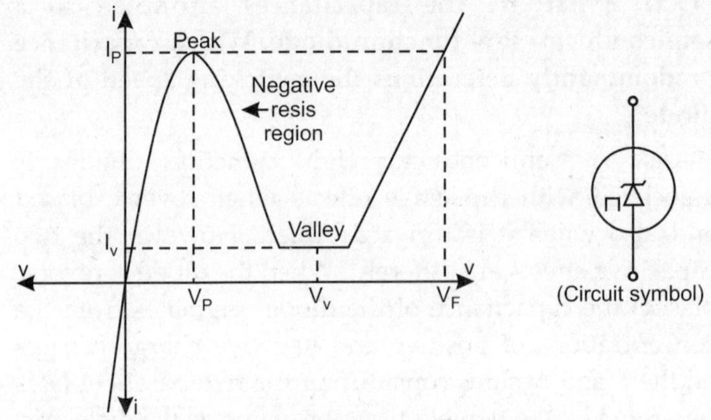

Fig. 9.13

At this point di/dv is zero. Afterwards, the current starts decreasing with the increase in voltage till a point (called valley) is reached where again, di/dv is zero. After this, current starts increasing again with increase in voltage and a current I_p may be reached at a voltage V_F.

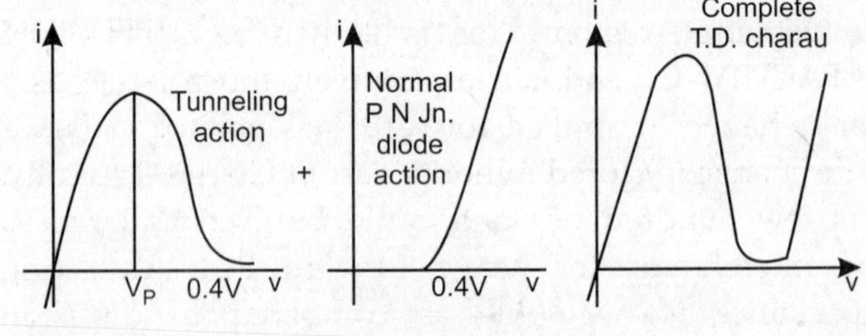

Fig. 9.14

These characteristics may be considered to be composed of following two characteristics, one due to normal *P-N* junction diode action and other due to tunnelling phenomenon (Fig. 9.14).

Due to tunnelling phenomenon, the decrease in current from I_p afterwards, may be explained on the basis that there are no more carriers available for tunnelling so that a reduction in current results with increase in voltage. The current approaches zero for a forward voltage of about say 0.4 or 0.5 V. But then, the normal P-N junction effect starts. The current due to normal P-N junction diode adds to the current due to the tunnelling to give a valley point as shown. The characteristics exhibit a negative resistance region from peak to valley.

Q.29. What are the materials used for making tunnel diodes? Which is the most commonly used one and why?

Ans. Tunnel diodes may be constructed either from Germanium or Galium Arsenide (GaAs). Silicon is not used. This is due to the reason that the ratio I_p/I_v is maximum in Galium Arsenide (=15) and comparatively smaller for Germanium (= 8), but very small for silicon (~3). In addition to this, Galium arsenide has maximum voltage swing, $V_F - V_P = (= 1\text{ V})$. A voltage swing of 0.45 V is common with Germanium tunnel diodes. Considering the above stated reasons, Gallium arsenide is the most commonly used semiconductor material used for making tunnel diodes.

Q.30. What is a negative resistance region, a negative resistance device?

Ans. According to Ohm's law, resistance is given by voltage divided by current. For same resistance, an increase in voltage should result in corresponding increase in current.

If the current shows a tendency to decrease when the voltage is increased, then the device is said to exhibit negative resistance.

A device, whose characteristics exhibit a negative resistance region, is called Negative Resistance Device. Tunnel diode is one of the examples of negative resistance devices.

Q.31. What is a tunnel rectifier?

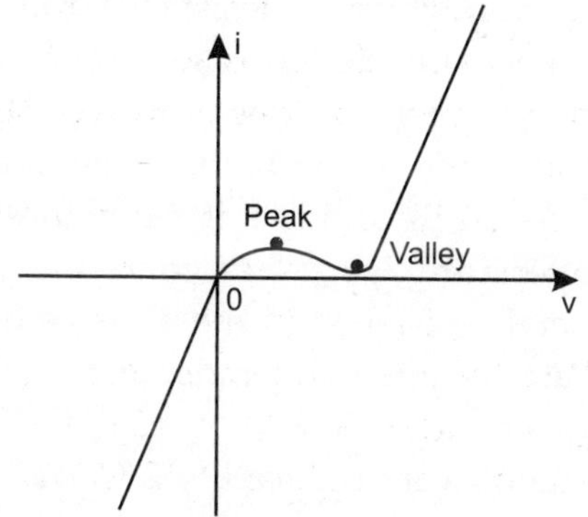

Fig. 9.15

Ans. A tunnel rectifier is a tunnel diode used as a rectifier. When used as rectifier, it is operated in the backward region and its forward characteristics are modified to have small value of peak current (almost zero). In other words, the peak is suppressed. The characteristics acc shown in Fig. 9.15.

Due to operation in the backward region, it is also called backward diode.

Q.32. What are the advantages of a tunnel rectifier over an ordinary semiconductor diode rectifier?

Ans. A tunnel rectifier responds with a huge current for a very small applied voltage in the reverse direction *i.e.*, the resistance offered is very low (of the order of 5 ohms or so).

An ordinary rectifier may start conducting after a voltage (= 0.6 V) has been applied. Thus, when a signal having an amplitude of say 0.1 V is applied to an ordinary rectifier for rectification, it would not start even throughout the cycle of applied input resulting into a poor rectification efficiency. On the other hand, a tunnel rectifier responds with a very high efficiency.

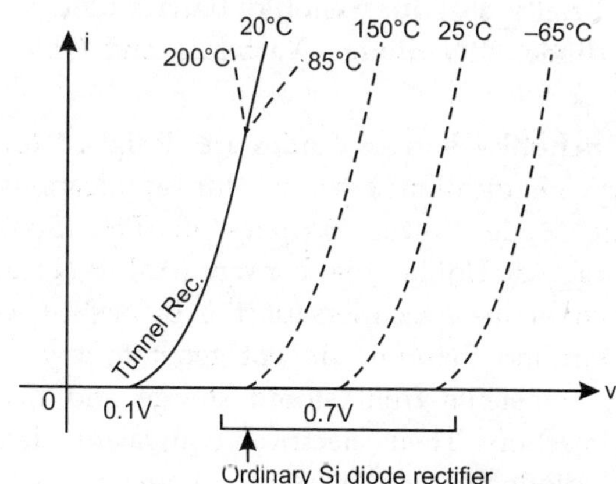

Fig. 9.16

In addition to this, tunnel rectifier undergoes a very small change in its characteristics due to change in temperature whereas an ordinary diode may undergo drastic changes in its characteristics with same change in temperature. Both of the above explained facts are illustrated diagrammatically in Fig. 9.16.

Q.33. 'Tunnel diode is a voltage controllable device' Comment.

Ans. From the tunnel diode characteristics, it is clear that the current is single valued function of voltage, *i.e.* for one value of voltage, there is only one current. That is -why it is called a voltage controllable device. Apart form being a single valued function of voltage, the curve is multivalued function of current, *i.e.* for one current, there are more than one voltages. For example, peak current I_p can be achieved at V_P as well as at V_F. Thus, a tunnel diode exhibits *V-I* characteristics that are single valued function voltage and a multivalued function of current.

There is another category of devices whose characteristics are single valued function of current and multivalued function of voltage. To distinguish them from the category of tunnel diodes, they are called current controllable devices UJTs, SCRS, Avalanche transistors etc. fall in this category.

Q.34. What are the advantages of tunnel diodes?

Ans. The advantages of tunnel diodes are :

(*i*) Low noise, low cost, simplicity.

(*ii*) It is immune to environmental changes.

(*iii*) It can be used as a high speed switch. Switching times of the order of a nanosecond are common. Switching times as small as 50 psec could be achieved.

(*iv*) Power consumption is low.

Q.35. Briefly describe Schottky barrier diodes, Point contact diode, PIN diodes, Varactors and Hot carrier diodes.

Ans. Schottky barrier diodes are designed for high efficiency rectification essential for applications like Switched Mode Power Supplies (SMPS), Switching regulators etc. Unlike the conventional P-N unction diode rectifier, these rectifiers function by majority carrier conduction and therefore do not generate any reverse recovery transients from stored charge and minority carrier injection. Their electrical equivalent circuit is an ideal diode in parallel with a known capacitance equivalent of junction capacitance resulting in a very fast recovery performance and low conduction voltage. Their use substantially improves the efficiency of the switching circuits. Its construction and operating principle is same as that of e hot carrier diode.

Point contact diodes are primarily intended for RF applications due to their extreme small internal capacitance considerably less than that of a junction diode designed for same circuit application?

A PIN diode is formed by diffusing heavily doped P^+ and N^+ regions into an almost intrinsically pure Silicon. In practice, it is impossible to obtain intrinsically pure material, the I-layer may to considered to be a lightly doped N-region. Characteristics of PIN diode are primarily determined by the semiconductor nature of the chip especially that of I-layer. PIN diodes are used in RF switches, attenuators and various types of phase shifting devices.

Varactors are used as voltage variable capacitors. They depend upon change in capacitance that occurs across their depletion layers when the reverse bias voltage is varied. Also mown as VARICAPS these are designed to provide various capacitance ranges from few *pF* to 100 *pF* or so. In different varactor types, the minimum and maximum capacitance may vary, the maximum to minimum capacitance ratio is typically 2.5 to 3 in commercially available types.

Hot carrier diode is a Metal-to-Semiconductor majority carrier conducting device with a single rectifying junction. The carriers are high mobility electrons in an *N*-type semiconductor material. These are particularly useful as mixers and detectors at VHF and higher frequency ranges. Notable features include its higher operating frequency and lower conduction voltage. It is both mechanically as well as electrically superior when compared to a point contact diode.

Q.36. Why are power diodes mounted on heat sinks for thermal management?

Ans. Power diodes are designed to operate at high power levels. The power dissipated in a P-N junction diode is the product of its forward biased voltage, which is 0.7 volt in case of Silicon diodes and 0.2 to 0.3 volt in the case of Germanium diodes, and the forward current. The P-N junction diode temperature is a function of the ambient operating temperature and junction-to-ambient thermal resistance. The junction-to-case and case-to-ambient thermal resistances depend upon diode construction and package. Higher the thermal resistances, higher will be the temperature of the diode junction for a given ambient temperature. By mounting the diode on a suitable heat sink, we are effectively reducing the case-to-ambient thermal resistance, which is the predominant component of overall thermal resistance from junction to ambient. The heat sink design is so chosen as to ensure that the junction temperature does not exceed the absolute maximum value even at the highest operating ambient temperature. As an illustration, if the highest operating ambient temperature were 50°C and the junction to ambient thermal resistance in the absence of heat sink were 10°C/Watt; the junction temperature for 20 watts of dissipated power will be $50 + 20 \times 10 = 250$°C. If the overall thermal resistance were reduced to 2°C/Watt by using a heat sink; the junction temperature would rise only to 90°C. If the power diode had absolute maximum junction temperature rating of 200°C. it would get damaged if used without heat sink.

Q.37. 'Transistors are current operated devices whereas tubes are voltage operated devices' Comment.

Ans. The amplification factor (μ) of a tube is the ability of the grid voltages and the plate voltage to produce an equal change in the output current. Thus. the voltage in the output is directly related to the input voltage. It is due to this reason that tubes are considered as voltage operated devices. Similarly, the characteristics of transistor are such that output current is a function of input current. That is why they are called current operated devices.

Q.38. What is a transistor? Show the transistor action for the N-P-N and P-N-P transistor.

Ans. A transistor may be thought of as a semiconductor device that is formed when two P-N junction diodes are tied together.

When a thin *P*-type material is sandwiched between two thick *N*-type materials, we get an *N-P-N* transistor. When a thin *N*-type material is sandwiched between two thick *P*-type materials, we get a *P-N-P* transistor.

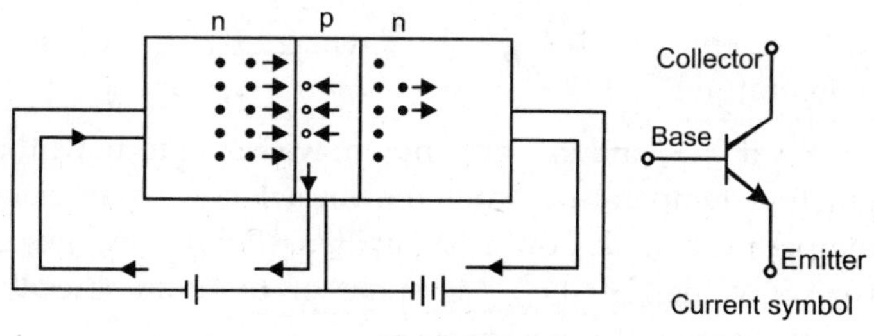

Fig. 9.17

An *N-P-N* transistor alongwith its circuit symbol is shown in Fig. 9.17.

When the emitter base junction is forward-biased, normal diode action takes place. Electrons (from *N*-type emitter) and holes (from *P*-type base) cross the junction and tend to recombine. Since base region is extremely thin, therefore, there are only a small number of holes available for recombination. All those electrons that do not recombine may be powerfully attracted towards *N*-type collector, if the second junction (between *P*-type base and *N*-type collector) is reverse biased (with *N*-type collector positive). This constitutes a current.

The action in a *P-N-P* transistor is exactly similar to that in the *N-P-N* transistor with the difference that it is the holes that act as charge carriers within the semiconductor material.

Q.39. How do you distinguish between

(*i*) β and α ; (*ii*) h_{FE} and h_{fe} (*iii*) I_{CBO} and I_{CEO}?

How do you calculate beta (β) if alpha (α) is given and I_{CEO} is known?

Ans. (*i*) **Beta** (β) is the current gain of the transistor from base to the collector. It is a general term for current gain. It may also be defined as the current gain of the transistors when used in common emitter (*CE*) configuration. **Alpha** (α) is a general term for current gain of the transistor when used in common base (*CB*) configuration. It is the current gain from emitter to collector terminals as in a common base configuration, input current flows in the emitter terminal and output current flows in the collector terminal.

(*ii*) h_{FE} is the DC current gain from base to collector whereas h_{fe} is the AC current gain from base to collector.

(*iii*) I_{CBO} is the leakage current from collector to base when emitter is open whereas

I_{CBO} is the leakage current from collector to emitter when base is open.

(β) and (α) are related to each other by the following expression :

$$\beta = \left(\frac{\alpha}{1-\alpha}\right)$$

I_{CEO} can be calculated from I_{CBO} only if h_{FE} at a base current equal to I_{CBO} is given. In that case,

$$I_{CEO} = (h_{FE} \times I_{CBO})$$

Q.40. Derive an expression for the power gain to be provided by transistor *Q* of Fig. 9.18.

Ans. Input power = Input voltage × Input current

Input voltage $= I_b R_{IN}$

Therefore, Input power $= I_b R_{IN} \times I_b = I_b^2 R_{IN}$

Output power = Output voltage × Output current

$$= I_C R_C \times I_C$$

$$= \beta I_b R_C \times \beta I_b$$

$$= \beta^2 I_b^2 R_C$$

$$\text{Power gain} = \frac{\text{Output}}{\text{Input power}}$$

$$= \left[\frac{\beta^2 I_b^2 R_C}{I_b^2 R_{IN}}\right]$$

$$= \beta^2 \frac{R_C}{R_{IN}}$$

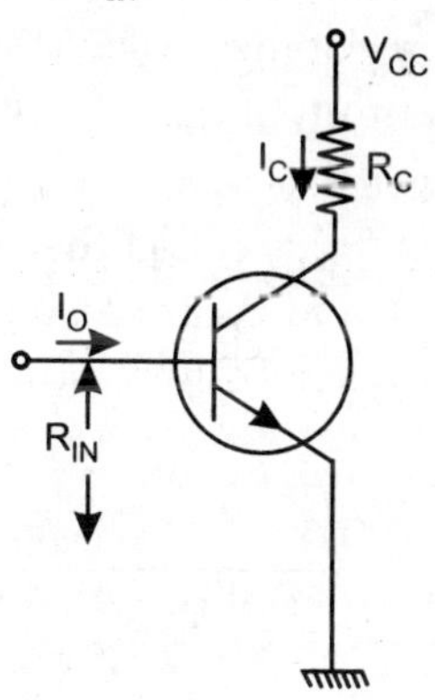

Fig. 9.18

Q.41. The input impedance, as illustrated in the previous question, affects the power gain provided by the transistor. How is it then that input impedance parameter does not figure in the transistor manuals?

Ans. The input impedance parameter does figure in

the usual list of parameters. However, it is not given as such as it comprises of a number of components. It is usually specified as a maximum base to emitter voltage (V_{be}) under specified input current conditions.

Q.42. What do you mean by (f_T) of a transistor? What is its significance. Why is it called unity gain crossover frequency?

Ans. The gain of a transistor falls at higher frequency values. The gain falls at a rate of 6 dB per octave. The frequency at which gain falls to unity (or 0 dB) is called unity gain crossover frequency and is denoted by f_T (Fig. 9.19). The gain bandwidth product at this frequency is (f_T). This gain bandwidth product remains unaltered throughout the 6 dB/octave slope as decrease in gain is accompanied by an increase in bandwidth. Thus (f_T) is the gain bandwidth product and is indicative of high frequency response of the transistor.

Q.43. What is the significance of thermal resistance of a power transistor? How do you define (θ_{J-C}) and (θ_{C-A}).

Ans. The thermal resistance is one of the parameters that indicates the power handling capability of a transistor. Quantitatively, it is given by the ratio of rise in transistor junction temperature with respect to some reference to the power dissipated. Obviously, thermal resistance is measured in °C/Watt.

There is another parameter called maximum junction temperature (more precisely, it is a transistor rating) which tells us as to what temperature of the transistor junction, the device will retain its properties. An estimate of transistor junction temperature can be made only when the thermal resistance parameter is also known in addition to the quantum of power dissipated.

One of the thermal resistances encountered with power transistors is (θ_{J-C}) called thermal resistance from junction to case. If θ_J and θ_C are respectively the junction and case temperatures in °C for a power dissipated equal to P watts, then

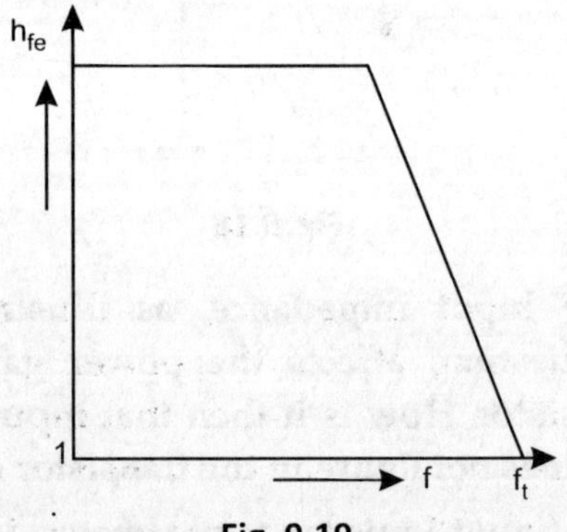

Fig. 9.19

$$\theta_{J-C} = \left[\frac{\theta_J - \theta_C}{P}\right]$$

The above expression can be used to obtain the value of (θ_J) from known values of (θ_{J-C}), (θ_C) and (P).

Another thermal resistance normally encountered is (θ_{C-A}). It is case-to-free air thermal resistance. Mathematically,

$$\theta_{C-A} = \left(\frac{\theta_C - \theta_A}{P}\right) \text{ ; where } \theta_A = \text{Ambient temperature.}$$

So, if θ_{C-A} and θ_{J-C} are known, we can determine the junction temperature from the knowledge of ambient temperature and power dissipated. Below are given typical values of θ_{C-A} for some of the commercially available transistor packages.

TO - 3 : 30°C/W   TO - 5 : 150°C/W

TO - 18 : 300°C/W   TO - 39 : 150 C/W

TO - 66 : 60°C/W   TO - 36 : 25°C/W

TO - 8 : 75°C/W.

Q.44. Is it theoretically possible to realise the maximum dissipation capability of a transistor. If so, under what conditions can we do so? What are the practical limitations.

Ans. First of all, it can be theoretically possible only if the ambient temperature is the one at which the transistor's maximum dissipation capability has been specified.

Secondly, there should be no temperature differential between the transistor case and the ambient. That is, the thermal resistance between transistor case and free air is zero which in turn necessitates that the transistor is mounted on an infinite heat sink. And that of course is practically impossible.

Consider the case of 2N3055 with junction temperature (max) rating of 200°C and maximum dissipation capability of 115 watts (at 25°C Ambient) and a thermal resistance from junction to case of I.5°C/watt, it is clear that 115 watts of power can be realised only if case temperature is at ambient temperature as

$$1.5 = \left(\frac{\theta_J - \theta_C}{115}\right) \text{ gives } \theta_C = 25°C$$

θ_J = Junction temperature

θ_C = Case temperature

Also θ_{C-A} = Thermal resistance from case to air

$$= \left(\frac{\theta_C - \theta_A}{115}\right)$$

If 115 watts of power is to be realised

where θ_A = Ambient temperature

Substituting for θ_C and θ_A in the above expression, $\theta_{C-A} = 0$.

Q.45. Define and interpret the following transistor ratings and specifications : V_{CEO}, V_{CBO}, V_{EB}, $I_{C(max)}$ $P_{D(max)}$ t_{ON}, t_{OFF}, C_{ob}.

Ans. V_{CEO} is defined has the maximum collector to emitter voltage with base open. It. is the worst case collector emitter breakdown voltage and is a very important specification particularly in switching transistors such as-those used for designing switched mode power supplies. This rating is often specified in different ways. Quite often V_{CER} (collector emitter breakdown voltage with a specified resistance usually 100 to 200 ohms connected between base and emitter) and V_{CES} (collector emitter breakdown voltage with base shorted to emitter) are also given in addition to V_{CEO} for switching transistors particularly designed for use in switching supplies. V_{CES} is the lowest of all and if design is such that V_{CES} rating is not exceeded, the possibility of collector emitter junction breaking down gets eliminated.

V_{CBO} is the collector to base breakdown voltage with emitter open. V_{CBO} determines the worst case output breakdown voltage when the transistor is connected in common base configuration. V_{EB} is the maximum base emitter voltage of the transistor.

$I_{C(max)}$ is the peak collector current rating of the transistor, either continuous or pulsed.

$P_D(max)$ is the maximum power dissipation capability of the transistor. It is usually specified at a given ambient temperature and it should be derated at higher temperatures as per the derating curve supplied by the manufacturer. Also, in the absence of the heat sink, the dissipation capability will be much less. The actual dissipation capability can be determined from the known values of maximum operating junction temperature, ambient temperature and the thermal resistances involved.

The turn-on time (t_{ON}) is the sum of delay time (t_d) and rise time (t_r). Delay time is the time taken by the collector current to change from its OFF-state value to 10 percent of its final ON-state value after the turn-on pulse is applied at the base. The rise time (t_r) is the time taken for the collector current to rise from 10 percent to 90 percent of its final ON-state value.

The turn-off time (t_{OFF}) is the sum of storage time (t_s) which is the time for the collector current to change from its final ON-state value to 90 percent of its final ON-state value and the fall Lime (t_f) which is the time for the collector current to change from 90 percent of its final ON-state value to 10 percent of its final ON-state value.

C_{ob} is the output capacitance of the transistor when connected in common base configuration. It is particularly useful in switching and high frequency applications.

Q.46. Sketch the typical input and output characteristics of a bipolar junction transistor when connected in (*i*) common emitter configuration and (*ii*) common base configuration, Also, give the figure of the circuit of NPN and PNP transistors for both for configurations.

Ans. Input Characteristics : The input characteristics of a bipolar junction transistor connected in common emitter configuration are the (I_B) versus (V_{BE}) characteristics. In case of common base configuration, they are a plot of (I_E) versus (V_{BE}). In both cases, the characteristics resemble those of a forward biased P-N junction diode with the only difference that the input characteristics of a junction transistor do depend to some extent on the collector-emitter voltage (in case of common emitter configuration) or collector-base voltage (in case of common base configuration). The input characteristics for the two cases are shown in Fig. 9.20 (*a*) and (*b*).

Output Characteristics : The output characteristics of a bipolar junction transistor when connected in common emitter configuration are a family of characteristic curves of (I_C) versus (V_{CE}) with (I_B) as the control parameter. The collector current (I_C) as a function of base current (I_B) is expressed by

$$I_C = \beta I_B + (\beta + 1) I_{CBO}$$

Fig. 9.20 (*c*) shows the family of output characteristics for a transistor in common emitter configuration.

When the transistor is connected in common base configuration, the family of output characteristics are a plot of (I_C) versus (V_{CB}) with (I_E) as the control parameter (Fig. 9.20 (*d*)) The collector current in this case is expressed by

$$I_C = \alpha I_E + I_{CBO}.$$

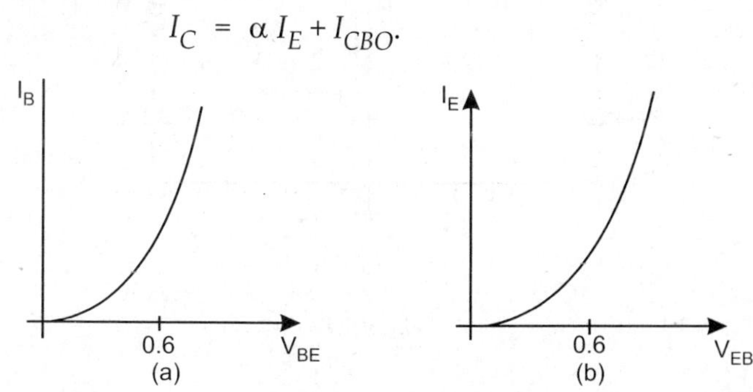

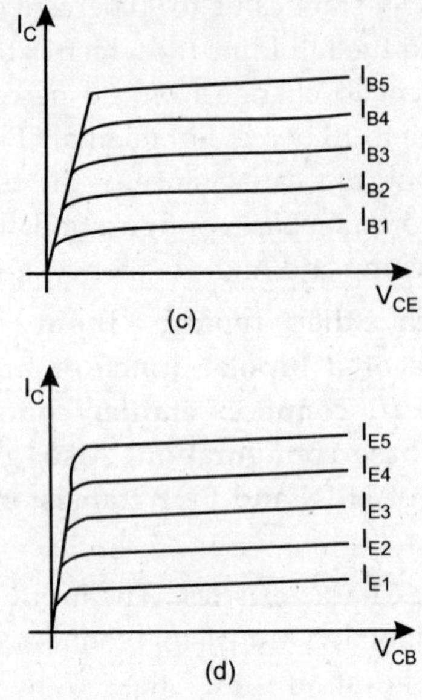

Fig. 9.20

Fig. 9.21 (*a*) and (*b*) shows the NPN and PNP transistor in the common base configuration respectively.

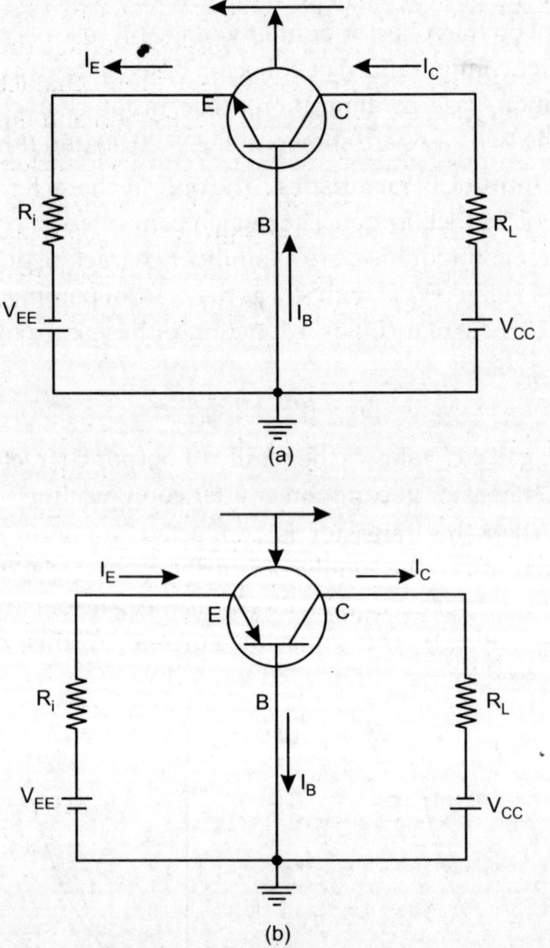

Fig. 9.21

Fig. 9.22 (*a*) and (*b*)shows the NPN and PNP transistor in the common emitter configuration respectively.

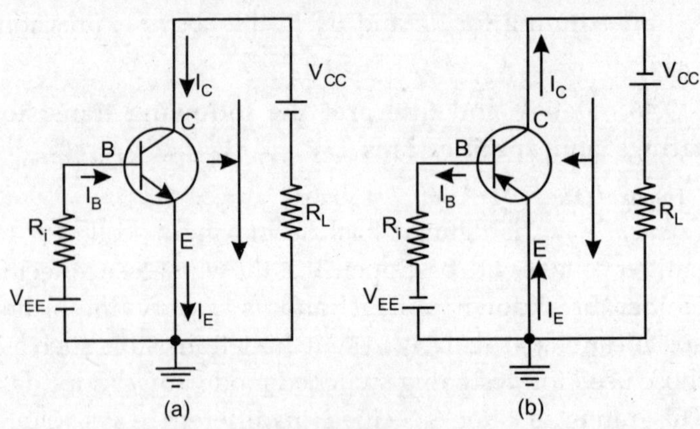

Fig. 9.22

Q.47. Describe the common-collector configuration with the circuit diagrams of NPN and PNP configurations. Also, sketch the typical input and output characteristics of a bipolar junction transistor when connected in common collector configuration.

Ans. In the common-collector configuration, the collector terminal is common to both the input and output sections. This configuration is also known as emitter follower configuration. The configuration is similar to common-emitter configuration that has been discussed before however, the output is taken from emitter terminal and not the collector terminal. Fig 9.23 (*a*) and (*b*) show common-collector configuration using NPN and PNP transistors respectively.

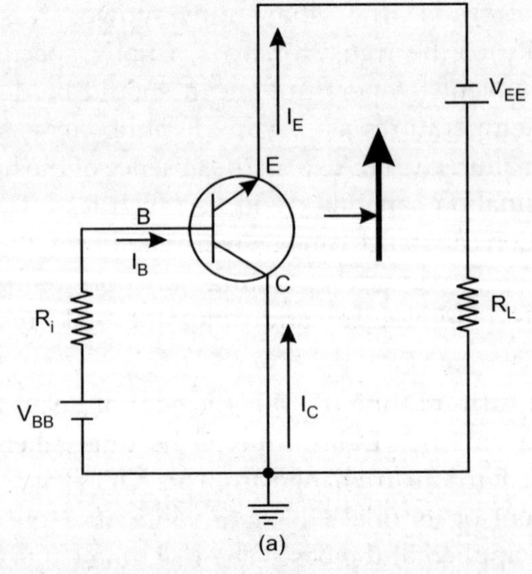

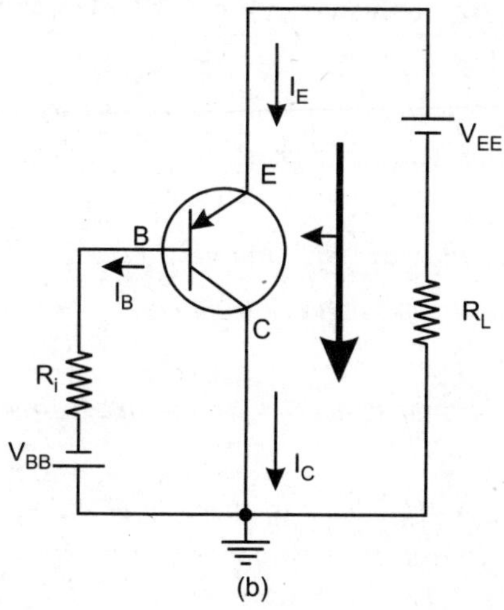

Fig. 9.23

Input characteristics of common-collector configuration relate the base current (I_B) to the base-collector voltage (V_{BC}) for different values of the emitter-collector voltage (V_{EC}) Fig 9.24 (*a*) shows the input characteristics of the common-collector configuration.

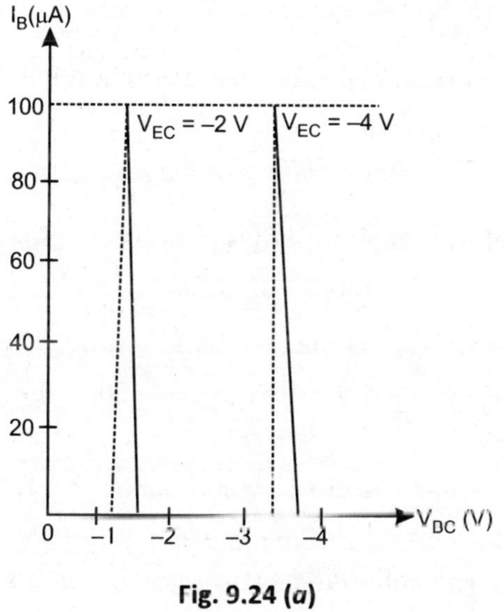

Fig. 9.24 (*a*)

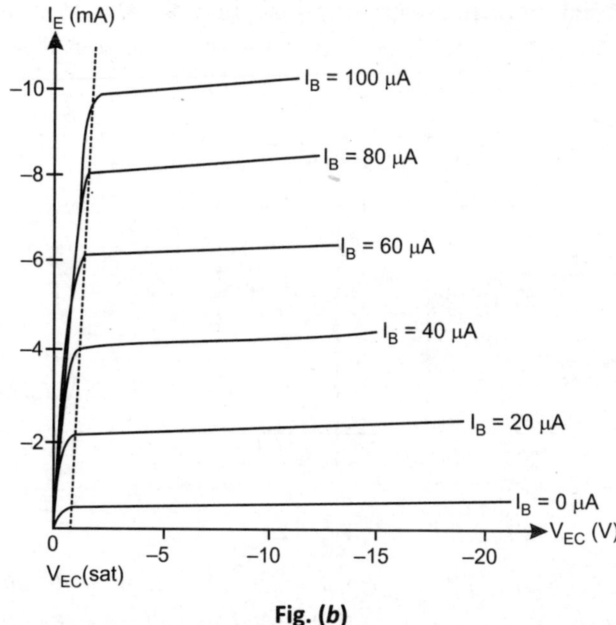

Fig. (*b*)

Output characteristics of the common collector configuration relate the emitter current (I_E) to the emitter-collector voltage (V_{EC}) for different values of base current (I_B). The characteristics are similar to that for the common-emitter configuration Fig 9.24 (*b*) shows the output characteristics of the common-collector configuration.

Gamma (γ) is the current gain in the common-collector configuration. The collector current in this case is expressed as:

$$I_C = \gamma I_B + \gamma I_{CO}.$$

$$\text{where } \gamma = (\beta + 1) = \frac{1}{(1 - \alpha)}$$

Q.48. Explain why a BJT in a common-collector configuration be used for impedance matching applications?

Ans. Common collector configuration offers high input impedance and low output impedance and hence it is used for impedance-matching applications, that is, for driving low-impedance load from a high-impedance source.

Solved Problems

Problem 9.1. *Draw the load line for the diode circuit of Fig 9.25 (a). The forward characteristic of the diode used are shown in Fig. 9.25 (b).*

Solution. The equivalent circuit of Fig. 9.25(*a*) is shown in Fig 9.26 (*a*). It has a Thevenin voltage (V_{th}) of $\left(3 \times \frac{100}{200}\right) = 1.5\ V$ and a Thevenin resistance (R_{th}) of $\left(\frac{100 \times 100}{200}\right) = 50\ \Omega$.

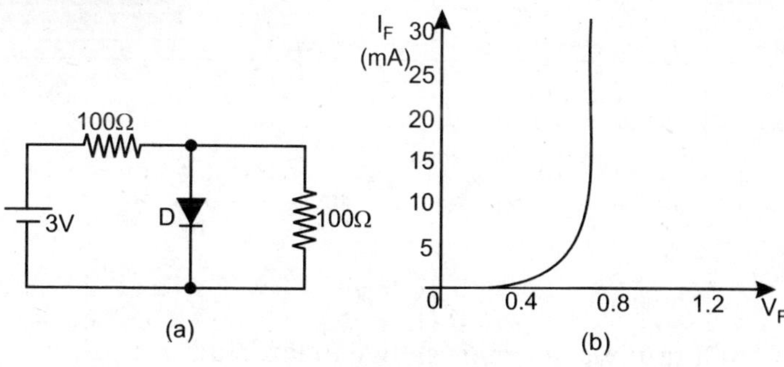

Fig. 9.25

The co-ordinates of the load line are (1.5 V, 0) and (0. 30 mA). The load line superimposed on the *I-V* characteristics is shown in Fig. 9.2 (*b*).

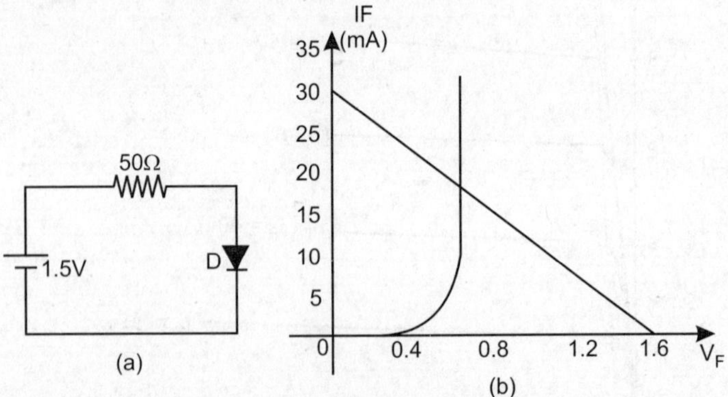

Fig. 9.26

Problem 9.2. *A semiconductor junction diode has a reverse saturation current of 0.5 μA at an ambient temperature of 25°C. What should be the approximate value of reverse, saturation current at 75°C ?*

Solution. The reverse saturation current approximately doubles for every 10°C rise in temperature. The temperature increases by 50°C in the present case. Thus the reverse saturation current increases by a factor of (2^5) = 32. The reverse saturation current at 75°C is therefore 32 × 0.5 = 16 *μA*.

Problem 9.3. *Sketch the output waveform when the input waveform of Fig. 9.27 a is fed to the zener diode circuits of Fig. 9.27 (b) and Fig. 9.27 (c).*

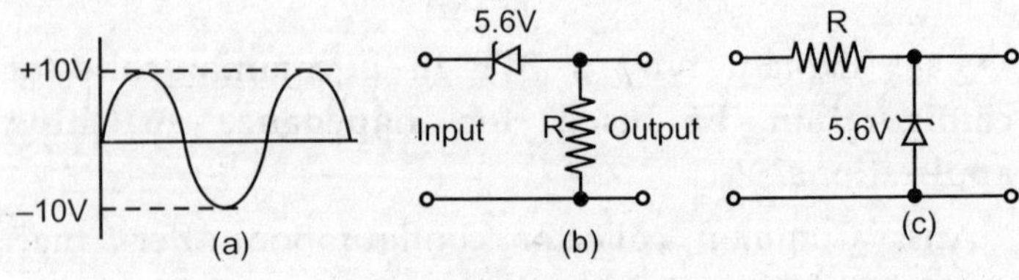

Fig. 9.27

Solution. For the circuit shown in Fig. 9.27 (*b*)

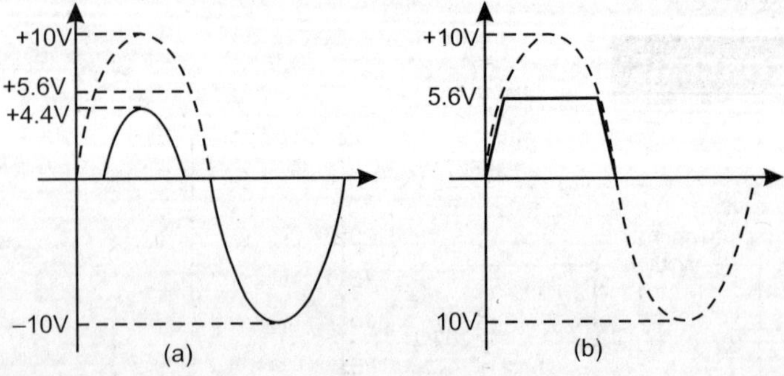

Fig. 9.28

Output waveform is shown in 9.28. (a)

For the circuit shown in Fig. 9.27 (*c*) output waveform is shown in Fig. 9.28 (*b*).

Problem 9.4. *A power transistor was used in conjunction with a heat sink. Various thermal resistances encountered were as follows :*

θ_{J-C} *(For the transistor in question) = 1.5°C / watt*

θ_{C-S} *(thermal resistance from case to heat sink) = 0.5°C/ watt*

θ_{S-A} *(thermal resistance from sink to free air) = 3°C / watt.*

If the power transistor is operated at an ambient of 40°C and if the maximum transistor junction temperature rating is 200°C, what will be

(*i*) *Maximum power dissipation capability of the transistor under steady state conditions.*

(*ii*) *Case temperature for the power dissipated equal to that calculated in (i) above.*

Solution. Maximum junction temperature = 200°C

Ambient temperature = 40°C

Maximum temperature differential that can be tolerated = 160° C

Total thermal resistance from junction to free air

$$= \theta_{J-C} + \theta_{C-S} + \theta_{S-A}$$

$$= 1.5 + 0.5 + 3 = 5\,°\text{ C/Watt}$$

Therefore, Maximum dissipation capability

$$= 160/5 = 32 \text{ watts}$$

(*b*) Now if (θ_C) is the case temperature. Then

$$\theta_{J-C} = \left(\frac{\theta_J - \theta_C}{P}\right) \text{ or } 1.5 = \frac{200 - \theta_C}{32}$$

or $\quad \theta_C = 200 \times 1.5 \times 32 = 200 - 48 = 120°\text{C}$

Problem 9.5. *According to data sheet of 2N 3055 power transistor, maximum dissipation capability at an ambient of 25° C is 115 watts. If junction to case thermal resistance of 2N 3055 is given to be 1.5°C / watt and thermal resistance from case to free air 30°C / watt (for TO-3 package), calculate the dissipation capability of the transistor in the absence of any heat sink. It is also given that 2N 3055 has a maximum junction temperature, rating of 200"C.*

Solution. Ambient temperature = 25°C

Case to air thermal resistance = 30°C/watt

Let maximum dissipation capability be = *x* watts

If θ_C is the case temperature, then we can write

$$\left(\frac{\theta_C - 25}{x}\right) = 30 \text{ or } 30x = \theta_C - 25$$

or $\quad x = \left(\frac{\theta_C - 25}{30}\right) \quad ...(1)$

Also, maximum acceptable θ_C can be computed from

$$= \left(\frac{200 - \theta_C}{x}\right) = 1.5$$

or $\quad x = \left(\frac{200 - \theta_C}{1.5}\right) \quad ... (2)$

Form (1) and (2), we get

$$\frac{\theta_C - 25}{30} = \frac{200 - \theta_C}{1.5} \text{ or } 1.5\,\theta_C - 37.5 = 6000 - 30\,\theta_C$$

or $\quad \theta_C = \frac{6037.5}{31.5} \approx 190°C$

Substituting the value of θ_C in (1) above

$$x = \frac{190 - 25}{30} = 5.5 \text{ watts.}$$

This example shows that absence of a suitable heat sink reduces the power handling capability of a transistor. And if there were no heat sink, it would be tremendously reduced. In the above problem, 115 watt transistor has been reduced to only a 5.5 watt in the absence of heat sink.

Problem 9.6. *Calculate the reverse current of a silicon diode at a temperature of 27°C if a forward voltage of 0.5 V causes a current of 1 A.*

Solution.

$$I = I_o\left[e^{\frac{5800V}{T}} - 1\right],\ T = 27°C = 300°\text{ K}$$

$$I = 1\,A,\ V = 0.5\,V$$

Substituting the values of T, I and V,

$$I_o\left[e^{\frac{5800 \times 0.5}{300}} - 1\right] = 1$$

or $I_o\,(e^{9.67} - 1) = 1$ or $I_o\,(15835 - 1) = 1$

This gives $I_o = 63\ \mu A$.

Problem 9.7. *A silicon diode operating at a temperature of 25°C and a forward biased-voltage of 0.6 V has a forward, current of 0.8 A flowing through it. If the current is held constant, what would be the voltage across the diode at an operating temperature of 75°C.*

Solution. For a silicon diode,

$$I = I_o\left[e^{\frac{5800}{T}} - 1\right]$$

Substituting the values of I, V and T

$$I_o\left[e^{\frac{5800 \times 0.6}{298}} - 1\right] = 0.8 \text{ or } I_o\,[117008 - 1] = 0.8$$

This gives $I_o = 6.837\mu A$

When the operating temperature becomes 75° C (= 348° K), The new voltage across the diode can be computed from:

$$I = I'_o\left[e^{\frac{5800\,V}{T}} - 1\right]$$

where $\quad T = 348°$ K ; I = held constant at 0.8 A

and $\quad I'_o$ = New value of reverse saturation current.

We know that (I_o) doubles for every 10°C rise in temperature.

Therefore,

$$I'_o = I_o\,(\text{at } 25°C) \times 2^5 = 32\,I_o = 32 \times 6.837$$

$$= 218.8\ \mu A$$

Substituting for I, I'_o and T,

$$0.8 = 218.8 \times 10^{-6} \times \left[e^{\frac{5800V}{348}} - 1\right]$$

or $\quad (e^{16.67V} - 1) = \frac{0.8}{218.8 \times 10^{-6}}$

$$= 3.656 \times 10^3 = 3656$$

or $e^{16.67}\,V = 3657$

This gives

$$V = \frac{8.2}{16.67} = 0.492 \text{ volt.}$$

Problem 9.8. *What should be the junction-to-case thermal resistance of a certain silicon transistor that has a maximum junction operating temperature of 200°C and that can safely dissipate a power of 300 watts at a case temperature of 25°C.*

Solution. The said transistor can safely dissipate a power of 300 watts at a case temperature of 25°C. Therefore, maximum allowable junction case temperature differential is (200 – 25) – 175°C. This gives required junction-to-case thermal resistance of:

$$\frac{175}{300} = 0.583°C/\text{ watt.}$$

Problem 9.9. *A silicon transistor has junction to case and junction-to-ambient thermal resistances (θ_{j-C}) and (θ_{j-A}) of 10°C / W and 100°C / W respectively. Determine*

(*a*) *the power dissipation capability of the transistor when operating at a case temperature of 50°C.*

(*b*) *the power dissipation capability of the transistor when the ambient temperature is 25°C.*

The maximum junction operating temperature is 200°C

Solution. (*a*) When the case temperature is 50°C, maximum allowable junction-case temperature difference is 150°C.

This gives power dissipation capability = $\frac{150}{10}$ = 15 watts

(*b*) Maximum allowable temperature difference between ambient and junction = 175°C Ambient-Junction thermal resistance = 100° C/W

This gives power dissipation capability = $\frac{175}{100}$ = 1.75 watts.

Problem 9.10. *A 12 V zener diode operating at a current of 10 mA has an observed dynamic resistance value of 10 ohms. What would be the change is zener voltage produced if the operating current changes by 100 μA ?*

Solution. The dynamic resistance is given by :

$$Z = \frac{\Delta V_z}{\Delta I_z}$$

where ΔV_z = change in zener voltage

ΔI_z = change in operating current

$$\Delta V_z = Z \times \Delta I_z = 10 \times 100 \times 10^{-6} = 1 \text{ mV}.$$

Problem 9.11. *A pentavalent impurity is added to an intrinsic Si semiconductor with one part in 10^8 parts. Find the resistivity in Ω cm for the intrinsic semiconductor. Given that the atomic weight of Si is 28.1, density is 2.33 g/cm^3, Avogadro's number is 6.023×10^{23}, effective mass of an electron is 1.08 × mass of an electron, mass of an electron is 9.11×10^{-31} kg, mobility of electron is 1300 cm^2/V.s, Boltzmann constant is 8.642×10^{-5} eV/K, Planck's constant is 6.626×10^{-34} J.s and temperature is 300 K.*

Solution: Concentration of donor atoms refers to the number of donor/cm^3 of the semiconductor material.

In 1 gram of Si , the number of atoms is $6.023 \times 10^{23}/28.1 = 2.14 \times 10^{22}$.

Number of atoms of Si in 1 cm^3 are $2.33 \times 2.14 \times 10^{22} = 4.996 \times 10^{22}$.

Given that there is one dopant atom per 10^8 Si atoms, therefore, the dopent concentration is $(4.986 \times 10^{22})/(1 \times 10^8) = 4.986 \times 10^{14}$ atoms/cm^3

The conductivity of an N-type semiconductor is given by,

$$\sigma = N_D \mu_n q = 4.986 \times 10^{14} \times 1300 \times 1.6 \times 10^{-10} = 0.103 \ (\Omega.\text{cm})^{-1}$$

Hence, resistivity = 1/0.103 = 9.642 Ω. cm

Problem. 9.12. *Find the concentration of holes (atoms/cm^3) in a P-type Si semiconductor at 300K if its conductivity is 1 (Ω.cm)$^{-1}$. (Given that the mobility of holes in Si is 500 cm^2/V.s)*

Solution: For a P-type semiconductor, the expression for conductivity is given by,

$$\sigma = N_A \mu_p q$$

therefore,

$$N_A = 1/(500 \times 1.6 \times 10^{-19}) = 1/(8 \times 10{-}17)$$
$$= 1.25 \times 10^{16} \text{ atoms/cm}^3$$

Problem. 9.13: *A Si diode has a forward voltage drop of 1 V for a forward DC current of 100 mA. It has a reverse current of 2 μA for a reverse voltage of 5 V. Find the bulk resistance of the diode in ohms.*

Solution: Bulk resistance of a diode is

$$\frac{V_F - V_B}{I_F} = \frac{1 - 0.7}{100 \times 10^{-3}} = 3\ \Omega$$

Problem. 9.14. *For the diode circuit shown in Figure 9.29 (a), find the operating point in (V, mA). The V-I characteristics of the diode are shown in Figure 9.29(b).*

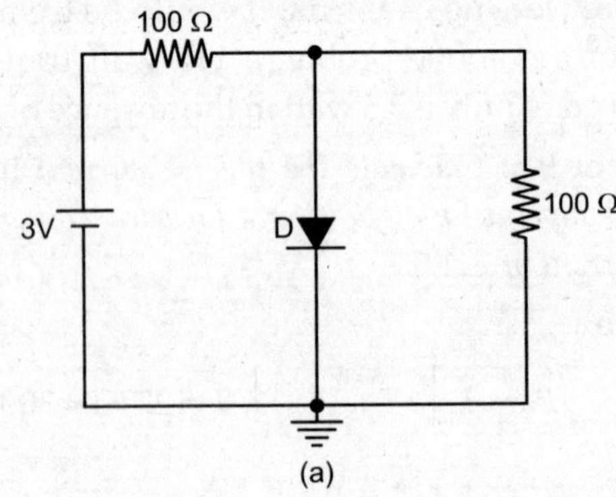

(a)

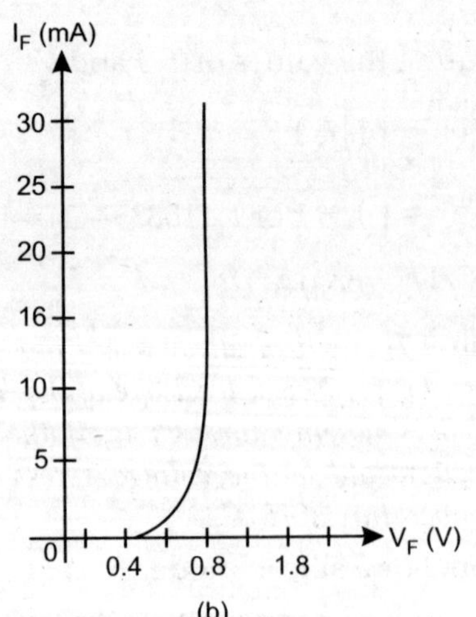

(b)

Fig 9.29

Solution: The circuit given can be simplified using the Thevenin's theorem.

Thevenin's equivalent voltage or V_{th} = (3 × 100/200) = 1.5 V.

Thevenin's equivalent resistance or R_{th} = (100 × 100/200) = 50 Ω.

Simplified equivalent circuit is shown in the Figure. 9.30 (*a*).

The load line is given by the equation

$$V_{th} = V_D + I_D \times R_{th}$$

$$1.5 = V_D + I_D \times 50$$

The co-ordinates of the load line on the *x* and *y* axis are (1.5 V, 0) and (0, 30 mA) respectively. The load line superimposed on the V-I characteristics of the diode are shown in Figure 9.30 (*b*).

The operating point is given by the intersection of the load line with the V-I characteristics of the diode. From Figure. 9.30 (*b*) we can see that the point is (0.8 V, 15 mA).

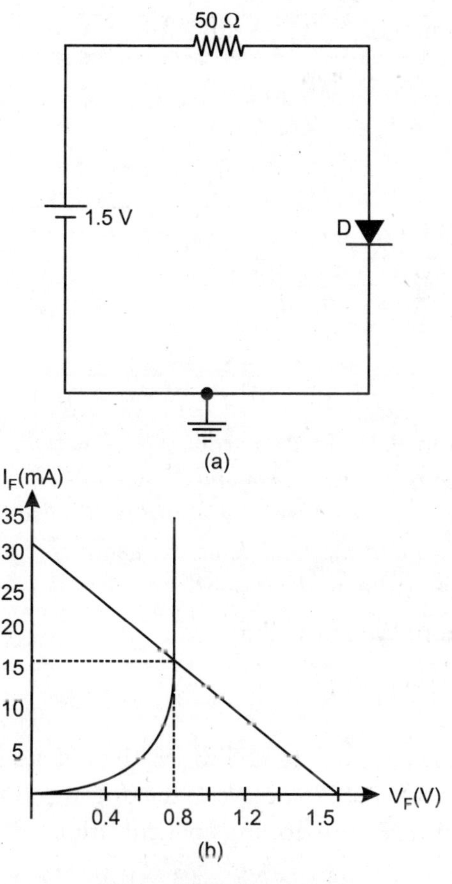

Fig .9.30

The line should cut *x*-axis at not symmetrical 1.5.

Problem. 9.15. *For the circuit shown in the Figure 9.31, find the value of the emitter current (I_E) in milliamperes. (Given that $\beta = 50$, $V_{BE} = 0.7$ V and $I_{CO} = 0$ μA)*

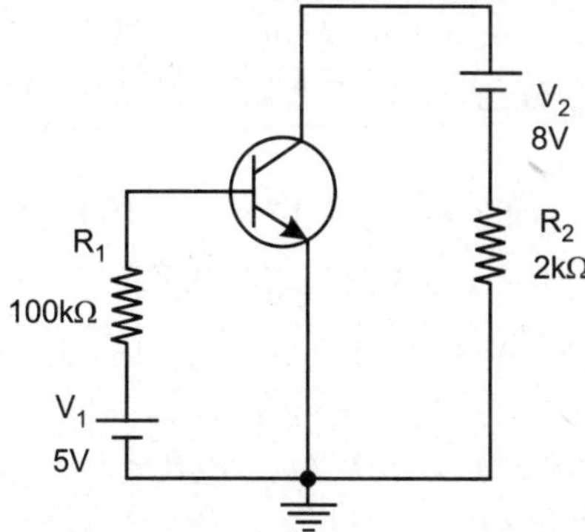

Fig. 9.31

Solution: The polarity of the voltage V_1 applied to the input section forward biases the emitter-base junction. Therefore, the transistor is either in active region or in the saturation region. Let us assume that the transistor is in the active region.

Applying Kirchhoff's voltage law to the input section, we get

$$5 - 100 \times 10^3 \times I_B - V_{BE} = 0$$

Substituting $V_{BE} = 0.7$V we get

$$I_B = \frac{(5-0.7)}{100} \times 10^3 \text{ A}$$

$$= 43 \ \mu A$$

Also,

$$I_C = \beta I_\beta$$

Hence,

$$I_C = 50 \times 43 \times 10^{-6} \text{ A} = 2.15 \text{ mA}.$$

Applying kirchhoff's voltage law to the output section, we get

$$8 - 2 \times 10^{-3} V_{CE} = 0$$

$$V_{CE} = 8 - 2 \times 10^3 \times 2.15 \times 10^{-3} = 8 - 4.3 - 3.7 \text{ V}$$

$$V_{CE} = V_{CB} - V_{BE}$$

Therefore,

$$V_{CB} = V_{CE} + V_{BE} = 3.7 + 0.7 = 4.4 \text{ V}$$

For the NPN transistor, positive value of V_{CB} represents a reverse-biased collector-base junction and hence the assumption that the transistor is in the active region is correct.

$$I_E = I_C + I_B = 2.15 \times 10^{-3} + 43 \times 10^{-6} = 2.193 \text{ mA}$$

Problem 9.16. *For a transistor, the value of α is specified to be 0.98 at a particular collector-base voltage. The value of α increases by 0.5% when the collector-to-base voltage is increased. Find the corresponding percentage change in the value of β.*

Solution: Original value of $\alpha = 0.98$.

Value of β for $\alpha = 0.98$ is $\beta = \alpha/(1-\alpha) = 0.98/(1-0.98) = 49$

New value of $\alpha = 0.98 + 0.5 \times 0.98/100 = 0.985$

Value of β for $\alpha = 0.985$ is $\beta = 0.985/(1-0.985) = 66$

Percentage change is $\beta = [(66-49)/49] \times 100 = 34.69\%$

Problem. 9.17. *The intrinsic carrier concentration of Si sample at 300 K is $1.5 \times 10^{16}/m^3$. If after doping, the number of majority carriers is $5 \times 10^{20}/m^3$, what is the minority carrier density?*

Solution: From the Law of mass action,

$$n_i^2 = np$$

Where n_i is the intrinsic concentration.

Therefore,

$$p = \frac{n_i^2}{n} = \frac{1.5 \times 10^{16} \times 1.5 \times 10^{16}}{5 \times 10^{20}} = 45 \times 10^{10}$$

$$= 4.5 \times 10^{11}/m^3$$

Problem. 9.18. *The electron concentration in a sample of uniformly doped N-type Si at 300 K varies linearly from $10^{17}/cm^3$ at $x = 0$ to $6 \times 10^{16}/cm^3$ at $x = 2\ \mu m$. Assume a situation that electrons are supplied to keep this concentration gradient constant with time. If electronic charge is 1.6×10^{-19} C and the diffusion constant $D_n = 35\ cm^2/s$, what is the current density in the Si, if no electric field is present.*

Solution: The current density for N-type semi-conductor is given by,

$$J_n = q\mu_n \eta \varepsilon + qD_n \frac{dn}{dx} \quad ...(1)$$

Given that no electric field is present, so $\varepsilon = 0$. Therefore, Eq. (1) reduces to

$$J_n = qD_n \frac{dn}{dx}$$

From the statement of the problem, we have

$$\frac{dn}{dx} = \frac{6 \times 10^{16} - 10 \times 10^{16}}{2 \times 10^{-6}} = -2 \times 10^{22}$$

Substituting in the above equation, we get

$$J_n = -1.6 \times 10^{-19} \times 35 \times 10^{-4} \times (-2 \times 10^{22})\ A/m^2$$

$$= 1120\ Acm^{-2}$$

Problem. 9.19. *A Si sample A is doped with 10^{18} atoms/cm^3 of Boron. Another sample B of identical dimensions is doped with 10^{18} atoms/cm^3 of Phosphorus. The ratio of electron to hole mobility is 3. What is the ratio of conductivity of the sample A to B?*

Solution: Conductivity of an N-type semiconductor is,

$$\sigma_n = nq\mu_n$$

Conductivity of a p-type seimconductor is,

$$\sigma_p = nq\mu_p$$

Therefore,

$$\frac{\sigma_p}{\sigma_n} = \frac{\mu_p}{\mu_n} = \frac{1}{3}$$

Problem 9.20. *A heavily doped N-type semiconductor has the following data: Hole-electron mobility ratio = 0.4, doping concentration = 4.2×10^8 atoms/m^3, intrinsic concentration = 1.5×10^4 atoms/m^3. What is the ratio of conductance of the N-type semiconductor to that of intrinsic semiconductor of same material and at the same temperature?*

Solution: For N-type semiconductor,

$$\sigma_n = nq\mu_n$$

For intrinsic semiconductor,

$$\sigma_1 = n_1 q(\mu_n + \mu_p)$$

Therefore,

$$\frac{\sigma_n}{\sigma_1} = \frac{n\mu_n}{n_1(\mu_n + \mu_p)}$$

$$= \frac{4.2 \times 10^8 \times \mu_n}{1.5 \times 10^4 \times \mu_n [1 + (\mu_p/\mu_n)]} = 20000$$

Problem 9.21. *In an abrupt P-N junction, the doping concentrations on the P side and N side are $N_A = 9 \times 10^{16}/cm^3$ and $N_D = 1 \times 10^{16}/cm^3$, respectively. The P-N junction is reverse biased and the total depletion width is 3 μm. What is the depletion width on the P side?*

Solution: We know that

$$\frac{W_N}{W_P} = \frac{N_A}{N_D}$$

Where, W_N is the depletion width of N-side, W_P is the depletion width of P-side; N_A is the doping concentration of P-side and N_D the doping concentration of N-side.

$$W_P = \frac{W_N \times N_P}{N_A} = \frac{(3 \times 10^{-6} - W_P) \times 10^{16}}{9 \times 10^{16}}$$

$$W_P = 0.3\ \mu m$$

Problem 9.22. *A Si P-N junction diode under reverse bias has depletion region of width 10 μm. The relative permittivity of Si, $\varepsilon_r = 11.7$ and the permittivity of free space $\varepsilon_0 = 8.85 \times 10^{-12}$ F/m. Calculate the depletion capacitance of the diode per square meter.*

Solution. Depletion capacitance of a diode is given by,

$$C = \frac{\varepsilon_0 \varepsilon_r A}{d}$$

Where, d is the width of the depletion region and A is the area.

The depletion capacitance of the diode per square metre is given by C/A. From the given data,

$$\frac{C}{A} = \frac{\varepsilon_0 \varepsilon_r A}{d} = \frac{8.85 \times 10^{-12} \times 11.7}{10 \times 10^6}$$

$$F = 10.35\ \mu F$$

Problem 9.23. *The I-V characteristics of the diode in the circuit in Figure 9.32 are*

$$i = \frac{V - 0.7}{500}\ A,\ V \geq 0.7\ V$$

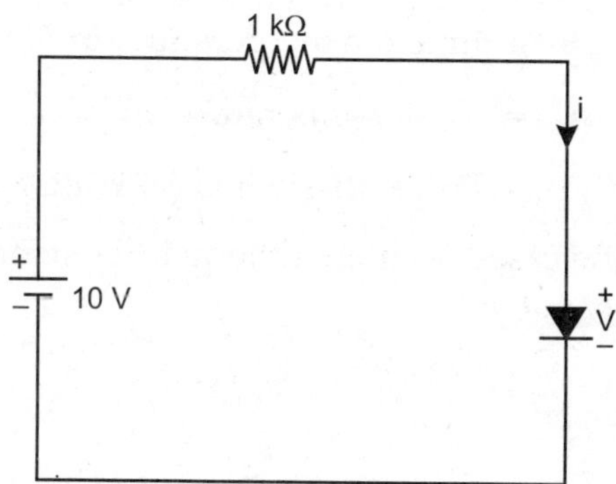

Fig. 9.32

Solution: Applying KVL to the given circuit, we get

$10 - 1000\, i - V = 0$

The diode in the given circuit is forward biased and the current through the diode is

$$i = \frac{V - 0.7}{500}\ A$$

Substituting this value of i in the above equation we have

$$10 - 1000 \times \frac{(V - 0.7)}{500} - V = 0$$

Solving the above equation, we get

$$V = \frac{11.4}{3} = 3.8\ V$$

Therefore, the current through the diode is

$$i = \frac{V - 0.7}{500} = \frac{3.8 - 0.7}{500} = 6.2\ mA$$

Problem. 9.24. *The DC current gain (β) of a BJT is 50. Assuming that the emitter injection efficiency is 0.995, what is the base transport factor?*

Solution. Given that,

$$\beta = 50$$

Therefore,

$$\alpha = \frac{\beta}{\beta + 1} = \frac{50}{51}$$

we know that,

$$\alpha = (\beta^*) y$$

Where, β^* = base transport factor and y = emitter injection efficiency.

Substituting values, we get

$$\beta^* = \frac{\alpha}{y} = \left(\frac{50}{51}\right)\frac{1}{0.995} = 0.9853$$

Summary of Important Formulas

1. The law of mass action is given by following expression.

$$np = n_i^2$$

where n = free electron concentration

p = hole concentration

ni = Intrinsic concentration

2. Drift current is given by the following expression.

$$J = (n\mu_e + p\mu_p)q\varepsilon$$

J = Current Density

n = Free electron concentration

p = Hole concentration

μ_e = Mobility of electron

μ_p = Mobility of hole

q = Charge of electron = 1.6×10^{-19} Coulamb

ε = Applied electric filed

3. The hole diffusion current density is expressed by:

$$J_p = -qD_p \frac{dp}{dx}$$

where J_p = Hole diffusion current density

D_p = Diffusion constant of holes

4. The electron diffusion current density expressed by

$$J_n = qD_n \frac{dn}{dx}$$

where J_n = Electron diffusion current density

D_n = Diffusion constant of electrons

5. The collector current in the case of common collector configuration given by:

$$I_C = \gamma(I_B + I_{CO})$$

where γ = Current gain in common collector configuration

$$= (\beta + 1) = \frac{1}{(1-\alpha)}$$

where β = Current in common emitter configuration

α = Current gain in common base configuration

6. A semiconductor diode's current-voltage relationship is expressed by:

$$I = I_o\left[e^{\frac{5800V}{T}} - 1\right] \text{ for Silicon diodes}$$

$$= I_o\left[e^{\frac{11600V}{T}} - 1\right] \text{ for Germanium diodes}$$

where V = Voltage across the diode in volts

T = Ambient temperature in degrees Kelvin

7. The generalised expressed for diode's current-voltage relationship is given by:

$$I = I_o\left[\frac{V}{e^{\eta V_T} - 1}\right]$$

where $V_T = \frac{T}{11600}$; a quantity dependent upon temperature and measured in volts

η = constant

= 1 for Germanium

= 2 for Silicon

8. α (current gain in common base configuration) and β (current gain in common emitter configuration) are interrelated by:

$$\beta = \frac{\alpha}{(1-\alpha)}$$

9. I_{CEO} (Leakage current from collector-to-emitter with base open) and I_{CBO} (leakage current from collector-to-base with emitter open) are interrelated by;

$$I_{CEO} = h_{FE} \times I_{CBO}$$

When h_{FE} is given at base current equal to I_{CBO}.

10. The junction-to-case thermal resistance (θ_{J-C}) is given by:

$$\theta_{J-C} = \frac{\theta_J - \theta_C}{P} \text{ °C/Watt}$$

θ_J = Junction temperature (in °C)

θ_C = Case temperature (in °C)

P = Power dissipated (in Watts)

11. The case-to-ambient thermal resistane (θ_{C-A}) is given by:

$$\theta_{C-A} = \frac{\theta_C - \theta_A}{P} \text{ °C/Watt}$$

θ_C = Case temperature (in °C)

θ_A = Ambient temperature (in °C)

P = Power dissipated (in Watts)

12. The output characteristics of a bipolar junction transistor when connected in common emitter configuration are expressed by:

$$I_C = \beta I_B + (\beta + 1)I_{CBO}$$

where the terms have their usual meaning

13. The output characteristics of a bipolar transistor when conncected in common base configuration are expressed by;

$$I_C = \alpha I_E + I_{CBO}$$

wher the terms have their usual meaning

SELF EVALUATION EXERCISE

MULTIPLE CHOICE QUESTIONS

1. In a semiconductor diode, P-side is grounded and N-side is applied a potential of -5V through a resistance of 100 ohms. The diode shall
   (*a*) conduct fully
   (*b*) not conduct
   (*c*) conduct partially
   (*d*) None of these
2. Cut-in voltage (Vγ) of a silicon diode is
   (*a*) 0.1 V (*b*) 0.2 V
   (*c*) 0.3 V (*d*) 0.6 V

3. The static resistance of a junction diode
   (*a*) is its opposition to the flow of DC currents
   (*b*) is its opposition to the flow of AC currents
   (*c*) is much smaller than its dynamic resistance
   (*d*) determines the diode's resistance when it is used as a rectifier
4. The junction diode's capacitance that is made use of in varactors is known as
   (*a*) Transition capacitance
   (*b*) Diffusion capacitance
   (*c*) Space charge capacitance
   (*d*) Inter-electrode capacitance
5. The semiconductor rectifier specification that indicates subcycle current rating of the rectifier is
   (*a*) Surge current (I_{FSM})
   (*b*) Peak Inverse Voltage (PIV)
   (*c*) $I^2 t$ (Ampere squared seconds)
   (*d*) Transition capacitance.
6. Identify the false statement
   (*a*) The dynamic resistance of a semiconductor junction diode varies inversely as the diode's forward current
   (*b*) The thermal runaway problem in transistor is primarily due to negative temperature co-efficient of base-emitter junction conduction.
   (*c*) A transistor when in the active region has: both its emitter and collector junctions forward biased.
   (*f*) when a transistor is driven to saturation, emitter and collector junctions are forward biased.
7. A power transistor has the following specifications : P_D = 100 watts. Maximum junction temperature = 200°C, case to ambient thermal resistance of 10°C/W and junction to case thermal resistance of 2°C/W. This transistor can dissipate a maximum power of.......... watts in the absence of any heat sink at an ambient temperature of 50°C.
   (*a*) 100
   (*b*) 50
   (*c*) 12.5
   (*d*) 22.5
8. Transistor's α and β are related by
   (*a*) $\beta = \left(\frac{\alpha}{1-\alpha}\right)$
   (*b*) $\alpha = \left(\frac{\beta}{1-\beta}\right)$
   (*c*) $\beta = \left(\frac{\alpha}{1+\alpha}\right)$
   (*d*) $\alpha = \left(\frac{1+\beta}{\beta}\right)$
9. When current through a zener diode increases by a factor of 2, voltage across its terminals
   (*a*) is halved
   (*b*) is doubled too
   (*c*) is practically unchanged
   (*d*) None of these
10. The dynamic resistance of a zener diode
    (*a*) increases with increase in current through it.
    (*b*) decreases with increase in current through it.
    (*c*) is independent of current passing through it.
    (*d*) None of these
11. Typical value of impurity concentration in a tunnel diode is
    (*a*) 1 part in 10^8 parts
    (*b*) 1 part in 10^6 parts
    (*c*) 1 part in 10^3 parts
    (*d*) None if these
12. A tunnel diode
    (*a*) is a current controllable device
    (*b*) exhibits characteristics that are represented by a single valued function of current
    (*c*) is a voltage controllable device
    (*d*) has a cut-in voltage of 0.3 V
13. The Backward diode
    (*a*) is a current controllable device
    (*b*) is a tunnel diode operated in negative resistance region of its V-I characteristics
    (*c*) is another name for a tunnel rectifier
    (*d*) None of these
14. Zener diodes have
    (*a*) a positive temperature co-efficient for their breakdown voltages
    (*b*) a negative temperature co-efficient for their breakdown voltages

(*c*) a breakdown voltage that is independent of temperature

(*d*) None of these

15. In Zener and Avlanche breakdown diodes, current flow is due to flow of

(*a*) minority carriers

(*b*) majority carriers

(*c*) minority as well as majority carriers

(*d*) minority carriers (in zener diodes) and majority carriers (in Avlanche diodes)

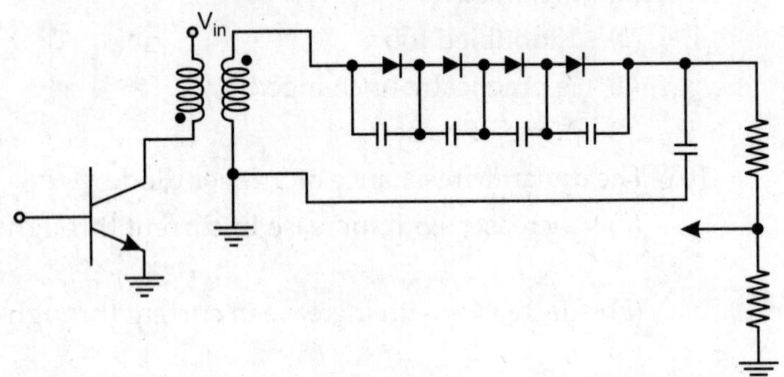

Fig. 9.33

16. N-type Si is obtained by doping Si with

(*a*) Germanium (*b*) Aluminium

(*c*) Boron (*d*) Phosphorus

17. Consider the following assertions:

S_1: For Zener effect to occur, a very abrupt junction is required.

S_2: For quantum tunnelling to occur, a very narrow energy barrier is required.

Which of the following is correct?

(*a*) Only S_2 is true.

(*b*) S_1 and S_2 are both true but S_2 is not a reason for S_1.

(*c*) S_1 and S_2 are both true and S_2 is a reason for S_1.

(*d*) Both S_1 and S_2 are false.

18. In a forward-biased P-N junction, what is the sequence of events that best describes the mechanism of current flow?

(*a*) Injection and subsequent diffusion and recombination of minority carriers

(*b*) Injection and subsequent drift and generation of minority carriers

(*c*) Extraction and subsequent diffusion and generation of minority carriers

(*d*) Extraction and subsequent drift and recombination of minority carriers

19. A Zener diode, when used in voltage stabilization circuits, is biased in

(*a*) Reverse-bias region below the breakdown voltage

(*b*) Reverse breakdown region

(*c*) Forward-bias region

(*d*) Forward-bias constant current mode

20. In a uniformly doped BJT, assume that N_E, N_B and N_C are the emitter, base and collector dopings in atoms/cm^3, respectively. If the emitter injection efficiency of the BJT is close to unity, which one of the following conditions is TRUE?

(*a*) $N_E = N_B = N_C$

(*b*) $N_E \gg N_B$ and $N_B > N_C$

(*c*) $N_E = N_B$ and $N_B < N_C$

(*d*) $N_E < N_B < N_C$

(II) Fig. 9.33 shows a portion of a switching supply circuit. The rectifier portion comprises of four diode rectifiers connected in series to get a higher overall PIV rating. What do you think why do we have to put capacitors across individual rectifiers. You have to make choice out of the following statements.

(*a*) These capacitors have been put to protect the diodes from the voltage transients.

(b) With the help of these capacitors, it is possible to operate the stack at higher frequencies.

(c) These capacitors have been put to improve upon the reverse recovery time.

(d) These capacitors ensure that the reverse voltage that appears across the rectifier stack is equally shared by all individual diodes in addition to improving the overall reverse recovery time of the rectifier stack.

(III) Two or more than two rectifiers may be connected in parallel to enhance their current handling capability. Fig. 9.34 shows a similar situation. What is the purpose of connecting small resistance (*r*) in series with each rectifier? Make the best choice out of the possible answers listed below.

(*a*) These are connected to ensure equal current sharing by individual diodes

(*b*) These are connected to match their reverse biased characteristics (*c*) These are connected to ensure equal reverse voltage across individual diodes

(*d*) These are connected to protect the diodes from transients.

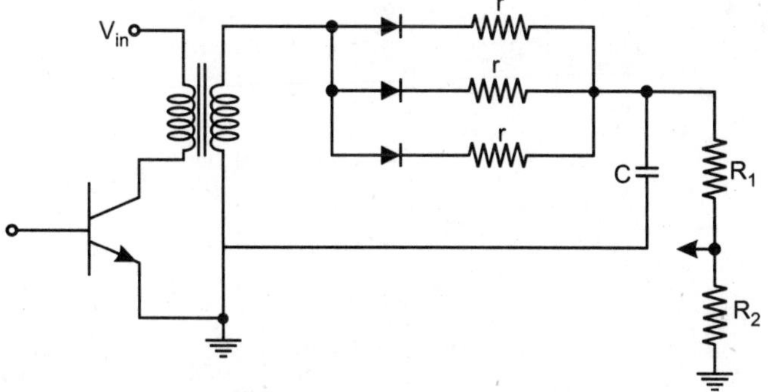

Fig. 9.34

ANSWERS TO SELF EVALUATION EXERCISE

1. (*a*)	2. (*d*)	3. (*a*)	4. (*a*) and (*c*)	5. (*c*)
6. (*c*)	7. (*c*)	8. (*a*)	9. (*c*)	10. (*b*)
11. (*c*)	12. (*c*)	13. (*c*)	14. (*a*)	15. (*a*)
16. (*d*)	17. (*b*)	18. (*a*)	19. (*b*)	20. (*b*)

(II) (*d*)

(III) (*a*)

Chapter 10

Field Effect Transistors

Learning Objectives

After completing the chapter, you will learn the following topics:

1. Difference between Field Effect Transistor (FET) and BJT
2. FET – A Unipolar Device
3. Construction of JFET
4. Characteristics of N-Channel JFET and P-Channel JFET
5. Characteristic parameters of JFETs
6. Insulated Gate Field Effect Transistor (IGFET)
7. Construction of MOSFET
8. Enhancement MOSFETs and Depletion MOSFETs
9. Construction of Depletion type MOSFET
10. Drain characteristics of Depletion type MOSFET and Enhancement type MOSFET
11. Transfer characteristics of Depletion type MOSFET and Enhancement type MOSFET
12. Characteristic parameters of MOSFETs
13. Difference between JFETs and MOSFETs
14. Dual-Gate MOSFET
15. VMOS devices
16. MESFET
17. Applications of JFETs and MOSFETs
18. Insulated Gate Bipolar Transistor (IGBT)

Q. 1. What is an FET and how does it differ from a conventional transistor ?

Ana. FET stands for Field-Effect-Transistor and it differs from a conventional transistor in the following respects :

(*i*) It is a voltage controlled device whereas current controls the operation of a conventional transistor- In this respect, FETs resemble tubes

(*ii*) It is a unipolar device, which means that the flow of current is due to one type of charge carriers (majority carriers in case of FETs), whereas conventional transistors are bipolar devices. which means that the flow of current is due to both types of charge carrier, majority carriers as well as minority carriers.

(*iii*) FETs have got exceptionally high input impedance (of the order of thousands of mega-ohms).

Q. 2. FET is a unipolar device'. Elaborate on the statement.

Ans. All semiconductor materials (*P*-type of *N*-type) have got excess of either holes or electrons. If in a semiconductor device (take conventional transistor). the current flow is due to majority carriers (electrons in *N*-type and holes in *P*-type). as well as minority carriers (electrons in *P*-type and holes in *N*-type). then the device is called a bipolar device. In an *N-P-N* transistor electrons and holes form the majority and minority carriers respectively. It is the opposite in *P-N-P* transistor.

If in a solid state device the current is contributed by majority carriers only, then it is termed as a unipolar device. FET (field effect transistor) is an example of this type.

Q.3. Show the constructional features of a junction field effect transistor and hence give the difference between an *N*-channel FET and a *P*-channel FET. Explain their operation in brief.

Ans. Fig. 10.1 shows the constructional features of a junction FET terminals are:

1. Source 2. Drain. 3. Gate

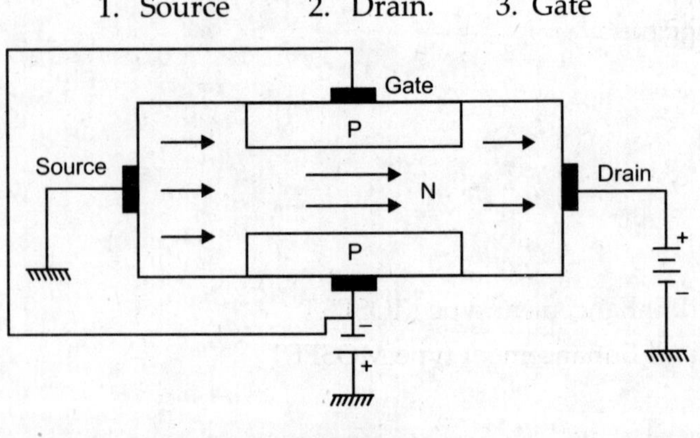

Fig. 10.1

The basic FET consists of a bar of an *N*-type or a *P*-type semiconductor material with a light doping level. On both sides of the bar, *P-N* junction is formed by diffusing the impurities of opposite character as that of the bar, *i.e.* forming *P*-type regions in an *N*-type bar and forming *N*-type regions in a *P*-type bar. Metal contacts on two ends to the bar serve as 'source' and 'drain' where as those on the diffused regions form the gate, as shown. In *N*-type bar, the majority carriers are the electrons. These electrons may travel from source to drain if a positive potential is applied at the drain.

FETs using *N*-type bar are called *N*-channel FETs whereas *P*-channel FETs are produced by using P-type bars.

Now if a negative potential (for N-channel FETs) or a positive potential (for *P*-channel FETs) is applied to the gate terminal, the junction formed by the gate terminal and the silicon bar is reverse biased A potential barrier exists at a reverse biased junction and this barrier can be represented by an equivalent voltage. Since there are two such junctions in an FET, two equivalent voltages appear which give rise to an electric field as shown in Fig. 10.2. This field controls the flow of carriers from source to drain. More the reverse bias, stronger the field is and lesser amount of carriers are able to reach the destination. Hence current in the drain circuit (contributed by flow of carriers) may be controlled by the voltage present at gate terminal. This phenomenon is exactly analogous to the control of output voltage with the voltage present at the grid in case of vacuum tubes.

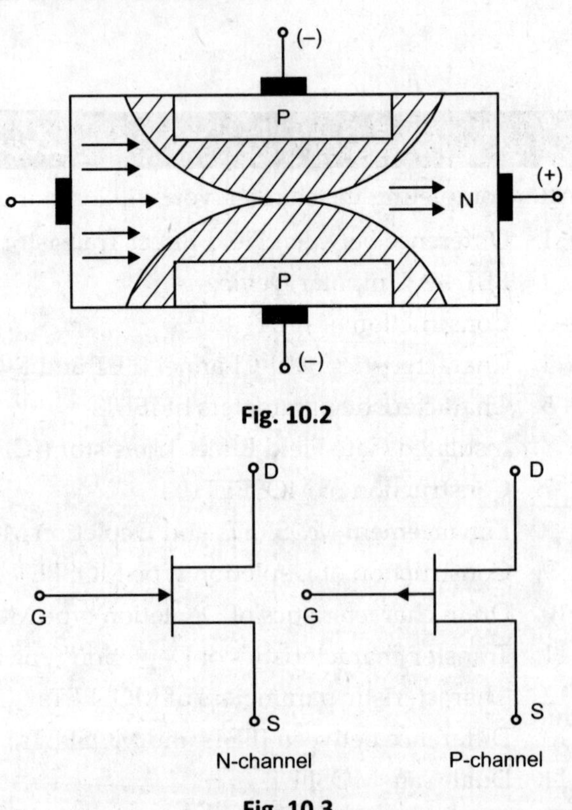

Fig. 10.2

Fig. 10.3

Q. 4. Draw the circuit symbols for an N-channel FET and a P-channel FET and also draw their characteristics.

Ans. The circuit symbols for N-channel and P-channel FETs are shown in Fig. 10.3. When there is no voltage at the gate terminal (*i.e.* for $V_{GS} = 0$), then the current increases as the drain to source potential is increased. The current saturates after some time for a particular value of V_{ds}. This increase in current with increase in is expected as a greater V_{ds} means a stronger force for the current carriers emitted by the source.

Now when a certain gate potential is present so as to reverse bias the gate source junction, it restricts the flow of carriers by the electric field produced due to *P-N* junctions being reverse biased. Thus the drain current saturates at a lower value. More the reverse bias, lesser the saturation value for drain current is. Such a behaviour is indicated in the characteristic curves shown in Fig. 10.4.

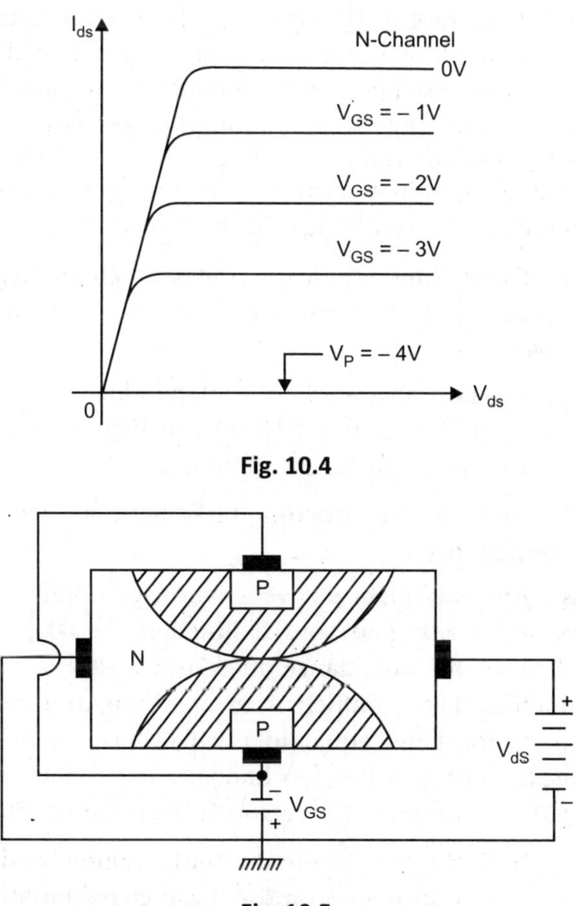

Fig. 10.4

Fig. 10.5

Q. 5. What is Pinch-Off voltage ?

Ans. It is that value of gate to source potential (V_{GS}) for which the drain current in zero. In other words, the electric field is strong enough to block the flow of electrons completely or pinch off the channel. The pinch-off voltage is indicated in Fig. 10.4 to be (– 4V) and Pinch-off phenomenon is shown in Fig. 10.5

Q.6. Draw the drain current versus drain voltage (V_{ds}) characteristics of a *P*-channel junction FET and show the pinch-off phenomenon.

Ans. The (I_{ds}) versus (V_{ds}) characteristics are shown in Fig. 10.6 (*a*). (I_{ds}) has been considered negative as the current in this case is constituted by holes and not electrons as is the case in an N-channel FET. The pinch-off phenomenon is shown in Fig. 10.6. (*b*).

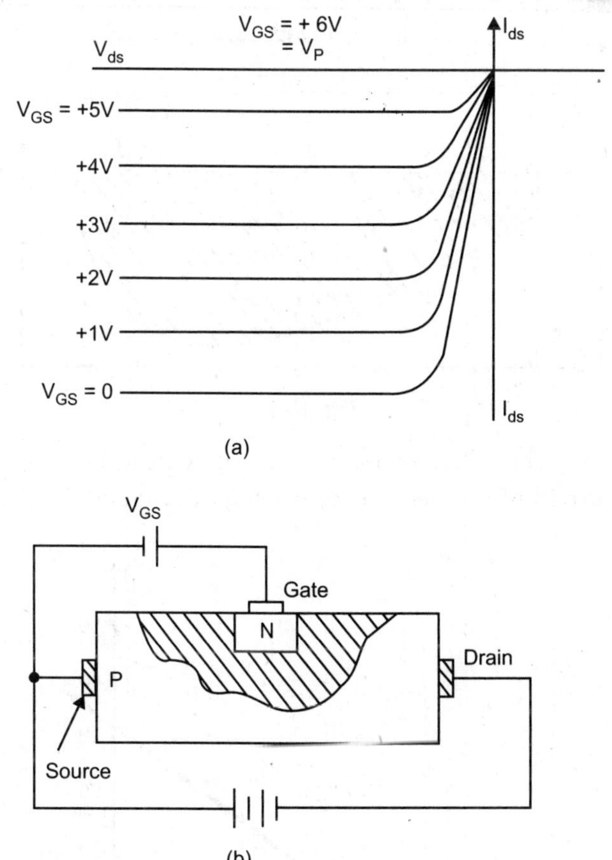

Fig. 10.6

Q. 7. Give one important reason for which N-channel FETs are preferred over P-channel FETs.

Ans. The mobility of electrons (which constitute the current in N-channel FETs) is much more than that of holes (which constitute the current in P-channel FETs). That is why N-channel FETs are preferred over P-channel FETs,

Mobility is the ease with which a charge carrier can be made to move. If we say that electrons have more mobility, this signifies that it is easier to make them mobile as compared to holes.

In other words, for the same applied drain source voltage, number of electrons reaching the drain (in N-channel FET) will be more than the number of holes (in P-channel FET) reaching the drain.

Q. 8. We can see from the drain characteristics (I_{ds} vs. V_{ds}) of the FET that the drain current (I_{ds}) tends to level off or saturate beyond a certain value of drain source voltage (V_{ds}) even if the gate source voltage is kept at a fixed value including V_{GS} = 0. Explain this phenomenon.

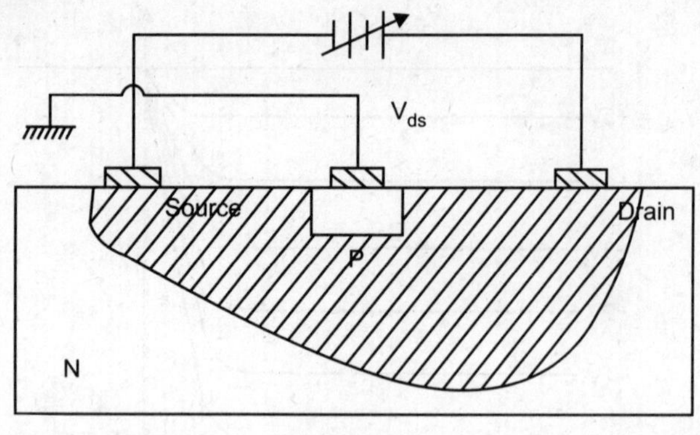

Fig. 10.7

Ans. The FET channel can be considered to be composed of a resistance between gate and source

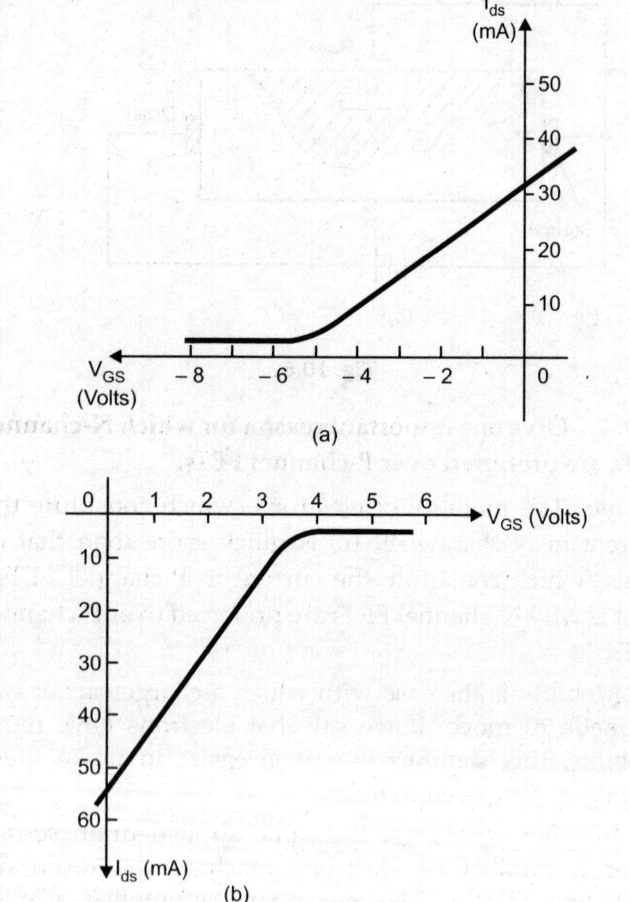

Fig. 10.8

terminals and another resistance between gate and drain terminals. This is being assumed of course strictly for the sake of explanation. The applied drain voltage (V_{ds}) presents a positive voltage on the *N*-side of the gate source junction in an *N*-channel FET and a negative voltage on the *P*-side of the gate source junction in a *P*-channel FET. In both cases, it tends to reverse bias the junction. The amount of this reverse bias increases with increase in (V_{ds}) even for a fixed (V_{GS}). A stage comes where an increase in (I_{ds}) resulting from an increase in (V_{ds}) is nullified by a greater channel constriction due to increase in reverse bias. This is where the saturation sets in. The channel constriction becomes larger as we move towards the drain as the amount of reverse bias which is the product of drain current and the channel resistance at that point is also increasing (Fig. 10.7.).

Q. 9. Draw the typical transfer characteristics of N-channel and P-channel junction field effect transistors.

Ans. The characteristics are shown in Fig. 10.8 (*a*) (for N-channel FETs) and Fig. 10.8 (*b*) (for P-channel FETs). The curves shown are self explanatory.

Q. 10. How do the junction FETs have a very high input impedance ?

Ans. Junction FETs as we have seen operate with a reverse biased gate source junction. So, the input impedance is the impedance offered by a reverse biased *PN* junction. For a sufficient reverse bias, the current drawn by the gate (equal to reverse biased junction's leakage current) may be a few nanoamperes, which yields an input impedance in the region of the order of 100 M.

Q. ll. Indicate the channel ohmic region and the channel pinch-off region on the drain characteristics of a junction FET. List the typical applications when you need to operate your FET in the channel ohmic region and channel pinch-off-region.

Ans. The channel ohmic region is the one where the drain current varies linearly with the drain voltage (Fig. 10.9) Also, the slope of this linear region depends upon the gate to source voltage with the slope for $V_{GS} = 0$ being widely different from what is at V_{GS} near the pinch-off. The FET ON-resistance may be as small as a few hundred ohms for $V_{GS} = 0$ and the FET OFF-resistance may be in hundreds of mega-ohms.

Thus for small values of (V_{ds}), FET can be used as a voltage variable resistor where the FET ON-resistance can be made to vary with a varying gate to source voltage. The applications based on the use of FET in the channel ohmic region (*i.e.* as a VVR) are the voltage controlled attenuator, automatic gain control circuit, FET as a switch in chopper circuits, FET voltameters etc.

FETs when used as amplifies and oscillators are operated in the channel pinch-off region.

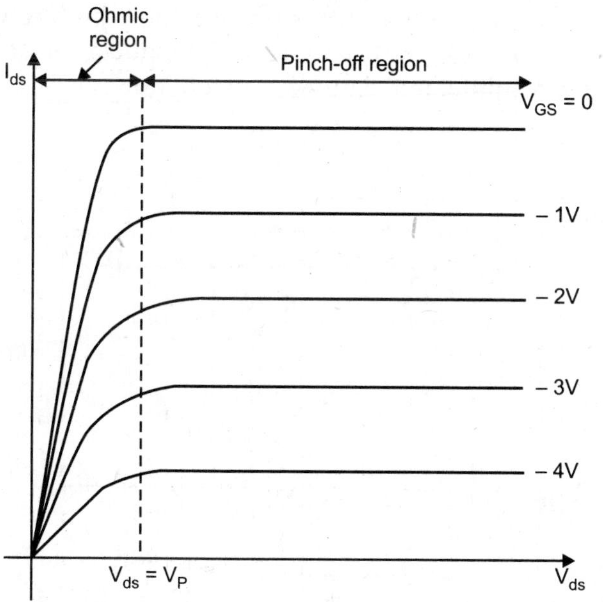

Fig. 10.9.

Q. 12. Define the Drain Resistance (r_d), the Transconductance (g_{fs}) and Amplification (μ) for junction field effect transistors.

Ans. The Drain Resistance (r_d) of a junction FET is defined as the ratio of a small change in V_{ds} (= ΔV_{ds}) to the corresponding change in I_{ds} (= ΔI_{ds}) (Fig.10.10 *a*). Drain resistance is usually specified for V_{GS} = 0. The change in V_{ds} is taken around V_{ds} somewhere in the channel pinch-off region for a constant V_{GS}

i.e. $$r_d = \left|\frac{\Delta V_{ds}}{\Delta I_{ds}}\right|, V_{GS} = \text{constant}$$

The drain resistance (r_d) is sometimes specified indirectly in terms of output conductance (Y_{oss}) in the specifications sheet. (Y_{oss}) also includes capacitive effects if any. At low frequencies, however, it is reciprocal of the drain resistance.

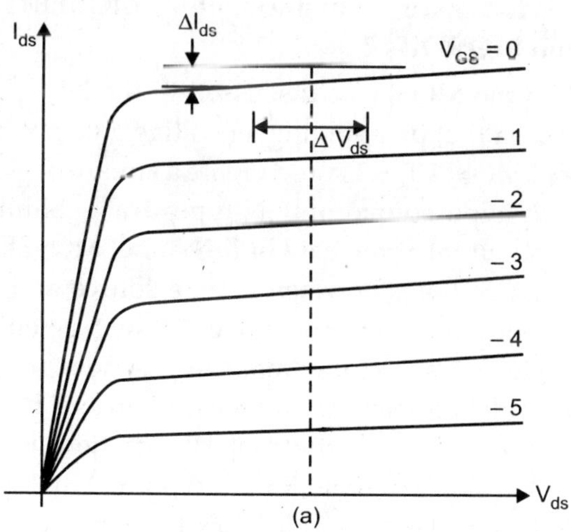

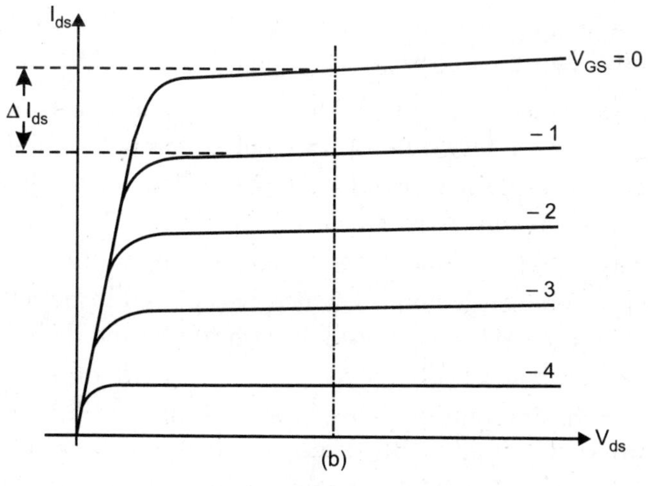

Fig. 10.10

Trabsconductance (g_{fs}) is the ratio of a change in I_{ds} (= ΔI_{ds}) produced to a small change in V_{GS} (= ΔV_{GS}) that produces it for a constant V_{ds} (Fig. 10.10 *b*).

That is, $$g_{fs} = \left|\frac{\Delta I_{ds}}{\Delta V_{GS}}\right| V_{ds} = \text{constant}$$

This measurement is also made for V_{ds} somewhere in the channel pinch-off region. The forward transconductance (g_{fs}) is also sometimes specified in terms of an admittance (Y_{rss}) which takes into account the capacitive effects of the FET. At low frequencies, however, $Y_{rss} = g_{fs}$.

The amplification factor (μ) is the ratio of change in V_{ds} (= ΔV_{ds}) to a change in V_{GS} (= ΔV_{GS}) to produce same change in I_{ds}.

That is, $$\mu = \left|\frac{\Delta V_{DS}}{\Delta V_{GS}}\right| I_{ds} = \text{constant, Also, } \mu = (g_{fs} \cdot r_d).$$

Q. 13. What is an Insulated Gate Field Effect Transistor (IGFET) ? Is there any difference between (IGFET) and Metal Oxide Semiconductor Field Effect Transistor (MOSFET) ?

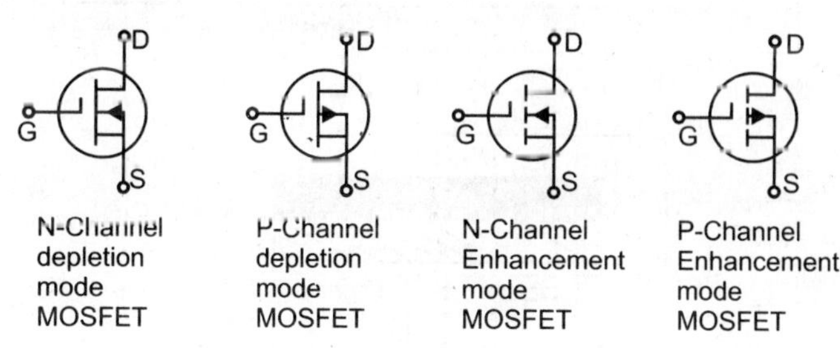

Fig. 10.11

Ans. IGFET is that FET in which the gate region is insulated from the current carrying semiconductor bar. If the insulating material used is a metal oxide, IGFET is called a MOSFET. Otherwise there is no other difference between a MOSFET and an IGFET. The MOSFETs are either of depletion type or of enhancement The circuit symbols for the two types are shown in Fig. 10.11.

Q. 14. Draw a diagram showing the constructional features of a MOSFET. Also label its parts. From the diagram, explain in brief how the voltage at the gate controls the flow of carriers.

Ans. An N-channel MOSFET is shown in Fig. 10.12. (*a*). P-type semiconductor, on which N-type semiconductor materials have been diffused to form source and drain, is called the base or substrate.

In the diagram, a positive voltage is applied between source and drain (with drain positive) and a positive voltage also appears at the gate terminal with respect to substrate.

In order to see the effect of (+ ve) voltage at gate, the gate region has been redrawn in an enlarged view in (Fig. 10.12 (*b*). A (+ ve) voltage at gate contact enables the electrons to break away from their atoms in the metal oxide and get attracted towards the gate contact, at the same time repelling the (+ ve) nuclii towards substrate end of the metal oxide. This positive charge on the substrate side of metal oxide attracts minority carriers from P-type substrate (minority carriers being electrons) as shown in (Fig. 10.12 (*a*)). These attracted electrons fill in the space between *N*-type source and *N*-type drain. This completes the channel and the current flow is enhanced.

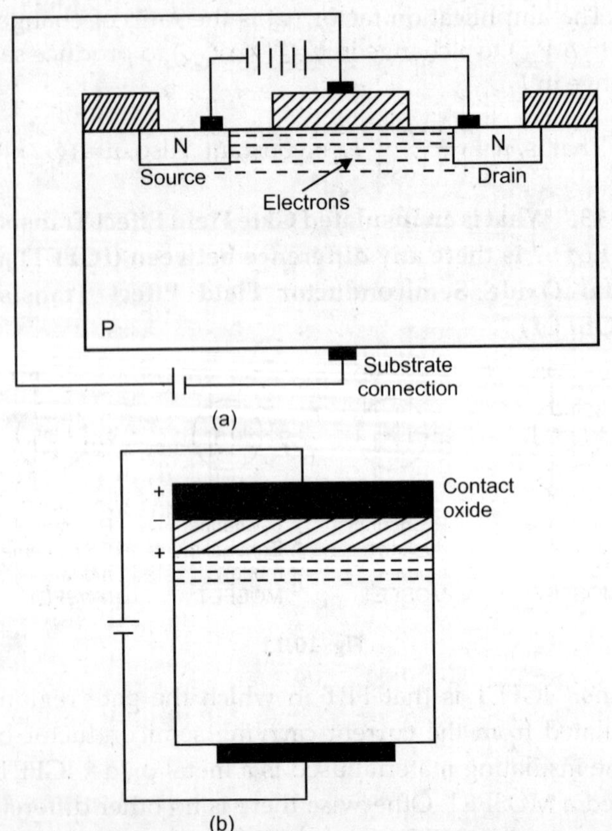

Fig. 10.12

More the positive voltage on gate, more the attracted electrons are and hence more the conductance of the channel resulting into a greater current flow.

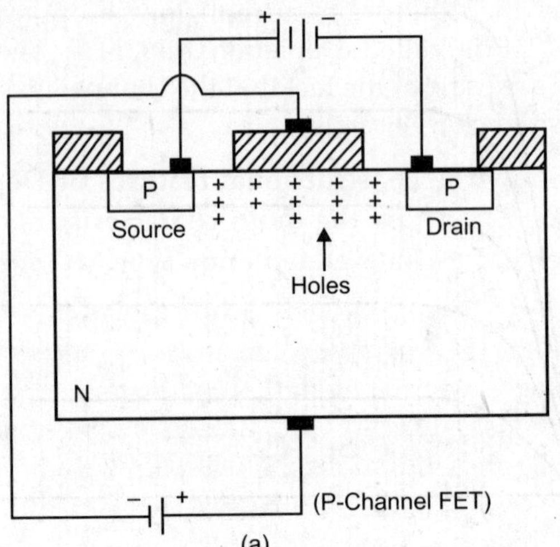

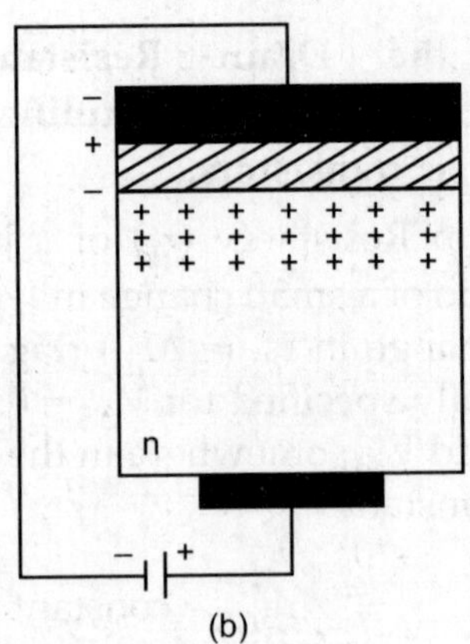

Fig. 10.13

The conditions for a *P*-channel MOSFET are completely reversed with holes constituting the current. The situation is illustrated in Fig. 10.3.

Q. 15. What are Enhancement MOSFETs and Depletion MOSFETs ?

Ans. In the MOSFETS discussed in the last question, we have seen that by applying a positive gate potential in *N*-channel MOSFET, we could create a stream of electrons between *N*-type source and *N*-type drain. Similarly, a negative potential at the gate of *P*-channel MOSFET gave us a stream of holes between *P*-type source and *P*-type drain. In both the cases, the current flow was enhanced because electrons are the majority carriers in *N*-type material and holes are the majority carriers in *N*-type material and holes are the majority carriers in *P*-type material. Thus, when a MOSFET is operated in this mode, it is called an Enhancement MOSFET.

On the other hand, if the gate potential is such that it creates holes in *N*-channel MOSFET and electrons in *P*-channel MOSFET to fill the space between source and drain, then the drain current flow is opposed and the MOSFET is called Depletion MOSFET. The name depletion arises from the fact that the channel is depleted of the desired charge carriers.

Q. 16. Show the constructional features of Depletion type MOSFETs. Draw the drain characteristics of both *N*-channel and *P*-channel depletion type MOSFETs.

Ans. The constructional features of *N*-channel depletion MOSFET and the *P*-channel depletion MOSFET are respectively shown in Figs. 10.14 (*a*) and (*b*).

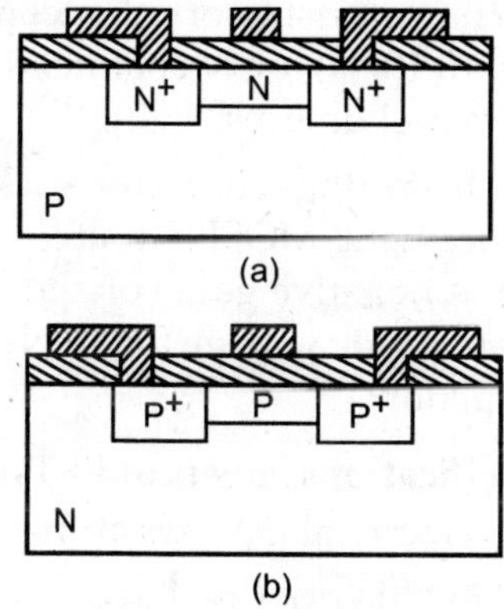

Fig. 10.14

As is, clear from the drain characteristics of depletion MOSFETs, these devices can be operated both in the depletion mode and the enhancement mode though the former is the usual mode of operation. The drain characteristics for *N*-channel and *P*-channel depletion MOSFETs are shown in Fig. 10.15 (*a*) and 10.15 (*b*) respectively.

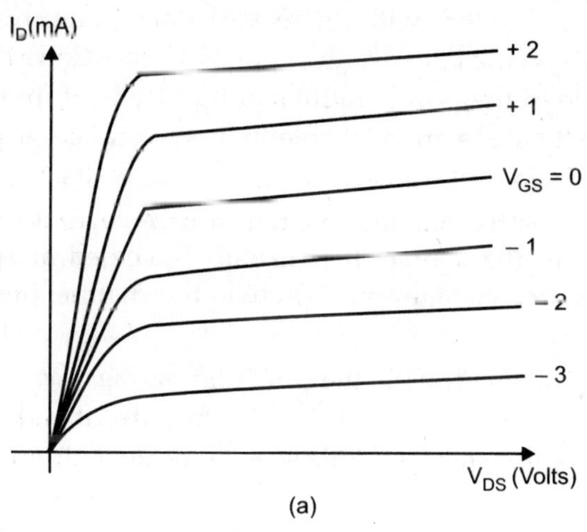

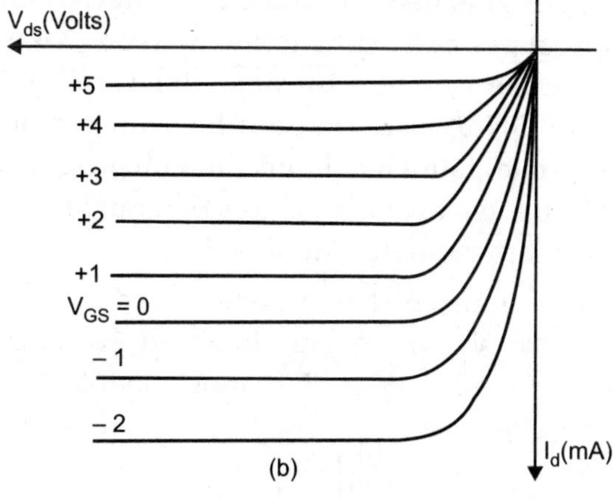

Fig. 10.15

Q. 17. Draw the drain characteristics of an Enhancement type MOSFET.

Ans. The drain characteristics of Enhancement type MOSFETs are shown in Fig. 10.16 (*a*) for *N*-channel and Fig. 10.16 (*b*) for P-channel devices.

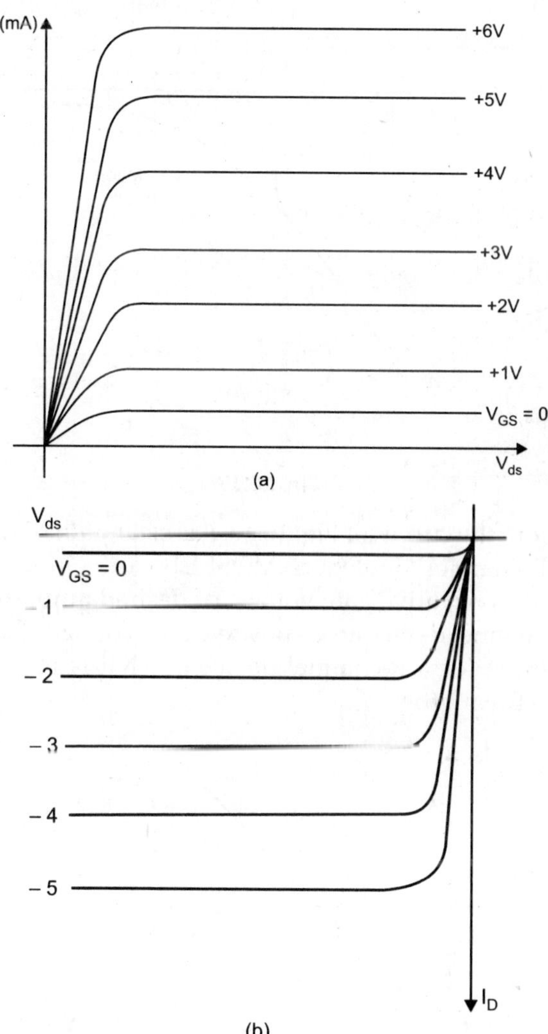

Fig. 10.16

Q. 18. With the help of transfer characteristics for depletion type and enhancement type MOSFETs, briefly describe as to how the depletion type MOSFETs are usually ON devices that can be switched off with an appropriate gate signal and the enhancement type MOSFETs are usually OFF devices that can be switched ON with an appropriate gate signal.

Ans. The transfer characteristics of depletion type MOSFETs are shown in Fig. 10.17 (*a*) for N-channel devices and Fig 10.17 (*b*) for *P*-channel devices.

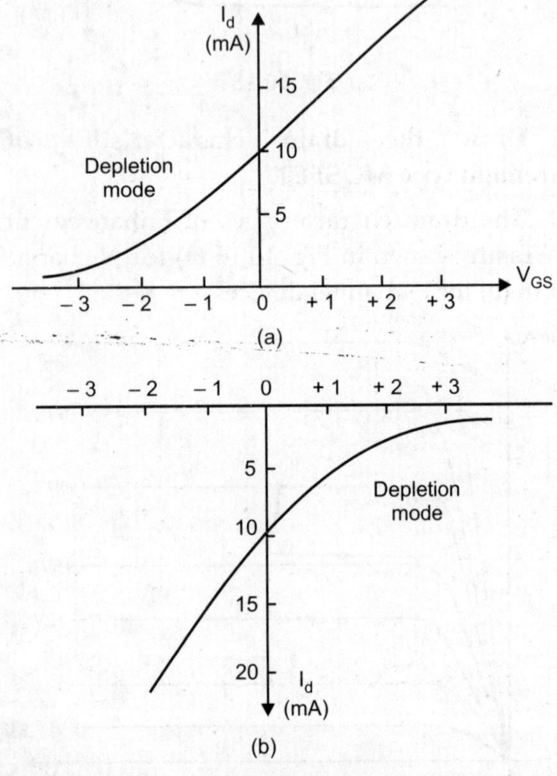

Fig. 10.17

From the curves of Fig. 10.17. it is clear both N-channel and P-channel depletion MOSFETs are On even for $V_{GS} = 0$. A negative gate voltage of desired amplitude for an *N*-channel device and a positive gate voltage of desired amplitude for a P-channel device is what is required to switch them off.

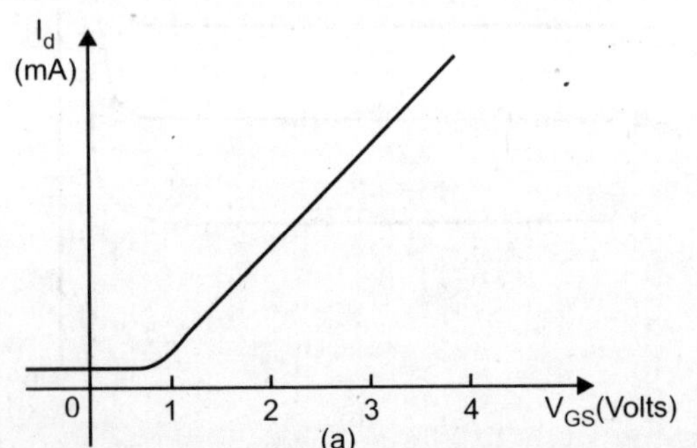

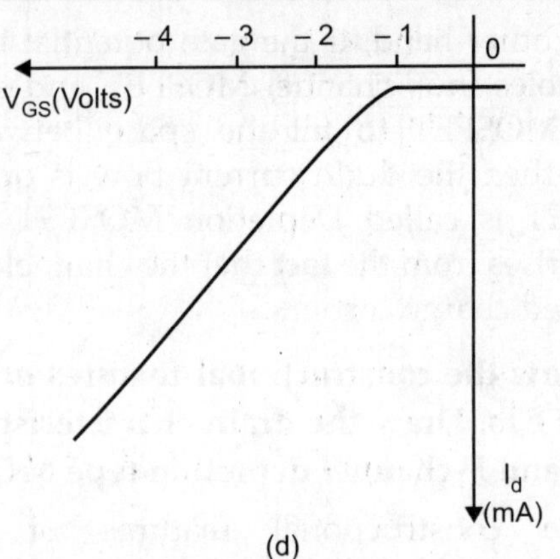

Fig. 10.18

The transfer characteristics of enhancement MOSFETs are shown in Fig. 10.18. (*a*) for N-channel devices and Fig. 10.18. (*b*) for *P*-channel devices.

As is clear from the characteristics shown both *N*-channel enhancement MOSFETs are usually in non-conducting state. A negative gate voltage for a *P*-channel device and a positive gate voltage for an N-channel device can initiate conduction.

Q. 19. The specifications sheets of FETs and MOSFETs contain host of electrical specifications. Some of the important ones are BV_{GS}, P_D, I_{DSS}, V_p (off) I_{GSS}, g_{fs}, g_{os}, r_{ds} (on), C_{iss}, C_{rss}, V_{GS} (th). Define there parameters giving the typical values wherever applicable.

Ans. (BV_{GS}) is the reverse gate source breakdown voltage. It is defined with either the drain open (BV_{GSO}) or the drain shorted to the source (BV_{GSS}). It is typically few tens of volts.

(P_D) is the power dissipation rating of the FET. It is the product of V_{ds} and I_{ds}. The power dissipation rating must be derated with increase in temperature. It is usually specified at an ambient of 25°C. Junction FETs are available in the power ratings of hundreds of milliwatts while MOSFETs are available in a wide range of power ratings from milliwatts to more than 100 watts.

(I_{DSS}) is the maximum drain source current for gate shorted to the source. It is usually specified at specific drain source voltage somewhere in the channel pinch-off region.

(V_p) or V_{GS} (off) is the pinch-off voltage *i.e.* the gate source voltage for which the drain current reduces to zero. It is also specified for a specific value of drain source voltage.

(I_{GSS}) is the gate to source leakage current with drain shorted to source. It is given for a specific value of gate source reverse bias voltage and is typically a few nanoamperes in junction FETs.

(g_{fs}) is forward transconductance of the FET and is usually given for specific values of V_{GS} and V_{ds}. It is typically few thousand micro-mhos to a few milli-mhos.

(g_{os}) is the output conductance. Its reciprocal gives the dynamic drain resistance (r_d). It is usually specified as an admittance parameter (y_{os}). (y_{os}) equals (g_{os}) at low frequencies.

r_{ds} (on) is the dynamic drain resistance of the FET, same as (r_d) defined earlier.

(C_{iss}) is the input capacitance of the FET when connected in the common source configuration. It is typically of the order of few pFs.

The overall input capacitance (C_i) in a common source amplifier configuration using a junction FET is larger than (C_{iss}) due to the presence of Miller component of gate-drain interelectrode capacitance (C_{gd} or C_{rss}). It is given by

$$C_i = C_{iss} + (I - A\upsilon)\, C_{rss}$$

(C_{rss}) is the reverse transfer capacitance for the common source configuration. It is of the order of few pF in junction FETs but only a small fraction of a pF in some MOSFETs.

(C_{oss}) is the output capacitance (drain source capacitance) for the common source configuration. This too is typically few pF.

V_{GS} (th) is the threshold voltage where the MOSFET begins to conduct. It is defined for enhancement mode MOSFETs only.

Q. 20: What are the important differences between JFETs and MOSFETs?

Ans: JFETs and MOSFETs are somewhat similar devices but there are quite a few differences between the two devices in terms of their principle of operation and the value of their characteristic parameters. These differences are as follows:

1. JFETs are operated in depletion mode only. DE-MOSFETs can be operated in both depletion and enhancement modes and E MOSFETs are operated in enhancement mode.
2. The input resistance offered by MOSFETs is much higher than that by JFETs. Input resistance for JFETs is greater than $10^9\ \Omega$ whereas that of MOSFETs is around $10^{13}\ \Omega$.
3. JFETs have higher drain resistance than MOSFETs and hence their characteristic curve is more flat than that of MOSFETs. Drain resistance for JFETs is in the range of 100 kΩ to 1 MΩ while that for MOSFETs is in the range of 1 to 50 kΩ.
4. The leakage gate current in MOSFETs is much smaller than that in JFETs. The gate current for MOSFETs is in the range of 100 nA to 10 pA whereas that for JFETs is in the range of 100 mA to 10 nA.
5. MOSFETs are easier to construct and are used more widely than JFETs.

Q. 21: Explain the construction and principle of operation of a dual-gate MOSFET. Also draw the transfer characteristics of an n-channel dual gate DE-MOSFET?

Ans: In a dual-gate MOSFET, an additional second insulated gate is provided as compared to a conventional MOSFET. The flow of current through the MOSFET is controlled by voltages at both the gate terminals. Since the control is exerted by two gates, the dual-gate MOSFET may be considered to be the counterpart of a tetrode. Fig. 10.19(*a*) shows the cross-section of an n-channel dual-gate DE-MOSFET and Fig. 10.19(*b*) shows the circuit symbol. The device acts if two MOSFETs are connected in series. The *n* + region in the middle acts as a drain for MOSFET-1 and a source for the MOSFET-2. The flow of current through the MOSFET is controlled by both gate voltages. For the dual-gate n-channel DE-MOSFET shown in the figure, the drain current decreases when the gate voltage at either of the two gate terminals is made negative. The gate terminal-1 provides higher transconductance as compared to the gate terminal-2. Owing to simultaneous control of two gate voltages, the device is used in applications such as AGC amplifier, mixers and demodulators. When it is used in AGC amplifier, the signal to be amplified is connected to Gate-1 and the voltage to control the gain is applied to Gate-2.

Figure 10.20 shows the transfer characteristics of a popular n-channel dual-gate DE-MOSFET. The drain characteristics are similar to that of a conventional n-channel DE-MOSFET.

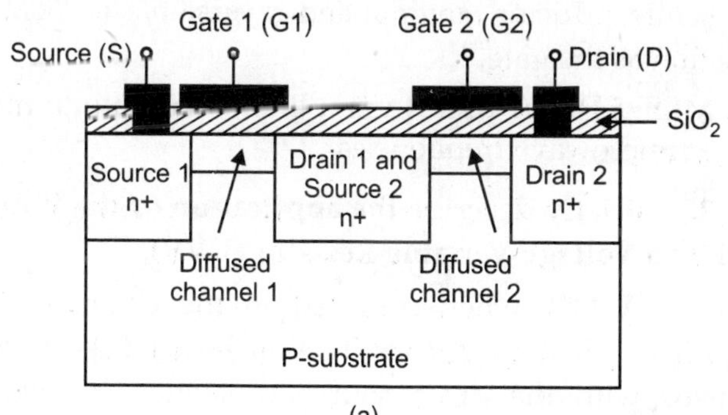

(a)

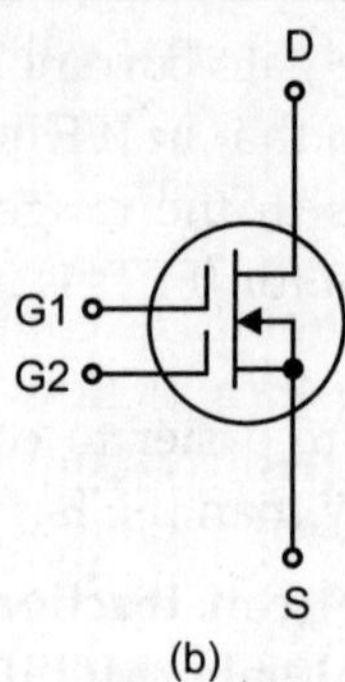

Fig. 10.19

Q. 22. Which characteristic of a DUAL GATE MOSFET gives it an edge over a junction FET as far as their use in designing amplifiers in the VHF range is concerned ?

Ans. A dual gate MOSFET has a much smaller value of drain gate capacitance (C_{gd}) typically 0.005pF against typically 5pF for JFET. This is the inter-electrode capacitance between drain and gate-1. Thus, if the input signal is applied at gate-1, the overall input capacitance that appears between gate-1 and source is the gate source inter-electrode capacitance. In contrast, in a junction FFT amplifier, the overall input capacitance that equals [C_{gs} + (1 – $A\upsilon$) C_{gd}] is quite large with (1 – A_v) C_{gd} contributing the most of it.

Q. 23. What are VMOS devices ?

Ans. VMOS stands for vertical MOS. It is a kind of MOSFET structure capable of higher current ratings than those possible with the conventional MOSFETs. These are basically power field effect transistors. Typical VMOS devices have a maximum drain current rating of a few tens of amperes and a drain-source voltage rating in the range of 200 to 300 volts.

Q. 24. What is MESFET? What are its typical applications?

Ans. MESFET stands for Metal Semiconductor FET. It is a field effect transistor made from Gallium arsenide as the semiconductor material and reverse biased Schottky junction as the gate.

MESFETS are used in amplifiers, oscillators, mixers etc. at microwave frequencies.

Q. 25. Briefly describe the application of the junction FET as a Voltage Variable Resistor (VVR).

Ans. A FET when operated in the channel ohmic region for smaller values of (V_{ds}) can be used as a variable resistor with the drain source resistance (r_d) varying almost linearly with gate source reverse bias voltage (Fig. 10.21. (*a*)). The basic circuit diagram is shown in Fig. 10.21. (*b*)). The output voltage (V_O) in this case can be made to vary from near zero to the maximum value as the gate source reverse bias is varied from zero to near pinch-off.

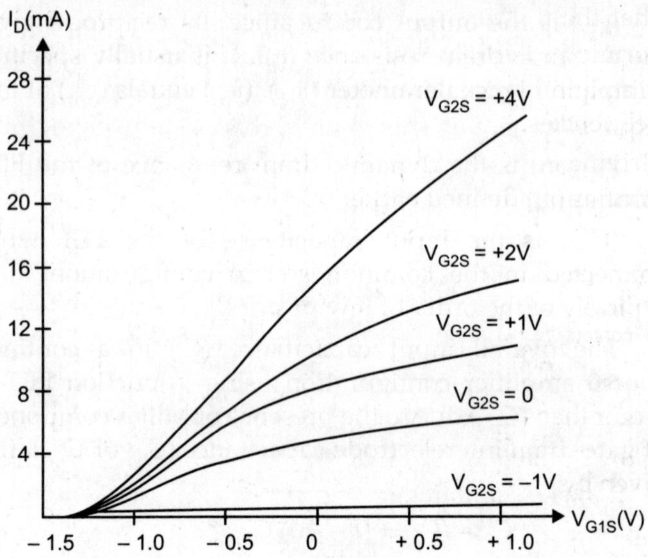

Fig. 10.20.

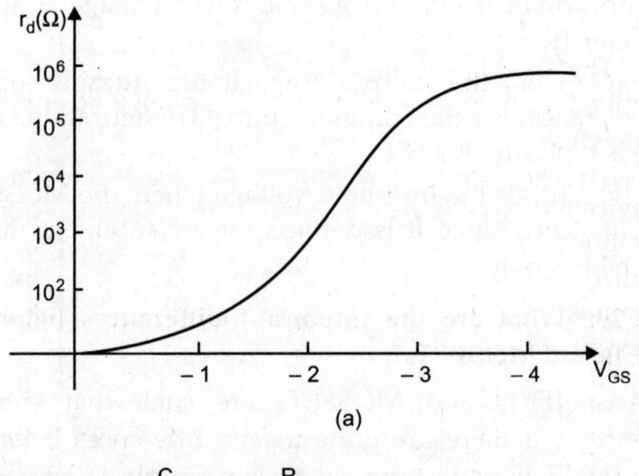

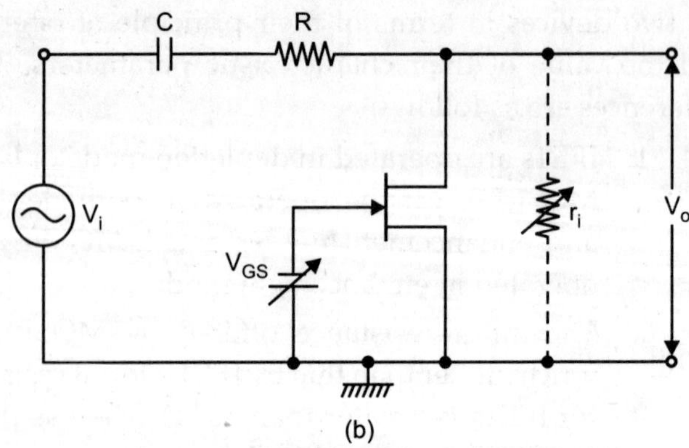

Fig. 10.21

Q. 26: Explain the construction of an IGBT with the help of a diagram. Also give the output characteristics of an n-channel IGBT device?

Ans: Figure 10.24 shows the cross-sectional view of an

n-channel Insulated Gate Bipolar Transistor (IGBT) cell. We can see from the figure that the construction of IGBT is very similar to that of power VMOS devices except that the n+ drain is replaced by p+ collector layer, thus forming a vertical PNP transistor. Figure 10.25 shows the simplified equivalent circuit for the n-channel IGBT. The equivalent circuit shows an n-channel power MOSFET driving a wide base PNP transistor in a Darlington configuration.

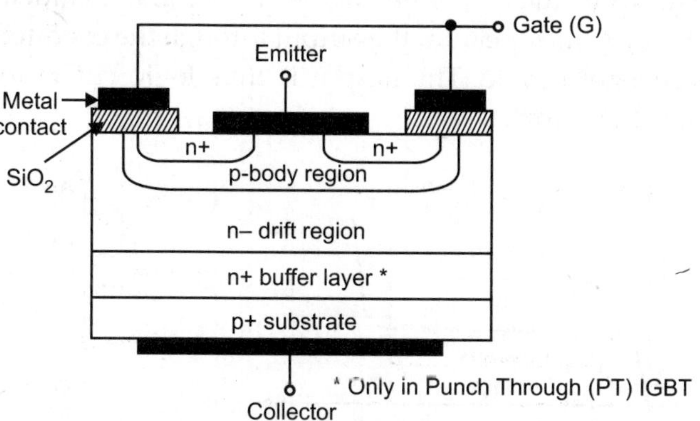

Fig. 10.24

The ON/OFF state of the device is controlled by the applied gate voltage w.r.t. to the emitter voltage. If the gate-emitter voltage is less that the threshold voltage (V_{TH}) of the MOSFET, no inversion layer is created and the device is in the OFF state. When the gate-emitter voltage is above the threshold voltage (V_{TH}), enough electrons will be drawn towards the gate to form a conductive channel across the body region, leading to the flow of current between the collector and the emitter terminals.

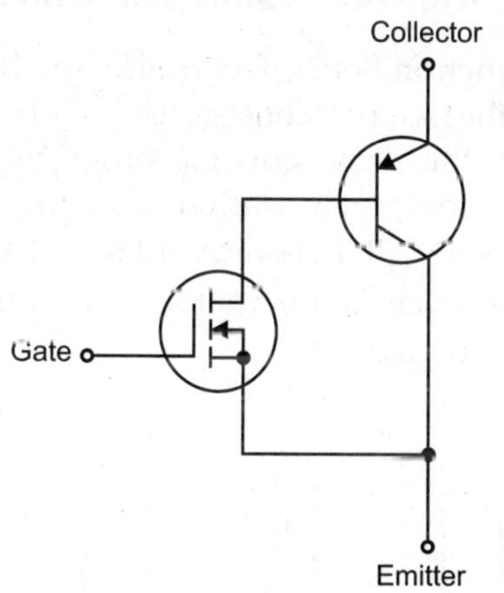

Fig. 10.25

Figure 10.26 shows the typical output characteristics of an n-channel IGBT device. The output characteristics are quite similar to that of an n-channel E-MOSFET. A noteworthy feature on the IGBT characteristics is the offset of approximately 0.7 V from the origin and the steep slope of the rising portion of the characteristics. The offset is because the ON-state voltage across the IGBT is one diode-drop higher than that of n-channel E-MOSFET. The steep slope of the transfer characteristics is attributed to the fact that current flow in an IGBT is due to flow of both electrons and holes as compared to an n-channel MOSFET where the current flow is due to the flow of electrons only. This reduces the effective resistance to current flow in the drift region. The resulting reduction in the ON-state voltage is the main advantage of IGBTs over power MOSFETs.

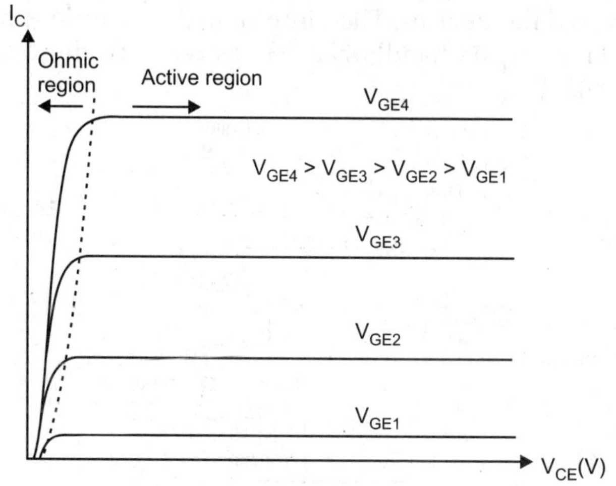

Fig. 10.26

Q. 27: "IGBT has positive attributes of both BJTs and MOSFETs." Elaborate the statement.

Or

How do you compare IGBT with a MOSFET or a BJT when used as a switch?

Ans. IGBTs are three-terminal power semiconductor devices having positive attributes of both BJTs and MOSFETs. IGBTs offer fast switching times similar to that of MOSFETs and lower ON state voltages and larger blocking voltages similar to that of BJTs. They are used in high efficiency and fast switching applications like switch mode power supplies, traction motor control and induction heating.

Q.28: What are the shortcomings of IGBT?

Ans: In an IGBT, holes are left in the drift region and they can only be removed by the process of recombination or by applying a voltage gradient. This causes a tail current in IGBTs during turn-off till all the holes are removed. An n+ buffer layer is added in some IGBTs to control the rate of recombination of holes by absorbing trapped holes during turn-off. IGBTs with n+ buffer layer are called punch-through (PT) IGBTs and those without the n+ buffer layer are called non-punch-through (NPT) IGBTs. PT IGBTs are also referred to as asymmetrical IGBTs and NPT IGBTs as symmetrical IGBTs.

Another problem associated with IGBTs is the occurrence of latch-up phenomenon. Latch-up refers to the failure mode where the IGBT can no longer be turned off by the gate voltage. Latch-up can be explained with the help of a more detailed equivalent circuit of the IGBT as shown in Fig. 10.27. The basic structure of an IGBT resembles a thyristor (series of PNPN junctions) and latch-up can occur if the IGBT were not properly used.

Like MOSFETs, IGBTs also are susceptible to gate insulation damage by the electrostatic discharge of energy through the devices. Therefore similar precautions must be taken while handling IGBTs as taken in the case of MOSFETs.

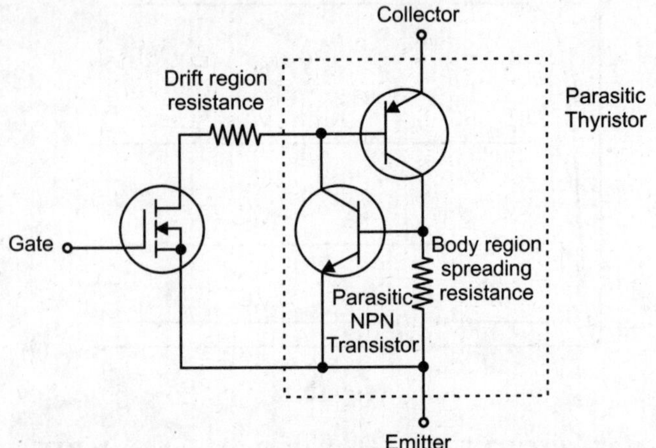

Fig. 10.27

Q. 29. The internal circuitries of CMOS digital integrated circuits make abundant use of *N*-channel and *P*-channel enhancement MOSFETs. The CMOS inverter is one of the most important building blocks of all these ICs. Draw the basic configuration of a CMOS inverter and briefly describe its functional principle.

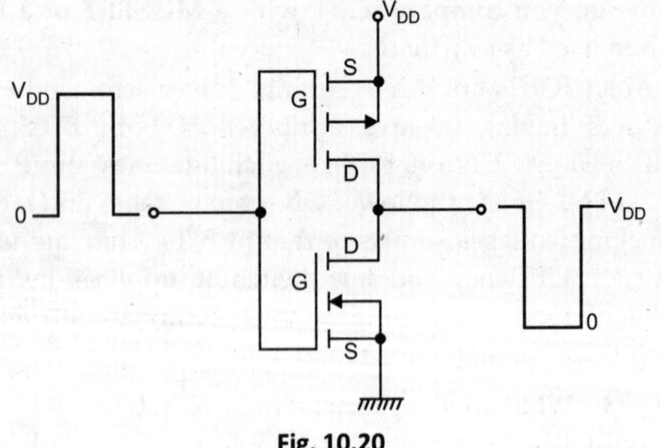

Fig. 10.20

Ans. Fig. 1022. shows the basic configuration. When the input to the circuit is a logic LOW, a GND at the gate of N-channel MOSFET forces it to the non-conducting state while the *P*-channel MOSFET goes into conduction. Thus, (V_{DD}) appears at the output through the conducting *P*-channel device. The output is thus logic HIGH for a logic LOW input.

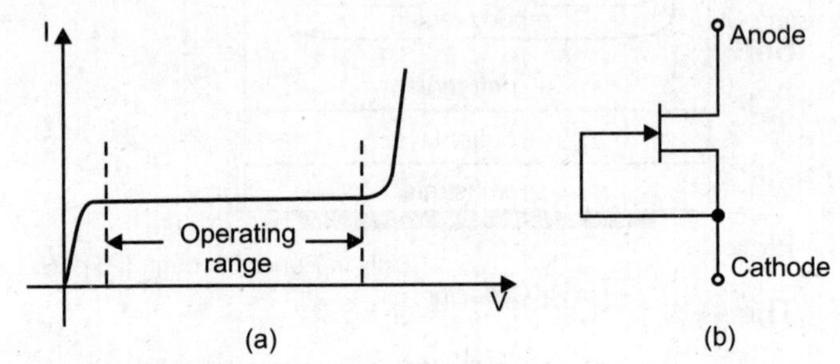

Fig. 10.23

When the input is a logic HIGH, a (V_{DD}) at the gate of *N*-channel MOSFET turns it ON while *P*-channel device goes to non-conducting state. The output goes to GND through the conducting *N*-channel MOSFET. The output is thus logic LOW for a logic HIGH input.

Q. 30. Current regulating diodes often comprise a junction FET inside the device. What is the configuration used and what are the terminals brought out ?

Ans. A junction field effect transistor (JFET) with gate shorted to the source constitutes a current regulating diode where the drain current varies by an extremely small amount for large variation in the drain voltage (Fig. 10.23. (*a*)) Junction FET is preferred over a MOSFET as the former has a much flatter characteristics thus providing better current regulation.

Solved Problems

Problem 10.1. *An N-channel junction FET has a pinch-off voltage (V_p) of – 6V. What would be the drain current (I_d) for a gate source voltage (V_{GS}) of – 3V given that the drain current for V_{GS} = 0 is 10 mA ?*

Solution. Drain current (I_d) is related to (V_{GS}) and (V_P) by

$$I_d = I_{DSS}\left[1-\left(\frac{V_{GS}}{V_P}\right)\right]^2$$

Also I_{DSS} = Drain current for $V_{GS} = 0 = 10\ mA$

Therefore, $I_d = 10\left[1-\left(\frac{-3}{-6}\right)\right]^2 = 10\,(1-0.5)^2 = 10 \times 0.25$ = 2.5 mA

Problem 10.2 *Figure 10.28 shows the transfer characteristics of an enhancement mode MOSFET. Determine the drain. current for $V_{GS} = 8$ V.*

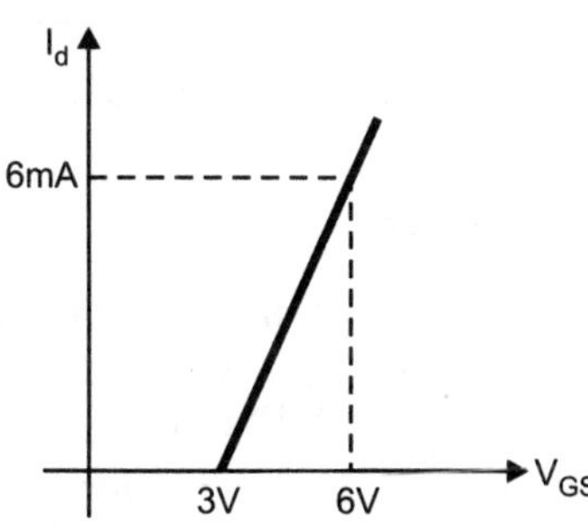

Fig. 10.28

Solution. For an enhancement mode MOSFET,

$$I_d = K[V_{GS} - V_{GS}(th)]^2$$

where K = constant

Here $I_d = 6\ mA$ for $V_{GS} = 6V$ and $V_{GS}(th) = 3V$

Therefore $6 = K(6-3)^2 = 9K$

$$K = \frac{2}{3}\ mA/V^2 = \frac{2}{3} \times 10^{-3}\ A/V^2$$

Now $$I_d = K[V_{GS} - V_{GS}(th)]^2 = \frac{2}{3} \times 10^{-3}(8-3)^2$$

$$= \frac{2}{3} \times 25 \times 10^{-3} = \frac{50}{3}\ \text{mA} = 16.7\ \text{mA}.$$

Problem 10.3. *A JFET is used as a common source AC amplifier. The drain resistance is 20 K and the amplifier feeds a load resistance of 20 K. The FET has the following specifications:*

$g_{fs} = 1000\ \mu$ mhos, $C_{gs} = 8$ pF, $C_{gd} = 4$ pF, $C_{ds} = 2$ pF. Determine the total input capacitance.

Solution. In capacitance, $C_i = C_{gs} + C_{gd}(1 + A_v)$ where A_v = Voltage gain

$= - g_{fs} \times$ (parallel combination of drain resistance and load resistance)

$$= -1000 \times 10^{-6} \times \frac{20 \times 20}{20 + 20} \times 10^3 = -10$$

This gives $C_i = 8 + 4[1 - (-10)] = 8 + 44 = 52\ pF$

Problem 10.4. *A certain junction FET's data sheet gives $I_{DSS} = 10$ mA and $g_{mo} = 5000$ micro-mho. Determine gate source cut-off voltage V_{GS} (off) for this JFET and also the value of (gm) for $V_{GS} = -3V$.*

Solution. Gate-source cut-off voltage V_{GS} (off) is given by:

$$V_{GS}(\text{off}) = \frac{-2I_{Dss}}{g_{mo}} = \frac{-2 \times 10 \times 10^{-3}}{5000 \times 10^{-6}} = -4\text{V}$$

Also, $$g_m = g_{mo}\left[1 - \frac{V_{GS}}{V_{GS}(off)}\right] = 5000\left(1 - \frac{-3}{-4}\right)$$

$$= 5000 \times (1/4) = 1250 \text{ micro-mho}$$

Problem 10.5 A *certain JFET has a maximum transconductance of 6000 μmho. Determine the drain current (I_d) for $V_{GS} = -3$ V if $I_{DSS} = 20$ mA. Also determine the magnitude of transconductance at this drain current.*

Solution. Maximum transconductance, $g_{mo} = 6000\ \mu$ mho

Also, I_{DSS} = 20 mA

Therefore, $$V_{GS}(\text{off}) = \frac{-2 \times 20 \times 10^{-3}}{6000 \times 10^{-6}} = -6.66\text{ V}$$

Now $$I_d = I_{DSS}\left[1 - \frac{V_{GS}}{V_{GS}(off)}\right]^2 = 20 \times 10^{-3}\left(1 - \frac{3}{6.66}\right)^2$$

$= 20 \times 10^{-3} \times (0.55)^2 = 20 \times 10^{-3} \times 0.3025 = 6$ mA

$$g_m = g_{mo}\left[1 - \frac{V_{GS}}{V_{GS}(off)}\right] = 6000\left(1 - \frac{3}{6.66}\right) = 6000 \times 0.55$$

$= 3300\ \mu$ ohm.

Problem 10.6. *Refer to the JFET circuits of Figs. 10.29. (a) and (b), The JFET used in the two cases has $I_{DSS} = 20$ mA and V_{GS} (off) = – 4 V. If the input signal is 100 mV peak to peak, determine the output signal for $V_{GS} = 0$ and also for $V_{GS} = -5V$.*

Solution. For Fig. 10.29. (*a*):

Drain current for $V_{GS} = 0 = I_{DSS} = 20$ mA

V_{GS} (off) $= -4V = -V_P$ (Pinch-off voltage)

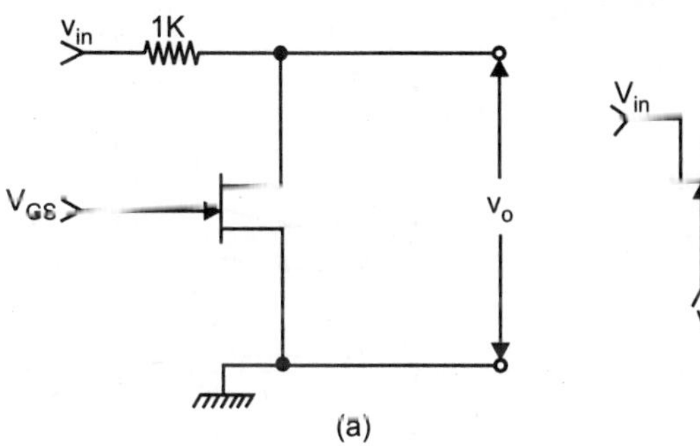

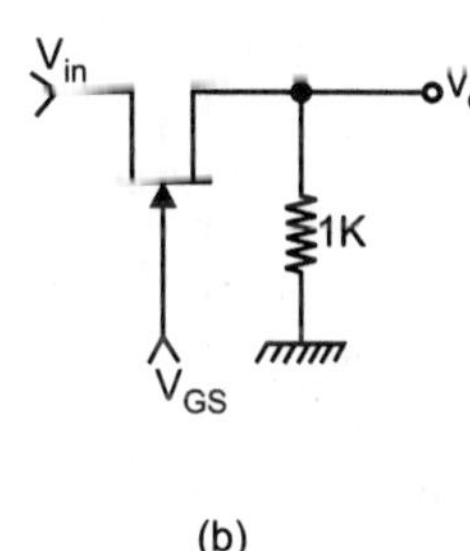

Fig. 10.29

Therefore,

$$V_P = 4\text{V}$$

$$R_{DS} = \frac{4}{20 \times 10^{-3}} = 200\ \Omega$$

For $V_{GS} = 0$, $$v_0 = \frac{v_{in} \times 200}{200 + 1000} = \frac{v_{in}}{6} = \frac{100}{6} = 16.67\text{ mV}$$ p-t-p

For V_{GS} = – 5 V, JFET is in cut-off

Therefore, $v_0 = v_{in}$ = 100 mV p-t-p

For Fig. 10.29. (*b*):

For V_{GS} = 0, $v_0 = v_{in}$ = 100 mV p-t-p

For V_{GS} = – 5 V, v_0 = 0 as JFET is in cut-off.

Problem 10.7. *Fig. 10.30(a) shows the transfer characteristics of a junction FET. Determine the drain current for V_{GS} = – 3V. Also determine the maximum value of transconductance.*

Solution. From the characteristics, I_{DSS} = 15 mA, V_{GS} (off) = – 6 *V*. The transfer characteristics of Fig. 10.30 (*a*) can be expressed by:

$$I_D = I_{DSS}\left[1 - \frac{V_{GS}}{V_{GS}(off)}\right]^2 = 15 \times 10^{-3}\left[1 + \frac{V_{GS}}{6}\right]^2$$

For V_{GS} = – 3 V, $I_D = 15 \times 10^{-3} (1 - 0.5)^2$
$= 15 \times 10^{-3} \times 0.25$ = 3.75 mA

Maximum value of transconductance, $g_{mo} = \dfrac{-2\, I_{DSS}}{V_{GS}(off)}$

$= \dfrac{I_{DSS}}{3}$ = 5 milli-mho.

Problem 10.8. *Figure. 10.30 (b) shows the drain characteristics of a junction FET. Determine (a) gate source cut-off voltage V_{GS} (off) (b) I_{DSS} (c) maximum value of drain-source voltage in the ohmic region (d) range of drain source voltage over which it acts as a constant current source (e) V_{GS} with a question mark.*

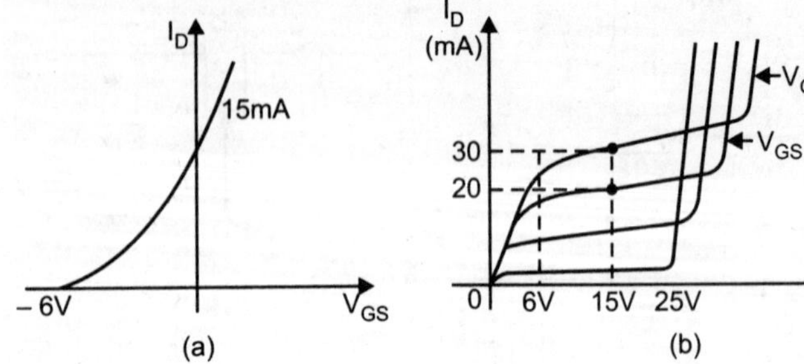

Fig. 10.30

Solution.

(*a*) V_{GS} (off) = – V_P = – 6 V (*b*) I_{DSS} = 30 mA

(c) 6 V (*d*) 6 V to 25 V

(*e*) Drain current (I_D) as a function of gate source voltage (V_{GS}) is given by:

$$I_D = I_{DSS}\left[1 - \frac{V_{GS}}{V_{GS}(off)}\right]^2 \text{ or } 20 = 30\left(1 + \frac{V_{GS}}{6}\right)^2$$

$$\text{or } \left(1 + \frac{V_{GS}}{6}\right)^2 = \frac{2}{3} \text{ or } \left(1 + \frac{V_{GS}}{6}\right) = 0.816$$

or V_{GS} = 6 (0.816 – 1) = – 1.1 V.

Problem 10.9. *Figure 10.31(a) shows the transfer characteristics of an enhancement mode MOSFET. Determine the drain current for V_{GS} = 5V. If this MOSFET is the type chosen for the circuit of Fig. 10.31. (b), then what would be the drain-source voltage ?*

Solution. The transfer characteristics of an enhancement MOSFET are expressed by:

$$I_D = K I_D \text{ (on)}$$

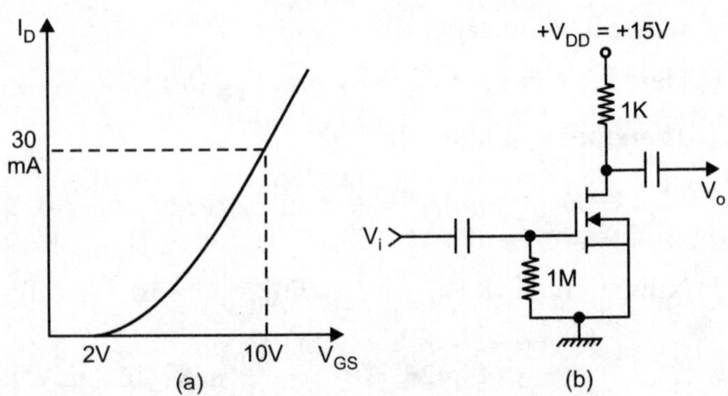

Fig. 10.31.

$$\text{where } K = \left[\frac{V_{GS} - V_{GS}(th)}{V_{GS}(on) - V_{GS}(th)}\right]^2$$

Here, V_{GS} (th) = 2 V, V_{GS} (on) = 10 *V*, I_D (on) = 30 mA

$$\text{Therefore, } K = \left(\frac{V_{GS} - 2}{10 - 2}\right)^2 = \left(\frac{V_{GS} - 2}{8}\right)^2$$

For V_{GS} = 5 V, $K = (3/8)^2 = 9/64$

Therefore, I_D (for V_{GS} = 5 *V*) = $\dfrac{9}{64} \times 30$ = 4.22 mA

In the circuit shown in Fig. 10.31 (*b*) V_{GS} = 0 which means that MOSFET is in cut-off.

Therefore, drain source voltage for V_{GS} = 0 = + 15 V.

Problem 10.10: *For the self-bias circuit in Fig. 10.32, determine the value of drain current (I_D) and gate-source voltage (V_{GS}).*

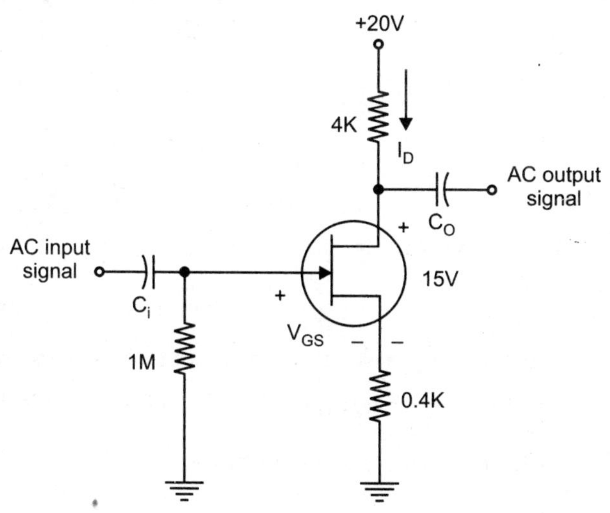

Fig. 10.32

Solution: The value of drain current is given by

$$I_D = \frac{V_{DD} - V_{DS}}{R_D + R_S}$$

Therefore, $I_D = (20 - 15)/(4 \times 10^3 + 0.4 \times 10^3) = 5/4.4 \times 10^3 = 1.14$ mA

The gate-source voltage is given by $V_{GS} = -I_D \times R_S = -1.14 \times 10^{-3} \times 0.4 \times 10^3 = 0.456$ V.

Drain current $I_D = 1.14\ mA$ and gate-source voltage $V_{GS} = 0.456$ V.

Problem 10.11: *For the circuit shown in Fig. 10.33, determine the value of drain-source voltage (V_{DS}). (Given that the gate-source voltage is – 0.8 V).*

Solution. The gate voltage is given by $V_G = 12 \times 100 \times 10^3/(100 \times 10^3 + \times 900 \times 10^3) = 1.2$ V

The gate-source voltage is equal to $V_{GS} = V_G - V_S$

Therefore $-0.8 = 1.2 - V_S$ or $V_S = 2.0$ V

The drain current is equal to $I_D = V_S/R_S = 2/(1 \times 10^3) = 2$ mA

The drain-source voltage is $V_{DS} = 12 - 2 \times 10^3 \times 2 \times 10^{-3} - 1 \times 10^3 \times 2 \times 10^{-3} = 12 - 4 - 2 = 6$ V

The drain-source voltage is equal to 6 V.

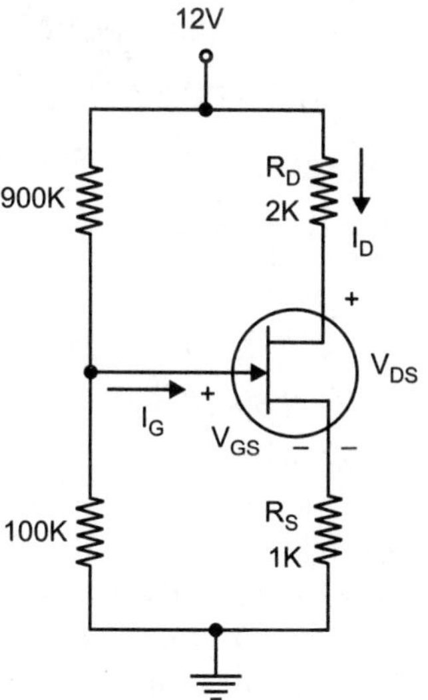

Fig. 10.33

Problem 10.12: *Figure 10.34 shows a circuit using E-MOSFET. Given that the threshold voltage for the MOSFET is 2 V and $I_{D(on)} = 6$ mA for $V_{GS(on)} = 5$ V, determine the value of the operating point.*

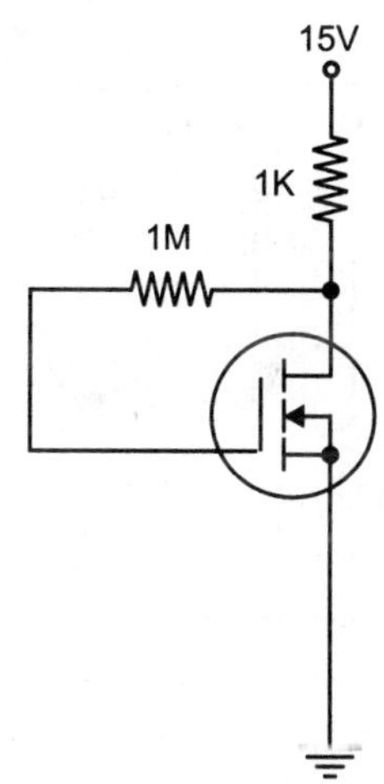

Fig. 10.34

Solution: Drain current in an E-MOFET is given by $I_D = K(V_{GS} - V_T)^2$

Therefore, $K = 6 \times 10^{-3}/(5-2)^2 = 0.67 \times 10^{-3}$ A/V^2 = 0.67 mA/V^2

Gate-source voltage in the feedback configuration is given by

$$V_{GS} = V_{DD} - I_D R_D$$

$$V_{GS} = 15 - I_D \times 1 \times 10^3 = 15 - 1000 I_D$$

Substituting this value of I_D in the expression $I_D = K(V_{GS} - V_T)^2$ we get

$$I_D = 0.67 \times 10^{-3}(15 - 1000 I_D - 2)^2$$

$$I_D = 0.67 \times 10^{-3}(13 - 1000 I_D)^2$$

$$3000I_D = 2\,(169 - 26000I_D + 10^6 I_D{}^2)$$

$$2 \times 10^6 I_D{}^2 - 55000I_D + 338 = 0$$

Solving for I_D we get I_D 9.3×10^{-3} A = 9.3 mA

The value of the drain-source voltage (V_{DS}) is given by

$$V_{DS} = V_{DD} - I_D R_D$$

$$V_{DS} = 15 - 9.3 \times 10^{-3} \times 1 \times 10^3$$

$$= 15 - 9.3 = 5.7 \text{ V}$$

The operating point is given by (9.3 mA, 5.7 V).

Problem 10.13: *Determine the values of voltages V_D and V_C for the circuit shown in Fig. 10.35, given that the transistor β is 100, V_{BE} = 0.7V, saturation drain current of JFET is 10 mA and the pinch-off voltage is –5 V.*

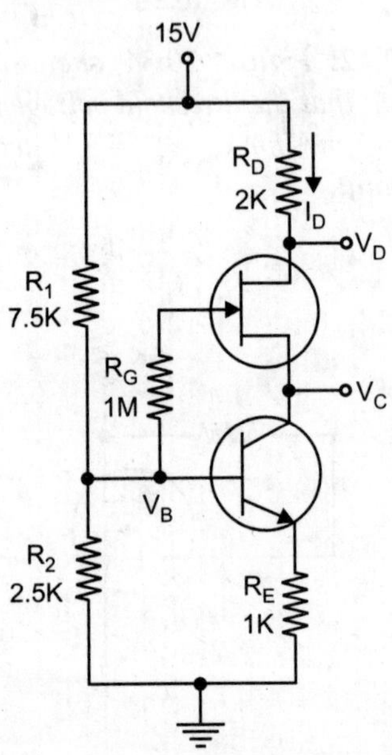

Fig. 10.35

Solution: The figure shows the voltage-divider transistor configuration with an additional JFET connected between the base and the collector terminals.

Value of $\beta \times R_E = 100 \times 1 \times 10^3 = 10^5$

Value of $10 \times R_2 = 10 \times 2.5 \times 10^3 = 2.5 \times 10^4$

As the value of ($\beta \times R_E$) is much large than ($10 \times R_2$), approximate analysis can be done to analyse the voltage-divider transistor configuration.

$$V_B = 15 \times 2.5 \times 10^3/(2.5 \times 10^3 + 7.5 \times 10^3) = 3.75 \text{ V}$$

$$V_E = V_B - V_{BE} = 3.75 - 0.7 = 3.05 \text{ V}$$

$$I_E = V_E/R_E = 3.05/(1 \times 10^3) = 3.05 \text{ mA}$$

$$I_C \cong I_E = 3.05 \text{ mA}$$

In the figure shown $I_D = I_S = I_C = 3.05$ mA

The drain voltage (V_D) is equal to $V_D = 15 - I_D \times 2 \times 10^3$

$$V_D = 15 - 3.05 \times 10^{-3} \times 2 \times 10^3 = 15 - 6.1 = 8.9 \text{ V}$$

As the current through the gate resistor is zero, therefore there is no voltage drop across the resistor R_G. The value of collector voltage (V_C) is equal to $V_C = V_B V_{GS}$

In a JFET the drain current is given by

$$I_D = I_{DSS}\left(1 - \frac{V_{GS}}{V_P}\right)^2$$

$$3.05 \times 10^{-3} = 10 \times 10^{-3}\,[1 - V_{GS}/(-5)]^2$$

$$V_{GS} = -2.24 \text{ V}$$

Therefore, $V_C = 3.75 - (-2.24) = -5.99$ V

Hence, $V_C = -5.99$ V, $V_D = 8.9$ V

Problem 10.14: *The circuit shown in Fig. 10.36 uses a JFET in an AGC circuit. The voltage V_C is the control voltage. Derive the expression for the ratio of the output voltage to the input voltage. Also find the value of the output voltage for V_{in} = 5 V when the control voltage V_C is equal to (a) 0 V (b) 0.5 V_P and (c) V_P (Given that R_L = 100 K, R = 10 K, the resistance of the JFET at zero gate-source voltage is 10 K.)*

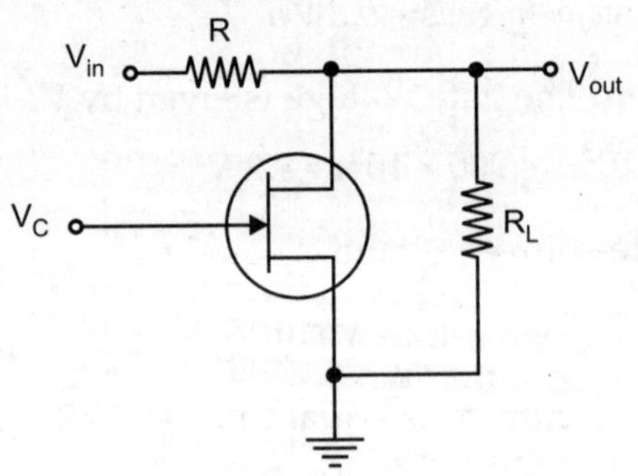

Fig. 10.36

Solution: If the value of the drain resistance of the FET is r_d then the ratio V_o to V_{in} is given by

$$\frac{V_o}{V_{in}} = \frac{r_d \parallel R_L}{(r_d \parallel R_L) + R} = \frac{r_d \times R_L}{(r_d \times R_L) + R(r_d + R_L)}$$

The value of the resistance (r_d) is given by the expression

$$r_d = \frac{r_0}{(1 - V_{GS}/V_P)^2}$$

where r_o is the resistance at $V_{GS} = 0$.

The control voltage is applied between the gate and the source terminals, therefore $V_C = V_{GS}$

When $V_C = 0$, $r_d = r_o = 10\ K$

$$V_o = \frac{5 \times 10 \times 10^3 \times 100 \times 10^3}{(10 \times 10^3 \times 100 \times 10^3) + 10 \times 10^3\,(10 \times 10^3 + 100 \times 10^3)}$$

$$= \frac{5 \times 1000 \times 10^6}{1000 \times 10^6 + 1100 \times 10^6} = 2.38\ \text{V}$$

When $V_C = 0.5\ V_P$, $r_d = 4\ r_o = 40\ \text{K}$

$$V_o = \frac{5 \times 40 \times 10^3 \times 100 \times 10^3}{(40 \times 10^3 \times 100 \times 10^3) + 10 \times 10^3\,(40 \times 10^3 + 100 \times 10^3)}$$

$$= \frac{5 \times 4000 \times 10^6}{4000 \times 10^6 + 1400 \times 10^6} = 3.78\ \text{V}$$

When $V_C = V_P$, r_d tends to ∞. Therefore

$$\frac{V_o}{V_{in}} = \frac{R_L}{R_L + R}$$

$$= 100 \times 10^3/(100 \times 10^3 + 10 \times 10^3) = 0.91$$

$$V_o = 5 \times 0.91 = 4.55\ \text{V}$$

Therefore,

$$\frac{V_o}{V_{in}} = \frac{r_d \times R_L}{(r_d \times R_L) + R(r_d + R_L)}$$

(*a*) V_o=2.38 V; (*b*) V_o = 3.78 V; (*c*) V_o = 4.55 V

SUMMARY OF IMPORTANT FORMULAS

1. Drain resistance of a junction FET is given by,

$$r_d = \left|\frac{\Delta V_{ds}}{\Delta I_{ds}}\right|,\ V_{GS} = \text{constant}$$

where r_d = Drain resistance

ΔV_{ds} = Change in voltage between drain and source

ΔI_{ds} = Corresponding change in current between drain and source

V_{GS} = Voltage between gate and source

2. Transconductance of a junction FET is given by,

$$g_{fs} = \left|\frac{\Delta I_{ds}}{\Delta V_{gs}}\right|,\ V_{DS} = \text{constant}$$

where g_{fs} = Transconductance

ΔI_{ds} = Change in current between drain and source

ΔV_{gs} = Change produced in voltage between gate and source

V_{DS} = Voltage between drain and source

3. Amplification factor of a junction FET is given by,

$$\mu = \left|\frac{\Delta V_{ds}}{\Delta V_{gs}}\right| I_{DS} = \text{constant}$$

Also,

$$\mu = (g_{fs} \times r_d)$$

where μ = Amplification factor

ΔV_{ds} = Change in voltage between drain and source

ΔV_{gs} = Change in voltage between gate and source

I_{DS} = Current between drain and source

SELF EVALUATION EXERCISE

1. In Table 10.1 column-1 are listed six types of FETs/MOSFETs. The circuit symbols, the physical construction and the V-I characteristics (Drain characteristics or Transfer characteristics) are given in a haphazard manner in columns 2, 3 and 4 respectively. You have to identify the relevant circuit symbol, the physical construction and the characteristics for each of the FET/MOSFET types listed in column-1. For instance, if you feel that N-channel JFET circuit symbol is the one given at number 2 from top in column-2, its physical construction is at number 3 from top in column-3 and the relevant characteristics are at number 4 from top in column = – 4, then your answer will be (1-2-3-4).

Table 10.1

Device	*Circuit symbol*	*Physical construction*	*Characteristics*
N-Channel	D, G, S	S, G, D; P; N; P	$V_{GS} = 0$, – 0.5V, – 1V, – 1.5V, – 2V; I_{DS}; $V_{DS}(V)$ →
P-Channel JFET	D, G, S	S, G, D; N+; N; N+; P	$V_{GS}(V)$ →; I_{DS} (mA)
N-Channel depletion mode MOSFET	D, G, S	S, G, D; N; N; P	← V_{DS}; 2, 1.5, 1, 0.5, $V_{GS} = 0$; I_{DS}
N-Channel enhancement mode MOSFET	D, G, S	S, G, D; P; P; N	← V_{GS}; – 1; I_{DS} (mA); 5
P-Channel Depletion mode MOSFET	D, G, S	S, G, D; N; P; N	5; I_{DS} (mA); 1; V_{GS} →
P-Channel enhancement mode MOSFET	D, G, S	S, G, D; P+; P; P+; N	I_{DS} (mA); 5; – 2, – 1, 1, 2; ← $V_{GS}(V)$

2. **Answer the following:**
   (*i*) Do we have enhancement mode JFETs also?
   (*ii*) What is the substrate semiconductor material in an N-channel JFET?
   (*iii*) What will be the effect of increased negative gate voltage on the drain current in an N-channel depletion MOSFET?
   (*iv*) Which one of the two, JFET or MOSFET, is likely to have lower gate drain inter-electrode capacitance?
   (*iv*) What do you think is the most significant advantage of a JFET or a MOSFET as compared to a bipolar transistor?
3. Recognize the FET circuit configuration shown in Fig. 10.37 (*a*), (*b*) and (*c*).

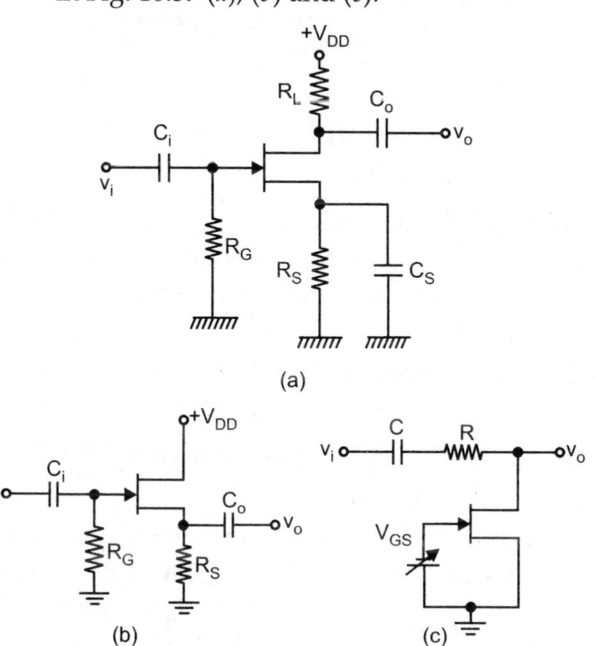

Fig. 10.37

4. Fig. 10.38 (*a*) shows the application of a JFET as a voltage variable attenuator Fig. 10.38 (*b*) shows the circuit diagram of a N-channel MOSFET used as a switch. There is something conceptually wrong with each of these configurations. What is it?

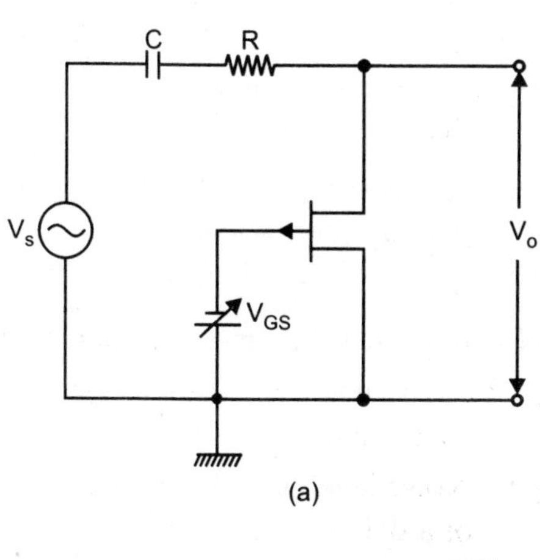

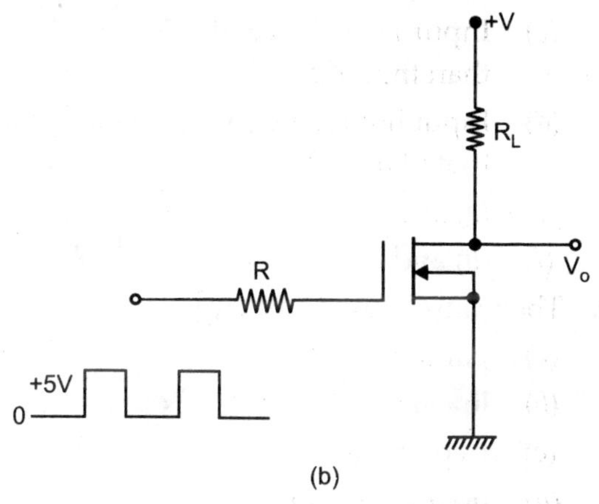

Fig. 10.38

Multiple Choice Questions

1. FETs are
   (*a*) voltage controlled devices with high input impedance
   (*b*) current controlled devices with low input impedance
   (*c*) voltage controlled devices with low input impedance
   (*d*) current controlled devices with high input impedance
2. For low values of drain-source voltages, the JFET acts as a
   (*a*) current source (*b*) voltage source
   (*c*) BJT (*d*) resistor
3. JFETs are also referred to as square-law devices because
   (*a*) the drain current varies as square of the drain-source voltage for fixed value of gate-source voltage
   (*b*) the drain current varies as square of the gate-source voltage for fixed drain source voltage

(*c*) the gate current varies as square of the drain-source voltage for fixed value of gate-source voltage

(*d*) the gate current varies as square of the gate-source voltage for fixed drain-source voltage

4. Which of the following devices has revolutionised the field of computers?
   (*a*) BJTs
   (*b*) enhancement MOSFETs
   (*c*) depletion MOSFETs
   (*d*) JFETs

5. Which of the following statements is/are false?
   (*a*) Noise level of a FET device is more than that of a BJT
   (*b*) Noise level of a FET device is less than that of a BJT
   (*c*) Input impedance of a MOSFET is greater than that of a JFET
   (*d*) Input impedance of a MOSFET is less than that of a JFET
   (*e*) (*a*) and (*c*)
   (*f*) (*b*) and (*c*)

6. The transconductance curve of a JFET is
   (*a*) parabolic
   (*b*) linear
   (*c*) hyperbolic
   (*d*) none of the above

7. Depletion MOSFETs can operate in
   (*a*) depletion mode only
   (*b*) enhancement mode only
   (*c*) both depletion and enhancement modes
   (*d*) none of the above

8. Dual-gate MOSFETs can be considered to be counterpart of
   (*a*) triodes
   (*b*) diodes
   (*c*) tetrodes
   (*d*) pentodes

9. The principle of operation of a VMOS device is similar to that of
   (*a*) enhancement MOSFET
   (*b*) depletion MOSFET
   (*c*) insulated gate bipolar transistor
   (*d*) junction FET

10. Which of the following statements is/are true?
   (*a*) IGBTs offer fast switching times similar to that of MOSFETs and lower on-state voltages and larger blocking voltages similar to that of BJTs
   (*b*) IGBTs offer fast switching times similar to that of BJTs and lower-on state voltages and larger blocking voltages similar to that of BJTs
   (*c*) The equivalent circuit of an IGBT is an n-channel. power MOSFET driving a wide base PNP transistor in a darlington configuration
   (*d*) The equivalent circuit of an IGBT is an n-channel power MOSFET driving a wide base PNP transistor in a cascode configuration
   (*e*) (*a*) and (*c*)
   (*f*) (*a*) and (*d*)
   (*g*) (*b*) and (*c*)

11. Which MOSFET allows the flow of drain current even with zero gate to source voltage just due to existence of channel between drain and source terminals?
   (*a*) Depletion MOSFET
   (*b*) Enhancement MOSFET
   (*c*) Depletion-Enhancement MOSFET
   (*d*) All of the above

12. The N-channel enhancement type basically termed as normally-OFF MOSFET works only with?
   (*a*) Large positive gate voltage
   (*b*) Large negative gate voltage
   (*c*) Large positive drain voltage
   (*d*) Large negative drain voltage

13. During the occurrence of latch-up condition in an IGBT, which of the following is true?
   (*a*) I_g is no longer controllable
   (*b*) I_c is no longer controllable
   (*c*) The device turns off
   (*d*) All of the above

14. Which of the following is the similarity between IGBT and BJT?
   (*a*) Low on-state power losses
   (*b*) High on-state power losses

(c) Low switching losses
(d) High input impedance

15. A MOSFET is sometimes called
    (a) Many gate JFET
    (b) Open gate JFET
    (c) Insulated gate JFET
    (d) Shorted gate JFET

16. The gate voltage in a JFET at which drain current becomes zero is called?
    (a) Cut-off voltage
    (b) Saturation voltage
    (c) Active voltage
    (d) Pinch-off voltage

Answers of Self Evaluation Exercise

1. (1-2-1-1) (2-1-5-3) (3-5-2-6) (4-6-3-5) (5-3-6-2) (6-4-4-4)
2. (i) No. JFETs are inherently depletion mode devices
   (ii) P-type material
   (iii) The drain current decreases
   (iv) MOSFET
   (v) It is the high input impedance, typically 10^6 to 10^7 ohms in JFETs and 10^{12} to 10^{15} ohms in MOSFETs.
3. Fig. 10.37 (a) : Common-Source amplifier
   Fig. 10.37 (b) : Source follower or Common drain amplifier
   Fig. 10.37 (c) : Voltage variable resistor
4. Fig. 10.38 (a) : The gate-source voltage polarity should have been positive as the JFET being used is a P-channel FET
   Fig. 10.38 (b) : The gate waveform should be negative going from zero to make N-channel depletion MOSFET switch between ON and OFF states.

Multiple Choice Questions

1. (a)	2. (d)	3. (b)	4. (b)	5. (e)	6. (a)
7. (c)	8. (c)	9. (a)	10. (e)	11. (c)	12. (a)
13. (b)	14. (a)	15. (c)	16. (d)		

Chapter 11

UJTs and Thyristors

Learning Objectives

After completing the chapter, you will learn the following topics:

1. Uni-Junction Transistor
2. Input diode characteristics of UJT
3. Occurrence of negative resistance
4. UJT as a current controllable device
5. UJT relaxation oscillator circuit
6. P-N-P-N diode
7. Characteristics of P-N-P-N diode
8. Silicon Controlled Switch (SCS)
9. SCS and four layer diode
10. Characteristics of SCS
11. Rate effect and its effect on P-N-P-N diode operation
12. Specifications and characteristic parameters of Thyristors
13. Thyristors as current controllable negative resistance devices
14. DIAC and TRIAC
15. Typical application circuits of Thyristors
16. Programmable UJT (PUT)

Q.1. What does UJT stand for ? Justify the name UJT.

Ans. UJT stands for Uni-Junction Transistor.

In fact, this device has got only one junction that is why the name unijunction, and since it has got three terminals like a transistor, hence the name unijunction transistor.

Q.2. Show the constructional features of a UJT diagrammatically and also, draw its circuit symbol.

Ans. The constructional features of a UJT are shown in Fig. 11.1 (*a*).

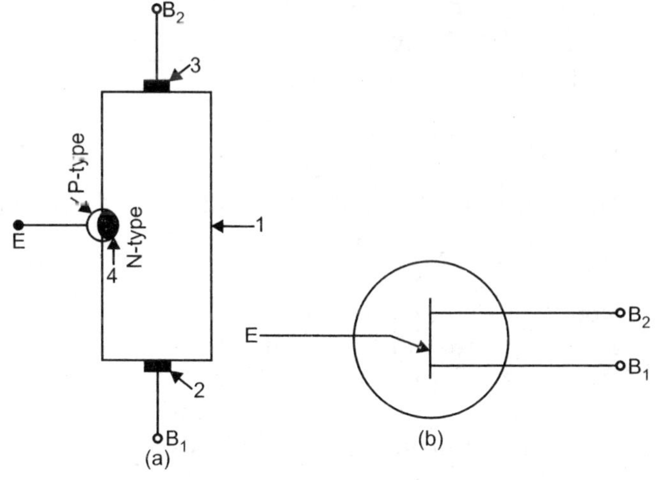

Fig. 11.1

1. N-type silicon bar 2. Ohmic contact, B_1
3. Ohmic contact, B_2 4. *P*-type material

Figure 11.1 (*b*) shows the circuit symbol.

Q.3. Draw the input diode characteristics of UJT and explain the occurrence of negative resistance in particular.

Ans. Fig 11.2 (*a*) shows the test circuit for determining the input characteristics. The characteristics are shown in Fig. 11.2 (*b*), It is essential to forward bias a *P-N* junction diode by a voltage $V_Y (= 0.5\,V)$, if it is to exhibit a very low resistance path. Thus, the voltage across the *P-N* junction diode has to be 0.5 *V* with *P*-side positive to make the conduction possible.

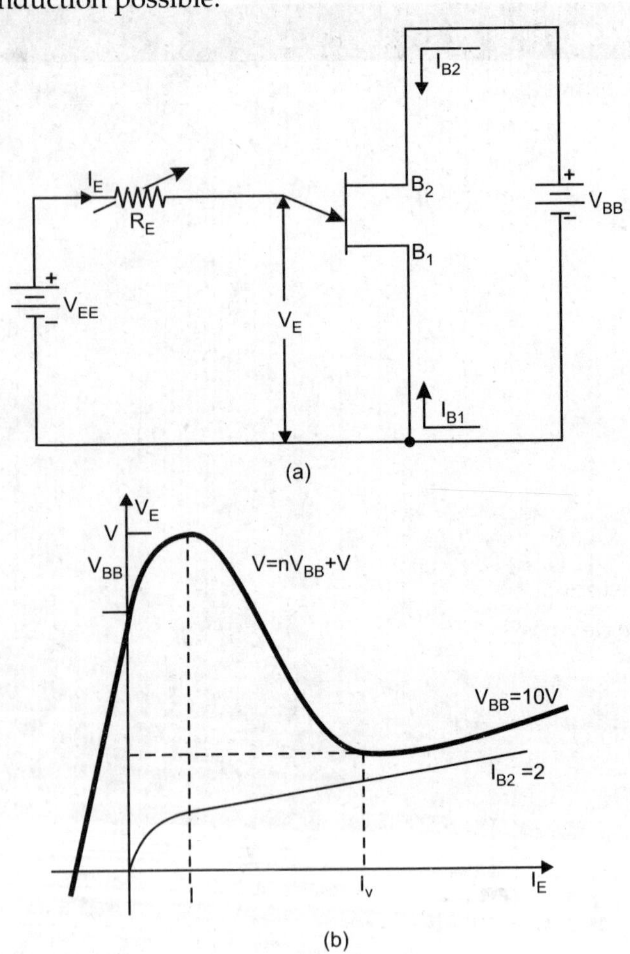

Fig. 11.2

The voltage appearing on *N*-side of the diode is equal to $\eta\, V_{BB}$, where $\eta = \dfrac{R_2}{R_1 + R_2}$.

Note that R_1 is the resistance of the semiconductor bar between *E* and B_1 and R_2 is the resistance between *E* and B_2. Now this input diode shall conduct only, when V_E exceeds 5.5 V ($V_E = \eta\, V_{BB} + V_\gamma$). The emitter voltage V_E at which diode starts conducting is termed as V_p.

From the characteristics, as the voltage is increased, the current remains negligibly small in magnitude till a voltage ($= V_p$) is reached at which diode is forward biased. After that current increases rapidly. It is seen that as the current increases, the voltage decreases giving a negative resistance region. Such a behaviour can be explained considering the fact that due to increase in forward current, the resistance of EB_1 region falls rapidly resulting in the reduction in the voltage. Finally, when I_E becomes very large, it may be considered to be much greater than I_{B2}. Under these conditions, the curve approaches asymptotically the curve for $I_{B2} = 0$ giving rise to a valley point. The curve for $I_{B2} = 0$ is the same as for ordinary *P-N* junction diode.

The region before peak point (*Vp, Ip*) is called cut-off region, as the input diode remains reverse biased in this region. The region after the valley point (V_v, I_v) is termed as saturation region, as there is not much increase in the emitter voltage for large changes in emitter current.

Q.4. Uni-Junction Transistor is a current controllable device. Comment.

Ans. As seen from the characteristics, the curve is a single valued function of current and a multivalued function of voltage. A single valued function of current implying that for each current value, there is a unique voltage. Such a behaviour is contrary to one seen in the tunnel diode characteristics which could be expressed by a single valued function of voltage. That is what made tunnel diode a voltage controllable device. Due to single valued function of current nature of its V-I characteristics, a UJT is called a current controllable device.

Q.5. Draw UJT relaxation oscillator circuit and explain its operation.

Ans. Fig 11.3 (*a*) shows the basic UJT relaxation oscillator circuit. The sawtooth waveform that appears across the capacitor is shown in Fig. 11.3(b). Initially UJT is in the cut-off region, *i.e.* the input diode is reverse biased. The capacitor starts charging from V_{CC} through *R*. When the voltage V_C across the capacitor becomes large enough to forward bias the input diode, the capacitor starts discharging through the low resistance between the emitter-base B_1 region. This discharge process continues until it reaches a point, where input diode is again reverse biased. At this point, the capacitor starts charging again. The process of charging through (*R*), comparatively a higher resistance, and discharging through low forward resistance of input diode continues and gives rise to a sawtooth signal as shown in Fig. 11.3 (*b*).

If the capacitor in the relaxation oscillator of Fig.11.3 (*a*) could be charged through a constant current source Fig.11.4 (*a*), the result will be a perfect sawtooth waveform as shown in Fig. 11.4 (*b*).

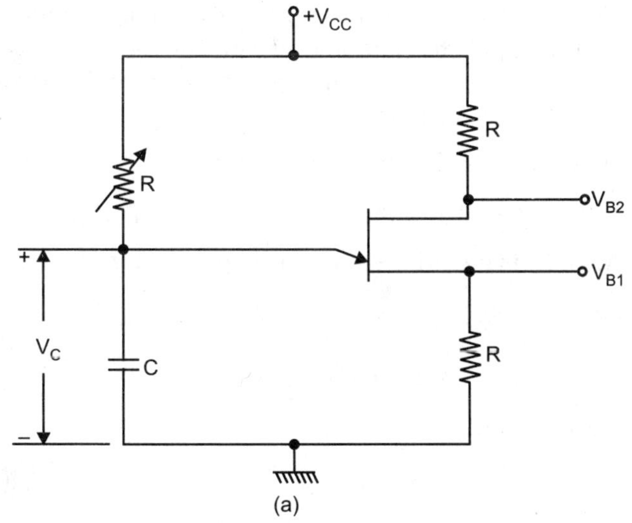

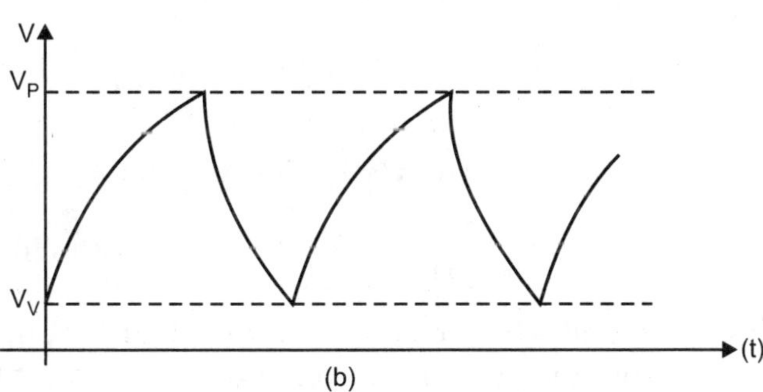

Fig. 11.3

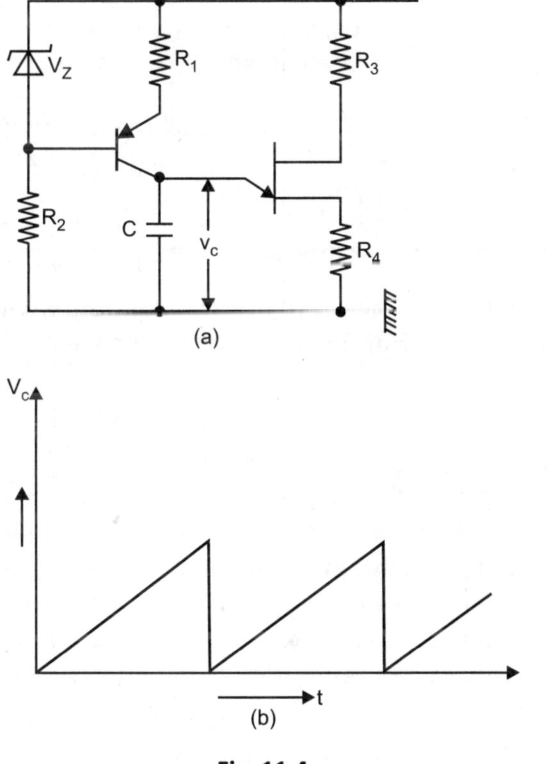

Fig. 11.4

Q. 6. Can you use the relaxation oscillator circuit discussed in the last problem to generate a train of pulses ?

Ans. If we see the voltage waveforms at B_1 and B_2, it is observed that we get a train of positive pulses at B_1 and a train of negative pulses at B_2. The pulses occur during the time UJT is ON. The waveforms are shown in Fig. 11.15.

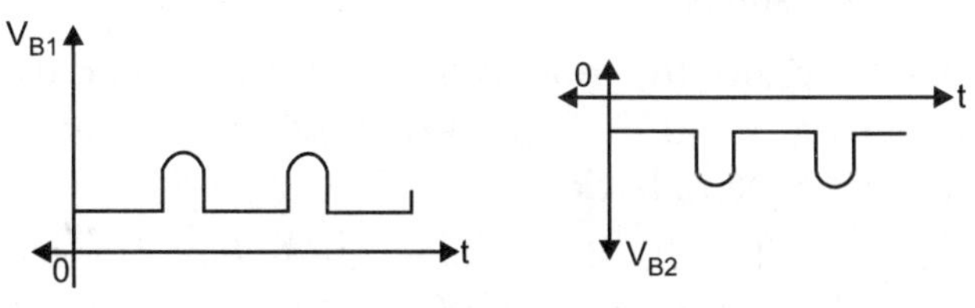

Fig. 11.5

Q.7. What is a *P-N-P-N* diode ? Draw its circuit symbol and explain in brief its operation.

Ans. It is a four-layer diode consisting of four alternate layers of *P*-type and *N*-type materials as shown in Fig. 11.6.

Two semiconductor regions at the corners (*P* and *N*) are called anode and cathode respectively.

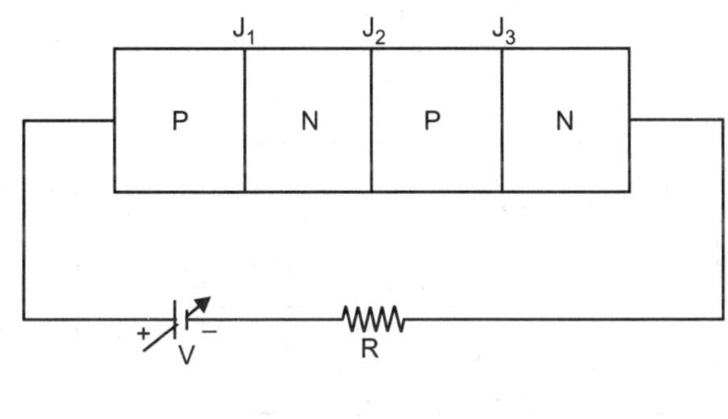

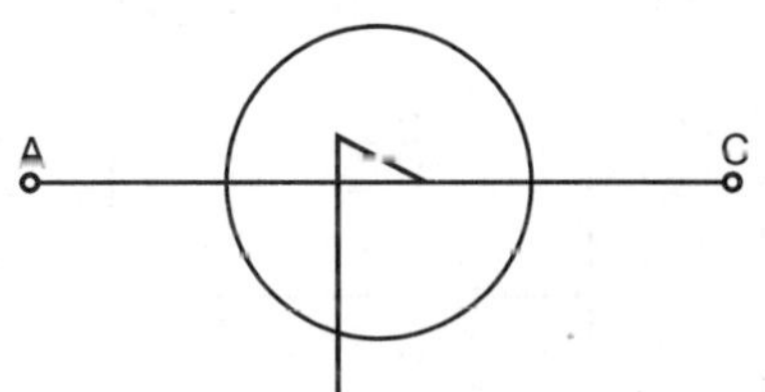

Fig. 11.6

If anode (*P*-type) is made positive with respect to the cathode (*N*-type) as shown above, then junction J_1 and J_3 are forward biased. This applied voltage then effectively appears across junction J_2, thus reverse biasing J_2. Now if the applied voltage is increased, a stage comes when breakover of J_2 occurs (from the reverse region characteristics of a *P-N* diode).

At this voltage, the current may increase all of a sudden from a very small value to a very large value. This increase in current is accompanied by reduction in the voltage giving rise to a negative resistance region. The reduction in voltage can be explained on the basis that after breakover occurs, all the junctions are effectively forward biased and the net voltage across anode and cathode is equal to sum of the three junction voltages.

One of the junction voltages is in a direction opposite to that of other two junction voltages so that the net voltage is equal to that across one of the junction diodes (= 0.6 V) for silicon diodes.

Q.8. Explain the operation of a four-layer diode considering it to be composed of two transistors connected back to back.

Ans. Fig. 11.7 illustrates how a *P-N-P-N* diode may be viewed as a composite of two transistors (One of them being *P-N-P* and other *N-P-N*) connected back to back. But we must always remember that two transistors connected back to back will not make a *P-N-P-N* diode.

Now the collector current of transistor in the active region is given by :

$$I_C = -\alpha I_E + I_{co} \quad ...(1)$$

where, I_c = Collector leurrent .

α = Short circuit gain of CB configuration

I_{co} = Reverse saturation current.

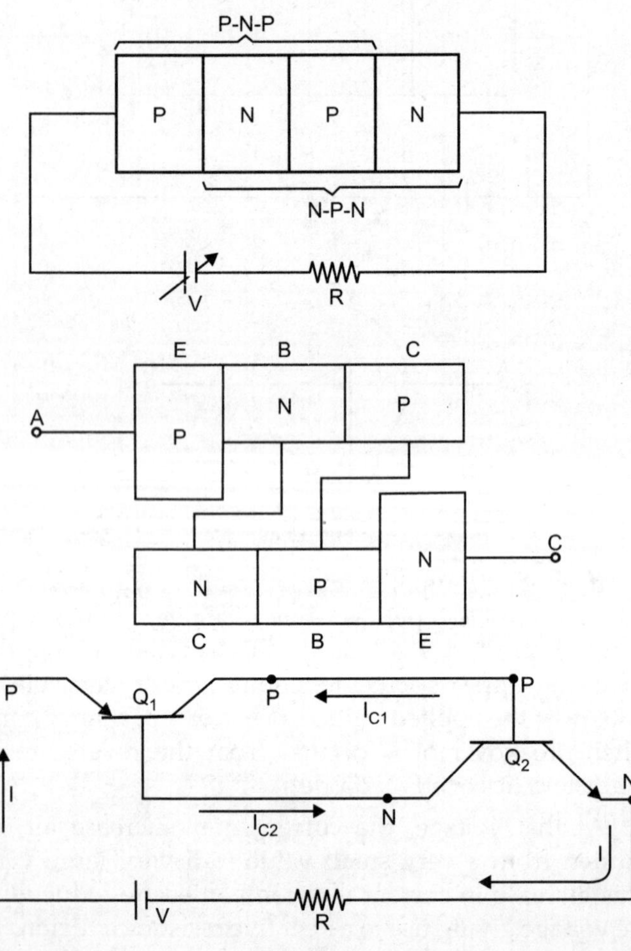

Fig. 11.7

Note that both the transistors will be in active region because the collector junction of both of them (*i.e.* J_2) is reverse biased and their emitter junctions (J_1 of *P-N-P* and J_3 of *N-P-N*) forward biased.

Note I_{E1} = Emitter current of $Q_1 = +I$

and I_{E2} = Emitter current of $Q_2 = -I$ ...(2)

Also, I_{co} is negative for Q_1 and positive for Q_2

So that

$$I_{CO2} = -I_{CO1} = -\frac{I_{CO}}{2} \quad ...(3)$$

From (1) and (2)

$$I_{C1} = -\alpha_1 I + I_{CO1} \quad ...(4)$$

and $$I_{C2} = \alpha_1 I + I_{CO2}$$

Combining all the equations (2), (3) and (4)

$$I = \frac{I_{CO2} - I_{CO1}}{I - (\alpha_1 + \alpha_2)}$$

$$I = \frac{I_{CO}}{1-(1)} \quad ...\text{form Eqn. 3}$$

Now α_1 and α_2 which are the forward current gains of two transistors under *CB* configuration shall have values lying between say 0 and 0.95 depending upon the currents flowing. Thus, when we are increasing the voltage, effectively we are doing nothing but increasing the values of α_1 and α_2, and when $(\alpha_1 + \alpha_2)$ reaches unity, there is a sudden increase in anode current as:

$$I = \frac{I_{CO}}{1-(1)} = \text{approaching infinity } (\infty)$$

for $(\alpha_1 + \alpha_2) \rightarrow 1$

Q.9. From the expression $I = \frac{I_{CO}}{1-(\alpha_1+\alpha_2)}$, it is obvious that when $(\alpha_1 + \alpha_2)$ exceeds unity, current *I* should reverse, but that does not happen actually. Why so ?

Ans. The moment $(\alpha_1 + \alpha_2)$ becomes unity or approaches unity, current becomes exceedingly large. This brings all of three junction diodes into saturation region. Or in other words, Q_1 and Q_2 go to saturation due to their collector junctions now forward biased. This saturation then decreases the values of α_1 and α_2 or limits their magnitudes so that $(\alpha_1 + \alpha_2)$ never exceeds unity. Put in another way, the transistor Q_1 and Q_2 enter saturation only upto the extent that $(\alpha_1 + \alpha_2)$ remains unity.

Q.10. Which semiconductor material is used for constructing *P-N-P-N* diodes ?

Ans. Silicon.

Q.11. Why cannot you use Germanium for making four-layer diodes? Is it impossible to make four-layer diodes with Germanium as the semiconductor material?

Ans. For making *P-N-P-N* diodes, we always use Silicon. Germanium is never used, as it does not have a stable off-state. To elaborate it further, with Germanium as the semiconductor material, the magnitudes of (α) may become large enough to give ($\alpha_1 + \alpha_2$) equal to unity for a very small applied voltage. That is, the device has tendency to switch to its 'ON' state from 'OFF' state very soon, thus giving an unstable 'OFF' state.

Q.12. Draw the characteristics of a *P-N-P-N* diode and mark the various regions. Also, give the typical values of resistances offered by the device when operating in various regions.

Ans. The *V-I* characteristics of a *P-N-P-N* diode are shown in Fig. 11.8. Resistance when operating in cut-off iegion equals hundreds of mega ohms.

Resistance in 'ON' state is few ohms.

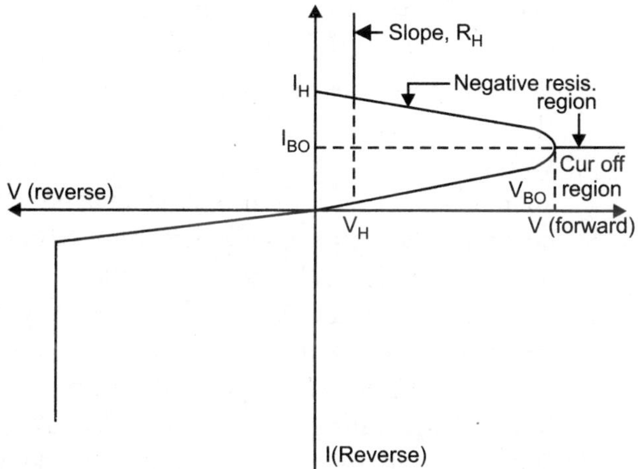

Fig 11.8

Q.13. What is SCS ? How does it resemble a four layer diode ?

Ans. SCS stands for silicon controlled switch. The construction of a silicon-controlled switch is exactly the same as that of a four layer diode with the only difference that the connections to the inner layers that are not accessible in a 4-layer diode are available in the case of SCS. Figure. 11.9 shows the junction diagram of an SCS.

N-type base near the anode is called anode gate and the *P*-type base near the cathode is called cathode gate. In some of the devices, both the connections are available but in most of the cases, only one connection is available, *i.e.* the cathode gate.

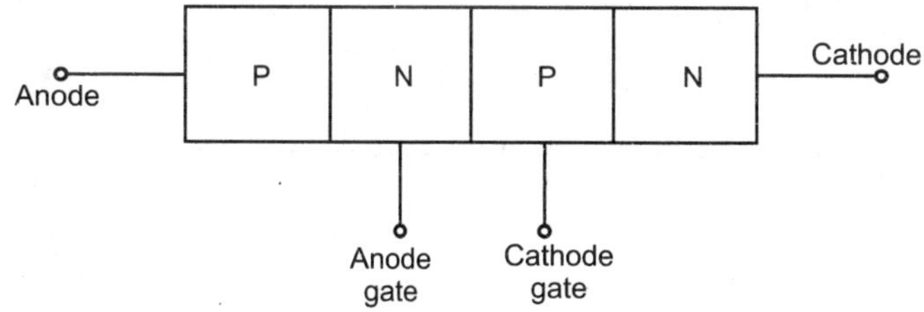

Fig. 11.9

Q.14. Justify the name Silicon Controlled Switch.

Ans. The *P-N-P-N* device conducts heavily, when the breakover voltage is reached. The breakover voltage further depends upon ($\alpha_1 + \alpha_2$) becoming unity. By inducing current at these gate terminals, ($\alpha_1 + \alpha_2$) may be made to approach unity at a reduced applied voltage. Or in other words, the breakover voltage may be reduced to an extent depending upon strengths of currents induced. This way, we have a sort of control over breakover voltage. Since the *P-N-P-N* diodes are always made from Silicon and these devices perform the switching operation, hence the name Silicon Controlled Switch.

Q. 15. Draw the circuit symbol and characteristics of a silicon controlled switch.

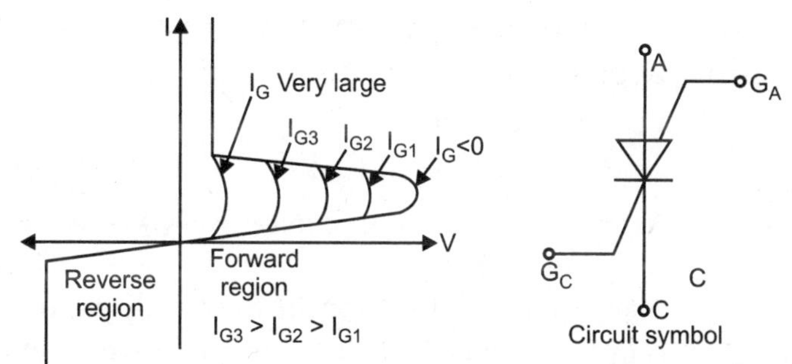

Fig. 11.10

Ans. It is clear from the characteristics that when gate current becomes very large and in a direction to forward bias the junction, the characteristics have the shape as that of an ordinary diode. Also, the breakover voltage may also be increased by injecting a negative current at the gate.

Q.16. What is rate-effect ? How does it affect the *P-N-P-N* diode operation ?

Ans. The effect of rate of change of applied voltage on the breakover voltage is termed as Rate-Effect.

We know that, in off state, the centre-diode is reverse biased and all the reverse biased diodes have some capacitance across them. A four layer diode represented by two forward biased and one reverse biased diode is shown in Fig. 11.11.

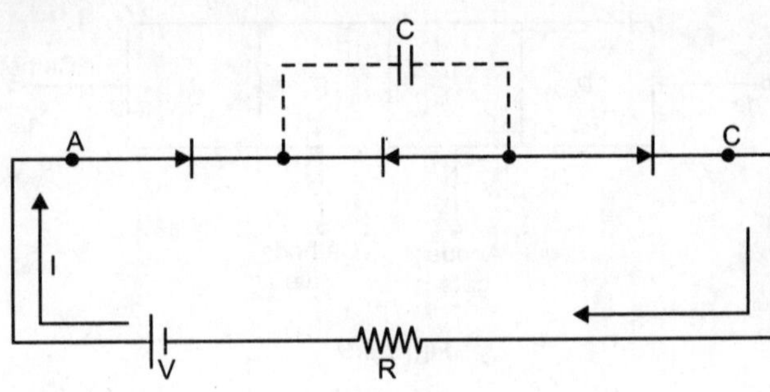

Fig. 11.11

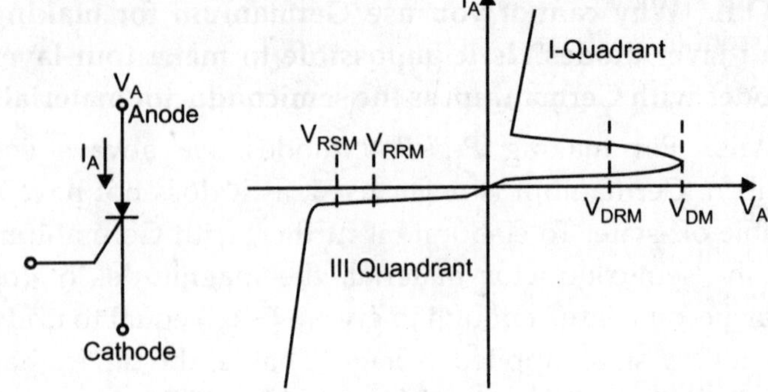

Fig. 11.12

For a lower rate of change of applied voltage, the capacitance C offers very high reactance and so the current through the capacitance can be ignored. When the rate of change of applied voltage becomes exceedingly large so that the current through C is not negligible, then the current may add to the total current and enable $(\alpha_1 + \alpha_2)$ reach unity at a reduced applied voltage. This higher rate of change of applied voltage may reduce the breakover voltage. This phenomenon is called rate-effect. Thus, rate effect may cause premature firing of a four-layer device.

Q.17. Briefly describe the following thyristor specifications:

(i) Repetitive peak reverse voltage, V_{RRM}
(ii) Non-repetitive peak reverse voltge, V_{RSM}
(iii) Repetitive peak off-state voltage, V_{DRM}
(iv) Peak off-state voltage, V_{DM}.

Ans.

(i) *Repetitive Peak Reverse Voltage* (V_{RRM}) is defined for reverse blocking mode type of thyristors like SCRs for operation in the third quadrant (Fig.11.12). It is the maximum repetitive reverse voltage (anode negative with respect to the cathode) that can be applied to the device safely and still keeping it in the blocked state. The current that flows through the device in this state is referred to as the Repetitive Peak Reverse Current (I_{RRM}).

(ii) *Non-Repetitive Peak Reverse Voltage* (V_{RSM}) is also defined in a similar way. For obvious reasons, this rating is little higher than the one discussed before. If the application is such that the reverse voltage does not appear across the device repetitively, may be because the device is being operated in the forward off-state, and if it is to be protected against any occasional reverse transients, then (V_{RSM}) and not the (V_{RRM}) is the rating we should look for. In practice, if we have not allowed any reverse voltage to exceed (V_{RRM}), we are very safe.

(iii) *Repetitive Peak_Off-State, Voltage* (V_{DRM}) is the maximum forward off-state voltage (anode positive with respect to cathode) that can be allowed to appear across the device when it is being operated in the first quadrant (in the forward blocking mode). This voltage is usually much less than the actual breakover voltage of the device which is basically a characteristic of the construction of the thyristor. The maximum value of current that can flow in the forward blocking state is (I_{DRM}).

(iv) *Peak Off-State Voltage* (V_{DM}) *Breakover Voltage* (V_{BO}) maximum off-state voltage. A forward voltage greater than this breaks the junction and switches the device to the ON-state even in the absence of any gate signal. As mentioned above, it does not depend upon the voltage grades of different thyristors in the same family. For instance, SCRs TY 505 F, TY 1005 F, TY 2005 F, TY 3005 F, TY 4005 F and TY 5005 F respectively have (V_{DRM}) rating of 50,100, 200, 300, 400 and 500 volts, whereas (V_{BO}) for all these devices belonging to the same class may be larger than the highest available (V_{DRM}) rating in this class. But it may be stressed once again that (V_{DRM}) rating should not be allowed to exceed in order that the device retains its listed characteristics.

Q.18. Define the 'Critical rate of rise of ON-state current' (*di/dt*) and 'Critical rate of rise of OFF-state voltage' (*dv/dt*) in case of thyristors. What is the significance of these specifications and what are the possible consequences, if these specifications are exceeded ?

Ans. When a thyristor is switched ON, initially, the thyristor can handle a very small current, as the current is concentrated in a small area of the device pellet. Gradually, the load current carrying capability increases to its specified rating. Critical rate of rise of ON-state current (*di/dt*) tells us about the maximum rate of change of ON-state current the device can handle safely. If this rating is exceeded during device switch ON, may be due to a faster switch ON, there could be a development of localized hot spots in the pellet at the time of switch ON and this could result in the device getting damaged, (*di/dt*) rating of 100*A*/μs is typical.

The critical rate of rise of OFF-state voltage (*dv/dt*) determines the maximum allowable rate of change off applied forward OFF-state voltage. When the applied forward voltage is a time varying one, a current (*C dv/dt*) flows through the reverse biased junction capacitance (C) of the thyristor. This current, if more than a certain value can cause premature firing of the thyristor due to decrease in its breakover voltage. Maintaining (*dv/dt*) within (*dv/dt*) rating of the device ensures that there is no change in the breakover voltage (V_{BO}) of the device, (*dv/dt*) rating of 100 *V*/ μ.s is quite common.

The (*dv/dt*) rating can be enhanced by connecting a low resistance between gate and cathode, thus providing a low resistance shunt path for the charging current mentioned above bypassing gate cathode junction. Majority of high (*dv/dt*) rating thyristors use what is called a 'Shorted Emitter Construction' which is nothing but an internally created low value resistance across gate cathode junction.

Q. 19. Define breakover voltage (V_{BO}) and breakover current (I_{BO}). Give their typical values.

Ans. The forward applied voltage at which the centre junction breaks over to give a sudden rise of current is called Breakover *Voltage.*

The current corresponding to breakover voltage (V_{BO}) is called Break over current. Both the parameters are indicated in the answer to a previous question.

Typical values of V_{BO} and I_{BO} are from tens of volts to hundreds of volts (for V_{BO},) of the order of hundreds of microamps (for I_{BO}).

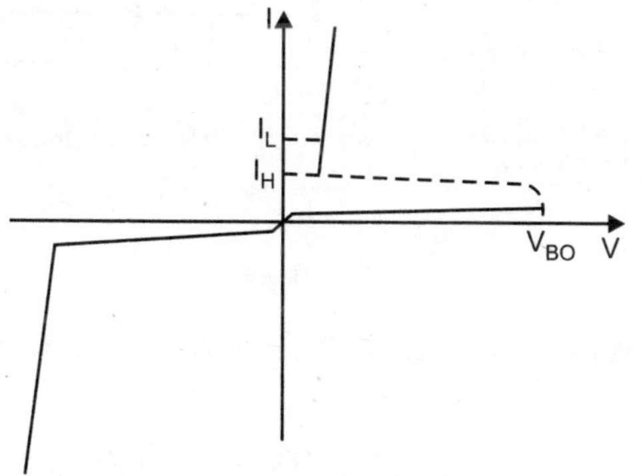

Fig. 11.13

These are also subject to changes with temperature.

Q. 20. What are 'Holding current' and 'Holding voltage'? Give their typical values.

Ans. *Holding current* and *Holding voltage* comes into picture, when the device is in the 'ON' state and has to be kept there only. Now if the voltage across the device is decreased so that the current decreases, then at a particular stage, the current may not be able to keep the device in the 'ON' state. So, the device goes to 'OFF'-state.

Holding current is the minimum current required to keep the device in the 'ON'-state once triggered. It is indicated in the characteristics shown in Fig 11.13

Voltage corresponding to holding current is termed as 'Holding voltage.'

The values of these parameters vary from type to type. Typical value of holding current may lie between several milliamps to several hundreds of milliamps. Holding voltage may be as small as 0.5 V for smaller units and as large as 20 V for larger devices. Furthermore, these values are subject to changes with temperature.

Q. 21. Distinguish between Holding Current (I_H) and Latching Current (I_L) in case of thyristors.

Ans. *Holding Current* (I_H) is the minimum ON-state current required to keep the thyristor in the low impedance state, once it has been triggered to the ON-state (Fig. 11.13). In order to switch off the thyristor, the anode current must be brought below the holding current value.

Latching current (I_L) is little larger than the holding current and ON-state current equal to or more than the latching current ensures that a switched ON device remains in the ON-state even after the gate signal is removed. Once the ON-state current has exceeded the latching current value after switch ON, the current then has to be brought below the holding" current to switch it off. To further illustrate the difference, let us assume that an SCR is triggered with a narrow gate trigger pulse and that the gate pulse has been withdrawn before the anode current of the switched ON device could reach the latching current amplitude. In such a case, the ON-state will not sustain and the device will come back to the forward blocking state immediately after the gate trigger pulse is withdrawn.

Q.22. What does the Amperes Squared Seconds (I^2t) rating of a thyristor signify ? Where do we make use of this rating ?

Ans. The Amperes Squared Seconds (I^2t) rating of a thyristor tells us about the surge current handling capability of the device for sub-cycle time periods, when the device is being used as a rectifier. (I^2t) rating is usually specified or it can be calculated from the maximum peak surge current that the device can sustain for a given time period which is usually one full cycle of a applied 50Hz/60 Hz waveform. (I^2t) rating of the thyristors can be used to calculate the surge current capability, when the thyristor conducts for only a part of the full cycle time period.

Q.23. Why are thyristors called current controllable negative resistance devices?

Ans. Thyristors are negative resistances devices, as a part of their *I-V* characteristics exhibits negative resistance where a reduction in the voltage is accompanied by an increase in current as shown in Fig. 11.14.

Thyristors are current controllable devices because their *I-V* characteristics can be expressed by a single valued function of current *i.e.* for every current value,

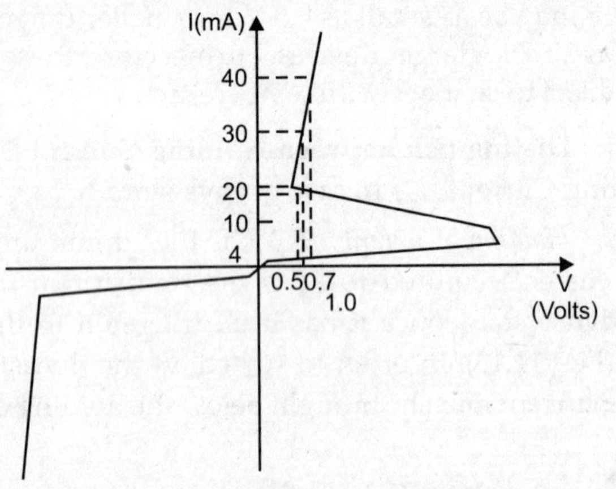

Fig. 11.14

there is one and only one corresponding voltage value (Fig. 11.14). For instance, for currents equal to 30 mA and 40 mA, the corresponding voltages are 0.5 V and 0.7 V respectively. On the other hand, these characteristics are representative of a (multivalued function of voltage) *i.e.* for every voltage value, there could be one or more than one corresponding current values. For instance, for a voltage of 0.7 volt, the current could be 4 mA or 20 mA or 40 mA. Of course, 20 mA point is in the unstable region and can not, thus sustain. It is because of 'single valued function of current' and 'multivalued function of voltage' nature of the I-V characteristics that thyristors are current controllable devices. UJTs also belong to the same class.

Q.24. When do we need to connect thyristors in series or in parallel? What are the basic considerations to be looked after while doing so?

Ans. Thyristors are connected in series to enhance the voltage ratings of the individual devices. For instance, two SCRs each having a peak repetitive OFF-state voltage rating of 1000 volts could be connected in series to get an overall (V_{DRM}) of 2000 volts provided, of course, that steps are taken to ensure that the applied voltage is equally divided between the individual devices. The devices to be connected in series should preferably have closely matched OFF-state characteristics like forward breakover voltage, OFF-state resistance and so on. One way to force an equal division amongst individual devices is to connect identical resistors across each one of the devices connected in series (Fig. 11-15 (*a*)). The value of (*R*) here should be a fraction of the forward OFF-state resistance so that any variations in this device parameter from device to device do not affect the voltage division. Capacitive dividers are also used as the resistors dissipate power and reduce the efficiency. But in case of capacitive dividers, a series resistance should be used along with each capacitor to damp the current pulses.

Thyristors are connected in parallel to enhance the current capability of the individual devices to be connected in parallel. Care should, however, be taken to ensure that the total ON-state current is equally divided amongst individual devices connected in parallel. To achieve this, the devices should preferably have closely matched forward conduction characterises like the ON-state dynamic impedance, ON-state voltage and so on. Equal division of current can be forced by connecting small resistors in series with each of the parallel connected thyristors (Fig. 11.15 (*b*)). (*R*) in this case is much larger than the ON-state dynamic impedance of the individual thyristors.

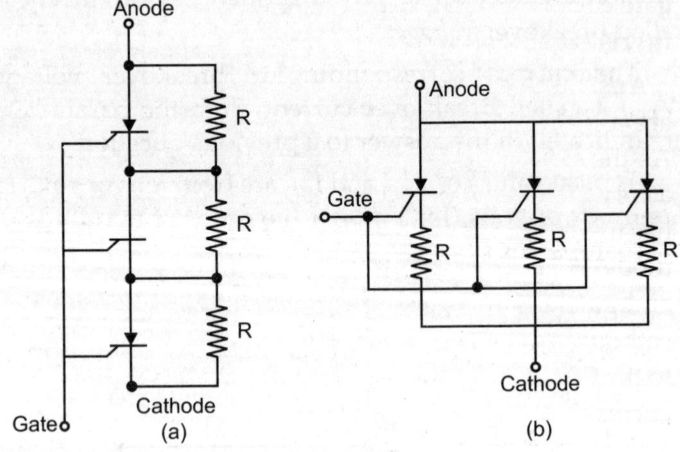

Fig. 11.15

Q.25. What is a TRIAC ? How does it differ from an SCR?

Ans. A TRIAC is a bi-directional thyristor. It can be considered to be equivalent to two *P-N-P-N* devices connected back to back (Fig. 1.16 (*a*)). The voltage-current characteristics of a triac are similar to those of an *SCR* with the difference that a triac exhibits identical characteristics both in the forward (first quadrant) and reverse (third quadrant) directions (Fig. 11.16 (*b*)). Like an *SCR*, a triac too is a three terminal device and the terminals are designated as Main Terminal-1 (*MT*-1), Main Terminal-2 (*MT*-2) and Gate (Fig. 11.16 (*c*)). Another popular bi-directional thyristor type is the *DIAC* which can be considered to be a triac without the gate contact.

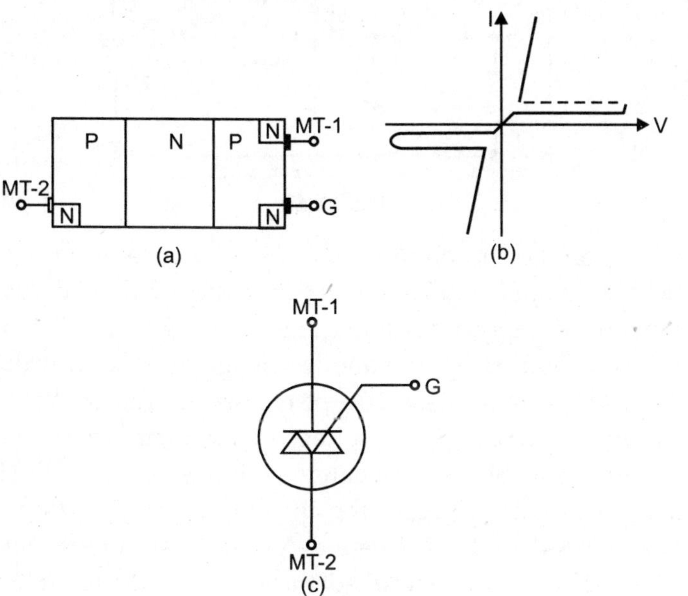

Fig. 11.16

Q.26. Briefly describe the gate triggering characteristics of a thyristor. What is the preferred combination of gate current and gate voltage for reliable triggering of a thyristor? Explain various limiting boundaries.

Ans. The gate triggering characteristics of a thyristor are a plot of gate voltage (V_G) versus gate current (I_G). As the gate cathode circuit of a thyristor is nothing but a forward based P-N junction, the characteristics curve has to resemble the one for a forward biased P-N junction diode. Before we do that, we must define the gate voltage and gate current. Here, the gate voltage (V_G) is the gate-cathode voltage just prior to the device getting triggered to the ON-state and gate current (I_G) is the forward biased current flowing in gate-cathode circuit. Also, we should also remember that the '*P*' and '*N*' materials constituting the gate and cathode respectively in a thyristor are lightly doped.

As a result of this light doping, there is a spread in the *V-I* characteristics even for a given type of thyristor. This spread is shown in Fig. 11.17 by characteristic curves '1' and '2'. While curve-1 represents the minimum voltage that must be applied to the gate to turn-on the thyristor, curve-2 tells us about the maximum possible value of the gate voltage that can be safely applied to the device. Similarly, there are minimum and maximum values of gate current too. Yet another requirement that needs to be fulfilled is in respect of the power dissipation. There is always a rated gate power dissipation specified for the thyristor, which should not be exceeded. This is also shown in Fig. 11.17. All these requirements together define the operating area for the gate trigger circuit. This operating area has been shown shaded in Fig. 11.17. The boundary specified by the power dissipation is for the average value of the gale power dissipation. A relatively higher power dissipation may be permitted for shorter pulse durations. For example, *SCR* type no. 2N3668 has P_G (*av*) rating of 0.5 W and a transient gate power dissipation P_{GM} specification of 40 W for a 10 μs trigger pulse. Now such a device could be triggered by a continuous current of 200 mA at a trigger voltage of 2.5 V. It could also be triggered by a 2 A trigger pulse at 20 V provided that trigger pulse was no wider than 10 μs.

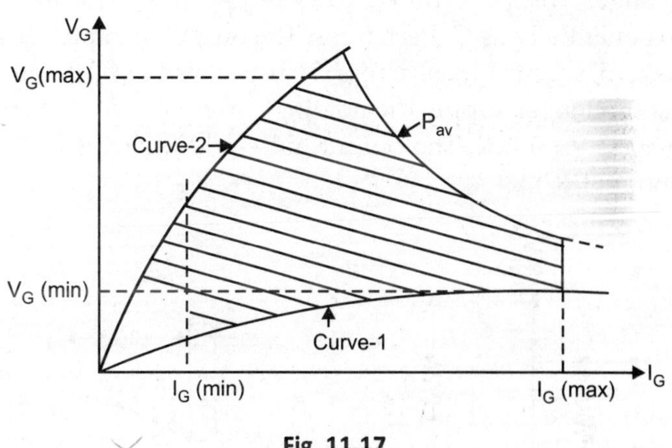

Fig. 11.17

Q.27. What are the advantages, if any, of triggering thyristors with high current pulses ?

Ans. Higher gate trigger currents yield faster turn on. But the fast turn on of the thyristor is also not without problems. If the rate of rise of the on-state current (*di*/*dt*) exceeds the critical value of (*di*/*dt*), the device could get damaged due to creation of localized hot spots.

Q.28. How does the required gate triggering current vary with temperature ?

Ans. The gate triggering characteristics vary with temperature. At lower temperatures, one requires higher gate trigger voltage and current. For example, for SCR type no. 2N3668, the maximum values of V_{GT} and I_{GT}

are 2 V and 40 mA respectively at a junction temperature of 25°C. At – 40°C, the same become 3V and 80mA respectively and at 100°C, the values are about 1.5 V and 30 mA.

Q.29. Draw the basic SCR circuit showing its use as pulse-generator. Explain in brief its operation.

Ans. Fig. 11.18 shows the use of SCR as a pulse generator. In the normal state, the output voltage would be at ground level, as the capacitor (C) would have been sitting charged to (*V*) volts. It may be mentioned that the breakover voltage of SCR must be greater than the voltage (*V*). When a triggering pulse is applied at the gate of the SCR, it fires and the capacitor (C) discharges through SCR's ON resistance (*r*) and resistance (*R*). It does so with a time constant (*RC*), if $r < R$. The capacitor can discharge only upto a voltage V_H (called Holding Voltage) as beyond that holding current of the SCR can no longer be supplied and the SCR turns-off. When the SCR fires, the voltage across SCR tends to change abruptly to V_H from *V*, a change of $(V - V_H)$. Since, voltage across a capacitor cannot change abruptly, the output (V_O) changes abruptly from 0 to $(V - V_H)$ and then decays exponentially as C discharges. The output abruptly goes to zero, when voltage across C has decayed to (V_H) which can no longer supply the holding current (I_H) and at that instant $V_O = I_H R$. The output pulse-width in this case is about (3*RC*).

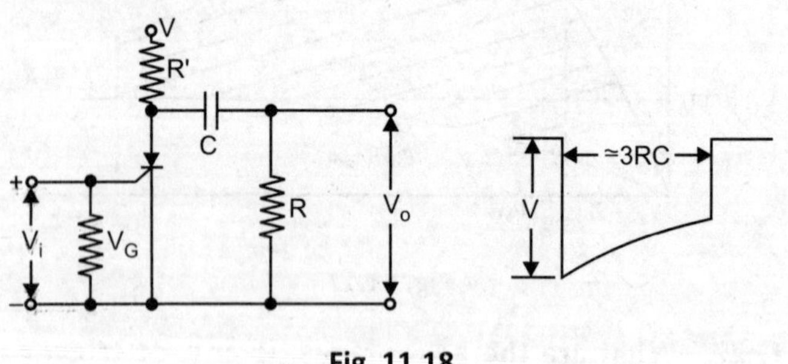

Fig. 11.18

Q.30. With the help of circuit diagram, explain in brief the use of an SCR as a bistable multivibrator.

Ans. Figure 11.19 shows the basic circuit arrangement of a bistable multivibrator using SCRs. In the circuit arrangement shown, if the supply voltage is less than the breakover voltage of the SCRs, the circuit can be used as a bistable multivibrator. If the supply voltage is greater than the breakover voltage, it becomes an astable multivibrator with the frequency of oscillation determined by R_1, R_2 and R_C.

In the bistable mode, the circuit functions as follows: Assume the SCR (1) is initially conducting and SCR (2) is cut-off. Obviously, the output voltage is low $(= V_H)$. Under these conditions, the current flowing through SCR (1) is limited by R_1 only.

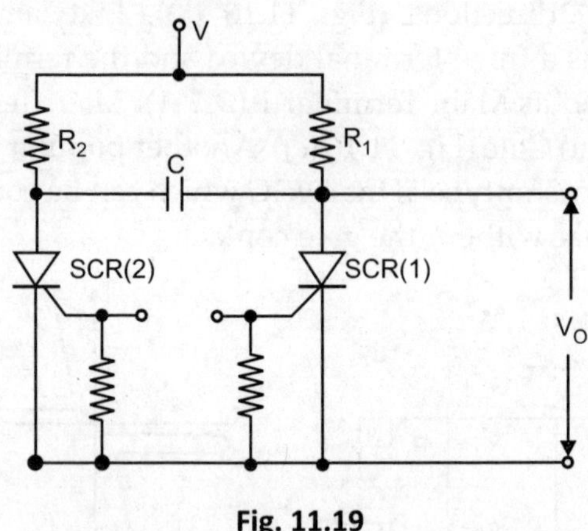

Fig. 11.19

R_1 can be chosen in such a way that current through SCR (1) is only slightly greater than the holding current. Now if we apply a positive trigger at the gate of SCR (2), it starts conducting, the anode voltage of SCR (2) drops from (*V*) to (V_H). This abrupt change is transmitted to the anode of SCR (1), as voltage across *C* cannot change instantaneously. This negative going step at SCR (1) anode turns it off. Also, the current that flows through *C* as a result of SCR (2) anode going to an almost zero potential has to be supplied through R_1. As mentioned above, R_1 cannot supply both the holding current for SCR (1) as well as current through C. As a result, current through SCR (1) falls below its holding current value and its turn-off is ensured. As SCR (1) goes to off-state, output goes high (= *V*). A positive trigger at SCR (1) gate can again change states.

Q.31. Draw the basic (Half Wave Rectifier) power control using SCR. Explain in brief its operation.

Ans. The basic circuit diagram is shown in Fig. 11.20 The point at which the gate trigger pulse is applied to the SCR gate during the positive half cycle of the AC source is controlled by a suitable circuitry in the box labelled *Phase Control*. During the negative half cycle, SCR remains reverse biased and there is no current through the load. Thus, input AC power appears across the load only for the period when SCR conducts. The AC power in the load, thus can be controlled by controlling the firing angle of SCR. Refer to waveforms of Fig.11.21 (*a*) and (*b*). Figure 11.21 (*a*) shows the waveform appearing across the load and Fig. 11.21 (*b*) shows SCR anode waveform.

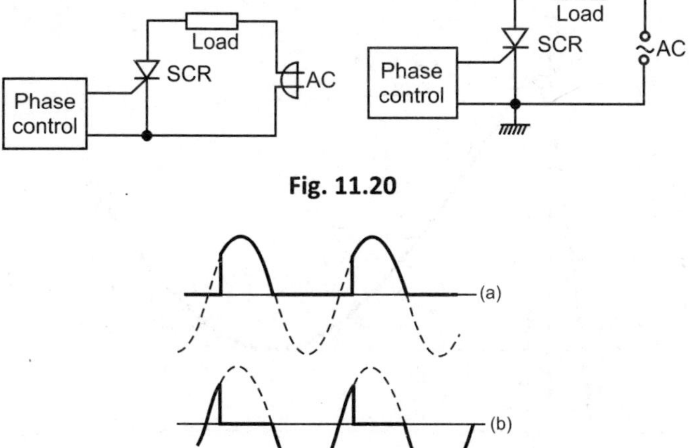

Fig. 11.20

Fig. 11.21

In the half-wave rectifier control of AC power using SCR, the firing angle can be anywhere between 0° and 180°.

Q.32. Show the basic SCR power control circuit that can accomplish full wave rectifier control.

Ans. The basic circuit arrangement is shown in Fig. 11.22 (*a*). Typical input and output waveforms are shown in Fig. 11.22 (*b*). Conduction angle is 90°.

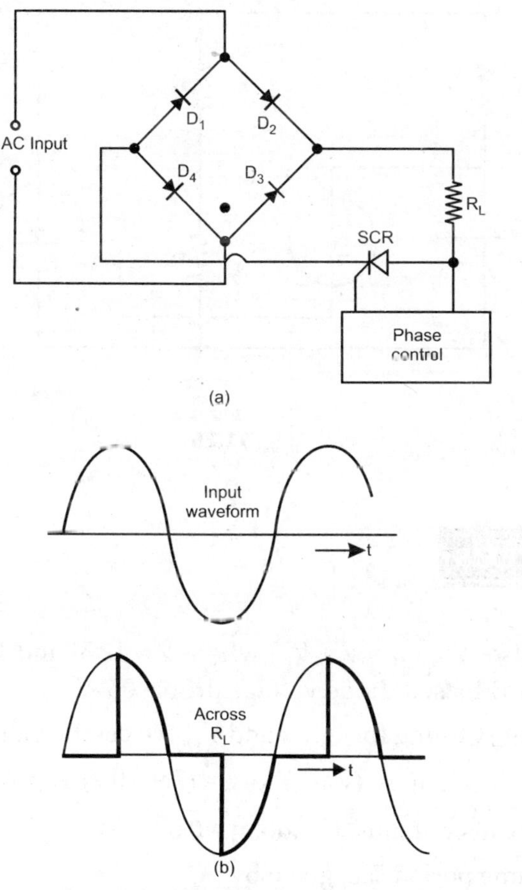

Fig. 11.22

Diode D_2 and D_4 conduct during positive half cycles of input, whereas diode D_1 and D_3 conduct during negative half cycles of input. The power control is provided by SCR.

Q.33. With the help of the basic circuit arrangement, explain in brief how a triac can be used for control of AC power in a load.

Ans. Fig. 11.23 shows basic circuit arrangement. The portion of input AC power that appears across the load is during the time triac is ON. The idea is similar to the one discussed in the case of an SCR with the triac having an advantage that it can control both positive as well as negative half cycles. In the circuit shown, phase angle is being controlled by a diac. During positive or negative half cycles, when the diac's breakover voltage is exceeded, the triac fires.

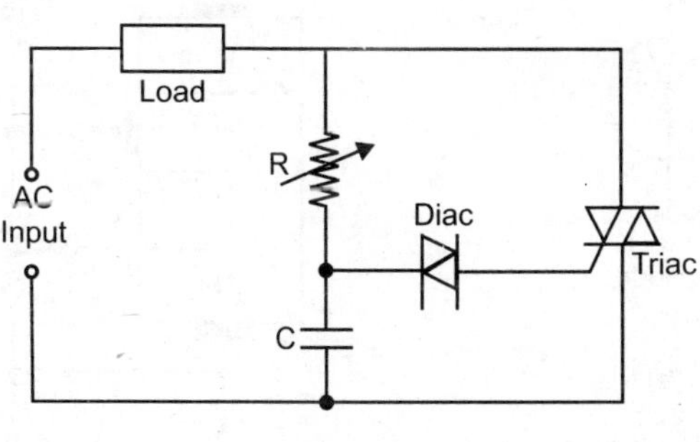

Fig. 11.23

The triac turns-off, when the current falls below its holding value. Another trigger during the negative half cycle turns the triac ON again. The period for which the triac conducts during positive or negative half cycles of AC input can be controlled by varying *R*. By controlling time constant (*RC*), we can control the time instant at which diac would fire which in turn triggers the triac.

Q.34. What is a Programmable unijunction transistor (PUT)? Explain the design of PUT. Comment on the statement "The name PUT is a misnomer".

Ans. Programmable unijunction transistor (PUT) behaves more or less like a UJT and has similar applications. A PUT is designed to enable the engineer to program UJT parameters, which include inter-base resistance (R_{BB}), intrinsic stand-off ratio (η), peak current (I_P) and valley current (I_V). Programming is possible with the help of two external resistors. Figure 11.24 shows the circuit symbol and constructional features of a PUT device. R_1 and R_2 are programming resistors. R_{BB} and η in the case of PUT are given by Eqns. (1) and (2). The peak current and valley current depend upon the equivalent resistance R_G of the gate control circuit. R_G is given by Eq. (3).

$$R_{BB} = R_1 + R_2 \quad ...(1)$$

$$\eta = R_1/(R_1 + R_2) \quad ...(2)$$

$$R_G = R_1 \times R_2/(R_1 + R_2) \quad ...(3)$$

As is obvious from Fig. 11.24 (*b*), the name PUT is a misnomer. PUT is a four-layer device like a thyristor with contacts, namely, anode (*A*) and cathode (*K*) made to the two extreme layers and a control contact called gate (*G*) made to one of the inner layers. In this case, the inner layer chosen is the one closer to the anode contact. The device is so called because it can perform functions similar to that of a UJT. Note that a PUT device is directly interchangeable with a UJT device.

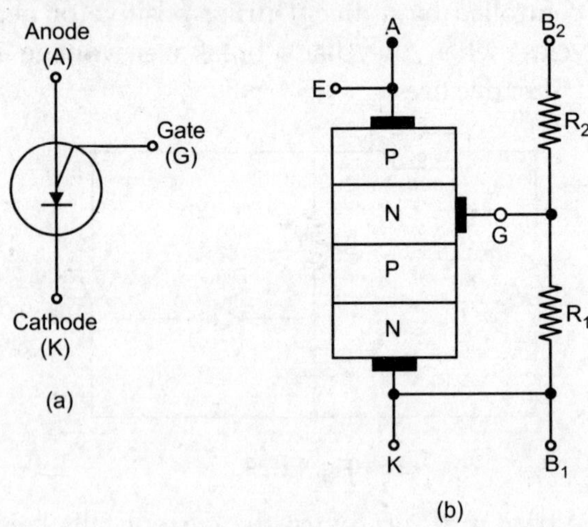

Fig. 11.24

Q.35. Draw the *V-I* characteristics of a Programmable unijunction transistor (PUT).

Ans. Figure 11.25 shows the *V* – *I* characteristics of a PUT device. The voltage V_S appearing at the control terminal is given by

$$V_S = V_B \times [R_1/(R_1 + R_2)] \quad ...(1)$$

where, V_B is the *DC* voltage applied at the B_2 terminal.

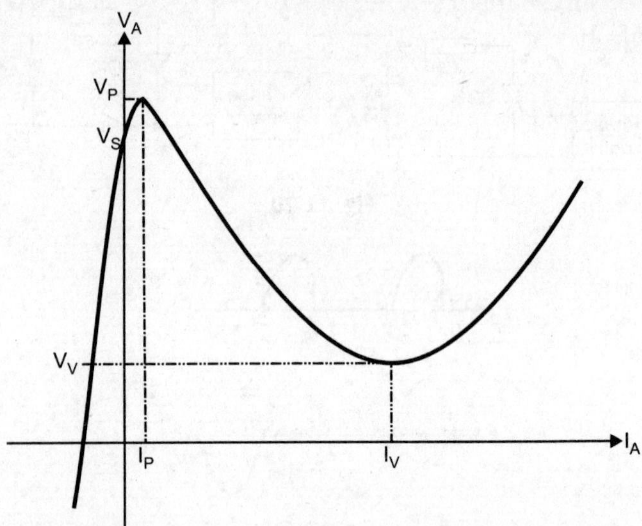

Fig. 11.25

As anode to cathode voltage exceeds V_S by one diode voltage drop, the device goes to conduction. We can notice the similarity in the *V* – *I* characteristics of a PUT and a UJT device. 2N 6027 and 2N 6028 are commonly used PUTs. Like UJT devices, the most common application of PUT devices also is in triggering of thyristors. Figure 11.26 shows one such circuit, which is self-explanatory.

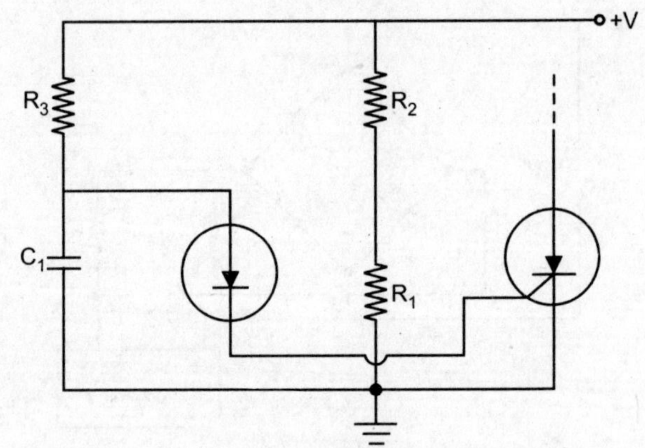

Fig. 11.26

SOLVED PROBLEMS

Problem 11.1. *Refer to the UJT-based relaxation circuit of Fig. 11.27(a). The waveform observed across capacitor C is shown in Fig. 11.27(b). Determine the intrinsic stand-off ratio η of the UJT used in the circuit. Also, determine the time period T of the waveform. (Given that the forward-biased diode voltage drop = 0.7 V.)*

Solution. From the waveform shown in Fig. 11.27(*b*), peak voltage $V_P = 7.9$ V.

Also, $V_P = \eta \times V + V_D$, where $V = 12$ V and V_D is the forward-biased diode voltage drop = 0.7 V.

Substituting for V_P, V and V_D, we get the value of η as

$$\eta = (V_P - V_D)/V = (7.9 - 0.7)/12 = 0.6$$

Intrinsic stand-off ratio, η = 0.6.

Time period *T* is given by

RC × In[1/(1– η)] = 10 × 10^3 × 0.1 × 10^{-6} × In[1/(1 – 0.6)] = 0.001 × ln 2.5 = 0.92ms

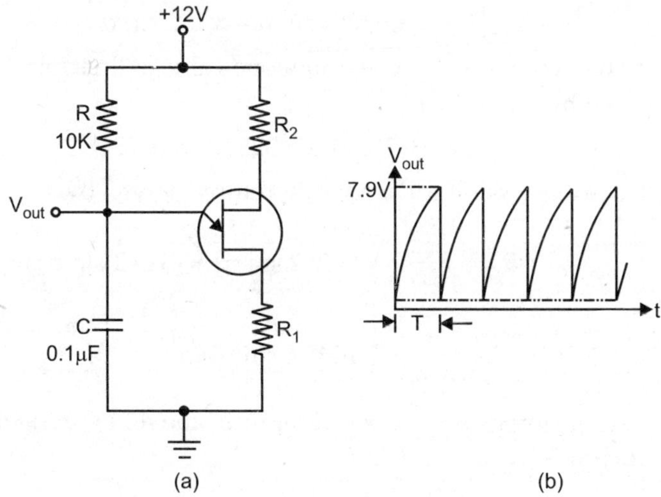

Fig. 11.27

Problem 11.2. *The PNPN diode used in the circuit of Fig. 11.28 (a) has a break-over voltage of 20 V and a holding current of 5 mA. The V – I characteristics of the device are shown in Fig. 11.28 (b). Determine (a) current flowing through the device and (b) the region of operation of the device, if the resistance R were increased to 10 K.*

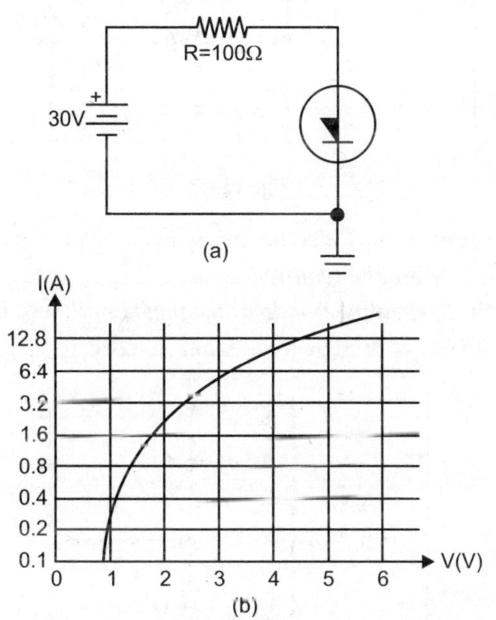

Fig. 11.28

Solution. The applied voltage is greater than the break-over voltage. The device is, therefore, in the saturation region. The current flowing through the device to first approximation is given by 30/100 = 0.3A.

For current equal to 0.3A, the voltage drop across the device as seen from the characteristic curve of Fig. 11.28 (*b*) is 1.0 V

Therefore, more exact value of current can be computed from (30 – 1)/100 = 0.29 A

When the resistance *R* is increased to 10 K, the current through the device reduces to 30/10000 = 3 mA

which is less than the holding current value. Therefore, the device switches back to the cut-off state.

Problem 11.3. Refer to the PNPN diode circuit of Fig. 11.29. The break-over voltage and holding current specifications of the device are 12 V and 4 mA, respectively. The variable voltage source is set in such a way that the device is conducting. If the knee voltage of the device is taken to be 0.7 V, determine the applied voltage at which the device will turn off.

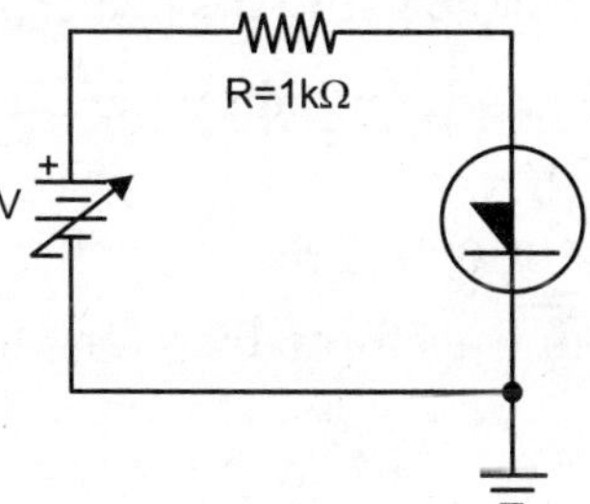

Fig. 11.29

Solution. The device will turn off, when the current falls below the holding current value.

If the applied voltage at that point is *V*, then

$(V – 0.7)/1 \times 10^3 = 4 \times 10^{-3}$

That is, $V = 4.7$ V

Problem 11.4. *Refer to the relaxation oscillator circuit of Fig. 11.30. The PNPN diode used in the circuit is 1N 5158 having a break-over voltage and holding current specifications of 10 V and 4 mA, respectively. Draw the waveform across the capacitor indicating its timing parameters.*

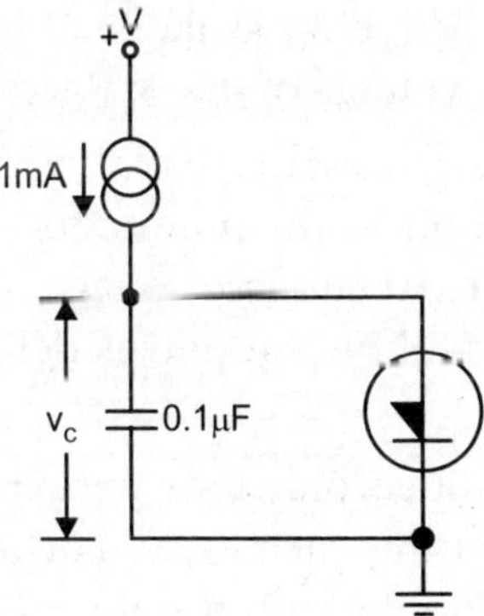

Fig. 11.30

Solution. The capacitor in this case is being charged from a constant current source of magnitude equal to 1 mA.

Therefore, the instantaneous voltage appearing across the capacitor is given by

V_C = (I/C) × t. Now I = 1.0 mA

and C = 0.1 μF.

Considering that the ON-resistance of the PNPN diode is negligible, the waveform appearing across the capacitor will be a sawtooth waveform as shown in Fig. 11.31. The capacitor voltage linearly rises to a voltage equal to the break-over voltage of the device from where it discharges almost instantaneously through the device. The voltage rise is linear due to constant current charging.

In the present case, the capacitor voltage rises to 10 V before the device switches to the ON-state.

The time period required by the capacitor to charge to 10 V is given by

$0.1 \times 10^{-6} \times 10/10^{-3} = 1$ m

The sawtooth waveform, therefore, has a time period of 1 ms.

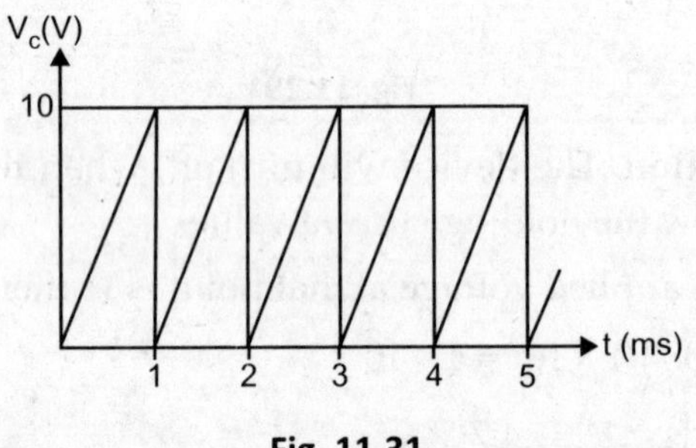

Fig. 11.31

Problem 11.5. Refer to the SCR-based half-wave power control circuit of Fig. 11.32. The 230 V AC source of supply feeds a 100 Ω load through a controlled rectifier. If the trigger circuit of the SCR was so adjusted as to start conduction at 30° after the start of each cycle, determine the total power delivered to the load by the AC supply. Assume holding voltage of the SCR to be zero.

Solution. The instantaneous value of the power drawn from the AC supply is the product of the instantaneous values of line current and line voltage. The average value of this product will be the power delivered by the AC supply.

The SCR conducts only for a period from 30°(π/6 rad) to 180°(π rad) during each cycle. For rest of the period, the SCR remains in the cut-off state.

Therefore, the current flows through the load only for a period from 30°(π/6 rad) to 180°(π rad).

The expression for instantaneous value of line current is given by

$$(230 \times \sqrt{2}/100) \sin \alpha = 3.252 \sin \alpha$$

The expression for instantaneous value of line voltage is given by

$$(230 \times \sqrt{2}) \sin \alpha = 325.2 \sin \alpha$$

The expression for power therefore is given by

$$P = \frac{1}{2\pi} \times \int_{\pi/6}^{\pi} (3.252 \sin \alpha) \times (325.2 \sin \alpha) d\alpha$$

$$= \frac{1}{2\pi} \times \int_{\pi/6}^{\pi} 1057.6 \sin^2 \alpha d\alpha$$

After integration operation, the above expression reduces to

$$P = \frac{1}{2\pi} \times |528.8\alpha - (528.8/2) \sin 2\alpha |_{\pi/6}^{\pi}$$

This simplifies to P = 256.78 W.

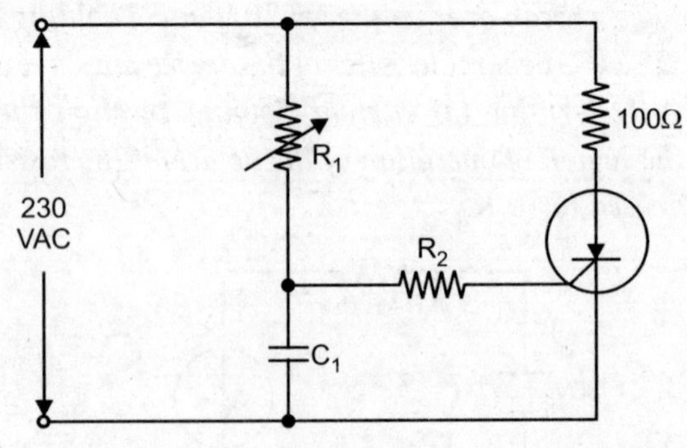

Fig. 11.32

Problem 11.6. *Refer to the SCR crowbar circuit of Fig. 11.33. Determine the required breakdown voltage of the Zener diode, if the load voltage could at the most be allowed to increase to 10 V. Take V_{GT} of SCR to be equal to 0.8 V.*

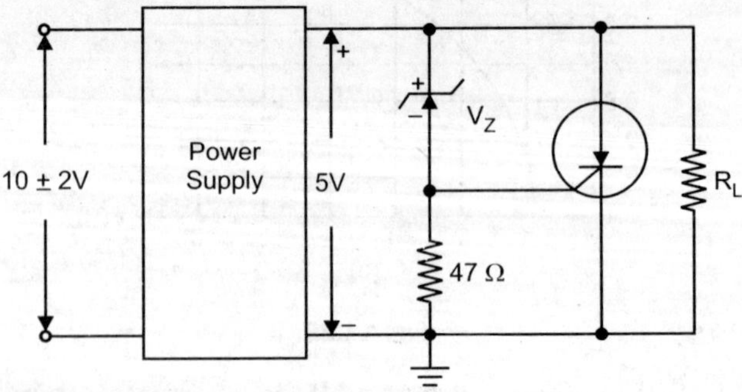

Fig. 11.33

Solution. In the present case, $V_Z + V_{GT} = 10$ V.

This gives $V_Z = 10 - V_{GT} = 10 - 0.8 = 9.2$ V.

SUMMARY OF IMPORTANT FORMULAS

1. Collector current of a bipolar transistor in the active region is given by:

$$I_C = -\alpha I_E + I_{CO}$$

Where, I_C = Collector current

α = Short circuit gain of CB configuration

I_{CO} = Reverse saturation current

2. Inter-base resistance in a programmable UJT (PUT) is given by:

$$R_{BB} = R_1 + R_2$$

Where, R_{BB} = Inter-base resistance

R_1 and R_2 = Programmable resistors

3. Intrinsic stand-off ratio (η) in a programmable UJT (PUT) is given by:

$$\eta = R_1/(R_1 + R_2)$$

Where, η = Intrinsic stand-off ratio

R_1 and R_2 = Programmable resistors

SELF-EVALUATION EXERCISE

1. Some of the devices in the thyristor family, namely the Silicon Controlled Rectifier (SCR), the Silicon Controlled Switch (SCS), the Silicon Unilateral Switch (SUS), the Silicon Bilateral Switch (SBS), the diac and the triac are listed in Table-11.1 from S.No. 1 to 6 with their circuit symbols, arrangement of different semiconductor layers/junctions and the *V-I* characteristics given in a haphzard manner in columns 3, 4 and 5 respectively. For each of the listed devices in column-2, identify the corresponding circuit symbol, the arrangement of semiconductor layers and the *V-I* characteristics. For example, if you feel that the device listed at S.No 1 has its circuit symbol, construction and its *V-I* characteristics respectively listed at S.No. 3, 6 and 4, then your answer for this part will be (1-3-6-4).

2. Answer the following

(*a*) Is thyristor a current controllable or a voltage controllable device ?

(*b*) Is UJT a current controllable or a voltage controllable device?

(*c*) Do we have SCR type devices constructed from Germanium?

(*d*) The *V-I* characteristics of an SCR and be expressed by (*i*) single valued function of current (*ii*) a single valued function of voltage.

(*e*) The *V-I* characteristics of a UJT can be expressed by (*i*) multivalued function of voltage (*ii*) multivalued function of current.

(*f*) Which SCR specification will have to be checked the most carefully when the input to the SCR is changing at fast rate (*i*) *dv*/*dt* rating (*ii*) Breakdown voltage (*iii*) Turn-on time (*iv*) Turn-off time.

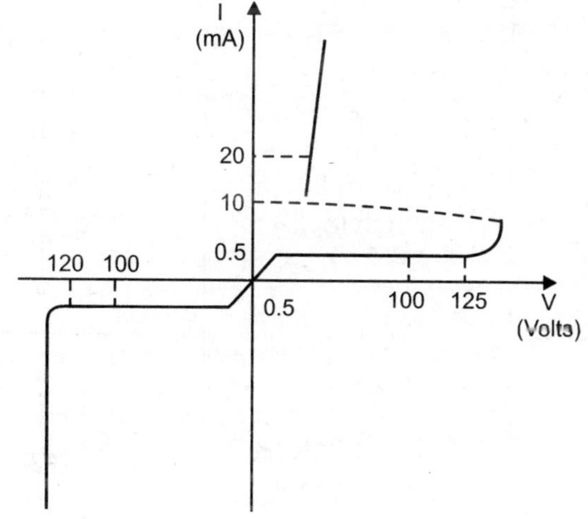

Fig. 11.34

(*g*) A triac is usually triggered by a diac or an SBS. Which of the two is capable of providing a larger peak triggering current.

Table 11.1

S.No.	*Device type*	*Circuit symbol*	*Arrangement of semiconductor layers*	*V–I Characteristics*
1	Silicon controlled rectifier (SR)	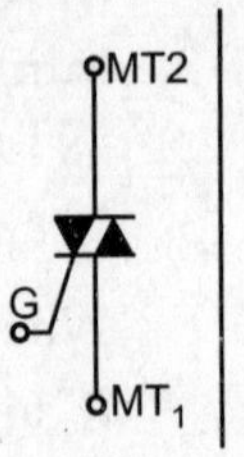 MT2, G, MT_1	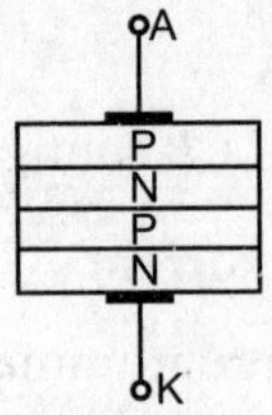 A, P, N, P, N, K	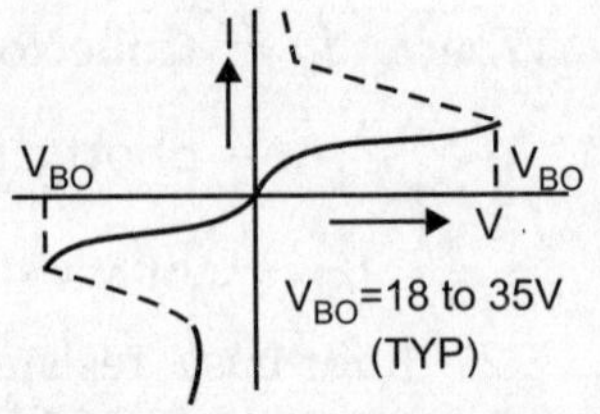 V_{BO}, V_{BO}, V, V_{BO}=18 to 35V (TYP)
2	Silicon controlled switch (SCS)	 K, A	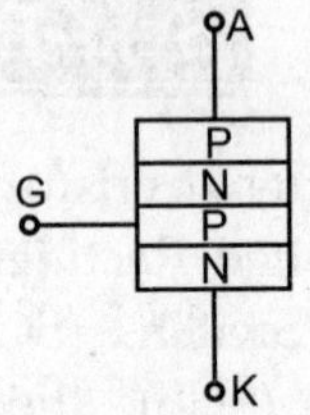 A, G, P, N, P, N, K	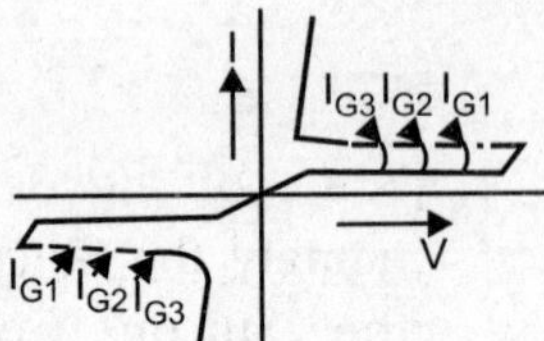 I_{G3} I_{G2} I_{G1}, I_{G1} I_{G2} I_{G3}, V
3	TRIAC	 A, G, K	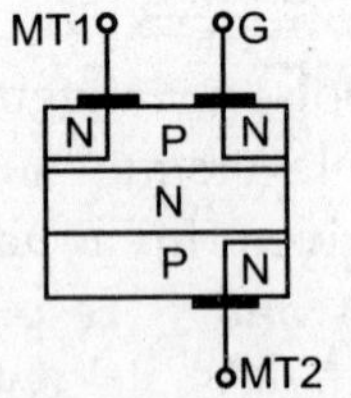 MT1, G, N, P, N, N, P, N, MT2	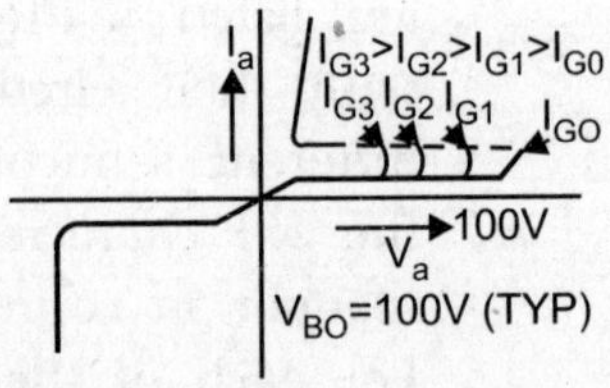 I_a, $I_{G3}>I_{G2}>I_{G1}>I_{G0}$, I_{G3} I_{G2} I_{G1} I_{GO}, 100V, V_a, V_{BO}=100V (TYP)
4	Silicon unilateral switch (SUS)	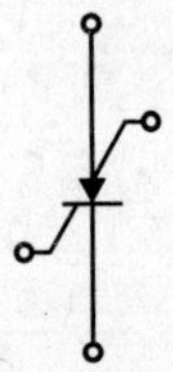	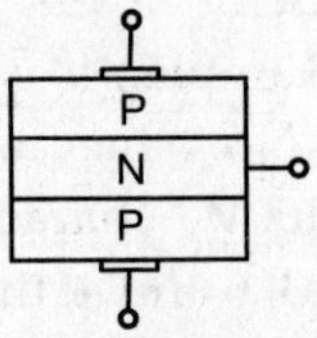 P, N, P	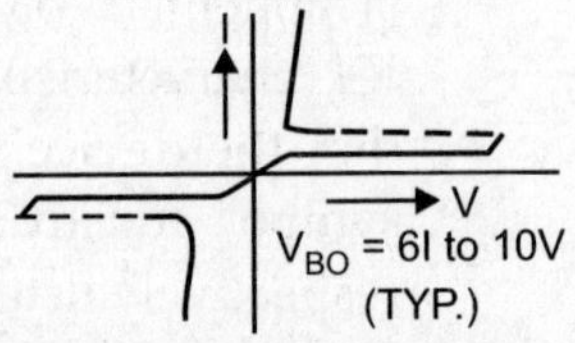 V, V_{BO} = 6I to 10V (TYP.)
5	Silicon bilateral switch (SBS)		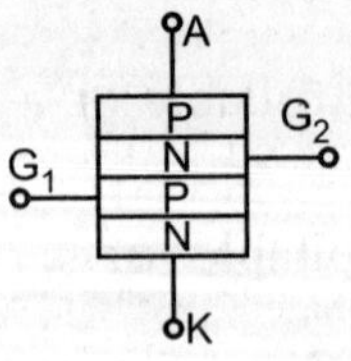 A, G_1, G_2, P, N, P, N, K	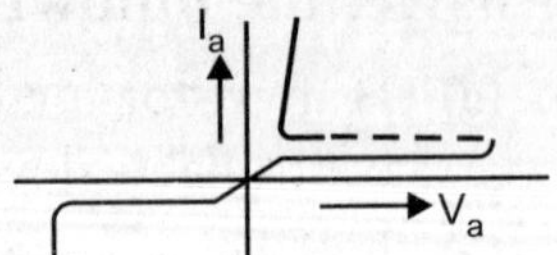 I_a, V_a
6	DIAC		 N, P, N, N, P, N	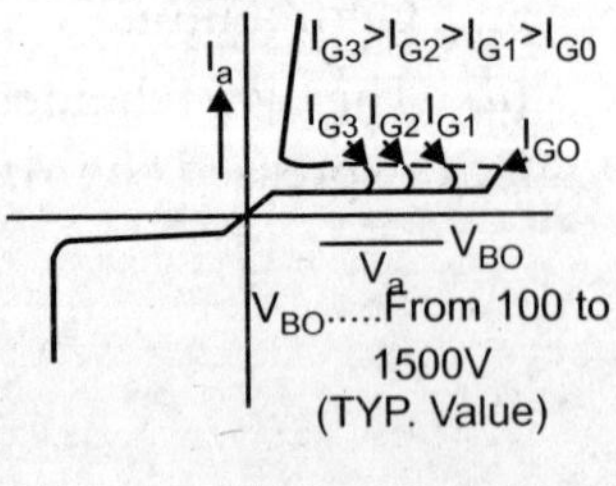 I_a, $I_{G3}>I_{G2}>I_{G1}>I_{G0}$, I_{G3} I_{G2} I_{G1} I_{GO}, V_{BO}, V_a, V_{BO}.....From 100 to 1500V (TYP. Value)

3. Figure 11.34 shows *V-I* characteristics of a certain SCR. Determine the following:

(*i*) Repetitive peak forward blocking (off-state) voltage

(*ii*) Non-repetitive peak reverse voltage

(*iii*) Holding current

(*iv*) Latching current

4. Recognize the thyristor application circuits shown in Fig. 11.35. Choose from Relaxation Oscillator, Astable Multivibrator, Controlled Half Wave Rectifier, Controlled Full Wave Rectifier.

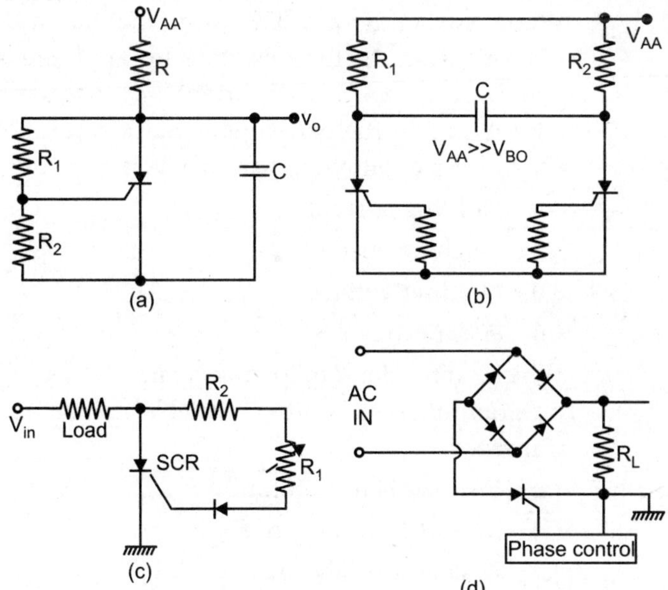

Fig. 11.35

Multichoice Questions

1. SCRs are connected in series to enhance
   (*a*) their overall *dv*/*dt* rating
   (*b*) their voltage ratings
   (*c*) their current handling capability
   (*d*) none of these
2. While connecting SCRs in series, the individual SCRs chosen for the purpose should have closely matched
   (*a*) *di*/*dt* rating
   (*b*) reverse biased characteristics
   (*c*) forward and reverse blocking characteristics
   (*d*) forward ON-state characteristics
3. SCRs are connected in parallel to enhance
   (*a*) voltage handling capability
   (*b*) current handling capability
   (*c*) *di*/*dt* rating
   (*d*) none of these
4. While connecting SCRs in parallel, the SCRs to be connected in parallel should have closely matched
   (*a*) forward ON-state characteristics
   (*b*) forward and reverse OFF-state characteristics
   (*c*) forward OFF-state and ON-state characteristics
   (*d*) SCRs cannot be connected in parallel
5. Once an SCR has been switched to the ON-state, the minimum value of the anode current required to keep the device in the ON-state is called
   (*a*) Latching current
   (*b*) Trigger current
   (*c*) Holding current
   (*d*) Break-over current
6. Which of the following Thyristors does not have a control terminal?
   (*a*) DIAC
   (*b*) TRIAC
   (*c*) SCR
   (*d*) GTO thyristor
7. The construction of a programmable unijunction transistor (PUT) is
   (*a*) similar to that of a conventional unijunction transistor except for the fact that both resistors forming the inter-base resistance R_{BB} are variable
   (*b*) similar to that of a four-layer PNPN diode
   (*c*) different from the conventional unijunction transistor in the sense that in the case of a PUT, the silicon bar is heavily doped and the P-type emitter is lightly doped
   (*d*) different from the conventional unijunction transistor in the sense that a PUT is made from Germanium and not Silicon

8. While switching an SCR from the forward-blocking state to the ON-state by applying a gate-trigger pulse, it is important for a successful switch on that the gate-trigger pulse is present till one of the following parameters is exceeded.
   (*a*) Holding current
   (*b*) Latching current
   (*c*) Holding voltage
   (*d*) *di*/*dt* rating
9. Which of the following parameters is programmable in a programmable unijunction transistor?
   (*a*) Intrinsic stand-off ratio
   (*b*) Inter-base resistance R_{BB}
   (*c*) Peak and valley currents
   (*d*) All the above
10. A UJT relaxation oscillator circuit produces a sawtooth like waveform
   (*a*) at B_1 terminal
   (*b*) at B_2 terminal
   (*c*) across the capacitor
   (*d*) none of these
11. UJT's *V-I* characteristics can be expressed by
   (*a*) a single valued function of voltage
   (*b*) a multivalued function of current
   (*c*) a single valued function of current
   (*d*) none of these
12. Pick the odd one out (Hint: One of these is not a thyristor)
   (*a*) 2N2646 (*b*) OE104
   (*c*) TY1005F (*d*) OE106
13. Pick the odd one out (Hint: Think of their *V-I* characteristics)
   (*a*) Tunnel diode (*b*) UJT
   (*c*) SCR (*d*) TRIAC
14. When we want to know the subcycle surge current capability of a thyristor, we should look for
   (*a*) *di*/*dt* rating
   (*b*) surge current, rating, I_{FSM}
   (*c*) I^2t rating
   (*d*) none of these
15. In half wave SCR power control circuit, if the firing angle is 30°C, then for one complete cycle of operation, the load gets power for
   (*a*) 60° (*b*) 150°
   (*c*) 330° (*d*) 30°

ANSWERS TO SELF EVALUATION EXERCISE

1. (1-3-2-6) (2-4-5-3) (3-1-3-2) (5-6-6-4) (6-5-4-1)
2. (*a*) Current controllable device
   (*b*) Current controllable device
   (*c*) No
   (*d*) Single valued function of current
   (*e*) Multivalued function of voltage
   (*f*) *dv*/*dt* rating
   (*g*) Diac
3. (*i*) 100 V (*ii*) 120 V
   (*iii*) 10 mA (*iv*) 20 mA
4. Figure 11.34 (*a*) Relaxation oscillator Fig. 11.34 (*b*): Astable multivibrator Fig. 11.34 (*c*) : Controlled half wave rectifier Fig. 11.34 (*d*): Controlled full wave rectifier.

MULTICHOICE QUESTIONS

1. (*b*)	2. (*c*)	3. (*b*)	4. (*a*)	5. (*c*)	6. (*a*)
7. (*b*)	8. (*b*)	9. (*d*)	10. (*c*)	11. (*c*)	12. (*a*)
13. (*a*)	14. (*c*)	15. (*b*)			

Chapter 12

Optoelectronic Devices

Learning Objectives

After completing the chapter, you will learn the following topics:

1. Operational principle of LEDs
2. Characteristic curves of a typical LED
3. LED parameters including Forward voltage, Candle power, Radiant power output, Peak spectral emission and Spectral bandwidth
4. Interfacing LED with digital logic
5. Photodiode – PN Junction, PIN and Avalanche
6. Different modes of operation of photodiodes – Photoconductive and Photovoltaic
7. Noise performance of photoconductive and photovoltaic photodiodes
8. Solar cells – Operational principle and characteristics
9. Phototransistors and Photodarlingtons
10. Photo SCR
11. Operational principle of bulk-type photoconductors
12. Photo FET
13. Image Sensors – CCD and CMOS
14. Electron Multiplying CCD
15. Full colour image in a CCD sensor
16. Comparison between CCD and CMOS sensors
17. Types of thermal Detectors and their operational principle
18. Operational principle of photoemissive type of photodetectors
19. Construction and features of Vacuum photodiodes
20. Image Intensifier - Operational principle and categories
21. Photosensor parameters including Responsivity, Detectivity, Dee-Star, Quantum efficiency, Spectral response and Response time
22. Different constituents of an optocoupler or optoisolator
23. Applications of optocoupler
24. Different devices used for fabricating displays
25. Different types of display fonts used with LED displays
26. Block diagram of a display subsystem

27. On board integrated circuit numeric displays (OBIC)
28. Stretched seven segment displays
29. Monolithic displays
30. Alphanumeric displays

Q. 1. What is an LED ? Briefly describe its operational principle. On what factors does the wavelength of the emitted photons depend and what are the practical materials used for fabricating LEDs?

Ans. A semiconductor P-N junction diode designed to emit light, when forward biased is called a light emitting diode (LED). Figure 12.1 shows its electrical symbol. Now how does an LED emit light ? When a P-N junction is forward biased, the electrons in the N-type material (forming the majority carriers) and the holes in the P-type material (again constituting the majority carriers) travel towards the junction. Some of these electrons and holes would tend to recombine with each other and in the process radiate energy in the form of light. The radiant energy can also be in the form of heat and the process named phonon emission. Phonon emission will not be discussed as that is not the one responsible for LED action. The wavelength of the emitted photons is a function of the band gap of energy in the semiconductor material. It is given by the empirical formula λ. (nm) = 1240/ΔE,where ΔE is the energy gap in eV. The above expression, thus relates the expected wavelength of photon emission to energy differential experienced by an electron undergoing recombination. The maximum possible energy of the emitted photons is determined by the band gap energy of the solid in which the P-N function is formed. There are numerous elements and elemental compounds which have band gap energies that lie in the region which could produce emission ranging from ultraviolet to infra-red However, very few of these materials are viable candidates for practical LED devices. Table 12.1 lists a few of the available materials, the associated band-gap energy and the emission wavelength to be useful for practical applications. Some have too low a conversion efficiency to be useful. Presently the only commercially available LED devices are manufactured using Gallium Arsenide (GaAs), Gallium phosphide (GaP) and the ternary compound Gallium Arsenide Phosphide Ga (As, P). These LEDs are available in numeral and alphabetic characters also measuring in height from 1/8 inch to more than an inch or so.

Fig. 12.1

Table 12.1.

Material	*Band gap Energy eV*	*Emission λ (nm)*
Ge	0.66	1880
Si	1.09	1140
GaAs	1.43	910
GaP	2.24	560
$GaAs_{60}P_{40}$	1.91	650
AlSb	1.60	775
InSb	0.18	6900
SiC	2.2–3.0	563 – 413

Q. 2. What are the characteristic curves of interest in a light emitting diode ? Sketch these curves for a typical LED.

Ans. The characteristics of interest in an LED are usually the:

(*a*) Volt-Ampere characteristics
(*b*) Spectral distribution
(*c*) Light output versus diode forward current
(*d*) Directional characteristics

These characteristics for a typical LED are shown in Fig. 12.2

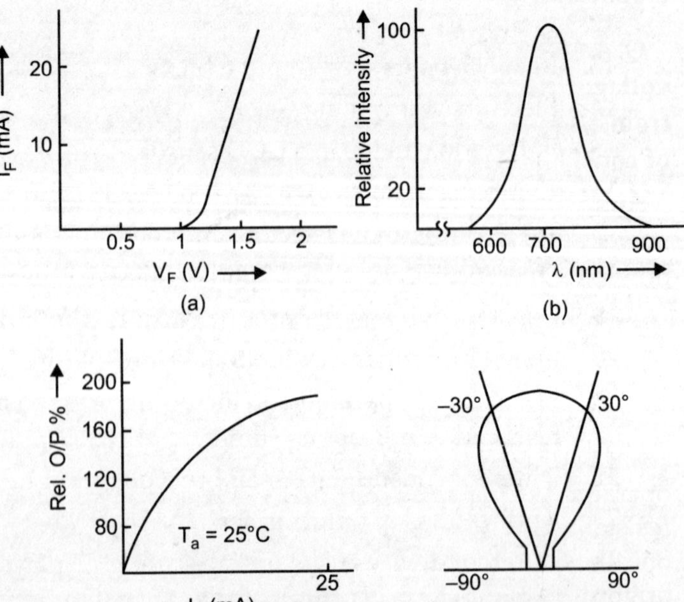

Fig. 12.2

While in operation, LED is forward biased like a junction diode. Fig. 12.2. (*a*) shows an example of Volt-Ampere characteristics of an LED. The threshold voltage is 1.5 V. Another curve of interest is the spectral distribution of light emnating from the LED. Such a curve is shown in Fig. 12.2 (*b*). The curve shows that the maximum intensity is obtained at 700 nm in case of the LED for which this curve has been drawn. The dependence of the brightness of the emitted light on the forward current is depicted in Fig. 12.2. (*c*). The curve of Fig. 12.2. (*d*) shows the distribution of light for different viewing angles.

Q. 3. Define the typical LED parameters.

Ans. The LED parameters of interest are the:

(*a*) Forward voltage (V_F)

(*b*) Candle power (*CP*)

(*c*) Radiant power output (P_o)

(*d*) Peak spectral emission (λ_{peak})

(*e*) Spectral bandwidth (BW)

Forward voltage (V_F) is the DC voltage across the LED, when it is ON.

Candle Power (CP) is a measure of the luminous intensity or the brightness of the emitted light.

Radiant power output (P_o) is the light power output of the LED.

Peak spectral emission (λ peak) is the wavelength of the brightest colour emitted by the LED.

Spectral bandwidth gives an indication as to how concentrated the brightest emitted colour is around its nominal wavelength.

Q. 4. How do you drive a LED from a regulated DC voltage? What are the shortcomings of driving LEDs from a constant voltage source. Briefly describe the use of constant current source for driving LEDs.

Ans. When an LED is driven from a regulated DC power supply, a resistor can be used to limit the current flowing through the LED. In Fig. 12.3. (*a*), the current that would flow through LED is given by

$$I = \left(\frac{V_{CC} - V_F}{R}\right), \text{ where } V_F \text{ is typically 1.5V.}$$

If you want to bias your LED at say 10 mA and your operating voltage is say 10 volts, you would require a resistance of 850 ohms. And if your supply is not regulated one, then of course the LED current and consequently the brightness would also vary directly as the supply voltage. If you are using a configuration like the one given in Fig. 12.3 (*b*), the LED current shall be given by

$$I_F = \frac{V_{CC} - V_F - V_{CE}(sat)}{R}$$

where, V_{CE} (*sat*) is the saturation collector to emitter voltage of the driver transistor. In such a case, we can write the following expression :

$$I_{F\,(max)} = [V_{CC\,(max)} - V_{F\,(min)} - VCE_{\,(sat)(min)}]/R_{(min)}$$

$$\text{and } I_{F\,(min)} = V_{CC\,(min)} - V_{F\,(max)} - [V_{CE\,(sat)}]_{max}/R_{(max)}$$

Keeping the above expression in mind, if you were asked to ensure a minimum of 10 mA of current through the LED when forward biased and connected in the configuration of Fig. 12.3 (*b*) and also when you have been told that the supply can vary from 10 volts to 15 volts and V_{CE} (sat) could be anywhere between 0.2 volt and 0.6 volt, then you can say that your current limiting resistance in no case could exceed 740 ohms, if the forward voltage drop across the LED can at the most be 2 volts. Of course, if V_{CC} is considerably larger than V_F + V_{CE} (sat), then small variations in V_F or V_{CE} (sat) will have a negligible effect on the LED current.

In some applications, it may be desirable that the LED lamp is driven from a current source. The current source can be used to regulate current through the LED regardless of the supply voltage variations or variations in V_F of LED from lamp to lamp. Figure 12.3 (*c*) gives you one example of LEDs being driven from a constant current source. For this circuit, current through LED string remains constant as long as V_{CC} is greater than $V_{CC\,(min)}$. The circuit uses an LED as voltage reference. If desired, it can be replaced by a Zener diode. Relevant mathematical expressions for the circuit of Fig. 12.3 (*c*) are as follows :

$$V_{REE} = 1.5\text{V (for a Ga AsP LED)}$$

$$R_1 = \frac{(V_{CC} - V_{REF})}{1mA}$$

(if LED – 1 is biased at 1 mA)

$$R_2 = (V_{REF} - V_{BE})/I_{CC}$$

$$V_{CC}\,(\text{min}) = V_{REF} + nV_F$$

(where *n* is the number of LEDs)

$$I_{CC} = (V_{REF} - V_{BE})/R_2$$

$$I = I_{CC} + (V_{CC} - V_{REF})/R_1$$

For the circuit shown in Fig. 12.3 (*d*)

$$I_{CC} = I_Z + I_F$$

$$I_Z = \frac{(V_{CC} - V_Z)}{R} \text{ and } I_F = (V_Z - V_F - V_{EB})/R'$$

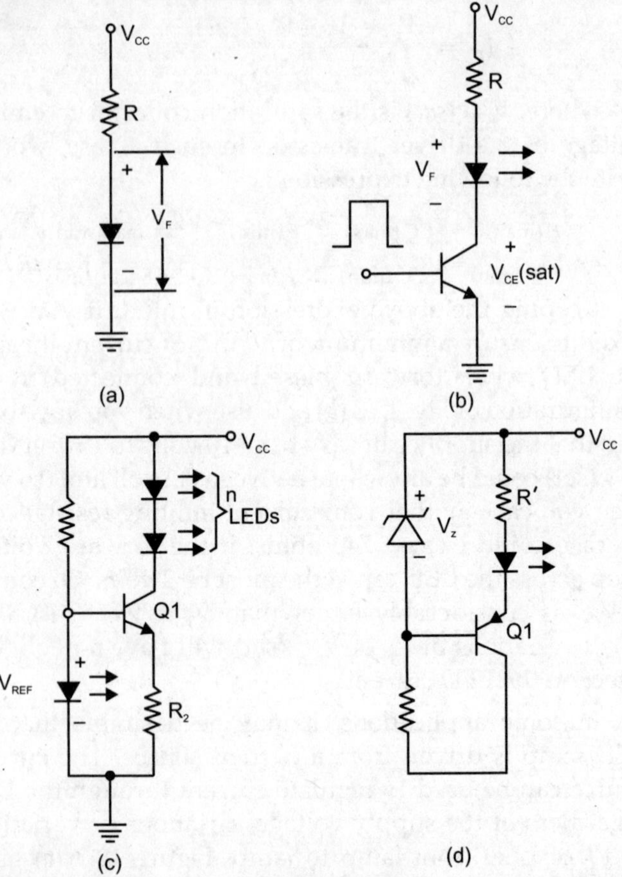

Fig. 12.3

Q. 5. Briefly describe how LEDs can be interfaced with digital logic? Also, outline the use of transistors and FETs for driving LEDs.

Ans. Since LED lamps operate at low voltages and currents, they can directly interface to most logic families Figure. 12.4 (*d*) depict some of the ways that an LED lamp can be used to interface to digital logic. Figure 12.4 (*a*) and (*b*) depict LED interface with logic gates with open collector outputs, whereas Figure 12.4 (*c*) and (*d*) show the interface circuits with active pull up totem pole gates. TTL logic families are guaranteed to sink a minimum of current which can drive most LED lamps without any additional buffering. If additional current is required, both TTL as well as CMOS buffers are available. Low cost transistors can also be used, when larger currents are required

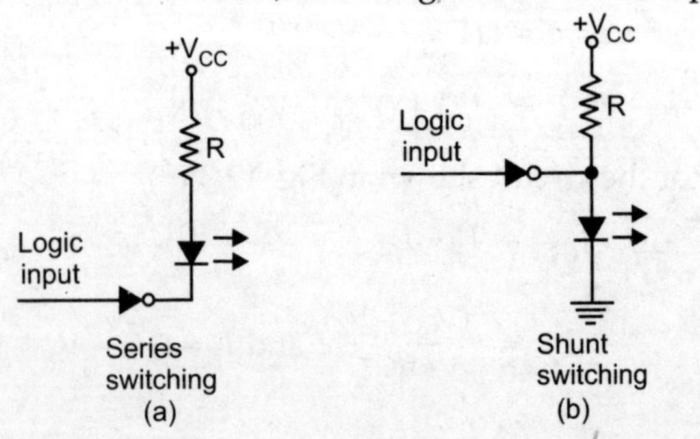

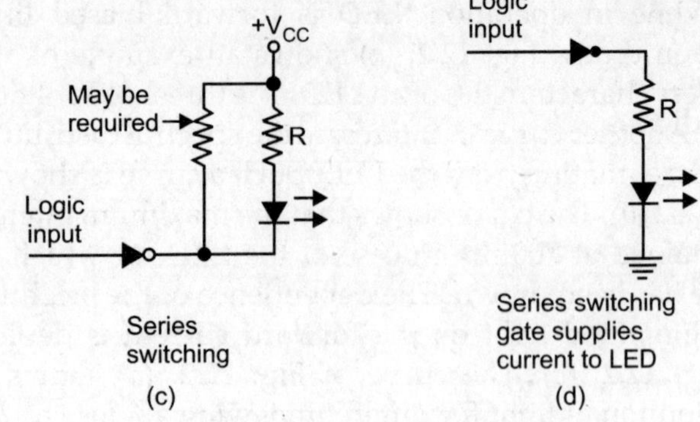

Fig. 12.4

than a logic gate can supply directly, Figure. 12.5 shows some of the commonly used drive schemes implemented with N-P-N and P-N-P transistors.

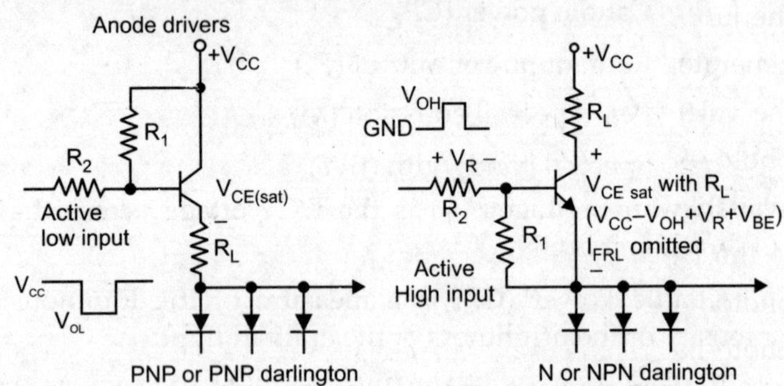

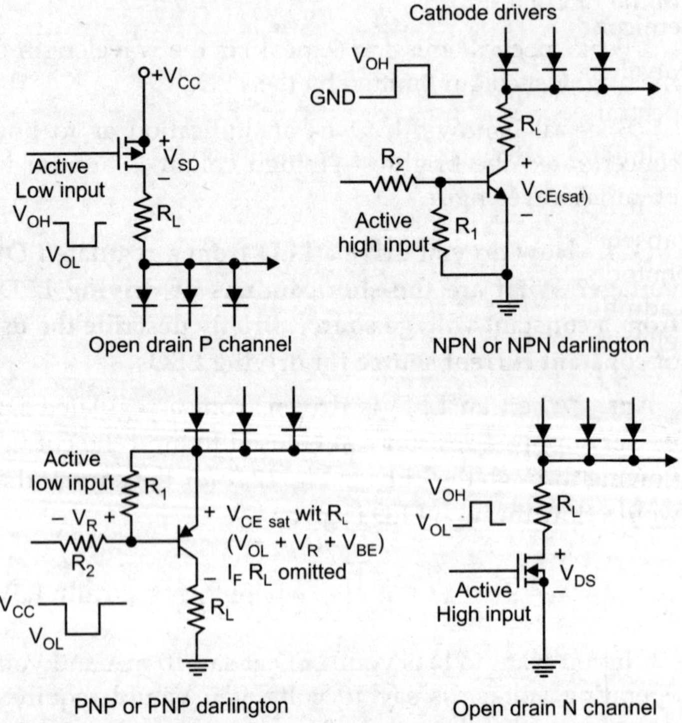

Fig. 12.5

Q. 6. What is a photodiode? Briefly describe the P-N junction photodiodes, the PIN photodiodes and the Avalancle photodiodes.

Ans. The photodiode is again a P-N junction diode through which a significant current flows, when light falls on it. There are three main types into which these photodiodes can be classified. (1) P-N junction photodiodes, (2) PIN Photodiodes, (3) Avlaaehe photodiodes (APD).

When the light is incident on a P-N photodiode, the impinging photons create electron-hole pairs on both sides of the junction. With no light applied, the reverse current is simply the reverse saturation current due to the minority carriers, holes in the N-type and electrons in the P-type. When light is applied, photo-induced electrons in the conduction band of the P-type will move across the junction down the potential hill with the thermally generated minority carriers. Similarly, holes produced in the valence band of the N-type (due to electrons being photo-excited into the conduction band of N-type) are available to add to the current flow by moving across to the P-side. Figure 12.6 (*a*) shows the photoeffect in a photodiode. Figure 12.6 (*b*) shows the variation of the photo-induced current as a function of irradiance.

The sensitivity of a photodiode made from a certain semiconductor material is different for different light wavelengths falling on it. This can be observed from the spectral distribution curve of the photodiode. The curve is similar to the one shown in Fig. 12.2 (*b*).

It is because, the sensitivity depends upon the band gap energy of the semiconductor used for making the photodiode. In addition to Germanium and Silicon, Cadmium Selenide, Cadmium sulphide and Cadmium Telluride are other commonly used materials.

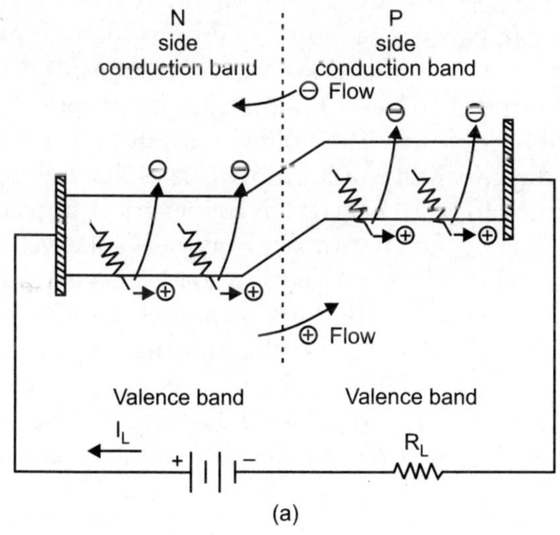

(a)

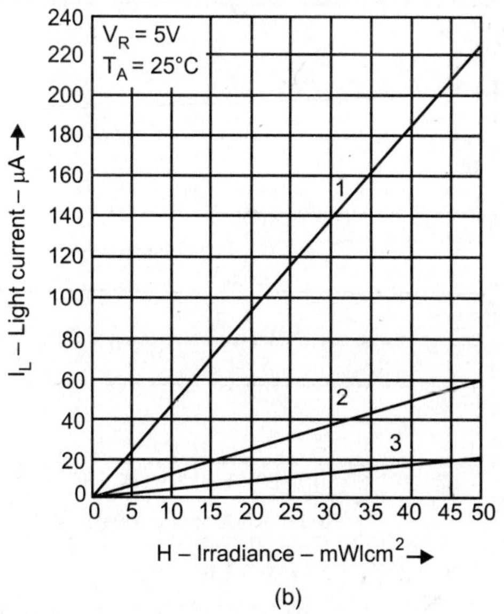

(b)

Fig. 12.6

In operation, the photodiode is reverse-biased. See Fig. 12.7. The active diameter of these devices is typically in the range of 2 to 15 mm^2 but are mounted in a standard TO-5 package with a window. They can operate at frequencies of the order of 1 MHz and are used in character recognition, optical communications equipment and encoders.

The second variety of photodiodes *i.e.* PIN photodiodes have a faster response. A PIN photodiode consists of a layer of intrinsic silicon between heavily doped *P* an N-type Silicon semiconductor materials. This has the effect of reducing the transit or the diffusion time of photo-induced electron-hole pairs. That is, carriers created by photons incident on the middle of the depletion region have less distance to travel than those generated on one side or the other side of the depletion layer. The relatively thick layer (2.5 microns) of high resistivity intrinsic silicon ensures the absorption of most of the photons. Figure 12.8 shows the electro-optical characteristics of typical PIN photodiode. You would notice that the device is essentially a constant current generator at a given light level.

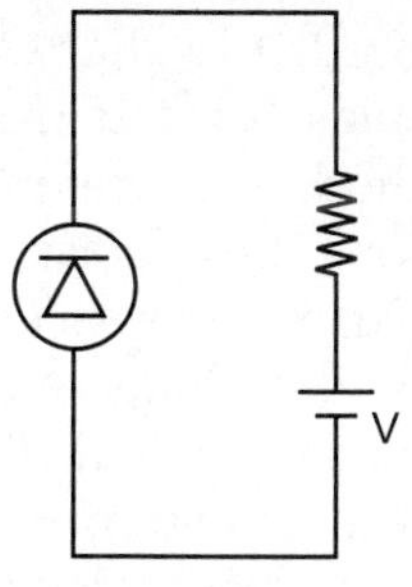

Fig. 12.7

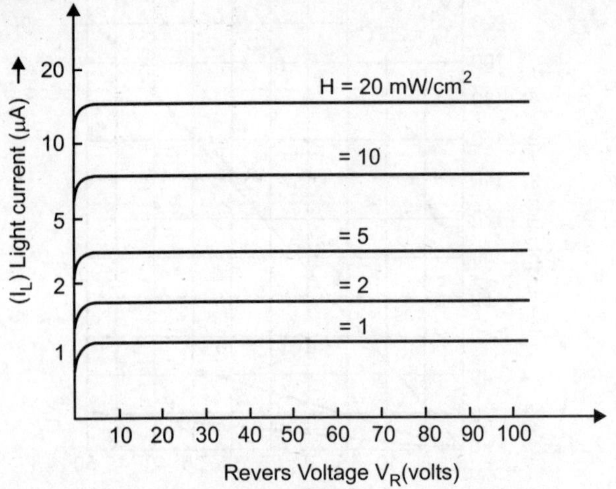

Fig. 12.8

A third variety is the i.e. Avalanche photo diode. It is a photodiode operated in the reverse breakdown or avlanche breakdown region. By doing like this, the current sensitivity may be increased 30 to 100 times. In the avalanche photodiode, the construction is such as to provide a very uniform junction that exhibits the avalanche effect at voltages typically between 30 V and 300 V. The electron-hole pairs that are generated by incident photons are accelerated by the high electric field to knock out more electrons in the valence band to make them available for conduction.

Q. 7. Distinguish between photoconductive and photovoltaic modes of operation of photodiodes. What are their preferred application areas?

Ans. A photoconductive photodiode operates on the principle of photoconduction, which says that electrical conductivity increases by generation of additional charge carriers, when light of sufficient energy falls on the material. In case of a photodiode, the sensitivity of the photo-induced current to incident light could be enhanced considerably, if the diode was operated in the reverse biased condition. The current that flows in the absence of any light in this case is the reverse saturation current of the diode known here as the dark current.

In the photovoltaic photodiode, an EMF is generated across the junction when illuminated. This photovoltaic EMF produces the photoinduced current, when the external circuit is closed. As a photovoltaic diode requires no external reverse bias, it is free from any dark current. Photovoltaic photodiodes are preferred in low noise, low frequency applications. Low noise mainly results from the absence of dark current effects. Due to higher value of junction capacitance and shunt resistance, photovoltaic photodiodes are not used beyond 100 kHz. Below 1 kHz, however, it exhibits a far superior signal to noise ratio than a photoconductive photodiode having same active area.

Photoconductive photodiodes are used in high frequency applications. Their frequency response easily extends beyond a few GHz. This is made possible due to reduced junction capacitance resulting from application of reverse bias. Photoconductive diodes also produce higher responsivities at longer wavelengths.

Q. 8. What are solar cells? Explain their operational principle and characteristics. Also, discuss about the efficiency of the solar cells.

Ans. Solar cells are devices whose operation is very similar to that of a photodiode in the photovoltaic mode. The operational principle of the basic solar cell is based on the photovoltaic effect. Due to the photovoltaic effect, there is a generation of an open circuit voltage across a P-N junction, when it is exposed to light, which is the solar radiation in case of a solar cell. This open circuit voltage leads to the flow of electric current through a load resistance connected across it as shown in Fig. 12.9.

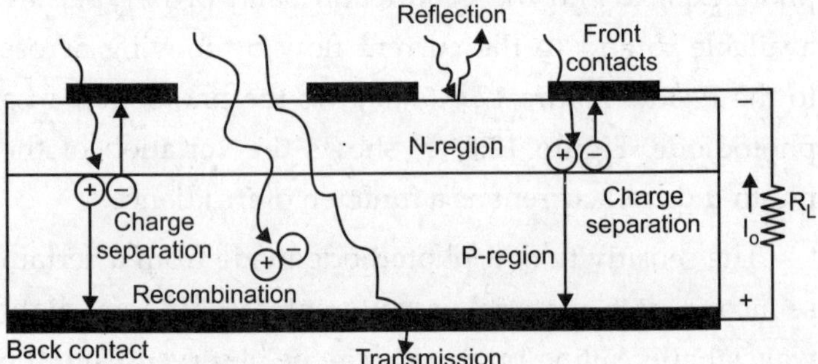

Fig. 12.9

As is evident from the figure, the impinging photon energy leads to generation of electron-hole pairs. The electron-hole pairs either recombine and vanish or start drifting in the opposite directions with electrons moving towards the N-region and holes moving towards the P-region. This accumulation of positive and negative charge carriers constitutes the open circuit voltage. This voltage can cause a current to flow through an external load or when the junction is shorted, the result is a short circuit current whose magnitude is proportional to input light intensity. The voltage output and the current delivering capability of an individual solar cell are very small for it to be of any use as an electrical power input to any system. As an example, a typical solar cell would produce 500 mV output with a load current capability of about 150 mA. The series-parallel arrangement of solar cells is done to get the desired output voltage with required power delivery capability. The series combination is used to enhance the output voltage while the parallel combination is used to enhance the current rating.

Figure 12.10 shows the current-voltage and power-voltage characteristics of a solar cell. As is evident from the figure, solar cell generates its maximum power at a certain voltage. The power-voltage curve has a point of maximum power called Maximum Power Point (MPP). The cell voltage and the corresponding current are less than the open circuit voltage (V_{OC}) and the short circuit current (I_{SC}) respectively at the maximum power point.

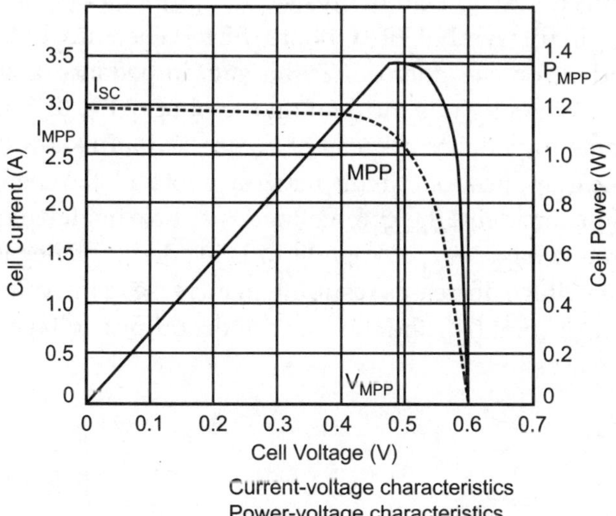

Fig. 12.10

Solar cell efficiency is the ratio of maximum electrical solar cell power to the radiant light power on the solar cell area. The efficiency figure for some crystalline solar cells is in excess of 20%. The most commonly used semiconductor material for making solar cells is silicon. Both crystalline and amorphous forms of silicon are used for the purpose. Another promising material for making solar cells is Gallium Arsenide (GaAs). Gallium Arsenide solar cells when perfected will be light weight and more efficient.

Q. 9. Compare the noise performance of photoconductive and photovoltaic photodiodes. What is Noise Equivalent Power (NEP)?

Ans. Noise in a photoconductive photodiode is a combination of shot noise (mainly due to the dark current) and the Johnson (or thermal) noise (due to series resistance, load resistance and channel resistance if any). The shot noise in photoconductive photodiode is expressed by

$$I_S \text{ (noise current)} = (2eI_d\,\Delta f)^{1/2}$$

where, e = Electronic charge = 1.6×10^{-19} Coulomb

I_d = Dark current (Amperes)

Δf = Noise Equivalent Bandwidth (Hertz)

The thermal noise contribution (I_t) is expressed by

$$I_t = \left(\frac{4KT\Delta f}{R}\right)^{1/2}$$

where, K = Boltzmann's constant = 1.38×10^{-23} J/°K

T = Temperature (°K)

Δf = Noise Equivalent Bandwidth (Hertz)

R = Resistance (series, load, channel)

Noise in a photovoltaic photodiode is mainly the thermal noise due to device's shunt resistance. It is expressed as,

$$I_t = \left(\frac{4KT\Delta f}{R_{sh}}\right)^{1/2}$$

where, R_{sh} = Shunt resistance (ohms)

Noise Equivalent Power (NEP) is the minimum incident power required to generate a photocurrent equal to the total RMS noise current of the photodiode. It is often considered as figure-of-merit for the photodiodes. It is expressed in watts per Hertz square root (watts/ $\sqrt{Hz}$) and can be determined from

$$\frac{\text{Noise current (in amperes)}}{\text{Responsivity (in amperes per watt)}}$$

It is always specified alongwith test conditions like source wavelength, test frequency and bandwidth.

Q. 10. How does a phototransistor differ from a photodiode? What are its advantages? What is a photodarlington?

Ans. A phototransistor differs from a photodiode in the sense that the photo induced current of the photodiode gets internally multiplied by transistor action resulting in a larger sensitivity. If (I_λ) is photoinduced current, then the collector current in a phototransistor would be I_λ (1 + h_{fe}). Some types of phototransistors have got a third lead (base lead) also brought out. This enables use of a phototransistor as a switching or bistable device, where incidence of a small amount of light can make the device switch from a low current level to a high current level.

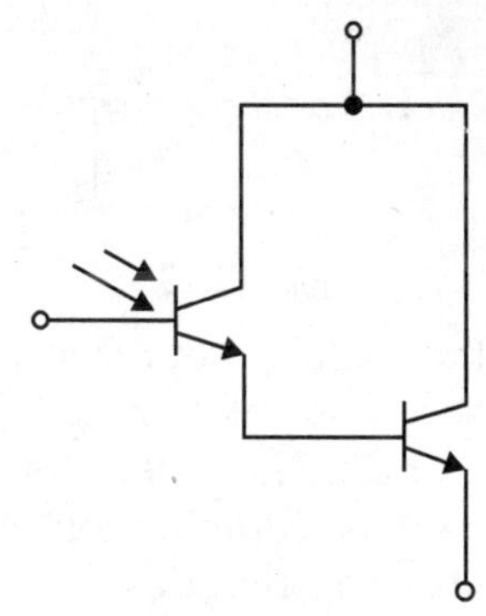

Fig. 12.11

A photodarlington (Fig. 12.11) is darlington configuration with the input transistor as light sensitive. The sensitivity of a photodarlington is an order of magnitude higher than even phototransistors.

Q. 11. What is a light activated SCR (or a photo SCR)?

Ans. It is an SCR that can be switched into conduction by light falling on gate cathode junction. Like a conventional SCR, once switched ON, the light loses control on the anode current. The LASCR can then be switched OFF only by bringing the anode current below the holding value.

Q. 12. What are bulk type photoconductors? Outline their principle of operation.

Ans. In case of bulk-type photoconductors, the conductivity of the device increases with increase in intensity of light impinging on the sensitive surface of the device. The conductivity increases due to release of electrons caused by impinging photon energy.

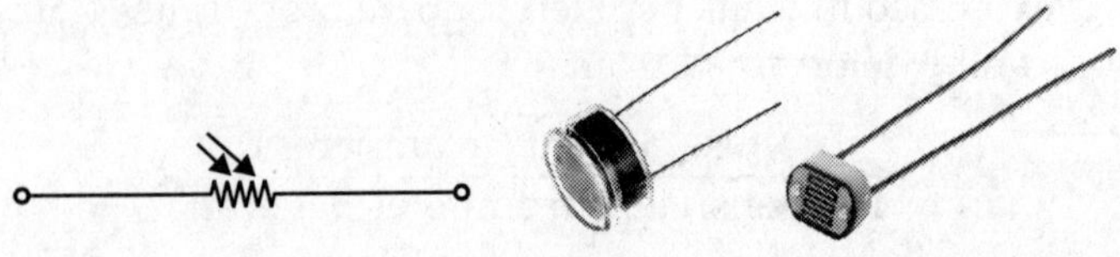

Fig. 12.12

The commonly used materials for fabrication of bulk type photoconductors are Cadmium Sulphide (CdS) that respond to visible band and Lead Sulphide (Pbs), Lead Selenide (PbSe) Silicon (Si) and Germanium (Ge) that respond to IR band. Figure. 12.12 shows the circuit symbol and also the pictures of typical bulk type photoconductors. Light Dependent Resistor (LDR) is another name for the same device. The electrical response of bulk type photoconductors is a current, when these are

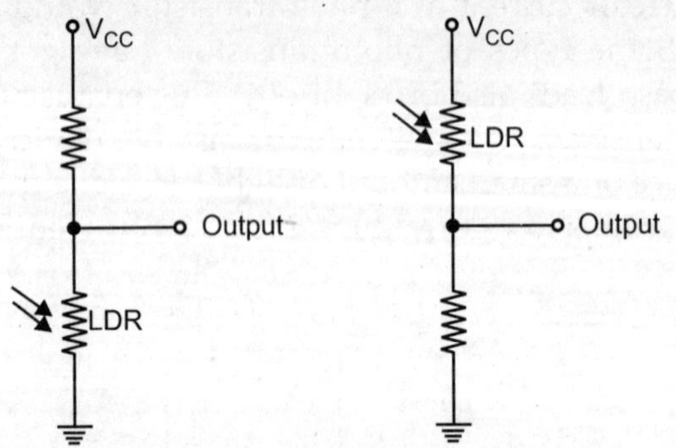

Fig. 12.13

biased by a constant voltage source and a voltage, when these are biased by a constant current. Figure 12.13 shows typical biasing schemes. These detectors offer a logarithmic Resistance versus Illumination response that is constant over almost four decades.

Q. 13 What is a photo FET? Draw the circuit symbol of a photo FET and a simple application circuit.

Ans. Photo FETs are light sensitive FET devices, wherein the diode formed by the reverse-biased gate-channel junction acts as a photodiode. Incident light generates additional photo carriers resulting in increased conductivity level. Gate current flows, if the gate is connected to an external resistor. Figure 12.14 (*a*) shows the circuit symbol of a photo FET. When no light is incident on the photo FET, the gate impedance is very high. When light is incident on the photo FET, the value of gate impedance decreases. Figure 12.14 (*b*) shows the typical application circuit using a photo FET. When no light is incident, the gate voltage is approximately equal to the voltage $-V_{GG}$. When light is incident, the negative gate voltage decreases resulting in increase in the value of drain current (I_D) and the value of the output voltage (V_o) decreases.

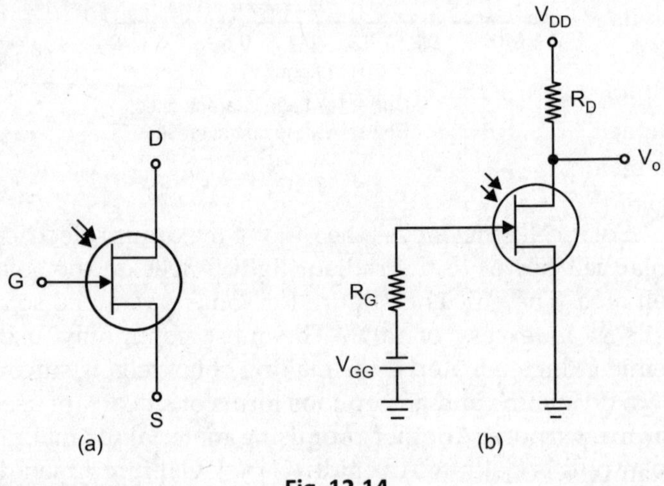

Fig. 12.14

Q. 14 What are image sensors? What are the two common types of image sensors?

Ans. Image sensors are used to store optical images in the form of electrical signals. An image sensor constitutes the heart of a digital camera. Two most common and widely used types of image sensors are the charge-coupled device (CCD) sensor and the CMOS sensor. Both types of sensors use a two-dimensional array of thousands to millions of discrete pixels. The amount of light falling on each of the pixels generates free electrons with the number of electrons and hence, the quantum of charge depending upon the intensity of impinging photons. The two types of sensors differ in the mode in which this charge is converted into voltage and subsequently read out of the chip for further processing.

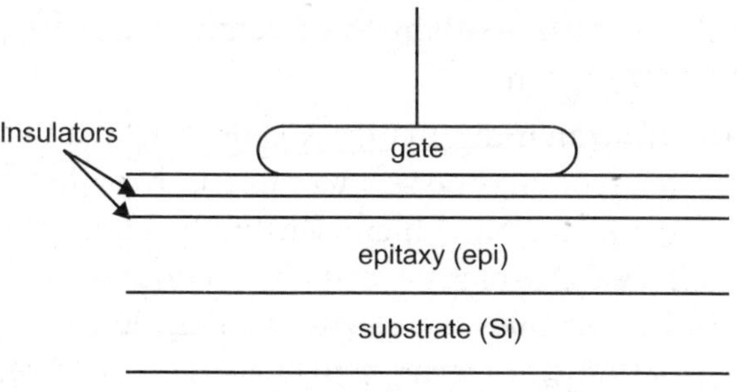

Fig. 12.15

Q. 15 What is a CCD sensor? Explain the construction and the operational principle of CCD sensor.

Ans. A charge coupled device (CCD) is basically an array of thousands to millions of light sensitive elements called pixels etched onto a silicon surface. Each of the pixels is a buried channel MOS capacitor. CCDs are typically fabricated on a p-type substrate and a buried channel is implemented by forming a thin n-type region on its surface. A silicon dioxide layer is grown on the top of the n-type region and an electrode, also called gate, on top of the insulating silicon dioxide completes the MOS capacitor. The electrode could be metal, but is more likely to be a heavily doped polycrystalline silicon conducting layer (Fig. 12.15).

The sensor is not actually flat, but has tiny cavities, like wells, that trap the incoming light and allow it to be measured. Each of these wells or cavities is a pixel. The size of the CCD is specified in mega-pixels. Mega-pixel value of the CCD can be computed by multiplying number of pixels in a row by number of pixels in a column. For example, 1000 pixels in a row and 1000 pixels in a column make 1.0 mega pixel CCD chip. When the light reflected off the target to be imaged is incident on this array of pixels; the impinging photons generate free electrons in the region underneath the pixels. In order to make sure that these free electrons don't combine with holes and disappear as heat energy; the electrons underneath each pixel are held there by applying a positive bias voltage to the pixels. If the sensor array were exposed to light for the same time; the number of electrons and hence, the quantum of charge held under a certain pixel would vary directly as the luminous intensity that particular pixel is exposed to. This charge pattern represents the light pattern falling on the device. The charge is read out by suitable electronics and then converted into a digital bit pattern that can be understood and stored in a computer. This digital bit pattern then represents the image.

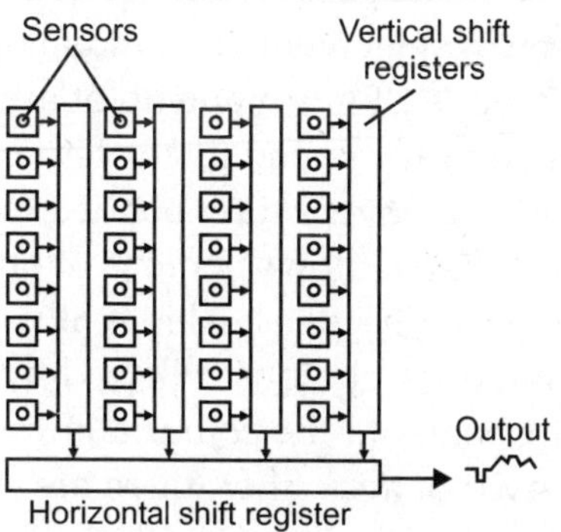

Fig. 12.16

The charge held in the bins corresponding to different pixels is read out, converted into equivalent analog voltages and then digitized with the help analog-to-digital converter. Charge on a CCD is shifted in two directions namely parallel (or horizontal) and serial (or vertical). While parallel shift occurs from the right to the left, the serial shift is performed from top to bottom and directs the charge packets to the measurement electronics. One way to make the read-out process faster is to split up into 2 or 4 different sections. Each section follows the process of parallel and serial shift. Figure 12.16 shows a typical CCD sensor with a single point read out. Figure 12.17 shows the photograph of a typical packaged CCD chip.

Fig. 12.17

Q. 16 How can you record an image in full colour using a CCD sensor?

Ans. The CCD sensor described earlier can only determine number of photons collected by each pixel and therefore, it carries no information about the wavelength or colour of those photons. Hence, the CCD sensor is capable of recording the image only in monochrome. In order to record image in full colour; a filter array is bonded to the sensor substrate. One such common colour filter array is the Bayer filter. Bayer's colour filter array (CFA) comprises an arrangement of red, green and blue filters to capture colour information. It has alternating Red/Green and Blue/Green arrangement and is sometimes called an RGBG filter. Bayer's CFA is made up of alternating rows of Red/Green and Blue/Green filters. Figure 12.18 shows a CCD array with a Bayer's filter bonded to its surface. A particular colour filter allows photons of only that colour to pass through to the pixel. This is illustrated in Fig. 12.19. The number of photons collected by each pixel in this case corresponds to the colour allowed by the filter above it. In Bayer pattern filter, there are twice as many green squares as there are red or blue. This is because the human eye is much more sensitive to green light than either red or blue, and has a much greater resolving power in that range.

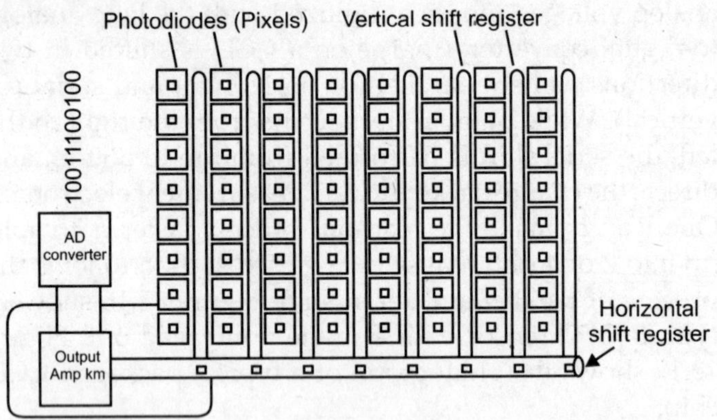

Fig. 12.18

There is space between adjacent light sensitive pixels and this space is used to house on-chip electronics. Micro lenses placed above the filter help in directing light into one or other of the adjacent pixels. As outlined earlier, a single pixel allows photons of one colour only. Full colour image is worked out by using a complex method called demosaicing. Simply stated, the camera treats each 2 × 2 set of pixels as a single unit thereby providing one red, one blue and two green pixels. The actual colour is estimated based on the photon levels in each of these four pixels.

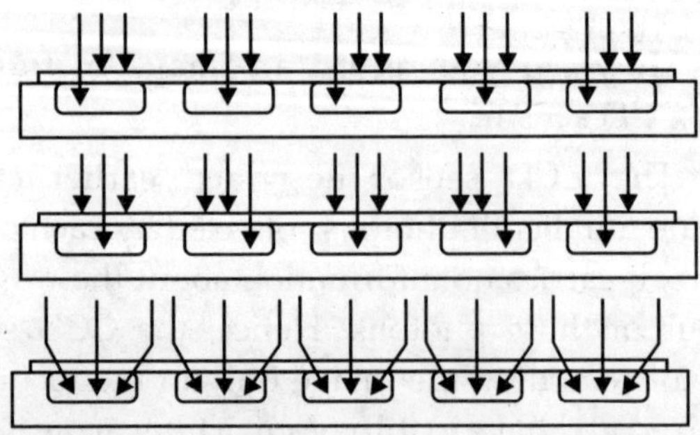

Fig. 12.19

Q. 17 Explain how an Electron Multiplying CCD overcomes the shortcomings of traditional CCD technology. Also, explain the operational principle of an EMCCD sensor.

Ans. Electron multiplying CCD (EMCCD) technology helps in overcoming the shortcoming of traditional CCD technology in offering high sensitivity at high speed. Traditional CCD sensors backed by readout noise figures of typically less than ten electrons offer high sensitivity. High sensitivity, however, comes at the cost of reduced speed of readout, which is typically less than 1 MHz. The speed constraint arises from the bandwidth limitation of the CCD charge amplifier. Increasing bandwidth, which is essential for high speed operation, increases noise too. That is, high speed amplifiers are relatively noisier than their slow speed counterparts. Electron multiplying CCD overcomes this limitation by building a unique electron multiplying structure into the chip. As a result of this, EMCCD as an image sensor is capable of detecting single photon events without an image intensifier. EMCCD sensors achieve high sensitivity, high speed operation by amplifying charge signal before the charge amplifier.

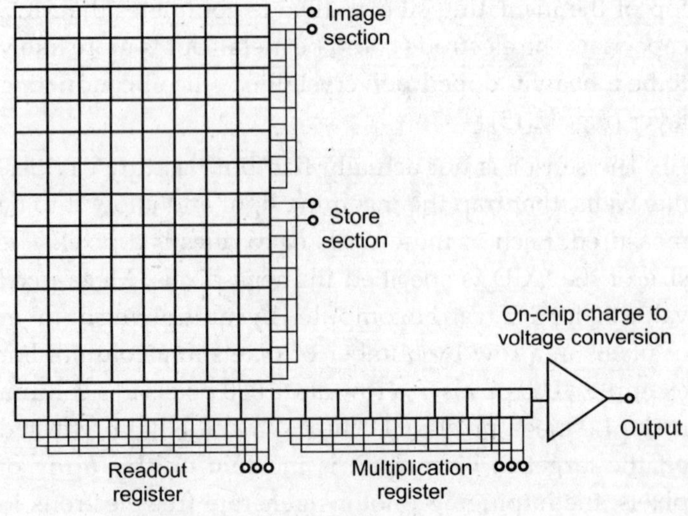

Fig. 12.20

Most Electron-multiplying CCDs utilize a Frame Transfer CCD structure shown in Fig. 12.20. In the case of frame-transfer CCD format, sensor area captures the image and storage area stores the image prior to read out. The storage area is normally identical in size to the sensor area and is covered with an opaque mask, normally made of aluminium. When the sensor area is exposed to light, an image is captured, which is automatically shifted downwards behind the masked region of the chip, and subsequently read out. During the readout process, another image is being captured by the sensor area. The aluminium mask therefore acts like an electronic shutter. As shown in Fig. 12.20, there is a multiplication register between the normal serial readout register and the charge amplifier. To readout the sensor the charge is shifted out through the readout register and through the multiplication register, where amplification occurs prior to readout by the charge amplifier. The multiplication register has several hundred stages or cells that use higher than normal clock voltages to achieve charge amplification. The amplification occurs in the multiplication register through a process known as clock induced charge or spurious charge that occurs naturally in CCDs. The multiplication register contains many hundreds of cells as outlined above and the amplification process occurs in each cell by harnessing clock-induced charge process. When clocking the charge through a register, there is a very small but finite probability (typically less than 2%) that the charges being clocked can create additional charges by a process known as impact ionization. Impact ionization is the process by which a charge having sufficient energy creates another electron-hole pair. Hence, a free electron charge in the conduction band can create another charge leading to amplification. The electron multiplication factor (M) may be computed from $M = (1+p)^N$, where (N) is number of cells and (p) is the probability value. If the multiplication register for instance had 512 cells or stages and the probability of secondary electron generation were 1.3%; the multiplication factor would be around 744. The electron multiplication prior to the output amplifier ensures that the readout noise introduced by the amplifier has negligible effect.

Major advantages of EMCCD sensors include high sensitivity in low light conditions, high speed imaging capability, good daytime imaging performance and reduced likelihood of sensor damage while viewing in bright conditions. By elevating photon-generated charge above the readout noise of the device, even at high frame rates, the EMCCD has the capability of meeting the needs of ultra-low-light imaging applications without the use of external image intensifiers. The disadvantage is its relatively higher power consumption due to need for active cooling of the CCD. The extreme low light capability of these EMCCDs lend themselves well to a range of applications including border and coastal surveillance, surveillance of ports and airports, protection of sensitive sites and critical assets and low light scientific imaging such as in astronomy.

Q. 18. What is the operational principle of CMOS sensors and how do they differ from the CCD sensors?

Ans. Both CCD and CMOS sensors depend for their operation on photoelectric effect to create electrical signal from light photons. CCD (charge coupled device) and CMOS (complementary metal oxide semiconductor) image sensors are two different technologies for capturing images digitally.

In a CCD sensor, quantum of charge held by different pixels is transferred through one or a very limited number of output nodes. Each pixel's charge is converted to a proportional voltage and after buffering sent off-chip as an analog signal. In a CCD sensor, all of the pixel can be devoted to light capture and the output's uniformity, which is a key factor in image quality, is high. On the other hand, in a CMOS sensor, each pixel has its own charge-to-voltage convertor, amplifier and a pixel select switch (Fig. 12.21). This is called active pixel sensor architecture in contrast to passive pixel sensor architecture used in a CCD sensor. Also, the sensor often also includes on-chip amplifiers, noise-correction, and analog-to-digital conversion circuits and other circuits critical to pixel sensors' operation. The chip in this case outputs digital bits. Inclusion of these functions reduces the area available for light capture. Also, with each pixel doing its own conversion, uniformity and consequently image quality is lower. While read out mechanism of a CCD sensor is serial; it is massively parallel in the case of a CMOS sensor allowing high total bandwidth for high speed.

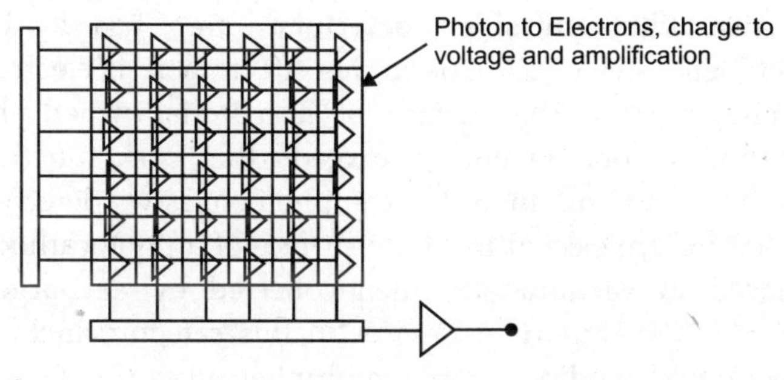

Fig. 12.21

Q. 19. What are some of the major differences between the two main types of sensors?

Ans. Fabrication of CMOS sensors involves standard CMOS technology well established for fabrication of integrated circuits. This also allows easy integration of required electronics on the same chip thereby resulting in devices that are compact and cost effective. On the other hand, CCD sensor fabrication involves dedicated and costly manufacturing processes.

In comparison with CCD sensors, CMOS sensors have relatively poor sensitivity and uniformity. The former is due to poor fill factor in the case of CMOS sensors and a large fill factor in CCD devices. The latter is due to use of separate amplifiers for different pixels in the case of CMOS sensors and a single amplifier for all pixels in the case of CCD sensors.

CMOS sensors have higher speed than CCD devices due to use of active pixels and having analog-to-digital converters on the same chip leading to lesser propagation delays. Low end CMOS sensors have low power requirements, but high speed CMOS cameras typically require more power than CCD devices.

Dynamic range of a CCD sensor is higher than that of a CMOS sensor.

As far as the noise performance is concerned, CMOS sensors score over CCD sensors, when it comes to temporal noise due to lower bandwidth of amplifiers at each pixel. But in terms of fixed pattern noise performance, CCD sensors are better due to single point charge-to-voltage conversion.

CMOS sensors allow on-chip incorporation of a range of functions such as automatic gain control, auto exposure control, image compression, colour encoding, motion tracking etc.

Due to absence of shift registers in CMOS sensors, they are immune to smearing around over-exposed pixels as it is caused by spilling of charge into the shift register.

Q. 20. What are photoemissive type of photodetectors? Briefly describe their principle of operation.

Ans. Photoemissive detectors are based on photoemissive effect. Due to this effect, there is electron emission from the surface of a material, when the impinging photon energy exceeds the work function of the material. In a typical photoemissive detector, electrons are ejected from the surface of a photocathode placed in vacuum and then collected by a charged anode. The important devices in this category include vacuum photodiodes, photomultiplier tubes (PMT) and microchannel multipliers (Fig 12.22).

Fig. 12.22 Photo emissive detectors

Q. 21. What are Vacuum photodiodes? Explain the construction and salient features of a photomultiplier tube.

Ans. Vacuum photodiode is the oldest photosensor. It comprises an anode and a cathode placed in a vacuum envelope. Cathode when irradiated, releases electrons that are attracted by the positively charged anode, thus producing a photocurrent proportional to the light intensity.

Photomultiplier tubes are extremely sensitive photo sensors operating in the ultraviolet, visible and near infrared spectrum. PMTs have internal gain of the order of 10^8 and can even detect single photon of light. They are constructed from a glass vacuum tube which houses a photocathode, several dynodes and an anode. When the incident photons strike the photocathode, electrons are produced as a result of the photoelectric effect. These electrons accelerate towards the anode and in the process electron multiplication taken place due to secondary emission process from the dynodes. PMTs require few kilo-volts of biasing voltages for proper operation. Figure 12.23 shows the cross-section of a photomultiplier tube. As we can see from the figure, the dynodes are given progressively increasing positive voltages with the dynode nearest to the cathode having the lowest voltage and the dynode nearest to the anode having the maximum voltage.

Salient features of photomultiplier tubes include low noise, high frequency response and large active area. Due to these features, PMTs are used in nuclear and particle physics, astronomy, medical imaging and motion picture film scanning. Avalanche photo diodes have replaced PMTs in some applications, but PMTs are still used in many cases. Figure 12.24 shows photographs of some representative PMTs.

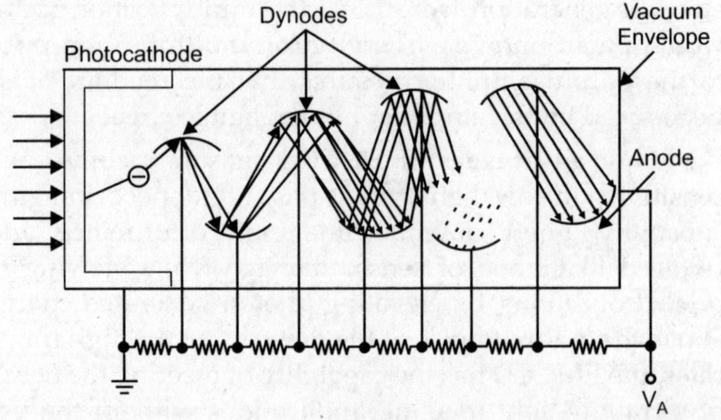

Fig. 12.23

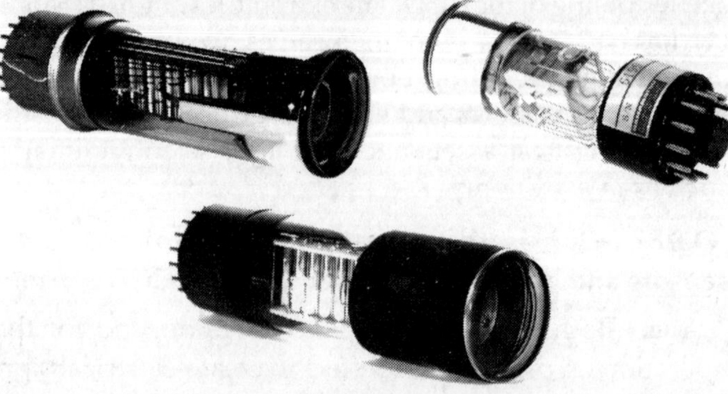

Fig. 12.24

Q. 22. Explain the operational principle of Image intensifiers. What are the categories that image intensifiers can be classified into?

Ans. Image intensifiers are devices that amplify visible and near infrared light from an image so that a dimly lit scene can be viewed by a camera or by human eye. Contemporary image intensifiers comprise an objective lens, vacuum tube with photocathode at one end, tilted micro channel plate (MCP) and a phosphor screen (as shown in Fig. 12.25). Objective lens focuses the image on to the photocathode. When the photons strike the photocathode, electrons are released due to the photoelectric effect. These photoelectrons are accelerated through around 4 – 5 kV into a tilted micro channel plate where secondary electron multiplication takes place. The electrons all move together due to the potential difference across the tube and for each photoelectron hundreds or even thousands of electrons are created. All these electrons hit the phosphor screen at the other end, releasing one photon for every electron. Thus, the screen converts the high energy electrons into photons, which corresponds to the input image radiation but with the incident flux being amplified many times.

Image intensifiers are classified into the following categories namely generation 0, generation 1, generation 2, generation 3 and generation 4. Generation 0 and generation 1 devices did not have the MCP and the stream of electrons generated by the photocathode was accelerated towards the phosphor screen by the applied potential. Generation 1 devices were a tremendous improvement upon generation 0 devices and had three times the photosensitivity of generation 0 devices. Generation 2 devices introduced the concept of micro channel plates. Generation 3 devices are the same as generation 2 devices except that the photocathode material in these devices is Gallium Arsenide (GaAs) whereas it was S-25 in case of generation 2 devices. Also, generation 3 devices had a better MCP. Generation 3 Ultra and generation 4 tubes are also available which offer slight improvement over generation 3 devices. Image intensifiers are used in night vision devices used for military applications.

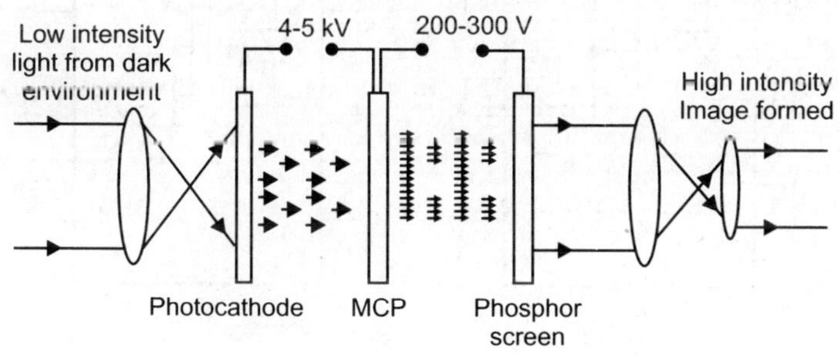

Fig. 12.25

Q. 23. What are thermal detectors ? What are different types of thermal detectors ? Briefly describe principle of operation of different types of thermal detectors.

Ans. Thermal detectors, all of them, absorb radiation and operate on the resulting temperature rise. There are three major categories of thermal detectors, namely:

1. Thermocouple (or Thermopile) type
2. Bolometric type
3. Pyroelectric detectors

Thermocouple (or thermopile) type of thermal detectors are based on Seebeck effect, which states that if a circuit is formed consisting of two dissimilar metals, a current flows in the circuit, if one of the junctions is at a higher temperature than the other. The spectral response of these detectors extends to about 40 μm and frequency response to about 20 Hz. Typical sensitivity of a thermocouple is 100 μV/°C. Thermopile is a series connection of more than one thermocouple, which enhances the sensitivity of the detector.

In case of a Bolometer, the sensing element is a resistance with a high temperature co-efficient of resistance. Different types include Metal bolometer, Thermistor bolometer and Semiconductor bolometer. Semiconductor bolometers with a response time of 0.5 ms or so are the fastest of the three types. Thermistor bolometers offer response times in the range of 1 to 100 ms depending upon sensitivity. Figure 12.26 shows a simple circuit for producing an electrical output with a bolometric sensor.

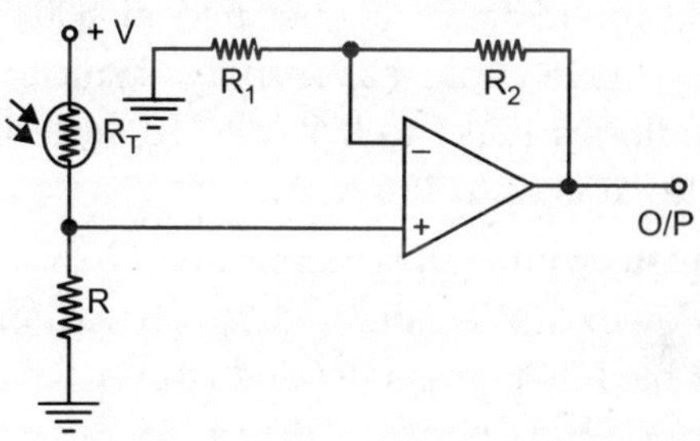

Fig. 12.26

Pyroelectric detectors depend for their operation on temperature dependent polarization change. As a result, pyroelectric detectors sense only changes in incident flux. Their spectral response extends upto 50 μm. Response time is in milli-seconds for high sensitivity versions and pico seconds in case of low sensitivity versions. Figure 12.27 shows a typical application circuit.

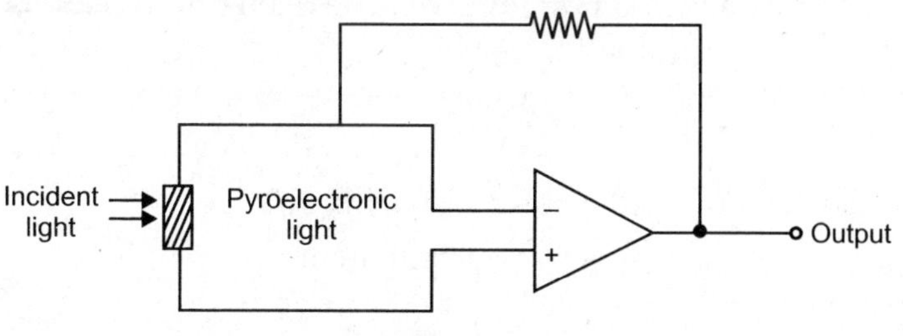

Fig. 12.27

Q. 24. Outline major characteristic parameters of photosensors.

Ans. Major characteristic parameters of photo-sensors include the following:

1. Responsivity
2. Noise Equivalent Power (NEP)
3. Sensitivity usually expressed as either Detectivity (D) or Dee-Star (D*)
4. Quantum Efficiency
5. Response time
6. Detector noise comprising of Johnson noise, Shot noise and Flicker noise
7. Spectral Response
8. Active area

Q. 25. Briefly describe:

(a) Responsivity **(b) Detectivity (D)**
(c) Dee-Star (D*) **(d) Quantum efficiency**
(e) Spectral Response **(f) Response time**

Ans. (a) **Responsivity** is defined as ratio of electrical output to radiant input. It is measured in Amperes per Watt (A/W) or Volts per Watt (V/W). It is specified at a particular wavelength, usually at the peak of response characteristics. The responsivity of a silicon PIN photodiode for instance is specified at 800 or 850 nm, where these have peak responsivity. Typical figures for PIN photodiodes is 0.4 to 0.6 A/W. In case of Avlanche photodiodes, it is upto 100 A/W.

(b) **Detectivity (D)** is reciprocal of Noise Equivalent Power (NEP) defined in an earlier question. Like NEP, it also depends upon active area and noise bandwidth.

(c) **Detectivity (D*)** is normalised detectivity and can be computed from:

$$D^* = D\sqrt{A \cdot \Delta f}$$

It is measured in cm $Hz^{1/2}$/W. It can be used to objectively compare all detectors.

(d) **Quantum Efficiency (η)** is the ratio of number of electrons released to number of photons absorbed at a given wavelength. It can be computed from :

$$\eta = \frac{1240\,R}{\lambda}$$

where, R = Responsivity in A/W
λ = Wavelength in nm

(e) **Spectral Response** is the variation of Responsivity as a function of wavelength.

(f) **Response time** is expressed as a rise time parameter in quantum detectors and as a time constant parameter in thermal detectors.

Q. 26. What is an optocoupler or an optoisolator? What are its different constituents and how do these constituents affect the performance of optocoupler? List typical circuit applications of optocouplers.

Ans. An optocoupler consists of a photon emitting device whose flux is coupled through an optically transparent insulation to some sort of photodetector. The photon emitting device may be an incandescent or neon lamp, or an LED. The transparent insulation may be air, glass, plastic or fiberoptic. The photodetector may be a photoconductor, a photodiode, phototransistor or an integrated combination of photodiode/amplifier. Various combinations of these elements result in a wide variety of input characteristics, output characteristics and coupled characteristics. The most commonly employed combination is an LED input, a thin layer of transparent insulation separating it from a solid-state photodetector. The schematic representations of different types of commercially available optoisolators are shown in Fig. 12.28.

The main concern in the choice of a photoemitter for an optocoupler is optimization of the coupling to the photodetector. The parameters to be optimised are gain, bandwidth, electrical characteristics. A low series resistance is also desirable; so a GaAs based photoemitter is the best choice. A low forward voltage is also desirable but not as important as optimisation of gain and bandwidth.

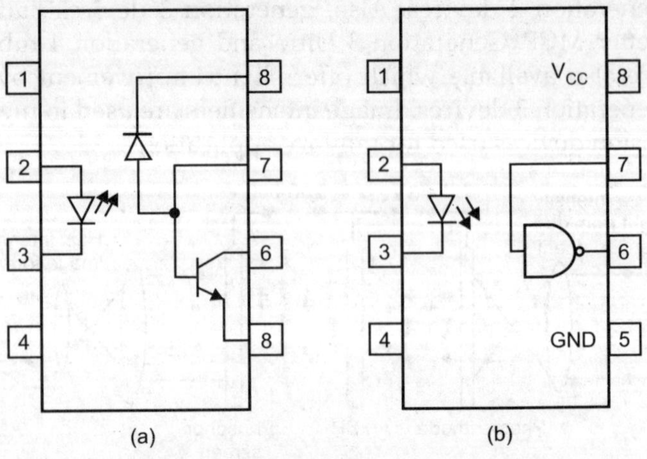

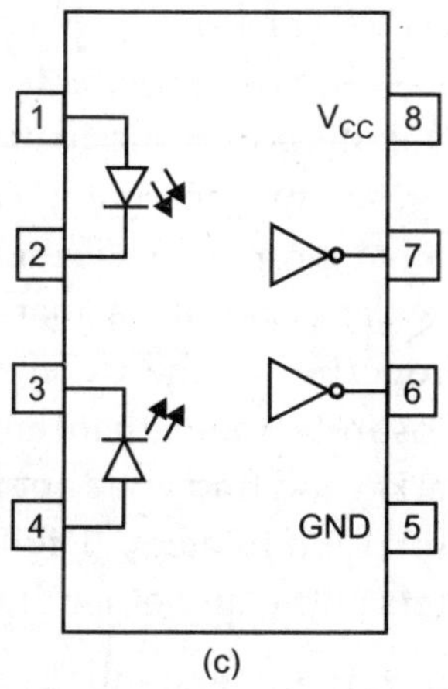

Fig. 12.28

In most of the optocouplers (especially those from Hewlett Packard, U.S.A.) the GaAs emitters used have an emitted wavelength of 700 nm.

In selection of the optical medium, insulation is a vital consideration if the emitter-to-detector distance is very small (e.g. "sandwich" construction). If the distance is very large such as through fiber optics, lenses or other medium (reflective or transmissive), the insulation is less important but spectral transmittance becomes significant, especially where plastics are involved. Most optocouplers use a transparent junction coating to reduce losses at the surfaces of the emitter and the detector. This also provides good insulation. The importance of having a coupling medium other than air is shown in Fig. 12.29. In the figure shown, we have considered an optocoupler having a GaAsP (with refractive index = 3.6) emitter and a silicon detector (with refractive index = 3.5). Whenever, light passes from one medium to another with refractive indices n_1 and n_2, the transmission is given by

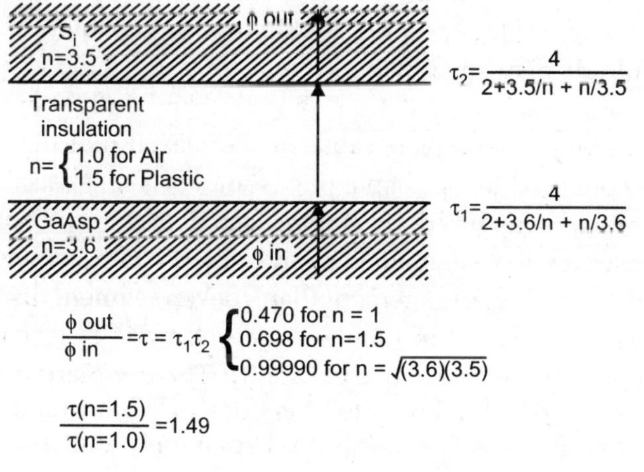

Fig. 12.29

$$T = \frac{4}{2 + (n_2/n_1) + (n_1/n_2)}$$

For Fig. 12.28 if T_1 and T_2 are transmittances at emitter-insulation and insulation-detector interfaces respectively, then the total transmittance from emitter to detector shall be

$$T = (T_1 T_2)$$

Using these expressions for GaAsP emitter, Silicon detectors and air insulator, we get $T = 0.470$. For a plastic with $n = 1.5$, we get $T = 0.7$ which gives about 50 percent more transmittance as compared to the first case.

Silicon photodiodes are very good detectors but usually require ancillary gain elements to produce adequate signal levels. To add gain with elements external to the optocoupler is an inconvenience, raises the component count and degrades performance. For these reasons, it is desirable to include the gain element within the optocoupler package. There are two ways to integrate a photodiode and an amplifier. One of these is the use of a phototransistor in which the photodetecting medium is the collector-base junction. The other is the use of a photodiode whose photocurrent is amplified by a transistor integrated separately on the same chip. Although, photo-transistor is less complicated to produce but it has some inherent performance disadvantages relative to the photodiode/transistor. The photo-transistor option gives poor linearity and low speed of response.

The amplifiers used in optocouplers are normally of two types viz; a single transistor with photodiode anode connected to the base and the cathode separated for reverse bias connection and the other type is called a split darlington in which the emitter of the first transistor is connected to the base of the other in the usual darlington manner but the collectors of the two transistors are separated to allow the output collector drop a lower V_{CE} in saturation. See Fig. 12.30.

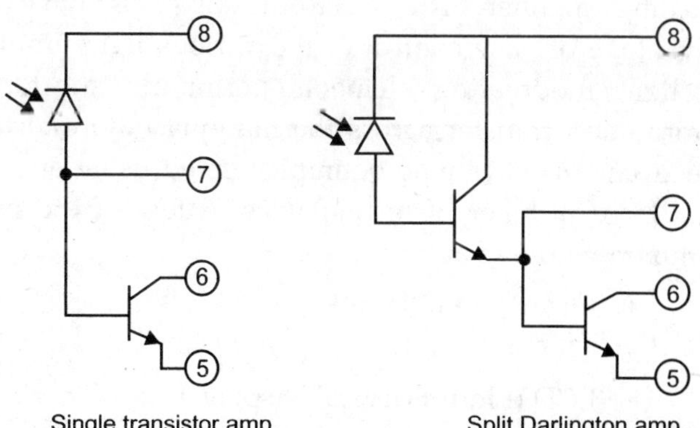

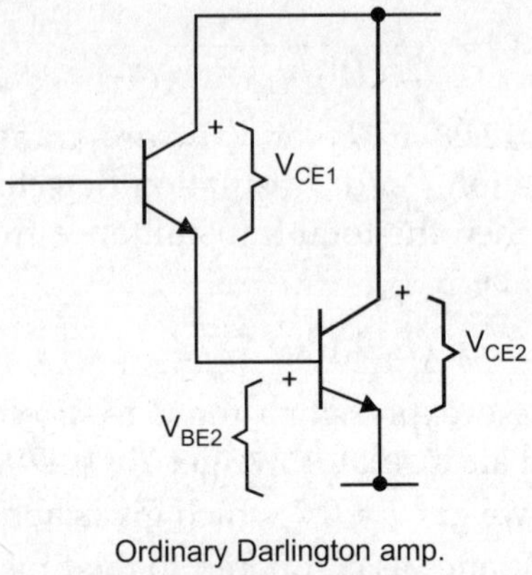

V_{CE1}
V_{CE2}
V_{BE2}
Ordinary Darlington amp.

Fig. 12.30

Applications: Optocouplers are useful for applications where analog or DC signals need to be transferred between two isolated systems in the presence of a large potential difference or induced noise. They are an inexpensive way to eliminate the shock hazard between an input transducer and an output circuit. Optically coupled isolators can reduce the common mode noise generated between an isolated input circuit and an output circuit. Potential applications include those in which large transformers, expensive instrumentation amplifiers or complicated A to D conversion schemes have been used. Examples include sensing circuits (thermocouples, transducers etc.), patient monitoring equipment, power supply feedback, high voltage current monitoring and audio or video amplifiers. In many applications, the coupler can transmit the analog signal directly. However, in applications where very high linearity and stability are critical, the analog signal can be converted into a digital form and then isolated.

Q. 27. What are the different basic devices used for fabricating displays? Which is the most popular and most commonly used light emitting device for making displays?

Ans. Digital display devices are largely used to present information in a clear, interesting and attention catching manner. The electronic displays have the advantage of easy control of the various light emitting or light modulating elements permitting substantial information transfer across the man-machine interface. Electronic displays may be implemented using any one of a large number of technologies. A few of the more commonly known are:

- Incandescent lamps
- CRT
- LCD (Liquid Crystal Displays)
- Light Emitting Diode Displays

Each of these technologies is characterised by distinct capabilities and requirements with respect to element density, colour, power dissipation, packing, unit size, drive voltages, environmental capabilities, cost, availability and other properties. The choice of any one display technology is generally a matter of optimising some of these properties at the expense of others. The LED displays constitute more than 80 per cent of the display device market and find wide application in Digital multimeters. Digital voltmeters, Digital LCR meters, Digital panel meters and many other digital instruments.

Q. 28. What do you understand by a display font? What are the different types of display fonts used with LED displays ?

Ans. The most prominent feature of any display is the physical arrangement of the display elements (individual LEDs, for instance). This arrangement, or font, as it is commonly termed, is important not only from the standpoint of the type of information which can be transmitted but also important in the sense that it dictates the nature and complexity of the support electronics required by the display.

Fig. 12.31 shows some of the commonly adopted display fonts. Fonts (*a*) and (*e*) are by far the most common in use in LED display technology. Both arrangements have some positive and some negative attributes. The seven-segment technique is easy to utilise from an electrical standpoint; however, it is limited to displaying numeric and a very small range of alphabetic information. The 5 × 7 dot matrix of (*e*) can display a wide range of numeric, alphabetic and other characters but involves some rather expensive electronic circuitry to implement other portions of the display subsystem. The sixteen segment approach of (*c*) has a full alphanumeric capability (*i.e.* it could display all alphabet from A to Z and could display all numerals from 0 to 9) but has had rather limited acceptance, perhaps because of the reason that this font is not capable of displaying some of the specialized characters required by sophisticated systems. Font (*f*) illustrates a 9 segment display which has a somewhat more pleasing arrangement than a seven segment display. It is because, it can display same numeric information and has more alphabetic capability. The dot-matrix fonts (*b*) and (*d*) are abbreviated versions of 35-dot matrix of (e) and are used primarily to display only numeric and hexadecimal information. Hexadecimal information implies display of numerals 0 to 9 and alphabet A to F.

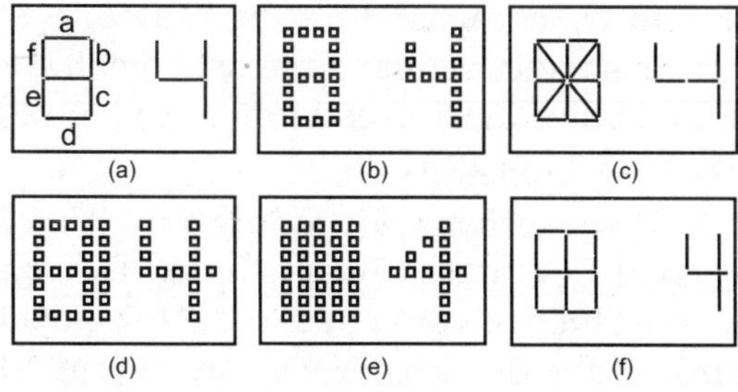

Fig. 12.31

Q 29. A complete display subsystem is often much more than LED matrix. With the help of a block diagram, briefly describe the functions of each one of the circuit blocks of a display subsystem.

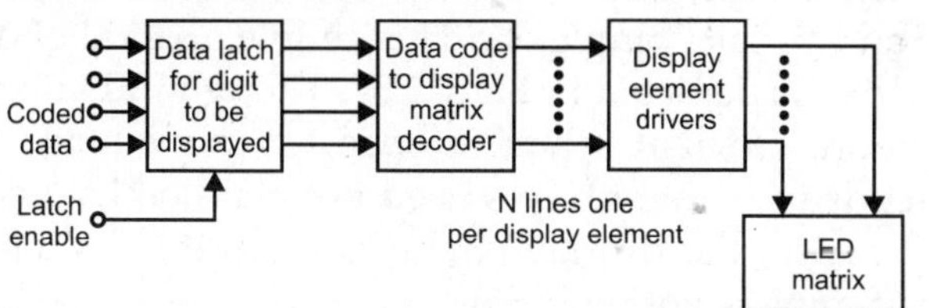

Fig. 12.32

Ans. The visible portion of a display (be it LED or some other technology) is only a minor portion of the electronic system necessary to convert electrically coded data into a form readily comprehended by a human viewer. Figure 12.32 is a block-diagram of what may be generally termed as display subsystem. A display user can purchase one, some or all of the functions in the block diagram as a single product. It is necessary, however, that all of the functions depicted be present in some part of any information display.

By way of example, a seven-segment decoder/driver may be combined with a data latch and a seven-segment diode matrix to form a display subsystem. Similarly, all four of the above functions may be purchased in a single package to perform the same end result. Often, the data latch depicted in the block diagram will be present as a part of the data source and doesn't have to be supplied as a portion of the display. Price, digit size, viewing conditions, power limitations, system architecture and other considerations must be taken into account, when deciding how to best design and partition a display subsystem.

Q. 30. Write short notes on the following:

(*a*) On board integrated circuit numeric displays (OBIC)

(*b*) Stretched seven segment displays

(*c*) Monolithic Displays

(*d*) Alphanumeric Displays

Ans. On-board integrated circuit numeric displays (OBIC): For many applications, it is desirable to have the display subsystem assembled into a compact configuration and packaged separately at some distance from the data source. One possible approach is to have the display PC board contain all the necessary components that comprise the display subsystem. However, a more efficient and in many cases less costly approach is to use a display device that has integrated circuit containing the subsystem logic functions of Data latch/Decoder/Driver packaged on the same substrate as the display matrix. Such a device is termed as an on board integrated circuit display (OBIC)

It has the advantages of reduced pin connections between the data source and the display assembly, significantly enhanced reliability due to reduction in circuit complexity and reduced component count, compactness etc. Also, a pleasing font may be used, such as modified 4 × 7 dot matrix, that need not conform to font codes available with discrete commercial decoders. The disadvantage of an OBIC display is that as more functions are added, the module becomes unique in its configuration and a second source is usually not available.

The Hewlett-Packard HP-5082-7300 display family belongs to OBIC category. This family offers a series of plastic or hermetic devices that display either decimal or hexa-decimal numeric information. The plastic devices are epoxy encapsulated and are designed for use in consumer and commercial equipment, such as control units for household appliances, electronic office equipment, computers and electronics measuring instruments. The hermetic devices, available in either industrial or military grade packages, are designed for use in high reliability applications involving the possible exposure to an adverse environment, such as the instrument panels of transportation vehicles, controllers in numerically controlled milling machines, chemical processing equipment, military and aerospace systems. Figure 12.33 depicts the constructional features of the hermetic OBIC display.

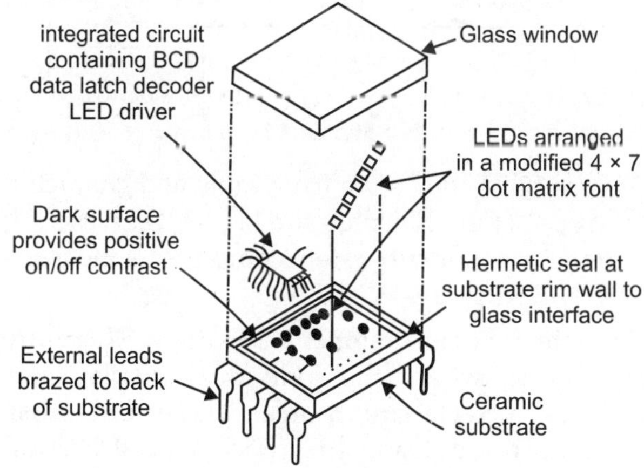

Fig. 12.33

Truth Table						
BCD Input				*Decimal Numeric*	*Hexi-Decimal*	*Over-range*
X_8	X_4	X_2	X_1			
L	L	L	L	0	0	+ 1
L	L	L	H	1	1	
L	L	H	L	2	2	
L	L	H	H	3	3	
L	H	L	L	4	4	
L	H	L	H	5	5	
L	H	H	L	6	6	
L	H	H	H	7	7	
H	L	L	L	8	9	
H	L	L	H	9	S	
H	L	H	L	8 2.	A	
H	L	H	H	(BLANK)	B	
H	H	L	L	(BLANK)	C	
H	H	L	H	....	D	
H	H	H	L	(BLANK)	E	
H	H	H	H	(BLANK)	F	

Fig. 12.34

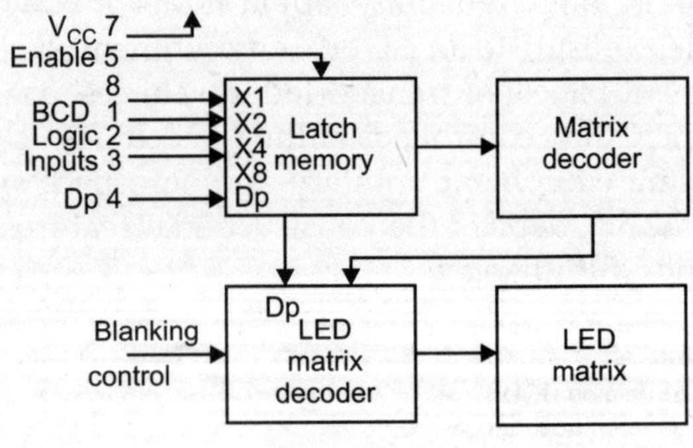

Fig. 12.35

Figure 12.34 shows the truth table and character font for this type of OBIC Display and Fig. 12.34 gives the block diagram of the four subsystem components incorporated into an OBIC display.

Stretched Seven Segment Displays: These are the display types where the individual display element illuminates a rectangular area. These LED displays are manufactured using the concept of stretching the light from an LED by diffusion and reflection. This is accomplished by encapsulating the LED into a special cavity which is a rectangular cone with small end down at the LED and the large top end forming a segment of the display as shown in Fig. 12.36

A typical logic system will produce output information in the form of BCD data (Binary Coded Decimal data). The BCD must then be converted to the 7-segment matrix code format of the display. Drivers are required to drive each segment of the display. The two methods of driving 7-segment displays are (l) direct DC drive of a single digit and (2) strobing a string of digits. The simplest method of driving seven-segment display is to have one decoder/driver for each bit. This is usually cost effective upto 4 or 5 digits depending upon trade-off between component cost and circuit complexity. This simple method of direct DC drive is depicted in Fig. 12.37. Though LED displays are more efficient when strobed, the advantage with direct drive is that drivers need not handle high current levels. Also, the display circuit complexity is simple as extra timing is not required.

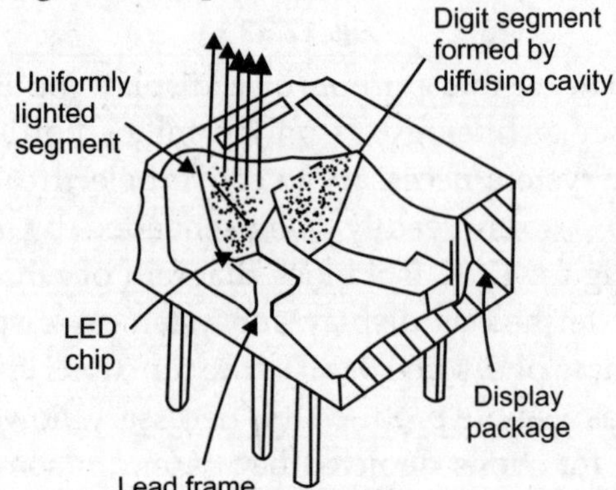

Fig. 12.36

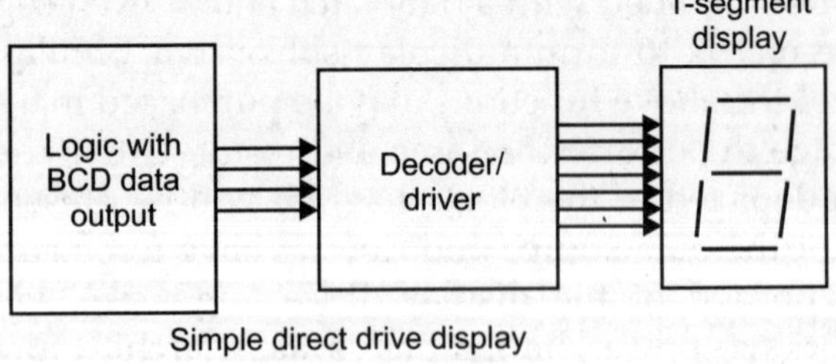

Fig. 12.37

The most efficient method of driving an LED display is to strobe it at high peak current and a low duly factor. This takes advantage of increased efficiency of LED at high peak currents and reduction in average power dissipation, when compared to DC operation. A strobed (multiplexed) display design becomes cost effective, when the decoder/driver and display PC board size and complexity required for DC drive become significant, when compared to the timing circuitry and digit driver cost and size of PC board needed for strobed operation. The power saving that is attainable with strobed operation may show up as a need for a smaller sized power supply, contributing to cost reduction. Fig. 12.38 shows a strobed six digit display system block diagram arrangement. The like segments of each digit in the display are tied to a common-segment bus-line which is connected to one of seven segment decoders. The scheme operates as follows: When first clock-pulse occurs, the multiplexing logic selects digit data buffer no. 1. This first digit BCD data is presented to the decoder/driver and the required segment lines are activated. However, the digit scanner enables only the digit driver no. 1 and the first number is displayed. When second clock pulse occurs, the multiplexing logic selects digit data buffer no. 2. This second digit data is presented to the decoder/driver and again required segment lines are activated. The digit scanner now enables only digit driver no. 2 and the second number is displayed. This process is continued through digits 3, 4, 5 and 6 and the whole process is repeated. The strobing rate to produce a flicker free display should be a minimum of 100 Hz.

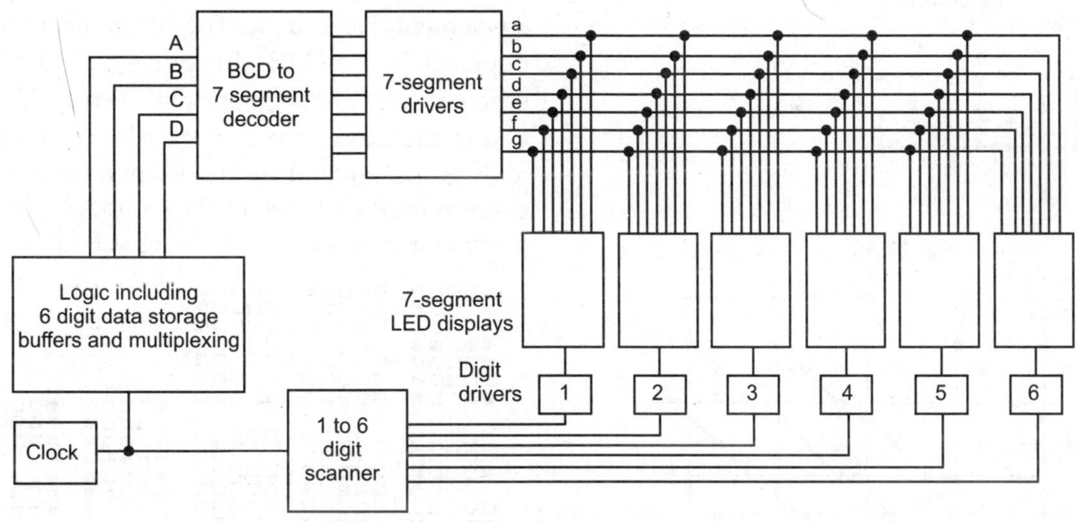

Fig. 12.38

Monolithic displays: The monolithic seven-segment displays represent one of the largest segment of the optoelectronics industry today. Monolithic LED displays are used because of their low cost, small size and low power requirements. Monolithic displays have a very high sterance which allows them to be used outdoors and in other areas of high ambient lighting. While monolithic displays have generally been used for low cost applications, improvements in character size and viewing angle have also made monolithic displays suitable for portable instruments and other high quality applications.

Monolithic displays differ from other types of LED displays in that the individual light emitting segments are formed by diffusing separate LED junctions on a single chip of GaAsP. Usually, GaAsP substrate is 'n' doped material and each LED junction is formed by a (*p*+) diffusion, so monolithic displays are normally of common cathode configuration. Although, most monolithic displays are constructed with the standard seven-segment format, almost any number of segments or character shapes can be diffused into GaAsP substrate. Figure 12.39 illustrates the construction of a monolithic GaAsP chip. Because GaAsP is relatively expensive, most monolithic displays are magnified to keep chip sizes small while attaining a viewable character size. In most cases, the monolithic display is magnified by an external lens. The resulting apparent character height is equal to the actual character height times the magnification of the lens. Monolithic displays are driven in the same way, as any other strobed common cathode display. The advantage of monolithic seven segment displays is that they can be driven at considerably lower average currents than large seven-segment displays for a satisfactory luminous intensity. The current requirement for comparable luminous intensities may be four to five times smaller.

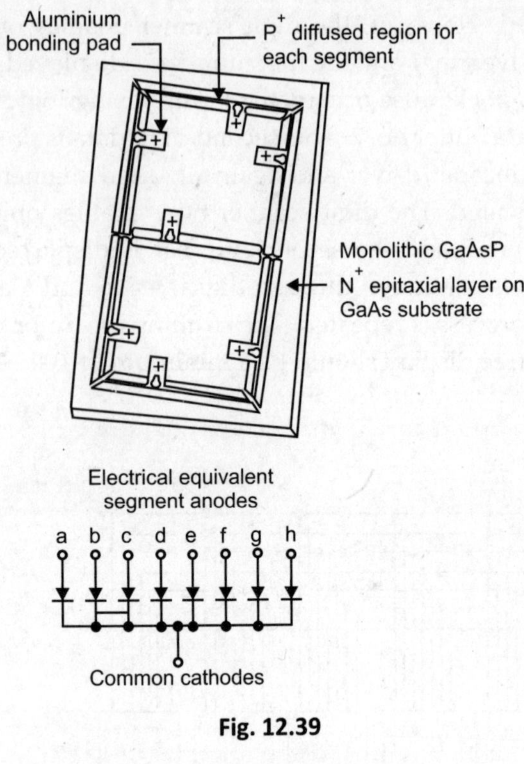

Fig. 12.39

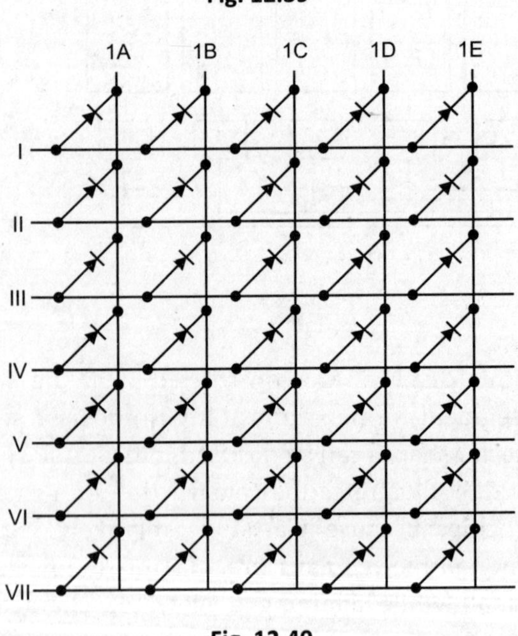

Fig. 12.40

Alphanumeric displays : Seven segment display devices are capable of transmitting only a limited number of useful different states of information. Theoretically, upto 49 different information states could be depicted utilising the 7-segment font, however, less than 1/3 of these have any meaning in normally utilised communications context. By increasing the number of elements in a display character from 7 to 9, it becomes possible to depict in rough form most of the English language symbols and numerals. Utilisation of a 14 or 16 segment character allows better symbol differentiation and use of some special symbols. However, the lower case alphabet cannot be represented and asthetic quality and readability are still poor. By increasing the number of segments still further, it is possible to add further refinement to the display quality and by utilising infinite number of elements, any two-dimensional character and character shape could be achieved. In practice, it is found that 35 elements in a 5 × 7 array or 63 elements in a 7 × 9 array offer satisfactory resolution for most alphabetic language communications. For LED alphanumeric dot matrix displays, the industry standard product is the 35-element 5 × 7 array. The 5 × 7 dot matrix array was depicted in Fig. 12.31 in font (*e*). Figure. 12.40 depicts schematically one digit of a (5 × 7) array. The row lines are tied to the anodes of all of the diodes of a particular row and are designated by the Roman numerals I -VII. The column lines tie together the cathodes of each diode of a particular column and are designated A-E.

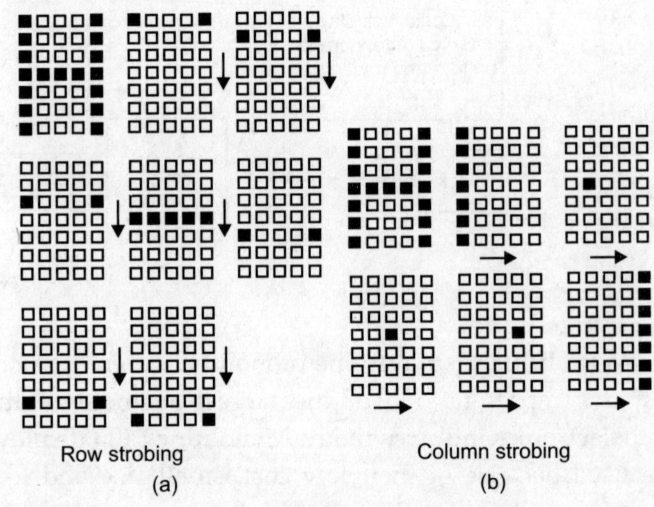

Fig. 12.41

There are two common methods for addressing the diodes in the array in order to generate character information. These are (1) Row strobing and (2) Column strobing. In Row-strobing, data for one row of the desired character is applied to the column lines and the row is then energized. The process is repeated for each of seven rows. In column strobing, data for one column of the desired character is supplied to row lines and the column is then energized. The process is repeated, varying the data, until all of the columns of the character have been selected. Figure 12.41. (*a*) and (*b*) depict the sequence of steps necessary to form the character (*H*) using ROW and COLUMN strobing techniques.

SOLVED PROBLEMS

Problem 12.1. *A photodiode has a noise current of 1fA, responsivity figure of 0.5A/W, active area of 1mm² and rise time of 3.5ns. Determine its (a) NEP, (b) detectivity, (c) D*, (d) quantum efficiency at 850nm.*

Solution. NEP = Noise current/Responsivity = $1 \times 10^{-15}/0.5 = 2 \times 10^{-15}$W = 2fW.

Detectivity = 1/NEP = $1/ 2 \times 10^{-15} = 0.5 \times 10^{-15}$W^{-1}

$$D^* = D \times \sqrt{(A \times \Delta f)}$$

$$A = 1\text{mm}^2 = 1 \times 10^{-2}\text{ cm}^2$$

Δf (in MHz) = $0.35/t_r$ (in μs) = $0.35/3.5 \times 10^{-3}$ = 100MHz = 1×10^8 Hz

Therefore, $D^* = 0.5 \times 10^{15} \times \sqrt{(1 \times 10^{-2} \times 1 \times 10^8)}$ = 5×10^{17} W^{-1} cm Hz$^{1/2}$

$$\text{Quantum efficiency, } \eta = \left[\frac{(1240 \times R)}{\lambda}\right]$$

$\eta = [(1240 \times 0.5)/850] = 0.729$

Problem 12.2. *An oscilloscope is used to measure the output of a photodiode. Determine the rise time of the pulse as seen on the oscilloscope, given that the rise time of the photodiode is 1 ns, rise time of the light pulse is 5 ns and the bandwidth of the oscilloscope is 350 MHz.*

Solution. Given that the rise time of the pulse is 5 ns and the rise time of the photodiode is 1ns.

Bandwidth of the oscilloscope = 350 MHz. Therefore, rise time of the oscilloscope is t_r(in μs) = 0.35/BW (in MHz) = $0.35/350 = 1 \times 10^{-3}$ μs = 1 ns.

Therefore, overall rise time

$$= \sqrt{(5 \times 10^{-9})^2 + (1 \times 10^{-9})^2 + (1 \times 10^{-9})^2}$$

$$= 5.19 \times 10^{-9} \text{ s} = 5.19 \text{ ns}$$

Problem 12.3. *Design a circuit using a photoconductor that generates a logic HIGH voltage when the light incident on it is above 200 lux, given that the photoconductor has a resistance of 10kΩ at light level of 100 lux, α = 0.5, power supply voltage (V_{CC}) = 10V and the reference voltage of the Zener diode i; 2.5V.*

Solution. The resistance of the photoconductor at 200 lux can be calculated using the expression

$$R_a = R_b \times \left(\frac{E_a}{E_b}\right)^{-\alpha}$$

Here, R_b = 10 kΩ, E_b = 100 lux, E_a = 200 lux, α = 0.5. Therefore,

$R_a = 10 \times 10^3 \times (200/100)^{-0.5} = 10 \times 10^3 \times (2)^{-0.5} = 7.07$ kΩ

Figure 12.42 shows one of the possible circuit configurations that can be used for the given application.

The comparator output will go high, when the voltage at positive terminal exceeds 2.5V.

The value of resistance R_L can be calculated using $[(V_{CC} \times R_L)/(R_L + R_{sen})] = 2.5$V.

As V_{CC} = 10V and R_{sen} at 200 lux is 7.07 kΩ.

$(10 \times R_L)/(7.07 \times 10^3 + R_L) = 2.5$

$$4R_L = 7.07 \times 10^3 + R_L$$

$$R_L = 2.36 \text{ k}\Omega$$

For a 10 mA current through the Zener diode, value of resistor R can be calculated using the expression $R = [(V_{CC} - V_Z)/i_Z] = (10 - 2.5)/(10 \times 10^{-3}) = 7.5$ kΩ.

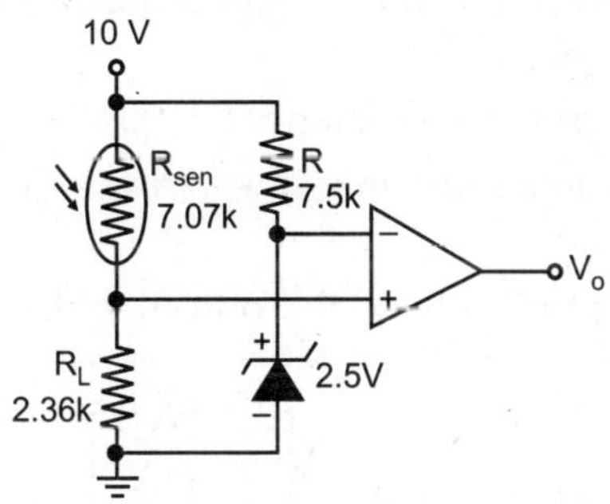

Fig. 12.42

Problem 12.4. *Determine the cut-off wavelengths for Silicon and Germanium photodiodes, given that their bandgap energies are 1.1 eV and 0.72 eV, respectively, at 25 °C. How will the cut-off wavelength change, when the operating temperature changes from 25 °C to 200 °C?*

Solution. The cut-off wavelength is given by the formula,

$$\lambda_c = \frac{1240}{E_g}$$

At 25 °C, for Silicon photodiode, E_g = 1.1 eV, therefore λ_c = 1240/1.1 = 1127.27nm

The temperature variation of the bandgap energy of silicon semiconductor is given by E_g (T) = 1.21 – 3.60 × $10^{-4}T$, where T is the temperature in Kelvin.

Therefore, bandgap energy of Silicon photodiodes at 200 °C is given by

$E_g = 1.21 - 3.60 \times 10^{-4} \times 473 = 1.21 - 0.17 = 1.04$ eV

The cut-off wavelength of Silicon photodiodes at 200 °C is given by λ_c = 1240/1.04 = 1192.31 nm

At 25 °C, for Germanium photodiode, E_g = 0.72 eV, therefore λ_c = 1240/0.72 = 1722.22 nm

The temperature variation of the bandgap energy of Germanium semiconductor is given by E_g (T) = 0.785 – 2.23 × 10^{-4} T, where T is the temperature in kelvin.

Therefore, bandgap energy of Germanium photodiodes at 200 °C is given by

$E_g = 0.785 - 2.23 \times 10^{-4} \times 473 = 0.785 - 0.105 = 0.68$ eV

The cut-off wavelength of Germanium photodiodes at 200 °C is given by $\lambda_c = 1240/0.68 = 1823.53$ nm

Problem 12.5. *For the circuit shown in Fig. 12.43, determine the amplitude of the output voltage pulse, when the light pulse having wavelength of 1000 nm, pulse width of 1 s and energy of 10 mJ is incident on the active area of the photodiode. The responsivity of the photodiode is 0.5 A/W at 1000 nm.*

Solution. The incident light pulse has an energy of 10mJ and a pulse width of 1 s.

Therefore, the input peak power = $10 \times 10^{-3}/1 = 10$mW.

Output current from the photodiode = $0.5 \times 10 \times 10^{-3}$ = 5mA.

Voltage across the resistance $R = 50 \times 5 \times 10^{-3} = 250$ mV.

Gain of the amplifier = $(1 + R_2/R_1) = [1 + (10 \times 10^3/1 \times 10^3)] = 11$.

Voltage amplitude of the output pulse = $250 \times 10^{-3} \times 11 = 2.75$V.

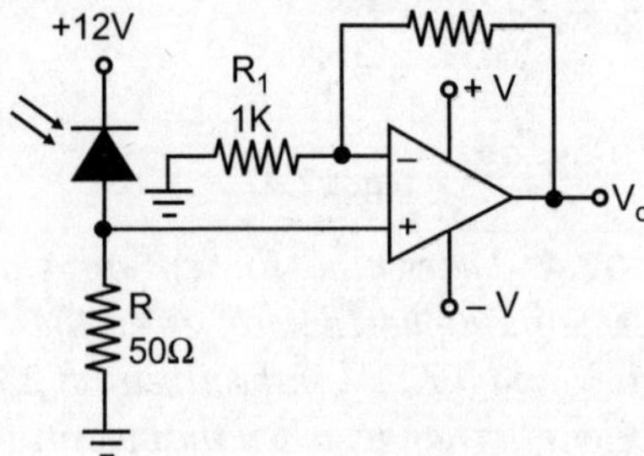

Fig. 12.43

Problem 12.6. *Determine the output voltage of the phototransistor circuit shown in Fig. 12.44 (a), when a Continuous Wave (CW) light radiation of 1 mW is incident on the active area of the phototransistor. The active area of the transistor is 10 mm² and its output characteristics are shown in Fig. 12.44 (b). The base-emitter voltage of the transistor Q_2 = 0.7 V, values of resistors R_1, R_C and R_E are 1 kΩ, 2.2 kΩ and 1 kΩ, respectively, and the supply voltage (V_{CC}) is 12 V.*

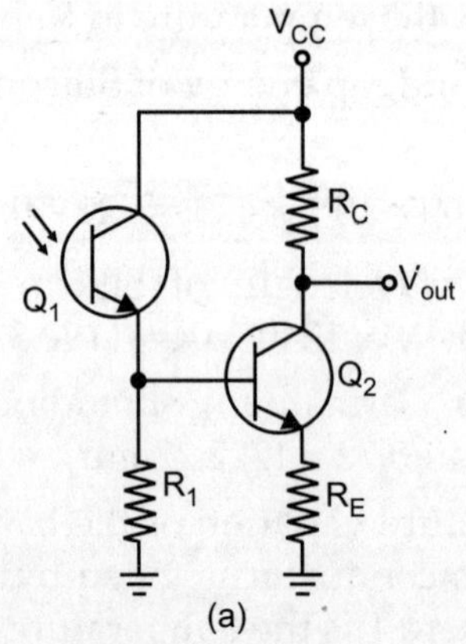

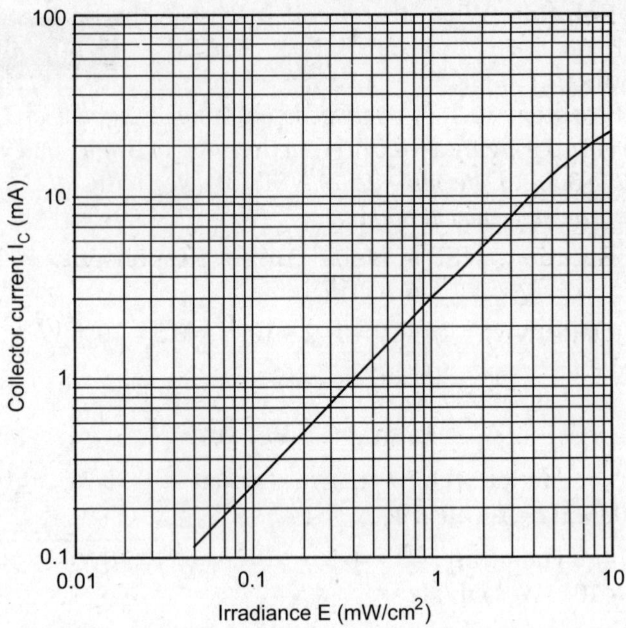

Fig. 12.44

Solution. The incident irradiance on the phototransistor = $1 \times 10^{-3}/(10 \times 10^{-1})^2$ W/cm² = 1mW/cm²

From output characteristics of the transistor, collector current generated is 3mA.

The voltage generated across the resistor R_1 is $3 \times 10^{-3} \times 1 \times 10^3 = 3$V

Therefore, the voltage applied to the base of the transistor Q_2 is 3V. So the transistor Q_2 goes into the conducting mode.

The voltage across resistor $R_E = 3 - 0.7$ V = 2.3 V.

The value of emitter current (I_E) is given by $I_E = 2.3/(1 \times 10^3) = 2.3$ mA

As the collector current (I_C) is approximately equal to the emitter current, therefore $I_C = 2.3$ mA.

The output voltage $V_{out} = V_{CC} - I_C \times R_C = 12 - 2.3 \times 10^{-3} \times 2.2 \times 10^3 = 12 - 5.06 = 6.94$V

Problem 12.7. *The circuit shown in Fig. 12.45 is kept in a dark room. Determine the output voltage (V_o). Also, determine the output voltage when a bright light is flashed on the circuit.*

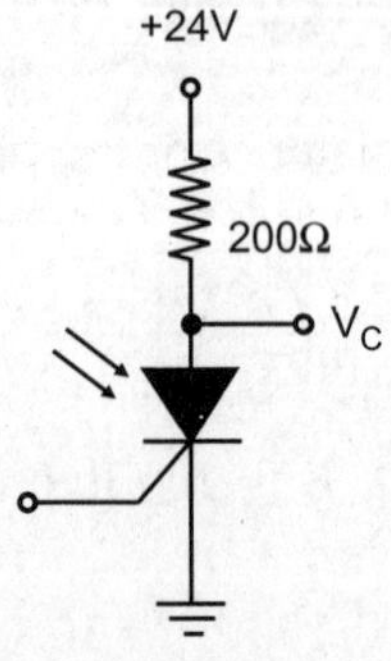

Fig. 12.45

Solution. When the circuit is kept in the dark room with no incident light, the photo SCR is in the non-conducting state. Therefore, the output voltage V_o is equal to the supply voltage, that is, $V_o = +24V$.

When a bright light is incident on the photo SCR, it goes into the conducting state and ideally, the output voltage is zero, that is, $V_o = 0V$.

Problem 12.8. *For the circuit shown in Fig. 12.46, determine the current through the LED when (a) $V_B = 0$ V and (b) $V_B = 10$ V. (V_{BE} of both the transistors is 0.7 V and $V_{CE(sat)}$ of both the transistors is 0.2 V.)*

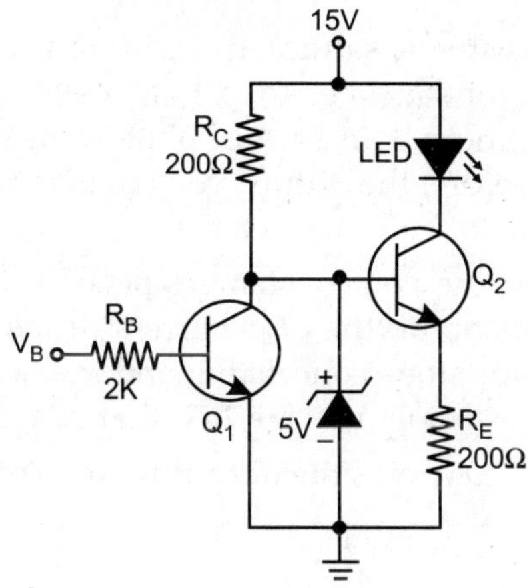

Fig. 12.46

Solution. When $V_B = 0V$, transistor Q_1 is in the cut-off region and there is no collector current flowing through the transistor.

Therefore, the base voltage of transistor Q_2 is equal to the reverse voltage of Zener diode Z_1. Therefore, $V_{B2} = 5V$.

Transistor Q_2 is in the conducting mode and the voltage at the emitter terminal of transistor $Q_2 = V_{B2} - 0.7$ $V = 5 - 0.7 = 4.3$ V.

The emitter current flowing through transistor $Q_2 = V_{E2}/R_E = 4.3/200 = 21.5$mA.

As the collector current is approximately equal to the emitter current, the value of $I_C = 21.5$ mA.

The current flowing through the LED is the same as the collector current of the transistor. Therefore, the current flowing through the LED is 21.5 mA.

When V_B = 10V, transistor Q_1 is in saturation. Therefore, the voltage $V_{CE1} = 0.2V$.

The base voltage of transistor Q_2, $V_{B2} = 0.2V$.

Hence, transistor Q_2 is in cut-off and the value of collector current is zero. Therefore, the current flowing through the LED is also zero.

Problem 12.9. *For the optocoupler circuit of Fig. 12.47(a), the voltage across the resistor R is 4V. Determine the value of the voltage V_B applied to the base of the transistor Q_1. The relationship between the input current flowing through the LED and the output phototransistor collector current is shown in Fig. 12.47(b). The base-emitter voltage of transistor Q_1 is equal to 0.7V.*

Solution. The current flowing through the resistor $R = 4/(4 \times 10^3) = 1 \times 10^{-3}A = 1mA$.

The collector current of the phototransistor I_C is equal to the current through resistor R.

Therefore, $I_C = 1mA$.

From the curve of Fig. 12.47(b), the input current flowing through the LED is approximately equal to 1.1mA.

The current flowing through the LED is equal to the collector current of transistor Q_1.

As the collector current of the transistor is approximately the same as its emitter current, therefore, the emitter current is equal to 1.1 mA.

The voltage drop across resistor $R_E = 800 \times 1.1 \times 10^{-3} = 0.88$ V.

The voltage $V_B = V_E + V_{BE} = 0.88 + 0.7 = 1.58$ V.

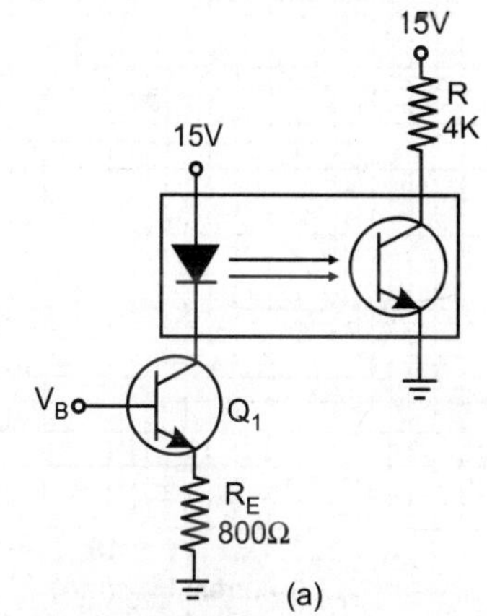

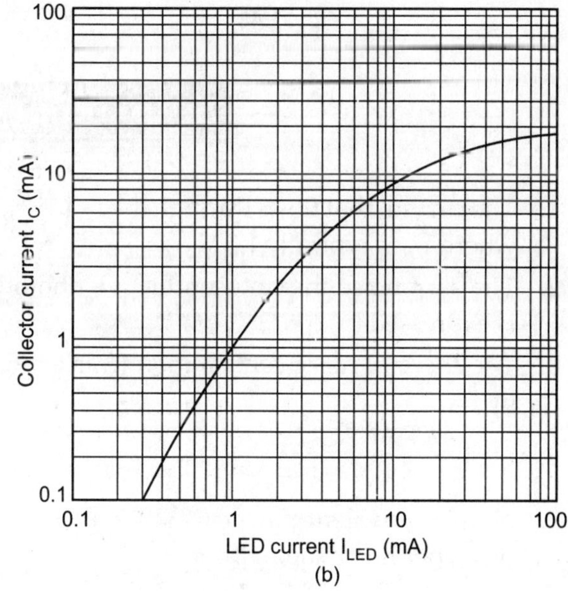

Fig. 12.47

Problem 12.10. *The optocoupler-based circuit of Fig. 12.48(a) is used to provide isolation from the power line and detect zero crossings of the line voltage. The relationship between the input LED current and the output current of the phototransistor for the optocoupler is shown in Fig. 12.48(b). Draw the output waveform $V_O(t)$. Assume the diodes to be ideal.*

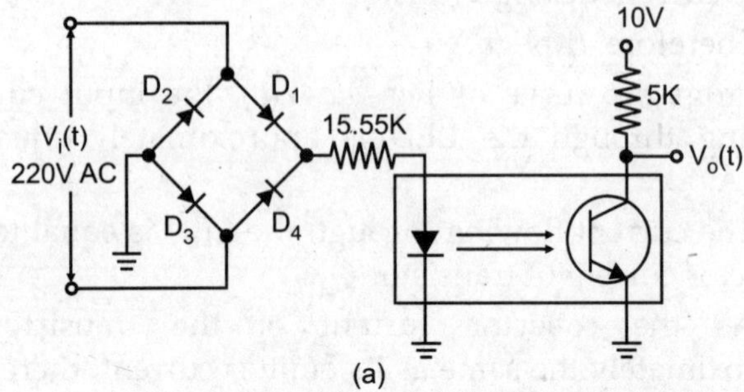

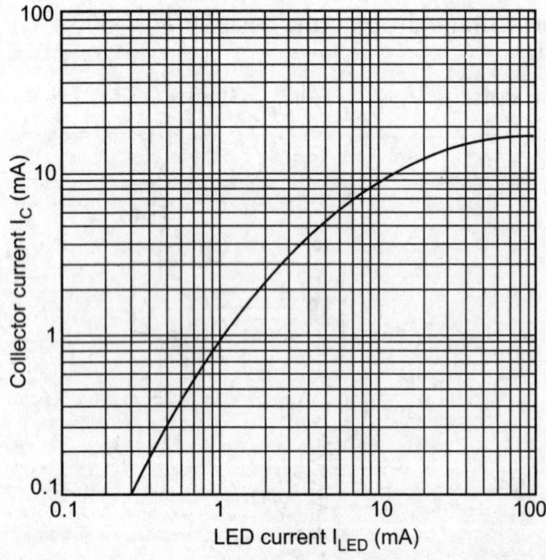

Fig. 12.48

Solution. The bridge rectifier produces a full-wave rectified output. Therefore, the current through the LED is also a full-wave rectified waveform. The peak current through the LED is $I_{LEDpeak} = (1.414 \times 220)/(15.55 \times 10^3) = 20 \times 10^{-3} A = 20mA$.

The value of collector current, when the phototransistor is in saturation $I_{Csat} = 10/(5 \times 10^3) = 2 \times 10^{-3} = 2mA$.

From Fig. 12.48(*b*), the collector current corresponding to the LED current of 20mA is 14mA. The collector current of the phototransistor never reaches 14mA as it saturates at 2mA.

The transistor is saturated during most of the cycle as the transistor saturates for all values of LED currents that produce more than 2mA of phototransistor collector current. Therefore, the output voltage is zero for most of the cycle.

When the line voltage changes polarity, that is, when zero crossings occur, the LED current drops to zero. The phototransistor stops conducting and the output voltage is equal to the supply voltage 10V, that is, $V_O(t) = 10V$.

Figure 12.49 shows the output waveform.

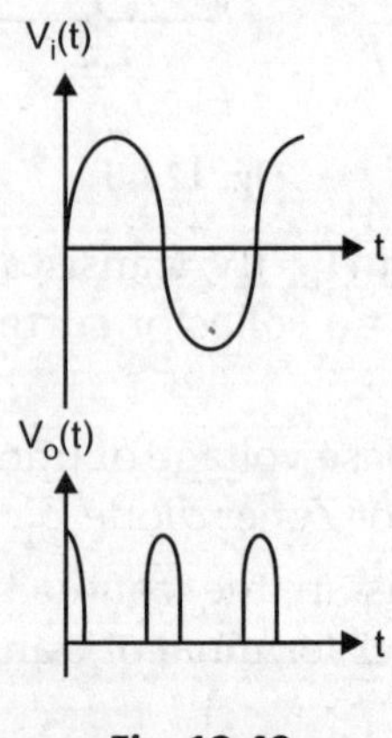

Fig. 12.49

Summary of Important Formulas

1. The current that flows through the $I = \left(\frac{V_{CC} - V_F}{R}\right)$ where V_F is typically 1.5 V.
2. The shot noise in photoconductive photodiode is expressed by $I_s = (2eI_d\,\Delta f)^{1/2}$
3. The thermal noise contribution (I_t) is expressed by
4. $I_t = \left(\frac{4KT\,\Delta f}{R}\right)^{1/2}$
5. Noise equivalent power (NEP) is expressed in watts per hertz square root
$$= \frac{\text{Noise curent (in amperes)}}{\text{Responsivity (in amperes per watt)}}$$
6. The electron multiplication factor (M) may be computed from $M = (1 + p)^N$, where (N) is number of cells and (p) is the probability value.
7. Normalised detectivity is given by
$$D^* = D\sqrt{A.\Delta f}$$
8. Quantum Efficiency is given by, $\eta = \frac{1240\,R}{\lambda}$, where R is Responsivity in A/W and λ is wavelength in nm

SELF EVALUATION EXERCISE

I. Multichoice Questions

1. What factors determine the type of photo sensor you select?
   (*a*) Speed of response
   (*b*) Minimum and maximum light levels
   (*c*) Spectral response
   (*d*) (*a*) and (*b*)
   (*e*) All the above
2. A photocell's resistance vs light can vary as much as
   (*a*) 10 : 1
   (*b*) 20 : 1
   (*c*) 50 : 1
   (*d*) Several orders of magnitude
3. LED's efficiency is a measure of the electrical energy required to produce a certain
   (*a*) Current
   (*b*) Resistance
   (*c*) Photometric efficiency
   (*d*) Light output
4. The resistance of a photoconductor
   (*a*) Decreases with increase in light intensity
   (*b*) Increases with increase in light intensity
   (*c*) Does not change with light intensity
   (*d*) Can increase or decrease with increase in light intensity
5. Increase in reverse bias voltage across the photodiode
   (*a*) Increases the rise time and decreases the width of the depletion region of the photodiode
   (*b*) Decreases the rise time and decreases the width of the depletion region of the photodiode
   (*c*) Decreases the rise time and increases the width of the depletion region of the photodiode
   (*d*) Increases the rise time and increases the width of the depletion region of the photodiode
6. The noise current and responsivity of a sensor are 10nA and 1A/W respectively. Determine its NEP?
   (*a*) 10nW (*b*) 1nW
   (*c*) 100nW (*d*) 0.1nW
7. The output wavelength of a LED is
   (*a*) Directly proportional to the band-gap energy of the semiconductor material
   (*b*) Inversely proportional to the band-gap energy of the semiconductor material
   (*c*) Proportional to the square of the band-gap energy of the semiconductor material
   (*d*) Inversely proportional to the square of the band-gap energy of the semiconductor material
8. The best parameter to compare the noise performance of any two sensors is
   (*a*) NEP (*b*) Dee-star
   (*c*) Detectivity (*d*) None of these
9. Which of the following statements is true?
   (*a*) Drive waveform for a LCD display is a DC voltage
   (*b*) Drive waveform for a LCD display is an AC waveform with zero RMS value
   (*c*) Drive waveform for a LCD display is an AC waveform with zero DC value
   d. Drive waveform for a LCD display can be a DC or an AC waveform
10. Typical value of CTR for a opto-coupler having a phototransistor at the output is
    (*a*) 10% to 50% (*b*) 1% to 10%
    (*c*) 60% to 90% (*d*) 10% to 70%
11. What is an LED made up of?
    (*a*) Phosphorescent material
    (*b*) Germanium
    (*c*) Silicon
    (*d*) Gallium Arsenide
12. In a phototransistor, the base current is
    (*a*) Set by a bias voltage
    (*b*) Directly proportional to light
    (*c*) Inversely proportional to light
    (*d*) Square to light intensity

13. A photodiode is used in reverse bias because
(*a*) Majority of electron-hole pairs swept are reversed across the junction
(*b*) Only one side is illuminated
(*c*) Reverse current is small compared to photocurrent
(*d*) Reverse current is large compared to photocurrent

14. The general connection for a phototransistor is
(*a*) Common base configuration
(*b*) Common collector configuration
(*c*) Common emitter configuration
(*d*) Darlington-pair configuration

15. When a photodiode is reverse biased and kept in a dark room, the current flowing through the device corresponds to
(*a*) Zero current
(*b*) Maximum current the device can hold
(*c*) Normal current that flows through the device
(*d*) Reverse saturation current

16. Emission of light in an LED results due to
(*a*) Emission of electrons
(*b*) Photovoltaic effect
(*c*) Generation of electromagnetic radiation
(*d*) Conversion of heat to electrical energy

17. The combined package of LED and a photodiode can be
(*a*) Optocoupler
(*b*) Opto Isolator
(*c*) Optically coupled isolator
(*d*) All of the above

18. Germanium photodiodes have dark currents of the order of
(*a*) 5 mA (*b*) 10 mA
(*c*) 25 mA (*d*) 10 A

19. Optocoupler is otherwise known as
(*a*) Laser
(*b*) Photodiodes
(*c*) Optoisolator
(*d*) Photoconductive cell

20. The advantages of opto-isolators are
(*a*) Easy to interface with logic devices
(*b*) Capable of wide band signal transmission
(*c*) Problems, such as noise transients, contact bounce are completely eliminated
(*d*) All of the above

21. To display the digit 0 in a seven segment display
(*a*) A must be lighted
(*b*) F must be off
(*c*) G must be on
(*d*) All segments except G should be lighted

22. One of the following photodetectors is a latching device
(*a*) Photodiode
(*b*) Photo SCR
(*c*) Photo FET
(*d*) Phototransistor

23. What does EMCCD offer to overcome the shortcoming of conventional CCD?
(*a*) Dynamic range
(*b*) Better noise performance
(*c*) High sensitivity at high speed
(*d*) Cost effective fabrication

24. Which of the following statement is false in the context of CMOS sensors?
(*a*) CMOS sensors have good sensitivity
(*b*) Dynamic range of CMOS is lower than CCD sensor
(*c*) CMOS sensors are immune to smearing
(*d*) CMOS technology is easy to integrate

25. Following is (are) the type(s) of Light sensor(s)
(*a*) Photo sensor
(*b*) Photo transistors
(*c*) Photo conductors
(*d*) All of the above

II. Figure. 12.50 shows the circuit symbols/package configurations of some popular opto-electronic devices and modules. You have to identify these devices and modules.

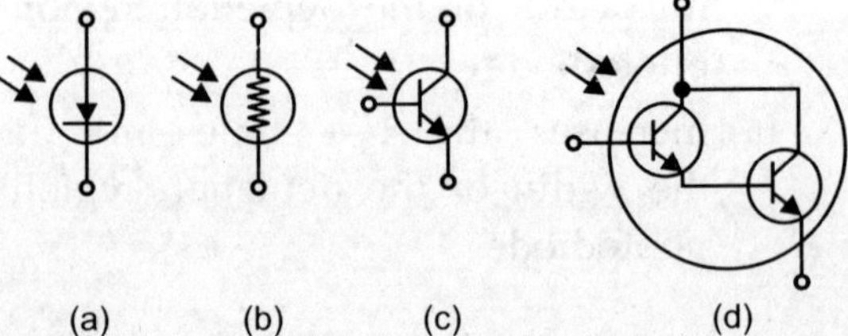

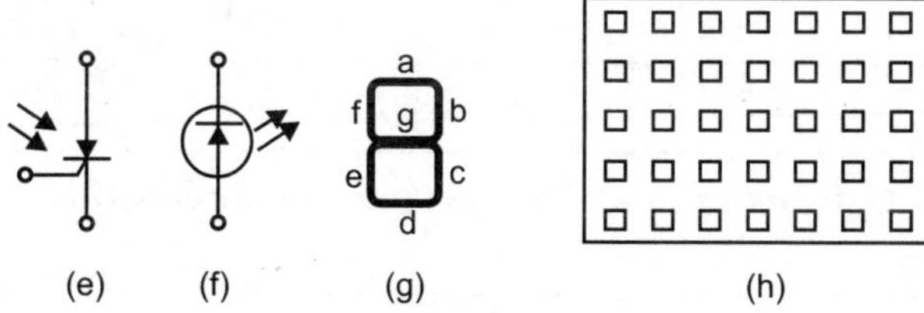

Fig. 12.50

III. Seven different optoelectronic devices are listed in column-1 of Table 12.2 In column-2 of this table are listed some well known optoelectronic effects while column-3 has only three entries. For each one of the devices in column-1, identify the photo-effect responsible for the basic operation of the devices concerned from column-2 and the broad class of detectors to which it belongs from column-3.

Device	*Effect*	*Class of Detector*
1. Photoresistor	1. Photoconductive effective	1. Quantum detectors
2. Photo diode	2. Photorvoltaic effect	2. Thermal detectors
3. Photo transistor	3. Photo emissive effect	3. Neither of the two
4. Photo FET	4. Charge coupled concept	
5. Photo thyristor	5. Pyroelectric effect	
6. Solar Cell		
7. Photomultiplier tube		

IV. Figure 12.51 shows V I characteristics of a photodiode when illuminated with a certain light intensity. The characteristics have been divided into three regions; Region AB, Region BC and Region CD. If this diode is being used in the photo-voltaic mode, which is the relevant part of the characteristics.

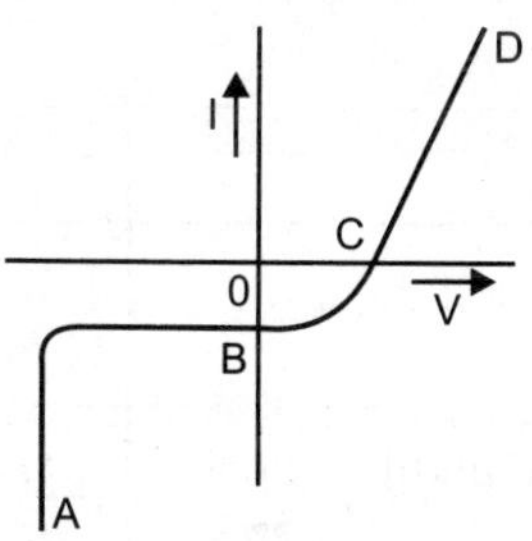

Fig. 12.51

V. **Do as directed :**

(*a*) How many free charge carriers (Electrons) will be generated when 2mW of 1μm radiation falls on a detector with a quantum efficiency of 80% for one second.

(*b*) Noise generated in a photodiode, when operated in photo-conductive mode is mainly the Johnson noise, whereas that generated in a photodiode operating in a photovoltaic mode is predominantly the shot noise and the Johnson noise, say TRUE or FALSE.

(*c*) When a photodiode is being operated in the photo-conductive mode, with increase in reverse bias voltage, the noise current...., the infrared responsivity.....and the rise and fall times.....(Decrease, Decreases, Increase, Increases)

(*d*) When you are looking for an ultra low noise performance and the frequency of operation is 1 kHz, you would opt for..... photodiode. (Photoconductive, Photovoltaic)

(*e*) A photodiode with a comparatively larger responsivity is....sensitive (More, Less)

VI. In column-1 of Table 12.3 are listed important photodiode parameters. In column-2 are given some statements, each having some relevance to one of the listed parameters in column-1. Match the parameters with the relevant statements.

Table 12.3

Parameter	Statement
1. Responsivity	1. It indicates the number of electrons created for each photon absorbed by the photodiode.
2. Noise Equivalent Power (NEP)	2. It is the Figure-of-Merit for the photodiode. It is measured in watts/$\sqrt{Hz}$. It can be calculated from the known values of detector noise current and responsivity for specified values of source wavelength, test frequency and noise bandwidth.
3. ShotNoiseinPhotoconductive Photodiode	3. It can be used to calculate the amount of photocurrent produced for a given incident light power.
4. Johnson noise current in photo-voltaic photo-diodes	4. It can be calculated from expression $I_n = \sqrt{(2q\, I_d\, \Delta f)}$ where, q = electronic charge I_d = Dark current Δf = Noise bandwidth
5. Quantum Efficiency	5. It exhibits a flat noise versus frequency characteristics.

Answer to Self Evaluation Exercise

I.

1. (*e*)	**2.** (*d*)	**3.** (*d*)	**4.** (*a*)	5. (*c*)
6. (*a*)	**7.** (*b*)	**8.** (*b*)	**9.** (*c*)	10. (*a*)
11. (*d*)	**12.** (*b*)	**13.** (*a*)	**14.** (*c*)	15. (*d*)
16. (*a*)	**17.** (*d*)	**18.** (*b*)	**19.** (*c*)	20. (*d*)
21. (*d*)	**22.** (*b*)	**23.** (*c*)	**24.** (*a*)	25. (*d*)

II. Fig. 12.50 (*a*) : Photodiode Fig. 12.50 (*b*) : Photoresistor Fig. 12.50 (*c*): Phototransistor
Fig. 12.50 (*d*): Photodarlington Fig. 12.50 (*e*) : LASCR Fig. 12.50 (*f*): LED
Fig. 12.50 (*g*): Seven Segment Display. Fig. 12.50 (*h*): 5 × 7 dot matrix array.

III. (1-1-1), (2-2-1), (3-2-1), (4-2-1), (5-2-1), (6-2-1); (6-3-1)

IV. Region BC

V. (*a*) 8×10^{15} (*b*) False (*c*) Increases, Increases, Decrease (*d*) Photovoltaic (*e*) More

VI. (1-3) (2-2) (3-4) (4-5) (5-1).

Chapter 13

Amplifiers

Basic Configurations–Biasing and Stability–Small Signal Analysis–Frequency Response–Large Signal Amplifiers

Learning Objectives

After completing the chapter, you will learn the following topics:

1. Basic Amplifier Configurations including CE, CB and CC configurations
2. Classification of Amplifiers as Voltage Amplifiers, Current Amplifiers, Transconductance Amplifiers and Transresistance Amplifiers
3. Amplifier Parameters including Current Gain, Voltage Gain, Power Gain, Input Impedance and Output Impedance, Dynamic Range, Noise Figure, Bandwidth and Slewing Rate
4. Hybrid Parameter (h-parameter) Model and Small Signal Analysis of Amplifiers
5. Hybrid-π Model of Bipolar Transistor
6. Hybrid-π Equivalent Circuits of CE, CB and CC Configurations
7. Amplifier Biasing and Bias Stability
8. Load Line, Operating Point (Q-point), Q-point Stabilization
9. Thermal Instability and Thermal Runaway
10. Frequency Response of Amplifiers including Low Frequency and High Frequency Responses
11. Classes of Amplifier Operation including Class-A, Class-B, Class-AB and Class-C Amplifiers
12. Large Signal Analysis and Power Amplifiers
13. Distortion in Amplifiers including Harmonic Distortion, Frequency Distortion, Phase Distortion and Intermodulation Distortion
14. Feedback Amplifiers
15. Types of Negative Feedback including Voltage Series, Current Series, Voltage Shunt and Current Shunt feedback topologies
16. Characteristics of Different Types of Negative Feedback in Amplifiers;
17. Cascade, Cascode, Darlington and Complementary Symmetry Amplifier Configurations

Q. 1. Briefly outline the three basic amplifier configurations and their salient features as regards current gain, voltage gain, input impedance and output impedance.

Ans. The three basic amplifier configurations are (*i*) The common-emitter (*CE*) configuration in which emitter is common to both input and output (*ii*) Common-base (*CB*) configuration in which base is common to both input and output and (*iii*) Common-collector (CC) configuration also known as emitter follower in which collector terminal is common to both input and output. Refer to Fig. 13.1 (*a*), (*b*) and (*c*) for the three basic circuit configuration built around bipolar-transistors.

The salient features of these amplifier configurations are tabulated in Table 13.1

Table 13.1

Parameter	*Common Emitter (CE)*	*Common Base (CB)*	*Common Collector (CC)*
Current Gain (A_i)	High (Typ. 50)	Less than unity (Typ. 0.98)	Highest (Typ. 100)
Voltage Gain (A_v)	High (Typ. 50)	Highest (Typ. 100)	Less than unity (Typ. 0.98)
Input impedance (Z_i)	Several Kilo-ohms	Lowest (Several ohms)	Highest (hundreds of K)
Output impedance (Z_o)	Several Kilo-ohms	Highest (Mega ohms)	Lowest (Several ohms)
Phase difference between input and output	180°	0°	0°

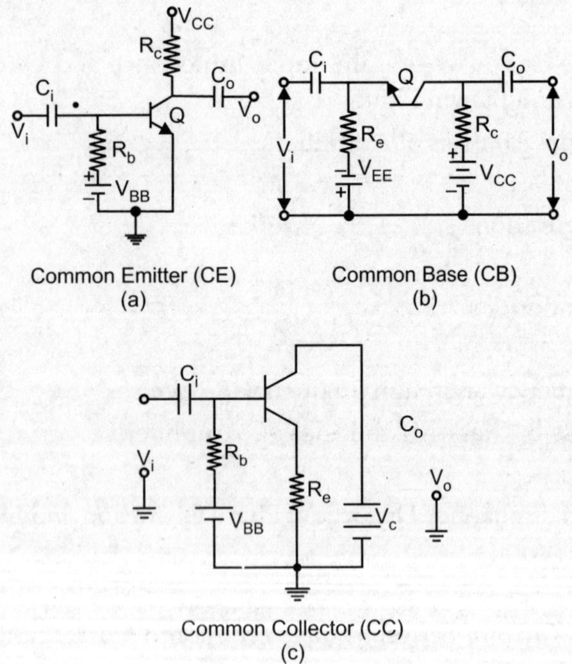

Fig. 13.1

Q. 2. In the last question, three basic amplifier configurations were shown using NPN transistors. Draw the same first using PNP bipolar transistors and then using junction FETs.

Ans. Figure. 13.2 (*a*), (*b*) and (*c*) respectively show the basic common-emitter, common-base and common-collector configurations using bipolar PNP transistors. You would notice that the only difference between these and the ones drawn in the immediately preceding question with NPN transistors is in the polarities of the DC supplies used. The polarities of the supplies used have been reversed. The idea is to forward bias base-emitter junction and reverse bias collector base junction.

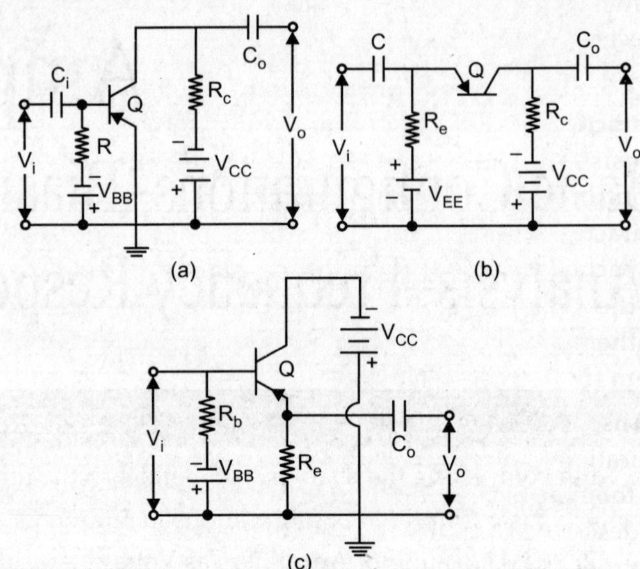

Fig. 13.2

When the bipolar transistors are replaced by field effect transistors, the configurations are known as common-source configuration (in place of common-emitter), common-gate configuration (in place of common base) and common drain configuration or source follower (in place of common-collector or emitter follower).

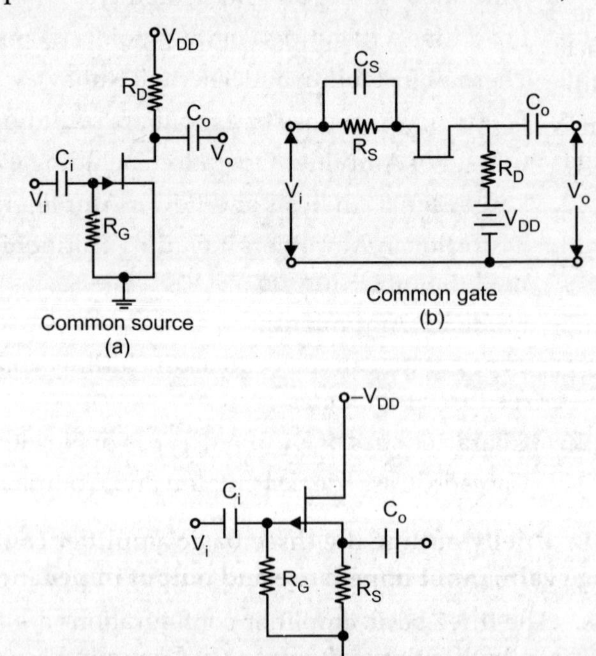

Fig. 13.3

Refer to Fig. 13.3 (*a*), (*b*) and (*c*) for basic circuit configurations. For a change, P-channel FET has been used in the common drain configuration. We can notice the similarities between the basic amplifier configurations drawn using bipolar transistors and then using field effect transistors.

Q. 3. In a previous question, we talked about various characteristics of amplifiers built around bipolar transistors. These included voltage and current gains, input and output impedances and so on. How would you characterise an FET amplifier? Briefly describe these in respect of the three FET amplifier configurations drawn in the previous question. Also, mention the relevant mathematical expressions without actually deriving them from first principles.

Ans. Operation of an FET has a lot of similarity to the operation of a vacuum tube. In fact, a junction FET can be thought of as a counterpart of a pentode tube. FET as described in detail in an earlier chapter is a voltage controlled device and is characterised by parameters like Drain resistance (r_d), Transconductance (g_m) and Amplification factor (μ). FET amplifiers are characterized by parameters like Voltage gain (A_v), which can be expressed in terms of the amplification factor (μ), Input and Output resistances. Another amplifier parameter that is of particular significance is the Input capacitance (C_i) as it affects the frequency response of the amplifier. In the following paragraphs, we shall discuss these characteristic parameters in respect of the three FET amplifier configurations.

In case of common-source amplifier (Fig. 13.4), the voltage gain is given by :

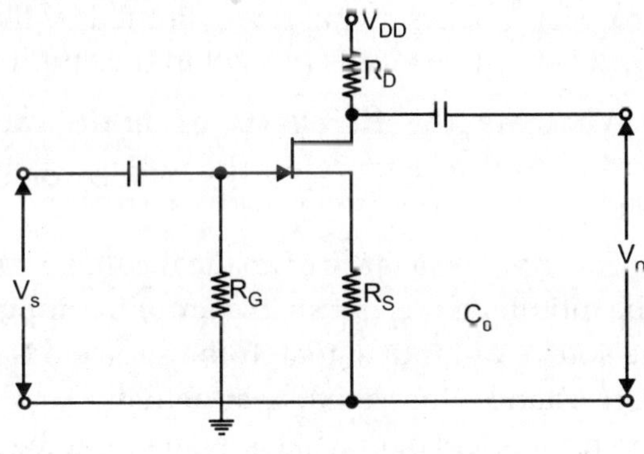

Fig. 13.4

$$A_V = \frac{-\mu R_D}{r_d + R_D} = -\frac{g_{fs}\, r_d\, R_D}{r_d + R_D}$$

For $r_d >> R_D$, the expression reduces to $A_V = -g_{fs} R_D$ (g_{fs}) is the transconductance.

The input resistance of this configuration, for all practical purposes, equals (R_G). To be precise, it is parallel combination of the input resistance of the FET and (R_G).

The output resistance of this configuration is given by parallel combination of (r_d) and (R_D) *i.e.* $r_d R_D/(R_D + r_d)$.

The input capacitance is given by $C_i = C_{gs} + (1 - A_v) C_{gd}$.

In case of common-gate configuration (Fig. 13.5), the parameters are given by :

Voltage gain, $A_V = \dfrac{(\mu + 1)R_D}{(r_d + R_D)} = \dfrac{(1 + r_d\, g_{fs})R_D}{(r_d + R_D)}$

Input Resistance, $R_i = \dfrac{r_d + R_D}{\mu + 1} = \dfrac{r_d + R_D}{(1 + g_{fs}\, R_D)}$

Output Resistance, $R_o = r_d + R_s\,(\mu + 1) = r_d + R_s\,(1 + g_{fs}\, R_D)$

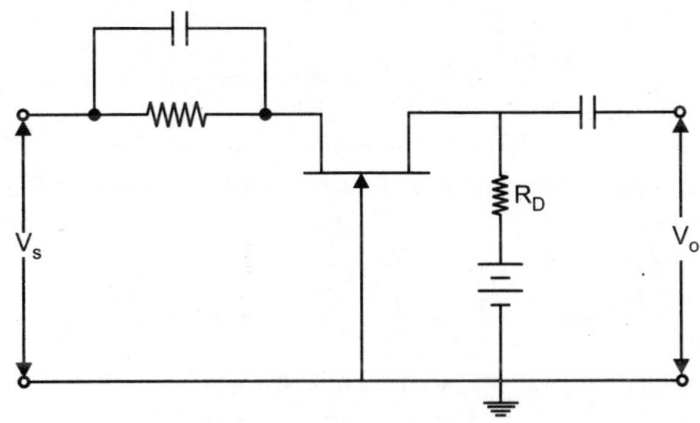

Fig. 13.5

In case of common-drain configuration (Fig. 13.6), various parameters are expressed by:

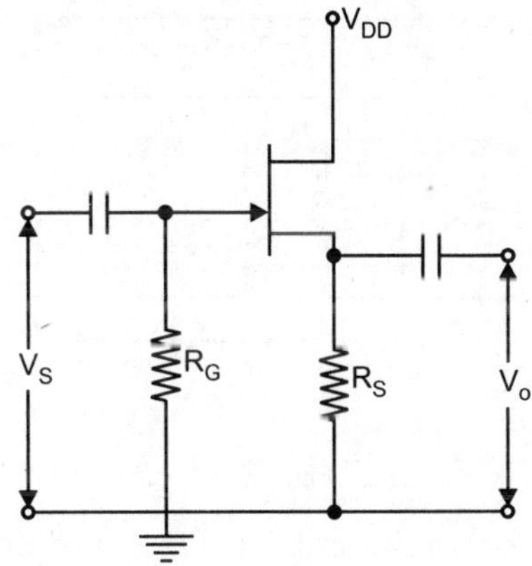

Fig. 13.6

Voltage $A_V \cong \dfrac{g_{fs}\, R_s}{1 + g_{fs}\, R_s}$

Input resistance, $R_i \cong R_G$

Output resistance, $R_o = R_s \left\| \dfrac{r_d}{1 + \mu} \right. \cong \dfrac{R_s}{(1 + g_{fs}\, R_s)}$.

Q. 4. The amplifiers are also sometimes classified as Voltage Amplifiers, Current Amplifiers, Transconductance Amplifiers and Transresistanee Amplifiers. How do you distinguish one type from the other?

Ans. To which of the above mentioned classes an amplifier belongs depends upon the nature of the amplification factor of that amplifier. If the amplification factor is voltage gain, then it is a voltage amplifier.

In other words, in a voltage amplifier, a small change in input voltage produces a large change in the output voltage. The amplifier's input and output circuits can then be represented by Thevenin's equivalent circuits (Fig. 13.7 (*a*))

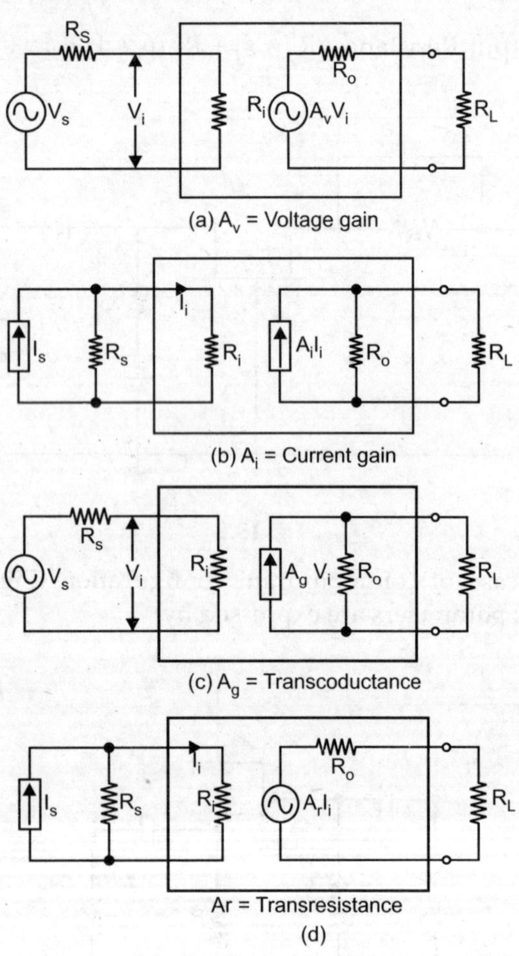

Fig. 13.7

In case of a current amplifier, the amplification factor is the current gain, The circuit is so optimised that a very small change in input current produces a very large change in output current. The input and output circuitries in this case can be represented by Norton's equivalent (Fig. 13.7 (*b*))

If the amplifier is such that a small change in input voltage produces a large change in output current, it is a transconductance amplifier. It is so called as the amplification factor here has the dimension of conductance. Such an amplifier can be represented by a Thevenin's equivalent circuit on the input side and a Norton's equivalent circuit on the output side (Fig. 13.7 (*c*)). The amplification factor in a transresistance amplifier is a resistance *i.e.* a small change in input current produces a large change in output voltage. Figuse. 13.7(*d*)) shows the equivalent circuit. Such a circuit is also referred to as a current-to-voltage converter.

Q. 5. Define the following with reference to amplifiers: Voltage Gain, Current Gain, Input Resistance, Output Resistance, Bandwidth, Power Gain, Slewing Rate.

Ans. The *voltage gain* of an amplifier is the ratio of output signal voltage to the input signal voltage. Mathematically, $A_v = (V_o/V_i)$

The current gain of an amplifier is the ratio of output signal current to the input signal current. Mathematically, $A_i = (I_o/I_i)$

The *input resistance* (or input impedance) of an amplifier is the resistance (or impedance) that effectively appears across the input terminals of the amplifier.

The *output resistance* (or output impedance) of an amplifier is the resistance (or impedance) that effectively appears across the output terminals of the amplifier.

The *bandwidth* of an amplifier is the range of frequencies over which the nominal gain or the mid frequency gain does not fall by more than 3 dB.

The *power gain* of an amplifier is the ratio of output signal power to input signal power. Mathematically, $A_p = (P_o/P_i)$.

The *slewing rate* of an amplifier is its ability to respond quickly to step change at its input. It is expressed as volts per microsecond. The slewing rate is limited by the input and output capacitive effects present in the amplifier.

Q. 6. What are the ill effects of finite values of input and output resistance on the voltage gain of an amplifier?

Ans. The input resistance of an ideal voltage amplifier should be infinite. If it is not so, the amplifier input loads on to the source of input signal. If the source has a finite internal resistance, the signal that actually reaches the amplifier input is less than what is available at the source. Consequently, the voltage gain reduces. However, if the source is an ideal one and has a zero internal resistance, then the amplifier's input resistance, no matter what it is, makes no difference to the overall voltage gain.

The- output resistance of an ideal voltage amplifier should be zero. If is not so, the load resistance would load on to the amplifier's output *i.e.*, the voltage actually appearing across the load would be less than the open circuited output voltage of the amplifier. Fig. 13.8 depicts the effects of input and output resistances on the voltage gain of the amplifier.

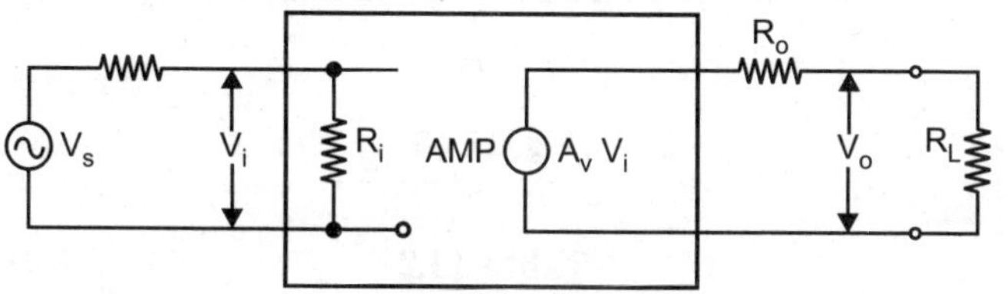

Fig. 13.8

Q. 7. Briefly describe the Dynamic Range and Noise Figure specifications of amplifiers.

Ans. The dynamic range of an amplifier can be defined as the ratio of its maximum usable output signal without exceeding the maximum acceptable specified distortion limit (usually 10%) to the minimum usable output signal having a certain specified minimum signal to noise ratio (usually 20 dB). It is expressed in decibels. Dynamic range of 50 dB is considered reasonably good while an amplifier with more than 70 dB dynamic range is considered to be exceptionally good.

The noise figure of an amplifier is defined as the ratio of signal to noise at the input to the signal to noise at the output. That is,

$$NF = \frac{(S_i/N_i)}{(S_o/N_o)}$$

It is expressed in decibels and is thus calculated from

$$NF = 10 \log \frac{(S_i/N_i)}{(S_o/N_o)}$$

An ideal amplifier (a noiseless amplifier) has a noise figure of zero decibel.

Q. 8. With the help of hybrid equivalent circuit of a transistor, briefly describe the different h-parameters. Why is the h-parameter model of a transistor suitable only for small signal, low frequency analysis of transistor amplifiers?

Ans. The hybrid equivalent circuit of a transistor has its input side represented by a Thevenin's circuit and the output side represented by a Norton's circuit (Fig. 13.9). h_{11}, h_{12}, h_{21} and h_{22} are the hybrid parameters and these parameters are inter-related by the following equations :

$$V_1 = h_{11}I_1 + h_{12}V_2 \quad ...(1)$$

$$I_2 = h_{21}I_1 + h_{22}V_2 \quad ...(2)$$

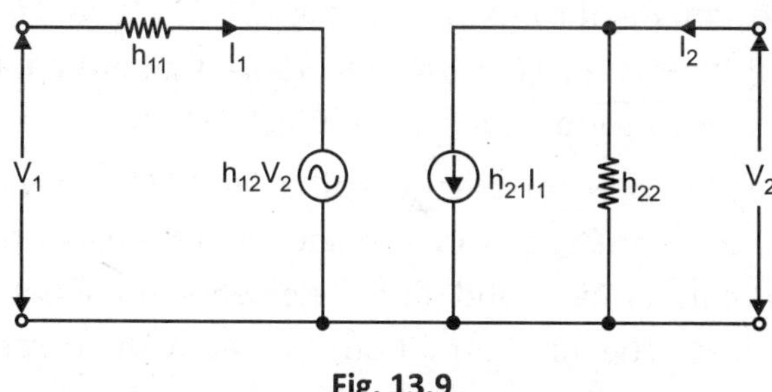

Fig. 13.9

Where V_1 and V_2 are input and output voltages while I_1 and I_2 are input and output currents respectively. These are called *h*-parameters or hybrid parameters as all have different dimensions, (h_{11}) is an impedance, (h_{12}) is a voltage ratio, (h_{21}) is a current ratio and (h_{22}) is a conductance.

Parameter (h_{11})

Substituting $V_2 = 0$ in equation (1), we get, $h_{11} = V_1/I_1$

This is the input impedance of the transistor with output shorted to *AC* signals, (h_{11}) is also written as (h_i). The *h*-parameter (h_i) for *CE*, *CB* and *CC* configurations is respectively denoted by h_{ie}, h_{ib} and h_{ic}. (h_{ie}) represents input impedance of transistor with output shorted, when it is connected in common-emitter configuration.

Parameter (h_{12})

Substituting $I_1 = 0$ in equation (1), we get, $h_{12} = V_1/V_2$

This is called the reverse voltage transfer ratio with input open circuited. (h_{12}) is written as (h_r). For *CE*, *CB* and *CC* configurations, (h_r) becomes (h_{re}), (h_{rb}) and (h_{rc}) respectively. (h_{re}) is thus the reverse voltage transfer ratio (ratio of input to output voltages) for a transistor in common emitter amplifier configuration with input open circuited (or input current = 0).

Parameter (h_{21})

From equation (2), substituting $V_2 = 0$, $h_{21} = I_2/I_1$

This is the forward current transfer ratio with output shorted to *AC* signals. For transistors, it is denoted as (h_f). For *CE*, *CB* and *CC* configurations, (h_f) becomes (h_{fe}), (h_{fb}) and (h_{fc}) respectively. (h_{fe}), for instance, is the forward current transfer ratio for a transistor in *CE* configuration with output short circuited. (h_{fe}) is thus, the short circuit current gain of transistor in CE configuration. It is same as (β) discussed earlier.

Parameter (h_{22})

Substituting $I_1 = 0$ in equation (2), we get, $h_{22} = I_2/V_2$

This is the output conductance also denoted by (h_o). For transistors connected in *CE*, *CB* and *CC* configurations, (h_o) becomes (h_{oe}), h_{ob}, (h_{oc}) respectively.

To sum up, the h-parameters are h_{ie}, h_{re}, h_{fe} and h_{oe} for CE-configuration, h_{ib}, h_{rb}, h_{fb} and h_{ob} for CB configuration and h_{ic}, h_{rc}, h_{fc} and h_{oc} for CC configuration.

The hybrid equivalent circuit of Fig. 13.9 is valid for any four terminal network provided that the network has a common terminal and all the elements in the network are linear. The first condition, is met with in case of transistor amplifier configurations, as we have emitter as the common terminal in CE-configuration, base as the common terminal in CB-configuration and collector as the common terminal in CC-configuration.

The second condition of the circuit having only linear elements is valid only, if the transistor is operated in the linear region of its characteristics *i.e.* far from the cut-off and saturation regions. And this is possible, only if the signals being handled are of smaller amplitudes.

Also, the hybrid equivalent circuit of a transistor does not take into account the inter-electrode capacitances and the input and output capacitive effects and therefore, it is not applicable to high frequency analysis. Due to these reasons, the h-parameter equivalent circuit of transistor amplifiers is used only for carrying out small signal low frequency analysis.

Q. 9. Draw the h-parameter equivalent circuits for a transistor connected in *CE*, *CB* and *CC* configurations.

Ans. The equivalent circuits are shown in Fig. 13.10 (*a*), (*b*) and (*c*) respectively for *CE*, *CB* and *CC* configurations. Table 13.2 gives approximate conversion formulae relating the h-parameters in the three configurations.

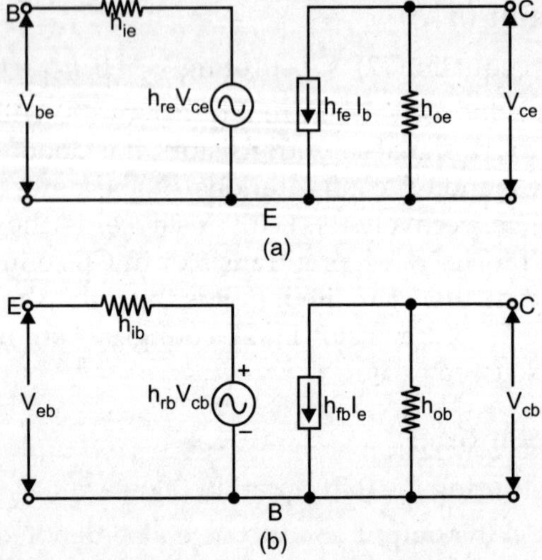

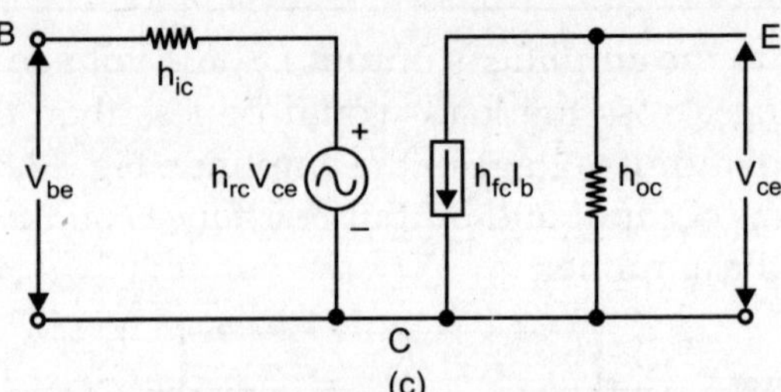

Fig. 13.10

Table 13.2

h-Parameter	*CE-Configuration*	*CB-Configuration*	*CC-Confuration*
h_{ie}	–	$\frac{h_{ib}}{(1+h_{fb})}$	h_{ic}
h_{re}	–	$\frac{h_{ib}\,h_{ob}}{1+h_{fb}} - h_{rb}$	$(1-h_{rc})$
h_{fe}	–	$-\frac{h_{fb}}{(1+h_{fb})}$	$-(1+h_{fc})$
h_{oe}	–	$\frac{h_{ob}}{(1+h_{fb})}$	h_{oe}
h_{ib}	$\frac{h_{ie}}{(1+h_{fe})}$	–	$-\frac{h_{ic}}{h_{fc}}$
h_{rb}	$\frac{h_{ie}\,h_{oe}}{(1+h_{fe})} - h_{re}$	–	$-\frac{h_{ic}\,h_{oc}}{h_{fc}}$
h_{fb}	$\frac{-h_{fe}}{(1+h_{fe})}$	–	
h_{ob}	$\frac{h_{oe}}{(1+h_{fe})}$	–	
h_{ic}	h_{ic}	$\frac{h_{ib}}{(1+h_{fb})}$	–
h_{re}	$(1 - \text{hre} \cong 1$	$\cong 1$	–
h_{fc}	$-(1+h_{fe})$	$-\frac{1}{(1+h_{fb})}$	–
h_{oc}	h_{oe}	$-\frac{h_{ob}}{(1+h_{fb})}$	–

Q. 10. Derive expressions for the amplifier's characteristics like current gain, voltage gain, input impedance and output impedance for a CE amplifier configuration fed from a source having an internal resistance of (R_s) and feeding a load (R_L) in terms of h-parameters. From the results obtained, briefly describe the effects of source resistance (R_s) and the load resistance (R_L) on the amplifier's characteristics.

Ans. The hybrid equivalent circuit for a CE-amplifier along with the source and the load is shown in Fig. 13.11

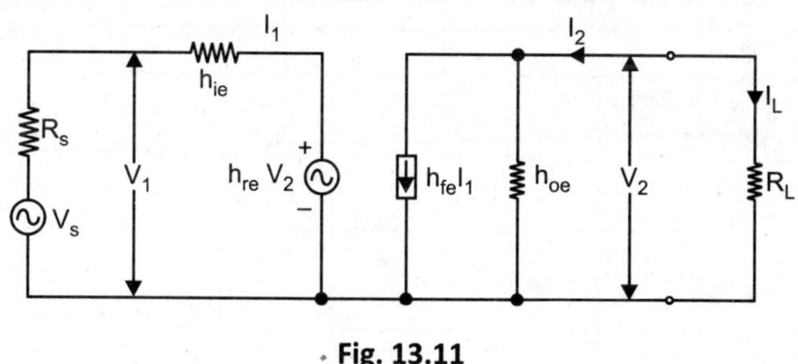

Fig. 13.11

CURRENT GAIN

Current gain $= \frac{I_L}{I_1} = -\frac{I_2}{I_1}$

From the equivalent circuit of Fig. 13.11.

$$I_2 = h_{fe} I_1 + h_{oe} V_2$$

Now $V_2 = I_L R_L = -I_2 R_L$

Therefore, $I_2 = h_{fe} I_1 + h_{oe}(-I_2 R_L) = h_{fe} I_1 - h_{oe} I_2 R_L$

or $I_2(1 + h_{oe} R_L) = h_{fe} I_1$ or $I_2/I_1 = h_{fe}/(1 + h_{oe} R_L)$

The current gain (A_i) is, therefore given by

$$A_i = \frac{-I_2}{I_1} = -\frac{h_{fe}}{(1 + h_{oe} R_L)}$$

INPUT IMPEDANCE (or INPUT RESISTANCE)

Input resistance, $R_i = \frac{V_1}{I_1}$

From the equivalent circuit of Fig. 13.11.

$$V_1 = h_{ie} I_1 + h_{re} V_2$$

or $\frac{V_1}{I_1} = h_{ie} + h_{re}\left[\frac{V_2}{I_1}\right]$

Now $V_2 = -I_2 R_L$

Therefore, $\frac{V_2}{I_1} = -\frac{I_2}{I_1} R_L = A_i R_L$

Substituting the value of (V_2/V_1), we get

Input resistance, $R_i = \frac{V_1}{I_1} = (h_{ie} + h_{re} A_i R_L)$

As $h_{re} \cong 0, R_i \simeq h_{ie}$

VOLTAGE GAIN

Voltage gain, $A_v = \frac{V_2}{V_1}$

From the equivalent circuit of Fig. 13.11

$$V_2 = -I_2 R_L = -\left[\frac{I_2}{I_1}\right] R_L I_1 = A_i R_L I_1$$

or $\frac{V_2}{V_1} = \frac{A_i R_L I_1}{V_1} = \frac{A_i R_L}{(V_1/I_1)} = \frac{A_i R_L}{R_i}$

or Voltage gain, $A_v = \frac{V_2}{V_1} = \frac{A_i R_L}{R_i} = -\frac{h_{fc}}{1 + h_{oe} R_L} \times \frac{R_L}{h_{ie}}$

$-\frac{R_L}{h_{ie}}$

OUTPUT IMPEDANCE (or OUTPUT RESISTANCE)

Output resistance, $R_O = \frac{V_2}{I_2}$

Writing mesh equations on input and output sides in the equivalent circuit of Fig. 13.11 we get

$$V_s = (R_s + h_{ie}) I_1 + h_{re} V_2 \quad ...(1)$$

$$I_2 = h_{fe} I_1 + h_{oe} V_2 \quad ...(2)$$

From equation (1) $I_1 = \frac{V_s - h_{re} V_2}{(R_s + h_{ie})}$

Therefore, from equation (2),

$$I_2 = \frac{V_s h_{fe}}{R_S + h_{ie}} - \frac{h_{fe} h_{re} V_2}{R_S + h_{ie}} + h_{oe} V_2$$

or $$I_2 = V_2\left[h_{oe} - \frac{h_{fe} h_{re}}{R_s + h_{ie}}\right] + \frac{V_s h_{fe}}{R_s + h_{ie}}$$

Now $$V_s = \frac{V_1 (h_{ie} + R_s)}{h_{ie}}$$

This gives

$$I_2 = V_2\left[h_{oe} - \frac{h_{fe} h_{re}}{R_s + h_{ie}}\right] + \frac{V_1 (h_{ie} + R_S) h_{fe}}{h_{ie}(R_S + h_{ie})}$$

$$= V_2\left[h_{oe} - \frac{h_{fe} h_{re}}{R_s + h_{ie}}\right] + \frac{V_1 h_{fe}}{h_{ie}}$$

or $I_2/V_2 = G_O (= 1/R_O) = \left(hoe - \frac{h_{fe} h_{re}}{R_s + h_{ie}}\right) + \frac{V_1 h_{fe}}{V_2 h_{ie}}$

$$\cong \left(h_{oe} - \frac{h_{fe} h_{re}}{R_s + h_{ie}}\right) \text{ as } \frac{V_1 h_{fe}}{V_2 h_{ie}} \cong 0$$

It is clear from the above expressions that the current gain (A_i), voltage gain (A_v), input resistance (R_i) are dependent upon the load resistance (R_L). Also, the output resistance (R_O) is a function of source resistance (R_s).

Table 13.3 lists the expressions for A_i A_v, R_i and R_0 for *CE*, *CB* and *CC* amplifier configurations. These are approximate expressions and it is assumed that $h_{oe} R_L$ and $h_{oe} R_e$ are less than or equal to 0.1. (R_e), here is the emitter resistor.

Table 13.3

Characteristic	*CE-config. with R_e bypassed*	*CE-config. with (R_e) unbypassed*	*CC*	*CB*
A_i	$-h_{fe}$	$-h_{fe}$	$1+h_{fe}$	$\frac{h_{fe}}{1+h_{fe}}$
R_i	h_{ie}	$h_{ie}+(1+h_{fe})R_e$	$h_{ie}+(1+h_{fe})$	$\frac{h_{ie}}{1+h_{fe}}$
A_v	$\frac{-h_{fe}R_L}{h_{ie}}$	$\frac{-h_{fe}R_L}{R_i}$	$1-\frac{h_{ie}}{R_i}$	$\frac{h_{fe}R_L}{h_{ie}}$
R_o	Infinity	Infinity	$\frac{R_s+h_{ie}}{1+h_{fe}}$	Infinity
$R'_O = R_O \parallel R_L$	R_L	R_L	$R_O \parallel R_L$	R_L

Q. 11. Outline qualitatively the small signal, low frequency amplifier characteristics in CE, CB and CC configurations.

Ans. The characteristics are tabulated in Table 13.4

Table 13.4

Characteristic	*CE*	*CB*	*CC*
Current Gain (A_i)	High	Low	Highest of all
Voltage Gain (A_v)	High	High	Low
Input Resistance (R_i)	Medium	Low	High
Output Resistance (R_o)	Medium	High	Low

Q. 12. Why is the h-parameter model of a transistor is not applicable at high frequencies? Draw a high frequency response for a common emitter configuration transistor and explain the circuit.

Ans. At high frequencies, the h-parameter model of a transistor is not applicable because at high frequencies, the transistor behaves in quite a different manner to what it does at low frequencies. At low frequencies, it is assumed that the transistor responds to the input voltage and current instantly, as the diffusion time of the carriers is very small as compared to the rise time of the input signal. However, at high frequencies this is not the case and hence the h-parameter model is not valid at high frequencies. A commonly used method at high frequencies is the hybrid-π model or the Giacoletto model. This model gives a fairly good approximation of the transistor's behaviour at high frequencies.

Figures 13.12 (*a*) and (*b*) show the circuit of a common-emitter NPN transistor and its hybrid-π model, respectively. The node B′ is an internal node and is not physically accessible. All the components, both capacitive as well as resistive, are assumed to be independent of frequency. They are dependent on the quiescent operating conditions, but under a given bias condition they do not vary much for small input signal variations.

Various circuit components are the base-spreading resistance (r_{bb}'), conductance between terminals *B′* and *E* ($g_{b'e}$), conductance between terminals *C* and *E* (g_{ce}), conductance between terminals *B′* and *C* ($g_{b'c}$), current source between terminals *C* and *E* ($g_m\ V_{b'e}$), collector-junction barrier capacitance (C_C) and diffusion capacitance between terminals *B′* and *C* (C_e). The ohmic base-spreading resistance (r_{bb}') is represented as a lump parameter between the external base terminal (*B*) and the node *B′*. The conductance ($g_{b'e}$) takes into account the increase in the recombination base current due to the increase in the minority carriers in the base region. g_{ce} is the conductance between the collector and the emitter terminals. The conductance ($g_{b'c}$) takes into account the feedback effect between the output and the input due to Early effect. Early effect results in modulation of the width of the base region due to varying collector–emitter voltage, which in turn causes a change in the emitter and the collector currents, as the slope of the minority-carrier distribution in the base region changes.

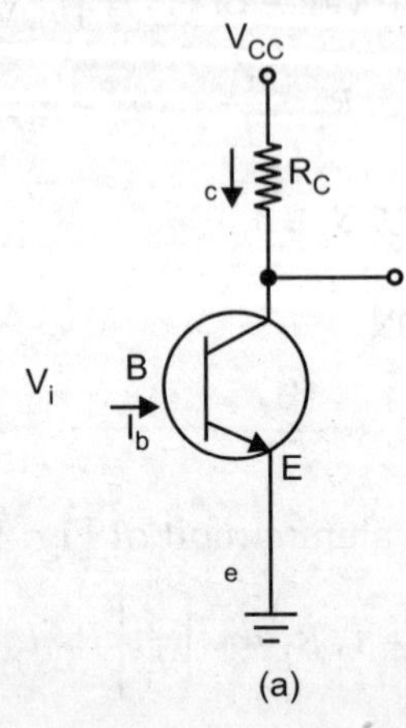

(a)

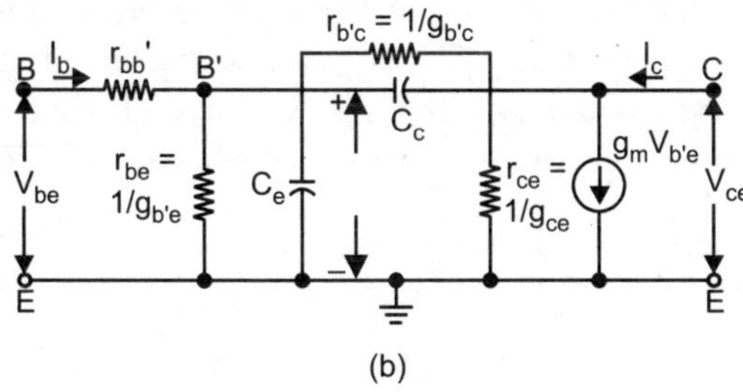

Fig. 13.12

Small changes in the value of voltage $V_{b'e}$ cause excess minority carriers, proportional to the voltage $V_{b'e}$, to be injected in to the base region. This results in small-signal collector current. Hence, the magnitude of the collector current for shorted collector and emitter terminals is proportional to the voltage $V_{b'e}$. The current generator $g_m V_{b'e}$ takes into account this effect. Note that g_m is the transconductance of the transistor and C_C is the collector-junction barrier capacitance. Sometimes, this capacitance is split into two parts, namely, the capacitance between C and B′ terminals and the capacitance between C and B terminals. The capacitance between C and B terminals is also referred to as the overlap-diode capacitance.

Q. 13 Draw a hybrid-π model for a common emitter amplifier at low frequencies. Derive the expressions for the components of hybrid-π model in terms of h-parameters.

Ans. Figure 13.13 (*a*) shows the hybrid-π model for a common-emitter transistor amplifier applicable at low frequencies and Fig. 13.13 (*b*) shows the h-parameter model for the same. As the hybrid-π model is drawn for low frequencies, the capacitive elements are considered as open circuit.

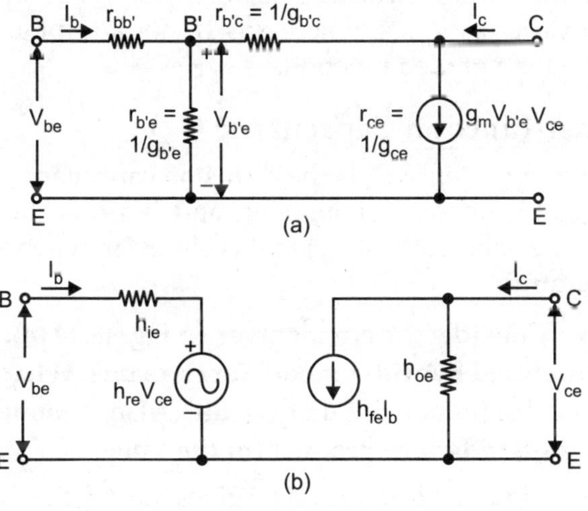

Fig. 13.13

Base-Spreading Resistance ($r_{bb'}$)

In the circuit shown in Fig. 13.13 (*b*), the value of input resistance when the output terminals are shorted, that is $V_{ce} = 0$, is equal to h_{ie}. Under these conditions for the circuit in Fig. 13.13 (*a*), the input resistance is given by

$$Z_i \,|\, V_{ce=0} = r_{bb'} + r_{b'e} \| r_{b'c} \quad ...(1)$$

Therefore,

$$h_{ie} = r_{bb'} + r_{b'e} \| r_{b'c} \quad ...(2)$$

As $r_{b'c} >> r_{b'e}$ therefore, Eq. (2) can be approximated as

$$h_{ie} = r_{bb}' + r_{b'e} \quad ...(3)$$

Conductance between Terminals *B′* and *C* or the Feedback Conductance ($g_{b'c}$)

For the circuit in Fig. 13.13 (*b*), if the input terminals are open-circuited, then h_{re} is the reverse voltage gain. In terms of the circuit in Fig. 13.13(*a*), the value of h_{re} is given by

$$h_{re} = \frac{V_{b'e}}{V_{ce}} = \frac{r_{b'e}}{r_{b'e} + r_{b'c}} \quad ...(4)$$

Arranging the terms in Eq. (4) we get,

$$r_{b'e}(1 - h_{re}) = h_{re}\, r_{b'c} \quad ...(5)$$

As the value of h_{re} is in the range of 10^{-4}, that is,h_{re} << 1, therefore, Eq.(5) can be approximated by

$$r_{b'e} = h_{re} r_{b'c} \text{ or } g_{b'c} = h_{re}\, g_{b'c} \quad ...(6)$$

The equation also verifies that the value of resistance $r_{b'c}$ is much larger than resistance ($r_{b'e}$), that is, $r_{b'c} >> r_{b'e}$.

Conductance between Terminals *C* and *E* (g_{ce})

For the circuit in Fig. 13.13 (*b*), if the input terminals are open circuit then

$$V_{b'e} = h_{re} V_{ce} \quad ...(7)$$

For the circuit in Fig. 13.13(*a*), with the input terminals open, that is with $I_b = 0$, the collector current I_c is given by

$$I_c = \frac{V_{ce}}{r_{ce}} + \frac{V_{ce}}{r_{b'e} + r_{b'c}} + g_m V_{b'e} \quad ...(8)$$

Value of h_{oe} is given by

$$h_{oe} = \left.\frac{I_c}{V_{ce}}\right|_{I_b=0} = \frac{1}{r_{ce}} + \frac{1}{r_{b'e} + r_{b'c}} + \frac{g_m V_{b'e}}{V_{ce}} \quad ...(9)$$

Substituting the value of $V_{b'e}$ given in Eq.(7) in Eq.(9) we get,

$$h_{oe} = \frac{1}{r_{ce}} + \frac{1}{r_{b'e} + r_{b'c}} + g_m h_{re} \quad ...(10)$$

Substituting the value of h_{re} as $g_{b'c}/g_{b'e}$, $1/r_{ce}$ as g_{ce}, $1/r_{b'c}$ as $g_{b'c}$ and assuming that $r_{b'c} >> r_{b'e}$, Eq.(10) can be rewritten as

$$h_{oe} = g_{ce} + g_{b'c} + g_m \frac{g_{b'c}}{g_{b'e}} \quad ...(11)$$

We will derive latter that the value of g_m is given by

$$g_m = h_{fe} g_{b'e} \quad ...(12)$$

Substituting the value of g_m given by Eq. (12) in Eq. (11), we get

$$h_{oe} = g_{ce} + g_{b'e} + g_{b'e} h_{fe} \quad ...(13)$$

Re-arranging the terms in the above equation we get

$$g_{ce} = h_{oe} - (1 + h_{fe}) g_{b'c} \quad ...(14)$$

As the value of $h_{fe} >> 1$, Eq.(14) can be approximated as

$$g_{ce} \cong h_{oe} - h_{fe} g_{b'c} \cong h_{oe} - g_m h_{re} \quad ...(15)$$

Conductance between Terminals B' and E or the Input Conductance ($g_{b'e}$)

In the circuit shown in Fig. 13.13 (*a*) as the value of resistance ($r_{b'c}$) is much greater than resistance ($r_{b'e}$), most of the current I_b flows into $r_{b'e}$ and the value of the voltage ($V_{b'e}$) is given by

$$V_{b'e} \cong I_b r_{b'e} \quad ...(16)$$

The short-circuit collector current (I_c) is given by

$$I_c = g_m V_{b'e} \cong g_m I_b r_{b'e} \quad ...(17)$$

As we have studied before the short-circuit current gain (h_{fe}) is defined as,

$$h_{fe} = \left.\frac{I_c}{I_b}\right|_{V_{ce} = const.} = g_m r_{b'e} \quad ...(18)$$

Re-arranging the terms in Eq.(18), we get,

$$r_{b'e} = \frac{h_{fe}}{g_m} \text{ or } g_{b'e} = \frac{g_m}{h_{fe}} \quad ...(19)$$

Transistor's Transconductance (g_m)

The transconductance of a transistor (g_m) is defined as the ratio of the change in the value of collector current to change in the value of voltage $V_{b'e}$ for constant value of collector–emitter voltage.

For a common-emitter transistor configuration, the expression for collector current is given by

$$I_c = I_{CO} + \alpha I_e \quad ...(20)$$

The short-circuit collector current is given by $g_m V_{b'e}$; therefore, the value of g_m is given by

$$g_m = \left.\frac{\partial I_c}{\partial V_{b'e}}\right|_{V_{CE} = const.} = \alpha \frac{\partial I_e}{\partial V_{b'e}} = \alpha \frac{\partial I_e}{\partial V_e} \quad ...(21)$$

The partial derivative of the emitter voltage w.r.t. to the emitter current (*i.e.*, $\partial V_e / \partial I_e$) can be represented as the emitter diode resistance (r_e). As we have studied before, the dynamic resistance of a forward-biased diode (r_d) is given as,

$$r_d = \frac{V_T}{I_D} \quad ...(22)$$

where, V_T is the volt equivalent of temperature and I_D is the diode current. Therefore, the value of g_m can be generalised as,

$$g_m = \frac{\alpha I_e}{V_T} = \frac{I_c - I_{CO}}{V_T} \quad ...(23)$$

As the value of $I_c >> I_{CO}$, therefore the value of g_m for an NPN transistor is positive. For a PNP transistor, the analysis can be carried out on similar lines and the value of g_m in the case of a PNP transistor is also positive. Therefore, the expression for g_m can be written as,

$$g_m = \frac{|I_c|}{V_T} \quad ...(24)$$

Q. 14. Discuss the two types of capacitances that you observe in a hybrid-π model.

Ans. In the hybrid-π model shown in Fig. 13.12 (*b*), there are two capacitances, namely the collector-junction barrier capacitance (C_C) and the emitter-junction diffusion capacitance (C_e).

Collector-Junction Capacitance (C_c)

The capacitance C_c is the output capacitance of the common-base transistor configuration with the input open ($I_E = 0$). It is also specified as C_{ob}. As the collector-base junction is reverse-biased, C_c is the transition capacitance and it varies as V_{CB}^{-n}, where n is 1/2 for abrupt junction and 1/3 for a graded junction.

Emitter-Junction Capacitance (C_e)

The capacitance C_e is the diffusion capacitance of the forward-biased emitter junction and is proportional to the emitter-bias current (I_E) and is almost independent of temperature.

Q. 15. Consider the circuit given in Fig. 13.14 (*a*). Draw the simplified hybrid-π model for the same. What is the β cut-off frequency for that circuit? Also, comment on the Gain-Bandwidth product for the same.

Ans. Figures 13.14 (*b*) and (*c*) gives the hybrid-π model for the circuit given in Fig. 13.14 (*a*).

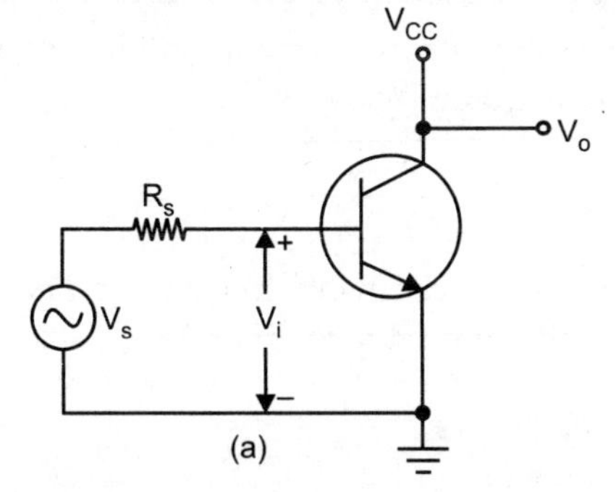

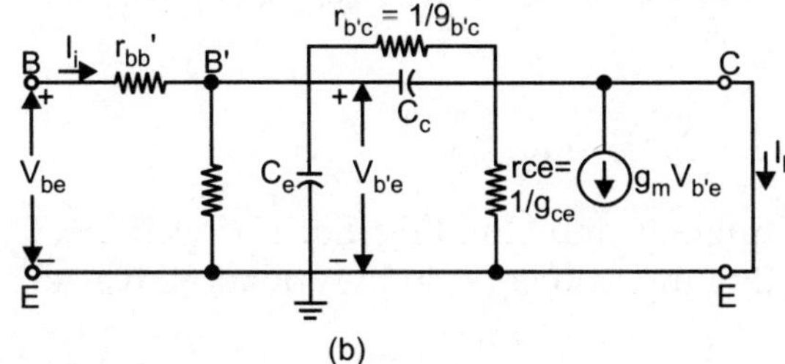

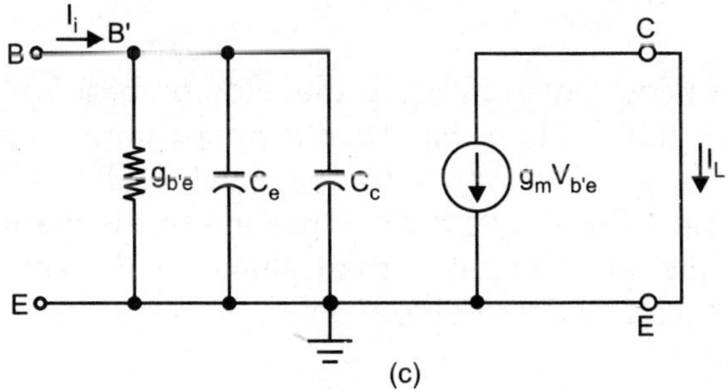

Fig. 13.14

For the circuit shown in Fig. 13.14 (*c*), the value of load current (I_L) is given by

$$I_L = -g_m V_{b'e} \qquad \text{...(1)}$$

The value of $V_{b'e}$ is given by

$$V_{b'e} = \frac{I_i}{g_{b'e} + j\omega(C_e + C_c)} \qquad \text{...(2)}$$

The value of current gain (A_i) under short-circuit condition is

$$A_i = \frac{I_L}{I_i} \qquad \text{...(3)}$$

Substituting the values of I_L and I_i given by Eqs.(1) and (2), respectively, in Eq.(3) we get,

$$A_i = \frac{-g_m}{g_{b'e} + j\omega(C_e + C_c)} \qquad \text{...(4)}$$

Rearranging the terms in Eq. (4), we get

$$A_i = \frac{-g_m / g_{b'e}}{1 + [j\omega(C_e + C_c)/g_{b'e}]} \qquad \text{...(5)}$$

Also, we know that

$$g_{b'e} = \frac{g_m}{h_{fe}}$$

Therefore, the value of current gain A_i is given by

$$A_i = \frac{-h_{fe}}{1 + [j\omega(C_e + C_c)/g_{b'e}]} = \frac{-h_{fe}}{1 + [j2\pi f\,(C_e + C_c)/g_{b'e}]} \qquad \text{...(6)}$$

At low frequencies the value of current gain A_i is given by

$$A_i = -h_{fe} \qquad \text{...(7)}$$

f_β is the frequency at which the value of current gain reduces to 0.707 times the gain value given in Eq. (7). Therefore,

$$A_i = \frac{-h_{fe}}{1 + j(f/f_\beta)} \qquad \text{...(8)}$$

Magnitude of current gain (A_i) is given by,

$$|A_i| = \frac{h_{fe}}{\sqrt{1 + (f/f_\beta)^2}}$$

The value of f_β is given by,

$$f_\beta = \frac{g_{b'e}}{2\neq(C_e + C_c)}\,\frac{g_{b'e}}{2\pi\,(C_e + C_c)} \qquad \text{...(9)}$$

As we can see from Eq. (7), the value of A_i is equal to $-h_{fe}$ at zero and low frequencies. Remember that h_{fe} is the low-frequency short-circuit current gain of the common-emitter configuration.

Gain-Bandwidth Product

The frequency f_T is the frequency at which the magnitude of the short-circuit current gain in the common-emitter amplifier becomes unity. As the value of $h_{fe} >> 1$, the magnitude of the current gain A_i becomes unity at the frequency given by the product of h_{fe} and f_β. Therefore, f_T is given by,

$$f_T \cong h_{fe} f_\beta \cong \frac{h_{fe} g_{b'e}}{2\pi\,(C_e + C_c)} \cong \frac{g_m}{2\pi\,(C_e + C_c)} \qquad \text{...(10)}$$

The parameter f_T is a strong function of the collector current of the transistor. The variation of f_T with collector current (I_c) is highlighted in Fig. 13.15.

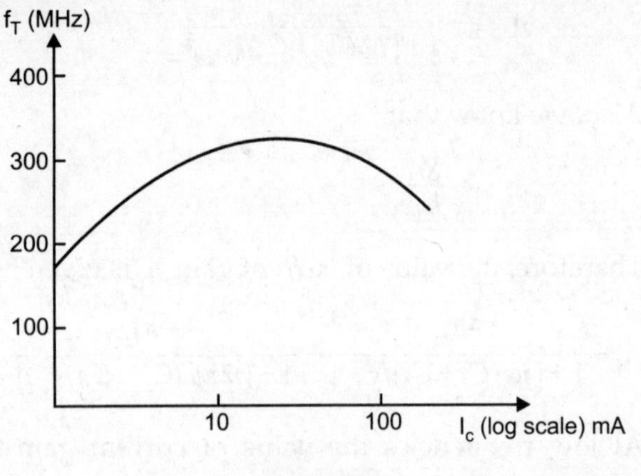

Fig. 13.15

The expression for current gain A_i can be written as

$$A_i \cong \frac{-h_{fe}}{1 + jh_{fe}(f / f_T)} \quad ...(11)$$

Figure 13.16 shows the variation of the current gain of a short-circuit common-emitter configuration with frequency. The bold dotted lines indicate the asymptotic curves while the regular bold line shows the actual curve. The two asymptotes intersect each other at $f = f_\beta$ and the gain thereafter decreases at a rate of 20 dB/decade. The actual gain at $f = f_\beta$ is 3 dB down (or 0.707 times) the value of mid-band gain.

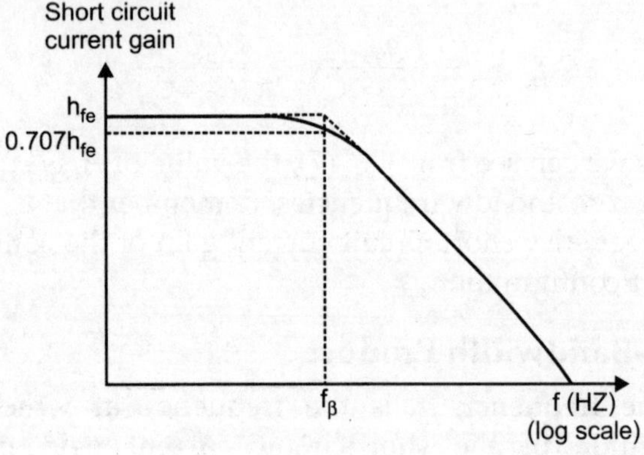

Fig. 13.16

Q. 16. What do you mean by α-cut-off frequency? Also, derive the expression for the α-cut-off frequency.

Ans. The α-cut-off frequency is the frequency at which the short-circuit current gain value in the common-base configuration drops by 3 dB to its value at low frequencies. It is represented as f_α. It may be mentioned here that the transistor used in common-base configuration has a much higher value of 3 dB frequency, as compared to the transistor used in common-emitter configuration. In other words, the value of f_α is much larger than the value of f_β.

The expression for the current gain in a common-base configuration is given by,

$$A_i = \frac{-h_{fb}}{1 + j(f / f_\alpha)} \quad ...(1)$$

where, $f\alpha$ is the α-cut-off frequency. The expression for the α-cut-off frequency is given by,

$$f_\alpha = \frac{1}{2\pi r_{b'e}(1 + h_{fb})C_e} \quad ...(2)$$

As the value of $h_{fe} \cong 1/(1 + h_{fb})$, Eq.(2) can be rewritten as

$$f_\alpha \cong \frac{h_{fe}}{2\pi r_{b'e} C_e} \quad ...(3)$$

Multiplying and dividing Eq. (3) by $(C_e + C_c)$ and substituting the value of f_β in the equation, we get,

$$f_\alpha \cong \frac{h_{fe} f_\beta (C_e + C_c)}{C_e} \quad ...(4)$$

The bandwidth offered by the common-base amplifier is much higher than that offered by a common-emitter amplifier, although the latter has much higher value of gain. Fig. 13.17 shows the comparison of the manner in which the short-circuit current gains for the common-emitter and common-base configurations vary with frequency.

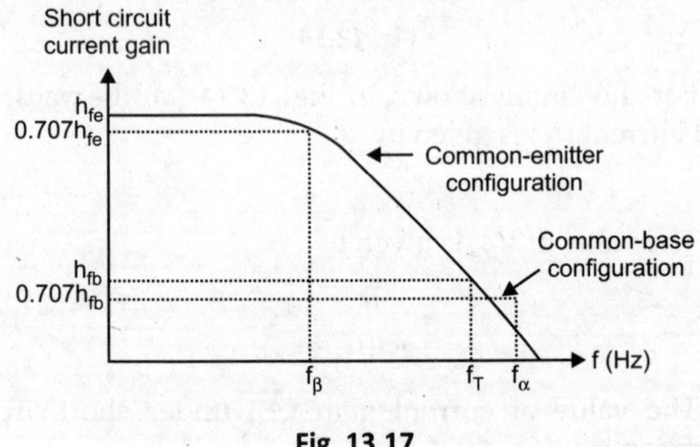

Fig. 13.17

Q. 17. What is the Miller theorem and where can it be useful? Derive the expression for input and output impedances using the theorem.

Ans. Consider the circuit shown in the Fig. 13.18 (*a*). As shown in the figure, it comprises three nodes, namely, input node 1, output node 2 and a ground node *G*. An impedance (*Z*) is connected between the input and the output nodes. This impedance is also referred to as the feedback impedance. This impedance has an effect on the functioning of the circuit. It is very difficult to analyse such a network, as the impedance affects the input and the output simultaneously. Miller's theorem helps to analyse such circuit configurations.

According to Miller's theorem, the circuit with feedback impedance can be replaced by an equivalent circuit such that the feedback impedance is split into two impedances: one between the input terminal and the ground (Z_{in}) and the other between the output terminal and the ground (Z_{out}). Figure 13.18 (*b*) shows the Miller's equivalent circuit of the network shown in Fig. 13.18 (*a*).

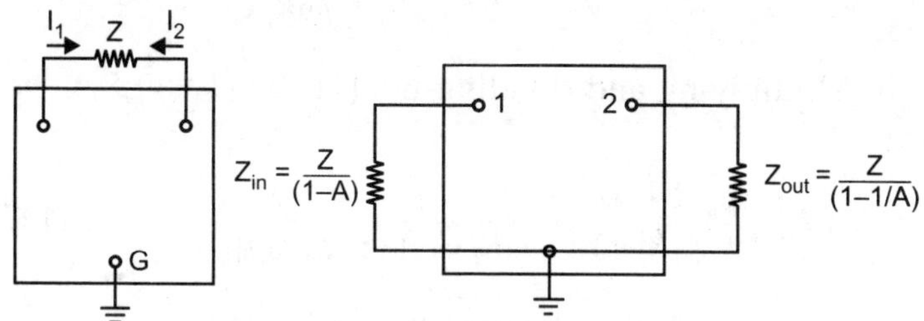

Fig. 13.18

Let the voltages at nodes 1 and 2 be V_1 and V_2, respectively. Let the ratio of V_2/V_1 be represented as A. Using Miller's theorem, the impedance connected between nodes 1 and 2 can be replaced by two impedances: one between node 1 and ground node *G* (Z_{in}) and the other between node 2 and ground node *G* (Z_{out}).

The value of current I_1 is given by

$$I_1 = \frac{V_1 - V_2}{Z} \quad ...(1)$$

The value of V_2 is given by $V_2 = AV_1$ Therefore current I_1 is given as,

$$I_1 = \frac{V_1 - AV_1}{Z} = \frac{V_1(1-A)}{Z} \quad ...(2)$$

From Eq. (2), the ratio of V_1/I_1 is given by,

$$\frac{V_1}{I_1} = \frac{Z}{1-A} \quad (3)$$

V_1/I_1 is the equivalent impedance of *Z*, as seen from the input side of the circuit. Therefore, Z_{in} is equal to the ratio V_1/I_1 and is given by,

$$Z_{in} = \frac{Z}{1-A} \quad ...(4)$$

The impedance Z_{in} appears in parallel with the input terminals of the network. The expression for output impedance Z_{out} can be derived as follows. The value of current I_2 is given by,

$$I_2 = \frac{V_2 - V_1}{Z} \quad ...(5)$$

Voltage V_1 is expressed in terms of voltage V_2 as

$$V_1 = \frac{V_2}{A}$$

Substituting this value of V_1 in Eq. (5) we get,

$$I_2 = \frac{V_2 - (V_2/A)}{Z} = \frac{V_2[1-(1/A)]}{Z} \quad ...(6)$$

The ratio V_2/I_2 is given by

$$\frac{V_2}{I_2} = \frac{Z}{(1-(1/A))} \quad ...(7)$$

The ratio V_2/I_2 is the equivalent impedance of *Z* as seen from the output terminals. It is therefore equal to Z_{out}. The expression for Z_{out} is given by

$$Z_{out} = \frac{Z}{1-(1/A)} \quad ...(8)$$

Q. 18. Draw a hybrid-π model for a common collector amplifier at high frequencies and derive the relevant equations using the Miller Theorem.

Ans. Figure 9.19 (*a*) shows the common-collector transistor amplifier. Capacitance C_L is included in parallel with the load resistor R_L, as the common-collector transistor due to its low output resistance is often used to drive capacitive loads. Figure 9.19(*b*) shows the hybrid-π equivalent model for the common-collector amplifier shown in Figure 9.19(*a*).

Applying Miller's theorem to the hybrid-π equivalent circuit of Figure 9.19(*b*), we get the equivalent circuit as shown in Figure 9.19(*c*). The parameter *K* is given by the ratio of voltages V_{ec} and $V_{b'c}$ (*i.e.*, $K = V_{ec}/V_{b'c}$). The input time constant t_{ic} is given by,

$$t_{ic} = \left[(R_s + r_{bb'}) \middle\| \left\{\frac{1}{g_{b'e}(1-K)}\right\}\right] [C_c + C_e(1-K)] \quad ...(1)$$

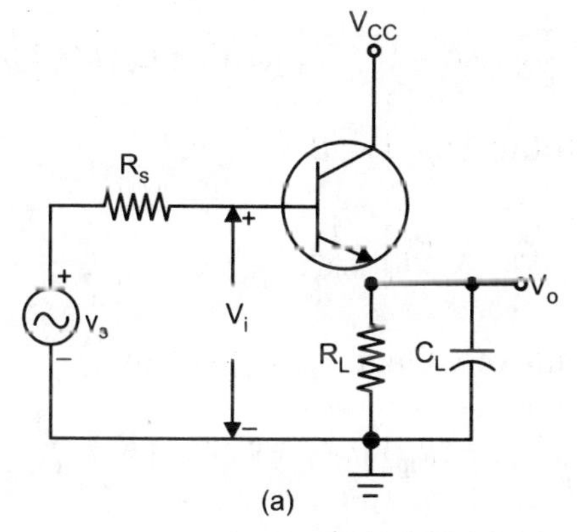

(a)

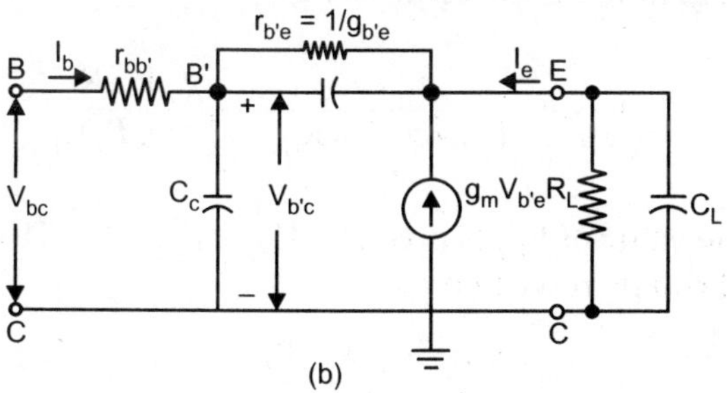

(b)

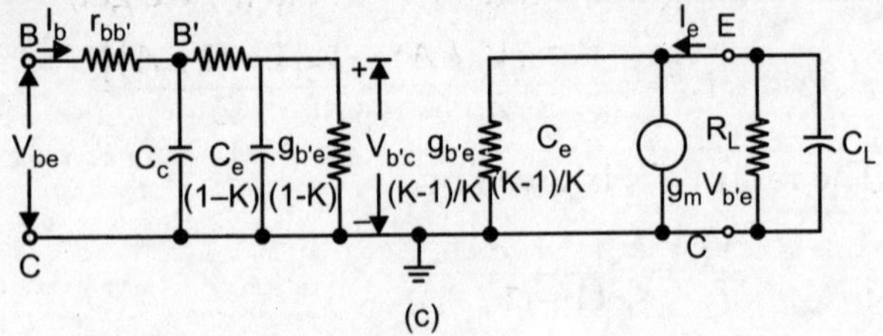

Fig. 13.19

As the low-frequency gain of the emitter–follower configuration is approximately equal to unity, therefore (1 – K) ≅ 0. Therefore, the expression for t_{ic} can be approximated by,

$$t_{ic} \cong (R_s + r_{bb'})C_c \quad ...(2)$$

The output time constant (t_{oc}) is given by,

$$t_{oc} = \left[R_L \left\| \left\{\frac{1}{g_{b'e}(K-1)/K}\right\}\right.\right] [C_L + C_e(K-1)/K] \quad ...(3)$$

As the value of (K – 1) ≅ 0, the above equation can be simplified as

$$t_{oc} \cong R_L C_L \quad ...(4)$$

As we have assumed that the output load is highly capacitive, therefore the value of $C_L >> C_c$. Hence,

$$R_L C_L >> (R_s + r_{bb'})C_c \quad ...(5)$$

This implies that the value of output time constant (t_{oc}) is much larger than the input time constant (t_{ic}). Hence, the upper 3 dB frequency is determined mostly by the output circuit alone.

The impedance of the output circuit (Z_o) is given by

$$Z_o = R_L \left\| \frac{1}{g_{b'e}(K-1)/K} \right\| \frac{1}{j\omega\{C_L + C_e(K-1)/K\}} \quad ...(6)$$

Substituting (K – 1) ≅ 0, we get,

$$Z_o \cong R_L \left\| \frac{1}{j\omega C_L} \right. \quad ...(7)$$

The value of voltage V_{ec} is given by

$$V_{ec} = g_m V_{b'e} Z_o \cong g_m V_{b'e}\left(R_L \left\| \frac{1}{j\omega C_L}\right.\right) \quad ...(8)$$

Re-arranging the terms we get,

$$V_{ec} = \frac{g_m V_{b'e}}{(1/R_L) + j\omega C_L} = \frac{g_m V_{b'e} R_L}{1 + j\omega R_L C_L} \quad ...(9)$$

The value of $V_{b'e}$ is given by $V_{b'e} = V_{b'c} - V_{ec}$. Therefore, Eq. (9) can be rewritten as,

$$V_{ec} = \frac{g_m (V_{b'c} - V_{ec}) R_L}{1 + j\omega R_L C_L}$$

The value of V_{ec} is given by

$$V_{ec} = \frac{g_m R_L V_{b'c}}{1 + g_m R_L + j\omega R_L C_L} \quad ...(10)$$

As $K = V_{ec}/V_{b'c}$, therefore the expression for K is given by

$$K = \frac{V_{ec}}{V_{b'c}} = \frac{g_m R_L}{1 + g_m R_L + j\omega R_L C_L} \quad ...(11)$$

Multiplying and dividing Eq. (11) by $(1 + g_m R_L)$, we get,

$$A = \left(\frac{g_m R_L}{1 + g_m R_L}\right) \frac{1}{1 + [j\omega R_L C_L/(1 + g_m R_L)]} \quad ...(12)$$

Equation (12) can be rewritten as Eq. (13)

$$A = \left(\frac{g_m R_L}{1 + g_m R_L}\right)\left(\frac{1}{1 + (jf/f_H)}\right) \quad ...(13)$$

where, $f_H = (1 + g_m R_L)/2\pi C_L R_L$. The value of f_H can be expressed as

$$f_H = \frac{1 + g_m R_L}{2\pi R_L C_L} \cong \frac{g_m}{2\pi C_L} \quad ...(14)$$

We already know that the value of unity gain bandwidth (f_T) is given by

$$f_T = \frac{g_m}{2\pi(C_e + C_c)} \quad ...(15)$$

As the value of C_e for a transistor is much larger than C_c, therefore f_T can be approximated by,

$$f_T \cong \frac{g_m}{2\pi C_e} \quad ...(16)$$

Substituting this value of f_T in Eq. (14), we get,

$$f_H \cong \frac{f_T C_e}{C_L} \quad ...(17)$$

Since the input impedance between terminals B' and C is much large as compared to $(R_s + r_{bb'})$, therefore K is approximately the overall voltage gain (A_{VS}), that is,

$$K \cong A_{VS} = \frac{V_{ec}}{V_s} \quad ...(18)$$

Q. 19. Compare the variations in the values of hybrid-π parameters with respect to collector current, collector-emitter voltage and temperature.

Ans. The variations in the values of hybrid-π parameters with change in collector current (I_C), collector–emitter voltage (V_{CE}) and temperature (T) are highlighted in Table 13.5.

Table 13.5 Variations in the Values of Hybrid-π Parameters

Parameter	$I_C\uparrow$	$V_{CE}\uparrow$	$T\uparrow$
g_m	Linear	Independent	Inverse
$r_{bb'}$	Decreases	Complex relation	Increases
$r_{b'e}$	Inverse	Increases	Increases
C_e	Linear	Decreases	Complex relation
C_c	Independent	Decreases	Increases

Q. 20. What do you understand by thermal instability of the operating point? What are its consequences? Define the overall stability factor for an amplifier.

Ans. When a particularly chosen operating point in an amplifier circuit shifts on its own due to change in temperature, the phenomenon is known as thermal instability. Shifting of operating point leads to distortions in the output waveform.

Let us take the case of common emitter amplifier to illustrate the phenomenon of shifting of operating point. Refer to Fig. 13.20 for the output characteristics, the DC load line and the location of operating point.

The collector current (I_c) for the CE-amplifier configuration can be written as,

$$I_C = \beta I_B + I_{CEO}$$

where, β = *DC* current gain

I_B = Base current

I_{CEO} = Reverse collector-emitter current with base open

Also, $I_{CEO} = (\beta + 1)\, I_{CBO}$

where, I_{CBO} = Reverse collector base current with emitter open.

This gives, $I_C = \beta I_B + (\beta + 1)\, I_{CBO}$ ...(1)

Equation (1) tells us that I_C is strongly dependent on β, I_B and I_{CBO}. And all the three parameters are observed to be strongly temperature dependent, particularly the I_{CBO}. In fact, I_{CBO} doubles for every 10°C rise in temperature for a silicon transistor. The effect becomes more severe as increase in I_{CBO} is getting multiplied by a large magnitude in (β + 1) to affect the collector current. Since (β) also increases with temperature, the overall effect sometimes becomes cumulative and leads to thermal runaway, if amplifier has not been designed to counter this thermal instability. The result is that the output characteristics shift upwards (as shown by dotted lines), when the temperature increases and the operating point which was chosen to be initially in the center may find itself somewhere near the saturation region of the characteristics.

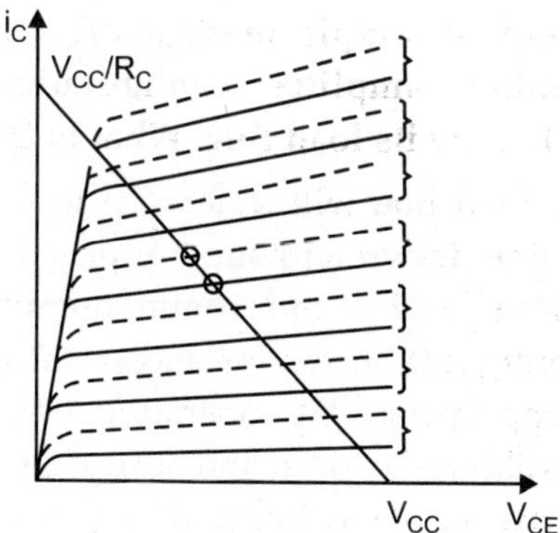

Fig. 13.20

The stability factor tells us about change in collector current due to changes in β, I_B (or V_{BE}) and I_{CBO} with temperature or even due to change of transistor (Transistor of even the same type number have large spread in their β values for instance). The overall stability factor is the combined effect of the stability factors resulting from all the three causes individually. It is expressed as,

$$\Delta I_C = S\,\Delta I_{CBO} + S'\,\Delta V_{BE} + S''\,\Delta\beta$$

where, ΔI_C = Total change in collector current

$$S = \frac{\Delta I_C}{\Delta I_{CBO}},\ \beta \text{ and } V_{BE} \text{ held constant}$$

$$S' = \frac{\Delta I_C}{\Delta V_{BE}},\ I_{CBO} \text{ and } \beta \text{ held constant}$$

$$S'' = \frac{\Delta I_C}{\Delta\beta},\ I_{CBO} \text{ and } V_{BE} \text{ held constant}$$

You would notice that smaller values for S, S' and S'' imply a smaller change in collector current due to variation in respective parameters. Thus, smaller the magnitude of stability factor, better the stability.

Q. 21. What do you think should be the design approach so that the amplifier does not have the thermal instability or the consequent thermal runaway problem?

Ans. If the amplifier circuit is so modified that the increase in $(\beta + 1)\, I_{CBO}$ term is compensated automatically by a decrease in (βI_B) term, the collector can be maintained at a constant value irrespective of the changes of β, I_B and I_{CBO}. Now, a reduction in the value of (β I_B) product can be brought out only by reducing I_B and that is what is precisely done. The circuit is modified in such a way that any tendency for the collector current to increase reduces

I_B, which in turn does not allow the collector current to increase. There are different types of bias stability schemes employed with common-emitter amplifier configuration. These are discussed in a subsequent question.

Q. 22. What is a load line with reference to amplifiers? With the help of simple mathematics and the basic common emitter amplifier configuration shown in Fig. 13.21 (*a*), draw its load line. What is Q-point?

Ans. The load line with reference to an amplifier is the straight line drawn and superimposed on its output characteristics (i_c *vs* V_{CE} in a common-emitter transistor amplifier configuration for instance) that enables us to choose an appropriate DC operating point (Operating point tells us about the input and output DC currents that flow in the absence of any input AC signal) also called the Q-point and also helps us in calculating output currents and output voltages for given input signal swing.

Figure 13.21 (*b*) shows the typical output characteristics for the basic common emitter amplifier circuit of Fig. 13.21 (*a*). From the circuit shown in (Fig. 13.21) (*a*),

$$V_{CC} = I_c R_c + V_{CE} \quad ...(1)$$

or
$$I_C = \left(-\frac{V_{CE}}{R_C} + \frac{V_{CC}}{R_C}\right) \quad ...(2)$$

This is the equation of *a* straight line of the type $y = mx + b$ with y, x, m and b representing I_C, V_{CE}, $(-1/R_C)$, and V_{CC}/R_C. This straight line called the load line has slope of $(-1/R_C)$ and a *y*-axis intercept of (V_{CC}/R_C).

The point of intersection of load line and the quiscent value of base current is called the Q-point. The choice of Q-point depends upon the desired characteristics. For instance, if you want the amplifier to operate in class-A, the Q-point is chosen somewhere around middle of output characteristics. Having chosen a proper operating point, the value of base resistance R_B to get the desired base current can be calculated from:

$$RB = \left[\frac{V_{CC} - V_{BE}}{I_B}\right]$$

(a) (b)

Fig. 13.21

Q. 23. Briefly describe and compare the operating point (Q-point) stabilization schemes commonly employed with CE-amplifier circuits, the basic configuration for which is shown in Fig. 13.22 (*a*).

Ans. The pre-dominant factor that leads to operating point instability is often considered as the stability factor (S) that tells us about change in collector current due to change in (I_{CBO}). Also, it is observed that an attempt to improve the stability factor (S) also improves upon the two stability factors (S') and (S''). That is, if you are able to modify your circuit in such a fashion that collector current becomes comparatively insensitive to changes in (I_{CBO}), it will automatically become insensitive to variations in (β) and (V_{BE}). The variations in (β) and V_{BE} could either be due to temperature changes or due to active device replacement.

Rewriting the expression for collector current for the CE-configuration, we have,

$$I_C = \beta I_B + (\beta + 1) I_{CBO}$$

or
$$\frac{dI_C}{dI_{CBO}} = S = (\beta + 1)$$

The stability factor is, thus very high for the basic configuration. The collector current change is (β + 1) times the change in I_{CBO}

Fig. 13.22 (*b*) shows the circuit of Fig. 13.22 (*a*) with a slight modification. In this circuit, as the collector current tends to increase, the quiscent collector voltage falls and so does the base current. As a result, the tendency of collector current to increase is checked. Similarly, you would notice that the tendency of the collector current to decrease will also be checked. So, the modification provides some kind of operating point stabilisation. This scheme goes by the name of collector-to-base bias. It has some shortcomings. Firstly, the degree of stabilisation achieved is not very high.

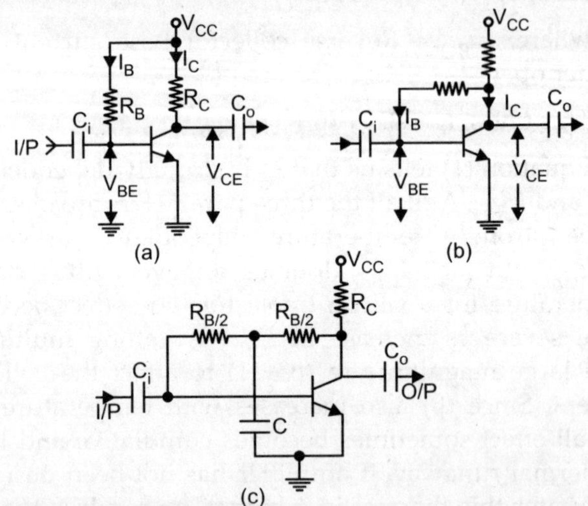

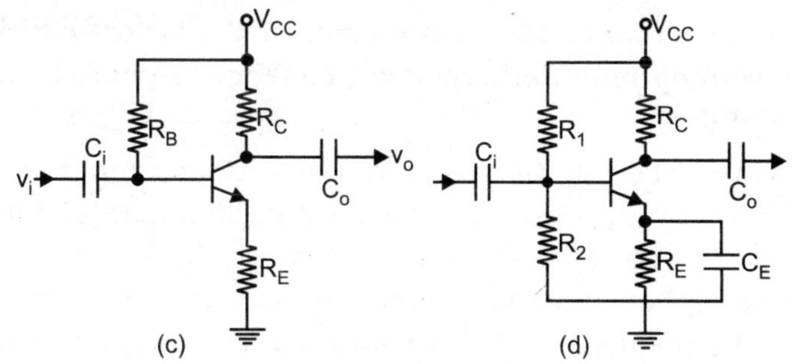

Fig. 13.22

Secondly, in order to get a certain base current, the values of R_B required in the collector-to-base bias circuit will be comparatively smaller. As a result, the base current becomes more sensitive to temperature changes. So, the compensating effect of base current when the collector current tends to increase is partly wasted due to base current itself becoming more sensitive. Thirdly, there will be a negative feedback from output to input through (R_B) resulting in degeneration. The last drawback can however be overcome by splitting (R_B) into two resistors as shown in Fig. 13.22 (*c*). The expression for stability factor (*S*) for the collector-to-base bias circuit is given by,

$$S = \frac{(\beta+1)}{1+\left(\frac{\beta R_C}{R_C+R_B}\right)} = \frac{(\beta+1)}{1+\left(\frac{\beta}{1+R_B/R_C}\right)}$$

In this expression, if $R_C << R_B$, $S \approx (\beta + 1)$. So, the scheme loses its stabilising capability in case the load resistance is very small. This is typically the scenario in case of transformer coupled amplifier stages where the load resistance is just the DC resistance of the primary winding.

All these problems are overcome to some extent in the circuit arrangement of Fig. 13.22 (*d*). This is known as emitter bias. When the collector current tends to increase, the DC voltage developed across the emitter resistor (R_E) reduces the base-emitter bias and thus the base current. The stability factor (*S*) in this case is given by,

$$S = \frac{(\beta+1)}{1-\left(\frac{\beta R_E}{R_E+R_B}\right)} = \frac{(\beta+1)}{1+\left(\frac{\beta}{1+R_B/R_E}\right)}$$

It is clear from the above expression that for this circuit to be effective R_B should be as small as possible and (R_C) should be as large as possible. It may not be possible to play with the value of (R_B), as it is more or less fixed as per base current requirement.

In the arrangement of Fig. 13.22 (*e*), the desired base current can be obtained with a much smaller base resistance and hence an improvement in the stability factor. This arrangment is known as voltage divider bias The effective base resistance in this case is $R_1 R_2/(R_1 + R_2)$.

The stability factor (*S*) is given,

$$S = \frac{\beta+1}{1+\left(\frac{\beta R_E}{R_E+R'_B}\right)}, \text{ where } R'_B = \frac{R_1 R_2}{R_1+R_2}$$

The emitter bypass capacitor has been used to prevent degeneration of AC signal.

Q. 24. Justify that the operating point in case of common-base and common collector (Emitter-follower) amplifier configurations is inherently stable.

Ans. In the case of common-base configuration (Fig. 13.23 (*a*) the expression for collector current is

$$I_C = \alpha I_E + I_{CBO}, \text{ where } \alpha = \frac{I_C}{I_E}$$

The stability factor, $S = dI_C/dI_{CBO} = 1$ which is the ideal value. This shows that the collector current (I_C) is insensitive to changes in (I_{CBO}).

The common collector configuration of Fig. 13.23 (*b*) is same as common-emitter configuration using voltage divider bias except that there is no bypass capacitor to prevent degeneration and the output is taken across emitter resistor. The stability characteristics of this configuration are as good as those of common-emitter configuration with voltage divider bias.

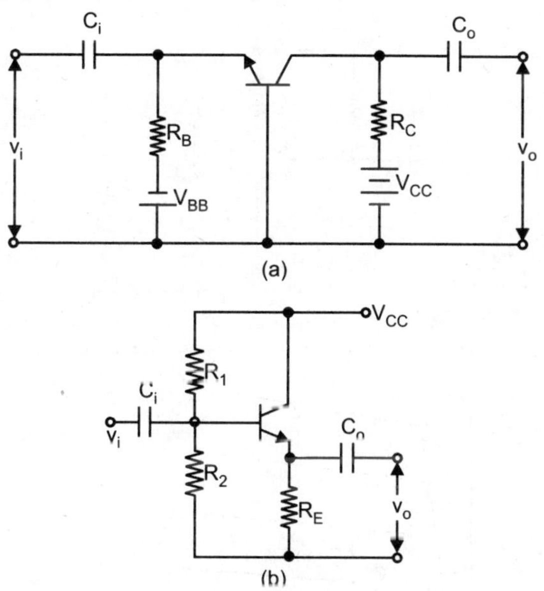

Fig. 13.23

Q. 25. What do you understand by AC load line ? How does it differ from the DC load line ?

Ans. As discussed earlier, the DC load line tells us about the quiscent values of the output voltage (V_{CE} in case of CE-amplifier, for instance) and the corresponding quiscent values of output current (I_C in case of CE-amplifier) that the amplifier can have. For a given input current value, the intersection of the DC load line and the output characteristic curve for that input current gives the values for output current and output voltage under DC conditions. As an illustration, refer to the output characteristics and the superimposed DC load line as shown in Fig. 13.24 (*b*) for the basic common emitter amplifier of Fig. 13.24 (*a*). This load line, as mentioned earlier, has a slope of $(-1/R_C)$.

Now if we connect a load resistance (R_L) across the output, the effective load resistance across the collector is now equal to a parallel combination of (R_C) and (R_L). Under AC conditions as capacitor (C_o) can be considered as a short at the frequency of operation. As a result, a new load line needs to be drawn on the output characteristics to be able to determine collector current and collector voltage for given input excitation. This line, called the AC load line, is drawn passing through the operating point and having a slope given by,

$$-\left[\frac{1}{\dfrac{R_L R_C}{R_L + R_C}}\right] = -\frac{(R_L + R_C)}{R_L R_C}$$

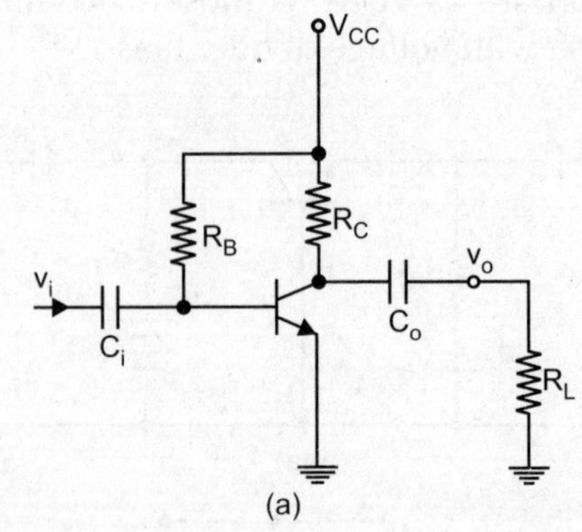

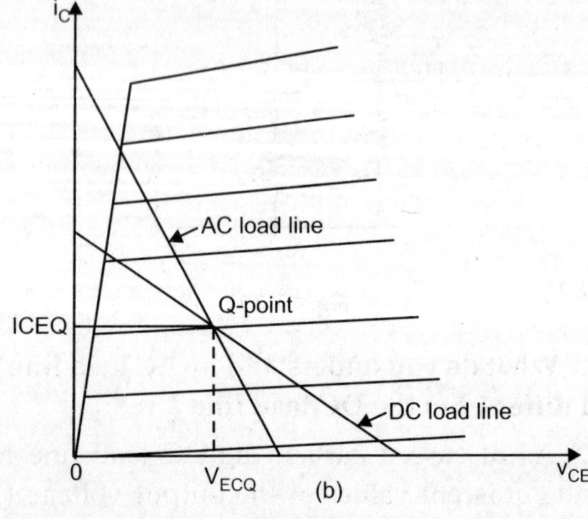

Fig. 13.24

Q. 26. What are the components or the factors which in your opinion limit the low frequency response of an amplifier?

Ans. The low frequency response of an amplifier is limited by presence of coupling capacitors and emitter bypass capacitors. These capacitors can be assumed to have negligible reactances only upto a certain frequency. As the frequency is further lowered, the amplifier gain drops, the reactances of coupling and emitter bypass capacitors increase with decrease in frequency. As a result, the signal coupled from the source to the input of the amplifier or from the output of one stage to the input of next (in case of cascaded arrangement) reduces (with increase of reactance of C_C) and also, there is an increasing degenerative feedback (due to increasing C_E reactance) which would not have been there in case C_W reactance was much smaller as compared to emitter resistance R_E with the result that the gain lowers.

Q. 27. Fig. 13.25 (*a*) shows a common emitter amplifier stage with the emitter resistance (R_E) bypassed by a capacitor (C_E). In the given amplifier circuit, how will you proceed to determine the lower 3 dB frequency $(f_L,$ in Fig. 13.25 (*b*)? You need not derive any mathematical expressions. You can make use of necessary mathematical results to arrive at the result.

Ans. If we assume that the size of the emitter bypass capacitor is large enough so as to have a negligible effect, on the low frequency response of the amplifier, then the lower 3 dB frequency considering the effect of coupling capacitor alone is given by,

$$f_L = \frac{1}{2\neq C_C (R_s + R)} \frac{1}{2\pi C_C (R_s + R)}$$

where, $R = R_i \,||\, R_1 \,||\, R_2 ......R_i$ is the input resistance of amplifier and is approximately equal to (h_{ie}).

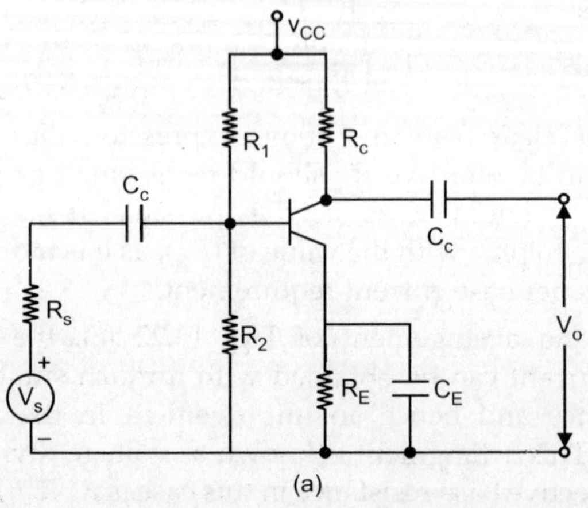

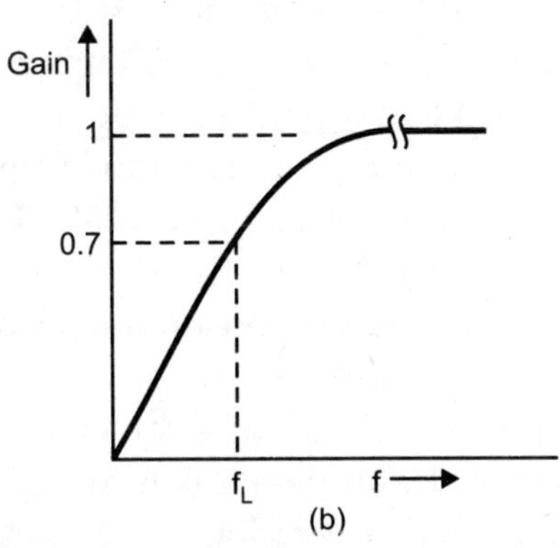

Fig. 13.25

Now if we ignore the effect of C_C assuming that its size is large enough so as to offer negligible reactance at the lowest frequency of operation, the lower 3 dB frequency due to the effect of C_e alone is given by,

$$f'_L = \left[\frac{(1+h_{fc})}{2\pi(2_s + h_{ie})C_E}\right]$$

After having determined (f'_L) and (f''_L), the lower 3 dB frequency due to the combined effects of both C_C and C_E is given by,

$$f_L = \frac{f_1}{\sqrt{2^{1/2}-1}}$$

where, (f_1) is the larger of the two frequencies (f'_L) and (f'_L)

Q. 28. What will be the lower 3 dB frequency, if two such stages of the type shown in Fig. 13.25 (*a*) were cascaded together?

Ans. We have seen in the last question that (f_L) is the over all lower 3 dB frequency of a single CE stage. The generalised expression for the lower 3 dB frequency when (n) identical stages are cascaded together is given by,

$$(f_{Ln}) = \frac{f_L}{\sqrt{2^{1/n}-1}} \quad \text{In our case, } n = 2$$

$$\text{Therefore, } (f_{Ln}) = \frac{f_L}{\sqrt{2^{1/2}-1}} \cong 1.55 f_L$$

In case various stages have unequal lower 3 dB frequencies, the value of (f_L) to be substituted in the above expression for calculating the overall lower 3 dB frequency is the one which is the largest of all.

Q. 29. Define lower 3 dB and upper 3 dB frequencies with reference to frequency response of an amplifier. These are also called half power points. Justify.

Ans. The lower 3 dB and upper 3 dB frequencies are the ones at which the voltage gain falls to 3 dB below its midband value. Expressed in other words, these are the frequencies at which gain falls to 70.7% of its midband value (Fig. 13.26). Since power is proportional to square of the voltage, when the voltage gain falls to 70.7% of midband value, and output power will fall to 50% of its midband value. That is why, 3 dB points are also referred to as half-power points. Mathematically also,

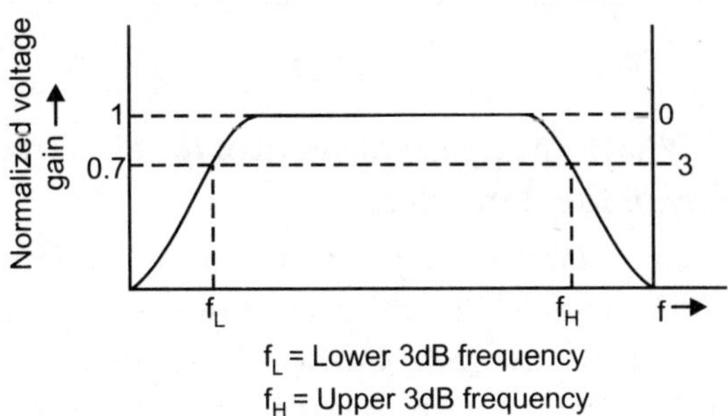

Fig. 13.26

Power gain = 10 log (P_o/P_i)

Let us assume (P_o/P_i) = 10 at midband

The power gain (in decibels) = 10 log 10 = 10 dB

Now when the power gain is 3 dB down *i.e.* 7 dB

Then, 10 log (P_o/P_i) = 7 *i.e.* (P_o/P_i) = Antilog 0.7 = 5

Thus, 3 dB reduction in power gain means power gain reduction to half of its original value. It may also be mentioned that 3 dB reduction in voltage gain implies that the voltage gain reduces to 70.7% of its original value.

Q. 30. How will an amplifier with a lower 3 dB frequency of (f_L) behave with a square-wave input of frequency (f)?

Ans. The flat top will get distorted. There will be an apparent sag in the flat portion of the input waveform. The sag expressed as percentage is given by,

$$\text{Percentage sag} = \left(\frac{\pi f_L}{f}\right) \times 100$$

Q. 31. What are primary factors responsible for the high frequency roll-off of a transistorised amplifier?

Ans. Unlike the low frequency response of a transistor amplifier, the high frequency response is primarily governed by the transistor's inherent characteristics. The high frequency response gets limited due to the capacitive effects of the transistor junctions. Therefore, the effects of these capacitances must be considered while studying the high frequency characteristics. These capacitances are represented by the hybrid π-model as depicted in Fig. 13.27.

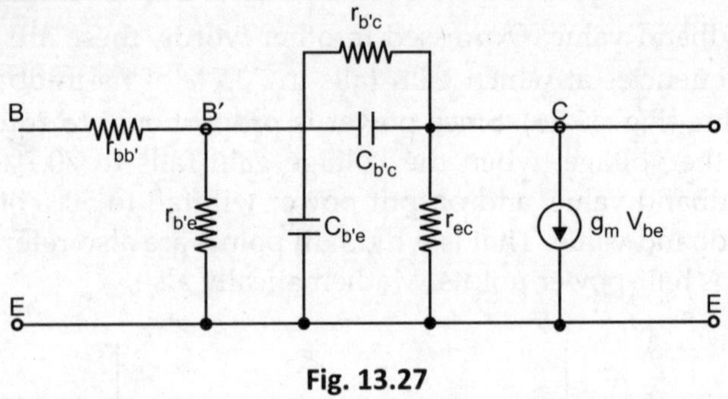

Fig. 13.27

Q. 32. What is the figure-of-merit of the high frequency capability of the transistor?

Ans. Gain-Bandwidth product is the figure of merit for the high frequency capability of transistors. Designated as (f_T), it is the unity-gain bandwidth product *i.e.*, it is the frequency at which the short circuit current gain falls to 0 dB or unity.

Q. 33. In transformer coupled multistage CE amplifier, on what factors do the low frequency and high frequency responses depend?

Ans. Low frequency response is limited by transformer's primary inductance, whereas the high frequency response is limited by the transformer's leakage inductance and stray capacitance.

Q. 34. How does the upper 3 dB frequency of a cascaded arrangement of (*n*) identical stages, each having an upper 3 dB frequency (f_H), depend upon (f_H)?

Ans. The overall upper 3 dB frequency is lower than the upper 3 dB frequency of a single stage and is given by,

$$f_{Hn} = f_H \sqrt{2^{1/n} - 1}$$

V_i Amp. (1) Amp. (2) Amp. (3) Amp. (N) V_o

Fig. 13.28

Fig. 13.28 shows the cascaded arrangement of (*n*) identical CE stages.

Q. 35. Briefly explain what do you understand by classes of operation of amplifiers? What would you like to comment on the active device dissipation and the amplifier's conversion efficiency (η) in each of these classes of operation?

Ans. The categorisation of amplifiers as class-A, class-B or class-C depends upon the time period for which current flows through the active device with reference to input signal period. If the current flows through the active device for full 360° of the input signal period, the amplifier is said to operate in class-A. If it flows for just one half of the input signal period, the amplifier is said to operate in class-B and if the current flows for a period less than one half of input signal period (typically 120° to 150°), the operation is termed as class-C. The portion of the input signal period for which current can flow through the active device depends upon amplifier's biasing which would decide the location of the input signal waveform vis-a-vis output versus input characteristics (Transfer characteristics) of the amplifier. Figure. 13.29 (*a*), (*b*) and (*c*) respectively depict operations in class-A, class-B and class-C.

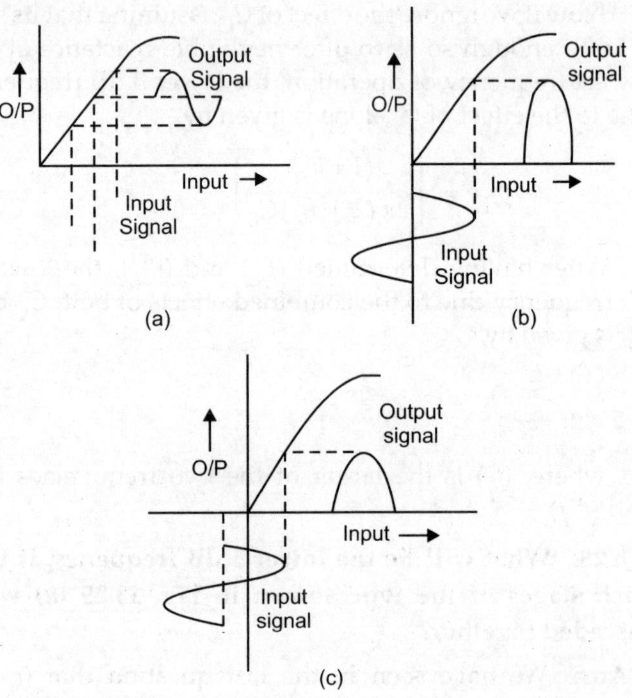

Fig. 13.29

Quite obviously, the active device dissipation will be the maximum in case of class-A and the minimum in case of class-C and for class-B, it is greater than that for class-C and less than that for class-A. An obvious inference from this is that a class-A amplifier has lowest conversion efficiency, class-B has comparatively higher efficiency and class-C has the highest efficiency of all the three classes of operation.

Q. 36. If class-C operation gives highest possible efficiency, why should amplifiers be operated at all in class-A or class-B? Explain.

Ans. In class-A operation, output current flows for the entire period of input signal cycle, and if the extremities of the input signal are confined within the linear region of the output versus input characteristics of interest, the output is going to be a true replica of the input which interpreted in another way implies least distortion. This is an essential requirement of all those amplifiers handling analog signals other than pure sinusoids. Audio and video signals are typical examples. Loss of signal intelligence or any distortion having crept into these signals is likely to cause severe impairment of the quality of sound or picture reproduced from received audio and video signals respectively.

In class-B operation (which has improved efficiency), the linear operation of class-A can be obtained by using two matched transistors in push-pull, each of them operating for half of the input signal period. The two halves are recombined in a transformer to reconstruct the amplified and undistorted version of input. Class-B operation, thus due to low distortion and comparatively higher efficiency has replaced class-A in quite a few places *e.g.*, large signal audio amplifiers.

Class-C amplifiers find application where a pure sinusoid is to be amplified. The output which is in the form of current pulses at a frequency equal to that of input sinusoidal signal gets amplified and appears in the form of amplified sinusoidal signal, if the load is a parallel tuned circuit resonant at that frequency.

Concludingly, we can say that each class of operation has its own area of applications. You should not expect a class-A amplifier in the carrier signal amplification stages of an AM transmitter and also you should not dare use a class-C amplifier in the audio amplifier stage of a radio receiver.

Table 13.6

Class of operation	*Current flows through active device for a period which is*	*Conversion Efficiency* (η)	*Distortion*	*Typical applications*
Class-A	Full 360° of the input signal period	Least	Least	Where intelligence is to be amplified; like audio and video amplifiers for example
Class-B	180° of the input signal period. Two transistors are needed at the output to reconstruct signal	Intermediate	Intermediate	Has replaced class-A in most of the applications
Class-C	Less than 180° of the input signal period. Load is a resonant circuit.	Maximum	Maximum	Where pure sinusoidal signals are to be amplified.

There is another class of operation known as class AB. In this, the active device biasing is such that output current flows for a period of input signal cycle which is more than 180° (It is 180° for class-B) and less than 360° (It is 360° in case of class-A).

Table 13.6 summarises the characteristics of three major classes of operation.

Q. 37. How does a transformer coupled class-A power amplifier differ from a direct coupled class-A power amplifier? Compare the two as regards (*i*) power dissipated in the transistor (*ii*) conversion efficiency and (*iii*) drain on the power supply.

Ans. The basic direct coupled class-A amplifier is shown in Fig. 13.30 (*a*). As it is a class-A amplifier, the operating point is chosen to be in the middle of the amplifier's output characteristics *i.e.* the output characteristics of the chosen transistor. The load line has a slope given by (– $1/R_L$). The centrally located operating point ensures a large voltage as well as a current swing without stretching the active device either towards cut-off or towards saturation (Fig. 13.30 (*b*).

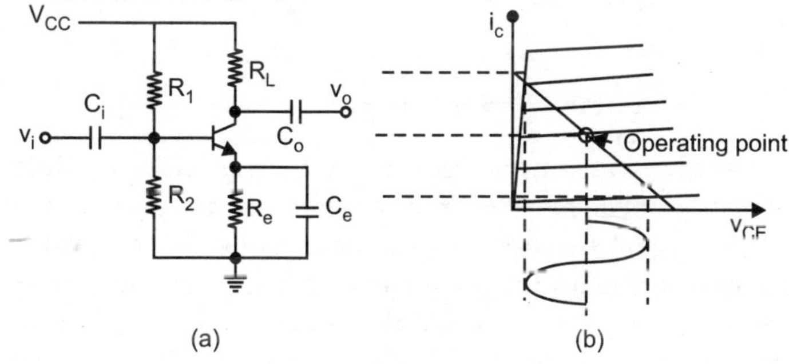

Fig. 13.30

During no signal present condition, 50 percent of the drain on the battery is dissipated in the transistor and the rest of 50 percent in the load (R_L). When the AC signal is present, 50 percent of the power being dissipated in the transistor gets converted to the useful AC power delivered to the load. That is, conversion efficiency of this type of amplifier circuit has a maximum value of 25 percent (Conversion efficiency denoted by 'η' is the ratio of AC power delivered to the load to the DC power drawn from the supply expressed as a percentage).

To sum up, for every watt drawn from the power supply, 500 mW each gets dissipated in the transistor and the load resistor under no signal conditions. Under AC signal amplification conditions, for every watt drawn from power supply, 500 mW are wasted in the load resistor (R_L), 250 mW are dissipated in the transistor and the remaining 250 mW appear in the load as useful AC power. In other words, if the AC power delivered to the load is (P_L), the maximum power dissipated in the transistor would be ($2P_L$) under quiscent conditions and drain on the power supply is always four times the useful AC power delivered to the load.

Figure. 13.31 (*a*) shows the transformer coupled load class-A amplifier. The AC load line in this case has a slope of $\left(\frac{-1}{R'_L}\right)$, where ($R'_L$) is ($R_L$) reflected onto the primary side of the coupling transformer. If the transformer has turns ratio of '*n*' (= N_P/N_S), then the load line slope is ($1/n^2R_L$). You can visualise that the DC load line in this case is almost vertical due to near zero DC resistance of primary winding and a small value of (R_c) if present (Fig. 13.31(*b*)

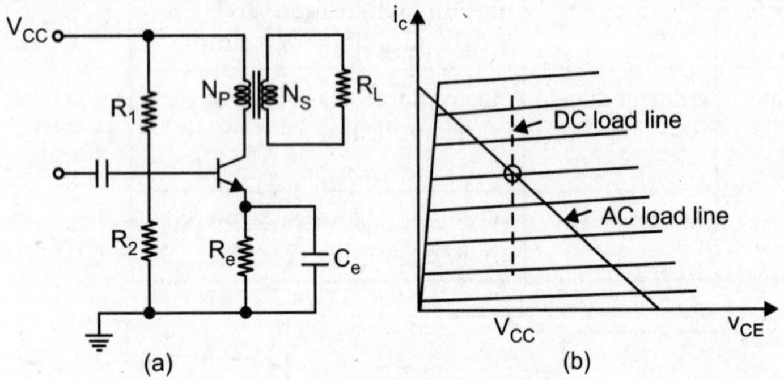

Fig. 13.31

As the transformer primary winding has very little DC resistance, no power is wasted in this part of the circuit. Again under no signal conditions, the transistor dissipates maximum power and this equals the power drawn from the supply. Under signal conditions, half of the power dissipated in the transistor is transferred to the load as useful AC power.

For a given transistor dissipation capability (P_T), the maximum useful AC power delivered to the load is ($P_T/2$) irrespective of whether the load is direct coupled or transformer coupled. Transformer coupled configuration has an advantage of much lower supply drain, one half of what it is in case of direct coupled configuration.

Q. 38. The single ended transformer coupled amplifier of the previous question has some distinct shortcomings. What are these and how are these overcome in class-A push-pull amplifiers?

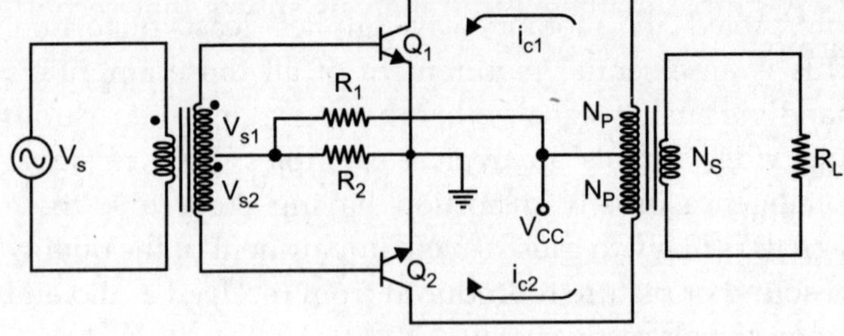

Fig. 13.32

Ans. The single ended transformer coupled amplifier has an objectionably high harmonic distortion with second harmonic causing the most of it. The harmonic distortion has also been found to vary with load resistance. The load resistance also partly determines the output power. Unfortunately, the same load resistance value does not optimise the both. (R_L) is often chosen to minimise distortion with the consequent reduction in the output power. Another drawback is the problems arising out of transformer saturation due to a large amount of DC current flowing in the transformer primary.

These problems are sorted out, to a great extent in a class-A push-pull amplifier (Fig. 13.32). It is class-A, as the current flows for full 360° of input cycle through the active device output. The circuit makes use of two phase splitting transformers; an input transformer that provides 180° out of phase signals to the two transistors' bases to ensure push-pull operation and an output transformer with a center tapped primary to combine the two halves of the amplified signal. Resistors (R_1), (R_2) provide the necessary forward bias for Q_1 Q_2 to ensure class-A operation.

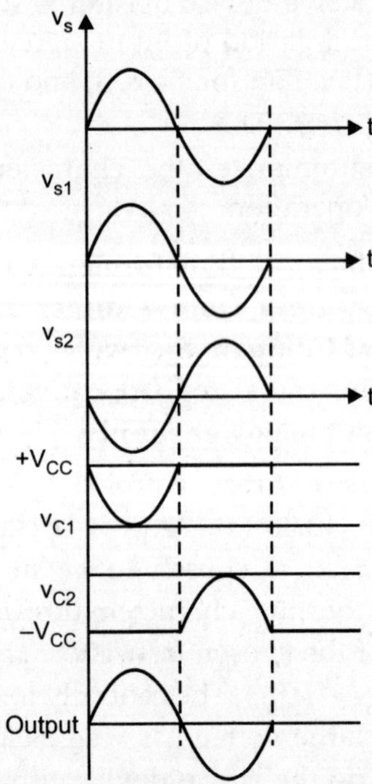

Fig. 13.33

A simple mathematical analysis shows that second harmonic gets eliminated in a push-pull amplifier. (In fact, all even harmonics get eliminated). This also allows us to choose the load resistance (R_L) as per maximum output power criterion. Transformer saturation problems are also eliminated as there is no DC component of current flowing through primary. The relevant waveforms are shown in Fig. 13.33.

Q. 39. Briefly describe the operation of a class-B push-pull amplifier. In what respects, is this amplifier configuration better than a class-A push-pull amplifier? What is the major drawback of class-B operation and how is this remedied?

Ans. The class-B push-pull amplifier configuration resembles a class-A push-pull amplifier circuit except that the former does not have any biasing resistors. Refer to Fig. 13.34 (*a*) for the basic circuit schematic. Both the transistors are biased at cut-off and as a result, transistors Q_1 and Q_2 conduct alternately. In the circuit shown, Q_1 conducts only during positive half cycles and transistor Q_2 conducts only during negative half cycles. The two halves are combined in the output transformer to deliver a full cycle to the load. As the transistors require some minimum base-emitter bias voltage to come into conduction (0.6 V in silicon and 0.2 V in germanium transistors), the class-B push-pull output waveform shows distortion near the cross-over point (Fig. 13.34 (*b*)) and this distortion goes by the name of cross-over distortion. This is the major drawback of a class-B push-pull amplifier.

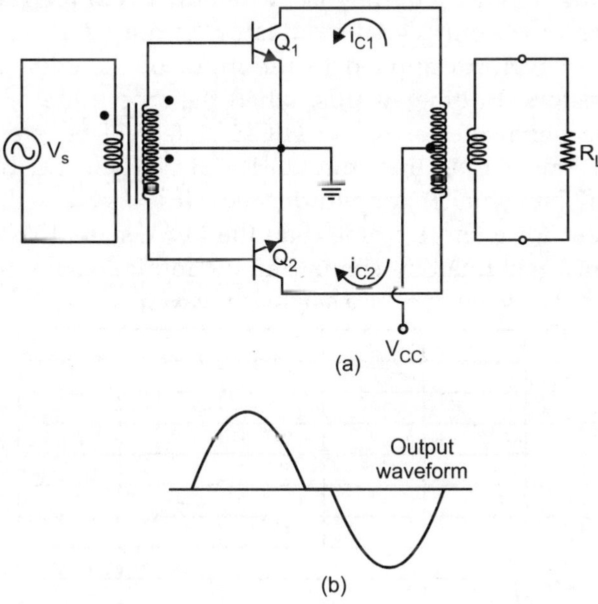

Fig. 13.34

As compared to a class-A push-pull amplifier, a class-B amplifier has a much higher conversion efficiency approaching 78.5 percent in a pure class-B amplifier resulting from the fact that the transistors do not dissipate any power in the standby.

If a class-B amplifier is slightly forward biased so that the collector current continues to flow, when the input signal approaches zero, the cross-over distortion can be eliminated (Fig. 13.35). As the transistors are not fully biased into conduction, this operation is somewhere between class-A and class-B. That is why, it is termed as class-AB. The maximum conversion efficiency of a class-AB amplifier is between 50 percent (maximum figure for class-A amplifier) and 78.5 percent (maximum figure for a class-B amplifier).

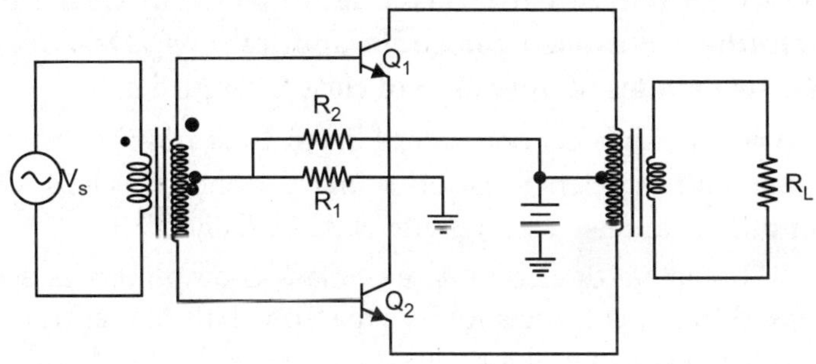

Fig. 13.35

Both class-B and class-AB push-pull amplifiers require that the push-pull transistors have identical input and output characteristics. Mismatch in the characteristics can lead to distortion.

In case the transistors are not fully matched, small resistors in series in the emitter leads of the two transistors can compensate for the mismatch. Though this reduction in distortion is at the cost of some reduction in the output power caused by degenerative feedback, yet it is worthwhile doing so.

Q. 40. Explain the configuration of a class AB amplifier. How are they able to solve the crossover distortion?

Ans. Class AB amplifiers do not suffer from the problem of crossover distortion, as in these amplifiers a small current flows even at zero input signal level. Figure 13.36 shows the configuration of a class AB push–pull amplifier. The circuit is essentially the same as that for class A amplifier. The voltage drop across resistor R_2 is approximately equal to the cut-in voltage of the transistor. When an AC signal is applied to the base, the collector current starts flowing immediately. This biases the transistor away from class *B* slightly towards class A. Resistors R_e are referred to as the emitter-stabilising resistors and reduce the amount of distortion produced. However, they cause a decrease in the output power due to the negative feedback effect.

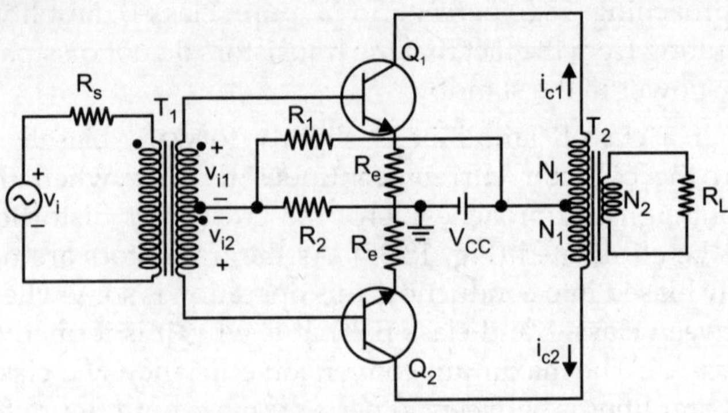

Fig. 13.36 Class AB amplifier

Q. 41. Briefly describe class C amplifiers. Why are these amplifiers not used for audio applications? Describe the two modes of operation of class C amplifier.

Ans. In applications where linearity is not the main issue and efficiency is of primary concern, class *C* amplifiers are used along with class *D*, *E* and *F*.

The active device in case of class *C* amplifiers is so biased that it conducts for less than one half of the input cycle, that is, it operates for less than 180° of the input cycle. Only a small portion of the input cycle is passed through the amplifier. This distorts the input signal and hence, class *C* amplifiers are not used for audio applications. They are widely used as *RF* and IF amplifiers. They offer very high efficiencies of the order of 90%.

Class C amplifiers operate in two modes, namely, the untuned mode and the tuned mode. However, mostly they are operated in the tuned mode. Figure 13.37 shows a tuned class C amplifier configuration. The tuned circuit provides a full cycle of the output signal at the fundamental or the resonant frequency of the tuned circuit. Hence, the unwanted frequencies are drastically suppressed. Other residual harmonics can be removed by using a filter. Class C amplifiers are generally used in RF transmitters and are primarily not intended for use as large signal or power amplifiers.

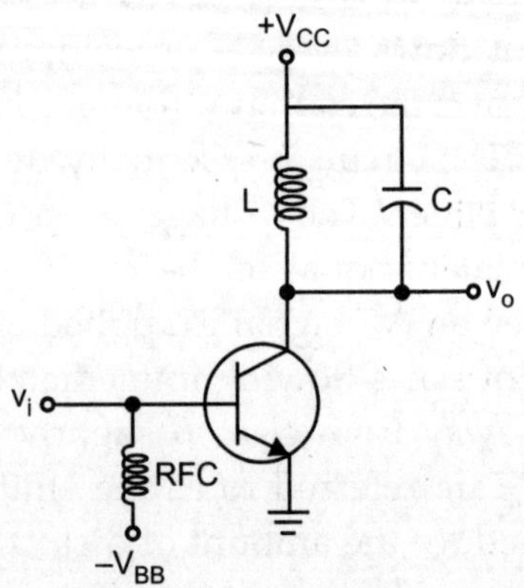

Fig. 13.37 Tuned class C amplifier

Q. 42. What are some of the advantages associated with class *D* amplifiers? Draw a graph to compare the performance of class *D* amplifier with a typical class AB amplifier and explain briefly, the operational principle of class *D* amplifiers.

Ans. Class *D* amplifiers operate with digital or pulsed signals. Class *D* amplifiers offer very high efficiency levels of the order of 90%; hence, they are quite desirable in power amplifiers and audio amplifiers. Figure 13.38 shows the typical efficiency versus output power curves for class AB and class *D* amplifiers. The theoretical maximum efficiency of class *D* amplifiers is 100% and over 90% is achievable in practice. High efficiency results in less power consumption for a given output power and reduces the heat-sink requirements drastically. Also, the loading on the power transformer is reduced by a substantial amount, resulting in the use of smaller transformers for the same output power.

The basic concept used in class *D* amplifiers is to convert the input signal into a pulsed waveform. In other words, the output transistors are operated as switches, that is, they operate either in the cut-off region (open switch) or in the saturation region (closed switch). When the transistor is in the cut-off region, the current through it is ideally zero and when it is in the saturation region, the voltage across it is very small, ideally zero. The power dissipation in both the cases is very low.

These amplifiers use PWM, pulse density modulation or sigma delta modulation to convert the input signal into a string of pulses. In the case, when the PWM technique is used, the input signal modulates the duty cycle of the drive waveform applied to the input of the switching transistors. In other words, when the amplitude of the input signal increases, the HIGH state will be present for a longer time than the LOW state. When the input signal has smaller amplitude, the HIGH state will be present for a shorter time than the LOW state. PWM is usually generated by comparing the input signal with a triangular waveform or a sawtooth waveform.

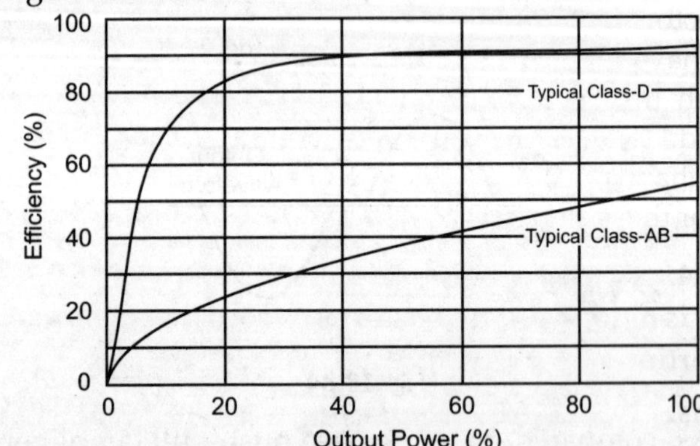

Fig. 13.38 Efficiency versus output power curves for class AB and class D amplifiers

Q. 43. What is the limiting factor in the power-handling capability of a transistor and how can it be enhanced?

Ans. Power transistors are used as active devices in power amplifiers. The maximum power that can be handled by an active device is related to its junction temperature, as the power dissipated by the device causes an increase in its junction temperature. The average power dissipated by the transistor can be approximated by the following equation:

$$P_D = V_{CE} I_C \qquad ...(1)$$

This power dissipation is allowed upto a certain temperature, above which the value of power dissipation capability decreases linearly with increase in temperature such that it reduces to zero at the maximum device case temperature. Larger the power handled by the transistor, higher is the case temperature. It may be mentioned here that the Silicon transistors offer larger maximum temperature ratings than Germanium transistors. Figure 13.39 shows the typical power derating curve for a Silicon power transistor.

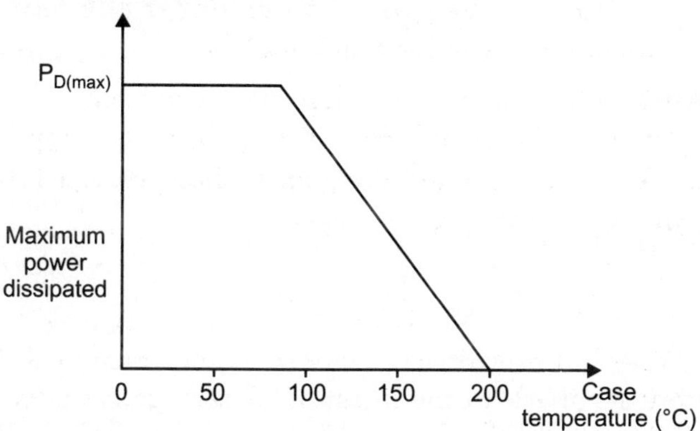

Fig. 13.39 Typical power derating curve for a silicon power transistor

Therefore, the limiting factor in the power-handling capability of a transistor is its maximum permissible collector junction temperature. Power transistors generally have large metal cases to provide a large area from which the heat generated by the device may be transferred. The power-handling capability of the device can be enhanced either by the use of heat sinks or by making use of thermoelectric coolers.

Q. 44. What is a non-linear distortion in an amplifier? Why is non-linear distortion also referred to as harmonic distortion?

Ans. Non-linear distortion causes amplitude distortion of the amplified signal. Due to non-linearity of the active device's transfer characteristics, the output not only contains the fundamental component but also its harmonics which are the root cause of amplitude distortion. That is why, non-linear distortion is also referred to as harmonic distortion.

Q. 45. Why do we talk about harmonic distortion in case of large-signal amplifiers only?

Ans. It is so because when the input signal amplitude is large, it is going to cover a large portion of active device's transfer characteristics which cannot be considered to be absolutely linear over the entire range. On the other hand, even if that characteristics have some non-linearity, and if the input signal amplitude is very small, one can safely assume that the region of characteristics being covered is linear. See Fig. 13.40 (*a*) and (*b*) for reference.

Fig. 13.40 (*a*) show the conditions of large signal input. The output in this case has undergone severe amplitude distortion. Figure. 13.40 (*b*) shows the case of small-signal input. In this case, the output is amplified version of input and has negligible amplitude distortion.

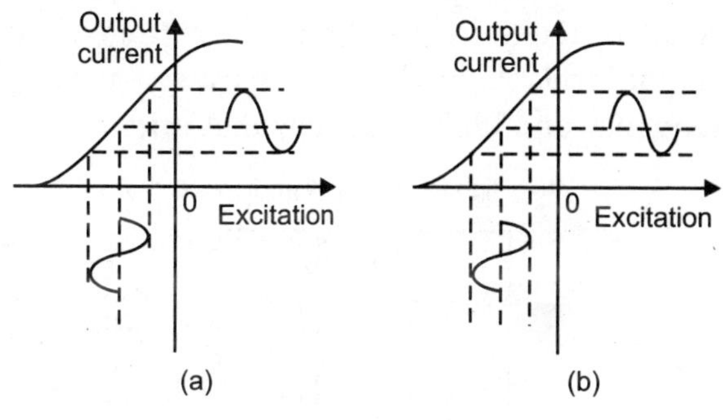

Fig. 13.40

Q. 46. What is intermodulation distortion. Why is it highly objectionable in case of audio amplifiers? What is transient intermodulation distortion?

Ans. When an amplifier has an input signal having more than one frequency component, the output signal in addition to containing the amplified versions of the frequency components of the input signal may also have sum and difference components. For example, if the input contains say (f_1) and (f_2), the output will contain (in the presence of intermodulation distortion) (f_1), (f_2), ($f_1 + f_2$) and ($f_1 - f_2$). It is highly objectionable in case of audio amplifiers as an-eudio signal is a complex signal composed of a number of frequency components. It is also caused due to non-linearity in the active device's transfer characteristics and thus, will be more troublesome in case of large signal amplifiers.

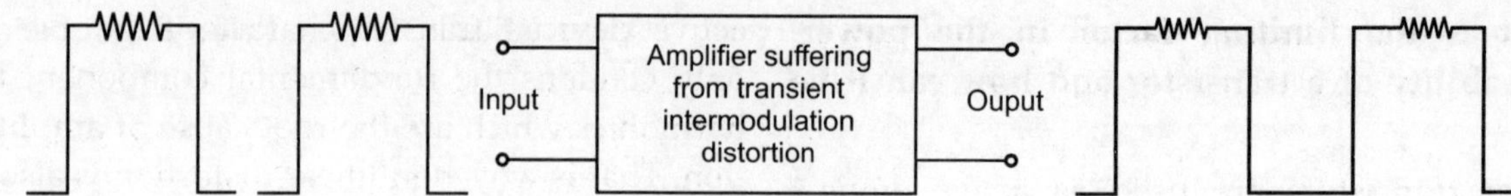

Fig. 13.41

When the input to the amplifier changes suddenly and if the amplifier cannot respond quickly, transient intermodulation distortion is said to be present. It can be checked with the help of a test signal comprising of a square-wave with a superimposed sine-wave of a much higher frequency (as compared to the frequency of the square-wave). In the output signal, high frequency components will be observed to be missing over a part of the square-wave, if transient intermodulation is present. Fig. 13.41 illustrates the point.

Q. 47. What do you understand by:

(i) Frequency distortion (*ii*) Phase distortion?

Ans. Frequency distortion is the result of tapering of the amplifier's frequency response on both the extremities. As a result, different frequency components get amplified to different levels. Figure 13.42 illustrates an example of low frequency and high frequency distortion.

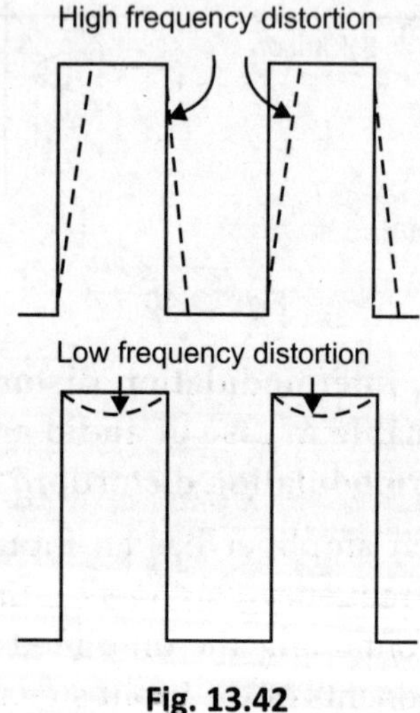

Fig. 13.42

Phase distortion is due to different frequency components undergoing different phase shifts in the process of amplification. For example, in a common emitter amplifier, we say that output is 180° out of phase with the input whereas in practice, this phase-difference may vary with frequency over as large arrange as 90° to 270°.

Q. 48. Why is phase distortion not objectionable in audio amplifiers? On the other hand, it is totally undesirable in the video amplifier of a television receiver. Comment.

Ans. Since ears are not sensitive to phase shifts, phase distortion is not objectionable in audio amplifiers. In video amplifiers, if the phase shift is proportional to frequency, the result will be a constant delay and hence, there is no visible distortion in the reproduced picture. If the phase shift varies with frequency in any other manner, different frequency components in the video signal undergo different delays and as a result, reproduced picture gets smeared.

Q. 49. What are the acceptable levels of total harmonic distortion and intermodulation distortion in a high quality audio amplifier?

Ans. Harmonic distortion is acceptable upto 0.7% to 1%. Intermodulation distortion is acceptable upto 2%.

Q. 50. How is the gain of an amplifier affected by introduction of negative feedback?

Ans. Introduction of negative feedback in an amplifier decreases its gain. The gain of the amplifier with feedback (A_{vf}) and the gain without feedback (A_v) are mathematically expressed as,

$$A_{vf} = \frac{A_v}{(1-\beta A_v)}$$

where, β = feedback factor. It is the fraction of the output feedback to the input. (β) has negative sign for negative feedback. As an illustration, if there is a 10% negative feedback, $\beta = -0.1$ and $A_{vf} = \frac{A_v}{1+0.1\,A_v}$. If A_v was 100, A_{vf} will be approximately equal to 9. Fig. 13.43 shows block schematic representation of a negative feedback amplifier.

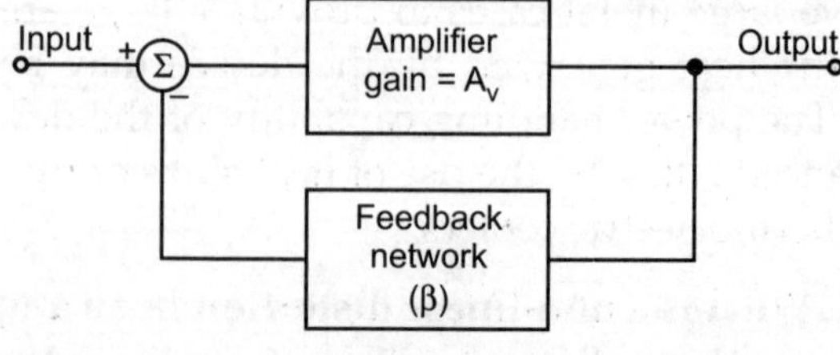

Fig. 13.43

Q. 51. If the negative feedback was to reduce the gain of the amplifier, what is the necessity of having negative feedback at all?

Ans. Introduction of negative feedback stabilises the amplifier gain against variations in transistor parameters like β, ageing effects and replacement of one transistor by another identical one which may not be truely identical. This can be verified mathematically also.

$$A_{vf} = \frac{A_v}{1-\beta A_v}, \text{ if } \beta A >> 1 \text{ then } A_{vf} \cong \frac{1}{\beta}$$

which implies A_{vf} is independent of amplifier parameters and is dependent on feedback network only. As long as feedback network is stable which is true as feedback network usually comprises of passive elements only, the amplifier gain will remain stable.

As an illustration,

Consider A_v= 100; β = 0.1 (10% feedback)

Then $A_{vf} = \frac{100}{1+0.1\times 100} = 9.1$

Let assume (A_v) changes to 150 due to changes in one or more of the amplifier parameters.

In that case. $A_{vf} = \frac{150}{1+0.1\times 150} = 9.37$

Thus, a 50% change in A_v (from 100 to 150) causes only a 3% change in A_{vf} (from 9.1 to 9.37).

In addition, there are other advantages with negative feedback *e.g.*, enhanced bandwidth, reduced noise distortion etc. Not only this, with proper choice of feedback network, one can have desired input and output impedances of the amplifier to suit a particular requirement.

Q. 52. "A fixed percentage of negative feedback reduces the gain by fixed percentage". Is this statement valid?

Ans. Gain and the amount of feedback have a nonlinear relationship and therefore a fixed percentage of feedback does not reduce the gain by fixed percentage, if the gain without feedback is different in the two cases.

As an illustration, let us assume that gain without feedback is 100. A 10% negative feedback reduces the gain to approximately 9. This is a reduction of about 91%. Now if the gain without feedback was 50,10% feedback would reduce the gain to 8.33 which is a reduction of about 83%.

Q. 53. What is the effect of negative feedback on the input resistance of amplifier?

Ans. The input resistance of the amplifier increases, when the negative feedback is introduced in series with the input and it decreases, when the feedback is introduced in shunt with the input.

Mathematically, $R_{if} = R_i(1-\beta A)$ for series feedback

$$R_{if} = \frac{R_i}{(1-\beta A)} \text{ .... for shunt feedback}$$

R_{if}= Input resistance of amplifier with feedback

R_i = Input resistance of amplifier without feedback

β = Feedback factor

A = Gain without feedback

Q. 54. Why does the input resistance of the amplifier increase with the introduction of the negative feedback in series with the input?

Ans. It is clear from Fig. 13.44 that introduction of negative feedback in series with input tends to reduce the input current that would flow in the absence of feedback. This implies an increase in the effective input resistance. Obviously, the extent to which input resistance increases depends upon the amount of negative feedback.

More the feedback, greater is the tendency of the input current to reduce and hence, larger is the increase in the effective input resistance.

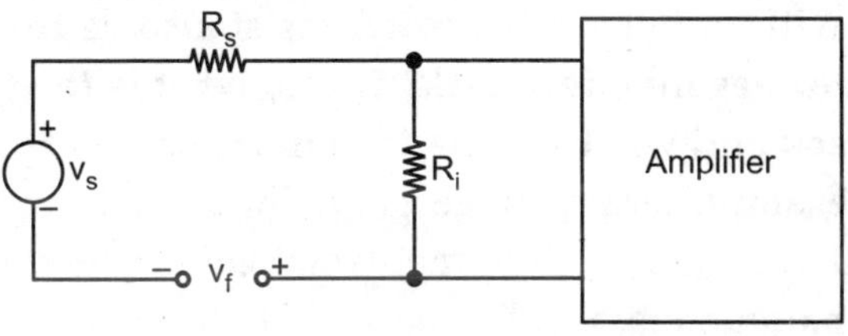

Fig. 13.44

Q. 55. How is input resistance of an amplifier affected in case of (*i*) voltage feedback (*ii*) current feedback?

Ans. The input resistance does not depend upon the type of feedback signal (voltage or current). It is affected by the mode in which it is feedback to the input, *i.e.* whether it has been fedback in series or in shunt with the input. The nature of feedback signal (whether it is voltage or current) affects the output resistance of the amplifier.

Q. 56. How will the output resistance of an amplifier be affected, if it has a voltage feedback?

Ans. The output resistance reduces when the feedback signal is voltage. Mathematically,

$$R_{of} = \frac{R_o}{(1-\beta A)}$$

where, R_{of} = Output resistance with feedback

R_o = Output resistance without feedback

A = Gain and β = Feedback factor.

Q. 57. Why does output resistance increase in case of current feedback?

Ans. Refer to Fig. 13.46. When output current is feedback, it indirectly appears as if there is an attempt to reduce the output current which implies an increase in the effective output resistance.

Mathematically $R_{of} = R_o (1 - \beta A)$.

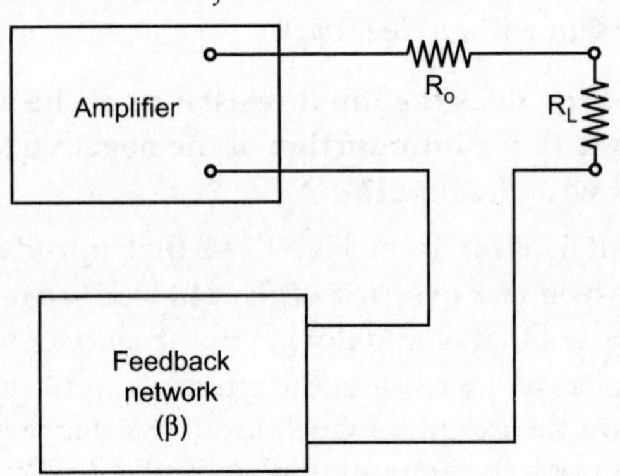

Fig. 13.45

Q. 58. Fig. 13.46 show a common-emitter amplifier stage with unbypassed emitter resistance. Is there any inherent negative feedback? If yes, what is the type of feedback and what will be the effects of this feedback on the amplifier's voltage gain, input resistance and output resistance? Give mathematical expressions for these parameters.

Ans. Yes, there is a negative feedback. The collector current flowing through R_e develops a voltage across it so as to reverse bias the base-emitter junction and thus oppose the input signal. The voltage across R_e is being applied in series with the input and also this voltage is proportional to the output current. Hence, it is a case of current series feedback. The feedback factor in this case will be,

$$-\frac{I_c R_e}{I_c} = -R_e$$

Gain with feedback is given by,

$$A_f = \frac{A}{1-\beta A} = \frac{A}{(1+R_e A)}$$

In the above expression, A is the transconductance of the amplifier without feedback and (A_f) is the transconductance with feedback. If we wish to find out how will the voltage gain be affected, then (β) will be taken as the fraction of the output voltage feedback to the input *i.e.*

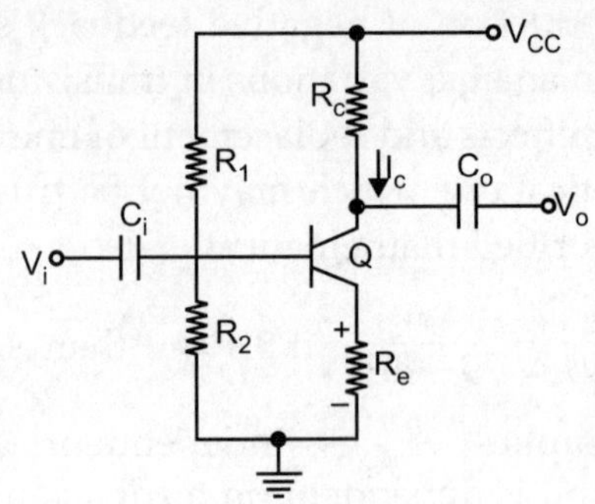

Fig. 13.46

$$\beta = \frac{-I_e R_e}{I_c R_c} = \frac{-(I_b + I_c) R_e}{I_c R_c} = -\frac{R_e}{R_c}\left\{1+\frac{I_b}{I_c}\right\}$$

$$= -\frac{R_e}{R_c}\left\{1+\frac{1}{h_{fe}}\right\} = -\frac{(1+h_{fe})}{h_{fe}} \times \frac{R_e}{R_c} \cong -\frac{R_e}{R_c}$$

$$\text{Therefore, } A_{vf} = \frac{A_v}{1-\beta A_v} = \frac{A_v}{1+\frac{R_e}{R_c}A_v}$$

$$\text{Now} \quad A_v \cong -\frac{h_{fe} R_c}{h_{ie}}$$

$$\text{Therefore, } A_{vf} = \frac{h_{fe} R_c / h_{ie}}{1+\frac{R_e}{R_c} \times \frac{h_{fe} R_c}{h_{ie}}} = \frac{h_{fe} R_c^2}{R_c h_{ie} + R_e R_c h_{fe}}$$

$$= \frac{-h_{fe} R_c}{h_{ie} + h_{fe} R_c}$$

The exact expression would have been

$$A_{vf} = \left[-\frac{h_{fe} R_c}{h_{ie} + (1+h_{fe}) R_e}\right].$$

$$\text{If we had substituted } \beta = \frac{(1+h_{fe})}{h_{fe}} \frac{R_e}{R_c}$$

$$R_{if} = h_{ie} + (1+h_{fe}) R_e$$

$$R_{of} = R_o\left[1+\frac{R_e}{h_{ie}}(1+h_{fe})\right].$$

Q. 59. Which of the basic amplifier configurations inherently has voltage series negative feedback? What is the effect of such a feedback on the voltage gain, input resistance and output resistance of the amplifier?

Ans. The common-collector configuration (emitter follower) is the one which has inherent negative feedback of the voltage-series type. In the figure shown (Fig. 13.47), the output is taken across R_e. The voltage that is fed back to the input is also the one developed across R_e. Therefore, the feedback is 100%. As a result, reduction in voltage-gain (due to negative feedback), increase in input resistance (due to feedback being in series with the input) and decrease in output resistance (due to voltage feedback), all are quite large (due to 100 per cent feedback). The voltage-gain feedback will infact be $\frac{A_v}{1+A_v}$ ≈ 1 (for $A_v >> 1$). The amplifier will have very high input resistance. That is how, it is an excellent configuration to be used as a buffer.

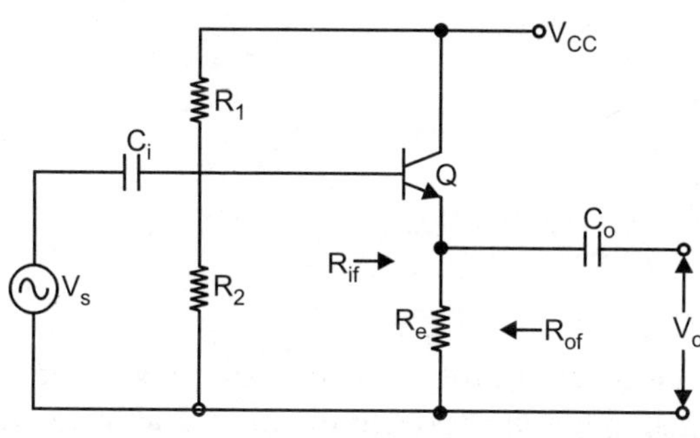

Fig. 13.47

Q. 60. Give an example of an amplifier configuration where there is a voltage shunt feedback. Which gain parameter of the amplifier is being stabilised by the feedback? Is it current gain, voltage gain or something else?

Ans. Figure 13.48 shows a common-emitter amplifier with a voltage shunt feedback provided by (R_B). The feedback factor (β) in this case equals ($1/R_B$). In this, a current that is proportional to the output voltage is being fed back to the input. Such a feedback results in a low input (due to shunt feedback) and output (due to voltage feedback) resistances.

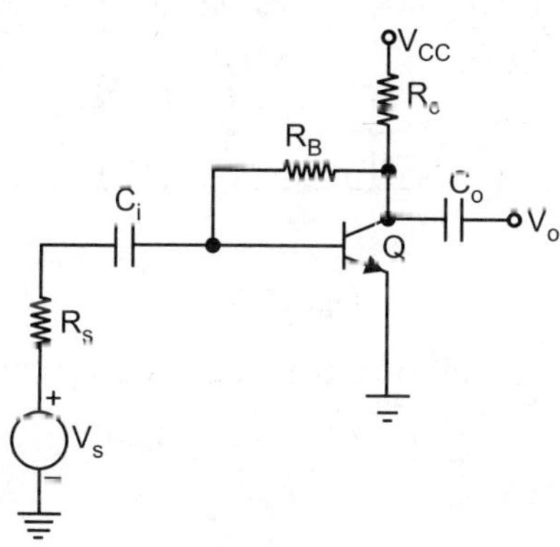

Fig. 13.48

Low output resistance implies that it is the output voltage that is significant, where low input resistance signifies that the amplifier input accepts the current signal. As a result of this, the amplifier becomes a sort of transresistance amplifier in which output voltage change divided by the corresponding input current change is the gain parameter. It has dimensions of resistance and is referred to as transresistance and it is this parameter which is being stabilised against variations in active device parameters etc.

Q. 61. According to you what should be the input and output resistances to make a good current amplifier and why?

Ans. A good current amplifier should have low input impedance and a high output impedance. It is so because lower the input impedance, higher is the fraction of input current flowing into the input of the amplifier (a desirable feature for a current amplifier). Also, higher the output impedance, higher is the fraction of output current flowing through the load. Refer to Fig. 13.49

In Fig. 13.49

For $R_i << R_s$

$I_i \cong I_s$ and if $R_o >> R_L$

Also, $I_L \cong I_o$

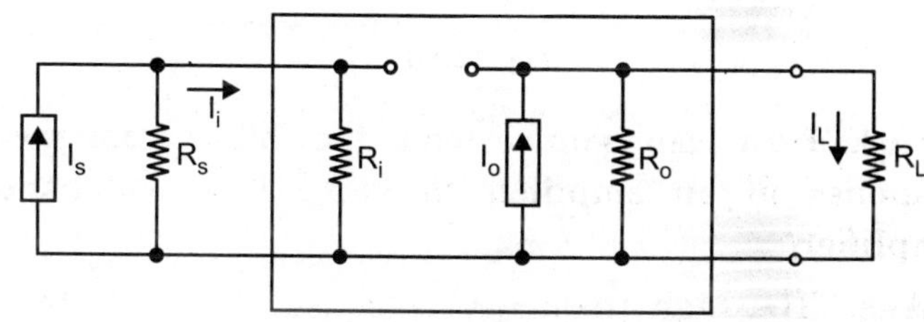

Fig. 13.49

Table 13.7 summarises the characteristic of four types of feedbacks encountered in amplifiers.

Table 13.7

Type of feedback	*Input Resistance*	*Output Resistance*	*Stabilized Gain parameter*	*Remarks*
Voltage series feedback	High	Low	Voltage Gain	We get a voltage amplifier
Current series feedback	High	High	Trans conduc-tance	We get a trans-conductance amplifier
Voltage shunt feedback	Low	Low	Trans-resistance	We get a Trans-resistance amplifier
Current shunt feedback	Low	High	Current gain	We get a current amplifier

Q. 62. Can you think of a schematic arrangement that would make a good current amplifier from basic circuit configurations?

Ans. A cascaded arrangement of transresistance amplifier provided by a common emitter amplifier configuration with a voltage shunt feedback followed by transconductance amplifier provided by a CE amplifier configuration with unbypassed emitter resistance to provide current series feedback will make a good current amplifier. First stage gives current to voltage amplification whereas second stage gives voltage to current amplification. Fig. 13.50 shows the arrangement,

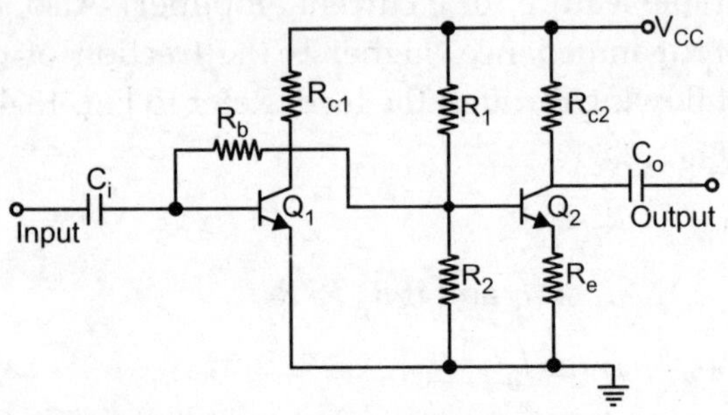

Fig. 13.50

Q. 63. How can you extend the high frequency response of an amplifier to make it a wideband amplifier?

Ans. The high frequency response of an amplifier is limited by shunt capacitance that may be present across the output. One of the methods to extend the high frequency response is to neutralize this capacitance by means of an inductance.

Another method is to use negative feedback. For example, if we have a common-emitter configuration with an emitter resistance bypassed by a capacitance (C_e) and if the value of (C_e) is so chosen that it is not large enough to completely bypass all frequencies of interest, in that case, higher frequencies will face lesser degeneration than the lower frequencies. As a consequence, amplifier gain at higher frequencies, where the gain would have otherwise fallen, will be higher than that at lower frequencies.

Thus, loss in gain at higher frequencies due to shunting effect of capacitances is compensated by increase in gain due to lesser degeneration at higher frequencies.

Figure. 13.51 (*a*) shows the method of enhancing the high frequency response by using a neutralising inductance. The circuit of Fig. 13.51 (*b*) relies on the bypass capacitor for increasing or more precisely extending the high frequency response.

In case of Fig. 13.51 (*a*), *L* is so chosen that L-R_C time constant is one half of the C_{sh} – R_C time constant.

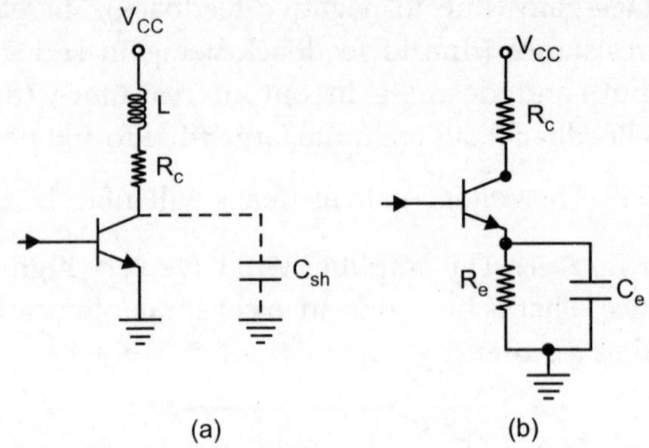

Fig. 13.51

$$\text{That is } \frac{L}{R_c} = 0.5\, R_c C_{sh}$$

$$L = 0.5\, R_C\, C_{sh}$$

Too large an inductance causes overshoots on fast transitions, whereas too small an inductance may not sufficiently extend the high frequency response.

Q. 64. Calculate the overall input impedance, output impedance and gain of a cascaded three-stage BJT amplifier.

Ans. Figure 13.52 shows a cascaded three-stage RC-coupled BJT amplifier. Let the h-parameters of the transistor Q_1 be h_{ie1}, h_{fe1}, h_{re1} and h_{oe1}; of transistor Q_2 be h_{ie2}, h_{fe2}, h_{re2} and h_{oe2} and that of transistor Q_3 be h_{ie3}, h_{fe3}, h_{re3} and h_{oe3}.

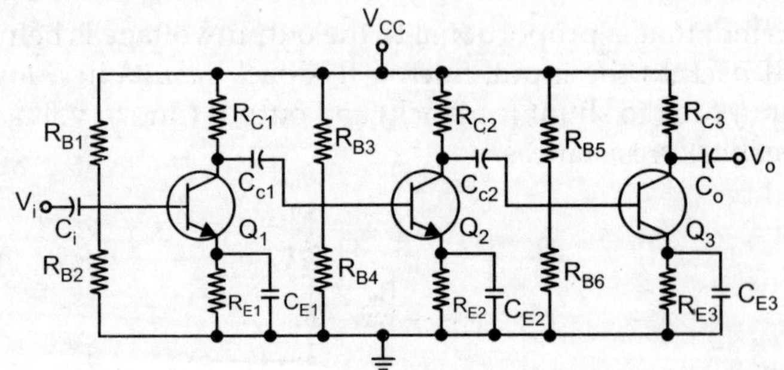

Fig. 13.52

The voltage gain of third stage is given by

$$A_{v3} = \frac{-h_{fe3} \times [R_{C3} \| (1/h_{oe3})]}{h_{ie3}} \quad ...(1)$$

Assuming $1/h_{oe} >> R_{C3}$, the equation can be simplified as

$$A_{v3} = \frac{-h_{fe3} \times R_{C3}}{h_{ie3}} \quad ...(2)$$

Similarly, the gain for the second stage A_{v2}, assuming $1/h_{oe2} >> R_{B5} \| R_{B6} \| R_{C2} \| h_{ie3}$ is given by

$$A_{v2} = \frac{-h_{fe2} \times (R_{B5} \| R_{B6} \| R_{C2} \| h_{ie3})}{h_{ie2}} \quad ...(3)$$

The gain for the first stage A_{v1}, assuming $1/h_{oe1} >> R_{B3} \| R_{B4} \| R_{C1} \| h_{ie2}$ is given by,

$$A_{v1} = \frac{-h_{fe1} \times (R_{B3} \| R_{B4} \| R_{C1} \| h_{ie2})}{h_{ie1}} \quad ...(4)$$

The overall voltage gain A_v is given by the product of the voltage gains of the three stages:

$$A_v = A_{v1} \times A_{v2} \times A_{v3} \quad ...(5)$$

The overall input impedance of the amplifier is the same as the input impedance of stage 1 and is given by,

$$Z_i = R_{B1} \| R_{B2} \| h_{ie1} \quad ...(6)$$

The output impedance of the amplifier is equal to the output impedance of the last stage and is given by,

$$Z_o = R_{C3} \| (1/h_{oe3}) \quad ...(7)$$

Q. 65. How does a darlington combination like the one shown in Fig. 13.53 give extremely high input resistance?

Ans. A darlington arrangement (Fig. 13.53) is nothing but two emitter follower stages in cascade, with the first emitter-follower (formed by Q_2 in this case) having infinite emitter resistance; represented by R_{e2} and is not actually connected in the circuit. Now, R_{i1} which is the input resistance looking into the base of Q_1 will be quite high and is mathematically given by $\{(h_{ie1} + (1 + h_{fe1} R_{e1})\}$. This becomes the effective emitter resistance for Q_2. As a result, the input resistance looking into the base of Q_2 will be still higher and approximately equal to $\{h_{ie2} + (1 + h_{fe}) R_{e2}\}$.where R_{e2} in this case is $\{h_{ie1} + (1 + h_{fe1}) R_{e1}\}$. Considering a typical example, where $h_{ie1} = h_{ie2} = IK$, $h_{fe1} - h_{fe2} = 100$ and $R_{e1} - 1K$, then R_{i2} is approximately equal to 10 MΩ.

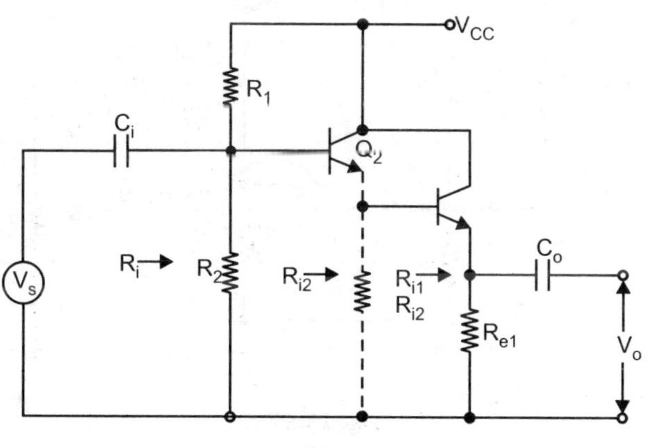

Fig. 13.53

Q. 66. In the preceding question, the overall input resistance of the cascaded arrangement including bias resistances (which you cannot ignore in practice) will be parallel combination of R_1 R_2 and R_{i2} which will be much less than R_{i2} as R_1 and R_2 cannot be too high due to bias requirements. Then what is the use of darlington arrangement to achieve high input resistance?

Ans. Yes, in an arrangement like the one shown in Fig. 13.53, the overall input resistance will not be as high as one would like to have in a Darlington combination. But the bootstrapping concept has been made use of to get a very high input resistance including the effect of bias resistances. Refer to Fig. 13.54. The value of bias resistance may be small due to bias requirements but its effective AC

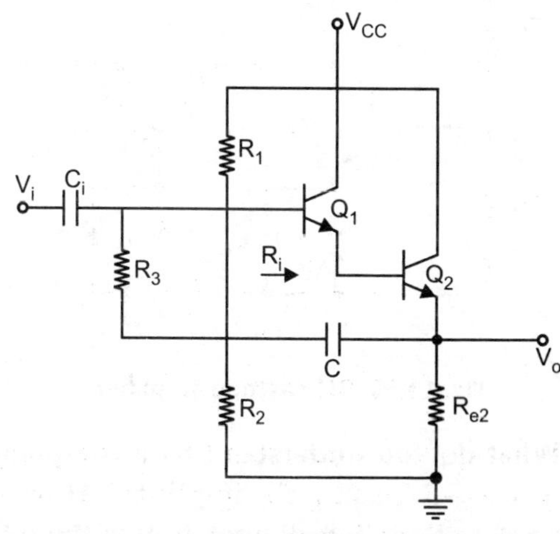

Fig. 13.54

resistance value is extremely high since C is large enough so as to offer a short circuit to the output signal at the lowest frequency of operation. Also, $v_o = v_i$ and v_o is in phase with v_i. As a result, there is a negligible current through R_3 which means increase in effective value of R_3.

Mathematically, R_3 (eff.) is given by,

$$R_3\,(\text{eff.}) = \frac{R_3}{(1 - A_v)} \quad ...(\text{Miller effect})$$

A_v is the overall voltage gain and in the present case can be anywhere between 0.95 and 1. For $A_v = 0.99$, R_3 (eff.) will be 100 times R_3.

Q. 67. Explain the design of a BJT cascode amplifier configuration. Why is it used for high-frequency operation?

Ans. Cascode amplifiers are two-stage amplifiers comprising a transconductance amplifier followed by a current buffer. They offer advantages like high input–output isolation, high input impedance and high output impedance. In a BJT cascode amplifier configuration, the common-emitter transistor is followed by a common-base connection. Figure 13.55 shows the circuit diagram of a cascode amplifier, wherein the transistor Q_1 is configured as common-emitter amplifier and Q_2 is configured as common-base amplifier. The common-base amplifier Q_2 offers large bandwidth but its input impedance is low. Because of the low input impedance of Q_2, the voltage gain of Q_1 is low. Therefore, the Miller component of capacitance of transistor Q_1 is small and the bandwidth of the cascode amplifier is wider than for the common-emitter stage. Hence, it is used for high-frequency operation.

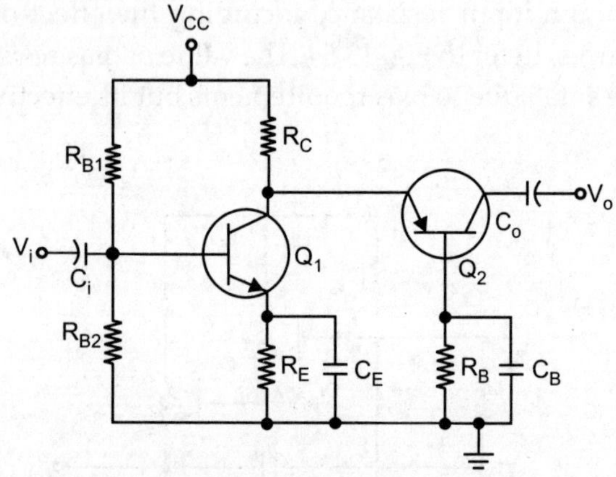

Fig. 13.55 BJT cascode amplifier

Q. 68. What do you understand by a complementary symmetry class-B push-pull amplifier? How does it give a true class-B push-pull operation without the use of center tapped input and output transformers?

Ans. Figure. 13.56(*a*) shows the basic complementary symmetry class-B amplifier. Use of an N-P-N and a P-N-P transistor ensures that one of the transistors conducts only during positive half cycles (N-P-N transistor in this case) and the other conducts only during negative half cycles (P-N-P transistor in this case). Thus there is no need for a phase splitting input transformer. Also, nowadays power transistors are available that have collector-emitter impedance of several ohms which alleviates the need for an output transformer needed to couple the output to a low impedance load usually a speaker. Biasing resistors have been added to prevent cross-over distortion. The transistors used should have matched characteristics. Matched pairs for audio applications for different power ratings are available commercially. These include CL 100 (*N-P-N*) and *CK* 100 (*P-N-P*), 2N5296 (*N-P-N*) and 2N6107 (*P-N-P*).

Single supply versions of complementary symmetry class-B amplifiers are more commonly used. Figure. 13.56 (*b*) shows the basic circuit.

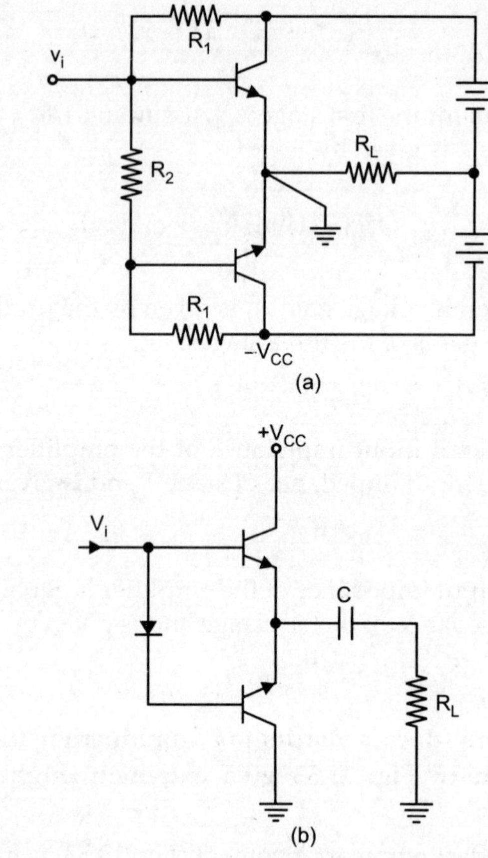

Fig. 13.56

Q. 69. What do you understand by a Quasi-complementary class-*B* amplifier? What are the advantages of such a design?

Ans. Another variation of the complementary-symmetry amplifier is the quasi-complementary class *B*

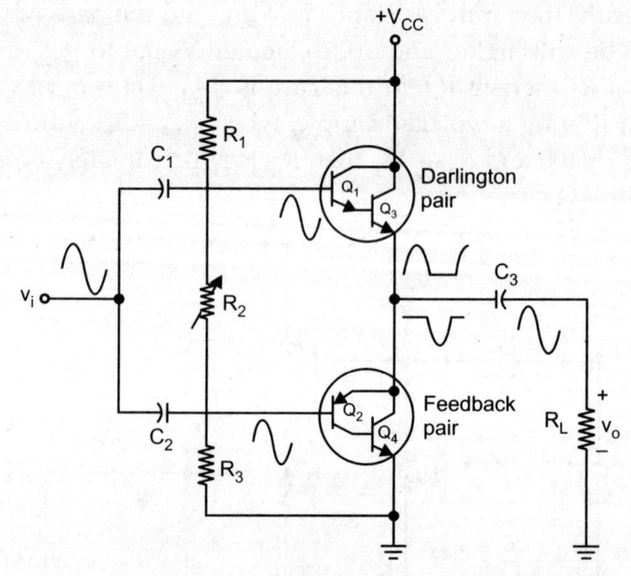

Fig. 13.57

amplifier shown in Fig. 13.57. The push–pull operation is achieved by using a pair of complementary transistors. Transistors Q_1 and Q_3 form a Darlington connection that provides output from a low-impedance emitter follower. Transistors Q_2 and Q_4 form a feedback pair and they provide a low-impedance drive to the load. The crossover distortion can be minimised by adjusting the value of the resistor R_2.

One of the main advantages of using the quasi-complementary configuration is that it employs matched NPN transistors as high current devices and does not require a high-power PNP transistor, as required in case of complementary-symmetry amplifiers. It may be mentioned here that quasi-complementary push–pull amplifier is the most popular form of power amplifiers.

SOLVED PROBLEMS

Problem 13.1. *Given that the hybrid parameters for the transistor are hie = 1.5kΩ, h_{fe} = 150, h_{re} = 1 × 10^{-4} and h_{oe} = 20 μmhos. Draw the hybrid equivalent circuit of the transistor in all the three configurations.*

Solution. Figure 13.58 (*a*) shows the hybrid equivalent circuit for the transistor in the common-emitter configuration.

The values of hybrid parameters in the common-collector configuration are $h_{ic} = h_{ie}$, $h_{rc} = 1$, $h_{fc} = -(1 + h_{fe})$ and $h_{oc} = h_{oe}$.

Therefore, h_{ic} = 1.5 kΩ, h_{rc} = 1, h_{fc} = – 151 and h_{oc} = 20 μmhos.

Figure 13.58 (*b*) shows the hybrid equivalent circuit of the transistor in the common-collector configuration.

The values of hybrid parameters in the common-base configuration are,

$$h_{ib} = \frac{h_{ie}}{1+h_{fe}}$$

$$= (1.5 \times 103)/151 = 9.93\ \Omega$$

$$h_{rb} = \frac{h_{ie}h_{oe}}{1+h_{fe}} - h_{re}$$

$$= (1.5 \times 10^3 \times 20 \times 10^{-6})/(151 - 1 \times 10^{-4}),$$

$$= 0.99 \times 10^{-4}$$

$$h_{fb} = -\frac{h_{oe}}{1+h_{fe}}$$

$$= -150/151 = -0.99$$

$$h_{ob} = \frac{h_{oe}}{1+h_{fe}}$$

$$= (20 \times 10^6)/151 = 0.13 \times 10^{-6}$$

Figure 13.58 (*c*) shows the hybrid equivalent of the transistor in the common-base configuration.

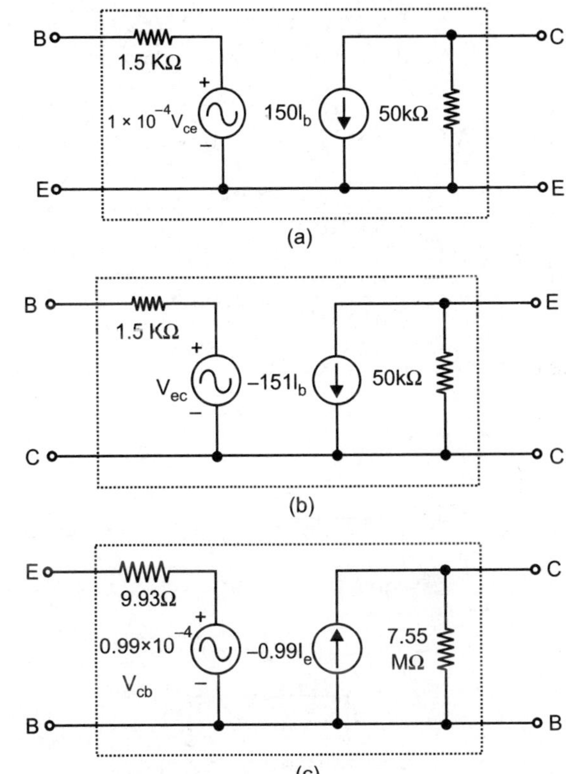

Fig. 13.58

Problem 13.2. *For the circuit shown in Fig. 13.59, determine the input impedance, overall voltage gain and current gain and output impedance. The values of the h-parameters of the transistor are h_{ie} = 1.5 kΩ, h_{fe} =100, h_{re} = 1 × 10^{-4} and h_{oe} = 25 mA/V.*

Solution. The complete hybrid equivalent model for the network is shown in Fig. 13.60 (*a*). Figure 13.60 (*b*) shows the simplified circuit of the network shown in Fig. 13.60 (*a*) with the input section being replaced by the Thevenin's equivalent network. As the resistor R_B >> R_S, therefore, the Thevenin's equivalent voltage V_{TH} is approximately equal to the source voltage V_S and the Thevenin's equivalent resistance R_{TH} is approximately equal to the resistor R_S.

The input impedance Z_i is equal to

$$Z_i = h_{ie} - \frac{h_{fe}h_{re}R_C}{1+h_{oe}R_C}$$

$$= 1.5\times10^3 - \frac{100\times1\times10^{-4}\times4\times10^3}{1+25\times10^{-6}\times4\times10^3}$$

$$= 1.5\times10^3 - \frac{40}{1.1}$$

$$= 1500 - 36.36 = 1.464\text{ k}\Omega$$

The voltage gain A_v is equal to

$$A_v = \frac{V_o}{V_i} = \frac{-h_{fe}R_C}{h_{ie}+(h_{ie}h_{oe}-h_{fe}h_{re})R_C}$$

$$= \frac{-100\times4\times10^3}{1.5\times10^3+(1.5\times10^3\times25\times10^{-6}-100\times1\times10^{-4})\times4\times10^3}$$

$$= \frac{-4\times10^5}{1.5\times10^3+0.0275\times4\times10^3}$$

$$= \frac{-4\times10^5}{1.61\times10^3} = -248.45$$

The current gain A_i is equal to

$$Ai = -\frac{h_{fe}}{1+h_{oe}R_C}$$

$$= \frac{-100}{1+25\times10^{-6}\times4\times10^3}$$

$$= -90.91$$

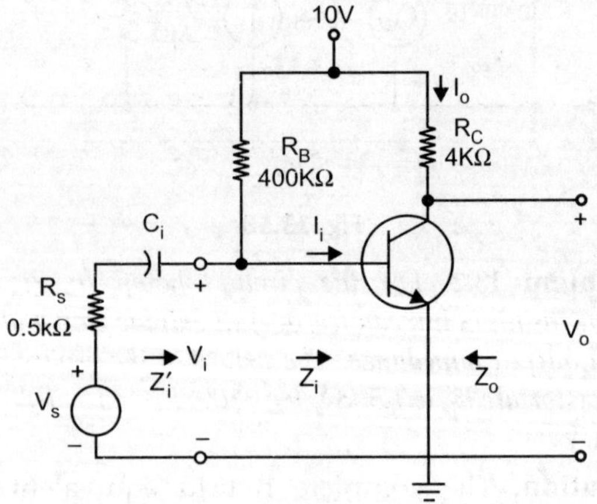

Fig. 13.59

The output impedance Z_o' is parallel combination of Z_o and R_C. Z_o is given by,

$$Z_o = \frac{1}{h_{oe}-[h_{fe}h_{re}/(h_{ie}+R_s)]}$$

$$= \frac{1}{25\times10^{-6}-[(100\times1\times10^{-4})/(1.5\times10^3+500)]}$$

$$= \frac{1}{25\times10^{-6}-5\times10^{-6}} = 50\text{ k}\Omega$$

The overall output impedance $Z_o' = 50\times10^3 \,||\, 4\times10^3 = 3.7\times10^3 = 3.7\text{ k}\Omega$

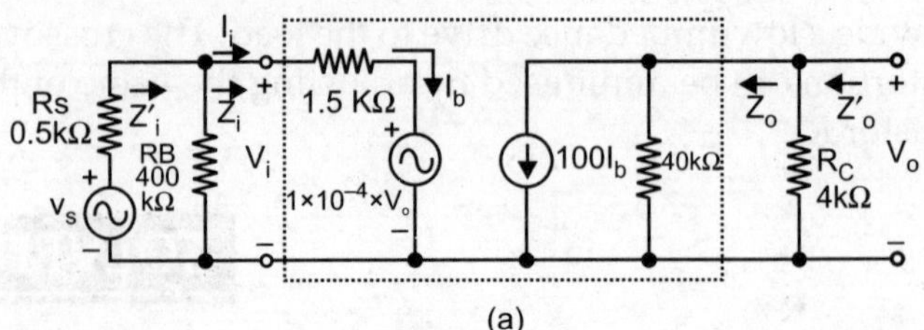

(a)

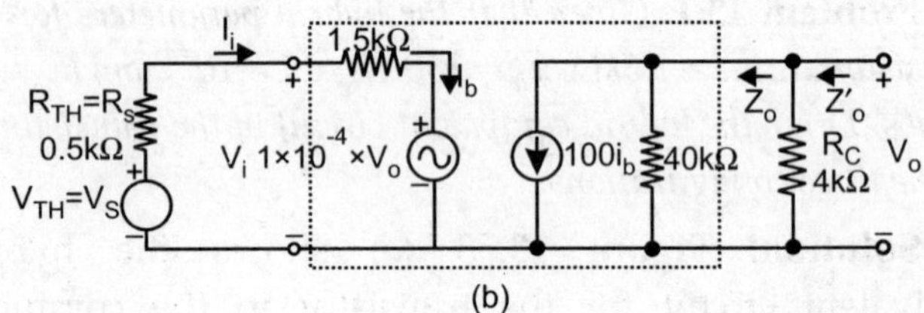

(b)

Fig. 13.60

Problem 13.3. *Determine the following parameters for the amplifier shown in Fig. 13.61. Given that the h-parameters of the transistor are h_{ie} = 1 kΩ, h_{fe} = 100, $h_{oe} = 40\times10^{-6}$ mhos.*

(a) Z_i *(b) Z_o*

(c) A_v *(d) A_i*

Fig. 13.61

Solution. The AC equivalent circuit for the amplifier in Fig. 13.61 is shown in Fig. 13.62.

Input impedance

$$Z_i = R_{F1} \,\|\, h_{ie}$$

$$= 300\times10^3 \,\|\, 1\times10^3$$

$$= 0.99\text{ k}\Omega$$

Output impedance (Z_o) is given by,

$$Z_o = R_{F2} \,\|\, R_C \,\|\, (1/h_{oe})$$

$$= 50 \times 10^3 \| 2 \times 10^3 \| 25 \times 10^3$$

$$= 1.79 \times \text{k}\Omega$$

Voltage gain A_v is given by,

$$A_v = \frac{-h_{fe} \times [R_{F2} \| R_C \| (1/h_{oe})]}{h_{ie}}$$

$$= \frac{-100 \times 1.79 \times 10^3}{1 \times 10^3}$$

$$= -1.79$$

The current gain A_i is given by,

$$A_i = \frac{h_{fe} R_{F1} [R_{F2} \| R_C \| (1/h_{oe})]}{(R_{F1} + h_{ie})[R_{F2} \| R_C \| (1/h_{oe}) + R_C]}$$

$$= \frac{100 \times 100 \times 10^3 \times (50 \times 10^3 \| 2 \times 10^3 \| 25 \times 10^3)}{(100 \times 10^3 + 1 \times 10^3) \times (50 \times 10^3 \| 2 \times 10^3 \| 25 \times 10^3 + 2 \times 10^3)}$$

$$= \frac{10^7 \times 1.79 \times 10^3}{101 \times 10^3 \times 3.79 \times 10^3} = 46.76$$

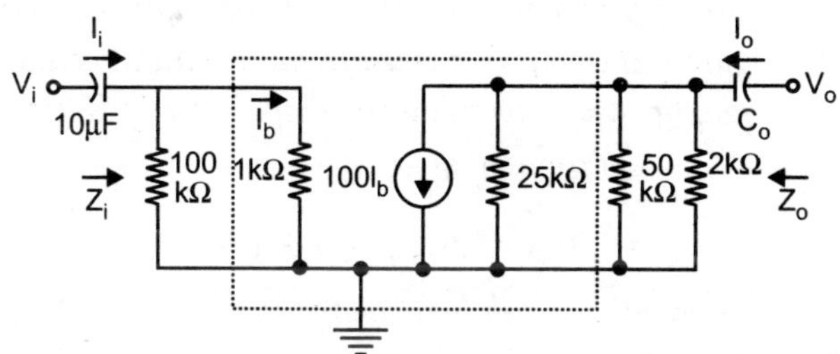

Fig. 13.62

Problem 13.4. *Figure 13.63 shows a two-stage CC-CB amplifier. The values of input and output impedances and the voltage gains for each stage are shown in the figure. All the values are for no-load conditions. However, the input impedance and the output impedance of the first stage are under loaded conditions. Determine*

(a) *The loaded voltage gain for each stage.*

(b) *The overall voltage gain of the amplifier.*

(c) *The total system voltage and current gain.*

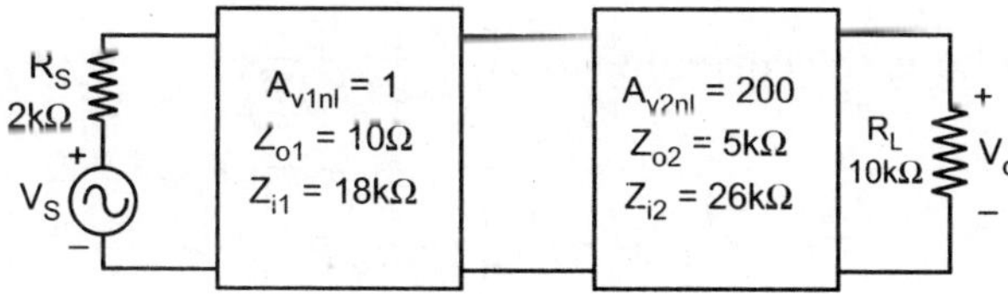

Fig. 13.63

Solution. The loaded voltage gain of the second-stage, that is, the CB amplifier is given by,

$$A_{v2} = A_{v2nl} \times [R_L / (R_L + Z_{o2})]$$

$$= 200 \times [(10 \times 10^3) / (10 \times 10^3 + 5 \times 10^3)]$$

$$= 133.33$$

The loaded voltage gain of the first stage, that is, the CC amplifier is given by,

$A_{v1} = A_{v1nl} \times [Z_{i2} / (Z_{i2} + Z_{o1})] = 1 \times [26/(26+10)] = 26/36 = 0.722$

The overall voltage gain of the amplifier $= A_{v1} \times A_{v2} = 0.722 \times 133.33 = 96.26$

The total system gain

$A_{vs} = A_v \times [Z_{i1}/(Z_{i1} + R_s)] = 96.26 \times [(18 \times 10^3)/(18 \times 10^3 + 2 \times 10^3)] = 86.63$

The value of the total system current gain

$A_{is} = -(A_{vs} \times Z_{i1})/R_L = (86.63 \times 18 \times 103)/(10 \times 10^3) = -155.93$

Problem 13.5. *Refer to basic common base amplifier configuration of Fig. 13.64 The h-parameter values for the transistor used in this circuit are as follows : $h_{ie} = 4K$, $h_{re} = 7 \times 10^{-4}$, $h_{fe} = 100$, $h_{oe} = 50\mu$ mhos.*

Determine the following :

(a) *The current gain taking source and load resistances into account*

(b) *The voltage gain taking source and. load resistances into account*

(c) *The power gain*

(d) *The input resistance*

(e) *The output resistance.*

Solution.

The h-parameter values given in the problem are for CE-configuration. As a first step, we convert the CE h-parameters to CB h-parameters.

$$h_{ib} = \frac{h_{ie}}{1+h_{fe}} = \frac{4000}{1+100} \cong 40 \text{ ohms}$$

$$h_{rb} = \frac{h_{ie} h_{oe}}{1+h_{fe}} - h_{re} = \frac{4000 \times 50 \times 10^{-6}}{1+100} \quad 7 \times 10^{-4}$$

$$= \frac{0.2}{101} - 0.0007 = 0.0013$$

$$h_{fb} = \frac{-h_{fe}}{1+h_{fe}} = \frac{-100}{101} \cong -1$$

$$h_{ob} = \frac{h_{oe}}{1+h_{fe}} = \frac{50 \times 10^{-6}}{1+100} \cong 0.5\mu \text{ mho.}$$

CURRENT GAIN

The current gain (A_i) without taking source resistance into account is given by

$$A_i = \frac{-h_{fb}}{1+h_{ob}R_L} = \frac{-(-1)}{1+0.5\times10^{-6}\times1000}$$

$$= \frac{1}{1.0005} \cong 1$$

The current gain (A_{is}) taking source resistance into account is then given by (Refer to Fig. 13.65 *(a)*

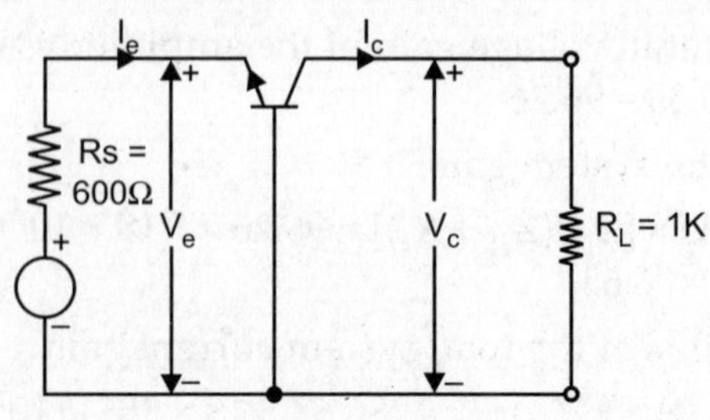

Fig. 13.64

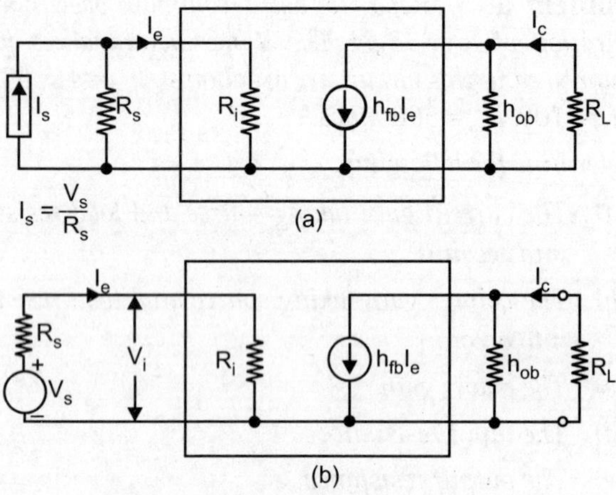

Fig. 13.65

$$A_{is} = \frac{A_i R_s}{R_s + R_i}$$

INPUT RESISTANCE

The input resistance (R_i) is given by,

$R_i = h_{ib} + h_{rb} A_i R_L = 40 + 0.0013 \times 1 \times 1000 = 41.3$ ohms.

The current gain (A_{is}) is therefore given by

$$A_{is} = \frac{1\times600}{600+41.3} = \frac{600}{641.3} \cong 0.93$$

VOLTAGE GAIN

The voltage gain (A_v) without taking source resistance into account is given by (Refer to Fig. 13.65 *(b)*)

$$A_V = \frac{A_i R_L}{R_i} = \frac{1\times1000}{641.3} \cong 1.5$$

Voltage gain (Avs) taking source resistance (R_s) into account is given by,

$$A_{vs} = \frac{A_v R_i}{R_i + R_s} = \frac{1.5\times41.3}{600+41.3} = \frac{1.5\times41.3}{641.3} \cong 0.09$$

POWER GAIN

Power gain (A_p) = Voltage gain × Current gain

$= 0.09 \times 0.93 = 0.084$

OUTPUT RESISTANCE

The output conductance (G_O), where $G_O = 1/R_O$, is given by,

$$G_O = h_{ob} - \frac{h_{fb}h_{rb}}{R_s+R_i} = 0.5\times10^{-6} - \frac{(-1)(0.0013)}{641.3}$$

$$= 0.5\times10^{-6} + \frac{0.0013}{641.3} = 2.5\times10^{-6} \text{ mhos}$$

The output resistance (R_o) is therefore given by,

$$R_o = \frac{1}{2.5\times10^{-6}} = 400K.$$

Problem 13.6. *A transistor having the following CE h-parameters is to be used as an emitter follower. Determine*

(a) the load resistance required to get an input impedance of 300K.

(b) the overall current gain, voltage gain and power gain if the amplifier is fed from the output of another amplifier having an output impedance of 100K.

$h_{ie} = 1000\ \Omega,\ h_{fe} = 100$

$h_{oe} = 50\ \mu$ mhos, $h_{re} = 0.001$

justify all assumptions made.

Solution.

First, we find the corresponding CC h-parameters from the given CE h-parameters.

$h_{ic} = h_{ie} = 1000\ \Omega$

$h_{rc} = (1 - h_{re}) = (1 - 0.001) \cong 1$

$h_{fc} = -(1 + h_{fe}) = -101$

$h_{oc} = 50 \times 10^{-6}$ mhos

(a) The input impedance (R_i) is given by,

$$R_i = h_{ic} + h_{rc} A_i R_L$$

where, the current gain, $A_i = \dfrac{h_{fc}}{1+h_{oc}R_L}$

or $$R_i = h_{ic} + h_{rc} R_L \left[\frac{-h_{fc}}{1+h_{oc}R_L}\right]$$

or $$-\frac{h_{fc}R_L}{(1+h_{oc}R_L)} = \frac{R_i - h_{ic}}{h_{rc}}$$

$$= \frac{300\times10^3 - 1000}{1}$$

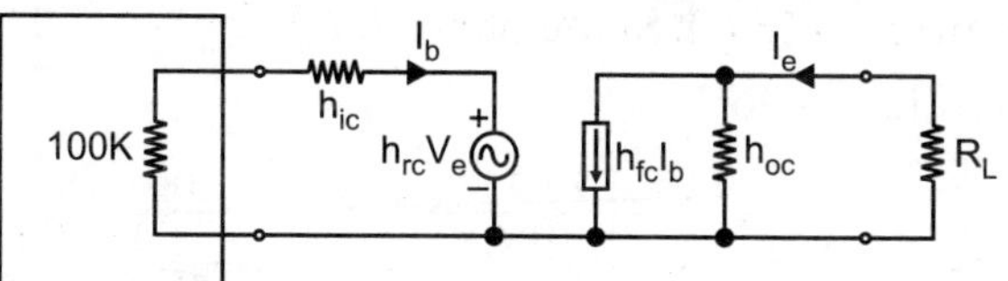

Fig. 13.66

or $\dfrac{R_L}{1+h_{oc}R_L} = \dfrac{299000}{101} = 2990$

or $\dfrac{R_L}{1+50\times10^{-6}R_L} = 2990$

or $R_L = 50\times10^{-6}\times2990\,R_L + 2990$

or $R_L(1-0.15) = 2990$

Therefore, $R_L = \dfrac{2990}{0.85} = 3517\ \Omega$

The load resistance required to give an input impedance of 300 K is 3.517 K.

(*b*) The current gain $A_i = -\dfrac{h_{fc}}{1+h_{oc}R_L}$

$$= -\frac{(-101)}{1+50\times10^{-6}\times3517}$$

$$= \frac{101}{1+0.175} = \frac{101}{1.175} = 86$$

The current gain (A_{is}) taking source resistance into account is given by,

$$A_{is} = \frac{A_i R_s}{R_s+R_i} = \frac{86\times100}{100+300} = \frac{86}{4} = 21.5$$

The voltage gain, $A_v = \dfrac{A_i R_L}{R_l} = \dfrac{86\times3.517}{300} \cong 1$

The voltage gain A_{vs} taking source resistance into account by,

$$A_{vs} = \frac{A_v R_i}{R_s+R_i} = \frac{1\times300}{100+300} = 0.75$$

Power gain = Current gain × Voltage gain

= 21.5 × 0.75 = 16.12 (= 12 dB).

Problem 13.7. *Determine the α- and β-cut-off frequencies for a transistor with the following specifications: g_m = 38 mmhos, $r_{b'e}$ = 5.9 kΩ, h_{ie} = 6 kΩ, $r_{bb'}$ = 100 Ω, $C_{b'c}$ = 12 pF, $C_{b'e}$ = 63 pF, f_T = 80 MHz and h_{fe} = 224 at 1 kHz. Also, determine the value of common-emitter short-circuit current gain at frequencies of f_β, f_T and fα.*

Solution.

$$f\alpha = \frac{h_{fe}}{2\pi r_{b'e}C_e}$$

$$= \frac{224}{2\times\pi\times5.9\times10^3\times10^{-12}} = 95.91\text{ MHz}$$

$$g_{b'e} = 1/r_{b'e} = 1/(5.9\times10^3) = 1.69\times10^{-4}$$

f_β=1.69 × 10^{-4}/[2 × π × (63 × 10^{-12} + 12 × 10^{-12})] = 358.63 kHz

Common-emitter short-circuit current gain is given by,

$$A_i = \frac{-h_{fe}}{1+j(f/f_\beta)}$$

For $f = f_\alpha$, A_i = – 224/[1 + j{(95.91 × 10^6)/(358.63 × 10^3)}] = – 224/(1 + 267. 43*j*)

$$|A_i| = 224/\sqrt{(1^2+267.43^2)} = 0.838$$

$$\angle A_i = 90° - \tan^{-1}(267.43) = 90° - 89.786° = 0.21°$$

For $f = f_\beta$, A_i = – 224/[1 + *j* {(358.63 × 10^3)/(358.63 × 10^3)}]

= – 224/(1 + *j*)

$$|A_i| = 224/\sqrt{2} = 158.39$$

$$\angle A_i = 90° - \tan^{-1}1 = 90° - 45° = 45°$$

$$f_T = h_{fe}f_\beta$$

$$f_T = 224\times358.63\times10^3 = 80.333\text{ MHz}$$

For $f = f_T$, A_i = – 224/[1 + j{(80333 × 10^6)/(358.63 × 10^3)}]

= – 224/(1 + 224*j*)

$$|A_i| = 224/\sqrt{1^2+224^2} = 1$$

$$\angle A_i = 90° - \tan^{-1}224 = 90°\ \ 89.744° - 0.256°$$

Problem 13.8. *For a certain common emitter amplifier configuration, assume V_{CC} = 24V, R_C = 10Ω., V_{BE} = 0. 7V. Fig. 13.67 shows the output characteristics of the amplifier. Draw the load line and calculate the value of base resistance required, if the quiscent base current is 100 mA.*

Solution. The load line is expressed by the straight line equation

$$i_C = \left(-\frac{1}{R_C}\right)V_{CE} + \frac{V_{CC}}{R_C}$$

For $i_C = 0$, $V_{CE} = V_{CC}$

and for

$$V_{CE} = 0,\ i_C = V_{CC}/R_C$$

The line joining the co-ordinates (V_{CC}. 0) and (0. V_{CC}/R_C) is the load line.

Quiscent base current = 100 mA

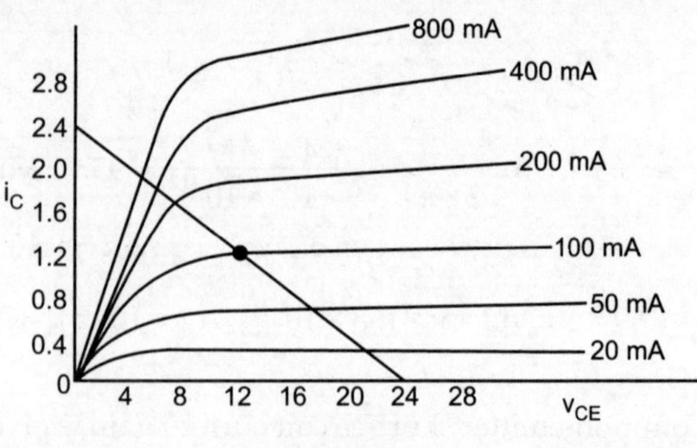

Fig. 13.67

Base resistance required to get this base current is given by,

$$R_B = \frac{V_{CC} - V_{BE}}{I_B} = \frac{24 - 0.7}{0.1} = \frac{23.3}{0.1} = 233\ \Omega.$$

Problem 13.9. *Determine the slopes of DC load lines for the common-emitter amplifier configurations of Fig. 13.68. Compare their stability factors, if all the three configurations have a (β) of 50 at the operating point.*

Solution. Figure. 13.68 (*a*) : Slope of DC load line = – 1/1100 amp/volt

$$\text{Stability} = \frac{(\beta + 1)}{1 + \left(\frac{\beta R_E}{R_E + R'_B}\right)}, \text{ where } R'_B = \frac{R_1 R_2}{R_1 + R_2}$$

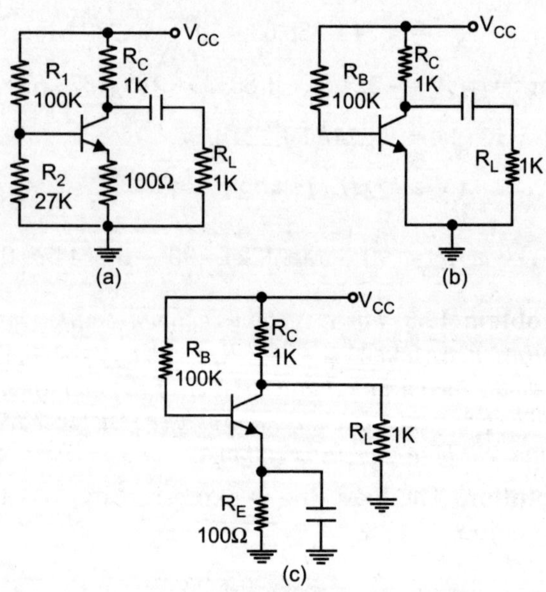

Fig. 13.68

$$= \frac{50 + 1}{1 + \left(\frac{50 \times 100}{100 + 2126}\right)} = \frac{51}{\left(1 + \frac{5000}{2226}\right)} \cong 15$$

Figure. 13.68 (*b*): Slope of DC load line = – 1/1000

Stability factor = β + 1 = 51

Figure. 13.68 (*c*): Slope of DC load line = – (1/1100)

Stability factor

$$= \frac{(\beta + 1)}{1 + \left(\frac{\beta R_E}{R_E + R_B}\right)} = \frac{(50 + 1)}{1 + \left(\frac{50 \times 100}{100000 + 100}\right)}$$

$$\approx \frac{51}{1 + \left(\frac{5000}{100000}\right)} = \frac{51}{21/20} = \frac{1020}{21} \cong 48.5.$$

Problem 13.10. *For the common-emitter amplifier circuit of Fig. 13.69 (a) and the output characteristics of the transistor 2N2222 used in the circuit of Fig. 13.69 (b), calculate the following:*

(a) Quiscent base current, I_B

(b) Quiscent collector current, I_{CEQ}

(c) Quiscent collector-emitter voltage, V_{CEQ}

(d) Slope of the DC load line.

(e) Co-ordinates of the AC load line.

Solution.

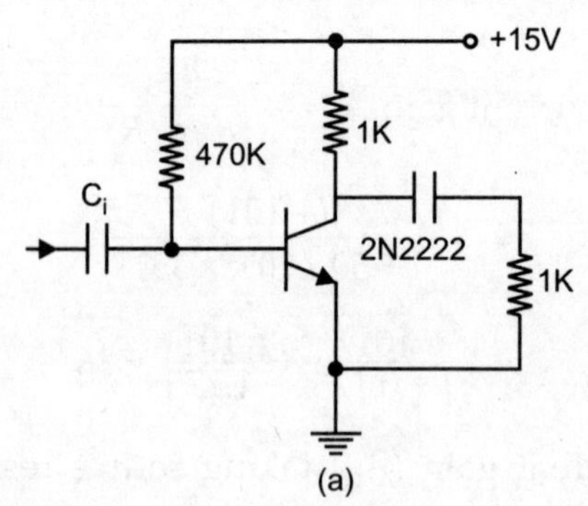

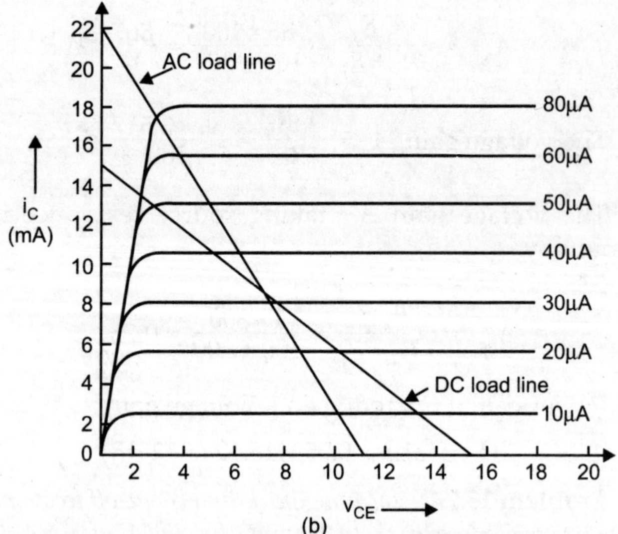

Fig. 13.69

(*a*) Quiscent base current = $\frac{15 - 0.7}{470}$ mA ≅ 30μA

(*b*) Quiscent collector current (I_{CEQ}) and quiscent collector voltage (V_{CEQ}) can be determined from the load line, as superimposed on the output characteristics. $I_{CEQ} = 8$ mA

(*c*) $V_{CEQ} = 7$V

(*d*) Slope of DC load line = – 1/1000

(*e*) Slope of AC load line = – 1/500

The equation for the AC load line can now be written as,

$$I_C - 8 \text{ mA} = -1/500\,(V_{CE} - 7)$$

for $I_C = 0$, $V_{CE} = 11$V

and for $V_{CE} = 0$, $I_C = 22$ mA.

The co-ordinates of AC load line are therefore, (11, 0) and (0, 22).

Problem 13.11. *A common-emitter (CE) amplifier with a lower 3 dB frequency of 70 Hz has been cascaded with another CE stage with a lower 3 dB frequency of 100 Hz. At what frequency will the voltage gain of the cascaded arrangement fall to 70.70% of the midband gain ? (Find out the frequency on the lower end of the frequency scale).*

Solution. The gain falls to 70.7% of midband value at a frequency on the lower end called lower 3 dB frequency. This frequency in this case is given by,

$$\frac{100}{\sqrt{(2^{1/2} - 1)}} = \frac{100}{\sqrt{0.414}} \cong 155 \text{ Hz.}$$

Problem 13.12. *The lower 3 dB frequency and upper 3 dB frequency of a common-emitter amplifier stage with a midband gain of 40 dB are 100 Hz and 100 kHz respectively. What is the amplifier gain at 100 Hz and 100 kHz?*

Solution. The,amplifier gain at these frequencies will be 40 – 3 = 37 dB.

Problem 13.13. *Fig. 13.70 shows the input and the output signal waveforms in case of a common-emitter amplifier obtained while doing square-wave testing. What is the lower 3 dB frequency of the amplifier?*

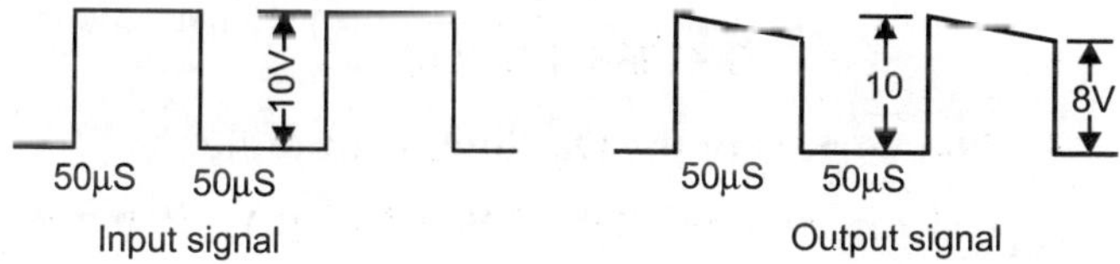

Fig. 13.70

Solution. The sag in the output waveform is given by

$$\left(\frac{\pi f_L}{f}\right)$$

where, f = signal frequency

and f_L = lower 3 dB frequency

$$\text{Also, sag} = \frac{10 - 8}{10}$$

$$\text{Therefore, } \frac{\pi f_L}{f} = 0.2; \text{ or } f_L = \left(\frac{0.2f}{\pi}\right)$$

Now $f = 10$ kHz.

$$\text{Therefore, } f_L = \frac{0.2 \times 10^4}{\pi} = \frac{2000}{\pi} \cong 636 \text{ Hz.}$$

Problem 13.14. *What will be the percentage sag in the output waveform, if the input signal in the last question had a frequency of 4 kHz instead of 10 kHz?*

Solution.

The lower 3 dB frequency = 636 Hz.

$$f = 4000 \text{ Hz.}$$

$$\text{Therefore, \% sag} = \frac{\pi f_L}{f} \times 100 = \frac{\pi \times 636}{4000} \times 100 = 50\%.$$

Problem 13.15. *Calculate the values of harmonic distortion components for an output signal having amplitude of 5 V at the fundamental frequency, second harmonic component of 0.5 V, third harmonic component of 0.2 V and fourth harmonic component of 0.05 V. Also, calculate the total harmonic distortion.*

Solution.

The second harmonic distortion component is given by,

$$D_2 = \left|\frac{A_2}{A_1}\right| \times 100\%$$

$$D_2 = \frac{0.5\,V}{5\,V} \times 100\% = 10\%$$

The third harmonic distortion component is given by,

$$D_3 = \left|\frac{A_3}{A_1}\right| \times 100\%$$

$$D_3 = \frac{0.2\,V}{5\,V} \times 100\% = 1\%$$

The fourth harmonic distortion component is given by,

$$D_4 = \left|\frac{A_4}{A_1}\right| \times 100\%$$

$$D_4 = \frac{0.05\,V}{5\,V} \times 100\% = 1\%$$

The total harmonic distortion (D) is equal to

$$D = \sqrt{D_2^2 + D_3^2 + D_4^2 + \ldots}$$

$$D = \sqrt{(10)^2 + (4)^2 + (1)^2} = \sqrt{117} = 10.8\%$$

Total harmonic distortion in percentage = 10.8%

Problem 13.16. *A class B amplifier provides a 20 V peak output signal to 15 Ω load. The system operates on a power supply of 25 V. Determine the efficiency of the amplifier.*

Solution. The peak load current $I_L(p) = 20/15 = 1.33$ A

The DC value of current drawn from the supply $I_{dc} = (2/\pi) \times I_L(p) = 0.847$ A

Input power delivered by the supply voltage $P_i = V_{CC} \times I_{dc} = 25 \times 0.847 = 21.175$ W

Output power delivered to the load $P_o = V_L(p)^2/2R_L = 20^2/(2 \times 15) = 400/30 = 13.33$ W

Efficiency $\eta = (P_o/P_i) \times 100\% = (13.33/21.175) \times 100\% = 62.95\%$

Problem 13.17. *Figure 13.71 shows a two-stage BJT cascaded amplifier. Calculate the voltage gain, input and output impedance of the amplifier, given that h_{ie} and h_{fe} parameters of both the transistors are 1 kΩ and 100, respectively. Assume that the effect of h_{oe} is negligible.*

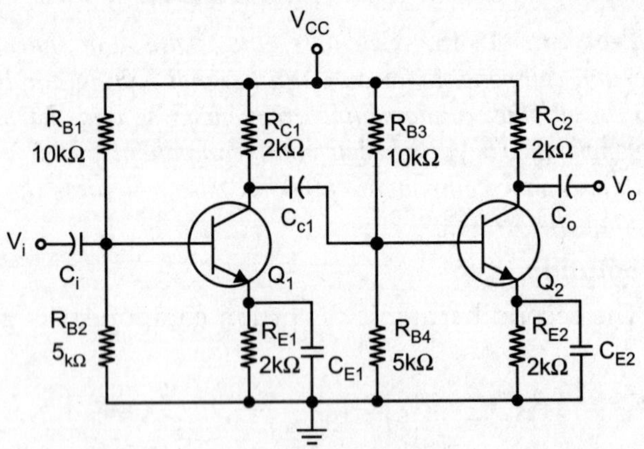

Fig. 13.71

Solution. Voltage gain of the second stage is given by,

$$A_{v2} = \frac{-h_{fe2} \times R_{C2}}{h_{ie2}}$$

$$= \frac{-100 \times 2 \times 10^3}{1 \times 10^3} = -200$$

Voltage gain of the first stage is given by,

$$A_{v1} = \frac{-h_{fe1} \times (R_{B3} \| R_{B4} \| R_{C1} \| h_{ie2})}{h_{ie1}}$$

$$= \frac{-100 \times (10 \times 10^3 \| 5 \times 10^3 \| 2 \times 10^3 \| 1 \times 10^3)}{1 \times 10^3}$$

$$= \frac{-100 \times 555.56}{1 \times 10^3} = -55.556$$

Overall voltage gain is given by, $A_v = A_{v1} \times A_{v2} = -200 \times -55.556 = 11111.2$

The input impedance of the amplifier Z_i is given by,

$$Z_i = R_{B1} \| R_{B2} \| h_{ie1}$$

$$= 10 \times 10^3 \| 5 \times 10^3 \| 1 \times 10^3 = 769.05\ \Omega$$

The output impedance of the amplifier is given by $Z_o = R_{C2} \| (1/h_{oe2})$. As the effect of hoe is negligible, therefore, $Z_o = R_{c2} = 2$ kΩ.

Problem 13.18 *Figure 13.72 shows a Darlington amplifier. The two transistors Q_1 and Q_2 are identical and the h-parameters for both the transistors are $h_{ie} = 1$ kΩ, $h_{fe} = 100$ and $h_{oe} = 40 \times 10^{-6}$ mhos. The values of the voltages $V_{CC} = 15$ V, $V_{BE1} = 0.7$ V and $V_{BE2} = 0.7$ V. Determine the following:*

(a) *DC voltage and currents*

(b) *Input impedance*

(c) *Output impedance*

(d) *Voltage gain*

(e) *Current gain*

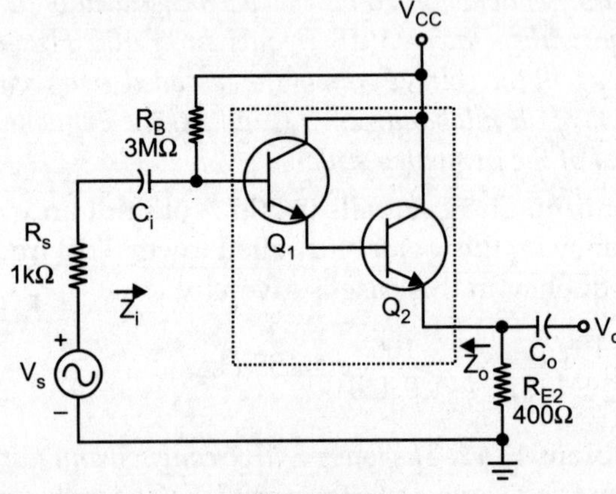

Fig. 13.72

Solution. The base bias current of the Darlington amplifier I_{B1} is given by,

$$I_{B1} = \frac{V_{CC} - V_{BE1} - V_{BE2}}{R_B + (1 + h_{fe})^2 R_E}$$

$$= \frac{15 - 0.7 - 0.7}{3 \times 10^6 + (101)^2 \times 400}$$

$$= \frac{13.6}{3 \times 106 + 4.08 \times 10^6} = 1.92\ \mu A$$

Emitter current of the Darlington amplifier is

$I_{E2} = (1 + h_{fe})^2 I_{B1} = (101)^2 \times 1.92 \times 10^{-6}\ \mu A = 19.6$ mA

The collector current $I_C \cong I_E = 19.6$ mA

The collector emitter voltage $V_{CE} = V_C - V_E = 15 - 19.6 \times 10^{-3} \times 400 = 7.16$ V

$$Z_i = R_B \left\| \frac{(1 + h_{fe})^2 R_{E2}}{1 + h_{oe} h_{fe} R_{E2}} \right.$$

$$= 3 \times 10^6 \left\| \frac{(101)^2 \times 400}{1 + 40 \times 10^{-6} \times 100 \times 400} \right.$$

$$= 3 \times 10^6 \| 1.57 \times 10^6 = 1.03\ M\ \Omega$$

$$Z_o \cong \frac{R_s + h_{ie}}{(1+h_{fe})^2} + \frac{h_{ie}}{1+h_{fe}}$$

$$= \frac{1\times 10^3 + 1\times 10^3}{(101)^2} + \frac{1\times 10^3}{101}$$

$$= 0.196 + 9.9$$

$$= 10.1\ \Omega$$

Voltage gain A_v is given by,

$$A_v = 1 - \frac{h_{fe}}{(1+h_{fe})R_{E2}}$$

$$= 1 - \frac{1\times 10^3}{101\times 400}$$

$$= 1 - 0.025$$

$$= 0.975$$

Current gain A_i is given by,

$$A_i = \frac{R_B}{Z_i + R_B} \times \left(\frac{(1+h_{fe})^2}{1+h_{oe}h_{fe}R_{E2}}\right)$$

$$= \frac{3\times 10^6}{3\times 10^6 + 1.03\times 10^6} \times \frac{(101)^2}{1 + 40\times 10^{-6}\times 100\times 400}$$

$$= 0.74 \times 3923.46 = 2903.36$$

Problem 13.19. *For the cascode amplifier circuit shown in Fig. 13.73, determine the values of resistors R_E, R_1 and R_2 such that the operating point is I_{CQ} = 10 mA and V_{CEQ} = 10 V. Given that the value of β = 100 and V_{BE} of each transistor is 0.7 V.*

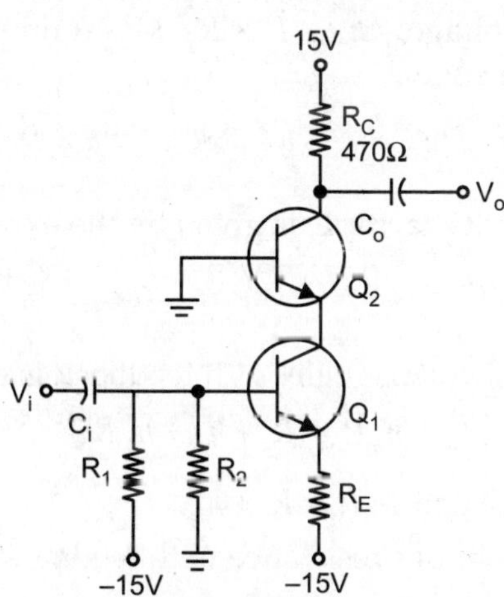

Fig. 13.73

Solution. DC voltage drop across the resistor R_C is given by 470 × 10 × 10^{-3} = 4.7 V

Therefore, the voltage drop across the resistor R_E is 15 – 4.7 – 10 – 10 – (– 15) = 30 – 24.7 = 5.3 V

Value of resistor R_E is approximately equal to 5.3/10 × 10^{-3} = 5.3 × 10^3 = 0.53 kΩ

Voltage at the base of transistor Q_1 is given by –15 + 5.3 + 0.7 = –9V

For good bias stability, current through resistor R_2 >> I_{BQ1}. Value of R_2 should not be so large such that this condition is not met and also, it should not be too small to have an undue load on the power supply.

Assume R_2 = 10 kΩ. Current through R_2 is 9/(10 × 10^3) = 0.9 mA

Current through resistor R_1 = 0.9 × 10^{-3} – 10 × 10^{-3}/100 = 9 × 10^{-3} – 10 × 10^{-5} = 0.89 × 10^{-3} = 0.89 mA

Voltage drop across R_1 is 15 – 9 = 6V

Value of resistor R_1 = 6/(0.89 × 10^{-3}) = 6.74 kΩ

Problem 13.20. *Calculate the efficiency of the amplifier circuit shown in Fig. 13.74. The input voltage applied is such that it produces a base-current of 10 mA peak. The base-emitter voltage of the transistor is equal to 0.7 V.*

Solution. The value of the base-current at the Q-point is given by I_{BQ} = (V_{CC} – 0.7)/[R_B + (β + 1)R_E] = (25 – 0.7)/(2 × 10^3 + 41 × 10) = 24.3/2410 = 10.08 × 10^{-3}A = 10.08 mA

The value of the collector current I_{CQ} is given by β × I_{BQ} = 40 × 10.08 × 10^{-3} = 403.3 × 10^{-3} A = 403.3 mA

V_{CEQ} = V_{CC} – $I_C R_C$ – $I_E R_E$ = 25 – 403.3 × 10^{-3} × 25 – 403.3 × 10^{-3} × 10 = 14.09 V

The input power P_i = V_{CC} × I_{CQ} = 25 × 403.3 × 10^{-3} = 10.08 W

The output power P_o = [$I_{C(p)}^2$ × R_C]/2

$I_{C(p)}$ = β × $I_{B(p)}$ = 40 × 10 × 10^{-3} = 400 × 10^{-3} = 400 mA

P_o = [(400 × 10)^2 × 25]/2 = 2W

Efficiency η = (P_o/P_i) × 100% = (2/10.08) × 100% = 19.84%

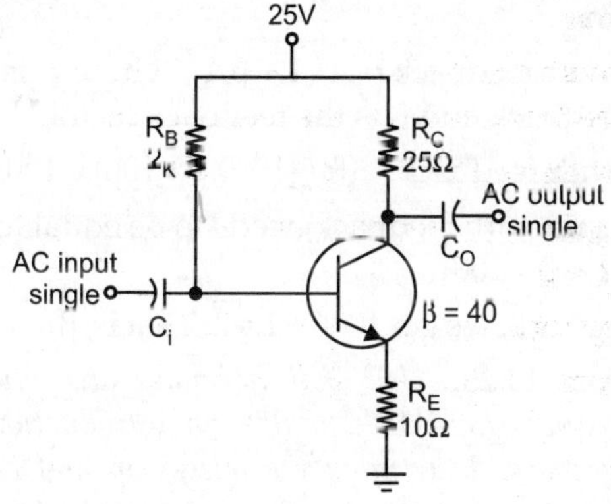

Fig. 13.74

Problem 13.21. *Refer to the non-inverting amplifier circuit of Fig. 13.75. The closed-loop gain of this amplifier is usually considered as being equal to [(1 + (R_2/R_1)]. Given that the open-loop gain of the op-amp is 10 000, values of R_1 and R_2 are 10 KΩ and 100 KΩ, respectively, and the value of* *the applied input voltage V_i is 1 V, determine the percentage error incurred in computing the output voltage by using this expression.*

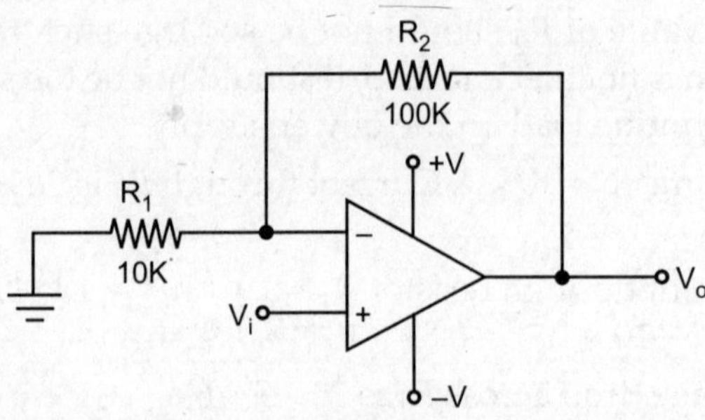

Fig. 13.75

Solution. Closed-loop gain ignoring the effect of finite value of open loop gain = $1 + (R_2/R_1)$.

Substituting for R_1 and R_2, it comes out to be 11.

The output voltage is, therefore, 11 V.

The true value of closed-loop gain is given by $[A/(1+ \beta A)]$. A is the open-loop gain and is equal to 10 000. β is the feedback factor and is given by, $[R_1/(R_1 + R_2)]$. This equals 1/11.

The true value of the closed-loop gain is, therefore, given by [10 000/{(1 + 10 000)/11}] = 10.988.

The true value of output voltage is, therefore, 10.988 V.

Percentage error = [(11 – 10.988)/11] × 100 = 0.11%

Problem 13.22. *An amplifier without feedback has a voltage gain of 100. The designer decides to use 10% negative feedback to bring the non-linear distortion to an acceptable level. Determine the gain of the amplifier in the presence of feedback. If the desired value of gain with feedback is 50, what should in that case be the feedback factor?*

Solution.

Gain with feedback = $A/(1 + \beta A)$, where A is the gain without feedback and β is the feedback factor.

Gain with feedback = 100/(1 + 0.1 × 100) = 100/11 = 9.09

If the gain with feedback needs to be equal to 50, then 50 = 100/(1 + β × 100)

Solving for β, we get 100β + 1 = 2. That is, β = 0.01 or 1%.

Problem 13.23. *The total harmonic distortion of an amplifier reduces from 10% to 1% on introduction of 10% negative feedback. Determine the open-loop and closed-loop gain values.*

Solution. Distortion (with feedback) = Distortion (without feedback)/(1 + βA), where A is the open-loop gain.

0.01 = 0.1/(1 + 0.1 A), which gives 1 + 0.1 A = 10 or A = 90.

Therefore, open-loop gain = 90.

Closed-loop gain = 90/(1 + 0.1 × 90) = 9.

Problem 13.24. *A negative-feedback amplifier has –20 dB of feedback. Determine the loop gain.*

Solution. Negative feedback in dB = 20 log (Gain with feedback/Gain without feedback) = –20.

Log (Gain with feedback/Gain without feedback) = – 1.

Gain with feedback/Gain without feedback = 0.1 = 1/ $(1 + \beta A)$.

This gives (1 + βA) = 10, or βA = 9.

Loop gain = –βA = –9.

Problem 13.25. *For the common-collector circuit of Fig. 13.76, determine expressions for voltage gain, input resistance and output resistance in the presence of negative feedback.*

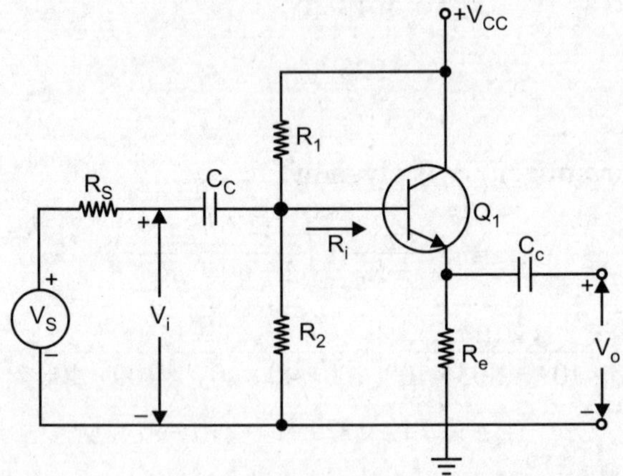

Fig. 13.76

Solution. Voltage gain, A_V = Output voltage/Input voltage

Output voltage = $h_{fe} \times I_b \times R_e$ (assuming $I_c \cong I_e$) and input voltage = V_s.

Therefore, $A_V = h_{fe} \times I_b \times R_e/V_s \cong h_{fe} \times R_e/(R_s + h_{ie})$ as $V_s/I_b = R_s + h_{ie}$.

Desensitivity factor D is given by the expression,

$D = 1 + \beta A_V = 1 + (h_{fe} \times R_e)/(R_s + h_{fe}) = (R_s + h_{ie} + h_{fe} R_e)/ R_s + h_{ie})$

Therefore, voltage gain with feedback is given by,

$$A_{Vf} = (h_{fe}R_e)/(R_s + h_{ie} + h_{fe}R_e)$$

Input impedance, $R_i = R_s + h_{ie}$.

Therefore, input resistance with feedback is given by,

$$R_{if} = R_i \times D = R_s + h_{ie} + h_{fe}R_e$$

We are interested in the resistance looking into the emitter. Hence R_e is considered as an external load. Therefore, $R_{of} = \infty$.

Since R_e is the load resistance in the present case, therefore, $R_o' = R_e$.

This gives $R_{of}' = R_o'/D = R_e \times (R_s + h_{ie})/(R_s + h_{ie} + h_{fe}R_e)$.

$$R_{of} = \lim_{Re \to \infty} R_{of}' = (R_s + h_{ie})/h_{fe}$$

Problem 13.26. *Refer to the op-amp-based inverting amplifier circuit of Fig. 13.75. Identify the type of negative feedback. Determine the trans-impedance gain, the input impedance and output impedance of the amplifier, given that trans-impedance, input impedance and output impedance parameters of the op-amp are 100 MΩ, 10 MΩ and 100 Ω, respectively.*

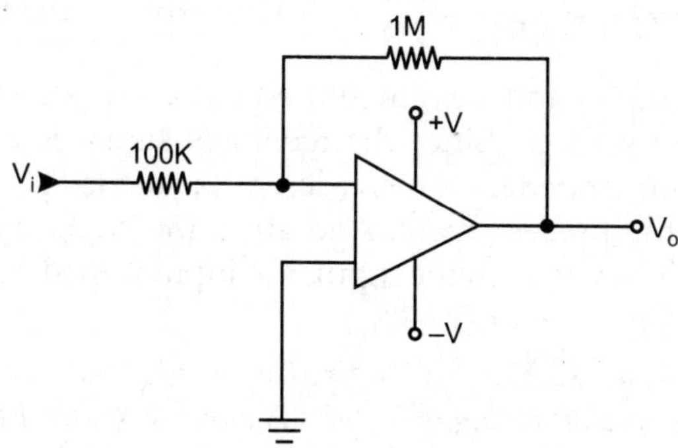

Fig. 13.77

Solution. The feedback is of voltage-shunt type.

Trans-impedance, $R_M = 100\ M\Omega = 100 \times 10^6\ \Omega = 10^8\ \Omega$.

Feedback factor, $\beta = 1/10^6\ (\Omega)^{-1} = 10^{-6}\ (\Omega)^{-1}$.

Therefore, trans-impedance with feedback,

$$R_{Mf} = R_M/(1 + \beta R_M)$$

$=10^8/(1 + 10^{-6} \times 10^8) = 10^8/101 = 0.99 \times 10^6\ \Omega = 990$ K

Input impedance, $R_{if} = R_i/(1 + \beta\ R_M) = 10^7/(1 + 10^{-6} \times 10^8) = 10^7/101 = 99$ KΩ.

As the output is seen across an open-circuit, $R_m = R_M$.

Output impedance, $R_{of} = R_o/(1 + \beta\ R_M) = 100/(1 + 10^{-6} \times 10^8) = 100/101 = 0.99\ \Omega$.

Problem 13.27. *Refer to the cascade arrangement of two CE amplifier stages as shown in Fig. 13.78. Prove that the circuit has current-shunt feedback. Derive an expression for the feedback factor β. What is the current gain with feedback?*

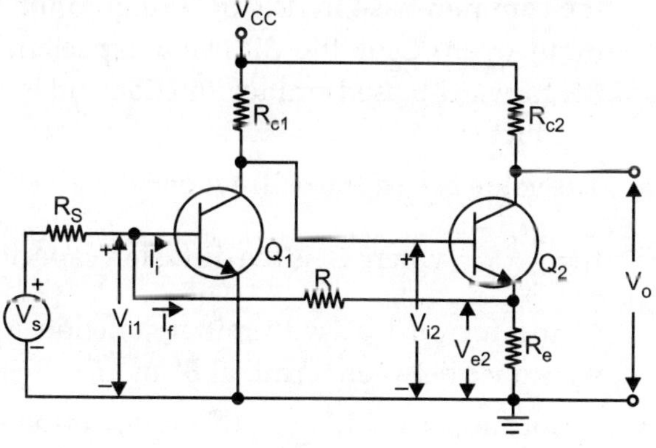

Fig. 13.78

Solution. Refer to the circuit of Fig. 13.78. As the output current is sampled and fed back in shunt with the input current I_i, it is a current-shunt feedback.

The feedback current I_f is given by $I_f = (V_{i1} - V_{e2})/R$.

Assuming $V_{e2} >> V_{i1}$, we get $I_f = - V_{e2}/R$.

$V_{e2} = (I_f - I_o) \times R_e$ considering base current of Q_2 to be negligible as compared to its collector current.

Therefore, $I_f = (I_o - I_f) \times (R_e/R)$.

This gives $I_f = (R_e/R) \times [R/(R + R_e)] \times I_o = [R_e/(R + R_e)] \times I_o = \beta \times I_o$ where, $\beta = R_e/(R + R_e)$.

Current gain with feedback A_{If} is approximately given by $1/\beta$. That is $A_{If} \cong (R + R_e)/R_e$.

Problem 13.28. *An amplifier in the absence of any feedback has following specifications :*

Gain, A	= 100
Input resistance, R_i	= 10 kΩ.
Output resistance, R_o	= 1 kΩ.

How will these specifications be affected on introduction of 10% negative feedback of the voltage series type?

Solution. The amplifier gain with feedback is given by,

$$A_f = \frac{A}{1 - \beta A},\ A = 100,\ \beta = -0.1$$

Therefore, $A_f = \dfrac{100}{1 + 10} \approx 9.$

The input resistance will increase (as it is a series feedback) and the new value is given by,

$$R_{if} = R_i\,(1 - \beta A) = 10\ (1 + 10) = 110\ \text{K}\Omega$$

The output resistance will decrease (as it is a voltage feedback) and the new value is given by,

$$R_{of} = \frac{R_o}{1 - \beta A} = \frac{1000}{1 + 10} = \frac{1000}{11} = 91\Omega.$$

Problem 13.29. *An amplifier has a midband gain of 100 and a bandwidth of 500 kHz. A negative feedback of 10 percent is introduced in the amplifier to stabilise the gain. What are the new gain and bandwidth values ? Also, determine the gain bandwidth product before and after feedback and comment on the results obtained.*

Solution.

Gain without feedback,	$A_v = 100$
Bandwidth without feedback,	$B = 500$ kHz
Feedback factor,	$k = 0.1$

Gain with feedback,

$$A_{vf} = \frac{A_v}{(1 - kA_v)} = \frac{100}{1 - (-0.1)(100)} = \frac{100}{11} \cong 9$$

Bandwidth with feedback, $B_f = B\ (1 - kAv) = 500\ (1 + 10) = 5500$ kHz

Gain-Bandwidth product before feedback = 100 × 500 = 50000 kHz = 50 MHz

Gain-Bandwidth product after feedback $= 9 \times 5500 = 50000$ kHz = 50 MHz

It can be seen from the expressions of grain with feedback and the bandwidth with feedback that the gain-bandwidth product is independent of feedback factor. Thus, gain bandwidth product remains constant. Enhanced bandwidth is achievable only at the cost of sacrifice in gain.

Problem 13.30. *Fig. 13.79 shows the basic common emitter amplifier with emitter bypass capacitor. The amplifier has a voltage gain of 50. If the emitter bypass capacitor is disconnected, the gain is observed to fall to about 33. How do you explain this? Does the disconnection of emitter bypass capacitor affect the input and output impedances of the amplifier. If yes, in what proportion, these impedances are affected?*

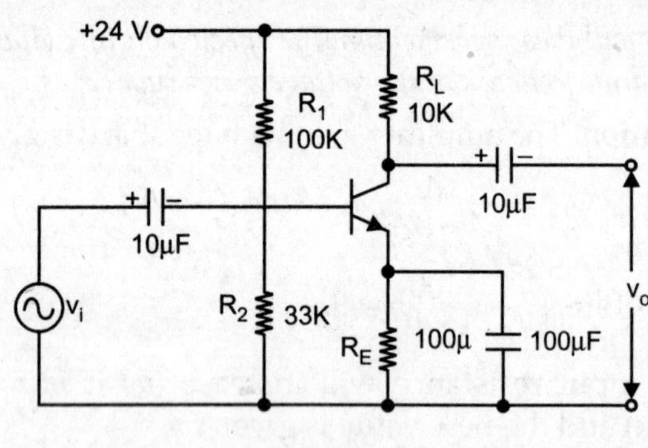

Fig. 13.79

Solution. In the presence of emitter bypass capacitor, there is no signal feedback from output to input as the output AC signal does not pass through the emitter resistor. When the bypass capacitor is disconnected, there is a signal feedback of the current series type with the feedback factor equal to (R_E/R_L).

The voltage gain without feedback $A_v = 50$

Feedback factor, $k = \frac{R_E}{R_L} = -\frac{100}{10000} = -0.01$

Voltage gain with feedback, $A_{vf} = \frac{A_v}{(1 - kA_v)}$

$= \frac{50}{1-(-0.01)(50)} = \frac{50}{1.5} \cong 33$ This expl the results.

The input and output impedances are affected by negative feedback. Since the feedback signal is current, the output impedance increases by a factor $(1 - kA_v)$ (= 1.5 in the present case). And since the signal is being fedback in series with the input, the input impedance also increases by a factor $(1 - kA_v)$.

Problem 13.31. *An audio transistor with maximum power dissipation rating of 5 watts is wired as a direct coupled load class-A amplifier. What is the maximum AC power the amplifier-can deliver to the load? What is the power drawn from the supply under these conditions? Repeat the same, if the transistor is wired as a single ended transformer coupled class-A amplifier.*

Solution.

Direct Coupled

Maximum power dissipation rating of the transistor, $P_T = 5$ W

Therefore, maximum *AC* power delivered to load = 5/2 = 2.5 W

Drain on power supply = 4 × 2.5 = 10 W

Transformer Coupled

Maximum AC power deliverable to load = 2.5 W

Drain on power supply = 2 × 2.5 = 5 W.

SUMMARY OF IMPORTANT FORMULAS

1. The voltage gain of an amplifier is given by $A_v = \frac{V_o}{V_i}$, where V_o is output signal voltage and V_i is output signal voltage.
2. The current gain of an amplifier is given by $A_i = \frac{I_o}{I_i}$, where I_o is output signal current and I_i is output signal current.
3. Noise Figure of an amplifier is given by N.F. = $\frac{(S_i/N_i)}{(S_o/N_o)}$ where (S_i/N_i) is input signal to noise ratio and (S_o/N_o) is output signal to noise ratio.
4. The value of f_β (β cut-off frequency) is given by $f_\beta = \frac{g_{b'e}}{2\pi(C_e + C_c)}$, where C_C is the output of the common-base transistor configuration with input open, C_e is the diffusion capacitance of the forward-biased emitter junction and is input conductance.
5. The value of f_α(α cut-off frequency) is given by $f_\alpha \cong \frac{h_{fe}}{2\pi r_{b'e} C_e}$, where C_e is the diffusion capacitance of the forward-biased emitter junction, $r_{b'e}$ is resistance between terminal B' and E.
6. According to Miller's theorem, impedance connected between two nodes can be replaced by two impedances, one between node-1 and ground and the other between node-2 and ground.

$Z_{in} = \frac{Z}{(1-A)}$, $Z_{out} = \frac{Z}{1-(1/A)}$, where Z is the impedance between 1 and 2 and A is the ratio of V_2/V_1.

7. Average power dissipated by the transistor is given by,

$P_D = V_{CE}I_C$, Where V_{CE} is voltage difference between collector and emitter and I_C is collector current.

8. For a three stage cascaded BJT amplifier the overall gain is given by $A_v = A_{v1} \times A_{v2} \times A_{v3}$

9. For a three stage cascaded BJT amplifier the overall output impedance is given by

$Z_i = R_{B1} \parallel R_{B2} \parallel h_{ie1}$

10. For a three stage cascaded BJT amplifier the overall input impedance is given by

$Z_o = R_{C3} \parallel (1/h_{oe3})$

11. For a cascaded amplifier, the lower 3 dB frequency is given by (fLn) $= \frac{f_L}{\sqrt{2^{1/n}-1}}$, where n is the number of stages of the cascade network and f_L is the overall lower 3dB frequency of a single is *CE* stage.

12. For a cascaded amplifier, the lower 3dB frequency is given by $(f_{Hn}) = f_H\sqrt{2^{1/n}-1}$, where n is the number of stages of the cascade network and f_H is the overall upper 3 dB frequency of a single CE stage.

13. Gain of an amplifier by introduction of negative feedback is given by $A_{vf} = \frac{A_v}{(1-\beta A_v)}$, where A_v is gain without feedback and β is feedback factor.

14. Input resistance of amplifier for series feedback is given by $R_{if} = R_i(1-\beta A)$

Where R_i is input resistance of amplifier without feedback, A is gain without feedback and β is feedback factor.

15. Input resistance of amplifier for shunt feedback is given by $R_{if} = \frac{R_i}{(1-\beta A)}$

Where R_i is input resistance of amplifier without feedback, A is gain without feedback and β is feedback factor.

16. Output resistance of amplifier for voltage feedback is given by $R_{of} = \frac{R_o}{(1-\beta A)}$

Where R_o is output resistance of amplifier without feedback, A is gain without feedback and β is feedback factor.

17. Output resistance of amplifier for current feedback is given by $R_{of} = R_o(1-\beta A)$

Where R_o is output resistance of amplifier without feedback, A is gain without feedback and β is feedback factor.

18. The common emitter *h*-parameters can be expressed in terms of common base and common collector *h*-parameters by following expressions:

$$h_{ie} = \frac{h_{ib}}{1+h_{fb}} = h_{ic}$$

$$h_{re} = \frac{h_{ib}h_{ob}}{1+h_{fb}} - h_{rb} = (1-h_{rc})$$

$$h_{fe} = -\frac{h_{fb}}{1+h_{fb}} = -(1+h_{fc})$$

$$h_{oe} = \frac{h_{ob}}{1+h_{fb}} = h_{oc}.$$

19. The common base *h*-parameter can be expressed in terms of common emitter and common collector *h*-parameters by following expressions:

$$h_{ib} = \frac{h_{ie}}{1+h_{fe}} = -\frac{h_{ic}}{h_{fc}}$$

$$h_{rb} = \frac{h_{ie}h_{oe}}{1+h_{fe}} - h_{re} = -\frac{h_{ic}h_{oc}}{h_{fc}}$$

$$h_{fb} = -\frac{h_{fe}}{1+h_{fe}}$$

$$h_{ob} = \frac{h_{oe}}{1+h_{fe}}$$

20. The common collector *h*-parameters can be expressed in terms of common emitter and common base *h*-parameters by following expressions:

$$h_{ic} = h_{ic} = \frac{h_{ib}}{1+h_{fb}}$$

$$h_{rc} = (1-h_{re}) \approx 1$$

$$h_{fc} = -(1+h_{fe}) = \frac{-1}{(1+h_{fb})}$$

$$h_{oc} = h_{oe} = \frac{h_{ob}}{1+h_{fb}}$$

21. In the case of CE, CB and CC configurations, the current gain A_I, the voltage gain A_V, the input resistance R_i the output resistance Ro (with no load) and R_o (with load resistance R_L) are given by following expressions:

Current gain, A_I = – h_{fe} (with R_e bypassed or unbypassed) – CE configuration

= $(1 + h_{fe})$ – CC configuration

= $\frac{h_{fe}}{1+h_{fe}}$ – CB configuration

Voltage gain, $A_v = -\frac{h_{fe}R_L}{h_{ie}}$

(with R_e bypassed or unbypassed) – CE configuration

= $1 - \frac{h_{ie}}{R_i}$ – CC configuration

= $\frac{h_{fe}R_L}{R_i}$ – CB configuration

Input resistance, $R_i = h_{ie}$ (R_e unbypassed) = $h_{ie} + (1 + h_{fe}) R_e$ (R_e bypassed)

= $h_{ie} + (1 + h_{fe}) R_L$ – CC configuration

= $\frac{h_{ie}}{1+h_{fe}}$ – CB configuration

Output resistance R_o = Infinity – *CE* configuration

= $\frac{R_s + h_{ie}}{1+h_{fe}}$ – CC configuration

= Infinity – CB configuration

Output resistance $R'_o = R_L$ – *CE* configuration

= $R_o \,||\, R_L$ – CC configuration

= R_L – *CB* configuration

Self Evaluation Exercise

I. Multichoice Questions

1. The parameter h_{oe} can be determined by

(*a*) taking the slope of the output characteristic curve at the operating point

(*b*) taking the slope of the input characteristic curve at the operating point

(*c*) cannot be determined using the input and output characteristic curves

(*d*) by taking the collector current increment for a fixed value of collector-emitter voltage

2. What is the unit of the output conductance parameter?

(*a*) Ohms (*b*) It is dimensionless

(*c*) Mhos (*d*) Ampere

3. Increase in the value of transistor's h_{fe} parameter results in

(*a*) decrease in the value of input impedance and increase in the value of current gain

(*b*) decrease in the values of both the input impedance and the current gain

(*c*) increase in the values of both the input impedance and the current gain

(*d*) increase in the value of input impedance and decrease in the value of current gain

4. Increase in the junction temperature of a transistor results in

(*a*) increase in the values of all the four h-parameters

(*b*) decrease in the values of all the four h-parameters

(*c*) increase in the value of h_{ie} and h_{fe} parameters and decrease in the values of h_{re} and h_{oe} parameters

(*d*) decrease in the value of h_{ie} and h_{fe} parameters and increase in the values of h_{re} and hoe parameters

5. Which of the following statement (*s*) is/are true?

(*a*) The low-frequency response of an amplifier is due to the bypass and the coupling capacitors.

(*b*) The high-frequency response of an amplifier is due to the bypass and the coupling capacitors.

(*c*) The low-frequency response of an amplifier is due to the junction capacitances and the stray-wiring capacitances.

(*d*) The high-frequency response of an amplifier is due to the junction capacitances and the stray-wiring capacitances.

(*e*) Both (*a*) and (*d*).

(*f*) Both (b) and (c).

6. The voltage gain of an amplifier decreases at 20 dB/decade above 100 kHz. If the mid-band frequency gain is 80 dB, what is the value of the voltage gain at 2 MHz?

(*a*) 60 dB (*b*) 52 dB

(*c*) 54 dB (*d*) 64 dB

7. In the *h*-parameter model the input and the output sections are modeled as
   (*a*) voltage sources
   (*b*) current sources
   (*c*) input section as voltage source and output section as current source
   (*d*) input section as current source and output section as voltage source

8. Larger the spacing between the curves of the output characteristics of a transistor
   (*a*) smaller is the value of h_{fe}
   (*b*) larger is the value of h_{fe}
   (*c*) h_{fe} is independent of the spacing
   (*d*) h_{fe} can increase or decrease depending upon the circuit configuration

9. Which of the following statement(s) is/are true?
   (*a*) In the case of an FET, there is no feedback from the output to the input, whereas in the case of a BJT there is feedback between the output and the input circuits through the parameter h_{re}.
   (*b*) In the case of a BJT, there is no feedback from the output to the input, whereas in the case of an FET there is feedback between the output and input circuits through the parameter g_m.
   (*c*) BJT is a more ideal amplifier as compared to an FET.
   (*d*) FET is a more ideal amplifier as compared to a BJT.
   (*e*) Both (*a*) and (*d*).
   (*f*) Both (*b*) and (*c*).

10. Which of the following statement(s) is/are false?
    (*a*) The parameters h_i and h_r are determined from the input characteristic curves of the transistors whereas the parameters h_f and h_o are determined from the output or the collector characteristics.
    (*b*) Cascode amplifiers are two-stage amplifiers comprising a transconductance amplifier followed by a current buffer.
    (*c*) Darlington connection refers to the connection of two bipolar junction transistors wherein their collector terminals are tied together and the emitter terminal of one transistor is connected to the base terminal of the other transistor.
    (*d*) Darlington connection refers to the connection of two bipolar junction transistors wherein their emitter terminals are tied together and the collector terminal of one transistor is connected to the base terminal of the other transistor.

11. The emitter diffusion capacitance for a transistor
    (*a*) is proportional to the emitter-bias current.
    (*b*) is inversely proportional to the emitter-bias current.
    (*c*) is independent of the emitter-bias current.
    (*d*) is proportional to the square of emitter-bias current.

12. The rise time of an amplifier is
    (*a*) inversely proportional to the upper 3 dB cut-off frequency.
    (*b*) directly proportional to the upper 3 dB cut-off frequency.
    (*c*) independent of the upper 3 dB cut-off frequency.
    (*d*) proportional to the square root of the upper 3 dB cut-off frequency.

13. The value of α-cut-off frequency is
    (*a*) smaller than the β-cut-off frequency.
    (*b*) greater than the β-cut-off frequency.
    (*c*) can be more or less than the β-cut-off frequency.
    (*d*) equal to the β-cut-off frequency.

14. The conductance ($g_{b'c}$) takes into account
    (*a*) the resistance between the emitter and the collector terminals.
    (*b*) the conductance between the base and collector due to flow of majority carriers.
    (*c*) the reduction in the flow of emitter current.
    (*d*) the feedback effect between the output and the input due to the Early effect.

15. The ohmic base spreading resistance is represented as
    (*a*) the increase in the recombination base current due to the increase in the minority carriers in the base region.

(b) the conductance between the collector and the emitter terminals.
(c) a lumped parameter between the external base terminal and the node B′.
(d) the feedback effect between the output and the input due to the Early effect.

16. Which of the following amplifier class suffers mainly from the problem of crossover distortion?
(a) Class A (b) Class B
(c) Class AB (d) Class C

17. Which of the following best describes a class A amplifier?
(a) High efficiency and high distortion
(b) Low efficiency and high distortion
(c) Low efficiency and low distortion
(d) High efficiency and low distortion

18. Heat sinks are used in power transistors to
(a) reduce the transistor power
(b) reduce the junction temperature
(c) increase the ambient temperature
(d) increase the collector current

19. In which class of amplifiers, it is theoretically possible to achieve 100% efficiency?
(a) Class D (b) Class A
(c) Class AB (d) Class C

20. In which of the following amplifier classes, the active device operates for the whole of the input signal cycle?
(a) Class A (b) Class AB
(c) Class B (d) Class C

21. The maximum possible efficiency in series fed class A amplifiers and transformer coupled class A amplifiers are
(a) 50% and 25% (b) 25% and 50%
(c) Both have 25% (d) Both have 50%

22. How much percentage of maximum power can be dissipated by a transistor for an operating temperature equal to its maximum junction temperature?
(a) 100% (b) 10%
(c) 50% (d) 0%

23. A 10 W class B push–pull amplifier should have transistors with individual collector dissipation of
(a) 2 W (b) 1 W
(c) 5 W (d) 10 W

24. Which of the following harmonic component has the main contribution in the total distortion in an amplifier?
(a) Third harmonic component
(b) Second harmonic component
(c) Fourth harmonic component
(d) Fifth harmonic component

25. Class C amplifiers are generally used as
(a) power amplifiers (b) RF amplifiers
(c) audio amplifiers (d) none of these

26. Introduction of negative feedback desensitises or stabilises the gain. The gain stability
(a) increases with increase in open-loop gain
(b) decreases with increase in open-loop gain
(c) decreases with increase in feedback factor
(d) increases with increase in loop gain

27. A voltage amplifier has an open-loop gain of 100. If 10% negative feedback were introduced in the amplifier, then an 11% change in open-loop gain would cause
(a) 1% change in closed-loop gain
(b) 11% change in closed-loop gain
(c) 1.1% change in closed-loop gain
(d) 0.1% change in closed-loop gain

28. In the case of a negative-feedback amplifier, which of the following is true?
(a) Closed-loop gain can be less than or more than the open-loop gain depending upon the type and amount of feedback.
(b) Closed-loop gain is always less than the open-loop gain.
(c) Closed-loop gain is always more than the open-loop gain.
(d) Closed-loop gain and bandwidth are always less than the corresponding open-loop values.

29. In which of the following feedback topologies, the input impedance increases with introduction of feedback.
(a) Voltage-shunt feedback
(b) Current-shunt feedback
(c) Voltage-series feedback
(d) None of these

30. In which of the following feedback topologies, the input impedance decreases with introduction of feedback.

(*a*) Voltage-shunt feedback
(*b*) Current-series feedback
(*c*) Voltage-series feedback
(*d*) None of these

31. In which of the following feedback topologies, the output impedance increases with introduction of feedback.
(*a*) Voltage-shunt feedback
(*b*) Current-shunt feedback
(*c*) Voltage-series feedback
(*d*) None of these

32. In which of the following feedback topologies, both input as well as output impedances decrease with introduction of feedback.
(*a*) Voltage-shunt feedback
(*b*) Current-shunt feedback
(*c*) Voltage-series feedback
(*d*) None of these

33. Which of the following amplifier configurations has an inherent current-series feedback.
(*a*) Emitter follower
(*b*) Common-base amplifier
(*c*) Common-emitter amplifier by bypassed emitter resistor
(*d*) Common-emitter amplifier with un-by-passed emitter resistor

34. It is desired to design a voltage controlled current source. What type of negative feedback should preferably be introduced to make it a stable source?
(*a*) Voltage-series feedback
(*b*) Current-series feedback
(*c*) Current-shunt feedback
(*d*) Voltage-shunt feedback

35. A voltage amplifier consists of cascade arrangement of two identical stages. Negative feedback is introduced from output of second stage to input of first stage to get a stable overall closed-loop gain of 45. In order that 10% variation in open-loop gain of each stage does not produce more than 1% variation in overall closed-loop gain, the minimum value of open-loop gain of each stage should be
(*a*) 30 (*b*) 900
(*c*) 45 (*d*) None of these

36. A voltage amplifier has 5% negative feedback. Its voltage gain is
(*a*) indeterminate from given data
(*b*) approximately 20
(*c*) approximately 5
(*d*) 100

37. For the expression of gain with feedback $A_f = A/(1 + \beta A)$ in the case of negative-feedback amplifier to be valid,
(*a*) feedback factor β should be independent of load and source resistances
(*b*) forward transmission through the feedback network should be zero
(*c*) reverse transmission through amplifier should be zero
(*d*) all the above should be true

38. Voltage-shunt feedback stabilises
(*a*) voltage gain
(*b*) current gain
(*c*) trans-resistance
(*d*) trans-conductance

39. An amplifier with open-loop input resistance of 100 *K* has – 20 dB of voltage-series feedback. Closed-loop input resistance would be
(*a*) 10 K (*b*) 1000 K
(*c*) 2000 K (*d*) 5 K

40. In a negative-feedback amplifier with a high open-loop gain, doubling the feedback factor
(*a*) doubles the closed-loop gain too
(*b*) has no effect on closed-loop gain
(*c*) reduces the closed-loop gain to one-fourth
(*d*) reduces the closed-loop gain to one-half

II. Figure 13.80 shows the AC load line drawn on the output characteristics of a certain CE amplifier circuit. Two choices for the operating point are given. Which is the preferred option?

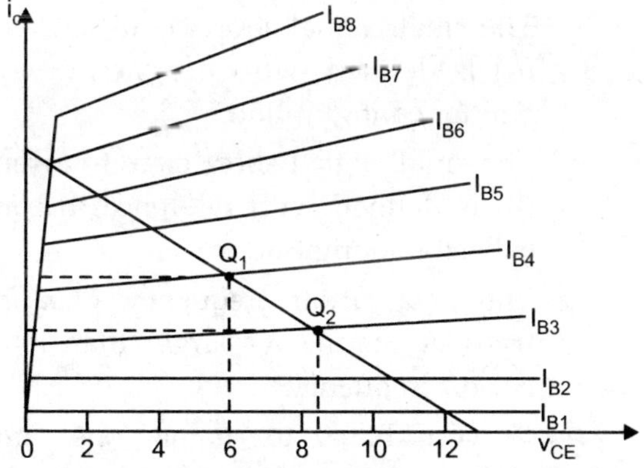

Fig. 13.80

III. Figure. 13.79 (*b*) shows the characteristics along with the DC load line for the CE amplifier configuration of Fig. 13.79 (*a*) Mark the end points of the load line and find out the base current at the quiscent point.

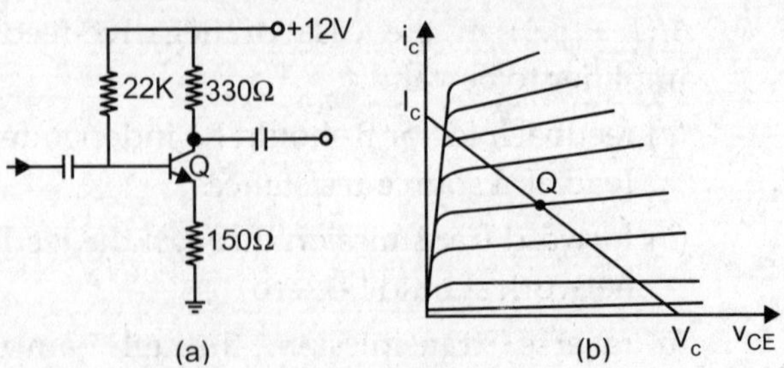

Fig. 13.81

IV. A single stage CE amplifier has a lower 3 dB cut-off 64 Hz and an upper 3 dB cut-off of 10 kHz. What will be the new values for these frequencies for two stage amplifier consisting of a cascaded arrangement of two identical stages of the type mentioned?

V. Each of the three basic amplifier configurations *i.e. CE, CC* and *CB* is characterised by its own set of electrical characteristics. Below are given some false statements with reference to these circuit configurations. For each of these false statements, write the correct one.

(*a*) The current gain per stage is the lowest in case of emitter-follower configuration.

(*b*) The voltage gain per stage is the highest in case of common-emitter configuration.

(*c*) Common-collector configuration is the only one that has both voltage gain as well as current gain greater than unity.

(*d*) Common-base configuration is used as a buffer stage to match a high impedance source to a low impedance load.

(*e*) The small signal short circuit current gain (α) is defined with reference to common emitter configuration.

(*f*) The small signal short circuit current gain (β) is defined with reference to common-collector configuration.

(*g*) The beta-cut-off frequency of a bipolar transistor is always larger than its alpha-cut-off frequency.

(*h*) If you have to name one amplifier configuration that has the widest application, then it is the common-collector configuration.

(*i*) The superiormost frequency capability of a transistor is achieved, when it is used in the CE-configuration.

(*j*) f_T is considered as figure-of-merit for a bipolar transistor. It is the frequency at which, the short circuit current gain becomes unity.

VI. Five amplifier relevant transistor connections are shown in Figs. 13.82 (*a*) to 13.82 (*e*). Each one of these connections can be identified with one of the five statements listed ahead. For each of the listed statements, identify and name the corresponding transistor connection diagram.

(*a*) This connection can be considered as two cascaded emitter followers with the first stage having an infinite emitter resistance.

(*b*) It is one of the most vital building blocks in all audio and video amplifiers available in IC package.

(*c*) It is most extensively used amplifier configuration for constructing power amplifiers.

(*d*) It is characterised by the fact that the output of one is fed to the input of the next.

(*e*) It is a popular configuration for constructing audio power amplifier, may be termed as a transformerless Class-B push-pull amplifier.

VII. Read the following statements and tell as to what type of distortion these statements are referring to Choose from : Harmonic Distortion, Intermodulation Distortion, Frequency Distortion, Phase (or Delay) Distortion.

(*a*) A 1 kHz tone at the amplifier input produces not only the amplified 1 kHz tone at the output, in addition, it also produces 2 kHz, 3 kHz and 4 kHz tones.

(*b*) The signal consisting of 1 kHz and 5 kHz components when applied to amplifier's input resulted in an output signal comprising of 1 kHz, 4 kHz, 5 kHz and 6 kHz components.

(*c*) Both these distortions result from the non-linearity in the characteristics of the active device making the amplifier.

(*d*) This type of distortion is not objectionable in an audio amplifier but could be troublesome in video amplifiers under certain conditions.

(*e*) This type of distortion is related to the frequency response of the amplifier.

Answers to Self Evaluating Exercise

I. Multichoice Questions

1. (*a*)	**2.** (*c*)	**3.** (*a*)	**4.** ()	**5.** (*e*)
6. (*d*)	**7.** (*c*)	**8.** (*b*)	**9.** (*e*)	**10.** (*d*)
11. (*a*)	**12.** (*a*)	**13.** (*b*)	**14.** (*d*)	**15.** (*c*)
16. (*b*)	**17.** (*c*)	**18.** (*b*)	**19.** (*a*)	**20.** (*a*)
21. (*b*)	**22.** (*d*)	**23.** (*a*)	**24.** (*b*)	**25.** (*b*)
26. (*d*)	**27.** (*a*)	**28.** (*b*)	**29.** (*c*)	**30.** (*a*)
31. (*b*)	**32.** (*a*)	**33.** (*d*)	**34.** (*b*)	**35.** (*a*)
36. (*b*)	**37.** (*d*)	**38.** (*c*)	**39.** (*b*)	**40.** (*d*)

II. Q_1 is the preferred choice.

III. I_C = 36 mA, V_C = 12V Base current at the quiscent point = 0.54 mA

IV. Lower 3 dB frequency = 100 Hz, Upper 3 dB frequency – 6.4 kHz

V. (*a*) There are three possible correct statements :

1. The current gain per stage is the lowest in case of common-base configuration.
2. The voltage gain per stage is the lowest in case of Emitter-follower configuration.
3. The current gain per stage is the highest in case of Emitter follower configuration.

(*b*) The voltage gain per stage is the highest in case of common-base configuration.

(*c*) Common-emitter configuration is the only one that has both current gain as well as voltage gain greater than unity.

(*d*) Common-collector configuration is used as a buffer stage to match a high impedance source to a low impedance load,

(*e*) The small signal short circuit current gain (α) is defined with reference to common-base configuration.

(*f*) The small signal short circuit current gain (β) is defined with reference to common-emitter configuration.

(*g*) The beta cut off frequency of a bipolar transistor is always smaller than its alpha cut-off frequency.

(*h*) If you have to name one amplifier configuration, that has the widest application, then it is the common-emitter configuration,

(*i*) The superiormost frequency capability of a transistor is achieved, when it is used in CB configuration.

(*j*) f_T is considered as a figure-of-merit for a bipolar transistor. It is the frequency at which the short circuit current gain, when used in CE configuration becomes unity,

VI. (*a*) Figure 13.82 (*b*): Darlington connection

(*b*) Figure 13.82 (*d*): Differential amplifier connection

(*c*) Figure 13.82 (*e*): Push-pull (Class-B) connection

(*d*) Figure 13.82 (*a*): Cascade connection

(*e*) Figure 13.82 (*c*): Complementary symmetry output connection.

VII. (*a*) Harmonic distortion

(*b*) Intermodulation distortion

(*c*) Harmonic and Intermodulation distortions

(*d*) Phase (or delay) distortion

(*e*) Frequency distortion

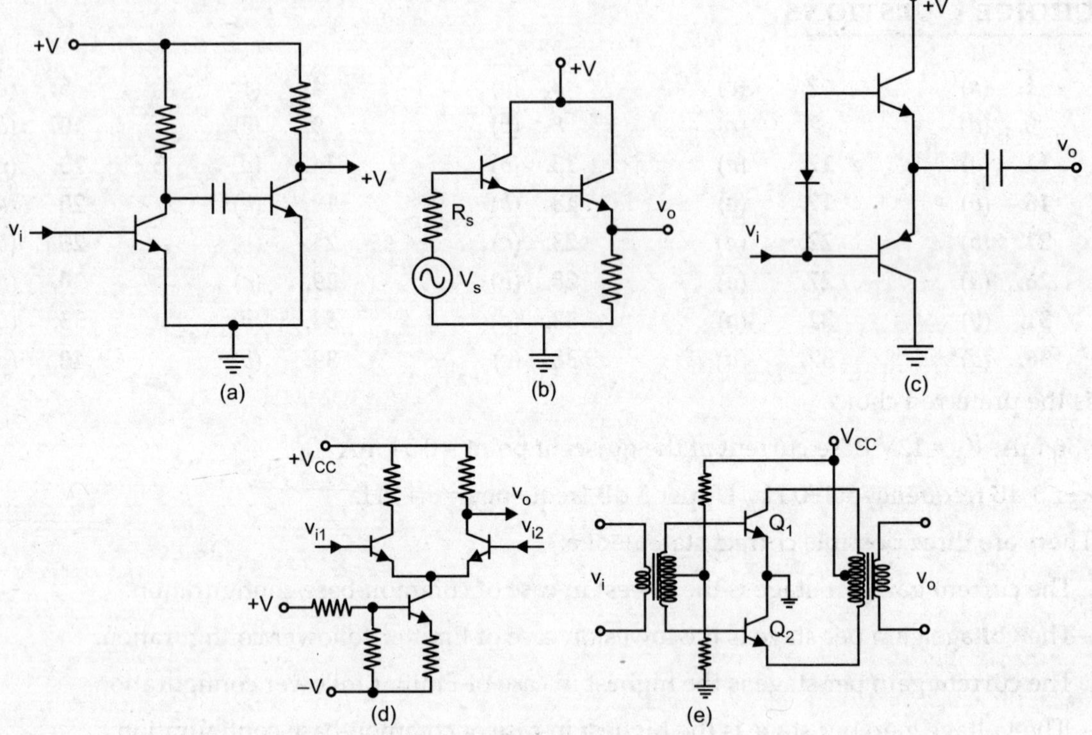

Fig. 13.82

Chapter 14

Oscillators

Sinusoidal and Non-sinusoidal

Learning Objectives

After completing the chapter, you will learn the following topics:

1. Barkhausen Criterion of Oscillations
2. RC Phase Shift Oscillator
3. Bubba Oscillator
4. Twin-T Oscillator
5. Wein-bridge Oscillator
6. Comparison between RC Phase Shift and Wein-bridge Oscillators
7. Armstrong Oscillator
8. Hartley Oscillator
9. Colpitt Oscillator
10. Clapp Oscillator
11. Comparison between RC and LC Oscillators
12. Crystal Oscillators
13. Piercing Oscillator
14. Multivibrators including Bistable, Monostable and Astable Multivibrators
15. Schmitt Trigger Circuit
16. Blocking Oscillators

Q. 1. How does an oscillator differ from an amplifier as for circuit configuration is concerned?

Ans. An oscillator is an amplifier with a positive feedback in which the amount of the output to be fed back is function of the amplifier gain.

Q. 2. What is the basic principle of operation of an oscillator? What is Barkhausen criterion for sustained oscillations?

Ans. Figure 14.1 is the block schematic form of an amplifier with a positive feedback. In an arrangement like this,if the feedback factor (β) is such that $|\beta A|$ equals unity and also if the feedback signal arrives in phase with the input (this is implied by positive feedback),the output will be there, even if the input is removed. In practice, the feedback network is a frequency selective network, it could be an *R-C* network or an *L-C* network or the one containing a quartz crystal. With such a circuit represented by block schematic of Fig. 14.1, due to some inevitable noise present at the amplifier input, there is an amplified output due to noise having all frequency components. But the conditions of $|\beta A| = 1$ and presence of positive feedback (characterised by zero loop phase-shift are satisfied at one frequency only. As a result, only that frequency component will be subsequently amplified and sustained. So, the oscillator out-put frequency will be the one selected by the feedback network. It is the frequency at which the loop phase-shift is zero or integral multiple of 360°. As per Barkhausen criterion, for oscillations to occur, the condition that must be satisfied is:

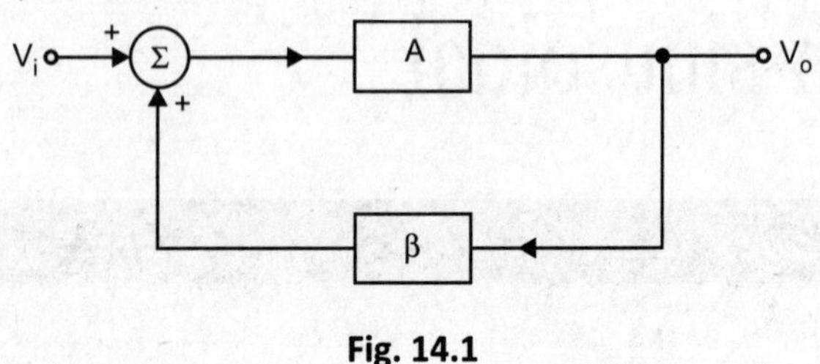

Fig. 14.1

$\beta . A = 1$, Which implies

(*i*) $|\beta A| = 1$ i.e. loop gain should be unity.

(*ii*) Arg $\beta A = 0$ or integral multiple of 360° *i.e.* loop phase shift should be zero or integral mulitple of 360°.

Q. 3. In practice, the loop gain is kept slightly greater then unity. Why is it so?

Ana. If for an oscillator, loop gain is precisely unity, no doubt, oscillation will occur but with change in parameters of the transistor or other active devices used in the amplifier part of the oscillator due to ageing or replacement,you may land up with gain less than unity and in that case, there will be no oscillations. So in practice, loop gain is kept sightly grater than unity to ensure that oscillator works, even if there is a slight change in the circuit parameters. Moreover, there is no harm in keeping loop gain slightly greater than unity as the output cannot increase infinitely as it appears because the output amplitude will be limited due to onset of non-linearity of active device used. However, if (β.A) is much larger than unity the oscillator output will have lot of distortion. So,it is recommended to keep the product (β.A) only slightly greater then unity.

Q. 4. If the feedback network of a certain oscillator provides 1% positive feedback, what should be the minimum gain for the amplifier section of the oscillator for sustained oscillation?

Ans. A feedback of 1% means $\beta = 0.01$. Therefore, amplifier's minimum gain is $1/\beta = 100$.

Q. 5. If in an oscillator, the amplifier portion is a two stage *CE* configuration, what should be the phase-shift to be introduced by the feedback network at the oscillation frequency for sustained oscillations?

Ans. Since a two-stage *CE* amplifier provides a phase-shift 2π radians, the feedback network must not introduce any more phase-shift in order to satisfy Barkhausen criterion for sustained oscillations.

Q. 6. Draw the basic phase shift oscillator and identify the amplifier and feedback portions. What should be the requirement on the amplifier gain for sustained oscillations and what will be the frequency of oscillations?

Ans. Figure. 14.2 shows *R-C* phase shift oscillator. R_1C_1, R_2C_2 and R_3C_3 constitute the feedback network, whereas transistor Q, biasing resistors, R_4, R_5 and R_6 and the bypass capacitor C_4 constitute the amplifier section. R_6 can be adjusted for the fine variation of amplifier gain so as to make it only slightly greater than the attenuation in the feedback network .

It can be mathematically verified that the above feedback network introduces an attenuation of (1/29) for $R_1 = R_2 = R_3 = R$ and $C_1 = C_2 = C_3 = C$. Therefore, the amplifier section must provide a gain of at least 29.

The oscillator output frequency which is the frequency at which overall phase-shift around the loop is zero or integr~l of 2π radians is given by,

$$f = \frac{1}{2\pi\sqrt{6}\,RC}$$

If the positions of (R) and (C) in the feedback network are interchanged, the frequency of oscillations would then be given by,

$$f = \frac{\sqrt{6}}{2\pi RC}$$

Fig. 14.2

Incidently,this is the frequency at which the phase-shift network introduces a phase shift of 180°. Another 180° is provided by *CE* amplifier stage to give an overall loop phase-shift of 360° or 2π radians.

Q. 7. Is it possible to have a tow-section phase-sift network instead of three-section network in an *R-C* phase-shift oscillator of the types discussed in the last question?

Ans. No, it is not. Two section network means each section will have to provide a phase-shift of 90°. A single *R-C* section can provide a phase shift of 90° only for $R = 0$. This means infinite attenuation which implies amplifier has to have a gain of infinity to produce sustained oscillations. All this not practical. Hence, it is not possible to construct an *R-C* phase-shift oscillator with a two section *R-C* phase-shift oscillator with a two section *R-C* feedback network.

Q. 8. Why is it difficult to have a variable frequency operation with an *R-C* phase shift oscillator?

Ans. Because, it is difficult to simultaneously vary three capacitors.

Q. 9. Why is it not feasible to vary three resistors instead of three capacitors to get variable frequency operation is an *R-C* phase shift oscillator?

Ans. Variation of resistance values will alter the loop gain of the oscillator circuit and there is likelihood of it not satisfying Barkhausen criterion for sustained oscillations.

Q. 10. What is the difference between Bubba oscillator and buffered RC phase shift oscillator?

Ans. Bubba oscillator is a slight variation of the buffered RC phase shift oscillator with the difference between the two being that the Bubba oscillator uses four RC sections in the feedback network with each RC section contributing a phase difference of 45°.

Q. 11. Draw and explain a twin-T oscillator circuit.

Ans. The twin-T oscillator employs a twin-T type of notch filter network as the frequency selective component in the feedback network. Figure 14.3 shows the basic circuit schematic of a twin-T oscillator. The circuit employs both positive as well as negative feedback. The positive feedback necessary to produce oscillations is provided by a voltage divider network of R_1 and R_2. The negative feedback is through the frequency selective twin-T network. Figure 14.4 shows the amplitude and phase response of the twin-T network as a function of normalised frequency (ω_c).

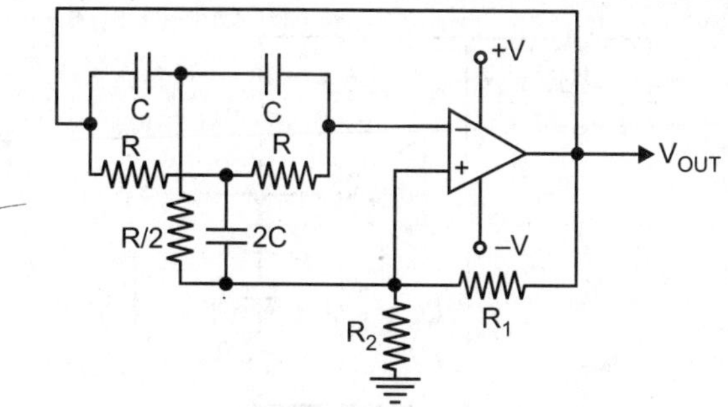

Fig. 14.3 Twin-T oscillator

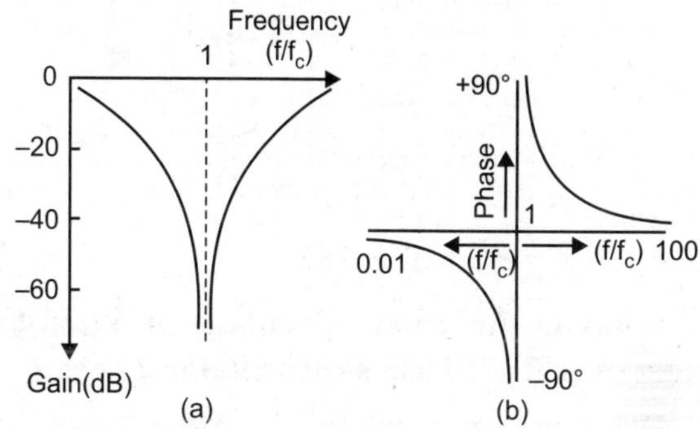

Fig. 14.4 (a) Magnitude and (b) phase response of twin-T network

Q. 12. Draw the basic Wien-bridge oscillator and explain why it is necessary to have a zero phase shift amplifier? As per Barkhausen criterion, what should be the amplifier gain for sustained oscillations? What will be the frequency of oscillations?

Ans. The feedback network in this case consists of series *R-C* branch in series with a parallel *R-C* branch introduces a zero phase shift at a frequency given by $f = \dfrac{1}{2\pi RC}$. Also, atenuation at this frequency is 1/3.

Therefore, for sustained oscillation as per Barkhausen criterion to be satisfied are (*i*) the amplifier gain should at least be 3. In practice,or should be slightly grater than 3 (*ii*) To satisfy the condition of zero loop phase-shift and amplifier must not introduce any phase-shift. Two stage *CE* amplifier as shown in Fig. 14.5 (*a*) and non-inverting op amp configuration as shown in Fig. 14.5 (*b*) are popular amplifier circuits for wien-bridge oscillators.

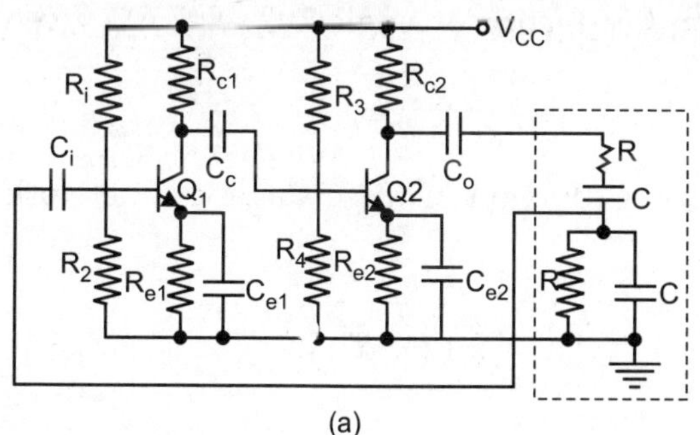

(a)

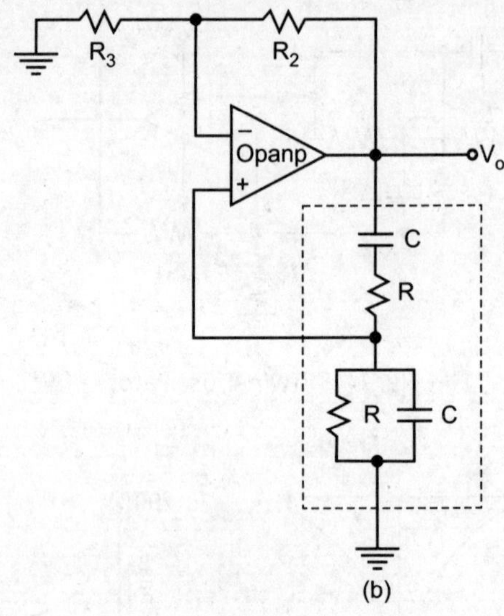

Fig. 14.5

Q. 13. What is the chief advantage of Wien-bridge oscillators over *R-C* phase-shift scillators?

Ans. Wien-bridge oscillator is more amenable to variable frequency operation, as it is easier to vary two capacitors simultaneously than varying three capacitors.

Q. 14. In an *R-C* phase-shift oscillator, each of three *R-C* sections has to provide phase shift of 60° satisfying loop phase shift condition of Barkhausen criterion and thus, produce desired oscillations. This means that oscillator frequency is given by $1/2\pi f RC = \tan 60° = \sqrt{3}$ or $f = \dfrac{1}{2\pi\sqrt{3}RC}$.

In practice the output frequency is observed to be $f = \dfrac{1}{2\pi\sqrt{6}RC}$. Why is that so?

Ans. While calculating the frequency of oscillations. the three *R-C* sections have been assumed to be identical in their circuit behaviour. In practice, it will not be so. In other words, if we derive an expression for the transfer function of this three stage network, we will notice that the resulting transfer function is not a cube of the transfer function of a single *R-C* section.

Transfer function of single *R-C* section is given by,

$$= \frac{s}{s + 1/RC}$$

Transfer function of three stage *R-C* network is not equal to,

$$\left\{\frac{S}{s + (1/RC)}\right\}^3$$

This is due to interaction of various stages and their dependence on one another. It can be mathematically proved in such cases that an overall phase shift of 180° to be introduced by feedback network will be provided at a frequency less the one calculated before by assuming the three sections independent of each other. This modified frequency expression is,

$$f = \frac{1}{2\pi\sqrt{6}\,RC}$$

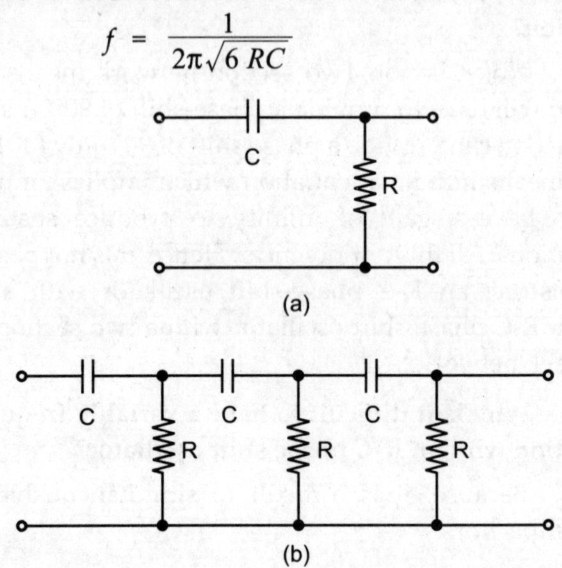

Fig. 14.6

Q. 15. Draw the basic circuit diagram of an *RC* phase shift oscillator using first an of operational amplifier and then a junction field effect transistor. Write the expressions for the frequency of oscillations and the minimum gain required for sustained or cillations in the two cases.

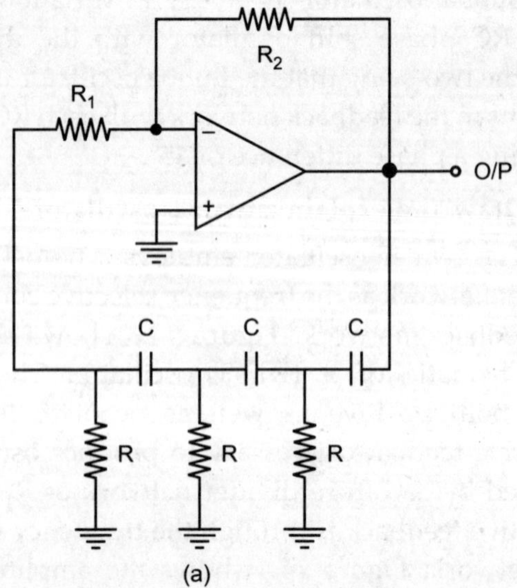

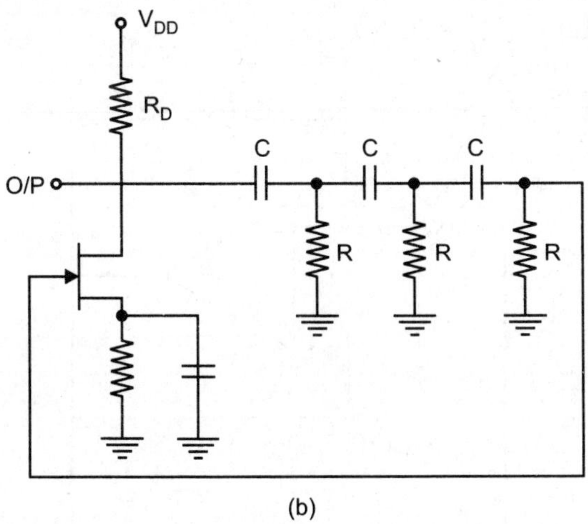

Fig. 14.7

Ans. Figure. 14.7 (*a*) shows an *R-C* phase shift oscillator using an operational amplifier. An op-amp in the amplifier part of the oscillator is always a better choice than a bipolar junction transistor as the former due to its high input impedance and low output impedance does not pose any loading problem.

The oscillation frequency is given by, $f = \dfrac{1}{2\pi\sqrt{6}\,RC}$

The minimum amplifier gain for sustained oscillations = 29

Gain in this case is given by-(R_2/R_1)

The *RC*- phase shift oscillator implemented with JEET is shown in Fig,14.7(*b*). Again the frequency of oscillation is given by $f = \dfrac{1}{2\pi\sqrt{6}\,RC}$

and the minimum gain for sustained oscillation is approximately 29. The gain in this case is given by g_m. (Total effective load resistance)]

$$= g_m(rd \,||\, R_D \,||\, R_L)$$

(*gm*) and (*rd*) are respectively the transconductance and drain resistance of the FET.

Q. 16. Explain the construction of Armstrong oscillator. Why is it also called a Tickler oscillator?

Ans. Armstrong oscillator also known as Meissner oscillator uses magnetic coupling as means of feeding part of output signal back to input to provide oscillations. It is also called a Tickler oscillator due to use of magnetic coupling between the tickler coil and the coupling coil. Tickler coil is the name given to a small coil connected in series with the plate circuit of a vacuum tube and coupled inductively to the grid circuit to provide feedback. In the case of a bipolar junction transistor or a FET, the tickler coil is placed in series with collector or drain circuit and is inductively coupled to the base or gate circuit. A capacitor is placed across either the tickler coil or the coupling coil to form a tank circuit that decides the operating frequency. Figures 14.8 (*a*) and (*b*) show the basic circuit arrangements in the two cases with N-channel junction FET used as the active device. Biasing components are omitted for the sake of simplicity.

The frequency of oscillation is primarily determined by the tank circuit and is given as follows:

$$f = \frac{1}{2\pi\sqrt{LC}}$$

(*L*) is the inductance of the coupling coil in the case of circuit shown in Fig .14.8 (*a*) and that of tickler coil in the case of circuit shown in Fig. 14.8 (*b*). In practice, the frequency of oscillations is slightly different from the one computed by using previous equation because of stray capacitances, loading of tank circuit etc. The feedback factor in this case is given by ratio of mutual inductance between the two coils to the inductance in the tank circuit. Minimum gain required to start oscillations is reciprocal of the feedback factor.

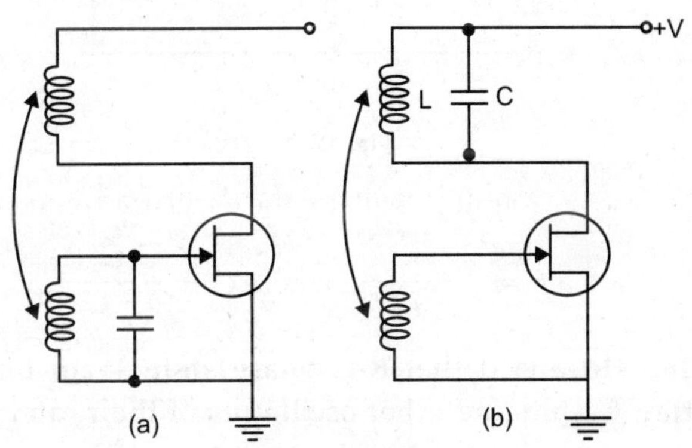

Fig. 14.8

Q. 17. How does the feedback network of Hartley oscillator differ from the feedback network of Colpit's oscillator? What will be the oscillator frequencies in the two cases?

Ans. The feedback of a Hartley oscillator consists of tapped coil with a capacitor connected across whereas in case of Colpitt's oscillator the feedback network is split capacitor with a coil connected across. See Fig. 14.9 (*a*) and (*b*). How a tapped coil or a split capaciton produce a phase-difference of 180°is obvious.

In case of Hartley oscillator, the oscillator frequency is,

$$f = \frac{1}{2\pi\sqrt{LC}}, \text{ where } L = L_1 + L_2$$

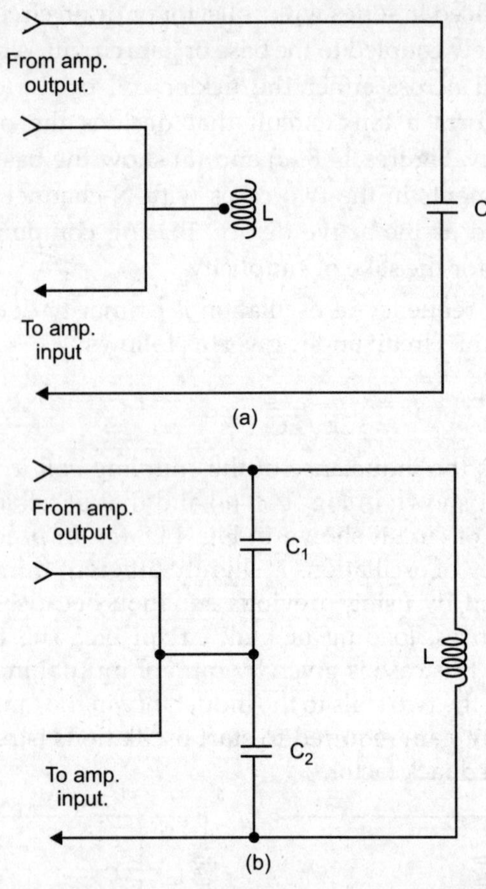

Fig. 14.9

In case of Colpitt's oscillator,the oscillator frequency is

$$f = \frac{1}{2\pi\sqrt{LC}}, \text{ where } C = \frac{C_1 C_2}{C_1 + C_2}$$

Q. 18. How is it that R-C phase shift, Wein bridge, Hartley, Calpitt and other oscillators of their categories produce sinusoidal output?

Ans. All these oscillators have frequency selective network in the feedback path. These feedback networks satisfy Barkhusen criterion as loop phase-shift only at a single frequency which implies that the output will be sinusoidal.

Q. 19. Draw the basic circuit implementaions of Hartley and colpitt's oscillators using op-amp as the active device. Write expressions for oscillation frequency and also mention the minimum gain required for sustained oscillations.

Ans. The circuit diagrams are shown in Fig. 14.10 (*a*) (for Hartley oscillator) and Fig. 14.10 (*b*) (for Colpitt oscillator)

In case of Hartey oscillator (Fig 14.10 (*a*)),

$$\text{Oscillation frequency, } f = \frac{1}{2\pi\sqrt{L_{eq}C}}$$

where $L_{eq} = L_1 + L_2$

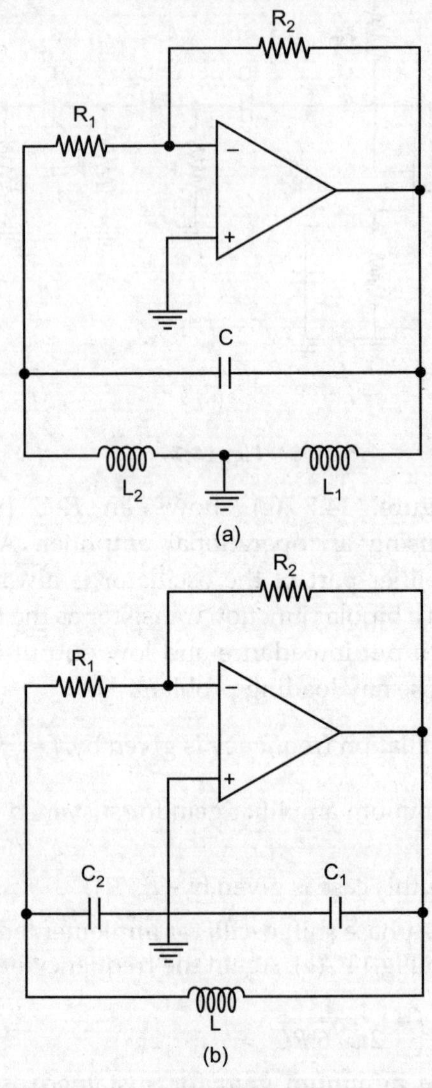

Fig. 14.10

The feedback network attenuates the signal by a factor (L_1/L_2).Therefore, the minimum amplifier gain sustained oscillator is (L_2/L_1).

In case of Colpitt oscillator (Fig. 14.10 (*b*))

$$\text{Oscillation frequency} = \frac{1}{2\pi\sqrt{LC_{eq}}}$$

$$\text{where, } C_{eq} = \frac{C_1 C_2}{C_1 + C_2}.$$

The feedback network attenuates the signal by a factor (C_2/C_1).Therefor, the minimum amplifier gain for sustained oscillations is (C_1/C_2).

Q. 20. What is a crystal oscillator? How does a crystal oscillator produce a highly stable frequency output?

Ans. A crystal oscillator is one in which a quartz crystal replaces the conventional *L-C* tank circuit in the feedback path. As a result, the oscillator frequency is completely determined by the crystal characteristics.

The frequency stability of an oscillator depends upon two factors (*i*) the frequency determining network should have very high *Q*-factor, so it has very high ($d\theta/d\omega$) value. A high value of ($d\theta/d\omega$), where θ represents phase, implies that even infinitesimally small change in (ω) will produce a sufficient change in (θ) to restore the frequency to the original value, (*ii*) the frequency determining network should be made up of elements with extremely high operational stability. On both these accounts, quartz crystal has no competitor. It has *Q*-factor approaching tens of thousands and as a component, it is extremely stable.

Q. 21. Why are *L-C* oscillators preferred over *R-C* oscillators at high frequencies, say in MHz range?

Ans. Firstly, because *L* and *C* have reasonable values at higher frequencies and secondly, because of comparatively mush better frequency stability at these frequencies, *L-C* oscillators are preferred over *R-C* oscillators.

Q. 22. How is Clapp oscillator different from the Colpitt oscillator?

Ans. Clapp oscillator circuit (Fig. 14.11) is a slight modification of the Colpitt oscillator circuit configuration. The feedback circuit in the case of Clapp oscillator uses an extra capacitor (C_3 in Fig. 14.11) in series with the coil. The function of C_3 is to minimise the effect of junction capacitance on the operating frequency.

The operating frequency (*f*) is given by equation (1),

$$f = \frac{1}{2\pi} \times \sqrt{\left[\frac{1}{L}\left\{\left(\frac{1}{C_1}\right)+\left(\frac{1}{C_2}\right)+\left(\frac{1}{C_3}\right)\right\}\right]} \quad (1)$$

If C_3 is chosen to be much smaller then either C_1 or C_2, then expression for frequency (*f*) simplifies to equation 2,

$$f = \frac{1}{2\pi\sqrt{LC_3}} \quad (2)$$

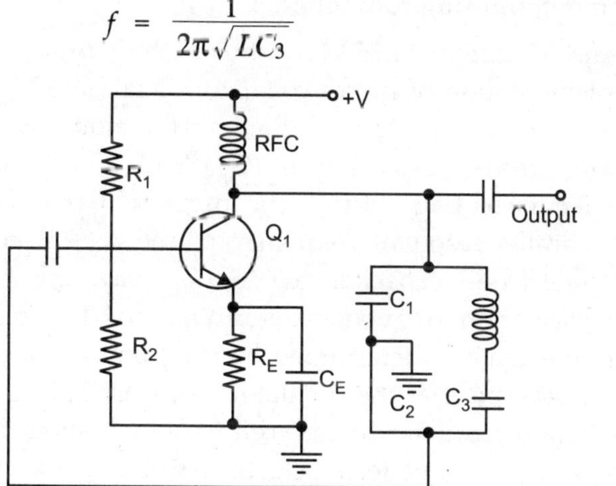

Fig. 14.11

Remember that we still need C_1 and C_2 to provide the required phase shift for regenerative feedback. Clapp oscillator is preferred over Colpitt oscillator for designing variable frequency oscillators. The tuning element (C_3 in Clapp oscillator) adjustment does not alter the attenuation factor in Clapp oscillator. The attenuation factor is decided by C_1 and C_2. In the case of Colpitt oscillator, any attempt to vary the frequency by varying either C_1 or C_2 might cause cessation of oscillations over a portion of desired frequency range.

Q. 23. Draw the three basic *L-C* oscillator circuits viz. Hartley, Colpitt, Clapp. Show the basic circuit configurations with respect to placement of feedback network using a *CE* amplifier configuration. Transistor biasing elements can be omitted.

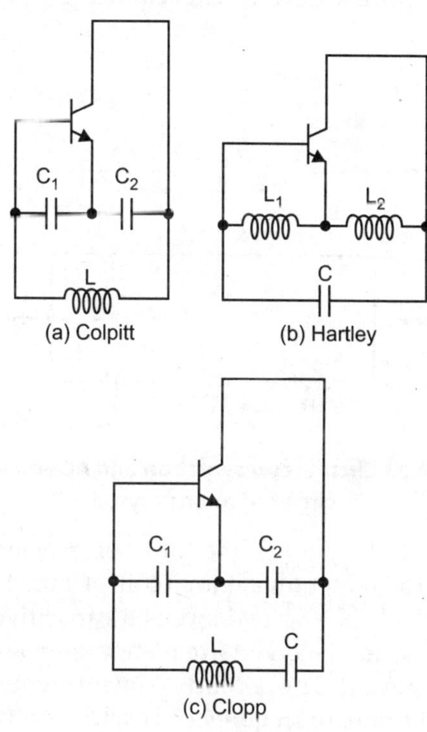

Fig. 14.12

Ans. The three oscillator configurations are shown in Fig. 14.12 (*a*) (Hartely), 14.12 (*b*) (Colpitt) and 14.12 (*c*)(Clapp).

Q. 24. Explain the construction of crystal oscillator. Construct the reactance versus frequency plot. Also, briefly describe the graph.

Ans. In the case of a crystal oscillator, a quartz crystal with the desired value of resonant frequency forms part of the frequency selective feedback network. Crystal oscillator is the natural choice when the accuracy and stability of oscillation frequency is of paramount importance.

Figure 14.13 shows the circuit representation and AC equivalent circuit of the quartz crystal. ***R***, ***L*** and C_S respectively represent the resistance, inductance and capacitance of the piezoelectric crystal element. C_M represents the mounting capacitance. It is in fact, the capacitance due to the parallel plate capacitor formed by the connecting electrodes and the piezoelectric element constituting the dielectric. Typically, R is in the range of few hundreds of ohms to a few kilo-ohms; *L* is of the order of few tens of milli-henries to few henries, C_S is a very small fraction of a pico-farad and C_M is few pico-farads. The *Q*-factor of the crystal is given by $Q = \omega L/R = 1/\omega C_S R$. The crystal exhibits two resonant frequencies. One is the series resonant frequency f_S. It is the frequency at which the inductive reactance of inductance *L* equals the capacitive reactance of capacitance C_S expressed by equation (1)

$$f_s = \frac{1}{2\pi\sqrt{LC_s}} \quad (1)$$

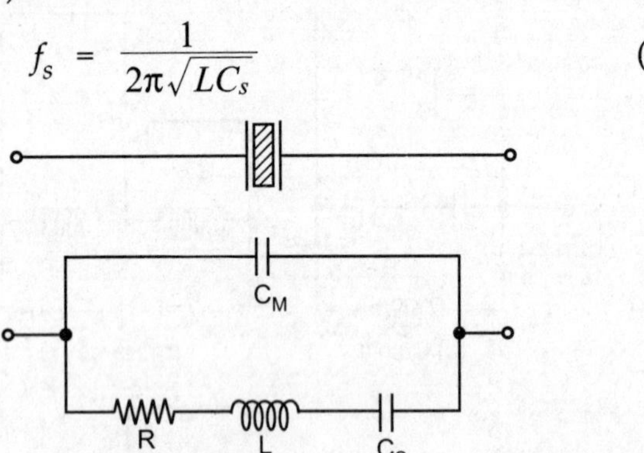

Fig. 14.13 Circuit representation and AC equivalent circuit of quartz crystal

Figure 14.14 shows the plot of reactance versus frequency for the equivalent circuit of Fig. 14.13. Quite understandably, the impedance is a capacitive reactance below the series resonant frequency and an inductive reactance above it. The second resonant frequency called the parallel resonant frequency f_P occurs at a value where the inductive reactance equals the capacitive reactance due to equivalent capacitance of the tank circuit. Parallel resonance occurs at a frequency where the circulating loop current is at its maximum. Since circulating loop current flows through series combination of C_S and C_M, therefore the equivalent capacitance of the parallel tuned circuit is given by equation (2).

$$C_P = \frac{C_M \times C_S}{(C_M + C_S)} \quad (2)$$

Parallel resonant frequency is given by equation 3.

$$f_p = \frac{1}{2\pi\sqrt{LC_p}}$$

Since C_S is much smaller than C_M, C_P is only marginally less than C_S with the result that f_S and f_P are very close to each other.

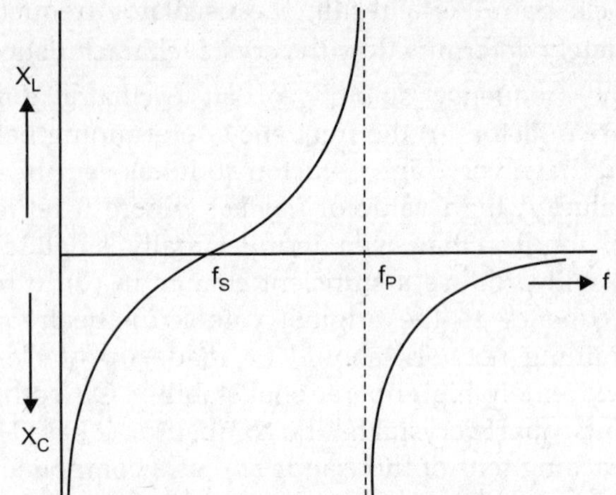

Fig. 14.14 Reactance versus frequency plot

The two resonant frequencies described in the previous paragraph are the fundamental resonant frequencies. Remember that the specified crystal frequency is between f_S and f_P. This area of frequencies between f_S and f_P is known as the area of usual parallel resonance or simply parallel resonance. A crystal can also resonate at harmonics of the fundamental frequency called overtones. The fundamental resonant frequency of the crystal is usually limited to less than 30 MHz due to the smallest physical dimension the crystal can be cut to. Operation in the overtone mode allows stable output at much higher frequencies.

Q. 25. One of the most popular crystal oscillator configurations is the crystal oscillator built around a FET and called 'Piercing oscillator'. Draw the basic circuit arrangement of this type of oscillator. Does it resemble any of the standard *L-C* oscillator configurations? If yes, which one does it resemble with? Briefly describe the operational principle and identify the frequency determining and the required amplifier gain determining components.

Ans. Figure. 14.15 (*a*) shows the basic circuit implementation of pierce oscillator. Dual gate MOSFET is the active device and the crystal along with the interelectrode capacitances (C_{gs}) and (C_{ds}) constitute the feedback (Fig. 14.15). The oscillation frequency is the parallel resonant frequency of the crystal and the inter-electrode capacitances do not play any role in the oscillation frequency determination. The required amplifier gain is determined by (C_{gs}) and (C_{ds}). In fact, it is observed that the oscillator stops oscillating as the oscillation frequency is reduced to below about 2 MHz due to insufficient feedback. In a situation like this, it becomes imperative to have external capacitors between gate source and drain-source. Pierce oscillator works very well in VHF and UHF frequency rages.

A closer look at the circuit configuration of Pierce oscillator will tell you that it resembles a Colpitt oscillator with the crystal replacing the inductance and inter-electrode capacitances providing the two capacitances. Purpose of putting RFC (Radio Frequency Choke) is to prevent the high frequency output form getting grounded through drain power supply.

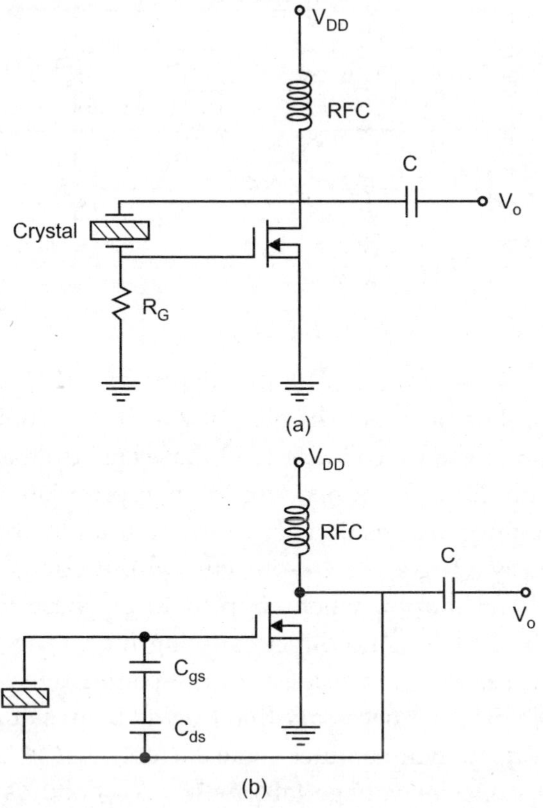

Fig. 14.15

Q. 26. Why are *R-C* phase-shift oscillators like Wien-bridge oscillators only suitable at low frequencies?

Ans. Fig 14.16 shown a Wienbridge oscillator using an opamp.

We know that the output developed across parallel *R C* branch is fed back to the input. If we calculate the function $(v_f/v_o)$and write an expression for phase angle (θ) between (v_f) and (v_o), we will calculate $\left(\frac{d\,\theta}{d\omega}\right)$ to be $\left\{\frac{2RC}{1+\omega^2R^2C^2}\right\}$. At the oscillation frequency $\omega = \frac{1}{RC}$, we find $\frac{d\theta}{R\omega}$ equals (RC). Now a higher frequency stability means higher $\frac{d\theta}{d\omega}$ and in the present case higher $\frac{d\,\theta}{d\,\omega}$ results from a higher value of product (RC). This implies that oscillator will have comaratively better output frequency stability at lower frequencies as $f = \frac{1}{2\pi RC}$.

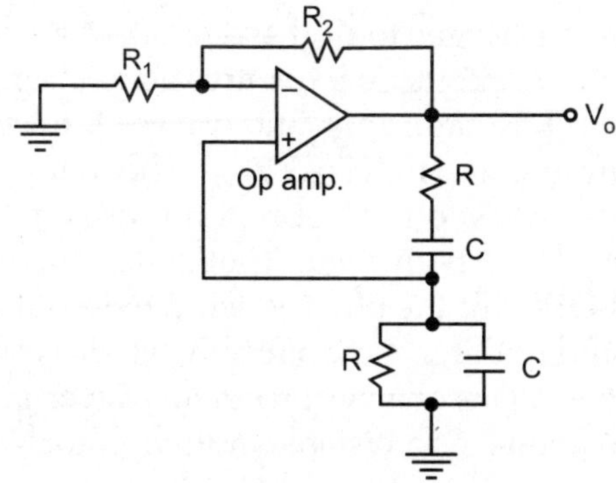

Fig. 14.16

Q. 27. Draw the basic bistable multivibrator circuit and explain in brief its operation.

Ans. The basic bistable multivibrator circuit is shown in Fig. 14.17. There are other varieties of bistable multivibrators like the self bias, binary or the emitter coupled binary. The one shown in Fig. 14.17 is the fixed bias one. The operational principles of all the types are however identical.

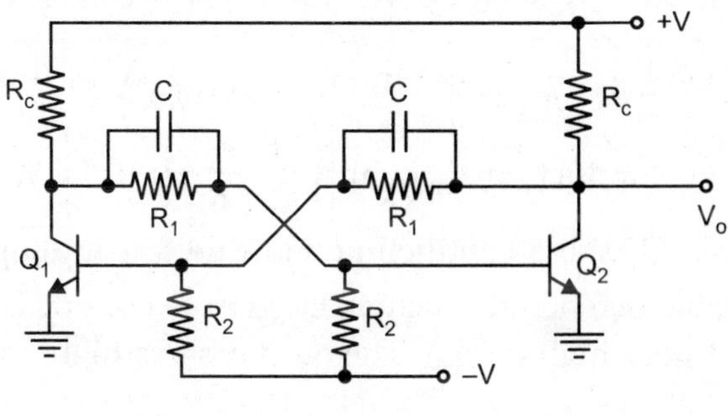

Fig. 14.17

In the circuit arrangement shown, it can be proved that both the transistors cannot be either simultaneously OFF or simultaneously ON. This is due to presence of regenerative feedback. Let us assume that both the transistors are conducting. Due to slight circuit imbalance, which is always there, collector current in one of the transistors will always be greater than in the other.

Let us assume $I_{c2} > I_{c1}$ lesser I_{c1} means larger V_{c1}. Now $V_{C1}$is coupled to Q_2-base and therefor rise in V_{c1} means rise in Q_2-base. This results in an increase in I_{c2} and a fall in V_{c2}. Fall in V_{c2} means fall in Q_1- base voltage which further lower I_{c1} with consequent increase in V_{c1}. Thus, a slight imbalance initiates this regenerative action which would culminate in transistor Q_2 going to saturation and transistor Q_1 going to cut-off. The conclusion is: whenever there is a tendency of one the transistors to conduct more than the other, it will end up with that transistor going to saturation and driving the other transistor to cut- off. Now if we take output form Q_2-collector, it will be LOW $(= V_{CE2(sat)})$ if Q_2 was initially in saturation. If we give a negative going trigger to Q_2-base to effect a decrease in its collector current, the regenerative action begins which finally takes Q_1 to saturation and drives Q_2 to cut- off. As a result output goes HIGH (= $+V_{CC}$). The output will stay high till we give another appropriate trigger to initiate a transition. Thus, both the output states, the one when output is LOW and the other when the output is HIGH, are stable and undergo a change only when a transition is induced by a means of an appropriate trigger pulse. That is why, it is known as a bistable multivibrator.

Q. 28. What is a Schmitt-trigger? Draw its basic circuit diagram and explain in brief how it works.

Ans. A Schmitt trigger (Fig. 14.18) is nothing but an emitter coupled bistable multivibrator. If you look at its basic circuit arrangement, you will notice that it resembles a binary shown earlier except that coupling form Q_2-collector to Q_1-base is missing. In fact, in this circuit, coupling is being provided by emitter resistance R_e. Let us assume V_{in}= 0. Therefore, Q_1 is in cut-off and coupling form Q_1-collector to Q_2-base ensures that Q_2 is in saturation. When Q_2 is in saturation the voltage across R_e equals $\left\{\frac{V_{cc}R_e}{R_e+R_{c2}}\right\}$ assuming $V_{CE2(sat)}$= 0. In order to make Q_1 conduct must equal $\left[\left\{\frac{V_{cc}R_e}{R_e+R_{c2}}\right\}+0.7\right]$. When it does so, Q_1 starts conducting and due to coupling and inevitable regenerative action, Q_2 goes to cut-off. That is, output goes high (= V_{cc}). The output stays high and the voltages across(R_e) changes to $\left\{\frac{V_{cc}R_e}{R_e+R_{c1}}\right\}$. Now when V_{in} is reduced and when it falls below a voltage equal to $\left[\left\{\frac{V_{cc}R_c}{R_e+c_1}\right\}+0.7\right]$, Q_1 tends to come out of saturation or conduct less heavily.

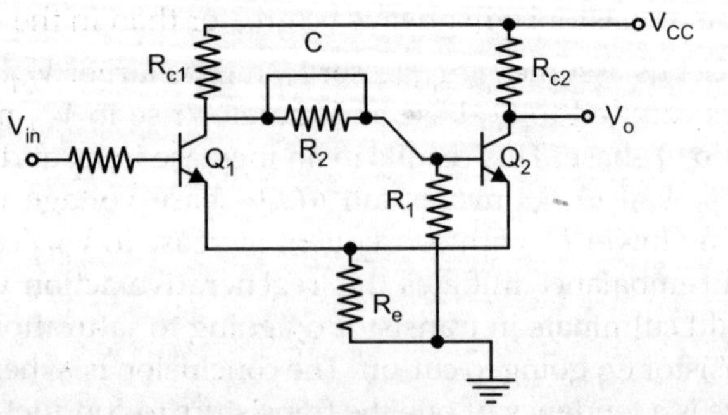

Fig. 14.18

The rest is done by the regenerative action and the result is Q_1 goes to saturation. So, the output goes LOW; thus the output state depends upon input state and the HIGH and LOW states correspond to two distinct input levels which depend mainly upon R_{c1}, R_{c2} Re and V_{cc}. The Schmitt trigger thus exhibits hysterisis.

Q. 29. Sketch the basic monostable multivibrator circuit and explain in brief its operation.

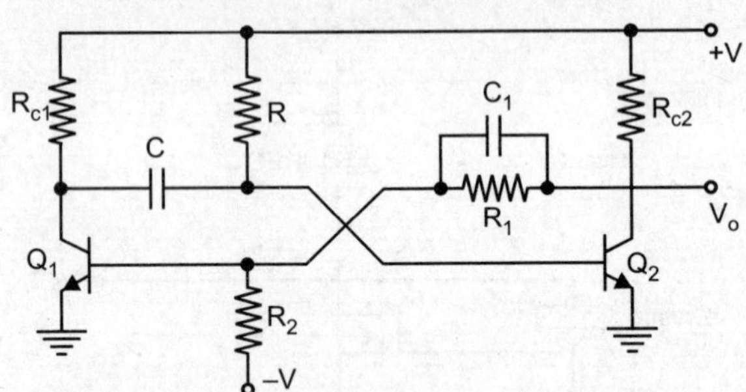

Fig. 14.19

Ans. Refer to Fig. 14.19. Initially transistor Q_2 will be in saturation, as it gets its base bias form V_{CC} though R. Coupling form Q_2-collector to Q_1-base ensures that Q_1 is in cut-off. Now What happens, if an appropriate trigger pulse induces a transition in Q_2 form saturation to cut-off. Obviously, Q_2 goes to cut-off with output going HIGH. This HIGH output, when coupled to Q_1- base turns it ON. Since there is no direct coupling form Q_1-collector to Q_2 base, which is necessary for a cumulative process to set-in, it is not necessary that Q_1 will be in saturation. However, it will conduct some,current, that is certain. The Q_1-collector voltage falls by (I_{C1} R_{C1}) and Q_2- base voltage falls by the same amount as voltage across a capacitor (C in this case) cannot change instantaneously. The conclusion is : the moment we applied the trigger, Q_2 went to cut-off and Q_1 started conducting. But now there is path for capacitor C to charge from V_{cc} through R and the conducting transistor. The polarity is such that Q_2-base potential rises. The moment, it exceeds Q_2-base cut-in voltage, it turns on Q_2 which due to coupling through R_1 turns off Q_1. And we are back to original state, the stable state. Whenever we trigger the circuit into the other state, it cannot stay there permanently and it returns back after a time period decided by R and C. Larger the time constant (RC), larger is the time for which it stays in the other state (also called quais-stable state). This is how a monostable multivibrator works.

Q. 30. Draw the basic astable multivibtaor circuit and explain how it functions as a free running square-wave oscillator?

Ans. Figure 14.20 shows the basic as table multivibrator circuit. Let us assume that transistor Q_2 is initially conducting *i.e.* output is LOW. Capacitor C_2 will charge through R_2 from V_{cc} and the moment Q_1- base potential exceeds its cut-in voltage, it is turned ON. Fall in the Q1-collector potential is transmitted to the Q_2-base, which thus, goes to cut-off and the output goes HIGH. But again, Q_2 cannot remain cut-of indefinitely as C_1 has started charging through R_1 and the conducting transistor Q_1 and the moment it charges to a voltage exceeding Q_2- base cut in voltage, Q_2 is again turned ON. This process continues and due to both the couplings being capacitive, nether of the states is stable. Both the states are quasi-stable. As a result, we get a square waveform at the output.

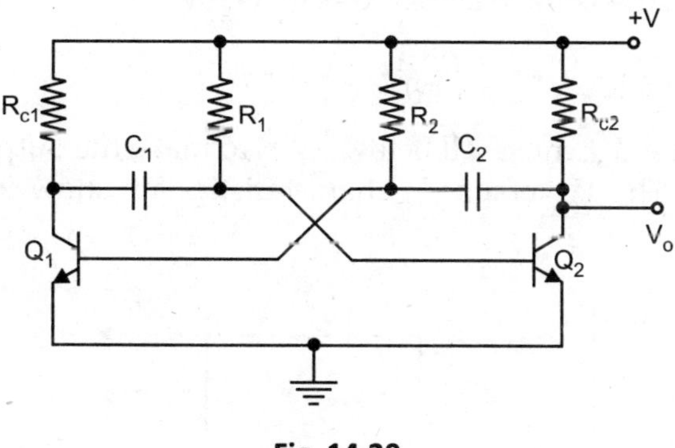

Fig. 14.20

Q. 31. In astable multivibrator of pervious question, if we want the high time of the output waveform to be greater than the low time, what should be relative values of the two time constants?

Ans. The high time (when Q_2 in cut off and Q_1 is in conduction) is obviously determined by R_1 C_1 time constant whereas the low time (when Q_2 is in conduction and Q_1 is in cut off) is determined by R_2 C_2 time constant. Therefore, for high time to be greater than the low time, $R_1 C_1 > R_2 C_2$.

Q. 32 What is a blocking oscillator? With the help of basic blocking oscillator configuration, briefly explain its principle of operation.

Ans. A blocking oscillator is basically a transformer coupled pulse generator which could work both in the monostable as well as astable mode. Fig. 14.21 shows basic blocking oscillator that works in the astable mode. Transistor Q gets its base bias though R. The transistor, which is connected in the *CE* configuration, provides phase inversion and if the transformer polarities are so chosen as to provide another phase inversion, there are conditions for positive feedback. When the transistor conducts, there are oscillations at the resonant frequency of primary winding. If the amount of positive feedback is very high, the oscillations will build up very rapidly. There is a burst of current in the base, which charges the capacitor in a direction to drive the transistor to cut- off. The capacitor starts discharging though R. After it has discharged to an extent that it cannot keep the transistor in cut-off the transistor conducts again followed by bust of base current, which charges the capacitor again to drive the transistor to cut- off and the process continues. The natural frequency of such an oscillator (also called Relaxation Oscillator) is given by $\frac{n+1}{RC}$.

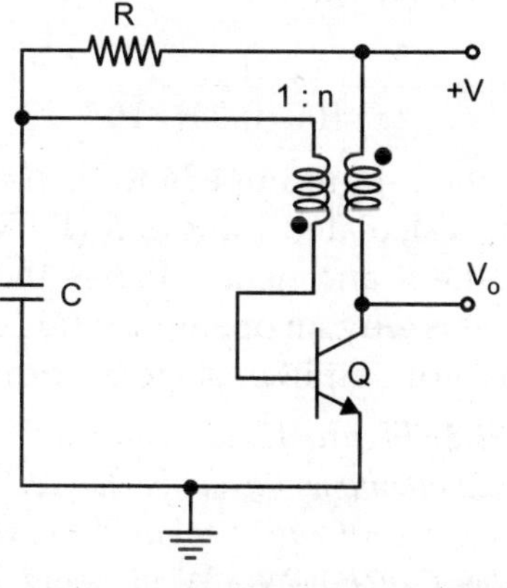

Fig. 14.21

Q. 33. What are synchronised blocking oscillators? What is their significance from application viewpoint?

Ans. In synchronised blocking oscillators, conduction of the transistor is synchronised with the synchronizing pulses. If in the blocking oscillator configuration of the previous question, the natural frequency is kept slightly lower than the frequency of the synchronising pulses, the blocking oscillator output will be synchronised with the applied sync pulse input.

Synchronised blocking oscillators are of considerable practical importance and are invariably used in the horizontal and vertical scanning sections of the television receivers. The horizontal deflection oscillator is synchronised with the horizontal sync and vertical deflection oscillator is synchronised with the vertical sync pulses.

Solved Problems

Problem 14.1. *The amplifier gain in an oscillator is 50. What should be percentage feedback for sustained oscillations?*

Solution. Amplifier gain = 50

Therefore, feedback factor $\beta = \frac{1}{50} = 0.02$

Percentage feedback = 2 %

Problem 14.2. *Calculate the values of R and C for 100 Hz output in wien bridge oscillator and explain why it is necessary to have an amplifier section with a very high input impedance?*

Solution. Frequency(f) in a wein-bridege oscillator is given by

$$f = \frac{1}{2\pi RC}$$

Assume $C = 0.001\mu F$

$$\therefore \quad R = \frac{1}{2\pi \times 100 \times 0.001 \times 10^{-6}} = 1.6\ M\Omega$$

If the amplifier, input impedance is not very high as compared to the calculated value of R (1.6 M in this case), it will load on the R and hence change the frequency of oscillations. That is why, an opamp or and FET input stage instead of transistor amplifier stage is normally used.

Problem 14.3. *Figure 14.22 shows an R-C phase shift oscillator using an opamp as the active device. The oscillator is oscillating at the correct frequency but the output is distorted severely near the signal peaks. What could be the possible causes?*

Solution. The minimum amplifier gain requirement is 29. In order to ensure that we get sustained oscillations and the device replacement and ageing have no effects, amplifier gain is chosen to be slightly more than what is demanded by Barkhausen criterion. It may exceed the minimum required gain value by 10 to 15 per cent. Too large a gain can cause signal clipping near the peaks.

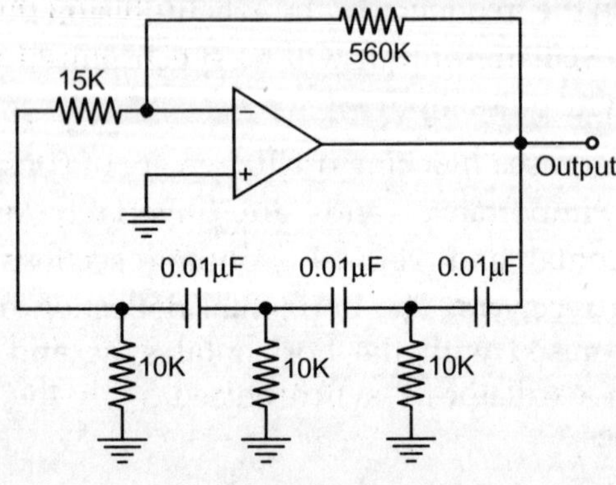

Fig. 14.22

The gain magnitude in the present case is given by,

$$\frac{560}{15} \cong 37.3$$

Which exceeds the required minimum by about 28 percent. This explains the distortion observed in the output.

Problem 14.4. *If in the Schmitt trigger circuit of Q. 28 (Fig. 14.18), $R_{c1} = 9\ k\Omega$, $R_e = 3\ k\Omega$, $R_{c2} = 3\ k\Omega$ and $V_{CC} = 12V$. Draw V_o vs. V_{in} characteristics. Assume $V_{CE(sat)} = 0$ for both the transistors.*

Solution. Initially, when Q_2 is in saturation, voltage across R_e is given by,

$$V_e = \frac{12 \times 3}{3+3} = 6V$$

Thus, the input must at least the be 6.7 volts for making the output go HIGH. When the output goes HIGH, the voltage across R_e is given by,

$$V_e = \frac{12 \times 3}{3+9} = 3V$$

Now V_m must fall below 3.7 V to make the output go LOW. The V_o versus V_{in} characteristics are show in Fig. 14.23.

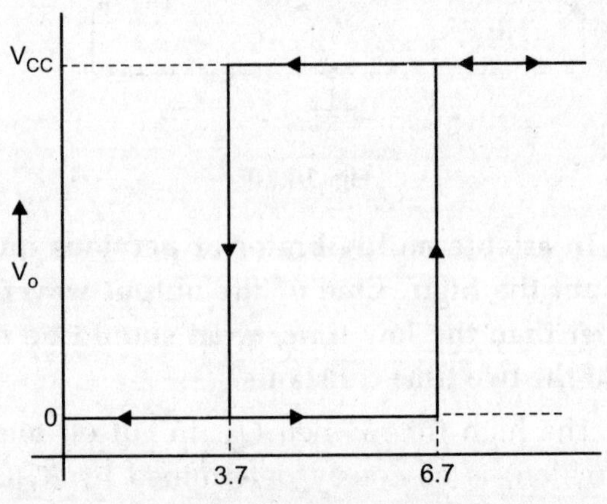

Fig. 14.23

Problem 14.5. *Determine the gain or phase shift as the case may be for the following oscillator circuits. If in an oscillator, the amplifier portion is a two-stage CE configuration, what should be the phase shift to be introduced by the feedback network at the oscillations frequency for sustained oscillations?*

Solution. A two-stage CE amplifier provides a phase shift of 2π radians.

Therefore, the feedback network must not introduce any more phase shift or introduce phase shift equal to multiples of 2π radians in order to satisfy Barkhausen criterion for sustained oscillations.

Problem 14.6. *Refer to the phase shift oscillator of Fig. 14.24. Determine the frequency at which this circuit would oscillate, if the loop gain criterion were met. Also, determine the maximum value of resistance (R_1) for sustained oscillations.*

How would the oscillation frequency change, if the positions of R and C in the feedback network were interchanged?

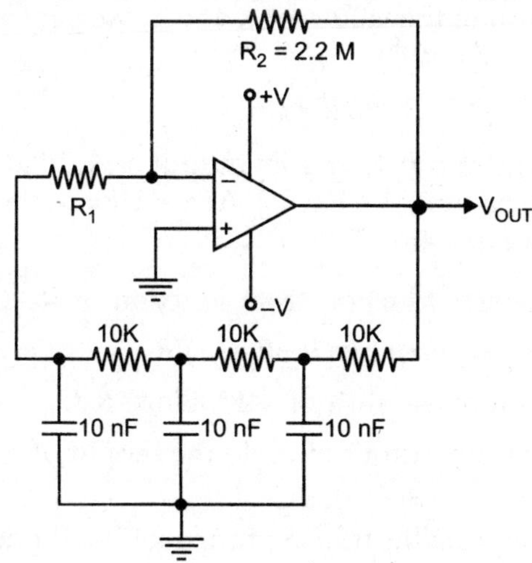

Fig. 14.24

Solution.

The oscillation frequency (f) is given by $f = \sqrt{6}/2\pi RC$. Therefore, $f = \sqrt{6}/(6.283 \times 10 \times 103 \times 10 \times 10^{-9}) = 3.9$ kHz

Minimum gain to be provided by the amplifier = 29

$|\text{Gain}| = R_2/R_1 = 29$

This gives $R_1 = 2.2 \times 10^6/29 = 75.86$ KΩ

R_1 should therefore be less than 75.86 KΩ

When the positions of R and C are interchanged, the oscillation frequency is given by $f = 1/2\pi\sqrt{6}RC$

Substituting the values of R and C, we get $f = 650$ Hz

Problem 14.7. *If in the RC phase shift oscillator circuit of Problem 11.6, the three RC sections were separated from each other by op-amp buffers. First buffer is connected between the first and second RC sections; second buffer is connected between second and third RC sections and the third buffer is connected between the third RC section and the op-amp input. Determine the new values of oscillation frequency and resistor (R_1) for sustained oscillations.*

Solution. The oscillation frequency (f) in this case is given by $f = \sqrt{3}/2\pi RC$. Therefore, $f = \sqrt{3}/(6.283 \times 10 \times 10^3 \times 10 \times 10^{-9}) = 2.758$ kHz

Minimum gain to be provided by the amplifier in this case is 8.

Therefore, R_1 should be less than $2.2 \times 10^6/8 = 275$ KΩ

Problem 14.8. *Refer to the oscillator of Fig. 14.25. Does the given oscillator circuit resemble any standard oscillator configuration? If yes, identify the oscillator configuration. Also, determine its frequency of oscillations and the required minimum value of amplifier gain.*

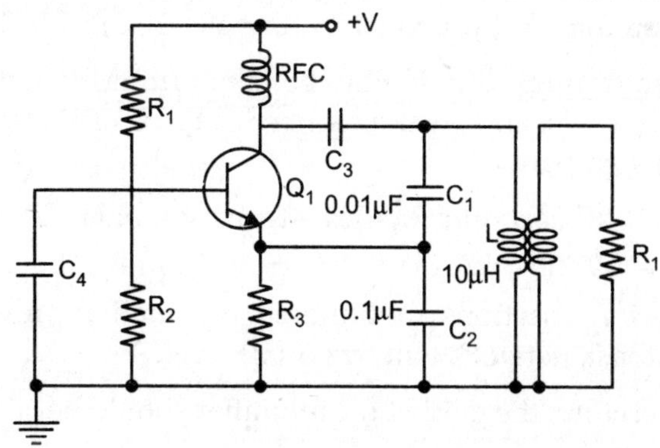

Fig. 14.25

Solution. If we carefully examine the given circuit, we find that it follows the Colpitt oscillator configuration.

The feedback circuit in this case is also a tank circuit comprising a pair of series connected capacitors C_1 and C_2 and an inductor L.

Also, the amplifier has been wired in common-base configuration. Note that the base terminal is effectively grounded for AC signal through capacitor C_4.

The output in this case appears across series combination of C_1 and C_2. It appeared across C_1 only in the case of Colpitt oscillator configured around common-emitter amplifier. The feedback signal appears across C_2 in both cases.

The feedback factor in this case is therefore given by $[C_1 \times C_2/(C_1 + C_2)]/C_2$, which simplifies to $C_1/(C_1 + C_2)$.

The required minimum value of amplifier gain is therefore, $(C_1 + C_2)/C_1$.

Frequency of oscillations can be computed from $f = 1/2\pi\sqrt{LC}$, where $C = C_1 \times C_2/(C_1 + C_2)$.

Now $C = 0.01 \times 10^{-6} \times 0.1 \times 10^{-6}/(0.01 \times 10^{-6} + 0.1 \times 10^{-6}) = 0.009$ μF

Therefore, $f = 1/2\pi\sqrt{(10 \times 10^{-6} \times 0.009 \times 10^{-6})} = 530.5$ kHz

Feedback factor $= 0.01 \times 10^{-6}/(0.01 \times 10^{-6} + 0.1 \times 10^{-6}) = 1/11$

Also, required minimum value of amplifier gain = 11

Problem 14.9. *Figure 14.26 shows a buffered RC oscillator circuit. Determine the following:*

(a) Frequency of oscillation

(b) Possible points in the circuit for quadrature outputs

(c) Value of resistance R_G (choose from 220 KΩ, 290 KΩ, 300 KΩ and 330 KΩ)

Solution. The circuit shown is that of Bubba oscillator. It employs four RC sections isolated from each other with op-amp buffers.

The frequency of oscillation is given by $f = 1/2\pi RC$ as each of the four sections contributes a phase shift of 45°.

Therefore, $f = 1/(6.283 \times 10 \times 10^3 \times 10 \times 10^{-9}) = 1.592$ kHz

Quadrature outputs may be taken from the outputs of op-amps A_2 and A_4. Remember outputs of A_2 and A_4 will be 90° apart.

Each RC section provides attenuation of $(1/\sqrt{2})$ at the operating frequency.

Since RC sections are buffered, attenuation provided by feedback network will be $(1/\sqrt{2})^4 = 1/4$.

Therefore, the gain of the amplifier should be slightly more than 4.

This implies that the input resistance R_G should be slightly less than 300 KΩ. That is, 290 KΩ is the correct choice.

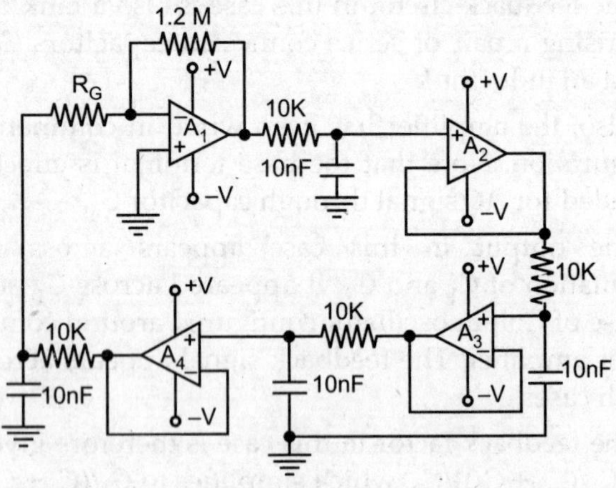

Fig. 14.26

Problem 14.10. *Figure 14.27 shows the circuit diagram of a quadrature oscillator. Determine the operating frequency. Also, determine the phase difference between the signals appearing at the outputs of opamps A_1 and A_2. If the peak amplitude of the signal appearing at the output of A_1 is 2 V, determine the peak amplitude of the signal at the junction of R_2 and C_2 and also, at the output of A_2.*

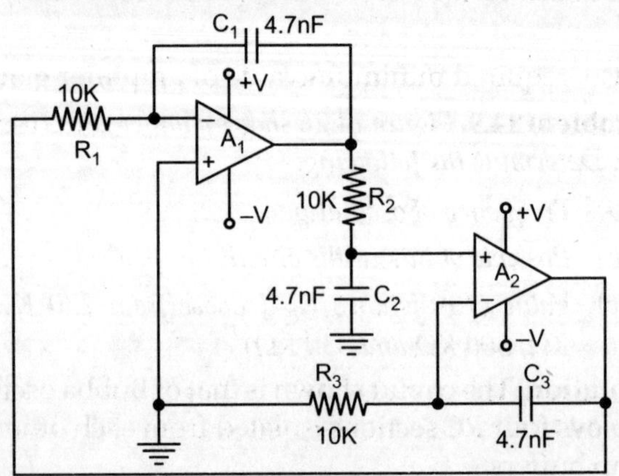

Fig. 14.27

Solution. The frequency of oscillation is given by $f = 1/2\pi RC$.

Substituting the values of R and C, we get $f = 1/(6.283 \times 10 \times 10^3 \times 4.7 \times 10^{-9})$

= 3.386 kHz

The transfer function from output of A_1 to junction of R_2–C_2 is given by $1/(1 + R_2C_2s)$. At $\omega = 1/R_2C_2$, it produces a phase shift of – 45°.

The transfer function from junction of R_2–C_2 to the output of A_2 is given by $(1 + R_3C_3s)/R_3C_3s$. At $\omega = 1/R_3C_3$, it produces a phase shift of –45°. Since $R_2C_2 = R_3C_3$, the phase shift from output of A_1 to the output of A_2 will be –90°.

It is clear from the transfer functions mentioned above that the two networks respectively provide attenuation and gain of $1/\sqrt{2}$ and $\sqrt{2}$.

Therefore, peak amplitude of signal at junction of $R_2 - C_2$ is $2/\sqrt{2} = \sqrt{2}$ V and that at the output of A_2 is $\sqrt{2} \times \sqrt{2} = 2$ V.

Problem 14.11. *Refer to the twin-T oscillator of Fig. 14.28. Determine the frequency of the output signal. How would the frequency change, if all component values in the twin-T network are doubled?*

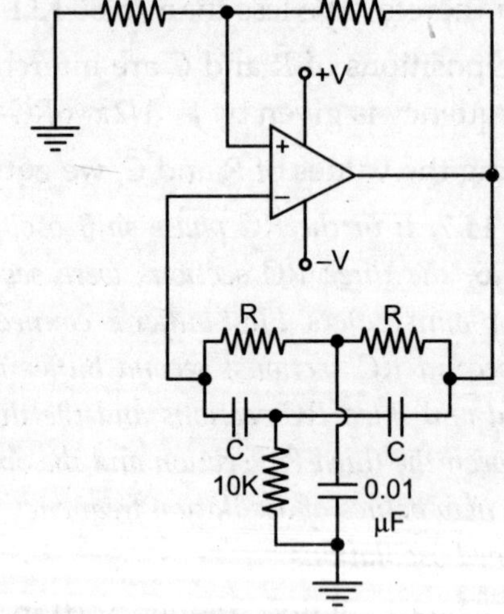

Fig. 14.28

Solution. The operating frequency (f) is given by $f = 1/2\pi RC$

If we compare the given twin-T network with the standard form of twin-T, we will find that $R = 2 \times 10 \times 10^3$ Ω = 20 KΩ and $C = 0.01 \times 10^{-6}/2 = 0.005$ μF

Therefore, $f = 1/(6.283 \times 20 \times 10^3 \times 0.005 \times 10^{-6}) = 1.592$ kHz

When all component values are doubled, the operating frequency will be reduced to one-fourth. That is, changed value of frequency = 1.592/4 = 398 Hz

Problem 14.12. *Refer to the Wien bridge oscillator circuit of Fig. 14.29. Determine the oscillator frequency and the preferred value of R_1.*

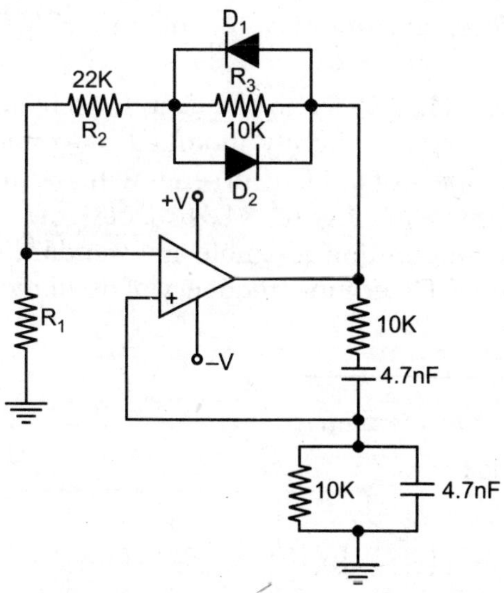

Fig. 14.29

Solution. Oscillator frequency is given by $f = 1/2\pi RC$ where $R = 10$ KΩ and $C = 4.7$ nF

Therefore, $f = 1/(6.283 \times 10 \times 10^3 \times 4.7 \times 10^{-9}) = 3.386$ kHz

Required minimum value of amplifier gain = 3

Amplifier gain = $1 + [(R_2 + R_3)/R_1]$

This gives $(R_2 + R_3)/R_1 = 2$, $R_1 = (22 \times 10^3 + 10 \times 10^3)/2 = 16 \times 10^3\ \Omega = 16$ KΩ

Since the required gain has to be slightly greater than 3, preferred value of R_1 therefore should be slightly less than 16 KΩ.

R_1 may be chosen to be 15 KΩ.

Problem 14.13. *Refer to the RC oscillator circuit of Fig. 14.30. Identify the type of oscillator. Also, determine the operating frequency and preferred value of R_3.*

Solution.

It is Wien bridge oscillator.

Operating frequency is given by $\omega = 1/\sqrt{(R_1R_2C_1C_2)}$.

Therefore, $\omega = 1/\sqrt{(10^3 \times 100 \times 10^3 \times 0.1 \times 10^{-6} \times 10^{-9})} = 10^4$ rad/s

$f = \omega/2\pi = 10^4/6.283$ Hz = 1.592 kHz

Attenuation provided by feedback network is given by,

$|\beta| = R_2C_1/(R_1C_1 + R_2C_2 + R_2C_1)$

$|\beta| = (100 \times 10^3 \times 0.1 \times 10^{-6})/(10^3 \times 0.1 \times 10^{-6} + 100 \times 10^3 \times 10^{-9} + 100 \times 10^3 \times 0.1 \times 10^{-6})$

$= 10^{-2}/(10^{-4} + 10^{-4} + 10^{-2}) = 1/1.02$

Therefore, minimum value of required gain = 1.02. That is, $1 + (R_4/R_3) = 1.02$, $R_4/R_3 = 0.02$

$R_3 = 2.2 \times 10^3/0.02 = 110$ KΩ.

Since the required gain has to be slightly greater than 1.02, preferred value of R_3 therefore should be slightly less than 110 KΩ.

R_3 may be chosen to be 100 KΩ.

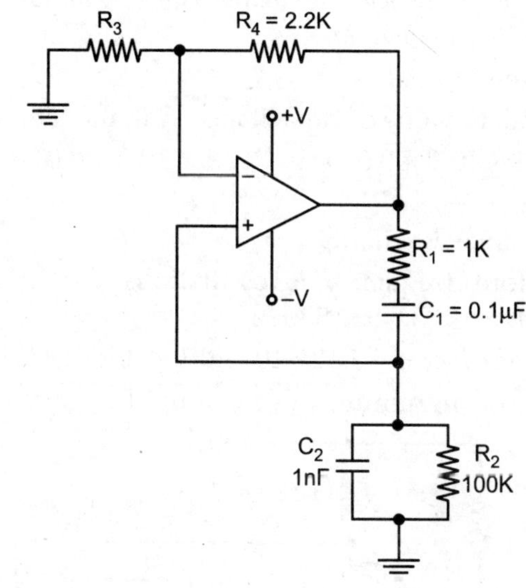

Fig. 14.30

Problem 14.14. *Refer to the Armstrong oscillator circuit of Fig. 14.31. Determine (a) frequency of oscillations and (b) minimum amplifier gain required to start oscillations. How would the oscillation frequency change, if the loaded Q-factor of the tank circuit were given to be 5?*

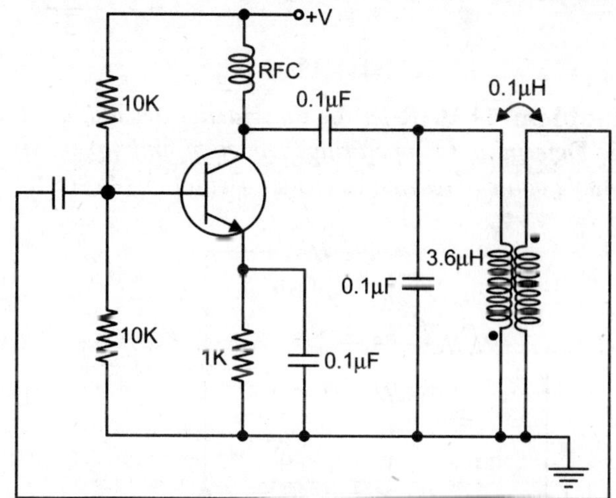

Fig. 14.31

Solution.

Frequency of oscillations, $f = 1/2\pi\sqrt{LC} = 1/(6.283\sqrt{(3.6 \times 10^{-6} \times 0.1 \times 10^{-6})}) = 265.3$ kHz

Feedback factor = $0.1 \times 10^{-6}/3.6 \times 10^{-6} = 1/36$

Therefore, minimum value of amplifier gain required to start oscillations = 36

Q-factor of the tank circuit = 5

Frequency of oscillation will reduce by a factor equal to $[\sqrt{Q^2/(Q^2+1)}]$.

Therefore, new frequency of oscillations will be 265.3 × 103 × $[\sqrt{5^2/(5^2+1)}]$ = 260.1 kHz

Problem 14.15 *Refer to the Colpitt oscillator of Fig. 14.32. Determine (a) frequency of oscillations and (b) minimum gain required to start oscillations.*

Solution.

Equivalent value of capacitance, *C* in the tank circuit

$C = 0.1 \times 10^{-6} \times 0.01 \times 10^{-6}/(0.1 \times 10^{-6} + 0.01 \times 10^{-6})$ = 0.009 μF

Inductance, L = 10 μH

Therefore, frequency of oscillations, $f = 1/(2\pi\sqrt{10 \times 10^{-6} \times 0.009 \times 10^{-6}})$ = 530.5 kHz

Feedback factor = $0.01 \times 10^{-6}/(0.1 \times 10^{-6})$ = 1/10

Therefore, minimum value of amplifier gain = 10

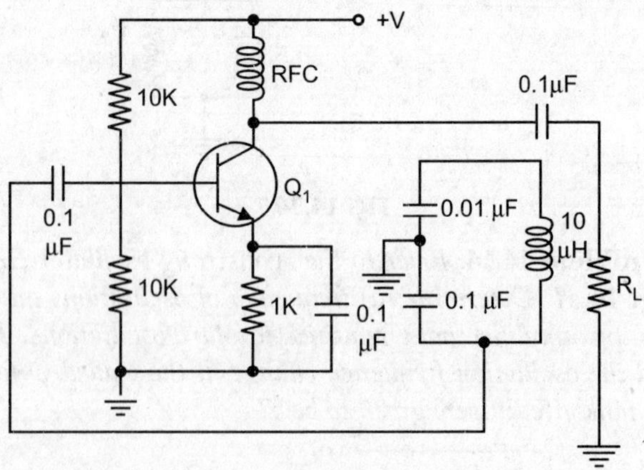

Fig. 14.32

Problem 14.16 *Refer to the Hartley oscillator of Fig. 14.33. Determine (a) operating frequency and (b) maximum acceptable value of resistance (R) for oscillations to start.*

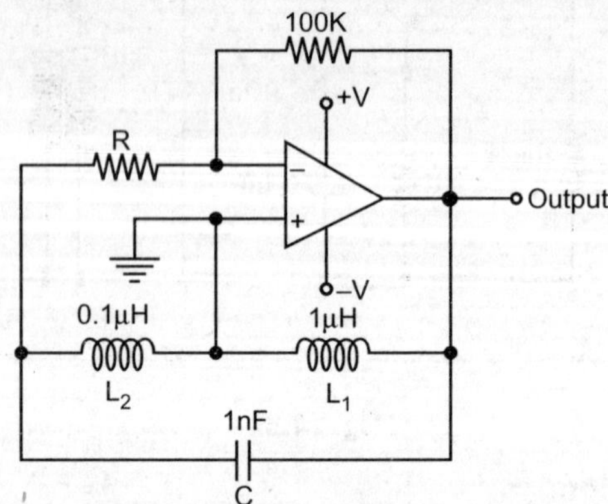

Fig. 14.33

Solution. Frequency of oscillations is given by $f = 1/2\pi\sqrt{LC}$.

$L = L_1 + L_2 = 1.0 \times 10^{-6} + 0.1 \times 10^{-6}$ = 1.1 μH and C = 1 nF

Therefore, $f = 1/(6.283 \times \sqrt{1.1 \times 10^{-6} \times 1 \times 10^{-9}})$ = 4.799 MHz

Feedback factor = $(0.1\times10^{-6})/(1.0 \times 10^{-6})$ = 0.1

Therefore, minimum required gain = 10.

Now, gain = $100 \times 10^3/R$.

Therefore, maximum value of $R = 10^5/10 = 10^4\Omega$ = 10 kΩ.

Problem 14.17 Refer to the Colpitt oscillator of Fig. 14.34. The circuit is slightly modified by connecting an additional capacitor of 1.0 nF in series with the inductance (*L*) as shown in Figure 14.34. Does the modified oscillator configuration resemble any standard oscillator configuration? Determine frequency of oscillations in this case.

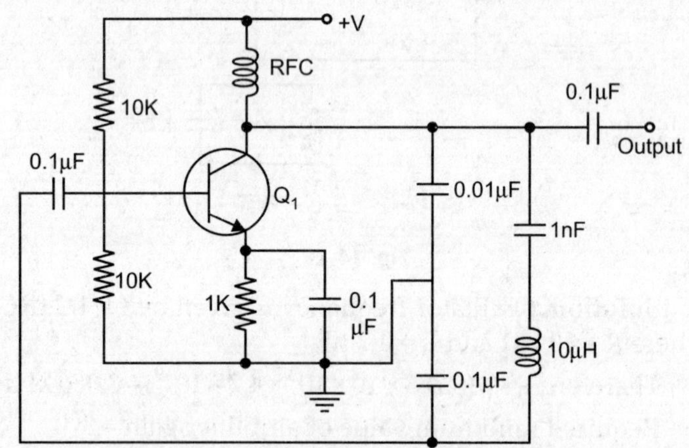

Fig. 14.34

Solution. The modified circuit is the Clapp oscillator configuration.

The frequency of oscillations is given by $f = 1/2\pi\sqrt{LC}$.

In the present case, C is given by $1/C = (1/C_1) + (1/C_2) + (1/C_3)$.

Substituting for C_1, C_2 and C_3, we get $1/C = 1/(0.01 \times 10^{-6}) + 1/(0.1 \times 10^{-6}) + 1/(1 \times 10^{-9})$.

This gives C = 0.0009 μF

Therefore, $f = 1/(6.283 \times \sqrt{10 \times 10^{-6} \times 0.0009 \times 10^{-6}})$ = 1.677 MHz.

Problem 14.17. *A quartz crystal is characterized by L = 2.5 H, R = 1 kΩ, C_S = 0.01 pF and C_P = 10 pF. Determine the series and parallel resonant frequencies of the crystal.*

Solution. Series resonant frequency, $f_s = 1/2\pi\sqrt{LC_s} = 1/(6.283\sqrt{2.5 \times 0.01 \times 10^{-12}})$ = 1.006614 MHz

Equivalent capacitance C_P for the parallel resonant circuit

$C_P = 0.01 \times 10^{-12} \times 10 \times 10^{-12}/(0.01 \times 10^{-12} + 10 \times 10^{-12})$ pF = 0.00999 pF

Therefore, parallel resonant frequency, $f_P = 1/2\pi\sqrt{LC_P}$.

Substituting the values, we get $f_P = 1/(6.283\sqrt{2.5 \times 0.00999 \times 10^{-12}})$ = 1.007117 MHz

Summary Of Important Formula

1. Barkhausen criterion of oscillations is $|\beta A| = 1$ and $\angle\beta A = 0°$ or integral multiple of 360°.
2. The oscillation frequency (ω) of RC phase shift oscillator with lag type of phase shift network is $\omega = \sqrt{6}/RC$.
3. Feedback factor (β) of RC phase shift oscillator with lag type of phase shift network is $|\beta|=1/29$
4. Oscillation frequency of phase shift oscillator using lead type of phase shift network is $\omega =1/(\sqrt{6}\,RC)$
5. Feedback factor (β) of phase shift oscillator using lead type of phase shift network is $|\beta| = 1/29$
6. The oscillation frequency of a buffered lag type RC phase shift oscillator is $f = \sqrt{3}/(2\pi RC)$
7. Minimum value of the amplifier gain for sustained oscillations for buffered lag type RC phase shift oscillator = 8.
8. For lead type of RC network, the oscillation frequency is $f = 1/(2\pi\sqrt{3}\,RC)$
9. For Twin-T oscillator, the oscillation frequency is $\omega = 1/RC$
10. For Wien bridge Oscillator, the frequency of oscillation is $\omega = \dfrac{1}{\sqrt{R_1 C_1 R_2 C_2}}$
11. In the Wien bridge oscillator when $R_1 = R_2 = R$ and $C_1 = C_2 = C$, magnitude of attenuation factor β is $|\beta|=1/3$
12. For Armstrong oscillator, the frequency of oscillation is $f = \dfrac{1}{2\pi\sqrt{LC}}$
13. For Hartley oscillator, frequency of oscillation is $\omega = \dfrac{1}{\sqrt{(L_1 + L_2)C}}$
14. For Harley oscillator, the magnitude of the feedback factor β is $|\beta| = \left(\dfrac{L_2}{L_1}\right)$
15. For Colpitt oscillator, at oscillation frequency is $\omega = \dfrac{1}{\sqrt{[LC_1 C_2/(C_1 + C_2)]}}$
16. For Clapp oscillator, the operating frequency is $|\beta| = \left(\dfrac{C_1}{C_2}\right)$
17. For Clapp oscillator , the operating frequency is $f = \dfrac{1}{2\pi} \times \sqrt{\left[\dfrac{1}{L} \times \left\{\left(\dfrac{1}{C_1}\right) + \left(\dfrac{1}{C_2}\right) + \left(\dfrac{1}{C_3}\right)\right\}\right]}$
18. For crystal oscillator, the series resonant frequency is $f_s = \dfrac{1}{2\pi\sqrt{LC_s}}$
19. For Crystal oscillator, parallel resonant frequency is $f_s = \dfrac{1}{2\pi\sqrt{LC_p}}$

SELF EVALUATION EXERCISE

Multichoice Questions

1. According to Barkhausen-Criterin for sustained oscillations:
   (*a*) $\beta A < 1$ (*b*) $\beta A > 1$
   (*c*) $\beta A = 1$ (*d*) $\beta A = 0$
2. The condition that decides the oscillator's output frequency is:
   (*a*) loop gain should at least he unity
   (*b*) loop phase-shift should be zero or integral multiple of 2π radians
   (*c*) loop gain should be precisely unity
   (*d*) loop phase-shift should be precisely zero radian
3. In a conventional transistor RC phase-shift oscillator using lead type R–C sections, frequency of oscillations is given by:
   (*a*) $f = \dfrac{1}{2\pi\sqrt{RC}}$ (*b*) $f = \dfrac{1}{2\pi RC}$
   (*c*) $f = \dfrac{1}{2\pi\sqrt{6}\,RC}$ (*d*) $f - \dfrac{1}{2\pi\sqrt{3}\,RC}$
4. In a Wien-bridge oscillator, frequency of oscillations is given by:
   (*a*) $f = \dfrac{1}{2\pi RC}$ (*b*) $f = \dfrac{1}{RC}$
   (*c*) $f = \dfrac{1}{2\pi\sqrt{6}\,RC}$ (*d*) $f = \dfrac{1}{2\pi\sqrt{3RC}}$
5. Most popular oscillator configuration for audio applications is:

(*a*) Hartley oscillator
(*b*) Coplitt s oscillator
(*c*) Wien-bridge oscillator
(*d*) R-C phase shift oscillator

6. One of the following oscillator types provides extremely stable output frequency:
(*a*) Hartley oscillator
(*b*) Wein brdge oscillator
(*c*) Crystal oscillator
(*d*) Clapp oscillator
(*e*) Tuned grid tuned plate oscillator

7. The multivibrator characterised by one quasi-stable state is :
(*a*) Bistable multivibrator
(*b*) Monostable multivibrator
(*c*) Astable multivibrator
(*d*) Schmitt trigger

8. The multivibrator circuit that has a typical application of converting sinusoidal input into a square wave output is:
(*a*) Emitter coupled binary
(*b*) Schmitt trigger
(*c*) Monoshot
(*d*) Free- running multivibrator

9. One of the following can be used to generate a pulse, whenever triggered:
(*a*) Flip-flop
(*b*) Monostable multivibrator
(*c*) Astable multivibrator
(*d*) Schmitt trigger

10. Pick the old-one out:
(*a*) Hartley oscillator
(*b*) Colpitt oscillator
(*c*) Clapp oscillator
(*d*) Wien-bridge oscillator

11. Figure 14.35 show the feedback network of one of the very popular types of sinusoidal oscillators. Which one is that?
(*a*) Wien bridge oscillator
(*b*) Hartley oscillator
(*c*) Colpitt s oscillator
(*d*) R-C phase shift oscillator

12. Figure 14.36 shows basic Hartley oscillator configuration. What do you think the polarities marked for the split coil are correct?
(*a*) Yes they are (*b*) No they are not
(*c*) It really doesn't matter

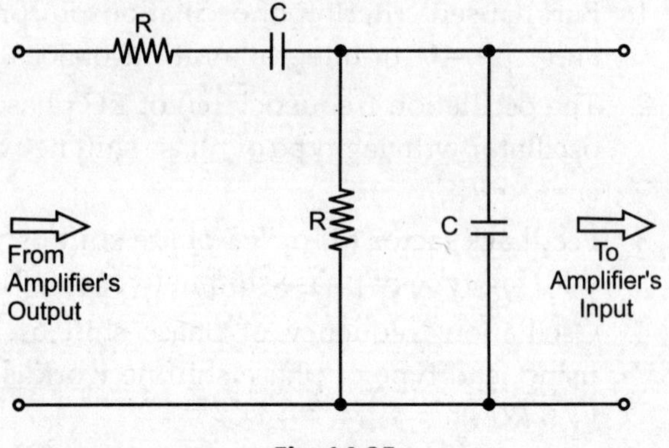

Fig. 14.35

Fig. 14.36

13. Of the following, the oscillator with the most stable oscillation frequency is
(*a*) Clapp Oscillator
(*b*) Colpitt Oscillator
(*c*) Hartley Oscillator
(*d*) Armstrong Oscillator

14. The series and parallel resonant frequencies of a quartz crystal are
(*a*) the same
(*b*) spaced far apart
(*c*) very close to each other
(*d*) in the range of tens of MHz

15. The feedback network of a conventional RC phase shift oscillator uses
(*a*) three lag-type RC sections
(*b*) three lead-type RC sections
(*c*) three RC sections of either lead or lag type
(*d*) twin-T notch filter

16. Pick the odd one out
    (*a*) Hartley oscillator
    (*b*) Colpitt's oscillator
    (*c*) Clapp oscillator
    (*d*) Wien-bridge oscillator

ANSWERS TO SELF EVALUATION EXERCISE

1. (*c*)	**2.** (*b*)	**3.** (*c*)	**4.** (*a*)
5. (*c*)	**6.** (*c*)	**7.** (*b*)	**8.** (*b*)
9. (*b*)	**10.** (*d*)	**11.** (*a*)	**12.** (*b*)
13. (*a*)	**14.** (*c*)	**15.** (*c*)	**16.** (*d*)

Chapter 15

Linear Power Supplies

Learning Objectives

After completing the chapter, you will learn the following topics:

1. Constituents of a Linear Power Supply
2. Rectifiers including Half-wave, Conventional Full-wave and Bridge Rectifiers for Positive Outputs
3. Rectifiers including Half-wave, Conventional Full-wave and Bridge Rectifiers for Negative Outputs
4. Rectifier Relevant Parameters including Ratio of Rectification, Transformer Utilization Factor, Ripple Frequency, Ripple Factor and Peak Inverse Voltage of Rectifier Diodes
5. Filters including Inductor, Capacitor and L-C Filters
6. CLC Filter (π-Filter) I
7. Regulated Power Supply Parameters including Line and Load Regulation, Output Impedance and Ripple Rejection Factor
8. Emitter Follower Regulator
9. Series Pass Regulator
10. Shunt Regulator

Q. 1. With the help of a schematic arrangement, briefly describe the function of each one of the constituents of a linear power supply.

Ans. A linear power supply essentially comprises a mains transformer, a rectifier, a filter and a regulator (Fig. 15.1). The transformer provides voltage transformation and produces across its secondary windings AC voltages required for producing desired DC voltages. It also provides electrical isolation between the power supply input *i.e.* AC mains and the DC output. The rectifier circuit changes the AC voltage appearing across transformer secondary to DC. Commonly used rectifier circuits include the half wave rectifier, conventional full wave rectifier requiring a tapped secondary and the bridge rectifier.

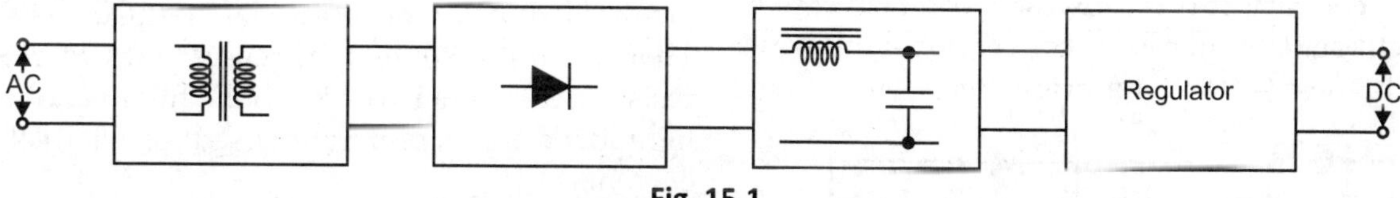

Fig. 15.1

The rectified voltage will always have some AC content known as power supply ripple. The filter circuit smoothes the ripple of the rectified voltage. The regulator circuit is a type of feedback circuit that ensures that the output DC voltage does not change from its nominal value due to changes in line voltage or load current. It is the nature of regulator circuit that distinguishes a linear power supply from a switching supply. In a linear regulator, the active device, usually a bipolar transistor is operated anywhere between cut-off and saturation, whereas in a switching regulator, it is either in cut-off or in saturation. In fact, complete regulator circuit is now available in an IC package in both fixed output voltage as well as variable output voltage varieties. They are popularly known as three terminal regulators.

Q. 2. Draw the basic circuit configurations of a half wave rectifier, a conventional full wave rectifier with center tapped secondary and a bridge rectifier for both positive as well as negative DC output voltages. Differentiate between the three types as regards (*i*) Ratio of rectification (*ii*) Transformer utilization factor (*iii*) Ripple frequency (*iv*) Ripple factor and (*v*) Peak inverse voltage of the diodes used.

Ans. Fig. 15.2 shows the three rectifier circuits for positive output voltages and Fig. 15.3 shows the same for negative output voltages.

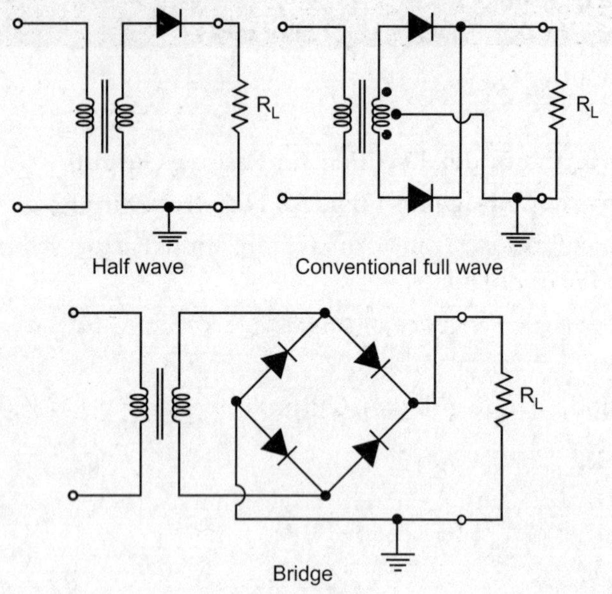

Fig. 15.2

Ratio of rectification is the ratio of DC power delivered to the load to the AC power input to the rectifier from the transformer secondary. It is 0.406 for half wave rectifier and 0.812 for both conventional full wave and bridge rectifier.

Transformer utilization factor is the ratio of DC power delivered to load to the AC power rating of the transformer secondary. It is the lowest (= 0.287) in case of half wave rectifier, 0.693 in case of conventional full wave rectifier and highest (= 0.812) for bridge rectifier.

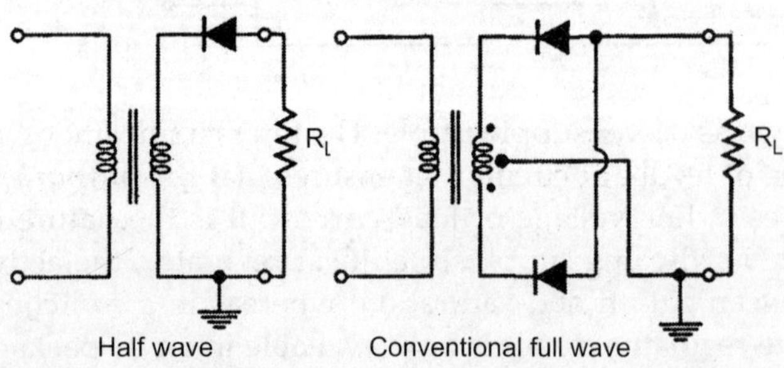

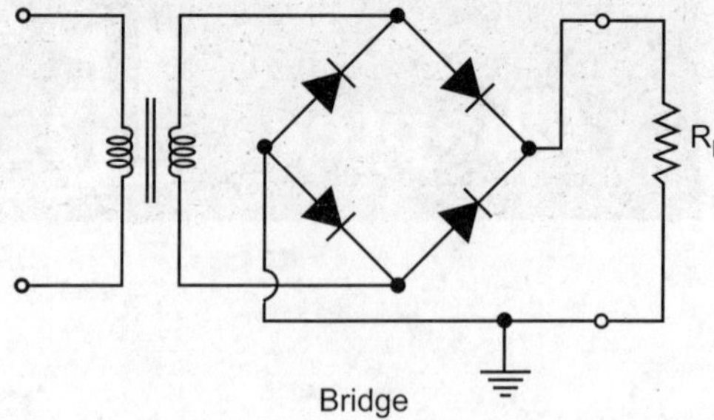

Fig. 15.3

Ripple frequency in case of half wave rectifier is (*f*), while in case of full wave rectifiers (conventional and bridge) it is (2*f*), where (*f*) is the frequency of AC mains.

Ripple factor is the ratio of RMS amplitude of AC components to the DC component. It is 1.21 for half wave and 0.482 for full wave rectifiers.

Peak inverse voltage that appears across the rectifier diodes in the three cases is (V_m) in a half wave rectifier, ($2V_m$) in a conventional full wave and (V_m) in bridge rectifier. Here, (V_m) is peak value of the AC voltage appearing across the secondary.

Q. 3. What is filtering process in a power supply? How does a choke connected in series with the load resistance (Fig. 15.4) provide filtering? How is filtering affected due to changes in load resistance for a fixed inductance value in a choke filter?

Ans. The filter in a power supply helps in reducing the ripple content (the amplitude of AC components) which in the rectified waveform is so large that the waveform can hardly be called a DC. The fact that an inductor offers high reactance to AC components is the basis of a choke providing filtering. The ripple factor (*r*) in this case is expressed as,

$$r = \frac{R_L}{3\sqrt{2}\,\omega L} \approx \frac{R_L}{1330L} \text{ for } f = 50 \text{ Hz}$$

As it is clear from the above expression, the ripple factor is directly proportional to load resistance (R_L). That is, ripple content increases with increase in load resistance. In other words, choke filter is not effective for light loads and is preferably used for relatively high load currents.

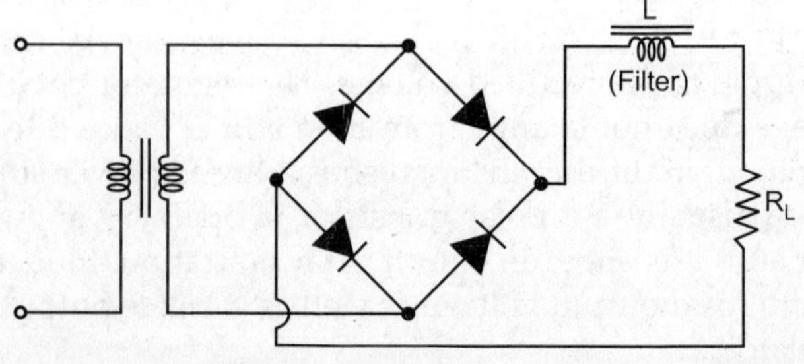

Fig. 15.4

Q. 4. How does a simple capacitor provide filtering action? Why is it suitable only at light loads ? How does the ripple factor vary with the load resistance?

Ans. The filtering action of a capacitor connected across the output of the rectifier comes from the fact that it offers a low reactance of AC components. The AC components are bypassed to ground through the capacitor and only the DC is allowed to go through the load. The capacitor charges to the peak value of the voltage waveform during the first cycle and as the voltage in the rectified waveform is on the decrease, the capacitor voltage is not able to follow the change, as it can discharge only at a rate determined by (CR_L) time constant. In case of light loads, when (R_L) is large, the capacitor would discharge only a little before the voltage in the rectified waveform exceeds the capacitor voltage, thus charging it again to the peak value (Fig. 15.5). The ripple is triangular in nature with the ripple content being inversely proportional to (C) and (R_L).

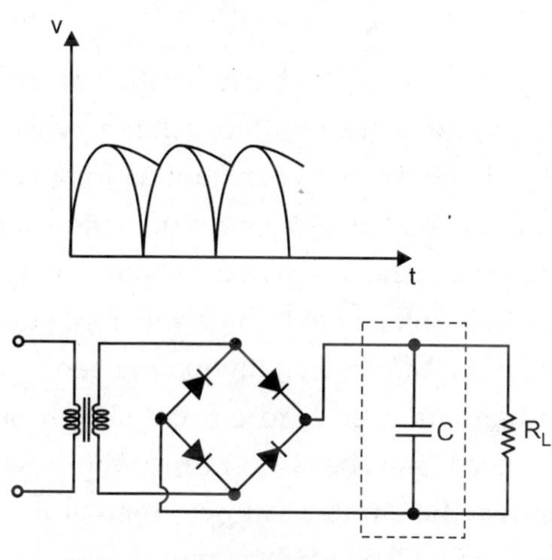

Fig. 15.5

Ripple can be reduced by increasing (C) for a given value of (R_L). For heavy loads, when (R_L) is small, even a large capacitance value may not be able to provide ripple within acceptable limits. Ripple factor (r) can be expressed by,

$$r = \left(\frac{2886}{CR_L}\right) \text{ for power line frequency of 50 Hz}$$

Where, C is in microfarads and R_L is in ohms

Q. 5. What are the advantages of choke input LC filter over simple single element inductor or capacitor filters? What is critical inductance?

Ans. We have seen that while a choke filter is effective only at heavy load currents, a capacitor filter provides filtering only for light loads. The performances of inductor and capacitor filters deteriorate fast as the load resistance is increased in case of former or decreased in case of the latter. Apparently, an appropriate combination of (L) and (C) could give us a filter that would provide adequate filtering over a wide range of (R_L) values. For all practical purposes, the ripple factor in a choke input LC filter (Fig. 15.6) is independent of (R_L).

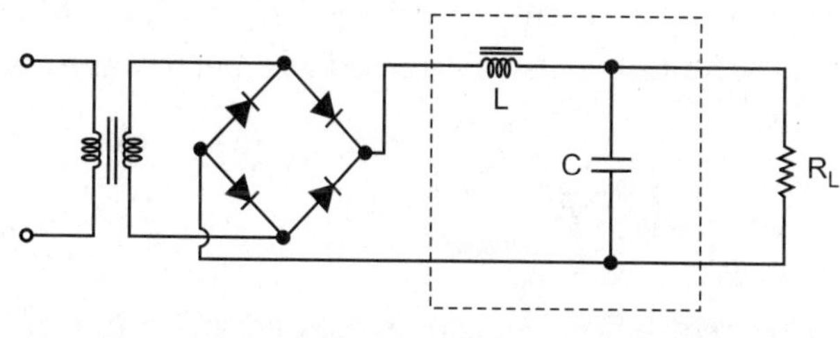

Fig. 15.6

$$r = 1.2/LC \text{ power line frequency of 50 Hz}$$

where, (L) is in henries and (C) is in microfarads.

The inductance value in the choke input LC filter should be such that the current (i) through the inductance never falls to zero which amounts to saying that the negative peak of the ripple current should never exceed the DC value of the current. The minimum inductance that achieves this is known as critical inductance. The chosen value of inductance should be greater than equal to the critical inductance. It can be expressed by the following empirical relationship.

$$L_C \geq \frac{R_L}{755}, \text{ where, } (R_L) \text{ is in ohms and } (L_C) \text{ is in henries.}$$

Q. 6. Draw the circuit diagram of a CLC filter and give the expression for the ripple factor. Also, mention some of the problems encountered by this filter.

Ans. Figure 15.7 shows the CLC filter, which is basically a capacitor filter followed by LC section. The ripple characteristics of this filter are similar to those of two-section LC filter and the expression for ripple factor can be written as,

$$\text{Ripple factor, } r = (\sqrt{2}) \times \left(\frac{X_C}{R_L}\right) \times \left(\frac{X_{C1}}{X_{L1}}\right)$$

The circuit, however, suffers from the problem of high diode peak currents, poorer regulation and a ripple that is dependent upon load resistance. In the case of very small load current, one may replace the inductance (L) with a resistance equal in value to the inductive reactance at the ripple frequency of 2ω.

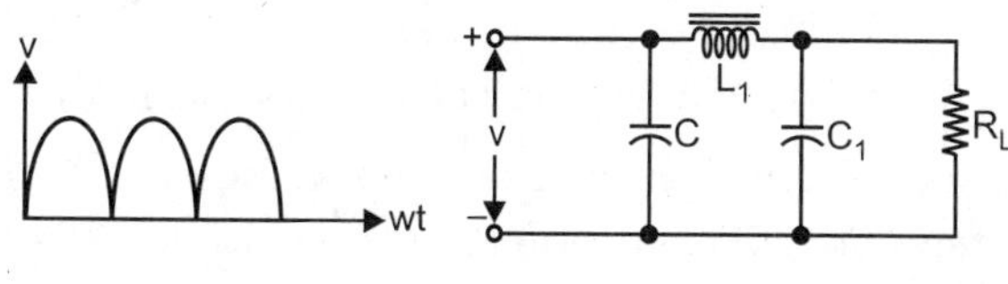

Fig. 15.7

Q. 7. Define line regulation and load regulation.

Ans. Line regulation is the ability of the power supply to maintain a constant output voltage despite variations in the input voltage. It is usually expressed as a percentage and line regulation of less than 0.01 per cent is typical. Load regulation is the ability of the power supply to maintain a constant output voltage over a wide variation in load current. It is also expressed as a percentage and is given by,

$$\left(\frac{V_{no\ load} - V_{full\ load}}{V_{full\ load}} \times 100\right)$$

Load regulation is also expressed in terms of the output impedance of the power supply. The output impedance is the ratio of a small change in the output voltage (ΔV_o) to the change in load current (ΔI_L) that caused the output voltage change.

Q. 8. What is the significance of the terms Output Impedance and Ripple Rejection Factor of a power supply?

Ans. Output impedance is an important parameter of a regulated power supply. It determines load regulation of the power supply. The regulated power supply may be represented by a Thevenin's equivalent circuit comprising of a voltage source equal to the open circuit voltage across power supply output terminals in series with impedance equal to the output impedance of the power supply. The voltage appearing across the load resistance is equal to the open circuit voltage minus drop across output impedance of the power supply. The voltage drop increases with increase in load current resulting in reduction of voltage across the load. Another way of explaining the same is that the output impedance of the power supply and the load resistance form a potential divider. The load voltage decreases with decrease in load resistance value. An ideal power supply has an output impedance of zero, which renders the output voltage independent of the load resistance value.

Practical power supplies very nearly approach the ideal condition because of emitter follower nature of regulator circuit characterised by low output impedance, which is further reduced by a factor of (1 + loop gain) due to voltage feedback. Loop gain is product of output voltage feedback factor and the gain of the error amplifier. Output impedance is typically of the order of milli-ohms.

Ripple rejection factor is defined as the ratio of ripple in the regulated output voltage to the ripple present in unregulated input voltage.

That is, ripple rejection factor $= \dfrac{V_{RIPPLE}(output)}{V_{RIPPLE}(input)}$

When expressed in decibels, ripple rejection

$$= 20\log\left[\frac{V_{RIPPLE}(output)}{V_{RIPPLE}(intput)}\right]\text{dB}$$

Ripple in unregulated input is nothing but a periodic variation in input voltage. It manifests at the output with a reduced value. Again, the factor by which ripple is reduced equals the de-sensitivity factor (1 + loop gain) due to negative feedback.

That is, $V_{RIPPLE}(\text{output}) = \dfrac{V_{RIPPLE}(input)}{(1 + loop\ gain)}$

Q. 9. Briefly describe the operational principle of an emitter follower regulator. Draw the basic circuit diagrams of positive output and negative output regulators.

Ans. Fig. 15.8 shows the basic positive output emitter follower regulator. The emitter voltage which is also the output voltage remains constant as long as the base voltage is held constant. A zener diode connected at the base ensures this. The regulated output voltage in this case is (V_Z – 0.6) volts. Due to high inherent current gain of the series pass transistor, a low power zener diode can be used to regulate high load current. The base current in this case need only be ($1/h_{FE}$) times the load current. Fig. 15.9 shows the emitter follower regulator circuit for negative voltages. If the load current is so large that it is beyond the capability of the zener diode to provide the requisite base current, a darlington combination can be used instead of a single transistor series pass element (Fig. 15.10 and Fig. 15.11).

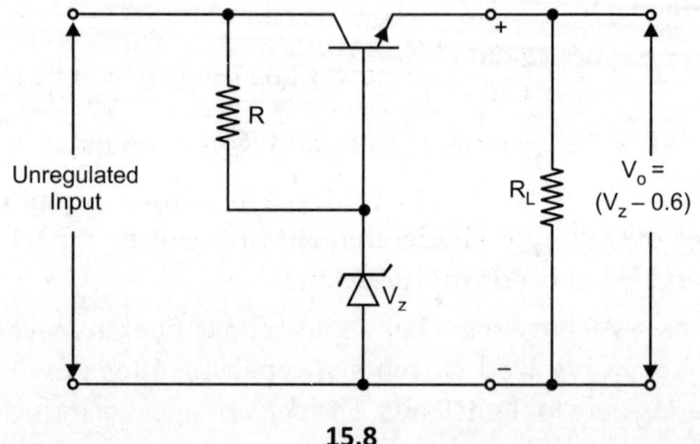

15.8

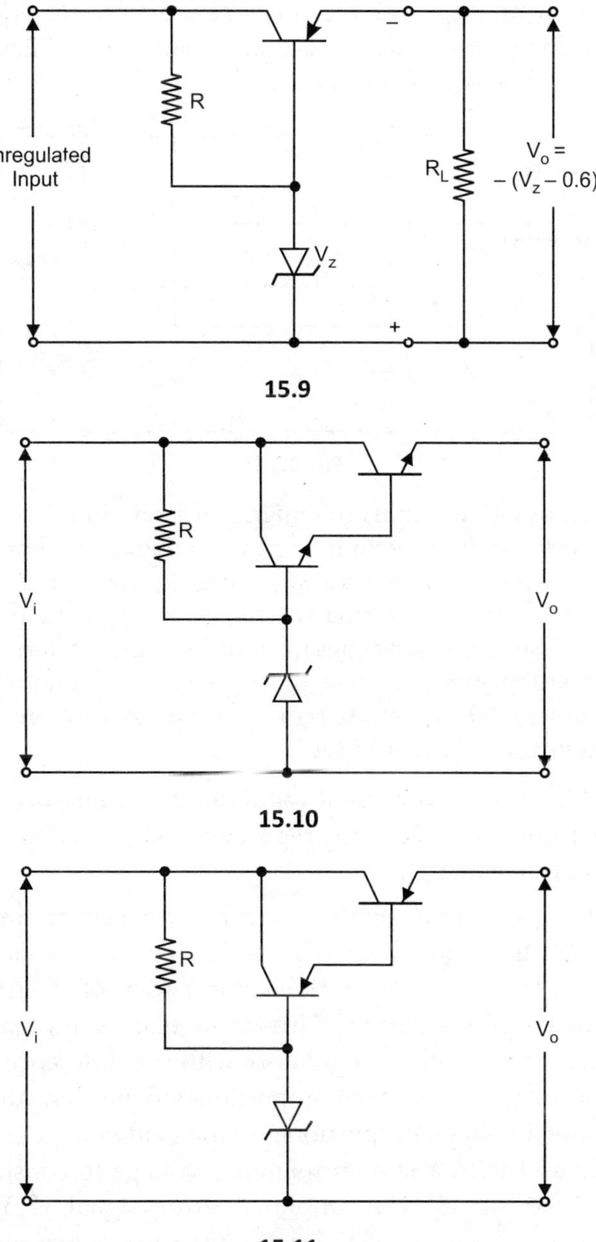

Q. 10. With the help of the basic circuit arrangement, briefly describe the operational principle of series pass element linear regulator.

Ans. Fig. 15.12 shows the basic constituents of a series pass type linear regulator. The series pass element, a bipolar transistor in the circuit shown, works like a variable resistance with the conduction of the transistor depending upon the base current. The regulator circuit functions as follows:

A small fraction of the output voltage is compared with a known reference DC voltage and their difference is amplified in a high gain DC amplifier. The amplified error signal is then fed back to the base of the series pass transistor to alter its conduction so as to maintain essentially a constant output voltage. The regulated output voltage is given by,

$$\frac{V_o R_2}{R_1 + R_2} = V_{ref} \text{ or } V_o = \frac{Vref(R_1 + R_2)}{R_2}$$

As the output voltage tends to decrease due to decrease in input voltage or increase in load current, the error voltage produced as a result causes the base current to increase. The increased base current increases transistor conduction, thus reducing its collector emitter voltage drop which compensates for the reduction in the output voltage.

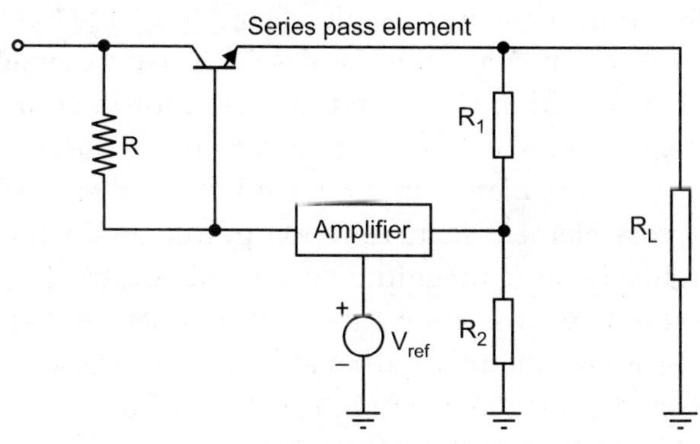

Fig. 15.12

Similarly, when the output voltage tends to increase due to increase in input voltage or decrease in load current, the error voltage produced as a consequence is of the opposite sense. It tends to decrease transistor conduction, thus increasing its collector emitter voltage drop again maintaining a constant output voltage. The regulation provided by this circuit depends upon the stability of the reference voltage and the gain of the DC amplifier. A typical series pass regulator circuit using a bipolar transistor as the error amplifier is shown in Fig. 15.13.

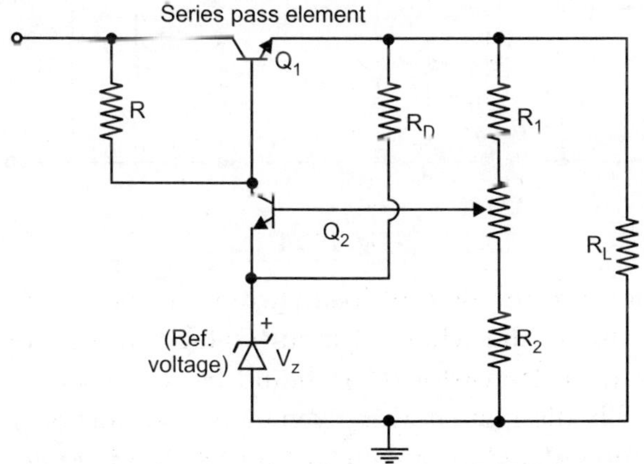

Fig. 15.13

Q. 11. Why is it desirable to incorporate current limiting circuitry in the voltage regulating type series pass element linearly regulated power supplies? Draw the typical series pass regulator with built-in overload current limiting feature. Briefly describe the functioning of the current limiting circuitry.

Ans. The power dissipated in the series pass transistor is the product of its collector-emitter voltage and the load current. As the load current increases or decreases within a certain range, the collector emitter voltage decreases or increases due to the feedback action keeping the output voltage as constant. The series pass transistor is so chosen that it can safely dissipate the power under normal load conditions. If there is an overload condition due to some reason or the other, the transistor is likely to get damaged, if such a condition is allowed to persist for long. In the worst case, if there is a short circuit on the output, the whole of unregulated input would appear across the series pass element increasing the power dissipation to prohibitively large magnitude eventually destroying the transistor. Even a series connected fuse does not help in such a case, as the thermal time constant of transistor is much smaller than that of fuse. Thus, it is always desirable to build overload protection or short circuit protection in the linearly regulated power supply design. One such configuration is shown in Fig. 15.14.

Under normal operating conditions, transistor (Q_3) is in saturation. Thus, it offers very little resistance to the load current path. In the event of an overload or a short circuit, diode (D_1) conducts, thus reducing the base drive to transistor (Q_3). Transistor (Q_3) offers an increased

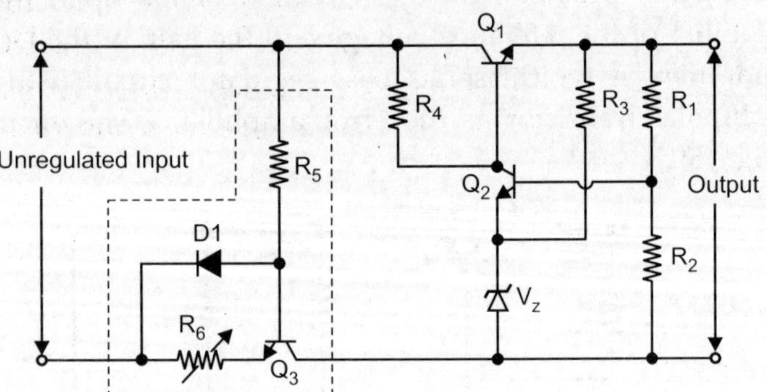

Fig. 15.14

resistance to the flow of load current. In the event of a short circuit, the whole of input voltage would appear across (Q_3). Transistor (Q_3), should be so chosen that it can safely dissipate power given by the product of worst case unregulated input voltage and the limiting value of current. Diode (D_1) and transistor (Q_3) should preferably be mounted on the same heat sink so that base-emitter junction of (Q_3) and diode's *P-N* junction are equally affected by temperature rise and the short circuit limiting current is as per the preset value.

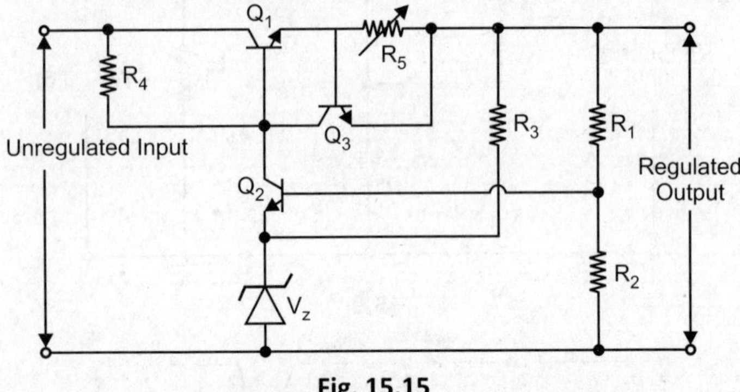

Fig. 15.15

Another circuit arrangement for providing overload or short circuit protection is shown in Fig. 15.15. For load currents less than the limiting value, the circuit regulates the output voltage normally. Transistor (Q_3) is then OFF. As the load current reaches the limiting value (determined by resistance R_5), transistor (Q_3) conducts and major portion of (Q_1) base current gets routed through (Q_3) virtually switching (Q_1) OFF.

Q. 12. What is a current regulator ? Briefly describe operation of a current regulator using series pass regulation concept.

Ans. A current regulator ensures a constant current through the load irrespective of line or load variations. Fig, 15.16 shows the V I characteristics of a current regulated power supply. Current regulators are usually similar to the voltage regulators with the difference that in this case, it is a voltage proportional to the load current and not a voltage proportional to the load voltage that is compared with a known reference voltage to constitute the error signal. The amplified error signal is again applied to the base of the pass transistor to maintain a constant load current. Fig 15.17 shows the block diagram representation of a simple current regulator using series pass regulation concept.

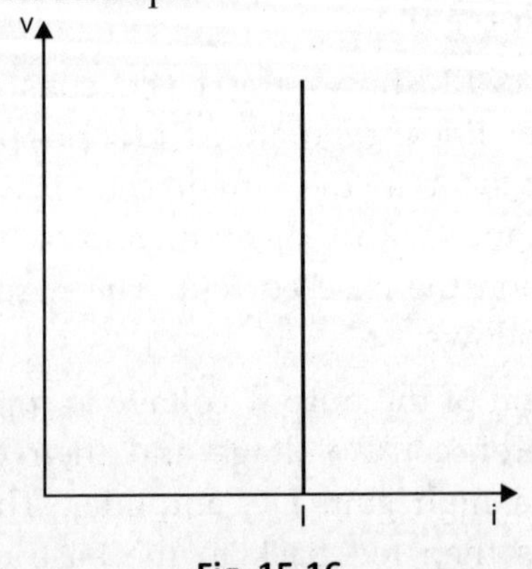

Fig. 15.16

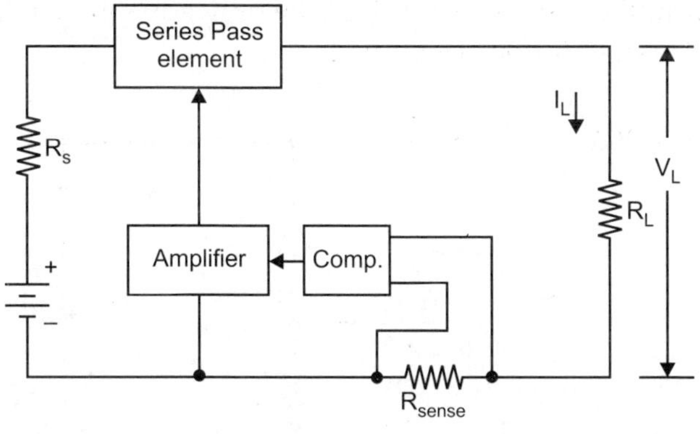

Fig. 15.17

Let us assume (I_L) tends to increase due to decrease in (R_L) value. Voltage (V_{Sense}) across (R_{sense}) increases. The drive to the base of the error amplifier increases and it conducts harder, thus reducing the base drive of the series pass transistor. This effectively increases the collector-emitter resistance of the series pass transistor to restore the load current to the nominal value. A similar chain of events occurs, when the load current tends to decrease. The error amplifier transistor in this case conducts less harder thereby increasing the base drive to the series pass element and effectively decreasing its collector emitter resistance. The load current is again restored to the nominal value.

Q. 13. What is foldback current limiting? What are its advantages?

Ans. Foldback current limiting is a form of over current protection, where the load current reduces to a small fraction of the limiting value of load the moment, the load exceeds this limiting value (Fig. 15.18). It is very commonly used on linear series pass type regulator circuits, as it would drastically reduce the dissipation of series transistors in case of short circuit conditions.

Q. 14. What do you understand by crowbaring in power supplies?

Ans. Crowbaring is a type of overvoltage protection for the power supplies. Whenever the output voltage reaches a certain preset limit, the power supply output is short circuited by an SCR which takes the short circuit current through it. Fig. 15.19 shows a simple arrangement. The output voltage is sensed and compared with a known reference. As (V_o) exceeds (V_{ref}), the SCR is triggered to the ON-state.

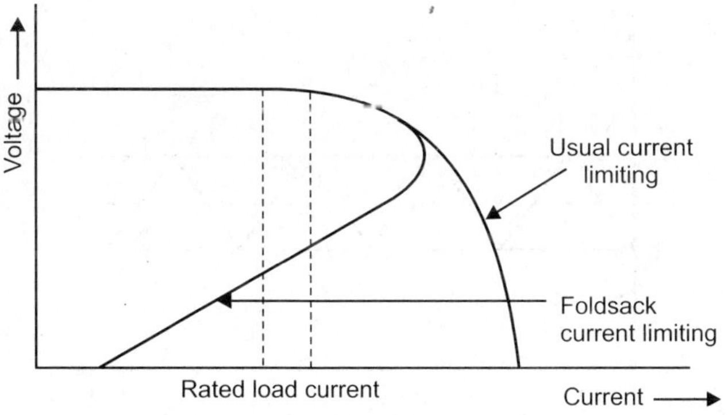

Fig. 15.18

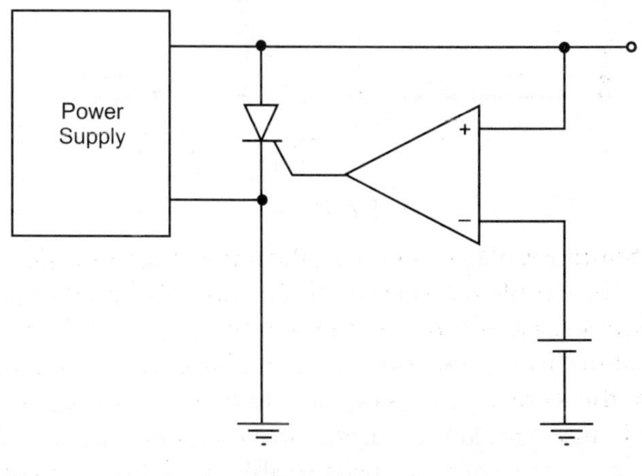

Fig. 15.19

Q. 15. With the help of the basic circuit diagram, briefly describe the operation of a shunt type linear regulator.

Ans. While in a series type linear regulator, the pass element is connected in series with the load and any decrease or increase in the output voltage is accompanied by a decrease or increase in the collector-emitter voltage of the series pass transistor, in a shunt type linear regulator (Fig. 15.20) regulation is provided by a change in the current through the shunt transistor to maintain a constant output voltage. The regulated output voltage in a shunt regulated linear power supply is the unregulated input voltage minus drop across a resistance (R_1). Now, the current through (R_1) is sum of load current (I_L) and current through shunt transistor (I_s). As the output voltage tends to decrease, the base current through the transistor reduces with the result that its collector current (1_s) decreases too. This reduces drop across and the output voltage is restored to its nominal value. Similarly, any tendency of the output voltage to increase is accompanied by an increase in current through the shunt transistor consequently increasing voltage drop across (R_1), thus maintaining a constant output voltage. The current capability of the shunt transistor can be enhanced by replacing it with a darlington combination (Fig. 15.21).

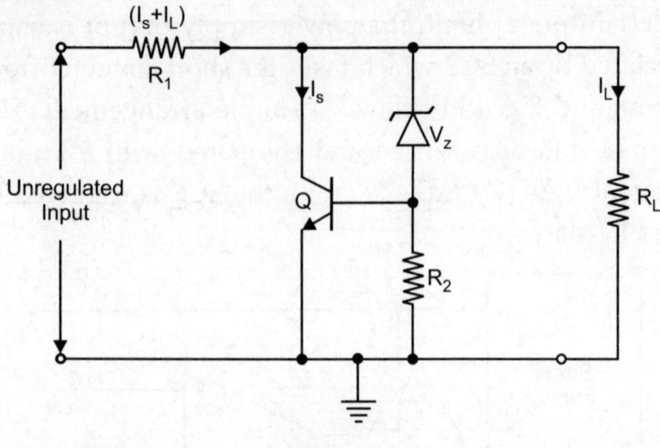

Fig. 15.20

Shunt regulator is not as efficient as a series regulator is for the simple reason that the current through the series resistor in case of shunt regulator is the sum of load current and shunt transistor current and it dissipates more power than the series pass regulator with same unregulated input and regulated output specifications. In a shunt regulator, the shunt transistor also dissipates power in addition to the series resistor. The only advantage with a shunt regulator is its simplicity.

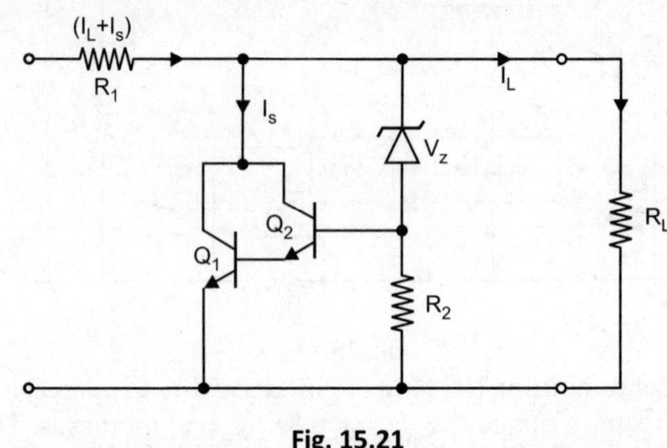

Fig. 15.21

Q. 16. What type of overcurrent protection technique is usually employed with shunt regulator?

Ans. A shunt regulator is automatically protected against over current due to presence of series resistance provided that the resistance can safely dissipate power due to the excess current under overcurrent conditions.

SOLVED PROBLEMS

Problem 15.1. *A full wave rectifier when connected to the secondary of a mains transformer produces a no load DC voltage of 12V. Determine the RMS value of the voltage across the secondary.*

Solution. If (V_m) is peak value of the voltage across transformer secondary, no load DC voltage at the output of full wave rectifier $= \dfrac{2V_m}{\pi}$.

$$\text{Therefore,} \quad \frac{2V_m}{\pi} = 12 \text{ or } V_m = \frac{12\pi}{2} = 18.84 \text{ volts}$$

$$V_{rms} = \frac{V_m}{\sqrt{2}} = \frac{18.84}{\sqrt{2}} = 13.32 \text{ volts}$$

Problem 15.2. *The no load DC output voltage of 15 volts in a power supply falls to 10 volts at the rated load current of 1 A. What is the internal resistance of the power supply?*

Solution. Internal resistance

$$= \frac{\text{No load voltage} - \text{full load voltage}}{\text{Full load current}}$$

$$= \frac{15-10}{1} = 5\Omega$$

Problem 15.3. *Determine the percentage ripple for power supply output waveforms of Fig. 15.22 (a) and (b).*

Solution. Figure 15.22 (*a*)

$$\text{Ripple factor} = \frac{\textit{RMS value of AC component}}{\textit{DC value of waveform}}$$

Peck to peak ripple = 2 Volts, Peak value = 1 volt

RMS value $= 1/\sqrt{2}$, DC value = 15 volts.

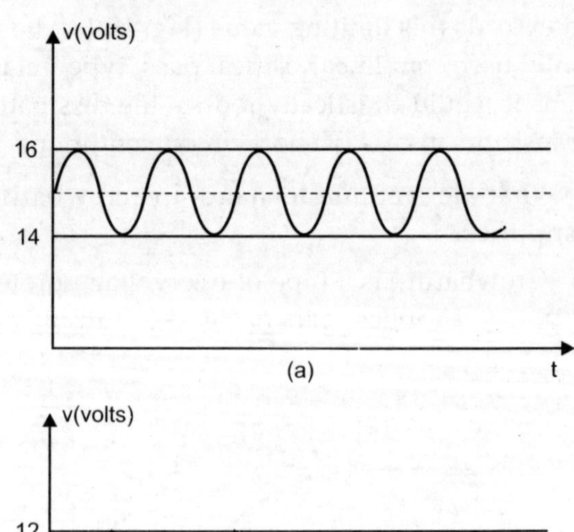

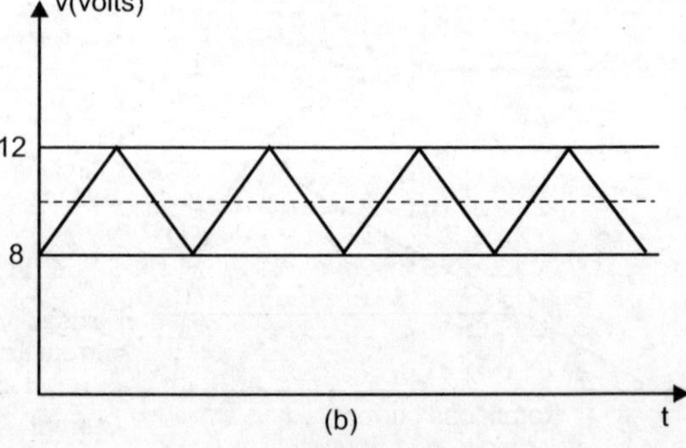

Fig. 15.22

$$r = \frac{1/\sqrt{2}}{15} = \frac{1}{15\sqrt{2}}, \quad \% r = \frac{100}{15\sqrt{2}} \approx 4.7\%$$

Figure 15.22 (*b*):

$$\text{RMS value of AC component} = \frac{\textit{Peak value}}{\sqrt{3}}$$

$$= \frac{2}{10\sqrt{3}} \text{ volts and DC value} = 10 \text{ volts}$$

$$r = \frac{2}{\sqrt{3}}, \quad \%r = \frac{200}{10\sqrt{3}} = 11.5\%.$$

Problem 15.4. *Refer to the emitter-follower regulator circuit of Fig. 15.23. Assume (β) of the transistor = 50, V BE (Q) = 0.7 V, forward voltage drop of diode D_1 = 0.7V, zener diode voltage (V_Z) = 12 V. What is the regulated output voltage?*

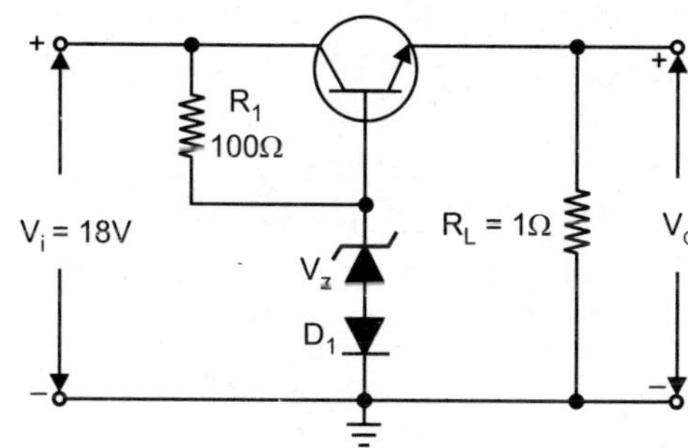

Fig. 15.23

Solution. Regulated output voltage,

$$V_O = V_Z + V_{D1} - V_{BE}(Q)$$
$$= 12 + 0.7 - 0.7$$
$$= 12 \text{ V}$$

Problem 15.5. *What is the current through the zener diode in the case discussed in the previous question?*

Solution. V_{CE} (Q) = 18 – 12 = 6V

Current through resistor R_1 = (18 – 12.7)/100 = 53 mA

Part of this current flows towards the base terminal of *Q* and the rest flows through the series combination of zener diode and diode D_1.

Now, load current = 12/1000 = 12 mA

Base current of transistor Q = 12 × 10^{-3}/(1 + β)

= 0.012/51 = 0.24 mA

Therefore, current through zener diode

$$= 53 \times 10^{-3} - 0.24 \times 10^{-3} = 52.76 \text{ mA}$$

Problem 15.6. *Figure 15.24 shows the basic shunt regulator circuit. Assume V_{BE} (Q_1) = 0.7 V. What is the value of the regulated output voltage? What is the maximum power dissipation in resistor R_S?*

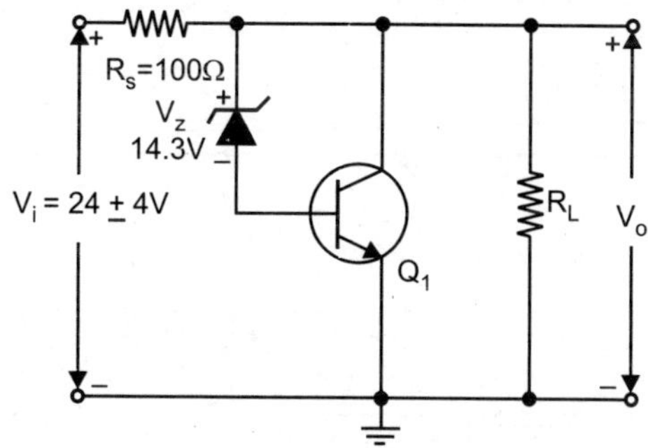

Fig. 15.24

Solution. Regulated output voltage,

$$V_O = V_Z + V_{BE}(Q_1)$$
$$= 14.3 + 0.7 = 15\text{V}$$

R_S dissipates maximum power, when the unregulated input voltage has maximum value.

Now, maximum unregulated input voltage = 28V

Therefore, maximum power dissipation

$$= (28 - 15)^2/10 = 16.9\text{W} \cong 17\text{W}$$

Problem 15.7. *A regulated power supply provides a ripple rejection of – 80 dB. If the ripple voltage in the unregulated input were 2V, what is the value of output ripple in mV?*

Solution. Ripple rejection in dB

$$= 20 \log \left[\frac{\text{output ripple}}{\text{input ripple}}\right] = -80\text{dB}$$

$$\text{Therefore, } \log \left[\frac{\text{output ripple}}{\text{input ripple}}\right] - 4 \text{ or}$$

output ripple/input ripple = 10^{-4}

Therefore, output ripple = 2 × 10^{-4} V = 0.2 mV

Problem 15.8. *Figure 15.25 shows load voltage versus load current characteristics of a regulated power supply. What is the output impedance of the power supply in ohms?*

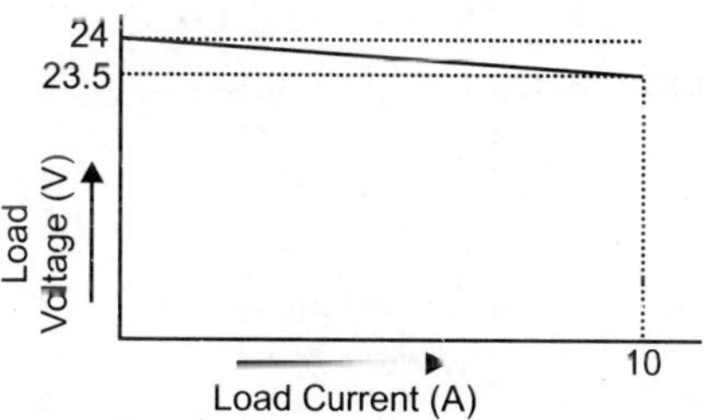

Fig. 15.25

Solution. Output impedance is given by ratio of change in output voltage for known change in load current. From the given characteristic curve, output impedance = (24 – 23.5)/(10 – 0) = 0.5/10 = 0.05 Ω = 50 mΩ.

Problem 15.9. *For the series pass regulator circuit of Fig. 15.26, what is the range (min, max) over which the regulated* *output voltage (in volts) is adjustable? Assume V_{BE} (Q_2) = 0.7 V.*

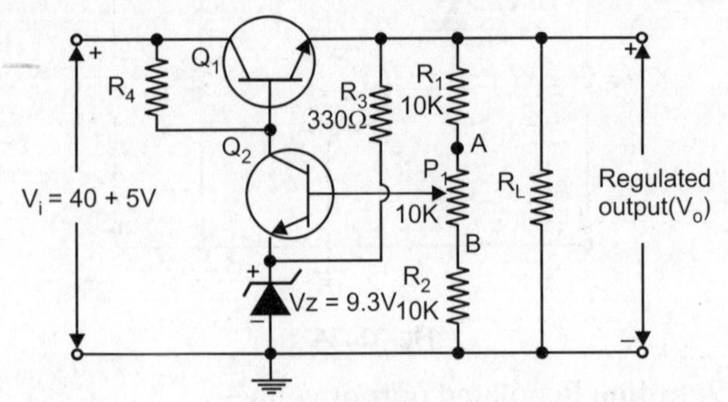

Fig. 15.26

Solution. Range over which the output voltage is adjustable is given by moving the position of the potentiometer P_1 from A to B

Regulated output voltage, when the potentiometer P_1 is at position A = $(9.3 + 0.7) \times [(10 \times 10^3 + 20 \times 10^3)/20 \times 10^3]$ = 15V

Regulated output voltage, when the potentiometer P_1 is at position B = $(9.3 + 0.7) \times [(10 \times 10^3 + 20 \times 10^3)/10 \times 10^3]$ = 30V

Therefore, regulated output voltage is adjustable from 15V to 30V

Problem 15.10 *In the voltage regulator shown in the Figure 15.27, the load current can vary from 100 mA to 500 mA. Assuming that the Zener diode is ideal (i.e., the zener knee current is negligibly small and Zener resistance is zero in the breakdown region), the value of R is*

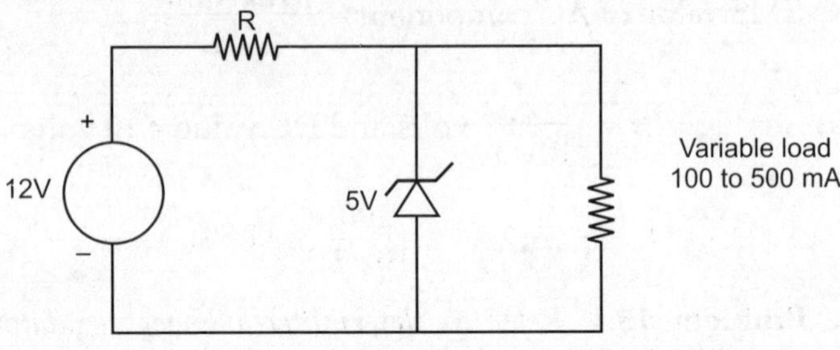

Fig. 15.27

Solution.

For satisfactory operation of the zener diode regulator

$$\frac{Vin - V_Z}{R} \geq I_{ZK} + I_L$$

Therefore, $(12 - 5)/R \geq I_L$ (as $I_{ZK} \approx 0$)

When I_L = 100 mA, $(12 - 5)/R \geq 100 \times 10^{-3}$. Therefore, $R \leq 70\ \Omega$

When I_L = 500 mA, $(12 - 5)/R \geq 500 \times 10^{-3}$. Therefore, $R \leq 14\ \Omega$

Therefore, $R = 14\ \Omega$, satisfies both the conditions

Summary of Important Formulas

1. Ripple factor for inductor filter is

$$r = \frac{R_L}{3\sqrt{2}\,(\omega L)}$$

2. Ripple factor for capacitor filter

$$r = \frac{V_r(RMS)}{V_{DC}} = \frac{2}{4\sqrt{3}\,fCR_L}$$

3. Ripple factor for LC filter,

$$r = \frac{\sqrt{2}\,X_c}{3X_L} = \left(\frac{\sqrt{2}}{3}\right)\times\left(\frac{1}{4\omega^2 LC}\right)$$

$$= \left(\frac{\sqrt{2}}{12\omega^2}\right)\times\left(\frac{1}{LC}\right)$$

4. Critical inductance for LC filter is $LC = \frac{R_L}{3\omega}$

5. For CLC Filter (π - Filter), ripple factor,

$$r = (\sqrt{2})\times\left(\frac{X_C}{R_L}\right)\times\left(\frac{X_{C1}}{X_{L1}}\right)$$

6. Ripple rejection factor

$$= \frac{V_{RIPPLE}(output)}{V_{RIPPLE}(input)} \text{ or } 20\log\left[\frac{V_{RIPPLE}(output)}{V_{RIPPLE}(input)}\right]$$

7. $V_{RIPPLE}(output) = \frac{V_{RIPPLE}(input)}{(1 + loop\ gain)}$

Self Evaluating Exercise

(I) Choose the correct answer

1. The output voltage at no load in an unregulated power supply is the same as its output voltage at rated load current. Its internal resistance is therefore

(*a*) extremely small (*b*) zero
(*c*) infinite (*d*) extremely large

2. The filter circuit that is effective for very large value load resistance is

(*a*) capacity filter (*b*) inductor filter
(*c*) choke input LC filter (*d*) π-type CLC filter

3. If the transformer utilization factor (T.U.F.) for a particular rectifier configuration is small, it implies that
   (*a*) for a given transformer rating, it would deliver a larger DC power to the load.
   (*b*) for a given transforer rating, it would deliver lesser DC power to the load.
   (*c*) The ratio of DC power delivered to load to the AC power available at the input of rectifier circuit from transformer secondary is small.

4. One of the following expressions is wrong:
   (*a*) $\text{T.U.F.} = \dfrac{\text{DC power delivered to the load}}{\text{AC rating of transformer secondary}}$
   (*b*) $\text{Load Regulation} = \dfrac{\text{No load output voltage–Full load output voltage}}{\text{No load output voltage}}$
   (*c*) $\text{Ripple factor}(r) = \dfrac{\text{RMS value of AC component}}{\text{DC value of the rectified wave}}$
   (*d*) $\text{Ratio of rectification} = \dfrac{\text{DC power delivered the load}}{\text{AC power available across transformer secondary}}$

5. One of the plots shown in Fig. 15.28 is for voltage regulating type linearly regulated power supply with foldback current limiting
   (*a*) Fig. 15.28 (*a*) (*b*) Fig. 15.28 (*b*)
   (*c*) Fig. 15.28(*c*) (*d*) Fig. 15.28 (*d*)

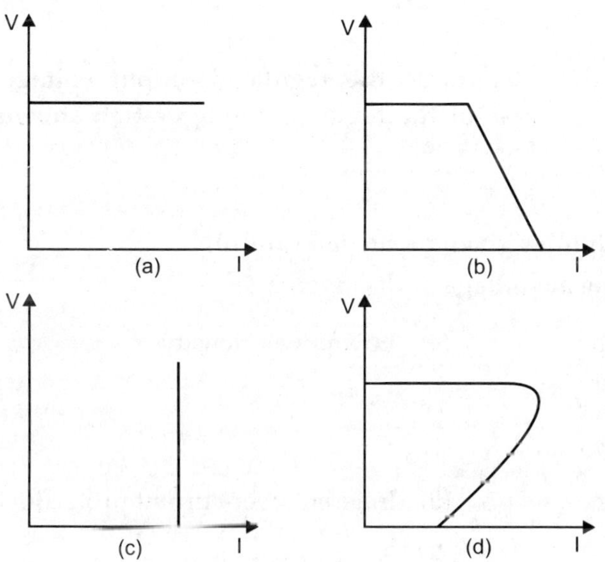

Fig. 15.28

6. The V-I plot that represents a current regulating type regulated power supply is shown in
   (*a*) Fig. 15.28 (*a*) (*b*) Fig. 15.28 (*b*)
   (*c*) Fig. 15.28 (*c*) (*d*) Fig. 15.28 (*d*)

7. Mark the rectifier circuit that produces the least ripple
   (*a*) Single rectifier, half wave, single phase
   (*b*) Four rectifiers, full wave, single phase
   (*c*) Two rectifiers, full wave, single phase
   (*d*) Six rectifiers, full wave, three phase.

8. The regulation of a linear regulator is normally........than that of a switching regulator.
   (*a*) better (*b*) worse
   (*c*) far better (*d*) far worse

9. In a series pass linear regulator, the voltage drop across the series pass element
   (*a*) is independent of changes in output voltage
   (*b*) changes directly with changes in output voltage
   (*c*) changes inversely with changes in output voltage
   (*d*) none of these

10. In a series pass linearly regulated power supply, regulated output voltage is 10 volts across a load of 1000 ohms. The unregulated input is 15V ±3V. The series pass transistor dissipation in the worst case will be
    (*a*) 0.08 watt (*b*) 0.8 watt
    (*c*) 8 watt (*d*) 0.18 watt
    (*e*) 0.15 watt

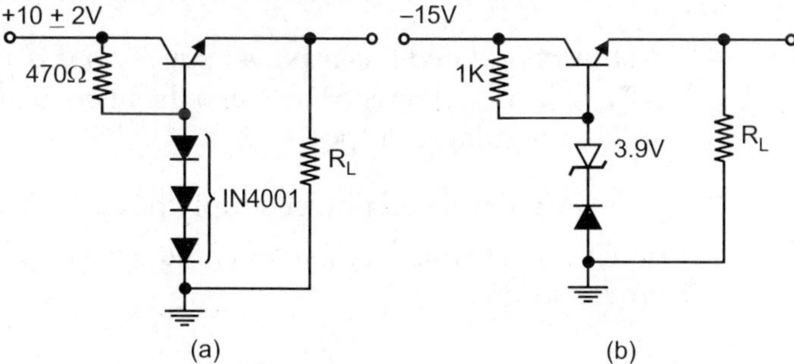

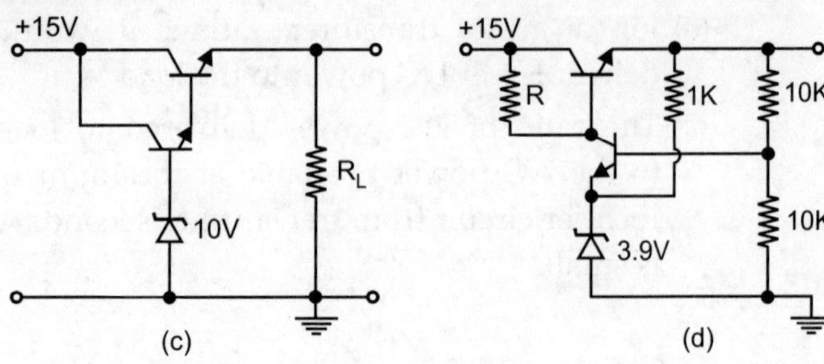

Fig. 15.29

11. The type of linear voltage regulator that is inherently immune to overload condition is
    (*a*) Emitter follower regulator
    (*b*) Series pass regulator with error amplifier in feedback loop
    (*c*) Shunt regulator
    (*d*) None of these

12. The output voltage of a regulated power supply drops by 1 volt from no load to rated full load of 1A. The no load output voltage of the power supply is 24 volts. What is the load regulation of the power supply?
    (*a*) 2.3% (*b*) 3.2%
    (*c*) 3.4% (*d*) 4.3%

13. The no load and rated load output voltage in a regulated power supply are the same. Its output impedance is therefore
    (*a*) Extremely small (*b*) Zero
    (*c*) Infinite (*d*) Extremely large

14. A voltage regulator provides a ripple rejection of – 60 dB. If the ripple in the unregulated input were 0.5 volt, the ripple in the regulated output would be
    (*a*) 0.5 mV (*b*) 60 mV
    (*c*) 1 mV (*d*) 5 mV

15. In a full-wave rectifier using two ideal diodes, V_{dc} and V_m are the dc and the peak values of the voltages respectively across a resistive load. If PIV is the peak inverse voltage of the diode, then the appropriate relationships for this rectifier are
    (*a*) $V_{dc} = \frac{V_M}{\pi}$, $PIV = 2\ V_m$
    (*b*) $V_{dc} = \frac{2V_m}{\pi}$, $PIV = 2V_m$
    (*c*) $V_{dc} = \frac{2V_m}{\pi}$, $PIV = V_m$
    (*d*) $V_{dc} = \frac{V_m}{\pi}$, $PIV = V_m$

(II) Determine the regulated output voltage in case of the regulator configuration shown in Fig. 15.29.

(III) Read the statements with reference to linear power supplies given on the left carefully. Match each one of these statements with the corresponding advantage or disadvantage.

1. In a series pass transistor regulated power supply, the power dissipated in the transistor is the product of load current and the voltage drop across the collector-emitter terminals and also the transistor is always conducting	(*a*) Low power density
2. A linear power supply operates at the power line frequency of 50 Hz, which leads to bulky.	(*b*) Inherent over-current protection
3. The regulator transistor is always operated in the active region of its output characteristics and it is not repetively switched ON and OFF at high frequencies	(*c*) Poor efficiency
4. A shunt regulated linear power supply has a series connected resistor between the unregulated input and the regulated output	(*d*) Low peak-to-peak noise in the regulated output
5. A linearly regulated power supply has a relatively much poorer efficiency as compared to a switched mode power supply	(*e*) Not preferred for high output power applications

ANSWERS TO SELF EVALUATION EXERCISE

(I)

1. (*b*)	**2.** (*a*)	**3.** (*b*)	**4.** (*b*)	**5.** (*d*)
6. (*c*)	**7.** (*d*)	**8.** (*a*)	**9.** (*b*)	**10.** (*a*)
11. (*c*)	**12.** (*d*)	**13.** (*b*)	**14.** (*a*)	**15.** (*b*)

(II)

Fig. 15.29 (*a*): 1.2 V **Fig. 15.29 (*b*):** –3.9 V **Fig. 15.29 (*c*):** 8.8 V **Fig. 15.29 (*d*):** 9 V

(III)

1. (*c*)	**2.** (*a*)	**3.** (*d*)	**4.** (*b*)	**5.** (*e*)

Chapter 16

Switched Mode Power Supplies

Learning Objectives

After completing the chapter, you will learn the following topics:

1. Linear Power Supplies versus Switched Mode Power Supplies
2. Power Inverter versus Power Converter
3. Constituents of a Switched Mode Power Supply
4. Types of Switched Mode Power Supplies
5. Forward Converters
6. Push-Pull Converters
7. Flyback Converters including Self-oscillating and Externally Driven Flyback Converters
8. Design Guidelines for Switched Mode Power Supplies
9. Switching Regulators including Buck, Boost and Inverting Regulators
10. Off-line Switched Mode Power Supplies
11. Connecting Power Converters in Series
12. Connecting Power Converters in Parallel

Q. 1. How do you differentiate a switched mode power supply from a linear power supply?

Ans. Depending upon the regulation technique used, power supplies are classified as either linear power supplies or switching power supplies. Conventional AC/DC power comprising a transformer, rectifier, filter and a regulator (series or shunt type) constitute the linear power supplies. In a linear power supply, the active device that provides regulation is always operated in the active (or linear) region of its characteristics. Any change in the output voltage due to a change in input voltage or load current results in a change in the voltage drop across the regulator transistor (in case of a series regulator) or a change in current through the regulator transistor (in case of a shunt regulator) so as to maintain a constant output voltage across the load.

On the other hand, in a switched mode power supply, the active device that provides regulation is always operated in a switched mode *i.e.* it is operated either in cut-off or in saturation. The input DC is chopped at high frequency (typically 10 kHz to 100 kHz) using an active device (bipolar transistor, SCR, power MOSFET, IGBT) and the converter transformer. The transformed chopped waveform is rectified and filtered. A sample of output is used as a feedback signal to usually adjust the duty cycle of the drive waveform for the switching transistor to achieve regulation.

Q. 2. Compare the linear and switched mode power supplies on the basis of major power supply specifications like line and load regulation, output ripple, efficiency and power density.

Ans. The strength of a linearly regulated power supply lies in its comparatively much better line and load regulation specifications. Line regulation for a linear supply may be anywhere between 0.02% to 0.05%, whereas the lower and upper limits for the line regulation in case of a switching supply are 0.05% and 0.1%. Load regulation in case of a linear supply typically varies from 0.02% to 0.1%. This figure could be as bad as 1% in a switching supply. Also, a linear power supply may have a worst case peak to peak ripple of 5 mV while the same for a switching supply could be as much as 100 mV.

The parameters, where a switching supply has a high degree of superiority over a linear supply are the efficiency and the power density. High efficiency in a switching supply (typically 70 to more than 90%) results from the switched mode operation of the active device. The switching transistor, one of the popular active device for switching supplies, dissipates the least power, when in cut-off or in saturation. Efficiency figures in a linear supply range from 40 to 50 percent. The power density indicates the size and weight vis-a-vis power delivering capability of the supply and it may be as high as 25 watts per cubic inch in a switching supply. Higher power density here arises from the reduced sizes of inductors, transformers and capacitors due to high frequency operation. The same in case of a linear power supply is typically 0.5 to 1 watt per cubic inch.

Q. 3. How do you distinguish between a power inverter and a power converter? Where do these circuits find application?

Ans. A power inverter converts DC power into AC power. Electronic systems operating from AC power can be made to operate on DC power with a power inverter.

A power converter converts DC power at one voltage level into a DC power at another voltage level. An electronic system may require a number of DC voltages for performing optimally and all these DC voltages may not be available from the primary source of DC power. DC to DC converters could be used to generate any DC voltage level from available DC voltage level. In fact a DC power converter (DC to DC) is an extension of power inverter (DC to AC) as the output of an inverter when rectified and filtered, gives us a DC to DC converter. Both DC to DC converters and Inverters, as said earlier in the introduction, belong to the class of switched mode power supplies.

Q. 4. What are the main constituents of a DC to DC converter? Explain in brief the role of each constituent part.

Ans. A DC to DC converter comprises a switching device, a transformer, a rectifier and a filter. The switching device, which is usually semiconductor transistor or a power MOSFET, acts like some kind of chopper and produces a waveform (from applied DC) acceptable to a transformer. The primary of the transformer is normally in series with the collector of the transistor or drain of a MOSFET. The transformer does the necessary power transformation. The transformer forms a very important part, as it decides the size of the converter, influences the frequency of operation, dictates the output voltage and affects the conversion efficiency. The output appearing across the secondary of the transformer is then rectified. The rectifier is either a halfwave-common with flyback converters — or a full wave one—common with push-pull forward converters. The rectified waveform is then filtered. The filter is almost invariably a capacitor filter. The size of the capacitor is determined by the load it has to deliver arid the frequency of operation. The detailed design of DC to DC converters shall be dealt with in subsequent questions.

Q. 5. What are various types of DC to DC converters? How do you differentiate between them from operational viewpoint?

Ans. DC to DC converters can be broadly classified as (*i*) Ringing choke or flyback converter and (*ii*) Forward converter.

In case of a flyback converter, the energy is stored in the magnetic field of the transformer primary during the ON time of the active device. The stored energy gets transferred to the secondary during the OFF time only. So, when the system is storing energy it is not transferring and vice-versa. Fig . 16.1 shows basic flyback converter.

Flyback converters can be further categorised as

(*i*) Externally driven flyback converter Fig. 16.1 (*a*)

(*ii*) Self-oscillating flyback converter Fig. 16.1 (*b*).

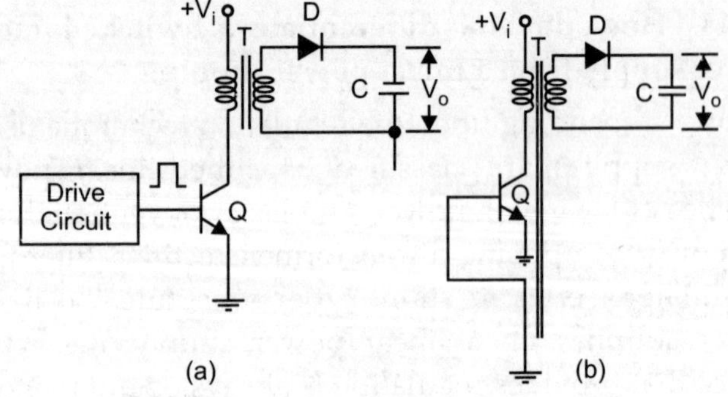

Fig. 16.1

In a self-oscillating type flyback mode DC to DC converter, the drive for the input of the active device (transistor Q as shown in the figure) is provided by one of the windings on the transformer core. In an externally driven converter like the one shown in Fig. 16.1 (*a*), the input drive is provided by a separate square wave oscillator.

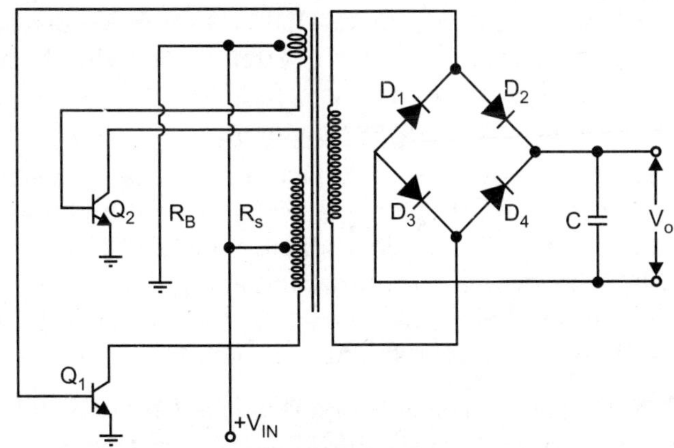

Fig,16.2

In case of a forward converter, the processes of energy storage and energy transfer take place simultaneously. That is, during both the ON time as well as the OFF time, the energy is getting stored as well as transferred. Push-pull converter is a typical case of a forward converter. Furthermore, push-pull converter can also be a self-oscillating one or an externally driven one. Fig. 16.2 shows basic self-oscillating push-pull DC to DC converter.

Q. 6. With the help of the basic circuit schematic, briefly describe the operational principle of a single ended forward converter with special reference to fundamental differences, if any between a single ended forward converter and a flyback converter.

Ans. Fig. 16.3 shows the basic circuit schematic for a single ended forward converter. In case of a forward converter, the current continues to flow through the load during both ON as well as OFF times of the switching transistor. This was mentioned in the last question too. There is a fundamental difference between a flyback converter discussed earlier and the conventional forward converter whose circuit schematic is shown in Fig. 16.3. Here, when the transistor is switched ON, the polarities of the transformer windings are so chosen as to forward bias diode D_1 and reverse bias diodes D_2 and D_3. As a result, most of the energy in a forward converter is stored in the output inductance rather than transformer primary used to store energy in a flyback converter. When the transistor switch is turned OFF, the magnetic field collapses, the polarities reverse and the diodes D_2 and D_3 are forward biased while diode D_1 gets reverse biased. As the current through an inductance cannot change instantaneously, the output current continues to flow through the output and the current path is provided by forward biased diode D_2. In the absence of the third winding and diode D_3, a good fraction of energy stored in the transformer primary is lost. The effect is more severe at higher switching frequencies. The third winding and the forward biased diode D_3 return the energy which would otherwise be lost and reset the transformer core after each operating cycle. This not only increases the converter efficiency but also makes the converter transformer core immune to saturation problems.

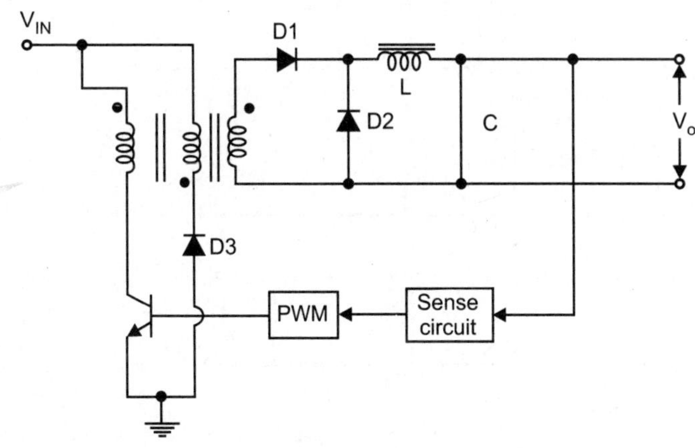

Fig. 16.3

Q. 7. With the help of a circuit diagram, explain in brief the principle of operation of an externally driven flyback converter. Also, draw the waveform at critical points.

Ans. The circuit diagram is shown in Fig. 16.4 (*a*).

The transistor *Q* acts as a chopper for input voltage V_{IN}. The driver circuit supplies a train of pulses to the input of the transistor *Q*. This train of pulses drives the transistor towards saturation (During the high state of the pulse-train) and cut off (during low state of the pulse train) alternately. During the conduction time of the transistor, the collector is effectively at ground potential and V_{IN} appears across primary winding of the transformer. The current through the primary inductance ramps upto a value given by

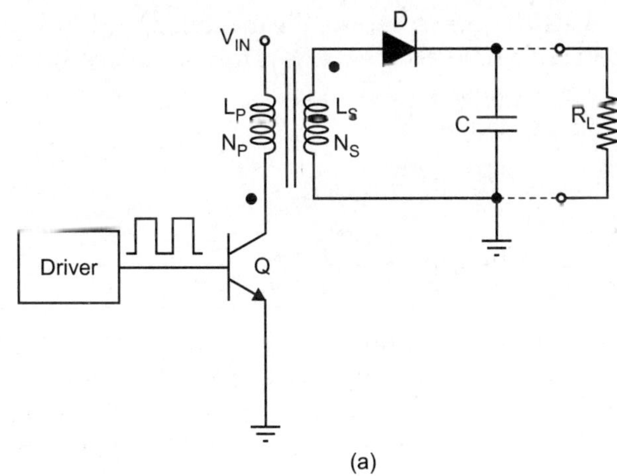

(a)

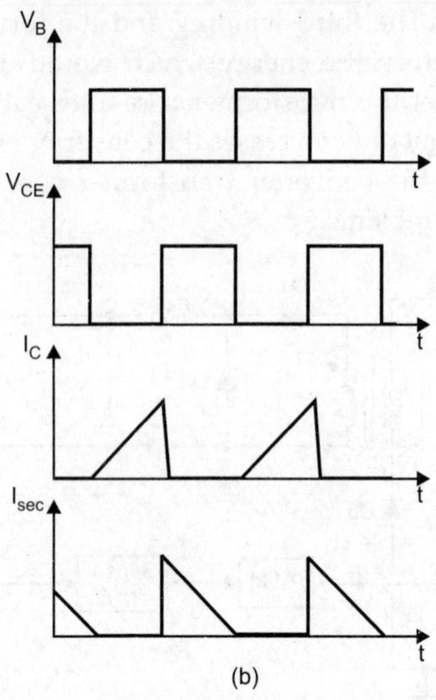

(b)

Fig. 16.4

$$V_{IN} = L_P \frac{dI_P}{dt}$$

It the current ramps up from zero to $\hat{I}_P$ in time (t_{ON}), then

$$\hat{I}_P = \left(\frac{V_{IN}\, t_{ON}}{L_P}\right)$$

Energy is stored in the magnetic field due to $\hat{I}_P$ flowing through primary of the transformer. The energy stored in given by,

$$\left(\frac{1}{2} LP\hat{I}_P^2\right)$$

Also, when the energy is getting stored, the diode in the secondary circuit is kept reverse biased by appropriately choosing the winding polarities. As a result, the energy cannot get transferred. However, when the transistor goes to cut-off, the magnetic field collapses, the polarities reverse, the diode is forward biased and the energy gets transferred. This energy charges the capacitor to a voltage V given by,

$$\frac{1}{2} CV^2 = \frac{1}{2} \hat{L}_P I_p^2$$

$$\text{or} \quad V = \left(\frac{L_P}{C} \hat{I}_P^2\right)^{1/2} = \sqrt{\frac{L_P}{C}}\, \hat{I}_P$$

This process continues. During every 'ON' time, the energy is stored and during following 'OFF' time, the same gets transferred, everytime the capacitor gets a quantum of voltage. Where does this process stop? It will stop at a stage, where the output voltages across the capacitor C or load R_L is such that V_0^2/R_L) equals the power for which the converter has been designed. That is, a flyback converter is a constant power output converter; if we try to draw a current which is more than its capability, the voltage will fall so as to keep the power unaltered.

In fact, diode does rectification and the capacitor does filtering. In a flyback converter, the filter capacitor exclusively supplies the load during the 'ON' time of the transistor as during that time, the energy is only being stored and not being transferred. The waveforms shown in Fig. 16.4 (*b*) are self-explanatory.

Q. 8. Since a flyback converter is a constant power converter, there is every likelihood of output voltage rising without any limit and thus damaging the transistor or the transformer or both if the load is accidentally taken off. What precautions do you suggest in this case? Also, suggest circuit modifications to take care of this.

Ans. A flyback converter cannot give us an output power greater than the one for which it has been designed. As a result, an increase in load current shall invariably result in a decrease in load voltage and vice-versa.

One way to take care of this drawback is to limit the output voltage to a value which is larger than the desired output voltage on load. This can be achieved by means of a resistance. Obviously, the value of resistance shall be larger than the load resistance, the value of this resistance can be calculatd from

$$R_{limit} = \left(\frac{V_{limit}^2}{P_o}\right)$$

V_{limit} should be chosen to be as high as possible to be limited only by the $V_{CEO(max)}$ rating of the transistor. This method however results in an unacceptable power loss, as the output power capability of the converter increases. In high power converters, some sort of voltage feedback loop is used instead. The output voltage is sensed; a fraction of it sampled and compared with a reference voltage. The comparator does the job of switching off the converter whenever the output voltage tends to exceed the desired output voltage. The two techniques discussed above are represented in block schematics in Figs. 16.5 (*a*) and (*b*).

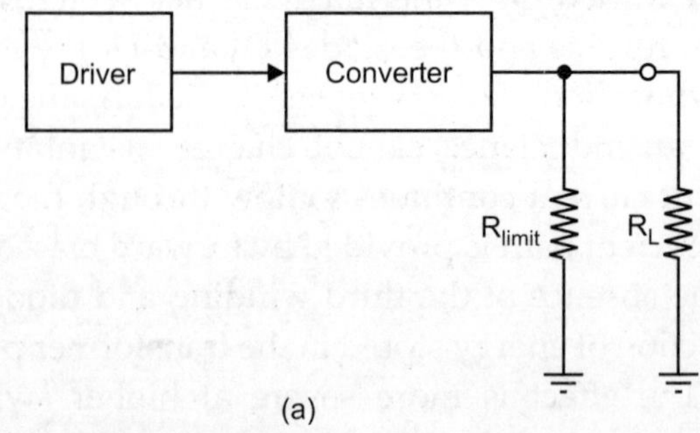

(a)

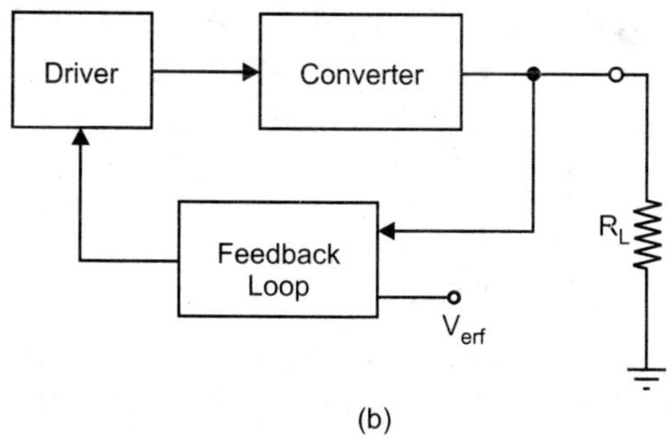

(b)

Fig. 16.5

Q. 9. With the help of a circuit diagram, explain the principle of operation of a self-oscillating flyback converter. Also, draw relevant waveforms.

Ans. Fig. 16.6 (*a*) shows a typical one transistor, one transformer self-oscillating flyback converter. Fig. 16.6 (*b*) shows relevant waveforms. During the conduction period of the transistor, the transistor is driven to saturation and V_{IN} appears across the primary of the transformer. The current through primary winding rises linearly and thus, stores energy in the magnetic field. This linearly rising current induces constant voltages across the base winding and secondary winding. However, the polarity of secondary voltage induced is such that diode *D* is reverse biased. The constant voltage induced across base winding sends a constant base current determined by this voltage and resistance R_B. If I_B is the base current, then I_C (or I_P) cannot rise beyond I_p given by (βI_B). After I_p is achieved, transistor starts coming out of saturation, I_p falls rapidly and the transistor is driven to cut-off. The magnetic field collapses because of rapidly falling I_P ; the induced voltages reverse polarities.

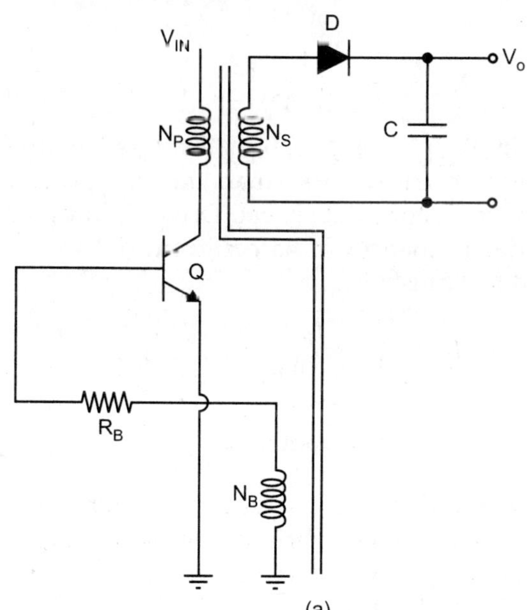

(a)

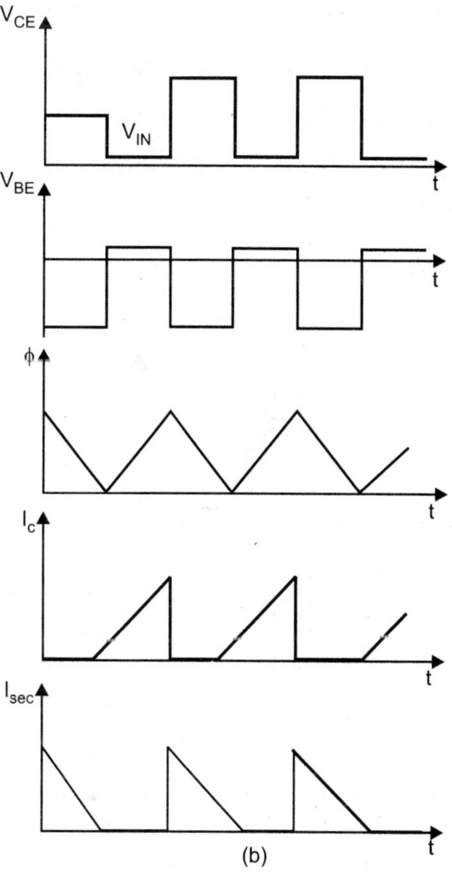

(b)

Fig. 16.6

The diode gets forward biased and the stored energy gets transferred to the output capacitor and the load. Since the capacitor tries to hold secondary voltage as constant, the secondary current decreases linearly. This turns the transistor 'ON' again, when the current reduces to zero and the cycle starts again. Magnetic flux in the core varies from zero to (ϕ_{max}) during the conduction period when primary current rises from zero to $\hat{I}_P$. The flux decreases from (ϕ_{max} to zero during 'OFF' time when $\hat{I}_P$ falls from $\hat{I}_P$ to zero. It may be mentioned that (ϕ_{max})is less than the flux corresponding to maximum flux density rating of the core material. The core should never be allowed to fully saturate.

Q. 10. Outline the design procedure of an externally driven flyback converter using a single transistor and a single transformer with special emphasis on the choice criterion for the transistor.

Ans. Refer to Fig. 16.7

Let V_{IN} = Input voltage in volts

P_o = Output power to be delivered to load R_L

V_o = Desired output voltage in volts

η = Expected conversion efficiency

f = Frequency of operation in Hz

t_{ON} = Conduction time in sec

If (L_p) is the primary inductance and $\hat{I}_P$ the peak current, then

$$V_{IN} = \frac{L_P \hat{I}_P}{t_{ON}} \quad ...(1)$$

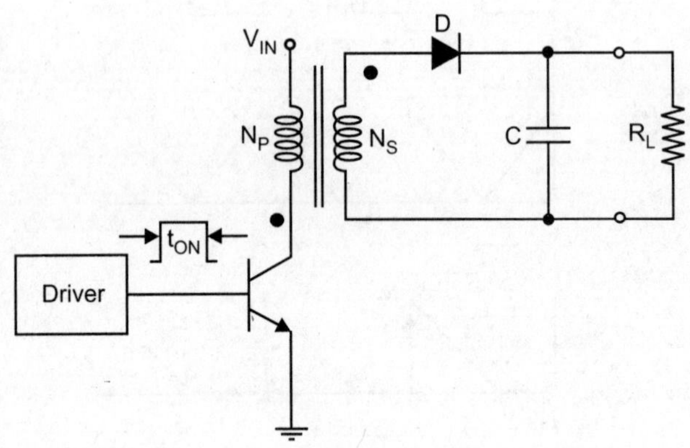

Fig. 16.7

Also, the P_o to be delivered power drawn form the input source is given by,

$$P_{IN} = P_0/\eta$$

Also $P_{IN} = \frac{1}{2} LP\hat{I}_P^2 f$

or $2P_0 = \eta L_P \hat{I}_P^2 f \quad ...(2)$

Equations (1) and (2) can be rewritten as,

$$V_{IN}\, t_{ON} = L_P \hat{I}_P \quad ...(3)$$

and $2P_0 = \eta L_P \hat{I}_P^2 f \quad ...(4)$

Solving(3) and (4) simultaneously,we get expres-sions for L_P and $\hat{I}_P$ as,

$$\hat{I}_P = \frac{2P_0}{(VI_N t_{oN} \eta f)}$$

$$L_P = \frac{(V_{IN}^2 t_{ON}^2 \eta f)}{2P_0}$$

As a second step, we can determine the number of primary turns which would not saturate the core.

Let A_e= Effective area of cross-section of the core in (m^2)

l_e = Effective magnetic path length in (m)

B_{max} = Maximum flux density of the more material Wb/m^2.

If N_P is the number of primary turns, then form first principles

$$V_{IN} = NP\frac{d\varphi}{dt} = \frac{N_P B_{max} A_e}{t_{ON}}$$

or $$N_P = \left(\frac{V_{IN}\, t_{ON}}{B_{max} A_e}\right) \quad (5)$$

B_{max} to be substituted in the above expression should be a little less then the B_{max} rating of the core material.

After $\hat{I}_P$, L_P and Np have been determined, we can determine the air gap size which could give (L_P) with (N_P) primary turns. We know that

$$L_P = \left(\frac{\mu_0 \mu_r N_P^2 A_e}{l_e}\right) \quad (6)$$

The above expression can be used to first find the effective permeability which would make (N_P) primary turns give (L_P) henries of primary inductance. Once (μ_e) is known, gap size (I_g) can be found from

$$l_g = (l_e/\mu_e)$$

As a third step, number of secondary turns can be determined from the step-up ratio, which in turn is governed by the transistor's $V_{CEO(max)}$ rating. During the OFF time, the voltage that appears across collector-emitter junction is given by,

$$V_{IN} + [V_0/\eta]$$

where, n is step -up ratio = $\left(\frac{N_s}{N_P}\right)$

Now $V_{CEO(max)} = \left(V_{IN} + \frac{V_o}{\eta}\right)$

or $$n = \frac{V_o}{[V_{CEO}(\max) - V_{IN}]}$$

or $$N_s = \frac{V_0 N_P}{[V_{CEO}(\max) - V_{IN}]}$$

Again V_{CEO} (max) to be substituted in the above expression should be less than the V_{CEO} (max) rating given in transistor's data sheet. Drive circuit parameters can be determined from base current (I_B) this circuit is required to furnish

$$I_B = \left(\frac{\hat{I}_P}{h_{FE}(\min)}\right)$$

Peak secondary current = $\hat{I}_P\left(\frac{N_P}{N_S}\right)$

Primary and secondary wire sizes can be determined from calculated values of primary and secondary currents. The currents to be considered should be the average ones and not the peak ones.

The diode D should be a fast recovery one. A fast recovery diode ensures that the diode is fully reverse biased and there is no leakage of power during the conduction time of the transistor. Its PIV should also be more than twice the desired output voltage.

Capacitor C should be such that (CR_L) time constant is at least ten times t_{ON}. This ensures that the capacitor maintains the output voltage as constant during the conduction time of the transistor when the capacitor is solely supplying energy to the load. Mathematically,

$$CR_L = 10t_{ON}$$

or $$C\left(\frac{V_0^2}{P_0}\right) = 10\, t_{ON}$$

or $$C = \left(10t_{ON}\frac{P_0}{V_0^2}\right)$$

One significant point to be noted here is that the transformation ratio does not depend upon ratio of V_0 and V_{IN}. In turn, it is governed by the transistor's V_{CEO} (max) rating. If a transistor with a higher V_{CEO} (max) rating is available, step-up ratio could be further reduced, thus reducing N_S.

Q. 11. Describe in brief the procedural steps you would follow in designing a self-oscillating flyback converter required to deliver a power of (P) watts to a load R_L. The converter is also required to furnish a load voltage of (V) volts from an available input voltage of (V_{IN}) volts.

Ans. Fig. 16.8 shows the circuit diagram of self-oscillating flyback converter. The design procedure is identical to the one described for an externally driven flyback converter in the preceding question.

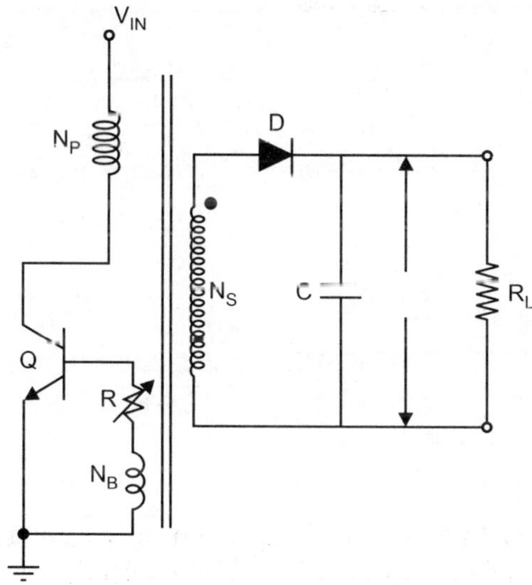

Fig. 16.8

The only difference is that in the present case, the base drive is being provided by another winding called the feedback winding on the switching transformer core. The number of base winding turns can be calculated from the criterion of producing a voltage equal *to* V_{BE} (max) for the chosen transistor. In fact, the voltage induced across the base winding is made to be 2 V_{BE} (max). Out of this, V_{BE} (max) is made to drop across (R) for the requisite base drive (I_B), (R) has been included to take care of the variation of V_{BE} (max) from transistor to transistor of the same type. (R) is varied to produce only V_{BE} (max) across base emitter junction.

Q. 12. With the help of a circuit diagram and relevant waveforms, explain the principle of operation of a Push Pull type forward converter. Which type of converter-Flyback type or Forward type-you would prefer in high power applications? Give reasons in support of your answer.

Ans. Fig. 16.9 (a) shows the circuit diagram and Fig. 16.9 (b) the relevant waveforms. The converter shown is a self-oscillating one with the transformer providing both the switching of the two transistors as well as the power transformation from primary to secondary. An externally driven converter like this could also be possible but is seldom used in practice. As we shall see a little later, the external drive circuit would be required to produce two drive signals (for Q_1 and Q_2) which are precisely 180° out of phase so that when Q_1 is ON, Q_2 is OFF and vice-versa.

In the circuit of Fig. 16.9 (a) when power is applied to the circuit, current starts flowing through Q_1 and Q_2. There can never be a perfect symmetry between Q_1, Q_2 and associated components. As a result, current through one of the transistors (say Q_1) is greater than that through Q_2. Depending upon the rate of rise of current, constant voltages are induced in the base windings. The polarities of these windings are so chosen that the current through Q_1 is aided, whereas that through Q_2 is impeded. Thus, the process of rise of current through Q_1 continues and so is the fall of current through Q_2. This regenerative action, however, cannot continue for very long. A stage comes, when collector current of Q_1 saturates the core of the transformer. It may be mentioned that flux density is directly proportional to the current. In this case,

$$B = \left(\frac{L_P I_C}{N_P A}\right)$$

The value of I_C, which makes $B = B_{max}$ of the core material, saturates the core. Once the core saturates, the inductance falls and collector current through Q_1 rapidly rises and in fact, tends to go to infinity.

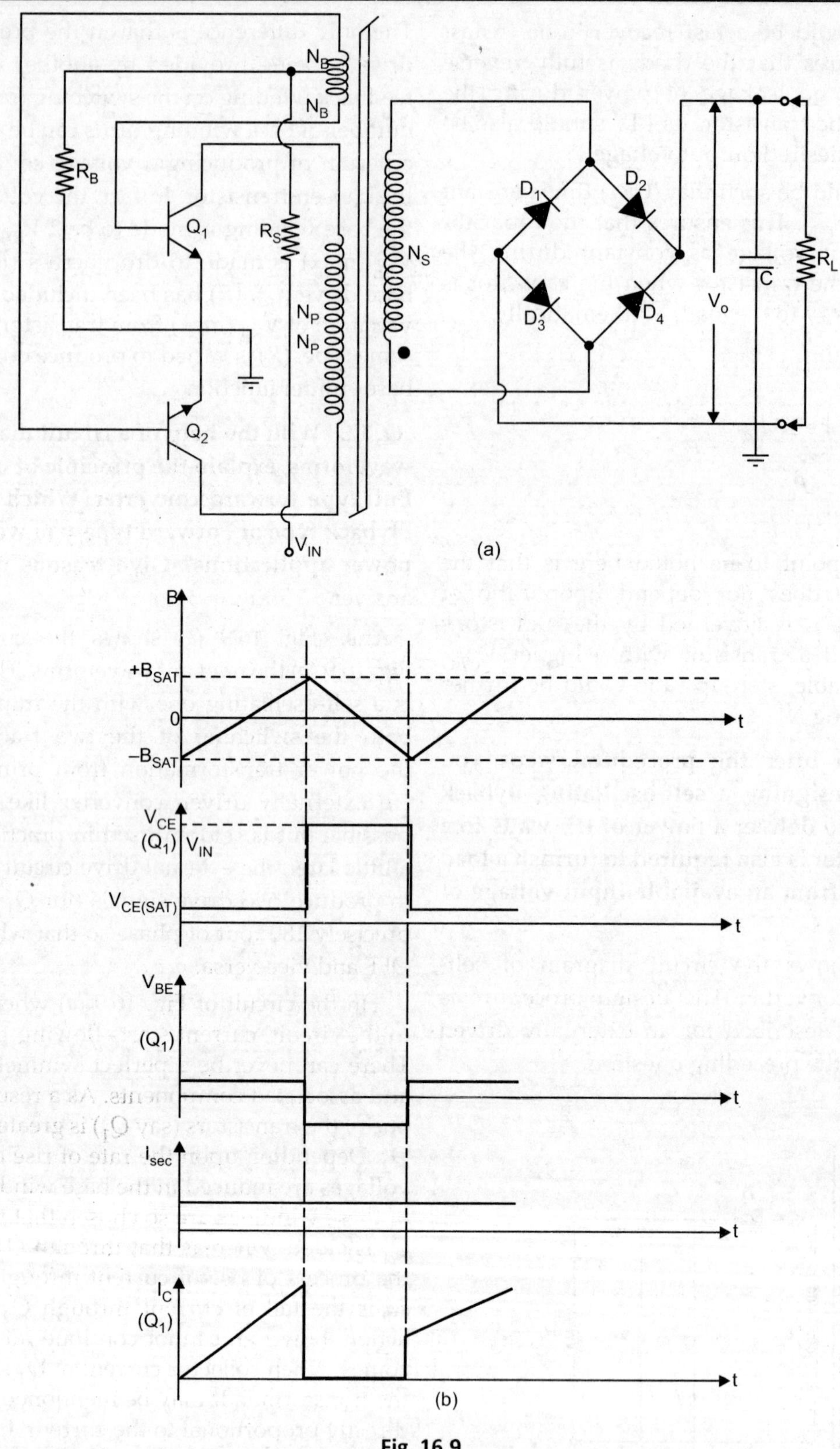

Fig. 16.9

But fortunately, it cannot do so. It is because collector current is limited by base drive current ($I_C = I_B h_{FE}$). There is no more rise in collector current and there is no more induced e.m.f. in base windings, as the magnetic field in the core is no more a changing one. Once, the cut-off voltage is removed from the base of Q_2, current starts flowing through its collector. The collector current induces voltages in base windings; but this time it aids current through its collector. The collector current induces voltages in base windings; but this time it aids current through Q_2 and impedes the current through Q_1. This process is further helped by a collapsing magnetic field due to current flowing through the primary winding of transistor Q_1. Again, a stage is soon reached where core saturates and we are back to the beginning again. This completes one cycle of operation. The process is repeated afterwards with Q_1 and Q_2 switching ON and OFF alternately. Thus, across the secondary we have a square waveform which is rectified by the bridge rectifier and then the rectified waveform is filtered by the capacitor. The waveforms shown in Fig. 16.9 (*b*) are self-explanatory.

Forward converters, especially of push-pull type, are highly suitable for high power applications because of the fact that the power is being delivered to the load continuously from the input source of power. It is different from ringing choke converters in the sense that there the power is stored in one-half of the cycle or a part of the cycle and transferred in the remaining part. In ringing choke converters, when the power is being stored, it is the capacitor filter which has to feed the load. A push-pull converter of the type discussed above has a higher efficiency and a better regulation as compared to that of ringing choke converter.

Q. 13. How do you design a push-pull type forward converter using a single saturable transformer ? Derive the design equations employed from the first principles.

Ans. Refer to Fig. 16.9 (*a*). From the first principles,

$$V_{IN} = N_P \frac{d\varphi}{dt} \quad ...(1)$$

N_p = Half of the primary turns

N_p is the number of primary turns in the collector circuit of either of the two transistors.

As shown in the waveforms in the preceding question (Fig. 16.9 (*b*), the flux varies from $-\varphi_{max}$ to φ_{max} or $+\varphi_{max}$ to $-\varphi_{max}$ when one of the transistors is conducting. Therefore, change in flux

$$d\varphi = -\varphi_{max} - (-\varphi_{max}) = (2\varphi_{max})$$

This change in flux occurs in a time equal to one-half of the total time period

or $\quad dt = 0.5\,T$

where, T = Time period for one complete cycle of operation

Also, $T = 1/f$, where (f) is the frequency of operation

From Equation (1), $V_{IN} = N_P . = (4\,N_P\,\varphi_{max} f)$

Now $\varphi_{max} = B_{max} \cdot A_e$

where, B_{max} = Maximum flux density of the core material

A_e = Effective area of cross-section of the core

Therefore, $V_{IN} = (4N_P\, B_{max}\, A_e f)$ ...(2)

Equation (2) can be used to find number of primary turns.

$$\text{From Equation (2), } N_P = \frac{V_{IN}}{4\,B\max\, A_e\, f} \quad ...(3)$$

Number of secondary turns are given by,

$$N_s = \left(N_P \frac{V_o}{V_{IN}}\right) \quad ...(4)$$

Once N_P is known, primary inductance(LP)can be determined from,

$$L_P = \left(\frac{\mu_0 \mu_r N_9^2 A_e}{l_e}\right) \quad ...(5)$$

The primary inductance (L_P) then determines the peak value of primary current can be determined from,

$$V_{IN} = L_P \frac{dI_P}{dt} = L_P \frac{2\hat{I}_P}{0.5T}$$

$$= 4Lp\,\hat{I}_P f$$

$$\text{or} \quad \hat{I}_P = \left(\frac{V_{IN}}{4L_P f}\right)$$

Maximum value of transistor collector current is then given by,

$$L_{Max} = \hat{I}_P + 1_{av} = \hat{I}_P + \frac{P_O}{\eta V_{IN}}$$

$$= \left(\frac{V_{IN}}{4L_P f} + \frac{P_o}{\eta V_{IN}}\right) \quad ...(6)$$

The transistors selected for the purpose should be capable of handling a collector current as given by the above expression. Another transistor parameter of interest is the V_{CEO} (max) rating. V_{CEO} (max) limits the maximum value of V_{IN} that can be used. Maximum V_{IN} that can be safely employed is given by,

$$V_{IN} \leq \left(\frac{V_{CEO}(\text{max}) - V_{BE}(\text{winding})}{2}\right) \quad ...(7)$$

R_B and N_B can be determined in the same fashion as in the case of ringing choke converter.

$$\text{That is, } N_B = \left(\frac{2V_{BE}(\text{max}) \times N_P}{V_{IN}}\right)$$

$$\text{and} \quad R_B = \frac{V_{BE}(\text{max})}{I_B} \quad ...(8)$$

Rs should be such that it produces a voltage of 0.6 V at the centre tap of feedback winding.

That is, $\frac{V_{IN} R_B}{(R_B + R_S)} = 0.6$

or $R_S = \left[\frac{(V_{IN} - 0.6)}{0.6}\right] R_B$ ...(9)

D_1 to D_4 are rectifier diodes. These diodes should have the requisite PIV and current carrying capability. Capacitor C should be such that CR_L time constant is at least 100 times the period T.

That is, $CR_L \geq 100$ T or $C \geq \frac{100T}{R_L} \geq \frac{100}{R_L f}$

Now $R_L = \frac{V_0^2}{P_0}$; $C \geq \left(\frac{100P_o}{V_0^2 f}\right)$ (10)

Equation 3, 4, 5, 6, 7, 8, 9, and 10 give the complete design information.

Q. 14. What are the preferable magnetic properties of the core material to be used in (*i*) a flyback converter and (*ii*) a push-pull type forward converter. Justify your answer.

Ans. In a push-pull converter, the transformer is operated in the saturating mode and it gives us a square wave output. Obviously, the core material should have square loop characteristics (Fig. 16.10) Not only this, the core material should also have a very high value of maximum flux density. In a flyback converter, which is operated in the non-saturating mode, the core material need not be square loop one. However, a high value of maximum flux density is preferable as higher the flux density, lesser will be the number of primary turns and consequently the secondary turns required.

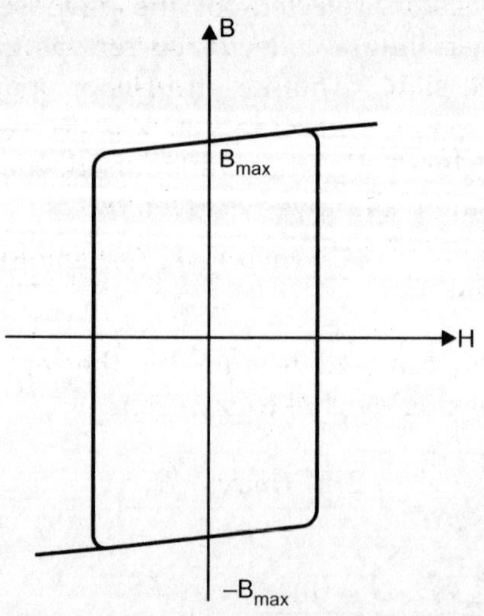

Fig. 16.10

Q. 15. Describe in brief the operation of a push pull converter using a non-saturable transformer for power transformation and a saturable transformer for switching. What is the main advantage of using this configuration over the conventional one transformer type push-pull type converter described in preceding question ?

Ans. Fig. 16.11 shows a push-pull type converter using two transistors and two transformers. One of the transformers (T_1) is operated in the non-saturating mode, whereas transformer (T_2) is a saturable transformer. T_1 performs power transformation, whereas T_2 does power switching. To explain the operation of this circuit, we assume that slight imbalance between Q_1 and Q_2 causes either Q_1 or Q_2 conduct more heavily than the other in the beginning. If Q_1 conducts more heavily, the voltage induced across one-half of the T_1 primary also appears across T_2 primary via feedback resistance R_f. Voltages induced across T_2 secondary (centre tapped) are in a polarity so that Q_1 is driven to saturation and Q_2 to cut-off. As transformer T_2 saturates, its primary draws a very large current. A large (T_2) primary current causes a large voltage drop across R_f so as to reduce the voltage appearing across T_2 primary to almost zero. The drive for transistor Q_1 disappears and it comes out of saturation and forcing Q_2 at the same time to go into conduction. The regenerative action forces Q_2 into saturation and Q_1 to cut-off. This completes one cycle of operation. And the process repeats afterwards. A square waveform appears across the (T_1) secondary which is rectified by a bridge rectifier constituted by diodes D_1 to D_4 and subsequently filtered by capacitor C.

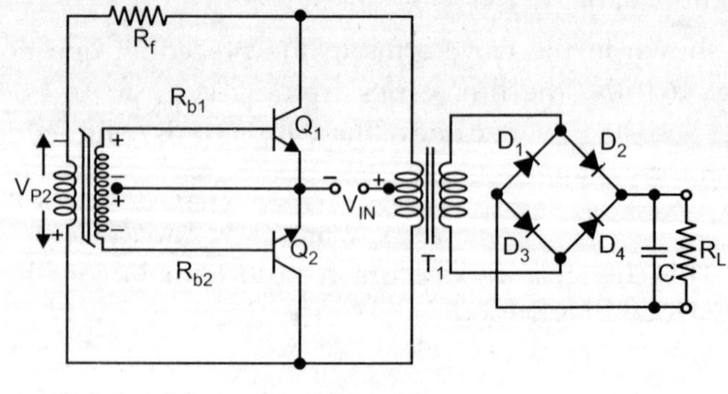

Fig. 16.11

The main advantage of this configuration arises from the fact that the output transformer in this case in a non-saturable type, which does not necessitate the use of a very expensive square-loop core material required for the transformer of earlier configuration. Also, the peak collector current in this case is a function of load, whereas in the one transformer configuration, it depended upon transistor's input characteristics and thus, could vary from transistor to transistor.

Q. 16. With the help of basic circuit configurations, briefly describe the operational principles of Buck, Boost and Inverting (also known as Buck-Boost) type of switching regulators?

Ans. Fig. 16.12 (*a*) shows the basic buck regulator. The **buck regulator** also called a step-down regulator resembles the conventional forward converter. A buck regulator differs from a forward converter in that there it does not make use of a transformer and there is no input-output isolation. Output voltage is always less than the input DC voltage and can be calculated from,

$$V_0 = D.V_{in} \text{ where, } D = \text{Duty Cycle} = \frac{T_{ON}}{T}$$

Output regulation is achieved by pulse width modulation of the transistor switch.

The boost regulator also known as step-up switching regulator (Fig. 16.12 (*b*)) is based on the flyback converter principle. It resembles the basic flyback converter, except that it is non-isolating type. The energy storage and transfer element in this case is inductor rather than a transformer. While the switching transistor conducts, the inductor stores energy in the form of magnetic field. The energy stored equals $\frac{1}{2} L_P I_P^2$. As the diode is reverse biased during this period, the energy cannot get transferred, as it is being stored. When the switching transistor is turned OFF, the diode gets forward biased and the stored energy is delivered to the load alongwith energy from DC input voltage. The voltage across the load equals the DC input voltage plus the voltage due to energy stored in the inductor. The output voltage in this case is given by,

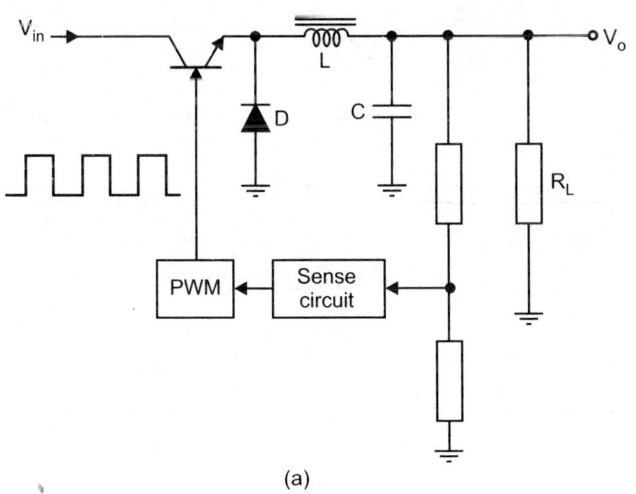

(a)

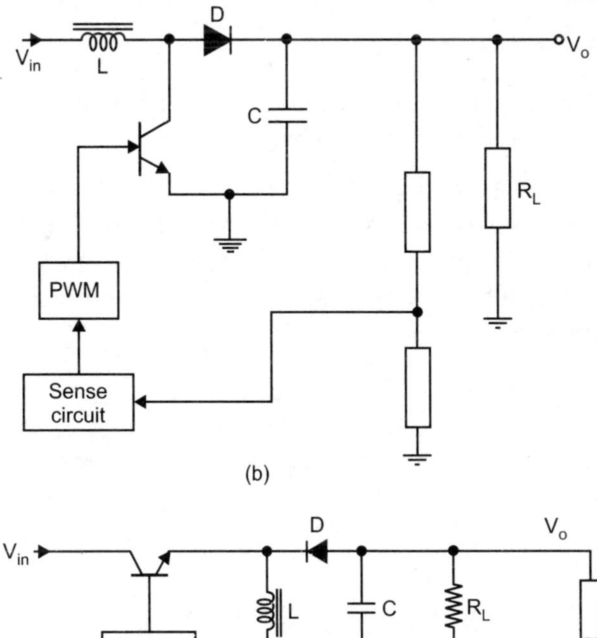

(b)

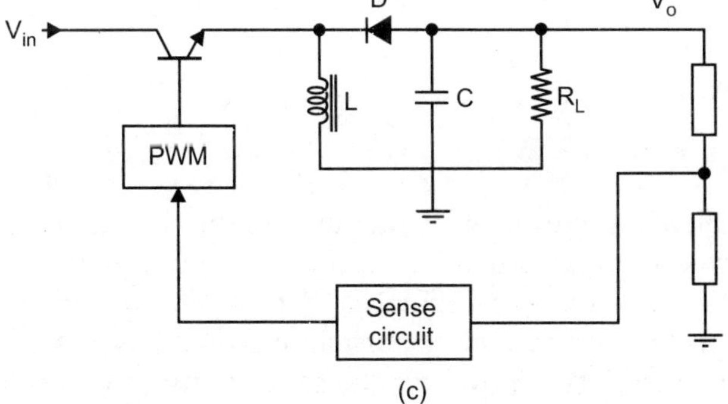

(c)

Fig. 16.12

$$V_o = \frac{V_{in}}{(1-D)} = V_{in}(T/T_{off})$$

$$\text{where, } D = \left(\frac{T_{on}}{T}\right) \text{ and } T = (T_{on} + T_{off})$$

The power output capability of boost regulator is $\frac{1}{2}$ ($L_P I_P^2 f$), where (*f*) is the operating frequency.

As the DC input voltage is not electrically isolated from the output, the output voltage in this case cannot be exclusively determined from the output power delivering capability of the circuit, as is the case in conventional flyback converter.

Inverting regulator (Fig. 16.12 (*c*)) is another circuit configuration based on the flyback converter principle. It converts an unregulated positive voltage into a regulated negative voltage. Energy is stored in the inductor (*L*) during the conduction time of the transistor. The stored energy is transferred during the OFF-time. The circuit delivers a constant power to the load. The output voltage is given by,

$$V_0 = -\sqrt{P_o R_L}$$

Regulation is achieved by controlling the duty cycle of the drive waveform. The output voltage can also be expressed by,

$$V_0 = -V_{in}\left(\frac{T_{on}}{T_{off}}\right)$$

The output voltage in this case can either be less than or greater than the input voltage.

Q. 17. What are the different constituents of an off-line switching power supply? Briefly describe the functions of each one of these constituents with particular reference to those building blocks not found in DC to DC converters.

Ans. Fig. 16.13 shows both the essential and desirable sections of an off-line switched mode ower supply. While the input voltage section, the chopper, the output rectifier the filter and the control section are essential features of an off-line switching supply. Inclusion of fuse, EMI filter and the current inrush control circuitry add to the safety and reliability of the power supply.

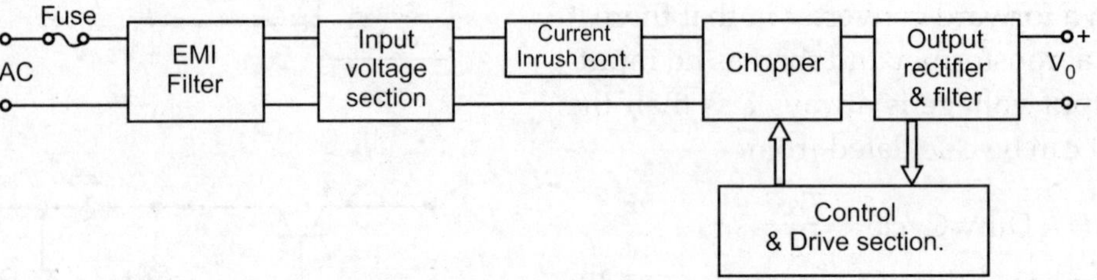

Fig. 16.13

FUSE is the most elementary method of providing overload protection to the power supply components, particularly the active devices. For an off-line switching supply line input fuse, a normal or slow blow fuse does the job very well. However, the continuous current rating of the fuse should be decided only after giving due consideration to the inrush current.

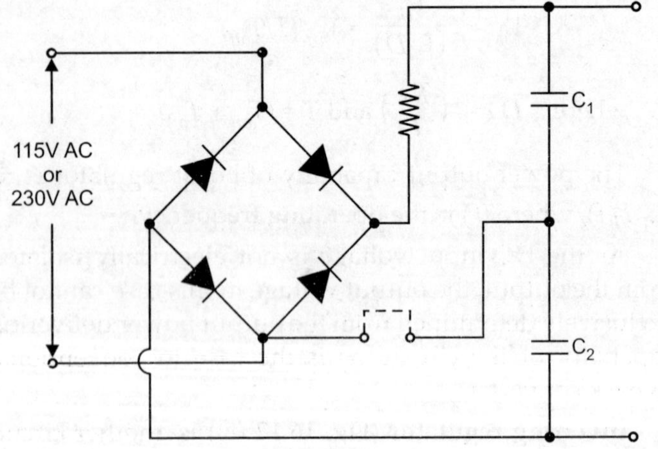

Fig. 16.14

Input voltage section provides input rectification and filtering circuitry to generate the DC for the converter input. Off-line switching supplies usually have a selectable AC input voltage of 115V (nominal) and 230V (nominal). Fig. 16.14 shows typical input voltage section.

Unlike linear power supplies, switching supplies are inherently a rich source of electromagnetic interference (EMI). A switching supply gives rise to EMI/RFI due to the high frequency switching action of the active device. With every switching action, a short burst of high frequency energy is generated and it is repeated at the switching frequency of the converter. EMI filter is used to suppress this interference. EMI appears in two different forms, namely the conducted EMI and the radiated EMI. Conducted EMI is the noise feedback from the power supply to the AC power line. Conducted EMI can be suppressed by using suitable filters, such as those shown in Fig. 16.15 (*a*) - used for differential mode conducted electromegnetic interference and Fig. 16.15 (b) - used for common mode conducted interference. Radiated interference is directly transmitted from the switching components, leads etc. Switching transformers using air

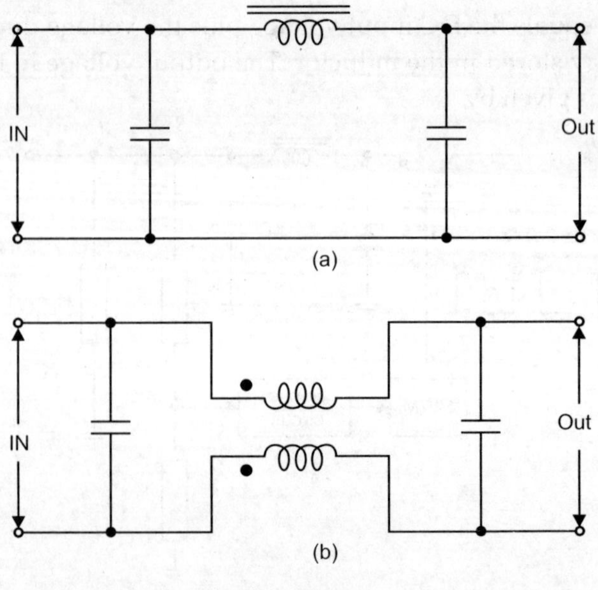

Fig. 16.15

gapped cores, long lead lengths carrying high current transients and ground loops are predominant sources of radiated EMI. It can be effectively suppressed by using twisted pairs, short leads, proper PCB layouts and Faraday shields around transformers, switching transistors, fast recovery rectifiers or by simply enclosing the power supply in a metallic enclosure.

In the off-line switching supplies, where the high voltage DC input to the converter is produced by directly rectifying AC mains with a bridge rectifier followed by a capacitive input filter, there is always a possibility of heavy inrush of current at the time of power on, if proper care is not taken to limit this current. Inrush current may be limited by using series resistors or NTC thermistors or even a resistor that gets shorted, when the currents have stabilised.

The other building blocks include the chopper, the output rectifier, the filter and the control section. The chopper represents the active device-converter transformer combination that chops the DC input at a high frequency. The output rectifier and filter combination convert the transformed chopped waveform back to DC and the control section not only provides the drive waveform/s for the chopper active device/s, it also provides regulation. Fig. 16.16 shows the typical circuit diagram for an off-line flyback type switching supply.

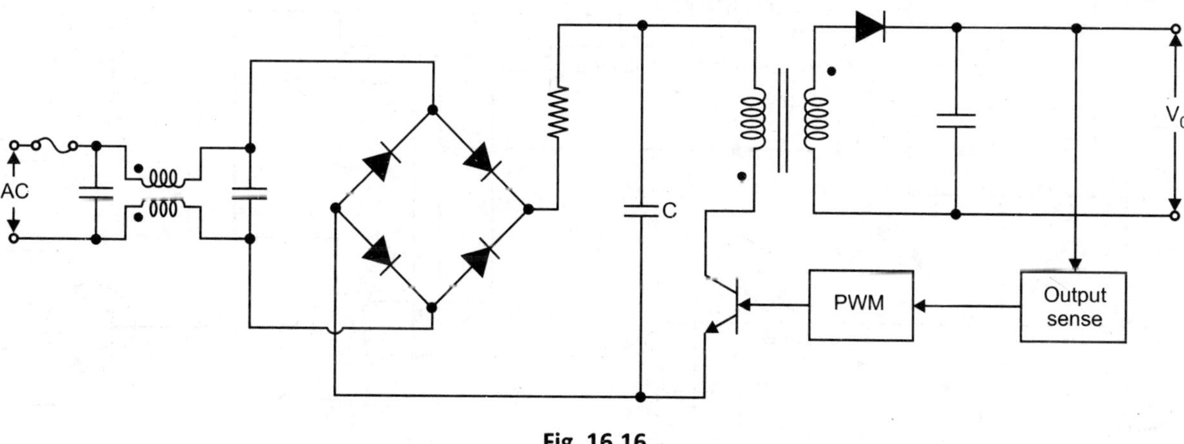

Fig. 16.16

Q. 18. Use of integrated circuits in the drive and control section of switching supplies is becoming very common. Outline the principal constituents of an IC based switching supply with the help of a block schematic arrangement. What are the advantages of IC switching regulator based supplies ?

Ans. Switching supply design today is dominated by integrated circuits. The drive and control circuitry functions in a switching supply invariably use an IC that is popularly known as a switching regulator IC. The state-of-the-art switching regulator ICs are internally equipped with all the required circuit blocks to implement almost any switching supply configuration.

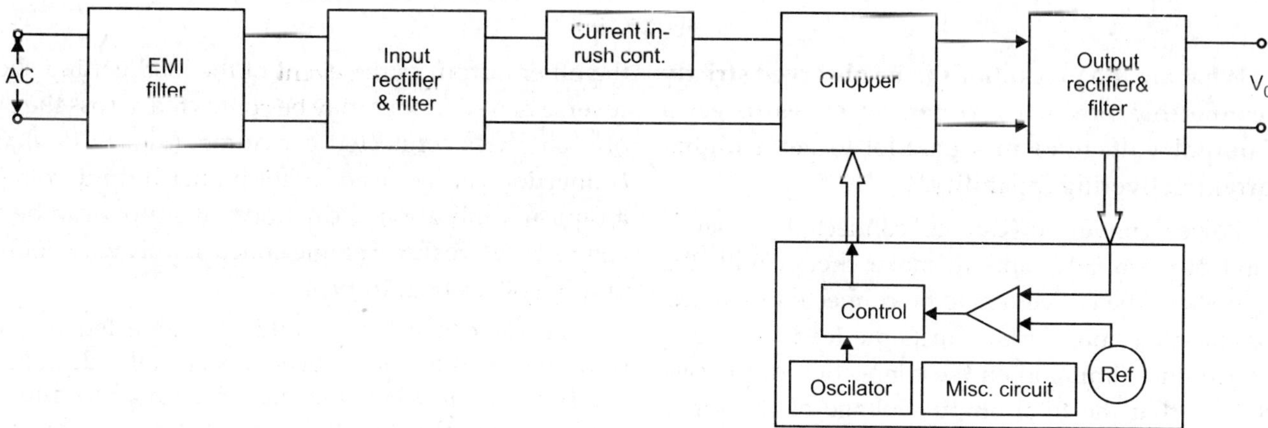

Fig. 16.17

Fig. 16.17. shows the basic IC based switching supply. The IC has all the building blocks to generate the desired waveform for the chopper active device, sense the output and compare it with a standard reference to produce an error signal which then modifies the drive waveform's duty cycle to maintain a constant output voltage. Some switching regulator ICs also have uncommitted internal transistor to give it a flexibility for use in a variety of switching supply configurations. These ICs also have a number of built-in protection features like overload protection, short circuit current limit, crowbaring, soft start etc.

The major advantages of switched mode power supplies, namely the compact size, high efficiency, greater reliability and low cost have been partly possible due to advances made in the development of discrete switching components but mainly due to introduction of a large number of drive and control ICs. Fig. 16.18 shows the internal schematic of a popular switching regulator IC.

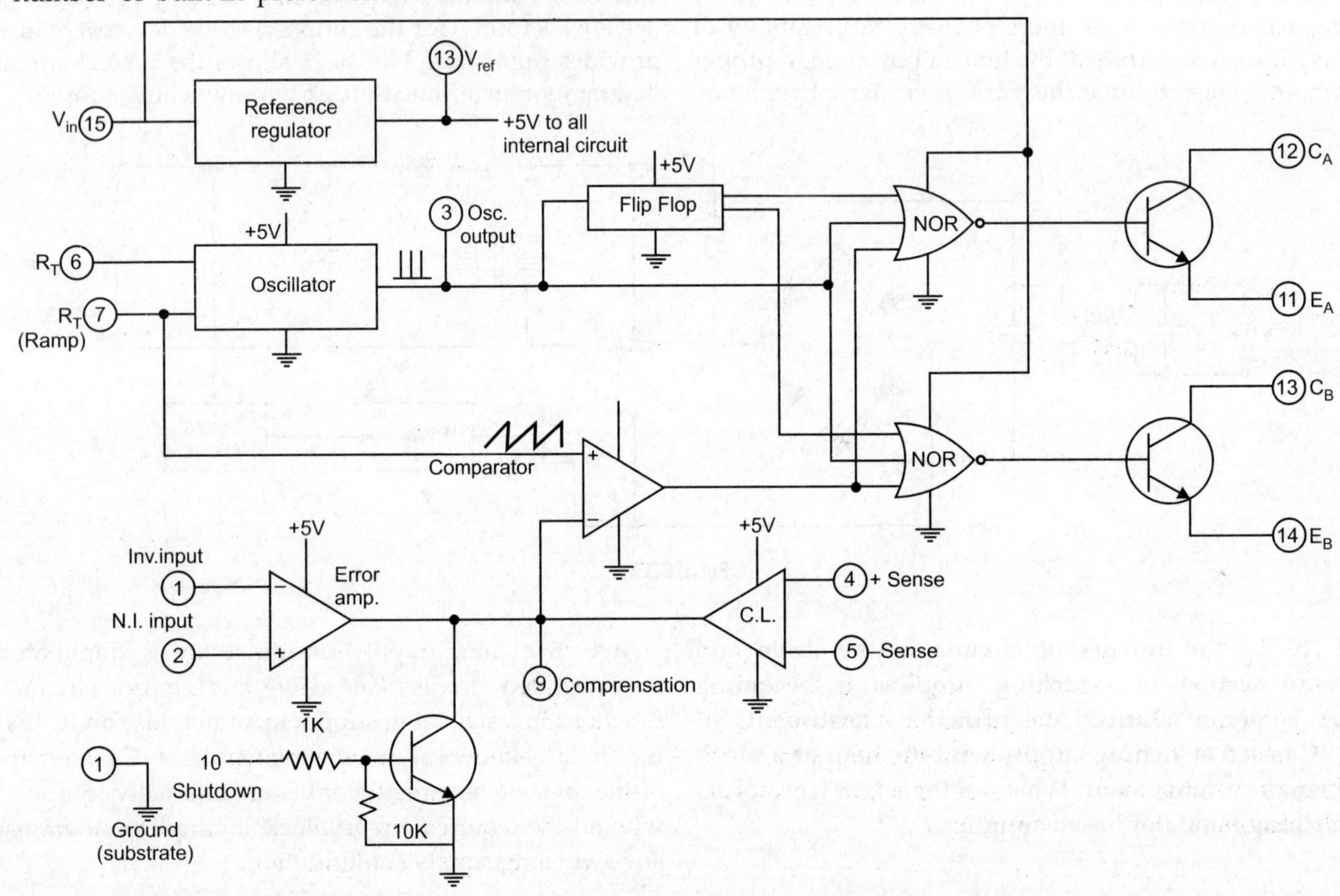

Fig. 16.18

Q. 19. What are the precautions to be observed strictly while connecting power converters in series to get a higher output voltage or in a parallel to get a higher load current delivering capability?

Ans Power converters can be connected in series in general but it is advisable to make checks with the specifications of the converters to be connected in series. It is possible that one output affects the feedback loop of other. Another limitaion on the connection of the two converters is that the total output voltage of the series connected converters should not exceed the working breakdown voltage of any one of power converters. In order to protect each output from the reverse voltage of the other output in the event of the load getting shorted, reverse biased diodes may be connected across the output of each series connected converters (Fig. 16.19 *a*). Series connection can be used to get higher output voltage. In a typical application, a dual output supply can be series connected to realise a single ended supply with double thy output voltage (Fig. 16.19 *b*).

Power converters should be connected in parallel only, when they have been specifically designed for the purpose or when the manufacturer recommends a parallel operation. The biggest problem in the parallel connection of two converters arises from unequal load sharing. Unequal load sharing is due to

(*i*) the output voltages not being equal

(*ii*) the converter with a greater output voltage tending to provide the entire load current.

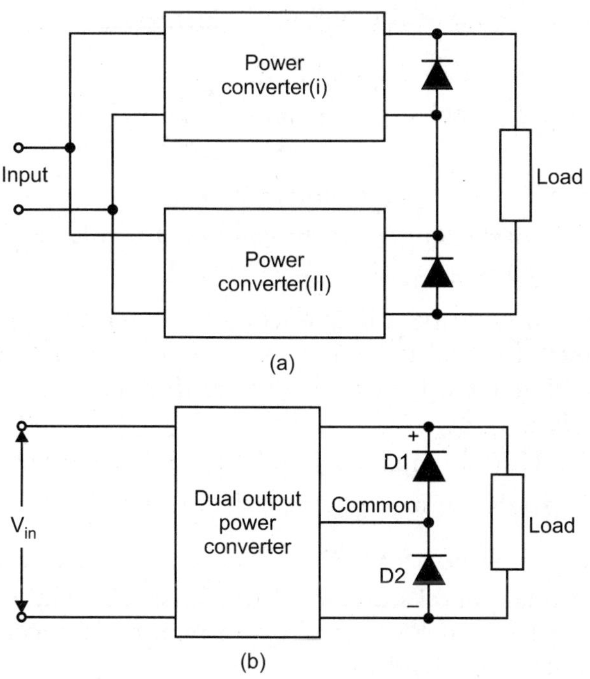

Fig. 16.19

Even if the output voltages are adjusted to be precisely equal, a difference in the output impedance and also, its drift with time and temperature will cause the loads to become unbalanced.

One method used to overcome this unequal load sharing is to use small individual series resistors (Fig. 16.20). While this type of parallel connection could be useful in a few applications, it must be borne in mind that the series resistors degrade the output regulation seriously. Also, the circuit will always have current imbalance. If the two converters in Fig. 16.20 had a nominal output voltage of 5V, a 50 mV difference in the output voltages may cause a current imbalance of 25%. In such a case, each converter should not be capable of providing just 50% of the load but 75% of it.

A good reason for parallel connection of power converter outputs may be connected in parallel through two diodes (Fig. 16.20 *b*). For 100 percent redundancy, each power converter must be capable of supplying the total load. In this case, it really does not matter whether each output provides at least a part of the load current. The diodes permit one output to fail without affecting the other which continues to supply power to the load. Such system is useful in applications, where power supply failure is not tolerable, or where a high degree of reliability is required.

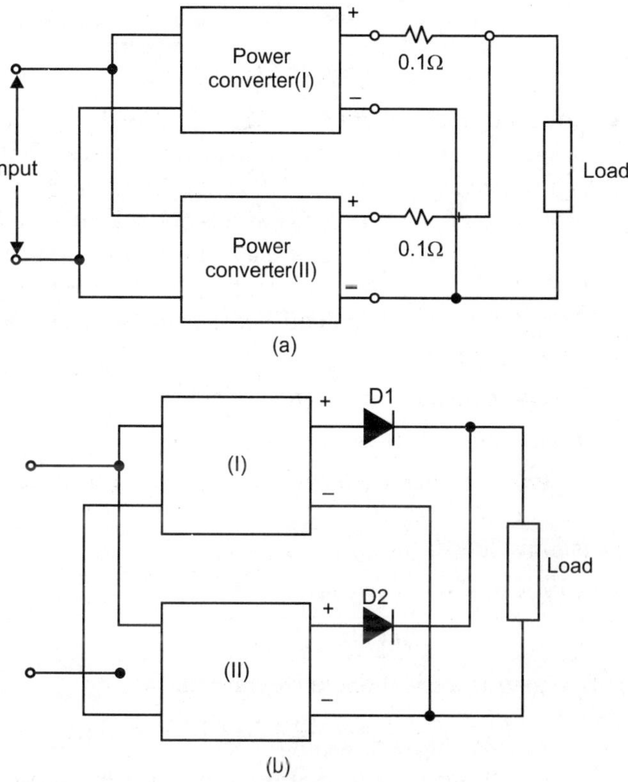

Fig. 16.20

SOLVED PROBLEMS

Problem 16.1. *Figure 16.21 shows the basic ringing-choke-type of flyback DC-to-DC converter along with the drive waveform across the feedback winding and the primary current waveform. From the data given in the circuit diagram, determine output voltage across the load resistance R_L, if the conversion efficiency were 80%.*

Solution. Power delivered by the converter of this type is given by the product of power stored in the primary of the switching transformer and the conversion efficiency. That is, power delivered = power stored×conversion efficiency.

Power stored in turn is equal to the product of energy stored and the switching frequency. That is, power stored $= (1/2) L_P I_P^2 f$

$$f = 1/(50 \times 10^{-6} + 50 \times 10^{-6}) \text{ Hz} = 10\text{kHz}$$

Therefore, power stored

$$= (1/2) \times 48 \times 10^{-6} \times 25 \times 25 \times 10 \times 10^3$$

$$= 150 \text{ W}$$

Power delivered to load resistance

$$R_L = 150 \times 0.8 = 120 \text{ W}$$

If V_0 were the voltage across R_L, then

$V_0^2/R_L = 120$, which gives $V_0 = \sqrt{(120\ R_L)} = \sqrt{(120 \times 100)} = 110$ V

Voltage across load resistance, $R_L = 110$ V

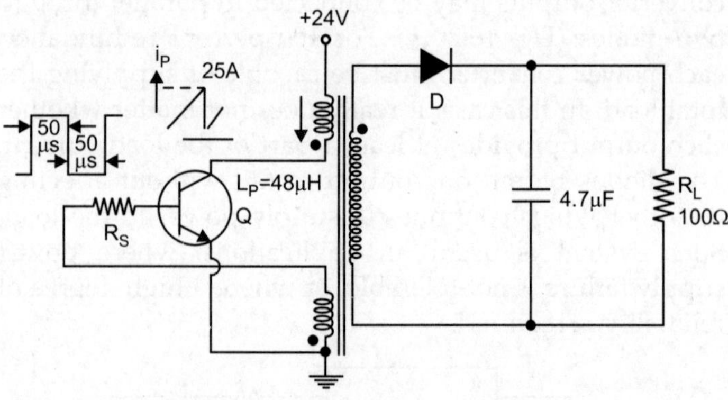

Fig. 16.21

Problem 16.2. *A flyback type DC-to-DC converter operating in discontinuous mode is to deliver a power of 50 W to a load resistance of 5000 Ω. The converter operates from a regulated input of 24 VDC and produces 500 V across the load resistance. The toroidal core chosen for the purpose is capable of delivering the desired output power at a switching frequency of 20 kHz and flux density of 3000 Gauss and has the following dimensional parameters:*

Effective magnetic pat length, $l_e = 6.2$ cm

Effective cross-sectional area, $A_e = 0.4\ cm^2$

Design the switching transformer assuming a conversion efficiency of 80% and the V_{CEO}(max) rating of the switching transistor as 100 V.

Solution. We know that,

$$f = 20 \text{ kHz}$$

For a symmetrical drive waveform,

$$t_{ON} = t_{OFF} = 1/(2 \times 20 \times 10^3)s = 25\ \mu s$$

As a first step, we will find the required values of primary inductance and the corresponding peak primary current. These can be determined as follows:

$$I_P = 2P_O/V_{IN}t_{ON}\ \eta f \text{ and } L_P = V_{IN}^2\ t_{ON}^2\ \eta f/2P_O$$

$$I_P = (2 \times 50)/(24 \times 25 \times 10^{-6} \times 0.8 \times 20 \times 10^3) = 100/9.6$$

$$= 10.4 \text{ A}$$

$$L_P = (24 \times 24 \times 25 \times 10^{-6} \times 25 \times 10^{-6} \times 0.8 \times 20 \times 10^3)/(2 \times 50) = 57.6\ \mu H$$

Number of primary turns can be determined as follows:

$$N_P = V_{IN}t_{ON}/B_{MAX}A_e$$

$$= 24 \times 25 \times 10^{-6})/(0.3 \times 0.4 \times 10^{-4}) = 50$$

The value of permeability required for 50 turns of primary winding producing a primary inductance of 57.6 µH can be calculated as follows:

$$L_P = \mu_0\mu_r N_P^2 A_e/l_e, \text{ which gives}$$

$$\mu_r = L_P l_e/\mu_0 N_P^2 A_e$$

$$= (57.6 \times 10^{-6} \times 6.2 \times 10^{-2})/(4 \times 3.14 \times 10^{-7} \times 50 \times 50 \times 0.4 \times 10^{-4})$$

$$= (357 \times 10^{-8})/(1.256 \times 10^{-7}) = 28.4$$

Size of the air gap can be computed as follows:

$$l_g = 6.2/28.4 \text{ cm} = 2.2 \text{ mm}$$

Assuming permissible collector-to-emitter voltage at turn-off equal to 75% of the V_{CEO}(max) rating of the switching transistor, the step-up ratio (n) can be computed as follows:

$$n = V_O/(V_{CEO}(\text{max}) - V_{IN}) = 500/(75 - 24) = 500/51 = 9.8$$

Therefore, number of secondary turns,

$$N_S = 50 \times 9.8 = 490$$

Primary and secondary wire sizes can be determined from calculated RMS values of primary and secondary currents, respectively.

Hence, Primary inductance $L_P = 57.6$ µH, number of primary turns = 50, permeability $\mu_r = 28.4$, size of the air gap = 2.2 mm, step-up ratio (n) = 9.8, number of secondary turns $N_S = 490$.

Problem 16.3. *An externally driven push–pull-type DC-to-DC converter is to be designed to deliver a power of 100 W to a load resistance of 10 K. The converter operates from a regulated input of 12 VDC and produces 1000 V across the load resistance. The toroidal core chosen for the purpose is capable of delivering the desired output power at a switching frequency of 20 kHz and flux density of 2000 gauss and has the following dimensional parameters:*

Effective magnetic pat length, $l_e = 8.3$ cm

Effective cross-sectional area, $A_e = 0.57\ cm^2$

Assuming core permeability and conversion efficiency to be 3000 and 80% respectively, design the transformer by determining primary and secondary number of turns and the relevant currents that would enable choose the right wire gauge for the primary and secondary windings. Assume two-transistor, single transformer topology and a symmetrical drive waveform.

Solution. As a first step, we will determine the number of primary turns, N_P.

$$N_P = V_{IN}/(4B_{max}A_e f)$$

$$= 12/(4 \times 0.2 \times 0.57 \times 10^{-4} \times 20 \times 10^3)$$

$$= 13.16$$

Primary winding can be taken as 14 – 0 – 14 center-tapped winding.

Number of secondary turns,

$$N_S = N_P(V_O/V_{IN})$$

$$N_S = 14 \times 1000/12 = 1167$$

The primary current in each half of the primary winding flows for half of the period of the switching waveform and is equal to the sum of magnetizing current and a DC component.

Primary current = $(V_{IN}/4L_P f) + (P_O/\eta V_{IN})$

$$LP = \mu_0\mu_r N_P^2 A_e/l_e$$

$$= 4 \times 3.14 \times 10^{-7} \times 3000 \times 14 \times 14 \times 0.57 \times 10^{-4}/8.3 \times 10^{-2} = 507\ \mu H$$

Therefore, maximum value of primary current for one of the period of switching waveform is given by,

$$[12/(4 \times 507 \times 10^{-6} \times 20 \times 10^3)] + [100/(0.8 \times 12)]$$

$$= 0.3 + 10.4 = 10.7\ A$$

RMS value of primary current can be determined from the maximum value of the current and the wave shape. In the present case, the waveform is almost a square waveform with a flat top and a duty cycle of 0.5 and the DC component is much larger than the magnetising component. RMS value of such a waveform is given by,

RMS value of primary current = $10.7 \times \sqrt{0.5} = 10.7 \times 0.707 = 7.56$ A

Secondary current is a DC given by V_O/R_L.

Secondary current = 1000/10000 = 0.1 A

Primary and secondary wire sizes can be chosen for 7.56 A and 0.1 A, respectively.

Problem 16.4. *Figure 16.22 shows the basic buck regulator configuration. It produces a regulated output voltage of +12 V. If the unregulated input voltage at a certain time is +24 V, determine the on-time of the drive waveform appearing at the base terminal of the switching transistor Q. Assume a switching frequency of 10 kHz.*

Solution. The output voltage in the case of buck regulator is given by,

$$V_O = V_{IN} \times (T_{ON}/T) = V_{IN} \times T_{ON} \times f$$

$$T_{ON} = V_O/(V_{IN} \times f) = 12/(24 \times 10^4) = 50\mu s.$$

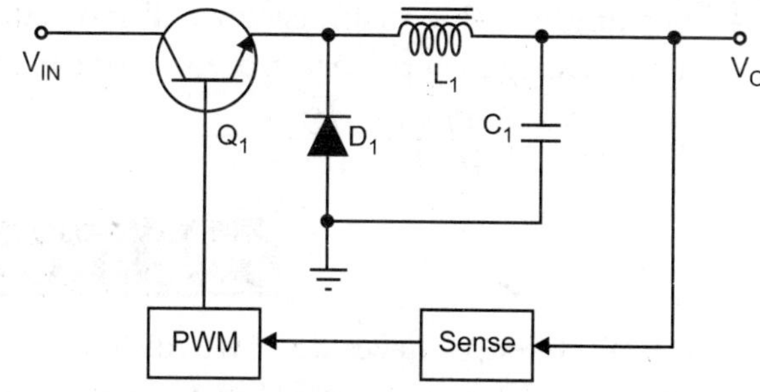

Fig. 16.22

Problem 16.5. *Figure 16.23 shows the basic boost regulator circuit using a pulse width modulated drive waveform control (V_{IN} = 12V). For the drive waveform shown in the circuit, determine the output voltage. Also, determine the changed on-time of the drive waveform, when the unregulated input voltage changes to + 18 V.*

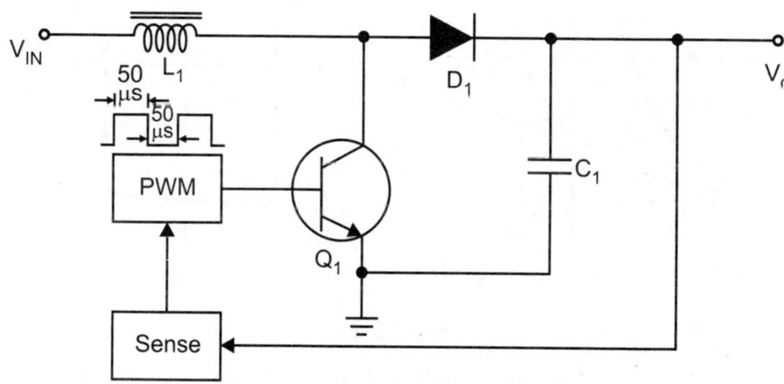

Fig. 16.23

Solution. From the given drive waveform, duty cycle,

$$D = (50 \times 10^{-6})/(50 \times 10^{-6} + 50 \times 10^{-6}) = 0.5$$

The output voltage V_O in the case of boost regulator configuration is given by,

$$V_O = V_{IN}/(1 - D) = 12/(1 - 0.5) = 24\ V$$

When the input voltage changes to +18 V, the new value of duty cycle D required to maintain the output voltage at + 24 V is given by,

$$24 = 18/(1 - D)$$

$$D = 1 - (18/24) = 0.25$$

Therefore, the changed value of on-time is given by $0.25 \times 100 \times 10^{-6} = 25\mu s$

Summary of Important Formulas

1. Voltage across capacitor in an externally driven flyback converter is $V = \sqrt{\frac{L_P}{C}}\hat{I}_P$
2. Value of capacitance of an externally driven flyback converter using a single transistor and a single transformer $C = \left(10t_{ON}\frac{P_o}{V_o^2}\right)$
3. Primary inductance in a push-pull type forward converter is given by $L_P = \left(\frac{\mu_o\mu_r N_P^2 A_e}{l_e}\right)$

4. Maximum value of transistor collector current in a push-pull type forward converter is given by, $I_{MAX} = \left(\frac{V_{IN}}{4L_{P}f} + \frac{P_{O}}{\mu V_{IN}}\right)$

5. Capacitor in a push-pull type forward converter is given by, $C \geq \left(\frac{100P_{O}}{V_{O}^{2}f}\right)$

SELF EVALUATION EXERCISE

1. The B-H loop curves for two different types of ferrite core used for making SMPS transformers are depicted in Figs. 16.29 (*a*) and (*b*). One of these is much better suited for making switching transformers of self-oscillating type push-pull DC to DC converters.

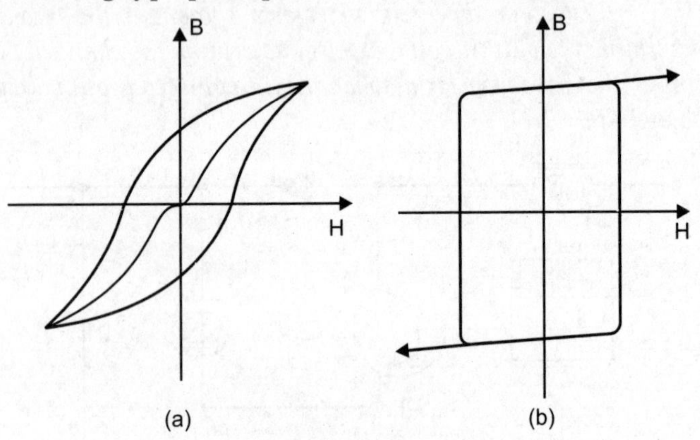

Fig. 16.29

2. One of the following statements is correct regarding the power handling capability of a transformer core.
   (*a*) It depends upon the maximum flux density rating of the core.
   (*b*) It depends upon core's cross-sectional area.
   (*c*) It depends upon the frequency of operation and the size of air gap.
   (*d*) It is directly proportional to the product of the core window area and the core cross-section.
3. Answer the following :
   (*a*) When DC/DC converter is normally referred to as a constant output power converter?
   (*b*) In which switching regulator configuration, the peak inductor current is independent of the load current?
   (*c*) Which DC/DC converter configuration is the most suitable one for low power, high output voltage applications?
   (*d*) Which DC/DC converter configuration you would for high output power applications?

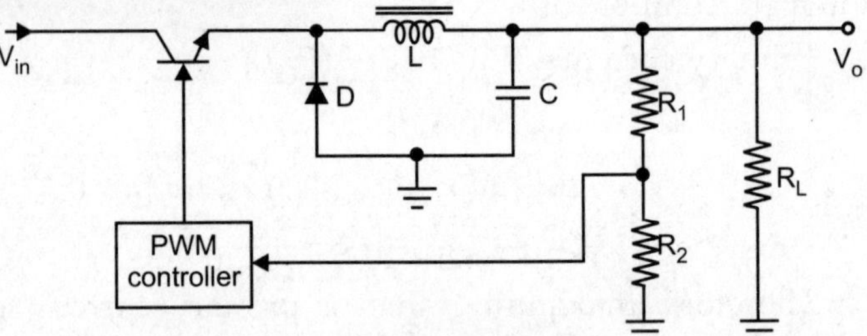

Fig. 16.30

4. In the typical switching regulator configuration of Fig. 16.30, the duty cycle of the transistor's base waveform $\left(\text{Duty Cycle} = \frac{T_{ON}}{T}\right)$ will as the output voltage tends to decrease. Fill in the missing blank
5. Figure 16.31 shows some basic DC/DC converter configurations of both the flyback and forward type. Identify the configurations with correct transformer winding polarities.

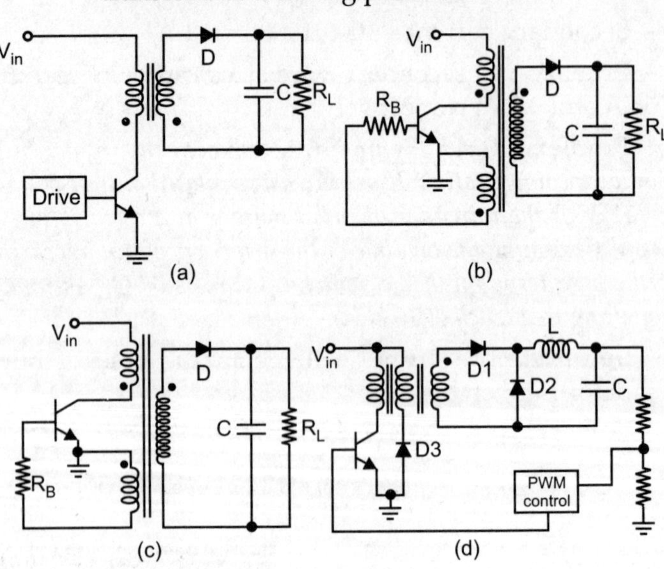

Fig. 16.31

6. A self-oscillating ringing choke type DC/DC converter designed to deliver an output power of 10 watts to a resistive load of 100 K is shown in Fig. 16.32 (*a*). Its collector emitter voltage waveform is shown in Fig. 16.32 (*b*). What do you think will happen, if this 100 K load

resistance is replaced by (*i*) 1K resistance (*ii*) IM resistance.

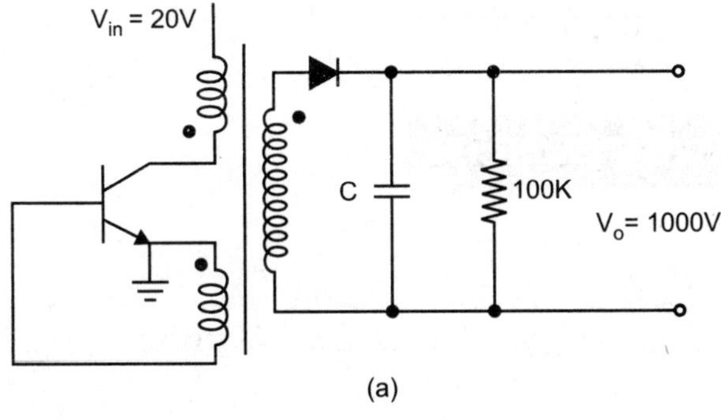

(a)

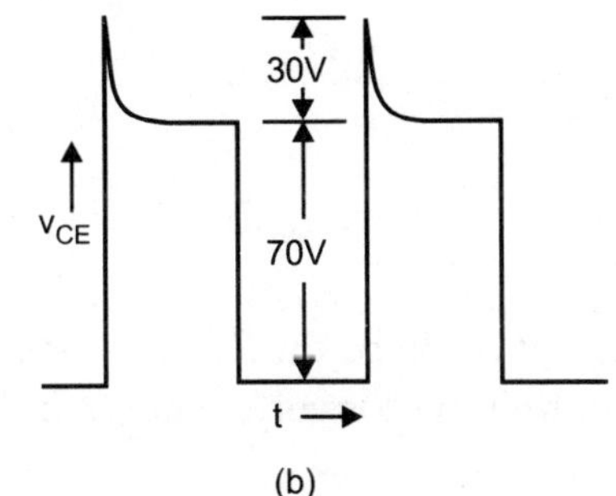

(b)

Fig. 16.32

7. In a flyback-type DC-to-DC converter, the energy is stored in the primary winding of the switching transformer during
   (*a*) turn-ON time of the switching device
   (*b*) turn-OFF time of the switching device
   (*c*) both turn-ON and turn-OFF times of the switching device
   (*d*) none of these
8. Which of the following statements is true in the case of ringing-choke-type power converter during turn-ON time of the switching transistor?
   (*a*) Energy stored, fast recovery rectifier forward-biased, filter capacitor charging
   (*b*) Energy transferred, fast recovery rectifier reverse-biased, filter capacitor charging
   (*c*) Energy stored, fast recovery rectifier reverse-biased, filter capacitor discharging
   (*d*) Energy transferred, fast recovery rectifier forward-biased, filter capacitor charging
9. Which of the following expressions can be used to determine the output voltage in the case of self-oscillating flyback-type DC-to-DC converter?
   (*a*) $V_O = \sqrt{(P_O \times R_L)}$
   (*b*) $V_O = n \times V_{IN}$ (*n* is the step-up ratio of the switching transformer)
   (*c*) $V_O = \eta \times V_{IN}$ (η is the conversion efficiency)
   (*d*) $V_O = \sqrt{(P_O/R_L)}$
10. A DC-to-DC converter having a conversion efficiency of 80% is delivering a power of 16 W to the load. If the converter were producing an output voltage of 400 V from an input of 20 V, what would be the current drawn from the 20 V source?
   (*a*) 1000 mA
   (*b*) 500 mA
   (c) 200 mA
   (*d*) cannot be determined from given data
11. Figure 16.33 shows the secondary current waveform in the case of
   (*a*) forward-type DC-to-DC converter
   (*b*) flyback-type DC-to-DC converter
   (*c*) Buck-type switching regulator
   (*d*) none of these

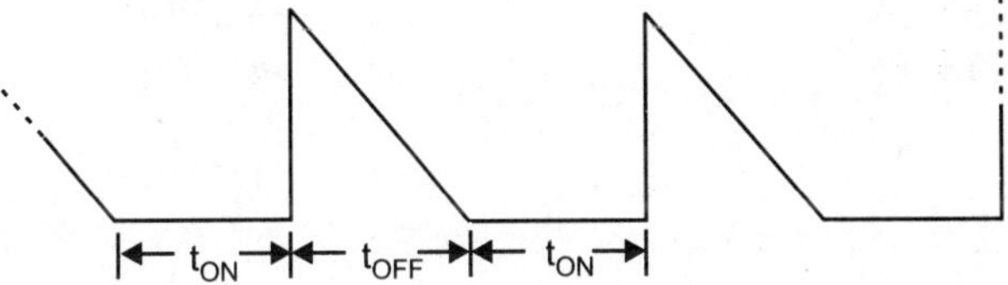

Fig. 16.33

12. The effect of core saturation on primary inductance is
   (*a*) constancy of primary inductance
   (*b*) sudden increase in primary inductance
   (*c*) sudden decrease in primary inductance
   (*d*) core saturation has no effect on primary inductance
13. Cutting an air gap in the magnetic path of a core
   (*a*) produces a relatively lower B for a given H
   (*b*) produces a relatively higher B for a given H
   (*c*) increases the core permeability
   (*d*) none of these
14. Which components' size is affected by frequency of operation of converter?
   (*a*) transformer and transistor
   (*b*) transformer and filter capacitor
   (*c*) switching transistor and filter capacitor
   (*d*) switching transistor, transformer, filter capacitor

15. The maximum value of primary current that can flow in a push–pull converter when either of the transistors is conducting is given by one of the following expressions:
(*a*) $(P_{OUT}/\eta V_{IN}) + (V_{IN}/4fL_P)$
(*b*) $P_{OUT}/\eta V_{IN}$
(*c*) $V_{IN}/4fL_P$
(*d*) None of these

Answer to Self Evaluation Exercise

1. Figure 16.24 (*b*)
2. (*d*)
3. (*a*) Self-oscillating flyback converter or an externally driven flyback converter without PWM controller in the feedback loop.
   (*b*) Step-up switching regulator
   (*c*) Flyback converter
   (*d*) Push-pull converter
4. Increase
5. Fig. 16.26 (*c*) and Fig. 16.26 (*d*)
6. In both cases, the converter is likely to get damaged. In the first case, it could be due to increased current and in the second case, it could be due to increased voltage. Remember that the converter shown in Fig. 16.27 (*a*) delivers a constant output power to the load.

7. (*a*)	8. (*c*)	9. (*a*)
10. (*a*)	11. (*b*)	12. (*c*)
13. (*a*)	14. (*b*)	15. (*a*)

Chapter 17

Wave Shaping Circuits

Learning Objectives

After completing the chapter, you will learn the following topics:

1. Basic R-C Low Pass Circuit and its response to Step and Pulse Inputs
2. R-C Low Pass Circuit as Integrator
3. R-C High Pass Circuit as Differentiator
4. R-L Integrator Circuits
5. R-L Differentiator Circuits
6. Diode Clipper Circuits
7. Positive Clamper Circuits
8. Negative Clamper Circuits

Q. 1. How does the basic RC circuit of Fig. 17.1 (*a*) respond to a STEP or a PULSE input of given time duration? Under what conditions can this circuit be called an integrator circuit, if at all it can be called one?

Ans. STEP INPUT

For a step input of 17.1 (*b*), the output voltage (V_o), which is also voltage across (C), rises exponentially towards the final value of (V) with a time constant (RC). The output voltage (V_o) is given by,

$$v_o = V(1 - e^{-t/RC})$$

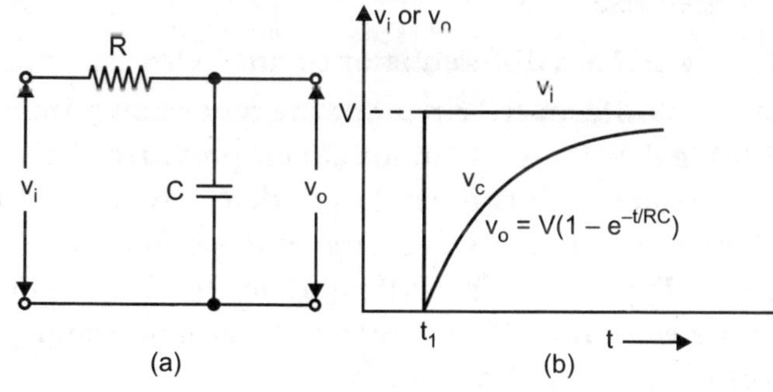

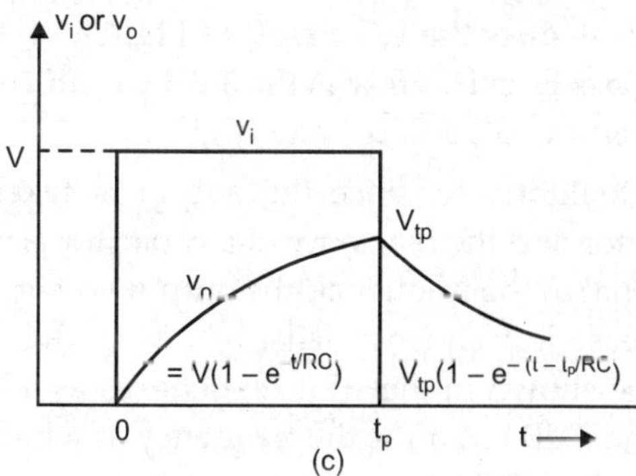

Fig. 17.1

This expression is valid only, when the capacitor is initially fully discharged. If the capacitor was initially charged to a voltage (V_o) less than (V), then the exponential charging equation would be,

$$v_o = [V - (V - Vo)e^{-t/RC}]$$

It this input step occurs at time $t = t_1$, then

$$v_o = V(1 - e^{-(t - t_1)/RC})$$

PULSE INPUT

The output (v_0) during the high time of the pulse is given by,

$$v_o = V(1 - e^{-t/RC})$$

For $t = t_p$

$$v_o \text{ (at } t = t_p) = (1 - e^{-t_p/RC}) = V_{(t_p)}$$

The output (V_o) during the low time of the pulse is given by,

$$v_o = V_{(tp)}[e^{-(t-t_p)/RC}]$$

If the circuit shown in Fig. 17.1 (*a*) is an integrator circuit, the output (v_o) should be integral of the input (v_i) *i.e.* $v_o = k \int v_i$ dt, where (*k*) is a constant. In the given *RC* circuit, if the product (*RC*) is much larger than the time period of the applied input, the capacitor voltage (or the output voltage in the present case) would change by only a very small amount as the input goes through a complete cycle. In such a case, it is not wrong to assume that whole of (v_i) appears across (*R*) only. As a result current (*i*) can be expressed as:

$$i = (V_i/R)$$

$$v_o = \frac{1}{C}\int i dt = \left(\frac{V_i}{R}\right)dt = \frac{1}{RC}\int v_i \, dt$$

Therefore, (v_o) is integral of (v_i) provided that $RC >> T$.

In fact, if $RC \geq 15\ T$, integration is near ideal.

Q. 2. How does the RC circuit of Fig. 17.1 (*a*) behave as a low pass circuit? How is the 3 dB cut-off frequency of this circuit related to (R) and (C)?

Ans. Qualitatively, since the output is taken across the capacitor and the reactance of a capacitor is inversely proportional to the frequency, the output voltage will fall with increase in frequency. That is how, an RC network of the type shown in Fig. 17.1 (*a*) behaves as a low pass circuit. The 3 dB cut-off is the frequency at which output amplitude is 0.707 times (or 3 dB below) the nominal maximum amplitude. The nominal maximum output amplitude is same as the input amplitude so that ratio {$v_o v_i$) equals 0.707 at 3 dB cut-off frequency.

When the resistance (*R*) equals capacitive reactance (X_C), the ratio (v_o v_i) indeed becomes 0.707 so that the cut-off frequency (f_C) is given by the condition,

$$R = X_C \text{ or } f_C = \frac{1}{2\pi RC}$$

Q. 3. When a pulsed waveform is applied to an RC low pass circuit of the type shown in Fig. 17.2, we are often interested in knowing the quality with which the leading and the trailing edges or the flat top portions will be passed. Such a situation arises particularly in the analysis of high frequency amplifiers, where the amplifier input has a finite capacitance and this capacitance alongwith source resistance constitutes the RC network. How is the 3 dB cut-off frequency of such a network, as determined in the last question related to the capability of the network to pass fast transitions ? On what factors will the faithful reproduction of flat top portion depend?

Ans. Refer to Fig. 17.2. The quality with which this network reproduces fast transitions is expressed by the magnitude of the rise time (t_r) which is the time taken by the output to change from 10% to 90% of the impressed transition or step. From the exponential charging relationship, it can be verified that, $t_r = 2.2\ RC$.

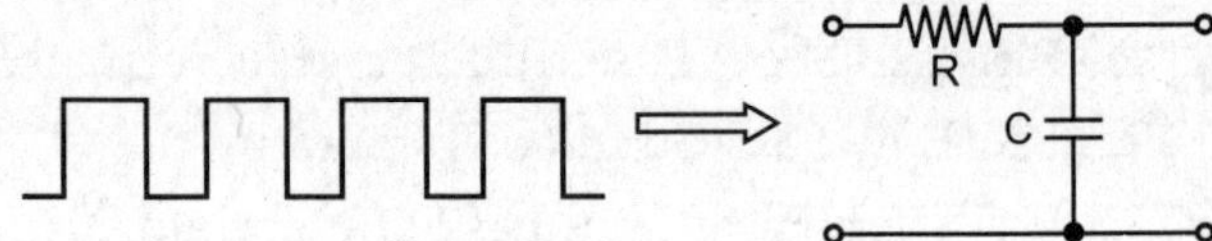

Fig. 17.2

Also, 3 dB cut- off frequency (f_c)is given by,

$$f_c = \frac{1}{2\pi RC} \text{ or } RC = \frac{1}{2\pi f_c}$$

$$\text{which gives } t_r = \frac{2.2}{2\pi f_c} = \left(\frac{0.35}{f_c}\right)$$

This expression indicates that higher the 3 dB cut-off, smaller is the rise time. Therefore, for faithful reproduction of fast transitions, (f_c) should be as high as possible. If the sharp transitions are reproduced, flat portions would automatically be passed undistorted. Infact, as long as 3dB cut-off frequency is equal to or larger than the frequency of the input square waveform or pulse train, flat portion will be passed with not too much of distortion. On the other hand, in order to allow the transitions to pass, the cut-off frequency should be much larger than the input frequency. In fact, the required (f_c) will be governed by the desired rise time.

Q. 4. What is a differentiator circuit ? Or

"While an *RC* circuit of the type shown in Fig. 17.3 (*a*) makes it an integrator circuit provided that the time constant (*RC*) is much larger than, say more than ten times, the input waveform time period), the *RC* circuit of Fig. 17.3 (*b*) is a differentiator circuit provided the time constant (*RC*) is much smaller than the input waveform time period." Comment !

Ans. A differentiator circuit is the one in which the response is proportional to differential of the excitation.

In other words, output is proportional to the slope of the input. The first part of the statement that the *RC* circuit of Fig. 17.3 (*a*) constitutes an integrator circuit under the specified conditions is true and has already been dealt with at length in the preceding questions. In case of the RC circuit of Fig. 17.3 (*b*), where the output is taken across (*R*), if the time constant (*RC*) is much smaller than the input waveform time period, it is safe to assume that whole of (v_i) appears across (*C*) only as the input goes through one complete cycle. Thus,

$$i = C\frac{dv_i}{dt}$$

$$v_o = i\,.\,R = RC\frac{dv_i}{dt} \text{ or } v_o \propto \frac{dv_i}{dt}$$

This explains the validity of the second part of statement.

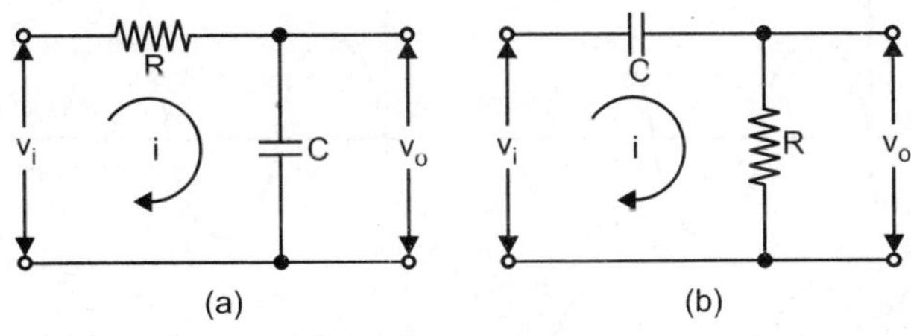

Fig. 17.3

Q. 5. Justify that RC differentiator circuit of Fig. 17.3 (*b*) is a high pass circuit. Determine the 3 dB cut-off frequency (also sometimes referred to as the lower 3 dB cut-off) for this network. Which portion of an input square waveform is directly affected by the magnitude of this cut-off frequency?

Ans. Qualitatively, since the reactance of a capacitor is inversely proportional to the frequency, the output voltage (v_o) $\left(v_o = \frac{v_i R}{\sqrt{R_2 + X_c^2}}\right)$ must fall with decrease in

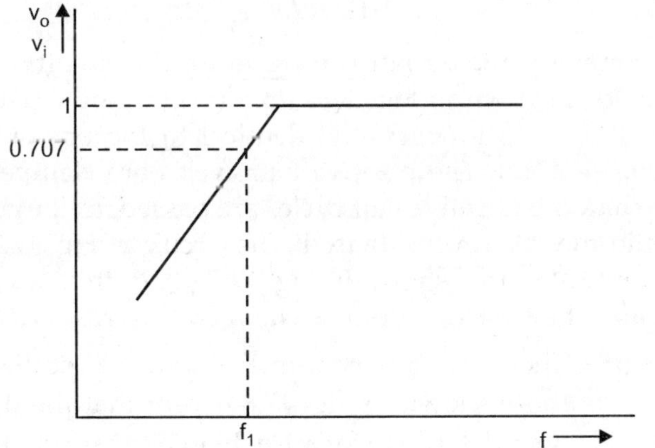

Fig. 17.4

frequency of the input waveform and hence, the high pass characteristics. The frequency, where the ratio(v_o/v_i) falls to 0.707 is known as the lower 3dB cut-off (Fig. 17.4).

Quite obviously,

$$\frac{v_o}{v_i} = \frac{R}{\sqrt{R^2 + X_c^2}} = 0.707$$

$$R = X_C \text{ or } f_1 = \left(\frac{1}{2\pi RC}\right)$$

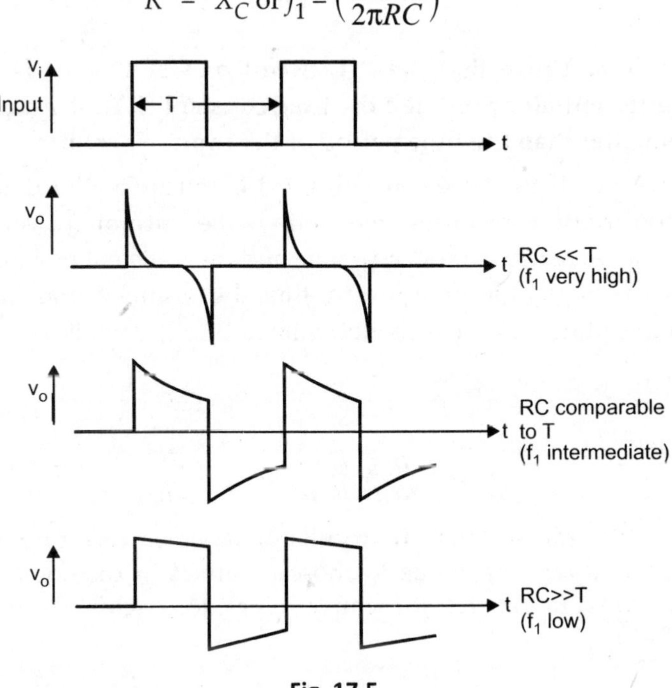

Fig. 17.5

where, f_1 is lower 3 dB cut-off frequency.

The lower 3 dB cut-off frequency affects the low frequency response due to the high pass nature of the circuit. Lower the lower 3 dB cut-off frequency, less severe is its effect on the flatter portions of the waveform. In a true differentiator circuit, where (f_1) is very high, the flat portion of square wave input may be completely missing.

The effects of different lower 3 dB cut-off frequencies in a high pass RC circuit on a square waveform input are depicted in Fig. 17.5

Q. 6. Prove that the *R-L* circuit of Fig. 17.6 (*a*) behaves as an integrator circuit provided the time constant (*L/R*) is much larger than the time period of the input waveform.

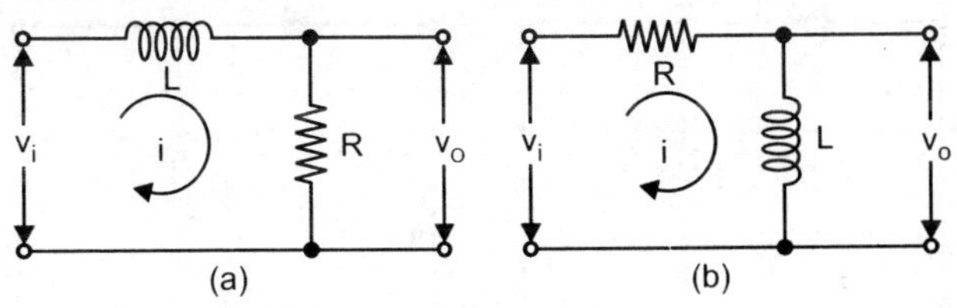

Fig. 17.6

Ans. If the time constant (L/R) is much larger than the time period of the input waveform, it is justified to assume that the current (i) is exclusively determined by (L) as the input goes through one complete cycle.

That is, $i = \frac{I}{L}\int v_o\, dt$

$$v_o = i\,.\,R = \frac{R}{L}\int v_i\, dt \text{ or } v_o \propto \int v_i\, dt.$$

Q. 7. Prove that the R-L circuit of Fig. 17.6 (*b*) is a differentiator provided the time constant (*L/R*) is much smaller than the time period of the input waveform.

Ans. If the time constant (L/R) is much smaller than the input waveform time period, the current (i) very quickly reaches the steady state and the transient portion takes negligible time. It can, thus, be assumed that the current (i) is exclusively determined by (R). That is,

$$i \propto \frac{v_i}{R}$$

$$v_o \propto L\frac{di}{dt} = \frac{L}{R}\frac{dv_i}{dt} \text{ or } v_o \propto \frac{dv_i}{dt}$$

We can see that the results obtained in case of R-L circuits are analogous to those obtained in case of RC circuits. In general, for simple integrator circuit (R–L or R–C).

$$v_o = \frac{1}{\tau}\int v_i\, dt$$

and for differentiator circuit (R – L or R–C),

$$v_i = \tau\left(\frac{dv_i}{dt}\right)$$

Where, τ = Time constant = RC (for R–C circuit) = $\frac{L}{R}$ (for R–L circuit).

Q. 8. Figure. 17.7 shows some diode clipper circuits. Why are these called non-linear waveshaping circuits? If these circuits are fed at the input with a sinusoidal signal. Sketch the clipped output waveform for each one of them when (*i*) diodes are ideal and (*ii*) the diodes have a forward biased voltage drop of 0.7 V and an infinite reverse bias resistance.

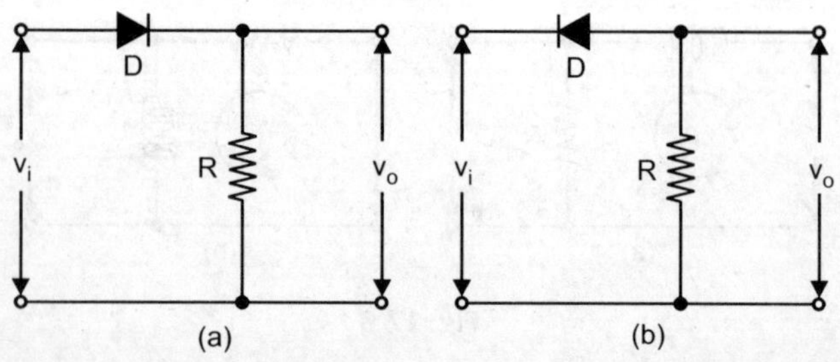

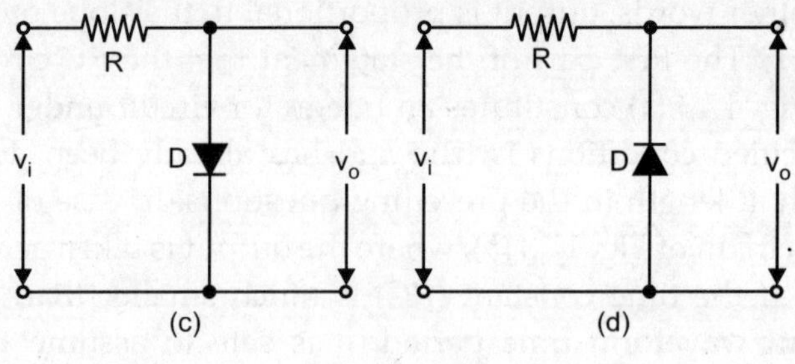

Fig. 17.7

Ans. The waveshaping circuits are called non-linear, when one or more than one elements have non-linear current-voltage characteristics. The non-linear element in the circuits of Fig. 17.7 is the diode.

(*i*) The input and output waveforms for the clipping circuit of Fig. 17.7 (*a*) are shown in Fig. 17.8 (*a*). It is self-explanatory.

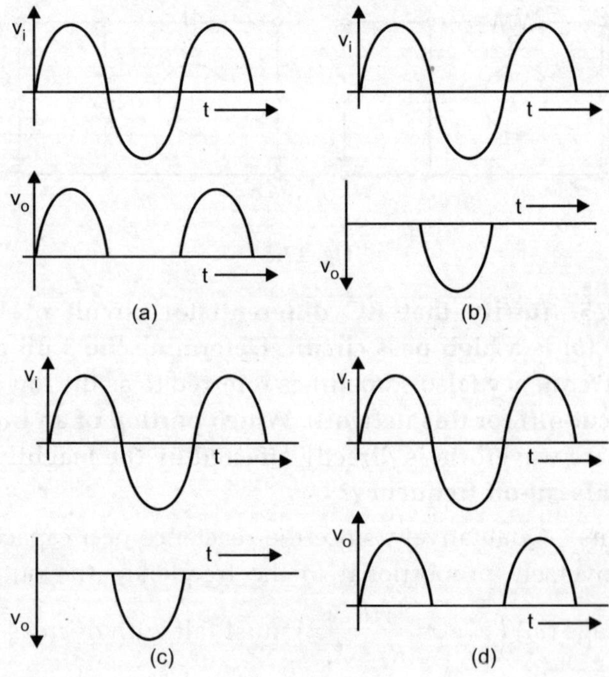

Fig. 17.8

Similarly, the output waveform for the circuit of Fig. 17.7 (*b*) is shown in Fig. 17.8 (*b*). The clipping circuit of Fig. 17.7 (*c*) is functionally identical to the one shown in Fig. 17.7 (*b*). The positive half cycles are clamped to zero and the negative half cycles are passed, as the diode would then be reverse biased. The circuit of Fig. 17.7 (*d*) is identical to that shown in Fig. 17.7 (*a*). In this case, the negative half cycles would be clamped to zero.

(*ii*) The output waveforms would be similar to those shown in Fig. 17.8 except that the diode would have a forward biased drop of 0.7 V instead of zero assumed in case of ideal diodes. The waveforms are shown in Fig. 17.9.

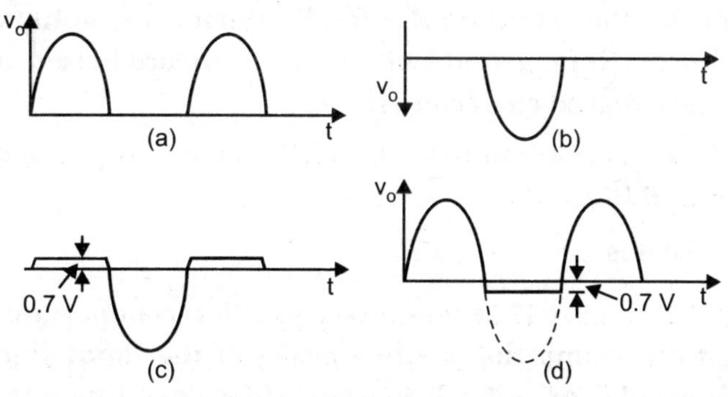

Fig. 17.9

Q. 9. Sketch the output waveform for the clipping circuits of Fig. 17.10. Assume the diodes to be ideal and the peak amplitude (V_m) of the input sinusoidal signal to be greater than the battery voltage (V_{BB}).

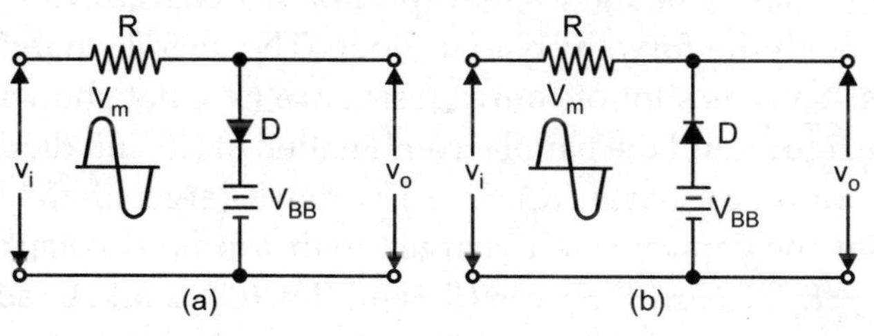

Fig. 17.10

Ans. The output waveforms are shown in Fig. 17.11 (*a*) for the circuit of Fig. 17.10 (*a*) and Fig. 17.11 (*b*) for the circuit of Fig. 17.11(b).

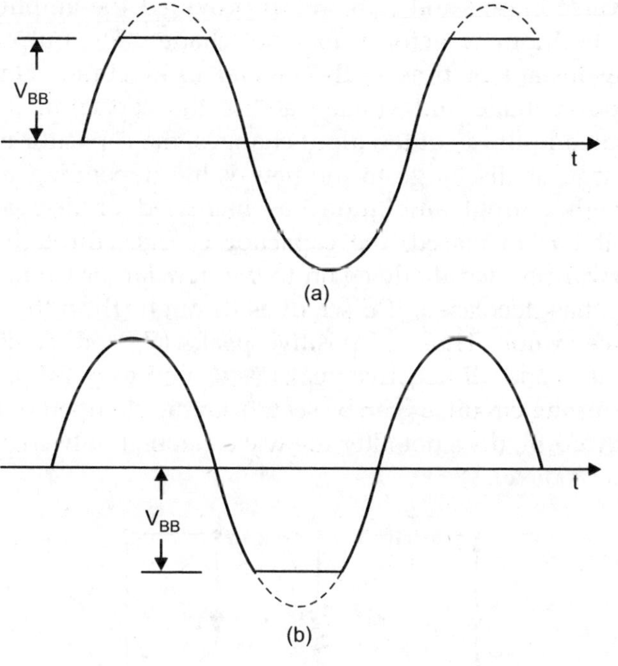

Fig. 17.11

In case of the circuit shown in Fig. 17.11(*a*), the diode gets forward biased, when the input exceeds (V_{BB}) and thus, the output remains clamped at (V_{BB}) as long as input remains greater than (V_{BB}). For $v_i < V_{BB}$, the diode is reverse biased and the output is same as the input. For the circuit of Fig. 17.10 (*b*), the diode is reverse biased for (v_i) greater than ($-V_{BB}$) and forward biased for (v_i) more negative than (V_{BB}).

Q. 10. Sketch the clipped output waveforms for the circuits of Fig. 17.12. Assume the diodes to be ideal and the peak amplitude of (v_m) of the input sinusoid to be greater than battery voltage (V_{BB}).

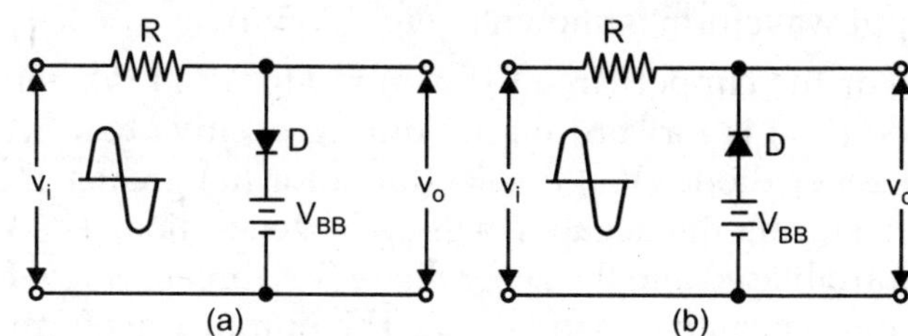

Fig. 17.12

Ans. For the circuit shown in Fig. 17.12 (*a*), the diode is forward biased for all input voltages except for the period, when (v_i) is more negative than the battery voltage ($-V_{BB}$), where it is reverse biased. For the circuit shown in Fig. 17.12 (*b*), the diode is forward biased for (v_i) less than equal to (V_{BB}) and reverse biased for (v_i) greater than (V_{BB}). The output waveforms for the two cases in Fig. 17.12 (*a*) and (*b*) are shown in Fig. 17.13 (*a*) and (*b*) respectively.

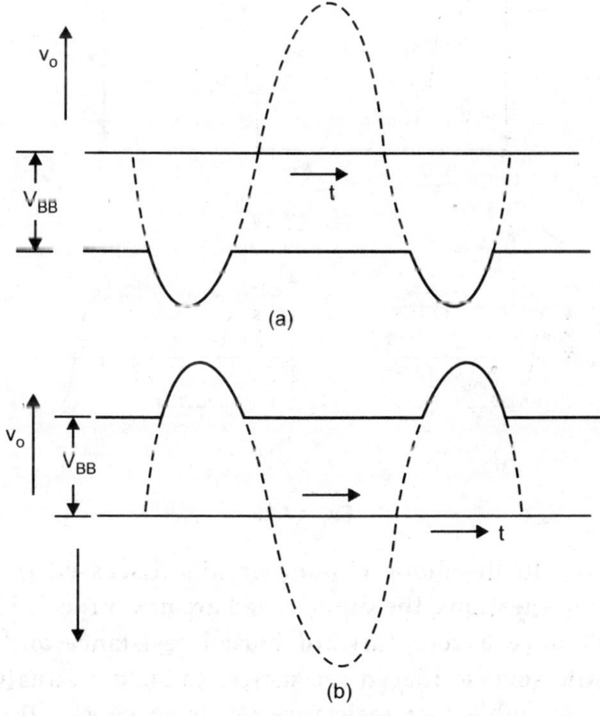

Fig. 17 .13

Q. 11. Sketch the clipped output waveforms for the double diode clipper circuits of Fig. 17.14 (*a*) and (*b*) Assume diodes D_1 and D_2 to be ideal and the forward biased voltage drops of the zener diodes to be zero.

Ans. For the circuit shown in Fig. 17.14 (*a*), during the positive half cycle, diode (D_2) remains reverse biased and the diode (D_1) gets forward biased for (v_i) greater than (V_{BB}). During the negative half cycle, diode (D_1) remains reverse biased and diode (D_2) is forward biased only for the period, where (v_i) is more negative than ($-V_{BB}$). The output waveform is shown in Fig. 17.15(a).

For the clipper circuit shown in Fig. 17.14 (*b*), zener diode (V_{z2}) is forward biased during positive half cycle and zener diode (V_{z1}) breaks down for (v_i) greater than (V_{z1}). During the negative half cycle, zener diode (V_{z1}) is forward biased and the zener diode (V_{z2}) breaks down for (v_i) more negative than ($-V_{z2}$). The output waveform is shown in Fig. 17.15 (*b*).

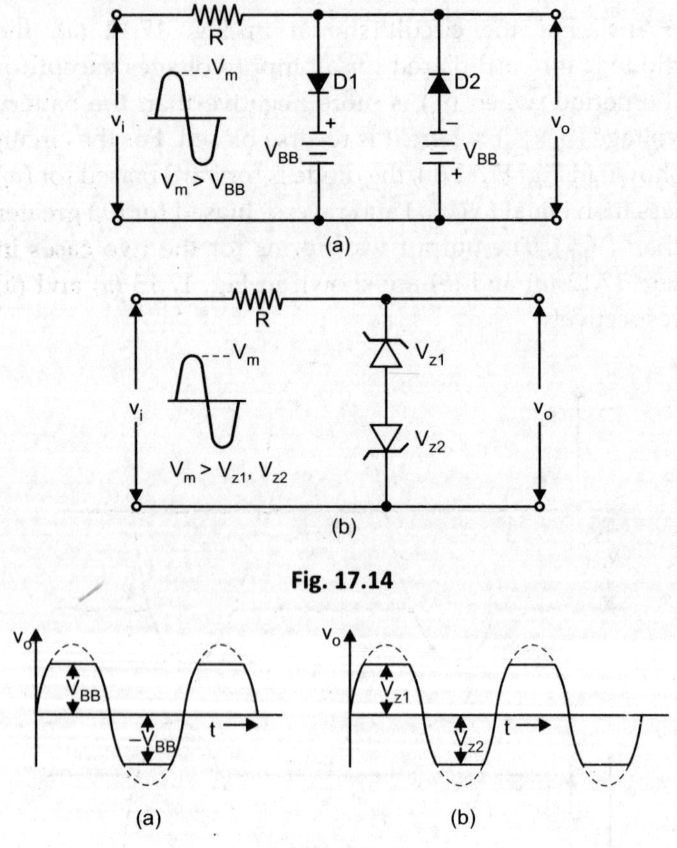

Fig. 17.14

Fig. 17.15

Q. 12. In the diode clipper circuits discussed in the earlier questions, the diodes used are never ideal. They don't have a zero forward biased resistance and an infinite reverse biased resistance. In such a situation, it is desirable that resistance (*R*) is so chosen that it satisfies the condition $R_f < R < R_r$. What is the optimum value of (*R*) for a diode with known forward biased and reverse biased characteristics ?

Ans. The optimum value of (*R*) is given by geometric mean of (R_f) and (R_r).

That is $R = \sqrt{R_f.R_r}$.

Q. 13. Figure 17.16 (*a*) shows a practical clamper circuit that can clamp the positive peaks of the input signal (sinusoidal or pulsed) to zero. How does this circuit function and what is the choice criteria for selecting different component values?

Ans. The clamped output waveform for a given sinusoidal input is shown in Fig. 17.16 (*b*). As the input (v_i) rises towards the positive peak (V_m) from zero in the first quarter of the cycle, capacitor (C) charges to (V_m) through the forward biased diode. The overall charging resistance is sum of source resistance (R_s), not shown in the figure, and the parallel combination of (*R*) and diode's forward resistance (R_f). If (*R*) is much larger than (R_f), then the capacitor (*C*) charges with a time constant of ($R_s + R_f$) C. Now for given (R_s) and (R_f), (*C*) should be such that it charges to (V_m) in a time which in no case is greater than one fourth of the time period of the input waveform. The total charging time may be considered to be equal to five times the time constant.

When the input starts decreasing, the diode becomes reverse biased and remains so provided the amplitude of the input waveform does not change. The output at any instant of time is then equal to algebraic sum of input voltage and voltage across the capacitor. If the peak amplitude of the input changes, the capacitor must charge or discharge to the new value depending upon whether input amplitude has increased or decreased. If it has increased, the capacitor charges through the forward biased diode again to the new larger value and if it has decreased, if discharges through (*R*) to the new lower value. Thus, all positive peaks (V_m) are clamped to zero and all negative peaks ($-V_m$) go to ($-2V_m$). The clamping circuit, as can be seen from the clamped output waveform, does not alter the waveshape, it only changes the DC level.

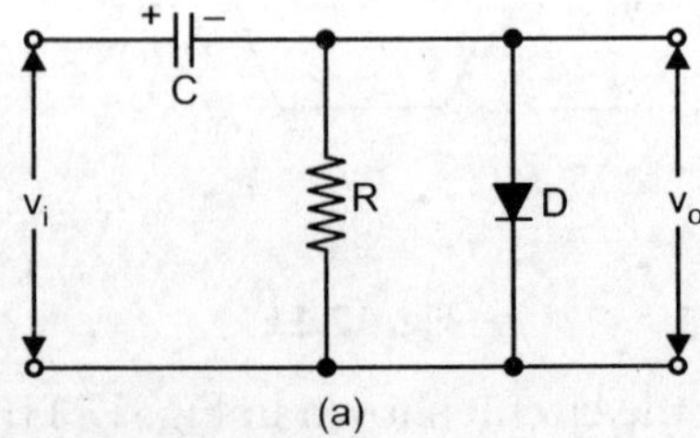

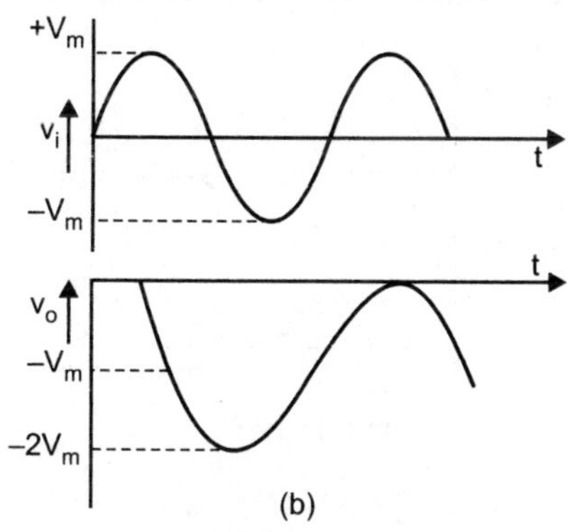

Fig. 17.16

In the absence of (R), the capacitor will be forced to discharge through the reverse biased diode resistance (R_r) which may be as large as 10 M. In that case, it may even be seconds before the capacitor discharges to new lower value and the output again clamped to the desired voltage.

As (R) is chosen to be much smaller than (R_r), this problem is not encountered and the clamping is usually restored within a few cycles of the input waveform at the most after the input undergoes a step change in amplitude. The most optimum value of (R) is given by geometric mean of (R_f) and (R_r) i.e. $R = \sqrt{R_f . R_r}$.

It may be mentioned here that the clamping circuit will function even in the absence of (R) provided that the input amplitude remains constant. In the event of input amplitude decreasing, one has to wait for a sufficiently long time depending upon {R_r C) time constant before normal clamping operation is restored.

After having calculated (R) for a given diode, (C) should be so chosen that it does not discharge appreciably during the time the diode remains reverse biased. The time constant $\left(R_s + \frac{R_r R}{R_c + R}\right)C$ of the discharge circuit should at least be 100 times the reverse biased time period. It is' more appropriate to choose (C) to meet extremely slow discharge requirement rather than fast charging requirement though (R_f). (R_f) is usually so small that (C) chosen by the discharge criterion almost always satisfies the charge criterion.

Q. 14. Draw a practical clamper circuit that clamps the negative extremities to zero.

Ans. The circuit is shown in Fig. 17.17 (*a*). The input and output waveforms are shown in Fig. 17.17 (*b*). The circuit functions in the same way as the one discussed in the last question except that the capacitor would charge to ($-V_m$) during the first negative half cycle, when the diode gets forward biased. Negative peaks are clamped to zero as for all negative peaks of the input waveform, the output (v_o) given by algebraic sum of (v_i) and (v_c) will be zero. (R) is again determined by,

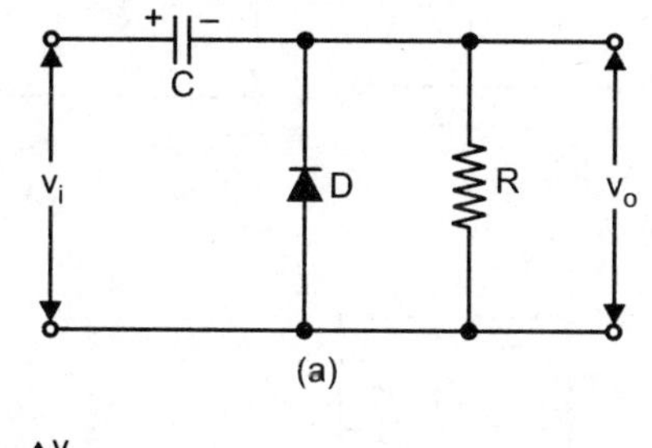

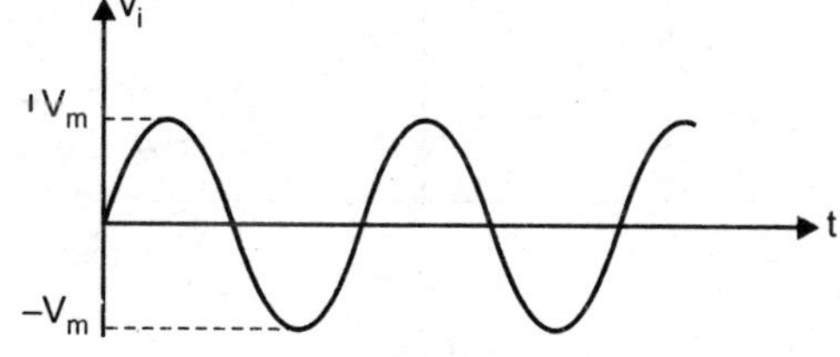

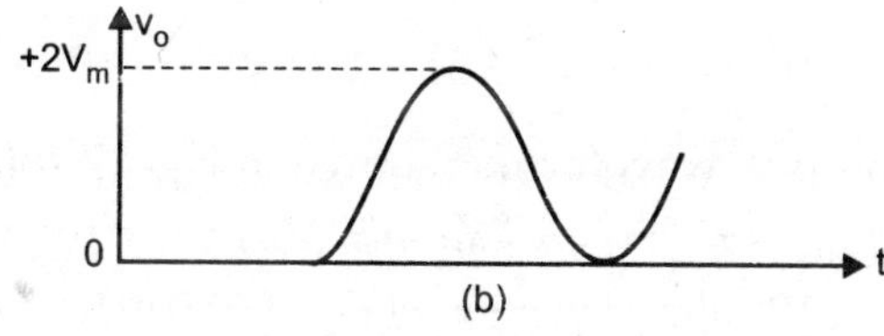

Fig. 17.17

$$R = \sqrt{R_f . R_r}$$

and (C) is determined form $\left(\frac{RR_r}{R + R_r}\right)C = 100\ T$

Where, (T) is time period of the input waveform.

SOLVED PROBLEMS

Problem 17.1. *Refer to Fig. 17.18. Determine the amplitude of (v_o) at the time of input pulse going low and also 1 ms after it has gone low.*

Solution. The time constant

$$= RC = 10 \times 10^3 \times 0.1 \times 10^{-6} = 10^{-3}\text{s}$$

Coincidentally, the time constant equals the pulse width. The output (V_o) will be 63.1 percent of the final value of 10 V *i.e.* $V_o = 6.31$. Volts at the time of pulse going low. In general, the charging process is governed by the expression,

$$v_o = V(1 - e^{-t/RC})$$

$$t = RC \quad \text{and} \quad v_o = 0.631\text{ V}$$

When the input pulse goes low, the capacitor starts discharging as per,

$$v_o = V'e^{-(t-t_p)/RC}$$

Here, $V' = 6.31$ V $t_p = 1$ ms

Again (v_o) will be 36.9 percent of (V) 1 ms after the capacitor starts discharging as 1 ms happens to be equal to circuit time constant.

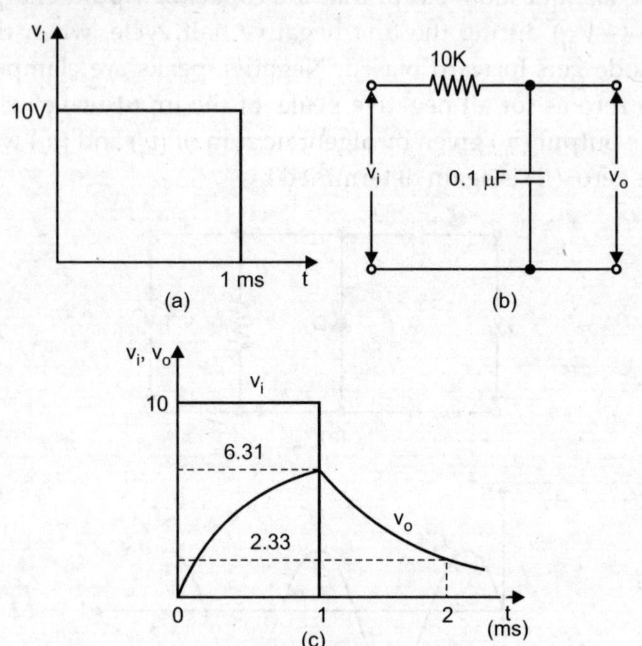

Fig. 17.18

$$v_o = 0.369 \times 6.31 = 2.33 \text{ V}$$

The output waveform is shown in Fig. 17.18(c)

Problem 17.2. *How will the circuit of Fig. 17.19(a) respond to a 10 volts step input of Fig. 17.19(b)?*

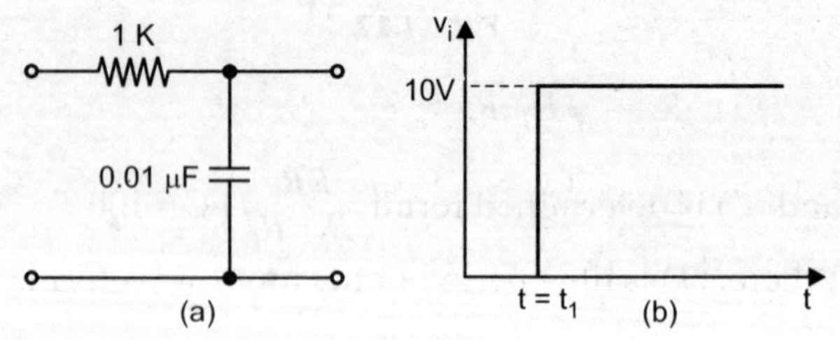

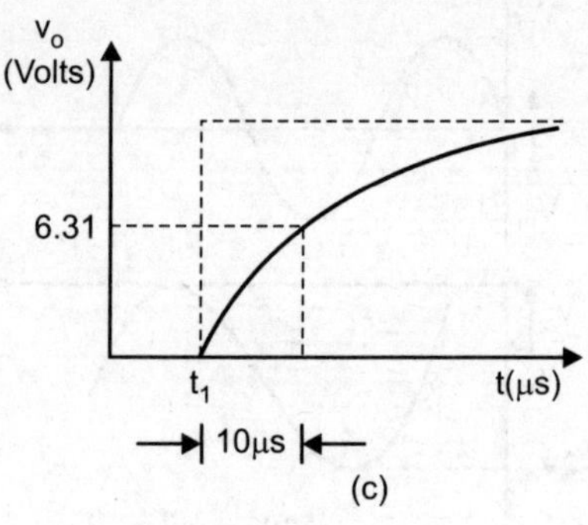

Fig. 17.19

Solution. Circuit time constant = RC = $10^3 \times 10^{-8}$ = µS

The response is governed by the equation,

$$v_o = (1 - e^{-t/10^{-5}}) = (1 - e^{-10^5 t})$$

For $t = 10\mu\text{s} = 10^{-5}\text{s}$

$$v_o = 10(1 - e^{-1}) = 0.631 \times 10 = 6.31 \text{ V}$$

The output (v_o) is shown in Fig. 17.19 (c).

Problem 17.3. *For the RC circuit of 17.19 (a) and the input waveform Fig. 17.19 (b), in what time will the output rise from 1 V to 9V?*

Solution. The time taken in this case would be to the rise time (t_r).

$$t_r = 2.2 \text{ RC} = 2.2\ 10^3 \times 10^{-8} = 22\mu\text{s}.$$

Problem 17.4. *The basic low pass RC circuit discussed in the previous problems has 3 dB cut-off frequency of 3.5 kHz. If this circuit is fed at the input with a 20 volts step, in what time will the output rise to 12.6 volts starting from the time of receiving the step?*

Solution. 3 dB cut-off = 3.5 KHz

$$\text{Rise time} = \frac{0.35}{f_c} = \frac{0.35}{3.5 \times 10^3} = 10^{-4} \text{ s}$$

$$\text{Therefore, } 2.2\ RC = \frac{10^{-4}}{2.2} = 45.5\ \mu\text{s}$$

The output will rise to 12.6 V (which is 63 per cent of the final value of 20 V) in 45.5 µs (= time constant).

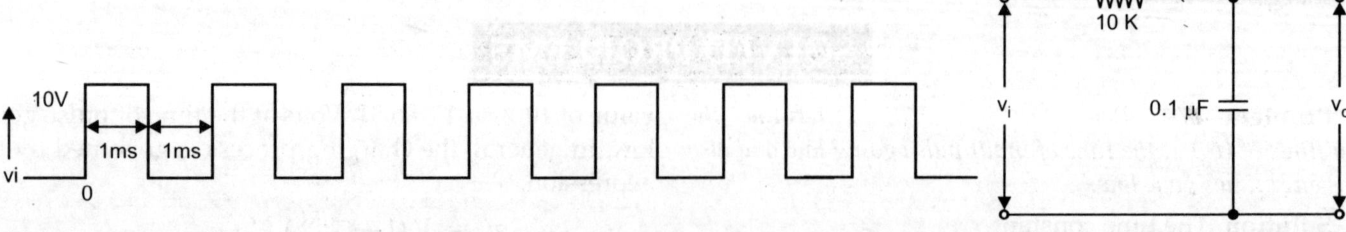

Fig. 17.20

Problem 17.5. *Refer to Fig. 17.20. Determine the output waveform (v_o) for at least 10 cycles of input waveform and comment on the results.*

Solution. With the leading edge of the first cycle, the capacitor starts charging and charging is governed by,

$$v_o = 10(1 - e^{-t/RC})$$

Since the pulse high time equals the time constant (= 1 ms), the output will rise to 6.3 V at the end of the first pulse.

The capacitor starts discharging then and the discharge equation is given by,

$$v_o = 6.3\, e^{-(t - 10^{-3})/RC}$$

In this case, the pulse low time also equals the time constant. Therefore, at t = 2 ms, v_o = 0.37 × 6.3 = 2.33 V.

With the beginning of the second pulse, the capacitor starts charging again but this time, it has an initial charge of 2.33 V and the charging equation now would be,

$$v_o = 10 - (10 - 2.33)\, e^{-(t - 2 \times 10^{-3})/10^{-3}}$$

for, t = 3 ms : $v_o = 10 - 7.67 \times e^{-1} = 7.18$ V

and for t = 4 ms : $v_o = 7.18 \times 0.37 = 2.65$ V

Similarly, at t = 5 ms : $v_o = 10 - (10 - 2.65)\, e^{-1} = 10 - 7.35/e = 7.3$ V

at t = 6 ms : $v_o = 7.3 \times 0.37 = 2.7$ V

at t = 7 ms : $v_o = 10 - (10 - 2.7)\, e^{-1} = 10 - 7.3/e = 7.32$ V

at t = 8 ms : $v_o = 7.32 \times 0.37 = 2.71$ V

at t = 9 ms : $v_o = 10 - (10 - 2.71)\, e^{-1} = 10 - 7.29/e = 7.32$ V

at t = 10 ms: $v_o = 7.32 \times 0.37 = 2.71$ V

In the subsequent cycles, the output will swing between 2.71 V and 7.32 V. It is clear from the above calculations that the steady state is achieved after five cycles. It can be verified that if the circuit time constant is much smaller than the input waveform time period, the steady state would be arrived at very quickly, most likely in the first cycle itself. If the time constant is much larger than the input waveform period, it would require comparatively larger number of cycles to reach steady state. Figure 17.21 shows the output waveform.

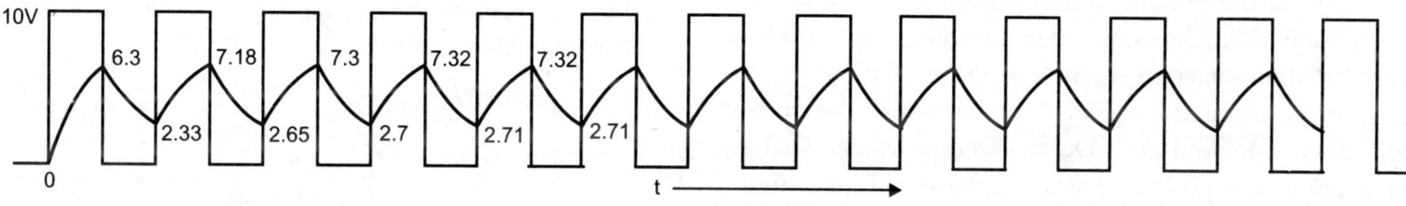

Fig. 17.21

Problem 17.6. *Refer to Fig. 17.22 Sketch the output waveform, preferably without doing any mathematical calculations.*

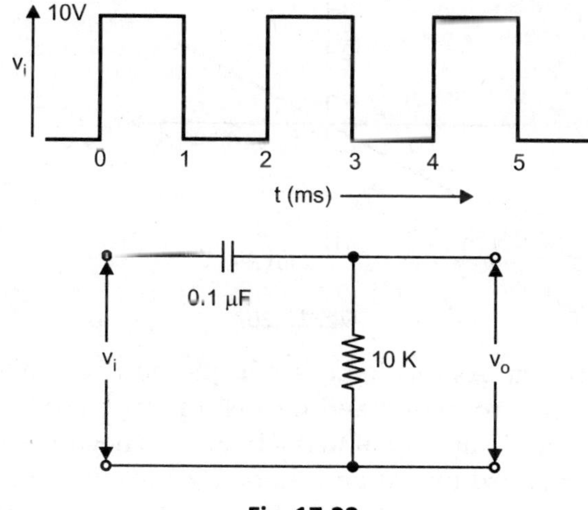

Fig. 17.22

Solution. The output waveform is shown in Fig. 17.23. The waveform is self-explanatory. The fact that the circuit time constant is 1 ms and so are the high and low times of the input waveform, all voltage levels can be determined without doing any calculations.

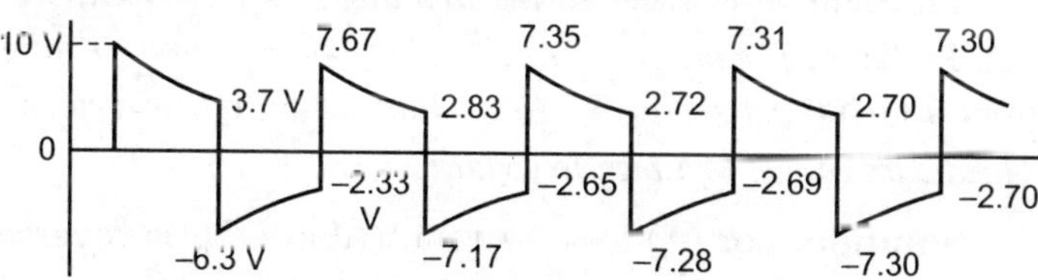

Fig. 17.23

Problem 17.7. *A 100 μs pulse is applied to the* **RC** *high pass circuit of Fig. 17.24 (a). Sketch the response waveform. Also, calculate the time taken by output pulse to go to near zero after the input pulse goes low.*

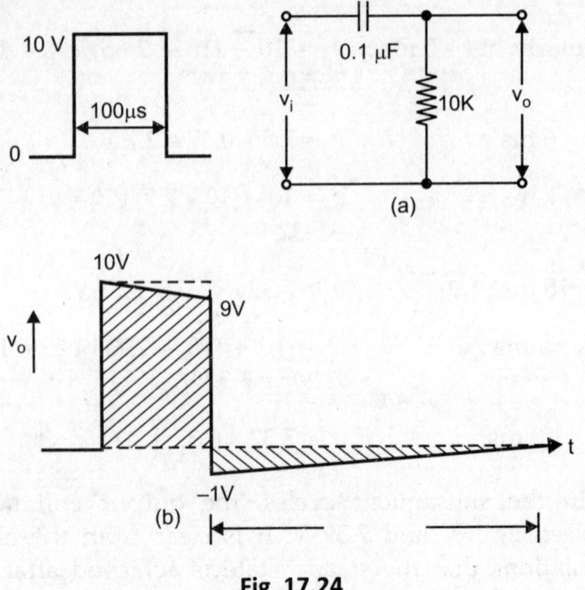

Fig. 17.24

Solution. The output voltage at the time of termination of input pulse can be calculated from,

$$V_o(\text{at } t = 100\ \mu s) = 10 - 10\left(1 - e^{-\frac{10^{-4}}{RC}}\right) = 10\ e^{-10^{-4}/10^{-3}}$$
$$= 10/e^{0.1} = 9\ V$$

As the input pulse goes to zero, the output goes to − 1 V, as the voltage across capacitor cannot change instantaneously. The output then gradually rises towards zero as the capacitor discharges. Since the input and output are isolated by a blocking capacitor, the output will always have a zero DC or average value. That is, area under the positive portion must equal area under the negative portion. Assuming the charge and discharge process to be linear which is a valid assumption, when the circuit time constant is much larger than the pulse width, we can calculate the required time for the output to decay to zero to be 2 ms. The response waveform is shown in Fig. 17.24 (*b*).

Problem 17.8. *Refer to the two diode clipper circuits of Fig. 17.25 (a). The input to this circuit is a linear ramp varying from 0 to 100 volts (Fig. 17.25 b). Plot the output waveform. The diodes can be assumed to be ideal.*

Solution. For (v_i) less than 35 V, diode D_1 is reverse biased and diode D_2 is forward biased with nodes *A* and *B* at a potential of 35 V. As (v_i) exceeds 35 volts, D_1 gets forward biased and from then onwards, potential at A is same as (v_i). Diode D_2 remains forward biased as long as (v_i) does not exceed 75 V. Thus, for (v_i) greater than 35 volts and less than 75 wits, (v_o) is same as (v_i).

As (v_i) exceeds 75 volts, D_2 is also reverse biased. From there onwards, the output is constant at 75 volts.

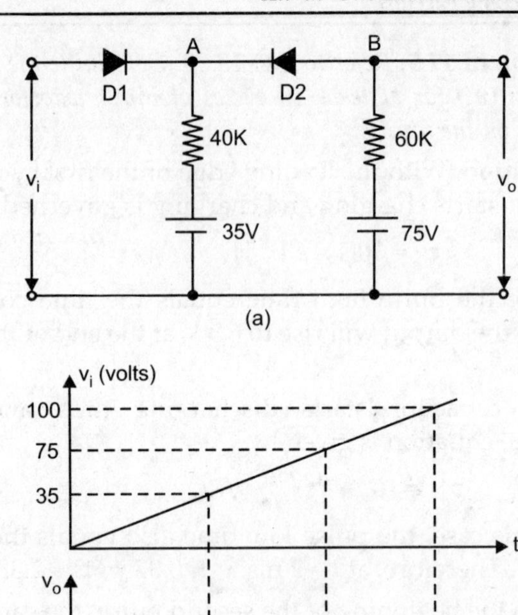

Fig. 17.25

Problem 17.9. *Draw the transfer characteristics (i.e. v_o vs. v_i) for the two diode clipper circuit of Fig. 17.26 (a). Assume the diodes to be ideal.*

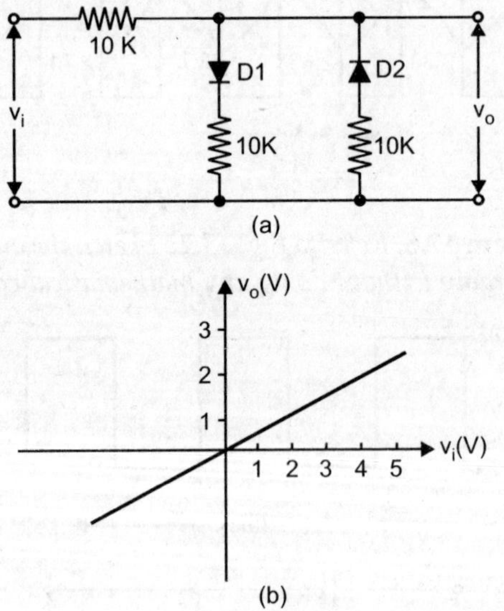

Fig. 17.26

Solution. As (v_1) increases in the positive direction, (D_1) is forward biased and (D_2) is reverse biased. (v_o) in this case is always equal to ($v_i/2$). For (v_i) negative, diode (D_2) is forward biased and diode (D_1) is reverse biased. Again the output is ($-v_i/2$). The transfer characteristics are shown in Fig. 17.26 (*b*).

Problem 17.10. *For the clamping circuit of Fig. 17.27 (a) and the input waveform of Fig. 17.27 (b), plot the output waveform for the first five cycles. Comment on the results obtained.*

Solution. As the input goes from 0 to + 10 V, the diode gets forward biased. As the voltage across the capacitor cannot change instantaneously, the output voltage (v_o) abruptly rises to only + 5V due to the potential divider arrangement of R_s (= 100 Ω) and R_f(= 100 Ω). The capacitor then starts charging towards + 10 V with a time constant ($R_f + R_s$) C.

The time constant = (100 + 100) × 10^{-6} = 200μs

The time period of the input waveform = 400μs (for = 2.5 kHz)

Therefore, during the first cycle, when the diode is forward biased, the capacitor would exponentially charge as per,

$$v_C = 10\,[1 - e^{-t/200 \times 10^{-6}}]$$

for t = 200 ms : v_C = 10 (1 – e^{-1}) – 6.3 V or $v_o = \frac{10 - 6.3}{2}$ = 1.85 V

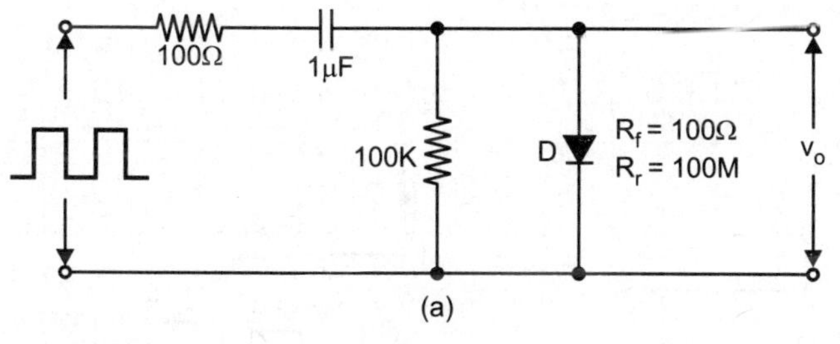

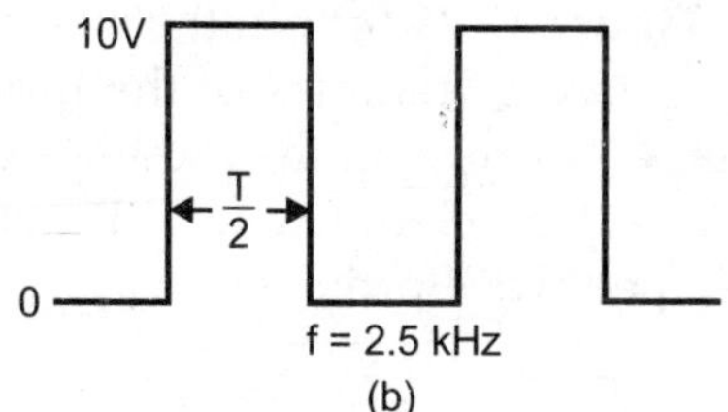

Fig. 17.27

As the input drops to zero, if (R) is much larger than (R_S) the output drops to (– 6.3) V (= capacitor voltage).

For the second half of the first cycle, when the input is zero, the output remains equal to the capacitor voltage of (– 6.3) V. Here, it is assumed that the capacitor does not discharge through (R) during this time which is quite valid as the time constant (RC) equal to 100 ms is 500 times the half cycle time of 200 us. The output returns to + 1.85 volts, as the input goes to + 10 V again with the beginning of the second cycle. The output starts decaying again with a time constant of 200 μs. It would decay to (1.85 × 0.37) = 0.68 V before the input drops to zero again during the second cycle.

The capacitor voltage at that instant of time is 10 – (2 x 0.68) = 8.64 V with the result that the output drops to (– 8.64) V. The output waveform progresses in the same fashion for the subsequent cycles and as seen from Fig. 17.28, the output is nearly clamped in the fourth cycle. From then onwards, the output has the positive extremities of the input waveform clamped to zero.

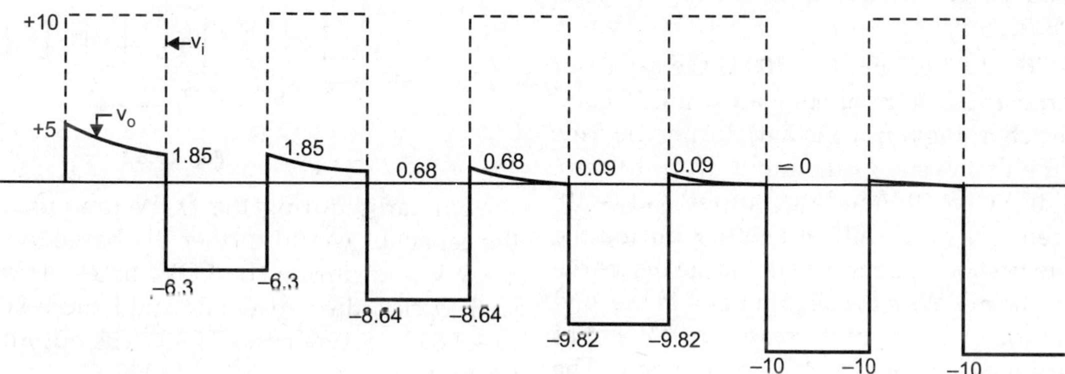

Fig. 17.28

Problem 17.11. *Sketch the steady state clamped output waveforms for the circuits shown in Fig. 17.29. Assume a zero forward biased voltage drop for diodes.*

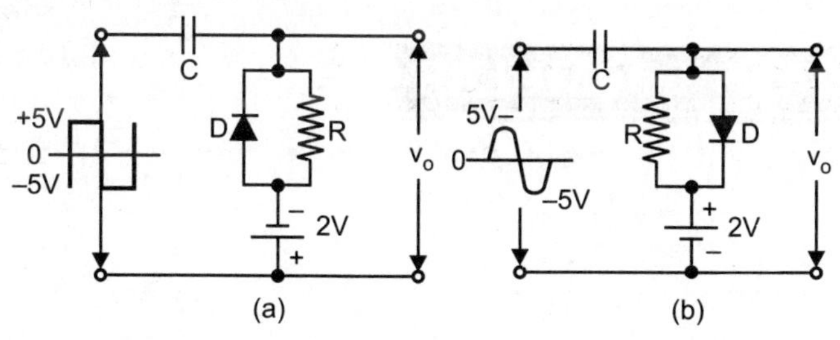

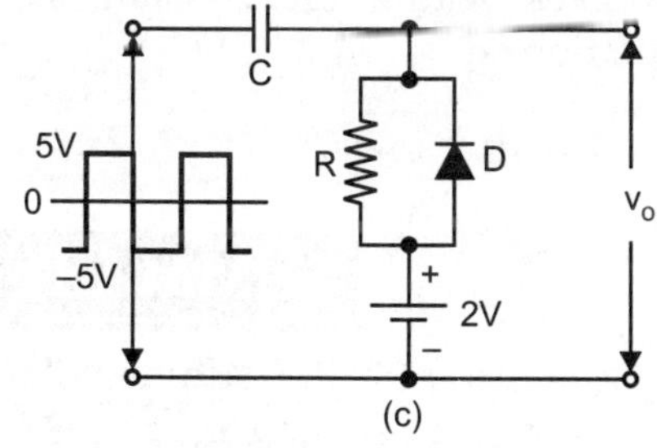

Fig. 17.29

Solution. Figure 17.29 (*a*) : In the absence of the 2 V battery, the negative extremities of the input waveform would be clamped to zero. In the presence of the battery, the waveform will be clamped to (–2V). Fig. 17.30 (*a*) shows the clamped waveform.

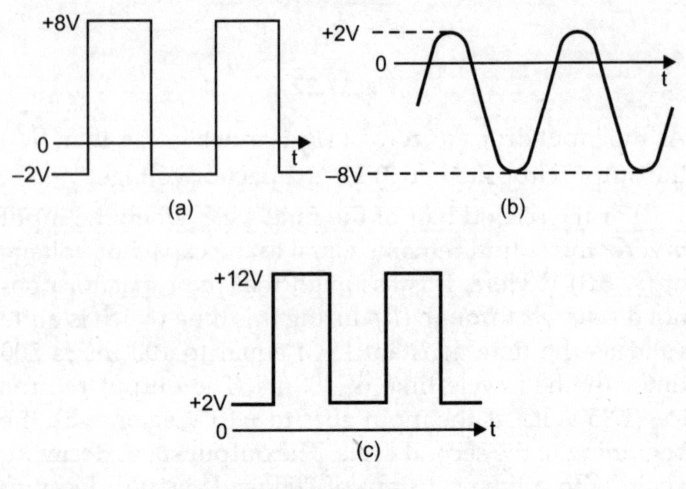

Fig. 17.30

Figure 17.29 (*b*): In this case, the positive peaks instead of being clamped to zero shall be clamped to (+ 2V). Fig. 17.30 (*c*) shows the waveform.

Figure 17.29 (*c*): In this case, the negative peaks instead of being clamped to zero, shall be clamped to + 2V. Fig. 17.30 (*c*) shows the waveform.

Problem 17.12. *Refer to the clamping circuit of Fig. 17.31 (a). The input waveform to this circuit is shown in Fig. 17.31 (b) for the first six cycles. Sketch the clamped output waveform for the first six cycles.*

Solution. With C = 1 μF and R_f = 100 Ω the capacitor (C) would charge to 10 V in about half a millisecond (100 μs being the charging time constant) during the 5 ms HIGH time of the first cycle. As the input drops to zero, the output too drops by 10 V and the output is at (– 10) V. The output remains practically at (– 10) V during the LOW time of the first cycle due to a very large discharge time constant of 100 ms. With the beginning of the second cycle, the amplitude of the input decreases by 4 V and as a result, the output can go upto (– 4)V and not zero. The capacitor starts discharging the moment input goes low in the second cycle with a time constant of 100 ms. The discharge equation is,

$$v_C = 10\, e^{\left(\frac{-t}{100\times10^{-3}}\right)}$$

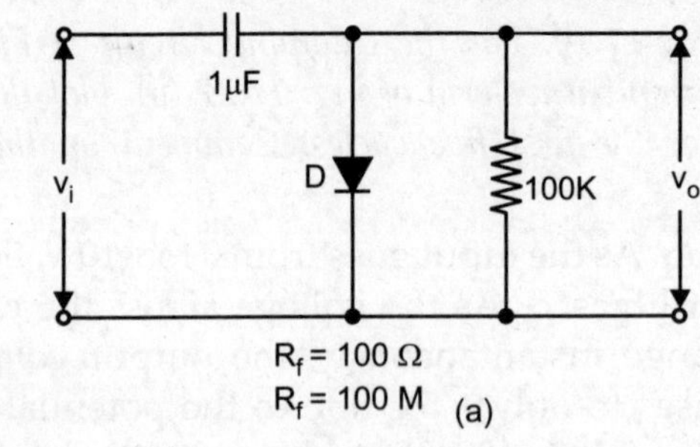

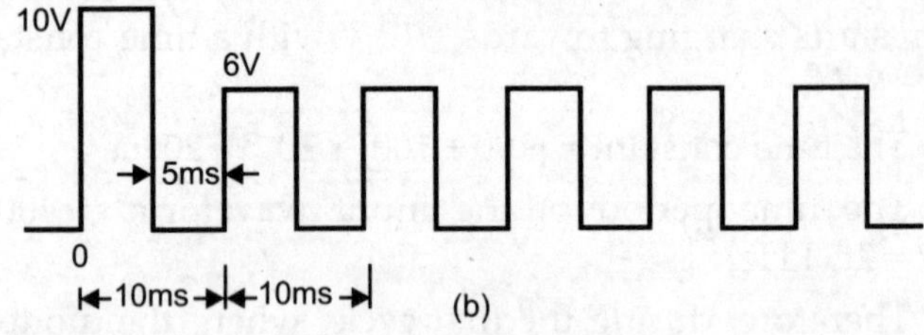

Fig. 17.31

In 5 ms, the new capacitor voltage will be

$$v_C = -10\, e^{\frac{-5\times10^{-3}}{100\times10^{-3}}} = -10\, e^{-1/20} = -9.52 \text{ V}$$

With the beginning of the third cycle, the output will be – 9.52 + 6 = – 3.52 V.

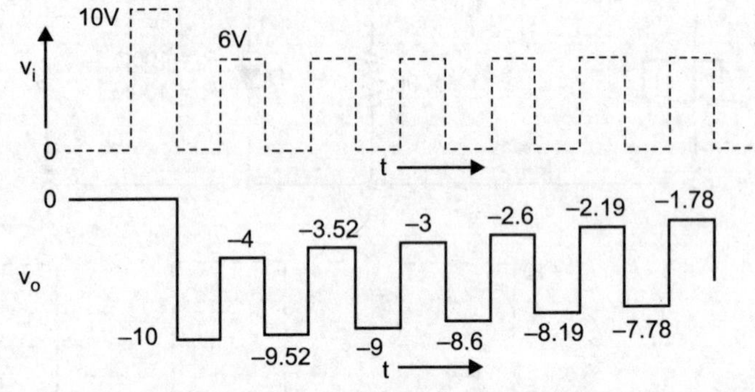

Fig. 17.32

Similarly, during the LOW time of the third cycle, the capacitor would further discharge to v_C = 9.52 $e^{-1/20}$ = – 9 V and during the LOW times of fourth, fifth and sixth cycles, the capacitor would successively discharge to – 8.6 V, – 8.19 V and– 7.78 V. The output waveform for the first six cycles is shown in Fig. 17.32.

It can be verified that the capacitor (C) would have discharged to (– 6) volts in eleven cycles. It is, therefore, ten cycles after the input amplitude drops that the output is properly clamped.

SUMMARY OF IMPORTANT FORMULAS

1. For a step input, the output voltage across a basic RC circuit is given by, $v_o = V\,(1 - e^{-t/RC})$

2. If the input step occurs at time $t = t_1$, then $v_o = V\,(1 - e^{-(t-t_1)RC})$

3. For an RC integrator circuit, $v_o = \frac{1}{RC}\int v_i\, dt$

4. 3 dB cut-off frequency for a low pass RC circuit, $f_c = \frac{1}{2\pi RC}$

5. For an RL differentiator circuit, $v_o = \frac{L}{R}\left(\frac{dv_i}{dt}\right)$

6. The optimum value of R in a clamper circuit for a diode with known forward biased and reverse biased characteristics is given by, $R = \sqrt{R_f . R_r}$

SELF EVALUATION EXERCISE

1. The input (v_1) to the *R-C* network of Fig. 17.33 (*a*) is shown in Fig. 17.33 (*b*). The output voltage waveform (v_o) for four different cases listed below are shown in a haphazard way in Figs. 17.33(*c*) to 17.33(*f*).

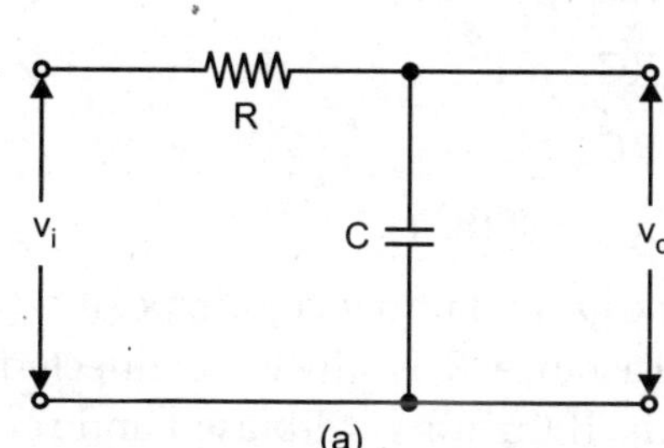

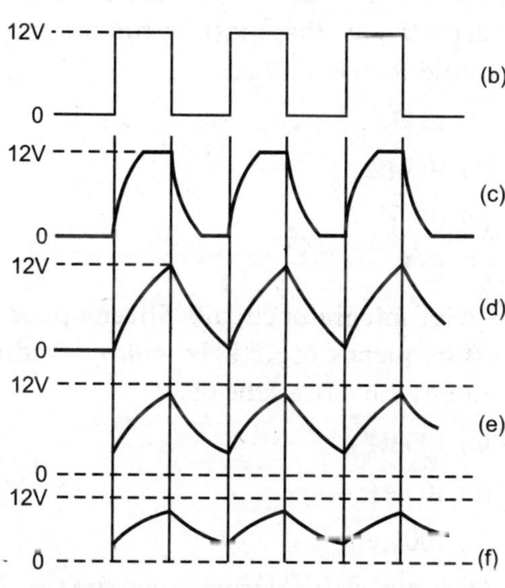

Fig. 17.33

Identify the correct output waveform for each one of the listed cases. Case (*i*) $RC = 2$ ms Case (*ii*) $RC = 20$ μs Case (*iii*) $RC = 5$ μs and Case (*iv*) $RC - 200$μs.

2. Fill in the blanks :

(*i*) An *RC* circuit to behave as a true differentiator for a symmetric square waveform of 1 kHz at the input should have a time constant of not greater than................................

(50μs, 1 ms, 20 ms)

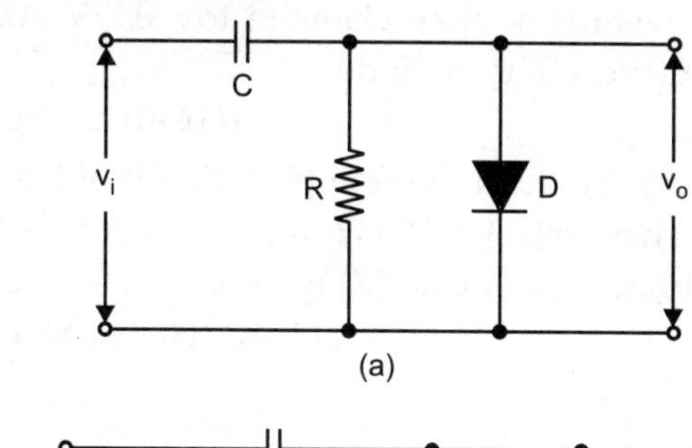

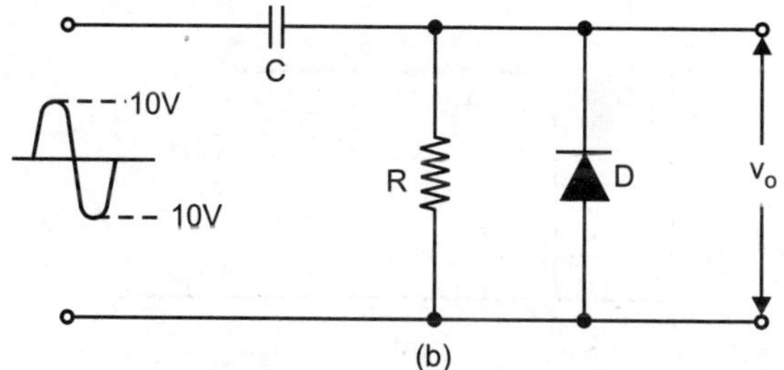

Fig. 17.34

(*ii*) If an *R-C* circuit is to behave as a true integrator circuit for a 1 kHz symmetric square waveform at the input, it must have a time constant not less than ....

(50 μs, 1 ms, 20 ms)

(*iii*) An integrator circuit is also a .... circuit.

(High pass, Low pass, Band pass)

(*iv*) A differentiator circuit is also a .... a circuit.

(High pass, Low pass, Band reject)

(*v*) An *R-C* differentiator circuit with a lower 3 dB cut-off of 3.5 kHz will respond to a step input with a rise time of ....

(100μs, practically nil value)

(*vi*) An *R-C* integrator circuit with an upper 3 dB cut-off of 3.5 kHz will respond to a step input with a rise time of ....

(100 μs, practically nil value)

(*vii*) The DC value of the waveform at the output of an *R-C* differentiator circuit is ....

(Zero, same as the DC value of the input waveform)

(*viii*) The positive extremities of the input waveform will be clamped to .... volt, when fed to the clamping circuit of Fig. 17.34 (*a*) Assume the diode to be a practical Germanium diode.
(0, 0.7, 0.2)

(*ix*) When a sinusoidal waveform with a peak amplitude of 10 V is fed to the clamping circuit of Fig. 17.34 (*b*) .... peaks of the waveform will be clamped to - 0.7 V. Assume practical Silicon diode.
(Positive, Negative)

(*x*) If the diode in the clamping circuit of Fig. 17.34 (*b*) has an (R_f) of 100 Ω and an (R_r) of 10 MΩ, the optimum value of (*R*) would be......................
(33 K, 100 K, 100 Ω 10 M)

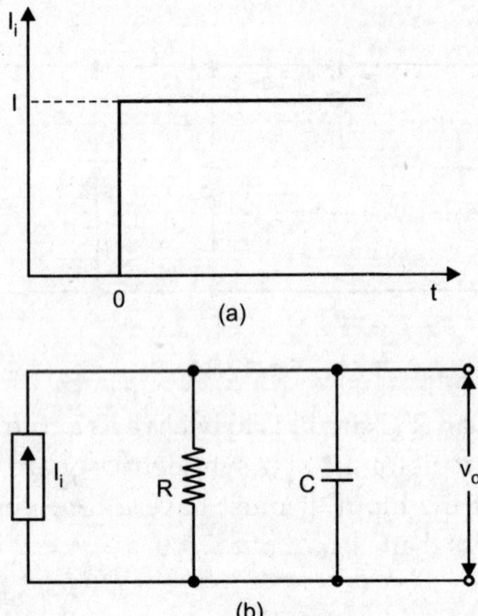

Fig. 17.35

3. The step current input of Fig. 17.35(*a*) is applied to the *R-C* circuit of Fig. 17.35(*b*). Sketch the output voltage waveform.
4. A low pass circuit can also possibly be
   (*a*) an integrator circuit
   (*b*) a differentiator circuit
   (*c*) either a differenttiaor or an integrator circuit
   (*d*) none of these
5. A high-pass circuit can also possibly be
   (*a*) an integrator circuit
   (*b*) a differentiator circuit
   (*c*) either a differentiator or an integrator circuit
   (*d*) none of these
6. A low-pass circuit with a relatively higher upper 3 dB cut-off frequency will
   (*a*) have relatively more sluggish step response
   (*b*) have relatively steeper step response
   (*c*) behave more like an integrator
   (*d*) none of these
7. A low-pass circuit is fed with a periodic waveform of time period T. For this circuit to function like an integrator, the necessary condition to be satisfied is
   (*a*) $RC = T$
   (*b*) $RC << T$
   (*c*) $RC >> T$
   (*d*) none of these
8. In the basic clamper circuit, positive or negative, a resistance *R* is always connected across the diode. If the forward-biased and reverse-biased resistances of the diode were 10 Ω and 10 MΩ, respectively, the most optimum value of *R* would then be
   (*a*) 10 Ω
   (*b*) 10 MΩ
   (*c*) 100 K
   (*d*) 10 K
9. An RC integrator circuit with an upper 3 dB cut-off frequency of 3.5 kHz will respond to a step input with a rise time of
   (*a*) 100 μs
   (*b*) 10 μs
   (*c*) 100 ms
   (*d*) indeterminate from given data
10. An RC differentiator circuit with a lower 3 dB cut-off frequency of 3.5 kHz will respond to a step input with a rise time of
   (*a*) 100 μs
   (*b*) 10 μs
   (*c*) 100 ms
   (*d*) practically nil value

ANSWERS TO SELF EVALUATION EXERCISE

1. Case (*i*) Fig. 17.33 (*f*)
   Case (*ii*) Fig. 17.33 (*d*)
   Case (*iii*) Fig. 17.33 (*c*)
   Case (*iv*) Fig 17.33 (*e*)
2. (*i*) 50 μs (*ii*) 20 ms (*iii*) Low pass (*iv*).High pass (*v*) Practically nil value (*vi*) 100 μs (*vii*) Zero (*viii*) 0.2 (*ix*) Negative (*x*) 33 K.
3. Fig 17.36 shows the output waveform.

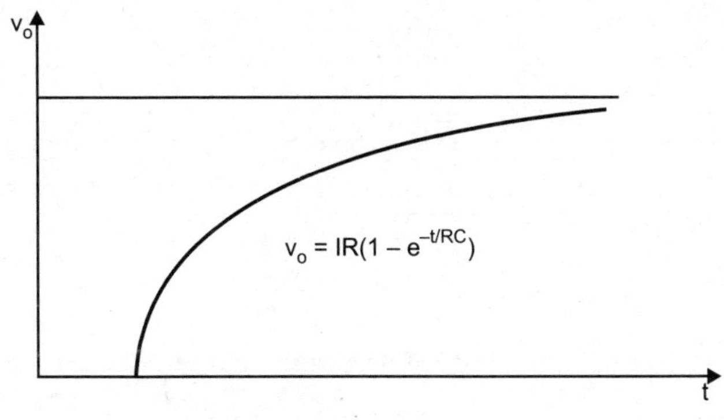

Fig. 17.36

4. (*a*) **5.** (*b*) **6.** (*b*) **7.** (*c*) **8.** (*d*) **9.** (*a*) **10.** (*d*)

Chapter 18

Operational Amplifiers

Fundamentals—Types—Characteristic Parameters

Learning Objectives

After completing the chapter, you will learn the following topics:

1. Internal Circuit of Operational Amplifier (Opamp)
2. Basic Characteristics of Ideal and Practical Operational Amplifiers
3. General purpose Opamps
4. Opamp Comparators
5. Norton Opamps
6. Low Noise Opamps
7. Video Opamps
8. Instrumentation Opamps
9. Isolation Opamps
10. Characteristic Parameters of Opamps
11. Relevance of Different Parameters to Different Application Circuits

Q.1. What is an opamp? What are its main attributes? Why it is called an operational amplifier?

Ans. An operational amplifier popularly known as an opamp is basically a high-gain differential amplifier capable of amplifying signals right down to DC. The capability of the opamp to amplify signals down to DC lies in the use of direct coupling mechanism in the internal architecture of the device. That is why, it is also called a direct-coupled or a DC amplifier. The other main attributes of an opamp are very high input impedance, very low output impedance, very large bandwidth, extremely high value of open-loop gain and so on. Figure-18.1 shows the circuit representation of an opamp. An opamp is usually a two input and one output device along with two power supply terminals one for positive supply and the other for negative supply with both having a common ground. The word "usually" is used here, as there are opamps with differential outputs. There are also opamps with single power supply terminal.

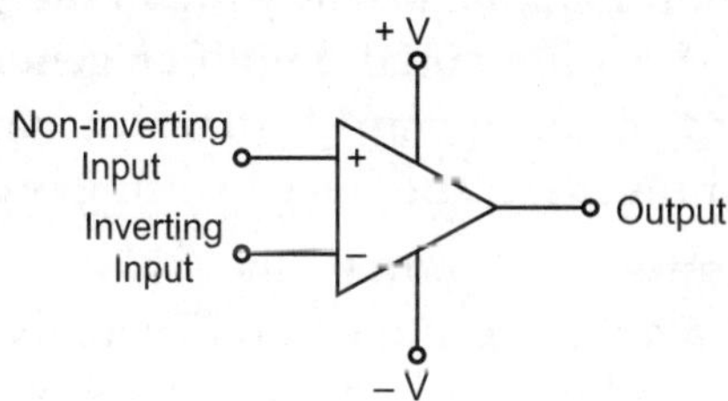

Fig. 18.1

It is called an operational amplifier, as it was originally conceived as an analog computation building block that could be used conveniently to perform mathematical functions like addition, subtraction, integration, differentiation and so on. Today opamps have unlimited applications. An opamp fits in any conceivable circuit application from analog computation to building amplifiers and oscillators, from active filters to phase shifters, from comparators to voltage regulators, from function generators to gyrators and so on and so forth.

Q.2. What is the basic circuit architecture found inside a general purpose opamp? Briefly describe the functioning of an opamp with the help of a block schematic representation.

Ans. The inside circuit of a typical opamp consists of a differential amplifier input stage that is followed by one or more differential amplifier gain stages. Generally, the differential input amplifier provides a differential output and the gain stages provide a single-ended output. The number of gain stages may be different for different opamps. The output of the final gain stage feeds the input of a class-B push–pull output stage through an emitter-follower buffer stage. The output stage acts like a level translator and output driver. Figure-18.2 shows the block schematic arrangement of the internal circuit of a typical opamp with a differential input and a single-ended output. Now,

$$V_{out} = A_{OL} \times V_{in}$$

A_{OL} is the open-loop voltage gain of the opamp, that is, voltage gain in the absence of any negative feedback.

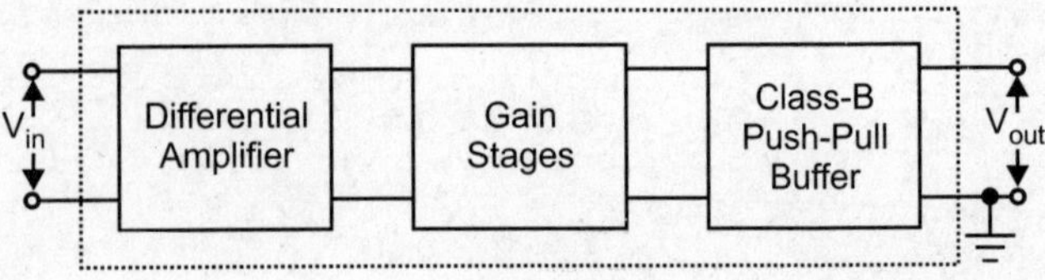

Fig. 18.2

Q.3. Why a differential amplifier stage is preferred as the input stage in an opamp? With the help of a typical circuit diagram, briefly explain the principle of operation of a differential amplifier generally found at the input of an opamp. Explain in particular the existence of inverting and non-inverting inputs.

Ans. Figure-18.3 shows the basic single-stage differential amplifier configuration. Note that the designers of IC amplifiers do not have the luxury of using coupling and bypass capacitors, as it is not practical to fabricate large-value capacitors (larger than 50 pF) on the IC chip. This is one of the reasons that differential amplifier which requires no coupling and emitter bypass capacitors is the designers' preferred choice as the opamp front-end circuit. Even large-value resistors are also not easy to fabricate in the IC form. Large-value resistances are simulated by suitable transistor configurations.

The differential amplifier configuration is also sometimes called a long tail pair, as the two transistors share a common-emitter resistor. The current through this resistor is called the tail current. Figure 18.3 shows the configuration of a differential input–differential output differential amplifier. In the circuit of Figure-18.3, if the output is single ended and is taken from the collector of transistor Q_2, then the base terminal of Q_1 is the non-inverting input and base terminal of transistor Q_2 is the inverting input. The output is in-phase with the signal applied at non-inverting input and out-of-phase with the signal applied at inverting input. This further implies that if two different input signals are applied to inverting and non-inverting inputs, the output is given by $A_d \times (V_1 - V_2)$.

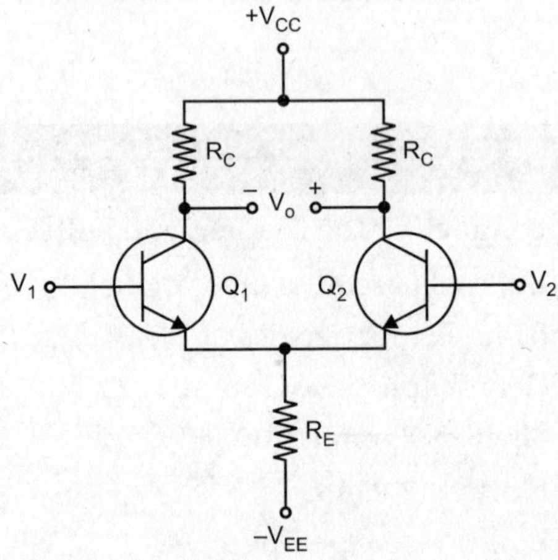

Fig. 18.3

For DC analysis, if we ignore the voltage drop across base-emitter junctions of the two transistors, the common-emitter point for all practical purposes is at ground potential. That is, the tail current is given by V_{EE}/R_E, which is constant. The tail current is divided into two separate paths, one through Q_1 and the other through Q_2. As the two halves are symmetrical, total current sharing depends upon the voltages applied to the two inputs.

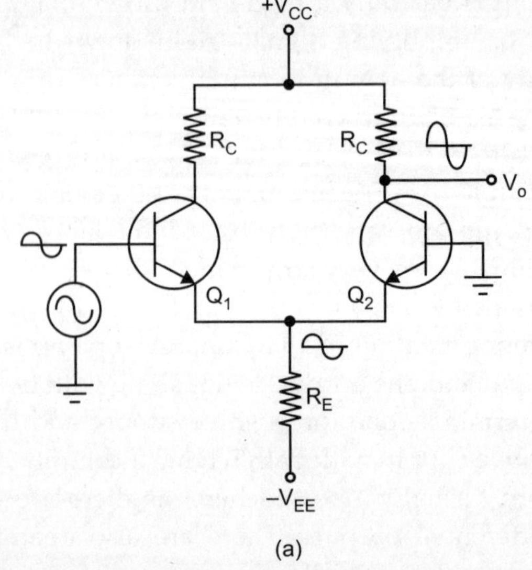

(a)

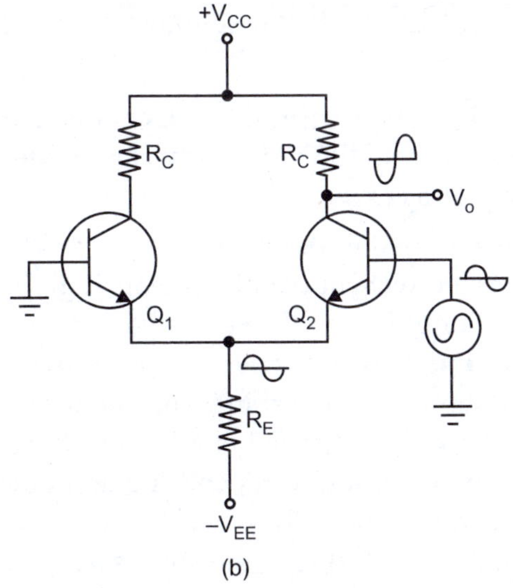

(b)

Fig. 18.4

Figure-18.4 shows the circuit diagram and signal wave shapes at relevant points, when the signal is applied to inverting and non-inverting inputs separately.

Q.4. A true differential amplifier should amplify only the differential input and not the common mode input. How does a differential amplifier stage achieve this? Or in other words, how would you modify the basic differential amplifier stage shown in Figure-18.3 to make it a true differential amplifier?

Ans. One of the key parameters of the differential amplifier is its ability to amplify differential input and its insensitivity to common mode input. It may be mentioned here that if the inverting and non-inverting inputs were applied DC voltages V_1 and V_2 respectively; then the differential input will be equal to $(V_1 - V_2)$ and common mode input will be equal to $(V_1 + V_2)/2$. This is expressed in terms of Common Mode Rejection Ratio (CMRR); which is nothing but ratio of differential gain to common mode gain. CMRR is described in detail in a subsequent question on opamp parameters. It can be verified that for the differential amplifier of Figure-18.3, differential gain is given by R_C/r_e' and the common mode gain is given by $R_C/2R_E$. CMRR therefore is given by $2R_E/r_e'$, where r_e' is the dynamic resistance of the base-emitter junctions of the two transistors.

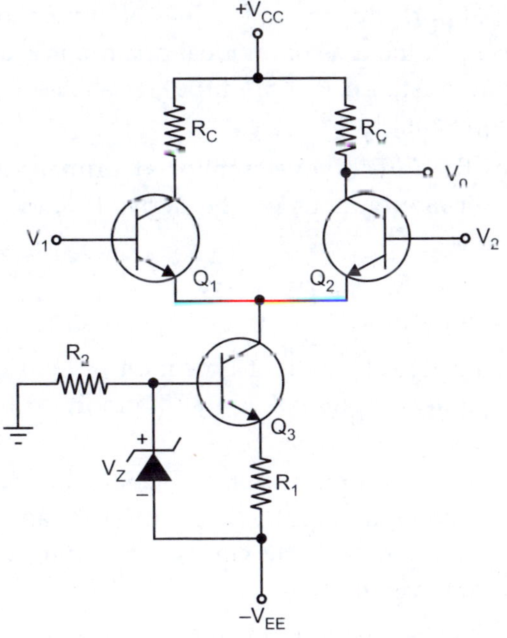

Fig. 18.5

It is evident from the expression for CMRR that the value of R_E should be as high as possible. That is why, in practical opamp circuits, R_E is replaced by a constant current source. A current mirror configuration is also used to implement a constant current source. Figures-18.5 and 18.6 respectively show differential amplifier circuits with a conventional constant current source and a current mirror configuration. We are familiar with the functioning of conventional constant current sources. The current mirror works as follows. If the diode connected across the base-emitter junction and the base-emitter junction diode of the transistor had matched current-voltage characteristics, the collector current of the transistor equals the current through resistor R. In other words, the collector current is a mirror image of the resistor current. As it is very easy to match the current-voltage characteristics of a diode and base-emitter junction diode of a transistor in ICs because of both being on same chip; current mirror configuration is commonly used as current source and active load in IC opamps. Figure-18.7 shows a differential amplifier stage with current mirrors being used instead of emitter resistor R_E and collector resistor R_C.

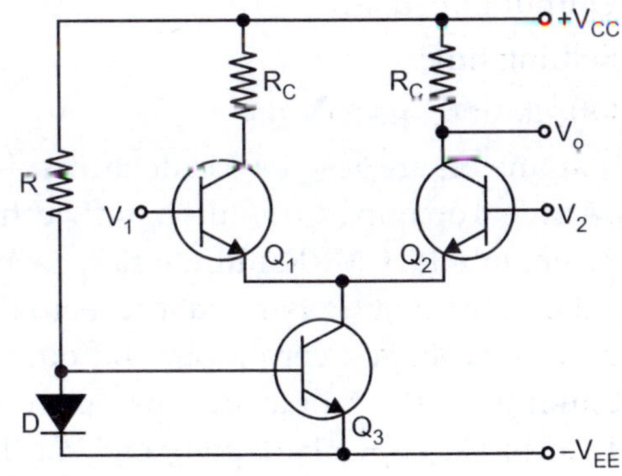

Fig. 18.6

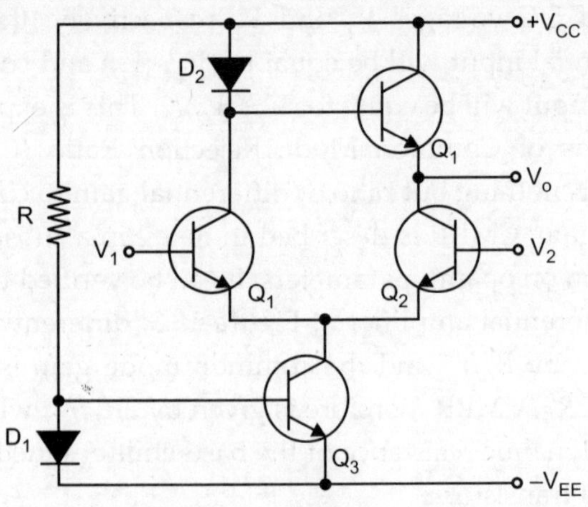

Fig. 18.7

Q.5. Which basic amplifier configurations are used as gain stages and the output stage in an opamp?

Ans. As outlined earlier, common-emitter amplifier and class-B push–pull amplifier constitute the gain and final stages of the opamp respectively. These have been described in detail in and earlier chapter on amplifiers (Chapter-13). Common-emitter amplifier stage or a cascade arrangement of more than one such stage provides most of the gain of the opamp. These stages when used inside an opamp, will have active loads.

Q.6. What are the important performance parameters of an opamp? How do you compare an ideal opamp with a practical opamp in terms of these parameters?

Ans. Important performance parameters of an opamp include the following:

1. Bandwidth
2. Slew rate
3. Open-loop gain
4. Common mode rejection ratio (CMRR)
5. Power supply rejection ratio (PSRR)
6. Input impedance
7. Output impedance
8. Settling time
9. Offsets and offset drifts

These parameters are described in detail in subsequent questions. An ideal opamp has infinite bandwidth; infinite open loop gain, infinite CMRR, infinite slew rate, infinite input impedance, zero output impedance, zero DC input and output offset voltages, zero input and output offset drifts and infinite PSRR. A practical opamp would have high bandwidth; high open loop gain, high CMRR, high input impedance, low output impedance, low DC input and output offset voltages, low input and output offset voltage drifts, high PSRR, low settling time and high slew rate.

Q.7. With the help of equivalent circuit of an opamp, briefly describe the difference between an ideal opamp and a practical opamp.

Ans. As outlined in the answer to an earlier question, an opamp is a direct-coupled high gain, high bandwidth differential amplifier with very high value of input impedance and very low value of output impedance. Figure-18.8 shows the Thevenin's equivalent circuit model of a generalised amplifier fed at the input from a source having a source resistance R_S and the amplified output feeding a load resistance R_L. Owing to finite values of R_i and R_o, the effective value of the gain is less than what it would have been had R_i and R_o been infinite and zero, respectively.

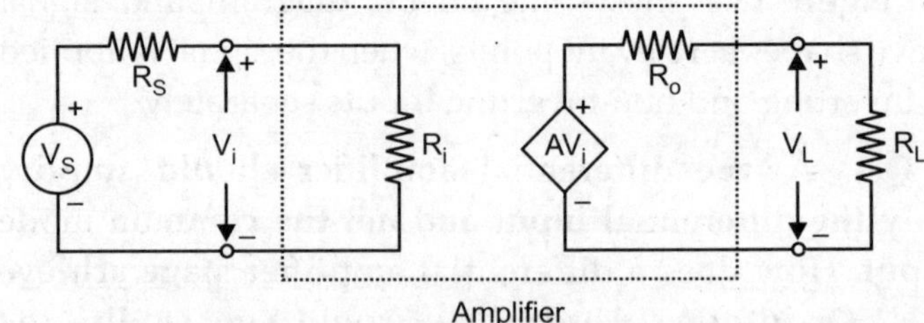

Fig. 18.8

Figure-18.9 shows the Thevenin's equivalent of the opamp. V_I and V_{NI} are, respectively, inverting and non-inverting inputs and A_d is the open-loop differential voltage gain. This is the equivalent circuit model of a practical opamp. There are loading effects at the input and output ports due to finite values of input and output resistances. In the case of an ideal opamp, it is assumed that Input resistance, $R_i = \infty$; Output resistance, $R_o = 0$ and Open-loop gain, $A_d = \infty$

From the three above-mentioned primary assumptions, other assumptions can be derived. These include the following:

1. Since $R_i = \infty$, $I_I = I_{NI} = 0$.
2. Since $R_o = 0$, $V_o = A_d \times V_d$
3. For linear mode of operation of opamp and a finite output voltage and infinite differential gain, $V_d = 0$.
4. Since output voltage depends only on differential input voltage, it rejects any voltage common to both inputs. Therefore, common mode gain = 0.
5. Bandwidth and slew rate are also infinite, as no frequency dependencies are assumed.

6. Drift is also zero as there are no changes in performance over time, temperature, power supply variations and so on.

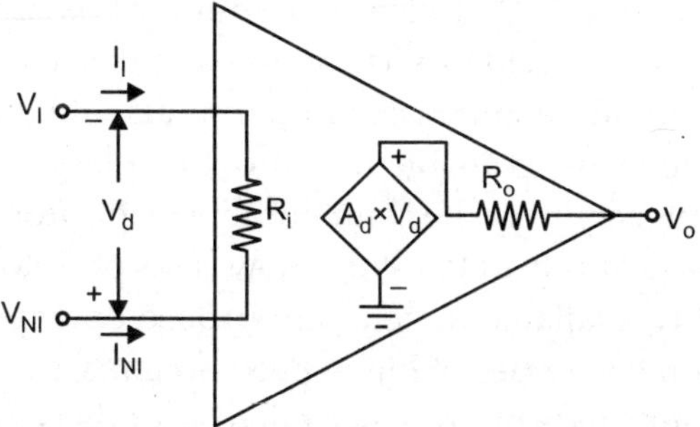

Fig. 18.10

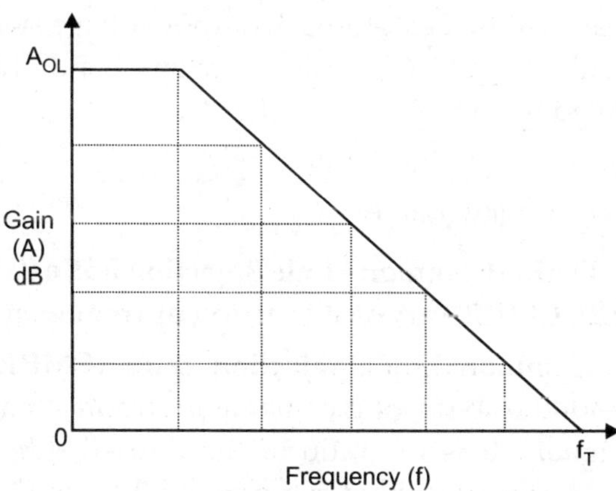

Fig. 18.11

Q.8. What do you understand by Open Loop Gain, Closed Loop Gain and Loop Gain? What is the significance of Loop Gain? How does the open loop gain of an opamp vary with frequency?

Ans. Open-loop gain is the ratio of single-ended output to the differential input. This parameter has a great bearing on the gain accuracy specification of the opamp wired as an amplifier. The **closed loop gain** is defined when the opamp is used with negative feedback, such as in inverting and non-inverting amplifier configurations. The ratio of the open-loop gain to the closed-loop gain (which depends upon the application circuit) is called the **loop gain.** Accuracy at any given frequency depends heavily on the magnitude of the loop gain at that frequency. The magnitude of loop gain at a given frequency depends directly on the value of the open-loop gain at that frequency, as the value of closed-loop gain for a given application circuit is fixed. Figure-18.11 shows the open-loop gain versus frequency curve of an opamp. As a thumb rule, the gain error at any given frequency is given by the ratio of the closed-loop gain to the open-loop gain. Thus a higher open-loop gain gives a smaller error for a given closed-loop gain.

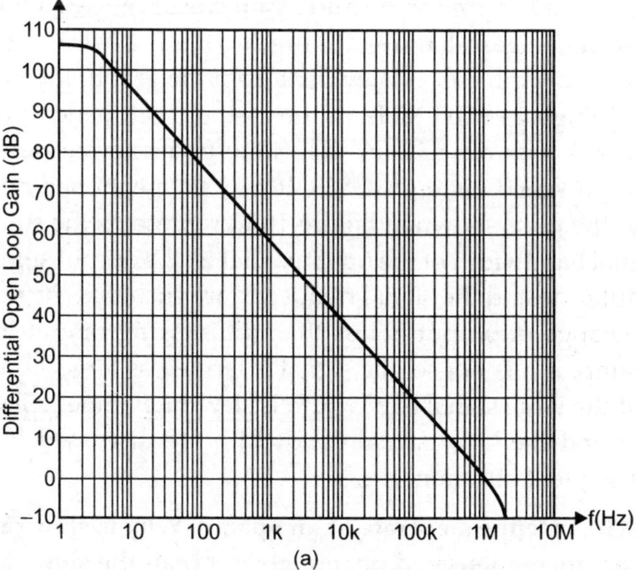

(a)

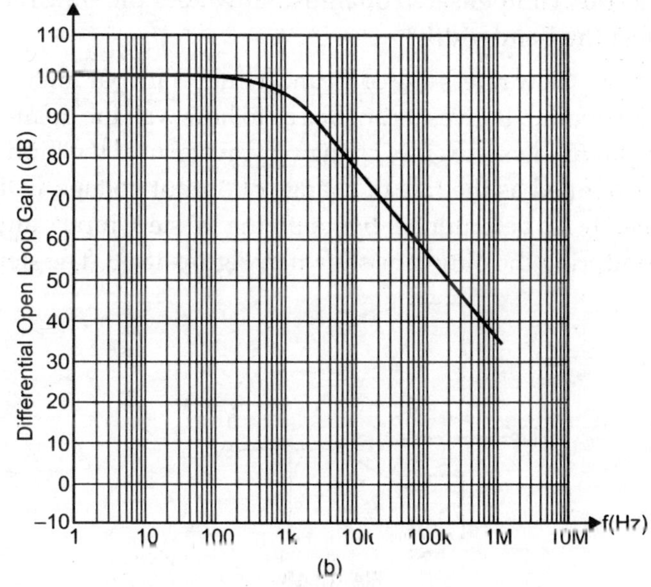

(b)

Fig. 18.12

Q.9. How do you define the bandwidth of an opamp? Can it be different for different types of opamps? What is unity gain cross-over frequency?

Ans. Bandwidth of an opamp tells us about the range of frequencies it can amplify for a given amplifier gain. The frequency response curve of a typical opamp looks like the graph of Figure-18.12 (*a*) shows the frequency response of the general-purpose opamp 741 and the graph has been reproduced from the data sheet of the said opamp. The high-frequency roll-off is attributed to capacitive effects appearing in shunt. Beyond cut-off, the frequency falls at a rate of 6 dB per octave or 20 dB per decade. When the opamp is used in the closed-loop mode; the bandwidth increases at the cost of the gain. The bandwidth is usually expressed in terms of the **unity gain crossover frequency** (also called **gain-bandwidth product**). It is 1 MHz in the case of opamp 741 as is evident from the graph. It could be as high as 1500 MHz in the case of high bandwidth opamps. Figure-18.12 (*b*) shows the frequency response of high-speed opamp AD829. It may be mentioned here that the plots shown in Figure-18.12 represents the small signal bandwidth of the opamp. That is, the output signal amplitude and the signal frequency are such that the rate of change of output is less than the slewing rate of the opamp. If it is not so, the signal is termed as large signal and the large signal bandwidth is slew-rate limited. Slew rate and the large signal bandwidth are discussed in a subsequent question.

Q.10. Define slew rate of an opamp. Why is slew rate of an uncompensated opamp greater than the slew rate of a fully compensated opamps? How does the slew rate affect the bandwidth?

Ans. Slew rate is one of the most important parameters of an opamp. It gives us an idea as to how well the opamp output follows a rapidly changing waveform at the input. It is defined as the rate of change of output voltage with time. It is determined by applying a step input and monitoring the output as shown in Figure-18.13. The step

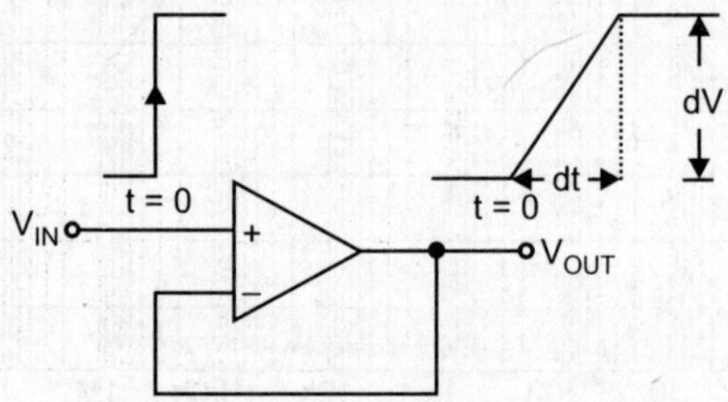

Fig. 18.13

input simulates the large signal conditions. The incapability of the opamp to follow rapidly rising and falling input is, respectively, due to the minimum charge and discharge times required by an internally connected capacitor across the output. This capacitor has a value that guarantees stable operation of the opamp down to a gain of unity. This implies that the amplifier will give a stable operation and will not get into oscillations for any gain value as unity gain happens to be the worst condition. Now if it is so, then the amplifier is said to be fully internally compensated. Opamp 741 is an example.

In the case of an uncompensated opamp, this capacitor needs to be connected externally. In that case, we have a control on the slew-rate specification. We can sacrifice stability to achieve a higher slew rate. For instance, if we know that we are never going to use our opamp for a gain less than 10, we could afford to connect a smaller capacitor and thus, get a higher slew rate. But all this is possible only, when we decide to use an uncompensated opamp. In an internally compensated opamp also, higher slew-rate versions are available that provide a large charging current internally for the compensation capacitor. Opamp 741 has a slew rate of 0.5 *V*/μs. Slew rates of upto 10 *V*/μs are usually available in general-purpose opamps. In the case of some varieties of high-speed opamps, slew rate as high as hundreds of volts per microseconds is available. EL 2444 (high-speed quad opamp from M/s Elantec) offers a slew rate of 325 *V*/μs. Slew rate limits the large signal bandwidth. Peak-to-peak output voltage swing for a sinusoidal signal ($V_{p\text{-}t\text{-}p}$), slew rate and bandwidth are interrelated by the following equation:

$$\text{Bandwidth (highest frequency, } f_{MAX}) = \text{Slew rate}/(\pi \times V_{p\text{-}t\text{-}p}).$$

Q.11. Define Common Mode Rejection Ratio (CMRR). How does CMRR vary as a function of frequency?

Ans. Common mode rejection ratio (CMRR) is a measure of the ability of the opamp to suppress common mode signals. It is the ratio of the desired differential gain (A_d) to the undesired common mode gain (A_c). The ratio CMRR is usually expressed as CMR given by 20 log (A_d/A_c) dB. A lower CMRR or CMR reflects in terms of smaller variation in the output of a differential amplifier due to variation in the common mode input, when the differential input stays put. The common mode input is the average value of the two inputs. The common mode input affects the bias point of the input differential pair. Owing to lack of perfect symmetry in the two halves of the input differential pair, change in bias point changes the offset voltage. This, in turn, changes the output voltage. CMRR is also defined as the ratio of the change in the common mode input to the corresponding change in the output offset voltage.

CMRR is always specified for a given input voltage range. Exceeding the input voltage range would degrade the CMRR specification. In some opamps, the input voltage range is specified separately which implies that the given value of CMRR is guaranteed over the listed input voltage range. CMRR as published in data sheet is a DC parameter. CMRR when graphed versus frequency falls with increase in frequency as shown in Figure 18.14. The graph shown in the figure is taken from data sheet of opamp 741.

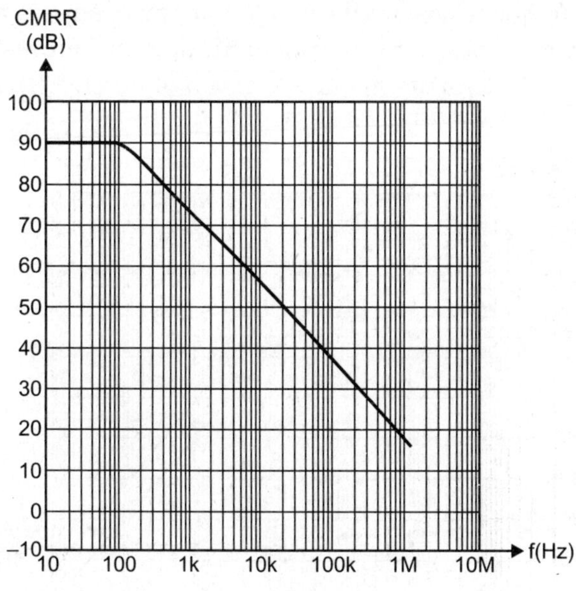

Fig. 18.14

Q.12. Define input impedance of an opamp. How does the open loop input impedance differ from input impedance of opamp in closed loop configuration?

Ans. Input impedance is the impedance looking into the input terminals of the opamp and is mostly expressed in terms of resistance only. The effective input impedance is, however, different from what is specified in the data sheets, when the opamp is used in the closed-loop mode. In the inverting amplifier configuration, the effective input impedance equals the input resistance connected externally from source of input signal to the inverting input terminal of the opamp. In the non-inverting amplifier configuration, it equals the product of loop gain and the specified input impedance.

Q.13. Define output impedance of an opamp. How does it affect the output signal amplitude?

Ans. Output impedance is defined as the impedance between the output terminal of the opamp and ground. Common-emitter (bipolar) and common-source (CMOS) output stages used in rail-to-rail output opamps have higher output impedance than emitter-follower output stages. Output impedance becomes a critical parameter, when using rail-to-rail output opamps to drive heavy loads. If the load were mainly resistive, it would decide how close the output can be to the rails. If the load were capacitive, the additional phase shift caused will erode the phase margin. Figure-18.15 shows the effect of output impedance on the output signal assuming that the output impedance is resistive in the case of a predominantly resistive load R_L [Figure-18.15 (*a*)] and a capacitive load C_L [Figure-18.15 (*b*)].

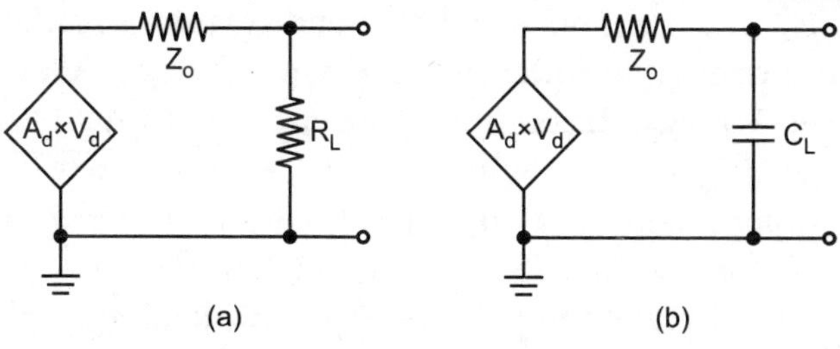

Fig. 18.15

In the case of resistive load, output signal amplitude is given by:

$$V_o = A_d \times V_d \times \left[\frac{R_L}{R_L + Z_o}\right]$$

In the case of capacitive load, output signal amplitude is given by:

$$V_o = A_d \times V_d \times \left[\frac{1}{j(f / f_o) + 1}\right]$$

Where, $f_o = 1/2\pi Z_o C_L$.

Q.14. Define settling time of an opamp? Does it depend upon closed loop gain? In which application circuits does settling time become critical?

Ans. Settling time is a parameter specified in the case of high-speed opamps or the opamps with a high value of gain-bandwidth product. It gives the response of the opamp to larger step inputs. It is expressed as the time taken by the opamp output to settle within a specified percentage of the final value, usually 0.1% or 0.01% of the final expected value, in response to a step at its input (Figure-18.16). The settling time is usually specified for

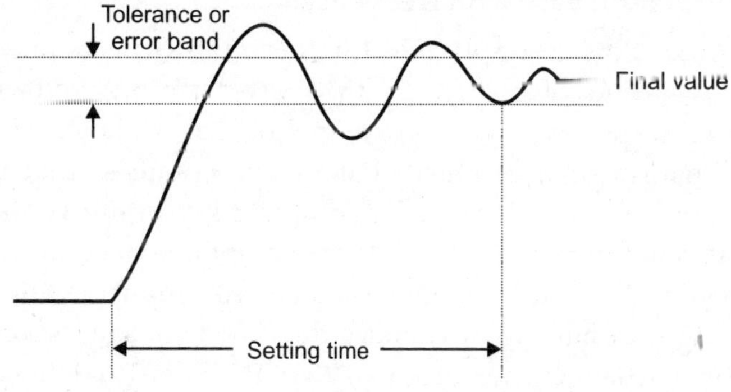

Fig. 18.16

opamp wired as a unity gain amplifier and it worsens for a closed-loop gain greater than one. Settling time is a design issue in data acquisition circuits, when signals are changing rapidly.

Q.16. What are different offset voltages and currents associated with opamps?

Ans. Different offset voltages and currents associated with opamps include **Input Offset Voltage, Output Offset Voltage, Input Bias Current** and **Input Offset Current. Offset** is a commonly used term with reference to opamps. An ideal opamp should produce a zero output for a zero differential input. But it is not so in the case of real opamps. It is observed that we need to apply a DC differential voltage externally to get a zero output. This externally applied input is then the **input offset voltage.** This parameter may be as large as 5 mV in general-purpose opamps and as small as 200 μV in low offset opamps. The input offset voltage is often a function of power supply voltages. This variation is expressed in terms of Power Supply Rejection Ratio (PSRR). **Output offset voltage** is the voltage at the output with both the input terminals grounded. Another offset parameter is the **input offset current**. It is the difference between the two bias currents flowing towards the inputs of the opamp. Yet another important opamp parameter is the **input bias current.** It is defined as the average of the two bias currents flowing into the two input terminals of the opamp. The listed value of input and output offset voltages and input offset and bias currents tend to drift with temperature. This drift is also specified in the data sheets. Some general-purpose opamps (741 for instance) have a provision for externally nullifying the input offset voltage by usually connecting the fixed terminals of a potentiometer of a given resistance value across the designated terminals and connecting the center terminal to the negative supply voltage.

Q.17. Define Power supply Rejection Ratio (PSRR). How does it vary with frequency?

Ans. Power supply rejection ratio (PSRR) is defined as the ratio of change in the power supply voltage to corresponding change in the output voltage. The mechanism that produces PSRR is the same as that is responsible for CMRR. The power supply affects the bias point of the input differential amplifier stage of the opamp. Owing to inherent mismatch in the input circuitry, changes in bias point changes the offset voltage, which in turn, changes the output voltage. PSRR is also defined as the ratio of the change in the input offset voltage to the change in one of the power supply voltages, with the other power supply voltage held constant.

PSRR like CMRR too is a DC parameter and its value falls with increase in frequency. The parameter is particularly significant, as switched mode power supplies can have noise in the frequency range of 20 kHz to 200 kHz and even higher. PSRR is almost zero at these frequencies with the result that power supply noise appears as noise at the output of the opamp. Figure-18.17 shows PSRR versus frequency graph of opamp type AD 829.

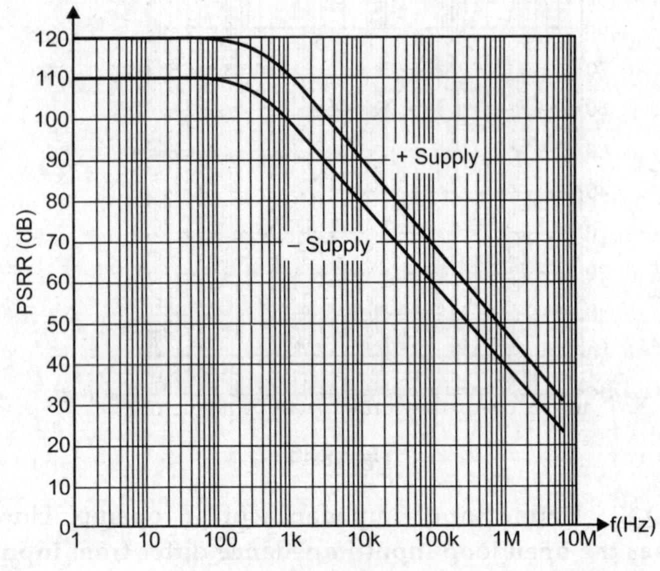

Fig. 18.17

Q.18. How do you choose a suitable opamp for your application circuit? How does the performance parameters described in previous questions influence the choice of a suitable opamp for a given application? Illustrate your answer with suitable examples.

Ans. Based on the circuit objectives, the first thing to be decided is the opamp category that would do the required job economically. Having made up our mind on the broad class of opamps, that is, we need a general-purpose opamp or a high-speed opamp or for that matter an instrumentation opamp, we can go through the specifications of the devices available in the chosen category to choose a device that suits our needs the best. We should give due attention to both AC as well as DC considerations. One of the major AC considerations, for instance, is the desired loop gain. As an example, if we wanted the opamp to yield an accuracy of 0.1% while amplifying an AC signal of 10 kHz by a gain of 10, then the opamp must have an open-loop gain of at least 10,000 at the operating frequency of 10 kHz. Opamp 741 will certainly not serve the purpose here. The open-loop gain for a given frequency can be verified from the gain versus frequency plot.

Similarly, slew rate must be high enough to follow the fastest changing input signal without causing distortion. It is always good as a thumb rule to choose an opamp that has a minimum slew rate of 25% larger than the fastest rate of change in the input signal.

In addition to open loop gain, input offset voltage and the input bias current are the other important parameters to be considered while choosing the right opamp in the precision category. Input bias current is particularly important, when the source of input signal has relatively higher impedance. FET-input opamps or instrumentation amplifiers deserve an attention in such cases.

Q.19. We have been talking so far about opamps in general. Opamps are often grouped into different categories based on their performance specifications and application areas. Briefly describe each one of these.

Ans. Opamps can be categorised on the basis of the performance specifications and consequently their application area and also, on the basis of their internal structure as General purpose opamps, High speed and High bandwidth opamps, Precision opamps, Instrumentation opamps, Norton opamps (also known as current differencing opamps), Opamp comparators and Isolation amplifiers.

General Purpose Opamps. The term General purpose opamp is generally used with reference to that category of opamps which have moderate or say reasonably good values for all the key parameters. In case of general purpose opamps, none of the major specifications, as outlined earlier can be said to be exceptionally good or state-of-the-art and also, none of the specifications is extremely poor. Within the broad category of general purpose opamps, some type numbers may have slightly better overall performance specifications and are, thus, called as high performance general purpose opamps. A general purpose opamp of the type 741 may have an open loop gain of typically 10^4 to 10^5, and input impedance of 0.5 to 1 megaohm, an output impedance of 50 to 100 ohms, a CMRR of the order of 70 to 100 dB, output offset voltage of the order of a few millivolts and an open loop bandwidth of 1 MHz. Those general purpose opamps that have a JFET based input differential amplifier stage inside the opamp have relatively much larger input impedance (typically 10^{12} ohms) and much better noise specifications. So, general purpose opamps due to overall good performance specifications are used in a wide range of applications in low cost designs.

JFET-input opamps (opamps with the input differential amplifier stage constructed with JFETs) are preferred in applications, where high input impedance and low noise are the main considerations.

High speed, Wide bandwidth opamps have high slew rate and bandwidth specifications. Today opamps with slew rate in hundreds of volts per microsecond and bandwidth in hundreds of MHz are commercially available. Precision opamps have extremely low offsets and a very high value of open loop differential gain. The output offset voltage in such opamps may be of the order of several tens of microvolts. The open loop gain approaches 120 dB.

Power Opamps. The power supply voltages and consequently the output voltage swing are no longer limited to ± 18 volts or so. Opamps with supply voltage ratings of the order of several hundred volts and current delivering capability of the order of several amperes are also available. However, these devices are very expensive and are manufactured by a few select manufacturers internationally. These are called **Power Opamps.**

Opamp Comparators are opamps specially designed for use in comparator applications. These have much faster response time (typically from a few nanoseconds to several tens of nanoseconds) than that of a conventional general purpose opamp (typically 1 microsecond). Comparator in which one of the inputs is normally a reference voltage, switch between two states at the output depending upon whether the other input is higher or lower than the reference. In case of general purpose opamps being used as comparators, the output switches between two levels that are fixed and depend upon supply voltages. These levels may not be compatible with the load requirements, which in most of the applications is a logic circuit. There might therefore be a need for the comparator output to be compatible to certain logic family like TTL or CMOS. The response time and the output compatibility happen to be the two most important specifications in case of opamp comparators. A comparator forms the basic building block in circuits like zero crossing detectors, level detectors, Schmitt triggers, square and triangle waveform generators.

A **Norton opamp** also known as the current differencing opamp differs from a conventional opamp in its internal circuit design. While in a conventional opamp, the input stage is a differential amplifier to achieve inverting and non-inverting input functions, in case of a norton opamp, the non-inverting input function is derived from the inverting input function by using a configuration called a current mirror. The non-inverting input current is derived from the one entering at the inverting input. While we talk about differential input voltage in case of a conventional opamp, it is differential input current in case of a norton opamp. Single supply operation and the fact that most of the general purpose opamp applications can be realised without significantly sacrificing performance characteristics plus low cost per opamp make norton opamp a very attractive choice. A norton opamp has a definite edge over a general purpose opamp (of 741 type) in single supply operation. When a general purpose opamp is used on a single supply, the minimum input voltage that can be applied while ensuring that the opamp responds to a differential input is limited by the minimum common mode input range (typically + 2V in general purpose opamps). This is rather large. In addition, general purpose opamps, when used on single supply have poorer output voltage swing. Norton opamps have no such limitations. A norton opamp is represented in a circuit by the circuit symbol of Fig. 18.18 (*b*). The conventional opamp is represented by the circuit symbol of Fig. 18.18 (*a*).

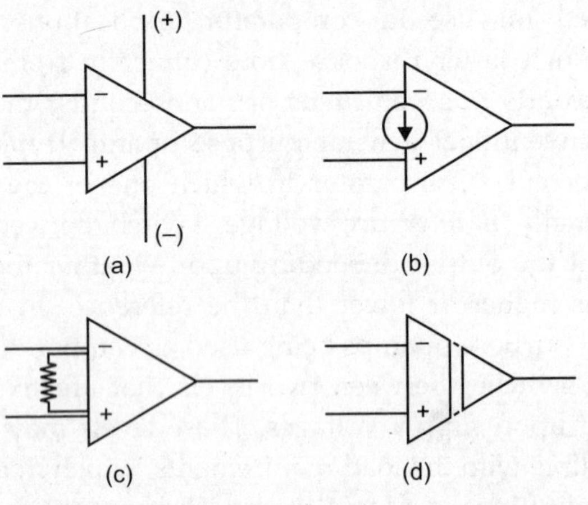

Fig. 18.18

An **Instrumentation Amplifier** is a differential amplifier with a very high input im-pedance, very large CMRR and extremely low values of offsets and offset drifts. Like an opamp of the type described in the earlier paragraphs, it is a differential input single ended output gain block. Gain in an instrumentation opamp, if we so call it, is usually set with a single resistor, whereas in a conventional opamp, the same is achieved with the help of two resistors. Instrumentation opamps are available in the single IC package. Fig. 18.18 (*c*) shows its circuit representation. Typical internal schematic of such a device is shown in Fig. 18.19. As is clear, it is a combination of three opamps. The output opamp has been wired as a differential amplifier with its non-inverting and inverting inputs fed from the outputs of the other two opamps wired as buffers (unity gain non-inverting amplifiers). The gain setting resistor is connected external to the device. Thus, the two major problems in a conventional opamp differential amplifier, namely the low effective input impedance loading on to the signal source with a comparatively high output impedance and the gain setting requiring simultaneous adjustment of two resistors leading to inaccuracies are totally overcome.

The essential characteristics of instrumentation amplifiers, as outlined above make them very useful in amplifying low level differential signals with a high common mode content. Transducer, such as thermocouples, biological probes, strain gauge bridges and current shunts produce small differential signals superimposed on common mode bias voltages. Instrumentation amplifier is an excellent choice in such cases for meaningful acquisition and analysis of data in addition to its use in general purpose applications requiring a precision high quality differential amplifier.

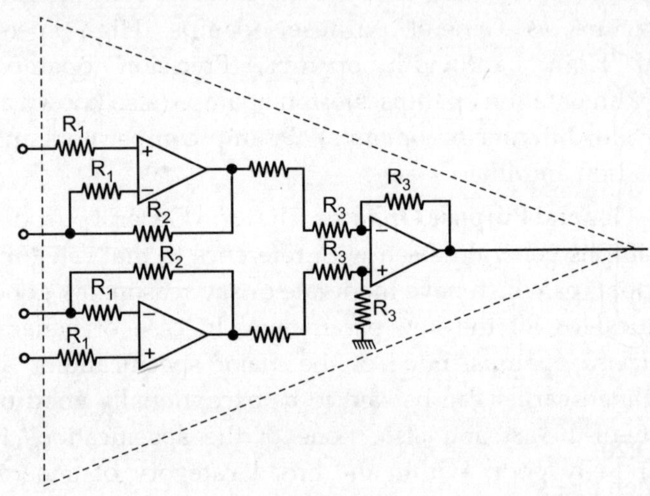

Fig. 18.19

An **Isolation Amplifier** is again a differential input signal ended output amplifier with its output electrically isolated from the input. Isolation impedance as high as 10^{12} ohms and an isolation voltage of about 1000 volts are common.

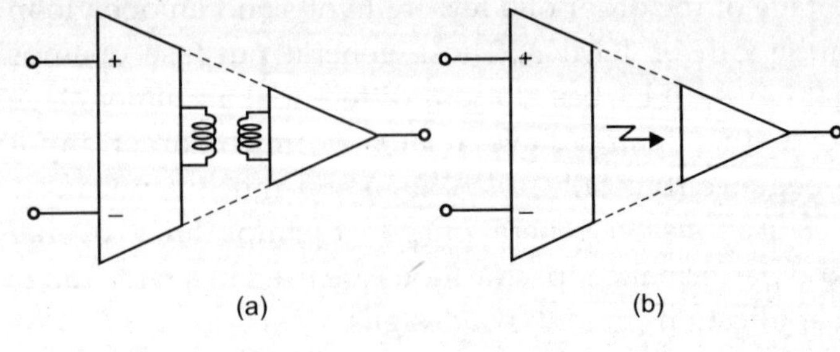

Fig. 18.20

The differential amplifier in an isolation amplifier may be an ordinary differential amplifier or an instrumentation amplifier. Fig. 18.18 (*d*) shows the circuit symbol. There are transformer coupled isolation amplifiers [Fig. 18.20 (*a*)] mainly used in applications where linearity, gain accuracy etc. are important and there are optically coupled isolation amplifiers [Fig. 18.20 (*b*)] used in applications, where speed and bandwidth are the criteria.

Isolation amplifiers are used for those applications, where electrical isolation between the source and the output is desirable. For instance, in medical instrumentation where the transducer or sensors are in physical contact with the patient, isolation amplifier is an excellent interface between the transducer and the equipment (Fig. 18.21). The isolation amplifier provides a floating input for the signal source output and need for source ground connections is completely eliminated.

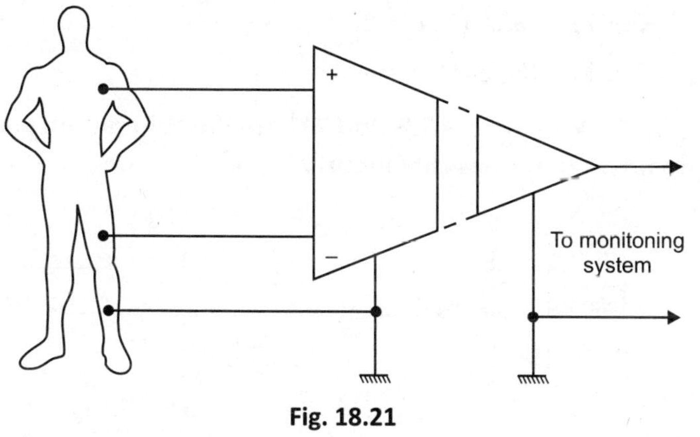

Fig. 18.21

Q.20. Justify the name unity gain crossover frequency, when the grain at this frequency is 0 dB instead of one dB.

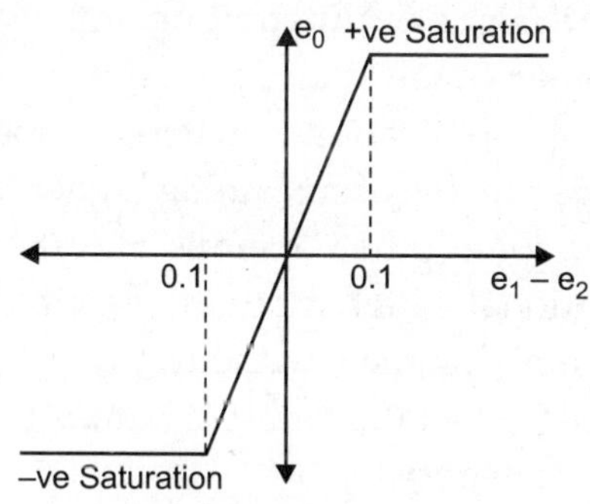

Fig. 18.22

Ans. Zero dB corresponds to a voltage ratio of unity, i.e. when the gain is expressed as a ratio of output voltage and the input voltage, then a ratio of unity gives zero, when converted into decibels (dB).

Q.21. Draw the voltage transfer characteristics of opamp and explain the shape of the curve.

Ans. For $(e_1 - e_2)$ positive, the output is +ve and if $(e_1 - e_2)$ is even slightly greater than 100 μV, the output goes to positive saturation.

For $(e_1 - e_2)$ negative, the output is (–ve) and if $(e_1 - e_2)$ is even slightly negative say by 100 μV, opamp goes to (–ve) saturation. Fig. 18.22 shows the transfer characteristics.

As an illustration, let $e_1 = 2$ mV and $e_2 = 1.9$ mV, so that $e_1 - e_2 = 0.1$ mV. If the gain is 100,000, then due to e_1 at (+ve input) we get $(2 \times 10^{-3}) \times (100{,}000) = 200\ V$ at the output. Due to 1.9 mV at the (–ve input), we get $-(1.9 \times 10^{-3}) \times 100{,}000 = -190$ V at the output.

Hence, when both e_1 and e_2 are present. Output voltage = 200 – 190 = + 10 V. Thus, as long as $e_1 > e_2$, e_0 goes to (+ ve) saturation. Now suppose e_2 exceeds e_1 by 0.1 mV, i.e. $e_2 = 2.1$ mV.

Voltage at the output due to e_2 is $-(2.1 \times 10^{-3}) \times (100{,}000) = -210$ V

Output voltage due to both e_1 and e_2 is 200 – 210 = – 10 V

Thus, as far as e_2 exceeds e_1, the output voltage goes to – ve saturation.

Q.22. How do you achieve offset nulling in the general purpose opamp type 741?

Ans. Refer to the data sheet of opamp 741. Offset nulling may be achieved by wiring a preset of l0K ohms between the null terminals (1 and 5) and joining the variable terminal to negative supply. The preset may be adjusted to give a zero output voltage with both the inputs grounded. Fig. 18.23 shows the circuit arrangement.

Q.23. From the transfer characteristics of an opamp (Fig. 18.24), it is clear that when $(e_1 - e_2)$ is slightly positive say by 0.1 mV or slightly negative say by 0.1 mV, the output goes to saturation. It is desired to increase the rate of change of output, i.e. output should go to saturation for a smaller value of $(e_1 - e_2)$. How would you achieve this?

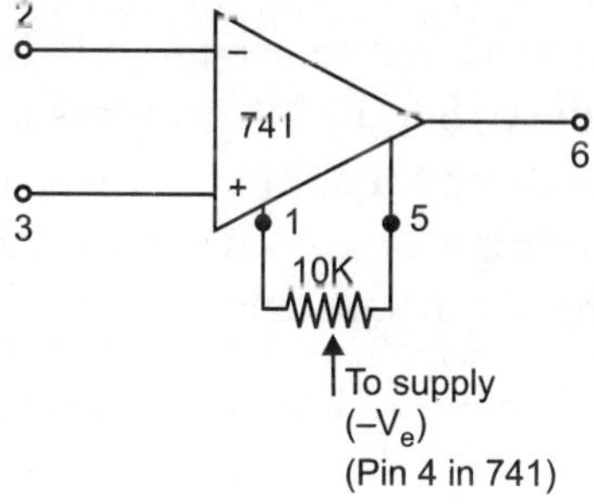

Fig. 18.23

Ans. We know that, $e_o = A(e_1 - e_2)$

where, A = open loop gain

Now $(e_1 - e_2) = e_o/A$

Thus, if the +ve saturation voltage or the –ve saturation voltage is kept at a lower level, then $(e_1 - e_2)$ required to achieve that saturation voltage decreases, as dictated by the equation above. Curve (B) in the figure shown has a high rate of change of output at the cost of reduced maximum output voltage.

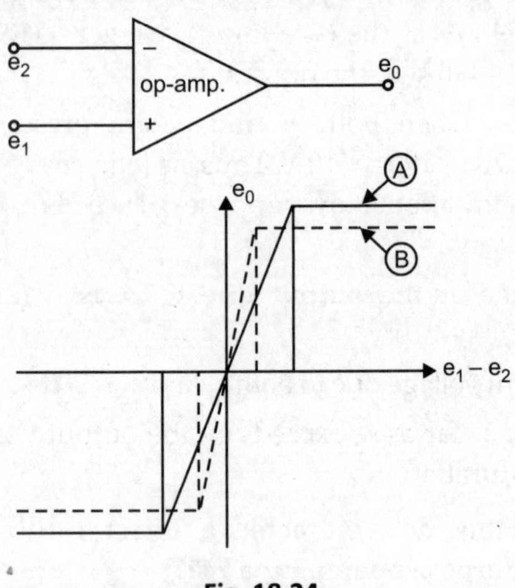

Fig. 18.24

Q.24. What are the applications, where the characteristics with a faster transition from negative to positive saturation and vice-versa are desirable ?

Ans. In all those applications, where the opamp is to be used as a comparator, the device must have the capability of causing a faster transition from one state to the other state.

Q.25. Which of tine two arrangements in Fig. 18.25. gives a higher output? Assume the CMRR to be infinite.

Ans. Both the arrangements give the same output because $e_0 = A(e_1 - e_2)$ and $(e_1 - e_2)$ is same in both the arrangement.

The important point to keep in mind is that it is the difference in the two inputs that carries any significance in differential amplifiers. Individual amplitudes do not carry any importance. This is of course with the assumption that CMRR is infinite.

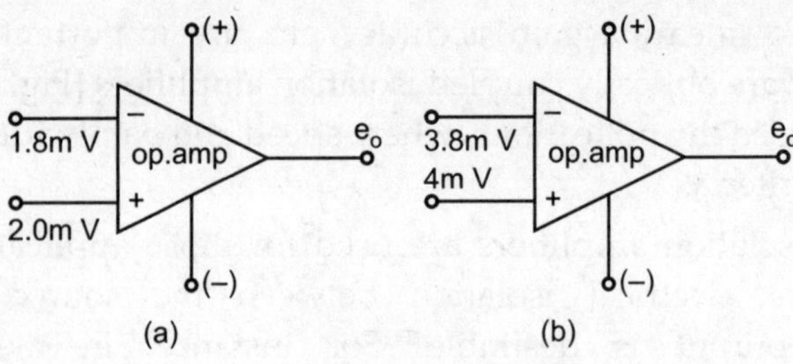

Fig. 18.25

Q.26. An opamp is having an open-loop gain of 100 dB. Find the magnitude of output for the arrangements shown in Fig. 18.26.

Ans. Voltage gain of 100 dB signifies gain ratio of 10^5.

For Fig. 18.26 (*a*), $e_o = 10$ V ;

For Fig. 18.26 (*b*), $e_o = -10$V.

Q.27. We say that 709 and 741 are the general purpose opams. What does that signify?

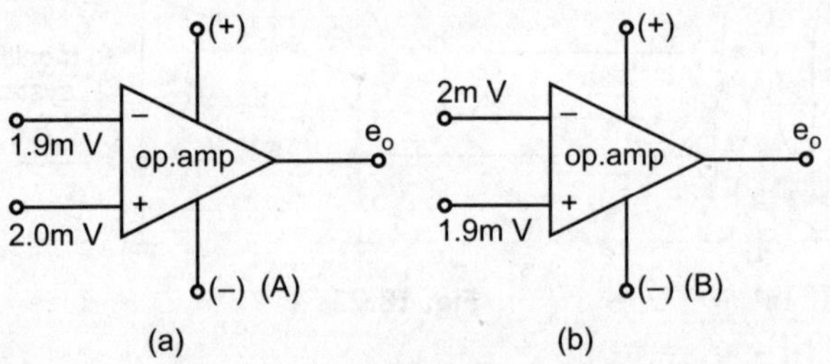

Fig. 18.26

Ans. The designs of different opamps vary from one type to the other. Some opamps are designed specifically for switching applications, whereas some may be designed to be capable of sustaining higher temperatures. Furthermore, some other opamps may be designed to exhibit exceptionally high input impedance. On the other hand, some are designed without any special purpose. They are called general propose opamps. (709) and (741) belong to this category.

SOLVED PROBLEMS

Problem 18.1. *Refer to the differential amplifier circuit of Figure-18.27. Determine the quiescent DC voltage at the collector terminal of each transistor assuming V_{BE} of the two transistors to be negligible. What will be the quiescent DC values, if V_{BE} is taken as 0.7 V?*

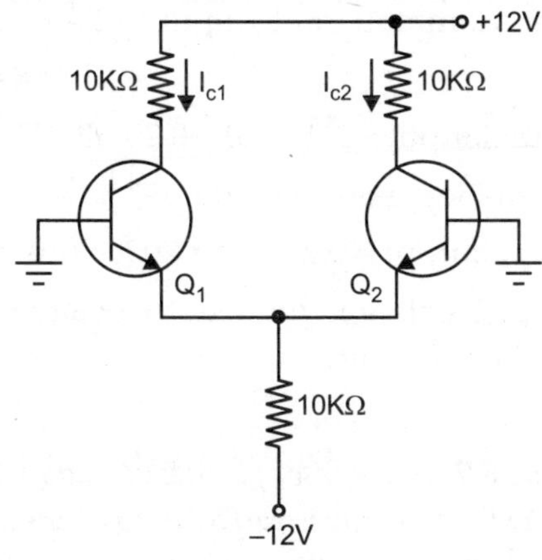

Fig. 18.27

Solution. Assuming V_{BE} to be negligible, the tail current $I_T = 12/10 \times 10^3 = 1.2$ mA

Therefore, emitter current of each transistor = $1.2 \times 10^{-3}/2 = 0.6$ mA

Collector current of each transistor is approximately equal to its emitter current

Therefore, collector current of each transistor = 0.6 mA

Quiescent DC voltage at the collector of each transistor $= 12 - 0.6 \times 10^{-3} \times 10 \times 10^3 = 6$ V

If $V_{BE} = 0.7\ V$, tail current = $(12 - 0.7)/(10 \times 10^3)$

$= 1.13$ mA

This gives emitter and hence collector current of each transistor as $(1.13 \times 10^{-3})/2 = 0.565$ mA

Quiescent DC voltage at each collector in that case equals $12 - 0.565 \times 10^{-3} \times 10 \times 10^3 = 6.35$ V

Problem 18.2. *Refer to the differential amplifier circuit diagram of Figure-18.28. Determine the differential voltage given that the transistors used in the circuit have $h_{ic} = 1\ k\Omega$ and $h_{fe} = 40$.*

Solution. Dynamic resistance r_e' of the base-emitter junction diode of the two transistors is given by h_{ie}/h_{fe}

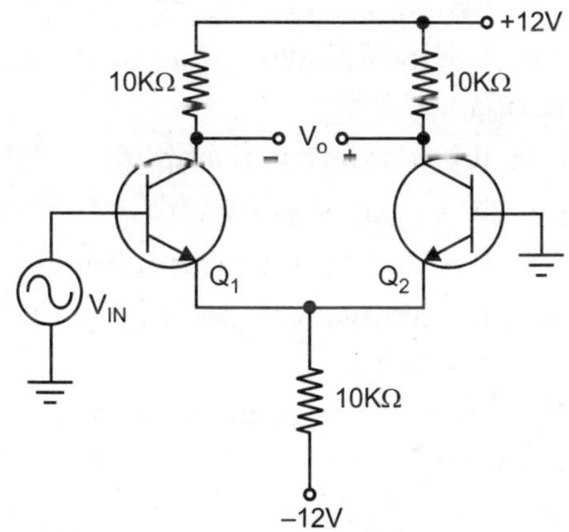

Fig. 18.28

Therefore, $r_e' = 1000/40 = 25\ \Omega$

Differential voltage gain is given by $Ad = R_c/r_e'$

Therefore, $A_d = 10000/25 = 400$

Problem 18.3. *Refer to the differential amplifier circuit of Figure-18.29. Determine the tail current I_T and also, the quiescent DC voltage across the output.*

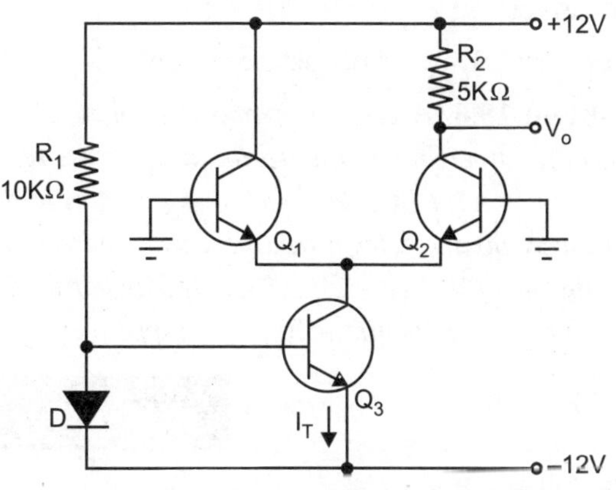

Fig. 18.29

Solution. Current I_R flowing through resistor R_1 is given by $I_R = (12 + 12 - 0.7)/(10 \times 10^3)$

$= 23.3/(10 \times 10^3) = 2.33$ mA

The tail current $I_T = I_R = 2.33$ mA

Current flowing through each transistor is equal to half of the total tail current

Therefore, current through each transistor = $2.33 \times 10^{-3}/2 = 1.165$ mA

Quiescent DC voltage across output = $12 - 5 \times 10^3 \times 1.165 \times 10^{-3} = 6.175$ V

Problem 18.4. *Opamp LM 741 is specified to have a slew rate of 0.5 V/μs. If the opamp were used as an amplifier and the expected peak output voltage were 10 V, determine the highest sinusoidal frequency that would get satisfactorily amplified. If the highest frequency component to be amplified were 15 kHz; what should be the minimum value of slew rate?*

Solution. Highest sinusoidal frequency f_{MAX} that would get satisfactorily amplified is given by f_{MAX} = Slew rate/$2\pi V_P$, where V_P is the expected peak output voltage

In the present case, slew rate = 0.5 V/μs and $V_P = 10$ V

Therefore, $f_{MAX} = (0.5 \times 10^6)/(2\pi \times 10) = 7.96$ kHz

For $f_{MAX} = 15$ kHz, desired value of slew rate = $0.5 \times 15/7.96 = 0.942\ V/\mu s$

Problem 18.5. *The differential voltage gain and CMRR of an opamp, when expressed in decibels are 110 dB and 100 dB, respectively. Determine the common mode gain expressed as a ratio.*

Solution. CMRR (in dB) = 20 log (A_d/A_{CM}) where A_d and A_{CM} are, respectively, the differential and common mode gain values.

CMRR (in dB) = 20 log A_d – 20 log A_{CM}.

That is, 20 log A_{CM} = 20 log A_d – CMRR = 110 – 100 = 10 dB.

This gives log A_{CM} = 10/20 = 0.5

Therefore, A_{CM} = Antilog 0.5 = 3.16

Problem 18.6. *A certain opamp is applied 2.0 volts and 2.2 volts at its inverting and non-inverting inputs respectively. If the opamp is used in open loop configuration and its open loop gain were 80 dB; determine differential and common mode inputs and output. Assume positive and negative saturation voltages to be +10 volts and – 10 volts respectively.*

Solution. Common mode input = (2.0 + 2.2)/2

= 2.1 volts

Differential input = 2.2 – 2.0 = 0.2 volt

Now open loop gain = 80 dB = 10^4

Therefore, output = 0.2 × 10^4 = 2000 volts

The output will get limited to positive saturation voltage equal to +10 volts

Problem 18.7. *In the case of a certain opamp, 0.5 V change in common mode input causes a DC output offset change of 5 μV. Determine CMRR in dB.*

Solution. CMRR = $\Delta V_{CM}/\Delta V_{OS}$ = 0.5/(5 × 10^{-6}) = 10^5

CMRR in dB = 20 log 10^5 = 100 dB

SUMMARY OF IMPORTANT FORMULAS

1. Common mode rejection ratio (CMRR) is given by:

   CMRR = 20 $\log_{10}$ (A_d/A_{CM})

   A_d = Differential gain

   A_{CM} = Common mode gain

2. CMRR is also expressed as ratio of change in common mode input to change in output offset voltage. That is,

   CMRR = $\Delta V_{CM}/\Delta V_{OS}$

3. Bandwidth (highest frequency, f_{MAX}) = Slew rate/ ($\pi \times V_{p\text{-}t\text{-}p}$)
4. Loop gain (in dB) = Open loop gain (in dB) – Closed loop gain (in dB)
5. Loop gain = Open loop gain/Closed loop gain

SELF EVALUATION EXERCISE

I. Below are given some characteristic parameters of opamps. Identify the type of opamp in each of the following cases. Choose from **High bandwidth opamp, Opamp comparator, General purpose opamp, Instrumentation opamp** and **Isolation Opamp:**

(*a*) Slew rate = 0.5 V/μs, Input impedance = 1M, unity gain cross-over frequency = 1 MHz

(*b*) Slew rate = 10 V/μs, Opamp produces logic compatible output

(*c*) CMRR = 120 dB, Input impedance = 10^{10} ohms, Input offset voltage = 10 μV

(*d*) Unity gain cross-over frequency : 2500 MHz

(*e*) Output stage is optically coupled or transformer coupled to input stage

II. MULTICHOICE QUESTIONS

1. An opamp can amplify
   (*a*) DC signals only
   (*b*) AC signals only
   (*c*) both AC as well as DC signals
   (*d*) neither DC nor AC signals
2. Which of the following statements best describes an opamp?
   (*a*) It is a differential amplifier.
   (*b*) It is a large bandwidth amplifier.
   (*c*) It is direct-coupled amplifier.
   (*d*) It is a direct-coupled high-gain differential amplifier.
3. The input stage of an opamp in most of the cases is a

(*a*) high-gain common-emitter amplifier
(*b*) differential amplifier
(*c*) emitter-follower buffer
(*d*) none of these

4. The common mode rejection ratio is
(*a*) ideally zero
(*b*) ratio of common mode gain to differential gain
(*c*) ratio of differential gain to common mode gain
(*d*) ratio of output DC offset voltage to change in common mode input

5. The input offset current is
(*a*) zero
(*b*) less than input bias current
(*c*) more than input bias current
(*d*) average of the two bias currents

6. One of the following opamp parameters limits the large signal bandwidth.
(*a*) Slew rate
(*b*) CMRR
(*c*) Settling time
(*d*) Unity gain bandwidth

7. The frequency response of an opamp shows the cut-off frequency to be equal to 100 Hz. If the mid-band voltage gain is 100 dB, then the unity gain cross-over frequency will be
(*a*) 1.0 MHz
(*b*) 100 MHz
(*c*) 10 MHz
(*d*) indeterminate from given data

8. An opamp has open-loop gain and unity gain bandwidth specifications of 120 dB and 10 MHz, respectively. The open-loop cut-off frequency will be
(*a*) 10 Hz
(*b*) 100 Hz
(*c*) 1000 Hz
(*d*) 1 Hz

9. The compensating capacitor connected externally in the case of uncompensated opamps is to
(*a*) increase large signal bandwidth
(*b*) prevent oscillations
(*c*) improve slew rate
(*d*) increase CMRR

10. Identify the opamp type in which the output stage is designed to be compatible with some logic family.
(*a*) Opamp comparator
(*b*) High speed opamp
(*c*) Isolation opamp
(*d*) Instrumentation opamp

11. Which of the following opamp internally comprises three opamps out of which two are wired as buffers and the third is a difference amplifier?
(*a*) Isolation opamp
(*b*) Norton opamp
(*c*) Instrumentation opamp
(*d*) Opamp comparator

12. In which of the following opamp types, the input stage is not a conventional differential amplifier:
(*a*) Norton opamp
(*b*) Instrumentation opamp
(*c*) Opamp comparator
(*d*) Isolation opamp

13. Which of the following opamp types is ideally suited to single supply operation?
(*a*) High bandwidth opamp
(*b*) Instrumentation opamp
(*c*) Norton opamp
(*d*) General-purpose opamp

14. In the front-end differential amplifier inside an opamp, the tail resistor is usually replaced by
(*a*) emitter-follower configuration
(*b*) current mirror configuration
(*c*) common-emitter amplifier configuration
(*d*) a junction diode

15. In the internal circuit of an opamp, collector resistors are usually replaced by active loads to
(*a*) increase differential gain
(*b*) decrease common mode gain
(*c*) increase bandwidth
(*d*) increase efficiency

ANSWER TO SELF EVALUATION EXERCISE

I.

(*a*) General Purpose opamp
(*b*) Opamp comparator
(*c*) Instrumentation opamp
(*d*) High bandwidth opamp
(*e*) Isolation opamp

II. MULTICHOICE QUESTIONS

1. (*c*)	2. (*d*)	3. (*b*)	4. (*c*)	5. (*b*)	6. (*a*)
7. (*c*)	8. (*a*)	9. (*b*)	10. (*a*)	11. (*c*)	12. (*a*)
13. (*c*)	14. (*b*)	15. (*a*)			

Chapter 19

Operational Amplifiers

Application Circuits

Learning Objectives

After completing the chapter, you will learn the following topics:

1. Opamp as Voltage Follower
2. Opamp as Inverting Amplifier
3. Opamp as Non-inverting Amplifier
4. Opamp as Inverting Summing Amplifier
5. Opamp as Non-inverting Summing Amplifier
6. Opamp as Subtractor
7. Opamp as Inverting Comparator
8. Opamp as Non-inverting Comparator
9. Opamp as Window Comparator
10. Opamp as Logarithmic Amplifier
11. Opamp as Anti-log Amplifier
12. Opamp as Peak Detector
13. Opamp as Half-wave and Full-wave Rectifiers
14. Active Low Pass, High Pass and Band Pass Filters using Opamps
15. Voltage Regulators using Opamps
16. Sinusoidal Oscillators using Opamps
17. Opamp as Instrumentation Amplifier
18. Opamps as Clipper and Clamper Circuits
19. Opamp as Current-to-Voltage and Voltage-to-Current Converters
20. Opamp as Averager Circuit

Q. 1: Briefly explain the concept of virtual earth in opamps.

Ans: The concept of virtual earth is applicable to opamp configurations with negative feedback. In all these circuit configurations, the differential input to the opamp is always zero. That is, voltage at inverting input is always equal to that at the non-inverting input. If the non-inverting input were grounded; the inverting input would also behave, as if it were grounded, though it is not physically grounded. Another way of explaining the concept of virtual earth is that the impedance (R_i) seen at inverting input terminal of opamp in this case would be equal to (R_f/A), where (R_f) is the feedback resistance and (A) is the open loop gain. Since (A) is very high; input impedance would be nearly zero (Figure-19.1).

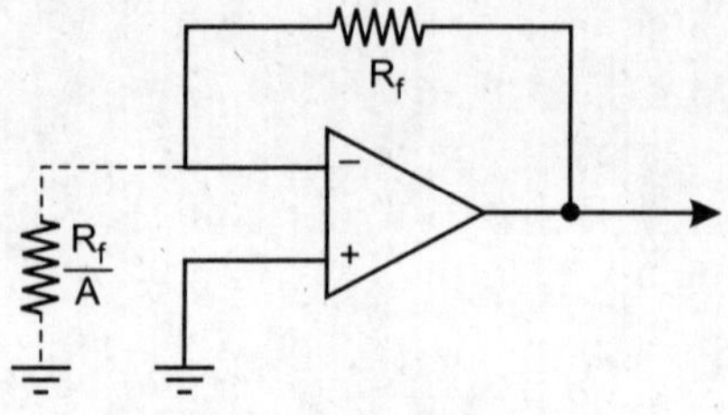

Fig. 19.1

Q. 2: With the help of circuit diagram, briefly describe the operation of voltage follower configured around an opamp. What are the main attributes of a voltage follower circuit?

Ans: Voltage follower is nothing but a non-inverting amplifier circuit with unity gain. Figure-19.1 shows the basic voltage-follower circuit. Due to extremely high open loop differential gain of the opamp and the presence of negative feedback, the differential input to the opamp is ideally zero. If the voltage at the inverting input tends to be slightly less than the voltage at the positive input; the output tends to increase rapidly towards positive saturation due to high open loop gain. Increase in output voltage is fed back to the negative input, which increases the inverting input voltage towards the voltage at non-inverting input. Similarly, if the voltage at inverting input tends to increase; the output is driven to negative saturation bringing the voltage at inverting input back towards the voltage at non-inverting input. The net result is that the differential input is always zero. That is voltage at the inverting input stays equal to the voltage at non-inverting input. In the circuit of Figure-19.2, the voltage at the output is equal to voltage at the inverting input, which is further equal to the voltage at non-inverting input and hence, the name voltage follower. The output voltage in this case follows the voltage applied at the input.

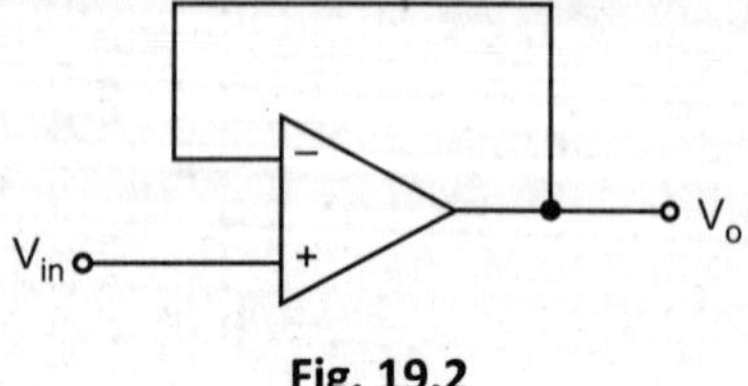

Fig. 19.2

Voltage-follower circuit has many advantages. Owing to its extremely high input impedance, extremely low output impedance and unity gain; it acts as an ideal interface between a high impedance source and a low impedance load. Also, unity closed-loop gain leads to a very high closed-loop bandwidth equal to the unity gain cross-over frequency. The output offset error is also very low because due to unity closed-loop gain, input errors are not amplified.

Q. 3: Draw the opamp based voltage follower circuit for AC applications.

Ans: Figure-19.3 shows the voltage follower circuit for *AC* applications. It is similar to the one shown in Figure-19.2 for *DC* applications except for the inclusion for capacitors in series with input and output. The capacitors block *DC* and allow only *AC* signals to appear at the input and then delivered to the output.

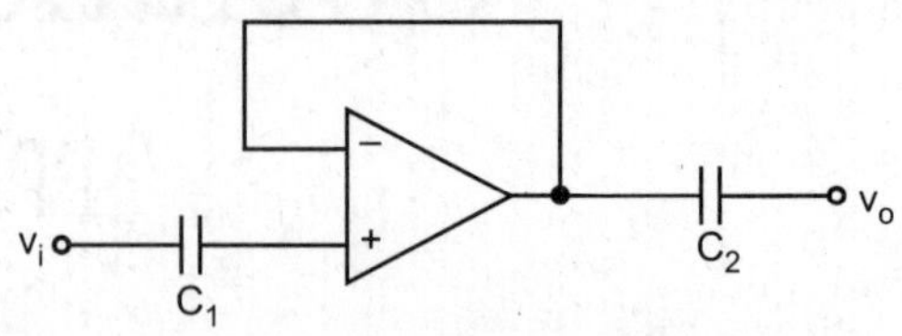

Fig. 19.3

Q. 4. How do you use an opamp as an inverting amplifier ?

Ans. An opamp inverting amplifier is shown in Fig. 19.4.

Assuming a zero inward current flow, same current flows through R_1 as well as R_2. Due to ground at (+ve) input, a virtual earth appears at $R_1 - R_2$ junction.

$$I = \frac{V_{in}}{R_1} = -\frac{V_o}{R_2}$$

so that $$\frac{V_o}{V_{in}} = \left(-\frac{R_2}{R_1}\right)$$

Minus sign indicates that output is the inverted version of input. By selecting proper values of R_1 and R_2 desired magnitude of gain can be achieved.

Q.5. Briefly describe the design guidelines for the inverting amplifier configuration of Figure. 19.4.

Ans. The inverting amplifier of Fig. 19.4 has an effective input impedance equal to (R_1), as it comes in parallel with the high differential input impedance of the opamp. For a given value of closed loop gain, a higher input impedance, which means a higher (R_1), leads to a higher value of (R_2). There is always an upper limit to the value of (R_2) for a given opamp. In fact, higher the input impedance of the opamp, larger is the maximum allowable value of (R_2). In fact, the problem starts, when the current flowing towards the negative terminal of the opamp becomes comparable to the current (*I*). Things will be more clear, if we look at the actual expression for gain that we would have arrived at, if we had not made any assumptions. The actual expression for gain for the amplifier circuit of Fig. 19.4. given by :

$$\text{Closed loop gain} = -\frac{A_{OL}\, R_2}{(R_1 + R_2 + A_{OL}\, R_1)}$$

$$= -\frac{R_2}{R_1 + (R_2/A_{OL})}$$

This implies that when ratio $\left(\frac{R_2}{A_{OL}}\right)$ is much smaller than (R_1), the gain expression reduces to the one that we are more familiar with *i.e.* closed loop gain

$$= \left(-\frac{R_2}{R_1}\right).$$

The input impedance of this circuit is same as the input resistance value *i.e.* R_1.

The output impedance of this circuit is approximated by :

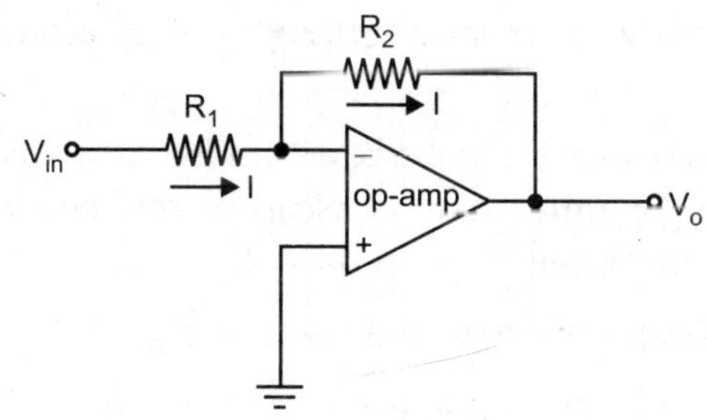

Fig. 19.4

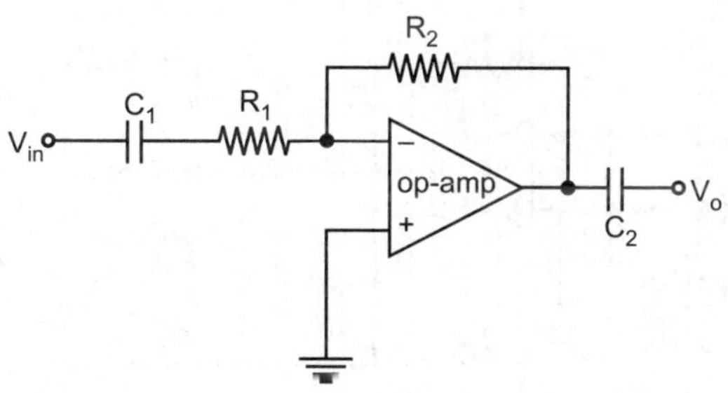

Fig. 19.5

$$R_o = \left[\frac{R_1 + R_2}{R_1 \cdot A_{OL}}\right]$$

where, R_{OL} = open loop output impedance.

Q. 6. Give the circuit arrangement of an inverting amplifier for *AC* applications.

Ans. Fig. 19.5 shows the circuit arrangement, which is self-explanatory.

Q. 7. Draw the circuit arrangements for a non-inverting amplifier for *DC* as well as *AC* applications. Derive the expression for gain. Outline basic design equations.

Ans. Fig. 19.6 (*a*) shows a non-inverting amplifier for *DC* applications. The same for *AC* applications is shown in Fig. 19.6 (*b*).

Following the same steps, as followed in case of an inverting amplifier.

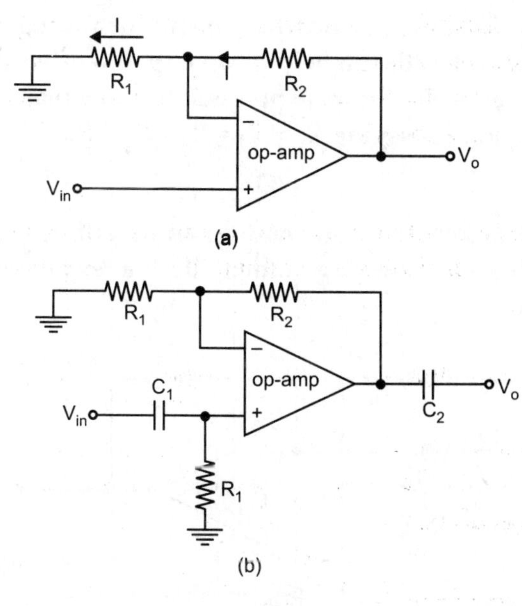

Fig. 19.6

Due to virtual earth, voltage of $R_1 - R_2$ junction equals (V_{in}). Also assuming that current flowing towards negative input of the opamp is zero, we can write,

$$I = \frac{V_{in}}{R_1} = \frac{V_o}{R_1 + R_2} : \text{Gain} = \frac{R_1 + R_2}{R_1}$$

Design Information

The actual gain expression in case of a non-inverting amplifier is given by :

$$\text{Closed loop gain} - \frac{A_{OL}\,(R_1 + R_2)}{R_1 + R_2 + A_{OL}\, R_1}$$

which reduces to $\frac{R_1 + R_2}{R_1 + (R_2/A_{OL})}$ for $A_{OL} >> 1$. Again if (R_2/A_{OL}) << R_1, the expression reduces to what we are all familiar with.

The input impedance of this configuration is given by:

$$\text{Input impedance} = R_i\, A_{OL}\left[\frac{R_1}{R_1 + R_2}\right]$$

$$= R_i\ (\text{Loop gain})$$

where, R_i = Open loop input impedance

$$\text{Loop gain} = \frac{\textit{Open loop gain}}{\textit{Closed loop gain}}$$

The output impedance can be computed from :

$$\text{Output impedance} = \frac{R_0}{Loop\ gain}$$

R_0 = open loop output impedance.

Q. 8. Give the circuit arrangement for adding various *DC* voltages with the help of opamp and other passive components. Derive an expression to show that $e_0 = -(e_1 + e_2 + e_3 + ...)$. Assume $R_1 = R_2 = R_3 = R_4 = R$.

Or

Give the circuit arrangement for an inverting summing amplifier or inverting adder. Derive expression for output.

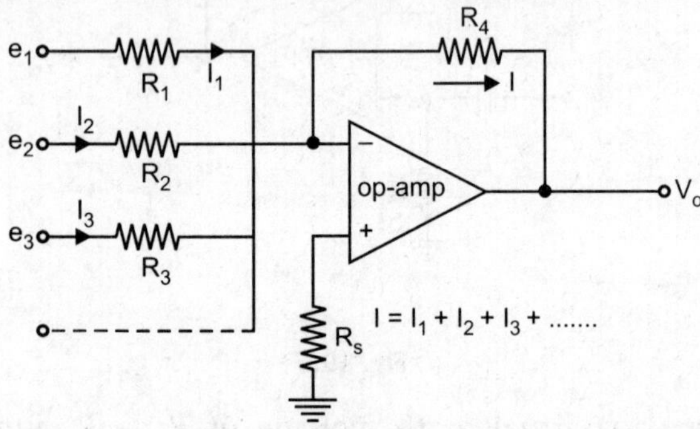

Fig. 19.7

Ans. Fig. 19.7. shows the opamp circuit for adding *DC* voltages. Assuming a virtual earth at negative input terminal.

$$I_1 = \frac{e_1}{R_1},\ I_2 = \frac{e_2}{R_2};\ I_3 = \frac{e_3}{R_3};\ I = I_1 + I_2 + I_3 = \frac{e_1}{R_1} + \frac{e_2}{R_2} + \frac{e_3}{R_3}$$

Also, $$I = -\frac{e_0}{R_4}$$

$$-\frac{e_0}{R_4} = \left(\frac{e_1}{R_1} + \frac{e_2}{R_2} + \frac{e_3}{R_3} + ...\right)$$

If $R_1 = R_2 = R_3 = R_4 = \text{R}$

Then $e_0 = -(e_1 + e_2 + e_3 + ...)$.

Q. 9. Design an adder circuit (with unity gain) to be used for addition of (say 3) audio voltages.

Ans. The basic adder circuit for *AC* applications can be obtained from its *DC* counterpart by wiring blocking capacitors in series with input and output and shown in Fig. 19.8.

Opamp number = 741

Now for unity gain, $R_1 = R_2 = R_3 = R_4 = R$

Select R = 100 K

$R_5 = R/2 = 50$ K

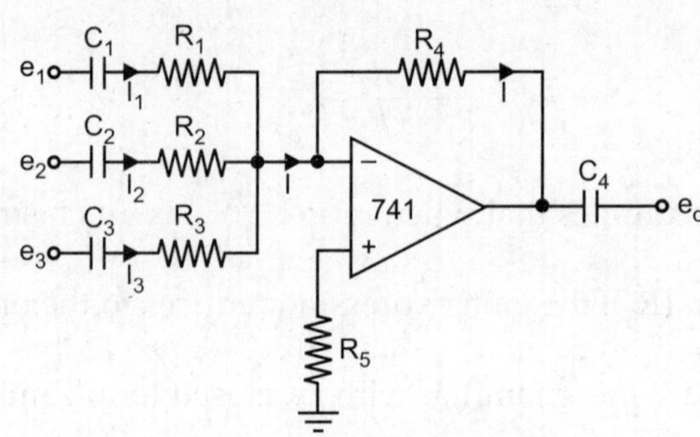

Fig. 19.8

Capacitors are to block *DC*. They must have a large value so that they allow the lowest frequency audio signal without any resistance.

Let $C_1 = C_2 = C_3 = C_4 = 1\ \mu F$.

Q. 10. Draw a non-inverting adder circuit using opamp.

Ans. A non-inverting adder can be constructed from its inverting counterpart by cascading it with a unity gain inverting amplifier.

The complete circuit is shown in Fig. 19.9.

$V_{01} = -(e_1 + e_2 + e_3)$ and $V_{02} = -V_{01} = (e_1 + e_2 + e_3)$

[Assuming $R_1 = R_2 = R_3 = R_4 = R$ and $R_5 = R_6 = R'$]

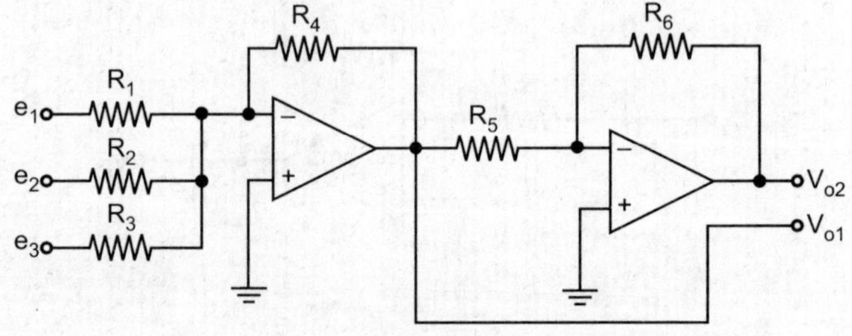

Fig. 19.9

Q. 11. Draw a basic *DC* phase-splitter circuit using opamp and explain its operation.

Ans. Fig. 19.10 shows the circuit arrangement. If

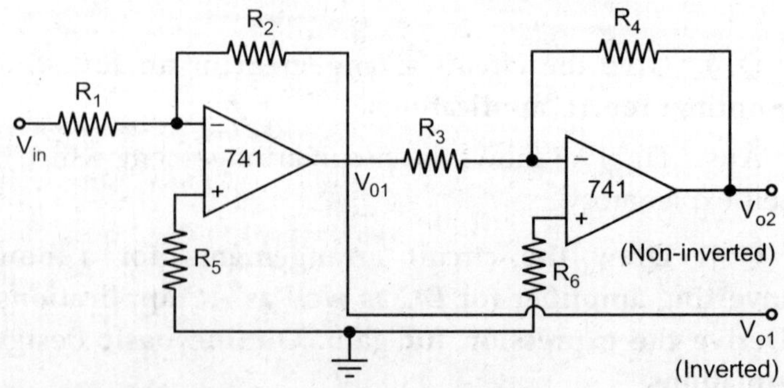

Fig. 19.10

$R_1 = R_2$ and $R_3 = R_4$ the circuit becomes a cascaded arrangement of two unity gain inverting amplifier sections. V_{01} (output of first section) is 180° out of phase with V_{in}, whereas V_{02} (output of second section) is 180° out of phase with V_{01} and hence in phase with V_{in}.

R_5 equals $R_1 \,||\, R_2$ and R_6 equals $R_2 \,||\, R_4$.

Q. 12. Show how an opamp may be used to build a subtracter. Derive an expression to show that $e_0 = (e_1 - e_2)$.

Ans. The subtractor circuit is shown in Fig. 19.11.

Consider that e_2 is grounded so that the circuit becomes a non-inverting amplifier with e_1 as input and gain equal to $\frac{R_2 + R_1}{R_1}$.

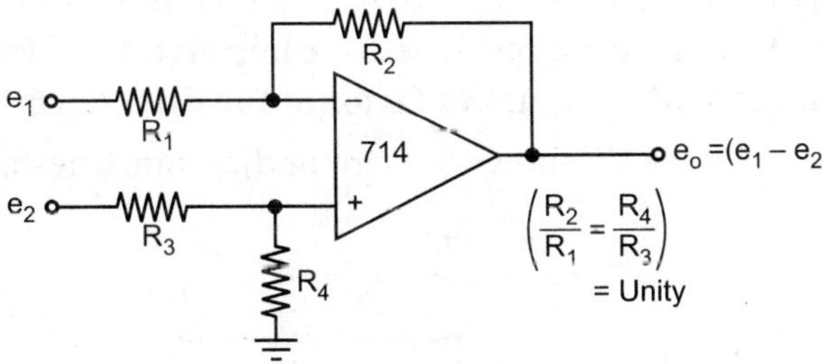

Fig. 19.11

But due to R_3 and R_4, effective input at +ve terminal is $\frac{e_1 \times R_4}{R_3 + R_4}$

so that non-inverting gain $\frac{R_2 + R_1}{R_1} \times \frac{R_4}{R_3 + R_4} =$

$\left(\frac{1 + R_2/R_1}{1 + R_3/R_4}\right)$, If $\frac{R_4}{R_3} = \frac{R_2}{R_1}$

Then non-inverting gain

$$= \frac{1 + R_2/R_1}{1 + R_1/R_2} = \frac{R_1 + R_2}{R_1} \times \frac{R_2}{R_2 + R_1} = \frac{R_2}{R_1}$$

which is the same as the gain, when e_1 is grounded and circuit becomes an inverting amplifier.

Thus, both the inputs provide a gain of R_2/R_1 but the outputs due to these inputs (e_1 and e_2) are 180° out of phase.

At the output, we get $\left(\frac{R_2}{R_1} . e_1\right)$ volts and $\left(\frac{-R_2}{R_1} e_2\right)$ volts so that, Output $= \frac{R_2}{R_1} (e_1 - e_2)$

and if $\frac{R_2}{R_1}$ = unity. Then $e_0 = (e_1 - e_2)$

Q. 13. Suggest some opamp circuit for non-linear amplification and explain its operation.

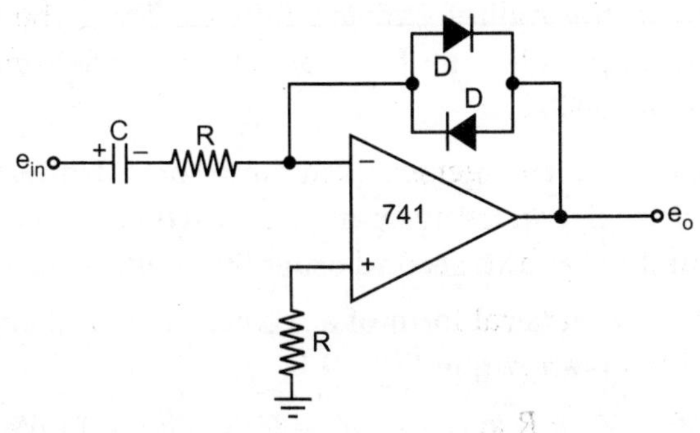

Fig. 19.12

Ans. Non-linear amplification can be achieved by having a non-linear device in the feedback path (Fig. 19.12). The circuit above uses ordinary silicon diodes for the purpose. The amplifier thus becomes a semi-log amplifier, as the forward current through silicon diodes varies as log of applied voltage.

With zero voltage, diodes act as open circuit and the gain is high due to minimum feedback; with large applied voltage, diodes offer very small resistance and thus, gain is low.

Such a circuit may give an output voltage change of 2 : 1 for an input change of 1000 : 1. This compression of output is decided by *R*. For higher *R*, compression is lesser, *i.e.* output may change in the ratio 3 : 1 for input change of 1000 : 1.

Q. 14. Where does this type of logarithmic amplifier find application ?

Ans. It finds application in *AC* bridge balance detectors. The output of bridge may vary over wide range around its null point. In order to achieve null, the output applied to an *AC* millivoltmeter whose sensitivity may have to be adjusted a number of times before a null is achieved. If the bridge output is applied to the semilog amplifier of the type described in the preceding question, the output of semilog amplifier varies only in a small range for a wide variation of bridge output (Fig. 19.13).

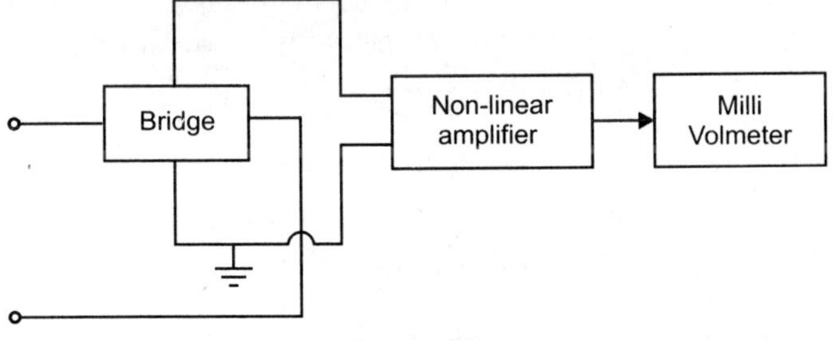

Fig. 19.13

A variation of 10000 : 1 in the bridge output may cause a small variation of 6 : 1 in amplifier output which is further applied to the milli-voltmeter, thus enabling the single range of milli voltmeter to accommodate variations over a range of 10000 :

Q. 15. Draw the generalised form of a Butterworth active filter circuit using opamp as an active device. Also, draw first order and second order low pass sections.

Ans. The general form of a second order butterworth active filter is shown in Fig. 19.14.

If $Z_1 = Z_2 = R$ and $Z_3 = Z_4 = C$, We get a LOW-PASS filter.

If $Z_1 = Z_2 = C$ and $Z_3 = Z_4 = R$, We get a HIGH-PASS filter.

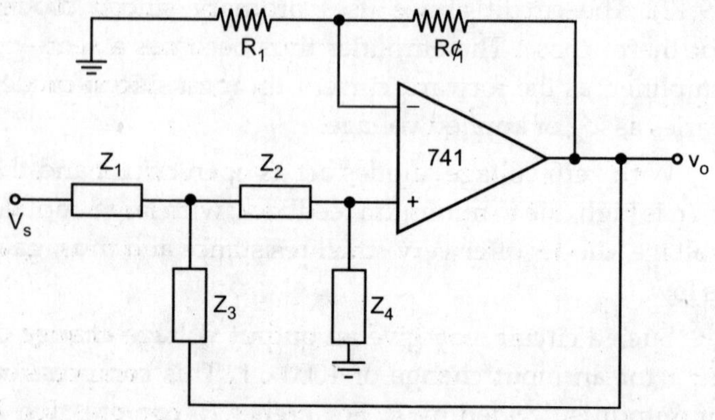

Fig. 19.14

The cut off frequency is given by, $f_0 = \dfrac{1}{2\pi RC}$

The values of R_1 and R'_1 can be determined from the relation.

$$A_0 = 3 - 2K$$

where, $2K$ is the co-efficient of s in the second order butterworth polynomial.

The first order and second order low-pass filters are shown in Figs 19.15 and 19.16 respectively.

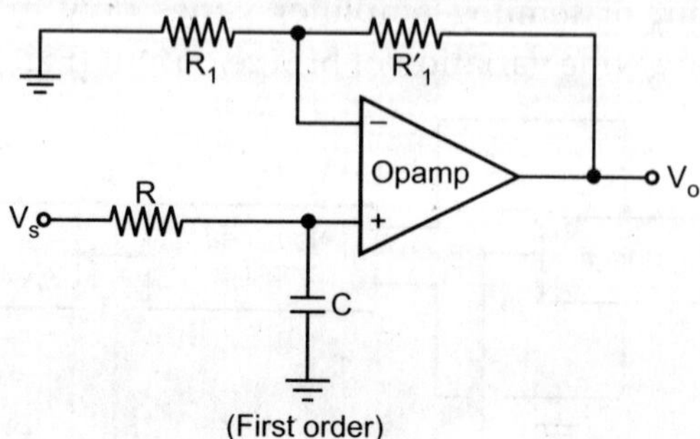

Fig. 19.15

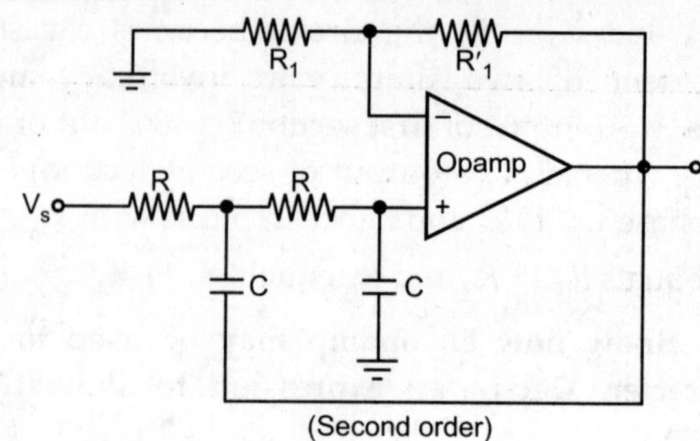

Fig. 19.16

Q. 16. One of the very popular active filter configurations is the second order band pass filter built around a single opamp. Draw the circuit diagram of a second order band pass filter using an opamp and briefly describe how this circuit behaves as a band pass filter? Also, give expressions for computation of center frequency and the Quality factor of the filter circuit.

Ans. Fig. 19.17. shows the circuit diagram. The circuit

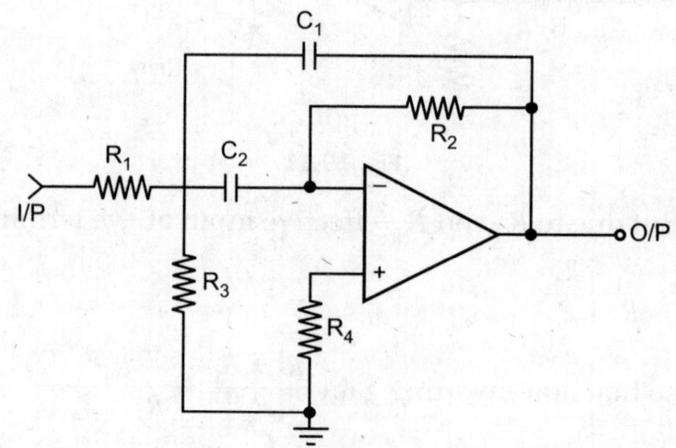

Fig. 19.17

behaves as a band pass filter as follows. A close examination of the circuit reveals that it is like a cascade arrangement of a low pass filter (constituted by R_1, C_1) followed by a high pass filter (constituted by R_2, C_2) If the LPF cut-off frequency were higher than the HPF cut-off frequency, the net result would be a band pass operation. That is what exactly happens in this circuit. Another way of qualitatively explaining the operation of this circuit is as follows. At very low frequencies, C_1 and C_2 would behave as very high impedances close to an open circuit. As a result, the input signal is prevented from reaching the output. At very high frequencies, the output is shorted to the inverting input, which converts the circuit to an inverting amplifier with zero gain. Again, there is no output. Thus, at both very low and very high frequencies, the output is zero. At some intermediate band of frequencies, the gain provided by the circuit offsets the loss due to the potential divider $R_1 - R_3$. Mathematical expressions governing the design of the filter circuit are as follows :

Resonant frequency, $f_R = 2Q/(2\pi R_2 C)$

where, Q = Quality factor

for $C = C_1 = C_2$

Quality factor, $Q = [R_1 . R_3/2R_3]^{1/2}$; Voltage gain,

$A_v = Q/(2\pi R_1 f_R C)$

Q. 17. With the help of circuit diagram, briefly describe the functioning of a band reject filter built around an opamp. List the important design expressions.

Ans. Fig 19.18. shows the circuit diagram. It uses a Twin-T network for frequency selection connected in series with the non-inverting input of the opamp. Similar circuits were explained earlier in the chapter with reference to frequency selective amplification. As mentioned earlier, a twin-T network offers a very high impedance at the resonance frequency and a very low impedance at frequencies off-resonance. This phenomenon explains the behaviour of the circuit. Another way of explaining the behaviour of the circuit is as follows. Very low frequency signals find their way to the output via the low pass filter formed by $R_1 - R_2 - C_3$. Very high frequency signals reach the output through the high pass filter constituted by $C_1 - C_2 - R_3$. In an intermediate band of frequencies, both filters pass the signal to some extent but the negative phase shift introduced by LPF is cancelled out by an identical positive phase shift by HPF with the result that at any instant, the net signal reaching the non-inverting input and hence the output is zero.

Component values of the twin-T network are chosen as :

$$R_1 = R_2 = R, R_3 = R/2 \text{ and } C_1 = C_2 = C, C_3 = 2C$$

$$0 \leq R4 \leq (R_1 + R_2); f_R = 1/(2\pi RC)$$

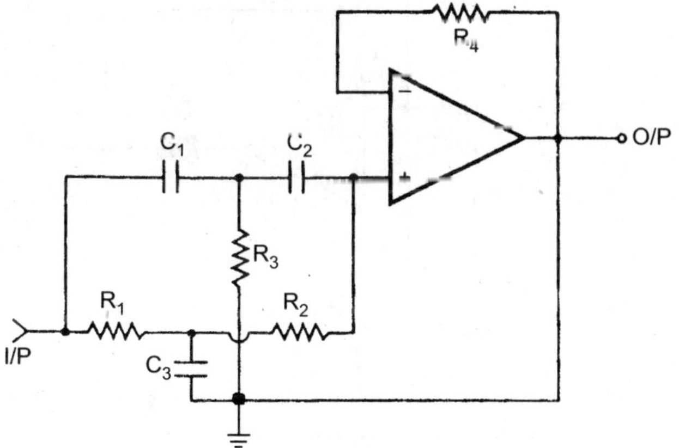

Fig. 19.18

Q. 18. Briefly describe the operation of an opamp based half wave rectifier circuit. What is the major advantage of such a rectifier circuit over the conventional circuit without the opamp ? How can we modify the circuit to get a full wave rectified output?

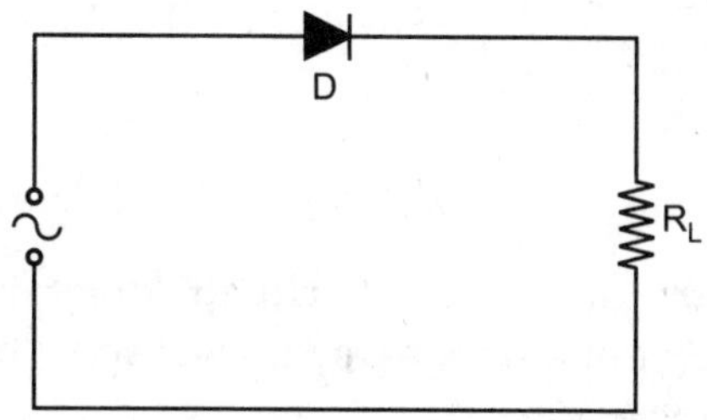

Fig. 19.19

Ans. The conventional rectifier circuits (Fig 19.19) do not produce at the output an ideal rectified waveform due to forward biased voltage drop across the diode, which is 0. 7 volt in case of silicon diodes. This problem is overcome in opamp based rectifier circuits. Fig. 19.20 shows the generalised rectifier circuit built around an opamp. The circuit functions as follows. Due to non-inverting input of the opamp being at ground potential, during positive half cycles, diode (D_1) is forward biased and the diode (D_2) is reverse biased.

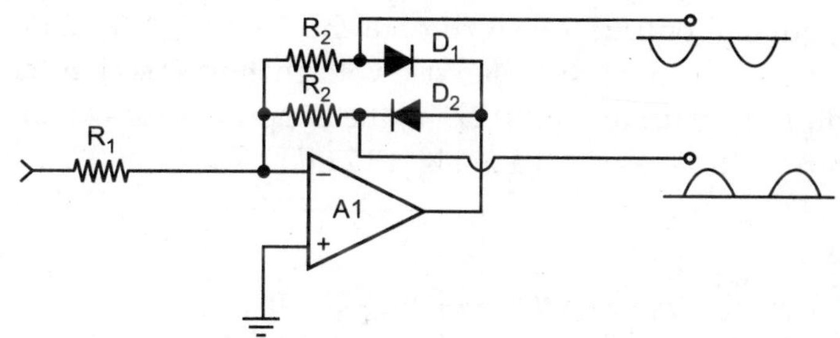

Fig. 19.20

The positive half cycles appear as negative half cycles due to phase inversion. Similarly, during negative half cycles, diode (D_1) is reverse biased and diode (D_2) is forward biased with the result that negative half cycles appear as positive half cyeles again due to phase inversion. Remember that the moment input increases by a few millivolts either in positive or negative direction, the output tends to go to negative or positive saturation respectively. It is, therefore, not necessary for the input to exceed the diode drop to produce the output. However, maximum values of peak positive output and peak negative output are (VSAT – 0.7) and (– VSAT + 0. 7) respectively. The two half wave rectified outputs can be summed up in another opamp stage to get a full wave rectified output, as shown in Fig. 19.21.

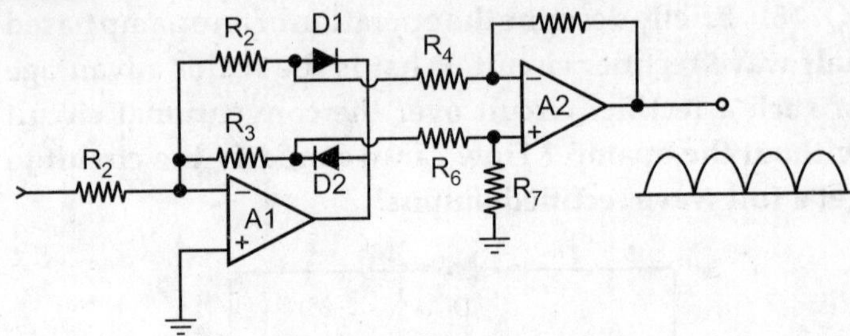

Fig. 19.21

Q. 19. With the help of circuit diagrams, briefly describe the operation of opamp-based positive and negative clipper circuits.

Ans. Fig. 19.22. (*a*) shows a positive clipper circuit, a clipper circuit that clips positive half cycles above a certain positive reference voltage. The circuit functions as follows. During positive half cycles, for input voltage less than or equal to reference voltage (V_{REF}), the opamp output goes to positive saturation and diode (D_1) is reverse biased with the result that the input appears as such at the output. The situation is same during negative half cycles. When the input voltage exceeds the reference voltage, the opamp output goes to negative saturation and the diode gets forward diased. The output gets shorted to inverting input and the out is clamped to (V_{REF}). The input and output waveforms are shown in Fig. 19.22. (*b*). If the polarity of the reference voltage is reversed in the clipper circuit of Fig. 19.22 (*a*), the clipping occurs below zero voltage, as shown in Fig. 19.22 (*b*).

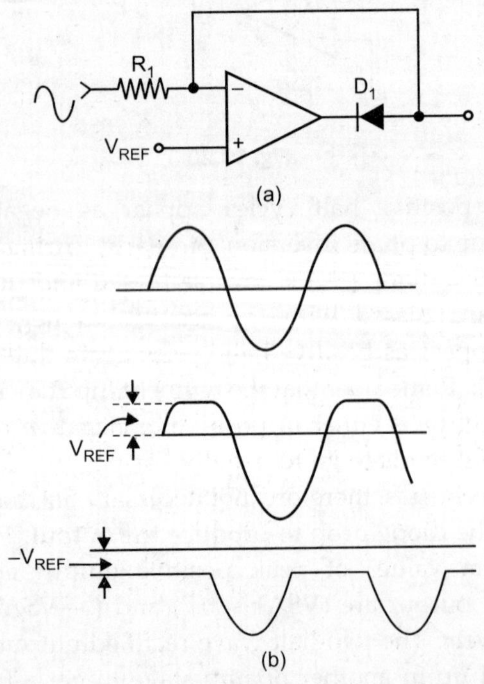

Fig. 19.22

Fig. 19.23 (*a*) shows a negative clipper circuit. The circuit functions as follows. During positive half cycles and also for input voltage less negative or equal to ($-V_{REF}$), diode (D_1) is reverse biased. The input appears as such at the output. For input voltage more negative than ($-V_{REF}$), diode is forward biased and the output gets clamped at the reference voltage. The input and output waveforms are shown in Fig. 19.23 (*b*). When the polarity of the reference voltage is reversed, clipping occurs above zero voltage. The output waveform in this case is also shown in Fig. 19.23 (*b*).

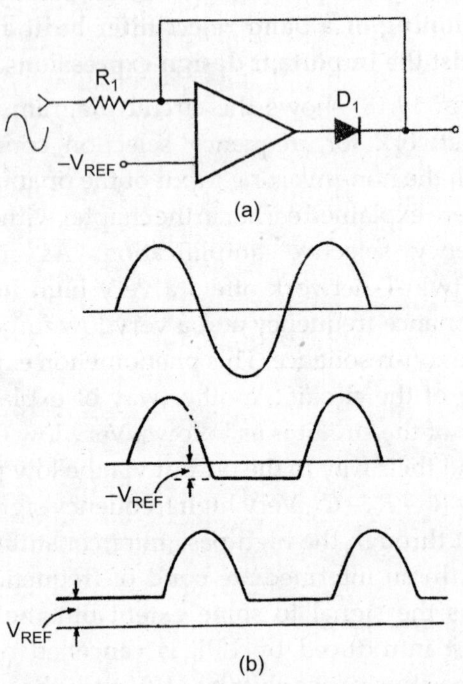

Fig. 19.23

Q. 20. With the help of a circuit diagram, briefly describe the operation of an opamp-based clamper circuit. What is the basic advantage of this active clamper circuit over the conventional passive clamper circuit?

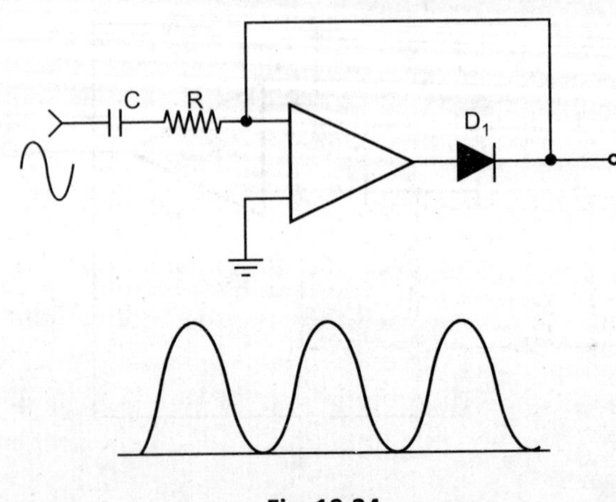

Fig. 19.24

Ans. Fig. 19.24. shows the clamper circuit that clamps the negative peaks to zero. The circuit operates as follows. During the first negative half cycle, the diode (D_1) gets forward biased and the capacitor charges through the resistance (R) and the forward biased diode to the peak of the negative half cycle. During the positive half cycle, the diode gets reverse biased. There is no rapid discharge path for the capacitor and in this case, the output equals the input voltage plus the voltage across the capacitor. The negative peaks are, thus, clamped to zero voltage. If the non-inverting input is applied a reference voltage (V_{REF}), positive or negative, the negative peaks are clamped at (V_{REF}) instead of zero. Fig. 19.25 shows the clamper circuit that clamps positive peaks to zero. Again, if the non-inverting input is applied a reference voltage (V_{REF}), positive or negative, the positive peaks are clamped at (V_{REF}) instead of zero.

The conventional clamper circuit can not function as a clamper, if the peak input signal is less than 0. 7 volt. The opamp-based clamper circuit has no such limitation. It functions, as if the diode were ideal. This implies that the circuit can be used to clamp even milli-volt signals.

Q. 21. With the help of circuit diagram, briefly describe the operation of an opamp-based peak detector circuit. Also, outline the factors that govern the choice of different component values.

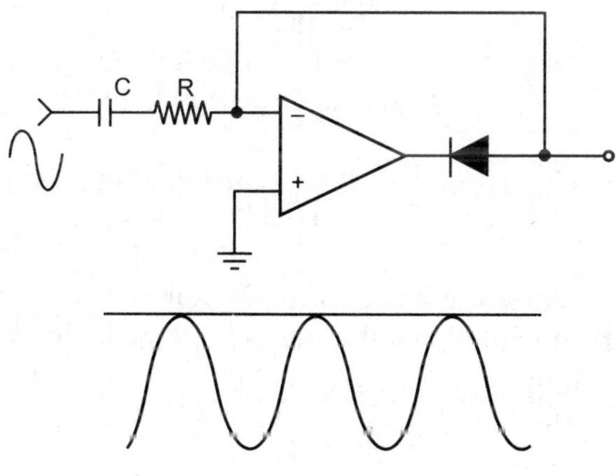

Fig. 19.25

Ans. Peak detector circuit produces a voltage at the output equal to peak amplitude (Positive or Negative) of the input signal. Fig. 19.26. shows a positive peak detector circuit. As we can see, it is essentially a clipper circuit with a parallel resistor-capacitor connected at its output. The clipper here reproduces the positive half cycles. During this period, the diode is forward biased. The capacitor rapidly charges to the positive peak from the output of the opamp through the ON resistance of the forward biased diode. As the input starts decreasing beyond the peak, the diode gets reverse biased thus isolating the capacitor from the output of the opamp. The capacitor can now discharge only through the resistor connected across it. The value of the resistor is much larger than the forward biased diode's ON resistance. The purpose of this resistor is to allow a discharge path so that the output can respond to changing amplitudes of the signal peaks, decreasing amplitudes of the signal peaks to be more precise. The buffer circuit connected ahead of the capacitor is to prevent any discharge of the capacitor due to loading effects of the following circuit. The circuit can be made to respond to the negative peaks by reversing the polarity of the diode. Rest of the circuit is the same.

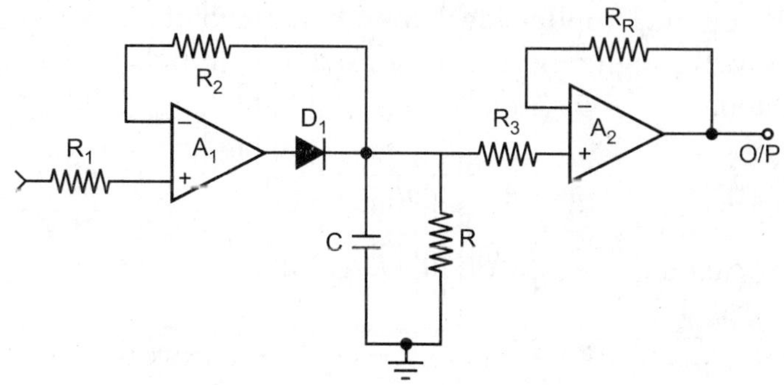

Fig. 19.26

The parallel R-C circuit time constant is typically 100 times the time period corresponding to the minimum frequency of operation. The R-C time constant also controls the response time. The response time is nothing but the time needed to respond to a decreasing amplitude. Surely, a large time constant would make the response more sluggish. An attempt to reduce the time constant to improve the response time increases the output ripple. The chosen time constant is compromise of the two conflicting requirements. Slew rate is the primary specification that needs to be looked into while choosing the right opamp for the clipper portion. The desired slew rate is such that the slew rate limited frequency, which is a function of peak-to-peak output swing and the slew rate is at least equal to the highest frequency of operation. The peak to peak voltage swing at the output of the opamp is equal to $[V_{pk} - (-V_{SAT})] = (V_{pk} + V_{SAT})$. V_{pk} is the maximum peak amplitude of the input signal.

Q. 22. Describe the operation of an integrator circuit using operational amplifier. Derive an expression to show that $v_o = K\int v_i\, dt$.

Ans. An integrator circuit using an opamp is shown in Fig. 19.27.

Since +ve input terminal has been grounded through a small resistance R_1, virtual earth appears at R-C junction.

Thus, the voltage e_0 effectively is the voltage across the capacitor C.

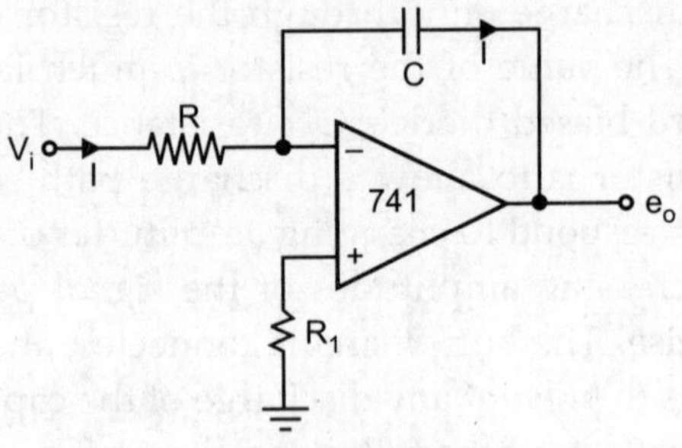

Fig. 19.27

Due to infinite impedance at the input terminals of the op-amp (under ideal conditions), whatever current flows through resistance R, same flows through capacitor C too.

i.e. $I = \frac{v_i}{R} = \frac{-C\,dv_o}{dt}$

So that, $v_o = \frac{-1}{RC}\int V_i\,dt = K\int V_i.\,dt$

Thus, $V_o = K\int V_i$ where, $K = -\frac{1}{RC}$ and hence the result.

Q. 23. Give some practical circuit which gives an output given by $e_0 = \int e_i\,dt$.

Give all the component values and underline design procedure in brief.

Ans. Basic circuit arrangement is shown in Fig. 19.28.

The circuit consists of two sections in cascade. The output of first section is given by,

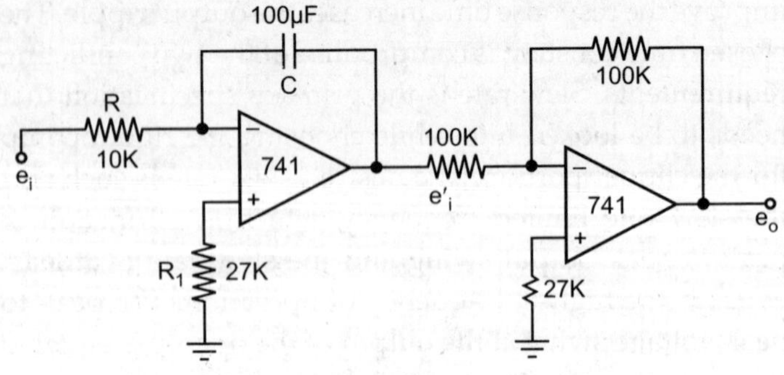

Fig. 19.28

$$e'_0 = K\int e_i\,dt \text{ where, } K = -\frac{1}{RC}$$

Let $R = 10\ K$, then for $|K| = \text{Unity}$

$C = 1/104\ F = 100\ \mu F$

Therefore, if $R = 10\ K$ and $C = 100\ \mu F$

Then $e'_0 = -\int e_i\,dt$ and $e_0 = \int e_i\,dt$.

Q. 24. It is desired to have an output which is sum of the integrals of the various inputs, i.e.,

$$e_0 = \int e_1\,dt + \int e_2\,dt + \int e_3\,dt + \ldots..$$

Given an appropriate circuit and prove the result.

Ans. The circuit is shown in Fig. 19.29.

From the circuit above $I = I_1 + I_2 + I_3 + \ldots. = -C\frac{de'_0}{dt}$

or $= \frac{e_1}{R} + \frac{e_2}{R} + \frac{e_3}{R} = -C\frac{de'_0}{dt}$

i.e. $e'_0 = -\frac{1}{RC}\left(\int r_1\,dt + \int e_2\,dt + \int e_3\,dt + \ldots\right)$

Selecting $R = 10\ K$ and $C = 100\ \mu F$, we get $RC = 1$

i.e. $e'_0 = -\left(\int e_1\,dt + \int e_2\,dt + \int e_3\,dt + \ldots.\right)$

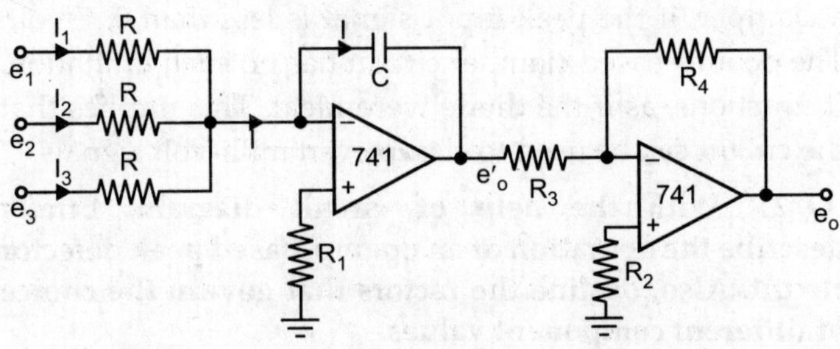

Fig. 19.29

If $\frac{R_4}{R_3} = \text{Unity (say } R_3 = R_4 = 100\ K)$

Then $e_0 = -e'_0 = \int e_1\,dt + \int e_2\,dt + e_3\,dt + \ldots.$ and hence, the result.

Q. 25. Draw the basic differentiator circuit using an opamp and show that the output voltage is differential of the input.

Ans. Circuit shown in Fig. 19.30 is the basic differentiator.

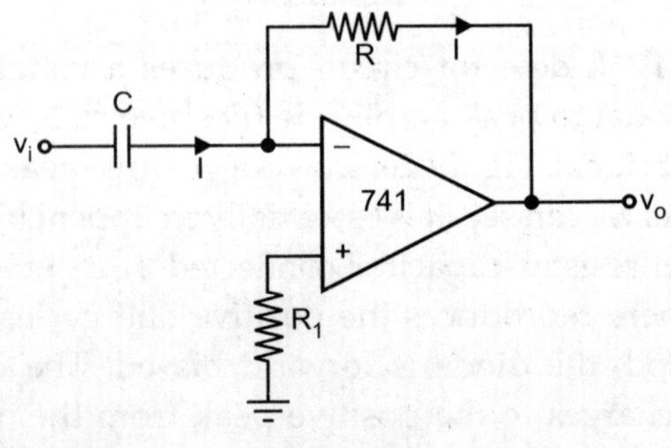

Fig. 19.30

Assuming a zero current towards (–ve) input terminal of opamp and also considering a virtual earth occurring at *R-C* junction we can write,

$$I = C\frac{de_i}{dt} = \frac{-v_o}{R} \text{ so that } v_o = -RC\frac{v_i}{dt}$$

$$V_o = \frac{K\,dv_i}{dt} = \text{where } K = -RC.$$

Q. 26. Draw an appropriate circuit which gives an output (e_0) given by $e_0 = \left(\frac{de_i}{dt}\right)$

Ans. The circuit is shown in Fig. 19.31. To make $e_0 = \frac{de_i}{dt}$ $RC = 1$

and $R_3 = R_4$; $R_2 = R_3 \parallel R_4$ and $R_1 = R$

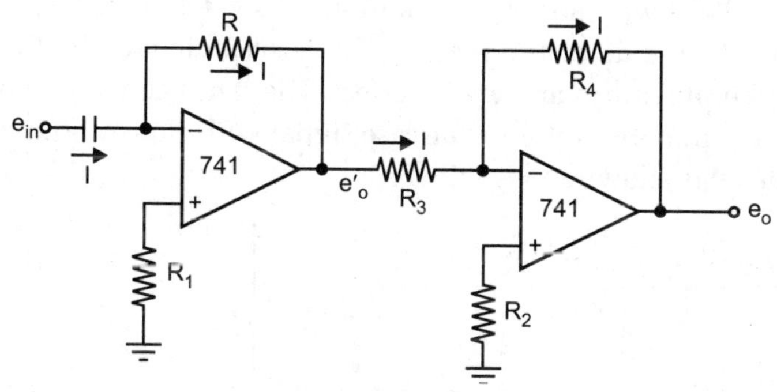

Fig. 19.31

Q. 27. Draw an appropriate circuit using opamp which gives an output equal to the sum of differentials of three inputs. Explain its operation by deriving an expression to prove the above stated result.

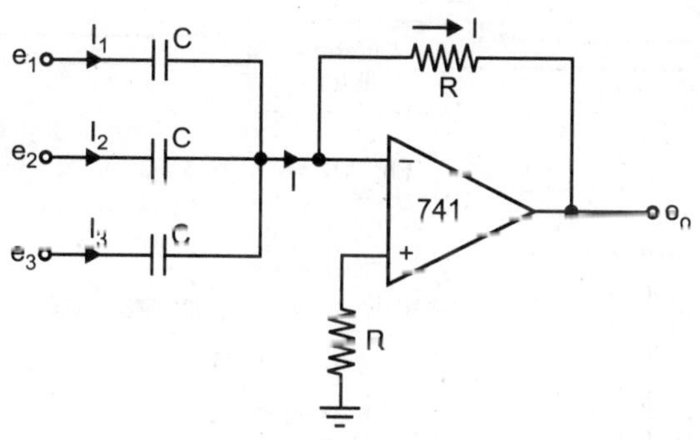

Fig. 19.32

Ans. An opamp circuit giving the required output is shown in Fig. 19.32. The negative input terminal being at virtual earth gives :

$$I_1 = \frac{Cde_1}{dt}, I_2 = \frac{Cde_2}{dt}, I_3 = \frac{Cde_3}{dt}$$

$$I = I_1 + I_2 + I_3 = C\left[\frac{de_1}{dt} + \frac{de_2}{dt} + \frac{de_3}{dt}\right]$$

Since currents flowing towards input terminals of an opamp are zero.

This gives current through $R = I = -\frac{e_0}{R}$

$$-\frac{e_0}{R} = C\left[\frac{de_1}{dt} + \frac{de_2}{dt} + \frac{de_3}{dt}\right]$$

$$e_0 = (-RC)\left[\frac{de_1}{dt} + \frac{de_2}{dt} + \frac{de_3}{dt}\right] \text{ and hence, the result.}$$

Q. 28. Design an opamp circuit giving an output e_0 given by $e_0 = \sin t - \cos t$ from the available inputs $e_1 = \sin t$ and $e_2 = \cos t$. Give all the component values and make the reasonable assumptions, wherever required.

Ans. The above stated output could be achieved from given inputs by a number of opamp circuits *e.g.* (*i*) Integrator (*ii*) Subtracter (*iii*) Differentiator.

Of all the three above mentioned designs, differentiator requires minimum number of circuit components as d/dt (sin t + cos t) = – (sin t – cos t) and when inverted gives desired output (Fig. 19.33). This inversion occurs automatically as the inputs are applied at inverting input. On the other hand, integrator's output – (sin t – cos t) has to be inverted by a unity gain inverting amplifier to get the desired result.

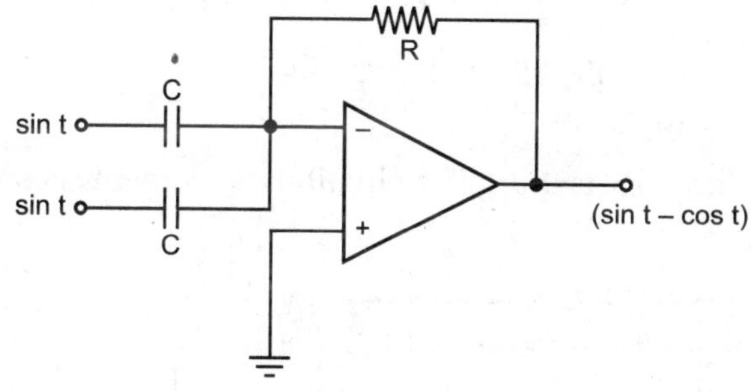

Fig. 19.33

Subtractor uses comparatively smaller number of elements. But the designer must not be tempted to use a differentiator circuit for the sake of compatibility of the circuit as it has certain disadvantages which far outweigh the savings in the circuit size. These disadvantages have been discussed in the next question.

Thus, either the 'integrator' or 'subtractor' is used for this design problem.

With Integrator: The basic circuit is shown in Fig. 19.34:

Form Integrator theory, output of first opamp

$$= -\frac{1}{RC}\int(\sin t + \cos t) = -\frac{1}{RC}(\sin t - \cos t)$$

$$= -(\sin t - \cos t) \text{ (If } RC = 1)$$

If $\frac{R_4}{R_3}$ = 1, then output of second opamp = (sin t – cos t) = Desired result.

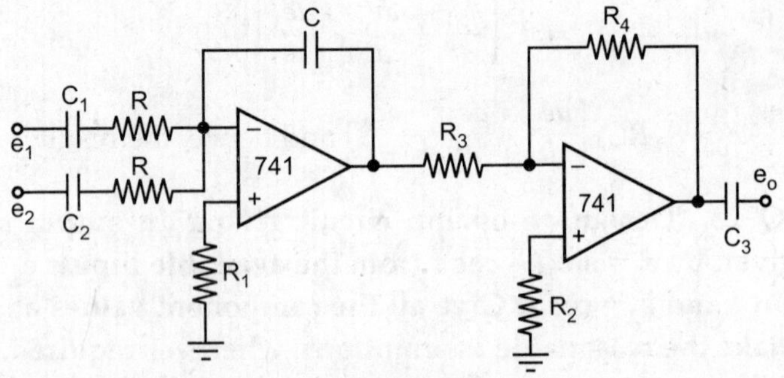

Fig. 19.34

Thus, design constraints are : $RC = 1, \frac{R_4}{R_3} = 1$

Let $R = 10$ K ; $C = 1/10{,}000 = 100$ μF

Choose $R_3 = R_4 = 100$ K

$C_1, = C_2, C_3$ are blocking capacitors.

Choose high value capacitors for C_1, C_2, C_3 so that they allow even minimum frequency AC signal.

Let $C_1 = C_2 = C_3 = 1000$ uF

$R_1 = R/2, R_2 = R_3 \parallel R_4$

Opamp No. = 741

With subtractor : The circuit arrangement is shown in Fig. 19.35:

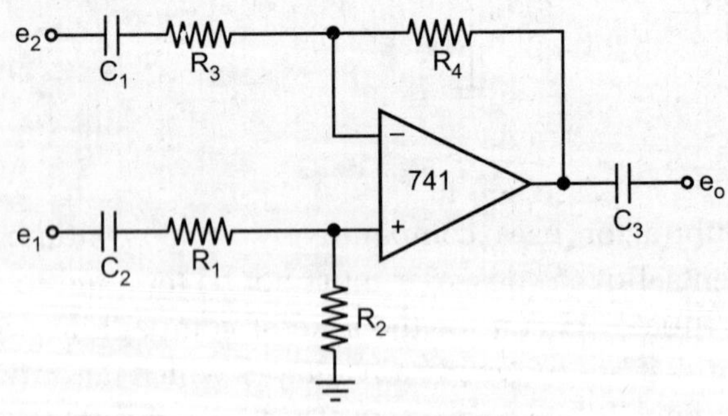

Fig. 19.35

From subtractor theory.

For $e_0 = e_1 - e_2$

$\frac{R_4}{R_3} = \frac{R_2}{R_1}$ = Unity

Let $R_3 = 100\ K$ Then, $R_4 = 100\ K$

Let $R_2 = 100\ K$ Then, $R_1 = 100\ K$

$C_2 = C_3 = 1000\ \mu F.$

Q. 29. Differentiator circuit may be avoided as far as possible and Integrator used instead. Use a Differentiator only, when it is a must and there is no other alternative." How far is the above statement true? Give illustrations in support of your answer.

Ans. Of all the types of signals used in the electronic systems, two most important ones are:

(*i*) Square Waveform (*ii*) Ramp Waveform.

Let us consider the output that will result when a ramp signal is applied to an integrator as well as a differentiator.

As the diagrams shown in Fig. 19.36. illustrate, a differentiator provides a constant output, whereas an integrator gives a hyperbolic variation.

This was an ideal situation. Now let us take the practical side of the picture. The electronic circuits that generate ramps are never perfect. The ramp generated by them may resemble a staircase signal with closely spaced steps as shown in Fig. 19.37.

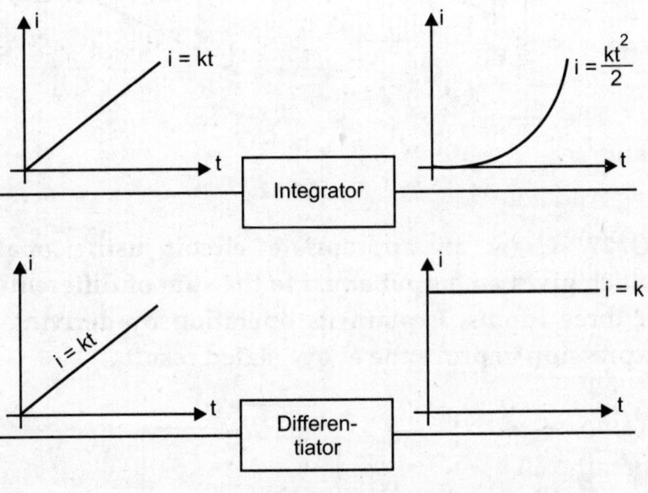

Fig. 19.36

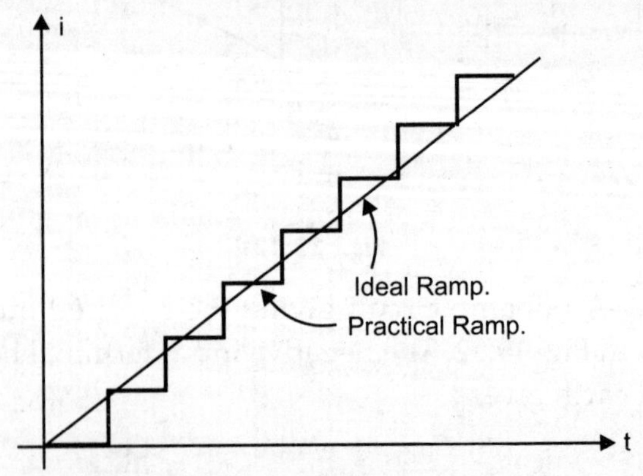

Fig. 19.37

When this type of ramp is applied to an integrator, the output varies slightly from what should have been in case of the input being an ideal ramp. But when it is applied to a differentiator circuit, we may get infinite amplitude pulses at abrupt transitions and no output during constant amplitude portion. Fig. 19.38. illustrates the phenomenon.

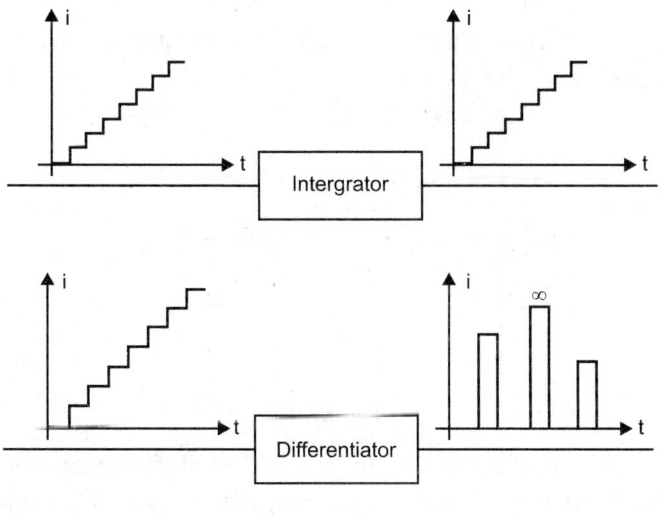

Fig. 19.38

The signal at the output of differentiator is, thus, quite senseless.

Another disadvantage of differentiator is that noise (having higher frequencies) may pass through the input capacitor and present itself at the output. On the other hand, integrators are much more immune to noise because of resistance being the first element and the circuit offering reduced gain at higher frequencies.

Q. 30. Give an arrangement to show how opamp adder circuit could be used for multiplying two voltages, .

Ans. The arrangement is shown in Fig. 19.39.

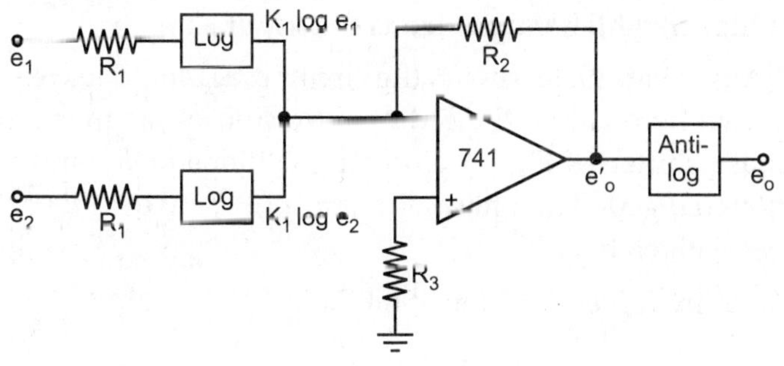

Fig. 19.39

$$e'_0 - - K_1 \log e_1 - K_1 \log e_2 = - K_1 (\log e_1 + \text{lag } e_2)$$
$$= - K_1 \log (e_1 . e_2)$$

Now $\quad e_0 = \text{Antilog } e'_0 = (\text{constant}) \times e_1 e_2$

Thus, $\quad e_0 = K e_1 e_2$

Q. 31. Give an arrangement to show how opamp adder circuit could be used to divide two voltages, .

Ans. The arrangement is shown in Fig. 19.40.

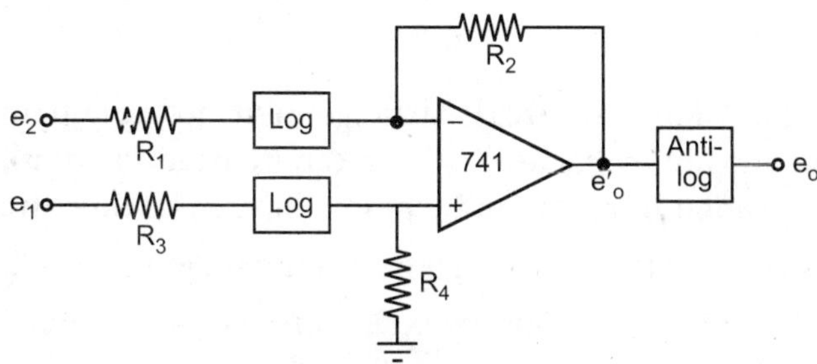

Fig. 19.40

$$e'_0 = K_1 (\log e_1 - \log e_2) = K_1 \log (e_1/e_2)$$

Therefore, $\quad e_0 = K (e_1/e_2)$

Q. 32. What is a Gyrator?

Ans. The word Gyrator means 'to turn round' or 'to invert'. The circuits which are used to simulate inductances with the help of opamps, resistors and capacitors are called Gyrators. In other words, Gyrator gives inductance without using any inductive coil. It does so by inverting or gyrating the frequency dependent characteristics of capacitive reactance ($Z_C = 1/\omega C$), into frequency dependent characteristics of inductive impedance ($Z_L = \omega L$) thus, justifying the name gyrator.

Q. 33. What is the purpose of using such a complicated arrangement to have inductive effects?

Ans. The fabrication of inductance in the integrated form requires a lot of space thus, prohibiting the use of inductance in the electronic circuits, if they are to be fabricated by *IC* technology. As the *IC* technology is gaining more and more importance every day due to its reliability, the compactness of circuit, longer life, lesser power consumption and many other advantages, it becomes essential to devise some method of simulating inductance easily by *IC* technology. The gyrator serves the purpose well, as the fabrication of opamps capacitors and resistors is very easy.

Q. 34. What do you mean by 'floating inductance' and 'grounded inductance'?

Ans. An inductance may be connected in a circuit either in series or in shunt (from one terminal to ground). First arrangement is called floating inductance, whereas in the latter case the inductance is termed as grounded inductance. Refer to Fig. 19.41. for illustration.

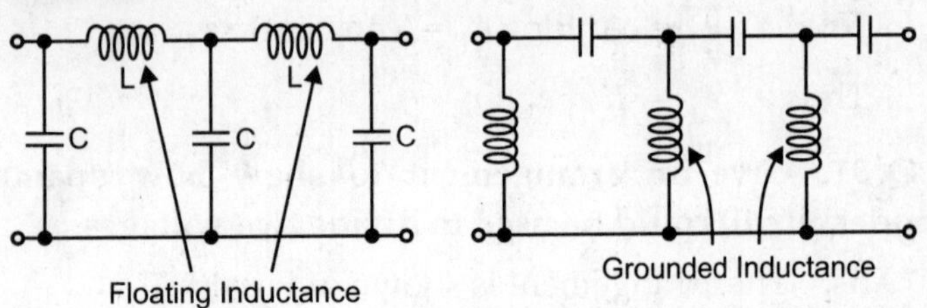

Fig. 19.41

Q. 35. Show a circuit arrangement using opamps resistors and capacitors, which can be used to simulate a grounded iron-core inductance. Explain its operation.

Ans. The circuit arrangement is shown in Fig. 19.42 (*a*)

In order to explain its operation, let us consider the electrical equivalent circuits of the choke at low and high frequencies as shown in Fig. 19.42 (*b*) and (*c*).

For $f = 0$, capacitor (*C*) becomes open so that V_{in} is effectively applied across R_1 [circuit (*a*)]; for $f = \infty$, C becomes short. The voltage at (+ve) input terminal becomes equal to V_{in} and same is at (–ve) input and hence, at the output.

Thus, voltage at the two ends of R_1 is V_{in} and hence, there is no current through R_1. That is, R_1 is effectively out of the circuit. Under these circumstances, V_{in} is appearing across R_2 only.

To summarise, at $f = 0$ and $f = \infty$ the opamp circuit shown behaves as in circuit (*b*) and circuit (*c*) respectively Fig. 19.42. (*b*) and (*c*) represent the equivalent circuits of a choke at $f = 0$ and $f = \infty$ respectively, hence opamp circuit shown in Fig. 19.42. (*a*) is equivalent to a choke.

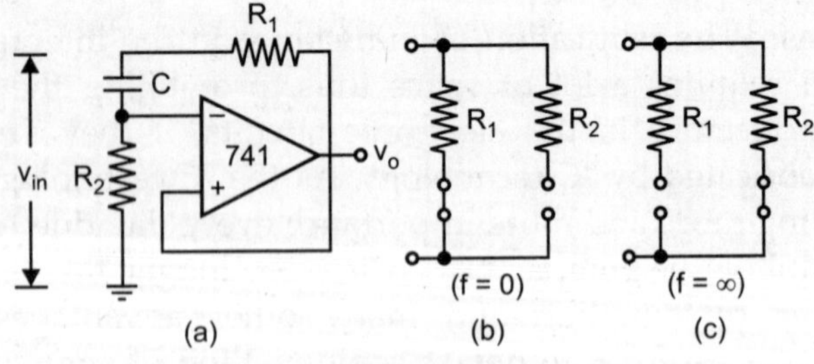

Fig. 19.42

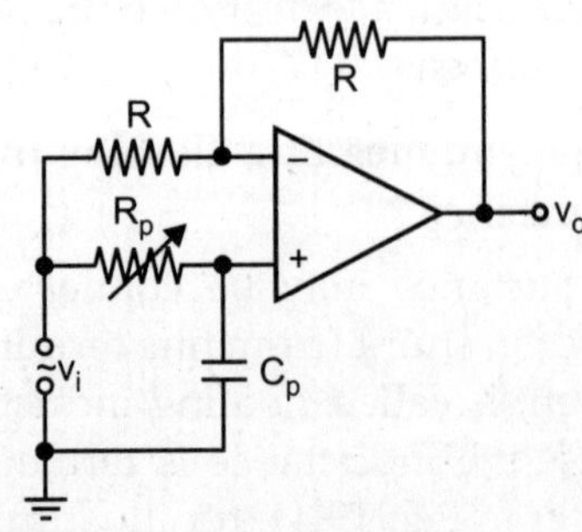

Fig. 19.43

Q. 36 With the help of circuit diagram, briefly describe the working of basic single opamp phase shifter circuit in which the output lags the input.

Ans. Fig. 19.43. shows the circuit diagram of single opamp based lagging type phase shifter circuit. The output lags the input by an angle (θ).

$$\theta \text{ (in degrees)} = -2 \tan^{-1} (\omega R_P C_P)$$

where, $\omega = 2\pi f$, where, (*f*) is the frequency of input signal.

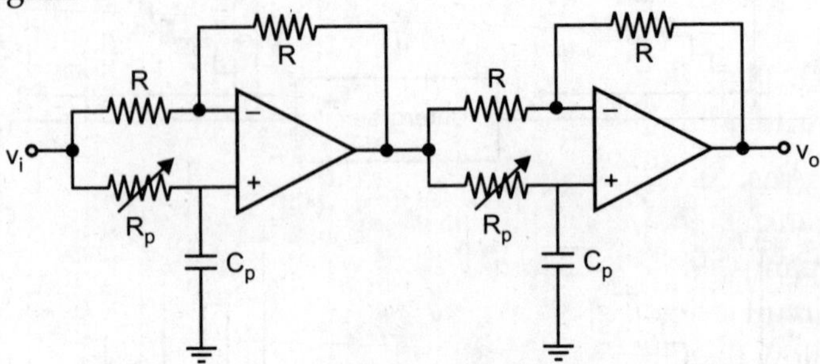

Fig. 19.44

The simple circuit can be used of shift the phase of the input signal over a wide range by varying (R_P) with 0° and – 180° being the extremes.

For $R_P << \dfrac{1}{\omega C_P}$, the phase shift is near zero. (It will be 0°, when $R_P = 0$ which is not practical).

For $R_P >> \dfrac{1}{\omega C_P}$ (θ) approaches – 180°

(θ = – 180° only for R_P = infinity which is again not feasible). For $R_P = \dfrac{1}{\omega C_P}$, θ = – 90°. Two such sections can be used in cascade to vary the phase shift over full 360°. Fig. 19.44. shows the circuit diagram.

Q. 37. Briefly describe the working of an opamp phase shifter in which output leads the input.

Ans. Fig. 19.45. shows the circuit diagram. The circuit shown here is just the redrawn version of lagging type phase shifter of Fig. 19.43. with positions of R_P and C_P interchanged. The phase difference (θ) between (v_i) and (v_0) is given by,

$$\theta \text{ (in degrees)} = 2 \tan^{-1} (\omega R_P C_P)$$

for $R_P = \dfrac{1}{\omega C_P}$; θ = 90°

for $R_P >> \dfrac{1}{\omega C_P}$; θ = 180°

for $R_P << \dfrac{1}{\omega C_p}$; θ = 0°

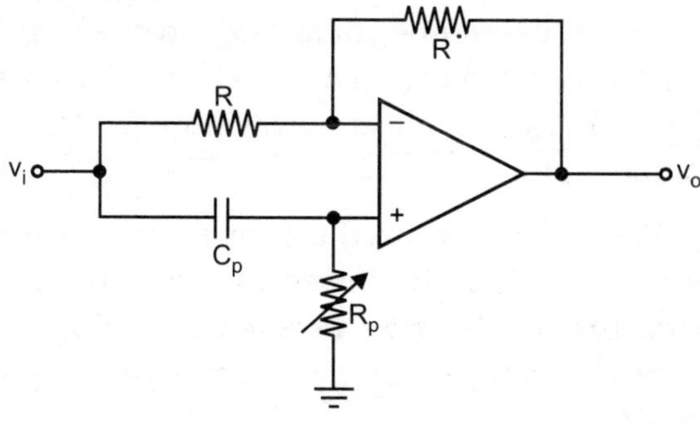

Fig. 19.45

Q. 38. Suggest a circuit using a Norton Opamp to build a non-inverting type high voltage DC amplifier.

Ans. Fig. 19.46.(*a*) shows the circuit diagram. The transfer characteristics are shown in Fig. 19.46. (*b*). The opamp used here is *LM* 3900. The voltage gain of this circuit is given by (R_2/R_1). Resistors (R_3) have been used to allow the output voltage (V_0) to go to zero for zero input. Typical value of (R_3) to be used in this configuration is 1*M*. This configuration is very useful for generating a variable high voltage *DC* output or a high voltage ramp, where the load current demand is very small.

Q. 39. Suggest a circuit using a Norton opamp to build an inverting type high voltage *DC* amplifier with the transfer characteristics as shown in Fig. 19.47. (*a*).

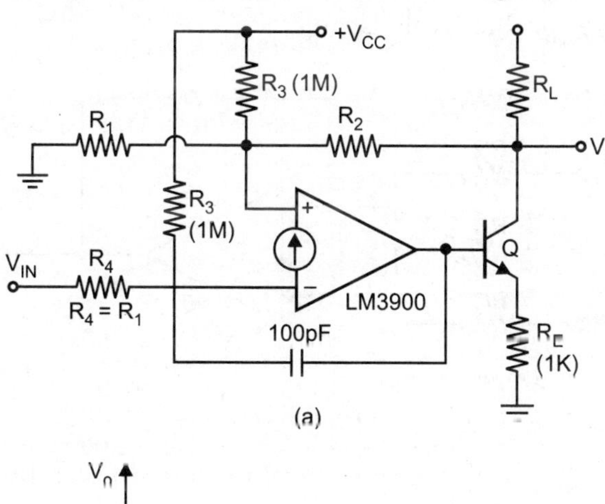

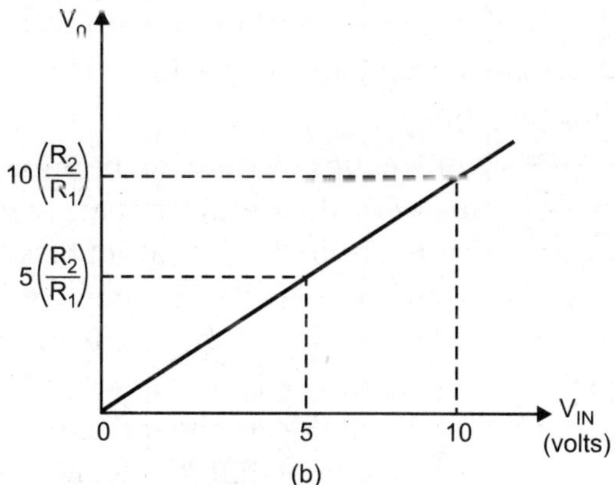

Fig. 19.46

Ans. Fig. 19.46. (*b*) shows the circuit diagram. The gain is again given by the ratio (R_2/R_1).

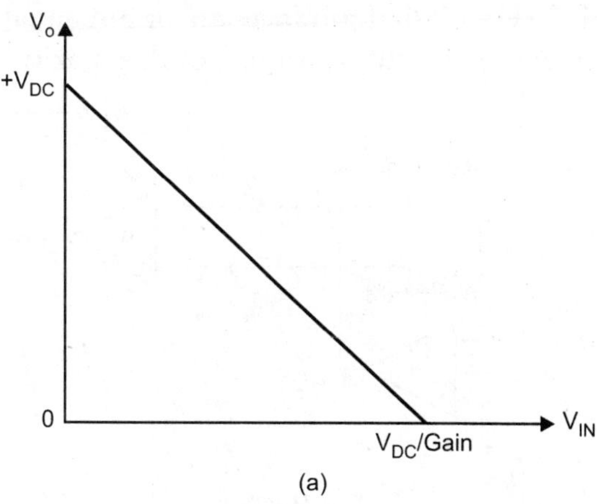

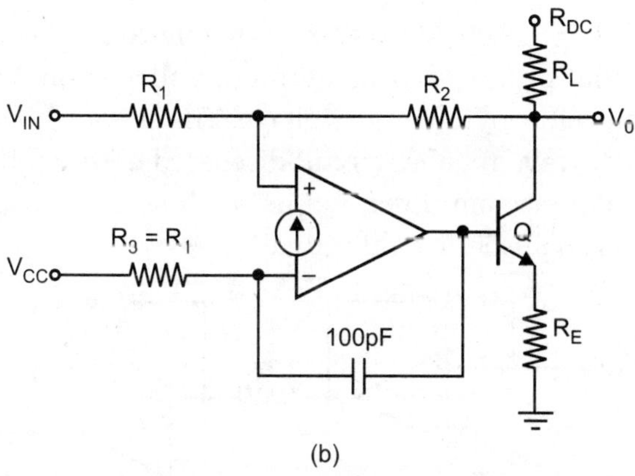

Fig. 19.47

Q. 40. With help of circuit diagram, explain how an opamp-based summer circuit can be used to build an averaging amplifier?

Ans. An Averaging Amplifier is the one that produces an output equal to the average of the multiple inputs. Fig. 19.48. shows the circuit that can produce at the output an average of (*N*) inputs. When all input resistors are equal (=*R*) and the feedback resistor equals (*R*/*N*), then it can be proved that the output is given by :

$$V_0 = (V_1 + V_2 + V_3 + \ldots\ldots\ldots\ldots\ldots + V_N)/N$$

Fig. 19.48

Q. 41. With the help of circuit diagram, explain how an opamp can be configured as an Absolute Value Circuit i.e. a circuit that produces an output equal to the magnitude of the input regardless of its polarity.

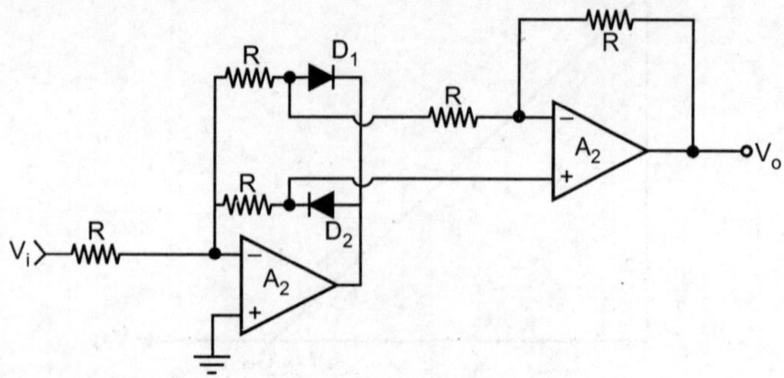

Fig. 19.49

Ans. Fig. 19.49. shows one possible opamp configuration that produces at its output a voltage equal to the absolute value of the input voltage. The circuit shown is a dual half wave rectifier circuit discussed earlier followed by a difference amplifier. Let us see how this circuit behaves as an absolute value circuit.

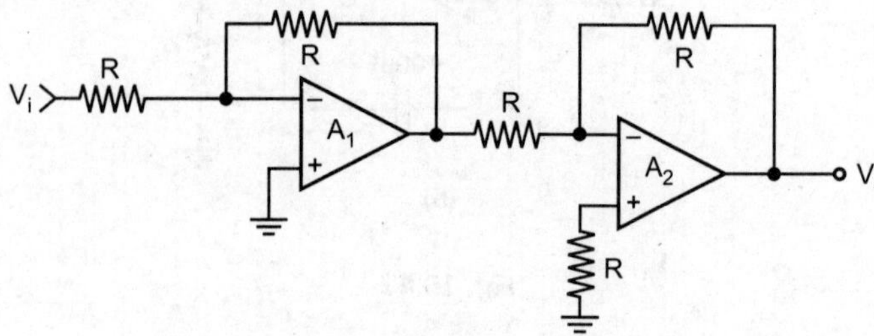

Fig. 19.50

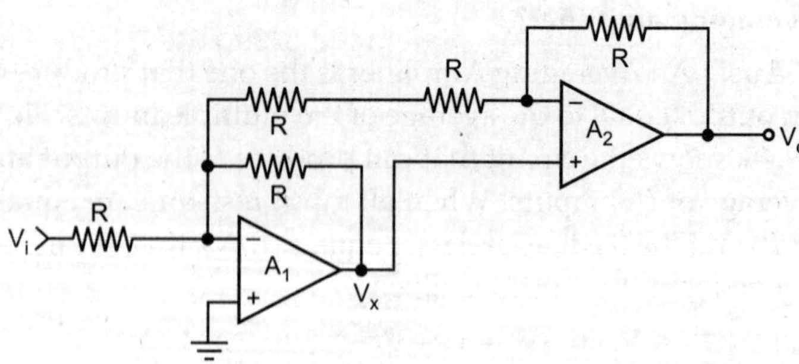

Fig. 19.51

When the applied input is of positive polarity (say +V), diode (D_1) is forward biased and diode (D_2) is reverse biased. The circuit reduces to what is shown in Fig. 19.50. Simple mathematics shows that output (– V_0) in this case is equal to (+ V). When the applied input is of negative polarity (say – V), diode (D_1) is reverse biased and the diode (D_2) is forward biased. The equivalent circuit in this case is shown in Fig 19.51. By applying Kirchoff's Current Law (KCL) at the inverting terminal of the first opamp, we can determine voltage (V_x) to be equal to (2/3) V. Also, (V_x) is related to (V_0) by $V_x = (2/3)\ V_0$. This implies that $V_0 = V$. Thus, the output always equals the absolute value of the input.

Q. 42. We are aware of the role of an opamp in a D/A converter, where it is used as a current to voltage converter. Briefly describe how an inverting summer opamp configuration can be used as a simple D/A converter ?

Ans. Fig. 19.52. shows how the basic inverting summer circuit can be used as a 4-bit D/A converter. As is clear from the circuit, the MSB has a gain of 8 (= 2^3), second MSB has a gain of 4 (= 2^2), third MSB has a gain of 2 (= 2^1) and the LSB has a gain of unity (= 2^0). If a digital '1' were represented by 1 volt and a digital '0' by zero volt, the analog output of this converter circuit will vary from 0 to 15 volts as the digital input varies from 0000 to 1111. However, it is not necessary for the analog output to be equal to decimal equivalent of the binary input. The full-scale analog output could as well be any other value. However, this basic circuit has the disadvantage that source of each digital bit sees a different input resistance, which keeps reducing as we go to higher order bits. Also, such a circuit becomes impractical for larger bit converters as the ratio of the smallest resistor and the largest resistor becomes prohibitively large.

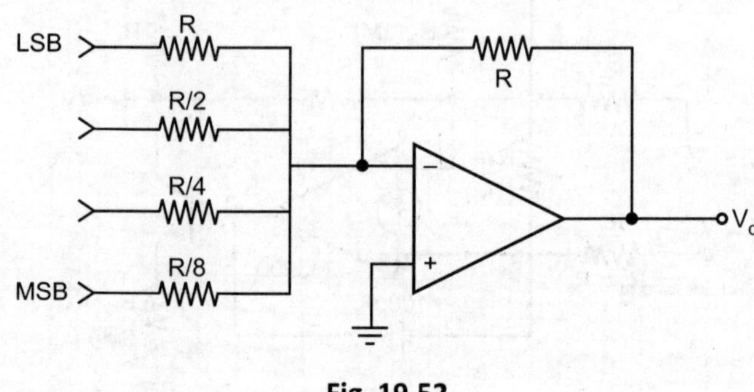

Fig. 19.52

Q. 43. Briefly describe how three opamps can be used to build an Instrumentation Opamp? Also, write the expression for gain.

Ans. An Instrumentation Opamp can be implemented with three opamps with the output opamp wired as a difference amplifier fed from the outputs of the other two opamps wired as buffers. Fig. 19.53. shows the circuit diagram. It can be verified that this circuit provides a differential gain of $[(2R/R_G) + 1]$. Resistance (R_G) can be used to control the gain of the amplifier. Instrumentation opamps are available as monolithic integrated circuits with performance parameters better than what are achievable with three-opamp configuration. Resistance (R_G) is connected external to the opamp.

Q. 44. There are opamp configurations that can simulate a stable voltage reference from a not too stable zener reference. With the help of circuit diagram, briefly describe the functional principle of a stable voltage reference built around an opamp.

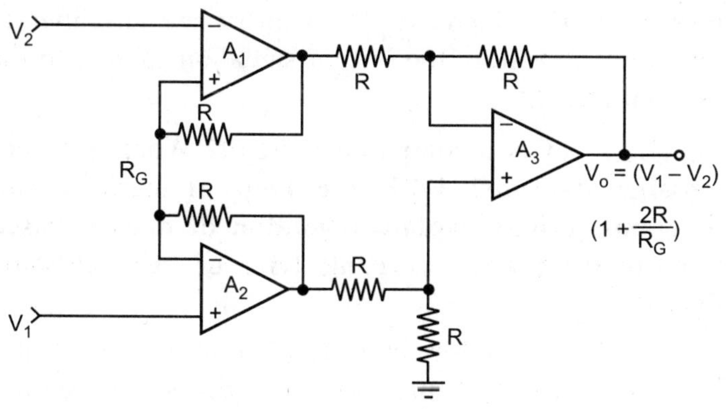

Fig. 19.53

Ans. Fig. 19.54. shows the circuit diagram. The opamp used here is a single supply opamp. Remember that the input stage of a single supply opamp is different from that of a conventional opamp. Without going into details, it may be mentioned that both inverting and non-inverting inputs behave as *PN* junctions connected from the terminal to ground. If this is true, then voltage across resistor (R_1) in Fig. 19.54. will be 0.6 volt which would

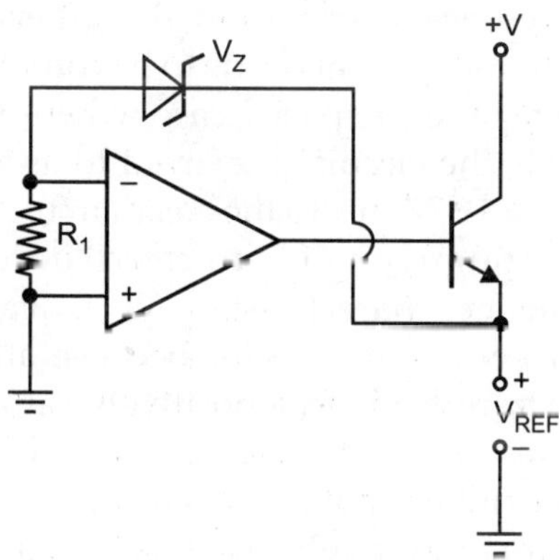

Fig. 19.54

ensure a constant current through the resistor. This constant current is also the bias current for the zener diode as the opamp input draws negligible current. A constant bias current through the zener diode leads to a stable zener reference voltage. The reference voltage available from the circuit is (V_Z + 0.6). The transistor acts as a current booster and it is there to meet the bias current requirements of the zener diode, which the opamp output usually would not be capable of. If the output reference voltage tends to change, it results in change in the input bias current to the opamp, as the voltage at the inverting input can not change. Change in input bias current to the opamp causes a change in opamp output voltage, which keeps the output reference voltage constant.

Q. 45. Suggest circuit configuration using a single opamp, which can be used as full wave rectifier.

Ans. Fig. 19.55 shows the circuit diagram. During the positive half cycles of the input signal, the diode is reverse biased and therefore the feedback resistor is disconnected from the output of the opamp. The positive half cycles appear as such. During negative half cycles, the diode is forward biased. The negative half cycles get inverted and again appear as positive half cycles. Thus, the output is a full wave rectified output.

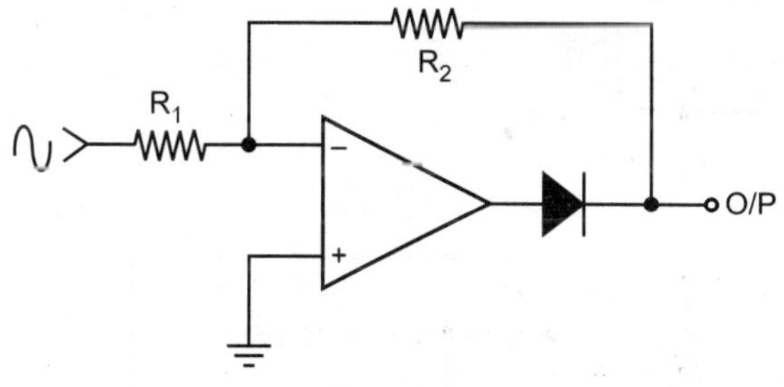

Fig. 19.55

Q. 46. One of the circuit applications to which an opamp adapts itself very well is the WEIN-BRIDGE type of *RC* phase shift oscillator. Briefly describe the operation of an opamp-based Wein-bridge oscillator. Also, outline the conditions for determining various component values for a given frequency of operation.

Ans. Wein-bridge oscillator is a type of *RC* phase shift oscillator in which the frequency selective network in the feedback path of the amplifier portion is the well-known Wein bridge network of Fig. 19.56. (*a*). Fig. 19.56 (*b*) shows the opamp-based Wein-bridge oscillator. As we can see that the resistive arms of the bridge network (R_1) and (R_2) along with the opamp constitute the amplifier portion of the oscillator. The amplifier in this case is a non-inverting amplifier configuration having a gain of $(1 + R_2/R_1)$. The amplifier in this case has got be a non-inverting one because the frequency selective network formed by series *R-C* and parallel *R-C* arms does not introduce any phase difference to the signal, as it travels from output to the non-inverting input of the opamp-based amplifier. This satisfies the phase shift condition of Barkhausen criterion. The frequency at which this phase shift condition is satisfied is given by:

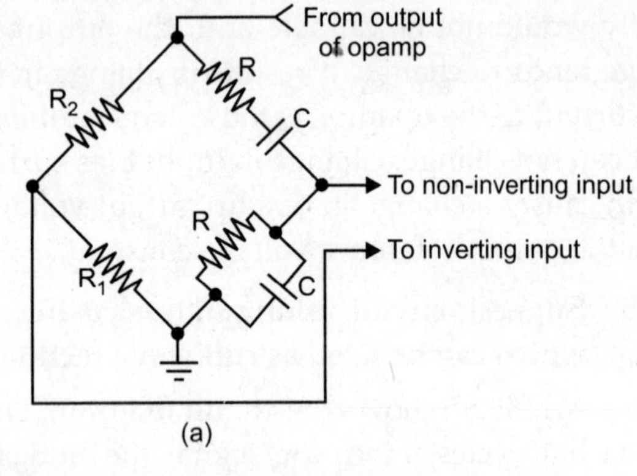

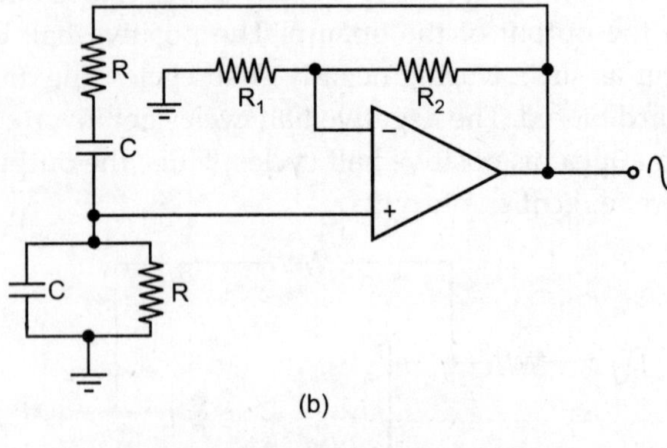

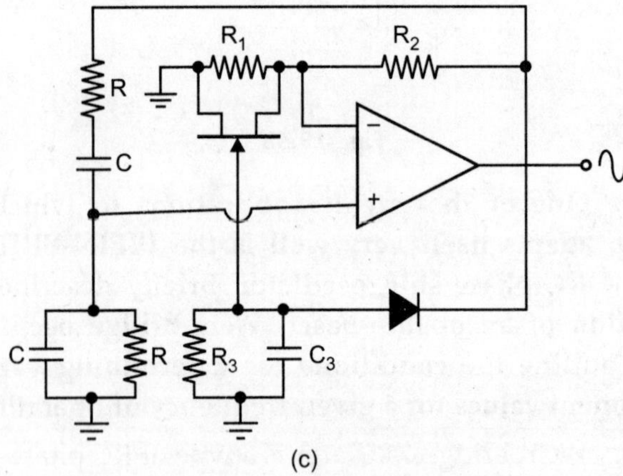

Fig. 19.56

$$\omega = 1/RC$$

The other condition to be satisfied for sustained oscillations is that of loop gain being unity. It can be mathematically verified that for $\omega_o = 1/RC$, the output signal is attenuated by a factor of 3 before it is fed back to the non-inverting input. This means that the amplifier portion of the oscillator must provide a gain of at least 3. This further implies that $R_2 = 2R_1$. In fact, R_2 can he slightly greater than $2R_1$.

To achieve amplitude stability of the output signal, a parallel combination of a fixed resistor and the drain-source of a junction FET often replace resistance (R_1) as shown in Fig. 19.56. (*c*). The gate terminal of the FET is driven by a negative *DC* voltage, which is produced from rectified output signal. An increase in output signal amplitude leads to an increased negative *DC* voltage at the FET gate. This increases the drain-source terminal resistance, which reduces the gain of the amplifier thus restoring the output amplitude. Similarly, when the output signal amplitude decreases, reduced negative *DC* voltage at the gate terminal decreases the drain-source resistance. This increases the gain of the amplifier to the extent to restore the output signal amplitude to the original value.

Q.47. What is a comparator circuit? What is a zero crossing detector? With the help of basic circuit diagrams, briefly explain operation of opamp based non-inverting and inverting type of zero crossing detectors.

Ans. A comparator circuit is a two-input, one-output building block that produces a high or low output depending upon the relative magnitudes of the two inputs. An opamp can be very conveniently used as a comparator, when used without negative feedback. Because of very large value of open-loop voltage gain, it produces either positively saturated or negatively saturated output voltage depending upon whether the amplitude of the voltage applied at the non-inverting input terminal is more or less positive than the voltage applied at the inverting input terminal.

One of the inputs of the comparator is generally applied a reference voltage and the other input is fed with the input voltage that needs to be compared with the reference voltage. In a special case, where the reference voltage is zero, the circuit is referred to as zero-crossing detector. Figure 19.57 shows the basic circuit arrangement of a non-inverting type of zero-crossing detector along with its transfer characteristics. It is called a non-inverting zero-crossing detector because an input more positive than zero leads to a positively saturated output voltage. Diodes D_1 and D_2 connected at the input are to protect the sensitive input circuits inside the opamp from excessively large input voltages. R is the current-limiting resistor.

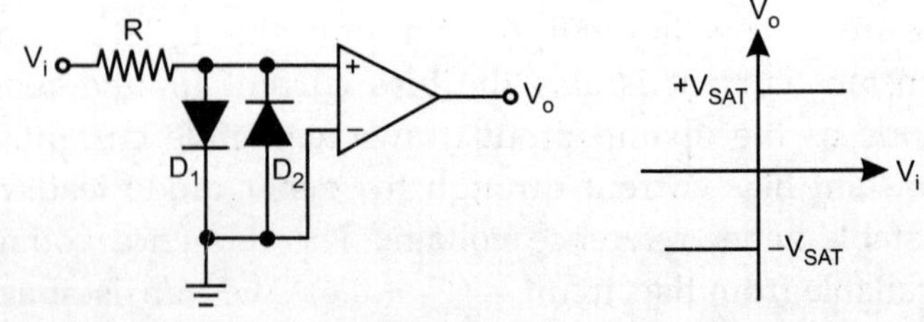

Fig. 19.57

Figure-19.58 shows the inverting type of zero-crossing detector along with its transfer characteristics. In this case, input voltage slightly more positive than zero produces a negatively saturated output voltage. One common application of zero-crossing detector is to convert sine wave signal to a square wave signal. Figures 19.59 (a) and (b) respectively show relevant waveforms for non-inverting and inverting type of zero-crossing detector circuits.

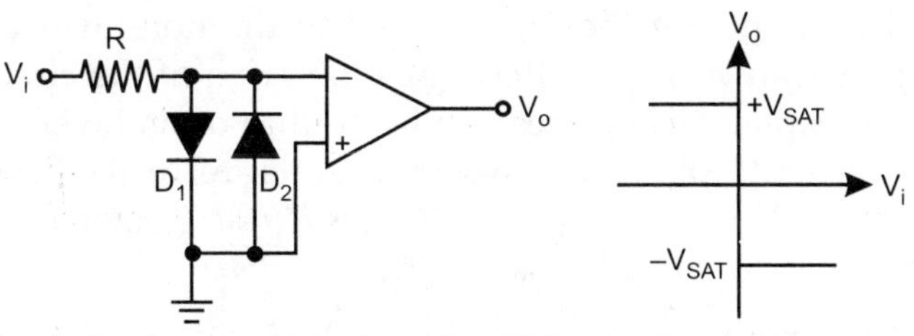

Fig. 19.58

Fig. 19.59

Q.48. In the generalised case of a voltage comparator, reference voltage may be a positive or a negative voltage. With the help of circuit diagram, briefly explain operation of non-inverting comparators with positive and negative reference voltages.

Ans. Fig. 19.60 (*a*) shows the circuit diagram of non-inverting comparator for a positive reference voltage. V_{REF} in this case is given by $+V_{CC} \times [R_2/(R_1 + R_2)]$. Fig. 19.60 (*b*) shows the circuit diagram of non-inverting comparator for a negative reference voltage. Reference voltage V_{REF} is given by $-V_{CC} \times [R_2/(R_1 + R_2)]$. Inverting comparators can be built in the same manner by applying the input to the inverting input of the opamp and the reference voltage to the non-inverting input of the opamp.

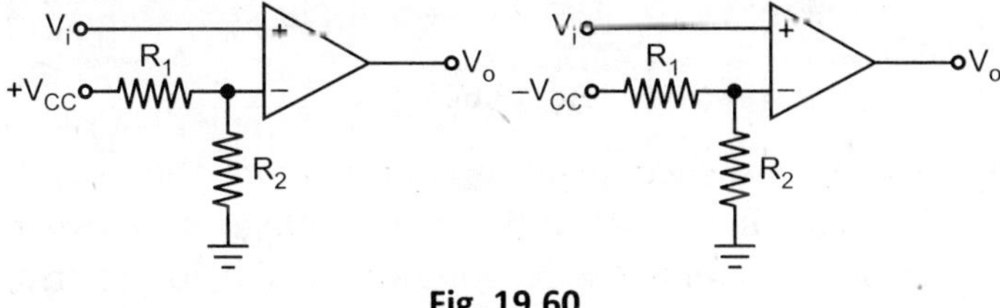

Fig. 19.60

Q.49. General purpose opamps are seldom used as voltage comparators. Opamp comparators are specially designed opamps optimised for comparator function. What are the limitations of general purpose opamps, when used as comparators and what are the chief attributes of opamp comparators?

Ans. In the preceding two questions, we have discussed use of general-purpose opamps as voltage comparators. General-purpose opamps when used as comparators, suffer from slew rate limitation. Relatively lower slew rate forces the transition time from one state to the other to be prohibitively large. Though this problem can be overcome by using a high-speed opamp with a higher slew rate specification; a better design approach to overcoming this limitation is by eliminating the compensation capacitor. It may be mentioned here that comparator works as a non-linear circuit and therefore, elimination of compensation capacitor has no derogatory effect on the performance. With capacitor removed, the only capacitance remaining is the stray capacitance across the output. Thus, slew rates can be very high.

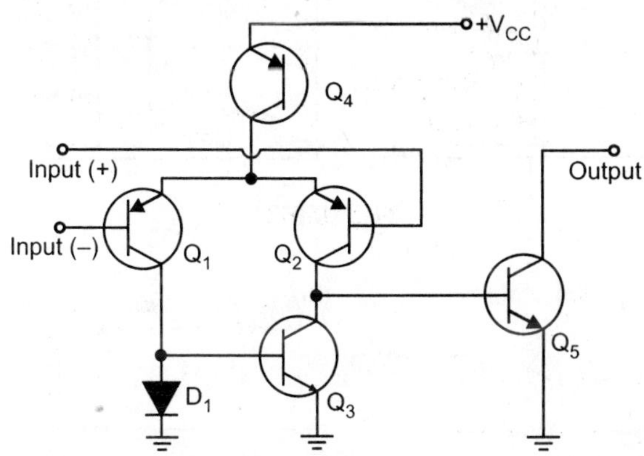

Fig. 19.61

Another important parameter of a comparator is its ability to operate from a single supply and interface conveniently with popular logic families. Input circuit of opamp comparator is tailored to meet these two requirements. Fig. 19.61 shows the simplified schematic diagram of opamp comparator. As seen in the figure, the output has an open-collector output stage. For the output stage to work properly, the output terminal needs to be connected to the positive supply voltage through an external resistor called pull-up resistor. It is called pull-up resistor as it pulls the output voltage to the supply voltage, when the output transistor Q_5 (in Fig. 19.61) is in cut-off state. Not all comparators have an open-collector output stage. In fact, pull-up resistor slows down the response time of the comparator. There are opamp comparators with active pull-up output stage that are capable of producing relatively much faster switching times. These comparators need dual power supplies.

Q.50. Why is there a need for the comparator circuit to have hysteresis, more so when the input signal to be compared is varying slowly? With the help of circuit diagrams, briefly describe the operation of non-inverting and inverting comparators with hysteresis.

Ans. When the input signal to comparator contains noise, transitions at the output around the trip point tend to become highly erratic. Instead of being smooth from one state to the other, transition around the trip point is a cluster of pulses with randomly varying pulse width. The problem becomes particularly severe, if the input signal were changing slowly. This phenomenon is demonstrated in Fig. 19.62. While Fig. 19.62 (*a*) illustrates the ideal condition; Fig. 19.62 (*b*) shows an erratic output in the presence of noisy signal. The two figures are self-explanatory.

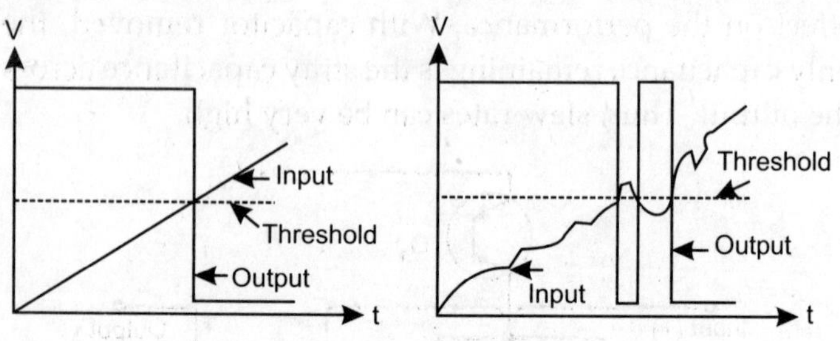

Fig. 19.62

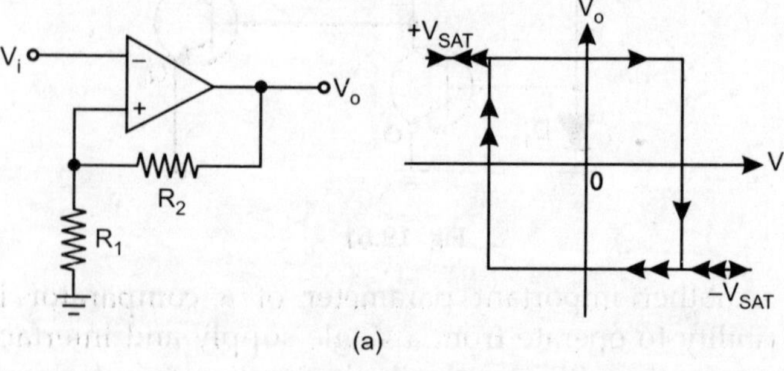

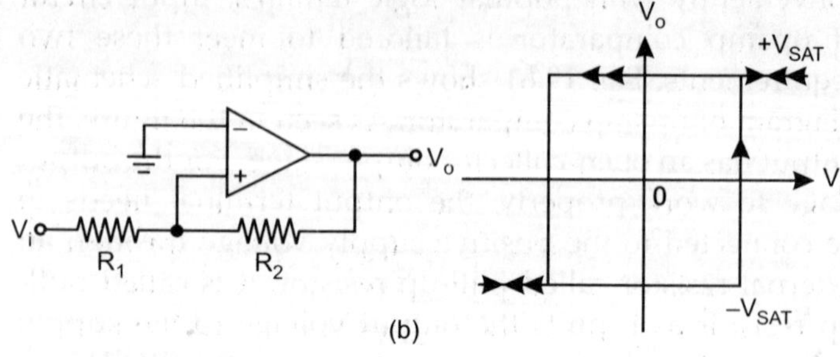

Fig. 19.63

Fig. 19.63 (*a*) shows the circuit schematic of an **inverting comparator** with hysteresis along with its transfer characteristics. The circuit functions as follows. Let us assume that the output is in positive saturation. Voltage at non-inverting input in this case is $V_{SAT} \times R_1/(R_1 + R_2)$. Now, the input signal needs to be more positive than this voltage for the output to go to negative saturation. Once the output goes to negative saturation, voltage fed back to non-inverting input becomes $-V_{SAT} \times R_1/(R_1 + R_2)$. The input signal amplitude needs to become more negative than this for the output to go to positive saturation. In this manner, the circuit offers a hysteresis of $2V_{SAT} \times R_1/(R_1 + R_2)$.

Non-inverting comparator with hysteresis can be built by applying the input signal to the non-inverting input as shown in Fig. 19.63 (*b*). Operation of the circuit can be explained on lines similar to that of its inverting counterpart. Upper and lower trip points are, respectively, given by $+V_{SAT} \times R_1/R_2$ and $-V_{SAT} \times R_1/R_2$. Hysteresis in this case is equal to $2V_{SAT} \times R_1/R_2$.

Q.51. What is a window comparator? With the help of circuit diagram, briefly explain the operation of opamp based window comparator.

Ans. In the case of conventional comparator, the output changes state when the input voltage goes above or below the preset reference voltage. In a window comparator, there are two reference voltages called the lower and the upper trip points. Output is in one state, when the input is inside the window created by the lower and upper trip points and in the other state, when it is outside the window. Fig. 19.64 (*a*) shows the basic circuit

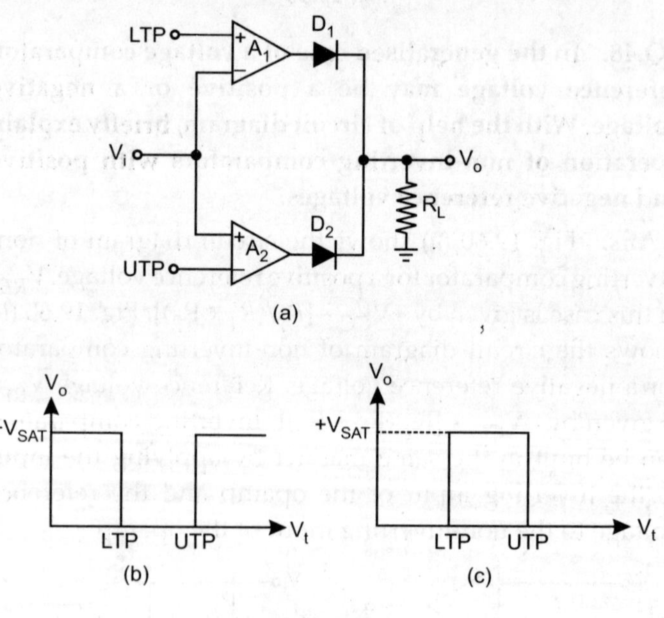

Fig. 19.64

diagram of one such window comparator. The circuit functions as follows. When the input voltage is less than the voltage reference corresponding to the lower trip point (LTP), output of opampA_1 is at $+V_{SAT}$ and that of opampA_2 is at $-V_{SAT}$. Diodes D_1 and D_2 are respectively forward- and reverse-biased. Consequently, output across R_L is at $+V_{SAT}$. When the input voltage is greater than the reference voltage corresponding to the upper trip point (UTP); the output of opampA_1 is at $-V_{SAT}$ and that of opampA_2 is at $+V_{SAT}$. Diodes D_1 and D_2 are respectively reverse-and forward-biased with the result that the output across R_L is again at $+V_{SAT}$. When the input voltage is greater than LTP voltage but lower than UTP voltage; the output of both opamps is at negative saturation with the result that diodes D_1 and D_2 are reverse-biased and the output across R_L is zero.

Q.52. With the help of circuit diagram, briefly describe the operation of opamp based current-to-voltage converter. Give all relevant design equations.

Ans. Current-to-voltage converter is nothing but a trans-impedance amplifier. An ideal trans-impedance amplifier makes a perfect current-to-voltage converter, as it has zero input impedance and zero output impedance. Opamp wired as trans-impedance amplifier very closely approaches a perfect current-to-voltage converter. Fig. 19.65 shows the circuit arrangement. The circuit is characterised by voltage shunt feedback. The expressions for output voltage, closed-loop input and output impedances are given as follows:

$$V_o = I_i \times R \times \left(\frac{A_{OL}}{1 + A_{OL}}\right)$$

For $A_{OL} >> 1$, the above equation simplifies to the following equation.

$$V_o = I_i \times R$$

$$Z_{in} = \frac{R}{1 + A_{OL}}$$

$$Z_o = \frac{R}{1 + A_{OL}}$$

Where, R_o is the output impedance of the opamp

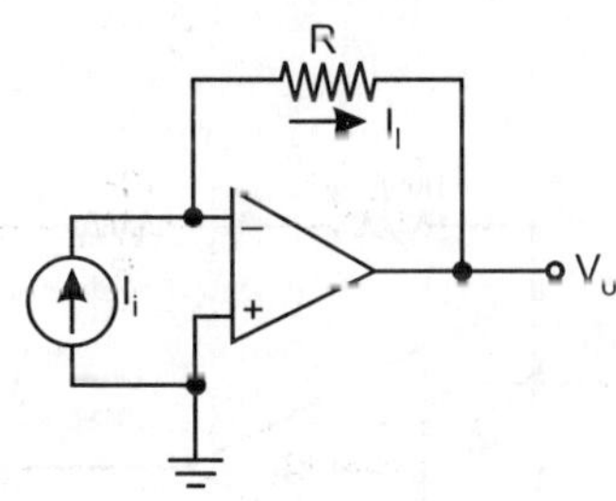

Fig. 19.65

Q.53. With the help of circuit diagram, briefly describe operation of opamp based voltage-to-current converter. Give all relevant design equations.

Ans. Voltage-to-current converter is a case of a trans-conductance amplifier. An ideal trans-conductance amplifier makes a perfect voltage-controlled current source or a voltage-to-current converter. Opamp wired as trans-conductance amplifier very closely approaches a perfect voltage-to-current converter. Fig. 19.66 shows the basic circuit arrangement. The circuit is characterised by current series feedback. Expressions for output voltage, closed-loop input and output impedances are given as follows:

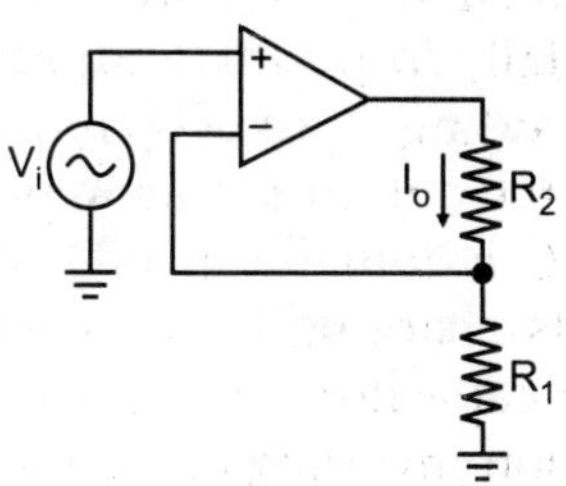

Fig. 19.66

$$I_o = \frac{V_i}{R_1 + [(R_1 + R_2)/A_{OL}]}$$

For $A_{OL} >> 1$, the above equation simplifies to the following equation:

$$I_o = \frac{V_i}{R_1}$$

Closed-loop input impedance is given by:

$$Z_{in} = R_i \times \left(1 + A_{OL} \times \frac{R_1}{R_1 + R_2}\right)$$

Where, R_i is the input impedance of the opamp.

Closed-loop output impedance is given by:

$$Z_o = R_1 \times \left(1 + A_{OL} \times \frac{R_1}{R_1 + R_2}\right)$$

Voltage to current converter of Fig. 19.66 operates with a floating load, which is not always convenient. Monolithic opamps specially designed as trans conductance amplifiers to feed single-ended load resistances are commercially available.

Q.54. What is a Relaxation Oscillator? With the help of a suitable circuit digram, briefly describe operation of opamp based relaxation oscillator.

Ans. Relaxation oscillator is an oscillator circuit that produces a non-sinusoidal output whose time period is dependent on the charging time of a capacitor connected as a part of the oscillator circuit. Opamps adapt very well to construction of relaxation oscillator circuits that produce a rectangular output. Fig. 19.67 shows the basic circuit arrangement of an opamp-based relaxation oscillator circuit.

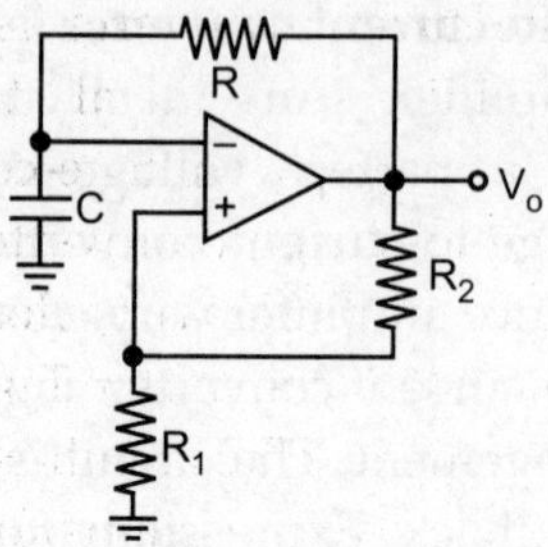

Fig. 19.67

The circuit functions as follows. Let us assume that the output is initially in positive saturation. As a result, voltage at non-inverting input of opamp is $+V_{SAT} \times R_1/(R_1 + R_2)$. This forces the output to stay in positive saturation as the capacitor C is initially in fully discharged state. Capacitor C starts charging towards $+V_{SAT}$ through R. The moment the capacitor voltage exceeds the voltage appearing at the non-inverting input, the output switches to $-V_{SAT}$. The voltage appearing at non-inverting input also changes to $-V_{SAT} \times R_1/(R_1 + R_2)$. The capacitor starts discharging and after reaching zero, it begins to charge towards $-V_{SAT}$. Again, as soon as it becomes more negative than the negative threshold appearing at non-inverting input of the opamp, the output switches back to $+V_{SAT}$. The cycle repeats thereafter. The output is a rectangular wave. The expression for time period of output waveform can be derived from the exponential charging and discharging process and is given by the following equation:

$$T = 2RC\ln\left(\frac{1+B}{1-B}\right)$$

Where, $B = R_1/(R_1 + R_2)$

Fig. 19.68 shows the relevant waveforms. The time period of output may be conveniently varied by varying the resistance value of R.

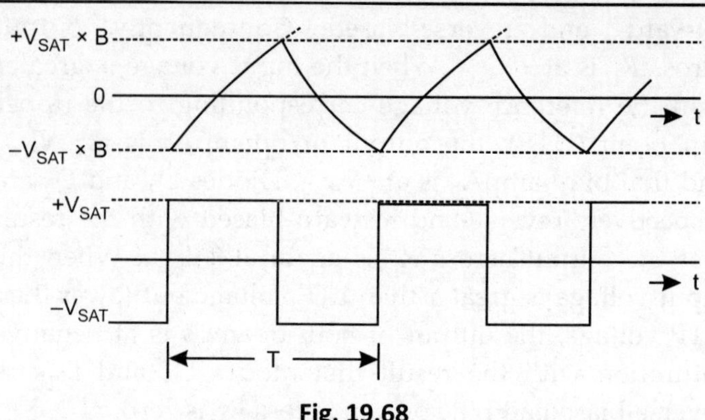

Fig. 19.68

Q.55. With the help of a suitable circuit diagram, briefly describe the operation of an opamp based average circuit that can be used to produce at its output a voltage equal to average of (*n*) input voltages.

Ans. An averager circuit produces an output that is equal to average of the amplitudes of the applied input signals. Fig. 19.69 shows the generalized form of an averager circuit for (*n*) inputs. The circuit configuration is similar to that of an inverting-type summing amplifier. The output voltage V_o is given by,

$$V_o = -\left(\frac{V_1 + V_2 + V_3 + \ldots\ldots + V_n}{n}\right)$$

A non-inverting averager may be built by connecting a unity gain inverting amplifier at the output of the circuit shown in Fig. 19.69.

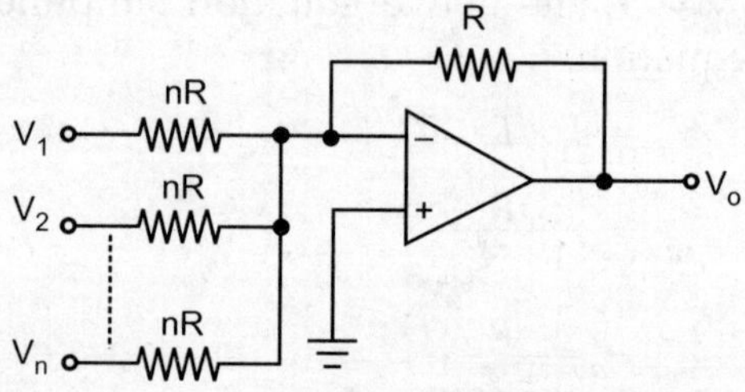

Fig. 19.69

Solved Problems

Problem 19.1 *Design an inverting amplifier whose gain is variable from unity to 100 for DC applications. Suggest some method to convert it into a non-inverting amplifier with the same gain variation. The input impedance of the amplifier should not be less than 10 K.*

Solution. The basic circuit arrangement of an inverting amplifier with variable gain is shown in Fig. 19.70

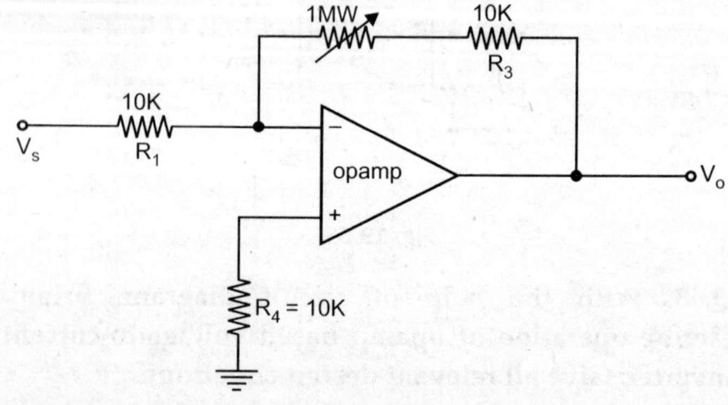

Fig. 19.70

From the theory of inverting amplifier, Magnitude of

Gain = $\frac{R_2 + R_3}{R_1}$

Desired gain = 1 to 100.

Thus $\frac{R_2 + R_3}{R_1}$ = Unity (for gain = 1)

and $\frac{R_2 + R_3}{R_1}$ = 100 (for gain = 100)

Select $R_1 = 10\ K\Omega$

The input impedance of inverting amplifier circuit is effectively equal to the input resistance used, R_1 in the present case. This explains the choice of 10 K as the input resistance.

$(R_2 + R_3) = 10\ k\Omega$ (for gain = 1)

and $(R_2 + R_3) = 1000\ K\Omega$ (for gain = 100)

Select $R_3 = 10\ K\Omega$

Then $R_2 = 0$ ohm (for gain = 1)

and $R_3 = 990$ K (for gain = 100)

Thus, R_o may be taken as 1 M variable resistance.

To convert it into a non-inverting amplifier, connect a unity gain inverting amplifier in cascade with the above designed section. The complete circuit arrangement is shown in Fig. 19.71

Choice of an opamp : Any general purpose opamp like 741, for instance, may be chosen.

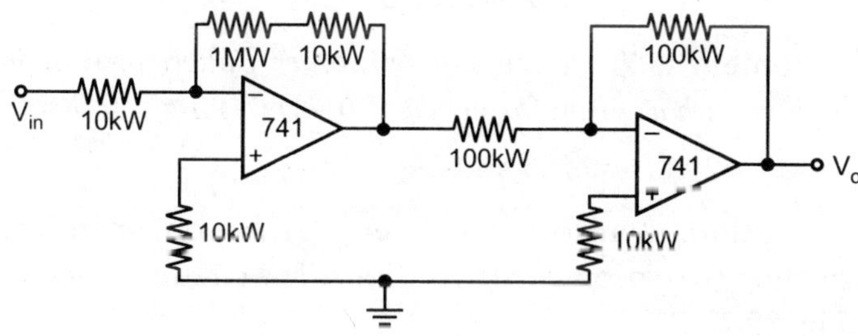

Fig 19.71

Problem 19.2. *Design a non-inverting adder circuit to be used for DC applications with a gain of 10, i.e.* $e_0 = 10\ (e_1 + e_2 + e_3)$.

Solution. Basic circuit is shown in Fig. 19.72

$$e_0 = -10\ (e_1 + e_2 + e_3),\ e'_0 = 10\ (e_1 + e_2 + e_3)$$

opamp number; 741

Rest of the design comprises of an inverting adder with gain = 10 and a unity gain inverting circuit. Gain of 10 can be achieved by selecting the ratio of

$$\left(\frac{\textit{Feedback resistor}}{(-ve)\ \textit{input resistor}}\right) = 10$$

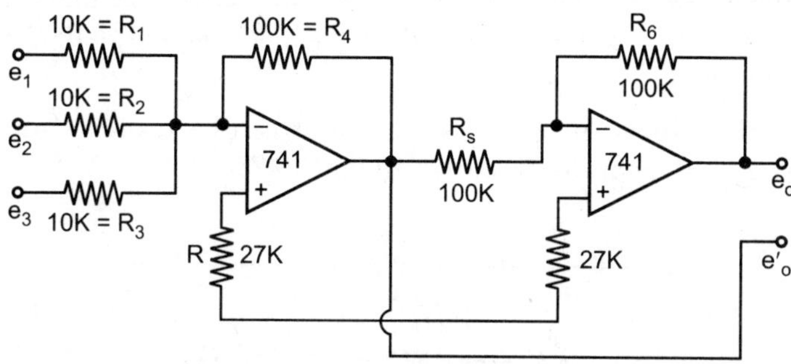

Fig. 19.72

Problem 19.3. *It is desired to have two outputs, one in phase with the input and amplified hy a factor of 100 and the other out of phase without any amplification. Give the circuit arrangement with appropriate component values and opamp number.*

Solution. The circuit is shown in Fig. 19.73. The design is self-explanatory.

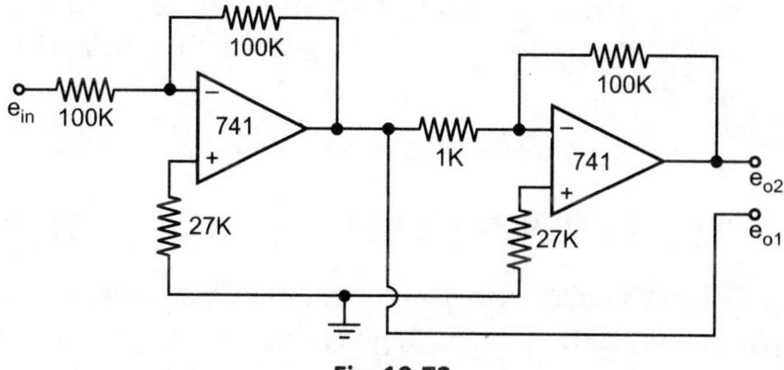

Fig. 19.73

Problem 19.4. *Give an opamp circuit that can be used as a subtractor with a gain of (× 10) with all the components values and opamp number.*

Solution. The circuit using opamp (741) is shown in Fig. 19.74. The gain is decided by the ratio R_2/R_1. Also, $R_4/R_3 = R_2/R_1$

$R_1 = 10$ K; $R_2 = 100$ K

$R_3 = 10$ K and $R_4 = 100$ K.

$e_0 = 10\ (e_2 - e_1)$

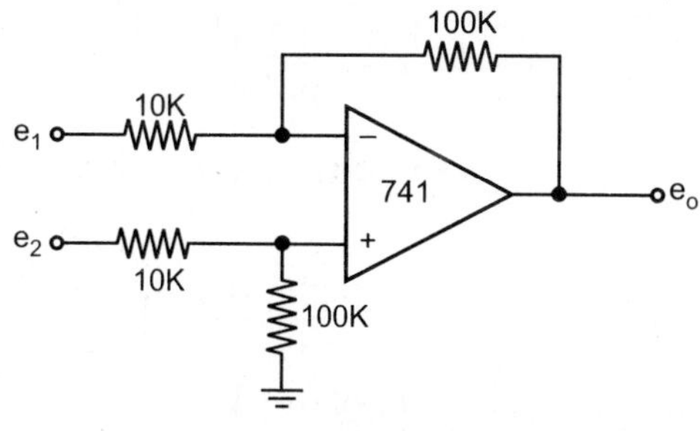

Fig. 19.74

Problem 19.5. *It is desired to have an amplifier which has such a response that it gives zero or minimum) amplification at a frequency of 1 kHz and maximum amplification at all other frequencies. Suggest some circuit with opamp and explain its operation.*

Solution. An opamp circuit that can be used to get the above response is shown in Fig. 19.75

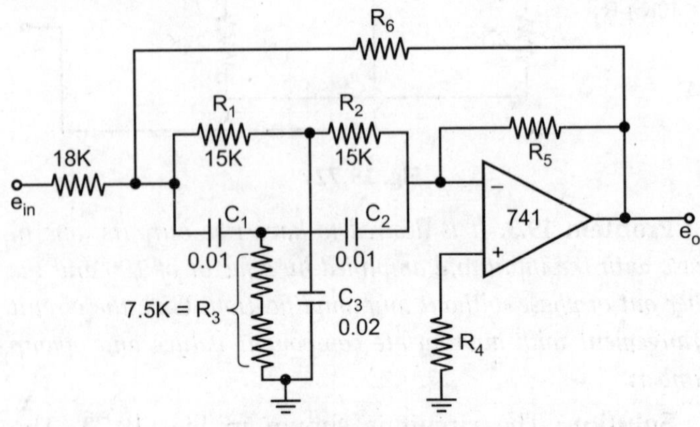

Fig. 19.75

In this circuit, a passive network consisting of R_1, R_2, C_1, C_2, R_3, C_3 has been incorporated in series with negative input.

$$R_1C_1 = R_2C_2 \text{ also } R_1 = R_2 = 2R_3$$

and $$C_1 = C_2 = \frac{C_3}{2}$$

This is called a Twin-T network. Its characteristics are such that at the frequency, f_0, given by,

$$f_0 = \frac{1}{2\pi R_2 C_2}$$

It exhibits infinite impedance (or maximum impedance) at f_0 and at all other frequencies, it gives minimum resistance. Thus, at centre frequency (f_0) amplifier gives minimum output and maximum output results at all other frequencies.

By increasing R_6 we can increase the magnitude of output at frequencies away for (f_0), whereas sharpness can be increased by increasing the value of R_5.

If a frequency of 1 kHz is to be rejected, then

$$f_0 = 1 \text{ kHz} = \frac{1}{2\pi R_2 C_2}$$

$$= \frac{1}{2\pi R_1 C_1} \text{ Now let } R_1 = R_2 = 15 \text{ K}$$

Then $$C_1 = C_2 = \frac{1}{2\pi \times 1000 \times 15 \times 1000}$$

$$= \frac{1}{30\pi} \mu F = 0.01 \mu F$$

$$R_3 = \frac{15}{2} = 7.5 \text{ k}$$

$$C_3 = 2 \times 0.01 = 0.02 \mu \text{ F.}$$

Problem 19.6. *Suggest some circuit that accepts 1 kHz and rejects all other frequencies.*

Solution. The circuit is shown in Fig. 19.76. The operation can be explained on similar basis as that of rejecter filter circuit.

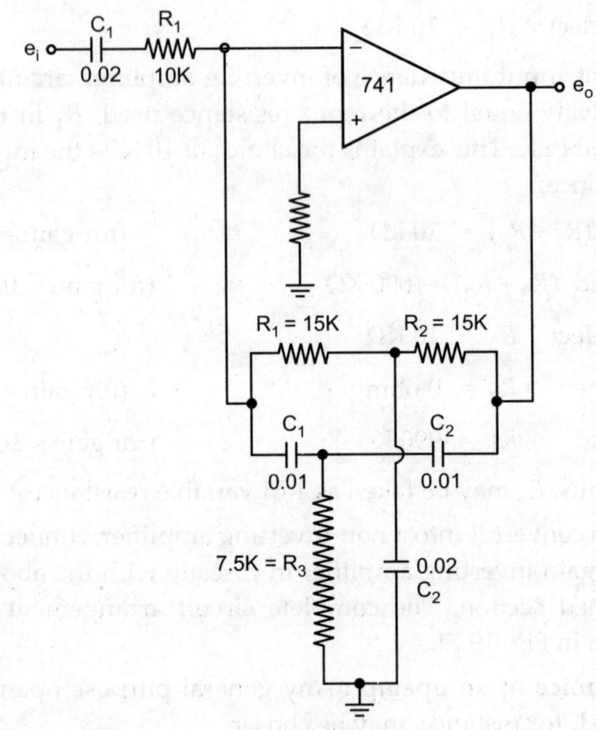

Fig. 19.76

Problem 19.7. *Design a fourth order Butterworth high pass-filter with a cut-off frequency of 10 kHz. It is given that*

$$2K1 = 0.765 \text{ and } 2K_2 = 1.848.$$

Solution. Fourth-order filter can be obtained by cascading two second order sections. The circuit is shown in Fig. 19.77.

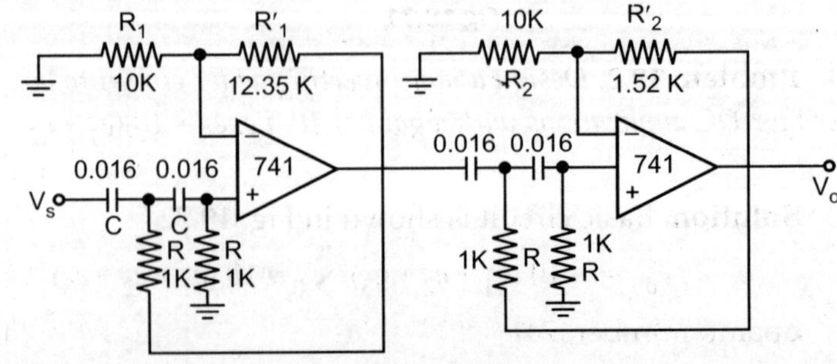

Fig. 19.77

$$A_{v1} = \frac{R'_1 + R_1}{R_1} = 3 - 2K_1 = 3 - 0.765 = 2.235$$

Selecting

$$R_1 = 10\text{ K}; R'_1 = 12.35\text{ K}$$

$$A_{v2} = \frac{R'_2 + R_2}{R_2} = 3 - 2K_2 = 3 - 1.848 = 1.152$$

Choosing

$$R_2 = 10\,K; R'_2 = 1.52\text{ K}$$

$$f_0 = \frac{1}{2\pi RC} = 10\text{ kHz}$$

Choosing R = 1 K,C = 0.016 μF.

Problem 19.8. *Design a low-pass fourth order butterworth Filter, with the same cut off frequency, i.e. 10 kHz. It is given that 2 K_1= 0.765 and 2K_2 = 1.848.*

Solution. The design procedure is entirely the same. The only change is in the position of R's and C's. By interchanging the positions of R's and C's, we get a low-pass filter.

Problem 19.9 *Design a second order band pass filter using opamps. The band of frequencies to be passed is from 1kHz to 10 kHz. It is given that 2K = 1.414*

Solution. A second order band pass filter (1 kHz to 10 kHz) may be obtained by cascading a low pass (second order) with cut-off frequency of 10 kHz with a second order high pass with cut-off frequency of 1 kHz. The circuit is shown in Fig. 19.78

$$A_{vl} = A_{v2} = 3 - 2K = 3 - 1.414 = 1.586$$

$$= \frac{R_1 + R'_1}{R_1}, \text{ Choosing R1} = 10\text{ K}$$

$$1 + \frac{R'_1}{R} = 1.586$$

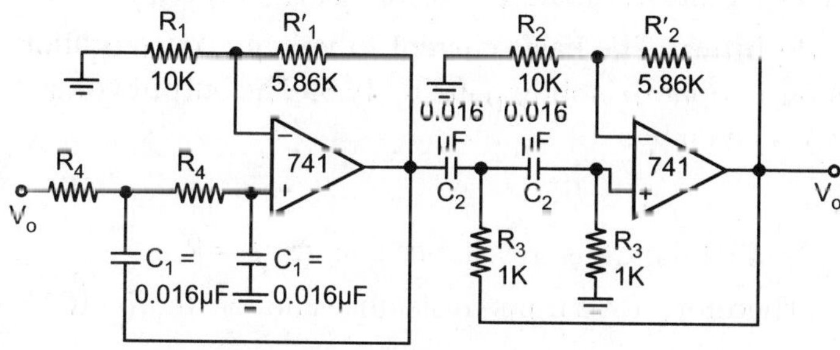

Fig. 19.78

$$R'_1 = 10 \times 0.586 = 5.86\text{ K}$$

Also, $(R'_2 + R_2)/R_2 = 1.586;$ (If $R_2 = 10$ K)

$$R'_2 = 5.86\text{ K}$$

$$\frac{1}{2\pi R_4 C_1} = 10 \times 10^3 \text{ Selecting } R_4 = 1\text{ K}$$

$$C_1 = \frac{1}{2 \times 1000 \times 10^4} F$$

$$= \frac{1}{2\pi}\,\mu F = \frac{1}{62.8}\,\mu F = 0.016\,\mu\text{F}$$

Also, $\frac{1}{2\pi R_3 C_2} = 10^3$, choosing $R_3 = 1$ K

$$C_2 = \frac{2}{2\pi \times 10^3 \times 10^3} = 0.16\,\mu\text{F}$$

The response of the filter circuit is shown in Fig. 19.79

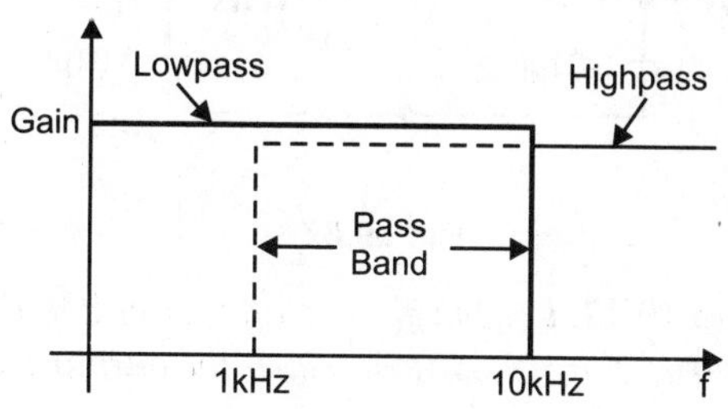

Fig. 19.79

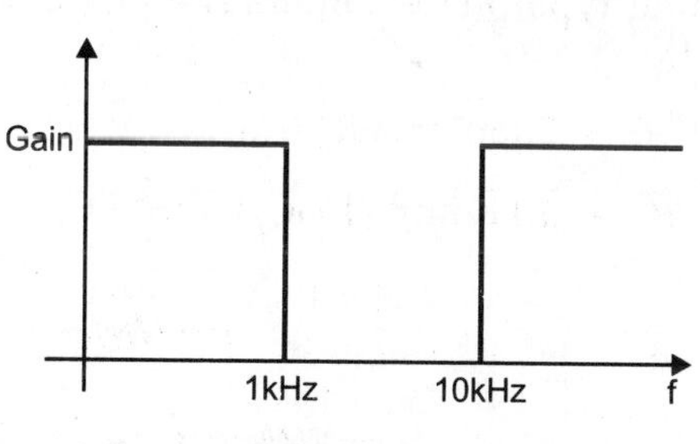

Fig. 19.80

Problem 19.10. *Design a band-reject filter (second order) to have a response of Fig. 19.80*

Solution. Second order butterworth polynomial is $(s^2 + 1.414\,s + 1) = 0$

Second order band reject filter can be designed using a second order low pass with cut off frequency equal to lowest frequency of the band to be rejected (= 1 kHz in this case) and a second order high pass with cut off frequency equal to highest frequency of the band to be rejected (10 kHz in this case) and then connecting them in the fashion shown in Fig. 19.81

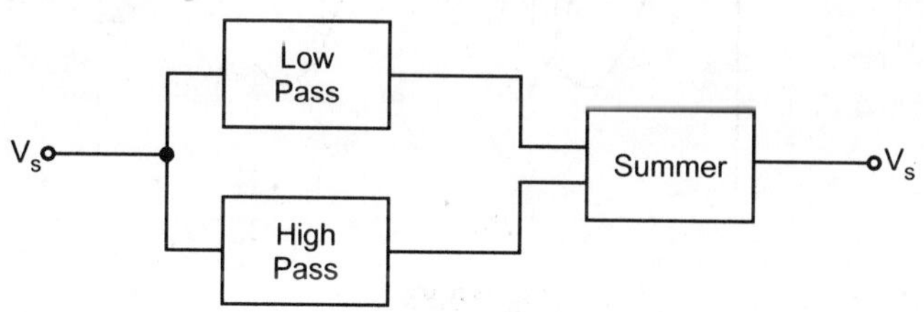

Fig 19.81

Problem 19.11. *Suggest a circuit using opamps to shift the phase of a 10 kHz sinusoidal signal by 180° with the output signal lagging the input signal. Give all component values,*

Solution. The circuit diagram is shown in Fig. 19.82 Each section provides a phase shift of (- 90°) as R_P has been chosen to be equal to $(1/\omega C_p)$ at the operating frequency. With a single section, it would not have been possible to get a phase shift of (– 180°) though a phase shift of little less than (– 180°) could be achieved as explained earlier.

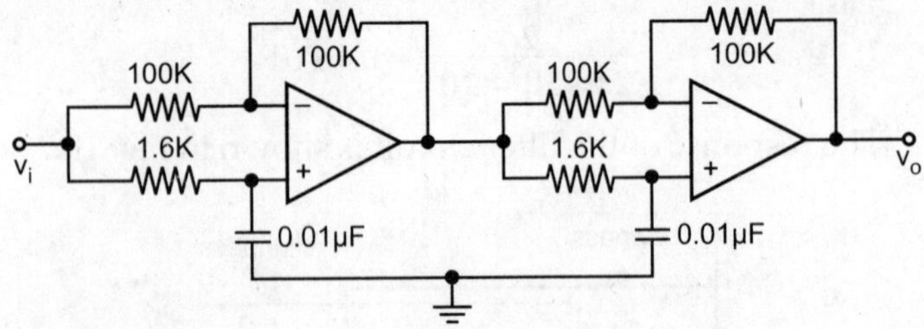

Fig. 19.82

Problem 19.12. *For the phase-shifter circuit and the given input waveform (Fig. 19.83 (a)), sketch the output waveform.*

Solution. The configuration shown in Fig. 19.94 (*a*) is that of leading type phase shifter. The phase shift angle (θ) is given by

$$\theta = 2 \tan^{-1} (\omega R_P C_P)$$

where, R_p = 10 K and $(1/\omega C_p) = 10^4\ \Omega$

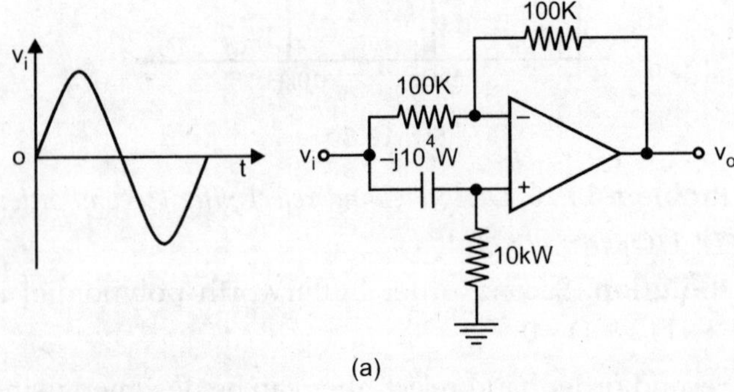

(a)

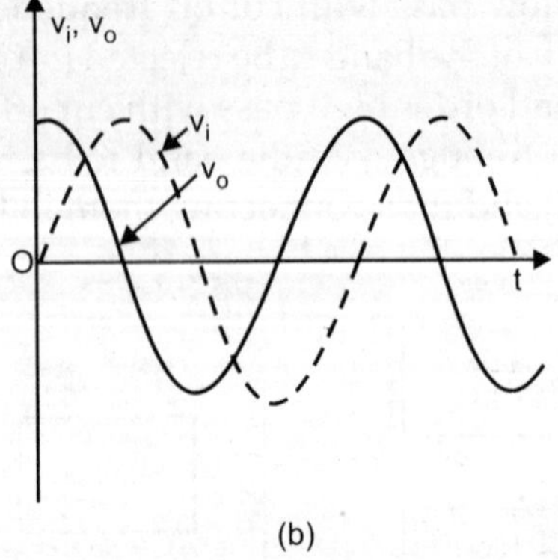

(b)

Fig. 19.83

Therefore, $\theta = 2 \tan^{-1} 1 = 90°$

The output thus leads the input by 90°. The output waveform is shown in Fig. 19.83 (*b*).

Problem 19.13. *Design an inverting amplifier around an opamp. The amplifier should have a gain of 10 and an input impedance not less than 10 K. If the input signal, a sinusoidal one, has a peak amplitude varying from 100 mV to 300 mV, determine the maximum possible input signal frequency that can be faithfully amplified, if the chosen opamp has a slew rate of 0.5 V/μs.*

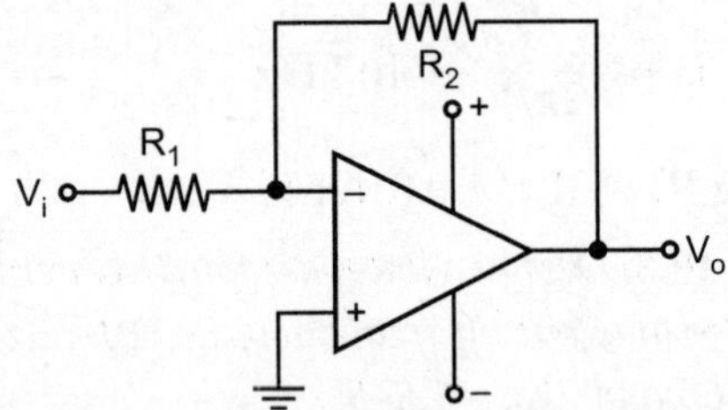

Fig. 19.84

Solution. Fig. 19.84 shows the inverting amplifier using an opamp. The gain of this amplifier configuration is $(-R_2/R_1)$ and its input impedance is (R_1). As it is desired that the input impedance is not to be less than 10K, therefore, we can safely choose (R_1) to be 10 K.

This gives $R_2 = 10 \times 10 = 100$ K

The largest input signal amplitude = 300 mV

Corresponding output signal amplitude = 300 × 10 = 3 V

Highest frequency that can be faithfully reproduced is given by :

$$f_{max} = \frac{\text{Slew Rate}}{2\pi V_o(\text{max})} = \frac{0.5}{\pi \times 3 \times 2} \text{ MHz}$$

$$= \frac{500}{6\pi} \text{ kHz} = 26.5 \text{ kHz.}$$

Problem 19.14. *Design a current to voltage converter having a trans-resistance gain of 10^5 using an opamp.*

Solution. The basic current to voltage configuration using an opamp is shown in Fig. 19.85 The output voltage (V_0) is given by:

$$v_o = IR$$

So that the trans- resistance gain $(v_o/I) = R$

Therefore, for a trans- resistance gain of 10^5, $R = 100$ K

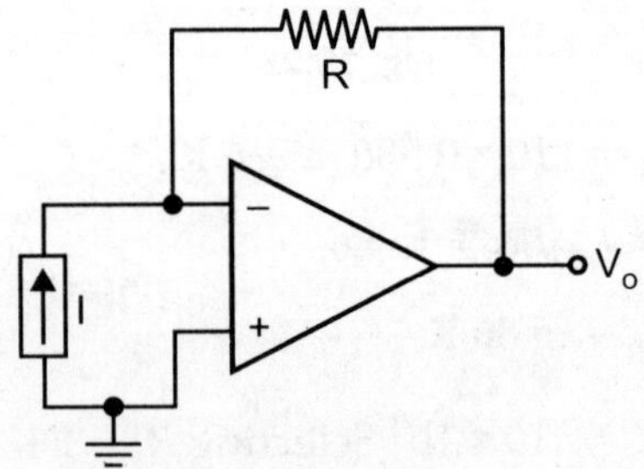

Fig. 19.85

Problem 19.15. *Design an opamp based non-inverting amplifier having a gain of 11. Determine the input impedance of this amplifier, if the chosen opamp has open loop gain of 100000 and open loop input impedance of 1 MΩ.*

Solution. Fig. 19.86 shows the basic non-inverting amplifier using an opamp. The gain of this amplifier is given by:

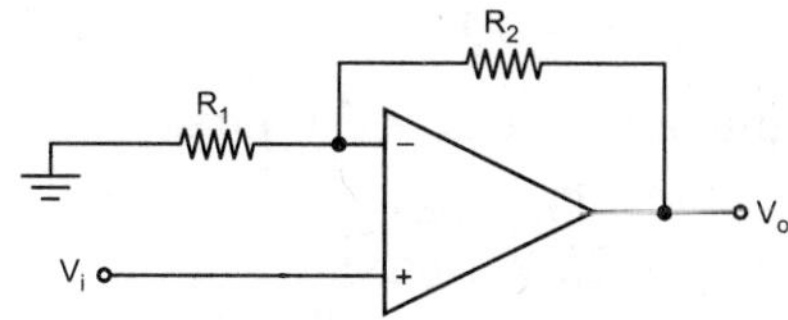

Fig. 19.86

$$\left(1+\frac{R_2}{R_1}\right)$$

Required gain = 11. Therefore, $R_2/R_1 = 11 - 1 = 10$

For $R_1 - 10\ K, R_2 - 100K$

Now, for the non-inverting amplifier shown in Fig. 19.86, the input impedance (R_i) is given by:

Open loop input impedance × loop gain

$$= \text{Open loop input impedance} \times \frac{\textit{open loop gain}}{\textit{closed loop gain}}$$

$$= 1 \times \frac{100000}{11}\ \text{M}\Omega$$

$$= \frac{100000}{11}\ \text{M}\Omega = 9091\ \text{M}\Omega$$

Problem 19.16. *You have to choose an appropriate opamp type number for your inverting amplifier configuration that has been designed for a gain of 10. The input is a sinusoidal signal with a peak to peak amplitude of 2 volts. If the highest expected input signal frequency is 50 kHz, what should be the slew rate of the chosen opamp?*

Solution. If the peak to peak amplitude of input signal is 2 volts, then the peak to peak amplitude of the output signal would be 20 volts.

Required Slew rate = πf_{max} × peak to peak output amplitude

$$= \pi \times 50 \times 10^3 \times 20\ V/s$$

$$= \frac{\pi \times 50 \times 10^3 \times 20}{10^6}\ \text{V/µs.}$$

$$= \pi\ \text{volts/µs} = 3.14\ \text{volts/µs}$$

Problem 19.17. *Design a non-inverting zero crossing detector with a hysteris of 100 mV. If the opamp has output saturation voltages of + 10 V and – 10 V, determine the highest input frequency that would yield an output waveform transition time not more than 10% of half of the input signal time period, if the chosen opamp has a slew rate of 10 V/ µs.*

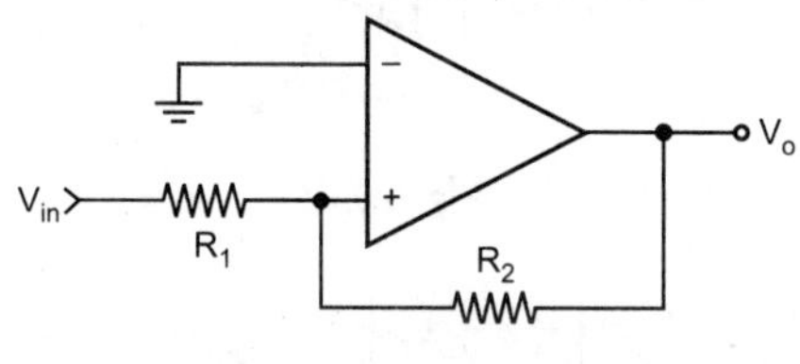

Fig. 19.87

Solution. Fig. 19.87 shows the basic opamp based zero crossing detector with hysterisis

Hysterisis is given by: $2\ V_{SAT}\left(\frac{R_1}{R_1+R_2}\right)$; Now, $V_{SAT} = 10\ V$

Therefore,

$$20\left(\frac{R_1}{R_1+R_2}\right) = 100 \times 10^{-3}$$

$$\text{or}\ \frac{R_1}{R_1+R_2} = 5 \times 10^{-3}$$

$$\text{or}\ 1+\frac{R_2}{R_1} = \frac{1}{5\times 10^{-3}} = 200$$

$$\text{or}\ \frac{R_2}{R_1} = 199$$

For $R_1 = 1$ K and $R_2 = 199$ K

For a slew rate of 10 V/µs, transition time from ($-V_{SAT}$) to + (V_{SAT}) is given by:

$$\frac{10-(-10)}{10} = 2\ \mu s$$

Half of time period at the highest input frequency = $1/2\,f_{max}$

10% of this time period = $\frac{1}{20\,f_{max}}$

Therefore,

$$\frac{1}{20\,f_{max}} = 2 \times 10^{-6}$$

$$f_{max} = \frac{1}{20 \times 2 \times 10^{-6}} = 25\ \text{kHz}$$

Problem 19.18. *Design an inverting voltage comparator with a reference voltage of 4 volts and a hysterisis of 0.4 volt. The input impedance of the comparator should not be less than 500 K. Assume that the chosen opamp has differential input impedance of 400 K and a* $|V_{SAT}| = 10$ *voits.*

Solution. Fig. 19.88 shows the basic inverting type voltage comparator with the reference input of 4 volts. The hysterisis in the circuit is given by :

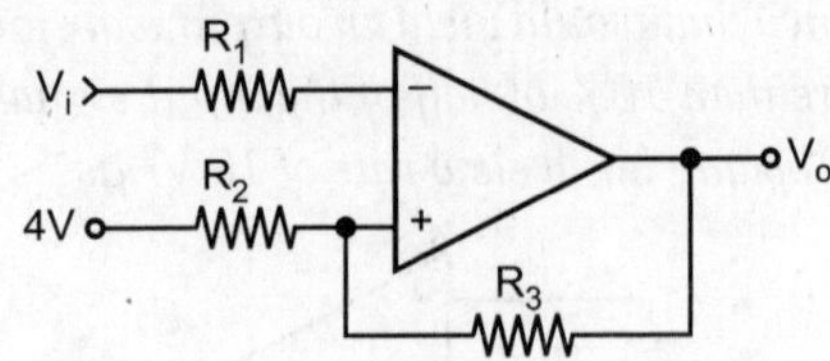

Fig. 19.88

$$2\,V_{SAT}\left[\frac{R_2}{R_2+R_3}\right]=2\times 10\left[\frac{R_2}{R_2+R_3}\right]$$

Desired hysterisis = 0.4 volt

Therefore,

$$\frac{20\,R_2}{R_2+R_3} = 0.4 \text{ which gives } R_3 = 49\,R_2$$

Usually, $R_1 = R_2$ which gives the input impedance as :

Input impedance = R_1 + Differential input impedance + $R_2 \| R_3$

If $\quad R_1 = R_2 = R$

Then Input impedance = R + 400 K + (R || 49 R) = R + 400 K + 0.98 R = 1.98 R + 400 K. Minimum desired input impedance = 500 K

Therefore, 1.98 R + 400 K = 500 K

or $\quad (R)_{min} = \dfrac{100}{1.98} = 50.5$ K

Therefore,

$$R_1 = R_2 = 50.5 \text{ K}$$

and $\quad R_3 = 49\,R_2 = 40 \times 50.5 = 2.474$ M

Problem 19.19. *Design an opamp based lagging type phase shifter that can shift the phase of an input sinusoidal signal by – 60° with a gain of unity. If the input signal has a peak amplitude of 5 volts and the highest input frequency is 50 kHz, what should be the slew rate of the chosen opamp so that it does not limit the bandwidth?*

Solution. Fig. 19.89 shows the circuit diagram. For unity gain, $R_1 = R_2$. The phase shift introduced by the circuit (θ) is given by :

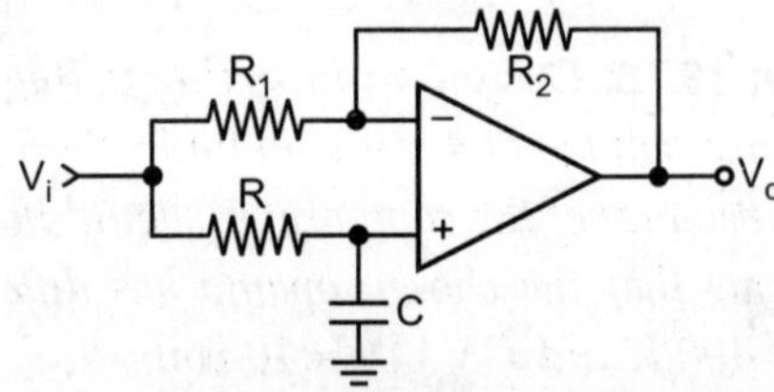

Fig. 19.89

$$\theta = -2\tan^{-1}(2\pi fRC)$$

or $\quad -60° = -2\tan^{-1}(2\pi \times 50 \times 10^3\, RC)$

or $10^5\,\pi \times RC = \tan 30° = 0.577$

or $\quad R_C = \dfrac{0.577}{10^5\pi} = 0.184 \times 10^{-5}$

For $\quad C = 0.001\ \mu$F

$$R = \frac{0.184 \times 10^{-5}}{10^{-9}} = 1840\ \Omega = 1.84 \text{ K}$$

Let $\quad R_1 = R_2 = 10$ K

Input signal amplitude = 5 V peak

$$f_{max} = \frac{Slew\ Rate}{2\pi\, V_o\,(\max)}$$

or Slew rate = $2\pi \times 5 \times 50 \times 10^3 = 31.4 \times 50 \times 10^3$

$= 157 \times 10^4$ V/s = 1.57 V/μs.

Problem 19.20. *Modify the phase shifter circuit of Fig. 19.89 so as to provide a leading phase shift of 60° with all other parameters remaining the same.*

Solution. The modified circuit can be obtained by interchanging the positions of (R) and (C) in Fig. 19.89. The leading phase shift is given by : $\theta = 2\tan^{-1}(2\pi f RC)$.

Problem 19.21. *A phase shifter network of the type shown in Fig. 19.89 provides a phase shift of – 45° at a certain frequency. With (R) and (C) remaining same, what phase shift will this circuit provide at twice the original frequency ?*

Solution. The phase shift is given by :

$$\theta = -2\tan^{-1}(2\pi f RC)$$

Therefore,

$$2\,\pi fRC = \tan(\theta/2) = \tan 22.5° = 0.4142$$

The new phase shift is given by :

$$\theta = -2\tan^{-1}(2\pi \times 2f \times RC)$$
$$= -2\tan^{-1}(2 \times 0.4142)$$
$$= -2\tan^{-1}(0.8284)$$
$$= -2 \times 39.638 = -79.27°$$

Problem 19.22. *Design an absolute value circuit using opamps which produces an output $v_o = |\, v_i \,|$ given that – 5V ≤ v_i < + 5V and the highest frequency likely to be encountered is 20 kHz. The absolute value circuit should have an input impedance not less than 15 K. Also, give the desired values for the slew rate and the unity gain cross-over specifications of the opamp or opamps used in the circuit.*

Solution. Fig. 19.90 shows the circuit diagram. For this circuit to give $v_o = |v_i|$, $R_1 = R_2 = R_3 = R_4 = R_5 = R$. The input impedance of this circuit is equal to (R_1).

Therefore,

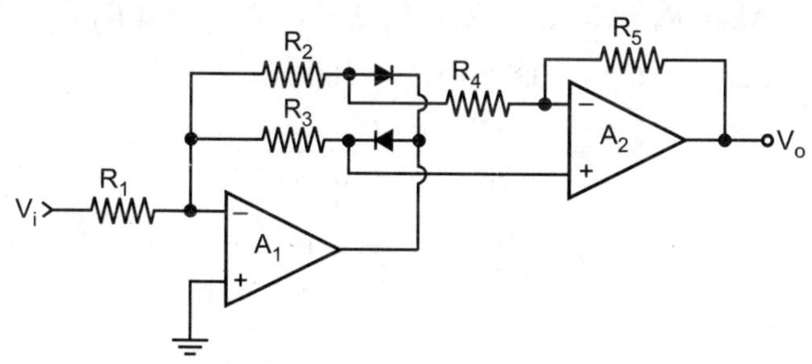

Fig. 19.90

$R_1 = R_2 = R_3 = R_4 = R_5 = 15K$

Maximum peak-to-peak value of the input signal = 10 V. Highest frequency = 20 kHz. Therefore, required slew rate = $\pi f_{max} \times 10 = 200 \times 10^3\ \pi$ volts/sec = 0.628 V/µs (A_1) and (A_2) in this circuit are respectively working at closed loop gains of unity and 1.5. Therefore, the unity gain cross-over frequency specification for (A_1) and (A_2) should be a minimum of 20 kHz and 30 kHz respectively.

Problem. 19.23 *Fig. 19.91 shows a second order low pass filter built around a single opamp. Calculate the values of R_1 R_2, C_1 C_2. R_3, if this filter has to have a cut-off frequency of 10 kHz, a Q-factor of 0.707 and. an input impedance not less than 10 K.*

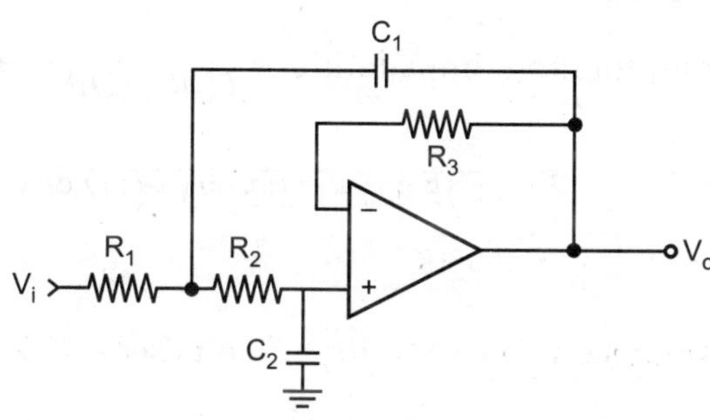

Fig. 19.91

Solution. For $R_1 = R_2 = R$, the cut-off frequency of the filter is given by :

$$f_c = \frac{1}{2\pi R\sqrt{C_1 C_2}}$$

The Q - factor is given by:

$$Q = \frac{1}{2}\sqrt{C_1/C_2}$$

For $\quad Q = 0.707, C_1 = 2\,C_2$

The circuit has a minimum input impedance of (R_1) which gives $R_1 = 10\ K = R_2$

$$f_c = \frac{1}{2\pi \times 10^4\sqrt{2C_2^2}}$$

$$= \frac{1}{2\sqrt{2}\,\pi \times 10^4\,C_2} = 10 \times 10^3$$

or $\quad C_2 = \dfrac{1}{2\sqrt{2} \times 3.14 \times 10^8}$

$$= \frac{1}{2\sqrt{2} \times 3.14 \times 10^2}\ \mu\text{F} = 0.0011\ \mu\text{F}$$

$$C_1 = 2C_2 = 0.0022\ \mu\text{F}$$

(R_3) is equal to ($R_1 + R_2$) in order to have equal DC resistance between each opamp input and ground.

Therefore, $R_3 = 20K$

Problem 19.24. *Fig. 19.92 shows a second order high pass filter built around an opamp. Calculate the values of R_1, R_2, R_3, C_1 and C_2, if the filter is to have following characteristics:*

(a) High frequency 3 dB cut-off = 10 kHz

(b) Quality factor = 0.707

(c) Input impedance = 20 K (mm)

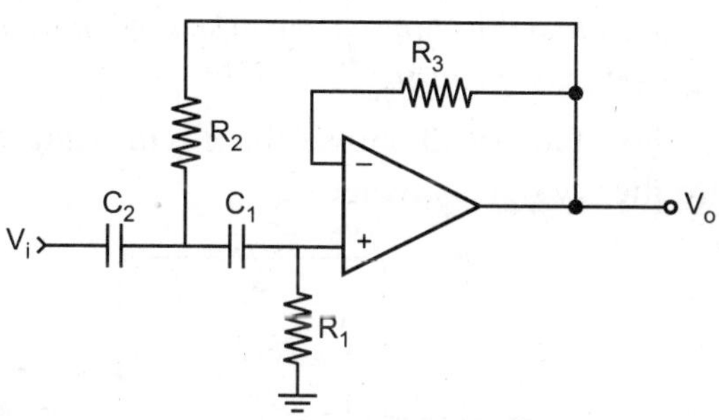

Fig. 19.92

Also, determine the required minimum unity gain cross-over frequency and the minimum slew rate of the opamp if the highest expected peak to peak output swing is 10 volts.

Solution. The Quality factor of the high pass filter of Fig. 19.92 is given by:

$$Q = \frac{1}{2}\sqrt{\frac{R_1}{R_2}} \quad \text{For } Q = 0.707, R_1 = 2\,R_2$$

For $C_1 = C_2 = C$, The cut-off frequency (fc) is given by:

$$f_c = \frac{1}{2\pi C\sqrt{R_1 R_2}}$$

For $\quad f_c = 10$ KHz, $2\pi C\sqrt{R_1 R_2} = 10^{-4}$

or $2\sqrt{2}\,\pi\, C\, R_2 = 10^{-4}$

or $\quad R_2C = \dfrac{10^{-4}}{2\sqrt{2}\pi}$

As the required minimum input impedance is 20 K, (R_1) can be chosen to be 22K (the nearest standard value). The input impedance varies inversely with input frequency. The limit is given by (R_1) in parallel with input impedance of the voltage follower. Therefore, for all practical purposes, limit is established by (R_1).

Now $\quad R_1 = 22K, R_2 = \dfrac{22}{2}\ 11\ K$

Which gives

$$C = \frac{10^{-4}}{2\sqrt{2}\ \pi \times 11 \times 10^3}$$

$$= \frac{100}{2\sqrt{2}\ \pi \times 11 \times 10^3}\ \mu F = 0.0y01\ \mu F$$

On similar lines as explained in case of low pass filter, $R_3 = R_1 = 22$ K

Minimum required unity gain cross-over frequency $= f_c = 10$ kHz

Minimum slew rate = πf_c × (peak to peak output swing) = n × 10^4 × 10 V/s = 0.314 V/μs

Problem 19.25. *Design an opamp based Twin-T band reject filter having a notch frequency at 100 kHz. Specify the small signal bandwidth of the opamp to be used in the circuit, if the highest expected frequency is 1 MHz.*

Solution. Fig. 19.93 shows the circuit diagram. The notch frequency (f_0) is given by :

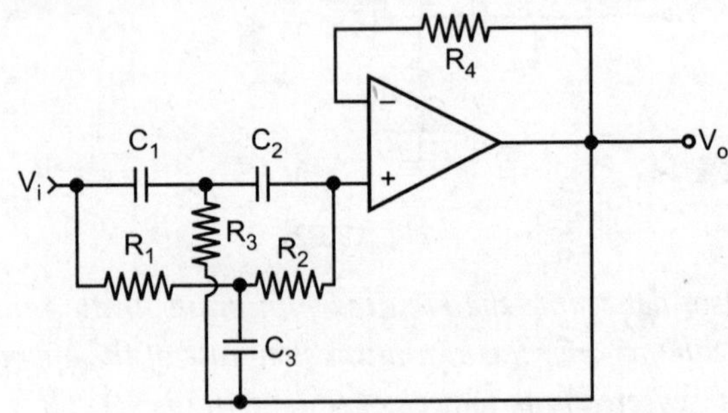

Fig. 19.93

$$f_0 = \frac{1}{2\pi R_1 C_1} = \frac{1}{2\pi\ R_2\ C_2} = \frac{1}{R\pi\ R_3\ C_3}$$

Also, $R_1 = R_2$, $C_1 = R_3 = R_1/2$, $C_3 = 2\ C_1$ and $R_4\ R_1 +$ R2

Let $C_1 = 0.0001\ \mu F\ (= 100\ pF)$

$$R_1 = \frac{1}{6.28 \times 100 \times 10^3 \times 10^{-10}}$$

$$= \frac{10^5}{6.28} = 15.92\ kW$$

Therefore,

$$C_1 = C_2 = 100\ \text{pF and } C_3 = 2 \times 100 = 200\ \text{pF}$$

$$R_1 = R_2 = 15.92\ \text{K and } R_3 = \frac{15.92}{2} = 7.96 k\Omega$$

$$R_4 = 15.92 + 1592 = 31.84\ K\Omega$$

Since the circuit provides a unity gam in the pass band, the small signal bandwidth equals the highest expected frequency.

Desired small signal bandwidth = 1 MHz

Problem 19.26. *Determine the minimum input impedance offered by the band, reject filter of problem 9.25. Also, determine the desired slew rate of the opamp if the peak to peak input is 5 volts.*

Solution.

$$\text{Minimum input impedanc e} = \frac{1}{(1/R_1)\,(1/R_2) + (1/R_3)}$$

$$= \frac{1}{(1/15.92) + (1/15.92) + (1/7.96)}$$

$$= 3.98\ K$$

Desired slew rate = π × 10^6 × 5 volts/sec = 15.7 V/μs

SUMMARY OF IMPORTANT FORMULAS

1. Voltage gain of an op-amp based inverting amplifier is given by (Refer to Figure-19.4):

$$A_{CL} = \frac{A_{OL}\ R_2}{R_1 + R_2 + A_{OL}\ R_1} \cong -\frac{R_2}{[R_1 + (R_2/A_{OL})]}$$

In the case of ideal opamp, $A_{CL} = -\left(\frac{R_2}{R_1}\right)$

Input impedance = R_1

Output impedance = $R_O = \left[\frac{R_1 + R_2}{R_1\ A_{OL}}\right] R_{OL}$

2. The voltage gain of a non-inverting amplifier is given by (refer to Figure-19.6 *a*):

$$A_{CL} = \frac{A_{OL}\ (R_1 + R_2)}{R_1 + R_2 + A_{OL}\ R_1}$$

In the case of ideal opamp

$$A_{CL} = 1 + \frac{R_2}{R_1}$$

Input impedance is given by:

$$R_i = R_{1L}\ A_{OL}\left(\frac{R_1}{R_1 + R_2}\right)$$

$$= R_{1L} \times \text{Loop gain}$$

Where, R_{1L} is the open-loop input impedance

The output impedance (R_O) is given by $R_O = R_{OL}$/Loop gain

Where, R_{OL} is the open-loop output impedance of the opamp

3. For an opamp based average circuit (Refer to Figure-19.69):

$$V_o = -\left(\frac{V_1 + V_2 + V_3 + \ldots + V_n}{n}\right)$$

4. For an opamp based integrator circuit (Refer to Figure-19.27):

$$V_o = -\frac{1}{RC}\int V_i dt = K\int V_i dt$$

5. For a differentiator circuit (Refer to Figure-19.30):

$$V_o = -RC\frac{dV_i}{dt} = K\frac{dV_i}{dt}$$

6. Output voltage in a current to voltage converter is given by (Refer to Figure-19.65):

$$V_o = I_i \times R \times \left(\frac{A_{OL}}{1 + A_{OL}}\right)$$

Where, AOL = Open loop gain of opamp

For $A_{OL} >> 1$, $V_o = I_i \times R$

Input and output impedances are expressed by:

$$Z_{in} = \frac{R}{1 + A_{OL}}$$

$$Z_o = \frac{R_o}{1 + A_{OL}}$$

Where, R_O os the output impedance of the opamp

7. Current I_O flowing through a floating load resistance R_2 in a voltage to current converter is given by (Figure 19.66):

$$I_O = \frac{V_i}{R_1 + [(R_1 + R_2)/A_{OL}]}$$

For $A_{OL} >> 1$, $I_O = \frac{V_i}{R_1}$

Closed-loop input impedance is given by:

$$Z_{in} = R_i \times \left(1 + A_{OL} \times \frac{R_1}{R_1 + R_2}\right)$$

Where, R_i is the input impedance of the opamp

Closed-loop output impedance is given by:

$$Z_o = R_1 \times \left(1 + A_{OL} \times \frac{R_1}{R_1 + R_2}\right)$$

8. The phase shift between output and input signals in the case of single stage lagging type phase shifter is given by (Refer to Figure 19.44)

θ (in degrees) = $-2\tan^{-1}(\omega R_P C_P)$

Where, $\omega = 2\pi f$, f being the frequency of the input signal.

9. The phase shift between output and input signals in the case of a single stage leading type phase shifter is given by (Refer to Figure 19.45)

θ (in degrees) = $2\tan^{-1}(\omega R_P C_P)$

Where, $\omega = 2\pi f$, f being the frequency of the input signal.

Self Evaluation Exercise

I. Multichoice Questions

1. The large signal bandwidth of an opamp is limited by its
   (*a*) Slew rate specification
   (*b*) Unity gain cross-over frequency specification
   (*c*) CMRR specification
   (*d*) Open loop gain.
2. Indicate the FALSE statement
   (*a*) Closed loop gain is always less than the open loop gain.
   (*b*) Input impedance is always greater than the output impedance.
   (*c*) The open loop gain falls to zero decibel at a frequency equal to unity gain crossover frequency.
   (*d*) Open loop bandwidth is always greater than the open loop bandwidth.
3. One of the following statements in reference to a unity gain noninverting buffer using an opamp is incorrect. Identify this statement -
   (*a*) It has extremely high input impedance, much higher than even its open loop input impedance
   (*b*) Its input impedance is very high almost approaching its open loop input impedance magnitude

(c) The voltage gain is unity
(d) The input, is applied to the non-inverting input.
(e) The feedback magnitude is 100 per cent.

4. One of the opamp configurations shown in Fig. 19.94 is the true voltage follower

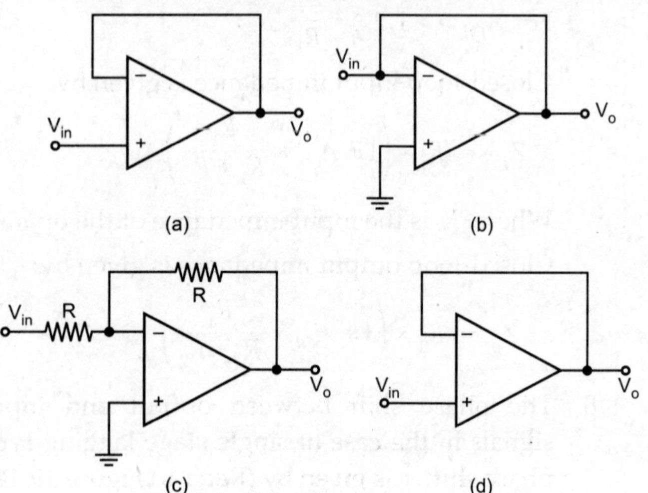

Fig. 19.94

(a) The one shown in Fig. 19.94 (a)
(b) The one shown in Fig. 19.94 (d)
(c) The one shown in Fig. 19.94 (c)
(d) The one shown in Fig. 19.94 (b)

5. The SLEW RATE specification of an opamp is usually measured in
(a) microvolts per second
(b) volts per microsecond
(c) decibels
(d) volts per microvolt.

6. The open loop gain of a certain opamp falls to zero decibel at a frequency of 10 MHz. Its transient response time is therefore,
(a) 35 ns (b) 350 ns
(c) 100 ns (d) none of these.

7. Write the expression for the output voltage for the integrator circuit of Fig 19.95 (a)

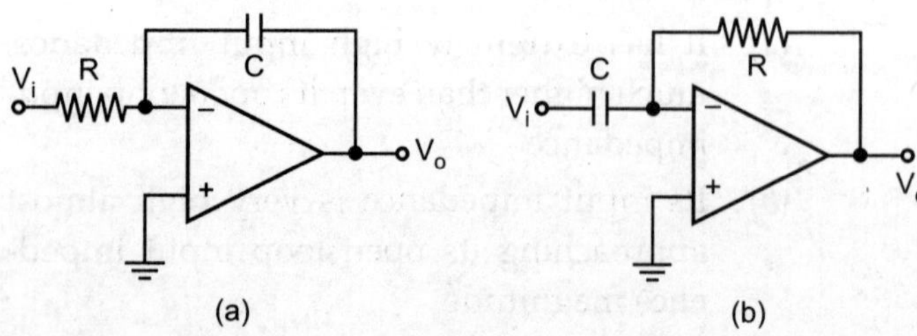

Fig 19.95

(a) $v_o = \frac{1}{RC}\int v_i dt$ (b) $v_o = -RC \int v_i dt$
(c) $v_o = -\frac{1}{RC}\int v_i dt$ (d) $v_o = -\int v_i dt$

8. Write the expression for the output voltage for the differentiator circuit of Fig. 19.95 (b).
(a) $v_o = -RC\frac{dv_i}{dt}$ (b) $v_o = -\frac{1}{RC}\frac{dv_i}{dt}$
(c) $v_o = \frac{dv_i}{dt}$ (d) $v_o = -\frac{dv_i}{dt}$

9. A given opamp has an open loop gain of 110 dB and a CMRR rating of 106dB. What should be the open loop common mode gain of this op-amp?
(a) It can not be determined from the given data
(b) 4 dB
(c) 216 dB
(d) 110 dB.

10. One of the following opamp type numbers is a comparator
(a) LM710 (b) LM741
(c) LM748 (d) LM747

11. One of the following type numbers is not an opamp
(a) LM748 (b) LM709
(c) MC1458 (d) LM 319
(e) LM 340

12. A certain general purpose opamp has a PSRR specification of 100 dB. This implies that a one volt change in the power supply will produce an input offset voltage change of
(a) 100μV (b) ImV
(c) 10 μV (d) 1 μV

13. A certain internally compensated opamp has an internally connected compensation capacitance of 30 pF. This compensation capacitor can be supplied with a charging current of 15 μA by the op-amp's internal circuitry. What opamp specification is being determined by this data? What is its value?
(a) Slew Rate, 0.5V/us
(b) Unity gain bandwidth, 1MHz
(c) CMRR, 10,0 dB
(d) Slew Rate, the value can not be determined from given data.

14. An. opamp has a unity gain bandwidth of 10 MHz. This opamp is wired as an inverting amplifier with a closed loop gain of 10. What is maximum usable input frequency, if the gain error is not allowed to exceed 1 per cent?

(*a*) lMHz (*b*) 100 kHz

(*c*) 10 kHz (*d*) None of these.

15. In a certain uncompensated opamp, it is recommended to externally connect a 50 pF capacitor between the specified terminals to ensure stability right upto the unity gain bandwidth value. By connecting a 25pF capacitor instead, the opamp performance is affected. Now what is it that gets affected.

(*a*) The slew rate and the opamp stability deteriorate

(*b*) The opamp capability to be immune to any tendency to oscillate deteriorates (it is stable upto a lower frequency of operation now) while the slew rate specification improves

(*c*) The slew rate deteriorates while the stability is not affected

(*d*) The slew rate and stability characteristic improve.

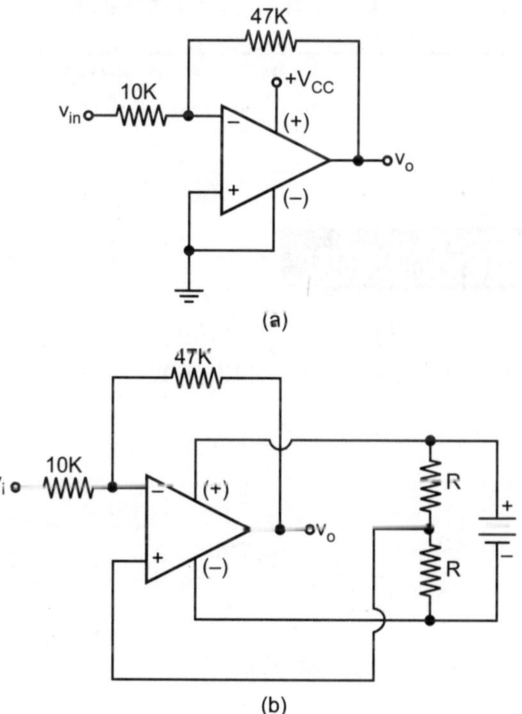

Fig. 19.96

II. Figs. 19.96 (*a*) and (*b*) show two inverting amplifier configurations intended to have identical gain and to be used with a single supply. Only one of the two configurations is correct. Identify

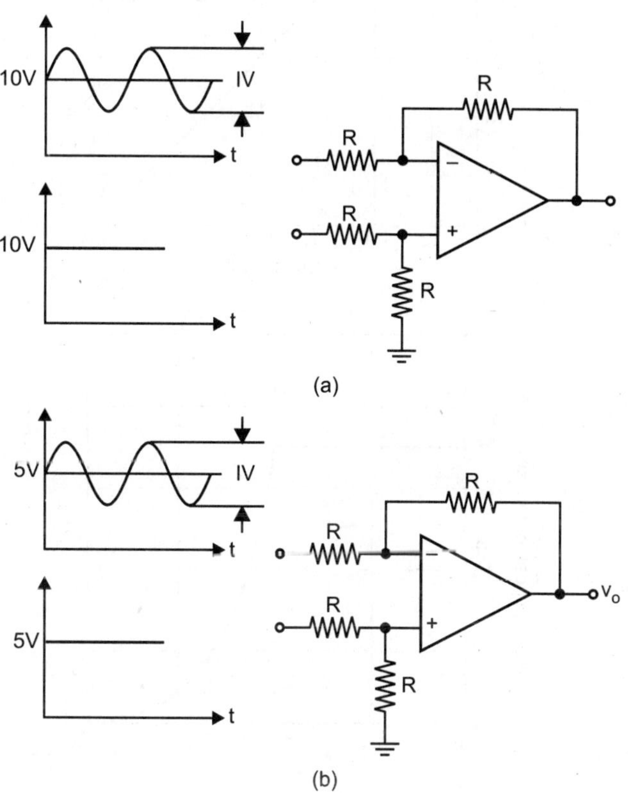

Fig 19.97

III. The output behaviour in case of two identical differential amplifier configurations shown in Figs. 19.97 (*a*) and (*b*) was observed to be slightly different, even though they were fed-with identical differential input. Which opamp specification is responsible for this ?

IV. Some of the performance characteristics of an opamp taken right from its data sheet are shown in Figs. 19.98 (*a*) to (*d*). These characteristics speak about one or more than one opamp specificatifflns. What specification or specifications can we determine for the given opamp from' each one of these characteristic curves.

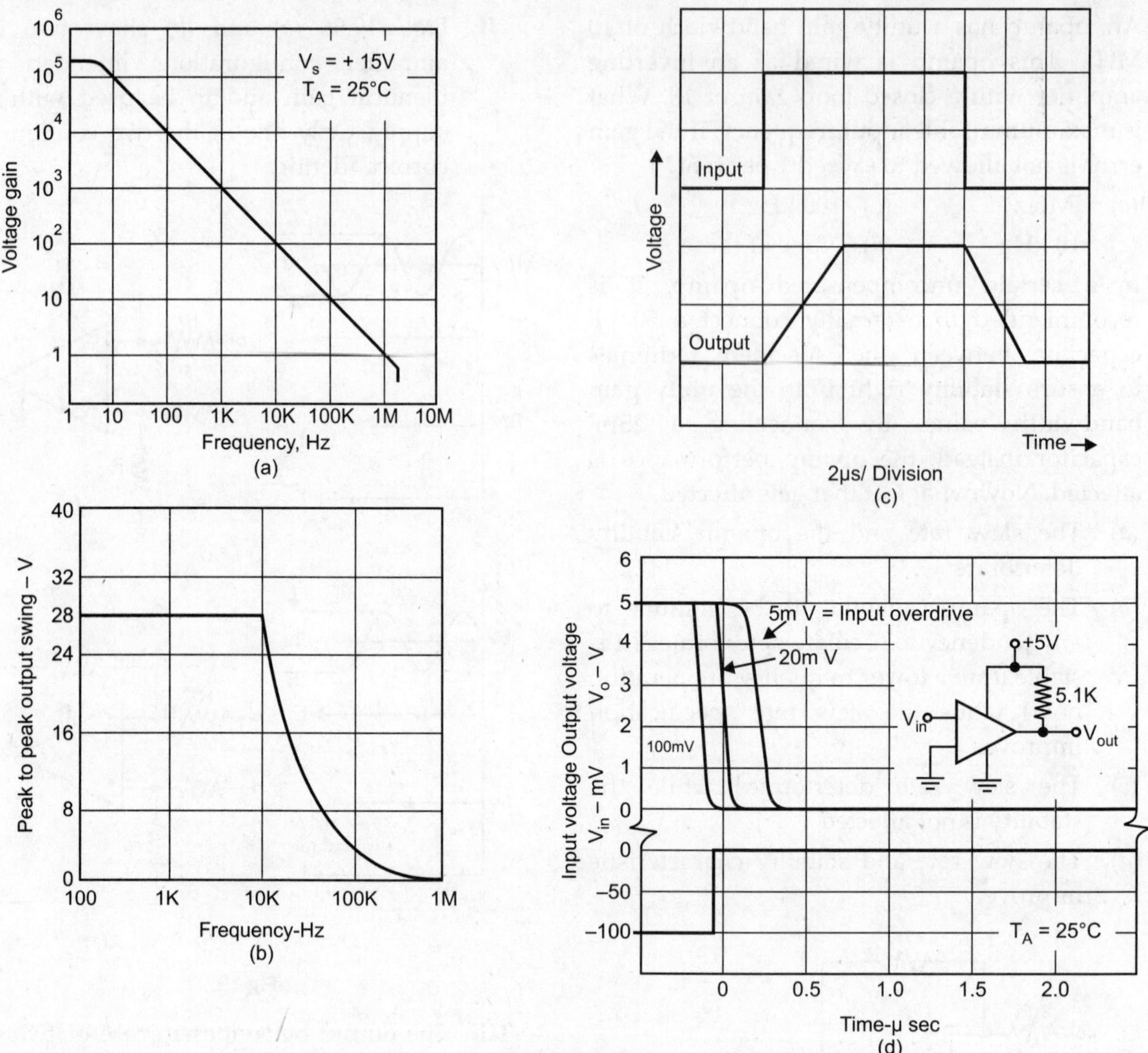

Fig. 19.98

Answers to Self Evaluation Exercise

I. 1. (*a*) 2. (*d*) 3. (*b*) 4. (*a*) 5. (*b*) 6. (*a*)
7. (*c*) 8. (*a*) 9. (*b*) 10. (*a*) 11. (*e*) 12. (*c*)
13. (*a*) 14. (*c*) 15. (*b*)

II. Fig. 19.6 (*b*)

III. Common mode Rejection Ratio (CMRR).

IV. Fig 19.98 (*a*):0 Open loop gain (AOL) and unity gain bandwidth
Fig 19.98 (*b*): Power Bandwidth
Fig. 19.98 (*c*): Slew rate
Fig. 19.98 (*d*): Response time (Normally defined with reference to comparators).

Chapter 20

IC Timers

Learning Objectives

After completing the chapter, you will learn the following topics:

1. Internal Architecture of IC Timer 555, Dual IC Timer 556 and Quad IC Timer 558
2. Internal Architecture of Programmable Timers 2240 and 7240
3. Different terminals of IC Timer 555 and their Significance
4. Different Astable Multivibrator Configurations Using IC Timer 555
5. Positive Edge-triggered Monostable Multivibrator Configuration Using IC Timer 555
6. Negative Edge-triggered Monostable Multivibrator Using IC Timer 555
7. IC Timer 555 as a Time Delay Generator
8. IC Timer 555 as a Frequency Divider
9. IC Timer 555 as a Pulse Width Modulator
10. IC Timer 555 as a Pulse Position Modulator
11. IC Timer 555 as a Sawtooth Waveform Generator
12. IC Timer 555 as a Positive to Negative DC Voltage Converter

Q.1. What is found inside of an integrated circuit timer 555? Outline the significance of each one of the IC timer terminals.

Ans. Fig. 20.1 shows the internal schematic arrangement of a timer IC 555. Basically, it consists of two comparators, a flip-flop, discharge transistor, an output stage and a potential divider arrangement that sets the comparators' reference levels at + $V_{CC}/3$ and + 2 $V_{CC}/3$, as the three resistances are identical. The comparators' outputs decide the fate of the flip-flop and subsequently the output. In the astable mode of operation, pins 2 and 6 are joined together and a capacitor connected from the junction to ground. There are two external resistances connected between pins (7) and (8) and between pins (7) and (2,6). Initially, comparator (2) sets the flip-flop and the output is HIGH. The output goes LOW, when capacitor charges to + 2Vcc/3 driving comparator (1) output HIGH and resetting the flip-flop. The capacitor also discharges through the now turned-on discharge transistor and the cycle repeats. In the monostable mode, trigger terminal is kept high and a negative trigger (from V_{CC} to ground) will initiate a pulse whose width is determined by external *RC* timing circuit.

Pin (1) of 555 is the **ground terminal.**

Pin (2) of 555 is the **trigger terminal.** If the voltage at this terminal is held greater than one-third of V_{CC}, the output remains LOW. A negative going pulse from V_{CC} to less than $V_{CC}/3$ triggers the output to go HIGH. The amplitude of the pulse should be able to make the comparator (inside the IC) change its state. However, the width of the negative going pulse must not be greater than the width of the expected output pulse.

Pin (3) is the **output terminal** of IC 555. There are two possible output states.

In the LOW output state, the output resistance appearing at pin (3) is very low (approximately 10Ω). As a result, the output will source zero current, if the load is connected from pin (3) to ground and sink a current I_{sink} (depending upon load), if the load is connected from

$+V_{CC}$ to pin (3). In the HIGH output state, there appears a resistance of approximately 10Ω between (V_{CC}) and pin (3). As a result, the output sources a current (depending upon load), if the load is connected from pin (3) to ground and sinks zero current, if the load is connected between + V_{CC} and pin (3).

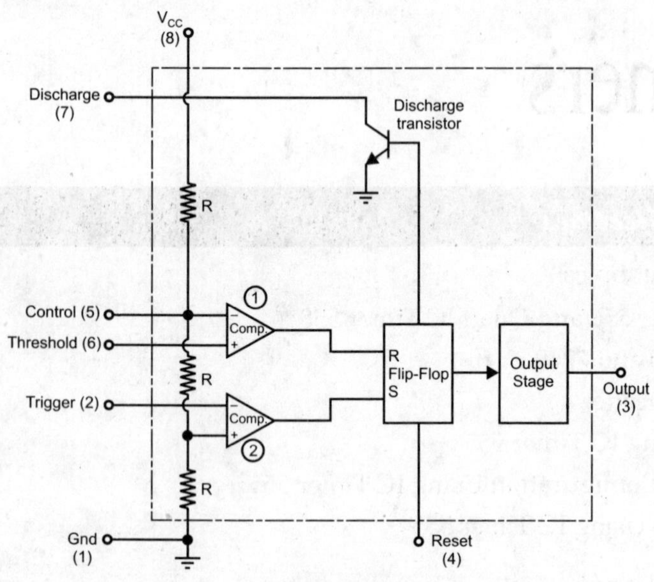

Fig. 20.1

Pin (4) is the **reset terminal**. When unused, it is connected to + V_{CC}. Whenever the potential of pin (4) is driven below 0.4V, the output is immediately forced to LOW state. The reset terminal enables the timer over-ride command signals at pin (2) of the IC.

Pin (5) is the **control voltage terminal.** This can be used to alter the reference levels at which the timer comparators change state. A resistor connected from pin (5) to ground can do the job. Normally, 0.01 μF capacitor is connected from pin (5) to ground. This capacitor bypasses supply noise and does not allow it to affect the threshold voltages.

Pin (6) is the **threshold terminal.** In both astable as well as monostable modes, a capacitor is connected from pin (6) to ground. Pin (6) monitors the voltage across the capacitor, when it charges from the supply and forces the already HIGH output to go LOW, when the capacitor reaches + 2/3 V_{CC}.

Pin (7) is the **discharge terminal.** It presents an almost open circuit, when the output is HIGH and allows the capacitor to charge from the supply through an external resistor or resistors (Fig. 20.2*a*) and presents an almost short circuit, when the output is LOW and allows the capacitor to discharge through it and an external resistor, if any (Fig. 20.2*b*).

Pin (8) is the + V_{CC} terminal. 555 can operate from any supply voltage from + 3 to + 18 V and is, thus compatible with both TTL logic and op-amps.

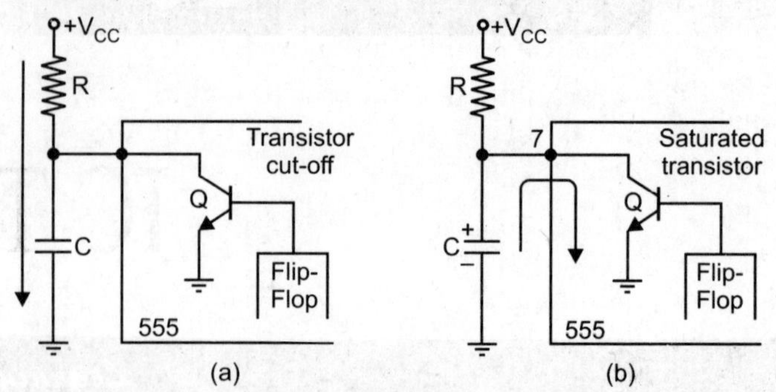

Fig. 20.2

Q.2. With the help of a circuit diagram, explain in brief the astable mode of operation of timer IC 555. Derive an expression for the frequency of the output waveform. Also, draw relevant waveforms.

Ans. Fig. 20.3 (*a*) shows basic 555 astable multivibrator configuration. When the output is HIGH, pin 7 presents an open circuit, the capacitor *C* charges from + V_{CC} through resistors R_1 and R_2 with a time constant determined by R_1, R_2 and C. When the voltage across capacitor C reaches 2/3 V_{CC}, the output goes LOW and pin 7 appears as a short circuit. *C* discharges through R_2 and the discharge transistor inside the *IC*. When the voltage across the capacitor falls below 1/3 V_{CC}, the output goes HIGH and the cycle repeats. Thus, the circuit behaves as an astable multivibrator with its ON-time governed by R_1, R_2 and *C* and OFF-time dependent on R_2 and *C*. The frequency of the output waveform is given by,

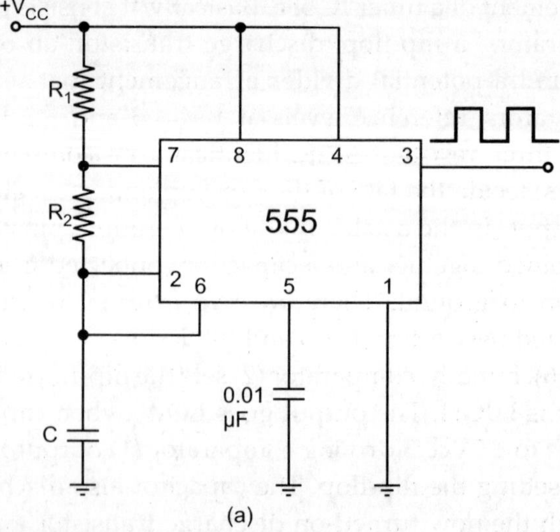

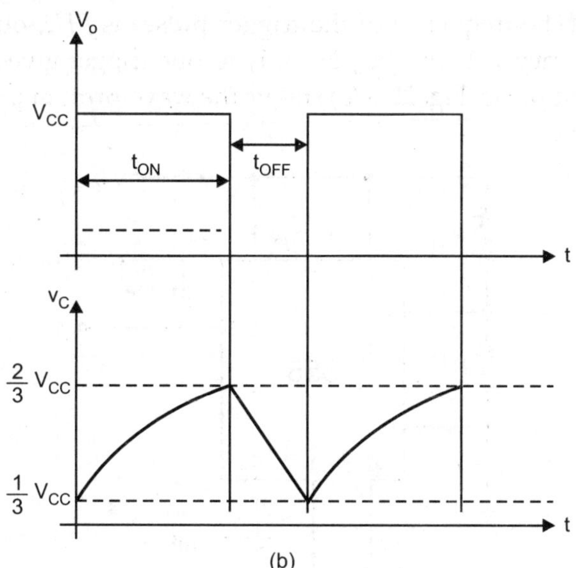

Fig. 20.3

$$f = \frac{1}{T}, \text{ where } T = (t_{ON} + t_{OFF})$$

Now t_{on} is governed by the charging process. The instantaneous value of the voltage across the capacitor is given by,

$$V_C = V_{CC}\left(1 - e^{-\frac{t}{(R_1 + R_2)C}}\right) \quad ...(1)$$

First of all, we find the time taken by the capacitor to charge from zero to $V_{CC}/3$

From equation (1),

$$\frac{V_{CC}}{3} = V_{CC}\left(1 - e^{\frac{-t}{(R_1 + R_2)C}}\right)$$

or $$\frac{-t}{(R_1 + R_2)C} \cdot \log_{10} e = \log_{10} 2/3$$

or $$t = \frac{-(R_1 + R_2)C \log_{10} 3/2}{\log_{10} e}$$

$$= -\left(\frac{\log_{10} 1.5}{\log_{10} e}\right)(R_1 + R_2)C$$

$$= 0.4\,(R_1 + R_2)C$$

Now, we find time taken by the capacitor to charge from zero to $2\,V_{CC}/3$

$$2/3 = \left(1 - e^{\frac{-t}{(R_1 + R_2)C}}\right)$$

or $$e^{\frac{-t}{(R_1 + R_2)C}} = \frac{1}{3} \text{ or } t = 1.09\,(R_1 + R_2)C$$

Now we can find out the time taken by the capacitor to charge from $1/3\ V_{CC}$ to $2/3\ V_{CC}$ and this time would be t_{ON}.

$$t_{ON} = 1.09\,(R_1 + R_2)\,C - 0.4\,(R_1 + R_2)\,C$$

$$= 0.69(R_1 + R_2)C$$

t_{OFF} would be the time taken by the capacitor to discharge from $2/3\ V_{CC}$ to $1/3\ V_{CC}$, $t_{OFF} = 0.69\ R_2C$.

Therefore, frequency of operation is given by,

$$f = \frac{1}{0.69\,(R_1 + R_2)\,C + 0.69\,R_2C}$$

Fig. 20.3 shows relevant waveforms.

Q.3. The astable circuit of Fig. 20.3 (*a*) cannot give us an output waveform with a duty cycle of 50% or less, as the charging resistance will always be greater than the discharge resistance and as a result t_{ON} is always greater than t_{OFF}. How do you modify this circuit to obtain duty cycles of 50 percent or less?

Ans. Fig. 20.4 (*a*) shows the modified circuit. Charging takes place through R_1, whereas discharge occurs through R_2, as the diode D will be reverse biased in that case.

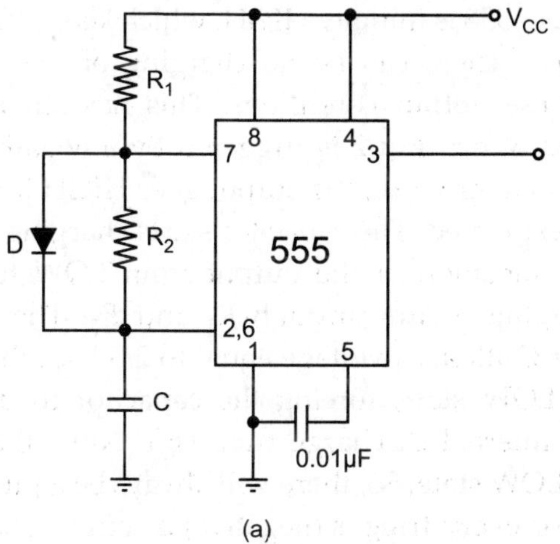

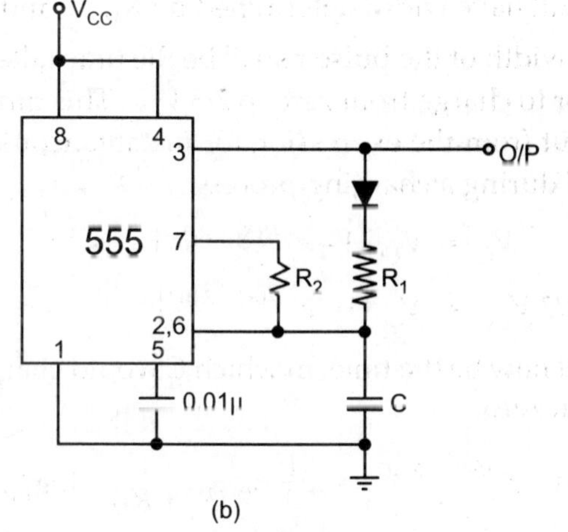

Fig. 20.4

Accordingly, $t_{ON} = 0.69\ R_1C$ and $t_{OFF} = 0.69\ R_2C$

So, charging and discharging times (t_{ON}) and (t_{OFF}) can be independently controlled. Frequency of output waveform,

$$f = \frac{1}{0.69\,(R_1 + R_2)\,C}$$

Another astable mode IC timer 555 circuit configuration, where ON and OFF time periods can be independently controlled is shown in Fig. 20.4 (*b*). The circuit relies on the fact that during the charging process, the output is HIGH. The capacitor, in this case, charges from the output. During LOW state of the output, the diode (*D*) prevents the capacitor to discharge through the conducting output transistor. Charging is through (R_1) and discharging is through (R_2).

Q.4. With the help of a circuit diagram, explain the operation of IC timer 555 in monostable mode. Derive an expression for the output pulse width. What would be the output frequency, if the frequency of trigger pulses is (*f*) Hz? Draw waveforms at (*i*) Pin 3 (*ii*) Pin 2 (*iii*) Pin 6.

Ans. Fig. 20.5 (a) shows the circuit diagram.

Pin 2 of 555 is initially HIGH, which keeps the output held LOW. There can be no charging of the capacitor under these output conditions. This has already been discussed. When Pin 2 is triggered by a negative going pulse (V_{CC} to ground), the output goes HIGH for reasons already explained. The capacitor starts charging with the positive transition of the output from LOW to HIGH. The charging occurs through R_1 and R_2. The moment capacitor *C* attains a voltage equal to 2/3 V_{CC}, the output goes to LOW state, forcing the capacitor to discharge through internal discharge transistor. Now, the output stays in LOW state. So, there will always be a pulse at the output for every trigger (negative) at pin 2. The output pulse shall have width determined by R_1, R_2 and *C*.

The width of the pulse τ shall be the time taken by the capacitor to charge from zero to 2/3 V_{CC}. This time can be found out from the expression for instantaneous voltage across *C* during a charging process.

$$V_C = V_{CC}[1 - e^{-t/(R_1 + R_2)C}]$$

$$\text{or } 2/3\ V_{CC} = V_{CC}[1 - e^{-t/(R_1 + R_2)C}]$$

t will now be the time, in which *C* would charge to 2/3 V_{CC} from zero.

$$\text{Now, } e^{-t/(R_1 + R_2)C} = \frac{1}{3} \text{ or } \frac{-t}{(R_1 + R_2)C}\log_{10}e = \log\frac{1}{3}$$

$$\text{or} \quad t = (R_1 + R_2)\,C\,\log_{10} 3/\log_{10} e$$

In general, if *R* is the total charging resistance, then pulse-width (τ) shall be given by,

$$\tau = 1.1\,RC$$

If the frequency of the trigger pulses is *f Hz*, output frequency will also be *f Hz* only as one trigger gives one output pulse. Fig. 20.5 (*b*) shows the waveforms at pins 2, 3 and 6.

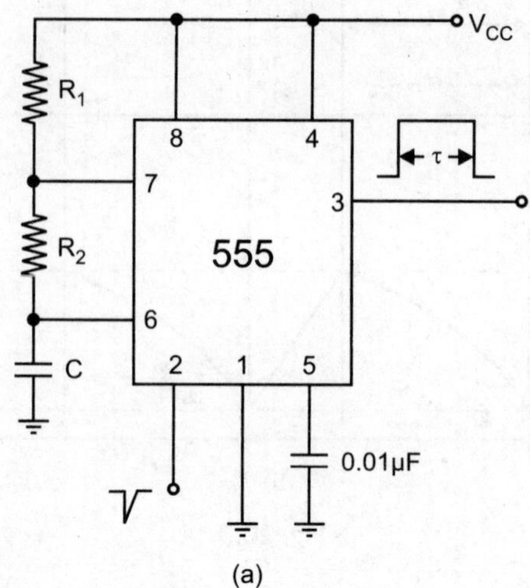

(a)

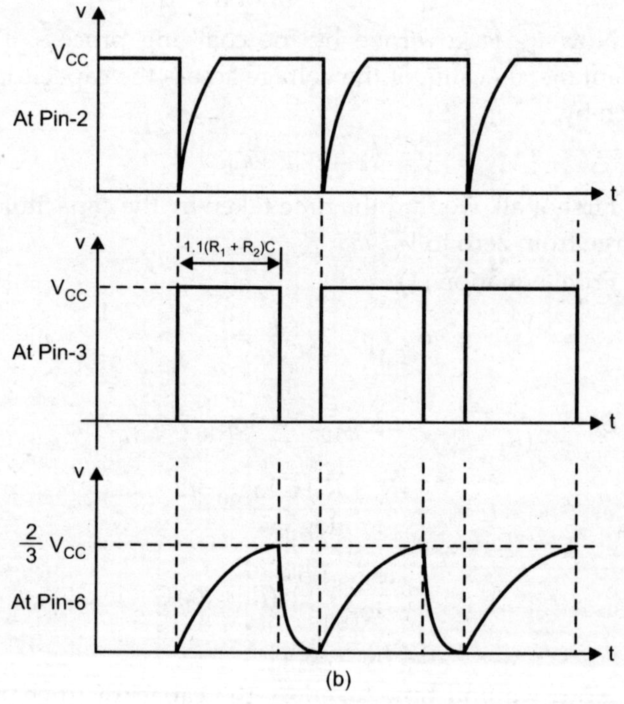

(b)

Fig. 20.5

Q.5. It is often desirable to trigger a 555 monostable multivibrator either on the leading edges or on the trailing edges of a continuous pulse train. Suggest 555 monoshot circuit for these two cases i.e. (*i*) When it is desired to trigger the monoshot on the leading edges, (*ii*) When it is desired to trigger the monoshot on the trailing edges.

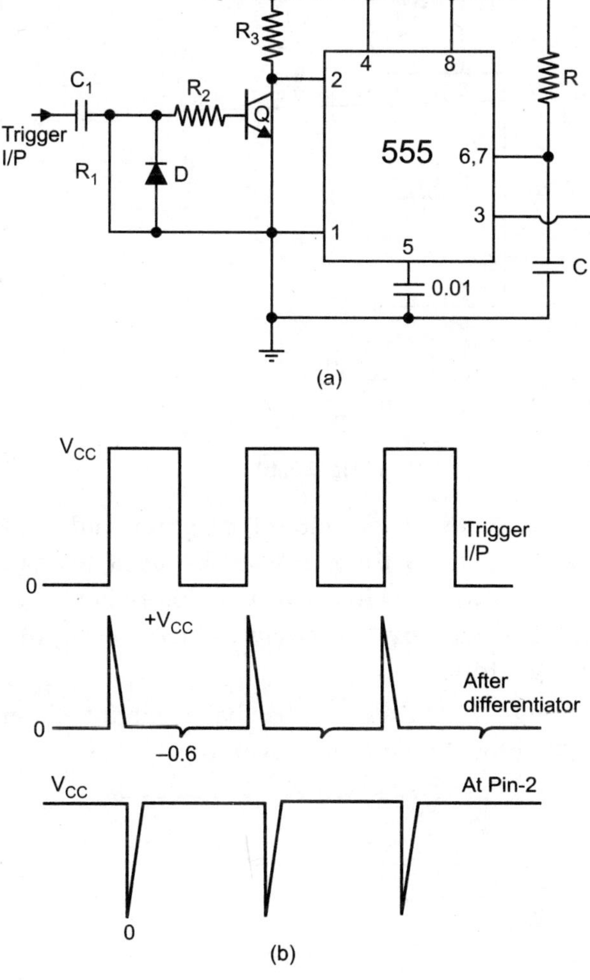

Fig. 20.6

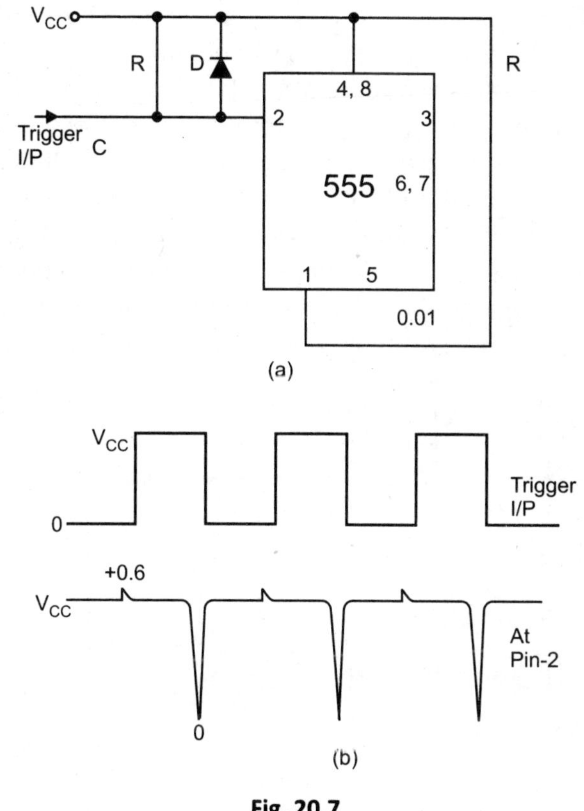

Fig. 20.7

Ans. Fig. 20.6 (*a*) shows the monoshot configuration that is triggered by the leading edges of the pulse train applied at trigger input. The pulse train applied at the trigger input is differentiated by $C_1 - R_1$ and the differentiated waveform is applied to the base of a transistor. The negative going spikes in the differentiated waveform are clipped by diode (*D*). The transistor switch then generates the desired trigger signal for pin-2 of 555. The relevant waveforms are shown in Fig. 20.6 (*b*). The differentiator time constant (R_1 C_1) should be much smaller than the ON-time of the pulse train for proper differentiation. Also, the differentiated pulse width should be smaller than the expected HIGH time of the monoshot output.

The monoshot configuration that triggers on the trailing edges along with the relevant waveforms is shown in Fig.20.7 (*a*) and (*b*). The circuit is self-explanatory.

Q.6. When a 555 timer is used in an astahle mode, after the reset is removed, the first pulse is almost twice as wide as the subsequent pulses. Why is it so ?

Ans. It is because of the reason that when the reset is pulled down in the astable mode of operation, the capacitor discharges completely so that it has to charge to 2/3 V_{CC} from ground, after the reset is removed. In the subsequent pulses, the capacitor has to charge to 2/3 V_{CC} from 1/3 V_{CC} only and the time taken to charge from ground to 2/3 V_{CC} is approximately twice as large as that taken by the capacitor to charge from 1/3 V_{CC} to 2/3 V_{CC}.

Q.7. What is the longest time-delay that one can get from timer IC 555 ?

Ans. Although it is theoretically possible to get time delays exceeding an hour, yet it is not advisable. It is because of the reason that very large value capacitors with low leakage specifications are pretty expensive. As a result, one would prefer using a different scheme to achieve time delays of this order. However, 30 to 35 minutes is the longest feasible time delay.

Q.8. Explain in brief, how a 555 timer IC can be used to generate a negative voltage from a positive voltage in a transformerless configuration?

Ans. Fig. 20.8 shows the basic circuit arrangement.

IC 555, resistors R_1, R_2 and capacitors C_1, C_2 constitute an astable multivibrator. The frequency of the output waveform is chosen to be a few kilohertz. Capacitor C_4 and diode D_1 form the clamping circuit. The positive peaks of the waveform get clamped to zero. Diode D_2 acts as a rectifier and capacitor C_3 as a filter. The output voltage tracks the input voltage linearly. However, the magnitude of output voltage is less than the positive input voltage by about 3 volts. Fig. 20.9 (*a*) to (*c*) show the

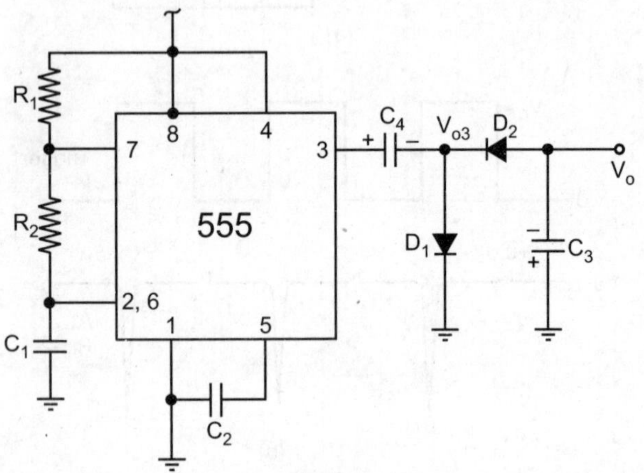

Fig. 20.8

waveforms at pin 3 of 555, at the anode of diode D_1 and across capacitor C_3 respectively. The circuit shown in Fig. 20.9 produces an unregulated negative voltage, *i.e.* the output voltage varies with change in input voltage or change in load conditions. The circuit can be modified to make it a source of regulated negative voltage. The output voltage can be sensed and compared with a reference voltage. The error signal can be utilized in conjunction with pin 4 of 555 to obtain regulation effect. Fig. 20.10 shows the circuit.

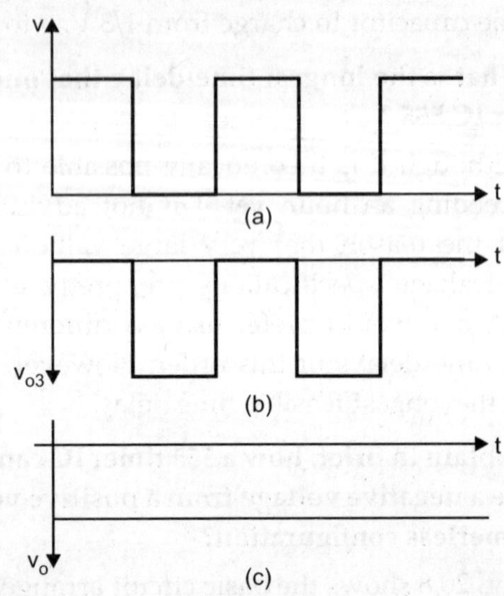

Fig. 20.9

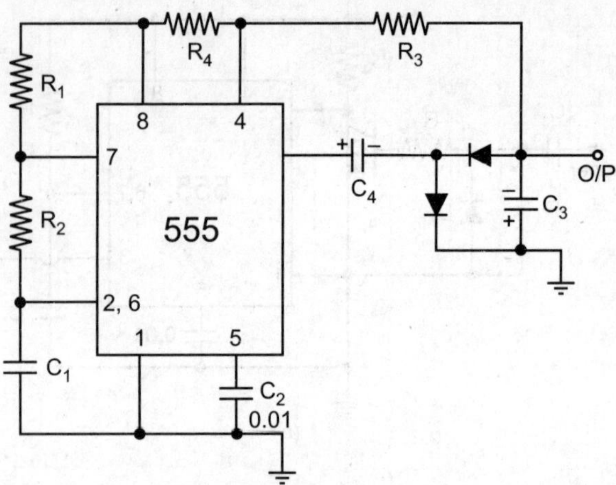

Fig. 20.10

Q.9. Draw the basic circuit diagram and explain in brief how a 555 timer IC can be used to make a frequency divider? Find cut the component values to make it divide by two circuit. Draw the input and output waveforms.

Ans. Fig. 15.11 shows the basic circuit diagram alongwith input and output waveforms.

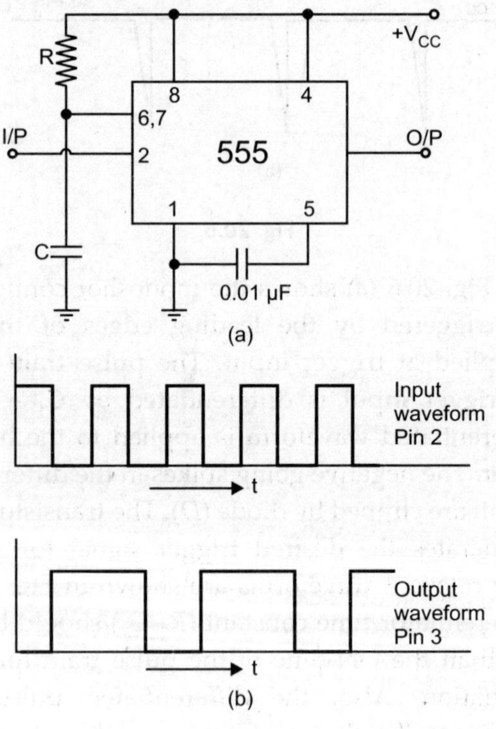

Fig. 20.11

The circuit shown is a simple monostable multivibrator circuit. It can be used as a frequency divider, if the length of the timing cycle is properly adjusted. Now, what happens is that output goes from LOW to HIGH on the negative going edge of the input waveform. To make it a divide by two circuit, the output should not respond to the succeeding negative transition of the input and also, the output should start a new cycle on the next negative transition of input waveform. This will be possible only, if the output is HIGH at the time of unwanted negative transitions and is zero at the time of desired negative transitions. Waveforms shown in Fig. 20.11 (*b*) further illustrate it. Suppose that we want to divide 10 kHz waveform by 2. The time period of a 10 kHz waveform is $(1/10^4) = 100\ \mu s$. Obviously, the length of the timing cycle should be anywhere between 100 μs and 200 μs to make it a divide by two circuit. Let it be 150 μs. Length of the timing cycle in a monoshot configuration is given by = 1.1 *RC*

$$1.1\ R_C = 150\ \mu s = 150 \times 10^{-6}$$

$$R_C = \frac{150 \times 10^{-6}}{1.1}, \text{ Let } C = 0.01\ \mu F$$

$$R = \frac{150 \times 10^{-6}}{1.1 \times 10^{-8}} = 15\ K.$$

Q.10. Explain in brief, how a 555 timer IC can be used as a pulse width modulator? Draw expected waveforms.

Ans. Fig. 20.12 (*a*) shows the circuit diagram and Fig. 20.12 (*b*) shows relevant waveforms.

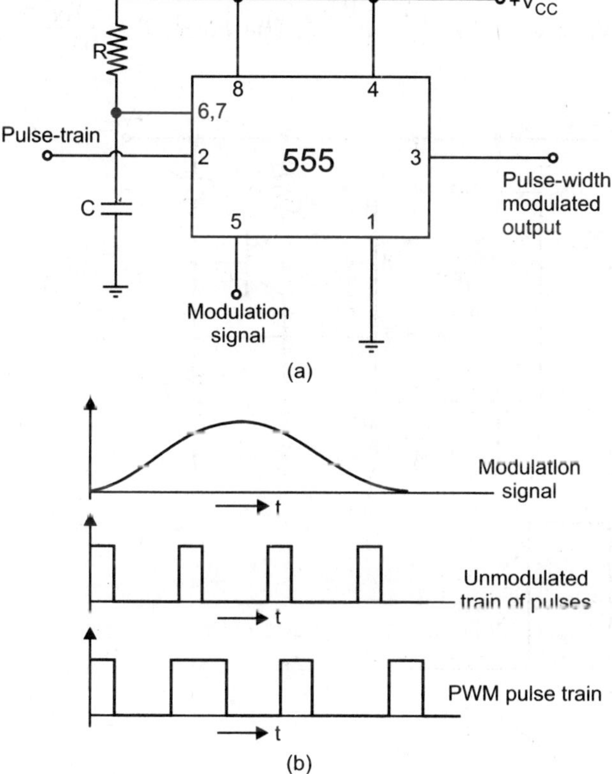

Fig. 20.12

The timer has been used in the monostable mode. It is being triggered by a continuous train of pulses.

Each pulse in the pulse-train initiates a pulse at the output. The width of the output pulse depends upon the time that elapses before capacitor charges to the upper threshold voltage. This upper threshold is 2/3 V_{CC} in the absence of any externally applied DC voltage at pin 5. In the circuit of Fig. 20.12 (*a*), the upper threshold varies with the amplitude of modulation signal and so does the pulse width in the pulse-train at the output. Fig. 20.12 (*b*) shows relevant waveforms.

Q.11. How do you use a 555 timer IC to make a DC to DC converter? Draw the basic circuit diagram and explain the principle of operation.

Ans. Fig. 20.13. shows a basic DC to DC converter using a 555 timer IC. In the circuit shown, the timer IC working in an astable multivibrator mode, changes the available input DC voltage V_{in} into a waveform that is acceptable to a transformer. The transformer does the amplitude conversion and the bridge rectifier rectifies the waveform appearing across the secondary of the transformer. The capacitor does the filtering.

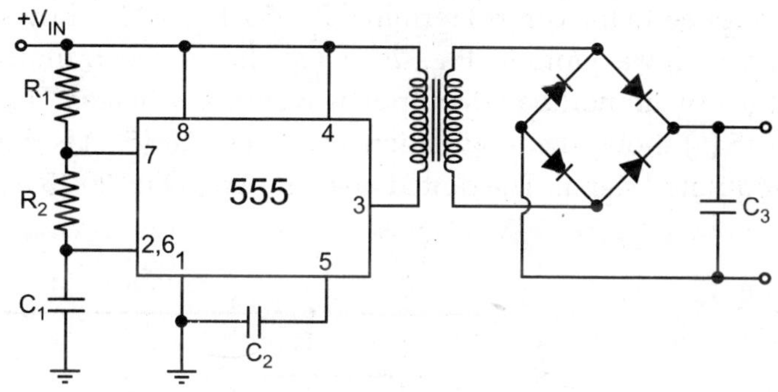

Fig. 20.13

Q.12. Explain in brief, the circuit application of a 555 timer IC as a pulse position modulator. Illustrate with relevant waveforms.

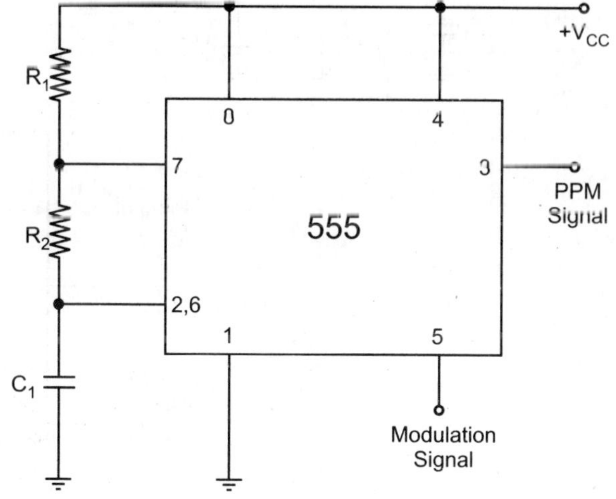

Fig. 20.14

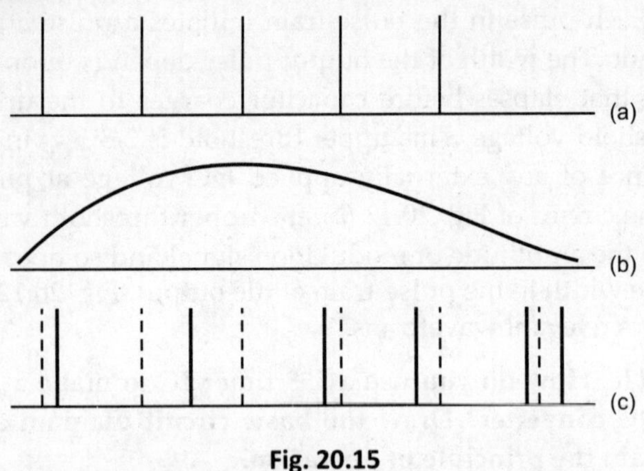

Fig. 20.15

Ans. Fig. 20.14 shows the circuit diagram. The timer has been connected in the free running mode. In the absence of any modulation signal, the output is a continuous train of pulses. In the pulse-train, which can be called an unmodulated one, the separation between the trailing edges of any two consecutive pulses is identical. The positions of these trailing edges can be made to alter in proportion to modulation signal amplitude, if the same is applied to the control terminal IC 555. Fig. 20.15 shows expected waveforms. Fig. 20.15 (*a*) shows the trailing edges of unmodulated output waveform, whereas Fig. 20.15 (*b*) shows the modulation signal. Fig. 20.15 (*c*) is the modulated signal. The dotted lines shown in Fig. 20.15 (*c*) constitute the unmodulated pulse train, the one shown in Fig. 20.15 (*a*).

Q.13. Briefly describe the use of IC timer 555 astable multivibrator as a sawtooth waveform generator.

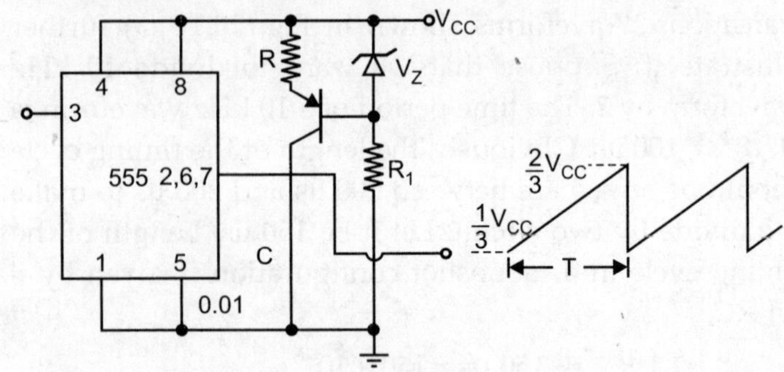

Fig. 20.16

Ans. Fig. 20.16 shows the circuit diagram. It is the basic astable multivibrator configuration with a slight modification in the capacitor charging circuitry. The capacitor, in this case, charges through a constant current source with the constant current value given by $[(Vz - 0.6)/R]$.

$$T = C\left[\frac{2/3\,V_{CC} - 1/3\,V_{CC}}{I}\right] = \frac{C\,V_{CC}}{3I}$$

Where, $I = \left(\frac{V_Z - 0.6}{R}\right)$. Therefore, $T = \frac{V_{CC}\,RC}{3(V_Z - 0.6)}$

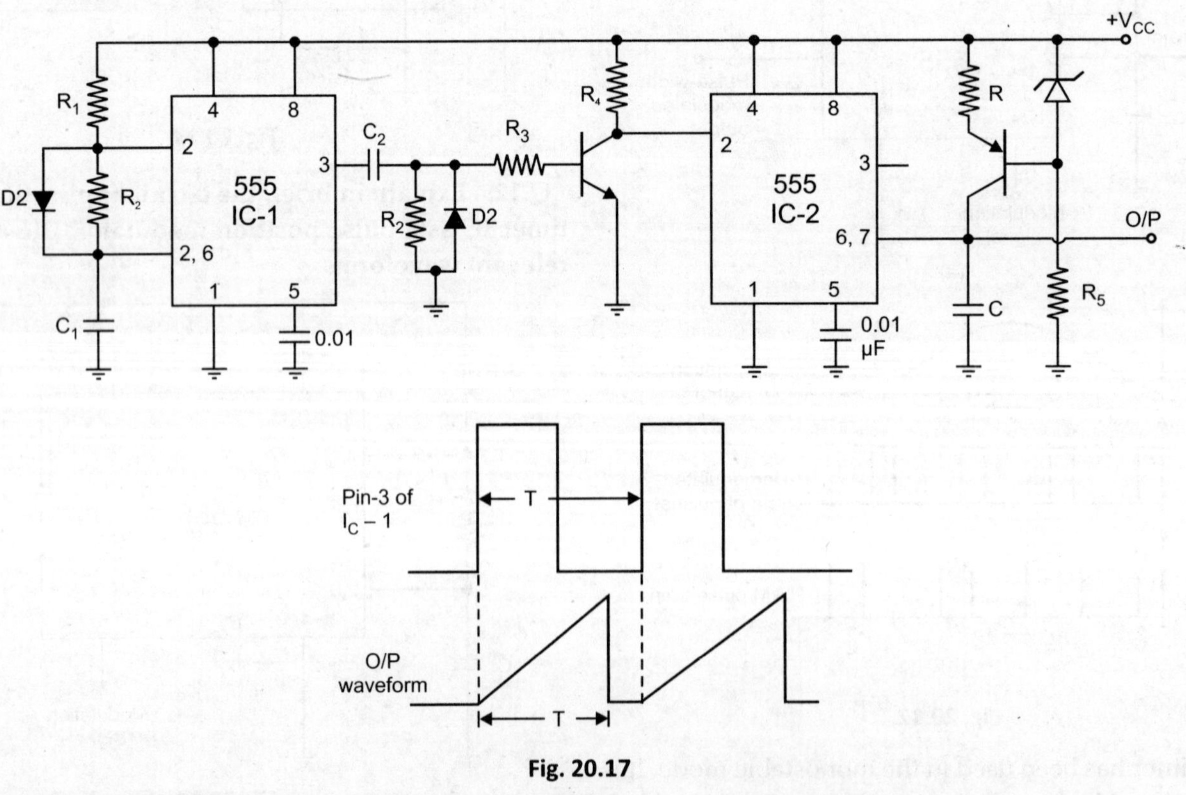

Fig. 20.17

The discharge is almost instantaneous. The lower and the upper voltage levels of the sawtooth waveform are 1/3 V_{CC} and 2/3 V_{CC}. A sawtooth waveform that ramps from zero to 2/3 V_{CC} can also be constructed by using a monoshot with constant current, charging with the trigger input of monoshot driven by the output of an astable multivibrator. The circuit diagram and the relevant waveform are shown in Fig. 20.17. The sawtooth time period is given by,

$$T = \left(\frac{2\,V_{CC}RC}{3(V_Z - 0.6)}\right)$$

The LOW time of the output sawtooth waveform $(T' - T)$. (T') can be adjusted to get desired sawtooth waveform.

Q.14. Suggest an IC timer 555 based circuit, whose output can drive a resistive load after a specified time delay (T_1) has elapsed after the receipt of the positive going trigger pulse. Everytime, there is a trigger pulse, the load should be driven for a specified time length (T_2).

Ans. Fig. 20.18 (*a*) shows the proposed circuit diagram. Both the timers have been wired as monostable multivibrator. The first or the input monoshot output goes HIGH on the leading edge of the trigger pulse. The output remains HIGH for a time length equal to (1.1 R_1C_1). The second or the output monoshot has been wired to trigger on the low going edge of the trigger pulse. The output pulse of the first monoshot acts as a trigger pulse for the second monoshot, which when triggered, goes HIGH for a time length equal to (1.1 R_2 C_2). Thus, the load (R_L) is driven for a time length of (1.1 R_2C_2). If R_1, C_1, R_2 and C_2 are so chosen that T_1 = 1.1 R_1 C_1 and T_2 = 1.1 R_2 C_2 are equal to the required time periods, the desired results are achieved. Also, the time constant R_3 C_3 should be much less than R_1 C_1 and time constant R_4 C_4 should be much smaller than R_2 C_2. The relevant waveforms are shown in Fig. 20.18 (*b*).

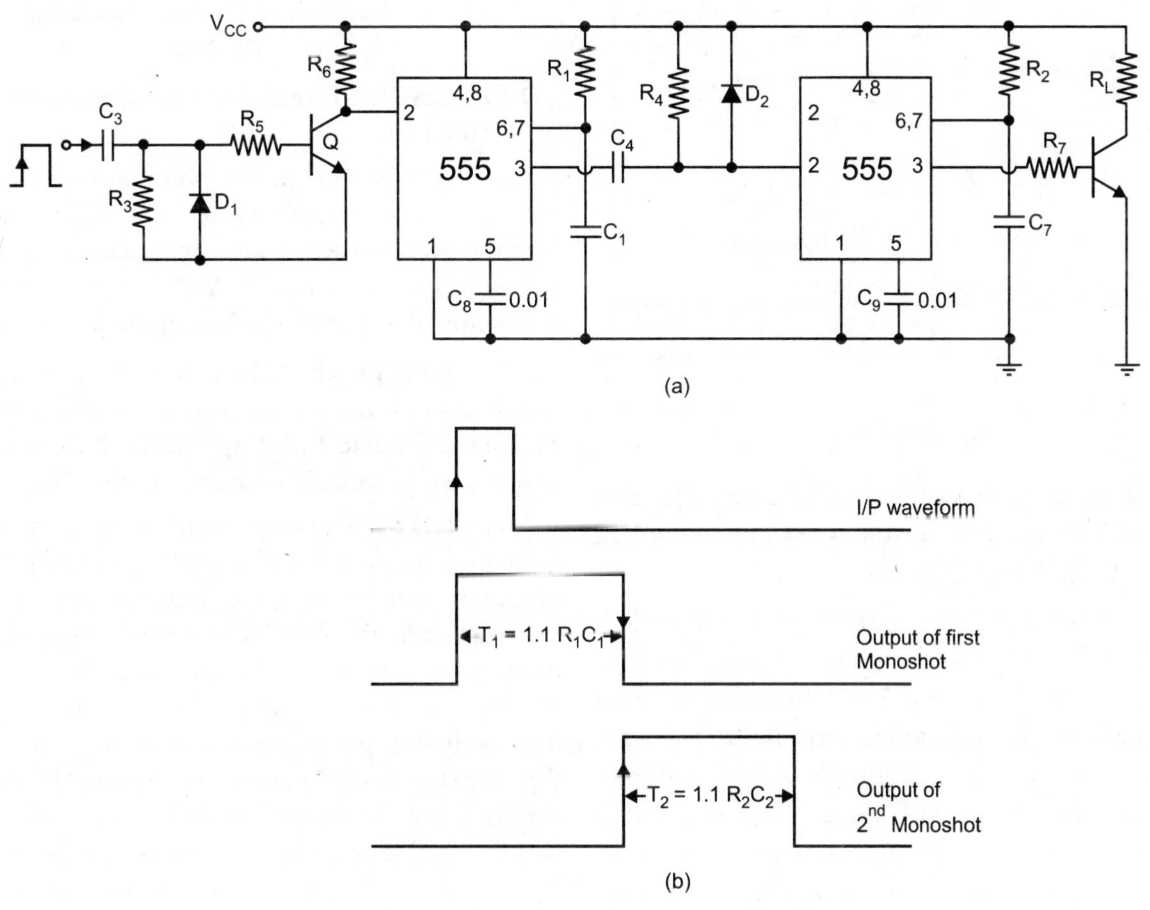

Fig. 20.18

Q.15. Is there a dual version of IC timer 555 also available? If yes, what is it and what are its salient features?

Ans. Yes, there is a dual version of IC timer 555, which is IC timer 556. Again like the timer IC 555, timing is provided by an external resistor and an external capacitor. The two timers operate independently sharing only the

V_{CC} and Ground. The timers can be triggered and reset on HIGH-to-LOW edges and the output structures can source or sink upto 200 mA. Typical applications include precision timing, pulse generation, sequential timing, time delay generation, pulse width modulation, pulse position modulation and linear ramp generator. These are also the typical applications of IC timer 555.

Q.16. Give the internal block schematic arrangement of IC timer 556.

Ans. As outlined in the previous question, IC timer 556 is a dual version of IC timer 555. Therefore, its internal architecture is also expected to be the same as that of IC timer 555. Figure-20.19 shows the internal block schematic of IC timer 556. The figure is self-explanatory.

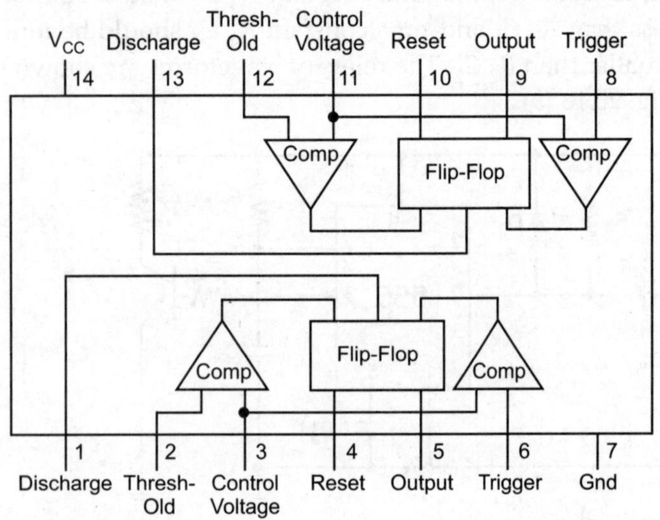

Fig. 20.19

Q.17. Is there a quad version of IC timer 555 also available? If yes, what is it and what are its salient features and typical applications?

Ans. Yes, there is a quad version of IC timer 555, which is IC timer 558. The IC timer 558 provides four independent timing functions. The timers can be used in a monostable mode to produce accurate time delays, from microseconds to hours. In the time delay mode of operation, the time is precisely controlled by one external resistor and one capacitor. The four timer sections in IC timer 558 are edge-triggered; therefore, when connected in tandem for sequential timing applications, no coupling capacitors are required. Each timing function can source or sink 100 mA. Typical applications include sequential timing, time delay generation, precision timing and industrial controls.

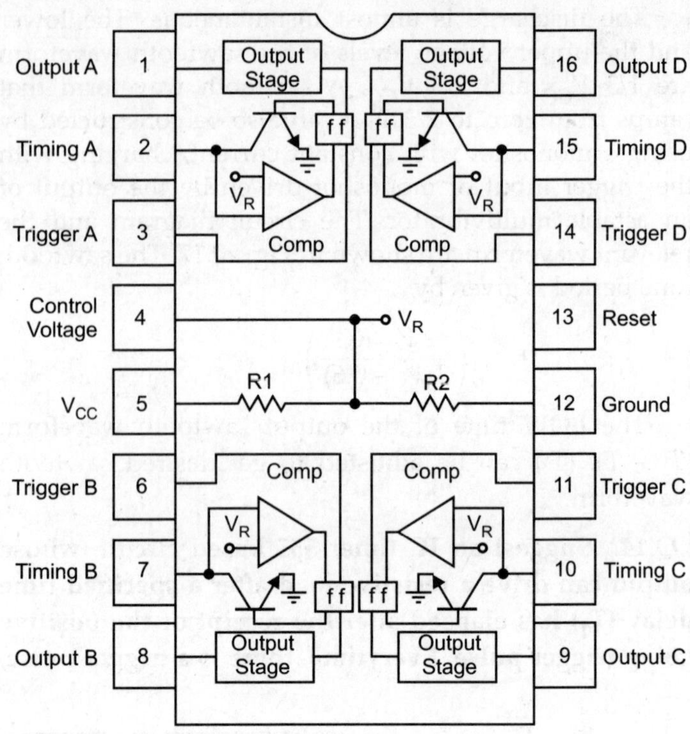

Fig. 20.20

Q.18. Give the internal block schematic arrangement of IC timer 558.

Ans. As outlined in the previous question, IC timer 558 is a quad version of IC timer 555. Therefore, its internal architecture is also expected to be the same as that of IC timer 555. Figure-20.20 shows the internal block schematic of IC timer 558. The figure is self-explanatory.

Q.19. What are the salient features of programmable timer 2240? With the help of its internal architecture, outline the basic building blocks 2240 comprises of. What is programmable timer/counter 7240?

Ans. IC 2240 is a programmable timer/counter capable of both monostable and astable operation. Monostable operation allows accurate time delays ranging from microseconds to about 120 hours. Long delays, upto three years, can easily be generated by cascading two timers. The basic building blocks of 2240 include a time-base oscillator, programmable 8-bit counterand a control flip-flop. Figure-20.21 shows the internal block schematic arrangement. An external resistor-capacitor (*RC*) network sets the oscillator frequency and allows delay times from 1 *RC* to 255 *RC*. In the astable mode of operation, a single *RC* network sets the base frequency. The frequencies of the square waves at the 8 outputs are each at different factors of two from the base frequency. If two or more of the outputs are shorted together, various pulse patterns can be generated. These frequencies or pulse pattern scan also easily be synchronised to an external signal. The trigger, reset and outputs are all TTL and CMOS compatible for easy interface with digital systems. The timer's high accuracy and versatility in producing a wide range of time delays makes it ideal as a direct replacement for mechanical or electromechanical devices. 7240 is CMOS version of the bipolar 2240.

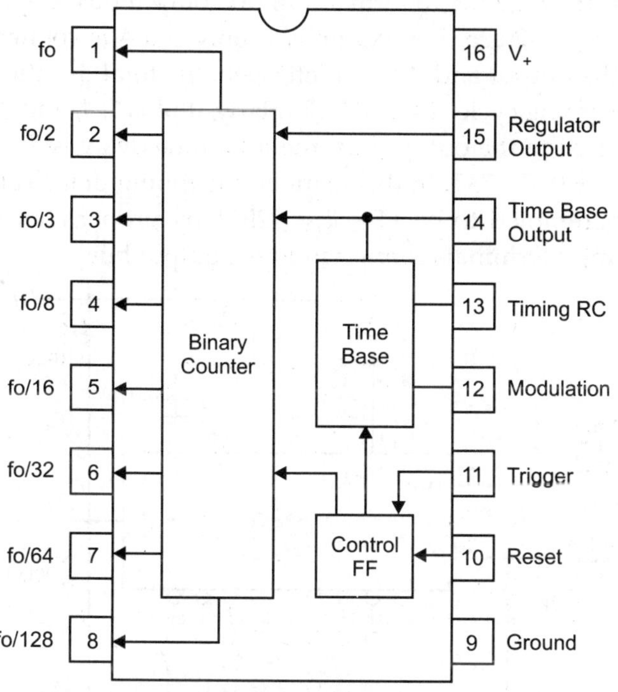

Fig. 20.21

Q.20. With reference to the block schematic representation of the internal architecture of programmable timer counter, briefly give the functional description of the device.

Ans. Refer to the internal block schematic arrangement of Figure-20.21. When power is applied to the timer counter IC 2240; with trigger or reset inputs not activated, the circuit starts with all outputs HIGH. Application of a positive going trigger pulse to the trigger pin-13 of the *IC* initiates the timing cycle. The trigger input activates the time-base oscillator thereby enabling the counter section. The counter outputs are set to LOW state. The time-base oscillator generates timing pulses at output of time-base oscillator, that is pin-14, with a period $T = 1\ RC$. These clock pulses are counted by the binary counter section. The timing sequence is completed, when a positive going reset pulse is applied to the RESET pin. Once triggered, the circuit is immune from additional trigger inputs until the timing cycle is completed or a reset input is applied. If both the RESET and trigger are activated simultaneously; the trigger takes precedence. Figure-20.22 gives the timing sequence of output waveforms at various circuit terminals after the counter section is triggered. When the circuit is in a RESET state, both the time-base and the counter sections are disabled, and the counter outputs are in HIGH state.

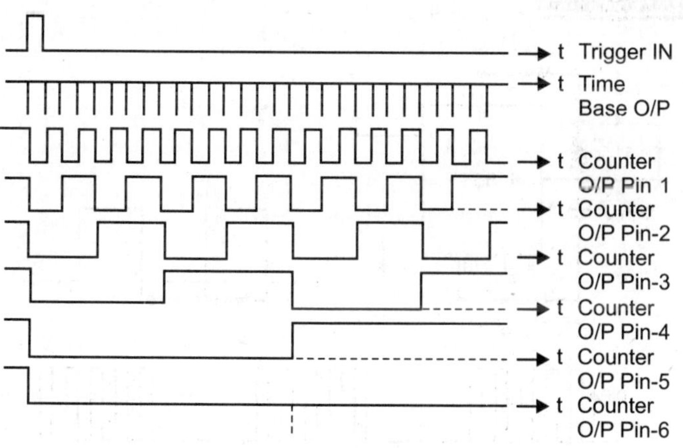

Fig. 20.22

Q.21. With the help of a suitable circuit diagram, briefly explain the operation of programmable timer/ counter 2240 in Astable mode of operation.

Ans. The Programmable Timer/Counter 2240 can be operated in its Astable or free running mode by disconnecting the reset terminal (pin 10) from the counter outputs. Two typical circuits are shown in Figures 20.23 and 20.24. The circuit in Figure-20.23 operates in its free

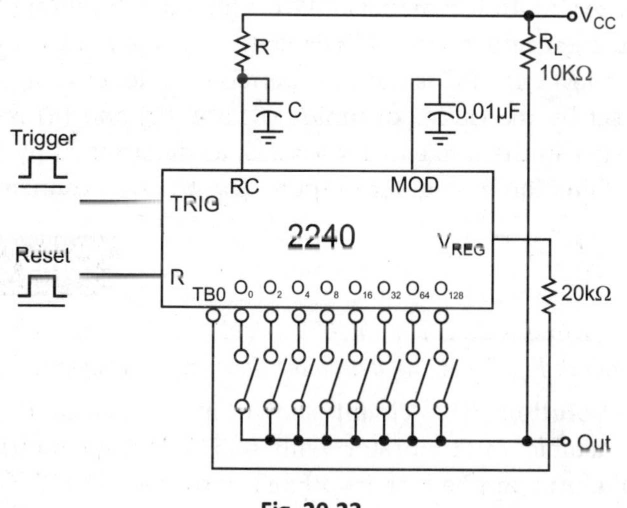

Fig. 20.23

running mode with external trigger and reset signals. It starts counting and timing, following a trigger input until an external RESET pulse is applied. Upon application of a positive going RESET signal to pin 10, the circuit reverts back to its RESET state. The circuit of Figure-20.24 is designed for continuous operation. It self-triggers automatically, when the power supply is turned on, and continues to operate in its free running mode indefinitely. In Astable or free running operation, each of the counter outputs can be used individually as synchronised oscillators, or they can be inter-connected to generate complex pulse patterns.

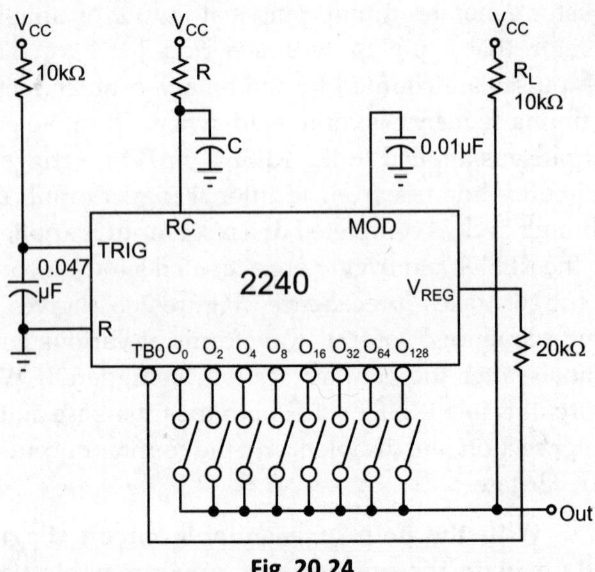

Fig. 20.24

Q.22. With the help of a suitable circuit diagram, briefly explain the operation of programmable timer/ counter in Monostable mode of operation.

Ans. Figure-20.25 shows the generalised circuit diagram of *IC* timer 2240 based Monostable multivibrator. The output is normally HIGH and goes LOW, following a trigger input. It remains LOW for the time duration (T_O) and then returns to a HIGH state. T_O is given by $T_O = nT = nRC$. $T = RC$ is the time period of time, base signal, as set by the choice of timing (*R*) and (*C*) and (*n*) is an integer in the range of $1 \leq n \leq 255$, as determined by the combination of counter outputs ($O_0 \ldots O_{128}$) connected to the output bus. The binary counter outputs, *O*0 . . . *O*128, pins 1 through 8, are open collector type stages and can be shorted together to a common pull-up resistor to form a wired-*OR* connection.The combined output will be LOW as long as any one of the outputs is LOW. The time delays associated with each counter output can, thus be added together. This is done by simply shorting the outputs together to form a common output bus as shown in Figure-20.25. For example, if only pin 4 is connected to the output and the rest left open, the total duration of the timing cycle, T_O, is $8T$. Similarly, if pins 1, 4, and 7 are shorted to the output bus, the total time delay is $T_O = (1 + 8 + 64)\ T = 73T$. In this manner, the timing delay can be programmed to be $1T \leq T_O \leq 255T$ by proper choice of counter terminals connected to the output bus.

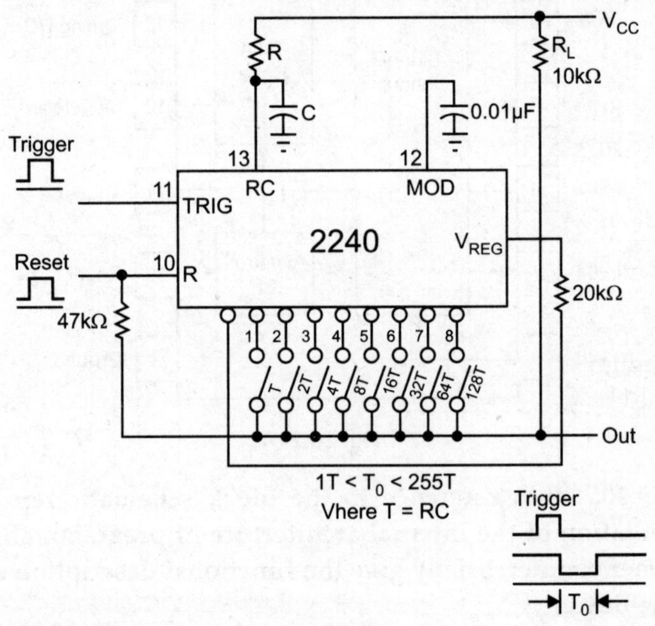

Fig. 20.25

Solved Problems

Problem 20.1. *For the 555 astable multivibrator circuit shown in Fig. 20.26 (a), determine the output waveform.*

Solution. The circuit shown in Fig. 20.26 (*a*) is an astable multivibrator with a 500 Hz symmetrical waveform applied at its RESET terminal. The RESET terminal is alternately HIGH and LOW for 1 ms. Thus, the astable output appears for 1ms (when the RESET is HIGH) and then the output is LOW for 1ms (when the RESET is LOW). The cycle repeats in the same fashion.

The astable multivibrator has an ON time of $0.69 \times 14.5 \times 10^3 \times 10^{-8}$ sec equal to 100 μs. The OFF time is also 100μs so that the astable has a frequency of 5 kHz or a time period of 200 μs.

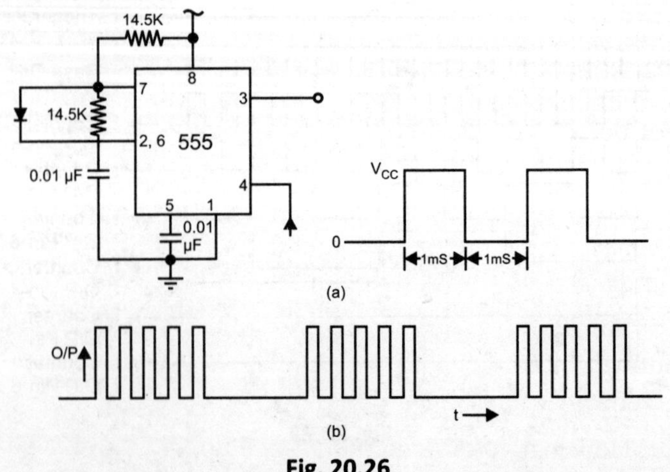

Fig. 20.26

As a result, each time the RESET terminal goes HIGH for 1ms, 5 complete cycles of astable multivibrator appear at the output.

The output waveform is shown in Fig. 20.26 (*b*)

Problem 20.2. *Identify the circuit configuration shown in Fig. 20.27 (a). Sketch the output waveform.*

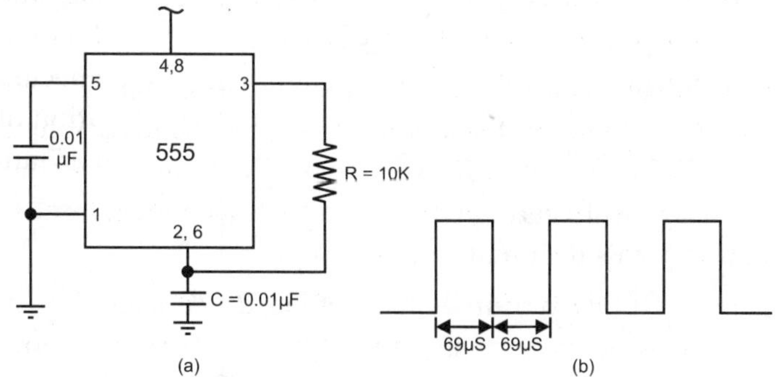

Fig. 20.27

Solution. The circuit shown in Fig. 20.27 (*a*) works like an astable multivibrator. When the output is HIGH, the capacitor charges from the HIGH output through the 10K resistance. In fact, there is about 10 ohms resistance between the (V_{CC}) terminal and the output terminal, when the output is HIGH. So, the exact charging path is from (V_{CC}) through that 10 ohms and then 10K. As 10 ohms is negligible as compared to 10K, the output HIGH time (or the charging time for the capacitor to charge from 1/3 V_{CC} to 2/3 V_{CC}) is given by 0.69 *RC* only.

That is, output HIGH time = $0.69 \times 10^4 \times 10^{-8} = 69$ μs

When the output goes LOW, there is an effective resistance of 10 ohms between output and ground and thus, the capacitor discharges through 10K and the ON-resistance of the conducting output transistor (which is about 10 ohms) from 2/3 V_{CC} to 1/3 V_{CC}.

That is, output LOW time = $0.69 \times 10^4 \times 10^{-8} = 69$ μs.

The output waveform is shown in Fig. 20.27 (*b*).

It should be remembered that we can not afford to use too small a value of (R) in this configuration. The current through the conducting output transistor can not be allowed to exceed its maximum current sinking capability of 200 mA (max.), when the output is LOW.

Problem 20.3. *Fig. 20.28 shows a 555 based monoshot alongwith the waveforms applied at pin-2 and pin-4. Sketch the output waveform.*

Solution. Everytime, the monoshot is triggered, the output should go HIGH and stay there for a time period given by 1.1 *RC*

Here, $R = 10K, C = 0.01$ μF

Therefore, 1.1 $RC = 1.1 \times 10^4 \times 10^{-8} = 110$ μs.

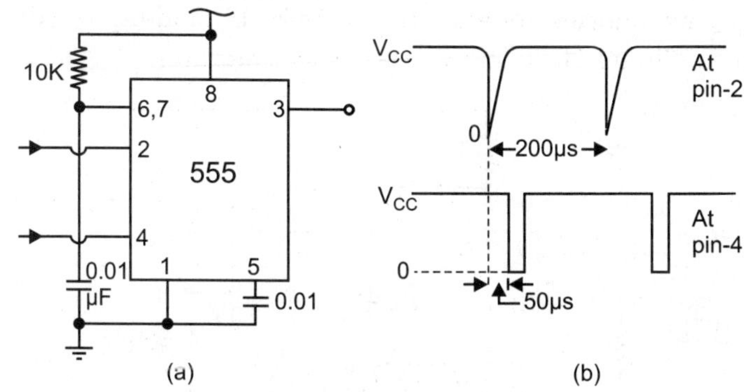

Fig. 20.28

That is, the output should stay HIGH for 110 μs everytime a genuine trigger pulse appears at pin-2. In the conventional monoshot configuration, the RESET terminal (pin-4) is tied to V_{CC}. As explained earlier, whenever RESET terminal is brought below 0.4 volt, the output goes LOW by overriding all other commands. Thus, in the present case, though the output should have stayed HIGH for 110 μs, if pin 4 was HIGH, it goes LOW 50 μs after occurrence of trigger pulse. The output waveform is shown in Fig. 20.29. The dashed portion in the output pulse shows the output waveform, if the RESET terminal is kept HIGH.

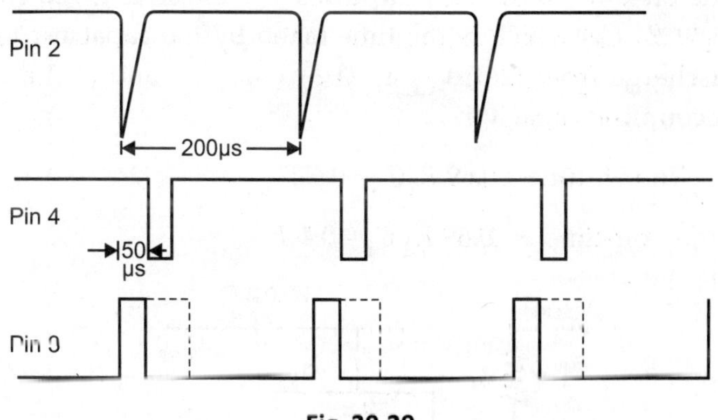

Fig. 20.29

Problem 20.4. *The monostable multivibrator circuit of Figure-20.25 is to be triggered by a clock signal of 10 kHz. Give the circuit diagram for generating a time delay of 25 ms.*

Solution. Clock frequency – 10 kHz. Clock time period, $T = 100$ μs

Let $R = 10K$, then $C = 0.01$ μ*F*

Desired time delay = 25 ms

Therefore, Desired time delay = $25 \times 10^{-3}/(100 \times 10^{-6})$ = 250T

Now, $250T = (128 + 64 + 32 + 16 + 8 + 2) \times T$

Therefore, counter outputs O_2, O_8, O_{16}, O_{32}, O_{64} and O_{128} are shorted together and outputs O_0 and O_4 are left open. Figure-20.30 shows the circuit diagram.

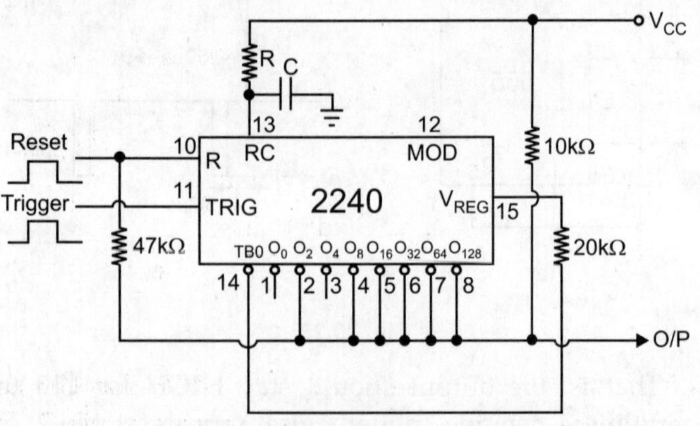

Fig. 20.30

Problem 20.5. *Design an astable multivibrator using 555 timer IC generating a 10 kHz pulsed output with a duty cycle of 0.6.*

Solution. Fig. 20.31 shows the basic astable multipvibrator circuit built around IC 555. The on-time of the output pulsed waveform in this case would be equal to (0.69 R_1 C_1) which is basically the time required by the capacitor (C_1) to charge from (1/3) V_{CC} to (2/3) V_{CC}. The off-time of the output pulsed waveform is given by (0.69 R_2 C_1) which is the time taken by the capacitor to discharge from (2/3) V_{CC} to (1/3) V_{CC}. (C_2) and (C_3) are decoupling capacitors.

The on-time = $0.69\,R_1C_1 = 0.6\,T$

off-time = $0.69\,R_2\,C_1 = 0.4\,T$

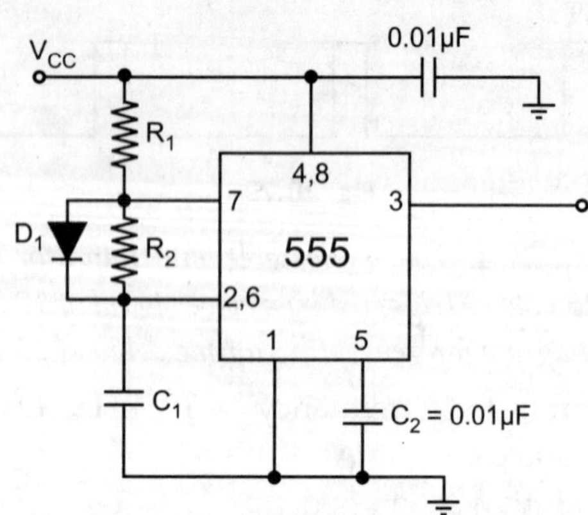

Fig. 20.31

where, T = time period $= \dfrac{1}{10^4}$

Let $C_1 = 0.01$ μF

Then $R_1 = \dfrac{0.6 \times 100 \times 10^{-6}}{0.69 \times 10^{-8}} = \dfrac{6000}{0.69} \cong 8.7$ K

and $R_2 = \dfrac{0.4 \times 100 \times 10^{-6}}{0.69 \times 10^{-8}} = \dfrac{4000}{0.69} \cong 5.8$ K.

Problem 20.6. *Modify the astable multivibrator circuit of problem-20.5, so as to generate at the output a 10 kHz burst of 10 cycles repeating at an interval of 10 ms.*

Solution. The RESET terminal (Pin-4) of IC 555 can be used to generate the required repetitive burst. When the RESET terminal of IC 555 is brought below 0.4 V, the output is reset to zero. In the normal astable multivibrator operation, this terminal is tied to (V_{CC}).

If the RESET terminal of 555 of Fig. 20.30 is fed with a pulsed waveform with an on-time of 1 ms (= time period occupied by 10 cycles) and an off-time of 9 ms, the output will be the desired signal. The modified circuit is shown in Fig. 20.32.

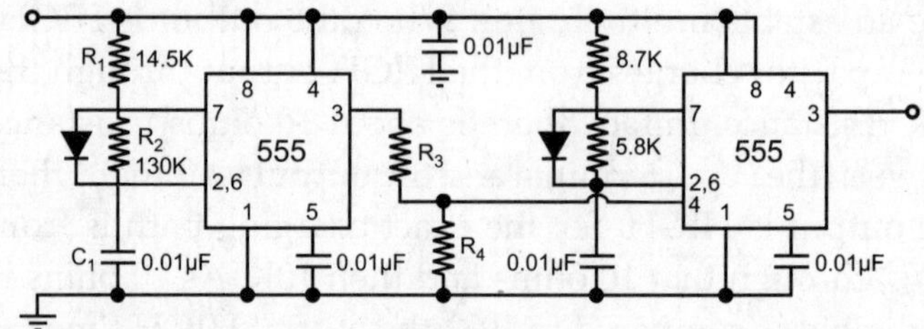

Fig. 20.32

The values of (R_1), (R_2) and (C_1) are calculated as 0.69 $R_1\,C_1 = 10^{-3}$

For $C_1 = 0.1$ μF,

$$R = \frac{10^{-3}}{0.69 \times 10^{-7}} \cong 14.5\,K \text{ and } 0.69\,R_2C_1 = 9 \times 10^{-3}$$

Therefore, $R_2 = \dfrac{9 \times 10^{-3}}{0.69 \times 10^{-7}} \cong 130\,K$

(R_3) and (R_4) constitute a potential divider to ensure that during the low time of the 555 output, pin-4 of the output 555 gets a voltage well below 0.8 V. R_3 = 1K and R_4 = 220 Ω giving a division of 6 is safe.

Problem 20.7. *Design a 555 based monostable multivibrator circuit that generates a 100 μs pulse at the output on the trailing edge of the trigger pulse, that is 1 ms wide.*

Solution. Fig. 20.33 shows the basic 555 timer based monostable multivibrator circuit. The width of the output pulse here is given by :

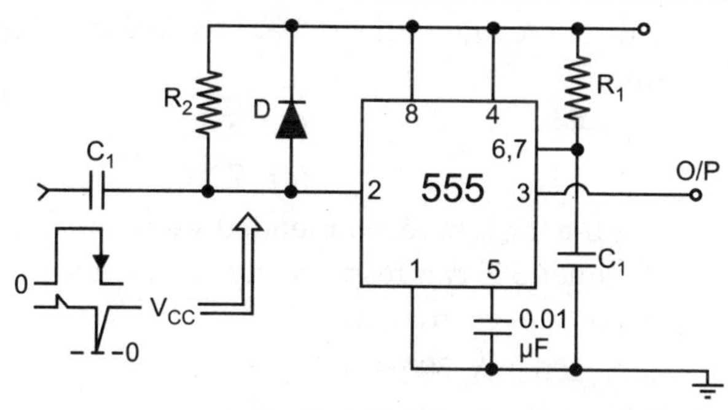

Fig. 20.33

$$T = 1.1\,RC$$

For $C_1 = 0.01\ \mu F,\ R_1 = \dfrac{100}{0.01 \times 1.1} = 9\ K$

This monostable circuit requires that the trigger pulse width at the trigger terminal (Pin-2) is smaller than the width of the expected output pulse. Also, the trigger terminal responds to HIGH-to-LOW transition. (R_2), (C_2) and diode (D) constitute a differentiator circuit with diode (D) disallowing the LOW-to-HIGH transition of the available trigger pulse to appear at pin-2. The differentiator time constant should be about one-tenth of the expected output pulse width.

Let $R_2 C_2 = 10\ \mu s$ (one-tenth of expected output pulse width)

For $C_2 = 0.001\ \mu F,\ R_2 = 10\ K$

Problem 20.8. *Modify the trigger portion in the monostable multivibrator circuit of problem 20.33, so that it produces an output pulse on the leading edge of the trigger pulse.*

Solution. The trigger portion for the leading edge trigger is shown in Fig. 20.34. The basic idea is that pin-2 should normally be at (V_{CC}) and it should go low at the leading edge of the trigger pulse for a period, which is smaller than the expected output pulse width.

The transistor 2N2222 and its associated base and collector resistances are acting like a switch that inverts the pulse present at the output of the differentiator circuit.

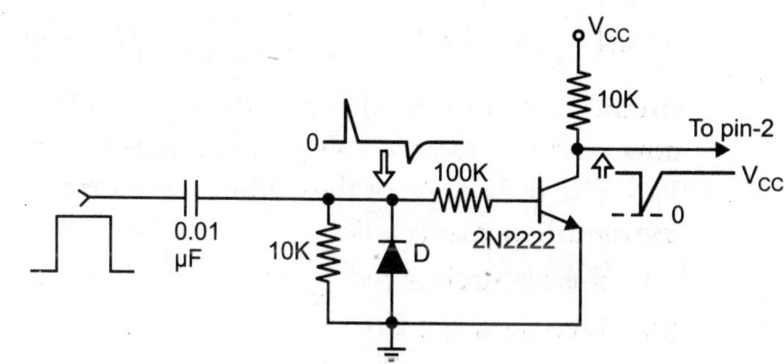

Fig. 20.33

SUMMARY OF IMPORTANT FORMULAS

1. In the case of IC timer 555 based Astable multibrator, the ON and OFF the periods of the output waveform are given by:

$$t_{ON} = 0.69\,R_1 C$$

and $$t_{OFF} = 0.69\,R_2 C$$

where, R_1 = Resistance of the charging path

R_2 = Resistance of the discharge path

C = Timing capacitance

The frequency of the output waveform is given by:

$$f = \frac{1}{0.69\,(R_1 + R_2)\,C}$$

2. In the case of IC timer 555 based Monostable multivibrator, the output pulse width is given by:

$$T = 1.1\,\text{RC}$$

Where, R = Resistance of the charging path

C = Timing

3. In the case of programmable Counter/Timer 2240, the time period of the time base of counter output is given by:

$$T = \text{RC}$$

Where, R, C are the externally connected resistance and capacitance to pin – 13 of the IC.

SELF-EVALUATION EXERCISE

I. MULTICHOICE QUESTIONS

1. With reference to astable mode of operation of IC 555, one of the following statements is true:

(*a*) Trigger terminal is always shorted to the threshold terminal

(*b*) Reset terminal is always grounded.

(*c*) Reset terminal is always at VQC

(*d*) Trigger terminal is left open.

2. If the total charging and discharging resistance in an astable mode IC 555 circuit is (R) each and the timing capacitor is (C), the frequency of operation (f) is given by:
   (a) $f = \frac{0.724}{RC}$
   (b) $f = 1.38\ RC$
   (c) $f = 0.69\ RC$
   (d) None of these.
3. In a 555 astable multivibrator operating from V_{CC} of 15 V, the output waveform duty cycle is 0.4 $\left(\text{Duty cycle} = \frac{\text{ON – time}}{\text{total time period}}\right)$. The circuit is slightly modified with the capacitor now charging from a voltage of 30 rather than V_{CC}. The duty cycle of the output waveform in the modified circuit will:
   (a) remain unchanged
   (b) become about 0.6
   (c) be approximately 0.3
   (d) None of these.
4. The pulse width in a 555 monostable multivibrator is given by:
   (a) $\tau = 0.69\ RC$
   (b) $\tau = 1.38\ RC$
   (c) $\tau = 1.1\ RC$
   (d) None of these
5. In the astable mode of operation of IC 555, the timing capacitor always charges between:
   (a) 0 and 1/3 V_{CC}
   (b) 1/3 V_{CC} and 2/3 V_{CC}
   (c) 0 and 2/3 V_{CC}
   (d) 0 and V_{CC}
6. The IC timer 555 terminal that can be used to modulate the pulse width is:
   (a) Trigger terminal
   (b) Reset terminal
   (c) Control terminal
   (d) Discharge terminal.
7. One of the following type numbers is not an IC timer:
   (a) 555 (b) XR-2240
   (c) 556 (d) 558
   (e) 565
8. One of the timer ICs can be classified as a digital timer:
   (a) 555 (b) 558
   (c) 556 (d) 7240.
9. For the highest recommended value of V_{CC}, the IC timer 555 can respectively source and sink a maximum current of:
   (a) 200 mA, 200 mA
   (c) 100 mA, 200 mA
   (b) 100 mA, 100 mA
   (d) 200 mA, 100 mA.
10. Identify the correct statement:
   (a) The IC timer 555 sinks current only, when the output is LOW and the load resistance is connected between the output and the V_{CC} terminals.
   (b) The IC timer 555 always sinks current, when the output is LOW irrespective of how the load is connected.
   (c) The IC timer 555 always sources current, when the output is HIGH irrespective of how the load is connected.
   (d) In the astable or monostable mode of operation, when the output is HIGH, the amplitude of this HIGH output is about 0.5 volt less than the V_{CC}.

II. The monoshot of Fig. 20.35 was designed by someone to generate a pulse at the output (the pulse width of course depeding upont R and C) whenever it was triggered by the available trigger pulse as shown. Is this configuration goint to work?

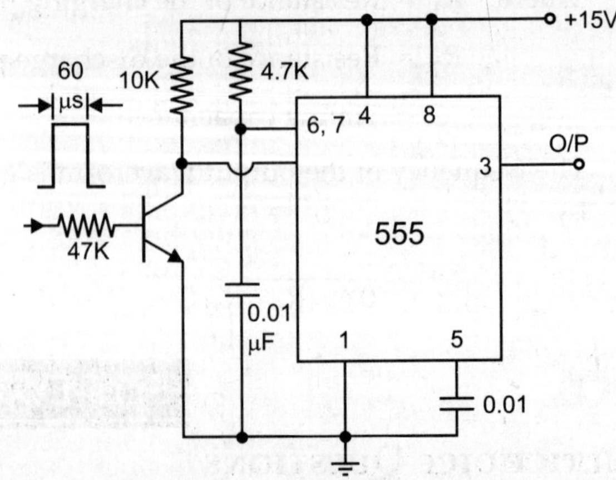

Fig. 20.35

III. Fig. 20.36 shows a conventinal astable multivibrator circuit feeding a resistive load. What is wrong with the set-up?

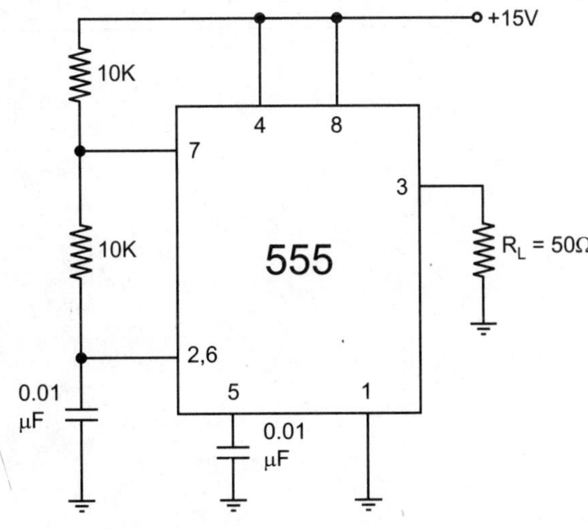

Fig. 20.36

Answers to Self-Evaluation Exercise

I. 1. (*a*) 2. (*a*) 3. (*c*) 4. (*c*) 5. (*b*) 6. (*c*)
7. (*e*) 8. (*d*) 9. (*a*) 10. (*a*)

II. It will not work. In the 555 monoshot configuration, the LOW going trigger pulse should not only go to less than (V_{CC}/3) from (V_{CC}), its width should also be less than the desired output pulse width. This second condition is not being complied with in the monoshot of Fig. 20.35.

III. 555 can source a maximum current of 200 mA for V_{CC} = 15 V. In the present case, we are trying to extract about 250 mA which the timer cannot source. The IC is likely to get damaged under these conditions. It may also be mentioned that the maximum current sourcing and sinking capability of timer 555 are (V_{CC}) depedent. For (V_{CC}) less than 15 V, the capability also goes down. Refer to electrical characteristics of IC timer 555.

Chapter 21

Phase Locked Loops

Learning Objectives

After completing the chapter, you will learn the following topics:

1. Phase Locked Loop (PLL) Fundamentals
2. Internal Architecture of Phase Locked Loop
3. Performance Parameters of Phase Locked Loops
4. Classification of Phase Locked Loops
5. Analog and Digital PLLs
6. Voltage Controlled Oscillator (VCO) in PLLs
7. Phase Comparator in PLLs
8. Low Pass Filter in PLLs
9. Application Data of Common PLLs
10. PLL as a Frequency Multiplier
11. PLL as a Frequency Divider
12. PLL as an Amplitude Demodulator
13. PLL as a Frequency Demodulator
14. PLL as a Narrowband Pass Tracking Filter
15. PLL for Frequency Synthesis
16. PLL for CW Carrier Recovery
17. PLL for Angle Modulated Signal Amplifier

Q. 1. What is a PHASE LOCKED LOOP ? With the help of a block diagram, explain in brief the operation of a PLL.

Ans. A phase locked loop is basically a feedback control system with a phase comparator and a low pass filter along with an error amplifier in the forward signal path and a voltage controlled oscillator (VCO) in the feedback path. Fig. 21.1 shows the basic phase locked loop, f_i and θ_i are the frequency and phase of the input signal, whereas f_0 and θ_0 are respectively the frequency and phase of the signal at the output of the voltage controlled oscillator. The phase comparator produces an error signal whose amplitude varies directly as the frequency difference and the phase difference between the input signal and the VCO output. The error voltage $V_e(t)$ passes through the low pass filter and is then amplified by the error amplifier before it is fed back to the control terminal of VCO. The operation of a basic PLL can be explained as follows:

In the absence of any input signal, there will be zero error voltage $V_e(t) = 0$ which gives $V_d(t) = 0$.

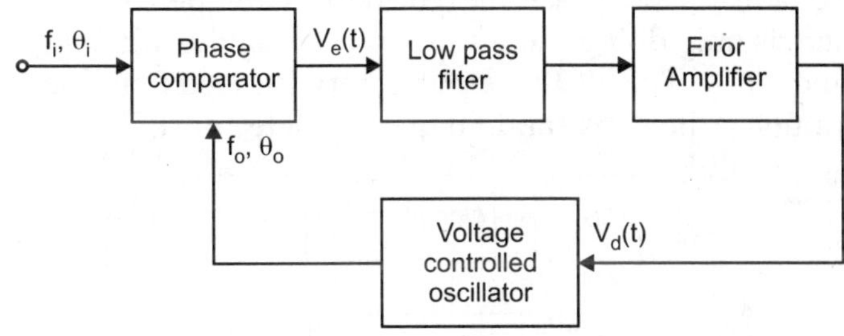

Fig. 21.1

As a result, the VCO runs at its free running frequency. When an input signal is applied, the phase comparator compares the two signals and produces an error voltage, whose magnitude is proportional to the phase difference and the frequency difference. This error voltage after amplification in the error amplifier appears at the control point of VCO. The polarity of the error voltage is such that it tends to reduce the difference between the input signal frequency and the VCO output. If the input frequency happens to be close to the VCO output, the VCO locks on to the input signal frequency. Once this lock is achieved, VCO becomes capable of tracking the input signal frequency. If the VCO output happens to be away from the input signal frequency by more than a certain predefined limit, the PLL may not be able to achieve lock. These aspects will be discussed in the subsequent questions.

Q. 2. What are the different classes of phase locked loops? Briefly explain the differences between different classes of PLLs.

Ans: There are four classes of phase locked loops as stated by Roland E. Best in his book titled *Phase Locked Loops: Theory, Design and Applications*.

1. Analog or Linear PLL (LPLL)
2. Digital PLL (DPLL)
3. All digital PLL (ADPLL)
4. Software PLL (SPLL)

The **Analog Phase Locked Loop,** also called **Linear Phase Locked Loop (LPLL)** is the classical form of PLL. All components in the LPLL operate in the continuous-time domain. The phase detector is typically some form of analog multiplier, either a double-balanced mixed (DBM) or an active four quadrant multiplier. Loop filter is an active or passive filter and it uses a voltage controlled oscillator. The loop filter typically results in either a first order or second order filter. Transfer functions in s-domain of the typical filter functions for analog PLLs include 1, $(1 + s\tau_2)/s\tau_1$, $(1 + s\tau_2)/ (1 + s\tau_1)$ and $1/ (1 + s\tau)$. Figure-21.2 shows the basic block diagram of an analog PLL. The diagram is self-explanatory. It may be mentioned here that the Frequency Divider block is not an essential element of the phase locked loop. A frequency divider here is only doing the job of multiplying the input signal frequency. The VCO output otherwise feeds one of the inputs of the Phase and Frequency Detector (PFD).

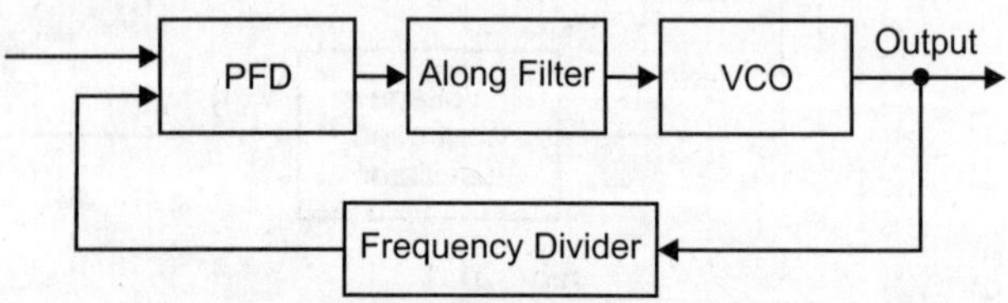

Fig. 21.2

The **Digital PLL** is nothing but an analog PLL with a digital phase detector. Figure-21.3 shows the block schematic arrangement of a Digital PLL. Commonly used digital phase detectors include Exclusive-OR gate (EXOR), Edge-triggered JK flip-flop and Phase Frequency Detector (PFD). Figure-21.3 shows the block schematic arrangement of a classical digital PLL. The DPLL is commonly used in frequency synthesis applications. In Figure-21.3, the optional digital divider and its different variations are used in frequency synthesis applications.

The **All-digital PLL (ADPLL)** is distinctly different from the first two PLLs mentioned, thus far. The ADPLL is a digital loop using all digital building blocks and also handling discrete time signals. The analog VCO in this

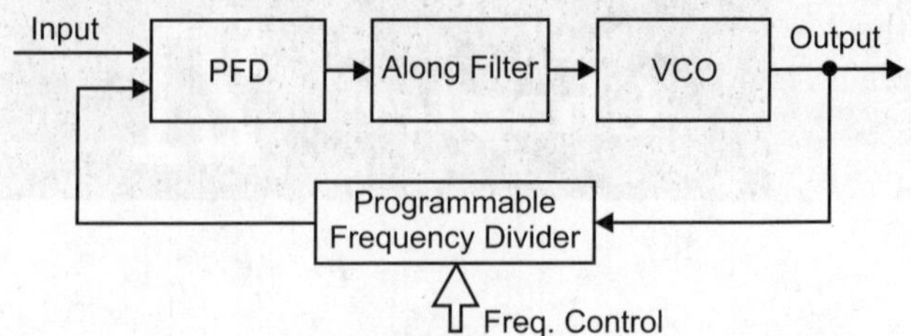

Fig. 21.3

case is replaced by a digitally controlled oscillator (DCO), also called a numerically controlled oscillator (NCO). IC 74HC/HCT297 of the pin-to-pin TTL compatible CMOS series is a common example of All-digital PLL. Figure-21.4 shows the internal architecture of the IC 74HC/HCT297. These devices contain all the necessary circuits, with the exception of the divide-by-n counter, to build first order phase-locked-loops. Both EXCLUSIVE-OR (XORPD) and edge-controlled (ECPD) phase detectors are provided for maximum flexibility. The length of the up/down K-counter is digitally programmable according to the K-counter function table. The phase detector generates an error signal waveform that at zero phase error, is a 50% duty factor square wave. At the limits of linear operation, the phase detector output will be either HIGH or LOW all of the time depending on the direction of the phase error $(\phi_{IN} - \phi_{OUT})$. Within these limits, the phase detector output varies linearly with the input phase error according to the gain k_d, which is expressed in terms of phase detector output per cycle or phase error. The output of the phase detector will be $k_d\phi_e$, where the phase error $\phi_e = \phi_{IN} - \phi_{OUT}$. The phase detector output controls the up/down input to the K-counter. The counter is clocked by input frequency Mf_c, which is a multiple M of the loop centre frequency f_c. When the K-counter counts up, it generates a CARRY pulse. Counting down generates a BORROW pulse. If the CARRY and the BORROW outputs are conceptually combined into one output, that is positive for a CARRY and negative for a BORROW, and if the K-counter is considered as a frequency divider with the ratio Mf_c/K, the output of the K-counter will equal the input frequency multiplied by the division ratio. Thus, the output from the K-counter is $(k_d\phi_e Mf_c)/K$. The CARRY and BORROW pulses go to the increment/decrement (I/D) circuit, which in the absence of any CARRY or BORROW pulses has an output that is 1/2 of the input clock (I/D_{CP}). The input clock is just a multiple, 2N, of the loop centre frequency. In response to a CARRY or BORROW pulse, I/D circuit will either add or delete a pulse at I/D_{OUT}. Thus, the output of I/D circuit will be $Nf_c + (k_d\phi_e Mf_c)/2K$.

The **Software PLL (SPLL)** can be viewed as a discrete-time implementation of either the linear PLL (LPLL) or the Digital PLL (DPLL). The implementation of the loop filter is typically a difference equation. The design/analysis of the loop filter is done using the z-transform and not Laplace transform, as is done in case of analog PLL. It has become feasible to simulate analog PLL designs using software tools. The SPLL is commonly used in digital communication applications.

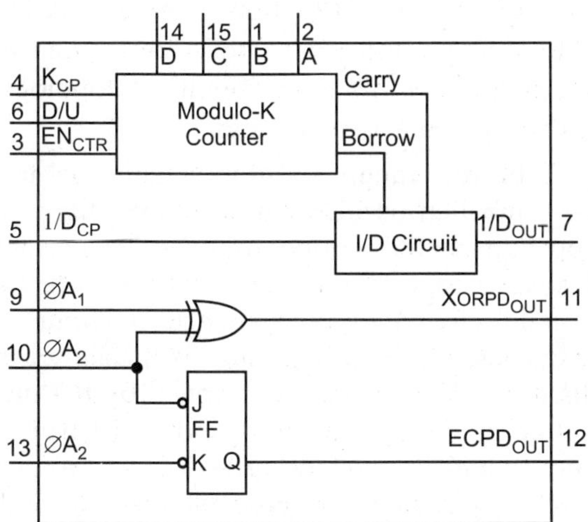

Fig. 21.4

Q. 3. Define the following terms with reference to Phase Locked Loops:

(a) Look-in-range or Synchronisation range

(b) Capture range or Acquisition range

(c) Free-running frequency

(d) Tracking range

(e) Lock-up time

(f) Loop gain

Also, briefly explain as to which PLL parameters, these are dependent on?

Ans.

(a) LOCK-IN-RANGE or SYNCHRONISATION RANGE is defined as the frequency range centered around VCO's initial frequency over which VCO can track the input signal once a lock has been achieved. Lock range depends on the maximum error voltage that can be generated and the corresponding deviation produced.

(b) CAPTURE RANGE is the frequency range over which the VCO output can acquire lock with the input frequency. It is also centered around the initial frequency of VCO output. Capture-range depends upon the response of the low pass filter, more precisely the band edge of the LPF response together with the closed-loop gain of the PLL. Capture range is less than the Lock-in range and can at best be equal to that. See Fig. 21.5 for reference.

(c) FREE RUNNING FREQUENCY is the VCO output frequency, when it is not in lock.

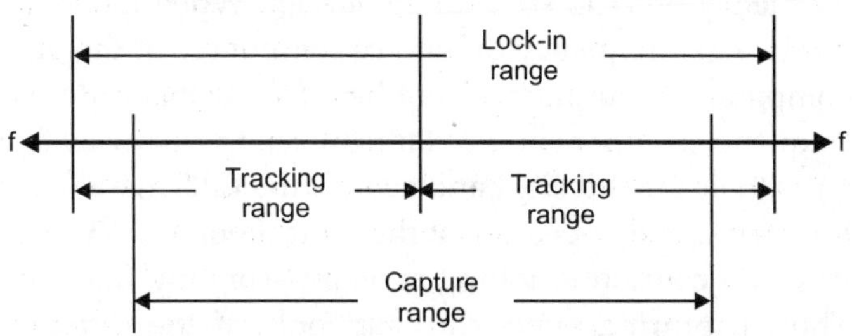

Fig. 21.5

(d) TRACKING RANGE is the frequency deviation on either side of the VCO free running frequency over which VCO can track the input once in lock. It is half of the lock-in range and depends upon the same parameters that decide the lock-in range.

(e) LOCK-UP TIME is the time taken up by the VCO output to achieve lock. Lock-up time is inversely proportional to the bandwidth of the low pass filter. Higher selectivity increases the lock-up time and at the same time, it also reduces the capture-range.

(f) LOOP GAIN is the product of the phase-comparator's conversion gain, VCO's conversion gain and the gain of low pass filter to DC. Phase comparator's conversion gain is the ratio of the output voltage to the phase difference between the input and the VCO signal under locking condition. It is measured in volts per radian. VCO conversion gain is the ratio of frequency shift (from f_0) to the error voltage. It is measured in radians per second per volt. Obviously units of measurement of loop gain would be $(\text{second})^{-1}$.

Q. 4. Explain in brief, why lock range is independent of the band edge of the low pass filter, whereas the capture-range is inversely proportional to the selectivity offered by the low pass filter?

Ans. Lock range has already been defined to be the frequency range centered around VCO free running frequency, over which VCO can track the input once lock has been achieved. Now under locking conditions, frequency difference is zero. Keeping in mind the fact that frequency is time derivative of phase ($\omega = d\theta/dt$), the output of phase comparator will be a DC which will always be passed by a low pass filter (LPF) irrepective of the band-edge of the LPF response. On the other hand, capture range as already defined signifies as to how close the input frequency should be to the free-running frequency of VCO so as to get locked. When the loop is not in lock, frequency difference component at the phase comparator's output will not be a DC. If this difference frequency component is so large (when f_i is far away from f_0) as to lie outside the band edge of the LPF, there will be no error signal appearing at the control point of VCO and it would continue running at its free-running frequency. Thus, capturing effect can start only, if the difference frequency component lies within the band edge of the LPF. In other words, position of the band edge determines the capture range. Fig. 21.6 illustrates it further. Fig. 21.6 (*a*) shows the LPF response curve for a smaller capture range PLL, whereas LPF response curve shown in Fig. 21.6 (*b*) is a representative of a larger capture range PLL.

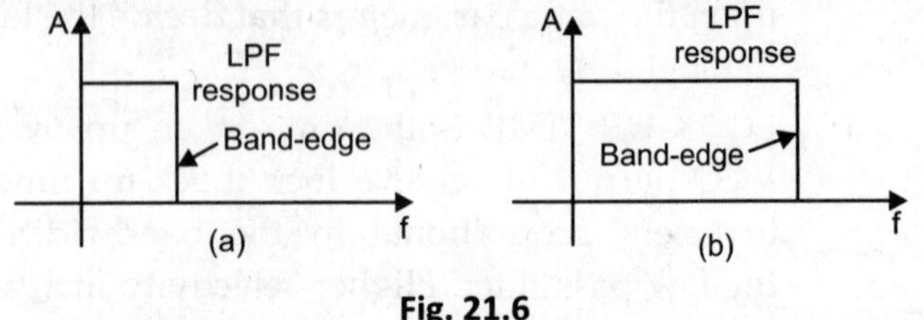

Fig. 21.6

Q. 5. What are the effects of Low Pass Filter Bandwidth (LPF BW) on the PLL performance ? Why can't we reduce the bandwidth beyond a certain limit ?

Ans. Following are the effects of LPF bandwidth on PLL performance :

1. Reducing the LPF bandwidth makes the capture process slower, *i.e.* it increase the lock-up time.
2. Capture range decreases with decrease in LPF Bandwidth.
3. If the low pass filter is highly selective, the interference rejection properties of the PLL improve as the error signal due to undesired interfering frequency gets further attenuated by LPF.
4. If the loop bandwidth is reduced, the transient response of the loop gets underdamped. From stability view point, the LPF bandwidth may not be allowed to become too small, so as to make the loop unstable. There is always a compromise between a very large bandwidth (giving a large capture range and a small lock-up time) and a very small bandwidth (giving good interference rejection properties).

Q. 6: What is its importance with reference to the operational performance of a PLL? What are the major requirement specifications that need to be addressed while designing a VCO for a PLL?

Ans: Voltage Controlled Oscillator (VCO) is an important building block of a phase locked loop, as it influences many of the major performance specifications of PLL frequency synthesizers. Performance parameters of a VCO of relevance to their use in design of a phase locked loop include VCO tuning range, VCO tuning gain, *f*-to-*V* Slope and phase noise performance. These characteristics also influence the circuit topology of the VCO to be used in a given design.

VCO tuning range should be such as to enable voltage controlled oscillator to tune over the range that the loop is expected to operate over. This requirement is not always easy to meet and may require the VCO or resonant circuit to be switched in some extreme circumstances. VCO tuning gain is measured in terms of Volts/Hz or Volts/MHz. As implied by the units, it is the change in VCO frequency produced corresponding to given change in voltage. The VCO f-to-V or V-to-f response curve can be seen to be relatively linear at lower frequencies flattening out at higher voltages, where the changes in capacitance from the varactor diodes reduce. The voltage controlled oscillator gain affects some of the overall loop design considerations and calculations. The voltage controlled oscillator used in a phase locked loop should have a monotonic voltage-to-frequency curve. It implies that it always changes in the same sense and the frequency typically increases for increasing voltage. The loop becomes unstable if it is otherwise. The phase noise performance of the voltage controlled oscillator is of particular importance, when PLLs are used as frequency synthesizers. Here, the phase noise performance of the VCO determines many of the overall phase noise performance characteristics of the overall loop and the overall synthesizer, if used in one.

Q. 7. What are the typical circuit topologies used for building VCOs of PLLs?

Ans: Two commonly used VCO circuit topologies for PLLs include Colpitt and Clapp oscillator circuits. Colpitt and Clapp oscillator circuits are similar and are discussed in detail, in the chapter on **Sinusoidal Oscillators.**

Q. 8. What is the role of phase detector in a phase locked loop? Briefly describe the different types of phase detector circuits used in phase locked loops.

Ans: The phase detector in a PLL is used to detect the phase difference between the VCO output and the input signal, and produce at its output an error signal representative of the phase difference. This error signal after passing through the low pass filter is used to control the frequency of the VCO thereby forcing the VCO output to have the same frequency as that of the input signal. There are two broad categories of phase detectors, namely *Phase Only Sensitive Detectors* and *Phase/Frequency Detectors.*

Phase Only Sensitive Phase Detectors

Phase detectors that are only sensitive to phase difference, produce an output that is proportional to the phase difference between the two signals. When the phase difference between the two incoming signals is steady, they produce a constant voltage. When there is a frequency difference between the two signals, they produce a varying voltage. Again, there are two major types of Phase Only Sensitive Detectors, namely Diode Ring Mixer Phase Detector and XOR Circuit Based Phase Detector.

Diode Ring Mixer Phase Detector is the simplest form of phase detector and is synthesised from a diode ring mixer. Figure-21.7 (*a*) shows the basic circuit of a diode ring mixer, indicating the RF, LO and IF ports. This circuit produces an output at the IF port that varies as cosine of the phase difference between the signals present at RF and LO ports. The IF port output produces a null for phase difference of zero and also odd multiples of (π/2). It produces a maximum output for phase difference of integral multiple of (π) and a minimum output for phase difference of integral multiple of (–π). Figure-21.7 (*b*) shows the phase detector response curve.

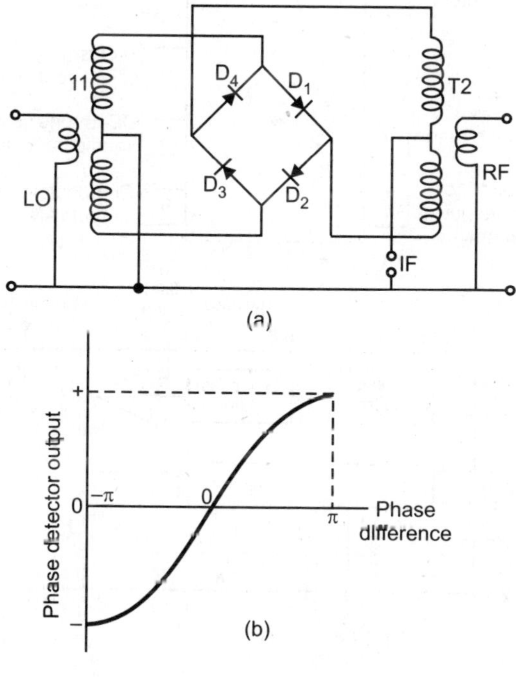

Fig. 21.7

The **Exclusive-OR gate** constitutes a simple phase detector circuit. The way in which an Exclusive-OR gate works as a phase detector is evident from the waveforms of Figure-21.8. As it is evident from the waveforms shown in Figure-21.8, the XOR output duty cycle varies as the phase difference between the two input signals.

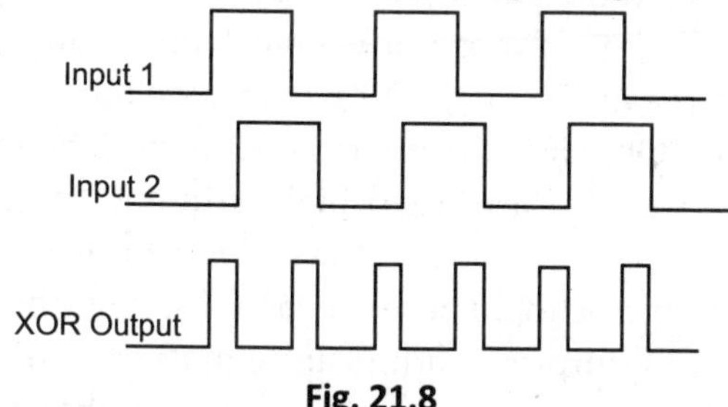

Fig. 21.8

Figure-21.9 shows the response curve for the XOR phase detector. As it is evident from the response curve; a null voltage is produced for a phase difference of (π/2) lagging or leading. Phase differences of zero and (π) respectively produce minimum and maximum voltage output. Unlike analog mixer phase detectors, XOR phase detector is independent of amplitude variations.

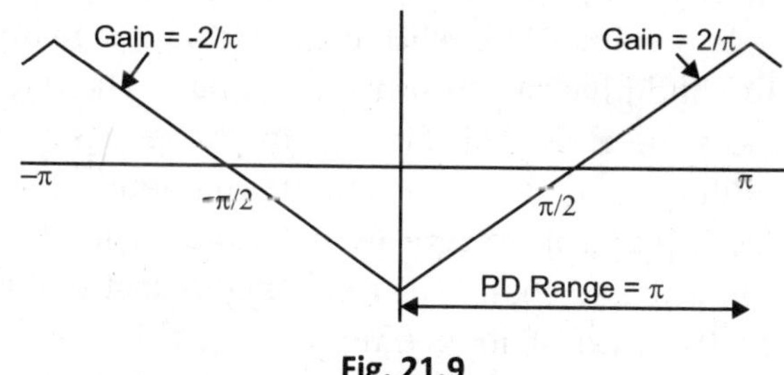

Fig. 21.9

Phase-frequency detectors

Phase-frequency sensitive detectors produce a voltage output proportional to the phase difference whilst the phase difference is between ± 180°. Beyond this, the circuit limits at one of the extremes. In this way no AC component is produced, when the loop is out of lock and the output from the phase detector can pass through the filter to bring the phase locked loop into lock. Two commonly used devices for building phase-frequency detectors include Edge-triggered J-K Flip-Flop and D-type Flip-Flop. In the case of J-K flip-flop, the VCO and reference signals are applied to the *J* and *K* inputs and the phase comparator output is taken from the Q-output of the flip-flop. The phase comparator is insensitive to waveform asymmetry, as it is edge triggered. In the case of D type flip-flop, two D-type flip-flops are used. The two inputs are used to clock the two flip flops whose outputs feed a NAND gate. The NAND gate feeds the RESET inputs of the two flip-flops and phase comparator output is taken from the Q-output of one of the flip-flops. The details of working of these phase comparators can be found in open literature.

Q. 9. What is the role of low pass filter in a phase locked loop? How does the cut-off frequency of low pass filter affect the PLL performance?

Ans: The low pass filter in a phase locked loop plays a crucial role in the performance of the PLL. It significantly influences phase noise performance and stability of the loop and the ability of the loop to change the VCO output frequency. One of the primary functions of the loop filter is to remove any unwanted high frequency components that might pass out of the phase detector and appear in the VCO control line. These unwanted signals then appear on the VCO output as spurious signals. As an example, let us take the case of a mixer used as a phase detector. The mixer produces both sum and difference signals. When the loop is in lock; the two signals, entering the phase detector, have the same frequency. In that case, the difference frequency is zero and a DC voltage is produced proportional to the phase difference as expected. The sum frequency is also produced and this will fall at a point equal to twice the frequency of the reference. If this signal is not attenuated, it will reach the control voltage input to the VCO and give rise to spurious signals. The filter also influences the ability of the loop to change frequencies. If the filter has a very low cut-off frequency; then the changes in control voltage occur slowly, and the VCO will not be able to change its frequency as fast as it would have in the case of loop filter, having a higher cut-off frequency. However, when using filters with high cut-off frequencies, care must be taken to ensure that unwanted frequencies are not passed along the control signal line resulting in generation of spurious signals. The loop filter also governs the stability of the loop. If the filter is not designed correctly, then oscillations can build up around the loop, and large signals will appear on the control line. The loop filter also decides the capture range of the PLL. This aspect has been discussed in an earlier question.

Q. 10. What are the most commonly used general purpose analog PLLs ? Briefly outline their preferred applications.

Ans. Most commonly used general purpose analog PLLs are 560, 561, 562, 564 and 565. All of them contain a voltage controllable oscillator, a phase comparator, a low pass filter and an error amplifier. PLL 560 is very well suited for signal reconstitution, FM demodulation, Tracking filters, Modems, FSK receivers and Data synchronisers. PLL 561 is preferably used for AM Demodulation. It is because of the reason that in addition to FM and oscillator outputs, PLL 561 also produces an output voltage, which is proportional to the voltage of incoming signal. 561 has been optimised for highly linear FM and AM demodulation. PLL 562 is especially designed for frequency synthesis applications. Also, it can perform all functions of PLL 560. PLL 564, finds application in high speed modems, FSK transmitters and receivers, etc. PLL 565 is preferred for low frequency FSK. It has been optimised to provide highly linear frequency to voltage transfer characteristics.

Q. 11. Briefly describe the salient features of PLL 560.

Ans. PLL 560 is a monolithic PLL comprising a phase comparator, a VCO, an amplifier and low pass filter. The centre frequency of the VCO can be set by an external capacitor and the low pass characteristics can be adjusted by two capacitors and two resistors to be connected externally again. Its salient features include (*i*) wide tracking range of ± 15% (*ii*) high linearity (*iii*) capability to provide FM demodulation without using tuned circuits and linear frequency to voltage characteristics. Pin connection diagram and internal block schematic are shown in Fig. 21.10.

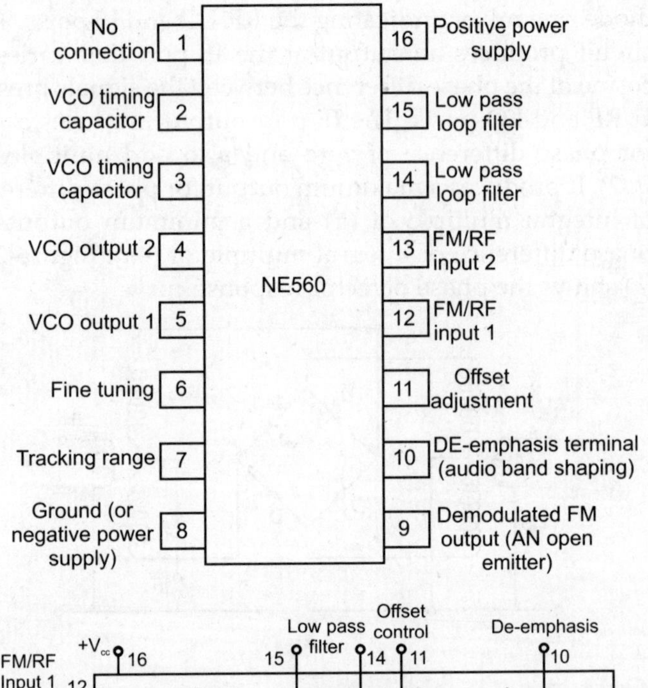

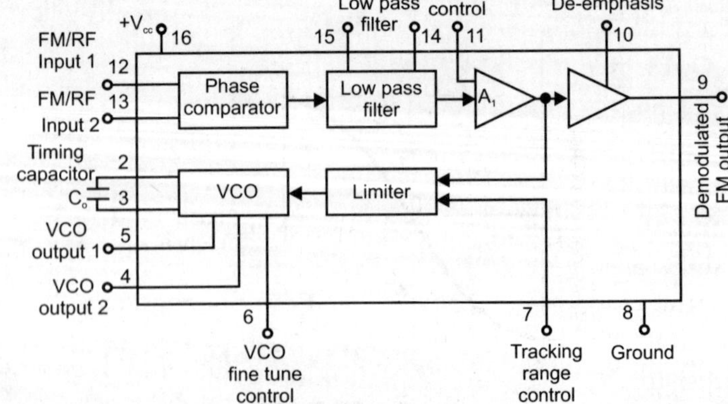

Fig. 21.10

Q. 12. Briefly outline the applications of PLL 561.

Ans. 561 like 560 is a general purpose PLL, comprising a VCO, a phase comparator, an amplifier and a low pass filter. In addition, it has an analog multiplier block built-in on the PLL chip, which enables it to be used as a synchronous AM demondulator in addition to all 560 applications. Pin connections and internal block schematic of 561 are shown in Fig. 21.11.

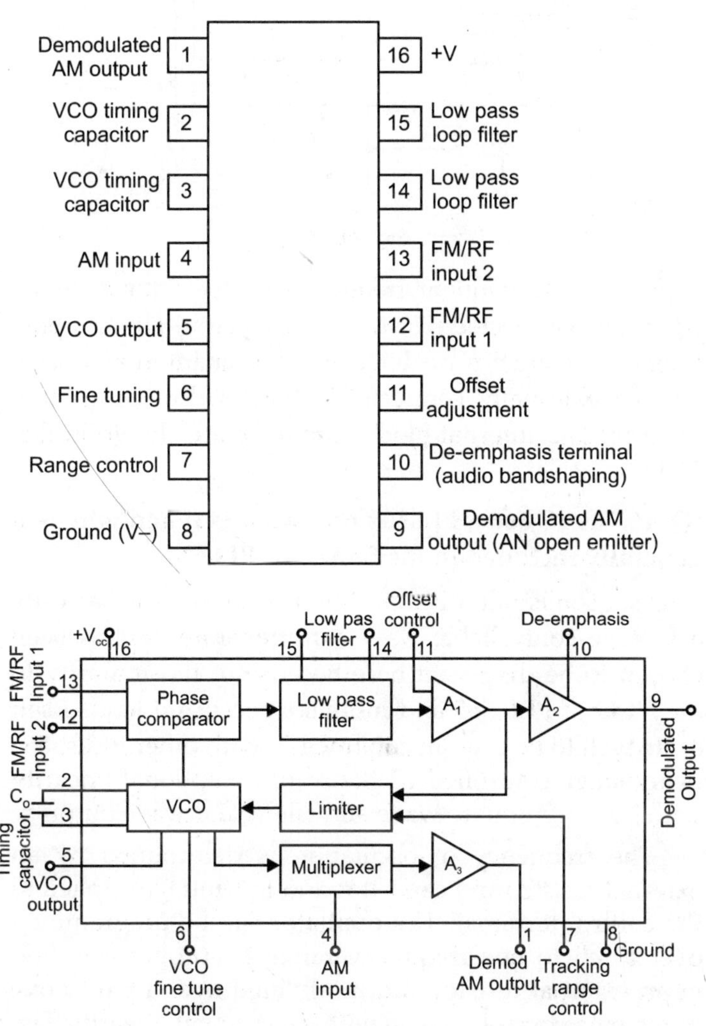

Fig. 21.11

Q. 13. Briefly outline the building blocks comprising PLL 562. What are its salient features?

Ans. PLL 562 is a general purpose monolithic PLL comprising a VCO, a phase comparator, a low pass filter and an error amplifier. While the VCO frequency can be set by an external capacitor (to be connected between pins 5 and 6), the low pass filter that determines the capture characteristics is constructed from two capacitors and two resistors (to be connected from pins 13 and 14). Its salient features include high linearity, adjustable tracking range to ± 15% and its capability to operate as FM demodulator without the use of tuned circuits. Fig. 21.12 shows its pin connections and internal block schematic.

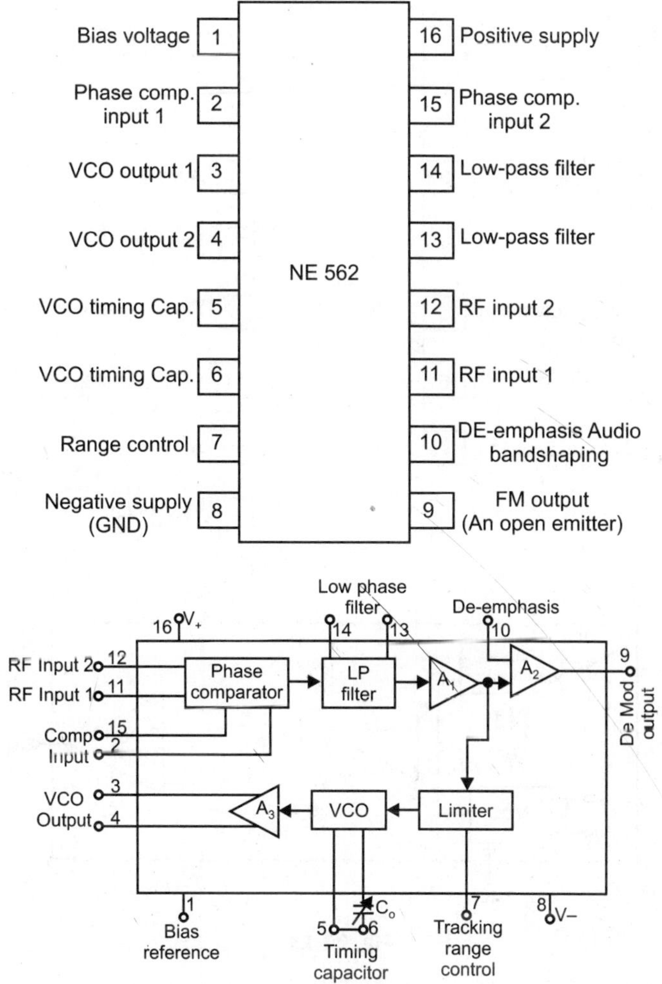

Fig. 21.12

Q. 14. Briefly describe the internal architecture of PLL 564. What are its typical applications?

Ans. PLL 564 is a high frequency phase locked loop with an operational frequency range upto 50 MHz. As it is clear from the internal block schematic shown in Fig. 21.13, it has some additional circuitry which goes by the name of Post Detection Processor that makes it truely versatile. PLL 564 comprises of a VCO that uses an emitter coupled oscillator construction to enable it go upto 50 MHz and whose center frequency can be set by an external capacitor, a PHASE COMPARATOR, which is basically a double balanced modulator with a limiter amplifier to improve AM rejection and a POST DETECTION PROCESSOR consisting of a unity gain transconductance amplifier and a Schmitt comparator with an adjustable hysterisis. The amplifier of the post detection processor can be used as a DC retriever, when the PLL is used as a FSK demodulator and as a post detection filter, when it is used as a linear FM demodulator. Pin connection diagram and internal block schematic of 564 are shown in Fig. 21.13

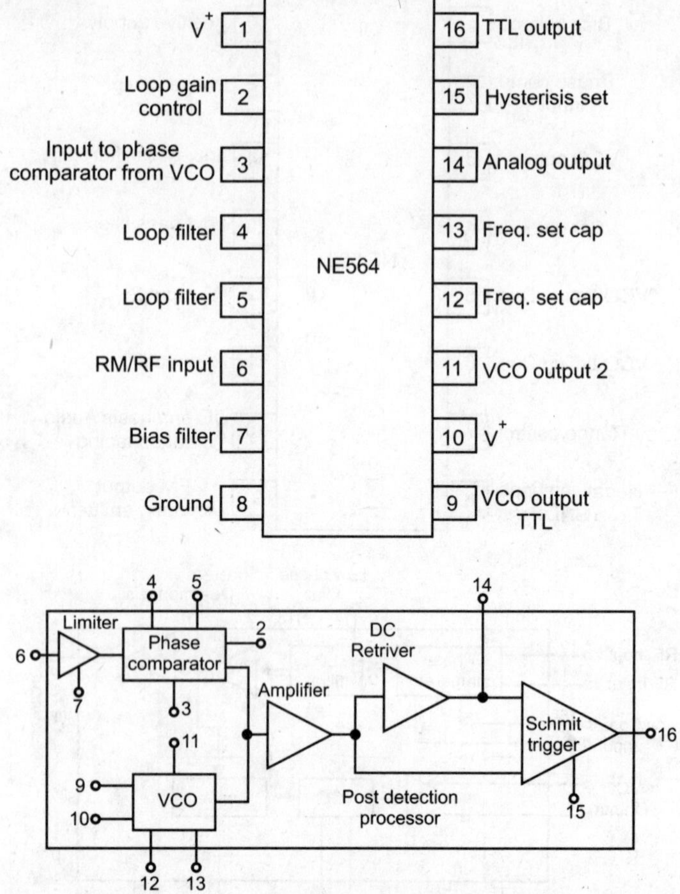

Fig. 21.13

Q. 15. Briefly describe PLL 565.

Ans. PLL 565 is a general purpose PLL, having an operating frequency range of 0.001 Hz to 500 kHz. Internally, the PLL has a VCO that has exceptional linearity and stability, a phase comparator, an error amplifier and a low pass filter (Fig. 21.14).

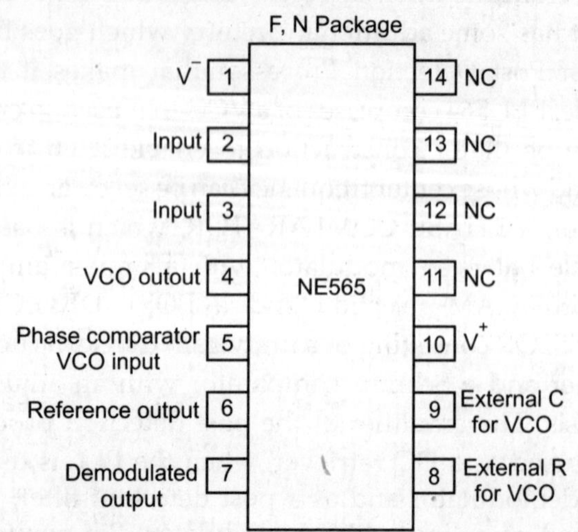

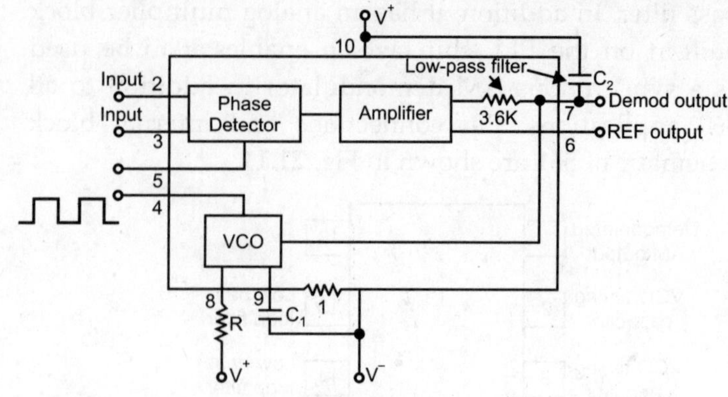

Fig. 21.14

The VCO free running frequency is determined by an externally connected resistor and capacitor. The low pass filter characteristics are determined by an internal resistor and an externally connected capacitor. Pin connection diagram and internal block schematic are shown in Fig. 21.14.

Q. 16. Is IC 566 a PLL? If not, what is it and why is it generally included in the family of PLLs?

Ans. 566 is not a PLL in the true sense. It is basically a voltage controllable waveform generator. It has been chosen to be discussed here because of its similarity to the VCO of PLL 565 and more so, because it lends itself quite well to be used in conjunction with other PLLs. 566 is a voltage controlled oscillator of exceptional linearity with buffered square wave and triangular wave outputs.

The frequency of oscillation is determined by an external resistor and capacitor and the voltage applied to the control terminal. The oscillator can be programmed over a ten to one frequency range by proper selection of an external resistor and modulated over a ten to one range by a control voltage with exceptional linearity. Fig. 21.15 shows the pin connections and the internal block schematic of 566.

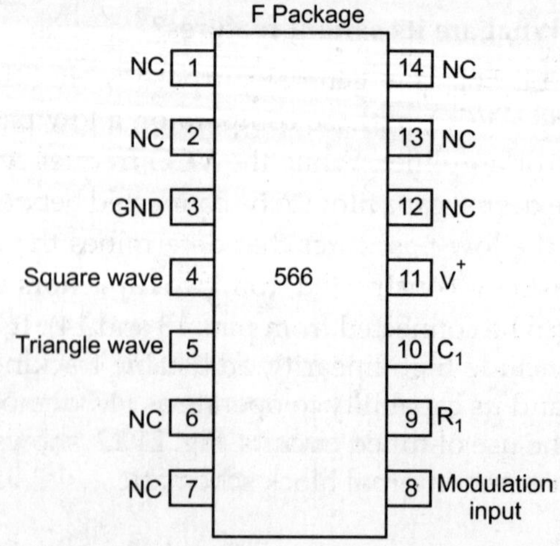

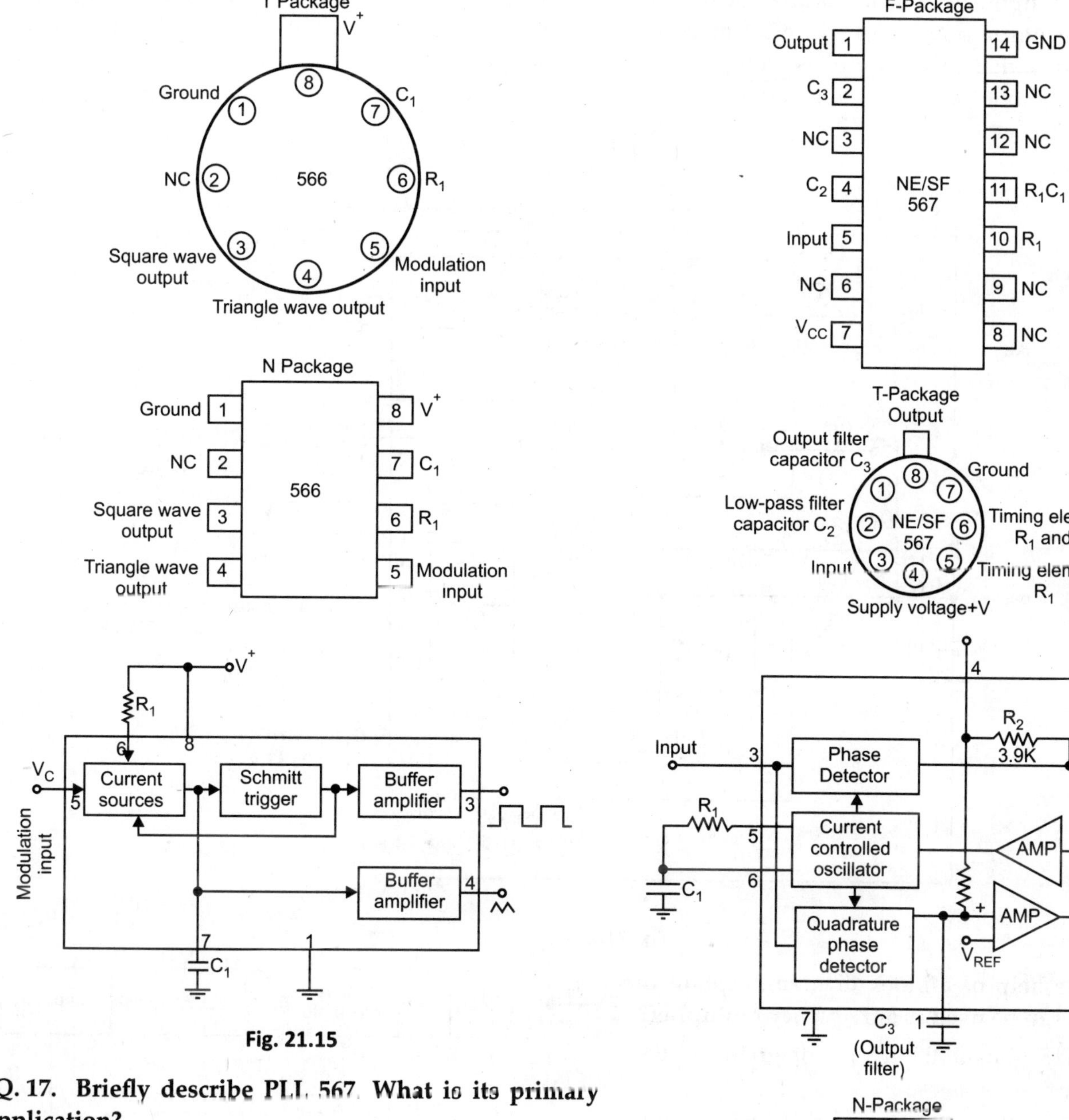

Fig. 21.15

Fig. 21.16

Q. 17. Briefly describe PLL 567. What is its primary application?

Ans. PLL 567 is a highly stable phase locked loop with synchronous AM lock detection and power output circuitry. It is primarily used as a tone and frequency decoder, where it is required to drive a load, whenever a sustained frequency within its detection band is present at its self biased input. The center frequency of the band and the output delay are independently determined by means of four external components. Pin connections and internal block schematic are shown in Fig. 21.16.

Q. 18. Briefly describe PLL 4046B What are its salient features?

Ans. PLL 4046B is a micropower phase locked loop consisting of a low power linear voltage controlled oscillator (VCO), two different phase comparators, having a common signal input amplifier and a common phase comparator input. Its salient features include (*i*) very low power consumption (*ii*) high VCO linearity (*iii*) option of using either of the two phase comparators (*iv*) on-chip zener diode for supply regulation (*v*) low VCO frequency drift with temperature (typically 0.04% per degree centigrade). Pin connections and internal schematic are shown in Fig. 21.17.

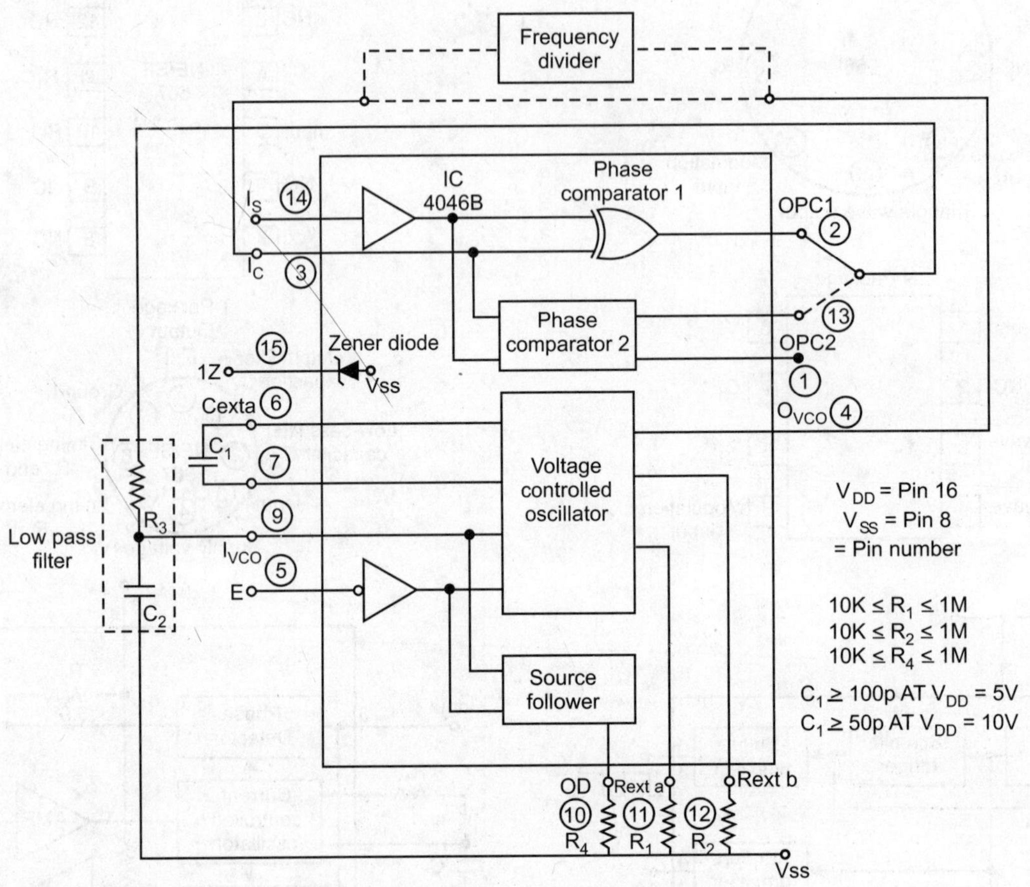

Fig. 21.17

Q. 19. With the help of a block diagram, explain in brief how a PLL can be used as a frequency multiplier?

Ans. Fig. 21.18 shows the block diagram of the frequency synthesiser.

The VCO runs at a multiple of the reference input. The counter inserted between the VCO and the phase comparator decides the multiplication factor.

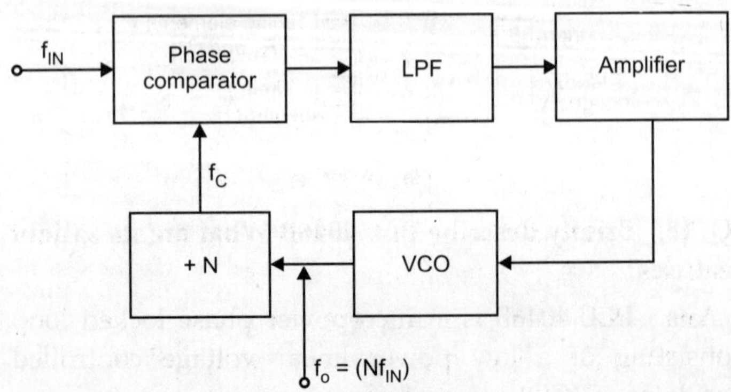

Fig. 21.18

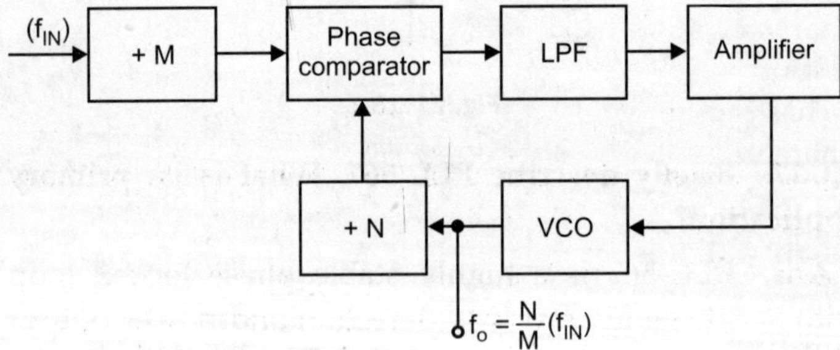

Fig. 21.19

That is, if we use a divide by (*N*) counter and an input reference of frequency (f_i), then the output frequency (output of VCO) will be (Nf_i). Another scheme for a frequency synthesiser is shown in Fig. 21.19. The arrangement shown has undoubtedly greater programme flexibility.

If one wishes to multiply the available reference input by 16/3, for instance, one can very well do it with the arrangement shown in Fig. 21.19 by having a divide-by-16 counter between VCO input and phase comparator input and divide-by-3 counter in series with input. This arrangement finds wide applications in frequency synthesisers.

Q. 20. With the help of a block schematic, explain in brief, how PLL can be used as an FM demodulator. Which general purpose PLL is suitable for FM demodulation applications, and why is it so?

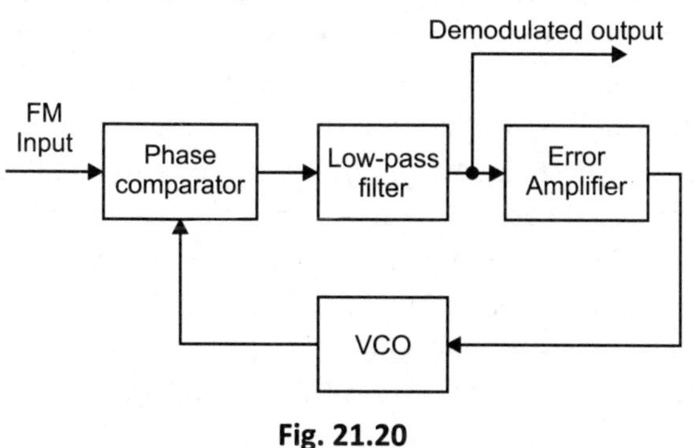

Fig. 21.20

Ans. Fig. 21.20 shows the block-schematic of a general purpose PLL. If the input to the PLL is a frequency modulated signal and the output is taken from low pass filter's output, it would represent demodulated FM signal. It is because VCO would always track the frequency variations of the input FM signal provided, these variations lie within the lock range of the PLL. To maintain these lock conditions, the error voltage produced at the output of phase comparator must represent these frequency variations. The filtered error voltage, which forces the VCO to maintain lock with the input signal is then the demodulated FM.

The linearity of demodulated FM signal depends upon the linearity of the voltage to frequency transfer characteristics of phase comparator of PLL. PLL 560 is normally used for FM demodulation purposes as it has sufficiently linear voltage to frequency transfer characteristics. PLL 562 is equally suitable.

Q. 21. Draw the basic PLL arrangement that can work as an AM demodulator. Explain in brief its operation. Which general-purpose PLL is the most suitable for this purpose ? Give reasons in support of your answer.

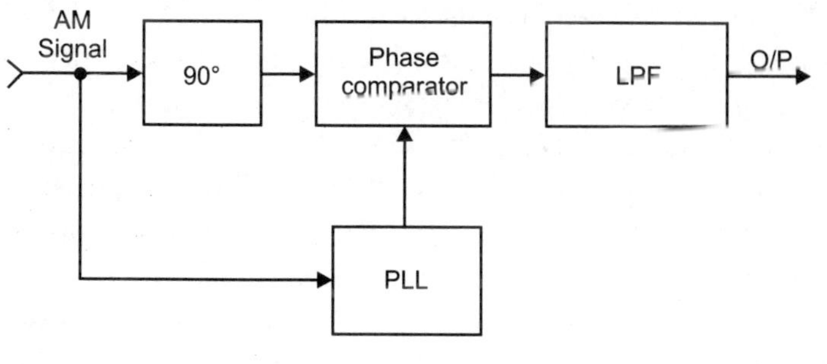

Fig. 21.21

Ans. Fig. 21.21 shows the block diagram. The input to the PLL (applied to the input of the phase-comparator inside the PLL) is the AM signal. The output from the PLL (taken from VCO output) will then be the input signal devoid of amplitude variations, *i.e.* it will be the carrier signal (under locking conditions). The PLL output is then applied to another phase comparator, which acts as a product demodulator. The inputs to the phase comparator are the carrier signal (appearing at PLL output) and the AM signal shifted in phase by 90°. This 90° phase-shift is intentionally introduced to ensure that the VCO signal and the AM signal fed to the phase comparator are in phase. It may be mentioned that in most of monolithic PLLs, the loop goes into lock with the input signals and the phase comparator being 90° out of phase. (However, there are certain PLLs, where the phase difference is zero under lock conditions. CMOS PLL type number 4046B is an example). It is because of this that an additional 90° phase shift is provided to ensure that the two signals on reaching the phase comparator (acting as a Product Detector) are in phase. The output filter removes all but the difference frequency component. The output of the phase comparator is directly proportional to the amplitude of the input signal. PLL 561 is the most suitable general purpose PLL for AM demodulation application. PLL 561 has a built-in quadrature phase detector (shown as a separate block in Fig. 21.21. This phase-detector forms a part of PLL 561.

Q. 22. With the help of a block schematic, explain in brief how a phase locked loop (PLL) can be used as a phase modulator?

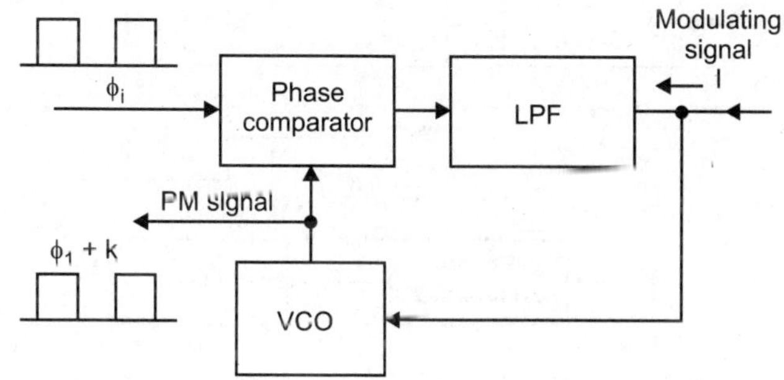

Fig. 21.22

Ans. Fig. 21.22 shows the block schematic. When the phase locked loop is locked on to the incoming signal at the free running frequency, the phase of VCO output with respect to the incoming signal will then be 90°. The basis of the circuit arrangement shown being a phase modulator is the fact that if the voltage at the control input of the VCO (or the output terminal of LPF) is changed externally, the phase of the signal at the output of the VCO will have to change so as to produce an opposing voltage at the output of phase comparator or output of LPF, which is necessary for maintaining locking condition. So, we can say that if a signal current (I) is injected externally as shown, the phase of the VCO output will change accordingly producing an opposing current at the output of LPF. This opposing current is of the same amplitude as that of one injected externally. Under such circumstances, externally injected current forms the modulating signal, the output of VCO forms the modulated signal and the carrier is the free-running frequency of the oscillator. If the modulating signal amplitude is low enough so that the loop frequency is limited by the phase comparator output rather than the VCO swing, then the phase of VCO output can vary from 0° to 180°.

Q. 23. How can a PLL be used as a frequency multiplier or as a frequency divider?

Ans. A PLL can be used as a frequency multiplier by inserting a divide-by-N circuit, such as a frequency counter between the VCO output and one of the inputs of phase comparator. In this case, VCO signal applied to the relevant phase comparator input is (1/N) times the VCO output. When the PLL is in lock; the VCO output will be (N) times the input signal frequency. Figure-21.23 (a) shows the arrangement. A PLL can be used as a frequency divider by inserting a divide-by-M circuit between the input signal and the relevant phase comparator input as shown in Figure-21.23 (b). In this case, VCO output frequency is (M) times the input signal frequency under lock conditions. Same purpose would be served, if a frequency multiplier with multiplication factor (M) were inserted between the VCO output and relevant phase comparator input. Figure-21.23 (c) shows the arrangement.

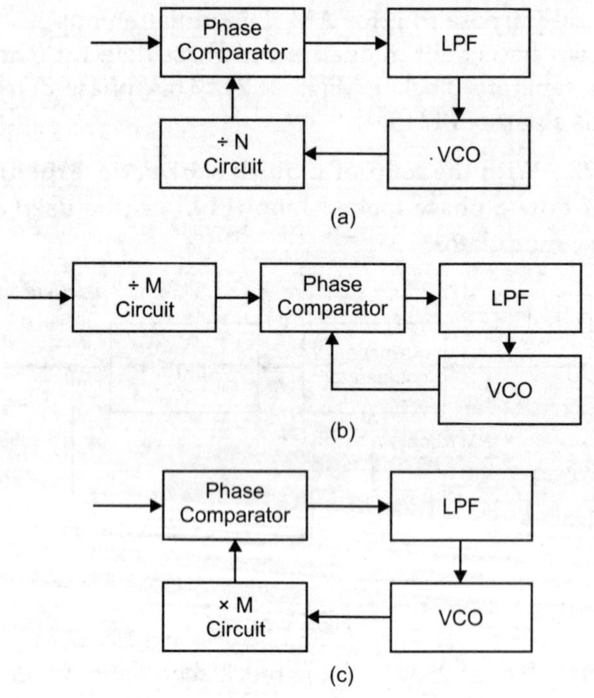

Fig. 21.23

Q. 24. How does a PLL work as a tracking narrow band pass filter? What are its advantages as compared to a conventional analog band pass filters?

Ans: A narrow band pass filter finds application in processing of angle modulated signals. Such narrow band pass filters, if implemented using conventional analog filter circuits, suffer because of the temperature dependence of their characteristics leading to shift in the carrier frequency. These filters can also not be used in applications, where the carrier frequency of angle-modulated signal is required to be varied. These limitations are overcome, if a PLL tracking the carrier is used as a band pass filter. The PLL also separates the spectrum of the angle-modulated signal from other interfering signals, or limits the transmitted spectrum to within specified bounds. A further advantage of PLL band pass tracking filter is that it rejects the amplitude modulation, that is, it may also be used as a limiter. Figure-21.24 shows the block schematic arrangement of a PLL based band pass tracking filter. If the loop parameters depend on the amplitude of the input signal, an AGC circuit must precede the phase detector in order to keep the filter parameters constant. The exercise of designing a PLL based narrow band pass tracking filter, in fact, gets reduced to the design and implementation of a baseband loop filter.

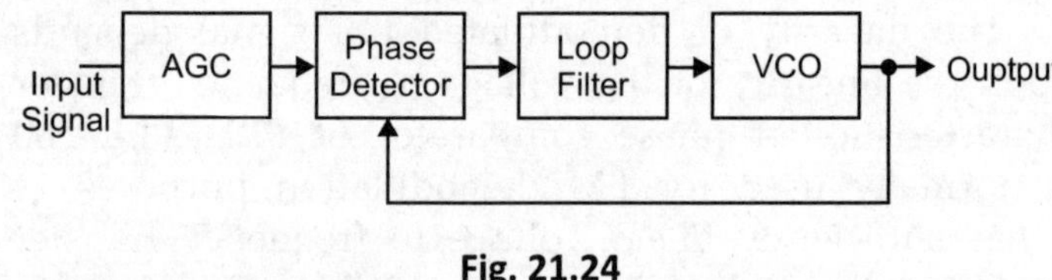

Fig. 21.24

Q. 25. How is a PLL used as a CW carrier recovery circuit?

Ans: Carrier recovery circuit using PLL is nothing but a PLL based narrow band pass tracking filter. The objective of CW carrier recovery is to retrieve the carrier and to suppress as much noise, modulation and interference as possible. The CW carrier recovery circuit is a narrowband band pass tracking filter implemented by a PLL, as shown earlier in Figure-21.24. The noise-free recovery of a carrier requires use of a very narrowband PLL. Unfortunately, the acquisition properties of narrowband PLLs are very poor. This problem may be eliminated by using a wide bandwidth during acquisition process and a narrow bandwidth in the steady-state after the phase-locked condition has been achieved.

Q. 26: How is PLL used as an amplifier of angle modulated signals?

Ans. The PLL may be used as an amplifier of angle modulated signal. In this case, the signal to be amplified is applied to the PLL input and the VCO output is the amplified signal. The gain is determined by the ratio of VCO output and PLL input powers. The amplification in this case is performed in the baseband providing it an edge over high frequency amplifiers, which are very expensive. In addition to amplification, the PLL also operates as a limiter and filter for the incoming angle-modulated signals. Figure-21.25 shows the basic block schematic arrangement of the PLL based amplifier. Due to the frequency multiplier placed in the feedback path, the VCO output frequency is f_i/M, where f_i is the input frequency. The input phase/frequency deviation is also divided by M. However, the modulating frequency remains unchanged. The output frequency multiplier following the power amplifier restores the original carrier frequency and its phase/frequency deviation.

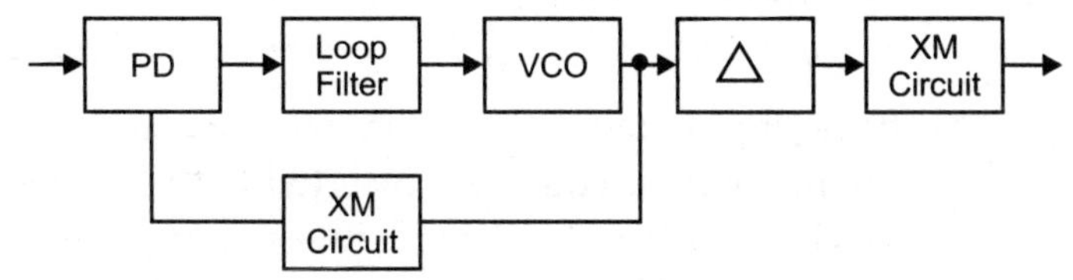

Fig. 21.25

SOLVED PROBLEMS

Problem 21. 1: *Draw an appropriate PLL arrangement, which can be used to generate a frequency of 27 MHz from an available crystal of 30 MHz.*

Solution. The block schematic is shown in Fig. 21.26. In the given arrangement, N and M are divided-by-9 and divided-by-10 circuit respectively. Under lock, output of VCO is 27 MHz.

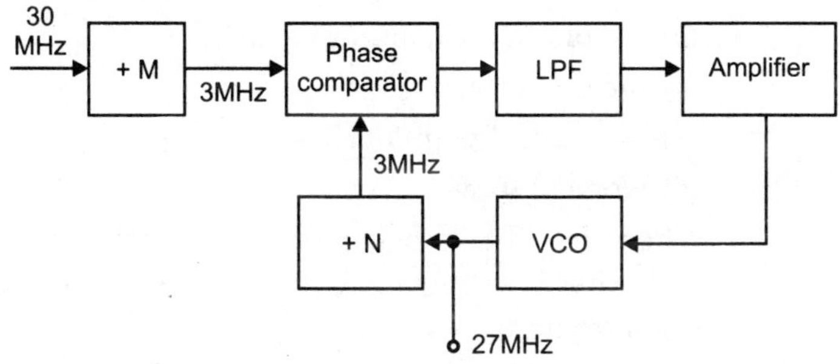

Fig. 21.26

Problem 21.2: *Give the basic block schematic arrangement of phase locked loop that can be used to synthesise 15 MHz signal from an available 35 MHz signal.*

Solution: Figure-21.26 shows the basic arrangement. A modulo-3 counter has been inserted between the VCO output and the relevant phase comparator input and the input signal is fed to the relevant phase comparator input through a modulo-7 counter.

Now, the input signal fed to one of the inputs of phase comparator is 5 MHz. Under phase lock, the other input to the phase comparator is also 5 MHz. Therefore, the VCO output signal frequency is 15 MHz.

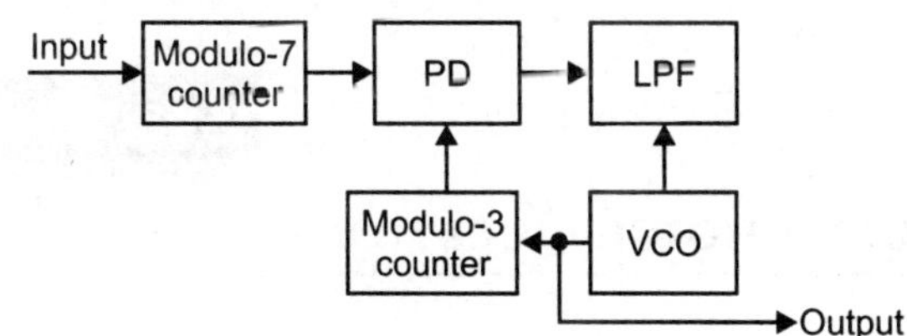

Figure. 21.26

Problem 21.3: *In a phase locked loop based frequency synthesiser application, (×4) frequency multiplier circuit is added between the VCO output and relevant phase comparator input. Also, input signal is fed to its relevant phase comparator input through a toggle flip-flop. What is VCO output frequency?*

Solution: The insertion of toggle flip flop, which is a divide-by-2 circuit, has the effect of dividing the VCO output frequency by two. The insertion of (×4) multiplier circuit has the effect of dividing the VCO output frequency. Therefore, VCO output frequency will be one-eighth the input signal frequency.

If the input frequency were (f_i); the signal frequency fed to relevant phase comparator input will be ($f_i/2$). Also, if (f_{VCO}) were the free running frequency of VCO; the signal frequency fed to relevant phase comparator input will be ($4 \times f_{VCO}$).

Now under phase lock, $4 \times f_{VCO} = f_i/2$

Which gives, $= f_{VCO} = f_i/8$

SUMMARY OF IMPORTANT FORMULAS

1. If the input signal frequency were (f_i) and the divide-by-M circuit were inserted between the signal source and relevant phase comparator input; the VCO output under phase lock conditions would be given by:

$f_{VCO} = f_i/M$

2. If the input signal frequency were (f_i) and the multiply-by-M circuit were inserted between the signal source and relevant phase comparator input; the VCO output under phase lock conditions would be given by:

$f_{VCO} = M \times f_i$

3. If the input signal frequency were (f_i) and the divide-by-N circuit were inserted between the VCO output and relevant phase comparator input; the VCO output under phase lock conditions would be given by:

$f_{VCO} = N \times f_i$

4. If the input signal frequency were (f_i) and the multiply-by-N circuit were inserted between the VCO output and relevant phase comparator input; the VCO output under phase lock conditions would be given by:

$f_{VCO} = f_i/N$

5. If the input signal frequency were (f_i); divide-by-N circuit were inserted between the VCO output and relevant phase comparator input; divide-by-M circuit were inserted between signal source and relevant phase comparator input; the VCO output under phase lock conditions would be given by:

$f_{VCO} = (N/M) \times f_i$

6. If the input signal frequency were (f_i); multiply-by-N circuit were inserted between the VCO output and relevant phase comparator input; multiply-by-M circuit were inserted between signal source and relevant phase comparator input; the VCO output under phase lock conditions would be given by:

$f_{VCO} = (M/N) \times f_i$

SELF EVALUATION EXERCISE

I. MULTICHOICE QUESTIONS

1. The lock range of a certain general purpose PLL with a free running frequency of 50 MHz is specified to be ± 10%. Its lock range in MHz is:
   (*a*) 10 MHz
   (*b*) 5 MHz
   (*c*) 15 MHz
   (*d*) indeterminate from given data.
2. The interference rejection properties of a PLL improve on:
   (*a*) reducing the LPF bandwidth
   (*b*) increasing the LPF bandwidth
   (*c*) increasing the gain of error amplifier
   (*d*) reducing the gain of error amplifier
3. The general purpose PLL, which is preferably used as an AM demodulator is:
   (*a*) 561 (*b*) 564
   (*c*) 567 (*d*) 565
4. When a PLL is being used as an FM demodulator, the demodulated signal appears at:
   (*a*) the output of phase comparator
   (*b*) the output of low pass filter
   (*c*) the VCO output
   (*d*) None of these.
5. When the PLL is being used as a phase modulator, the modulating signal is injected at:
   (*a*) the phase comparator input
   (*b*) the LPF input
   (*c*) the control point of VCO
   (*d*) None of these
6. When the PLL is being used as a frequency multiplier or a frequency divider, the output is taken from:
   (*a*) the VCO input
   (*b*) the output of phase comparator
   (*c*) the VCO output
   (*d*) LPF output
7. Greater LPF bandwidth in a PLL signifies:
   (*a*) larger capture range and small lock-up time
   (*b*) smaller capture range and larger lock-up time
   (*c*) larger lock-in range
   (*d*) larger capture range and lock-up time.
8. The essential building blocks of a phase locked loop are:
   (*a*) the VCO, the phase comparator and the error amplifier
   (*b*) the VCO, the phase comparator and the low pass filter

(c) the VCO, the low pass filter and the error amplifier

(d) the phase comparator, the low pass filter and the error amplifier

(e) the phase comparator and the VCO

9. In a typical frequency multiplier application using a PLL, a decade counter was inserted between the VCO output and the relevant phase comparator input. If the input frequency is 10 kHz, the VCO output frequency will be:

(a) 10 kHz

(b) 1 kHz

(c) 100 kHz

(d) indeterminate from given information.

10. A 1 MHz signal is fed to the input of phase locked loop through a T-flip-flop. The loop between VCO output and the relevant phase comparator input is broken and a Modulo-6 counter inserted, the output of the VCO when the PLL is in lock, is:

(a) 3 MHz (b) 6 MHz

(c) 1 MHz (d) 133.33 kHz

11. The phase locked loop in which all components operate in continuous time domain is:

(a) Analog or linear ALL

(b) Digital PLL

(c) Software PLL

(d) None of these

12. Identify the characteristic of PLL influenced by loop filter:

(a) Loop stability

(b) Ability to reject unwanted high frequency components

(c) Capture range

(d) All of above

13. If a frequency multiplier cerate were added between the VCO output and the reterant phase comparator input, the PLL would act as a:

(a) Frequency multiplier

(b) Frequency divider

(c) Frequency multiplier or divider

(d) None of these

14. When the PLL is used as a narrow band pass tracking filter, the loop filter:

(a) has a high cut-off frequency

(b) has a low cut-off frequency

(c) is a narrow band pass filter

(d) None of these

15. CMOS IC CD 4046 B constitutes one elements of a PLL. What can it be:

(a) VCO (b) Loop filter

(c) Phase comparator (d) None of these

II. Fill in the Blanks

1. A PLL is analogous to........................feedback control system.
(Positive, Negative)

2. If the PLL is a feedback system, it will obviously have some circuit blocks in the forward signal path and some in the feedback path. The ones in the forward signal path are .... (i)................and..................and the ones in the feedback path are (ii)............
(VCO, Phase comparator, LPF, Error amplifier)

3. The VCO output in a monolithic PLL is always a........................waveform.
(sinusoidal, square, sinusoidal or square)

4. Out of the following, the specifications you would attach highest priority to when you are looking for a PLL to be used as an FM demodulator is........................
(VCO voltage to frequency conversion linearity, LPF bandwidth, phase comparator conversion gain, Lock-up time)

5. When the PLL is being used as a tracking filter, the output is taken from..................output.
(VCO, LPF, Phase comparator)

Answers to Self Evaluation Exercise

I.

1. (a)	2. (a)	3. (a)	4. (b)
5. (c)	6. (c)	7. (a)	8. (b)
9. (c)	10. (a)	11. (a)	12. (d)
13. (b)	14. (b)	15. (c)	

II. 1. Negative

2. (i) Phase comparator, LPF, Error Amplifier
(ii) VCO

3. Square

4. VCO voltage to frequency conversion linearity

5. VCO.

Chapter 22

Digital Electronics Fundamentals

Number Systems and Codes–Binary Arithmetic

Boolean Algebra and Minimisation Techniques

Learning Objectives

After completing the chapter, you will learn the following topics:

1. Different Types of Data and Data Representation
2. Different Number Systems Including Binary, Octal, Decimal and Hexadecimal Number Systems
3. 1's and 2's Complements
4. Representation of Positive and Negative Numbers
5. Conversion from One Number System to Another Including Binary-to-Decimal and Decimal-to-Binary, Binary-to-Octal and Octal-to-Binary, Binary-to-Hex and Hex-to-Binary, Octal-to-Decimal and Decimal-to-Octal, Octal-to-Hex and Hex-to-Octal, Decimal-to-Hex and Hex-to-Decimal
6. Number System Codes Including 8421 BCD Code, Gray Code, EBCDIC and ASCII Codes, Hollerith Code, 7-bit Selectric Code and 7-segment Display Code
7. BCD Code-to-Binary and Binary-to-BCD Code Conversion, Binary-to-Gray and Gray-to-Binary Conversion
8. Parity Bits
9. Binary Multiplication and Division
10. Half Adder and Full Adder, Half Subtractor and Full Subtractor
11. Laws and Theorems of Boolean Algebra Including Commutative, Associative Laws, Distributive and Involution Laws, Absorption Law and DeMorgan's Theorems
12. Sum-of-Products and Product of Sums Boolean Expressions
13. Simplifying Boolean Expressions Using Quine-McCluskey Tabular Method and Karnaugh Maps

Q. 1. Distinguish between the following terms-BIT, BYTE, NIBBLE, COMPUTER WORD and WORD LENGTH.

Ans. BIT is an abbreviation of the words Binary Digit and is the smallest unit of information. It is either '0' or '1'. A string of four bits or a binary representation of four bits is called a NIBBLE and that of 8-bits usually a BYTE. Byte is the basic unit of data operated on as a single unit in computers. COMPUTER WORD is again a string of bits whose size (number of bits in the string) called the WORD LENGTH or WORD SIZE is fixed for a specified computer, though it may vary from computer to computer. The word 'length' may be 1 Byte, 2 Bytes or 4 Bytes or even larger. The illustration shown ahead provides a good comparison of these terms for a 16-bit machine.

X...BIT

XXXX..NIBBLE

XXXXXXXX.................................BYTE

XXXXXXXXXXXXXXXX.............COMPUTER WORD,

where X is either a '0' or a '1'.

Q.2. What is DATA ? Outline the different types of data used in computers.

OR

What are the different data types used in computers to perform arithmetic computations, logical operations and data processing?

Ans. Data is basically binary numbers or other forms of binary coded information operated on as a unit in computers to perform arithmetic computations, data processing and logic operations. Different data types used in digital computers are classified as under :

(*a*) Numeric Data

(*b*) Character or Alphabetic Data.

(*c*) Alphanumeric Data

(*d*) Logic Data

NUMERIC DATA consists of Integers and Real numbers. Integers are the set of whole numbers including '0'. Positive integers may or may not be preceded by a positive sign but the negative integers must be preceded by a (–) sign. A real number consists of an integral part and a fractional part. Nevertheless, all integers are also real numbers but in the language of computers, presence or absence of decimal points distinguishes a real number from an integer. For computer operations, 98 is an integer while 98.0 is a real number although both have same magnitude. 98.768, however, is a real number and not an integer. Real numbers are represented either in the fixed point representation form or in the floating point/ exponential representation form.

CHARACTER or ALPHABETIC DATA consists of English alphabet, both upper case and lower case, written in Roman script. For example, if you are asked to arrange the given list of the names of the countries in the Asian subcontinent in an alphabetic order, it would constitute a form of character data.

ALPHANUMERIC DATA, necessitated by the need of both numeric as well as character data in computers, comprises of alphabetic characters A to Z and the numerics 0 to 9.

LOGICAL DATA is the one that can take only one of the two possibilities *i.e.* TRUE or FALSE. Such data is made use of in computers to make logical decisions like comparison etc. It is also used in operations like OR, AND, NOT and so on. Such data is also usually in the form of statements having character data or numeric data or both alongwith some special characters like $(<)$, $(>)$, $(< >)$, $(\leq)$, $(\geq)$ etc. The statement $X + Y = 5$ can be either TRUE or FALSE, depending upon the magnitudes of X and Y and is, thus a form of logical data.

Q. 3. Define RADIX/BASE of a number system. What is the radix for Binary, Octal, Decimal and Hexadecimal number systems? Assuming an arbitrary number system of radix-3, write down the first ten numbers in this number system.

Ans. RADLX/BASE of a number system is defined as the number of different symbols used in the number system. Binary number system uses two different symbols '0' and '1' and thus, is a BASE-2 system or has a radix of 2. Similarly, Octal number system has a radix of 8, decimal number system has a radix of 10 and the hexadecimal number system has a radix of 16.

Let us assume that the arbitrary number system has the symbols 0, 1, 2. In that case, the first ten numbers in this assumed system will be 0, 1, 2, 10, 11, 12, 20, 21, 22, 100.

Q. 4. What is Decimal system of number representation? How can a decimal number be represented as a sum of different digits multiplied by their respective place values or weights ?

Ans. The decimal number system has ten different symbols or digits *i.e.* 0, 1, 2, 3, 4, 5, 6, 7, 8, and 9 and all higher numbers in this system are represented in terms of these ten digits only. The process of writing the decimal numbers after '9' consists of writing the second digit first and then following it up with other digits one by one to get the next ten numbers from 10 to 19. The next ten numbers from 20 to 29 are obtained by writing the third digit first and then following it up with the digits '0' to '9'. This process continues till we have exhausted all two digit combinations and we have reached '99'. Then, we begin with three digit combinations starting with '100' and the process goes on endlessly.

The decimal system has a radix of 10 or it is a base-10 number system as it has ten different digits to work with. It is a positional system *i.e.*, each digit in a given decimal number has a weight or a value. The value or magnitude of a given decimal number can be expressed as a sum of various digits multiplied by their place values or weights. If the given number has both an integer part as well as a fractional part, the two portions should be dealt with separately. The weights or place values of different digits in a mixed decimal number are listed below:

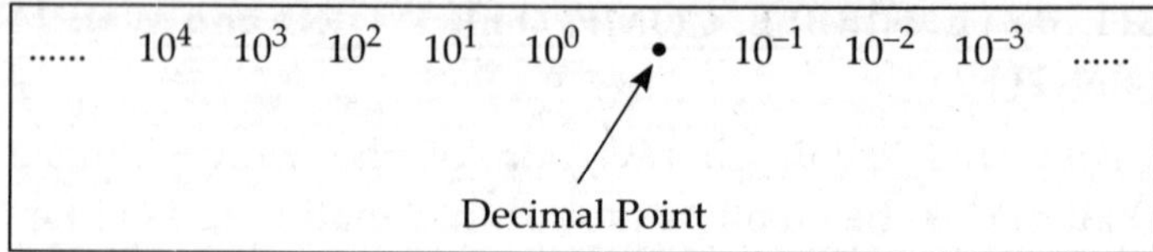

Let us take the case of decimal number 5395.8742. The decimal number and the weights of different digits are tabulated below :

Decimal Number	5	3	9	5	•	8	7	4	2
Weight	10^3	10^2	10^1	10^0		10^{-1}	10^{-2}	10^{-3}	10^{-4}

The Integer part of the decimal number can also be written as :

$$5395 = 5 \times 10^3 + 3 \times 10^2 + 9 \times 10^1 + 5 \times 10^0$$
$$= 5000 + 300 + 90 + 5 = 5395$$

And the fractional part can be expressed as :

$$8 \times 10^{-1} + 7 \times 10^{-2} + 4 \times 10^{-3} + 2 \times 10^{-4}$$
$$= 0.8 + 0.07 + 0.004 + 0.0002 = 0.8742$$

Thus, the value of a decimal number clearly depends upon the weights of the different digits in addition to the magnitudes of these digits. Also, weights are a function of the radix of the number system and the position of the digit. In fact the concept of each digit having a weight, depending upon the position of the digit and the radix of the number system are equally valid in other number systems that are of much greater interest to a computer engineer.

Q. 5. What is a BINARY NUMBER SYSTEM? Illustrate with the help of an example as to how can we convert a binary number into its decimal equivalent?

Ans. The binary number system is a base-2 number system (it has a radix of 2), as it has only two different symbols or digits. These are 0 and 1 and all binary numbers are expressed in terms of these two digits only. Table 22.1 lists first twenty numbers in the binary number system alongwith their decimal equivalents.

Table 22.1

Binary	*Decimal*	*Binary*	*Decimal*
0	0	1011	11
1	1	1100	12
10	2	1101	13
11	3	1110	14
100	4	1111	15
101	5	10000	16
110	6	10001	17
111	7	10010	18
1000	8	10011	19
1001	9	10100	20
1010	10		

Each one of the digits in a binary number has a place value or weight on the same pattern, as we have in case of decimal number system. The binary weights in a mixed binary number are listed below:

.....	2^4	2^3	2^2	2^1	2^0	•	2^{-1}	2^{-2}	2^{-3}	2^{-4}	.....
					Binary	↑	Point				

The decimal equivalent of a given binary number is obtained as a sum of various digits multiplied by their respective weights. In case of mixed numbers, the integer and fractional parts should be handled separately. This can be best illustrated with the help of an example. Conversion of a binary number 11011.1011 into its decimal equivalent is illustrated ahead.

Binary Number	1	1	0	1	1	•	1	0	1	1
Weight	2^4	2^3	2^2	2^1	2^0	•	2^{-1}	2^{-2}	2^{-3}	2^{-4}

Integer Part = $11011 = 1 \times 2^0 + 1 \times 2^1 + 0 \times 2^2 + 1 \times 2^3 + 1 \times 2^4$

$$= 1 + 2 + 0 + 8 + 16 = 27$$

The decimal equivalent of the fractional part is given by:

$$1 \times 2^{-1} + 0 \times 2^{-2} + 1 \times 2^{-3} + 1 \times 2^{-4}$$
$$= 0.5 + 0 + 0.125 + 0.0625 = 0.6875$$

The decimal equivalent of 11011.1011 is, therefore, 27.6875.

Q.6. What do you understand by l's COMPLEMENT and 2's COMPLEMENT of binary numbers? Illustrate with the help of examples as to how positive and negative numbers are represented in l's and 2's complement form?

Ans. l's complement of a binary number is obtained by complementing all its bits *i.e.* by replacing 0s by 1s and 1s by 0s. 2's complement of a binary number is obtained by adding 1 to its l's complement. Thus, for finding 2's complement, we first find its l's complement and then add 1 to it.

As an illustration, 1's complement of 110101 is 001010 and its 2's complement is 001011.

Note that 001011 = 001010 + 1

l's Complement Notation. The positive binary numbers remain unchanged, when represented in the 1's complement form. For instance, + 7 and + 18 in the 1's complement will be 00000111 and 00010010 (in 8-bit notation) respectively. The negative numbers in 1's complement form are obtained by taking l's complement of its positive counterpart. – 7 and – 18 are respectively represented as 11111000 and 11101101 (in 8-bit notation).

2's Complement Notation. In the 2's complement representation of binary numbers, the MSB represents the sign. '1' in the MSB position represents a (–) sign and a 0 in the MSB position represents a (+) sign. The remaining bits give the magnitude of the number. The positive amplitudes are represented in the same way like we do in case of binary numbers and the negative magnitudes are represented by the 2's complement of their positive counterparts. Another important thing to remember here is that in 2's complement representation of binary numbers, the magnitude of a number (positive or negative) is given by number of bits one less than the total number of bits, as the MSB gives the sign. The whole procedure can be best explained with the help of an example. Let us say that we want to have 8-bit 2's complement representation + 15 and – 15. Now in the 8-bit 2's complement representation, only seven bits will represent the magnitude and the eighth bit (MSB) provides the sign information. Thus, +15 is represented as :

0	0001111
Sign	Magnitude

For – 15 sign bit is 1 and the magnitude is represented by the 2's complement of 0001111. Thus, – 15 in the 8-bit 2's complement form is expressed as :

1	1110001
Sign	Magnitude

Also, one can notice that in this notation, if 2's complement of the magnitude of + 15 gives the magnitude of –15, then the reverse process is also true *i.e.,* 2's complement of magnitude of –15 gives the magnitude of + 15. It is because 2's complement of 2's complement of a binary number, is the same binary number. Also, 1's complement representation of a positive binary number is same as its 2's complement representation.

Q.7. What are the maximum positive and minimum negative numbers that can be represented by n-bit 1's complement and 2's complement notations?

Ans. The maximum positive decimal number that can be represented using n-bit 1's complement notation is $(2^{n-1} - 1)$ and the minimum negative decimal number that can be represented using n-bit l's complement notation is also $-(2^{n-1} - 1)$.

These maximum positive and minimum negative numbers in case of n-bit 2's complement notation are $(2^{n-1} - 1)$ and -2^{n-1} respectively.

Table 22.2 lists all possible positive and negative numbers represented in 4-bit 1's and 2's complement notation.

Q.8. What is Sign magnitude system of representing binary numbers ?

Ans. In the Sign Bit Magnitude system of representing binary numbers, the MSB represents the sign with '0' denoting (+) sign and '1' denoting a (–) sign. The remaining bits give the magnitude. The magnitudes for the positive as well as the negative magnitudes are represented in the same way as we do in case of binary numbers. For example, in 4-bit representation in this system, + 6 is represented as 0 110 and – 6 is written as 1110. Table 22.2 lists all possible positive and negative 4-bit numbers in sign bit magnitude representation.

Table. 22.2

Decimal Number	*Sign-bit Magnitude*	*1's Complement*	*2's Complement*
+ 7	0111	0111	0111
+ 6	0110	0110	0110
+ 5	0101	0101	0101
+ 4	0100	0100	0100
+ 3	0011	0011	0011
+ 2	0010	0010	0010
+ 1	0001	0001	0001
+ 0	0000	0000	0000
– 0	1000	1111	1000
– 1	1001	1110	1111
– 2	1010	1101	1110
– 3	1011	1100	1101
– 4	1100	1011	1100
– 5	1101	1010	1011
– 6	1110	1001	1010
– 7	1111	1000	1001
– 8	—	—	1000

Q. 9. Compare the advantages and disadvantages of using Sign-Bit Magnitude and 2's complement notations for representing positive and negative numbers.

Ans. **SIGN-BIT MAGNITUDE** notation has the advantage of simplicity. However, the arithmetic circuits required to perform addition and subtraction of binary numbers represented in sign-bit magnitude notation are very complex. This notation is, therefore, used in only those places where we don't have to perform addition, subtraction operations on the binary data. One such application is in Analog to Digital converters.

2's COMPLEMENT notation is commonly used to represent positive and negative numbers, as it is very easy to generate 2's complement of a binary number and as a result binary addition and subtraction operations become very simple.

Q. 10. With the help of an illustration, explain the method of converting a given decimal number having both an Integer as well as a Fractional part into corresponding binary number.

Ans. The first thing to bear in mind while converting a given decimal number into an equivalent binary number is that the integer part and the fractional part should be handled separately.

DOUBLE-DABBLE METHOD is the most commonly used technique for converting decimal numbers into their binary equivalents. The decimal number (or the integer part of the decimal number, in case it has both integer as well as fractional parts, is progressively divided by 2 and the remainders after each division noted till the quotient becomes '0'. The remainders, written in the reverse order, constitute the binary equivalent.

For the conversion of the fractional part, the fraction is progressively multiplied by 2 and the carry recorded till the result of the multiplication becomes '0' with a carry of '1' or the desired number of digits have been obtained, whichever is earlier. The CARRYS written in the forward order then constitute the binary equivalent.

Let us find the binary equivalent of the decimal number 13.25.

Integer Part

	Integer Part	Remainder	
2	13	–	↑
	6	1	
	3	0	
	1	1	
	0	1	

The remainders written in the reverse order constitute 1101, which is binary equivalent of 13.

Fractional Part

$0.25 \times 2 = 0.5$ with a CARRY of 0

$0.50 \times 2 = 0$ with a CARRY of 1

The CARRYS written in the forward order constitute 01.

The binary equivalent of 13.25 is, therefore, 1101.01

Q. 11. What is the OCTAL SYSTEM of number representation? With the help of an illustration, explain as to how can we make Octal- Decimal-Octal conversions?

Ans. The octal number system has a radix or base of 8 *i.e.* It has 8 distinct digits and all higher digits are expressed as a combination of these on the same pattern as the one followed in case of binary and decimal number systems. These different digits are 0, 1, 2, 3, 4, 5, 6 and 7. The next digit after 7 in this system will be 10 followed by 11,12,13,14 and so on. In fact, if we omit the numbers having digits 8 and 9 in the decimal numbers, we end up with the octal number system.

Each digit in the octal number system has a place value. As it is a base–8 system, the place values or weights starting from LSB towards MSB will be 8^0, 8^1, 8^2, 8^3, ... for the integer part and 8^{-1}, 8^{-2}, 8^{-3}, 8^{-4} for the fractional part.

....	X	X	X	X	X	X	•	X	X	X	X	X
....	8^5	8^4	8^3	8^2	8^1	8^0	•	8^{-1}	8^{-2}	8^{-3}	8^{-4}	....

Radix ⇑ Point

Here, 'X' is an Octal Digit.

The given octal number can be converted into its decimal equivalent by taking the sum of all the digits multiplied by their respective place values. For instance, 72 in the octal number system equals 58 in the decimal system as,

$$2 \times 8^0 + 7 \times 8^1 = 58$$

Also,

$$(63.25)_8 = [(3 \times 8^0 + 6 \times 8^1).(2 \times 8^{-1} + 5 \times 8^{-2})]_{10}$$
$$= (51.32)_{10}$$

Integer part of the decimal number can be converted into its octal equivalent by progressively dividing the decimal number by 8 and noting down the remainders till we get a quotient of '0'. The remainders taken in the reverse order constitute the octal number integer. Converson of $(58)_{10}$ to its octal equivalent is illustrated below:

	Integer Part	*Remainder*	
8	58	–	↑
	7	2	
	0	7	$(58)_{10} = (72)_8$

Fractional part of the decimal number can be converted into its equivalent octal fraction by progressively multiplying the decimal fraction by 8 and recording the CARRY till the result of multiplication is '0' with a certain carry or when the desired number of digits have been obtained. $(0.35)_{10}$ is converted into its octal fraction as follows :

$0.35 \times 8 = 0.80$ with a carry of 2

$0.80 \times 8 = 0.40$ with a carry of 6

$0.40 \times 8 = 0.20$ with a carry of 3

$(0.35)_{10}$, therefore, equals $(0.263)_8$

The CARRYS written in the FORWARD ORDER constitute octal fraction.

Q. 12. What is the HEXADECIMAL SYSTEM of numbers? List all the sixteen symbols or digits of this system and illustrated with the help of an example as to how can convert a hexadecimal number into its decimal equivalent and vice versa ?

Ans. HEXADECIMAL NUMBER SYSTEM is a base-16 or radix -16 number system, as it has 16 different digits. These digits are 0,1, 2, 3, 4, 5, 6, 7, 8, 9, A, B, C, D, E, F, The PLACE VALUES or WEIGHTS of different digits in hexadecimal number are as under:

...X	X	X	X	X	X	X	X	X	X	X....
16^4	16^3	16^2	16^1	16^0	•	16^{-1}	16^{-2}	16^{-3}	16^{-4}	16^{-5}

Radix ↑ Point

Here, *X* is any hexadecimal digit.

Decimal equivalent of a hexadecimal number is equal to the sum of different digits multiplied by their respective place values or weights.

A decimal number can be converted into an equivalent hexadecimal number by progressively dividing the INTEGER PART of the decimal number by 16 and noting the remainders till we reach 0 and arranging the remainders in reverse order and by progressively multiplying the DECIMAL FRACTION by 16 and recording the CARRYS till the result of multiplication is 0 or desired number of digits have been obtained and then arranging these carrys in the forward order.

Illustration: Let us see, as to how do we convert $(235.25)_{10}$ into an equivalent hexadecimal number.

Integer Part

16	235	Remainder
	14	11 (B hexadecimal)
	0	14 (E in hexadecimal)

↑

Refer to Table 22.3 for the basic hexadecimal numbers and their decimal number equivalents.

$$(235)_{10} = (EB)_{16}$$

Fractional Part

$0.25 \times 16 = 0$ with a CARRY of 4

Therefore, $(235.25)_{10} = (EB.4)_{16}$

Illustration. : Let us see as to how do we convert a given hexadecimal number $(FB5.35)_{16}$ into its decimal equivalent.

$$(FB5.35)_{16} = \{(5 \times 16^0 + 11 \times 16^1 + 15 \times 16^2) .$$
$$(3 \times 16^{-1} + 5 \times 16^{-2})\}_{10}$$
$$= (4021.207)_{10}$$

Table 22.3. Basic Hex Numbers and Their Decimal Equivalents

Decimal Number	*Hexadecimal Equivalent*	*Decimal Number*	*Hexadecimal Equivalent*
0	0	8	8
1	1	9	9
2	2	10	A
3	3	11	B
4	4	12	C
5	5	13	D
6	6	14	E
7	7	15	F

Q. 13. With the help of an example, outline the steps required for converting an octal number into an equivalent binary and a binary number into an equivalent octal number.

Ans. Octal-Binary Conversion.

Octal binary conversion is straight forward. As the base of the octal number system is third power of the base of the binary number system (*i.e.* $8 = 2^3$), an octal number can be converted by replacing each octal digit by its three bit binary equivalent. All we have to then remember is the binary equivalents of the basic digits of the octal number system (Refer Table 22.4).

As an illustration.

$$(36.42)_8 = (011110.100010)_2$$
$$= (11110.10001)_2$$

0s on the extreme left position of the integer part and extreme right position of the fractional part are meaningless and, therefore, can be omitted.

Binary-Octal Conversion

Binary number can be converted into an equivalent octal number by splitting the INTEGER and FRACTIONAL parts in groups of three bits starting from the binary point both sides. 0s can be added to complete the outside groups, if needed. Examples ahead further illustrate the concept.

Table. 22.4.

Octal	*Binary*
0	000
1	001
2	010
3	011
4	100
5	101
6	110
7	111

$(11010011.011011)_2$

$= (0\,1\,1\,0\,1\,0\,0\,1\,1\,.\,0\,1\,1\,0\,1\,1)_2$

'0' added

$= (3\,2\,3\,.\,3\,3)8$

Also, $(1001101.1101)_2$

$= (001\ 001\ 101\ .110\ 100)_2$

Two 0s added on both sides

$= (115.64)_8$

Q. 14. With the help of examples, outline steps to convert a hexadecimal number into its binary equivalent and a given binary number into its hexadecimal equivalent.

Ans. Hexadecimal-Binary Conversion

Hexadecimal binary conversion is straight forward. It is done by replacing each hexadecimal digit by its four bit binary equivalent. Four bits are used as the base of the hexadecimal system is fourth power of the base of the binary system. Thus, all that we have got to remember is the 4-bit binary equivalent of the 16 basic digits of the hexadecimal system. Examples ahead further illustrate the concept.

$$\begin{array}{cccccccc} (2 & B & F & . & 2 & 9 & B)_{16} \\ \downarrow & \downarrow & \downarrow & & \downarrow & \downarrow & \downarrow \\ (0010 & 1011 & 1111 & . & 0010 & 1001 & 1011)_2 \end{array}$$

$= (0\,0\,1\,0\,1\,0\,1\,1\,1\,1\,1\,1\,.\,0\,0\,1\,0\,1\,0\,0\,1\,1\,0\,1\,1)_2$

$= (1\,0\,1\,0\,1\,1\,1\,1\,1\,1\,.\,0\,0\,1\,0\,1\,0\,0\,1\,1\,0\,1\,1)_2$

0s at the extreme left positions of integer part and extreme right positions of the fractional part are meaningless and thus, can be omitted.

Binary-Hexadecimal Conversion

A binary number can be converted into an equivalent hexadecimal number by splitting the integer and fractional parts in groups of 4 starting from the binary point towards both sides and completing the outside groups by adding 0s, if needed. Following example illustrates the concept further.

$(110110111\,.\,1011110)_2$

$= (\underline{000}1\ 1011\ 0111\,.\,1011\ 110\underline{0})_2$

$= (1\,B\,7\,.\,B\,C)_{16}$

Q. 15. Where does the hexadecimal number system find actual use in computers?

Ans. Hexadecimal number system provides a condensed and short hand way of representing binary digits stored and processed inside the computer. Also, the addresses of various memory locations are more convenient to express in hexadecimal notation. Let us take up the case of a machine having a memory with 64K memory locations, each location capable of storing 16 bits of information. Such a memory requires 64 K ($= 2^{16}$ or 65536) different addresses. These addresses can be designated as 0 to 65535 in the decimal number system and (00000000 00000000) to (11111111 11111111) in the binary number system. Decimal number system is not used in computers and the binary numbering appears too cumbersome and inconvenient to handle. In the hexadecimal form, it is reduced to (0000) to (FFFF). Thus, contents of the memory, when expressed in the hexadecimal form are much less cumbersome to represent.

Q. 16. What is the BCD number system? Illustrate with the help of examples.

Ans. BCD stands for Binary Coded Decimal. In the BCD number system, decimal numbers are represented with each decimal digit replaced by its four bit binary equivalent. Some examples are given below:

683 = 0110 1000 0011

59.54 = 0101 1001 . 0101 0100

Table. 22.5.

Hex Number	Binary Equivalent	Hex Number	Binary Equivalent
0	0 0 0 0	9	1 0 0 1
1	0 0 0 1	A	1 0 1 0
2	0 0 1 0	B	1 0 1 1
3	0 0 1 1	C	1 1 0 0
4	0 1 0 0	D	1 1 0 1
5	0 1 0 1	E	1 1 1 0
6	0 1 1 0	F	1 1 1 1
7	0 1 1 1		
8	1 0 0 0		

Q. 17. Outline the procedure for making Octal-Hexadecimal and Hexadecimal-Octal conversions.

Ans. Octal-Hexadecimal Conversion

The given octal number may first be converted into an equivalent binary number and the binary number is then transformed to its hexadecimal equivalent. The other option is to first convert the octal number into its decimal equivalent and then convert the decimal number into its hexadecimal equivalent. The former approach is more convenient and straightforward.

As an illustration:

$(57.61)_8 = (101\ 111\ .\ 110\ 001)_2$

$= (101111\ .\ 110001)_2$

$= (00101111\ .\ 11000100)_2$

$= (2F\ .\ C4)_{16}$

Therefore, $(57.61)_8 = (2F\ .C4)_{16}$

Hexadecimal-Octal Conversion

The given hexadecimal number is first converted into its binary equivalent and the binary number, thus found is then transformed into the octal number. We can also go from hexadecimal to decimal and then onwards to octal but the former method is more convenient and straight forward.

As an illustration:

$(2F\ .\ C4)16 = (0010\ 1111\ .\ 1100\ 0100)_2$

$= (00101111\ .\ 11000100)_2$

$= (101111\ .\ 110001)_2$

$= (101\ 111\ .\ 110\ 001)2 = (57.61)8$

Therefore, $(2F,\ C4)_{16} = (57.61)_8$

To summarise, the following four axioms must always be remembered to convert a given number in one number system to its equivalent in another:

I. Whenever it is desired to find the DECIMAL EQUIVALENT of a given number in another number system, it is given by the SUM of all the digits multiplied by their respective PLACE VALUES or WEIGHTS. Do not forget to handle the integer and fractional parts separately, Starting from the radix point, the weights of different digits in the binary number system are, 2^0, 2^1, 2^2, 2^3....... for the integer part and 2^{-1}, 2^{-2}, 2^{-3}, ... for the fractional part. Similarly, the weights for the Octal number system are 8^0, 8^1, 8^2 ...... for the integer part and 8^{-1}, 8^{-2}, 8^{-3} for the fractional part and for the hexadecimal number system these are 16^0, 16^1, 16^2 ..... for the integer part and 16^{-1}, 16^{-2}, 16^{-3}...... for the fractional part.

II. To convert a given MLXED DECIMAL NUMBER into an equivalent in another number system, as a first step, HANDLE THE INTEGER PART AND FRACTIONAL PART SEPARATELY, The integer part is progressively divided by (*r*) and the remainders noted till the result of division yields a zero quotient. THE REMAINDERS WRITTEN IN THE REVERSE ORDER CONSTITUTE THE EQUIVALENT IN THE OTHER SYSTEM, (*r*) here is the base or radix of the transformed number system, *i.e.* in order to find the binary equivalent, you would progressively divide by 2. The fractional part is progressively multiplied by (*r*) and the CARRYS recorded till the result of multiplication gives zero or when the desired number of bits have been obtained, THE CARRYS WRITTEN IN THE FORWARD ORDER CONSTITUTE THE EQUIVALENT. For finding the octal equivalent fraction of a decimal fraction, you would progressively multiply by 8.

Table 22.6. First Sixteen Numbers in Different Number Systems and their complements

Binary			*Decimal*			*Hexadecimal*			*Octal*		
No.	*1's Comp.*	*2's Comp.*	*No.*	*9's Comp.*	*10's Comp.*	*No.*	*15's Comp.*	*16,s Comp*	*No.*	*7's Comp*	*8's Comp.*
0000	1111	10000	0	9	10	0	F	10	0	7	10
0001	1110	1111	1	8	9	1	E	F	1	6	7
0010	1101	1110	2	7	8	2	D	E	2	5	6
0011	1100	1101	3	6	7	3	C	D	3	4	5
0100	1011	1100	4	5	6	4	B	C	4	3	4
0101	1010	1011	5	4	5	5	A	B	5	2	3
0110	1001	1010	6	3	4	6	9	A	6	1	2
0111	1000	1001	7	2	3	7	8	9	7	0	1
1000	0111	1000	8	1	2	8	7	8	10	67	70
1001	0110	0111	9	0	1	9	6	7	11	66	67
1010	0101	0110	10	89	90	A	5	6	12	65	66
1011	0100	0101	11	88	89	B	4	5	13	64	65
1100	0011	0100	12	87	88	C	3	4	14	63	64
1101	0010	0011	13	86	87	D	2	3	15	62	63
1110	0001	0010	14	85	86	E	1	2	16	61	62
1111	0000	0001	15	84	85	F	0	1	17	60	61

III. The Octal-Binary conversion and the reverse process and also the Hexadecimal-Binary conversion and the reverse process are straight forward. For Octal-Binary conversion, replace each digit in the octal number by its 3-bit binary equivalent (3-bit equivalent because octal system base is third power of the binary system base). For Hexadecimal Binary conversion, replace each Hex digit by its 4-bit binary equivalent (4-bit equivalent because hex system base is fourth power of the binary system base). For Binary-octal conversion, split the binary number in the groups of 3 bits starting from the binary point and if needed, complete the outside groups by adding 0s and then write the octal equivalent of these 3-bit groups. For Binary-Hexadecimal conversion, split the binary number in the groups of 4-bits starting from the binary point and, if needed, complete the outside groups by adding 0s and then write the Hex equivalents of these 4-bit groups.

Table 22.7

Binary Equivalent	*Decimal Number*
0000	0
0001	1
0010	2
0011	3
0100	4
0101	5
0110	6
0111	7
1000	8
1001	9

IV. For Octal-Hexadecimal conversion, we can go from octal number to its binary equivalent and then from binary equivalent to hex equivalent.

For Hexadecimal Octal-conversion, we can go from hexadecimal number to its binary equivalent and then from binary number to the octal equivalent.

Q. 18. What is floating point notation? What are the advantages of floating point representation? With the help of illustrations, briefly describe floating point representation of decimal and straight binary numbers.

Ans. Floating-point notation can be used conveniently to represent both large as well as small fractional or mixed numbers. This makes the process of arithmetic operations on these numbers relatively much easier. Floating-point representation greatly increases the range of numbers, from the smallest to the largest, that can be represented using a given number of digits. Floating-point numbers are in general expressed in the form $N = m \times b^e$, where m is the fractional part, called the significand or mantissa, (e) is the integer part, called the exponent, and (b) is the base of the number system or numeration. Fractional part (m) is a p-digit number of the form ± (d.dddd…..dd), with each digit (d) being an integer between 0 and (b – 1) inclusive. If the leading digit of (m) is nonzero, then the number is said to be normalised.

Floating point representation of decimal, hexadecimal and binary number systems is written as follows:

Decimal system: $N = m \times 10^e$, Hexadecimal system: $N = \text{m} \times 16^e$, Binary system: $N = \text{m} \times 2e$

For example, decimal numbers 0.0003754 and 3754 will be represented in floating-point notation as 3.754×10^{-4} and 3.754×10^3 respectively. A hex number 257. ABF will be represented as 2.57 ABF $\times 16^2$. In the case of normalised binary numbers, the leading digit, which is the most significant bit, is always '1' and thus, does not need to be stored explicitly. Also, while expressing a given mixed binary number as a floating-point number, the radix point is so shifted as to have the most significant bit immediately to the right of the radix point as a '1'. Both the mantissa and the exponent can have a positive or a negative value. The mixed binary number $(110.1011)_2$ will be represented in floating-point notation as .1101011 $\times 2^3$ = 1101011e + 0011. Here, .1101011 is the mantissa and e+0011 implies that the exponent is +3. As another example, $(0.000111)_2$ will be written as .111e–0011, with .111 being the mantissa and e–0011 implying an exponent of –3. Also, $(-\ 0.00000101)_2$ may be written as $-.101 \times 2^{-5}$ = –.101e–0101, where –.101 is the mantissa and e–0101 indicates an exponent of –5. If we wanted to represent the mantissas using eight bits, then .1101011 and .111 would be represented as .11010110 and .11100000.

Q. 19. What determines the range of numbers and the precision with which numbers can be represented by a given number system?

Ans. The range of numbers that can be represented in any machine depends upon the number of bits in the exponent, while the fractional accuracy or precision is ultimately determined by the number of bits in the mantissa. The higher the number of bits in the exponent, the larger is the range of numbers that can be represented. For example, the range of numbers possible in a floating-point binary number format using six bits to represent the magnitude of the exponent would be from 2^{-64} to 2^{+64}, which is equivalent to a range of 10^{-19} to 10^{+19}. The precision is determined by the number of bits used to represent the mantissa. It is usually represented as decimal digits of precision. The concept of precision as defined with respect to floating-point notation can be explained in simple terms as follows. If the mantissa is stored in (n) number of bits, it can represent a decimal number between 0 and $(2^n - 1)$ as the mantissa is stored as an unsigned integer. If (M) is the largest number such that $10^M - 1$ is less than or equal to $2^n - 1$, then (M) is the precision expressed as decimal digits of precision. For example, if the mantissa is expressed in 20 bits, then decimal digits of precision can be found to be about 6, as $(2^{20}-1)$ equals 1 048 575, which is a little over $(10^6 - 1)$.

Q. 20. Briefly explain why 8421 BCD code cannot be used for addition of decimal number, where the addition of two decimal digits produces a number greater than 9?

Ans. BCD code numbers cannot be used to perform addition of decimal numbers, where the addition involves adding two decimal digits to produce a number that exceeds 9. It is so because in 8421 BCD code, there is no code for a number greater than 9. It can best be explained with the help of an example.

12	0001	0010	38	0011	1100
+ 73	+ 0111	0011	+ 45	+ 0100	0101
+ 85	1000	0101	83	0111	1101

Remember, 0 + 0 = 0, 1 + 0 = 1, 0 + 1 and 1 + 1 = 0 with a carry of 1.

Thus, we have seen that although the decimal addition of 12 and 73 when performed in the BCD notation, produces the correct result, addition of 38 and 45 yields an incorrect answer. It is so, because we cannot decode (01111101); the 4-bit group (1101) in the 8421 BCD code notation does not have a decimal equivalent. In fact, all six 4-bit binary numbers following 1001 *i.e.* 1010, 1011, 1100 1101, 1110 and 1111 are forbidden numbers in 8421 BCD code.

Q. 21. With the help of an example, explain as to how can we convert a given binary number into its BCD equivalent and also, a given BCD number into its binary equivalent.

Ans. Binary-8421 BCD Conversion. The given binary number is first converted to its equivalent decimal number which is then converted to its BCD equivalent, The example below illustrates the procedure:

Binary Number: 110101.101

Equivalent Decimal number : (1 + 4 + 16 + 32). (0.5 + 0.125) = 53.625

Equivalent BCD number = (01010011. 0110 0010 0101)

BCD-Binary Conversion. The given BCD number is first converted into an equivalent decimal number which is straightforward and then the resulting decimal number is transformed into its binary equivalent.

Illustration: As an illustration, let us see as to how do we convert the BCD number (0011 0111. 0111 0101) to its binary equivalent.

BCD-Decimal Conversion

BCD Number	=	0011	0011	.	0111	0101
Equivalent decimal number	=	3	7	.	7	5

Decimal Binary Coversion

Integer Part

2	37	Remainder
	18	1
	9	0
	4	1
	2	0
	1	0
	0	1

$(37)_{10} = (100101)_2$

Fractional Part

0.75 × 2 = 0.5 with a carry of 1

0.5 × 2 = 0 with a carry of 1

$(0.75)_{10} = (.11)_2$

therefore, $(37.75)_{10} = (100101.11)_2$ or in other words

$(0011\ 0111\ .\ 0111\ 0101)_{BCD} = (100101.11)_2$

Q.22. What is Excess-3 Code? With the help of an example, outline the steps for converting given decimal number to its Excess-3 code equivalent. Also, list the excess-3 code equivalents for the first fifteen decimal numbers.

Ans. The Excess-3 code, another important BCD code, is an unweighted code. The Excess-3 code is particularly significant for arithmetic operations. When the sum of two decimal digits exceeds 9, they cannot be added using 8421 BCD code. Excess-3 code has no such limitation and considerably simplifies arithmetic operations. Table 22.8 lists first sixteen entries for the excess-3 code.

Table 22.8

Decimal	*Excess-3*		*Decimal*	*Excess-3*	
0	0011	0011	8	0011	1011
1	0011	0100	9	0011	1100
2	0011	0101	10	0100	0011
3	0011	0110	11	0100	0100
4	0011	0111	12	0100	0101
5	0011	1000	13	0100	0110
6	0011	1001	14	0100	0111
7	0011	1010	15	0100	1000

A given decimal number can be converted into its Excess-3 equivalent as follows :

1. Add 3 to each decimal digit in the given decimal number.
2. Replace each digit of the newly formed decimal number by its 4-bit binary equivalent.

Illustration:

(*i*) Given decimal number = 43

After adding + 3 to each digit,

the new decimal number = 76

The Excess-3 code equivalent of 43 is, therefore, 0111 0110

(*ii*) Given decimal number = 269

After adding + 3 to each digit,

2	6	9
+3	+3	+3
5	9	12

The Excess-3 equivalent of 269 = 0101 1001 1100. Remember not to take any carry forward in case by adding 3 to a digit, you get a number more than 9.

Q. 23. What is GRAY CODE? Outline the procedure for converting a given binary number to its Gray code equivalent and a given Gray code number to its binary equivalent.

Ans. It is another important unweighted BCD code and its application arise from the fact that each number in the grey code differs from it immediately preceding number by one bit only.

Table. 22.9

Decimal No.	*Binary No.*	*Gray Code Equivalent*	*Decimal No.*	*Binary No.*	*Gray Code Equivalent*
0	0000	0000	8	1000	1100
1	0001	0001	9	1001	1101
2	0010	0011	10	1010	1101
3	0011	0010	11	1011	1110
4	0100	0110	12	1100	1010
5	0101	0111	13	1101	1011
6	0110	0101	14	1110	1001
7	0111	0100	15	1111	1000

As a result of this property, the maximum error that can creep into a system using binary Gray code to encode data is much less than the worst case error, if straight binary code was used instead.

Binary-Grey Code Conversion

Following steps should be followed for converting a given binary number to its Gray code equivalent:

1. Begin with the most significant Bit (MSB). The MSB of the Gray code equivalent is the same as the MSB of given binary number.
2. The second most significant bit (adjacent to MSB) in the Gray code number is obtained by adding the MSB and the second MSB of the binary number and ignoring carry, if any.
3. The third most significant bit in the Gray code number is obtained by adding the second MSB and third MSB in the binary number and ignoring carry, if any.

This process continues till we obtain the LSB.

Illustration:

1. Binary Number : 1101
   ↓
   Gray Code Number : 1
2. Binary Number : 1101
   Gray Code Number : 10
3. Binary Number : 1101
   Gray Code Number : 101
4. Binary Number : 1101
   Gray Code Number : 1011

Grey-code Binary Conversion

Following steps should be followed for converting a given Gray code number to its binary equivalent:

1. Begin with the most significant (MSB). The MSB of the binary number is the same as the MSB of the Gray code number.
2. The bit next to the MSB in the binary number is obtained by adding the MSB in the binary number to the second MSB in the Gray code number.
3. Similarly, the third MSB in binary number is obtained by adding second MSB in the binary number to the third MSB of Gray code number.

The process continues till we obtain the LSB. The conversion procedure is best illustrated with the help of the following illustration :

1. Gray Code Number : 1010
   Binary Number : 1
2. Gray Code Number : 1010
   Binary Number : 11
3. Gray Code Number : 1010
   Binary Number : 110
4. Gray Code Number : 1010
   Binary Number : 1100

The Binary equivalent of Gray Code Number 1010 is 1100

Q. 24. What is a PARITY BIT and what are the two types of parity bits ? How does the parity bit help in detecting errors that are likely to creep in the computer words, while they are being operated on or being stored or moved from one place to another inside the computer?

Ans. PARITY BIT is an extra bit added to a computer word in order to detect any error that might have crept into, it while it was being stored or processed and moved from one place to another inside the computer. Any such error if not detected and subsequently corrected can be disastrous. For instance, if we are adding two decimal numbers 12 and 25 in the Excess-3 form, the two BCD numbers to be added will be (0100 0101) and (0101 1000). Their addition yields the Excess-3 equivalent of 37. Let us assume that one of the two numbers (say 25) gets a one bit error, when it was being taken from memory to the arithmetic logic unit and it becomes (1101 1000). The result of addition in this case is Excess-3 equivalent of 17 (the correct answer being 37).

We have an EVEN PARITY, when an extra bit is added to a word so that the total number of 1s becomes an even number (this added bit could be a '0' or a '1') and an ODD PARITY, where an extra bit is added to a word to make number of 1s as odd (this added bit could be '0' or a *T*). For instance, if we have to add an even parity bit to 4-bit number 1101, it will a '1' and if we have to add an odd parity bit to the same number, it would be a '0'. The parity check can be made at different points to look for any possible single bit error. One can appreciate that for a given word, the even parity bit is complement of the odd parity bit. We will also appreciate that addition of a single parity bit (even or odd) cannot be used to detect 2-bit errors. It may, however, be mentioned that in normal operation, chances of 2-bit errors are very remote. Any noise or transient is likely to cause only single bit errors. Nevertheless, there are some data storage media, magnetic tapes for instance, that are highly prone to 2-bit errors.

Q. 25. What do you mean by DOUBLE PARITY ? How does it help detect 2-bit errors in digital data stored on magnetic tape?

Ans. Double parity is often used with magnetic tapes to detect any possible 2-bit errors. On a magnetic tape, digital data is stored in blocks. Figure 22.1 shows a portion of the tape alongwith one block of digital data consisting of 9 rows and 9 columns. First 8-bits in each row indicate a binary number and the 9th bit its parity bit. The entire 9th row contains the parity bits. Thus the block of digital data contains eight 8-bit binary numbers. In the arrangement shown, we have used odd parity check. The solid dots represent the magnetised spots (or say binary 1s) and hollow dots represent unmagnetised spots (or binary 0s). All nine rows and nine columns have an odd parity. Let us see as to what happens, when first two bits in the binary number represented by first row change state. In such a case, the horizontal parity of this number is not changed, whereas the vertical parity is disturbed.

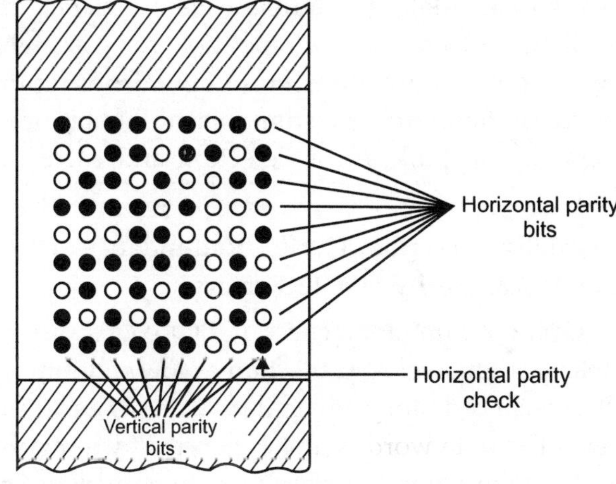

Fig. 22.1

The number was originally. (10110101) with a '0' parity bit. After the errors have crept into this number, it becomes (01110101) with a '0' parity bit. Both the numbers have an odd parity and the errors would have gone unnoticed, had there been no column parity check.

The first column which originally read 100101011 becomes 000101011 thus spoiling odd parity (the number of 1s now is even). The magnetic tape when read electronically, can be checked for horizontal and vertical parity and any possible errors detected.

Q. 26. Briefly describe the features of Repetition Code as error detection and detection code.

Ans. The repetition code makes use of repetitive transmission of each data bit in the bit stream. In the case of threefold repetition, '1' and '0' would be transmitted as '111' and '000' respectively. If, in the received data bit stream, bits are examined in groups of three bits, the occurrence of an error can be detected. In the case of single-bit errors, '1' would be received as 011 or 101 or 110 instead of 111, and a '0' would be received as 100 or 010 or 001 instead of 000. In both cases, the code becomes self-correcting, if the bit in the majority is taken as the correct bit. There are various forms in which the data are sent using the repetition code. Usually, the data bit stream is broken into blocks of bits, and then each block of data is sent some predetermined number of times. For example, if we want to send eight-bit data given by

11011001; it may be broken into two blocks of four bits each. In the case of threefold repetition, the transmitted data bit stream would be 110111011101100110011001. However, such a repetition code, where the bit or block of bits is repeated '3'times are not capable of correcting two-bit errors, although it can detect the occurrence of error. For this, we have to increase the number of times each bit in the bit stream needs to be repeated. For example, by repeating each data bit '5' times, we can detect and correct all two-bit errors. The repetition code is highly inefficient and the information throughput drops rapidly as we increase the number of times each data bit needs to be repeated to build error detection and correction capability.

Q. 27. Briefly describe Cyclic Redundancy Check Code and its capability to detect errors.

Ans. Cyclic redundancy check (CRC) codes provide a reasonably high level of protection at low redundancy level. The cycle code for a given data word is generated as follows. The data word is first appended by a number of 0s equal to the number of check bits to be added. This new data bit sequence is then divided by a special binary word, whose length equals n+1, n being the number of check bits to be added. The remainder obtained as a result of modulo-2 division is then added to the dividend bit sequence to get the cyclic code. The code word so generated is completely divisible by the divisor used in the generation of the code. Thus, when the received code word is again divided by the same divisor, an error-free reception should lead to an all '0' remainder. A non-zero remainder is indicative of the presence of errors.

The probability of error detection depends upon the number of check bits, (n), used to construct the cyclic code. It is 100 % for single-bit and two-bit errors. It is also 100 %, when an odd number of bits are in error and the error bursts have a length less than (n + 1). The probability of detection reduces to $(1 - (1/2)^{n-1}$ for an error burst length equal to $n + 1$, and to $1 - (1/2)^n$ for an error burst length greater than (n + 1).

Q. 28. What is Hamming Code? Briefly describe algorithm for writing generalized Hamming code.

Ans. We have seen, in the case of the error detection and correction codes described above, how an increase in the number of redundant bits added to message bits can enhance the capability of the code to detect and correct errors. If we have a sufficient number of redundant bits, and if these bits can be arranged such that different error bits produce different error results, then it should be possible not only to detect the error bit but also to identify its location. In fact, the addition of redundant bits alters the 'distance' code parameter, which has come to be known as the Hamming distance. The Hamming distance is nothing but the number of bit disagreements between two code words. For example, the addition of single-bit parity results in a code with a Hamming distance of at least 2. The smallest Hamming distance in the case of a threefold repetition code would be 3. Hamming noticed that an increase in distance enhanced the code's ability to detect and correct errors. Hamming's code was, therefore, an attempt at increasing the Hamming distance and at the same time having as high an information throughput rate as possible.

The algorithm for writing the generalised Hamming code is as follows:

1. The generalised form of code is $P_1\, P_2\, D_1\, P_3\, D_2\, D_3\, D_4\, P_4\, D_5\, D_6\, D_7\, D_8\, D_9\, D_{10}\, D_{11} P_5$........, where P and D respectively represent parity and data bits.
2. We can see from the generalised form of the code that all bit positions that are powers of 2 (positions 1, 2, 4, 8, 16, ...) are used as parity bits.
3. All other bit positions (positions 3, 5, 6, 7, 9, 10, 11, ...) are used to encode data.
4. Each parity bit is allotted a group of bits from the data bits in the code word, and the value of the parity bit (0 or 1) is used to give it certain parity.
5. Groups are formed by first checking (N – 1) bits and then alternately skipping and checking (N) bits following the parity bit. Here, (N) is the position of the parity bit; 1 for P_1, 2 for P_2, 4 for P_3, 8 for P_4 and so on. For example, for the generalized form of code given above, various groups of bits formed with different parity bits would be $P_1\, D_1\, D_2\, D_4\, D_5$........, $P_2\, D_1\, D_3\, D_4\, D_6\, D_7$........, $P_3\, D_2\, D_3\, D_4\, D_8\, D_9$........., $P_4\, D_5\, D_6\, D_7\, D_8\, D_9\, D_{10}\, D_{11}$........and so on. To illustrate the formation of groups further, let us examine the group corresponding to parity bit P_3. Now, the position of P_3 is at number 4. In order to form the group, we check the first three bits (N – 1 = 3) and then follow it up by alternately skipping and checking four bits (N = 4).

The Hamming code is capable of correcting single-bit errors on messages of any length. Although the Hamming code can detect two-bit errors, it cannot give the error locations. The number of parity bits required to be transmitted along with the message, however, depends upon the message length, as shown above. The number of parity bits n required to encode m message bits is the smallest integer that satisfies the condition $(2^n - n) > m$.

Table 22.14: Generation of Hamming Code

	P_1	P_2	D_1	P_3	D_2	D_3	D_4
Date Bits (without Parity)			0		1	1	0
Data Bits with Parity Bit P_1	1		0		1		0
Data Bits with Parity Bit P_2		1	0			1	0
Data Bits with Parity Bit P_3				0	1	1	0
Data Bits with Parity	1	1	0	0	1	1	0

Q. 29. Briefly describe the Hamming code with code word length of seven bits with four message bits and three parity bits. Illustrate with the help of a suitable example, how Hamming code helps in detecting and correcting single bit errors?

Ans. The most commonly used Hamming code is the one that has a code word length of seven bits with four message bits and three parity bits. It is also referred to as the Hamming (7, 4) code. The code word sequence for this code is written as $P_1 P_2 D_1 P_3 D_2 D_3 D_4$, with P_1, P_2 and P_3 being the parity bits and D_1, D_2, D_3 and D_4 are the data bits. We will illustrate step-by-step the process of writing the Hamming code for a certain group of message bits and then the process of detection and identification of error bits with the help of an example. We will write the Hamming code for the four-bit message 0110 representing numeral '6'. The process of writing the code is illustrated in Table 22.14, with even parity. Thus, the Hamming code for 0110 is 1100110. Let us assume that the data bit D1gets corrupted in the transmission channel. The received code in that case is 1110110. In order to detect the error, the parity is checked for the three parity relations mentioned above. During the parity check operation at the receiving end, three additional bits X, Y and Z are generated by checking the parity status of $P_1 D_1 D_2 D_4$, $P_2 D_1 D_3 D_4$ and $P_3 D_2 D_3 D_4$ respectively. These bits are a '0' if the parity status is okay and a '1, if it is disturbed. In that case, Z Y X gives the position of the bit that needs correction. The process can be best explained with the help of an example. Examination of the first parity relation gives $X = 1$ as the even parity is disturbed. The second parity relation yields $Y = 1$ as the even parity is disturbed here too. Examination of the third relation gives $Z = 0$, as the even parity is maintained. Thus, the bit that is in error is positioned at 011 which is the binary equivalent of '3'. This implies that the third bit from the MSB needs to be corrected. After correcting the third bit, the received message becomes 1100110, which is the correct code.

Q. 30. What are Alphanumeric codes ? Briefly describe ASCII and EBCDIC alphanumeric codes.

Ans. The binary codes that can be used to represent all the letters of alphabet, numbers, mathematical symbols and punctuation marks are known as the ALPHANUMERIC CODES. Alphanumeric codes enable us interface the input-output devices like keyboard, puched cards, printers, video display units, etc. with the computer. One such alphanumeric code that is extensively used in microcomputers for interfacing microcomputers with keyboards, printers, video display units is the ASCII (American Standard Code for Information Interchange) code pronounced as "ask-ee". It has become a world standard alphanumeric Code for microcomputers. It is a 7-bit code that can be used to represent 26 upper case letters (A to Z), 26 lower case letters (a to z), 10 numerals (0 to 9), 33 special characters and symbols and 33 control characters. It comes to a total of 128 characters and that is what a 7-bit code at the most can represent ($2^7 = 128$). An 8-bit version of ASCII code known as USASC-II-8 or ASCII-8 has also been developed. The 8-bit version can represent a maximum of 256 characters. Table 22.11 lists the ASCII code for all the 128 characters. A '0' has been added in the MSB position to make it a BYTE, as a Byte is conveniently handled in digital systems. Also the 8-bit representation has been split into two 4-bit numbers so as to facilitate an easy remembrance of the code. For instance, all the numerals from 0 to 9 have the first four bit number as 1011, while the second number varies from 0000 (binary equivalent of 0) to 1001. The table contains the binary as well as the hex equivalent of various characters in the ASCII code. It is very simple to work with Table 22.11. For instance, if you want to find out the ASCII code for the character '&', the four most significant bits for this are 0010 (Hex equivalent = 2) and four least significant bits are 0110 (Hex equivalent = 6), which gives '&' in the ASCII code as 00100110 or 26 (in Hex).

Table 22.11

			Least significant digit (or nibble)														
	Binary		0000	0001	0010	0011	0100	0101	0110	0111	1000	1001	1010	1011	1100	1101	1110	1111
		Hex	0	1	2	3	4	5	6	7	8	9	A	B	C	D	E	F
Most Significant digit (or nibble)	0000	0	NUL	SOH	STX	ETX	EOT	ENQ	ACK	BEL	BS	HT	LF	VT	FF	CR	SO	SI
	0001	1	DLE	DC1	DC2	DC3	DC4	NAK	SYN	ETB	CAN	EM	SUB	ESC	FS	GS	RS	US
	0010	2	SP	!	"	#	$	%	&	'	(	)	*	+	,	–	.	–
	0011	3	0	1	2	3	4	5	6	7	8	9	:	;	<	=	>	?
	0100	4	@	A	B	C	D	E	F	G	H	I	J	K	L	M	N	O
	0101	5	P	Q	R	S	T	U	V	W	X	Y	Z	[	/	]	^	–
	0110	6	,	a	b	c	d	e	f	g	h	i	j	k	l	m	n	o
	0111	7	p	q	r	s	t	u	v	w	x	y	z	[	1	]	–	DEL

Quite often a parity bit is added to the 7-bit ASCII code, when ASCII coded digital data is transmitted over telephone lines or other such media, where there is a likelihood of 1-bit errors.

Table 22.12

			Least significant digit (or nibble)														
	Binary		0000	0001	0010	0011	0100	0101	0110	0111	1000	1001	1010	1011	1100	1101	1110	1111
		Hex	0	1	2	3	4	5	6	7	8	9	A	B	C	D	E	F
Most Significant digit (or nibble)	0000	0	NUL	SOH	STX	ETX	PF	HT	LC	DEL			SMM	VT	FF	CR	SO	SI
	0001	1	DLE	DC1	DC2	TM	RES	NL	BS	IL	CAN	EM	CC	CU1	1FS	IGS	IRS	IUS
	0010	2	DS	DOS	FS		BYP	LF	ETB	ESC			SM	CU2		ENQ	ACK	BEL
	0011	3			SYN		PN	RS	UC	EOT				CU3	DC4	NAK		SUB
	0100	4	SP										[	;	<	(	+	1!
	0101	5	&										!]	$	*	)	;	A
	0110	6	–	/									1	,	%	.	>	?
	0111	7										'	;	#	@	'	=	"
	1000	8		*a*	*b*	*c*	*d*	*e*	*f*	*g*	*h*	*i*						
	1001	9		*j*	*k*	*l*	*m*	*n*	*o*	*p*	*q*	*r*						
	1010	A		-	*s*	*t*	*u*	*v*	*w*	*x*	*y*	*z*						
	1011	B																
	1100	C	{	*A*	*B*	*C*	*D*	*E*	*F*	*G*	*H*	*I*						
	1101	D	}	*J*	*K*	*L*	*M*	*N*	*O*	*P*	*Q*	*R*						
	1110	E	/		*S*	*T*	*U*	*V*	*W*	*X*	*Y*	*Z*						
	1111	F	0	1	2	3	4	5	6	7	8	9						

EBCDIC (Extended Binary Coded Decimal Interchange Code) pronounced as "eb-si-dik" is another widely used alphanumeric code mainly popular with larger computer systems. It is an 8-bit code and thus can accommodate upto 256 characters. Table 22.12 gives the listing of the characters in binary as well as Hex form in EBCDIC. The arrangement is similar to one adopted in Table 22.11 for the ASCII code.

Q. 31. Briefly describe Seven Segment Display Code.

Ans. SEVEN SEGMENT DISPLAY CODE

SEVEN SEGMENT DISPLAYS are now-a-days are very common and are found everywhere from pocket calculators, Digital clocks, Laboratory equipment to Petrol pumps. Fig. 22.2 (a) shows one such display with its seven segments designated as *a, b, c, d, e, f* and *g*. Each such display usually a DOT POINT (DP). An LED type seven segment display (we have the LCD type also) is nothing but a combination of LEDs arranged as shown in Fig. 22.2 (*b*), known as COMMON CATHODE TYPE or Fig. 22.2 (*c*) known as COMMON ANODE TYPE. This display can be used to generate numerals 0 to 9 and letters to A to F. A '1' lights a segment in common cathode type display and a '0' lights the segment in case of common anode type display. The binary code for the common anode type and common cathode type displays is given in Table 22.13.

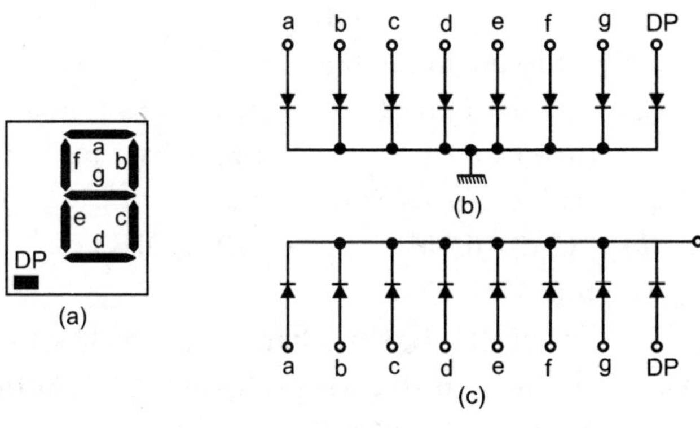

Fig. 22.2

Table 22.13

	Common Cathode Type									*Common Anode Type*							
	'1' means ON									*'0' means ON*							
	a	*b*	*c*	*d*	*e*	*f*	*g*	DP		*a*	*b*	*c*	*d*	*e*	*f*	*g*	DP
0	1	1	1	1	1	1	0			0	0	0	0	0	0	1	
1	0	1	1	0	0	0	0			1	0	0	1	1	1	1	
2	1	1	0	1	1	0	1			0	0	1	0	0	1	0	
3	1	1	1	1	0	0	1			0	0	0	0	1	1	0	
4	0	1	1	0	0	1	1			1	0	0	1	1	0	0	
5	1	0	1	1	0	1	1			0	1	0	0	1	0	0	
6	1	0	1	1	1	1	1			0	1	0	0	0	0	0	
7	1	1	1	0	0	0	0			0	0	0	1	1	1	1	
8	1	1	1	1	1	1	1			0	0	0	0	0	0	0	
9	1	1	1	0	0	1	1			0	0	0	1	1	0	0	
A	1	1	1	1	1	0	1			0	0	0	0	0	1	0	
B	0	0	1	1	1	1	1			1	1	0	0	0	0	0	
C	0	0	0	1	1	0	1			1	1	1	0	0	1	0	
D	0	1	1	1	1	0	1			1	0	0	0	0	1	0	
E	1	1	0	1	1	1	1			0	0	1	0	0	0	0	
F	1	0	0	0	1	1	1			0	1	1	1	0	0	0	

Q. 32. Outline the basic rules that can be used for carrying out ADDITION and SUBTRACTION of binary numbers with larger number of bits.

OR

Justify the following:

(a) **(i) 0 + 0 = 0** **(ii) 0 + 1 = 1**
(iii) 1 + 0 = 1 **(iv) 1 + 1 = 10**
(v) 1 + 1 + 1 = 11

(b) **(i) 0 – 0 = 0** **(ii) 1 – 0 = 1**
(iii) 1 – 1 = 0
(iv) 0 – 1 = 1, with a borrow of 1 or 10 – 1 = 1

Ans. Following are the basic rules of binary addition:

(i) 0 + 0 = 0 (ii) 0 + 1 = 1 (iii) 1 + 0 = 1
(iv) 1 + 1 = 0 with a carry of 1 (or 10)
(v) 1 + 1 + 1 = 1 with a carry of 1 (or 11).

These equations have been written in the same way like we do in case of well known decimal number system. The fact that adding '0' to a certain digit produces the same digit as the answer (*e.g.* 7 + 0 = 7, 0 + 6 = 6...) explains the first three equations. When we add '1' to a number or digit in the decimal number system, we get the next higher digit or number as the answer. For example, 7 + 1 in decimal equals '8' because '8' immediately follows '7' in decimal number system or 7 + 1 in octal equals '10' as in the octal number system, the next adjacent higher number after 7 is 10. This explains equation-iv. Also, 1 + 1 + 1 = 11 as 1 + 1 + 1 = 10 + 1 = 11.

Following are the basic rules of binary subtraction :

(i) 0 – 0 = 0 (ii) 1 – 0 = 1
(iii) 1 – 1 = 0
(iv) 0 – 1 = 1 with a borrow of 1 (or 10 – 1 = 1)

The above equations are justified, if we recall the rules for subtracting decimal numbers. Subtracting '0' from any digit or number produces the same digit or number as the result (e.g. 7 – 0 = 7, 9 – 0 = 9). Subtracting 1 from any digit or number in decimal produces the immediately preceding digit or number in decimal as the answer. 7 – 1 in decimal gives 6 as 6 immediately precedes 7 in decimal number system. Similarly, 1 – 1 in binary equals 0 and 10 – 1 in binary equals

Q.33. With the help of an illustration, outline the procedure for performing addition of larger bit straight binary numbers.

Ans. The addition of larger binary integers, fractions or mixed binary numbers is performed column-wise in just the same way as we do in case of decimal numbers although in this case, we have to remember the basic laws governing the addition of two or three binary digits.

Consider two generalised 4-bit binary numbers (A3 A2 Al A0) and (B3 B2 Bl B0) with A3, B3 representing the MSBs and A0, B0 the LSBs of the two numbers. The addition of these two numbers is performed as follows :

Starting from the LSB position, in the first step, we add the bits A0, B0 (LSBs of the two numbers) and record the sum S0 below these bits in the same column and take the carry C0 forward to the next column of bits. For instance, if A0 = 1 and B0 = 1, then S0 = 0 and C0 = 1. Next we add the bits Al, Bl and the carry C0 from the previous addition. Same procedure is followed till we reach the MSB. The four steps are shown below :

```
(i)              (C0)                (ii)          (C1)  (C0)
      A3   A2    A1    A0                A3   A2    A1    A0
    + B3   B2    B1    B0              + B3   B2    B1    B0
    ----------------------             ----------------------
                       S0                           S1    S0
    ----------------------             ----------------------

      (C2) (C1)  (C0)                         (C2)  (C1)  (C0)
(iii) A3   A2    A1    A0    (iv)        A3   A2    A1    A0
    + B3   B2    B1    B0              + B3   B2    B1    B0
    ----------------------             ----------------------
           S2    S1    S0        (C3)  S3     S2    S1    S0
```

C0, C1, C2 and C3 are CARRYS, if any, produced as a result of adding first, second, third and fourth column additions respectively starting from LSB and proceeding towards MSB. A similar procedure is followed when the given numbers have both the integer as well as fractional parts.

Q.34. Outline the steps for carrying out subtraction of straight binary numbers with larger number of bits.

Ans. Subtraction is also done column wise in the same way like we do in case of decimal number system. Let us see how do we subtract $(1011)_2$ from $(1100)_2$.

In the first step, we subtract the LSBs. As the SUBTRAHEND in the given example is larger than MINUEND; we borrow from the next adjacent higher bit position having a 1. In this case, 1 is borrowed from the second MSB position leaving a 0 in that position. The borrow is first brought to the third MSB position to make it (10). Out of (10) in this position, 1 is taken to the LSB position to make (10) there leaving a 1 in the third MSB position. (10-1) in the LSB column gives 1, (1-1) in the third MSB column gives 0, (0-0) in the second MSB column gives 0 and (1-1) in the MSB also gives 0 to complete subtraction. Subtraction of mixed numbers is also done in *i.e.*, same manner. All these steps are depicted below :

```
(i)    1   1   0   0     (ii)    1   1   0   0
     – 1   0   1   1           – 1   0   1   1
     ---------------           ---------------
                   1                     0   1
     ---------------           ---------------
```

(*iii*)	1	1	0	0	(*iv*)	1	1	0	0
	– 1	0	1	1		– 1	0	1	1
		0	0	1		0	0	0	1

Q. 35. Outline the steps for carrying out EXCESS-3 ADDITION of two given 8421-BCD numbers.

Ans. Excess-3 addition of two given 8421-BCD numbers is performed as per the following guidelines:

1. The given 8421-BCD numbers are written in the Excess-3 form by adding '0011' to each 4-bit group.
2. Carry out the addition following the basic laws of binary addition.
3. Subtract '0011' from the 4-bit group that does not produce a carry.
4. Add '0011' to the 4-bit group that produces a carry. The result thus obtained is in the Excess 3 form.

Q. 36. With the help of examples, explain as to how can we use the method of 2's COMPLEMENT to perform the addition of two decimal numbers for the following four cases: (*i*) when the given decimal numbers are positive (*ii*) when the smaller of the two numbers is negative (*iii*) when larger of the two decimal numbers is negative (*iv*) when both the numbers are negative.

Ans. 2's complement is the most commonly used code for processing positive and negative binary numbers. It forms the basis of arithmetic circuits is modern computers. The following examples testify that when the two decimal numbers to be added are expressed in 2's complement form, addition of these numbers following the basic laws of binary addition gives correct results:

(*i*) Consider the decimal numbers + 26 and + 39

2's complement representation of + 26 is 00011010

2's complement representation of + 39 is 00100111

Addition of two numbers gives :

	0	0	0	1	1	0	1	0	....	+26
+	0	0	1	0	0	1	1	1	....	+39
	0	1	0	0	0	0	0	1	....	+65

(*ii*) Consider the two decimal numbers + 39 and –26

2's complement representation of + 39 is 00100 111

2's complement representation of – 26 is 11100 110

Addition of two numbers gives :

	0	0	1	0	0	1	1	1	....	+39
+	1	1	1	0	0	1	1	0	....	–26
	0	0	0	0	1	1	0	1	....	+13

The final carry in the 9th bit position is disregarded in 8-bit arithmetic.

(*iii*) Consider the two decimal numbers + 26 and – 39

2's complement representation of + 26 is 00011010

2's complement representation of – 39 is 11011 001

The addition of two numbers gives :

	0	0	0	1	1	0	1	0	....	+26
+	1	1	0	1	1	0	0	1	....	–39
	1	1	1	1	0	0	1	1	....	–13

It may be mentioned here that the result of binary addition is also in 2's complement form. It is immaterial, when the result of addition is a positive number as 2's complement representation of a positive decimal number is the same as its straight binary representation. If the result is a negative number, we must first take the 2's complement of the magnitude part of the result before we find its decimal equivalent. In the above example, 11110011 would represent a negative number as the sign bit is 'T' and its magnitude would be the decimal equivalent of 2's complement of 1110011. 2's complement of 1110011 is 0001101 and its decimal equivalent is 13. Thus, 11110011 represents – 13 in 2's complement form.

(*iv*) Consider the two decimal number – 26 and – 39

– 26 is expressed in 2's complement form as 11100110

– 39 is expressed in 2's complement form as 11011001

Their addition gives :

	1	1	1	0	0	1	1	0	....	–26
+	1	1	0	1	1	0	0	1	....	36
	1	0	1	1	1	1	1	1	....	–65

Final carry in the 9th bit position is disregarded. A '1' in the sign bit position indicates a negative number. The magnitude of this number is given by 2's complement of 0111111. 2's complement of 0111111 is 1000001 and its decimal equivalent is 65.

Q. 37. Prove that 8-bit 2's complement arithmetic cannot be used to perform addition of two decimal numbers, when the expected result lies outside the range of – 128 to + 127.

Ans. Let us consider the addition of + 80 and + 75 using 8-bit 2's complement arithmetic.

+ 80 in the 8-bit 2's complement form is expressed as 01010000

+ 75 in the 8-bit 2's complement form is expressed as 01001011

The addition of the two numbers gives :

```
  0 1 0 1 0 0 0 0
  0 1 0 0 1 0 1 1
  ---------------
  1 0 0 1 1 0 1 1
```

A '1' in the sign bit position indicates a negative number which is not correct. Addition + 80 and + 75 yields + 155. Let us see as to what happens, when we add – 56 and – 84 again using 8-bit 2's complement arithmetic.

– 56 is expressed as 11001000 and – 84 is expressed as 10101100 and their addition yields :

```
  1 1 0 0 1 0 0 0
  1 0 1 0 1 1 0 0
  ---------------
  0 1 1 1 0 1 0 0
```

With the final carry disregarded

'0' in the sign bit position indicates a positive number which is also not correct.

Thus, we have considered two cases, where the expected decimal addition result lies outside the range – 128 to + 127 and we have seen that 8-bit 2's complement arithmetic cannot be used to perform these additions. It is because 8-bit 2's complement notation cannot represent decimal numbers outside the range of – 128 to + 127. In general, n-bit 2's complement notation has a range of (-2^{n-1}) to $(2^{n-1}-1)$.

Q. 38. Outline the steps for carrying out subtraction of binary numbers for the following cases: (*i*) both Minuend and Subtrahend are positive (*ii*) Subtrahend is negative and smaller in magnitude (*iii*) Subtrahend is negative and greater in magnitude (*iv*) both minuend and subtrahend are negative.

Ans. The subtraction is similar to the addition process. Adding 2's complement of the subtrahend to the minuend does the job. The steps to be followed will be still more clear after going through the examples to follow.

Case (*i*) Let us subtract + 16 from + 25

+ 25 (the minuend) is expressed as 00011001

+ 16 (the subtrahend) is expressed as 00010000

2's complement of the subtrahend (*i.e.* + 16) is 11110000

Therefore, (+ 25) – (+ 16) is given by:

```
    0 0 0 1 1 0 0 1
  + 1 1 1 1 0 0 0 0
  -----------------
    0 0 0 0 1 0 0 1
```

with the final carry disregarded

The decimal equivalent of 00001001 is + 9 which is the correct answer.

Case (*ii*) Let us subtract – 16 from + 25

+ 25 (the minuend) is expressed as 00011001

– 16 (the subtrahend) is expressed as 11110000

2's complement of subtrahend (*i.e.,* – 16) is 00010000

Therefore, (+ 25) – (– 16) is given by:

```
    0 0 0 1 1 0 0 1
  + 0 0 0 1 0 0 0 0
  -----------------
    0 0 1 0 1 0 0 1
```

The decimal equivalent of 00101001 is + 41 which is the correct answer.

Case (*iii*) Let us subtract – 25 from + 16

+ 16 (the minuend) is expressed as 00010000

– 25 (the subtrahend) is expressed as 11100111

2's complement of subtrahend (*i.e.,* – 25) is 00011001

Therefore, (+ 16) – (– 25) is given by:

```
    0 0 0 1 0 0 0 0
  + 0 0 0 1 1 0 0 1
  -----------------
    0 0 1 0 1 0 0 1
```

The decimal equivalent of 00101001 is + 41 which is the correct answer.

Case (*iv*) Let us subtract – 16 from – 25

– 25 (the minuend) is expressed as 11100111

– 16 (the subtrahend) is expressed as 11110000

2's complement of subtrahend (*i.e.* – 16) is 00010000

Therefore, (– 25) – (– 16) is given by:

```
    1 1 1 0 0 1 1 1
  + 0 0 0 1 0 0 0 0
  -----------------
    1 1 1 1 0 1 1 1
```

A '1' in the sign bit position indicates a negative number. The magnitude of this number is the decimal equivalent of 2's complement of 1110111. 2's complement of 11110111 is 0001001 and its decimal equivalent is 9. Thus, 11110111 in 2's complement form represents – 9, which is the correct answer.

Q. 39. "8-bit 2's complement arithmetic cannot be used to perform either addition or subtraction, if the result of such an operation lies outside the range of – 128 to

+ 127." What do you do to overcome this OVERFLOW problem?

Ans. It is not a hardware problem. On the contrary, it is a software problem. The person who is writing the program to carry out certain addition or subtraction operation (Subtraction can be considered alongwith addition, as it is basically addition of 2's complement of subtrahend to the minuend) should include in his program some instructions to check, if there is any change in sign bit. In case, it is found out that the sign bit has changed, indicating the occurrence of an overflow, the program must tell the computer to switch over to 16-bit arithmetic.

Q. 40. With the help of an illustration, briefly describe the process of REPEATED LEFT SHIFT AND ADD to multiply two given binary numbers. How do you determine the sign of the product?

Ans. In the REPEATED LEFT SHIFT AND ADD method, the end product is the sum of several partial products (the number of partial products is equal to number of bits in the multiplier binary number) with the second partial product shifted 1-bit position to the left relative to the first partial product and the third partial product shifted 1-bit position to the left relative to the second partial product and so on. We begin with the LSB of the multiplier to obtain the first partial product. If the LSB is a '1', a copy of the multiplicand forms the first partial product and it is an all 0 sequence, if the LSB is a '0'. We proceed towards the MSB of the multiplier and obtain various partial products. We stop, when all multiplier bits have been exhausted. Addition of all partial products gives the final answer. If the multiplicand and multiplier have different signs, the end result has a negative sign otherwise it is positive. Following example further illustrates the concept:

$(26)_{10}$	....	11010	Multiplicand
$(3)_{10}$	....	×011	Multiplier
		11010	First Partial Product
		11010	Second Partial Product
		00000	Third Partial Product
		1001110	....$(78)_{10}$

Microprocessors and microcomputers, however, follow what they call it REPEATED ADD AND RIGHT SHIFT algorithm to do binary multiplication, as it is comparatively much more convenient to implement than the Repeated Left Shift and Add algorithm.

Q. 41. With the help of an example, briefly describe the use of Repeated Add and Right Shift algorithm (or approach) to perform binary multiplication.

Ans. Consider the binary multiplication $(10110)_2 \times (1101)_2$

Multiplicand	10110
Multiplier	1101

The sequence of operations needed to do the above multiplication is as follows :

10110	Multiplicand
×1101	Multiplier
00000 +10110	START
10110	Result of first addition
01011 +00000	0 (Result of addition shifted 1-bit to the right)
01011	Result of second addition
00101 +10110	10 (Result of addition shifted 1-bit to the right)
11011	Result of third addition
01101 +10110	110 (Result of addition shifted 1-bit position to the right)
100011	Result of fourth addition
10001	1110

Thus, $(10110)_2 \times (1101)_2 = (100011110)_2$

At the outset, we write an all '0s' sequence with number of 0s equal to the number of bits in the multiplicand. We begin with the LSB of the multiplier, if it is a '1', a copy of the multiplicand is added to the all 0s sequence. If the LSB is a '0', an all 0s sequence is added to the already written all 0s sequence. The result of addition is shifted 1-bit position to the right in the second step and the next higher bit in the multiplier is examined. If it is a '0', an all 0s sequence and if it is a '1', a copy of the multiplicand is added to the 1-bit shifted result of the previous addition. The process is repeated in the third step and the other subsequent steps till we have exhausted all the bits in the multiplier. The sign of the multiplication result depends upon the signs of the multiplicand and the multiplier. The result is positive, when the multiplicand and the multiplier have same signs. It is negative otherwise.

We can verify the correctness of the result by doing equivalent decimal multiplication.

$(10110)_2 \times (1101)_2 = (100011110)_2$

Now $(10110)_2 = (22)_{10}$ and $(1101)_2 = (13)_{10}$

$22 \times 13 = 286 = (100011110)_2.$

Q. 42. With the help of an illustration, describe in brief the method of REPEATED SHIFT AND SUBTRACT to divide two given binary numbers. How do you determine the sign of the quotient?

Ans. Let us consider the division $(10011010)_2$ by $(1100)_2$

Divident: 10011010

Divisor: 1100

The sequence of operations needed to carry out the above division is tabulated ahead.

At the outset, we begin with as many number of bits (starting from MSB end) in the dividend as the number of bits in the divisor and check whether the divisor is smaller or greater than the selected group of bits in the dividend. If it happens to be greater, we record '0' in the quotient. If it is smaller, we subtract the divisor bits from the dividend bits and record a '1' in the quotient. If it is greater and we have already recorded a '0' in the quotient, then as a second step, we include the next adjacent bit in the dividend bits, shift the divisor to the right by 1-bit position and again make a similar check like the one made in the first step. If it is smaller and we have done the subtraction, then in the second step, we append the next MSB of the dividend to the remainder, shift the divisor to the right by 1-bit position and again make a similar check. The options are again the same. The process continues till we have exhausted all the bits in the dividend.

First Step	Quotient 0	1 0 0 1 1 0 1 0 1 1 0 0	Dividend Divisor
Second Step	1	1 0 0 1 1 – 1 1 0 0	First 5 MSBs of dividend Divisor right shifted
		0 1 1 1	First subtraction Remainder
Third Step	1	0 1 1 1 0 – 1 1 0 0	Next MSB added from Dividend, Divisor right shifted
		0 0 0 1 0	Second subtraction Remainder
Fourth Step	0	0 0 0 1 0 1 – 1 1 0 0	Next MSB added Divisor right shifted
Fifth Step	0	0 0 0 1 0 1 0 – 1 1 0 0	Next MSB added Divisor right shifted Divisor Still greater

Quotient : 01100 Remainder : 1010

In the example shown, 10011010 is to be divided by 1100. In the first step, we compare 1100 with first four MSBs (as the divisor has four bits) of the dividend i.e. 1001. Since 1100 is greater than 1001, we record a '0' in the quotient column. That is all about the first step. In the second step, we append the next MSB to the dividend bits to get 10011, shift the divisor by 1-bit position to the right and again make a check. Now the divisor is smaller than the dividend and as a consequence, we have recorded a '1' in the quotient column and also, at the same time, we have subtracted the divisor from the dividend to get 0111 as remainder. In the third step, we have appended to it the next MSB from the dividend bits to get 01110, shifted the divisor 1-bit position to the right and made a cheek. Again the divisor is smaller and thus, we have recorded a '1' in the quotient column. The subtraction yields 00010 as remainder. In the fourth step, the next MSB from dividend, when appended to the remainder makes it 000101. The divisor is shifted as usual and a check made. The divisor this time happens to be greater; so a '0' is recorded for the quotient. In the fifth step, the next MSB, which is also the LSB in the present example, is appended to make the remainder as 0001010. The divisor is shifted and the two compared. The divisor is still larger. Therefore, another '0' is recorded in the quotient column to get the final quotient as 01100 while 1010 is the final remainder. The remainder can, however, be further processed by successively adding 0 and subtracting the shifted divisor to get the fractional part of the quotient. The sign of the quotient is determined by the signs of the dividend and the divisor. When the dividend and the divisor have the same sign, the quotient is positive; otherwise it is negative.

Q. 43. With the help of an illustration, briefly describe the REPEATED SUBTRACT AND LEFT SHIFT ALGORITHM to divide two given binary numbers.

Ans. Let us consider the division (10011010) by (1100)

Dividend: (10011010)

Divisor: (1100)

The steps needed to perform the division are outlined as follows: We begin with the first four MSBs of the dividend (four because the divisor is 4-bit long). We subtract divisor from the dividend, if the subtraction requires a borrow in the MSB position, enter a '0' in the quotient column. If it doesn't, enter a '1'. In the example under consideration, there exists a borrow in the MSB position and so, there is a '0' in the quotient column. In case there is a borrow, divisor is added to the result of subtraction to restore the shifted group of bits (while doing so, final carry, if any is ignored). Next MSB is appended to the result of first subtraction (in case there wasn't a borrow) or the result of subtraction restored by adding the divisor (in case there was a borrow).

Appending the next MSB amounts to shifting the remaining bits of the dividend 1-bit position to left. It is again compared with the divisor and the process repeated. It goes on till we have exhausted all the bits of dividend. The final remainder can be further processed by successively appending 0s and trying subtraction to get fractional part bits of the quotient.

Quotient	1001 −1100	1010
0	1101 +1100	Borrow exists
	1001 10011 −1100	Final carry ignored Nex MSB appended
1	0111 01110 −1100	No borrow Next MSB appended
1	00010	No borrow Next MSB appended
0	1001 +1100	Borrow exists
	0101 01010 −1100	Final carry ignored Next MSB appended
0	1110 +1100	Borrow exists
	1010	Restored
	Quotient = 01100	Remainder = 1010

Q. 44. Outline the steps to be followed to perform addition and subtraction of floating point numbers.

Ans. The undermentioned steps should be followed, while doing either an addition or a subtraction operation of two given floating point numbers:

(*i*) In both addition as well as subtraction operations, the first step is to scale the operands so that they have the same exponents. For instance, if you wanted to add the two floating point decimal numbers 0.5321×10^3 and 0.3068×10^2, then the second number must be scaled and rewritten as 0.03068×10^3 before the addition operation is carried out. As the numbers are assumed to be initially normalised, scaling implies that the smaller of the two numbers be shifted to the right until its exponent equals to that of larger number.

(*ii*) Addition is then done by adding the two mantissas and subtraction is carried out by subtracting the subtrahend mantissa from the minuend mantissa. The exponents are left untouched

(*iii*) When two normalised numbers are added, there might be an overflow digit in the result. It must be corrected and the mantissa restored to a fractional number by shifting it one place to the right and incrementing the exponent value by 1. For instance, addition of 0.65×10^3 and 0.47×10^3 yields 1.12×10^3, which can be expressed as 0.112×10^4.

(*iv*) Similarly, when the two normalised numbers are subtracted, there might be an underflow. An underflow means occurrence of one or more than one 0s in MSB position. The mantissa in such a case must be shifted by as many placed to the left, as there are 0s and the exponent decremented by a number equal to number of places shifted. For instance, subtraction of 0.58×10^4 from 0.63×10^4 yields 0.05×10^4, which is rewritten as 0.5×10^3.

Q. 45. How do you carry out MULTIPLICATION and DIVISION with the floating point numbers ? Briefly describe the procedure to be followed.

Ans. Multiplication of two given floating point numbers is carried out by multiplying their mantissas and adding the exponent values. For intance, 0.6×10^3 multiplied by 0.5×10^4 gives 0.3×10^7 ($0.3 = 0.6 \times 0.5$ and $7 = (4 + 3)$). Similarly, (0.110×2^3) multiplied by 0.10×2^2 gives 0.01100×2^5, which after normalisation becomes 0.1100×2^4.

Division of two given floating point numbers is carried out by dividing the dividend mantissa by the divisor mantissa and subtracting their exponent values. For instance, 0.65×10^5 divided by 0.13×10^2 gives 5×10^3 which after normalisation becomes 0.5×10^4.

Q. 46. Distinguish between the HALF ADDER and a FULLER ADDER circuit block with reference to their capabilities to perform binary addition.

Ans. A HALF ADDER is an ARITHMETIC CIRCUIT BLOCK that can be used to add only two bits. Such a circuit has, thus two inputs that represent the two binary digits (or bits) to be added and two outputs with one output giving the SUM and the other indicating the CARRY. All possible input combinations and their corresponding outputs (*i.e.* SUM AND CARRY) are shown in tabular form (Fig. 22.3). If we recall the procedure of adding larger binary numbers whether it is a straight binary addition or a 2's complement addition, in both cases, we have to finally add two given binary numbers. In case of 2's complement addition, of course, the numbers that are actually added inside the computers are the 2's complement of the numbers to be added and the result of addition is also in 2's complement form. Coming back to the addition process, we begin with the addition of LSBs of two numbers and record the sum under LSB column and take the carry (if any) forward to the next higher column bits. As a result, when we add the bits of the next higher adjacent column, we would be required to add three bits, if there was a carry from the previous addition. We have a similar situation for the other higher column bits also till we reach the MSB.

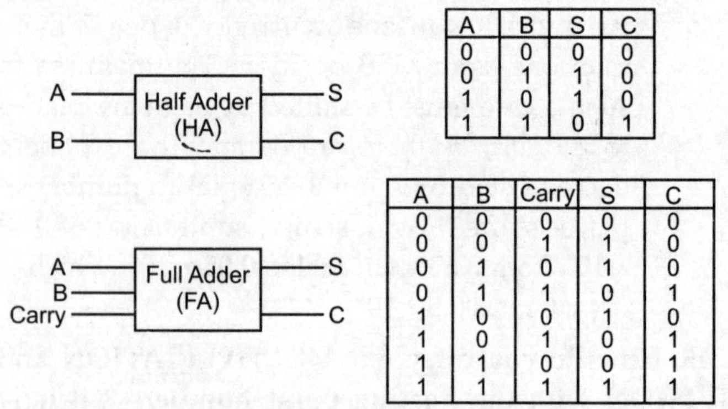

A	B	S	C
0	0	0	0
0	1	1	0
1	0	1	0
1	1	0	1

A	B	Carry	S	C
0	0	0	0	0
0	0	1	1	0
0	1	0	1	0
0	1	1	0	1
1	0	0	1	0
1	0	1	0	1
1	1	0	0	1
1	1	1	1	1

Fig. 22.3

Thus, in the hardware implementation of an adder circuit capable of adding larger binary numbers, although we could do with a half adder circuit as far as addition of LSBs is required, we will certainly have to find a circuit capable of adding three bits for the higher order column bits. The arithmetic circuit block that is capable of adding three bits to produce a sum and carry output is the FULL ADDER. The FULL ADDER CIRCUIT, therefore, has three inputs and two outputs (Figure. 22.3). The two bits to be added and carry from the previous edition are the three inputs. The two output are the sum and carry outputs. The accompanying table lists all possible inputs and their corresponding outputs for a full adder.

Q. 47. What are COMMUTATIVE LAWS of Boolean Algebra ?

Ans. The **Commutative Laws** of Boolean algebra are expressed by :

$$(A + B) = (B + A) \text{ and } A.B = B.A$$

Where, A and B are binary variables.

The first commutative law implies that the order in which the binary variables are added (or ORed) is immaterial. That is, the result of *A* OR *B* is the same as the result of *B* OR *A*.

The second commutative law implies that the order in which the binary variables are multiplied (or ANDed) is immaterial. That is, the result of A AND B is the same as the result of *B* AND *A*. These laws hold good in ordinary algebra too.

Q. 48. State and explain ASSOCIATIVE LAWS of Boolean Algebra.

Ans. The **Associative Laws** of Boolean algebra are expressed by:

$$(A + B) + C = A + (B + C)$$

and $(A. B).C = A . (B . C)$

where, *A*, *B* and *C* are binary variables.

The first associative law implies that when three binary variables are being added (or ORed), it is immaterial whether you do this by adding the result of addition of *A* and *B* to *C* or by adding *A* to the result of addition of *B* and *C*. Addition in the mathematics of logic means OR-operation Similarly, multiplication of *A*, *B* and *C* can be done either by ANDing the result of *A* and *B* to *C* or by ANDing the result of *B* and *C* to *A*. Multiplication in logic means AND-operation.

Q. 49. State and explain the DISTRIBUTIVE LAW of Boolean Algebra.

Ans. The **Distributive Law** of Boolean algebra is expressed by:

$$A . (B + C) = A . B + A . C$$

Distributive Law implies that the Boolean expressions can always/be expanded term by term and also, in case of Boolean expression being a sum of two or more than two terms having a common variable, the common variable can be taken common like we do in case of ordinal algebra. The proof of Distributive law is obvious in the form of the accompanying truth table Comparison of columns representing $A . (B + C)$ and $A . B + A . C$ establishes the truth of Distributive Law.

A	B	C	B + C	A.(B + C)	A.B	A.C	A.B + AC
0	0	0	0	0	0	0	0
0	0	1	1	0	0	0	0
0	1	0	1	0	0	0	0
0	1	1	1	0	0	0	0
1	0	0	0	0	0	0	0
1	0	1	1	1	0	1	1
1	1	0	1	1	1	0	1
1	1	1	1	1	1	1	1

Q. 50. State and prove INVOLUTION LAW of Boolean Algebra.

Ans. According to **Involution Law**, if (*x*) is a binary variable, then complement of complement of (x) is (*x*), *i.e.* $\overline{\overline{x}} = x$.

Truth Table

x	$y = \overline{x}$	$\overline{y} = \overline{\overline{x}} = x$
0	1	0
1	0	1

Q. 51. What do you understand by the IDEMPOTENT LAW of Boolean Algebra ?

Ans. The **Idempotent Laws** of Boolean algebra can be expressed by :

$x + x + x + x \ldots\ldots\ldots + x = x$

$x \,.\, x \,.\, x \,.\, x \,.\, x \ldots\ldots\ldots\ldots x = x$

The first part of the Idempotent Law is a direct outcome of the ORing operation. The output of an OR-gate is logic '0', when all its inputs are in logic '0' state and logic '1', state when all its inputs are in logic '1' state (from the truth table of an OR-gate). Refer Fig. 22.4 (*a*).

The second part is an outcome of AND-gate operation. From the truth table of an AND gate, its output is logic '0', when all its inputs are logic '0' and its output is logic '1', when all its inputs are logic '1'. Refer Fig. 22.4 (*b*).

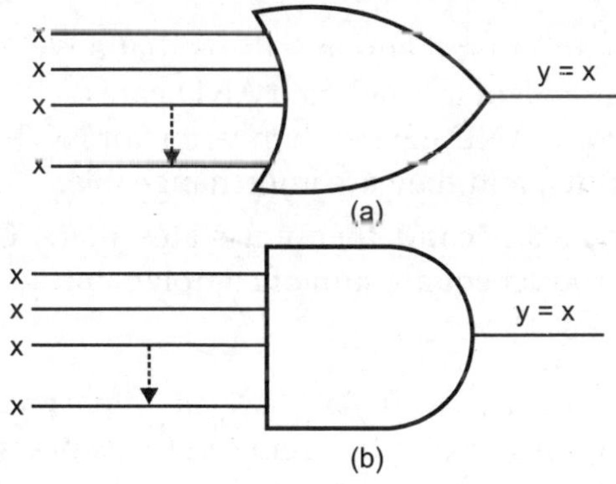

Fig. 22.4

An interesting corollary of the Idempotent Law is as follows :

"When all the inputs of an OR-gate or an AND-gate are shorted to get a one input one output gate, the logic state of the input is same as the logic state of the output."

Q. 52. State and explain ABSORPTION LAW of Boolean algebra. Also, justify your answer with the help of truth table.

Ans. According to **Absorption Law :**

$A \,.\, (A + B) = A$

where, *A* and *B* are binary variables.

L.H.S. $= A \,.\, (A + B)$

$= A \,.A + A.B$ Distributive Law

$= A + A.B$ Idempotent Law

$= A \,.\, (1 + B)$ Distributive Law

$= A \,.\, 1 \ldots\ldots\ldots\ldots (1 + B = 1)$

$= A$ (R.H.S.)

The truth table below illustrates the proof of Absorption Law.

A	B	A + B	A.(A + B)
0	0	0	0
0	1	1	0
1	0	1	1
1	1	1	1

Q. 53. How do you find the DUAL of a given Boolean expression ?

Ans. The DUAL of a given Boolean identity can be found out by replacing (+) or OR signs by (.) or AND signs, (.) or AND signs by (+) or OR signs and complementing 0's and l's. For instance, dual of a Boolean identity ($A.B + A.B$) is $(A + B) \,.\, (A + \overline{B})$ and dual of another Boolean identity ($\overline{A}.B + A.\overline{B}$) is $(\overline{A} + B).\ (A + \overline{B})$.

Q. 54. What is the significance of principle of duality? Give illustrations to prove your point.

Ans. Duality principle is very useful in simplifying logic expressions. This can be best explained with the help of some illustrations :

From Idempotent Laws of Boolean Algebra, we know that if (*x*) is a binary variable, then

$x + x + x + \ldots\ldots\ldots\ldots\ldots\ldots x$

$x \,.\, x \,.\, x \,.\, x \,.\, \ldots\ldots\ldots\ldots\ldots x$

If we closely examine the two laws, we will notice that forms of Indempotent one is dual of the other.

Another established law of boolean algebra says that

$A + (B + C) = (A + B) + C \ldots\ldots\ldots\ldots\ldots\ldots$Associative law.

Taking duals of both the sides, we get another valid logic equation *i.e.*

$A.(B.C) = (A.B).C$.........Another form of associative law

Similarly, when $(A+\overline{A}) = 1$ is a valid equation, its dual $A.\overline{A} = 0$ is equally valid.

There are many more such logic equations, whose duals lead to many other useful valid laws of Boolean algebra.

Q. 55. Find the duals of following Boolean identities :

(a) $(A+B)(B+C)(C+D)$

(b) $A.B+B.C+C.D$

(c) $A(B+C)+B.(C+A)+C.(A+B)$

(d) $1.(A+B+C)$ (A, B, C and D are binary variables)

Ans. Changing (+) signs to (.) signs and vice-versa and complementing 0s and 1s, we get,

(a) Dual of $(A+B).(B+C).(C+D) = A.B+B.C+C.D$

(b) Dual of $A.B+B.C+C.D = (A+B)(B+C)(C+D)$

(c) Dual of $A.(B+C)+B.(C+A)+C.(A+B) = (A+B.C).(B+C.A).(C+A.B)$

(d) $0+(A.B.C)$

Q. 56. Below are given some established laws of Boolean algebra. Write their duals.

(a) $A+\overline{A}=1$ **(b) $A+1=1$**

(c) $A.A.A.=A$ **(d) $A.(A+B)=A$**

Ans.

(a) $A.\overline{A}=0$ (b) $A.0=0$

(c) $A+A+A=A$ (d) $A+A.B=A$

Q. 57. Draw the logic circuit for the Boolean expression $y=(A+B).(C+D)$ using the basic logic gates. Write the dual for this Boolean identity and draw the logic circuit for the dual. What conclusions do you draw from this exercise?

Ans. The logic circuit for the given Boolean expression is shown in Fig. 22.5.

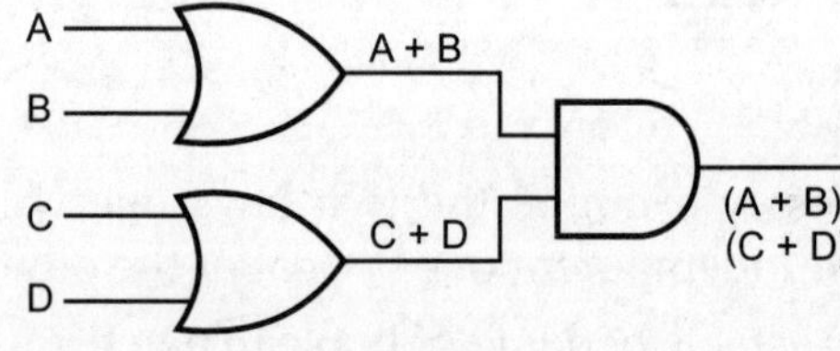

Fig. 22.5

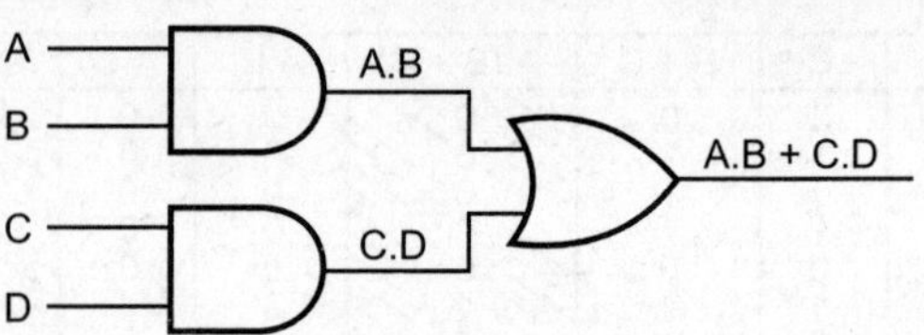

Fig. 22.6

The dual expression can be written as : $y = A.B+C.D$

The logic implementation for this Boolean expression is shown in Fig. 22.6.

If in the logic circuit of Fig. 22.5, OR gates are replaced by AND gates and the AND gates are replaced by OR gates, we get the logic circuit of Fig. 22.6. Thus, for finding the dual, OR operators should be replaced by AND operators and the AND operators should be replaced by OR operators. This is in fact, another way of saying that for finding the dual, (+) signs are to be replaced by (.) signs and vice-versa. This is of course in addition to complementing 0's and 1's.

Q. 58. STATE and explain DeMORGAN'S THEOREMS of Boolean Algebra.

Ans. Demorgan's first theorem states that "The complement of a sum equals the product of the complements." It is expressed as :

$$\overline{(A+B)} = \overline{A}.\overline{B}$$

Where, A and B are binary variables. In general, if X_1 X_2, X_3, X_4 .... X_n are binary variables, then according to DeMorgan's first theorem,

$$\overline{X_1+X_2+...+X_n} = \overline{X_1}.\overline{X_2}\,\overline{X_3}...\overline{X_n}$$

Fig. 22.7 shows the logic implementation or graphic expression of DeMorgan's first theorem.

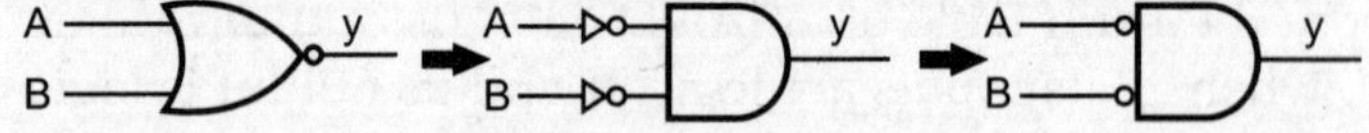

Fig. 22.7

Demorgan's first theorem tells us that a NOR gate is logically equivalent to a bubbled AND gate (A BUBBLED AND gate is an AND gate with inverters or NOT gates at all of its inputs) and they are interchangeable.

Demorgan's second theorem states that, "Complement of a product equals sum of complements."

i.e., $\overline{A.B} = (\overline{A}+\overline{B})$ ...(2)

In general, if X_1; X_2, X_3, X_4 .... X_n are binary variables, then DeMorgan's second theorem can (be expressed as :

$$\overline{(X_1.X_2.X_3.......X_n)} = \overline{X_1}+\overline{X_2}+\overline{X_3}+..........\overline{X_n}$$

Fig. 22.8 shows the logic representation of DeMorgan's second theorem which tells us that a NAND gate is equivalent to a bubbled OR gate. The two are interchangeable.

The truth table below proves the two DeMorgan's theorems. The truth table is self-explanatory.

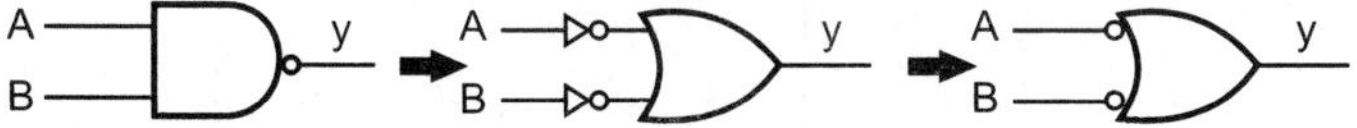

Fig. 22.8

A	B	$\overline{A+B}$	$\overline{A.B}$	$\overline{A}$	$\overline{B}$	$\overline{A}.\overline{B}$	$\overline{A}+\overline{B}$
0	0	1	1	1	1	1	1
0	1	0	1	1	0	0	1
1	0	0	1	0	1	0	1
1	1	0	0	0	0	0	0

Q. 59. How do you implement a 2-input OR gate using NAND gates only?

Ans. Desired Boolean expression is:

$$Y = A + B \text{ (OR-gate)}$$
$$= \overline{\overline{A + B}} \text{ (Involuation Law)}$$
$$= \overline{(\overline{A}\,.\,\overline{B})} \text{ (Demorgan's theorem)}$$

$\overline{(\overline{A}\,.\,\overline{B})}$ can be simulated by feeding $\overline{A}$ and $\overline{B}$ to the input of NAND gate. $\overline{A}$ and $\overline{B}$ can in turn be obtained from A and B respectively by using NAND gates with shorted inputs. Fig. 22.9 shows the logic implementation.

Q. 60. Implement a 2-input AND gate using 2-input NOR gate only.

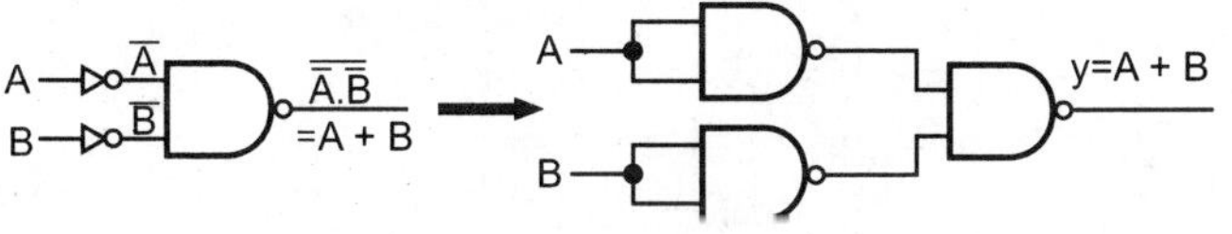

Fig. 22.9

Ans. Desired Boolean expression

$$Y = A\,.\,B \quad \text{(AND gate)}$$
$$= \overline{\overline{A}}\,.\,\overline{\overline{B}} \quad \text{(Involution Law)}$$
$$= \overline{(\overline{A} + \overline{B})} \quad \text{(DeMorgan's Law)}$$

Fig. 22.10. shows the logic implementation

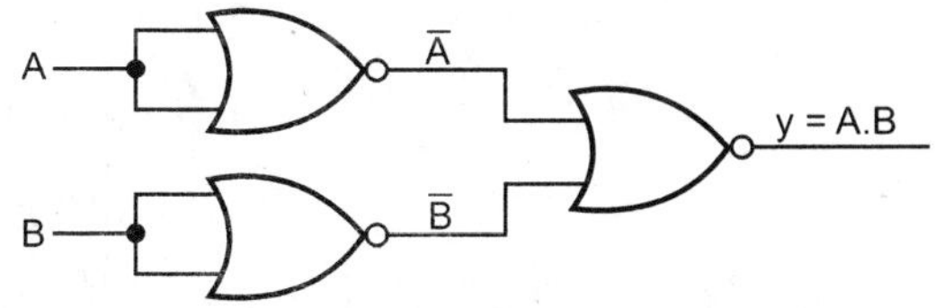

Fig. 22.10

Q. 61. Implement a 2-input EX-OR gate using (*i*) 2-input NAND gates only (*ii*) 2-input NOR gates only.

Ans. (*i*) Desired Boolean expression:

$$Y = \overline{A}\,.\,B + A\,.\,\overline{B}$$

Let $\overline{A}\,.\,B = C$

and $A.\,\overline{B} = D$

C and D are binary variables.

Boolean expression now can be rewritten as:

$$Y = C + D$$
$$= \overline{(\overline{C}\,.\,\overline{D})} \text{.......DeMorgan's theorem.}$$

Substituting the values of C and D, equation (1) becomes:

$$Y = \overline{(\overline{\overline{A}\,.\,B}\,.\,\overline{A\,.\,\overline{B}})}$$

Fig 22.11 shows the logic implementation

(*ii*) $\overline{A}\,.\,B = \overline{(A + \overline{B})}$ and $A\,.\,\overline{B} = \overline{(\overline{A} + B)}$ DeMorgan's theorem

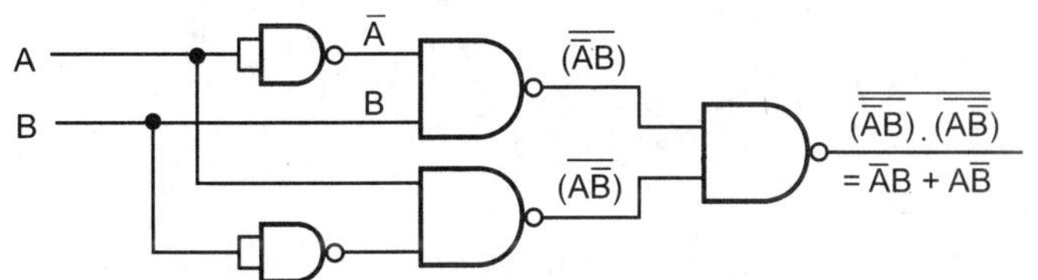

Fig. 22.11

Therefore, $Y = \overline{(A+\overline{B})} + \overline{(\overline{A}+B)}$

Fig. 22.12 shows the logic implementation.

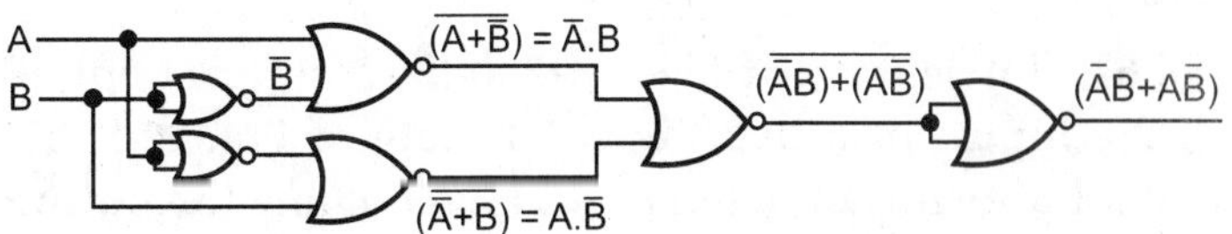

Fig. 22.12

Q. 62. Implement the following Boolean expression:

y = A.B.C. + A . $\overline{B}$. $\overline{C}$ + $\overline{A}$. $\overline{B}$. C + $\overline{A}$. B . $\overline{C}$ with minimum number of gates.

Ans. The given Boolean expression s:

$$= A.B.C. + A\,.\,\overline{B}\,.\,\overline{C} + \overline{A}\,.\,\overline{B}\,.\,C + \overline{A}\,.\,B\,.\,\overline{C}$$
$$= A\,.\,(B\,.\,C + \overline{B}\,.\,\overline{C}) + \overline{A}\,.\,(\overline{B}\,.\,C + B\,.\,\overline{C})$$

(Distribution Law)

$$= A\,.\,(B\,.\,C + \overline{B}\,.\,\overline{C}) + \overline{A}\,.\,(\overline{B}\,.\,C + B\,.\,\overline{C})$$
$$= A\,.\,\overline{((\overline{B} + \overline{C}).(B + C))} + \overline{A}\,.\,(\overline{B}\,.\,C + B\,.\,\overline{C})$$
$$= A\,.\,\overline{(\overline{B}C + B.\overline{C})} + \overline{A}\,.\,(\overline{B}\,.\,C + B\,.\,\overline{C})$$

Let $D = (\overline{B}\,.\,C + B\,.\,\overline{C})$ so that $y = \overline{A}\,.\,D + A\,.\,\overline{D}$

From the above expressions, it is clear that with A and D as the inputs of a 2-input *EX-OR* gate, its output will be desired (y). Also, (D) will be the output of another EX-OR gate, if B and C are its inputs. Refer to Fig. 22.13 (*a*) to (c). Fig. 22.13 (*c*) shows the implementation of the given Boolean expression, while Figs. 22.13 (*a*) and (*b*) represent two steps towards achieving the final result.

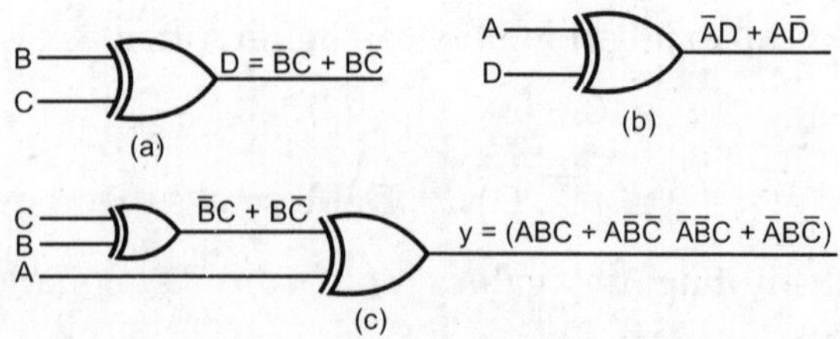

Fig. 22.13

Q. 63. Name the basic logic gate represented by each of the logic notations/diagrams shown in Fig. 22.14.

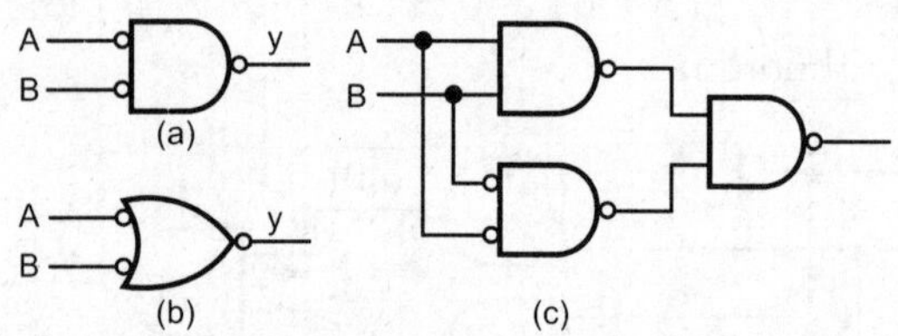

Fig. 22.14

Ans.

Fig. 22.15 (*a*) - OR gate

Fig. 22.15 (*b*) - AND gate

Fig. 22.15 (*c*) - EX-OR gate

Q. 64. Pulse train of Fig. 22.15. (*a*) are in turn applied to the inputs of a bubbled AND gate of Fig. 22.15 (*b*) and a bubbled OR gate Fig. 22.15. (*c*). Draw the output waveforms in the cases. Numbers 1 to 10 are nothing but the time instances.

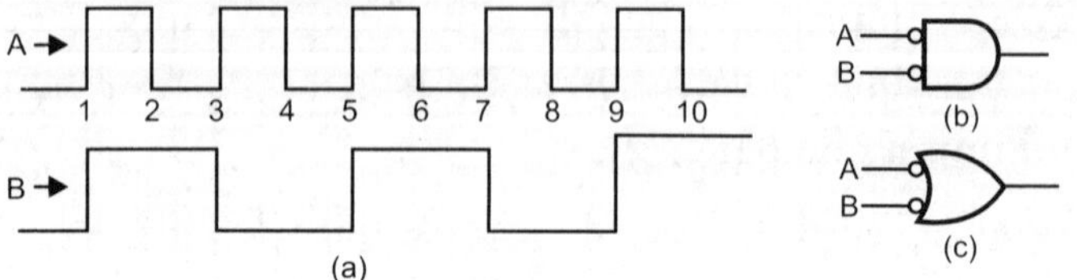

Fig. 22.15

Ans. A bubbled AND gate is nothing but a NOR gate and a bubbled OR gate is the same as a NAND gate. Keeping in mind the truth table of a NOR gate, while drawing the waveforms in case of the bubbled AND gate and following the truth table of a NAND gate while drawing the output waveform in case of the bubbled OR gate is all that is required to solve this problem. The output waveforms for the two cases are shown in Fig. 22.16 (*a*) - for the bubbled AND gate and 22.16 (*b*) - for the bubbled OR gate.

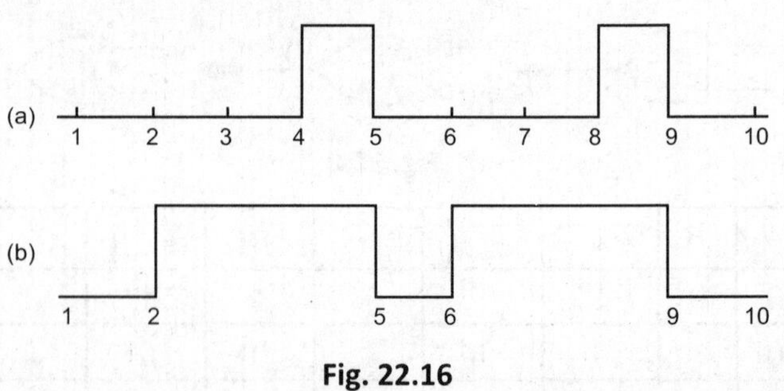

Fig. 22.16

Q. 65. Use basic laws and theorems of Boolean algebra to simplify the following logic expression:

(*i*) $z = \overline{\overline{(X + \overline{Y})} + \overline{(X\,.\,\overline{Y})} + \overline{(\overline{X}\,.\,Y)}}$

(*ii*) $z = \overline{\overline{(B\,.\,\overline{C}\,.\,D)} + \overline{(\overline{A}\,.\,C\,.\,D)}}$

(*iii*) $z = \overline{(\overline{A} + B)}$

(*iv*) $z = \overline{(A\,.\,B + B\,.\,C)\,.\,(B\,.\,C + C.\,D)\,.\,(C\,.\,D + A\,.\,D)}$

(*v*) $z = (A + B)\,.\,(A + \overline{B})\,.\,(\overline{A} + B)$

Ans. (*i*) $Z = \overline{\overline{(X + \overline{Y})} + \overline{(X\,.\,\overline{Y})} + \overline{(\overline{X}\,.\,Y)}}$

$= \left(\overline{\overline{X + \overline{Y}}}\right).\overline{\overline{(X\,.\,\overline{Y}}}\,.\overline{\overline{(\overline{X}\,.\,Y)}}$ (DeMorgan's)

$= (X + \overline{Y})\,.\,(X\,.\,\overline{Y})\,.\,(\overline{X}\,.\,Y)$

(Involution Law)

$= (X + \overline{Y})\,.\,X\,.\,\overline{X}\,.\,Y\,.\,\overline{Y}$

$= (X + \overline{Y})\,.\,0\,.\,0$

$= 0$

(*ii*) $Z = \overline{\overline{(B\,.\,\overline{C}\,.\,D)} + \overline{(\overline{A}\,.\,C\,.\,D)}}$

$= \overline{\overline{(B\,.\,\overline{C}\,.\,D)}} + \overline{\overline{(\overline{A}\,.\,C\,.\,D)}}$

(DeMorgan's Theorem)

$= (\overline{B} + \overline{C}\,.\,D)\,.\,(\overline{A}\,.\,C\,.\,D)$

(DeMorgan's Theorem)

$= \overline{B}\,.\,\overline{C}D + \overline{A}\,.\,C\,.\,D$

(*iii*) $Z = \overline{(\overline{A} + B)}$

$= \overline{\overline{A}}\,.\,\overline{B}$ (DeMorgan's theorem)

$= A\,.\,\overline{B}$ (Involution Law)

$Z = \overline{(A\,.\,B + B\,.\,C)\,.\,(B\,.\,C + C\,.\,D)\,.\,(C\,.\,D + A\,.\,D)}$

$$= \overline{(A.B.B.C + A.B.C.D + B.C.B.C + B.C.C.D).(C.D + A.D)}$$

$$= \overline{(A.B.C + A.B.C.D + B.C + B.C.D)}.\overline{(C.D + A.D)}$$

$$= \overline{(A.B.C.C.D + A.B.C.A.D + A.B.C.DC.D + A.B.C.D.A.D + B.C.C.D + B.C.A.D + B.C.D}$$

$$\overline{C.D. + B.C.D.A.D)}$$

$$= \overline{A.B.C.D + A.B.C.D + A.B.C.D} + \overline{A.B.C.D + B.C.D} + \overline{A.B.C.D + B.C.D + A.B.C.D}$$

(*v*) $$Z = \overline{B.C.D}$$

$$Z = (A + B).(A + \overline{B}).(\overline{A} + B) = (A + B)$$

$$= (A + B).(A.B + \overline{A}.\overline{B})$$

$$= A.B + A.B \quad \text{(Idempotent Law)}$$

$$= A.B$$

Q. 66. What is a Sum-of-Products Boolean expression? What type of logic building blocks or arrangement can be used to implement such an expression ? What is a Minterm expression?

Ans. Sum-of-Products expression contains sum of different terms with each term being either a single literal or a product of more than one literals. It can be obtained from the truth table directly by considering those input combinations which produce a '1' at the output. Each such input combination produces a term. Different terms are given by product of corresponding literals. Sum of all terms gives the expression. For the truth table given below, the Boolean function representing this truth table is given by:

$$y = \overline{A}\,\overline{B}\,\overline{C} + \overline{A}\,BC + A\,\overline{B}\,\overline{C} + ABC$$

Inputs			Output
A	*B*	*C*	*Y*
0	0	0	1
0	0	1	0
0	1	0	0
0	1	1	1
1	0	0	1
1	0	1	0
1	1	0	0
1	1	1	1

Considering the first term, the output is '1', when $A = 0$, $B = 0$, $C = 0$. This is possible only, when $\overline{A}$, $\overline{B}$ and $\overline{C}$ are ANDed. Also, for the second term, the output is '1' only when $\overline{A}$, B and C are ANDed. The other terms can be explained similarly. A sum-of-products expression is also known as minterm expression. Such an expression can be implemented by using AND-OR logic.

Q. 67. What is a Product-of-Sums Boolean expression? How is it different from a Sum-of-Products expression? What is Maxterm expression?

Ans. Product-of-sums expression contains product of different terms with each term being either a single literal or a sum of more than one literal. It can be obtained from the truth table by considering those input combinations that produce a '0' at the output. Each such input combination gives a term and product of all such terms gives the expression. Different terms are obtained by taking sum of corresponding literals. Here, 0 and 1 respectively mean the uncomplemented and complemented variables unlike the case of sum-of-products expressions, where 0 and 1 respectively meant complemented and uncomplemented variables. To illustrate it further, consider the truth table referred to earlier in case of sum-of-products expression. Since each term is going to be sum of literals, it implies that it is going to be implemented using OR-operation.

An OR-gate produces a '0' only, when all its inputs are '0' which means that first term is going to be $(A + B + \overline{C})$. Similarly, the other terms would be $(A + \overline{B} + C)$, $(\overline{A} + B + \overline{C})$ and $(\overline{A} + \overline{B} + C)$. The product-of-sums Boolean function is given by :

$$y = (A + B + \overline{C})(A + \overline{B} + C)(\overline{A} + B + \overline{C})$$
$$(\overline{A} + \overline{B} + C)$$

A Product-of-sums expression is also called a maxterm expression. It can be implemented by using OR-AND logic.

Q. 68. With the help of a suitable expression, briefly describe the procedure for transforming a Product-of-Sums expression into an equivalent Sum-of-Products expression and vice-versa.

Ans. Transforming a product-of-sums expression into an equivalent sum-of-products expression is a straight forward process. Multiplying out the given expression and carrying out the obvious simplifications provides the equivalent sum of-products expression.

As an illustration, $(A+B+\overline{C})(A+\overline{B}+C)(\overline{A}+B+\overline{C})(\overline{A}+\overline{B}+C)$

$$= (A + A\overline{B} + AC + AB + BC + A\overline{C} + \overline{B}\,\overline{C})$$
$$(\overline{A} + \overline{A}\,\overline{B} + \overline{A}C + \overline{A}B + BC + \overline{A}\,\overline{C}$$
$$+ \overline{B}\,\overline{C})$$
$$= (A + BC + \overline{B}\,\overline{C})(\overline{A} + BC + \overline{B}\,\overline{C}) = BC + \overline{B}\,\overline{C}$$

A given sum-of-products expression can be transformed into an equivalent product-of-sums expression by (*i*) taking dual of given expression (*ii*) multiplying out different terms to get the sum-of-products form (*iii*) removing redundancy and (*iv*) taking a dual to get the equivalent product-of-sums expression.

As an illustration, let us find the equivalent product-of-sums expression of the given sum-of-products expression

$$AB + \overline{A}\,\overline{B}$$

Dual of $AB + \overline{A}\,\overline{B} = (A+B)(\overline{A}+\overline{B})$

$$= A\overline{A} + A\overline{B} + \overline{A}B + B\overline{B} = A\overline{B} + \overline{A}B$$

Dual of

$$A\overline{B} + \overline{A}B = (A+\overline{B})(\overline{A}+B)$$

Therefore,

$$(AB + \overline{A}\,\overline{B}) = (A+\overline{B})(\overline{A}+B)$$

Q. 69. What is the EXPANDED form of a Boolean expression ? With the help of suitable examples, explain how can you obtain the expanded form of given Sum-of-Products or Product-of-Sums Boolean expression ?

Ans. Expanded sum-of-products and product-of-sums forms of Boolean expressions are useful not only in analysing them, they are also used in the application of minimisation techniques, such as Quine-McCluskey tabular method and Karnaugh's mapping method for simplifying a given Boolean expression.

The expanded form, sum-of-products or product-of-sums, is obtained by including all possible combinations of missing variables.

As an illustration, consider a sum-of-products expression :

$$\overline{A}B + A\overline{B}C + B\overline{C}$$

It is a 3-variable expression. The expanded versions of different terms would be :

$$\overline{A}B = \overline{A}B(C+\overline{C}) = \overline{A}BC + \overline{A}B\overline{C}$$

$A\overline{B}C$ is a complete term and has no missing variable.

$$B\overline{C} = B\overline{C}(A+\overline{A}) = AB\overline{C} + \overline{A}B\overline{C}$$

The expanded sum-of-products expression is therefore given by:

$$\overline{A}BC + \overline{A}B\overline{C} + A\overline{B}C + AB\overline{C} + \overline{A}B\overline{C} =$$
$$\overline{A}BC + \overline{A}B\overline{C} + A\overline{B}C + AB\overline{C}$$

As an another illustration, consider the product-of-sums expression:

$$(A+\overline{B})(\overline{A}+C)$$

$(A+\overline{B})$ expands to $(A+\overline{B}+C)(A+\overline{B}+\overline{C})$

and $(\overline{A}+C)$ expands to $(\overline{A}+B+C)(\overline{A}+\overline{B}+C)$

The expanded product-of-sum expression is therefore given by:

$$(A+\overline{B}+C)(A+\overline{B}+\overline{C})(\overline{A}+B+C)(\overline{A}+\overline{B}+C)$$

As another illustration, $(A+B)$ in $(A+B)(\overline{A}+B+C+D)$ will expand to

$$(A+B+C+D)(A+B+C+\overline{D})(A+B+\overline{C}+D)(A+B+\overline{C}+\overline{D})$$

An expanded form of Boolean expression, where each term contains all Boolean variables in their true or complemented form, is also known as the CANONICAL FORM of the expression.

As an illustration, $f(A, B, C) = \overline{A}\,\overline{B}\,\overline{C} + \overline{A}B\overline{C} + \overline{A}BC + AB\overline{C}$ is a Boolean function of three variables expressed in canonical form. This function after simplification reduces to,

$$f(A, B, C) = B\overline{C} + \overline{A}B + \overline{A}\,\overline{C}$$

and loses its canonical form.

Q. 70. With the help of examples, explain Π and 'Σ' notations of representing Boolean expressions. How can we get one from the other ?

Ans. Σ and Π notations are respectively used to represent sum-of-products and product-of-sums Boolean expressions. These notations are better explained with the following illustrations.

Let us see how we can represent a Boolean function

$$f(A, B, C, D) = ABC + A\overline{B}\,\overline{C} + \overline{A}B\overline{C}D + ABC\overline{D} + \overline{A}\,\overline{B}\,\overline{C}\,\overline{D}$$

using Σ nomenclature

The first step is to write expanded sum-of-products given by :

$$f(A, B, C, D) = ABCD + ABC\overline{D} + A\overline{B}\,\overline{C}D + A\overline{B}\,\overline{C}\,\overline{D} + \overline{A}B\overline{C}D + ABC\overline{D} + \overline{A}\,\overline{B}\,\overline{C}\,\overline{D}$$

$$= ABCD + ABC\overline{D} + A\overline{B}\,\overline{C}D + A\overline{B}\,\overline{C}\,\overline{D} + \overline{A}B\overline{C}D + \overline{A}\,\overline{B}\,\overline{C}\,\overline{D}$$

Different terms are then arranged in the ascending order of the binary numbers represented by various terms with true variables representing a '1' and a complemented variable representing a '0'. The expression becomes

$$f(A, B, C, D) = \overline{A}\,\overline{B}\,\overline{C}\,\overline{D} + \overline{A}B\overline{C}D + A\overline{B}\,\overline{C}\,\overline{D} + A\overline{B}\,\overline{C}D + ABC\overline{D} + ABCD$$

with different terms representing 0000, 0101, 1000, 1001, 1110 and 1111.

The decimal equivalent of these terms enclosed in the Σ then gives the Σ notation for the given Boolean function.

That is, $f'(A, B, C, D) = \Sigma\ 0, 5, 8, 9, 14, 15$

The complement of $f(A, B, C, D)$ *i.e.* $f'(A, B, C, D)$ can be directly determined from Σ notation by including the left out entries from the list of all possible numbers for a four variable function.

That is, $f(A, B, C, D) = \Sigma\ 1, 2, 3, 4, 6, 7, 10, 11, 12, 13$

Let us now take the case of a product-of-sums Boolean function and its representation in Π-nomenclature.

Consider $f(A, B, C, D) = (A + \overline{B} + \overline{C} + D)\ (A + B + C + \overline{D})\ (B + \overline{C} + \overline{D})$

The expanded product-of-sums Boolean function is given by :

$$f(A, B, C, D) = (A + \overline{B} + \overline{C} + D)\ (A + B + C + \overline{D})\ (A + B + \overline{C} + \overline{D})\ (\overline{A} + B + \overline{C} + \overline{D})$$

The binary numbers represented by different sum terms are 0110, 0001, 0011 and 1011 (True and complemented variables here represent 0 and 1 respectively). These numbers, when arranged in ascending order, are 0001, 0011, 0110, and 1011.

Therefore, $f(A, B, C, D) = \Pi\ 1, 3, 6, 11$

and $f'(A, B, C, D) = \Pi\ 0, 2, 4, 5, 7, 8, 9, 10, 12, 13, 14, 15$

An interesting corollary of what we have discussed above is that if a given Boolean function $f(A, B, C)$ is given by :

$$f(A, B, C) = \Sigma\ 0, 1, 4, 7$$

then $f(A, B, C) = \Pi\ 2, 3, 5, 6$

and $f'(A, B, C) = \Sigma\ 2, 3, 5, 6 = \Pi\ 0, 1, 4, 7$

Optional combinations can also be incorporated into Σ and Π nomenclature using suitable identifiers. 'ϕ' or '*d*' are used as identifiers.

For example, if $f(A, B, C) = \overline{A}\,\overline{B}C + \overline{A}B\overline{C} + A\overline{B}\,\overline{C} + AB\overline{C}$ and $\overline{A}\,\overline{B}C$, $\overline{A}BC$ are optional combinations, then

$$f(A, B, C) = \Sigma\ 0, 2, 4, 6 + \sum_{\phi} 1, 3$$

$$\Sigma\ 0, 2, 4, 6, + \sum_{d} 1, 3$$

Also, $f(A, B, C) = \Pi\ 5, 7 + \prod_{\phi} 1, 3$

or $\Pi\ 5, 7 + \prod_{d} 1, 3$

Q. 71. Describe various steps involved in simplifying a given Boolean expression using Quine-McCluskey tabular Method.

Ans. Quine-McCluskey tabular method of simplification is based on the complementation theorem according to which $XY + X\overline{Y} = X$, where '*X*' represents not only a variable but a term or an expression also and '*Y*' is a variable. This theorem implies that if a Boolean expression contains two terms that differ only in one variable, then they can be combined together and replaced by a term that is smaller by one literal. Same procedure is applied for the other pairs of terms, wherever such a reduction is possible. All these terms reduced by one literal are further examined to see, if they can be reduced further. The process continues till the terms become irreducible. The irreducible terms are called PRIME IMPLICANTS. An optimum set of prime implicants that can account for all the original terms, then constitutes the minimised expression. The technique can be applied equally well for minimising sum-of-products and product-of-sums expressions and is particularly useful for Boolean functions having more than six variables, as it can be mechanised and run on a computer. On the other hand, Karnaugh's Mapping method, to be discussed after this, is a graphical method and becomes very cumbersome, when the number of variables exceeds six.

The step-by-step procedure for application of tabular method for minimising Boolean expressions, both sum-of-products and product-of-sums, is illustrated below :

1. The Boolean expression to be simplified is expanded, if it is not in the expanded form already. In a four variable expression with *A*, *B*, *C*, *D* as variables, a term *AB* for instance would expand to :

$$AB\ (\overline{C}\,\overline{D} + \overline{C}D + C\overline{D} + CD) = AB\overline{C}\,\overline{D} + AB\overline{C}D + ABC\overline{D} + ABCD$$

ABC would expand to : $ABC\ (D + \overline{D}) = ABCD + ABC\overline{D}$

A would expand to : $A(\overline{B}\,\overline{C}\,\overline{D} + \overline{B}\,\overline{C}D + \overline{B}C\overline{D} + \overline{B}C\overline{D} + B\overline{C}\,\overline{D} + B\overline{C}D + BC\overline{D} + BCD)$

$$= A\overline{B}\,\overline{C}\,\overline{D} + A\overline{B}\,\overline{C}D + A\overline{B}C\overline{D} + A\overline{B}CD + AB\overline{C}\,\overline{D} + AB\overline{C}D + ABC\overline{D} + ABCD$$

2. Different terms in the expression are divided into groups depending upon number of 1's, they have. True and complemented variables in a sum-of-products expression mean '1' and

'0' respectively. Reverse is the case in case of a product-of-sums expression.

For instance, $\overline{A}BCD$ stands for 0111, whereas $(\overline{A} + B + C + D)$ means 1000. Groups are then arranged in the ascending order, that is, beginning with a group having least number of 1's in its included terms. Terms within the same group are also arranged in the ascending order of the decimal numbers represented by these terms.

As an illustration, consider an expression $ABC + \overline{A}BC + A\,\overline{B}\,\overline{C} + A\,\overline{B}\,C + \overline{A}\,\overline{B}\,\overline{C}$. The grouping of different terms and arrangement of different terms within the group are shown below:

	$A\,B\,C$	
$\overline{A}\,\overline{B}\,\overline{C}$	0 0 0	First group
$A\,\overline{B}\,\overline{C}$	1 0 0	Second group
$\overline{A}\,B\,C$	0 1 1	Third group
$A\,\overline{B}\,C$	1 0 1	
$A\,B\,C$	1 1 1	Fourth group

As another illustration, consider a product-of-sums expression given by:

$(A + B + C + D)\,(\overline{A} + B + C + D)(A + \overline{B} + C + D)(\overline{A} + \overline{B} + C + D)(\overline{A} + B + \overline{C} + D)(A + B + \overline{C} + D)(\overline{A} + \overline{B} + \overline{C} + \overline{D})(A + B + \overline{C} + D)(A + B + C + \overline{D})(\overline{A} + \overline{B} + \overline{C} + D)$.

Formation of groups and arrangement of terms within different groups are illustrated in the table below:

	$A\,B\,C\,D$	
$\overline{A}\,\overline{B}\,\overline{C}\,\overline{D}$	0 0 0 0	First group
$\overline{A}\,\overline{B}\,\overline{C}\,D$	0 0 0 1	Second group
$\overline{A}\,\overline{B}\,C\,D$	0 0 1 1	
$\overline{A}\,B\,\overline{C}\,D$	0 1 0 1	Third group
$A\,B\,\overline{C}\,\overline{D}$	1 1 0 0	
$\overline{A}\,B\,C\,D$	0 1 1 1	
$A\,\overline{B}\,C\,D$	1 0 1 1	Fourth group
$A\,B\,\overline{C}\,D$	1 1 0 1	
$A\,B\,C\,\overline{D}$	1 1 1 0	
$A\,B\,C\,D$	1 1 1 1	Fifth group

Optional terms are also considered, while forming groups. This completes the first table.

3. The terms of the first group are successively matched with those in the next adjacent higher order group to look for any possible matching and consequent reduction. The terms are considered matched, when all literals except one match. The pairs of matched terms are replaced by a single term, where the literals that don't match are replaced by '—'. These new terms formed as a result of the matching process find a place in the second table. The terms in the first table that don't find a match are called the prime implicants and are marked with an asterik sign (*). The matched terms are ticked (√).

4. Terms in the second group are compared with those in the third group to look for possible match. Again, terms in the second group that don't find a match become the prime implicants.

5. The process continues till we reach the last group. This completes first round of matching. The terms resulting from the matching in the first round are recorded in the second table.

6. The next step is to perform matching operations in the second table. While comparing the terms for a match, it is important that a dash (—) is also treated like any other literal, that is the dash signs also need to match. The process continues onto the third, fourth tables till the terms become irreducible any further.

7. An optimum selection of prime implicants to account for all the original terms constitute the terms for the minimised expression. Although, optional (or don't care) terms are considered—or matching, they don't have to be accounted for once prime implicants have been identified.

The process can be best explained with the help of an example. Let us consider simplification of sum-of-products expression:

$$\overline{A}BC + \overline{A}\,\overline{B}D + A\overline{C}D + B\overline{C}\,\overline{D} + \overline{A}B\overline{C}D$$

In the first step, we write the expanded version of the given expression given by :

$\overline{A}BCD + \overline{A}BC\overline{D} + \overline{A}\,\overline{B}CD + \overline{A}\,\overline{B}CD + A\,\overline{B}CD + A\overline{B}\,\overline{C}D + AB\,\overline{C}\,\overline{D} + \overline{A}B\overline{C}\,\overline{D} + \overline{A}\,B\,\overline{C}D$

Formation of groups, placement of terms in different groups and first round matching are all demonstrated below:

A	B	C	D
0	0	0	1
0	0	1	1
0	1	0	0
0	1	0	1
0	1	1	0
0	1	1	1
1	0	0	1
1	1	0	0
1	1	0	1

A	B	C	D	
0	0	0	1	√
0	1	0	0	√
0	0	1	1	√
0	1	0	1	√
0	1	1	0	√
1	0	0	1	√
0	1	1	1	√
1	1	0	1	√

A	B	C	D	
0	0	—	1	√
0	—	0	1	√
—	0	0	1	√
0	1	0	—	√
0	1	—	0	√
—	1	0	0	√
0	—	1	1	√
0	1	—	1	√
—	1	0	1	√

0	1	1	—	√
1	—	0	1	√
1	1	0	—	√

In the second round of matching, each term in the first group is compared with every term in the second group. For instance, first term in the first group (0 0–1) matches with the second term in the second group (01 – 1) to yield 0 – –1, which is recorded in the next table as shown below. The process continues till all terms have been compared for a possible match. Since this new table has only one group, the terms contained therein are all prime implicants. In the present example, the terms in the first and second tables have all found a match. But that is not always the case.

A	*B*	*C*	*D*	
0	—	—	1	*
—	—	0	1	*
0	1	—	—	*
—	1	0	—	*

The next table is what is known as prime implicant table. The prime implicant table contains all the original terms in different columns and all the prime implicants recorded in different rows as shown below :

0001	0011	0100	0101	0110	0111	1001	1100	1101		
√	√		√		√				0 – – 1	$P \rightarrow \overline{A}D$
√			√			√		√	– – 0 1	$Q \rightarrow \overline{C}D$
		√	√	√	√				0 1 – –	$R \rightarrow \overline{A}B$
		√	√				√	√	– 1 0 –	$S \rightarrow B\overline{C}$

Each prime implicant is identified by a letter. Each prime implicant is examined one by one and the terms it can account for are ticked as shown. The next step is to write a product-of-sums expression using the prime implicants to account for all the terms. For the table above, this expression is given by:

$$(P + Q)\,(P)\,(R + S)\,(P + Q + R + S)\,(R)\,(P + R)\,(Q)\,(S)\,(Q + S)$$

Obvious simplification reduces this expression to *PQRS*, which can be interpreted to mean that all prime implicants, that is, *P*, *Q*, *R* and *S* would be needed to account for all the original terms.

Therefore, the minimised expression $= \overline{A}D + \overline{C}D + \overline{A}B + B\overline{C}$. What has been described above, is the formal method of determining the optimum set of prime implicants. In most of the cases where the prime-implicant table is not too complex the exercise can be done even intuitively. The exercise begins with identification of those terms that can be accounted for by only a single prime implicant. In the present example, 0011,0110,1001,1100 are such terms. As a result *P*, *Q*, *R* and *S* become the essential prime implicants. The next step is to find out, if any terms have not been covered by the essential prime implicants. In the present case, all terms have been covered by essential prime implicants. In fact, all prime implicants are essential prime implicants in the present example.

As another illustration, let us consider a product-of-sums expression given by:

$$(A + B + C + D)(A + B + C + \overline{D})\ (A + B + \overline{C} + D)(\overline{A} + B + C + D)(\overline{A} + B + \overline{C} + D)$$

The procedure is similar to what has been described in case of simplification of sum-of-products expressions. The resulting tables leading to identification of prime implicants are given below:

A	B	C	D
0	1	0	1
0	1	1	1
1	1	0	1
1	1	1	0
1	1	1	1

A	B	C	D	
0	1	0	1	√
0	1	1	1	√
1	1	0	1	√
1	1	1	0	√
1	1	1	1	√

A	B	C	D	
0	1	–	1	√
–	1	0	1	√
–	1	1	1	√
1	1	–	1	√
1	1	1	–	*

A	B	C	D	
–	1	–	1	*

The prime implicant table is constructed after all prime implicants have been identified to look for optimum set of prime implicants needed to account for all the original terms. The table is shown below. Here, again, both the prime implicants turn out to be essential ones.

0101	0111	1101	1110	1111	Prime implicants
			√	√	111–
√	√	√		√	–1–1

Minimised expression = $(A + B + C)\ (B + D)$

Q. 72. Define a Karnaugh map. Which type of Boolean expressions can be represented on a Karnaugh map? Give different styles of representing two, three and four variable expressions on Karnaugh maps.

Ans. Karnaugh map is a graphical representation of the logic system. It can be drawn directly from either minterm (sum-of-products) or maxterm (product-of-sums) Boolean expressions. Drawing a Karnaugh map from truth table involves an additional step of writing the minterm or maxterm expression depending upon whether it is desired to have minimised sum-of-products or a minimised product-of-sums expression.

An n-variable Karnaugh map has (2^n) squares and each possible input is alloted a square. A '1' is placed in all those squares for which the output is desired, that is, the output is '1' and '0' is placed in all those squares for which the output is not desired, that is, the output is '0' in use of a minterm Karnaugh map. 0's are omitted for simplicity. An 'X is placed in squares corresponding to 'Don't care' conditions. In case of a maxterm Karnaugh map, a '1' is placed in all those squares for which the output is '0' and a '0' is placed for input entries corresponding to a '1' output. Again 0's are omitted for simplicity and a 'X' is placed in squares corresponding to 'Don't' care conditions.

The choice of terms identifying different rows and columns of a Karnaugh map is not unique or a given number of variables. The only condition to be satisfied is that the designation of adjacent rows and adjacent columns should be the same except for one of the literals being complemented. Also, the extreme rows and extreme columns are considered adjacent. Some of the possible

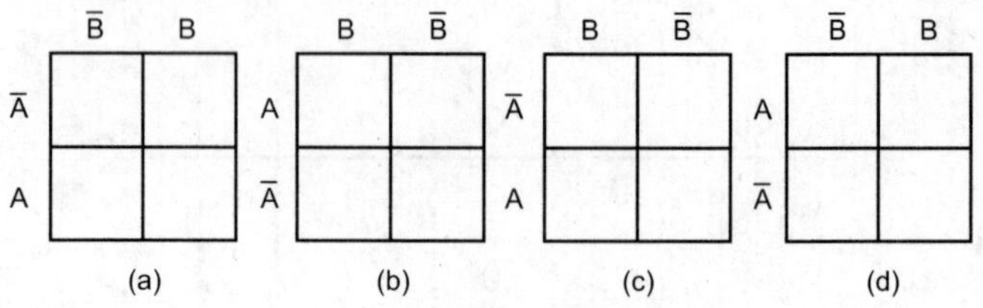

Fig. 22.17

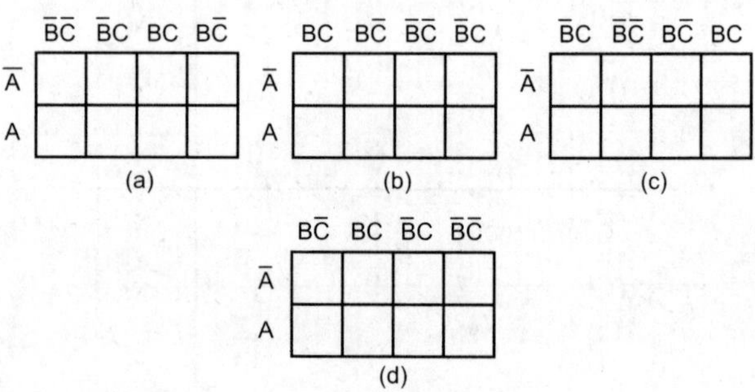

Fig. 22.18

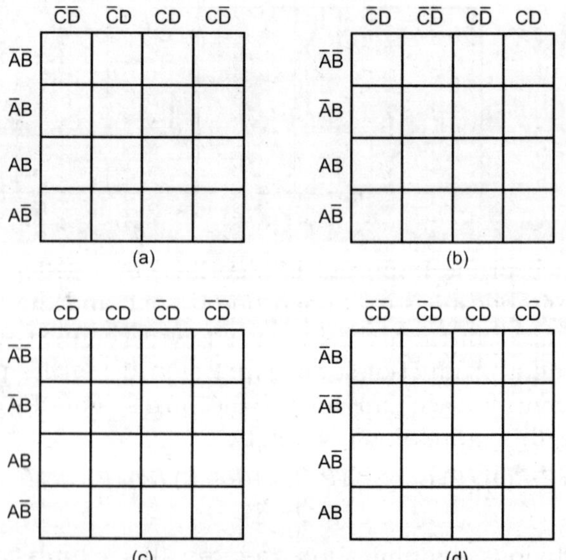

Fig. 22.19

designation styles for 2, 3 and 4 variable minterm Karnaugh maps are given in Figs. 22.17, 22.18 and 22.19 respectively.

The style of row identification need not be the same as that of column identification as long as it meets the basic requirement with respect to adjacent terms. It is, however, a practice to adopt a uniform style of row and column identification. Also, the style shown in Figs. 22.17 (*a*), 22.18 (*a*) and 22.19 (*a*) are more commonly used.

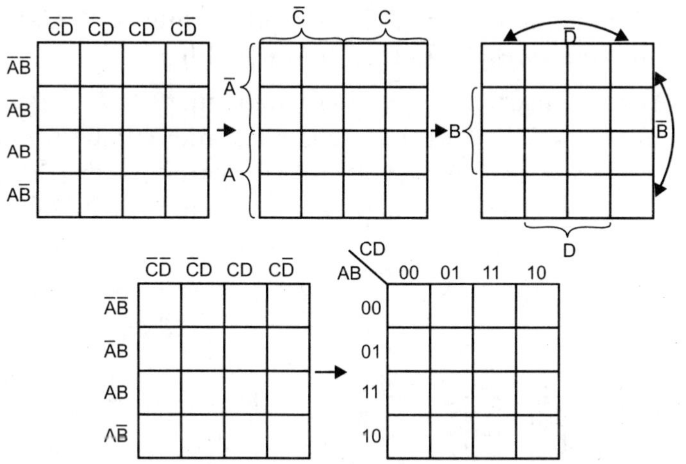

	$\overline{C}\,\overline{D}$	$\overline{C}D$	CD	$C\overline{D}$
$\overline{A}\,\overline{B}$				
$\overline{A}B$				
AB				
$A\overline{B}$				

→

AB \ CD	00	01	11	10
00				
01				
11				
10				

Fig. 22.20

Some more styles are shown in Fig. 22.20. A similar discussion applies for maxterm Karnaugh maps.

Q. 73. With the help of suitable examples, describe step-by-step procedure to simplify a given Boolean expression by Karnaugh map method.

Ans. The Karnaugh method of simplifying Boolean expression will be explained with the help of following three expressions :

Truth table

A	B	Y
0	0	0
0	1	1
1	0	1
1	1	1

	$\overline{B}$	B
$\overline{A}$		1
A	1	1

Sum-of-products K-map

	$\overline{B}$	B
$\overline{A}$		
A		1

Product-of-suns K-map

Fig. 22.21

1. $y = \overline{A}B + A\overline{B} + AB$......2-variable expression
2. $y = \overline{A}\,\overline{B}\,\overline{C} + \overline{A}B\,\overline{C} + A\,\overline{B}\,\overline{C} + AB\,\overline{C}$....3-variable expression
3. $y = \overline{A}\,\overline{B}\,\overline{C}\,\overline{D} + \overline{A}\,\overline{B}\,\overline{C}\,D + \overline{A}\,B\,\overline{C}\,\overline{D} + \overline{A}B\,\overline{C}D + AB\,\overline{C}\,\overline{D} + AB\,\overline{C}D + A\overline{B}\,\overline{C}D$.... 4-variable expression.

1. $y = \overline{A}\,B + A\,\overline{B} + AB$ (sum-of-products). Corresponding product-of-sums expression is

   $y = A + B$ (product-of-sums). Fig. 22.21 shows the truth table and K-maps

2. $y = \overline{A}\,\overline{B}\,\overline{C} + \overline{A}B\,\overline{C} + A\,\overline{B}\,\overline{C} + AB\overline{C}$. Corresponding Product-of-sums expression is

   $y = (A + B + \overline{C})\,(A + \overline{B} + \overline{C})\,(\overline{A} + B + \overline{C})\,(\overline{A} + \overline{B} + \overline{C})$.

   Fig. 22.22 shows the truth table and the K-maps

A	B	C	Y
0	0	0	1
0	0	1	0
0	1	0	1
0	1	1	0
1	0	0	1
1	0	1	0
1	1	0	1
1	1	1	0

	$\overline{B}\overline{C}$	$\overline{B}C$	BC	$B\overline{C}$
$\overline{A}$	1			1
A	1			1

Sum-of-products K-map

	$\overline{B}+\overline{C}$	$\overline{B}+C$	$B+C$	$B+\overline{C}$
$\overline{A}$	1			1
A	1			1

Product-of-sums K-map

Fig. 22.22

Truth Table

A	B	C	D	Y
0	0	0	0	1
0	0	0	1	1
0	0	1	0	0
0	0	1	1	0
0	1	0	0	1
0	1	0	1	1
0	1	1	0	0
0	1	1	1	0
1	0	0	0	1
1	0	0	1	1
1	0	1	0	0
1	0	1	1	0
1	1	0	0	1
1	1	0	1	1
1	1	1	0	0
1	1	1	1	0

	$\overline{C}\,\overline{D}$	$\overline{C}D$	CD	$C\overline{D}$
$\overline{A}\,\overline{B}$	1	1		
$\overline{A}B$	1	1		
AB	1	1		
$A\overline{B}$	1	1		

Sum-of-products K -map

	$\bar{C}+\bar{D}$	$\bar{C}+D$	$C+D$	$C+\bar{D}$
$\bar{A}+\bar{B}$	1	1		
$\bar{A}+B$	1	1		
$A+B$	1	1		
$A+\bar{B}$	1	1		

Product-of-suns K-map

Fig. 22.23

2. $y = \bar{A}\,\bar{B}\,\bar{C}\,\bar{D} + \bar{A}\,\bar{B}\,\bar{C}\,D + \bar{A}B\,\bar{C}\,\bar{D} + \bar{A}B\,\bar{C}\,D + A\bar{B}\,\bar{C}\,\bar{D} + A\bar{B}\,\bar{C}D + AB\bar{C}D$. Corresponding product-of-sums expression is

$$y = (A + B + \bar{C} + D)\,(A + B + \bar{C} + \bar{D})\,(A + \bar{B} + \bar{C} + D)\,(A + \bar{B} + \bar{C} + \bar{D})\,(\bar{A} + B + \bar{C} + D)\,(\bar{A} + B + \bar{C} + \bar{D})\,(\bar{A} + \bar{B} + \bar{C} + D)\,(\bar{A} + \bar{B} + \bar{C} + \bar{D})$$

Fig. 22.23 shows the truth table and the K-maps.

Having drawn the K-map, the next step is to form groups of l's as per the following guidelines:

1. Each square containing a '1' must be considered at least once, though it can be considered as often as desired.
2. The objective should be to account for all the marked squares in the minimum number of groups.
3. The number of squares in a group must always be a power of 2, that is, groups can have 1, 2, 4, 8, 16.....squares.
4. Each group should be as large as possible which means that a square should not be accounted for by itself, if it can be accounted for by a group of two squares; a group of two squares should not be made, if the involved squares can be included in group of four squares and so or on.
5. Optional entries can be used in accounting for all of 1-squares to make optimum groups. Optional entries are marked 'X' in the corresponding squares. It is, however, not necessary to account for all optional entries. Only those optional combinations that can be used to advantage should be used.

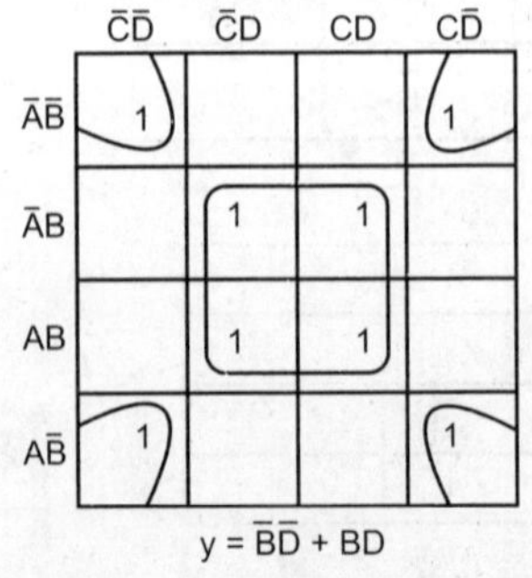

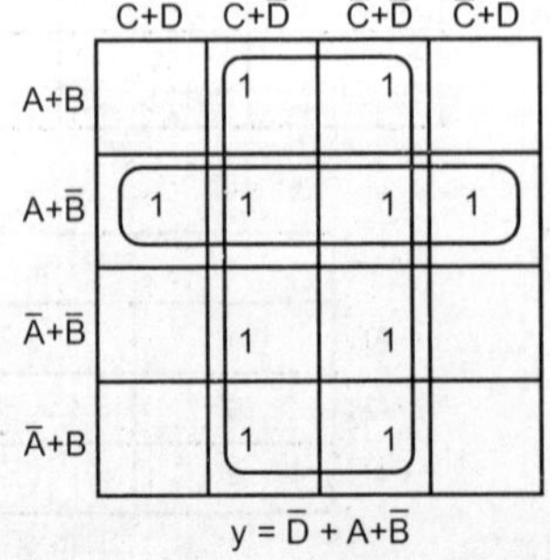

Fig. 22.24

The minimum 'sum-of-products' or the 'product-of-sums' expressions can be written directly from Karnaugh map after groupings have been done. The process is illustrated below in Fig, 22.24 for 4-variable minterm and maxterm Karnaugh maps.

Q. 74. Briefly describe the procedure for using Karnaugh map method for simplification of Boolean expressions with more than four variables.

Ans. Construction of Karnaugh maps for larger number of variables is a complex and cumbersome exercise though manageable upto six variables. Five and six variable Karnaugh maps are shown in Figs. 22.25 (*a*) and (*b*) respectively.

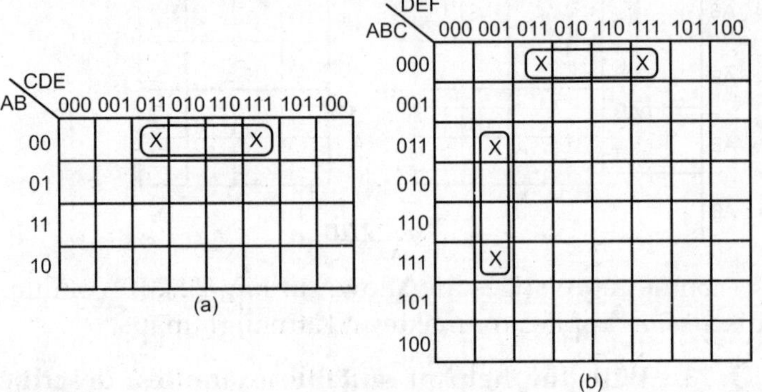

Fig. 22.25

One important point to remember while forming groups in the Karnaugh maps involving more than four variables is that terms equidistant from central, horizontal and vertical lines are considered adjacent. These lines are shown thicker in Figs. 22.25 (*a*) and (*b*). Squares marked 'X' in Figs. 22.25 (*a*) and (*b*) are adjacent and therefore can be grouped.

The Boolean expressions with more than four variables can also be represented by more than one four variable maps. Five, six, seven and eight variable Boolean expressions can be represented by two, four, eight and sixteen 4-variable maps respectively. In general, an n-variable Boolean expression can be represented by 2^{n-4} 4-variable maps. In such multiple maps, groups are made as before except that in addition to adjacencies discussed earlier, corresponding squares in two adjacent maps are also considered adjacent and can therefore be grouped. Examples below illustrate the procedure :

Consider the 5-variable Boolean expression given by:

$$A\bar{B}\,\bar{C}DE + A\bar{B}CDE + \bar{A}BC\,\bar{D}E + AB\bar{C}DE + \bar{A}\,\bar{B}CD\bar{E} + \bar{A}BCDE + ABCD\bar{E} + A\,\bar{B}CD\bar{E} + \bar{A}BC\bar{D}\,\bar{E}$$

The multiple Karnaugh map for this 5-variable expression is shown in Fig. 22.26. The construction of the Karnaugh map and formation of groups are self-explanatory.

The minimised expression is given by:

$$CD\bar{E} + \bar{A}BC\bar{D} + A\bar{C}DE + A\bar{B}DE$$

As another illustration, consider a six-variable expression given by:

$$\bar{A}BC\bar{D}\,\bar{E}\,\bar{F} + AB\bar{C}D\bar{E}F + \bar{A}\,\bar{B}\,\bar{C}\,\bar{D}\,\bar{E}\,\bar{F} + ABCDEF + A\bar{B}CDE\bar{F} + \bar{A}\,\bar{B}\,\bar{C}\,\bar{D}\,\bar{E}\,F + \bar{A}\,B\,\bar{C}\bar{D}E\bar{F}$$

The Karnaugh map for this expression comprising of four 4-variable k-maps is given in Fig. 22.27 alongwith formation of groups.

The minimised expression is given by:

$$\bar{A}\,\bar{B}\,\bar{C}\,\bar{D}\,\bar{E} + \bar{A}\,B\,\bar{C}\bar{D}\,\bar{F} + A\bar{B}CDE\bar{F} + AB\bar{C}D\bar{E}F + ABCDEF$$

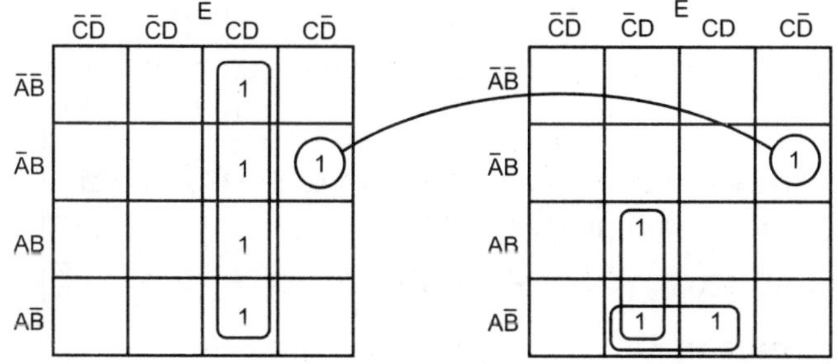

Fig. 22.26

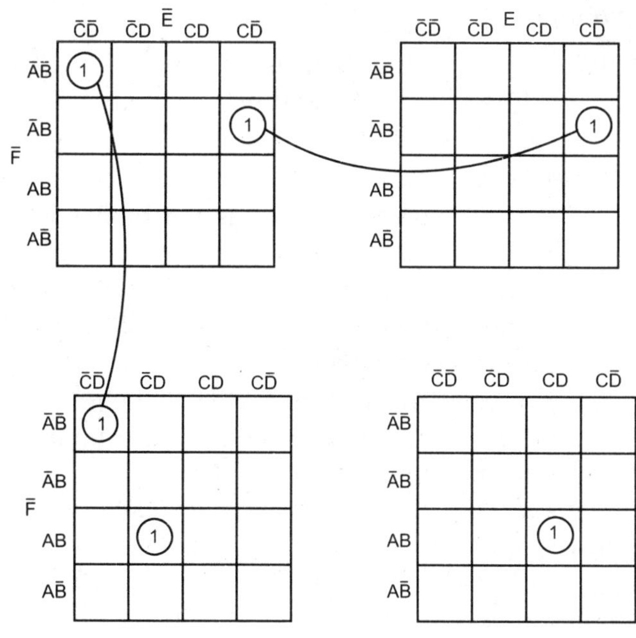

Fig. 22.27

Solved Problems

Problem 22.1. *Find the BINARY WEIGHTS for the bits shown within squares for the mixed binary numbers shown below :*

(*a*) 1 1 [1] 1 1 . 1 1 [1]

(*b*) 1 0 1 0 [1] . 1 [0] 0

(*c*) [1] 0 0 0 0 0 0 0 0 . 0 0 0 [1]

Solution.

(*a*) For the integer part starting from the LSB (Least Significant Bit) position and proceeding towards left, the binary weights are 2^0 (= 1), 2^1 (= 2) 2^2 = (4), 2^3 (= 8) and so on. The bit within the square is third from right and its binary weight is $2^2 = 4$. For the fractional part, starting from the leftmost position, the binary weights are 2^{-1}, 2^{-2}, 2^{-3}, 2^{-4}.........The binary weight for the bit within the square is therefore (2^{-3}).

(*b*) Integer part: $2^0 = 1$, Fractional part: $2^{-2} = 0.25$

Integer part: 2, Fractional part: $2^{-4} = 0.0625$.

Problem 22.2. *Find the decimal equivalents of the binary numbers of Problem 22.1.*

Solution.

(*a*) Binary Number : 11111. 111

Decimal equivalent of the integer part is given by:

$$1 \times 2^0 + 1 \times 2^1 + 1 \times 2^2 + 1 \times 2^3 + 1 \times 2^4$$

$$1 + 2 + 4 + 8 + 16 = 31$$

Decimal equivalent of the fractional part is given by :

$$1 \times 2^{-1} + 1 \times 2^{-2} + 1 \times 2^{-3}$$

$$= 0.5 + 0.25 + 0.125 = 0.875$$

The decimal equivalent of the given binary number is therefore, 31.875

(*b*) Binary Number : 10101.100

Decimal equivalent of the integer part is given by:

$$1 \times 2^0 + 0 + 1 \times 2^2 + 0 + 1 \times 2^4 = 1 + 4 + 16 = 21$$

Decimal equivalent of the fractional part is given by :

$$1 \times 2^{-1} + 0 + 0 = 0.5$$

Decimal equivalent of 10101.100 is therefore, 21.5

(*c*) Binary Number = 100000000 . 0001

Decimal equivalent of the integer part is given by:

$$0 + 0 + 0 + 0 + 0 + 0 + 0 + 0 + 1 \times 2^8 = 256$$

Decimal equivalent of the fractional part is given by:

$$0 + 0 + 0 + 1 \times 2^{-4} = 0.03125$$

Decimal equivalent of the given binary number is therefore, 256.03125.

Problem 22.3. *Below are given some binary numbers in 2's complement notation. Write their decimal equivalents, (i) 01101 (ii) 10011.*

Solution.

(*i*) It is a positive number as the MSB position shows a '0'. The magnitude of this number is

$1 \times 2^0 + 0 \times 2^1 + 1 \times 2^2 + 1 \times 2^3 = 13$

Thus, 0 1101 represents + 13 in 2's complement form

(*ii*) It is a negative number as the MSB position shows a '1'. To find the magnitude of this number, we find the 2's complement of this number to know the magnitude bits for the positive counterpart of this number.

2's complement of 0011 is 1101 and 1101 represents 13. Thus 1 0011 in 2's complement notation represents –13.

Problem 22.4. *Find the decimal numbers represented by following binary numbers given in Sign-Bit Magnitude notation, 0001, 0011,1101, 1111.*

Solution.

0001 represents + 1

0011 represents + 3

1101 represents – 5

1111 represents – 7.

Problem 22.5. *Convert the following decimal numbers into their corresponding binary equivalents.*

(*i*) 31

(*ii*) 16.73.

Solution.

Binary equivalent of 31 is 11111

(*ii*) Integer Part

	Integer	Remainder
2	16	–
	8	0
	4	0
	2	0
	1	0
	0	1

Binary equivalent of 16 = 1 0 0 0 0

Fractional Part

0.73 × 2 = 0.46 with a CARRY of 1

0.46 × 2 = 0.92 with a CARRY of 0

0.92 × 2 = 0.84 with a CARRY of 1

0.84 × 2 = 0.68 with a CARRY of 1

0.68 × 2 = 0.36 with a CARRY of 1

The binary equivalent of 0.73 = 10 111

and the binary equivalent of 16.73 is therefore, 10000.10111.

Problem 22.6. *(a) Write the first twenty numbers in the octal number system, (b) Convert $(169)_{10}$ into an equivalent octal number, (c) Convert $(203)_8$ into an equivalent decimal number.*

Solution.

(*a*) 0, 1, 2, 3, 4, 5, 6, 7, 10, 11, 12, 13, 14, 15, 16, 17, 20, 21, 22, 23.

(*b*)

	169	Remainder
8	21	1
	2	5
	0	2

The remainders written in the reverse order give the equivalent octal number as $(251)_8$

(*c*) $(203)_8 = 3 \times 8^0 + 0 \times 8^1 + 2 \times 8^2 = (131)_{10}$

Problem 22.7. *Do the following conversions:*

(*a*) $(532.78)_{10} = (?)_8$

(*b*) $(234.55)_8 = (?)_{10}$

Solution.

(*a*) Integer Part.

	532	Remainder
8	66	4
	8	2
	1	0
	0	1

Writing the remainders in the reverse order, the octal equivalent of $(532)_{10}$ is $(1024)_8$

Fractional Part

0.78 × 8 = 0.24 with a CARRY of 6

0.24 × 8 = 0.92 with a CARRY of 1

0.92 × 8 = 0.36 with a CARRY of 7

Writing the CARRY in the forward order, the octal equivalent of the decimal fraction $(0.78)_{10}$ is $(0.617)_8$. Therefore, combining the two results, we get,

$$(532.78)_{10} = (1024.617)_8$$

(*b*) $(234.55)_8$

$= (4 \times 8^0 + 3 \times 8^1 + 2 \times 8^2) . (5 \times 8^{-1} + 5 \times 8^{-2})$

$= (4 + 24 + 128) . (0.625 + 0.078) = (156.703)_{10}$.

Problem 22.8. *Do the following number conversions :*

(*a*) $(1BF.25)_{16} = (?)_{10}$ (*b*) $(65.35)_{10} = (?)_{16}$.

Solution.

(*a*) $(1BF)_{16} = 15 \times 16^0 + 11 \times 16^1 + 1 \times 16^2$.........($F \rightarrow 15$ and $B \rightarrow 11$)

$= 15 + 176 + 256 = 447$

$(.25)_{16} = .(2 \times 16^{-1} + 5 \times 16^{-2})_{10}$

$= .(2/16 + 5/256)_{10} = (.14)_{10}$

Therefore,

$(1BF.25)_{16} = (447.14)_{10}$

(*b*) Integer Part

	65	Remainder
16	4	1
	0	4

Therefore, $(65)_{10} = (41)_{16}$

Fractional Part

$0.35 \times 16 = 0.60$ with a CAERY of 5

$0.60 \times 16 = 0.60$ with a CARRY of 9

Therefore, $(0.35)_{10} = (0.59)_{16}$

Combining the results of the integer and fractional parts, $(65.35)_{10} = (41.59)_{16}$

Problem 22.9. *Mark the forbidden 4-bit groups in 8421 BCD code in the following. 1001, 0110, 0101, 0000, 1111, 1101, 1011.*

Solution.

The forbidden groups are 1111,1101 and 1011. The reasons are given in the text.

Problem 22.10. *Write the decimal equivalents of the following 8421 BCD numbers*

(*a*) 00111001.01110101

(*b*) 1000000

(*c*) 10000001.0001

Solution.

(*a*) BCD Number = 00111001.01110101

= 0011 1001 . 0111 0101

Equivalent Decimal number = 39.75

(*b*) BCD Number = 10000000

= 1000 0000

Equivalent Decimal number = 80

(*c*) BCD Number = 10000001 . 0001

= 1000 0001 . 0001

Equivalent Decimal number

= 8 1 .1

Problem 22.11. *Identify the invalid BCD numbers in the following:*

01011101 . 0010, 01001001 . 0100,

1101111 . 0111, 01111000.0110

Solution.

The invalid BCD numbers are :

01011101.0010 as 01011101.0010 → 01011101.0010 and 1101 has no decimal equivalent in BCD notation.

and 11011111. 0111 as 11011111. 0111 → 11011111. 0111 and 1101,1111 are forbidden groups.

Problem 22.12. *Write the 8421 BCD equivalents of the following binary numbers :*

(*a*) 110101

(*b*) 10111.011.

Solution.

(*a*) The decimal equivalent of the given binary number is given by :

$1 \times 2^0 + 0 \times 2^1 + 1 \times 2^2 + 0 \times 2^3 + 1 \times 2^4 + 1 \times 2^5 = 53$

The 8421 BCD equivalent = 0101 0011

(*b*) Binary-decimal conversion

Integer part: $(10111)_2 = 1 \times 2^0 + 1 \times 2^1 + 1 \times 2^2 + 0 \times 2^3 + 1 \times 2^4 = 23$

Fractional part: .011

$= 0 + 0.25 + 0.125 = 0.375$

therefore, $(10111.011)_2 = (23.375)_{10}$

Decimal-BCD Conversion

$(23.375)_{10} = (0010\ 0011\ .\ 0011\ 0111\ 0101)_{BCD}$

$= (00100011\ .\ 001101110101)_{BCD}$

Problem 22.13. (*i*) Find the Gray numbers for the binary numbers 101101 and 101001101.

Solution.

101101 (Binary) = 111011 (Gray) → 1 0 1 1 0 1
1 1 1 0 1 1

101001101 (Binary) – 111101011 (Gray) → 1 0 1 0 0 1 1 0 1
1 1 1 1 0 1 0 1 1

(*ii*) Find the Binary numbers for the Gray Numbers 110101101 and 101101011

Solution.

110101101 (Gray) = 100110110 (Binary) → 1 1 0 1 0 1 1 0 1
1 0 0 1 1 0 1 1 0

101101011 (Gray) = 110110010 (Binary) → 1 0 1 1 0 1 0 1 1
1 1 0 1 1 0 0 1 0

Problem 22.14. *Find the Gray Code equivalent of the decimal number 26.*

Solution.

The straight binary equivalent of 26 is first determined as follows:

	26	Remainder
2	13	0
	6	1
	3	0
	1	0
	1	1
	0	1
		$(26)_{10} = (11010)_2$

The Gray Code equivalent of $(11010)_2$ is then found out as follows:

Binary = 1 1 0 1 0

Gray = 1 0 0 1 1

The Gray Code equivalent of $(26)_{10}$ is, therefore, 10011.

Problem 22.15. *Write the Excess-3 equivalent of the following decimal numbers with each decimal number expressed as three 4-bit groups:*

8,17, 263

Solution.

8 can also be expressed as 008. The Excess-3 equivalent of 008 is : 0011 0011 1011.

Similarly, Excess-3 equivalent of 17 is given by 0011 0100 1010 and Excess-3 equivalent of 263 is (0101 1001 0110).

Problem 22.16. *List all possible 4-bit binary gray code numbers along with parity bits to have (i) odd parity and (ii) even parity.*

Solution. Refer to Table 22.14.

Table 22.14

Grey Code	*Parity Bit (ODD)*	*Parity Bit (EVEN)*	*Grey Code*	*Parity Bit (ODD)*	*Parity Bit (EVEN)*
0000	1	0	1100	1	0
0001	0	1	1101	0	1
0011	1	0	1111	1	0
0010	0	1	1110	0	1
0110	1	0	1010	1	0
0111	0	1	1011	0	1
0101	1	0	1001	1	0
0100	0	1	1000	0	1

Problem 22.17. *Determine the floating-point representation of $(-142)_{10}$ using the IEEE single-precision format.*

Solution.

As a first step, we will determine the binary equivalent of $(142)_{10}$. Following the procedure outlined in an earlier part of the chapter, the binary equivalent can be written as:

$$(142)_{10} = (10001110)_2$$

$$(10001110)_2 = 1.000\,1110 \times 2_7 = 1.0001110e + 0111$$

The mantissa = 0001110 00000000 00000000

The exponent = 00000111

The biased exponent = 00000111 + 01111111 = 10000110

The sign of the mantissa = 1

Therefore, $(-142)_{10}$ = 11000011 00001110 00000000 00000000

Problem 22.18. *Add the following straight binary numbers and verify the results by showing equivalent decimal addition:*

(*a*) $(110101)_2 + (100100)_2$

(*b*) $(1010)_2 + (111111)_2$

(*c*) $(1011.110)_2 + (1100.01)_2$

Solution.

```
          (1)
     1  1  0  1  0  1  ............   5 3
(a) +1  0  0  1  0  0  ............ + 3 6
  1  0  1  1  0  0  1                 8 9

  (1) (1) (1) (1)
     0  0  1  0  1  0  ............   1 0
(b) +1  1  1  1  1  1  ............ + 6 3
  1  0  0  1  0  0  1                 7 3
```

One of the numbers here has four digits while the other has six bits. As the given numbers are integers, the 4-bit number can be expressed as a 6-bit number by adding 0s towards left without making any difference.

```
    (1) (1) (1) (1)    (1)
     1   0   1   1  .  1  1  0  .....    1 1 . 75
(c) +1   1   0   0  .  0  1  0  .....  + 1 2 . 25
  1  1   0   0   0  .  0  0  0           2 4 . 00
```

The fractional part of one of the numbers has only two bits while the other had three bits. 2-bit fractional part can be expressed as a 3-bit fraction by adding 0s towards right without making any difference.

Problem 22.19. *Subtract $(10110.011)^2$ from $(11010.10)^2$ and verify the result by showing equivalent decimal subtraction.*

Solution.

The minuend is first modified to have same number of bits in the integer and fractional parts as the subtrahend by adding 0s as explained in problem 1.

```
  1 1 0 1 0 . 1 0 0 .....   26 . 500
  1 0 1 1 0 . 0 1 1 .....   22 . 375
  -----------------        ---------
  0 0 1 0 0 . 0 0 1          4 . 125
```

Problem 22.20. *Add (0010 0011 0110) and (0100 0101 1001) and verify the result by showing equivalent decimal addition.*

Solution.

The Excess-3 equivalents of (0010 0011 0110) and (0100 0101 1001) are (0101 0110 1001) and (0111 1000 1100) respectively.

```
   0 1 0 1     0 1 1 0     1 0 0 1
  +0 1 1 1     1 0 0 0     1 1 0 0
  --------------------------------
   1 1 0 0     1 1 1 1     0 1 0 1
   0 0 1 1    -0 0 1 1    +0 0 1 1
  --------------------------------
   1 0 0 1     1 1 0 0     1 0 0 0
```

Therefore, (1001 1100 1000) is Excess-3 representation of the result of addition

Addition of Decimal Equivalents

Decimal equivalent of 0010 0011 0110 is $(236)_{10}$

Decimal equivalent of 0100 01011001 is $(459)_{10}$

Their addition yields $(695)_{10}$

The excess-3 BCD representation of $(695)_{10}$ is (1001 1100 1000)

Which agrees with the result found above.

Problem 22.21. *Can you use 16-bit 2's complement arithmetic to add 15690 and 18310?*

Solution.

16-bit 2's complement arithmetic has a range of (-2^{15}) to $+(2^{15}-1)$ *i.e.* – 32768 to + 32767. Addition of decimal numbers 15690 and 18310 gives 34000. Thus, 16-bit 2's complement notation, if used to perform this addition will produce incorrect result.

Problem 22.22. *Add – 110 and – 32 first using 8-bit 2's complement arithmetic and then using 16-bit 2's complement arithmetic. Comment on the results.*

Solution.

– 110 in 8-bit 2's complement form is 10010010

– 32 in 8-bit 2's complement form is 11100000

```
  10010010
  11100000
  --------
  01110010   Final carry disregarded
```

The carry in the 9th bit position is ignored, as we are dealing with 8-bit arithmetic.

– 110 in 16-bit 2's complement form is 1111111110010 010

– 32 in 16-bit 2's complement form is 1111111111100000

```
  1111111110010010
  1111111111100000
  ----------------
  1111111101110010
```

The carry in the 17th bit position is ignored, as we are dealing with 16-bit arithmetic.

01110010 (the result obtained in 8-bit solution) represents +114 in 2's complement form and 1111111101110010 (the result obtained in 16-bit solution) represents -142 in 2's complement form.

The 16-bit 2's complement arithmetic has produced the correct result, as the expected result lies within the range of 16-bit 2's complement notation.

Problem 22.23. *Subtract (i) – 64 from +32 (ii) 29H from 4FH ('H' only tells that it is hexadecimal number) using 8-bit 2's complement arithmetic.*

Solution.

(*i*) + 32 is expressed as 00100000

– 64 is expressed as 11000000

2's complement of-64 is given by 01000000

(+32) – (– 64) is therefore,

```
   00100000
 + 01000000
 ----------
   01100000
```

Decimal equivalent of 01100000 is + 96

(*ii*) 4FH = $(79)_{10}$ and 29H = $(41)_{10}$

+79 is expressed as 01001111

+41 is expressed as 00101001

2's complement of + 41 is 11010111

(+ 79) – (+41) is therefore

```
   01001111
 + 11010111
 ----------
   00100110   with final carry disregarded.
```

Decimal equivalent of 00100110 is +38 and its hex equivalent is 26H. Therefore, 4FH – 29E = 26H.

Problem 22.24. *Subtract (i) 80H from C0H (ii) $(52)_8$ from $(65)_8$.*

Solution.

(*i*) C0H = 11000000

80 H = 10000000

2's complement of 10000000 is 10000000 only.

Therefore, C0H – 80H is given by:

```
  11000000
+ 10000000
  01000000  with final carry disregarded.
```

01000000 = 40 H = $(64)_{10}$

(*ii*) $(65)_8$ = 110101 = 00110101

$(52)_8$ = 101010 = 00101010

(0s have been added to make it 8-bit long)

2's complement of $(52)_8$ is 11010110

$(65)_8 - (52)_8$ is therefore

```
  00110101
+ 11010110
  00001011  with final carry disregarded.
```

00001011 is equivalent to 001011 (two 0s at the extreme left have been omitted to make it 6-bit long so that it can be directly converted into an octal number). 001011 is equivalent to $(13)_8$.

Therefore, $(65)_8 - (52)_8 = (13)_8$.

Problem 22.25. *Multiply the following using ADD AND RIGHT SHIFT algorithm.*

(*i*) $(10000)_2 \times (100)_2$

(*ii*) *7BH × 3H*

Verify the results by doing equivalent decimal multiplication.

Solution. (*i*)

10000 × 100	Multiplicand Multiplier
00000 +00000	START
00000	First Addition
00000 +00000	0 (shifted 1-bit to right)
00000	Second Addition
00000 +10000	00 (Shifted 1-bit to right)
10000	Third Addition
01000	000 (shifted 1-bit to right)

Multiplication Result = 01000000

$(10000)_2 = (16)_{10}$ and $(100)_2 = (4)_{10}$

$16 \times 4 = 64\ (1000000)_2$

(*ii*) 7B H = 01111011 and 3 H = 0011 = 11

01111011 × 11	Multiplicand Multiplier
00000000 +01111011	START
+01111011 +01111011	First Addition →1
10111000 01011100	Second →01

All Multiplier bits exhausted

Multiplication Result: 0101110001 = 000101110001 = 171H

(Remember HEX-BINARY Conversion)

Let us verify the result by doing equivalent decimal multiplication.

7BH = $(123)_{10}$, 3 H = $(3)_{10}$

$123 \times 3 = 369 = 171$ H.

Problem 22.26. *Use "Repeated Right Shift and Subtract" algorithm to dividend (11010110 by 1011. Determine both the integer as well as fractional parts of the quotient. The fractional^ part may be determined upto four bit places.*

Solution.

The sequence of operations is tabulated. The operations are self-explanatory.

1	11010110 −1011	Dividend Divisor
0	00100 −1011	
0	001001 −1011	
1	0010011 −1011	
	0001000	
1	00010000 −1011	
	00000101	
0	000001010 −1011	

Cont...

1	0000010100 −1011	
1	00000010010 −1011	
	00000000111	
1	1110 −1011	
	0011	

Quotient = 10011.0111

Problem 20.27. *Use "Repeated Subtract and Left Shift" algorithm to divide (100011) by (100) to determine both integer and fractional parts of quotient. Verify the result by showing equivalent decimal division.*

Solution.

The sequence of operations as listed ahead are self-explanatory.

Quotient 1	100 −100	011	Divided Divisor
	000		
0	000 −100	11	
	100 +100	11	
	000	11	
0	0001 −100	1	
	101 +100	1	
	001	1	
0	0011 −100		
	111 +100		
1	0110 −100		
	0010		
1	100 −100		
	000		

Quotient = $1000.11 = (8.75)_{10}$

Now $100011 = (35)_{10}$ and $100 = (4)_{10}$, 35 divided by 4 gives 8.75.

Problem 22.28. *In the problems solved so far, we have not talked about the sign of numbers entering the division. We have only mentioned that the sign of the quotient would depend upon the signs of the dividend and the divisor. Divide AFH by 09H using the method of repeated right shift and subtract keeping in mind the signs of the numbers.*

Solution.

Dividend = AFH

As it is a negative hexadecimal number, the magnitude of this number is determined by its 2's complement (or more precisely by its 16's complement in the hexadecimal, numbers language).

16's complement of AFH = 51H ...... (H only signifies that it is a hex number)

The binary equivalent of 51H = 01010001 = 1010001

Divisor = 09 H

It is a positive number. Binary equivalent of 09H = 00001001

As the dividend is a negative number and the divisor a positive number, the quotient will be a negative number.

1	01010001 −1001	Divisor less than Dividend
	0001	
0	00010 −1001	Divisor greater than dividend
0	000100 −1001	Divisor still greater than Dividend
1	0001001 −1001	Divisor less than dividend
	0000000	

Quotient = 1001 = 09 H

As the quotient should be a negative number, it must be represented by 16's complement of 09H *i.e.* F7H.

Therefore, AFH divided by 09 H = F7H

Problem 22.29. *Add* 0.110101×2^3 *and* 0.101011×2^4.

Solution.

First 0.110101×2^3 is scaled to have an exponent value of 4, *i.e.*, $110101 \times 2^3 = 0.0110101 \times 2^4$, also one '0' is added to the extreme right of the mantissa of the other number so that both the numbers have same number of bits in their fraction mantissas. Adding '0' to the extreme right of a fractional part makes no difference to its value. (0.101011×2^4) can be written as (0.1010110×2^4).

Adding the two mantissas gives :

$$\begin{array}{r} 0.0110101 \\ 0.1010110 \\ \hline 1.0001011 \\ \hline \end{array}$$

Therefore, $(0.110101 \times 2^3) + (0.101011 \times 2^4)$

$= 1.0001011 \times 2^4$

$= 0.10001011 \times 2^5$

Problem 22.30. Subtract (*i*) 0.11011×2^4 from 0.110111×2^5 and (*ii*) $0.6\text{ B} \times 16^2$ from $0.7\text{A} \times 16^3$. Verify the results by showing equivalent decimal subtraction.

(*i*) $0.11011 \times 2^4 = 0.011011 \times 2^5$

Therefore, $(0.110111 \times 2^5 - 0.011011 \times 2^5)$

$= 0.011100 \times 2^5$

(Subtracting the two Mantissas)

$= 0.111000 \times 2^4$

$0.11011 \times 2^4 = 1101.1 = (13.5)_{10}$

$0.110111 \times 2^5 = 11011.1 = (27.5)_{10}$

Therefore, subtraction of the two numbers should give $(14)_{10}$. Also, $111000 \times 2^4 = 1110.00 = (14)_{10}$ and hence the proof.

(*ii*) $0.6\text{B} \times 16^2 = 0.06\text{ B} \times 16^3$

$0.110111 \times 25 = 11011.1 = (27.5)_{10}$

Also, $0.7\text{A} \times 16^3 = 0.7\text{A}0 \times 16^3$

Therefore, $(0.7\text{A} \times 16^3 - 0.6\text{B} \times 16^2)$

$= (0.7\text{A}0 \times 16^3 - 0.06\text{B} \times 16^3)$

$= 0.735 \times 16^3$

$0.6\text{B} \times 16^2 = 6\text{BH} = (107)_{10}$

$0.7\text{A} \times 16^3 = 7\text{A}0\text{H} = 0 + 16 \times 10 + 256 \times 7 = (1952)_{10}$

Therefore, subtraction of the two numbers should give $(1845)_{10}$

Also, $0.735 \times 16^3 = 735\text{ H}$

$= 5 \times 16^0 + 3 \times 16^1 + 7 \times 16^2 = (1845)_{10}$ and hence, the proof.

Problem 22.31. *Find the dual of:*

(*a*) $(A + \overline{B})(\overline{A} + B)(\overline{A} + \overline{B})$

(*b*) $A\overline{B} + B\overline{C} + C\overline{D}$.

Solution.

(*a*) The dual of $(A + \overline{B})\ (\overline{A} + B)\ (\overline{A} + \overline{B})$ is given by $A\,.\,\overline{B} + \overline{A}\,.\,B + \overline{A}\,.\,\overline{B}$

(*b*) The dual of $A\overline{B} + B\overline{C} + C\overline{D}$ is given by $(A + \overline{B})\ (B + \overline{C})\ (C + \overline{D})$.

Problem 22.32. *Simplify (a)* $(A + \overline{B} + C)\ \overline{A}B\overline{C}D$ *(b)* $(C + \overline{D} + \overline{C}D)\overline{E}$.

Solution.

(*a*) Since $\overline{A}B\overline{C}$ is complement of $(A + \overline{B} + C)$, therefore, $(A + \overline{B} + C)\overline{A}B\overline{C} = 0$ which gives

$(A + \overline{B} + C)\ \overline{A}B\overline{C}D = 0$

(*b*) Since $\overline{C}D$ is complement of $(C + \overline{D})$, therefore, $C + \overline{D} + \overline{C}D = 1$ which gives

$(C + \overline{D} + \overline{C}D)\overline{E} = 1\,.\,\overline{E} = \overline{E}$.

Problem 22.33. Prove that $(\overline{L}\ \overline{M}\ \overline{N} + \overline{A})(A + L + M + N) = 1$ and $(\overline{L}\ \overline{M}\ \overline{N} + A)\ [\overline{A}\,.\,(L + M + N)] = 0$ given that $\overline{(L + M + N)} = A$.

Solution.

$(\overline{L}\,\overline{M}\,\overline{N} + \overline{A})(A + L + M + N)$

$= (A + \overline{A})\ (A + \overline{A}) = 1.1 = 1\ldots.\ \overline{(L + M + N)} = \overline{L}\,\overline{M}\,\overline{N} = A$

$(L + M + N) = A$

$(\overline{L}\,\overline{M}\,\overline{N} + A)\ [\overline{A}\,.\,(L + M + N)]$

$= (A + A)(\overline{A}\,.\,\overline{A}) = A\,.\,\overline{A} = 0.$

Problem 22.34. Prove that $ABCD + AB\overline{C}\,\overline{D} + ABC\overline{D} + AB\overline{C}D + ABCDE + AB\overline{C}\,\overline{D}\,\overline{E} + AB\overline{C}DE$ equals AB.

Solution.

$ABCD + AB\overline{C}\,\overline{D} + ABC\overline{D} + AB\overline{C}D + ABCDE + AB\overline{C}\,\overline{D}\,\overline{E} + AB\overline{C}DE$

$= ABCD + AB\overline{C}\,\overline{D} + ABC\overline{D} + AB\overline{C}D$

$= AB\,(CD + \overline{C}\,\overline{D} + C\overline{D} + \overline{C}D) = AB$

$ABCD$ appears in $ABCDE$, $AB\overline{C}\,\overline{D}$ appears in $AB\overline{C}\,\overline{D}\,\overline{E}$ and $AB\overline{C}D$ appears in $AB\overline{C}DE$ with the result that $ABCDE$, $AB\overline{C}\,\overline{D}\,\overline{E}$, and $AB\overline{C}DE$ are redundant. Also, C, D appear in all possible combinations and therefore, redundant.

Problem 22.35. *Prove that* $AB\overline{C} + (\overline{A} + \overline{B})D + \overline{C}D + (\overline{B} + C)\ D + AD = AB\overline{C} + D$.

Solution.

$AB\overline{C} + (\overline{A} + \overline{B})D + \overline{C}D + (\overline{B} + C)D + AD$

$= AB\overline{C} + \overline{A}D + \overline{B}D + \overline{C}D + \overline{B}D + CD + AD$

$= AB\overline{C} + D(A + \overline{A}) + D(C + \overline{C}) + \overline{B}D$

$= AB\overline{C} + D + D + \overline{B}D\ldots.\ A + \overline{A} = 1,\ C + \overline{C} = 1$

$= AB\overline{C} + D + \overline{B}D\ \ldots.\ D + D = D$

$= AB\overline{C} + D\ \ldots.\ D$ appears in $\overline{B}D$, which makes $\overline{B}D$ redundant.

Another possible approach is given as under.

$(\overline{A} + \overline{B})$ is complement of (AB). Therefore, from $AB\overline{C}$ and $(\overline{A} + \overline{B})D$, $\overline{C}D$ would be redundant. Also, $(\overline{B} + C)$ is complement of $B\overline{C}$. Therefore, from $AB\overline{C}$ and $(\overline{B} + C)D$, AD would be redundant. The given expression then reduces to

$$AB\overline{C} + (\overline{A} + \overline{B})D + (\overline{B} + C)D$$

$$= AB\overline{C} + D(\overline{A} + \overline{B} + \overline{B} + C)$$

$$= AB\overline{C} + D(\overline{A} + \overline{B} + C)$$

Now, $(\overline{A} + \overline{B} + C)$ is complement of $AB\overline{C}$, therefore, the expression reduces to $AB\overline{C} + D$.

Problem 22.36. *Prove the following :*

(*a*) $L(M + \overline{N}) + \overline{L}\,\overline{P}\,Q = (L + \overline{P}\,Q)(\overline{L} + M + \overline{N})$

(*b*) $(A + \overline{B}\,\overline{C})\overline{D} + D\,\overline{E}(F + \overline{G}) = [D + (A + \overline{B}\,\overline{C})][\overline{D} + \overline{E}(F + \overline{G})]$

(*c*) $[A\,\overline{B} + \overline{C} + \overline{D}]\,[D + (E + \overline{F})G = D(A\,\overline{B} + \overline{C}) + \overline{D}\,G(E + \overline{F})$

Solution.

(*a*) From transposition theorem, $XY + \overline{X}Z = (X + Z)(\overline{X} + Y)$

If we consider $X = L$, $Y = (M + \overline{N})$ and $Z = \overline{P}Q$, then the given expression reduces to the above form.

Therefore, $XY + \overline{X}Z = L(M + \overline{N}) + \overline{L}\,\overline{P}\,Q$

$$= (L + \overline{P}Q)\,(\overline{L} + M + \overline{N})$$

(*b*) If we consider, $X = \overline{D}$, $Y = A + \overline{B}\,\overline{C}$ and $Z = \overline{E}(F + \overline{G})$, then the given expression reduces to $(XY + \overline{X}Z)$. Applying transposition theorem,

$$XY + \overline{X}Z = (X + Z)(\overline{X} + Y)$$

$$= [\overline{D} + \overline{E}(F + \overline{G})][D + A + \overline{B}\,\overline{C}]$$

(*c*) If we consider, $X = \overline{D}$, $Y = A\,\overline{B} + \overline{C}$ and $Z = (E + \overline{F})G$, then the given expression reduces to $(X + Y)(\overline{X} + Z)$. Applying transposition theorem,

$$(X + Y)(\overline{X} + Z) = XZ + \overline{X}Y$$

$$= \overline{D}G\,(E + \overline{F}) + D(A\overline{B} + \overline{C}).$$

Problem 22.37. *Using McCluskey's tabular method, minimise the Boolean function represented by :*

$f(A, B, C, D) = \overline{A}\,\overline{B}\,\overline{C}\,\overline{D} + \overline{A}\,\overline{B}\,C\overline{D} + \overline{A}\,\overline{B}\,CD + \overline{A}B\,\overline{C}\,\overline{D} + \overline{A}B\overline{C}D + \overline{A}BCD + A\overline{B}\,\overline{C}\,\overline{D} + A\overline{B}CD + AB\overline{C}D + +ABC\overline{D}$

Input equal to $A\overline{B}C\overline{D}$ is a don't care condition.

Solution.

Formation of groups, first round of matching, second round of matching, identification of prime implicants and the prime implicant table are all given below. Different steps to obtaining these tables have already been discussed in detail.

A	*B*	*C*	*D*	
0	0	0	0	√
0	0	1	0	√
0	1	0	0	√
1	0	0	0	√
0	0	1	1	√
0	1	0	1	√
1	0	1	0	√
0	1	1	1	√
1	0	1	1	√
1	1	0	1	√
1	1	1	0	√

A	*B*	*C*	*D*	
0	0	–	0	√
0	–	0	0	*
–	0	0	0	√
0	0	1	–	√
–	0	1	0	√
0	1	0	–	*
1	0	–	0	√
0	–	1	1	*
–	0	1	1	√
0	1	–	1	*
–	1	0	1	*
1	0	1	–	√
1	–	1	0	*

A	*B*	*C*	*D*	
–	0	–	0	*
–	0	1	–	*

0000	0010	0011	0100	0101	0111	1000	1011	1101	1110	Prime Implicants	
√			√							0 – 00	L
			√	√						010 –	M
		√			√					0 – 11	N
				√	√					01 – 1	P
				√				√		– 101	Q
									√	1 – 10	R
√	√					√				– 0 – 0	S
	√	√					√			– 01 –	T

A product-of-sums expression that tells about the combination of prime implicants required to account for all the terms is given by :

$$(L + S)(S + T)(N + T)(L + M)(M + P + Q)(N + P)\,(S)\,(T)\,(Q)\,(R)$$

Obvious simplification reduces the expression to

$$STQR\,(L + M)(N + P) = LNRSTQ + LPRSTQ + MNRSTQ + MPRSTQ$$

Now, this sum-of-products expression logically states that all the input combinations can be accounted for by the prime implicants (*L, N, R, S, T, Q*) or (*L, P, R, S, T, Q*) or (*M, N, R, S, T, Q*) or (*M, P, R, S, T, Q*). Since all the options are of the same complexity, any of the options can be considered. If we consider *LPRSTQ*, then the minimised expression is given by :

$$f(A, B, C, D) = \overline{A}\,\overline{C}\,\overline{D} + \overline{A}\,B\,D + A\,\overline{C}\,\overline{D} + \overline{B}\,\overline{D} + \overline{B}\,C + B\,\overline{C}\,D$$

Problem 22.38. *Using Quine-McCluskey's tabular method, find the minimum sum. of products for f(A, B, C, D) =* $\Sigma(1, 2, 3, 9, 12, 13, 14) + \sum_{\phi} 0, 7, 10, 15.$

Solution.

Different steps are tabulated below. Eight prime implicants have been identified. These prime implicants along with the inputs (not including optional inputs) constitute the prime implicant table that is then processed to find the desired prime implicants.

A B C D	
0 0 0 0	√
0 0 0 1	√
0 0 1 0	√
0 0 1 1	√
1 0 0 1	√
1 0 1 0	√
1 1 0 0	√
0 1 1 1	√
1 1 0 1	√
1 1 1 0	√
1 1 1 1	√

A B C D	
0 0 0 —	√
0 0 — 0	√
0 0 — 1	√
— 0 0 1	*
0 0 1 —	√
— 0 1 0	*
0 — 1 1	*
1 — 0 1	*
1 — 1 0	*
1 1 0 —	√
1 1 — 0	√
— 1 1 1	*
1 1 — 1	√
1 1 1 —	√

A B C D	
0 0 — —	*
1 1 — —	*

0001	0010	0011	1001	1100	1101	1110	*Prime Implicants*					
√			√				—	0	0	1	→	L
	√						—	0	1	0	→	M
		√					0	—	1	1	→	N
			√		√		1	—	0	1	→	P
						√	1	—	1	0	→	Q
							—	1	1	1	→	R
√	√	√					0	0	—	—	→	S
				√	√	√	1	1	—	—	→	T

The product-of-sums expression that tells about the combination of prime implicants required to account for all the terms is given by:

$$(L + S)\ (M + S)(N + S)\ (L + P)(T)(P + T)(Q + T)$$

which after obvious simplification reduces to

$$T(L + S)\ (M + S)\ (N + S)\ (L + P)$$

$$= T(LM + LS + MS + S)\ (LN + PN + LS + PS)$$

$$= T(LM + S)(LN + PN + LS + PS)$$

$$= T(LMN + LMPN + LMS + LMP + LNS + PNS + LS + PS)$$

$$= T(LMN + LMP + LS + PS)$$

$$= TLMN + TLMP + TLS + TPS$$

Now, this sum-of-products expression states that all the input combinations can be accounted for by the prime implicants (*T, L, M, N*) or (*T, L, M, P*) or (*T, L, S*) or (*T, P, S*).

The most optimum expression would result from either *TLS* or *TPS*.

Therefore, minimised Boolean function is given by:

$$f(A, B, C, D) = AB + \overline{B}\,\overline{C}D + \overline{A}\,\overline{B},$$

or

$$f(A, B, C, D) = AB + \overline{A}\,\overline{B} + A\overline{C}D.$$

Problem 22.39. *For a Boolean function f(A, B) = Σ 0, 2 prove that f(A, B) = Π, 1, 3 and f′(A, B) = Σ 1, 3 = Π 0, 2.*

Solution.

$$f(A, B) = \Sigma\, 0, 2 = \overline{A}\,\overline{B} + A\overline{B} = \overline{B}\,(A + \overline{A}) = \overline{B}$$

$$\text{Now } \Pi\, 1, 3 = (A + \overline{B})(\overline{A} + \overline{B}) = A\overline{A} + A\overline{B} + \overline{A}\,\overline{B} + \overline{B}\,.\,\overline{B}$$

$$= A\overline{B} + \overline{A}\,\overline{B} + \overline{B} = \overline{B} = f(A, B)$$

$$\text{Now } \Sigma\, 1, 3 = \overline{A}B + AB = B(A + \overline{A}) = B$$

$$\text{and } \Pi\, 0, 2 = (A + B)\ (\overline{A} + B) = A\overline{A} + AB + \overline{A}B + B\,.\,B$$

$$= AB + \overline{A}B + B = B$$

Therefore, $\Sigma\, 1,3 = \Pi\, 0, 2$

Also, $f(A, B) = \overline{B}$

Therefore, $f'(A, B) = \overline{\overline{B}} = B$

$$f'(A, B) = \Sigma\, 1,3 = \Pi\, 0, 2$$

Problem 22.40. *Minimise using Karnaugh map the Boolean function, f(L, M, N, P) given by:*

$$f(L, M, N, P) = \overline{L}\,\overline{M}N + LP + \overline{L}\,\overline{P} + \overline{P}(M + N) + L\overline{N}$$

$$= \overline{L}\,\overline{M}N + LP + \overline{L}\,\overline{P} + M\overline{P} + N\overline{P} + LN.$$

Solution.

The expanded version of *f*(*L, M, N, P*) is given by :

$$\overline{L}\,\overline{M}NP + \overline{L}\,\overline{M}N\overline{P} + L\overline{M}\,\overline{N}P + L\overline{M}NP + LMNP + LM\overline{N}P$$

$$+ \overline{L}\,\overline{M}\,\overline{N}\,\overline{P} + \overline{L}\,\overline{M}N\overline{P} + \overline{L}MN\overline{P} + \overline{L}M\overline{N}\,\overline{P} + \overline{L}MN\overline{P}$$

$$+ LMN\overline{P} + LM\overline{N}\,\overline{P} + \overline{L}MN\overline{P} + LMN\overline{P} + LM\overline{N}\,\overline{P} + LMN\overline{P}$$

$$+ L\overline{M}N\overline{P} + L\overline{M}\,\overline{N}\,\overline{P} + L\overline{M}\,\overline{N}P + LM\overline{N}P + LM\overline{N}\,\overline{P}$$

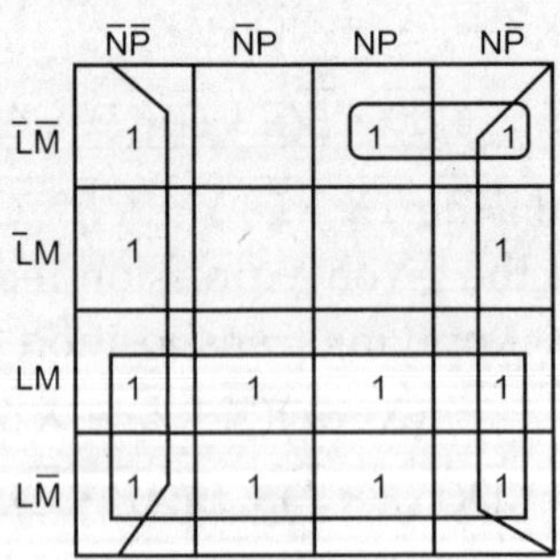

Fig. 22.28

Removing the duplicate terms,

$$f(L, M, N, P) = \overline{L}\,\overline{M}\,\overline{N}\,\overline{P} + \overline{L}\,\overline{M}N\overline{P} + \overline{L}\,\overline{M}NP + \overline{L}M\overline{N}\,\overline{P}$$

$$+ \overline{L}MN\overline{P} + LMN\overline{P} + LMNP + L\overline{M}N\overline{P} + L\overline{M}NP + LM\overline{N}\,\overline{P}$$

$$+ LM\overline{N}P + LMN\overline{P} + LMNP$$

The Karnaugh map is shown in Fig. 22.28. The minimised expression is given by :

$$f(L, M, N, P) = L + \overline{P} + \overline{L}\,\overline{M}N.$$

Problem 22.41. *Minimise the following Boolean function using the mapping method in both the minimised sum-of-products as well as minimised product-of sums forms.*

$$f(A, B, C) = \Sigma\, 0, 1, 3, 5 + \sum_{\phi} 2, 7.$$

Solution.

$$f(A, B, C) = \Sigma\, 0, 1, 3, 5 + \sum_{\phi} 2, 7$$

That is, $f(A, B, C) = \overline{A}\,\overline{B}\,\overline{C} + \overline{A}\,\overline{B}\,C + \overline{A}BC + A\overline{B}C$ in the sum-of-products form.

Also, $f(A, B, C) = \Sigma\, 0,1, 3,5 + \sum_{\phi} 2,7 = \Pi\, 4, 6 + \prod_{\phi} 2,7$

That is, $f(A, B, C) = (\overline{A} + B + C)\,(\overline{A} + \overline{B} + C)$ in the product-of-sums form.

Don't care input combinations for sum-of-products expression $= \overline{A}B\overline{C}$, ABC and don't care input combinations for product-of-sums expression $= (A + \overline{B} + C)$, $(\overline{A} + \overline{B} + \overline{C})$. The Karnaugh maps for the two cases are shown in Figs. 22.29 (*a*) and (*b*).

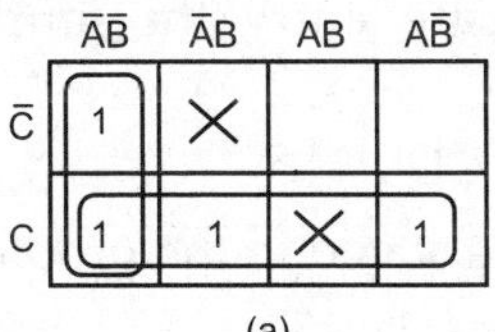

(a)

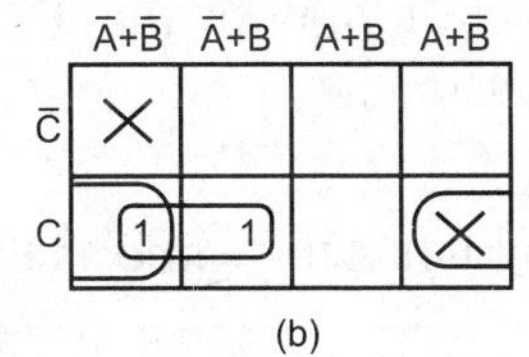

(b)

Fig. 22.29

The minimised sum-of-products Boolean function is given by :

$$f(A, B, C) = C + \overline{A}\,\overline{B}$$

The minimised product-of-sums Boolean function is given by :

$$f(A, B, C) = (\overline{A} + C)(\overline{B} + C).$$

Problem 22.42. For the Karnaugh map shown in Fig. 22.30 below, draw the truth table. Also write the Boolean function in Σ-nomenclature.

AB \ CD	00	01	11	10
00		1	1	
01		1	1	
11	1	X	X	1
10	1			1

Fig. 22.30

Solution.

The truth table for the given Karnaugh map is shown below.

A	B	C	D	Y
0	0	0	0	0
0	0	0	1	1
0	0	1	0	0
0	0	1	1	1
0	1	0	0	0
0	1	0	1	1
0	1	1	0	0
0	1	1	1	1
1	0	0	0	1
1	0	0	1	0
1	0	1	0	1
1	0	1	1	0
1	1	0	0	1
1	1	0	1	0
1	1	1	0	1
1	1	1	1	0

The Boolean function, $y = \Sigma\, 1, 3, 5, 7, 8, 10, 12, 14 + \sum_{\phi} 13, 15$.

Problem 22.43. *Write the simplified Boolean expression given by the Karnaugh map shown in Fig. 22.31.*

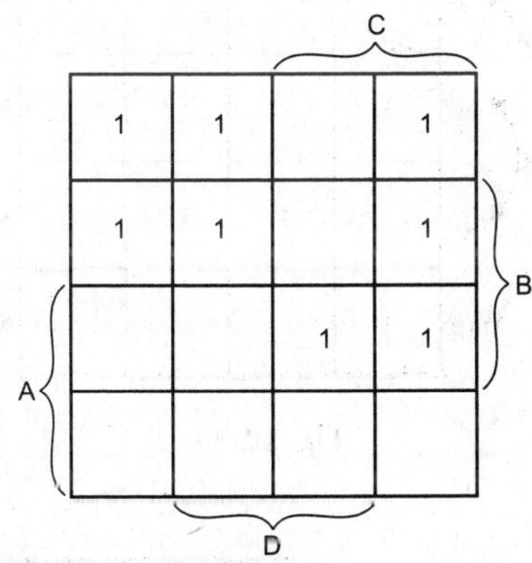

Fig. 22.31

Solution.

The Karnaugh map with groupings is shown in Fig. 22.32.

The group of four 1's on the top left of the map yields a term $\overline{A}\,\overline{C}$.

The group of four l's (2 on the extreme left and 2 on the extreme right) yields a term $\overline{A}\,\overline{D}$.

The third group of two l's is in the third row of the map. The third row corresponds to intersection of *A* and *B* as is clear from the map. Therefore, this group yields a term *ABC*.

The simplified Boolean expression is given by:

$$y = \bar{A}\,\bar{C} + \bar{A}\,\bar{D} + ABC.$$

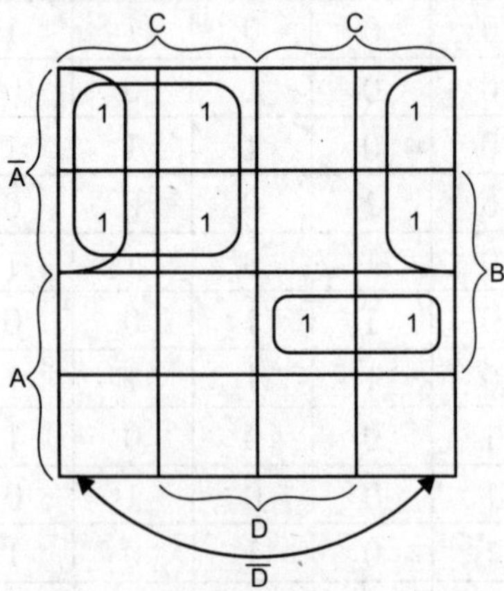

Fig. 22.32

Problem 22.44. *Minimising a given Boolean expression using Quine-McCluskey tabular method yields the following prime implicants : — 0—0, — 1 —1, 1 — 10 and 0 — 00. Draw the corresponding Karnaugh map.*

Solution.

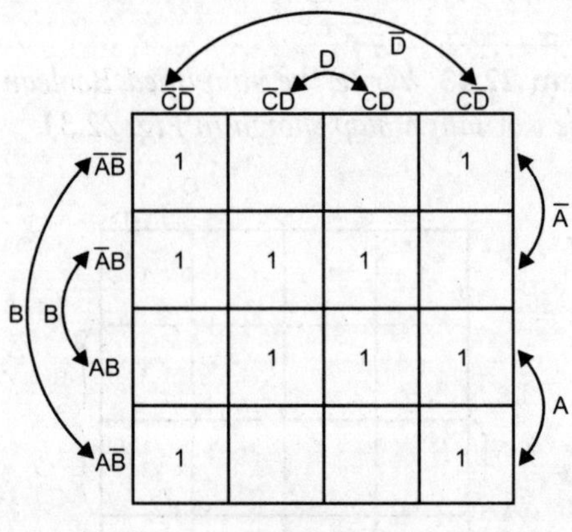

Fig. 22.33

As is clear from the prime implicants, the expression has four variables. If the variables are assumed as *A*, *B*, *C* and *D*, then the given prime implicants correspond to $\bar{B}\,\bar{D}$ (–0–0), BD (–1–1), $AC\bar{D}$ (1–10) and $\bar{A}\,\bar{C}\,\bar{D}$ (0–00). The Karnaugh map can now be drawn, as shown in Fig. 22.33.

Problem 22.45. *($\bar{A}B$ + CD) is a simplified Boolean expression of the expression $ABCD + \bar{A}\,\bar{B}CD + \bar{A}B$. Determine, if there are any 'Don't Care' entries.*

Solution.

The expanded version of the given expression is :

$$ABCD + \bar{A}\,\bar{B}CD + \bar{A}B\,(\bar{C}\,\bar{D} + \bar{C}D + CD + C\bar{D})$$

$$= ABCD + \bar{A}\,\bar{B}CD + \bar{A}B\,\bar{C}\,\bar{D} + \bar{A}B\bar{C}\,D + \bar{A}BCD + \bar{A}BC\bar{D}$$

The Karnaugh map for this Boolean expression is shown in Fig, 22.34.

Now, if $\bar{A}B$ + CD is to be simplified version of the given expression, then the lowermost square in the 'CD' column should not be empty. This implies that there is a Don't care entry $A\bar{B}CD$. This has been reflected in the map by putting, 'X', in the relevant square. With the groups farmed alongwith the 'Don't Care' entry, the simplified expression is :

$$\bar{A}B + CD$$

which is the same as the one stated in the problem.

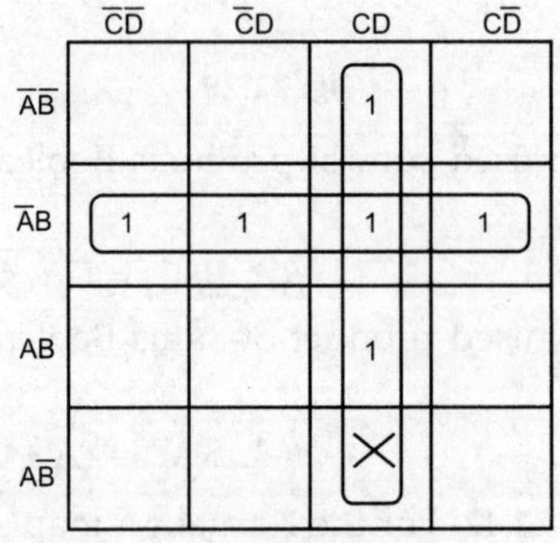

Fig. 22.34

SUMMARY OF IMPORTANT FORMULAS

1. Radix of:

   Binary number system = 2

   Octal number system = 8

   Decimal number system = 10

   Hexadecimal number system = 16

2. Maximum numbers that can be represented by an n-bit number in a radix-r number system = r^n
3. Maximum positive decimal number that can be represented by n-bit binary number in 1's complement notation = $(2^{n-1} - 1)$
4. Maximum negative decimal number that can be represented by n-bit binary number in 1's complement notation = $-(2^{n-1} - 1)$
5. Maximum positive decimal number that can be represented by n-bit binary number in 2's complement notation = $(2^{n-1} - 1)$
6. Maximum negative decimal number that can be represented by n-bit binary number in 2's complement notation = -2^{n-1}
7. Basic rules of binary addition:

0 + 0 = 0; 0 + 1 = 1; 1 + 0 = 1, 1 + 1 = 0 with CARRY of 1 and 1 + 1 + 1 = 1 with CARRY of 1

8. Basic rules of binary subtraction:
   0 – 0 = 0; 1 – 0 = 1; 0 – 1 = 1 with a BORROW of 1 and 1 – 1 = 0
9. Commutative laws of Boolean algebra:
   $A + B = B + A; A . B = B . A$
10. Associative laws of Boolean algebra:
    $A + (B + C) = A + (B + C); A.(B.C) = A.(B.C)$
11. Distributive law of Boolean algebra:
    $A . (B + C) = A . B + A . C$
12. Involution law of Boolean algebra:
    $X'' = X$
13. Idempotent laws of Boolean algebra:
    $X + X + X + \ldots\ldots + X = X$ and $X.X.X\ldots\ldots X = X$
14. Absorption law of Boolean algebra:
    $A . (A + B) = A$
15. DeMorgan's theorems of Boolean algebra:
    $(X_1 + X_2 + X_3 + \ldots\ldots + X_N)' = (X_1'. X_2'. X_3' + \ldots\ldots X_N')$
    $(X_1. X_2 . X_3 + \ldots\ldots X_N)' = (X_1' + X_2' + X_3' + \ldots\ldots + X_N')$

Self Evaluation Exercise

I. Multichoice Question

1. Number (–39) when represented in 2's complement form is
   (*a*) 11011001
   (*b*) 10100111
   (*c*) 11011000
   (*d*) none of these
2. Number (–39) when represented in Sigh-Bit magnitude form is
   (*a*) 11011001
   (*b*) 10100111
   (*c*) 11011000
   (*d*) none of these
3. Number (–13) when represented in l's complement form is
   (*a*) 11011001 (*b*) 10100111
   (*c*) 11110000 (*d*) 00001111
4. The numbers that come immediately after and before hex number $(FFEF)_{16}$ are respectively
   (*a*) $(FFF0)_{16}$ and $(FFE0)_{16}$
   (*b*) $(FFFE)_{16}$ and $(FFEE)_{16}$
   (*c*) $(FFF0)_{16}$ and $(FFFE)_{16}$
   (*d*) $(FFF0)_{16}$ and $(FFEE)_{16}$
5. The numbers that immediately follow $(777)_8$ and $(1111)_2$ are respectively
   (*a*) $(1000)_8$ and $(10000)_2$
   (*b*) $(7770)_8$ and $(11110)_2$
   (*c*) $(7777)_8$ and $(11111)_2$
   (*d*) none of these
6. The numbers that immediately precede $(777)_8$ and $(1111)_2$ are respectively
   (*a*) $(776)_8$ and $(1110)_2$
   (*b*) $(7770)_8$ and $(1110)_2$
   (*c*) $(776)_8$ and $(11110)_2$
   (*d*) none of these
7. The binary number $(1101.111)_2$ when represented by floating point notation is
   (*a*) $(.1101111 \times 2^4)$ (*b*) $(1101111)_2$
   (*c*) $(.1101111)_2$ (*d*) none of these
8. The decimal number $(839.57)_{10}$ when represented by floating point notation is
   (*a*) (.83957) (*b*) $(.83957 \times 2^3)$
   (*c*) $(.83957 \times 8^3)$ (*d*) $(.83957 \times 10^3)$
9. The hex number $(FFF.FF)_{16}$ in floating point notation is written as
   (*a*) $(.FFFFF \times 16^2)$
   (*b*) $(.FFFFF \times 16^3)$
   (*c*) (.FFFFF)
   (*d*) none of these
10. l's complement of $(11101)_2$, $(FFF)_{16}$, $(FIEA)_{16}$ and $(732)_8$ respectively are
    (*a*) $(00010)_2$, $(000)_{16}$, $(0E15)_{16}$ and $(045)_8$
    (*b*) $(00010)_2$, $(000)_{16}$, $(0E15)_{16}$ and $(156)_8$
    (*c*) $(00010)_2$, $(111)_{16}$, $(lF26)_{16}$ and $(045)_8$
    (*d*) none of these
11. An arbitrary number system has a radix of 32 with '0' to '9' and 'A' to 'V' as its 32 basic digits, '0' being the first and 'V' being the 32nd. Decimal number $(128)_{10}$ in this arbitrary number system.

(*a*) can be expressed as $(40)_{32}$
(*b*) is $(128)_{32}$
(*c*) is $(100000)_{32}$
(*d*) cannot be determined from given data

12. The number immediately following 'V' in the arbitrary number system of Q-11 will be
(*a*) $(V0)_{32}$ (*b*) $(10)_{32}$
(*c*) $(VI)_{32}$ (*d*) none of these

13. Decimal equivalent of BCD 8421 number (10000000.1001) is
(*a*) 80.9 (*b*) 128.9
(*c*) 80.25 (*d*) 128.25

14. Amongst the following 4-bit groups, pick the one that is forbidden in 8421 BCD code: 1001, 1000, 1010, 0100, 0101, 0110
(*a*) 1001 (*b*) 0110
(*c*) 1010 (*d*) 0100

15. The decimal number 78 in 5421 BCD code will be
(*a*) 0 1111000 (*b*) 10101011
(*c*) 1001110 (*d*) none of these

16. Each of the following Boolean expressions can be implemented using a single logic gate except one of them
(*a*) $Z = \overline{(x + y)}$
(*b*) $Z = (\overline{x} . \overline{y})$
(*c*) $Z = x . \overline{y} + \overline{x} . y$
(*d*) $Z = x . y + \overline{x} . y + x . \overline{y}$

17. All but one of the following Boolean expressions can be implemented using a single 2-input NAND-gate
(*a*) $y = \overline{x}$ (*b*) $y = \overline{A} + \overline{B}$
(*c*) $y = \overline{C . D}$ (*d*) $y = \overline{A} . \overline{B}$

18. Binary number $(1110101)_2$ when converted into an equivalent decimal number gives
(*a*) 117 (*b*) 58
(*c*) 116 (*d*) 57

19. Binary number $(1101)_2$ when added to another binary number $(101)_2$ yields
(*a*) 10011 (*b*) 10010
(*c*) 11010 (*d*) 11001

20. 64 in decimal is equivalent to
(*a*) 64 in octal too (*b*) 100 in octal
(*c*) 24 in octal (*d*) 80 in octal

21. 128 in decimal is equivalent to
(*a*) 80 in hexadecimal (*b*) 180 in octal
(*c*) 1000000 in binary (*d*) none of these

22. $(.11)_2$ is equivalent to
(*a*) $(0.75)_{10}$ (*b*) $(0.24)_{10}$
(*c*) $(0.375)_{10}$ (*d*) none of these

23. One's complement of $(11)^2$ in 8-digit arithmetic is
(*a*) 00000000 (*b*) 11111100
(*c*) 11111111 (*d*) none of these

24. Two's complement of $(10\,1)_2$ in 8-digit arithmetic is
(*a*) 00000011 (*b*) 11111010
(*c*) 11111011 (*d*) none of these

25. 9's complement of $(2903)_{10}$ is
(*a*) 7096
(*b*) 2903
(*c*) There is nothing like 9's complement
(*d*) none of these

26. Decimal number (3) when converted into Gray code gives
(*a*) 0011 (*b*) 0100
(*c*) 0010 (*d*) none of these

27. Gray code (1011) converted into an equivalent binary gives
(*a*) 1101 (*b*) 0101
(*c*) 0111 (*d*) none of these

28. Three Boolean operators are
(*a*) NAND, NOR, OR (*b*) OR, AND, NOT
(*c*) NOR, NAND, NOT (*d*) none of these

29. According to Boolean algebra, A+A+A + ... + A is
(*a*) nA (*b*) 0
(*c*) 1 (*d*) A

30. According to Boolean algebra, $A . A . A$ ....+ A is
(*a*) nA (*b*) 0
(*c*) 1 (*d*) A

31. A.l according to Boolean algebra is
(*a*) A (*b*) 1
(*c*) 0 (*d*) none of these

32. Mark the incorrect Boolean expression
(*a*) $1 + A = A$ (*b*) $1 + A = 1$
(*c*) $A + \overline{A} = 0$ (d) $A + AB = A$

33. The Boolean expression $y = \overline{A} + \overline{B}$ should preferably be implemented by
    (a) a single 2-input NAND
    (b) two NOT gates and one 2-input OR
    (c) one 2-input AND and one NOT
    (d) none of these
34. One of the following Boolean expressions represents De Morgan's theorem
    (a) $A + \overline{A} = 1$ (b) $A \cdot \overline{A} = 0$
    (c) $\overline{(A + B)} = \overline{A} \cdot \overline{B}$ (d) $1 + A = 1$
35. "A NOR-gate is equivalent to a bubbled AND-gate and a NAND-gate is equivalent to a bubbled OR-gate". This statement is derived from
    (a) Idempotent law
    (b) Involution law
    (c) De Morgan's theorems
    (d) Absorption law

II. Fill in the Blanks

1. The .... BCD code is an example of a weighted code (8421, Grey, Excess-3)
2. The .... code is a BCD code (Excess-3, Grey)
3. The .... code is usually used in arithmetic operations (Excess-3, Grey)
4. The parity bit helps in .... errors occurring during data transmissions (correcting, detecting)
5. ASCII is an abbreviation for (i) .... and EBCDIC is an abbreviation for (ii) ....
6. The .... is a 12-bit code used for recording information on punched cards (Selectric, Hamming, Hollerith)
7. The 7-bit .... code that has become an industry standard for microcomputer input/output interface (ASCII, EBCDIC, BCDIC, Selectric)
8. (1000 1101) is .... 8421 BCD number (a valid, an invalid)
9. The binary number 110010111 has an .... parity (Odd, Even)
10. A 7-bit alphanumeric code can be used to represent a minimum of .... characters (64, 128, 256)
11. In 2's complement addition or subtraction, the result of addition or subtraction is .... in.
    (2's complement form, l's complement form, straight binary form)
12. 8-bit 2's complement arithmetic can be used to add or subtract binary numbers, when the expected result lies in the decimal range .... (i)......... to ....(ii)......... .
    (– 128, – 256, + 255, + 256, + 128, + 127)
13. Binary multiplication in computers is a process of ....(i)......... and ....(ii)......... operations.
    (Compare, Subtract, Add, Shift)
14. Binary division in computers is a process of ....(i)......... and ....(ii)......... operations. (Shift, Add, Subtract, Complement)
15. A half adder circuit can be used to add .... binary digits only (2, 3)
16. 0101 when divided by 1010 produces a quotient ....(i)......... and a remainder ....(ii)......... (.01, .10, 010,0000)
17. If 00001101 is a binary number represented in 2's complement form, then the magnitude of its decimal equivalent is .... (115, 13, 243)
18. 11010010 is .... a binary number (necessarily, not necessarily)
19. Adding 2H to FEH yields .... (FFH, FE1H, FF1H, 100H)
20. Subtracting $(3)^8$ from $(100)^8$ yields .... (75, 97, 86)
21. If (X) is a binary variable, then X + X + X equals ....(i)......... and X.X.X equals ....(ii)......... (X, X, 3X, X^3)
22. If (X) is a binary variable, then $X + \overline{X}$ equals .... (i)......... and $X.\overline{X}$ equals ....(ii)......... (X, $\overline{X}$, 1, 0).
23. If (A) is a binary variable, then 1 + A equals .... (i)......... and 1. A equals ....(ii)......... (1, A, 0)
24. According to Boolean algebra, $ABCD + \overline{ABCD}$ equals .... (0, 1)

III. Say True or False

1. 823.01 is not a valid octal number.
2. –8 cannot be represented in 2's complement form using 4 bits.
3. A 16-bit data contained in a certain location of computer memory can be expressed in terms of four hexadecimal digits only.

4. Range covered by a floating point number is dependent upon the number of bits in the mantissa.
5. Adding 2's complement of a binary number (X) to another binary number (Y) achieves ($Y - X$).
6. Adding 2's complement of a binary number (X) to 2's complement of another binary number (Y) achieves ($X + Y$).
7. Addition of (+ 75) and (+ 85) is possible using 8-bit 2's complement representation.
8. 2's complement arithmetic can be used to carry out addition and subtraction of decimal numbers irrespective of the signs of the numbers involved.
9. While 1101.1011 is a fixed point binary representation, $.11011011 \times 2^4$ is a floating point representation of the same number.
10. 16's complement of Hex number $(EEE1)_{16}$ is 111F.

Answers to Self Evaluation Exercise

I.

1. (*a*)	2. (*b*)	3. (*c*)	4. (*d*)	5. (*a*)	6. (*a*)
7. (*a*)	8. (*d*)	9. (*b*)	10. (*a*)	11. (*a*)	12. (*b*)
13. (*a*)	14. (*c*)	15. (*b*)	16. (*d*)	17. (*d*)	18. (*a*)
19. (*b*)	20. (*b*)	21. (*a*)	22. (*a*)	23. (*b*)	24. (*c*)
25. (*a*)	26. (*c*)	27. (*a*)	28. (*b*)	29. (*d*)	30. (*d*)
31. (*a*)	32. (*a*)	33. (*a*)	34. (*c*)	35. (*c*)	

II.

1. 8421 2. Excess–3 3. Excess–3 4. detecting
5. (*i*) American Standard Code for Information Interchange
(*ii*) Extended Binary Coded Decimal Interchange Code
6. Hollerith 7. ASCII 8. an invalid 9. Even 10. 128 11. 2's complement form
12. (*i*) – 128 (*ii*) + 127 13. (*i*) Add (*ii*) Shift 14. (*i*) Shift (*ii*) Subtract
15. 2 16. (*i*) .10 (*ii*) 0000 17. 243 18. not necessarily
19. 100H 20. 75 21. (*i*) X (*ii*) X 22. (*i*) 1 (*ii*) 0 23. (*i*) 1 (*ii*) A 24. 1

III.

1. True (8 is an invalid octal digit) 2. False 3. True 4. False 5. True
6. True 7. False 8. True 9. True 10. True

Chapter 23

Logic Gates, Logic Families and Combinational Logic Circuits

Learning Objectives

After completing the chapter, you will learn the following topics:

1. Positive and Negative Logic Systems
2. Logic Gates Including AND, OR, NOT, NAND, NOR, Exclusive-OR, Exclusive-NOR and INHIBIT gates
3. Implementing AND, OR, NOT, Exclusive-OR and Exclusive-NOR gates Using NAND and NOR Gates
4. Logic Families
5. Transistor-Transistor Logic (TTL) and its Sub-families
6. Complementary Metal Oxide Semiconductor (CMOS) Logic Family and its Sub-families
7. Emitter Coupled Logic (ECL) Logic Family and its Variants
8. NMOS and PMOS Logic Families
9. I^2L Family
10. Interface Between Logic Gates of Different Logic Families
11. Buffers and Transceivers
12. Multiplexers and De-multiplexers
13. Encoders and Decoders
14. Priority Encoders
15. Arithmetic Circuits Including Half Adder, Full Adder, Half Subtractor, Full Subtractor
16. Controlled Inverters
17. Magnitude Comparators
18. Parity Generators and Checkers
19. Read Only Memory (ROM)
20. Programmable Logic Array (PLA) Devices
21. Programmable Array Logic (PAL) Devices

Q. 1. What are POSITIVE LOGIC and NEGATIVE LOGIC systems? Give examples to illustrate your answer.

Ans. The binary variables as we know can have either of the two states *i.e.* the logical '0' state and the logical '1' state. These logic states in digital systems, like computers, for instance, are represented by two different voltage levels or two different current levels. If the more positive of the two voltage or current levels represents a logic '1' and the less positive of the two levels represents a logic '0', then the logic system is referred to as the Positive Logic System. If the more positive of the two voltage or current levels represents a logic '0' and the less positive of the two levels represents a logic '1', then the logic system is termed as the Negative Logic System. Following examples further illustrate the concept:

1. If the two voltage levels are 0V and +5V, then in the positive logic system, 0V represents logic '0' and +5V represents logic '1'. In the negative logic system, 0V represents logic '1' and +5V would represent a logic '1'.
2. If the two voltage levels are 0V and –5V, then in the positive logic system, 0V represents logic '1' and –5V represents a logic '0'. In the negative logic system, 0V would represent a logic '0' and –5V would represent a logic '1'. As we go through the basics of logic gates, we would have an interesting observation vis-a-vis positive and negative logic systems and that is, and AND-gate in positive logic systems becomes an OR-gate in negative system and vice versa. Similarly, a NAND-gate in positive logic system becomes a NOR-gate in negative logic system and vice versa.

Q. 2. What is a TRUTH TABLE ? What is its significance in logic operations ?

Ans. A TRUTH TABLE lists all possible combinations of input binary variables and the corresponding outputs of a logic system. The logic system output can be found out from the logic expression, often referred to as the Boolean expression, that relates the output with the inputs of that very logic system. Refer to Fig. 23.1 (*a*). In this case, there can be four possible inputs *i.e.* X = 0, Y = 0 (*ii*) X = 0, Y = 1 (*iii*) X = 1, Y = 0 (*iv*) X = 1, Y = 1. Let us assume that this logic circuit is such that the outputs for the above four cases respectively are 0, 1, 1 and 1, then this entire information can be put together in a tabular form as shown in Fig. 23.1 (*b*). This is then known as the Truth Table for this logic system.

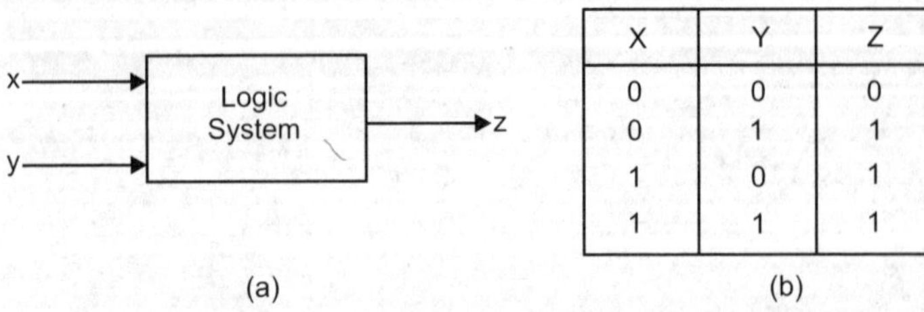

X	Y	Z
0	0	0
0	1	1
1	0	1
1	1	1

(b)

Fig. 23.1

Q. 3. A particular logic system has three inputs and one output. How many rows will the truth table of this logic system have ?

OR

How many rows will be there in the truth table of a logic system having (*n*) input binary variables?

Ans. When the number of input binary variables is only one, then there are only two possible inputs *i.e.* '0' and '1'. In case number of inputs is two, there can be four possible input combinations *i.e.* 00, 01, 10 and 11. Similarly, for three binary inputs, the number of possible input combinations goes upto eight *i.e.* 000,001,010,011,100,101,110 and 111. This statement can be generalised to say that if a logic circuit has (*n*) binary inputs, its truth table will have (2^n) possible input combinations or in other words (2^n) rows. Fig. 23.2 shows the truth table for a 3-input logic circuit and it has ($8 = 2^3$) rows. The output column is left blank as the entries in that column will depend upon the type of the logic circuit represented by this truth table.

A	B	C	Y
0	0	0	–
0	0	1	–
0	1	0	–
0	1	1	–
1	0	0	–
1	0	1	–
0	1	0	–
1	1	1	–

Fig. 23.2

Q. 4. What are Logic Gates? Name the three elementary or basic logic gates.

Ans. Logic gates are the most basic and the most important building blocks of any digital system including computers. Each one of the basic logic gates is a piece of hardware or an electronic circuit that can be used to implement the most basic logic expressions (often referred to as the Boolean expressions). While laws of Boolean algebra could be used to do manipulation with binary variables and simplify the logic expressions, these are actually implemented inside the computer with the help of electronic circuits called logic circuits. And logic Gates can be said to be the most elementary logic circuits. Three basic logic gates are the OR-gate, the AND-gate and the NOT-gate.

Q. 5. What is an OR-gate? Explain in brief, the functioning of an OR-gate with the lip of its truth table.

Ans. An OR-gate is a logic circuit with two or more than two inputs and one output. The output of an OR-gate is logic '0' only when all of its inputs are at logic '0'. For all other possible input combinations, the output is logic '1'. Fig. 23.3 shows the circuit symbol and the truth table of a 2-input OR-gate. The operation of an OR-gate is explained by the logic expression:

$$Y = A + B$$

A	B	Y
0	0	0
0	1	1
1	0	1
1	1	1

(a) (b)

Fig. 23.3

This logic expression is not read as Y equals A Plus B as is done in case of ordinary algebra; instead it is read as Y equals A OR B. As an illustration, if we have say four binary variables A, B, C and D and we want to know the result of (A + B + C + D), then the result would be the output of a 4-input OR-gate with A, B, C and D as its inputs.

Q. 6. Draw the circuit symbols of 3-input and 4-input OR gates alongwith the logic expressions that explain their operation. Also, draw the truth table of 3-input OR-gate.

Ans. The circuit symbol of 3-iriput OR-gate is shown in Fig. 23.4 (*a*). Fig. 23.4 (*b*) shows the circuit symbol of 4-input OR-gate.

Logic expression for 3-input OR-gate is given by: Y = A + B + C

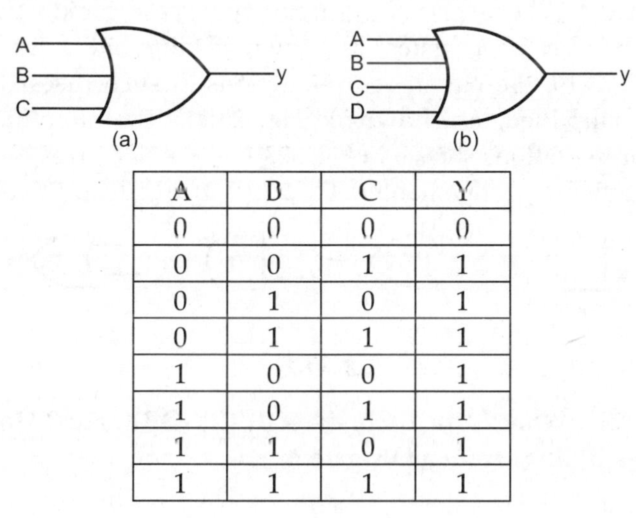

A	B	C	Y
0	0	0	0
0	0	1	1
0	1	0	1
0	1	1	1
1	0	0	1
1	0	1	1
1	1	0	1
1	1	1	1

(c)

Fig. 23.4

Logic expression explaining the functioning of 4-input OR-gate is given by:

$$Y = A + B + C + D$$

Fig. 23.4. (*c*) shows the truth table of 3-input OR-gate.

Q. 7. In case of logic configuration shown in Fig. 23.5, write the input-output logic expression.

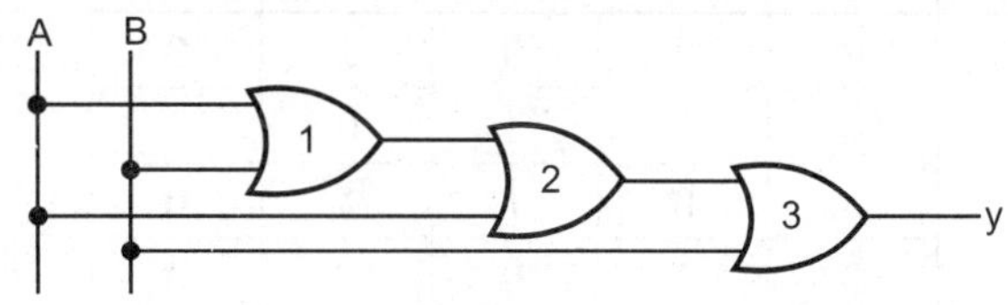

Fig. 23.5

Ans. The output of gate-1 is $(A + B)$ which is one of the inputs to the second gate, the other input being A, the output of the second OR-gate is $(A + B + A)$. The output of the third OR-gate will then be $(A + B + A + B)$

Now $A + A = A$ and $B + B = B$. Therefore, $Y = A + B$

This is same as the logic expression for a two-input OR-gate. Thus, in this case, what three OR gates are doing could have been done with a single 2-input OR-gate.

Q. 8. What is an AND-gate ? Explain in brief, the functioning of an AND-gate with the help of its truth table.

Ans. An AND-gate is a logic circuit having two or more than two inputs and one output. The output of a AND-gate is logic '1' only, when all of its inputs are in logic '1' state. In all other cases, the output is logic '0'. Figure 23.6 shows the logic symbol and truth table of a 2-input And -gate.

$$Y = A.B$$

A	B	Y
0	0	0
0	1	0
1	0	0
1	1	1

(a) (b)

Fig. 23.6

Q. 9. Draw the circuit symbols of 3-input and 4-input AND gates. Give respective logic expressions. Also, draw the truth table for 4-input AND-gate.

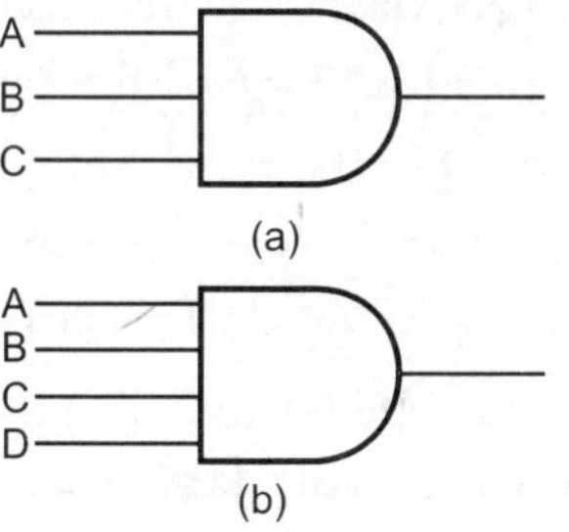

A	B	C	D	Y
0	0	0	0	0
0	0	0	1	0
0	0	1	0	0
0	0	1	1	0
0	1	0	0	0
0	1	0	1	0
0	1	1	0	0
0	1	1	1	0
1	0	0	0	0
1	0	0	1	0
1	0	1	0	0
1	0	1	1	0
1	1	0	0	0
1	1	0	1	0
1	1	1	0	0
1	1	1	1	1

Fig. 23.7

Ans. 3-input AND-Gate

Refer to Fig. 23.7(*a*) for the circuit symbol. Its logic expression is : $Y = A.B.C$

4-inputAND-Gate

Refer to Fig. 23.7 (*b*) for the circuit symbol. Its logic expression is : $Y = A.B.C.D$

Fig. 21-3 (*c*) shows the truth table of a 4-input AND-gate.

Q. 10. Write the input-output logic expression for the logic configuration shown in Fig. 23.8.

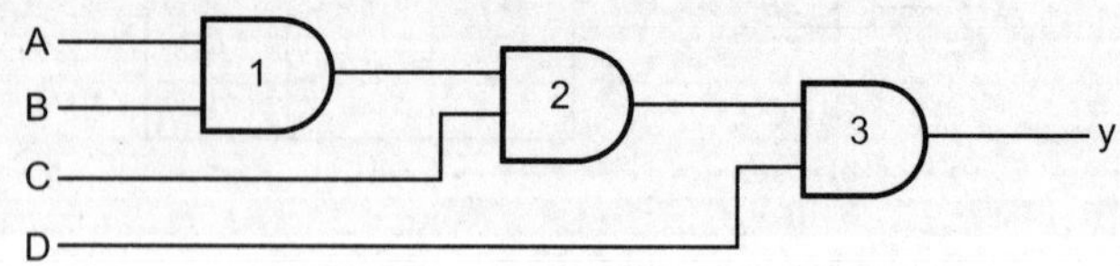

Fig. 23.8

Ans. Output of the first AND-gate = *AB*

Inputs to the second AND-gate are *A.B* and *C*

Therefore, output of the second AND-gate = *A.B.C*

Inputs to the third AND-gate are *A.B.C* and *D*

Therefore, output of the third AND-gate, $Y = A.B.C.D$

It is interesting to note here how three 2-input AND-gates have been used to build a 4-input AND-gate.

Q. 11. What is a NOT-gate? Explain in brief, its operation and draw its truth table.

OR

"A NOT-Circuit complements the input". Explain.

Ans. A NOT-circuit is a one input one output logic gate whose output is always the complement of the input. That is, a logic '0' at the input produces a logic '1' at the output and vice-versa. It is also known as a **'Complementing Circuit'** or an **'Inverting Circuit'** or a **Negating Circuit.** Fig. 23.9 shows its circuit symbol and the truth table.

NOT operation on a binary variable (x) is denoted as ($\bar{x}$) or (x'). That is, if x is the input to a NOT gate, then its output z is given by :

$$z = \bar{x} \text{ read as } z = \text{NOT } x$$

Thus, if $x = 0$, $z = 1$ and if $x = 1$, $2 = 0$.

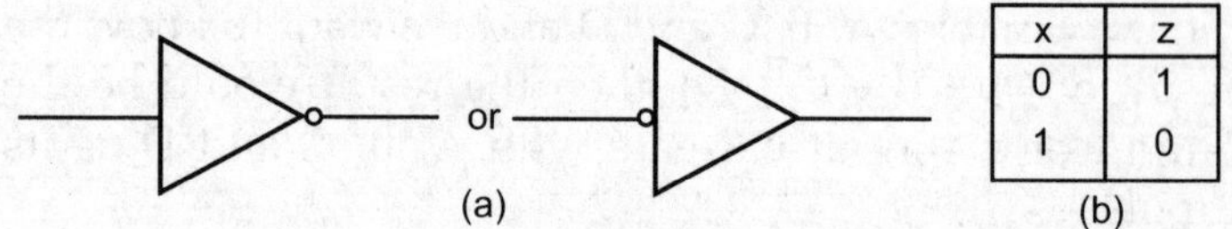

Fig. 23.9

Q. 12. Draw the alternative logic circuit representations for the logic gate arrangements shown in Fig. 23.10.

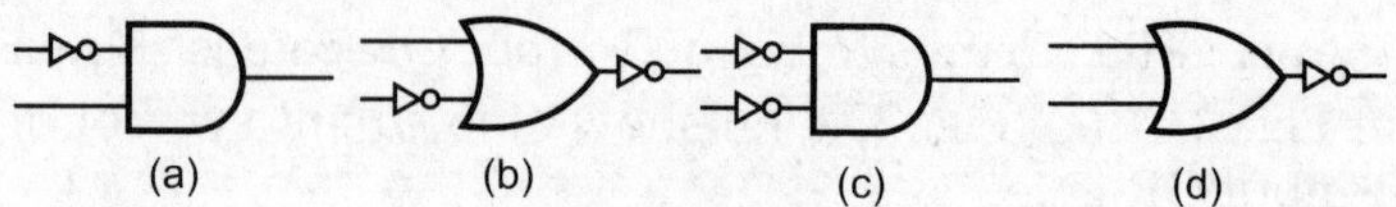

Fig. 23.10

Ans. Whenever a NOT circuit appears in conjunction with other basic gates, as shown in different cases of Fig. 23.10, the circuit representations in such cases can be simplified, as shown in Fig. 23.11. The simplified representations shown in Figs. 23.11 (*a*) to (*d*) correspond to the logic arrangements of Fig. 23.10 (*a*) to (*d*) respectively.

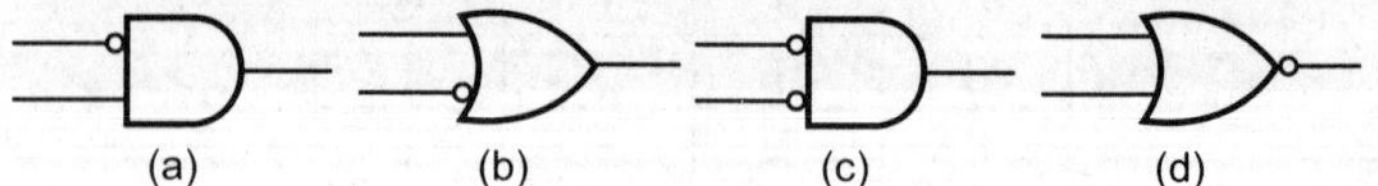

Fig. 23.11

Q. 13. What is an Exclusive-OR (EX-OR) gate ? How does it differ from a OR-gate ?

Or

Draw the logic symbol for an EX-OR gate and explain in brief, its logic operatic with the help of its truth table.

Ans. EX-OR stands for Exclusive-OR. Fig. 23.12 compares the truth tables for a 2-input OR and a 2-input EX-OR. EX-OR gate differs from an OR gate in only one of the entries in the truth table. The output of an EX-OR gate, when both its inputs are logic '1' is logic '0' instead of logic '1' in case of an OR gate. This entry has been highlighted in the truth table. The operation of a EX-OR gate can also be explained by the statement that "The output is logic '0' for like inputs and the output is logic '1' for unlike inputs. Its input-output logic expression is given by:

$$z = \bar{x}\,y + x\,\bar{y} \text{ written as } z = x \oplus y$$

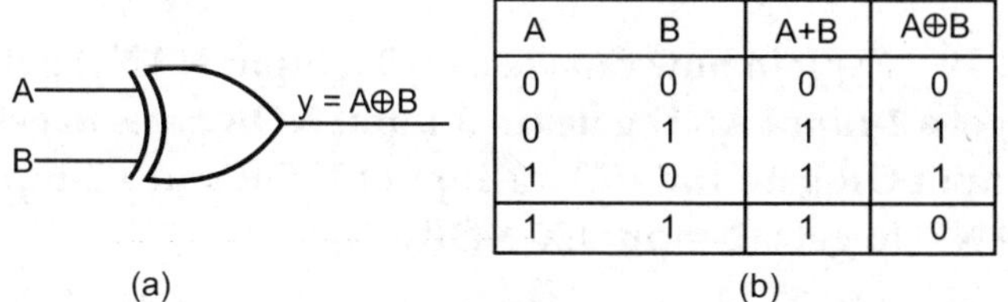

A	B	A+B	A⊕B
0	0	0	0
0	1	1	1
1	0	1	1
1	1	1	0

(b)

Fig. 23.12

In other words, this expression implies that z (output) is logic '1' when either $\bar{x}\,.\,y = 1$ OR $x.\bar{y} = 1$ (from the basic definition of an OR gate). Now $\bar{x}.y$ equals logic '1' only, when both $\bar{x}$ as well as y are logic '1' *i.e* $x = 0$ $y = 1$. Similarly, $x.\bar{y} = 1$ implied $x = 1$ and $y - 0$. Thus, either of the two conditions *i.e.* (*i*) $x = 0.$ $y = 1$ (*ii*) $x = 1, y = 0$ should, be satisfied to get a logic '1' at the output of an EX-OR gate.

Q. 14. Draw the circuit symbol of a NAND gate and explain in brief, its logic operation with the help of its truth table.

Or

With the help of truth table of a NAND gate, explain in brief the functioning of a NAND gate.

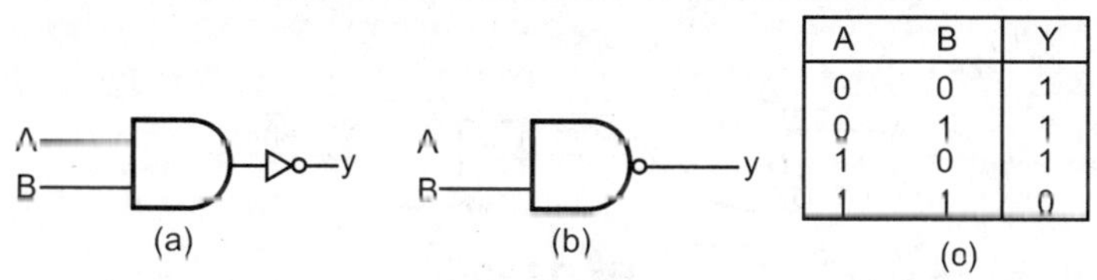

A	B	Y
0	0	1
0	1	1
1	0	1
1	1	0

(c)

Fig. 23.13

Ans. NAND stands for NOT-AND. An AND gate followed by a NOT circuit makes it a NAND gate (Refer to Figs. 21.13(*a*) and 21.13(*b*). The truth table of a 2 input NAND gate is shown in Fig. 21.13(*c*). Truth table of a NAND gate is obtained from the truth table of an AND gate by complementing the output entries.

The output of a NAND gate is logic '0' when all its inputs are logic '1'. For all other cases, the output is logic '1'. NAND gate operation is logically expressed by: $Y = \overline{A.B}$.

Q. 15. What is a NOR-gate ? How does it differ from an OR-gate ? Explain in brief, the operation of a NOR-gate with the help of its logic expression and truth table.

Ans. NOR stands for NOT-OR. Thus NOR gate is obtained by complementing the output of an OR-gate. The circuit symbol and the truth table for 2-input NOR-gate are respectively shown in Figs. 23.14(*b*) and (*c*). The truth table of a NOR-gate is obtained from the truth table of an OR-gate by replacing the output entries by their complements.

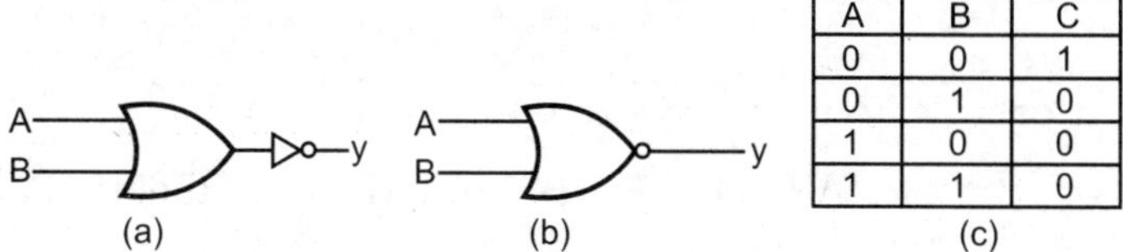

A	B	C
0	0	1
0	1	0
1	0	0
1	1	0

(c)

Fig. 23.14

NOR-gate is logically expressed by the expression (Boolean Expression): $Y = \overline{A+B}$

The output of a NOR-gate is logic '1' when all its inputs are logic '0'. For all other cases, the output is logic '0'. In general, the Boolean expression for a NOR-gate with more than two inputs can be written as:

$$Y = \overline{(A+B+C+D+.........)}$$

Q. 16. Why are NAND and NOR gates called UNIVERSAL GATES ?

Or

While OR, AND and NOT gates are the basic logic gates, NAND and NOR gates are referred to as the Universal Gates. Comment.

Ans. OR, AND and NOT gates are the basic logic gates, as they together can be used to construct the logic circuit for any given logic or Boolean expression. The NAND and NOR gates have a special property that these gates can individually be used to construct logic circuit for any Boolean expression. That is, it is possible to use either only the NAND gates or only the NOR gates to implement any Boolean expression. It is so because a combination of NAND gates or a combination of NOR gates can be used to perform the functions of any of the basic logic gates. It is because of this reason that NAND and NOR gates are called **Universal Gates.**

Q. 17. What is an EX-NOR gate ? With the help of its truth table and logic expression, explain in brief, its operation.

Ans. EX-NOR means NOT EX-OR *i.e.* the logic gate that we get by complementing the output of an EX-OR gate. Fig. 23.15 shows its circuit symbol alongwith its truth table.

The truth table of EX-NOR gate is obtained from the truth table of an EX-OR gate by complementing the output entries.

Logically, $Y = \overline{A \oplus B}$

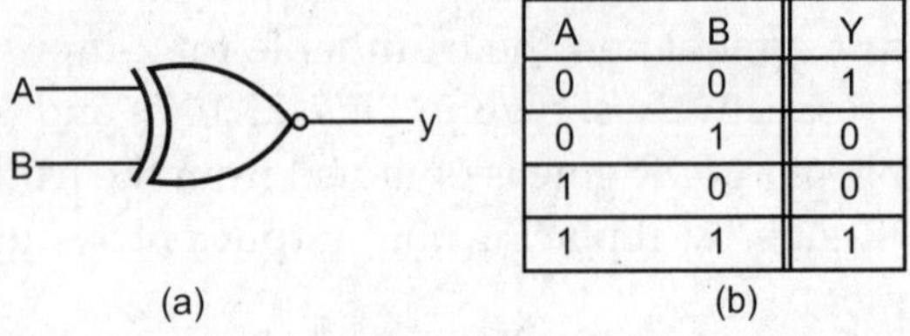

A	B	Y
0	0	1
0	1	0
1	0	0
1	1	1

Fig. 23.15

or $\quad Y = \overline{(\overline{A}.B + A.\overline{B})}$

$= (A + \overline{B}).(\overline{A} + B)$ .... DeMorgan's theorems

$= A.B + \overline{A}.\overline{B}$

The output of an EX-NOR gate is logic '1', when the inputs are like and logic '0' when the inputs are unlike.

Q. 18. What logic functions do you get when you short both the inputs of a NAND or a NOR to get a one input, one output logic circuit ?

Ans. Fig. 23.16 shows a 2-input NAND with its inputs shorted alongwith the truth table of a 2-input NAND. When the input is logic '0', both A and B inputs will be logic '0'. From the truth table, $y = 1$. When the input is logic '1', both the inputs will be logic '1'. From the truth table, output $y = 0$. The output is thus complement of the input. A NAND-gate with shorted inputs is, therefore, a NOT-gate. The statement can be generalised for a NAND gate with more than two inputs to say that when all the inputs of a NAND are shorted, it becomes a NOT-gate.

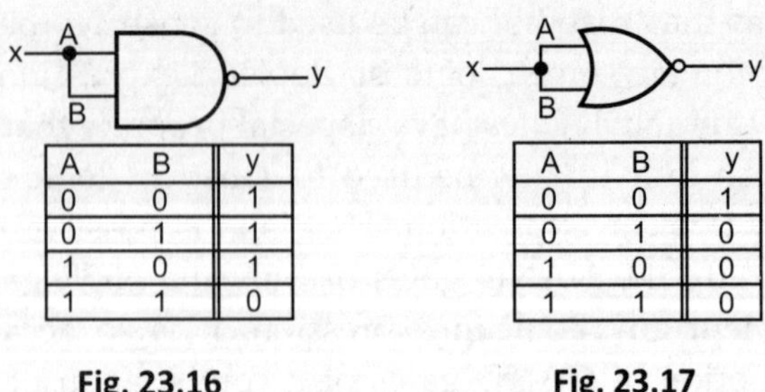

A	B	y
0	0	1
0	1	1
1	0	1
1	1	0

Fig. 23.16

A	B	y
0	0	1
0	1	0
1	0	0
1	1	0

Fig. 23.17

Logically, $Y = \overline{A.B.C.D.......}$ (NAND-Gate)

When all the inputs are shorted, then

$A = B = C = D = x$(say)

where, (x) is some binary variable.

In that case, $Y = \overline{x}$ (x.x.x.......$= x$, Idempotent Law)

Same is true for a NOR-gate, when all its inputs are tied together to get a one input one output logic circuit. To prove this consider a two-input NOR with its inputs shorted. When the input is logic '0' and from the truth table of a 2-input NOR, the output is logic '1'. Similarly, when the input is logic '1', the output is logic '0'. Thus the output is complement of the input Fig. 23.17 shows the NOR-gate and its truth table.

Logically, $Y = \overline{A + B + C + D + ...}$ (NOR-Gate)

when $A = B = C = D$ .....$= x$ (say), where (x) is some binary variable,

they $y = \overline{x}\,(x + x + x + x \text{ ....} = x$, Idempotent Law).

Q. 19. Explain how can we use (*i*) 2-input NAND gates to get a 2-input AND gate (*ii*) 3-input NOR gates to get a 3-input OR gate and (*iii*) a 2-input EX-OR and a 2-input NAND to get a 2-input EX-NOR?

Ans. Fig 23.18 shows the use of two 2-input NAND gates to get an AND gate. The second NAND has been wired as a NOT circuit.

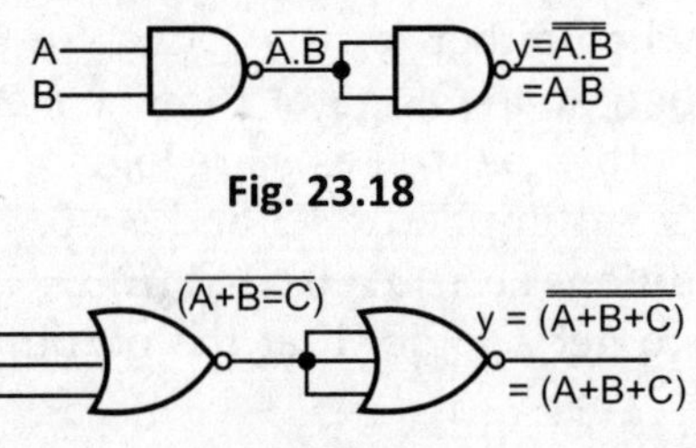

Fig. 23.18

Fig. 23.19

Fig. 23.19 depicts the use of two 3-input NOR gates to implement a 3-input OR-gate. The second NOR gate has been wired as a NOT circuit.

2-input EX-OR followed by a 2-input NAND wired as an INVERTER (NOT circuit) by shorting its inputs makes the combined arrangement an EX-NOR gate. Refer to Fig. 23.20.

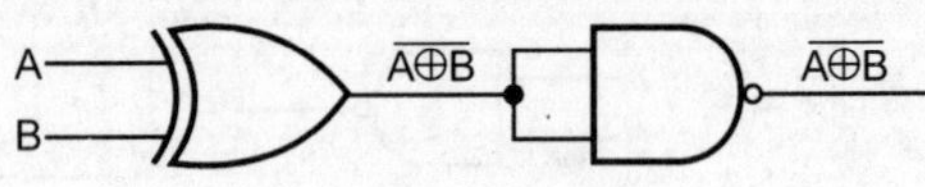

Fig. 23.20

Q. 20. Write Boolean Expressions for OR, AND, NAND, NOR, EX-OR and EX-NOR gates. Write generalised expressions assuming that the gates have more than two inputs. Also, draw the combined truth table for the above gates to show a comparison of their logic operations. For the purpose of the truth table, you can assume all the gates to be 2-input ones.

Ans. The generalized Boolean expression for an OR-gate can be written as :

$$Y = A + B + C + D + ...$$

where, *A*, *B*, *C*, *D*.........are input binary variables and *Y* is the output.

Fer an AND-gate, $Y = (A.B.C.D.......)$

For a NOR-gate, $Y = (\overline{A + B + C + D + ...})$

For an EX-OR gate, $Y = \overline{A}\ .B + A\ .\overline{B}$

For an EX-NOR gate, $Y = A.B + \overline{A}\ .\ \overline{B}$

It may be mentioned here that EX-OR and EX-NOR are usually 2-input gates.

The truth table showing a comparison of various gates is given in Fig. 23.21.

Truth table

A	B	A + B OR	A . B AND	$\overline{A+B}$ NOR	$\overline{A.B}$ NAND	A ⊕ B EX-OR	$\overline{A \oplus B}$ EX-NOR
0	0	0	0	1	1	0	1
0	1	1	0	0	1	1	0
1	0	1	0	0	1	1	0
1	1	1	1	0	0	0	1

Fig. 23.21

Q. 21. For the logic circuit shown in Fig. 23.22, write the input output Boolean expression and draw its truth table.

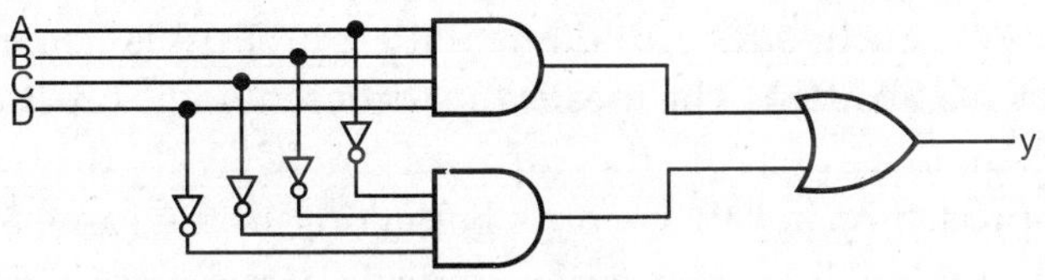

Fig. 21.22

Ans. The input output Boolean expression is : $Y = A.B.C.D + \overline{A}.\overline{B}.\overline{C}.\overline{D}$.

You would notice that the output is logic '1' when all inputs are like.

Truth table

A	B	C	D	Y	A	B	C	D	Y
0	0	0	0	1	1	0	0	0	0
0	0	0	1	0	1	0	0	1	0
0	0	1	0	0	1	0	1	0	0
0	0	1	1	0	1	0	1	1	0
0	1	0	0	0	1	1	0	0	0
0	1	0	1	0	1	1	0	1	0
0	1	1	0	0	1	1	1	0	0
0	1	1	1	0	1	1	1	1	1

Fig. 23.23

Q. 22. What is FAN-OUT with reference to logic gates?

Ans. It is a common occurrence in logic circuits where the output of one logic gate feeds the inputs of several others. It is not that we can drive unlimited number of logic gates' inputs from the output of a single logic gate. It is limited by the current sourcing capability of the output, when it is HIGH and the current sinking capability of the output, when it is LOW and the requirements of the logic gates' inputs in the two states. Number of logic gate inputs that can be driven from the output of a single logic gate with reliability (*i.e.* without causing any false output) is called the FAN-OUT. It is the characteristic of the family to which the IG belongs. It is 10 for the logic gates belonging to the standard TTL family and 50 for CMOS family logic gates.

Q. 23. Why is that the FAN-OUT of standard TTL logic gates is 10 and not more than 10 or less than 10 ?

Ans. When the output of a logic gate belonging to a standard TTL family is LOW, it can sink a maximum current of 16 mA. The LOW-state current of standard TTL logic gate input is 1.6mA which implies that 10 such inputs could be connected to the output of a single gate. Also, when the output of the logic gate is HIGH, it can source a maximum current of 0.4 mA. The requirement of TTL logic gate input in the High state is 40 μA. Again, a maximum of 10 gates can be driven from the output of a single logic gate belonging to TTL family. It leads one to the conclusion that the TTL logic gates have a FAN-OUT of 10. It may be mentioned here that there are sub-groups of the TTL family and we are referring to the standard TTL.

Q. 24. What is an INHIBIT-gate ? Draw the circuit symbol of a 4-input INHIBIT-gate along with its truth table.

Ans. It is basically an AND gate with one of its inputs negated by an inverter. Negated input acts to inhibit the gate or in other words, the gate shall give a logic '1' output only, when the negated input is driven to logic '0'. Fig. 23.24 shows the logic symbol of a 4-input INHIBIT gate. Fig. 23. 25 shows the truth table.

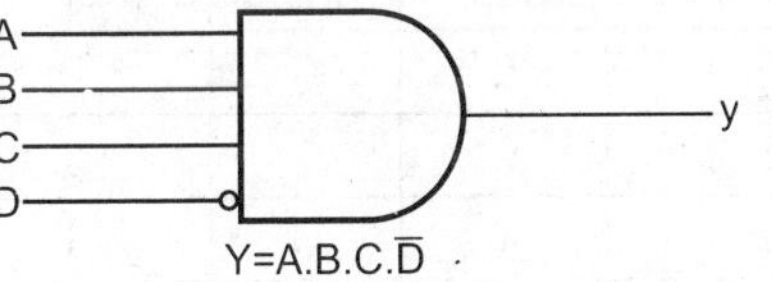

Fig. 23.24

A	B	C	D	Y
0	0	0	0	0
0	0	0	1	0
0	0	1	0	0
0	0	1	1	0
0	1	0	0	0
0	1	0	1	0
0	1	1	0	0
0	1	1	1	0
1	0	0	0	0
1	0	0	1	0
1	0	1	0	0
1	0	1	1	0
1	1	0	0	0
1	1	0	1	0
1	1	1	0	1
1	1	1	1	0

Fig. 23.25

Q. 25. About a 2-input EX-OR gate, we say that "Like inputs produce a '0' output and unlike inputs produce a '1' output." How do you explain the operation of an EX-OR gate, if it had more than two inputs? Draw the truth table of a 4-input EX-OR and a 4-input EX-NOR.

Ans.

A	B	C	D	EX-OR	EX-NOR
0	0	0	0	0	1
0	0	0	1	1	0
0	0	1	0	1	0
0	0	1	1	0	1
0	1	0	0	1	0
0	1	0	1	0	1
0	1	1	0	0	1
0	1	1	1	1	0
1	0	0	0	1	0
1	0	0	1	0	1
1	0	1	0	0	1
1	0	1	1	1	0
1	1	0	0	0	1
1	1	0	1	1	0
1	1	1	0	1	0
1	1	1	1	0	1

Fig. 23.26 (a)

Fig. 23.26 (*a*) shows the truth table. In general, if the number of l's in the input sequence is odd, the output is '1'. If the number of l's in the input sequence is even, the output is '0'.

Q. 26. If 3-input EX-OR gates are not available, then how do you implement a 3-input EX-OR gate using 2-input EX-OR gates?

Ans. Two of the three inputs are fed to a 2-input EX-OR gate and its output is then fed to the input of another EX-OR gate whose second input is the third binary variable. The output of this second EX-OR gate is then the final output. Fig. 23.26 (*b*) shows the logic implementation.

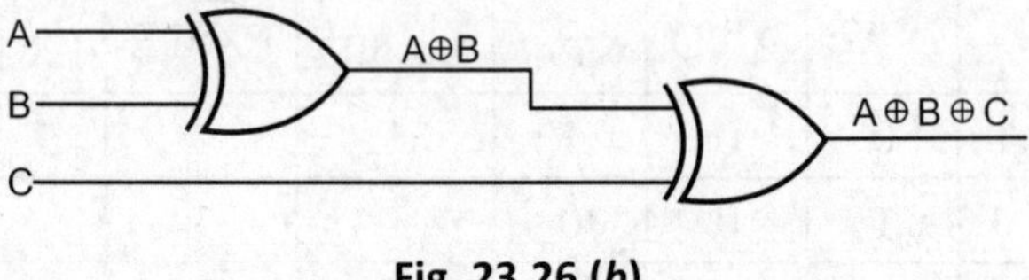

Fig. 23.26 (*b*)

Q. 27. Briefly explain what do you understand by the term LOGIC FAMILY ? What features are generally used to characterise a certain logic family?

Ans. There are a variety of circuit configurations or more appropriately various approaches on which different digital integrated circuits available today are based. Each such fundamental approach is called a LOGIC FAMILY. The idea is that different logic functions, when fabricated in the form of an IC with the same approach or in other words belonging to the same logic family shall have identical electrical characteristics. These characteristics include supply voltage range, speed of response, power dissipation, input and output logic levels etc. In other words, the set of digital ICs belonging to the same logic family are electrically compatible with each other. A digital system in general comprises of digital ICs performing different logic functions and choosing these ICs from the same logic family guarantees that different ICs are compatible with respect to each other and that the system as a whole performs the intended logic function. In case output of an IC belonging to a certain family feeds the inputs of another IC belonging to a different family, established interface techniques should be used to ensure compatibility.

Q. 28. What are the different types of logic families? What in your opinion are the most commonly used ones? What are sub-families ?

Ans. The entire range of digital ICs is fabricated using either the BIPOLAR DEVICES or the MOS DEVICES. Different logic families falling in the first category are called BIPOLAR FAMILIES and some of these are the

Resistance Transistor Logic (RTL), the Diode Transistor Logic (DTL), Transistor Transistor Logic (TTL), Emitter Coupled Logic (ECL) also known as Current Mode Logic (CML), Integrated Injection Logic (I^2L) and High Threshold Logic (HTL). The logic families that use MOS devices as basic devices are known as MOS families and the prominent members belonging to this category are the PMOS family (using P-channel MOSFETs), NMOS-family (using N-channel MOSFETs) and CMOS-family (Using both N and P-channel devices.)

The two most popular logic families are the TTL logic family and the CMOS logic family. The TTL logic family has several subfamilies, namely:

(*a*) Standard TTL (*b*) Low power TTL
(*c*) High power TTL (*d*) Low power Schottky TTL
(*e*) Schottky TTL (*f*) Advanced low power Schottky TTL
(*g*) Advanced Schottky TTL

The ICs belonging to TTL family are designated as 74 or 54 (for standard TTL), 74L or 54L (for low power TTL), 74H or 54H (for high power TTL), 74LS or 54LS (for low power Schottky TTL) and 74S or 54S (for Schottky TTL). An alphabetic code preceding this indicates the name of the manufacturer (DM for National Semi-conductors, SN for Texas Instruments and so on).

A two, three or four digit numerical code tells the logic function performed by the IC. It may be mentioned that 74-series devices and 54-series devices are identical except for their operational temperature range. 54-series are MIL-qualified (operational temperature range of – 55°C to + 125°C) versions of corresponding 74-series ICs (Operational temperature range of 0°C to 70°C). For example, 7400 and 5400 are both Quad 2-input NAND gates.

The popular CMOS subfamilies include:

(*a*) 4000 A CMOS family
(*b*) 4000 B CMOS family
(*c*) 4000 UB CMOS family
(*d*) 54/74C family
(*e*) 54/74HC family
(*f*) 54/74HCT family
(*g*) 54/74AC family
(*h*) 54/74ACT family.

Q. 29. Briefly describe different sub-families belonging to TTL logic family with respect to the typical internal circuitry and performance characteristics.

Ans. TTL as outlined above stands for TRANSISTOR-TRANSISTOR LOGIC. It is a logic family implemented with bipolar process technology that combines or integrates NPN transistors, PN junction diodes and diffused resistors in a single monolithic structure to get the desired logic function. NAND-gate is the basic building block of this logic family. Figs. 23.27 to 23.33 show the circuit schematics of the TTL NAND-gate in case of standard TTL, Low power TTL, High power TTL, Low power Schottky TTL, Schottky TTL, Advanced low power Schottky TTL and Advanced Schottky TTL subfamilies respectively. Table 23.1 gives a comparison of the major performance specifications of the TTL logic subfamilies. Different terms and definitions used in Table 23.1 have been described in Table 23.2.

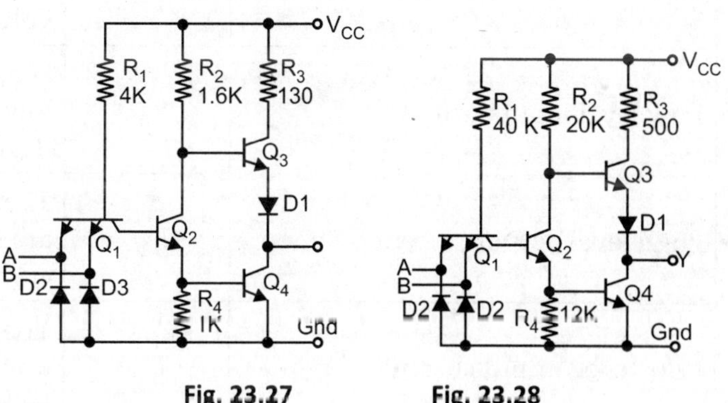

Fig. 23.27 Fig. 23.28

Table 23.1

Logic Family	Designation	Supply Voltage Min. (V)	Supply Voltage Max. (V)	Min. V_{IH} (V)	Max. V_{IL} (V)	Min. V_{OH} (V)	Max. V_{OL} (V)	Max. I_{IH} (μA)	Max. I_{IH} (mA)	Max. I_{OH} (mA)	Max. I_{OL} (mA)	Prop. Delay (nS)
Standard TTL	74	4.75	5.25	2.0	0.8	2.4	0.4	40	1.6	0.4	16	22
Low power TTL	74L	4.75	5.25	2.0	0.7	2.4	0.4	10	0.18	0.2	3.6	60
High power TTL	74H	4.75	5.25	2.0	0.8	2.4	0.4	50	2.0	20	20	10
Low power Schottky TTL	74LS	4.75	5.25	2.0	0.8	2.7	0.5	20	0.4	0.4	8	15
Schottky TTL	74S	4.75	5.25	2.0	0.8	2.7	0.5	50	2.0	1.0	20	5
Advanced low power Schottky TTL	74ALS	4.5	5.5	2.0	0.8	2.5	0.5	20	0.1	0.4	8	11
Advanced Schottky TTL	74 AS	4.5	5.5	2.0	0.8	2.5	0.5	20	0.5	2.0	20	4.5

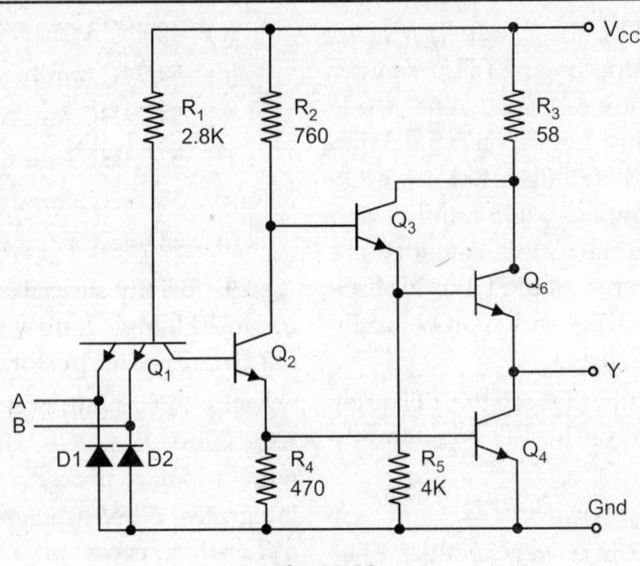

Fig. 23.29

Table 23.2

Parameter	Notation	Description
Maximum high level input current	I_I	When the input is applied the maximum voltage as specified by the concerned logic family, the current flowing into the inputs is then the maximum high level input current.
High level input current	I_{IH}	It is the current flowing into an input, when it is applied a high level input voltage equal to the minimum high level output voltage specified for the family.
Low level input current	I_{IL}	When the voltage applied at the input is equal to the maximum low level output voltage specified for the family, the current flowing out of the logic input is the low level input current (I_{IL}).
High level output current	I_{OH}	The current flowing out of an output, when the input conditions are such that the output is in logic HIGH state. It tells about the current sourcing capability of the output.
Low level output current	I_{OL}	The current flowing out of an output, when the input conditions establish a logic LOW at the output. It tells about the current sinking capability of the output.
High level input voltage	V_{IH}	It is the minimum positive voltage level applied at the input to be recognised as a legal HIGH level for the specified family.
Low level input voltage	V_{IL}	It is the maximum positive voltage level applied at the input to be recognised as a legal LOW level for the specified family.
High level output voltage	V_{OH}	It is the output voltage, when the input conditions establish a logic HIGH at the output for the specified family.
Low level output voltage	V_{OL}	It is the output voltage, when the input conditions establish a logic LOW at the output for the specified family.
Supply Current	I_{CC}	The supply current, when the output is HIGH, LOW and in high impedance state is respectively designated as I_{CCH}, I_{CCL} and I_{CCZ}.

Supply Current	I_{CC}	The supply current, when the output is HIGH, LOW and in high impedance state is respectively designated as I_{CCH}, I_{CCL} and I_{CCZ}.
Rise Time	t_r	It is the time that elapses between 10% and 90% of the final signal level, when signal is making a transition from a logic LOW to a logic HIGH.
Fall Time	t_F	It is the time that elapses between 90% and 10% of the signal level, when the signal is making a HIGH to LOW transition.
Propagation Delay (Low to High)	t_{pLH}	It is the time delay between the specified voltage points on the input and output waveforms with the output changing from LOW to HIGH.
Propagation Delay (High to Low)	t_{pHL}	It is the time delay between the specified voltage points on the input and output waveforms with the output changing from HIGH to LOW.
Maximum Clock frequency	f_{max}	It is the maximum clock frequency that can be used with a logic circuit without causing any false change of logic states both at input and output.
Speed-Power Product		The speed of a logic circuit can be increased (*i.e.* propagation delay can be reduced) at the expense of power dissipation. The Speed-Power dissipation product (usually measured in pico-joules) is a useful figure-of-merit for evaluating different logic families. It is the product of power dissipated per gate in milliwatts and the propagation delay measured in nano-seconds to give the product in pico-joules. Smaller the value of this product, better the logic family is.

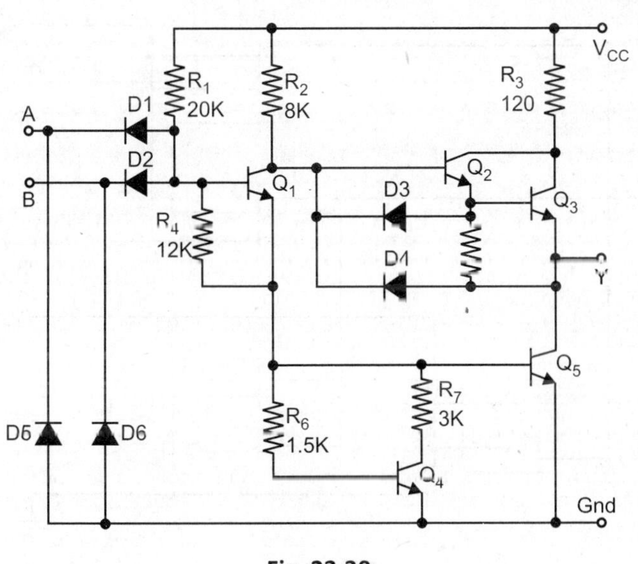

Fig. 23.30

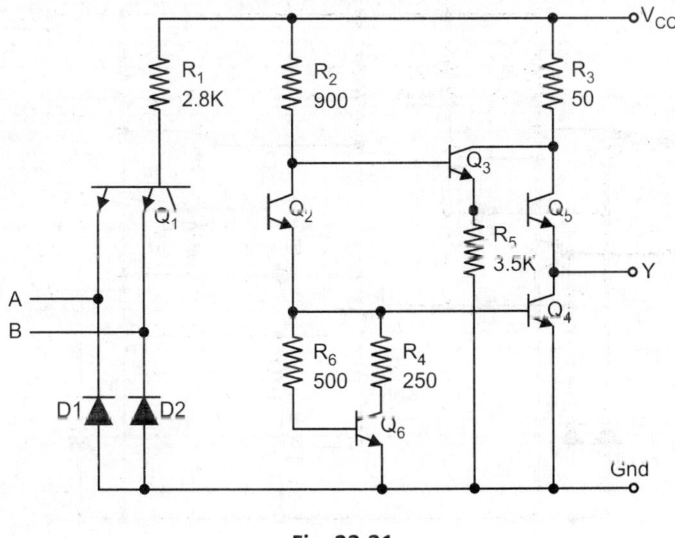

Fig. 23.31

Q. 30. Briefly describe the basic difference between different CMOS logic subfamilies.

Ans. The basic difference between different CMOS logic subfamilies, such as 4000A series, 4000B series, 4000UB series, 74C series etc. is in the fabrication process used and not in the design of the circuits used to implement the intended logic function Figs. 23.34 to 23.36 show the circuit schematics for the basis CMOS inverter, NAND-gate and NOR-gate respectively.

4000A-series CMOSICs were the first to arrive the scene from the CMOS logic family. 4000A CMOS subfamily is obsolete now and has been replaced by 4000B and 4000UB subfamilies. 4000B series is a high voltage version of4000A series and also, all the outputs in this series are buffered. 4000UB series is also a high voltage version of 4000A series but here the outputs are not buffered. 74C CMOS subfamily offers pin to pin replacements of 74 series TTL logic functions. For instance, if 7400 is a Quad 2-inputNAND in standard TTL, then 74C00 is a Quad 2-input NAND with same pin connections in CMOS. 74HC series is the high speed CMOS version of the 74 series logic functions. This is achieved using silicon gate CMOS technology rather than metal gate CMOS technology used in earlier CMOS subfamilies. 74 HCT series is only a process variation of 74HC series. The parameters like V_{IH}, V_{IL}, V_{OH} and V_{OL} in 74HCT series are compatible with those in bipolar TTL logic families. 74AC is presently the fastest CMOS logic family. This logic family has the best combination of high speed, low power consumption and high output drive capability. Again 74ACT is only a process variation of 74AC. The input thresholds (V_{IH} V_{IL}) and output voltage levels (V_{OH}, V_{OL}) in 74ACT are compatible with bipolar TTL families. Table 23.3 gives a comparison of different CMOS logic subfamilies.

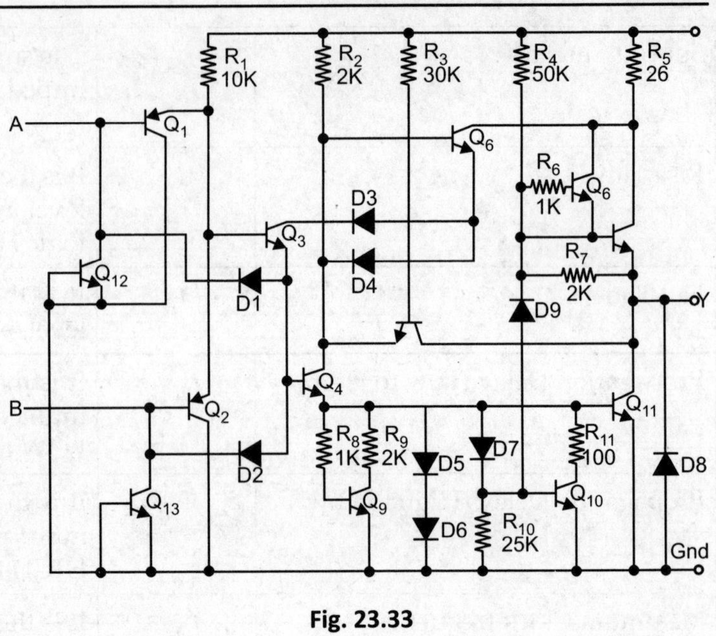

Fig. 23.33

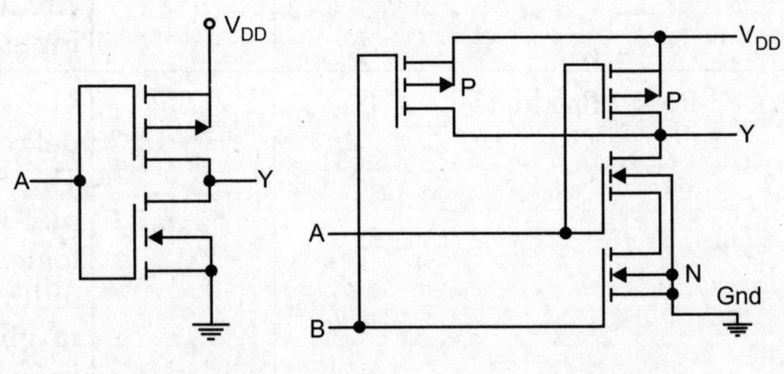

Fig. 23.34 Fig. 23.35

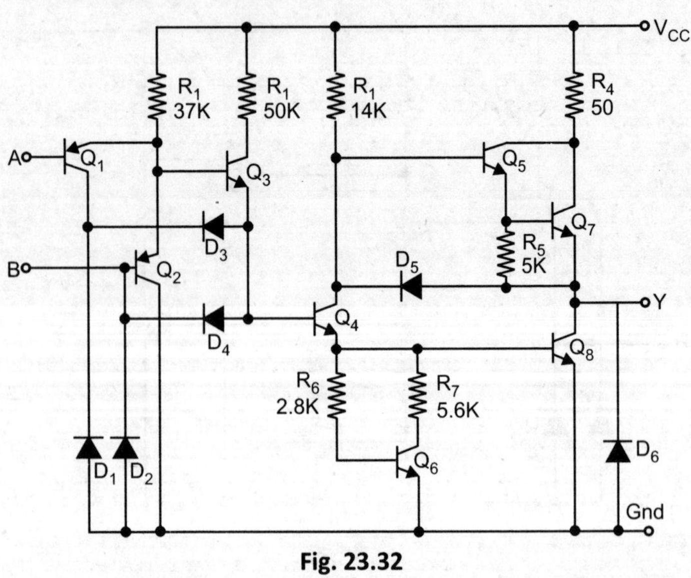

Fig. 23.32

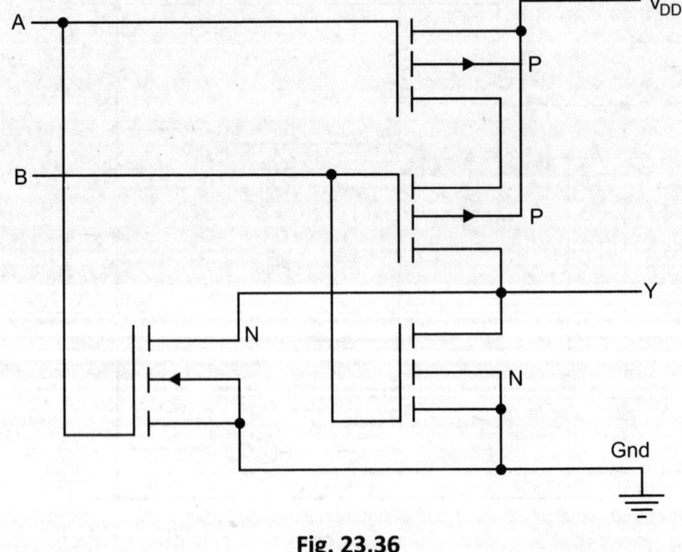

Fig. 23.36

Table 23.3

Logic Family	Supply Voltage		Min. VIH	Max. VIL	Min. VOH	Max. VOL	Max. IIH	Max. IIL	Max. IOH	Max. IOL	Prop. Delay
	Min. (V)	Maxx. (V)	(V)	(V)	(V)	(V)	(μA)	(μA)	(MA)	(mA)	(ns)
4000A	3	12	3.5	1.5	4.95	0.05	1	1	0.24	0.24	120
4000B	3	18	3.5	1.5	4.95	0.05	1	1	0.42	0.42	250
4000UB	3	18	4	1	4.95	0.05	1	1	0.42	0.42	120
74C	3	15	3.5	1.5	4.5	0.5	1	1	0.01	0.01	90
74HC	2	6	3.15	0.9	3.84	0.33	1	1	4	4	23
74HCT	4.5	5.5	2	0.8	3.84	0.33	1	1	4	4	25
74AC	2	6	3.15	1.35	3.76	0?37	1	1	24	24	8.5
74ACT	2	6	2	0.8	3.76	0.37	1	1	24	24	12.5

Q. 31. Briefly describe the ECL family. Outline the major characteristics of MECL 10K and MECL/PECL III logic families.

Ans. ECL (Emitter Coupled Logic) family is the fastest logic family and it belongs to the group of bipolar logic families. The high speed or the short propagation delay time of the emitter coupled logic is the result of several factors. These are: (*i*) It is non-saturating logic *i.e.* the transistors in this logic are always operated in the active region while in conduction (*ii*) The logic swing *i.e.* the difference in the voltage levels corresponding to logic LOW and HIGH states is kept small (typically 1) with the result that the output capacitance needs to be charged and discharged by a relatively much smaller voltage differential. (*iii*) The circuit currents are relatively high and the output impedance is low with the result that the output capacitance can be charged and discharged quickly. Fig. 23.37 shows the circuit schematic of 2-input OR/NOR gate in emitter coupled logic. The popular ECL subfamilies are:

(*i*) MECL 10,000 (or 10K) family (*ii*) MECL-III family (*iii*) MECL 10KH family

MECL 10KH family is an improved and high speed version of MECL 10K family.

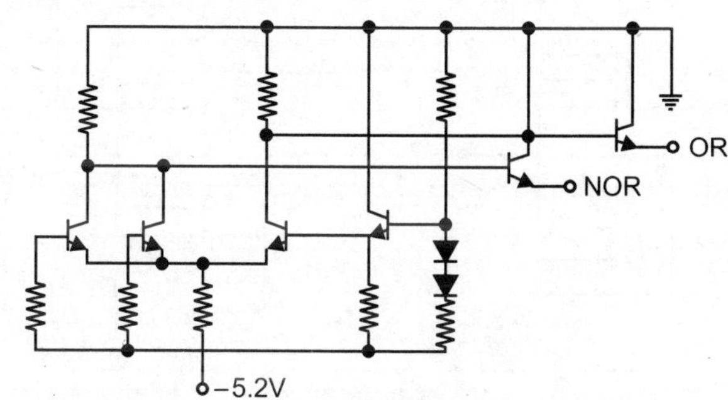

Fig. 23.37

Table 23.4.

Logic Family	Supply Voltage		Min. V_{IH} (V)	Min V_{OH} (V)	Max. V_{IL} (V)	Max. V_{OL} (V)	Max. I_{IL} (μA)	Min. I_{OL} (mA)	Max. I_{IH} (μA)	Min. I_{OH} (mA)	Max operating frequency (MHz)
	Min (V)	Max. (V)									
MECL 10,000	– 4.7	– 5.7	– 1.105	– 0.980	– 1.630	– 1.475	0.5	50	265	50	200
MECL III or PECL III	– 4.7	– 5.7	– 1.105	– 0.980	– 1.485	– 1.600	350	40	350	40	1000

Q. 32. Briefly describe NMOS and PMOS logic families. Why are these more suitable for implementing digital functions with large scale integration as compared to TTL, CMOS and ECL families ?

Ans. The logic families discussed so far are the ones that are commonly used for implementing discrete logic functions, such as logic gates, flip flops, counters, multiplexers, demultiplexers etc. in relatively less complex digital ICs belonging to the class of Small Scale Integration (SSI) and Medium Scale Integration (MSI) level of inner circuit complexities. The TTL, the CMOS and the ECL logic families are not suitable for implementing digital ICs that have Large Scale Integration (LSI) and above level of inner circuit complexity. The competitors for LSI class of digital ICs are the PMOS, the NMOS and the Integrated Injection Logic (I^2L). The first two are briefly discussed in this question and the third is discussed in the next question.

The PMOS logic family uses P-channel MOSFETS. Fig. 23.38. (*a*) shows an inverter circuit using PMOS logic. MOSFET Q_1 acts as an active load for the MOSFET switch Q_2. For the circuit shown, (GND) and $(-V_{DD})$ respectively represent logic '1' and logic '0' for a positive logic system. When the input is grounded (*i.e.* logic '1'), Q_2 remains in cut-off and $-V_{DD}$ appears at the output through the conduction Q_1. When the input is at $(-V_{DD})$ or near $(-V_{DD})$, Q_2 conducts and the output goes to near zero potential (*i.e.* logic '1'). Fig. 23.38 (*b*) shows the PMOS logic based 2-input NOR gate. As is clearly seen from the arrangement shown in Fig. 23.38 (*b*), the output goes to a logic '1' state (*i.e.* ground potential) only, when both Q_1 and Q_2 are conducting. And this is possible only when both the inputs are at logic '0' (*i.e.* $-V_{DD}$). For all other possible input combinations, the output is at logic '0' (*i.e.* $-V_{DD}$) because with either Q_1 or Q_2 non-conducting, the output is nearly $-V_{DD}$, which is logic '0', through the conducting Q_3. The circuit of Fig. 23.38 (*b*), thus, behaves like 2-input NOR gate in the positive logic.

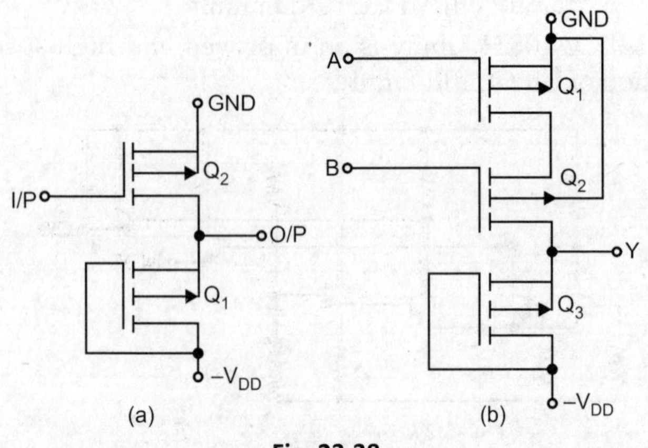

Fig. 23.38

It may be mentioned here that the MOSFET being used as a load (Q_1 in Fig. 23.38 *a* and Q_3 in Fig. 23.38*b*) is so designed as to have an ON-resistance that is much greater than the total ON-resistance of the MOSFETs being used as switches (Q_2 in Fig. 23.38 *a* and Q_1, Q_2 in Fig. 23.38 *b*).

NMOS logic family uses N-channel MOSFETS. N-channel MOS devices require a smaller chip area per transistor as compared to P-Channel devices with the result that NMOS logic offers a higher density. Also, due to greater mobility of charge carriers in N-channel devices, NMOS logic family offers higher speed too. It is because of this reason that most of the MOS memory devices and microprocessors employ NMOS logic or some variation of it, such as VMOS, DMOS, HMOS etc. VMOS, DMOS and HMOS are only structural variations of NMOS aimed at further reducing the propagation delay. Figs. 23.39 (*a*) and (*b*) respectively show a 2-input NOR and a 2-input NAND using NMOS logic. The arrangements shown are self-explanatory.

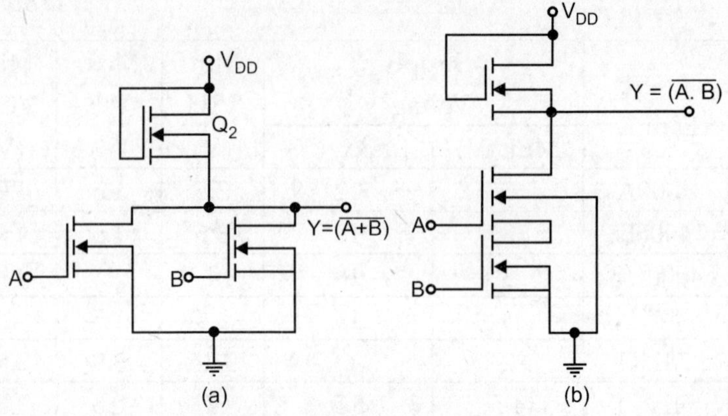

Fig. 23.39

The circuits shown can be extended to accommodate more inputs. For instance, Fig. 23.40 shows a 3-input NOR gate using NMOS logic.

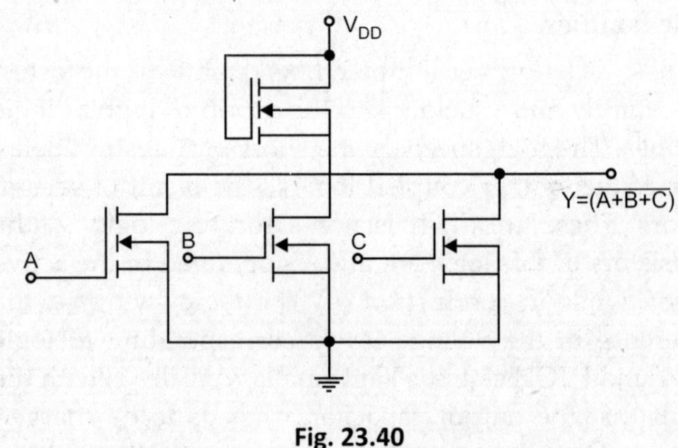

Fig. 23.40

Q. 33. Briefly describe operational principle and characteristics of I^2L family with particular reference to its suitability to implementing LSI and VLSI logic functions.

Ans. Integrated Injection Logic (I^2L) also known as current injection logic, family is well-suited to implementing LSI and VLSI digital functions and is a close competitor to NMOS logic family. Fig. 23.41 shows the basic I^2L family building block. Transistors (Q_3) and (Q_4) constitute current sources. The magnitude of current, known as injection current, depends upon externally connected (*R*) and applied (+V). If the input (A) is HIGH, the injection current through (Q_3) flows through base emitter junction of (Q_1). (Q_1) saturates and its collector drops to a low voltage, typically 50mV to 100mV. When (A) is LOW, the injection current is swept away from the base emitter junction of (Q_1). (Q_1) becomes open and the injection current through (Q_4) saturates (Q_2) with the result that (Q_1) collector potential equals base emitter saturation voltage of (Q_2), typically 0.7 *V*.

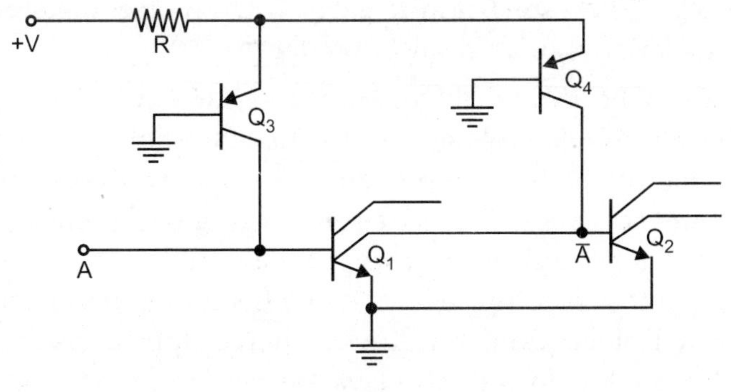

Fig. 23.41

The speed of I^2L family devices is a function of injection current (I) and improves with increase in current as a higher current allows a faster charging of capacitive loads present at bases of transistors. The programmable injection current feature is made use of in I^2L family of digital ICs to choose the desired speed depending upon intended application. The logic '0' level is V_{CE} (sat) of the driving transistor (Q_1 in the present case) and the logic '1' level is V_{BE} (sat) of the driven transistor (Q_2 in the present case) and the logic '1' level is V_{BE} (sat) of the driven transistor (Q_2 in the present case). Typically, logic '0' and logic '1' levels are respectively 0.1V and 0.7V. Speed-power product of I^2L family is typically under 1 pJ.

Q. 34. What are logic gates with open collector outputs with reference to TTL logic family devices ? Mention one important advantage of logic gates with open collector outputs.

Ans. The TTL family logic gates that do not have the usual totem pole output stage and instead have the output transistor with an open collector are referred to as the open collector gates. Fig. 23.42 shows the internal diagram of a TTL NAND with an open colleges tor output. An externally connected resistor called the pull-up resistor ensures that the circuit functions like a NAND gate. Typical value of pull-up resistor is IK.

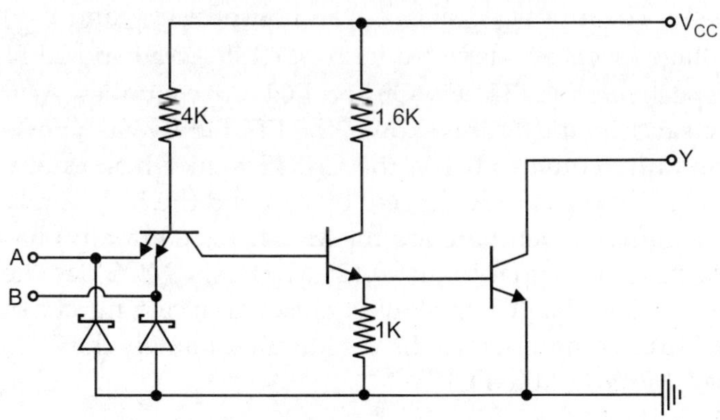

Fig. 23.42

The advantage of using open collector gates lies in their capability to provide ANDing operation, when outputs of several open collector NANDs are tied together through a common pull-up resistor without having to use an AND gate for the purpose. Refer to Fig. 23.43. the output in this case would be given by:

$$Y = \overline{AB} \,.\, \overline{CD} \,.\overline{EF}$$

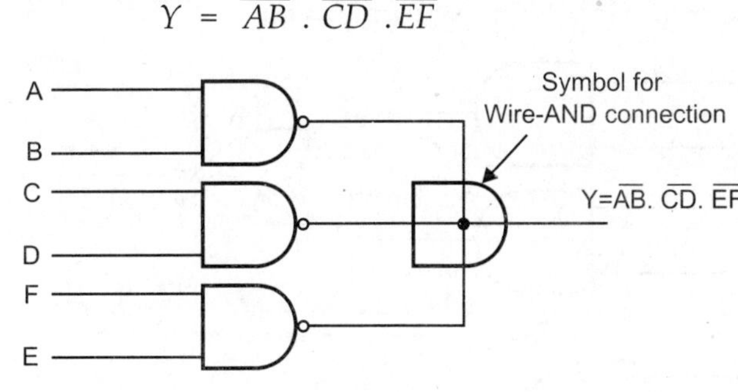

Fig. 23.43

Such a connection is also referred to as WIRE-AND connection. However, open collector gates are noisier and slower than their totem pole output counter parts.

Q. 35. What are tri-state logic gates ?

Ans. The tristate output logic gates have three possible output states *i.e.* logic '1' state, logic '0' state and a high impedance state. The high impedance state is controlled by an external enable input.

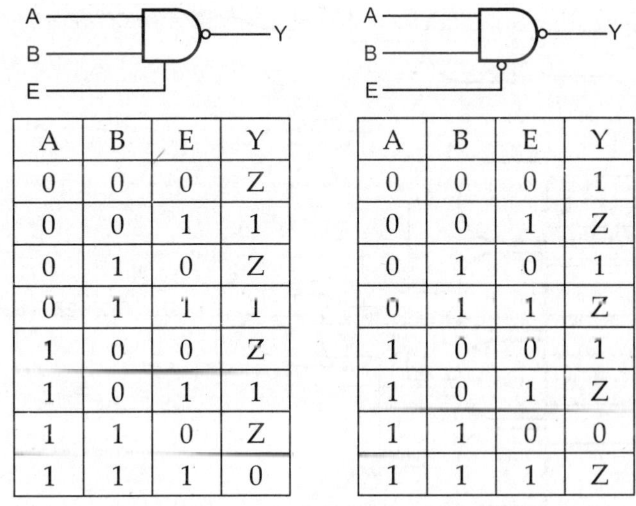

A	B	E	Y
0	0	0	Z
0	0	1	1
0	1	0	Z
0	1	1	1
1	0	0	Z
1	0	1	1
1	1	0	Z
1	1	1	0

A	B	E	Y
0	0	0	1
0	0	1	Z
0	1	0	1
0	1	1	Z
1	0	0	1
1	0	1	Z
1	1	0	0
1	1	1	Z

Fig. 23.44

The enable input decides whether the gate is active or is in the high impedance state. When active, it can be '0' or '1' depending upon input conditions. The outputs and inputs of tristate state be gates can be connected to a common bus line. Fig. 23.44 (*a*) shows the circuit symbol of a tristate NAND with active HIGH Enable input alongwith its truth table. The one shown in Fig 23.44 (*b*) has an active LOW Enable input.

Q. 36. What are AND-OR-INVERT gates? How are these gates usually designated ?

Ans. AND-OR and OR-AND gates can be usefully employed to implement sum-of-products and product-of-sums Boolean expressions respectively. Figs. 23.45(*a*) and 23.45(*b*) respectively show the symbols of AND-OR-INVERT and OR-AND-INVERT gates.

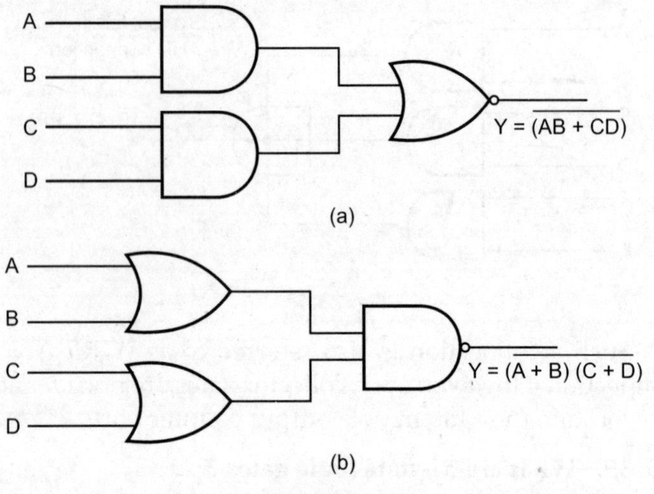

Fig. 23.45

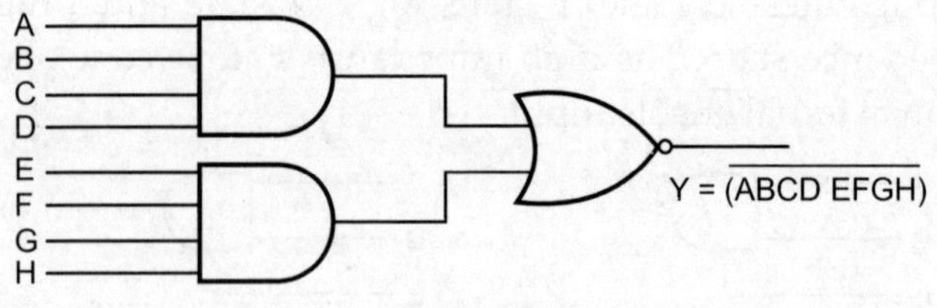

Fig. 23.46

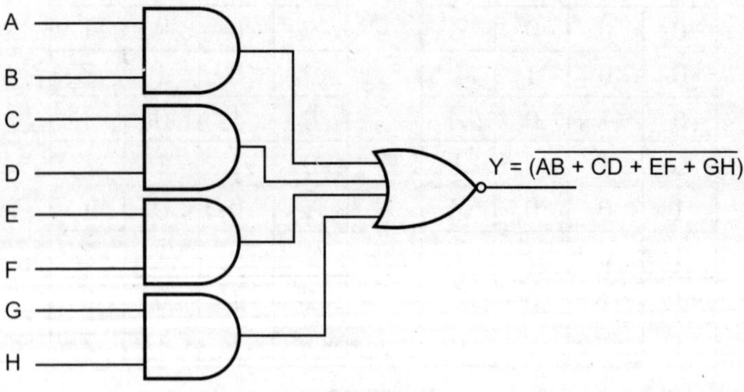

Fig. 23.47

Another method of designating the gates shown in Fig. 23.45 is to call them 2-wide, 2-input AND-OR-INVERT or OR-AND-INVERT gates, as the case may be. The gate is 2-wide as there are two gates at the input and 2-input as each gate has 2 inputs. Other varieties, such as 2-wide, 4-input AND-OR-INVERT (Fig. 23.46), 4-wide, 2-input AND-OR-INVERT (Fig. 23.47) etc. are also available in IC form.

Q. 37. What are Schmitt gates ? Draw the transfer characteristics of a Schmitt inverter.

Ans. The logic gates discussed so far have a single input threshold voltage level. This threshold voltage is same for both Low-to-High as well as High-to-Low output transitions. There are some logic gate varieties in NAND gates and Inverters that are available with built-in schmitt action. These have two different input voltage threshold levels, one for Low-to-High output transition and the other for High-to-Low output transition. Figs. 23.48 (*a*) and (*b*) respectively show circuit symbols of Schmitt NAND and Schmitt inverter. Fig. 23.48 (*c*) shows the typical transfer characteristics for such a device. These characteristics have been taken from the data sheet of 74LS132, which is a Quad 2-input Schmitt NAND belonging to low power Schottky TTL family. The difference between the two threshold levels is called systerisis.

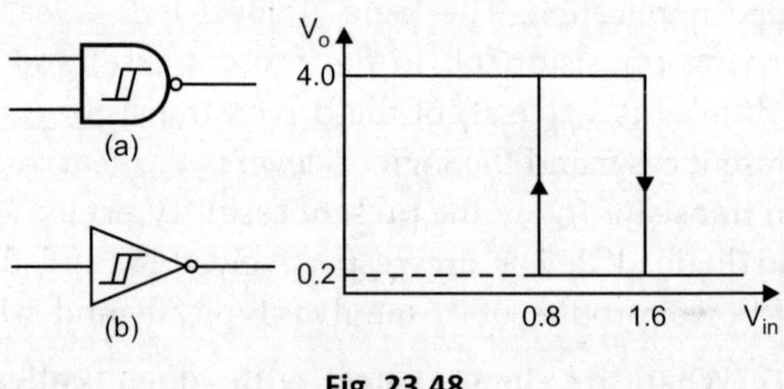

Fig. 23.48

Q. 38. What is NOISE IMMUNITY with reference to logic gates ? Compare TTL, CMOS and ECL families on the basis of noise immunity.

Ans. Noise immunity of a circuit is basically its insensitivity to undesired electrical signals In the context of logic circuits, the noise immunity tells us about the maximum quantum of noise that can be tolerated without causing any false change of logic state. The noise immunity of a digital IC is a characteristic of the logic family to which the IC belongs. It can be computed from $[V_{OL(\max)} - V_{IL\,(\max)}]$ or $[V_{OH\,(\min)} - V_{IH\,(\min)}]$ whichever is less. The input (V_{IH} and V_{IL}) and output (V_{OH} and V_{OL}) voltage levels are depicted in Figs. 23.49, 23.50 and 23.51 respectively for TTL, CMOS and ECL logic families. As it is clear from the figures shown, the TTL family has a noise immunity equal to 0.4 V, the CMOS family has a noise immunity equal to 30 percent of V_{DD} and the ECL family (the figures given here are for MECL 10,000 family) has an immunity equal to just 0.125 V. Thus, CMOS has the highest and the ECL the lowest noise immunity figures of the three commonly used logic families, namely the TTL, the CMOS and the ECL.

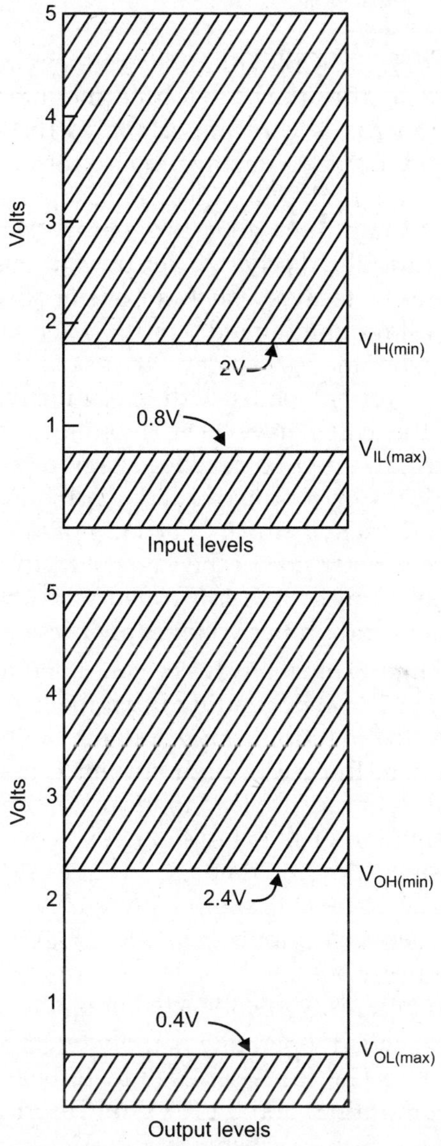

Fig. 23.49

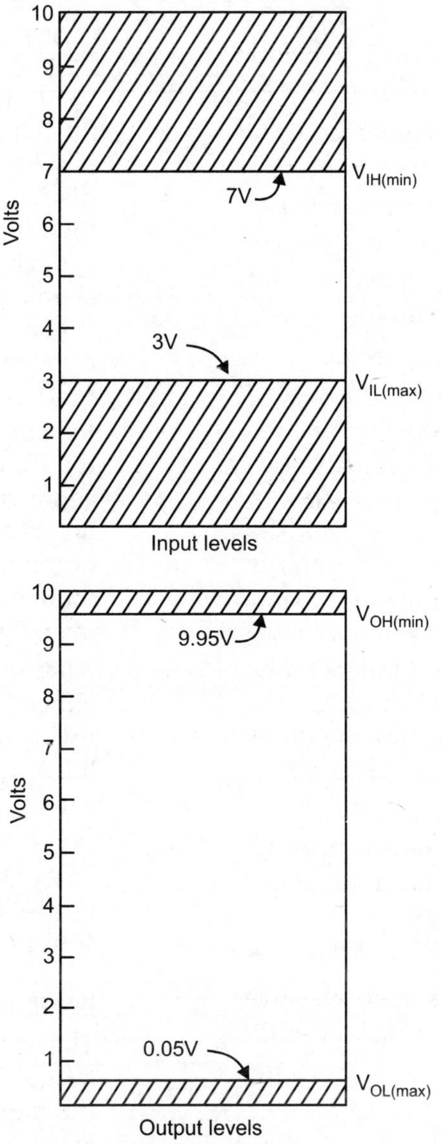

Fig. 23.50

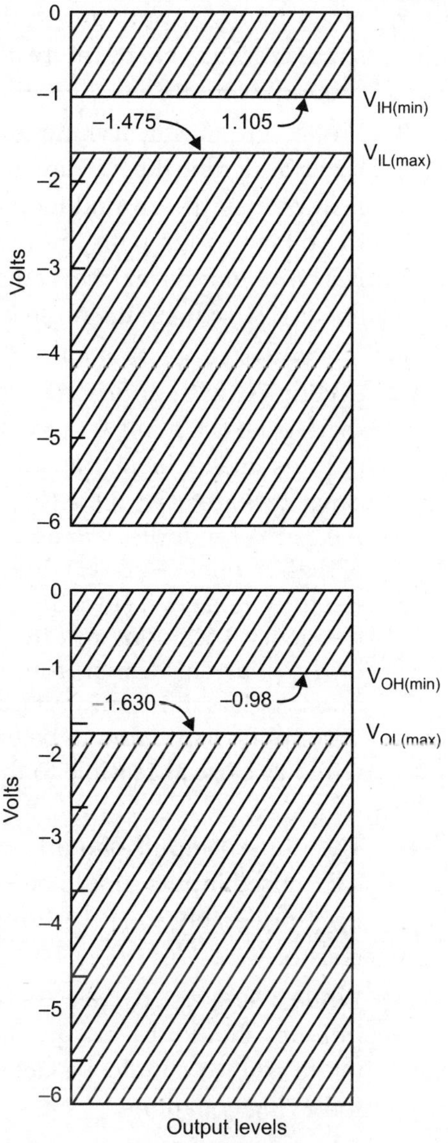

Fig. 23.51

Q. 39. Briefly outline the guidelines to be followed while using TTL logic family devices.

Ans. 1. Compatibility while using different TTL Subfamilies. Replacing a TTL IC of one TTL subfamily with the one belonging to another subfamily (the type numbers remaining the same) should not be done blindly. The designer should ensure that the replacement device is compatible with the existing circuit with respect to parameters like output drive capability, input loading, speed and so on. As an illustration, let us assume that you are using 74S00 (Quad 2-input NAND), whose output drives twenty different NAND inputs implemented using 74S00 again (Fig. 23.52). This circuit works well, as the Schottky TTL family has a fan out of 20 with an output HIGH drive capability of 1mA and input HIGH current requirement of 50 μA. If you try replacing 74S00 that is driving others with a 74LS00 the circuit fails to work, as 74LS00 NAND has an output HIGH drive capability of 0.4 mA only. It can not feed 20 NAND input loads implemented using 74S00.

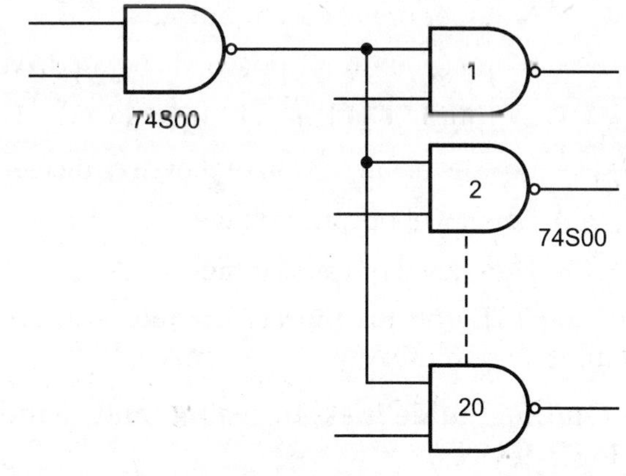

Fig. 23.52

2. None of the inputs and outputs of TTL ICs should be driven by more than 0.5 V below ground reference.
3. Proper grounding techniques should be used while designing the PCB layout. If the grounding is improper, the ground loop currents give rise to voltage drops with the result that different ICs will not be at the same reference. This effectively reduces the noise immunity.
4. The power supply rail must always be properly decoupled with appropriate capacitors so that there is no drop in (V_{CC}) rail as the inputs and outputs make logic transitions. Usually, two capacitors are used at the (V_{CC}) point of each IC. A 0.1 μF ceramic disc should be used to take care of high frequency noise, while typically a 100 μF electrolytic is good enough to eliminate any low frequency variations resulting from variations in (I_{CC}) current drawn from (V_{CC}), depending upon logic states of inputs and outputs. The decoupling capacitors to be effective should be wired as close as feasible to the (V_{CC}) pin of the IC.
5. The unused inputs should not be left floating. All unused inputs should be tied to a solid logic level preferably to (V_{CC}) through a 1 K to 5 K resistor.
6. While using open collector devices, resistive pull-up should be used. The value of pull-up resistance should be determined from the following equations:

(*i*) $$R_X = \frac{V_{CC(max)} - V_{OL}}{I_{OL} - N_{2(LOW)} \times 1.6\, mA}$$ and

(*ii*) $$R_{(max)} = \frac{V_{CC\,(max)} - V_{OL}}{N_1 \times I_{OH} + N_{2\,(HIGH)} \times 40\, \mu A}$$

where, R_x = External pull-up resistor

N_1 = Number of wired-OR outputs

N_2 = Number of unit input loads being driven

I_{OH} = Output HIGH leakage current

I_{OL} = Low level fan out current of driving element

V_{OL} = Low level output voltage

V_{OH} = High level output voltage.

Also, one TTL unit load in HIGH state = 40 μA and one TTL unit load in LOW state = 1.6 mA.

Q. 40. Outline guidelines to using and handling CMOS logic devices.

Ans.

1. Proper handling of CMOS ICs before they are used and also, after they have been mounted on the PC boards is very important, as these ICs are highly prone to damage to electrostatic discharge. Although all CMOS ICs have inbuilt protection networks to guard them against electrostatic discharge, yet precautions should be taken to avoid such an eventuality While handling unmounted chips, potential differences should be avoided. It is a good practice to cover the chips with a conductive foil. Once the chips have been mounted on the PC board, it is a good practice again to put conductive clips or conductive tape on the PC board terminals. Remember that PC board is nothing but an extension of the leads of the ICs mounted on it unless it is integrated with the overall system and proper voltages are present.
2. All unused inputs must always be connected to either (Y_{SS}) or (V_{DD}) depending upon the logic involved. A floating input can result in a faulty logic operation. In case of high current types, such as buffers, it can also lead to the maximum power dissipation of the chip being exceeded, thus, causing device damage. A resistor (typically 220 K to 1M) should preferably be connected between input and the (V_{SS}) or (V_{DD}), if there is a possibility of device terminals becoming temporarily unconnected or open.
3. The recommended operating supply voltage ranges are 3 to 12V for A-series (3 to 15 being the maximum rating) and 3 to 15V for B-series and UB-series (3 to 18V being the maximum). CMOS IC application circuits that are operated in a linear mode over a portion of voltage range, such as RC or crystal oscillators, a minimum (V_{DD}) of 4V is recommended.
4. Input signals should be maintained with the power supply voltage range $V_{SS} \leq V_i \leq V_{DD}$ ($-\,0.5\ V \leq V_i \leq V_{DD} + 0.5$ V being the absolute maximum). If the input signal exceeds the recommended input signal range, the input current should be limited to ± 100 μA.
5. CMOS ICs like active pull-up TTL ICs cannot be connected in WIRE-OR configuration. Paralleling of inputs and outputs of gates is also recommended for ICs in the same package only.
6. Majority of CMOS clocked devices have maximum rise and fall time ratings of normally 5 to 15 microseconds. The device may not function properly with larger rise and fall times. The restriction however does not apply

to those CMOS ICs having in built Schmitt trigger shaping in the clock circuit.

7. Whenever two sequential CMOS ICs (flip flops for instance) are cascaded and clocked synchronously, the maximum rise time of the clock should be limited, as per

Maximum t_2 (clock) $= \frac{0.8\, V_{DD}}{1.25} \times t_{p\,(nS)}$ for A–series CMOS ICs.

$= \frac{0.8\, V_{DD}}{1.15} \times t_{p\,(nS)}$ for B–series CMOS ICs.

where, $t_p = t_{pHL}$ or t_{pLH} whichever is less of the first input device as per the device's data sheet at specified value of (V_{DD}) and loading conditions.

Q. 41. Briefly describe the requirements when it comes to interfacing a logic device of one logic family with a logic device belonging to another for the following cases :

(*a*) CMOS-TTL interface (*b*) TTL-CMOS interface

(*c*) TTL-ECL interface (*d*) ECL-TTL interface

(*e*) CMOS-ECL interface (*f*) ECL-CMOS interface

Ans. CMOS and TTL are the two most widely used logic families *i.e.* majority of digital ICs in use today belong to either of these two families. Though ICs belonging to the same logic family have no special interface requirements, that is, output of one can directly feed the input of the other, same is not the case, if we have to interconnect digital ICs belonging to different logic families. Incompatibility of ICs belonging to different families mainly arises from different voltage levels and current requirements associated with LOW and HIGH logic states at the inputs and outputs. Here, we shall discuss simple interface techniques that can be used for CMOS-to-TTL, TTL-to-CMOS CMOS-ECL, ECL-CMOS and TTL-ECL, ECL-TTL interconnections.

CMOS-to-TTL Interface. First possible type of CMOS-to-TTL interface is the one where both ICs are operated from a common supply. We have read in the earlier questions that TTL family has a recommended supply voltage of 5V, whereas CMOS family devices can operate over a wide supply voltage range of 3V to 18V. In the present case, both ICs would operate from 5V. As far as the voltage levels in the two logic states are concerned, the two have become compatible. CMOS output has a $V_{OH(min)}$ of ≈ 5V (for V_{CC} = 5V) and, $V_{OL\,(max)}$ ≈ 0 V, which is compatible with $V_{IH(min)}$ and $V_{IL(max)}$ requirements of approximately 2V and 0.8V respectively of TTL family devices. In fact, in a CMOS-to-TTL interface, with the two devices operating on the same (V_{CC}) the voltage level compatibility is always there. It is the current level compatibility that needs attention. That is, in the LOW state, the output current sinking capability of CMOS IC in question must at least equal the input current sinking requirement of the TTL IC being driven. Similarly, in the HIGH state, the HIGH output current drive capability of CMOS IC must equal or exceed the HIGH level input current requirement of TTL IC. For a proper interface, both the above conditions must be met with. As a thumb rule, a CMOS IC belonging to 4000 B family (the most widely used CMOS family) can feed one LS TTL or two Low power TTL unit loads. When a CMOS IC needs to drive a standard TTL or a Schottky TTL device, a CMOS buffer (4049B or 4050B) is used. 4049B and 4050B are hex buffers of inverting and non-inverting type respectively with each buffer capable of driving two standard TTL loads. Fig. 23.53 shown CMOS-to-TTL interface with both devices operating from 5V supply and the CMOS IC driving a low power TTL or a low power Schottky TTL device. Fig. 23.54 shows the CMOS-to TTL interface where the TTL device in use is either a standard TTL or a Schottky TTL.

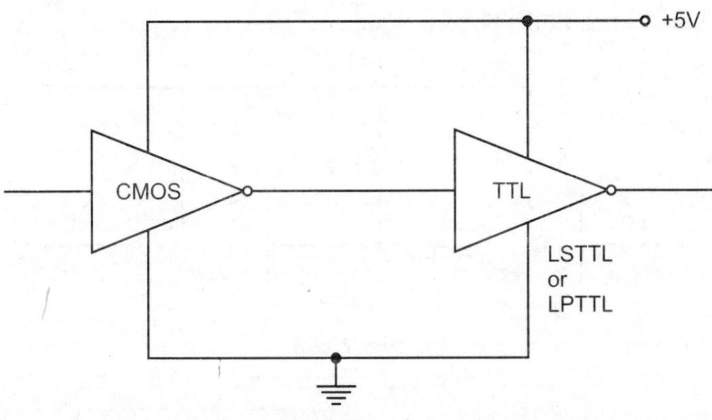

Fig. 23.53

CMOS-to-TTL interface, when the two are operating on different power supply voltage can be achieved in several ways. One such scheme is shown in Fig. 23.55. As can be seen from the arrangement, there is both voltage level as well as current level compatibility.

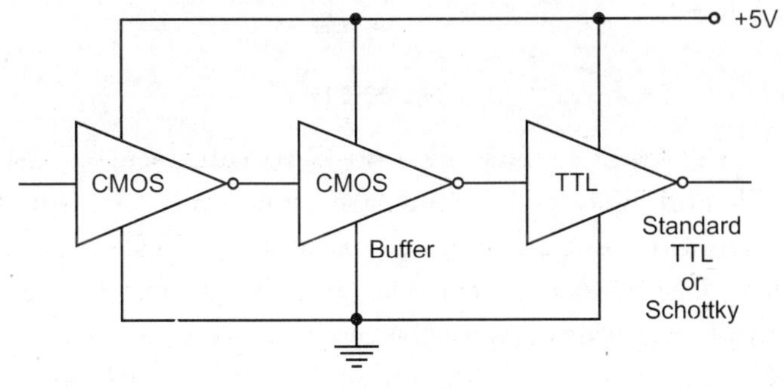

Fig. 23.54

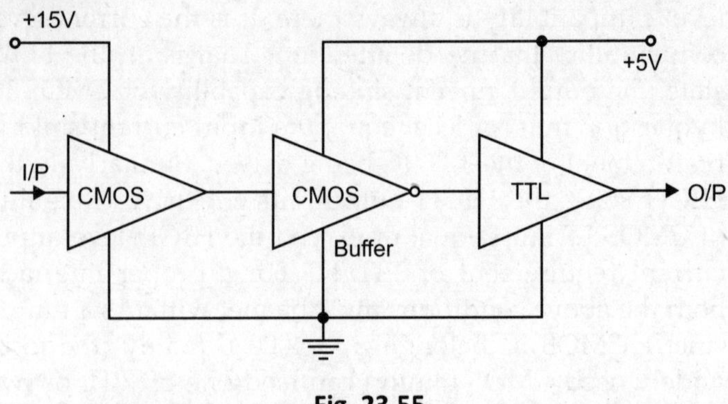

Fig. 23.55

TTL-to-CMOS Interface. In the TTL-to-CMOS interface, current compatibility is always there. The voltage level compatibility in the two states is a problem. When the two devices are operating on the same power supply voltage *i.e.* 5V, a pull-up resistor of 10K achieves compatibility (Fig 23.56). When the two are operating on different power supplies, one of the simplest interface techniques is to use a transistor (as a switch) in between the two, as shown in Fig. 23.57. Another technique is to use an open-collector type TTL buffer.

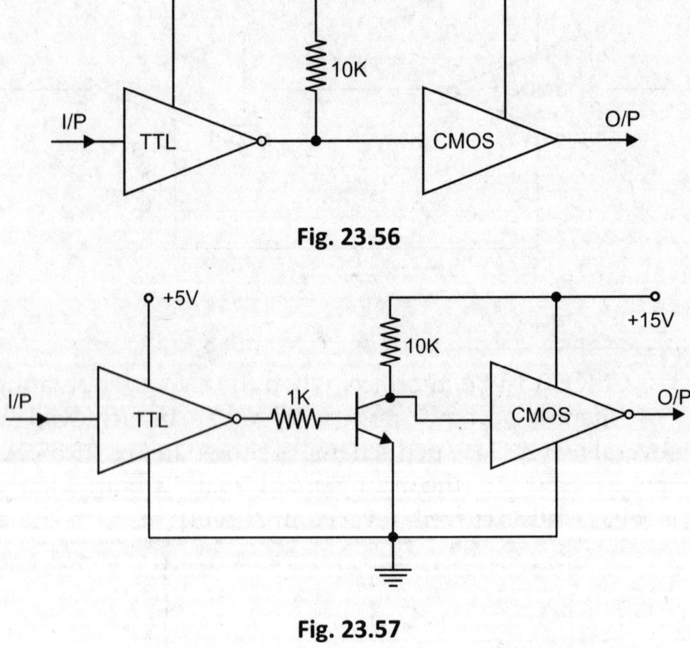

Fig. 23.56

Fig. 23.57

TTL-to-ECL and ECL-to-TTL Interface. TTL-to-ECL and ECL-to-TTL interface connections are not as straightforward as they are in case of TTL-to-CMOS and CMOS-to-TTL due to widely different power supply requirements for the two types and also, because ECL devices have differential inputs and differential outputs. Nevertheless, special chips are available that can take care of all these aspects. These are known as level translators. MC10124 is a quad TTL-to-ECL level transistor. That is, there are four independent single input and complementary output translators inside the chip. Fig. 23.58 shows TTL-to-ECL interface using MC10124. MC10125 is another level translator for ECL-to-TTL interface, MC10125 has differential inputs and single ended output. 23.59 shows a typical interface schematic using MC10125. You would notice that in the interface schematics of Figs. 23.58 and 23.59 only one of the available four translators has been used.

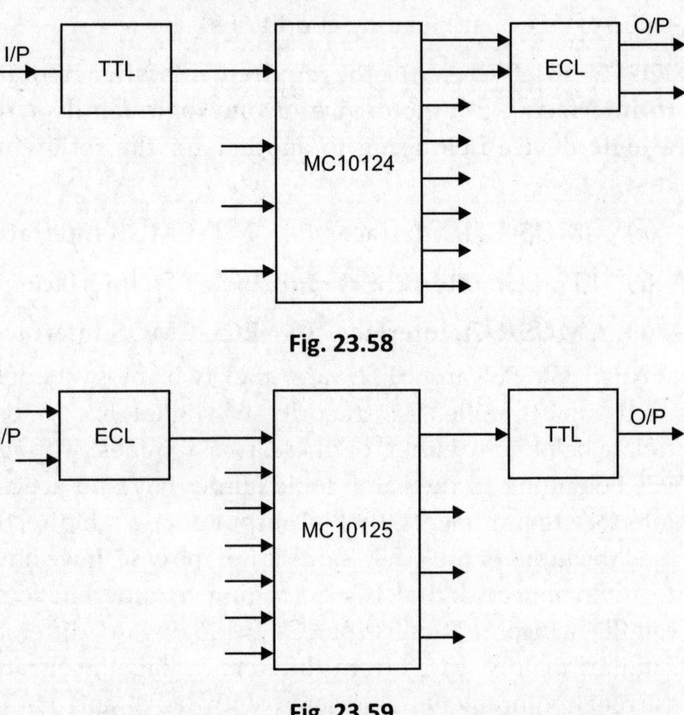

Fig. 23.58

Fig. 23.59

CMOS-to-ECL and ECL-to-CMOS Interface. CMOS -to-ECL and ECL-to-CMOS Interfaces are similar to the one described in case of TTL-to-ECL and ECL-to-TTL interfaces. Again dedicated level translators are available. MC10352 for instance is a quad CMOS-to-ECL level translator chip. CMOS-to-ECL interface is also possible by having first a CMOS-to-TTL interface followed by a TTL-to-ECL interface using MC10124 or a similar chip. Fig. 23.60 shows the arrangement. Similarly, ECL-to-CMOS interface is possible by having a ECL-to-TTL interface using MC10125 or a similar chip followed by TTL-to-CMOS interface. Fig. 23.61 shows the typical interface schematic.

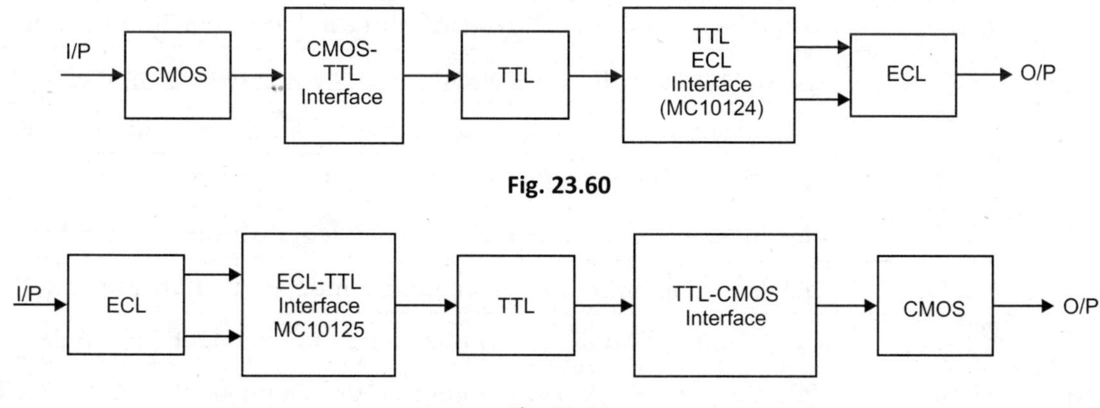

Fig. 23.60

Fig. 23.61

Q. 42. How do you classify digital integrated circuits on the basis of complexity of the circuit that has been integrated on the chip?

Ans. We are all familiar with terms like SSI, MSI, LSI, VLSI being used with reference to digital integrated circuits. These terms refer to groups in which digital ICs have been divided based on the complexity of the circuitry that has been integrated on the chip. It is a common practice to consider the complexity of a logic gate as a reference for defining the complexities of the other digital IC functions. A broadly accepted definition of different groups of ICs mentioned above is as follows :

A VLSI (Very Large Scale Integration) chip contains circuitry equivalent in complexity to 1000 or more gates.

An LSI (Large Scale Integration) chip is the one that contains circuitry equivalent in complexity to 100 or more gates.

An MSI (Medium Scale Integration) chip is the one that contains circuitry equivalent in complexity to 10 or more gates.

An SSI (Small Scale Integration) chip has a circuit complexity less than that of an MSI chip.

Table 23.5

IC No.	*Logic Family*	*Description*
7400	TTL	Quad 2-input NAND i.e. It contains four independent 2-input NAND-gates.
7410	TTL	Triple 3-input NAND i.e. It contains three independent 3-input NAND-gates
7420	TTL	Dual 4-input NAND i.e. It contains two independent 4-input NAND-gates.
7430	TTL	Single 8-input NAND i.e. It contains one independent 8-input NAND-gates.
4011B	CMOS	Quad 2-input NAND i.e. It contains four independent 2-input NAND-gates.
4012 B	CMOS	Dual 4-input NAND i.e. It contains two independent 4-input NAND-gates.
4023 B	CMOS	Triple 3-input NAND i.e. It contains three independent 3-input NAND-gates.
4081B	CMOS	Quad 2-input AND gate i.e. It contains four independent 2-input AND gates.
7408	TTL	Quad 2-input AND i.e. It contains four independent 2-input AND gates.
7411	TTL	Triple 3-input AND i.e. It contains three independent 3-input AND gates.
7421	TTL	Dual 4-input AND i.e. It contains two independent 4-input AND gates.
7432	TTL	Quad 2-input OR i.e. It contains four independent 2-input OR gates.
4071B	CMOS	Quad 2-input OR i.e. It contains four independent 2-input OR-gates.
4001B	CMOS	Quad 2-input NOR i.e. It contains four independent 2-input NOR-gates.
4002B	CMOS	Dual 4-input NOR i.e. It contains two independent 4-input NOR gates.

Cont...

4025B	CMOS	Triple 3-input NOR i.e. It contains three independent 3-input NOR gates.
7402	TTL	Quad 2-input NOR i.e. It-contains four independent 2-input NOR gates.
7427	TTL	Triple 3-input NOR i.e. It contains three independent 3-input NOR gates.
74260	TTL	Dual 5-input NOR i.e. It contains two independent 5-input NOR gates.
7404	TTL	Hex Inverter i.e. It contains six independent inverting/NOT gates.
4049 UB	CMOS	Hex Inverting Buffer i.e. It contains six independent inverting/NOT gates.
7486	TTL	Quad 2-input EX-OR i.e. It contains four independent 2-input EX-OR gates.
4030B	CMOS	Quad 2-input EX-OR i.e. It contains four independent 2-input EX-OR gates.
4070B	CMOS	Quad 2-input EX-OR i.e. It contains four independent 2-input EX-OR gates.
74266	TTL	Quad 2-input NOR i.e. It contains four independent 2-input NOR gates.

Q. 43. What are Buffers and Tranceivers ? Why are these usually tristate devices?

Ans. A BUFFER has a larger load driving capability than a logic gate. It could be an inverting or a non-inverting buffer with a single input and single output, a NAND buffer, a NOR buffer an OR buffer or an AND buffer. Driver is another name for a buffer. A Driver is sometimes used to designate a circuit that has even larger drive capability than a buffer. Buffers are usually tristate devices to facilitate their use in bus oriented systems. Fig. 23.62 shows the symbols and functional tables of inverting and non-inverting buffers of the tristate type.

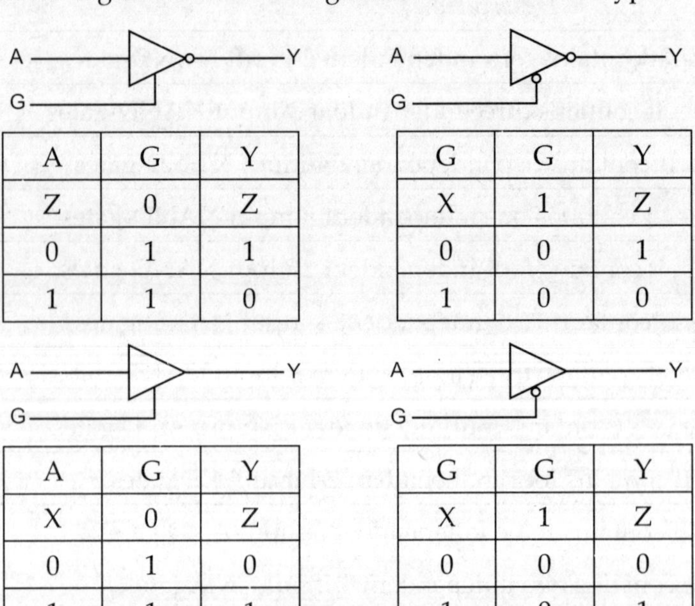

A	G	Y
Z	0	Z
0	1	1
1	1	0

G	G	Y
X	1	Z
0	0	1
1	0	0

A	G	Y
X	0	Z
0	1	0
1	1	1

G	G	Y
X	1	Z
0	0	0
1	0	1

Fig. 23.62

A TRANSCEIVER is a bi-directional buffer with additional direction control and Enable inputs. It allows flow of data in both directions depending upon the logic status of control inputs, transceivers also are usually tristate devices to make them compatible with bus oriented systems. Figs. 23.63 (*a*) and (*b*) respectively show the circuit symbols of inverting and non-inverting transceivers.

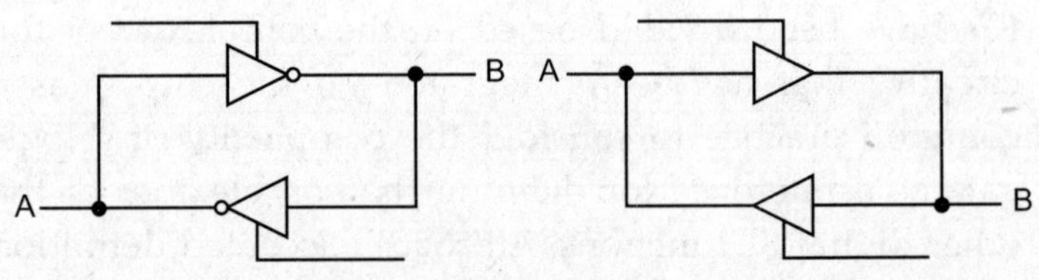

Fig. 23.63

Fig. 23.64 shows a typical logic circuit arrangement of a tristate non-inverting transceiver along with its functional table.

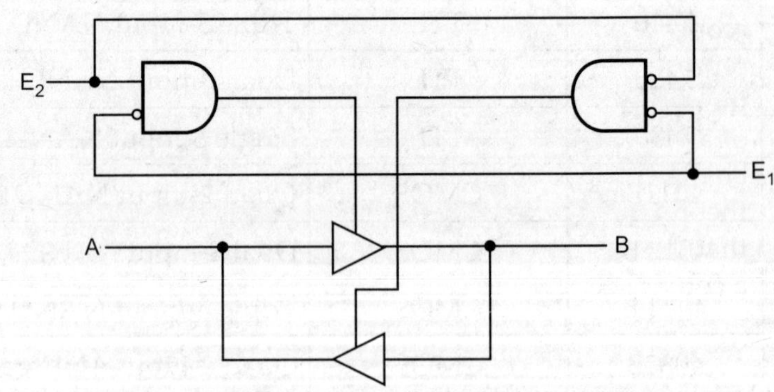

E_1	E_2	Operation
L	L	*Data flow from B to A*
L	H	*Data flow from A to B*
H	X	*Isolation*

Fig. 23.64

Q. 44. What is a Half Adder ? Beginning with the Boolean expressions for SUM and SCARRY outputs,

show how a half adder circuit can be implemented by using either only NAND or only NOR gates.

Ans. A **Half Adder** is an arithmetic circuit block that can be used to add only two bits. Such a circuit, thus, has two inputs that represent the two bits to be added and two outputs with one output giving the SUM and the other giving the CARRY. Figure 23.65 shows the block schematic of a Half Adder alongwith its truth table showing all possible input combinations and the corresponding outputs.

The Boolean expressions for the two outputs are :

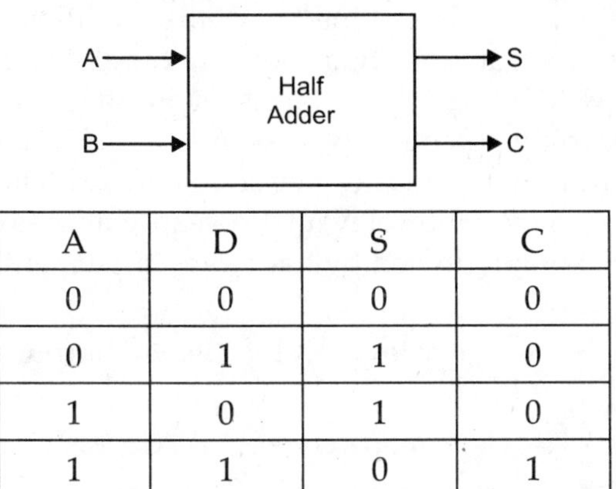

A	D	S	C
0	0	0	0
0	1	1	0
1	0	1	0
1	1	0	1

Fig. 23.65

$$S = \overline{A} \cdot B + A \cdot \overline{B} \text{ and } C - A \cdot B$$

Fig. 23.66 shows the simplest form of hardware implementation of a half adder.

A look at the two expressions tells that there is no scope for further simplification. While the first one representing the SUM output is that of an EX-OR gate, the second one representing the CARRY output is that of an AND gate. But these two expressions can certainly be represented in different forms using various laws and theorems of Boolean algebra just to illustrate the flexibility that the designer has to implement in hardware as simple a combinational function as that of a half adder.

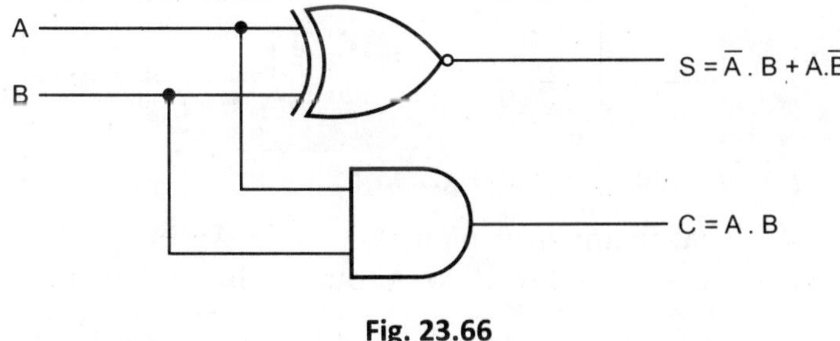

Fig. 23.66

Considering the Boolean expression for SUM output, this can be written in many ways, such as the following :

$$\overline{A} \cdot B + A \cdot \overline{B} = [\overline{\overline{(\overline{A} \cdot B)} \cdot \overline{(A \cdot \overline{B})}}]$$

The hardware implementation of this expression is shown in Fig. 23.67 using only NAND gates.

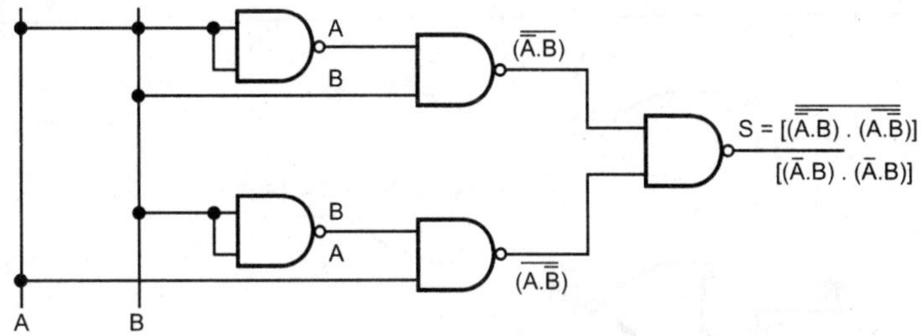

Fig. 23.67

Also, $\overline{A} \cdot B + A \cdot \overline{B} = A \cdot (\overline{A} + \overline{B}) + B \cdot (\overline{A} + \overline{B}) = A \cdot \overline{(A \cdot B)} + B \cdot \overline{(A \cdot B)}$

$$= [\overline{\overline{\{A \cdot \overline{(A \cdot B)}\}} \cdot \overline{\{B \cdot \overline{(A \cdot B)}\}}}]$$

The hardware implementation of this transformed expression requires only four NAND gates and is shown in Fig. 23.68.

Also, $\overline{A} \cdot B + A \cdot \overline{B} = [\overline{\overline{(A \cdot \overline{B}) + (\overline{A} \cdot B)}}]$

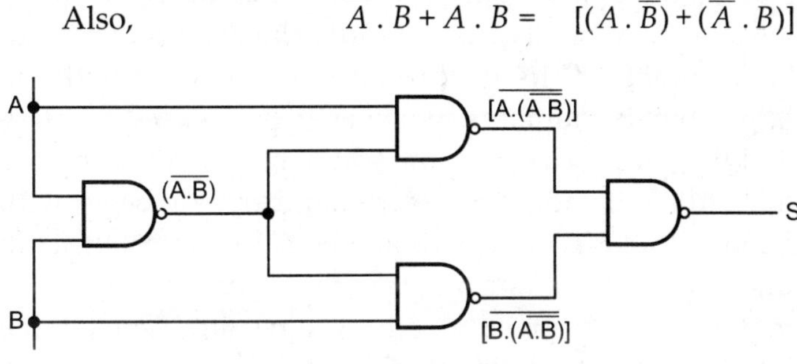

Fig. 23.68

The hardware implementation of this expression using only the NOR gates is shown in Fig 23.69.

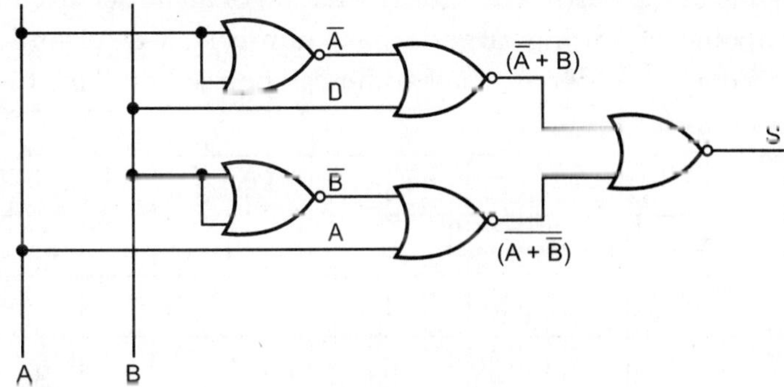

Fig. 23.69

The CARRY output can similarly be implemented by using either a single AND gate (Fig. 23.70 (*a*)) or two NAND gates, as shown in Fig. 23.70 (*b*) or three NOR gates as shown in Fig 23.70 (*c*).

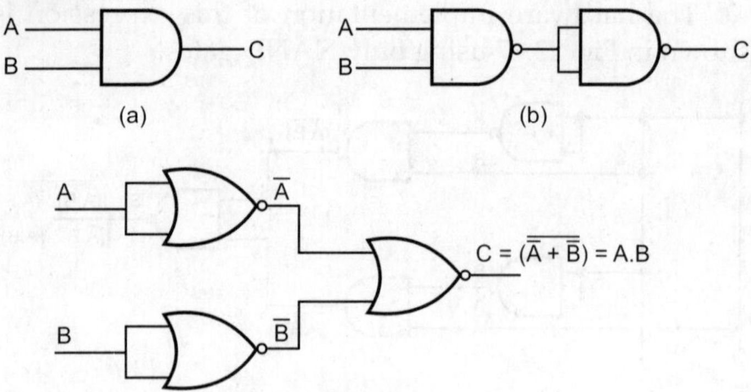

Fig. 23.70

Q. 45. Briefly describe the functioning of a full adder circuit. How does it differ from a half adder? Also, give the block schematic arrangement of 4-bit binary adder using Full-Adder building blocks.

Ans. A **Full Adder** circuit is an arithmetic circuit block that is capable of adding three binary digits to produce a SUM and a CARRY output. Such a building block comes very handy, when it comes to adding binary numbers with large number of bits. If we recall the procedure for adding larger binary numbers, whether it is a straight binary addition or 2's complement addition, in both cases, we have to finally add two binary numbers. We begin with addition of LSBs of the two numbers. We record the sum under the LSB column and take the carry, if any, forward to the next higher column bits. As a result, when add the bits of the next adjacent higher column bits, we would be required to add three bits if there was a carry from the previous addition. We have a similar situation for the other higher column bits also till we reach the MSB. Thus in the hardware implementation of an adder circuit capable of adding larger binary numbers, we certainly need a full adder as a half adder can be used only for the addition of LSBs.

A	B	Cin	SUM (S)	Cout
0	0	0	0	0
0	0	1	1	0
0	1	0	1	0
0	1	1	0	1
1	0	0	1	0
1	0	1	0	1
1	1	0	0	1
1	1	1	1	1

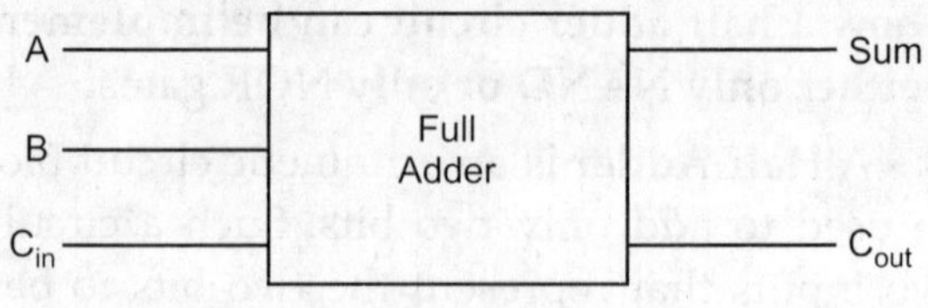

Fig. 23.71

Fig. 23.71 shows the block schematic representation of a full adder circuit alongwith it truth table showing all possible input combinations and corresponding output entries.

The combinational logic implementation of a full adder can be determined by first writing Boolean expressions for the two output variables, *i.e.* SUM and CARRY outputs, in terms of input variables and then simplifying these expressions by using any of the simplification techniques described in the earlier part of the chapter. The relevant Boolean expressions, their simplification followed by hardware implementation are given below.

The Boolean expressions for the two output variables are :

$$S = \overline{A}\,.\,\overline{B}\,.\,C_{in} + \overline{A}\,.\,B\,.\,\overline{C}_{in} + A\,.\,\overline{B}\,.\,\overline{C}_{in} + A\,.\,B\,.\,C_{in}$$

and $C_{out} = \overline{A}\,.\,B\,.\,C_{in} + A\,.\,B\,.\,\overline{C}_{in} + A\,.\,B\,.\,\overline{C}_{in} + A\,.\,B\,.\,C_{in}$

The Karnaugh maps for the two expressions are given in Fig. 23.72 *(a)* for SUM output and 23.72. *(b)* for C_{out} output. As it is clear from the two maps, the expression for SUM output can not be simplified any further whereas simplified Boolean expression for (C_{out}) is given by :

$$C_{out} = B\,.\,C_{in} + AB + AC_{in}$$

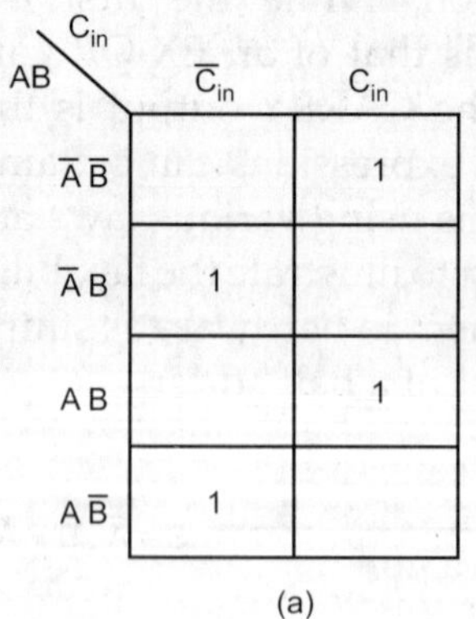

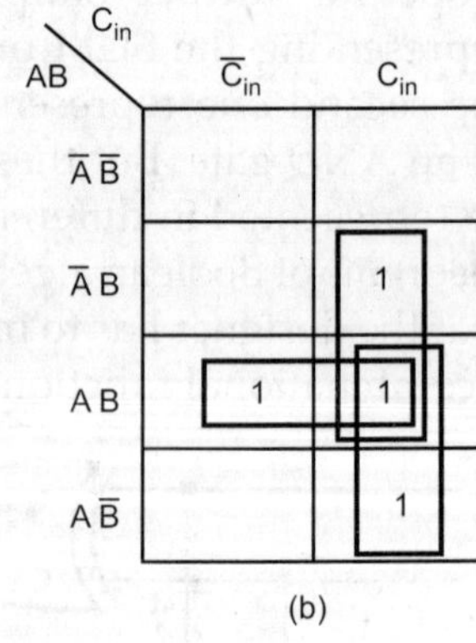

Fig. 23.72

The hardware implementation is shown in Fig. 23.73. The expression for the SUM output can be rewritten as :

$$S = \overline{C}_{in}\,(\overline{A}\,.\,B + A\,.\,\overline{B}) + C_{in}\,(\overline{A}\,.\,B + A\,.\,\overline{B})$$

$$= \overline{C}_{in}\,(\overline{A}\,.\,B + A\,.\,\overline{B}) + C_{in}\overline{(\overline{A}\,.\,B + A\,.\,\overline{B})}$$

Similarly, expression for carry out can be rewritten as:

$$C_{out} = A\,.\,B + B\,C_{in}\,(A + \overline{A}) + AC_{in}\,(B + \overline{B})$$

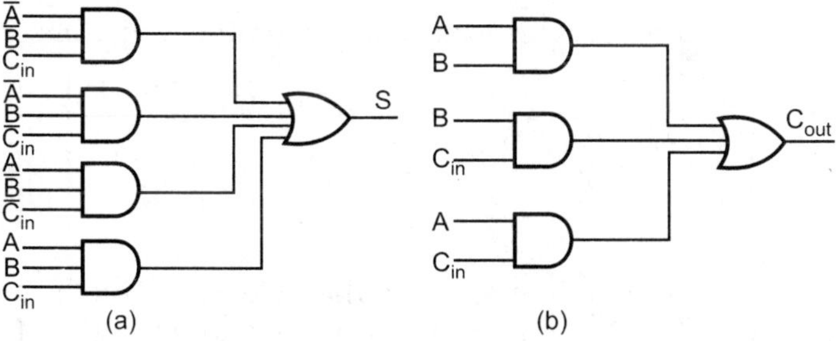

Fig. 23.73

$$= A \,.\, B + A \,.\, B \,.\, C_{in} + \overline{A} \,.\, B.\, C_{in} + A \,.\, B.\, C_{in} + A \,.\, \overline{B} \,.\, C_{in}$$

$$= A \,.\, B + A \,.\, B \,.\, C_{in}\, \overline{A} \,.\, B \,.\, C_{in} + A \,.\, \overline{B} \,.\, C_{in}$$

$$= A \,.\, B(1 + C_{in}) + C_{in}(\overline{A} \,.\, B + A \,.\, \overline{B})$$

$$= A \,.\, B + C_{in}(\overline{A} \,.\, B + A \,.\, \overline{B})$$

If we attempt to implement in hardware these re-written Boolean expressions for SUM and CARRY outputs, we get a very familiar logic circuit, as shown in Fig. 23.74. If we take a closer look at this circuit, we find that it comprises of two half adders and an OR-gate. The logic circuit of Fig. 23. 74 is re-drawn in Fig. 23.75, where the portion of the circuit representing a half adder has been replaced by a block.

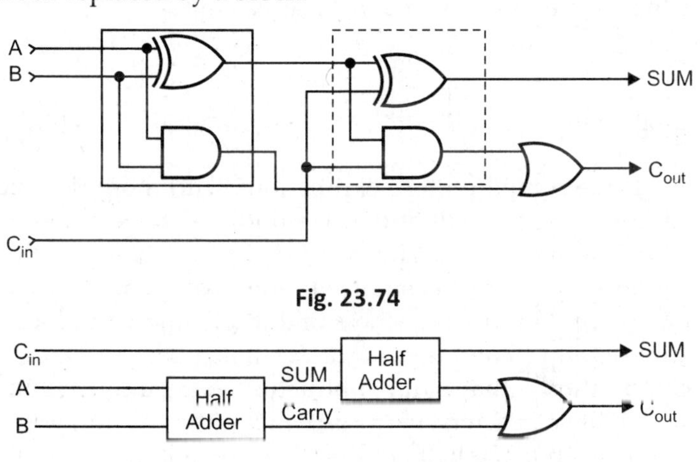

Fig. 23.74

Fig. 23.75

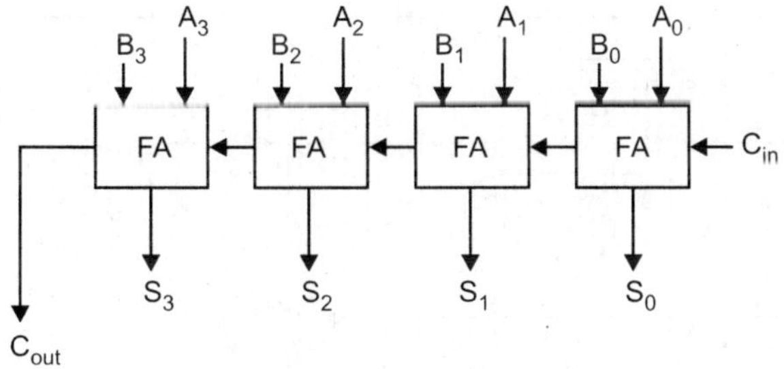

Fig. 23.76

The FULL ADDER of the type described above forms the basic building block for constructing binary adders. The full adder, as we have seen above, is capable of adding only 1-bit binary numbers. These adders can be added in cascade to form adders capable of adding binary numbers with larger number of bits. For example, a 4-bit binary adder would require four full adders of the type shown in Fig. 23.75 to be connected in cascaded. Fig. 23.76 shows the arrangement. Here, ($A_3\ A_2\ A_1\ A_0$) and ($B_3\ B_2\ B_1\ B_0$) are the two binary numbers to be added) with A_0 and B_0 representing LSBs of the two numbers.

Q. 46. What is a Half Subtractor? Briefly describe its functioning principle with the help of relevant truth table and hardware.

Ans. The subtraction of two given binary numbers can be achieved, as outlined earlier, by adding 2's complement of the subtrahend to the minuend. With this, the subtraction operation can be hardware implemented with adder circuits. However, we do have counterparts of half adder and full adder in half subtractor and full subtractor for direct implementation of subtraction operation using logic gates.

A **Half Subtractor** is a combinational circuit that can be used to subtract one binary digit from another to produce a difference output and a borrow output. The borrow out here specifies whether a '1' has been borrowed to perform the subtraction. The truth table of a half subtractor as shown in Fig. 23.77, explains it further.

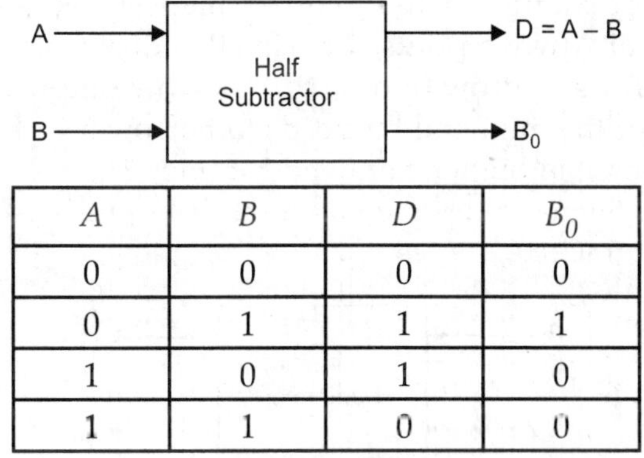

A	B	D	B_0
0	0	0	0
0	1	1	1
1	0	1	0
1	1	0	0

Fig. 23.77

The Boolean expression for the two output are:

$$D = \overline{A} \,.\, B + A \,.\, \overline{B}$$

$$B_0 = \overline{A} \,.\, B$$

A

B

D = A – B

B_0

Fig. 23.78

As it is clear from the two expressions, there is no scope for any further simplification. While the expression for the difference output is that of an EX-OR gate, the expression for Borrow (B_0) is that of an AND gate with input (A) complemented before it is fed to the gate. Fig. 23.78 shows the logic complementation of a half subtractor.

We can notice the similarities between a half subtractor and a half adder. The expressions for SUM and DIFFERENCE outputs are just the same. The expression for BORROW is also similar to what we have in case of CARRY. If the input (A), that is the minuend, is complemented, an AND-gate can be used to implement the BORROW output.

Q. 47. Briefly describe the functional principle of a full-subtractor. Show the hardware implementation of the same and also, explain how the basic full-subtractor building blocks can be cascaded to perform subtraction of larger bit binary numbers?

Ans. A **Full Subtractor** performs subtraction operation on two bits, a minuend and a subtrahend and also takes into consideration the practical fact whether a '1' had already been borrowed by the next adjacent lower minuend bit or not. As a result, there are three bits to be handled at the input of a full subtractor namely the two bits to be subtracted and a borrow bit designated as (B_{in}). There are two outputs, namely the difference output (D) and the Borrow output (B_o). Borrow output bit tells whether the minuend bit needs to borrow a '1' from the next possible higher minuend bit. Fig. 23.79 shows the block schematic representation and the truth table of a full subtractor.

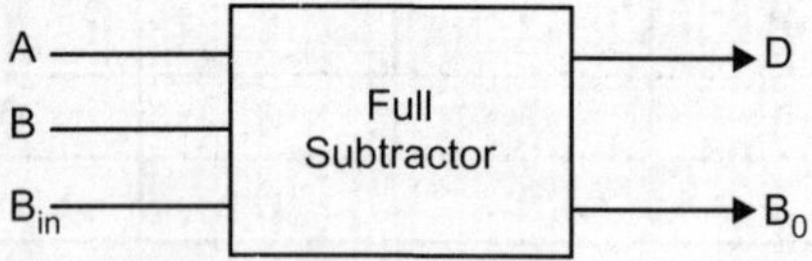

Minuend (A)	Subtrahend (B)	Borrow In (B_{in})	Difference (D)	Borrow Out (B_0)
0	0	0	0	0
0	0	1	1	1
0	1	0	1	1
0	1	1	0	1
1	0	0	1	0
1	0	1	0	0
1	1	0	0	0
1	1	1	1	1

Fig. 23.79

The Boolean expressions for the two output variables are :

$$D = \overline{A}.\overline{B}.B_{in} + \overline{A}.B.\overline{B}_{in} + A.B.B_{in} + A.B.$$
$$B_{in}$$

$$\text{and } B_0 = \overline{A}.\overline{B}.B_{in} + \overline{A}.B.\overline{B}_{in} + \overline{A}.B.B_{in} + A.B.$$
$$B_{in}$$

The Karnaugh maps for the two expressions are given in Fig. 23.80 (a) for difference output (D) and (b) for borrow output (B_0).

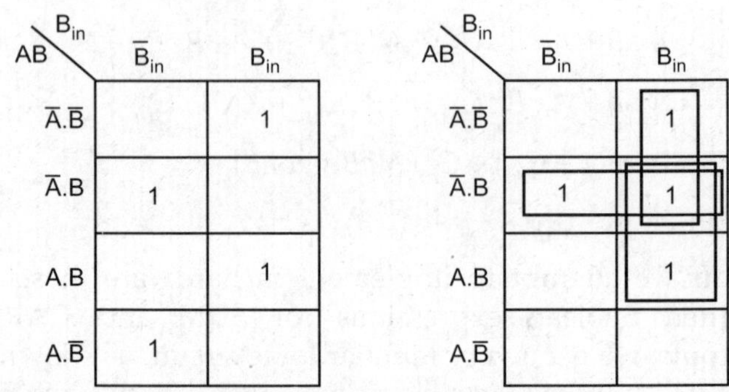

Fig. 23.80

As it is clear from the two Karnaugh maps, no simplification possible for the difference output (D). The simplified expression for (B_o) is given by :

$$B_o = \overline{A}.B + \overline{A}.B_{in} + BB_{in}$$

$$\text{Also, } D = \overline{A}.\overline{B}.B_{in} + \overline{A}.B.\overline{B}_{in} + A.\overline{B}.\overline{B}_{in} + A.$$
$$B.B_{in}$$

If we compare these expressions with those derived earlier in case of full adder, we find that the expression for (D) is same as that for SUM output. Also, expression for borrow out (B_o) is similar to the expression for carry out (C_o). They become the same if (A) input in case of subtractor is complemented. By a similar analysis, it can be shown that a full subtractor can be implemented with half subtracters, in the same way as a full adder was constructed using half adders. The relevant diagrams are shown in Figs. 23.81 and 23.82.

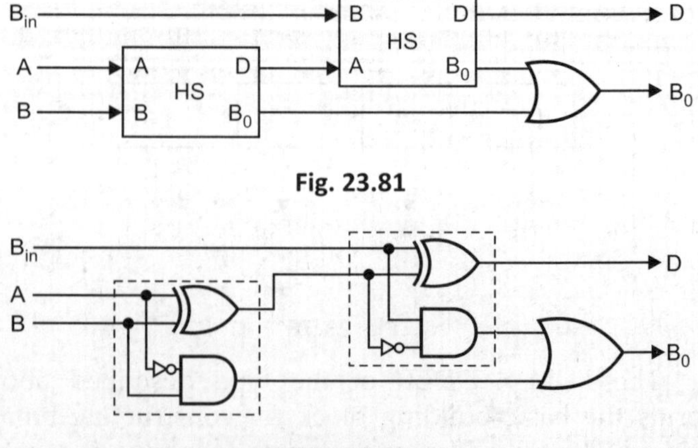

Fig. 23.81

Fig. 23.82

Again, more than one full subtractors can be connected in cascade to perform subtraction on two larger binary numbers. As a illustration, Fig. 23.83 showsa 4-bit subtractor.

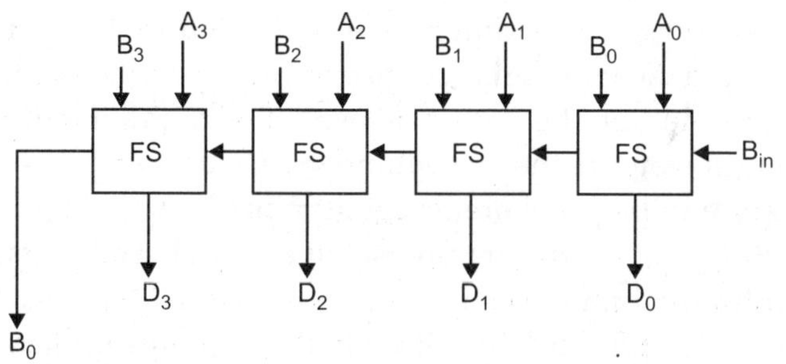

Fig. 23.83

Q. 48. What are CONTROLLED INVERTERS ?

Ans. The subtraction operation is usually executed with the help of adders, the underlying principle being the fact that subtraction is nothing but addition of 2's complement of subtrahend to the minuend. Thus the first step, towards practical implementation of a subtractor is to determine the 2's complement of the subtrahend. And for this, one needs to first find 1's complement. A CONTROLLED INVERTER does exactly this. A 1-bit controlled inverter is nothing but a 2-input EX-OR gate with one of its inputs treated as a control input, as shown in Fig. 23.84. When the control input is LOW, the input bit is passed as such to the output. (Recall the truth table of an EX-OR gate). When the control input is HIGH, the input bit gets complemented at the output *i.e.* output is complement of the input. Fig. 23.85 shows an 8-bit controlled inverter of this type.

Fig. 23.84

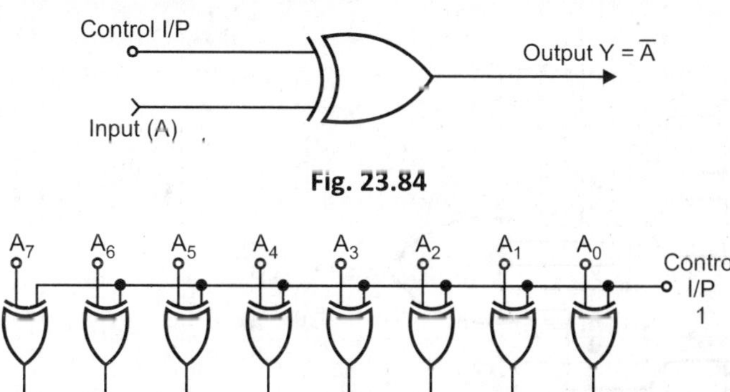

Fig. 23.85

When the control input is LOW, the output ($Y_7\ Y_6$ ......Y_0) is same as the input ($A_7\ A_6$......A_0). When control input is HIGH, the output is 1's complement of the input. As an example, (1011 0110) at the input would produce (0100 1001) at the output, when control input is in logic '1' state.

Q. 49. With the help of a suitable diagram, briefly describe the operation of a 4-bit Adder-Subtractor circuit.

Ans. Subtraction of two binary numbers, as mentioned earlier, can be accomplished by adding 2's complement of the subtrahend to the minuend. The final carry, if any, is disregarded. After disregarding the final carry (or over flow), if MSB bit is a '0', then the result of addition is the correct answer. If the MSB bit is a 1, this implies that the answer has a negative sign. The true magnitude in this case would be determined by taking 2's complement of the answer. To sum up, adding 2's complement of subtrahend to minuend accomplishes subtraction any way.

As outlined earlier, full adders can be used to perform subtraction provided, we can cater for taking 2's complement of the subtrahend and disregard the final carry or overflow. Fig. 23.86 shows a hardware arrangement. Let us see how it can be used to perform subtraction of two 4-bit binary numbers. A close look at the diagram would reveal that it is the hardware arrangement for a 4-bit binary adder with the slight change that one of the binary numbers' bits are fed through controlled inverters. As said before, when the SUB input (which is the control input in Fig. 23.85) is in Logic 1 state, the four bits of one of the numbers ($B_3\ B_2\ B_1\ B_0$ in this case) get complemented. If the same '1' is also fed to the CARRY IN of the LSB full adder, what we finally achieve is the addition of 2's complement and not l's complement. Thus, in an adder arrangement of Fig. 23.86 we are basically adding 2's complement of ($B_3\ B_2\ B_1\ B_0$) to ($A_3\ A_2\ A_1\ A_0$) which means we are accomplishing operation ($A - B$). The final carry (*i.e.* carry out of MSB full adder) gets ignored once not displayed.

For implementing an 8-bit adder-subtractor, one would require eight full adders and eight 2-input EX-OR gates. 4-bit full adders and Quad 2-input EX-OR gates are individually available in integrated circuit form. IC 7483 is, for instance, a 4-bit binary adder-subtractor circuit implemented with 7483 and 7486. An 8-bit adder-subtractor circuit would require two 7483's and two 7486's.

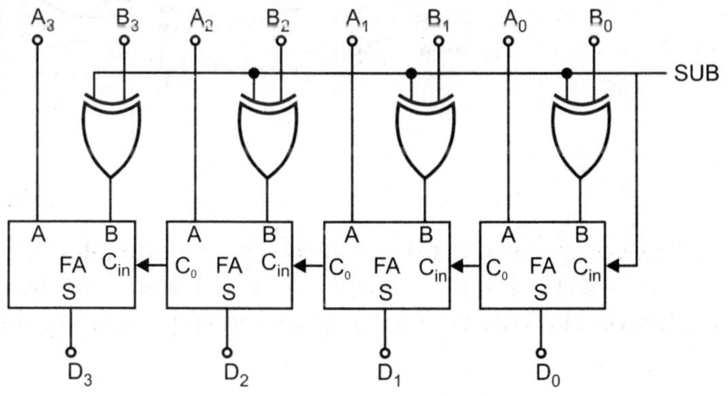

A	D	C_{in}	SUM (S)	C_{out}
0	0	0	0	0
0	0	1	1	0
0	1	0	1	0
0	1	1	0	1
1	0	0	1	0
1	0	1	0	1
1	1	0	0	1
1	1	1	1	1

Fig. 23.86

Q. 50. What are MAGNITUDE COMPARATORS ? Show the hardware implementation of a typical 4-bit magnitude comparator.

Ans. A **Magnitude Comparator** is a combinational circuit that compares two given numbers and determines whether one is equal to, less than or greater than the other. The output is in the form of three binary variables, representing the conditions $A = B$, $A > B$ and $A < B$, if A and B are the two numbers. Depending upon the relative magnitudes of the two numbers, the relevant output changes state. If the two numbers, let us say, are 4-bit binary numbers and are designated as (A_3 A_2 A_1 A_0) and (B_3 B_2 B_1 B_0), the two numbers will be equal, if all pairs of significant digits are equal *i.e.* $A_3 = B_3$, $A_2 = B_2$, $A_1 = B_1$ and $A_0 = B_0$. In order to determine whether A is greater than or less than B, we inspect the relative magnitudes of pairs of significant digits starting from most significant position. The comparison is done by successively comparing the next adjacent lower pair of digits, if digits of the pair under examination are equal.

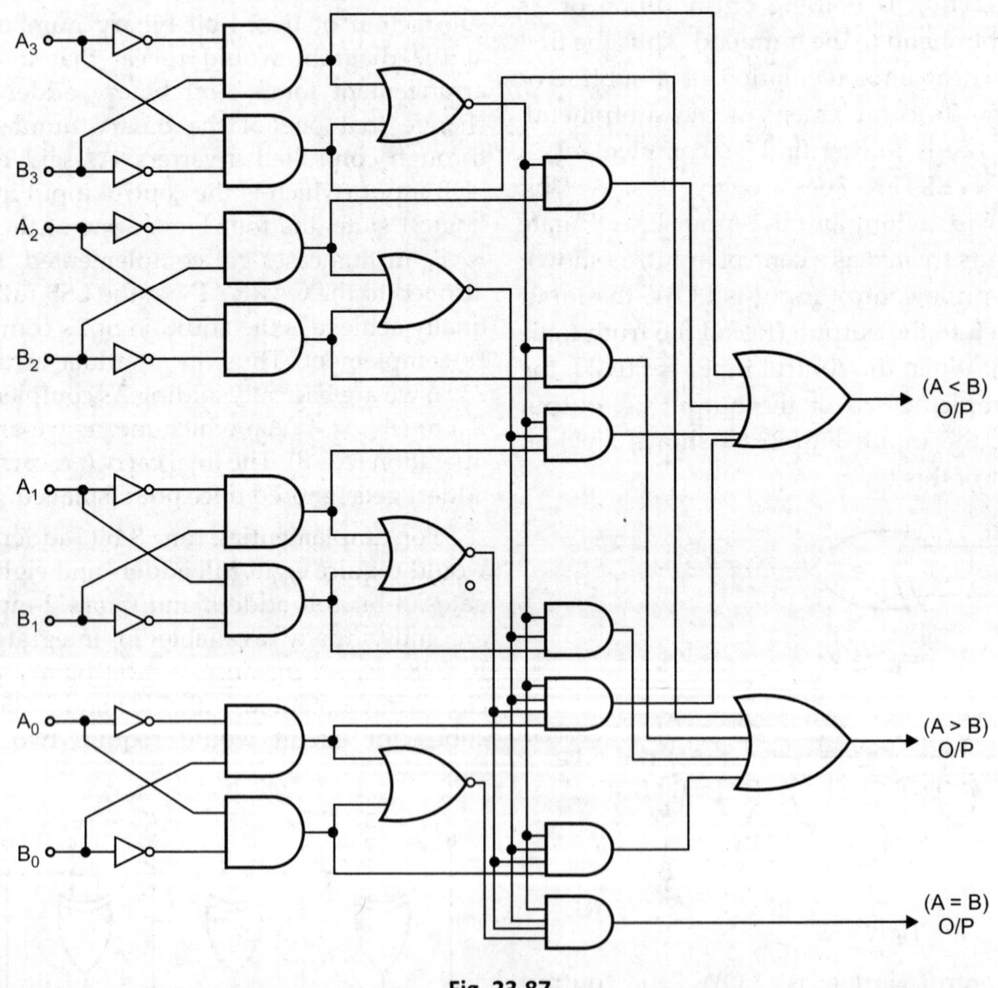

Fig. 23.87

The comparison continues until a pair of unequal digits is reached. In the pair of unequal digits, if $A = 1$, $B = 0$, then $A > B$ and if $A = 0$, $B = 1$, then $A < B$. If X, Y and Z are three variables respectively representing $A = B$, $A > B$ and $A < B$ conditions, then the Boolean expressions representing these conditions are :

$$X = x_3 . x_2 . x_1 . x_0$$

where, $x_i = A_i Bi + \overline{A}_i \overline{B}_i, i = 0, 1, 2, 3$

$$Y = A_3 . \overline{B}_3 + x_3 . A_2 \overline{B}_2 + x_3 . x_2 . A_1 . \overline{B}_1 + x_3 . x_2 . x_1 . A_0 \overline{B}_0$$

$$Z = \overline{A}_3 . B_3 + x_3 . \overline{A}_2 . B_2 + x_3 . x_2 . \overline{A}_1 . B_1 + x_3 . x_2 . x_1 . \overline{A}_0 . B_0$$

Refer to first of three equations given above, (x_3) will be '1' only, when both A_3 and B_3 are equal. Similarly, (x_2) will be '1', when both A_2 and B_2 are equal and (x_1) will be '1', when both A_1 and B_1 are equal and finally (x_0) will be '1', when both A_0 and B_0 are equal. ANDing of x_3, x_2, x_1, x_0 ensures that (X) will be '1', when all x_0, x_1, x_2 and x_3 are in logic '1' state. Thus, $x = 1$ means that $A = B$. On similar lines, it can be visualised that second and third equations respectively represent $A > B$ and $A < B$ conditions.

Fig. 23.87 shows hardware implementation of 4-bit magnitude comparator.

Q. 51. How can combinational circuits be used for generation and checking of parity bits?

Ans. The significance of adding a parity bit to a string of data has been discussed earlier. Both even and odd parity were discussed. To summarise what is already said in the even parity method, the parity bit is so chosen as to make the number of 1's even. In odd parity selection, the parity bit makes number of 1's as odd. For example, an even parity bit for a 4-bit code group '1001' will be a '0'. It will be a '1', if an odd parity is desired.

Another important thing that is worth mentioning again here is, that addition of parity bit can be used to detect only single bit errors that occur during transmission of data from one location to another. The parity bit, odd or even, becomes a part of code group. Also, there is no way for the receiving side to determine which bit is in error, as it does not know what the code group is supposed to be.

Parity generation and checking can be accomplished by means of simple combinational circuits. Fig. 23.88 (*a*) shows a circuit that can be used generate an even parity bit for a 4-bit code group. The logic circuit shown here does nothing but an EX-OR operation. That is, if D_0, D_1, D_2, D_3 are the four bits, the output is given by :

$$P = D_0 \oplus D_1 \oplus D_2 \oplus D_3$$

Such a circuit would generate a '0' where number of 1's in the 4-bit code group is even and a '1', if the number of 1's is odd. That is what is required for an even parity bit generation. The parity bit generated here always makes total number of 1's as even . The circuit shown in Fig. 23.88 (*a*) can be easily adapted for any number of bits. Also, similar circuits comprising of EX-OR and equivalence logic functions can be used for generation of odd parity bit. Fig. 23.88 (*b*) shows an odd parity generation logic circuit for 4-bit code groups.

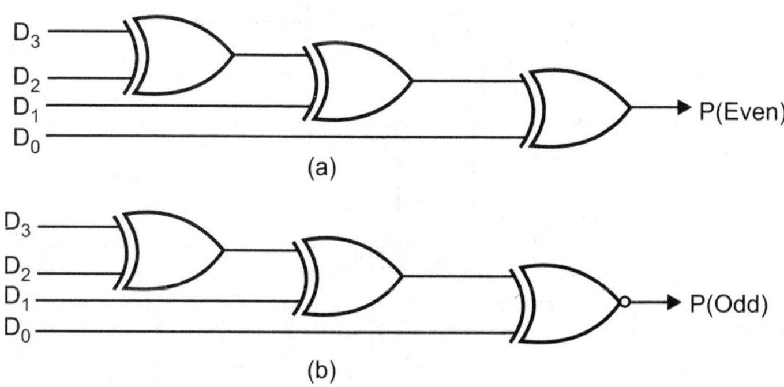

Fig. 23.88

The circuits used for parity check are also similar. Fig. 23.89 shows the logic circuit that can be used for even parity check. It produces a '0' in case of no error and a '1' in case of 1-bit error.

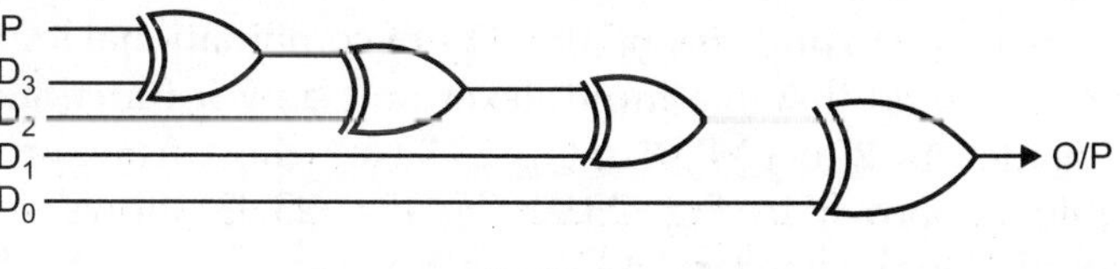

Fig. 23.89

Fig. 23.90 shows another logic circuit that can be used for odd parity check. Again a '0' at the output means no error, whereas a '1' indicates a single bit error.

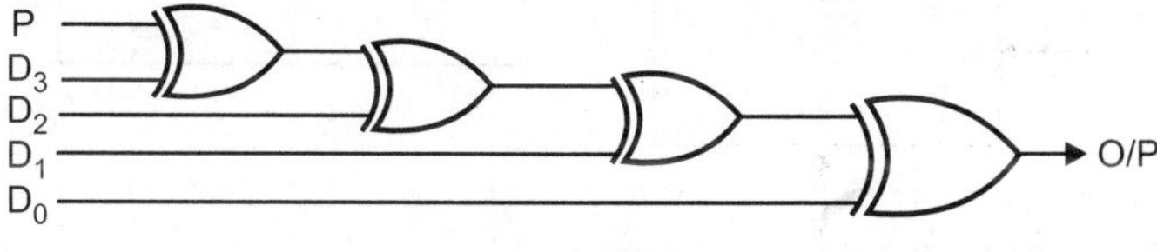

Fig. 23.90

The logic circuits shown in Figs. 23.88, 23.89 and 23.90 are self-explanatory.

Q. 52. What is a Multiplexer circuit ? Briefly describe the operation of a Multiplexer circuit with the help of relevant hardware and truth table.

Ans. A **Digital Multiplexer** is a combinational circuit having more than one input lines, one output line and more than one selection line. It selects binary information present on any one of the input lines and routes it to the output line. The logic status of the selection lines decides which input line is selected. If there are (n) selection lines, then the number of maximum possible input lines is (2^n) and the multiplexer is referred to as 2^n-to-1 multiplexer. Figs. 23.91 (*a*) and (*b*) respectively show the circuit representation and truth table of 4-to-1 multiplexer.

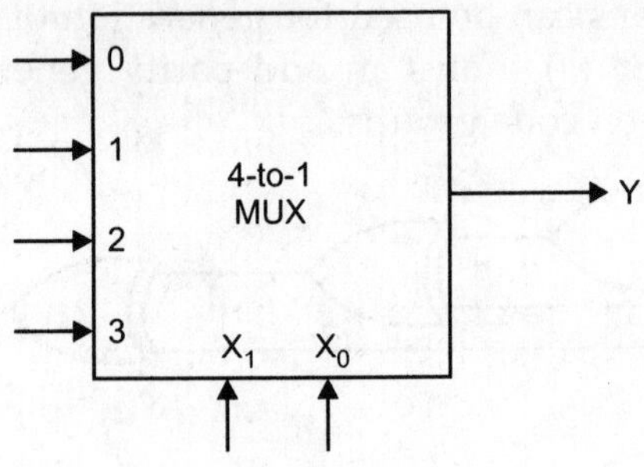

Fig. 23.91

X_1	X_0	Y
0	0	I_0
0	1	I_1
1	0	I_2
1	1	I_3

In order to understand the type of combinational logic circuit found inside a multiplexer and how it functions, consider the 2-to-1 MUX of Fig. 23.32 (*a*) whose functional table is shown in Fig. 23.92 (*b*). Fig. 23.93 shows the possible logic diagram of this multiplexer.

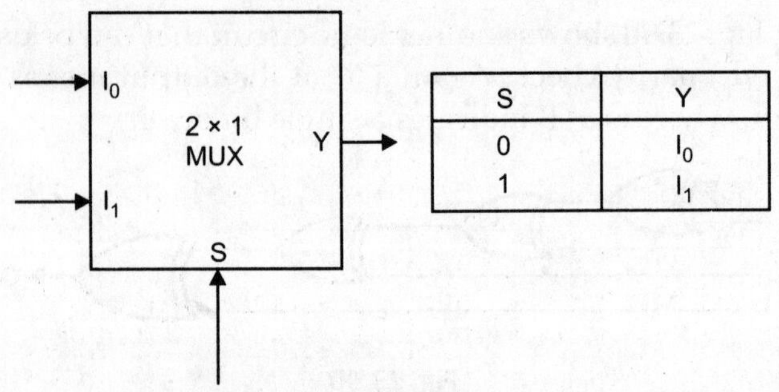

Fig. 23.92

The circuit function as follows:

For $S = 0$, the Boolean expression for the output becomes $y = I_0$

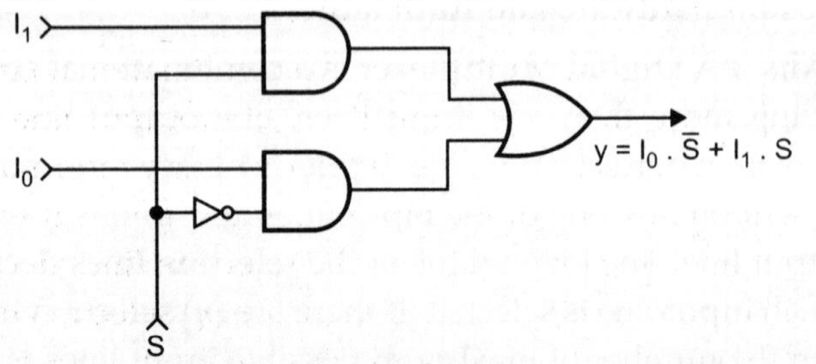

Fig. 23.93

For $S = 1$, the Boolean expression for the output becomes $y = I_1$

Thus, inputs I_0 and I_1 ate respectively switched to the output for $S = 0$ and $S = 1$. Extending the concept further, Fig. 23.94 shows the logic diagram of a 4 × 1 MUX. The input combinations 00, 01,10 and 11 on the select lines respectively switch I_0, I_1, I_2 and I_3 to the output.

The operation of the circuit is governed by the Boolean expression,

$$Y = I_0 \bar{S}_0 \bar{S}_1 + I_1 S_0 \bar{S}_1 + I_2 \bar{S}_0 S_1 + I_3 S_0 S_1$$

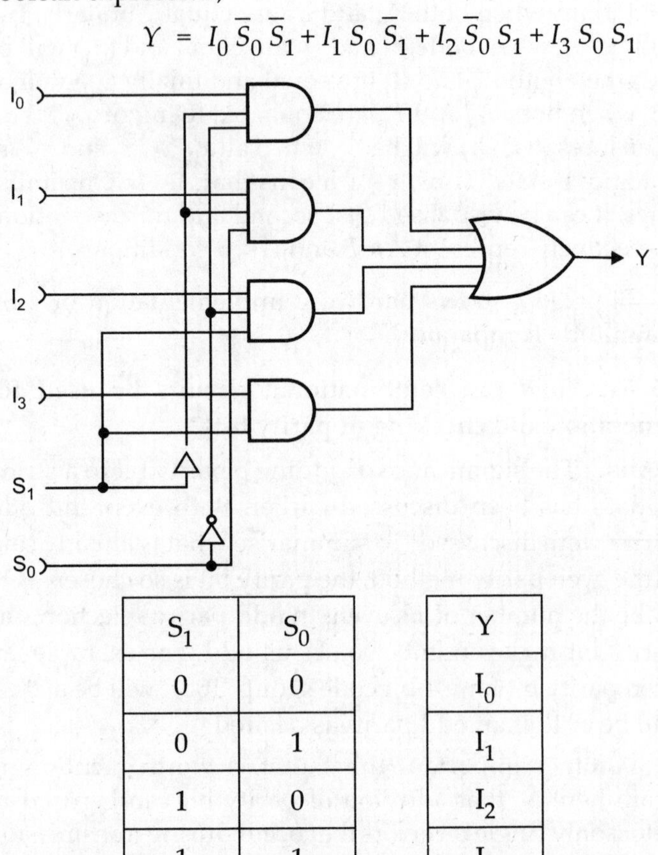

S_1	S_0	Y
0	0	I_0
0	1	I_1
1	0	I_2
1	1	I_3

Fig. 23.94

The multiplexers usually have an ENABLE input that can be used to control the operation. When this input is enabled, that is, when it is in logic '1' or logic '0' state depending upon whether ENABLE input is active HIGH or active LOW respectively, the outputs are disabled. The MUX functions normally when ENABLE input is inactive. Fig. 23.95 shows how 2 × 1 MUX of Fig. 23.95 can be modified to include an ENABLE input. The functional table of this modified MUX is also shown in Fig. 94. The ENABLE input here is active HIGH. Some IC packages have more than one multiplexer. In that case, ENABLE input and Selection inputs are common to all multiplexers within the same IC package. Fig. 23.96 shows 4 × 1 MUX with an active LOW ENABLE input.

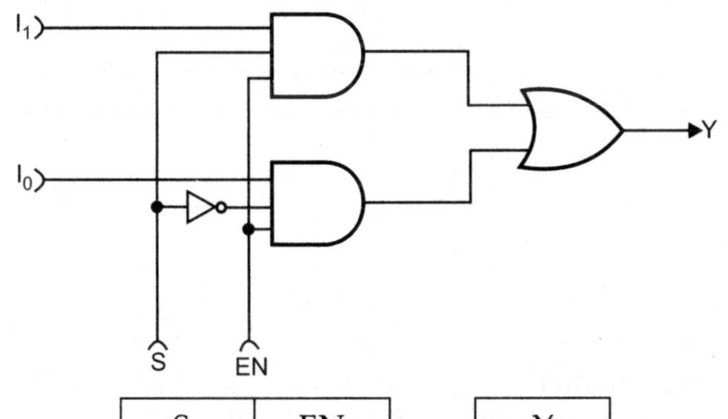

S_1	EN
X	1
0	0
1	0

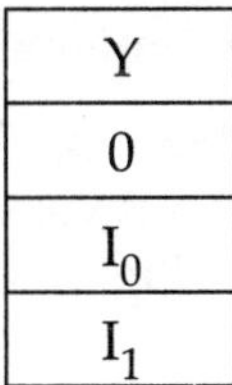

Y
0
I_0
I_1

Fig. 23.95

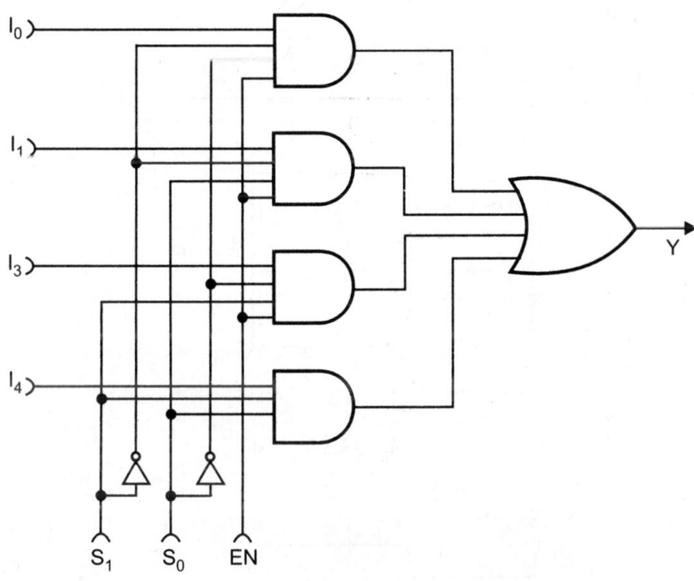

S_1	S_0	EN	Y
X	X	1	0
0	0	0	I_0
0	1	0	I_1
1	0	0	I_2
1	1	0	I_3

Fig. 23.96

Q. 53. One of the main applications of a multiplexer circuit is to implement complex Boolean expressions. How can a multiplexer be used to do that ? Explain with the help of suitable examples.

Ans. One of the most common applications of a multiplexer is the implementation of a Boolean function. The simplest technique to do so is to employ a ($2^n \times 1$) MUX to implement an (n) variable Boolean function. The input lines corresponding to each of the minterms present in the Boolean function are made equal to logic '1' state. The remaining minterms that are absent in the Boolean function are disabled by making their corresponding input lines equal to logic '0'. As an example, Fig. 23.97 shows the use of (3 × 1) MUX for implementing the Boolean function,

$$F(A, B, C) = \Sigma\, 2, 4, 7$$

which can be written in the expanded form as,

$$F(A, B, C) = \overline{A}\, B\, \overline{C} + A\, \overline{B}\, \overline{C} + ABC$$

However, there is a better technique available to do the same. In this, an (n × 1) MUX can be used to implement a Boolean function with (n + 1) variables. The procedure is as follows. Out of (n + 1) variables, (n) are connected to the (n) selection lines of (2^n × 1) MUX. The left over variable is used with the input lines. The input lines are made equal to either the left over variable or its complement or '0' or '1'.

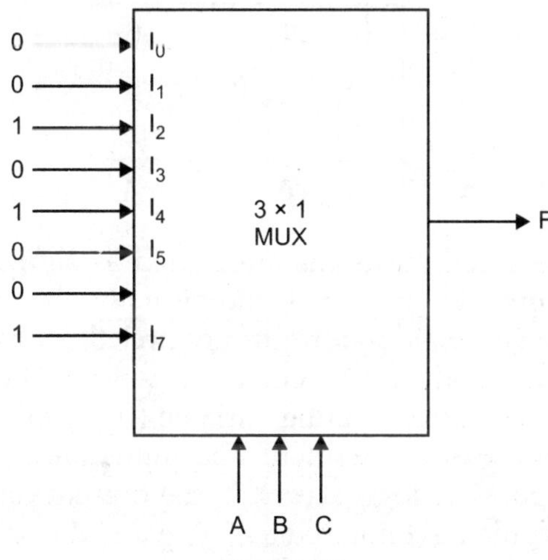

Fig. 23.97

Which line is given what logic status can be easily determined with the help of a simple procedure. The complete procedure is illustrated for the Boolean function considered earlier and given by:

$$F(A, B, C) = \Sigma\, 2, 4, 7$$

It is a three variable expression and we shall see how this can be implemented with a (4 × 1) MUX. This MUX has two selection lines. The first step here is to determine the truth table of the given Boolean function. This is shown below :

Minterm	*A*	*B*	*C*	*F*
0	0	0	0	0
1	0	0	1	0
2	0	1	0	1

Cont...

3	0	1	1	0
4	1	0	0	1
5	1	0	1	0
6	1	1	0	0
7	1	1	1	1

In the next step, connect the remaining variables to the selection lines leaving out the left most variable, 'A' in this case. The higher order variable is connected to the higher order selection line in the same order. For instance, in the present case, variables B and C would respectively be connected to selection lines S_1 and S_0. In the third step, construct a table of the type shown below. Under the inputs to the MUX, minterms are listed in two rows as shown. The first row list those terms where 'A' is complemented and second row lists those terms, where 'A' is uncomplemented.

	I_0	I_1	I_2	I_3
$\overline{A}$	0	1	②	3
A	④	5	6	⑦
	A	0	$\overline{A}$	A

The required minterms are circled as shown. Now, each column is inspected individually. If both the minterms of a certain column are not circled, a '0' is written below that. If both are circled, a '1' is written. If only one is circled, the corresponding variable (complemented or uncomplemented) is written. The input lines are then given appropriate logic status. In the present case, I_0, I_1, I_2 and I_3 would be connected to A, 0, $\overline{A}$ and $\overline{A}$ respectively. Fig. 23.98 shows the logic implementation.

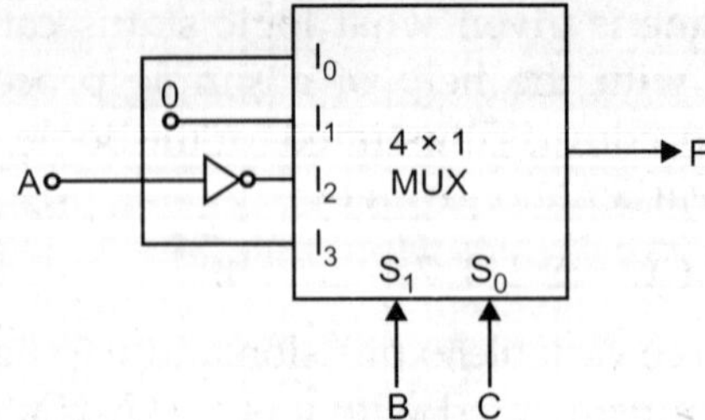

Fig. 23.98

It is not necessary to choose only the left most variable in the sequence to be used as input to MUX. Any of the variables can be used provided the implementation table is constructed accordingly. In the problem illustrated above, 'A' was chosen as the variable for the input lines and accordingly, the first row of the implementation table contained those entries where 'A' was complemented and second row contained entries, where 'A' was uncomplemented. If we consider 'C' as the left out variable, the implementation table will be as shown.

	I_0	I_1	I_2	I_3
$\overline{C}$	0	④	4	6
C	①	3	5	⑦
	0	$\overline{C}$	$\overline{C}$	A

Fig. 23.99 shows MUX implementation. In case of 'B' being the left out variable, the (implementation table is as shown below. Fig. 23.100 shows the MUX implementation.

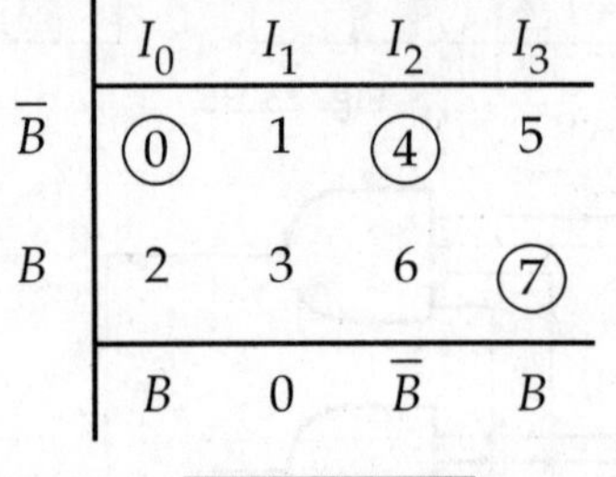

	I_0	I_1	I_2	I_3
$\overline{B}$	⓪	1	④	5
B	2	3	6	⑦
	B	0	$\overline{B}$	B

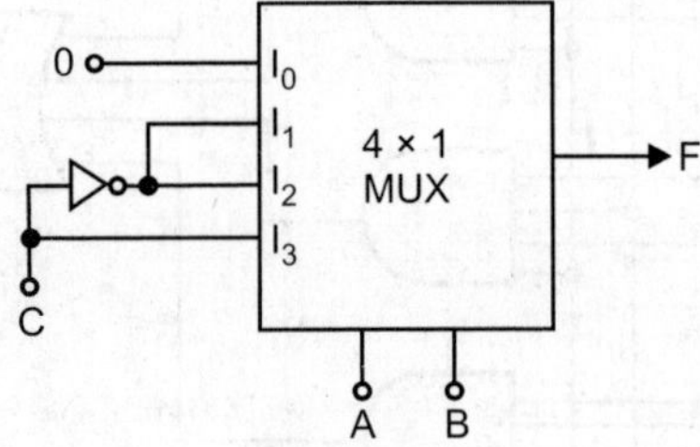

Fig. 23.99

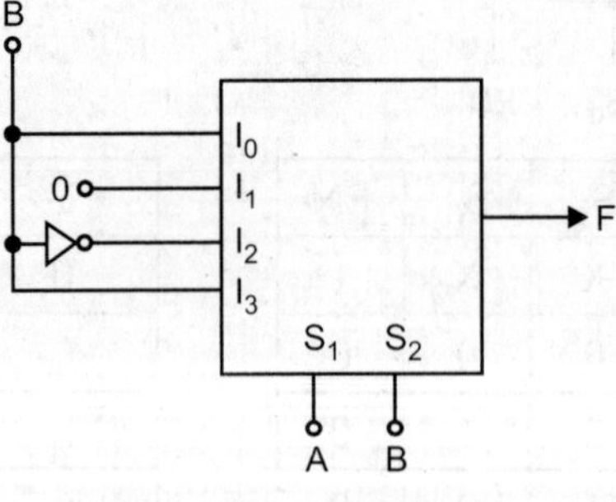

Fig. 23.100

Q. 54. Briefly explain with the help of a suitable example how a multiplexer can be used for conversion of parallel data to serial data?

Ans. Though the data is processed parallely in many digital systems due to the processing being faster, when it comes to transmitting this data to relatively large distances, it is done so serially. The parallel arrangement in this case is highly undesirable, as it would require a large number of transmission lines. Multiplexer can possibly be used for parallel-to-serial conversion. Fig. 23.101. shows one such arrangement, where an (8 × 1) MUX has been used to convert 8-bit parallel binary data to serial form. A 3-bit MOD-8 counter controls the selection inputs. As the counter goes through 000 to 111, the multiplexer output goes through X_0 to X_7. The conversion process takes a total of 8 clock cycles.

Q. 55. What is a De-multiplexer? How does a decoder differ from a de-multiplexer? Briefly describe the operation of a decoder.

Ans. A **Demultiplexer** is a combinational logic circuit with an input line, (2^n) output lines and (n) select lines. It routes the information present on the input line to any of the output lines. The output line that gets the information present on the input line is decided by the bit status of the selection lines. A **Decoder** is a special case of a demultiplexer without the input line. Fig. 23.102 (*a*) shows the circuit representation of a (1 × 4) Demultiplexer. Fig. 23.102 (*b*) shows the truth table of the demultiplexer when the input line is held HIGH.

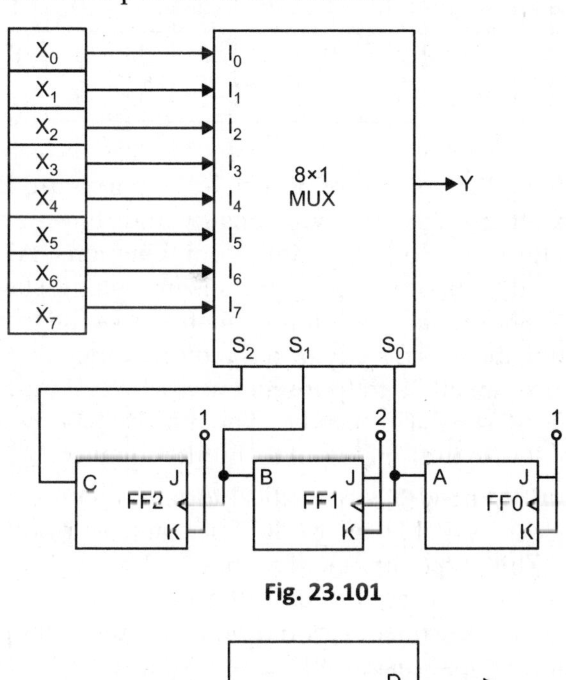

Fig. 23.101

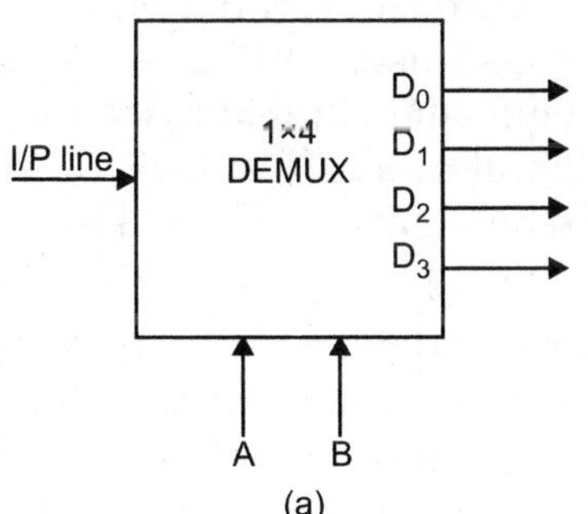

(a)

I/P	Select		O/P			
	A	B	D_0	D_1	D_2	D_3
1	0	0	1	0	0	0
1	0	1	0	1	0	0
1	1	0	0	0	1	0
1	1	1	0	0	0	1

(b)

Fig. 23.102

A decoder, as mentioned earlier, is a combinational circuit that decodes the information on (n) input lines to a maximum of (2^n) unique output lines. In case, there are some unused or don't care combinations in the n-bit code, then there will be less than (2^n) output lines. As an illustration, if there are 3 input lines, it can have a maximum of 8 unique output lines. In case, in the 3-bit input code, the only used 3-bit combinations are 000, 001, 010, 100, 110 and 111 (011 and 101 being either unused or don't care combinations), then this decoder will have only six output lines. In general, if (n) and (m) are respectively number of input and output lines, then

$$m < 2^n$$

A decoder can generate a maximum of (2^n) possible minterms with an n-bit binary code. Fig. 23.103 shows symbols for 2-to-4, 3-to-8 and 4-to-16 line decoders.

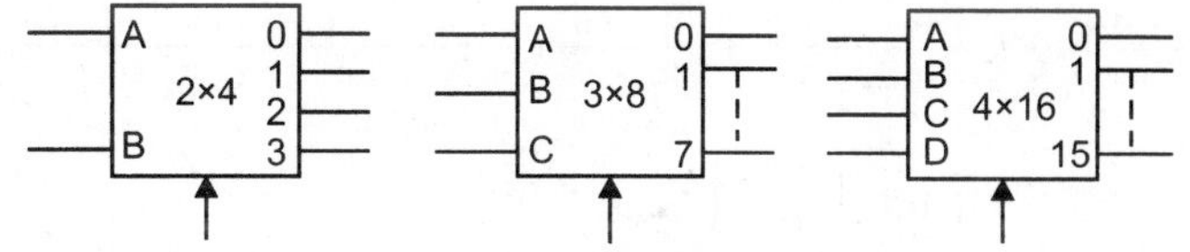

Fig. 23.103

In order to illustrate further the operation of a decoder, consider the logic circuit diagram of Fig. 23.104. This logic circuit, as we shall see, implements 3-to-8 line decoder function. This decoder has 3 inputs A, B, C and 8 outputs D_0, D_1 .... D_7. The truth table of such a circuit is also shown alongwith the figure. As is clear, for any given input combination, only one of the eight outputs is HIGH. Thus, each output produces a certain minterm. The output line that is in logic '1' state represents the minterm equivalent of the binary number currently present at the input. In the present case, D_0, D_1, D_2, D_3, D_4, D_5, D_6 and D_7 respectively represent $\overline{A} . \overline{B} . \overline{C}$, $\overline{A} . \overline{B} . C$, $\overline{A} . B . \overline{C}$, $\overline{A} . B . C$, $A . \overline{B} . \overline{C}$, $A . \overline{B} . C$, $A . B . \overline{C}$ and $A . B . C$. A decoder can be very conveniently used to implement a given Boolean function. The decoder is used here to generate the required minterms and an external OR-gate is used to produce sum of minterms. In general, an n-to-2^n decoder and (m) external OR gates can be used to implement any combinational circuit with (n) inputs and (m) outputs.

It can be visualised here that a Boolean function that has a large number of minterms, if implemented with a decoder and an external OR-gate would require an OR-gate with number of inputs equal to number of minterms. In all those cases, where number of minterms in a given Boolean function (F) with (n) variables is greater than $2^n/2$ (or 2^{n-1}), then the complement Boolean function ($\overline{F}$) will have fewer minterms than ($\overline{F}$). In that case, it would be more advantageous to sum the minterms of (F) using a NOR gate whose output then will be nothing but the Boolean function (F).

Inputs			Outputs							
A	B	C	D_1	D_1	D_2	D_3	D_4	D_5	D_6	D_7
0	0	0	1	0	0	0	0	0	0	0
0	0	1	0	1	0	0	0	0	0	0
0	1	0	0	0	1	0	0	0	0	0
0	1	1	0	0	0	1	0	0	0	0
1	0	0	0	0	0	0	1	0	0	0
1	0	1	0	0	0	0	0	1	0	0
1	1	0	0	0	0	0	0	0	1	0
1	1	1	0	0	0	0	0	0	0	1

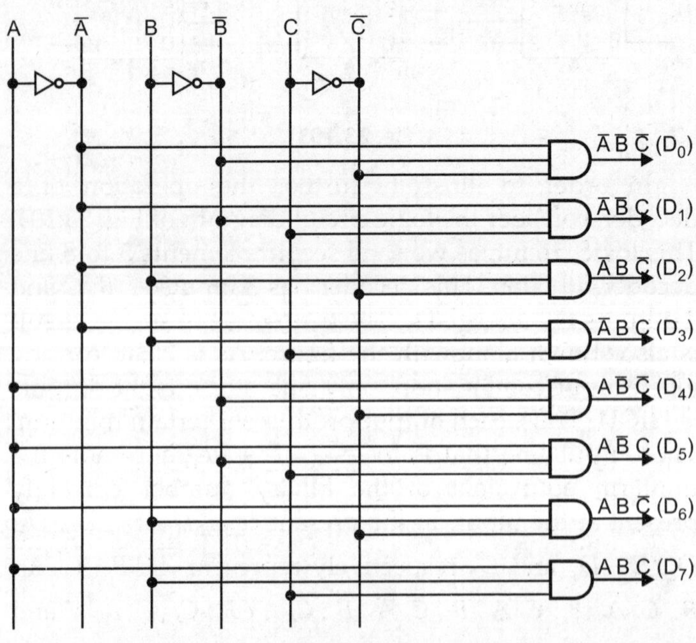

Fig. 23.104

Q. 56. What is an Encoder? How does a Priority Encoder differ from a conventional encoder?

Ans. An ENCODER is a multiplexer without the single output line of a multiplexer. It is a combinational logic function that has 2^n (or less) input lines and (n) output lines, which were (n) selection lines in a multiplexer. The (n) output lines generate the binary code for the possible (2^n) input lines. Let us take the case of an octal-to-binary encoder. Such an encoder would have 8 input lines, each representing an octal digit and 3 output lines representing the 3-bit binary equivalent. The truth table of such an encoder is shown below. In the truth table, D_0 ....D_7 represent octal digits 0 to 7. A, B, C represent the binary digits.

Inputs								Outputs		
D_0	D_1	D_2	D_3	D_4	D_5	D_6	D_7	A	B	C
1	0	0	0	0	0	0	0	0	0	0
0	1	0	0	0	0	0	0	0	0	1
0	0	1	0	0	0	0	0	0	1	0
0	0	0	1	0	0	0	0	0	1	1
0	0	0	0	1	0	0	0	1	0	0
0	0	0	0	0	1	0	0	1	0	1
0	0	0	0	0	0	1	0	1	1	0
0	0	0	0	0	0	0	1	1	1	1

The 8 input lines would have $2^8 = 256$ possible combinations. But in case of an octal-to-binary encoder only 8 of these 256 combinations would have any meaning. Rest are don't care input combinations. Also at a time, only one of the input lines is in logic '1' state. Fig. 23.105. Shows hardware implementation of the truth table shown above. This circuit has a discrepancy that it also produces an all 0's output sequence, when all inputs are in logic '0' state. This can be overcome by having an additional line to indicate an all 0's input sequence.

A **Priority Encoder** is a practical form of an encoder. The encoders available in the IC form are all priority encoders. This type of encoder has a known input priority, which ensures that only the line with highest input priority is encoded. As an example, if the octal-to-binary encoder just discussed has an input priority for higher order digits, then in that case, if for example D_2, D_4, D_7 are all simultaneously in logic '1" state, only (D_7) will be encoded and output will be 111.

The truth table of such a priority encoder will then be as shown ahead. Looking at the last row of the table, it implies that if $D_7 = 1$, then irrespective of the logic status of other inputs, output is 111 as (D_7) will only be encoded.

As another example, Fig. 23.106 shows the logic symbol and truth table of a 10-line decimal to 4-line

BCD encoder. In the functional table shown, the input line with highest priority having a LOW on it is encoded irrespective of logic status of other input lines. The functional table shown in Fig. 23.106 is that of IC 74147, a Decimal-to-BCD priority encoder with active LOW inputs and outputs.

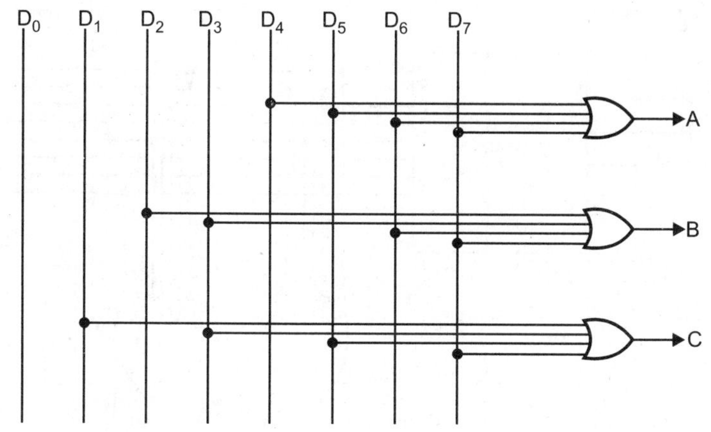

Fig. 23.105

Inputs								*Outputs*		
D_0	D_1	D_2	D_3	D_4	D_5	D_6	D_7	*A*	*B*	C
1	0	0	0	0	0	0	0	0	0	0
X	1	0	0	0	0	0	0	0	0	1
X	X	1	0	0	0	0	0	0	1	0
X	X	X	1	0	0	0	0	0	1	1
X	X	X	X	1	0	0	0	1	0	0
X	X	X	X	X	1	0	0	1	0	1
X	X	X	X	X	X	1	0	1	1	0
X	X	X	X	X	X	X	1	1	1	1

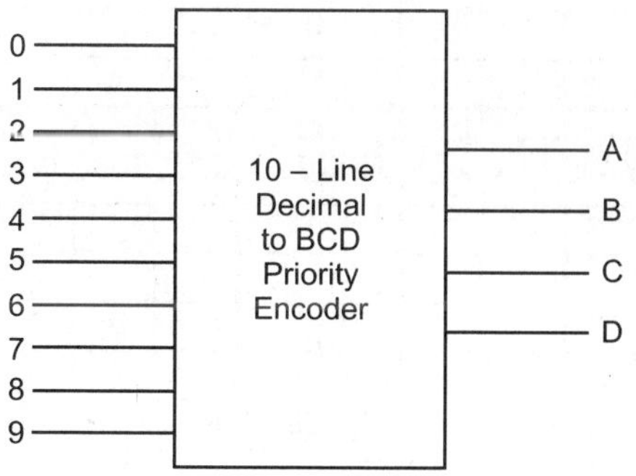

Inputs										*Output*			
0	1	2	3	4	5	6	7	8	9	*D*	*C*	*B*	*A*
X	X	X	X	X	X	X	X	X	0	0	1	1	0
X	X	X	X	X	X	X	X	0	1	0	1	1	1
X	X	X	X	X	X	X	0	1	1	1	0	0	0
X	X	X	X	X	X	0	1	1	1	1	0	0	1
X	X	X	X	X	0	1	1	1	1	1	0	1	0
X	X	X	X	0	1	1	1	1	1	1	0	1	1
X	X	X	0	1	1	1	1	1	1	1	1	0	0
X	X	0	1	1	1	1	1	1	1	1	1	0	1
X	0	1	1	1	1	1	1	1	1	1	1	1	0
0	1	1	1	1	1	1	1	1	1	1	1	1	1

Fig. 23.106

Q. 57. Briefly describe the functional principle of a ROM. How is a ROM used to implement a Boolean expression?

Ans. A **Read Only Memory (ROM)** is essentially a memory device that can be used to store a certain fixed set of binary information. These devices have certain inherent links which can be fused (or made) or broken to store any user specified binary information in the device Once a given pattern is formed, it remains as such even, if power is turned off and on. There are types of ROMs, where the user can erase the data already stored on the ROM chip and load it with fresh data. Different types of ROMs and memory related issues of ROMs are discussed in detail in chapter-27 dealing with microprocessors and microcomputers. In the present section, we shall discuss use of ROMs for combinational logic implementation which is one of the most widely exploited applications of ROMs.

A ROM in general has (n) input lines and (m) output lines, (n) input lines constitute the address lines to give (2^n) possible unique addresses and an equal number of memory locations. (m) output lines indicate that each location stores a binary word of (m) bits. Fig. 23.107 shows the representation of a ($2^n \times$ m) ROM. Out of (2^n) possible m-bit binary words stored on the chip, the word available on the output lines at any given time depends upon the address input. Each input address has a unique corresponding word. Internally, a ROM is a combinational circuit with input AND gates wired as a decoder and having OR gates equal to number of outputs. A ROM with 5 input lines and 4 output lines, for instance, would have an equivalent of (5 × 32) decoder at the input that would generate 32 possible minterms and 4 OR gates at the output. Each of these 4 OR gates would be 32-input gates fed from 32 outputs of the decoder through fusible links. Fig. 23.108 shows the internal architecture of a generalised ROM. Very complex combinational functions can be generated with ROMs by suitably making or breaking these links. To sum up, for implementing an n-input or variable, m-output combinational circuit, one would need a ($2^n \times m$) ROM. As an illustration, let us see how ROM's internal architecture would be like when it is used to implement the following Boolean function with two outputs given by:

$$F_1(A, B, C) = \Sigma\, 0, 1, 5$$

and

$$F_2(A, B, C) = \Sigma\, 1, 3, 4$$

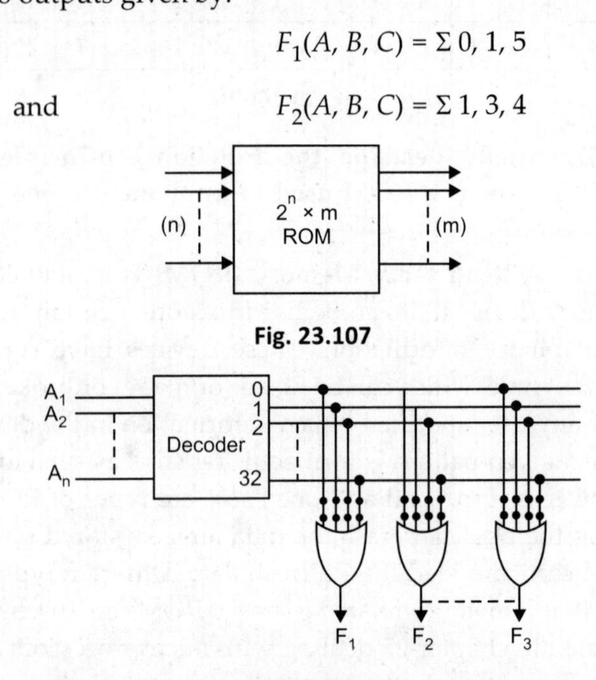

Fig. 23.107

Fig. 23.108

Implementation of this Boolean function would require a (8 × 2) ROM. The internal architecture of ROM in this case would be as shown in Fig. 23.109.

It may be mentioned here that in practice, ROM would not be used to implement as simple a Boolean function as illustrated above. The purpose here was to tell the readers how a ROM implements a Boolean function. In actual practice, ROMs would be used only in case of very complex Boolean functions.

Another noteworthy point here is that when it comes to implementing Boolean functions with ROMs, it is not a economical to use ROM for those Boolean functions which have a large number of, 'Don't Care Conditions'. In case of a ROM, each 'Don't Care Condition' would be represented by an address and the location corresponding to that address input would have either all 0's or all l's. In other words, the space on the chip is not optimally utilised.

A more appropriate device for such situations would be a Programmable Logic Array (PLA). A PLA is a device similar in concept to a ROM and is described in the next question.

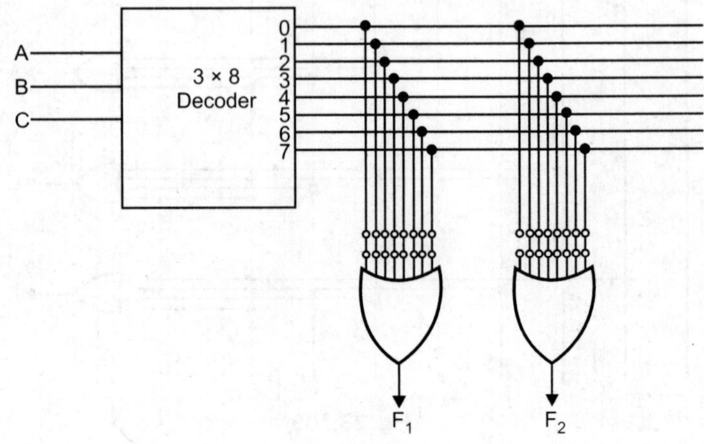

Fig. 23.109

Q. 58. What is a Programmable Logic Array (PLA)? How does it differ from a ROM?

Ans. A Programmable Logic Array (PLA) enables logic functions expressed in sum-of-products form to be implemented directly. It is similar in concept to a ROM. But unlike ROM, the PLA does not provide full decoding of the variables and does not generate all possible minterms, as is the case in a ROM.

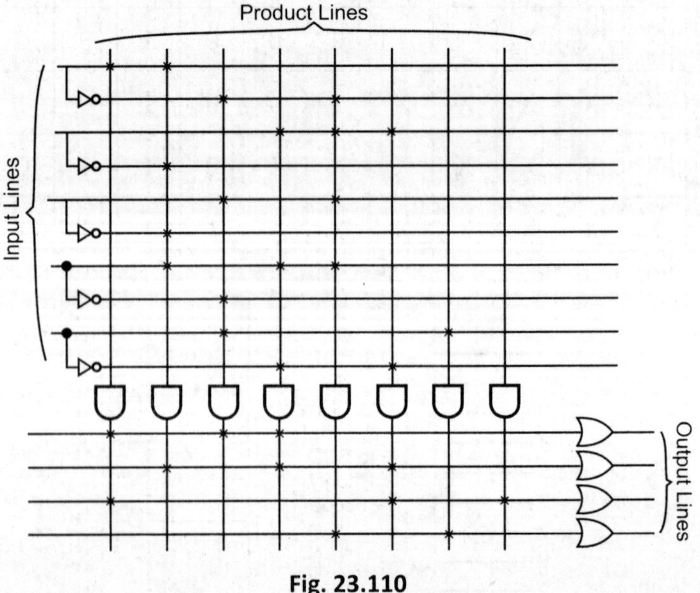

Fig. 23.110

While a ROM has a fixed AND gate array at the input and a programmable OR gate array at the output, a PLA has programmable AND gate array at the input and a programmable OR gate array at the output. In a PLA, each of the product terms of the given Boolean function is generated by an AND gate, which can be programmed to form the AND of any subset of inputs or their complements. For instance, if A, B, C are input variables, each AND gate in array could be programmed to produce any of the following product terms namely $A, \bar{A}, B, \bar{B}, C, \bar{C}, \bar{A}\,\bar{B}, \bar{A}B, A\bar{B}, AB, \bar{A}\,\bar{B}\,\bar{C}, \bar{A}\,\bar{B}\,C, \bar{A}\,B\,\bar{C}, \bar{A}\,B\,C, A\,\bar{B}\,\bar{C}, A\,\bar{B}\,C, A\,B\,\bar{C}$ and ABC. Thus, we have a programmable AND gate array at the input. The product terms so produced can be summed up in an array of programmable OR gates. Thus, we have a programmable OR gate array at the output. The input and output gates are constructed in the form of arrays with input lines orthogonal to product lines and the product lines being orthogonal to output lines. As an example, Fig. 23.110 shows a PLA structure with 5 input lines, 8 product lines and 4 output lines. The cross (X) indicates a connection. The structure shown in Fig. 23.110 can also be represented by the schematic form of Fig. 23.111. PLAs usually have inverters at the output of OR gates to enable PLAs implement a given Boolean function in either AND-OR or AND-OR-INVERT form.

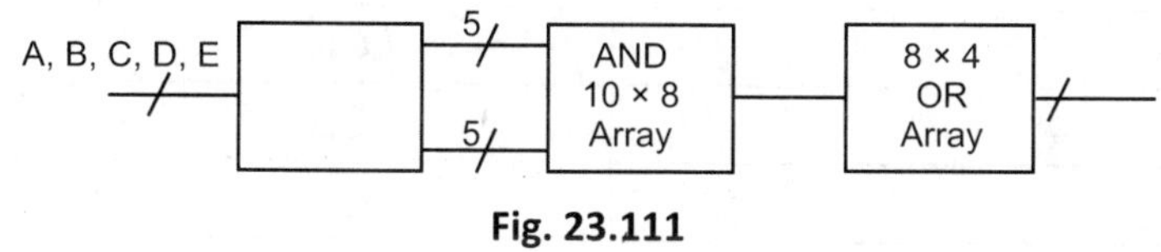

Fig. 23.111

Q. 59. What is a Programmable Array Logic (PAL) device ? How does it differ from PLA device?

Ans. A variant of PLA devices is the one which has programmable AND gate array at the input and a fixed OR gate array at the output. There are many applications, where product term sharing capability of the PLA is not fully utilised and is wasted. This device is called Programmable Array Logic (PAL), a trade mark of Advanced Microdevices Inc. A PAL based solution is more economical. This implies that if programmability of the OR array in a PLA is dispensed with, the new device, which is a PAL, will have a large number of applications. PAL devices are, however, less flexible than PLA devices. The flexibility of a PAL device can be enhanced by adding controllable tristate drivers as shown in Fig. 23.112. These drivers allow those pins to which they are connected to act as either inputs or outputs on both. This feature allows the device to be used in a wide range of applications than would be possible with a device with fixed input and output allocations. Mask programmed version of PAL is HAL (Hard Array Logic). A HAL device is pin to pin compatible to its PAL counterpart.

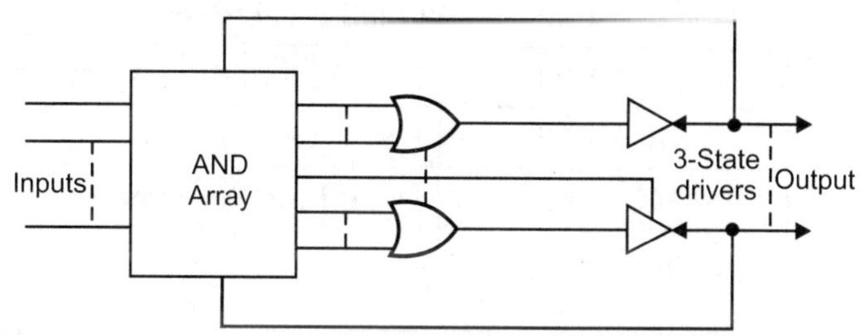

Fig. 23.112

Table 23.6: Buffers and Transceivers

IC Type No.	*Function*	*Logic Family*
4049B	Hex inverting Buffer	CMOS
4050B	Hex Non-Inverting Buffer	CMOS
40097B	3-state Hex Non-Inverting Buffer	CMOS
40098B	3-state Hex Inverting Buffer	CMOS
74125	Quad 3-state Buffer (LOW Enable)	TTL
74126	Quad 3-state Buffer (High Enable)	TTL
74240	Octal 3-state Inverting Bus/Line Driver	TTL
74241	Octal 3-state Bus Line Driver	TTL
74242	Quad 3-state Inverting Bus Transceiver	TTL
74243	Quad 3-state Non-Inverting Bus Transceiver	TTL
74244	Octal 3-state Non-Inverting Driver	TTL
74245	Octal 3-state Non-Inverting Bus Transceiver	TTL
74365A	Hex 3-state Buffer with Common Enable	TTL
74366A	Hex 3-state Inverter with Common Enable	TTL

Cont...

74367A	Hex 3-state Buffer, 4-Bit and 2-Bit	TTL
74368A	Hex 3-state Inverter Buffer, 4-Bit and 2-Bit	TTL
74465	Octal 3-state Buffer, Gated Enable-Inverted	TTL
74540	Octal 3-state Inverting Buffer/Line Driver	TTL
74541	Octal 3-state Buffer/Line Driver	TTL
74640	Octal 3-state Bus Transceiver	TTL
74641	Octal 3-state Non-Inverting Bus Transceiver	TTL
74645	Octal 3-state Non-Inverting Bus Transceiver	TTL
MC 10114	Triple Line Receiver	ECL
MC 10115	Quad Line Receiver	ECL
MC 10116	Triple Line Receiver	ECL
MC 10123	Triple 4-3-3 Input Bus Driver	ECL
MC 10128	Bus Driver	ECL
MC 10129	Quad Bus Receiver	ECL
MC 10188	Hex Buffer with Enable	ECL
MC 10192	Quad Bus Driver	ECL
MC 10194	Dual Simultaneous Transceiver	ECL
MC 10195	Hex Inverter/Buffer	ECL
MC 1692	Quad Line Receiver	ECL

Table 23.7: Arithmetic Circuits

IC Type No.	*Function*	*Logic Family*
7483	4-Bit Full Adder	TTL
7485	4-Bit Magnitude Comparator	TTL
74181	4-Bit ALU	TTL
74182	Look Ahead Carry Generator	TTL
74183	Dual Carry Save Full Adder	TTL
74283	4-Bit Binary Adder	TTL
4008	4-Bit Binary Full Adder	CMOS
4527	BCD Rate Multiplier	CMOS
40085	4-Bit Magnitude Comparator	CMOS
40181	4-Bit Arithmetic Unit	CMOS
40182	Look Ahead Carry Generator	CMOS
10179	Look Ahead Carry Block	ECL
10180	Dual High Speed 2-Bit Adder/Subtractor	ECL
10181	4-Bit Arithmetic Logic Unit/Function Generator	ECL
10182	4-Bit Arithmetic Logic Unit/Function Generator	ECL
10183	4 × 2 Multiplier	ECL

Table 23.8: Multiplexers and Demultiplexers

IC Type No.	*Function*	*Logic Family*
74151	8-input Multiplexer	TTL
74152	8-input Multiplexer	TTL
74153	Dual 4-input Multiplexer	TTL
74157	Quad 2-input non-inverting Multiplexer	TTL
74158	Quad 2-input Inverting Multiplexer	TTL
74251	8-input 3-state Multiplexer	TTL
74253	Dual 4-input 3-state Multiplexer	TTL
74257	Quad 2-input 3-state Multiplexer	TTL
74258	Quad 2-input 3-state Multiplexer	TTL
74298	Dual 2-input Multiplexer with output latches	TTL
74353	Dual 4-input multiplexer	TTL
74398	Quad 2-input Multiplexer with output register	TTL
74399	Quad 2-input Multiplexer with output register	TTL
7442	l-of-10 Decoder	TTL
74138	l-of-8 Decoder/Demultiplexer	TTL
74139	Dual l-of-4 Decoder/Demultiplexer	TTL
74145	l-of-10 Decoder/Driver (open collector)	TTL
74154	4-to-16 Decoder/Demultiplexer	TTL
74155	Dual l-of-4 Decoder/Demultiplexer	TTL
74156	Dual l-of-4 Decoder/Demultiplexer (open collector)	TTL
74256	Dual-4-Bit Addressable Latch	TTL
74259	8-bit Addressable Latch	TTL
74147	10-lineyto 4-line Priority Encoder	TTL
74148	8-line to 3-line Priority Encoder	TTL
74348	8-line to 3-line Priority Encoder (3-state)	TTL
74247	BCD-to-7 segment Decoder/Driver (open collector)	TTL
74248	BCD-to-7 segment Decoder/Driver with Pull-ups	TTL
4019	Quad 2-input Multiplexer	CMOS
4512	8 input 3 state Multiplexer	CMOS
4539	Dual 4-input Multiplexer	CMOS
4028	l-of-10 Decoder	CMOS
4511	BCD to 7-segment Latch/Decoder/Driver	CMOS
4514	l-of-16 Decoder/Demultiplexer with input Latch	CMOS
4515	l-of-16 Decoder/Demultiplexer with input Latch	CMOS
4543	BCD to 7-segment. Latch/Decoder/Driver for LCD displays	CMOS
4555	Dual l-of-4 Decoder/Demultiplexers	CMOS
4556	Dual l-of-4 Decoder/Demultiplexers	CMOS
4723	Dual 4-bit Addressable Latch	CMOS
4724	8-bit Addressable Latch	CMOS
40147	10-line to 4-line BCD Priority Encoder	CMOS
4532	8-line to 3-line Priority Encoder	CMOS

Cont...

10132	DualLMultiplexes with Latch and Common Reset	ECL
10134	Dual Multiplexer with Latch	ECL
10158	Quad 2-input Multiplexer (Non-inverting)	ECL
10159	Quad 2-input Multiplexer (Inverting)	ECL
10164	8-line Multiplexer	ECL
10173	Quad 2-input Multiplexer	ECL
10174	Dual 4-to-l Multiplexer	ECL
10165	8-input Encoder	ECL
10161	Binary to 1-8 (LOW)	ECL
10162	Binary to 1-8 (HIGH)	ECL
10171	Dual Binary to 1-4 (LOW)	ECL
10172	Dual Binary to 1-4 (HIGH)	ECL

Solved Problems

Problem 23.1. *A and B inputs of a 2-input NAND gate are applied continuous tram of pulses as shown in Fig. 23.113. 0V represents logic '0' and + 5V represents logic '1'. Draw the output waveform.*

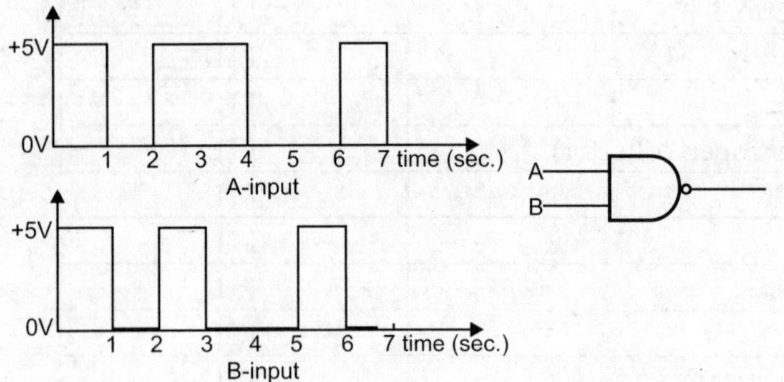

Fig. 23.113

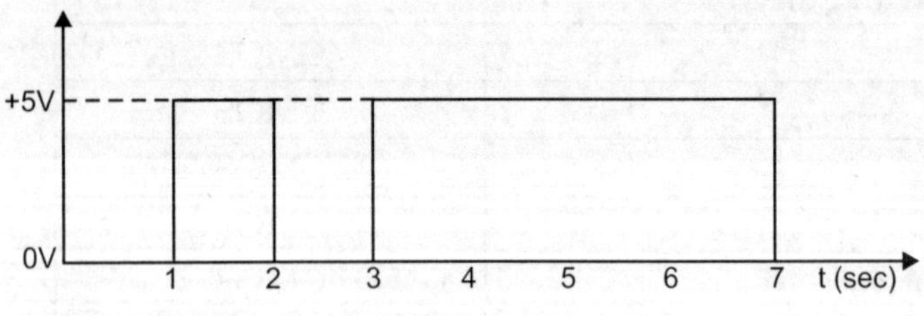

Fig. 23.114

Solution.

The output pulse train can be easily drawn keeping the truth table of a 2-input NAND in mind. The waveform is shown in Fig. 23.114.

Problem 23.2. *How would you hardware implement a 4-input OR gate using 2-input OR gates only.*

Solution.

Fig. 23.115 (*a*) shows one possible arrangement of 2-input OR gates that simulates a 4-input OR gate. *A, B, C* and *D* are the logic inputs and *Y* is the output. Fig. 23.115 shows another possible arrangement.

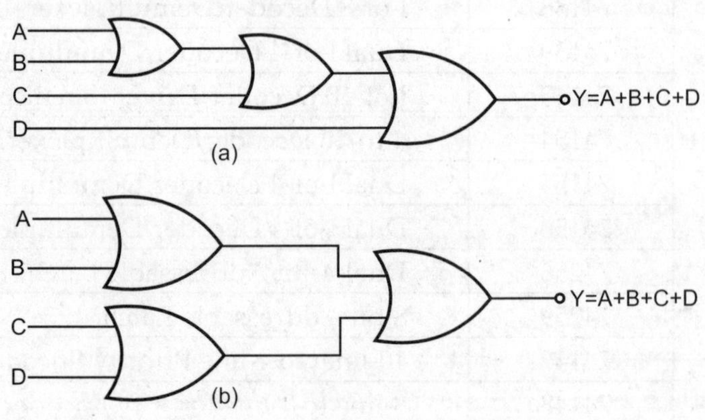

Fig. 23.115

Problem 23.3. *Show the logic arrangements for implementing*

(*a*) *a 4-input AND gate using 2-input AND gates*

(*b*) *a 4-input NAND gate using 2-input AND gates and NOT circuits.*

(*c*) *a 3-input NAND gate using 2-input NAND gates.*

(*d*) *a NOT gate using 2-input NAND gates.*

Solution.

(*a*) Fig. 23.116 (*a*) shows hardware implementation of a 4-input AND gate using 2-input AND gates.

(*b*) Fig. 23.116 (*b*) shows the required logic arrangement. It is the same as the one shown in Fig. 23.116 (*a*) except for the additional NOT circuit at the output.

(*c*) Fig. 23.116 (c) shows the logic arrangement

(*d*) A NAND gate with all its inputs shorted becomes a one input one output NOT circuit using a 2-input NAND gate.

The logic arrangements shown in Figs. 23.116 (*a*) to (*d*) are self-explanatory.

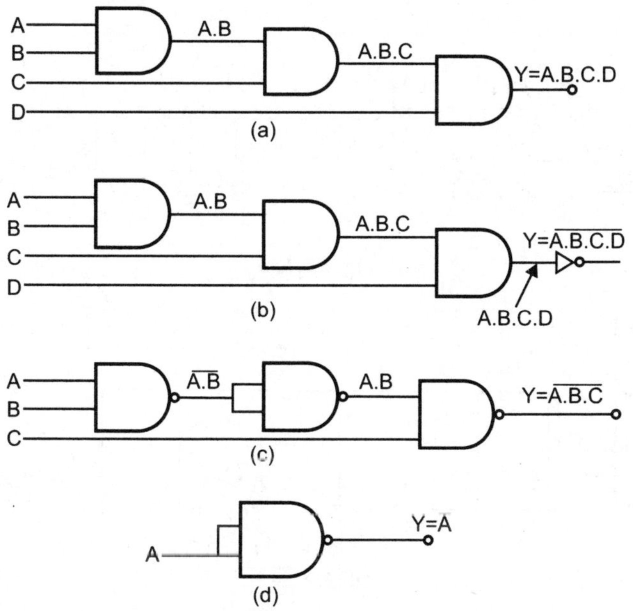

Fig. 23.116

Problem 23.4. *Show the hardware implementation of a NOT circuit using (a) 2-input EX-OR gates (b) 2-input EX-NOR gates.*

Solution.

Refer to the truth tables of 2-input EX-OR and EX-NOR gates reproduced here for the purpose or explaining the intended hardware.

A	B	Y
0	0	0
0	1	1
1	0	1
1	1	0
EX-OR		

A	B	Y
0	0	1
0	1	0
1	0	0
1	1	1
EX-NOR		

Fig. 23.117

Fig. 23.118

It is clear from the EX-OR truth table that if one of the inputs is permanently tied to a logic '1' level, then the other input and the output perform the function of a NOT circuit. In case of an EX-NOR gate, the same is true, if one of the inputs is permanently tied to the logic '0' level. Figs. 23.117 (*a*) and (*b*) show the desired implementation.

Problem 23.5. *Refer to the INHIBIT gate of Fig. 23.118 (a). The waveform shown in Fig. 23.118 (b) is applied to the inhibit input. Draw the waveform at the output.*

Solution.

Since all the other inputs of the gate have been permanently tied to a logic '1' level, a logic '0' at the inhibit input would produce a logic '1' at the output and a logic '1' at the inhibit input would produce a logic '0' at the output.

The output waveform is, therefore, the inversion of the input waveform and is shown in Fig. 23.119.

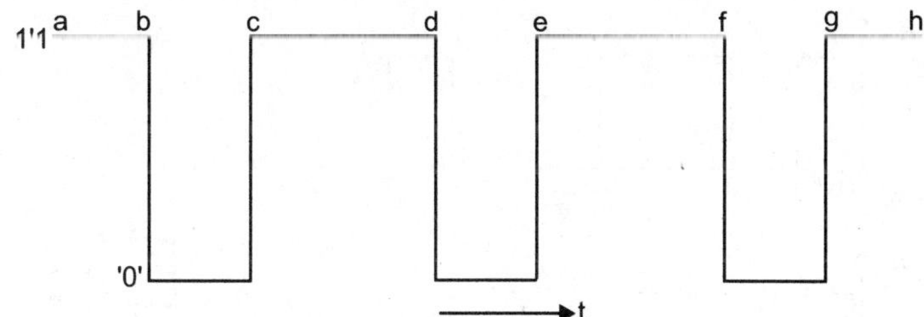

Fig. 23.119

Problem 23.6. *Show the logic implementation of (a) 3-input OR gate using 3-input NOR gates (b) NOT gate using an 8-input INHIBIT gate.*

Solution.

(*a*) Fig. 23.120 shows the logic implementation. The arrangement is self-explanatory.

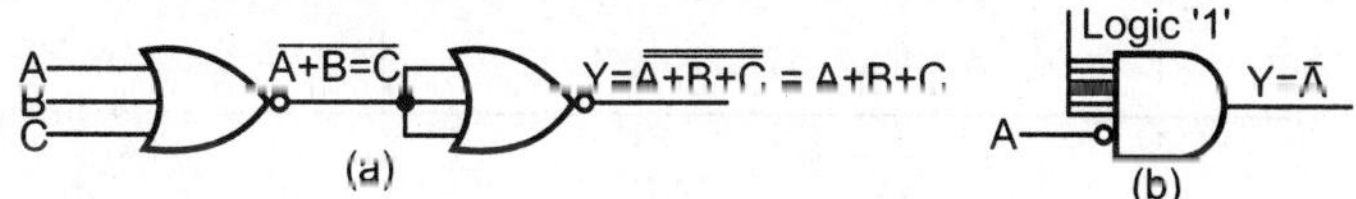

Fig. 23.120

(*b*) Fig. 23.120 (*b*) shows the logic implementation. The inhibit input and the output constitute the NOT circuit, if all other inputs of the gate are tied to a logic '1' level.

Problem 23.7. *How do you implement a 3-input EX-OR function using 2-input EX-OR gates. Also, show the logic implementation of a 3-input EX-NOR gate using 2-input EX-NOR gates.*

Solution.

Fig. 23.121 snows the implementation of a 3-input EX-OR gate using 2-input EX-OR gates. The 3-input EX-NOR gate implemented with 2-input EX-NOR gates is depicted in Fig. 23.121 (*b*) Both the circuits are self-explanatory.

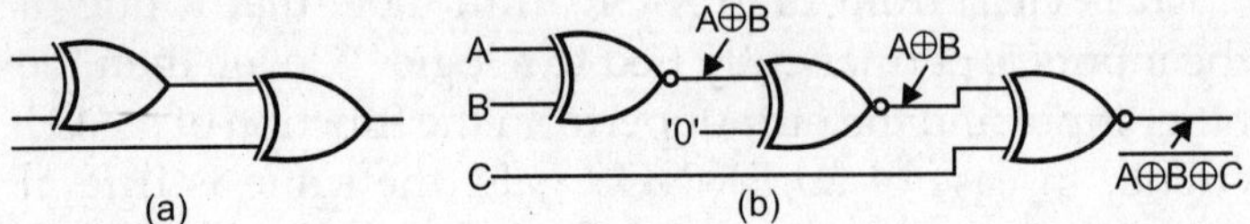

Fig. 23.121

Problem 23.8. *A and B inputs of 2-input NAND are applied continuous train of pulses, as shown in Fig. 23.122. 0V represents logic '0' and +5V represents logic '1'. Sketch the output waveform.*

Solution.

The output pulse train can be drawn keeping in mind the truth table of a 2-input NAND gate. In other words, the output of a NAND is logic '0' only when both its inputs are in logic '1' state. For all other input combinations, the output is logic '1'. The output waveform is shown in Fig. 23.123.

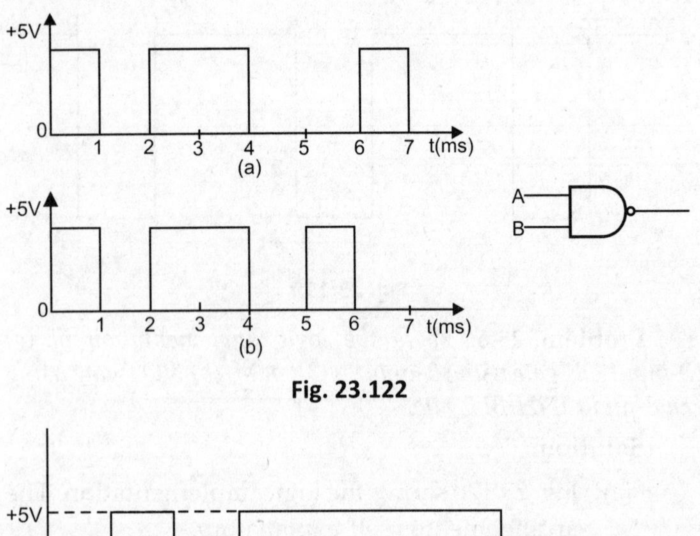

Fig. 23.122

Fig. 23.123

Problem 23.9. *Refer to the INHIBIT gate shown in Fig. 23.124 (a) and the inhibit input waveform shown in Fig. 23.124 (b) Sketch the output waveform.*

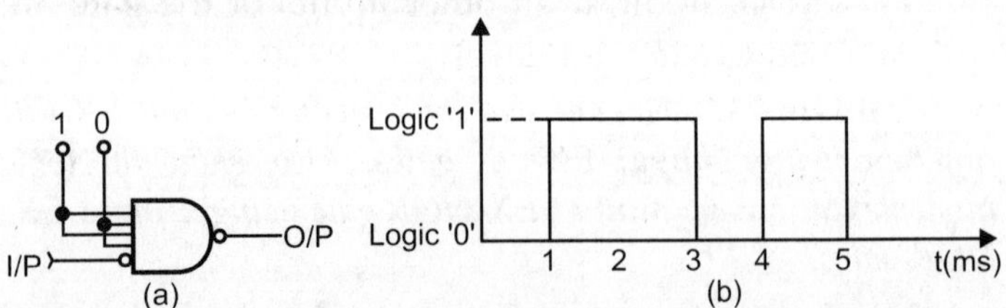

Fig. 23.124

Solution.

The output will always be logic '0' as two of the inputs of the logic gate have been permanently tied to logic '0'.

Problem 23.10. *Refer to the circuit shown in Fig. 23.125 (a). Sketch the output waveform for the input waveform shown in Fig. 23.125 (b).*

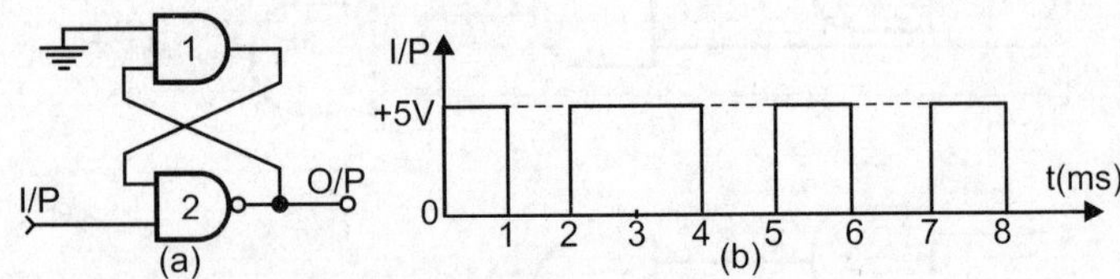

Fig. 23.125

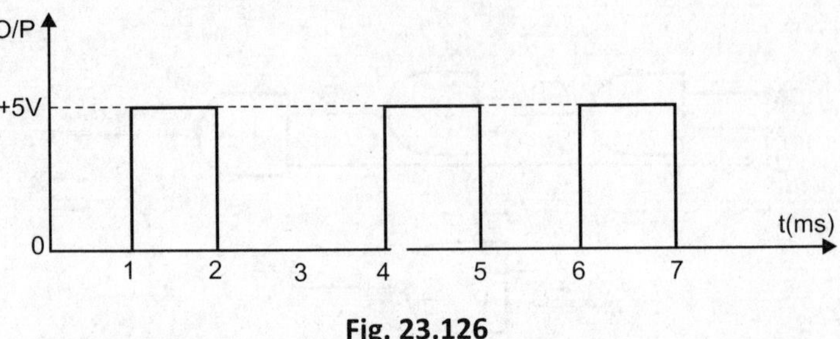

Fig. 23.126

Solution.

One of the inputs to the NAND gate (1) is grounded (*i.e.* logic '0'). Therefore, the output of NAND gate (1) is permanently HIGH (logic '1'). The output of NAND (1) has been fed back to one of the inputs of NAND gate (2). Therefore, a logic '0' at the other input produces a logic '1' at the output of NAND (2) and a logic '1' produces a logic '0' at the output *i.e.* the output is inverted version of the input. The output waveform is shown in Fig. 23.126.

Problem 23.11. *Sketch the output waveform. The NAND used in the circuit has transfer characteristics of Fig. 23.127 (b).*

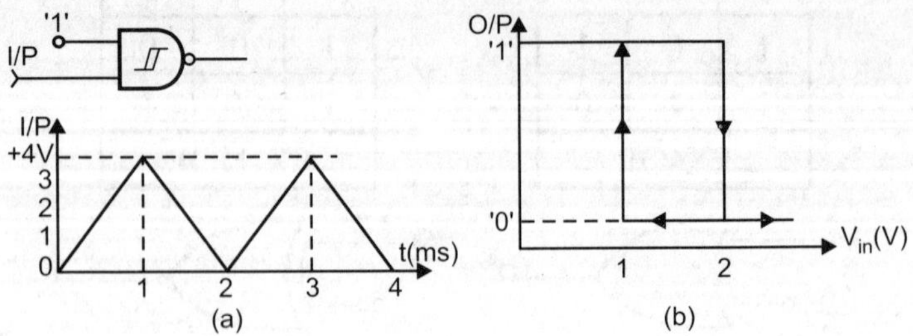

Fig. 23.127

Solution.

The output waveform is shown in Fig. 23.128. The output is initially in logic '1' state. It goes from logic '1' to logic '0', as the input exceeds 2V. The output goes from logic '0' to logic '1' again, as the input drops below 1V.

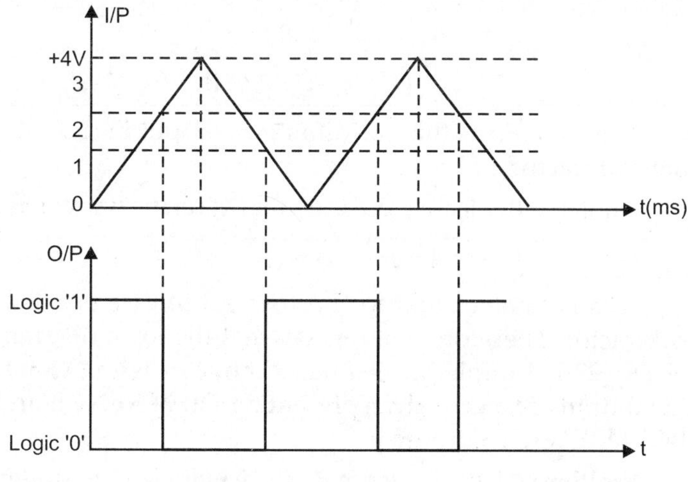

Fig. 23.128

Problem 23.12. *For the half adder of Fig. 23.129 (a), the inputs applied at (A) and (B) are as shown in Fig. 23.129 (b). Plot the corresponding SUM and CARRY outputs on the same scale.*

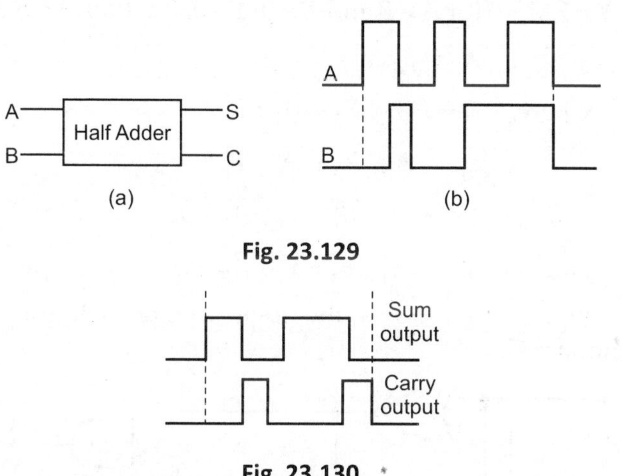

Fig. 23.129

Fig. 23.130

Solution.

The SUM and CARRY waveforms can be plotted from our knowledge of the truth table of the half adder. All that we need to remember to solve this problem is that (0 + 0) yields a '0' as SUM output and '0' as CARRY. (0 + 1) or (1 + 0) yield '1' as SUM output and '0' as CARRY (1 + 1) produces a '0' as SUM output and '1' as CARRY. The output waveforms are as shown in Fig. 23.130. The figures are self-explanatory.

Problem 23.13. *Draw the logic diagram of a half adder using only NAND gates. Show the steps involved therein to arrive at the final configuration starting with the basic half adder logic expressions.*

Solution.

The logic expressions for the SUM and CARRY outputs of a half adder are as follows:

SUM, $S = \overline{A}.B + A.\overline{B}$ and CARRY, $C = A.B$

where A and B are the input binary variables

Now, $S = \overline{A}.B + A.\overline{B} = \overline{A}.B + B.\overline{B} + A.\overline{B} + A.\overline{A}$

(Addition of $B.\overline{B}$ and $A.\overline{A}$ does not make any difference as $A.\overline{A} = B.\overline{B} = 0$.

$$S = B.(\overline{A} + \overline{B}) + A.(\overline{A} + \overline{B})$$

$$= B.(\overline{A.B}) + A.(\overline{A.B})$$

or $$S = [\overline{\overline{\{B.(\overline{A.B})\}}.\overline{\{.\overline{A.B}\}}}]$$

Also, $$C = A.B = \{\overline{(\overline{A.B})}\}$$

These expressions (for S and C) when implemented in hardware lead to the logic circuit of Fig. 23.131.

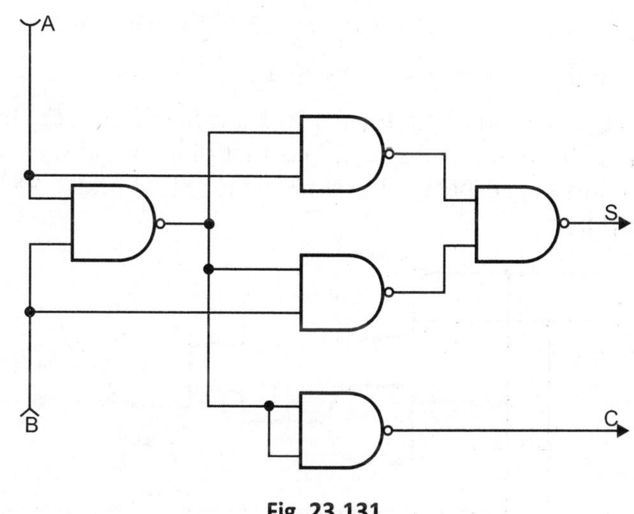

Fig. 23.131

Problem 23.14. *Refer to the logic diagram of Fig. 23.132. Write down simplified Boolean expressions for DIFFERENCE and BORROW outputs.*

Solution.

If (X) and (Y) were the inputs to the half subtractor (HS), then the Difference and Borrow outputs could be written as :

Difference $= X \oplus Y = \overline{X}.Y + X.\overline{Y}$

and Borrow $= \overline{X}.Y$

If (X) is considered as SUM output of the adder and (Y) is considered as (C), then

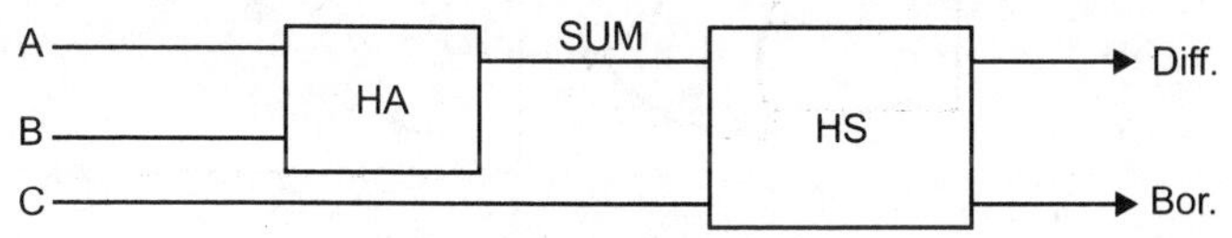

Fig. 23.132

$$X = A \oplus B = \overline{A}.B + A.\overline{B}$$

and $$Y = C$$

Substituting these values, we get

$$\text{Difference} = \overline{(\overline{A}.B.+A\overline{B})}.C+(\overline{A}.B.+A.\overline{B}).\overline{C}$$

$$= (A.B+\overline{A}.\overline{B}).C+(\overline{A}.B+A.\overline{B}).\overline{C}$$

$$= A.B.C+\overline{A}.\overline{B}.C+\overline{A}.B.\overline{C}.+A.\overline{B}.\overline{C}$$

$$\text{Borrow} = (A.B.+\overline{A}.\overline{B}).C$$

$$= A.B.C+\overline{A}.\overline{B}.C$$

Problem 23.15. *The logic diagram of Fig. 23.133 performs the function of a very common arithmetic building block. Recognize that. Derive expressions, if needed, to justify your answer.*

Solution. Writing Boolean expressions for (X) and (Y),

$$X = \overline{\{\overline{(\overline{A}.B)}.\overline{(A.\overline{B})}\}}$$

$$= [\overline{\overline{(\overline{A}.B)}}]+[\overline{\overline{(A.\overline{B})}}] = \overline{A}.B+A.\overline{B}$$

and $$Y = \overline{(\overline{A}+\overline{B})} = A.B$$

It is obvious from the expressions for (X) and (Y) that the logic circuit of Fig. 23.133 is that of a half adder with (X) and (Y) respectively representing SUM and CARRY outputs.

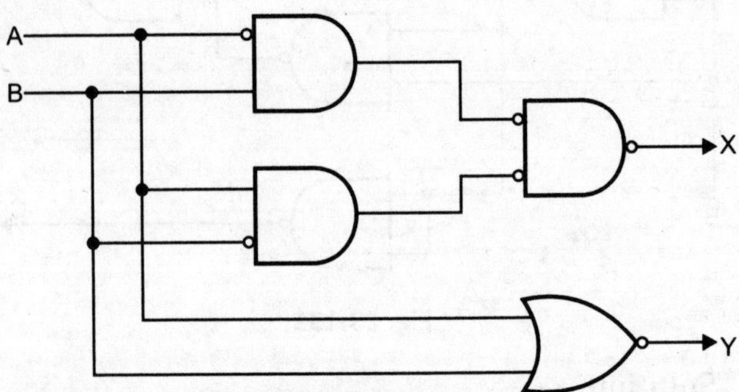

Fig. 23.133

Problem 23.16. *Prove that the logic diagram of Fig. 23.134 represents a half subtractor provided that (Y) represents the DIFFERENCE and (X) represents the BORROW OUT.*

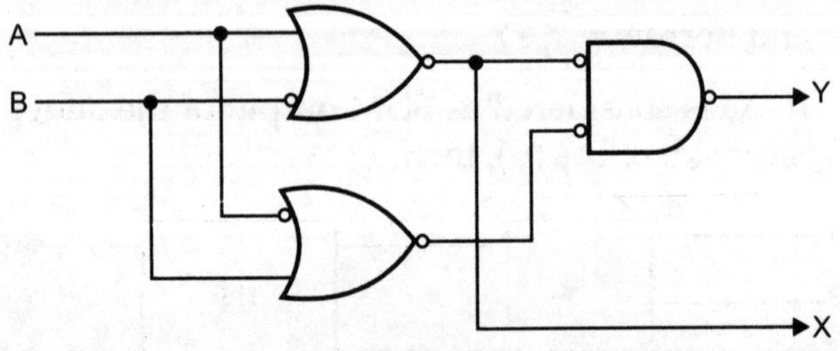

Fig. 23.134

Solution. The Boolean expression for (Y) can be written as :

$$Y = \overline{[\{\overline{(\overline{A}+B)}\}.\{\overline{(A+\overline{B})}\}]}$$

$$= (\overline{A}+B).(A+\overline{B})$$

$$= \overline{A}.B+A.\overline{B} = A\oplus B$$

This is same as that for difference output in case of a half subtractor.

Similarly, Boolean expression for (X) can be written as :

$$X = \overline{(A+\overline{B})} = \overline{A}.(\overline{\overline{B}}) = \overline{A}.B$$

This is same as that for Borrow out in case of a half subtractor. Therefore, we can say that the logic diagram of Fig. 23.114 implements a half subtractor with (Y) and (X) outputs respectively representing DIFFERENCE and BORROW OUT outputs.

Problem 23.17. *Design a 2-bit Magnitude Comparator. Also, write relevant Boolean expressions for the three possible outcomes of magnitude comparison.*

Solution. Let A (A_1 A_0) and B (B_1 B_0) be the two numbers. If X, Y and Z represent the conditions $A = B$, $A > B$ and $A < B$ respectively *i.e.* X = 1, Y = 0, Z = 0 for $A = B$, X = 0, Y = 1, Z = 0 for $A > B$ and X = 0, Y = 0, Z = 1 for $A < B$, then

$$X = x_1 \cdot x_0$$

where, $$x_1 = A_1 \cdot B_1 + \overline{A_1} \cdot \overline{B_1},$$

$$x_0 = A_0 \cdot B_0 + \overline{A_0} \cdot \overline{B_0}$$

$$Y = A_1 \cdot \overline{B_1} + x_1 \cdot A_0 \cdot \overline{B_0}$$

and $$Z = \overline{A_1} \cdot B_1 + x_1 \cdot \overline{A_0} \cdot B_0$$

Fig. 23.135 shows the logic diagram of the 2-bit comparator.

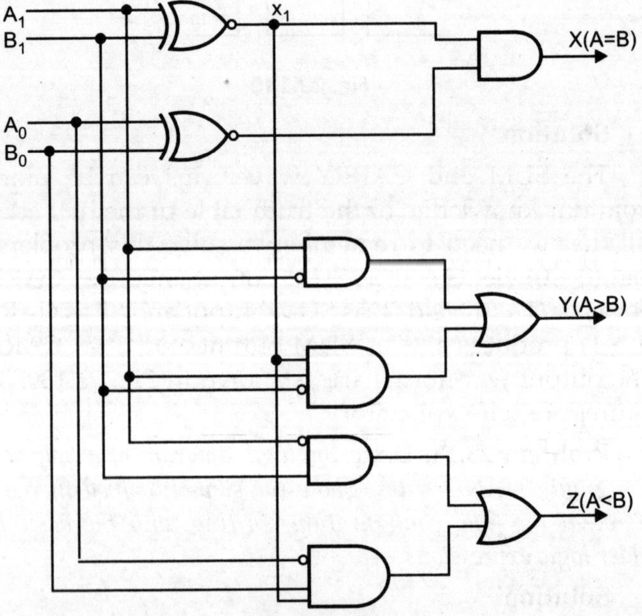

Fig. 23.135

Problem 23.18. *Determine the number of HA and FA blocks required to construct (a) an 8-bit binary parallel adder*

(b) a 16-bit binary parallel adder (c) an 8-bit Adder/Subtractor in addition to logic gates required, if any.

The objective is to have minimum possible hardware.

Solution. (*a*) In case of an 8-bit adder,

Number of Half Adders = 1

Number of Full Adders = 7

Additional logic gates = Nil

(*b*) In case of a 16-bit adder,

Number of Half Adders = 1

Number of Full Adders = 15

Additional logic gates = Nil

(*c*) In case of 8-bit Adder/Subtractor,

Number of Half Adders = Nil

Number of Full Adders = 8

Additional logic gates = Eight 2-input EX-OR gates

We would recall that in case of an Adder/Subtractor, even the LSB adder block needs to be a full adder. A half adder cannot be used here.

Problem 23.19. Implement the following Boolean function with a multiplexer:

$F(A, B, C, D) = S(0,1, 3, 5, 7, 9, 15)$.

Solution. This is a four variable Boolean function and can be implemented with a multiplexer having three selection lines. The procedure begins with the truth table shown below.

Truth table

D	*C*	*B*	*A*	*F*
0	0	0	0	1
0	0	0	1	1
0	0	1	0	0
0	0	1	1	1
0	1	0	0	0
0	1	0	1	1
0	1	1	0	0
0	1	1	1	1
1	0	0	0	0
1	0	0	1	1
1	0	1	0	0
1	0	1	1	0
1	1	0	0	0
1	1	0	1	0
1	1	1	0	0
1	1	1	1	1

The next step is to construct the implementation table. We choose variables *A*, *B*, *C* for the selection lines. The implementation table is shown below:

	I_0	I_1	I_2	I_3	I_4	I_5	I_6	I_7
$\overline{D}$	(0)	(1)	2	(3)	4	(5)	6	(7)
D	8	(9)	10	11	12	13	14	(15)
	$\overline{D}$	1	0	$\overline{D}$	1	$\overline{D}$	0	1

Fig. 23. 136 shows the hardware implementation.

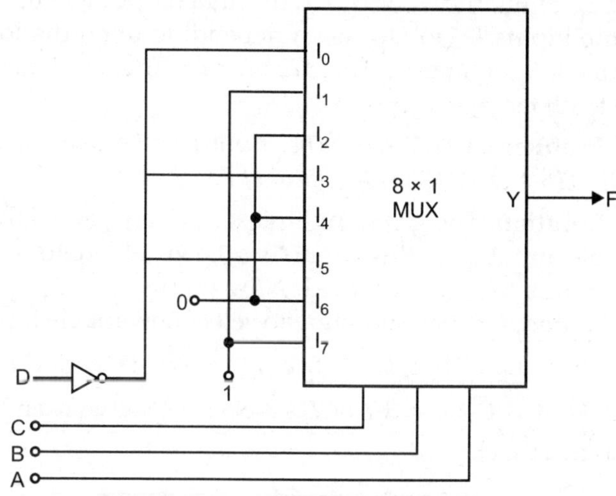

Fig. 23.136

Problem 23.20. *Design a (16 × 1) MUX using two (8 × 1) multiplexers having an Active LOW enable input.*

Solution. A (16 × 1) MUX can be constructed from two (8 × 1) MUX having an ENABLE input. The ENABLE

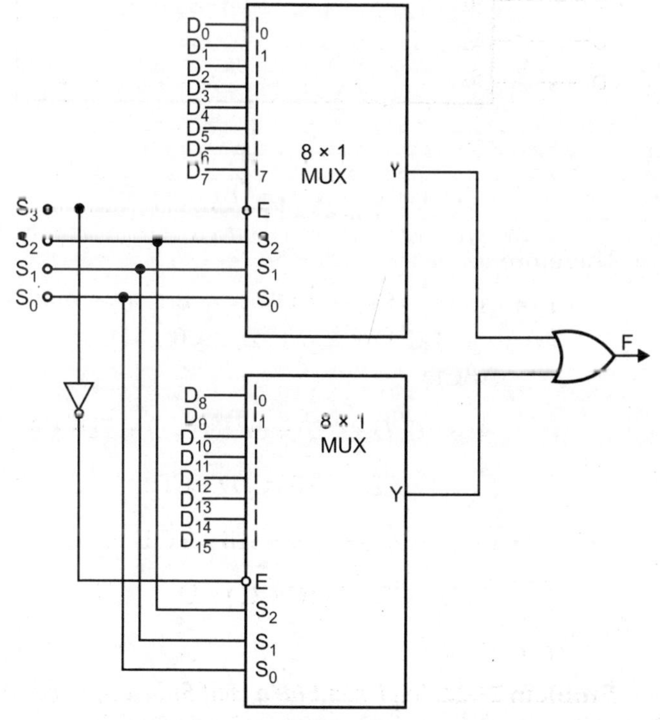

Fig. 23.137

input is taken as the fourth selection variable occupying the MSB position. Fig.23.137. shows the complete logic circuit diagram. IC 74151 can be used to implement (8 × 1) MUX.

When (S_3) is logic '0', upper MUX is enabled and the lower one is disabled. If we recall the truth table of a 4-variable Boolean function, (S_3) would be '0' for the first eight entries and '1' for the remaining eight entries. Therefore, when $S_3 = 0$, the final output will be any of the inputs D_0 to D_7 depending upon the logic status of S_2, S_1 and S_0. Similarly, when $S_3 = 1$, the final output will be any of the inputs D_8 to D_{15} again depending upon the logic status of S_2, S_1 and S_0. The circuit, therefore, implements the truth table of a (16 × 1) MUX.

Problem 23.21. *Write the Boolean expression implemented by the (16 × 1) MUX arrangement of Fig. 23.138.*

Solution. The given multiplexer has an active HIGH Enable input and this input has been added to + 5V. Therefore, the IC is enabled. Now inputs I_1 I_3, I_5, I_7, I_9, I_{11}, I_{13} and, I_{15} corresponding to selection variable inputs of $\overline{A}\,\overline{B}\,\overline{C}\,D$, $\overline{A}\,\overline{B}\,C\,D$, $\overline{A}\,B\,\overline{C}\,D$, $\overline{A}\,B\,C\,D$, $A\,\overline{B}\,C\,D$, $A\,B\,\overline{C}\,D$, $A\,\overline{B}\,\overline{C}\,D$, and $ABCD$ respectively have been tied to logic HIGH.

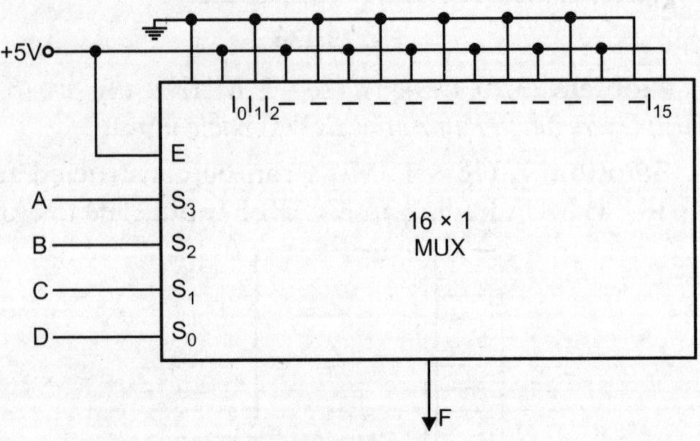

Fig. 23.138

Therefore,

$$F = \overline{A}\,\overline{B}\,\overline{C}\,D + \overline{A}\,\overline{B}\,CD + \overline{A}\,B\,\overline{C}\,D + \overline{A}\,BCD + A\,\overline{B}\,\overline{C}\,D + A\,\overline{B}\,C\,D + A\,B\,\overline{C}\,D$$
$$ABCD$$
$$= \overline{A}\,\overline{B}\,(\overline{C}\,D + CD) + \overline{A}\,B\,(\overline{C}\,D + CD) + A\,\overline{B}$$
$$(\overline{C}\,D + CD) + AB\,(\overline{C}\,D + CD)$$
$$= (\overline{A}\,\overline{B} + \overline{A}\,B + A\,\overline{B} + AB)\,.\,(\overline{C}D + CD) = \overline{C}$$
$$\overline{C}\,.\,D + CD = D\,(C + \overline{C}) = D$$

$$F = D.$$

Problem 23.22. *Implement (a) a Half Subtracter and (b) a Full Adder circuit separately using a (3 × 8) Decoder.*

Solution. A decoder with an OR-gate at the output can be used to implement a given Boolean function. The decoder should at least have as many input lines as the number variables in the Boolean function to be implemented.

(*a*) The truth table of the half subtractor is given below :

A	B	D	B_0
0	0	0	0
0	1	1	1
1	0	1	0
1	1	0	0

Difference output, $D = \overline{A}\,B + A\overline{B} = \Sigma\, 1,2$

Borrow output, $B_0 = \overline{A}\,B$

Though, these expressions (for D and B) can be implemented with a (2 × 4) decoder implementation with a (3 × 8) decoder requires that MSB input line is permanently disabled.

Fig. 23.139 (*a*) shows the diagram.

(*b*) The truth table of a Full adder is shown below:

A	B	C	S	C_0
0	0	0	0	0
0	0	1	1	0
0	1	0	1	0
0	1	1	0	1
1	0	0	1	0
1	0	1	0	1
1	1	0	0	1
1	1	1	1	1

Output, $S = \Sigma\, 1, 2, 4, 7$

Output $C_0 = \Sigma\, 3, 5, 6, 7$

Fig. 23.139 (*b*) shows the hardware implementation.

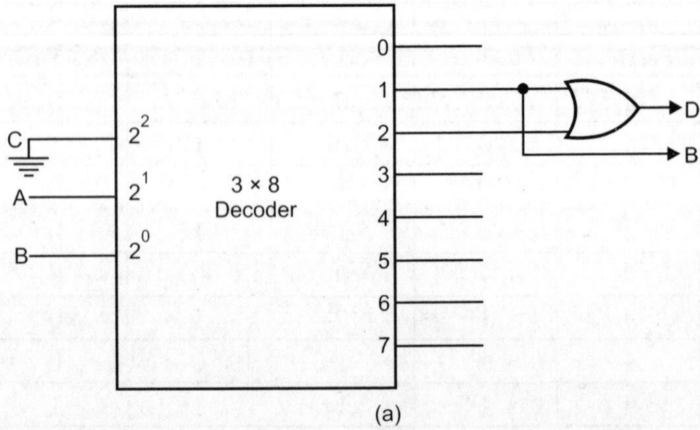

(a)

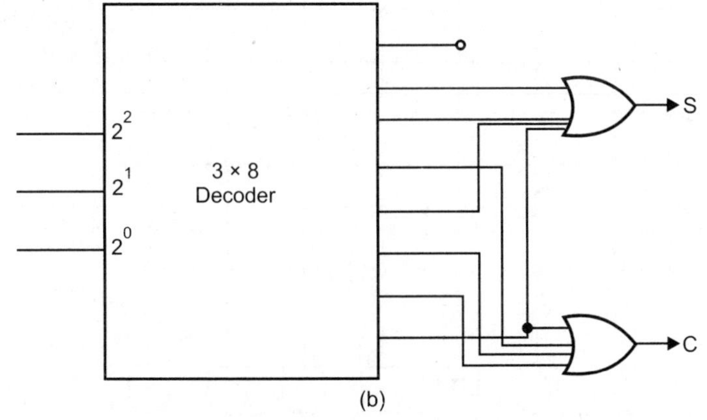

Fig. 23.139

Problem 23.23. *Design a 4-line to 2-line priority encoder with active HIGH inputs and outputs.*

Solution. The truth table for such a priority encoder is given below with D_0, D_1, D_2, D_3 as inputs and X, Y as outputs:

INPUTS				OUTPUTS	
D_0	D_1	D_2	D_3	X	Y
1	0	0	0	0	0
X	1	0	0	0	1
X	X	1	0	1	0
X	X	X	1	1	1

The Boolean expressions for outputs X Y can be written as:

$$X = D_2 \overline{D_3} + D_3 = D_2 + D_3$$

$$Y = D_1 \overline{D_2}\ \overline{D_3} + D_3 = D_1 \overline{D_2} + D_3$$

Fig. 23.140 shows the hardware implementation.

Problem 23.24. *Show the structures or logic arrangement of both a ROM as well as a PLA required to implement a binary full adder.*

Solution. The truth table of a full adder is given below:

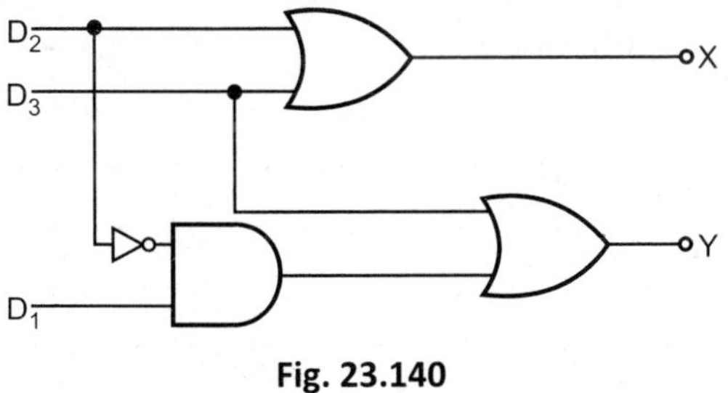

Fig. 23.140

A	B	C_{in}	S	C_0
0	0	0	0	0
0	0	1	1	0
0	1	0	1	0
0	1	1	0	1
1	0	0	1	0
1	0	1	0	1
1	1	0	0	1
1	1	1	1	1

The Boolean expressions for Sum (S) and Carry Out (C_0) can be written as :

$$S = \Sigma\, 1, 2, 4, 7$$

and $$C_0 = \Sigma\, 3, 5, 6, 7$$

Fig. 23.141. shows the implementation with a (8 × 2) ROM.

If we make an attempt to simplify the Boolean expressions for sum and carry outputs, we shall find that the expression for sum output cannot be simplified any further and also that the expression for carry out can be simplified to three product terms with fewer literals. If we examine even the existing expressions, we find that we would need seven AND gates in the PLA implementation. And if we use simplified expression, even then we would require the same number of AND gates. So, the simplification here would not help as far as its implementation with a PLA is concerned. Fig. 23.142 shows the structure.

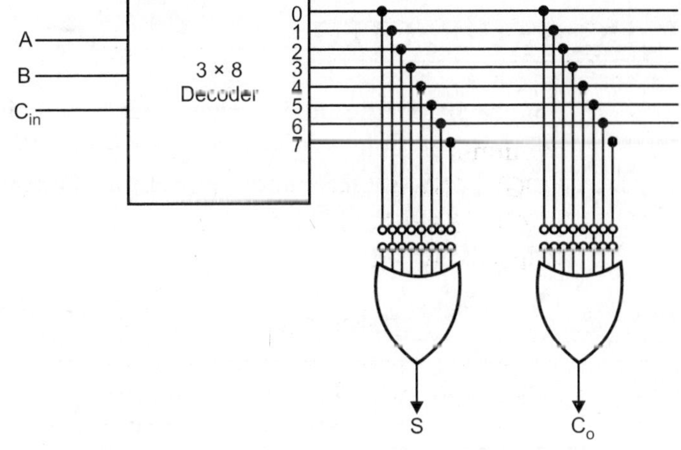

Fig. 23.141

Problem 23.25. *Determine the size of ROM required to implement:*

(*a*) *A binary multiplier that multiplies two 4-bit numbers* (*b*) *Dual 8-to-l multiplexer with common selection inputs*

(*c*) *A BCD Adder / Subtractor with a control input for selection of operation.*

Solution. (*a*) The number of inputs required here would be 8. The result of multiplication would be in 8 bits.

Therefore, size of ROM = $2^8 \times 8 = 256 \times 8$

(*b*) The number of inputs = 8 + 8 + 3 = 19.....(No. of selection inputs = 3)

Number of outputs = 2

Therefore, size of ROM = $2^{19} \times 2 = 512\text{ K} \times 2$

(*c*) Number of inputs = 4 (Augend bits) + 4 (Addend bits) +1 (Carry In) +1 (Control Input) = 10

Number of outputs = 4 (Sum or subtraction output bits) + 1 (Carry or Borrow bit) = 5

Size of ROM = $2^{10} \times 5 = 1024 \times 5$.

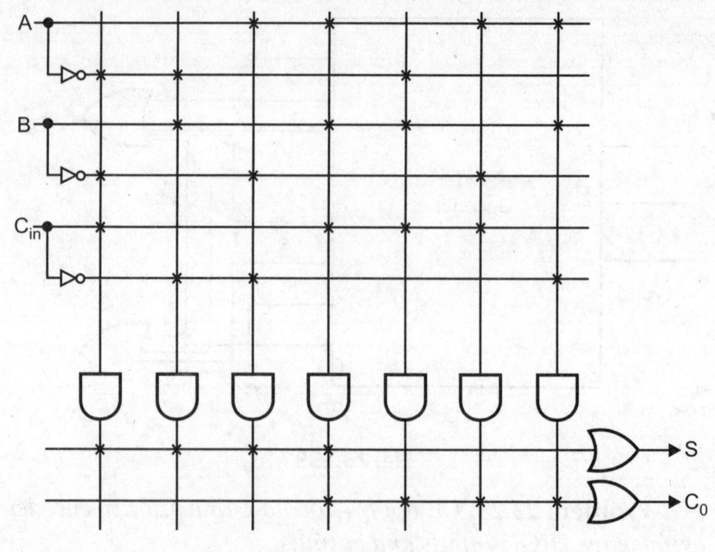

Fig. 23.142

SUMMARY OF IMPORTANT FORMULAS

1. In the case of a 2^n – to – 1 multiplexer, if (n) is the number of data inputs and (m) are teh number of control inputs, then

$$2^m = n$$

2. In the case of a de-multiplexer, if (n) is the number of data outputs and (m) is the number of control inputs, then

$$2^m = n$$

3. In the case of a decoder, if (n) and (m) are the number of inputs and decoded output lines respectively; then

$$m \leq 2^n$$

SELF EVALUATION EXERCISE '

I. DO AS DIRECTED

1. For each of the logic gates mentioned in column-2 of Table. 23.6, locate the corresponding LOGIC SYMBOLS and the Input-Output LOGIC EXPRESSION from columns 3 and 4 respectively.

Table 23.6

S.No.	Logic Gate	Logic Symbol	Input-output expression
1.	AND-Gate		$z = x + y$
2.	QR-Gate		$z = x.y$
3.	NAND-Gate		$z = \overline{(x + y)}$
4.	NOR-Gate		$z = \overline{(x . y)}$
5.	EX-OR-Gate		$z = \overline{x} . y + x. \overline{y}$

2. For the following logic expressions, state whether they are TRUE or FALSE:

(*a*) 0.0 = 0
(*b*) 0 + 1 = 0
(*c*) 1.0 = 0
(*d*) 1.1.1 = 1
(*e*) 1 + 1 + 1 + 1 = 1
(*f*) 0 + 0+ 0 + 1 = 1
(*g*) 1.1.1.0 = 1
(*h*) 1 + 1 = 1
(*i*) $0 \oplus 1 = 1$
(*j*) $x + 1 = x$
(*k*) $\overline{x \oplus x} = 1$
(*l*) $x . \overline{x} = 1$
(*m*) $x + \overline{x} = 0$
(*n*) $x . x = x$
(o) $x \oplus x = 0$

3. We have a 2-input logic gate whose output is logic '0' for all possible input combinations except when both the inputs are logic '0'. Can you identify the gate ?
4. The truth table of 2-input AND-gate (Fig. 23.143.) has some missing entries. Complete the truth table by filling in the missing places.
5. Identify the logic gates from the truth tables given in Figs. 23.144. (*a*) and 23.144 (*b*).
6. The basic logic gates *i.e.* AND, OR, NOR, NAND, NOT can be implemented in terms of a combination of logic gates other than the one being implemented. Some such logic implementations are shown in Fig. 23.145. You are required to tell us the logic gate represented by each one of these logic circuits.
7. For the logic circuit shown in Fig. 23.146, write the input-output Boolean expression.
8. Carefully examine the logic circuit shown in Fig. 23.147. It has two inputs and one output and behaves exactly like one of the popular logic gates. Can you name this logic gate?

A	B	C	Z
0	0	0	–(i)
0	0	1	0
0	1	0	0
0	1	1	–(ii)
1	0	0	0
1	0	1	–(iii)
1	1	–(iv)	0
1	1	1	–(v)

Fig. 23.143

Z	Y	Z
0	0	1
0	1	1
1	0	1
1	1	0

(a)

X	Y	Z
0	0	0
0	1	1
1	0	1
1	1	0

(b)

Fig. 23.144

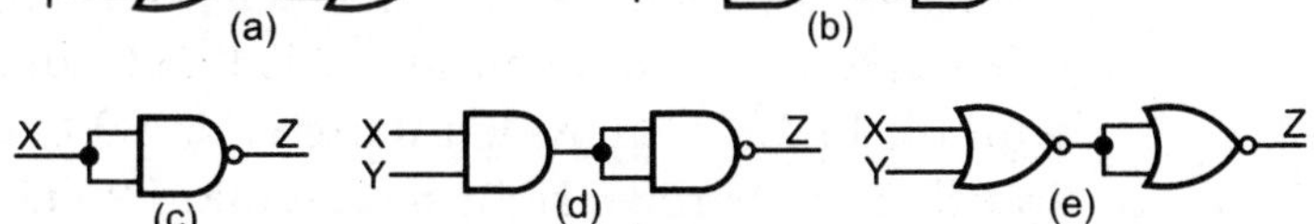

Fig. 23.145

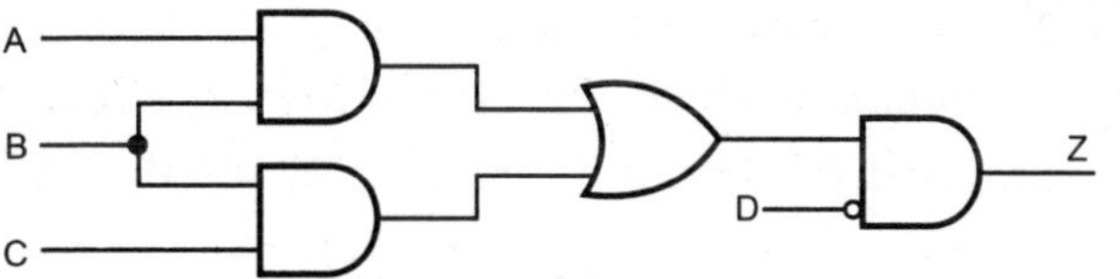

Fig. 23.146

9. **Fill in the blanks:**
   (*a*) The NOT gate can have—input variable(s). (one, many)
   (*b*) The IC7400 is from-----family. (TTL, CMOS)
   (*c*) IC4011 is a quad 2-input NAND gate belonging to...family of ICs. (TTL, CMOS)
   (*d*) There are three basic logic gates. Two of them are the AND and the OR. The third is -----(NAND, NOR, NOT)
   (*e*) NAND and —gates are often referred to as the UNIVERSAL gates. (NOR, AND, NOT)

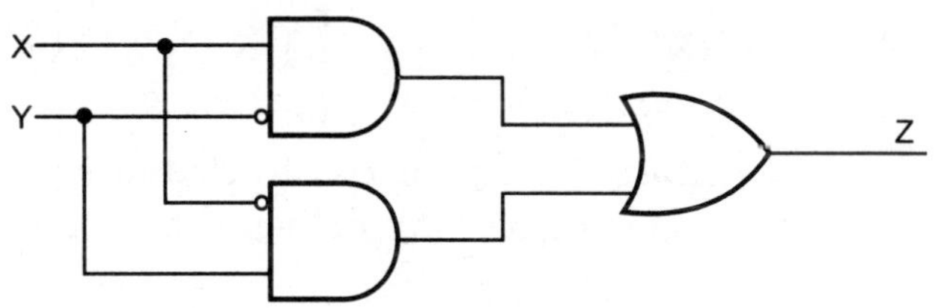

Fig. 23.147

   (*f*) "One can not think of an X-OR or an X-NOR logic function with more than two inputs." This statement is—(TRUE, FALSE)
   (*g*) The FAN-OUT of standard TTL ICs is----- (10, 50)
   (*h*) In an OR-gate, irrespective of the number of inputs, only one of the output entries in the truth table is logic '0' and all other entries are logic '1'. May be for a different input combination, there is one more gate having only one of the output entries in the truth table as logic '0'. It is------(NAND-gate, AND-gate, NOR-gate)

II. If You Can Pick the Most Correct Answer

1. "The output is '1' for like inputs and '0' for unlike inputs." This statement is representative of which logic gate ?
   (*a*) AND (*b*) OR
   (*c*) NAND (*d*) EX-NOR

2. "If we short all the inputs of a NAND-gate to get a one-input one-output logic circuit, it is the basic NOT-circuit. "Is this statement true ?
   (*a*) YES (*b*) NO
   (*c*) NOT ALWAYS
3. "Shorting all the inputs of a NOR-gate makes it a NOT-circuit. Is this statement true ?
   (*a*) YES (*b*) NO
   (*c*) NOT ALWAYS
4. In general, the logic gate whose all output entries are logic '0' except for one entry that" logic '1' is
   (*a*) AND, NOR (*b*) OR, NAND
   (*c*) NAND, NOR (*d*) NOR
5. In general, logic gates whose all the output entries are logic '1' except for one entry that logic '0' are
   (*a*) AND, OR (*b*) NAND, OR
   (*c*) NOR, NAND (*d*) EX-OR
6. A half-adder can be constructed using two 2-input logic gates. One of them is an AND-gate the other is
   (*a*) OR (*b*) NAND
   (*c*) NOR (*d*) EX-OR
7. Each of the following Boolean expressions can be implemented using a single logic gate except one of them. Which is that ?
   (*a*) $z = \overline{(x+y)}$
   (*b*) $z = \overline{(x.y)}$
   (*c*) $z = x.\overline{y} + \overline{x}.y$
   (*d*) $z = x.y + \overline{x}\ y + x.\overline{y}$
8. A logic gate with four inputs can have
   (*a*) 16 possible input combinations.
   (*b*) 4 possible input combinations,
   (*c*) 8 possible input combinations.
   (*d*) 2 possible input combinations.
9. A data selector is also called a
   (*a*) De-multiplexer
   (*b*) Priority encoder
   (*c*) Multiplexer (*d*) Decoder
10. An arithmetic digital building block that has one data input, more than one data outputs and more than one control inputs is the
   (*a*) Data Selector (*b*) Multiplexer
   (*c*) Decoder (*d*) Demultiplexer
11. A decoder is nothing but a demultiplexer without
   (*a*) control inputs (*b*) data input
   (*c*) enable input (*d*) none of these
12. In a l-to-16 de-multiplexer, the number of control inputs will be
   (*a*) 4 (*b*) 1
   (*c*) 2 (*d*) 16
13. In a l-to-8 demultiplexer, the control inputs are designated as ABCD with A as the LSB and D as the MSB, the outputs are designated as 0 to 15. If the data input is steered to the output designated 7 when clocked, then the status of control inputs in the order DCB A just prior to clocking is
   (*a*) 1110 (*b*) 0111
   (*c*) 1000 (*d*) none of these
14. A multiplexer has (*n*) data inputs, (*m*) control inputs and one output, then
   (*a*) $2^m = n$ (*b*) $2^n = m$
   (*c*) $m^n = 2$ (*d*) $n^2 = m$
15. A de-multiplexer has one data input, (*m*) control inputs and (*n*) outputs, then
   (*a*) $2^n = \text{m}$ (*b*) 2^m
   (*c*) $n^m = 2$ (*d*) $m^n = 2$
16. A BCD-to-Decimal decoder has
   (*a*) 10 inputs and 10 outputs
   (*b*) 4 inputs and 10 outputs
   (*c*) 10 inputs and 4 outputs
   (*d*) none of these
17. If the arithmetic and logic operations are to be done on larger bit numbers and they are to be done with high speed, a digital building block that is invariably used is
   (*a*) the look ahead carry generator
   (*b*) a RAM
   (*c*) a ROM
   (*d*) an ALU (Arithmetic Logic Unit)
18. There is 10-line decimal designated 0, 1, 2, 3, 4, 5, 7, 8, 9 to 4-line BCD priority encoder with a LOW deciding the priority. The encoder outputs are active LOW and are designated as D (MSB) C B A (LSB). If the input bit stream is 1010111011 with the bit on extreme left representing 0-line and that on extreme right representing 9-line, then the encoded output is

(*a*) 0111 (*b*) 1000
(*c*) 1001 (*d*) 0000

19. A half adder circuit has two inputs and
(*a*) one output (*b*) two outputs
(c) three outputs (*d*) none of these

20. A full adder circuit has
(*a*) two inputs and one output
(*b*) two inputs and two outputs
(*c*) two inputs and three outputs
(*d*) three inputs and two outputs

21. A half adder is constructed from two-input logic gates. One of them is a 2-input-EX-OR gate. The other is a
(a) 2-input AND-gate
(*b*) 2-input NAND-gate
(c) 2-input OR-gate
(*d*) 2-input NOR-gate

22. A full adder is constructed from five logic gates. There are two 2-input EX-OR gates, two 2-input AND gates and a
(*a*) 2-input OR-gate
(*b*) 2-input EX-NOR gate
(*c*) 2-input AND-gate
(*d*) 2-input NAND-gate

23. A 4-bit binary adder could be constructed with a minimum of
(*a*) 3 1-bit full adders and a half adder
(*b*) 4 1-bit full adders
(*c*) 4 half adders
(*d*) none of these

24. The half subtractor has
(*a*) 2 inputs and 2 outputs
(*b*) 3 inputs and 2 outputs
(*c*) 2 inputs and 1 output
(*d*) none of these

25. The half subtractor can be constructed from a 2-input EX-OR, a 2-input AND and
(*a*) a 2-input OR
(*b*) a 2-input EX-NOR
(*c*) an inverter
(*d*) none of these

26. A full subtractor can be constructed from two half subtractors and a
(*a*) 2-input NAND (*b*) 2-input NOR
(*c*) 2-input OR (*d*) 2-input AND

27. Given that 7483 is a 4-bit parallel adder chip, how do you build a 16-bit parallel adder circuit
(*a*) By cascaded arrangement of 16 7483's
(*b*) By a cascaded arrangement of 4 7483's
(*c*) 16-bit parallel adder cannot be built using 7483's
(*d*) none of these

28. It is said that full adder can be used as a subtractor. Let us say, you want to build a 4-bit parallel subtractor. All that you need is one 7483 (which is a 4-bit full adder) and
(*a*) four inverters
(*b*) four 2-input AND gates
(*c*) sixteen 2-input NAND-gates
(*d*) eight 2-input NAND gates

29. The programmable logic device (PLD) having a programmable AND-array at the input and a programmable OR-array at the output is called a
(*a*) Programmable Logic Array (PLA)
(*b*) Programmable Array Logic (PAL)
(*c*) Programmable Gate Array (PGA)
(*d*) Application Specific Integrated Circuit (ASIC)

30. A Programmable Array Logic (PAL) device has a
(*a*) programmable AND-array at the input and a fixed OR-array at the output
(*b*) fixed AND-array at the input and a programmable OR-array at the output
(*c*) fixed AND-array at the input and a fixed OR-array at the output
(*d*) none of these

31. A masked programmed version of a PAL device that is pin compatible to its PAL counterpart is known as
(a) a PLA device (*b*) HAL device
(*c*) a PLS device (*d*) PGA device

32. It is desired to implement a binary multiplier capable of multiplying two 3-bit binary numbers. Size of ROM would be
(*a*) 64 × 4 (*b*) 1024 × 6
(*c*) 64 × 8 (*d*) 128 × 6

33. A ROM is usually not preferred to implement those Boolean functions which
(*a*) are very complex
(*b*) can otherwise be implemented using a PLA
(*c*) have a large number of 'Don't Care Conditions
(*d*) have large number of outputs

Answer to Self Evaluation Exercise

I. 1. (1-3-2) (2-5-1) (3-1-4) (4-2-3) (5-4-5)

2. (*a*) TRUE (*b*) FALSE (*c*) TRUE
(*d*) TRUE (*e*) TRUE (*f*) TRUE
(*g*) FALSE (*h*) TRUE (*i*) TRUE
(j) FALSE (*k*) TRUE (*l*) FALSE
(*m*) FALSE (*n*) TRUE (*o*) TRUE

3. NOR-gate

4. (*i*) 0 (*ii*) 0 (*iii*) 0 (*iv*) 0 (*v*) 1

5. Fig. 23.144 (*a*) EX-OR gate Fig. 23.144 (*b*) NAND gate

6. Fig. 23.145 (*a*) NOR Fig. 23.145 (*b*) AND Fig. 23.145 (*c*) NOT Fig. 23.145 (*d*) NAND Fig. 213.145 (*e*) OR

7. $z = (A.B + B.C).\ \overline{D}$

8. EX-OR gate

9. (*a*) one (*b*) TTL (*c*) CMOS (*d*) NOT (*e*) NOR (*f*) FALSE (*g*) 10 (*h*) NAND

II.

1. (*d*)	2. (*a*)	3. (*a*)	4. (*a*)
5. (*b*)	6. (*d*)	7. (*d*)	8. (*a*)
9. (*c*)	10. (*d*)	11. (*b*)	12. (*a*)
13. (*b*)	14. (*a*)	15. (*b*)	16. (*b*)
17. (*a*)	18. (*a*)	19. (*b*)	20. (*d*)
21. (*a*)	22. (*a*)	23. (*a*)	24. (*a*)
25. (*c*)	26. (*c*)	27. (*b*)	28. (*a*)
29. (*a*)	30. (*a*)	31. (*b*)	32. (*a*)
33. (*c*)			

Chapter 24

Sequential Logic Circuits

Flip-Flops, Counters and Registers

Learning Objectives

After completing the chapter, you will learn the following topics:

1. Sequential Logic Circuits versus Combinational Logic Circuits
2. Flip-Flops
3. R-S Flip-Flop
4. J-K Flip-Flop
5. Race Problem and Use of Master-Slave Configuration
6. D-type Flip-Flop
7. T-type Flip-Flop
8. Level Triggered and Edge Triggered Flip-Flops
9. Important Parameters of Flip-Flops
10. Ripple Counter versus Synchronous Counter
11. Functional Description of Ripple and Synchronous Counters
12. Decade and BCD Counters
13. UP, DOWN and UP/DOWN Counters
14. Modulo Counters
15. Presettable Counters
16. Designing Counters
17. Shift Registers
18. Types of Shift Registers Including Serial-In Serial-Out, Serial-In Parallel Out, Parallel-In Serial Out and Parallel-In Parallel Out Shift Registers
19. Universal Shift Register
20. Ring and Shift Counters
21. Decoding Counters

Q. 1: What are sequential logic circuits? How do these circuits differ from combinational logic circuits?

Ans: In the case of a sequential logic circuit, the output logic state depends upon not only the present input but also on the past input or/and past output. This is uunlike combinational logic circuits that change state depending upon the actual signals being applied to their inputs at that time. Sequential logic circuits have some form of inherent memory that allows the sequential logic circuits to take into account their previous input state as well as those actually present.

Q. 2: With the help of a simple block schematic representation, briefly explain the operation of a sequential logic circuit.

Ans: Figure-24.1 shows the generalized block schematic representation of a sequential logic circuit. The diagram is self-explanatory. Sequential logic circuits are generally termed as Bistable devices, whose output or outputs are initially set in one of the logic states, that is, logic '0' or logic '1'. These remain indefinitely in this current state until they are triggered to change the state. In a sequential logic circuit, it is the clock signal that controls the operation. Bistable latches and flip-flops are the basic building blocks of a sequential logic circuit. Flip-flops themselves are constructed from universal NAND and NOR gates configured in a certain combinational way.

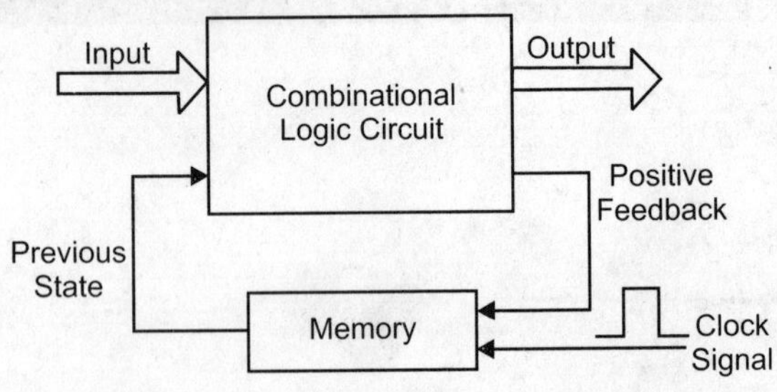

Fig. 24.1

Q. 3. What is a flip-flop? Show the logic implementation of a SET-RESET flip-flop also known as *R-S* flip-flop with the help of NAND gates and describe in brief, its operational principle.

Ans. A flip-flop is basically a bistable multivibrator having two states. The flip-flop can be made to change state from one logic level to another with an appropriate trigger command, Bistable multivibrators using discrete components have already been discussed in an earlier chapter. Fig. 24.2 (*b*) shows the logic symbol of an *R-S* flip-flop.

An *R-S* flip-flop implemented with NAND gates is shown in Fig. 24.2. (*a*). The output of each NAND has been fed back to one of the inputs of the other NAND. Operation of this flip-flop is explained by the truth table of Fig. 24.2. (*c*).

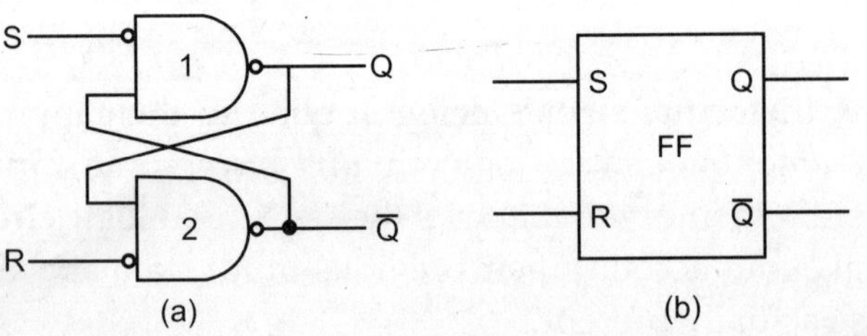

Operation Mode	S	R	Q_{n+1}
No Change	0	0	Qn
SET	1	0	1
RESET	0	1	0
Prohibited	1	1	–

(c)

Fig. 24.2

The subscript (n + 1) in this truth table indicates the state of the flip-flop after it has been triggered by an appropriate pulse at either R or S input. (n) indicates the state of the flip-flop before it is triggered. The truth table is self-explanatory. When $S = 1$ and $R = 1$, the output (taken from output of NAND gate-1) appears to be simultaneously at logic 0 as well as logic 1 and that is why, such an input is not permissible in this arrangement.

When $S = 1$ and $R = 0$, the output

$Q_{n+1} = 1$ and the flip-flop is said to be SET. When S = 0 and $R = 1$,

$Q_{n+1} = 0$ and the flip-flop is said to be RESET. When $R = S = 0$, the flip-flop remains in the state it was in prior to the application of the trigger pulse.

What we have described above is the R-S flip-flop with active HIGH inputs *i.e.* the inputs are active, when HIGH. The logic implementation using NAND gates, the circuit representation and the truth table for an *R-S* flip-flop with active LOW inputs are respectively shown in Fig. 24.3 (*a*), (*b*) and (*c*).

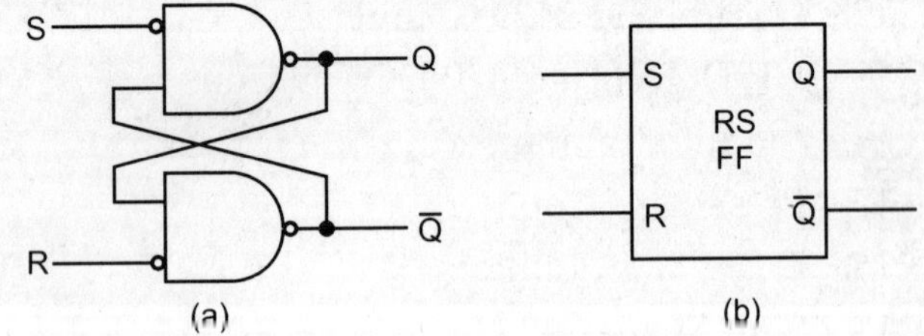

Operation Mode	S	R	Q_{n+1}
No Change	1	1	Q_n
SET	0	1	1
RESET	1	0	0
Prohibited	0	0	–

(c)

Fig. 24.3

Q. 4. Draw the NOR-gate counterparts of active HIGH input and active LOW input RS flip-flops of Figs. 24.2 (*a*) ana 24.3 (*a*) respectively.

Ans. Active HIGH input RS flip-flop built around NOR gates is shown in Fig. 24.4 (*a*). The same with active LOW inputs is shown in Fig. 24.4 (*b*).

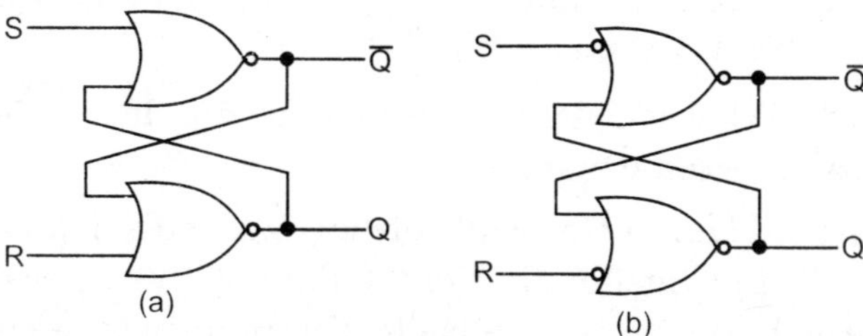

Fig. 24.4

Q. 5. What is the need for having a CLOCKED Flip-Flop? Draw the logic representation, the circuit symbol and the truth table of a clocked *R-S* flip-flop.

Ans. In majority of digital application circuits like counters, registers or a combination of these, it is necessary that operation of all flip-flops being used in the overall system be synchronised by a master clock. The flip-flops then change their states as per the inputs only on the occurrence of the clock pulse. Such flip-flops are known as the clocked flip-flops.

Fig. 24.5 (*a*) shows the logic representation of a clocked R-S flip-flop. The truth table of the clocked R-S flip-flop as given in Fig. 24.4 (*c*) is self-explanatory. When the clock is LOW, the AND gates are disabled and as a result the output cannot respond to the input. When the clock goes HIGH, the AND gates are enabled and the *R* and *S* inputs are passed on to the inputs of the NAND gates that constitute the flip-flop. The output changes state as per the logic states of the *R* and *S* inputs.

Q. 6. Differentiate between EDGE TRIGGERED flip-flops and the LEVEL TRIGGERED flip-flops. Draw the circuit symbols for the following types of flip-flops.

(*i*) Positive edge triggered clocked R-S flip-flop with active LOW inputs.

(*ii*) Level triggered R-S flip-flop with active HIGH inputs.

(*iii*) Negative edge triggered clocked R-S flip-flop with active HIGH inputs.

(*iv*) Level triggered clocked R-S flip-flop with active LOW inputs.

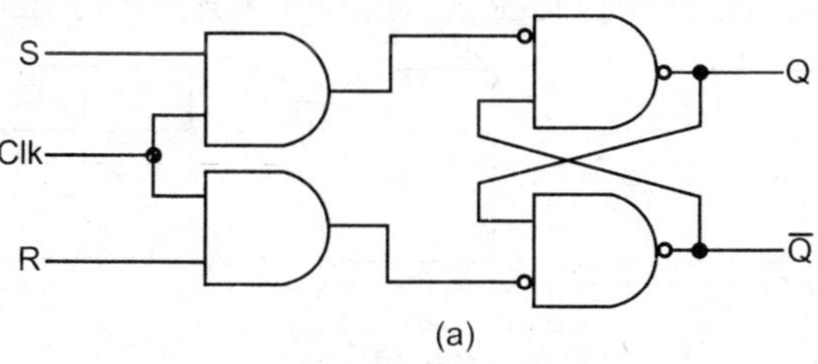

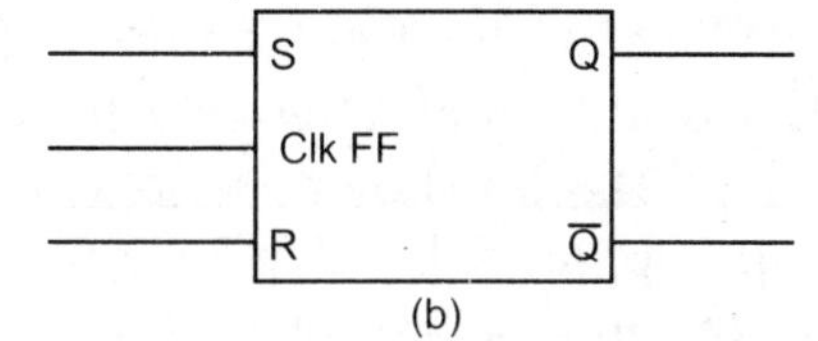

S	R	*Clk*	Q_{n+1}
0	0	0	Q_n
0	0	1	Q_n
0	1	0	Q_n
0	1	1	0
1	0	0	Q_n
1	1	1	1
1	0	0	Q_n
1	1	1	*Not permitted*

(c)

Fig. 24.5

Ans. In an EDGE TRIGGERED flip-flop, the output responds to the data at the inputs only on the LOW to HIGH (or HIGH to LOW) transition of the clock pulse. Any changes in the input during the time the clock pulse is HIGH (or LOW) do not have any effect on the output. In case of all flip-flops, it will be known whether the flip-flop responds to the positive going transitions or the negative going transitions.

In a LEVEL TRIGGERED flip-flop, the output responds to the data present at the inputs during the time, when the clock pulse level is HIGH (or LOW). That is, any changes at the input during the time, the clock is HIGH (or LOW) are reflected at the output as per the truth table.

For circuit symbols of different flip-flops mentioned from (*i*) to (*iv*) above, refer to Fig. 24.6 (*a*) to (*d*).

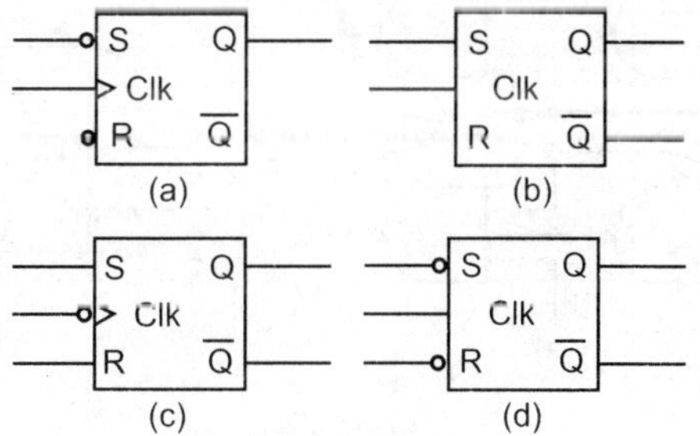

Fig. 24.6

Q. 7. What is a clocked *J-K* flip-flop ? What improvement does it have over the clocked R-S flip-flops? What are the PRESET and CLEAR inputs normally available in the clocked *J-K* flip-flops in the IC form?

Ans. Fig. 24.7 shows how we can use a clocked R-S flip-flop with two additional AND gates to get a clocked J-K flip-flop. Fig. 24.8 (*a*) shows the logic symbol of a clocked J-K flip-flop. A J-K flip-flop behaves in the same fashion as an R-S flip-flop except for one of the entries in the truth table. In case of an R-S flip-flop, the input combination $S = 1$, $R = 1$ (in case of a flip-flop with active HIGH inputs) and the input combination $S = 0$, $R = 0$ (in case of a flip-flop with active

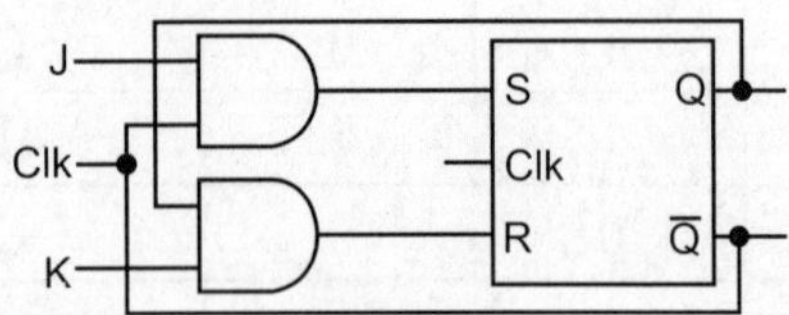

Fig. 24.7

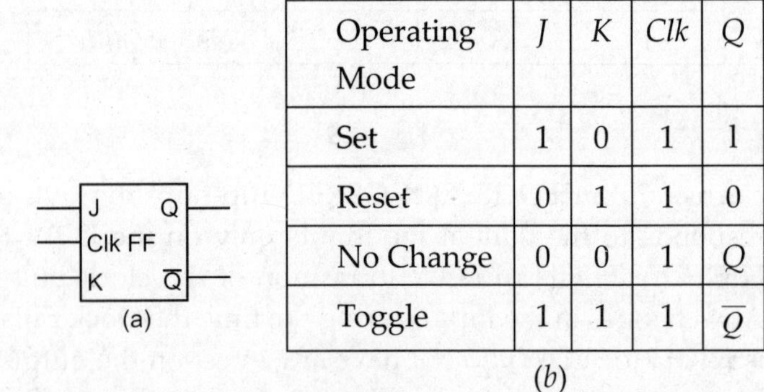

Operating Mode	*J*	*K*	*Clk*	*Q*
Set	1	0	1	1
Reset	0	1	1	0
No Change	0	0	1	Q
Toggle	1	1	1	$\overline{Q}$

(*b*)

Fig. 24.8

LOW inputs) are prohibited. In case of a J-K flip-flop with active HIGH inputs, the output of the flip-flop TOGGLES (*i.e.* it goes to the opposite logic state) for J = 1, K = 1. The output toggles for the input J = 0, K = 0 in case of the flip-flop having active LOW inputs. Thus, the problem of a prohibited input entry in an R-S flip-flop is not their in case of J-K flip-flop. Fig. 24.8 (*b*) shows the truth table of a clocked J-K flip-flop.

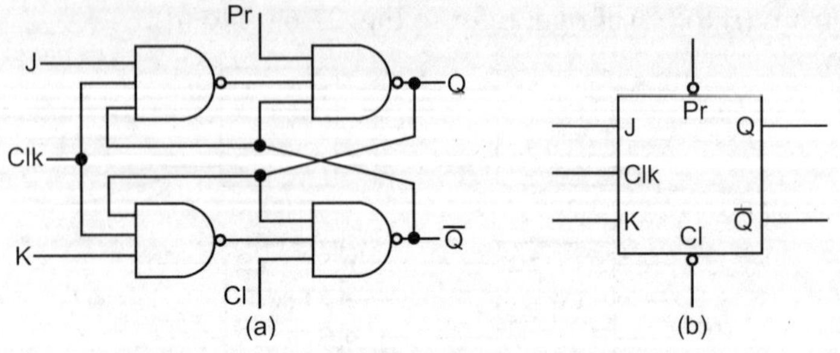

Fig. 24.9

It is often necessary to clear a flip-flop to logic '0' state ($Q_n = 0$) or preset it to a logic '1' state ($Q_n = 1$). An example of how this is realised is illustrated in Fig. 24.9. The circuit shown in self-explanatory. The flip-flop is cleared (*i.e.* $Q_n = 0$), whenever CLEAR input is 0 and PRESET input is 1. The flip-flop is preset to logic '1' state, whenever PRESET input is 0 and the CLEAR input is 1. In both the cases, the clock is assumed to be in logic '0' state. Quite clearly, here the CLEAR and PRESET inputs are active, when LOW. Fig. 24.9 (*b*) shows the logic symbol for a Clearable - Presettable clocked J-K flip-flop.

Q. 8. What is RACE PROBLEM in flip-flops ? How do you overcome that ?

Ans. Whenever the width of the trigger pulse is greater than the propagation time of the flip-flop, the change of state at the output is not reliable. This problem is referred to as RACE PROBLEM. As the propagation delays are normally very small, there is a reasonable likelihood of race condition existing in flip-flops. One technique to get over this problem in flip-flops is to use a MASTER-SLAVE flip-flop. A master-slave flip-flop constructed with two J-K flip-flops is shown in Fig. 24.10. The first flip-flop is called the master flip-flop and the second is called the slave. The clock to the slave flip-flop is the complement of the clock to the master flip-flop. When the clock pulse is present, the master flip-flop is enabled while the slave flip-flop is disabled. As a result, the master flip-flop can change state while the slave flip-flop cannot. When the clock goes LOW, the master flip-flop gets disabled while the slave flip-flop is enabled. Therefore, the slave J-K flip-flop changes state as per the logic states at its J and K inputs. The contents of the master flip-flop are, therefore, transferred to the slave flip-flop, the master flip-flop being disabled can acquire new inputs without affecting the output.

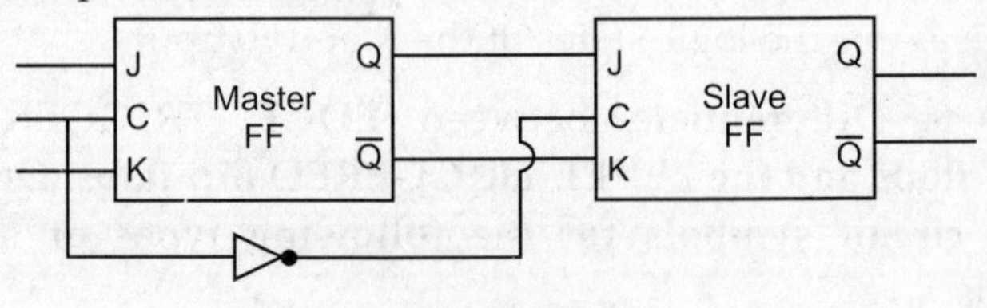

Fig. 24.10

Q. 9. What is a T-flip-flop ? Explain its operation with the help of its truth table.

Ans. The output of T-flip-flop called TOGGLE FLIP-FLOP changes state every time, it is triggered at its T-input called the TOGGLE input. That is, the output becomes '1' if it was '0' and '0' if it was '1' A positive edge triggered T-flip-flop (Fig. 24.11 *a*) responds to the positive going edges of the clock waveform, whereas a negative

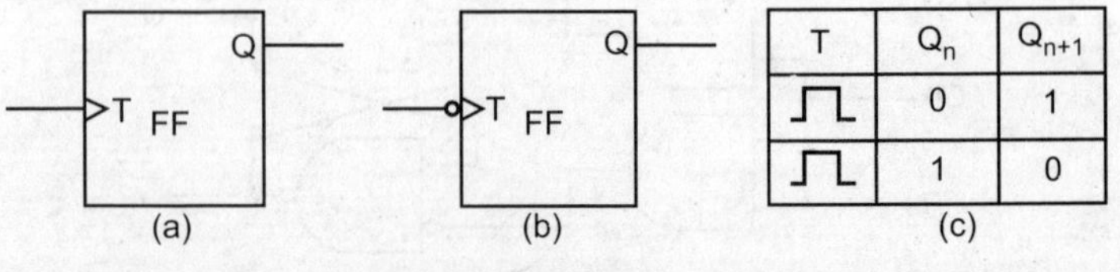

T	Q_n	Q_{n+1}
⎍	0	1
⎍	1	0

(c)

Fig. 24.11

triggered flip-flop (Fig. 24.11 *b*) responds to the negative going edges of the clock waveform. Fig. 24.11 (*c*) shows its truth table.

Q. 10. Briefly explain operation of a D-flip-flop with the help of its truth table.

Ans. A D-flip-flop also called DELAY Flip-flop can be used to provide temporary storage of one bit of information. Fig. 24.12 (*a*) shows its circuit symbol and Fig. 24.12 (*b*) the truth table. When the clock is active, the data bit present at the D-input is transferred to the *Q*-output. In the *D*-flip-flop whose circuit symbol is shown in Fig. 24.12 (*a*), this data transfer will take on the negative going edge of the clock input. D inputs can acquire new input data, when the clock is inactive. Thus, a D-flip-flop can provide a maximum delay equal to the clock period. Digital integrated circuits with two D-type flip-flops or four *D*-type flops integrated within the same IC package are commercially available. A Quad *D*-flip-flop, for instance, can provide a temporary storage of 4-bit information. The IC numbers for different types of flip-flops are given in Table 24.1.

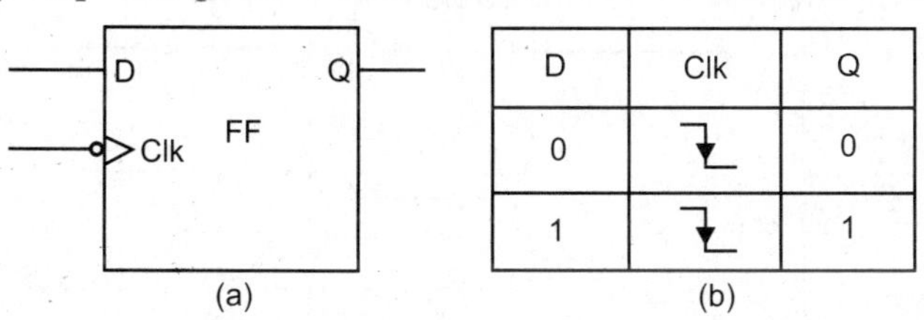

D	Clk	Q
0	↓	0
1	↓	1

Fig. 24.12

Q. 11. What is D-type Latch ? How does it differ from a D-type flip-flop ?

Ans. In a D-type latch, output *Q* follows the D-input as long as the clock input (also called the ENABLE input) is HIGH or LOW depending upon which clock level it responds to. When the clock goes to the inactive level, the output holds on to the logic state during the entire time period, the clock or ENABLE is inactive.

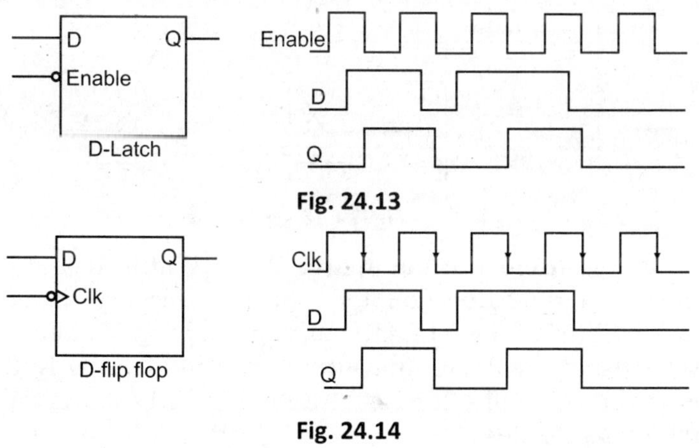

Fig. 24.13

Fig. 24.14

In a D-flip-fiop, the data on D-input is transferred to the *Q*-output on the positive on the positive or negative going transition of the clock waveform depending upon the flip-flop and this logic state is held at the output till we get the next effective clock transition. Figs. 24.13 and 24.14 depict the two cases.

Q. 12. Why is a J-K flip-flop called a UNIVERSAL Flip-flop ? Show the logic implementation of a T-flip-flop and a D-flip using J-K flip-flop.

Ans. A J-K flip-flop can be used to construct almost any other flip-flop. That is why it is called a UNIVERSAL flip-flop. Fig. 24.15 depicts the use of J-K flip-flops to construct different flip-flops.

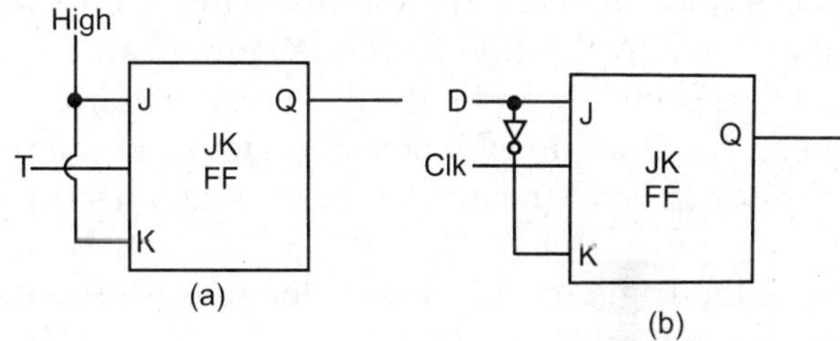

Fig. 24.15

Q. 13. What do you understand by Synchronous and Asynchronous inputs with reference to flip-flops ?

Ans. Synchronous inputs are those whose effect on the flip-flop output is synchronised with the clock input. *R*, *S*, *J*, *K*, *D* inputs are all synchronous inputs. Asynchronous inputs are those which operate independently of the synchronous inputs and the input clock signal. These are in fact override inputs as their status overrides the status of all synchronous inputs and also the clock input and place the flip-flop in one state or the other. PRESET and CLEAR inputs are examples of asynchronous inputs. When active, the PRESET and CLEAR inputs place the flip-flop's Q-output in '1' and '0' state respectively. Usually, these are active LOW inputs When it is desired that the flip-flop functions as per its synchronous inputs' status, the asynchronous inputs are kept in their inactive state. Also, both the asynchronous inputs, if available on a given flip-flop, are not made active simultaneously.

Q. 14. List important timing parameters of a flip-flop. Briefly describe each one of these.

Ans. You would find certain timing parameters listed in the specification sheet of a flip-flop. Some of these parameters, as we shall see in the paragraphs to follow, are specific to the logic family the flip-flop belongs to. There are some parameters which have different values for different flip-flops belonging to the same broad logic family. It is, therefore, important that one considers these timing parameters before using a certain flip-flop in a given application. Some of the important ones are:

1. Set-up and Hold times
2. Propagation delay
3. Clock pulse HIGH and LOW times
4. Asynchronous input active pulse width
5. Clock transition time
6. Maximum clock frequency

Set-up and Hold Times. The SET-UP TIME is the minimum time period for which the synchronous input (*R, S, J, K, D*) must be stable prior to the active clock transition for the flip-flop output to respond reliably at the clock transition. It is usually denoted by t_s (min). As an example, if in a JK flip-flop (*J*) and (*K*) inputs are to go to '1' and '0' respectively and if the flip-flop is negative edge triggered, the situation is depicted in Fig. 24.16 (*a*). Here, time periods (t_1) and (t_2) must at least equal t_s (min). The set-up time in case of 74ALS109A, which is a Dual J-K positive edge triggered flip-flop belonging to Advanced Low power Schottky TTL logic family, is 15ns. Also, the asynchronous inputs, such as PRESET and CLEAR, if there should be inactive prior to the clock transition for a certain minimum time period, if the outputs have to respond as per synchronous inputs. The set-up time may, therefore, be separately defined for synchronous and asynchronous inputs. In case of 74ALS109A, the asynchronous input set-up time is 10 ns. The asynchronous input set-up time for active low PRESET and CLEAR inputs is shown in Fig. 24.16 (*b*) assuming a positive edge triggered flip-flop.

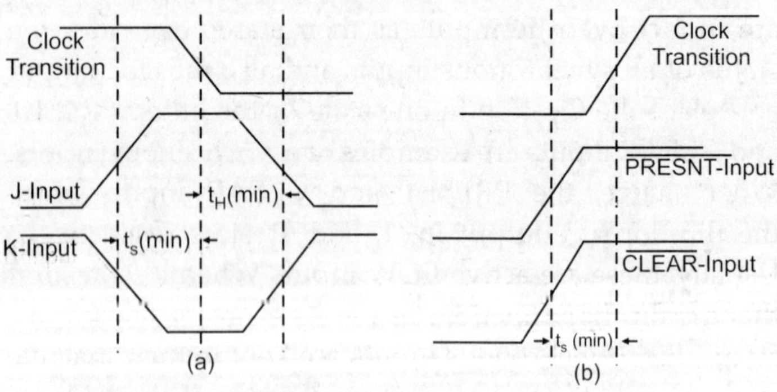

Fig. 24.16

The HOLD TIME t_H (min) is the minimum time period for which the synchronous inputs (*R, S, J, K, D*) must remain stable in the desired logic state after the active clock transition for the flip-flop to respond reliably. The same is depicted in Fig. 24.16 (*a*), if the desired logic status for (*J*) and (*K*) inputs is '1' and '0' respectively and the flip-flop is negative edge triggered. The Hold Time for flip-flop 74ALS109A is specified to be zero. To sum up, for a flip-flop to respond properly and reliably at the active clock transition, the synchronous inputs must be stable in their intended logic states and the asynchronous inputs in their inactive states for at least a time period equal to specified minimum set-up times prior to the clock transition and the synchronous inputs must be stable for a time period equal to at least the specified minimum hold time after the clock transition.

Propagation Delay. There is always a time delay, known as PROPAGATION DELAY from the time instant the signal is applied to the time the output makes the intended change. The flip-flop data sheet usually specifies propagation delays for both High-to-Low output transition (t_{pHL}) as well as for Low-to-High output transition (t_{pLH}). Propagation delays are measured between 50% points on input and output waveforms and are usually specified for all types of inputs including synchronous and asynchronous inputs. Propagation delays for High-to-Low and Low-to-High output transitions for a positive edge triggered flip-flops are shown in Fig. 24.17. For flip-flop 74ALS109A, the (t_{pHL}) and (t_{pLH}) for clock input to output are respectively 14 ns and 16 ns. The same *i.e.* (t_{pHL}) and (t_{pLH}) from asynchronous inputs to output for this flip-flop or 15 ns and 13 ns respectively.

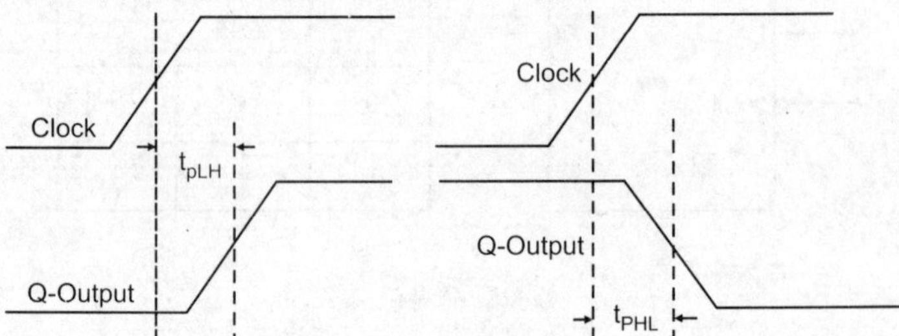

Fig. 24.17

Clock Pulse HIGH and LOW Times. Clock pulse HIGH time, t_w (*H*) and clock pulse LOW time, t_w (*L*) are respectively the minimum time durations for which the clock signal should remain HIGH and LOW. Failure to meet these requirements can lead to unreliable triggering. Fig. 24.18 depicts these timing parameters. t_w (*H*) and t_w (*L*) for 74ALS109A are 4 ns and 5.5 ns respectively.

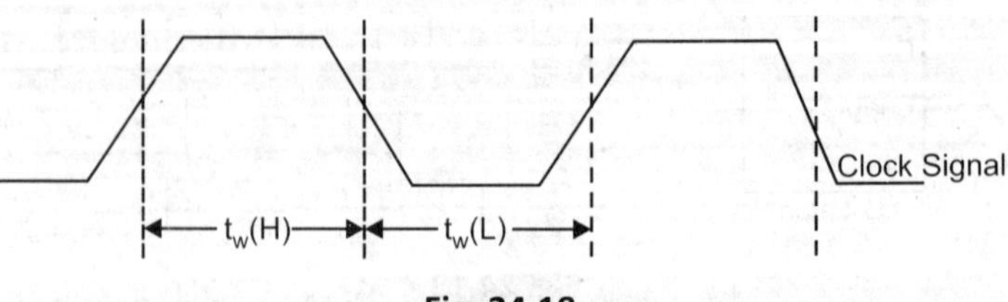

Fig. 24.18

Asynchronous input Active Pulse Width. It is the minimum time duration for which the asynchronous input (PRESET or CLEAR) must be kept in its active state, usually LOW, for the output to respond properly. It is 4ns in case of flip-flop 74ALS109A. Fig. 24.19 shows the timing parameter.

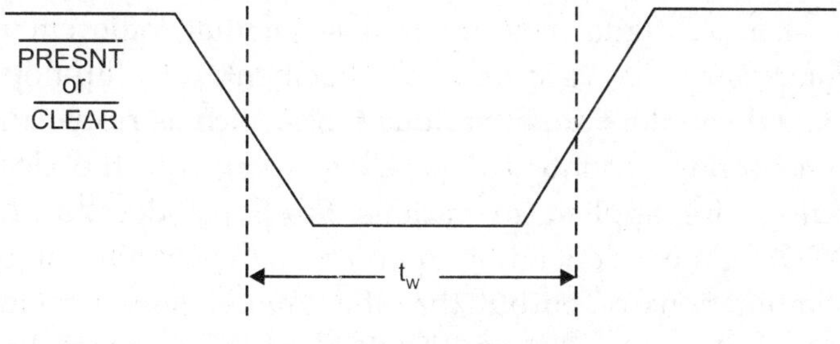

Fig. 24.19

Clock Transition Times. The manufacturers specify the maximum transition times, Rise time and Fall time, for the output to respond properly. If these specified figures are exceeded, the flip-flop may respond erratically or even may not respond at all. This parameter is logic family specific and is not specified for individual devices. The allowed maximum transition time for TTL devices would be much smaller than that for CMOS devices. Also, within the broad TTL family, it would vary from one subfamily to another.

Maximum Clock Frequency. This is the highest frequency that can be applied to clock input. If this figure is exceeded, there is no guarantee that the device would work reliably and properly. This figure may vary slightly from device to device of even the same type number. What the manufacturer usually specifies is a safe value. If this specified value is not exceeded, the manufacturer guarantees that the device will trigger reliably. For 74ALS109A, this figure, for instance, is 34 MHz.

Q. 15. How can we use a flip-flop for debouncing of mechanical switches ?

Ans. Due to switch bounce phenomenon, the mechanical switch cannot be used as such to produce a clean voltage transition. Refer to Fig. 24.20 (*a*). When the switch is moved from position-1 to position-2, what is desired at the output is a clean voltage transition from '0' to + *V*, as shown in Fig. 24.20 (*b*). What actually happens in depicted in Fig. 24.20 (*c*). The output makes several transitions between '0' and + *V* for a few milliseconds due to contact bounce before it finally settles at + *V*. Similarly, when it is moved from position-2 back to position-1, it makes several transitions before coming to rest at '0'. Though this random behaviour lasts only for a few milliseconds, it is unacceptable for many digital circuit applications. A NAND or a NOR latch can solve this problem and provide a clean output transition. Fig. 24.20 (*d*) shows a typical switch debounce circuit built around a NAND latch. The circuit functions as follows :

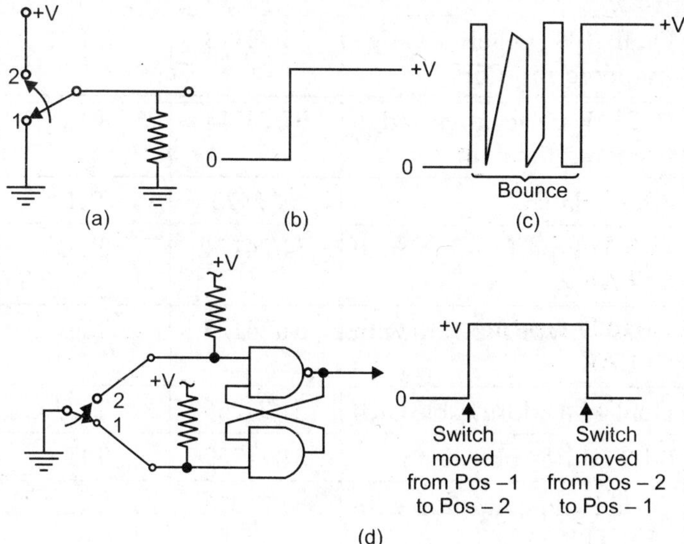

Fig. 24.20

When the switch is in position-1, the output is at '0' level. When it is moved to position, the output goes to '1' level within a few nanoseconds (depending upon the propagation delay of the NAND gate) after its first contact with position-2. When the switch contact bounces, it makes and breaks contact with position-2 before it finally settles at the intended position. Making of contact always leads to a '1' level at the output and breaking of contact also leads to a '1' level at the output due to the fact that the contact break produces a '1' level at both inputs of the latest, which forces the output to hold its existing logic state. On similar lines, it can be explained than when the switch is brought back position-1, the output makes a neat transition to '0' level.

Table 24.1. Functional Index of Flip-Flops and Latches

Function	*IC Number*	*Logic Family*
Dual J-K negative edge triggered flip-flop	54/7473	TTL
Dual D-type positive edge triggered flip-flop	54/7474	TTL
Dual J-K flip-flop with SET and CLEAR	54/7476	TTL
Dual J-K flip-flop with RESET	54/7478	TTL
Dual J-K flip-flop with CLEAR	54/74107	TTL
Dual J-K positive edge triggered flip-flop	54/74109	TTL
Dual J-K negative edge triggered flip-flop	54/74112	TTL

Cont...

Dual J-K negative edge triggered flip-flop	54/74113	TTL
Dual J-K negative edge triggered flip-flop	54/74114	TTL
4-bit D-latch	54/7475	TTL
Hex D-type flip-flop with CLEAR	54/74174	TTL
Quad D-type flip-flop with CLEAR	54/74175	TTL
Dual 4-bit addressable latch	54/74256	TTL
8-bit addressable latch	54/74259	TTL
Octal D-type flip-flop with MASTER RESET	54/74273	TTL
Quad SET-RESET latch	54/74279	TTL
Octal transparent latch (3-state)	54/74373	TTL
Octal D-type flip-flop (3-state)	54/74374	TTL
Octal D-type flip-flop with common enable	54/74377	TTL
Hex D-type flip-flop with enable	54/74378	TTL
4-bit D-type flip-flop with enable	54/74379	TTL
Octal D-type flip-flop (3 state)	54/74534	TTL
Octal D-type latch (3 state)	54/74573	TTL
Octal D-type flip-flop (3 state)	54/74574	TTL
Quad D-type latch	4042 B	CMOS
Quad R-S latch (3-state)	4044 B	CMOS
Dual 4-bit addressable latch	4723 B	CMOS
8-bit addressable latch	4724 B	CMOS
Hex D-type flip-flop	40174 B	CMOS
Quad D-type flip-flop	40175 B	CMOS
Dual D-type flip-flop	4013 B	CMOS
Dual J-K flip-flop	4027 B	CMOS
Quad D-type flip-flop (3-state)	4076 B	CMOS

Q. 16: What is a digital counter? What is it generally used for?

Ans: A digital counter is a sequential logic circuit comprising a cascaded arrangement of flip-flops. Digital counters are of various types, such as ripple and synchronous counters depending upon how the clock pulses are applied to various flip-flops; decade and BCD counters depending upon the nature of ten unique combinations of output that the counter goes through before it resets; UP and DOWN counters depending upon whether the counter increments or decrements each time, it is clocked and so on. These are discussed in subsequent questions. Counters are manufactured as integrated circuits (ICs) belonging to different logic families. Digital counters are widely used for counting the number of events occurring in a given time interval or for measurement of time interval between two specified events. Counters are also used as frequency dividers.

Q. 17. Distinguish between a RIPPLE COUNTER and a SYNCHRONOUS COUNTER.

Ans. In a RIPPLE COUNTER also known as a SERIAL COUNTER, the clock input is applied to only the first flip-flop, the flip-flop that represents the LSB. The clock input to any subsequent flip-flops is the output of its immediately preceding flip-flop. For instance, clock input to second flip-flop will be the output of the first flip-flop and the clock input to the third flip-flop will be the output of the second flip-flop and so on. As a natural consequence of this, all the flip-flops do not change state at the same time. In a typical case of a counter having a cascaded arrangement of four flip-flops, if the fourth flip-flop was to change state for a particular clock pulse, it would do so only after a time delay which is equal to sum of propagation delays of all the four flip-flops.

On the other hand, in a SYNCHRONOUS COUNTER also known as PARALLEL COUNTER, all the flip-flops change state synchronously or at the same time. The clock in this case is simultaneously applied to the clock inputs of different flip-flops. The delay involved in this case is equal to the propagation delay of one flip-flop only irrespective of number of flip-flops used to construct the counter.

Ripple and Synchronous counters will be more clear, as we go through their operational details in the subsequent questions.

Q. 18. Draw the basic logic circuit arrangement of a 3-bit ripple counter using flip-flops and briefly describe its operational principle. Also, draw the timing diagram to illustrate how a 3-bit counter can be used as a Divide-By-8 circuit.

Ans. Fig. 24.21 (*a*) shows the 3-bit ripple counter implemented with negative edge triggered J-K flip-flops.

All the three flip-flops are used in the toggle mode. The output of first flip-flop feeds the clock input of the second and the output of second flip-flop feeds the clock input of the third. The outputs of various flip-flops are designated as Q_0 (LSB), Q_1 and Q_2 (MSB). Let us assume that all the flip-flops are initially reset to logic '0' state. The clock input waveform and the waveforms that would appear at the outputs of various flip-flops as the counter counts through different clock cycles are shown in Fig. 24.21 (*b*).

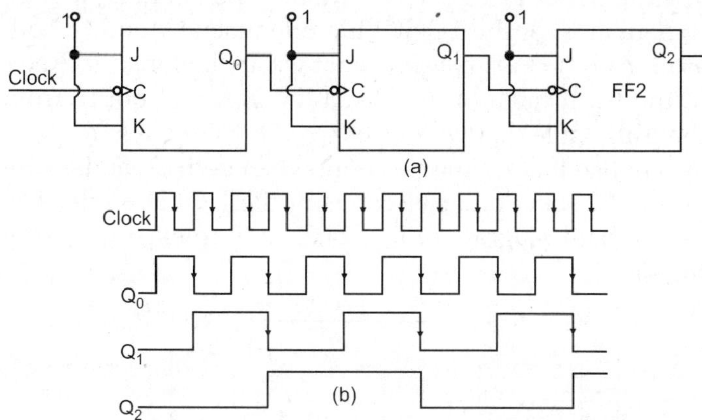

Fig. 24.21

On the negative going edge of the first clock pulse, Q_0 (LSB) goes from 0 to 1. As the flip-flops used here are the negative edge triggered ones, FF1 and FF2 remain in their initial states. Therefore, at the end of first clock cycle, $Q_0 = 1$, $Q_1 = 0$, $Q_2 = 0$. On the negative going edge of the second clock pulse. Q_0 goes from 1 to 0. This 1 to 0 transition at Q_0 forces Q_1 change from 0 to 1. FF2 output remains unchanged again, as its clock input does not receive a 1 to 0 transition. At the end of second clock cycle, therefore, $Q_0 = 0$, $Q_1 = 1$ and $Q_2 = 0$. This process continues and at the end of the seventh clock pulse, $Q_0 = 1$, $Q_1 = 1$, $Q_2 = 1$. On the negative going transition of the eighth clock pulse, Q_0 goes from 1 to 0, Q_1 also goes from 1 to 0 as its clock input also receives a 1 to 0 transition and Q_2 also goes from 1 to 0 for the same reasons. Thus, we are back to where we started from *i.e.* 000 state. Thus, with the arrangement of three flip-flops, we were able to count eight different states from 000 to 111. The results are tabulated ahead.

If we had a cascaded arrangement of four flip-flops, we could have gone through 16 different states from 0000 to 1111. In general, if we had (*n*) flip-flops, we could count upto 2^n pulses before the counter resets to the initial state.

You would notice from the timing diagram that FF0 toggles on every clock pulse, FF1 toggles on every alternate clock pulse and FF2 changes state on every fourth clock pulse. As each flip-flop must toggle twice to complete one cycle of waveform at its output, the frequencies of the waveforms present at the outputs of FF0, FF1 and FF2 are respectively (*f*/2), (*f*/4) and (*f*/8) where (*f*) is the frequency of input clock waveform. Thus this counter circuit of Fig. 24.21 (*a*) can be used as a divide by 8 circuit, when the pulse train to be divided is applied at the clock input of FF0. The divided waveform appears at FF2 output.

At the end of	Q_2	Q_1	Q_0
First clock cycle	0	0	1
Second clock cycle	0	1	0
Third clock cycle	0	1	1
Fourth clock cycle	1	0	0
Fifth clock cycle	1	0	1
Sixth clock cycle	1	1	0
Seventh clock cycle	1	1	1
Eighth clock cycle	0	0	0

Q. 19. Define the MODULUS of a counter.

Ans. The MODULUS of a counter is the total number of logic states through which the counter can progress. For instance, the modulus of the counter of Q. 18 is 8. Such a counter is called a Mod – 8 counter.

Q. 20. If (*n*) is an integer, then a simple cascaded arrangement of (*n*) flip-flops can be used to construct a binary ripple counter having a modulus of 2^n. Such a counter will count from 000....0 to 111...1 without skipping any of the states in its natural count sequence. How do you construct a ripple counter having a modulus other than 2^n, *i.e.* modulus that is not an integral power of 2.

Ans. The basic cascaded arrangement of flip-flops as shown earlier in Fig. 24.21 can be modified to skip some counts and reset to all 0s without going upto the final count by using feedback techniques. This forms the basis of constructing ripple counters with a modulus that is not precisely equal to an integral power of 2. A three flip-flop counter with an appropriate feedback circuit can be used to construct a counter with any modulus upto 8. Similarly, four flip flops could be used to construct a ripple counter of any modulus upto 16.

Let us see how do we construct a Mod-6 counter using a 3 flip-flop arrangement. Fig. 24.22 (*a*) shows the natural count sequence of a Mod-8 counter. If somehow we can skip the last two states (*i.e.* 110 and 111), we could get a Mod-6 counter that would count from 000 to 101 and then reset to 000 to restart the count sequence. Fig. 24.22 (*b*) shows the arrangement that achieves this. The flip-flops used here, should have CLEAR facilities. As the counter counts from 000 to 101, the NAND output remains HIGH keeping the CLEAR inputs inactive. During the sixth clock transition when the counter tends to go to 110, the NAND produces a 0 at the output, thus, clearing all the flip-flops to 0 state. The counter thus instead of going to 110 gets reset to 000. The seventh clock transition again takes it to 001 and the process continues.

Q_2	Q_1	Q_0
0	0	0
0	0	1
0	1	0
0	1	1
1	0	0
1	0	1
1	1	0
1	1	1

(a)

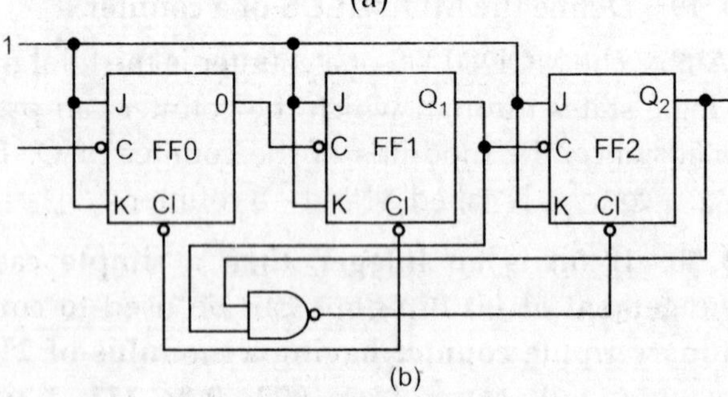

(b)

Fig. 24.22

Q. 21. What is the most significant advantage of a synchronous counter over a ripple counter. Draw the logic diagram of a 3-bit synchronous counter using flip-flops and briefly describe its operational principle. Also, sketch the waveforms present at the outputs of various flip-flops as the clock goes through one complete count cycle.

Ans. The ripple counter discussed in the preceding questions is an asynchronous counter, where various flip-flops are not clocked simultaneously. The total propagation delay in such counters is equal to sum of individual delays and becomes prohibitibly large in a counter with a large count. On the other hand, in a synchronous counter, all flip-flops are clocked simultaneously and thus, change their states at the same time. Due to the propagation delays of different flip-flops, there will always be a limit to the maximum frequency at which the ripple counters can be operated without giving any erroneous counts. Such a problem starts making its presence felt, when the propagation delay of one flip-flop becomes a significant part of the clock cycle time period. Synchronous counters do not have these problems.

Fig. 24.23 (*a*) shows the logic diagram of a 3-bit synchronous counter. The first flip-flop representing the LSB has been wired as a TOGGLE flip-flop. The second flip-flop can operate in either of the two modes *i.e.* the HOLD mode when Q_0 is 0 and the TOGGLE mode, when Q_0 is 1. The third flip-flop representing the MSB can also operate in either the HOLD mode when either Q_0 or Q_1 or both are 0 or the TOGGLE mode when Q_0 and Q_1 both are 1. And that is precisely what we are looking for (Refer to the truth table of Fig. 24.23 (*b*). You will notice from the truth table that whenever Q_0 is 1, Q_1 toggles. You will also notice that Q_2 toggles only when both Q_0 and Q_1 are in logic 1 state, otherwise it is in the HOLD mode. The logic circuit diagram of Fig. 24.23 (*a*) implements exactly the same.

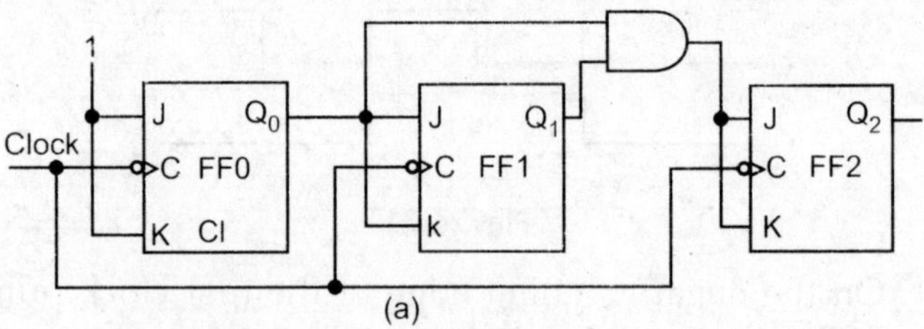

(a)

Q_2	Q_1	Q_0
0	0	0
0	0	1
0	1	0
0	1	1
1	0	0
1	0	1
1	1	0
1	1	1
0	0	0

(*b*)

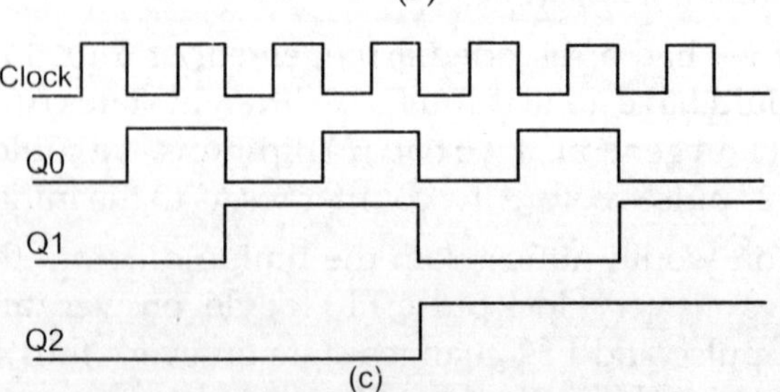

(c)

Fig. 24.23 Shows the timing diagram

Q. 22. Define the following:

1. **UP-Counter** 2. **DOWN-Counter**
3. **UP-DOWN Counter** 4. **Decade Counter**
5. **BCD Counter** 6. **Modulo Counter.**

Ans.

1. An UP-COUNTER or a Forward Counter is the one that counts upwards everytime it is pulsed. For instance, a typical decade counter counting in the UP-count mode will progress as 0000, 0001, 0010, 0011, 0100, 0101, 0110, 0111,1000,1001, 0000, 0001 and so on.
2. A DOWN-COUNTER counts downwards everytime it is clocked. A typical decade counter in the DOWN-count mode will count as 0000, 1001, 1000, 0111, 0110, 0101, 0100, 0011, 0010, 0001, 0000,1001, and so on.
3. An UP-DOWN COUNTER is a bidirectional counter and can be made to count upwards as well as downwards. In some of the UP-DOWN Counter ICs like 74192 (Presettable UP-DOWN Decade Counter), for instance, there are separate clock input terminals for the UP and DOWN count modes. Whether the counter counts upwards or downwards depends upon the clock input that is pulsed. In some other types of UP-DOWN counter ICs like 74190 (Presettable UP-DOWN Decade Counter) for instance, there is one clock input and a UP/DOWN mode control terminal. The count mode is selected from this terminal.
4. A DECADE COUNTER is the one which goes through 10 unique combinations of outputs and then resets, as the clock proceeds. We may use some sort of a feedback in a 4-bit binary counter to skip any six of the sixteen possible output states from 0000 to 1111 to get a decade counter. A decade counter does not necessarily count from 0000 to 1001, it could count as 0000, 0001, 0010, 0011,1000, 1001, 110, 111, 1110,1111, 0000, 0001 and so on.
5. A BCD COUNTER is a special case of a decade counter in which the, counter counts from 0000 to 1001 and then resets. The output weights of the flip-flops in these counters are in accordance with 8421-code. For instance, at the end of seventh clock pulse, the output sequence will be 0111 (Decimal equivalent of 0111 as per 8421-code is 7). These counters will, thus, be different from other decade counters that provide the same count by using some kind of forced feedback to skip some of the natural binary counts. Fig. 24.24 shows a decade counter of the BCD type.
6. A MODULO COUNTER is the one in which the number of unique output states the counter goes through in one complete count cycle is not an integral power of 2. That is, the modulus of the counter is other than 2^n, where (n) is an integer. A decade counter can also be termed as a modulo-10 counter.

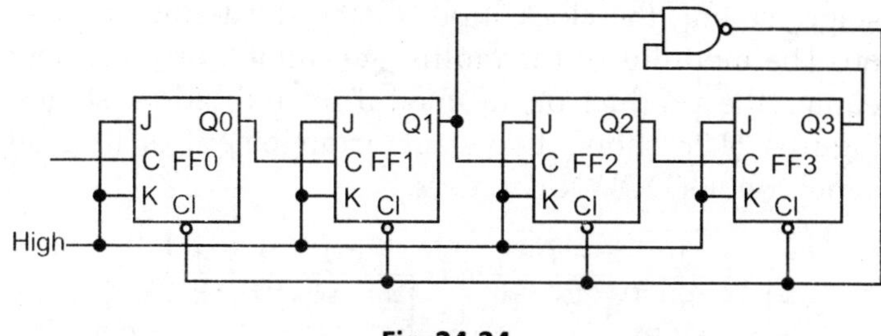

Fig. 24.24

Q. 23. What are PRESETTABLE COUNTERS ?

Ans. PRESETTABLE COUNTERS are those which can be preset to any starting count either asynchronously (independent of clock signal) or synchronously (with the active transition of the clock signal). The presetting function is accomplished with the help of PRESET and CLEAR (or MASTER RESET) inputs available on flip-flops. Presetting operation is also known as 'Preloading' or simply 'Loading' operation.

Presettable counters can be UP-counters, DOWN-counters or UP/DOWN-counters. Additional inputs/outputs available on presettable UP/DOWN counters usually include PRESET inputs from where any desired count can be loaded, PARALLEL LOAD (PL) input which when active allows the PRESET inputs to be loaded onto counter outputs and TERMINAL COUNT (TC) outputs which become active, when the counter reaches the terminal count.

Q. 24: Why do we need to use a cascaded arrangement of counters? Briefly explain the operation of cascaded arrangement of binary counters.

Ans: A cascade arrangement of counters is used to build counters with a higher modulus than is possible with a single stage. The terminal count outputs allow more than one counter to be connected in a cascade arrangement.

Figure-24.25 shows the cascaded arrangement of two four-bit synchronous binary UP counters. In order to construct a multistage UP-counter, all counter stages are connected in the count-UP mode. The clock is applied to the clock input of a lowest-order counter, the terminal count UP (TCU), also called the carry-out (C_o), of this counter is applied to the clock input of the next higher counter stage and the process continues. If it is desired to build a multistage DOWN counter, all counters are wired as DOWN counters, the clock is applied to the clock input of the lowest-order counter and the terminal count DOWN (TCD), also called the borrow-out (B_o), of the lowest-order counter is applied to the clock input of the next higher counter stage. The process continues in the same fashion, with the TCD output of the second stage feeding the clock input of the third stage and so on. The modulus of the multistage counter arrangement equals the product of the moduli of individual stages. Figures-24.26 shows two-stage arrangement of four-bit synchronous DOWN counters.

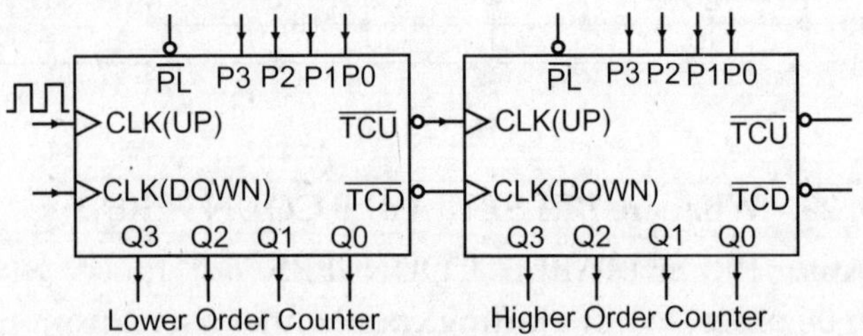

Fig. 24.25

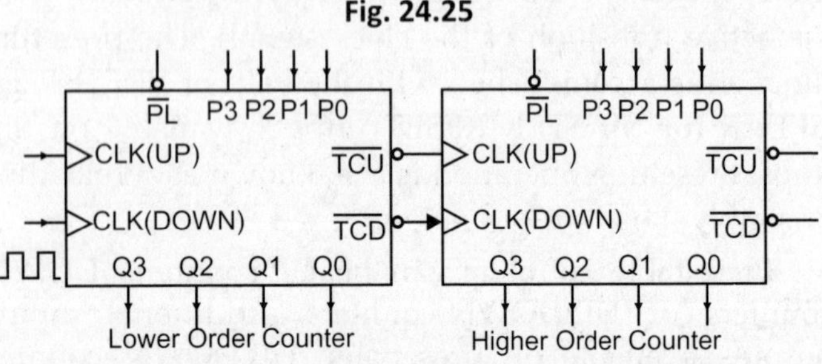

Fig. 24.26

Q. 25: Why do we need to cascade BCD counters? What decides the number of BCD counter stages required for a certain application? Briefly explain the functioning of cascade arrangement of multiple BCD counters.

Ans: We know that the BCD counters are used, when the application involves counting of pulses and the result of counting is to be displayed in decimal. A single stage BCD counter counts from 0000 (decimal equivalent '0') to 1001 (decimal equivalent '9') and thus, is capable of counting upto a maximum of nine pulses. Output in a BCD counter is in binary coded decimal (BCD) form. BCD output needs to be decoded appropriately before it can be displayed. More than one BCD counter stages need to be used in a cascade arrangement in order to be able to count up to a larger number of pulses. The number of BCD counter stages to be used equals the number of decimal digits in the maximum number of pulses we want to count upto. A maximum count of 9999 or 3843, both would require a 4-stage BCD counter arrangement with each stage representing one decimal digit.

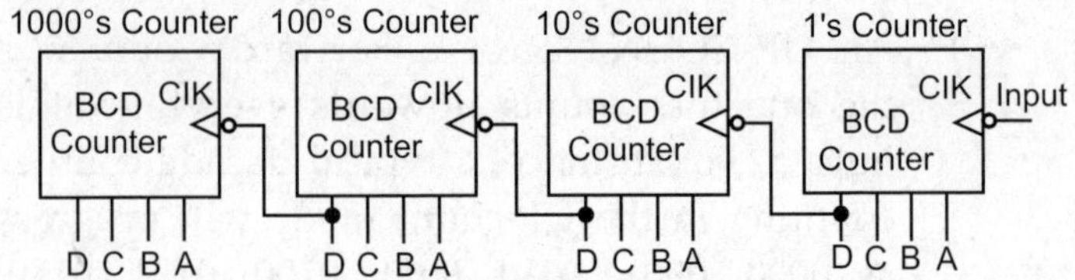

Fig. 24.27

Figure-24.27 shows a cascade arrangement of four BCD counter stages. The arrangement works as follows. Initially, all four counters are in all 0's state. The counter representing the decimal digit of one's place is clocked by the pulsed signal that needs to be counted. The successive counters are clocked by the MSB of the immediately previous counter stage. The first nine pulses take the one's place counter to (1001). The tenth pulse resets it to (0000) and 1-to-0 transition at MSB of one's place counter clocks ten's place counter. Ten's place counter gets clocked on every tenth input clock pulse. On the hundredth clock pulse, the MSB of ten's counter makes a 1-to-0 transition which clocks the 100's place counter. This counter gets clocked on every successive hundredth input clock pulse. On thousandth input clock pulse, the MSB of 100's counter makes 1-to-0 transition for the first time and clocks the 1000's place counter. This counter is clocked thereafter on every successive thousandth input clock pulse. With this background, we can always tell the output state of the cascade arrangement. For example, immediately after 7364th input clock pulse, the state of 1000's, 100's, 10's and 1's BCD counters would respectively be 0111, 0011, 0110 and 0100.

Q. 26: What do you understand by Excitation Table of a flip-flop? Explain with the help of examples.

Ans: The excitation table lists the present state, the desired next state and the flip-flop inputs (J, K, and D etc.) required for achieving that. The same for a J-K flip-flop and a D-flip-flop are shown in Tables 24.2 and 24.3 respectively. Referring to Table-24.2, if the output is in logic '0' state and it is desired that it goes to logic '1' state on occurrence of clock pulse, the J-input must be in logic '1' state while K-input can be either in logic '0' or logic '1' state. This is true as for a '0' to '1' transition; there are two possible input conditions which can achieve this. These are J=1, K=0 (SET-mode) and J=K=1 (Toggle mode), which further leads to J=1, K=X (either 0 or 1). The other entries of the excitation table can be explained on similar lines. In case of D-flip-flop (Table-24.3), the D-input is the same as the logic status of the desired next state. This is true as in case of a D-flip-flop, the D-input is transferred to the output on the occurrence of the clock pulse. It is irrespective of the present logic status of the Q-output.

Table. 24.2: Excitation table of J-K flip-flop

Present State (Q_n)	Next State (Q_{n+1})	J	K
0	0	0	X
0	1	1	X
1	0	X	1
1	1	X	0

Table-24.3: Excitation table of D- flip-flop

Present State (Q_n)	Next State (Q_{n+1})	D
0	0	0
0	1	1
1	0	0
1	1	1

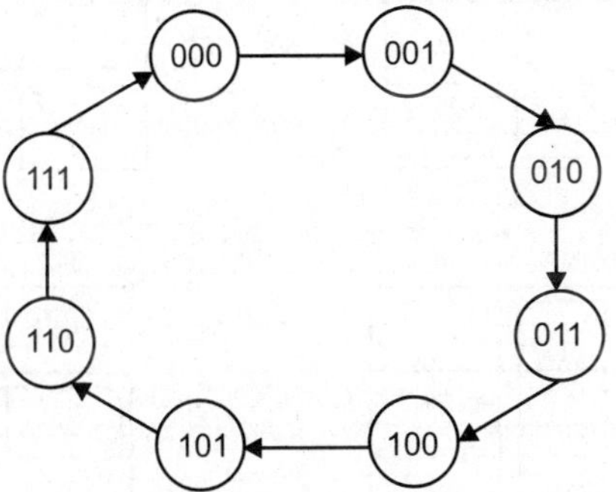

Fig. 24.28

Q. 27: What do you understand by State Transition Diagram of a sequential circuit?

Ans: The STATE TRANSITION DIAGRAM is a graphical representation of different states of a given sequential circuit and the sequence in which these states occur in response to a clock input. Different states are represented by circles and the arrows joining them tell the sequence in which different states occur. As an example, Figure-24.28 shows the state transition diagram of MOD-8 binary counter.

Q. 28: Outline different steps required to design a counter with an arbitrary count sequence.

Ans: So far, we have discussed different types of synchronous and asynchronous counters. A large variety of synchronous and asynchronous counters are available in the IC form, some of which were even mentioned and discussed in the previous questions. The counters discussed in the previous questions count in either the normal binary sequence with a modulus of (2^N) or with slightly altered binary sequences where one or more of the states were skipped. The latter type of counters had a modulus of less than (2^N), (N) being the number of flip-flops used. But nevertheless, even these counters had a sequence that was either upwards or downwards and not an arbitrary. There are applications where a counter is required to follow a sequence that is arbitrary and not binary. As an example, a MOD-10 counter may be required to follow a sequence that is 0000, 0010, 0101, 0001, 0111, 0011, 0100, 1010, 1000, 1111, 0000, 0010 and so on. In such cases, the simple and seemingly obvious feedback arrangement with a single NAND gate discussed in the earlier sections of this chapter for designing counters with a modulus of less than (2^N) cannot be used.

There are several techniques for designing counters that follow a specified arbitrary sequence. In the present section, we shall discuss in detail a commonly used technique for designing synchronous counters using J-K flip-flops or D-flip-flops. The design of the counters basically involves designing a suitable combinational logic circuit that takes its inputs from the normal and complemented outputs of the flip-flops used and decodes the different states of counter to generate the correct logic states for the inputs of the flip-flops such as J, K, and D etc.

We shall illustrate the design procedure with the help of an example. We shall do it for a MOD-6 synchronous counter design, which follows the count sequence as 000, 010, 011, 001, 100, 110, 000, 010......

1. Determine the number of flip-flops required for the purpose. Identify the undesired states. In the present case, number of flip-flops required is 3 and the undesired states are 101 and 111.
2. Draw the state transition diagram showing all possible states including the ones that are not desired. The undesired states should be depicted to be transiting to any of the desired states. We have chosen (000) state for the purpose. It is important to include the undesired states to ensure that if the counter accidentally gets into any of these undesired states due to noise or power-up, the counter will go to a desired state to resume the correct sequence on application of next clock pulse. Figure-24.29 shows the state transition diagram.

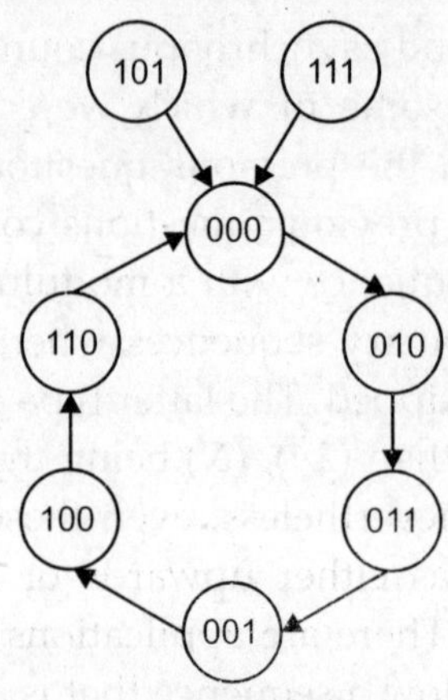

Fig. 24.29

3. Draw the excitation table for the counter listing the present states, next states corresponding to present states and the required logic status of flip-flop inputs (J, K inputs in case counter is to be implemented with J-K flip-flops). The excitation table is shown in Table-24.4.

Table-24.4: Excitation table

Present State			*Next State*			*Inputs*					
C	B	A	C	B	A	J_C	K_C	J_B	K_B	J_A	K_A
0	0	0	0	1	0	0	X	1	X	0	X
0	0	1	1	0	0	1	X	0	X	X	1
0	1	0	0	1	1	0	X	X	0	1	X
0	1	1	0	0	1	0	X	X	1	X	0
1	0	0	1	1	0	X	0	1	X	0	X
1	0	1	0	0	0	X	1	0	X	X	1
1	1	0	0	0	0	X	1	X	1	0	X
1	1	1	0	0	0	X	1	X	1	X	1

The circuit excitation table can be drawn very easily once we know the excitation table of the flip-flop to be used for building the counter. For instance, let us look at the first row of the excitation Table-24.4. The counter is in (000) state and it is to go to (010) on application of a clock pulse. That is, the normal outputs of *C*, *B* and *A* flip-flops have to undergo 0 to 0, 0 to 1 and 0 to 0 transitions respectively. Referring to the excitation table of a *J-K* flip-flop, the desired transitions can be realised, if the logic status of J_A, K_A, J_B, K_B, J_C and K_C is as shown in the excitation table.

4. The next step is to design the logic circuits for generating J_A, K_A, J_B, K_B, J_C and K_C inputs from available A, $\overline{A}$, B, $\overline{B}$, C and $\overline{C}$ outputs. This can be done by drawing Karnaugh maps for each one of the inputs, minimising them and then implementing the minimised Boolean expressions. The Karnaugh maps for J_A, K_A, J_B, K_B, J_C and K_C are respectively shown in Figures-24.30 (*a*), (*b*), (*c*), (*d*), (*e*) and (*f*). The minimised Boolean expressions are given in equations are given as under:

$$J_A = B\overline{C}$$

$$K_A = \overline{B} + C$$

$$J_B = \overline{A}$$

$$K_B = A + C$$

$$J_C = A\overline{B}$$

$$K_C = A + B$$

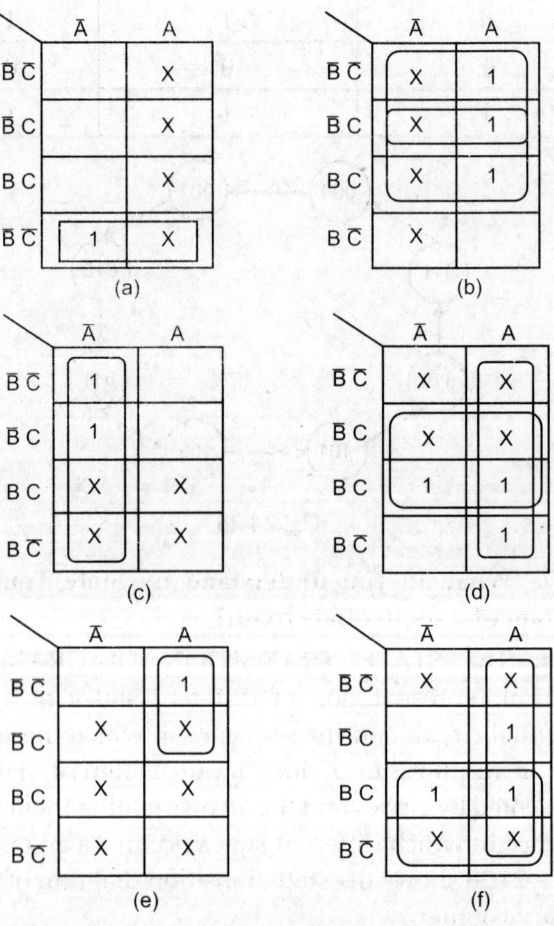

Fig. 24.30

The above expressions can now be used to implement combinational circuits to generate J_A, K_A, J_B, K_B, J_C and K_C

inputs. Figure-24.31 shows the complete counter circuit. The design procedure illustrated above can be used to design a synchronous counter for any given count sequence with the condition that no state occurs more than once in the given count sequence's one complete cycle as the design cannot handle a situation, where a particular present state has more than one future states.

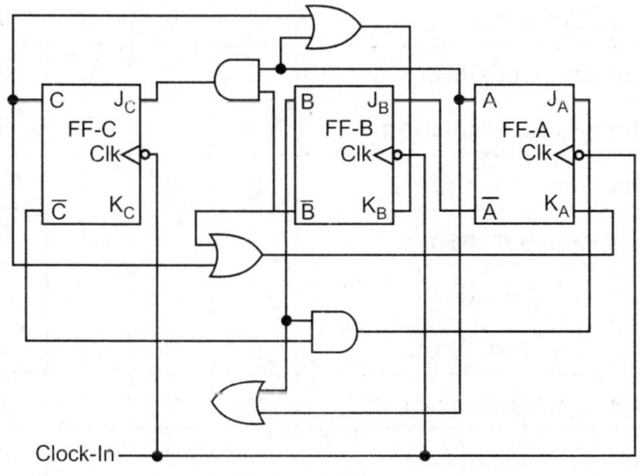

Fig. 24.31

Q. 29. What is a Shift Register ? What does it consist of and where do such devices find application ?

Ans. A SHIFT REGISTER is a storage device that is used to momentarily store binary data. The data to be stored could be the data appearing at the output of an encoding matrix before it is fed to the main digital system for processing or it might be the data present at the output of a microprocessor before it is fed to the driver circuitry of the output devices. The shift register thus forms an important link between the main digital systems and the input/output channels. The shift registers can also be configured to construct some special types of counters that can be used to perform a number of arithmetic operations like Subtraction, Multiplication, Division, Complementation etc. The basic building block in all shift registers is the flip-flop, mainly a D-type flip-flop. Although in many of the commercial shift register ICs, their internal circuit diagram might indicate the use of (*R-S*) flip-flops, a careful examination will reveal that these (*R-S*) flip-flops have been wired as D-type flip-flops only.

Q. 30. Classify shift registers.

Ans. Based on the method by which data can be loaded onto and read from shift registers, they are classified as : (Refer Fig. 24.32)

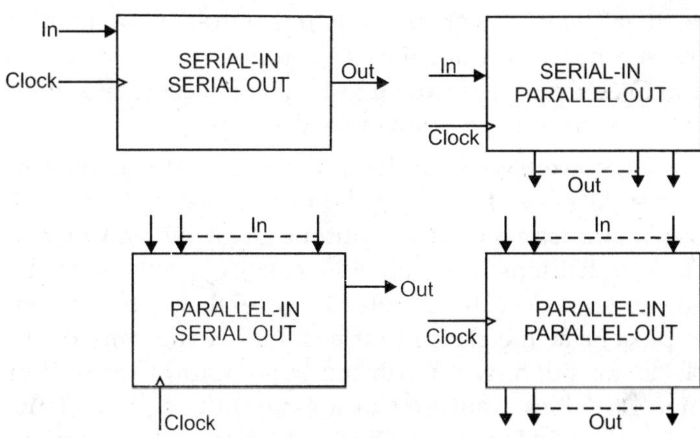

Fig. 24.32

1. Serial-In Serial-Out shift registers
2. Serial-In Parallel-Out shift registers
3. Parallel-In Serial-Out shift registers
4. Parallel-In Parallel-Out shift registers

Q. 31. Define a UNIVERSAL SHIFT REGISTER.

Ans. Registers are made in IC form in a variety of types. Those shift registers which can be used in serial-serial, shift right, shift left, serial-parallel, parallel-serial and parallel-parallel data register transfers are usually called universal registers. IC 74194 is a universal shift register.

Q. 32. With the help of logic diagram and the timing waveforms, describe the functioning of a Serial-in Serial-Out shift register.

Ans. Fig. 24.33 shows the basic 4-bit serial-in serial-out shift register implemented using D-type flip-flops. The circuit functions as follows:

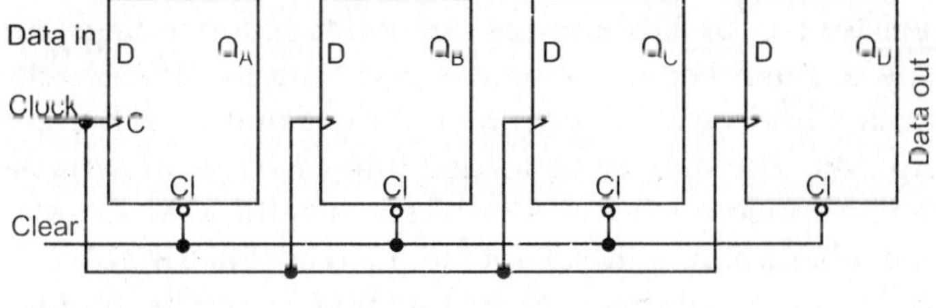

Fig. 24.33

A reset pulse applied to the CLEAR input of all the flip-flops resets their *Q* outputs to 0s. Refer to the timing waveforms of Fig. 24.34. The waveforms shown include the clock pulse train, the waveform representing the data to be loaded onto the shift register and the *Q*-outputs of different flip-flops.

The flip-flops shown respond to the LOW-to-HIGH transition of the clock pulses as indicated by their logic symbols. During the first clock transition, Q_A output goes from 0 to 1. The outputs of other three flip-flops remain 0 as their D inputs were 0 at the time of transition. During the second clock transition, Q_A-output goes from 1 to 0 and Q_B-output goes from 0 to 1 again in accordance with laws governing operation of D-type flops.

Thus, we have seen that a Logic '1' that was present at the data input prior to the occurrence of first clock transition has reached the output Q_B at the end of two clock transitions. This bit will reach Q_D-output at the end of four clock transitions. In general, a data bit, that is present at the data input terminal at the time of nth clock transition, will reach the Q_D-output at the end of $(n + 4)$th clock transition in a 4-bit shift register of the type shown in Fig. 24.33. During the fifth and subsequent clock transitions, data bits continue to shift to the right and at the end of eighth clock transition, the shift register is again reset to all 0s. Thus, in a 4-bit serial-in serial-out shift register, it takes four clock cycles to load the data bits and another four cycles to read the data bits out of the register. The contents of the register, as it is clocked eight times are summarised below.

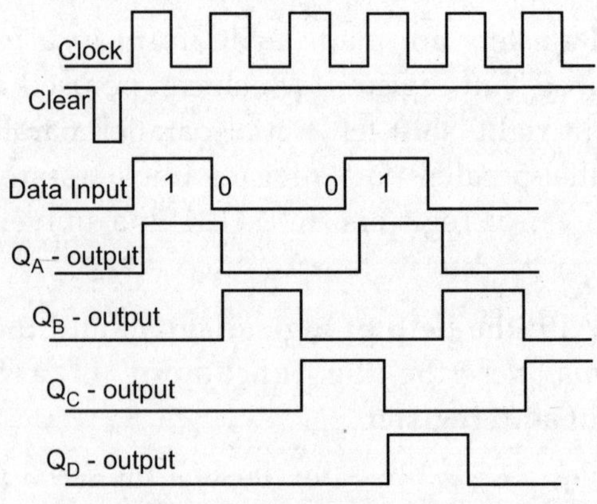

Fig. 24.34

IC 7491 is a popular 8-bit serial-in serial-out shift register Fig. 24.35 shows its internal functional diagram. Due to the inverter between R and S inputs of the data input flip-flop, it is functionally the same as a D-type flip-flop. The data to be loaded into the register serially can be applied either at A or B inputs of the NAND gate. The other input is then kept HIGH to enable the NAND. In that case, data present at A or B gets complemented, as it appears at the NAND output. Another inversion provided by the inverter however restores the original status so that for a '1' at data input, there is '1' at SET input of the flip-flop and a '0' at the RESET input of the flip-flop and for a '0' at the data input, there is a '0' at the SET and a '1' at the RESET input of the flip-flop. The NAND gate provides only a gating function and if it is not required, the two inputs of the NAND can be shorted to have a single line data input. The shift register responds to the LOW to HFGH transitions of the clock pulses.

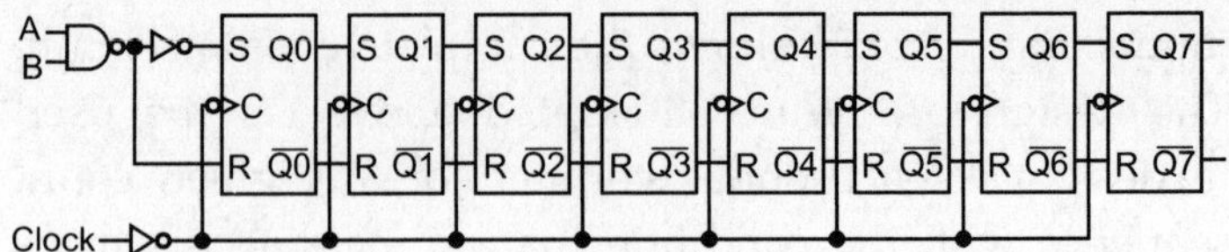

Fig. 24.35

Clock	Q_A	Q_B	Q_C	Q_D
Initial Contents	0	0	0	0
After first transition	1	0	0	0
After second transition	0	1	0	0
After third transition	0	0	1	0
After fourth transition	1	0	0	1
After fifth transition	0	1	0	0
After sixth transition	0	0	1	0
After seventh transition	0	0	0	1
After eighth transition	0	0	0	0

Q. 33. With the help of the functional diagram, describe the operation of a SERIAL-PARLLEL-OUT shift register.

Ans. A serial-in parallel-out shift register is constructionally identical to a serial-in serial-out shift register except that in case of former, all flip-flop outputs are also brought out on the terminals. Fig. 24.36 shows the functional diagram of IC 74164, a popular 8-bit serial-in parallel-out shift register. Here also, data can be entered serially through one of the two inputs (A or B). The unused input must be tied to HIGH or both inputs should be shorted to get a single a data input.

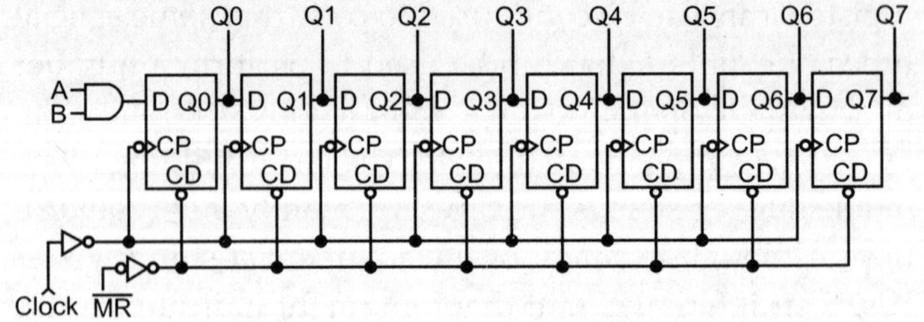

Fig. 24.36

Q. 34. With the help of a functional diagram, describe the operational principle of 4-bit PARALLEL-IN SERIAL-OUT shift register.

Ans. Fig. 24.37 shows the functional diagram. The shift register shown consists of a Parallel In or Serial-In, Serial-Out 4-bit shift register with gated clock inputs and an overriding CLEAR input. The parallel-in or serial-in modes are controlled by SHIFT/LOAD input. It may mentioned here that the circuit shown is a part of the functional diagram of IC 74166 which is an 8-bit shift register of this type. If the circuit shown is extended on the right and another four flip-flops added alongwith the input gate circuitry, as included in case of the flip-flops shown the diagram, we get the functional diagram of IC 74166.

When the SHIFT/LOAD input is held HIGH, the serial data input AND gates are enabled and the circuit behaves like a serial-in serial-out shift register. When the SHIFT/LOAD input is held LOW, parallel data input AND gates are enabled and data is loaded parallel in synchronism with the next clock pulse. Clocking is accomplished on the LOW to HIGH transition of the clock pulse via a 2-input NOR gate. Holding one of the inputs of the NOR gate HIGH in hibits the clock applied to the other input. Holding an input LOW enables the clock applied to the other input. An active LOW CLEAR inputs overrides all the inputs including the clock and resets all flip-flops to '0'.

The timing waveforms shown in Fig. 24.38 for the register of Fig. 24.37 explain both serial-in saurel-out as well as parallel-in serial-out operations.

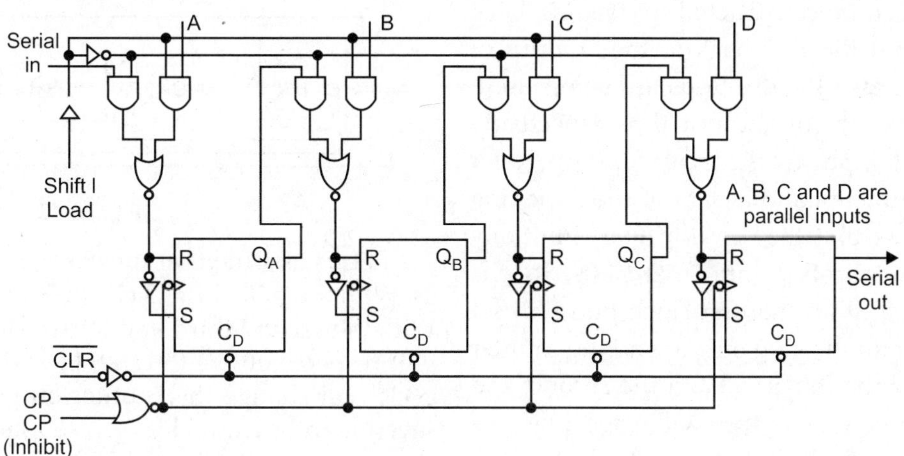

Fig. 24.37

Q. 35. Briefly describe the operation of a PARALLEL-IN, PARALLEL-OUT shift egister.

Ans. The hardware qf a PARALLEL-IN, PARALLEL-OUT shift register is similar to that a PARALLEL-IN SERIAL-OUT shift register. If in a parallel-in, serial-out shift register, outputs of different flip-flops are brought out, it becomes a parallel-in, parallel-out shift register. In fact, the functional diagram of one of the popular 8-bit parallel-in, parallel-out shift register is identical to the functional diagram of an 8-bit parallel-in serial-out shift register IC 74166 mentioned in the last question except that the flip-flop outputs have been brought out on the terminals.

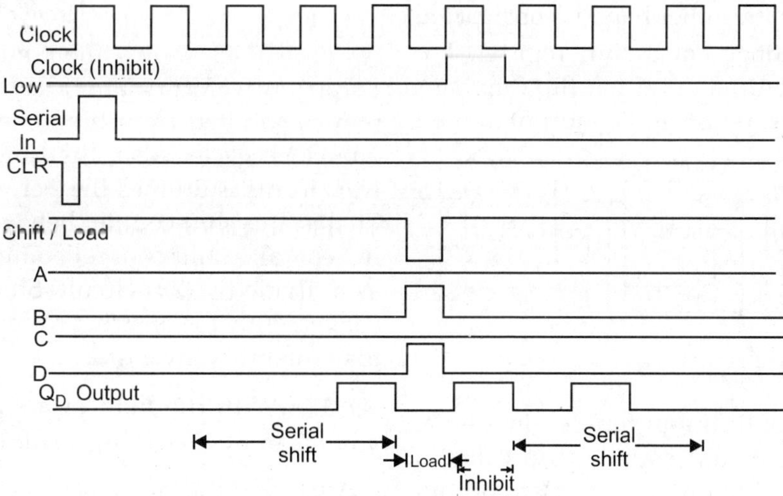

Fig. 24.38

Q. 36. With the help of functional diagrams, briefly explain how shift register could be modified to construct a (*i*) Ring counter and a (*ii*) shift counter also known as Johnson counter. Also, describe the operational principle of these counters with the help of respective timing diagrams.

Ans. A RING COUNTER is obtained from a shift register by directly feeding back the true output of the output flip-flop to the data input terminal of the input flip-flop. In case, D-type flip-flops are being used to construct the shift register, the ring counter also called a circulating register can be constructed by feeding back *Q*-output of the output flip-flop back to the D-input of the input flip-flop. In case J-K flip-flops are being used, *Q*-output and $\overline{Q}$-output of the output flip respectively fed back to the *J* and *K* inputs of the input flip-flop. Refer to Fig. 24.39 for the functional diagram of the 4-bit ring counter. Let us assume that the flip-flop FF0 is initially set to '1' and all other flip-flops are reset to 0s. So, the counter output is 1000. With the first clock pulse, this 1 gets shifted to the second flip-flop output and the counter output becomes 001 and similarly. With the second and third clock pulses, the counter output will respectively be 0010 and 0001 due to feedback, with the fourth clock pulse, the counter output will again become 1000. The count cycle repeats in the subsequent clock pulses. Circulating registers of this type find wide application in the control section of microprocessor based system, where one event should follow the other. The timing waveforms for the circulating register of Fig. 24.39, as shown in Fig. 24.40 further illustrate their utility as a control element in a digital system to generate control pulses that must occur one after the other sequentially.

A SHIFT COUNTER on the other hand is constructed by having an inverse feedback in a shift register. For instance, if we connect Q-output of output flip-flop back to the K input of the input flip-flop and $\overline{Q}$-output of the

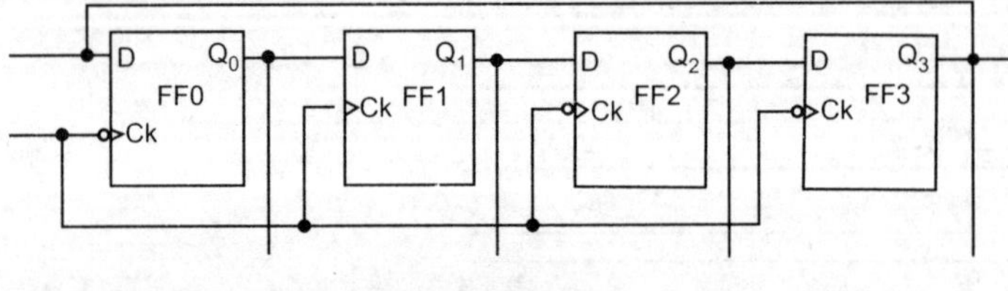

Fig. 24.39

output flip-flop to J input of the input flip-flop in a serial shift register, the result is a shift counter also called a Johnson counter. In case R-S flip-flops are used, output goes to the R input and the Q-output is conneeled to the S input. Fig. 24.41 shows the basic 4-bit shift counter.

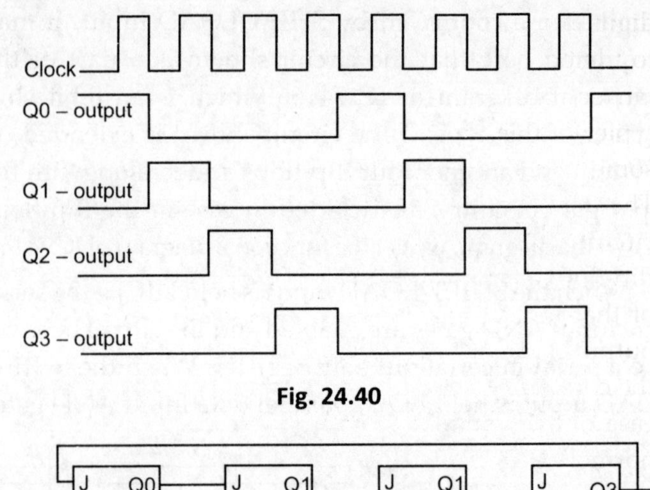

Fig. 24.40

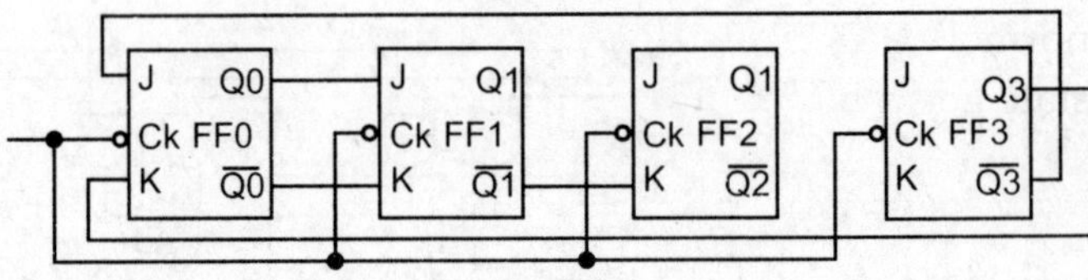

Fig. 24.41

Let us assume that the counter is initially reset to all 0s. With the first clock cycle, the output will become 0001. With the second third and fourth clock cycles, the output will respectively be 0011, 0111 and 1111. The fifth clock cycle will change the counter output to 1110. The sixth, seventh and eighth clock pulses successively change the output to 1100, 1000 and 0000. Thus, one count cycle is completed in eight cycles.

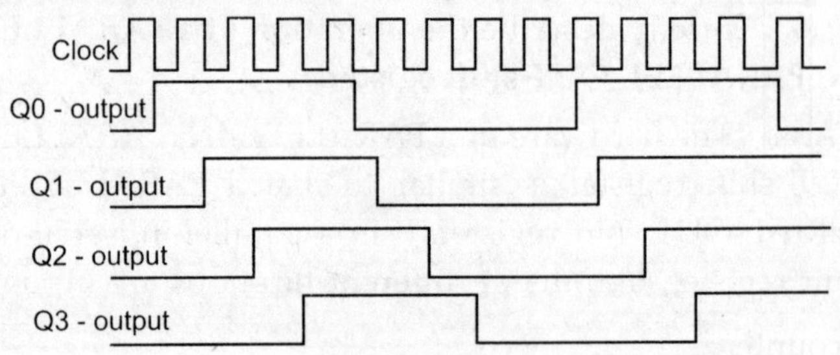

Fig. 24.42

Fig. 24.42 shows the timing waveforms. Different output waveforms are identical except the fact that they are shifted from the inmediately preceding one by one clock cycle. Also, the time period of each of these waveforms is 8 times the period of the clock waveform. That is, this shift counter behaves as a divide-by-8 circuit. In general, a shift counter comprising of (*n*) flip-flops acts as a divide by (2*n*) circuit. Shift counters can used very conveniently to construct counters having modulus other than infegral power of 2.

Q. 37. With the help of a suitable example, what do you mean by Decoding a counter?

Ans. The output state of a counter at any time instant, as it is being clocked, is in the form of a sequence of binary digits. For a large number of applications, it is important to detect or decode different states of the counter whose number equals the modulus of the counter. One typical application could be a need to initiate or trigger some action after the counter reaches a specific state. The decoding network, therefore, is going to be a logic circuit that takes its input from the outputs of different flip-flops constituting the counter and then makes use of that data to generate as many outputs. The decoder outputs are normally LOW and for a given counter state, the corresponding decoder output goes to HIGH state. In case of Active-LOW decoding, the decoder outputs are normally HIGH and the output representing the counter state goes to the LOW state.

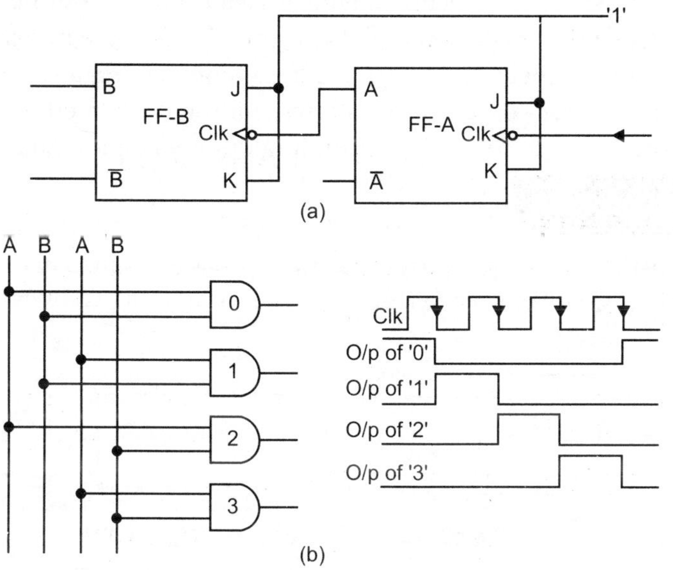

Fig. 24.43

We shall further illustrate the concept of decoding a counter with the help of an example. Consider the 2-stage MOD-4 ripple counter of Fig. 24.43 (*a*). This counter has four states in 00, 01, 10 and 11, which need to be decoded. Let us now look at the arrangement of four 2-input AND gates as shown in Fig. 24.43 (*b*) and how the outputs look like as the counter clock goes through first four pulses. We should also keep in mind, before proceeding further that our decoder network has four AND gates, which equals the number of counter states to be decoded, which further equals modulus of the counter. Also, each AND-gate has two inputs, which, equals the number of flip-flops used in the counter. We can see from the waveforms of Fig. 24.43. (*b*) that when the counter is in (00) state, the AND-gate designated '0' is in HIGH state and the outputs of other gates designated '1', '2' and '3' are in LOW state. Similarly, for 01, 10 and 11 states of the counter, the outputs of gates 1,2 and 3 are HIGH. This is incidentally Active-High decoding. We can visualise that if the AND-gates were replaced by NAND-gates with the inputs to the gates remaining the same, we shall get an Active-LOW decoder. For a counter that uses '*N*' flip-flops and has a modulus of '*X*', the decoder shall have '*X*' number of (*N*)-input AND or NAND gates depending upon whether we want an active-HIGH or active-LOW decoder.

Table 24.5 Functional Index of Counters and Registers

Function	*IC Number*	*Logic Family*
Decade counter	54/7490	TTL
4-bit binary counter	54/7493	TTL
BCD decade counter with asynchronous RESET	54/74160	TTL
4-bit binary counter with asynchronous RESET	54/74161	TTL
BCD decade counter with synchronous RESET	54/74162	TTL
4-bit binary counter with synchronous RESET	54/74163	TTL
UP/DOWN decade counter	54/74168	TTL
UP/DOWN decade counter	54/74169	TTL
Presettable BCD decade UP/DOWN counter 4-bit	54/74190	TTL
Presettable/binary UP/DOWN counter	54/74191	TTL
Presettable BCD decade UP/DOWN counter	54/74192	TTL
Presettable 4-bit binary UP/DOWN counter	54/74193	TTL
Decade counter	54/74290	TTL
4-bit binary counter	54/74293	TTL
Dual decade counter	54/74390	TTL
Dual 4-bit binary counter	54/74393	TTL
8-bit shift register, serial-in serial-out	54/7491	TTL
4-bit shift register	54/7495	TTL
8-bit shift register (serial-in parallel-out)	54/74164	TTL
8-bit shift register (Parallel-in, Serial-out)	54/74166	TTL
4-bit Right/Left shift register	54/74194	TTL
4-bit shift register	54/74195	TTL

Cont...

5-stage Johnson counter	4017B	CMOS
4-stage divide by 9 Johnson counter	4022B	CMOS
UP/DOWN decade counter	4510B	CMOS
Dual decade counter	4518B	CMOS
Dual 4-bit binary counter	4520B	CMOS
4-bit programmable BCD DOWN counter	4522B	CMOS
4-bit programmable binary DOWN counter	5426B	CMOS
4-bit synchronous programmable counter	40161B	CMOS
Presettable UP/DOWN binary counter	40163B	CMOS
8-stage static shift register (Parallel or serial input, serial output)	4014B	CMOS
Dual 4-stage static shift register (Serial input parallel output)	4015B	CMOS
8-bit static shift register (Asynchronous parallel input or synchronous serial input, serial output)	4021B	CMOS
4-bit universal shift register	4035B	CMOS

Cont...

Q. 38: What is a bi-directional shift register?

Ans: A bi-directional shift register allows shifting of data either to the left or to the right. This is made possible with the inclusion of some gating logic having a control input. The control input allows shifting of data either to the left or to the right, depending upon its logic status.

SOLVED PROBLEMS

Problem 24.1. *Fig. 24.44 shows an edge triggered flip-flop with active LOW inputs alongwith the pulse waveforms appearing at S, R and clock inputs. Draw the Q-output waveform.*

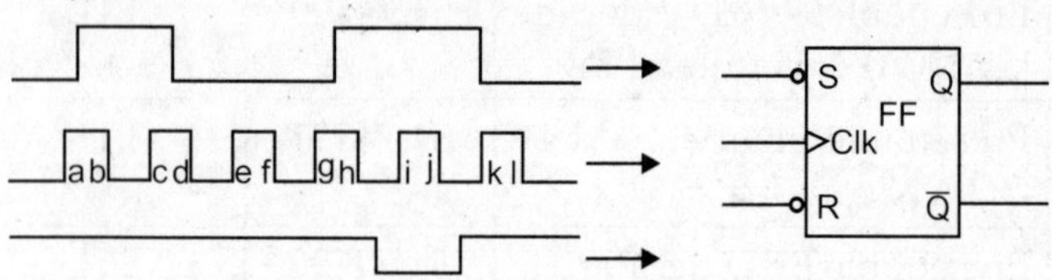

Fig. 24.44

Solution. The circuit symbol indicates that the flip-flop responds to LOW-to-HIGH transition of the clock waveform. Also since the circuit has active LOW inputs, $S = 0, R = 0$ combination is prohibited. $S = 0, R = 1$ SETS the flip-flop ; $S = 1, R = 0$ RESETS it and $S = 1$ $R = 1$ causes no change. All these changes at the output take place only on the LOW-to-HIGH transition of the clock pulse waveform. With this background, the *Q*-output waveform can be drawn as shown in Fig. 24.45.

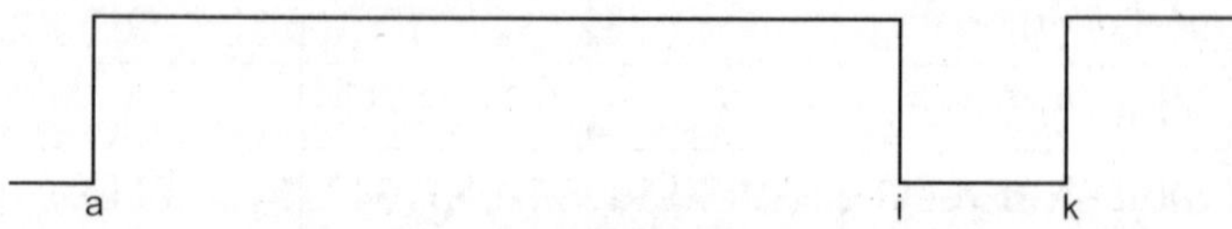

Fig. 24.45

Problem 24.2 *If the S, R and clock waveforms of problem 24.1 are applied to the flip-flop of Fig. 24.46. what will be the Q-output waveform ?*

Solution. The flip-flop shown is a level triggered flip-flop with active LOW inputs. With the operational principle of a level triggered flip-flop as explained earlier, the Q-output waveform can be drawn as shown in Fig. 24.47.

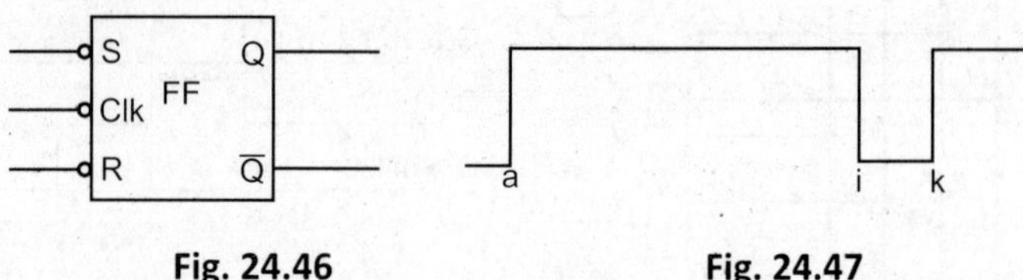

Fig. 24.46 Fig. 24.47

Problem 24.3. *Draw the logic circuit implementation of a basic R-S latch with active LOW inputs using 2-input NOR gates only.*

Solution. The logic implementation is shown in Fig. 24.48. The circuit shown is self-explanatory.

Problem 24.4. *Identify the type of clocked R-S flip-flop from the logic circuit diagram of Fig. 24.49 in respect of (i) whether the flip-flop has active LOW or active HIGH inputs (ii) Whether it is EDGE triggered or LEVEL triggered (iii) if it is edge triggered, will it trigger on the LOW-to-HIGH transition or HIGH-to-LOW transition and (iv) if it is level triggered, which clock pulse level (HIGH or LOW) will the flip-flop respond to?*

Solution. It is apparent from the circuit diagram of Fig. 24.49 that the clock pulse is being used to either enable or disable the input AND gates which means that the *S* and *R* inputs are passed onto the NAND latch portion of the circuit shown as long as the clock is HIGH. When the clock goes LOW, the AND gates are disabled and any changes on *S* and *R* inputs can not reach the latch inputs. Therefore, the logic circuit represents a level triggered flip-flop and also the flip-flop responds, when the clock level is HIGH. The truth table of this logic circuit is shown in Fig. 24.50. It is obvious from the truth table that the flip-flop is SET when $S = 0$, $R = 1$ and it is RESET, when $S = 1$, $R = 0$ provided the clock level is HIGH. The entries in the Q-output column, when the clock level is LOW, are marked (–) and it indicates here that when the clock is LOW, the flip-flop output holds the stage it was in just before the clock level going LOW. To conclude, the flip-flop has active LOW inputs.

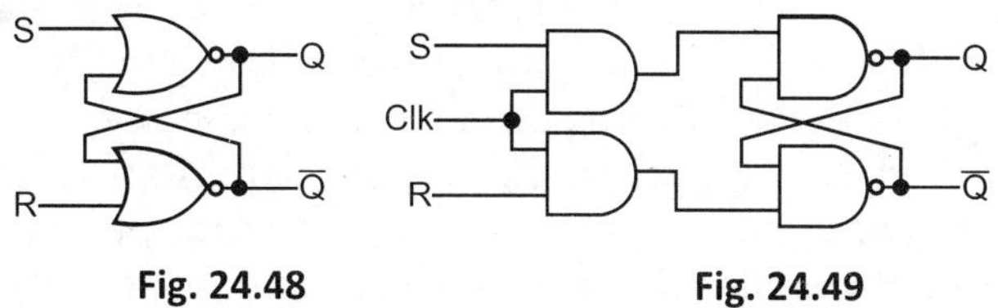

Fig. 24.48 **Fig. 24.49**

Problem 24.5. *Draw the logic circuit diagrams to implement the following clocked R-S flip-flop functions:*

(*i*) *HIGH level triggered with active HIGH inputs*

(*ii*) *HIGH level triggered with active LOW inputs.*

Solution.

(*i*) The logic diagram is shown in Fig. 24.51 (*a*). The cross coupled NOR gates at the output constitute the *R-S* latch with active LOW inputs. When the clock is HIGH, the *R* and *S* inputs are complemented by the input NAND gates and as a result, the active LOW inputs become the active HIGH inputs.

(*ii*) The logic diagram is shown in Fig. 24.51(*b*). The operation of the circuit can be explained on same lines as in the case of part (*i*) of this problem.

S	R	CLK	Q_{n+1}
0	0	0	–
0	0	1	Not permitted
0	1	0	–
0	1	1	1
1	0	0	–
1	0	1	0
1	1	0	–
1	1	1	No change

Fig. 24.50

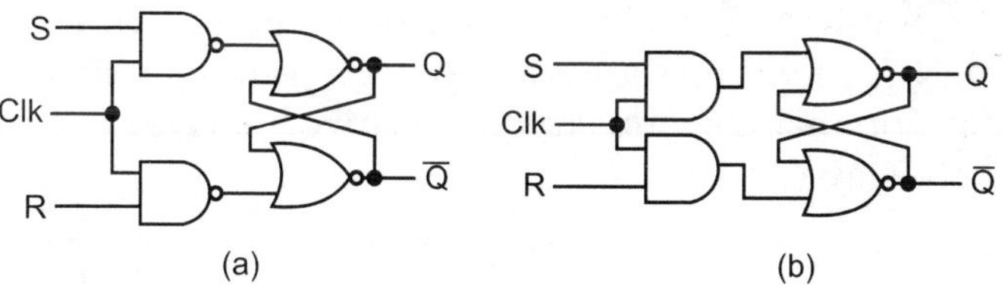

Fig. 24.51

Problem 24.6. *What will be the Q-output logic state for the J-K flip-flop of Fig. 24.52. for the following input conditions immediately after it is clocked ?*

(*a*) $J = 0, K = 0, Pr = 1, Cl = 1$

(*b*) $J = 1, K = 0, Pr = 1, Cl = 1$

(*c*) $J = 0, K = 1, Pr = 1, Cl = 1$

(*d*) $J = 1, K = 1\ Pr = 1, Cl = 1$

(*e*) $J = 1, K = 1, Pr = 0, Cl = 1$

(*f*) $J = 0, K = 1, Pr = 0, Cl = 1$

(*g*) $J = 1, K = 0, Pr = 1, Cl = 0$

(*h*) $J = 0, K = 1, Pr = 1, Cl = 0.$

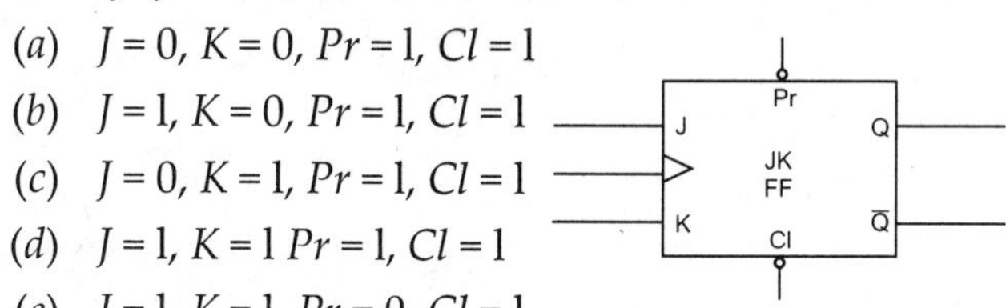

Fig. 24.52

Solution.

(*a*) It will be the same as it was prior to clocking, that is, if it was '1', it will continue to be '1' and if it was '0', it will continue to be '0'.

(*b*) It will be '1'

(*c*) It will be '0'

(*d*) It will toggle. That is, if it was '1', it will become '0' and if it was '0', it will become '1'.

(*e*) It will be '1'. In this case, the output will be preset to logic '1' state, as the PRESET input is logic '0' and PRESET input in this case has active LOW input. *J* and *K* inputs in such a case have no effect on the output.

(*f*) The output shall again be logic '1'. The reasons have been explained in case of (*e*) above.

(*g*) The output will be logic '0'. As the CLEAR input is, '0', the output will be cleared to '0' and the *J* and *K* inputs do not have any effect on the output.

(*h*) The output will be logic '0' for the reasons explained above.

Problem 24.7. *Draw the Q-output and $\overline{Q}$-output for the J-K flip-flop and the input waveforms shown in Fig. 24.53.*

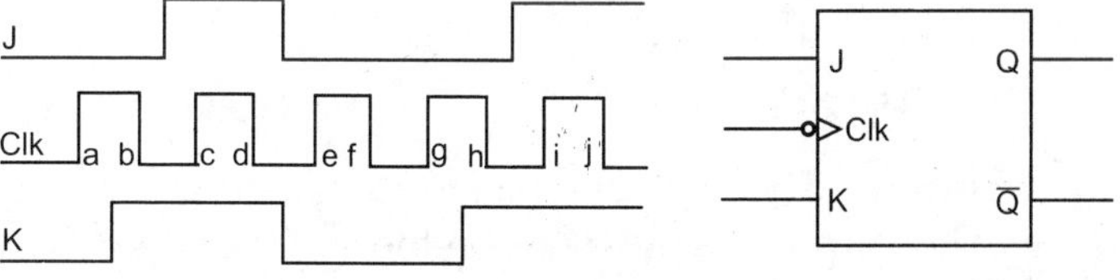

Fig. 24.53

Solution. The output waveforms are shown in Fig. 24.54 The flip-flop shown in Fig. 24.53 is an edge triggered flip-flop that responds to the negative going edge of the clock waveform.

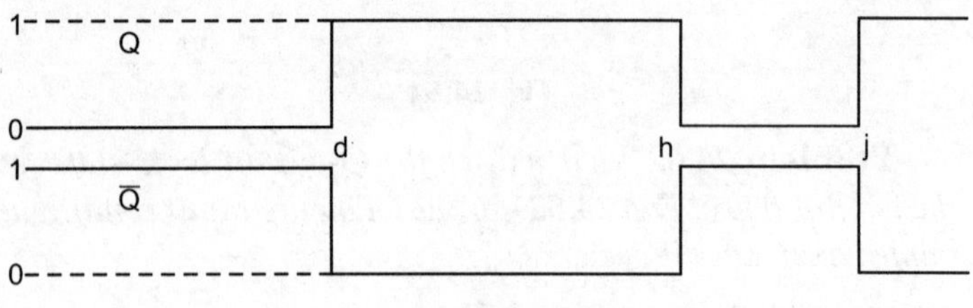

Fig. 24.54

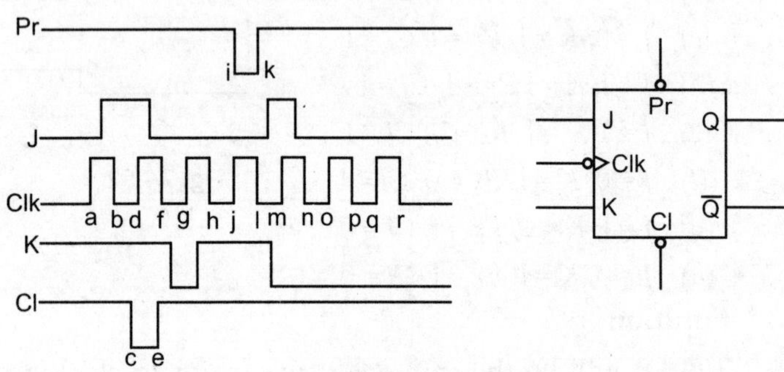

Fig. 24.55

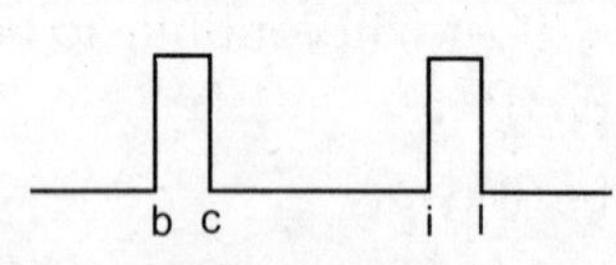

Fig. 24.56

Problem. 24.8. Fig. 24.55 shows a presettable, clear-able, negative edge triggered *J-K* flip-flop alongwith the waveforms at its various inputs. Draw the *Q*-output waveform.

Solution. The *Q*-output waveform is shown in Fig. 24.56 As is clear from the circuit symbol the PRESET and CLEAR inputs are active LOW inputs and therefore, the *Q*-output will be set to logic '1' state whenever PRESET input is LOW and cleared to logic '0', whenever the CLEAR input is LOW irrespective of *J* and *K* inputs.

Problem 24.9. *If the J, K, Clock, Preset and Clear waveforms of Fig. 24.55 were applied to a M/S J-K flip-flop of Fig. 24.57, what would be the Q-output waveform?*

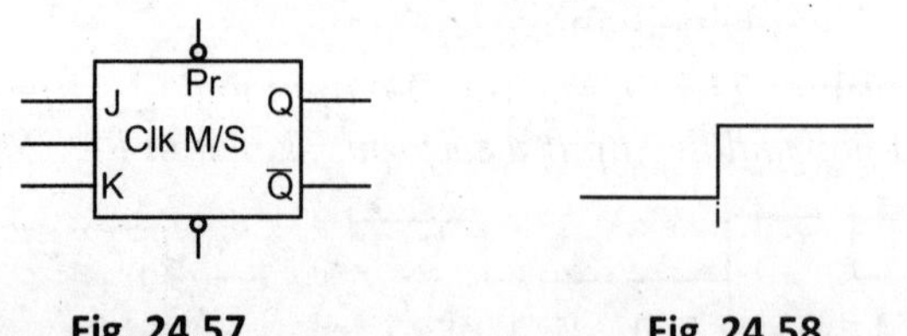

Fig. 24.57 Fig. 24.58

Solution. The output waveform is shown in Fig. 24.35. The output changes state on the leading edge of the clock waveform.

Problem 24.10. *Draw the output waveform for the cascaded arrangement of T-flip-flops shown in Fig. 24.59.*

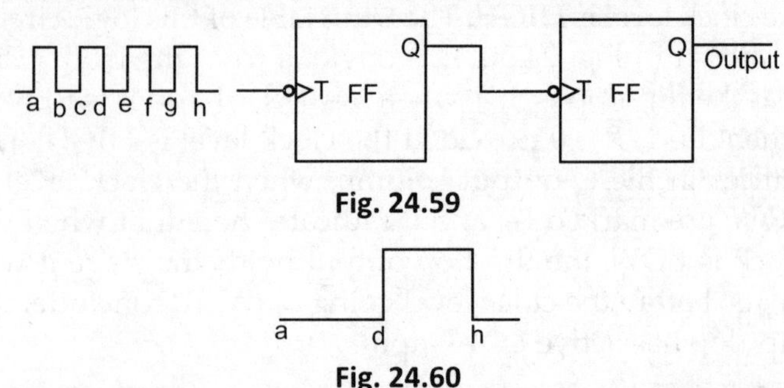

Fig. 24.59

Fig. 24.60

Solution. The output waveform is shown in Fig. 24.60 Output of first T-flip-flop changes state for every negative going edge of the input clock waveform. Its frequency is, therefore, half of the frequency of clock pulses. The output of first flip-flop acts as clock input for the second T-flip-flop as shown in the Fig. 24.59. The second flip-flop output toggles for every negative going edge of the waveform appearing at its clock input *i.e.* the waveform appearing at the output of the first T-flip-flop. The final output thus has a frequency that is one-fourth of the input close pulses.

Problem 24.11. *Draw the output waveform for the D-flip-flop and the input waveforms shown in Fig. 24.61 (a).*

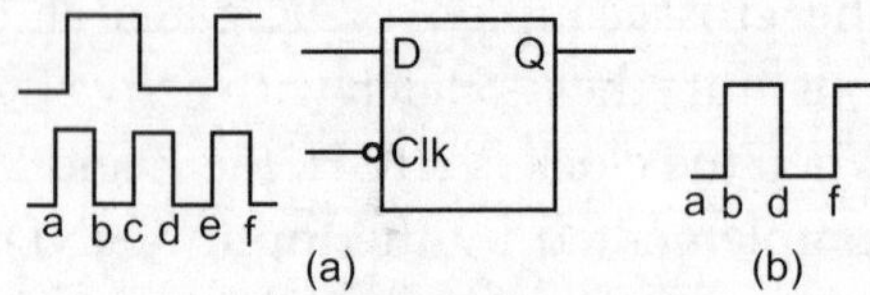

Fig. 24.61

Solution. The *Q*-output waveform is shown in Fig. 24.61 (*b*).

Problem 24.12. *Fig. 24.62 shows the internal logic circuit diagram of one of the four D-type latches of a 4-bit D-latch in IC 7475. Give argument to prove that the Q-output will track the D-input only, when the ENABLE input is HIGH.*

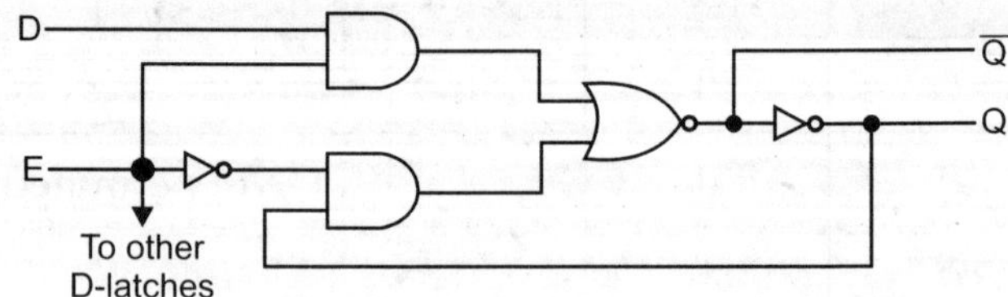

Fig. 24.62

Solution. When the ENABLE input is HIGH, the upper AND-gate is enabled while the lower AND-gate is disabled. As a result, output of upper AND is '*D*' and the output of lower AND is '0'. The output of NOR gate in this case will be $\overline{D}$ which implies that $Q = D$.

It can also be proved that the *Q*-output holds its value it had just before the ENABLE input went LOW during the time the ENABLE input is LOW.

Problem. 24.13 *In the 3-bit counter of Fig. 24.21. (a) what would be the output of FF0, FF1 and FF2 at the end of 13 clock cycles?*

Solution. $Q_0 = 1$ $Q_1 = 0$ $Q_2 = 1$.

Problem 24.14. *The input clock frequency to the ripple counter of Fig. 24.63. is 1 MHz. What is the frequency of FF2 output ? What will be the logic status of FF1 output at the end of 27th clock cycle?*

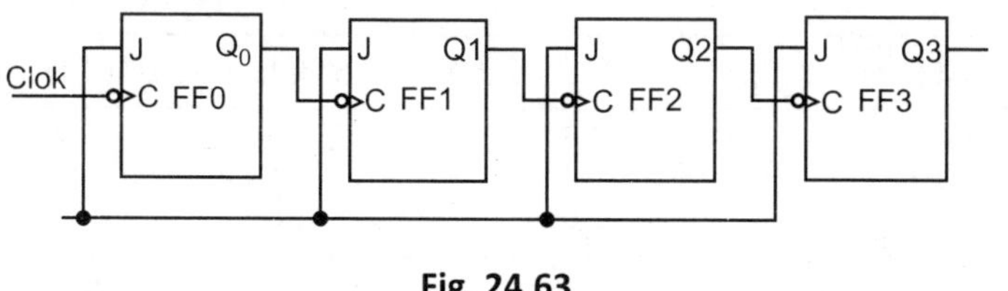

Fig. 24.63

Solution. The frequency of FF2 output would be 125 kHz (= lMHz/8).

At the end of 16th clock cycle, the output will be 0000. The 17th clock cycle will make it 0001. The counter progresses in the same fashion and at the end of 27th clock cycle, the output will be 1011. The logic status of FF1 at the end of 27th clock pulse will, therefore, be '1'.

Problem 24.15. *What is the hardware required to construct a Mod-3 counter that counts in the natural count sequence using flip-flops and gates? Draw the circuit arrangement.*

Solution. The hardware required includes two *J-K* flip-flops with CLEAR facilities and a 2-input NAND gate.

The circuit arrangement is shown in Fig. 24.64. The feedback is such that the counter skips the state 11 and is reset to 00 on the third clock transition.

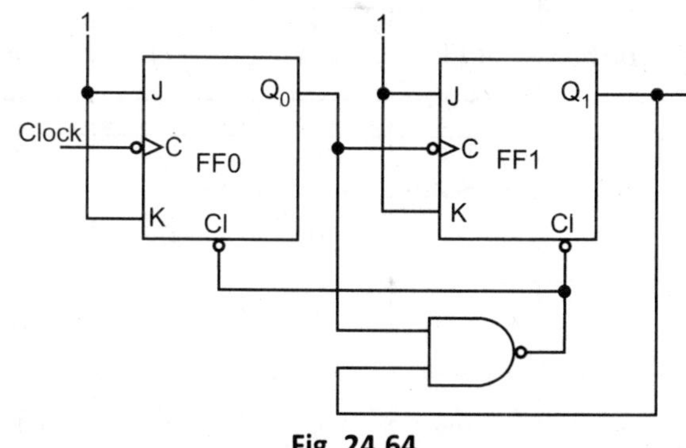

Fig. 24.64

Problem 24.16. *Draw the logic circuit diagram of a Mod–5 counter that counts in its natural count sequence from 000 to 100 using clearable negative edge triggered J-K flip-flops and NAND gates.*

Solution. Fig. 24.65 shows the circuit diagram that is self-explanatory.

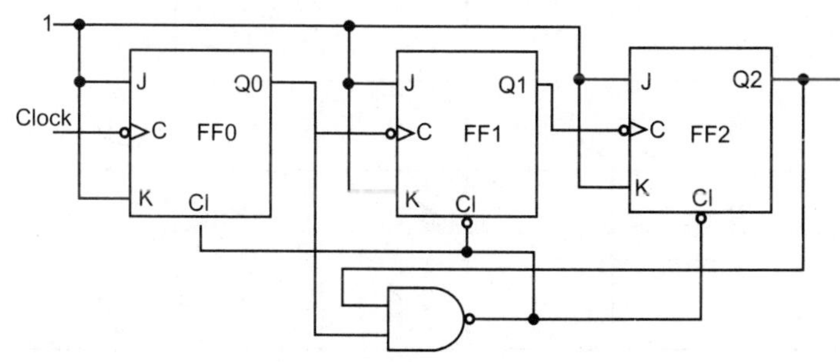

Fig. 24.65

Problem 24.17. *Fig. 24.66 shows some logic circuit arrangement of four J-K flip-flops and an AND gate. Find out if this arrangement represents any counter circuit by drawing its truth table and the timing waveforms.*

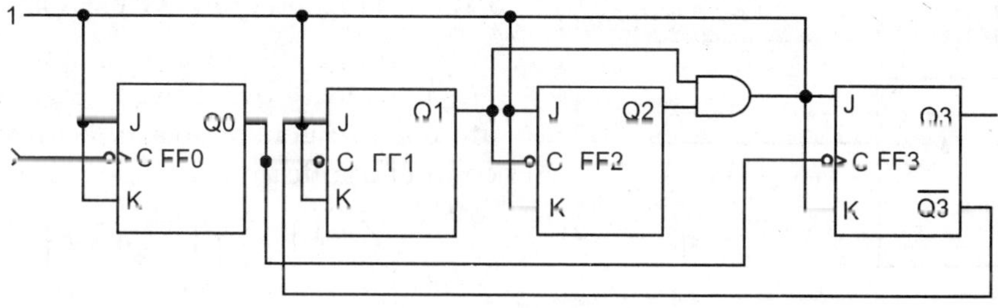

Fig. 24.66

Solution. It is obvious from the circuit arrangement that:

Flip-flop FF0 toggles for every negative transition of the clock pulse.

Flip-flop FF1 toggles, whenever Q0 = 1 and Q3 = 0.

It is reset to 0 whenever Q0 = 1 and Q3 = 1.

It holds its existing state, whenever Q0 = 0 irrespective of the state of Q3.

Flip-flop FF2 toggles, whenever Ql goes from 1 to 0. For all other conditions of Ql, Q2 holds.

Flip-flop FF3 toggles whenever Ql = Q2 = 1 and Q0 goes from 1 to 0. For all other conditions of Ql and Q2. FF3 is reset to 0, whenever Q0 goes from 1 to 0.

With the behaviour of different flip-flops as explained above, the truth table for the circuit of Fig. 24.66 is as shown in Fig. 24.67 (*a*). The timing waveforms are shown in Fig. 24.67 (*b*).

Count Number	*Q3*	*Q2*	*Q1*	*Q0*
0	0	0	0	0
1	0	0	0	1
2	0	0	1	0
3	0	0	1	1
4	0	1	0	0
5	0	1	0	1
6	0	1	1	0
7	0	1	1	1
8	1	0	0	0
9	1	0	0	1
0	0	0	0	0

(a)

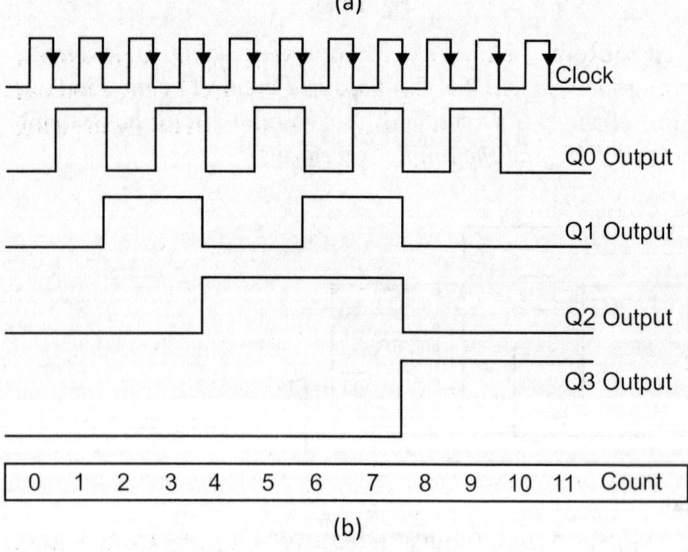

(b)

Fig. 24.67

Problem 24.18. *Fig. 24.68 shows the application of IC 74190 (Presettable BCD/Decade Counter) as a decade counter. The mode select table for this IC is also given. Draw the waveforms that would appear at Q0, Q1, Q2 and Q3, as the clock goes through one complete count cycle.*

Mode Select Table				
INPUTS				
$\overline{PL}$	$\overline{CE}$	$\overline{U}/D$	*CP*	***MODE***
H	L	L	↑	Count UP
H	L	H	↑	Count Down
L	X	X	X	Preset (Async.)
H	H	X	X	No Change (Hold)

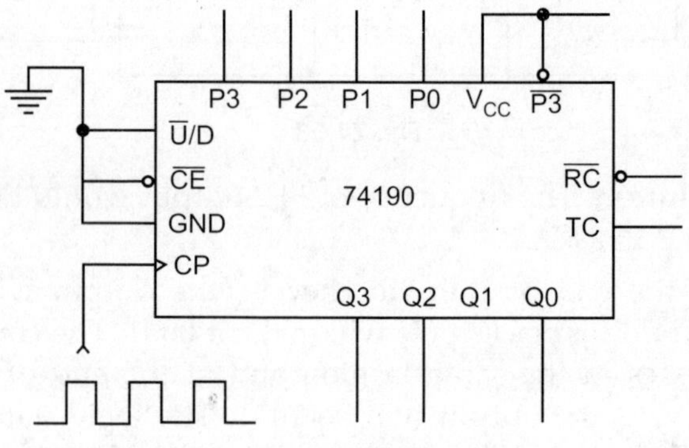

Fig. 24.68

Solution. From the interpretation of different terminals of 74190, we will know that the IC has been wired to count in the upward sequence.

$\overline{CE}$ is the COUNT ENABLE terminal with active LOW input. Its connection to ground ensures that the counter is enabled. $\overline{U}$/D is the $\overline{UP}$/Down count control input. When LOW, it counts upwards.

$\overline{PL}$ is the Parallel Load Control input that is active when LOW. When it is LOW, the data present at P3, P2, P1, P0 is loaded onto Q3, Q2, Ql, Q0.

A Terminal Count (TC) output and a Ripple clock ($\overline{RC}$) output provide overflow/underflow indication and are used to generate carry/borrow signals in multistage counter applications.

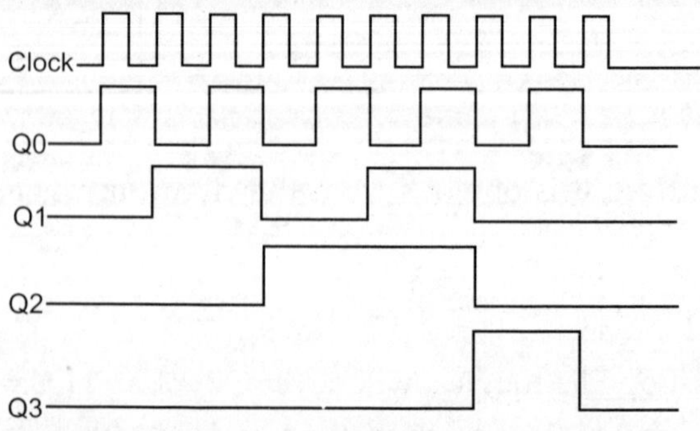

Fig. 24.69

Problem 24.19. *Draw the timing waveforms for the counter of Fig. 24.68, when the $\overline{U}$/D terminal is connected to + V_{cc}, other connections remain unchanged.*

Solution. The counter now counts downwards. Fig. 24.70 shows the timing waveforms.

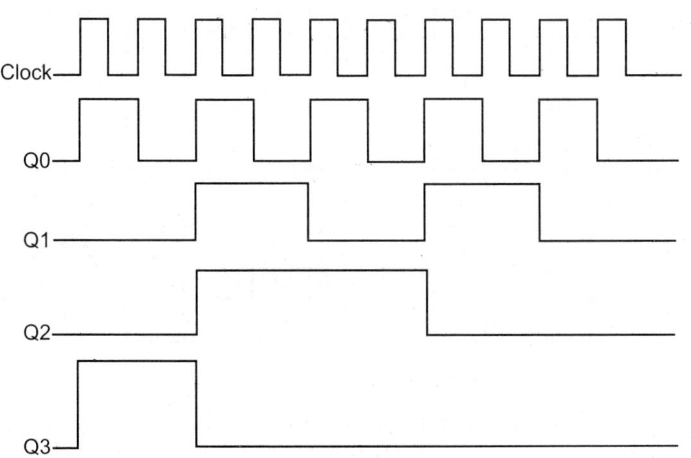

Fig. 24.70

Problem 24.20. *Fig. 24.71 shows a cascaded arrangement of two 74190s. Bath the Up / Down Decade counters have been wired as up counters. What will be the outputs designated A, B, C, D, E, F, G, H after 27th clock pulse.*

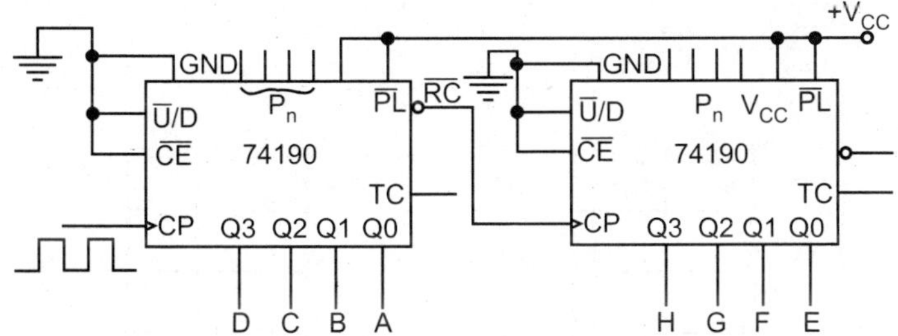

Fig. 24.71

Solution. The cascaded arrangement shown forms basically a BCD counter that can count from 0 to 99. The counter 74190 shown on the left forms l's counter while the one down on the right forms the 10's counter. Every time I's counter counts from 0000 to 1001, the 10's counter counts one pulse, as it is driven by ripple clock (RC) output that is internally enabled by terminal count (TC). Therefore, at the end of tenth clock pulse, A, B, C, D will all be 0 and E, F, G, H will respectively be 1, 0, 0, 0. Similarly, at the end of 27th clock pulse, A, B, C, D, E, F, G and H will respectively be 1, 1, 1, 0, 0, 1, 0 and 0

Problem 24.21. *Refer to Fig. 24.72 showing application of IC 7491 alongwith the waveform at A.*

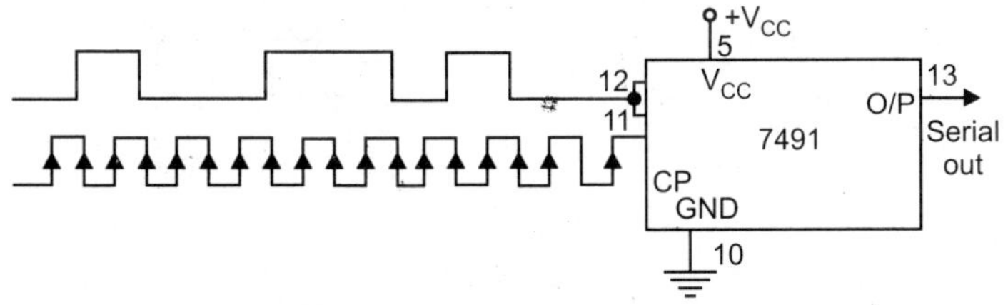

Fig. 24.72

(*i*) *What will be the data bit present at the output at the end of 11th LOW to HIGH transition of the clock waveform?*

(*ii*) *If there is a '1' at the end of nth LOW to HIGH clock transition at the Q3 output, what will be Q5 output at the end of (n + 2) the transition?*

Solution.

(*i*) At the end of 8th LOW to HIGH clock transition, the data bits loaded into the register will be 10110010 with 0 on the extreme right appearing at the output Q7 (Refer to Fig. 24.35). The 9th clock transition will shift this 0 out of the register and the next adjacent bit (*i.e.* 1) will take its place on Q7. Each subsequent clock pulse will shift the bits one step towards right. So, at the end of 11th clock transition, the Q7, output will-be a logic 0.

(*ii*) It will be 1 only. The Q3 output will be shifted 2 bit positions to right by two clock transitions.

Problem 24.22. *Refer to the functional diagram of IC74164 shown in Fig. 24.36. Terminals B and $\overline{MR}$ are tied HIGH and terminal A is applied the waveform shown in Fig. 24.73 (a). Draw the Q3 output waveform as the clock goes through 8 cycles.*

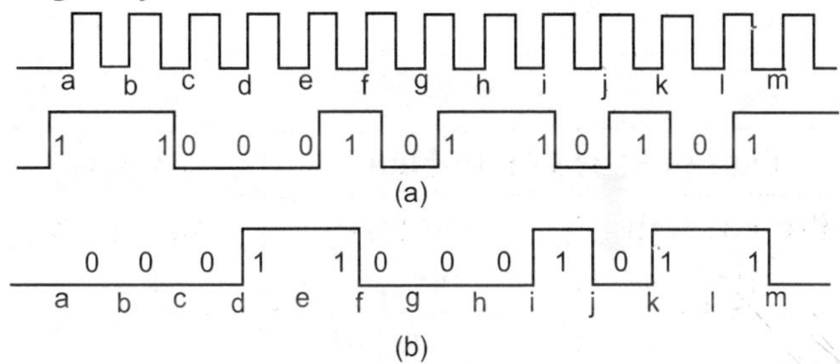

Fig. 24.73

Solution. By tying B and $\overline{MR}$ HIGH, the input AND gate is enabled so that the data bits to be loaded can reach the D-input of the first flip-flop and the CLEAR input is made inactive. The Q3 output waveform is shown in Fig. 24.73. (*b*):

Problem 24.23. *With the help of the functional diagram, explain how can a 3-bit shift counter be used as a modulus-5 counter?*

Solution. Fig. 24.74 (*a*) shows the functional diagram. Due to feedback from the true output of flip-flop FF1 to the *K*-input of flip-flop FF0, the counter skips the state 111. As a result, the counter has 5 distinct states instead of 6 that it would normally have in case it had the conventional feedback arrangement of a shift counter. The truth table giving the different states through which the counter passes as the clock goes through one complete count cycle is shown in Fig. 24.74. (*b*)

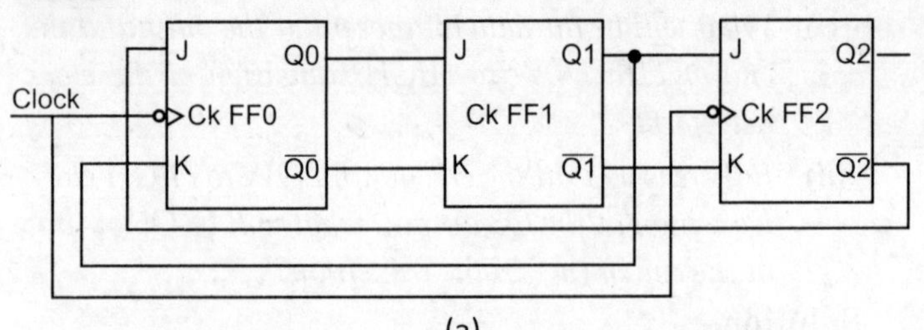

(a)

Clock Pulse	*Q2*	*Q1*	*Q0*
1	0	0	1
2	0	1	1
3	1	1	0
4	1	0	0
5	0	0	0

(b)

Fig. 24.74

Problem 24.24. *Table-24.6 gives excitation table of a certain flip-flop having (X1) and (X2) as its inputs. Draw the circuit excitation table of a MOD-5 synchronous counter using this flip-flop for the count sequence 000, 001, 011, 101, 110, 000...... If the present state is an undesired one, it should transit to (110) on application of clock pulse. Design the counter circuit using the flip-flop whose excitation circuit is given in Table-24.6.*

Table-24.6

Output State of Flip-Flop		**Inputs**	
Present State (Q_n)	**Next State (Q_{n+1})**	**X1**	**X2**
0	0	0	0
0	1	0	1
1	0	1	X
1	1	X	1

Solution: The circuit excitation table is shown in Table-24.7. The number of flip-flops required is 3. X1 (A), X2 (A) are the inputs of the flip-flop (A), which is also the LSB flip-flop. X1 (B), X2 (B) represent inputs to B-flip-flop.

Table-24.7

Present State			*Next State*			*Inputs*					
C	B	A	C	B	A	X1 (A)	X1 (B)	X2 (A)	X2 (B)	X3 (A)	X3 (B)
0	0	0	0	0	1	0	1	0	0	0	0
0	0	1	0	1	1	X	1	0	1	0	0
0	1	0	1	1	0	0	0	X	1	0	1
0	1	1	1	0	1	X	1	1	X	0	1
1	0	0	1	1	0	0	0	0	1	X	1
1	0	1	1	1	0	1	X	0	1	X	1
1	1	0	0	0	0	0	0	1	X	1	X
1	1	1	1	1	0	1	X	X	1	X	1

X = don't care condition

X1 (C) and X2 (C) are the inputs to the C-flip-flop, which is also the MSB flip-flop. The excitation table is self-explanatory. Next step is to draw Karnaugh maps, one each for different inputs to the three flip-flops. Figures-24.75 (a) to (f) show the Karnaugh maps for X1 (A), X2 (A), X1 (B), X2 (B), X1 (C) and X2 (C) respectively. The minimised expressions are given in the following equations:

$X_1(A) = A, X_2(A) = A + \overline{B}\,\overline{C}, X_1(B) = B, X_2(B) = A + B + C, X_1(C) = C, X_2(C) = B + C$

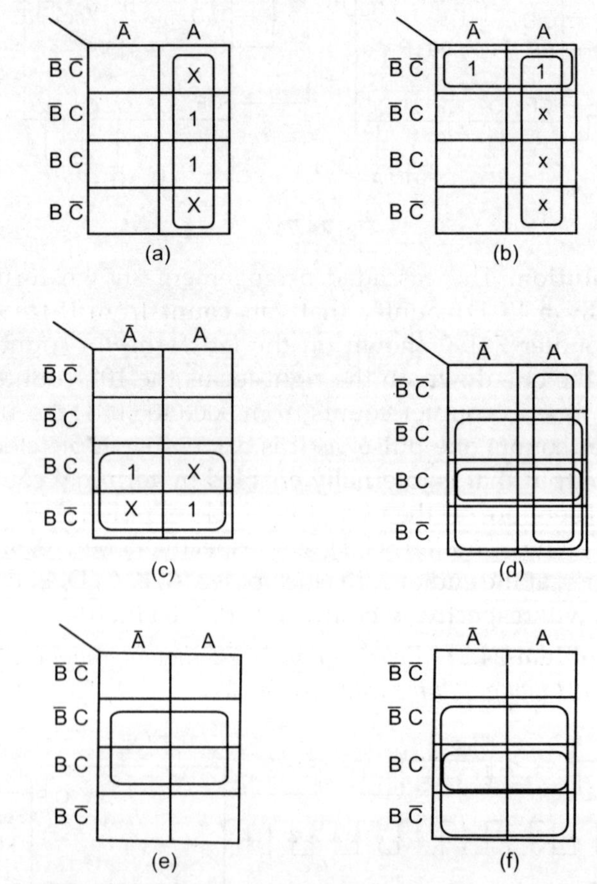

Fig. 24.75

Fig. 24.76 shows the final circuit.

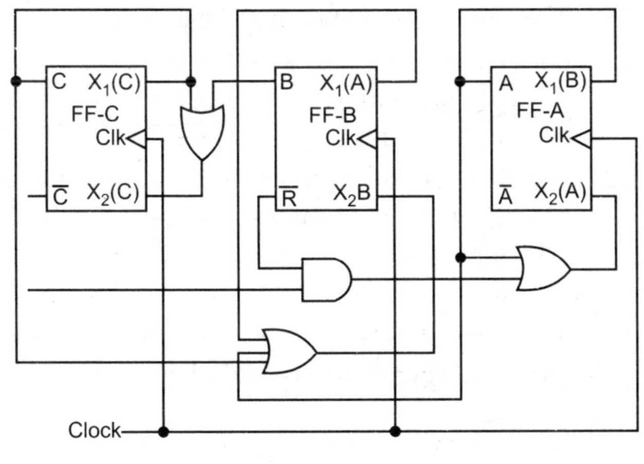

Fig. 24.76

Problem 24.25. *Determine the number of flip-flops required to construct (a) a MOD-10 ring counter and (b) a MOD-10 Johnson counter. Also, write the count sequence in the two cases.*

Ans: (*a*) The modulus of a ring counter is the same as the number of bits (or flip-flops). Therefore, the number of flip-flops required=10. The count sequence is 1000000000, 0100000000, 0010000000, 0001000000, 0000100000, 0000010000, 0000001000, 0000000100, 0000000010, 0000000001 and back to 1000000000.

(*b*) The modulus of a Johnson counter is twice the number of flip-flops. Therefore, the number of flip-flops = 5. The count sequence is 00000, 10000, 11000, 11100, 11110, 11111, 01111, 00111, 00011, 00001 and back to 00000.

Summary of Important Formulas

1. The modulus of an n-bit counter is given by:
   Modulus, $M \leq 2^n$
2. Minimum number of flip-flops (*N*) required to build a modulo-M counter is given by:
   N = An interger $\geq 3.33 \log M$
3. The overall propagation delay of an n-bit binary ripple counter is given by:
   Propagation Delay, $t_p(n) = n \times t_p$
   Where t_p = Propagation delay of one flip-flop
4. The overall propagation delay of an n-bit synchronous counter is given by:
   Propagation delay, $t_p(n) = t_p$
   Where t_p = propagation delay of one flip-flop
5. In Modulus-N shift or Johnson Counter, number of flip-flop required is given by:
   Number of flip-flops = N/2
6. In a Modulus-N Ring Counter, number of flip-flops required is given by:
   Number of flip-flops = N

Self Evaluation Exercise

I. Multichoice Questions

1. In an R-S flip-flop with active LOW inputs, when R = S = 0, the output *Q* when the flip-flop is clocked is
   (*a*) 0
   (b) 1
   (*c*) Indeterminate
   (*d*) 0 if it was 1 earlier and 1 if it was 0 earlier.
2. In a presettable, clearable positive edge triggered J-K flip-flop, the *Q* output when the flip-flop is clocked for *J* = 1, *K* = 0, Preset = 1 and Clear = 0 will be
   (*a*) 1
   (*b*) 0
   (*c*) Indeterminate
   (*d*) Complement of what it was at the time of clocking (Assume all the inputs to be active LOW inputs).
3. The *Q*-output in a presettable, clearable *J-K* flip-flop with active LOW inputs is logic 1 at a certain time. The output when the flip-flop is clocked for J = 1, K = 1, Pr = 0, CI = 1 will
   (*a*) become 0 (*b*) toggle
   (*c*) remain 1 (*d*) none of these.
4. For one of the following conditions, clocked J-K flip-flop can be used as DIVIDE BY 2 circuit when the pulse train to be divided is applied at clock input.
   (*a*) J = 1, K = 1 and the flip-flop should have active HIGH inputs
   (*b*) J = 1, K = 1 and the flip-flop should have active LOW inputs

(*c*) J = 0, K = 0 and the flip-flop should have active HIGH inputs

(*d*) J = 1, K = 1 and the flip-flop should be a negative edge triggered one.

5. We have two negative edge triggered J-K flip-flops with active LOW inputs. The *J* and *K* inputs of both the flip-flops have been permanently tied to logic 0. The Q-output of first flip-flop (Q_1) feeds the clock input of the second flip-flop. What will be the logic status of outputs Q_1 and Q_2 at the end of five clock cycles, if the two outputs were cleared to logic 0 each before start.

(*a*) $Q_1 = 1, Q_2 = 0$ (*b*) $Q_1 = Q_2 = 1$

(*c*) $Q_1 = 0, Q_2 = 1$ (*d*) $Q_1 = 0, Q_2 = 0$.

6. A 4-bit circulating register is initially set to 0001 (1 is the true output of the first flip-flop and represents LSB). 4-bit number preset in the register at the end of 16 clock pulses is

(*a*) 0001 (*b*) 0010

(*c*) 0100 (*d*) 1000

7. A shift counter comprising of 5 flip-flops with an inverse feedback from the output of the MSB flip-flop to the input of the LSB flip-flop is a

(*a*) Divide by 32 counter

(*b*) Divide by 10 counter

(*c*) 5-bit shift register

(*d*) Modulus-5 counter

8. Minimum number of flip-flops needed to construct a BCD decade counter is

(*a*) 4 (*b*) 3

(*c*) 10 (*d*) none of these

9. A 4-bit UP/DOWN binary ripple counter is initially reset to 0000. The UP/DOWN mode select terminal designated as $\overline{U}/D$ on the pin connection diagram of the IC is tied to HIGH input. What will be the counter output status after the first clock pulse ?

(*a*) 0001 (*b*) 1000

(*c*) 1111 (*d*) none of them

10. A 4-bit binary number 1010 is to be loaded in a 4-bit serial-in serial-out shift register feeding in first the MSB. What will be the logic status of the true output of the output flip-flop at the end of 6 clock cycles ?

(*a*) 0 (*b*) 1

(*c*) Indeterminate

11. The shift registers belong to the class of

(*a*) sequential logic circuits

(*b*) combinational logic circuits

(c) neither of the two

12. Number of D-flip-flops required to confute Modulo-12 shift counter would be

(*a*) 5 (*b*) 10

(*c*) 6 (*d*) 12

13. Number of D-flip-flops required to build model-12 ring counter would be

(*a*) 12 (*b*) 6

(*c*) 10 (*d*) 5

14. In a J.K flip-flop with active-HIGH input, J and K inputs are tied logic '1' state. The flip-flop would behave as a

(*a*) D-flip-flop (*b*) Toggle flip-flop

(*c*) JK flip-flop only (*d*) None of these

15. A Modulo-6 counter would divide the input clock frequency by

(*a*) 6 (*b*) 3

(*c*) 12 (*d*) None of these

II. Fill in the Blanks

1. "D-type latch is another name of a D-type flip-flop". This statement is................(True, False)

2. In a negative edge triggered D-flip-flop, the data at D-input is transferred to the Q-output during............transition of the clock input. (L to H, H to L)

3. The clocked R-S flip-flop is said to operate......... (Synchronously, Asynchronously)

4. The RACE PROBLEM does not exist in............. (R-S flip-flop, J-K flip-flop, M/S J-K flip-flop)

5. The flip-flop is the fundamental building block of...........logic circuits. (Sequential, Combinational)

III. Name the Flip-Flops Whose Circuit Symbols are Shown in Fig. 24.77.

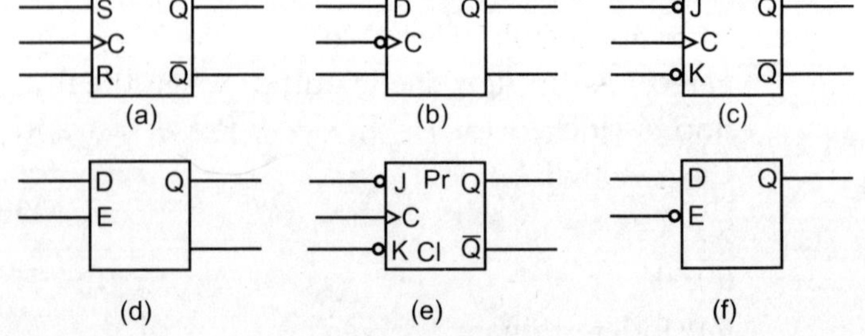

Fig. 24.77

S = Low to high transition of the clock

X = Irrelevant

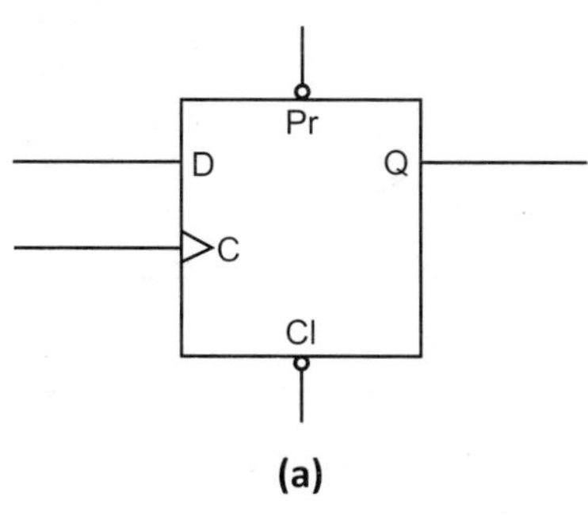

(a)

Mode of operation	*Inputs*		*Output*			
	Pr	*Clr*	*Clk*	*D*	*Q*	$\overline{Q}$
Asynchronous SET	(i)	1	X	X	1	0
Asynchronous RESET	1	0	X	X	0	1
Prohibited	0	(ii)	X	X	1	1
SET	1	1	↑	1	1	0
RESET	1	1	↑	0	0	(iii)

(b)

Fig. 24.78

IV. Fig. 24.78 (*a*) shows a D-type flip-flop with preset and clear facilities. An incomplete truth table for this flip-flop is shown in Fig. 24. 78 (*b*). Fill in the missing places.

V. Identify the type of counters and registers represented by the functional diagrams of Fig. 79.

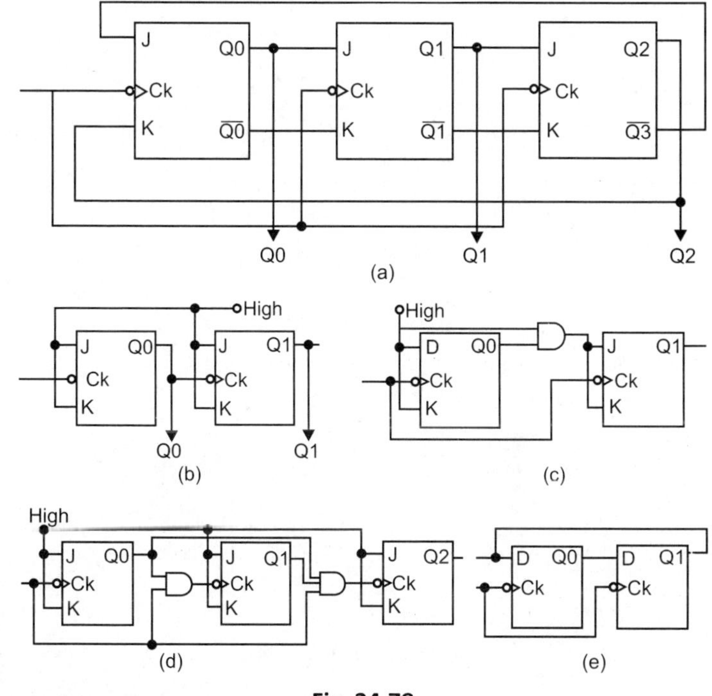

Fig. 24.79

Answer to Self Evaluation Exercise

I. 1. (*c*) 2. (*b*) 3. (*c*) 4. (*a*)
5. (*a*) 6. (*a*) 7. (*b*) 8. (*a*)
9. (*c*) 10. (*b*) 11. (*a*) 12. (*e*)
13. (*a*) 14. (*b*) 15. (*a*)

II. 1. False 2. H to L 3. Synchronously
4- M/S J/K flip-flop 5. Sequential

III. Fig. 24.77 (*a*) - Positive edge triggered R-S flip-flop with active HIGH inputs

Fig. 24.77 (*b*) - Negative edge triggered D-type flip-flop

Fig. 24.77 (*c*) - Positive edge triggered J-K flip-flop with active LOW inputs

Fig. 24.77 (*d*) - D-type latch with active HIGH Enable input.

Fig. 24.77 (*e*) - Presettable, Clearable positive edge triggered J-K flip-flop with all inputs active, when LOW

Fig. 24.77 (*f*) - D - type latch with active LOW enable input,

IV. (*i*) 0 (*ii*) 0 (*iii*) 1.

V. Fig. 24.79 (*a*)-3-bit shift counter Fig. 24.79 (*d*)-3-bit synchronous counter

Fig. 24.79 (*b*)-2-bit binary ripple counter Fig. 24.79 (*e*)-2-bit circulating register or 2-bit shift register

Fig. 24.79 (*c*)-2-bit synchronous counter

Chapter 25

D/A and A/D Converters

Learning Objectives

After completing the chapter, you will learn the following topics:

1. Resistive Networks for D/A Conversion
2. Binary Ladder Network for D/A Conversion
3. D/A Converter Specifications Including Resolution, Accuracy, Non-linearity, Differential Non-linearity, Dynamic Range and Monotonicity
4. Types of D/A Converters Including Mutiplying, Bipolar Output, BCD Input and Companding D/A Converters
5. Modes of Operations of D/A Converters Including Current Steering Mode and Voltage Switching Mode
6. Common Applications of D/A Converters
7. A/D Converter Specifications Including Accuracy, Resolution, Non-linearity, Differential Non-linearity, Conversion Time, Aperture and Acquisition Time, Code width, Sampling Rate, Gain and Offset Errors and Gain and Offset Drifts
8. A/D Converter Terminology
9. Types of A/D Converters Including Flash or Simultaneous A/D Converters, Half Flash A/D Converters, Counter Type A/D Converters, Successive Approximation A/D Converters, Single Slope, Dual Slope and Multi Slope A/D Converters, Tracking Type A/D Converters and Sigma-Delta A/D Converters.
10. Common Applications of A/D Converters

Q. 1. How do you convert a digital signal into an equivalent analog signal ? Explain in brief the process of D/A conversion in both the Resistive-Divider network and the modified Binary Ladder network.

Ans. The digital information available to us in the form of 1s and 0s can be converted into an equivalent analog signal by means of simple resistive networks. One such resistive network that can convert a 3 bit digital signal into an equivalent analog signal is shown in Fig. 25.1. This network, however, can be extended further to enable it to perform digital to analog conversion of larger number of bits. In Fig. 25. 1, if (R_L) is much larger than (R), it can be proved with simple electrical circuit theorems that the output analog signal is given by:

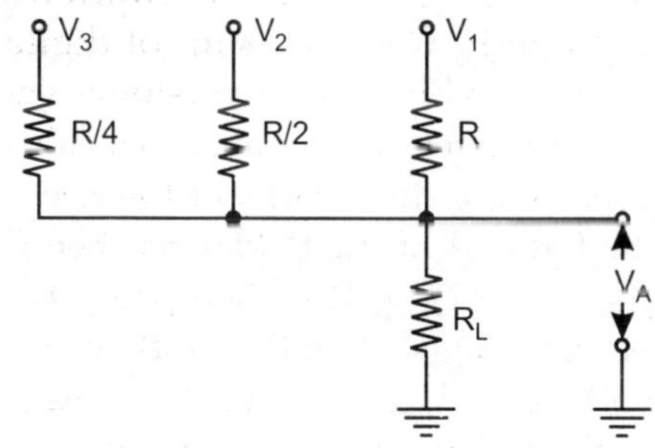

Fig. 25.1

$$V_A = \frac{\left(\frac{V_1}{R}\right)+\left(\frac{V_2}{R/2}\right)+\left(\frac{V_3}{R/4}\right)}{\left(\frac{1}{R}\right)+\left(\frac{1}{R/2}\right)+\left(\frac{1}{R/4}\right)} = \frac{\left(\frac{V_1}{R}\right)+\left(\frac{2V_2}{R}\right)+\left(\frac{4V_3}{R}\right)}{\left(\frac{1}{R}\right)+\left(\frac{2}{R}\right)+\left(\frac{4}{R}\right)}$$

$$V_A = \frac{V_1 \times 2^0 + V_2 \times 2^1 + V_3 \times 2^2}{(2^3 - 1)}$$

If we had four bits, then

$$V_A = \frac{V_1 \times 2^0 + V_2 \times 2^1 + V_3 \times 2^2 + V_4 \times 2^3}{(2^4 - 1)}$$

We can generalise this expression for the case of the n-bit binary information,

$$V_A = \frac{V_1 \times 2^0 + V_2 \times 2^1 + V_3 \times 2^2 ... + V_n \times 2^{n-1}}{(2^n - 1)}$$

If we want to convert (010) into equivalent analog signal, let us see what does the above expression give us,

We have $V_1 = 0$, $V_2 = V$ (say) and $V_3 = 0 = V_4 = V_5 = ... = V_n$

Therefore, $V_A = \frac{0 \times 2^0 + V \times 2^1 + 0 \times 2^2 + ...}{7} = \frac{2V}{7}$

Assume $V = 7$ volts, then $V_A = 2$ volts. Similarly for (110), we get $V_A = 6$ volts. ,

Thus, if we take $V = 7$ volts, the output analog signal is just the decimal equivalent of input binary number. Even if V is not equal to 7 volts, the analog output progresses in the same fashion as the input binary number.

This resistive divider network has two serious drawbacks. The first drawback arises from the fact that each resistor in the network is of a different value. Since these dividers are constructed using precision resistors, the added expense becomes unattractive. Secondly, the resistor used for the most significant bit (MSB) is required to handle a much larger current than the LSB resistor. For instance, in a 10-bit network, current through MSB resistor will be about 500 times the current through LSB resistor. To overcome these drawbacks, a second type of network called a LADDER is usually used in practice. The binary ladder too is a resistive network whose output voltage is a properly weighted sum of digital inputs. A binary ladder for a 4-bit D/A conversion is shown in Fig. 25.2. It is made up of resistors of only two values and thus, overcomes one of the drawbacks of the resistive divider network. The left end of the ladder has been terminated in a resistance of $(2R)\ \Omega$. It can be proved with simple mathematics that a logic '1' in the MSB position provides an output voltage of (V/2), a logical '1' in the second MSB position provided an output voltage of (V/4), the third MSB provides a voltage of (V/8) and so on. Table 25.1 summarizes the results. In general, analog output (V_A) is given by,

$$V_A = \frac{V_0 2^0 + V_1 2^1 + V_2 2^2 + V_3 2^3 + ... + V_{n-1}\, 2^{n-1}}{2^n}$$

Table 25.1

Bit Position	*Binary weight*	*Output voltage*
MSB	$\frac{1}{2}$	V/2
2nd MSB	$\frac{1}{4}$	V/4
3rd MSB	$\frac{1}{8}$	V/8
4th MSB	$\frac{1}{16}$	V/16
.	.	.
.	.	.
.	.	.
.	.	.
.	.	.

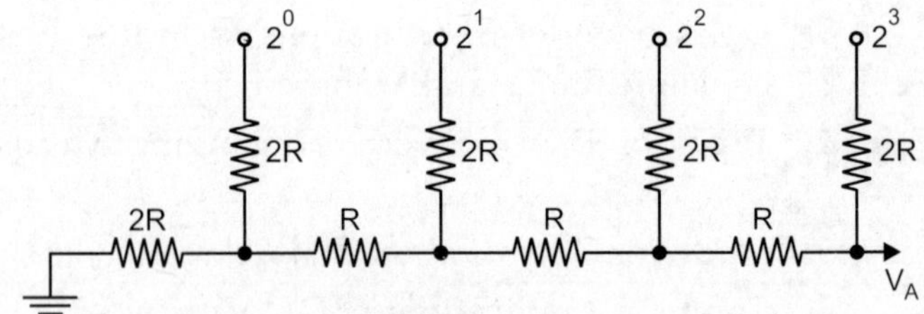

Fig. 25.2

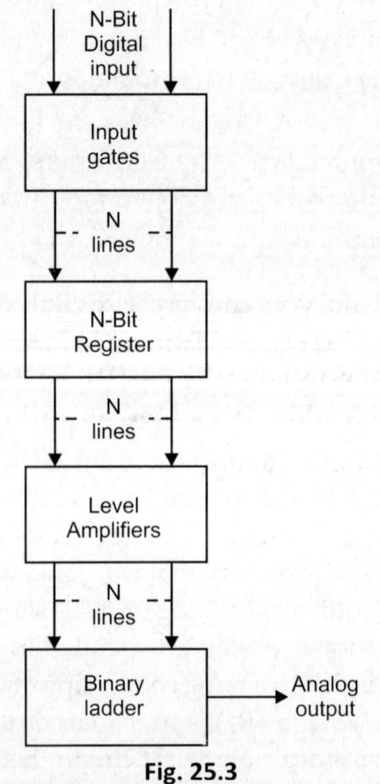

Fig. 25.3

Either the resistive divider or the binary ladder can be used as the basis for a D/A converter. It is in this network that actual digital to analog conversion takes place. There is, however, need for additional circuitry to complete the design of a D/A converter. As an integral part of a D/A converter, there must be a register which can be used to store digital information temporarily. There must be level amplifiers between register and resistive network to make sure that digital signals presented to the resistive network are all of the same level. Fig. 25.3 shows the block diagram representation of a complete N-Bit D/A converter.

D/A converter of different sizes (8-bit, 12-bit, 16-bit etc.) are available in the form of integrated circuits. In these ICs, all the component blocks just discussed are integrated within the same IC chip.

Q. 2. The important criteria for selecting a good D/A converter are the RESOLUTION, the ACCURACY and the SPEED. Briefly describe each one of these.

Ans. The RESOLUTION of a D/A converter is the number of states (2^n) that the full scale range is divided or resolved into. Here (n) is the number of bits in the input digital word. Higher the number of bits, better the resolution. For instance, a 12-bit D/A converter has a resolution of 0.024% (a 12-bit word has 4096 different states), whereas an 8-bit D/A converter has a resolution of 0.39% (an 8-bit D/A converter has only 256 different states).

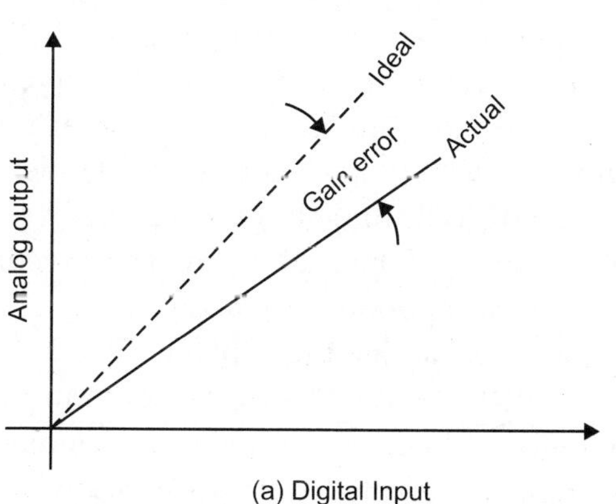

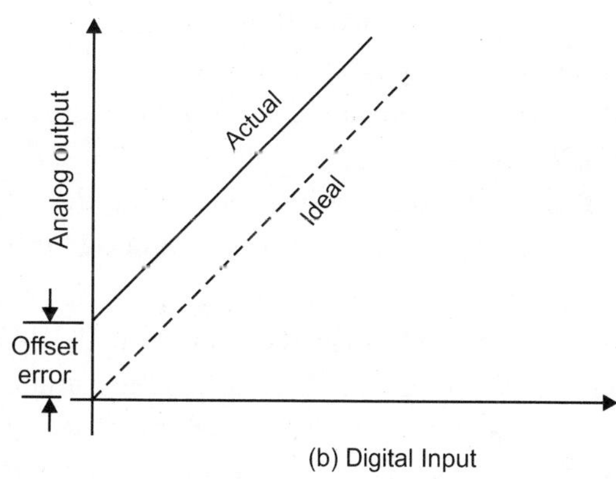

Fig. 25.4

The ACCURACY of a D/A converter is the difference between the actual analog output and ideal expected output, when a given digital input is applied. Sources of error include the Gain Error (or Full Scale Error). Offset Error (or Zero Scale Error), Non-linearity errors and a drift of all these factors. Gain error is difference between the actual and the ideal analog output voltage expressed as a percentage of full scale range or in terms of LSB (Fig. 25.4*a*).

The offset error is the error at analog zero with the data converter operating in bipolar mode (Fig. 25.4*b*). NON-LINEARITY is the maximum deviation of the analog output voltage from a straight line drawn between the end points expressed as a percentage of the full scale range or in terms of LSBs. DIFFERENTIAL NON-LINEARITY (DNL) is the worst case deviation of any adjacent analog outputs from the ideal 1 LSB step size.

SPEED of a D/A converter is expressed in terms of its settling time. Settling time is the elapsed time for the analog output to reach its final value within a specified error band after a digital input code change. General purpose D/A converters have a settling time of several microseconds while some of the ultra-high speed D/A converters have a settling time that is of the order of few nanoseconds. D/A converter Type No. AD 9768 from Analog Devices U.S.A. for instance, has as a settling time of 5 ns.

Q. 3. What do you understand by Dynamic Range of a D/A converter?

Ans. This is the ratio of the largest output to the smallest output, excluding zero, expressed in dB. For linear D/A converters it is $20 \times \log 2^n$, which is approximately equal to $6n$. Companding-type D/A converters, discussed in a subsequent question,have larger dynamic range than the other types of D/A converters.

Q. 4. Distinguish between Non-linearity (*NL*) and Differential Non-linearity (DNL), specifications of a D/A converter.

Ans. Non-linearity (*NL*) is the maximum deviation of analog output voltage from a straight line drawn between the end points expressed as a percentage of the full scale range or in terms of LSBs.

Differential Non-linearity (DNL) is the worst case deviation of any adjacent analog output from the ideal 1

LSB step size. The deviation of actual step size from the ideal step size corresponding to 1 LSB is called DNL error.

Q. 5. What do you understand by

(*a*) Multiplying D/A Converters?

(*b*) Bipolar Output D/A Converters?

(c) Companding D/A Converters ?

Ans. (*a*) Multiplying D/A Converters: In a MULTIPLYING TYPE D/A CONVERTER, the converter multiplies an analog reference by a digital word (Fig. 25.5). Some D/A converters can multiply only positive digital words by a positive reference. This is known as single quadrant operation (QUAD-I). Two quadrant operation (QUAD I and III) can be performed by a D/A converter that usually operates in Quadrant I-by configuring the output for bipolar output operation. This is accomplished by offsetting the output by a negative MSB (1/2 of full scale range) so that MSB becomes the sign bit. Multiplying D/A converter is particularly useful, when we are looking for digitally programmable attenuation of an analog input signal.

(*b*) **Bipolar Output D/A Converters:** In BIPOLAR OUTPUT D/A CONVERTERS, the analog output signal range includes both positive and negative values. The transfer characteristics of an ideal 2-quadrant bipolar output D/A Converter are shown in Fig. 25.6.

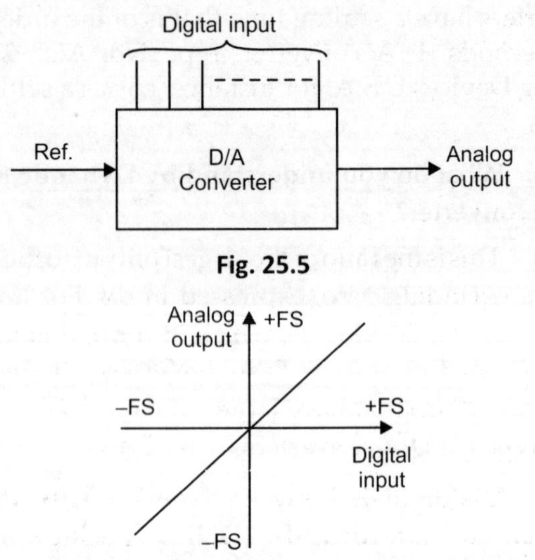

Fig. 25.5

Fig. 25.6

(*c*) **Companding D/A Converters:** Companding type D/A converters are so constructed that the more significant bits of the digital input have a larger than binary relationship to the less significant bits. This decreases the resolution of the more significant bits which increases the analog signal range. The effect of this is to compress more data into more significant bits.

Q. 6. Briefly describe the CURRENT STEERING MODE of operation of a D/A Converter. What will be the analog output voltages in a 4-bit converter, if V_{ref} = + 15 volts and the feedback resistance in the output opamp equals the ladder resistance R for the following digital inputs (*i*) 0101 (*ii*) 1010 (*iii*) 1111.

Ans. In the CURRENT STEERING MODE of operation of a D/A converter, the analog output is a current that is equal to the product of the fractional binary value (D) of the digital word applied at the input and a reference voltage V_{ref}. The output current is often converted into a voltage using an external opamp (Fig. 25.7). Some D/A converter types have an in-built opamp for current to voltage conversion. If the feedback resistance R_{FB} connected externally equals the ladder resistance R, the analog output voltage is given by($- D\ V_{ref}$). The arrangement of Fig. 25.7 can be conveniently used to explain the operation of a digital to analog converter of the R-2R ladder type in the current steering mode. The R-2R ladder divides the input current (I) due to a reference voltage V_{ref} applied at reference voltage pin into binary weighted currents as shown. These currents are then steered to either the output designated OUT-1 or the output designated OUT-2 by the current steering switches. The positions of current steering switches are controlled by the digital input word. A logic '1' steers the corresponding current to OUT-1, whereas a logic '0' steers it to OUT-2. For instance, a logic '1' in MSB position will steer the current (I/2) to OUT-1. OUT-2 here is the ground terminal. In the simplified arrangement of a 4-bit D/A converter of Fig. 25.7, the maximum analog output voltage will be – 15/16 V_{ref}. In general, the maximum output is given by $-(1-2^{-n})\ V_{ref}$, where (n) is the number of bits.

(*i*) Digital word = 0101

Fractional binary value, $D = \frac{5}{16}$

Analog output voltage is, therefore, $-\frac{5}{16} \times 15 = -\frac{75}{16}$ volts

(*ii*) Digital word = 1010

Therefore, $D = \frac{10}{16} = \frac{5}{8}$

Analog output voltage $= -\frac{5}{8} \times 15 = -\frac{75}{8}$ volts

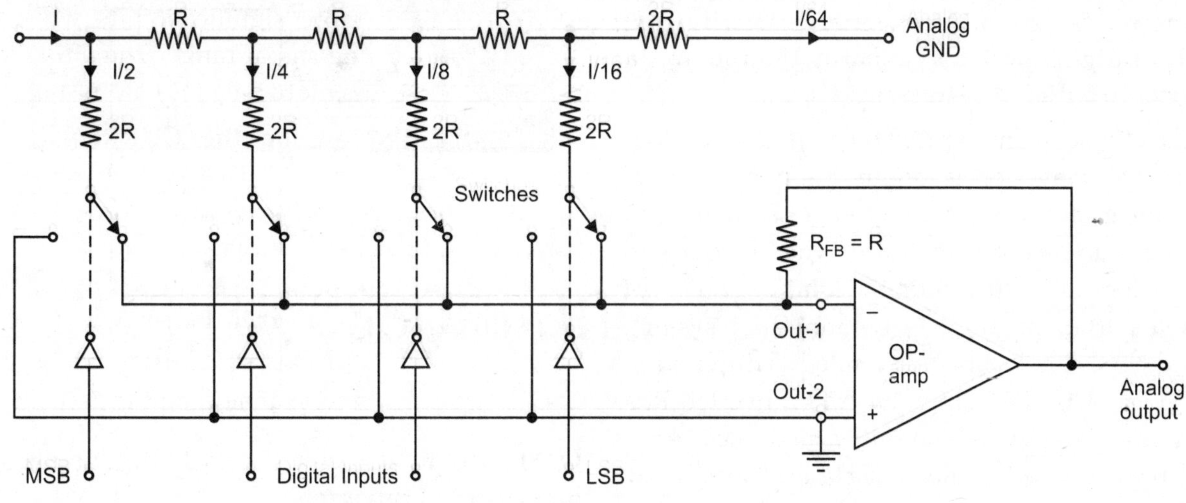

Fig. 25.7

(*iii*) Digital input work = 1111

Therefore, $D = \frac{15}{16}$

$$\text{Analog output Voltage} = -\frac{15}{16} \times 15$$

$$= -\frac{225}{16} \text{ volts.}$$

Q. 7. Briefly describe the voltage switching mode of operation of D/A converters. What are their salient features?

Ans. In the VOLTAGE SWITCHING MODE of operation of R-2R ladder type D/A converters, the reference voltage is applied at OUT-1 terminal and the output is taken from reference voltage terminal (Fig. 25.8) OUT-2 is joined to analog ground.

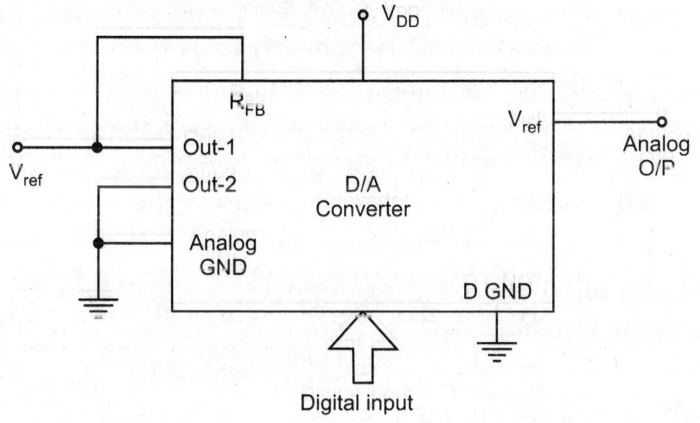

Fig. 25.8

The output voltage is the product of fractional binary value of the digital input word and the reference voltage applied at OUT-1 terminal. That is, analog output voltage = $D.V_{ref}$.. As the positive reference voltage produces a positive analog output voltage, the voltage switching mode of operation is possible with a single supply. Also, the analog output is a voltage without the use of any feedback resistance and opamp. However, the reference voltage applied to OUT-1 terminal in this case will see different input impedances for different digital codes. This necessitates that the input source be buffered.

Q. 8. What do you mean by MONOTONICITY of a D/A converter ? Under what conditions is a D/A converter Monotonic ?

Ans. In an ideal D/A converter, the analog output should increase by an identical step size for every subsequent 1 LSB increment in the digital input word. When such a converter is fed at its input from the output of a counter, the converter output is a perfect staircase waveform (Fig. 25.9). This is the perfect monotonicity. However, a D/A converter is said to be monotonic, if its analog output either increases or remains the same but does not decrease as the digital input code advances in 1 LSB steps.

A D/A converter is monotonic, if its differential non-linearity error (DNL) is less than or equal to twice the worst case non-linearity (NL) error.

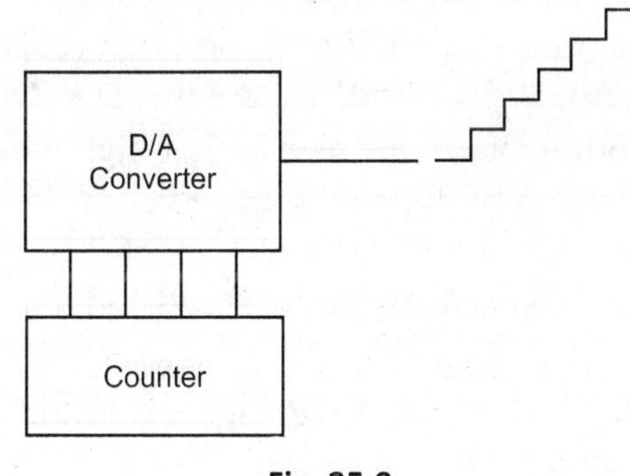

Fig. 25.9

Q. 9. Where does an A/D converter primarily find application in digital systems? What is the role of an A/D converter in a digital system?

Ans. Like digital-to-analog (A/D) converters, analog-to-digital (A/D) converter is another important data conversion integrated circuit component known. A/D converter forms an essential interface, when it comes to analysing analog data with a digital computer. It is an indispensable part of any digital communication system, where the analog signal to be transmitted is digitised at the sending end with the help of an A/D converter. It is invariably used in all digital read-out test and measuring equipment. Be it a digital voltmeter or a laser power meter, or for that matter even a pH meter, an A/D converter is the heart of all of them. The A/D converter takes at its input an analog voltage and after a certain amount of time produces a digital output code representing the analog input.

The A/D conversion process is generally more complex than the D/A conversion process. There are various techniques developed for the purpose of A/D conversion, and these techniques have different advantages and disadvantages with respect to one another, which have been utilised in the fabrication of different categories of A/D converter ICs. A D/A converter circuit, as we will see in the following paragraphs, forms a part of some of the types of A/D converter.

Q. 10. What is a BCD input D/A converter? How does it function? How do you define percentage resolution of such a converter?

Ans. A BCD input D/A converter accepts BCD equivalent of decimal digits at its input. A 2-digit BCD. input D/A converter, for instance, is an 8-bit converter. Fig. 25.10 shows circuit representation of an 8-bit BCD input D/A converter. Such a device would have 99 steps and accept decimal digits 00 to 99 at its input. A 12-bit converter would similarly have 999 steps. The weight of different bits in the Least Significant Digit (LSD) will be 1 (for A_0), 2 (for B_0), 4 (for C_0) and 8 (for D_0). The weight of corresponding bits in the next higher digit would be 10 times the weights of the bits in the adjacent lower digit. For the converter of Fig. 25.10, the weights of A_1, B_1, C_1 and D_1 would respectively be 10, 20, 40 and 80.

In general, an n-bit D/A converter of BCD input type will have $(10^{n/4} - 1)$ steps. Percentage resolution of such a converter is given by:

$$\frac{1}{[10^{(n/4)} - 1]} \times 100$$

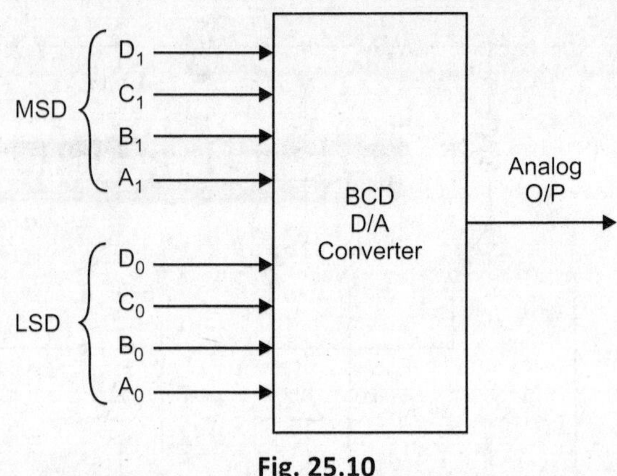

Fig. 25.10

Q. 11. Briefly describe the undermentioned specifications of A/D Converters:

(*i*) Resolution

(*ii*) Accuracy

(*iii*) Gain and Offset Errors

(*iv*) Gain and Offset Drifts

(*v*) Quantization Error

(*vi*) Non-linearity

(*vii*) Differential Non-linearity

(*viii*) Conversion Time

(*ix*) Aperture and Acquisition Time

(*x*) Code Width.

Ans.

(*i*) **Resolution:** of an A/D converter is the amount of input analog voltage change required to increment its digital output between one code change and the next code change. An n-bit converter can resolve 1 part in 2^n. It may be expressed as percentage of full scale or in bits. Resolution of an 8-bit converter can be expressed as 1 part in 256 or as 0.4% of FS or simply as 8-bit resolution. If such a converter has a full scale range of 10V, it will be able to resolve a 40 mV change in its input.

(*ii*) **Accuracy** specification describes the maximum sum of all the errors, both from analog sources (mainly the comparator and the ladder resistors) as well as the digital sources (Quantization error) of the A/D converter. These errors mainly include the GAIN ERROR, the OFFSET ERROR and the QUANTIZATION ERROR. The accuracy would describe the actual analog input and full-scale weighted equivalent of the output code corresponding to actual analog input. Accuracy specification is rarely provided on the datasheets and quite often several sources of errors are listed separately.

(iii) Gain and Offset Error: GAIN ERROR is the difference between the actual full-scale transition voltage and the ideal full-scale transition voltage. It is expressed either as percentage of full-scale range (% of FSR) or in LSBs. OFFSET ERROR is the error at analog zero for an A/D converter operating in the bipolar mode. It is measured in % of FSR or in LSBs.

(iv) Gain Drift, Offset Drift: GAIN DRIFT is the change in the full-scale transition voltage measured over the entire operating temperature range. It is expressed in FS per degree celcius or PPM of FS per degree celcius or LSBs. OFFSET DRIFT is the change with temperature of analog zero for an A/D converter operating in the bipolar mode. It is generally expressed in PPM of FS per degree celcius or LSBs.

(v) Quantization Error: An A/D converter having an n-bit output can only identify (2^n) output codes while there exist an infinite number of analog input values adjacent to the LSB of the A/D converter, which are assigned the same output code. For instance, if we are digitising an analog signal with a peak value of 7V using 3 bits, then in that case all analog voltages equal to or more than 5.5V and less than or equal to 6.5 V will be represented by same output code *i.e.* 110 (if the output coding is in straight binary form). The error is ± 0.5 V or ± 1/2 LSB as 1 LSB change in the output corresponds to an analog change of 1V. The ± 1/2 LSB limit to resolution is known as the fundamental quantization error. Expressed in percentage, the quantization error in an 8-bit converter is 1 part in 256 or 0.4%.

(vi) Non-linearity: NON-LINEARITY specification (also referred to as the Integral Non-linearity, INL by some manufacturers) of an A/D converter describes its departure from a linear transfer curve. Non-linearity error does not include Gain, Offset and Quantization errors. It is expressed as a % of FS or in LSBs.

(vii) Differential Non linearity: It indicates the worst case difference between the actual analog voltage change and the ideal 1 LSB voltage change. DNL specifications are as important as the INL specifications, as an A/D converter having a good INL specification may have a poor quality transfer curve, if DNL specification is poor. DNL is also expressed in % of FS or in LSBs. DNL infact explains the smoothness of the transfer characteristics and is, thus, of great importance to the user.

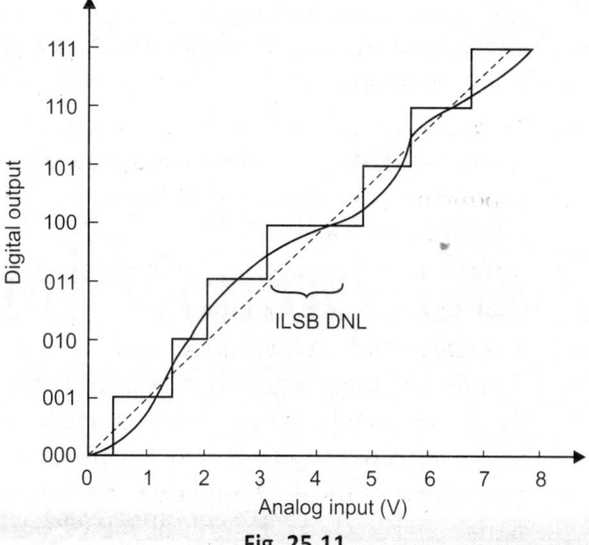

Fig. 25.11

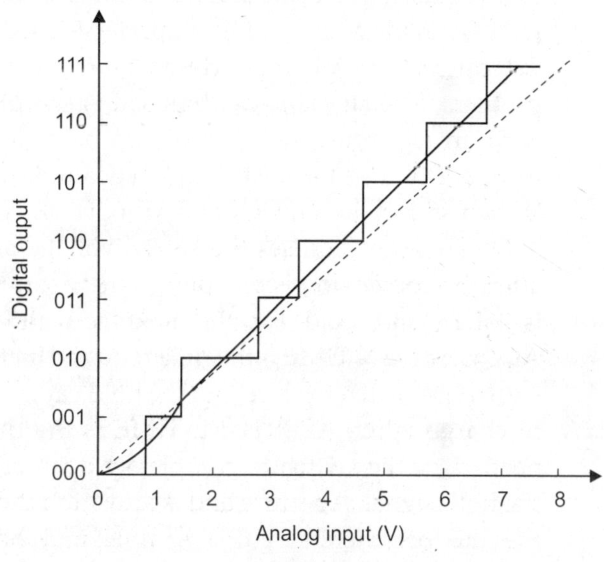

Fig. 25.12

Fig. 25.11 shows the transfer curve for 3-bit A/D converter with 7V full-scale range, 1/4LSB NL and 1 LSB DNL. Fig. 25.12 shows the same for 7V FS range, 1LSB NL and 1/4 LSB DNL. Although the former has much better NL specifications, the latter with better DNL specifications, has a much better and smoother curve and may, thus, be preferred. Too high a value of DNL may even grossly degrade the converter resolution. In a 4-bit converter with ± 2LSB DNL, the 16-step transfer curve may be reduced to a 6-step curve. DNL specification should in no case be ignored unless the NL specification is tight enough to guarantee the desirable DNL.

(viii) Conversion Time: It is the time that elapses from the time instant of the start conversion signal until the conversion complete signal occurs. It ranges from a few nanoseconds for FLASH A/D converters to a few microseconds for SUCCESSIVE APPROXIMATION TYPE

A/D converters and may be as large as tens of milliseconds for DUAL SLOPE INTEGRATING A/D converters.

Note. Fig. 25.11 has 1/4LSB NL as about 5.75 V produces 6 LSB change in the output against 6V producing 6LSB change. 0.25 V corresponds to 1/4LSB in this example. DNL is 1LSB as in the worst case 2V change in the input produces a 1LSB change in the output.

(*ix*) **Aperture and Acquisition Time:** When a rapidly changing signal is digitised, the input signal amplitude would have changed even before the conversion is complete with the result that output of the A/D converter represents the signal amplitude at the end of the conversion cycle rather than at the start. A SAMPLE AND HOLD circuit with a buffer amplifier is used at the input of the A/D converter to overcome this problem. The signal to be digitised is sampled with an electronic switch that can be rapidly turned ON and OFF. The sampled amplitude is then stored on the HOLD CAPACITOR. The A/D converter digitises the stored voltage and after the conversion is complete, a new sample is taken and held for the next conversion. ACQUISITION TIME is the time required by the electronic switch to close and the hold capacitor to charge while APERTURE TIME is the time needed by the switch to completely open after the hold signal. Aperture and Acquisition times are the parameters of the sample and hold circuit. Ideally, both the times should be zero. The maximum sampling frequency is, thus, determined by the Aperture and Acquisition time in addition to the conversion time.

(*x*) **Code Width:** The amount of input voltage change which occurs between the output code transitions expressed in LSBs of full-scale is the CODE WIDTH. CODE WIDTH UNCERTAINTY is the dynamic variation or JITTER in the code width due to noise.

Q. 12. Distinguish between unipolar and bipolar modes of operation of A/D converters.

Ans. In Unipolar mode of operation, the analog input to the A/D converter varies from 0 to full-scale voltage of one polarity only. A/D converter configured to convert both positive as well as negative analog input voltages is said to be operating in bipolar mode.

Q. 13. What do you understand by CODING with reference to A/D converters?

Ans. Coding defines the nature of A/D converter output data format. Commonly used formats include straight binary, offset binary, complementary binary, 2's complement, low byte and high byte.

Q. 14. Briefly explain Low Byte, High Byte, Right Justified and Left Justified Data formats in A/D converter data formats.

Ans. In A/D converters with resolution greater than 8-bits, some products are offered in high byte or low byte format to simplify their interface with 8-bit microprocessor systems. The *Low Byte* output contains the least significant bit and some or all of the lower 8 bits of the A/D converter output. In the *High Byte*, output contains the MSB and some or all of the upper 8 bits. In the *Right Justified Data* output format, data bit sets shorter than 8 bits are placed starting with the right side of the data output transfer register. This could apply to upper or lower byte. For example, a 12-bit ADC will have 4 extra bits which could be right justified. In the *Left Justified Data*, data bit sets shorter than 8-bits are placed starting with the left side of data output transfer register. This could apply to the lower or the upper byte. For example, a 12-bit ADC will have 4 extra bits, which could be left justified.

Q. 15. What are command register and status register in A/D converter?

Ans. *Command Register* is an internal register of the A/D converter that can be programmed by the user to select various modes of operation, such as unipolar or bipolar mode selection, range selection, data output format selection etc. *Status Register* indicates the current status of the analog-to-digital conversion with a 'Busy' or 'Conversion Complete' signal.

Q. 16. What are control lines in an A/D converter? Name some.

Ans. Digital input/output pins that activate/monitor and control ADC operation are called control lines. Some of the examples are Chip Select, Write, Start Convert, Conversion Complete, etc.

Q. 17. Explain in brief the principle of operation of an A/D converter based on simultaneous conversion technique. What do you have to say regarding speed of such converters? What is the main drawback of this technique?

Ans. The simultaneous conversion method of A/D conversion is based on using a number of comparators. The number of comparators needed in N-bit A/D conversion is $(2^N - 1)$. One such system capable of converting an analog input signal into a 2-bit digital output is shown in Fig. 25.13. The analog signal to be digitised serves as one of the inputs to each of the comparators. The second input for each of the comparators is a reference input, different for each comparator.

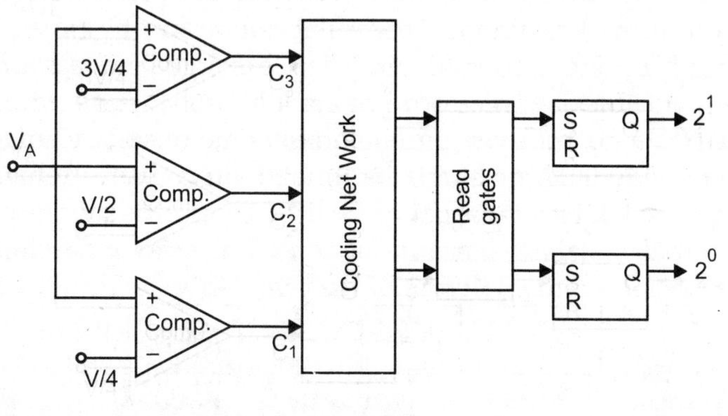

Fig. 25.13

The reference voltages to be used for comparators are in general $\frac{V}{2^N}, \frac{2V}{2^N}, \frac{3V}{2^N}$ and so on. V is the maximum analog signal that the converter can digitise and (N) is the number of bits in the digital output. In the present case of 2-bit A /D converter, the reference voltages for the three comparators will be $\frac{V}{4}, \frac{V}{2}$ and $\frac{3V}{4}$. If we wanted a 3-bit output, the reference voltages would have been V/8, V/4, 3V/8, V/2, 5V/8, 3V/4 and 7V/8. Referring to Fig. 25.13, the output status of various comparators depends upon the input analog signal V_A. For instance, when the input V_A lies between V/4 and V/2, output C_1 is HIGH, whereas both C_2 and C_3 outputs are LOW. The results are summarised in Table 25.2. The three comparator outputs can then be fed to a coding network (comprising of logic gates etc.) to provide two bits, which are digital equivalent of input analog voltage. The bits at the output of the coding network can then be entered into a flip flop register, for storage.

The construction of a simultaneous A/D converter also known as Flash A/D converter is quite straightforward and relatively easy to understand. However, as the number of bits in the desired digital signal increases, the number of comparators required to perform A/D conversion increases very rapidly and it may not be feasible to use this approach once the number of bits exceeds 6 or so. The greatest advantage of this technique lies in its capability of extremely fast conversion rates.

Table 25.2

Input voltage (VA)	C_1	C_2	C_3	2^1	2^0
0 to V/A	Low	Low	Low	0	0
V/4 to V/2	High	Low	Low	0	1
V/2 to 3V/4	High	High	Low	1	0
3V/4 to V	High	High	High	1	1

Q. 18. What are Half-Flash A/D converters? What are the advantages of these converters? How do they compare with Flash or Simultaneous A/D converters?

Ans. Half flash A/D converter, also known as Pipeline A/D converter, is a variant of flash type converter that overcomes to a large extent the primary disadvantage of requirement of a prohibitively large number of comparators in high resolution full flash converters without significantly degrading its high speed conversion performance. When compared to a full flash converter of certain resolution, while the number of comparators and associated resistors reduce drastically in a half flash converter; conversion time increases approximately by a factor of 2. For an n-bit flash converter, number of comparators required is 2^n [$(2^n - 1)$ for encoding of amplitude and one comparator for polarity], the same for an equivalent half flash converter would be $2 \times (2^{n/2})$. In case of an 8-bit converter, the number is 32 (for half flash) against 256 (for full flash). How it is achieved, is explained in the following paragraphs considering the example of an 8-bit half flash converter.

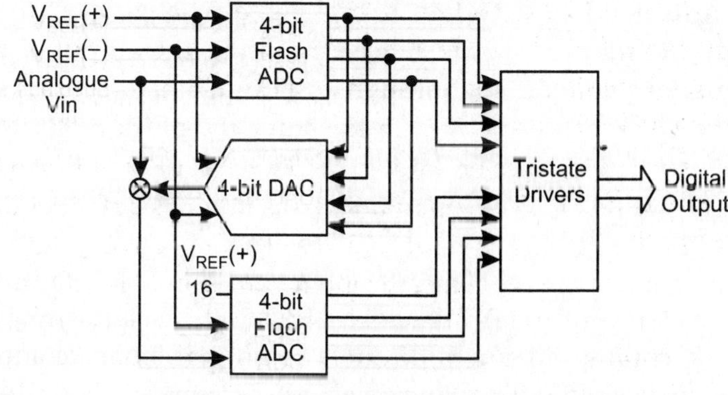

Fig. 25.14

A half flash converter uses two full flash converters with each full flash converter, having a resolution equal to half the number of bits of the half flash converter. That is, an 8-bit half flash converter uses two 4-bit flash converters. In addition, it uses a 4-bit D/A converter and an 8-bit latch. Figure-25.14 shows the basic architecture of such a converter. The timing and control circuitry is omitted for the sake of simplicity. The circuit functions as follows. The most significant 4-bit A/D converter converts the input analogue signal into a corresponding 4-bit digital code, which is stored in the most significant 4 bits of the output latch. This 4-bit digital code, however, represents the low resolution sample of the input. Simultaneously, it is converted back into an equivalent analogue signal with a 4-bit D/A converter. The approximate value of the analog signal so produced is then subtracted from the sampled value and the difference is converted into digital code using least significant 4-bit A/D converter. Least significant A/D converter is referenced to 1/16th (= $1/2^4$) of the reference voltage used by most significant A/D converter. The new 4-bit digital output is stored in least significant 4-bits of the output latch. The latch now contains the 8-bit digital equivalent of the analog input. The digitized output is the same as would be produced by an 8-bit full flash converter. The only difference is that the conversion process takes a little longer. It may also be mentioned here that 8-bit half flash converter can be either used as a 4-bit full flash converter or 8-bit half flash converter. Some half flash converters use a single full flash converter and reuse it for both conversions. This is achieved by using additional sample and hold circuitry.

Q. 19. Explain in brief with the help of block schematic the counter method of A/D conversion.

Ans. Higher resolution A/D converters using only one comparator could be constructed, if a variable reference voltage were available. This reference voltage could then be applied to the comparator and when this becomes equal to the analog input voltage, conversion would be complete. This variable reference voltage must correspond to a certain digital signal changing in the desired fashion. For instance, it could be the output of a digital counter going through its natural binary count, as the clock proceeds. Fig. 25.15 shows the block diagram of such a converter known as COUNTER-TYPE A/D CONVERTER. The operation is as follows : To begin with, the counter is reset to all 0s. Then a convert signal appears on the start-line, the input gate is ENABLED and the clock pulses are allowed to the counter's clock-input. The counter advances through its normal binary count sequence, the staircase waveform is generated at the output of the binary ladder constituting a D/A converter. This staircase waveform forms one of the inputs of the comparator whose other input is the analog input signal. Whenever the binary ladder output exceeds the analog input voltage, the comparator changes state, the gate is DISABLED and the counter stops. The counter output is then the required digital output corresponding to analog input signal.

The counter-type A/D converter provides a very good method for digitising to a high resolution. This method is much simpler than the simultaneous method for higher resolution A/D converters. The drawback with this converter is that the required conversion time is longer. Since the counter always begins from all 0s position and counts through its normal binary sequence, it may require as many as 2^n counts before conversion is complete. The average conversion-time can be taken to be $2^n/2 = 2^{n-1}$ counts. One clock cycle gives one count. As an illustration, if we have a 4-bit converter and a 1 MHz clock, the average conversion time would be 8μS. It would be as large as 0.5 ms for a 10-bit converter of this type at 1 MHz clock-rate.

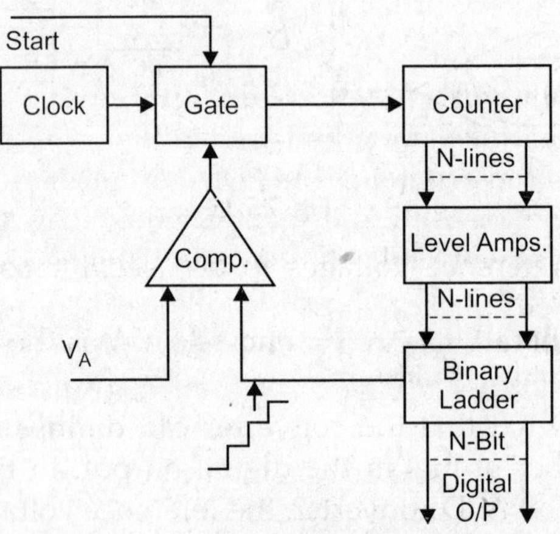

Fig. 25.15

Q. 20. What is a Tracking A/D converter? How does it overcome the primary shortcoming of the counter type A/D converter?

Ans. In the counter type A/D converter described in the previous question, the counter is reset to zero at the start of each new conversion. The D/A converter output staircase waveform always begins at zero and increases in steps till it reaches a point, where analog output of D/A converter exceeds the analog input to be digitised. As a result, counter type A/D converter of the type discussed above, is slow. Tracking type A/D converter, also called Delta encoded A/D converter, is a modified form of counter type converter that overcomes to some extent the shortcoming of the latter. In the modified arrangement, the counter, which is primarily an UP-counter, is replaced by an UP/DOWN counter. It counts in upward sequence whenever D/A converter output analog voltage is less than the analog input voltage to be digitised and it counts in the downward sequence, whenever D/A converter output analog voltage is greater than analog input voltage. In this type of converter, whenever a new conversion is to begin, the counter is not reset to zero; in fact it begins counting either up or down from its last value depending upon the comparator output. The D/A converter output staircase waveform contains both positive going and negative going staircase signals that track the input analog signal.

Q. 21. Explain in brief the operation of a 'Successive Approximation Type' A/D converter. How do you compare its speed with that of a counter type A/D converter?

Ans. The development of A/D converters has progressed in a quest to reduce the conversion time. SUCCESSIVE APPROXIMATION TYPE OF A/D CONVERTER aims at approximating the analog signal to be digitised by trying only one bit at a time. The process of A/D conversion by this technique can be illustrated with the help of an example. Let us take a 4-bit successive approximation type A/D converter. Initially, the counter is reset to all 0s. Then the start pulse sets the MSB. That is, the flip flop representing the MSB is set. The counter output is converted into equivalent analog signal and then compared with the analog signal to be digitised. A decision is then taken whether the MSB is to be left-in (*i.e.* flip flop representing MSB remains set) or it is to be taken out (*i.e.* flip flop is reset), when the first clock pulse sets the seconds MSB. Once the second MSB is set, again a comparison is made and a decision taken whether the second MSB is to remain set or not, when the subsequent clock pulse sets the third MSB. The process continues till we go down to LSB. We will notice that every time we make a comparison, we will tend to narrow down the difference between the analog signal to be digitised and the analog signal representing the counter count. Refer to the operational diagram of Fig. 25.17. It is clear from the diagram that to reach any count from 0000 to 1111, the converter requires 4 clock cycles. In general, the number of clock cycles required for each conversion will be (n) for n-bit A/D converter of this type. Fig. 25.18 shows the block-schematic representation of a successive approximation type of A/D converter. Since only one flip flop (in the counter) is operated upon at one time, a ring counter which is nothing but a circulating register (serial shift register with outputs Q and $\overline{Q}$ of the last flip flop connected to J and K inputs respectively of the first flip flop) is used to do the job. Referring to Fig. 25.17, the dark lines show the sequence in which the counter arrives at the desired count, assuming that 1001 is the desired count.

This type of A/D converter is much faster than the counter-type A/D converter previously discussed. In an n-bit converter, the counter-type A/D converter on an average would require (2^{n-1}) clock cycles for each conversion whereas successive approximation type converter requires only (n) clock cycles.

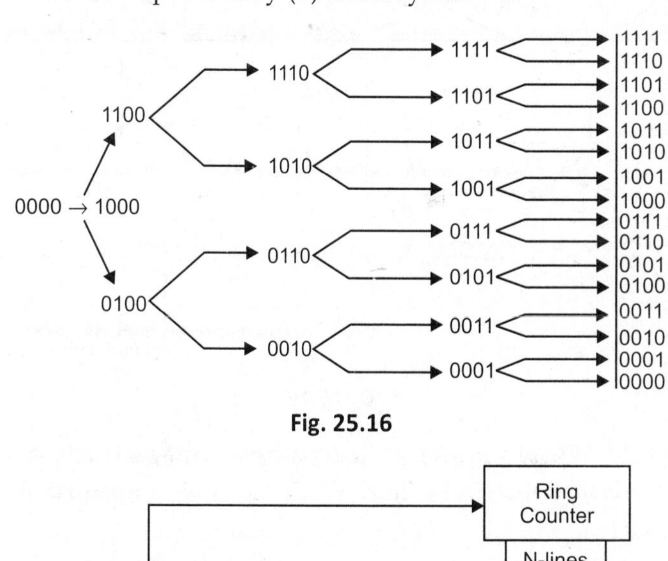

Fig. 25.16

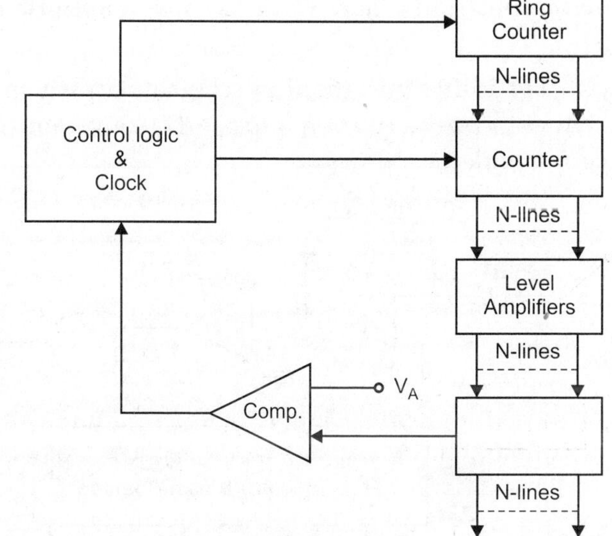

Fig. 25.17

Q. 22. Briefly explain the operation of Single Slope A/D Converter.

Ans. Figure 25.18 shows the block schematic representation of a single slope A/D converter. In this type of converter, one of the inputs to the comparator is a ramp of fixed slope while the other input is the analog input to be digitised. The counter and the ramp generator are initially reset to 0's. The counter starts counting with the first clock cycle input. The ramp is also synchronised to start with the first clock input. The counter stops, when the ramp amplitude equals the analog input. In this case, the counter count is directly proportional to the analog signal. It is a low cost, reasonably high accuracy converter but it suffers from the disadvantage of loss of accuracy due to changes in the characteristics of the ramp generator. This shortcoming is overcome in dual slope integrating type A/D converter.

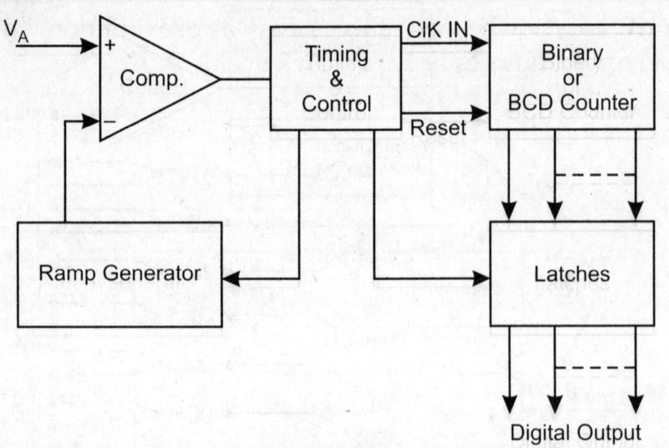

Fig. 25.18

Q. 23. How does a dual-slope integrating A/D converter function? Where are these converters normally used?

Ans. Fig. 25.19 shows the block schematic arrangement of a dual-slope integration type A/D converter. The converter works as follows :

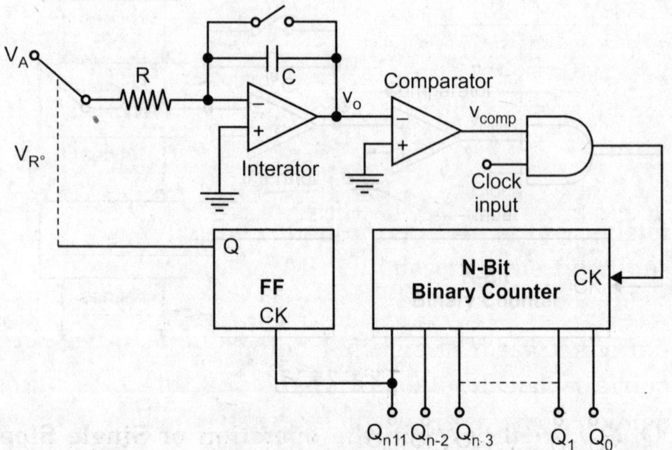

Fig. 25.19

Initially, switch S is connected to the analog voltage (V_A) to be digitised. The output of the integrator is mathematically given by,

$$v_0 = -\frac{1}{RC}\int_0^t V_A\, dt = (-V_A/\text{RC})\, t$$

The moment v_0 tends to go below zero, clock pulses reach the clock input terminal of the counter, which is initially clear. The counter begins counting from 0000 ....0. At the (2^n)th clock pulse, the counter is again cleared, the 1 to 0 transition of MSB of the counter sets a flip flop which controls the state of switch S, which now connects the integrator input to a reference voltage $(-V_R)$. The integrator output moves in the positive direction; the counter has again started counting after it got reset (at say $t = T_1$). The moment, the integrator output tends to exceed zero, the counter stops as the clock pulses no longer reach the counter's clock input. The counter output at this stage (say at $t = T_2$) is proportional to the analog input. Mathematically, it can be proved that $n = \left(\frac{V_A}{V_R}\right) 2^n$, where ($n$) here is the count recorded in the counter as $t = T_2$. Refer to Fig. 25.20 for further illustration. This type of A/D converter is very popular in digital voltmeters due to its good conversion accuracy and low cost.

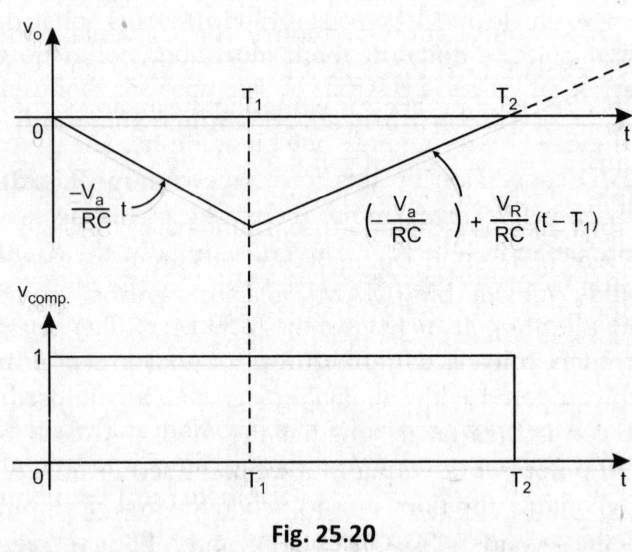

Fig. 25.20

Q. 24. What are triple slope and quad slope A/D converters?

Ans. There are multi-slope converter architectures too aimed at further enhancing the performance of integrating A/D converters. For example, the triple-slope architecture is used to increase the conversion speed at the cost of added complexity. Increase in conversion speed is accomplished by carrying out integration from reference voltage at two distinct rates, a high speed rate and a low speed rate. The counter is also divided into two sections, one for MSB bits and the other for LSB bits. A properly designed triple slope converter achieves increased conversion speed without compromising the inherent linearity, differential linearity and stability characteristics of dual slope converter. Bias currents, offset voltages and gain errors associated with operational amplifiers, used as integrator and comparator do introduce some errors. These can be cancelled by using additional charge/discharge cycles and then using the results to correct the initial measurement. One such A/D converter is quad-slope converter that uses two charge/discharge cycles as compared to one charge/discharge cycle in case of dual slope converter. Quad-slope A/D converters have a much higher accuracy than dual slope counterparts.

Q. 25. What is a sigma-delta A/D converter? Briefly explain the concept behind the operation of a Sigma-Delta converter. Where does this type of converter find application?

Ans. Sigma-delta A/D converter employs a different concept from what has been discussed in some of the previous questions in the case of various types of A/D converters. While the A/D converters covered so far rely on sampling of analog signal at the Nyquist frequency and encode the absolute value of the sample, in the case of sigma-delta converter, as explained in the following paragraphs, the analog signal is over sampled by a large factor (*i.e.* sampling frequency is much larger than the Nyquist value) and also it is not the absolute value of the sample but the difference between the analog values of two successive samples that is encoded by the converter.

In the case of A/D converters sampled at Nyquist rate (f_s), the RMS value of the quantization noise is uniformly distributed over the Nyquist band of DC to ($f_s/2$), as shown in Figure-25.21 (*a*). The signal to noise ratio for a full scale sine wave input in this case is given by (S/N) = (6.02n + 1.76) dB, (*n*) being the number of bits. The only way to increase the signal to noise ratio is by increasing number of bits. On the other hand, a sigma-delta converter attempts to enhance signal to noise ratio by over sampling the analog signal, which has the effect of spreading the noise spectrum over a much larger bandwidth, and then filtering out the desired band. If the analog signal were sampled at a rate of (Kf_s), the quantization noise gets spread over DC to ($Kf_s/2$), as shown in Figure-25.21 (*b*). (*K*) is a constant referred to as oversampling ratio. Enhanced S/N ratio means higher resolution, which is achieved by other types of A/D converters by way of increasing number of bits.

It may be mentioned here that if we simply use over sampling to improve resolution, it would be required to oversample by a factor of (2^{2N}) to achieve an N-bit increase in resolution. Sigma-delta converter does not require to be oversampled by such a large factor because it not only limits the signal pass band but also shapes the quantization noise in such a way that most of it falls outside this pass band, as shown in Figure-25.21 (c). In the following paragraphs is explained the operational principle of sigma-delta A/D converter.

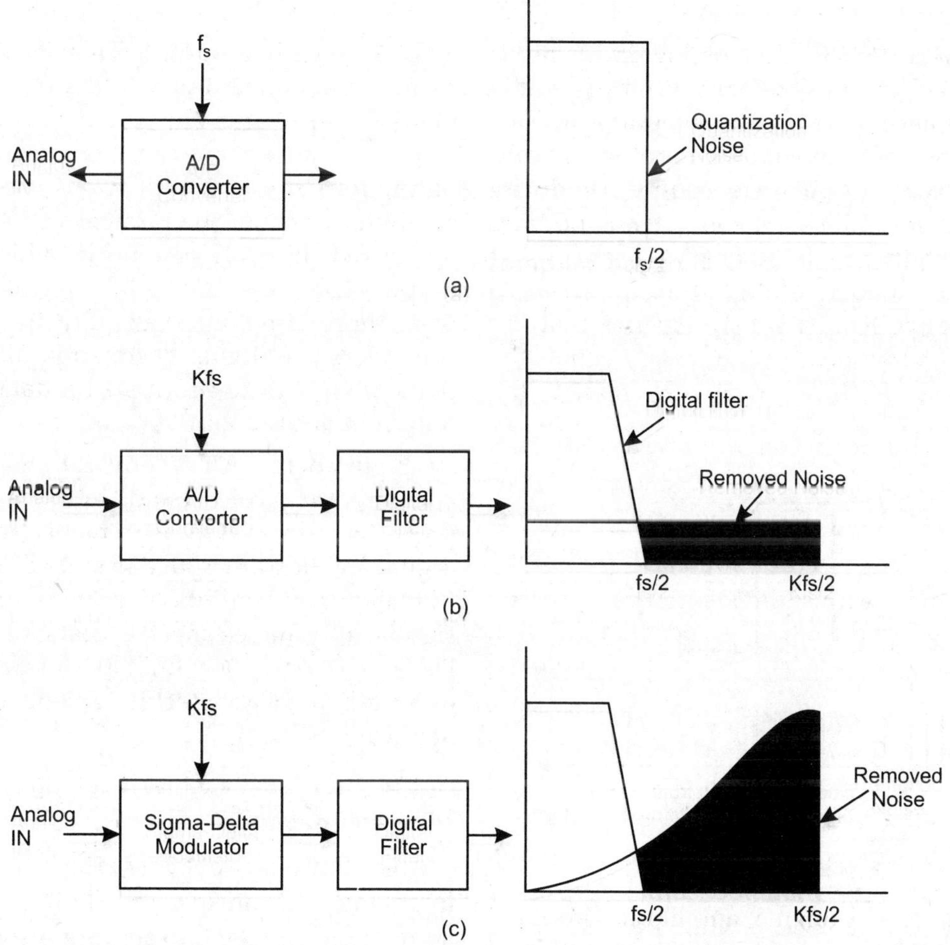

Fig. 25.21

The heart of sigma-delta converter is the delta modulator. Fig. 25.22 shows the block schematic representation of delta modulator, which is basically a 1-bit quantizer of the flash type (single comparator). Output of delta modulator is a bit stream of 1's and 0's with number of 1's relative to number of 0's over a given number of clock cycles indicating the amplitude of analog signal over that time interval. An all 1's sequence over a given interval corresponds to maximum positive amplitude and an all 0's sequence indicates maximum negative amplitude. Equal number of 1's and 0's indicates zero amplitude. Other values between positive and negative maximum are indicated by a proportional number of 1's relative to number of 0's. This is further illustrated in Fig. 25.23.

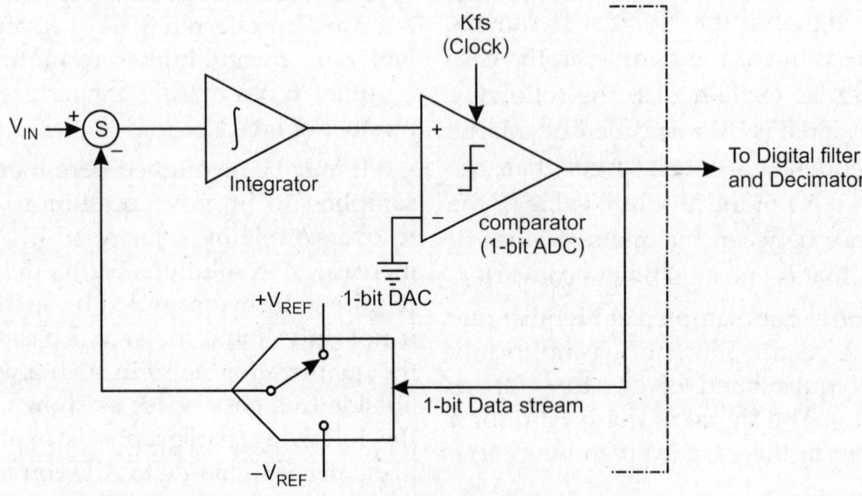

Fig. 25.22

Coming back to the delta modulator (Fig. 25.22), input to the 1-bit quantizer, which is basically a comparator, is from the output of an integrator. The integrator in turn is fed from difference between analog input signal and the analogue equivalent of quantized output produced by 1-bit D/A converter. 1-bit D/A converter is nothing but a two-way switch that feeds either ($+V_{REF}$) or ($-V_{REF}$) to the summing point depending upon the bit status at its input. The negative feedback loop ensures that the average value of D/A converter output nearly equals the analog input so as to produce a near zero input to the integrator.

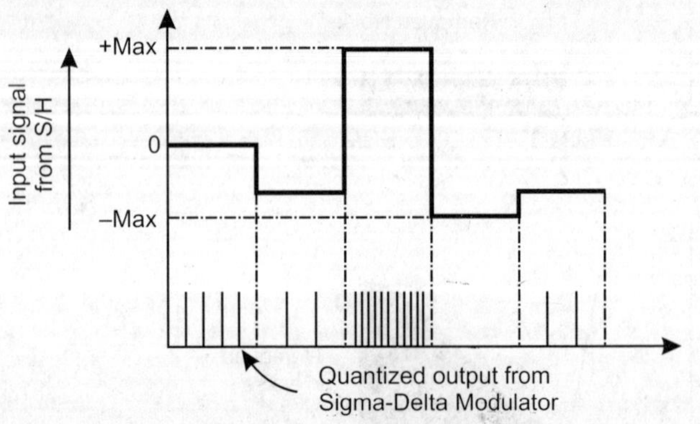

Fig. 25.23

An increase in analog signal amplitude produces a larger number of 1's at the quantizer output and consequently a higher average value of the analog signal at D/A converter output. This means that number of 1's in the quantizer output bit stream over a given time interval, represents the analog signal amplitude. The single bit data stream can then be encoded into desired output format. One simple way to do this could be to use a counter to count the number of 1's in the data stream over fixed intervals of time in which case the counter output represents the digital equivalent of analog signal over those time intervals. Practical sigma-delta A/D converters use digital decimation filter at the output of delta modulator to process 1-bit data stream to produce output in desired format.

Sigma-delta A/D converter is widely used for contemporary voice band, audio and high resolution precision industrial measurement applications. Its highly digital architecture is ideally suited for such applications, as it allows easy addition of digital functionality without significantly increasing the cost. AD 1871 from Analog Devices is one such high performance A/D converter of sigma-delta architecture intended for digital audio applications.

Q. 26. On what factors does the accuracy of an A/D converter depend ?

Ans. Since an A/D converter is a closed loop system involving both analog and digital systems, the overall accuracy must take into account errors from both analog as well as digital portions of the A/D converter. The source of digital error is the "quantization error". It is generally ± 1 bit. This error is minimised by having more number of bits in the counter.

The main source of analog error in an A/D converter is the comparator. Other sources of analog error are the resistors in the ladder, the reference voltage, supply ripple and noise. The source of error in the comparator, which is the primary source of analog error, is the variation in the offset, gain, amplifier linearity with input voltage levels and temperature.

Q. 27. How can you use a D/A converter to build a digitally programmable gain element ? Write the expression for the gain.

Ans. If the feedback resistance is used as the input resistor and the D/A converter is connected as a feedback element, the circuit acts as a divider or a programmable gain element. Fig. 25.24 shows the circuit configuration. The output is given by $V_0 = -(V_{in}/D)$. For smaller values of digital fraction (D), the output increases and one should ensure that the amplifier does not saturate under these circumstances.

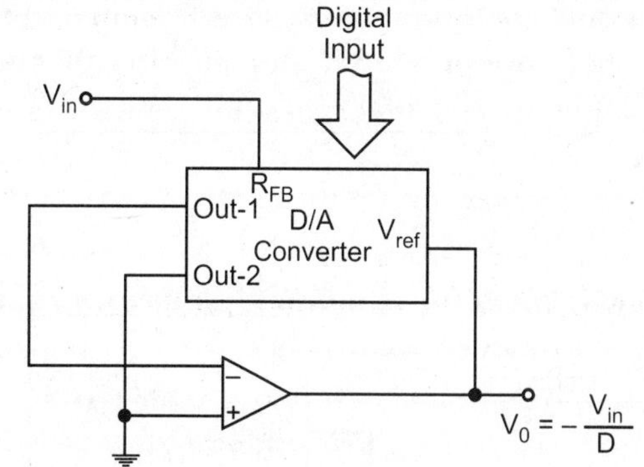

Fig. 25.24

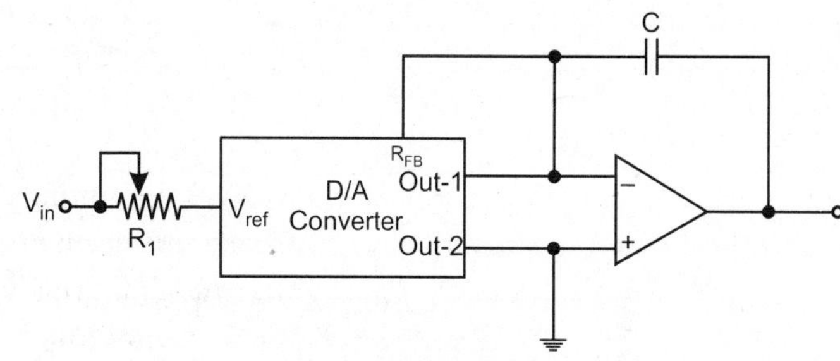

Fig. 25.25

Q. 28. Briefly describe how a D/A converter can be incorporated into an opamp integrator circuit to make it a programmable integrator ?

Ans. Fig. 25.25 shows an inverting type of programmable integrator whose output is expressed by:

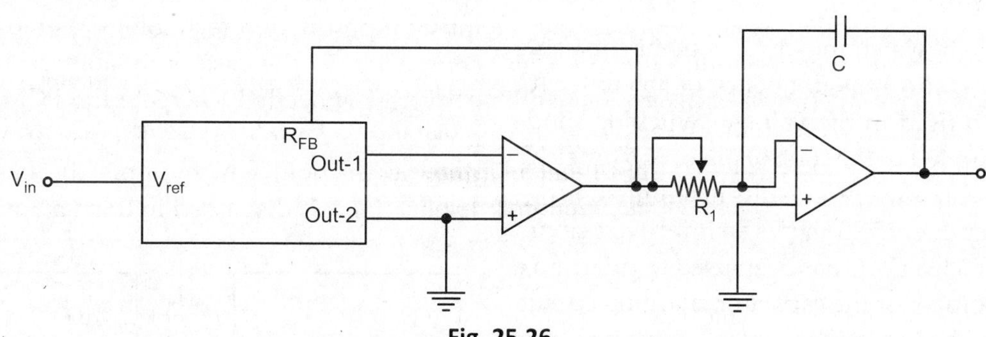

Fig. 25.26

$$V_0 = -\frac{1}{C(R_{DAC} + R_1)} \cdot D \int V_{in}\, dt$$

(R_{DAC}) is the input resistance of the D/A converter at V_{ref} terminal. Resistance (R_1) has been used to get an appropriate value of integrator time constant for full scale value of (D) integrator time constant given by $C\,(R_{DAC} + R_1)/D$ is the largest, when the input digital code is near zero and the smallest, when (D) has the full scale value. The non-inverting type of programmable integrator is shown in Fig. 25.26. The output in this case is given by :

$$V_0 = \frac{D}{CR_1} \int V_{in}\, dt$$

Fig. 25.27

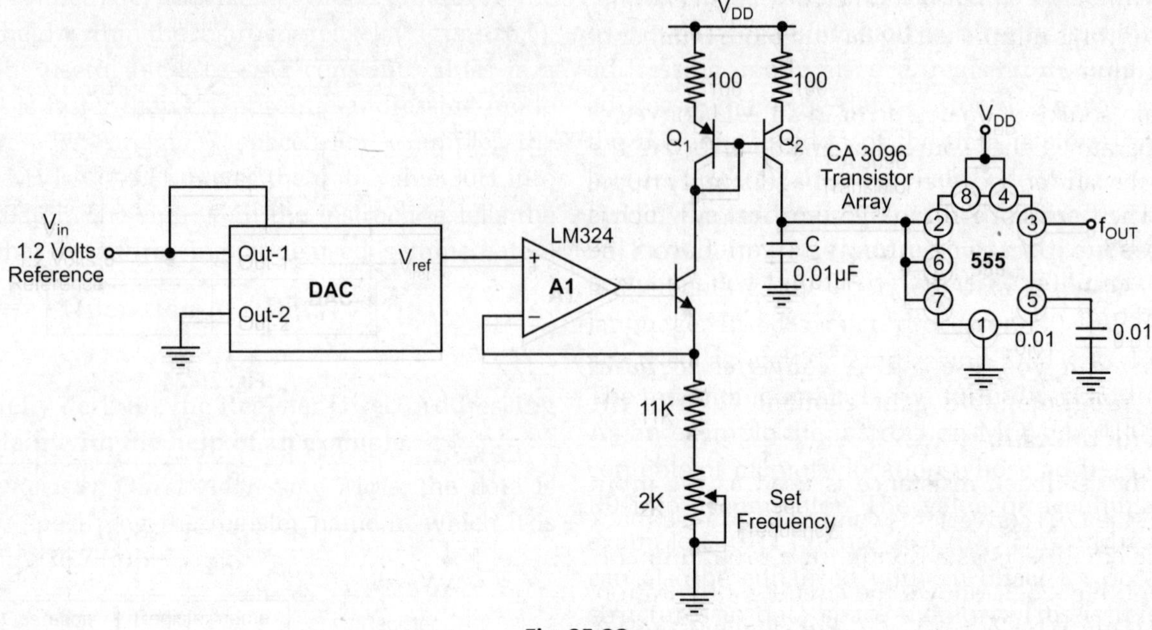

Fig. 25.28

Programmable integrator forms the basis of a number of medium frequency function generators.

Q. 29. With the help of a suitable circuit, briefly describe how a D/A converter can be used to digitally control the frequency of a 555-timer based astable multivibrator circuit?

Ans. Fig. 25.27 shows simple arrangement for using a D/A converter to control the frequency of 555 timer IC. The DAC has been used in the voltage switching mode and its output is applied to the control pin of 555 timer IC. Such an arrangement, however, results in a limited range of output frequency. Fig. 25.28 shows an improved circuit arrangement in which a DAC has been used to program a constant current source for the capacitor charging circuit. The DAC again connected in the voltage switching mode supplies a voltage that determines current in the constant current circuit. This current mirrored by the transistor pair Q_1, Q_2 to supply charging current for the timing capacitor '*C*'.

Q. 30. Outline important areas of application of A/D converters.

Ans. Like D/A converters, A/D converters have numerous applications. A/D converters are used in virtually all those applications where the analog signal is to be processed, stored or transported in digital form. They form an essential interface, when it comes to analysing analog data with a digital computer, the process being known as 'data acquisition'. They are an indispensable component of any digital communication system, where the analog signal to be transmitted is digitised at the sending end with an A/D converter. They are invariably used in all digital readout test and measuring equipment, such as digital multimeters (DMMs), digital storage oscilloscopes (DSOs), etc. Also, A/D converters are integral to contemporary music reproduction technology, as most of it is done on computers. In the case of analog recording too, an A/D converter is needed to create the PCM data stream that goes onto a compact disc. The use of A/D converters for data acquisition, which forms the basis of most other applications, is discussed in the next question.

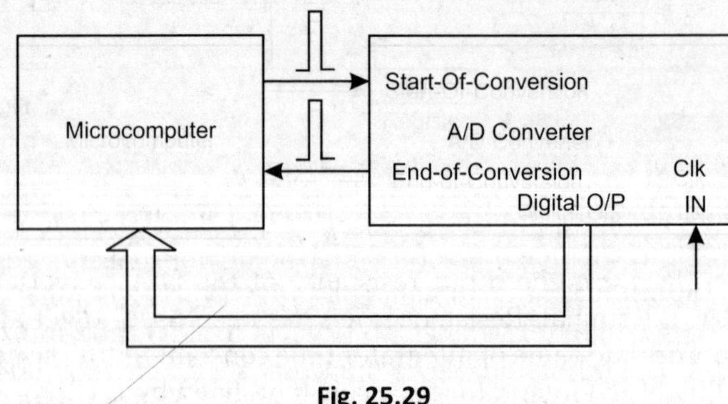

Fig. 25.29

Q. 31. With the help of a suitable block schematic representation, briefly explain use of A/D converter in a data acquisition block.

Ans. There are a large number of applications, where an analog signal is digitised to be subsequently stored or processed in a digital computer. The computer may store the data to be later passed on to a D/A converter to reconstruct the original signal, as in a digital-storage oscilloscope. It may process the digitised signal to generate the desired outputs in a process control application. Figure-25.29 shows the basic data acquisition building block. The computer generates a start-of-conversion signal. At the time instant of occurrence of the end-of-conversion signal generated by the A/D converter, the computer loads the digital output of the A/D converter onto its memory.

Solved Problems

Problem 25.1. *An 8-bit D/A converter produces an analog output voltage of 62.5 mV for a digital input of 00000001. What will-be the analog output for a digital input of 10000000?*

Solution. It is given that a '1' in LSB position produces an output of 62.5 mV. This implies that a '1' in the successive higher places as we move towards MSB would produce an output of 125 mV, 250 mV, 500 mV, 1V, 2V, 4V and 8V respectively.

The analog output for a digital input of 10000000 will be 8V.

Problem 25.2. *A 4-bit D/A converter produces a full scale output of 15 V. Determine the resolution of this converter. Also, determine the analog output for a digital input of 1010.*

Solution. The full scale output is given by : V_A (full scale) $= \left(\frac{2^n - 1}{2^n}\right)$ V

where n = Number of bits V = Voltage corresponding to a logic '1'.

Now, $\frac{(2^n - 1)V}{2^n} = 15$ This gives, $V = \frac{15 \times 2^4}{(2^4 - 1)} = 16V$

Resolution is the analog output produced by a '1' in LSB position and is given by,

$$\frac{1}{2^n} V = \frac{16}{16} = 1 \text{ volt}$$

Analog output for a digital input of (1010) is given by,

$$\frac{0 + 16 \times 2^1 + 0 + 16 \times 2^3}{16} = 10 \text{ volts.}$$

Problem 25.3. *An 8-bit D/A converter has a step size of 20 mV. Determine the full scale output and percentage resolution.*

Solution. The step size is same as the resolution.

Therefore, $\frac{1}{2^n} \times V = 20$ mV

or $V = 20 \times 2^8 = 20 \times 256 = 5.12$ volts

['V' is voltage corresponding to a logic '1']

$$\text{Full scale output} = \left(\frac{2^n - 1}{2^n}\right) V = \left(\frac{2^8 - 1}{2^8}\right) \times 5.12$$

$$= \frac{255}{256} \times 5.12 = 5.1 \text{ volts}$$

$$\text{Percentage resolution} = \frac{1}{2^n - 1} \times 100 = \frac{100}{255} = 0.392\%$$

Resolution can also be determined from,

$$\text{Percentage Resolution} = \frac{\text{Step size}}{\text{Full scal output}} \times 100$$

$$= \frac{20 \times 10^{-3}}{5.1} \times 100 = 0.392\%.$$

Problem 25.4. *An 8-bit BCD D/A converter produces a full scale output of 9.9 V. Determine the following:*

(*a*) *Total number of steps*

(*b*) *Resolution or step size*

(*c*) *Percentage resolution*

Solution. An 8-bit BCD D/A converter will have two decimal digits at the input. The input will vary from 00 to 99.

(*a*) Total number of Steps = 99

(*b*) Resolution = Step size $= \frac{\text{Full Scale Output}}{\text{Number of Steps}}$

$= \frac{9.9}{99} = 0.1$ volt

(*c*) Percentage Resolution $= \frac{1}{99} \times 100 = 1.01\%$

It can also be computed from,

$$\text{Percentage Resolution} = \frac{\text{Step Size}}{\text{Full Scale Output}} \times 100$$

$$= \frac{0.1}{9.9} \times 100 = 1.01\%$$

Problem 25.5. *How many bits should a current output D/A converter have so that its full scale output is 10 mA and its resolution is better than 50 μA?*

Solution. Resolution $= \frac{\text{Full Scale Output}}{\text{Number of Steps}}$

If (n) is number of bits, then number of steps = $(2^n - 1)$

Therefore, $50 \times 10^{-6} = \frac{10 \times 10^{-3}}{(2^n - 1)}$

or $2^n - 1 = \frac{10 \times 10^{-3}}{50 \times 10^{-6}} = 200$

or $2n = 201$ or $n = \dfrac{\log 201}{\log 2} = \dfrac{2.3032}{0.3010} = 7.65$

For resolution to be better than 50 μA, $n > 7.65$. Since (n) has to be a positive integer, therefore, $n = 8$.

Problem 25.6. *Compare the step size and percentage resolution of a D/A converter with 12-bit binary input with a D/A converter having a 12-bit BCD input. Both have a full scale output of 9.99 V.*

Solution. Step size $= \dfrac{\text{Full Scale Output}}{\text{Number of Steps}}$

In case of binary input, step size $= \dfrac{9.99}{(2^{12}-1)}$

$= \dfrac{9.99}{4095} = \dfrac{9990}{4095} = 2.44$ mV

Percentage Resolution $= \dfrac{1}{4095} \times 100 = 0.0244\%$

In case of BCD input, step size $= \dfrac{9.99}{999}$

$= \dfrac{9990}{999} = 10$ mV

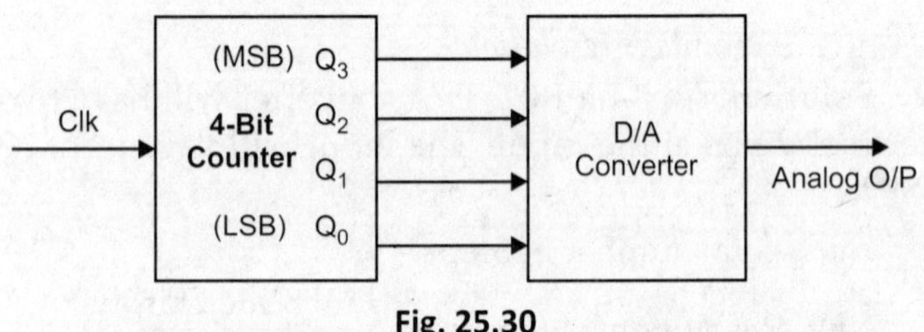

Fig. 25.30

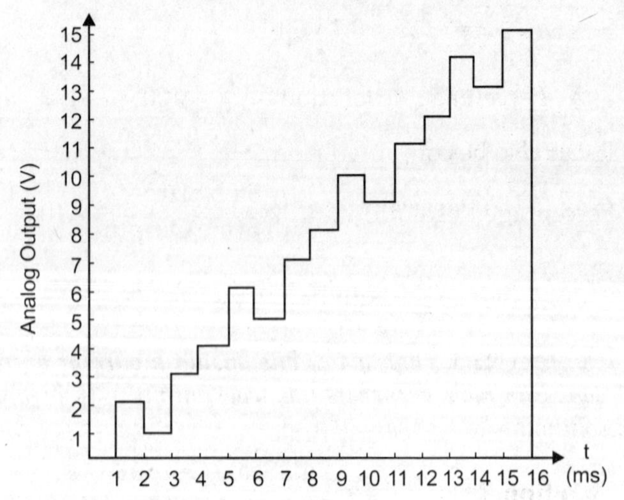

Fig. 25.31

Percentage Resolution $= \dfrac{1}{999} \times 100 = 0.1\%$.

Problem 25.7. *An experimenter connects a 4-bit ripple counter to a 4-bit D/A converter to perform a stair-case test using a 1kHz clock as shown in Fig. 25.30. The output staircase waveform is shown in Fig. 25.31 The cause of incorrect staircase signal is later determined to be wrong connection between counter output and D/A converter input. What is it?*

Solution. The correct staircase waveform would be generated at the output of D/A converter, if the LSB output of the counter is fed to LSB bit of D/A converter and also, the remaining bits of the counter are connected to the corresponding bits of D/A converter in the same order. If we recall the sequence in which counter advances, it can be visualised that the given staircase waveform would result, if the interconnections of LSB and the next adjacent higher bit of the counter output and D/A converter input are interchanged.

Problem 25.8. *An 8-bit D/A converter produces an analog output of 12.5 mV for a digital input of 0000010. Determine the analog output for a digital input of 00000100.*

Solution: The LSB in this case would produce an analog output of 12.5/2 = 6.25 mV. Starting with the LSB, different binary digits would produce analog outputs of 6.25 mV, 12.5 mV, 25 mV, 50 mV, 100 mV, 200 mV, 400 mV and 800 mV.

Therefore, analog output voltage for 00000100 would be 25 mV.

Problem 25.9. *A 12-bit D/A converter has a resolution of 2.44 mV. Determine its analog output for a digital input of 111111111111.*

Solution. If (V) is the voltage corresponding to logic '1', then $(1/2^n) \times V = 2.44 \times 10^{-3}$

This gives, $V = 4096 \times 2.44 \times 10^{-3} = 10$V

Analog output for 111111111111= Full scale output = $[(2^n - 1)/2^n] \times V = (4095/4096) \times 10 = 10$V

Problem 25.10. *How many bits should a current output D/A converter have so that its full scale output is 20 mA and its resolution is better than 25 μA?*

Solution. In n-bit converter, number of steps = $2^n - 1$

Resolution = Full scale output/Number of steps

$25 \times 10^{-6} = 20 \times 10^{-3} / (2^n - 1)$

$2^n - 1 = 20000/25 = 800$; $2^n = 801$

This gives $n = (\log 801/\log 2) = 9.6465$

Since (n) is an integer, $n = 10$

Problem 25.11. *Determine the resolution of a 12-bit A/D converter having a full scale analog input voltage of 5V.*

Solution. A 12-bit A/D converter resolves analog input voltage into $(2^{12} - 1)$ levels. Resolution is nothing but the step size.

Resolution $= \dfrac{5}{(2^{12}-1)} = \dfrac{5000}{(4096-1)} = \dfrac{5000}{4095} = 1.22$ mV.

Problem 25.12. *Determine the conversion time of an 8-bit A/D converter of the counter type shown earlier in Fig. 25.15 for an input clock frequency of 10 MHz.*

Solution. The counter type A/D converter shown in Fig. 25.15 has a variable conversion time which is maximum, when the input analog voltage is just below the full scale analog input voltage. An average conversion time equal to half of the maximum conversion time is usually defined in case of such converters.

Now maximum conversion time = Time taken by $(2^8 - 1)$ cycles of clock input.

Clock time period = $(1/10^7) = 0.1$ µs

Therefore, maximum conversion time = 255 × 0.1 = 25.5 µs

Average conversion time = (25.5/2) = 12.75 µs.

Problem 25.13. A 10-bit A/D converter of the successive approximation type has a resolution (or quantization error) of 10 mV. Determine the digital output for an analog input of 3.325 Volts.

Solution. In case of a successive approximation type of A/D converter, the final analog output of its D/A converter portion always settles at a value below the analog input voltage to be digitised within the resolution of the converter.

Analog input voltage = 3.325 V, Resolution = 10 mV

$$\text{Number of steps} = \frac{3.325 \times 10^3}{10} = 332.5.$$

Step number 332 will produce a D/A converter output of 332 × 10 = 3.32V and step number 333 will produce a D/A converter of 3.33 V. The A/D converter will settle at step 332. The digital output will be the binary equivalent of $(332)_{10}$ which is 0101001100.

Note. When this converter actually performs the transformation or conversion, in the tenth clock cycle, the LSB will be set to '1' initially. This would produce a D/A converter output of 3.33 V which exceeds the analog input voltage of 3.325 V. The comparator changes state which in turn resets the LSB to '0' bringing the D/A converter output of 3.32 V. This is how a converter of this type settles at a value, where D/A converter output settles at a value that is one step below the value that makes it exceed the analog input to be digitised.

Problem 25.14. *An 8-bit successive approximation type A/D converter works at a clock frequency of 1 MHz. Determine conversion time for an analog input of (a) 2.32 V (b) 3.95 V. The converter has a step size of 20 mV.*

Solution. In case of an A/D converter of the successive approximation type, the conversion time is independent of analog input voltage. It is the time taken by (n) clock cycles in case of an n-bit converter.

In the present case, for both analog inputs,

Conversion time = 8 × time period of clock input = 8 x 1 µs = 8 us.

Problem 25.15. *Compare the average conversion time of an 8-bit counter type A/D converter with that of an 8-bit successive approximation type A/D converter, if both are working at 1 MHz clock frequency.*

Solution. Clock time period = 1 µs

Average conversion time in case of counter type A/D converter is given by :

$$\frac{(2^8 - 1)}{2} \times 1 = 127.5 \text{ µs}.$$

Conversion time in case of successive approximation type A/D converter is given by:

$$8 \times 1 = 8 \text{ µs}.$$

Problem 25.16. *Determine the conversion time of a 12-bit A/D converter of the counter type A/D converter for an input clock frequency of 1 MHz.*

Solution: The counter type A/D converter has a variable conversion time depending upon magnitude of analog input voltage. It is the maximum, when the input analog voltage is just below the full scale analog input voltage. An average conversion time equal to half of the maximum conversion time is usually defined in case of such converters.

Maximum conversion time = Time taken by $(2^{12} - 1)$ = 4095 cycles of clock input

Clock time period = $1/(1 \times 10^6) = 1$ µs

Therefore, maximum conversion time = 4095 × 1 = 4095 µs = 4.095 ms

Average conversion time = (4.095/2) = 2.047 ms

Problem 25.17. *The A/D converter of Problem 25.16 has a full scale analog input of 10 volts. Determine the digital output produced by this converter for an analog input of 3.5 volts.*

Solution: Full scale analog input = 10 volts

Size of each step (or resolution) = 10/4095 = 2.442 mV

Therefore, number of steps required for an analog input of 3.56 V is given by:

$3.56/(2.442\times10^{-3}) = 1457.8 = 1458$ steps

The digital output will therefore be equal to binary equivalent of 1458

Digital Output = 010110110010

Problem 25.18. *The D/A converter of a counter type AID converter produces a staircase output having a step size of 10 mV. The A/D converter has a 10-bit resolution and is specified to have a quantization error of ± 1/2 LSB. Determine the digital output for an analog input of 4.012 volts. Assume that the comparator has a comparison threshold of 1 mV.*

Solution: The comparator has a comparison threshold of 1 mV. It implies that for the comparator to change state, the voltage at the relevant input should be 1 mV more than the voltage at the other input. Now, one of the inputs to the comparator is the analog input voltage (= 4.012 V in the present case). The other input to the comparator is a voltage that is equal to sum of D/A converter output voltage and a fixed voltage corresponding ½ LSB. This is the case, when the quantization error of A/D converter is specified to be ± ½ LSB. In case the quantization error is 1 LSB, the D/A converter directly feeds the other input of comparator.

In the present case, 1 LSB corresponds to 10 mV

Therefore, ½ LSB corresponds to 5 mV.

For an analog input of 4.012 V, the voltage at the other input needs to be 4.013 V (due to comparator threshold of 1 mV). This implies that D/A converter output needs to be 4.008 V.

Therefore, number of steps = $4.008/(10\times10^{-3}) = 400.8 = 401$

Digital output is binary equivalent of $(401)_{10}$, which equals 0110010001

SUMMARY OF IMPORTANT FORMULAS

1. For a n-bit D/A converter using binary ladder network, full scale analog output voltage, (V_A) is given by:
   $[(2^n - 1)/2^n]V$
   Where, V = Voltage corresponding to logic '1' state
2. In general, for an n-bit D/A converter using binary ladder network,
   $$V_A = \frac{V_1 \times 2^0 + V_2 \times 2^1 + V_3 \times 2^2 + ... + V_n \times 2^{n-1}}{2^n}$$
3. Percentage Resolution of an n-bit D/A converter is given by:
   Percentage Resolution = $[1/(2^n - 1)] \times 100$
4. Dynamic range of a linear D/A converter is given by:
   Dynamic Range = $20 \log 2^n$
5. Percentage resolution of BCD-input D/A converter is given by:
   Percentage resolution = $[1/(10^{n/4} - 1)] \times 100$
6. Percentage resolution of a D/A converter can also be determined from:
   Percentage Resolution = (Step Size/Full Scale Output) × 100
7. Percentage resolution of n-bit A/D converter is given by:
   Percentage Resolution = $(1/2^n) \times 100$
8. The number of comparators needed for n-bit simultaneous or flash A/D convertor is (2^n), $(2^n - 1)$ for amplitude and one for polarity
9. The number of comparators needed for n-bit half flash A/D convertor is $2 \times (2^{n/2})$
10. For an n-bit counter type A/D converter, maximum conversion time= $2^n \times T$
    Where, T = Clock period
11. For an n-bit counter type A/D converter, average conversion time = $(2^n/2) \times T$
    Where, T = Clock period
12. For an n-bit successive approximation type A/D converter, conversion time = $n \times T$
    Where, T = Clock period

SELF EVALUATION EXERCISE

I. Fill in the Blanks

(*a*) A one step increase in the digital input word of a 6-bit D/A converter increases the analog output by......of its maximum analog output value. (1/64,1/63,1/32)

(*b*) A D/A converter with 1.6 percent resolution has a......resolution than a D/A converter with 3.2 percent resolution. (Higher, Lower)

(*c*) A parallel or flash type 6-bit A/D converter requires......comparators at the input. (63, 32, 7)

(*d*) The A/D converter type that is widely used in digital voltmeter applications is the...... (Successive approximation type, Counter type, Dual slope integrating type)

(*e*) A 4-bit A/D converter has a percentage resolution of......(1.58, 3.2, 6.6)

(*f*) The resolution in volts of a 4-bit D/A converter of the multiplying type for a reference voltage input of 16 volts will be......(0.1, 0.01,1)

(*g*) The conversion time for a 6-bit successive approximation type A/D converter in the 1 MHz clock would be......µs. (32, 64, 6)

(*h*) The average conversion time in case of a 6-bit counter type A/D converter with 1 MHz clock is......µs. (32, 64, 6)

II. Multichoice Questions

1. Number of comparators needed to build a 6-bit simultaneous A/D converter is
   (a) 63 (b) 64
   (c) 7 (d) 6
2. Number of comparators needed in addition to other circuit blocks to construct a 6-bit counter type A/D converter is
   (a) 1 (b) 63
   (c) 64 (d) 6
3. The average conversion time in a 4-bit counter type A/D converter run by a 10 MHz clock shall be
   (a) 8μs
   (b) 400 ns
   (c) 800 ns
   (d) None of these
4. Out of the following, the fastest A/D converter type is
   (a) Counter type A/D converter
   (b) Successive Approximation type A/D converter
   (c) Dual Slope Integrating type A/D converter
5. A counter does not constitute a building block in one of the following A/D converter types.
   (a) Successive Approximation Type A/D converter.
   (b) Tracking type A/D converter.
   (c) Dual Slope Integrating type A/D converter.
   (d) Flash or simultaneous A/D converter.
6. The dynamic range of an n-bit linear D/A converter is expressed by:
   (a) Log 2^n
   (b) 20 log 2^n
   (c) 20 log2^{n-1}
   (d) None of these
7. Percentage solution of a 16 bit BCD input D/A converter would be
   (a) 0.1 (b) 1
   (c) 0.01 (d) None of these
8. An 8-bit BCD input D/A converter has a full scale output of 9.9 volts. Percentage resolution is
   (a) 0.1 (b) 0.01
   (c) 0.001 (d) 1
9. Minimum number of bits the D/A converter must have for a full scale output of 50 mA and resolution better than 50 μA one
   (a) 10 (b) 8
   (c) 12 (d) 16
10. One of the following is a characteristic of sigma-Delta type A/D converter.
    (a) It is the absolute value of analog signal that is sampled
    (b) The analog signal is sampled at Nyquist rate
    (c) In the case of Sigma-Delta A/D converter is a comparator
    (d) In the case of Sigma-Delta A/D converter, the analog signal is sampled at a rate much higher than Nyquist rate

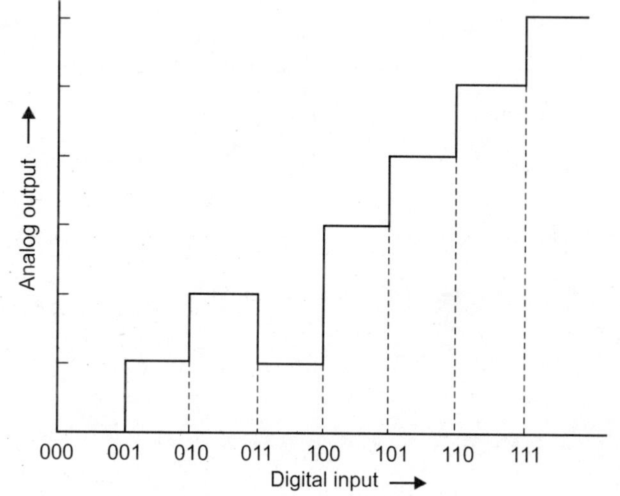

Fig. 25.32

III. Say TRUE or FALSE

(a) An 8-bit A/D converter with a full scale analog input of 10 volts can resolve an analog input change of about 40 mV.
(b) Accuracy of an A/D converter depends upon factors like gain error, offset error, quantization error etc.
(c) The sampling rate in an A/D converter is determined by conversion time.
(d) If the Differential Non-linearily (DNL) error in a D/A converter is ± 1LSB, the converter is monotonic.

(*e*) While checking the monotonicity of a D/A converter with a frequency counter, if the analog output waveform looks like the waveform of Fig. 25.32 the converter can be said to be monotonic.

Answers to Self Evaluation Exercise

I.

(*a*) 1/64 (*b*) Higher (*c*) 63
(*d*) Dual Slope Integrating
(*e*) 6.6 (*f*) 1 (*g*) 6 (*h*) 32

II.

1. (*a*)	2. (a)	3. (*c*)	4. (*b*)
5. (*d*)	6. (*b*)	7. (*c*)	8. (*a*)
9. (*a*)	10. (*d*)		

III.

(a) True (*b*) True (*c*) False
(*d*) True (*e*) False

Chapter 26

Microprocessors and Memory Devices

Learning Objectives

After completing the chapter, you will learn the following topics:

1. Introduction to Microprocessors
2. Microprocessor Architecture Including Arithmetic Logic Unit, Register File and Control Unit
3. Basic Microprocessor Instructions
4. Microprocessor Addressing Modes
5. 8085 Microprocessor Including Architecture, Instruction Set, Registers and Addressing Modes
6. RISC versus CISC Microprocessors
7. Peripheral Devices Including Programmable Timer/Counter, Programmable Peripheral Interface, Programmable Interrupt Controller, DMA Controller, Programmable Communication Interface, Math Co-processor, Programmable Keyboard/Display Interface, Programmable CRT Controller, Clock Generator, Floppy Disk Controller and Octal Bus Transceiver
8. Primary Memory Devices
9. Random Access Memory (RAM)
10. SRAM and DRAM
11. RAM Applications
12. Read Only Memory (ROM)
13. Types of ROM Including PROM, EPROM and Flash Memory
14. ROM Applications
15. Expanding Memory Capacity
16. Secondary Storage Devices

Q. 1: What is a microprocessor? What is its role in a digital system? What are the basic constituents of a microprocessor based system?

Ans: A microprocessor is a programmable device that accepts binary data from an input device and processes the data according to the instructions stored in the memory and provides results as output. In other words, microprocessor executes the program stored in the memory and transfers data to and from the outside world through I/O ports. Any microprocessor based system essentially comprises three parts, namely, microprocessor, memory and peripheral I/O devices. The microprocessor is generally referred to as the heart of the system, as it performs all the operations and also controls the rest of the system. The three are interconnected by the data bus, the address bus and the control bus.

The memory stores the binary instructions and data for the microprocessor. Memory can be classified as primary or main memory and secondary memory. Read/write memory (R/WM) and read only memory (ROM) are the examples of primary memory and are used for executing and storing programs. Magnetic disks and tapes are examples of secondary memory. They are used to store programs and results after the completion of program execution. Microprocessors do not execute programs stored in the secondary memory directly. Instead, they are first copied on to the R/W primary memory.

The input/output devices are means through which microprocessor interacts with the outside world. The commonly used input devices include keyboards, A/D converters, cameras, scanners, microphones and so on. LEDs, seven-segment displays, LCD displays, printers and monitors are some of the commonly used output devices. A bus is basically a communication link between the processing unit and the peripheral devices. It is a group of wires that carry information in the form of bits.

Q. 2: Briefly describe different types of buses in use in a microprocessor based system.

Ans: There are three types of buses in a microprocessor based system. These include the Address Bus, the Data Bus and the Control Bus. The *Address Bus* is unidirectional and is used by the central process unit (CPU) to send out address of the memory location to be accessed. It is also used by the CPU to select a particular input or output port. It may consist of 8, 16, 20 or even more number of parallel lines. Number of bits in the address bus determines the maximum number of data locations in the memory that can be accessed. A 16-bit address bus, for instance, can access 2^{16} data locations. It is labelled as $A_0 \ldots A_{n-1}$, where n is width in bits of the address bus.

The *Data Bus* is bidirectional, that is, data flow occurs both to as well as from microprocessor and peripherals. The data bus size has a considerable influence on the computer architecture as the parameters like the word length and the quantum of data that can be manipulated at a time are determined by size of the data bus. There is an internal data bus, which may not be of the same width as the external data bus that connects the microprocessor to I/O and memory. The size of the internal data bus determines the largest number that can be processed by the microprocessor in a single operation. The largest number processed, for instance, by a 16-bit internal data bus is 65535. The data bus is labelled as $D_0 \ldots D_{n-1}$, where (n) is the data bus width in bits.

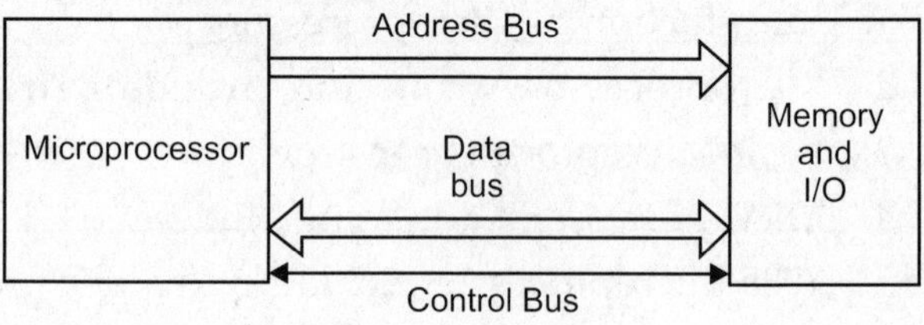

Fig. 26.1

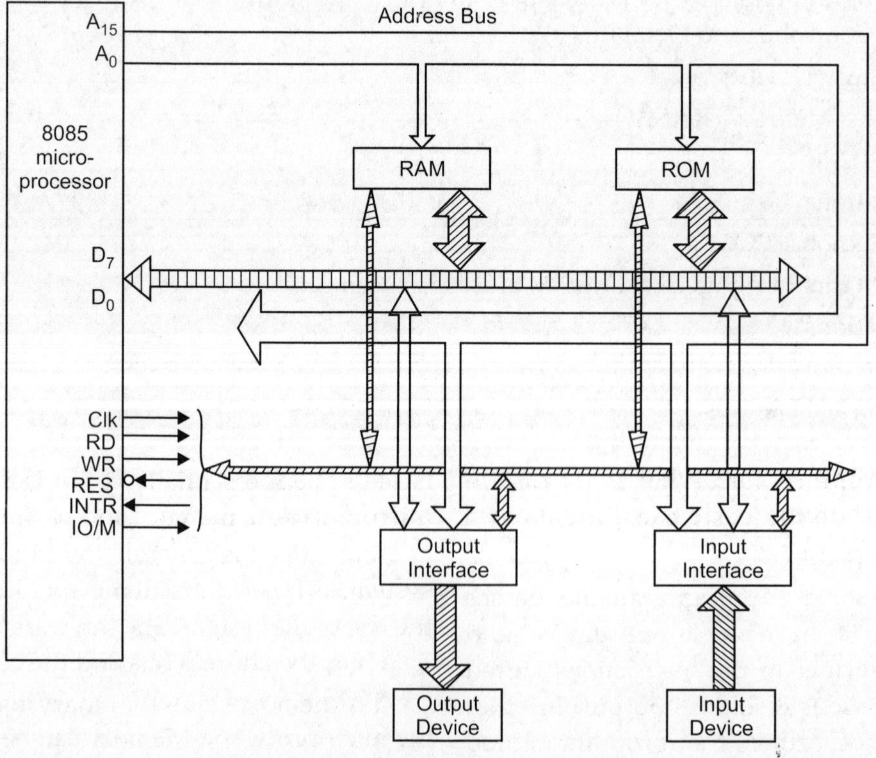

Fig. 26.2

The *Control Bus* contains a number of individual lines carrying synchronising signals. 'Bus' here would normally imply a group of lines working in unison. The control bus (if we call it a bus) sends out control signals to memory, I/O ports and other peripheral devices to ensure proper operation. It carries control signals, such as memory read, memory write, read input port, write output port, hold, interrupt, etc. For instance, if it is desired to read the contents of a particular memory location, the CPU first sends out address of that very location on the address bus and a 'memory read' control signal on the control bus. The memory responds by outputting data stored in the addressed memory location onto the data bus. 'Interrupt' tells the CPU that an external device needs to be read or serviced. 'Hold' allows a device, such as direct memory access (DMA) controller to take over the address and data busses. Figure 26.1 shows the bus interface between the microprocessor and its peripheral devices. The microprocessor in this figure is an eight-bit microprocessor, such as Intel 8085. Figure 26.2 shows a more descriptive schematic representation of Figure 26.1.

Q. 3. With the help of a suitable block schematic arrangement, briefly describe the basic functional blocks of a microprocessor.

Ans. Figure 26.3 shows a simplified typical schematic arrangement of a microprocessor. The figure shown is a generalised one and is not the actual structure of any of the commercially available microprocessors. The important functional blocks of a microprocessor include the following:

1. Arithmetic logic unit (ALU)
2. Register file
3. Control unit

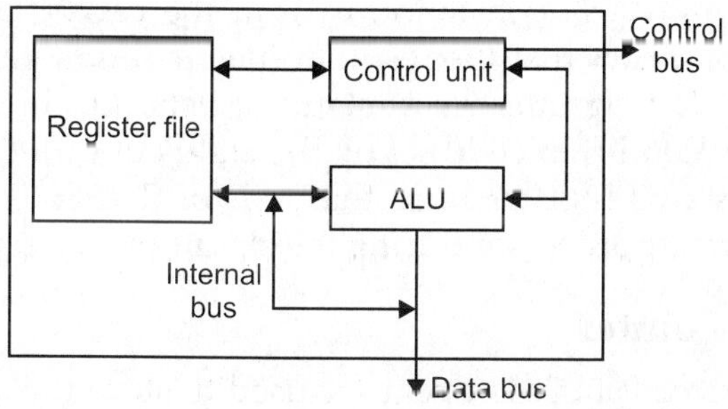

Fig. 26.3

Arithmetic Logic Unit

Arithmetic Logic Unit (ALU) is the core component of all microprocessors. It performs the entire integer arithmetic and bit-wise logical operations of the microprocessor. ALU is a combinational logic circuit and has two data input lines, a data output line and a status line. It gets data from the registers of the microprocessor, processes the data according to the instructions from the control unit and stores the results in its output registers. All modern ALU's use binary data in two's complement format. The integer arithmetic operations performed by the ALU include the addition and subtraction. It performs AND, OR, NOT and EX-OR logical operations. Some 16-bit, 32-bit and 64-bit microprocessors also perform the multiplication and division operations. In other microprocessors, the multiplication and division operations are performed by writing algorithms using addition and subtraction operations. ALU also performs the bit shifting operations and the comparison of data operations.

Register File

Register File comprises various registers used primarily to store data, addresses and status information during the execution of a program. Registers are sequential logic devices built using flip-flops. Some of the commonly found registers in most of the microprocessors include program counter (PC), instruction registers, buffer registers, status register, stack pointer, general purpose registers and temporary registers.

Control Unit

Control Unit governs and co-ordinates the activities of different sections of the processor and I/O devices. It is responsible for controlling the cycle of fetching machine instructions from memory and executing them. It also co-ordinates the activities of the input and output devices. It is undoubtedly the most complex of all functional blocks of the microprocessor and occupies most of the chip area. Control unit is a sequential logic circuit, which steps the processor through a sequence of synchronized operations. It sends a stream of control signals and timed pulses to the components and external pins of the microprocessor. As an illustration, to execute an instruction from the memory, the control unit sends out a Read command to the memory and reads the instruction (or data) that comes back on the data bus. The control unit then decodes the instruction and sends appropriate signals to the ALU, the general purpose registers, the multiplexers, the de-multiplexers, the program counter and so on. If the instruction was to store data in the memory, the control unit sends out address of the memory location on the address bus, the data to be stored on the data bus and a 'write' command on a control line. Figure-26.4 shows descriptive block diagram of a microprocessor. Multiplexers and de-multiplexers here don't represent primary functions and are there to facilitate flow of data between different blocks and also between different blocks and the outside world.

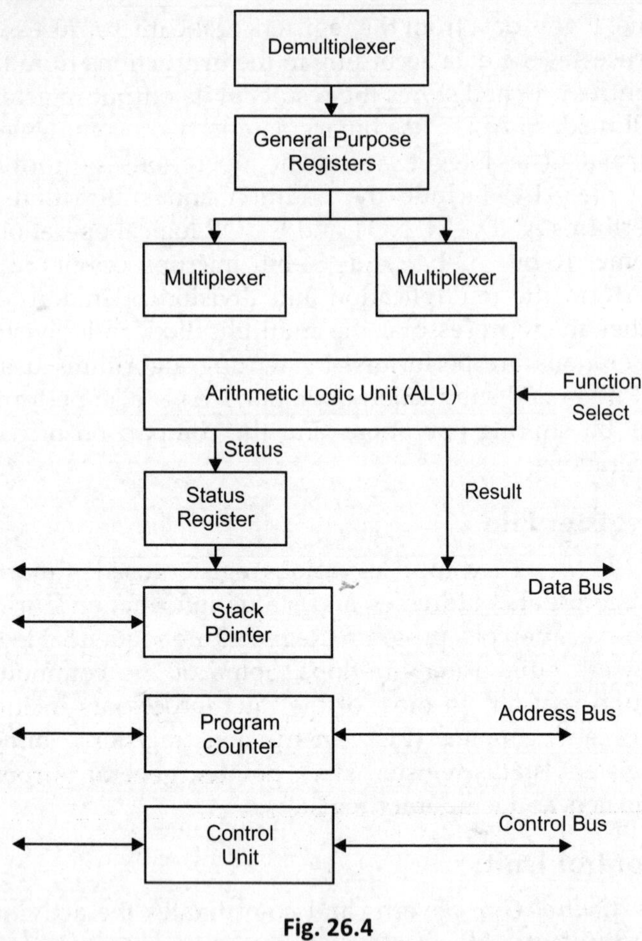

Fig. 26.4

Q. 4. What are the common registers in the register file of a microprocessor? Briefly describe the functions of these registers.

Ans. Commonly found registers in most of the microprocessors include program counter (PC), instruction registers, buffer registers, status register, stack pointer, general purpose registers and temporary registers.

Program Counter (PC)

The *Program Counter* is a register that stores address of the next instruction to be executed and hence, plays a central role in controlling the sequence of machine instructions that the processor executes. After the instruction is read into the memory, the PC is automatically incremented by '1'. This is of course with the assumption that the instructions are executed sequentially. Its contents are affected by JUMP and call instructions. In the case of JUMP instruction, the PC is first loaded with the new address and then incremented thereafter until another JUMP instruction is encountered. When the microprocessor receives an instruction to begin a subroutine, the contents of the program counter are incremented by '1' and are saved in the stack. The PC is loaded with the address of the first instruction of the subroutine. Its contents are incremented by '1' until a return instruction is encountered. The saved stack contents are then loaded into the PC and the program continues, executing each instruction sequentially until another JUMP instruction or a subroutine call is encountered. The interrupt process also alters the contents of program counter.

Instruction Register

An *Instruction Register* stores the code of the instruction currently being executed. Control unit extracts the operation code from the instruction register, which determines the sequence of signals necessary to perform the processing required by the instruction.

Buffer Register

Buffer Registers interface the microprocessor with its memory system. The two standard buffer registers are memory address register (MAR) and memory buffer register (MBR). MAR is connected to the address pins of the microprocessor and holds the absolute memory address of the data or instruction to be accessed. MBR, also known as memory data register, is connected to the data pins of the microprocessor. It stores all data written to and read from memory.

Status Register

The *Status Register* stores the status outputs of the result of an operation and gives additional information about the result of an ALU operation. The status of bits stored in the status register tells about occurrence or non-occurrence of different conditions and one or more bits may be updated at the end of an operation. Each bit is a Boolean flag representing a particular condition. The most common conditions are the carry, overflow, zero and negative. For instance, a '1' in the carry status bit position shows that the result of the operation generates a carry. The significance of status register lies in the fact that the condition code set by the status of different bits in the status register form the basis of decision-making by the microprocessor during the execution of a program.

Stack Pointer

A *Stack Pointer* is a register used to store the address of a memory location belonging to the most recent entry in the stack. A stack infact is a block of memory locations designated for temporary storage of data. It is used to save data of another general purpose register during execution of a subroutine or when an interrupt is serviced. The data is moved from a general register to the stack by a PUSH instruction at the beginning of a subroutine call and back to the general register by a POP instruction at the end of the subroutine call. Microprocessors use a stack because it is faster to move data using PUSH and POP instructions than moving data to/from memory using MOVE instruction.

General Purpose Registers

There is a set of registers for general purpose use designated as *General Purpose Registers*. They are used explicitly to store data and address information. Data registers are used for arithmetic operations, while the address registers are used for indexing and indirect addressing. These enhance processing speed of the microprocessor by avoiding large number of external memory read/ write operations while an ALU operation is being performed, as it is much easier and faster to read from or write into an internal register than to read from or write into an external memory location. Earlier microprocessors had only one register called accumulator for ALU operations. It needed at least four assembly language instructions to perform a simple addition including carrying data from an external memory location to the accumulator, adding contents of accumulator to those of another memory location, storing the result in the accumulator and transferring the contents of the accumulator back to the external memory location. With the availability of more number of general purpose registers, it would be possible to perform many ALU operations without even a need to store data in external memory.

Temporary Registers

They are used, when data has to be stored during the execution of a machine instruction. They are completely hidden from the user of the microprocessor.

Q. 5. Briefly describe the hard-wired and micro-coded types of control units used in microprocessors.

Ans. Control units are categorised into two types depending upon the way they are built. These include hard-wired and micro-coded control units. Hard-wired controllers are sequential logic circuits, the states of which correspond to the phases of the instruction execution cycle. In the case of hard-wired controllers, there is an electronic circuitry in the control unit to generate control signals for each instruction. They are very compact and fast, but are difficult to design. This design is also known as reduced instruction set computer (RISC) design. Micro-coded control units are easy to design and execution of an instruction in this case involves executing a micro-program consisting of a sequence of micro instructions. This design is also known as complex instruction set computer (CISC) design. Micro-coded control units offer more flexibility as compared to hardwired control unit but they are comparatively slower than hard-wired control units.

Q. 6. Outline the different basic operations performed by a microprocessor.

Or

What are the different instructions used by microprocessor?

Ans. Microprocessors perform the following basic operations:

1. Data transfer instructions
2. Arithmetic instructions
3. Logic instructions
4. Control transfer or Branch or Program Control instructions
5. Machine control instructions

Data Transfer Instructions transfer data from one location designated as the source location to another location designated as destination. Arithmetic Instructions performed by microprocessors include addition, subtraction, multiplication, division, comparison, negation, increment and decrement. *Logic Instructions* are used to perform all the logic operations that are performed by hard-wired logic, which include AND, OR, NOT and EXCLUSIVE-OR operations. Other logic operations include SHIFT and ROTATE operations. Microprocessors execute machine codes from one memory location to the next, that is, they execute instructions in a sequential manner. *Control Transfer* or *Branch Instructions* change the flow of the program either unconditionally or under certain test condition. Branch instructions include JUMP, CALL, RETURN, and INTERRUPT. *Machine Control Instructions* include HALT and NOP instructions.

Q. 7. What are Data Transfer Instructions? Briefly describe data transfer instructions of 8085 microprocessor.

Ans. As outlined in the answer to the previous question, data transfer instructions transfer data from one location designated as the source location to another location designated as destination. The data transfer could take place from one register to another, from one memory location to another memory location, from a memory location to a register or from a register to a memory location and so on. In fact, they are more correctly referred to as data movement operations as the contents of the source are not transferred, but are copied into the destination register without modifying the contents of the source. It may be mentioned here that these operations do not affect the flags. Data transfer operations of 8085 microprocessor are of three types, namely MOVE, LOAD and STORE.

MOV destination, source	Copy data from the source to the destination location
LDA address	Copy the data byte at memory location specified by 16-bit address into the accumulator
STA address	Copy the data from the accumulator to the memory location specified by 16-bit address

Q. 8. What are Arithmetic Instructions? What are the different arithmetic operations supported by 8085 microprocessor?

Ans. Arithmetic instructions performed by microprocessors include addition, subtraction, multiplication, division, comparison, negation, increment and decrement. It may be mentioned here that most of the eight-bit microprocessors do not support the multiplication and the division operations. These operations are supported by the 16-bit and 32-bit microprocessors. The arithmetic operations supported by the 8085 microprocessor are addition, subtraction, increment and decrement operations. Examples are given as follows:

ADD R	Adds the contents of the register to accumulator
ADI eight-bit	Adds the eight-bit data to accumulator
SUB R	Subtracts the contents of register from accumulator
SUI eight-bit	Subtracts eight-bit data from the contents of accumulator
INR R	Increments the contents of the register
DCR R	Decrements the contents of the register

Q. 9. What are logic Instructions with reference to a microprocessor? Give examples.

Ans. Microprocessors can perform all the logic functions of hard-wired logic. The basic logic operations performed by all microprocessors are AND, OR, NOT and EX-OR. The other logic operations include 'shift' and 'rotate' operations. All these operations are performed on a bit-for-bit basis on bytes or words. For instance, 11111111 AND 10111010 equals 10111010 and 11111111 OR 10111010 equals 11111111. Some microprocessors also perform bit-level instructions, such as 'set bit', 'clear bit' and 'complement bit' operations. It may be mentioned that logic operations always clear the carry and overflow flags, while the other flags change to reflect the condition of the result.

The basic shift operations are the 'shift left' and 'shift right' operations. In the shift left operation also known as arithmetic shift left, all bits are shifted one position to the left with the rightmost bit set to '0' and the leftmost bit-transferred to the carry position in the status register. In the shift right operation also known as logic shift right, all bits are shifted one bit position to the right with the leftmost bit set to '0' and the rightmost bit transferred to the carry position in the status register. If in the shift right operation, leftmost bit is left unchanged, it is called arithmetic shift right. In a 'rotate' operation, the bits are circulated back into the register. Carry may or may not be included. As an illustration, in a 'rotate left' operation without carry, the leftmost bit goes to the rightmost bit position and in a 'rotate right' with carry included, the rightmost bit goes to the carry position and the carry bit takes the position of left most bit. Examples of logic instructions performed by the 8085 microprocessor include the following:

ANA R/M	Logically AND the contents of Register/ memory with the contents of accumulator
ANI eight-bit	Logically AND the eight-bit data with the contents of the accumulator
ORA R/M	Logically OR the contents of Register/ memory with the contents of accumulator
ORI eight-bit	Logically OR the eight-bit data with the contents of the accumulator
XRA R/M	Logically EX-OR the contents of Register/ memory with the contents of accumulator
XRI eight-bit	Logically EX-OR the eight-bit data with the contents of the accumulator
CMA	Complement the contents of the accumulator
RLC	Rotate each bit in the accumulator to the left position
RRC	Rotate each bit in the accumulator to the right position

Q. 10. What are Control Transfer or Branch or Program Control Instructions? Briefly describe different branch instructions.

Ans. Microprocessors execute machine codes from one memory location to the next, that is, they execute instructions in a sequential manner. Branch instructions change the flow of the program either unconditionally or under certain test condition. Branch instructions include JUMP, CALL, RETURN, and INTERRUPT.

JUMP instructions are of two types, namely, (1) UNCONDITIONAL JUMP instructions and (2) CONDITIONAL JUMP instructions. If the microprocessor is so instructed as to load a new address in the program counter and start executing instructions at that address, it is termed as an UNCONDITIONAL JUMP. In the case of a CONDITIONAL JUMP, the program counter is loaded with a new instruction address only, if and when certain conditions are established by the microprocessor after reading the appropriate status register bits. CALL instructions transfer the flow of the program to a sub-routine. The CALL instruction differs from JUMP instruction as CALL saves a return address (address of program counter plus one) on the stack. The RETURN instruction returns control to the instruction whose address was stored in the stack, when CALL instruction was encountered. INTERRUPT is a hardware-generated CALL (externally driven from a hardware signal) or a software-generated CALL (internally derived from the execution of an instruction or by some internal event). Examples of transfer control instructions of 8085 microprocessor are the following:

MP 16-bit address	Change program sequence to location specified by 16-bit address
JZ 16-bit address	Change program sequence to location specified by 16-bit address, if zero flag is set
JC 16-bit address	Change the program sequence to the location specified by 16-bit address, if carry flag is set
JNZ 16-bit address	Change the program sequence to the location specified by 16-bit address, if zero flag is reset
JNC 16-bit address	Change the program sequence to the location specified by 16-bit address, if carry flag is reset
CALL 16-bit address	Change the program sequence to the location of the subroutine specified by the 16-bit address,
RET	Return to the calling program

Q. 11. What are Machine Control Instructions?

Ans. *Machine control instructions* include HALT and NOP instructions. Machine control instructions performed by 8085 microprocessor include the following:

HLT	Stop processing and wait
NOP	No operation

Q. 12. What are Addressing Modes with reference to microprocessors? Name common used addressing modes.

Ans. Microprocessors perform operations on data stored in register or memory. This data is specified in the operand field of the instruction. The data can be specified in various ways as direct data value or stored in some register or memory location and so on. These are referred to as the addressing modes of the microprocessor. In other words, addressing mode as expressed in the instruction tells us as to how and from where the microprocessor can get the data to act on. Addressing modes are of direct relevance to compiler writers and to programmers writing the code in assembly language. Different microprocessor architectures provide a variety of addressing modes. RISC microprocessors have far less addressing modes than CISC microprocessors. The most commonly used addressing modes are absolute, immediate, register direct, register indirect, indexed, PC relative, implicit and relative addressing modes. They account for more than 90 percent of the total addressing modes.

Q. 13. Briefly describe the Absolute Addressing Mode. Explain with the help of an example.

Ans. In *Absolute Addressing Mode,* the data is accessed by specifying its address in the memory [Figure-26.5 (a)]. This mode is useful for accessing fixed memory locations, such as memory mapped I/O devices. For example, the instruction MOV A, 30H in 8085 microprocessor moves the contents of memory location 30H into accumulator [Figure-26.5(b)]. In this case accumulator has the value 07H.

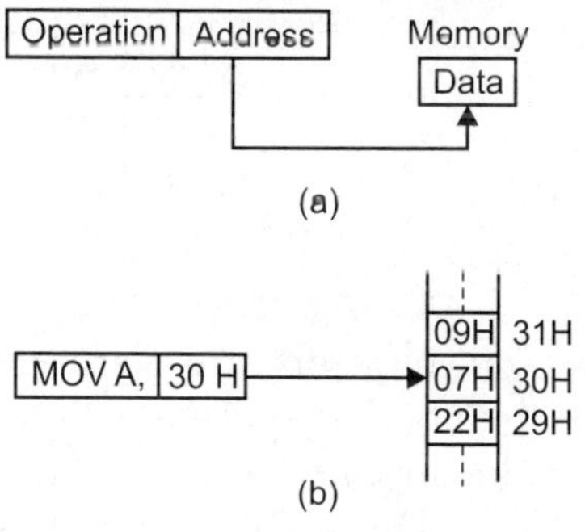

Fig. 26.5

Q. 14. Briefly describe the Immediate Addressing Mode. Explain with the help of an example.

Ans. In immediate addressing mode, value of the operand is held within the instruction itself (Figure-26.6). This mode is useful for accessing constant values in a program. It is faster than the absolute addressing mode and requires less memory space. For example, the instruction MVI A, #30H moves the data value 30H into the accumulator. The sign '#' in the instruction tells the assembler that the addressing mode used is immediate.

Operation	Data

Fig. 26.2

Q. 15. Briefly describe the Register Direct Addressing Mode. Explain with the help of an example.

Ans. In *Register Direct Addressing Mode*, the data is accessed by specifying the register name in which it is stored [Figure 26.7 (a)].

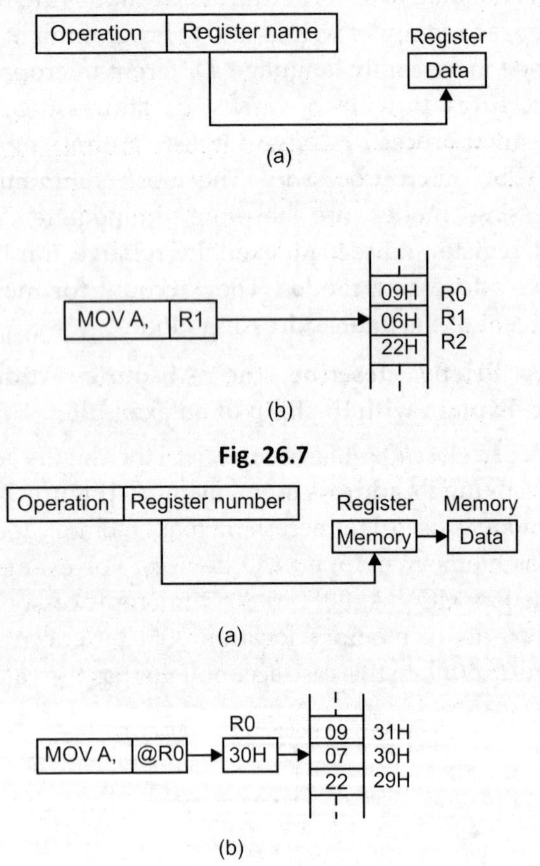

Fig. 26.7

Fig. 26. 8

Operations on registers are very fast and hence instructions in this mode require less time than absolute addressing mode instructions. As an example, the instruction MOV A, R1 in 8051 microprocessor moves the contents of register R1 into accumulator [Figure 26.7 (b)]. The contents of accumulator after the instruction are 06H.

Q. 16: Briefly describe the Register Indirect Addressing Mode. Explain with the help of an example.

Ans: In all modes discussed so far, either the value of the data or its location is directly specified. Indirect addressing mode uses a register to hold the actual address, where the data is stored. That is, in this case the memory location of the data is stored in a register [Figure-26.8 (a)]. In other words, in indirect addressing mode, the address is specified indirectly and has to be looked up. This addressing mode is useful, when implementing the pointer data type of high-level language. In 8085 microprocessor, R0 and R1 registers are used as eight-bit index and DPTR as 16-bit index. The mnemonic symbol used for indirect addressing is @. As an example the instruction MOV A, @R0 moves the contents of memory location whose address is stored in R0 into accumulator. The value of accumulator in this example is 07H [Figure-26.8 (b)]. This addressing mode can also be enhanced with an offset for accessing data structures in data space memory. This is referred to as register indirect with displacement. As an example, the instruction MOVC A,@A+DPTR copies the code byte at the memory address formed by adding the contents of A and DPTR to A.

Q. 17. Briefly describe the Indexed Addressing Mode. Explain with the help of an example.

Ans. In an *Indexed Addressing Mode*, the address is obtained by adding the contents of a register to a constant (Figure-26.9). The instruction 'Move the contents of Accumulator A to the memory location, whose address is given by the contents of register 1 plus 5' is an example of indexed addressing.

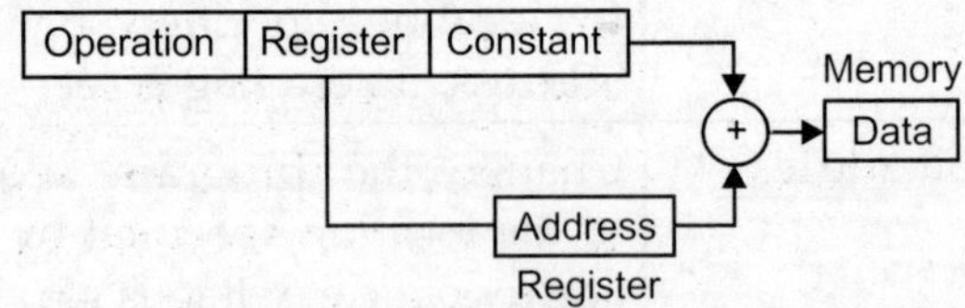

Fig. 26.9

Indexed addressing mode is useful, whenever the absolute location of the data is not known till the program is running. This addressing mode is used to access a continuous table or array of data items stored in memory. The content of the constant gives the starting address while the contents of the register determine the element of the array or table to be accessed. If program counter is used in the indexed addressing mode, it is known as PC relative addressing mode.

Q. 18. Briefly describe the Implicit and Relative Addressing Modes. Give examples.

Ans. In an Implicit Addressing Mode, no operand is used in the instruction and the location of the operand is obvious from the instruction itself. Examples include 'Clear carry flag', 'Return from sub-routine' and so on. The Relative Addressing Mode is used for JUMP and BRANCH instructions only. In this, a displacement is added to the address in the program counter and the next instruction is fetched from the new address in the program counter. This mode is particularly useful in connection with conditional JUMPs.

Q. 19. How do we program microprocessors? Briefly explain the use of assembly language and high level languages for programming microprocessors.

Ans. Microprocessors execute programs stored in the memory in the form of sequence of binary digits. Programmers do not write the program in binary form but write it either in the form of a text file containing assembly-language source code or using a high-level language.

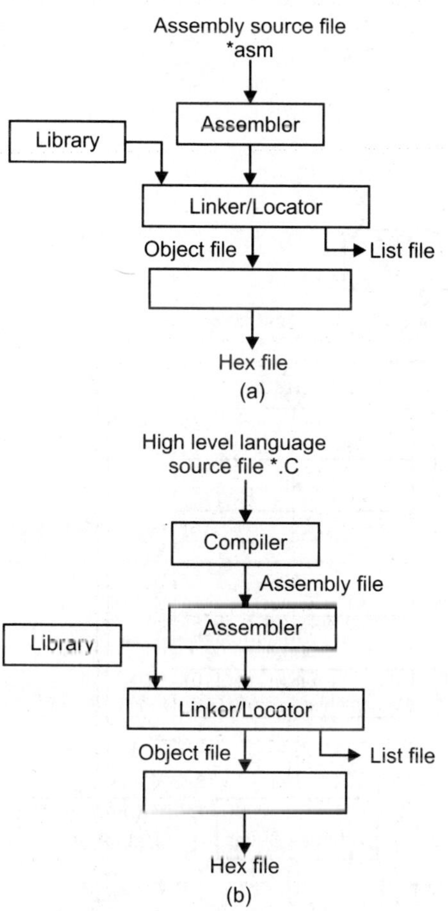

Fig. 26. 10

Programs, such as editor, assembler, linker and debugger enable the user to write the program in assembly language, covert it into binary code and debug the binary code. Editor is a program that allows the user to enter, modify and store a group of instructions or text under a file name. The assembly language source code is translated into object code by a program called assembler. Linker converts the output of the assembler into a format that can be executed by the microprocessor. The debugger is a program that allows the user to test and debug the object file.

Programming in assembly language produces code which is fast and takes up little memory. However, it is difficult to write large programs using assembly language. Another disadvantage of assembly language programming is that it is specific to a particular microprocessor. High-level language programming overcomes these problems. Some of the popular high-level languages used include C, C++, Pascal, etc. Compiler programs are primarily used to translate source code from a high-level language to a lower level language (e.g., assembly language or machine language). Figures-26.10 (a) and (b) show the various steps involved in executing assembly language programs and programs written in high-level languages, respectively.

Q. 20. Briefly describe CISC and RISC microprocessor with particular reference to their salient features.

Ans. The primary goal of complex instruction set computer (CISC) architecture is to complete a task in as few lines of assembly as possible. This is achieved by building processor hardware that is capable of understanding and executing a series of complex operations. In this case, each instruction can execute several low-level instructions. One of the primary advantages of this system is that the compiler has to do very little work to translate a high-level language statement into assembly. Because the length of the code is relatively short, very little RAM is required to store instructions. In nutshell, the emphasis is to build complex instructions directly into the hardware. Examples of CISC processors include the following: CDC 6600, System/360, VAX, PDP-11, and Motorola 68000 family, and Intel and AMD × 86 CPUs.

Reduced instruction set computer (RISC) is a microprocessor that emphasises simplicity and efficiency. RISC designs start with a necessary and sufficient instruction set. The objective of any RISC architecture is to maximise speed by reducing clock cycles per instruction. Almost all computations can be done from a few simple operations. The goal of RISC architecture is to maximise the effective speed of a design by performing infrequent operationsin software and frequent functions in hardware, thus, obtaining net performance gain.

To understand this phenomenon, let us consider any assembly-level language program. It has been observed that it uses MOV instruction much frequently as compared to MUL. So, if the architectural design implements MOV in hardware and MUL in software, then there will be considerable gain in speed, which is the basic feature of RISC technology. Examples of RISC processors include Sun's SPARC, IBM and Motorola's PowerPC, and ARM-based processors.The salient features of a RISC processor are as follows:

1. The microprocessor is designed using hardwired control. For example, one bit can be dedicated for one instruction. Generally, variable length instruction formats require microcode design. All RISC instructions have fixed formats, so no microcode is required.
2. RISC microprocessor executes most of the instructions in a single clock cycle. This is due to the fact that they are implemented in hardware.
3. The instruction set typically includes only register-to-register load and store.
4. The instructions have simple format with few addressing modes.
5. RISC microprocessor has several general-purpose registers and large cache memories, which support very fast access of data.
6. RISC microprocessor processes several instructions simultaneously and so includes pipelining.
7. Software can take advantage of more concurrency.

Q. 21. With the help of a suitable diagram indicating different blocks inside 8085 microprocessor, briefly describe the functions of different pins of the microprocessor.

Ans. Figure-26.11 shows the architecture of 8085 microprocessor. Functions of different pins of the microprocessor are briefly described in Table-26.1.

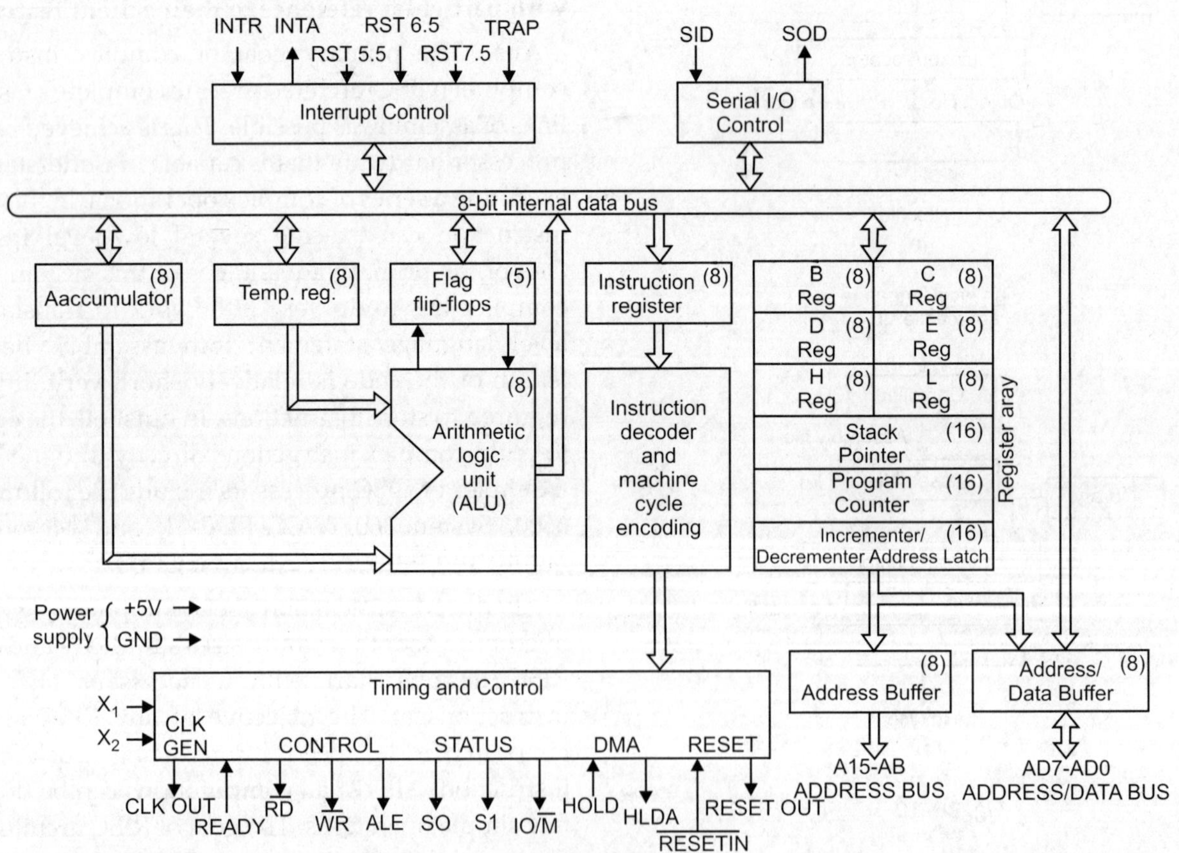

Fig. 26. 11

Q. 22. Briefly describe the Register Set of 8085 microprocessor.

Ans. The 8085 microprocessor has the following registers:

1. **Accumulator:** 8085 microprocessor has an eight-bit accumulator.
2. **Flag register:** 8085 has aeight-bit flag register containing five one-bit flags, namely, sign, zero, auxiliary carry, parity and carry.
3. **B and C registers:** 8085 has eight-bit B and C registers which can be used as one 16-bit BC register pair.
4. **D and E registers:** 8085 has eight-bit D and E registers which can be used as one 16-bit DE register pair.
5. **H and L registers:** 8085 has eight-bit H and L registers which can be used as one 16-bit HL register pair.
6. **Stack pointer:** 8085 has a 16-bit stack pointer.
7. **Program counter:** 8085 has a 16-bit Program counter.

Q. 23. Name the Addressing Modes of 8085 microprocessor.

Ans. 8085 microprocessor has four addressing modes. These are:

1. Register addressing
2. Register indirect addressing
3. Direct addressing mode
4. Immediate addressing mode

Table-26.1

Single	Description
Address bus (12–19, 21 – 28)	A 16-bit address. The lower eight bits are multiplexed with the data bus. The most significant eight bits of the memory address (or I/O address) are denoted by $A_{8.}$ – A_{15}. The lower eight bits of the memory address (or I/O address) appear on the multiplexed address/data bus (AD_0 – AD_7) for the first clock cycle of the machine cycle. It then becomes the data bus during the second and third clock cycles.
Data bus (12 – 19)	Eight-bit data bus is multiplexed with lower eight bits of the address bus ($AD_0 - AD_7$)
Control and Status Signals	
ALE (Address) Latch Enable (30)	It is a positive-going during the first clock state of the machine cycle that indicates that the bits on $A_0 - AD_7$ are address bits. It is used to latch the low-order address on the on-chip latch from the multiplexed bus
READ ($\overline{RD}$) (32)	A LOW on ($\overline{RD}$) indicates that the selected memory or I/O device is ready to be read and the data bus is available for data transfer
WRITE ($\overline{WR}$) (31)	A LOW on ($\overline{WR}$) indicates that the data on the bus are to be written into a selected memory or I/O location. Data is set up at the trailing edge of the ($\overline{WR}$) signal
IO/$\overline{M}$ (34)	This is a status signal that is used to differentiate between I/O and memory operations
S_1 and S_0 (29, 33)	These are status signal and can identify various operations
Power Supply and Clock Frequency	
V_{CC} (40)	+ 5 V
V_{SS} (20)	Ground
X_1, X_2 (1, 2)	A crystal, LC or RC network is connected at these two pins to drive the internal clock generator. X1 can also be an external clock input from a logic gate. The frequency is internally divided by 2 to give the internal operating frequency of the processor. The crystal frequency must be at least 1 MHz and must be twice the desired internal clock frequency.

Cont...

CLK OUT–Clock output	This output signal can be used as a system clock for devices on the board. The period of CLK is twice the X_1, X_2 input period
Interrupts and other Operations: 8085 has Five Interrupt Signals	
INTR: Interrupt Request (10)	This is a general-purpose signal. The microprocessor issues acknowledge signal (INTA), when the interrupt is requested
RST 7.5 (7) RST 6.5 (8) RST 5.5 (9)	These are restart interrupts. These are vectored interrupts and transfer program control to specific memory locations
TRAP (6)	It is a non-maskable interrupt and has the highest priority

In addition to these interrupts RESET, HOLD and READY pins accept externally initiated signals as inputs

HOLD (39) HLDA (38)	A HOLD signal indicates that another master device is requesting the use of data and address buses. The microprocessor, upon receiving the HOLD request, will relinquish the use of the bus after completion of the current bus transfer. It sends the HOLD ACKNOWLWDGE (HLDA) signal, indicating that it will relinquish the bus in the next clock cycle

Q. 24: Distinguish between Instruction, Instruction Cycle, Instruction Set and Machine Cycle.

Ans: An *Instruction* is a binary pattern designed inside a microprocessor to perform a specific function. The entire group of instructions a microprocessor can perform is referred to as its *Instruction Set*. *Instruction Cycle* is defined as the time required completing the execution of an instruction. The 8085 instruction cycle consists of one to six machine cycles. *Machine Cycle* is defined as the time required for completing one operation of accessing memory, I/O and so on. This will comprise of three to six T-states, which is defined as one sub-division of the operation performed in one clock period.

Q. 25: Name the common peripheral devices used along with a microprocessor in a microcomputer system.

Ans: Microprocessors and peripheral devices provide a complete solution in increasingly complex application environments. A peripheral device typically belongs to the category of MSI logic devices. Common peripheral devices used along with a microprocessor in a microcomputer system include programmable counter/timer, Programmable peripheral interface (PPI), EPROM, RAM, programmable interrupt controller (PIC), DMA controller, programmable communication interface-universal synchronous asynchronous receiver transmitter (USART), math co-processor, programmable keyboard/display interface, CRT controller, floppy disk controller, clock generators and transceivers.

Q. 26: Briefly describe the function of the following peripheral devices. Give typical type numbers of ICs used to perform each of these functions:

1. **Programmable Timer/Counter**
2. **Programmable Peripheral Interface (PPI)**
3. **Programmable Interrupt Controller (PIC)**
4. **DMA Controller**
5. **Programmable Communication Interface**
6. **Math Coprocessor**
7. **Programmable Keyboard/Display Interface**
8. **Programmable CRT Controller**
9. **Floppy Disk Controller**
10. **Clock Generator**
11. **Octal Bus Transceiver**

Ans: Programmable Timer/Counter

Programmable timer/counter is used for generation of accurate time delay for event counting, rate generation, complex waveform generation applications and so on. Examples of programmable timer/counter devices include Intel's 8254 and 8253 family of devices. Intel 8254 contains three 16-bit counters< which can be programmed to operate in several different modes. Some of the functions common to microcomputers and implementable with 8254 are real time clock, event counter, digital one shot, programmable rate generator, square wave generator, binary rate multiplier, complex wave form generator and complex motor controller.

Programmable Peripheral Interface (PPI)

PPI devices are used to interface the peripheral devices with the microprocessors. 8255 PPI is a widely used programmable parallel I/O device. 8255 can be programmed to transfer data under various conditions, from simple I/O to interrupt I/O. It can function in bit reset (BSR) mode or I/O mode. In I/O mode, it has three ports, namely, port A, port B and port C. The I/O mode is further divided into three different modes, namely mode 0, mode 1, mode 2. In mode 0, all ports function as simple I/O ports. Mode 1 is a handshake mode, whereby port A and/or B use bits from port C as handshake signals. In mode 2, port A can be set up for bidirectional data transfer using handshake signals from port C and port B can be set up either in mode 0 or in mode 1. In BSR mode, individual bits in port C can be set or reset.

Programmable Interrupt Controller (PIC)

A *PIC* is a device that allows priority levels to be assigned to its interrupt outputs. It functions as an overall manager in an interrupt driven system environment. When the device has multiple interrupt outputs, it will assert them in the order of their relative priority. Common modes of a PIC include hard priorities, rotating priorities and cascading priorities. PICs often allow the cascading of their outputs to inputs between each other. Intel 8259 is a family of PICs designed and developed for use with the Intel 8085 and Intel 8086 microprocessors.

DMA Controller

In *Direct Memory Access* (DMA) data transfer scheme, data is transferred directly from an I/O device to memory or vice versa without going through the CPU. *DMA Controller* is used to control the process of data transfer. Its primary function is to generate, upon a peripheral request, sequential memory address, which will allow the peripheral to read or write data directly to or from memory. One of the popular known programmable DMA controllers is Intel's 8257.

Programmable Communication Interface

Programmable Communication Interface (PCI) is an interface device that is used for data communication applications with microprocessors. They basically convert the data from the microprocessor into a format acceptable for communication and also convert the incoming data into a format understood by the microprocessor. 8251 is a PCI device designed for Intel's 8085, 8086 and 8088 microprocessors and is used in serial communication applications.

Math Co-processor

Math co-processors are special purpose processing units that assist the microprocessor in performing certain mathematical operations. The arithmetic operations performed by the co-processor are floating-point operations, trigonometric, logarithmic and exponential functions and so on. The examples include Intel's 8087, 80287, etc. The 8087 Numeric coprocessor provides the instructions and data types needed for high performance numeric application, providing upto 100 times the performance of a CPU alone. Another widely used math co-processor is 80287. The 80287 Numeric processor extension (NPX) provides arithmetic instructions for a variety of numeric data types in 80286 systems. It also executes numerous built-in transcendental functions (e.g., tangent and log functions).

Programmable Keyboard/Display Interface

Programmable keyboard/display Interfaces are the devices used for interfacing the keyboard and the display to the microprocessor. The keyboard section of the device debounces the keyboard entries and provides the data to the microprocessor in the desired format. The display section converts the data output of the microprocessor into the form desired by the display device in use. 8279 is general purpose programmable keyboard and display I/O interface device designed for use with Intel microprocessors. The keyboard portion can provide a scanned interface to a 64-contact key matrix. The keyboard entries are debounced and strobed in an eight-character FIFO. If more than eight characters are entered, overrun status is set. Key entries set the interrupt output line to the CPU. The display portion provides a scanned display interface for LED, incandescent, and other popular display technologies. Both numeric and alpha-numeric segment displays may be used. The 8279 has 16 × 8 display RAM.

Programmable CRT Controller

Programmable CRT Controller is a device to interface CRT raster scan displays with the microprocessor system. Its primary function is to refresh display by buffering the information from main memory and keeping track of the display position of the screen. One of the commonly used programmable CRT controllers is Intel's 8275H. It allows simple interface to almost any raster scan CRT display with a minimum of external hardware and software overhead. The number of display characters per row and the number of character rows per frame are software programmable.

Floppy Disk Controller

A *Floppy Disk Controller* is used for disk drive selection, head loading, issue of read/write commands, data separation, and serial-to-parallel and parallel-to-serial conversion of data. Some examples of floppy disk controllers include Intel 82078, Intel 82077 and Intel 8272.

Clock Generator

A clock generator is a circuit that produces timing signal for synchronisation of circuit's operation. Examples of clock generators used in microprocessor systems include 8284 and 82284. 8284 generates the system clock for the 8086 and 8088 processors. It requires a crystal or a TTL signal source for producing clock waveforms. It provides local READY and MULTIBUS READY synchronisation.

Octal Bus Transceiver

Bus Transceivers are devices with high output drive capability for interconnection with data buses. In a microprocessor based system, they provide interface between the microprocessor bus and the system data bus. 8286 is eight-bit bipolar transceiver with three-state output used in a wide variety buffering applications in microcomputer systems. It is packaged in 20-pin DIP package.

Q. 27: What are the different types of memory devices in a microcomputer system? Distinguish between primary memory and secondary storage devices. What are volatile and non-volatile memory devices?

Ans: A computer memory refers to components, devices, chips and recording media that are used for temporary, semi-permanent and permanent storage of data. There are several types of memory devices used in a computer. These include RAM, ROM, cache, flash memory, hard disk, floppy disk, CDs, etc. The memory devices can be broadly classified into two types, namely, the primary memory and the secondary storage. Figure-26.12 shows the various types of memory devices present in a typical computer system. It may be mentioned here that in computer terminology memory usually refers to RAM and ROM and the term 'storage' refers to hard disk, floppy disk and CDs.

Primary memory also called the main memory is directly accessible by the CPU to store and retrieve information. It is also called a volatile memory, as the data is not stored permanently and gets lost, when the power is switched off. Different types of RAMs and ROMs are examples of primary memory. Secondary memory or storage devices are also known as external memory or non-volatile memory. These are called external memory devices, as the CPU does not access these memories directly and instead they are accessed via input-output routines. Contents of secondary memories are first transferred to main memory, and then CPU can access it. These are called non-volatile memory devices, as these are used for storing data/information permanently and the stored data doesn't get lost, when the power is switched off. Common examples of secondary storage include disks, magnetic tapes, CD-ROM, DVD etc.

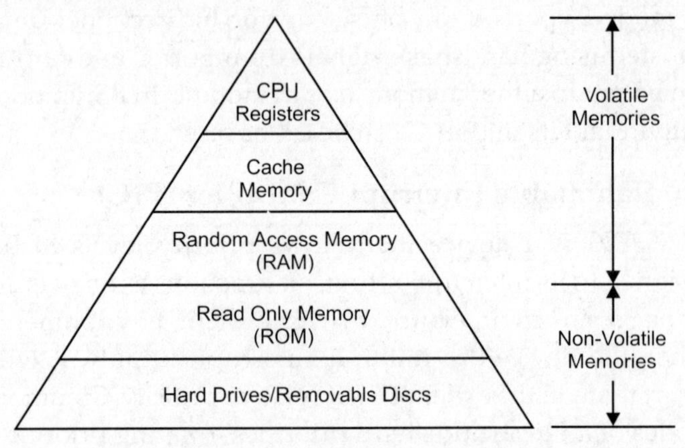

Fig. 26.12

Q. 28: RAM, ROM and Cache memory are different constituents of primary of a computer system. Briefly distinguish between these three types of primary memory devices.

Ans: The primary memory holds the program instructions for the program to be executed, the input data to be processed and the intermediate results of any calculations, when processing is being done. Primary memory is also used for storing BIOS and start-up programs. When a program and data are entered into a computer, the control unit directs them to the primary memory. Each program instruction and each data item is stored in a memory location that has a unique address. These data and instructions are held till new data items and instructions are written over them. So, same data can be accessed repeatedly, if so desired and same instructions can be executed repeatedly, if so required. This is what is called stored program concept. The primary memory of a computer further comprises process registers, random access memory (RAM), and cache memory and read only memory (ROM). Process registers are memory cells built into the CPU that contain the specific data needed by the CPU. Cache memory is basically a type of RAM memory.

RAM is a 'read/write' memory, where the data can be read from or written into any of the memory locations regardless of the order in which they are arranged. Therefore, all the memory locations in a RAM can be accessed at the same speed. RAM is used to store data, program instructions and the results of any intermediate calculations during the execution of a program. Also, same data can be read any number of times and different data written into same memory location with every fresh data item over-writing the existing one. It is typically used for short-term data storage, as it cannot retain data when the power is turned off. RAM is available in the form of ICs as well as in form of plug-in modules. The plug-in modules are small circuit boards containing memory ICs and having input and output lines connected to an edge connector. They are available as single in-line memory modules (SIMMs) and dual in-line memory modules (DIMMs). More than one memory ICs (or chips) can be used to build the RAM for larger systems. The capacity or size of a RAM is measured in bytes. RAM chips are available in the memory capacities ranging from 2 KB to as much as 32 MB. One KB of memory equals $2^{10} = 10^{24}$ bytes and one MB memory equals 2^{20} bytes. The words kilo (K) and mega (M) have been used as 2^{10} and 2^{20} are approximately equal to 1000 and 1000000, respectively. As an illustration, a microcomputer with a 64 KB of RAM has $64 \times 2^{10} = 2^6 \times 2^{10} = 2^{16} = 65536$ bytes of memory. The two categories of RAM are static RAM (SRAM) and the dynamic RAM (DRAM).

In the case of ROM, instructions can be written into the memory only once at the manufacturer's premises. These instructions can, however, be read from a ROM as many times asdesired. Once it is written, a ROM cannot be written into again. The contents of a ROM can, thus, be accessed by a CPU but cannot be changed by it. The instructions stored on a ROM vary with the type of application it is made for. The ROM for a general purpose microcomputer, for instance, would contain system programs such as those designed to handle operating system instructions. In the case of some special types of ROMs, it is possible for the user to have his own instructions stored on the ROM as per his requirements. Such ROM chips are called programmable read only memories (PROMs). PROM contents once programmed cannot be changed. However, there are some special types of PROMs, whose contents can be erased and then reprogrammed. These are known as erasable programmable read only memories (EPROMs).

Q. 29. What are the basic building blocks of an SRAM (Static Random Access Memory)? How do you classify RAMs on the basis of nature of memory cell?

Ans. RAM has three basic building blocks, namely an array of memory cells arranged in rows and columns with each memory cell capable of storing either a '0' or a '1', an address decoder and a read/write control logic. Depending upon the nature of memory cell used, there are two types of RAM, namely static RAM (SRAM) and dynamic RAM (DRAM). In SRAM, the memory cell is essentially a latch and can store data indefinitely as long as the DC power is supplied. DRAM on the other hand, has a memory cell that stores data in the form of charge on a capacitor. Therefore, DRAM cannot retain data for long and hence, needs to be refreshed periodically. SRAM has higher speed of operation than DRAM but has lesser storage capacity.

Q. 30. Briefly describe the memory cell of a static RAM.

Ans. As mentioned in the previous question, the basic element of SRAM is a latch memory cell. Figure-26.13 shows a basic SRAM memory cell. The memory cell is selected by setting the 'select' line active. The data bit is written in to the cell by placing it on the 'data in' line and is read from the 'data out' line.

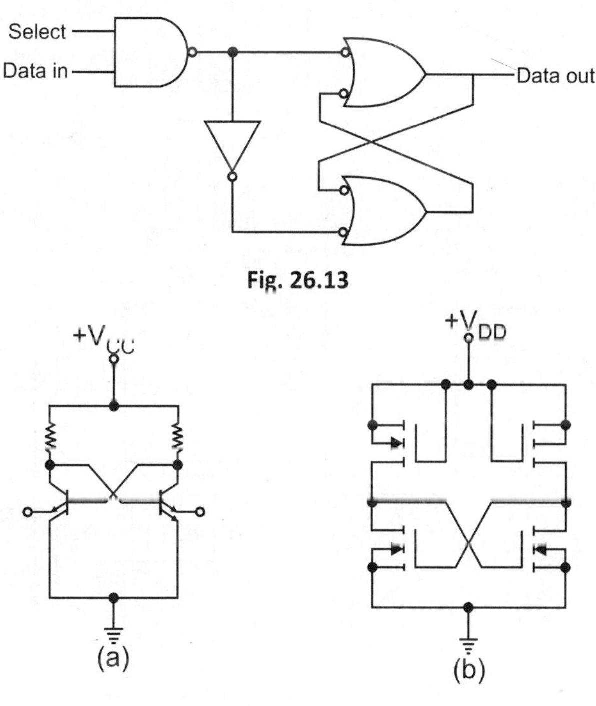

Fig. 26.13

Fig. 26.14

SRAMs can be broadly classified as asynchronous SRAM and synchronous RAM. Asynchronous SRAMs are those, whose operations are not synchronised with the system clock, that is, they operate independent of the clock frequency. In these RAMs, 'data in' and 'data out' are controlled by address transition. Synchronous RAMs are those whose timings are initiated by clock edges. Also, 'address', 'data in', 'data out' and all other control signals are synchronised with the clock signal. Synchronous RAMs normally have an address burst feature, which allows the memory to read and write at more than one location using a single address. Both synchronous and asynchronous SRAMs are available in bipolar, MOS and BiCMOS technologies. While bipolar SRAM offers relatively higher speed of operation, MOS technology offers higher capacity and reduced power consumption. Figure-26.14 (a) and (b) respectively show the basic bipolar memory cell and the MOS (NMOS more specifically) memory cell.

Q. 31. With the help of a suitable block diagram and the relevant timing diagram, describe the operation of an asynchronous SRAM.

Ans. Figure-26.15 shows the typical architecture of 64 × 8 asynchronous SRAM. It is capable of storing 64 words of eight bits each. The main blocks include six-line (as 64 = 2^6) to 64-line address decoder, I/O buffers, 64 memory cells and control logic for read/write operations. The memory cells in a row are represented as a register.

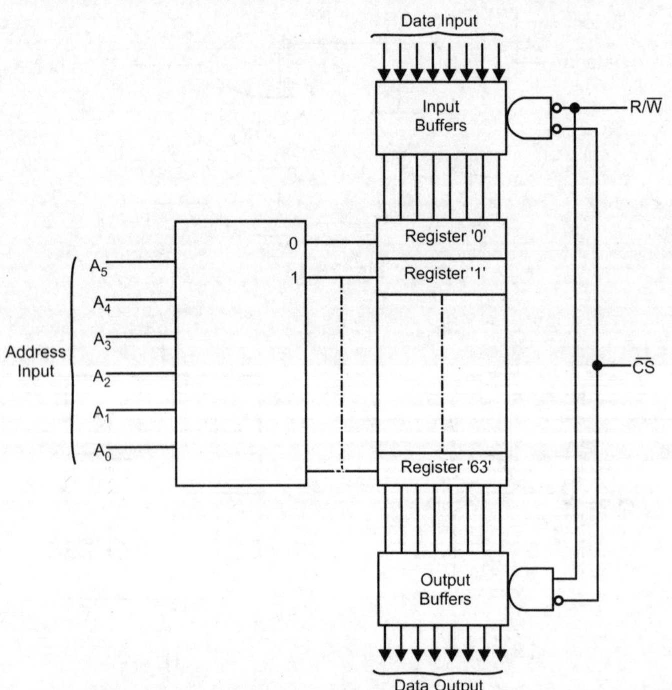

Fig. 26.15

Each register is an eight-bit register and can be read from as well as written into. As can be seen from the figure, all the cells inside the same register share the same decoder output line, also referred to as 'row line'. The control functions are provided by $R/\overline{W}$ (read/write) and the $\overline{CS}$ (chip select) inputs $R/\overline{W}$ and $\overline{CS}$ inputs are also referred to as (write enable) and (chip enable) inputs, respectively. The 'data input' and 'data output' lines are usually combined by using common 'input/output' lines in order to conserve the number of pins on the IC package.

The memory is selected by making $\overline{CS} = 0$. During the 'Read' operation, the status of $R/\overline{W}$ and $\overline{CS}$ pins are '1' and '0', respectively, while during the 'write' operation, it is '0' and '0', respectively. During the 'read' operation, the input buffers are disabled and the contents of the selected register appear at the output. During the 'write' operation, the input buffers are enabled and the output buffers are disabled. The contents of the input buffers are loaded into the selected register whose previous data is overwritten by the new data. The output buffers being tri-state are in high impedance state during the 'write' operation. $\overline{CS}$ =1 deselects the chip and both the input and output data buffers get disabled and go to the high impedance state. The contents of memory in this case remain unaffected. The 'chip select' inputs are particularly important, when more than one RAM memory chip is combined to get larger memory capacity.

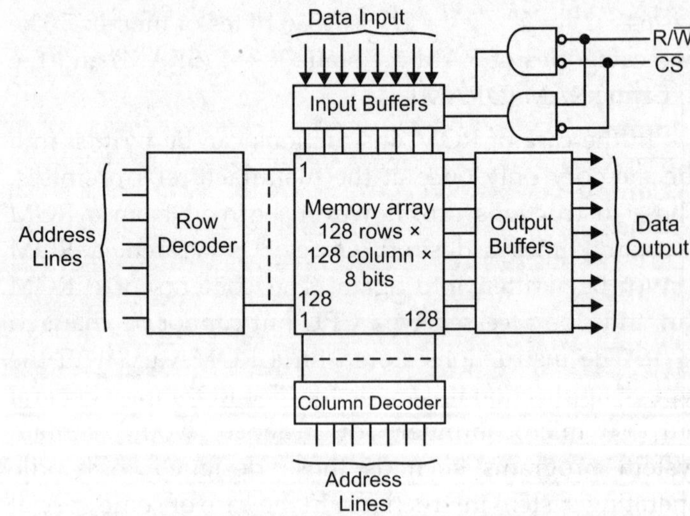

Fig. 26. 16

In the case of larger SRAM memories, there are two address decoders, namely, one for rows and one for columns. They are referred to as row decoders and column decoders, respectively. A part of the address lines are fed to the row decoder and the rest of the address lines are fed to the column decoder. Figure-26.16 shows the architecture of a typical 16K × 8 asynchronous SRAM. The memory cells are arranged in eight arrays of 128 (=2^7)

rows and 128 (=2^7) columns each. Memories with a single address decoder are referred to as two-dimensional memories and those with two decoders are referred to as three-dimensional memories.

Figures-26.17(*a*) and (*b*) show the timing diagrams during 'read' and 'write' operations respectively. The diagrams are self-explanatory. Read and write cycle time intervals of few nano-seconds to a few tens of nanoseconds are common in the case of asynchronous SRAMs. The different timing intervals shown in the diagram are defined as follows:

Complete read cycle time (t_{RC})	It is defined as the time interval for which a valid address code is applied to the address lines during the 'read' operation.
RAM access time (t_{ACC})	It is defined as time lapse between application of new address input and appearance of valid output data.
Chip enable access time (t_{CO})	It is defined as the time taken by RAM output to go from Hi-Z state to a valid data level, once $\overline{CS}$ is activated.

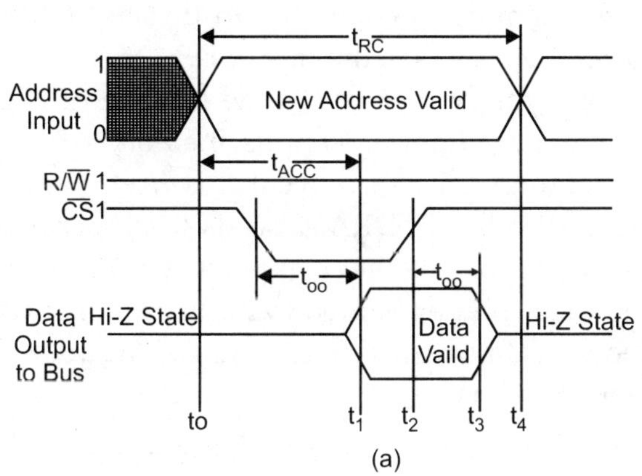

(a)

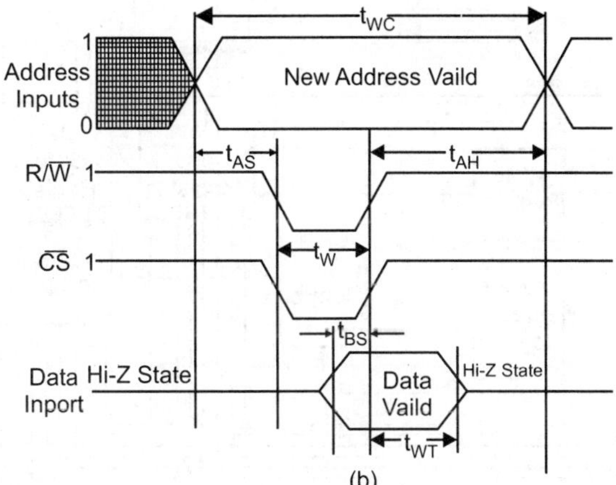

(b)

Fig. 26.17

Chip disable access time (t_{OD})	It is defined as the time taken by RAM to return to Hi-Z state after CS is inactivated.
Complete write cycle time (t_{WC})	It is defined as the time interval for which a valid address code is applied to the address lines during 'write' operation.
Write Pulse Width (t_W)	It is the time for which $R/\overline{W}$ is held 'LOW' during 'write' operation.
Address set-up time (t_{AS})	It is the time interval between appearance of a new address and $R/\overline{W}$ going 'LOW'.
Data set-up time (t_{DS})	It is defined as the time interval for which $R/\overline{W}$ must remain 'LOW' after valid data is applied to the data inputs.
Data hold time (t_{DH})	It is defined as the time interval for which valid input data must remain on the data lines after the $R/\overline{W}$ input goes 'HIGH'.
Address hold time interval (t_{AH})	It is defined as the time interval for which the valid address must remain on the address lines after the $R/\overline{W}$ input goes 'HIGH'.
Address hold time interval (t_{AS})	It is defined as the required time interval after which $R/\overline{W}$ can go 'LOW' after a valid address appears on the address lines.

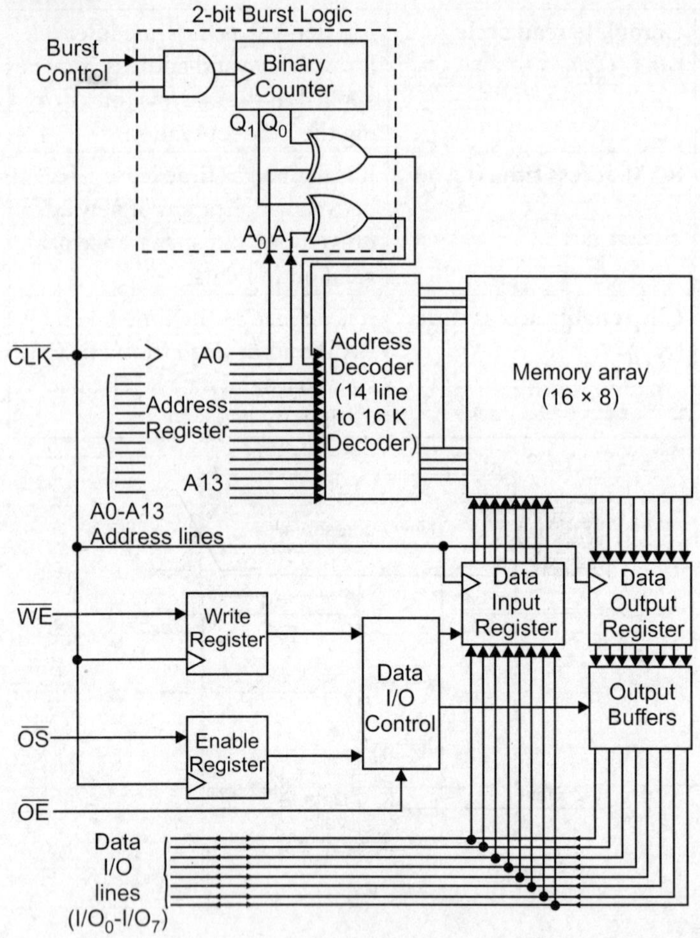

Fig. 26.18

Q. 32. With the help of a suitable block schematic, briefly describe the architecture of a synchronous RAM.

Ans. Synchronous RAM, as mentioned before, is synchronised with the system clock. In the case of a computer system, it operates at the same clock frequency at which the microprocessor operates. This synchronisation of the microprocessor and memory ensures faster execution speeds. The basic difference between the architecture of synchronous and asynchronous RAMs is that synchronous RAM makes use of clocked registers to synchronise 'address', $R/\overline{W}$, $\overline{CS}$ and 'data in' lines to the system clock. Figure-26.18 shows the basic architecture of a 32K × 8 synchronous SRAM with burst feature. As we can see from the figure the memory array block, the address decoder block and $R/\overline{W}$ and $\overline{CS}$ are the same, as in the case of asynchronous SRAM. As mentioned before, most synchronous RAMs have address burst feature. In this case, when an external address is latched to the address register, a certain number of lowest address bits are applied to the burst logic. Burst logic comprises of a binary counter and EX-OR gates. The output of the burst logic which basically produces a sequence of internal addresses is fed to the address bus decoder. In the case of an eight-bit burst logic, the internal address sequence generated is given by A_1A_0, $A_1\overline{A}_0$, $\overline{A}_1A_0$, $\overline{A}_1\overline{A}_0$, where A_0 and A_1 are the address bits applied to the burst logic. The burst logic shown in Fig.26.18 is also a two-bit logic.

Q. 33. What is DRAM? Briefly describe the memory cell of a DRAM.

Ans. The memory cell in the case of a DRAM comprises of a capacitor and a MOSFET. The cell holds a value of '1', when the capacitor is charged and '0', when discharged. The main advantage of this type of memory is its higher density or more bits per package as compared to SRAM. This is because the memory cell is very simple as compared to that of SRAM. Also, the cost per bit is less in the case of DRAM. The disadvantage of this type of memory is the leakage of charge stored on the capacitors of various memory cells, when they are storing a '1'. To prevent this from happening, each memory cell in a DRAM needs to be periodically read, its charge (or voltage) compared with a reference value and then charge restored to the capacitor. This process is known as 'memory refresh' and is done approximately in every 5–10 ms.

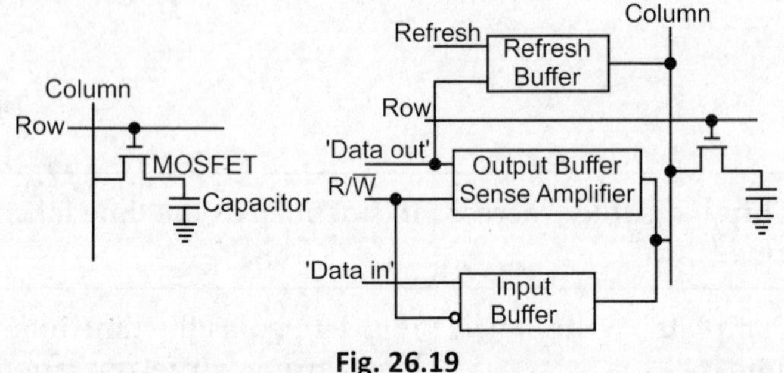

Fig. 26.19

Figure 26.19 shows the basic memory cell of DRAM and its basic principle of operation. The MOSFET acts like a switch. When in the 'write' mode ($R/\overline{W}$ = 0), the input buffers are enabled while the output buffers are disabled. When '1' is to be stored in the memory, 'Data in' line must be in 'HIGH' state and the corresponding 'row line' should also be in 'HIGH' state so that the MOSFET is switched ON. This connects the MOSFET to the 'data in' line and it charges to a positive voltage level. When '($R/\overline{W}$ = 1)' needs to be stored, 'data in' line is 'LOW' and the capacitor also acquires the same level. When the 'Row line' is taken to 'LOW' state, the MOSFET is switched OFF and disconnects the 'MOSFET' from the bit line. This traps the charge on the capacitor. When in 'read' mode ( ), the output buffers are enabled while the input buffers are disabled. When the 'row line' is taken to 'HIGH' logic, the MOSFET is switched ON and connects the capacitor to the 'data out' line through the output buffer. Refresh operation is performed by setting and by enabling the refresh buffer. There are two basic modes of refreshing the memory, namely, the burst refresh and distributed refresh mode. In burst refresh mode, all rows in the memory array are refreshed consecutively during the refresh burst cycle. In distributed refresh mode, each row is refreshed at intervals interspaced between 'read' and 'write' operations.

Q. 34. With the help of a suitable diagram and timing diagram, describe the architecture and operation of DRAM. How does architecture of DRAM differ from that of SRAM?

Ans. The architecture of DRAM memory is somewhat different from that of SRAM memory. Row and column address lines are usually multiplexed in a DRAM. This is done to reduce the number of pins on the package. Row address select (RAS) and column address select (CAS) inputs are used to indicate whether a row or a column is to be addressed. Address multiplexing is particularly attractive for higher capacity DRAMs. A 4M RAM, for instance, would require to have 22 address inputs (2^{22} = 4M).

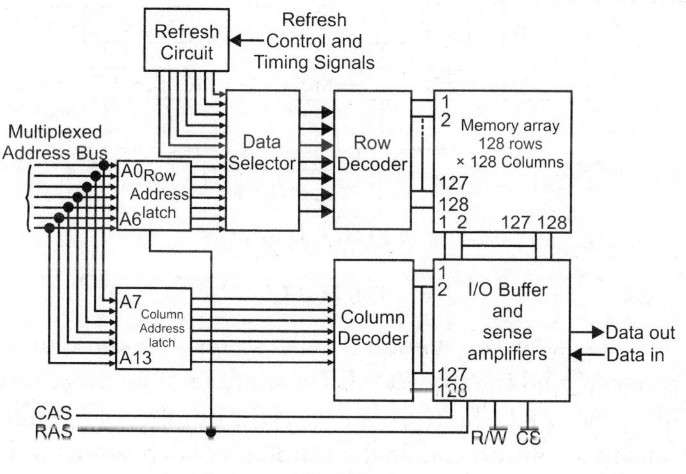

Fig. 26.20

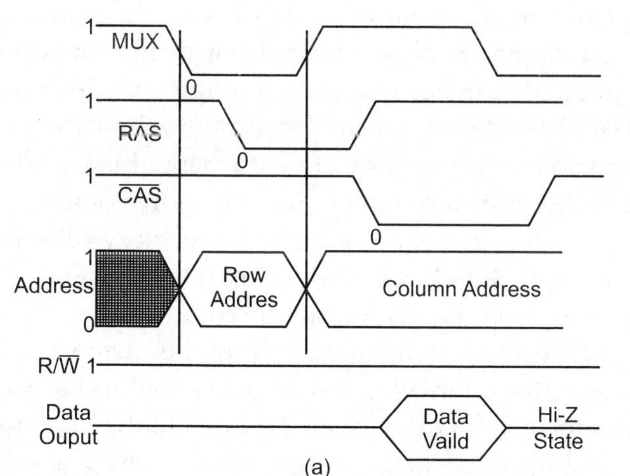

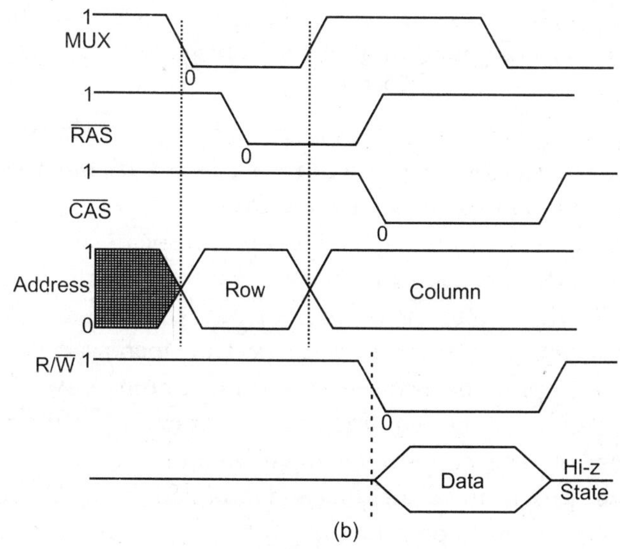

Fig. 26.21

Figure-26.20 shows the architecture of a 16K × 1 DRAM. The heart of a DRAM is an array of single bit memory cells. Each cell has a unique position vis-à-vis row and column. Other important blocks include address decoders (row decoder and column decoder), refresh control, address latches (row address latch and column address latch).As can be seen from the figure, seven address lines are time multiplexed at the beginning of the memory cycle by the RAS and CAS lines. First, the seven bit address (A_0–A_6) is latched into the row address latch and then the seven-bit address is latched to the column address latch (A_7–A_{13}). They are then decoded to select the particular memory location. Larger word sizes can be achieved by combining more than one chip. This is discussed in the latter part of the chapter.

Figures 26.21 (*a*) and (*b*), respectively, show the timing diagrams for read and write operations. The diagrams are self-explanatory. DRAMs are relatively slower than SRAMs. Typical access time is in the range of 100–250 ns.

Q. 35. Name different types of DRAM.

Ans. DRAM memories can be classified as follows.

1. Fast page mode (FPM) DRAM
2. Extended data output (EDO) DRAM
3. Burst extended data output (BEDO) DRAM
4. Synchronous (S) DRAM.

In FPM DRAM, the row address is specified only once for access to several successive column addresses. Hence, the read and write times are reduced. EDO DRAM is similar to FPM DRAM with the additional feature that a new access cycle can be started while keeping the data output of the previous cycle active. BEDO DRAM is an EDO DRAM with address burst capability. All the DRAMs discussed till now are asynchronous DRAMs and their operation is not synchronised with the system clock. SDRAM, as the name suggest, is a synchronous DRAM whose operation is synchronised with the system clock.

Q. 36. What is the role of cache memory in a computer? Briefly describe cache memory and distinguish between L1 and L2 levels of cache memory.

Ans. Advances in microprocessor technology and also, the software have enhanced manifold the application potential of present day computers. These enhanced performance features and increased speed can be optimally utilised to the maximum only, if the computer has the required capacity of main (or internal) memory. The computer's main memory, as we know, stores program instructions and data that the CPU needs during normal operation. In order to get the maximum out of the system, this would normally require that all of system's main memory has speed comparable to that of CPU. It may not be economical to have all of main memory as a high speed one in many systems. It is where cache memory comes in.

Cache memory is a block of high speed memory located between main memory and the CPU. The cache memory block is the one that communicates directly with CPU at high speed. It stores the most recently used instructions or data. When the processor needs data, it checks in high speed cache to see, if the data is there. If it is there, called a 'Cache hit', the CPU accesses the data from the cache. If not there, called a 'Cache Miss', then the CPU retrieves it from relatively slower main memory. Cache memory mostly uses SRAM chips but it can also use DRAM.

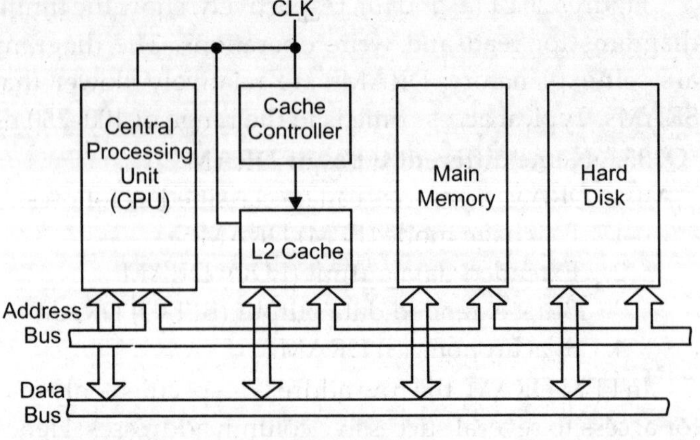

Fig. 26.22

There are two levels of cache memory. First is Level-1 cache (L1 or primary or internal cache). It is physically a part of microprocessor chip. Second is Level-2 cache (L2 or secondary or external cache). It is in the form of memory chips mounted external to the microprocessor. It is larger than L1 cache. L1 and L2 cache memories are of the sizes in the range of 2KB to 64KB and 256KB to 2MB, respectively. Some systems have higher level caches L3, L4 etc. but L1 and L2 are most common. Figure-26.22 shows the use of L1 and L2 cache memories in a computer system.

Q. 37. With the help of representative schematic and timing diagram, briefly describe ROM architecture and memory READ operation.

Ans. The internal structure or architecture of a ROM comprises of three basic parts, namely, the array of memory cells, address decoder the output buffers. The address decoder comprises of a single decoder in the case of small memories. In the case of large memories, it comprises of two decoders referred to as row and column decoders. The operation of a ROM can be best explained with the help of the simplified representation of a 32 × 8 ROM, as shown in Figure-26.23.

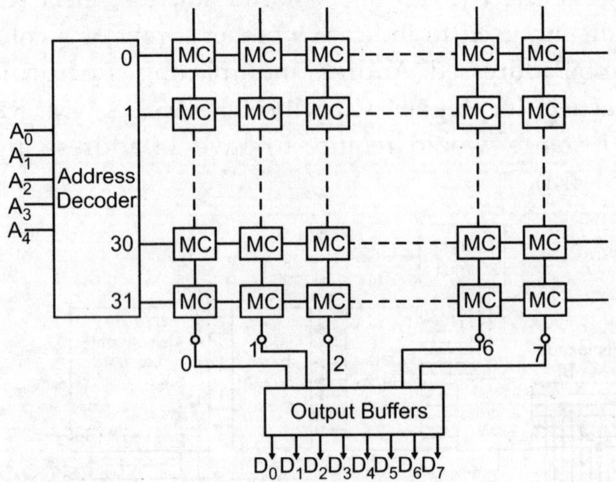

Fig. 26.23

The array of memory cells stores the data to be programmed into the ROM. The number of memory cells in a row equals the word size and the number of memory cells in a column equals the number of such words to be stored. In the memory shown in Fig.28.23, the word size is eight bits and the number of words is 32. The data outputs of each of the memory cells in the array are connected to an internal data bus that runs through the entire circuit. The address decoder, 1-of-32 decoder in this case sets the corresponding 'row' line HIGH, when a binary address is applied at its input lines. Five-bit address code (A4 A3 A2 A1 A0) is needed to address 32 memory cells. As an illustration, an address code of 10011 will identify 19th row. The output is read from the column lines. The data placed on the internal data bus by the memory cells is fed to output buffers. $\overline{CS}$ is an active low input used to select the memory device. In the case of larger memories, the address decoder comprises of row as well as column decoders. Let us consider a 2K bit ROM device with 256 × 8 organisation. The memory is arranged in the format of 32 × 64 matrix instead of 256 × 8 matrix. Five of the address lines are connected to the row decoder and rest three of the lines are connected to the column decoder. Row decoder is a 1-of-32 decoder and it selects one of the 32 rows. The column decoder comprises of eight 1-of-8 decoders. It selects eight of the total 64 columns. Thus, eight-bit word appears on the data output, when the address is applied and the $\overline{CS} = 0$.

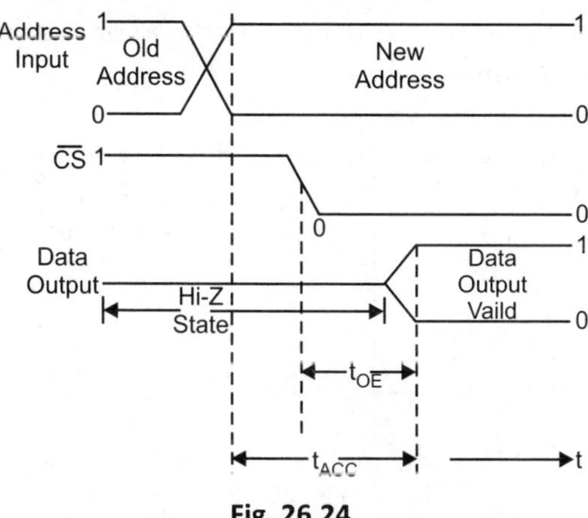

Fig. 26.24

Figure-26.24 shows the typical diagram of a ROM read operation. It shows that there is a time delay that occurs between the application of an address input and the availability of corresponding data at the output. It is this time delay that determines ROM's operating speed. This time delay is known as access time, t_{ACC}. Another useful timing parameter is the output enable time, t_{OE} which is the time delay between application of input and appearance of valid data output.

The typical bipolar ROMs have access times in the range of 30–90ns. In the case of NMOS devices, they have access times in the range of 35–500 ns. The output enable time (t_{OE})in the case of bipolar ROMs is in the range of 10 20ns. For MOS based ROMs, the same is in the range of 25–100 ns.

Q. 38: How do you classify ROMs on the basis of methodology of programming, erasing and reprogramming information into ROMs?

Ans: Depending upon the methodology of programming, erasing and reprogramming information into ROMs, they are classified as follows:

1. Mask-programmed ROMs
2. Programmable ROMs (PROM)
3. Erasable programmable ROMs (EPROM)
   (*a*) Ultraviolet erasable programmable ROMs (UV-EPROM)
   (*b*) Electrically erasable programmable ROMs (EEPROM)

Mask-Programmed ROM

In the case of a mask-programmed ROM, the ROM is programmed at the manufacturer's site according to the specifications of the customer. A photographic negative called a mask is used to store the required data on the ROM chip. A different mask would be needed for storing each different set of information. As preparation of a mask is an expensive proposition; mask-programmed ROM is economical only, when manufactured in large quantities. The limitation of such a ROM is that once programmed, it cannot be reprogrammed.

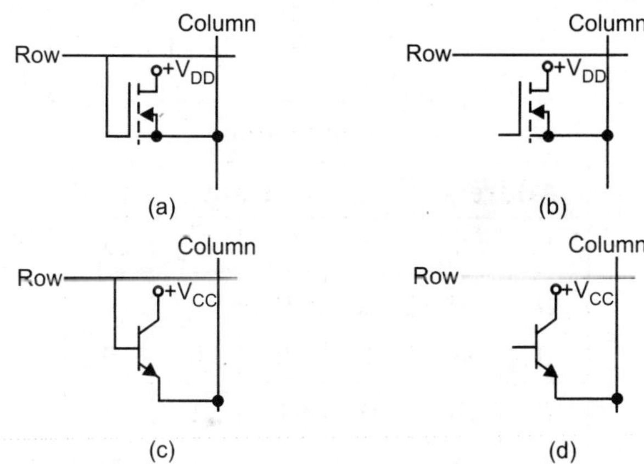

Fig. 26.25

The basic storage element is an NPN bipolar transistor connected in common-collector configuration or a MOSFET in common drain configuration. Figures-26.25(a) and (b) show a MOSFET based basic cell connection when storing a '1' and '0' respectively. As is clear from the figure, connection of the 'row line' to the gate of the MOSFET stores '1' at the location, when the 'row line' is set to level '1'. A floating gate connection is used to store '0'. Figures 26.25(c) and (d) show the basic bipolar memory cell connection, when storing a '1' and '0' respectively.

As an illustration, Figure-26.26 shows the internal structure of a 4 × 4 bipolar mask programmed ROM. The data programmed into the ROM is given in the adjoining truth table. The transistors with an open base store a '0', whereas those with their bases connected to the corresponding decoder output store a '1'. In the present case, transistors Q30, Q20, Q10 and Q00in Row-0 store 1, 0, 1 and 0 respectively. The stored information in a given row is available at the output when the corresponding decoder is enabled and that 'row line' is set to level '1'. The output of the memory cells appear at the column lines. For example, when the address input is '11', Row-3 is enabled and the data at the output is 0110.

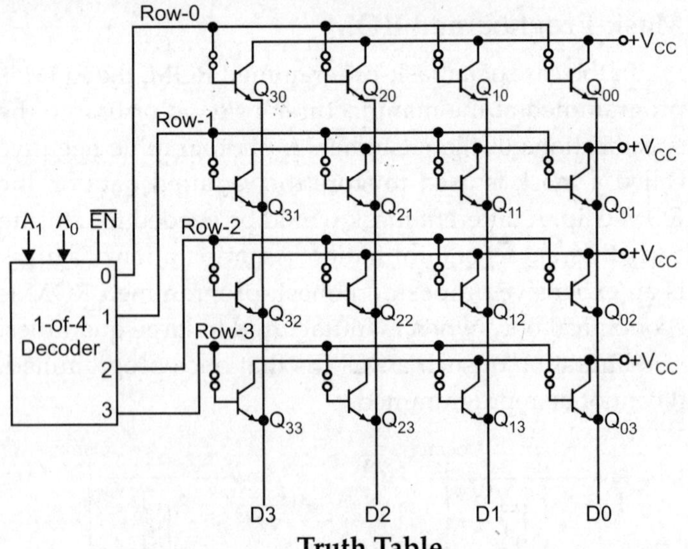

Truth Table

Address		Data			
A_1	A_0	D_3	D_2	D_1	D_0
0	0	1	0	1	0
0	1	1	0	0	0
0	0	0	1	1	1
1	1	0	1	1	0

Fig. 26.26

In the ROM architecture shown in Figure-26.26, number of memory cells in a row represents the word size. The four memory cells in a row here constitute a four-bit register. There are four such registers in this ROM. In a 16 × 8 ROM of this type, there will be 16 rows of such transistor cells with each row having eight memory cells. The decoder in that case would be a 1-of-16 decoder.

Programmable ROM (PROM)

In the case of PROMs, the programming instead of being done at the manufacturer's premises during the manufacturing process is done by the customer with the help of a special gadget called PROM programmer. Since the data once programmed, cannot be erased and reprogrammed, these devices are also referred to as one time programmable ROMs.

The basic memory cell of PROM is similar to that of a mask programmed ROM. Figures 26.27(a) and (b) show MOSFET based memory cell and bipolar memory cell respectively. In the case of a PROM, each of the connections that were left either intact or open in the case of mask programmed ROM are made with a thin fusible link, as shown in Figure-26.27. Basic fuse technologies used in PROMs are metal links, silicon links and PN junctions. These fusible links can be selectively blown off to store desired data. A sufficient current is injected through the fusible link to burn it open to store '0'. The programming operation, as said earlier, is done with PROM programmer. The PROM chip is plugged into the socket meant for the purpose. The programmer circuitry selects each address of the PROM one by one, burns in the required data and then verifies the correctness of the data before proceeding to the next address. The data is fed to the programmer from a keyboard or a disk drive or from a computer.

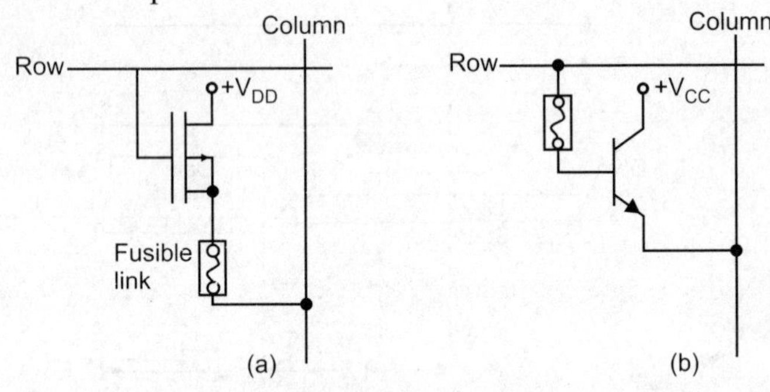

Fig. 26.27

PROM chips are available in various word sizes and capacities. 27LS19, 27S21, 28L22, 27S15, 24S41, 27S35, 24S81, 27S45, 27S43 and 27S49, respectively,are 32 × 8, 256 × 4, 256 × 8, 512 × 8, 1K × 4, 1K × 8, 2K × 4, 2K × 8, 4K × 8 and 8K × 8 PROMS. Typical access time in the case of these devices is in the range of 50 to 70 ns. MOS PROMs are available with much greater capacities than bipolar PROMs. Also, the power dissipation is much lower in MOS PROMs than it is in the case of bipolar PROMs with similar capacities.

Erasable PROM (EPROM)

EPROM can be erased and reprogrammed as many times as desired. Once programmed, it is non-volatile, that is it holds the stored data indefinitely. There are two types of EPROM, namely, the ultraviolet erasable PROM (UV-EPROM) and electrically erasable PROM (EEPROM).

The memory cell in a UV-EPROM is MOS transistor with a floating gate. In the normal condition, the MOS transistor is OFF. It can be turned ON by applying a programming pulse (in the range of 10–25V) that injects electrons in the floating gate region. These electrons remain trapped in the gate region even after removal of programming pulse. This keeps the transistor ON once it is programmed to be in that state even after removal of power. The stored information can, however, be erased by exposing the chip to ultra-violet radiation through a transparent window on the top of the chip meant for the purpose. The photo current, thus, produced removes the stored charge in the floating gate region and brings the transistor back to OFF state. The erasing operation takes about 15–20 min and the process erases information on all the cells of the chip. It is not possible to carry out any selective erasure of memory cells. Intel's 2732 is a (4K × 8) UV-EPROM hardware implemented with NMOS devices. Type numbers 2764, 27128, 27256 and 27512 have capacities of (8K × 8), (16K× 8), (32K × 8) and (64K × 8), respectively. The access time is in the range of 150–250 ns. UV-EPROMs suffer from disadvantages, such as need for removing the chip from the circuit if it is to be reprogrammed; non-feasibility of carrying out selective erasure and reprogramming process takes several tens of minutes. These are overcome in EEPROMs and flash memories discussed in the following sections.

The memory cell of an EEPROM is also a floating gate MOS structure with a slight modification that there is a thin oxide layer above the drain of the MOS memory cell. Application of a high voltage programming pulse between gate and drain induces the charge in the floating gate region which can be erased by reversing the polarity of the pulse. Since the charge transport mechanism requires very low current, erasing and programming operations can be carried out without removing the chip from the circuit. EEPROMs also have another advantage that it is possible to erase and rewrite data in the individual bytes in the memory array. The EEPROMs, however, have lower density (bit capacity per square mm of silicon) and higher cost as compared to UV-EPROMs.

Q. 39. What is a Flash Memory? Why is it called a flash memory? What does its basic memory cell comprise of?

Ans. Flash memories are high density non-volatile read/write memories with higher density. Flash memory combines the low cost and high density features of an UV-EPROM and in-circuit electrical erasability feature of EEPROM without compromising on the high speed access of both. Structurally, the memory cell of a flash memory is like that of an EPROM. The basic memory cell of a flash memory is shown in Figure-26.28. It is a stacked gate MOSFET with a control gate and floating gate in addition to drain and source. The floating gate stores charge, when sufficient voltage is applied to the control gate. '0' is stored, when there is more charge and '1' when there is less charge. The amount of charge stored on the floating gate determines if the MOSFET is turned ON or not.

It is called a flash memory because of its rapid erase and write times. Most of the flash memory devices use a 'bulk erase' operation in which all the memory cells on the chip are erased simultaneously.

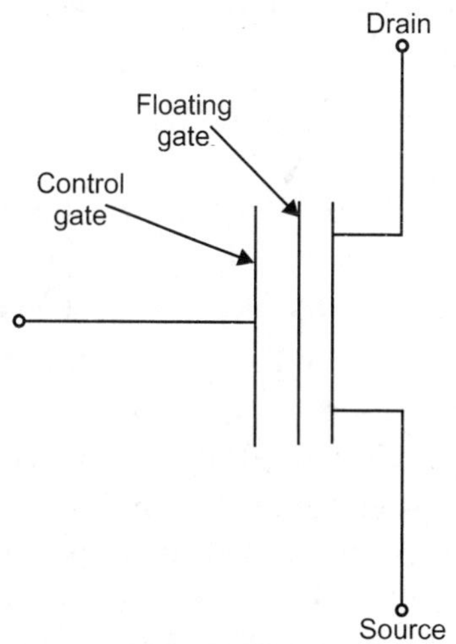

Fig. 26.28

Some flash memory devices offer a 'Sector Erase' mode in which specific sectors of the memory device can be erased at a time. This mode comes handy, when only a portion of the memory needs to be updated.

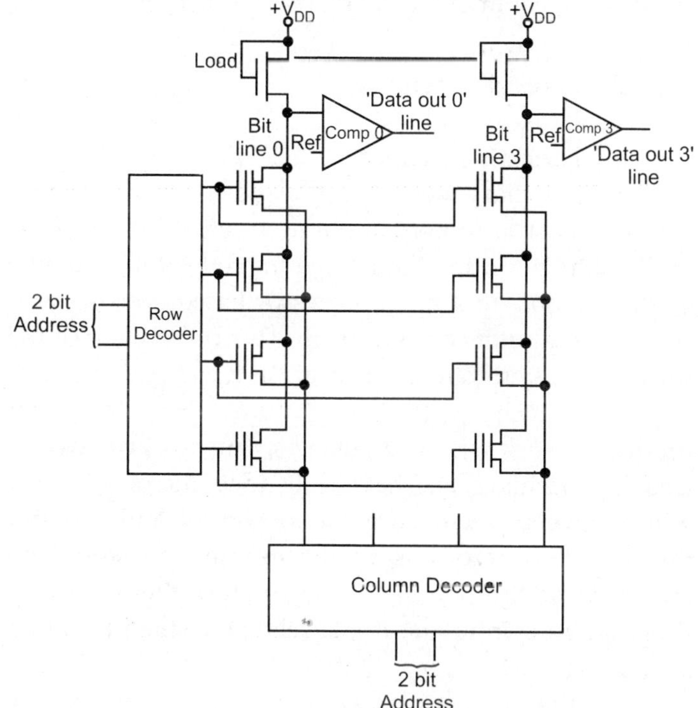

Fig. 26.29

Figure 26.29 shows the basic array of 4 × 4 flash memory. Similar to the case of earlier memories, there is an address decoder which selects the row. During the read operation, for a cell containing '1', there is current through the bit line which produces a voltage drop across the active load. This is compared with the reference voltage and the output bit is '1'. In case the memory cell has '0', there is very little current in the bit line. USB memory stick is an example of flash memory. Flash memory is at the core of solid state drives, which are fast replacing hard disk drives. Memory sticks are available in a wide range of capacities. The largest capacity memory stick available at the time of writing is 64 GB. The capacity will surely go up in future. Memory Stick PRO is reported to have a maximum theoretical capacity of 2 TB.

To sum up, while PROMs are least complex and low cost, they cannot be erased and reprogrammed. UV-EPROMs are little more complex and costly, but then they can be erased and reprogrammed by taking them out of the circuit. Flash memories are in-circuit electrically erasable either in sector wise or bulk mode. The most complex and most expensive are the EEPROMs, but then they offer byte-by-byte electrical erasability in-circuit.

Q. 40. What are common areas of applications of ROMs? Briefly describe each one of them.

Ans. Majority of ROM applications originate from the need for non-volatile storage of data or program codes. Some of the common application areas include:

1. Firmware
2. Bootstrap memory
3. Look-up tables
4. Function generators
5. Auxiliary memory

The most common application of ROM chips is in the storage of data and program codes that must be made available to a microprocessor based system, such as microcomputers on power-up. This component of the software is referred to as firmware as it comes embedded in the hardware with the machine. Even consumer products, such as CD players, microwave ovens, washing machines, etc., have embedded microcontrollers which have a microprocessor to control and monitor the operation according to the information stored on the ROM. ROMs are also used to store the 'bootstrap program' in computers. It is a relatively small program containing instructions that will cause the CPU to initialise the system hardware after it is powered on. The bootstrap program then loads the operating system programs stored in the secondary memory into its main internal memory. The computer then begins to execute operating system program. This start-up operation is also called 'booting operation'.

ROMs are frequently used as 'look-up tables'. There are two sets of data, one constituting the address and the other corresponding to the data stored in various memory locations of the ROM. The corresponding to each address input, there is a unique data ouput. One typical application is that of code conversion. As an illustration, a ROM can be used to build a binary to BCD converter, where each memory location stores BCD equivalent of corresponding address code expressed in binary.

A ROM can be an important building block in a waveform generator. In a typical waveform generation set-up, ROM is used as a look-up table with each of its memory location storing a unique digital code corresponding to a different amplitude of the waveform to be generated. The address inputs of the ROM are fed from the output of a counter. The data outputs of ROM feed a D/A converter, whose output constitutes the desired analog waveform. This concept is also utilised in speech synthesisers, where the digital equivalent of speech waveform values is stored in the ROM.

Q. 41. There are always situations where the required memory capacity (RAM or ROM) is larger than what is available on a single chip. How can we use more than one memory chip to increase the memory capacity both in terms of number of memory location and/ or the word size? Illustrate your answer with the help of examples.

Ans. When a given application requires a RAM or ROM with a capacity that is larger than what is available on a single chip, more than one such chip can be used to achieve the objective. Now, the enhancement in capacity required could be either in terms of increasing the word size or increasing the number of memory locations. How this can be achieved is illustrated in the following sections with the help of examples.

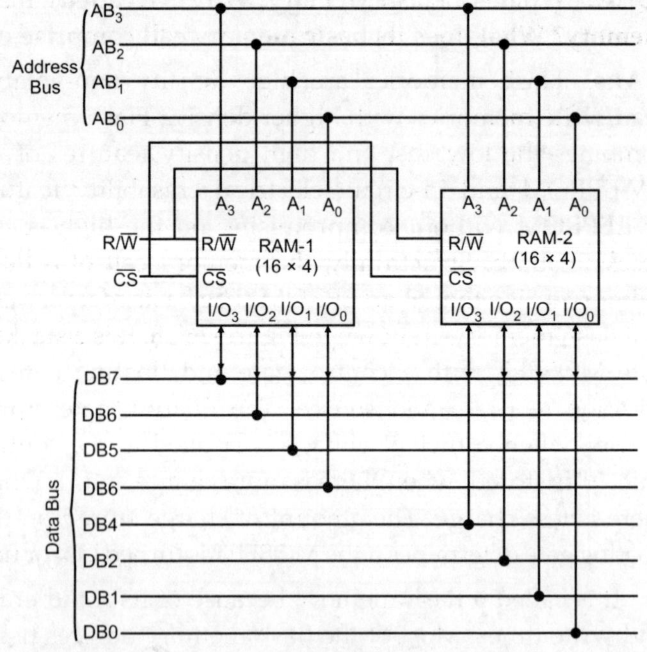

Fig. 26.30

Word Size Expansion

Let us take up the task of expanding the word size of an available 16 × 4 RAM chip from 4 bits to 8 bits. Figure-26.30 shows the diagram, where two such RAM chips have been used to achieve the desired word size. The arrangement is straightforward. Both chips are selected or deselected together. Also, input that determines whether it is a 'read' or 'write' operation is common to both chips. That is, both chips are selected for 'read' or 'write' operation together. The address inputs to the two chips are also common. The memory locations corresponding to various address inputs store higher order four bits in the case of RAM-1 and lower order four bits in the case of RAM-2. In essence, each of the RAM chips stores half of the word. Since the address inputs are common, same location in each chip is accessed at the same time.

Memory Locations Expansion

Figure-26.31 shows how more than one memory chips can be used to expand the number of memory locations. Let us consider use of two 16 × 8 chips to get a 32 × 8 chip. A 32 × 8 chip would need 5 address input lines. Four of the five address inputs other than the MSB address bit are common to both 16 × 8 chips. The MSB bit feeds input of one chip directly and input of the other chip after inversion. Input to the two chips is common.

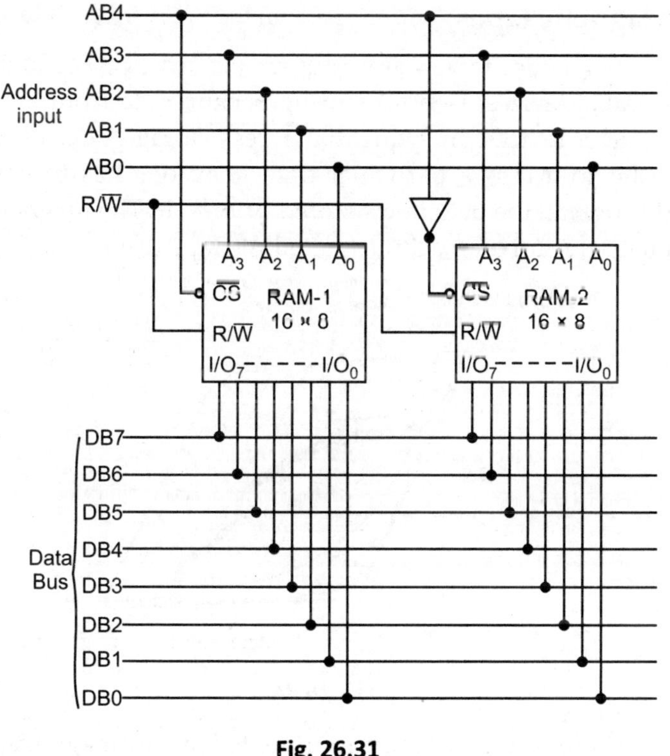

Fig. 26.31

Now, for first half of the memory locations corresponding to address inputs 00000 to 01111 (a total of 16 locations), the MSB bit of the address is '0' with the result that RAM-1 is selected and RAM-2 is deselected. For the remaining address inputs of 10000 to 11111 (again a total of 16 locations), RAM-1 is deselected, while RAM-2 is selected. Thus, the overall arrangement offers a total of 32 locations, 16 provided by RAM-1 and 16 provide by RAM-2. The overall capacity is thus (32 × 8).

Q. 42. What are secondary storage device generally used for? Name commonly used secondary storage devices.

Ans. Secondary storage devices are used for mass non-volatile storage of data and programs. It is often not practical to build a very large sized primary memory to meet all the storage requirements of the system, as it will increase the size and cost. That is where secondary storage is useful. Usually, it is located physically outside the machine. Though, it is not an essential component in theoretical terms, the secondary storage is almost indispensable, if one wants to exploit the full potential of a computer. Secondary storage devices are also referred to as auxiliary storage devices.

The primary storage due to its semiconductor nature can be accessed much faster than any of the storage media used for secondary storage. The secondary storage on the other hand is economical as far as cost per unit data stored is concerned and has an unlimited storage capacity. It is also safe from getting tampered with by any unauthorised persons. Commonly used secondary storage devices include magnetic, magneto-optical and optical storage devices. Another very common secondary storage device is the USB flash drive.

Q. 43. Name common magnetic storage devices and briefly describe each of them with particular reference to their structure and performance.

Ans. Common magnetic storage devices include Magnetic hard disks, Floppy disks and Magnetic tapes.

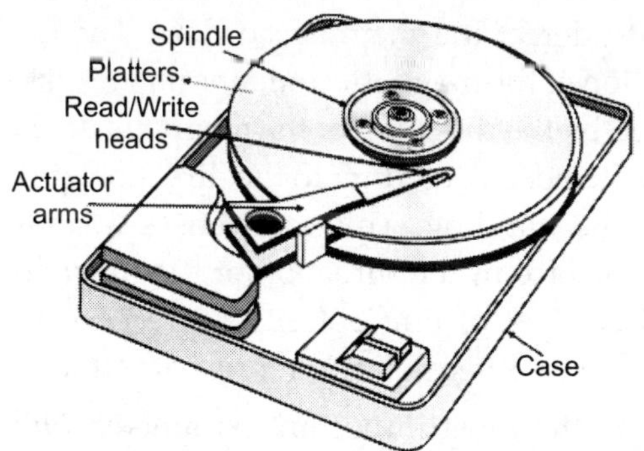

Fig. 26.32

Magnetic hard disks

Hard disks are non-volatile random access secondary data storage devices i.e. the desired data item can be accessed directly without actually going through or referring to other data items. They store the data on the magnetic surface of hard disk platters. Platters are made of aluminum alloy or a mixture of glass and ceramic covered with a magnetic coating. Figure-26.32 shows the internal structure of a typical hard disk. As can be seen from the figure, there are a few (two or more) platters stacked on top of each other on a common shaft. The shaft rotates these platters at speeds of several thousand rpm. Each platter is organised into tracks and sectors (Figure-26.33) both having a physical address used by the operating system to look for the stored data. Tracks are concentric circles used to store data. Each track is further subdivided into sectors so that total number of sectors per side of the magnetic disk is the product of the number of tracks per side and number of sectors per track. And if it is a double-sided disk, the total number of sectors gets further multiplied by two. From known values of total number of sectors and the number of bytes stored per sector, the storage capacity of the disk in bytes can then be computed.

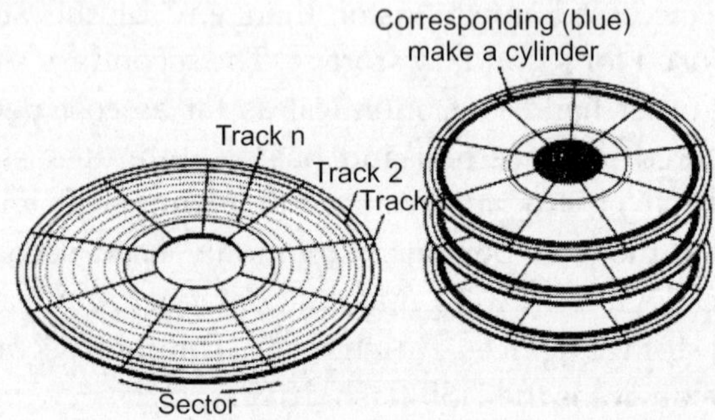

Fig. 26.33

There is a read/write head on one or both sides of the disk depending upon whether it is a single sided or a double sided disk. The head does not physically touch the disk surface; it floats over the surface and is close enough to detect the magnetised data. The direction or polarisation of the magnetic domains on the disk surface is controlled by the direction of the magnetic field produced by the write head according to the direction of the current pulse in the winding. This magnetises a small spot on the disk surface in the direction of the magnetic field. A magnetised spot of one polarity represents a binary '1' and that of the other polarity represents a binary '0'.

One of the most important parameters defining the performance of the hard disk is the size of the disk. They are available in various sizes ranging from 20GB to as large as 80GB. Other parameters defining hard disk's performance includes seek time and latency time. Seek time is defined as the average time required by the read/write head to move to the desired track. Latency time is defined as the time taken by the desired sector to spin under the head once the head is positioned over the desired track.

Floppy Disks

Floppy disks are removable disks made of flexible polyester material with magnetic coating on both sides. Important parts of a floppy disk are shown in Figure-26.34. Floppy disks are also organized in the form of tracks and sectors similar to a hard disk. A floppy disk drive unit is required to read data from or write data into a floppy disk. A read/write head that forms a part of the drive unit does this job. During read or write operation, the disk rotates to the appropriate position and the head makes a physical contact with the disk to do the desired operation. Earlier floppy disks were available in 5.25 inch size, having storage capability of 360KB. They were known as Double Sided Double Density (DSDD) floppy disks. They have been superseded by 3.5 inch floppy disks, having capability of 1.44 MB. Floppy disks have been replaced by CD disks and USB drives.

Magnetic tapes

Magnetic tapes are sequential access secondary storage devices used for storing backup data from mass storage devices. In sequential access storage devices, in order to access a particular data item, one has to pass through all the data items stored prior to it. The magnetic tapes are run on machines called the tape drive units.

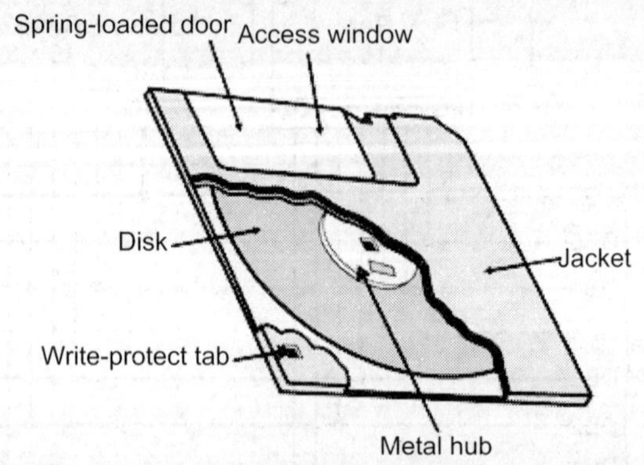

Fig. 26.34

The data on the tape is represented by tiny magnetised spots with the presence of a spot representing the binary digit '1' and its absence representing a binary digit '0'.

A simple and commonly used method of representing data on tapes is by using EBCDIC code. Magnetic tape is available in the form of reels, cassettes and cartridges. Reels are the most popular type.

Magneto-optical storage devices

Magneto-optical storage devices use a combination of magnetic and optical technologies for data storage. Magnetic coating, used in case of these devices, requires heat to alter the magnetic polarisation making them extremely stable at ambient temperatures. For data write operation, a laser beam having sufficient power is focused onto a tiny spot on the disk. This raises the temperature of the spot. Then the magnetic field generated by the write head changes the polarisation of the magnetic particles of that spot depending upon whether a '1' or '0' needs to be stored. For the read operation, a laser beam with less power is used. It makes use of the 'Kerr effect', where the polarity of the reflected beam is altered depending upon the polarisation of the magnetic particles of the spot.

Optical storage devices

One of the most significant developments in the field of storage media has been that of optical storage devices. Having arrived on the scene in the form of CD-Audio (Compact Disk-Audio) in early 1980's, since then optical disks have undergone tremendous technological development. These are available in various forms, namely CD-ROM (Compact Disk Read Only Memory), WORM disks (Write Once Read Many), CD-R (Compact Disk Read), CD-RW (Compact Disk Read/Write) and DVD-ROM (Digital Versatile Disk Read Only Memory).

An optical disk differs from a conventional hard disk (solid magnetic disk) in the method by which information is stored and retrieved. While hard disks use a magnetic head to read and write data; in the case of an optical disk, it is done with a laser beam. The high storage density of optical disks primarily results from the ability of the coherent laser beam to be focused onto a very tiny spot. The main advantages of optical disks include their vast storage capacity, immunity to illegal copying and their easy removability. Also, they do not transfer virus from one user to the next.

CD-ROM: CD-ROM is a disk comprising of three coatings namely polycarbonate plastic on the bottom, a thin aluminum sheet for reflectivity and a top coating of lacquer for protection. It can store upto 660 MB of data. It is formatted into a single spiral track having sequential sectors. CD-ROMs are prerecorded at the factory and store data in the form of pits and lands.

These are classified by the Access time and Data transfer rate. The performance of CD-ROM disks is enhanced by spinning them faster to achieve higher transfer rate and faster access time. These are rated as 2X, 4X, 6X, 16X, 24X and so on. A 16X CD-ROM drive will be 16 times faster than the original drives. The spinning rate of the drive is the number of revolutions per minute. Its seek time is the time the drive takes to locate a track, where desired data is stored. The time for which the drive has to wait for data to rotate under it is the latency. Sum of seek time and latency is the access time.

The read operation is performed by using a low power laser beam. The laser beam is focused on to pits and lands. Laser light reflected form a pit is 180 degrees out of phase with the light reflected from land. This light is detected by a photodiode followed by a processing circuitry. As the disk rotates a series of pits and lands are sensed and the data stored in them is read.

WORM: It is a type of optical storage device, where the data can be written once by the user which cannot be erased later but can be read many times. A low power laser beam is used to burn microscopic pits on the disk surface. Burned surfaces represent a binary '1' and unburned areas represent a binary '0'.

CD-R: It is a type of WORM which allows multiple write sessions to different areas of the disk. In this case, a laser is used to write data in the form of microscopic pits in organic dye layer. Presence and absence of a bump indicate a '1' and '0' respectively.

CD-RW: In this case the data can be recorded, erased, rewritten and read many times. The recording of data is by way of changing the state of material from a well structured crystalline state to a less ordered amorphous state.

DVD-ROM: Originally the term DVD was an abbreviation of Digital Video Disk but today it is used for referring to Digital Versatile Disks. It has much higher storage densities as compared to CD-ROMs. This is because the pit-size is smaller in case of DVD-ROMs as compared to CD-ROMs.

CD-ROMs are single-side storage devices whereas DVD-ROMs are available in single-side as well as double-side formats. As against 660 MB storage capacity of a CD-ROM, single sided DVD of same size offers 4.7GB in a single layer. A double layer or double sided DVD would offer 9.4 GB of storage capacity and a double sided, double layer DVD would have upto 17 GB which is about 30 times the storage capacity available on CD-ROM.

DVD-R and DVD-RAM are the counterparts of CD-R and CD-RW. Figure-26.35 shows the use of laser beam for read operation.

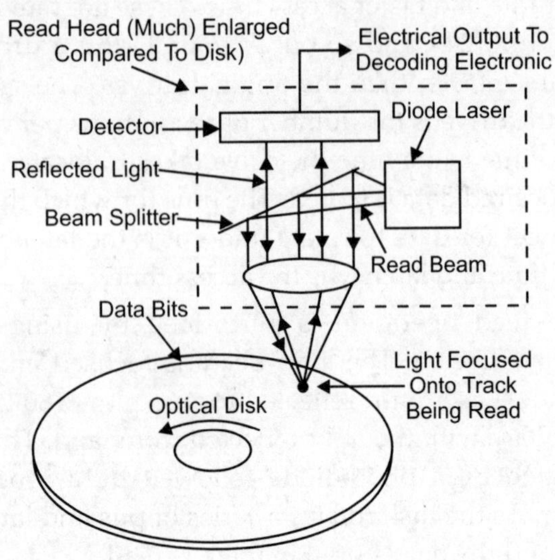

Fig. 26.35

USB flash drive

USB flash drives are plug and play flash memory data storage devices integrated with the USB interface. They are lightweight, rewritable, erasable storage devices. Flash memories have been discussed in the answer to an earlier question.

Q. 44. What is a serial port? What is it used for? Name common serial ports and briefly describe each one of them.

Ans. Serial port is a physical communication interface through which the information transfer takes place one bit at a time. Serial ports are used to connect mouse, keyboard and modems to the computer. Some of the commonly used serial standards include RS-232C port, PS/2, Fire Wire and USB.

RS-232C port: RS-232 is one of the oldest and the most well-known standard for serial interface approved by the Electronic Industries Association (EIA). It was developed to interface the Data Terminal Equipment (DTE) with the Data Communication Equipment (DCE). RS-232C, a variant of the RS-232 standard, is the most relevant for the computer world. RS-232C is mostly used to connect modem and other communication devices to the computer. In this case, the computer is referred to as the DTE and the attached device as DCE. RS-232C standard specifies 25 communication lines between the DTE and the DCE. Hence, the standard RS-232C connector is a 25-pin connector (DB-25). For personal computer applications, not all the 25 pins are required. Hence, most personal computers have a nine pin connector (DE-9). Figure-26.36 shows DE-9 connector along with its pin assignments.

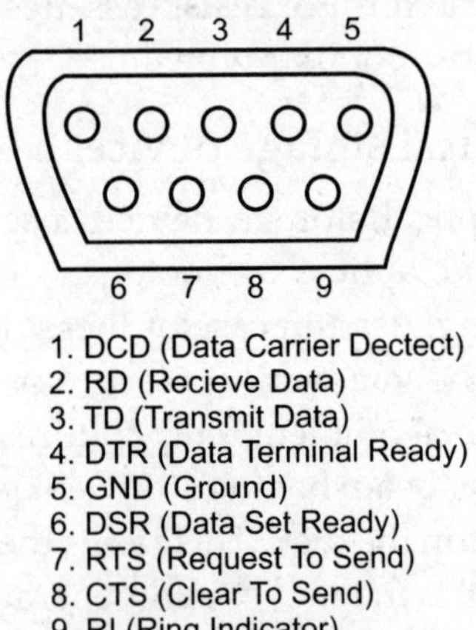

Fig. 26.36

The maximum specified cable length for RS-232C interface is 50 feet for a data transmission rate of 20 Kbaud. As the cable length increases the transmission rate decreases. RS-422 and RS-423 standards have higher transmission speeds as compared to RS-232C. They also support larger cable lengths. But, RS-232C still remains the most commonly used serial port.

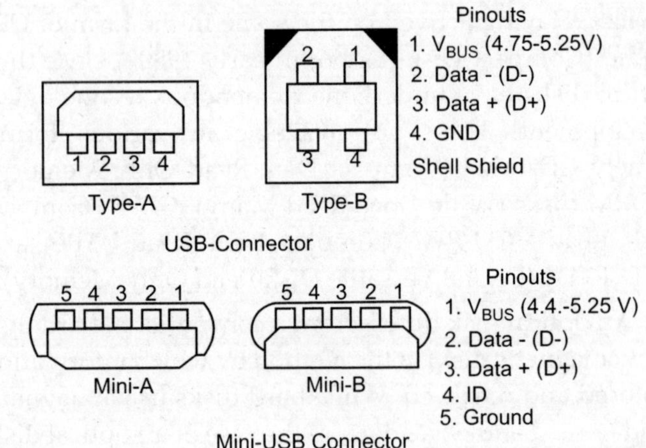

Fig. 26.37

Fire Wire: Fire Wire is the name of the interface specified by the IEEE standard 1394. This high speed serial bus standard is used for interfacing graphics and video peripherals, such as digital cameras and camcorders to the computer. Fire Wire can be used to connect upto 63 devices in a cyclic topology. It supports both plug-and-play and hot swapping. It is available in two versions, namely the Fire Wire 400 and Fire Wire 800. Fire Wire 400 hardware is available in 6-pin and 4-pin connectors and can support data rates of 100, 200 and 400 Mb/s. The 4-pin connector is used mostly in consumer electronic goods and the 6-pin connector is used in computers. Fire Wire

800 is based on IEEE 1394b standard and supports data rate of 786.432 Mb/s. It has a 9-wire connection.

Universal Serial Bus (USB): USB port was introduced in the year 1997 and is used to connect printers, mouse, scanners, digital cameras and external storage devices to the computer. Different versions of the USB standard include 0.9, 1.0, 1.1 and 2.0 with USB 2.0 being the latest. Another variant of the USB standard is the radio spectrum based USB implementation, known as Wireless USB. A USB port can be used to connect 127 devices. It supports two data rates of 1.5 Mb/s (Low speed) and 12 Mb/s (Full speed). Most of the USB 2.0 devices also support data rates of 480 Mb/s (Hi speed). USB is a 4 wire connection and is available in two standard types referred to as Type A and Type B. Miniature versions of the USB connector are also available namely Mini-A and Mini-B. Figure-26.37 shows different types of USB connectors along with their pin details.

PS/2 connector: PS/2 connector is used for connecting keyboards and mouse to a personal computer. The PS/2 mouse and PS/2 keyboard connectors are similar to each other except for that the PS/2 keyboard connector has an open-collector output. PS/2 mouse and keyboard connectors have replaced the DE-9 and 5-pin DIN connectors respectively. Figure-26.38 shows the PS/2 connector with the pin details.

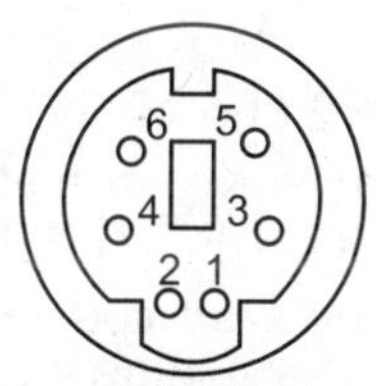

Pinouts

1. + DATA
2. Reserved
3. GND
4. V_{CC} (+5VDC)
5. + CLK
6. Reserved

Fig. 26.38

Q. 45. What is a parallel port? What is it used for? Name common parallel ports and briefly describe each one of them.

Ans. Parallel ports send multiple bits at the same time over a set of wires. They are used to connect printers, scanners, CD burners, and external hard drives etc. to the computer. Commonly used standard parallel ports include IEEE-488, Small Computer System Interface (SCSI) and IEEE 1284.

IEEE-488: IEEE-488 is a short-range parallel bus standard widely used in test and measurement applications. It is also referred to as General-Purpose Interface Bus (GPIB). IEEE 488 standard specifies a 24-wire connection for transferring eight data bits simultaneously. Other connections include eight control signals and eight ground lines. The maximum data rate is 1MB/s in the original standard and about 8 MB/s with the modified standard (HS-488). Figure-26.39 shows the pin connections and pin details.

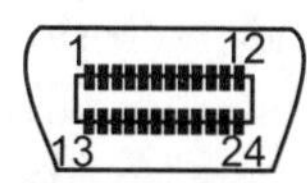

1EEE 488 Connector

Pinouts

1. Data 1/O1
2. Data 1/O2
3. Data 1/O3
4. Data 1/O4
5. EOI (End or Identity)
6. DAV (Data Valid)
7. NRFD (Not Ready for Data)
8. NDAC (Not Data Accepted)
9. IFC (Interface Clear)
10. SRQ (Service Request)
11. ATN (Attention)
12. Sheild Ground
13. Data 1/O5
14. Data 1/O6
15. Data 1/O7
16. Data 1/O8
17. REN (Remote Enable)
18. P/O Twisted Pair with 6
19. P/O Twisted Pair with 7
20. P/On Twisted Pair with 8
21. P/O Twisted Pair with 9
22. P/O Twisted Pair with 10
23. P/O Twisted Pair with 11
24. Signal Ground

Fig. 26.39

Small Computer System Interface (SCSI): SCSI is a widely used standard for interfacing personal computers and peripherals. SCSI is a standard given by the American National Standards Institute (ANSI). There are several variations of this standard and one variant may not be compatible with the other standard. Some of the SCSI versions include SCSI-1, SCSI-2, Wide SCSI, Fast SCSI, Fast Wide SCSI, Ultra SCSI, SCSI-3, Ultra SCSI-2 and Wide Ultra SCSI-2. Description of all these interfaces is beyond the scope of the book.

IEEE-1284: IEEE 1284 is a standard that defines bi-directional parallel communications between computers and other devices. It supports a maximum data rate of 4 Mb/s. It supports three types of connectors: DB-25 (type-A) for the host connection, Centronics 36 pin (type B) for the printer or device connection and Mini Centronics 36 pin (type C), a smaller alternative for the device connection. IEEE 1284-I devices use IEEE 1284-A and IEEE 1284-B connectors, while IEEE 1284-II devices use IEEE-1284-C connectors. Type C connector is not very popular.

Q. 46. What are the Internal Buses in a computer used for? Briefly describe Local Bus, PCI Bus and ISA Bus.

Ans. Internal buses in a computer are used to carry address, data and control signals between the CPU, cache memory, SRAM, DRAM, disk drives, expansion slots and other internal devices. Internal buses are of three types, namely the Local bus, PCI bus and the ISA bus.

Local bus: Local bus connects the microprocessor to the cache memory, main memory, coprocessor and the PCI bus controller. This bus includes the data bus, address bus and the control bus. It is also referred to as the primary bus. This bus has high throughput rates, which is not possible with buses using expansion slots.

PCI bus: Peripheral Control Interconnect (PCI) bus is used for interfacing the microprocessor with external devices like hard disks, sound cards etc. via expansion slots. It has VESA local bus as the standard expansion bus. Variants of the PCI bus include PCI 2.2, PCI 2.3, PCI 3.0, PCI-X, PCI-X 2.0, mini PCI, Card bus, Compact PCI and PC/104-Plus bus. PCI bus will be supersided by the PCI Express bus. PCI originally had 32 bits and operated at 33 MHz. Different variants have different bits and data transfer rates.

ISA bus: Industry Standard Architecture (ISA) bus is a computer standard bus for IBM compatible computers. It is available in 8-bit and 16-bit versions. VESA local bus was designed to solve the bandwidth problem of the ISA bus. It worked alongside the ISA bus, where it acted as a high-speed conduit for memory-mapped I/O and DMA, while the ISA bus handled interrupts and port mapped I/O. Both these buses have been replaced by the PCI bus.

Solved Problems

Problem 26.1. *In case of a 64KB Static RAM, determine the following:*

(*a*) *Number of registers in the array*

(*b*) *Size of each register*

(*c*) *Number of address inputs*

(*d*) *Type of address decoder*

Solution. (*a*) Number of registers = 64K = 64 × 1024 = 65536

(*b*) Size of each register = 8-bits

(*c*) Number of address inputs = 16 (as 64K= $2^6 \times 2^{10} = 2^{16}$)

(*d*) Type of decoder = 1-of-16 decoder

Problem 26.2. *A certain ROM is capable of storing 16KB of data. If the internal architecture of the ROM uses a square matrix of registers, determine the following:*

(*a*) *Number of registers in each row*

(*b*) *Number of registers in each column*

(*c*) *Total number of address inputs*

(*d*) *Type of row decoder*

(*e*) *Type of column decoder*

Solution. (*a*) ROM capacity = 16K = 16 × 1024 = 16384 bytes

Therefore, total number of registers = 16384

Since the registers are arranged in a square matrix,

Number of rows = Number of columns

Number of registers in each row = √16384 = 128

(*b*) Number of registers in each column = 128

(*c*) Total number of memory locations = 16384 = 214

Therefore, total number of address inputs = 14

(*d*) 1-of-7 decoder

(*e*) 1-of-7 decoder

Problem 26.3. *Determine the total number of memory cells, each capable of storing either a '0' or a '1', and the type of decoder used in the architecture of a 32K × 8 ROM.*

Solution. Total number of memory cells is given by:

$$32K \times 8 = 32 \times 1024 \times 8 = 2{,}62{,}144\ (= 2^{18})$$

This memory can store 32768 (= 2^{15}) 8-bit words. Therefore, type of decoder needed to address the ROM is 1-of-15 decoder.

Problem 26.4. *Determine the minimum size of a ROM required to convert a 4-bit straight binary code into a Gray code equivalent. Also, write data to be programmed in various memory locations of the ROM.*

Solution. Table-26.2 shows the 4-bit straight binary numbers and their gray code equivalents. It is clear from the table that the MSB of the straight binary number is same as the MSB of the gray code equivalent. This can, therefore, be passed on as such to the output. In that case, each memory location of the ROM need to store only a 3-bit data as the 4th bit is available as such from the input. The required size of the ROM is, therefore Gray Code (16 × 3).

Table 26.2

Binary Code				Gray Code			
A_3	A_2	A_1	A_0	D_3	D_2	D_1	D_0
0	0	0	0	0	0	0	0
0	0	0	1	0	0	0	1
0	0	1	0	0	0	1	1
0	0	1	1	0	0	1	0
0	1	0	0	0	1	1	0
0	1	0	1	0	1	1	1
0	1	1	0	0	1	0	1
0	1	1	1	0	1	0	0
1	0	0	0	1	1	0	0

1	0	0	1	1	1	0	1
1	0	1	0	1	1	1	1
1	0	1	1	1	1	1	0
1	1	0	0	1	0	1	0
1	1	0	1	1	0	1	1
1	1	1	0	1	0	0	1
1	1	1	1	1	0	0	0

The 3-bit data to be programmed into 16 different memory locations of the ROM corresponding to address inputs of (0000) to (1111) in the same order would be (000), (001), (011), (010), (110), (111), (101), (100), (100), (101), (111), (110), (010), (011), (001) and (000). Figure-26.40 shows this in ROM representation.

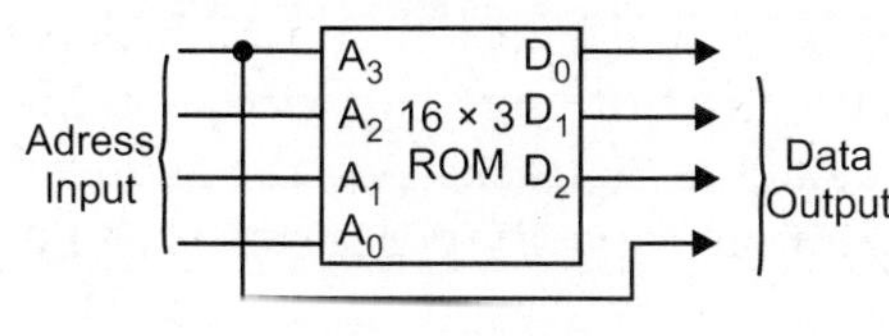

Fig. 26.40

Problem 26.5. *Two 16 MB RAMs are used to build a RAM capacity of 32 MB. Show the configuration and also, state the address inputs for which the two RAMs will be active. The two RAMs have common I/O pins, a Write Enable Input that is active LOW and a chip select input that is active HIGH.*

Solution.

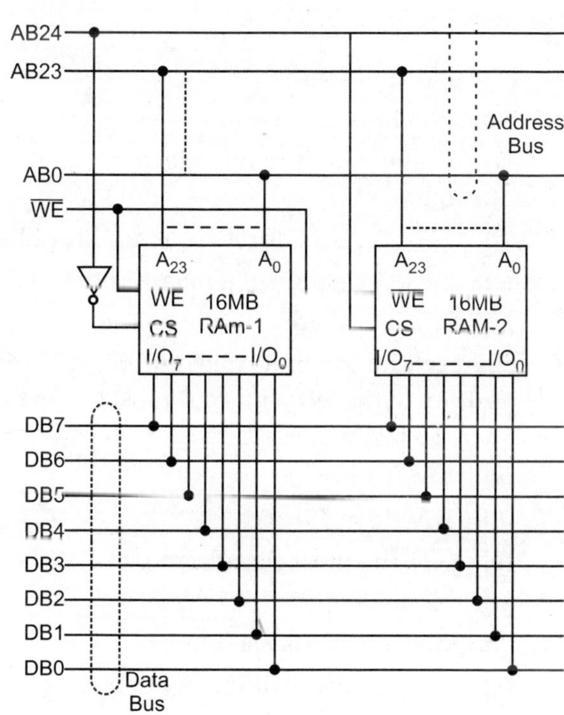

Fig. 26.41

Solution. Figure 26.41 shows the arrangement. Since the overall RAM capacity is 32 MB, it will have 25 address inputs (AB0 to AB24) as 32M = 2^{25}. For address inputs $(0000000)_{Hex}$ to $(0FFFFFF)_{Hex}$, which account for 16 M (= 2^{24}) memory locations, RAM-1 is enabled and 16 M locations of RAM-1 are available. RAM-2 is deselected for these address inputs. For address inputs $(1000000)_{Hex}$ to $(1FFFFFF)_{Hex}$, the total number of address in this group being equal to 16M, RAM-2 is selected and RAM-1 is deselected. 16 M locations of RAM-2 are available. Thus out of 32 MB, 16 MB is stored in RAM-1 and 16 MB is stored in RAM-2.

Problem 26.6. *What is available is a 1K × 8 chip of the type shown in Figure 26.42. This chip, as 'HIGH'. Show how to such ROMs can be connected to get 2K × 8 ROM without using any additional logic?*

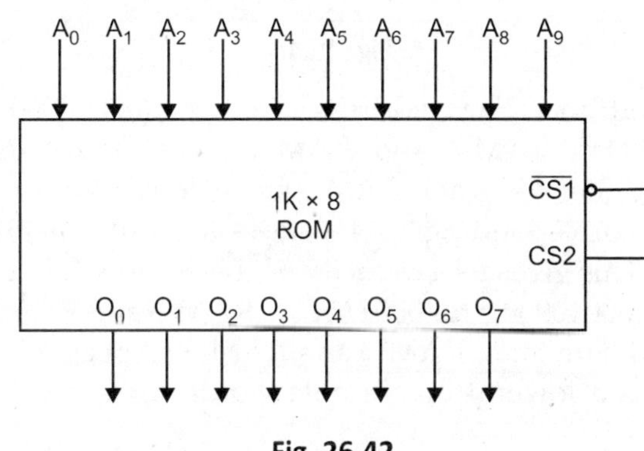

Fig. 26.42

Solution. Figure-26.43 shows the arrangement. The address bit AB_{10} is low for the first 1024 address inputs (from 00000000000 to 01111111111) and ROM-1 is selected. For the remaining 1024 address inputs (from 10000000000 to 11111111111), AB_{10}- bit is 'HIGH' thus enabling ROM-2.

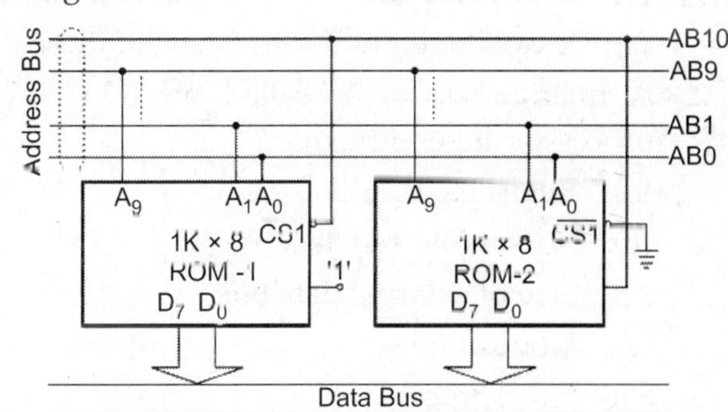

Fig. 24.43

Problem. 26.7. *Figure 26.44 shows an arrangement of four memory chips, each 16 × 4 RAM with an active LOW chip select input. Determine the total capacity and the word size. Which RAMs will put data on the data bus, when the address input is 00001101. Also, determine the address input range for which RAM-1 and RAM-2 will be active.*

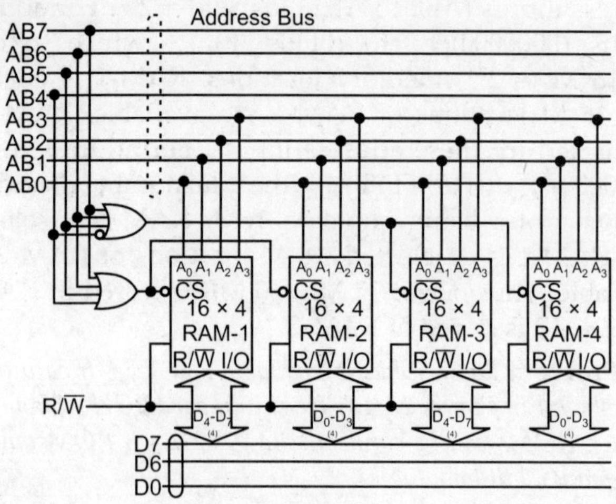

Fig. 26.44

Solution. For address inputs $(00000000)^2$ to $(00001111)^2$, RAM-1 and RAM-2 are selected. RAM-1 stores higher 4 bits and RAM-2 stores lower 4-bits of data words corresponding to 16 address inputs mentioned above. This gives us a capacity of 16 x 8. Now for address inputs $(00010000)^2$ to $(00011111)^2$, RAM-3 and RAM-4 are selected. Similarly, RAM-3 and RAM-4 respectively store upper and lower 4 bits of data words corresponding to these address inputs. This again gives a capacity of 16 × 8. Thus, the overall capacity is 32 × 8. The word size is 8.

For an address input 00001101, RAM-1 and RAM-2 will be selected. The address input range for which RAM-1 and RAM-2 are active is $(00000000)^2$ to $(00001111)^2$.

Problem 26.8. *Compute the storage capacity of a DSDD (Double sided double density) floppy diskette with 40 tracks per side. The disk has 9 sectors per track and each sector can store 512 bytes.*

Solution. Now number of sectors per side = 9 × 40 = 360

Total number of sectors on both sides = 2 × 360 = 720

Therefore, number of bytes that can be stored = 720 × 512 = 360kB

Problem 26.9. We know that a high density double sided diskette has a storage capacity of 1.2 MB. A high density diskette has 80 tracks per side and each track has 15 sectors. Storage capacity per sector is again 512 bytes. Verify the data.

Solution. Total number of sectors = Number of sides × Number of tracks per side × Number of sectors per track = 2 × 80 × 15 = 2400

Therefore, storage capacity = 2400 × 512 = 1.2 MB

Self Evaluation Exercise

I. Multichoice Questions

1. Two operands can be checked for equality using
   (*a*) OR operation (*b*) AND operation
   (*c*) EX-OR operation (*d*) None of these
2. A microprocessor is called an n-bit micro-processor depending upon
   (*a*) Registers' length
   (*b*) Size of internal data bus
   (*c*) Size of external data bus
   (*d*) None of these
3. 8085 microprocessor is a
   (*a*) Zero address microprocessor
   (*b*) One address microprocessor
   (*c*) Two address microprocessor
   (*d*) None of these
4. Setting contents of a microprocessor to zero can be efficiently done by
   (*a*) MOV Immediate instruction using zero as immediate data
   (*b*) AND Immediate instruction using zero as immediate data
   (*c*) XORing register with itself
   (*d*) None of these
5. Stack memory is used to
   (*a*) Provide additional memory to the base memory
   (*b*) Save return addresses of a subroutine
   (*c*) Save the status of a microprocessor
   (*d*) None of these
6. 'Shift left'instruction causes all bits shifted one position to the left with rightmost bit set to zero. The effect is to
   (*a*) Multiply by 2
   (*b*) Divide by 2
   (*c*) SET the most significant bit
   (*d*) None of these
7. Program counter is used to
   (*a*) Store address of the next instruction to be executed
   (*b*) Store temporary data to be used in arithmetic operations
   (*c*) Store the status of the microprocessor
   (*d*) None of these

8. The set of commands which give directions to the assembler during the assembly process but are not translated into machine instructions are called
   (*a*) Mnemonics (*b*) Identifiers
   (*c*) Directives (*d*) Operands
9. With reference to 8085 microprocessor, ANA R/M is
   (*a*) A logic instruction
   (*b*) An arithmetic instruction
   (*c*) Data transfer instruction
   (*d*) Control instruction
10. The following signal is used, when a peripheral device requests the microprocessor to have a DMA operation.
   (*a*) $I_0/\overline{M}$
   (*b*) READY
   (*c*) HOLD and HLDA
   (*d*) $R\overline{D}$ and $W\overline{R}$
11. Identify the primary memory device(s).
   (*a*) Registers built into CPU
   (*b*) RAM and ROM
   (*c*) Cache memory
   (*d*) All of these
12. SRAM devices are made using
   (*a*) Bipolar, MOS or BiMOS technologies
   (*b*) MOS technology
   (*c*) Bipolar technology
   (*d*) BiMOS technology
13. Two 1K×8 ROM chips can be used to configure
   (*a*) One 2k × 8 ROM
   (*b*) One 1k × 16 ROM
   (*c*) Both (*a*) and (*b*)
   (*d*) None of these
14. The basic memory cell in a DRAM is a
   (*a*) Capacitor
   (*b*) MOS switch and a capacitor
   (*c*) MOSFET
   (*d*) Flip flop
15. Which one of the following types of RAM needs periodic refreshing?
   (*a*) Asynchronous SRAM
   (*b*) DRAM
   (*c*) Synchronous SRAM
   (*d*) All types of RAM need periodic refreshing
16. The synchronisation between microprocessor and memory is done by
   (*a*) ALE signal (*b*) HOLD signal
   (*c*) READY signal (*d*) None of these
17. RISC processor is characterised by
   (*a*) Hardwired control design with no microcodes and fixed instruction format
   (*b*) Executing most of the instructions in a single clock cycle
   (*c*) Several general purpose registers and large cache memories that support very fast access to data
   (*d*) All of the above
18. It is a type of microprocessor instruction in which the contents of the source are copied into destination register without modifying the contents of the source.
   (*a*) Arithmetic instructions
   (*b*) Data transfer instructions
   (*c*) Control transfer instructions
   (*d*) Machine control instructions
19. Identify the true statement with reference to immediate addressing mode.
   (*a*) In this mode, data is accessed by specifying the register name in which it is stored
   (*b*) In this mode, value of operand is held within instruction itself
   (*c*) In this mode, memory location of data is stored in a register
   (*d*) This mode is useful for accessing fixed memory locations, such as memory mapped I/O devices
20. With reference to a 2K bit ROM organised as 256×8 array of memory cells, which one of the following statements is true?
   (*a*) It uses 256 rows of eight cells each
   (*b*) It uses 2048 memory cells and 8-line to 256-line address decoder
   (*c*) Both (*a*) and (*b*) are correct
   (*d*) None of these
21. With reference to 16K × 8 asynchronous SRAM with memory cells arranged in array of 128 rows and 128 columns,
   (*a*) There will be 8-line to 128-line row and column address decoders
   (*b*) There will be 7-line to 128-line row and column address decoders

(*c*) There will be 1, 31, 072 memory cells
(*d*) Both (b) and (c) are true

22. The memory device that communicates with the CPU and has relatively much higher speed than the main memory is
(*a*) DVD ROM
(*b*) Cache memory
(*c*) Primary memory
(*d*) Hard disk

23. With reference to level-1 and level-2 cache memory, one of the following statements is true.
(*a*) Level-1 cache has relatively higher storage capacity.
(*b*) Level-2 cache is part of microprocessor itself
(*c*) Level-2 cache has relatively lower storage capacity and is mounted external to the microprocessor.
(*d*) Level-1 cache has relatively lower storage capacity and is part of microprocessor itself

24. A type of memory device in which data is stored in the form of charge on a capacitor is
(*a*) Asynchronous SRAM
(*b*) Synchronous SRAM
(*c*) DRAM
(*d*) All of these

25. Identify the characteristic feature or features of a random access memory.
(*a*) It is a serial access memory.
(*b*) Data can be read from or written into any of the memory locations regardless of the order in which they are arranged
(*c*) All memory locations can be accessed with same speed.
(*d*) Both (*b*) and (*c*) are correct

26. A certain memory is specified as 16K × 8. Number of bits in each word would be
(*a*) 8
(*b*) 16
(*c*) Can be either 8 or 16
(*d*) None of these

27. In the memory specified in Q-26, number of words stored and number of memory locations would respectively be
(*a*) 16, 384 and 16, 384
(*b*) 1, 31, 072 and 16, 384
(*c*) 16, 384 and 1, 31, 072
(*d*) None of these

28. A certain memory is specified as 32K × 8. The number of address input lines and data input lines respectively would be
(*a*) 8 and 15 (*b*) 15 and 8
(*c*) 14 and 8 (*d*) 32 and 8

29. In the memory specified in Q-28, type of decoder would be
(*a*) 1-of-15 decoder (*b*) 1-of-32 decoder
(*c*) 8-of-15 decoder (*d*) None of these

30. It is desired to construct 64K × 16 RAM from available 16K × 8 RAM chips. The number of RAM chips required would be
(*a*) 16 (*b*) 4
(*c*) 8 (*d*) 2

II. Fill in the Blanks

1. A keyboard is a/an ....device. (Input, Output, Processing)
2. Cache memory is a type of .... (RAM, ROM)
3. RS-232C is a ....port. (Serial, Parallel)
4. IEEE-488 is a ....port. (Serial, Parallel)
5. MICR and OCR devices are .... devices (Input, Output)
6. Primary memory is a .... memory. (Volatile, Non-volatile)
7. USB flash memory is a ....storage device. (Primary, Secondary)
8. In a microprocessor system, data transfer from I/O device and memory and vice-versa is controlled by .... (DMA Controller, Programmable Communication Interface, Programmable Interrupt Controller)
9. Intel's 8257 is .... (Programmable DMA Controller, Programmable Interrupt Controller, Programmable Communication Interface)
10. Primary memory is ....than secondary memory. (Faster, Slower)

Answer to Self Evaluation Exercise

I. Multichoice Question

1. (c)	2. (b)	3. (a)	4. (c)	5. (b)
6. (a)	7. (a)	8. (c)	9. (a)	10. (d)
11. (d)	12. (a)	13. (c)	14. (b)	15. (b)
16. (c)	17. (d)	18. (b)	19. (b)	20. (c)
21. (d)	22. (b)	23. (d)	24. (c)	25. (d)
26. (a)	27. (c)	28. (b)	29. (a)	30. (c)

II. Fill in the blanks

1. Input 2. RAM 3. Serial 4. Parallel
5. Input 6. Volatile 7. Secondary
8. DMA Controller
9. Programmable DMA Controller
10. Faster

Chapter **27**

Programmable Logic Devices

PROMs–PALs – Microcontrollers – CPLDs – FPGAs

Learning Objectives

After completing the chapter, you will learn the following topics:

1. Difference between fixed and programmable logic
2. Advantages and disadvantages of fixed and programmable logic devices
3. Using PROM as a PLD
4. Programmable Array Logic (PAL) and its variants including PLAs and GALs
5. Difference between a microprocessor and a microcontroller
6. General architecture of a microcontroller
7. Important parameters of a microcontroller
8. Communication interfaces for microcontrollers
9. Power saving modes in microcontrollers
10. Simple Programmable Logic Devices (SPLDs) and Complex Programmable Logic Devices (CPLD)
11. CPLD architecture
12. CPLD applications
13. Field Programmable Gate Array (FPGA)
14. FPGA architecture
15. FPGA applications
16. Programmable interconnect technologies
17. Introduction to PLD programming languages

Q. 1. Distinguish between fixed and programmable logic devices. Illustrate with the help of suitable examples.

Ans. There are two broad categories of logic devices, namely fixed logic devices and programmable logic devices. While a fixed logic device, such as logic gate or multiplexer or flip flop performs given logic function that is known at the time of device manufacture; a programmable logic device on the other hand can be configured by the user to perform a large variety of logic functions. In terms of internal schematic arrangement of two types of devices, the circuits or building blocks and their interconnections in a fixed logic device are permanent and cannot be altered after the device is manufactured.

Programmable logic devices offer to the user a wide range of logic capacity in terms of digital building blocks, which can be configured by the user to perform the intended function or set of functions. This configuration can be modified or altered any number of times by the user by reprogramming the device. Figure-27.1 shows a simple logic circuit comprising of four 3-input AND gates and a 4-input OR-gate. This circuit produces an output that is the sum output of a full adder. Here, A and B are the two bits to be added and C is the carry-in bit. It is a fixed logic device, as the circuit is unalterable from outside due to fixed interconnections between various building blocks.

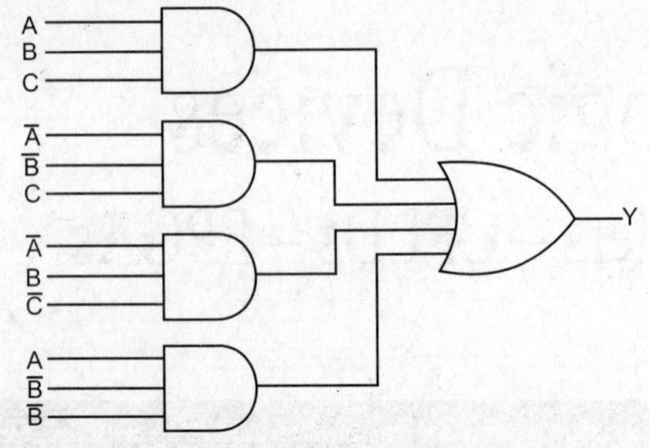

Fig. 27.1

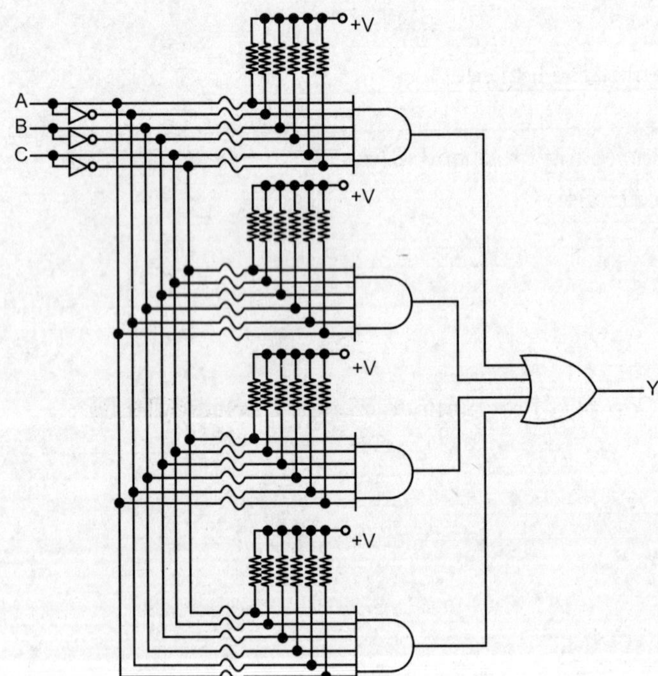

Fig. 27.2

Figure 27.2 shows the logic diagram of a simple programmable device. The device has an array of four 6-input AND gates at the input and a 4-input OR gate at the output. Each AND-gate can handle three variables and thus, can produce a product term of three variables. The three variables (*A, B* and *C* in this case) or their complements can be programmed to appear at the inputs of any of the four AND gates through fusible links called anti-fuse. This means that each AND gate can produce the desired three-variable product term. It may be mentioned here that an anti-fuse performs a function that is opposite to that performed by a conventional electrical fuse. A fuse has a low initial resistance and permanently breaks an electrically conducting path, when current through it exceeds a certain limiting value. In case of anti-fuse, initial resistance is very high and it is designed to create a low resistance electrically conducting path, when voltage across it exceeds a certain level. As a result, this circuit can be programmed to generate any three-variable sum-of-products Boolean function having four minterms by activating desired fusible links. For example, the circuit could be programmed to produce the sum output resulting from addition of three bits (sum output in case of a full adder) or to produce difference output resulting from subtraction of two bits with a borrow-in (difference output in case of a full subtractor).

We can visualise that the logic circuit of Figure-27.2 has a programmable AND array at the input and a fixed OR gate at the output. Incidentally, this is the architecture of programmable logic devices called Programmable Array Logic (PAL). Practical PAL devices have much larger number of programmable AND gates and fixed OR gates to have enhanced logic capacity and performance capability. PAL devices are discussed in detail in the latter part of the chapter.

Q. 2. Compare fixed logic devices and programmable logic devices vis-a-vis their merits and demerits.

Ans. The two types of logic devices are compared and contrasted in the following paragraphs:

1. If we want to build a fixed logic device to perform a certain specific function, time required from design to the final stage, when the manufactured device is actually available for use could easily be several months to a year or so. PLD based design requires much less time from design cycle to production run.
2. In case of fixed logic devices, the process of design validation followed by incorporation of changes, if any, involve substantial Non-Recurring Engineering (NRE) costs, which leads to enhanced cost of initial prototype device. In case of PLDs, inexpensive software tools can be used for quick validation of designs. Programmable feature of these devices allows quick incorporation of changes and also a quick testing of the device in actual application environment. In this case, device used for prototyping is the same as the one that would qualify for use in the end equipment.
3. In case of programmable logic devices, the user can change the circuit as often as he wants to

until the design operates to his satisfaction. PLDs offer to the users much more flexibility during design cycle. Design iterations are nothing but changing the programming file.

4. Fixed logic devices have an edge for large volume applications as they can be mass produced more economically. They are also the preferred choice in applications requiring highest performance level.

Q. 3. Briefly describe the following programmable logic devices:

(*a*) Programmable ROM

(*b*) Programmable Logic Array (PLA)

(*c*) Programmable Array Logic (PAL)

(*d*) Generic Array Logic (GAL)

(*e*) Complex Programmable Logic Device (CPLD)

(*f*) Field Programmable Gate Array (FPGA)

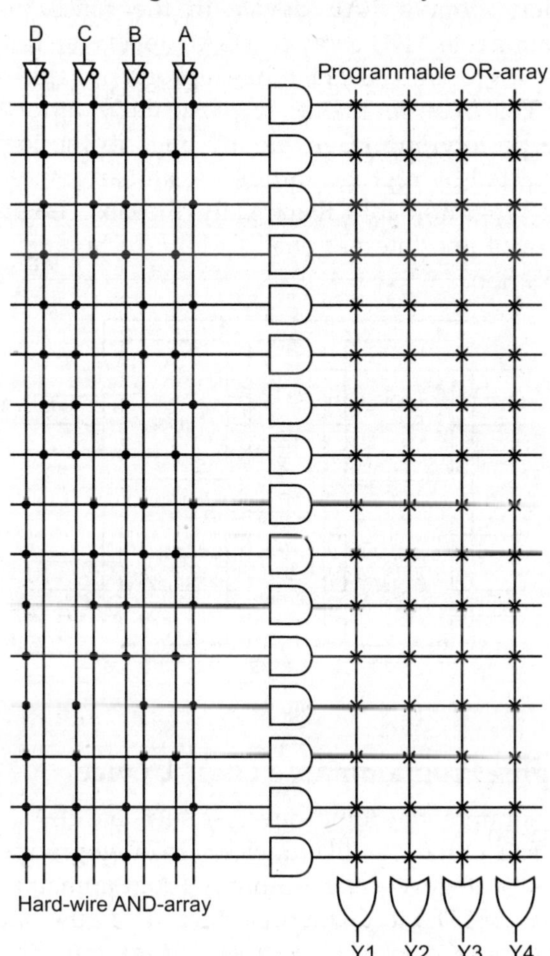

Fig. 27.3

Ans. The programmable logic devices outlined in the question differ from one another in terms of architecture, logic capacity, programmability and certain other specific features. Each one of them is briefly described in the following paragraphs:

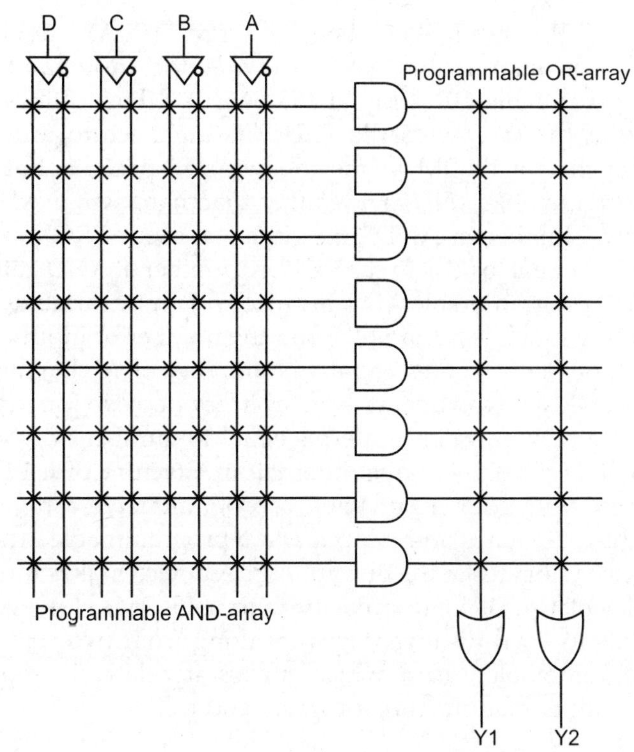

Fig. 27.4

Programmable ROMs

PROM (Programmable Read Only Memory) and EPROM (Ultraviolet Erasable Programmable Read Only Memory) can be considered to be predecessors to PLDs. The architecture of a programmable ROM allows the user to hardware implement an arbitrary combinational function of a given number of inputs. When used as a memory device, n-inputs of the ROM (called address lines in this case) and m-outputs (called data lines) can be used to store 2^nm-bit words. When used as a PLD, it can be used to implement (m) different combinational functions with each function being a chosen function of (n) variables. Any conceivable n-variable Boolean function can be made to appear at any of the m-output lines. A generalised ROM device with (n) inputs and (m) outputs has (2^n) hard-wired AND gates at the input and (m) programmable OR gates at the output. Each AND-gate has (n) inputs and each OR-gate has (2^n) inputs. Thus, each OR-gate can be used to generate any conceivable Boolean function of (n) variables and this generalised ROM can be used to produce (m) arbitrary n-variable Boolean functions. Figure-27.3 shows internal architecture of a PROM having four input lines, a hard wired array of 16 AND gates and a programmable array of four OR gates. A cross (×) indicates an intact (or unprogrammed) fusible link or interconnection and a dot (•) indicates a hard wired interconnection.

Programmable Logic Array

A Programmable Logic Array (PLA) device has a programmable AND array at the input and a programmable OR array at the output, which makes it one of the most versatile PLDs. Its architecture differs from that of PROM in the following respects. It has a programmable AND array rather than hard wired AND array. Number of AND gates in an m-input PROM are always equal to (2^m). In case of PLA, number of AND gates in the programmable AND array for (n) input variables is usually much less than (2^m) and the number of inputs of each of the OR gates equals the number of AND gates. Each OR-gate can generate an arbitrary Boolean function with a maximum of minterms equal to number of AND gates. Figure-27.4 shows internal architecture of a PLA device with four input lines, a programmable array of eight AND gates at the input and a programmable array of two OR gates at the output. A PLA device makes more efficient use of logic capacity than a PROM. However, it has its own disadvantages resulting from two sets of programmable fuses, which makes it relatively more difficult to manufacture, program and test.

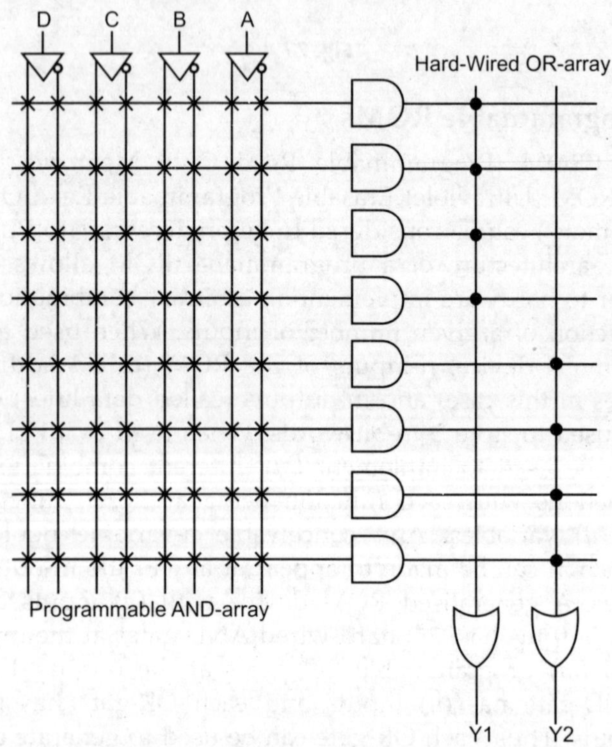

Fig. 27.5

Programmable Array Logic

Programmable Array Logic (PAL) architecture has a programmable AND array at the input and a fixed OR array at the output. Programmable AND array of a PAL device is similar to that of PLA device. That is, number of programmable AND gates is usually fewer than the number required to generate all possible minterms of given number of input variables. The OR array is fixed and the AND outputs are equally divided between available OR gates. For instance, a practical PAL device may have eight input variables, 64 programmable AND gates and four fixed OR gates with each OR gate having 16 inputs. That is, each OR gate is fed from 16 of the 64 AND outputs. Figure 27.5 shows the internal architecture of a PAL device that has four input lines, an AND array of eight AND gates at the input and two OR gates at the output to introduce the readers to the arrangement of various building blocks inside a PAL device to allow them a comparison between different programmable logic devices.

Generic Array Logic

Generic Array logic (GAL) device is similar to PAL device and was invented by Lattice Semiconductor. It differs from a PAL device in the sense that the programmable AND array of a GAL device can be erased and reprogrammed. Also, it has reprogrammable output logic. This feature makes it particularly attractive in device prototyping stage, as any bugs in the logic can be corrected by reprogramming. A similar device called PEEL (Programmable Electrically Erasable Logic) was introduced by International CMOS Technology (ICT) Corporation.

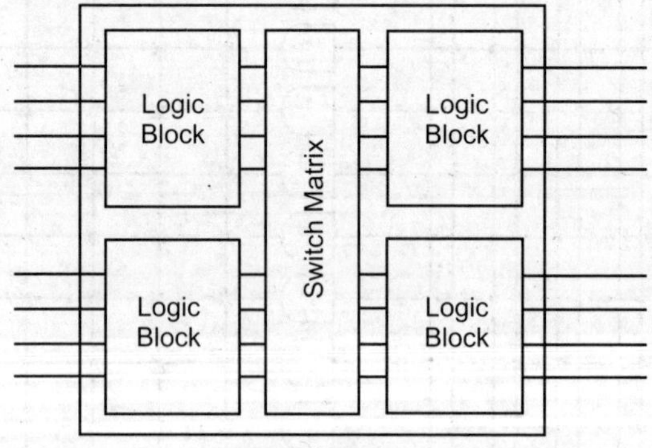

Fig. 27.6

Complex Programmable Logic Device

Programmable logic devices, such as PLAs, PALs, GALs and other PAL-like devices are often grouped into a single category called Simple Programmable Logic Devices (SPLDs) to distinguish them from the ones that are far more complex. Complex Programmable Logic Device (CPLD), as the name suggests, is a much more complex device than any of the programmable logic devices discussed so far. A CPLD may contain circuitry equivalent of several PAL devices linked to each other by programmable interconnections. Figure 27.6 shows the internal structure of a typical CPLD. Each of the four logic blocks is equivalent of a PLD, such as a PAL device. Number of logic blocks in a CPLD could be more or less than four. Each of the logic blocks has programmable interconnections. Switch matrix is used for logic block to logic block interconnections. Also, switch matrix in a CPLD may or may not be fully connected. That is, some of the possible connections between logic block outputs and inputs may not be supported by a given CPLD. While complexity of a typical PAL device may be of the order of few hundred logic gates; a CPLD may have complexity equivalent to tens of thousands of logic gates. When compared to FPGAs, CPLDs offer predictable timing characteristics due to their less flexible internal architecture and thus, ideal for critical control applications and other applications, where high performance level is required. Also, because of their relatively much lower power consumption and lower cost, CPLDs are an ideal solution to battery operated portable applications, such as mobile phones, digital assistants and so on. CPLDs can be programmed either by using a PAL programmer or by feeding it with a serial data stream from a PC after soldering it on the PC board. A circuit on CPLD decodes the data stream and configures it to perform the intended logic function.

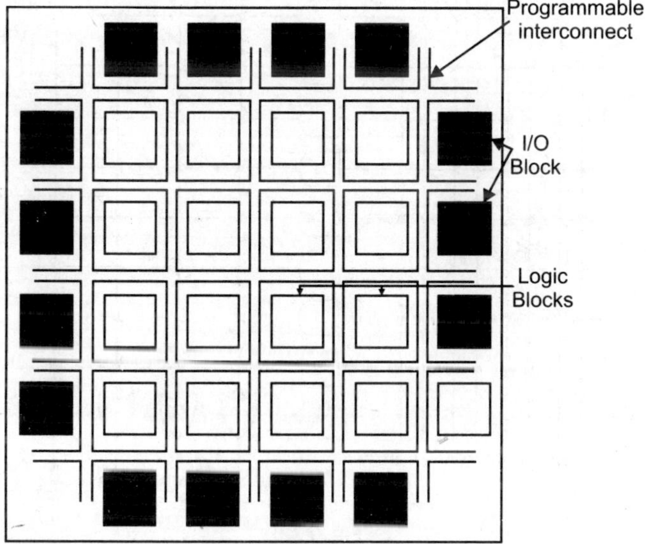

Fig. 27.7

Field Programmable Gate Array

Field Programmable Gate Array (FPGA) uses an array of logic blocks, which can be configured by the user. The word field programmable here signifies that the device is programmable outside the factory, where it is manufactured. Internal architecture of an FPGA device has three main parts, namely array of logic blocks, programmable interconnects and I/O blocks. Figure 27.7 shows architecture of a typical FPGA. Each of the I/O blocks provides individually selectable input, output or bidirectional access to one of the general purpose I/O pins on the FPGA package. Logic blocks in an FPGA are no more complex than a couple of logic gates or a look-up table feeding a flip flop. Programmable interconnects connect logic blocks to logic blocks and also, I/O blocks to logic blocks.

FPGAs offer a much higher logic density and much larger performance features as compared to CPLDs. Some of the contemporary FPGA devices offer logic complexity equivalent to that of eight million system gates. Also, these devices offer features like built-in hardwired processors, large memory, clock management systems and support for many of the contemporary device-to-device signaling technologies. FPGAs find extensive use in a variety of applications, which include data processing and storage, digital signal processing, instrumentation and telecommunications.

FPGAs are also programmed, like CPLDs after they are soldered on to the PC board. In case of FPGA, the programmed configuration is usually volatile and therefore, needs to be reloaded whenever power is applied or a different functionality is required.

Q. 4. With the help of a suitable example, explain how a PROM can be used to implement a given Boolean function. Is PROM implementation of Boolean functions with large number of 'Don't Care Conditions' an optimal solution?

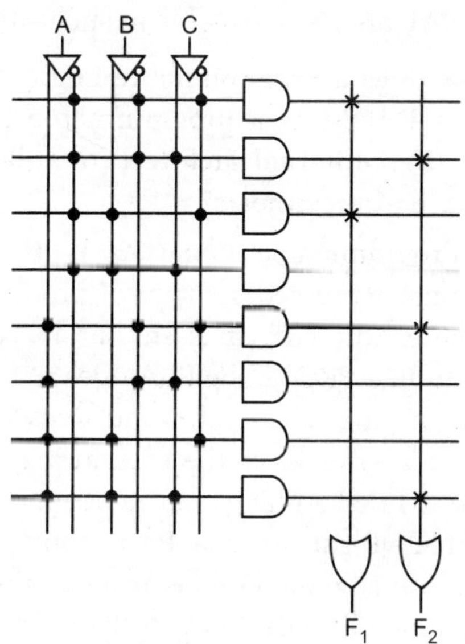

Fig. 27.8

Ans. A PROM in general has (n) input lines and (m) output lines and is designated as ($2^n \times m$) ROM. Looking at the internal architecture of a PROM device, it is a combinational circuit with the AND gates wired as a decoder and having OR gates equal to number of outputs. A PROM with five input lines and four output lines, for instance, would have an equivalent of (5 × 32) decoder at the input that would generate 32 possible minterms or product terms. Each of these four OR gates would be a 32-input gate fed from 32 outputs of the decoder through fusible links. As an illustration, let us see how PROM can be used to implement the following Boolean function with two outputs given by following two minterm equations:

$F_1(A, B, C) = \Sigma 0, 2$

$F_2(A, B, C) = \Sigma 1, 4, 7$

Implementation of this Boolean function would require a (8×2) PROM, '8' because the two outputs are three-variable functions and '2' because there are two outputs. The internal logic diagram of PROM in this case after it is programmed would be as shown in Figure-27.8.

It may be mentioned here that in practice, PROM would not be used to implement as simple a Boolean function as illustrated above. The purpose here was to tell the readers how a PROM implements a Boolean function. In actual practice, PROMs would be used only in case of very complex Boolean functions.

Another noteworthy point, which was highlighted earlier too, is that when it comes to implementing Boolean functions with PROMs, it is not an economical to use ROM for those Boolean functions, which have a large number of 'Don't Care Conditions'. In case of a PROM, each 'Don't Care Condition' would have either all 0's or all 1's. In other words, the space on the chip is not optimally utilised. Other programmable logic devices, like PLA or PAL are more suitable in such situations.

Q. 5. How does a Programmable Logic Array (PLA) differ from a PROM as a programmable logic device in terms of their internal architecture and number of programmable interconnects?

Ans. A Programmable Logic Array (PLA) enables logic functions expressed in sum-of-products form to be implemented directly. It is similar in concept to a PROM. But unlike PROM, the PLA does not provide full decoding of the input variables and does not generate all possible minterms as is the case in a PROM. While a PROM has a fixed AND gate array at the input and a programmable OR gate array at the output; a PLA device has a programmable AND gate array at the input and a programmable OR gate array at the output. In a PLA device, each of the product terms of the given Boolean function is generated by an AND gate, which can be programmed to form the AND of any subset of inputs or their complements. The product terms so produced can be summed up in an array of programmable OR gates. Thus, we have a programmable OR gate array at the output. The input and output gates are constructed in the form of arrays with input lines orthogonal to product lines and the product lines being orthogonal to output lines.

Figure 27.9 shows internal architecture of a PLA device with four input lines, eight product lines and four output lines. That is, programmable AND-gate array has eight AND gates. Each of the AND gates here has eight inputs, four corresponding to four input variables and their complements. Input to each of the AND gates can be programmed to be any of the possible 16 combinations of four input variables and their complements. Four OR gates at the output can generate four different Boolean functions, each having a maximum of eight minterms out of 16 minterms possible with four variables. The logic diagram depicts the unprogrammed state of the device. PLAs usually have inverters at the output of OR gates to enable PLAs implement a given Boolean function in either AND-OR and AND-OR-INVERT form. Figure 27.10 shows generalised architecture of a PLA device having (n) inputs, (m) outputs and (k) product terms with (n), (m) and (k) respectively representing number of input variables, number of OR gates and number of AND gates. The number of inputs to each OR gate and each AND gate are (k) and ($2n$) respectively.

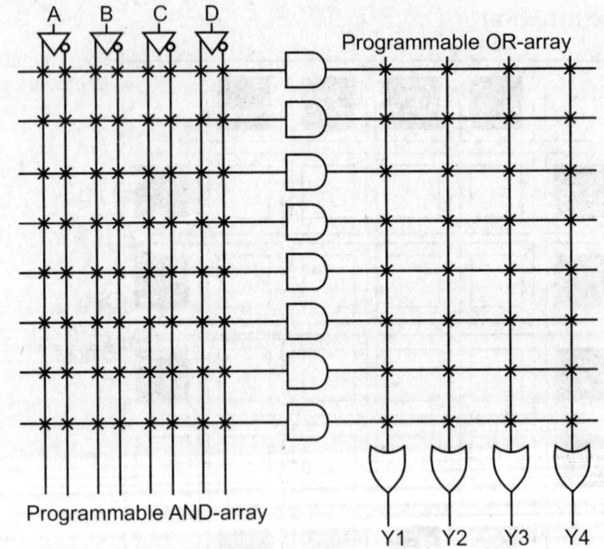

Fig. 27.9

As is clear from the description given in the preceding paragraph, the PLA would have a total of ($2kn + km$) programmable interconnections. A ROM with same number of input and output lines would have ($2^n \times m$) programmable interconnections.

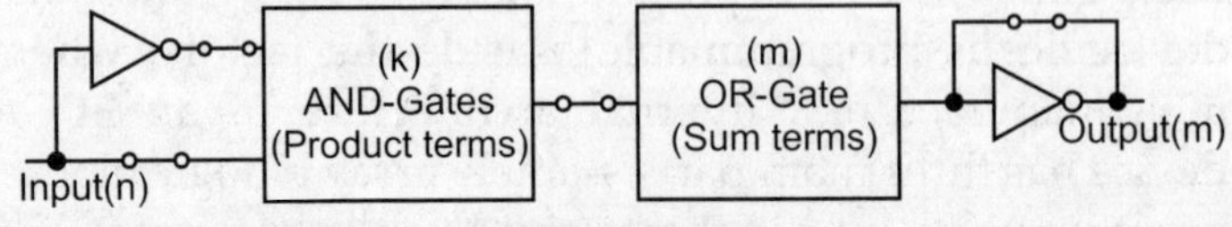

Fig. 27.10

Q. 6. What is a PAL device? How does it differ from a PLA device? What is a HAL (Hard Array Logic) device?

Ans. Programmable Array Logic (PAL) device is a variant of PLA device. While a PLA device has a programmable AND gate array at its input and a programmable OR gate array at its output; a PAL device has a programmable AND gate array at the input and a fixed OR gate array at the output. The idea to have a fixed OR-array at the output and make the device less complex originated from the fact that there were many applications, where product term sharing capability of the PLA was not fully utilised and thus, wasted. PAL device is a trade mark of Advanced Micro Devices Inc. PAL devices are, however, less flexible than PLA devices. The flexibility of a PAL device can be enhanced by having different output logic configurations including availability of both OR (also called active HIGH) and NOR (also called active LOW) outputs, bidirectional pins that can act both as inputs as well as outputs, having clocked flip flops at the outputs to provide what is called registered outputs. These features allow the device to be used in a wide range of applications than would be possible with a device with fixed input and output allocations. Mask programmed version of PAL is known as a HAL (Hard Array Logic) device. A HAL device is pin to pin compatible to its PAL counterpart.

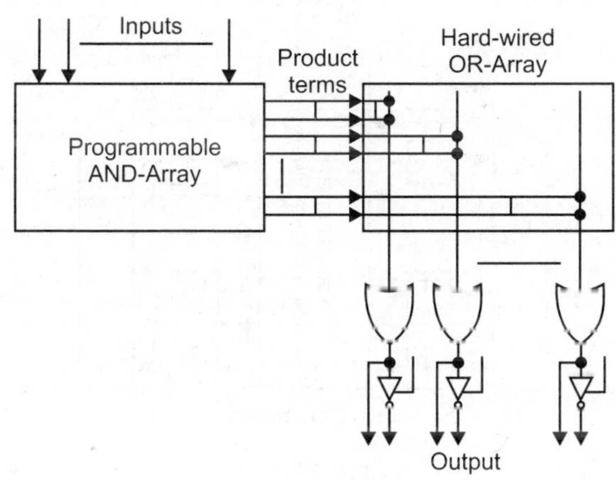

Fig. 27.11

Q. 7. With the help of a suitable diagram, briefly explain the internal architecture of a PAL device.

Ans. Figure 27.11 shows the block schematic representation of generalised architecture of a PAL device. As we can see from the arrangement shown, the device has a programmable AND-gate array that is fed with various input variables and their complements. Programmable input connections allow any of the input variables their complements to appear at the inputs of any of the AND gates in the array. Each of the AND gates generates a minterm of user defined combination of input variables and their complements. As an illustration, Figure 27.12 gives an example of generation of minterms.

Outputs from programmable AND array feed an array of hard wired OR gates. Here, output of each of the AND gates does not feed the input of each of the OR gates. Each OR-gate is fed from a subset of AND gates in the array. This implies that the sum-of-product Boolean functions generated by each of the OR gates at the output shall have only a restricted number of minterms depending upon the number of AND gates, it is being fed from. Outputs from the PAL device, as is clear from the generalised form of representation shown in Figure-27.11, are available both as OR outputs as well as complemented (or NOR) outputs.

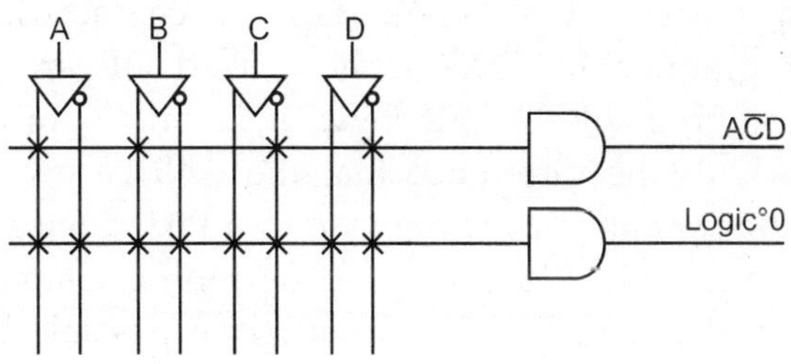

Fig. 27.12

Q. 8. Briefly describe various logic arrangements available at the output of practical PAL devices to enhance their flexibility?

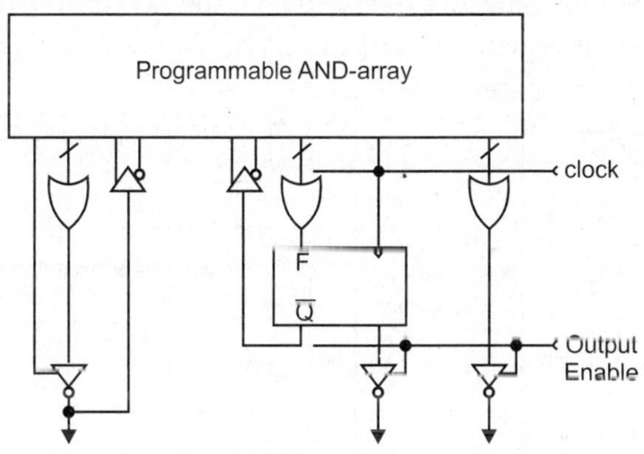

Fig. 27.13

Ans. Practical PAL devices offer various output logic arrangements. One of them, of course, is the availability of both OR and NOR outputs, as mentioned in the answer to the previous questions. Another feature available with many PAL devices is that of registered outputs. In case of registered outputs, OR-gate output drives the D-input of a D-type flip flop, which is loaded with the data on either LOW-to-HIGH or HIGH-to-LOW edge of a clock signal. Yet another feature is availability of bidirectional pins, which can be used both as outputs and inputs. This facility allows the user to feed a product term back to the programmable AND array. It helps particularly in those multi-output function logic circuits that share some common minterms. Some of the common output logic arrangements available with PAL devices are shown in Figure 27.13. Some PAL devices offer an EX-OR gate following the OR gate at each output. One of the inputs to the EX-OR gate is programmable, which allows the user to configure it either as an inverter or a non-inverting buffer or as a two-input EX-OR gate. This feature is particularly useful, while implementing parity and arithmetic operations.

Q. 9 What is a GAL (Generic Array Logic) device? How does it differ from a PAL device?

Ans. Generic Array Logic (GAL) is characterised by a reprogrammable AND array, a fixed OR array and a reprogrammable output logic. It is similar to a PAL device with the difference that the AND array is not just programmable as is the case in a PAL device but is reprogrammable. That is, it can be reprogrammed any number of times. This has been made possible by the use of electrically erasable PROM cells for storing the programming pattern. The other difference is in use of reprogrammable output logic, which provides more flexibility to the designer. GAL devices employ Output Logic Macro Cells (OLMC) at the output, which allows the designer to configure the outputs either as combinational outputs or registered outputs.

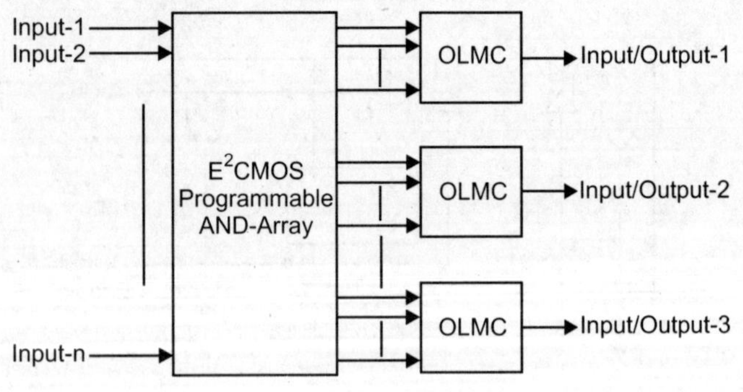

Fig. 27.14

Q. 10. With the help of a suitable block schematic, briefly describe the architecture of a GAL device.

Ans. Figure 27.14 shows the block schematic representation of a GAL device and Figure 27.15 shows the architecture of a typical OLMC used with GAL devices. The OLMC of the type shown in Figure 27.15 can be configured to produce four different outputs depending upon the selection inputs. These include the following:

1. S_1S_0 = 00: Registered mode with active LOW output
2. S_1S_0 = 01: Registered mode with active HIGH output
3. S_1S_0 = 10: Combinational mode with active LOW output
4. S_1S_0 = 11: Combinational mode with active HIGH output

We can see that two of the four inputs to 4-to-1 multiplexer are combinational outputs and the other two are the registered outputs. Also, of the two combinational outputs, one is the active HIGH output, while the other is an active LOW output. Same is the case with registered outputs. Of the four inputs to the multiplexer, the one appearing at the output depends upon selection inputs. 2-to-1 multiplexer ensures that the final output is also available as feedback to the programmable AND array.

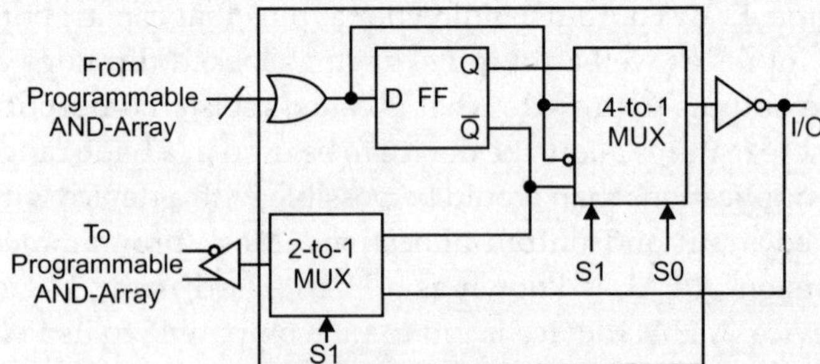

Fig. 27.15

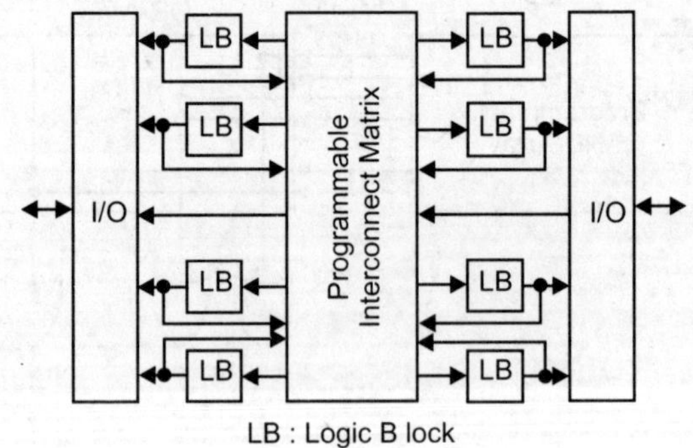

Fig. 27.16

Q. 11. Briefly describe the standard PAL numbering system. Give examples in support of your answer.

Ans. Standard PAL numbering system uses an alphanumeric designation comprising of a two-digit number indicating number of inputs followed by an alphabet that tells about the architecture/type of logic output. Another number following the alphabet indicates number of outputs. In case of PAL devices offering a combination of different types of logic outputs, the rightmost number indicates the number of output type implied by the alphabet used in the designation. For example, a PAL device designated PAL-16L8 shall have 16 inputs and eight active LOW outputs. Another PAL device designated PAL-16R4 has 16 inputs and four registered outputs. Also, number of inputs as given by the number designation includes dedicated inputs, user programmable inputs accessible from combinational I/O pins and any feedback inputs from combinational and registered outputs. For example, PAL-16L8 has 10 dedicated inputs and six inputs accessible from I/O pins.

In addition to the numbering system described above, an alphanumeric designation on the extreme left may be used to indicate the technology used. 'C' stands for CMOS, '10H' for 10KH ECL and '100' for 100K ECL. TTL is represented by a blank. An alphabet on the extreme right may be used to indicate power level with 'L' and 'Q' respectively indicating low and quarter power levels and blank representing full power.

Q. 12. What is a Complex Programmable Logic Device (CPLD)? Why is there a need to have a CPLD, when there already are programmable logic devices, such as PROMs, PLAs and PALs?

Ans. CPLD is nothing but integration of multiple PLDs, programmable interconnect matrix and I/O control block on a single chip. If we examine the internal architecture of simple programmable logic devices (SPLDs), such as PLAs and PALs; we find that it is not practical to increase their complexity beyond a certain level. It is because of the reason that the size of the programmable plane (such as programmable AND plane in a PLA or PAL device) increases too rapidly with increase in number of inputs to make it a practically viable device. One way to increase the logic capacity of simple programmable logic devices was to integrate multiple SPLDs on a single chip with a programmable interconnects between them. These devices have the same basic internal structure that we see in case of SPLDs and are grouped together in the category of Complex Programmable Logic Devices (CPLDs). Typically, CPLDs may offer a logic capacity equivalent to that of about 50 SPLDs.

Q. 13. With the help of suitable diagram/s, briefly describe the internal architecture and operation of a CPLD.

Ans. As outlined in the previous paragraph, CPLD is nothing but integration of multiple PLDs, programmable interconnect matrix and I/O control block on a single chip. Each of the identical PLDs is referred to as Logic Block or Function Block. Figure-27.16 shows the architecture of a typical CPLD. As is evident from the block schematic arrangement, the programmable interconnect matrix is capable of connecting input or output of any of the logic blocks to any other logic block. Also, input and output pins connect directly to both the interconnect matrix as well as logic blocks.

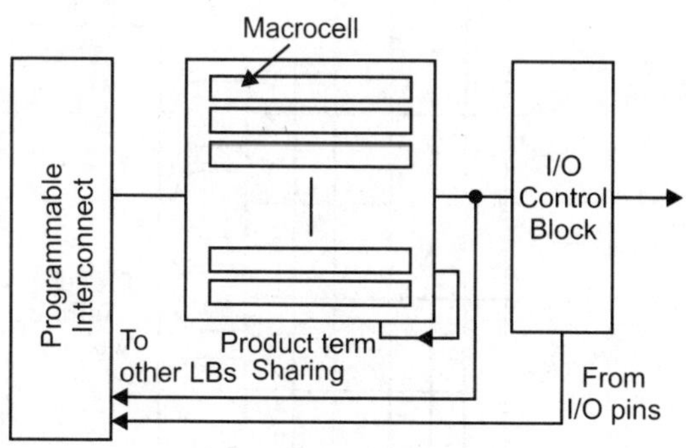

Fig. 27.17

Logic blocks may further comprise of smaller logic units called Macrocells, where each of the macrocells is a subset of PLD-like logic block. Figure 27.17 shows the structure of a logic block along with its interconnections with the programmable interconnect matrix and I/O block. The horizontal gray coloured bars inside the logic block constitute array of macrocells. Typically, each macrocell comprises of a set of product terms generated by subset of programmable AND array and feeding a configurable output logic. The output logic typically comprises of an OR-gate, an Exclusive-OR gate and a flip flop. The flip flop in case of most of contemporary CPLDs is configurable as D-type, J-K, T, R-S flip flop or can even be transparent. Also, the OR-gate can be fed with any or all of the product terms generated within the macrocell. Most of the contemporary CPLDs also offer an architecture, where the OR gate can also be fed with some additional product terms generated within other macrocells of the same logic block. For example, logic block in case of MAX-7000 series of CPLDs from Altera offers this product term flexibility, where OR-gate of each macrocell can have upto 15 additional product terms from other macrocells in the same logic block apart from a maximum of five product terms from within the same macrocell.

Figure 27.18 shows the logic diagram of a macrocell typical of macrocells in the logic blocks of most of contemporary CPLDs. The diagram is self-explanatory. There may be minor variations in devices from different manufacturers. For example, macrocells in XC-7000 series CPLDs from Xilinx have two OR gates fed from a two-bit Arithmetic Logic Unit (ALU) and its output feeds configurable flip flop.

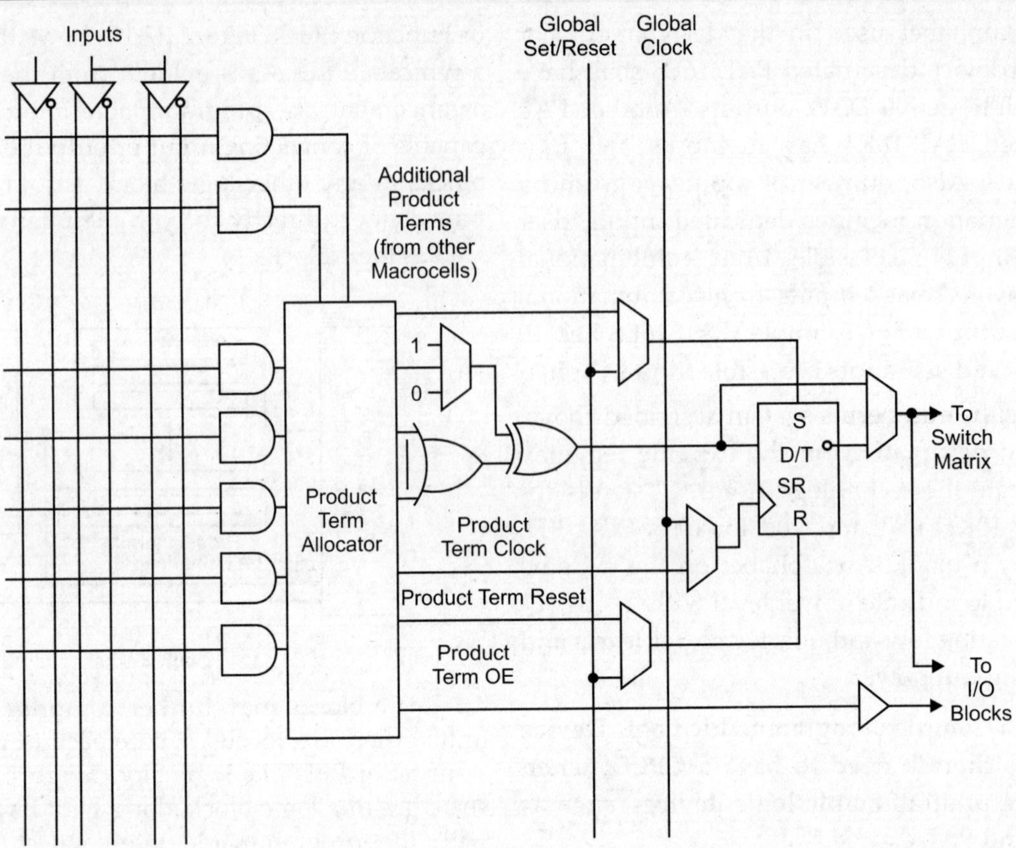

Fig. 27.18

Q. 14. Briefly outline salient features and applications of CPLDs.

Ans. Due to their less flexible internal architecture leading to predictable timing performance, high speed and a range of logic capacities; CPLDs find extensive use in a wide assortment of applications. These include implementation of random glue logic to prototyping small gate arrays, implementing critical control designs, such as graphics controller, cache control, UARTs, LAN controller and many more.

CPLDs are fast replacing SPLDs in complex designs. Complex designs using a large number of SPLDs can be replaced by a CPLD-based design with much smaller number of devices. This is particularly attractive in portable applications, like mobile phones, digital assistants and so on. CPLD architecture particularly suits those designs that exploit wide AND/OR gates and do not require a large number of flip flops.

Reprogramming feature of CPLDs makes incorporation of design changes very easy. With the availability of CPLDs having in-circuit programming feature, it is possible to even reconfigure the hardware without power down. Changing protocol in a communication circuit could be one such example.

One of the most significant advantages of CPLD architecture comes from its simple SPLD-like structure, which allows the design to partition naturally into SPLD-like blocks. This leads to a highly predictable timing or speed performance than would be possible, if the design were split into many pieces and mapped into different areas of the chip.

Q. 15. What is an FPGA? How does it differ from a CPLD?

Ans. An FPGA consists of an array of uncommitted configurable logic blocks, programmable interconnects and I/O blocks. Basic architecture of FPGA is shown in Figure-27.19. The basic difference between a CPLD and an FPGA lies in their internal architecture. CPLD architecture is dominated by relatively smaller number of programmable sum-of-products logic arrays feeding small number of clocked flip flops, which makes the architecture less flexible but with more predictable timing characteristics. On the other hand, FPGA architecture is dominated by programmable interconnects and the configurable logic blocks are relatively simpler. Logic blocks within an FPGA can be as small as the macrocells in a PLD called fine-grained architecture or larger and more complex called coarse-grained architecture. But they are never as large as the entire PLD like the logic blocks of a CPLD. This feature makes these devices far more flexible in terms of range of designs that can be implemented with these devices. Contemporary FPGAs have on-chip presence of higher level embedded functions, and embedded memories. Some of them even come with an on-chip microprocessor and related peripherals to constitute what is called complete 'System on a programmable chip'. Virtex-II Pro and Virtex-4 FPGA devices from Xilinx are examples. These devices have one or more PowerPC processors embedded within FPGA's logic fabric.

Q. 16. With the help of suitable diagrams, briefly describe typical logic blocks of FPGAs.

Ans. Figure 27.20 shows a typical logic block of an FPGA. It consists of a four-input look-up table (LUT) whose output feeds a clocked flip flop. The output can either be a registered output or the unregistered LUT output. Selection of output takes place in the multiplexer. LUT is nothing but a small one-bit wide memory array with its address lines representing the inputs to the logic block and one-bit output acting as the LUT output. A LUT with (n) inputs can realise any logic function of (n) inputs by programming the truth table of the desired logic function directly into the memory. Logic blocks can have more than one LUTs and flip flops also to give them capability to realise more complex logic functions. Figure-27.21 shows an LUT-based architecture that uses multiple LUTs and flip flops. This logic block can be configured either as four four-input LUTs or two five-input LUTs or one six-input LUT.

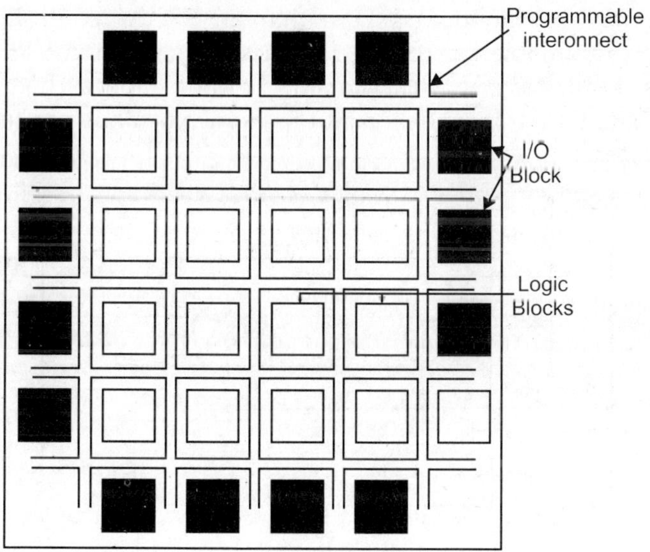

Fig. 27.19

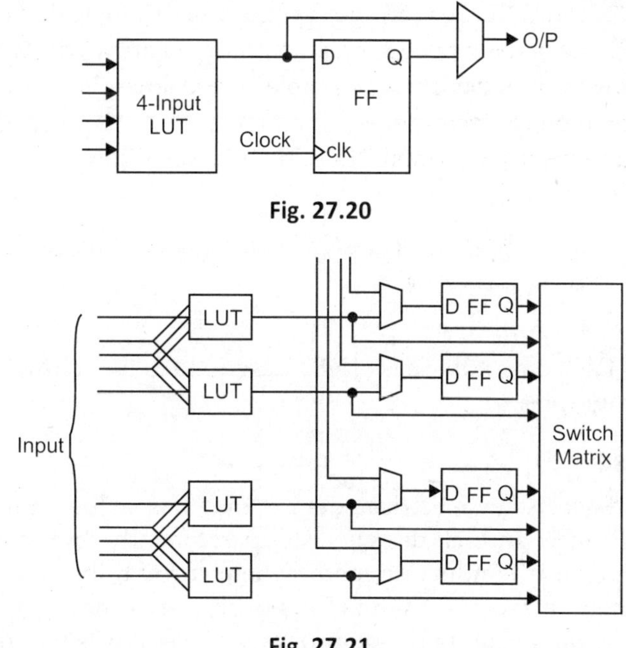

Fig. 27.20

Fig. 27.21

Q. 17. Briefly outline common application areas of FPGA devices.

Ans. In the early days of their arrival on the scene, FPGAs began as competitors to CPLDs for applications as glue logic for PCBs. With increase in their logic capacity and capability, availability of large embedded memory, higher level embedded functions, such as adders and multipliers, emergence of hybrid technologies combining the logic blocks and interconnects of traditional FPGAs with embedded microprocessors and facility of full or partial in-system re-configuration have immensely widened the scope of applications of FPGAs. FPGAs today offer complete system solution on a single chip, though very complex systems might be implemented with more than one FPGA devices.

Some of the major application areas of FPGA devices include Digital signal processing, Data storage and processing, Software-defined radio, ASIC prototyping, Speech recognition, Computer vision, Cryptography, Medical imaging, Defence systems, Bioinformatics, Computer hardware emulation and Re-configurable computing. Re-configurable computing also called customised computing involves using the programmable parts to execute software rather than compiling the software to be run on a regular CPU. This has rendered possible due to in-system re-configuration, which allows the internal design to be altered on-the-fly.

Q. 18. Briefly describe different interconnect technologies used in programmable logic devices.

Ans. Programmable features of every PLD, be it simple programmable logic devices (SPLDs), like PLAs,

PALs and GALs, or complex programmable logic devices (CPLDs) or even field programmable gate arrays (FPGA), come from their programmable interconnect structure. Interconnect technologies that have evolved over the years for programming PLDs include the following:

1. Fuse
2. EPROM or EEPROM floating gate transistors
3. Static RAM
4. Anti-fuse

Each one of these is briefly described in the following paragraphs.

Fuse

Fuse is an electrical device that has a low initial resistance and is designed to permanently break an electrically conducting path when current through it exceeds a specified limit. It uses bipolar technology, is non-volatile and one time programmable. It was the first user-programmable switch developed for use in PLAs. They were earlier used in smaller PLDs and are now being rapidly replaced by newer technologies.

Floating Gate Transistor Switch

This interconnect technology is based on the principle of placing the floating gate transistor between two wires in such a way as to facilitate a wire-AND function. This concept is used in EPROM and EEPROM devices and that is why, floating gate transistor is sometimes referred to as EPROM or EEPROM transistor. Figure 27.22 shows the use of floating gate transistor interconnects in the AND- plane of a CPLD or SPLD. All those inputs that are required to be part of a particular product term are activated to drive the product wire to a logic '0' level through the EPROM transistor. For inputs that are not part of the product term, relevant transistors are switched off.

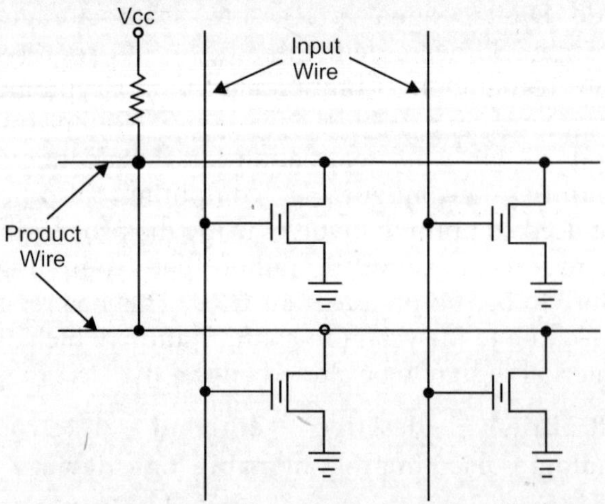

Fig. 27.22

This technology is commonly used in SPLDs and CPLDs. Floating gate transistor based switch matrix, however, requires a large number interconnects and therefore transistors. For example, a CPLD with 128 macrocells with four inputs and one output each would require as many as 65536 interconnects for 100% routability. Large number of interconnects also adds to the propagation delay. Use of multiplexers can reduce this number significantly and can also address the problem of increased propagation delay. MUX-based interconnect matrix is being used in CPLDs. CPLD type XPLA3 from Xilinx is an example.

Static RAM Controlled Programmable Switches

Static RAM (SRAM) is basically a semiconductor memory and the word 'static' implies that it is a non-volatile memory. That is, memory retains its contents as long as power is on. SRAM with (m) address lines and (n) data lines is referred to as ($2^m \times n$) memory and is capable of storing (2^m) n-bit words. Figure 27.23 shows the basic SRAM cell comprising of six MOSFET switches with four of them connected as cross-coupled inverters. Basic SRAM cell can store one bit of information. Reading operation is carried out by pre-charging both the bit lines (BL and $\overline{BL}$) to logic '1' and then asserting WL-line. Writing operation is done by giving desired logic status to BL-line and its complement to line and then asserting WL-line.

Figure 27.24 shows the use of SRAM controlled switches. SRAMs are used to control not only the gate nodes but also the select inputs of multiplexers that drive the logic block inputs. As an illustration, it shows the routing scheme for feeding the output of one logic block to the input of another via SRAM controlled pass transistor switches and SRAM controlled multiplexer. It may be mentioned here that an SRAM controlled programmable interconnect matrix does not necessarily use both, the pass transistors and multiplexers. Whether it uses pass transistors or multiplexers or both is product specific.

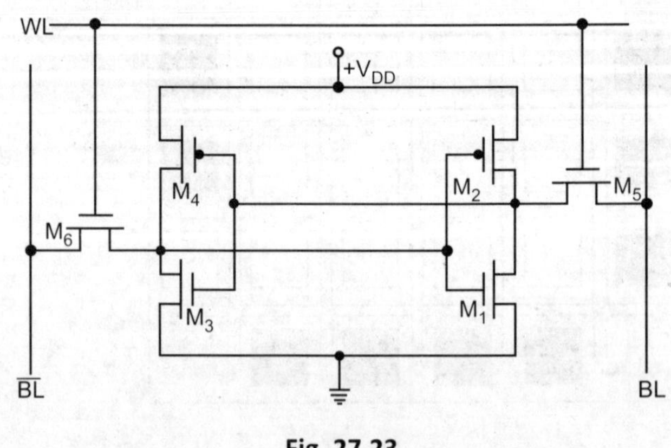

Fig. 27.23

Anti-fuse

Anti-fuse is an electrical device with a high initial resistance and designed to permanently create an electrically conducting path typically, when voltage across it exceeds a certain level. Anti-fuse uses CMOS technology, which is one of the main reasons for its wide use in PLDs, FPGAs in particular. A typical anti-fuse consists of an insulating layer sandwiched between two conducting layers. In the unprogrammed state, insulating layer isolates top and bottom conducting layers. When programmed, the insulating layer gets transformed to a low resistance link. Typically, it uses metal for conductors and amorphous silicon as insulator. Application of high voltage across amorphous silicon permanently transforms it into poly-crystalline silicon-metal alloy having a low resistance. There are other anti-fuse structures too, such as that used in Actel's anti-fuses. This anti-fuse known as PLICE uses poly-silicon and (n+) diffusion as conductors and ONO (Oxide-Nitride-Oxide)as insulator. This type of anti-fuse is usually triggered by a small current of the order of few milliamperes. High current density produced in the thin insulating layer produces heat, thus, melting the insulating layer and creating an irreversible resistive silicon link.

Anti-fuses are widely used as programmable interconnects in PLDs (Figure 27.25). Anti-fuse PLDs are one-time programmable in contrast to SRAM controlled interconnect based PLDs, which are re-programmable. It may be mentioned here that re-programmable feature helps the designers fix logic bugs or add new functions. Anti-fuse PLDs have advantages of non-volatility and usually higher speeds. Anti-fuses may also be used in PROMs. In that case, each bit contains both a fuse and an anti-fuse. The device is programmed by triggering one of the two.

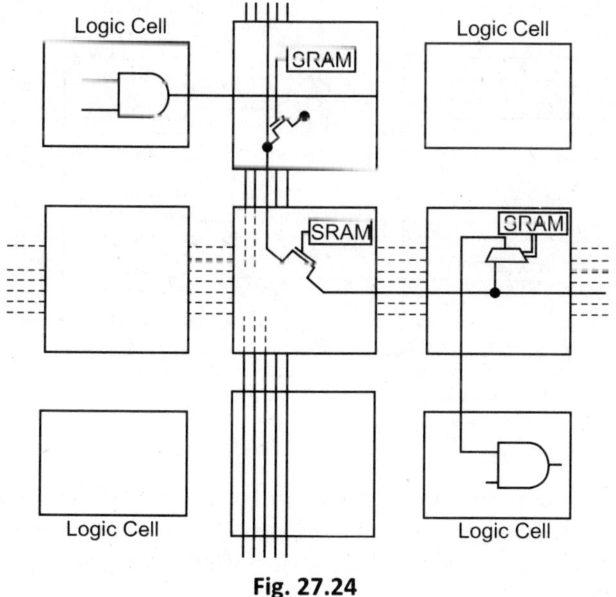

Fig. 27.24

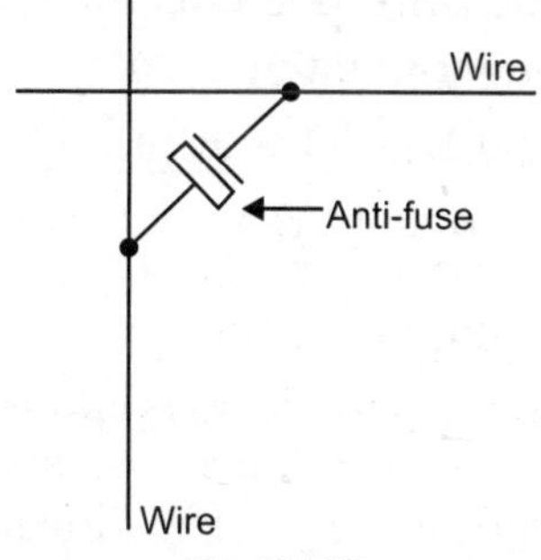

Fig. 27.25

Q. 19. Briefly outline the software programs used for programming PLDs.

Ans. During the PLD development cycle, from design entry to generation of bit stream that can be loaded onto the chip using some kind of electronic programming system, two types of software programs are needed to perform two different functions. First is a Hardware Description Language (HDL), which is needed at the design entry stage. HDL is a software programming language that is used to model or describe the intended operation of a piece of hardware. In the present case, it is the function that the PLD chip is intended to perform after it is programmed. It may be worth mentioning here that modern computer languages including both hardware description languages and high level programming languages almost invariably contain declarative and executable statements, and the hardware description languages are particularly rich in the former. If we compare the results of a high level programming language, such as (C++) and a HDL, it will be an executable program in case of former and hardware in case of latter. Hardware description languages that have evolved over the years include ABEL-HDL, VHDL (VHSIC HDL), Verilog and JHDL (Java HDL). VHSIC stands for Very High Speed Integrated Circuit. Second is a computer program, called logic compiler, used to transform a source code written in HDL into a bit stream. Logic compilers are available from manufacturers or third party vendors. In the paragraphs to follow, we shall briefly describe each of the hardware description languages mentioned above.

Q. 20 What are the common hardware description languages?

Ans. Common hardware description languages include the following:

1. ABEL – HDL Hardware Description Language
2. VHDL – VHSIC Hardware Description Language
3. Verilog
4. Java – HDL

Q.21. Briefly describe each of the hardware description languages mentioned in Q-20.

Ans. ABEL-Hardware Description Language

ABEL-HDL from DATA I/O was intended for relatively simpler PLD circuit designs that could be implemented on SPLDs. ABEL allows the designers to describe the digital circuit designs expressed in the form of truth tables, Boolean functions, state diagrams or any combination of these. It also allows the designer to optimise the design through design validation without specifying a device. In other words, ABEL-HDL facilitates writing hardware independent programs and it is only after the design verification and optimisation, have taken place that the PLD device is chosen. The source code written in ABEL environment is in standard format to have interface compatibility with other tools.

VHDL-VHSIC Hardware Description Language

VHDL is the most widely used hardware description language used for the purpose of describing complex digital circuit designs that would be implemented on CPLDs and FPGAs. VHDL was originally developed to document the behaviour of ASICs used by various manufacturers in their equipment. It was subsequently followed by development of logic simulation and synthesis tools that could read VHDL files and output a definition of the physical implementation of the circuit. With modern synthesis tools capable of extracting various digital building blocks, like counters, RAMs, arithmetic blocks etc. and implement them as specified by the user, same VHDL code could be synthesised differently for optimum performance.

VHDL is a strongly typed language. One of the key features of VHDL is that it allows the behaviour of the intended hardware to be described and then verified with the help of synthesis tools before the design is translated into actual hardware. Another feature of VHDL that makes it attractive for digital system design is that it allows description of a concurrent system.

Verilog

Verilog like VHDL supports design, design validation and subsequent implementation of analog, digital and mixed signal circuits at various levels of abstraction. Verilog-based design consists of a hierarchy of modules, whose behaviour is defined by concurrent and sequential statements. Sequential statements are placed inside a 'Begin/End' block and sequential statements contained inside the block are executed sequentially. All concurrent statements and all 'Begin/End' blocks in the design are executed in parallel. A subset of statements in Verilog is synthesisable. Therefore, if in a given design, different modules use only synthesisable statements; the design can be translated into a netlist, which can further be translated into a bit stream.

Verilog has some similarities and dissimilarities with C-language. It has similar pre-processor, major control keywords like 'if', 'while' etc. and also similar formatting mechanism in the printing routines and language operators. Dissimilarities include use of 'Begin/End' instead of curly braces to define a block of code and also, that Verilog does not have structures, pointers and recursive subroutines. Also, definition of constants in Verilog requires bit width along with their base.

Java HDL

Java HDL (JHDL) was developed in the Configurable Computing Laboratory of Brigham Young University (BYU). It is a low level hardware description language; which primarily uses object oriented approach to build circuits.

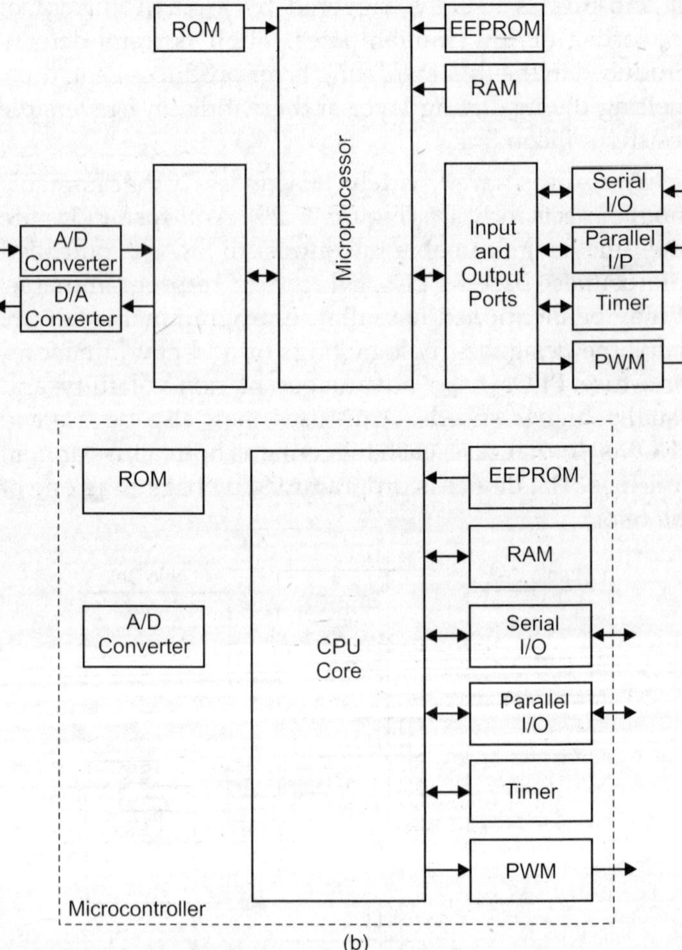

(b)

Fig. 27.26

It was developed primarily for the design of FPGA-based hardware and developers have paid particular attention to support Xilinx series of FPGA chips.

Q. 22. What is a microcontroller? What distinguishes a microcontroller from a microprocessor?

Ans. Microcontroller may be considered as a specialised computer-on-a-chip or a single-chip computer. The word 'micro' suggests that the device is small and the word 'controller' suggests that the device may be used to control one or more functions of objects, processes or events. It is also called an embedded controller, as microcontrollers are often embedded in the device or system that they control.

It contains a simplified processor, some memory (RAM and ROM), I/O ports and peripheral devices, such as counters/timers, analog-to-digital converters, etc., all integrated on a single chip. It is this feature of the processor and peripheral components available on a single chip that distinguishes it from a microprocessor based system. A microprocessor is nothing but a processing unit with some general purpose registers. A microprocessor based system also has RAM, ROM, I/O ports and other peripheral devices to make it a complete functional unit but all these components are external to the microprocessor chip. While a microprocessor based system is a general purpose system that may be programmed to do any of the large number of functions, it is capable of doing; microcontrollers are dedicated to one task and run one specific program. This program is stored in ROM and generally does not change.

Figure 27.26 further illustrates the basic difference between a microprocessor based system and a microcontroller. As is evident from the two block schematics shown in the figure; while a microprocessor based system needs additional chips to make it a functional unit; in a microcontroller, functions of all these additional chips are integrated on the same chip.

Q. 23. Outline common areas of application of microcontrollers.

Ans. Microcontrollers are embedded inside a large number of product categories including automobiles, entertainment and consumer products, test and measurement equipment, desk top computers to name some prominent ones. Any device or system that measures, stores, controls, calculates or displays information is sure to have an embedded microcontroller as a part of the device or system. In automobiles, one or more microcontrollers may be used for engine control, car cruise control and anti-lock brakes and so on. Test and measurement equipment such as signal generators, multimeters, frequency counters, oscilloscopes etc. make use of microcontrollers to add features, such as the ability to store measurements, display messages and waveforms and to create and store user routines. In desk top computers, microcontrollers are used in peripheral devices, such as keyboards, printers, modems etc. Consumer and entertainment products, such as TV, video recorders, camcorders, microwave ovens, washing machines, telephones with caller ID facility, cellular phones, air conditioners, refrigerators and many more products make extensive use of microcontrollers to add new control and functional features.

Q. 24. Name the building blocks found inside a microcontroller. Briefly describe each one of these.

Ans. Figure 27.27 shows the block schematic arrangement of various components of a microcontroller. As outlined earlier, a microcontroller is an integrated chip with an on-chip CPU, memory, I/O ports and some peripheral devices to make a complete functional unit. A typical controller as depicted in Figure 27.27 has the following components:

1. Central processing unit (CPU)
2. Random access memory (RAM)
3. Read only memory (ROM)
4. Special function registers
5. Peripheral components including Serial and/or parallel ports, Timers and counters, Analog-to-digital (A/D) converters and Digital-to-analog (D/A) converters

Central Processing Unit processes the program. It executes the instructions stored in the program memory pointed to by the program counter in synchronization with the clock signal. The processor complexity could vary from simple 8-bit processors to sophisticated 32-bit or even 64-bit processors.

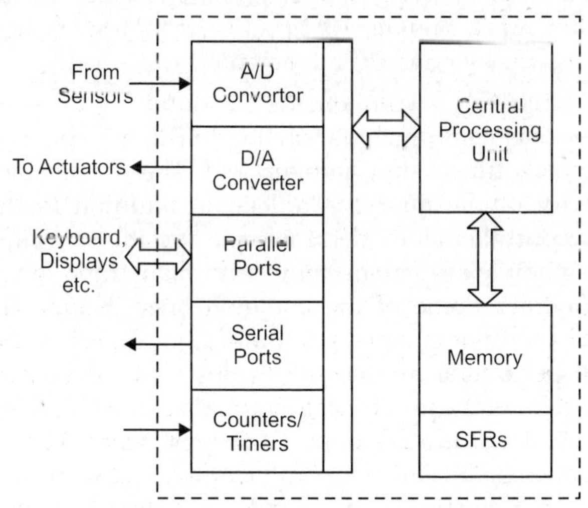

Fig. 27.27

Random Access Memory (RAM) is used to hold intermediate results and other temporary data during the execution of the program. Typically, microcontrollers have few hundreds of bytes of RAM. As an example, microcontroller type numbers 8XC51/80C31, 8XC52/80C32 and 68HC12 respectively have 128 bytes, 256 bytes and 1024 bytes of RAM.

Read Only Memory (ROM) holds the program instructions and the constant data. Microcontrollers use one or more of the following memory types for the purpose. These include ROM (Mask programmed ROM), PROM (one time programmable ROM, which is not field programmable), EPROM (field programmable and usually UV erasable), EEPROM (filed programmable, electrically erasable, byte erasable) and flash (similar to EEPROM technology). Microcontroller type numbers 8XC51, 8XC51FA, 8XC52 have 4K, 8K and 16K of ROM. As another example, 68HC12 16-bit microcontroller has 32 K of flash EEPROM, 768 bytes of EEPROM and 2 K of erase protected boot block.

Special Function Registers control various functions of a microcontroller. There are two categories of these registers. First type includes those registers that are wired into the CPU and do not necessarily form part of addressable memory. These registers are used to control program flow and arithmetic functions. Examples include status register, program counter, stack pointer etc. These registers are, however, taken care of by compilers of high level languages and therefore programmers of high level languages such as C, Pascal etc. do not need to worry about them. The other category of registers is the one that is required by peripheral components. Contents of these registers could for instance set a timer or enable serial communication and so on. As an example, special function registers available on 80C51 family of microcontrollers (80C51, 87C51, 80C31) include program counter, stack pointer, RAM address register, program address register and PC incrementer.

Peripheral components found on most microcontrollers include analog-to-digital converters, I/O ports, timers and counters etc. These available on majority of the microcontrollers. In addition to these, microcontrollers intended for some specific or relatively more complex functions come with many more on-chip peripherals. Some of the common ones include pulse width modulator, serial communication interface (SCI), serial peripheral interface (SPI), inter-integrated circuit (I^2C) two-wire communication interface, RS-232 (UART) port, IrDA (Infrared port), USB port, controller area network (CAN) and local interconnect network (LIN). These peripheral devices are briefly described in the following paragraphs.

Q. 25. Briefly describe the role of analog-to-digital converters on a microcontroller.

Ans. Analog-to-digital and digital-to-analog converters provide interface with the analog devices. For example, analog-to-digital converter provides interface between the microcontroller and the sensors that produce analog electrical equivalents of the actual physical parameters to be controlled. Digital-to-analog converter, on the other hand, provides an interface between the microcontroller and the actuators that provide the control function. As an example, both 68HC11 and 68HC12 from Freescale Semiconductor have 8-channel, 8-bit analog-to-digital converters. Digital-to-analog converter function in microcontrollers is provided by a combination of PWM (Pulse width modulator) followed by a filter. As an example, 68HC12 has an on-chip 16-bit/2-channel pulse width modulator.

Q. 26. What is the function of I/O ports in a microcontroller?

Ans. I/O ports provide interface between the microcontroller and the peripheral I/O devices, such as keyboard, display etc. 80C51 family of microcontrollers have four 8-bit I/O ports. Microcontroller 68HC11 offers 38 general purpose I/O pins including 16 bidirectional I/O pins, 11 input-only pins and 11 output-only pins.

Q. 27. What are the usual functions performed by counter/timer block inside a microcontroller? With the help of suitable block diagrams, briefly explain the operation of a typical timer subsystem of a microcontroller.

Ans. Counters/timers usually perform the following three functions. They are used to keep time and/or measure time interval between events, count the number of events and generate baud rates for the serial ports. Microcontroller 68HC11 has a 16-bit timer system comprising of three input capture channels, four output compare channels and one additional channel that can be configured as either input or output channel. Another popular microcontroller type number, PIC 16F84, has an 8-bit timer/counter with 8-bit prescaler.

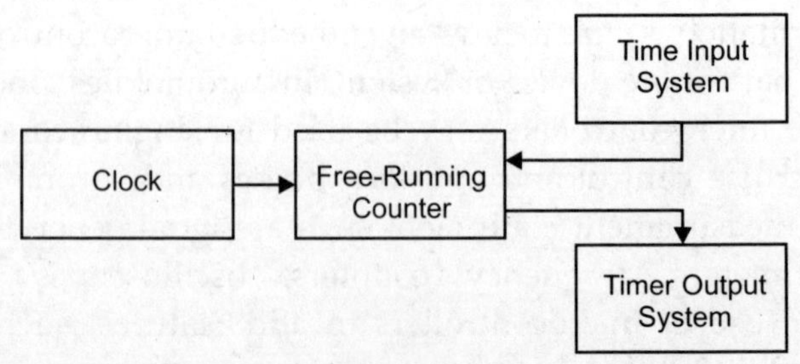

Fig. 27.28

Figure 27.28 shows a generalised block schematic representation of the timer subsystem of a microcontroller. Clock signal controls all timing activities of microcontroller. Counter is used to both capture external timing events (accomplished by timer input block) as well as to generate timing events for external devices (accomplished by timer output block). While the former process is typically used to measure frequency and time interval of periodic signals, the latter generates control signals for external devices.

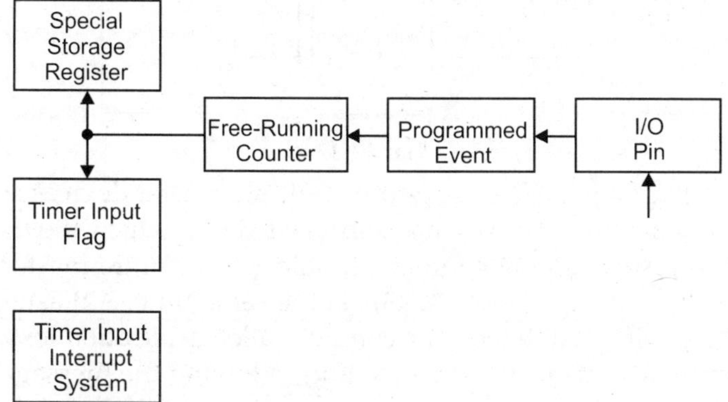

Fig. 27.29

It may be mentioned here that a timing event to be captured or generated is nothing but change of logic status on one of the microcontroller I/O pins configured as an input pin, in case the event is to be captured and an output pin, if it were to be generated. Figure 27.29 shows the block schematic arrangement of the timer input block of Figure 27.28. As shown in the figure, the counter captures the input time event in the form of its contents at the time of occurrence of the event. In fact, the counter captures the relative time of the event as the counter is free running. Absolute timing values can be computed from the relative system clock values. As an example, consider a microcontroller with a 10 MHz clock and a 16-bit counter/timer subsystem. This counter will take 6.5536 ms to count from 0000 to FFFF (hex notation). Let us assume that it is desired to find the frequency of a periodic signal whose successive rising or falling edges are observed to occur at 0010 and 0150. 0010 and 0150 respectively correspond to 16 and 336 in decimal. Therefore, the time interval between the two successive edges equals $320 \times 0.1 = 32$ μs. Signal frequency is therefore (1/32) MHz = 31.25 kHz.

Figure 27.30 shows the block schematic arrangement of the timer output block of Figure 27.28. The diagram is self-explanatory. Again free running counter values can be used to synchronise the time of desired logic state changes on the output pin. This feature can also be used to generate an aperiodic pulse or a periodic signal of any desired duty cycle.

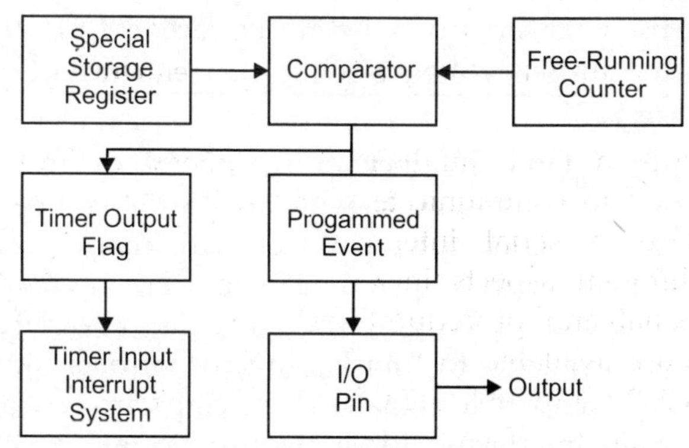

Fig. 27.30

For timer's input and output operations, microcontroller needs to set up some special registers. For timer input operation, as shown in Figure 27.29, registers are required to program the event (logic HIGH or logic LOW), configure the physical I/O pin as an input pin and also set up parameters for the related interrupt, if used. Another register is used to capture the counter value at the time of occurrence of the event. For time output operation, as shown in Figure 27.30, the physical I/O pin is to be configured as an output pin, event is to be programmed and the timing value is to be set in the special register to tell, when the programmed event should appear on the output pin. Output timer system also has an associated interrupt that can be utilized, if needed.

Q. 28. Briefly describe the different types of serial communication interfaces. Name some of the serial communication interfaces common with microcontrollers.

Ans. There are two types of serial communication interfaces, namely asynchronous communication interface and synchronous communication interface. Asynchronous communication interface uses a start and stop bit protocol to synchronise the transmitter and receiver. Start and stop bits are embedded in each data byte. As compared to synchronous communication interface, it offers lower data transmission rates. It is also referred to as the UART (Universal Asynchronous Receiver Transmitter) or the SCI (Serial Communication Interface). Synchronous communication interface uses a synchronized clock to transmit and receive each bit. Synchronization of transmitter and receiver clocks is usually accomplished by using an additional clock line linking the transmitter and the receiver. It is not recommended for long distance communication. It is also referred to as SPI (Serial Peripheral Interface). Microcontroller 68HC11 offers asynchronous non-return-to-zero serial communication interface and also a synchronous serial peripheral interface. 80C51 family of microcontrollers offers a full duplex enhanced UART interface.

Since a large number of peripheral devices are equipped to communicate with an RS-232 compatible interface, a serial interface standard that specifies the different aspects including electrical, mechanical, functional and procedural specifications; a variety of chips are available to translate microcontroller signals to RS-232 compatible signals. These chips are equipped to provide interfacing for a two-way communication system. Other serial communication interfaces common with microcontrollers include I^2C (Inter-Integrated Circuit) bus, CAN (Controller Area Network) bus and LIN (Local Interconnect Network) bus.

Q. 29. Briefly describe Inter-Integrated Circuit (I^2C) bus and its salient features.

Ans. Inter-Integrated Circuit (I^2C) Bus is a two-wire, low-to-medium speed serial communication interface developed by Philips Semiconductors in early 1980s for chip-to-chip communications. The two wires in the I^2C bus are called clock (SCL) and data (SDA). The SDA wire carries data, while SCL wire synchronises the transmitter and receiver during data transfer.

It is a proven industry standard communication protocol used in a variety of electronic products, which is particularly facilitated by its low cost and powerful features. It is supported by a large number of semiconductor and system manufacturers, who offer a variety of electronic products including input and output devices, different types of sensors, memory devices, displays, data entry devices etc. Some of the important features offered by I^2C devices are briefly described in the following paragraphs.

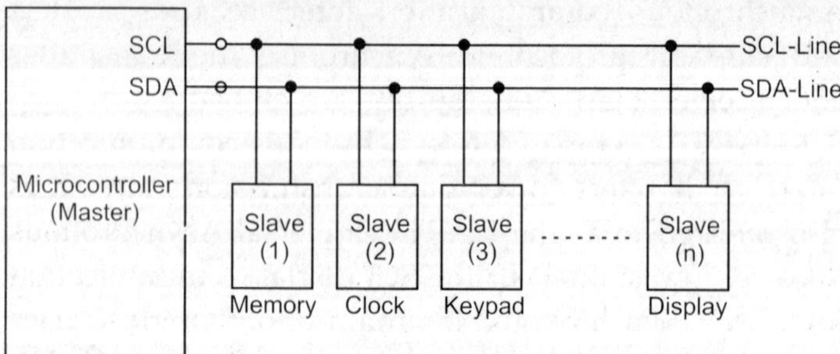

Fig. 27.31

I^2C devices offer master-slave hierarchy. These are classified as either master (device that initiates the message) or slave (device that responds to the message). The device can be either master-only or slave-only or can be switched between master and slave depending upon the application requirement. One possible master-slave configuration is the one, where one master (*e.g.* microcontroller) is connected to many chips connected as slaves as shown in Figure-27.31. Each of the I^2C slave devices is identifiable by a unique address. When the master device sends a message, it includes the address of the intended slave device at the beginning of the message.

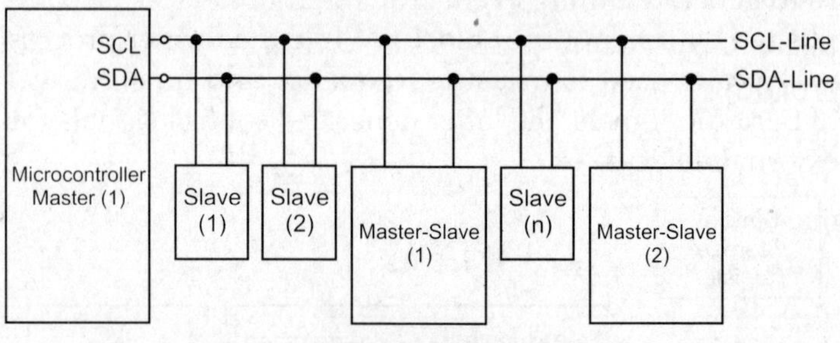

Fig. 27.32

I^2C interface also supports multiple master devices at the same time. The bus has a special feature, which allows it to resolve signal conflicts, should two or more master devices try to talk on the bus at the same time. A master I^2C device that detects the conflict, called arbitration loss, terminates its use of the bus, thus, allowing the message sent by another master to cross the bus unharmed. Figure-27.32 shows one such multi-master support arrangement.

Q. 30. Briefly describe Controller Area Network (CAN) protocol.

Ans. Controller area network (CAN) bus is a rugged serial communication interface used in a broad range of embedded as well as automation control applications. It was introduced by Bosch in 1986 for in-vehicle networks in automobiles. The CAN protocol was internationally standardised in 1993 as ISO-11898-1 and comprises the data link layer of the seven layer ISO/OSI reference model. The protocol provides two communication services namely data frame transmission (sending of a message) and remote transmission request (requesting of a message). All other services, such as error signaling, automatic retransmission of erroneous frames etc. are performed by CAN chips. Some of the important features of CAN protocol include the following. It provides multi-master hierarchy. This allows the user to build intelligent and redundant systems. It uses broadcast communication method. A sender of message transmits to all devices connected to the bus. All devices read the message and decode, if intended for them. This feature guarantees data integrity. Data integrity is also ensured by error detection mechanisms and automatic retransmission of faulty messages. CAN protocol is capable of providing low speed, fault-tolerant transmission at the rate of 125 kbps upto 40 meters, which can function over one wire, if there is a short. Transmission without fault tolerance is provided at the rate of 1 Mbps up to 40 meters. Transmission rates of 50 kbps are achievable upto a distance of 1 km.

Q. 31. Briefly describe Local Interconnect Network (LIN).

Ans. Local Interconnect Network (LIN) bus is a broadcast serial network that is used as a low cost sub-network of a CAN bus to integrate intelligent sensors or actuators in modern automobiles. It comprises of one master (typically a moderately powerful microcontroller) and up to 16 slaves (less powerful, cheaper microcontrollers or ASICs). It does not offer collision detection feature and therefore, all messages are initiated by master with at the most one slave replying to a given message identifier. Multiple such LIN networks may all be linked to a CAN upper layer network through their respective masters.

Q. 32. Briefly describe Von Neumann and Harvard architectures used by microcontroller processor to access memory.

Ans. Von Neumann architecture and Harvard architecture are the two fundamental architectures used by the processing units to access memory. The two architectures are briefly described in the following paragraphs:

Von Neumann Architecture

Von Neumann architecture uses a single memory to hold both program instructions and data. There is one common data and address bus between processor and memory (Figure-27.33). Instructions and data are fetched in sequential order, thus, limiting the operation data transfer rate or the throughput. This phenomenon is commonly referred to as Von Neumann bottleneck. The throughput is very small as compared to the size of the memory. In present day machines, the throughput is also very small as compared to the rate at which the processor itself can work. In the condition, where the processor is required to perform minimal processing on large amounts of data, the processor is forced to wait for vital data to be transferred from or to memory. Microcontroller type number 68HC11 uses Von Neumann architecture.

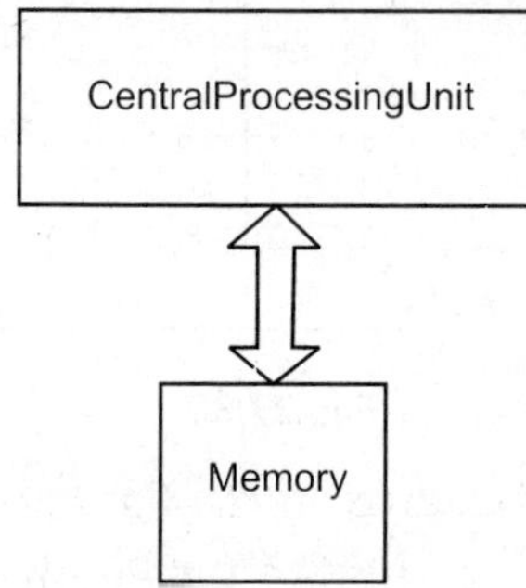

Fig. 27.33

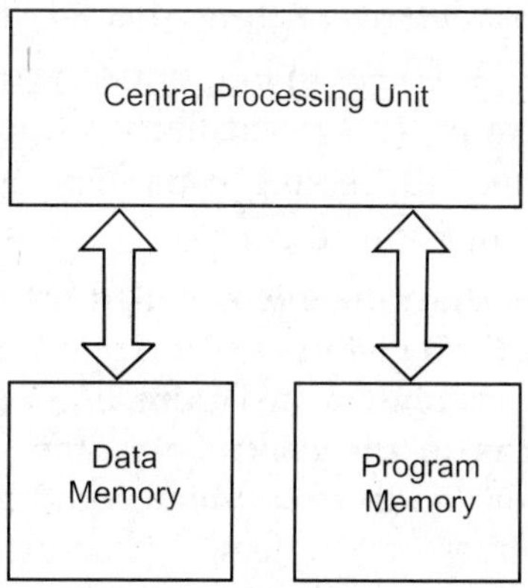

Fig. 27.34

Harvard Architecture

Harvard architecture uses physically separate memories for program instructions and data. It requires therefore separate buses for program and data, as shown in Figure 27.34. In such architecture, instructions and operands can be fetched simultaneously, which makes microcontrollers using this architecture much faster as compared to the ones using Von Neumann architecture. Also, different data and program bus widths are possible; which allows program and data memory to be better optimised to architectural requirements. In fact, the word width, timing, implementation technology and memory address structure can be different in the two cases. Program memory is usually much larger than the data memory, which implies that the address bus for the program memory is wider than the address bus for the data memory.

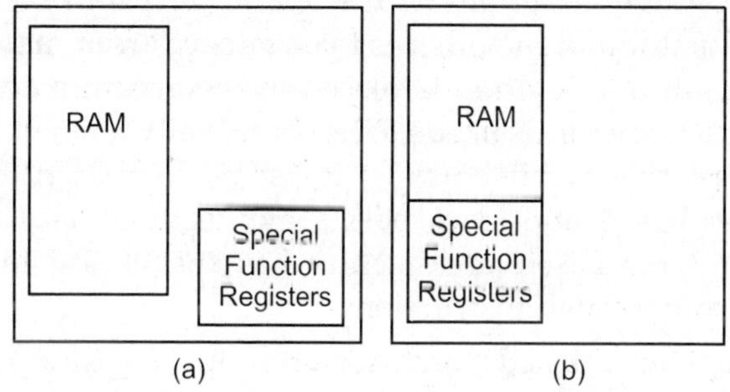

Fig. 27.35

Q. 33 Briefly describe the fundamental architectures used by microcontrollers for mapping special function registers into memory space.

Ans. There are again two fundamental architectures used for mapping special function registers into the memory space. In the first type of arrangement, I/O space and memory space are separated, as shown in Figure 27.35 (*a*). I/O devices have a separate address space, which is accomplished by either an extra I/O pin on the

CPU's physical interface or through a dedicated I/O bus. As a result of this, access to I/O control registers requires special instructions. It is particularly attractive in CPUs, having a limited addressing capability. It is generally found on Intel microprocessors.

In the second arrangement, called memory mapped I/O, I/O control registers are mapped into memory address space, as shown in Figure 27.35 (*b*). Read and write operations to the control registers are done via absolute memory addresses, which could be variables at absolute addresses or pointers to absolute addresses in high level languages. In this case, no special instructions are needed to access I/O control registers. Memory-mapped I/O uses same bus to address both memory and I/O devices. CPU instructions used to read from or write to memory are also used in accessing I/O devices.

Q. 34. Name different architectures used by processors of microcontrollers and briefly describe each one of these.

Ans. Processor architecture can be categorised as one of the following:

1. Accumulator based architecture
2. Register based architecture
3. Stack based architecture
4. Pipeline architecture

Accumulator based architecture: In *Accumulator Based Architecture,* as shown in Figure-27.36, instructions begin and end in accumulators (ACC A and ACC B in Figure-27.36), which are specially designated registers. In a typical operation, one of the operands is found in the accumulator and the other is to be fetched from memory. The result of operation is placed in the accumulator. As one of the operands needs to be continually fetched from memory, this architecture is slower as compared to the register based and stack based architectures. However, accumulator based architecture has the ability to run fairly complicated instructions.

Register based architecture: In *Register Based Architecture,* as shown in Figure-27.37, both operands are stored in registers and the result of operation is also stored in a register. Registers are typically collocated with the processor. Since processor and registers operate at the same speed, this architecture is much faster than the previously discussed accumulator based architecture. The contents of the register are read from and written to memory using background operation.

Stack based architecture: In the *Stack Based Architecture,* both operands as well as to-be-performed operation are stored on the stack, which could be configured around dedicated registers or a special portion of RAM. The result of operation is placed back on the stack. Figure 27.38 shows typical block schematic arrangement of this type of architecture.

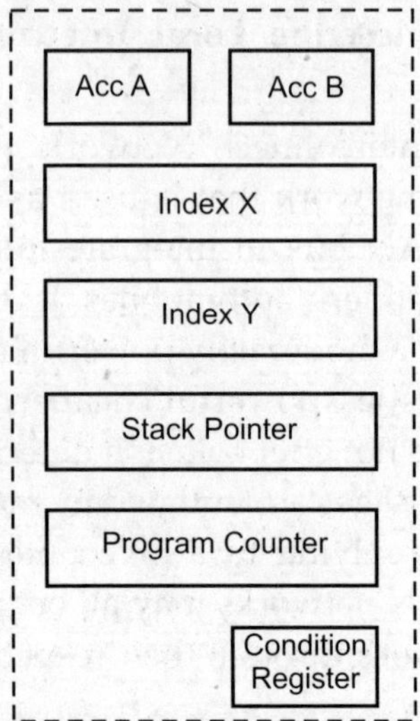

Fig. 27.36

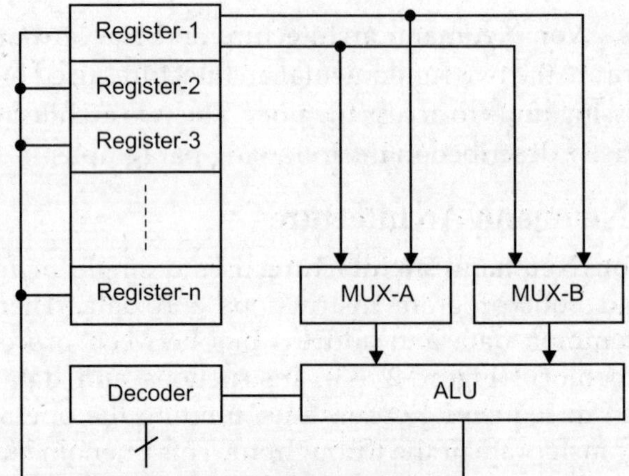

Fig. 27.37

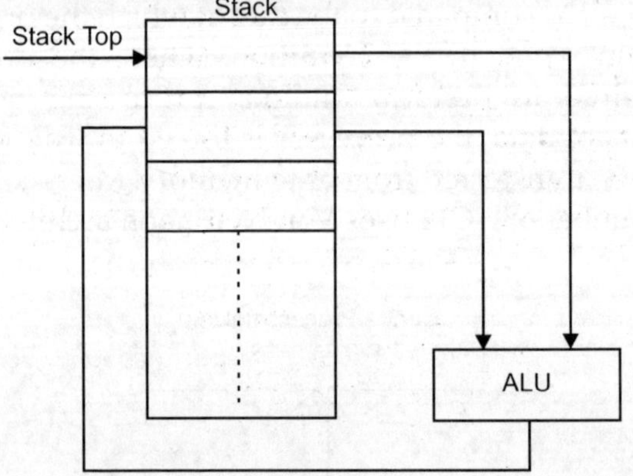

Fig. 27.38

Pipeline architecture: In the *Pipelined Architecture,* as shown in Figure 27.39, there are separate hardware stages for execution of different steps involved in execution of an instruction. These different steps include fetching an instruction from memory, decoding the instruction, fetching instruction operands from memory or registers, executing the instruction and then finally placing result back on the memory. Pipelining allows these stages to overlap and perform with parallelism. Overall result is that execution of an instruction is completed on every clock cycle. Instructions in a pipelined architecture are usually simple instructions that can be implemented within a single stage. These simple instructions act as building blocks for more complex instructions.

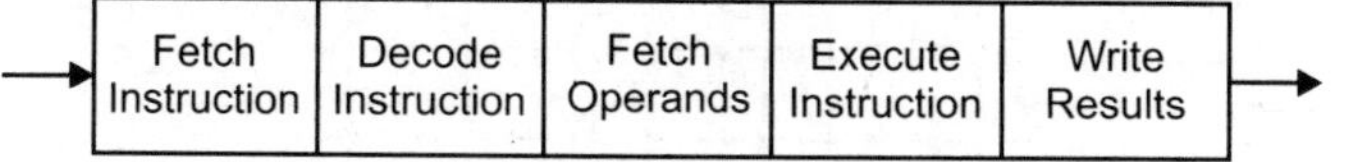

Fig. 27.39

Q.35. What do the microcontrollers use power saving modes for? Briefly describe the power saving modes typical to microcontrollers. Explain with the example of a practical microcontroller chip.

Ans: Power consumption is one of the important issues in battery powered devices. Most microcontrollers come with various power saving features. For a given application requirement, designers use these features to keep the power consumption down to an optimum value without compromising the operational requirements of the device. It may be mentioned here that not all modes are for power saving. Some microcontrollers support in-circuit debugging. Stop Clock Mode, Idle Mode, Power down Mode and LPEP (Low Power EPROM) operation are the power saving modes available in 80C51 family of microcontrollers. These are briefly described in the following paragraphs:

Stop Clock Mode allows the clock oscillator to be stopped or clock speed to be reduced to as low as 0MHz. When the oscillator is stopped, special function registers and RAM retain their values. This mode allows reduced power consumption by lowering the clock frequency to any value.

Idle Mode is another power saving mode available with 80C51 family of microcontrollers. In this mode, the processor puts itself to sleep, while all on-chip peripheral components stay active. The processor contents, the on-chip RAM and all special function registers remain intact during the idle mode. The instruction that invokes this mode is the last instruction executed in the normal operating mode before the idle mode gets activated. Idle mode can be terminated by either any enabled interrupt or by a hardware reset. By an enabled interrupt, the process is picked up at the interrupt service routine and continued. Hardware reset starts the processor in the same manner as it does on a power-on reset.

Power Down Mode is recommended for the lowest power consumption. When this mode is enabled, the oscillator stops and the instruction that invokes the power down mode is the last instruction executed. Special function registers and on-chip RAM retain their values down to V_{CC} amplitude of 2.0 volts. V_{CC} must be brought to minimum specified operating voltage before this mode is deactivated. Either a hardware reset or external interrupt can be used to terminate power down mode. While hardware reset redefines all the special function registers and retains on-chip RAM values, an external interrupt allows both special function registers and the on-chip RAM to retain their values. For proper termination of the power down mode, reset or external interrupt should not be executed, unless V_{CC} is restored to its normal operating level and also has been held active long enough for the oscillator to start and stabilise

Yet another mode available with 80C51 family of microcontrollers that helps in power saving is the *LPEP*. The EPROM array contains some analog circuits that are not required for V_{CC} less than 4.0 volts. This feature can be used to save power by setting the LPEP bit resulting in reduced supply current. This mode should be used only for applications that require V_{CC} of less than 4.0 volts.

SOLVED PROBLEMS

Problem 27.1. *Determine the size of PROM required for implementing the following logic circuits:*

(*a*) *Binary multiplier that multiplies two 4-bit numbers*

(*b*) *Dual 8-to-1 multiplexer with common selection inputs*

(*c*) *Single digit BCD Adder/Subtractor with a control input for selection of operation*

Solution: (*a*) The number of inputs required here would be 8. The result of multiplication would be in eight bits.

Therefore, size of PROM= $2^8 \times 8 = 256 \times 8$

(*b*) Number of inputs = 8+8+3 = 19...... (Number of selection inputs = 3)

Number of outputs = 2

Therefore, size of PROM = $2^{19} \times 2 = 512K \times 2$

(*c*) Number of inputs = 4 (Augend bits) + 4 (Addend bits) + 1 (Carry in) + 1 (Control input) = 10

Number of outputs = 4 (Sum or subtraction output bits) +1 (Carry or Borrow bit) = 5

Size of PROM = $2^{10} \times 5 = 1024 \times 5 = 1K \times 5$

Problem 27.2: *It is desired to find 'Binary Square' of a four-bit binary number using a PROM. Determine the size of PROM required for the purpose.*

Solution: Apparently, size of PROM should be $2^4 \times 8$, that is, 16×8.

This will be true, if none of the output bits is common with any of the input bits and also, if none of the output bits is either all 0's or all 1's. If A_0, A_1, A_2 and A_3 are input bits and B_0, B_1, B_2, B_3, B_4, B_5, B_6 and B_7 are output bits, it can be verified that when binary square is computed; the result reveals $A_0 = B_0$ and $B_1 = 0$, which means that only six of the output bits need be computed by PROM. Therefore, size of PROM = 16×6.

Problem 27.3: *Determine the number of programmable interconnections in the following programmable logic devices:*

(a) *1K × 8 PROM*

(b) *2K × 4 PROM*

(c) *PLA device with eight input variables, 64 AND gates and four OR gates*

Solution: (*a*) Number of OR gates = 8

Number of inputs to each OR-gate = 1024

Since a PROM device has a programmable OR-gate array, number of programmable interconnections = 1024×8 = 8192

(*b*) On similar lines, number of programmable interconnections = 2048 × 4 = 8192

(c) PLA device has programmable AND-gate array at the input and a programmable OR-gate array at the output.

Table 27.1

A	*B*	*Carry-in* (C_i)	*Sum* (*S*)	*Carry-out* (C_o)
0	0	0	0	0
0	0	1	1	0
0	1	0	1	0
0	1	1	0	1
1	0	0	1	0
1	0	1	0	1
1	1	0	0	1
1	1	1	1	1

Number of input variables = 8

Therefore, number of inputs to each AND-gate in the array = 16

Number of AND gates = 64

Therefore, programmable interconnections on the input side = 16 × 64 = 1024

Number of OR gates = 4

Number of inputs to each of the OR gates in the array = 64

Therefore, programmable interconnections on the output side = 64 × 4 = 256

Total number of programmable interconnections = 1024 + 256 = 1280

Problem 27.4. *Show the logic arrangement of both a PROM as well as a PLA required implementing a binary full adder.*

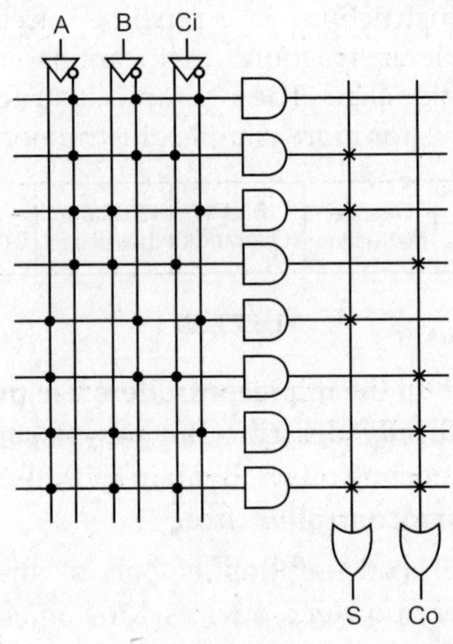

Fig. 27.40

Solution: The truth table of a full adder is given in Table-27.1.

The Boolean expressions for Sum (*S*) and Carry out (C_o) can be written as given in the following two equations.

$S = \Sigma 1, 2, 4, 7$

$C_o = \Sigma 3, 5, 6, 7$

Figure-27.40 shows the implementation with a (8 × 2) PROM.

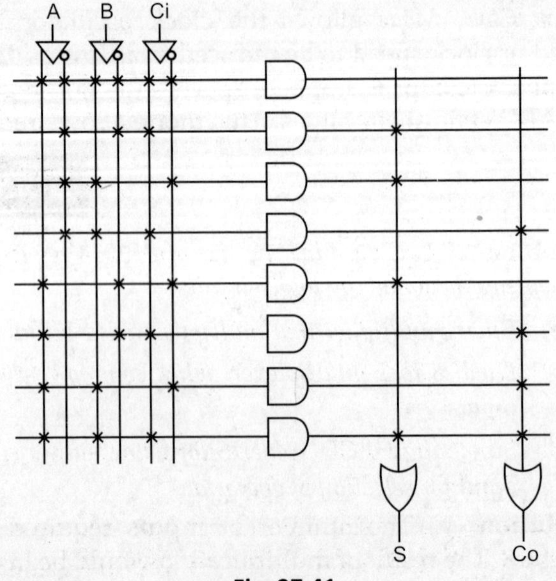

Fig. 27.41

If we simplify the Boolean expressions for sum and carry outputs, we shall find that the expression for sum output cannot be simplified any further and also that the expression for carry out cannot be simplified to three product terms with fewer literals. If we examine even the existing expressions, we find that we would need seven AND gates in the PLA implementation. And if we use simplified expression, even then we would require the same number of AND gates. So, the simplification here would not help, as far as its implementation with a PLA is concerned. Figure-27.41 shows the implementation of full adder with a PLA device.

Problem 27.5. *We have two two-bit binary numbers (A_1A_0) and (B_1B_0). Design PLA device to implement magnitude comparator to produce outputs for (A_1A_0) being 'Equal to', 'Not Equal to', 'Less than' and 'Greater than' (B_1B_0).*

Solution: Table 27.2 shows the function table with inputs and desired outputs. Boolean expressions for the desired outputs are given in the following four equations:

Output-1 (Equal to) = $\overline{A_1}\overline{A_0}\overline{B_1}\overline{B_0} + \overline{A_1}\, A_0\, \overline{B_1}\, B_0 + A_1A_0B_1B_0 + A_1\, \overline{A_0}\, B_1\, \overline{B_0}$

Table 27.2

A_1	A_0	B_1	B_0	Output 1	Output 2	Output 3	Output 4
0	0	0	0	1	0	0	0
0	0	0	1	0	1	1	0
0	0	1	0	0	1	1	0
0	0	1	1	0	1	1	0
0	1	0	0	0	1	0	1
0	1	0	1	1	0	0	0
0	1	1	0	0	1	1	0
0	1	1	1	0	1	1	0
1	0	0	0	0	1	0	1
1	0	0	1	0	1	0	1
1	0	1	0	1	0	0	0
1	0	1	1	0	1	1	0
1	1	0	0	0	1	0	1
1	1	0	1	0	1	0	1
1	1	1	0	0	1	0	1
1	1	1	1	1	0	0	0

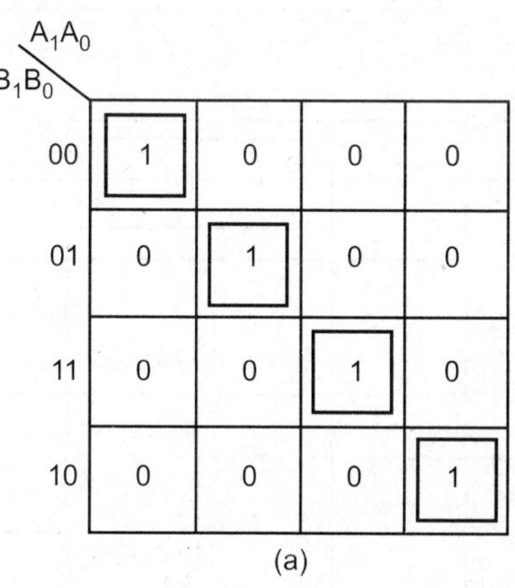

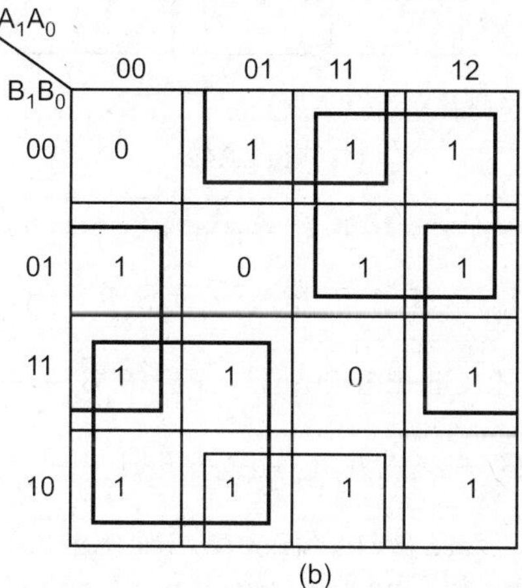

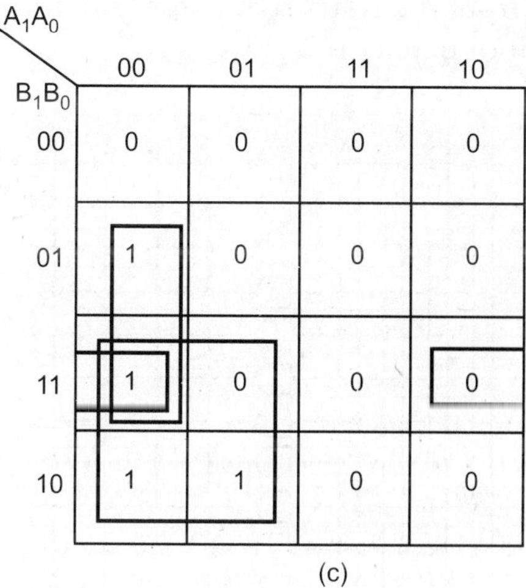

Output-2 (Not equal to) =

$$\overline{A_1}\overline{A_0}\overline{B_1}\overline{B_0} + \overline{A_1}\overline{A_0}\, B_1\, B_0 + \overline{A_1}\overline{A_0}B_1B_0 + A_1A_0\overline{B_1}\overline{B_0} +$$

$$\overline{A_1}A_0B_1\overline{B_0} + \overline{A_1}A_0B_1B_0 + A_1\overline{A_0}\overline{B_1}\overline{B_0} +$$

$$A_1\overline{A_0}\overline{B_1}B_0 + A_1\overline{A_0}B_1B_0 + A_1A_0\overline{B_1}\overline{B_0} + A_1A_0\overline{B_1}B_0 +$$

$$A_1A_0B_1\overline{B_0}$$

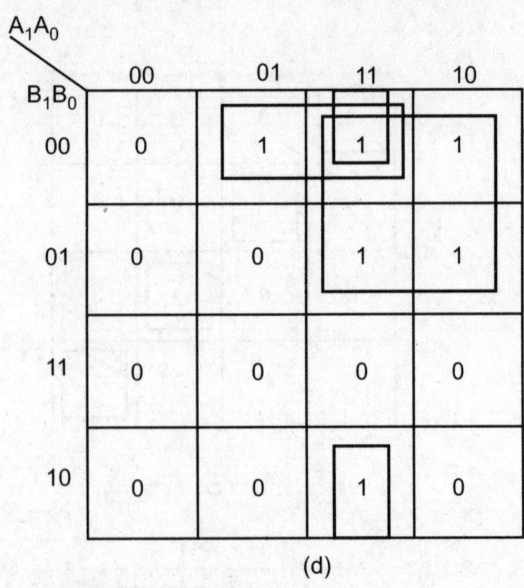

Fig. 27.42

Output-3 (Less than) = $\overline{A_1}\,\overline{A_0}\,\overline{B_1}\,\overline{B_0} + \overline{A_1}\,\overline{A_0}B_1\overline{B_0} +$

$\overline{A_1}\,\overline{A_0}B_1B_0 + \overline{A_1}A_0B_1\overline{B_0} + \overline{A_1}A_0B_1B_0 + A_1\overline{A_0}B_1B_0$

Output-4 (Greater than) = $\overline{A_1}A_0\overline{B_1}\,\overline{B_0} + A_1\overline{A_0}\,\overline{B_1}\,\overline{B_0} +$

$A_1\overline{A_0}\,\overline{B_1}B_0 + A_1A_0\overline{B_1}\,\overline{B_0} + A_1A_0\overline{B_1}B_0 + A_1A_0B_1\overline{B_0}$

Figures 27.42 (*a*) to (*d*) show the Karnaugh maps for the four outputs. The minimised Boolean expressions can be written from the Karnaugh maps and are given in the following four equations:

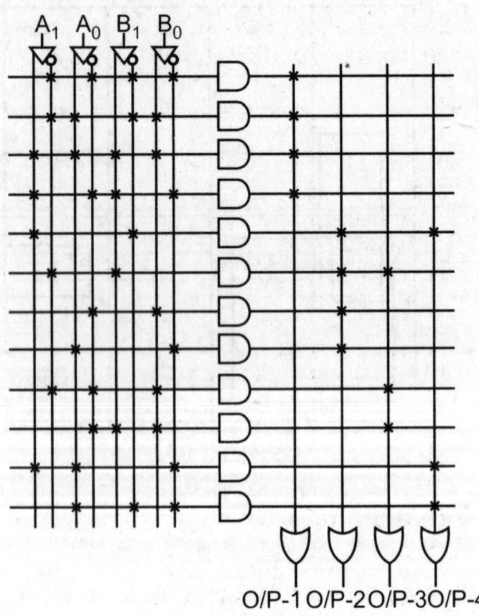

Fig. 27.43

Output-1 (Equal to) = $\overline{A_1}\,\overline{A_0}\,\overline{B_1}\,\overline{B_0} + \overline{A_1}A_0\overline{B_1}B_0$

$+ A_1A_0B_1B_0 + A_1\overline{A_0}B_1\overline{B_0}$

Output-2 (Not Equal to) = $\overline{A_1}B_1 + A_1\overline{B_1} + \overline{A_0}B_0$

$+ A_0\overline{B_0}$

Output-3 (Less than) = $\overline{A_1}B_1 + \overline{A_1}\,\overline{A_0}\,\overline{B_0} + \overline{A_0}B_1B_0$

Output-4 (Greater than) = $A_1\overline{B_1} + A_1A_0B_1 + A_0\overline{B_1}\,\overline{B_0}$

Examination of minimised Boolean expressions reveals that there are 12 different product terms to be accounted for. Therefore, a PLA device with 12 AND gates shall meet the requirement. Also, since there are four outputs; we need to have four OR gates at the output. Figure-27.43 shows the programmed PLA device.

Problem 27.6: *Table 27.3 shows the function table of a BCD to Gary code converter. Starting with the Boolean expressions for the four Gray code outputs, minimize them using Karnaugh maps and then implement in hardware this code converter with a suitable PLD with PAL architecture.*

Table 27.3

A	*B*	*C*	*D*	*P*	*Q*	*R*	*S*
0	0	0	0	0	0	0	0
0	0	0	1	0	0	0	1
0	0	1	0	0	0	1	1
0	0	1	1	0	0	1	0
0	1	0	0	0	1	1	0
0	1	0	1	1	1	1	0
0	1	1	0	1	0	1	0
0	1	1	1	1	0	1	1
1	0	0	0	1	0	0	1
1	0	0	1	1	0	0	0
1	0	1	0	X	X	X	X
1	0	1	1	X	X	X	X
1	1	0	0	X	X	X	X
1	1	0	1	X	X	X	X
1	1	1	0	X	X	X	X
1	1	1	1	X	X	X	X

Solution: From the given function table of BCD to Gray code converter, we can write the Boolean expressions for the four outputs as given in the following four equations:

$$P = \overline{A}.B.\overline{C}.D + \overline{A}.B.C.\overline{D} + \overline{A}.B.C.D + A.\overline{B}.\overline{C}.\overline{D}$$

$$+ A.\overline{B}.\overline{C}.D$$

$$Q = \overline{A}.B.\overline{C}.\overline{D} + \overline{A}.B.\overline{C}.D$$

$$R = \overline{A}.\overline{B}.C.\overline{D} + \overline{A}.\overline{B}.C.D. + \overline{A}.B.\overline{C}.\overline{D} + \overline{A}.B.\overline{C}.D.$$
$$+ \overline{A}.B.C.\overline{D} + \overline{A}.B.C.D$$

$$S = \overline{A}.\overline{B}.\overline{C}.D + \overline{A}.\overline{B}.C.\overline{D} + \overline{A}.B.C.D + A.\overline{B}.\overline{C}.\overline{D}$$

Karnaugh maps for the four outputs (*P*, *Q*, *R* and S) are respectively shown in Figures 27.44 (*a*) to (*d*). The minimised Boolean expressions are given in the following four equations:

$$P = B.D + B.C + A$$

$$Q = B.\overline{C}$$

$$R = B + C$$

$$S = \overline{A}.\overline{B}.\overline{C}.D + B.C.D + A.\overline{D} + \overline{B}.C.\overline{D}$$

Next step is to choose a suitable PAL device. Since there are four output functions; we shall need a PAL device with at least four OR gates at the output. Since each of the OR gates is to be hard wired to only a subset of programmable AND array and also, because one of the output functions has four product terms; we shall need an AND array of 16 AND gates. Since there are four input variables, we need each AND-gate in the array to have eight inputs to cater for four variables and their complements. To sum up, we choose a PAL device that has eight inputs, 16 AND gates in the programmable AND array and four OR gates at the output. Each OR-gate has four inputs.

Figure 27.45 shows the architecture of the programmed PAL device. We can see that P-output has only three product terms. Fourth input to the relevant OR-gate needs to be applied logic '0' input. This is achieved by feeding the inputs of corresponding AND-gate with all four variables and their complements. Logic 0's, wherever required are implemented in the same manner. Note that in the programmed PAL device of Figure 27.45, unprogrammed interconnection indicated by (×) is a 'Make' connection.

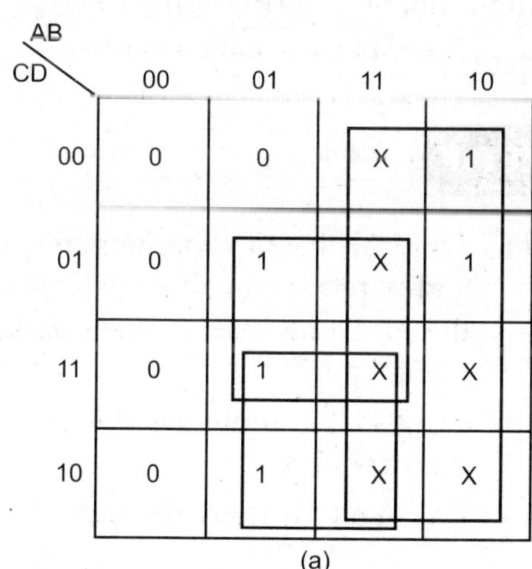

(a)

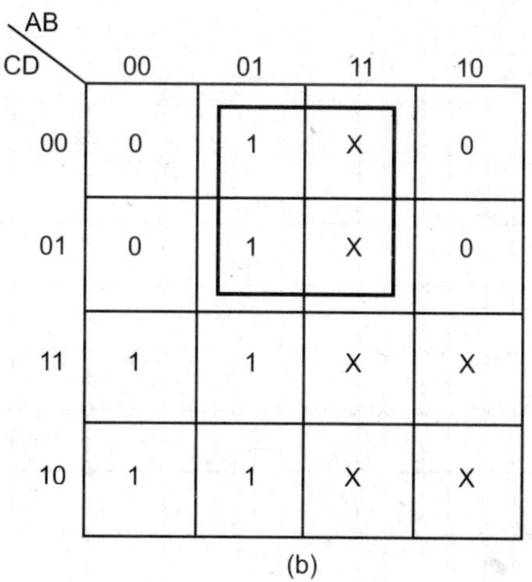

(b)

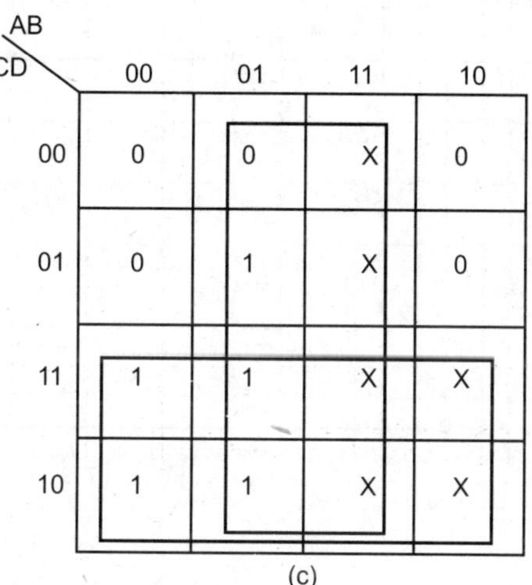

(c)

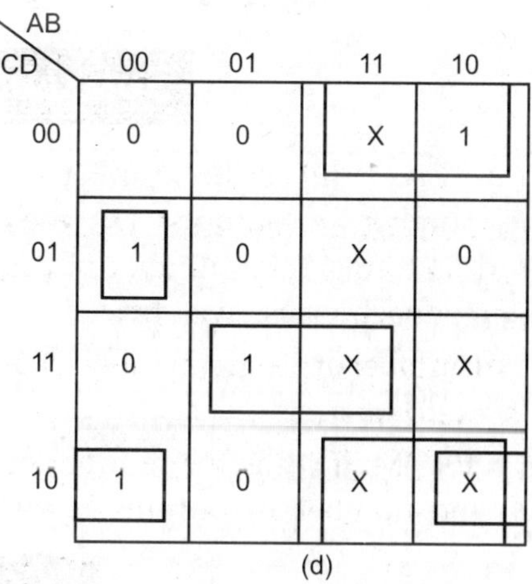

(d)

Fig. 27.44

Problem 27.7. *A certain microcontroller has an on-chip 16-bit counter/timer system. It is used to measure the width of an input pulse. Microcontroller has been programmed to measure the time of occurrence of rising and falling edges of an input pulse on a certain I/O pin. If the microcontroller uses 8 MHz clock and the count values observed at the time of occurrence of rising and falling edges of the input pulse are 001F and 00F1 (in hex), determine the pulse width as measured by the microcontroller.*

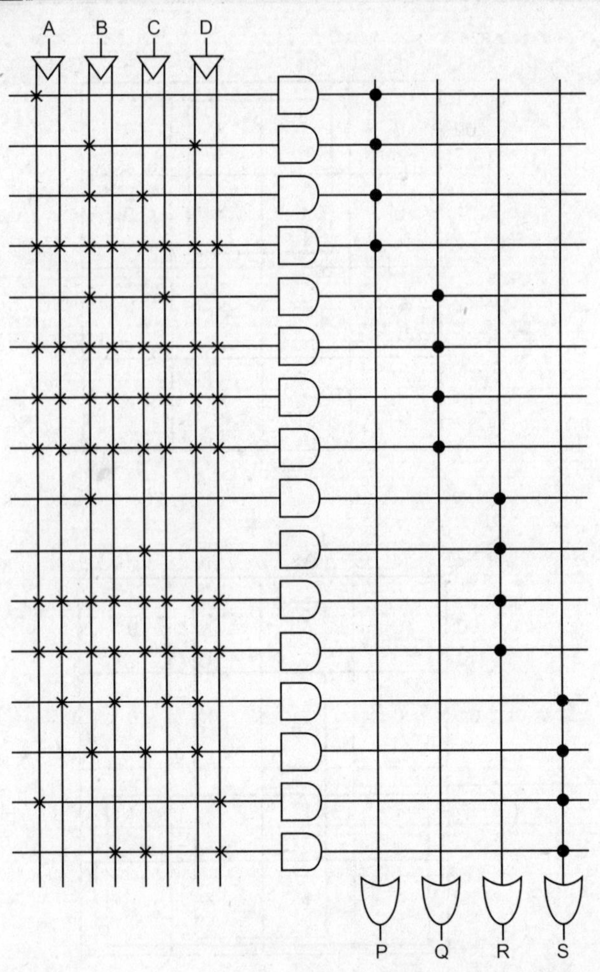

Fig. 27.45

Solution: Since the microcontroller uses a 16-bit counter, it counts from 0000 to FFFF (in hex) or 0 to 65536 in decimal. The rising edge of the input pulse occurs at 001F, whose decimal equivalent is 31. The falling edge occurs at 00F1, whose decimal equivalent is 241.

Therefore, the input pulse width accounts for (241–31) = 210 clock cycles.

Clock signal time period = 1/8 = 0.125 μs

Therefore, time period corresponding to 210 cycles = 210×0.125 = 26.25 μs

Pulse width measured by microcontroller = 26.25 μs

Problem 27.8. *It is desired to design a microcontroller based periodic signal generator with minimum and maximum time period specifications of 125 ns and 100 ms. What should be the system clock frequency?*

Solution: Minimum time period that can be generated by the microcontroller equals the time period corresponding to one clock cycle.

Therefore, one clock cycle time period = 125 ns

Clock frequency = (1/125) GHz = (1000/125) MHz = 8 MHz

Summary of Important Formulas

1. In a PLA, if (n) is the number of inputs; (m) the number of outputs; (K) the number of product terms; total number of programmable interconnections is given by:
   Total number of programmable interconnections $= K \times (2n + m)$
2. In a PROM, if (n) is the number of inputs and (m) the number of outputs; total number of programmable interconnections is given by:
   Total number of programmable interconnections $= 2^n \times m$
3. Minimum time period that can be generated by a microcontroller is given by:
   Minimum time period (μs) = $1/f_{CLOCK}$
   f_{CLOCK} = Clock frequency in MHz

Self Evaluation Exercise

Multichoice Questions

1. The difference between a PLA and a PAL is:
   (*a*) The PLA has a programmable OR plane and a programmable AND plane, while the PAL only has a programmable AND plane.
   (*b*) The PAL has a programmable OR plane and a programmable AND plane, while the PLA only has a programmable AND plane.
   (*c*) The PAL has more possible product terms than the PLA.
   (*d*) PALs and PLAs are the same thing.

2. PALs tend to execute ________ logic.
   (a) SAP (b) SOP
   (c) PLA (d) SPD
3. A GAL is essentially a ________.
   (a) Non-reprogrammable PAL
   (b) PAL that is programmed only by the manufacturer
   (c) Very large PAL
   (d) Reprogrammable PAL
4. What is an OTP device?
   (a) Optical transporting port
   (b) Octal transmitting pixel
   (c) Operational topical portable
   (d) One-time programmable
5. .... are used at the inputs of PAL/GAL devices in order to prevent input loading from a large number of AND gates.
   (a) Simplified AND gates
   (b) Fuses
   (c) Buffers
   (d) Latches
6. A programmable logic device .... consists of a programmable array of AND gates that connects to a fixed array of OR gates and is usually One Time Programmable (OTP).
   (a) GAL (b) CPLD
   (c) PAL (d) SPLD
7. SPLDs, CPLDs, and FPGAs are all which type of device?
   (a) PAL (b) PLD
   (c) EPROM (d) SRAM
8. A (n) .... is a section of embedded logic that is commonly found in FPGAs.
   (a) LUT (b) Core
   (c) DSP (d) PI
9. ASIC stands for:
   (a) Advanced speed integrated circuit.
   (b) Advanced standard integrated circuit.
   (c) Application specific integrated circuit.
   (d) Application speedy integrated circuit.
10. Most look-up tables in field-programmable gate arrays (FGPAs) use ________ inputs, resulting in .... possible outputs.
    (a) 4,16 (b) 8,16
    (c) 4,12 (d) 6,12
11. CLB is the acronym for ....
    (a) Configurable Logic Block
    (b) Configurable Logic Buffer
    (c) Critical Logic Buffer
    (d) Constant Logic Buffer
12. A circuit that implements a combinational logic function by storing a list of output values that correspond to all possible input combinations is ....
    (a) Output logic macrocell
    (b) Look-up table
    (c) Parallel logic expander
    (d) Logic element
13. By adding an OR gate to a simple programmable logic device (SPLD) the foundation for a (n) .... is made possible.
    (a) PAL (b) PLA
    (c) CPLD (d) EEPROM
14. PIA is an acronym for ....
    (a) Programmable Interface Array
    (b) Post Integrated Array
    (c) Programmable Input Array
    (d) Programmable Interconnect Array
15. Field-programmable gate arrays (FGPAs) use ....memory technology, which is ....
    (a) DRAM, non-volatile
    (b) SRAM, non-volatile
    (c) SRAM, volatile
    (d) RAM, volatile
16. The macrocells in a PAL/GAL are located ....
    (a) After the programmable AND arrays
    (b) Ahead of the programmable AND arrays
    (c) At the input terminals
    (d) At the output terminals
17. What does a dot mean when placed on a PLD circuit diagram?
    (a) A point that is programmable
    (b) A point that cannot change
    (c) An intersection of logic blocks
    (d) An input or output point
18. What gives a GAL its flexibility?
    (a) Its speed
    (b) Its reprogrammable EPROM

(c) Its large logic arrays
(d) Its programmable OLMCs

19. What programmable technology is used in FPGA devices?
(a) SRAM (b) FLASH
(c) Anti-fuse (d) All of the above

20. A macrocell basically contains ....
(a) A programmable AND-OR gate array and some input buffers
(b) An OR-gate array and some output logic
(c) An AND-OR gate array and some output logic
(d) Licensed programming

21. Which of the following statements are true?
(a) Both VHDL and Verilog are used as language to develop code targeted for FPGA device
(b) Syntax in Verilog programming is like C language
(c) Both (a) and (b)
(d) None of the above

22. Which of the following statements are FALSE?
(a) Verilog and VHDL both are concurrent languages.
(b) Any hardware descriptive language is concurrent in nature.
(c) None of the above
(d) Both (a) and (b)

23. A programmable interconnect device that has high initial resistance and can be permanently programmed to a highly conducting path is known as
(a) Fuse
(b) Floating gate transistor
(c) Anti-fuse
(d) None of these

24. The most widely used hardware description language for describing digital designs on CPLD and FPGA devices is
(a) VHDL
(b) Java-HDL
(c) Verilog
(d) ABEL

25. A PAL device is marked PAL-16L8. It implies the following.
(a) It has 16 active LOW inputs
(b) It has 8 active LOW outputs
(c) It has 16 inputs and 8 active LOW outputs
(d) It has 8 inputs and 16 active LOW outputs

26. A masked programmed version of a PAL device that is pin compatible to its PAL counterpart is known as
(a) PLA device
(b) HAL device
(c) FPGA device
(d) None of these

27. A PLA like architecture that can implement a full adder should at least have
(a) Eight 6-input programmable AND gates and two-8-input programmable OR gates
(b) Seven 6-input programmable AND gates and two 4-input programmable OR gates
(c) Eight 6-input programmable AND gates and two 4-input programmable OR gates
(d) Sixteen 3-input programmable AND gates and two 8-input programmable OR gates

28. The total number of programmable interconnections in a PLA device with (n) inputs, (m) outputs and (K) product terms will be
(a) $K \times (2n + m)$ (b) $K \times (n + m)$
(c) $n \times (K + m)$ (d) $m \times (K + n)$

29. The total number of programmable interconnections in a PROM device with (n) input and (m) output lines will be
(a) $2^m \times n$ (b) $2^n \times m$
(c) $m \times n$ (d) None of these

30. Identify the synchronous communication interface.
(a) UART
(b) Inter-Integrated Circuit (I^2C) Bus
(c) Serial Peripheral Interface (SPI)
(d) Controller Area Network (CAN)

31. Identify the asynchronous communication interface.
(a) UART
(b) Inter-Integrated Circuit (I^2C) Bus
(c) Controller Area Network (CAN)
(d) All of these

32. Minimum time period that a microcontroller can measure depends upon
(*a*) Clock frequency
(*b*) Size of counter/timer
(*c*) Both (*a*) and (*b*)
(*d*) None of these

33. The processor architecture that uses same memory to store program instructions and data is
(*a*) Von Neumann architecture
(*b*) Harvard architecture
(*c*) Accumulator based architecture
(*d*) None of these

34. Identify the peripheral function that is provided in microcontrollers by a pulse width modulator followed by a filter.
(*a*) A/D converter
(*b*) D/A converter
(*c*) Timer
(*d*) Counter

35. A PAL is numbered PAL-16R8. Identify the correct statement.
(*a*) It has 8 registered inputs and 16 active LOW outputs.
(*b*) It has 8 inputs and 16 registered outputs
(*c*) It has 16 active LOW inputs and 8 outputs.
(*d*) It has 16 inputs and 8 registered outputs

ANSWERS TO SELF EVALUATION EXERCISE

MULTICHOICE QUESTIONS

1. (*a*)	2. (*b*)	3. (*d*)	4. (*d*)	5. (*c*)
6. (*c*)	7. (*b*)	8. (*b*)	9. (*c*)	10. (*a*)
11. (*a*)	12. (*b*)	13. (*a*)	14. (*d*)	15. (*c*)
16. (*a*)	17. (*b*)	18. (*d*)	19. (*d*)	20. (*b*)
21. (*c*)	22. (*d*)	23. (*c*)	24. (*a*)	25. (*c*)
26. (*b*)	27. (*b*)	28. (*a*)	29. (*b*)	30. (*c*)
31. (*d*)	32. (*a*)	33. (*a*)	34. (*b*)	35. (*d*)

Chapter 28

Feedback and Control Systems

Learning Objectives

After completing the chapter, you will learn the following topics:

1. Open loop and closed loop control systems, Linear and Non-linear control systems, Continuous time and discrete time control systems, Time varying and time invariant control systems and Causal systems
2. Free and forced response, Steady state, transient and total response and Unit, step and impulse response
3. Second order control system
4. Transfer function
5. Concept of stability and relative stability
6. Stability criterion including Routh stability criterion, Hurwitz stability criterion and continued fraction stability criterion
7. Block diagram algebra including block diagram reduction techniques
8. Canonical form of representation of feedback control system
9. Signal flow graphs including signal flow graph terminology, computation of transfer function
10. Feedback control system classification including Type 'l' systems
11. Error constants including position, velocity and acceleration error constants
12. Steady state errors corresponding to unit step, unit ramp and unit parabolic input functions
13. Sensitivity parameters
14. Control system controllers including Proportional (P), Proportional-Integral (PI) and Proportional-Integral-Differential (PID) controllers
15. Control system compensators including Lead, Lag and Lag-Lead compensators
16. Root locus analysis including angle and magnitude criteria; construction of root locus; gain and phase margin; determining damping ration and determining closed loop transfer function
17. Nyquist analysis including introduction to polar and Nyquist stability plots, Nyquist stability criterion, gain and phase margin and gain factor compensation
18. Bode analysis including magnitude and phase plots, computation of gain and phase margin
19. State variable analysis including state equation representation of LTI systems, state transition matrix and its properties, controllability and observability of linear systems and Eigen values

Q. 1. What do you understand by a control system ? Define the inputs and outputs of a control system.

Ans. The word CONTROL usually implies an action that directs, commands or regulates. A CONTROL SYSTEM may, therefore, be considered to be an arrangement of physical components related to each other in such a manner as to direct, regulate or command itself or another system. A regulated power supply, where the output voltage is held constant irrespective of the changes in the load current or line voltage is an example of a control system.

The excitation applied to a control system from an external source is the INPUT. The input is applied usually to produce a specified response from the control system. For example, in a control system represented by a thermostatically controlled heater, the input to the system is a reference temperature usually specified by appropriately setting a thermostat. The output, which is the actual temperature, in this case is regulated at a temperature specified by the input.

The OUTPUT is the actual response obtained from the control system. It is not necessarily equal to the specified response implied by the input. Control systems can have more than one inputs or outputs.

Q. 2. Distinguish between an OPEN LOOP and a CLOSED LOOP control system.

Ans. In an OPEN LOOP CONTROL SYSTEM, the control action is independent of the output. Their ability to perform accurately is determined by their calibration. An automatic toaster that is controlled by a timer is an example of an open loop control system. Once having set the time, the control over the quality of the toast, which is the output in this case, is lost. An automatic washing machine and a voltmeter are also examples of open loop control systems. Open loop control systems do not usually have instability problems.

In a CLOSED LOOP CONTROL SYSTEM, also known as a FEEDBACK CONTROL SYSTEM, the control action in some way or the other is dependent upon the output. The feedback control systems are characterised by increased accuracy, reduced non-linearity and distortion and enhanced bandwidth.

Thermostatically controlled heaters, ovens, furnaces, refrigerators are all examples of closed loop (or feedback) control systems. A regulated power supply, a phase locked loop, a common emitter amplifier with unbypassed emitter resistor are examples of electronic circuits that too represent a closed loop control system.

Q.3. Distinguish between a REGULATING TYPE control system and a SERVOMECHANISM.

Ans. A regulating type control system is the one in which the reference input is constant for long periods of time. It could be constant even for the entire operational life of the system. A regulated power supply is a typical example.

A Servomechanism is a special type of feedback control system where the controlled variable is a mechanical position or a time derivative of position, such as velocity or acceleration.

A regulating type feedback control system differs from a servomechanism in that the primary function of the former is usually to maintain a constant controlled output while that of the letter is to make the output often follow a varying input.

Q. 4. Distinguish between a PARTIAL DIFFERENTIAL EQUATION and an ORDINARY (or TOTAL) DIFFERENTIAL EQUATION.

Ans. A partial differential equation is an equality that involves one or more than one dependent variables, two or more independent variables together with the partial derivatives of the dependent variables with respect to independent variables.

An **ordinary differential equation** is an equality that involves one or more than one dependent variables, one independent variable and one or more derivatives of the dependent variables with respect to the independent variable.

$\frac{dy}{dt} + \frac{dx}{dt} + x + y = 0$ where $x = x\ (t)$ and $y = y(t)$ is an ordinary differential equation with (x) and (y) as dependent variables and (t) as the independent variable.

The diffusion equation $\frac{\partial T}{\partial x} = K\frac{\partial T}{\partial t}$ is a partial differential equation where $T = T(x,\ t)$ is the dependent variable and (x), (t) are the two independent variables.

Q. 5. Distinguish between a TIME VARIABLE DIFFERENTIAL EQUATION and a TIME INVARIANT DIFFERENTIAL EQUATION.

Ans. In a **time variable differential equation,** one or more terms depend explicitly on the independent variable time (t). For example, the differential equation $\frac{t\,dy}{dt} + y = x$ is a time variable differential equation because the term $\left(\frac{t\,dy}{dt}\right)$ explicitly depends upon the independent variable (t).

In a **time invariant differential equation**, none of the terms depends explicitly on the independent variable time (t). Any differential equation of the form

$$\sum_{i=0}^{n} a_i \frac{d^i y}{dt^i} = \sum_{i=0}^{m} b_i \frac{d^i x}{dt^i}$$

where, $a_0, a_1, a_2 \ldots\ldots a_n, b_0, b_1, b_2 \ldots. b_m$ are constants, is time-invariant as the above equation depends implicitly on (t) through the dependent variables (x) and (y) and their derivatives. The control systems that are represented by time-invariant differential equations do not vary with time.

Q. 6. Distinguish between LINEAR and NON-LINEAR differential equations.

Ans. A linear differential equation comprises of linear terms only. A linear term is the one which is first degree in dependent variables and their derivatives.

The differential equation $\frac{d^2y}{dt^2} + \frac{dy}{dt} + y = x$ is a linear differential equation as the terms $\left(\frac{d^2y}{dt^2}\right), \left(\frac{dy}{dt}\right)$, y and x are all first degree terms.

Differential equations with one or more higher degree terms are non-linear differential equations.

The differential equation, $\frac{d^2y}{dt^2} + x\frac{dy}{dt} + y = xy$ is a non-linear one, as the terms $\left(\frac{x\,dy}{dt}\right)$ and (xy) are second degree.

Q. 7. Distinguish between linear and non-linear control systems. Do linear control systems exist in practice?

Ans. A *linear control system* is the one in which the principle of superposition applies. This happens, when magnitudes of signals in a control system are such that the components of the system exhibit linear characteristics. According to the superposition principle, when applied to a linear control system, the system response due to more than one inputs acting simultaneously is equal to sum of the responses due to various inputs acting alone. If $y(t)$ is the system response due to inputs $x_1(t)$, $x_2(t)$, $x_3(t) \ldots. x_n(t)$ acting simultaneously, then $y(t) = y_1(t) + y_2(t) + y_3(t) \ldots.. y_n(t)$

Where, $y_1(t)$ = system response due to $x_1(t)$ alone

$y_2(t)$ = system response due to $x_2(t)$ alone

$y_3(t)$ = system response due to x_3 (t) alone

:

$y_n(t)$ = system response due to x_n (t) alone

In the case of a linear control system, $y(ax) = a\, y\, (x)$

It may also be mentioned here that the concepts of linearity and superposition are equivalent and any system which satisfies the principle of superposition is linear. The above properties don't apply to a non-linear control system, where the magnitudes of signals force the components of the control system operate outside the range of linear operation. It may be mentioned here that linear control systems don't exist in practice. All physical systems are non-linear to some extent.

A linear control system is represented by a linear differential equation. A linear differential equation comprises of linear terms only. A linear term is the one which is first degree in dependent variables and their derivatives. The differential equation $\frac{d^2y}{dt^2} + \frac{dy}{dt} + y = x$ is a linear differential equation as the terms $\left(\frac{d^2y}{dt^2}\right), \left(\frac{dy}{dt}\right)$, y and x are all first degree terms.

A non-linear control system is represented by a nonlinear differential equation. Differential equations with one or higher degree terms are non-linear differential equations. The differential equation $\frac{d^2y}{dt^2} + x\frac{dy}{dt} + y = xy$ is a non-linear one, as the terms $\left(\frac{x\,dy}{dt}\right)$ and (xy) are second degree.

Q. 8. Distinguish between continuous time control system and discrete time variant control system.

Ans. In the case of a continuous time control system, all system variables are dependent on a continuously variable time variable. On the other hand, in the case of discrete time variant control systems, one or more variables are known only at certain discrete intervals of time.

Q. 9. Distinguish between time varying control system and time invariant control system.

Ans. In the case of a **time varying control system,** irrespective of the fact whether input and output are functions of time or not, system parameters vary with time. In the case of a time invariant control system, control system parameters are stationary with respect to time during the operation of the system. Practical systems are seldom ideally time invariant, as most physical systems have elements that drift with time. For example, control system of a guided missile, where mass of missile decreases with time with the consumption of fuel and a DC motor in which winding resistance increases, as it heats up with time are examples of time varying control systems. A time varying control system is represented by a time variable differential equation. In a time variable differential equation, one or more terms depend explicitly on the independent variable time (t). For example, the differential equation $\frac{t\,dy}{dt} + y = x$ is a time variable differential equation because the term $\left(\frac{t\,dy}{dt}\right)$ explicitly depends upon the independent variable (t).

A ***time invariant control system*** is represented by a time invariant differential equation. In a *time invariant differential equation,* none of the terms depends explicitly on the independent variable time (t). Any differential equation of the form,

$$\sum_{i=0}^{n} a_i \frac{d^i y}{dt^i} = \sum_{i=0}^{m} b_i \frac{d^i x}{dt^i}$$

$a_0, a_1, a_2, \ldots, a_n, b_0, b_1, b_2, \ldots,$ are constants, is time-invariant as the above equation depends implicitly on (t) through the dependent variables (x) and (y) and their derivatives. The control systems that are represented by time-invariant differential equations do not vary with time.

Q. 10. Under what conditions can a system be called CAUSAL ?

Ans. Causality is relevant only in case of those control systems where time is the independent variable. A system in which time is the independent variable is called a causal system if its output depends only on the past and present values of the inputs. In other words, a causal system can not anticipate its future inputs.

Q. 11. What are characteristic equation and characteristic polynomial with reference to a linear constant co-efficient differential equation?

Ans. In case of a linear constant co-efficient equation given by

$$\frac{d^n y}{dt^n} + a_{n-1}\frac{d^{n-1} y}{dt^{n-1}} + a_{n-2}\frac{d^{n-2}}{dt^{n-2}} + \ldots\ldots + a_1\frac{dy}{dt} + a_0 y = x$$

the characteristic equation is given by

$$D^n + a_{n-1}D^{n-1} + a_{n-2}D^{n-2} + \ldots\ldots + a_1 D + a_0 = 0$$

where, $D \equiv \frac{d}{dt}$ or $D^n \equiv \frac{d^n}{dt^n}$

and $D^n + a_{n-1}D^{n-1} + a_{n-2}D^{n-2} + \ldots\ldots + a_1 D + a_0$ is the characteristic polynomial.

For example, the characteristic equation and characteristic polynomial of a linear, constant co-efficient differential equation given by,

$$\frac{3\,d^2 y}{dt^2} + \frac{5\,dy}{dt} + 11y = x$$

are respectively $3D^2 + 5D + 11 = 0$ and $(3D^2 + 5D + 11)$.

Q. 12. With reference to control systems, briefly describe the following:

1. Free response 2. Forced response
3. Steady state response 4. Transient response
5. Total response

Ans. The FREE RESPONSE of a control system is its response, when the input to the system is zero. It is the solution of the differential equation representing the control system when the input is zero. The response of such a system then depends only on the initial conditions.

The FORCED RESPONSE of a control system is the solution of the differential equation representing the control system, when all the initial conditions are zero. The forced response thus depends only on the input $x(t)$.

The STEADY-STATE RESPONSE is that part of the total response which does not approach zero, as the time (t) approaches infinity.

The TRANSIENT RESPONSE is that part of total response which approaches zero, as the time (t) approaches infinity.

The TOTAL RESPONSE is equal to the sum of free response and forced response. It is also equal to the sum of steady-state response and transient response.

Q. 13. Define the UNIT STEP, UNIT RAMP and UNIT IMPULSE response of a control system.

Ans. The UNIT STEP RESPONSE of a control system is its output $y(t)$, when the input to the system is a unit step function $u(t)$ and all initial conditions are zero. Fig. 28.1 (*a*) shows a unit step function.

The UNIT RAMP RESPONSE of a control system is its output $y(t)$, when the input to the system is a unit ramp function $r(t)$ and all initial conditions are zero. Fig. 28.1 (*b*) shows a unit ramp function.

The UNIT IMPULSE RESPONSE of a control system is its output $y(t)$, when the input to the system is a unit impulse and all initial conditions are zero. Fig 28.1 (*c*) shows an impulse function.

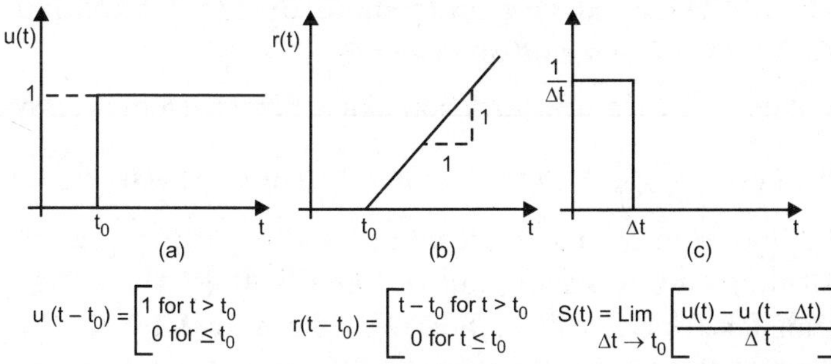

Fig. 28.1

Q. 14. The linear constant co-efficient second order differential equation of the form $\frac{d^2y}{dt^2} + 2\zeta w_n \frac{dy}{dt} + \omega_n^2 y = \omega_n^2 x$ is very important from the viewpoint of analysing a given control system performance as even the higher order control systems can often be approximated by second order systems. What do (ζ), (ω_n) stand for in the above equation? How can we deduced the values of damping co-efficient α, the system time constant (τ) and the damped natural frequency (ω_d) from the above equation?

Ans. (ζ) in the above equation is called the DAMPING RATIO and (ω_n) is called the UNDAMPED NATURAL FREQUENCY. The product ($\zeta\,\omega_n$) is the DAMPING CO-EFFICIENT (α). The system time constant (τ) is nothing but the reciprocal damping co-efficient (α). The DAMPED NATURAL FREQUENCY (ω_d) can be determined from $\omega_d = \omega_n\sqrt{1-\zeta^2}$.

Q. 15. Plot the unit step response of a second order control system represented by a generalised second order differential equation of the type mentioned in the previous question for different values of (ζ). Make a qualitative assessment of the family of curves.

Ans. The family of unit step responses of a generalised second order control system for different values of damping ratio (ζ) are shown in Fig. 28.2. As we can see from the family of plots, the response is a damped sinusoid before it comes to the steadystate for (ζ) less than 1. We can also notice that the frequency of this damped sinusoid decreases with increase in the value of (ζ) and is zero for $\zeta = 1$. Such an observation is evident from the relationship

$$\omega_d = \omega_n\sqrt{1-\zeta^2}$$

(ω_d) equals (ω_n) for $\zeta = 0$. As the system time constant (τ) decreases with increase in the value of (ζ), the sinusoid damps faster as expected. The output settles to within about 1.8% of its steadystate value in a time duration equal to four time constants. Thus, the system settles faster for decreasing values of time constant (or increasing value of damping ratio, ζ). The settling time increases with decrease in the value of (ζ).

Another important characteristic of this unit step response is the time (which represents the rise time) and the magnitude (which represents the overshoot) of the first peak of the response. These parameters can

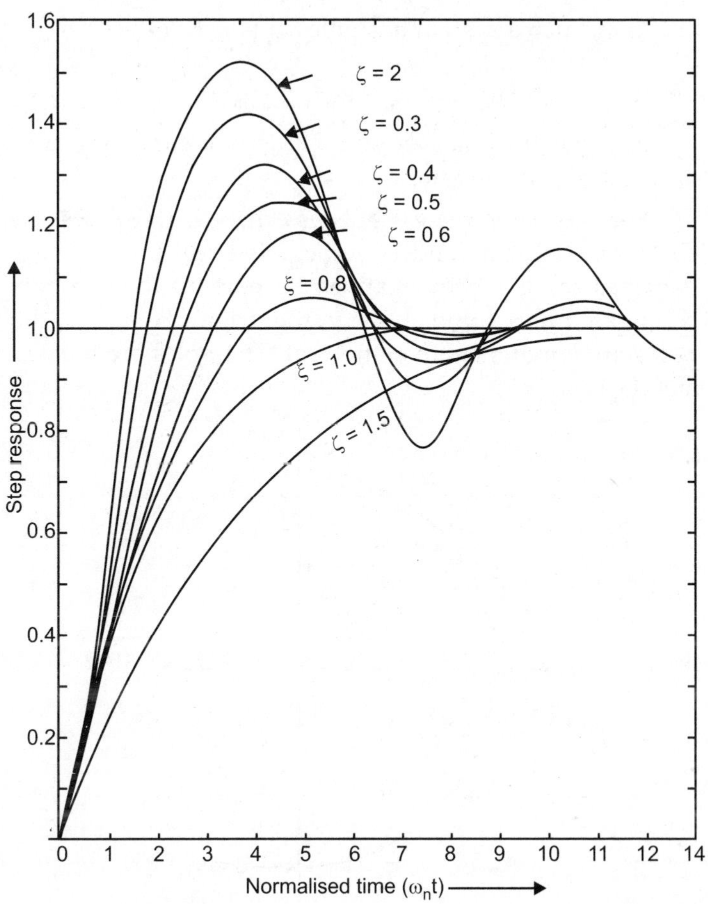

Fig. 28.2

be determined by putting differential of the response function equal to zero. The time for the first peak is given by,

$$t_p = \frac{\pi}{\omega_n\sqrt{1-\zeta^2}}$$

The overshoot which is equal to the difference of the magnitude of the first peak and the final steady state value is given by,

$$\text{Exp}\left[-\left(\frac{\zeta\pi}{\sqrt{1-\zeta^2}}\right)\right]$$

The overshoot, as is evident from the above equation, is a function of the damping ratio only and is independent of (ω_n).

Q. 16. Briefly describe the behaviour of location of poles of a linear constant co-efficient second order control system as the damping ratio (ζ) varies from 0 to 1.

Ans. If the second order system is considered by a second order differential of the form

$$\frac{d^2y}{dt^2} + 2\zeta\omega_n \frac{dy}{dt} + \omega_n^2 y = \omega_n^2 x$$

then, such a system has poles at $(-\alpha \pm j\,\omega_d)$

where, $\alpha = \zeta\omega_n$ and $\omega_d = \omega_n\sqrt{1-\zeta^2}$

For $\zeta = 0$, the poles are at $\pm j\,\omega_n$ *i.e.* poles are imaginary and complex conjugate.

For ζ greater than 0 but less than 1, the poles are complex conjugate and have negative real parts. As (ζ) becomes smaller, the negative real parts of the complex conjugate poles come closer to the origin. For ζ = 1, the poles are equal, negative and real. The poles are located at $(-\omega_n)$.

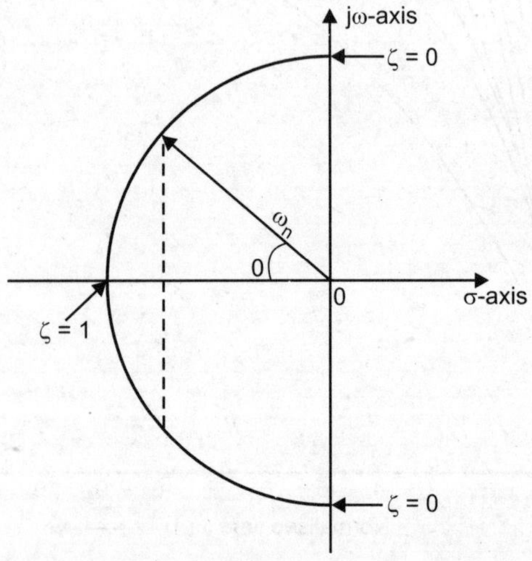

Fig. 28.3

For ζ, greater than 1, the poles are negative and real.

For ζ, less than 0, the poles lie in the R.H.P. of the s-plane.

If we plot the locus of the poles as a function of damping ratio (ζ) in the s-plane, we get a semicircle of radius (ω_n), as shown in Fig. 28.3. The angle (θ) that represents the locus is given by,

$$\theta = \cos^{-1}\zeta$$

Lower the damping ratio (ζ), higher the angle (θ) and closer to the imaginary axis of the s-plane are the system poles. As will be evident from the questions and answers covered in the latter of the chapter, the location of system poles have a direct bearing on the stability of the system.

Q. 17. What do you understand by the TRANSFER FUNCTION of a control system?

Ans. The transfer function of a control system is given by, $P(s) = \frac{Y(s)}{X(s)}$, where $Y(s)$ and $X(s)$ are respectively the Laplace transforms of output $y(t)$ and input $x(t)$ ignoring all those terms arising out of initial values. The system transfer function can, thus, be determined by finding Laplace transform of system differential equation and ignoring initial values. The transfer function of a system is also given by the Laplace transform of its impulse response. It is so because the Laplace transform of an impulse function $\delta(t)$ is 1.

Q. 18. How can you determine system poles, system zeros and the characteristic equation from the system transfer function?

Ans. The roots of the numerator of the transfer function give system zeros and roots of the denominator of the transfer function give system poles. The denominator of the transfer function set equal to zero gives the characteristic equation.

Q. 19. What do you understand by STABILITY of a control system? What is RELATIVE STABILITY? What is the relationship between the system stability and the roots of the characteristic equation describing the system?

Ans. A system is considered stable, if it remains at rest for all time unless it is excited by an external source of input. Such a system would return to its state of rest, when the excitation is removed. The stability can be defined in several ways. One way to express system stability is to determine the impulse response of the system. If the system's impulse response approaches zero as the time approaches infinity, the system is considered stable. Another definition says that the system is stable, if every bounded input (an input whose magnitude is less than some finite value for all values of t) produces a bounded output. It is important to note here that if a system produces a bounded output for a certain specific bounded input, it does not guarantee stability.

RELATIVE STABILITY is a measure of how close or far a stable system is from becoming unstable. A given system may be highly stable or only marginally stable. Relative stability of a control system is expressed in terms of parameters like Gain Margin, Phase Margin (to be discussed later).

The roots of the characteristic equation tell a lot about the stability of the control system. If the roots of the characteristic equation have negative real parts, the system is stable. Negative real parts ensure that the impulse response will decay exponentially with time. A system, whose characteristic equation has one or more positive real parts, is unstable. If the system characteristic equation has some roots with real parts equal to zero, the system is said to be marginally stable. Marginally stable systems are considered unstable. Incidentally, roots of the characteristic equation are basically the poles of the system.

Q. 20. Briefly describe the application of Routh stability criterion for determining the stability of a control system.

Ans. Routh stability criterion basically tries to establish whether all the roots of the system characteristic equation have negative real parts or not. Determining the roots of a second order characteristic equation is a simple exercise and may not necessitate the use of Routh or any other similar stability criterion. However, when the characteristic equation is of a higher order, the Routh criterion greatly simplifies the procedure.

Routh stability criterion can be used for determining the stability of a system that has (n^{th}) order characteristic equation of the form

$$a_n s^n + a_{n-1} s^{n-1} + a_{n-2} s^{n-2} + \ldots\ldots + a_1 s + a_0 = 0$$

The Routh stability criterion involves construction of Routh table from the known characteristic equation. The Routh table is given by,

s^n	a_n	a_{n-2}	$a_{n-4}\ldots\ldots$
s^{n-1}	a_{n-1}	a_{n-3}	$a_{n-5}\ldots\ldots$
s^{n-2}	b_1	b_2	$b_3\ldots\ldots\ldots$
.	c_1	c_2	$c_3\ldots\ldots\ldots$
.	:		
.	.		
s^0	.		

$$\text{Where, } b_1 = \frac{a_{n-1}\times a_{n-2} - a_n \times a_{n-3}}{a_{n-1}},$$

$$b_2 = \frac{a_{n-1}\times a_{n-4} - a_n \times a_{n-5}}{a_{n-1}},$$

$$b_3 = \frac{b_1 \times a_{n-3} \quad a_{n-1}\times b_2}{b_1},$$

$$c_2 = \frac{b_1 \times a_{n-5} - a_{n-1}\times b_3}{b_1}$$

It should be noted that while constructing the Routh table, any row can be multiplied by a constant before the next row is computed without affecting the result. The table is continued horizontally and vertically until only zeros are obtained. All the roots of the characteristic equation have negative real parts, if and only if all the elements of the first column of the table have same sign. Number of changes of sign give the number of roots or system poles with positive real parts.

Q. 21. Briefly describe the application of Hurwitz stability criterion to determine the stability of a control system.

Ans. Hurwitz stability criterion like Routh criterion determines the system stability by finding, if the roots of the system characteristic equation have negative real parts. It does so through the use of determinants formed from the co-efficients of different terms of the charac-teristic equation. The number of determinants to be evaluated equals the order of the charac-teristic equation. If the characteristic equation is of the form

$$a_n s^n + a_{n-1} s^{n-1} + a_{n-2} S^{n-2} + \ldots\ldots + a_1 s + a_0 = 0$$

Then, the first determinant, Δ_n =

$$\begin{vmatrix} a_{n-1} & a_{n-3} & a_{n-5}\ldots\ldots 0 \\ a_n & a_{n-2} & a_{n-4}\ldots\ldots 0 \\ 0 & a_{n-1} & a_{n-3}\ldots\ldots 0 \\ 0 & a_n & a_{n-2}\ldots\ldots 0 \\ \vdots & & \\ \vdots & & \\ \vdots & & \\ 0\ldots\ldots & \ldots\ldots & \ldots\ldots a_0 \end{vmatrix}$$

The remaining $(n-1)$ determinants are formed as the principal minor determinants of this determinant.

$$\Delta_1 = a_{n-1}$$

$$\Delta_2 = \begin{vmatrix} a_{n-1} & a_{n-3} \\ a_n & a_{n-2} \end{vmatrix}$$

$$\Delta_3 = \begin{vmatrix} a_{n-1} & a_{n-3} & a_{n-5} \\ a_n & a_{n-2} & a_{n-4} \\ 0 & a_{n-1} & a_{n-3} \end{vmatrix} \text{ and so on}$$

The system is stable, if and only if all determinants have a positive value.

Q. 22. Briefly describe the concept of CONTINUED FRACTION STABILITY CRITERION to establish whether all the roots of system characteristic equation have negative real parts.

Ans. The continued fraction stability criterion is applied by forming a continued fraction from the even and odd portions of the characteristic polynomial.

For the characteristic polynomial of the form

$a_n s^n + a_{n-1} s^{n-1} + a_{n-2} s^{n-2} + \ldots\ldots + a_1 s + a_0$

the two polynomials are

$a_n s^n + a_{n-2} s^{n-2} + \ldots\ldots$

and $a_{n-1} s^{n-1} + a_{n-3} s^{n-3} + a^{n-5} s^{n-5} + \ldots\ldots$

A continued fraction is formed from the fraction given by the ratio of the two polynomials as follows :

$$\frac{a_n s^n + a_{n-2} s^{n-2} + a_{n-4} s^{n-4} + \cdots}{a_{n-1} s^{n-1} + a_{n-3} s^{n-3} + a_{n-5} s^{n-5} + \cdots}$$

$$= \frac{a_n s}{a_{n-1}} + \frac{\left(a_{n-2} - \frac{a_n a_{n-3}}{a_{n-1}}\right) s^{n-2} + \left(a_{n-4} - \frac{a_n a_{n-5}}{a_{n-1}}\right) s^{n-4} + \cdots}{a_{n-1} s^{n-1} + a_{n-3} s^{n-3} + a_{n-5} s^{n-5} + \cdots}$$

$$= h_1 s + \cfrac{1}{h_2 s + \cfrac{1}{h_3 s + \cfrac{1}{\begin{matrix}\vdots \\ \vdots \\ \frac{1}{h_n s}\end{matrix}}}}$$

The roots of the characteristic polynomial shall have negative real parts if $h_1, h_2, h_3 \ldots\ldots h_n$ are all positive. The application procedure of the continued fraction stability criterion will be further illustrated in the solved problems to follow.

Q. 23. Comment on the following statements :

(*a*) If any of the co-efficients of the characteristic polynomial representing a control system is zero, the system is unstable.

(*b*) If all the co-efficients of the characteristic polynomial representing a control system do not have the same sign, the system is unstable.

(*c*) Routh, Hurwitz and Continued fraction stability criteria can be applied to only those systems which can be represented by finite polynomials in (*s*). They can not be applied to analyse systems which contain time delays.

(*d*) Routh, Hurwitz and Continued fraction stability criteria tell whether the system is stable or unstable. They do not tell anything about the relative stability of the system.

Ans.

(*a*) The statement is true. It can be mathematically proved that for any given finite polynomial in (*s*), the only way any of the co-efficients of the polynomial can be zero is for one or more than one of the roots to have zero or positive real parts. In both cases, the system would be unstable. For example, while the roots of $(s^2 + 3s + 2)$ are $s = -2$ and $s = -1$, the roots of $(s^2 + 2)$ are $\pm\sqrt{2}\,j$. Thus, omitting the (*s*) term (*i.e.* zero co-efficient for *s*) has yielded roots with zero real parts.

(*b*) The statement is true. It can be proved that for generalised polynomial

$a_n s^n + a_n - 1\, s^{n-1} + a_{n-2} s^{n-2} + \ldots\ldots\ldots + a_1 s + a_0$

if all its roots have negative real parts, all the co-efficients should have same sign. If any of the co-efficients has a different sign from that of the (a_n), it indicates one or more than one roots having positive real parts. For example, while $(s^2 + 3s + 2)$ has roots (-1) and (-2), the roots of $(s^2 - 3s + 2)$ are $(+1)$ and $(+2)$.

(*c*) The statement is true. The characteristic polynomial representing a system that contains time delay strictly speaking has infinite roots. However, if an approximation is used for the term representing time delay so as to give the characteristic polynomial the desired form, one of the stated criteria could be used to get some information about system stability. The information thus obtained is not very accurate.

(*d*) The relative stability must tell us as to how close the system is to becoming unstable, if any criterion declares it stable. The relative stability is usually expressed in terms of Gain and Phase margins (to be discussed later). The stated stability criteria tell whether the system is stable or not.

BLOCK DIAGRAM ALGEBRA AND SIGNAL FLOW GRAPHS

Q. 28. What do you understand by CANONICAL FORM of feedback control system?

Ans. The block diagram representation of a closed loop feedback control system, as shown in Fig. 28.4 is known as the canonical form of the feedback control system. In the figure shown, (*G*) represents the transfer function of the forward signal path, (*H*) is the transfer function of the feedback signal path.

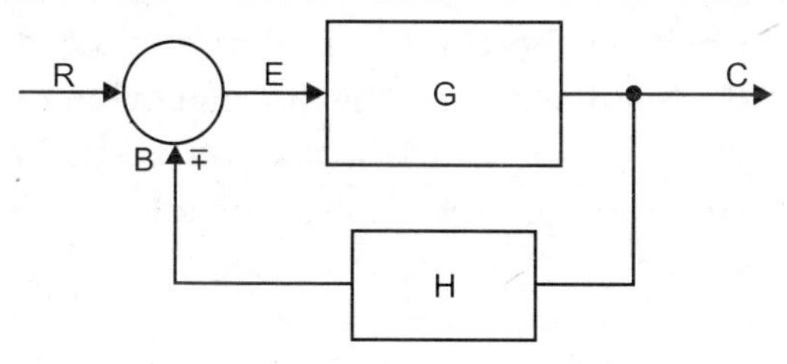

Fig. 28.4

The closed loop transfer function (C/R) is $\frac{C}{R} = \frac{G}{1 \pm GH}$

The other relevant expressions are $\frac{E}{R} = \frac{1}{1 \pm GH}$,

$\frac{B}{R} = \frac{GH}{1 \pm GH}$

Where, GH = open loop transfer function.

The characteristic equation of this feedback control system is given by,

$$1 \pm GH = 0$$

If $N(s)$ is the numerator polynomial and $D(s)$ is the denominator polynomial of the open loop transfer function, the characteristic equation is given by,

$$D(s) \pm N(s) = 0$$

Q. 25. What is a UNITY FEEDBACK control system? How do we express a given canonical form of a control system into an equivalent unity feedback type configuration?

Ans. In a unity feedback control system, the transfer function of the feedback block is unity. That is, the signal fed back to the input is of the same magnitude as the output signal. It is a situation of 100 percent feedback. Fig. 28.5 shows the unity feedback equivalent of the canonical form, as shown in Fig. 28.4 in the previous question.

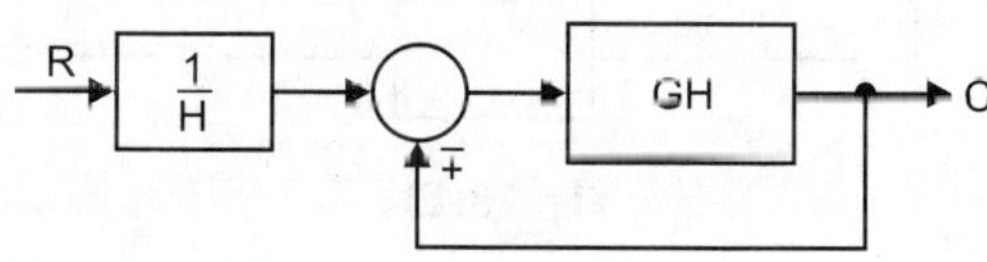

Fig. 28.5

Q. 26. Distinguish between a negative feedback and a positive feedback control system.

Ans. In a negative feedback system, the feedback signal is 180° out of phase with input. In a positive feedback system, the feedback signal is in phase with the input.

The transfer function of a negative feedback system (referring to the canonical form of Fig. 28.4) is given by,

$$\frac{C}{R} = \frac{G}{1 + GH}$$

The transfer function of a positive feedback system is given by, $\frac{C}{R} = \frac{G}{1 - GH}$.

Q. 27. How do we simplify (*a*) cascade arrangement of multiple blocks (*b*) multiple blocks in parallel?

Ans. Simplification of cascade arrangement of multiple blocks: A cascade arrangement of blocks is equivalent to a single block with its gain equal to multiplication of all individual gains. Figure 28.6 illustrates it further.

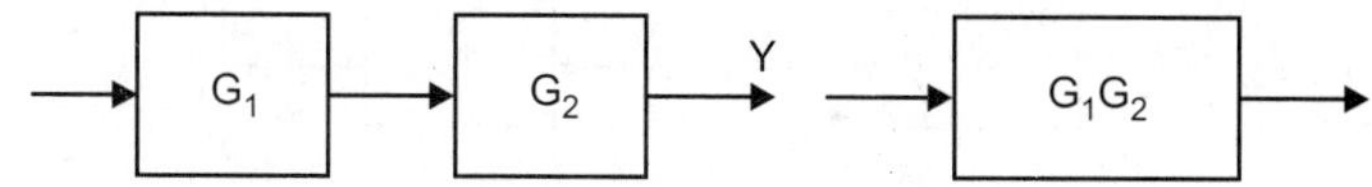

Fig. 28.6

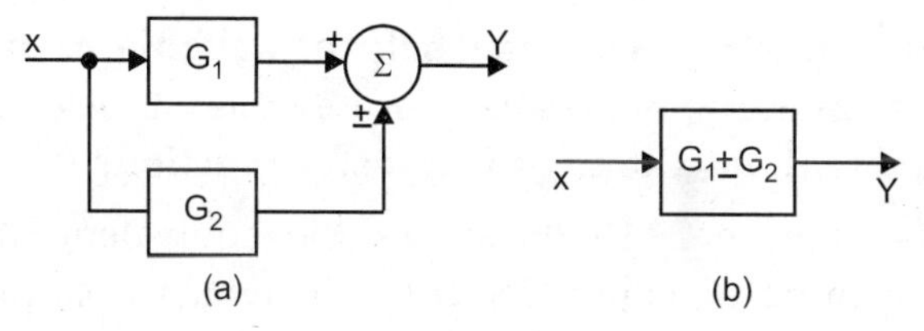

Fig. 28.7

Simplification of blocks in parallel: The relevant equation is $Y = G_1X \pm G_2X$. Figure 28.7 (*b*) shows simplification of the block schematic arrangement of Figure 28.7 (*a*).

Q. 28. Refer to the block diagram of Fig. 28.8 (*a*). Draw the equivalent block diagram by removing the block from the feedback loop.

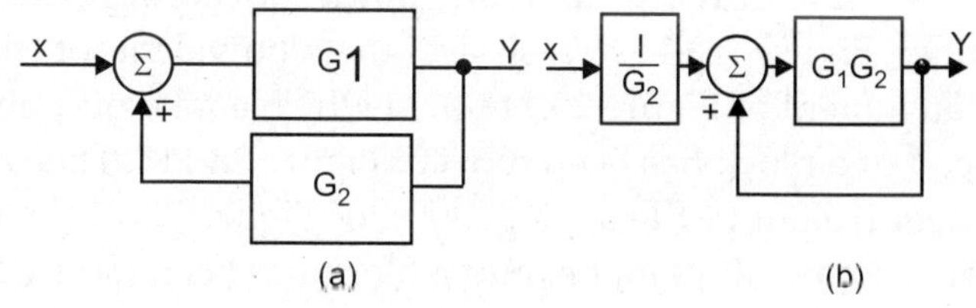

Fig. 28.8

Ans. Figure 28.8 (*b*) shows the equivalent of the block schematic arrangement of Figure 28.8 (*a*).

Q. 29. With the help of a suitable illustrative block diagram, explain how the summing points in a given block diagram can be rearranged?

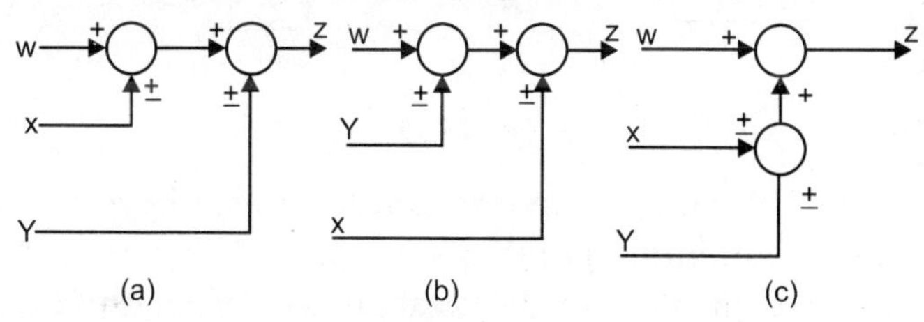

Fig. 28.9

Ans. Refer to the block diagram of Figure 28.9 (*a*). Figures 28.9 (*b*) and (*c*) show two possible equivalent block diagrams for the arrangement of Fig. 28.9 (*a*).

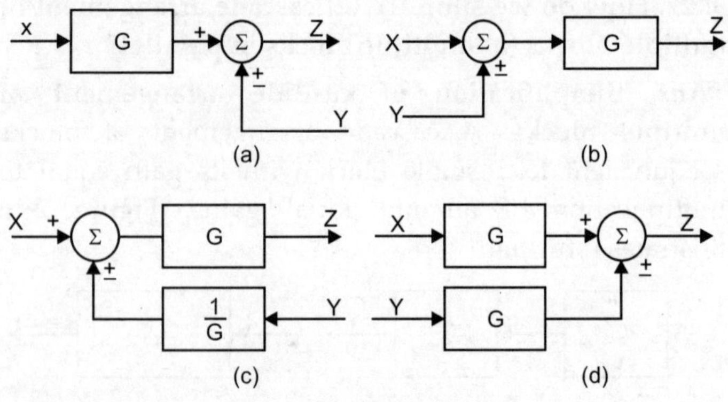

Fig. 28.10

Q. 30. Illustrate with the help of suitable examples how a summing point ahead or beyond a block can be moved without altering the transfer function?

Ans. Figure 28.10 (*c*) shows the equivalent of the arrangement of Figure 28.10 (*a*), in which a summing point ahead of a block has been moved. Figure 28.10 (*d*) shows the equivalent of the arrangement of Figure 28.10 (*b*), in which a summing point beyond a block has been moved.

Q. 31. Illustrate with the help of suitable examples how (*a*) take-off point ahead/beyond a block can be moved (*b*) take-off point ahead/beyond a summing point can be moved?

Ans. (*a*) Moving take-off point ahead/beyond a block: Figure 28.11 (*c*) shows the equivalent of the arrangement of Figure 28.11 (*a*), in which a take-off point ahead of a block has been moved. Figure 28.11 (*d*) shows the equivalent of the arrangement of Figure 28.11 (*b*), in which a take-off point beyond a block has been moved.

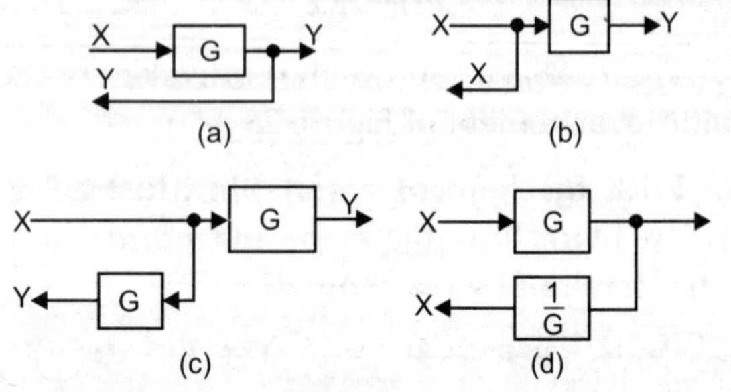

Fig. 28.11

(*b*) Moving take-off point ahead/beyond a summing point: Figure 28.12 (*c*) shows the equivalent of the arrangement of Figure 28.12 (*a*) in which a take-off point ahead of a summing point has been moved. Figure 28.12 (*d*) shows the equivalent of the arrangement of Figure 28.12 (*b*), in which a take-off point beyond a summing point has been moved.

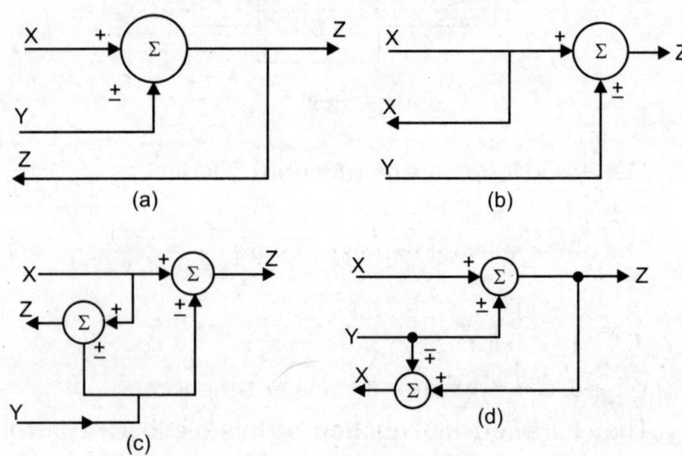

Fig. 28.12

Q. 32. Some control systems have more than one inputs applied at different points in the system. How do we find the response of such systems using block diagram algebra? Illustrate your answer with the help of a simple example.

Ans. In case of multiple inputs, one input is considered at a time with all other inputs put qual to zero. The given system block diagram is reduced to its canonical form and the response determined for each of the inputs acting alone. The overall response is then given by algebraic sum of all individual responses.

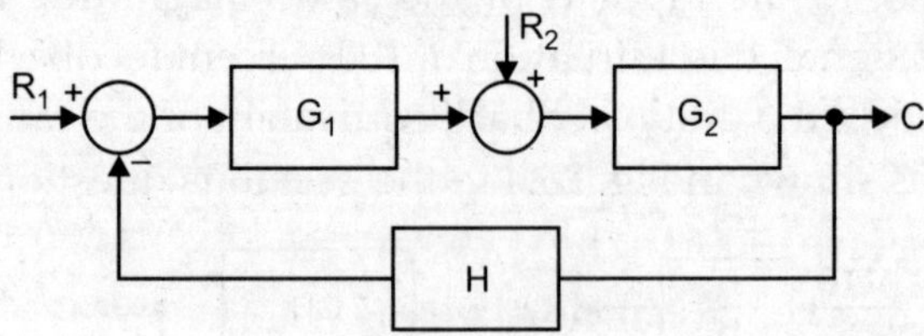

Fig. 28.13

As an illustration, consider the system block diagram of Fig. 28.13. Thy system has two inputs (R_1) and (R_2). Considering $R_2 = 0$, the given block diagram can be reduced to the canonical form Fig. 28.14 (*a*) and considering $R_1 = 0$, the given block diagram can be reduced to the canonical form of Fig. 28.14 (*b*). The transfer functions in the two cases are given by,

$$\frac{C_1}{R_1} = \left(\frac{G_1 G_2}{1 + G_1 G_2 H}\right); C_1 = \left(\frac{G_1 G_2 R_1}{1 + G_1 G_2 H}\right)$$

and
$$\frac{C_2}{R_2} = \left(\frac{G_2}{1 + G_1 G_2 H}\right); C_2 = \left(\frac{G_2 R_2}{1 + G_1 G_2 H}\right)$$

The total response is therefore given by,

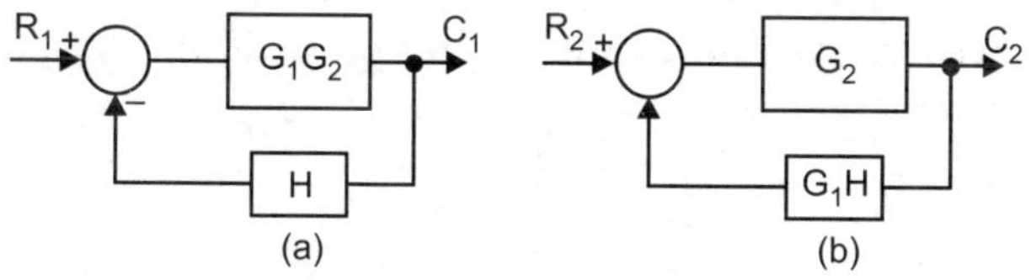

Fig. 28.14

$$C = C_1 + C_2 = \frac{G_1 G_2 R_1}{1 + G_1 G_2 H} + \frac{G_2 R_2}{1 + G_1 G_2 H}$$

$$= \frac{G_2}{1 + G_1 G_2 H}(G_1 R_1 + R_2)$$

Q. 33. Prove that in a closed loop feedback system, if the open loop gain is very large, the system transfer function depends only the feedback transfer function. In other words, the system characteristics depend largely on the nature of feedback.

Ans. If (G) is the forward path transfer function and (H) the feedback path transfer function, then the system transfer function, referring to the canonical form, is given by,

$$\frac{C}{R} = \frac{G}{1 \pm GH} = \frac{G/GH}{(1/GH) \pm 1} = \frac{1/H}{(1/GH) \pm 1}$$

In the limit, when the open loop gain (GH) tends to infinity $\lim_{GH \to \alpha}\left(\frac{C}{R}\right) = \pm 1/H$ and hence, the result.

Operational amplifier due to its extremely large open loop gain is ideally suited to a very large variety of application circuits using feedback configuration. It is because, the behaviour of the circuit configuration is then largely governed by the feedback components.

Q. 34. What is a SIGNAL FLOW GRAPH ? How do we draw a signal flow graph from a given system block diagram representation or a set of simultaneous equations?

Ans. A **signal flow graph** is a graphical representation of the set of simultaneous equations that describe the system. The signal flow graph of the system represented by an arrangement of blocks having non-interacting transfer functions can be drawn very easily by directly referring to the block diagram. Each variable is assigned a node and each block becomes a branch. The signal flow graph for the block diagram of Fig. 28.15 (*a*) is drawn in Fig. 28.15 (*b*). Extending the concept a little further, the signal flow graph for the canconical from of feedback control system of Fig. 28.16 (*a*) is shown in Fig. 28.16. (*b*).

The signal flow graph can be similarly drawn for a set of simultaneous equations describing the system.

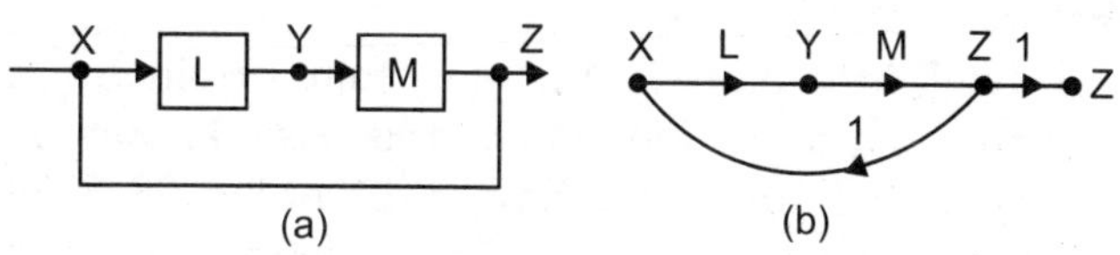

Fig. 28.15

As an illustration, let the set of simultaneous equations be

$$I_1 = 2V_1 + 3V_2$$

$$I_2 = 3V_1 + 5V_2$$

The signal flow graph for this set of equation is shown in Fig. 28.17.

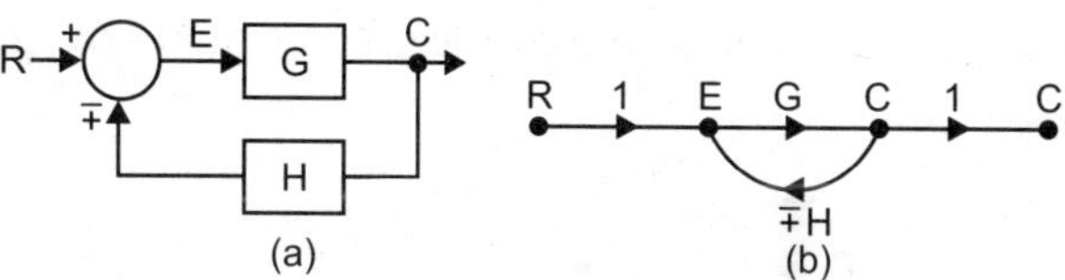

Fig. 28.16

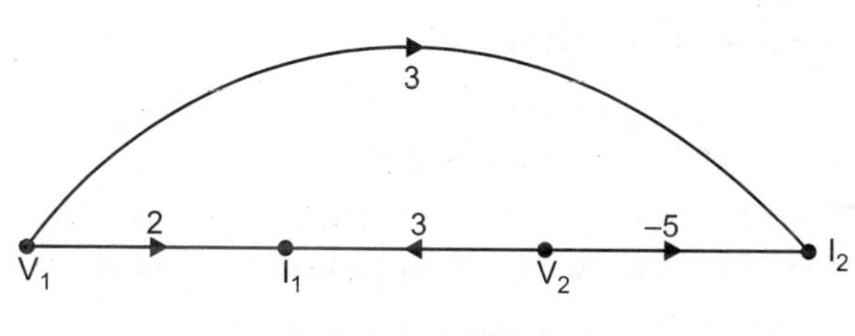

Fig. 28.17

Q. 35. Define the terms INPUT NODE, OUTPUT NODE, PATH, FORWARD PATH, FEEDBACK PATH, SELF LOOP AND LOOP GAIN with reference to signal flow graphs.

Ans. An **input node** is a node that has only outgoing branches. It is also known as source. If in a signal flow graph, the input node appears to have one or more incoming node, a dummy node is created with a branch gain of unity to satisfy the requirement. In the signal flow graph of Fig. 28.16 (*b*), a dummy input node has been created because the existing input node had one incoming branch.

An **output node** has only incoming branches. It is also known as sink. If in a signal flow graph, the output node appears to have one or more outgoing nodes, a dummy node with a unity branch gain is created to satisfy the requirement. In Fig. 28.16 (*b*), a dummy output node has been created because the existing output node had one outgoing branch.

A **path** is continuous, unidirectional succession of branches along which no node is passed more than once.

A **forward path** is the path from input node to the output node. There can be more than one forward paths too.

A **feedback path** is the one that originates and terminates on the same node. It is also known as a feedback loop. In the signal flow graph of Fig. 28.16 (*b*), *E* to *C* and then back to *E* constitutes of feedback loop.

Self loop is a feedback loop consisting of a single branch.

Loop gain is the product of the branch gains of the loop. For example, in Fig. 28.16 (*b*), the loop gain of the loop $(E - C - E)$ is $\mp GH$.

Q. 36. How do you find the system transfer function from the given system's signal flow graph?

Ans. The generalised expression for determining the system transfer function from the given signal flow graph is given by,

$$T = \frac{\sum_i P_i \Delta_i}{\Delta}$$

where, P_i = The *i*th forward path gain ...... (*i*) is the number of forward paths.

Δ = 1 – (sum of all loop gains) + (sum of all gain products of two non-touching loops)

– (sum of gain products of three non-touching loops) + (sum of gain products of four non-touching loops)....

Δ_i = (Δ) evaluated with all those loops touching P_i eliminated.

To make the matters simple, if a particular signal flow graph has two possible forward paths, the expression for transfer function reduces to

$$T = \frac{P_1\Delta_1 + P_2\Delta_2}{\Delta}$$

In case of a single forward path, $T = \left(\frac{P_1\Delta_1}{\Delta}\right)$

In the signal flow graph of Fig. 28.16 (*b*), there is only one forward path with a path gain of (*G*).

Therefore, $P_1 = G, \Delta = 1 - (\mp GH) = 1 \pm GH$

Δ_1 = 1 as the only loop in the graph is touching the forward path

This gives $T = \left(\frac{G}{1 \pm GH}\right)$

The methodology will be further illustrated in the solved problems that follow.

Q. 37. Outline the mathematical expressions for mass, spring and damper elements of a translational mechanical system. Also, derive expression for the transfer function of generalised translational mechanical system, as shown in Figure 28.19.

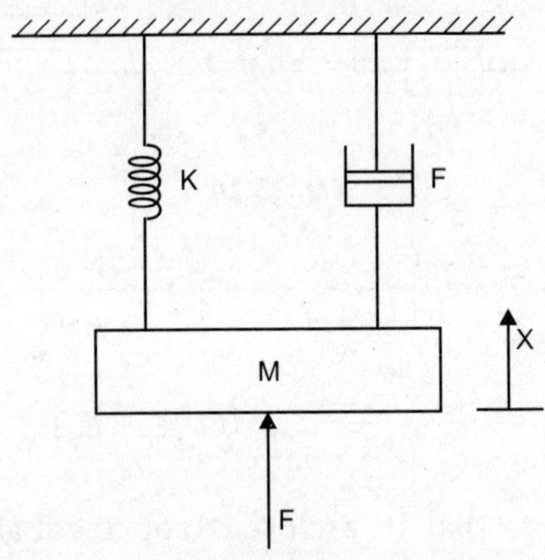

Fig. 28.18

Ans. Common elements of translation and rotation mechanical systems include mass, spring and damper (elements of translational mechanical systems) and inertia, torsional spring and damper (elements of rotational mechanical systems).

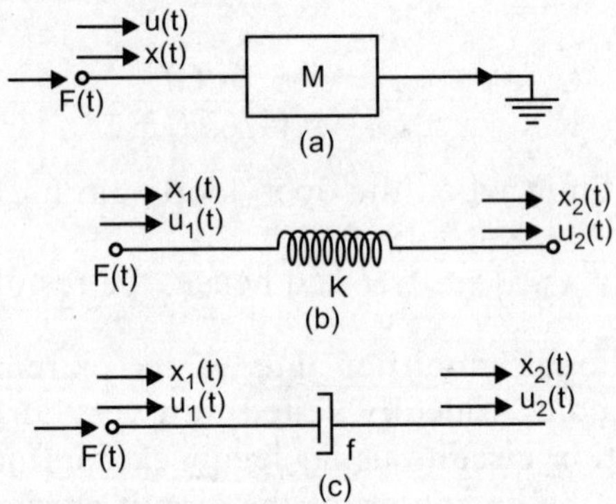

Fig. 28.19

Relevant mathematical expression for mass element (Fig. 28.19*a*) is given by:

$$F(t) = M.du\,(t)/dt = M.d^2x\,(t)/dt^2$$

Relevant mathematical expression for spring element (Fig. 28.19*b*) is given by:

$F(t) = K \,.\, (x_1 - x_2) = K \,.\, x(t)$................. K = Spring constant

Relevant mathematical expression for damper element (Fig. 28.19*c*) is given by:

$F(t) = f.\,[u_1(t) - u_2(t)] = f.[dx_1dt - dx_2/dt]$.....*f* = damping constant

As an illustration, one can write the mathematical expression describing the mechanical system of Figure 28.18 as follows:

$$F = M.\,d^2x\,(t)/dt^2 + f\,.\,dx(t)/dt + K.\,x(t)$$

The transfer function for this system can be written as follows:

Transfer function $= X(s)/F(s) = 1/(Ms^2 + fs + K)$

Q. 38. Outline the mathematical expressions for inertia, torsional spring and damper elements of a rotational mechanical system.

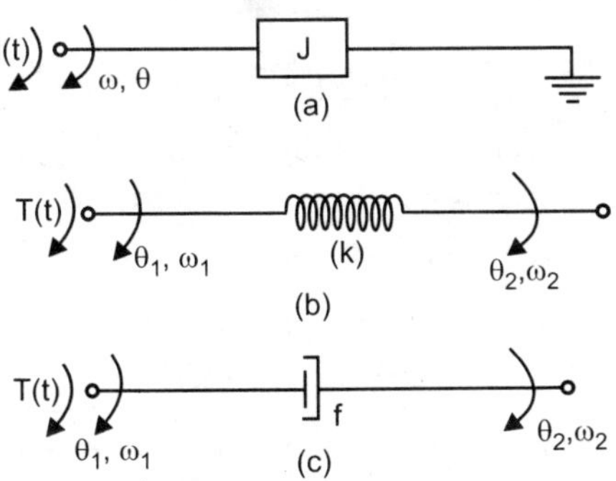

Fig. 28.20

Ans. Relevant mathematical expression for inertia element (Figure 28.20*a*) is given by:

$I(t) = J.d\omega(t)/dt = J.\ d^2\theta(t)/dt^2$........... J = Inertia constant

Relevant mathematical expression for torsional spring element (Figure 28.20*b*) is given by:

$T(t) = K.(\theta_1 - \theta_2) = K.\theta(t)$.............. K = Torsional spring constant

$= K.\int(\omega_1 - \omega_2)dt = K.\int\omega\,(t)dt$

Relevant mathematical expression for torsional damper element (Figure 28.20*c*) is given by:

$T(t) = f.(\omega_1 - \omega_2) = f.[d\theta_1(t)/dt - d\theta_2\,(t)/dt]$.................... f = Torsional damping constant

STABILITY ANALYSIS TECHNIQUES

Q.39. How do you classify feedback control systems by type. What is a Type-1 canonical feedback system ?

Ans. A canonical feedback system whose open loop transfer function can be expressed by,

$$GH = \frac{Ks^a \prod_{i=1}^{m-a}(s+z_1)}{s^b \prod_{i=1}^{n-b}(s+p_i)} = \frac{K\prod_{i=1}^{m-a}(s+z_1)}{S^l \prod_{i=1}^{n-a-l}(s+p_i)}$$

$$= \frac{KB_1(s)}{s^l B_2(s)}$$

where, a = number of zeros at origin

b = number of poles at origin

m = number of zeros

n = number of poles

$l = b - a$

K = constant, $b \geq a,\ m \leq n$

$(-z_i)$ and $(-p_i)$ are non-zero finite zeros and poles.

Q. 40. The POSITION ERROR CONSTANT (K_p), the VELOCITY ERROR CONSTANT (K_v) and the ACCELERATION ERROR CONSTANT (K_a) are the indicators of the effectiveness of feedback in stable, type '*l*' feedback systems. Define these parameters for a stable unity feedback system with an open loop transfer function *G*(*s*).

Ans. The POSITION ERROR CONSTANT (K_p) is a measure of the steady state error between the input and the output, when the input is a unit step function. It is defined by,

$$K_p = \lim_{s\to 0} G(s)$$

The VELOCITY ERROR CONSTANT (K_v) is a measure of the steadystate error between the input and the output, when the input is a unit ramp function. It is defined by,

$$K_v = \lim_{s\to 0} sG(s)$$

The ACCELERATION ERROR CONSTANT (*Ka*) is a measure of the steadystate error between the input and the output, when the input is a unit parabolic function. It is defined by,

$$K_a = \lim_{s\to 0} s^2 G(s)$$

Q. 41. How is the steadystate error in a unity feedback stable system related to the error constants ?

Ans. The steadystate error is related to the three error constants K_p, K_v and K_a by the following expressions :

$$\text{Steadystate error} = \frac{1}{1+K_p} = \frac{1}{K_v} = \frac{1}{K_a}$$

Q. 42. Comment on the following statements :

(*a*) The steadystate error in a unity feedback stable system for a unit step input is zero for a type *l* system for $l > 0$.

(*b*) The steadystate error in a stable unity feedback system for a unit ramp input is infinity for a type '0' system.

(c) The steadystate error in a stable unity feedback system for a unit parabolic input is infinity for type '0' and type '1' systems.

Ans. (*a*) The steadystate error is given by $[1/(1 + K_p)]$

where, (K_p) in general is given by $K_p = \underset{s \to 0}{Lim} \dfrac{KB_1(s)}{s^l B_2(s)}$

For $l > 0$, (K_p) is infinity which implies that steadystate error is zero. The statement is, therefore, true.

(*b*) The steadystate error in this case is given by $(1/K_v)$, where (K_v) is in general given by,

$$K_v = \underset{s \to 0}{\text{Lim}} \frac{KB_1(s)}{s^{l-1} B_2(s)}$$

This gives $K_a = 0$ for $l = 0$, which means that the steadystate error is infinity. The statement is, therefore, true.

(*c*) The steadystate error in this case is given by $(1/K_a)$, where (K_a) is in general given by,

$$K_a = \underset{s \to 0}{\text{Lim}} \frac{KB_1(s)}{s^{l-2} B_2(s)}$$

This given $K_a = 0$ for $l = 0$, 1 which means that the steadystate error is infinity. The statement is, therefore, true.

Q. 43. How do you express the steady error in terms of position, velocity and acceleration constants for a generalised stable system? How are these newly defined error constants related to the error constants defined earlier in case of stable unity feedback systems?

Ans. If K_s, K_r and K_{pa} are the step, ramp and parabolic error constants in case of a generalised system, then the steadystate error for the three cases is given by,

$$\text{Steadystate error} = \frac{1}{K_s} = \frac{1}{K_v} = \frac{1}{K_{pa}}$$

These constants are related to their counterparts defined earlier in case of stable unity feedback systems by the following expressions :

$$K_s = (1 + K_p),\ K_r = K_v,\ K_{pa} = K_a$$

Q. 44. What do you understand by the SENSITIVITY of a feedback control system? What are Magnitude Sensitivity and Phase Angle Sensitivity?

Ans. The SENSITIVITY of a feedback system is defined as a measure of the amount by which the system transfer function differs from its nominal value, when one of its parameters changes from its nominal value. Sensitivity is, therefore, defined with respect to a certain parameter. The sensitivity figure could be different for different system parameters. The sensitivity of a system with a transfer function (T) with respect to a parameter (K) is mathematically given by,

$$S_K^{T(k)} = \frac{dT\,(k)}{dk} \cdot \frac{K}{T\,(k)}$$

As the transfer function can be expressed in the polar form,

$$T(K) = |\,T(K)\,|\,e^{j\phi_T}$$

where, $|T(K)|$ is the magnitude of $T(K)$ and (ϕ_T) its phase. Then, the sensitivity of magnitude of $T(K)$ with respect to (K) is written as,

$$S_K^{|T(K)|} = \frac{d|\,T\,(K)\,|}{dk} \cdot \frac{K}{|T(K)|}$$

and the sensitivity of phase (ϕ_T) with respect to (K) is written as,

$$S_K^{\phi_T} \frac{d\phi_T}{dk} \cdot \frac{K}{\phi_T}$$

$$\text{Also, } S_K^{T(K)} = S_K^{|T(K)|} + j\,\phi_T S_k^{\phi_T}$$

Q. 45. Two very important parameters that indicate relativity stability of a feedback control system are the GAIN MARGIN and the PHASE MARGIN. Define these parameters. Indicate them on Gain vs. Frequency and Phase vs. Frequency plots.

Ans. GAIN MARGIN is defined as the reciprocal of the magnitude of open loop transfer function at the phase crossover frequency. The phase crossover frequency is the one at which the phase of open loop transfer function is (– 180°). That is,

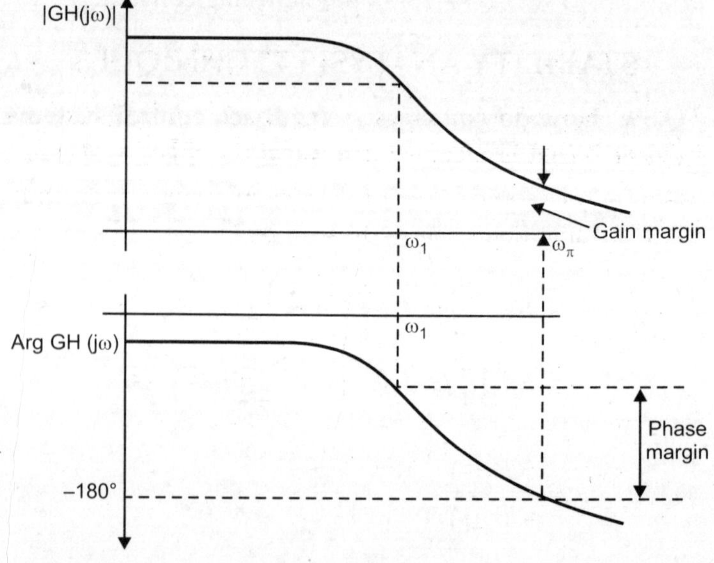

Fig. 28.21

Gain Margin $= \dfrac{1}{|GH(j\omega_\pi)|}$ where ω_π = Phase crossover frequency

PHASE MARGIN is defined as 180° plus a phase angle (ϕ_1). (ϕ_1) is the phase angle of open loop transfer function evaluated at a frequency (ω_1), where (ω_1) is the gain crossover frequency- frequency at which $|GH(j\omega)| = 1$. The two parameters are illustrated in Fig. 28.21.

Phase Margin = 180° + Arg $GH(j\,\omega_1)$

Q. 46. What are industrial controllers? Briefly describe Proportional (*P*), Integral (*I*), Derivative (*D*), Proportional-Integral (*PI*) and Proportional-Integral-Derivative (*PID*) controllers.

Ans. An **Industrial Controller** is a feedback control mechanism widely used in industrial controllers where it is used to the output of the process. It does so by taking error between the process variable and a reference value as the input and generates an output that is used to control the process variable to minimise the error value. We have *proportional controllers* called *P*-controllers, where the controller output depends upon the present value of the error input and is proportional to the error input; *integral controllers* called I-controllers, where the controller output accumulates past errors and is integral of the error input and *derivative controllers* called D-controllers, where the controller output predicts the future values of error input and is equal to derivative of error input. Figures 28.22, 28.23 and 28.24 respectively show block diagram representations of proportional, integral and derivative controllers. It may be mentioned here that in case of derivative controller, the control signal consists of proportional error signal summed up with the derivative of the error signal.

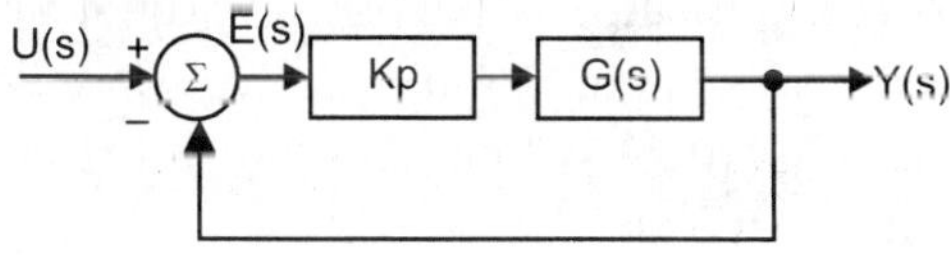

Fig. 28.22

In addition to *P*, *I* and D-controllers, we also have combination of more than one basic controller elements to achieve desired controller characteristics. For example, we have PI (Proportional-Integral) controllers, PD (*Proportional-Derivative*) controllers and PID (*Proportional-Integral-Derivative*) controllers. *PI* and PID are the most commonly used industrial controllers with the latter being even more popular of the two. *PI* controllers are fairly common, since derivative action is sensitive to measurement noise, whereas the integral term may allow the system reach its target value due to the control action. *PID* controllers due to their functional simplicity and robust performance in a wide range of operating conditions are referred to as universal controllers.

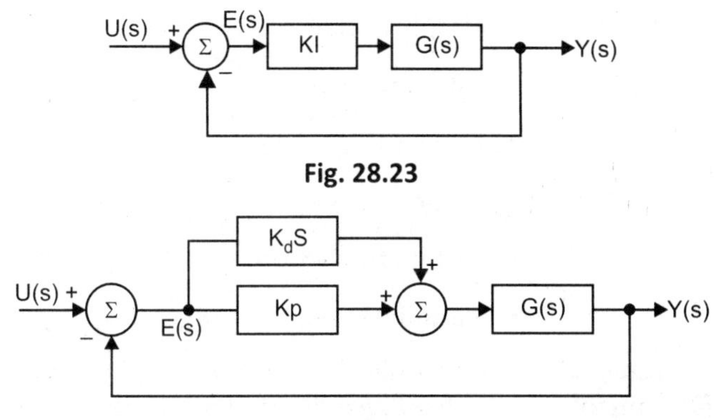

Fig. 28.23

Fig. 28.24

In a *Proportional Controller*, as outlined earlier, the control signal is proportional to the error signal. In the case of a P-controller, large controller gain improves steady state error. It also gives larger control loop bandwidth but higher sensitivity to measurement noise. Proportional controllers cannot be used to stabilise higher order processes. In case a constant steady state error is acceptable in a process; P-controller can be used. Derivative control mode is used, when prediction of the error can improve control or when it is necessary to stabilise the system.

PI-controller offers advantages over *on-off* control and proportional control by eliminating forced oscillations encountered in the case of former and steady state error in the case of the latter. However, introducing integral mode has a negative effect on speed of the response and overall stability of the system. PI-controller, however, has the disadvantage of being slow, which is expected since PI-controller does not have means to predict error in future. This problem, however, can be addressed by introducing a derivative element. This is what is done in the case of a PID (Proportional-Integral-Differential) controller, which combines all the good features of proportional, integral and derivative controllers.

Q. 47. With the help of a suitable block schematic representation, briefly describe the operation of PID controller outlining the role of proportional, integral and differential elements.

Ans. The PID control scheme is named after its three elements, namely proportional, integral and derivative elements, whose sum constitutes the PID controller output. The proportional, integral, and derivative terms are summed to calculate the output of the PID controller. The controller output $u(t)$ in this case is given by the following expression:

$$u(t) = K_p.e(t) + K_i \int e(\tau)\, d\tau + K_d\, d/dt\, e(t)$$

Taking Laplace transform, we get,

$$U(s) = Kp.E(s) + K_i.E(s)/s + K_d.sE(s)$$

K_p = Proportional gain

K_i = Integral gain

K_d = Derivative gain

e = Error

t = Instantaneous time

τ = Variable of integration

Figure 28.25 shows block schematic of one form of PID controller in a feedback loop. This is the parallel or non-interacting form of PID control. There are other alternative forms not discussed here. The basic principles however remain the same.

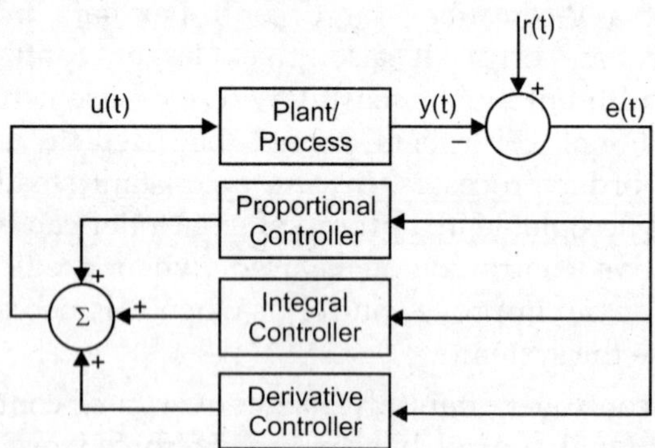

Fig. 28.25

Proportional Element: The proportional component in the PID controller produces an output that is proportional to the current error value. The constant of proportionality is called proportional gain constant (K_p). (K_p) can be varied to adjust the proportional response of the controller. The proportional element output (P_{out}) is given by the following expression.

$$P_{out} = K_p e(t) \text{ or } P_{out}(s) = K_p.E(s)$$

The system is prone to becoming unstable, in the case the proportional gain is too high. If the proportional gain is too low, the control action may be too small, when responding to system disturbances and the controller becomes less sensitive. Tuning theory and industrial practice indicate that the proportional term should contribute the bulk of the output change. A proportional controller generally operates with a steady state error, as a zero error would produce a zero controller output and a non-zero error would be required to drive it. The steady state error is inversely proportional to proportional gain constant and directly proportional to process gain. Higher process gain produces larger error while a higher proportional gain produces smaller error. This error can however be mitigated by adding a bias term to the set point. A better method is to add an integral element described in the following paragraph.

Integral Element: The contribution from the integral element of the PID controller is proportional to both the magnitude and the duration of the error. The integral element in a PID controller is the sum of the instantaneous error over time and gives the accumulated offset. The accumulated error is then multiplied by the integral gain (K_i) and added to the controller output. The integral term is given by the following expression. The integral element output (I_{out}) is given by the following expression:

$$I_{out} = K_i \int_0^t e(\tau)\, d\tau$$

$$I_{out}(s) = (K_i/s).\, E(s)$$

The integral term accelerates the movement of the process towards set point and practically reduces the residual steady state error to zero that is otherwise some finite value with a pure proportional controller.

Derivative Element: The derivative element produces an output that is proportional to rate of change of error. The derivative element output (K_d) is given by the following expression:

$$D_{out} = K_d\, d/dt\, e(t) \text{ or } D_{out}(s) = K_d\, s\, E(s)$$

Derivative action improves settling time and stability of the system due to its capability to predict system behaviour. An ideal derivative is not causal, so that implementations of PID controllers include an additional low pass filtering for the derivative term, to limit the high frequency gain and noise.

Q. 48. What is the purpose of using control system compensators?

Ans. A **compensator** makes some adjustments in the feedback control system to make up for some deficiencies in the system. Control system compensators are used for the following reasons:

1. To obtain the desired performance of the system. Compensating networks are applied to the system in the form of feed forward path gain adjustment.
2. Compensate a unstable system to make it stable.
3. To minimise overshoot.
4. To increase the steady state accuracy of the system. An important point to be noted here is that the increase in the steady state accuracy brings instability to the system.

5. To introduce poles and zeros in the system thereby causing changes in the transfer function of the system. Due to this, performance specifications of the system change.

Q. 49. Briefly describe a Lead Compensator. How does it affect the control system performance?

Ans. The generalised transfer function of a lead compensator is given by the following expression:

$$P(\text{Lead}) = \left(\frac{s+a}{s+b}\right),\ a < b$$

The lead network provides compensation by virtue of its phase lead property in low to medium frequency range and negligible attenuation in the high frequency region. More than one lead networks may be cascaded to get a larger phase lead. Lead compensation generally increases the system bandwidth.

The lead compensator may be used to improve upon the transient performance of the system for a given value of the gain factor. In some systems, it may even be used with appropriate choice of (*a*) and (*b*) to allow an increase in the value of gain factor, thus, providing greater accuracy without adversely affecting transient performance. Figure 28.26 (*a*) and (*b*) respectively show the pole zero plot and polar plot of the generalised lead compensator.

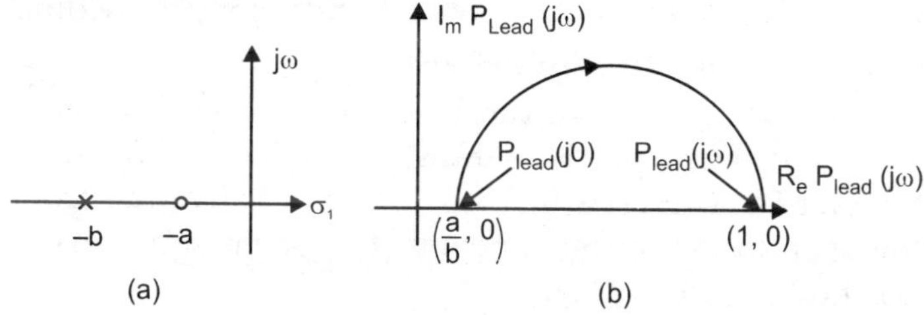

Fig. 28.26

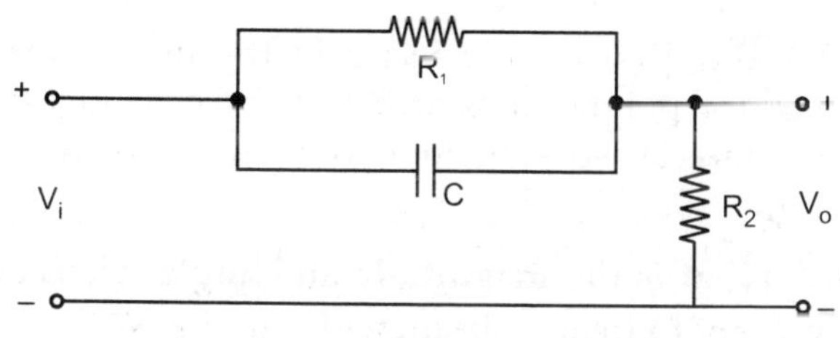

Fig. 28.27

Figure 28.27 shows an R-C network mechanization of a lead compensator. The transfer function of this network is given by the following expression:

$$P_{\text{lead}} = (Cs + 1/R_1)/(Cs + 1/R_1 + 1/R_2)$$

$$= (s+a)/(s+b)$$

$$a = 1/R_1C \text{ and } b = 1/R_1C + 1/R_2C$$

Q. 50. Briefly describe a Lag Compensator. How does it affect the control system performance?

Ans. The lag compensator provides compensation by virtue of its attenuation property in the high frequency region of its polar plot. The generalized transfer function of a lag compensator is given by the following expression:

$$P(\text{Lag}) = \frac{a}{b}\left(\frac{s+b}{s+a}\right),\ a < b$$

More than one phase lag compensators can be cascaded to provide even larger high frequency attenuation. The lag compensator usually decreases system bandwidth. The system, however, becomes more sluggish due to increase in the system time constant. It improves upon the steady state performance for a given relative stability specification or improves upon the relative stability for a given value of error constant. Figure 28.28 (*a*) shows the pole zero plot of a lag compensator. Figure 28.28 (*b*) shows polar plot. Figure 28.29 shows an R-C network implementation of a lag compensator. The transfer function of this network is given by the following expression:

$$P_{\text{lag}} = (R_2 + 1/Cs)/(R_1 + R_2 + 1/Cs)$$

$$= [a(s+b)]/[b(s+a)]$$

$$a = 1/[(R_1 + R_2)C] \text{ and } b = 1/R_2C$$

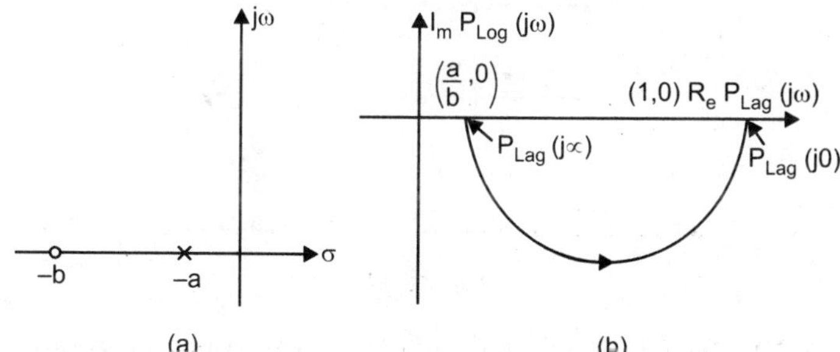

Fig. 28.28

Fig. 28.29

Q. 51. Briefly describe a Lag-Lead Compensator. How does it affect the control system performance?

Ans. The lag-lead compensator has a transfer function given by the following expression:

Figure 28.30 (*a*) shows the pole zero plot of such a control system compensator. Figure 28.30 (*b*) shows the polar plot of the same. A lag-lead compensator has all the desirable features of a lead and a lag compensator without having much of their undesirable characteristics. For example, a lag-lead compensator can be used to achieve many of system specifications, such as desired transient response, required accuracy etc. without excessive bandwidth (a characteristic of a lead compensator) or an excessive time constant (a characteristic of a lag compensator). Figure 28.31 shows the R-C network implementation of a lag-lead compensator. The transfer function of this network is given by the following expression:

$$[(s + 1/R_1C_1)\ (s + 1/R_2C_2)]\ /\ [s^2 + (1/R_1C_1 + 1/R_2C_2 + 1/R_2C_1)s + 1/R_1C_1R_2C_2]$$

$$a_1 = 1/R_1C_1,\ b_2 = 1/R_2C_2,\ a_1b_2 = b_1a_2,$$

$$b_1 + a_2 = a_1 + b_2 + 1/R_2C_1$$

Fig. 28.30

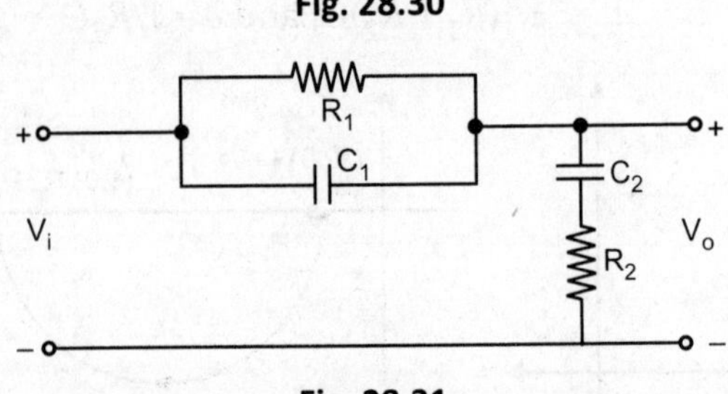

Fig. 28.31

Q. 52. Briefly distinguish between Series Compensation, Feedback Compensation and Load Compensation.

Ans. Connecting compensating circuit between error detector and the plant is known as series compensation. Figure 28.32 shows the block schematic representation of a control system employing series compensation.

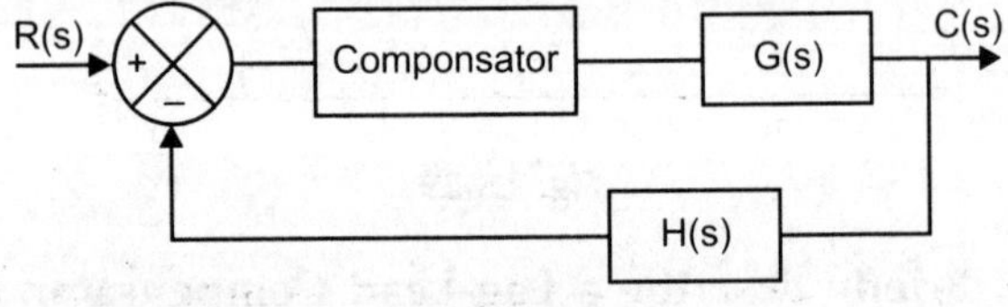

Fig. 28.32

When a compensator is used in a feedback manner, its called feedback compensation. Figure 28.33 shows the block schematic representation of a control system employing feedback compensation.

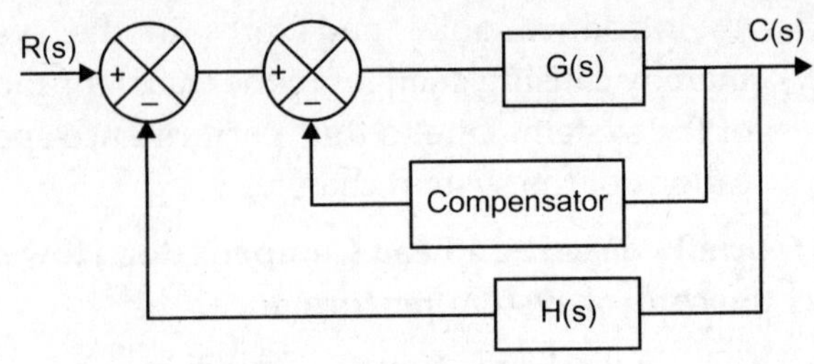

Fig. 28.33

A combination of series and feedback compensator is called load compensation.

Q. 53. What is root locus with reference to control systems? What is root locus method used for? Can we use the root locus analysis, if we knew only the poles and zeros of open loop transfer function?

Ans. Root locus is the locus of the roots of the characteristic equation plotted as a function of the gain factor of the open loop transfer function in s-plane. Now the roots of the characteristic equation are nothing but the poles of the closed loop transfer function.

Thus, root locus technique is a method of displaying the poles of the closed loop system in the s-plane as a function of the gain factor of the loop transfer function.

Root locus analysis is used for determining both absolute as well as relative stability of control systems. Root locus analysis permits accurate computation of the system's time domain response in addition to yielding frequency response information.

Root locus analysis can be carried out even if poles and zeros of the open loop transfer function are known as the system characteristic equation can be deduced from the open loop transfer function. The open loop transfer function $GH(s)$ is given by:

$$GH(s) = K\frac{N(s)}{D(s)}$$

$N(s)$ and $D(s)$ are respectively the numerator and denominator polynomials and K is the gain factor. The system characteristic equation is then given by $D(s) + KN(s) = 0$

Q. 54. What is the magnitude and angle criteria for a certain point to lie on a branch of root locus?

Ans. (s_1) must be a root of the system characteristic equation if a branch of the root locus for that system is to pass through that point. If the open loop transfer function is expressed as:

$GH(s) = K\dfrac{N(s)}{D(s)}$, then the system characteristic equation is given by $D(s) + KN(s) = 0$

If (s_1) lies on the root locus, then

$D(s_1) + KN(s_1) = 0$

$D(s_1) + KN(s_1) = -1$, which implies

$|K| = |D(s_1)/N(s_1)|$ and

$$\text{Arg}\left[\frac{N(s_1)}{D(s_1)}\right] = (2l+1)\pi \text{ Radians for } K > 0$$

$= 2l\pi$ Radians for $K < 0$ where $l = 0, \pm 1, \pm 2$ ....

To sum up, if a point (s_1) is to lie on a branch of the root locus, it must satisfy the *Magnitude Criterion* given by:

$$|K| = \left|\frac{D(s_1)}{N(s_1)}\right|$$

Angle Criterion given by:

Arg $[N(s_1)/D(s_1)] = (2l+1)\pi$ Radians for $k > 0$ and $2l\pi$ Radians for $k<0$

Q. 55. Briefly outline different steps to be followed to construct root locus from the given open loop transfer function of a control system. Illustrate different steps with the help of an example.

Ans. Following are the steps to be followed for construction of the root locus from the given open loop transfer function:

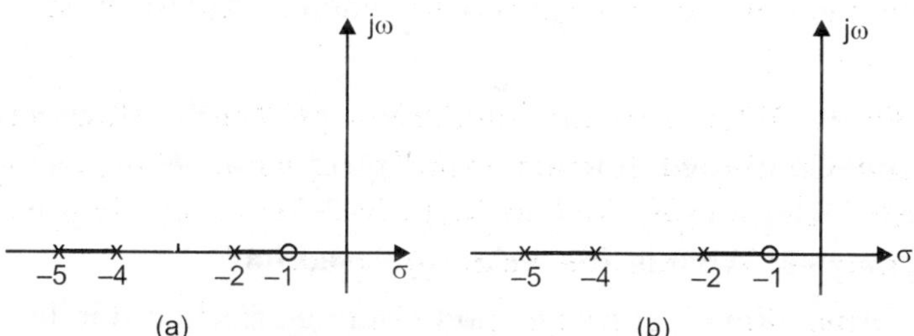

Fig. 28.34

1. The first step is to find out the number of branches (also known as the *loci*) of the root locus. The number of branches equals the number of poles of the open loop transfer function. Real axis loci can be determined as follows. For the gain factor (*K*) greater than zero ($K > 0$), points of the root locus on the real axis lie to the left odd number of finite poles and zeros. For the gain factor less than zero ($K < 0$), points of the roots locus lie to the left of an even number of finite poles and zeros. As an illustration, if the open loop transfer function of a certain closed loop system is given by:

$$GH(s) = \frac{K(s+1)}{(s+2)(s+4)(s+5)}$$

then the root locus is going to have three branches in all. The real axis loci for $K > 0$ and $K <$ 0 are respectively shown in Figure 28.34 (*a*) and (*b*).

2. *Asymptotes* play a vital role in the construction of root locus. The branches of the root locus approach a set of straight line asymptotes for large distances from the origin in the s-plane. These asymptotes originate from a point on the real axis known as center of asymptotes given by:

$$\sigma_c = \frac{\sum_{i=1}^{n} p_i - \sum_{i=1}^{m} z_i}{(n-m)}$$

(n) and (m) are respectively the number of poles and zeros of the open loop transfer function. (p_i) and (z_i) are respectively the poles and zeros of the open loop transfer function. The center of asymptotes, for instance, for the open loop transfer function considered above is given by:

$$\sigma_c = \frac{(2+4+5)-1}{(3-1)} = 5$$

3. The *Angles Between the Asymptotes* and the real axis are given by:

$$\beta = \begin{cases} \dfrac{(2l+1)\pi}{n-m} & \text{for } K > 0 \\[2ex] \dfrac{(2l)\pi}{n-m} & \text{for } K > 0 \end{cases}$$

$$l = 0, 1, 2, \ldots\ldots\ldots\ldots (n-m-1)$$

4. In the next step, we determine *Breakaway* Points if any. A breakaway point (σ_b) is a point on the real axis where two or more branches of the root locus depart from or arrive at the real axis. The location of the breakaway point (σ_b) can be determined from the equation

$$\sum_{i=1}^{n} \frac{1}{(\sigma_h + pi)} = \sum_{i=1}^{m} \frac{1}{\sigma_h + zi}$$

$(-p_i)$ and $(-z_i)$ are poles and zeros of the open loop transfer function.

Determination of the location of the breakaway point involves factorisation of a polynomial of the order $(n - m - 1)$. The correct location of the breakaway point can be determined from more than one value of (σ_b) that we get as a result of solving the polynomial by considering these locations in relevance to the real axis loci. Remember that the breakaway points lay on the real axis loci.

5. The last thing to consider while constructing a root locus is the *Departure and Arrival Angles.* These are relevant only to those systems which have complex poles and zeros. The Departure *Angle* of the root locus from a complex pole is given by:

$$\theta_D = 180° - \text{Arg}\,(GH')$$

Arg (*GH'*) is the phase angle of the open loop transfer function (*GH*) computed at the complex pole in question. While computing the phase angle, the contribution of the complex pole in question is ignored. The angle of arrival is relevant to complex zeros. It is given by:

$$\theta_A = 180° - \text{Arg}(GH'')$$

Arg (*GH''*) is the phase angle of the open loop transfer function (*GH*) at the complex zero, but ignoring the contribution of that particular zero.

Q. 56. How do we compute the Gain Margin and Phase Margin from the root locus plot?

Ans. *Gain Margin* can be determined from the root locus plot from:

$$\text{Gain Margin} = \frac{\text{Value of }(K)\text{ at imgainary cross - over}}{\text{Desing value of }(K)}$$

The Phase Margin can be determined from:

$$\text{Phase Margin} = 180° + \text{Arg}\,GH\,(j\omega_1)$$

(ω_1) is the gain cross-over frequency given by $|GH(j\omega_1)| = 1$

Q. 57. How do we determine Damping Ratio from the root locus plot?

Ans. If (θ) is the angle that a line starting from origin makes with the negative real axis, then the expression $\theta = \cos^{-1}(\zeta)$ can be used to determine (*K*) for a known (ζ) or (ζ) for a desired (*K*). For example, if the gain factor (*K*) is given, a point corresponding to the given value of (*K*) is located on the root locus plot. This point is joined with origin and the angle (θ) measured. Damping ratio (ζ) is then given by ($\cos\theta$). If (ζ) is known, a line is drawn from origin making an angle (θ) with the negative real axis, where $\theta = \cos^{-1}\zeta$. The value of (*K*) is then given by the point, where this line cuts the root locus plot.

Q. 58. How do we determine the closed loop transfer function and the system's time domain response from the root locus plot?

Ans. Root locus technique can be used to determine closed loop transfer function and thereby the time domain response for a given input. Closed loop transfer function (*C/R*) for a control system represented in a canonical unity feedback form is given by:

$$C/R = G/(1+G)$$

where, (*G*), the open loop transfer function, is further represented by $G = K\,N(s)/D(s)$.

N(*s*) and *D*(*s*) are polynomials whose roots respectively represent zeros ($-z_i$) and poles ($-p_i$). (*G*) is further expressed as:

$$G = \frac{K(z+z_1)\,(s+z_2)\,.....\,s+(z_m)}{(s+p_1)\,(s+p_2)\,.....\,(s+p_n)}$$

where, $m \le n$

$$C/R = KN/(D+KN)$$

It is evident that for $K = 0$, *G* and *C/R* have same poles and zeros. Location of closed loop poles and zeros is by definition determined directly from root locus plot for a specified value of open loop gain (*K*). In case the system is not a unity feedback system,

$$C/R = G/(1+GH)$$

Here, the closed loop poles may be determined directly from root locus for a given value of (*K*) but the closed loop zeros are not equal to open loop zeros.

Once the transfer function (*C/R*) has been determined from the root locus plot, time domain response *c*(*t*) may be determined for a given *r*(*t*) by taking Laplace inverse transform.

Q. 59. What are the limitations of Routh, Hurwitz and Continued fraction stability criteria, when used for determining system stability? What is Nyquist analysis? What is this technique used for ?

Ans. Routh, Hurwitz and Continued fraction stability criteria, with few exceptions, can only be used for determining absolute stability. These criteria can only be used on those systems whose characteristic equation is a finite polynomial in (*s*). If the system has a delay component somewhere in the loop, these techniques can be used only after using approximation for the term representing delay. Under these circumstances, these techniques yield only approximate stability information.

Nyquist analysis is a frequency response method and is essentially a graphical procedure, where a polar plot of sinusoidal open loop transfer function *GH*($j\omega$) with imaginary part of *GH*($j\omega$) taken along *y*-axis and real part of *GH*($j\omega$) taken along *x*-axis can give information about absolute and relative stability of the feedback system, once it has been put in the canonical form.

The Nyquist technique can be applied to systems with time delays without resorting to any approximations and thus, yields exact results about absolute and relative stability of the system.

Q. 60. What is Nyquist path?

Ans. The **Nyquist path** is a closed contour drawn on s-plane enclosing the entire right half plane of s-plane. Fig. 28.35 shows the generalised Nyquist path. Small semicircles along the imaginary axis or at the origin of the transfer function to be analysed are required in the path, if the said transfer function has poles at the origin or on the $j\omega$-axis. In the Nyquist path shown in Fig. 28.35, the transfer function has a pole at the origin and a pair of complex conjugate poles $\pm j\omega_0$. Different portions of the Nyquist path can be expressed as :

Path ab : $s = j\omega$ for $0 < \omega < \omega_0$

Path be : $s = \lim\limits_{p \to 0} (j\omega_0 + \rho e^{j\theta})$ for $-90° \le \theta \le +90°$

Path cd : $s = j\omega$ for $\omega_0 \le \omega \le \infty$

Path def: $s = \lim\limits_{p \to \infty} Re^{j\theta}$ for $+90° \le \theta \le -90°$

Path fg : $s = j\omega$ for $-\infty < \omega < -\omega_0$

Path gh : $s = \lim\limits_{\rho \to 0} (-j\omega_0 + \rho e^{j\theta})$ for $-90° \le \theta \le +90°$

Path hi : $s = j\omega$ for $-\omega_0 < \omega_0 < 0$

Path ija : $s = \lim\limits_{\rho \to 0} \rho e^{j\theta}$ for $-90° \le \theta \le +90°$

Q. 61. What is a Nyquist stability plot ? State Nyquist stability criterion.

Ans. The **Nyquist stability plot** is an extension of the polar plot. It maps the entire Nyquist path into the plane of the transfer function. In case of Nyquist analysis, the transfer function considered is the open loop transfer function $GH(j\omega)$ with imaginary part of $GH(j\omega)$ taken along the y-axis and the real part of $GH(j\omega)$ taken along the x-axis.

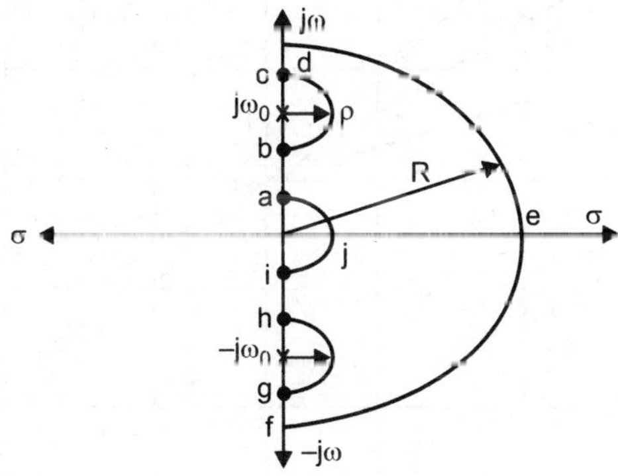

Fig. 28.35

According to the NYQUIST STABILITY CRITERION, the closed loop system is stable, if and only if

$$N = -P_0 \le 0$$

where, P_0 = number of poles of GH in the RHP

N = total number of clockwise encirclements of the point $(-1, 0)$ in the GH (s) plane. $(-1, 0)$ is the point where $GH(s) = -1$

If $N > 0$, then the number of zeros of $(1 + GH)$ or poles of the closed loop system in the RHP is given by $(N + P_0)$. Also, $N \le 0$ implies that the point $(-1, 0)$ is not enclosed by the Nyquist stability plot. **(N) is determined by traversing the stability plot in the prescribed direction and shading the region to the right of the contour. The shaded region does not contain (– 1, 0) point if $N \le 0$.**

Q. 62. How can we use Nyquist stability plot to determine relative stability 'parameters' such as Gain Margin and Phase Margin for a given closed loop system?

Ans. The **gain margin**, as described in an earlier question, is given by:

$$\text{Gain Margin} = \left| \frac{1}{GH(j\omega_\pi)} \right|$$

where, ω_π is phase crossover frequency.

The phase crossover frequency (ω_π) is the frequency at which the phase angle of $GH(j\omega)$ is $-180°$. It is the frequency at which the Nyquist plot crosses the negative real axis. The phase crossover frequency and the gain margin are illustrated in Fig. 28.36.

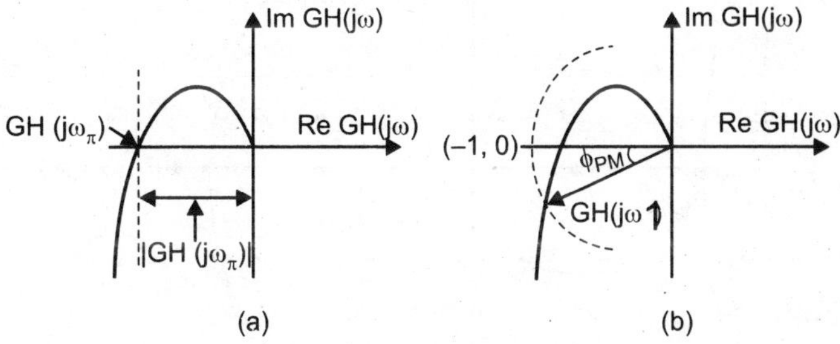

Fig. 28.36

The **phase margin**, as described in an earlier question, is given by,

$$\phi_{PM} = [180° + \text{Arg}GH(j\omega_1)]$$

where, ω_1 is Gain crossover frequency

Gain crossover frequency is the frequency at which $|GH(j\omega)| = 1$. (ω_1) and (ϕ_{PM}) are illustrated in Fig. 28.36 (b). The phase margin may also be defined as an angle by which the Nyquist stability plot may be rotated to make it pass through $(-1, 0)$ point.

Q. 63. Illustrate with the help of an example how the gain factor can destabilise a stable system or make an unstable system stable using a Nyquist stability plot.

Ans. Let us take the case of a control system represented by :

$$GH(s) = \frac{1}{s(s+1)(s+0.5)}$$

$$GH(j\omega) = \frac{1}{\omega\sqrt{(\omega^2+1)(\omega^2+0.25)}}$$

$$\angle -90° - \tan^{-1}\omega - \tan^{-1}2\omega$$

Now $GH(j\omega) = \infty\angle -90°$ for $\omega \to 0$

and $GH(jw) = 0\angle +90°$ for $\omega \to \infty$

The Nyquist stability plot is shown in Fig. 28.37

At the phase crossover frequency, ω_π

$$|GH(j\omega_\pi)| = \frac{4}{3}$$

Therefore, point (– 1, 0) is enclosed, as shown in Fig. 28.37 too. The given transfer function has a gain factor of '1' we compare this with a generalised transfer function of this type given by,

$$GH(s) = \frac{K}{s(s+p_1)(s+p_2)}$$

Now, if we reduce from 1 to 0.5, $|GH(j\omega_\pi)| = 0.66$

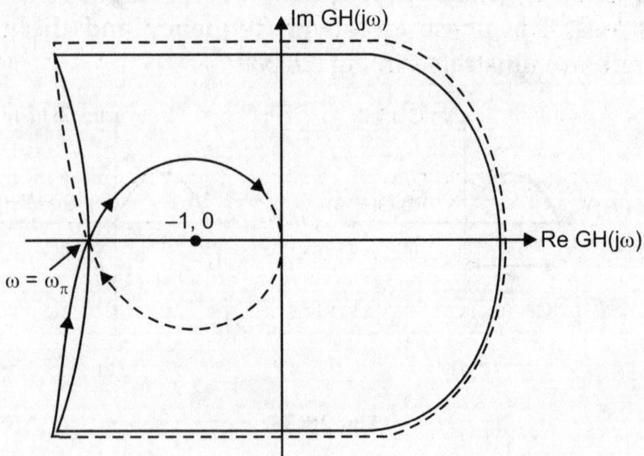

Fig. 28.37

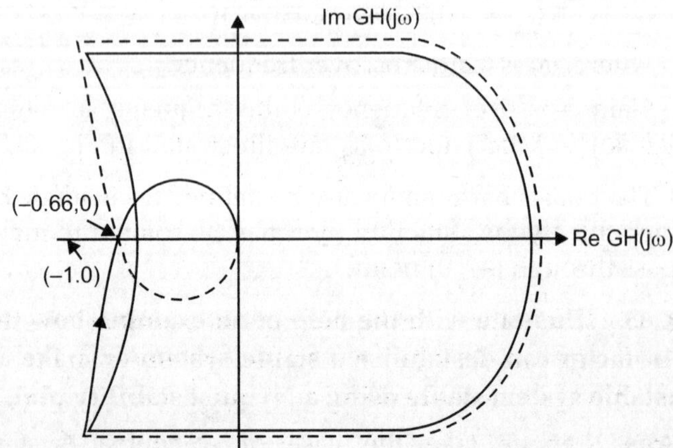

Fig. 28.38

and the Nyquist stability plot gets modified to what is shown in Fig. 28.38. The system has become stable now. The technique of changing the gain factor for altering the stability characteristic of a system is known as GAIN FACTOR COMPENSATION.

Q. 64. What type of plots are made use of in Bode analysis of closed loop control systems? Which closed loop control system parameters are the easiest to determine in the Bode analysis approach?

Ans. There are two types of Bode plots. The first is a plot of magnitude of $GH(j\omega)$ *vs.* radian frequency (ω). Logarithmic scales are used for both $|GH(j\omega)|$ and (ω). The second is a plot of phase angle of $GH(j\omega)$ vs. radian frequency (ω).

Gain margin and phase margin of a closed loop control system can be determined from the Bode plots with minimum of computational effort.

Q. 65. With the help of suitable examples, briefly describe how Bode magnitude and phase plots would look like for different types of control systems, where (*a*) open loop transfer function is a constant (*K*) (*b*) open loop transfer function has (*l*) poles at origin (*c*) open loop transfer function has (*l*) zeros at origin (*d*) open loop transfer function is given by [*p*/(s + p)] with *p* > 0 (*e*) open loop transfer function is given by [(*s* + *z*)/*z*] with *z* > 0 and (*f*) open loop function is given by $GH(s) = \omega_n^2/(s^2 + 2\zeta n_s + \omega_n^2)$

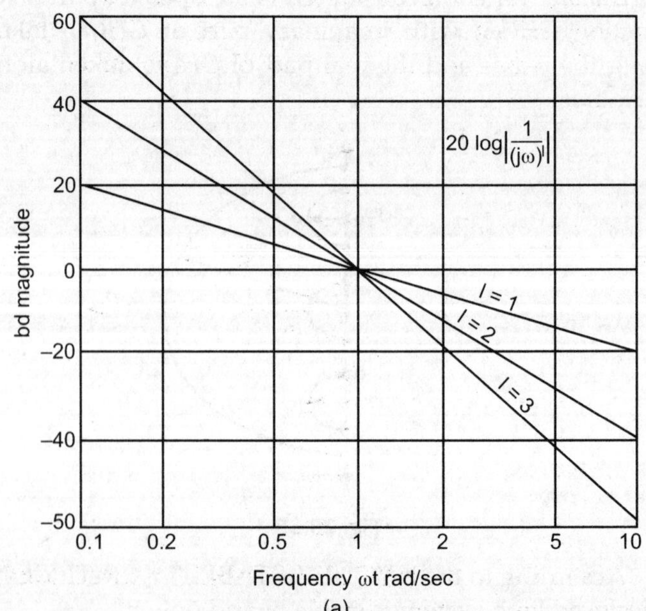

(a)

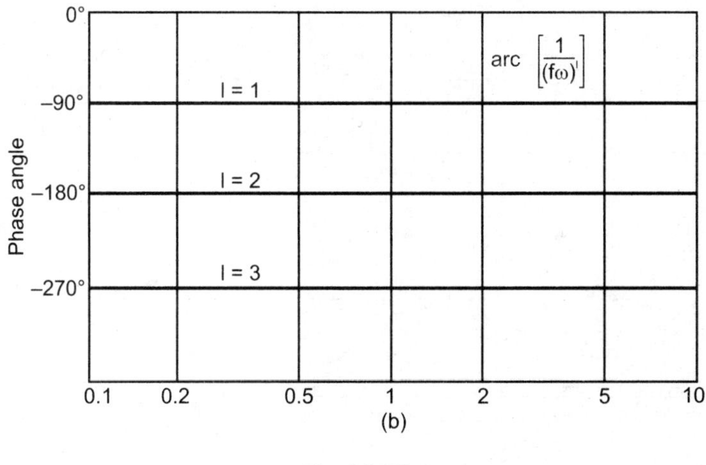

(b)

Fig. 28.39

Ans. Bode magnitude and phase plots for different types of control systems are explained as under:

(*a*) Magnitude plot of a control system whose open loop transfer function is a constant (K) is a straight line with magnitude equal to 20 $\log_{10} |K|$ and parallel to x-axis representing the frequency. The phase plot is also a straight line parallel to x-axis with magnitude equal to 0° for a positive (K) and – 180° for a negative (K).

(*b*) The magnitude and phase Bode plots for a control system whose open loop transfer function has (l) poles at origin are respectively given in Figures 28.39 (*a*) and (*b*).

(*c*) The magnitude and phase Bode plots for a control system whose open loop transfer function has (l) zeros at origin are respectively given in Figures 28.40 (*a*) and (*b*).

(*d*) The magnitude and phase Bode plots for a control system whose open loop transfer function is given by $GH(s) = p/(s + p)$ with $p > 0$ are respectively given in Figures 28.41 (*a*) and (*b*).

(*e*) The magnitude and phase Bode plots for a control system whose open loop transfer function is given by $GH(s) = (s + z)/z$ with $z > 0$ are respectively given in Figures 28.42 (*a*) and (*b*).

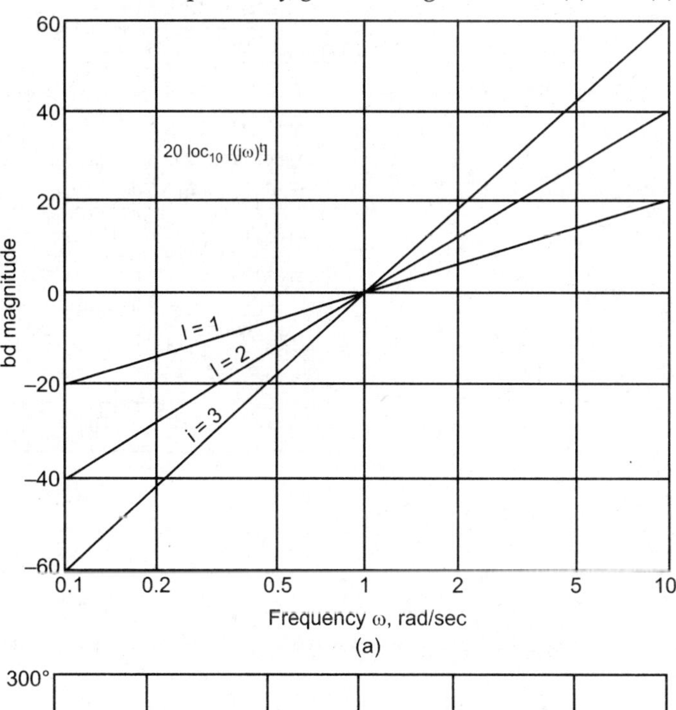

(a)

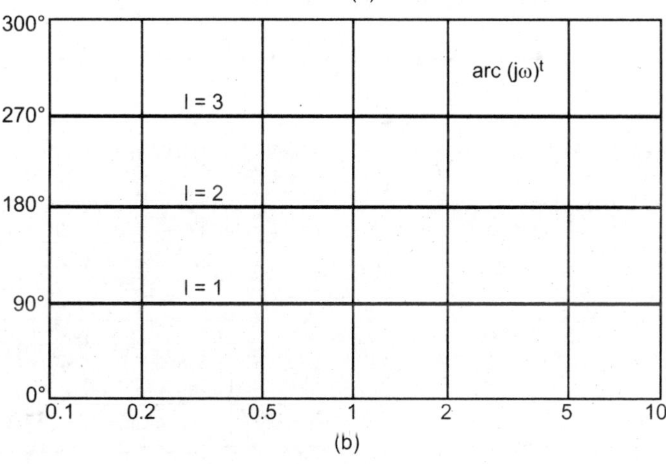

(b)

Fig. 28.40

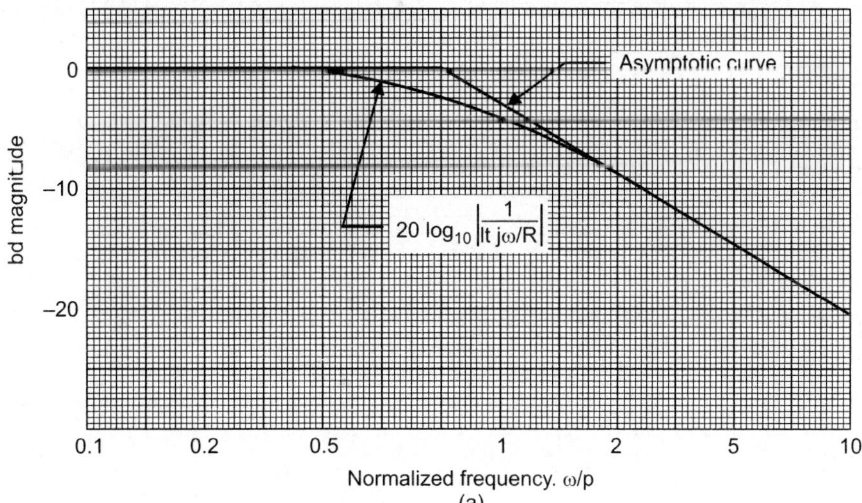

(a)

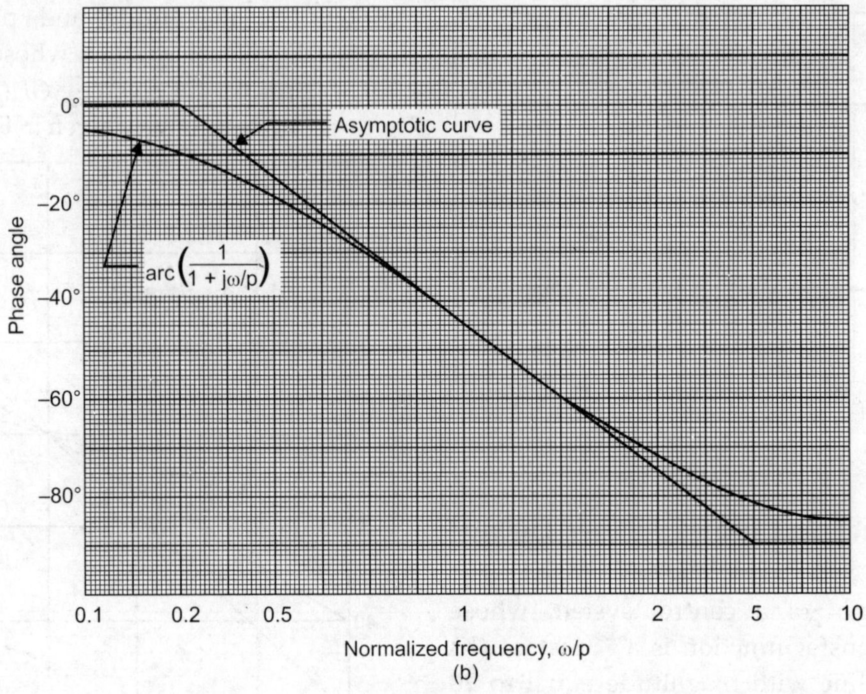

(b)

Fig. 28.41

(*f*) The Bode plots for the second order system with open loop transfer function given by $GH(s) = \omega_n^2/(s^2 + 2\zeta\omega_n{}^2 + \omega_n^2)$ or $GH(j\omega) = \omega_n^2/(-\omega^2 + 2j\zeta\omega_n + \omega_n^2)$ are shown in Figures 28.43 (*a*) and (*b*).

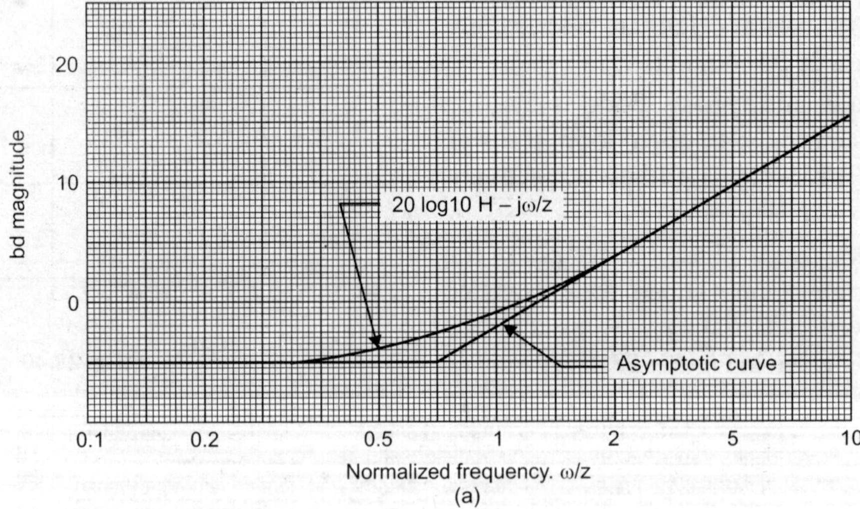

(a)

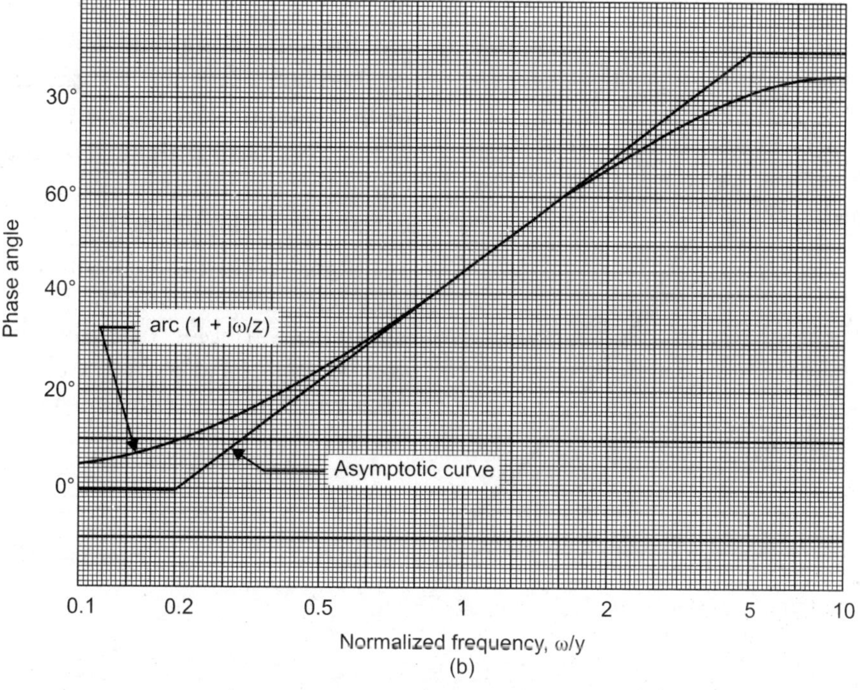

(b)

Fig. 28.42

Q. 66. How do you determine GAIN MARGIN and PHASE MARGIN from Bode plots ?

Ans. According to the definition of gain margin,

Gain margin = $\frac{1}{|GH(j\omega_n)|}$, where (ω_n) is phase cross-over frequency.

Now zero dB corresponds to a magnitude of 1. Therefore, gain margin is the number of dBs that $|GH(j\omega_\pi)|$ is below 0dB. Gain margin computation is illustrated in Fig. 28.43.

According to the definition of phase margin :

Phase margin = 180° + Arg $GH(j\omega_1)$

where, (ω_1) is the gain cross-over frequency i.e. frequency at which $|GH(j\omega)|$ is unity (or 0 dB). The phase margin is the number of degrees Arg $GH(j\omega_1)$ is above – 180°. Phase margin computation from Bode plots is also illustrated in Fig. 28.43.

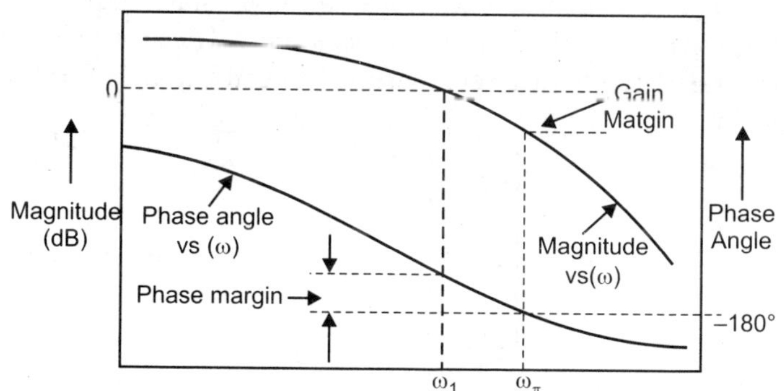

Fig. 28.43

Q. 67. What is a NICHOLS CHART? What is NICHOLS CHART PLOT?

Ans. A Nichols chart is a MAGNITUDE (in dB) vs. PHASE ANGLE plot of the loci of the constant dB magnitude and phase angle of the closed loop transfer function $\left[\frac{C}{R}(j\omega)\right]$ graphed as $|G(j\omega)|$ versus Arg $G(j\omega)$.

A Nichols chart plot is a dB magnitude-phase angle plot of a frequency response function $\left[\frac{C}{R}(j\omega)\right]$ superimposed on a Nichol's chart.

Q. 68. State variable analysis method of analyzing feedback control systems overcomes certain limitations experienced in the case of Root locus technique and frequency response techniques, such as Nyquist analysis and Bode plots. Briefly describe these shortcomings.

Ans. The techniques such as root locus method and frequency response methods, such as Nyquist analysis and Bode plots discussed in earlier questions for analysis of feedback control systems have certain limitations. For example, transfer function method is applicable to linear time invariant systems only. Also, these methods are used for single input, single output (SISO) control systems. These methods don't give any information regarding the state variables of the system and tell us only about system's output for a given input.

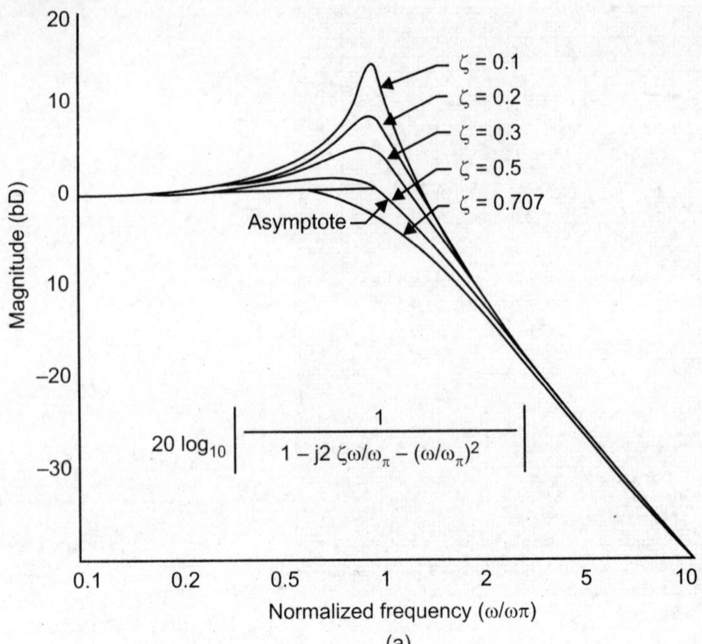

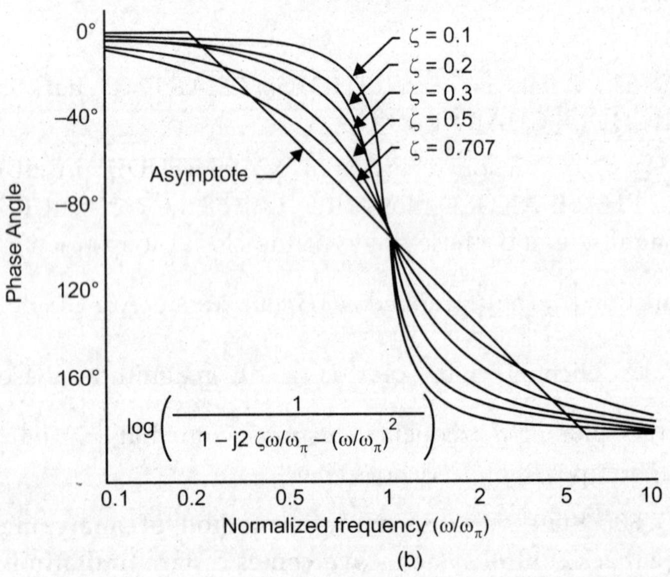

Fig. 28.44

State variable analysis on the other hand provides information about state variables at some predetermined points along signal flow in addition to giving information on the output. State variable analysis is applicable to both time invariant and time variant systems and also to single input, single output as well and multi-input, multi-output (MIMO) control systems. State variable analysis allows easy and convenient solutions through the use of digital computers. State variable approach also allows analysis of both continuous time and discrete time systems.

Q. 69. Define 'State Variables' and 'State Vector'.

Ans. State Variables: The state variables of a system are defined as a minimal set of variables $x_1(t)$, $x_2(t)$, x_3 (t).......... $X_n(t)$. The knowledge of these variables at any time (t_0) along with information on input excitation applied subsequently is sufficient to determine the state of the system at any time $t > t_0$. State variables should not be confused with system outputs. In general, output variables are usually defined as a function of state variables and input variables.

State Vector: If (n) number of state variables are necessary to determine the behaviour of a given system, these (n) variables can then be considered to be (n) components of a vector called state vector. The state vector in this case is a $(n \times 1)$ column matrix written as follows:

$$\begin{matrix} x_1(t) \\ x_2(t) \\ x_3(t) \\ \cdot \\ \cdot \\ \cdot \\ \cdot \\ x_n(t) \end{matrix}$$

Q. 70. How can a linear time invariant (LTI) control system be represented by state variable equations?

Ans. First, we shall discuss matrix representation of state equations. For an nth order system, n state equations can be written as follows:

$$\frac{dx_i(t)}{dt} = f_i[x_1(t), x_2(t), \ldots, x_n(t), r_1(t), r_2(t), \ldots, r_p(t)]$$

Where, $i = 1, 2, 3, \ldots, n$ and $r_1(t), r_2(t), \ldots, r_p(t)$ are the p input variables.

In general, the output variables are functions of the state variables and the input variables. If $c_1(t), c_2(t), \ldots, c_q(t)$ be the q output variables, then the output equations of a dynamic system can be expressed as follows:

$$c_j(t) = g_j[x_1(t), x_2(t), \ldots, x_n(t), r_1(t)\, r_2(t), \ldots, r_p(t)]$$

The q output equations and n state equations together are called *dynamic equations*. These equations can be conveniently represented in vector matrix form as follows:

$$x(t) = \begin{bmatrix} x_1(t) \\ x_2(t) \\ x_3(t) \\ \vdots \\ x_n(t) \end{bmatrix} \qquad \text{(this is } n \times 1 \text{ matrix)}$$

$$r(t) = \begin{bmatrix} r_1(t) \\ r_2(t) \\ r_3(t) \\ \vdots \\ r_p(t) \end{bmatrix} \quad \text{(this is } p \times 1 \text{ matrix)}$$

$$c(t) = \begin{bmatrix} c_1(t) \\ c_2(t) \\ c_3(t) \\ \vdots \\ c_q(t) \end{bmatrix} \quad \text{(this is } q \times 1 \text{ matrix)}$$

The $n \times 1$ column matrix $x(t)$ is called *state vector; p* $\times$ 1 column matrix is called *input vector* and $q \times 1$ column matrix is called *output vector*. The n state equations can be expressed as follows:

$$\frac{dx(t)}{dt} = f[x(t), r(t)]$$

Where, f is an $n \times 1$ column matrix that contains the function $f_1, f_2, f_3, \ldots, f_n$ as elements.

The q output equations can be expressed as follows:

$$c(t) = g[x(t), r(t)]$$

Where, f is a $q \times 1$ column matrix that contains the function $g_1, g_2, g_3, \ldots, g_n$ as elements.

A linear time invariant system may be represented in standard state equation form by a set of coupled first-order differential equations given as follows:

$$\frac{dx(t)}{dt} = Ax(t) + Br(t)$$

$$c(t) = Dx(t) + Er(t)$$

Where, x, r and c, respectively, represent state, input and output variables. The first part on the RHS of the state equation is called homogeneous part and the second part involving $r(t)$ is called *forcing function*. Here, A is an $n \times n$ co-efficient matrix with constant elements given by:

$$A = \begin{bmatrix} a_{11} & a_{12} & \cdots & a_{1n} \\ a_{21} & a_{22} & \cdots & a_{2n} \\ \vdots & \vdots & \cdots & \vdots \\ a_{n1} & a_{n1} & \cdots & a_{nn} \end{bmatrix}$$

Also, B is an $n \times p$ coefficient matrix with constant elements given by:

$$B = \begin{bmatrix} b_{11} & b_{12} & \cdots & b_{1p} \\ b_{21} & b_{22} & \cdots & b_{2p} \\ \vdots & \vdots & \cdots & \vdots \\ b_{n1} & b_{n2} & \cdots & b_{np} \end{bmatrix}$$

Also, D is a $q \times n$ coefficient matrix with constant elements given by:

$$D = \begin{bmatrix} d_{11} & d_{12} & \cdots & d_{1n} \\ d_{21} & d_{22} & \cdots & d_{2n} \\ \vdots & \vdots & \cdots & \vdots \\ d_{q1} & d_{q2} & \cdots & d_{qn} \end{bmatrix}$$

E is a $q \times p$ coefficient matrix with constant elements given by:

$$E = \begin{bmatrix} e_{11} & e_{12} & \cdots & e_{1p} \\ e_{21} & e_{22} & \cdots & e_{2p} \\ \vdots & \vdots & \cdots & \vdots \\ e_{q1} & e_{q2} & \cdots & e_{qp} \end{bmatrix}$$

Q. 71. Briefly describe the State Transition Matrix.

Ans. The homogeneous state equation can be written as follows:

$$\frac{dx(t)}{dt} = Ax(t)$$

The matrix that satisfies the linear homogeneous state equation is known as the *state transition matrix*. If $\phi(t)$ is the $n \times n$ matrix that represents the state transition matrix, then,

$$\frac{d\phi(t)}{dt} = A\,\phi(t)$$

If $x(0)$ denotes the initial state at $t = 0$, then $\phi(t)$ can also be defined by the state equation:

$$x(t) = \phi(t)\,.\,x(0)$$

This is the solution of homogeneous state equation for $t \geq 0$. Without going into elaborate mathematics, two possible expressions for state transition matrix are given as under:

$$\phi(t) = L^{1}[(sI - A)^{1}] \ldots t \geq 0 \text{ and}$$

$$\phi(1) = e^{At} = 1 + At + \left(\frac{1}{2}\right)!\,A^2t^2 + \left(\frac{1}{3}\right)!\,A^3t^3 + \ldots$$

Q. 72. With the help of a suitable example, describe the concepts of state variables, state equation and state transition matrix.

Ans. We shall illustrate the concepts of state variables, state equation and state transition matrix with the help of finding solution to a simple RL series network shown in Figure 28.44. The state equation of this network for time $t \geq 0$ can be written as follows:

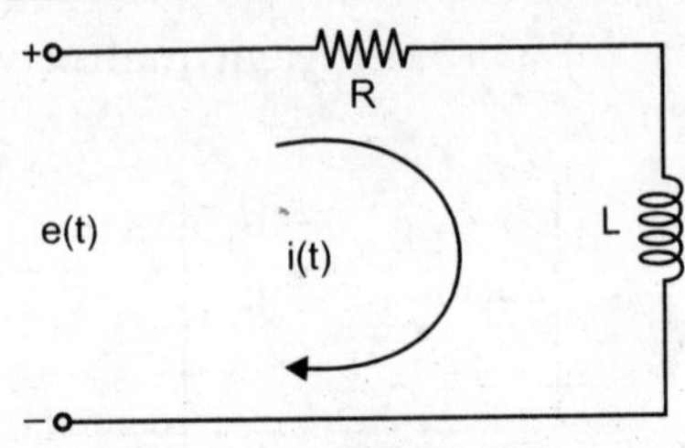

Fig. 28.44

$$\frac{di(t)}{dt} = \left(\frac{R}{L}\right)i(t) + \left(\frac{I}{L}\right)e(t)$$

Here, $i(t)$ is the state variable and $e(t)$ is the input. The solution to this equation can be found by taking Laplace transform on both sides, solving for $I(s)$ and then taking inverse Laplace transform. The solution is given by the following expression:

$$i(t) = e^{-(R/L)t}\,i(0) + \frac{E}{R}[1 - e^{-(R/L)t}]$$

It is evident that $i(t)$ satisfies the basic requirements as a state variable as it is the energy storage capability of the inductor element that holds the history of the system. The same is valid for the voltage across a capacitor. The first term on the RHS of the above equation is the homogeneous state equation, that is, state equation with zero input. The state transition matrix, therefore, is given by:

$$\phi(t) = e^{At} = e^{-(R/L)t},\text{ where } A = -\left(\frac{R}{L}\right)$$

Q. 73. Briefly describe the important characteristics of state transition matrix.

Ans. State transition matrix represents the free response of the system, as it satisfies the homogeneous state equation. It dictates the system response that is triggered by initial conditions only. Since the state transition matrix is dependent only on A; it is also referred to as state transition matrix of A. The state transition matrix completely defines the transition of states from time $t = 0$ to any time t for zero input. Following are the important characteristics of state transition matrix:

1. $\phi(0) = I$, where I is identity matrix.
2. $\phi^{-1}(t) = \phi(-t)$.
3. $\phi(t_2 - t_1)\phi(t_1 - t_0) = \phi(t_2 - t_0)$ for any t_0, t_1 and t_2.
4. $[\phi(t)]^k = \phi(kt)$, where k is an integer.

Q. 74. Describe different steps involved in finding solution to linear time invariant state equation using Laplace transform method.

Ans. The generalised form of linear time invariant state equation is given by:

$$\frac{dx(t)}{dt} = Ax(t) + Br(t)$$

The above equation can be solved by either of the following two methods:

1. Classical method of solving differential equations.
2. Laplace transform method.

In the case of more commonly employed Laplace transform method, the first step is to take Laplace transform on both sides. This gives:

$$sX(s) - x(0) = AX(s) + BR(s)$$

In the second step, solving for $X(s)$ gives the following expression:

$$X(s) = (sI - A)^{-1}x(0) + (sI - A)^{-1}BR(s)$$

In the third step, the state equation can then be obtained by taking inverse Laplace transform on both sides to get the following expression for $x(t)$:

$$x(t) = L^{-1}[(sI - A)^{-1}]\,x(0) + L^{-1}[(sI - A)^{-1}BR(s)]$$

Using the definitions of state transition matrix that says that $\phi(t) = L^{-1}[(sI - A)^{-1}]$ and also the convolution integral, the above equation can be written as follows:

$$x(t) = \phi(t)\,x(0) + \int_0^t \phi(t-\tau)\,Br(\tau)\,d\tau \qquad (t \geq 0)$$

The above equation is valid only if initial time were defined to be $t = 0$. In general, if the initial time was defined to be t_0 and the input $r(t)$ was applied at $t = 0$, then the above equation gets transformed to the following equation:

$$x(t) = \phi(t - t_0)x(t_0) + \int_{t_0}^t \phi(t-\tau)\,Br(\tau)\,d\tau$$

Having determined the state equation, the output vector can be determined as a function of initial condition and the applied input vector by substituting $x(t)$ into output equation given earlier:

$$c(t) = D\phi(t - t_0)x(t_0) + \int_{t_0}^t \phi(t-\tau)\,Br(\tau)\,d\tau + Er(t)$$

Q. 75. What do you understand by Controllability of a process? Explain with the help of a suitable example.

Ans. A process G is said to be completely controllable, if each of the state variables of the process can be transformed from an initial state to a desired final state in a finite time interval by some unconstrained control input. Even if there were one state variable that was independent of the control input; the system is said to be not completely controllable. A linear time invariant system as described earlier is defined by the following expressions:

1. $dx/dt = Ax(t) + Bu(t)$
2. $c(t) = Dx(t) + Eu(t)$

where, $x(t)$ is the $n \times 1$ state vector, $u(t)$ is the $r \times 1$ input vector, $c(t)$ is the $p \times 1$ output vector, A is the $n \times n$ co-efficient matrix, B is the $n \times r$ co-efficient matrix, D is the $p \times n$ co-efficient matrix and E is the $p \times r$ co-efficient matrix.

The state $x(t)$ is considered controllable at $t = t_0$ if there exists a control input $u(t)$ that can drive the state to any final state in a finite time $t_f - t_0 \geq 0$. If every state $x(t_0)$ of the system is controllable in a finite time interval, the system is said to be state controllable.

The condition of controllability depends upon coefficient matrices A and B of the system. The necessary and sufficient condition for system being completely state controllable is that the matrix,

$$[Q_c] = [B \; AB \; A^2B \; A^3B \; ... \; A^{n-1} \; B]$$

has a rank of n and is non-singular. The following example illustrates the method of determining controllability of a linear time invariant system: Let us take the case of a linear system defined by the following matrix representation of state equations:

$$\begin{bmatrix} dx_1(t)/dt \\ dx_2(t)/dt \end{bmatrix} = \begin{bmatrix} 0 & 1 \\ -1 & -2 \end{bmatrix} \begin{bmatrix} x_1 \; (t) \\ x_2 \; (t) \end{bmatrix} + \begin{bmatrix} 0 \\ 1 \end{bmatrix} u(t)$$

and $c = x_1(t)$. From the given state equations,

$$A = \begin{bmatrix} 0 & 1 \\ -1 & -2 \end{bmatrix}, B = \begin{bmatrix} 0 \\ 1 \end{bmatrix} \text{ and } AB = \begin{bmatrix} 1 \\ -2 \end{bmatrix}$$

Therefore, the controllability matrix is

$$[Q_C] = [B \; AB] = \begin{bmatrix} 0 & 1 \\ 1 & -2 \end{bmatrix}$$

Matrix $[Q_C]$ is non-singular and therefore, the system is controllable.

Q. 76. What do you understand by Observability of a process? Explain the conditions of observability with the help of a suitable example.

Ans. Observability is similar in concept to controllability and allows observing or estimating state variables from the measurements made on input and output variables. If a state variable cannot be observed from the measurements of the outputs; it is said to be unobservable. For a linear time invariant system, a state variable $x(t)$ at $t = t_0$ is considered to be observable, if for any given input $u(t)$, there exists a finite time $t_f \geq t_0$ such that knowledge of $u(t)$ for the time interval $t_0 \leq t \leq t_f$, the matrices A, B, C and D and the output $c(t)$ for $t_0 \leq t \leq t_f$ are sufficient to determine $x(t_0)$. The system is said to be completely observable, if every state of the system is observable for finite time t_f.

Condition of observability depends on the coefficient matrices A and D of the system. The necessary and sufficient condition for the system to be completely observable is that the $(n \times np)$ matrix has a rank of n, that is,

$$[Q_O] = [D' \; A'D' \; (A')^2D' \; ... \; (A')^{n-1} \; D']$$

If the system had one output, D would be a $1 \times n$ matrix and Q_O would be an $n \times n$ matrix. The system is observable, if Q_O is non-singular. We shall illustrate the procedure with help of an example. Let us consider the system expressed by state equations as follows:

$$\begin{bmatrix} dx_1(t)/dt \\ dx_2(t)/dt \end{bmatrix} = \begin{bmatrix} -2 & 0 \\ 0 & -1 \end{bmatrix} \begin{bmatrix} x_1(t) \\ x_2(t) \end{bmatrix} + \begin{bmatrix} 2 \\ 1 \end{bmatrix} u(t) \text{ and}$$

$$c = [2 \; 0] \begin{bmatrix} x_1(t) \\ x_2(t) \end{bmatrix}$$

From the given information, we get,

$$D = [2 \; 0] \text{ and } D' = \begin{bmatrix} 2 \\ 0 \end{bmatrix}$$

Also,

$$A = \begin{bmatrix} -2 & 0 \\ 0 & -1 \end{bmatrix}$$

Therefore,

$$A'D' = \begin{bmatrix} -4 \\ 0 \end{bmatrix}$$

This gives

$$[Q_o] = [D' \; A' \; D'] = \begin{bmatrix} 2 & -4 \\ 0 & 0 \end{bmatrix}$$

Since $[Q_O]$ is singular, the system is unobservable.

Q. 77. What are Eigen values with reference to state variable approach of analysing linear systems? Name one important property of Eigen values.

Ans. The characteristic equation plays a significant role in the study of linear systems. In the state variable approach of analysing linear systems, the transfer function can be written as follows:

$$G(s) = \frac{C(s)}{R(s)} = D(sI - A)^{-1} \; B + E$$

Now,

$$(sI - A)^{-1} = \frac{Adj(sI - A)}{|sI - A|}$$

This gives

$$G(s) = D\left[\frac{Adj(sI - A)}{|sI - A|}\right]B + E$$

$$= \frac{D\{AdJ[(sI - A)B] + (sI - A)E]\}}{|sI - A|}$$

The characteristic equation is obtained by setting denominator to zero. That is, $|sI - A| = 0$ is the characteristic equation of $G(s)$. The roots of the characteristic equation are referred to as Eigen values of the matrix A. An important property of the characteristic equation and Eigen values is that they remain unchanged during a non-singular transformation. That is, if the matrix A is transformed by a non-singular transformation $x = Py$ so that $A' = P^{-1} AP$, then the characteristic equation and the Eigen values of A' are the same as those of A.

Solved Problems

Problem 28.1. *Classify the following differential equations according to whether they are ordinary or partial, time variable or time invariant, linear or non-linear.*

(i) $\frac{d^2x}{dt^2} + 2x = 0$

(ii) $\frac{y\,dy}{dt} + 2y = 0$

(iii) $\frac{d}{dt}(t^2y) = 0$

(iv) $\frac{d^2y}{dt^2} + y\frac{dy}{dt} + y^2 = 0$

(v) $\frac{\partial T}{x} + k\frac{\partial T}{\partial t} = 0$

(vi) $\frac{\partial f}{\partial x} + \frac{\partial f}{\partial y} + 4x + 5y = 0$, where $f = f(x, y)$

Solution.

(i) Ordinary, Time invariant, Linear
(ii) Ordinary, Time invariant, Non-linear
(iii) Ordinary, Time variable, Linear
(iv) Ordinary, Time invariant, Non-linear
(v) Partial, Time invariant, Linear
(vi) Partial, Time invariant, Linear

Problem 28.2. *Determine the damping ratio (ζ), undamped natural frequency (ω_n), damped natural frequency (ω_d), time constant (τ), settling time (t_s) and the percentage overshoot in case of a second order control system represented by the differential equation.*

$$\frac{3d^2y}{dt^2} + \frac{6\,dy}{dt} + 12y = 12x$$

Solution.

The given differential equation can be written as

$$\frac{d^2y}{dt^2} + \frac{2dy}{dt} + 4y = 4x \text{ or } \frac{d^2y}{dt^2} + \frac{2dy}{dt} + (2)^2y = (2)^2x$$

Comparing the above equation with the generalised second order control system differential equation,

$$\frac{d^2y}{dt^2} + 2\,\zeta\,\omega_n\frac{dy}{dt} + \omega_n^2 y = \omega_n^2 x$$

we get, $2\zeta\,\omega_n = 2, \omega_n = 2$

That is, $2\zeta = \frac{2}{\omega_n} = \frac{2}{2} = 1, \tau = 0.5$

Also, damped natural frequency, $\omega_d = \omega_n\sqrt{1 - \zeta^2}$

$$= 2\sqrt{1 - (0.5)^2}$$

$$= 2\sqrt{0.75} \cong 1.73 \text{ radians/sec}$$

Time constant, $t = \frac{1}{\zeta\,\omega_n} = \frac{1}{0.5 \times 2} = 1$ sec

Settling time $t(s) = 4 \times$ time constant $= 4$ secs

Overshoot $= \text{Exp}\left[-\left(\frac{\zeta\pi}{\sqrt{1 - \zeta^2}}\right)\right]$

$$= \text{Exp}\left(-\frac{0.5\,\pi}{\sqrt{0.75}}\right)$$

$$= \text{Exp}(-0.58\pi) \cong 0.17 = 17\%$$

Problem 28.3. Find the percentage overshoot in case of the second order control systems represented by.

(i) $4\frac{d^2y}{dt^2} + 36y = 36x$ (ii) $\frac{d^2y}{dt^2} + 4\frac{dy}{dt} + 4y = 4x$

Solution.

(i) Comparing the given differential equation with the generalised form, we get

$$\omega_n = 3, \zeta = 0$$

Substituting the value of (ζ) in the expression for determining overshoot,

$$\text{Overshoot} = \text{Exp}\left[-\left(\frac{\zeta\pi}{\sqrt{1 - \zeta^2}}\right)\right] = 1 = 100\%$$

(*ii*) Comparing the given differential equation with the generalised form,

$\omega_n = 2$, $\zeta = 1$, ζ equal to unity gives an overshoot of 0.

Problem 28.4. *Determine the settling time for the second order systems of problem 28.3.*

Solution.

(*i*) System time constant, $\tau = \dfrac{1}{\zeta\,\omega_n} = \dfrac{1}{0}$ = infinity

Settling time is also therefore infinity. Ideally therefore, such a control would always be oscillating with a frequency of 3 rad/sec and

(*ii*) System time constant, $\tau = \dfrac{1}{\zeta\omega_n} = \dfrac{1}{2} = 0.5$ sec

Settling time = $4t = 4 \times 0.5 = 2$ secs

Problem 28.5. *The location of the poles of a second order system is shown in Fig. 28.45. Determine the damping ratio (ζ), percentage overshoot and the settling time.*

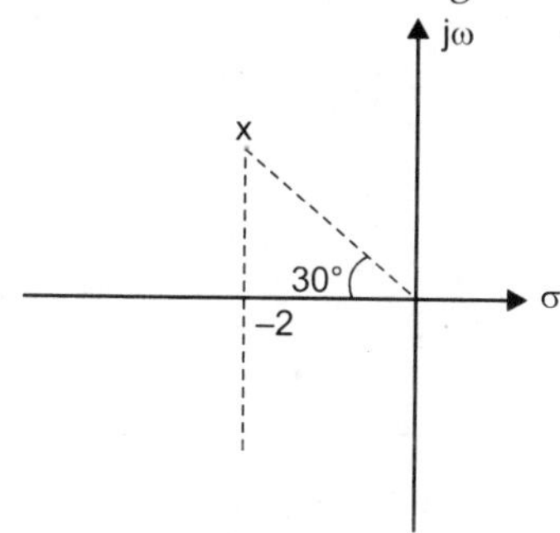

Fig. 28.45

Solution.

Damping ratio is given by

$$\zeta = \cos\theta = \cos 30° = \sqrt{3}/2 = 0.866$$

Percentage overshoot is given by

$$\text{Exp}\left[-\left(\frac{\zeta\pi}{\sqrt{1-\zeta^2}}\right)\right] \times 100$$

$$= \left[-\frac{0.866 \times 3.14}{\sqrt{0.25}}\right] \times 100$$

$$= \text{Exp}\,[-0.866 \times 3.14 \times 2] \times 100$$

$$= \text{Exp}\,(-5.43) \times 100 \cong 1\%$$

Settling time $\cong$ 4 × time constant = $\dfrac{4}{\zeta\omega_n}$

From Fig. 28.45, $\omega_n = \dfrac{2}{\cos 30°} = \dfrac{2}{\sqrt{3/2}} = \dfrac{4}{\sqrt{3}}$

Therefore, settling time = $\dfrac{4}{\left[\dfrac{\sqrt{3}}{2} \times \dfrac{4}{\sqrt{3}}\right]}$ = 2 secs

Problem 28.6. *The time domain total response of a second order system is given by $y(t) = (8e^{-t} - 7e^{-2t} + 3t - 4)$. Determine its steady state response and transient response.*

Solution.

The steadystate response is that part of the total response that does not approach zero as the time approaches infinity.

Obviously, steadystate response is given by $y_s(t) = 3t - 4$

The transient response is that part of the total response that approaches zero as the time approaches infinity.

Therefore, transient response is given by $y_T(t) = 8e^{-t} - 7e^{-2t}$.

Problem 28.7. *Determine the free response of a second order control system described by the differential equation, $\dfrac{d^2y}{dt^2} + 4\dfrac{dy}{dt} + 3y = x$ with initial conditions $y(0) = 0$ and $\left.\dfrac{dy}{dt}\right|_{t=0} = 2$.*

Solution.

The free response is determined by the initial conditions. In general, free response is given by

$$y_a(t) = c_1 e^{-D_1 t} + c_2 e^{-D_2 t}$$

where, c_1 and c_2 are unknown constants depending upon the initial conditions; D_1 and D_2 are the roots of the system's characteristic equation. The characteristic equation of the given system is

$D^2 + 4D + 3 = 0$ which gives, $D_1 = 1$ and $D_2 = 3$

Therefore, $y_a(t) = c_1 e^{-t} + c_2 e^{-3t}$

The first initial condition gives $c_i + c_2 = 0$

The second initial condition gives

$$\left.\frac{dy}{dt}\right|_{t=0} = \left.\frac{d}{dt}\left(c_1 e^{-t} + c_2 e^{-3t}\right)\right|_{t=0} = -c_1 - 3c_2 = 2$$

Solving for c_1 and c_2, we get $c_1 = 1$, $c_2 = -1$

Therefore, free response, $ya(t) = (e^{-t} - e^{-3t})$.

Problem 28.8. *Determine the forced response of the second order system of problem 38.7 for $x(t) = 1$.*

Solution.

The forced response is determined by putting all initial conditions equal to zero. The forced response depends only on the input $x(t)$. The forced response of a control system described by a linear, constant co-efficient, ordinary differential equation can be written in terms of a convolution integral

$$\int_0^t \omega(t-\tau)\,x(\tau)\,d\tau$$

where, $\omega(t - \tau)$ is the weighting function of the differential equation. The weighting function $\omega(t)$ is the free response of the system.

The forced response, $y_b(t) = \int_0^t \omega(t-\tau)x(\tau)\,d\tau$

$$y_b(t) = \int_0^t \omega(t-\tau)\,d\tau$$

Now $\omega(t) = (e^{-t} - e^{-3t})$

Therefore, $\omega(t-\tau) = e^{-(t-\tau)} - e^{-3(t-\tau)} = e^{-t}.e^{\tau} - e^{-3t}.e^{3\tau}$

$$y_b(t) = e^{-t}\int_0^t e^{\tau}d\tau - e^{-3t}\int_0^t e^{3\tau}d\tau$$

$$= e^{-t}\left|e^{\tau}\right|_0^t - e^{-3t}\left|e^{3\tau/3}\right|_0^t$$

$$= e^{-t}(e^t - 1) - e^{-3t}\left(\frac{e^{3t}}{3} - \frac{1}{3}\right)$$

$$= 1 - e^{-t} - \frac{1}{3} + \frac{1}{3}e^{-3t} = \left(\frac{2}{3} - e^{-t} + \frac{e^{-3t}}{3}\right)$$

$$= \frac{1}{3}(2 - 3e^{-t} + e^{-3t})$$

Problem 28.9. *Find the total response of the system described by the differential equation of problem 28.7.*

Solution.

The total response is given by sum of free and forced responses.

Total response $= y_a(t) + y_b(t) = e^{-t} - e^{-3t} +$

$$\frac{1}{3}(2 - 3e^{-t} + e^{-3t})$$

$$= \frac{2}{3} - \frac{2}{3}e^{-3t} = \frac{2}{3}(1 - e^{-3t})$$

Problem 28.10. *Determine the steadystate and the transient response for the system described the differential equation of problem 28.7.*

Solution.

The total response as determined in solution to problem 28.9 is

$$y(t) = \frac{2}{3}(1\ e^{-3t}) = \frac{2}{3} - \frac{2}{3}e^{-3t}$$

The steadystate part, therefore, is $y_s(t) = \frac{2}{3}$ and the transient part, $y_T(t) = -\frac{2}{3}e^{-3t}$

Problem 23.11. *Determine the transfer function of the system represented by the differential equation.*

$$\frac{d^2y}{dt^2} + \frac{dy}{dt} + 3y = \frac{dx}{dt} + 2x$$

Solution.

Taking Laplace transform of both sides of the given differential equation and assuming all initial conditions to be zero, we get,

$$(s^2 + s + 3)Y(s) = (s + 2)X(s)$$

Therefore, $P(s) = \dfrac{Y(s)}{X(s)} = \dfrac{(s+2)}{(s^2+s+3)}$

Problem 23.12. *Determine the system differential equation from the system transfer function.*

$$P(s) = \frac{(s+1)(s+3)}{s(s+2)(s+4)}$$

Solution.

$$P(s) = \frac{(s+1)(s+3)}{s(s+2)(s+4)} = \frac{(s^2+4s+8)}{s(s^2+6s+8)}$$

$$= \frac{(s^2+4s+3)}{s^3+6s^2+8s}$$

$$\frac{Y(s)}{X(s)} = \frac{s^2+4s+3}{s^3+6s^2+8s}$$

$$y(t) = \left(\frac{D^2+4D+3}{D^3+6D^2+8D}\right)x(t), \text{ where } D \equiv \frac{d}{dt}$$

or $D^3y + 6D^2y + 8Dy = D^2x + 4Dx + 3x$

$$\frac{d^3y}{dt^3} + 6\frac{d^2y}{dt^2} + 8\frac{dy}{dt} = \frac{d^2x}{dt^2} + 4\frac{dx}{dt} + 3x$$

Problem 28.13. *A particular system is expressed by a differential equation,* $\frac{dy(t)}{dt} + 3y(t) = x(t - T)$. *Determine its transfer functions.*

Solution.

Taking Laplace transform of both sides of the system differential equation.

$$sY(s) + 3Y(s) = e^{-sT}X(s)$$

or $$Y(s)[s+3] = e^{-st}X(s)$$

or $$P(s) = \frac{Y(s)}{X(s)} = \frac{e^{-sT}}{s+3}$$

Problem 23.14. *The impulse response of a certain system is Sin 2t. Determine the system transfer function.*

Solution.

The Laplace transform of the impulse response of the system gives the transfer function.

Therefore, $P(s) = \frac{2}{s^2+4}$ ......Laplace transform of sin ωt is $\left(\frac{\omega}{s^2+\omega^2}\right)$.

Problem 23.15. *The step response of a certain control system is e^{-t} sin t. Determine its transfer function.*

Solution.

The impulse function is derivation of step function. Therefore, impulse response of the system is given by,

$$\frac{d}{dt}(e^{-t}\sin t) = -e^{-t}\sin t + e^{-t}\cos t = e^{-t}(\cos t - \sin t)$$

The transfer function is nothing but Laplace transfer of impulse response.

$$P(s) = \text{Laplace transform of } (e^{-t}\cos t - e^{-t}\sin t)$$

$$= \text{Laplace transform of } e^{-t}\cos t - \text{Laplace transform of } e^{-t}\sin t$$

$$= \frac{(s+1)}{(s+1)^2+1} - \frac{1}{(s+1)^2+1} = \frac{(s+1)-1}{(s+1)^2+1}$$

$$= \frac{s}{(s+1)^2+1} = \frac{s}{(s^2+2s+2)}$$

The transfer function can also be determined as follows:

$$y(t) = e^{-t}\sin t$$

$$Y(s) = \frac{1}{(s+1)^2+1} = \frac{1}{s^2+2s+2}$$

$$x(t) = u(t)$$

$$X(s) = 1/s$$

Therefore, $P(s) = \frac{Y(s)}{X(s)} = \frac{1}{(s^2+2s+2)} \times \frac{1}{1/s}$

Problem 28.16. *Comment on the stability of the control systems represented by the conditions outlined below:*

1. *The system has poles at – 2 and – 5 and zeros at 2 and – 3.*
2. *The system characteristic equation is (s + 1) (s + 3) (s – 2) = 0.*
3. *The system's impulse response is given by h(t) = 1.*
4. *The system's impulse response is given by e^{-2t} sin 3t.*
5. *The system is described by differential equation $\frac{dy}{dt} = x$.*
6. *A step input applied to the system yields an output y = t.*

Solution.

1. Since the system poles have negative real parts, the system is stable.
2. The system has poles at – 1, – 3 and + 2. It is unstable.
3. The impulse response does not decay to zero as the time approaches infinity. The system, therefore, is unstable.
4. The impulse response decays to zero as the time approaches infinity. The system is stable.
5. The system characteristic equation is D = 0. So, the system has a pole at the origin. It is not stable.
6. The output is unbounded for a bounded input. The system is unstable.

Problem 28.17. *A second order control system is represented by the differential equation, $\frac{d^2y}{dt^2} + 3\frac{dy}{dt} + 4y = 4x$. With the help of Routh stability criterion, determine whether the system is stable or not.*

Solution.

The system characteristic equation (in s-domain) is given by, $s^2 + 3s + 4 = 0$ The routh table is constructed as follows:

s^2	1	0
s^1	3	0
s^0	4	0

The elements of the first column (1, 3, 4) have all positive signs. The system is, therefore, stable.

Problem 28.18. *Comment upon the stability of the system whose characteristic equation is given by $s^5 + s^4 - s^3 + s - 1 = 0$. Use Routh stability criterion.*

Solution.

The characteristic equation can be re-written as

$s^5 + s^4 - s^3 + 0s^2 + s - 1 = 0$

The Routh table is constructed as :

s^5	1	–1	1	0
s^4	1	0	–1	0
s^3	–1	2	0	
s^2	2	–1	0	
s^1	$\frac{3}{2}$	0		
s^0	–1	0		

The elements of the first column (1, 1, – 1, 2, 3/2, – 1) show change in sign at three places. The system is unstable and has three poles with positive real parts.

Problem 28.19. *The characteristic equation of a given system is $s^3 + (K+4)s^2 + 6s + 12 = 0$. Determine the values of (K) for which the system would be stable.*

Solution.

The Routh table for this system is

$$\begin{array}{c|cc} s^3 & 1 & 6 \quad 0 \\ s^2 & (K+4) & 12 \quad 0 \\ s^1 & \dfrac{6(K+4)-12}{(K+4)} & 0 \\ s^0 & 12 & \end{array}$$

For the system to be stable, $K + 4 > 0$ or $K > -4$

Also, $\dfrac{6(K+4)-12}{K+4} > 0$ or $6(K+4) - 12 > 0$

or $(K+4) > 2$ or $K > -2$

The condition of $K > -2$ also satisfies the condition $K > -4$. Therefore, the system is stable for $K > -2$.

Problem 28.20. *Find the range of values of gain factor (K) for which the control system with characteristic equation $s^2 + K(2+s) - 1 = 0$ is stable. Use Hurwitz stability criterion.*

Solution.

The characteristic equation can be re-written as

$$s^2 + Ks + (2K-1) = 0$$

The Hurwitz determinants are

$$\Delta_2 = \begin{vmatrix} K & 0 \\ 1 & 2K-1 \end{vmatrix} \text{ and } \Delta_1 = K$$

$$|\Delta_2| = (2K-1)K = 2K^2 - K$$

for $|\Delta_1|$ and $|\Delta_2|$ to be positive, $K > 0$ and $(2K-1) > 0$

or $2K > 1, K > 1/2$

Thus, the system is stable, if $K > 1/2$.

Problem 28.21. *For what values of (K), the system with characteristic polynomial of $s^3 + (4+K)^2 + 6s + 12$ is stable? Use Hurwitz stability criterion.*

Solution.

The Hurwitz determinants are

$$\Delta_3 = \begin{vmatrix} (4+K) & 12 & 0 \\ 1 & 6 & 0 \\ 0 & (4+K) & 12 \end{vmatrix}$$

$$\Delta_2 = \begin{vmatrix} (4+K) & 12 \\ 1 & 6 \end{vmatrix} \text{ and } \Delta_1 = (4+K)$$

Now $|\Delta_3| = 72(4+K) - 12 \times 12 = 72(4+K) - 144$

$|\Delta_2| = 6(4+K) - 12$

$|\Delta_1| = 4 + K$

For all the determinants to be positive,

$4 + K > 0$ or $K > -4$

$6(4+K) - 12 > 0$ or $6(4+K) > 12$ or $4 + K > 2$ or $K > -2$

$72(4+K) - 144 > 0$ or $72(4+K) > 144$ or $4 + K > 2$

or $K > -2$

Summarising the three conditions, we can say that the system is stable, when $K > -2$.

Problem 28.22. *Using continued fraction stability criterion, determine whether the system with a characteristic polynomial of $s^3 + 6s + 5s + 6$ is stable or not.*

Solution.

The characteristic polynomial = $s^3 + 6s + 5s + 6$

Let $Q_1(s) = s^3 + 5s$

$Q_2(s) = 6s + 6$

The continued fraction is formed as under

$$\frac{Q_1(s)}{Q_2(s)} = \frac{s^2 + 5s}{6s^2 + 6} = \frac{1}{6}s + \frac{4s}{6s^2 + 6}$$

$$= \frac{1}{6}s + \cfrac{1}{\frac{2}{3}s + \cfrac{1}{\frac{4s}{6}}}$$

The co-efficients of (s) are $\frac{1}{6}, \frac{3}{2}, \frac{4}{6}$. They are all positive. Therefore, the system is stable.

Problem 28.23. *Using continued fraction stability criterion, determine whether the system with the characteristic polynomial $s^3 - 2s^2 + 4s + 6$, is stable or unstable.*

Solution.

The characteristic polynomial = $s^3 - 2s^2 + 4s + 6$

Let $Q_1(s) = s^3 + 4s$

$Q_2(s) = -2s^2 + 6$

The continued fraction is formed as under

$$\frac{Q_1(s)}{Q_2(s)} = \frac{s^3 + 4s}{-2s^2 = 6} = -\frac{1}{2}s + \frac{7s}{-2s^2 + 6}$$

$$= -\frac{1}{2}s + \cfrac{1}{-\frac{2}{7}s + \cfrac{1}{\frac{7}{6}s}}$$

The co-efficients are $-\frac{1}{2}, -\frac{2}{7}, \frac{7}{6}$

Since the co-efficients are not all positive, the system is unstable.

Problem 28.24. *For what values of (K) will the control system represented by the characteristic polynomial $s^4 + 7s^3 + 24s^2 + 35s + K$ is stable. Use continued fraction stability criterion.*

Solution.

Let $Q_1(s) = s^4 + 24s^2 + K$

$Q_2(s) = 7s^3 + 35s$

The continued fraction is formed as under

$$\frac{Q_1(s)}{Q_2(s)} = \frac{s^4+24s^2+K}{7s^3+32s} = \frac{1}{7}s = \frac{19s^2+K}{7s^3+35s}$$

$$= \frac{1}{7}s + \cfrac{1}{\frac{7}{19}s + \cfrac{1}{\left(\frac{19}{35-\frac{7}{19}K}\right)s + \cfrac{1}{\left(\frac{35-\frac{7}{19}K}{K}\right)s}}}$$

The co-efficients are $\frac{1}{7}, \frac{7}{19}, \left(\frac{19}{35-\frac{7K}{19}}\right)$ and $\left(\frac{35-\frac{7K}{19}}{K}\right)$

For the system to be stable, $\frac{19}{35-\frac{7K}{19}}$ and $\frac{35-\frac{7K}{19}}{K}$ should positive

That is $35 - \frac{7K}{19} > 0$ or $-\frac{7K}{19} > -35$

or $\frac{7K}{19} < 35$ or $K < 95$

Also, $\frac{35-\frac{7K}{19}}{K} > 0$

That is either both $\left(35 - \frac{7K}{19}\right)$ and K should be negative or both $\left(35 - \frac{7K}{19}\right)$ and K should be But $\left(35 - \frac{7K}{19}\right)$ should be positive from the first condition.

Therefore, $K > 0$

To summarise the results, the necessary condition for the system to be stable is $0 < K < 95$.

Problem 28.25. *Fig. 28.46 shows some simple block diagram representations. Find out their equivalents from the block diagrams shown in Fig. 28.47.*

Solution.

The diagram shown in Fig. 28.46 (*a*) is the canonical form of a negative feedback system. Its transfer function is given by,

$$\frac{C}{R} = \left(\frac{G}{1+GH}\right)$$

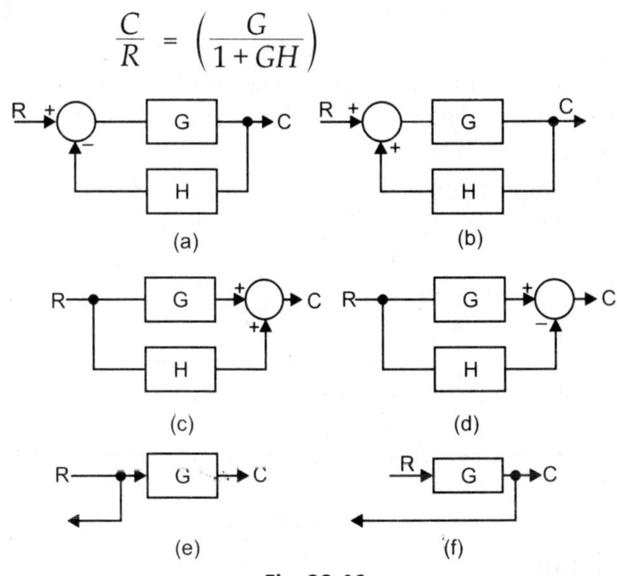

Fig. 28.46

The unity feedback equivalent of this circuit is shown in Fig. 28.47 (*c*). Conversion of canonical form to the unity feedback form has been discussed earlier. The diagram shown in Fig. 28.46 (*b*) is also a canonical form of a positive feedback system. Its transfer function is given by,

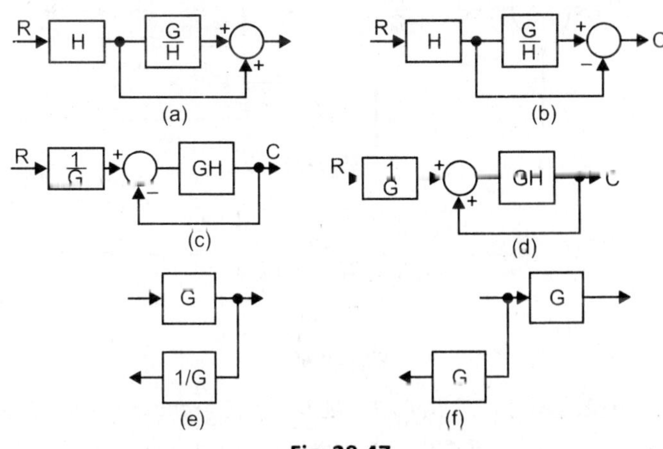

Fig. 28.47

$$\frac{C}{R} = \left(\frac{G}{1-GH}\right)$$

Fig. 28.47 (*d*) shows its unity feedback equivalent. The transfer function for the block diagram of Fig. 38.47 (*c*) is given by,

$$\frac{C}{R} = (G+H)$$

If we remove the block with transfer function (H) from the forward signal path, the resulting block diagram is shown in Fig. 28.47 (*a*). Therefore, the block diagrams of Figs. 28.46 (*c*) and 28.47 (*a*) are identical. On similar lines, the block diagram of Fig. 28.47 (*b*) is the equivalent of the block diagram of Fig. 28.46 (*d*).

In Fig. 28.46 (*e*), if we shift the take-off point from ahead of the block to after the block, we must introduce in this path a block with the transfer function equal to (1/G). The resulting block diagram is shown in Fig. 28.17 (*f*). On similar lines, the equivalent of block diagram of Fig. 28.46 (*f*) is shown in Fig. 28.47 (*e*).

Problem 28.26. *Determine the transfer function of the system represented by the block diagram of Fig. 28.48.*

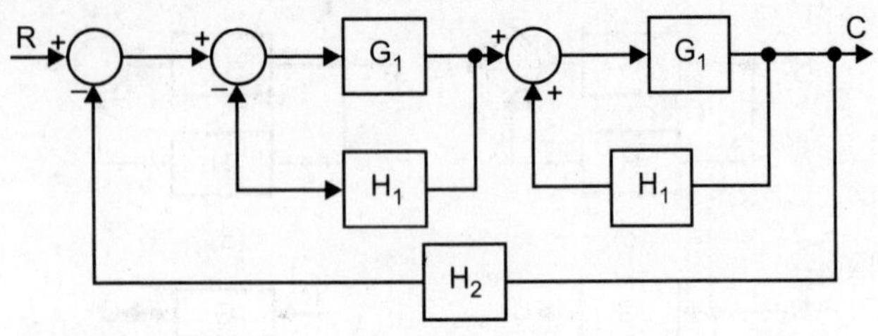

Fig. 28.48

Solution.

As a first step, we shall reduce the inner loops. The resulting diagram is shown in Fig 28.49 (*a*). The block diagram is now more or less in the canonical form. Applying the transformation again, we get the block diagram of Fig. 28.49 (*b*).

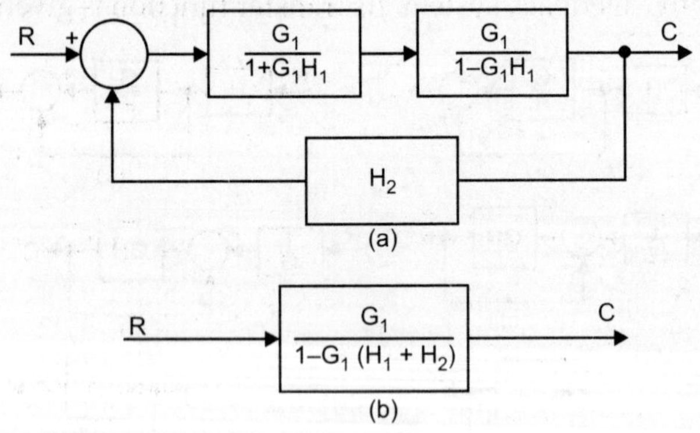

Fig. 28.49

Problem 28.27. *Refer to the block diagram representation of Fig. 28.50. Determine the system characteristic equation. Is the system represented by this block diagram stable?*

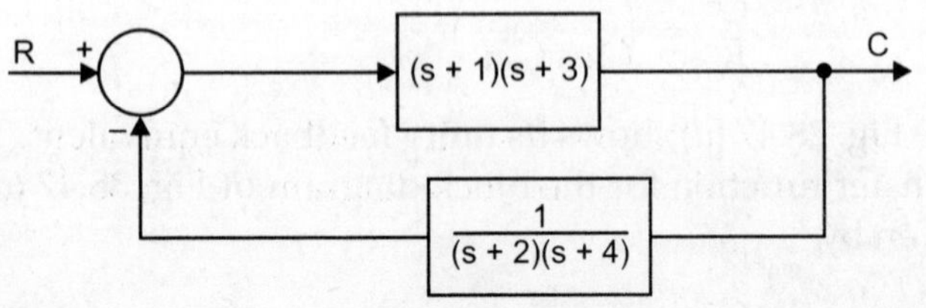

Fig. 28.50

Solution.

The open loop gain of the system is given by

$$GH(s) = \frac{(s+1)(s+3)}{(s+2)(s+4)}$$

The system characteristic equation is given by

$$(s+1)(s+3) + (s+2)(s+4) = 0$$

or $\quad s^2 + 4s + 3 + s^2 + 6s + 8 = 0$ or $2s^2 + 10s + 11 = 0$

or $\quad s^2 + 5s + 5.5 = 0$ or $(s+1.6)(s+3.4) = 0$

The roots of the characteristic equation, – 1.6 and – 3.4, have negative real parts. Therefore, the system is stable.

Problem 28.28. *Find out the overall transfer function of the system represented by the block diagram of Fig. 28.51.*

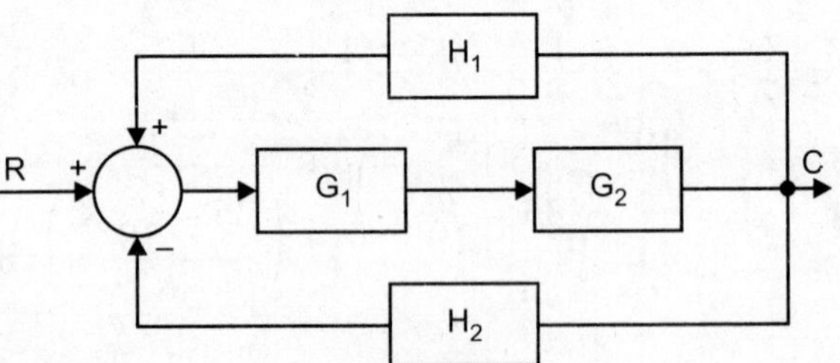

Fig. 28.51

Solution.

The block diagram Fig. 28.51 is equivalent to the one shown in Fig. 28.52 (*a*) which is further reduced to the single block representation of Fig. 28.52 (*b*). The transfer function is given by,

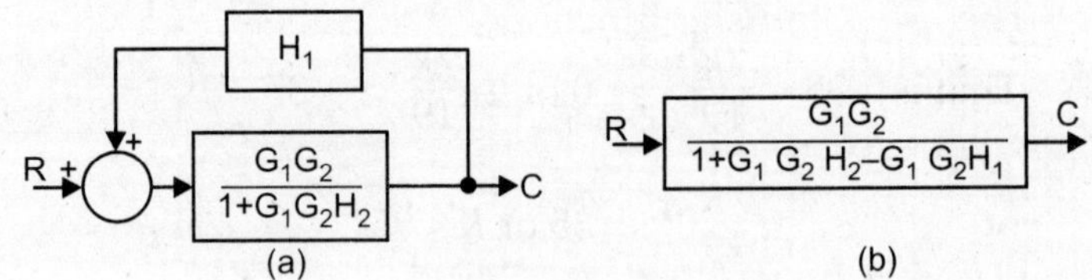

Fig. 28.52

$$\frac{C}{R} = \frac{\dfrac{G_1G_2}{1+G_1G_2H_2}}{1-\left(\dfrac{G_1G_2H_1}{1+G_1G_2H_2}\right)}$$

$$= \frac{G_1G_2}{1+G_1G_2H_2 - G_1G_2H_1}$$

Problem 28.29. *Draw the signal flow graph for the block diagram of Fig. 28.53.*

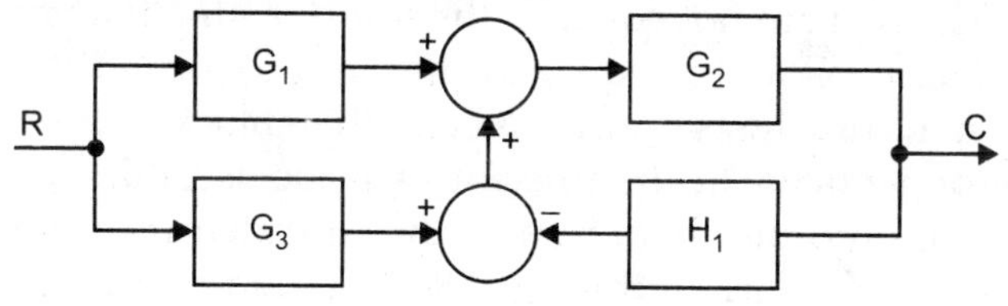

Fig. 28.53

Solution.

The signal flow graph is shown in Fig. 28.53. The graph is self-explanatory.

Problem 28.30. *Determine the transfer function (C/R) for the system of problem 28.29 from its signal flow graph shown in Fig. 28.54.*

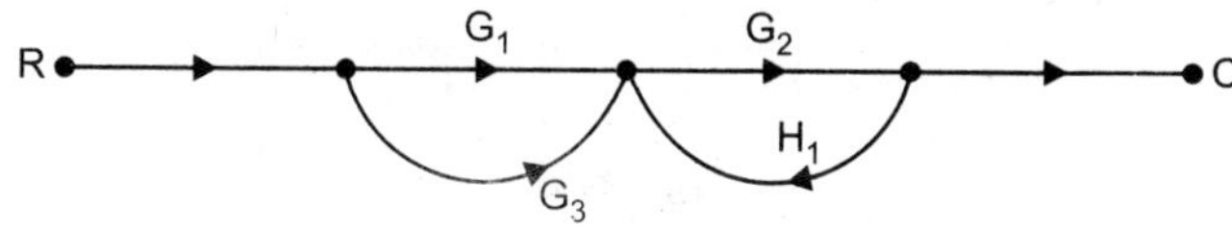

Fig. 28.54

Solution.

The signal flow graph of Fig. 28.54 has two forward paths with

$$P_1 = \text{path gain of the first path} = G_1G_2$$

$$P_2 = \text{path gain of the second path} = G_3G_2$$

$\Delta = 1 - (\text{sum of all loop gains})$..... there is only one feedback loop

$$= 1 - (-G_2 H_1) = 1 + G_2H_1$$

$$\Delta_1 = 1, \Delta_2 = 1$$

$$\text{Therefore, transfer function} = \frac{P_1 \Delta_1 + P_2\Delta_2}{\Delta}$$

$$= \frac{G_1 G_2 + G_3 G_2}{1 + G_2H_1} = \frac{G_2(G_1 + G_3)}{1 + G_2 H_1}$$

Problem 28.31. *Draw the signal flow graph for the block diagram of Fig. 28.55. Determine the system transfer function (C/R).*

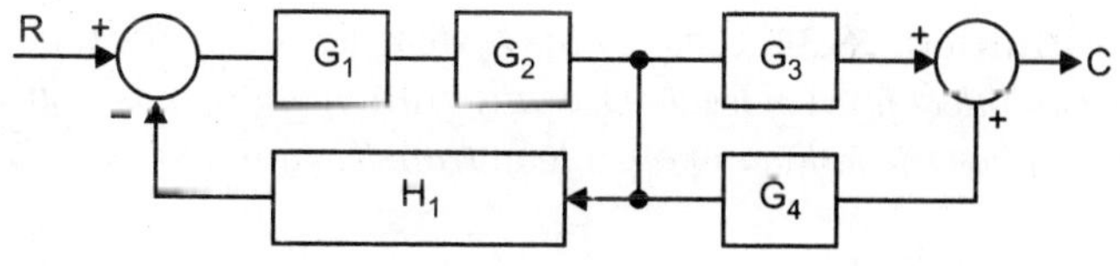

Fig. 28.55

Solution.

The signal flow graph is shown in Fig. 28.56. The graph is self-explanatory.

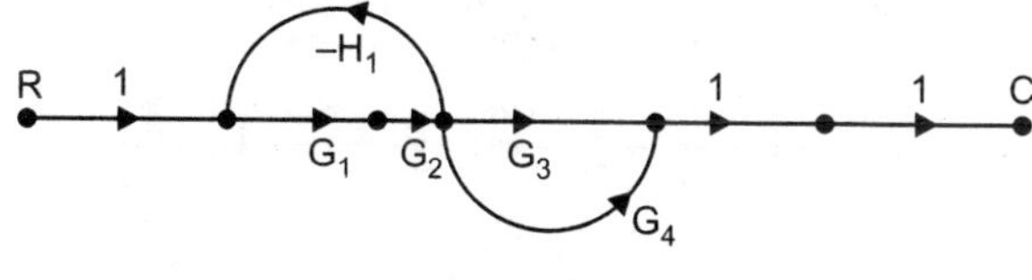

Fig. 28.56

The signal flow graph has two forward paths with

$$P_1 = \text{path gain of the first path} = G_1\, G_2\, G_3$$

$$P_2 = \text{path gain of the second path} = G_1\, G_2\, G_4$$

$$\Delta = 1 - (-G_1\, G_2\, H_1) = (1 + G_1\, G_2\, H_1)$$

$\Delta_1 = 1$, $\Delta_2 = 1$.......the only feedback loop is touching both the forward paths

$$\text{Therefore, transfer function} = \frac{P_1\, \Delta_1 + P_2\, \Delta_2}{\Delta}$$

$$= \frac{G_1\, G_2\, G_3 + G_1\, G_2\, G_4}{1 + G_1\, G_2\, H_1}$$

$$= \frac{G_1\, G_2\,(G_3 + G_4)}{1 + G_1\, G_2 H_1}$$

Problem 28.32. *Draw the signal flow graph for the simple potential divider network of Fig. 28.57 (a). Determine its transfer function. Then draw the signal flow graph for the network of Fig. 28.57 (b) which is nothing but two stages of the network of Fig. 28.57 (a) cascaded together. Determine the transfer function of the cascaded network and explain why it is not equal to the square of the transfer function determined in the first case.*

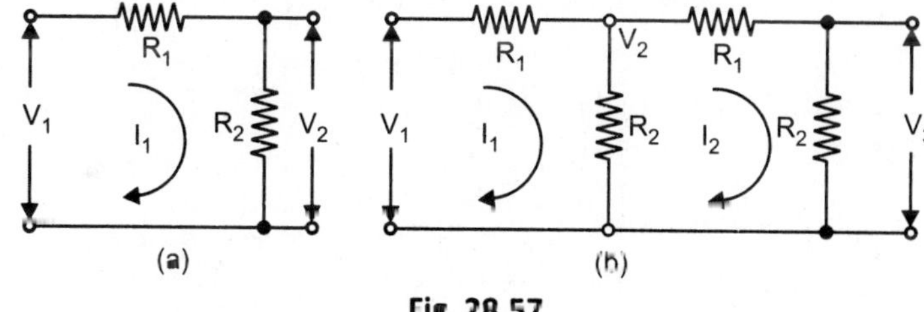

Fig. 28.57

Solution.

For the network of Fig. 28.57 (*a*), the relevant equation are

$$I_1 = \frac{V_1 - V_2}{R_1} = \frac{V_1}{R_1} - \frac{V_2}{R_1} \text{ and } V_2 = V_1\, R_2$$

The signal flow graph for this network is shown in Fig. 28.58 (*a*).

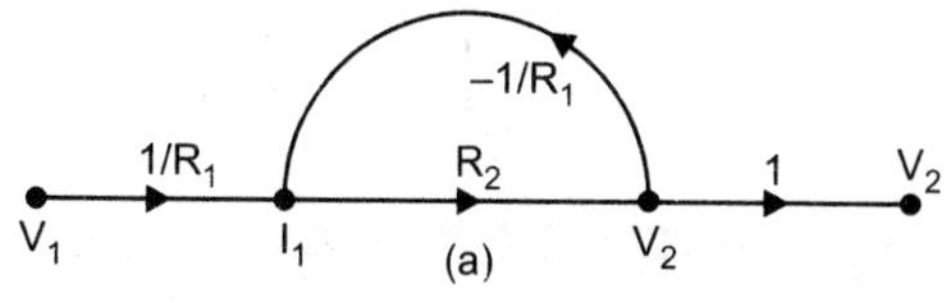

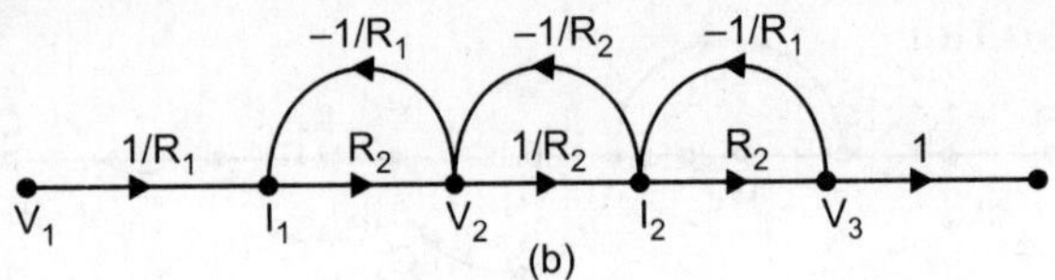

Fig. 25.58

$$I_1 = \frac{V_1 - V_2}{R_1} = \frac{V_1}{R_1} - \frac{V_2}{R_1},$$

$$I_2 = \frac{V_2 - V_3}{R_1} = \frac{V_2}{R_1} - \frac{V_3}{R_1}, \quad V_3 = I_2 R_2$$

and $V_2 = (I_1 - I_2) R_2 = I_1 R_2 - I_2 R_2$

The signal flow graph for the cascaded network is shown in Fig 28.58 (*b*).

From Fig. 25.58 (*a*), the transfer function is given by

$$\text{Transfer Function} = \frac{P_1 \Delta_1}{\Delta} = \frac{(R_2/R_1) \times 1}{1 - (-R_2/R_1)}$$

$$= \frac{(R_2/R_1)}{1 + R_2/R_1} = \frac{R_2}{R_1 + R_2}$$

For the network of Fig. 28.57 (*b*), the transfer function as determined from the signal flow graph of Fig. 28.58 (*b*) is given by

$$\text{Transfer Function} = \frac{P_1 \Delta_1}{\Delta} \text{ Now } P_1 = \left(\frac{R_2^2}{R_1^2}\right)$$

$$\Delta = 1 - \left[\left(-\frac{R_2}{R_1}\right) + \left(-\frac{R_2}{R_1}\right) + \left(\frac{-R_2}{R_1}\right)\right] + \left[\left(\frac{-R_2}{R_1}\right)^2\right]$$

$$= 1 + \frac{3R_2}{R_1} + \frac{R_2^2}{R_1^2} = \frac{R_1^2 + 3R_1 R_2 + R_2^2}{R_1^2}$$

$$\Delta_1 = 1$$

$$\text{Therefore, transfer function} = \frac{R_2^2/R_1^2}{(R_1^2 + 3R_1 R_2 + R_2^2)/R_1^2}$$

$$= \frac{R_2^2}{(R_1^2 + 3R_1 R_2 + R_2^2)}$$

The square of the transfer function of the single stage is given by

$$\frac{R_2^2}{(R_1 + R_2)^2} = \frac{R_2^2}{R_1^2 + 2R_1 R_2 + R_2^2}$$

The two are not equal as the transfer function of two blocks in cascade is equal to the product of individual transfer functions only, when the blocks are not interdependent. In the present case the second stage is loading onto the first stage in the cascaded connection. For instance, current through (R_2) is (I_1) in case of single stage, it becomes ($I_1 - I_2$) in the cascaded arrangement.

Problem 28.33. *Classify the system according to the 'type' for the following cases. All the cases refer to the canonical form of feedback control system of the type shown in Fig. 28.59.*

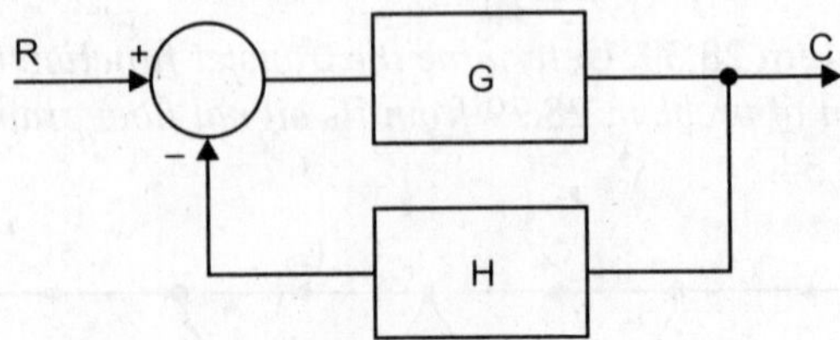

Fig. 28.59

(a) $G(s) = s(s+1)(s+3)$

$H(s) = 1/[s^2 (s+2)(s+5)]$

(b) $G(s) = (s+3)$

$H(s) = 1/[s(s+1)(s+4)]$

(c) $G(s) = (s+1)(s+3)/(s+2)(s+8)$

$H(s) = 1$

Solution.

(a) $GH(s) = \dfrac{s(s+1)(s+3)}{s^2 (s+2)(s+5)}$ which means $l = 2 - 1 = 1$.

Therefore, it is a type '1' system

(*b*) $GH(s) = \dfrac{(s+3)}{(s+1)(s+4)}$ which means $l = 1 - 0 = 1$.

Therefore, it is a type '1' system

(*c*) $GH(s) = \dfrac{(s+1)(s+3)}{(s+2)(s+8)}$ which gives $l = 0$.

Therefore, it is a type '0' system.

Problem 28.34. *Classify the system according the type for the canonical form of feedback configuration of Fig. 28.59, if the system has the following open loop transfer function.*

$$GH(s) = \frac{25}{5s^2 (s+1)(s+3)}$$

Solution.

$$GH(s) = \frac{25}{5s^2 (s+1)(s+3)} = \frac{5}{s^2 (s+1)(s+3)}$$

Which gives $l = 2$

Therefore, it is a type '2' system.

Problem 28.35. *Classify the system represented by the block diagram of Fig. 28.60 by 'type', determine the position error constant K_p velocity error constant K_v and the acceleration error constant K_a for this system.*

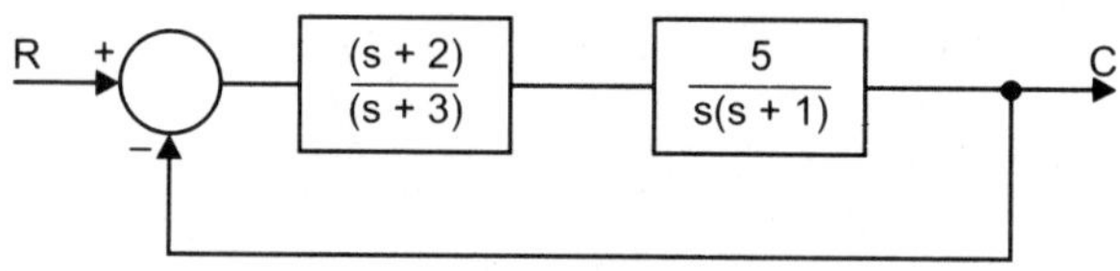

Fig. 28.60

Solution.

$$GH(s) = \frac{5(s+2)}{s(s+1)(s+3)}$$

For this system, $H(s) = 1$

$$GH(s) = \frac{5(s+2)}{s(s+1)(s+3)} \text{ which gives } l = 1$$

It is, therefore, a type '1' system.

Position error constant, $K_p = \lim_{s \to 0} G(s)$

$$= \lim_{s \to 0} \frac{5(s+2)}{s(s+1)(s+3)} = \infty$$

Velocity error constant, $K_v = \lim_{s \to 0} s\,G(s)$

$$= \lim_{s \to 0} \frac{5s(s+2)}{s(s+1)(s+3)}$$

$$= \lim_{s \to 0} \frac{5(s+2)}{(s+1)(s+3)} = \frac{10}{3}$$

Acceleration error constant, $K_a = \lim_{s \to 0} s^2 (G)(s)$

$$= \lim_{s \to 0} \frac{5s^2(s+2)}{s(s+1)(s+3)}$$

$$= \lim_{s \to 0} \frac{5s(s+2)}{(s+1)(s+3)} = 0$$

To summarise the results, $Kp = s$, $K_v = \frac{10}{3}$ and $K_a = 0$.

Problem 28.36. *Determine the steady state error for (a) unit step input (b) unit ramp input and (c) unit parabolic input for the system of problem 28.35.*

Solution.

The steadystate error for a unit step input $\frac{1}{1+K_p} = \frac{1}{\infty} = 0$

The steadystate error for a unit ramp input $\frac{1}{K_v} = \frac{1}{10/3} = 0.3$

The steadystate error for a unit parabolic input $= \frac{1}{K_a} = \frac{1}{0} = \infty$

Problem 28.37. *Determine the sensitivity of the system represented by the transfer function $T(s) = \frac{K}{s^2+3s+7}$, with respect to parameter (K).*

Solution.

$$T(s) = \frac{K}{s^2+3s+7}$$

$$S_k^{T(K)} = \frac{dT(K)}{dk} \cdot \frac{k}{T(K)}$$

$$\frac{d}{dk} = \left(\frac{K}{s^2+3s+7}\right) \cdot \frac{K}{K/(s^2+3s+7)}$$

$$\frac{d}{dk} = \frac{K}{(s^2+3s+7)} \cdot (s^2+3s+7) = 1.$$

Problem 28.38. *Determine the Gain and Phase margins for a system with the open loop transfer function given by $GH = \frac{2}{(s+1)}$.*

Solution.

$$GH = \frac{2}{s+1}$$

$$GH(j\omega) = \frac{2}{1+j\omega}$$

The gain crossover frequency (ω_1) is given by

$$\frac{2}{\sqrt{1+\omega_1^2}} = 1 \text{ or } (1+\omega_1^2) = 4,\ \omega_1 = \sqrt{3}$$

The phase crossover frequency (ω_π) is given by

$$-2\tan^{-1}(\omega_\pi) = -\pi$$

$$\tan^{-1}(\omega_\pi) = \frac{\pi}{2} \text{ or } \omega_\pi = \text{infinity}$$

$$\text{Gain Margin} = \frac{1}{|GH(j\omega_\pi)|} = \frac{1}{0} = \text{infinity}$$

Phase Margin = $180° + \text{Arg}\, GH(j\omega_1) = 180° - 120°\ 60°$

Problem 28.39. *Draw the Nyquist stability plots for the closed loop systems with open loop transfer functions given by*

(a) $GH(s) = \dfrac{1}{s(s+1)}$

(b) $GH(s) = \dfrac{1}{s(s-1)}$

(a) $GH(s) = \dfrac{1}{s^4(s+3)}$

(b) $GH(s) = \dfrac{1}{(s+4)(s+5)}$

Solution.

(a) $GH(s) = \dfrac{1}{s(s+1)}$

Substituting $s = j\omega$, we get

$$GH(j\omega) = \frac{1}{j\omega(1+j\omega)} = \frac{1}{\omega\sqrt{1+\omega^2}} \angle -90° - \tan^{-1}\omega$$

$$\text{Now } GH(j\omega) = \frac{1}{\omega\sqrt{1+\omega^2}} \angle -90° + \tan^{-1}\omega$$

represents the Nyquist stability plot for those portions of Nyquist path, where $s = j\omega$. Since the open loop transfer function has one pole at the origin, for the small semicircle around origin in the Nyquist path Fig. 28.61 (a).

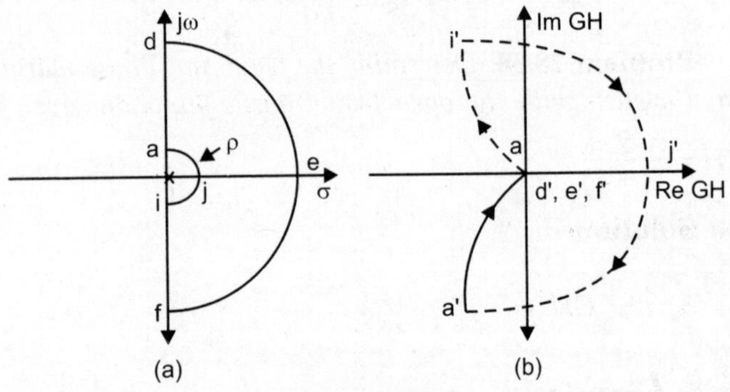

Fig. 28.61

$$s = \lim_{\rho \to 0} \rho e^{j\theta} \text{ for } -90° \le \theta \le 90°$$

$$\text{This gives } \lim_{\rho \to 0} GH(\rho e^{j\theta}) = \lim_{\rho \to 0}\left[\frac{1}{\rho e^{j\theta}(\rho e^{j\theta}+1)}\right]$$

$$= \lim_{\rho \to \infty}\left[\frac{1}{\rho e^{j\theta}}\right] = \infty\, e^{-j\theta} = \infty \angle -\theta$$

The Nyquist stability plot for the different regions of Nyquist path can now be plotted. For region (ad), the extreme points are :

$$\lim_{\omega \to 0} GH(j\omega) = \infty \angle -90° \text{ and } \lim_{\omega \to \infty} GH(j\omega)$$

$$= 0 \angle -180°$$

The stability plot of the region (*fi*) is the mirror image of the plot for the region (ad). The region (def) which is a semicircle with radius equal to infinity maps into origin in the stability plot. The points in the stability plot corresponding to points *i*, *j* and *a* in the Nyquist path are given by $\infty \angle 90°$, $\infty \angle 0°$ $\infty \angle -90°$ respectively. The resulting Nyquist stability plot is shown in Fig. 28.61 (*b*)

(b) $GH(s) = \dfrac{1}{s(s-1)}$

$$GH(j\omega) = \frac{1}{j\omega(j\omega-1)} = \frac{1}{\omega\sqrt{1+\omega^2}}$$

$$\angle -90° + \tan^{-1}\omega$$

On the lines discussed in part (*a*) of the problem, the Nyquist stability plot for this case shown in Figs. 28.62. The extreme points for $\omega = 0$ and $\omega = \infty$ are given by,

$$\lim_{\omega \to 0} GH(j\omega) = \infty \angle +90°$$

$$\lim_{\omega \to \infty} GH(j\omega) = 0 \angle -180°$$

(c) $GH(s) = \dfrac{1}{s^4(s+3)}$

$$GH(j\omega) = \frac{1}{(j\omega)^4(j\omega+3)} = \frac{1}{\omega^4\sqrt{\omega^2+9}}$$

$$\angle -\tan^{-1}\frac{\omega}{3}$$

The extreme points are given by:

$$\lim_{\omega \to 0} GH(j\omega) = \infty \angle 0°$$

$$\text{and } \lim_{\omega \to \infty} GH(j\omega) = 0 \angle -90°$$

The Nyquist stability plot is shown in Fig. 28.63. Note four semicircles in the stability plot, while mapping the region around origin in the Nyquist path (Nyquist path for this part is not shown) and also that the open loop transfer function has four poles at origin.

(d) $GH(s) = \dfrac{1}{(s+4)(s+5)}$

$$GH(j\omega) = \frac{1}{(j\omega+4)(j\omega+5)}$$

$$= \frac{1}{(\omega^2+16)(\omega^2+25)} \angle -\tan^{-1}\frac{\omega}{4}$$

$$-\tan^{-1}\frac{\omega}{5}$$

The extreme points are given by,

$$\lim_{\omega \to 0} GH(j\omega) = \frac{1}{20}\angle 0°$$

$$\lim_{\omega \to 0} GH(j\omega) = 0 \angle -180°$$

New Nyquist stability plot is shown in Fig. 28.64.

Probliem 28.40. *From the Nyquist stability plot of Fig. 28.62 function* $GH(s) = \frac{1}{s(s-1)}$, *determine whether the system is stable or not. If not, what are the number of closed loop poles in the right half plane.*

Solution.

The Nyquist stability plot of Fig. 28.62 reproduced in Fig. 28.65 with the region to the right of the contour shaded when the contour is traversed along its direction. Clearly, point (– 1, 0) is enclosed by the shaded region.

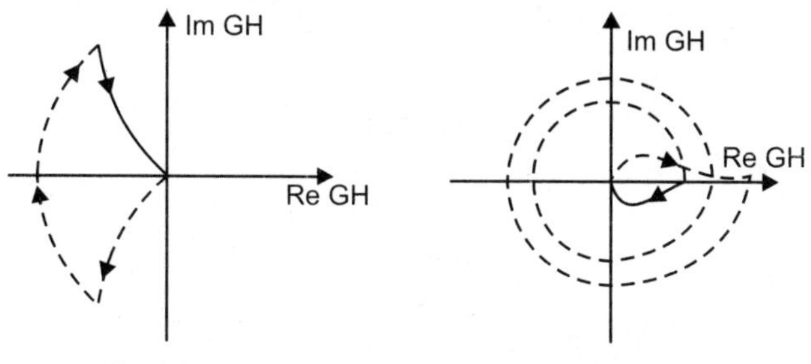

Fig. 28.62

Fig. 28.63

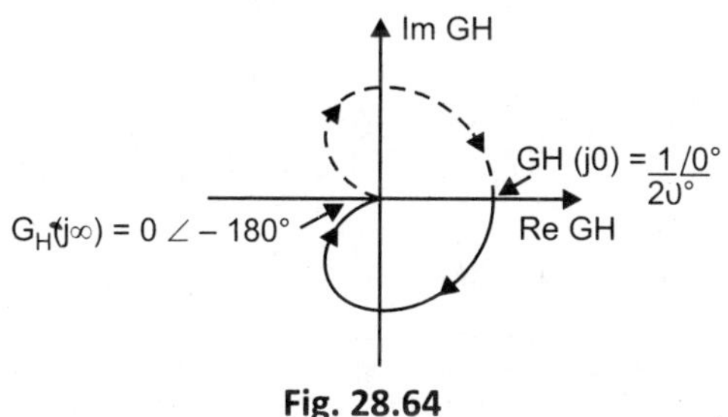

Fig. 28.64

Therefore, $N = 1$

Also, $P_0 = 1$ as the GH has one pole in the RHP. This gives the number of zeros of $(1 + GH)$ in the RHP as

$$Z_0 = N + P_0 = 2$$

which means that the closed loop system has two poles in the RHP. Thus, the system is unstable and it has two poles in the right help plane.

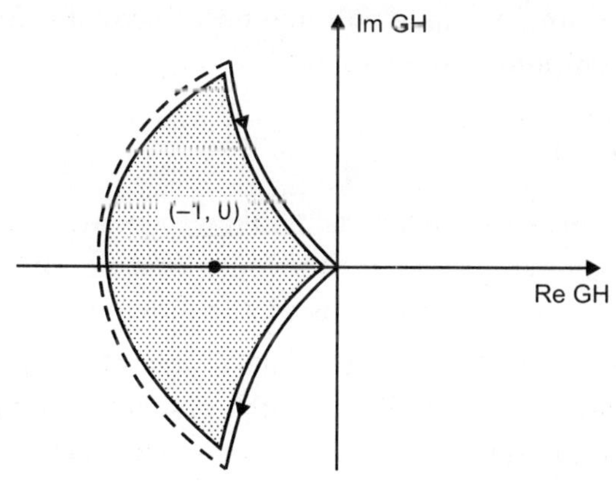

Fig. 28.65

Problem 28.41. *With the help of Nyquist stability plot, determine whether a closed loop system represented by an open loop transfer function of* $GH(s) = \frac{1}{s^3(s+3)}$ would be stable.

Solution.

$$GH(s) = \frac{1}{s^3(s+3)}$$

$$GH(j\omega) = \frac{1}{(j\omega)^3(j\omega+3)}$$

$$= \frac{1}{\omega^3\sqrt{\omega^2+9}} \angle 90° - \tan^{-1}\frac{\omega}{3}$$

Now, $GH(j\omega) = \infty \angle 90°$ for $\omega \to 0$

and $GH(j\omega) = 0 \angle 0°$ for $\omega \to \infty$

As the open loop transfer function $GH(s)$ has three poles at the origin, the Nyquist stability plot would have three infinite radius semicircles. The stability plot is shown in Fig. 28.40.

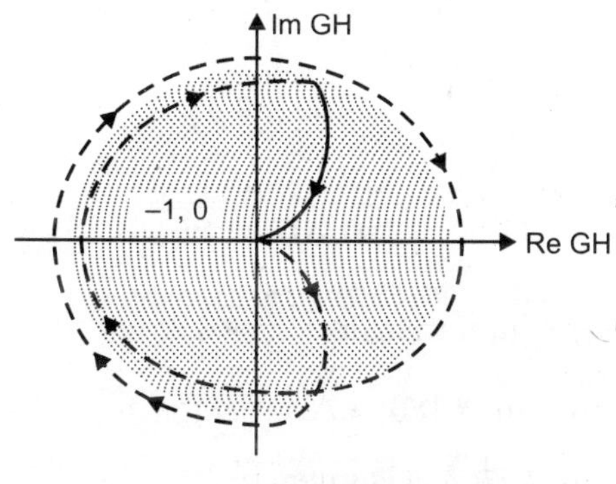

Fig. 28.66

If the region to the right of contour is shaded, it would be clear that the point (– 1, 0) is enclosed. Also, P_0 (number of poles of GH in the RHP) is zero in this case.

Therefore, $N \neq P_0$

Hence, the system is unstable.

Problem 28.42. *A closed loop system is represented by an open loop transfer function of* $GH(s) = \frac{1}{s(s+1)(s+0.5)}$. *Determine the Grain margin and Phase margin of the system.*

Solution.

$$GH(s) = \frac{1}{s(s+1)(s+0.5)}$$

$$GH(j\omega) = \frac{1}{j\omega(j\omega+1)(j\omega+0.5)}$$

$$= \frac{1}{\omega\sqrt{(\omega^2+1)(\omega^2+0.25)}}$$

$$\underline{/-90° - \tan^{-1}\omega - \tan^{-1} 2\omega}$$

At the phase crossover frequency, ω_π,

$$-\frac{\pi}{2} - \tan^{-1}\omega_\pi - \tan^{-1} 2\omega_\pi = -\pi$$

or $$\frac{-\pi}{2} - \tan^{-1}\left(\frac{3\omega\pi}{1-2\omega^2\pi}\right) = -\pi$$

or $$\frac{\pi}{2} = \tan^{-1}\left[\frac{3\omega\pi}{1-2\omega^2\pi}\right]$$

or $$\left(\frac{3\omega\pi}{1-2\omega^2\pi}\right) = \tan\frac{\pi}{2} = \infty \quad \text{or } \omega_\pi = 0.707$$

Also, if (ω_1) is the gain crossover frequency, then

$$\frac{1}{\omega_1\sqrt{(\omega_1^2+1)(\omega_1^2+0.25)}} = 1$$

or $$\omega_1 = 0.8$$

$$\text{Gain margin} = \frac{1}{|GH(j\omega_\pi)|}$$

$$|GH(j\omega_\pi)| = \frac{1}{0.707\sqrt{(1.5)(0.75)}} = \frac{4}{3}$$

Therefore, gain margin $= \frac{3}{4} = 0.75$

Phase margin = 180° + Arg $GH(j\omega_1)$

Now, Arg $GH(j\omega_1)$ is given by

$$-\frac{\pi}{2} - \tan^{-1}\omega_1 - \tan^{-1} 2\omega_1 = \frac{-\pi}{2} - \tan^{-1} 0.8 - \tan^{-1} 1.6 = -186.5°$$

Therefore, phase margin = 180° – 186.5° = – 6.5°

Problem 28.43. *Draw the root locus plot for the system represented by the open loop transfer function* $GH(s) = \frac{K}{s(s+1)(s+3)}$ *for* $K > 0$.

Solution.

The pole-zero plot and the real axis loci for the given open loop transfer function are shown in Fig. 28.67. The center of asymptotes is given by,

$$\sigma_c = \frac{(0+1+3)-0}{3-0} = -1.33$$

There are three asymptotes located at angles of

$$\frac{(2l+1)\pi}{n-m}; l = 0, 1, 2 \,.......(n-m-1), \text{ i.e., } 60°, 180° \text{ and } 300°.$$

The breakaway point (σ_b) is given by

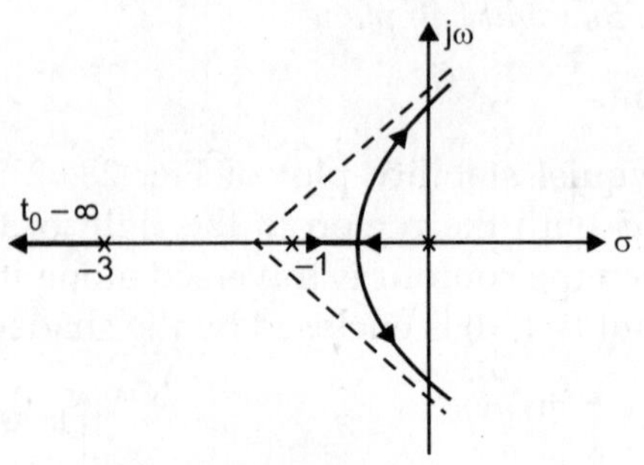

Fig. 28.67

$$\frac{1}{\sigma_b} + \frac{1}{(\sigma_b+1)} + \frac{1}{(\sigma_b+3)} = 0$$

or $$(\sigma_b+1)(\sigma_b+3) + \sigma_b(\sigma_b+3) + \sigma_b(\sigma_b+1) = 0$$

or $$3\sigma_b^2 + 8\sigma_b + 3 = 0$$

or $$\sigma_b = \frac{-8 \pm \sqrt{64-36}}{6} = \frac{-8 \pm \sqrt{28}}{6} = -2.2, -0.45$$

The correct value of σ_b is (– 0.45) as (– 2.2) does not lie on the real axis loci.

The root locus which can be considered as an approximate root locus is then drawn, as shown in Fig. 28.67. The exact root locus plot can be drawn by applying angle and magnitude criteria at different points.

Problem 28.44. *Sketch the root locus lot for the transfer function given in Problem 28.42 for* $K < 0$.

Solution.

The real axis loci in this case are as shown in Fig. 28.43. The center of asymptotes remains unchanged at (– 1.33). The angles made with the negative real axis by the three asymptotes are given by,

$$\frac{2l\pi}{(n-m)}, l = 0, 1, 2, \ (n-m-1) \text{ i.e. } 0°, 120°, \text{ and } 240°.$$

The correct value of the breakaway point in this case would be (– 2.2), as this is the value that lies on the real axis. The resulting root locus plot is shown in Fig. 28.68.

The portion of real axis from 0 to ∞ has been shown as the real axis locus for $K < 0$, as the system has no pole at infinity and therefore, this locus lies to the left of '0' pole at infinity.

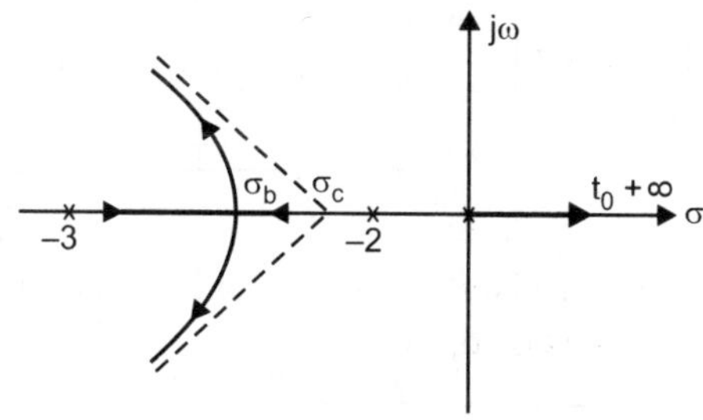

Fig. 28.68

Problem 28.45. *Fig. 28.69 shows the root locus plot of a certain closed loop control system. From the given root locus plot, determine the following:*

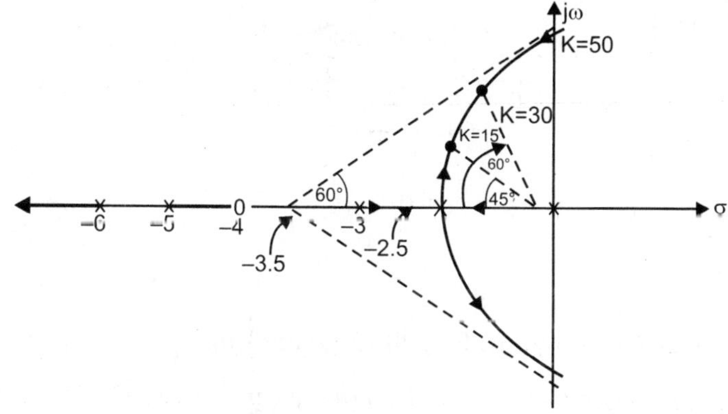

Fig. 28.69

(*i*) *Gain margin, if the design value of (K) is 10*

(*ii*) *The breakaway point and the center of asymptotes*

(*iii*) *The gain factor (K) for which the damping ratio is 0.707.*

(*iv*) *Whether the given root locus plot is for K > 0 or for K < 0.*

Solution.

(*i*) Gain margin = $\frac{50}{10} = 5$

Here, the value of (*K*) where the root locus crosses the (*j*ω) axis is 50

(*ii*) The breakaway point = – 2.5

center of asymptotes = – 3.5.

(*iii*) The damping ratio (ζ) is given by

$$\theta = \cos^{-1}\zeta = \cos^{-1} 0.707 = 45°$$

The line drawn from origin and making an angle of 45° with the negative real axis cuts the root locus at a point corresponding to *K* = 15.

(*iv*) The given root locus plot is for *K* > 0. You would notice that the real axis loci lie to the left of an odd number of finite poles and zeros.

Problem 28.46. *Sketch plots for (i)* $GH(s) = \frac{1}{s^2}$ *(ii)* $GH(s) = \frac{4}{(s+4)}$ *(iii)* $GH(s) = s^3$

Solution.

$$(i) \quad GH(s) = \frac{1}{s^2};\ GH(j\omega) = \frac{1}{(j\omega)^2} \quad |GH(j\omega)| = \frac{1}{\omega^2}$$

or

$$\text{Log}\,|GH(j\omega)| = -2 \log \omega;\ 20 \log$$

$$|GH(j\omega)| = -40 \log \omega$$

$$\text{Arg}\ GH(j\omega) = -180°$$

The Bode plots are shown in Figs. 28.70 (*a*) and (*b*)

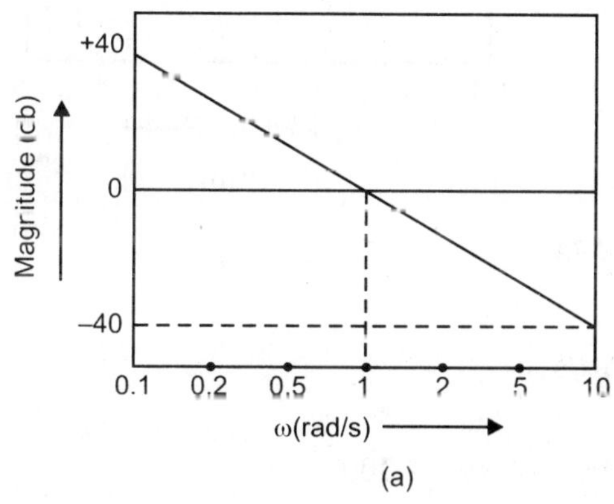

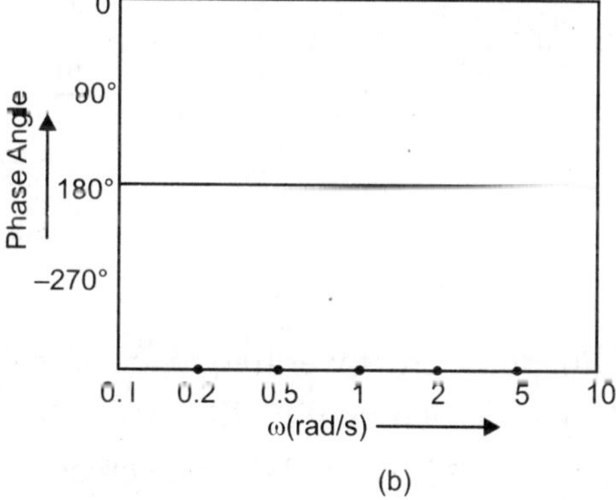

Fig. 28.70

$$(ii) \quad GH(s) = \frac{4}{(s+4)} = \frac{1}{(1+s/4)}$$

$$GH(j\omega) = \frac{1}{|1+j\omega/4|}$$

$$\text{Magnitude in dB} = 20 \log |GH(j\omega)| = 20 \log \left| \frac{1}{(1 + j\omega/4)} \right| = -20 \log |1 + j\omega/4|$$

$$\text{Arg } GH(j\omega) = -\tan^{-1}\left(\frac{\omega}{4}\right)$$

The Bode plots are shown in Figs. 28.71 (*a*) and (*b*). In these plots, on the x-axis, the frequency has been normalised to (ω/4) to simplify computation. The dotted lines show linear approximations of the curves.

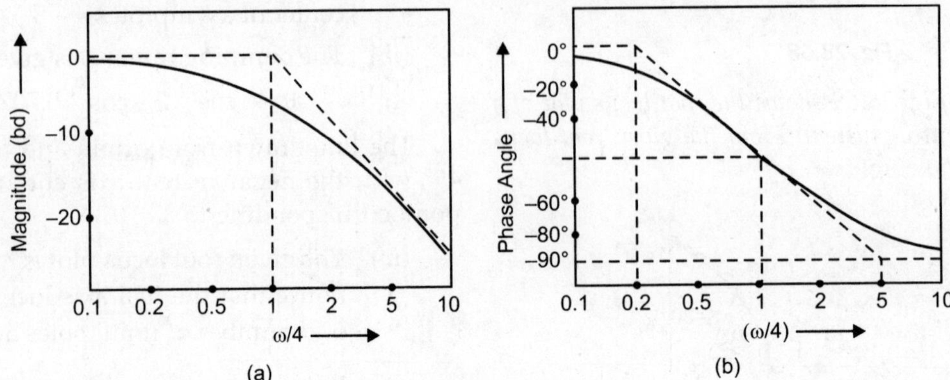

Fig. 28.71

(*iii*) $GH(s) = s^3$; $GH(j\omega) = (j\omega)^3$

Magnitude in dB = 20 log |*GH*(*j*ω)| = 60 log ω. The Bode plots are shown in Fig. 28.72 (*a*) and (*b*).

Problem 28.47. *Figs. 28.73 (a) and (b) show Bode plots for a certain closed loop system. Determine gain and phase margins. Also, determine how much we could afford to increase the gain factor without making the system unstable, if a phase margin of 60° or better was acceptable.*

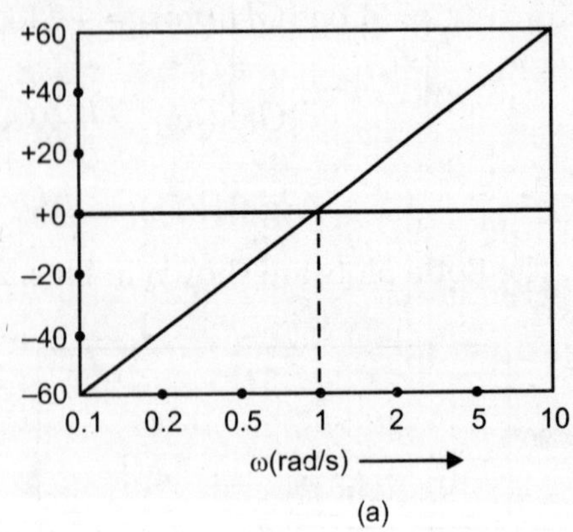

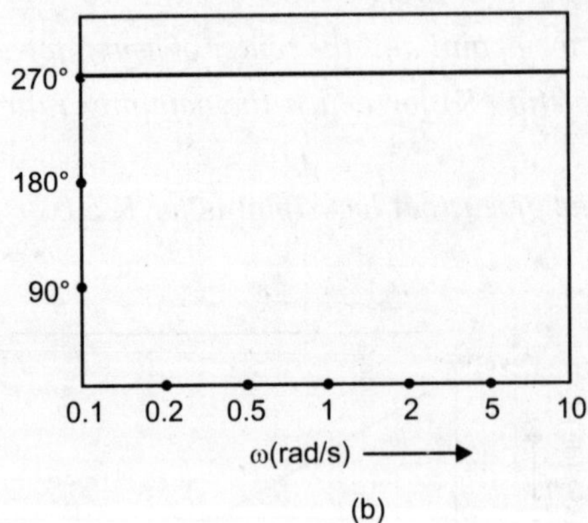

Fig. 28.72

Solution.

From Fig. 28.73 (*b*), phase crossover frequency, $\omega_\pi = 5$ rad/s

From Fig. 28.47 (*a*), Gain Margin = 15 dB

Also, from Fig. 28.73 (*b*), phase angle at gain crossover frequency = – 100°

Therefore, phase margin = 180° – 100 = 80°

If a phase margin of 60° is acceptable, the corresponding phase angle is – 120°. From Fig. 28.73 (*b*), the new gain crossover frequency is 2 rad/s.

Now at ω = 2 rad/s, from Fig. 28.73 (*a*), the new gain margin = 5 dB.

That is, the gain factor could now be increased by factor of 1.78.

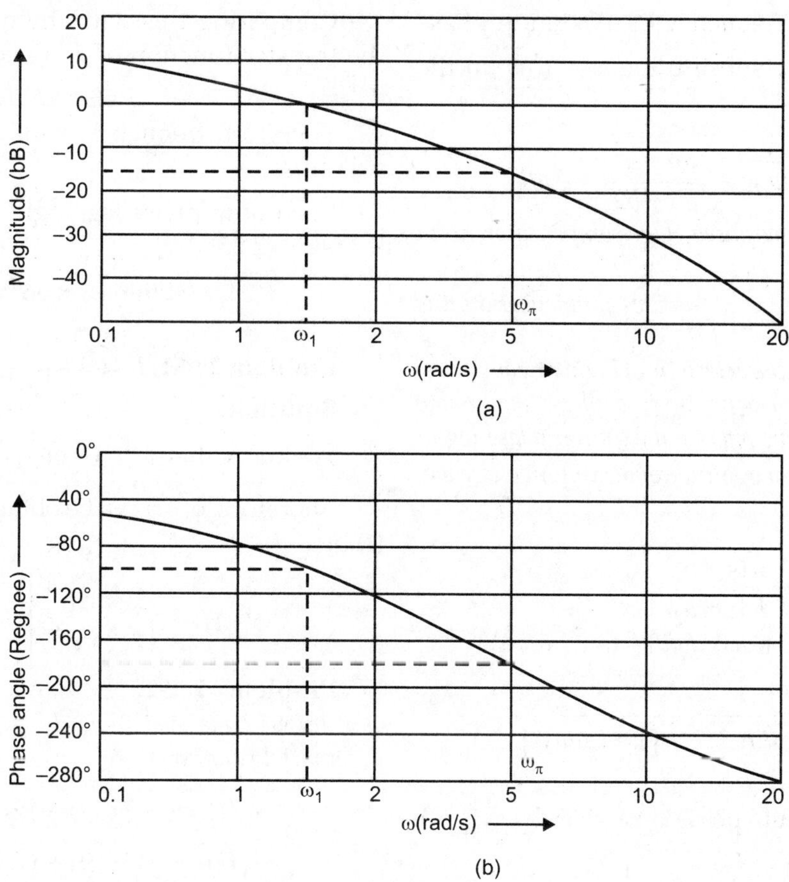

Fig. 28.73

Problem 28.48. *Prove that the maximum phase lead for the lead compensator having a generalised transfer function of* $\frac{(s+a)}{(s+b)}$ *occurs at a frequency (ω) given by* $\sqrt{ab}$. *Also, determine the value of this maximum phase lead and the magnitude attenuation at this frequency of maximum phase lead.*

Solution.

The transfer function of a lead compensator as also given in the statement of the problem is expressed by,

$$\frac{(s+a)}{(s+b)} = \frac{(j\omega+a)}{(j\omega+b)}$$

The phase angle (ϕ) for the lead compensator can therefore be expressed by,

$$\phi = \tan^{-1}(\omega/a) - \tan^{-1}(\omega/b)$$

Now $$\frac{d\phi}{d\omega} = \frac{1}{a\left[1+\left(\frac{\omega}{2}\right)^2\right]} - \frac{1}{b\left[1+\left(\frac{\omega}{b}\right)^2\right]}$$

The condition for which (ϕ) becomes (ϕ_{max}) can be determined by setting $\left(\frac{d\phi}{d\omega}\right) = 0$

That is, $$\frac{1}{a\left[1+\left(\frac{\omega}{\alpha}\right)^2\right]} = \frac{1}{b\left[1+\left(\frac{\omega}{b}\right)^2\right]}$$

or $$\frac{a^2+\omega^2}{a} = \frac{\omega^2+b^2}{b}$$

or $$\omega^2(b-a) = ab^2 - a^2b \text{ or } \omega^2 = ab,\ \omega = \sqrt{ab}$$

Themaximumphaselead= $\tan^{-1}\left(\frac{\sqrt{ab}}{a}\right) - \tan^{-1}\left(\frac{\sqrt{ab}}{b}\right)$

or $$\omega^2(b-a) = ab^2 - a^2b \text{ or } \omega^2 = ab,\ 2 =$$

$$= \tan^{-1}\sqrt{b/a} - \tan^{-1}\sqrt{a/b}$$

Nowtan$^{-1}\sqrt{\frac{b}{a}} = \pi/2 - \tan^{-1}\sqrt{a/b}$

Therefore, maximum phase lead

$$= 90° - \tan{-1}\sqrt{\frac{a}{b}} - \tan^{-1}\sqrt{\frac{a}{b}}$$

$$= 90° - 2\tan^{-1}\sqrt{ab}$$

The magnitude of the transfer function of the lead compensator is given by

$$\frac{\sqrt{\omega^2+a^2}}{\sqrt{\omega^2+b^2}}$$

The magnitude at the frequency of maximum phase lead can be determined by substituting $\omega = \sqrt{ab}$ in the above expression.

$$\text{Magnitude (for } \omega = \sqrt{ab}\text{)} = \frac{\sqrt{ab + a^2}}{\sqrt{ab + b2}} = \sqrt{a/b}.$$

Problem 28.49. *A generalised lag compensator with transfer function,* $P(s) = \dfrac{a(a+b)}{b(s+a)}$ *provides compensation due to its magnitude attenuation characteristic at higher frequencies. This type of compensation also produces a phase lag. Show that a compensator of this type offers a maximum phase lag at* $\omega = \sqrt{ab}$. *Also, determine the magnitude of maximum phase lag.*

Solution.

$$P(s) = \frac{a(s+b)}{b(s+a)}$$

$$\text{Phase angle, } \phi = \tan^{-1}(\omega/b) - \tan^{-1}(\omega/\alpha)$$

$$\frac{d\phi}{d\omega} = \frac{1}{b[1 + (\omega/b)^2]} - \frac{1}{a[1 + (\omega/a)^2]}$$

The condition of maximum phase lag is given by $\dfrac{d\phi}{d\omega} = 0.$

That is,
$$\frac{1}{b[1 + (\omega/b)^2]} = \frac{1}{a[1 + (\omega/a)^2]}$$

or
$$a[1 + (\omega/a)^2] = b[1 + (\omega/b)^2]$$

or
$$\frac{\omega^2 + a^2}{a} = \frac{\omega^2 + b^2}{b}$$

or
$$\omega^2(b - a) = ab^2 - a^2b \text{ or } \omega = \sqrt{ab}$$

Magnitude of maximum phase lag

$$= \tan^{-1}\left(\frac{\sqrt{ab}}{b}\right) - \tan^{-1}\left(\frac{\sqrt{ab}}{a}\right)$$

$$= \tan^{-1}\sqrt{ab} - \tan^{-1}\sqrt{b/a}$$

$$= \frac{\pi}{2} - \tan^{-1}\sqrt{b/a} - \tan^{-1}\sqrt{b/a}$$

$$= (90° - 2\tan^{-1}\sqrt{b/a})$$

Problem 28.50. *A control system uses a phase lag compensator having a transfer function of* $P(s) = \dfrac{4(s+9)}{9(s+4)}$. *Determine the radian frequency at which the compensator offers maximum phase lag. Also, determine the magnitude of this maximum phase lag.*

Solution.

$$P(s) = \frac{4(s+9)}{9(s+4)}$$

Comparing this with the generalised expression for the transfer function of a lag compensator, we get $a = 4$, $b = 9$

Therefore, frequency of maximum phase lag $= \sqrt{4 \times 9}$ $= 6$ rad/s

Maximum phase lag angle $= 90° - 2\tan^{-1}\sqrt{9/4} = 90° - 2\tan^{-1}(2.25)$

$$= 90° - 2 \times 66° = -42°.$$

Problem 28.51. *If* $\phi(t) = \begin{bmatrix} e^{-t} & e^{-2t} \\ 2e^{t} & e^{2t} \end{bmatrix}$, *determine* $\phi^{-1}(t)$

Solution.

We know that $\phi^{-1}(t) = \phi(-t)$

Therefore, $\phi(-t)$ can be obtained from $\phi(t)$ by replacing (t) by $(-t)$

$$\phi(-t) = \begin{bmatrix} e^{t} & e^{2t} \\ 2e^{-t} & e^{-2t} \end{bmatrix}$$

Problem 28.52. *A linear second-order single-input continuous-time system is described by the following set of differential equations:*

$$\dot{x}_1(t) = -2x_1(t) + 4x_2(t)$$

$$\dot{x}_1(t) = -2x_1(t) + x_2(t) + u(t)$$

Where, $x_1(t)$ and $x_2(t)$ are the state variables and $u(t)$ is the control variable. Determine whether the system is controllable and stable.

Solution.

We have

$$\begin{bmatrix} \dot{x}_1 \\ \dot{x}_2 \end{bmatrix} = \begin{bmatrix} -2 & 4 \\ +2 & -1 \end{bmatrix}\begin{bmatrix} x_1 \\ x_2 \end{bmatrix} + \begin{bmatrix} 0 \\ 1 \end{bmatrix} u(t)$$

The controllability matrix is

$$[S] = [B\ AB]$$

Where,

$$AB = \begin{bmatrix} -2 & 4 \\ +2 & -1 \end{bmatrix}\begin{bmatrix} 0 \\ 1 \end{bmatrix} = \begin{bmatrix} 4 \\ 1 \end{bmatrix} \text{ and } S = \begin{bmatrix} 0 & 4 \\ 1 & -1 \end{bmatrix}$$

and

Therefore, $|s| = 0 - 4 = -4 \neq 0$

Hence, the given system is controllable.

Now, the characteristic equation is given by $|s - A| = 0$

$$\begin{bmatrix} s+2 & -4 \\ -2 & s+1 \end{bmatrix} = 0$$

That is, $(s + 2)(s + 1) - 8 = 0 \Rightarrow s^2 + 3s - 6 = 0$

or $\qquad s = -4.37, 1.37$

Since one pole lies on the RHS of s-plane, the given system is unstable.

Problem 28.53. *A linear time invariant system is described by the state variable model*

$$\begin{bmatrix}\dot{x}_1\\ \dot{x}_2\end{bmatrix} = \begin{bmatrix}-1 & 0\\ 0 & -2\end{bmatrix}\begin{bmatrix}x_1\\ x_2\end{bmatrix} + \begin{bmatrix}0\\ 1\end{bmatrix}u$$

$$Y = [1 \;\; 2]\begin{bmatrix}x_1\\ x_2\end{bmatrix}$$

Determine whether the system is observable.

Solution.

We have

$$A = \begin{bmatrix}-1 & 0\\ 0 & -2\end{bmatrix} \text{ and } B = \begin{bmatrix}0\\ 1\end{bmatrix}$$

$$AB = \begin{bmatrix}-1 & 0\\ 0 & -2\end{bmatrix}\begin{bmatrix}0\\ 1\end{bmatrix} = \begin{bmatrix}0\\ 2\end{bmatrix}$$

The controllable matrix [QC] is given by

$$Q_C = [B \quad AB]$$

That is,

$$Q_C = \begin{bmatrix}0 & 0\\ 1 & -2\end{bmatrix}$$

Hence, $|Q_C| = 0 - 0 = 0$

Therefore, the given system is not completely controllable. Now,

$$A^T = \begin{bmatrix}-1 & 0\\ 0 & -2\end{bmatrix}$$

$$C = [1 \;\; 2]$$

$$C_T = \begin{bmatrix}1\\ 2\end{bmatrix}$$

$$A^T C^T = \begin{bmatrix}-1 & 0\\ 0 & -2\end{bmatrix}\begin{bmatrix}1\\ 2\end{bmatrix} = \begin{bmatrix}-1\\ -4\end{bmatrix}$$

The observability matrix Q_O is given by:

$$Q_O = [C^T \;\; A^T C^T]$$

That is, $$Q_O = \begin{bmatrix}1 & -1\\ 2 & -4\end{bmatrix}$$

Hence, $|Q_O| = -4 - (2) = -2 \neq 0$

So, the given system is completely observable.

Problem 28.54. A certain linear time invariant system has the state and the output equations given below.

$$\begin{bmatrix}\dot{x}_1\\ \dot{x}_2\end{bmatrix} = \begin{bmatrix}1 & 1\\ 0 & 1\end{bmatrix}\begin{bmatrix}\dot{x}_1\\ \dot{x}_2\end{bmatrix} + \begin{bmatrix}0\\ 1\end{bmatrix}u$$

$$y = [1 \;\; 1]\begin{bmatrix}x_1\\ x_2\end{bmatrix}$$

Determine $dy/dt \mid t = 0$

Solution.

We have

$$\dot{x}_1 = x_1 - x_2$$

That is,

$$\dot{x}_1(0) = x_1(0) - x_2(0)$$

$$\dot{x}_1(0) = 1 - (-1) = 2$$

$$\dot{x}_1(0) = x_2(0) - 1$$

Now, $$y = x_1 + x_2$$

Therefore,

For $t = 0$, we get $\dfrac{dy}{dt} = 2 + (-1) = 1$

Summary of Important Formulas

1. A linear constant coefficient second-order control system is represented by

$$\frac{d^2y}{dt^2} + 2\zeta\omega_n\frac{dy}{dt} + \omega_n^2 y = \omega_n^2 x$$

Where, ζ is the damping ratio and ω_n is the undamped natural frequency.

2. The damped frequency is

$$\omega_d = \omega_n\sqrt{1-\zeta^2}$$

3. The damping factor is

$$\alpha = \zeta\omega_n$$

4. The time constant is

$$t = \frac{1}{\zeta\omega_n}$$

5. A linear second-order control system has its poles at

$$-\alpha + j\omega_d$$

6. In the case of a linear second-order control system, the time for first peak is given by

$$t_p = \frac{\pi}{\omega_n\sqrt{1-\zeta^2}}$$

7. In the case of a linear second-order control system, overshoot is given by

$$\text{Exp}\left[-\left(\frac{\zeta\pi}{\sqrt{1-\zeta^2}}\right)\right]$$

8. In the case of a linear control system,

$$y(t) = y_1(t) + y_2(t) + y_3(t) + ... + y_n(t)$$

where, $y_1(t)$ is the system response due to $X_1(t)$ alone, $y_2(t)$ is the system response due to $X_2(t)$ alone, $y_3(t)$ is the system response due to $X_3(t)$ alone and $y_n(t)$ is system response due to $X_n(t)$ alone.

9. In the case of a linear control system,

$$y(ax) = a \,.\, y(x)$$

10. The transfer function is

$$\frac{C(s)}{R(s)} = \frac{G(s)}{1 + G(s) \,.\, H(s)}$$

11. Mason's Formula:

$$T = \frac{\sum_i P_i \Delta_i}{\Delta}$$

Where, P_i = the ith forward path gain; (i) is the number of forward paths;

Δ = 1 – (Sum of loop gains) + (Sum of all gain products of two non-touching loops) – (Sum of gain products of three non-touching loops) + (Sum of gain products of four non-touching loops) – ... Δ_i = (Δ) (Evaluated with all those loops touching P_i eliminated)

12. The transfer function of canonical form of feedback control system:

$$\frac{G}{1 + GH} \quad \text{(For negative feedback)}$$

$$\frac{G}{1 - GH} \quad \text{(For positive feedback)}$$

13. Mass element:

$$F(t) = M \,.\, \frac{du(t)}{dt} = M \,.\, \frac{d^2 x(t)}{dt^2}$$

14. Spring element:

$$F(t) + K \,.\, (x_1 - x_2) = K \,.\, x(t)$$

Where, K is the spring constant.

15. Damping element:

$$F(t) = f \,.\, [u_1(t) - u_2(t)] = f \,.\, \left(\frac{dx_1}{dt} - \frac{dx_2}{dt}\right)$$

Where, f is the damping constant

16. Inertia element:

$$I(t) = J \,.\, \frac{d\omega(t)}{dt} = J \,.\, \frac{d^2\theta(t)}{dt^2}$$

Where, J is the inertia constant

17. Torsional spring element:

$$T(t) = K \,.\, (\theta_1 - \theta_2) = K \,.\, \theta(t)$$

$$= K \,.\, \int(\omega_1 - \omega_2)dt = K \,.\, \int\omega(t)dt$$

Where, K is the torsional spring constant

18. Torsional damping element:

$$T(t) = f \,.\, (\omega_1 - \omega_2) = f \,.\, \left(\frac{d\theta_1(t)}{dt} - \frac{d\theta_2(t)}{dt}\right)$$

Where, 'f' is the torsional damping constant

19. The position error constant K_p is given by:

$$K_p = \lim_{s \to 0} G(s)$$

20. The steady-state error for a unit step input is

$$\frac{1}{1 + K_p}$$

21. The velocity error constant K_v is given by

$$K_v = \lim_{s \to 0} sG(s)$$

22. The steady-state error for a unit ramp input is

$$\frac{1}{K_v}$$

23. The acceleration error constant K_a is given by

$$K_a = \lim_{s \to 0} s^2 G(s)$$

24. The steady-state error for a unit parabolic input is

$$\frac{1}{K_a}$$

25. The sensitivity of magnitude of $T(K)$ with respect to K is written as

$$S_K^{|T(K)|} = \frac{d|T(K)|}{dk} \cdot \frac{K}{|T(K)|}$$

26. The sensitivity of phase with respect to K is written as

$$S_K^{\phi_T} = \frac{d\phi_T}{dk} \cdot \frac{K}{|T(K)|}$$

27. $S_K^{T(K)} = S_K^{|T(K)|} + j\phi_T S_k^{\phi_T}$

28. For a generalised feedback control system, the step error constant is

$$K_s = \frac{1}{\lim_{s \to 0}\left(T_d - \frac{C}{R}\right)}$$

The steady-state error for step input is

$$\frac{1}{K_s}$$

29. For a generalized feedback control system, the ramp error constant is

$$K_r = \frac{1}{\lim_{s \to 0}\frac{1}{s}\left(T_d - \frac{C}{R}\right)}$$

The steady-state error for a ramp input is $\frac{1}{Kr}$

30. For a generalised feedback control system,

$$K_{pa} = \frac{1}{\lim_{s \to 0}\frac{1}{s^2}\left(T_d - \frac{C}{R}\right)}$$

The steady-state error for a unit parabolic input is $(1/K_{pa})$

31. For a generalised feedback control system with the transfer function of the form

$$T = \frac{A_1 + kA_2}{A_3 + kA_4}$$

The sensitivity of the system with respect to parameter k is given by:

$$S_k^T = \frac{dT}{dk} \cdot \frac{k}{T} = \frac{k(A_2A_3 - A_1A_4)}{(A_1 + kA_2)/(A_3 + kA_4)}$$

32. Integral controller:

$$I_{out}(s) = \left(\frac{K_i}{s}\right)E(s)$$

33. Derivative controller:

$$D_{out}(s) = K_d s\, E(s)$$

34. PID controller:

$$U(s) = K_p E(s) + K_i\frac{E(s)}{s} + K_d sE(s)$$

35. Lead compensator:

$$P_{Lead} = \left(\frac{s+a}{s+b}\right),\ a < b$$

36. Lag compensator: ,

$$P_{Lag} = \frac{a}{b}\left(\frac{s+b}{s+a}\right),\ a < b$$

37. Lag–lead compensator:

$$P_{Lag-Lead} = \left(\frac{s+a_1}{s+b_1}\right)\left(\frac{s+b_2}{s+a_2}\right),\ \frac{b_1}{a_1} = \frac{b_2}{a_2} > 1$$

38. Magnitude criterion is

$$|K| = \left|\frac{D(s_1)}{N(s_1)}\right|$$

39. Angle criterion is

$$\text{Arg}\left[\frac{N(s_1)}{D(s_1)}\right] = (2l+1)\pi \text{ rad, (for } K > 0)$$
$$= 2l\pi \text{ rad, (for } K < 0)$$

40. Number of loci is equal to number of poles of open-loop transfer function.

41. Center of asymptotes is

$$\sigma_c = \frac{\sum_{i=1}^{n} p_i - \sum_{i=1}^{m} z_i}{n - m}$$

Where, n and m, respectively, are the number of poles and zeros of the open-loop transfer function $-p_i$ and $-z_i$, respectively, are the poles and zeros of the open-loop transfer function. Angles between the asymptotes and the real axis are given by:

$$\beta = \begin{cases} \frac{(2l+1)\pi}{n-m} & \text{for } K > 0 \\ \frac{(2l)\pi}{n-m} & \text{for } K > 0 \end{cases}$$

Where, $l = 0, 1, 2, \ldots, (n - m - 1)$.

42. Breakaway points can be computed from

$$\sum_{i=1}^{n}\frac{1}{\sigma_b + p_i} = \sum_{i=1}^{m}\frac{1}{\sigma_b + z_i}$$

and p_i are z_i poles and zeros of the open-loop transfer function.

43. Departure and arrival angles can be computed from the following:

Departure angle, $\theta_D = 180° - \text{Arg}(GH')$

Arg (GH') is the phase angle of the open-loop transfer function GH computed at the complex pole in question.

$$\theta_A = 180° - \text{Arg}(GH'')$$

Arg (GH'') is the phase angle of the open-loop transfer function GH at the complex zero, but ignoring the contribution of that particular zero.

44. Gain and phase margin can be computed from the following:

Gain margin =

$$\frac{\text{Value of } K \text{ at imgainary cross - over}}{\text{Design value of } K}$$

Phase margin = 180° + Arg [GH($j\omega_1$)]
Where, ω_1 is the gain cross-over frequency and is given by |GH ($j\omega_1$)| = 1

45. Damping ratio (ζ) is related to angle θ as follows:

$$\theta = \cos^{-1}\zeta$$

Where, 'θ' is the angle made by the line drawn from origin to the point on root locus corresponding to gain factor K of interest with the negative real axis.

46. Gain margin is

$$\frac{1}{GH(j\omega_\pi)}$$

Where, 'ω_π' is phase crossover frequency

47. Phase margin is

$$\phi_{PM} = [180° + \text{Arg } GH(j\omega_1)]$$

ω_1 is gain crossover frequency.

48. Nyquist stability plot: Number of infinite semicircles in 'type l' system is l.

49. Nyquist stability criterion: A closed-loop control system is stable only if

$N = -P_0 < 0.$

P_0 = number of poles of GH is the RHP

N = total number of clockwise encirclements of the point (– 1, 0) in the $GH(s)$ plane.

50. State equation of linear time-invariant system:

$$\frac{dx(t)}{dt} = Ax(t) + Br(t)$$

$$c(t) = Dx(t) + Er(t)$$

Where, x, r and c, respectively, represent state, input and output variables.

51. Homogeneous state equation:

$$\frac{dx(t)}{dt} = Ax(t)$$

52. State transition matrix:

$$\phi(t) = L^{-1}[(sI - A)^{-1}] \ldots t > 0$$

and $\phi(I) = e^{At} = 1 + \left(\frac{1}{2}\right)!A^2t^2 + \left(\frac{1}{3}\right)!A^3t^3 + \ldots$

53. Properties of state transition matrix:

(a) $x(t) = I$, where I is identity matrix.

(b) $\phi^{-1}(t) = \phi(-t)$.

(c) $\phi(t_2 - t_1)\,\phi(t_1 - t_0) = \phi(t_2 - t_0)$ for any t_0, t_1 and t_2.

(d) $[\phi(t)]^k = \phi(kt)$, where k is an integer.

54. Linear time-invariant system:

(a) $x(t) = L^{-1}[(sI - A)^{-1}]\, x(0) + L^{-1}\,[(sI - A)^{-1}\, BR(s)]$

(b) $x(t) = \phi(t)x(0) + \int_0^t \phi(t - \tau)\, Br(\tau)\, d\tau \quad (t \geq 0)$

(c) $x(t) = \phi(t - t_0)x(t_0) + \int_{t_0}^t \phi(t - \tau)\, Br(\tau)\, d\tau$

(d) $c(t) = D\phi(t - t_0)x(t_0) + \int_{t_0}^t \phi(t - \tau)\, Br(\tau)\, d\tau + Er(t)$

55. Controllability matrix:

$$[Q_C] = [B\ AB\ A^2B\ A^3B \ldots A^{n-1}\ B]$$

56. Observability matrix:

$$[Q_O] = [D'\ A'D'\ (A')^2D'\ (A')^3D' \ldots (A')^{n-1}D']$$

57. Eigen values of $[A]$: The roots of characteristic equation given by

$$|sI - A| = 0$$

Self Evaluation Exercise

I. Multichoice Questions

1. A negative feedback control system is characterised by

(a) reduced bandwidth
(b) increased accuracy
(c) increased non-linearity
(d) reduced distortion

2. In a linear feedback control system, the characteristic equation is of the form

(a) $D^n + a_{n-1}D^{n-1} + a_{n-2}D^{n-2} + \ldots\ldots + a_1D + a_o = 0$
(b) $(D^n + a_{n-1}D^{n-1} + a_{n-2}D^{n-2} + \ldots\ldots + a_1D + a_o)x = 0$
(c) $(D^n + a_{n-1}D^{n-1} + a_{n-2}D^{n-2} + \ldots\ldots + a_1D + a_o)y = 0$
(d) none of these.

3. Free response of a control system is the one characterised by

(a) input approaching infinity with time
(b) input being a finite value
(c) zero input
(d) all initial conditions being zero

4. Transient response of a control system is the one

(*a*) which approaches zero when the time approaches infinity
(*b*) which does not depend upon time
(*c*) which is time dependent
(*d*) none of these

5. A causal system is the one in which the output depends upon

(*a*) input as well as output
(*b*) present and past inputs
(*c*) present as well as future inputs
(*d*) present and past inputs and past outputs

6. A control system is represented by $y(t) = (t + T)$ with $T > 0$. Is it causal?

(*a*) Yes (*b*) No
(*c*) Not necessarily (*d*) none of these

7. Can $y(t) = (c^{-2t} + e^{-t} + 1)$ possibly be the transient response of a certain control system?

(*a*) Yes
(*b*) No
(*c*) only under special circumstances

8. Area under a unit impulse function is

(*a*) infinity (*b*) zero
(*c*) unity (*d*) none of these

9. Unit impulse response of a system is the one when

(*a*) the input to the system is a unit impulse
(*b*) the input to the system is a unit impulse and all initial conditions are zero
(*c*) the output is a unit impulse for all possible inputs
(*d*) none of these

10. A control system is represented by the second order differential equation $\frac{d^2y}{dt^2} + 2\zeta\,\omega_n\left[\frac{dy}{dt}\right] + \omega_n^2 y = \omega_n^2 x$. The damping co-efficient (α) of such a system is

(*a*) $\alpha = \zeta\omega_n$ (*b*) $\alpha = \frac{1}{\zeta\omega_n}$
(*c*) $\alpha = \frac{\zeta}{\omega_n}$ (*d*) none of these

11. Damped natural frequency (ω_d) for the system of Q.10 is given by

(*a*) $\omega_d = \zeta\omega_n$ (*b*) $\omega_d = \omega_n(1 - \zeta^2)$
(*c*) $\omega_d = \omega_n\sqrt{1-\zeta^2}$ (*d*) $\omega_d = \sqrt{\omega_n(1-\zeta^2)}$

12. "Laplace transform of the impulse response of a system gives its transfer function" comment on the statement.

(*a*) It is true because transfer function $P(s)$ is $Y(s)/X(s)$ where $Y(s)$ and $X(s)$ are Laplace transforms of output and input respectively and in case of impulse response, $X(s) = 1$ giving $P(s) = Y(s)$
(*b*) This statement is baseless as the transfer function has nothing to do with the impulse response
(*c*) It is true only under some conditions
(*d*) none of these

13. Transfer function of a lead compensator is given by

(*a*) $\left(\frac{s+a}{s+b}\right)$ with $a > b$ (*b*) $\left(\frac{s+a}{s+b}\right)$ with $a < b$
(*c*) $\left(\frac{s-a}{s+b}\right)$ with $a < b$ (*d*) $\left(\frac{s+a}{s-b}\right)$ with $a > b$

14. Stability of a control system can be determined from one of the following responses :

(a) Impulse response
(*b*) Step response
(*c*) Ramp response
(*d*) Steadystate response

15. One of the following methods can be used to determine only the absolute stability of a control system

(*a*) Routh stability criterion
(*b*) Root locus analysis
(*c*) Bode plots
(*d*) Nichols charts and plots

16. A canonical feedback system is the one with a

(*a*) unity feedback
(*b*) negative feedback
(*c*) positive feedback
(*d*) none of these

17. A feedback control system has a transfer function given by $P(s) = \frac{(s+1)(s+3)}{s^2(s+2)(s+4)}$. It is a

(*a*) type-5 system (*b*) type-2 system
(*c*) type-2 system (*d*) type-4 system

18. If (ω_n) is the phase crossover frequency *i.e.* the frequency (ω) at which Arg $GH(j\omega)$ is 180°, then the gain margin can be expressed as

(*a*) Gain margin = $|GH(j\omega_\pi)|$

(*b*) Gain margin = $\dfrac{1}{|GH(j\omega_\pi)|}$

(*c*) Gain margin = $\dfrac{-1}{|GH(j\omega_\pi)|}$

(*d*) none of these

19. Gain margin can be determined from the root locus plot by

(*a*) $\dfrac{\text{Value of gain factor } (K) \text{ at imaginary crossover}}{\text{Design value of } (K)}$

(*b*) $\dfrac{\text{Design value of gain factor } (K)}{\text{Value of gain factor } (K) \text{ at imaginary crossover}}$

(*c*) $\dfrac{\text{Value of gain factor } (K) \text{ at real axis crossover}}{\text{Desing value of } (K)}$

(*d*) none of these

20. If (ω_1) is the gain crossover frequency, *i.e.*, the frequency (ω) at which $|GH(j\omega)|$ = unity, then the phase margin can be expressed as

(*a*) Phase margin = [180° + Arg $GH(j\omega_1)$] degrees

(*b*) Phase margin = [180° – Arg $GH(j\omega_1)$] degrees

(*c*) Phase margin = Arg $GH(j\omega_1)$

(*d*) none of these

21. A higher value of gain factor (*K*) at the imaginary axis crossover of the root locus plot of a control system

(*a*) brings it closer to instability

(*b*) takes it farther from instability

(*c*) is always undesirable

(*d*) none of these

22. Routh table was constructed from the characteristic equation of a control system. The first column of the table contained the following integers : 1, 4, 5, 6, 13. The system is

(*a*) Stable (*b*) Unstable

(*c*) Marginally stable (*d*) none of these

23. A stable unity feedback system has one pole at the origin. A unit parabolic input will give a steadystate error of

(*a*) zero (*b*) Infinity

(*c*) 1 (*d*) none of these

24. For drawing the root locus plot of a certain closed loop control system, one needs to know

(*a*) only the open loop transfer function of the system

(*b*) the closed loop transfer function

(*c*) the closed loop system pole-zero map

(*d*) the open loop transfer function, gain and phase margins

25. Roots of system characteristics equation are nothing but

(*a*) the closed loop system poles

(*b*) the open loop system poles

(*c*) the closed loop system poles and zeros

(*d*) the closed loop system zeros

26. The homogeneous part of the state equation $dx(t)/dt = Ax(t) + Br(t)$ is given by

(*a*) $dx(t)/dt = Ax(t)$

(*b*) $dx(t)/dt = Br(t)$

(*c*) $dx(t)/dt = Ax(t) + Br(t)$

(*d*) $dx(t)/dt = 0$

27. The state transition matrix represents

(*a*) Forced response of the system

(*b*) Free response of the system

(*c*) Transient response of the system

(*d*) None of these

28. One of the following expressions relating state transition matrix is correct.

(*a*) $\phi(t) = L^{-1}[(sI - A)^{-1}]$

(*b*) $\phi(t) = [(sI - A)^{-1}]$

(*c*) $\phi(t) = L^{-1}[sI - A]$

(*d*) none of these

29. One of the following expressions is correct regarding state transition matrix.

(*a*) $\phi^{-1}(t) = -\phi(t)$ (*b*) $\phi^{-1}(t) = \phi(-t)$

(*c*) $\phi^{-1}(t) = \phi(t)$ (*d*) none of these

30. For a second-order system, the state transition matrix for $t = 0$ is given by:

(*a*) $\begin{bmatrix} 1 & 0 \\ 0 & 1 \end{bmatrix}$ (*b*) $\begin{bmatrix} 1 & 1 \\ 1 & 1 \end{bmatrix}$

(*c*) $\begin{bmatrix} 0 & 1 \\ 1 & 0 \end{bmatrix}$ (*d*) None of these

31. The forcing function part of the state equation $dx(t)/dt = Ax(t) + Br(t)$ is given by:

(a) $dx(t)/dt = Ax(t)$
(b) $dx(t)/dt = Br(t)$
(c) $dx(t)/dt = Ax(t) + Br(t)$
(d) $dx(t)/dt = 0$

32. One of the following expressions is correct for the characteristic equation of a linear system in state variable method of analysis

(a) $|sI - A^2| = 0$ (b) $|s^2I - A| = 0$
(c) $|sI - A| = 0$ (d) none of these

33. One of the following expressions is correct regarding state transition matrix:

(a) $\phi(0) = \infty$
(b) $\phi^{-1}(t) = \phi(t)$
(c) $\phi(t_2 + t_1)\phi(t_1 + t_0) = \phi(t_2 + t_0)$ for any t_0, t_1 and t_2
(d) $[\phi(t)]^k = \phi(kt)$, where k is an integer

34. For the third-order system, the state transition matrix for t = 0 is given by

(a) $\begin{bmatrix} 1 & 0 & 0 \\ 0 & 1 & 0 \\ 0 & 0 & 1 \end{bmatrix}$ (b) $\begin{bmatrix} 1 & 1 & 1 \\ 1 & 1 & 1 \\ 1 & 1 & 1 \end{bmatrix}$

(c) $\begin{bmatrix} 0 & 0 & 0 \\ 0 & 0 & 0 \\ 0 & 0 & 0 \end{bmatrix}$ (d) $\begin{bmatrix} 1 & 0 & 1 \\ 0 & 1 & 0 \\ 1 & 0 & 1 \end{bmatrix}$

35. One of the following expressions is correct for controllability matrix $[Q_C]$.

(a) $[Q_C] = [B\ AB\ A^2B\ A^3B\ A^4B\ ...\ A^{n-1}\ B]$
(b) $[Q_C] = [A\ AB\ A^2B\ A^3B\ A^4B\ ...\ A^{n-1}\ B]$
(c) $[Q_C] = [A\ AB\ AB^2\ AB^3\ AB^4\ ...\ AB^{n-1}]$
(d) $[Q_C] = [B\ AB\ A^2B\ A^3B\ A^4B\ ...\ A^n\ B]$

36. Observability matrix $[Q_O]$ is given by one of the following expressions:

(a) $[Q_O] = [D\ AD\ A^2D\ A^3D\ ...\ A^{n-1}\ D]$
(b) $[Q_O] = [D'\ A'D'\ (A')^2D'\ (A')^3D'\ ...\ (A')^n\ D']$
(c) $[Q_O] = [D'\ A'D'\ (A')^2D'\ (A')^3D'\ ...\ (A')^{n-1}\ D']$
(d) $[Q_O] = [D'\ AD'\ A^2D'\ A^3D'\ ...\ A^{n-1}\ D']$

II. Identify TRUE and FALSE Statements

1. A servomechanism is a feedback control system where the controlled variable is a mechanical position or its time derivative.
2. A negative feedback system is always characterised by increased bandwidth and reduced gain.
3. The transient response of a feedback control system is that part of the total response that does not approach zero as the time approaches infinity.
4. The magnitude of the open loop transfer function of a closed loop system with a gain margin of 10 evaluated at its phase cross-over frequency is 0.1.
5. While a lead compensator reduces system bandwidth, a lag compensator on the other hand increases the system bandwidth.

Answers to Self Evaluation Exercise

I. 1. (b) and (d)	2. (a)	3. (c)	4. (a)	5. (b)	6. (b)
7. (b)	8. (c)	9. (b)	10. (b).	11. (c)	12. (a)
13. (b)	14. (a)	15. (a)	16. (d)	17. (b)	18. (b)
19. (a)	20. (a)	21. (b)	22. (a)	23. (b)	24. (a)
25. (a)	26. (a)	27. (b)	28. (a)	29. (b)	30. (a)
31. (b)	32. (c)	33. (d)	34. (a)	35. (a)	36. (c)

II. 1. True 2. True 3. False 4. True 5. False

Chapter 29

Electrical Measurements and Instrumentation

Learning Objectives

After completing the chapter, you will learn the following topics:

1. Basic Moving Coil Meter Movement
2. Basic Moving Iron Movement
3. Hot Wire Ammeter
4. Thermocouple Meter
5. Linear and Square Law Scales in Meters
6. Damping in Meters
7. Sensitivity, Accuracy and Ohms Per Volt Rating of Meters
8. Use of Shunts to Measure Higher Currents
9. Use of Series Resistors to Measure Voltages
10. Ammeter-Voltmeter Method, Substitution Method and Wheatstone Bridge for Measuring Resistances
11. DC and AC Bridges and Null Condition
12. AC Bridges including Maxwell's Bridge, Maxwell-Wein bridge, Hay's Bridge, De-Saughty Bridge, Schering Bridge
13. Clip-On Meters
14. Electronic Meters including Electronic Voltmeter, Electronic Ammeter and Electronic Ohmmeter
15. Digital Multimeters
16. Analog Oscilloscopes and Important Performance Parameters
17. Digital Oscilloscopes including Sampling Oscilloscopes and Digital Storage Oscilloscopes
18. Digital Storage Oscilloscope Parameters
19. Oscilloscope Probes
20. Frequency Counters including Universal Counters
21. Continuous Count and Reciprocal Counters
22. Frequency Counter Parameters
23. Microwave Frequency Counters
24. Frequency Synthesizers
25. Synthesized Function Generators
26. Arbitrary Waveform Generators
27. Logic Probes
28. Logic Analysers

Q. 1. How can a current meter measure voltage and resistance ?

Ans. Voltage, current and resistance are inter-related by Ohm's law which states that:

$$I = \frac{V}{R}; V = IR \text{ and } R = \frac{V}{I}$$

Thus, if R is kept constant, then $V \propto I$ (Principle of voltmeter). If V is kept constant, the $R \propto (1/I)$ (Principle of ohm meter).

Q. 2. What are the two basic types of current meters?

Ans. There are two basic types of current meters:

(*i*) **Electromagnetic Current Meter:** It makes use of magnetic field around a current carrying wire to measure the current.

(*ii*) **Thermal Current Meter:** It makes use of heat produced by the current flowing through a wire to measure the current.

Q. 3. If the current through a coil is increased, what happens to the magnetic field?

Ans. The magnetic field strength increases as the current increases.

Q. 4. Draw the basic moving coil meter movement.

Ans. The moving coil meter, as shown in Fig. 29.1, operates on electro-magnetic effect. In its simplest form, the moving coil meter uses a coil of very fine wire wound on a light aluminium frame placed in the magnetic field of a permanent horse–shoe magnet. When the current to be measured flows through the coil; the coil is magnetised and polarity of this magnetism (or magnetic field) is such that coil repels the field due to permanent magnet. The distance travelled by coil depends upon the magnitude of current flowing through the coil. Therefore, by attaching a pointer to a coil frame and by adding a scale calibrated in units of current; the magnitude of current flowing through the meter can be measured. It is also called **d'Arsonval meter.**

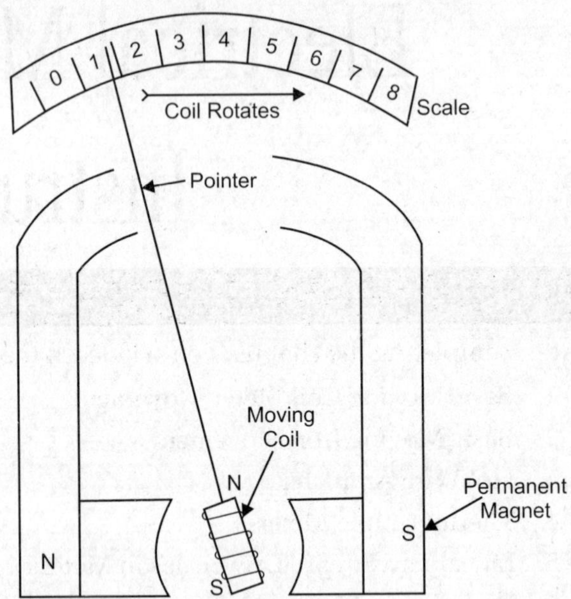

Fig. 29.1

Q. 5. What are various moving iron type movements?

Ans. There are three basic types of moving iron movements including (*i*) Radial Vane Type, (*ii*) Concentric Vane Type. (*iii*) Plunger Type.

All moving iron meters consist of two soft iron pieces placed inside a coil which is fixed. One of the soft iron pieces is fixed while other is free to move. When the current is passed through the coil; it produces magnetic field whose strength depends upon magnitude of current. The iron pieces are magnetised with same polarity and they repel each other with a force depending upon the strength of the magnetic field in which they are placed. This field, in turn, depends upon current under measurement. The distance travelled by moving piece is the indicator of current.

If iron pieces are rectangular and movable piece moves on one edge (radially), then the movement is called radial vane-type. If iron pieces are concentric and semicircular in shape and above vane moves about center, then it is concentric vane type. In plunger type movement, a soft iron core is placed partially within a fixed coil.

Q. 6. Draw a hot wire ammeter and explain its principle.

Ans. A wire becomes heated, when current passes through it. The hot wire ammeter makes use of this effect. A wire expands, when heated. More the wire is heated, more it expands.

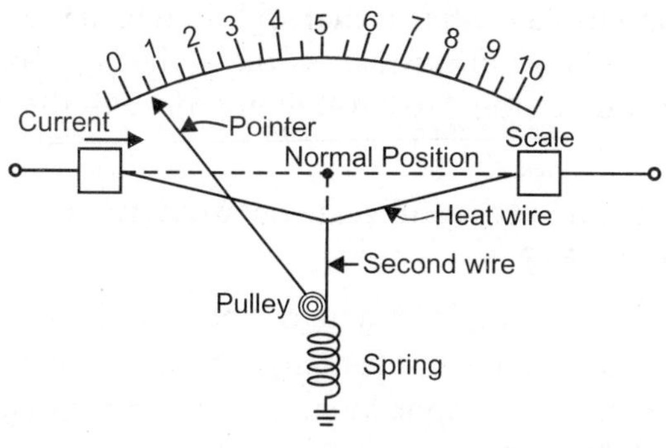

Fig. 29.2

Therefore, greater the current, more is the expansion. If a second wire and spring are attached to the original current carrying wire; they will pull the original current carrying wire, whenever it expands. If a pointer is attached to the second wire; it will move as the current carrying wire expands. The distance, the pointer moves, is used to indicate the magnitude of current through the wire. The constructional details are shown in Fig. 29.2.

Q. 7. What is a thermocouple?

Ans. A thermocouple consists of dissimilar metals which when joined, produce an e.m.f., when their junction is heated.

Q. 8. Briefly describe a thermocouple meter.

Ans. In this meter movement, usually a hot wire is also attached to the junction of the thermocouple. The current to be measured is passed through the thermocouple meter 'hot wire'. As in the case of hot wire ammeter, the wire heats upto a temperature that depends upon magnitude of current flow. The wire then heats the thermocouple junction and thermocouple, in turn, generates a small DC voltage. This voltage causes a current to flow in moving coil meter movement to indicate magnitude of current flowing through the hot wire.

Constructional details are shown in Fig. 29.3.

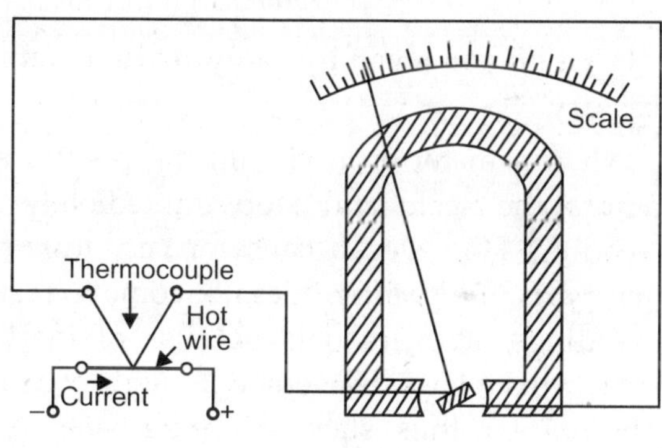

Fig. 29.3

Q. 9. Name the basic parts common to most of the meter movements.

Ans. The basic parts common to almost all meters are (*i*) coil (*ii*) pointer (*iii*) scale (*iv*) pivots, bearings, springs (*v*) retaining pins (*vi*) zero adjusting screw and (*vii*) a damping mechanism.

Q. 10. What are pole pieces? Why are they used?

Ans. Pole pieces are soft-iron pieces attached to the poles of horse shoe permanent magnet in moving coil meter. They are used to concentrate the lines of force between the magnetic poles. Refer to Fig. 29.4 for reference.

Q. 11. What is the difference between the coils used in moving coil type and moving iron type current meters?

Ans. The coil used in moving iron type meter consists of a number of turns so as to produce a strong magnetic field because it has to move an iron-vane, which is comparatively heavier.

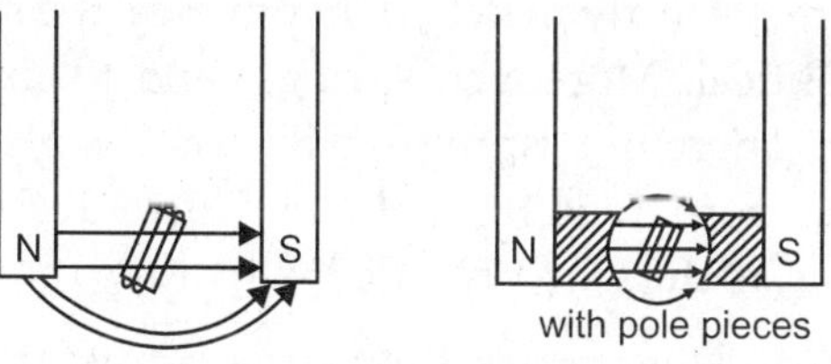

Fig. 29.4

Since the coil of a moving coil meter has itself to move, therefore, it must be as light as possible. And it is due to this reason that coil of a moving coil type of meter consists of several turns of fine wire wound on a light aluminium frame.

Q. 12. What is the purpose of springs?

Ans. When a meter is connected into a circuit, the pointer must indicate the magnitude of current flowing in the circuit and when it is disconnected, it must read zero. This action is controlled by the springs in the meter.

Q. 13. How do the springs affect the accuracy of the meter?

Ans. It is the springs that control the rotary motion of pointer. If the springs are not precisely manufactured, then there is every possibility that they become defective with age and meter pointer may be caused to rest at an incorrect reading or it may not read zero, when the circuit is disconnected.

Q. 14. Why do we use two oppositely wound springs?

Ans. Two springs are used with each of meter movement and these springs are oppositely wound (Fig. 29.5). This is done because metals expand on heating and contract as temperature falls. If both springs are wound in same directions, then as the temperature changes; this would cause the pointer to rotate off zero without any current flowing. When they are oppositely wound, when one expands; it tries to rotate the coil in one direction, whereas other tries to rotate it in opposite direction on expansion and hence their effects cancel and pointer remains at zero. Similar phenomenon occurs, when temperature falls.

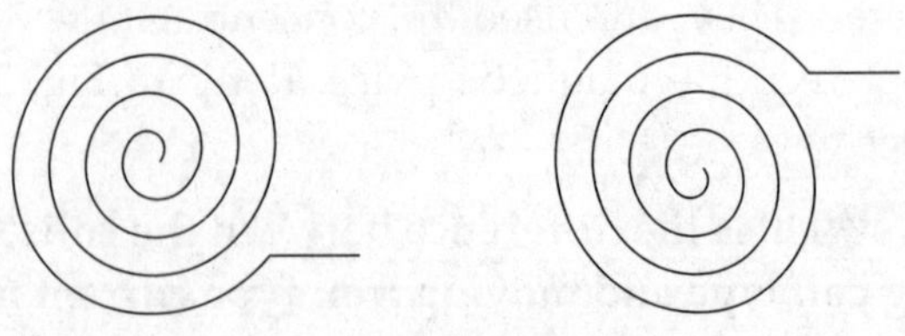

Fig. 29.5

Q. 15. What is the function of zero adjust screw?

Ans. The pointer shall remain at zero only, when two springs are perfectly matched. In practice, the springs are never identical. Moreover, springs tend to lose tension with age. Therefore, springs will not always keep the pointer at zero reading with no current flowing. To correct this, meters are provided with a zero adjust screw.

Q. 16. How does zero adjust screw adjust the zero?

Ans. As this screw, which is located on the front panel of the meter, is turned; it increases or decreases the tension on one of the springs, depending upon direction in which it is turned.

By turning the screw, the tension may be adjusted to be equal to that of other spring and when tension on both the springs are equal, the pointer reads zero.

Q. 17. What is a linear scale and a square law-scale?

Ans. A linear scale is that scale in which same incremental changes are represented by same spacings. The distance that the pointer defects across the scale is directly proportional to the magnitude of current flowing through the coil. See Fig. 29.6.

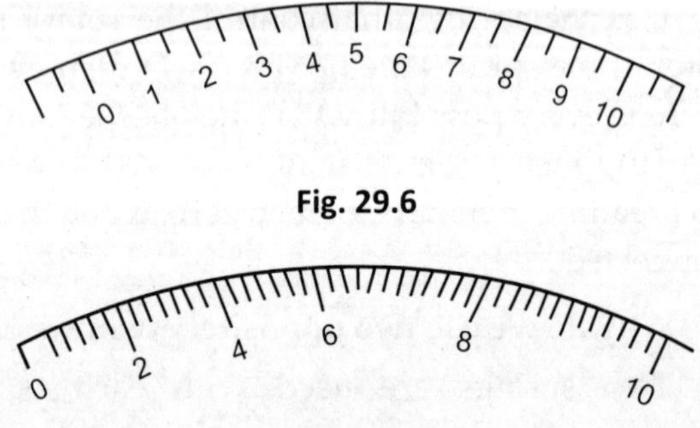

Fig. 29.6

Fig. 29.7

A 'Square-law' scale is that scale in which the distance that the pointer deflects across the scale, is proportional to the square of the current flowing through the coil. See Fig. 29.7.

Q. 18. Which type of scale is used in moving coil meter and why?

Ans. In moving coil meter, linear scale is used. In moving coil meter movement, the pointer travels a distance that is proportional to the deflection of coil (pointer is attached to the coil frame) and deflection of coil is proportional to the strength of magnetic field around the coil, which in turn depends upon amount of current flowing through the coil. Thus, the distance travelled by pointer is directly proportional to the magnitude of current flowing through the coil and hence linear scale.

Q. 19. Which type of scale is used in moving iron meters and why?

Ans. In moving iron meters, square-law scale is used. In moving iron meters, the distance travelled by the movable vane depends upon force of attraction (attraction type) or force of repulsion between the two iron-vanes (movable and fixed). In attraction type, there is only one vane that is attracted by the coil. Now this force of attraction or repulsion depends upon product of their pole strengths.

$$f \propto \frac{m_1 m_2}{r^2}$$

The pole strength induced in two vanes depends upon magnetic field, in which they are placed and magnetic field (provided by coil) is directly proportional to magnitude of current flowing.

$$m_1 \propto I, m_2 \propto I \text{ or } f \propto I^2$$

Thus, force of attraction depends upon square of the current under measurement and hence the square law-scale.

Q. 20. What do you mean by damping in meters? Why is it necessary?

Ans. When a meter is in circuit; the pointer should move across the scale and stop immediately at the correct reading. However, because of very little friction of rotating parts, the pointer does not come to rest at the correct readings, it overshoots because of inertia and then spring pulls it back, it overshoots slightly again and so on. The pointer, thus, starts vibrating back and forth before coming to rest.

To overcome this problem, the meter movement must be damped. Damping may be thought of to be a braking action on rotating parts. Damping also eliminates another problem. When circuit is de-energised, the pointer may come back so quickly that it bends, if there is no damping.

Damping may apply a braking action to the meter rotating parts, when the pointer comes back to zero.

Q. 21. How do you achieve damping in moving coil meters ?

Ans. Moving coil meters make use of aluminium frame on which coil is wound to provide damping. Since aluminium is a conductor, the frame acts as a one turn coil. When the coil assembly and the pointer rotate; the frame also cuts the lines of flux of the permanent magnet. Small currents called eddy currents are induced in the frame which set up their own magnetic field. The polarity of this magnetic field is opposite to that about the coil. This fact reduces the overall field about the coil so that coil along with the pointer swings more slowly. As soon as coil assembly comes to rest; the opposing force disappears.

When the meter is disconnected from the circuit, then also the field set up due to eddy currents opposes the motion of coil assembly to provide damping.

Q. 22. How do you achieve damping in moving iron meters ? Can you use the same method as used in moving coil meters?

Ans. Since in case of moving iron instruments, the coil remains stationary; the method used for providing damping in moving coil meters cannot be used in moving iron instruments.

In moving iron instruments, air damping is used. In this method, a light aluminium vane (damping vane) is attached to the same shaft as the pointer and moving vane and is enclosed in an air tight damping chamber. As the moving vane is repelled; the damping vane rotates with it and compresses the air ahead of it, as it moves. This decreases the speed at which the vane moves and hence decreases the speed of moving vane and pointer. When pointer comes to rest; no more air is compressed and damping action ceases.

Q. 23. What type of scale is used in thermocouple meters and why ?

Ans. In a thermocouple meter, there is a square law scale. Thermocouple consists of a moving coil meter and a thermocouple. Deflection in moving coil meter is proportional to the voltage generated by thermocouple, which depends upon heat at the junction and heat generated is proportional to square of the current under measurement.

Q. 24. What do you mean by meter sensitivity ?

Ans. Sensitivity of a meter is defined as the magnitude of current required to give full scale deflection.

Q. 25. Define 'Accuracy' of a meter.

Ans. Accuracy is defined as the percentage error at full scale deflection. If in a 100 mA meter, the accuracy is ± 2%, that means at 100 mA, the actual reading may be anywhere between 98 to 102 mA. Another important point is that whatever is the error at full scale, same is at other readings less than full scale. That is at a reading of 50 mA also, the actual reading may be anywhere between 48 and 52 mA.

Q. 26. Why is it preferable to use meters that give full scale deflection?

Ans. At full scale deflection, the percentage error is less, whereas it increases as the reading decreases.

Q. 27. What type of current meters are used for measurement of DC only? Which is the most suitable one?

Ans. The current meters which are capable of measuring d.c. are :

(1) Moving coil (2) Moving iron (3) Thermocouple.
Moving coil is the most suitable one.

Q. 28. Can you use moving coil for measuring AC currents? If not, why?

Ans. Moving coil meter's deflection depends upon field produced in the coil. If the current is alternating, the pointer shall be swinging forward and backward with change in polarities of AC, so it cannot measure AC.

Q. 29. Can a thermocouple measure AC, if yes, how?

Ans. Thermocouple is capable of measuring AC of high frequencies very accurately. It is based on the principle of generation of heat due to current. Heat generated does not depend upon frequency. Hence, thermocouple meter movement is insensitive to frequency.

Q. 30. What is the main application of thermocouple meters?

Ans. It is mainly used to measure radio frequency currents, having frequencies ranging from hundreds of megahertz to thousands of megahertz.

Q. 31. Can a moving iron instrument measure AC?

Ans. Yes, it can. Its action depends upon induction of magnetism in iron vanes which does not depend upon the type of signal, whether AC or DC.

Q. 32. Can you use moving coil meter for measuring AC with the help of some other circuit?

Ans. If we convert the AC to be measured into DC first and then apply it to moving coil meter, then it shall measure AC. Thus, moving coil meter along with a rectifier can measure AC. This type of instrument is called rectifier-meter.

Q. 33. What is a shunt? Where is it used?

Ans. Shunt is a resistance to be used in parallel with the coil of the basic meter movement (Fig. 29.8). The purpose of shunt is to enable the meter to measure large currents. Maximum value of current that a basic meter movement can measure is about 10 mA. If we want to measure larger currents, then rest of the current has to be delivered through the resistance, shunt allowing only 10 mA to pass through coil. Thus, in large range meters, shunts are provided inside the meter itself. These are also called internal shunts.

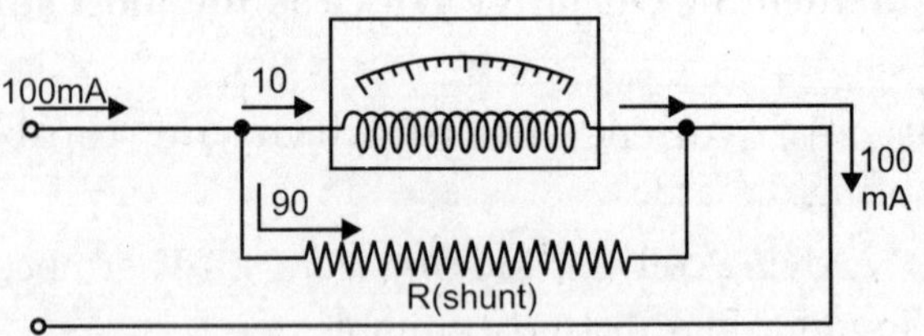

Fig. 29.8

Q. 34. What are external shunts? Why are they used?

Ans. Suppose in a particular application, current to be measured is 100 A, where meter available is capable of measuring only 1 A. Then we may provide an external resistance in parallel with meter to divert 99A of current. This is called external shunt.

Q. 35. How do you calculate the value of shunt resistance ?

Ans. Since shunt is in parallel with coil, therefore, voltage across shunt is same as that across coil (Fig. 29.9).

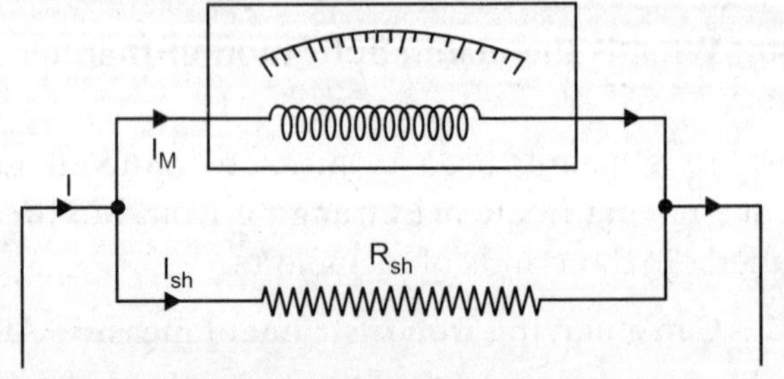

Fig. 29.9

$$I_M R_M = I_{sh} R_{sh} \text{ or } R_{sh} = \frac{I_M R_M}{I_{sh}}$$

Q. 36. Shunts have (*i*) high resistance or (*ii*) low resistance. Why?

Ans. Shunts have low resistance. Shunts are used to divert most of the current through them to enable the basic meter to measure larger currents. The current through them shall be large only, if their resistance is relatively small as compared to coil resistance In the previous example also, R_{sh} comes out to be equal to 3 ohms.

37. What is the arrangement inside a multi-range current meter?

Ans. Different values of shunts can be employed for different ranges.

Q. 38. How do you connect a current meter in a circuit?

Ans. A current meter is always connected in series with the circuit.

Q. 39. If we connect the current meter in parallel with the power source; what will happen?

Ans. Since current meter is a low resistance device; we may create almost a short circuit. If the meter is put across the power; this results in burn out of coil of the meter.

Q. 40. What are the precautions that must be observed while using current meter on DC?

Ans. Polarities must be taken into account while measuring DC. Positive of meter must be connected to positive of circuit and negative to negative of circuit.

Q. 41. Is there any DC current meter where polarity consideration is not important?

Ans. Those current meters which are having their scale such that zero is at the center of scale, need not consider polarities.

Q. 42. How can a current meter be used to measure voltage?

Ans. We know that, $V = IR$.

or $$I = \frac{V}{R} = V \times \frac{1}{R}$$

Thus, current is proportional to voltage for constant resistance. Suppose there is a current meter of rating $0 \rightarrow 1$ mA and resistance 1000 ohms. Now at full scale deflection, 1 mA is flowing through the coil, therefore, voltage developed across the coil resistance will be,

$$1 \times 1000 \text{ mV} = 1\text{V}$$

And if $I = 0.5$ mA

$$V = 0.5 \text{ V}$$

Thus, this current meter can measure voltage upto 1 V.

Q. 43. How do you measure larger voltages with basic current meter which is capable of measuring only small voltages (say of the order of *I V*)?

Ans. This can be done by increasing the series resistance of meter because

$$V = I \times R$$

For same *I*, if *R* is large, *V* will be large. Suppose we put a resistance of 1000 ohms in series with meter discussed above.

Then for full scale deflection $V = 1\text{ mA} \times (1000 + 1000) = 2\text{ V}$

Thus, the range is doubled. These series resistance are also called multiplier resistances.

Q. 44. How do you find the value of multiplier resistor for extending the range?

Ans. Let E_{max} → Maximum value of voltage to be measured

I_{max} → Current that meter can handle

Then total resistance, $R_{TOT} = \frac{E_{max}}{I_{max}}$

$R_{MULT} = R_{TOT} - R$

Q. 45. What is ohms/volt rating of a voltmeter? What is its significance?

Ans. The ohms/volt rating is given by $(R_M + R_{MULT})$ required for full scale deflection.

For example, 1 mA, 1000 ohms meter movement indicates full scale deflection for 1 *V*. Therefore, its ohms/voltage rating is

$$\frac{1000}{1} = 1000 \text{ ohms/volt.}$$

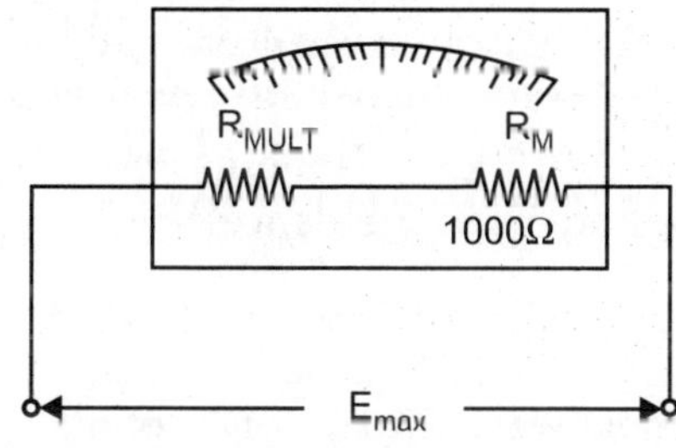

Fig. 29.10

It indicates the sensitivity of the voltmeter. Higher the ohms/volt rating, better the voltmeter is.

Q. 46. How will you connect a voltmeter in a circuit and why?

Ans. It is connected in parallel with a power source, as it is a high resistance device. If connected in series, it will disturb the circuit conditions.

Q. 47. What is the principle of measurement of power?

Ans. Power $= V_I = I^2 R = V^2/R$

If *V* and *I* or *I* and *R* or *V* and *R* are known, *P* can be determined.

Wattmeters are based on the principle that when voltage is applied across one coil and load current through another, magnetic field produced due to the two coils interact and the pointer deflects in proportion to *VI*, thus measuring power.

Q. 48. What is an energy meter?

Ans. An energy meter is capable of measuring energy consumed in a circuit. Since energy is given by power multiplied by time for which power is consumed, therefore, energy can be measured with a modified watt meter.

Q. 49. How do you classify resistances?

Ans. The resistances are classified into three categories:

(*i*)	Low resistances	1 ohm or less
(*ii*)	Medium resistances	Between 1 ohm and 100 kilo ohms
(*iii*)	High resistances	More than 100 kilo ohms.

Q. 50. Justify that method used for measurement of medium resistances will not be suitable for measurement of low resistances.

Ans. Suppose resistance to be measured is having a value of about 0.01 ohm. The contact resistance of the apparatus is having a value of about 0.001 ohm. So, it will cause a great error, whereas error will be negligible, if the resistance to be measured is having a value of 100 K. Hence, the answer.

Q. 51. How do you decrease the error in voltmeter ammeter method ? Give an example.

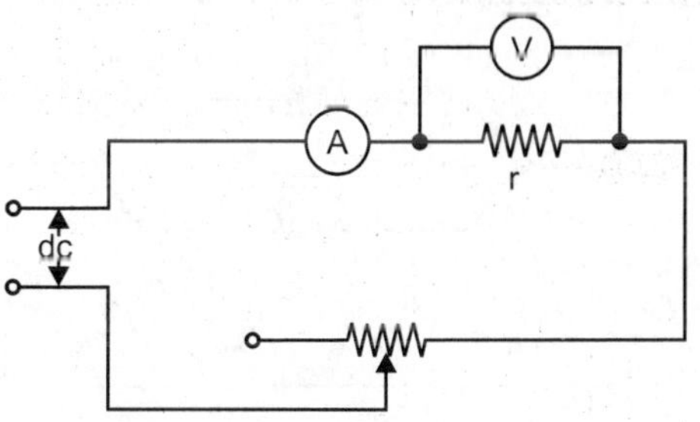

Fig. 29.11

Ans. If we make R_v = 500 ohms in the previous question and if voltmeter can read upto 10 V, then

Voltmeter reading $= r \times 0.1$

where, $\frac{1}{r} = \frac{1}{500} + \frac{1}{1} = \frac{501}{500} \approx 1$

$$r = \frac{500}{501} \approx 1 \text{ ohm.}$$

Voltmeter reading = 0.1 V

$$\text{Measured } R = \frac{0.1}{0.1} = 1 \text{ ohm.}$$

So, we have seen that error is reduced to zero. Thus, by making current carrying capacity of resistance under measurement large, the error in voltmeter-ammeter method can be decreased.

Q. 52. What are the common methods used to measure medium resistances?

Ans. The methods are :

(*i*) Ammeter-voltmeter method

(*ii*) Substitution method.

(*iii*) Wheatstone bridge

Q. 53. Draw the basic circuit diagram for substitution method and explain its working in brief.

Ans. Initially, switch *S* is in position 2, the reading on the deflection instrument is noted accurately, then *S* is shifted to 1 and *R* is varied (*R* is standard known variable resistance) till we get same deflection gain. Then, *R* gives directly the value of unknown resistances. See Fig. 29.12 for reference.

Q. 54. On what factors does accuracy of substitution method depend?

Ans. It depends upon :

(*i*) Constancy of applied voltage

(*ii*) Accuracy of indicating instrument

(*iii*) Accuracy of known variable resistance

(*iv*) Observation errors.

Q. 55. Which is the most accurate and commonly used method for measurement of medium resistances?

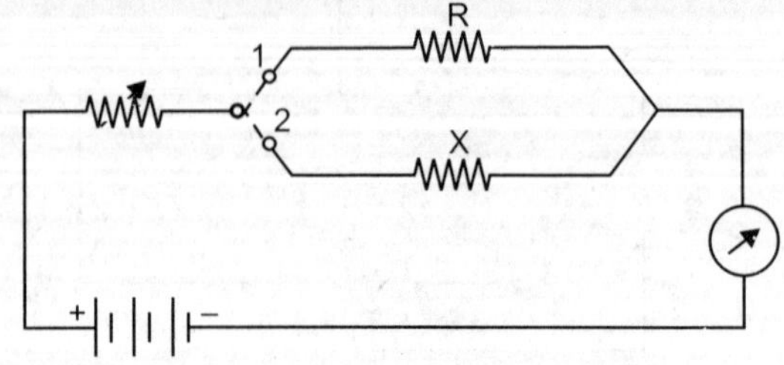

Fig. 29.12

Ans. Wheatstone bridge

Q. 56. Give the basic circuit arrangement of Wheatstone bridge and explain in brief its working.

Ans. Fig. 29.13 shows the basic Wheatstone bridge. *P* and *Q* are ratio arms. *R* is unknown resistance, *S* is a known resistance.

S is adjusted till galvanometer shows no reading. Then *B* and *D* are at the same potential.

$$\text{or} \quad V_{AB} = V_{AD}, V_{BC} = V_{CD}$$

$$\text{or} \quad \frac{V_{AB}}{V_{BC}} = \frac{V_{AD}}{V_{CD}}$$

Since whatever current flows through *AB*, also flows through *BC* and since whatever current flows through *AD*, also flows through *DC*.

$$\frac{P}{Q} = \frac{R}{S} \text{ or } R = \frac{P}{Q} \times S.$$

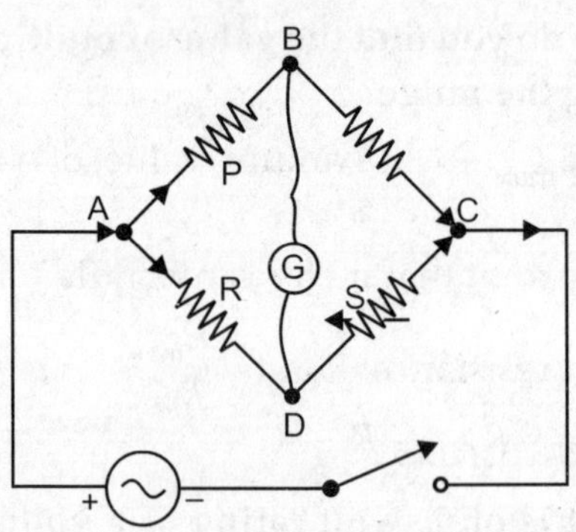

Fig. 29.13

Q. 57. How do you make an accurate measurement of medium resistances with Wheatstone bridge? Assume the resistance to be measured to be equal to 60 ohms.

Ans. Let *P* = 10 ohms, *Q* = 10 ohms.

Adjust *S*, till you get balance. Let us say we get balance for *S* between 57 and 58 ohms. This means that if *S* = 57 ohms, *G* shows deflection to left (Say), then if *S* = 58 ohms, it shows deflection towards right indicating that balance point lies somewhere between 57 and 58.

Now make *P* = 10, *Q* = 100 ohms and say setting of *S* is between 576 and 577 (for balance).

$$\text{Then} \quad R = \frac{10}{100} \times 576 = 57.6 \text{ ohms.}$$

Then *R* lies between 57.6 and 57.7 ohms.

Make *Q* = 1000 ohms keeping *P* same and obtain balance by varying *S*. Let us say that setting of *S* is between 5763 and 5764 ohms.

$$\text{Then} \quad R = \frac{10}{1000} \times 5763 = 57.63 \text{ ohms.}$$

R lies between 57.63 ohms and 57.64 ohms.

$$R = 57.635 \text{ ohms.}$$

Thus, *R* has been determined to a fair degree of accuracy.

Q. 58. What is bridge sensitivity?

Ans. It is the minimum change or alteration required under balance conditions to give full seale deflection in galvanometer connected across B and D. Smaller the change required, greater the sensitivity.

Q. 59. On what factors does sensitivity of Wheatstone bridge depend?

Ans. It depends upon the sensitivity of detector used. More sensitive the detector, more is the bridge sensitivity.

Q. 60. What is the condition for maximum sensitivity?

Ans. It is $P = Q = R = X = G$, where G is galvanometer resistance.

Q. 61. While operating a Wheatstone bridge, battery key is pressed before galvanometer key. Why?

Ans. The resistances used in different bridge arms are never perfect. They are always associated with certain inductances and capacitances. If the galvanometer key is pressed before the battery key, the galvanometer shall give a sudden kick due to transients in inductances and capacitances in the beginning. Thus, battery key is pressed first so that when the transients die out and steady state values are reached, then only galvanometer key is pressed.

Q. 62. Is there any provision for avoiding the resistance of lead contacts in Wheatstone bridge? If no, can you name any bridge where there is any provision of avoiding the effect of resistance of lead contacts?

Ans. No. In Kelvin's double bridge, there is provision of avoiding resistance of lead contacts. It is based on the same principle as Wheatstone bridge.

Q. 63. It is a common practice to use a reversible key in battery circuit of a Wheatstone bridge as shown in Fig. 29.14 to make two separate measurements of unknown resistance, first measurement with positive connected to A and second with positive connected to C. Why?

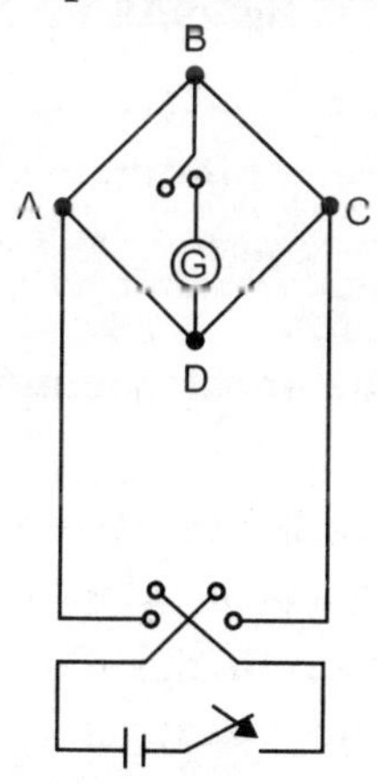

Fig. 29.14

Ans. When a current is passed through a resistance, thermo-electric emf's are produced at the junction depending upon the direction of flow of current. If a two-way key is used and two measurements are made with either direction of current, then error due to thermo-electric emf can be eliminated.

Q. 64. Using Thevenin's theorem, prove that circuit shown in Fig. 29.15 (*a*) can be replaced by circuit in Fig. 29.15 (*b*).

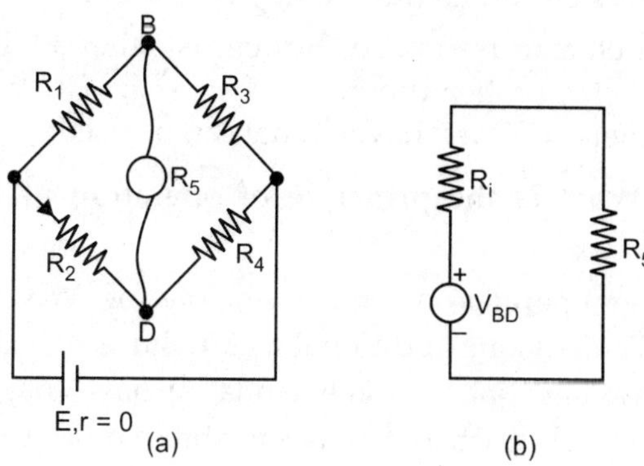

Fig. 29.15

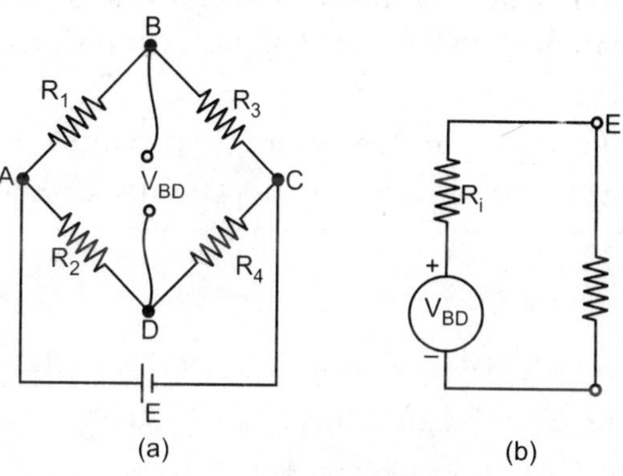

Fig. 29.16

Ans. If R_5 is removed from B to D, then circuit after applying Thevenin's theorem is shown in Fig. 29.16 (*b*).

$$R_i = \frac{R_1 R_3}{R_1 + R_3} + \frac{R_2 R_4}{R_2 + R_4}$$

Potential drop across $AB = \dfrac{ER_1}{R_1 + R_3}$

Similarly, potential drop across

$$AD = \frac{ER_2}{R_2 + R_4}$$

$$V_{BD} = \frac{ER_2}{R_2 + R_4} - \frac{ER_1}{R_1 + R_3}$$

Thevenin's equivalent circuit is as shown in Fig. 29.17(*b*). where V_{BD} and R_i have been calculated.

Q. 65. What is the most commonly used type of method for measurement of inductances and capacitances?

Ans. By AC bridges.

Q. 66. What is the difference between an AC bridge and a DC bridge?

Ans. A DC bridge employs real impedances (resistance) in its arms and the source of current is DC. In AC bridges, the bridge arms have resistances, capacitances, inductances etc. and supply used is alternating.

The detector used in AC bridges is either a vibration galvanometer or headphones, whereas in DC bridges, galvanometer (*D' Arsonaval Type*) can be used.

Q. 67. What is the principle of operation of an AC bridge ?

Ans. It operates on the principle of Wheatstone bridge. There is one additional factor. Since *AC* has both magnitude and phase; it is essential to have magnitude balance as well as phase balance to obtain null reading in the detector.

Q. 68. What do you mean by magnitude balance and phase balance? What are balance conditions in AC bridges?

Ans. In an AC bridge having impedances Z_1 Z_2, Z_3 and Z_4 in place of *P*, *Q*, *R*, and *S*, according to Wheatstone bridge principle, the balance condition is

$$Z_1 Z_3 = Z_2 Z_4$$

Magnitude balance implies $|Z_1| \times |Z_3| = |Z_2| \times |Z_4|$

where $|Z_1|$ = Magnitude of Z_1 and so on.

Phase balance implies $\varphi_1 + \varphi_3 = \varphi_2 + \varphi_4$.

Q. 69. If *Z* is an impedence due to an inductance *L* having a resistance *R*, find magnitude and phase.

Ans. Then

$$|Z| = \sqrt{R^2 + \omega^2 L^2}$$

$$\text{Phase } \varphi = \tan^{-1} \frac{\omega L}{R}.$$

Q. 70. If *Z* is an impedance due to a capacitance *C* having a resistance *R*, find $|Z|$ and φ.

Ans.

$$|Z| = \sqrt{R^2 + 1/\omega^2 C^2} \quad \varphi = \tan^{-1}\left(-\frac{1}{\omega RC}\right).$$

Q. 71. In the circuit shown in Fig. 29.17, find voltage across resistance when the circuit is in resonance? (*v*) is an alternating source of voltage.

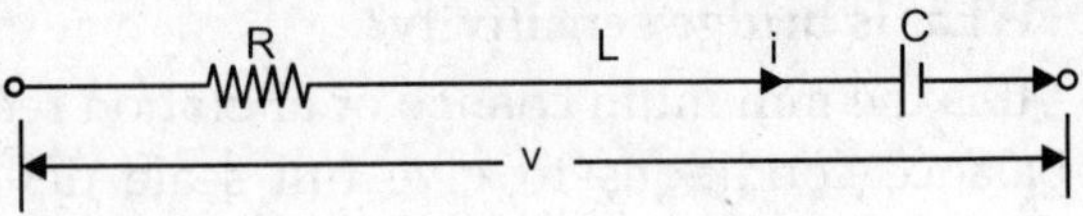

Fig. 29.17

Ans. Voltage across $R = v$

It is because voltage across *L* leads (*i*) by 90° and voltage across capacitance lags (*i*) by 90°. Thus, they cancel each other. Hence, whole of the voltage appears across *R*.

In other words at resonance, the resistance of series *LC* circuit is zero and hence, only element left is resistance *R*.

Q. 72. Is the bridge shown in Fig. 29.18 valid for AC operation?

Ans. No, because for obtaining balance,

$$\angle \varphi_1 + \angle \varphi_3 = \angle \varphi_2 + \angle \varphi_4$$

Here, $\angle \varphi_2 + \angle \varphi_4 = 0$

Here, $\varphi_1 \rightarrow$ + ve, Inductances introduce + ve phase

$\varphi_3 \rightarrow$ + ve

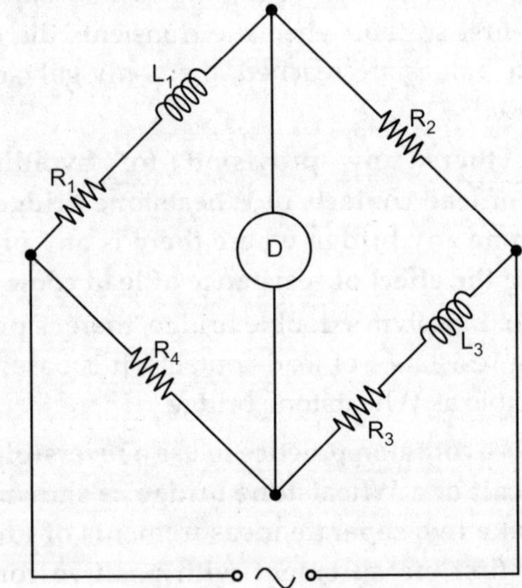

Fig. 29.18

Thus, the condition.

$\angle \varphi_1 + \angle \varphi_3 = 0$ cannot be attained which means null point cannot be achieved and hence, the bridge is not valid.

Q. 73. What are the most commonly used type of bridges?

Ans. (*i*) **Maxwell's Bridge**–For inductance measurement in terms of standard known inductance.

(*ii*) **Maxwell-Wein Bridge**–For inductance measurement in terms of capacitance.

(*iii*) **Hay's Bridge**–For inductance measurement in items of capacitance.

(*iv*) **De Saughty Bridge**–For capacitance measurement in terms of capacitance.

(*v*) **Schereing Bridge**–For capacitance measurement and for finding loss angle of capacitors.

Q. 74. Give the basic circuit arrangement for Maxwell's Bridge and Maxwell-Wein Bridge.

Ans. Fig. 29.19 shows Maxwell Bridge, whereas Fig. 29.20 shows Maxwell-Wein Bridge.

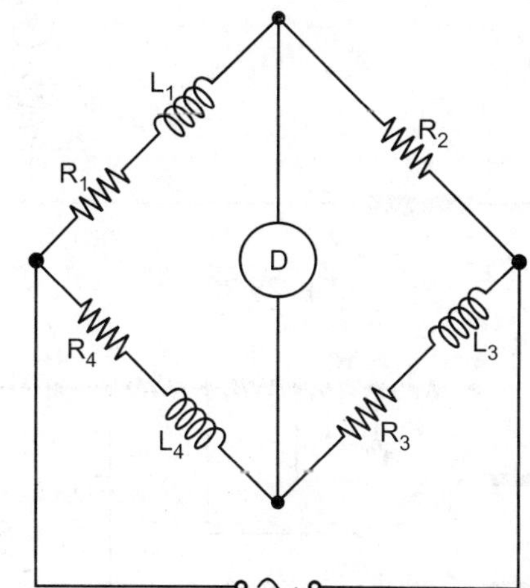

Fig. 29.19

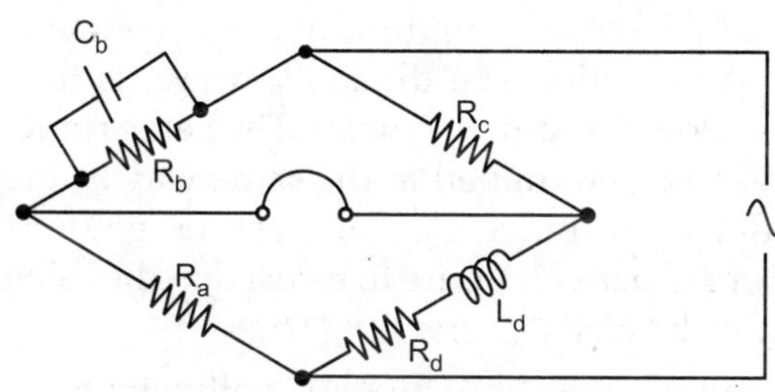

Fig. 29.20

Maxwell's Bridge

$L_1 = L_4 \times R_2/R_3, R_1 = R_4 \times R_2/R_3$

Maxwells-Wein Bridge

$$L_d = R_a R_b R_c, R_d = \frac{R_a}{R_b R_c}, Q_d = \frac{\omega L_d}{R_d} = \omega C_b R_b.$$

Q. 75. Give the basic circuit arrangement for (*i*) Hay's bridge (*ii*) De-saughty bridge and (*iii*) Schering bridge. Also, write relevant expressions for balance conditions.

Ans. The circuit arrangements alongwith balance conditions for Hay's bridge, De-saughty bridge and Schering bridge are respectively given in Fig. 29.21, 29.22 and 29.23.

Q. 76. What does the basic Analog Ohmmeter comprise of? What is its basic principle of operation?

Ans. A power source and a variable resistance connected in series with the basic meter movement, as shown in Fig. 29.24, converts it into an ohmmeter. When the leads are shorted, representing zero resistance, the current through the meter is maximum and is so adjusted

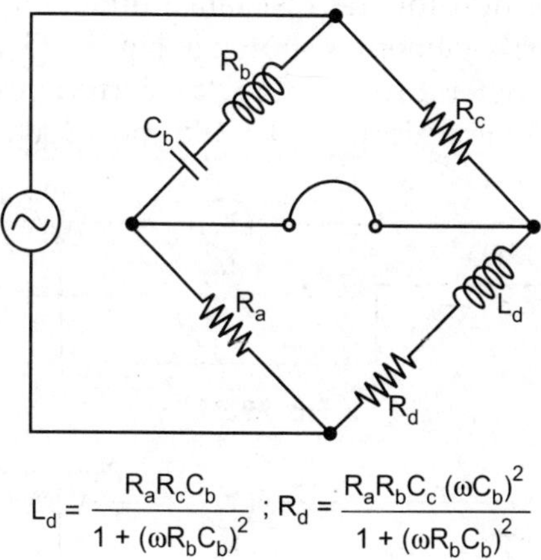

Fig. 29.21

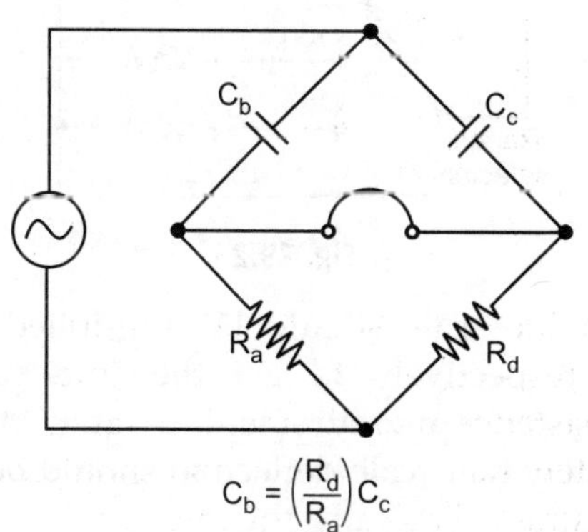

Fig. 29.22

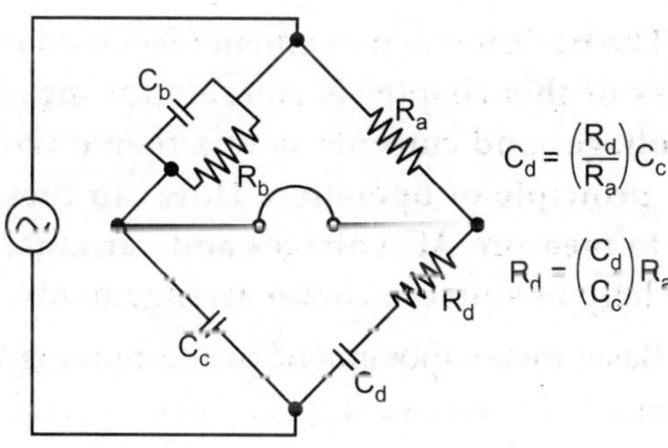

Fig. 29.23

with the variable resistance as to read zero located on the R.H.S. of the scale. The current flowing through the meter and hence, the deflection reduces, when an external unknown resistance is introduced in the circuit. The deflection reduces by an amount depending upon the value of resistance introduced. The scale is calibrated accordingly. Due to the non-linear nature of the current-external resistance relationship, the scale is crowded towards the L.H.S of the scale. The extreme L.H.S. reads infinity implying zero current and no deflection. Such a meter exhibits highest accuracy in the mid range. So, it is preferred to have an arrangement that gives near mid range deflection for the resistance under measurement. One such arrangement is shown in Fig. 29.25. It is a multi-range ohmmeter. In × 1 range, the correct reading is the one read on the scale. In × 10, × 100 and × 1000 ranges, the

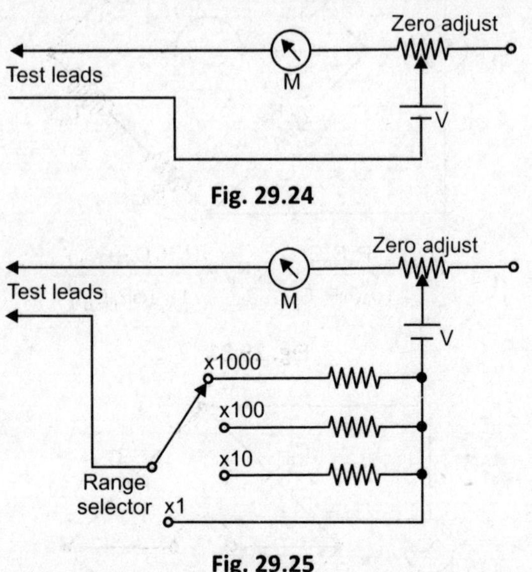

Fig. 29.24

Fig. 29.25

reading on the scale should be multiplied by 10, 100 and 1000 respectively to get the true value. While making resistance measurement, a range that leads to approximately half scale deflection should be chosen for highest accuracy. Typically, the mid scale reads 2K. So, if the resistance under measurement is expected to be of the order of 200K or so, x 100 range should be selected.

Q. 77. The basic meter movement discussed in the first few pages of this chapter is suitable for measurement of DC voltages and currents as was then obvious from its basic principle of operation. How can these meters be used to measure AC voltages and currents? Explain with the help of suitable circuit arrangements.

Ans. Basic meter movement of the moving coil type is not suitable for measuring AC currents and voltages for reasons explained earlier. To enable the meter do so, the AC input is rectified and an equivalent DC generated which is then fed to the meter movement. For frequencies greater than 15 Hz, the meter movement responds to the average value of the rectified current pulses. Since the relationship between the applied AC input voltage and average current pulses is well established, the scale can be calibrated to read either RMS or peak or peak-to-peak value.

One such meter arrangement using half wave rectifier is shown in Fig. 29.26. If I_{fs}, V_{fs} and R_{in} are the full scale DC current, full scale RMS voltage and input resistance of the meter shown, then

$V_{fs} = 2.22 \times I_{fs} \times R_{in}$ which gives $R_{in} = 0.45 \times V_{fs}/I_{fs}$ and $R_{in} = R_M + R_s$

In other words, ohms per volt rating of this AC voltmeter is given by $(0.45/I_{fs})$.

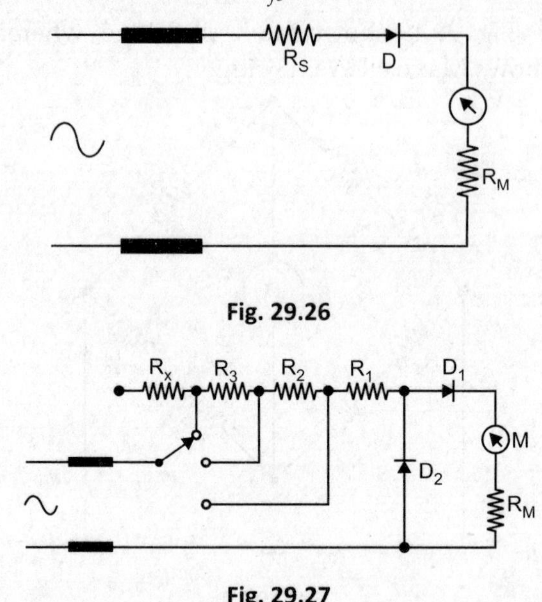

Fig. 29.26

Fig. 29.27

Fig. 29.27 shows a multi-range AC voltmeter using a half wave rectifier. The diode D_2 prevents the reverse current to flow through the meter. The values of R_1, R_2, R_3 and R_x can be determined in the same way as they were in case of a multi-range DC voltmeter except for the fact that the ohms per volt figure to be used in the calculations would now be $(0.45/I_{fs})$ and not $(1/I_{fs})$

The full wave rectifier type AC voltmeter as shown in Fig. 29.28 has enhanced AC sensitivity. It is twice that of half wave sensitivity. The input resistance of the meter is doubled too, thus reducing loading on the circuit under test. But it doubles the diode's forward voltage drop too and because of this, its use is limited to relatively large signal AC measurements.

AC current meters would have a rectifier and an appropriate shunt. Fig. 29.29 shows a typical AC current meter using a half wave rectifier.

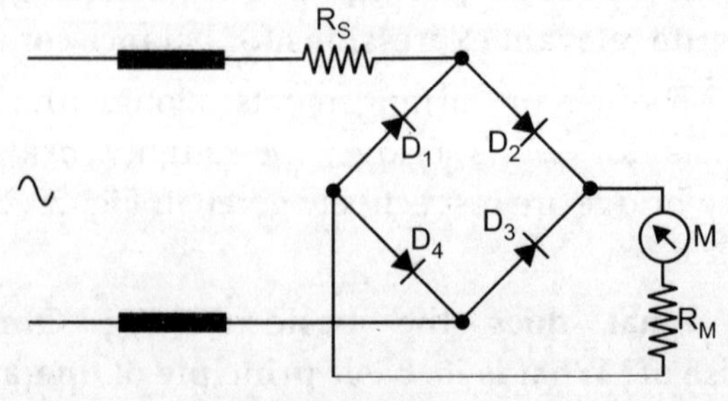

Fig. 29.28

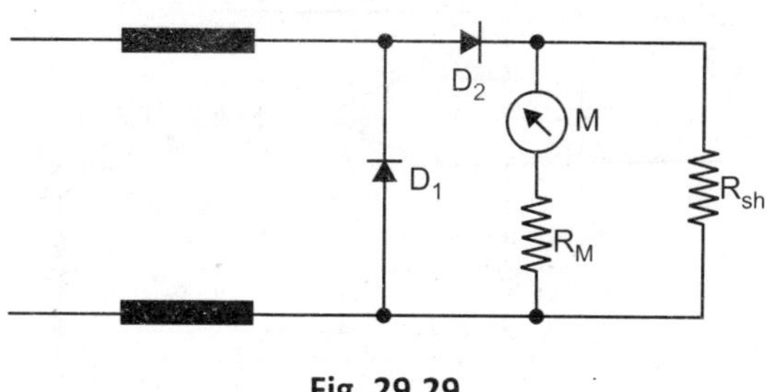

Fig. 29.29

Q. 78. What are clip-on meters? What is the basic principle of operation of these meters?

Ans. Clip-on meters are the non-invasive type AC current meters. These can be used to measure AC currents without breaking the circuit and are, thus primarily useful for measuring relatively higher currents, where it is impractical to break the circuit to insert the meter. Based on electromagnetic induction, the basic meter comprises of a coil wound on a ferrite core as shown in Fig. 29.30. The alternating current under measurement sets up a proportional magnetic field in the core which then induces a voltage in the coil. The voltage is then measured on a meter.

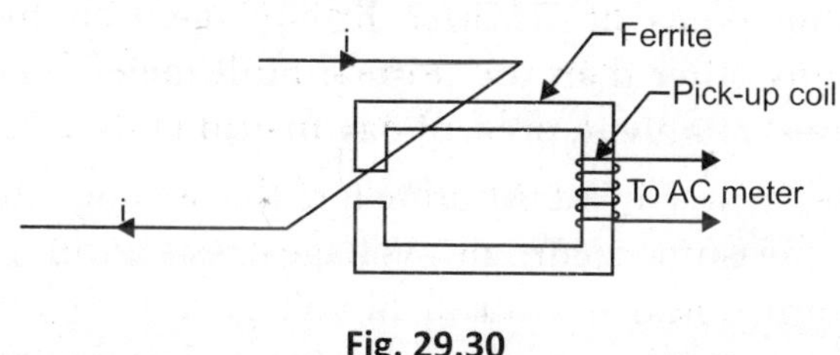

Fig. 29.30

Q. 79. What are Electronic Meters? How do they overcome the limitations of analog meters built around meter movements and passive components?

Ans. Electronic meters make use of active devices to overcome most of the limitations encountered in analog meters built around meter movements and passive components and discussed in the earlier paragraphs. These limitations are primarily overcome by using an amplifier with a very high input impedance at the front end in case of voltage measurements. The rectifier to be used in case of AC voltage measurements is also built around active devices, usually opamps, offering near zero rectifier forward voltage drop. The high input impedance eliminates the loading problem.

In case of current measurements, the front end is usually a current to voltage converter. This configuration offers near zero input impedance to the current source, thus eliminating any loading effect due to the meter.

The resistance measurements are also done by injecting a known current into a current to voltage configuration and then developing a proportional voltage across the unknown resistance. Electronic meters invariably use an opamp, preferably an FET input opamp as the active device due to its high input impedance, high open loop gain, high power supply rejection ratio (PSRR), low noise and so on.

Q. 80. With the help suitable basic circuit arrangements, briefly describe the functioning of:

(a) Electronic Voltmeter

(b) Electronic Ammeter

(c) Electronic Ohmmeter

(d) Electronic AC Voltmeter.

Ans. Electronic Voltmeter : Fig. 29.31 shows the basic multi-range electronic DC voltmeter. The opamp is wired as non-inverting amplifier. There is a different potential divider at the input for each range. Resistors R_1, R_2, R_3 and R_6 are so chosen as to produce the same voltage at the non-inverting input of the opamp for maximum voltage input in any of the ranges and also that the meter has the desired input impedance. The amplifier gain is chosen to make use of the maximum possible output voltage without sacrificing linearity. For an opamp with ± 12V supply voltages, the maximum opamp output may be taken as ± 10 V. The meter movement and the associated series resistance then should give full scale deflection for 10 V input. Following mathematical expressions govern the functioning and design of the meter:

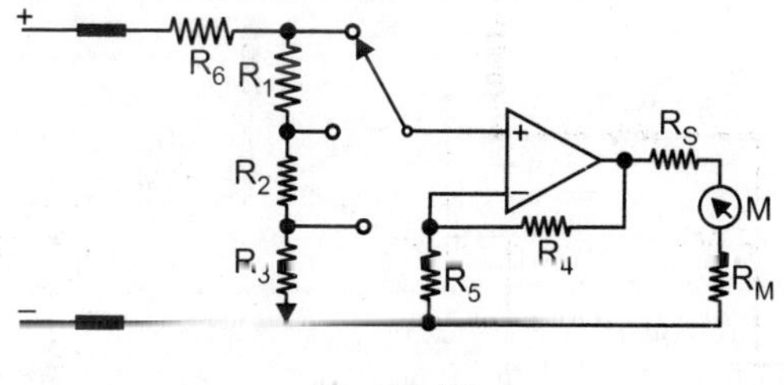

Fig. 29.31

1. $R_{in} = R_1 + R_2 + R_3 + R_6$
2. $V_L \times (R_1 + R_2 + R_3)/R_T = V_M \times (R_2 + R_3)/R_T = V_H \times R_3/R_T = V_{in}$ (+)
3. V_{in} (+) × Gain = 10
4. Gain = $1 + R_4/R_5$ (for ± 12 V supply)

R_s is so chosen as to give full scale deflection for 10 V input to the meter. As an illustration, for V_L = 2V, V_M = 20V, V_H = 200V, desired R_{in} = 10M and available meter movement of 100μA, 1K:

R_1 = 4.5M, R_2 = 450K, R_3 = 50K, R_s = 99K,

R_4 = 90 K and R_5 = 10K

Electronic Ammeter : Fig. 29.32 shows the basic electronic ammeter. If I_{max} is the maximum current measuring capability of the meter, then R can be chosen to have I_{max}. R = 10 Volts (for ± 12 V supplies)

$R_{in} = R/A_{OL}$, where A_{OL} is the open loop gain of the opamp. This expression underlines the importance of the opamp to have as high an open loop gain as possible. The multi-range ammeter will have selectable feedback resistors.

Electronic Ohmmeter : Fig. 29.33 shows the basic electronic ohmmeter. The unknown resistance is connected in place of the feedback resistance. A known current (= V/R) is injected into the circuit. The voltage at the opamp output equals ($V.R_x/R$). For given values of V and R, the output voltage is proportional to R_x, which can be read on the meter. The meter has a linear scale as against a non-linear one in case of passive meter discussed earlier.

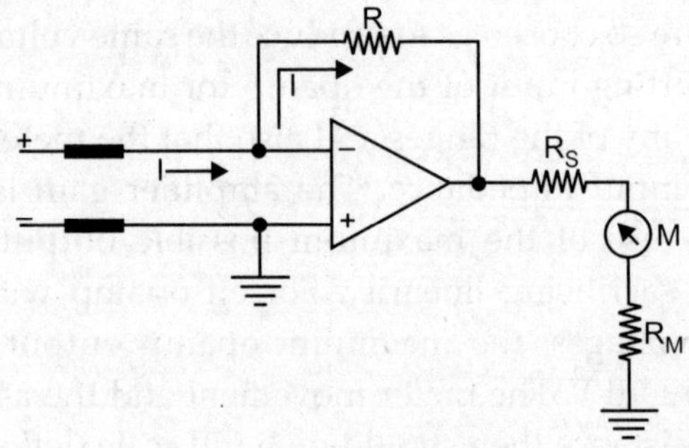

Fig. 29.32

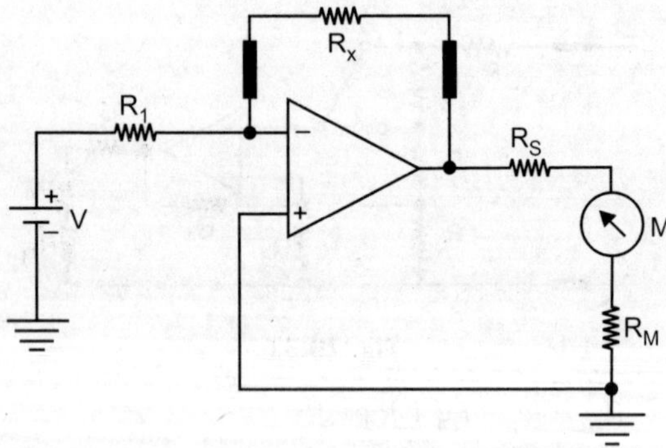

Fig. 29.33

Electronic AC Voltmeter: The AC voltemeter uses an active rectifier circuit as shown in Fig. 29.34. The forward voltage drop of this rectifier circuit comprising of opamp and diode D is given by ($0.7/A_{OL}$), where 0.7 volt has been taken to be the forward voltage drop of diode D, R and C constitute a high pass circuit and block DC. The sensitivity of this meter can be enhanced by building some gain around the opamp.

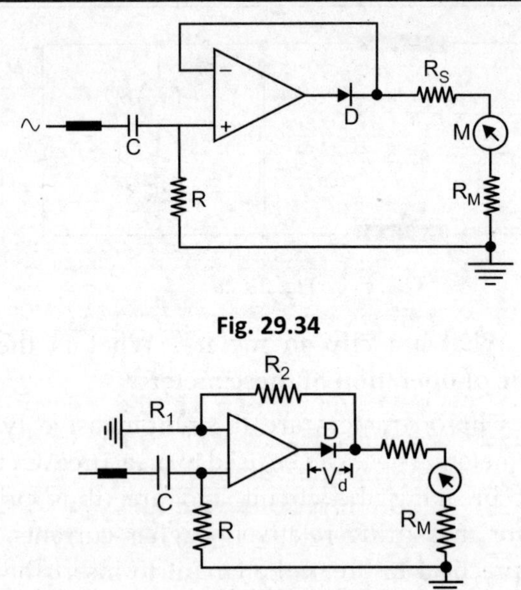

Fig. 29.34

Fig. 29.35

Fig. 29.35 shows the circuit. This is basically an amplifier-rectifier configuration. The gain equals $(1 + R_2/R_1)$.

Q. 81. Briefly describe operational principle of a Digital Multimeter (DMM). What are the commonly available types of DMMs? Briefly mention the test functions other than the normal multimeter functions that are available in present day instruments.

Ans. In a Digital Multimeter, the analog quantity to be measured (current, voltage, resistance) is first transformed into an equivalent voltage, if the parameter to be measured is current or resistance. The transformed analog voltage is then digitised using an analog to digital converter (ADC). To be more precise, the analog voltage is converted into a pulse train whose frequency depends upon voltage magnitude. The pulses are counted over a known gating period in a counter. The counter outputs are decoded and displayed. The displayed totalise count represents the magnitude of the parameter under measurement.

In another approach, which is also in common use, the input analog voltage is compared with a ramp from a ramp generator. The comparator generates a gating pulse whose width equals the time interval between the ramp amplitude rising from zero to the analog voltage under measurement. The counter in the ADC counts clock pulses of a known frequency over this gating interval and the counter count is decoded and displayed. So, while in the former method, there is a V/F conversion and the equivalent frequency representing the analog voltage is counted over a fixed gating interval, in the latter, a fixed frequency is counted over a variable gating interval with the gating interval being proportional to the analog voltage.

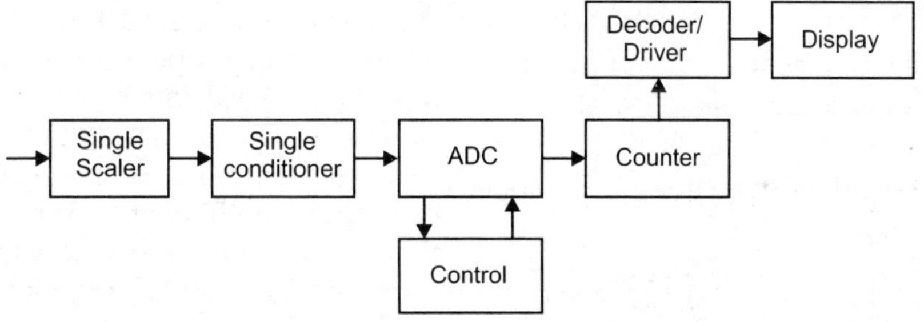

Fig. 29.36

The digital multimeters have the advantages of offering unambiguous display with no allowance for any human error, improved accuracy (± 0.1% against ± 3% in analog meters) and improved resolution (± 0.1% against 1% in analog meters). The other advantages include easy incorporation of features, such as Autoranging, Automatic polarity, Diode and Transistor test and so on. The cost advantage that used to exist in favour of analog meters has narrowed down to a small amount with advances in IC technology and DMMs are fast replacing Analog meters even for routine measurements. However, Analog meters are relatively immune to noise and are preferred in electrically noisy environment.

Fig. 29.36 shows the schematic arrangement of a digital meter. The signal scaler at the input is basically an attenuator/amplifier block and is partly used for range selection function. In autoranging meters, the input signal level is sensed on application and the signal scaler gain selected accordingly. Signal conditioner generates a DC voltage proportional to the input signal. The ADC used in DMM applications is usually the integrating type ADC (single slope or dual slope) with the latter being the preferred one because of its higher accuracy, insensitivity to changes in integrator parameters and low cost. All the building blocks depicted in Fig. 29.36 except for the display are available on a single chip ICL 7106/7107 is an example.

DMMs invariably have a display that has an additional 1/2-digit. We have $3\frac{1}{2}$, $4\frac{1}{2}$, $5\frac{1}{2}$ digit DMMs rather than 3, 4, 5 digit DMMs. While the usually so called full digits can display all digits from 0 to 9, a 1/2 digit can display either a '0' or '1'. Addition of a 1/2-digit in MSB position of the display preserves the resolution of the multimeter upto a higher range. For instance, a 3-digit multimeter has a resolution of 0.1 V upto 99.9 volts. A 31/2- digit meter with practically no additional hardware would give you a resolution of 0.1 V upto 199.9 volts. This increase in resolution range comes with the addition of one additional seven segment display and no change in hardware complexity. The display resolution is also sometimes expressed in terms of counts.

Fig. 29.37

The $3\frac{1}{2}$ digit DMM has 2000 count resolution. DMMs with 4000 count resolution referred to $3\frac{3}{4}$ digit meters are also commercially available. These meters will also have four numbers seven segment displays but have a little additional hardware.

Digital multimeters are made in a large variety of sizes, shapes and performance specifications ranging from pen type $3\frac{1}{2}$ digit DMM to $7\frac{1}{2}$ digit high resolution bench top versions Hand held versions are available typically upto $4\frac{1}{2}$ digit resolution. Majority of them have in-but Diode test, Transistor test and Continuity check. Some of them even offer *L-C* measurement frequency measurement without any significant change in price. Fig. 29.37 shows the photograph of one such multimeter. It is a high end multimeter with a 40,000 counts display, and in-built continuity check, Diode test, Frequency measurement, Capacitance measurement etc. in addition to conventional functions.

Q. 82. Distinguish between an Analog and a Digital oscilloscope from the operational viewpoint.

Ans. Analog oscilloscopes display the signal directly and enable us to see the waveform shape in real time. The signal update rate in them is the fastest possible, as there is only the beam retrace timing and the trigger-rearm between the two successive sweeps. Consequently, an analog oscilloscope has a much higher probability of capturing the desired event than any other class of oscilloscope. Analog oscilloscopes find wide application for viewing both repetitive as well as single shot events upto a bandwidth of about 500 MHz. Fig. 29.38 shows the photograph of a 400 MHz bandwidth analog oscilloscope (Model LA314 from LeCroy).

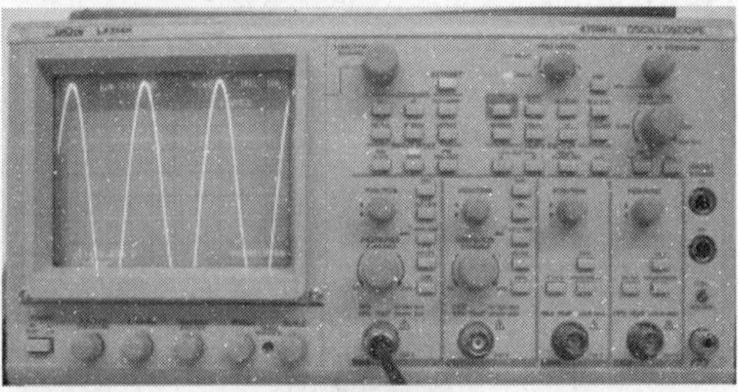

Fig. 29.38

Analog oscilloscopes do not give a desirable display, when viewing very low frequency repetitive signals or single shot events. In such cases, the display is nothing but a bright dot moving slowly across the screen to trace the waveform. Such a waveform is not at all convenient to analyse and needs some kind of photographic memory.

A CRT storage type analog oscilloscope overcomes this problem by using a special type of CRT. In one such type, the phosphor dots have higher persistence. As a result, the moving dot leaves behind a visible trail, as it sweeps across the screen even at much lower sweep speeds. There are two main types of storage modes currently in use for these oscilloscopes-the bistable storage mode capable of storing signals for many hours and the more popular variable persistence storage mode that can store signals for a maximum of ten minutes. Majority of commercially available CRT storage oscilloscopes have the option of both the above mentioned storage modes.

CRT storage type oscilloscope is an excellent choice for slowly changing signals. As the writing rate is faster than the conventional analog oscilloscopes, it is extremely good for viewing fast transient events. These can be used to store both repetitive and single shot signals having a bandwidth of 500 MHz or so. Oscilloscope type 7934 from Tektronix for instance has bandwidth of 500 MHz and a maximum writing speed of 4000 cm/μs. Even the handheld versions of these scopes with a reasonably good writing speed are available. Analog storage oscilloscope technology is fast being replaced by digital storage oscilloscope technology due to far more superior performance features of the latter.

Digital Oscilloscopes. In a Digital oscilloscope, the signal to be viewed is first digitised inside the scope using a fast A/D converter. The digitised signal is stored in a high speed semiconductor memory to be subsequently retrieved from the memory and displayed on the oscilloscope screen. There are two digitising techniques—real time sampling and equivalent time sampling. The digital storage oscilloscopes (DSO) use real time sampling as shown in Fig. 29.39, so that they can capture both repetitive and single shot signals. In real time digital storage oscilloscopes, the digitiser samples the entire input waveform with a single trigger. Sampling oscilloscopes use equivalent time sampling and are limited to capturing repetitive signals. Some digital storage, oscilloscopes also use equivalent time sampling to extend their useful frequency range for capturing repetitive signals. The equivalent time sampling technique is, thus applicable to only stable repetitive signals and can be implemented in at least three different ways, namely, as sequential single sample, sequential sweep and random interleaved sampling (RIS).

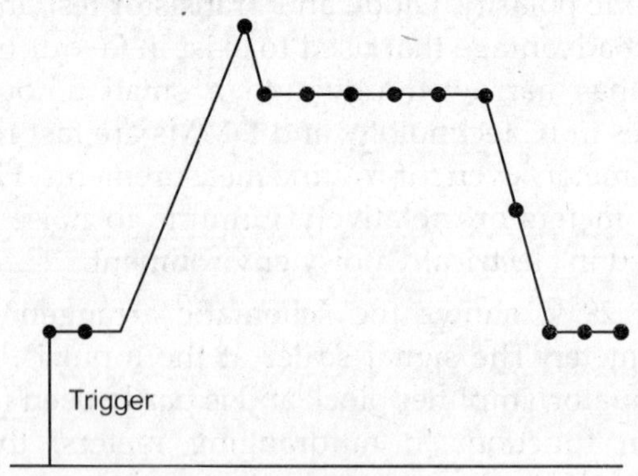

Fig. 29.39

In **sequential single sample technique** (Fig. 29.40), the digitiser acquires a single sample with the first trigger pulse. It then waits for the second trigger and on receipt of second trigger, a time delay equal to the reciprocal of the desired sampling rate is executed and then the second sample is acquired. The trigger-to-acquisition delay is incremented by the desired intersample period (Δt) for each subsequent acquisition. The resulting capture has, thus, an equivalent sample rate of $1/\Delta t$. Clearly, this method is slow as (N) trigger cycles would be needed to gather (N) samples and the scopes using this type of digitising technique cannot provide real time operation.

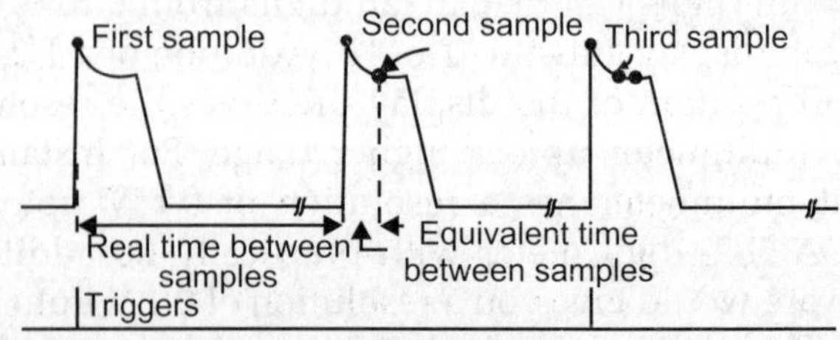

Fig. 29.40

In **sequential sweep equivalent time sampling** (Fig. 29.41), a sweep of samples spanning the desired display time range is acquired for each trigger. Here, (N) samples are acquired in (M) trigger cycles, where N = kM. Upon receipt of each trigger, (k) sequential samples are acquired at sample rate, f_S. These are stored in every Mth location of the acquisition memory allocated for (N) samples. The first sweep (k) samples are acquired directly on receipt of trigger with delay increments being equal to ($1/Mf_S$) with reference to trigger detection in order to give an apparent sample rate of Mf_S.

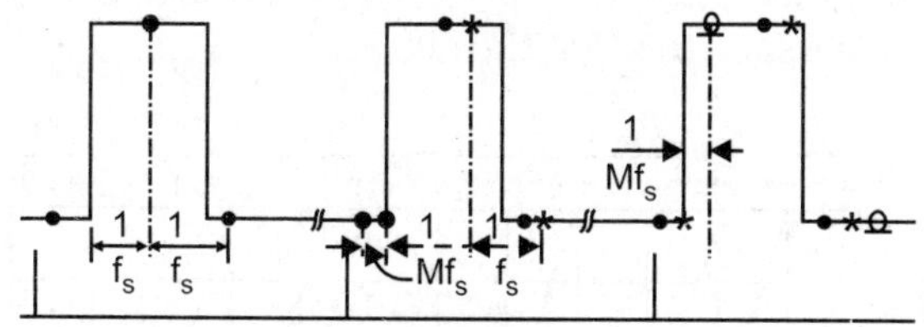

Fig. 29.41

Random interleaved sampling (RIS) technique uses a memory distribution scheme that is philosophically similar to that of sequential sweep equivalent time sampling with the difference that the samples are random with respect to the trigger. Sampling in this case occurs on both sides of trigger points which gives this technique the 'pretrigger view' capability not available in the first two equivalent time sampling techniques, as both methods gather signals only following the receipt of trigger.

If we want to view a 1 GHz signal, the sweep speed requirement would be enormous. Even if we are successful in achieving this high speed, the beam would almost be invisible. We have often noticed that as the time base setting is made faster, we are forced to adjust the intensity control to maintain an acceptable intensity level setting. Another major problem in designing real time oscilloscope for viewing very high frequency signals (in GHz range) is the difficulty to build such a high bandwidth in the vertical amplifier. Sampling oscilloscope using any of the equivalent time sampling techniques outlined above is an answer to all these problems. In such scopes, it is not imperative to take a sample or a group of samples from each cycle of the signal to be viewed. The next adjacent sample or group of samples may be 10,000 cycles away. As a result, the bandwidth of vertical amplifier can be afforded to be much lower than the frequency of the signal.

Another type of sampling oscilloscope, though not very common in use, is the analog sampling oscilloscope, where a conventional sample/hold circuit consisting of an electronic switch and a capacitor is used for signal acquisition (Fig. 29.42). It can be used to view high frequency repetitive signals in the non-storage mode unlike the digital sampling scopes, where the signal is sampled digitally and then stored in semiconductor memory for subsequent retrieval, and used for viewing high frequency repetitive signals in the storage mode, though not in real time (Fig. 29.43).

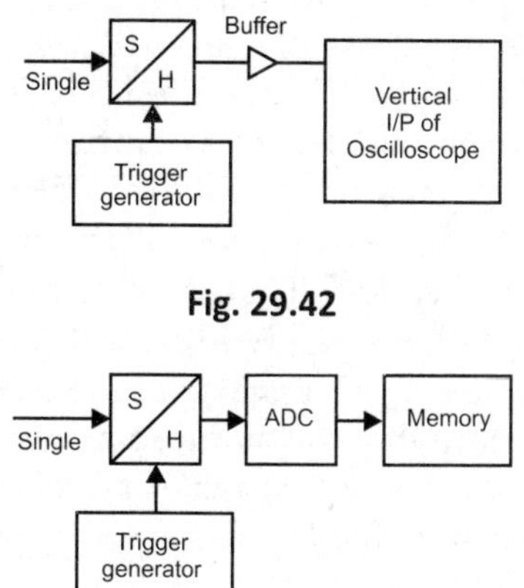

Fig. 29.42

Fig. 29.43

Almost all oscilloscopes available today use one or a combination of the technologies discussed above. Each technology has its own benefits and shortcomings. While signal manipulation and its consequent benefits is the strong point of the digital technology, extremely fast update rates coupled with low cost is a feature associated with analog scopes. In fact, many state-of-the-art oscilloscopes are not simple analog or digital. The offer advantages of both the technologies.

Q. 83. Define key specifications of Bandwidth (or Rise time), Vertical sensitivity and Accuracy with reference to an Analog oscilloscope.

Ans. Though oscilloscopes are characterised by scores of performance specifications, not all of them are important. Bandwidth (or rise time), vertical sensitivity and accuracy are the key analog oscilloscope specifications. Other features like triggering capabilities, display modes, sweep speeds etc. are secondary in nature.

Bandwidth and rise time: Bandwidth and rise time specifications of an oscilloscope are related to one another. Each can be calculated from the other.

$$\text{Bandwidth (in MHz)} = \frac{350}{t_r\,(in\ ns)}$$

Bandwidth is the most important specification of any oscilloscope. It gives us a fairly good indication of the signal frequency range that can be viewed on the oscilloscope with an acceptable accuracy. If we try to view a signal with a bandwidth equal to the oscilloscope's bandwidth, the measurement error may come to as large as 40 per cent (Fig. 29.44). As a rule, the oscilloscope bandwidth should be three to five times the highest frequency one is likely to encounter in order to keep the measurement error to less than 5 per cent.

Vertical sensitivity : Vertical sensitivity specification tells us about the minimum signal amplitude that can fill the oscilloscope screen in the vertical direction. 5 mV/div sensitivity is quite common. Oscilloscopes with sensitivity specifications of 1 mV/div are also available. Sensitivity and bandwidth are often trade-offs. Though a higher bandwidth enables us to capture high frequency signals, there is a good possibility of unwanted high frequency noise being captured, if the oscilloscope has a higher sensitivity too. That is why most of the high sensitivity scopes have bandwidth limit controls to enable a clear view of low level signals of moderate frequencies. It is also important that the oscilloscope we choose has an adequate volts/div range to make possible a full screen or near full screen display for a wide range of signal amplitudes.

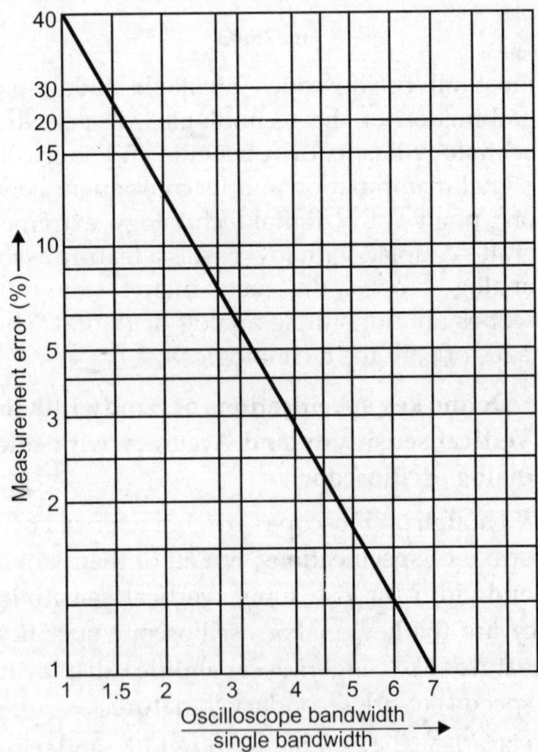

Fig. 29.44

Accuracy : Accuracy specification indicates the degree to which our measurement conforms to a true and accepted standard value. Accuracy of ± 1 per cent to ± 3 per cent is typical. Almost all oscilloscopes are provided with a magnification of 5X in the volts/div selector switch. This alters the nominal vertical deflection scale from say 5m V/div-5V/div to 1 m V/div-lV/div. It may be mentioned here that the accuracy suffers with the magnifier pull. Most of the manufacturers list both the accuracy specifications for the oscilloscopes manufactured by them.

Q. 84. Define WRITING SPEED with reference to an Analog storage oscilloscope.

Ans. With the CRT storage-type oscilloscope, the stored writing speed is usually the main criterion for choosing the instrument. The speed of a CRT storage scope depends on the speed of the input signal (signal frequency) and size of the trace it draws. The chart of Fig. 29.45 shows trace size in centimeters on the vertical axis and the speed of the input signal along the horizontal axis. The signal speed has been expressed both in terms of rise time and frequency. For sine waves, use frequency and for pulses, use rise time to determine the writing speed necessary to store the signal.

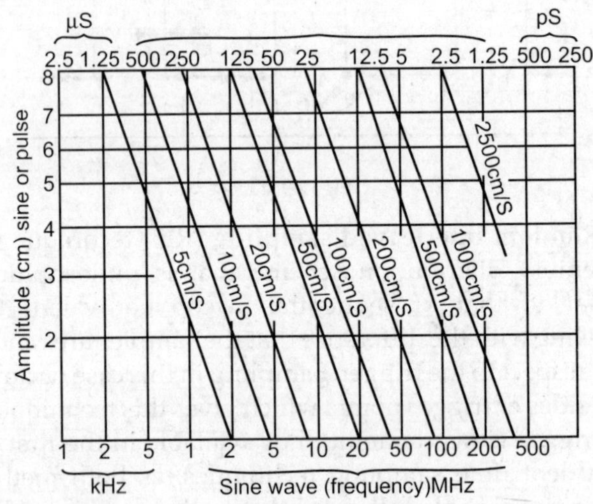

Fig. 29.45

Q. 85. Briefly describe the following specifications of a Digital Storage Oscilloscope.

(a) Bandwidth

(b) Sampling Rate

(c) Vertical Resolution and Accuracy

(d) Acquisition Memory.

Ans. *Bandwidth and sampling rate* : Bandwidth is an important specification of digital oscilloscopes just as it is for analog oscilloscopes. Bandwidth which is primarily determined by the frequency response of input amplifiers and filters, must exceed the bandwidth of the signal, if the sharp edges and peaks are to be accurately recorded.

Sampling rate is another vital digital scope specification. In fact, the sampling rate determines the true usable bandwidth of the scope. While the bandwidth is associated with the scope's analog front end (amplifiers, filters etc.) and is specified in Hz, sampling rate is associated with the digitising process and, if it is not adequate, it degrades the bandwidth. A clear understanding of sample rate specification is, thus important, when it comes to establishing adequacy of a particular sample rate to achieve a given bandwidth.

Digital oscilloscope specification sheets often contain two sample rates-one for single shot events and the other for repetitive signals. In some cases, both repetitive and single shot events are sampled at the sample rate through the oscilloscope's bandwidth capability for the two cases is different. It is lower in case of single shot events.

Theoretically, Nyquist criteria holds true and this criteria states that at least two samples must be taken for each cycle of the highest input frequency. In other words, the highest input frequency (also called Nyquist frequency) cannot exceed half the sample rate. Given this scenario, a (sin x/x) interpolation algorithm can exactly reproduce digitised signal. Interpolation algorithm is the mathematical function used by a scope to join two successive sample points while reconstructing the signal. The (sin x/x) interpolation has a tendency to amplify noise in the signal, particularly when each cycle is sampled only twice. With (sin x/x) interpolation, four samples per cycle are found to be quite adequate. The additional sample points effectively enhance the signal to noise ratio for (sin x/x) interpolation. With straight line interpolation, at last ten samples are required per cycle for good results.

For repetitive signals, however, even a smaller sample rate does the job, as explained in case of sampling oscilloscopes. Thus, it becomes important to look into sample rate specification together with the interpolation algorithm used. For instance, in a digital storage oscilloscope with a single shot sample rate of 400 MS/s using (sin x/x) interpolation technique, can give us a single shot bandwidth of 100 MHz, while the same sample rate will provide a bandwidth of only 40 MHz, if straight line interpolation algorithm is used instead.

Thus, the single shot bandwidth capability of a digital storage oscilloscope must always be gauged by its single shot sample rate. The sample rate in samples per second should be at least twice the highest frequency component or four times the highest frequency component for good results, or anywhere between 2 and 4, assuming (sin x/x) interpolation. For repetitive signals, if it is not a real time DSO, the sampling rate could be smaller.

Memory length : Memory length is a vital digital oscilloscope specification and should not be considered to be an insignificant one. Not only does it affect the sample rate and consequently the single shot bandwidth, longer memories also have many more peripheral benefits. The sample rate as quoted by the manufacturer always refers to the maximum digitising rate attainable in the single shot mode. Interestingly, the quoted sample rate figure does not hold true for the entire range of time base settings. For a given memory length, the attainable sample rate is observed to reduce, as the time base is made slower. Some manufacturers offer record length, which is nothing but the size of the memory used, while displaying the signal.

Suppose a particular DSO has a memory length of 1 K and a quoted sample rate specification of 100 MS /s. In the limit when the record length equals the memory length, we can store approximately 1000 samples. At the given sample rate, the displayed waveform will cover a time span of 10 μs, *i.e.* a time base setting of 1 Ms/div, if the waveform is to cover the full screen in the horizontal direction. If the time base setting is changed to 10 μs/div, the effective sample rate would be limited to only 10 MS/s, thus reducing the single shot bandwidth. The only method to maintain the sample rate at the quoted value for a larger time base setting range is to have a longer acquisition memory.

The effect of memory length on the single shot bandwidth as a function of time base setting is expressed by,

Sample rate = Memory length/(10 × time base setting)

where, 10 represents the total number of divisions in the horizontal direction.

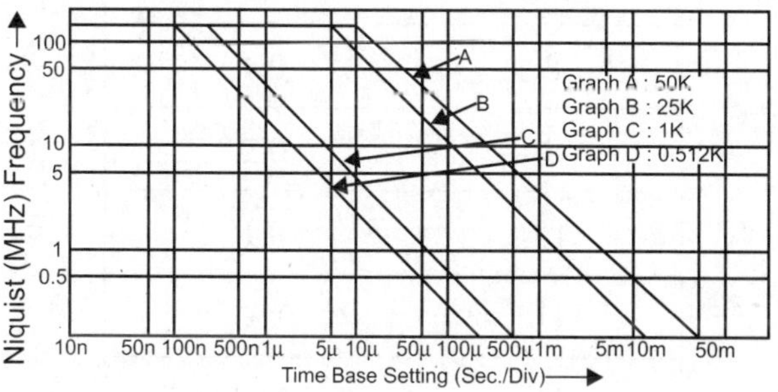

Fig. 29.46

Fig. 29.46 shows the changes in sample rate as a function of time base setting for digital oscilloscopes of different memory lengths. Given two oscilloscopes with identical sample rate and single shot bandwidth specifications, the one with the longer acquisition memory has a decisive edge. Hence, it must be accorded its due importance while choosing one to meet your requirements. For a given time resolution, longer memory enables events of longer duration to be recorded. For instance, a DSO with 1 K memory can record a 1 sec transient with a time resolution of 1 ms, whereas a DSO with 10 K memory can record a 10 sec long event with the same time resolution. In other words, for same transient duration, longer memories give enhanced time resolution. Long memories also help in acquiring hard-to-catch signals and also minimise signal reconstruction distortion.

Vertical accuracy and resolution: The accuracy specifications tells us how closely the measurement matches the actual value. Accuracy of a DSO is affected by various sources of error including gain and offset errors, differential non-linearity, quantisation error and so on. Quantisation error indirectly indicates vertical resolution, *i.e.* uncertainty associated with any reading or the scope's ability to see small changes in amplitude measurements. Choosing a scope with less than 8 bits of resolution is not recommended. Resolution specification must not be considered in isolation from accuracy specification. For instance, larger than 8 bits of resolution is meaningless, when the overall accuracy itself is ± 1 per cent. 8-bit resolution gives a ± 0.4 per cent uncertainty which is fairly acceptable if the overall accuracy is ± 1 per cent (Table 29.1). Also, digital scopes with more than 7 bits of resolution can resolve signal details better than visual measurements made with analog scopes.

Table 29.1

No. of Bits	*Uncertainty (%)*
6	1.6
7	0.8
8	0.4
9	0.2
10	0.1
11	0.05
12	0.02

Q. 86. What is purpose of an oscilloscope probe? What are 10 X and 100 X probes?

Ans. Oscilloscope probe acts as a sort of interface between the circuit under test and the oscilloscope input. The signal that we intend to view on the oscilloscope screen is to be fed to the vertical input (designated as the Y-input) of the oscilloscope. An appropriate probe ensures that the circuit under test is not loaded by the input impedance of the oscilloscope's vertical amplifier. This input impedance is usually 1-megohm in parallel with a capacitance of 10 to 50 pF.

The most commonly used general-purpose probes are *X*1, *X*10 and *X*100 probes. These probes respectively provide attenuation by factors of 1 (*i.e.* no attenuation), 10 and 100. That is, if we are measuring a 10 volts signal with a *X*10 probe, the signal being actually fed to the oscilloscope input will be 1 volt. *X*10 and *X*100 probes are quite useful for measuring high amplitude signals. Another significant advantage of using these probes is that the capacitive loading on the circuit under test is drastically reduced.

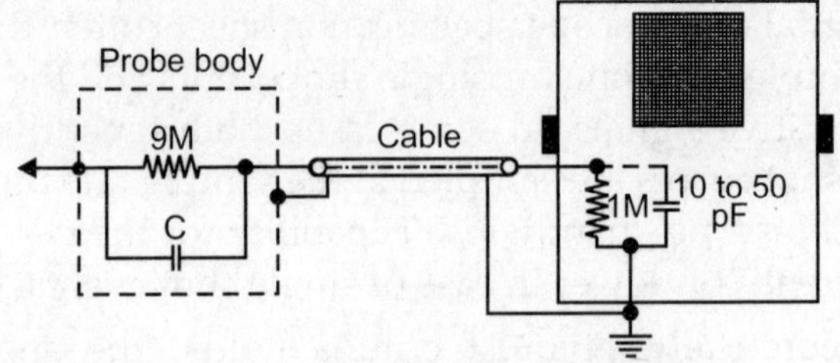

Fig. 29.47

Refer to the internal circuit of a *x*10 probe as shown in Fig. 29.47. The probe's RC time constant equals the scope's input RC time constant. Since the probe resistance is nine times the scope's input resistive component, in order to provide attenuation by factor of ten, the probe capacitance has got to be smaller than the scope's input capacitance by the same amount. As a result, the circuit under test with a probe of *X*10 will never see a capacitance of more than 5 pF.

Q. 87. What do you understand by PROBE COMPENSATION? How do you check whether a given probe is compensated or not?

Ans. The probe is compensated, when its RC time constant equals the RC time constant of the scope's input. With this, what we see on the scope's screen is what we are trying to measure independent of the frequency of the input signal. If the probe is not properly compensated, the signal will be attenuated more than the attenuation factor of the probe at higher frequencies due to reduction in the effective input impedance of the scope's vertical input.

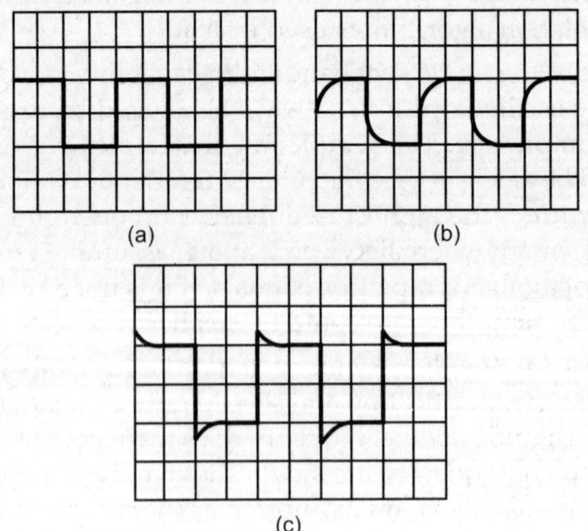

Fig. 29.48

To check for probe compensation, the probe can be used to see the calibration signal (CAL position on front panel) available on the scope. If the probe is properly compensated, the CAL signal will appear in the perfect rectangular shape (Fig. 29.48 *a*) with no rounding-off of edges (Fig. 29.48 *b*) or any spikes on fast transitions (Fig. 29.48 *c*) Rounding-off of edges indicates too little a probe capacitance while spikes indicate too large a probe capacitance. The probe capacitance can be adjusted by turning a screw or rotating the probe barrel after loosening the locking nut (in some probes) to get a perfect calibration signal.

Q. 88. What is a Frequency Counter? Which signal parameters can be measured with a modern frequency counter?

Ans. The most basic function of a frequency counter is the measurement of an unknown frequency. Modern frequency counters, however, offer much more than just the frequency measurement. The other related parameters such as *Time Period*, which is reciprocal of frequency, *Time Interval* between two events and *Totalise Count* which is nothing but cumulative count over a known period are the other functions that are available with the present day frequency counters. These instrument offering a variety of measurement options particularly two channel functions, such as time interval measurement are usually referred to as *Universal Counters*. One such counter is shown in Fig. 29.49. This is a programmable Counter/Timer (Model 776 from M/s Keithley).

Q. 89. Describe different functional modes of a Universal Counter.

Ans. The functions available with modern universal counters other than measurement of an unknown frequency are (*i*) Time interval measurement (*ii*) Period (*iii*) Time interval average (*iv*) Totalise (*v*) Frequency ratio A/B (*vi*) Phase A rel. to B (*vii*) Pulse width etc.

Time Interval Measurement

This mode measures the time that elapses between the occurrence of two events.

Fig. 29.49

One of the events called the start signal is usually fed into one of the channels, while the other called the stop signal feeds the second channel. Resolution of measurement is typically 10 ns or better. A typical application of this measurement mode is in determination of propagation delay in logic circuits. Variations of this mode can be used to measure pulse width, rise and fall times.

Time Interval Average

This mode can be used to improve measurement resolution in the time interval measurement mode for a given clock frequency. The resolution improves as square root of the number of measurements. That is, average of 100 measurements would give a 10-fold improvement in resolution.

Period

In this mode, the time period of the input signal is measured by counting clock pulses between two successive leading or trailing edges of the input signal. Again period average function can be used to improve upon measurement resolution for a given clock. For instance, if the measurement is done for 100 periods instead of one period for a given clock frequency, the measurement resolution would also improve by a factor of 100.

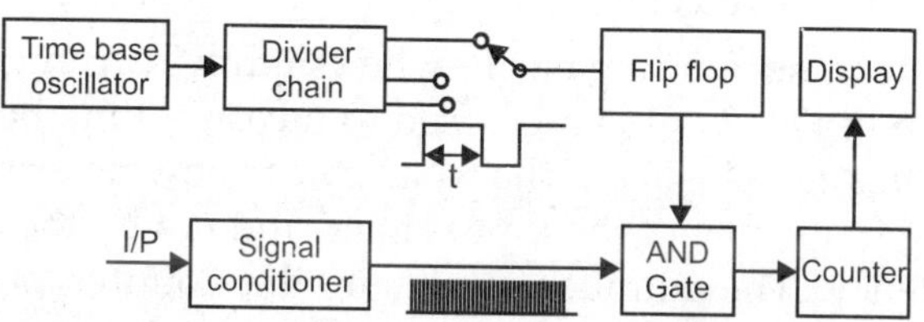

Fig. 29.50

Totalise : Totalise mode gives a cumulative count of events over a known time period.

Frequency Ratio A/B : This gives the ratio of the frequencies of signals fed to A and B channels. This feature can be used to test the performance of prescalers and frequency multipliers.

Phase A rel to B

It compares the phase delay between the signals with similar frequencies.

The counter/timer instrument shown in Fig. 29.50 has a DC coupled frequency range of 0 to 225 MHz, which can be extended to 2.4 GHz with an optional accessory. It has selectable impedance (50 Ω or 1 MΩ), sensitivity of 25 mV, time interval measurement of ± (1LSD + 10ps), single shot pulse width measurement range of 5 ns to 2000 s and so on. It offers measurement functions, such as Frequency, Pulse width, Time interval, Time interval averaged, Phase A to B (A and B are two inputs), Totalise, Ratio A to B etc.

Q. 90. With the help of block schematic arrangement, briefly describe the functional principle of measurement of frequency and time period.

Ans. Fig. 29.50 shows the architecture of a frequency counter, when it is being used in the frequency measurement mode. The oscillator section comprising of a crystal based oscillator and a frequency divider chain generates the clock pulses. The clock pulses are used to trigger a flip flop whose output serves to enable or disable the AND gate. When the AND gate is enabled, the input signal after passing through the signal conditioning section comprising of level shifting amplifiers, comparators etc. reaches the counter. In the simplest case, if the AND gate is enabled for 1 s (which is the case when the flip flop clock input is 1 Hz), then the counter count will represent the signal frequency. The measurement resolution in this case would be 1 Hz. The measurement resolution can be improved by enabling the AND gate for a longer time. For instance, a 0.1 Hz clock at the flip flop input would give a 10 s gate time and a consequent 0.1 Hz resolution. Similarly, a shorter measurement for a gate time of 0.1 s (corresponding to clock of 10 Hz) gives a measurement resolution of 10 Hz.

The same building blocks when slightly rearranged as shown in Fig. 29.51 can be used to measure time period. Opening and closing of the AND gate is now determined by the frequency of the input signal and not by the clock frequency. The number stored in the counter here is proportional to the number of clock pulses that reach the counter during the period of the input signal. The same set-up can be used for time interval (*TI*) measurement by having two input signal channels with one, enabling the AND gate by say setting the flip flop and the other disabling the same by resetting the flip flop.

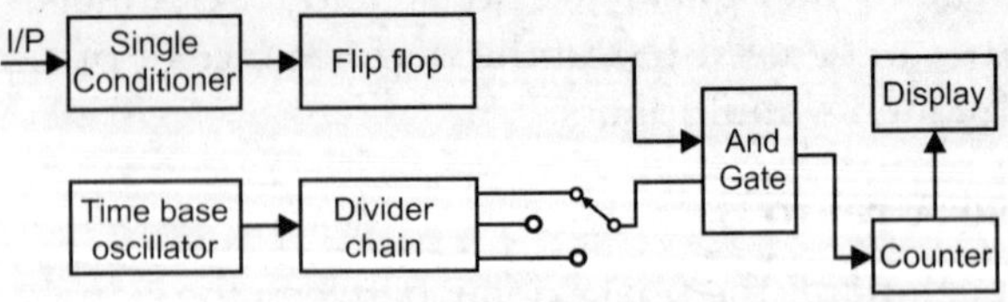

Fig. 29.51

Q. 91. With the help of a block schematic arrangement, describe the functional principle of a Reciprocal Counter. How does a reciprocal counter overcome resolution limitation of a conventional counter ?

Ans. The reciprocal counter overcomes some of the limitations of the basic counter described above, the most important of all being inadequate resolution, particularly when measuring low frequencies. The basic counter had a resolution of 1 Hz for a gate time of Is and resolution could be enhanced only by increasing the gate time. If the gate time could not be increased in a certain application, the resolution gets restricted to 1 Hz. The basic counter measures frequency in terms of event count. Depending upon the gate time which is 1s or any other decade value, such as 10, 100 etc. the decimal point appropriately placed in the count gives frequency. What is important to note here is that computation of frequency involves computation of event count only. The frequency which is given by event count divided by time taken is calculable from event count itself, if the time is 1s, 100s etc.

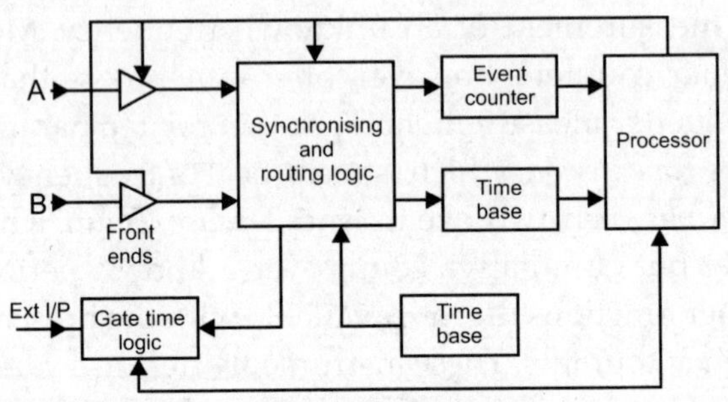

Fig. 29.52

In the reciprocal counter, both events as well as time are computed and the ratio of the two gives the frequency. Advent of reciprocal counter was made possible due to availability of digital logic that could perform arithmetic division economically and with precision. Fig. 29.52 shows the reciprocal counter hardware. The processor is the heart of the counter hardware and controls almost every other building block. The synchronising logic block routes the *A* and *B* channel inputs and the time base signal to the Event and Time counters. The routing is determined by the measurement function. The computations are done in the processor block.

As a matter of comparison, let us see as to how the two counters having an internal clock of 10 MHz would respond to measurement of a signal frequency 50.38752 Hz. The basic counter will display 50 Hz assuming a gate time of 1s as the event count will be 50. The reciprocal counter will also have an even count of 50 but it will also measure time with a resolution of 100 ns (for 10 MHz clock) equal to 0.9923328 s. The measured frequency will therefore be 50.38752 Hz. The frequency resolution offered for 10 MHz clock is 7 digits equal to 0.000005 Hz in the present case for a Is gate. The resolution could be further enhanced by increasing the clock frequency. Since clock frequencies upto 500 MHz are practical, a reciprocal counter would give a resolution of 2 ns for a 1s gate time.

The resolution of reciprocal counters can be further enhanced by using a technique called interpolation. It is possible to achieve a 9-digit resolution with a 10 MHz clock using interpolation techniques which otherwise would require a clock of 1 GHz. Higher resolution is important, when we are locking for a given resolution in shorter gate times. The details of interpolation techniques are beyond the scope of this text.

Q. 92. Briefly describe Continuous Count Counters.

Ans. The counter architectures discussed in the preceding paragraphs had a counter that counted for a known period equal to the gate time. These counters have a dead time, when the gate is disabled. Such counters could miss vital information that could be important to the measurement. The continuous count counter architecture is based on the fact that if different measurements of a certain parameter of a signal are not disjoint and the relationship that they have is made use of, the measurement resolution can be significantly enhanced by applying what we call as curve fitting algorithms. These counters have all the attributes of reciprocal counters with the additional ability of reading the event, the time and the counter without having to disable the gates.

Q. 93. List important specifications of frequency counters. Briefly describe each one of them.

Ans. The important specifications of a counter include Sensitivity, Bandwidth, Resolution, Accuracy and Throughput.

Sensitivity : It refers to the smallest signal that the instrument can measure and is usually expressed as mV (rms) or p-t-p. Sensitivity of 10 mV(rms) to 20 mV(rms) is typical. In majority of measurement situations, sensitivity is not the issue.

Bandwidth: The bandwidth of the counter is its front end bandwidth and is not necessarily the same as the maximum frequency that the counter is capable of measuring. Measuring a signal frequency higher than the instrument's bandwidth only reduces its sensitivity specification and requires a larger minimum input signal. For instance, the counter of Fig. 29.49 has a sensitivity of 25 mVrms (sine) upto a signal frequency of 100 MHz. It degrades to 50 mVrms (sine) for signal frequency in the range of 100 MHz to 225 MHz. However, the bandwidth does affect the measurement accuracy in case of some parameters, rise time being one such parameter. Thus, it is always preferable to choose a counter with as high a bandwidth as possible. Bandwidth is not explicitly mentioned in the specifications. However, it can be guessed by looking at variation in sensitivity across the frequency range of the instrument.

Resolution : Resolution refers to the minimum frequency increment in case of frequency measurement and time increment in case of time interval measurement that can be resolved. The resolution is usually very close to the least significant digit and is often ± 1 count of LSD. Noise in the input signal, noise in the front end and input signal slew rate are some of the factors that affect resolution.

Accuracy: Accuracy is related to resolution but not the same as resolution. Factors like Time base (or clock) accuracy and Trigger accuracy must be considered alongwith resolution specification to determine ultimate accuracy of frequency measurement. Time base error affects measurement accuracy according to:

Frequency accuracy = Resolution ± Time base error × Frequency

Trigger level accuracy is the precision with which trigger level can be set. If there is an error in the trigger level setting, the trigger timing is changed, thus affecting measurement accuracy.

Throughput: Throughput is related to resolution. For instance, increasing the gate time of a certain frequency measurement increases the measurement resolution by the same factor but it slows down the throughput by almost the same amount. Other factors affecting the throughput are more related to the speed of the microprocessor and the interface system. Two factors to watch here are the number of measurements the counter can deliver through the interface and the speed with which the counter can switch between different functions or set-ups. If short gate times are being used and/or measurements are being switched between different functions repeatedly, these factors become important.

Q. 94. How does the architecture of a Microwave frequency counter differ from that of a conventional counter ?

Ans. The counter architectures discussed in the preceding paragraphs (conventional, reciprocal, continuous count) are usually good enough upto 500 MHz or so. Counters meant for carrying out measurements at frequencies at RF beyond 500 MHz and microwave frequencies employ a different architecture. There are two types of architectures in use for building microwave counters. One uses a prescaler, while the other is based on down conversion.

Prescaler counters use a prescaler placed between the front end and the gating circuitry of the counter. Infact, prescalers are available inside the counters as an optional channel to extend the frequency range of measurement. Extension upto 3 GHz is typically available with a prescaler. Prescalers are not used with pulsed microwave counters due to their tendency to self oscillate. When used with basic counter, prescaler causes degradation of resolution. It is because frequency resolution of basic counter is dependent upon contents of the event counter and due to location of the prescaler before the gating circuitry, its contents can not be read. The resolution is not affected when the same is used in a reciprocal counter.

In the down converter based microwave counter, the input signal frequency is down converted and an IF produced. IF, which is the difference between the input signal frequency and the local oscillator frequency is then counted. The actual frequency is then computed from (*LO* + *IF*). Covering a frequency range of tens of GHz for an LO is an expensive proposition. The solution is to use a relatively lower frequency LO (approx. 200 MHz). The LO drives a step recovery diode that produces a sharp pulse with usable harmonics upto the desired range. This pulse drives a sampler which samples points of the input signal. The resulting IF is low pass filtered and counted. The actual input frequency is then given by ($N \times LO + IF$), where (*N*) tells us which harmonic of *LO* goes through mixing operation. One of the methods to determine (*N*) is to measure IFs at two slightly different *LO* frequencies. (*N*) is then given by $(IF_1 - IF_2)/ (LO_1 - LO_2)$. But all this is instrument's headache and may take several tens of milliseconds only. Fig. 29.53 shows the architecture of a microwave counter.

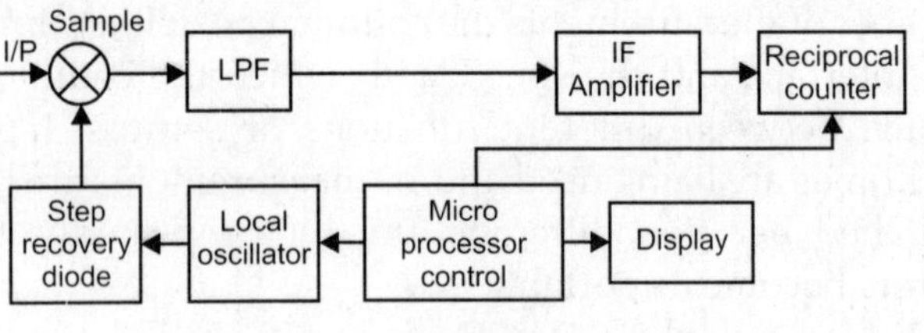

Fig. 29.53

Pulse microwave counters use the architecture described above with the additional gating circuitry to position the gate comfortably within the pulse.

Q. 95. What are frequency synthesisers and synthesised function Generators or Synthesised Signal Generators? What is common between these different categories of instruments?

Ans. Frequency synthesisers generate sinusoidal signals of extremely high frequency stability and exceptional output level accuracy. Frequency synthesisers and similar instruments, such as Synthesised function/ signal generators are used to provide test signals for characerisation of devices, subsystems and systems. Synthesised function generators in addition to providing spectrally pure and accurate CW sinusoidal signals also provide other waveforms, such as ramp, triangle, square and pulse. Synthesised signal generators in addition to providing spectrally pure and accurate CW signals, also have modulation capability and can be used to generate AM, FM, PM and pulse modulated signals. There is another class of synthesised function generators called synthesised arbitrary waveform generators. Majority of synthesised function generators have a limited arbitrary waveform generation capability built into them. However, these are available as individual instruments also. All the above mentioned instruments have one thing in common and that is synthesis of the signal which lends ultrahigh frequency stability and amplitude accuracy to the generated waveform. They, therefore, have more or less similar architecture for a given technique used for frequency synthesis.

Q. 96. Various circuit topologies have been used for frequency synthesis in frequency synthesiser instruments. These include :

(*a*) Direct Frequency synthesis,

(*b*) Indirect synthesis

(*c*) Direct Digital synthesis or Sample Sine Synthesis.

Describe each one of these topologies.

Ans. Direct Frequency Synthesis: Frequency synthesiser in its basic form uses reference oscillator, which is an ultrastable crystal oscillator, and other signal process circuits to multiply the oscillator frequency by a fraction M/N (where, *M* and *N* are integers), generate the desired output frequency. One such arrangement is shown in Fig. 29.54 comprises of an assortment of frequency multipliers and dividers, mixers and band pass filter (BPFs).

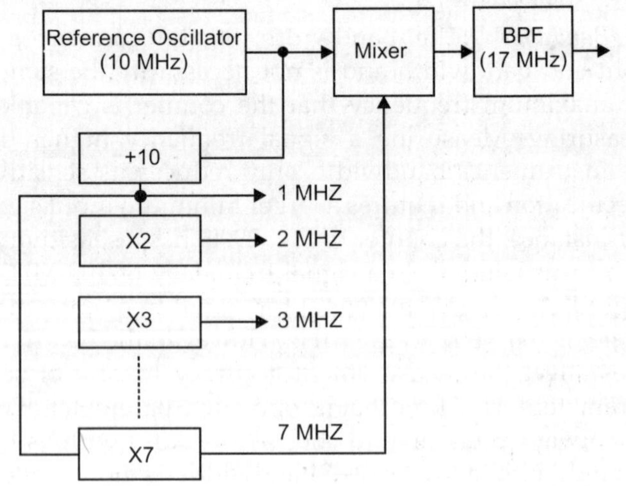

Fig. 29.54

The diagram shows the use of this architecture to generate 17 MHz. In this arrangement, if the BPF had a passband centered around 3 MHz, the output would have been 3 MHz as the mixer produces both sum and difference components. This method of frequency synthsesis has several disadvantages, more prominent being the technique being highly hardware intensive and therefore expensive, loss of phase continuity while switching frequencies with the result that this technique has not found favour with the designers of frequency synthesisers.

Indirect Synthesis: In the indirect synthesis, the output is not directly derived from quartz crystal based reference oscillator. Instead, the reference oscillator is used in a phase locked loop wired as a frequency multiplier to generate an output frequency that is M/N times the reference oscillator frequency. The output is taken from the VCO of the phase locked loop Fig. 29.55 shows the basic arrangement. If we insert a divide-by-N circuit between the reference oscillator and phase detector's signal input and a divide-by-M circuit between the VCO output and phase detector's VCO input, then the loop will lock with the VCO output as f_{ref} × (M/N). The frequency resolution of this architecture is × (f_{ref}/N). f_{ref} is the reference oscillator output frequency. The loop frequency switching speed is of the order of ten times the period of reference frequency input to loop phase detector. That is, if we desired a frequency resolution of 1 Hz, the switching time is going to be of the order of 10s, which is highly unacceptable. Another disadvantage of this architecture is that frequency multiplier loops also multiply noise at the phase detector, which manifests itself in the form of noise sidebands at VCO output. This restricts the maximum multiplication factor to few thousands in this arrangement with the result that resolution gets limited. In case finer resolution is needed, sequences of multiplication, division and addition are used involving more than one phase locked loops. One such arrangement is shown in Fig. 29.56. The synthesiser output in this case is given by :

$$f_{ref} \times [(M/N_1 \times N_2) + 1]$$

This technique can be extended to get any desired resolution. Since the multiplication numbers are low and the loop frequency is high, the output will have low noise sidebands. Also, the synthesiser is capable of fast frequency switching.

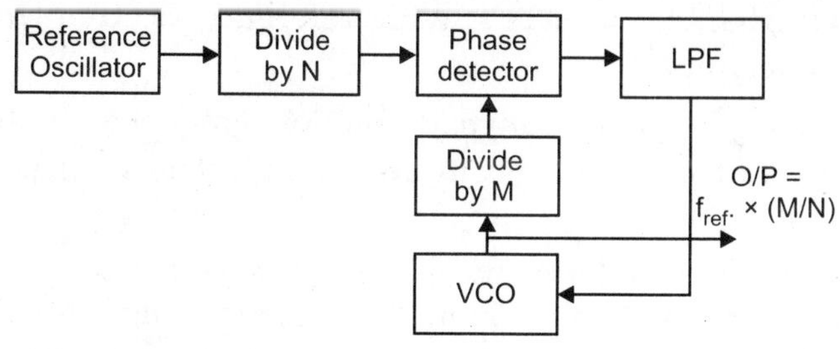

Fig. 29.55

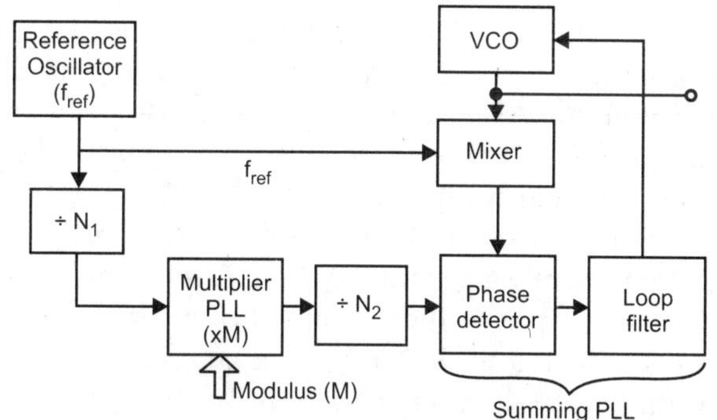

Fig. 29.56

Another popular method of indirect synthesis is Fractional *N* synthesis where a single PLL is made to lock to the non-integer multiple of loop reference. This technique can be used to achieve frequency resolution of the order of microhertz at switching speed of the order of a millisecond or so. Fig. 29.57 shows the basic architecture. The configuration functions as follows:

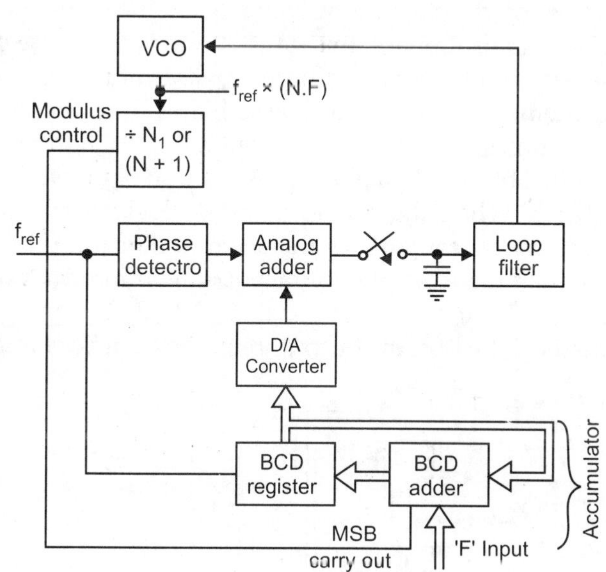

Fig. 29.57

The integer part of the desired multiplier is supplied to the digital divider placed between VCO output and phase detector in the form of its dividing factor. The fractional part is supplied to the accumulator. The accumulator is clocked by the reference source derived from crystal oscillator. The quantum of fractional input is added to the accumulator contents every clock cycle. The VCO output is *N* × *F* times the reference input, when the loop is locked. The circuit functions in such a way that the contents of the accumulator predict the expected phase detector output resulting from frequency difference of the two phase detector input signals. The D/A convertor is then so scaled and polarised that the D/A output waveform cancels the phase detector output waveform. The two waveforms are added in the analog adder, sampled and filtered to provide the oscillator control voltage. Also, to keep the phase detector output within its linear range, whenever the phase difference between the two inputs to the phase detector tends to become 360°, which is the maximum the phase detector can tolerate without going out of range, the phase of the divider output (which is ahead of reference input in phase) is retarded by 360° by either changing the divider modulus to ($N + 1$) momentarily or by any other means. In the architecture shown, the modulus is changed to accomplish this on receiving a command from the BCD adder at the time of accumulator overflow.

Sampled Sine Synthesis (Direct Digital Synthesis) : This method of frequency synthesis is based on generating the waveform of desired frequency by first producing the samples, as they would look like, if the desired waveform was sampled or digitised according to the Nyquist sampling theorem, and then interpolating amongst these samples to construct the waveform. As the frequency is rate of change of phase, this information is made use of to generate samples. Sine of different phase values are stored in a memory which is addressed by phase increment information stored in an accumulator. Fig. 29.58 shows the simplified block schematic representation of direct digital synthesis. When the accumulator is clocked at a fixed frequency, the contents of the accumulator jump by the phase increment whose digital equivalent information is stored in phase increment register (PIR). By changing the contents of the PIR, the output frequency can be changed.

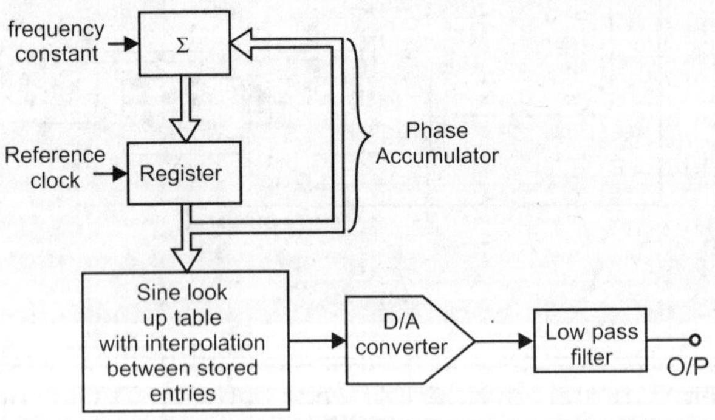

Fig. 29.58

The rate at which the look up table in the memory is addressed is given by clock frequency and phase increment during one clock period, as given by PIR contents. For instance, if the contents of PIR represented a phase angle of 36°, then the digital samples present at the output of the memory would correspond to phase differences of 36°, 72°, 108°, 144°, 180°, 216°, 252°, 288°, 324° and 360° to complete one cycle of output waveform. The ten samples will be produced in ten clock cycles. Therefore, the output frequency will be one-tenth of the clock frequency. In general, output frequency is given by :

$[\phi/2\pi] \times f$ clock; where ϕ = phase increment in radians

The digital samples are converted into their analog counterparts in *D/A* convertor and the analog samples interpolated to construct the waveform. The interpolator here is a low pass filter. The relevant waveforms are shown in Fig. 29.59.

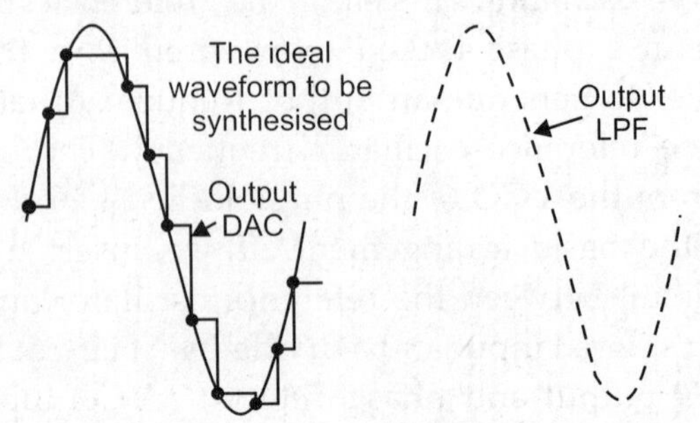

Fig. 29.59

This method of synthesis derives its accuracy from the fact that both the phase increment information and the time in which phase increment occurs can be computed to a very high degree of accuracy. With frequency being rate of change of phase, the resulting waveform is highly stable. The most important feature of this technique, however, is its capability to provide instantaneous switching. This is possible because size of angle increments between the two consecutive table look-ups may be changed instantaneously. The limitations of this technique are Quantisation noise and Aliasing inherent to any sampled data system. Another serious disadvantage is that of presence of spurious components due to imperfections and inaccuracies in D/A convertor. The highest frequency that can be synthesised is limited by the maximum speed of the available digit logic. The usable frequency range of direct digital synthesis output may be extended by a variety of techniques. However, depending upon the technique used, some of the advantages of the DDS may be lost. Just as with more conventional synthesisers, the DDS output may be doubled, mixed with other fixed sources or used as a reference inside a PLL.

Q. 97. Briefly describe FREQUENCY RANGE, RESOLUTION, FREQUENCY SWITCHING SPEED and SIGNAL PURITY with reference to frequency synthesisers.

Ans. Frequency range, Resolution, Frequency switching speed and Signal purity are the important synthesiser specifications.

While considering the *Frequency Range*, it is important to note whether the claimed frequency range is being covered in a single band or a series of contiguous bands. This aspect is significant from the viewpoint of noise performance which may be different in different bands, in case the frequency range is covered in more than-one band. This often leads to a larger transient, when the frequency switching involves switchover of the band also. *Frequency resolution* is usually the same throughout the range. It is typically 0.1 Hz though resolution as fine as 1 μ Hz is also available in some specific instruments.

Frequency Switching Speed is a measure of the time required by the source to stabilise at a new frequency after a change is initiated. In the PLL based synthesisers, it depends upon loop's transient response characteristics. Switching time is typically several hundreds of microseconds to tens of milliseconds in PLL based synthesisers and a few microseconds in instruments using direct digital synthesis (DDS) technique.

Signal Purity tells how well the output signal approximates the ideal single spectral line. Phase noise is one parameter that affects signal purity. It refers to the sidebands which result from phase modulation of carrier by noise. It is specified as the total sideband power in dB with respect to carrier (dBc). Presence of spurious signals resulting from undesired coupling between different circuits within the instrument and distortion products in the signal mixers also spoil signal purity.

Q. 98. Write brief notes on

(a) Synthesised function Generators and (*b*) Arbitrary waveform Generators

Ans. Synthesised function Generators are function generators with the frequency precision of a frequency synthesiser. The hardware of a synthesised function generator is similar to that of a frequency synthesiser with additional circuitry to produce pulse, ramp, triangle and square functions. These instruments with additional modulation capability are referred to as Synthesised Signal Generators.

Direct digital synthesis described in the earlier pages of this chapter is almost invariably used in synthesised function/signal generator design. Advances in digital technology have made DDS based synthesised function/ signal generators truely versatile. Synthesised sine wave output upto 30 MHz and other functions, such as pulse, ramp, triangle etc., upto 100 kHz, all with a resolution of 1 microhertz, are available in contemporary synthesised function generators.

Fig. 29.60 shows one such Synthesised function generator from M/s Hewlett Packard (Model HP 33120 A) that employs direct digital synthesis for achieving a high level of stability. It offers sine and square outputs upto 15 MHz. Triangle and ramp outputs are available upto 100 kHz. It also offers Noise (Gaussian) output with a bandwidth of 10 MHz. All outputs are available with a frequency resolution of 10 μHz.

Fig. 29.60

The **Arbitrary Waveform Generator (AWG)** is a signal source that is used to generate user specified custom analog waveforms. Using a custom stimulus waveform and measuring the response waveform provides realistic characterisation of the device or system under test. The Contemporary AWG allows generations of almost any conceivable waveform.

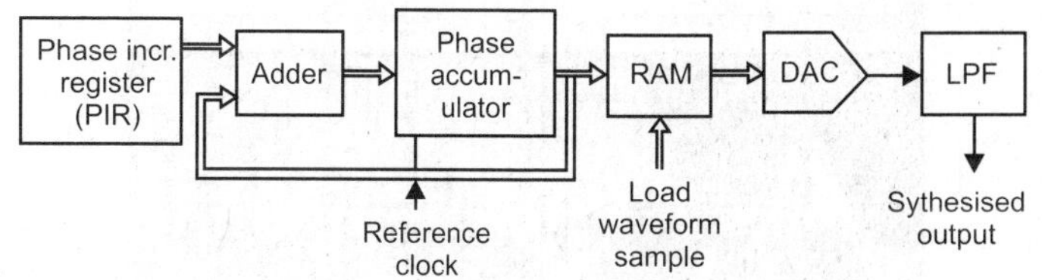

Fig. 29.61

Direct digital synthesis again is the heart of an arbitrary waveform generator. Fig. 29.61 shows the hardware. It looks very much similar to the one shown in Fig. 29.58. The sequential amplitude values of the waveform to be generated are stored in the RAM. The size of RAM decides the number of samples that can be stored in turn deciding the maximum number of Samples that one period of the desired waveform can be divided into. These sample values can be entered into RAM from the keyboard. Once the sample values are loaded into the RAM, they can be stepped through at a repetition rate governed by the Frequency Word input to the phase accumulator in the same way, as explained in case of a frequency synthesiser. The complexity of that waveform that can be synthesised by this process is limited by the size of RAM. As a thumb rule, a minimum of about 3 to 4 samples per cycle of the highest frequency in the waveform should be used. This is intended to eliminate aliasing.

Fig 29.62 shows an Arbitrary Waveform Generator from M/s Hewlett Packard (Model HP-8175 A) having a frequency range of DC to 25 MHz. Models with higher frequency range are also available. Fig. 29.64 shows some of the arbitrary waveforms possible with AWG of Fig. 29.63.

Fig. 29.62

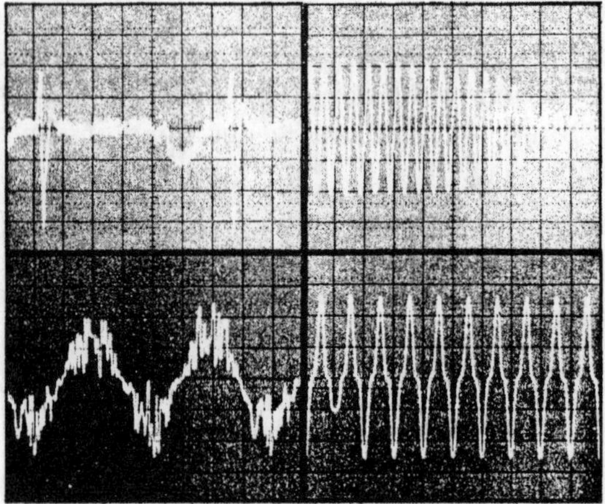

Fig. 29.63

Q. 99. What is a LOGIC PROBE and for what it is used?

Ans. A **Logic Probe** is the most basic tool used for troubleshooting of digital circuits. It is a small, hand held pen-like test instrument with a metallic tip on one end (Fig. 29.64). The instrument can be used to know the logic status of various points of interest such as pins of digital integrated circuits in a digital circuit. The logic status is indicated by a glowing LED. There may be typically three LEDs to indicate logic HIGH, logic LOW and Hi-Z states. Depending upon the actual logic status, where the probe is touched, the corresponding LED comes on. The probe can for be used to determine open and short circuits. Also, the probe has settings for different logic families to accommodate different acceptable voltage level ranges for logic LOW and HIGH status for different for logic families.

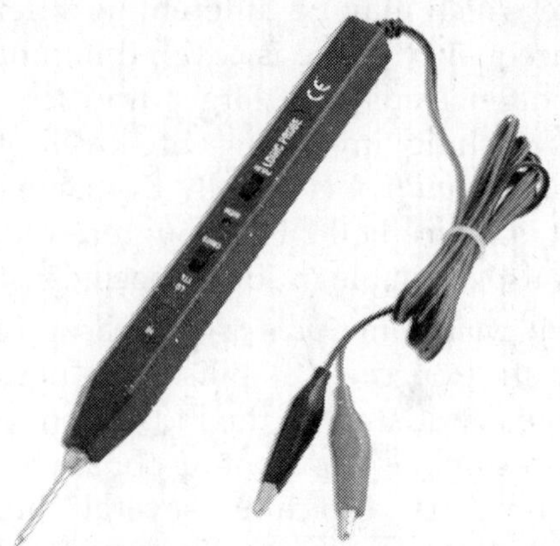

Fig. 29.64

Q. 100. What are LOGIC ANALYSERS? What are they used for? Briefly describe the synchronous and Asynchronous modes of operation of Logic Analysers.

Ans. Logic analysers are used for performance analysis and fault diagnosis of digital systems. Logic analysers have become a very relevant and indispensable diagnostic tool in the present day instrumentation scenario with the whole gamut of electronic instruments being centered around micro-processor/micro-computer based digital architecture. In addition, most logic analysers can be configured to format their outputs as a sequence of microprocessor instructions which makes them useful for debugging software too.

The logic analyser works in either of the following two modes of operation, namely (*a*) Asynchronous timing mode and (*b*) Synchronous state mode. A brief description of each of the two modes follows:

Asynchronous Timing Mode: In this mode of operation, the signals being probed are recorded either as a logic '0' or '1' and the logic analyser provides the time base, referred to as 'Internal Clock'. The time base determines, when data values are clocked into analyser's memory. On screen, the asynchronous mode display looks similar to an oscilloscope display except for the number of channels that can be displayed, which is much larger in case of a logic analyser.

Synchronous State Mode: In this mode of operation, samples of signals are stored in the memory on clock edge referred to as external clock supplied by the system under investigation. The logic analyser samples new data values or states only, when directed by the clock signal. On a given clock edge, the logic states of various signals constitute a group. The logic analyser display in this mode shows progression of states represented by these groups.

Q. 101. With the help of block schematic arrangement, briefly describe the functioning of a logic analyser.

Ans. Fig. 29.65 shows the block schematic arrangement of a logic analyser. All logic analysers have the following constituents, as shown in the figure. These are :

(*a*) Probes (*b*) Memory
(*c*) Trigger generator (*d*) Clock generator
(*e*) Storage qualifier (*f*) User interface

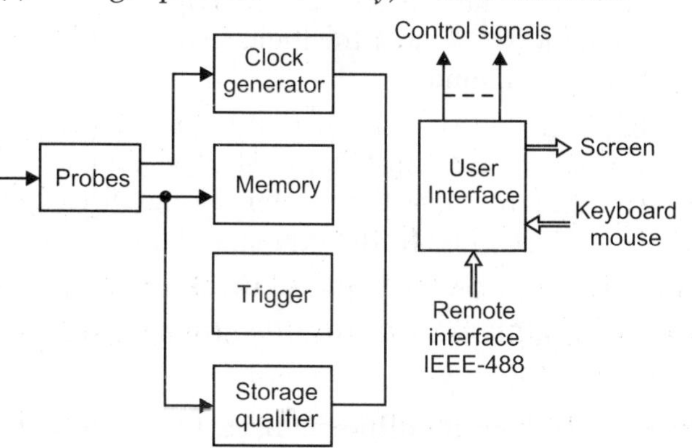

Fig. 29.65

Probes : Probes are used to provide physical connection to the circuit under test without causing any loading effects so that the logic signal of interest is not unduly affected and its timing integrity is maintained. The probes usually operate as voltage dividers. By this, the comparators inside the probe are asked to handle the lowest possible voltage slew rate which enables higher speed signals to be captured. These comparators have adjustable threshold to make the probes compatible with different logic families, as different families have different voltage thresholds. The comparators transform the input signals into logic 1's and 0's.

Memory: The memory stores the sampled logic values, the address for given sample being supplied internally. In a typical measurement using logic analyser, the user is interested in observing the logic signals around some event called the measurement trigger and the samples have a timing relationship with this trigger event. These samples are arbitrarily placed in the memory depending upon the instantaneous value of the internally supplied address.

Trigger: Logic analysers have both a combinational (or word recognised) trigger mode as well as an external trigger mode. In the combinational trigger mode, the trigger circuitry compares the incoming data with a word programmed by the user from the front panel. A trigger signal is generated, when the incoming data matches with the programmed word. Data is being sampled and stored in the memory by either an internal or an external clock. On occurrence of a trigger, the stored data samples are displayed on the screen.

Clock Generator: As stated earlier, the clock is either internal or external depending upon whether the selected operational mode is the asynchronous timing mode or the synchronous state mode. The two modes were described briefly earlier in the chapter. Again, in the timing mode, there are two commonly used approaches. Some logic analysers offer both the approaches.

In the first approach, called continuous storage mode, the clock is generated at the selected rate irrespective of the activity occurring on the input signals. The logic status of the input signal is stored in the memory on every clock cycle [Fig. 29.66. (*a*)]. In the second approach, called transitional timing mode, the input signals are again sampled at the selected rate but clock generator circuitry allows the samples to be stored in the memory only, if one or more signals change their logic status. Thus, the memory storage locations are used only, if inputs change leading to more efficient use of memory. For each sample, however, a time marker is recorded as shown in Fig. 29.66 (*b*). This approach offers distinct advantage, when long time records of infrequent or bursts of finely timed events are to be recorded.

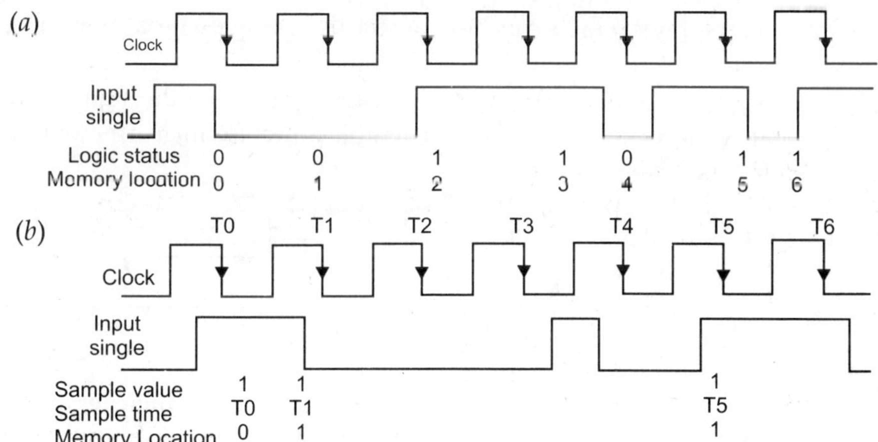

Fig. 29.66

Storage Qualifier: The function of the storage qualifier is to determine which data samples are clocked into the memory. The storage qualifier block looks at the sampled data and tests them against a criterion. If the criterion is met, the clocked sample is stored in the memory. This feature is particularly useful in troubleshooting microprocessor architectures. For instance, if the circuit under test is a microprocessor bus, this function can be used to separate bus cycles to a specific *I/O* port from cycles to all other ports or from instruction cycles.

User Interface : Bench top logic analysers typically use a dedicated keyboard and CRT display. Graphical user interfaces (GUI) similar to those available on personal computers are also available with many products. Also, interfaces like RS-232, IEEE-488 or Local Area Network (LAN) enable the use of the instrument from a personal computer or a work station. Remote interfaces are important in manufacturing applications of logic analysers. LAN interfaces have emerged as critical links in research and development activities, where these instruments can be tied to project databases.

Q. 102. Outline major performance specifications of a logic analyser. Briefly describe each one of these.

Ans. Some of the important specifications of logic analysers include :

(*a*) Sample rate (*b*) Set-up and Hold times

(c) Probe loading (*d*) Memory depth

(*e*) Channel count

Sample Rate in the timing mode determines the minimum time interval which can be resolved. Since the relationship of the sample clock and the input signal transition is random, two edges of the same signal can be measured to an accuracy of two sample periods. Measuring a transition on one signal with respect to a transition on another signal can also be done with an accuracy of two sample periods plus whatever skew exists between the channels. In the state mode, sample rate determines the maximum clock rate, which can be measured in the target state machine.

Set-up and Hold time specification in case of logic analysers also is similar to what we have in case of flip flops, registers and memory elements. Like these devices, a logic analyser also needs stable data a specified time before the clock becomes active. This specified time is the Set-up time. The Hold time is the time interval, for which the data must be held after the clock's active transition on enable data capture. Hold time is typically zero for logic analysers. It is desired that the target system is not perturbed by probe loading. Logic analysers with a sampling rate of equal to or less than 500 MHz have probe specifications of typically 100K, 6 to 8 pF. Analysers having greater than 1 GHz sample rate usually come with SPICE models for their probes so as to enable the users know the true impact of probe loading on signal integrity.

Channel count is the number of available input channels. Together with maximum rate, channel count determines the cost of instrument.

Memory Depth determines the maximum time window that can be captured in the timing mode or the total number of states or bus cycles that can be captured in the state mode. Most of the logic analysers offer 4 *K* to 1 *M* samples of memory.

Trigger resources, availability of preprocessors/ inverse assemblers, non-volatile storage and ability of logic analyser to store time value alongwith captured data are the other key features.

Q. 103. What is a LOCK-IN AMPLIFIER ? What is it used for ? Briefly describe the operational principle of a lock-in amplifier.

Ans. A **lock-in amplifier** is used to measure a very weak signal that is almost completely obscured by noise. With a lock-in amplifier, it is possible to measure a signal as small as a few nanovolts burried in noise that is thousands of times stronger than the signal. It may be mentioned here that use of amplifiers to enhance the signal or filters to reject noise really don't help much, when the signal is neck deep in noise. A typical lock-in amplifier measurement situation is depicted in Fig. 29.67. A system or an experiment is excited by a signal. The response of the system, that is, the signal present at the output of the system in response to input stimulus is obscured by noise. This is the signal desired to be measured. This forms the input to the lock-in amplifier. The input stimulus or its equivalent forms the reference input. The lock-in amplifier then detects the input signal at the frequency of the reference input. The heart of a lock-in amplifier is a phase sensitive detector (PSD) which can detect the desired signal with a very narrow bandwidth approaching zero in the theoretically ideal case thus increasing manifold the signal to noise ratio which makes possible a precise measurement of signal amplitude.

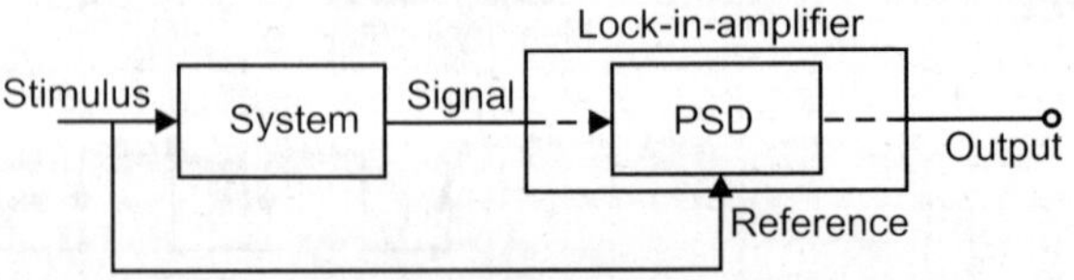

Fig. 29.67

Operational Principle: The heart of a lock-in amplifier as mentioned above is a phase sensitive detector followed by a low pass filter whose bandwidth equals the detection bandwidth. The phase sensitive detector is basically a multiplier circuit which multiplies the signal input with a reference and generates at its output sum and difference frequency components. If the reference frequency is same as the signal frequency, and also if the phase difference between the two is time invariant, then the difference component is a DC which is unaffected, as it passes through the low pass filter (LPF), even if the filter bandwidth is the narrowest possible. On the other hand, the sum component and all other sum and difference components resulting from noise frequencies and the reference fall well outside the LPF band edge and thus, get removed. Thus, if the reference to the multiplier is at the desired signal frequency, and if the phase relationship's maintained, the output of the LPF is proportional to the signal amplitude.

The reference signal that is used as an excitation to the system is seldom used as the reference input for the lock-in phase sensitive detector. Lock-in amplifier generates its own internal reference signal usually by a phase locked loop (PLL) locked to the external reference (Fig. 29.68). The external reference is applied to the PLL inside the lock-in amplifier as an input signal and the PLL locks the internal reference oscillator to this external reference. Since the PLL actively tracks this external reference, any changes in the external reference that constitutes the stimulus to the experiment do not affect the measurement.

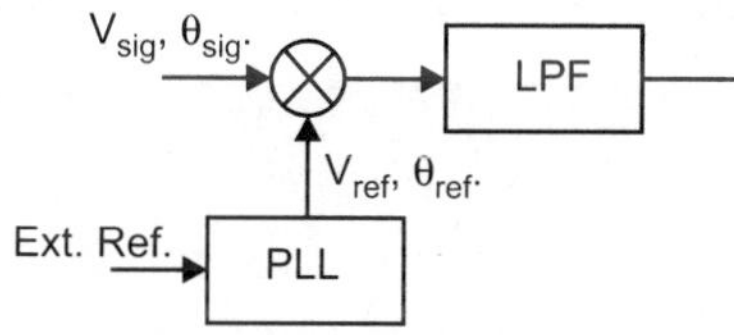

Fig. 29.68

The PSD and the following LPF detect only those signals whose frequencies are very close to the lock-in reference frequency. How far the signal frequency can be to the reference frequency and still get detected depends upon LPF bandwidth. A narrower LPF bandwidth will remove noise sources very close to the reference frequency. Only the signal at the reference frequency will result in a true DC output and be unaffected by LPF. The output of the PSD for signal frequency equal to reference frequency is proportional to :

$$V_{sig} \cos (\theta_{sig} - \theta_{ref})$$

Thus, the output is phase dependent. By making $\phi = (\theta_{sig} - \theta_{ref}) = 0°$, the output can be made proportional to V_{sig}. For $\phi = 90°$, there will be no output at all. This phase dependency can be eliminated by having a second PSD that multiplies the signal and a reference signal that is reference signal to PSD – 1 shifted 90° in phase. The output of PSD – 2 is then proportional to :

$$V_{sig} \sin (\theta_{sig} - \theta_{ref}) \text{ or } Vsig \sin \phi$$

If the outputs of PSD – 1 and PSD – 2 are designated as X and Y,

then, $X = V_{sig} \cos \phi$ and $Y = V_{sig} \sin \phi$

X and Y represent the signal as a vector quantity relative to lock-in reference oscillator. X is called the in-phase component and Y is called the quadrature component. Magnitude and phase of this vactor which appear at the output of a dual phase lock-in of this type are given by:

$$R = \sqrt{X^2 + Y^2} \text{ and } \phi = \tan^{-1} (Y/X)$$

The concept is demonstrated in Fig. 29.69

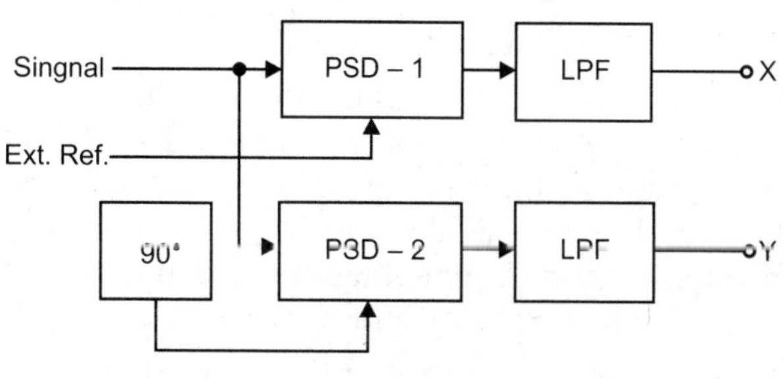

Fig. 29.69

Q. 104. Briefly describe the two types of Lock-in amplifiers namely (*a*) Analog Lock-in (*b*) Digital Lock-in.

Ans. There are two broad categories of lock-in amplifiers based on the technique of doing phase sensitive detection. These are :

(*a*) Analog lock-in, and (*b*) Digital lock-in.

Analog Lock-in: In a traditional lock-in amplifier, the signal and the reference are analog in nature. The PSD is an analog multiplier which multiplies the two analog signals. The result of multiplication is filtered with one or more stages of filters. Fig. 29.70 shows a typical block schematic arrangement of an analog lock-in amplifier. The input signal is amplified by a low noise front end which could be a differential amplifier for voltage input or a current to voltage converter for a current input. The signal is then selectively filtered to eliminate line frequency related interference and other unwanted signals. The conditioned signal forms one of the inputs of the analog multiplier. The other input *i.e.* the reference input, is generated by phase locking the internal reference oscillator to the external reference in a PLL.

The signal input also feeds another PSD (in case of a dual phase lock-in) as shown, whose reference input is the reference signal for PSD-1 90° phase shifted. The outputs of PSD-1 and PSD-2 are then low pass filtered and DC amplified to provide $R \cos \phi$ and $R \sin \phi$ outputs designated as X and Y in an earlier paragraph.

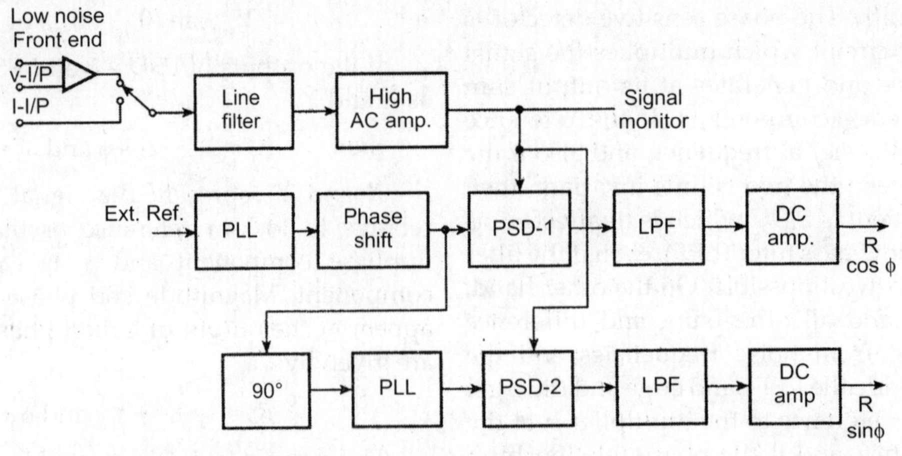

Fig. 29.70

Digital Lock-in : A digital lock-in amplifier differs considerably from its analog counterpart. Fig. 29.71 shows the basic architecture of a digital lock-in. Although the front end of both analog and digital lock-in is a low noise amplifier and the line filters are also present in both the cases but the similarity ends there.

A digital lock-in multiplies the signal and reference digitally. The input signal is digitised using an analog to digital converter before it feeds the digital multiplier. The ADC is preceded by an anti-aliasing filter for obvious reasons. The digital PSD multiplies the digitised input signal with a digitally computed sine wave at the reference frequency. In case of an external reference, a PLL locked to the external reference serves as a source of reference signal. The multiplied output is filtered in multiple cascaded stages of digital LPFs. The output from LPF section is passed through a digital to analog converter to provide front panel outputs. There is another identical detection channel, where the signal is multiplied with another digitally computed reference sine wave whose phase is 90° shifted from that of the reference sine wave of the first channel. The two channels provide in-phase and quadrature outputs.

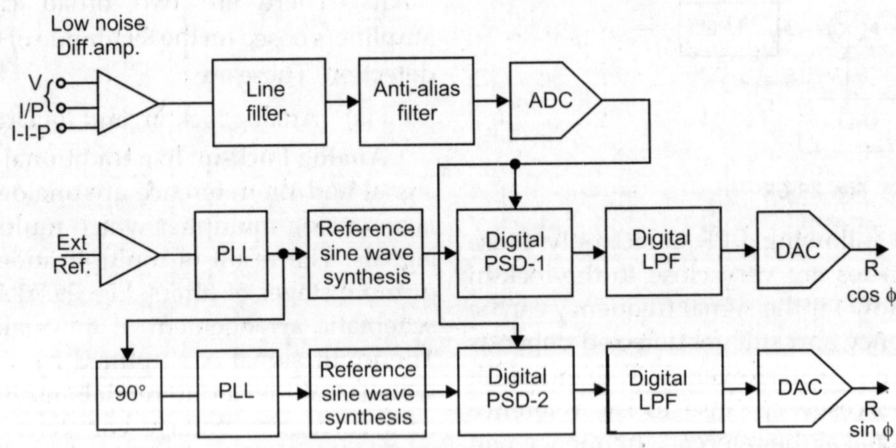

Fig. 29.71

Q. 105. What do you understand by DYNAMIC RESERVE with reference to Lock-in amplifiers ? How do analog and digital lock-in amplifiers compare in respect of dynamic reserve specification?

Ans. Dynamic Reserve is the most important performance specification of a lock-in amplifier, analog or digital. The dynamic reserve of a lock-in amplifier at a given full scale input voltage is the ratio (in dB) of the largest interfering signal to the full scale input voltage. The largest interfering signal here is defined as the amplitude of the largest signal at any frequency that can be applied to the input before the lock-in can not measure a signal with its specified accuracy. Digital lock-in has a superior dynamic reserve specification than their analog counterpart having a dynamic reserve in excess of 100 dB against a maximum of 80 dB in case of an analog lock-in.

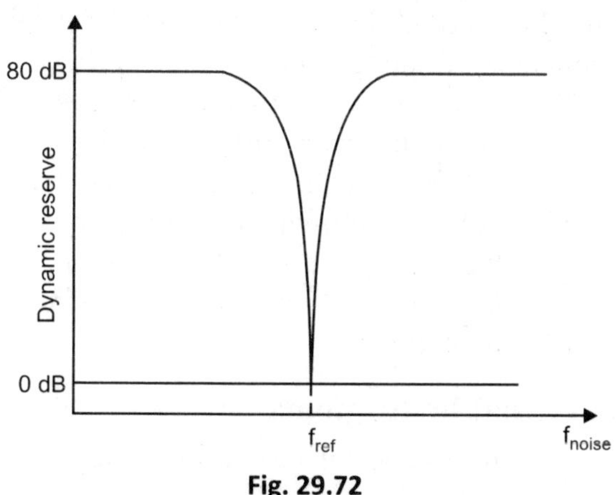

Fig. 29.72

It is always important to keep in mind each manufacturer's definition of dynamic reserve, when comparing specification. Some manufacturers define dynamic reserve as the largest interfering signal that can be applied without overloading the instrument. However, long before the instrument is overloaded, its accuracy may be severely degraded making this figure an overly optimistic estimate of the useful dynamic reserve. Similarly, some manufacturing quote dynamic reserve for interfering signals that are outside the dynamic range of the instrument and are, therefore, easier to reject. These aspects need to be paid attention to while comparing dynamic reserve specifications of two lock-in amplifiers.

The dynamic reserve of a lock-in also depends upon the noise frequency. Quite obviously, noise at the reference frequency will appear at the output unattenuated. So, the dynamic reserve for noise frequency equal to reference frequency is 0 dB (Fig. 29.72). As the noise frequency moves away from the reference frequency, the dynamic reserve increases. The dynamic reserve increases according to the low pass filter roll-off. When the noise frequency is far away, the reserve is limited by the gain distribution and the overload level of each gain element.

Q. 106. Why do we need computer-instrument interface? Briefly describe IEEE-488 interface standard.

Ans. Quite often, in complex measurement situations, more than one instrument are required to measure a parameter. In another situation, the system may require a large number of parameters to be measured simultaneously with each parameter being measured by a dedicated instrument. In such measurement situations, management of different instruments becomes very crucial. This has found a solution in the automated measurement set-ups, where various instruments are controlled by a computer. Another reason for which instruments are placed into such automated measurement set-ups is to achieve the capabilities that the individual instruments do not have. If there was a single instrument that did all the measurements the user required, automated them and compiled all the data in the required format, probably there would be no need for an integrated system. The probability of a single system doing all this is extremely remote, when there are a large number of different measurements to be made. Yet another reason for having a computer controlled instrument system is that it enables the user to make measurements faster and free of any human error.

In an integrated measurement set-up, there has to be transfer of data back and forth between different instruments and also between individual instruments and computer. Different interface standards have involved to allow transfer of data. The IEEE-488 interface is the most commonly used one for instrument-computer interface. This and some of the other popular interface standards are briefly discussed in the following paragraphs:

IEEE-488 Interface: The IEEE-interface has evolved from the Hewlett Packard Interface Bus (HP-IB) also called General Purpose Interface Bus (GP-IB). Presently, it is the standard interface bus used internationally for interconnecting programmable instruments in an automated measurement set-up.

Figs. 29.73 (*a*) and (*b*) show the general interface and bus structure of IEEE-488/HP-IB. Fig. 29.74 (*a*) shows interconnection of different types of programmable devices, such as talkers, listeners, controllers etc. A listener is an instrument that can only receive data from other instruments. A printer is an example of a listener type of instrument. A talker, such as a frequency counter is capable of transmitting data to the other instruments connected to the bus. There are some instruments which may perform both the functions. In the listening mode, they receive instructions to carry out certain measurements and in the talking mode, they transmit the results of measurements. A controller is supposed to manage the interface bus *i.e.* to decide what and when a particular instrument should do.

The interface bus has 16 lines and uses a 24-pin connector. A maximum of 15 devices/instruments can be connected to this interface bus in parallel. Typical data rate is 250 to 500 kilobytes per second over the full transmission path that ranges from 2 to 20 m. Maximum data rate is 1 M byte per second. There are 8 lines dedicated for data transfer (D-101 to D-108) in bit parallel format. There are three lines for data byte transfer control (DAV, NRFD, NDAC) and five lines for general interface management (IFC, ATN, SRQ, REN, EOI). Different lines in the interface bus carry addresses, program data, measurement data, universal commands and status bytes to and from the devices interconnected in the system. The data lines are held at +5V for a '0' and pulled to ground for a '1'.

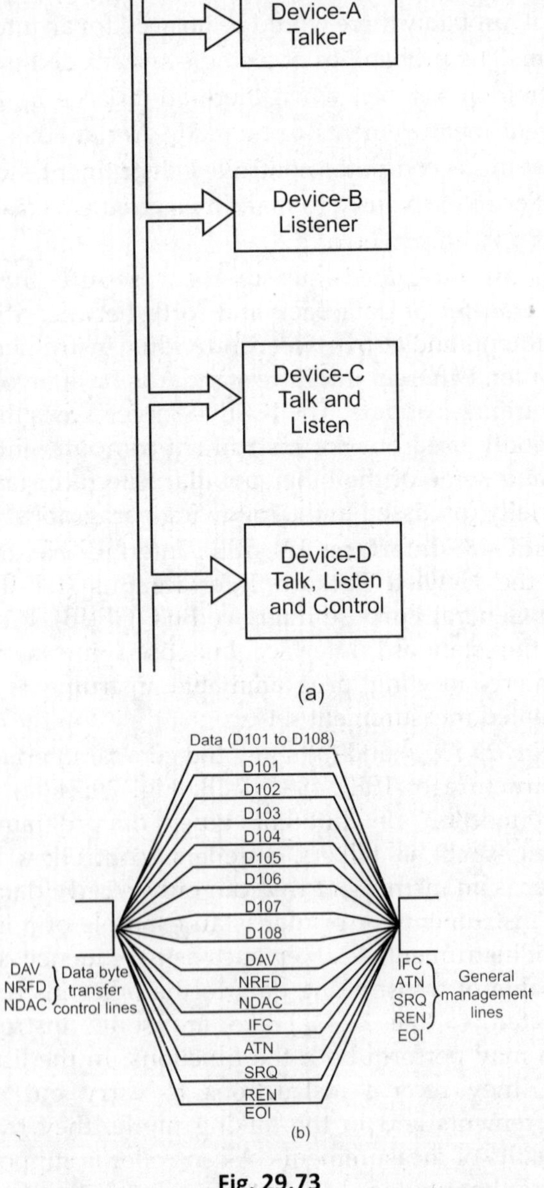

Fig. 29.73

The other popular instrument interface buses that allow interconnection of stand alone instruments and computers are the VXI-bus, the PCI bus and the MXI-bus. These interface buses are more relevant to fast growing concept of virtual instrumentation and therefore, are discussed in the following questions on virtual instruments.

Q. 107. What do you understand by VIRTUAL INSTRUMENTATION? What are the different types of virtual instrumentation set-ups? Briefly describe each one of these.

Ans. Advances in software development and rapid increase in the functional capabilities available on PC platform have changed the traditional instrumentation scenario. The scene is fast changing from the box like conventibnal stand alone instruments to printed circuit cards offering various instrument functions. These cards are inserted either in a card or in a PC slot. These acquire the measurement data which is then processed in the computer and subsequently displayed on the monitor in a format as required by the user. Such an instrumentation concept is commonly referred to as Virtual Instrumentation.

Use of Virtual Instruments

There are four types of virtual instrumentation set-ups. These include :

(*a*) A set of instruments used as virtual instrument
(*b*) Software graphical panel as a virtual instrument
(*c*) Graphical programming techniques as virtual instrument
(*d*) Reconfigurable building blocks as virtual instrument

A Set of Instruments as a Virtual Instrument

In complex measurement situations, usually more than one instrument are required to do the intended measurement. An instrumentation set-up that is used to qualify various sub-systems and systems for electromagnetic compatibility (EMC) is an example. In such a set-up, as shown in Fig. 29.74 the computer receives measurement data from all the standalone instruments, works on the data and then displays the measurement results. Another similar set-up that has been customised to perform a certain test on a certain specific product however would not be classified as a virtual instrument.

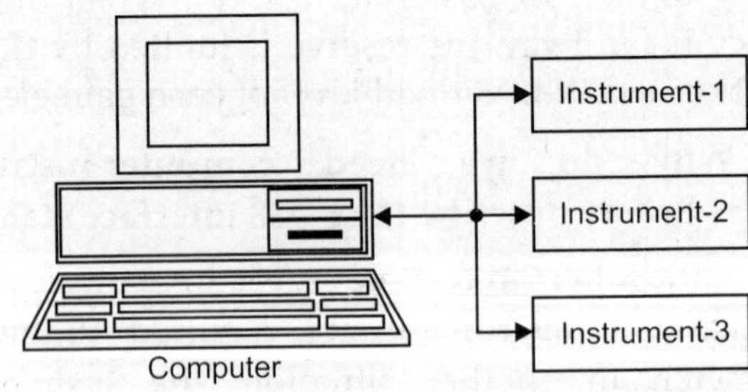

Fig. 29.74

Software Graphical Panel as a Virtual Instrument

In this type of virtual instrumentation set-up, the instrumentation hardware is controlled by a personal computer from a keyboard or a mouse. The PC screen is used to display the measurement results (Fig. 29.75). The instrumentation hardware could be a traditional box like instrument or a PC card offering the desired measurement function. The computer control of the instrument is through an interface bus, such as IEEE-488 (Fig. 29.76).

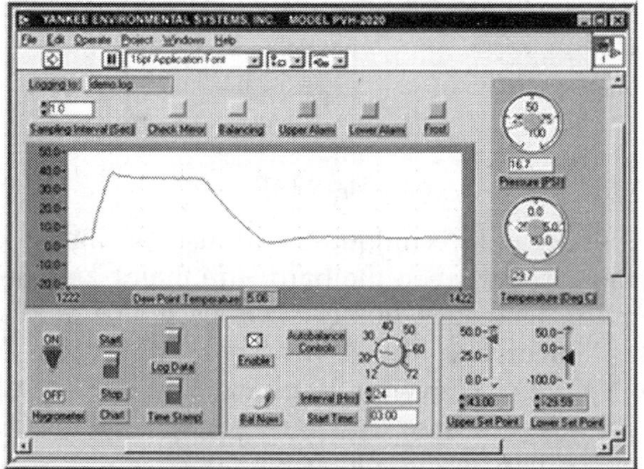

Fig. 29.75

Graphical Programming Technique as Virtual Instrument : In a typical computer controlled instrument set-up, the software to do the job is written using a textual programming language, such as C, BASIC, Pascal, Fortran. Due to constant increase in computer power and instruments' capabilities, the development of software to make full use of instrumentation setup has become a tedious and time consuming job, if it is done using one of the available textual programming languages. There has been a distinct trend to move away from the conventional 'programming languages and to move towards graphical programming languages. A graphical programming equivalent of a program is a set of interrelated icons (graphical objects) joined by lines and arrows. Use of graphical programming language leads to drastic reduction in programming time, sometimes by a factor as large as ten.

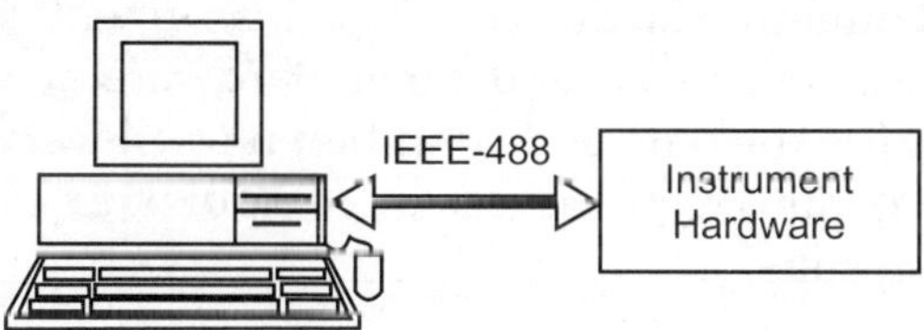

Fig. 29.76

Having written a graphical program for a certain test, all icons appear on the screen with programmed interactions. It may be mentioned here that with graphical language, the instrument control as well as the program flow and execution are determined graphically. A graphical programming product lists the interface buses and instruments that are supported by it. Graphical programming languages are typically used, where one wants to decrease the effort needed to develop a software for instrument systems. However, they require substantial computing power and size of these programs can reduce the speed of application in some cases.

Reconfigurable Building Blocks as Virtual Instrument: If one tends to look into the building blocks of various instruments, one is sure to find a lot of commonality. The building blocks such as front ends, ADCs, DACs, DSP modules, Memory modules etc. are the commonly used ones. One or more of these building blocks are invariably found in voltmeters, oscilloscopes, spectrum analysers, waveform analysers, counters, signal generators and so on. In an instrumentation set-up comprising of more than one instrument functions, there is going to be a lot of redundant hardware.

A fast emerging concept is to have instrument hardware in the form of building blocks, which can be configured from a graphical user interface (GUI) to emulate the desired instrument function. These building blocks could be reconfigured at will to become voltmeters, oscilloscopes, spectrum analysers, waveform recorders and so on. A graphical panel would represent each virtual instrument.

Q. 108. What are different components of a virtual instrument? Briefly describe each one of these with suitable illustrations or examples.

Ans. The basic components of a virtual instrument as shown in Fig. 29.77 are :

(*a*) Computer and Display (*b*) Software

(*c*) Bus structure (*d*) Instrument hardware

Computer and Display: Majority of virtual instruments are built around personal computers or work stations with a high resolution monitor. The chosen computer should meet the system requirements, as dictated by the software packages.

Software: Software is the brain of any virtual instrument set-up. The software uniqly defines the functional capabilities of the instrument set-up, and in most of the cases, is designed to run industry standard operating systems for personal computers and workstations.

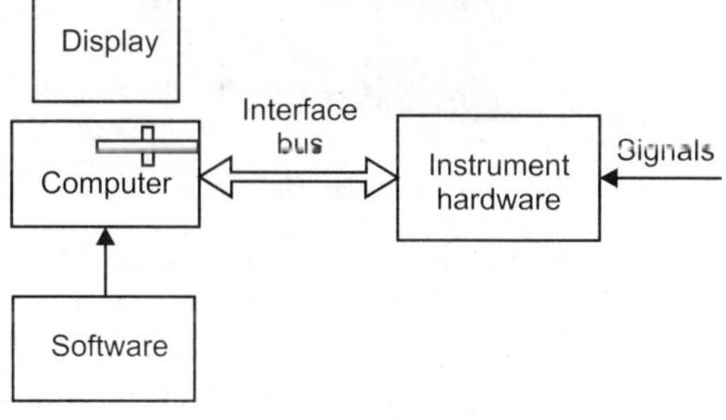

Fig. 29.77

Interface Bus Structure: The commonly used interface bus structures for computer-instrument interface are the

IEEE-488, described in the earlier pages, PC-bus and the VXI-bus. The other two are described here:

In a **PC-bus virtual instrument** set-up, the instrument function available on a printed circuit card (Fig. 29.78) is inserted directly into a vacant slot in the personal computer. Since the cards are plugged directly into the computer back plane and contain no embedded command interpreter as found in IEEE-488 instruments, these cards are invariably delivered with drive software so that they can be operated from the computer. PC-bus instruments offer a low cost solution to building a data acquisition system. Due to limited printed circuit space and close proximity to sources of electromagnetic interference, PC-bus instruments offer a lower performance level than their IEEE-488 counterparts.

VXI-bus instruments are plug-in-instruments that are inserted into specially designed can cages called mainframes (Fig. 29.79) Mainframe contains power supplies, air cooling etc. that are common to all the modules. VXI-instruments combine the advantages of computer back plane buses and IEEE-488. A VXI-bus instrument has high speed communication as offered by computer back plane buses (such as VME-bus) and a high quality EMC environment that allow high performance instrumentation similar to that found in IEEE-488 instruments.

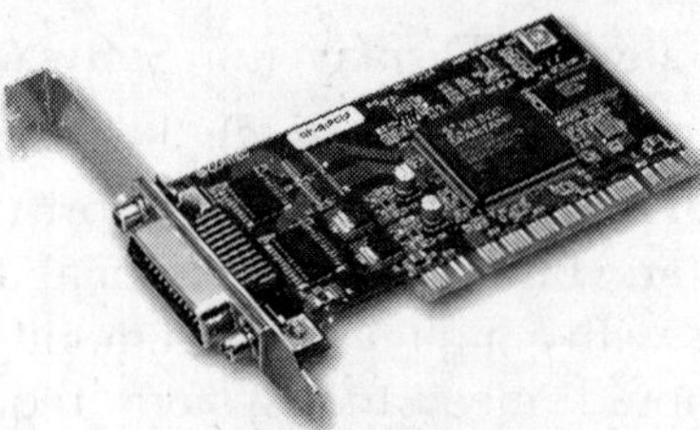

Fig. 29.78

Fig. 29.79

One of the methods to communicate with VXI instruments is *via* IEEE-488 as shown in Fig. 29.80 In this case, an IEEE-488 to VXI-bus converter module is plugged into VXI-bus nainframe. The mainframe then interfaces with IEEE-488 interface card in the computer using he standard interface cable. The set-up is easy to program but the overall speed is limited by IEEE-488 data transfer rate.

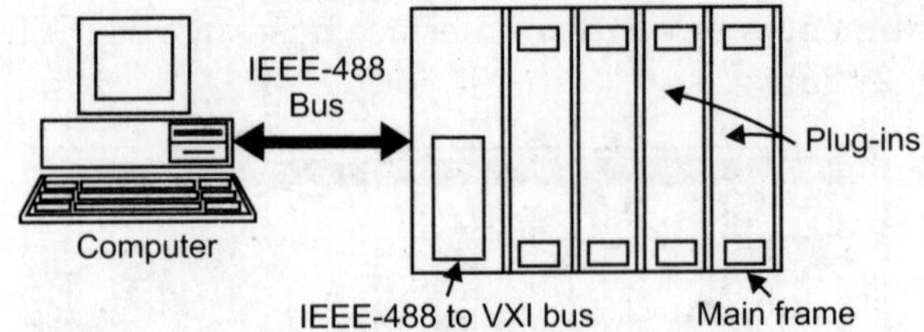

Fig. 29.80

The second technique is to use a higher speed interface bus between the hardware mainframe and the computer. One such bus is MXI-bus which is basically an implementation of VXI-bus on a flexible cable. In this case, VXI-MXI converter is plugged in the mainframe and an MXI-interface card with software is installed in computer. This set-up allows the use of off-the-shelf PCs to communicate with VXI instruments at speeds much faster than IEEE-488 instruments.

Yet another approach is to insert a powerful VXI-bus computer in the hardware mainframe to take full advantage of the VXI-bus instruments. The disadvantage of such a set up is that due to low volume requirement of VXI computers, these may not be able to match the industry standard personal computers on the price performance criteria. The set-up is shown in Fig. 29.81

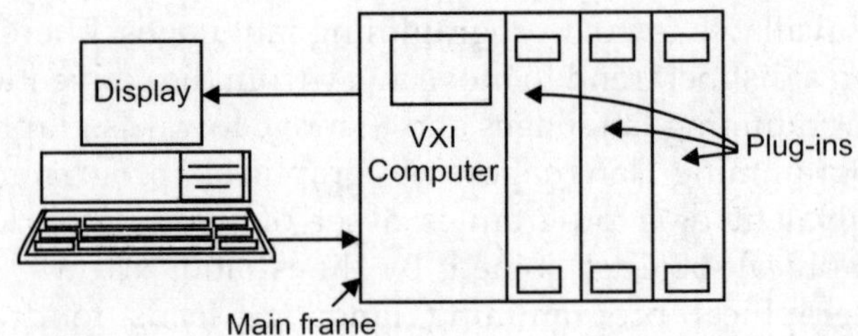

Fig. 29.81

Instrument Hardware : The instrument hardware comprises of sensors and other hardware components that acquire the data and condition it to a level and form so that it can be processed in computer to extract the desired results.

Q. 109. What are strip chart recorders and X-Y recorders?

Ans. A **strip-chart recorder** essentially consists of a drive motor, chart paper, stylus, servomotor, follow-up potentiometer and amplifier. The drive motor pulls a roll of chart paper past a stylus at a uniform rate. The stylus is moved across the chart paper (at right angles to the chart paper line of motion) by servo which receives its input from amplifiers output. Fig. 29.82 shows the basic constructional features of a strip chart recorder.

Strip chart recorders are particularly useful in recording data over a long period of time such as monitoring industrial processes. For instance, if it is desired to record some pressure variations, if any, over a 24 hour long period, the pressure is converted first into an electrical voltage by a suitable transducer which is then fed to the input of the amplifier in the recorder. The stylus records voltage variations or indirectly the pressure variations.

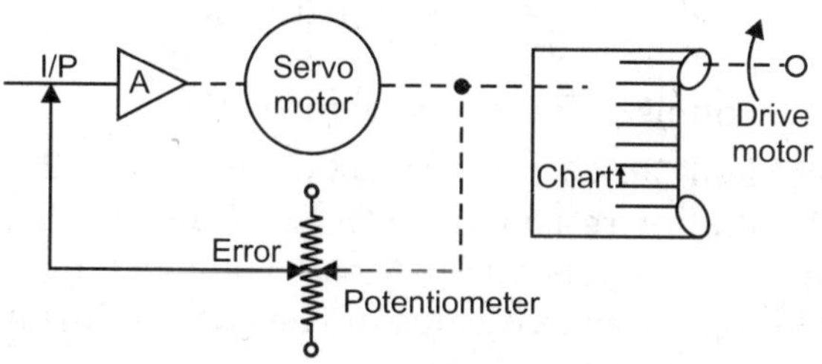

Fig. 29.82

The **X-Y recorder** (also referred to as X-Y plotter) is a two axes instrument that records a one shot display on a single non-moving chart or graph. Both the vertical and horizontal axes are identical or nearly identical and are similar in operation to the single axis of a strip chart recorder. However, in an X-Y recorder, the stylus is moved by two servomotors (horizontal and vertical) which in turn receive their input from corresponding amplifiers. Refer to Fig. 29.83.

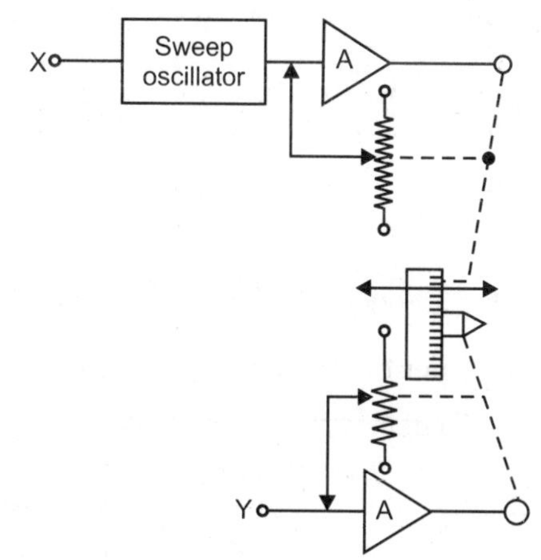

Fig. 29.83

Often, the horizontal axis servo system will include a sweep oscillator similar to that in an oscilloscope. The sweep oscillator produces a sawtooth voltage that moves the stylus horizontally at a fixed rate or it can even be bypassed so that horizontal system will respond to sine waves. In general, stylus constantly maintains a position that corresponds to the two input signals Thus, the display of an X-Y recorder is similar to that of an oscilloscope in X-Y mode except that is recorded on paper.

SOLVED PROBLEMS

Problem 29.1. *Meter A needs 500 μA for full scale deflection while meter B takes 1 mA for fulls scale deflection. Which is more sensitive? Why?*

Solution. Meter *A* is more sensitive because *A* requires lesser current for same deflection or in other words, for same current, deflection in *A* is more.

Problem 29.2. *If the current in a 500 μA meter is*

(i) 1 mA (ii) 10 mA, what will happen?

Solution.

(*i*) The pointer goes off the scale and may bend at the right hand retaining pin.

(*ii*) The coil may burn out along with the bend pointer due to large current.

Problem 29.3 *In a meter, the maximum current that can be measured is 500 mA. If the accuracy is 2%, find percentage error when reading is 100 mA, 200 mA and 500 mA. Also, find sensitivity.*

Solution. Sensitivity = 500 mA

$$\text{Error at 500 mA} = \pm 2 \times \frac{500}{100} = \pm 10 \text{ mA}.$$

Thus, for 500 mA, actual value lies between 490 and 510 mA

Error at 200 mA = ± 10 mA (Same as that at 500 mA)

Reading lies between 190 and 210 mA.

Percentage error = ± 5%

Error at 100 mA = ± 10 mA

Reading lies between 90 and 110 mA

Percentage error = ± 10%

Thus, percentage error increases as reading decreases.

Problem 29.4. *Find the value of shunt required to extend the range of 1 mA to 10 mA, when meter movement has a resistance of 27 ohms.*

Solution. $R_{sh} = \dfrac{I_M R_M}{I_{sh}}$

Now maximum current to be measured =10 mA

A maximum current that meter can measure = 1 mA

$$I = 10 \text{ mA}, I_{sh} = 9 \text{ mA}, I_M = 1 \text{ mA}$$

$$R_{sh} = \frac{27 \times 1}{9} = 3 \text{ ohms}.$$

Problem 29.5. *Find the value of shunts for a current meter whose range is to be extended from 0 → 10 mA to 0 → 100 mA, 0 → 1A and 0 → 10A. The meter has resistance of 9 ohms.*

Solution. For 0 → 100 mA,

$$R_{sh} = \frac{I_M R_M}{I_{sh}} = \frac{9 \times 10}{90} = 1 \text{ ohm}.$$

For $0 \to 1A$,

$$R_{sh} = \frac{I_M R_M}{I_{sh}} = \frac{9 \times 10}{990} = 0.091 \text{ ohm} = 0.1 \text{ ohm}.$$

For $0 \to 1A$,

$$R_{sh} = \frac{9 \times 10}{9990} \approx 0.009 \text{ ohm}.$$

The arrangement is shown in Fig. 29.84

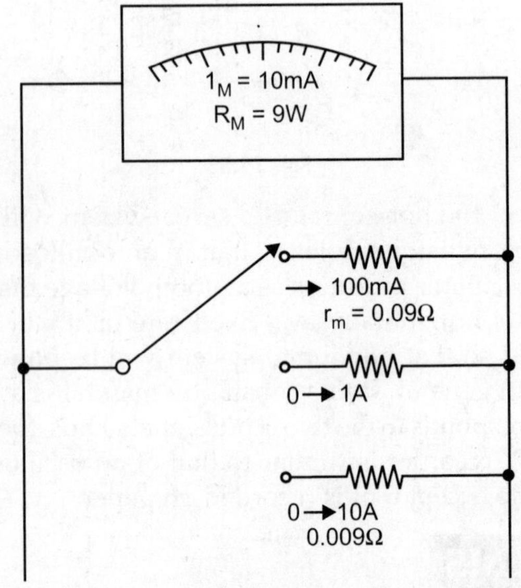

Fig. 29.84

Problem 29.6. *Calculate the value of R_{MULT} for enabling a current meter (0 → 10 mA, 1000 ohms) to measure voltage upto (i) 10 V (ii) 25 V.*

Solution. (*i*) $R_{TOT} = \frac{10}{10 \times 10^{-3}} = 1 \text{ K ohm}.$

$$R_{MULT} = 1000 - 1000 = 0 \text{ ohm}$$

Thus 10 V can be measured without any multiplier resistor.

(*ii*) $R_{TOT} = \frac{25}{10 \times 10^{-3}} = 2.5\ K = 2500 \text{ ohms}$

$$R_{MULT} = 2500 - 1000 = 1500 \text{ ohms}.$$

Problem 29.7. *There is a current meter, 1 mA, 1000 ohms and its range is extended to measure upto 10 V. What is ohms / volt rating for 0 → 10 V case?*

Solution. It is same as for basic current meter (*i.e.* 1000/1 = 1000 ohms/volt). Ohms/volt rating is the inherent characteristic of meter and remains same for different ranges.

For example, if range is extended to 10 V, then a 9000 ohms resistance is to be put in series.

$$\text{Ohms/volt} = \frac{(9000 + 1000)}{10} = 1000 \text{ ohms/volt}.$$

Problem 29.8. *Find ohms/volt rating of a 50 μA meter.*

Solution. The resistance required for

$$1\text{V} = \frac{1}{50 \times 10^{-6}} = 20{,}000 \text{ ohms}.$$

Thus, ohms/volt = 20,000 ohms/volt.

Problem 29.9. In the circuit show in Fig. 29.11, the actual value of resistance to be measured is 1 ohm. If current meter reads 100 mA and voltmeter reads upto 100 mV, find measured value of resistance, if voltmeter is having a resistance of 5 ohms.

Solution. Measured current = 0.1 A

Measured voltage = $r \times 0.1\ A$

where, $\frac{1}{r} = \frac{1}{5} + \frac{1}{1} = \frac{6}{5}$

$$r = 0.833 \text{ ohm}.$$

Measured voltage = 0.0833 V

$$\text{Measured resistance} = \frac{0.0833}{0.100} = 0.833 \text{ ohm}.$$

So, there is an error of 0.166 ohm which is approximately 16.6%. This method of measuring resistances is known as Ammeter-Voltmeter method.

Problem 29.10. *In the circuit shown in Fig. 29.85, find the ammeter reading.*

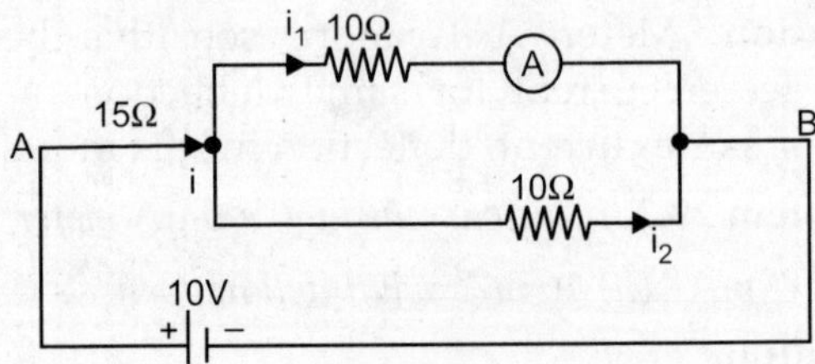

Fig. 29.85

Solution. Total resistance between *A* and *B*

$$= 15 + \frac{10 \times 10}{10 + 10} = 15 + 5 = 20 \text{ ohms}.$$

$$\text{Total current } i = \frac{10}{20} = \frac{1}{2}\text{A}$$

Therefore, current through ammeter, $i_1 = \frac{0.5 \times 10}{20}$ = 0.25 A.

Problem 29.11. *In the circuit shown in Fig. 29.86. Find current through 3 ohms resistance.*

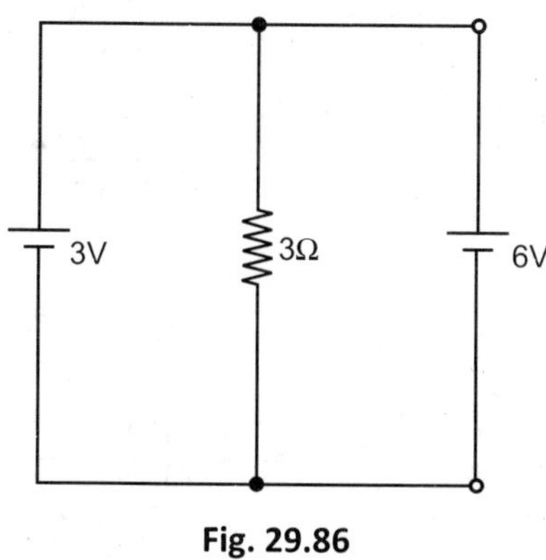

Fig. 29.86

Solution. Current = 2A

6 V in parallel with 3 V gives 6 V if the batteries are ideal.

Problem 29.12. *A square waveform with a time period of 10 μs and a rise time of 100 ns is seen on a 20 MHz bandwidth oscilloscope using a 100 MHz bandwidth probe. What will be the rise lime of the waveform as seen on the oscilloscope?*

Solution. Bandwidth of oscilloscope = 20 MHz

Corresponding rise time, τ_{osc} = 350/20 = 17.5 ns

Bandwidth of probe = 100 MHz

Corresponding rise time, τ_p = 350/100 = 3.5 ns

Rise time of the input waveform, τ_s = 100 ns

Rise time as observed on oscilloscope,

$$\tau = \sqrt{\tau_s^2 + \tau_p^2 + \tau_{osc}^2} = \sqrt{(100)^2 + (3.5)^2 + (17.5)^2}$$

$$= \sqrt{10000 + 12.25 + 306.25} = \sqrt{10318.5} = 101.6 \text{ ns.}$$

Problem 29.13. *If the same set up of probe and oscilloscope is used to view a 50 MHz signal having a rise time of 2 ns, what would be the observed rise time?*

Solution.

$$\text{Observed rise time} = \sqrt{2^2 + (3.5)^2 + (17.5)^2}$$

$$= \sqrt{4 + 12.25 + 306.25} = \sqrt{322.5} = 18 \text{ ns}$$

Problem 29.14. *A digital storage oscilloscope (DSO) has a 50K points memory. The specified sampling rate is 1 GS/s. Determine the slowest time base setting which can utilise the specified, maximum sampling rate of 1 GS/s, if the record is to fill all the 10 divisions of the oscilloscope, screen horizontally.*

Solution.

$$\text{Sampling Rate} = \frac{\text{Memory size}}{\text{Tme base setting} \times \text{Number of divisions}}$$

$$\text{or Time base setting} = \frac{50000}{10^9 \times 10} = 5 \times 10^{-6} \text{ s/div}$$

$$= 5 \text{ μs/div.}$$

Problem 29.15. *An amplitude modulated signal when observed on a spectrum analyzer looked like the spectrum shown in Fig. 29.87. Determine the percentage of modulation for this AM signal. Also, determine the modulating frequency.*

Solution. The carrier level = 20 dB

Therefore carrier level (in volts) = Antilog (20/20) = Antilog 1 = 10

The sideband level = 14 dB

Therefore sideband level (in volts) = Antilog (14/20) = Antilog 0.7 = 5

Percentage of modulation

$$= \frac{2 \times \text{sideband level in volts}}{\text{Carrier level in volts}} \times 100 = \frac{2 \times 5}{10} \times 100$$

$$= 100\%$$

Modulating frequency = 1 kHz

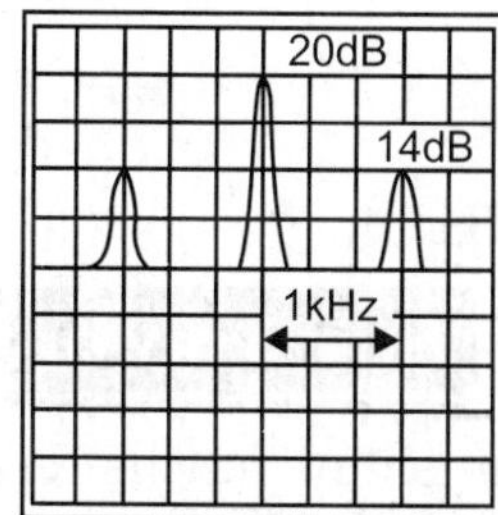

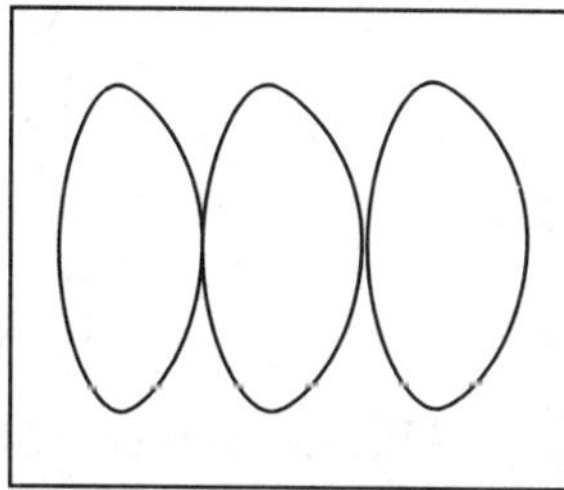

Fig. 29.87 Fig. 29.88

Problem 29.16. *Lissajous method is used to determine an unknown frequency. The Lissajous pattern thus obtained by feeding the unknown signal to the horizontal input and a known signal of 10 kHz to the vertical input of the oscilloscope is shown in Fig. 29.88. Determine the unknown frequency.*

Solution.

If (f_1) is the unknown frequency and (f_2) is the known frequency, then,

$$\frac{f_2}{f_1} = \frac{\text{Number of points of tangency to the horizontal line}}{\text{Number of points of tangency to the vertical line}}$$

$$= \frac{3}{1} \text{ which gives}$$

$$f_1 = \frac{f_2}{3} = \frac{10}{3} = 3.33 \text{ kHz}$$

Problem 29.17. *In a frequency synthesiser application using direct digital synthesis, the phase angle stored in the phase increment register is (π) radians. It the clock frequency is 10 MHz; determine the frequency of the output signal.*

Solution: In the case of direct digital synthesis of synthesising signals, the output signal frequency is given by:

$$f_{out} = \left(\frac{\phi}{2\pi}\right) \times f_{CLOCK}$$

where, φ = phase angle increment in radians

Here $\phi = \pi$ radians

$f_{CLOCK} = 10$ MHz

Therefore, fout = $\left(\frac{\neq}{2\neq}\right)\left(\frac{\pi}{2\pi}\right) \times 10$

= 5 MHz

Problem 29.18. *A certain frequency synthesiser is represented by the block schematic arrangement of 29.89. If $N_1 = 3$ and $N_2 = 5$ and $M = 15$; determine frequency of output signal for $f_{REF} = 50$ MHz.*

Solution:

$$\text{fout} = f_{REF} \times \left[\frac{M}{N_1 \times N_2} + 1\right]$$

$$= 50 \times \left[\frac{15}{3 \times 5} + 1\right]$$

$$= 50 \times 2 = 100 \text{ MHz}$$

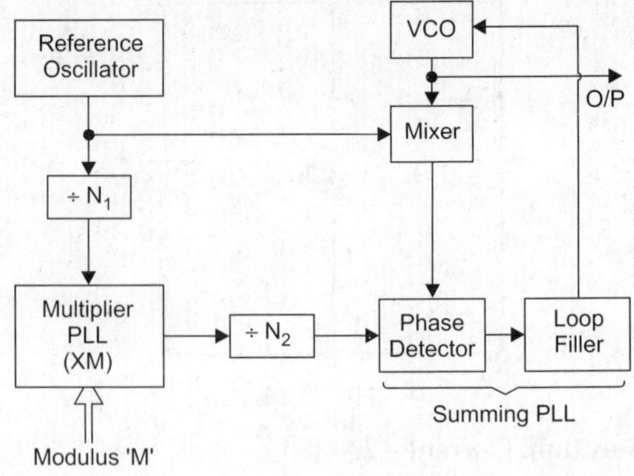

Fig. 29.89

Summary of Important Formulas

1. The resistance value of shunt resistance (R_{sh}) required to increase the currant measuring capillarity of the basic meter movement having resistence (R_M) and maximum current rating (I_M) can be computed from:

$$R_{sh} = \frac{I_M \times R_M}{I_{sh}}$$

where, $I_{sh} + I_M = I$ = Maximum desired current rating

2. The resistance ratue of the multiplier resistor (R_{MULT}) required to extend the voltage range to (V_{max}) can be computed from:

$$R_{MULT} = R_{TOT} - R_M$$

where, $R_{TOT} = \frac{V_{max}}{I_{max}}$

R_M = Meter resistance

I_{max} = Maximum current that meter can handle

3. The unknown resistance (R) in a Wheatstone bridge can be computed from:

$$R = \left(\frac{P}{Q}\right) \times S$$

where, P, Q = Ratio arm resistances

S = Known resistance

4. In the case of Maxwell's bridge (Fig. 29.19), the balance conditions are given by:

$$L_1 = L_4 \times \left(\frac{R_2}{R_3}\right)$$

$$R_1 = R_4 \times \left(\frac{R_2}{R_3}\right)$$

5. In the case of Maxwell-Wein bridge (Fig. 29.20), the balance conditions are given by:

$$L_d = R_a R_b R_c$$

$$R_d = \frac{R_a R_b}{R_c}$$

$$Q_d = \frac{\omega L_d}{R_d} = \omega C_b R_b$$

6. In the case of Hay's bridge (Fig. 29.21), the balance conditions are given by:

$$L_d = \frac{R_a R_c C_b}{1 + (\omega R_b C_b)^2}$$

$$R_d = \frac{R_a R_b R_c (\omega C_b)^2}{1 + (\omega R_b C_b)^2}$$

7. In the case of De-saughty bridge (Fig. 29.22), balance conditions are given by:

$$C_b = \left(\frac{R_d}{R_a}\right) C_c$$

8. In the case of Schering bridge (Fig. 29.23), balance conditions ave given by:

$$C_d = \left(\frac{R_b}{R_a}\right) C_c$$

$$R_d = \left(\frac{C_b}{C_c}\right) R_a$$

9. Bandwidth and Rise time specifications of an oscilloscope ave interrelated by:

$$\text{Bandwidth (in MHz)} = \frac{350}{t_r\,(in\ ns)}$$

10. In a digital storage oscilloscope, the effect of memory on the single shot bandwidth as a function of time base setting is expressed by:

$$\text{Sample Rate} = \frac{\text{Memory Length}}{\text{Time base setting} \times 10}$$

11. The output frequency in a frequency synthesiser employing dived direct synthesis is given by:

$$f_{OOT} = \left[\frac{\phi}{2\pi}\right] \times f_{CLOCK}$$

where, ϕ = Phase increment in radians

f_{CLOCK} = Clock frequency

Self Evaluation Exercise

I. Multichoice Questions

1. The permanent magnet moving coil instrument has .... scale. (Uniform, Non-uniform)
2. If the instrument has a square law response *i.e.* the pointer deflection is proportional to square of current or voltage under measurement, it can be used for measurement of
   (*a*) AC only
   (*b*) both AC as well as DC
   (*c*) DC only
   (*d*) None of these
3. One of the following instruments is suitable for measuring AC only
   (*a*) Permanent magnet moving coil type
   (*b*) Moving iron type
   (*c*) Dynamometer type
   (*d*) Induction type
4. A basic permanent magnet moving coil meter has markings 0–10 mA, 5 ohms. Its ohms per volt rating is
   (*a*) 100 (*b*) 5
   (*c*) 1000 (*d*) 10,000
5. The range of the meter of Q-4 is extended to 1A by using an external shunt resistance. Its ohms/volt rating
   (*a*) increases to 5000 (*b*) increases to 500
   (*c*) remains the same (*d*) none of these
6. In order to convert a 50 microamperes, 2000 ohms permanent magnet moving coil instrument into an ammeter capable of measuring upto 5A,
   (*a*) Connect a resistance of about 20 milliohms across the meter
   (*b*) Connect a resistance of about 20 milliohms in series with the meter.
   (*c*) replace the 50 microamperes scale with a 5A scale.
   (*d*) Connect a resistance of 20 megaohms across the meter.
7. A basic permanent magnet moving coil instrument gives full scale deflection for 1 mA while another gives full scale deflection for 100 mA.
   (*a*) The former is more sensitive
   (*b*) The latter has a higher sensitivity
   (*c*) Both the meters are equally sensitive as sensitivity is independent of full scale deflection current
   (*d*) None of these
8. A permanent magnet moving coil meter based voltmeter has an ohms per volt rating of 1000 ohms per volt. If the meter gives full scale deflection for 100 volts DC, it has a total series resistance (including the coil resistance) of
   (*a*) 100K (*b*) 1000 ohms
   (*c*) 10K (*d*) none of these
9. In a Wheatstone bridge, Null is obtained when
   (*a*) $P/Q = R/S$ (*b*) $P/Q = S/R$
   (*c*) $P/R = S/Q$ (*d*) None of these
10. One of the following methods will be used for measuring a resistance value likely to be between 0.01 ohm and 0.1 ohm.
    (*a*) Wheatstone bridge
    (*b*) Kelvin double bridge
    (*c*) Loss of charge method
    (*d*) Meggar
11. All analog meters used for measuring currents, voltages and resistances are basically
    (*a*) Current meters (*b*) Voltmeters
    (*c*) Ohm-meters (*d*) Multimeters
12. An ammeter is connected in .... (*i*) .... with the component to measure .... (*ii*) .... (Series, parallel, current, voltage)

13. The scale used with a moving coil meter is
(a) a linear one
(b) a non-linear one
(c) a square law scale
(d) none of these

14. Accuracy of a meter is ± 5%. When the meter is used to measure current approximately half the meter rating, the percentage error in the reading may be
(a) ± 10% (b) ± 5%
(c) ± 2.5% (d) none of these

15. The sensitivity of Wheatstone bridge is maximum when
(a) $P/Q = R/X$ (b) $R = X, P = X$
(c) $P = Q = R = X = G$ (d) none of these

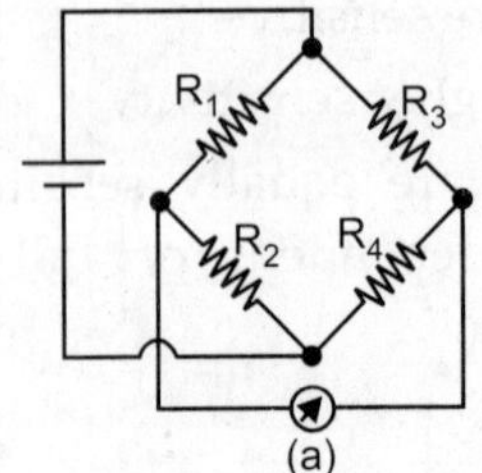

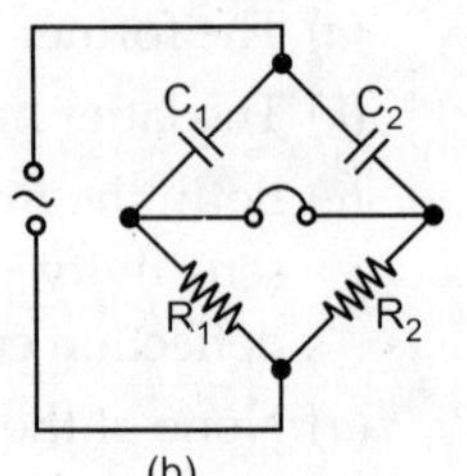

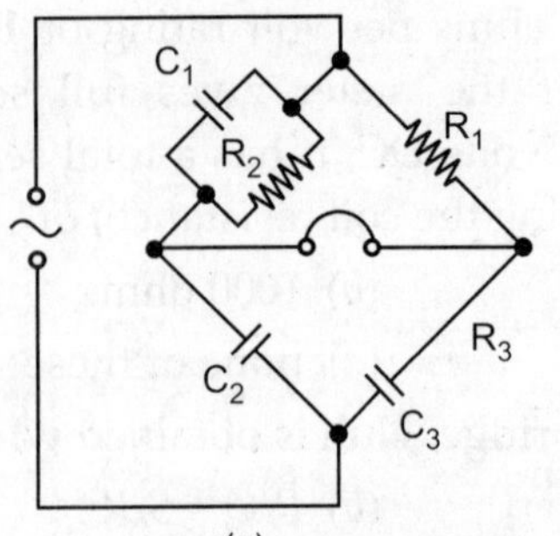

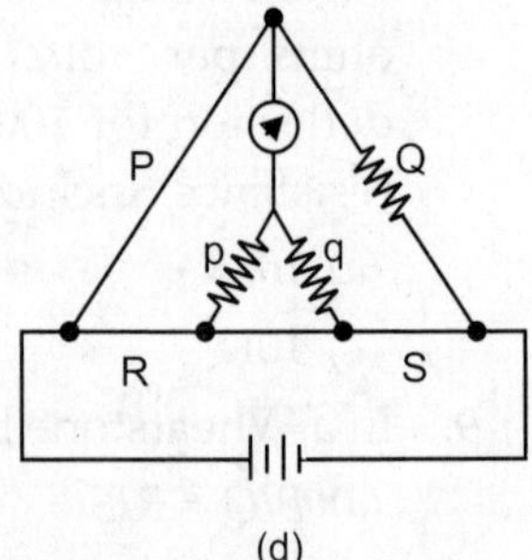

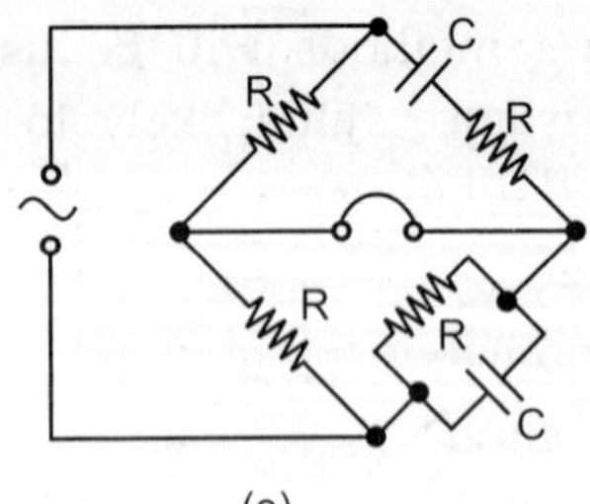

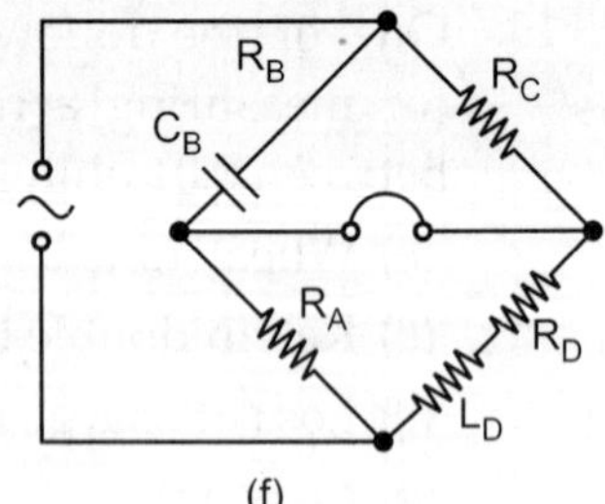

Fig. 29.90

16. Digital multimeters usually come with $3\frac{1}{2}$, $4\frac{1}{2}$, $5\frac{1}{2}$ digit displays rather than 3, 4, 5, digit displays. Addition of this 1/2 digit effectively
(a) increases accuracy of the meter
(b) increases the resolution of the meter
(c) increases the maximum voltage measuring capability of meter
(d) allows the meter to measure both positive as well as negative voltages

17. The heart of any digital multimeter is
(a) an A/D conversion building block
(b) a D/A conversion building block
(c) a voltage-to-frequency circuit
(d) a voltage-to-current and current-to-voltage converter

18. A digital multimeter with a 4000 count resolution would preserve 0.1 volt resolution upto
(a) 400 volts (b) 1000 volts
(c) 999.9 volts (d) 399.9 volts

19. The A/D converter type popular with digital voltmeter applications
(a) Successive Approximation type
(b) Dual Slope Integratimg type
(c) Counter type
(d) Simultaneous type

20. Bandwidth and Risetime specification of an oscilloscope are interrelated by
(a) Bandwidth (in MHz) = $\frac{350}{t_r\ (in\ ns)}$
(b) Bandwidth (in GHz) = $\frac{350}{t_r\ (in\ ns)}$
(c) Bandwidth (in MHz) = $\frac{0.35}{t_r\ (in\ ns)}$
(d) Bandwidth (in MHz) = $\frac{350}{t_r\ (in\ \mu s)}$

21. When the signal bandwidth equals the oscilloscope bandwidth, the amplitude measurement error could be as much as
(a) 10% (b) 30%
(c) 5% (d) 40%

B A digital storage oscilloscope has a single shot sampling rate of 400 MS/second. It uses $\left(\frac{\sin x}{x}\right)$ interpolation. It would have a single shot bandwidth of
(a) 100 MHz (b) 200 MHz
(c) 400 MHz (d) none of these

23. For a given time base setting, a higher acquisition memory means
(a) a lower effective sample rate

(b) a higher effective sample rate
(c) capability to store larger duration signals
(d) none of these

24. While using 10 X probe, as compared to a 1 X probe, the circuit under test will
(a) see a higher capacitance
(b) see a lower capacitance
(c) be severely loaded
(d) lead to higher measurement error

25. A universal counter is used to measure
(a) unknown frequency
(b) time interval (c) time period
(d) time interval average
(e) frequency ratio (f) all the above

26. In case of a conventional frequency counter architecture, the frequency measurement resolution improves with
(a) decrease in gate time
(b) increase in gate time
(c) increase in clock frequency
(d) none of these

27. The signal synthesis technique used almost invariably in synthesised function generators is
(a) Direct frequency synthesis
(b) Indirect (PLL-based) synthesis
(c) Direct digital synthesis
(d) Crystal controlled synthesis

28. The logic probe can be used to
(a) know the logic status (High, Low or High impedance) at a given point in a digital circuit
(b) analyse the flow of digital data in a digital circuit
(c) study the transient behaviour of digital data
(d) study counter states

29. One of the following instruments can be used to recover signal that is almost completely obscured by noise.
(a) Lock in amplifier
(b) Digital storage oscilloscope
(c) Logic analyser
(d) Programmable filter

30. The standard interface bus used for interconnecting programmable instruments in an automated measurement set-up is
(a) RS-232 interface
(b) IEEE-488 interface
(c) 1553 interface
(d) none of these.

II. Identify the Bridge circuits of Fig. 29.90. Choose from WHEATSTONE BRIDGE, WEIN BRIDGE, KELVIN DOUBLE BRIDGE, HAY'S BRIDGE, SCHERING BRIDGE, DE-SAUGHTY BRIDGE.

Answers to Self Evaluation Exercise

I.

1. Uniform	2. (c)	3. (d)	4. (a)	5. (c)	6. (a)
7. (b)	8. (a)	9. (a)	10. (b)	11. (a)	12. (i) series (ii) current
13. (a)	14. (a)	15. (c)	16. (b)	17. (a)	18. (d)
19. (b)	20. (a)	21. (d)	22. (a)	23. (b)	24. (b)
25. (f)	26. (b)	27. (c)	28. (a)	29. (a)	30. (b)

II. Fig. 29.90 (a) – Wheatstone bridge
Fig. 29.90 (b) – De-saughty bridge
Fig. 29.90 (c) – Schering bridge
Fig. 29.90 (d) – Kelvin double bridge
Fig. 29.90 (e) – Wein bridge
Fig. 29.90 (f) – Hay's Bridge

Chapter 30

Sensors and Transducers

Learning Objectives

After completing the chapter, you will learn the following topics:

1. Difference Between Sensor, Transducer and Actuator
2. Types of Actuators
3. Important Parameters of Transducers
4. Strain gauges including Resistive, Semiconductor and Vibrating wire Strain Gauges
5. Potentiometer Transducers
6. Capacitive Transducers
7. Inductive Transducers
8. Induction-type Transducers
9. Hall Effect Transducers
10. Linear Variable Differential Transformer (LVDT)
11. Thermocouples and Thermopiles
12. Thermocouple Instruments
13. Thermistors
14. Resistance Thermometry
15. Thermocouple Thermometry versus Resistance Thermometry
16. Synchro Transducers
17. Shaft Encoders including Absolute and Incremental Shaft Encoders
18. Hot Wire Gauges
19. PIRANI Gauge

Q. 1. What are Sensors, Transducers and Actuators? Briefly distinguish between the three. Give representative examples.

Ans. A sensor may be considered as a device that senses a physical quantity, such as temperature, pressure, fluid flow, speed, humidity etc. and converts it into an analog quantity that can be measured electrically in some form, such as voltage, current, capacitance, inductance, and ohmic resistance and so on. Thermocouples, Resistance Temperature Detectors (RTD) and Thermistors used to measure temperature and Photodiodes, Light Dependent Resistors (LDR), Photomultiplier tubes, all used to measure optical intensity are some representative examples of sensors.

A transducer on the other hand is a device that is used to transform one kind of energy to another. While sensors are always some or the other kind of transducers; the reverse is not true. Loudspeakers that converts audio signal into sound energy; Microphones that convert sound into an electrical signal;Antennas that convert electrical current into electromagnetic waves or electromagnetic waves into electrical current and Thermometers are some representative examples of transducers.

Actuators convert energy into motion. All actuators require a source of energy and a control signal. The source of energy in most cases is electrical, mechanical, hydraulic or pneumatic. It could be manual too. The control signal is a relatively much lower energy signal and could also be manual, electrical, hydraulic or pneumatic in nature. On receipt of control signals, the actuator responds by converting the energy into motion. The motion typically can be linear or rotary. A Rack and Pinion arrangement, a Screw Jack and an Electric Motor are some representative examples of actuators.

Q. 2. What are the different types of commonly used sensors? Give suitable examples.

Ans. Sensors are mainly classified based on the nature of measurand *i.e.* the quantity they measure. They may also be classified on the basis of application. Common types of sensors include the following:

1. **Acoustic and Sound Sensors,** Examples: Microphone, Hydrophone
2. **Acceleration Sensors,** Example: Accelerometers
3. **Speed Sensors,** Examples: Speedometer, Radar gun
4. **Distance or Displacement Sensors,** Examples: Laser range finder, Triangulation sensors, Linear Variable Differential Transformer (LVDT)
5. **Chemical/Bio Sensors,** Examples: PH sensor, LIDAR sensors
6. **Electric and Magnetic Sensors,** Examples: Galvanometer, Hall effect sensor
7. **Environmental Sensors,** Examples: Rain gauge, Snow gauge, Moisture sensor
8. **Optical Sensors,** Examples: Photo diode, Photo Multiplier Tube (PMT), Wave front sensor
9. **Mechanical Sensors,** Examples: Strain gauge, Potential meter
10. **Temperature Sensors,** Examples: Thermocouple, Thermopile, Thermistor
11. **Proximity Sensors,** Examples: Laser proximity sensor, Radar
12. **Inertial Sensors,** Example: Gyroscopes
13. **Image Sensors,** Example: CCD and CMOS sensors
14. **Flow Sensors,** Examples: Coriolis mass flow meters, Differential pressure flow meters, Ultrasonic flow meters

Sensors may also be broadly classified as Active sensors and Passive sensors depending upon whether they need a source of electrical energy for their operation or not. Examples of active sensors include LIDAR sensors and Photoconductive photodiodes. Radiometer is an example of a Passive sensor. Based on the nature of output, sensors are classified as Analog sensors and Digital sensors.

Q. 3. Actuators are generally classified on the basis of source of energy. Briefly describe common types of actuators illustrated with suitable examples.

Ans. As outlined in the answer to an earlier question, an actuator converts energy into motion. It takes hydraulic fluid, electric current or other sources of power and converts the energy to facilitate the motion. Most actuators produce either linear (straight line), rotary (circular) or oscillatory motion. Based on the source of energy, actuators are classified as (*a*) Electric actuators (*b*) Hydraulic actuators (*c*) Pneumatic actuators and (*d*) Mechanical actuators.

Electric actuators convert electrical energy to mechanical torque. The electrical energy is used to create motion. A typical application of electric actuators is in engines, where they are used to open and close different valves. Electrical actuators are considered to be one of the cleanest and readily available forms of actuators as they don't directly involve oil or other fossil fuels. There are many designs of electric actuators and this depends on their function in the engine that they are installed in. We have *Electric Linear Actuators* that convert electrical energy into linear displacement through mechanical transmission. These are used primarily in automation applications requiring controlled movement of a machine component, tool etc. Figure 30.1 shows the photograph of a representative electric linear actuator. We also have *Electric Rotary Actuators* that are used to convert electrical energy into rotary motion. These are primarily used in applications where a gate or valve requires controlled movement to a particular rotational position. Figure 30.2 shows the photograph of one form of a electric rotary actuator.

Fig. 30.1

Fig. 30.2

Hydraulic actuators consist of a cylinder or fluid motor that utilises hydraulic power to facilitate linear, rotary or oscillatory motion. Since liquids are nearly incompressible, they take longer to gain speed and power but are capable of exerting great force. They also allow for very precise control of the movement produced. A typical linear hydraulic actuator comprises a hollow cylinder that contains a liquid, usually oil, and a piston that is inserted in it. It operates on applying pressure either manually, such as in hydraulic car jack or through a hydraulic pump seen in construction equipment, such as cranes or excavators. The term single acting is used, when the fluid pressure is applied to just one side of the piston. The piston in this case can move in only one direction and a spring being frequently used to give the piston a return stroke. In the case of double acting hydraulic actuator, pressure is applied on each side of the piston. The direction of movement of piston is decided by nature of differential pressure between the two sides of the piston. Figure. 30.3 shows schematic of a typical linear hydraulic actuator.

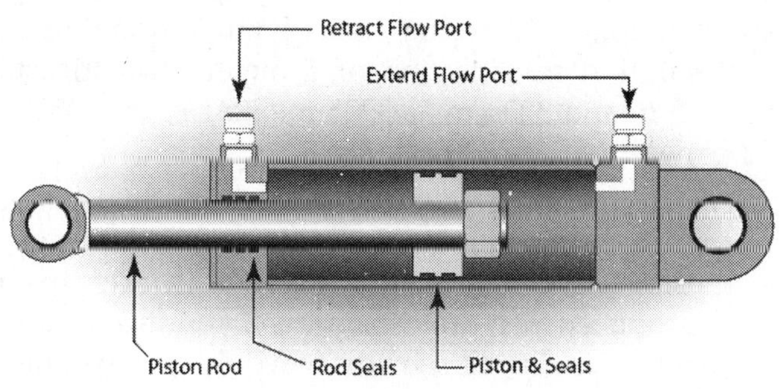

Fig. 30.3

Pneumatic actuators are similar to hydraulic actuators in concept as both use fluids for storage of energy. While hydraulic actuators use liquids; compressed gas is used instead in the case of pneumatic actuators. The energy in the compressed gas is converted into linear or rotary motion, depending on the type of actuator. Pneumatic actuators are preferred for applications, such as main engine controls because of their quick response in both starting as well as stopping. Pneumatic actuators are preferred over hydraulic actuators in places, where cleanliness is important, since the fluid in hydraulic actuators might leak and contaminate the surroundings. Pneumatic actuators, however, take up a lot of space and create a lot of noise. They are also difficult to transport once installed in a place.

Fig. 30.4

Mechanical actuators function through conversion of one kind of motion, such as rotary motion into another kind, such as linear motion. Mechanical actuators employ combination of structural components, such as gears and rails or pulleys and chains. Some of the simple mechanisms used to convert motion are screws, where the rotation of the actuator's nut causes the screw shaft to move in a straight line, the wheel and axle, where the rotating motion of a wheel causes a belt or something similar to move in a linear motion and a rack and pinion arrangement (Figure. 30.4) comprising a pair of gears to convert rotational motion into linear motion.

Q. 4. Sensors and transducers are also classified on the basis of material or basic component used and also whether they need any external power for their operation. What are these categories? Give examples. What are their typical application areas?

Ans. Transducers are broadly classified as passive and active transducers, depending upon whether they need or don't need any external power for their operation. Passive transducers need external power for their operation. Resistive, inductive, capacitive, voltage and current transducers are some representative examples of passive transducers. Common examples of resistive transducers include Resistive Strain Gauge (used for measurement of force, torque, displacement), Potentiometer transducers (used for measurement of pressure, displacement), Pirani Gauge (used for measurement of gas flow, gas pressure), Light Dependent Resistors (used for measurement of light intensity), and Thermistors (used for measurement of temperature). Capacitance microphone (used for speech, music, noise), Dielectric Gauge (used for measurement of thickness, liquid level) and Variable Capacitance Pressure Gauge (used for measurement of pressure, displacement) are some common examples of capacitive transducers. Differential Transformers (used for measurement of pressure, force), Magnetic Circuit Transducers (used for measurement of pressure, displacement), Eddy Current Gauge (used for measurement of thickness, displacement) and Reluctance Pick-up (used for measurement of vibrations, pressure, displacement) are some representative examples of inductive transducers. In the category of voltage and current transducers, we have Hall-effect Transducers (used for measurement of magnetic flux, current, power), Photo Emissive Cells and Photo Multiplier Tubes (used for measurement of light and radiation).

Active transducers need no external power for their operation. These are also known as self-generating transducers. In this category of transducers, auxiliary power supply constitutes major part of output signal power. The contribution of input signal is insignificant. Some representative examples of active transducers include Thermocouples and Thermopiles (used for measurement of temperature, radiation, and heat flow), Piezoelectric Pick-up (used for measurement of vibrations, acceleration, pressure changes, sound level) and Photo Voltaic Cells (used for measurement of light intensity).

Q. 5. What are static characteristics of a measurement device? What are the main static characteristic parameters? Briefly describe each one of them.

Ans. Static characteristics of a measuring instrument are used to measure the static or unvarying process conditions. Major static characteristic parameters include the following. The values of these parameters, in addition to the type of sensor or transducer used, will also depend upon associated circuitry:

1. Range and span
2. Accuracy
3. Precision
4. Resolution
5. Sensitivity
6. Repeatability and Reproducibility
7. Hysteresis
8. Dead Zone
9. Dead Time
10. Backlash
11. Bias
12. Tolerance
13. Drift
14. Zero Drift
15. Uncertainty

Range specifies the minimum and maximum values of the intended parameter, the measuring instrument is designed for. The temperature measurement range of a certain thermocouple instrument may be specified to be 0°C to 200°C. The distance measuring capability of a laser range finder may be specified to be 300 m to 20 km. Span is the difference between the maximum and minimum values of the quantity to be measured. In the above examples, the span would be 200 °C and 19.7 km respectively.

Accuracy is indicative of how closely the measured value matches the true value of the quantity being measured. Accuracy may be specified as percentage of full scale reading or percentage of true value or even as percentage of span. When the accuracy is specified as percentage of true value; it varies with change in true value.

Precision refers to repeatability and consistency of measurements, when the measurements are carried out under identical conditions at short intervals of time. Precision measurements may not necessarily be accurate. As an illustration, if the true value of temperature of a certain hot body were 90°C; and the measured values in a short span of time were 80°C, 81°C, 79°C, 80°C and 81°C; these measurements are precise but not accurate.

Resolution is specified as the minimum change in the amplitude of the measured parameter that can cause an increment in the output. In other words, it is the minimum resolvable change in the value of the parameter being measured. As an illustration, a laser range finder sensor with resolution specification of 5 meters will measure both 3233 m and 3234 m as 3235 m. 3231 m and 3332 m will be measured as 3230 m.

Sensitivity is the ratio of the infinitesimal change in output signal to infinitesimal change in input signal. It is the slope of the calibration curve. It is indicative of output produced per unit of input. If the input-output relationship were linear; sensitivity would be constant for all values of input. Also, if the measuring instrument had number of elements in series; then the overall sensitivity would be equal to product of individual sensitivities.

Repeatability indicates precision. As mentioned earlier, it is indicative of closeness of agreement among, multiple consecutive measurements of a given value of input with measurements made under identical operating conditions. **Reproducibility** is nothing but repeatability over a period of time. Good reproducibility indicates that calibration once done, would not drift over a long period of time.

Hysteresis is the difference in the output for a given input, when the input is approached from opposite directions. As a result of this, the output – input curves don't coincide for ascending and descending values of input. In the case of mechanical systems, it would imply that loading and unloading curves would not coincide.

Dead zone is the largest range of values of measure and which the measuring instrument does respond to. **Dead time** is the time period that elapses before the output begins to change after a change input has occurred.

Backlash is defined with respect to mechanical systems. It is the lost motion or clearance of mechanical elements, such as gears, linkages etc. It can be defined as the maximum distance or angle through which any part of a mechanical system may be moved in one direction without applying appreciable force or motion to the next part in mechanical sequence. It is the mechanical form of deadband.

Bias is the constant error that is present over the full range of measurement of a measuring instrument. **Tolerance** is the maximum allowable error in the measurement.

Drift is defined as the change in the output for a given input over a period of time. **Zero drift** is the changes occurring in the output for a zero input and is usually expressed as a percentage of full scale output.

Uncertainty is a non-negative parameter characterising the dispersion of the values attributed to a measured quantity. A measurement result is complete only, when the associated uncertainty is also specified.

Q. 6. What are dynamic characteristics of a measurement device? What are the main dynamic characteristic parameters? Briefly describe each one of them.

Ans. Dynamic characteristics represent dynamic response of the measuring instrument. The dynamic response tells the behaviour of the instrument under time varying input and output conditions. Dynamic characteristics are further categorised as periodic steady state and transient. In the case of former, the magnitude of output has a definite repeating cycle with time. In the case of latter, the magnitude of output does not repeat with time. The dynamic behaviour of a measuring system is characterised by following characteristic parameters:

1. Over shoot
2. Time constant
3. Rise time
4. Response time
5. Settling time
6. Types of input

Over shoot is the peak value of the output beyond the steady state value. **Time constant** is an indication of how fast the output responds to change in input and is inversely proportional to the response time. **Response time** is the time period taken by the measuring instrument to produce a specified percentage of the expected steady state value. **Rise time** is the time taken by the measuring instrument for the output to change from 10 per cent to 90 per cent of its steady state value. **Settling time** is the time period that elapses before the output settles to within specified percentage of its steady state value. The **input parameters** used to determine the behaviour of the measuring instrument include step, ramp, impulse and sinusoidal input.

Q. 7. What is a Resistance Strain Gauge? How does it look like? Describe in brief the principle of operation of using a resistance strain gauge for measurement of displacement.

Ans. It is the most commonly used transducer for measurement of displacement. A resistance strain gauge consists of a grid of fine wire cemented to a thin paper sheet or a very thin bakelite sheet and covered with a protective covering. To carry out the displacement measurement, the strain gauge paper sheet is bonded to the structure under study with the help of an adhesive. Fig. 30.5 (*a*) and (*b*) show two types of resistance strain gauges. The use of strain gauge is based on the following principle :

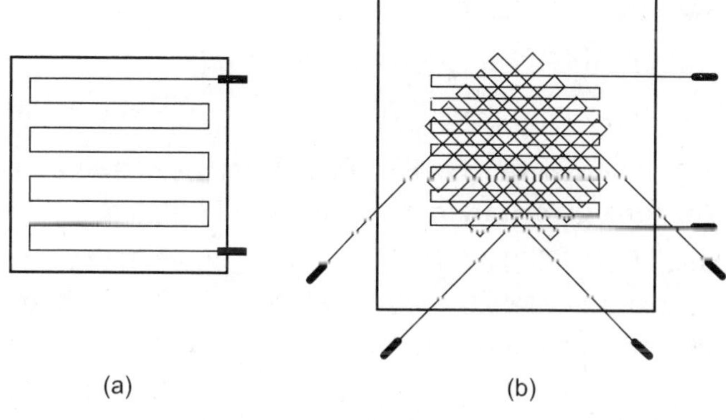

Fig. 30.5

When the structure under study is strained, the resistance of the wire changes owing to the change in the length and the diameter of the wire grid of the gauge (Resistance is proportional to LENGTH/AREA). This resistance change, which is a measure of the structure displacement or strain can be measured by any of the standard electrical methods, Wheatstone bridge being the most commonly used one. The strain that we have to consider is both due to the change in length as well as the diameter of the wire as the total change in resistance, is obviously due to change in length as well as change in area of cross section of the strain gauge grid wire.

Q. 8. From the basic principles, derive an expression to show the relationship of total change in resistance of the resistance strain gauge with the longitudinal strain (*e*), Poisson's Ratio (μ) and the unstrained resistance (*R*).

Ans. Change in resistance due to change in length is given by :

$$\Delta R = \frac{\text{R(Change in length)}}{\text{Original length}} = \text{R.e.}$$

where, e = Longitudinal Strain

Change in resistance due to change in area is given by:

$$\frac{\text{R(Change in Area)}}{\text{Original Area}}$$

$$\text{Change in area} = \frac{\pi}{4}d^2 - \frac{\pi(d - \mu ed)^2}{4}$$

$$= \frac{\pi}{4}d^2(2\mu e - \mu^2 e^2)$$

where, μ = Poisson's Ratio

d = Unstrained wire diameter

e = Longitudinal strain

Therefore, change in resistance due to change in area is given by:

$R(2\mu e - \mu^2 e^2) = R \,.\, e\,(2\mu - \mu^2 \,.\, e)$

Total resistance change $= \text{R.e}\,(2\mu - \mu^2.\, e + 1) \cong R\,.\,\text{e}\,(1 + 2\mu)$.

Q. 9. Define GAUGE FACTOR. What does the Gauge Factor for a given strain gauge material indicate?

Ans. GAUGE FACTOR is a very important strain gauge parameter and is defined as the ratio of unit change in resistance to unit change in length. It is given by $(1 + 2\mu)$.

Quite obviously, gauge factor indicates the sensitivity of the strain gauge. Larger the gauge factor, better the sensitivity. It is also referred to as the SENSITIVITY FACTOR.

Table 30.1 lists gauge factors for different commonly used wire materials for making resistance strain gauges. In practice, gauge factors are observed to vary from (–12) for some Nickel alloys to (+5) for Irridium-Platinum gauges.

Table 30.1

Material	*Gauge Factor*	*Material*	*Gauge Factor*
Nickel	– 12.10	Platinum	4.8
Manganin	0.47	Carbon	20.0
Nichrome	2.0	Soft Iron	4.2
Constantan	2.1	Doped Crystals	100 to 5000

Q. 10. Briefly mention typical applications of a resistance strain gauge for measurement of non-electrical quantities?

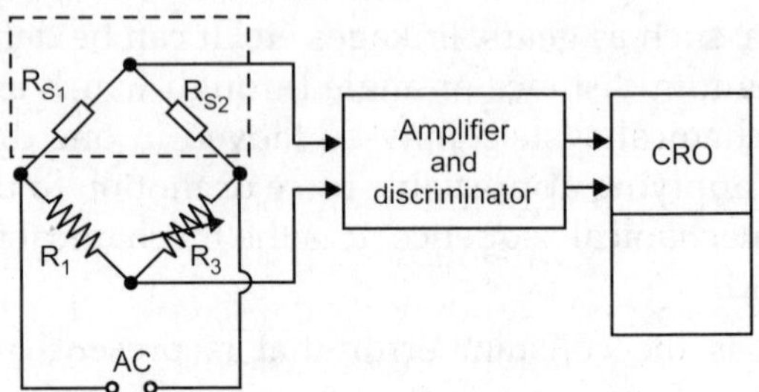

Fig. 30.6

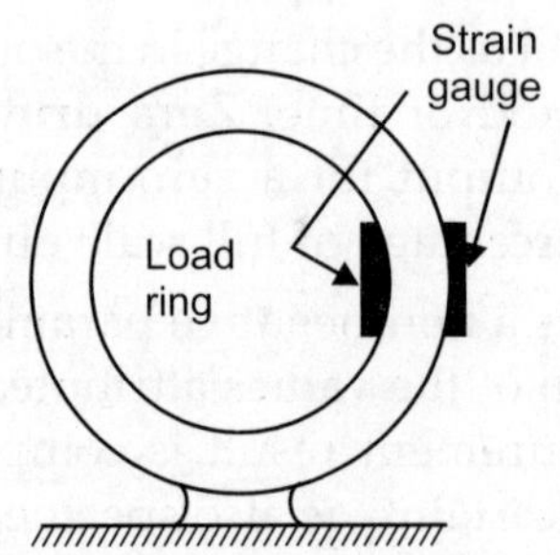

Fig. 30.7

Ans. 1. Strain gauges are extensively used for analysing dynamic strains in complex structures, such as Stress and strain in bridges, automobiles, roads etc. The strain gauge is used in a bridge connection and the output from the bridge can be recorded with an oscilloscope calibrated to read instantaneous strain or stress (Fig. 30.6).

2. Resistance wire strain gauges are very useful for measurement of Tension, Torque, Force and Stresses in structures. They are particularly useful, where the presence of the transducer should cause minimum of disturbance as in wind tunnel measurements.
3. Strain gauges are used in load cells and load rings to measure force by the strain produced. Fig. 30.7 shows a load ring which is usually made of steel. The strain produced in the load ring is measured with a strain gauge. The strain is a measure of the force applied to the load ring. It may be mentioned here that strain

gauges are used for measurement of very small displacements of the order of 0.025 mm or so.

Q. 11. What are potentiometer transducers? What are these used for? Briefly describe the principle of operation of a potentiometer transducer.

Ans. A potentiometer transducer makes use of a potentiometer, which is used to transform a linear or rotary movement to a varying resistance. The changing resistance value is then measured as a changing output voltage for fixed excitation input voltage. We have linear potentiometers and rotary potentiometers. Figures 30.8 and 30.9 respectively show constructional features of linear and rotary potentiometers. In the case of linear potentiometer, the resistance between the output terminals varies as a function of linear movement of slider arm. In the case of rotary potentiometer, the resistance between wiper terminal (W) and one of the fixed terminals (A or B) varies as a function of angular movement of the wiper.

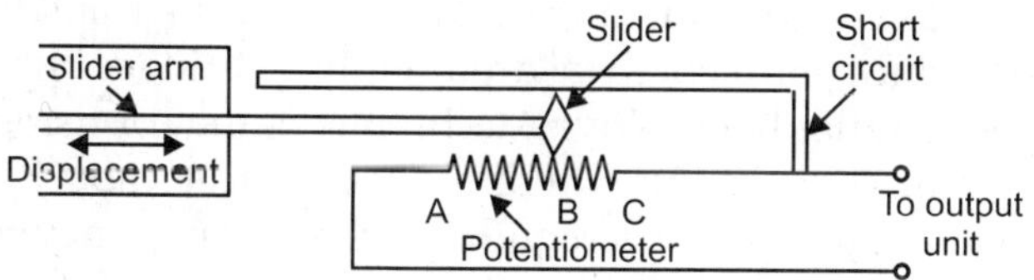

Fig. 30.8

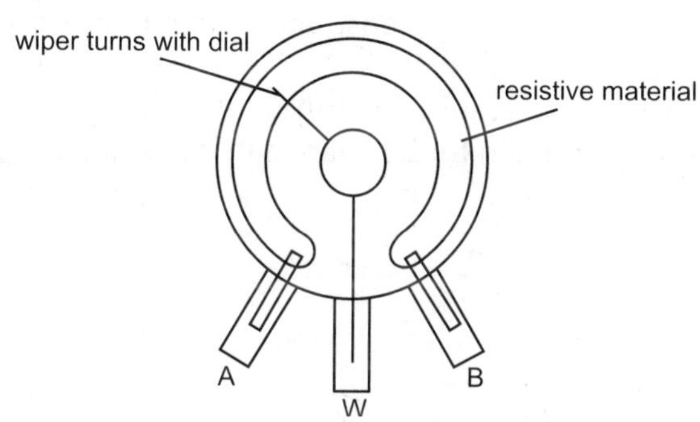

Fig. 30.9

Figure 30.10 explains the principle of operation of the potentiometer transducer. If (R_1) and R_2) are the resistance values between the variable terminal and the two fixed terminals; assuming that the load resistance is infinite, the output voltage for a fixed input voltage of (V_{IN}) would be given by:

$$V_O = V_{IN} \times [R_2/(R_1 + R_2)]$$

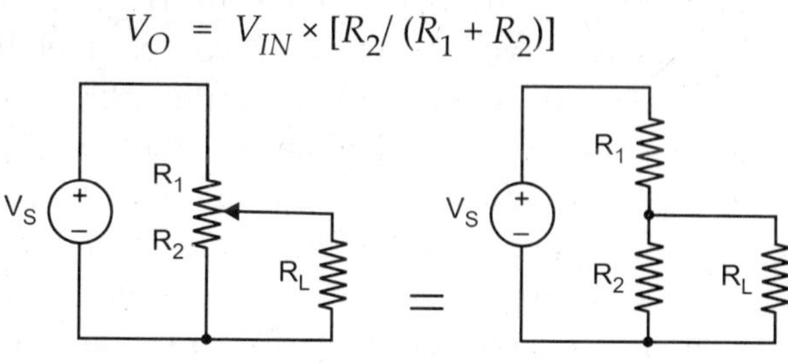

Fig. 30.10

In case, (R_L) is finite, then

$V_O'=V_{IN} \times [R_2'/(R_1 + R_2')]$ where $R_2' = R_2 \times R_L/[(R_2 + R_L)]$

With the help of simple mathematics, we can prove that the error produced due to finite value of load resistance is given by:

$V_O - V_O' = V_{IN} \times [(R_1 + R_2)/R_L] \times (x^2 - x^3)$, where $x = R_2/(R_1 + R_2)$

Q. 12. Explain in brief the principle of operation of PARALLEL PLATE CAPACITOR TYPE Capacitive transducer. Which non-electrical quantities can be effectively measured using these transducers?

Ans. This type of capacitive transducers could be pretty useful for measurement of linear displacement and pressure. Linear displacement or pressure is made to vary the capacitance of the transducer, which in turn can be measured with some appropriate electrical means.

Fig. 30.11 shows one such transducer. The transducer is basically a parallel plate capacitor one of whose plates is a cantilever type. A linear displacement causes plate separation to change, thus changing the capacitance (Capacitance of a parallel plate capacitor is given by $k \cdot A/d$, where 'k' is the dielectric strength.

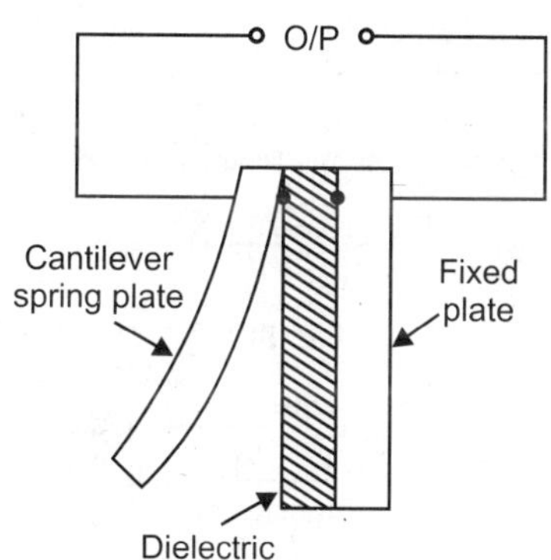

Fig. 30.11

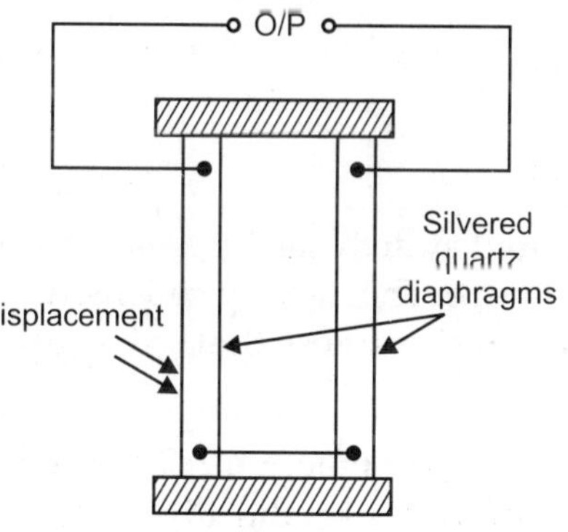

Fig. 30.12

'A' is the plate cross-section and 'd' is plate separation. This arrangement is fundamentally non-linear as the capacitance does not vary linearly with plate separation. However, linearity can be closely approximated for smaller displacements or by having a medium of high dielectric strength.

Another transducer based on similar principle is shown in Fig. 30.12. It consists of two silvered quartz diaphragms separated by an insulation. The displacement of these diaphragms varies the capacitance.

The capacitance variation is typically 0.4 pF for a displacement of 1 micron. This arrangement can also be used for measurement of pressure.

Q. 13. Explain in brief the functional principle of a capacitive transducer that can be used for measurement of rotational displacement.

Ans. A capacitive transducer that can be used to measure rotational displacement is shown in Fig. 30.13. The capacitance of the capacitor formed by rotor and stator is a function of the position of the rotor.

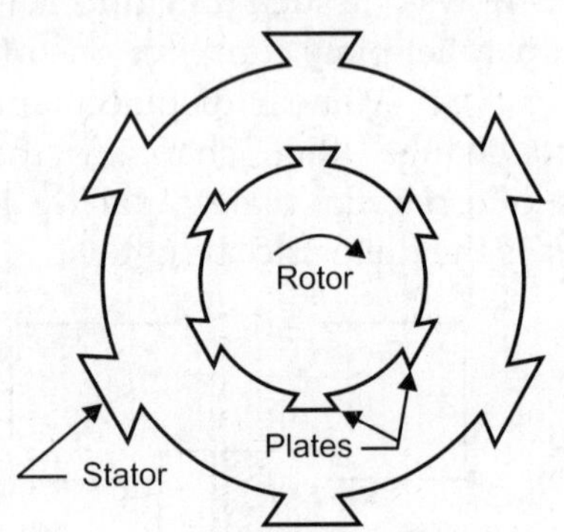

Fig. 30.13

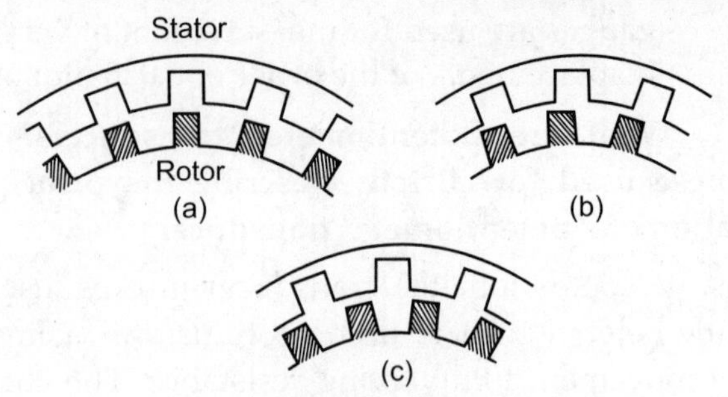

Fig. 30.14

Different rotor positions are depicted in Figs. 30.14 (*a*), (*b*) and (*c*). In the rotor position depicted in Fig. 30.14 (*a*), the capacitance is the least. It is intermediate in case of the position shown in Fig. 30.14 *b* and maximum in case of rotor position depicted in Fig. 30.14 (*c*). The change in capacitance is, thus, a measure of rotational displacement. The capacitance varies linearly over a limited range. The non-linearity outside this range is caused by fringing effects. This type of transducer is also quite useful for measurement of torque. The change in transducer capacitance can be measured with a bridge but the more commonly employed technique is that of using an oscillator (Fig. 30.15), where the transducer capacitance forms a part of the frequency determining network. Oscillators '*A*' and '*B*' operate at a frequency difference of 465 kHz. Under this condition, output of DC amplifier is zero. Variations in transducer capacitance vary the frequency of oscillator '*A*'. The mixer output varies in frequency and as result, the output of DC amplifier varies in direct proportion to the change in capacitance.

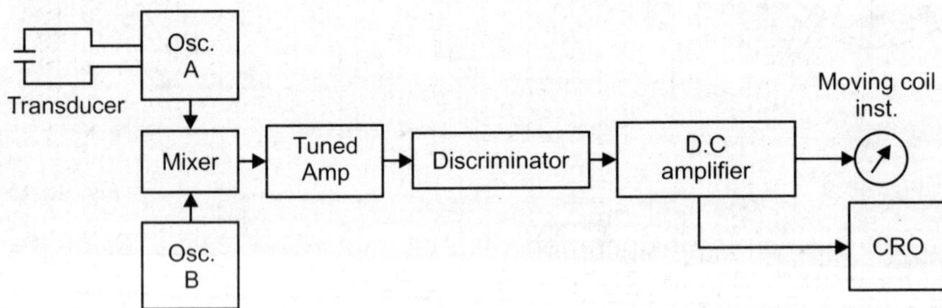

Fig. 30.15

The readout system can be a moving coil instrument or an oscilloscope. The moving coil instrument is used for static studies, while an oscilloscope is used for dynamic studies.

Q. 14. What is a push-pull type of displacement sensor? Briefly describe the principle of operation of such a sensor.

Ans. Push-pull type of displacement sensor is a variable distance capacitive transducer. The transducer comprises three parallel plates with two fixed outer plates and free or movable center plate (Figure. 30.16). The transducer has, therefore, two parallel plate capacitors; one between center plate and upper plate and the other between the center plate and the lower plate. The movement of center plate as a result of displacement to be measured increases the capacitance value of one capacitor and decreases the capacitance value of the other. If the plate moves downwards; capacitance values are given by:

$C_1 = \varepsilon_0\varepsilon_r A/(d+x)$ and $C_2 = \varepsilon_0\varepsilon_r A/(d-x)$

In the case of upward movement of center plate,

$C_1 = \varepsilon_0\varepsilon_r A/(d-x)$ and $C_2 = \varepsilon_0\varepsilon_r A/(d+x)$

Here, (A) is the area of overlap between the two plates, (d) is inter-plate distance, (x) is displacement of center plate, (ε_0) is free space permittivity and (ε_r) is relative permittivity of dielectric material between plates.

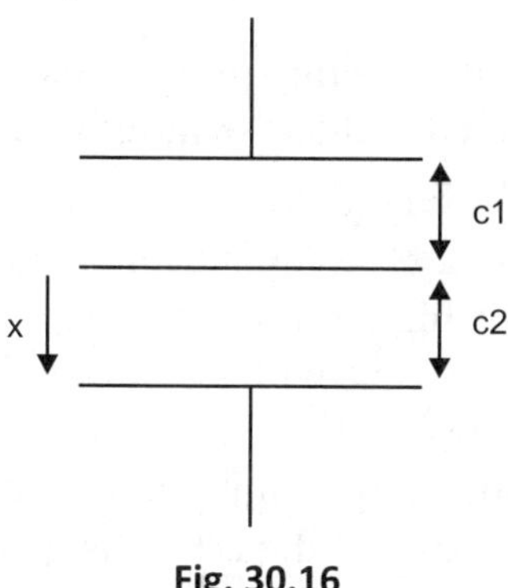

Fig. 30.16

Q. 15. Describe in brief the basic construction and operational principle of different types of INDUCTIVE TRANSDUCERS used for measuring linear displacement.

Ans. There are two basic types of variable inductance transducers. These are (*i*) the ones in which variation in inductance as a function of displacement is made use of for measuring the displacement and (*ii*) the ones in which displacement is measured in terms of induced e.m.f. in a coil. Fig. 30.17 (*a*) shows transducer of the first class. The transducer comprises of a coil wound on a ferromagnetic yoke with a ferromagnetic armature separated from the yoke by an air gap. The displacement under measurement is applied to the armature. The applied displacement decreases the air gap between the ferromagnetic armature and the yoke.

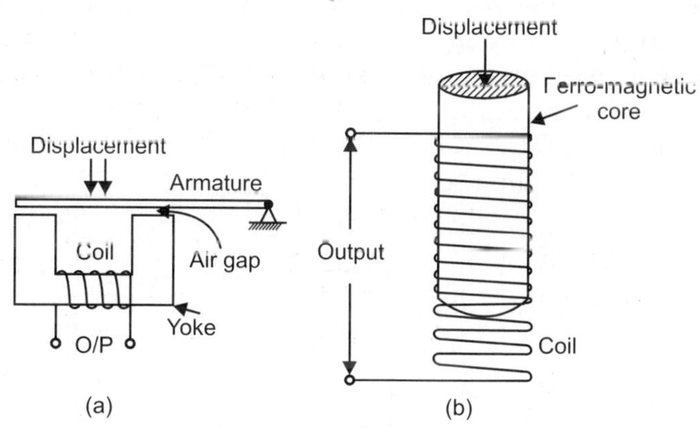

Fig. 30.17

The decrease in air gap increases the inductance. The inductance is a function of air gap length and hence, the displacement. Fig. 30.17 (*b*) shows another arrangement based on the same principle. The displacement changes the position of the core with respect to the coil resulting in a change of inductance of the coil.

The DIFFERENTIAL TRANSFORMER falls into the second category of inductive transducers; Fig. 30.18 (*a*) shows a simple form of differential transformer. It consists of an input or a primary winding wound on the central part of the tube. The output or secondary winding is divided into two equal parts wound on each end of the tube. The two parts of the secondary winding are connected in opposition. A ferromagnetic core is placed in the center of the tube. When the ferro-magnetic core is similarly placed with respect to the two parts of the secondary winding, the flux linkages of both the parts are equal. This results in equal induced e.m.fs in the two parts and since they are connected in opposition, the net induced e.m.f. is zero. If a displacement causes the core to move up, more flux links the upper part of the secondary winding than it does with the lower part. The induced e.m.fs are, thus unequal and there is a net voltage across the output terminals. This net voltage is a function of the displacement and the phase of this net voltage is the same as that of the induced e.m.f. in the upper part. Similarly, when the displacement causes the core to move down, there is again an induced e.m.f., but this time the phase of the induced e.m.f. is 180° out of phase with the phase of the induced e.m.f. in the upper part in the first case. In other words, magnitude and phase of the induced e.m.f. represent the magnitude and direction of the displacement.

Fig. 30.18 (*b*) shows another form of a differential transformer. The primary winding is put on the central limb while two parts of the secondary winding are wound on the outer limbs.

These two parts are similar and are wound in opposition. The iron armature is on the top of the core. The armature can revolve around an axis at the center of the central limb.

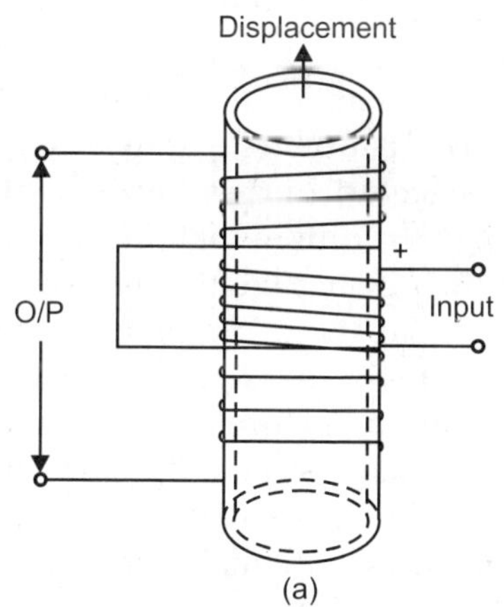

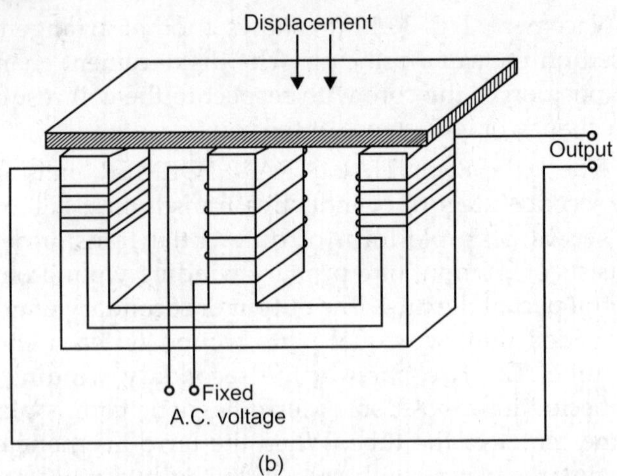

Fig. 30.18

When no displacement is applied to the armature, the magnetic circuits of the two parts of the secondary windings are similar. The flux linked with the two are equal and the net voltage is zero. When a displacement is applied say on the left side, the length of the air gap decreases on this side and increases on the right hand side. Thus, the flux linking with the left hand part of secondary winding is more than that linking with the right hand part. An output voltage that is a function of the displacement thus appears across the output.

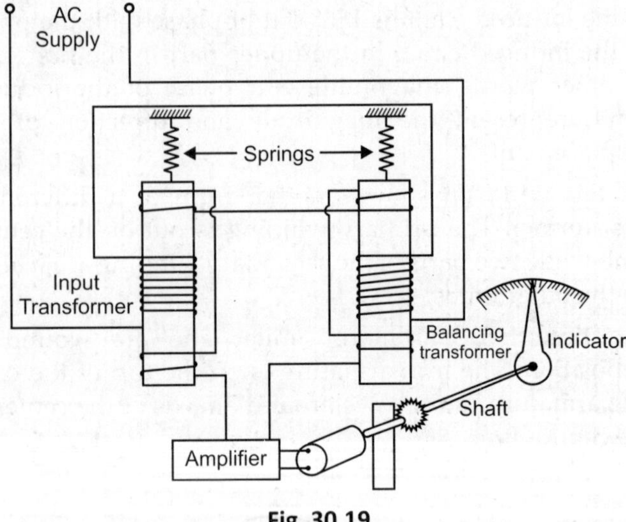

Fig. 30.19

Differential transformers are often used in servo motor operated null systems. In these systems (Fig. 30.19), the output voltage of the differential transformer is amplified and supplied to one phase of the two phase servo motor. The rotation of the servo motor is determined by the position of the core of the input differential transformer called a balancing transformer. The output voltage of this transformer opposes the output voltage of the input transformer and the motor will turn until the outputs of the two transformers are equal. The displacement can be directly read from the indicator on the motor shaft.

Fig. 30.20

Q. 16. What is a semiconductor strain gauge? How does it compare with the resistance strain gauge?

Ans. Semiconductor strain gauge employs a semiconductor material, commonly silicon. As the resistivity of semiconductor is higher, it is possible to have a sensitive element that is shorter, wider and simpler. The resistivity of semiconductors is very sensitive to strain and as a result, semiconductors strain gauges have gauge factors many times greater than those for the gauges made out of simple metals. But they tend to have higher temperature sensitivity and are less linear. They can be made integral with the structural components and are used in this mode for measurement of pressure.

Q. 17. What is the basic principle of operation of vibrating-wire strain gauge? What are its major applications and strong points?

Ans. In a vibrating-wire strain gauge, the wire is put under tension and its natural frequency of vibration in its first mode is

$$f = 1/2\,l\,\sqrt{T/m}$$

where, (l) is the length, (T) is tension and (m) is mass per unit length.

The fixed points of the wire are bonded to the material whose strain changes are to be measured. Due to any strain changes or more significantly the changes in the tension (T), strain can then be determined by monitoring the natural frequency. The most common circuit arrangement is to excite the wire with a single current pulse and then measuring the frequency of the damped vibrations. Fig. 30.21 shows the constructional features of a vibrating-wire strain gauge.

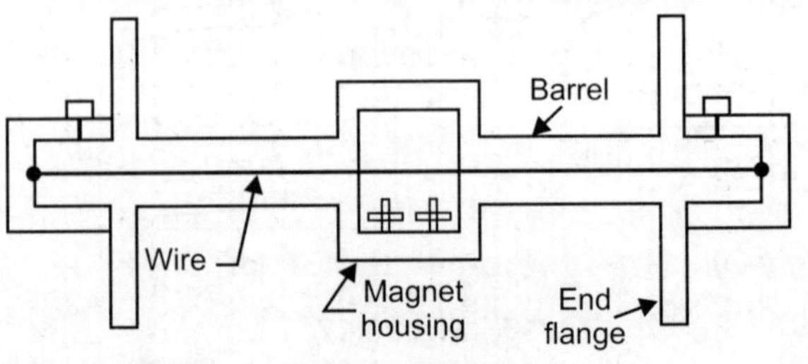

Fig. 30.21

These are comparatively larger than the bonded resistance strain gauges. Also, because of large force needed to stretch it, the gauge must be properly mounted. These gauges find wide application in the measurement of internal strains in concrete, where both the above mentioned factors have been taken care of. It has been observed that vibrating-wire strain gauges can be very stable.

Q. 18. All force and displacement capacitive transducers depend for their functioning on the variation in the capacitance value of the transducer due to displacement or force or pressure under measurement. Now this capacitance variation may take place in any of the three ways (*i*) Variation in the conductor separation (*ii*) Variation in conductor's area (*iii*) varying the dielectric strength of the dielectric separating the two conductors. Briefly explain the principle of operation of capacitive transducers of each of the above three types.

Ans. The capacitive transducer that relies on change in distance between the two capacitor conductors is basically a two plate capacitor. The capacitance in this case is a function of distance and is mathematically expressed by $C = f(d)$. The response characteristic is hyperbolic and is linear over a very small range of displacement. The sensitivity of such a transducer as a function of distance is given by:

$$S = \frac{\Delta C}{\Delta d} = -\frac{\varepsilon a}{d^2}$$

Quite often, it is the fractional change in the capacitance value and not the capacitance variation that is measured and in such cases, the sensitivity factor is given by :

$$S' = \frac{\Delta C/C}{\Delta d} = -\frac{1}{d}$$

The response characteristic in this case is linear.

Another type of change-of-distance capacitive transducer is the DIFFERENTIAL CAPACITOR shown in Fig. 30.22. Such a system has three plates. Two fixed and one movable plate form two capacitors in series. When the movable plate is in the centre, the two capacitors have equal values. When the movable plate is moved by a small amount in response to an applied mechanical displacement, one of the constituent capacitances decreases, while the other increases. This difference in the two capacitors' values or their ratio can then be used to generate a voltage or current that indicates the applied displacement.

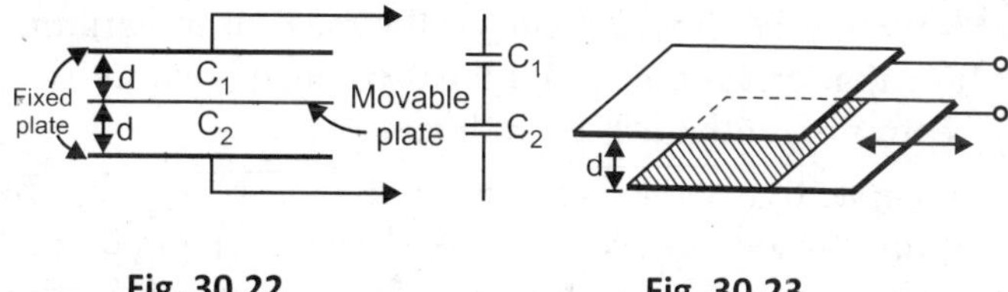

Fig. 30.22 Fig. 30.23

Fig. 30.23 shows a type of CHANGE-OF-AREA capacitive transducer for longitudinal movements. Fig. 30.24 shows a similar transducer for rotational movements. In the transducer of Fig. 30.23 displacement of one of the plates in the direction of the arrow causes a variation of the overlapping area of the two plates to effect a change in the value of capacitance. The capacitance in the transducer arrangements of Fig. 30.23 and 30.24 changes linearly with the displacement but the characteristic can be modified by appropriately shaping one of the two plates. The sensitivities, $\Delta C/\Delta L$ in case of the transducer of Fig. 30.23 and ($\Delta C/\Delta \theta$) in case of the transducer of Fig. 30.24 can be increased by having multiplate arrangement.

Such a transducer is also called a push–pull displacement sensor. It has been described in an earlier question.

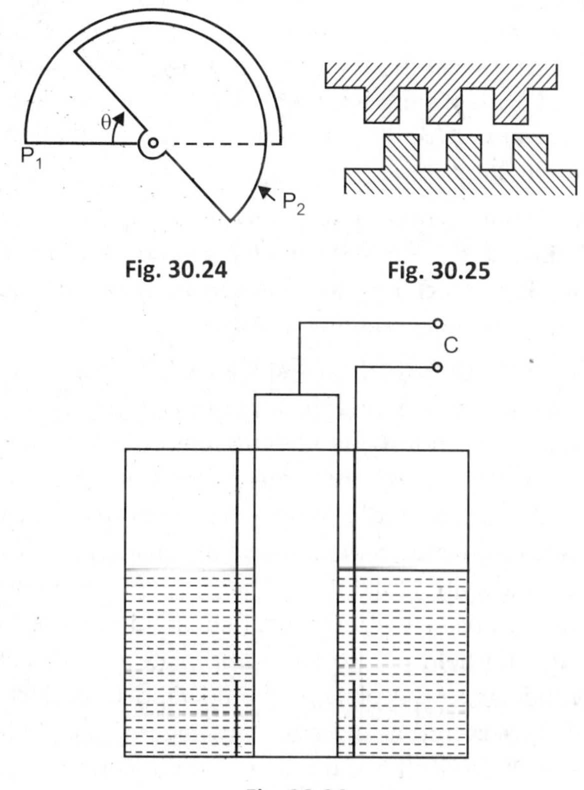

Fig. 30.24 Fig. 30.25

Fig. 30.26

The serrated capacitive transducer (Fig. 30.25) is another change-of-area transducer used for measurement of linear displacement. Such a transducer has a minimum of capacitance, when a tooth in one plate is opposite a hole in the other. Shifting the relative position of the plates by one tooth width causes the maximum variation in the capacitance value. 1 pF/0.0001 inch sensitivity is very common with such transducers.

A capacitive transducer type that is based on the variation of dielectric constant is shown in Fig. 30.26. A cylindrical capacitor is immersed in a non-conducting liquid. The variation in the liquid level causes a change in the value of capacitance. Such transducers can be used for chemical analysis of binary mixtures of non-conducting fluids, where both components differ in their dielectric constant. The dielectric constant of the mixture is measured and it indicates the relative concentration of both components.

In cases, where the dielectric constant is observed to vary with pressure, the capacitive transducers can be used to measure pressure. For instance, in case of Benzene, the dielectric constant varies by as much as 0.5 per cent as the pressure changes from 1 atm. to 100 atm. Dielectric constants of some of the crystals are also found to vary with pressure. These include *MgO. LiF, NaCl, KCl, KBr* etc. Variation of dielectric strength with humidity in some of the dielectric materials has been used to advantage.

The change is pretty small but is measurable. The dielectric strength of dry air at 45°C is 1.000247, whereas the dielectric strength of air at the same temperature but saturated with water is 1.00593. This is one of the most accurate and effective methods for determination of moisture in gases.

Q. 19. What is the basic principle of operation of a HALL EFFECT TRANSDUCER? Explain in brief as to how Hall Effect Transducers can be made use of in measuring displacement or position?

Ans. A HALL EFFECT TRANSDUCER is based on the effects of magnetic fields on moving charges. Refer to Fig. 30.27. A current (*I*) is passing lengthwise as shown through a plate made from a semiconducting material like Indium antimonide, *N*-type Germanium or Indium arsenide. Two conductive contacts are made at *X-X* and these contacts are so adjusted that the voltage between them in the absence of any magnetic field is zero. Now if a magnetic field (*B*) traverses the plate in a direction perpendicular to the plate, it would tend to deflect the charge carriers and as a result, a voltage (E_H) called the HALL VOLTAGE appears across two contacts. This voltage is given by:

$$E_H = K_H \cdot \frac{I.B}{t}$$

Where, (*t*) is the plate thickness and (K_H) is the Hall coefficient whose value depends upon the semiconductor material used.

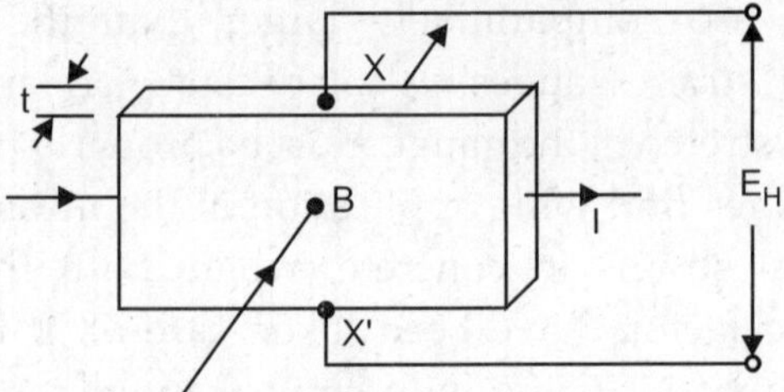

Fig. 30.27

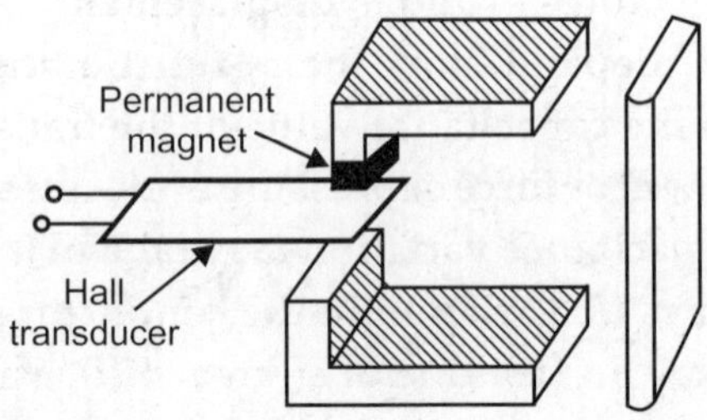

Fig. 30.28

Fig. 30.28 shows one method of using Hall effect transducers for measurement of location of displacement of a structural element. The transducer in this case consists of a ferromagnetic structure, having a permanent magnet. A Hall effect transducer is located within a gap. The output voltage in this Hall effect transducer depends upon the strength of the magnetic field in the gap. Any displacement in the ferromagnetic structure as shown, changes the magnetic field and hence, the output voltage. This method can be used to measure displacements down to 0.001 inch.

Q. 20. What are the other possible applications of Hall effect transducers ?

Ans. 1. It can be used as a magnetic to electric transducer, where it can be used to measure the strength of a magnetic field in terms of an electrical voltage. The system has the advantage that it requires very little space in the field direction and as a result, can be inserted in narrow gaps. It has the disadvantages that it is sensitive to temperature variations and that each plate needs individual calibration as Hall co-fficient could vary from plate to plate.

2. It can also be used to measure current without the need for cutting the conductor and without making any electrical connections between the conductor circuit and the meter. In its basic operation, (Fig. 30.29), the Hall effect transducer is used to measure the magnetic field caused by the current (*DC* or *AC*) under measurement. Such a system can be used to measure currents right from milliamperes to thousands of amperes.

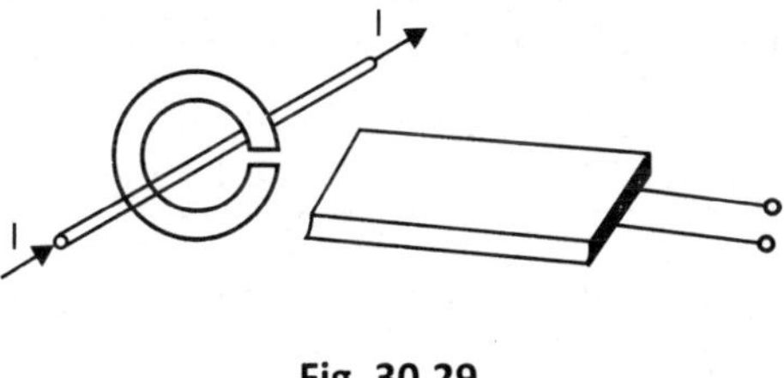

Fig. 30.29

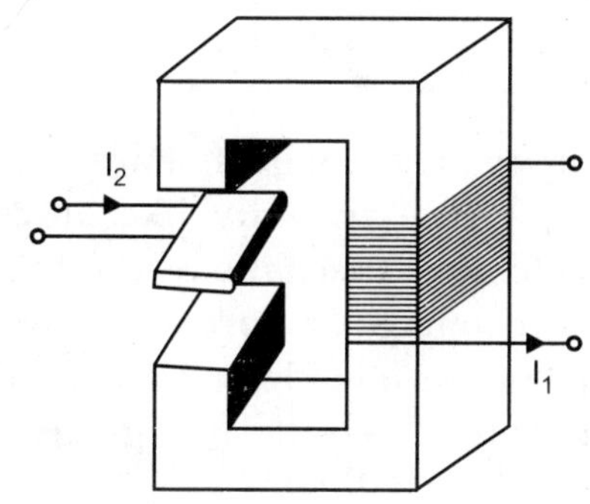

Fig. 30.30

3. Fig. 30.30 shows the application of a Hall effect transducer as an analog multiplier. If a current (I_1) is passing through the coil and a current (I_2) is passing through the plate of the Hall effect transducer, then the Hall effect voltage is given by:

$$E_H = (\text{constant}) \, . \, K_H \, . \, I_1 \, I_2 \, / t$$

The output voltage is, thus, proportional to the product of two currents I_1 and I_2. If I_1 and I_2 are both positive or negative, output voltage is positive, if I_1 is positive and I_2 is negative or vice-versa, the output voltage is negative. Thus, the multiplier operates in all the four quadrants. It can be used for both, DC as well as AC operations.

Q. 21. What is the basic principle of operation of INDUCTION TYPE transducers? What are the different types of induction type transducers?

Ans. Induction type transducers are all based on the famous Faraday's laws of electromagnetic induction, which talk about an e.m.f. being induced in a conductor placed in a magnetic field, whenever there is a rate of change of flux linking the conductor. It may be mentioned that there can be rate of change of flux both, when the conductor is stationary and the field is changing or when the field is constant and the conductor is moving.

Figs. 30.31 (*a*) to (*d*) show several induction type transducers. All of them produce an output proportional to the velocity of the moving member. They can also be used to measure the displacement by integrating the output or in such cases, where there is a known relationship between displacement and velocity. All diagrams are self-explanatory. Fig. 30.31 (*a*) shows a transducer that can be used to measure small movements as in ballistocardiography which is the measurement of recoil movement human body caused by opposite movement of body and food stream. Fig. 30.31 (*b*) shows a ribbon microphone. The metallic ribbon moves due to impinging sound waves. The output voltage is proportional to the velocity of the ribbon. Fig. 30.31 (*c*) shows a phonograph cartridge useful for making velocity or displacement measurements in the laboratory. Fig. 30.31 (*d*) depicts a magnetic tachometer. In this case, the output is proportional to the rotational velocity of the moving member.

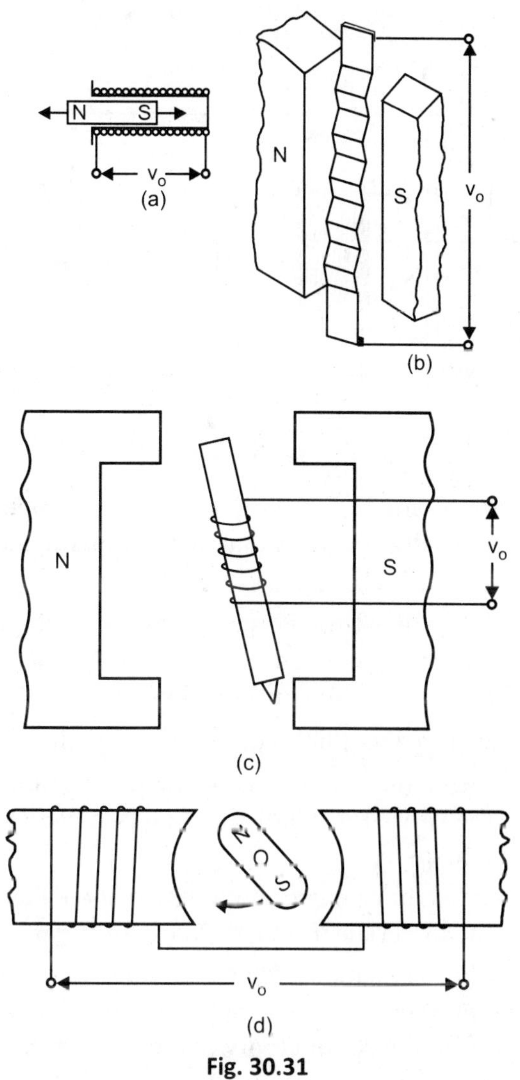

Fig. 30.31

Q. 22. Explain in brief as to how can one use an induction type transducer for proximity measurement or as a metal locator ?

Ans. Refer to Fig. 30.32. A high frequency alternating current passes through the coil '*L*'. It produces an alternating magnetic field surrounding the coil. If a metal piece is brought into this field in the vicinity of the coil, eddy currents are induced in the metal piece. These eddy currents create their own magnetic field that opposes the very field that was responsible for producing them. Closer the metal piece to the coil, more is the opposition. The opposing magnetic field reduces the effective inductance of the coil which causes a change of the magnitude and phase of the alternating current. This information can be used in many ways. It can be used as a proximity transducer or as metal locator. It can also be used as a thickness gauge for a metallic or non-metallic coating on a base metal provided, the coating has an electrical conductivity that is substantially different from that of base metal.

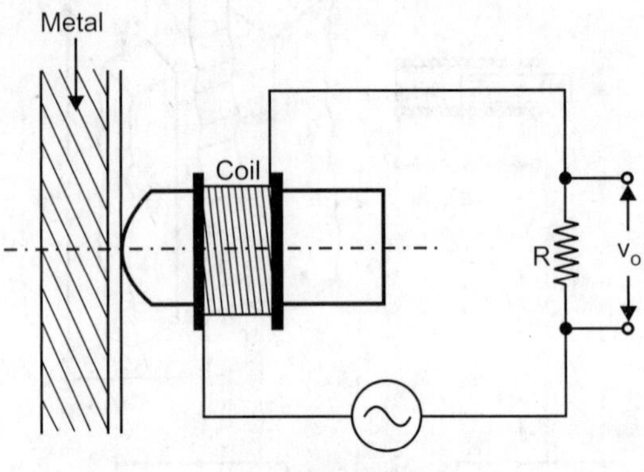

Fig. 30.32

Q. 23. What are the typical physical characteristics that give an edge to the capacitance transducers over resistive and inductive transducers?

Ans. Capacitive transducers offer a combination of desirable physical characteristics which are usually not available with other transducers. These are :

1. Its force requirements are very small.
2. These have a very good dynamic response, as the capacitor plates themselves have very little mass.
3. These have a very high dimensional stability, if the materials used are Quartz or Invar.
4. The capacitance whose change indicates the parameter to be measured, does not depend upon the conductivity of metal electrodes or its variation with temperature.
5. The capacitance transducers can be shielded against the effect of external electric stray fields. It is not affected by magnetic stray fields.
6. In applications like measurement of displacement or vibration, there is no need to connect any mechanical parts to the moving member and all that is required is to bring the capacitor plate in the vicinity of the moving member.
7. The temperature co-efficient of the dielectric constant of commonly used dielectric materials in capacitive transducers is several orders of magnitude smaller than the temperature co-efficient of materials used in resistive transducers or that of magnetic permeability of ferrous materials used in the construction of inductive transducers.

Q. 24. Whenever a capacitive transducer is used to measure a physical quantity like Force, Pressure, Displacement etc., the capacitance variations have got to be converted into voltage or current variations by some kind of a signal processing block, named MODIFIER in Fig. 30.33. Suggest some circuit arrangement that can be effectively used as a Modifier.

Ans. Fig. 30.34 shows a typical circuit arrangement (a Twin-T type network) that converts capacitance variations into voltage or current variations. The high frequency alternating signal source generates a symmetrical wave

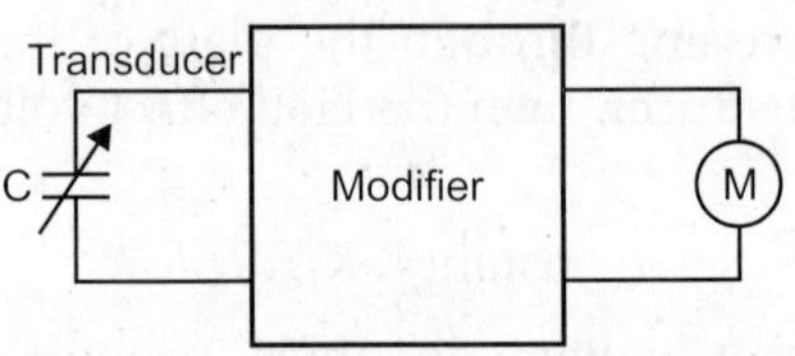

Fig. 30.33

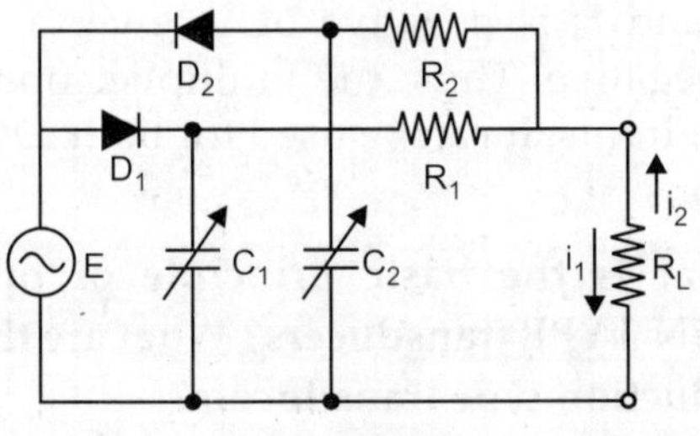

Fig. 30.34

(usually a sine wave) of amplitude 'E', Diode 'D_1' conducts during the positive half cycle enabling capacitor 'C_1' (one of the component capacitors in the differential capacitor) to change. During the negative half cycle, capacitor 'C_1' discharges through 'R_1' and 'R_L' (R_L could be the meter or recorder load) and through 'R_2' and 'D_2' (D_2 is forward biased during the negative half cycle). Also, during the negative half cycle. C_2 gets charged through R_2 and D_2. During the subsequent positive half cycle, C_2 discharges through R_2 and R_L and through R_1 and D_1. Thus, C_1 and C_2 charge during positive and negative half cycles respectively and discharge during negative and positive half cycles respectively.

If $C_1 = C_2$ and $R_1 = R_2$ and also assuming that diodes D_1 and D_2 have identical characteristics, currents i_1 (when C_1 discharges) and i_2 (when C_2 discharges) are equal and average current through R_L is zero. A meter connected across R_L or in place of R_L shows zero deflection. Any variation of C_1 or C_2 causes a deflection in the meter.

Q. 25. Resistance wire strain gauges, both bonded as well as unbonded, are frequently used as the basic transducer for measurement of any linear displacements or strains. The resistive change that is representative of any mechanical strain is converted into an equivalent voltage or current for measurement or recording purposes with the help of an appropriate circuit arrangement. Give a suitable circuit for this purpose and explain in brief its functioning.

Ans. A WHEATSTONE BRIDGE is the most frequently used circuit arrangement for converting the resistance changes of a strain gauge into an equivalent voltage signal. Refer to Fig. 30.35 The strain gauge resistance (R_x) forms the resistance of one of the arms of the bridge. R_s is a known standard resistance.

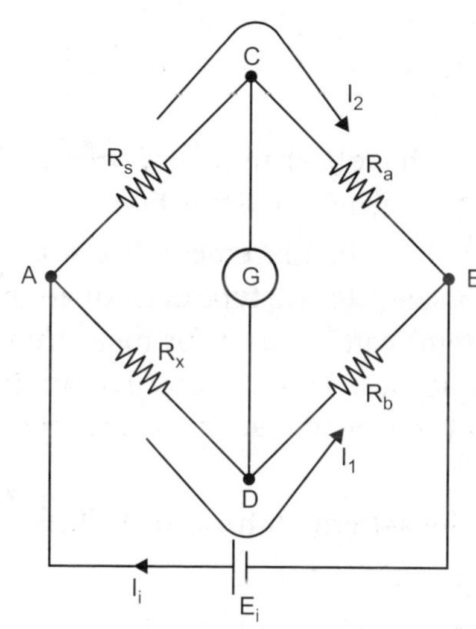

Fig. 30.35

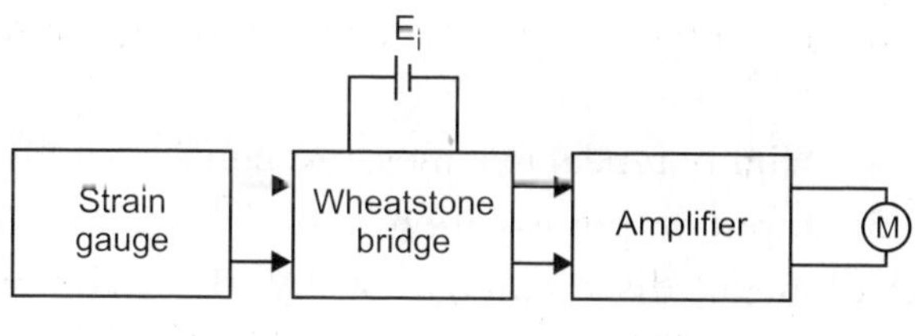

Fig. 30.36

Initially in the absence of any strain, the bridge is balanced by choosing appropriate value for (R_a/R_b) and R_s. In the balanced condition, $R_s . R_b = R_a . R_x$ or $R_x = (R_s . R_b/R_a)$ and the galvanometer shows zero deflection.

Any resistance change ΔR_x in R_x due to strain unbalances the bridge and there appears a voltage across the terminals C and D. This voltage indicates change in resistance.

The resistance change can also be calculated by restoring the balance by either changing R_s or ratio R_b/R_a. In that case,

$$\Delta R_x = \Delta R_s . R_b/R_a \text{ or } \Delta R_x = R_s . \Delta (R_b/R_a)$$

The complete instrumentation required for measuring strains using a strain gauge may look like the arrangement of Fig. 30.36.

Q. 26. The sensitivity of such an instrument depends upon the sensitivity of the basic transducer (indicated by the gauge factor), bridge sensitivity and the amplifier gain. What are the optimum conditions for getting optimum bridge sensitivity?

Ans. The voltage and current sensitivities are respectively expressed by : $\delta E_g/\delta R_x$ and $\delta I_g/\delta R_x$)

The sensitivities depend upon a number of factors including 'E_i' and also various resistances and also whether we have a constant 'E_i' or a constant 'I_i' or a constant current through R_x. The optimum conditions for maximum sensitivity can be derived by taking partial differentials of bridge equations with respect to 'R_x'. The derivations are quite cumbersome. These are summarised in Table 30.2.

Table 30.2

Bridge Output	*Bridge Input*	*Optimum Conditions*
Maximum Current Sensitivity	E_i = constant I_i = constant I_1 = constant	$R_a << R_b$ $R_s >> R_b$ $R_a << R_b << R_s$
Maximum Voltage Sensitivity	E_i = constant I_i = constant I_1 = constant	$R_b = R_s$ $R_a >> R_b$ & $R_s >> R_b$ $R_s >> R_b$

Q. 27. Explain the basic principle of operation of a Linear Variable Differential Transformer (LVDT). Draw the typical transfer curve for a LVDT and define SENSITIVITY.

Ans. Fig. 30.37 shows the schematic diagram of a LINEAR VOLTAGE DIFFERENTIAL TRANSFORMER (LVDT). It comprises of a primary winding (P) and two secondary windings (S_1) and (S_2) and a magnetic core (C). A steady AC driving signal is applied to the primary winding. The frequency of the applied AC signal is typically between 60 Hz and 20 kHz and its amplitude is typically several volts. The induced e.m.fs in the two secondaries are nearly equal, when the core is symmetrically located with respect to the two windings with the result that the output is zero. The displacement can be monitored by measuring the difference between the voltages induced in the two secondaries.

Fig. 30.38 shows the typical transfer curve. The output changes its phase by 180°, as the core is moved through zero position. The non-zero output, even when the core is at zero position is due to incomplete electric and magnetic balance. It is typically 1% of the output voltage (max.) in the linear range. It can be reduced by balancing the individual output circuits using resistive and capacitive shunts and by adequate grounding.

Sensitivity in case of an LDVT is output volts/ displacement/input volts. For a displacement of 0.001 inch/V applied to the primary coil, the output is of the order of 0.5 mV to 3 mV. Under standard operating conditions (6.3 V, 50 Hz input), sensitivity is between 3 mV and 20 mV for a 0.001 inch displacement. The sensitivity and the output voltage can be increased by increasing the primary voltage but this can lead to heating of the primary coil and other associated drawbacks. Fig. 30.39 shows the typical construction of an LVDT.

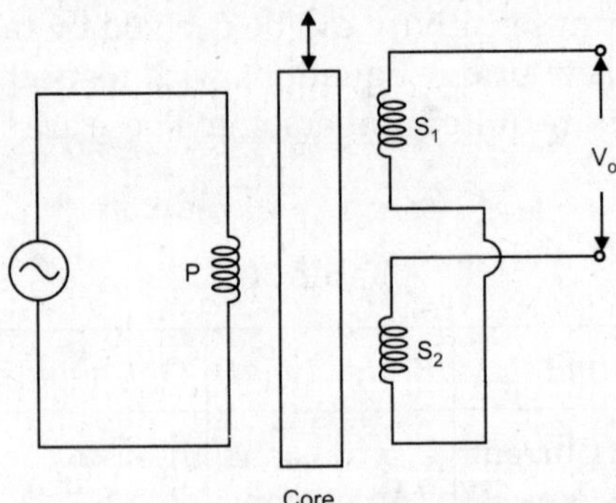

Fig. 30.37

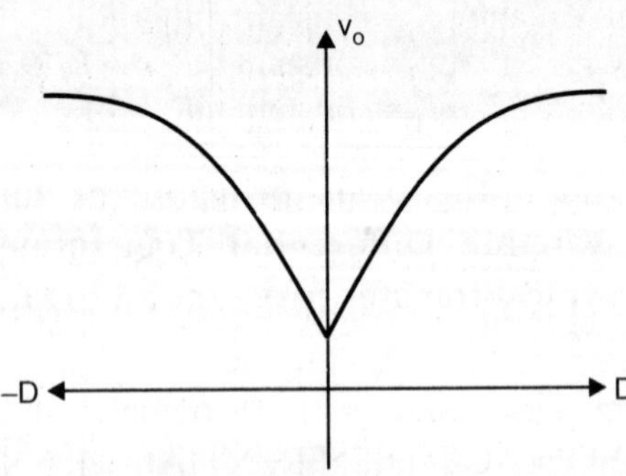

Fig. 30.38

Q. 28. Briefly mention major applications of LVDTs?

Ans. Although LVDT is a very important and a highly exploited transducer for measurement of mechanical displacement, it is being used for measurement of other physical parameters also due to easy availability of reliable means by which these parameters can be easily translated into displacement. The ruggedity, the reliability and .their capability to operate in an hostile environment of extreme pressure, temperature and shock have made these devices highly popular in industry. Their applications include:

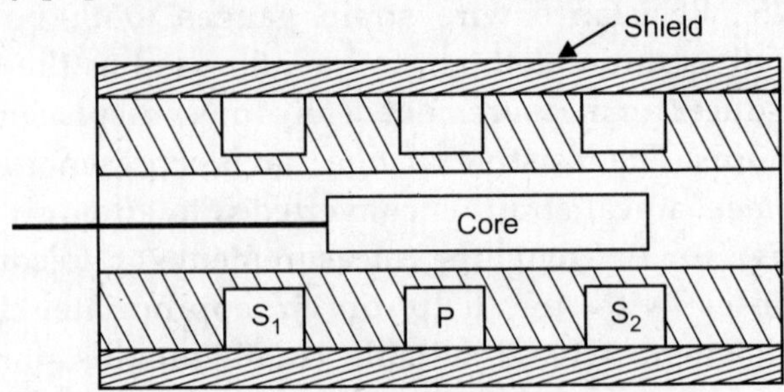

Fig. 30.39

1. These are being used for either direct displacement measurement or as a secondary sensor of a parameter which has been translated to displacement. Examples include on-time thickness measurement, gauging machine profiles, surface roughness etc.
2. LVDTs have also been used for force measurement.
3. LVDTs have been used for pressure measurement by measuring the deflection of a diaphragm.
4. LDVTs as displacement transducers in Force-Balance system (a type of feedback measurement system) can be used for measurement of force, torque, acceleration etc. Discussion of Force-Balance transducers is beyond the scope of this book.

Q. 29. List the salient features of LVDTs.

Ans. These are:

1. Frictionless measurement
2. Long mechanical life as a result of frictionless measurement
3. Infinite resolution (dependent on measurement S/N ratio only)
4. Highly repeatable measurements
5. Input output isolation

One major disadvantage of LVDT is the hazard of induced noise in an electromagnetically noisy environment. However, this can be overcome with proper shielding.

Q. 30. What is a Thermocouple? How does it function?

Ans. A THERMOCOUPLE is formed, when two dissimilar metallic conductors are joined to form junctions.

Whenever there is a temperature difference between the two junctions, an e.m.f. is induced across the thermocouple. The magnitude of induced e.m.f depends upon the temperature difference and types of metals used for making the thermocouple. The operation of a thermocouple is based on the SEEBECK EFFECT, which states that if a circuit is formed consisting of two dissimilar metallic conductors, a current flows in the circuit, if one of the junctions is at a higher temperature than the other.

Q. 31. We know that the operation of thermocouple is based on SEEBECK EFFECT. But how do you find the direction of flow of thermoelectric current or the polarity of thermoelectric e.m.f. for a given thermocouple?

Ans. As per the Seebeck effect, if a closed circuit is formed of two metals and the metal junctions, formed thereby are at different temperatures, an electric current flows in the circuit. The magnitude of this thermoelectric current as a result of thermoelectric e.m.f. is proportional to the temperature difference between the two junctions. For instance, if the two metals are copper and iron, the current flows from copper to iron at the hot junction and from iron to copper at the cold junction (Fig. 30.40).

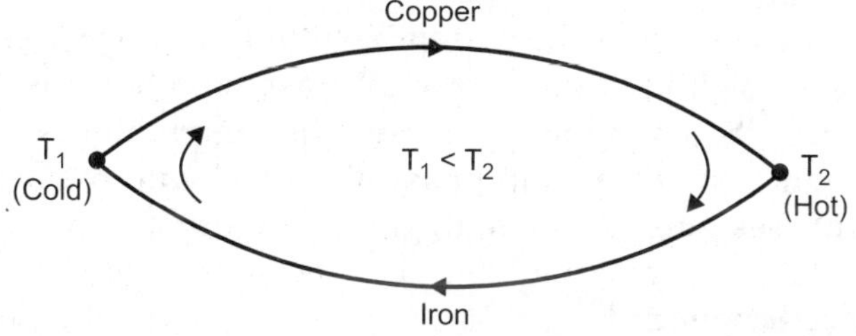

Fig. 30.40

There are a number of pairs of metals which exhibit this phenomenon. These metals were arranged in a series by Seebeck so that in a circuit made up of any two of the metals, the direction of the current at the hot junction is from the one mentioned earlier in the series to the one mentioned later in the series.

The metals were arranged like :

Bi-Ni-Co-Pd-Pt-U-Cu-Mn-Ti-Hg-Pb-Sn-Cr-Mo-Rh-Ir-Au-Zn-W-Cd-Fe-As-Sb-Te

For example: If the thermocouple was made up of ZINC and IRON, the thermoelectric current will flow from Zinc to Iron at the hot junction and from Iron to Zinc at the cold junction.

Q. 32. What is a Thermopile?

Ans. When a number of thermocouples are connected in series, the overall transducer becomes capable of measuring small temperature rises. Such a transducer is called a THERMOPILE. In a thermopile, all hot junctions are on the object, whose temperature is to be measured and all the cold junctions are kept at a constant known temperature. Typical applications of thermopiles are in infrared radiation measurement.

Q. 33. What do you mean by COLD JUNCTION COMPENSATION in case of thermocouples?

Ans. It is usually impractical to maintain the cold junction at 0°C as is the requirement. In case the cold junction is at the ambient, the measured e.m.f. is equivalent to having cold junction at 0°C. This means that an e.m.f. must be added to the measured value to get the required signal.

or $$E_{0-t} = E_{a-t} + E_{0-a}$$

This is known as *Cold Junction Compensation* and the voltage E_{0-a} as the cold junction compensation voltage.

E_{a-t} = Measured thermoelectric e.m.f., when the cold junction is at ambient

E_{0-t} = The thermoelectric e.m.f., when the cold junction is at 0°C

E_{0-a} = Cold junction compensation voltage

Q. 34. Write down the expression for the thermally induced e.m.f. in a thermocouple as a function of the junction temperatures.

Ans. The e.m.f. induced in a thermocouple circuit is given by :

$$e = a(T_2 - T_1) + b(T_2 - T_1)^2$$

where, e = induced e.m.f.

T_1 = Temperature of reference junction

T_2 = Temperature of hot junction

a, b = constants

Thus, the e.m.f. induced in a thermocouple is a parabolic function of temperature difference of the two junctions. The approximate values for the constants a, b are :

a = 40 to 50 microvolts per degree centigrade of temperature difference

b = few tenths or hundredths of a microvolt per degree centigrade square

Quite obviously, 'b' can be neglected with respect to 'a'.

Therefore, $e = a(T_2 - T_1)$

It is clear from this expression that a 100°C of temperature differential will produce a thermal e.m.f. of 5 mV.

Since the induced e.m.f. depends upon the temperature difference of the hot junction and the reference junction, the temperature of the latter should remain absolutely constant in order that the induced e.m.f. gives a correct idea of the temperature being measured.

Table 30.3. Commonly used thermocouples along with their temperature ranges

Thermocouple	*Temperature Range*
Copper-Constantan	– 200°C to 550°C
Iron-Constantan	– 20°C to 750°C
Chromel-Alumel + 10% Rhodium	300°C to 1000°C
Platinum-Platinum + 13% Rhodium	700°C to 1600°C

Q. 35. In order to make an absolute measurement of temperature using thermocouples, it is essential to keep one of the junctions at a constant known temperature, usually 0°C. The other junction known as the hot junction is then made use of to measure the unknown temperature. What are the various techniques by which the thermocouples can be applied to do the job?

Ans. Following application techniques ar available for this purpose :

(*a*) SURFACE CONTACT : In this, the hot junction of the thermocouple is brought in physical contact with the body whose temperature is to be measured.

(*b*) IMMERSION : In this, the hot junction is inserted or immersed in the body (Gas or Liquid) whose temperature is to be measured.

(*c*) SUCTION PYROMETER : In this technique, the hot gas being extracted is made to flow continuously over the hot junction of the thermocouple as is the case in a suction pyrometer.

(*d*) RADIATION PYROMETER : An optical system focuses the radiation into a definite solid angle so that it falls on the sensitive element which responds to the temperature rise caused by the radiation. In such a case, the sensitive element is mounted in an evacuated bulb.

Q. 36. What are the essential components of a Thermocouple instrument that can be used for measurement of an unknown current or voltage? How does such an instrument function?

Ans. Fig. 30.41 shows the essential components of a thermocouple instrument. These are:

(*i*) A HEATER ELEMENT that carries the current to be measured (in case of voltage measurement, it carries a current proportional to the voltage to be measured). It is important that the resistance of the heater wire remains constant, even when the wire gets heated due to the current flowing through the wire.

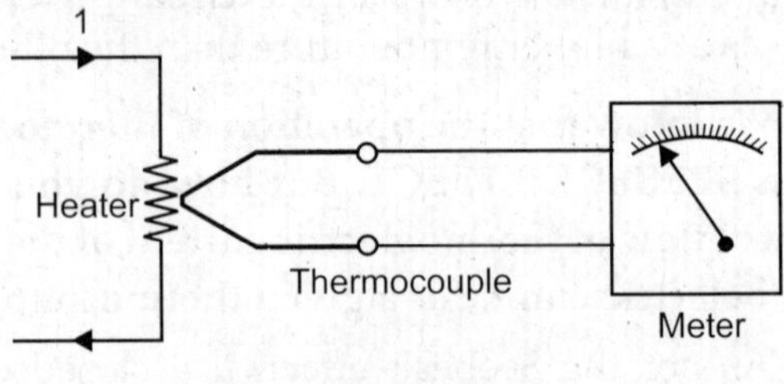

Fig. 30.41

(*ii*) A THERMOCOUPLE having its hot junction in contact with the heater and its cold junction near or at room temperature.

(*iii*) A sensitive PERMANENT MAGNET MOVING COIL METER that operates on the thermal e.m.f. generated in the thermocouple.

The instrument operates as follows :

The current or the current resulting from the voltage to be measured flows through the wire and thus, heats it causing its temperature to rise. The temperature of the hot junction which is in physical contact with the heater wire rises and as a result, there is thermal e.m.f. The needle of the moving coil meter deflects in response to this thermal e.m.f.

Q. 37. Derive an expression to show that a thermocouple ammeter or a voltmeter has a square law response i.e. the meter deflection is proportional to the square of the current or voltage under measurement ?

Ans. For a current (I) flowing through the heater wire, the heat produced is $I^2 R$, where (R) is the resistance of the wire. The temperature difference between the hot and cold junctions can then be expressed as :

$\Delta T = K_1 . I^2 . R$ where K_1 is a constant.

This temperature difference produces a thermal e.m.f. (E) which is proportional to the amount of temperature difference in a fashion already discussed. That is :

$$E = a(\Delta T) + b(\Delta T)^2$$

This e.m.f is fed to a permanent magnet moving coil meter, whose deflection is proportional to the e.m.f.

That is deflection. $\theta = K_2 . E$, where K_2 is another constant. Substituting the value of (E).

$$\theta = K_2 . [a . (\Delta T) + b (\Delta T)^2.$$

Substituting for (T), $\theta = K_2 \cdot (a \cdot K_1 \cdot I^2 \cdot R + b \cdot K_1^2 \cdot K_2 \cdot R^2 \cdot I^4)$

$$= a \cdot K_1 \cdot K_2 \cdot R \cdot I^2 + b \cdot K_1^2\, K_2\, R^2 \cdot I^4$$

Constant (b) is very small as compared to (a) as discussed earlier

Therefore, $\theta = a \cdot K_1 \cdot K_2 \cdot R \cdot I^2$

$$= K_3 \cdot I^2$$

where, $K_3 = a \cdot K_1 \cdot K_2$ is another constant.

In case of voltage measurement, (I) in the expression should be replaced by (V/R).

Therefore, for voltage measurement,

$$\theta = a \cdot K_1 \cdot K_2 \cdot R \cdot (V/R)^2 = a \cdot K_1 \cdot K_2 \cdot V^2/R$$

$$= K_4 \cdot V^2$$

where, $K_4 = a \cdot K_1 \cdot K_2/R$ is another constant.

Thus, we get a square law response. In fact, we don't get an exact square law response, as the resistance R of the heater element does not remain constant but varies with temperature rise as the element will have a positive or negative coefficient of resistance. Also, in reality, the thermal e.m.f. is not proportional to the first power of the temperature difference and indeed it is a parabolic function of the temperature rise. These effects are usually very small so that if the air gap field of the permanent magnet moving coil meter is uniform, the scale of the instrument may be laid-off approximately with the square of the heater current.

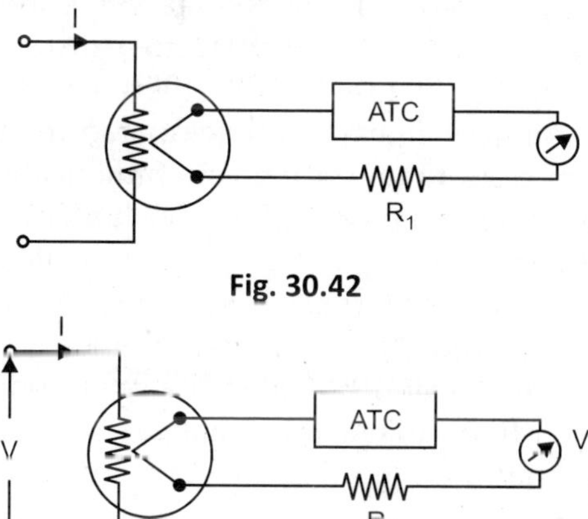

Fig. 30.42

Fig. 30.43

This basic thermocouple instrument can be used as an ammeter (Fig. 30.42), where the calibration of the instrument is adjusted with R_1. In some instruments, an automatic temperature compensation circuit (*ATC*) is used to correct for changes in ambient temperature. The instrument can be converted into a voltmeter by connecting a resistance in the heater circuit (Fig. 30.43).

Q. 38. What are various advantages and disadvantages of thermocouple instruments used for measuring currents and voltages?

Ans. The thermocouple instrument have the following advantages:

1. The thermocouple instrument correctly indicates the R.M.S. value of current or voltage irrespective of the waveform.
2. These instruments are not affected by stray magnetic fields.
3. These instruments are free from frequency errors and hence can be used over a wide frequency range. In fact, these are quite useful for *RF* current measurements.
4. These have a high sensitivity.
5. Due to their square law response, these can be used to measure both DC as well AC.

The thermocouple instruments have the following disadvantages:

1. The overload capacity of thermocouple instruments is small as compared to that of other instruments. It is because, the temperature attained by the heater is proportional to the square of the current passing through it. If the temperature of the heater element is, say 300°C at its rated capacity, then it would shoot upto 1200°C when twice the rated current flows through it.

It is obvious that this square law rate could bring the heater to its burn-out temperature. Thus, it is imperative that heater is protected against these damaging overloads.

The fuses don't provide any protection as the heater wire may burn out before the fuse blows off. The typical overload capacity of a thermocouple instrument is only 150% of the full scale current.

Q. 39. What is THERMOELECTRIC INVERSION? Define NEUTRAL TEMPERATURE.

Ans. Thermoelectric inversion means reversal of the polarity of thermoelectric e.m.f. As the temperature of the hot junction is increased above the temperature of the cold junction (maintained at a constant value), the thermoelectric e.m.f. increases proportionally. After a certain temperature for the hot junction, the e.m.f. starts decreasing with increase in temperature. This transition temperature is called the NEUTRAL TEMPERATURE. The e.m.f. keeps reducing till it becomes zero. Any increase in temperature after that results in an inversion of the polarity of the thermoelectric e.m.f. That is, the current would flow from iron to zinc at the hot junction and not from zinc to iron, as is normally the case in a Zinc-Iron thermocouple. Fig. 30.44 shows the variation of thermoelectric e.m.f. with hot junction temperature. The inversion of the thermoelectric e.m.f. is also depicted.

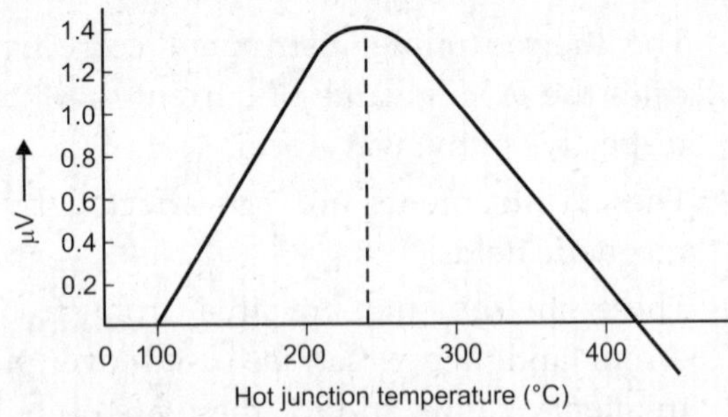

Fig. 30.44

Q. 40. What do you understand by J, K, R, S, T types of thermocouples?

Ans. The thermocouples are designated as J, K, R, S, T types and so on depending upon the materials used in making the thermocouple, their operational temperature range etc. and so on.

Q. 41. What is a Thermistor? Distinguish between NTC and PTC thermistors as regards their construction and Resistance/Temperature characteristics.

Ans. A thermistor is a device whose resistance varies with temperature.

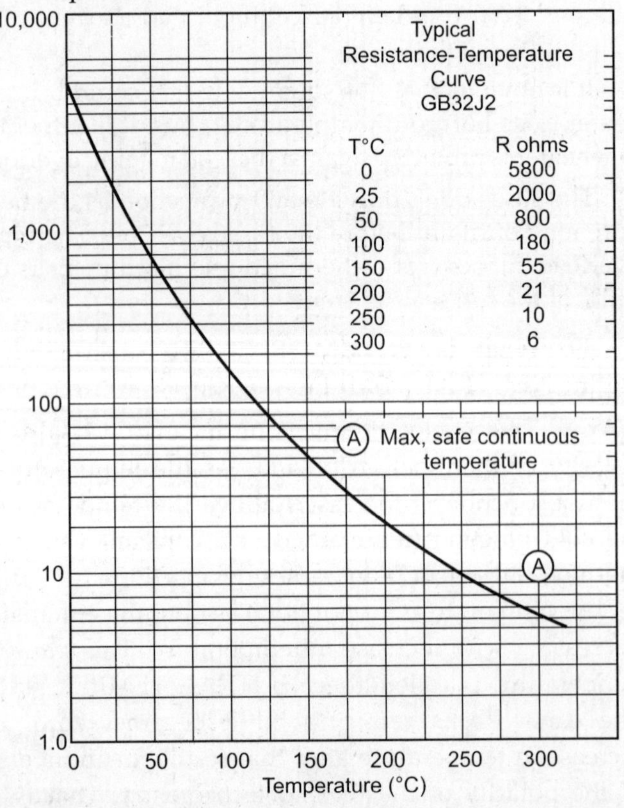

Fig. 30.45

These devices have a very high temperature co-efficient of resistance. The NTC THERMISTORS have a negative temperature co-efficient of resistance *i.e.* their resistance decreases with increase in temperature (Fig 30.45). These are basically ceramic materials made by sintering different mixtures of metallic oxides of Maganese, Nickel, Cobalt, Copper, Iron, Uranium etc. and are semiconductors. These are available in the form of Beads, Discs, Washers, Rods, Probes etc. Their electrical characteristics depend upon the type of oxide used as well as on the physical size and configuration of the device.

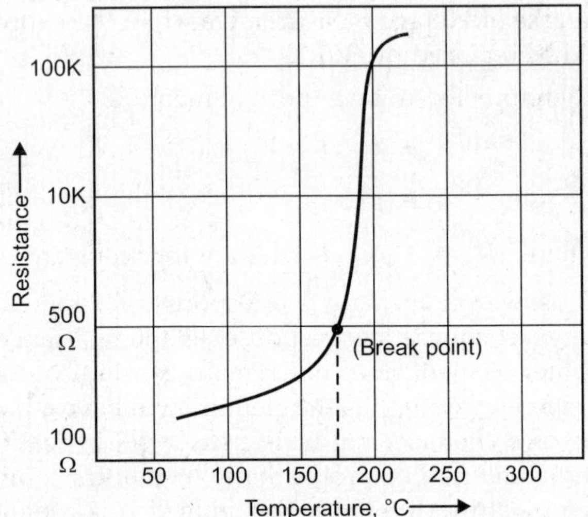

Fig. 30.46

PTC THERMISTORS on the other hand have a positive temperature co-efficient of resistance i.e. their resistance increases with temperature (Fig. 30.46).

It can be noticed from the Resistance vs. Temperature characteristics of NTC and PTC thermistors that the resistance change in case of NTC thermistors is gradual and more or less linear with temperature. In case of PTC thermistors, the change is rather abrupt. PTC thermistors exhibit a more or less constant low resistance upto a certain critical point called the break point after which the resistance increases quite abruptly. These are manufactured from doped polycrystalline and semiconductor barium titanate. The break point of PTC thermistors can be adjusted by changing the composition of the material. Fig. 30.47 shows the circuit symbols for thermistors.

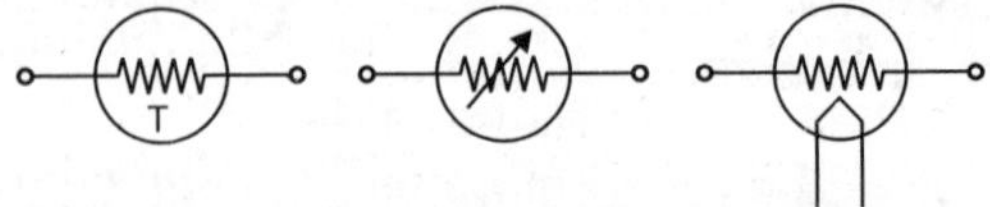

Fig. 30.47

Q. 42. Define the terms DISSIPATION CONSTANT and TIME CONSTANT with reference to NTC thermistors and BREAK POINT with reference to PTC thermistors.

Ans. DISSIPATION CONSTANT is the amount of power which if dissipated by the thermistor, will raise its temperature by 1°C. Typical values are 10 mW/°C, if the measurement is made by suspending the thermistor in a well stirred oil bath and 1 mW/°C, if it is suspended in still air.

The TIME CONSTANT is the time taken by the thermistor to indicate 63% of the newly impressed step temperature input. Again it may be several seconds for thermistors suspended in a well stirred oil bath and a few tens of seconds for thermistors suspended in still air.

THE BREAK POINT of a PTC thermistor is a point of the rapidly rising portion of its Resistance vs. Temperature characteristics after the curie temperature, where its resistance is approximately five times its base resistance. Refer to Fig. 30.46

Q. 43. With the help of typical VOLT-AMPERE characteristics of NTC thermistor, explain the two distinctly different modes of operation of these thermistors. Mention briefly the applications arising out of these operational modes.

Ans. Fig. 30.48 shows the typical characteristics of an NTC thermistor. The portion of the curve upto (*A*) indicates that if the thermistor current is very small so that contribution to the temperature rise from the power dissipated by the thermistor, as a result of the current flowing through it, is negligibly small, then (*V*) and (*I*) obey ohm's law and the device exhibits an approximately constant resistance. Under these circumstances, the variation in resistance will depend upon changes in temperature of the ambient. The resistance change with temperature could be shown on the *V-I* curve and would consist of almost parallel curves displaced from the curve, for say 25°C by an amount dependent on the temperature. The curves would move upwards towards the left for a drop in temperature and downwards towards right for an increase in temperature as indicated by diagonal lines. It is this LINEAR REGION of the curve (associated with small currents and low power dissipation) that is used for TEMPERATURE MEASUREMENT applications. In such an application, the thermistor forms one of the arms of a Wheatstone bridge and an unbalance due to a temperature change is indicated by a suitably calibrated ammeter/voltmeter.

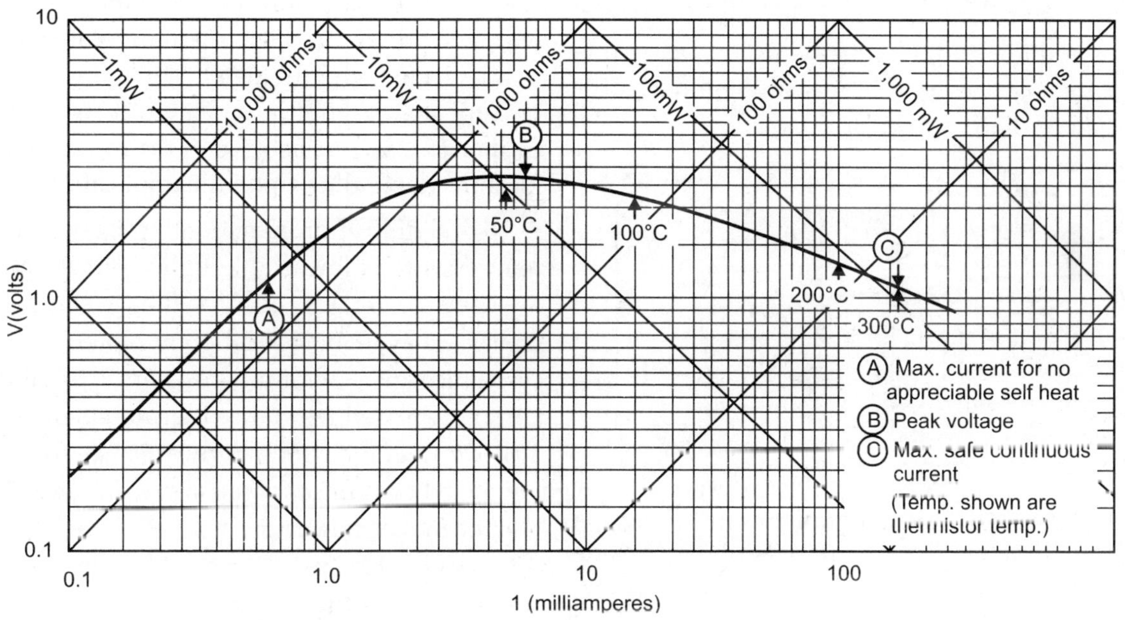

Fig. 30.48

It can be seen from the curve that an increase in current beyond point (*A*), causes a reduction in thermistor resistance. This is due to the heat developed in the thermistor itself as a result of current flowing through it. At point (*B*), the self heating of the thermistor is so much that any increase in current after that results in a decreased thermistor voltage *i.e.* the device exhibits negative resistance (Region B to C). The current continues to increase until a steady state is reached, where the thermistor dissipates as much power as is supplied to it. A series resistor is often used to limit the thermistor temperature to a safe value. Thus, in this self heating mode of operation, the thermistor resistance is dependent on the power dissipated as long as the thermistor temperature due to power dissipation is considerably greater than the ambient.

The thermistor is very sensitive to the rate of heat removal in this mode of operation and it may be used as a flow measurement device capable of measuring very small flow rates as small as 0.001 cubic cm/minute. Such a flow measurement device utilising a thermistor probe in the fuel line of an automobile could be used as an instantaneous fuel consumption indicator or if combined with a voltage proportional to speed could give instantaneous GALLONS PER MILE measurement. Another interesting application in the self-heating mode is found in a voltage regulating circuit that can regulate AC voltages in the face of line or load variations. It is possible as the thermistor is not polarity sensitive.

Q. 44. With the help of typical circuit diagrams explain as to how a thermistor can be used for the measurement of absolute temperature and differential temperature and for temperature control applications.

Ans. In a conventional temperature measurement application, the thermistor forms one of the arms of a Wheatstone bridge (Fig. 30.49). Any change in the thermistor resistance as a result of temperature change is reflected in the readout device. This method provides very accurate information about temperature under measurement. Accuracy in most applications is limited by the readout device only. As the lead length between the thermistor and the bridge is not a limiting factor, this basic arrangement can be used to measure temperatures at several locations. Refer to Fig. 30.50.

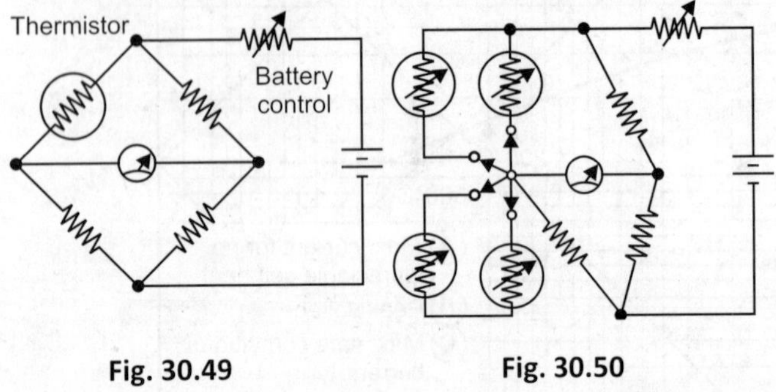

Fig. 30.49 **Fig. 30.50**

Fig. 30.51 shows as to how a thermistor can be used as differential thermometers. The circuit is similar to the one previously described. Two thermistors are placed in two arms of the wheatstone bridge circuit.

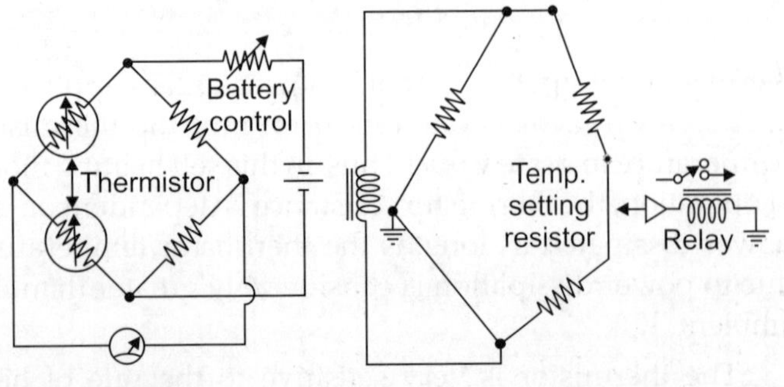

Fig. 30.51 **Fig. 30.52**

Initially the bridge is balanced and unbalance due to any temperature differential is indicated by the readout meter.

A temperature control system can be designed using a thermistor with known Temperature/Resistance characteristics to form one leg of the Wheatstone bridge and a variable resistor calibrated in temperature to form another leg. The thermistor is constantly monitoring the temperature of the body, whose temperature is to be controlled. The desired temperature is set from the variable resistance. In case, the temperature of the body is different from the desired value, there is a bridge unbalance. The output of the bridge is fed to an amplifier which actuates a relay to provide a source of heating or cooling. Bridge balance occurs when the desired temperature is attained. The relay gets de-energised and the source of heating or cooling is turned off. Refer to Fig. 30.52 for the circuit arrangement.

Q. 45. What is the principle of operation of a RESISTANCE THERMOMETER? How does it compare with a thermistor gauge and what are the major precautions to observe while using resistance thermometers?

Ans. The variation in resistance of a metal with temperature is the basis of temperature measurement in a resistance thermometer. The metal used is Platinum or Tungsten. Platinum or Tungsten wire is wound on a former to give a resistance in the range of 10 ohms to 10 kilohms depending upon the application.

To a first approximation, resistance variation with temperature is linear although more complex equations are used for greater accuracy. In fact, standard calibration charts are available for different materials. The change in resistance in a platinum resistance thermometer (most commonly used resistance thermometer) is very small; it only doubles for a temperature change from 0°C to 250°C. On the other hand, a thermistor, which also depends upon change in resistance with temperature for its operation, is much more sensitive. For the same temperature change, resistance in a thermistor shows a variation of 1000 : 1. However, the calibration of thermistor is relatively more difficult due to their response being somewhat non-linear.

Fig. 30.53 shows a commonly used circuit technique for making temperature measurements using resistance sensors. A constant current is made to flow through the sensor, thus converting resistance variation into voltage variation. The measurement can also be done with help of bridge or potentiometric methods, as is done in case of other sensors.

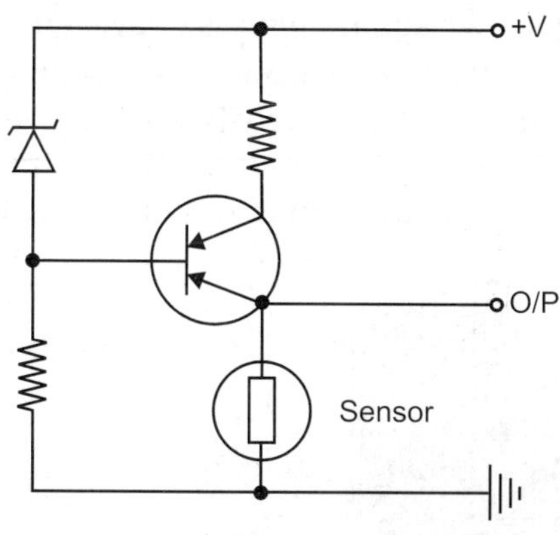

Fig. 30.53

until the two rotors are at the same angle. The receiver thus follows the transmitter position. Voltages induced in stator windings uniquely define the rotor angle.

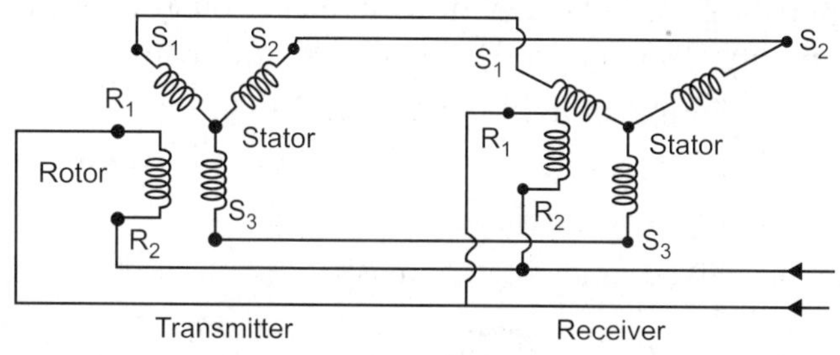

Fig. 30.54

Care should be taken regarding lead lengths to the sensor. Operating current must be kept low or I^2R heating will distort the reading. Sensor package should be such that the probe is genuinely measuring the desired temperature, as many probes conduct heat away down the connection leads causing the actual sensor temperature to be considerably lower than the temperature being measured.

Q. 46. What are the relative advantages and disadvantages of using THERMOCOUPLE THERMOMETRY and RESISTANCE THERMOMETRY?

Ans. Thermocouple Thermometry

Advantages. These have a wide temperature range, are versatile and are simple in application. These are available in a variety of package configurations. It is available in an insulated cable housing or ultra fine wires.

Disadvantages. Thermocouples need a temperature reference for their operation. They need great attention where high accuracy is needed. They need extension cables for long runs.

Resistance Thermometry

Advantages. These provide potentially the most accurate method for temperature measurement. These have a simple installation and need only copper cables for long runs.

Disadvantages. These need energising current. Sensor types are very limited. These are larger in size as compared to thermocouple junctions.

Q. 47. What is a SYNCHRO TRANSDUCER? How does it function?

Ans. Synchro is a classic POSITION MEASURING DEVICE. Fig. 30.54 shows the arrangement of a typical synchro transmitter and receiver link. If the rotors of the transmitter and receiver synchros are not at the same angular position, current flows in the stator leads and there is a torque on the receiver shaft. Rotors are energised with AC (usually 400 Hz). The torque will rotate the shaft

Q. 48. What are shaft encoders ? Distinguish between an INCREMENTAL SHAFT ENCODER and an ABSOLUTE SHAFT ENCODER.

Ans. Shaft Encoders, also called Rotary Encoders, are ANGULAR MEASUREMENT devices that make use of optical or magnetic techniques to measure the shaft angle in terms of a digital number. Shaft encoders essentially consist of a disc and a readout head, the disc is coded to provide a digital readout for discrete (very small) displacement increments. Based on the type of reading mechanism, the shaft encoders are classified as (*i*) Photoelectric Encoders and (*ii*) Magnetic Encoders. In photoelectric encoders, the reading head consists of light source assembly on one side of the disc and corresponding light sensor assembly facing it on the other side of the disc. The coded pattern in this case is partly translucent and partly opaque. Magnetic encoders on the other hand have a magnetic head and the coded pattern is partly magnetic and partly non-magnetic.

INCREMENTAL ENCODERS. Incremental Encoders have simple ON and OFF coded pattern. The output is in the form of number of counts between the start and end of displacement. Thus, start position must be known in order to know the end position in absolute terms.

ABSOLUTE ENCODERS. These have a coded pattern that gives a unique digital output for each discrete displacement increment. Different codes are practiced for this purpose, such as Binary, Binary coded decimal and Gray code.

The most critical selection criteria for shaft encoders are the Range, Resolution, Starting force, Over travel, Type and Magnitude of the full scale output. Accuracy and dynamic characteristics should also be considered.

Q. 49. What are the basic constituents of an optical shaft or rotary encoder? Brief describe different technologies used in the operation or functioning of an optical encoder.

Ans. As outlined in the previous question, an optical encoder is a type of rotary encoder that uses a sensor to identify position change, as light passes through a patterned encoder wheel or disk. Basic constituent parts of an optical shaft encoderinclude the following:

1. Source of light (usually an LED)
2. Photo sensor
3. Moveable disk or code wheel
4. Fixed or stationary mask

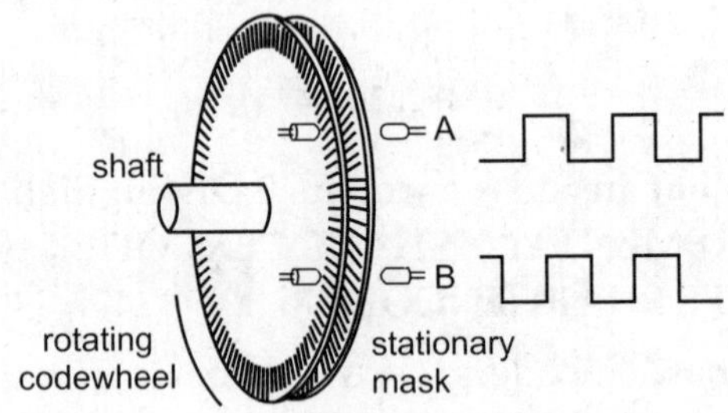

Fig. 30.55

The LED shines through one side of the optical shaft encoder. The encoder wheel or disk has a series of tracks on it, similar to the concentric grooves in an *LP*. For every track on the on the disk, there is a corresponding track on the mask. Small perforations, called windows, are cut along the tracks in the mask, which are either covered or open, as the disk moves depending upon the angular position of the moving disk. Each arc in the rotation indicates a different position and has a different pattern of open/closed windows. The sensor behind the mask identifies the optical encoders' current pattern. In this case, the light radiation is received by the photo sensor, when it encounters an open window. In the case of reflective mode, the photo sensor receives light radiation, when it encounters cover window.

Usually, optical shaft encoders have either a single channel output or a dual channel output. Single channel output encoders have only one LED-photo diode pair. Therefore, when it moves over the encoder disk, it generates square wave which is used for encoding. The number of HIGH-to-LOW and LOW-to-HIGH transitions is used to encode the position or distance. The rate of transitions can be used to compute the instantaneous speed of the system. The single output encoder has a shortcoming that it lacks sense of direction, which makes it highly unreliable in environments with vibrations. This is where the dual channel output becomes important. A dual channel optical rotary encoder has two detectors which are kept at a finite pre-determined distance from each other. The two square wave signals generated by the two channels will have a constant phase shift between the outputs because of the finite distance between the two detectors. The characteristics of the two outputs are used to decide upon the direction of motion of the encoder.

Fig. 30.56

What we have explained in the previous paragraphs is the basic mask optical encoder. Another type of encoder similar to basic mask encoder is the phased-array optical rotary encoder. Phased-array optical encoders use multiple signal outputs to average together to create a single signal. By using averages instead of a single reading, phased array-optical encoders produce much more stable signals that make them a preferred choice in less stable environments. Figure-30.56 shows photograph of an integrated dual channel optical shaft encoder.

Q. 50. Briefly describe the constructional features and operational principle of a magnetic encoder.

Ans. The operational principle of a magnetic shaft recorder is similar to that of an optical shaft encoder. While optical encoder uses light; a magnetic recorder does so using magnetic fields rather than light. In the case of magnetic encoder, a large magnetised wheel spins over a plate of magneto-resistive sensors. The spinning wheel produces predictable responses in the sensor, depending upon the strength of magnetic field. The sensor output feeds a signal conditioning circuit. The number of magnetised pole pairs on the wheel, the number of sensors, and the type of electrical circuit, all work together to determine the resolution of the magnetic encoder. For example, 1024 positions (or 10-bit resolution) can be achieved with four sensors and 128 poles. As compared to optical encoders, magnetic encoder signal is relatively immune to dust, moisture, shock and extreme temperatures.

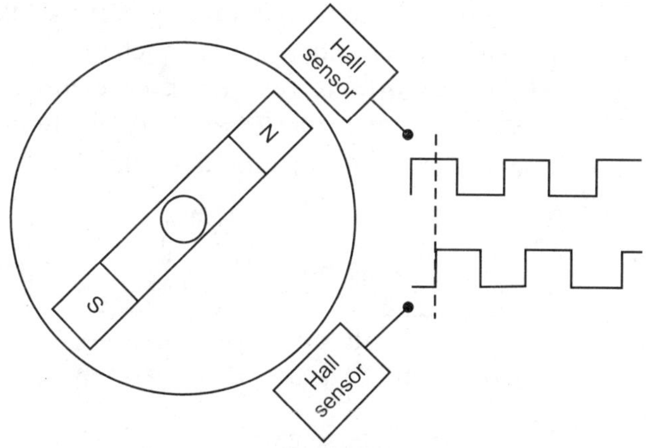

Fig. 30.57

Fig. 30.58

As outlined in the previous paragraph, magnetic encoders use a combination of permanent magnets and magnetic sensors to detect angular position and movement. A typical magnetic encoder has three main constituent parts namely the magnetised disk or wheel attached to a shaft and having number of pole pairs around its circumference; multiple magnetic sensors that are either Hall-effect devices that sense a change in voltage or magneto-resistive sensors that sense change in magnetic field and a signal conditioning circuit that extracts desired information on position and speed from signals produced by the sensors. The sensors detect changes in the magnetic field, as the alternating poles of the magnetised wheel pass over it. The simplest configuration would have a single magnet, with its north and south poles on opposite edges of the wheel, and a single sensor. Such a device would produce a sine wave output, whose frequency equals the rotational speed of the shaft attached to the wheel. Use of a second sensor located at 90° from the first, as shown in Figure 30.57 enables generation of quadrature signals *i.e.* Sine and cosine signals. This allows detection of not only the direction of rotation but also to interpolate the absolute position of the shaft. For incremental encoders, the sinusoidal outputs from the sensors are converted to square waves so the resulting quadrature waveforms can only be encoded to one of four possible angular positions. Figure 30.58 shows the photograph of a modular magnetic encoder showing the magnetised wheel and the signal conditioning card with magnetic sensors.

Q. 51. What is the principle of operation of HOT WIRE GAUGES ? What for are they used ?

Ans. HOT WIRE GAUGES depend for their operation on the thermal conductivity of the heated wire due to change in pressure. These are normally used to measure pressure of gases and vapours in the range of 0.1 Pa to 1000 Pa. However, separate calibration against an absolute gauge is required for each gas. Also, known as THERMAL CONDUCTIVITY GAUGES, the prominent members in the family of hot wire gauges include THERMOCOUPLES GAUGE, PIRANI GAUGE and THERMISTOR GAUGE.

Q. 52. With the help of a schematic arrangement, explain in brief the operation of a typical thermocouple gauge for the detection of leaks in vacuum systems.

Ans. Refer to Fig. 30.59. It consists of an electrically heated wire operating at a temperature of approximately 320°K mounted inside a glass or metal envelope that is connected to the vacuum system where a leak and consequent rise in pressure is to be measured. Any pressure charge causes a temperature change of the heated wire and this temperature change is sensed with the help of thermocouple.

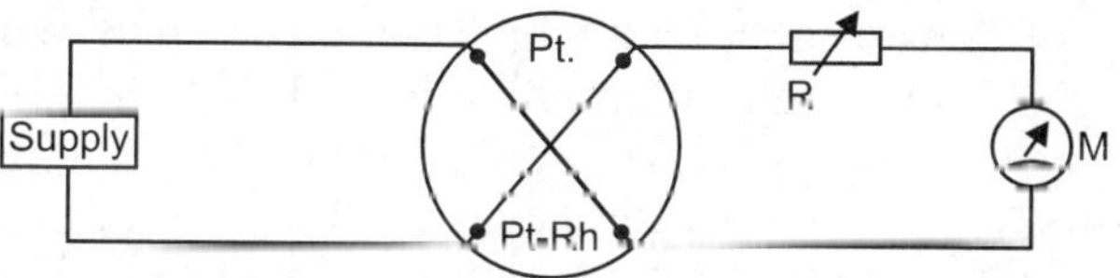

Fig. 30.59

Four parallel lead through wires pass through one end of the envelope, the two noble metal thermocouple wires are fixed across diagonal pairs. The wires are welded at their intersection so that one dissimilar pair forms the hot wire and the other dissimilar pair forms the thermocouple. The hot wire is brought to the requisite temperature with the help of a stabilised supply. Variable resistance '*R*' is adjusted to get a zero on the millivoltmeter. A rise in pressure increases the heat loss from the wire causing a fall of temperature. This results in a reduction of the thermocouple output, which can be calibrated to measure the pressure directly.

Q. 53. What is the basic principle of operation of a PIRANI GAUGE? How is it connected in the Wheatstone bridge arrangement for measurement of low pressures or vacuum?

Ans. It consists of an electrically heated platinum or tungsten wire mounted along the axis of a glass or metal tube (Fig. 30.60). The wire is heated to a temperature of approximately 320° K. The glass or metal tube is connected to the vacuum apparatus whose pressure is to be measured. Changes in the pressure inside the system cause a temperature change of the wire due to change in thermal conductivity associated with the transfer of heat from the hot wire to the surroundings.

Change in resistance due to change in temperature is a measure of the pressure. The wire temperature may also be affected by the changes in the room temperature. These effects can be compensated by using an identical dummy gauge sealed off at low pressure (Fig. 30.61).

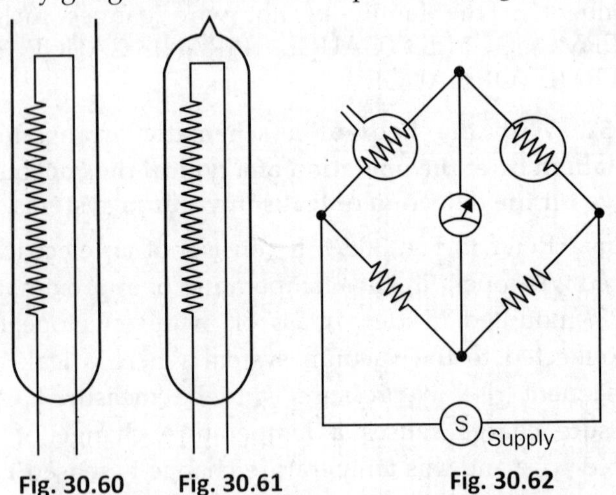

Fig. 30.60 Fig. 30.61 Fig. 30.62

The gauge and the dummy head form the adjacent arms of the wheatstone bridge (Fig. 30.62). The wire is brought to the operating point and the bridge is initially balanced. A rise in pressure results in an increase in thermal conductivity and hence, a fall in temperature of the hot wire. Similarly, a fall in pressure causes a decrease in thermal conductivity and thereby a rise in temperature of the hot wire. Change in resistance accompanying change in temperature causes bridge unbalance, which is indicated by meter deflection. The meter can be calibrated to read the pressure directly. This transducer, which is also known as Knudsen Hot Wire Manometer, is quite effectively used for measurement of very low pressures in the range of 0.1 Pa to 1000 Pa.

Q. 54. What is the acceleration sensor used for? Briefly describe the basic principle of operation of an accelerometer.

Ans. An accelerometer is one of the most commonly used inertial sensors. It is a dynamic sensor capable of measuring acceleration, velocity and displacement in a range of application scenarios. Accelerometers that can measure acceleration in one, two, or three orthogonal axes are commercially available. They are typically used in one of three modes, which include the following:

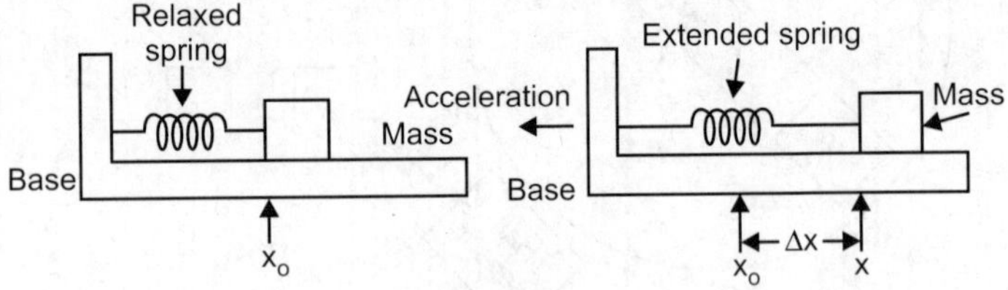

Fig. 30.63

1. As an inertial sensor for measurement of velocity and position
2. As a sensor of inclination, tilt, or orientation in two or three dimensions and
3. As a vibration or shock sensor

The basic principle of operation of an accelerometer is based on Newton's second law of motion and Hooke's law of elasticity. If a mass (m) is undergoing acceleration (a); it exerts a force (F). This force (F) is opposed by restraining effect of a spring having stiffness (k). Mass (m) experiences a displacement (Δx). Acceleration (a) can then be computed from the following equation.

$$\text{ma} = k\Delta x$$

Which gives $a = \dfrac{K\Delta x}{m}$

Fig. 30.64

Figure. 30.63 illustrates the behaviour of such a system. If the system had a damper with damping co-efficient (c); then the damping force is expressed by (cv) where (v) is velocity (v) experienced by mass (m) and (c) is damping coefficient. The behaviour of this second order system, as shown in Figure. 30.64 and representing an accelerator is expressed by the following equation.

$k\Delta x + cv = ma$, which gives $a = (k\Delta x + cv)/m$

Q. 55. Briefly describe capacitive and piezoelectric types of accelerometers.

Ans. There are several different principles used in operation of different types of accelerometers.

Two common types are based on the principles of capacitive sensing and the piezoelectric effect to sense the displacement of the proof, mass proportional to the applied acceleration. Both types are briefly described in the following paragraphs:

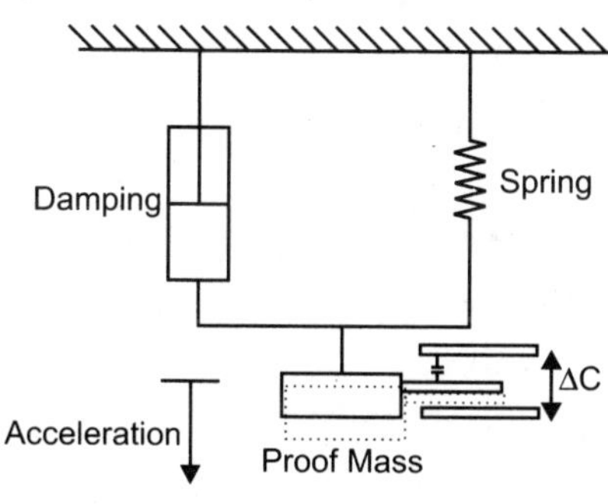

Fig. 30.65

Capacitive Accelerators

A *capacitive accelerometer* is also based on the spring-mass working principle as described in the previous question. The sensing element in this case is a parallel plate capacitor whose capacitance is made to vary by the force experienced by proof mass due to acceleration. Figure. 30.65 shows schematic representation of a typical capacitive accelerometer. The sensing electrodes are in stationary state, and the diaphragm which is attached to the seismic mass is sandwiched in between the two sensing electrodes creating two capacitors. The force experienced due to acceleration causes the proof mass to move. The movement of the diaphragm causes a capacitance shift by altering the distance between the two parallel plates, with the diaphragm itself being one of the plates. The motion of the mass leads to the capacitance change of the sensing electrodes so as to determine the acceleration. Change in capacitance is measured as an output voltage. Capacitive accelerometers are characterised by high accuracy and stability; high immunity to noise and varying temperature. These sensors typically dissipate less power, and can have larger bandwidths due to internal feedback circuitry.

Piezoelectric

Piezoelectric accelerometers use a piezoelectric quartz crystal that accumulates charge, which is further translated into proportional voltage or current as a consequence of accelerative force. Figure. 30.66 shows constructional features of a piezoelectric accelerometer. Due to the very nature of its operational principle, that is, the piezoelectric effect; these accelerometers respond only to AC phenomenon, such as vibration or shock and don't respond to DC phenomenon, such as acceleration due to gravity. That is why, piezoelectric accelerometers are commonly known as vibration sensors. They are characterised by a wide dynamic range and are best suited to measure vibration or shock rather than acceleration due to gravity.

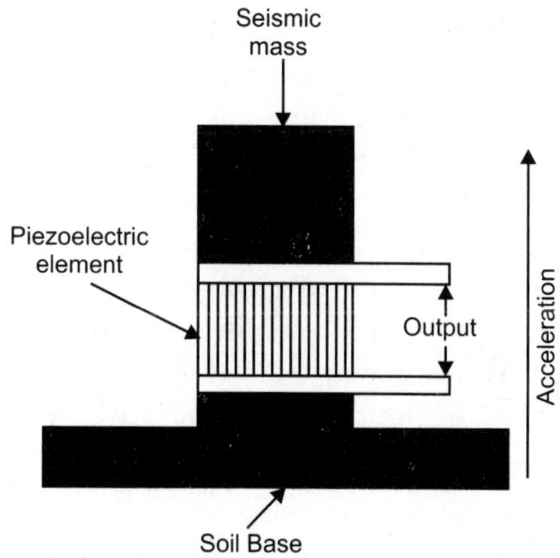

Fig. 30.66

Q. 56. Briefly describe the construction, operational principle and application areas of potentiometric accelerometers.

Ans. The working principle of *potentiometric accelerometer* is also based on the spring-mass system. Figure. 30.67 illustrates the constructional features f potentiometric accelerometer. It comprises a mass (seismic mass), a spring, a damper and a resistive element. The mass is connected between a spring and a dashpot. The wiper of the potentiometer is connected to the mass.

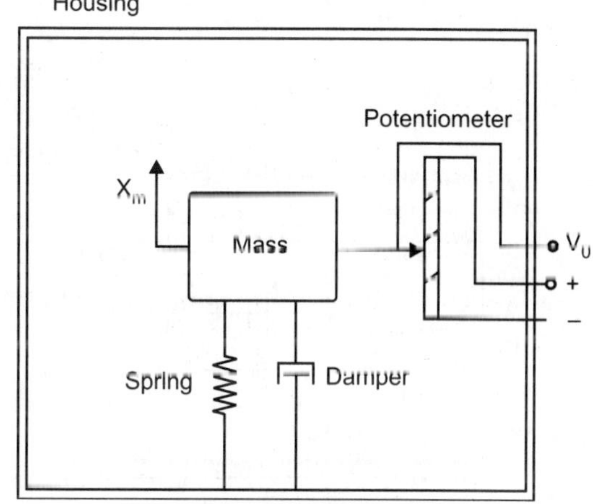

Fig. 30.67

The accelerometer measures the motion of the seismic mass by attaching the wiper arm to the spring-mass system. The displacement of the seismic mass produced due to accelerative force changes the position of the wiper of the potentiometer, thereby changing the resistance of the resistive element. Change in resistance is measured as an electrical signal. Potentiometric accelerometers are used in low frequency vibration measurements as its natural frequency is generally less than 30 Hz.

Q. 57. Briefly describe the construction and operational principle of Hall-effect accelerometer.

Ans. *Hall-effect accelerometer* is also based on the working principles of spring-mass system. The output voltage varies according to a change in magnetic field from the magnet attached to a seismic mass. The mass deflects because of the forces due to acceleration. The output Hall voltage varies due to changes in magnetic field. The Hall voltage is calibrated in terms of acceleration. Figure. 30.68 shows constructional features of Hall-effect accelerometer. The figure is self-illustrative.

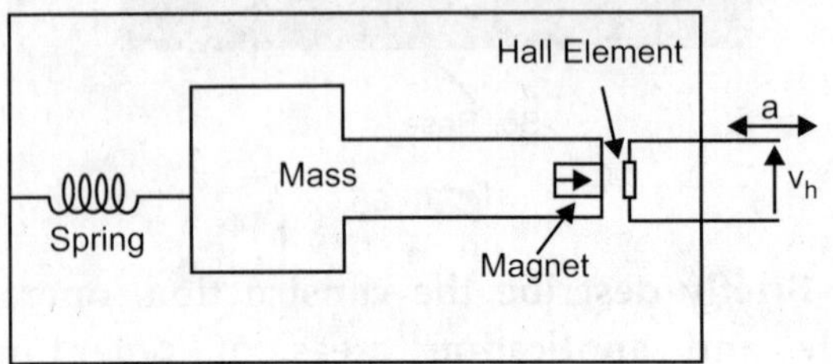

Fig. 30.68

Q. 58. Briefly describe the construction and operational principle of Piezoresistive accelerometers.

Ans. Unlike the capacitive, potentiometric and Hall-effect accelerometers that depend for their operation on spring-mass system; Piezoresistive accelerometers do not use a spring. Instead of that, the mass is attached to cantilever beam, which in turn, is sandwiched in between strain gauges.

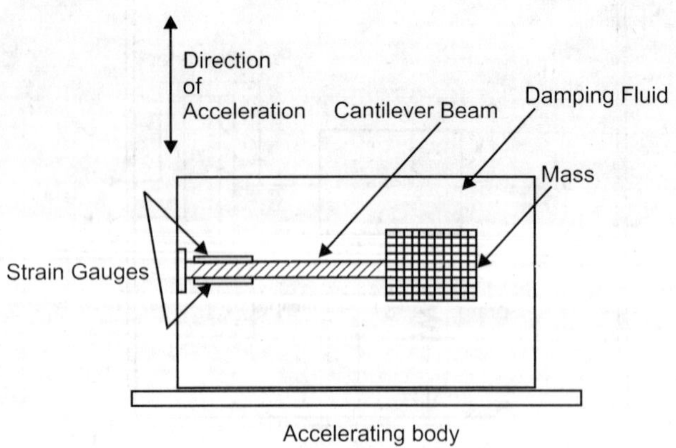

Fig. 30.69

The force exerted by the seismic mass changes the resistance of the strain gauges. Change in resistance of strain gauge is measured as an electrical output. Figure. 30.69 depicts the constructional features of a Piezoresistive accelerometer. Piezoresistive accelerometers are used in high shock applications capable of measurements upto ± 1000g. They have low frequency response down to zero Hz but have limited high frequency response.

Q. 59. Name important performance specifications of an accelerometer. Briefly describe each one of them.

Ans. Important specifications of accelerometers include the following:

1. Type of output, Analog/Digital
2. Number of axes
3. Sensitivity (voltage output per '*g*')
4. Dynamic range
5. Bandwidth
6. Amplitude stability
7. Grounding
8. Mass

Other specifications include zero '*g*' offset (voltage output at zero value of '*g*'), Noise (sensor minimum resolution), Temperature range, Bias drift with temperature (effect of temperature on voltage output at 0 *g*), Sensitivity drift with temperature (effect of temperature on voltage output per *g*) and Power consumption.

Type of Output: Accelerometers produce either analog or digital output. Analog accelerometers output a variable DC voltage depending on the magnitude of acceleration applied. Older digital accelerometers output a pulse width modulated (PWM) output, where the acceleration being measured is proportional to the pulse width (or duty cycle) of the PWM signal. Newer digital accelerometers are more likely to output their value using multi-wire digital protocols such as I^2C or SPI.

Number of axes: Accelerometers capable of measuring acceleration along one, two, or three axes are available. The most familiar type of accelerometer measures across two axes. However, three-axis accelerometers are increasingly common and inexpensive.

Sensitivity: Sensitivity specifies change in output signal produced for a given change in acceleration. A sensitive accelerometer will be more precise and probably more accurate. Accelerometers typically fall into two categories, those producing either 10 mV/g or 100 mV/g. Low output accelerometers are used to measure high vibrational levels, while high output accelerometers are used to measure low level vibrations.

Dynamic range: It is the range between the smallest acceleration detectable by the accelerometer to the largest before the output signal gets clipped or distorted. The dynamic range of the accelerometer should be greater than the expected vibration amplitude range of the sample.

Bandwidth: The bandwidth specifies the frequency response of the accelerometer. Low frequency cut-off and high frequency cut-off are the related parameters. Low Frequency cut-off is the frequency, where the output starts to fall off below the stated accuracy. High frequency Limit is the frequency, where the output exceeds the stated output deviation. It is typically governed by the mechanical resonance of the accelerometer.

Amplitude stability: This is not a specification in itself. It describes sensor's change in sensitivity depending on its application; for instance over varying temperature or time.

Grounding: There are two types of signal grounding in accelerometers leading to Case grounded Accelerometers and Ground Isolated Accelerometers. Case Grounded accelerometers have the low side of the signal connected to their case. As the case is part of the signal path and may be attached to a conductive material; care must be used, when using this type of accelerometer to avoid noise from the ground plain. Ground isolated accelerometers are much less susceptible to ground-induced noise, as they have their electrical components isolated from the case.

Mass: The mass of the accelerometer should be significantly smaller than the mass of the system to be monitored so that it does not change the characteristic of the object being tested.

Q. 60. What is thermal anemometry and what is it used for? What features of Hot Wire Anemometers make them popular?

Ans. Thermal anemometry depends for its operation on the convective heat loss to the surrounding fluid from an electrically heated sensing element or probe. It is the most commonly employed method for measurement of instantaneous velocity of fluids. The heat loss is interpreted as a measure of fluid velocity. This technique has been in use for a long time in the study of laminar, transitional and turbulent boundary layer flows; for measurement of unsteady flows, such as those arising behind rotating blade rows, when the flow is viewed in the stationary frame of reference.

There are two fundamentally different types of sensors, namely cylindrical sensors (hot wires and hot films) and flush sensors (hot films). While the former are most commonly used to measure the fluid velocity; flush sensors are employed to measure the wall shear stress. While hot-wire sensors are made from short lengths of resistance wire and are circular in section; hot-film sensors consist of a thin layer of conducting material that has been deposited on a nonconducting substrate. Hot-film sensors may also be cylindrical but may also take other forms, such as those that are flush-mounted.

Thermal anemometry, due to the use of very small probes, offers very high spatial resolution and excellent frequency response characteristics. The basic principles of the technique are relatively straightforward and the probes are difficult to damage if reasonable care is taken. Most sensors are operated in the constant temperature mode. They are operated with variable current so as to maintain a constant temperature. We also have constant current hot wire anemometers.

Q. 61. Briefly describe the operational principle of a hot wire anemometer.

Ans. The **hot wire anemometer** measures the instantaneous velocity of a fluid at a point. Figure. 30.70 shows the constructional features of a representative hot wire anemometer probe. It comprises a small sensing element placed in the flow field at the point of measurement. The sensing element is a short thin wire of platinum or tungsten, connected to a suitable electronic circuit. The hot wire anemometer functions as follows. The electrical resistance of the wire is a function of its temperature, which in turn depends upon surrounding fluid density. Also, the rate of heat transfer increases with increase in velocity of flow past the wire. To summarise, the wire resistance depends upon its temperature, which depends upon rate of convective heat transfer. The rate of convective heat transfer further is a function of fluid density around the wire, which in turn, depends upon fluid flow velocity. Thus, the fluid velocity at the location of hot wire anemometer probe is interpreted in terms of change in resistance. By maintaining either a constant voltage across the wire or a constant current though the wire by a suitable circuit; the change in current or voltage respectively becomes a measure of velocity pasts the hot wire. It is mainly used for measuring the velocity of gases. It is not preferred for liquids because bubbles and smaller solid particles tend to collect on the wire and spoil the calibration.

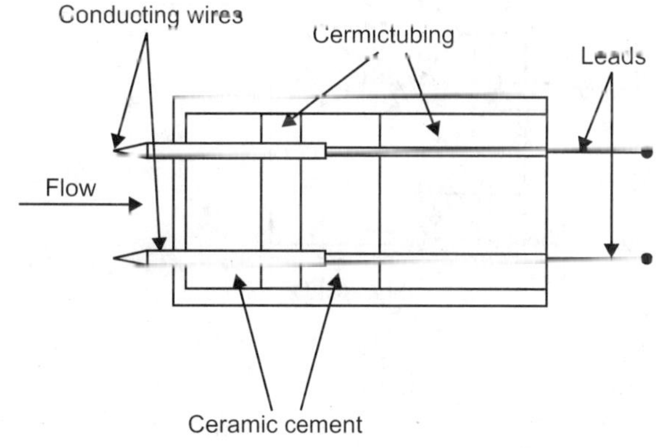

Fig. 30.70

Q. 62. Briefly describe constant current and constant temperature methods of measurement in hot wire anemometry.

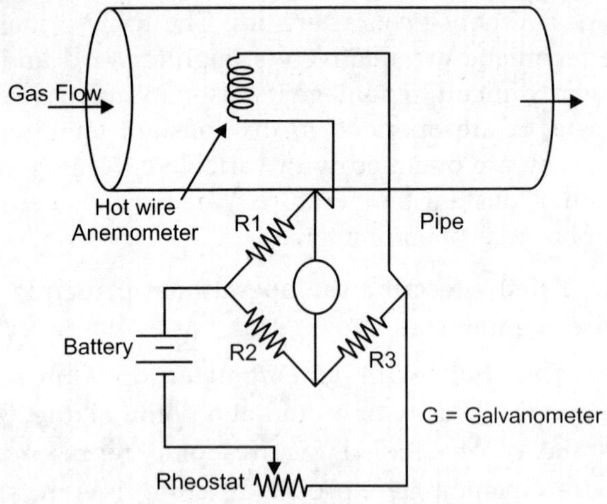

Fig. 30.71

Ans. Figure. 30.71 illustrates the constant current method of measuring the fluid flow rate using hot wire anemometer probe. The diagram is self-illustrative. A constant current is passed through the sensing wire. Initially, in the absence of any fluid flow, the galvanometer is brought to null position. In the presence of fluid flow, heat transfer takes place from the sensing wire to the flowing fluid. The temperature of sensing wire changes due to convective heat transfer to the fluid, which in turn, depends upon flow rate. Due to this, the galvanometer, which was initially at zero position, deflects and this deflection of the galvanometer becomes a measure of flow rate of the gas, when calibrated.

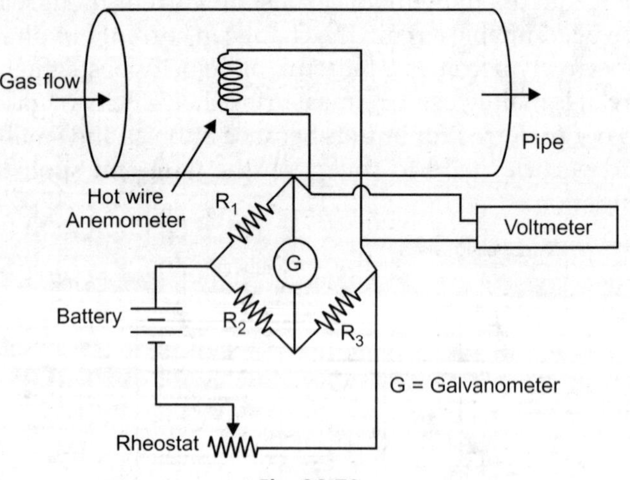

Fig. 30.72

Figure. 30.72 shows the circuit arrangement in the case of constant temperature method. A current is initially passed through the wire. In the presence of fluid flow, the resistance of the sensing wire tends to change due to change in temperature of the wire. In this case, however, the temperature and resistance of the sensing wire are maintained at a constant level by changing the current through the sensing wire. The electrical current required bringing the resistance and hence, the temperature of the wire back to their initial levels becomes a measure of fluid flow rate, when calibrated.

Q. 63. What are proximity sensors? Based on operational principle, what are different types of proximity sensors?

Ans. Proximity sensors are used to detect the presence or absence of target objects using electromagnetic fields, light, and sound. Based on the operational principle, proximity sensors are classified as inductive, capacitive, Photoelectric and ultrasonic. *Inductive sensors* are non-contact type proximity sensors capable of detecting proximity to ferrous targets. They do so by detecting the presence of ferrous objects in the magnetic field, produced by the sensor coil. *Capacitive sensors* are capable of detecting both ferrous as well as non-ferrous objects. They do so by detecting increase in capacitance occurring due to proximity to the object. *Photoelectric sensors* detect proximity by transmitting a light beam towards the target and detecting reflected radiation. One method is to detect the magnitude of received signal and establish proximity, when the received signal is greater than a specified threshold. In another method, the receiver determines distance to the target on a continuous basis and generates a pulse, when the distance to the target reaches specified value. Both laser time-of-flight and triangulation principles have been put to use for the purpose. While laser time-of-flight proximity sensors offer relatively much larger ranges; laser triangulation has much higher accuracy. Ultrasonic sensors are similar to photoelectric sensors. They use sound waves instead of light.

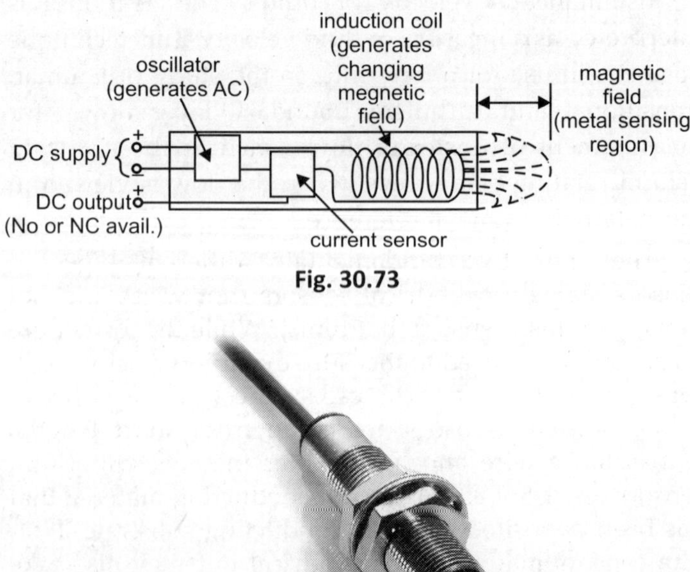

Fig. 30.73

Fig. 30.74

Q. 64. Briefly describe the principle of operation of inductive proximity sensors.

Ans. As outlined in the answer to the previous question, inductive sensors can detect only ferrous targets. An inductive proximity sensor primarily comprises a coil, an oscillator that is used to generate an oscillating magnetic field by feeding the coil with an alternating current, a detection circuit that produces an output, depending upon the amplitude of magnetic field and an output circuit. Figure 30.73 shows the basic schematic arrangement of inductive proximity sensor. Figure 30.74 shows the photograph of typical inductive proximity sensor. When a ferrous target enters this magnetic field; eddy currents are induced on the metal target's surface. This changes the reluctance of the magnetic circuit; which in turn, reduces the oscillation amplitude (Figure 30.75). The magnetic field continues to shrink further as more and more metal enters the magnetic field eventually getting completely collapsed. The detection circuit responds to these amplitude changes, thereby changing its output state. The circuit begins to oscillate again when the metal target moves out of sensor's range. As the amplitude of oscillations crosses the other threshold; the detection circuit returns the sensor to its previous output state. Depending upon type of sensor, a NC, NO or NO + NC (complementary) output states are available. The output states can be read by a suitable controller to extract the use able information. Because of magnetic field limitations, inductive sensors have a relatively narrow sensing range from fraction of a millimeter to few tens of millimeters.

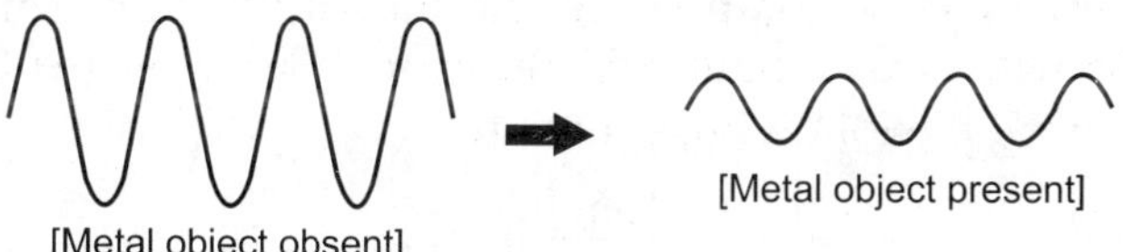

Fig. 30.75

Q. 65. Do we also have inductive proximity sensors that can detect non-ferrous objects? If yes, briefly describe the principle of operation of such a sensor.

Ans. Yes, we do have inductive proximity sensors that can detect non-ferrous metal materials, such as copper, aluminium etc. These sensors are categorised as high frequency oscillation type sensors. The sensor detects increase in frequency of oscillations caused by an approaching non-ferrous material. When a non-ferrous-metal target, such as aluminium or copper approaches the sensor; the oscillation frequency increases (Figure. 30.76). On the contrary, the oscillation frequency decreases, when a ferrous-metal target, such as iron approaches the sensor.

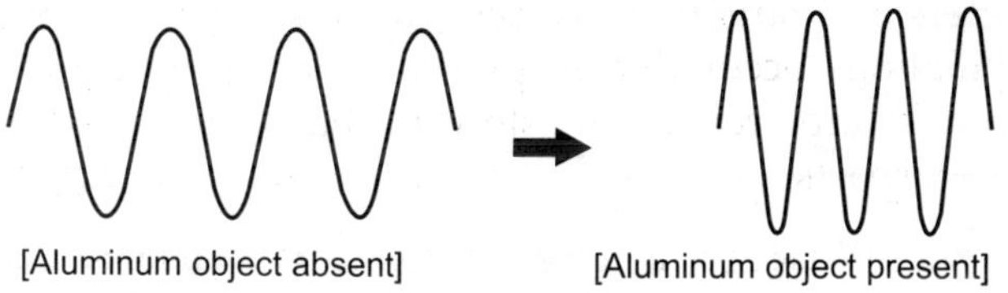

Fig. 30.76

Q. 66. Briefly describe the principle of operation of capacitive proximity sensors.

Ans. While inductive sensors use magnetic field; capacitive proximity sensors use electric field for proximity detection. They can detect metallic, non-metallic and non-ferrous targets. The capacitive proximity sensor too has an oscillator, a detection circuit used to detect the amplitude of electric field and an output circuit. The coil of inductive sensor is replaced by a capacitor in the case of capacitive sensor. In capacitive sensors, the two conduction plates are constituted by a plate housed in the sensing head and the target acting as the other plate (Figure 30.77). The capacitor operates like an open capacitor with air acting as an insulator. In the absence of any target material, there is little capacitance offered by the capacitor and there are practically no oscillations.

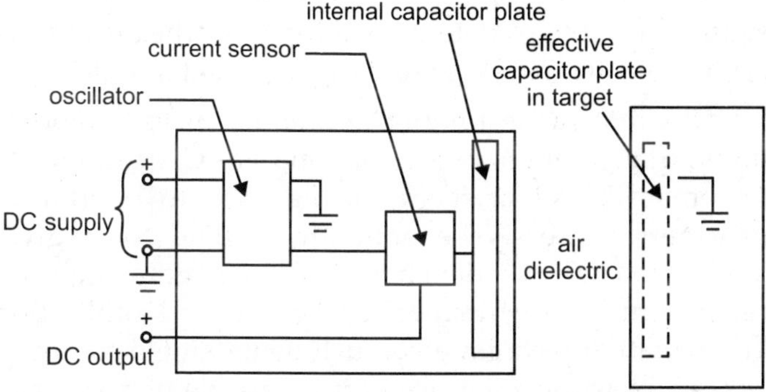

Fig. 30.77

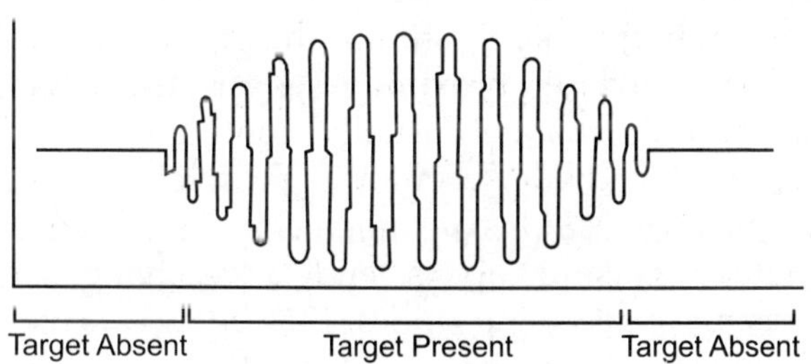

Fig. 30.78

As a target enters the sensing zone; the capacitance increases, causing a change in oscillation amplitude as shown in Figure. 30.78, which in turn, changes the output state of detection circuit. The basic difference between the operation of inductive and capacitive sensors is that while in the case of former, the oscillations collapse in the presence of target; in the case of capacitive sensors, the oscillations occur due to presence of target. Capacitive sensors are relatively slower than their inductive counterparts.

Q. 66. What is a laser proximity sensor? Briefly describe the operational principle of triangulation type proximity sensor.

Ans. Laser proximity sensor is a type of laser range finder designed to accurately measure relatively smaller distances to target as compared to those encountered in conventional laser range finders used in tactical battlefield applications, such as observation and surveillance and fire control. While a conventional battlefield laser range finder measures distance to target and the range information is displayed to the observer, when the device is used in observation and surveillance role or fed to a fire control computer in a fire control application; in the case of a proximity sensor, the processing circuitry generates a command signal, when the distance to target is equal to the preset distance value within a certain specified tolerance. The command signal in turn could be used to perform a variety of control functions. In military parlance, the most common application of a proximity sensor is in a laser proximity fuze, where the command signal is used to initiate detonation of warheads of large artillery shells, aviation bombs and guided missiles.

Different range finding techniques include time-of-flight, triangulation, phase shift and FM-CW techniques. Theoretically, all of them can be used to build laser proximity sensors. The optimum method for a given application however depends on the maximum range and range accuracy requirement of the intended application. Triangulation technique is much better suited to design of laser proximity sensors as it is capable of measuring shorter distances upto few meters with high accuracy. The range accuracy however falls off rapidly with increasing distance. Figure. 30.79 shows the basic principle of operation of a laser proximity sensor. The sensor has a transmitting channel usually configured around a semiconductor diode laser or a passively Q-switched diode pumped solid state microchip laser along with associated transmit optics and a receiving channel comprising of receiving optics, PIN or APD sensor and range processing circuitry.

As shown in the figure, transmitted laser beam reflected from the target located at three different distances produces the beam images that are displaced across the active area of the photo sensor. The receiver is designed to produce the focused laser beam spot at the center of the active area for preset value of proximity distance. There will be a distance both longer and shorter than the desired distance, where the focused beam spot falls just outside the active area.

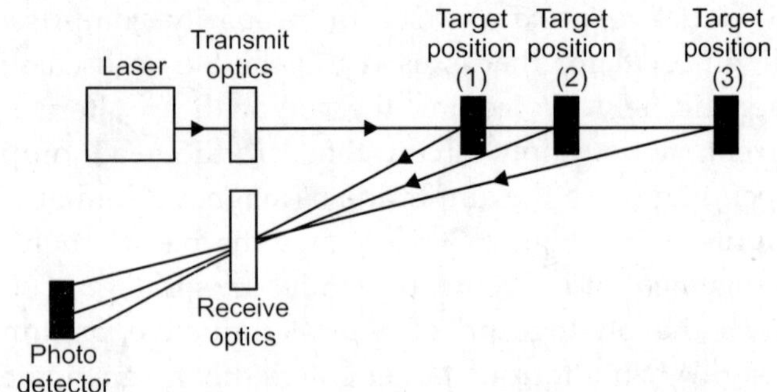

Fig. 30.79

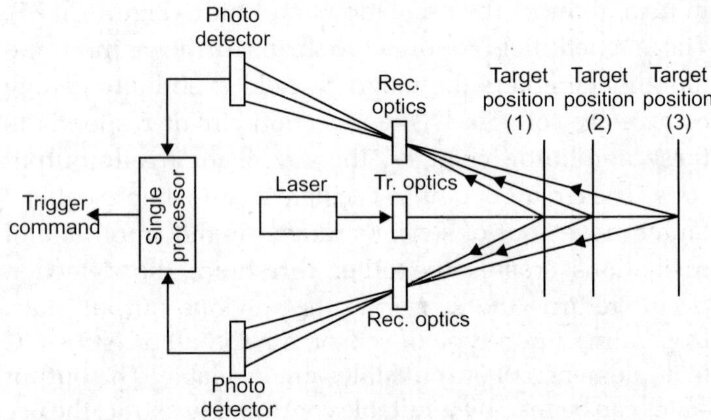

Fig. 30.80

The distance-to-target (d) from laser transmitter location can be computed from known values of length (l)between transmitter and receiver and angles (α) and (β) using the following equation:

$$d = l \times \frac{\sin \beta}{\sin (\alpha + \beta)}$$

Where α = Angle made by line joining transmitter and receiver and the line joining transmitter and target and β = Angle made by line joining transmitter and receiver and the line joining receiver and target.

This forms the basis of operation of laser proximity sensor in general and laser proximity fuze in particular. The performance of laser proximity sensor can be enhanced by using an axially symmetric arrangement of multiple aperture photo sensors as shown in Figure. 30.80. The design offers better ballistics due to centre of gravity being located on longitudinal axis of the ammunition round and higher S/N ratio due to averaging of multiple return signals.

SOLVED PROBLEMS

Problem. 30.1. *A certain temperature measuring system comprises of an RTD type temperature transducer, a Wheatstone bridge, an amplifier and a pen type chart recorder. The sensitivities of these individual elements are 0.25 ohm/°C (transducer), 0.01 Volt/ohm (Wheatstone bridge), 100 Volts/Volt (Amplifier) and 1.0 mm/Volt (Recorder). Determine (a) overall sensitivity (b) temperature change corresponding to recorder pen movement of 20 mm.*

Solution: (*a*) Overall sensitivity is the product of individual sensitivities.

Therefore, overall sensitivity = 0.25 × 0.01 × 100 × 1 = 0.25 mm/°C

(*b*) Temperature change = 20/0.25 = 80°C

Problem. 30.2. *A quartz piezoelectric crystal transducer used in a pressure sensor having a voltage sensitivity of 0.05 Volt-m/N is subjected to a pressure of 1.4 × 10^6 N/m^2. Determine the voltage output, if the thickness of crystal were 2.5 mm.*

Solution: Output voltage = Voltage sensitivity × Pressure × Thickness

= 0.05 × 1.4 × 10^6 × 2.5 × 10^{-3} = 175 Volts

Problem. 30.3. *A rotary potentiometer is used to measure rotational position of a shaft. The potentiometer has 1000 turns of wire. The input and output ranges of the potentiometer transducer respectively are – 150° to + 150° and 0 to 12 Volts. Determine (a) Span of potentiometer (b) Sensitivity and (c) Average resolution.*

Solution: (*a*) Span = 150 – (–150) = 300°

(*b*) Sensitivity = (12 – 0)/300 = 0.04 Volt/degree

(*c*) Average resolution = 12/1000 = 0.012 Volt

Problem. 30.4. *A milli-ammeter has a uniform scale with 100 divisions and full scale reading of 100 mA. Determine resolution of ammeter in Amperes, if one could measure 1/10th of a division with fair degree of certainty.*

Ans. Each division corresponds to 100/100 = 1 mA 1/10th division is, therefore, corresponds to 0.1 mA = 0.0001 Ampere

Problem. 30.5. *A certain potentiometer transducer used for measurement of linear displacement produces an output voltage of 10 Volts. If the total winding resistance of potentiometer were 10 K and the maximum displacement range were 6 cm; determine the displacement under measurement, if the power dissipation in the potentiometer winding were not to exceed 90 mW for maximum displacement.*

Solution: Voltage corresponding to maximum displacement that the potentiometer transducer can measure = $\sqrt{(90 \times 10^{-3} \times 10 \times 10^3)}$ = 30 Volts

30 Volts corresponds to 6 cm

Therefore, displacement corresponding to 10 Volts = 6 × 10/30 = 60/30 = 2 divisions

Problem. 30.6. *A resistance strain gauge having a gauge factor of 3.0 is bonded to a steel structural member. The structural member is subjected to a stress of 100 Mega-Newton/m^2. Determine percentage change in strain gauge resistance due to applied stress, if the Young's modulus of elasticity of steel were 250 Giga-Pascal.*

Solution: strain produced due to applied stress = 100 × 10^6/250 × 10^9 = 4 × 10^{-4}

Gauge factor = Percentage change in resistance/Percentage change in length

= Percentage change in resistance/Strain

Percentage change in resistance = Gauge factor × Strain = 3 × 4 × 10^{-4} = 0.0012 = 0.12%

Problem. 30.7. *A resistive strain gauge with unstrained length of 200 mm and cross-section of 5 cm^2 is bonded to a beam of steel. If the gauge factor is 2 and unstrained resistance of the strain gauge is 200 ohms; determine the change in length and the applied force, if the resistance of strain gauge changes by 10 milli-ohms. Assume Young's modulus of steel to be 200 × 10^9 Pa.*

Solution: Gauge factor = percentage change in resistance/Percentage change in length (Both taken as fraction)

Gauge factor = ($\Delta R/R$)/ ($\Delta L/L$), which gives

ΔL = ($\Delta R/R$) × L/Gauge Factor

Now $\Delta R/R$ = 0.01/200 = 0.00005

Therefore, ΔL = 0.00005 × 0.2/2 = 0.000005 m = 5 × 10^{-6} m

Applied stress = 200 × 10^9 × 5 × 10^{-6} /0.2 = 5 × 10^6 N/m^2

Applied force = Applied stress × Area of cross-section = 5 × 10^6 × 5 × 10^{-4} = 2500 N

Problem. 30.8. *The output of an LVDT sensor produces a current of 5 mA when its core moves through a distance of 1 mm. The current output is connected to a current meter through an amplifier having a gain of 100. The current meter has full scale reading of 10 Ampere. It has 100 divisions and each division has 10 sub-divisions. Determine resolution of LVDT with which the displacement can be measured.*

Solution: Sensitivity of LVDT = 3 × 10^{-3}/1 = 3 × 10^{-3} Ampere/mm

Sensitivity of measurement set-up = Sensitivity of LVDT × Amplifier Gain = 3 × 10^{-3} × 100 = 0.3 Ampere/mm

Each sub-division of current meter corresponds to 10/(100 × 10) = 0.01 Ampere

Therefore, resolution = 0.01/0.3 = 0.033 mm = 33 micron

Problem. 30.9. *A capacitive displacement sensor comprises a parallel plate capacitor; each plate having an area of 1200 cm^2. If the separation between two plates were 1.5 mm; determine displacement sensitivity of the sensor. Assume permittivity of dielectric between the two plates as 2. Permittivity of free space = 8.85×10^{-12} F/m.*

Solution: Capacitance, $C = \varepsilon_0 \varepsilon_r A/D$ where terms have their usual meaning.

Displacement sensitivity = $dC/dD = \varepsilon_0 \varepsilon_r A \times (-1/D^2) = -\varepsilon_0 \varepsilon_r A/D^2$

Substituting the values of different parameters, we get

$dC/dD = -8.85 \times 10^{-12} \times 2 \times 0.12/(1.5 \times 10^{-3})^2$

$= -8.85 \times 10^{-6} \times 0.24/2.25 = 0.944 \times 10^{-6}$ F/m = – 944 nF/m

Negative sign indicates that the capacitance increases with decrease in distance and vice-versa.

Problem. 30.10. *Determine change in length of an aluminium rod when its temperature is changed from 0 to 100°C. The length of the rod is measured to be 10 m at 25°C using a bimetallic strip. Assume co-efficient of thermal expansion of aluminium to be 25×10^{-6}/°C.*

Ans: The generalised expression for length as a function of temperature is given by:

$$I = I_0 \times [1 + \alpha \Delta T]$$

Length at 0°C can be computed from:

$L = 10 \times [1 + 25 \times 10^{-6} \times (0 - 25)] = 10 \times (1 - 625 \times 10^{-6})$ = 9.99375 m

Length at 120°C can be computed from:

$L = 10 \times [1 + 25 \times 10^{-6} \times (120 - 25)] = 10 \times (1 + 2375 \times 10^{-6})$ = 10.02375 m

Therefore, change in length = 10.02375 – 9.99375 = 0.03 m = 3 cm

Problem. 30.11. *A platinum resistance thermometer is used to measure the temperature of a gas. The thermometer has a resistance of 100 ohms at 0°C and 150 ohms at 100°C. The resistance increases to 300 ohms, when the thermometer comes in contact with the gas. Assume co-efficient of thermal expansion for platinum to be 0.0039/°C.*

Solution: Temperature (T_2) of the gas can be computed from the following expression.

$$T_2 = T_1 + [(R_2 - R_1)/\alpha R_0]$$

$$= 100 + [(300 - 150)/0.0039 \times 100]$$

$$= 100 + (150/0.39) = 484.6°C$$

Problem 30.12. *A soft iron wire of small diameter has a gauge factor of 4. Determine the Poisson's ratio of the wire.*

Solution. Gauge factor is given by $(1 + 2\mu)$ where 'μ' is the Poisson's ratio. Therefore, $(1 + 2\mu) = 4$ or $\mu = 1.5$.

Problem 30.13. *Strain in a beam is being measured with the help of a strain gauge attached to the beam. The resistance of the strain gauge having a gauge factor of '3' changes from 150 ohms to 151 ohms, when strained. Determine the strain in the beam to which the gauge has been attached.*

Solution. If ΔR and R are respectively the change in resistance and unstrained resistance, ΔL and L are respectively the change in length and the unstrained length, then the gauge factor can also be expressed as:

Gauge Factor = $\dfrac{\Delta R/R}{\Delta L/L}$, Substituting the given values,

Strain (given by $\Delta L/L$)

$$= \frac{\text{Change in resis tan ce/Unstrained resis tan ce}}{\text{Gauge Factor}}$$

$$= \frac{1/150}{3} = 0.0022$$

Problem 30.14. A strain gauge is being used to determine strain in a beam. The strain gauge is used in a bridge configuration, as shown in Fig. 30.81. $R_1 = R_2$ = 100 ohms. Initially, when the beam is unstrained, the bridge gets balanced at R_3 = 120.63 ohms and when the beam is strained, R_3 has to be adjusted to 120 ohms to restore balance. If the strain gauge has a gauge factor of 2.04, determine the strain (e) in the beam.

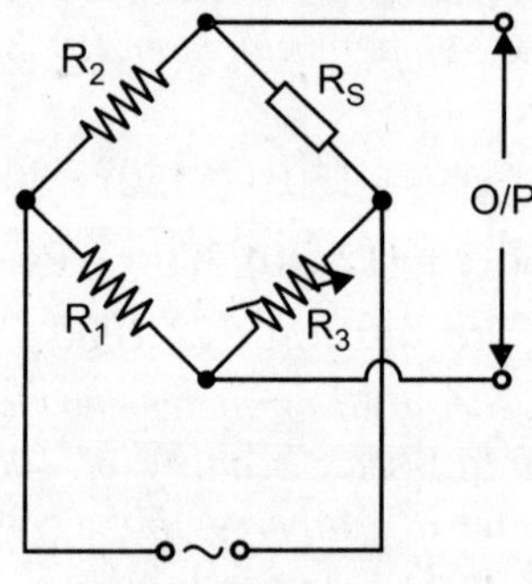

Fig. 30.81

Solution.

The balanced condition of the bridge is given by : $R_1/R_2 = R_3/R_s$ *i.e.* $R_s = R_2 R_3/R_1$

Therefore, unstrained resistance of the strain gauge

$$= \frac{100 \times 120.63}{100} = 120.63 \text{ ohms.}$$

When strain (*e*) is applied, the change in resistance R_3 needed to balance the bridge is given by : (120.63 – 120) = 0.63 ohm.

Therefore, change in resistance of the strain gauge is given by: R_2/R_1 (change in R_3) = (100/100) × 0.63 = 0.63 ohm.

Now, gauge factor = 2.04

$$= \frac{\text{(Unit change in strain gauge resis tan ce)}}{\text{Strain}}$$

Unit change in strain gauge resistance = 0.63/120.63 = 0.0052

Therefore, strain = 0.0052/2.04 = 0.00254

Problem 30.15. *A strain gauge bridge as shown in Fig. 30.81 has two fixed resistances R_1 and R_2 of 100 ohms and 150 ohms respectively. It has a variable resistance which produces a null at 100 ohms when the straingauge is unstrained. It has to be changed to 101 ohms to produce the null, when the gauge is strained. Determine the strain. (Gauge Factor = 3)*

Solution. Unstrained strain gauge resistance is given by:

$(R_2/R_1) \times R_3$ = (150/100) x 100 = 150 ohms

Strained strain gauge resistance is given by:

= (150/100) × 101 = 151.5 ohms

Therefore, change in strain gauge resistance due to strain is given by ;

= 151.5 – 150 = 1.5 ohms

Unit change in resistance = 1.5/150 = 0.01

Gauge Factor = 3.0 = 0.01/strain.

Therefore strain = 0.01/3 = 0.0033

Problem 30.16. *A pressure measuring capacitive transducer consists of two silvered quartz diaphragms. The diaphragms are 2.5 cm² each in cross-section and are 3 mm apart. A pressure of 10 Newtons/m produces 0.3 mm deflection in the diaphragm. If the unstrained capacitance of the transducer is 300 pF, what should be the capacitance of the transducer under strain?*

Solution. The unstrained capacitance = 300 pF

$$= \frac{k.A}{d} = \frac{k.(2.5 \times 10^{-4})}{0.003}$$

$$\text{Therefore } k = \frac{0.003 \times 300 \times 10^{-12}}{2.5 \times 10^{-4}} = 3.6 \times 10^{-9}$$

Capacitance under strain

$$= \frac{3.6 \times 10^{-9} \times 2.5 \times 10^{-4}}{2.7 \times 10^{-3}} = 333 \text{ pF}$$

Problem. 30.17. *In a range measurement experiment using triangulation method, the transmitter and receiver are separated by 10 m. The angles that the lines of sight between target and transmitter and target and receiver make with the line joining transmitter and receiver are respectively measured as 85º and 90º. Determine the measured distance to the target.*

Solution: The distance to target, d can be computed from

$$d = l \times \frac{\sin \beta}{\sin (\alpha + \beta)}$$

Given that, α = 85º, β = 90º and l = 10 m.

Therefore, d = 10 × sin 90° / sin (85º + 90°) = 10/sin 175° = 10/0.087 = 115 m

SUMMARY OF IMPORTANT FORMULAS

1. Change in resistance (ΔR) due to change in length (longitudinal strain) is given by:

   $\Delta R = R \times e$

   Where, R = Unstrained resistance and e = Longitudinal strain = $\Delta L/L$

2. Change in resistance ($\Delta R'$) due to change in area of cross-section is given by:

   $\Delta R' = R \times e \times (2\mu - \mu^2 \times e)$

3. Total resistance change due to change in length and change in area of cross-section is given by $R \times e \times (1 + 2\mu)$, where (μ) is Poisson's ratio.

4. Gauge factor of a resistance strain gauge = (1 + 2μ)

5. For a vibrating-wire strain gauge, the natural frequency of vibration in its first mode is given by:

   $f = 1/2l \times \sqrt{(T/m)}$

   Where, m = Mass per unit length

   T = Tension

   l = Length

6. The sensitivity (S) of a capacitive transducer is given by:

   $S = \varepsilon \times A/d$

   Where, ε = Permittivity = $\varepsilon_0 \times \varepsilon_r$

   A = Area of each conductor

   d = Separation between conductors

   In case, it is fractional change in capacitance that is measured and not the change in capacitance value; then sensitivity (S') is given by:

   $S' = -(1/d)$

7. Thermally induced EMF in a thermocouple is given by:

   $e = a \times (T_2 - T_1) + b \times (T_2 - T_1)^2$

   Where, e = Induced EMF

   T_1 = Temperature of reference junction

T_2 = Temperature of hot junction

a, b = Constants

Constant (b) is very small. This reduces the above expression to $e = a \times (T_2 - T_1)$

8. Distance-to-target (d) in a triangulation sensor is given by:

$$d = l \times \frac{\sin \beta}{\sin (\alpha + \beta)}$$

Where, α = Angle made by line joining transmitter and receiver and the line joining transmitter and target

β = Angle made by line joining transmitter and receiver and the line joining receiver and target

I = Distance between transmitter and receiver

SELF EVALUATION EXERCISE

Multiple Choice Questions

1. The Gauge Factor of a resistance wire strain gauge is a measure of:
   (*a*) Sensitivity of the gauge
   (*b*) Dynamic Range of displacement measurement capability
   (*c*) Resolution
   (*d*) none of the above
2. One of the following materials' constants determines the gauge factor of resistance wire strain gauge :
   (*a*) Young's modulus of elasticity
   (*b*) Poisson's ratio
   (*c*) Resistivity
   (*d*) Thermal conductivity
3. A resistance wire strain gauge having a gauge factor of 2 was strained due to a certain displacement causing its original length to change from 100 cm to 101 cm. The percentage change in resistance will be :
   (*a*) 1% (*b*) 0.5%
   (*c*) 20% (*d*) 2%
4. The operation of a thermocouple is based on :
   (*a*) Seebeck Effect (*b*) Peltier Effect
   (c) Thomson Effect (*d*) Faraday's Laws
5. If in the Seebeck series of metals, Platinum comes earlier than Rhodium, then in a Pt-Rh thermocouple if one the junctions is at a higher temperature than other, the current flows (thermoelectric current):
   (*a*) from Platinum of Rhodium at the hot junction
   (*b*) from Rhodium to Platinum at the hot junction
   (*c*) from Platinum to Rhodium at the cold junction
   (*d*) None of the above
6. For all practical purposes, the e.m.f. produced in a thermocouple is :
   (*a*) directly proportional to temperature difference between the two junctions
   (*b*) inversely proportional to the temperature difference between the two junctions
   (*c*) proportional to square of the temperature difference between the two junctions
   (*d*) None of the above
7. In a Thermocouple Ammeter or Voltmeter, meter deflection is proportional to :
   (*a*) current or voltage being measured
   (*b*) inverse of current or voltage being measured
   (*c*) square of current or voltage being measured
   (*d*) none of the above
8. The VOLT-AMPERE characteristics of a thermistor exhibit a negative resistance region. Is this true?
   (*a*) Yes, it is (*b*) No, it is not
   (*c*) Yes, it is in case of NTC thermistors
   (*d*) Yes, it is in case of PTC thermistors.
9. The thermistor associated with an abrupt change in resistance with temperature is :
   (*a*) NTC thermistor
   (*b*) PTC thermistor
   (*c*) Neither of the two. Both types show a gradual change in resistance with temperature
   (*d*) None of the above
10. Which thermistor type is ideally suited for temperature measurement application :
   (*a*) NTC type (*b*) PTC type

11. At what point of thermistor V-A characteristics would you like to operate while making absolute temperature measurements ?
   (*a*) In the negative resistance region
   (*b*) At the transition point where the resistance changes from positive to negative
   (*c*) At a point where the excitation current is very small so that the temperature rise due to this current is much smaller than the ambient
   (*d*) None of the above
12. All Induction type transducers are based on:
   (*a*) Faraday's Laws (*b*) Seebeck Effect
   (*c*) Peltier Effect (*d*) Thomson Effect.
13. One of the following transducers is used for very small linear displacement measurements:
   (*a*) LDVT (*b*) Pirani Gauge
   (*c*) Thermocouple (*d*) Shaft Encoder
14. Pick the odd one out:
   (*a*) Pirani Gauge
   (*b*) Thermocouple Gauge
   (*c*) Thermistor Gauge
   (*d*) Resistance wire strain gauge
15. One of the following transducers is very popular for measurement of rotational displacement:
   (*a*) Shaft Encoder
   (*b*) Differential Capacitor
   (*c*) LDVT
   (*d*) Strain Gauge
16. Recognize the transducer type from the schematic shown in Fig. 30.82

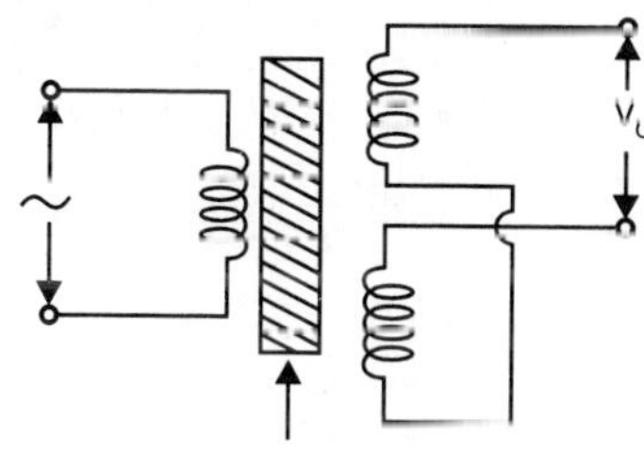

Fig 30.82

   (*a*) Induction type transducer
   (*b*) Differential transformer
   (*c*) LDVT
   (*d*) Synchro.
17. A transducer converts:
   (*a*) Mechanical energy into electrical energy
   (*b*) Electrical energy into mechanical energy
   (*c*) Acoustic energy into electrical energy
   (*d*) One form of energy into another form
18. Self-generating transducers are:
   (*a*) Active transducers
   (*b*) Passive transducers
   (*c*) Primary transducers
   (*d*) Secondary transducers.
19. Precision of an instrument refers to one of the following features:
   (*a*) Accuracy of measurements
   (*b*) Resolution with which measurements can be made with the instrument
   (*c*) Exactness of measurements
   (*d*) None of these
20. A set of voltage measurements at a particular node of a circuit read 13V, 13.1V, 12.9V, 13V and 13.1V. If the true value of voltage were 10.5V; the measurements indicate:
   (*a*) High accuracy and high precision
   (*b*) Low accuracy and low precision
   (*c*) High precision and low accuracy
   (*d*) Low precision and high accuracy
21. Sensitivity of instrument refers to:
   (*a*) Accuracy
   (*b*) Resolution
   (*c*) Ratio of output to input
   (*d*) None of these
22. A Hall-effect transducer is used for measurement of:
   (*a*) Displacement (*b*) Power
   (*c*) Current (*d*) All the above
23. One of the following is a shortcoming of Hall-effect transducers:
   (*a*) Poor resolution
   (*b*) Sensitivity to temperature variations
   (*c*) Variation in Hall's coefficient from plate to plate
   (*d*) Both (*b*) and (*c*)
24. A milli-ammeter has a uniform scale with 100 divisions. The resolution of the milli-ammeter assuming that full scale reading is 50 mA and $1/10^{th}$ of a division can be read with fair degree of certainty would be:
   (*a*) 50 mA (*b*) 25 mA
   (*c*) 100 mA (*d*) 5 mA

25. The output of an LVDT sensor is connected to an amplifier with a gain of 10. The amplifier outputs 5 mA for a core displacement of 0.8 mm. The sensitivity of LVDT is:
(*a*) 6.25 mA/mm (*b*) 0.625 mA/mm
(*c*) 0.0625 mA/mm (*d*) None of these

26. A triangulation displacement sensor is characterised by:
(*a*) High accuracy for longer distances
(*b*) Poor accuracy for shorter distances
(*c*) High accuracy for shorter distances
(*d*) None of these

27. Pirani gauge is used for measuring:
(*a*) Atmospheric pressure
(*b*) High pressure
(*c*) Low pressure
(*d*) None of these

28. A potentiometric type accelerometer can be used for measurement of frequencies:
(*a*) Lower than 50 Hz
(*b*) 10 Hz to 100 Hz
(*c*) Higher than 10 kHz
(*d*) Lower than 10 kHz

29. LVDT cannot be used for measuring:
(*a*) Temperature (*b*) Pressure
(*c*) Displacement (*d*) Velocity

30. Thermopile is a combination of:
(*a*) A number of thermistors connected in series or parallel
(*b*) A number of thermocouples connected in series or parallel
(*c*) A number of RTDs connected in series-parallel arrangement
(*d*) None of these

ANSWERS TO SELF EVALUATION EXERCISE

1. (*a*)	**2.** (*b*)	**3.** (*d*)	**4.** (*a*)	**5.** (*a*)
6. (*a*)	**7.** (*c*)	**8.** (*a*)	**9.** (*b*)	**10.** (a)
11. (*c*)	**12.** (*a*)	**13.** (*a*)	**14.** (*d*)	**15.** (*a*)
16. (*c*)	**17.** (*d*)	**18.** (*a*)	**19.** (*c*)	**20.** (*c*)
21. (*c*)	**22.** (*d*)	**23.** (*d*)	**24.** (*a*)	**25.** (*b*)
26. (*c*)	**27.** (*c*)	**28.** (*c*)	**29.** (*a*)	**30.** (*b*)

Index

D

P

Q

R

S

U

V

W

ALL-IN-ONE ELECTRONICS SIMPLIFIED

DR. ANIL K. MAINI | NAKUL MAINI

FEEDBACK FORM

1. Which topics from your examination point of view are inadequately or not covered in this book?

2. Is there any misprints/ mistakes/ factual inaccuracy in this book? If so, please specify.

3. What is your assessment of this book as regards the presentation of the subject-matter, expression, precision and price in relation to the other recommended books available on this subject?

4. Which competing books you regard as better than this? Please mention their author and publishers.

5. Any further suggestions/comments you would like to make for the improvement of the book?

PS: *If needed, please attach a separate sheet.*

Name: ____________________

College: ____________________

Address: ____________________

Email: ____________________

Mobile: ____________________

Please mail the feedback form at:
KHANNA BOOK PUBLISHING CO. (P) LTD.
4C/4344, Ansari Road, Daryaganj, New Delhi-110002

You can also email this feedback form at **contact@khannabooks.com** or **Whatsapp** at +91 – 9910909320

BEST FEEDBACK WILL BE REWARDED SUITABLY